Thermal Spray Fundamentals

Maher I. Boulos · Pierre L. Fauchais
Joachim V. R. Heberlein

Thermal Spray Fundamentals

From Powder to Part

Second Edition

Springer

Maher I. Boulos
Département de génie chimique et
de genie biotechnologique
Université de Sherbrooke
Sherbrooke, QC, Canada

Pierre L. Fauchais
Sciences des Procédés Céramiques et
de Traitements de Surface (SPCTS)
Université de Limoges
LIMOGES, France

Joachim V. R. Heberlein (deceased)
Department of Mechanical Engineering
University of Minnesota
Minneapolis, MN, USA

ISBN 978-3-030-70673-9   ISBN 978-3-030-70672-2   (eBook)
https://doi.org/10.1007/978-3-030-70672-2

This Springer imprint is published by the registered company Springer Nature Switzerland AG
The registered company address is: Gewerbestrasse 11, 6330 Cham, Switzerland

Foreword

Thermal spraying was invented and applied for the first time close to a century ago. In 1909, the Swiss engineer Max Ulrich Schoop got a first patent for the application of lead and zinc as protective coatings. Initially, he used a combustion process as a heat source for melting of metal wires. A high velocity flow of air or inert gas served to atomize the liquefied metal and to propel the formed droplets toward the substrate, where it solidified as a layer. Later on, he also used an electrical arc for this purpose.

These inventions represent, in principle, the beginning of the flame and wire arc spray technology. Over the following decades, thermal spraying evolved in terms of both technology and the great variety of the materials that could be sprayed. The driving force has been mostly economical since it offered means of modifying surface properties of parts and components where only the surface has to be made of expensive materials. Due to the relatively low process temperature of flame spraying, the selection of materials which could be processed at that time was limited and hence the variety of producible coatings and intended applications. In the middle of the last century, the need for coatings with materials of higher quality and melting temperature increased notably and with the development of the first plasma spray torches in mid-1950s by Thermal Dynamics Corporation, Metco, and Plasmadyne. These devices allowed for the generation of considerably higher process temperatures, which in turn extended the range of materials and applications that could be coated, and improved the overall quality and economics of the products.

The stringent requirements of the aerospace industry were strong technology drivers at these times. Based on the first plasma torch designs, new spray devices have been developed which continued up to now. For instance, besides the standard torches with plasma generation by a direct current (DC) arcs struck between two electrodes, electrodeless devices such as inductively coupling radio frequency (RF) plasma torches have been developed and are increasingly used for powder processing and the deposition of near net-shaped parts. Several process modifications also came into application, such as plasma spraying in vacuum or low pressure (VPS) or controlled atmospheric spraying. Also, in the lower temperature range improvements and new processes could be attained, like spraying with high velocity flames (HVOF and HVAF) and eventually the so-called "Cold Spraying." Concerning the precursor materials, the spectrum increased considerably. In addition to the conventional wires and powders of metals, ceramics, and polymers, suspensions and liquids can now be used by new spraying methods. In particular, with suspension spraying, the grain size of powders suitable for processing can be extended into the sub-micron range, for nanostructured coatings, giving rise to considerable improvement of coating qualities.

With these new process modifications and the enlarged spectrum of precursor materials, the variety of coatings and their industrial applications increased to an impressive extent. New surface qualities and interesting combinations of substrate and surface properties could be made, which otherwise were not possible, at least not with acceptable price and effort. These include coatings for protection against corrosion or wear, for electrical or thermal insulation, for high mechanical stability and hardness, and for specific electrical, optical, electrochemical, or catalytic properties, which could now be produced on an industrial scale.

A main precondition for improving quality, economy, and reproducibility, and for the acceptance in new application fields, was the need to gain a fundamental knowledge about the basic physics of the whole process chain concerning:

- The interaction of the deposition material with the heat source
- The transient melting and accelerating phase of the material
- The interaction with the surrounding atmosphere
- The material deposition, where surface conditions of the substrate has a critical impact on the resulting coating quality

In its early stages, thermal spraying could be considered as an "art," where empiricism and experience together with a "hot hand" of the operator for selecting the right set of the numerous process parameters guaranteed the success. Over the past three decades, thermal spraying evolved into a mature science. This important step was greatly helped by improved fast diagnostics methods, mainly based on lasers and computers and the use of numerical modeling for the prediction of the influence of the different process parameters on the coating quality and its performance.

The three authors, leading international scientists in the field of thermal spray, have presented in this book a large volume of knowledge and experience gathered over many years of successful research and development work. This book is by far one of the most comprehensive reviews of the state of the art in this field. As indicated in its title, it covers the field of thermal spray from the powder all the way to the final part. On the one hand, a systematic survey is given of the different types of power sources and process characteristics in thermal spraying for a large spectrum of applications, together with information about the necessary components like materials, equipment, control units, and applied diagnostic methods. On the other hand, the analysis of process fundamentals is not restricted to a phenomenological description, but is supported by extended theoretical considerations to understand the basic phenomena involved behind the whole processes. Separate chapters are also devoted to feedstock preparation, whether in the form of powders, wires, cords, or rods, surface preparation, coating formation and characterization, online diagnostics and control, process integration, and industrial applications. An impressive number of figures and extensive list of references facilitate considerably the understanding of the material and allow for going further into details, whenever necessary.

This impressive book is addressed to newcomers in the field of thermal spray as well as experienced researchers and engineers wanting to know more about the scientific details, either to improve the quality of their own products, and of their process efficiency, or to have guidance in a decision phase, when new applications or processes for increasing the product portfolio are the goal.

Dr. Ing. Rudolf Henne
Retired from German Aerospace Establishment (DLR)
Institute of Technical Thermodynamics
Stuttgart, Germany

Preface

Following the publication of our first edition of this textbook and based on the very position reception it received from the general scientific community, the need to keep it updated in this rapidly developing field became increasingly obvious. Moreover, as we used the book on a regular basis, we realized that with such a broad coverage of the field, certain improvements could be made in its overall structure to make it more "'reader friendly" and facilitate its use as

- *Textbook* for teaching purposes at the graduate school level
- *Reference book* by research scientists in their pursuit of further developments in this field by providing access to a huge volume of scientific literature
- *Manual* for practicing engineers for the purpose of system or process design that hopefully would help in the understanding of some of the basic concepts behind thermal spray technologies

One of the main changes made in this edition is the subdivision of the material presented into four distinct parts.

Part I, Basic Concepts, Chaps. 1, 2, 3, 4 and 5, is dedicated to a review of thermal spray coating and its position in the broader field of surface modification technologies. While not being a textbook on combustion science or thermal plasma fundamentals, a brief review is presented on basic concepts in these two fields. Considering that a large part of thermal spray technologies is based on the in-flight heating and melting particles as they are injected into a hot, combustion, or plasma spray stream, the material covering plasma or combustion gas/particle momentum and heat transfer has been split into two chapters. One dealing with basic transport phenomena taking place, in-flight, between a hot fluid stream and a single particle. The following chapter is dedicated to a discussion of gas and particle dynamics in thermal spray processes including the important concept of gas-particle interactions under dense loading conditions.

Part II, Thermal Spray Technologies, Chaps. 6, 7, 8, 9, 10, 11 and 12, is dedicated to the description of advances made in each of the thermal spray technologies including cold spray, which, while strictly not a thermal spray process, is well positioned in this field and used in a number of important applications. One change made in this group of technologies is the splitting of the chapter on DC plasma spraying into two reasonably sized chapters, one dealing with DC plasma spraying fundamentals and the other with DC plasma process technologies.

Part III, Coating Formation and Characterization, Chaps. 13, 14, 15, 16 and 17, is dedicated to the presentation of some of the broad topics in this field, including the characteristics and main manufacturing routes of powders, wires, and cords as feed material for different thermal spray processes. Surface preparation and coating formation through the layering of successive splats are covered in the following three chapters. Special attention is given to the area of nano-crystalline and nano-structured coatings and the growing interest in the field of solution and suspension plasma spraying. An overview of basic coating characterization techniques is discussed in a separate chapter in this part without attempting to be a reference book in this field, which is widely covered in the material science literature.

Part IV, Process Integration and Industrial Applications, Chaps. 18 and 19, has also been restructured with the chapter on process diagnostics and online monitoring and control integrated into the chapter on process integration and control to avoid any duplication due to the importance of instrumentation and control in modern thermal spray processes. This chapter is also expanded by introducing a wider range of examples of "spray booth" designs commercially available in the field. The last chapter of this book, dedicated to industrial applications of thermal spray technologies, was revamped, including a comparative analysis of different thermal spray processes, which can be helpful in selecting the appropriate process for a given application. This is followed by a review of thermal spray process applications grouped in terms of either "its process objectives," or its area of use "by industry" or by "country." As with our last edition, a brief economic analysis is presented at the end of this chapter without attempting either to cover this topic in an exhaustive fashion, being already the subject of numerous textbooks dedicated to economic analysis in the chemical process industries.

Such an in-depth revision of this textbook was not an easy task, requiring hundreds of hours of hard work which thanks to COVID-19, was involuntarily available to us over the year 2020. The task was not simplified either by the death of our dear colleague Professor Joachim V.R. Heberlein, who passed away in February 2014. His memory is cherished by us and many of his friends and colleagues in this field for his valuable contribution to the first edition of this book. We would like to thank our life partners Alice Boulos and Paulette Fauchais for their continued support and encouragement, without which this formidable task would not have been possible.

Sherbrooke, QC, Canada   Maher I. Boulos
Limoges, France   Pierre L. Fauchais

Preface (First Edition)

This book is based on a series of continuing education courses which have been offered by the authors across the world in conjunction with the International Symposium on Plasma Chemistry (ISPC) as part of its summer school over the period 1995–2011. A similar course, though more oriented toward thermal spray technology, was also offered by the authors in conjunction with ASM International Thermal Spray Conferences (ITSC) over the period 1998–2010. Both courses were offered to graduate students, practicing engineers, and researchers actively involved in the field of thermal plasmas. Their emphasis was on the fundamentals behind plasma processing and thermal spray technology, and the aim was to provide a grassroots understanding of the basic phenomena involved, necessary for taking the technology over the crucial step from being an "art" based on operator talent and experience to a mature science with quantitative predictive capabilities.

This step did not come easily and without the intense involvement of many leading researchers in this field across the world. The three determining factors which were of critical importance to the evolution of this field over the past three decades are as follows:

- Major improvement in process diagnostics and online controls
- The fast and significant development of numerical modeling and computing capabilities
- Major development in materials science and materials characterization techniques

In the process of preparation of the manuscript for this book, which spans many years, the authors were confronted with the critical need to strike a good balance between the need to be concise in the overall presentation of the subject and being inclusive in stressing the fundaments without overlooking the important applications which were the economical driver of the technology. New technologies were also developed over this period which, while not being "plasma technologies," were relevant to the overall field of surface treatment and coating. These were accordingly included in the book such as the combustion-based technologies and "cold spray."

We have no pretensions about having covered every aspect of this technology or exhaustively reported on every relevant publication in this field. Exhaustive lists of references are given at the end of each chapter. For those who were not cited, our apologies, it was not intentional. A book of this size and scope could not have been possible without the extensive help of students, research assistants, colleagues, and associates. Our sincere thanks to all who have helped make this book a reality. Particular thanks are due to Dr. Rudolf Henne who so generously gave his time in the process of reviewing the manuscript of the book in its final preparation stage. We also appreciate his willingness to write the foreword for this book which reflects his long and broad experience in the field of thermal spray. The financial assistance of the numerous government and private funding agencies and industrial partners who have also supported the basic and applied research behind this technology in our respective research laboratories is gratefully acknowledged. Our sincere thanks to our respective families and life partners, Paulette Fauchais, Yuko Heberlein, and Alice Boulos who had to cope with the long hours of intense personal efforts that were needed to complete this book.

Limoges, France   Pierre L. Fauchais
Minneapolis, MN, USA   Joachim V. R. Heberlein
Sherbrooke, QC, Canada   Maher I. Boulos

Contents

Part II   Thermal Spray Technologies

Part IV Process Integration and Industrial Applications

Basic Concepts

Abbreviations

APS	Atmospheric Plasma Spraying
DC	Direct current
HVAF	High-velocity air fuel
HVOF	High-velocity oxy fuel
ICP	Inductively coupled plasma
IPS	Induction plasma spraying
LPPS	Low pressure plasma spraying
OEM	Original Equipment Manufacturer
PTA	Plasma transferred arc
RF	Radio Frequency
RF-TPS	Radio Frequency Induction Plasma Spraying
WAS	Wire arc spraying
YSZ	Yttria stabilized zirconia

1.1 Introduction

The motto of the Olympic games *"citius, altius, fortius"* (faster, higher, stronger) entices athletes to continuously establish new records. Similarly, there is a continuous push in every part of industry to set new performance standards for the improvement of one part of human life. These performance improvements can be summarized as better functional performance, longer component life, and lower component cost. Besides the geometrical design, it is the choice of materials that will determine the performance and cost of the component. Advanced materials such as specialty steels, super alloys, and advanced ceramics have been developed for exceptional functional performance in a large number of applications. The increasing demands for combined functional requirements such as the high-temperature resistance to corrosive atmospheres in addition to abrasive wear resistance and the added difficulty in machining some of the specialty alloys to the relatively complex final forms, while keeping the final cost of the part at an acceptable level, led to the ever-increasing demand for surface modification technologies which allows the decoupling of the surface properties of a part from its bulk and structural properties. Surface modification can generally be achieved through either:

- Surface transformation through physical or chemical treatment
- Surface coating with a compatible metallic or ceramic material

This book is devoted to a review of thermal spray technologies which is one of the leading surface coating technologies. In this chapter, a brief introduction is given to this rapidly developing fields giving highlights of the technology, its historical development over the past century, and examples of typical industrial applications. An outline of the book content is presented at the end of the chapter in order to guide the reader through its different parts and provide easy access to information and pertinent references.

1.2 Needs for Coatings

The motivation for coating structural parts can be summarized by the following needs:

- Improve functional performance by allowing, for example, tolerance to higher temperature exposure using thermal barrier coatings
- Improve component life by reducing wear due to abrasion, erosion, and corrosion
- Extend the component life by rebuilding worn parts to their original dimensions avoiding the need for replacing the entire component, e.g., a shaft or axle.
- Reduce component cost by improving functionality of a low-cost material with an appropriate coating

© Springer Nature Switzerland AG 2021
M. I. Boulos et al. (ed.), *Thermal Spray Fundamentals*, https://doi.org/10.1007/978-3-030-70672-2_1

Coating technologies can be divided into two broad groups based on coating thickness. These are:

Thin films have thicknesses of less than 20 μm, offering excellent enhancement of surface properties. These are mostly obtained using atomic level deposition technologies such as chemical vapor deposition (CVD) or physical vapor deposition (PVD) which can provide surfaces with unparalleled hardness or corrosion resistance [Goto and Katsui (2015)], [Ogawa et al. (2018)]. The results can be further improved when assisted by plasma [Harder et al. (2017), Thull and Grant (2001)], laser [Katsui and Goto (2017)], or electron beam [Singh and Wolfe (2005)]. Most of thin film technologies, however, require a reduced pressure environment and, therefore, are more expensive and impose a limit on the size and shape of the part to be coated. The exceptions are the super thin surface modification films of large polymer sheets which are made possible using atmospheric pressure non-equilibrium discharges to manufacture materials such as nanocarbon [Hatakeyama (2017)] or Teflon-like layers on cellophane surfaces [Cruz-Barba et al. (2003)].

Thick films have generally a thickness greater than 20–30 μm, up to several millimeters. They are required when the functional performance depends on the layer thickness. Thermal barrier coatings are a typical example where the level of thermal protection of the substrate depends on the coating thickness. Wear resistance coatings exposed to strong erosion and corrosion conditions are another example where the component life depends on the layer thickness. Thick coatings in excess of a few millimeters may also be required for such applications as the rebuilding of worn parts to their original dimensions. Thick film deposition methods include chemical/electrochemical plating, brazing, weld overlays, and thermal spray. Each of these methods offers certain advantages and has their limitations. Wet chemical methods suffer from environmental hazards; brazing and weld overlays have limitations regarding the materials that can be deposited and the shape of the substrate, while thermal spray may require a post-deposition treatment to reach full density or eliminate all open porosity.

1.3   Thermal Spraying

The definition of thermal spraying is given in the thermal spray terminology compendium [Hermanek (2001)] as: "Thermal Spraying comprises a group of coating processes in which finely divided metallic or nonmetallic materials are deposited in a molten or semi-molten condition to form a coating. The coating material may be in the form of powder, ceramic rod, wire, or molten materials."

A block diagram of the principal components of a thermal spray system is illustrated in Fig. 1.1. The system is centered around the "spray torch" in which different sources of energy whether chemical (combustion) or electrical (electric discharge), depending on the spray process, are converted into a stream of hot gas in which the material to be sprayed is injected. The material introduced into the torch, whether as powder, wire, or other forms, is heated, melted, and entrained by the high-temperature, high-velocity gas stream toward the substrate on which the particle/molten droplet impact generating splats which piles up on the substrate forming the coating. The splats formed either through the lateral flow of the melted material or, as in cold spray process, the ductile deformation of the particle represent the building block for the formation of the coating. The form and properties of these splats depend on the nature of the material sprayed, particle/droplet size distribution, particle/droplet temperature, and velocity prior to its impact on the surface of the substrate as well as on the surface properties of the substrate.

A schematic representation of a typical atmospheric pressure DC plasma spray system is given in Fig. 1.2a. It consists essentially of five principal components:

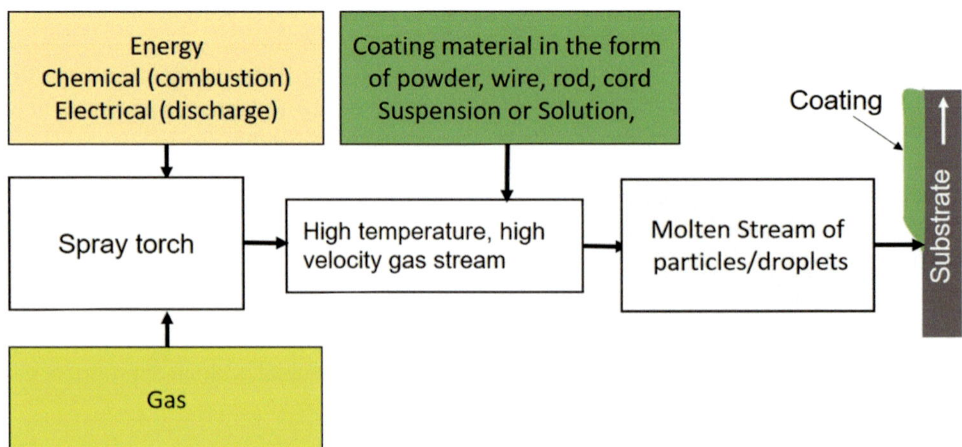

Fig. 1.1 Block diagram of a generalized thermal spray system

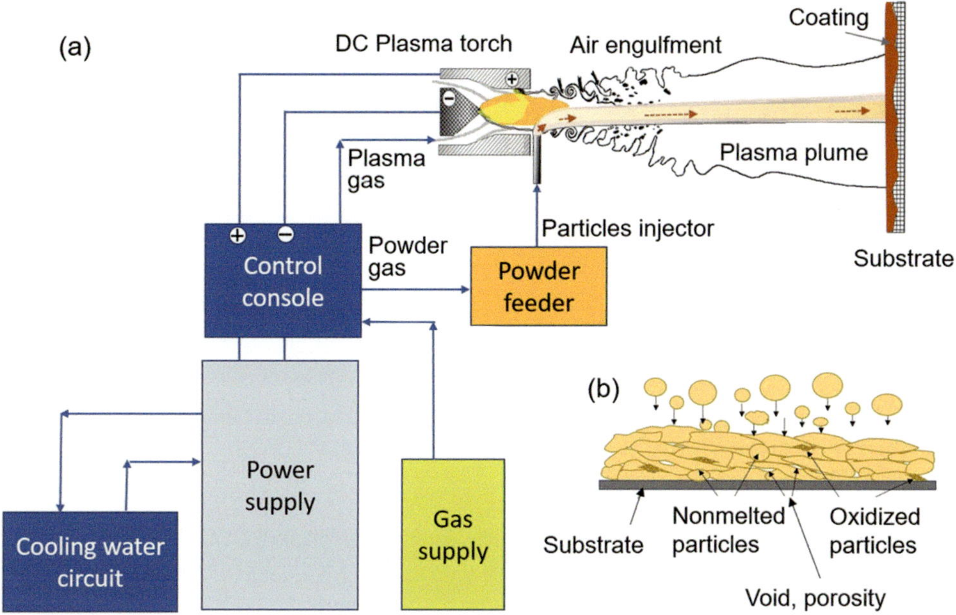

Fig. 1.2 (**a**) Schematic of a conventional DC plasma spray system. (**b**) Illustration of the coating building mechanism through the compilation of successive splat formation

Energy source, in the form of a DC electrical power supply

Gas supply, necessary for the operation of the plasma torch and the transport and injection of the material to be sprayed in powder form into the hot plasma stream

Powder feeder for the controlled feeding of the material into the plasma jet

Spraying torch and substrate manipulator, necessary for the movement of the spray torch and the substrate relative to each other to insure the control of the coating thickness

Control console, responsible for the monitoring and control of the different system components which are required for implementation of a reliable and safe coating processes

To these, it is necessary to add a broad range of ancillary equipment's and infrastructure services such as a cooling water circuits required for the protection of the different system components exposed to high-temperature plasma flow, exhaust gas cleaning and evacuation system, substrate preparation infrastructure which are necessary for optimal control of the spraying conditions, and, in certain cases, tooling for the post deposition finishing of the coating.

As illustrated in Fig. 1.2b, the coating is formed through the successive piling up of individual splats formed on impact of the particles/molten droplets on the surface of the substrate. The properties of the individual splats depend on the temperature and velocity of the particles/droplets prior to their impact on the substrate. The chemistry of the coating environment, the angle of projection of the particles/droplets on the substrate, and substrate surface preparation and its temperature can also be responsible for the creation of key features, or defects, in the coating as indicated in Fig. 1.2b. These could include non-melted or partially melted particles, oxidized particles, or poorly spread splats giving rise to the formation of voids/pores in the coating which, in turn, would affect the coating properties and its performance.

Substrate preparation has a critical impact on the quality of the coating since the adhesion of the coating to the substrate is directly related to its cleanliness, roughness, and sometimes the proper machining of the substrate. The presence of adsorbed contaminants on the substrate has also been reported to reduce the adhesion of the splat to the substrate. The effect is attributed to the heating and evaporation of the adsorbates and condensates present on the substrate during the splat formation step creating a localized high-pressure zone under the flattening particle and thus decreasing its contact with the substrate [Li and Li (2004)]. According to Fukumoto et al. (2006), such a situation can be avoided if the substrate is preheated over a critical transition temperature, T_t, which is in the range of 200–400 °C depending on the substrate material. To illustrate this point, micrographs given in Fig. 1.3. show splats of 58 μm diameter alumina particle plasma sprayed on a stainless steel substrate (ASI304) with a relative velocity of 138 m/s and an impact angle of 30°. The in-flight particle temperature prior to its impact on the substrate was reported as 2400 K. Micrograph given in Fig. 1.3a, obtained with a cold substrate, without substrate preheating, shows an elliptical shape with numerous fingers immerging from it, while that given in Fig 1.3b, which was obtained on a preheated substrate, has a pure elliptical shape with no visible fingers or splats. As will be discussed later in Part III of this book, the shape and properties of these splats have a significant impact on the overall properties of the coating.

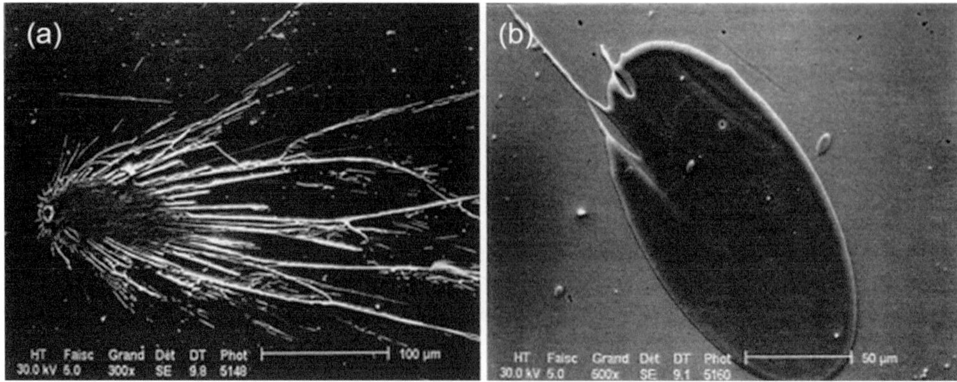

Fig. 1.3 Typical splats of plasma-sprayed alumina particles on an inclined stainless steel (ASI304) substrate at an impact angle of 30°. (**a**) Substrate at room temperature and (**b**) substrate preheated above transition temperature [Bianchi et al. (1997)]

Fig. 1.4 (**a**) YSZ feedstock particle formed by the agglomeration (spray-drying) of individual nanosized particles (**b**) higher magnification micrographs of the constituent nanosized particles (30–130 nm) [Lima and Marple (2007)]

Another area where significant developments in the field of thermal spray have been made over the past few decades is the spraying of nanostructured coatings with the objective of improving the toughness of plasma sprayed coatings [Fauchais et al. (2011)]. McPherson in 1973 was one of the first to identify nanosized features in conventional thermal-sprayed alumina coatings. Subsequent studies in this area aiming at producing nanostructured coatings using nanopowders were greatly hindered by the difficulty of feeding them into a flame or plasma stream using carrier gas-based conventional powder feeding techniques. The challenges being the extremely small mass of the nanopowders and the difficulty to provide them with the momentum necessary to have them penetrate a fast-moving plasma stream. One approach proposed by Lima and Marple (2007) was to agglomerate the nanoparticles into micron-sized particles and feed them into the plasma using conventional techniques. Micrographs of yttria stabilized zirconia (YSZ) particles produced using this technique are given in Fig. 1.4a, with individual nanosized constituents integrated in the agglomerated shown in Fig. 1.4b. The thermal diffusivity of the nanostructured plasma-sprayed YSZ coating obtained using these agglomerated powders was found to be lower than those obtained using conventional YSZ coatings up to temperatures of 1200 °C (heating and cooling steps). These nanostructured YSZ coating had a thermal shock resistance 2–4 times higher than that of the conventional YSZ coatings. The challenge, however, remained in the fact that great care has to be exercised in the use of such agglomerated powders since their temperature during the spray process need to be maintained just below the melting temperature of the material in order to avoid their complete melting which would erase irreversibly all of their nanostructure features.

The growing need to manufacture coatings with enhanced properties for a wide range of applications [Fauchais et al. (2015)] has led to the emergence of alternate powder feeding techniques that could be used with nanopowders that would not require the particle agglomeration step. *Suspension* and *solution* plasma-spraying techniques are two such novel powder injection approaches which have been gaining wide acceptance in the scientific community.

Suspension plasma spraying was used for the first time by Bouyer et al. (1997), Gitzhofer et al. (1997) for the feeding of hydroxyapatite nanopowders in an RF inductively coupled plasma (RF-IPS)-spraying system. In this approach, the nanoparticles are maintained in a suspension which is injected and atomized into the plasma stream. The formed droplets of the suspension containing the nanoparticles are further fragmented in-flight and the liquid evaporated without affecting the nanostructure of the powder. The dried agglomerated powder particles entrained by the relatively colder plasma stream, integrating the evolved liquid vapors, are heated to a sufficiently high temperature to adhere to the substrate on impact without causing them to melt completely and lose their nanostructure feature. Occasionally the produced agglomerated particles will disintegrate releasing individual nanoparticles which are entrained by the plasma flow and deposited on the substrate. A micrograph of typical micro-splats formed using this approach is given in Fig. 1.5 for the suspension plasma-sprayed YSZ nanopowder using ethanol as liquid medium for the formation of the suspension.

The substrate used in this case was smooth stainless steel, preheated over its transition temperature. Splats of molten particles are noted to be well flattened with a relatively low mean flattening degree of 2.1 (ratio of splat diameter to the original particle/droplet diameter) reflecting their relatively low impact velocity on the substrate.

Solution spraying, on the other hand, is a technique in which the precursor material to be deposited is dissolved in an appropriate liquid, injected and atomized in the form of fine liquid droplet mist. As these droplets are entrained by the plasma stream, they are heated and evaporated producing fine nanoparticles, depending on the solute concentration in the liquid. The formed nanoparticles are further heated and partially or completely melted before their impact on the surface of the substrate forming corresponding micron-sized splats. The main advantage of solution spraying compared to suspension spraying lies in the efficient mixing at the molecular level of the chemical constituents allowing for an excellent chemical homogeneity of the feedstock material [Ravi et al. (2006)]. The success in forming the phase required for a given system depends on the decomposition characteristics of the different precursors. Once the solid particles have been formed, the situation is the same as that with suspensions. Several liquid precursors, such as complexes solutions/sols/polymerics, have been evaluated for different oxide systems [Gell et al. 2008)].

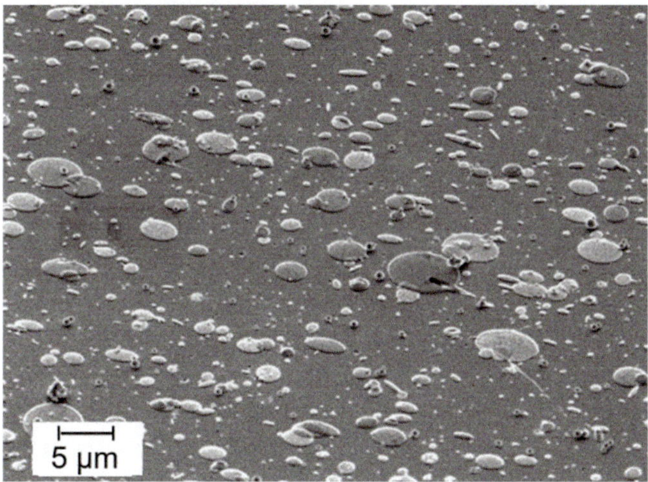

Fig. 1.5 SEM photography of YSZ splats resulting from suspension plasma spraying of nanosized particles on a preheated steel substrate [Fauchais et al. (2016)]

1.4 Classification of Thermal Spray Processes

A preliminary classification of thermal spray technologies according to the source of energy used in the process for the generation of the high-temperature, high-velocity gas stream is given in Fig. 1.6. These are grouped into two broad categories:

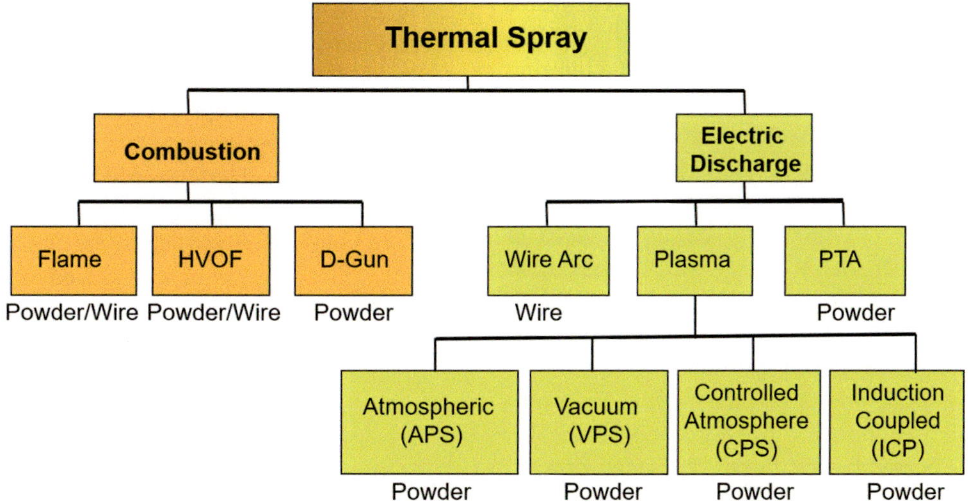

Fig. 1.6 Classification of thermal spray coating technologies

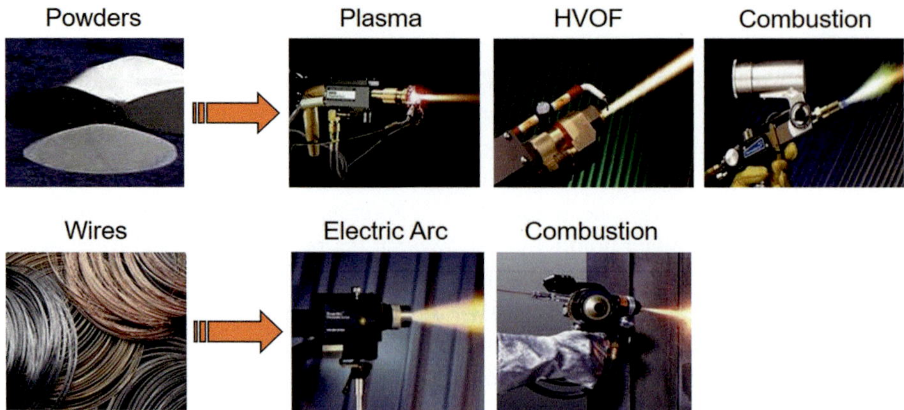

Fig. 1.7 Illustration of thermal spray processes using powders and wires. [Reproduced with kind permission of Oerlikon Metco Corp]

Combustion-based thermal spray processes, which historically represent the origin of the thermal spray technology, include *flame spraying, high-velocity oxy fuel (HVOF)*, or *high-velocity air fuel (HVAF)* and *detonation-gun*, or *(D-gun)*, spraying processes. HVOF, in which the sprayed particle velocities can reach up to 650 m/s with temperatures in the range of 2000 °C, are used to spray metals, cermets, and a few ceramics. Attention must be given to avoid in-flight particle oxidation or partial decarburization which can significantly affect the quality of the coating. This can be achieved using HVAF due to the higher particle velocities and lower particle temperatures achieved compared to HVOF. The increased particle kinetic energy gives rise to increased particle plastic deformation with harder substrates compensating for the lower particle temperatures.

The second group, which evolved rapidly over the past few decades, is *electric discharge-based* technologies include *atmospheric and vacuum plasma spraying (APS and VPS)*, wire arc spraying (WAS), and *plasma transferred arc (PTA)* deposition. The central part of these system is either a direct current (DC) or a radiofrequency (RF) induction plasma torch, which makes use of electrical energy for the generation of a high-temperature, high-velocity plasma stream which serves for the heating, melting, and atomizing the feed material in the spraying process.

Cold spray processes, which make use of a supersonic cold gas flow for the entrainment and acceleration of the spray material in the form of fine ductile particles to velocities between 300 and 1500 m/s, are also covered in this book. While this technology strictly does not belong to either the combustion or electric discharge groups, nor does it conform to the definition of thermal spraying, it is traditionally included as part of thermal spray technologies.

A second level for classification of the thermal spray processes is according to the form in which the coating material is introduced into the energy source, as powder, wires, or rods. As noted in Fig. 1.6 and illustrated in Fig. 1.7, the great

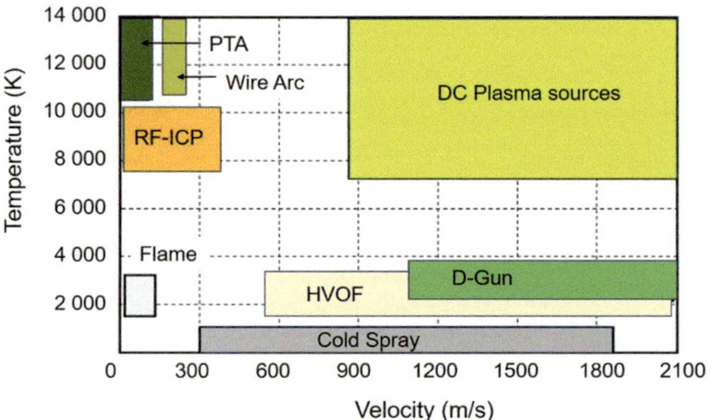

Fig. 1.8 Gas temperatures Vs velocity mapping associated with different thermal spray processes

majority of the thermal spray processes require that the material be sprayed in powder form. Wire arc spraying, developed in the 1960s for the protection of steel structures against corrosion [Steffens et al. (1990), Marantz and Marantz (1990)], is an exception in this regard since it is limited to the spraying of materials in the form of a wire with the tip of these wires molten by an electric arc struck between two wires, or between a single wire and a fixed hot electrode. A high-velocity gas flow across the arc constantly removes the molten material from the wire tips, breaks down larger droplets into smaller ones in a secondary atomization, and propels them toward the substrate. The wires need to be continuously fed to maintain a constant arc gap. Flame spraying and HVOF, which are combustion processes, are flexible enough to be able to use either powder or wire as feed material.

A further classification of thermal spray processes can be made according to the predominant gas velocity and gas temperature levels in the energy sources used. These are illustrated in the velocity–temperature diagram given in Fig. 1.8. This shows DC plasma sources distinguished by their highest-velocities and highest-temperature range among all thermal spraying processes. Combustion sources, on the other hand, are temperature limited to 3000–400 K while being able to achieve velocities as high as 2000 m/s. RF induction plasma sources offers high temperatures in the 8000–10,000 K, with relatively lower velocities generally below 200 or 300 m/s. With the relatively large volume of RF discharges, they offer high residence time of the particles in the plasma region (of the order 10–15 ms), compared to less than 1 ms for DC sources, which allows for the heating and melting of relatively large particles (>100–200 μm) of most refractory materials (T_m > 2000 K). Wire arc spray and PTA are also recognized as being high-temperature sources (T > 12,000 K) with relatively low-gas velocities.

One must keep in mind, however, that:

- Different materials require different deposit conditions
- Specific coating properties (high density or desired porosity) may require specific particle velocity/temperature characteristics
- The heat fluxes to the substrate vary for the different coating methods and for some substrates the heat flux need to be minimized to eliminate or reduce residual stresses in the substrate
- Substrate preheating and temperature control during spraying strongly influence coating adhesion and residual stresses
- Frequently a trade-off needs to be made between coating quality and process economics

A comparative review of the different thermal spraying processes as part of the broad field of surface modification technologies is presented in Chap. 2. It is clear, however, that every process has some unique features that make it particularly suited for a specific coating application. This makes different thermal spray processes to a large extent complementary rather than competitive for optimum product performance.

1.5 Historical Evolution of Thermal Spray Technology

The invention of thermal spray is credited to M Schoop and his collaborators, who received several patents on various aspects of thermal spray process and successfully commercialized the technology on an industrial scale [Schoop (1910), Schoop 1911, Schoop (1915)]. His impact on thermal spray is well described by Berndt (2001) and Knight (2005) who point out his unique vision of this field at that time and anticipated future developments. In his early patents, Schoop M. U. (1910) describes the atomization of liquid metals by a high-pressure air or inert gas stream, the use of metallic or ceramic powders heated by a flame [Schoop (1911)], the use of wires or rods of various materials in a patent by [Morf (1912)], or a description of the twin wire arc spray process [Schoop (1915)]. It should also be mentioned that Schoop was the first to recognize the importance of the droplet velocity and temperature on the coating formation, e.g., by commenting in anticipation of HVOF and cold spray: "At a gas pressure of 20 atmospheres, for example, coatings are produced, the density and hardness of which are above normal values; at a gas pressure of only 5 atmospheres the density and hardness are below normal values" [Schoop (1910)]. While thermal spray technology was developed in the beginning of the twentieth century, the technology was limited essentially to flame spraying. In the mid-1950s, the first plasma spray torch was developed by Thermal Dynamics Corporation, followed by Metco and Plasmadyne, who developed the basis of the torch designs currently used in this field. Figure 1.9 [courtesy of Oerlikon Metco] illustrates

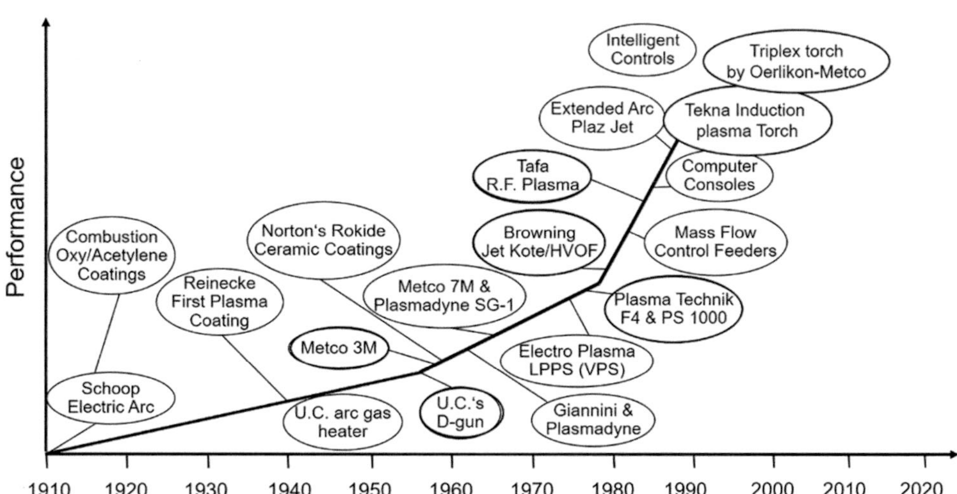

Fig. 1.9 Milestones in the development of the thermal spray industry. [Reproduced with kind permission of Oerlikon Metco Corp]

the evolution of thermal spray technology turning points in plasma torch designs used in the thermal spray industry over the past 100 years.

The requirements for new materials in the aeronautical and space industries led to a rapid development of thermal spray technology in the 1960s. Since then there has been an accelerating pace of developments for spray coating devices and processes, materials, process diagnostics and controls, and new applications. The 1970s saw the development of low-pressure plasma spraying and of induction plasma spraying, of HVOF, and of many process improvements. A trend was pursued to extend the process conditions to higher particle velocities and lower particle temperatures, through the introduction of HVAF and eventually that of cold spray. The use of coatings spread to more and more industries, and the increased requirements for improved process reliability led to an increase in research and development activities. At the same time, robotic systems were developed for the coating of complex shapes. Thermal spray technology became a truly interdisciplinary field, with research in plasma diagnostics, fluid dynamic modeling, new developments in describing plasma heat transfer, awareness of transient phenomena (instabilities), diagnostics for describing the in-flight properties of individual particles, and lastly the description of the coating formation. This research stimulated the development of more robust torches, as well as sensors, which can characterize properties of particles and of the substrate in the harsh environment of spray booths. These sensors not only allowed unprecedented gathering of information on the influence of process parameters on the particle states; they also set the stage for developing effective online control systems. Among the new torch developments, the torches with multiple cathodes [Zierhut et al. (1998), Barbezat and Landes 2000] or multiple anodes should be mentioned [Dzulko et al. (2005)] as well as the central injection torch [[Moreau et al. (1995), Burgess (2002)]. Induction plasma spraying, or as more commonly known as vacuum induction plasma spraying (VIPS), has attracted increasing attention over the past three decades. At the same time, cold spray technology became part of the family of industrially accepted coating technologies [Gärtner et al. (2006), Irissou et al. (2008)], and strong developments of this technology are continuing. This low temperature deposition process avoids the formation of oxides, and the properties of the deposited films are close to those of the wrought material. The discovery of the special properties of materials consisting of nanometer-sized grains (nanophase materials or nanomaterial) has led to developments of depositing coatings of such materials, either by using powders consisting of nanoparticle agglomerates or by using newly developed techniques of suspension spraying, where droplets of a suspension of nanoparticles are injected into the plasma, or solution spraying, where precursors of nanoparticles are injected as a liquid into the

plasma where nanoparticles then nucleate [Fauchais et al. (2011)]. A separate chapter in this book is devoted to these new developments.

1.6　Thermal Spray Applications

A survey of thermal spray applications in Europe in 2002 is presented in Fig. 1.10 [Ducos and Durand (2001)]. This shows that the earlier domination of aerospace applications (around 50%) is challenged by a significant increase of the use of thermal spray in the automotive and chemical process industries. Such an extension from high added value products, to high-volume production items, is a clear indication of the maturity of the technology and its competitiveness for the opening new markets [Vardelle et al. (2015)]. While such diversification of applications is likely to increase, the main pillars of users of thermal spray are expected to remain the aerospace, automotive, power, and chemical industries [Vardelle et al. (2016)].

One of the main areas in which thermal spray offers significant advantages is substitutions of steel with lighter weight materials such as Al or Mg. The use of aluminum engine blocks in automobiles to reduce the weight is an example for how thermal spray technology can respond to an industrial/societal need. Spray torches have been developed to coat the inside walls of the Al alloy cylinder blocks with materials withstanding the high temperatures and corrosive environment of the internal combustion chamber [McCune Jr. et al. (1993) and Barbezat (2001)]. Another strong driver for expanding thermal spray technology is the need for replacement of wear-resistant hard chrome coatings [Wasserman et al. (2001)]. The main advantages of thermal spray coatings being:

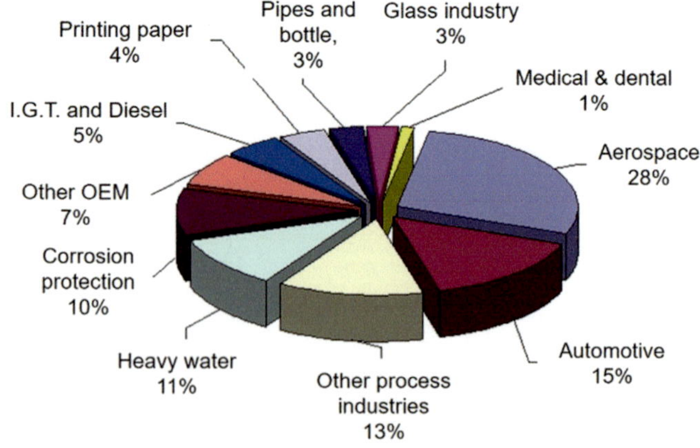

Fig. 1.10 Industrial applications of thermal spray technology in Europe in 2001, after Ducos and Durand (2001). Reprinted with permission of ASM International. All rights reserved

- Reduction of environmental concerns associated with exposure to hexavalent chromium
- Allowing for the deposition of thicker coatings
- Offering improved corrosion resistance and higher fatigue life

cost savings because of higher process productivity, and wider range of materials that can be deposited, having no limitations on the size of the part to be coated, and providing longer life of the coated part.

A comparison of wear resistance of chrome-plated surfaces with surfaces coated by HVOF or plasma spraying demonstrated the advantages of the thermally sprayed coatings in applications in the paper and textile industries. However, it is the large variety of materials that can be chosen for the coatings that are responsible for the continued diversification of the thermal spray market. Ceramic coatings can offer, besides good wear resistance, good tribological properties reducing friction, or can have high acceptance of color ink for printing presses [Wewel et al. (2002)]. Hip or knee or dental implants rely on biocompatible or bioactive coatings that increase the speed of connecting with the bone tissue.

According to Read (2003), the thermal spray market in 2003 was estimated at 3.5 billion USD. Details market distribution given in Table 1.1 show that the service of providing the coatings constitutes 40% of that market, with OEMs another 40% and 20% of the market related to powder sales. It should be pointed out, however, that in several applications, the coating is the "enabling technology" for components with a much larger market value than its actual cost.

Dorfman (2018) point out that a key feature of thermal spray coatings is that they can provide a functional surface to protect or modify the behavior of a substrate material and/or component. A substantial number of the world's industries utilize thermal spray for many critical applications. Key application functions include restoration and repair; corrosion protection; various forms of wear such as abrasion, erosion, and scuff; heat insulation or conduction; oxidation and hot corrosion; electrical conductors or insulators; near-net-shape manufacturing; seals, engineered emissivity; abradable coatings; decorative purposes; and more. Figures 1.11, 1.12, 1.12, 1.13, 1.14, 1.15 and 1.16 provide examples of the diversity of the technology and its industrial applications. Figures 1.11 and 1.12 show the multitude of parts that are coated using thermal spray technology in a jet engine and the automobile industry [Walser (2003)]. Figure 1.13 shows the spray deposition of a coating on a turbine blade and the assembly of coated blades in a gas turbine. Typically, such a blade is coated with at least two layers of different materials. Figure 1.14 illustrates the coating of a large drum used in the paper manufacturing industry, and Fig. 1.15 shows spray-coated spindles used in the textile industry, both coated for reducing wear. Figure 1.16 shows a plasma-coated hip implant which represents a rapidly

Table 1.1 Estimated thermal spray market size in 2003 as compiled by J. Read (2003), reproduced with kind permission of ITSA-Spraytime

OEM/end users	1400 M$ US
Large coating service companies	800 M$ US
Small coating companies	600 M$ US
Powder/equipment sales	700 M$ US
Estimated total market	3500 M$ US

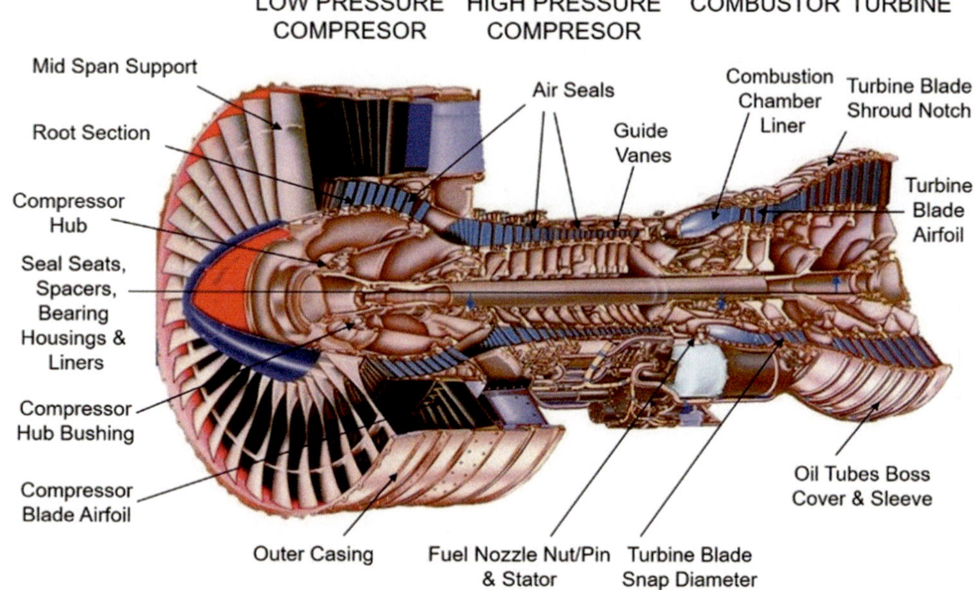

Fig. 1.11 Thermal-sprayed coatings on aircraft turbine engine parts. [Reproduced with kind permission of Oerlikon Metco Corp]

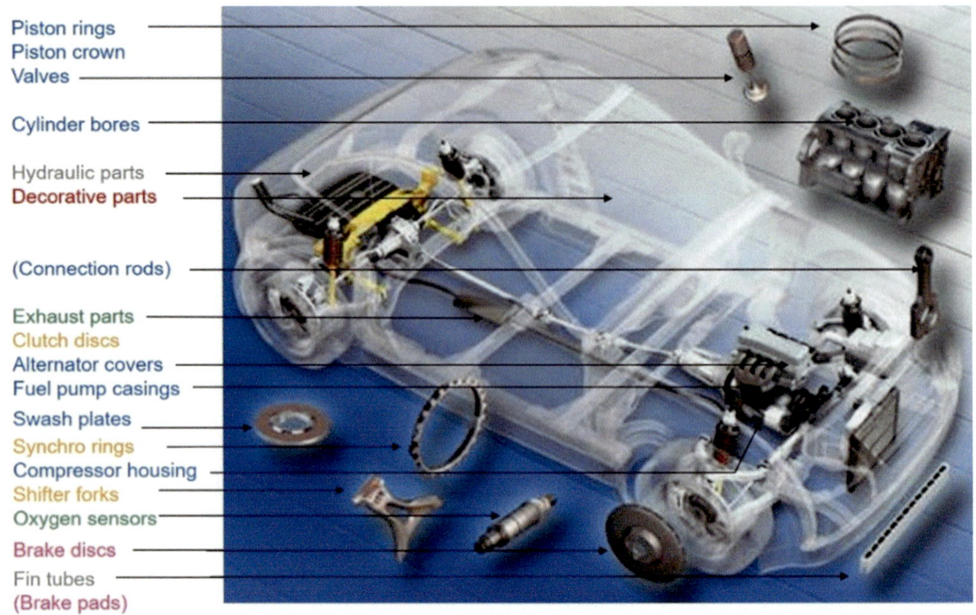

Fig. 1.12 Thermal spray components in the automotive industry. [Reproduced with kind permission of Oerlikon Metco Corp]

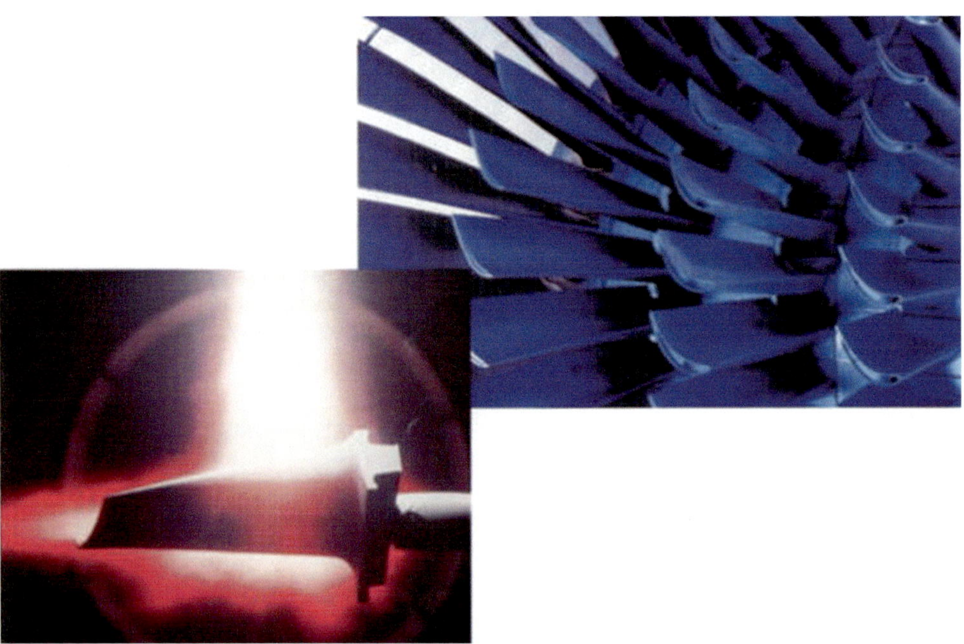

Fig. 1.13 Plasma-sprayed turbine blade coating. [Reproduced with kind permission of Oerlikon Metco Corp]

growing market. A detailed discussion of industrial applications of thermal spray and process economics is presented in Part IV of this book.

1.7   Overview of Book Content

Optimal integration of a thermal spray coating process into an industrial production line requires a fundamental understanding of the basic phenomena involved in the different stages of the process from the initial preparation of the coating material as a powder or wire to the final finishing step of the coated part. In this book, all issues related to producing a part that has the desired surface properties and performance characteristics are discussed. The book is divided into four distinct parts.

Part I covers essentially the fundamental aspects of the technology, with Chaps. 1 and 2 providing respectively brief introduction to thermal spraying and an overview of the vast field of surface modification technologies. Chapter 3 is

Fig. 1.14 Plasma spray applications in the paper and printing industry. [Reproduced with kind permission of Oerlikon Metco Corp]

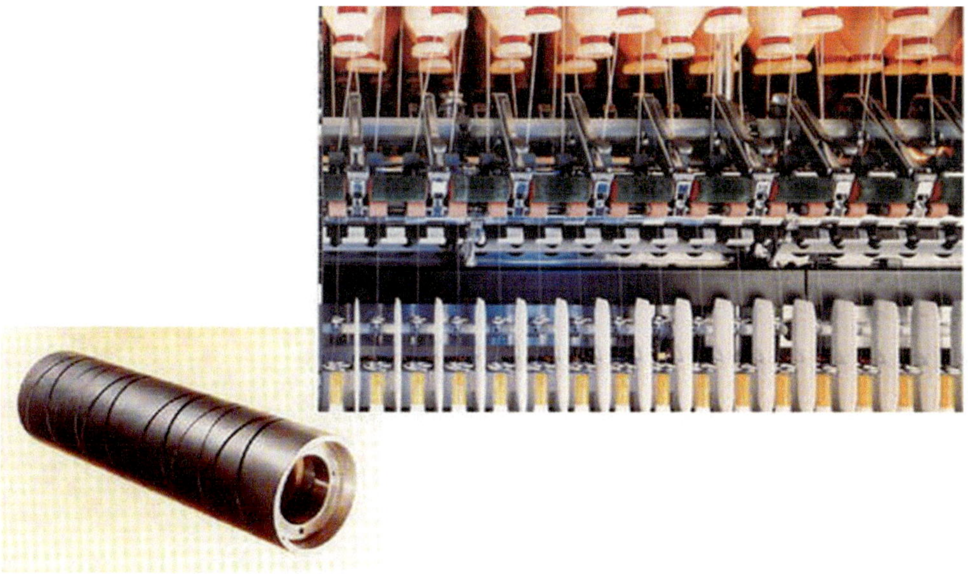

Fig. 1.15 Applications of plasma spray technology for the coating of spools and high wear parts in the textile industry. [Reproduced with kind permission of Oerlikon Metco Corp]

devoted a review of the fundamentals of combustion and thermal plasma generation and its thermodynamic and transport properties. Chapters 4 and 5 deal with the important field of transport phenomena under plasma conditions and plasma–particle interactions which are at the core of the thermal spray process in order to insure the proper entrainment and in-flight heating and melting of the coating materials in the combustion or plasma stream prior to its impact on the substrate. Examples are provided to understand and control the particle trajectories and their thermal histories in a flame or plasma-spraying operations.

Part II of this book is devoted to a detailed description of the different thermal spray deposition techniques. In Chap. 6, an overview is given of the "cold spray process" which while not being a "thermal spray process" is included in this book because of its complementarity to other thermal spray coating processes. Combustion-based thermal spray processes are described in Chap. 7, while atmospheric and vacuum DC and RF induction plasma-spraying technologies are described in great details in Chaps. 8, 9, and 10. Wire arc spraying (WAS) and plasma transferred arc (PTA) deposition are presented in Chaps. 11 and 12, respectively.

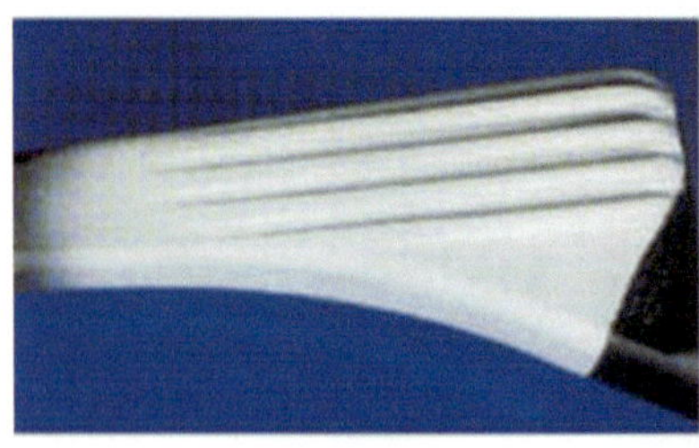
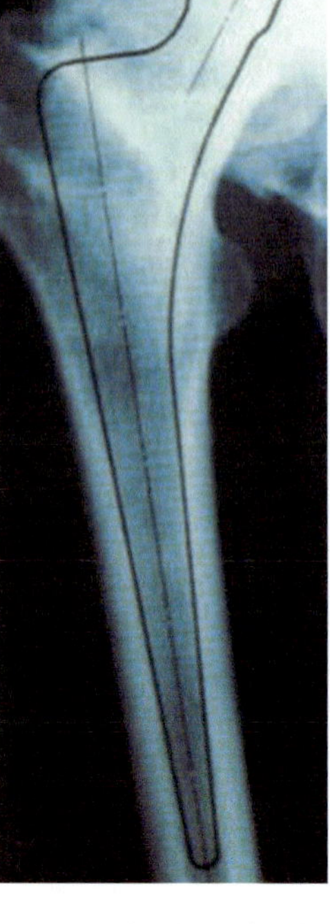

Fig. 1.16 Medical applications of plasma spray technology for the coating of hip and dental implants. [Reproduced with kind permission of Oerlikon Metco Corp]

Part III is devoted to a detailed discussion of coating formation and characterization, with Chaps. 13 and 14 dealing, respectively, with the important issues of preparation of the coating material to be used in the spraying process and substrate surface preparation. Over the past decade, significant progress has been made in understanding the details of the coating formation process, through modeling and sophisticated experimental studies. A detailed review of process fundamentals is provided in Chap. 15 including splat formation and coating buildup. The relatively new area of the spraying of nanostructured coatings using either agglomerated nanoparticles, suspension, or solution plasmaspraying techniques is discussed in Chap. 16. Chapter 17 is devoted to a review of various analysis techniques used to characterize the coatings and to describe both their material and functional properties.

Part IV, the final part of this book, is devoted to process integration and industrial applications. A description of ancillary equipment's such as spray booth design, powder feeders, and robotics for spray torch and substrate manipulation needed for the safe and efficient operation of the process in an industrial environment is presented in Chap. 18. This chapter also covers the important topic of health and operator safety concerns and a brief discussion of coating post treatment and finishing steps. Process instrumentation and online controls which have been the subject of significant developments in recent years are discussed in Chap. 19, while Chap. 20 is devoted to an overview of present and potential industrial applications of thermal spray technology including criteria for the selection of the coating process and materials as well as preliminary process economic analysis.

References

Barbezat, G. 2001. The internal plasma spraying on powerful technology for the aerospace and automotive industries. In *Proceedings of the International Thermal Spray Conference, (ITSC) Singapore, 2001*, ed. C.C. Berndt, K.A. Khor, and E. Lugscheider, 135–139. Materials Park, OH: ASM International.

Barbezat, G., and K. Landes. 2000. Plasma technology TRIPLEX for the deposition of ceramic coatings in the industry. In *Proceedings of the 1st. International Thermal Spray Conference (ITSC), Montreal, Canada*, ed. C.C. Berndt, 881–885. Materials Park, OH: ASM International.

Berndt, C.C. 2001. The origin of thermal spray literature. In *Proceedings of the International. Thermal Spray Conference, (ITSC) Singapore, 2001*, ed. C.C. Berndt, K.A. Khor, and E.F. Lugscheider, 1351–1360. Materials Park, OH: ASM International.

Bianchi, L., A. Denoirjean, F. Blein, and P. Fauchais. 1997. Microstructural investigation of plasma sprayed ceramic splats. *Thin Solid Films* 299: 125–135.

Bouyer, E., F. Gitzhofer, and M.I. Boulos. 1997. Suspension plasma spraying for hydroxyapatite powder preparation by RF plasma. 25 (5): 1066–1072.

Burgess, A. 2002. Hastelloy C-276 parameter study using the axial III plasma spray system. In *Proceedings of the International Thermal Spray Conference, (ITSC) Essen, Germany*, ed. E. Lugscheider, 516–551. ASM International.

Cruz-Barba, L.E., S. Manolache, and F. Denes. 2003. Generation of Teflon-likelayers on cellophane surfaces under atmospheric pressure non-equilibrium SF6-plasma environments. *Polymer Bulletin* 50: 381–387.

Dorfman, M.R. 2018. Thermal spray coatings. In *Handbook of environmental degradation of materials*, 3rd ed., 469–488.

Ducos, M., and J.P. Durand. 2001. Thermal coatings in Europe: A business perspective. In *Proceedings of the International Thermal Spray Conference, (ITSC) Singapore, 2001*, ed. C.C. Berndt, K.A. Khor, and E.F. Lugscheider, 1267–1271. Materials Park, OH: ASM International.

Dzulko, H., G. Forster, K.D. Landes, J. Zierhut, and K. Nassenstein. 2005. Plasma torch developments. In *Proceedings of the International. Thermal Spray Conference, (ITSC) Basel, Switzerland, DVS, Düsseldorf, Germany, 2005*. unpaginated CD.

Fauchais, P., G. Montavon, R.S. Lima, and B.R. Marple. 2011. Engineering a new class of thermal spray nano-based microstructures from agglomerated nanostructured particles, suspensions and solutions: An invited review. *Journal of Physics D: Applied Physics* 44: 093001. (53pp).

Fauchais, P., M. Vardelle, S. Goutier, and A. Ardella. 2015. Specific measurements of in-flight droplet and particle behavior and coating microstructure in suspension and solution plasma spraying. *Journal of Thermal Spray Technology* 24 (8): 1498–1505.

Fauchais, P., M. Vardelle, and S. Goutier. 2016. Latest researches advances of plasma spraying: From splat to coating formation. *Journal of Thermal Spray Technology* 25 (8): 1534–1553.

Fukumoto, M., H. Nagai, and T. Yasui. 2006. Influence of surface character change of substrate due to heating on flattening behavior of thermal sprayed particles. *Journal of Thermal Spray Technology* 15 (4): 759–764.

Gärtner, F., T. Stoltenhoff, T. Schmidt, and H. Kreye. 2006. The cold spray process and its potential for industrial applications. *Journal of Thermal Spray Technology* 15 (2): 223–232.

Gell, E., H. Jordan, M. Teicholz, B.M. Cetegen, N.P. Padture, L. Xie, D. Chen, X. Ma, and J. Roth. 2008. Thermal barrier coatings made by the solution precursor plasma spray process. *Journal of Thermal Spray Technology* 17 (1): 124–135.

Gitzhofer, F., E. Bouyer, and M.I. Boulos. 1997. *Suspension plasma spraying*. US Patent 5 609 921.

Goto, T., and H. Katsui. 2015. Chapter 9: Chemical vapor deposition of Ca–P–O film coating. In *Interface oral health science 2014*, ed. K. Sasaki et al., 8–9. https://doi.org/10.1007/978-4-431-55192.

Harder, B.J., D. Zhu, M.P. Schmitt, and D.E. Wolfe. 2017. Microstructural effects and properties of non-line-of-sight coating processing via Plasma Spray-Physical Vapor Deposition (PS-PVD). *Journal of Thermal Spray Technology* 26: 1052–1061.

Hatakeyama, R. 2017. Nanocarbon materials fabricated using plasmas. *Reviews of Modern Plasma Physics*: 1–7.

Hermanek, F.J. 2001. *Thermal spray terminology and company origins*. Materials Park: ASM International.

Irissou, E., J.-G. Legoux, A.N. Ryabinin, B. Jodoin, and C. Moreau. 2008. Review on cold spray process and technology: Part I—Intellectual property. *Journal of Thermal Spray Technology* 17 (4): 495–516.

Katsui, H., and T. Goto. 2017. Bio-ceramic coating of Ca–Ti–O system compound by laser chemical vapor deposition. In *Interface oral health science 2016*, ed. K. Sasaki et al. https://doi.org/10.1007/978-981-10-1560-1-4.

Knight, R. 2005. Thermal spray: Past, present and future. In *Proceedings of the ISPC-17, International Symposium on Plasma Chemistry, Toronto, Canada*.

Li, C.J., and J.-L. Li. 2004. Evaporated-gas-induced splashing model for splat formation during plasma spraying. *Surface and Coatings Technology* 184: 13–23.

Lima, R.S., and B.R. Marple. 2007. Thermal spray coatings engineered from nano-structured ceramic agglomerated powders for structural, thermal barrier and biomedical applications: A review. *Journal of Thermal Spray Technology* 16: 40–63.

Marantz, D., and D.R. Marantz. 1990. State of the art arc spray technology. In *Proceedings of the Thermal Spray Research and Applications, Proceedings of the 3rd. National Thermal Spray Conference, (NTSC) Long Beach, California, 1990*, ed. T.F. Bernecki, 113–118. Materials Park, OH: ASM International.

McCune, R.C., Jr., L.V. Reatherford, and M. Zaluzec. 1993. *Thermally spraying metal/solid lubricant composites using wire feedstock*. US Patent No. 5,194,304.

Moreau, C., P. Gougeon, A. Burgess, and D. Ross. 1995. Characterization of particle flows in an axial injection plasma torch. In *Proceedings of the 8th. National Thermal Spray Conference, (NTSC) Houston, Texas*, ed. C.C. Berndt and S. Sampath, 141–147. Materials Park, OH: ASM International.

Morf, E. 1912. *A method of producing bodies and coatings of glass and other substances*. UK Patent 28,001.

Ogawa, F., C. Masuda, and H. Fujii. 2018. In situ chemical vapor deposition of metals on vapor-grown carbon fibers and fabrication of aluminum-matrix composites reinforced by coated fibers. *Journal of Materials Science* 53: 5036–5050.

Ravi, B.G., S. Sampath, R. Gambino, P.S. Devi, and J.B. Parise. 2006. Plasma spray synthesis from precursors: Progress, issues, and considerations. *Journal of Thermal Spray Technology* 15 (4): 701–707.

Read, J. 2003. *Keynote address at the China International Thermal Spray Conference*. Proc., Dalian, China, 2003.

Schoop, M.U. 1910. *Improvements in or connected with the coating of surfaces with metal, applicable also for soldering or uniting metals and other materials*. UK Patent 5,712.

———. 1911. *An improved process of applying deposits of metal or metallic compounds to surfaces*. UK Patent 21,066.

———. 1915. *Apparatus for spraying molten metal and other fusible substances*. US Patent 1,133.

Singh, J., and D.E. Wolfe. 2005. Nano and macro-structured component fabrication by electron beam-physical vapor deposition (EB-PVD). *Journal of Materials Science* 40: 1–26.

Steffens, H.D., Z. Babiak, and M. Wewel. 1990. Recent developments in arc spraying. *IEEE Transactions on Plasma Science* 18 (6): 974–979.

Thull, R., and D. Grant. 2001. Chapter 10: Physical and chemical vapor deposition and plasma-assisted techniques for coating titanium. In *Titanium in medicine*, ed. D.M. Brunette et al. Berlin/Heidelberg: © Springer.

Vardelle, A., C. Moreau, N.J. Themelis, and C. Chazelas. 2015. A perspective on plasma spray technology. *Plasma Chemistry and Plasma Processing* 35: 491–509.

Vardelle, A., C. Moreau, et al. 2016. The 2016 thermal spray roadmap. *Journal of Thermal Spray Technology* 25 (8): 1376–1440.

Walser, B. 2003. The importance of thermal spray for current and future applications in key industries. *Spraytime* 10 (4): 1–7.

Wasserman, C., R. Boeckling, and S. Gustafsson. 2001. Replacement of hard chromeplating in printing machinery. In *Proceedings of the International. Thermal Spray Conference, (ITSC) Singapore, 2001*, ed. C.C. Berndt, K.A. Khor, and E. Lugscheider, 69–74. Materials Park, OH: ASM International.

Wewel, M., G. Langer, and C. Wassermann. 2002. The world of thermal spraying – Some practical applications. In *Proceedings of the International Thermal Spray Conference, (ITSC) Essen, Germany, 2002*, ed. E. Lugscheider, 161–164. Düsseldorf, Germany: DVS.

Zierhut, J., P. Haslbeck, K.D. Landes, B.G.M. Muller, and M. Schutz. 1998. TRIPLEX – An innovative three-cathode plasma torch. In *Proceedings of the International. Thermal Spray Conference, (ITSC) Nice, France, 1998*, ed. C. Coddet, 1374–1379. Materials Park, OH: ASM International. 1440.

Abbreviations

ALD	Atomic Layer Deposition
ALR	Atomizing gas and Liquid mass Ratio
APS	Atmospheric Plasma Spraying
CAPS	Controlled Atmosphere Plasma Spraying
CVD	Chemical Vapor Deposition
DC	Direct Current
D-Gun	Detonation Gun
DLC	Diamond-Like Carbon
EB-PVD	Electron Beam Physical Vapor Deposition
HF	High Frequency
HVOF	High-Velocity Oxy-Fuel
IPS	Induction plasma Spraying
LPPS	Low-Pressure Plasma Spraying
PA-CVD	Plasma Assisted Chemical Vapor Deposition
PE-CVD	Plasma Enhanced Chemical Vapor Deposition
PA-PVD	Plasma Assisted Physical Vapor Deposition
PLD	Pulsed Laser Deposition
PSZ	Partially Stabilized Zirconia
PTA	Plasma-Transferred Arc
PVD	Physical Vapor Deposition
RF	Radio Frequency
RGS	Ratio of Gas to Suspension volume flow rates
slm	Standard Liter per Minute
SMT	Surface Modification Technologies
TBC	Thermal Barrier Coating
VPD	Vapor Phase Deposition
VPS	Vacuum Plasma Spraying
WAS	Wire Arc Spraying
YSZ	Yttria Stabilized Zirconia

2.1 Introduction

Thermal Spray is one of the corner stones of the broad field of Surface Modification Technology. Its primary objective is the independent improvement of the surface properties of materials without compromising their thermophysical or structural properties. The technology allows the design-engineer a great flexibility in choosing materials of construction for their optimal mechanical and physical properties independent of their surface properties that could be subsequently adapted to process requirements through appropriate surface treatments. Numerous scientific papers, textbooks, and international conference proceedings have been published on Surface Modification Technologies (SMT). The textbook edited by Bach et al. (2006) titled *Modern Surface Technology* provides a particularly good overview of the field and its science base. As stated by Tillmann and Vogli (2006), in Chap. 1 of this book, "It is necessary that developers already consider surface requirement during concept phases, directly after taking down customer and market demands." Four fundamental aspects are identified for consideration in choosing a material and its surface properties.

- *Function*: What are the functional characteristics of the surface and existing requirement?
- *Purpose*: What are the objectives of surface modification, and which properties need to be maximized and what needs to be minimized?
- *Limitations*: What are the principal technical, economic, and environmental constraints that need to be met during life cycle of the part?
- *Options*: Which technologies are available that can respond to the design and engineering requirements while respecting economic and environmental limitations?

In most design situations, it is unlikely to have a perfect match of both bulk material and its surface properties with design requirements and economic constraints. Surface modification represents in these cases a viable alternative allowing for the independent modification of the surface properties without altering the bulk properties of the material.

© Springer Nature Switzerland AG 2021
M. I. Boulos et al. (ed.), *Thermal Spray Fundamentals*, https://doi.org/10.1007/978-3-030-70672-2_2

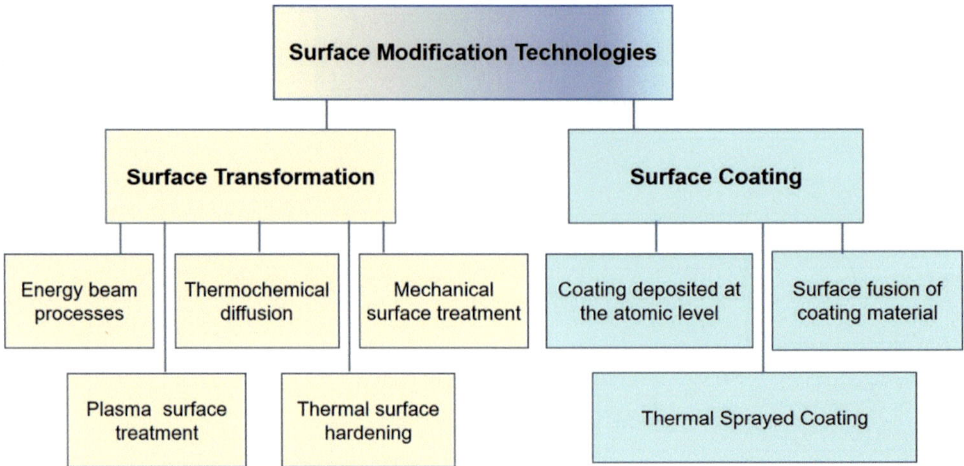

Fig. 2.1 Classification of surface modification technologies

A classification of the broad range of available surface modification technologies is given in Fig. 2.1. These can be classified under two broad groups of technologies: (i) *surface transformation*, through physical or chemical treatment, and (ii) *coating*, through the addition of a different material in the form of a thin, or thick, coating on the surface of the part.

Surface Transformation *Mechanical surface treatment* or *strain hardening*, in which the materials are submitted to plastic deformation processes by shot peening or water jet penning, rolling, or impact loading. The depth of hardened layer is typically below 1 mm for penning and rolling processes, but it can reach as high as 20 mm for impact loading.

- *Thermal surface hardening*, such as flame hardening, induction hardening, and case hardening, which is mostly applied to steels, through the heating of the steel to its austenitizing temperature followed by its rapid cooling at above a critical cooling rate. The depth of hardened layer is typically between 0.5 and 5 mm.
- *Thermochemical diffusion*, resulting in the diffusion of an external element such as carbon, nitrogen, boron, or niobium into the surface structure of the material at elevated temperatures. In the normal carburizing cycle, the thickness of the carburized layer can typical be in the range of 1–2 mm, depending on the carburization time, temperature, and carburizing gas (CO, CH_4, etc.).
- *Plasma and energy beam surface treatment*, using lasers, electron, and ion beams and low-pressure cold plasmas for the polymerization, carburization, or the nitridation of the surface of the material.

Coatings

- *Coating deposited at the atomic level*, often referred to as *thin coatings*, including plating, whether chemical or electroplating, Physical Vapor Deposition (PVD), and Chemical Vapor Deposition (CVD).
- *Thick coating built through the spraying of solid particles, molten droplets, or dispersed liquid*, against the substrate including cold spray and thermal spray processes such as Flame Spraying (FS), Plasma Spraying (PS), and Wire Arc Spraying (WAS).
- *Coating built through the surface fusing of a thick layer of the coating* material to the substrate including cladding, build-up brazing, and Plasma-Transferred Arc (PTA) deposition.

Considering that *surface transformation technologies* are well distant from the core field of this book, in this chapter we will be limited to a review of the main features of *coating techniques* used for surface modification and identify the main characteristics of each of these techniques with emphasis on atomic-level coating technologies used for the deposition of *thin coating*, *thermal-sprayed* coatings, and *fused coatings*, which are generally used for the deposition of relatively *thick coatings*.

The motivation for coating structural parts can be summarized by one or more of the following needs:

- Improve functional performance by allowing, for example, higher temperature exposure using Thermal Barrier Coatings (TBC).
- Improve component life by reducing wear due to abrasion, erosion, or corrosion.

- Extend the component life by rebuilding the worn parts to their original dimensions avoiding the need for replacing the entire component, for example, shafts or axles.
- Reduce component cost by improving functionality of a low-cost material with a high-performance coating.

A wide range of coating technologies have been developed in order to respond to these needs. These vary widely in terms of the following parameters:

- Coating material used, whether metals and alloys, ceramics, cermets, or polymers.
- Coating chemistry modification/contamination during the deposition process.
- Coating thickness—thin (<20 or up to 30 μm) or thick coatings (>50 μm).
- Coating density—fully dense or porous, depending on application.
- Coating adhesion to the substrate (chemical, mechanical, or fused).
- Tolerated substrate temperature (cold, mild, or high).
- Minimal post-deposition machining or treatment.

2.2 Coating Deposited at the Atomic Level

Thin film coatings with thicknesses of less than 20 or up to 30 μm are mostly obtained using atomic-level deposition technologies such as:

- Plating
- Physical Vapor Deposition (PVD)
- Chemical Vapor Deposition (CVD)

PVD and CVD belong to the general field of Vapor Phase Deposition (VPD) technologies. Considering that most *thin film* coating technologies require a reduced pressure environment, they are generally expensive and impose limits on the size and shape of the part to be coated. The exceptions are the use of super thin films for the surface modification of large polymer sheets using atmospheric pressure nonequilibrium discharges [Hatakeyama (2017)] or Teflon-like layers on cellophane surfaces [Cruz-Barba et al. (2003)]. In the following, a brief review is presented of the most widely used thin film coating techniques. Extensive reviews on the subject are available in the open literature [Bunshah et al. (1982), Lang et al. (1983), Dobkin and Zuraw (2003)], and more recently in the *Handbook of Thin Film Technology* by Frey and Khan (2013).

2.2.1 Plating

Plating is one of the classical surface modification technologies that have been used for centuries for the deposition of thin metallic films on the surface of metals and polymers for functional or decorative purposes. These can be grouped under two broad categories, *electroless* and *electroplating*.

2.2.1.1 Electroless Plating
Electroless plating is one of the oldest of these techniques that can be traced back to ancient civilizations for the deposition of a continuous thin films of precious metals, such as silver and gold, on a nonconductive surface without the use of an electric current. It is based on a pure chemical treatment in a liquid, aqueous, or nonaqueous solution, with the deposited material reduced from its ionic state by means of a chemical reducing agent. An excellent review on the subject is given by Djokic and Cavallotti (2010). The part to be coated generally needs to be pretreated with chemicals to remove oils and other contaminants. It is then activated with an acid etch or an equivalent treatment. The application of anti-oxidation chemicals completes the process, rendering the surface of the component resistant to corrosion and wear. Alternately, in phosphating procedure, an aqueous solution of a heavy metal salt, primarily phosphate, plus free phosphoric acid is reacted with a metal surface to produce an adherent coating of an insoluble complex phosphate on the metal surface.

Two main mechanisms are used for the electroless deposition of metals: *displacement deposition* and *autocatalytic deposition*. The former, also known as galvanic or immersion plating, involves the immersion of a less noble metal in a solution containing ions of a more noble metal. The less noble metal acts as reducing agent with the more noble metal acting as oxidizing agent. The transformation leads to the reduction of the more noble metal and its deposition in the form of a thin dense coating on the surface of the less noble metal. Typical examples of this type of deposition are Ag/Zn, Au/Ni, Au/Ag, Cu/Fe, Cu/Al, Pd/Ni, and so on. A*utocatalytic deposition*, on the other hand, makes use of various reducing agents, such as formaldehyde, hydrazine, polyhydroxy alcohols, and hydrogen, for the deposition process. The nature of the reducing agent can have a significant influence on the reaction kinetics, the surface morphology of the coating, and its physicochemical properties.

Electroless deposition processes have been the subject of extensive studies for numerous industrial applications. Nickel, copper, silver, gold, and related alloys are among the most investigated elements for their possible use in electronics, electromagnetic shielding, magnetic materials, energy devices, and biomedical applications. According to Djokic and Cavallotti (2010), almost all the metals and/or alloys that can be electrodeposited from aqueous solutions can also be electrolessly deposited under proper conditions. Recent developments in this field involves the co-deposition of particulate matter or substance within the growing film, which led to a new generation of electroless composite coatings, many of which possess excellent wear and corrosion resistance properties [Argarwala R. and V. Argarwala (2003)].

Hot dip coating [Maas et al. (2011)], on the other hand, is based on the dipping of the part to be coated into a molten bath of the coating material. Metals used, generally to protect steels from corrosion, are low-melting temperature ones such as zinc, zinc alloys, aluminum, aluminum alloys, and lead–tin alloys. Zinc is the most widely used material for the production of "galvanized steel." Metal coatings obtained with hot dipping techniques have thicknesses of the order of 10 µm or more. Their advantages are that they are omnidirectional, that is, all surfaces exposed to the liquid are coated, including blind holes where the liquid can penetrate; the substrate is kept at a relatively low temperature; and their associated capital and operation costs are relatively low.

2.2.1.2 Electroplating

Electroplating is an alternate thin film deposition technique that represents a natural extension of electroless techniques. The principal difference is in the use of a direct electric current for the reduction/oxidation reaction in the electrolyte, which is achieved through the immersion of two electrodes in the electrolytic cell. As described by Schwartz (1982), the electrical transfer in the solution is carried out by the ions. Positive ions (cation) travel toward the negative electrode (cathode) while negative ions (anion) travel toward the positive electrode (anode). As represented by Eqs. 2.1 and 2.2, respectively, the anodic reactions are oxidation reactions by which electrons are liberated and the valence state of the ions increase, while the deposition reactions at the cathode are characterized as reduction reactions since electrons are consumed and the valence state of the ions involved are reduced.

$$\text{at the anode}\quad M^{o} \;\rightarrow\; M^{n+} + ne \qquad (2.1)$$

$$\text{at the cathode}\quad M^{n+} + ne \;\rightarrow\; M^{o} \qquad (2.2)$$

Where, n is the valence state of the metal ion. For the case of electroplating with nickel or copper in an acidic solution, the corresponding ions are Ni^{2+} and Cu^{2+} giving rise to the value of $n = 2$ in the above equations. Each set of reactions represents half-cell reactions and proceeds independently of the other, limited by a condition of material balance. The process involves the desolation in the acidic electrolyte of the

coating material at the anode, and its deposition from the electrolyte on the surface of the part to be coated, which acts as the cathode.

Electroplating from solutions in which the metallic ions are combined with other complex ions involves a more complex mechanism. Cyanide-containing electrolytes represent one of the most common electrolytes used in electroplating. Deposition of the metal ion at the cathode occurs in this case directly from the complex ions. Examples for such reduction reactions for the deposition of copper, silver, and gold can be represented by the following reactions:

$$\text{Copper}\quad \left[Cu(CN)_3\right]^{2-} + e \;\rightarrow\; Cu + 3\ CN^{-} \qquad (2.3)$$

$$\text{Silver}\quad \left[Ag(CN)_2\right]^{-} + e \;\rightarrow\; Ag + 2\ CN^{-} \qquad (2.4)$$

$$\text{Gold}\quad \left[Au(CN)_2\right]^{-} + e \;\rightarrow\; Au + 2\ CN^{-} \qquad (2.5)$$

It should be noted that the above-listed complex ions are anionic, and accordingly would migrate toward the anode during the electrolysis. Yet, the reduction reactions listed above (Eqs. 2.3, 2.4 and 2.5) takes place at the cathode, which implies convective and diffusive transfer of the anions toward the cathode.

A relatively recent review of the basic principles, processes, and practice of electroplating is given by Kanani (2004). Electrodeposits are applied to metal substrates for decorative purposes or the enhancement of corrosion resistance, wear resistance, electrical properties, magnetic properties, or solderability [Fayyad et al. (2018), Wu L.P. et al. (2018)]. It is to be pointed out, however, that the main disadvantage of electroplating in general, and particularly hard chrome electroplating technology, is due to serious environmental concerns, and health-endangering problematic issues resulting from the use of hexavalent chromium in the plating bath [Von Burg, R., Liu, D. (1993), Costa, M. Dayan (1997), Dayan A.D., Paine, A. J (2001)]. For this reason, the use of hard chrome plating technology is increasingly subject to stricter rules and restrictions, which make it more challenging and costly [Directive 2002/95/EC of the European Parliament (2003)], with alternate approaches such as thermally sprayed wear-resistant coatings gaining wider acceptance.

2.2.2 Physical Vapor Deposition

Physical Vapor Deposition (PVD) is one of the main Vapor Phase Deposition (VPD) technologies used for the deposition of thin film with thicknesses varying from a few Angstrom up to hundreds of micrometers. The technology is very versatile enabling the deposition of virtually any inorganic material including metals, alloys, compounds, and their mixtures as well a limited number of organic materials for a wide range of applications mostly in the microelectronics industry, and advanced surface coating/modification. It is a line-of-sight process operated at low pressures typically in the 10^{-3} to 10^{-4} Pa range, which also implies relatively low depositing rates in the range of 0.1–10 µm/min. The process involves essentially the following three principal steps:

- Generation of the depositing species in the vapor phase by either evaporation or sputtering.
- Transport/migration of the species from the source to the substrate.
- Deposition of the film onto the substrate by nucleation and growth processes.

A wide range of technologies was developed varying in terms of the source used for the generation of the depositing species in vapor phase, and the mechanism of its migration to the substrate. These include:

- Evaporation by resistive heating
- Electron beam evaporator
- Ion beam evaporator (ion plating)
- Sputtering
- Plasma-Assisted Physical Vapor Deposition (PA-PVD)
- Pulsed Laser Deposition (PLD)

Highlight of some of these techniques are described. These are limited to a description of the basic concepts involved and its principal characteristics.

2.2.2.1 Evaporation by Resistive Heating

Vapors are produced by heating the precursor material in the form of chunks, wires, or foils in an open boat or suspended to a heated wire as illustrated schematically in Fig. 2.2. These types of evaporating sources have a limited capacity and small thermal mass allowing for the rapid heating-up and cooling-down of the material reducing the time of the deposition cycle. The absence of careful preheating and degassing of the evaporator does not allow for the close control of the quality of the film deposited. The coating formed on a plane surface from a single source can have significant variations of its thickness depending on the distance between the source and the substrate and the angle of incidence of the vapor stream toward the substrate surface. An example of the principal components in a typical evaporation PVD resistive-heated system is given in Fig. 2.3. This constitutes mainly of a vacuum chamber equipped with an appropriate vacuum pumping system and controls, an access door, one or more types of evaporators, a substrate holder, and a rate monitoring device for process monitoring and control. A shutter system, not illustrated, is often used to cover the vapor source and the

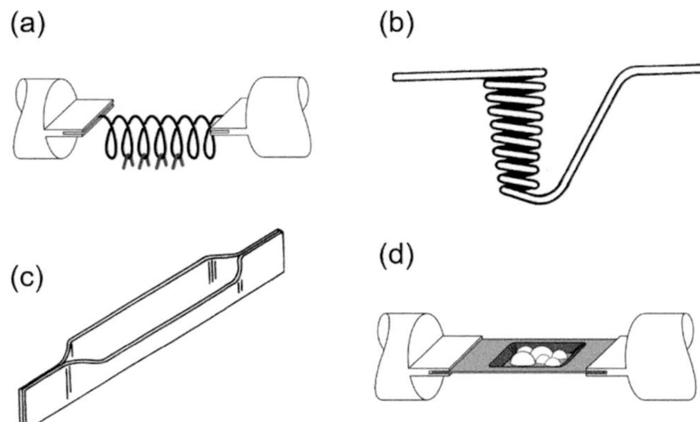

Fig. 2.2 Schematics of typical open-source evaporators: (**a**) wire-coils; (**b**) boat

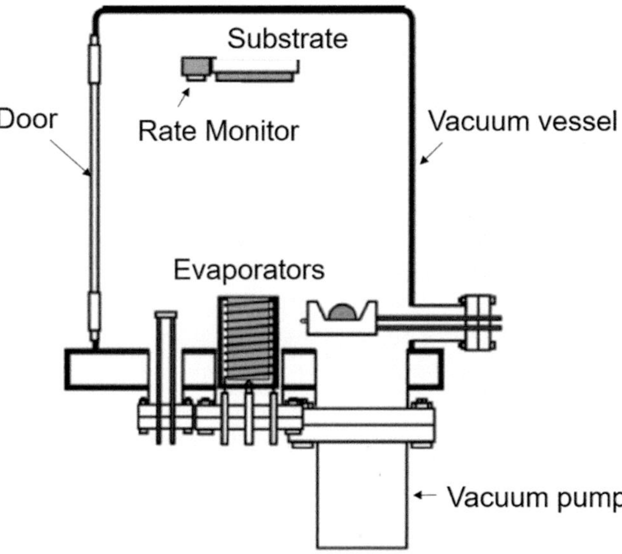

Fig. 2.3 Principal components of a typical thin film evaporation/deposition system

substrate during the stabilization period of the evaporation conditions.

A detailed analysis of the vapor flux distribution from a cylindrical evaporation cell, reported by Luscher et al. (1979), is given in Fig. 2.4. Details of the evaporator cell design is shown in Fig. 2.4a. The corresponding atomic flux distribution as function of the cell diameter and the level of the material in the cell is given in Fig. 2.4b. The results show that for a completely full cell, evaporated material leaves the surface in a roughly cosine distribution, which implies that the vapor flux density distribution relative to the surface normal varies according to the cosine of the angle from that normal. It may be noted that at $60°$ to the normal, the vapor flux density drops to 50% of its maximum value at $\theta = 0$. To achieve homogeneous thickness several sources must be used, or the substrates have to be subjected to an oscillatory

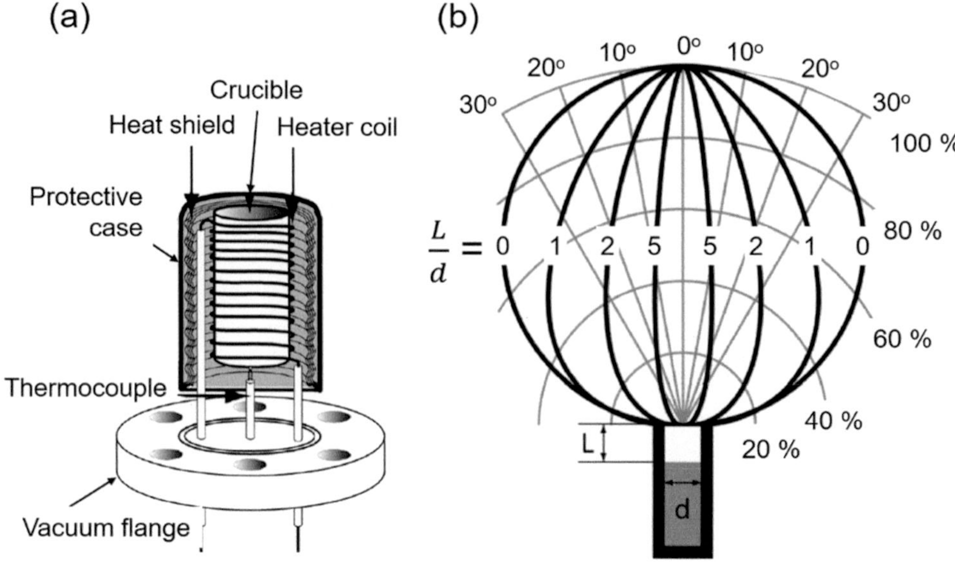

Fig. 2.4 (**a**) Simplified schematic of a typical evaporation cell. (**b**) Flux distribution of atoms as function of the cell diameter and depth [Luscher et al. (1979)]

movement. Resistively heated open-source evaporation cells are generally used for high-rate, low-volume coatings with little sensitivity to either microstructure or contamination and where the coating thickness is not critical.

2.2.2.2 Electron and Ion Beam Vacuum Evaporator/Coating Systems

The basic components of an electron beam vacuum evaporating system used in PVD applications are illustrated in Fig. 2.5. Superposed on the figure is the vapor flux profile at 200 mm from the source. The observed large variations in vapor flux are directly responsible for relatively important variations of the deposit thickness with distance from center-line of the source with the maximum deposit thickness directly above the source. The problem is generally overcome by impairing a relative motion to the substrate with respect to the source in order to even out the vapor flux distribution on the surface to be coated. Alternately, a gas stream is introduced in the chamber causing the dispersion of the vapor species that undergo multiple collisions during their path from the source to the surface of the substrate. Obviously in larger systems, the use of multiple evaporation sources strategically placed in the vacuum chamber would also help in achieving a relatively uniform vapor flux distribution toward the substrate.

In a corresponding ion beam deposition process illustrated in Fig. 2.6, often referred to as *ion plating*, the material is evaporated in a manner similar to the electron beam evaporator with the exception that the vaporized material passes through a gaseous glow discharge on the way to the substrate thus ionizing some of the vapor atoms. The glow discharge is produced by admitting a gas, usually argon, into the

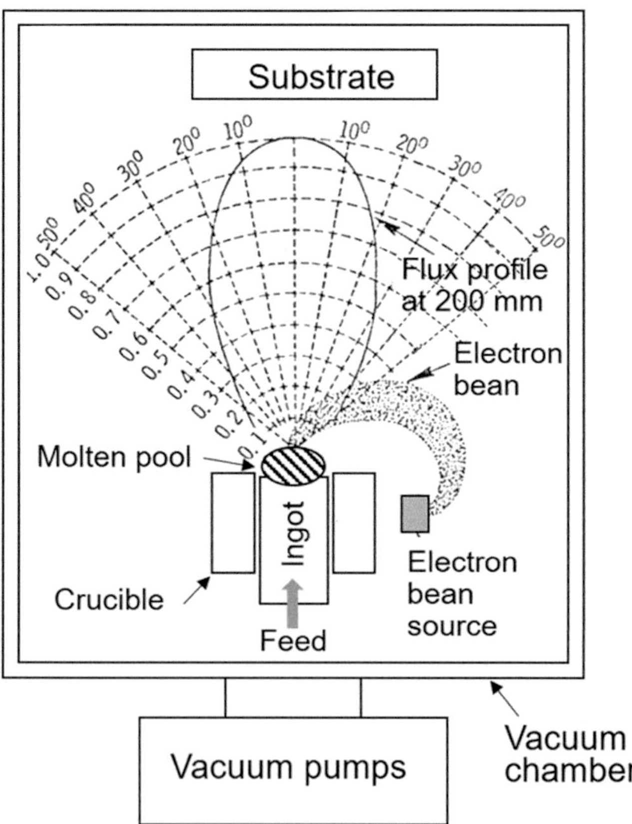

Fig. 2.5 Principal components of a typical electron beam evaporation/deposition system [Bunshah RF (1982)]

evaporation chamber and biasing the substrate to a high negative potential of 2–5 kV. Such a system offers the main advantage of continuous cleaning of the substrate as a result

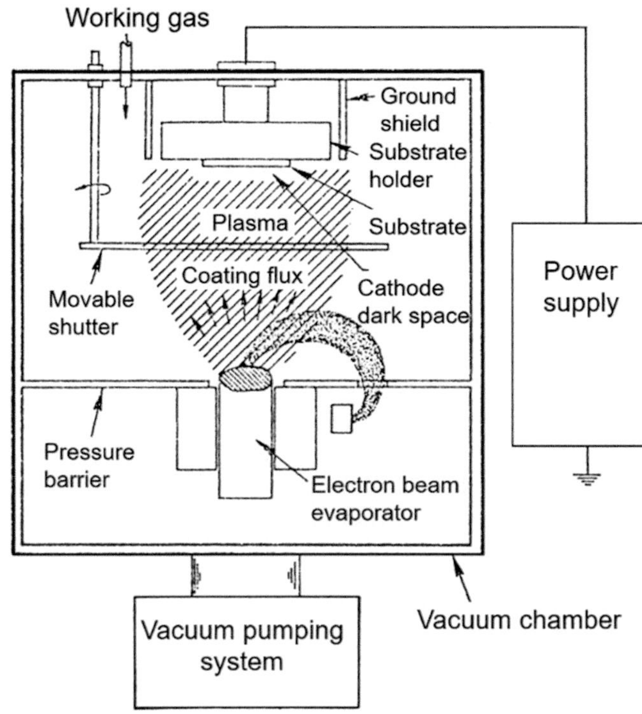

Fig. 2.6 Principal components of a typical ion-plating evaporation/deposition system [Bunshah RF (1982)]

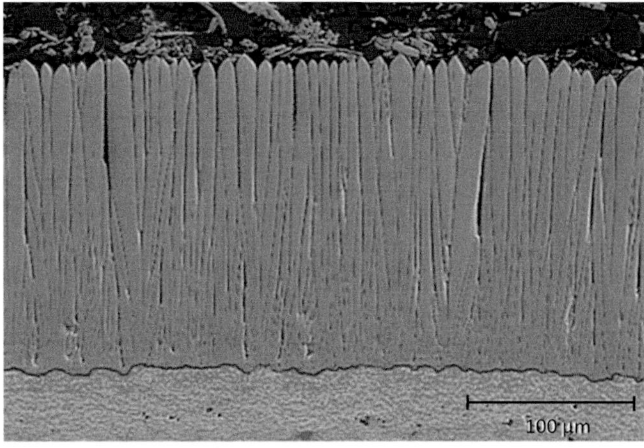

Fig. 2.7 Typical EBPVD 7YSZ TBC microstructure in cross section (courtesy of U. Schulz, DLR, Inst. of Materials Research, Cologne, Germany) [Niessen and Gindrat (2011)]

of its bombardment by the high-energy gas ions, giving rise to a better adhesion and lower impurity content of the deposit on the substrate. On the other hand, ion bombardment can also cause microstructure modification and residual stresses in the coating, undesired heating of the substrate, and a reduction in the deposition rate due to the sputtering-off of some of the formed coating.

The preparation of Diamond-Like Carbon (DLC) thin films by Shielded Arc Ion Plating (SAIP) was demonstrated by Taki et al. (1997). These are remarkably hard a-C:H films formed of hydrogenated amorphous carbon (a-C:H) network where both sp^2 and sp^3 carbon microdomains are randomly linked, with the sp^3 domains predominant. Such films are very useful because of their significant properties, such as high hardness, electrical resistivity, optical transparency, and chemical inertness. In their study, Taki et al. (1997) inserted a shielding plate between the target and the substrate. SAIP is simpler than filtered cathodic arc evaporation to prepare DLC films.

Electron Beam Physical Vapor Deposition (EB-PVD) of relatively thick zirconia–yttria Thermal Barrier Coatings (TBC) was developed in the early 1970s at Pratt & Whitney. The first tests on turbine blades were published in 1976. According to the reviews of Miller (1987, 1997), the burner rig lives of early EB-PVD coatings were reported to have been far better than those of dense TBCs obtained by standard Atmospheric Plasma Spraying (APS). According to

Sing et al. (2002, 2004); Mauer et al. (2013a,b, 2015) and Niessen et al. (2010), (2011), ZrO_2–6 to 8 wt% Y_2O_3 ceramic layers deposited by EB-PVD on soft-vacuum, plasma-sprayed, MCrAlY (M = Ni, or Co or NiCo) bond coats, showed considerably better performance compared to standard APS deposited on the same bond coat. The columnar microstructure specific to EB-PVD coatings (Fig. 2.7) (Niessen et al. 2011) seems to impar excellent strain tolerance to the material and the possibility to support frequent cycling. Moreover, these coatings presented an excellent adherence to smooth surfaces, had a relatively smooth surface finish, and the ability not to clog fine cooling holes when deposited. EB-PVD coatings also allowed operation at higher engine temperatures, of the order of 1200 °C at the surface of the ceramic coating [Vidal-Setif et al. (2012)], which represented the best results for TBCs. The relatively high investment cost, however, of EB-PVD installations combined with the limitations on the size of the part to be coated, compared to standard APS coatings systems, represents an important challenge for the general acceptance of EB-PVD on an industrial scale.

2.2.2.3 Sputtering

In contrast to resistive and EB-PVD processes, target evaporation in sputtering process, as schematically illustrated in Fig. 2.8, is carried out by positive ions (usually argon) generated in a glow discharge at a pressure in the range of 0.13–13 Pa. These are used to bombard the negatively biased target material, which acts as cathode, dislodging groups of atoms that pass into the vapor phase and deposit on the substrate. The energy of particles ejected from the target and impacting the substrate can reach hundreds of eV (up to 1000 eV) creating adherent and dense coatings. The densities of sputtered deposits are dependent on substrate temperature and deposition rate. The introduction of a magnetic field under the target intensifies the sputtering process by increasing the electrons' density and thus that of ions in the area where the magnetic field is parallel to the target surface: magnetron sputtering.

Since sputtering processes cannot be used with nonconductive materials such as refractory carbides, nitrides, and oxide, reactive Direct Current (DC) or Radio Frequency (RF) sputtering must be used. In this case sputtered metallic particles are made to react with the corresponding gas to form the desired ceramic (for example, Al particles sputtered react with oxygen). To improve the reaction an RF coil is disposed close to the reactive gas injection close to the substrate. However, collision frequencies must be increased to promote reactions and for that the chamber pressure must be increased (from 0.1 to 1 Pa). To increase the deposition rate, arcs are also used. A laser flash or a high-voltage flashover arc initiates cathodic arc PVD between the cathode (the material to be evaporated) and an anode (the chamber wall in many cases). A stream of electrons (a few tens to hundreds A) is generated and flows through the cathode, melting locally at numerous cathodic spots on the cathode surface. Cathodic arc PVD is used for the evaporation of metals (Ti, Al, Cr, and their alloys) but also carbon for hard DLCs. The addition of reactive gases permits deposition of nitrides and carbides. This technique is increasingly used in industry to coat all types of parts including turbine blades, automobile engine components, hydraulic components, medical tools and instruments, as well as a wide range of decorative coatings. Figure 2.9 presents schematic diagrams of PVD coating systems using either magnetron sputtering or evaporation [Baptista et al. (2018). Magnetron sputtering technique was shown to produce smooth surfaces using lower temperatures and presenting excellent mechanical and tribological properties together with good adhesion to the substrate.

According to Baptista et al. (2018), one of the industry concerns is to optimize the energy consumption of the PVD process, which is relatively high compared to alternate coating techniques. Their studies show that the reduction in the energy consumption in the magnetron processes must focus

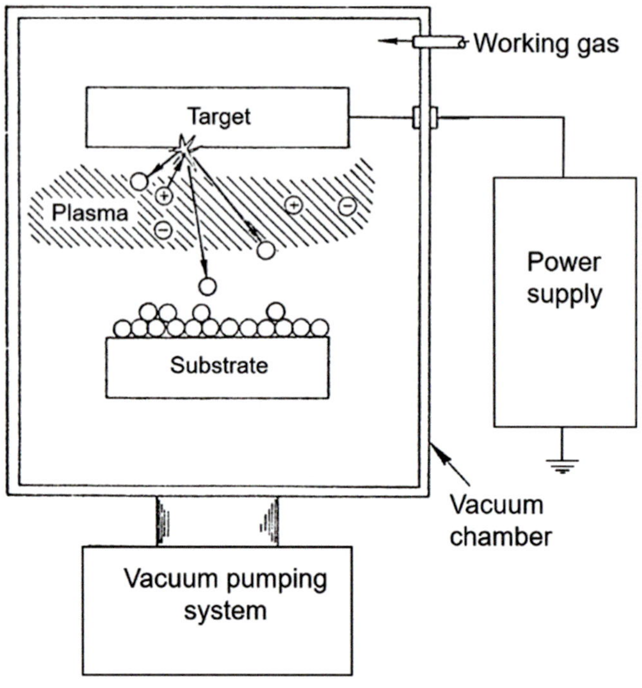

Fig. 2.8 Principal components of a basic sputtering system [Bunshah RF (1982)]

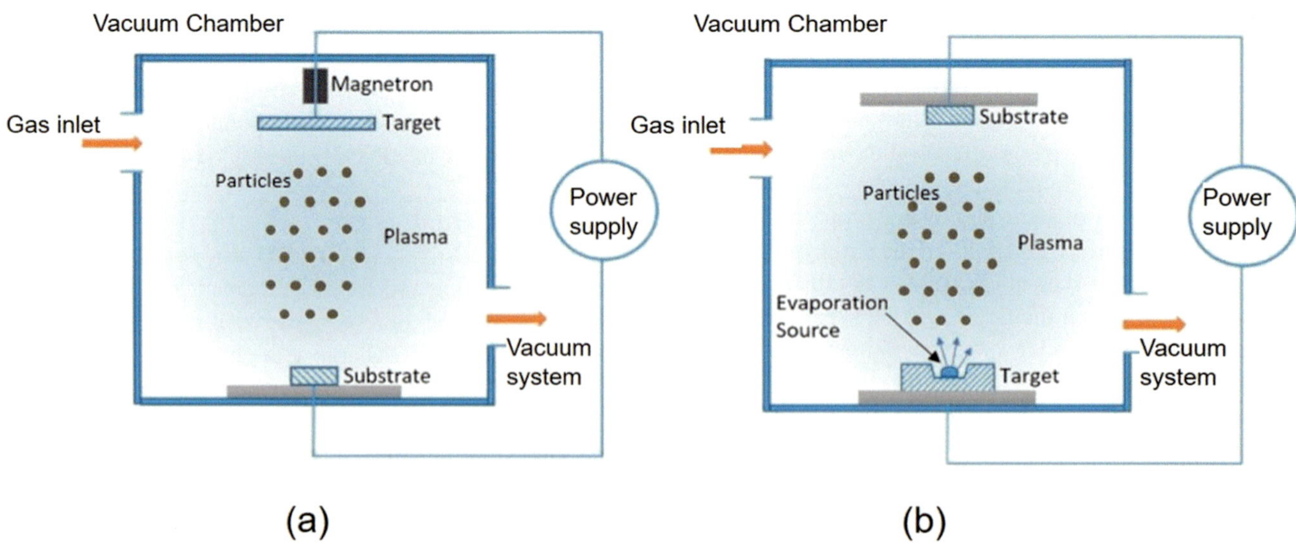

Fig. 2.9 Schematic diagram of the PVD by (**a**) sputtering and (**b**) evaporation method [Baptista et al. (2018)]

on the coating step since it is one of the largest energy-consuming steps in the whole process. DC sources continue to be one of the most widely used sources in the sputtering magnetrons. RF sources are less used although the appearance of the Medium Frequency (MF) source has brought new opportunities, allowed the combination of the two sources, and led to the appearance of new techniques such as Dual Anode Sputtering (DAS) that allow for reducing the heat load in the substrate.

Bobzin et al. (2018) point out that in PVD processes the control of the substrate temperature has a major impact on the morphology of the coating and its adhesion to the substrate. Substrate temperature is limited, on the other hand, by the maximum permissible value T_{max}. Recently developed industrial coating units, such as CC800/9 HPPMS (CemeCon AG, Würselen, Germany), have a built-in substrate temperature monitoring system that allows for a close control of the heating power depending on the measured substrate temperature, which allows in turn for the successful deposition of CrAlN6/3/4 and CrAlN8/8/4 coatings.

2.2.2.4 Pulsed Laser Deposition

Pulsed laser deposition (PLD) [Gladush et al. (2011)] uses a high-power laser beam (mostly an excimer laser with a wave length in the range of $200 < \lambda < 400$ nm) focused periodically onto a target material with laser pulses of a few tens of ns and frequencies of a few tens of Hz. Locally the target material is heated by the absorbed laser energy; it evaporates or sublimes, and if the laser energy flux is sufficient, the gas is transformed into a dense plasma inside the plume of the evaporated material. Resulting atoms, electrons, and ions are driven away from the target at high speeds into the vacuum under controlled conditions and deposit on the surface of the substrate leading through nucleation and growth to the formation of a thin film with the same chemical composition as that of the evaporated material. Additionally, gases can be introduced inside the reaction chamber to react with the atoms or molecules from the plume and modify the chemical composition of the material deposited on the substrate surface. As already pointed out in PVD processes the increase in the pressure in the deposition chamber confines the vapor plume expansion allowing for a homogeneous deposition over larger areas of the substrate surface. The process, being a line of sight with a cone angle of about $10°$ above the laser impact point, is not well adapted for coating large surfaces and it is generally limited to electronics or optical applications with coating thicknesses below a few μm. Nelea et al. (2002) have investigated the ability of PLD to produce crystalline and mechanical resistant calcium phosphate, including hydroxyapatite films.

2.2.3 Chemical Vapor Deposition

Contrary to PVD processes in which the precursor is evaporated from its solid elemental state using intense energy sources, such as resistive heating and electron or laser beams, Chemical Vapor Deposition (CVD) [Dobkin et al (2003)] is based on the vapor phase decomposition of a volatile form of the precursor such as organo-halogenic compounds, such as chlorides, fluorides, bromides, iodides, or phosphates, hydrocarbons, and ammonia complexes. The gaseous precursor is thermally decomposed to produce the coating on the component surface, which is usually at high temperatures (800–1100 °C), thus limiting the choice of the substrate material. When exothermic chemical reactions occur at temperatures lower than the decomposition temperature of the coating precursor, the substrate surface temperature is correspondingly lowered. The process, working at pressures between 13 and 100 kPa, is omnidirectional and gaseous mixture can easily penetrate into blind holes and deposit the coating, even on the internal surfaces of porous bodies. When using organo-metallic precursors, the part temperature can be lowered down to 400–600 °C and the pressure range varies from 0.13 and 105 Pa. Pure metals, alloys, intermetallics, borides, silicides, carbides, oxides, and sulfides can be deposited using CVD. The control of the coating microstructure depends on the temperature and the supersaturation of the precursor (ratio of the local effective concentration of the deposited material to its equilibrium concentration). With the high temperature of the substrate, diffusion between coating and substrate occurs easily resulting in a good diffusion bonding of the coating to the substrate.

During this process, thin film coatings are formed as a result of reactions between various gaseous phases and the heated surface of substrates within the CVD reactor. As different gases are transported through the reactor, distinct coating layers are formed on the tooling substrate. For example, TiN is formed as a result of the following chemical reaction:

$$TiCl_4 + N_2 + 3H_2 \xrightarrow{1000°C} TiN + 4\,HCl + H_2 \qquad (2.6)$$

Titanium Carbide (TiC) is formed as a result of the following chemical reaction:

$$TiCl_4 + CH_4 + 3H_2 \xrightarrow{1030°C} TiC + 4\,HCl + H_2 \qquad (2.7)$$

The final product of these reactions is a hard, wear-resistant coating that exhibits a chemical and metallurgical bond to the substrate.

To coat small components (3 μm to 3 mm), fluidized beds are used. The bulk density of the bed can be varied between 1800 and 19,000 kg/m^3 and it can be easily adjusted to that of the parts to be treated, and the reactive gases circulate in the bed kept in a furnace at the right temperature for the CVD reaction to occur.

To work at lower temperatures (25–400 °C), the decomposition of the precursor is achieved using Plasma-Enhanced Chemical Vapor Deposition (PE-CVD), or plasma-assisted chemical vapor deposition (PA-CVD) [Glocker et al. (2002), Mahan (2000)], which allows using all types of substrates including polymers with deposition rates higher than those of conventional thermal CVD. The energy necessary for the activation of the decomposition reaction of the precursor is supplied by means of plasmas produced by either glow discharges (DC voltage, pulsed DC voltage, radio frequency) or microwaves. For example, it is possible to achieve hard coatings of amorphous carbon (DLC: Diamond-Like Coatings) at temperatures below 200 °C with pulsed or RF discharges using acetylene as precursor. The pressure in these cases has to be lowered to 0.13 Pa and 13 kPa thus reducing significantly the deposition rate and thus restricting the use of technology to thin films (below 50 μm).

A typical example of a CVD reactor is shown in Fig. 2.10a, which they used to develop the so-called Atomic Layer Deposition (ALD), which is also a chemical gas phase thin film deposition technique, but unlike CVD it utilizes "self-limiting" adsorption reactions to control the thickness of the deposited films. For example, a typical CVD reaction is

$$\text{surface} + X + Y \;\rightarrow\; \text{surface} : XY \qquad (2.8)$$

where X and Y precursors (reactants) enter into the chamber as gases in which they react on a heated substrate to produce a film and/or can also react just above surface. To produce a film, ALD breaks up the CVD chemical reaction into two self-limiting "half reactions," A and B:

$$\text{Reaction (A):} \qquad \text{surface} + X \rightarrow \text{surface} : X \qquad (2.9)$$

$$\text{Reaction (B):} \qquad \text{surface} : X + Y \rightarrow \text{surface} : XY \qquad (2.10)$$

Each "half reaction" is sequenced one at a time followed by a purge cycle that pumps away any unreacted or excess reactant in the chamber. This allows for cycling of the half reactions: pulse A, purge A, pulse B, purge B, and then repeated for N amount of cycles. This sequencing results in thin film deposition. ALD cycle times range from sub-second to many seconds depending on sample geometry and chemistry. Figure 2.10b shows a bright field Cross-Sectional Transmission Electron Micrograph (XTEM) of a ~140 nm thick ALD ZnO/ZrO$_2$ formed of 8-bilayer nanolaminate coating on silicon substrate with its ~2 nm thick native SiO$_2$ layer. Platinum was deposited to protect the coating from cross-sectional milling.

2.2.3.1 Low-Pressure Chemical Vapor Deposition

Low-Pressure Chemical Vapor Deposition (LP-CVD) is a chemical vapor deposition technology that uses heat to initiate a decomposition reaction of the precursor gas on the solid substrate forming the solid phase thin coating. Low Pressure is used to decrease any unwanted gas phase reactions and improve the coating uniformity over the substrate. Low-pressure CVD can also be used for the control of the grain structure of the coating through proper optimization of the growth parameters. For example, controlled growth of

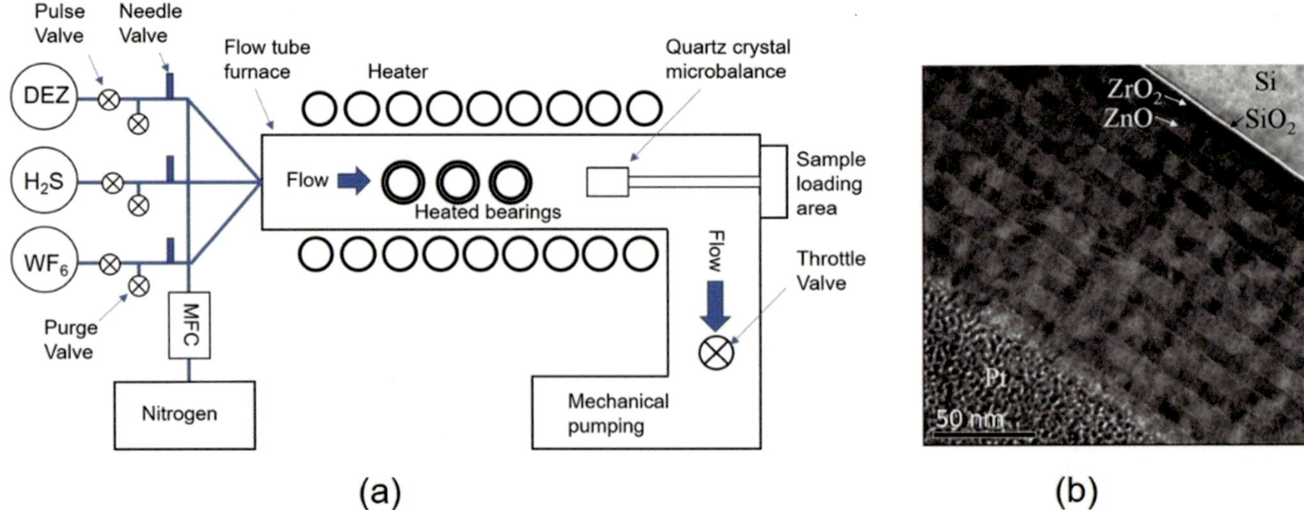

Fig. 2.10 (a) Schematic of viscous flow reactor used to grow ALD WS$_2$ coatings on rolling element bearings. (b) Cross-sectional transmission electron micrograph of ALD ZnO/ZrO$_2$ formed of 8 bilayer nanolaminate coating grown at 200 °C [Doll et al (2010)]

single-crystal, high-quality, "track-and-field ground" shaped graphene domains and the morphological evolution from hexagonal to hexagram graphene domain, even square and circular graphene domain, has been achieved on solid copper substrate by precise tuning of the deposition parameters. The etching reaction of graphene has also been studied, and results show that a low flow rate of high-purity hydrogen (99.999%) is favorable to form hexagonal structure for the etching reaction of graphene due to the absence of oxygen or oxidizing impurities in hydrogen gas commonly used [Jiang et al. (2018)].

2.2.3.2 Plasma-Enhanced Chemical Vapor Deposition

Plasma-enhanced chemical vapor deposition has been used for the thin films on copper foil at temperatures as low as 25–400 °C. In a typical PE-CVD reactor shown in Fig. 2.11, monolayer graphene films were synthesized on Cu foil using various ratios of hydrogen and methane in a gaseous mixture. The in-situ plasma emission spectrum was measured to elucidate the mechanism of graphene growth in a plasma-assisted thermal CVD system. According to this process, a distance must be maintained between the plasma initial stage and the deposition stage to allow the plasma to diffuse to the substrate. Raman spectra revealed that a higher hydrogen concentration promoted the synthesis of a high-quality graphene film. The results demonstrate that plasma-assisted thermal CVD is a low-cost and effective way to synthesize high-quality graphene films at low temperature for graphene-based applications.

Low-pressure plasma technologies have also been widely and successfully utilized for the production of a large variety of organic–inorganic nanocomposite thin films consisting of metal or metal oxide. Fanelli et Fracassi (2014) studied the production of a large variety of organic–inorganic nanocomposite thin films consisting of metal or metal oxide nanoparticles embedded in a polymer matrix. Recently, the deposition of this class of coatings has been also accomplished by these authors at atmospheric pressure with cold plasmas using aerosol-assisted processes in which a dispersion containing preformed inorganic nanoparticles and the liquid precursor of the polymeric component is atomized and injected in aerosol form in the atmospheric plasma. The deposition of organic–inorganic NC coatings by AP cold plasmas has been accomplished utilizing an aerosol-assisted process in which a dispersion containing inorganic NanoParticles (NPs) and a liquid organic precursor is atomized and injected in aerosol form in the atmospheric plasma. Even if this methodology is not yet mature enough, it seems to be very convenient for the deposition of multicomponent coatings. The most appealing advantage is the fact that, due to the aerosol utilization, in principle any type of preformed NPs can be injected into the plasma and incorporated in the coating, provided that a good dispersion of the NPs in a suitable solvent is obtained.

It is also possible to deposit with PE-CVD hard coatings of amorphous carbon (DLC: Diamond-Like Coatings) at temperatures below 200 °C using acetylene as precursor and pulsed or RF discharges. The process being monodirectional allows achieving coatings with rather complex architectures. Plasmas with low fractional ionization are of great interest for materials processing because electrons are very light, compared to atoms and molecules, thus energy exchange between the electrons and neutral gas is very inefficient. Consequently, the electrons can be maintained at very high equivalent temperatures (tens of thousands of Kelvins), while the neutral atoms remain at the ambient temperature. These energetic electrons can induce many processes that would otherwise be very improbable at low temperatures, such as dissociation of precursor molecules and the creation of large quantities of free radicals.

Another benefit of deposition within a discharge arises from the fact that electrons are more mobile than ions. As a consequence, the plasma is normally more positive than any object it is in contact with, as otherwise a large flux of electrons would flow from the plasma to the object. The difference in voltage between the plasma and the objects in its contacts normally occurs across a thin sheath region. Ionized atoms or molecules that diffuse to the edge of the sheath region feel an electrostatic force and are accelerated toward the neighboring surface. Thus, all surfaces exposed to the plasma receive energetic ion bombardment. The potential across the sheath surrounding an electrically isolated object (the floating potential) is typically only 10–20 V, but much higher sheath potentials are achievable by adjustments in reactor geometry and configuration. When a high-density plasma is used, the ion density can be high enough for the occurrence of significant sputtering of the deposited films; this sputtering can be employed to help planarize the film and fill trenches or hole. Plasma deposition is often used in semiconductor manufacturing to deposit films conformally

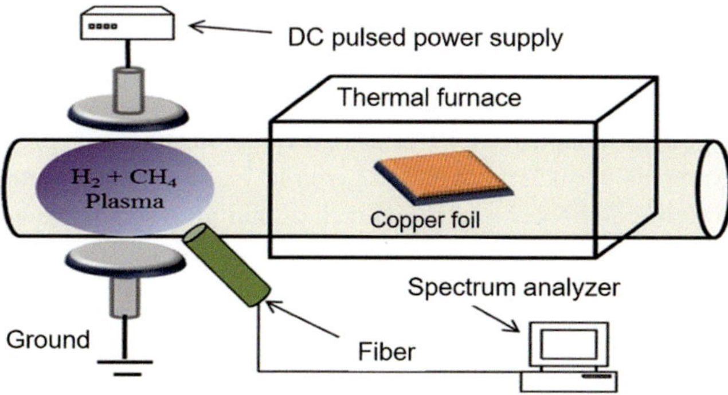

Fig. 2.11 . Apparatus for plasma-assisted thermal CVD and emission diagnostic system [Chan et al. (2013)].

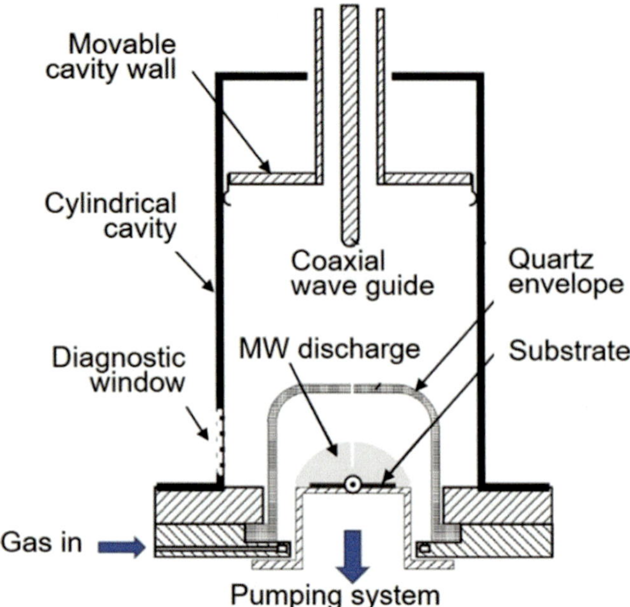

Fig. 2.12 Schematic of a microwave PE-CVD reactor used for the deposition of high-quality, single-crystal diamond films [Vikharev et al. (2007)]

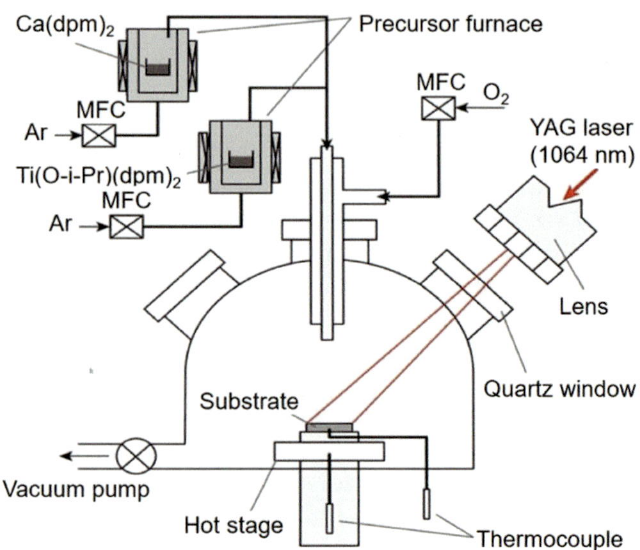

Fig. 2.13 Schematic of laser-enhanced CVD apparatus used by Katsui and Goto (2017)

(covering sidewalls) and onto wafers containing metal layers or other temperature-sensitive structures. PE-CVD also yields some of the fastest deposition rates while maintaining film quality (such as roughness, defects/voids), as compared with sputter deposition and thermal/electron beam evaporation, often at the expense of uniformity.

Microwave PE-CVD was used by Vikharev et al. (2007) for the deposition of high-quality single-crystal diamond films in a CVD reactor. A schematic of the reactor used is given in Fig. 2.12. The conditions used for this study were based on the results of homo-epitaxial growth of polycrystalline diamond films on diamond substrates and on numerical simulation of the microwave discharge in a CVD reactor. A high-quality single-crystal diamond layer was synthesized on a synthetic, type 1b diamond substrate. The properties of the obtained monolayer were studied by means of Raman and X-ray diffraction spectroscopy as well as optical and atomic-force microscopy.

2.2.3.3 Laser-Enhanced Chemical Vapor Deposition

In conventional CVD thin film deposition processes, the chemical reaction at the interface between the gas and substrate surface is driven by thermal energy. Katsui and Goto (2017) point out that laser irradiation can accelerate the chemical reactions and enable deposition at lower temperatures to avoid degradation and corrosion of the substrate materials. Figure 2.13 shows a schematic of a laser-enhanced CVD apparatus used for the deposition of well-crystallized Ca–Ti–O films with various crystal phases and

microstructures obtained at high deposition rates. The $CaTiO_3$ films were highly (011)-, (101)-, and (121)-oriented, forming cauliflower-like, granular, and faceted morphologies. These various preferred orientations and morphologies affected the solubility, regeneration of calcium phosphate films, and the bio-inertness of $CaTiO_3$ films. For the Ca-rich compositions, $Ca_{n+1}TinO_{3n+1}$ films with a Ruddlesden–Popper-type crystal structure were formed and exhibited promising bioactivity for calcium phosphate regeneration.

2.2.4 Thin Film Coating Technologies in Industry

The development of thin film coating technologies whether PVD or CVD, or variations of these two approaches such as EB-PVD, sputtering, and plasma- or laser-enhanced CVD, has had a major impact on industrial-scale developments in the fields of microelectronics, semiconductors, and green technologies such as solar cells, sensors, and light emitting devices. This has manifested itself by the rapid development of a wide range of industrial-scale, thin film deposition chambers adapted to PVD or CVD applications. Typical examples of such reactors are given in Fig. 2.14. According to Carter and Norton (2013), much of thin films deposited by PVD or CVD are carried over from the semiconductor industry and is often not thought of as ceramic processing. However, without thin films of Si oxide and Si nitride, "Silicon Valley" might still be green fields.

Bhattacharya and Meyer (2003) attempted to investigate the interactions between science and technology in this field by analyzing patent citations, publications, and patent outputs

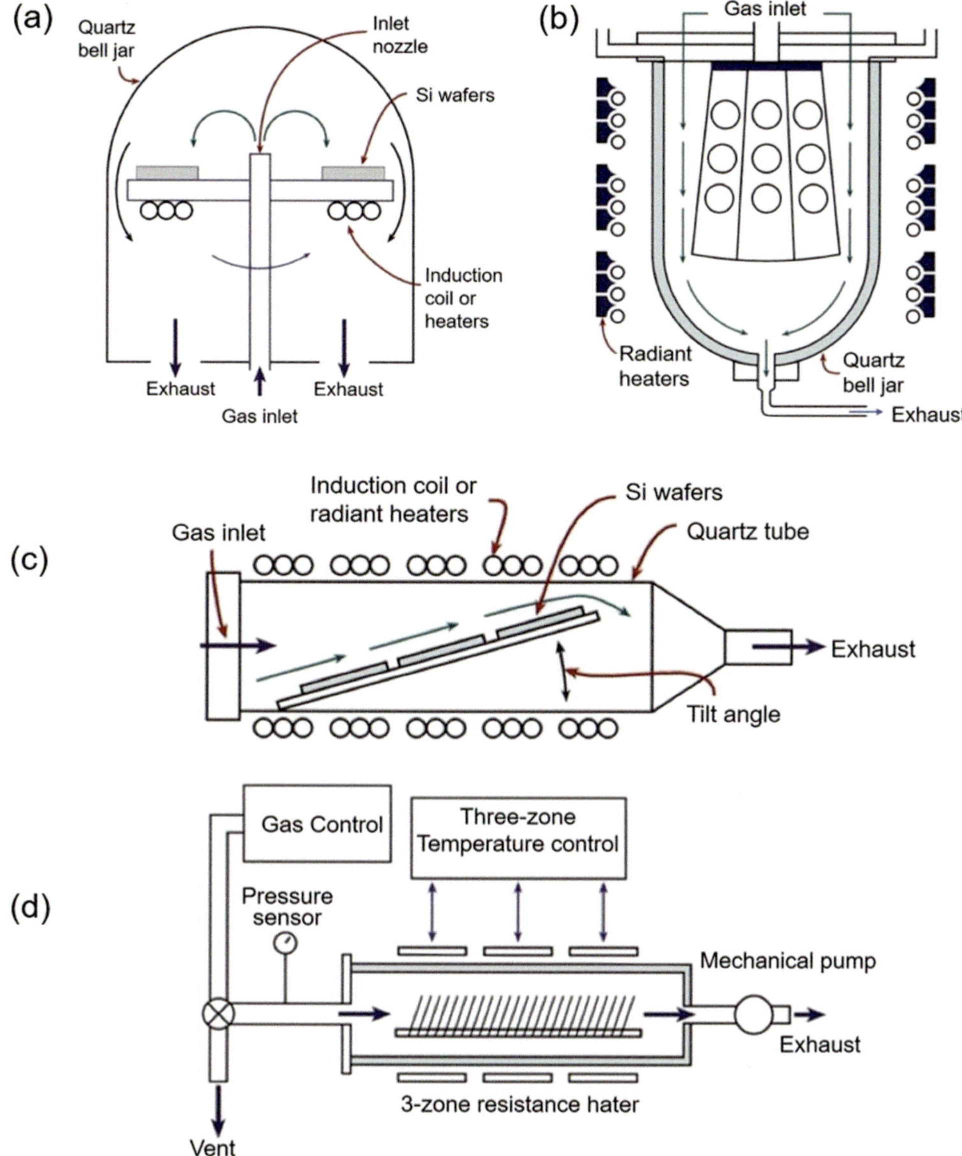

Fig. 2.14 Examples of industrial-scale Chemical Vapor Deposition (CVD) reactor configurations: (**a**) pancake, (**b**) barrel, (**c**) horizontal, (**d**) LP-CVD [Carter and Norton (2013)]

of MultiNational Corporations (MNCs) in thin film technology. MNCs have played a major role in the development of this technology as seen from the patenting activity in this field. Five years of patenting activity (1993–1997) in this area have resulted in 1739 patents with 506 patents (29% of the total) contributed by the top 10 MNCs, and 253 patents ($\approx$15%) by the next 10 MNCs. Our investigations are for the period 1997. Reasons for choice of this year were guided by two factors. In 1997, activities of MNCs show significant enhancement compared to previous years. Also, for this period, MNCs are not restricted to only the USA and Japan.

2.3 Thermal-Sprayed Coatings

2.3.1 Basic Concepts

Thermal spray is a generic term for a group of coating processes where the coating is deposited on *a prepared substrate* by applying *a stream of molten particles/droplets* of a *metallic or nonmetallic material that flatten upon their impact on the substrate forming more or less platelets, called splats, which are the building blocks of the coating.* Thermal spraying can be used exceptionally for the deposition of thin coatings with a coating thickness of 50 μm. The great majority of thermally sprayed coatings, however, tend to be relatively thick with total coating thickness ranging between 100 μm and a few 100 μm or more. These can be composed of a single material, a mixture in a single layer, or be multilayered, including a "bond coat," an "intermediate coat," and a "finishing coat."

As illustrated in the block diagram given in Fig. 2.15, a thermal spray system is centered around the "spray torch" in which different sources of energy whether chemical (combustion) or electrical (electric discharge), depending on the spray process, are converted into a stream of hot gases in which the material to be sprayed is injected. The material introduced into the torch, whether in the form of powder, suspension, solution, wire, or other forms, is heated, melted, and entrained by the high-temperature, high-velocity gas stream toward the substrate on which the particles/molten droplets impact generating the splats that pile up on the substrate forming the coating. The form and properties of these splats depend on the nature of the material sprayed, particle/droplet size distribution, and particle/droplet temperature and velocity prior to their impact on the surface of the substrate as well as on the temperature and surface properties of the substrate.

A schematic of typical DC plasma spray torch is given in Fig. 2.16. Operation in this case is in an open air with radial powder injection into the plasma jet either internally in the exit nozzle channel, or externally at the base of the plasma jet nozzle exit. Included in this figure is a representation of the coating formation through the compilation of successive splats indicating potential sources of defects in the coating resulting from the inclusion of unmolten particles and pores into the coating. Micrographs of typical shapes of splats formed with different materials under different thermal

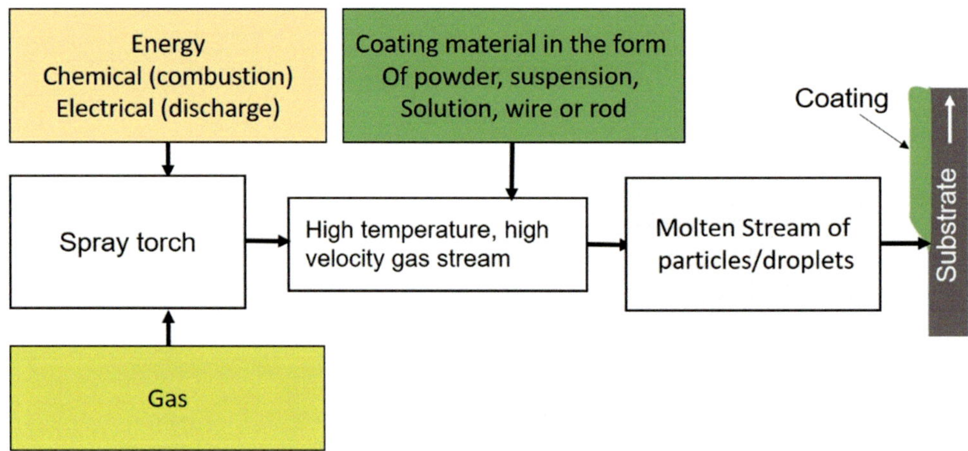

Fig. 2.15 Block diagram of a generalized thermal spray coating system

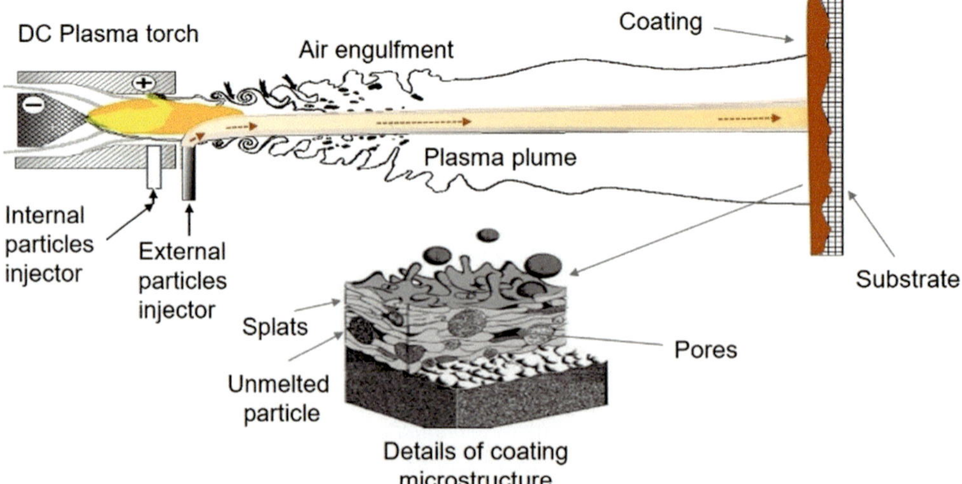

Fig. 2.16 Schematic of a conventional DC plasma spraying showing the coating building mechanism through the compilation of successive splat on the surface of the substrate

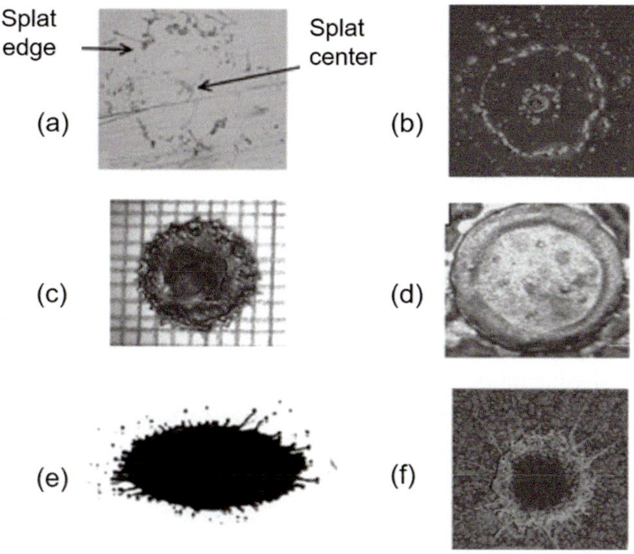

Fig. 2.17 Examples of splat formation in DC plasma spraying for different materials and deposition conditions: (**a**) alumina on cold 304-SS [Goutier S. et al. (2011)], (**b**) Ni on 304 -SS pre-oxidized at 150 °C [Dhiman R. et al. (2007)], (**c**) Cu on AISI-304-SS [Fukumoto M et al. (2011)], (**d**) alumina on low carbon steel preheated at 200 °C [Sabiruddina K. et al. (2011)], (**e**) alumina on SS preheated at 400 °C [Goutier S. et al. (2011)], and (**f**) Ni on pre-oxidized SS at 650 °C [Dhiman R. et al. (2007)]

spray operating conditions are given in Fig. 2.17. Such variations in splat morphology combined with deposition parameters, such as the use of a cross flow "cross wind" for the elimination of unmolten particles from the molten droplet flux to the substrate, can also have a significant impact on the cross-sectional microstructure of the coating as shown in Fig. 2.18 for Yttria Stabilized Zirconia (YSZ) coating deposited on a SS substrate.

A preliminary classification of thermal spray technologies according to the source of energy used in the process for the generation of the high-temperature, high-velocity gas stream is given in Fig. 2.19.

These are grouped into two broad categories:

- Combustion-based processes
- Plasma-based processes

To these one should add "cold spray," which while not being a "thermal spray" process as such bears significant similarity to the basic phenomena involved with the important exception that the energy provided into the gas stream is in the form of a high-pressure cold gas flow rather than a high-temperature flow. As feed material is injected into this high-velocity gas stream, the solid particles of the feed

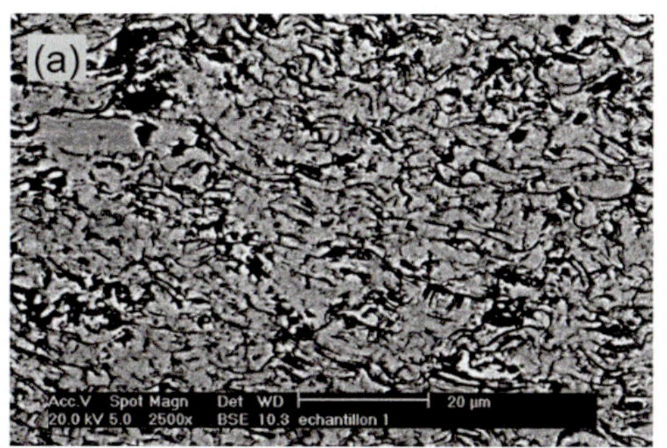

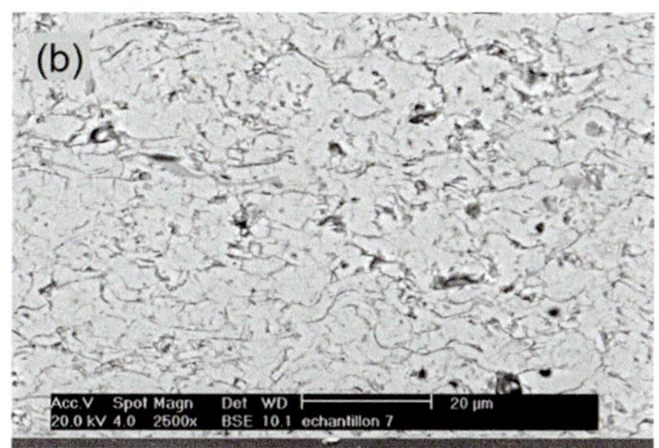

Fig. 2.18 YSZ spray coatings obtained with different operating and spray conditions: (**a**) spraying performed without wind jet; (**b**) with wind jet [Fauchais et al. (2010)]

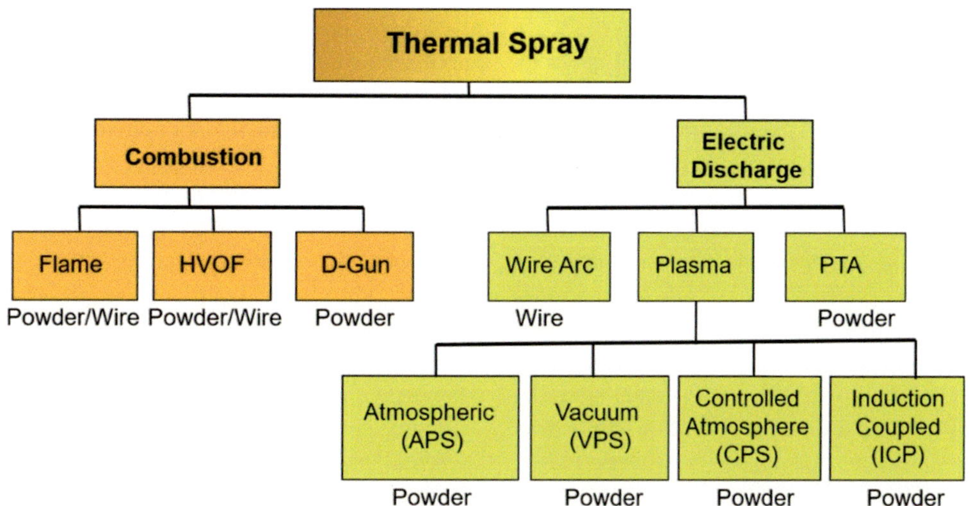

Fig. 2.19 Classification of thermal spray coating technologies

material are entrained to very high velocities prior to their impact on the surface of the substrate. On impact with the substrate, the kinetic energy of the particles is converted into a thermal energy that contributes to the softening of the material and its plastic deformation producing a splat with excellent adhesion to the substrate that is the building bloc of the coating.

2.3.1.1 Combustion-Based Processes

- *Flame Spraying* (FS) of powders, wires, or rods, the High-Velocity Oxy-Fuel (HVOF) with gaseous fuels, or more recently with liquid fuels [Song et al. (2017)].
- *High-Velocity Air Fuel* (HVAF) or *HighVelocity Oxy-Fuel* (HVOF) processes with significantly higher upstream pressures and a Laval nozzle allowing attainment of supersonic speeds, using mostly powders.
- *Detonation Spray* (DS) process where an explosive mixture is ignited upstream of a Laval nozzle to create a detonation with high pressure and temperature for heating and accelerating the powders toward the substrate.

These are historically the first group of thermal spray technologies that was developed in the early twentieth century and remain among the most widely used spray coating processes because of their very favorable economics, which make them a viable option for a wide range of applications The principal limitation of combustion spray processes lies, however, in the relatively low gas temperature given by the adiabatic flame temperature of the combustion mixture, which reduces its value for spraying of refractory metals and ceramics. Moreover, the fact that the spraying gas constitutes mainly of the combustion products leaves open the possibility of in-flight chemical reaction between the sprayed material and the combustion products resulting in coating contamination with oxides or other debris, which can be detrimental to the coating performance.

2.3.1.2 Plasma-Based Processes

- Atmospheric and controlled atmosphere DC plasma spraying
- RF Induction Plasma Spraying (IPS)
- Wire Arc Spraying (WAS)
- Plasma-Transferred Arc (PTA) deposition

In the plasma-based processes the gas is heated by an electrical discharge, that is, an arc, or a radio frequency inductively coupled discharge. These are by far the most versatile of all thermal spray processes because they have few limitations on the materials that can be sprayed, or on the substrate properties, its size, and shape. The coating quality is in general higher than that obtained with flame spraying. While the majority of the plasma spray processes are carried out in an open-air environment such as Atmospheric Plasma

Spraying (APS), the coating quality, that is, density, uniformity, and reproducibility, can generally be enhanced by spraying in a controlled environment, such as Controlled Atmosphere Plasma Spraying (CAPS), Low-Pressure Plasma Spraying (LPPS), or vacuum plasma spraying (VPS). The improved coating quality is, however, associated with a significant increase in the required capital investment and operating cost. High Frequency (HF) or Radio Frequency (RF) Induction Plasma Spraying (IPS) process is an alternate CAPS process where the electrical discharge is generated through the electromagnetic coupling of the energy between the induction coils surrounding the discharge cavity. Being an electrodeless process, the technology is particularly suited for the deposition of high-purity materials such as silica for preform building in the fiber optics industry or the coating of sputtering targets for the electronic industry. By offering a larger plasma volume and longer residence time of the particles in flight in the plasma prior to their impact on the substrate, IPS is particularly suitable for the melting and spraying of relatively coarse powders with a particle size in the hundreds of micrometers compared to DC plasmas that are generally limited to use with powders of particle size less than 100 μm.

The use of wires is listed separately as a different process because the material melting in this case is accomplished directly by an arc struck between two wires, or between a wire and a dummy electrode. As the tip of the wire melts, a high-velocity gas stream is used to atomize the molten metal and propel the formed droplets toward the substrate on which they impact forming the coating. This process is more economical because of the use of wires that are less expensive compared to powders, though the choice of materials is limited because the wires need to be ductile and electrically conducting. Wire Arc Spraying (WAS) has usually a higher porosity than those of the other plasma-based processes.

The PTA deposition process, which is also part of the weld overlay family of coating technologies, is not a thermal spray process because of the melting of the substrate at the point of deposition of the coating and the fact that coating material in the form of powder is injected into an arc struck between the auxiliary cathode and the substrate or workpiece, which acts as the anode. Often, part or all of the coating material can also be fed directly into the molten metal pool formed in the substrate in which the material is heated, melted, and fully blended before solidifying into a coating. PTA coatings are known for their very high coating densities and excellent bonding to the substrate. The transferred arc deposition can also use wires as the coating material; however, in this case the process is essentially a weld overlay process.

It is important to point out that each of the above-listed processes offers distinctly different characteristics and spraying conditions. Accordingly, they are best adapted to different materials and different coating objectives depending

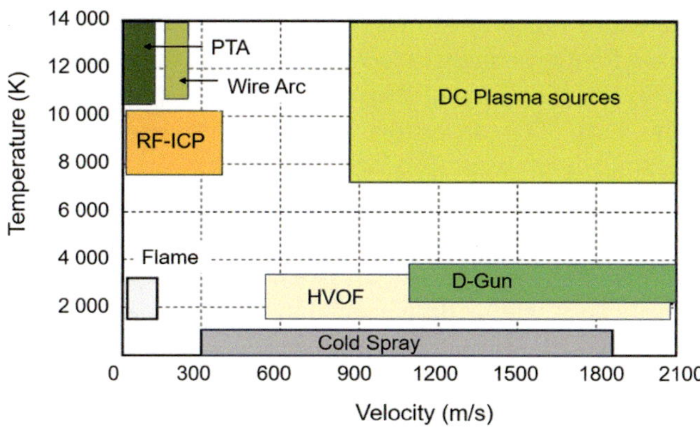

Fig. 2.20 Gas temperatures vs velocity mapping associated with different thermal spray processes

on process needs. In a way they are all complimentary processes rather than competing for the same application. A classification of thermal spray processes can be made according to the predominant gas velocity and gas temperature levels in the energy sources used. These are illustrated in the velocity–temperature diagram given in Fig. 2.20. It shows DC plasma sources distinguished by their highest velocities and highest temperature range among all thermal spraying processes. Combustion sources, on the other hand, have temperatures limited to 3000–4000 K, while being able to achieve velocities as high as 2000 m/s. Induction plasma sources offer high temperatures in the range of 8000–10,000 K, with relatively lower velocities generally below 100–300 m/s. On the other hand, with their relatively large volume of the discharges, they offer longer residence time of the particles in the plasma region (of the order 10–15 ms), compared to less than 1 ms for DC sources, which allows for the heating and melting of relatively large particles (>100–200 μm) of most refractory materials (T_m > 2000 K). Wire arc spray and PTA are also recognized as being high-temperature sources (T > 12,000 K) with relatively low gas velocities.

2.3.2 Energetic Gas Flow Generation

2.3.2.1 Cold Spray

As mentioned earlier, *cold spray* is "a kinetic spray process" utilizing supersonic jets of compressed gas to accelerate the powder particles to ultra-high velocities (up to 1500 m/s). On their impact on the substrate, the powder particles plastically deform and consolidate with their substrate to create a coating (Gärtner et al. 2006; Champagne 2007; Alkhimov et al. n. d.; Stoltenhoff et al. 1994). The high-velocity gas jets are generated using convergent–divergent Laval nozzles with upstream pressure between 2 and 2.5 MPa for a nozzle throat with an internal diameter in the range of 2–3 mm. Gases used

are He, N_2, or their mixtures at very high flow rates (up to 5 m³/min). Typically, the mass flow rate of the gas must be close to ten times that of the entrained powder. This implies that for a powder feed rate of 6 kg/h, a volumetric flow rate of 336 m³/h of helium and 52.3 m³/h for nitrogen will be necessary. Gases introduced (nitrogen or helium) are often preheated up to 700–800 °C to avoid their liquefaction under expansion and increase their velocity. This implies that the heating device must be capable of heating 90 m³/min of a He/N_2 mixture from room temperature up to 700–800 °C. As particles are injected upstream of the nozzle throat, the powder feeder has to be at a slightly higher pressure compared to the upstream gas pressure in the chamber. When spraying with He, cost constrains require that the spray operation be carried out within an enclosure in order to allow for full gas recycling.

For particles to adhere to the substrate, their impact velocity has to be above a critical value between 500 and 900 m/s depending on the nature of the sprayed material. The spray pattern typically covers an area of roughly 20–60 mm², at spray rates between 3 and 6 kg/h. Feed stock particle sizes are typically between 1 and 50 μm and deposition efficiencies reach easily 70–90%. Only ductile metals or alloys can be sprayed (Zn, Ag, Cu, Al, Ti, Nb, Mo, Ni–Cr, Cu–Al, Ni alloys, MCrAlYs, and polymers), owing to the impact-fusion coating build-up. Blends of ductile materials (>50 vol.%) with brittle metals or ceramics are also used. It is also important to underline that the substrate is not heated by the gas flow remaining typically at temperatures below 200 °C.

A portable, lower-pressure version of the cold-spray torch, using air instead of helium as spray gas, was developed [Kashirin et al. (2002a, b), Shkodkin et al. (2006) and Kashirin et al. (2007)], where the upstream pressure is below 1 MPa, and a much lower gas consumption of about 0.4 m³/min. Because of the lower gas velocities, 300–400 m/s, the particle size needed to achieve a coating is significantly smaller than conventional cold-spray systems. Mixing the spray powder with larger size particles, which can be ceramics, enhances the spray process. While the large particles mostly rebound upon impact, their role is to "press" the small metal particles and the previously deposited layers onto the substrate in a manner equivalent to shot peening. Powder recuperation is hardly possible for multi-component powder mixtures. While the deposition efficiency is much lower than with high-pressure systems, the process is still attractive for short production runs because of the considerably lower investment and running cost.

2.3.2.2 Flame Spray

Flame spray is the simplest and among the oldest spray processes developed by Dr. Schoop in 1912. The jet of high-temperature gas used in the spraying operation is generated through the combustion of an adequate gaseous or liquid fuel using either air or pure oxygen as oxidizer. The adiabatic flame temperature and its specific enthalpy are determined by the fuel composition (C_xH_y), and the fuel and oxidizer (oxygen or air) flow rates [Glassmann I. (1977)]. The material to be sprayed is fed in powder form with a simple powder hopper, or a more sophisticated powder feeder, through the central bore of the burner nozzle where the powder is heated by the oxy-fuel flame and entrained by the hot gases, melted in flight, and projected on the substrate or workpiece. The process takes place at atmospheric pressure. It is important to point out when using air as oxidizer that one mole of air contains only 0.2 moles of O_2, which is consumed in the combustion process; the balance 0.8 moles of N_2 and other inert gases, while they do not contribute to the combustion process, must still be heated thus reducing the overall specific enthalpy of the combustion products, and a reduction in the bulk flame temperature. At atmospheric pressure, the maximum temperature achieved with the stoichiometric combustion of acetylene ($2C_2H_2 + 5O_2$) is 3410 K. Flame velocities depend on the combustible gas flow rates and are generally below 100 m/s. Typically a 40-kW burner will require 18 standard liter per minute (slm) C_2H_2 + 30 slm of O_2. The specific mass of hot gas is about one-tenth that of the cold gas at room temperature.

In spite of the fact that flame-sprayed coatings exhibit generally high porosity (>10%) and relatively low adhesion (<30 MPa) to the substrate, the process is widely used on an industrial scale because of its simplicity, the wide range of materials that can be sprayed, and its overall low cost. Typical applications of flame spraying include the spraying of self-fluxing alloys, such as NiCrBSi, which contains boron and silicon as fluxing agents limiting oxidation. Metallurgical bonding between the coating and the substrate can be improved by heating the coating after spraying to its melting temperature provided that the melting temperature of the substrate is higher than that of the coating. Flame spraying is used for depositing wear-resistant coatings under low-load conditions, for spraying thermoplastics, and the rebuilding worn parts. The deposition of composite coatings used against abrasive wear (friction, erosion) and corrosion (cold or hot) is also possible. Flame spraying can also be used with feed stocks in the form of a wire or a rod or a cord. Typically, the wire, rod, or cord is fed axially and continuously at a velocity such that its tip is melted in the flame. A stream of compressed air, surrounding the flame, is used to atomize the molten material and generate a continuous stream of droplets that are projected toward the substrate. Flame-spray guns are generally lightweight and easy to use for hand-spraying of complicated shapes.

Detonation spraying is a flame combustion process in which the controlled explosion of a mixture of fuel, oxygen, and powdered coating material is utilized to melt and propel the material to the workpiece [Hermanek FJ (2001)]. First developed in Russia, it was introduced in western hemisphere in the early 1950s by Gfeller and Baiker working for Union Carbide. The detonation gun (D-Gun), as shown in Fig. 2.21, is composed essentially of 1-m long barrel, closed at one end, in which the fuel and oxygen combustion mixture is introduced. The ignition of the mixture by a spark plug close to the closed end generates a detonation [Kadyrov et al. (1995)]. Pressures around 2 MPa are generated at the closed end of the tube that propagates toward the open-end entraining particles injected around the middle section of the tube. The injected particles are accelerated and heated in flight, melting them in most cases. Pressures from the detonation close the gas feed valves until the chamber pressure is equalized. When this occurs, the tube is flushed with nitrogen, filled with combustion gases, and the cycle is repeated (between 4 and 8 times per second or more with the recent guns). In contrast to most other spray processes, the specific

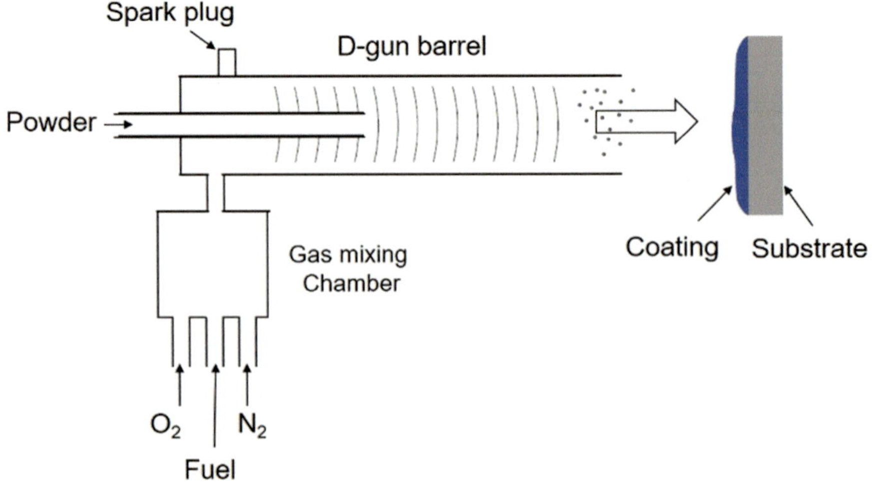

Fig. 2.21 Schematic of the D-Gun [Kadyrov et al. (1995)]

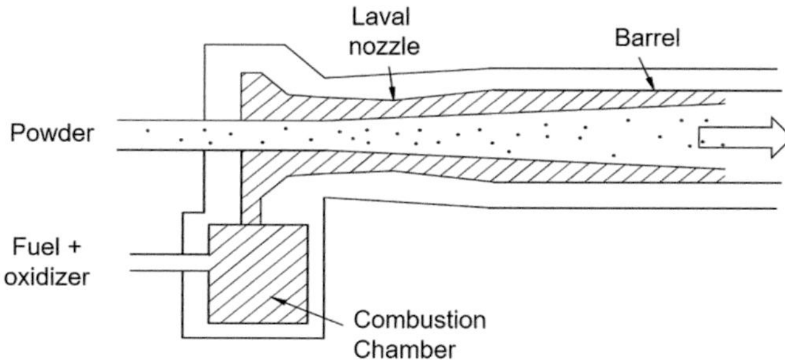

Fig. 2.22 Schematic of the HVOF process based on the Jet Kote gun [Thorpe et al. (1992)]

mass of the gas that accelerates and propels particles toward the substrate is higher than that of the cold gas (about five times) resulting in very high particle velocities. The detonation speed is the sound velocity in the combustion products [Glassmann I. (1977)]. The velocity of the gases behind the wave front, as defined by Kadyrov et al. (1995), is between 1000 and 3000 m/s. They depend strongly on the composition of the combustible gas mixture and its pressure. The D-Gun produces premium coatings, especially metallic and cermet ones, with properties that have been the goal of almost all of other spraying processes to reproduce, that is, thick high-density coatings with almost no oxidation, higher hardness, bonding and cohesive strength, smooth as sprayed surfaces, improved corrosion barrier, and wear resistance.

"High-velocity flame spraying" using "high-velocity oxy-fuel" (HVOF) or "high-velocity air fuel" (HVAF) guns represent a significant improvement over the standard "flame spray" process [Thorpe et al. (1992)]. Combustion is achieved in this case in a pressurized chamber, followed by a Laval-type nozzle as schematically represented in Fig. 2.22. The combustion at pressures higher than atmospheric pressure (between 0.2 and 1 MPa) increases slightly the combustion temperature (see Chap. 3). For example, the adiabatic temperature for a stoichiometric combustion mixture of methane with oxygen at atmospheric pressure is 3030 K, while that at 2 MPa it is 3733 K [Glassmann I (1977)]. The main advantage of the gun design is in the high gas velocities, which, as a result of the hot gas expansion in the Laval nozzle, can be as high as 2000 m/s (Fig. 2.20). Such velocities are supersonic as noted by shock diamonds observed downstream of the nozzle. Two types of spray guns exist characterized by the chamber pressure. The low-pressure HVOF uses a combustion at pressures between 0.24 and 0.6 MPa with heat inputs below 600–700 MJ, while the high-pressure guns are operated in the pressure range of 0.62–0.82 MPa, generally using kerosene as fuel either with oxygen or air as oxidizer with heat inputs over 1 GJ. These guns are often termed "hyper velocity guns." Generally, lower-pressure guns are fueled with hydrogen, propylene, methane, propane, heptane, kerosene, and a few trademark

gases together with oxygen or air as oxidizer. The first HVOF gun was introduced in the early 1980s by Browning and Witfield. Particles are injected either axially upstream of the nozzle (a pressurized powder feeder is required), as shown in Fig. 2.22, or radially downstream of the nozzle (with a conventional powder feeder). Compared to conventional flame spraying guns, HVOF guns are fed with high fuel gas flow rates of 60–120 slm, and oxygen flow rates ranging from 280 to 600 slm, generating power levels of a few hundreds of kW. With kerosene as fuel, the fuel feed rate is between 20 and 30 liter/min with oxygen flows in the order of 1 m^3/min, or air up to 5 m^3/min. The combination of high temperature and high velocity of the plasticized particles gives rise to very dense coatings achieved using this technology. The mixing of the flame with surrounding air can be delayed allowing the particles to be kept accelerating by extending the gun nozzle using a barrel. For the spraying of wires, special HVOF guns have been developed [Smith et al. (1997)]. The principle is the same as that of wire flame spraying. The flame (propane–oxygen) is positioned at the nozzle face and the molten tip of the wire is atomized by the pressurized flame and the airflow. Under such conditions, the gas temperature reaches values of 3100 K, and its velocity reaches values of up to 1600 m/s. Corresponding particle velocities are much higher than with standard wire flame spraying. The HVOF guns were very successful for spraying dense, wear-resistant, WC–Co coatings. Dense coatings of a wide range of metals, alloys, and cermets were also obtained with good adhesion to the substrate and low oxide content, compared to conventional flame spraying operations. They compare favorably with high-energy plasma-sprayed coatings. Typically, WC + Co, WC + NiCr, or WC + CrCo can have hardness values between 1100 and 1400 ± 150 HV_{5N}.

2.3.2.3 Plasma Spraying

The energy source in *plasma spraying torches* is an *electric discharge* that can be generated by either a nontransferred DC arc sources (DC plasma spraying) or an inductively coupled, electrodeless Radio Frequency (RF) discharge (induction plasma spraying). Plasmas are electrically conducting gases composed of a mixture of molecules, atoms, ions, in their fundamental and excited states, electrons, and photons in a state of local electrical neutrality. This implies that the number of negatively charged species in a plasma, mostly electrons, is equal to the positively charged one, ions. For gases used in plasma spraying, the plasma state exists as soon as the gas temperature reaches the threshold of ionization, which is usually higher than 7000–8000 K at atmospheric pressure. The large majority of plasma-forming gases used in plasma spraying operations are composed of a gas mixture containing a primary heavy gas, such as Ar or N_2, mixed with a secondary gas such as H_2 or He, which

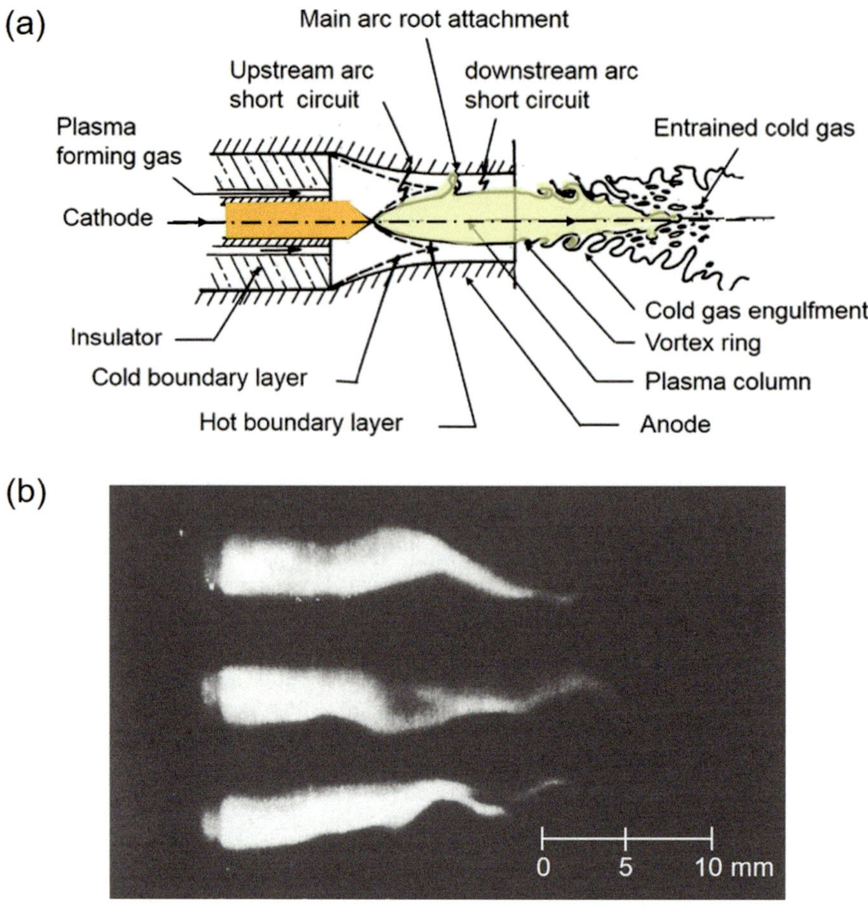

Fig. 2.23 (**a**) Schematic of a typical DC plasma torch used for plasma spraying. (**b**) High-speed photographs of the oscillation of the plasma jet at the exit of the torch nozzle

serves to increase the heat conductivity of the plasma (for details see Chap. 3).

A schematic of a typical *DC plasma spray torch* is given in Fig. 2.23a. This is mainly composed of a rod-type, thoriated tungsten cathodes (2 wt% ThO_2), not permitting traces of oxygen or water vapor in the plasma-forming gas (tungsten oxidation starts at 1800 K, while the cathode tip temperature is typically above 3600 K). The cathode is concentrically placed, but electrically isolated from a water-cooled cylindrical copper anode. The plasma gas is injected into the annular region between the anode and cathode, typically with a vortex motion. An arc is struck between the cathode and the anode, typically with an arc current of a few hundred amperes, and an arc voltage that can vary between 20 and 40–60 Volts or more depending on the composition of the plasma gas. Power levels are between 20 and 80 kW. Through the design of the vortex motion of the gases in the discharge cavity, the arc root point of attachment on the inner surface of the anode is kept in continuous motion over the surface of the anode, thus distributing the local heat load on the anode and reducing anode wear, which has a direct impact on anode life time. The plasma gases through their

contact with the arc heat to relatively high temperatures and emerge from the anode nozzle as a high-velocity, high-temperature, mostly turbulent jet with temperatures in the range of 8000–14,000 K. Gas velocities for conventional anode nozzle with internal diameters between 6 and 8 mm range between 500 and 2600 m/s (subsonic velocities at these temperatures). Gas flow rates are between 0.8 and 2 g/s, and it must be emphasized that the mass of the secondary gas, generally helium or hydrogen, is generally negligible compared to that of the primary gas (argon or nitrogen). The continuous fluctuations of the arc root attachment, and accordingly of the arc length and power, are reflected by the continuous, high-frequency (k Herz) oscillation of the plasma jet at the exit of the torch as shown by the high-speed photographs given in Fig. 2.23b. These, however, average out with time allowing for the injection, melting, and spraying powders at feed rates up to 3–6 kg/h.

Button-type cathodes made of thoriated tungsten or other refractory metals embedded in a copper holder are also used in DC plasma spray torches, the arc being centered on it by a strong vortex flow of the plasma-forming gases. Typical design configurations of the cathode anode and plasma gas

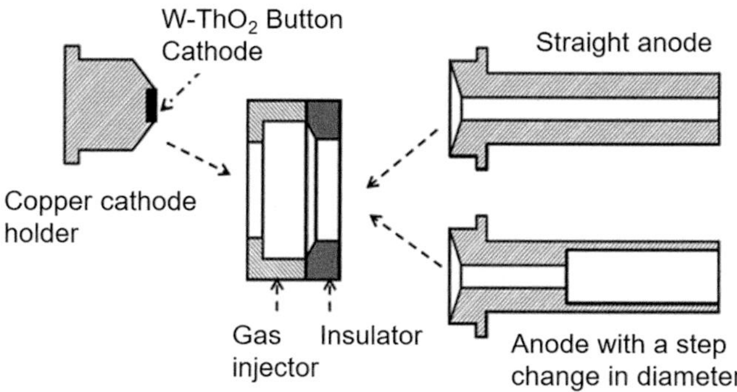

Fig. 2.24 Schematic of a DC plasma torch with a button-type hot cathode [Morishita T (1991)]

injector are given in Fig. 2.24 [Morishita T (1991)]. Plasma-forming gas flow rates are (5–7 g/s) higher than those with a stick-type cathode, with anode nozzle internal diameter of 8–10 mm. The bulk temperature of the plasma jet is in the 8000–10,000 K range, with velocities below 2000 m/s. These velocities can be supersonic at temperatures below 9000–10,000 K. Power levels are 200–250 kW. Typical powder flow rates are up to 20 kg/h. These types of torches are used mainly for high-power/high-deposition rate applications, where the cathode attachment has high current densities. Of course other types of plasma torches exist, which are described in Chap. 7, with more detail about the plasma spray process.

When operating under soft-vacuum conditions, the length of the plasma jet increases and so does the plasma velocity at the base of the torch exit nozzle. This is usually accompanied by a significant increase in the spray particle velocity and considerable improvement in the deposit quality in terms of deposit density and adhesion to the substrate. To capitalize on such improvement, the profile of the anode nozzle is modified through the addition of a Laval nozzle in order to limit the expansion of the jet and reduce the formation of shock waves in the flow. Typical examples of such a low *Vacuum Plasma Spraying (VPS)* nozzle design and the corresponding changes in the length of the plasma jet at different chamber pressures are given in Fig. 2.24a and b, respectively. It is important to point out that VPS has the added advantage of avoiding possible powder oxidation and contamination when spraying metallic coats. The improvement in coating quality is associated, however, with a significant increase in associated investment and operating cost of the system. (Fig. 2.25)

Induction Plasma Spraying (IPS) is an alternate plasma spray technology in which the energy is transferred through electromagnetic coupling from an RF power supply into the discharge. It is an electrodeless, large-volume discharge, which can be operated with inert, reducing, or oxidizing gases, at atmospheric or low-pressure conditions. The oscillator frequency is typically in the range of 1–4 MHz, with

nominal power levels in the range of 5–100 kW or more. A schematic of an induction plasma torch is given in Fig. 2.26 together with a section view of an industrial torch by Tekna Plasma Systems Inc. These show the water-cooled induction coil embedded into a polymer-matrix composite that houses the central plasma confinement ceramic tube. By applying an oscillating high-frequency current to the coil, an oscillating magnetic field is generated in the central region of the coil, which on ignition generates an induced annular current that heats the gas in the discharge cavity and maintains it under partially ionized plasma conditions. An internal gas flow introduced through the gas distributor head ensures the shielding of the internal surface of the ceramic tube from the intense heat to which it is exposed. The hot plasma gases exit the torch cavity through the downstream flange-mounted nozzle. The easy access to the discharge cavity allows for the axial introduction of the spray powder into the torch where the individual powder particles are heated, melted, and projected toward the substrate building-up the coating.

For plasma torches with RF power ratings between 15 and 100 kW (mostly used in plasma spraying) and plasma gas flow rates of 50–100 slm, the bulk plasma temperature is between 8000 K and 10,000 K, and the mean exit plasma velocity is generally below 100 m/s. The exit plasma velocity can be significantly increased by operating the discharge under sub-atmospheric pressure conditions leading eventually to the generation of supersonic plasma jets at velocities up to 2000 m/s. As shown in Fig. 2.26, induction plasma spray torches allow for easy axial injection of the powder or precursor, into the center of the discharge using a water-cooled powder injection probe. Induction plasma torches are well suited for materials processing when a long dwell time of particles in the plasma is advantageous [Boulos M (1992)]. They also offer the possibility to use reactive gases to alter or modify the particles in flight. Being an electrodeless torch, the technology is also well suited for the deposition of high-purity materials. It is a niche technology, with a big commercial success for spray powder preparation such as particle densification and spheroidization (not exclusively for spraying), as illustrated in Fig. 2.27. For more details about the process, see Chap. 8.

Wire arc spraying is a plasma spray process in which an arc is struck between two consumable wire electrodes of the coating material to be strayed. As the wire tips are melted by the arc, a compressed high-velocity gas stream is used to atomize the formed liquid metal and propel it to the substrate. Figure 2.28a represents schematically such an electrode arrangement with one of the wires acing as cathode and the other as anode. Both wires have to be continuously fed at a well-controlled speed in order to compensate for the wire tip melting and to maintain the proper distance between the wire tips and consequently a stable arc. High-speed photographs given in Fig. 2.28b show the wire melting and droplet

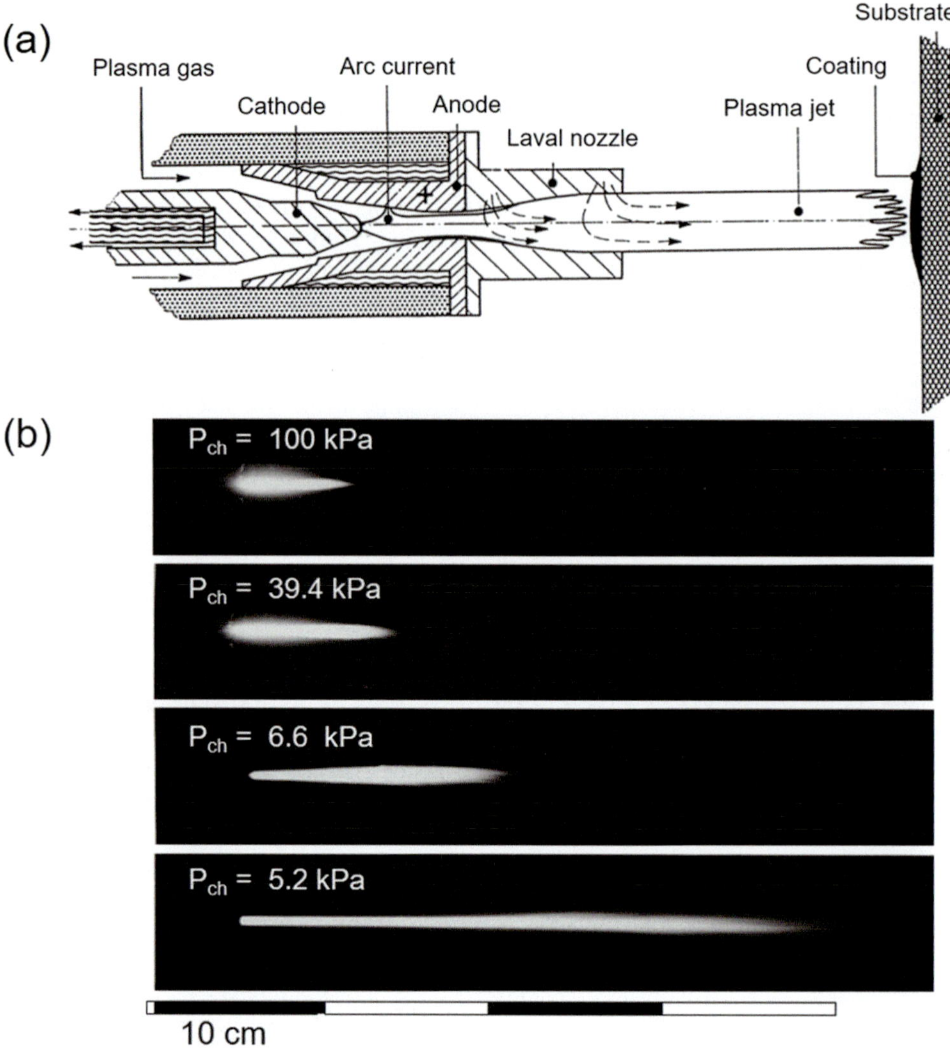

Fig. 2.25 (**a**) Schematic of a modified DC vacuum plasma spray torch incorporating a Laval nozzle. (**b**) Photographs of DC plasma jets under controlled atmosphere at different chamber pressures varying from P_{ch} = 100 down to 5.2 kPa

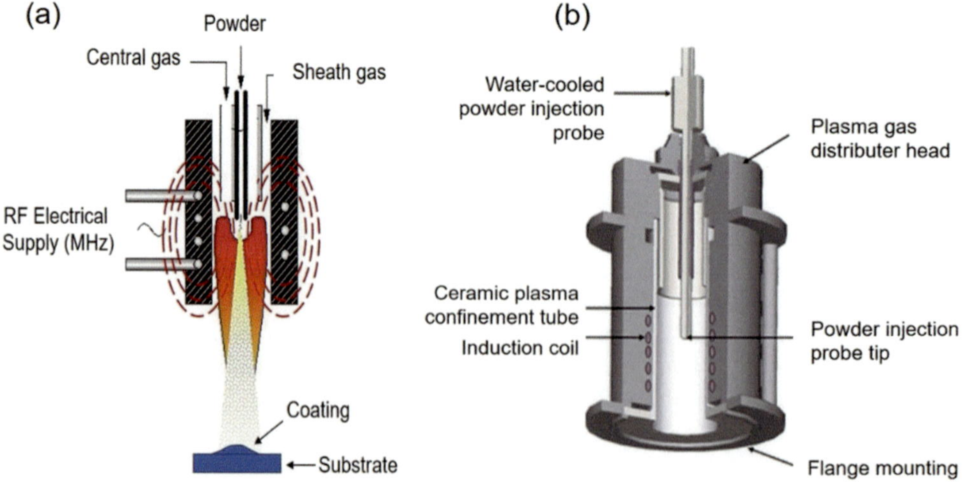

Fig. 2.26 (**a**) Schematic of RF induction plasma spray torch. (**b**) Sectional view of the Tekna induction plasma torch (courtesy of Tekna Plasma Systems Inc., Sherbrooke, Québec Canada)

formation by the atomizing gas flow. Arc power levels are generally between 2 and 10 kW, and atomizing gas flow rates (generally air) are in the range of 0.8 to about 3 m³/min. Gas velocities are a few hundreds of m/s and in spite of the fact that the arc temperature is around 20,000 K, the volume of the arc is too small for any meaningful heating of the atomizing gas flow, which allows the substrate temperature to be maintained at low temperatures without cooling. This is one of the main advantages of the process, which makes it particularly convenient for low-melting point substrates such as polymers. The process has also the advantage of being

capable of high rate deposition primarily for corrosion resistance coatings (zinc, aluminum, zinc–aluminum), with arcs operated up to 1500 A (with 400 A or less in conventional guns). Coatings are rather porous, and contain solidified particles and oxides (up to 25 wt%). Soft coatings can be densified by online shot peening just after deposition. Zinc is particularly efficient as cathodic protection against corrosion, especially when impregnated with epoxy paints in order to seal the porosity in the coating. This process is extensively used for the protection of steel bridges. For more details about the process, see Chap. 9.

2.3.2.4 Plasma-Transferred Arc Deposition

Plasma-Transferred Arc (PTA) deposition is more of a plasma welding rather than a true plasma spraying process. It is treated in this section because of the fact that its principal objective is still one of surface modification through plasma treatment. The process as illustrated in Fig. 2.29 involves establishing a transferred arc between a tungsten rod, which acts as cathode, and the part to be treated, which acts as the anode, necessarily a metal or an alloy. A water-cooled nozzle constricts the arc with a relatively low flow rate of less than 0.5 g/s plasma-forming gas (mostly argon). The transferred arc stability is often improved by using an auxiliary arc between the cathode and the torch nozzle, the nozzle thus becoming an auxiliary anode at a lower potential than that of the substrate. In both cases, however, the energy released at the anode attachment to the substrate results in the localized

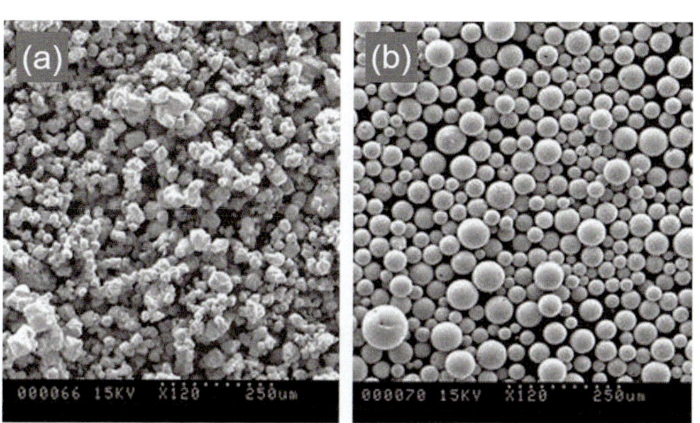

Fig. 2.27 Spheroidization of tungsten particles. (**a**) Initial particle diameter, $d_{50} = 78$ μm; apparent density, $\rho_o = 7400$ kg/m³. (**b**) After plasma treatment, $d_{50} = 69$ μm; $\rho_o = 11,600$ kg/m³ [courtesy of Tekna Plasma Systems Inc. Sherbrooke, Québec, Canada]

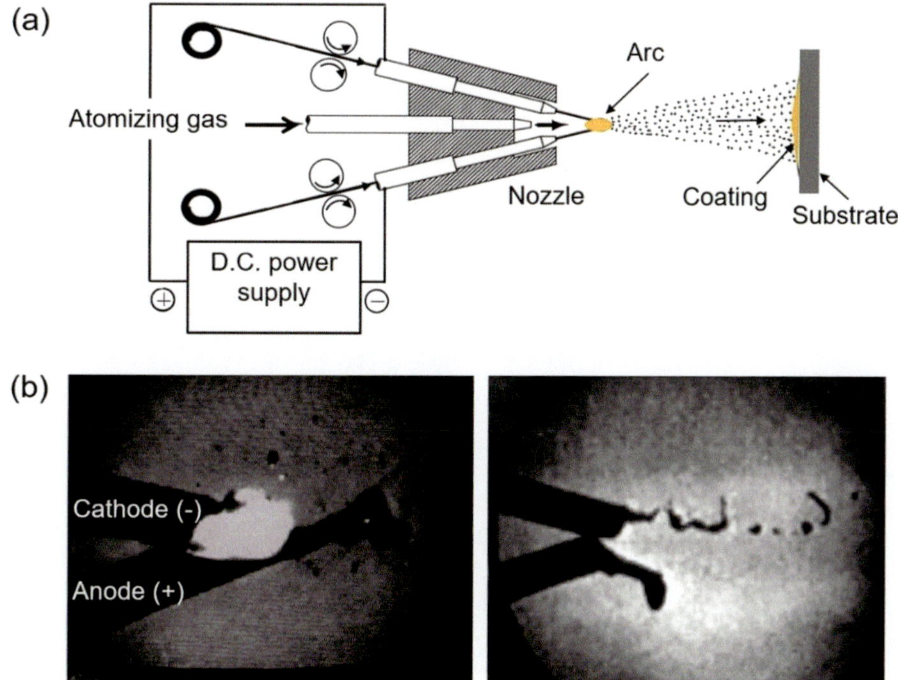

Fig. 2.28 (**a**) Schematic of twin wire arc spraying setup. (**b**) High-speed images of liquid metal droplet formation with 100 ns exposure time [Hussary and Heberlein (2007)]

melting of the surface of the substrate giving rise to the formation of a molten metal pool. The coating material in the form of powder is introduced into the arc where the individual particles are heated, though not melted, and projected toward the substrate in the molten metal pool where they complete their melting and mixing with molten substrate material. Then they stick onto the substrate where the transferred arc melts them. Depending on the power level of the arc, high powder flow rates can be deposited (up to 30 kg/h). At higher power levels and higher velocities, the process allows reducing drastically the mixing of the molten materials of the substrate and the coating. It is also possible to inject two different types of powders: a metal powder close to the nozzle exit and heated below their melting temperature, and a ceramic one close to the molten bath to be included unmolten within the coating. Whenever necessary, a nozzle surrounding the arc is added to the system allowing the introduction of a shielding gas flow in order to protect the molten metal from oxidation. The constriction of the arc favors the formation of a columnar-shaped arc, minimizing the effect of arc length variation on energy density. Regular steel parts with an appropriate PTA coating can exhibit superior corrosion and wear behavior even compared to specialty alloys. Mixing various metal alloy and ceramic powders and controlling their mix proportion can produce the coating with optimal properties. Moreover, as the metallic bond is formed between the coating and the base metal, the coating exhibits excellent peeling resistance. Thick coatings can be easily produced compared to other plasma coating methods. PTA coating can be used in various industries since it can achieve a high hardness at high temperatures and excellent wear resistance, burning resistance, and corrosion resistance. Among the metals often used in PTA are stainless steels, nickel alloys, and stellite family. An example of such a coating on the tooth of an excavator is presented in Fig. 2.30a, with a cross section of the coating (Fig. 2.30b) showing a good dispersion of WC irregular particles (with no evidence of melting and spheroidizing). For more details about the process, see Chap. 10.

2.3.3 Material Preparation and Injection

The precursor material used in different Thermal Spray Coating (TSC) technologies varies widely in terms of chemical composition and morphology depending on the process needs. As illustrated in Fig. 2.31, the large majority of these processes are best adapted to use precursors in powder form pneumatically injected into the combustion flame or plasma discharge. Alternately, the precursor material can also be used in the form of wires or cords, as in Wire Arc Spraying (WAS) or Plasma-Transferred Arc (PTA) deposition. The use of cords is of particular interest because of its adaptability for different materials. As shown in Fig. 2.32a, cords are made of wrapped metallic foil forming a ductile envelope that is filled with the coating material in the form of a fine powder.

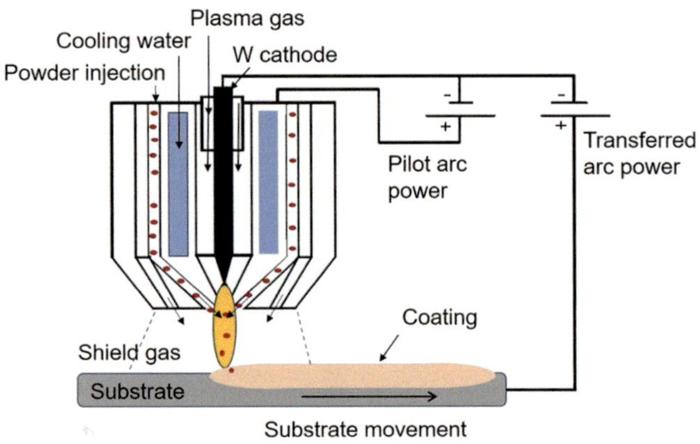

Fig. 2.29 Schematic of plasma-transferred arc (PTA) deposition [Wilden et al. (2006)]

Fig. 2.30 (**a**) PTA-coated tooth of excavator with Ni-base coating and (**b**) cross section of the coating [courtesy of Castolin]

Fig. 2.31 Precursor materials used in thermal spray coating applications

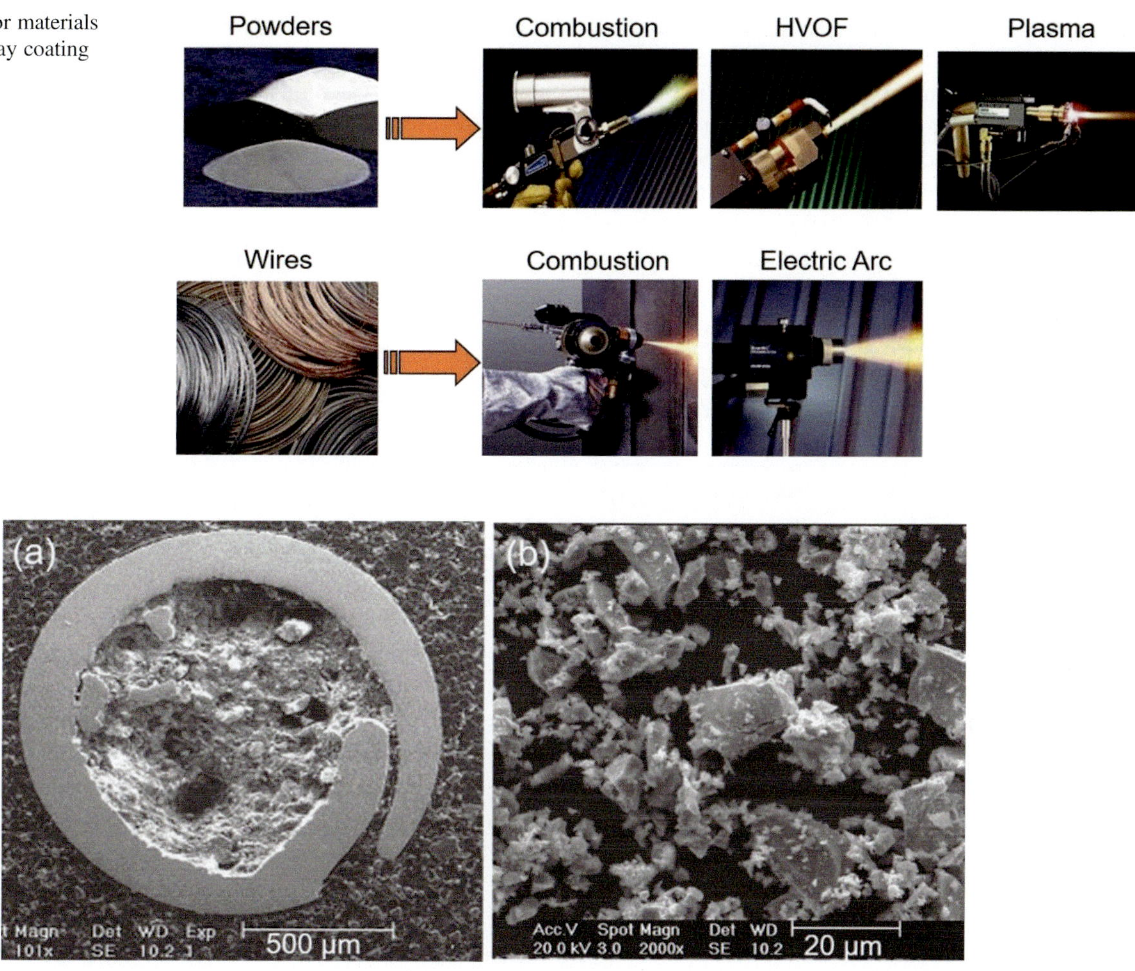

Fig. 2.32 (**a**) Cross section of a cord wire used in wire arc spraying. (**b**) Micrographs of powder filling used in the cords 1.5 µm < d_p < 192 µm [Courtesy of Orlikon Metco Corp]

Obviously, the metal used for the foil should be tolerated in the coating. The powder filling, which represents the bulk of the feed material, can be a metal, alloy, ceramic, or a mixture of them (Fig. 2.31b). More recently the use of precursors in the form of a liquid suspension or solution has attracted increasing attention with very encouraging results in terms of the quality of coating obtained. A detailed discussion of powder characteristics, their principal manufacturing techniques, and the influence of their injection conditions on their trajectories and in-flight particle heating and melting is given in Chaps. 4 and 5.

2.3.3.1 Powder Injection

Powders used in most thermal spray coating applications have particle sizes between 10 and 110 µm, with the exception of induction plasma spraying and PTA deposition, which are capable of using powders with larger particle size of 200 µm or more. The choice of the proper powder size distribution is a key issue that has a major impact on the coating quality. Powders must be chosen according to the

material sprayed, especially the difficulty of melting it, with a size distribution as narrow as possible in order to limit excessive particle dispersion in the flow. It is important to keep in mind that the inertia of a particle is proportional to its mass, which, in turn, is a function of the cube of the particle diameter!

Particle injection into the hot gas stream, whether combustion or plasma flow, is carried out either internally or externally to the spray torch; see Figs. 2.16, 2.21, 2.22, and 2.26. Most powder injectors used are cylindrical (in most cases a straight tube, with an internal diameter between 1.5 and 2.5 mm). During the transport of the particles by the carrier gas from the powder feeder to the spray torch, individual particles collide between themselves, as well as with the walls of the injector, resulting in a significant loss in their momentum with a corresponding reduction in their velocity to almost half that of the transport gas [Bronet et al (1990)]. At the exit of the injector, end of transport line, the particle-transport gas flow assumes a diverged conical trajectory with a half-angle between 5° and 10°. The divergence angle increases with the decrease in the particle diameter below 20 μm, and the decrease in their specific mass below 6000 kg/m^3. The further drop in the particle velocity below that of the transport gas is accentuated in curved sections of the injector because of the increased friction between the particles and the walls of the injectors in the curved section.

The proper control of the particle trajectory in DC thermal spray coating applications is of critical importance in order to achieve the right level of heating and melting of the particles prior to their impact on the substrate. Figure 2.33a illustrates typical trajectories of particle feed of the same dimension, and mass m_p, injected into a DC plasma jet at different initial velocities, $v_{p,i}$. It may be noted that of the three illustrated trajectories, only the one with the injection velocity $v_{p,2}$ represent optimal conditions with a maximum residence time of the particle in the central hot zone of the plasma jet. Particles injected at lower velocities, $v_{p,1}$, or higher ones, $v_{p,3}$, will not achieve the same degree of heating and melting prior to their impact on the substrate and accordingly will give rise to lower-quality coating compared to the one that could be achieved with the central trajectory. It is to be noted that optimal trajectory in this hypothetical simplified case would correspond to a trajectory with an angle of about 3.5–4° with the torch axis as shown in Fig. 2.33a. The corresponding particle dispersion cone in the jet is illustrated in Fig. 2.33b. It is also important to note that the position of the injector relative to the nozzle exit (external or internal injection) and the torch axis plays a key role in determining the particle trajectory [Vardell et al. (2001)]. Particle acceleration must be higher for an injection internal to the nozzle (the jet has not yet expanded). It can be reduced with external injection, this reduction increasing with the distance from the nozzle exit due to the jet expansion. The injector must not be too close to the jet to avoid its overheating, which could induce premature particle melting inside the injector and clogging. On the other hand, if the injector is too far away from hot gases, the injected particles will bypass the plasma or flame jet. The simultaneous injection of powders of different materials such as ceramic and metals and particle size distributions is also possible using multiple injectors located at different positions. This is illustrated in Fig. 2.34 representing the injection of two different powders with two opposite

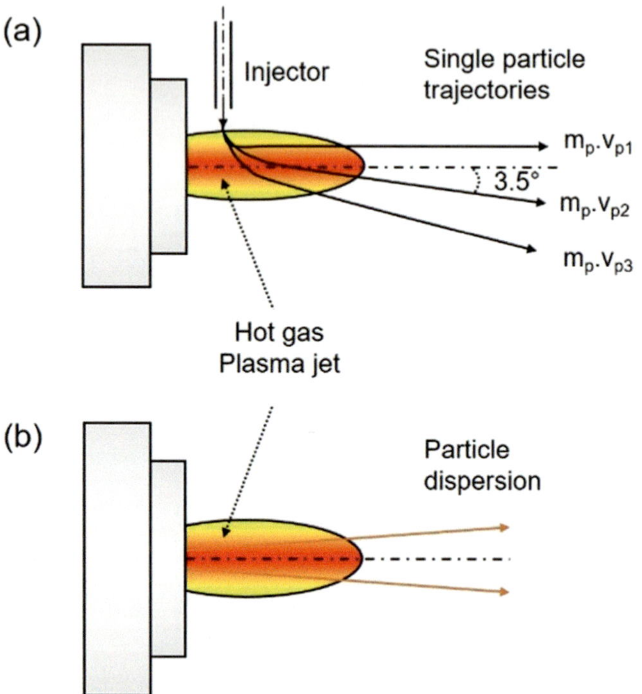

Fig. 2.33 (**a**) Trajectories of single particle injected radially with the same mass but with different injection velocities. (**b**) Dispersion cone of particles in the plasma jet

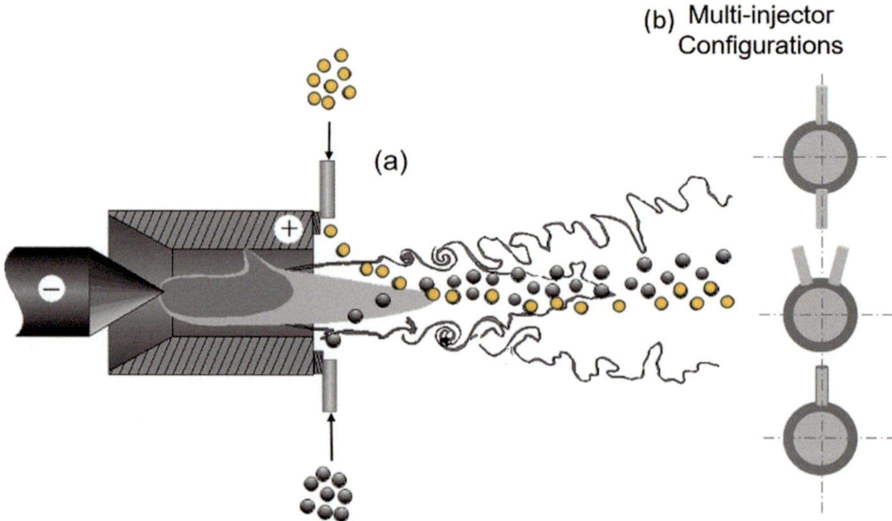

Fig. 2.34 (**a**) Injection of two different powders with opposite radial injectors. (**b**) Illustration of possible configurations of radial injectors

injectors, thus allowing the adaptation of the carrier gas flow rates to each powder. In this figure, different configurations of multi-injectors are also shown.

It is important to point out that the above analysis is obviously an oversimplified presentation of a rather complex problem of powder injection in combustion and plasma flows. An intensive research effort has been devoted over the past four decades to the study of in-flight particle conditions during thermal spray operations leading to the development of elaborate mathematical models supported by extensive diagnostic measurements for model validation. These will be discussed separately in Chaps. 4 and 5 as well as in subsequent chapters of this book dedicated to different thermal spray technologies.

2.3.3.2 Wire, Rod, or Cord Injection

Wires and cords, as those shown in Figs. 2.31 and 2.32, are necessarily made of ductile materials. They are supplied wound on plastic spools. Their diameters are typically in the range between 1.2 and 4.76 mm. Larger, stiffer wires ($d > 2$ mm) are more difficult to feed as uncoiling requires more force and larger wire tension. Nonductile materials can also be used in cored wires, which consist of an outer sheath made of a ductile material surrounding one or more powdered nonductile metals or ceramics. Rods and cords are mainly used for ceramic materials. Rods of ceramic materials 3.16, 4.75, 6.35, or 7.94 mm in diameter have a limited length of 608 mm, which corresponds to spray times of a few minutes. Cords made of a cellulosic or plastic casing are filled with ceramic particles and an appropriate organic binder that starts decomposing at 250 °C and would completely decompose at about 400 °C. Compared to rods, cords with diameters between 3.16 and 6.35 mm are supplied on plastic spools in lengths of 120 m allowing for uninterrupted spraying time for hours.

Wires and cords are introduced into the combustion or plasma torch by simple mechanical devices consisting of electrical or air-driven motors, drive rollers, and associated speed controllers. It is of primary importance to maintain a very stable feeding rate of the material into the flame, plasma, or wire arc. The torque of the motors must be high enough to overcome the friction of the system. Drive rollers must grip the wire, rod, or cord sufficiently to prevent slipping, while not deforming the wire or crushing the rod or cord. The feeding of wires can be achieved by systems, which pull only, push only, or push and pull, the latter being more complex because of the need to synchronize the two drive motors. Besides the wire guides and drive wheel geometry, critical feed issues include wire tension, wire straightness, wire diameter consistency, and wire surface characteristics.

The steady motion of the wire, rod, or cord must be adapted to their diameter and composition, as well as gas composition, flow rate, and plasma power. If the wire

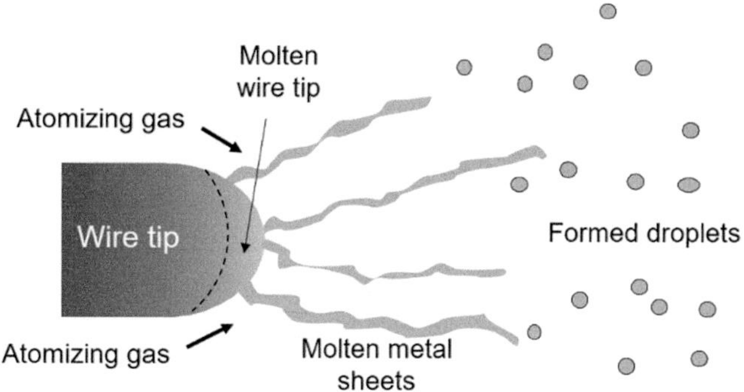

Fig. 2.35 Schematic of droplet detachment from the tip of a molten wire by the atomizing gas [Hussary et al. (2001)]

velocity is too high, the tip cannot melt because its residence time in the flame is too short. When using low-velocity spray guns, such as a flame gun, the molten material is atomized and accelerated by an auxiliary gas fed into the spray gun. The compressed atomizing gas (generally air), fed around the flame (see Chap. 5), creates liquid metal sheets, which are disintegrating, and their flapping motion is responsible for creating showers of drops. The size distribution of drops and their divergence angle depend strongly on the atomizing gas nozzle design and the atomizing gas flow rate. When using HVOF, instead of flame, the melted wire tip can be atomized directly by the hot gas flow.

Compared to powder feeding, the advantage of devices working with wires, rods, and cords is that *only fully melted drops are directed toward the substrate*. The way the molten wire tip form droplets has been the subject of extensive study for wire arc spraying [Hussary et al. (2001)]. The atomizing gas deforms the molten tips into sheets of molten metal that become self-aligned in the flow, the instabilities creating droplets at their extremities (see Fig. 2.35).

2.3.3.3 Liquid Injection

For deposition of coatings with nanometer-sized structure, liquids (suspensions or solutions) are injected into the hot gases in the thermal spray processes. These require, however, that sufficient energy be available to compensate for the cooling of the hot gases by the liquid evaporation process, and for the subsequent heating of the generated vapors, processes where the gases have a sufficient enthalpy are used, such as HVOF or DC or RF induction plasma spraying. The liquid injected into the plasma is atomized either by gas atomization or mechanical injection.

Gas Atomization   Very often coaxial atomization is used. This process consists of injecting a low-velocity liquid inside a nozzle where the liquid is fragmented by a gas (usually Ar because of its high specific mass) expanding within the nozzle bore. The quality and degree of atomization is a function of the "ratio of the gas to suspension volume flow rates" (RGS, generally over 100) or the "atomizing gas and liquid suspension mass ratio" (ALR, below 1), the relative velocity between the liquid–gas, the nozzle design, and the properties of the liquid (density, surface tension, dynamic viscosity). The limitations of this atomization approach include the space, droplet diameter, typically 5–100 μm for an air cap atomizer, divergence of the atomized droplet jet, and droplet velocity distribution.

Mechanical Injection The simplest and most common mechanical atomization of liquids or suspensions is to feed the liquid at very high pressures to an atomization nozzle that breaks down the liquid stream through strong shear forces developed in the nozzle at its atomization exit level. Alternately, a mechanical vibrating device, such as a magnetostrictive rod, is placed at the back side of the nozzle, which superimposes pressure pulsations at variable frequencies (up to a few tens of kHz).

In either of these liquid/suspension injection methods, the key issue is to control the size and the velocity of the droplets. Mechanical injection offers a controlled velocity and a constant size, although it may not be simple to achieve drop diameters below 100 μm. This control is by far more difficult with gas atomization. Besides having a relatively broad distribution of droplet sizes, droplets have a more or less broad velocity distribution and the trajectories show different spray angles.

2.3.4 Substrate Preparation

Substrates for coating applications can be metals, ceramics, composites, glasses, woods, plastics, etc. All of these materials have very different melting or decomposition temperatures, as well as different oxidation rates generating oxide layers that can vary from thin (<50 nm) to thick (>100 nm). This means that with most spray processes, except cold spray and wire arc spraying, the substrate temperature must be controlled during spraying. Whatever the substrate may be, its preparation prior to spraying is a critical step for the bonding and adhesion of the coating. This preparation comprises essentially three steps:

- *Cleaning the surface* to eliminate contamination, particularly from oil or grease. Solvent rinsing or vapor degreasing are commonly used processes to remove oil, grease, and other organic compounds. For porous materials, baking is sometimes necessary.
- *Roughening the surface* to provide asperities or irregularities to enhance coating adhesion and provide a larger effective surface. The substrate roughening in most

cases is achieved by grit blasting. It is also possible to use water jets with pressures between 200 and 400 MPa. In both cases, however, compressive stress, which can reach up to 2000 MPa, is induced in the first tenths of mm below the substrate-roughened surface. This stress contributes to the final stress distribution within the coating and the substrate. Some substrates (e.g., composites, plastics) require special preparations such as acid pickling.

- *A second-cleaning step* after grit blasting; it is often necessary to remove as much as possible of grit particle residues (generally made of a ceramic material) that are embedded into the substrate surface and will create a defect into the coating. The problem is amplified when the coating is exposed to temperature variations because of the expansion mismatch between substrate and grit materials. The removal of such grit residue can be achieved using compressed air jets, or ultrasonic baths or both. Obviously, in this respect, water roughening has the main advantage of not involving the use of any grit and consequently would not require such a second-cleaning step.

2.3.5 Coating Formation

As mentioned earlier in Sect. 2.3.1, the building block of thermal spray coatings is the formation of single splat as a result of the impact of a molten particle on the surface of the substrate. The shape of the splat and its adhesion to the substrate are directly dependent on a large number of parameters including the particle conditions prior to their impact on the substrate (such as particle diameter, velocity, and temperature), substrate condition (substrate temperature and surface roughness), as well as the particle and substrate thermophysical material properties. A schematic representation of the splat formation process is given in Fig. 2.36. This shows that as the particle touches the substrate surface, the molten liquid flows radially outward through kinetic energy conservation while its surface pulls itself inward toward the

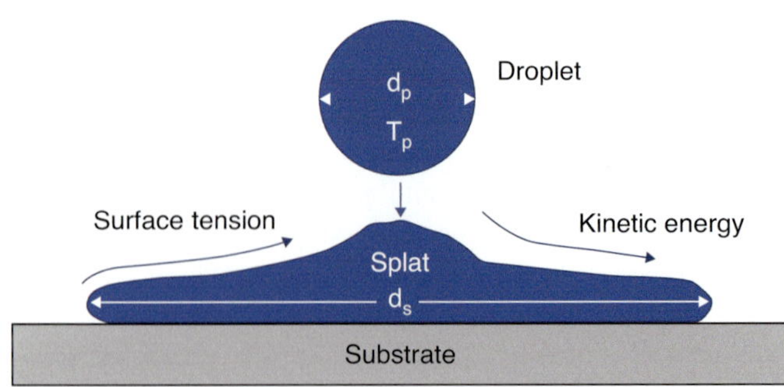

Fig. 2.36 Schematic diagram of splat formation on a smooth surface

center of the splat through surface tension effects resulting in the upper curling of the edges of the splat around its rim. The lower surface of the splat, through heat losses to the substrate, starts to solidify leading to the formation of the final shape of the splat. One of the first papers in this area as summarized by Dykhuizen (1994) reports that the flattening degree of a splat, ξ, defined as the ratio of the splat diameter, D_s, to that of the impacting droplet, d_p, can be expressed as a function of the particle Reynolds number, Re_p, that quantifies the viscous dissipation of the inertia forces by a relationship of the type:

$$\xi = \frac{D_s}{d_p} = C . Re_p^{\alpha} \tag{2.11}$$

According to different authors, C varies between 0.8 and 1.3, while different values of α were reported between 0.125 and 0.2. Typical values reported for ξ were ($2 < \xi < 6$).

Splat formation for different materials and under different plasma spraying conditions has been at the center of numerous studies in this field. An excellent review paper on the subject published by Fauchais et al. (2004) highlights studies carried out on splat formation on smooth substrates. These show that the substrate temperature as well as the angle of incidence of the droplet with respect to the substrate can have major impact on the form of the splat. The concept of a critical transition temperature, T_t, defined and introduced by Fukumoto et al. (1995a) as the temperature at which the splat shape changes from "splash splat" to "disc splat," is presented. Supporting data by Fukumoto et al. (1995b) given in Fig. 2.37 for the impact of nickel droplets on AISI304 SS substrate show such a switch to take place at a temperature around 550 K, which was identified as transition temperature. The change in the form of the splats was accompanied by a major improvement in the cohesion and adhesion of the splats to the substrate.

Similar splash-to-disc transition was also reported for the spraying of alumina on stainless steel. The effect was more pronounced with the deviation of the angle of incidence of the droplet trajectory with respect to the orthogonal direction to the substrate, as shown respectively in Figs. 2.38 and 2.39.

The transition from "splash splats" to "disc splats" was later confirmed with mathematical modeling work by Pasandideh-Fard et al. (2002) who carried out simulation of the splat formation for a 60 μm nickel droplet heated at 1600 K impacting at 73 m/s on a stainless-steel substrate. Computations were carried out for substrate temperatures of 563 K and 673 K. The corresponding contact resistance used in these calculations were 10^{-7} m^2 K/W and 10^{-6} m^2 K/W, respectively. Results given in Fig. 2.40 show a distinct change in the shape of the splat obtained in each of these two cases with transition temperature around 600 K somewhere between these two temperature limits. For most sprayed materials on different substrates, the transition temperature is below $0.3T_m$ (T_m being the particle melting

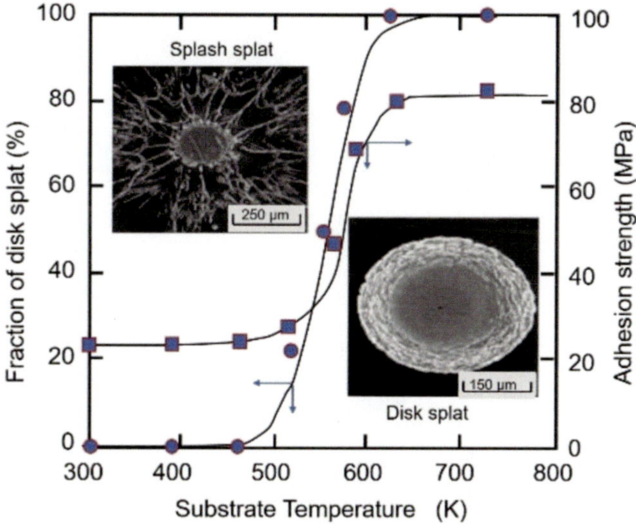

Fig. 2.37 Variation of the adhesive strength of the coating with substrate temperature (Ni-sprayed material with a size distribution 10–44 μm; stainless steel AISI304 substrate) [Fukumoto et al. (1995b)]

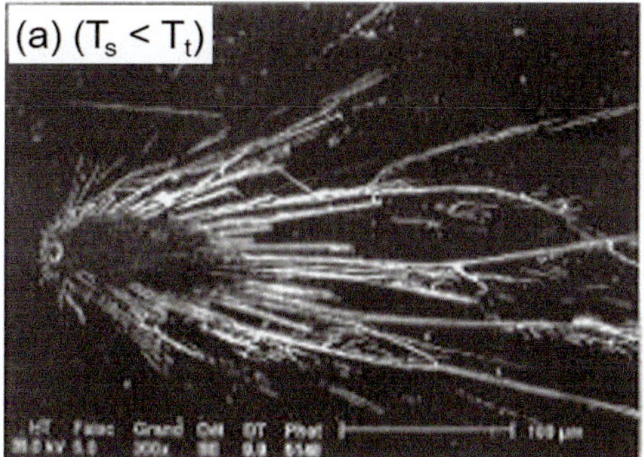

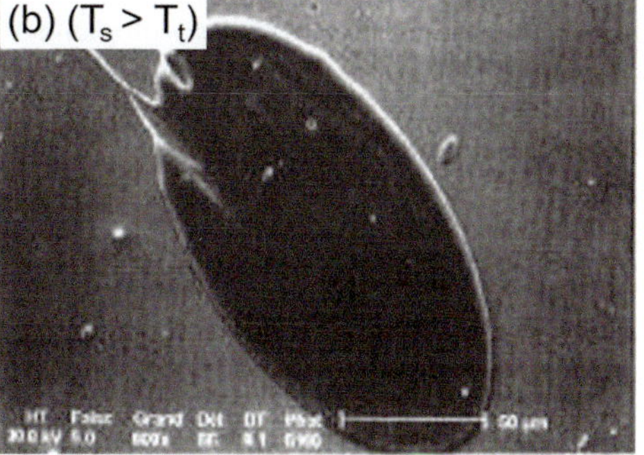

Fig. 2.38 Impact of alumina plasma-sprayed particles ($d_p = 58$ μm, $v_p = 138$ m/s, $T_p = 2400$ K) onto an inclined ASI304 stainless steel (impact angle 30°). (**a**) Room temperature substrate. (**b**) Preheated substrate over transition temperature [Bianchi et al. (1977)]

temperature). Thus, the cooling rate of the flattening particle is practically not affected by the substrate preheating at the transition temperature. The flattening behavior of the drop above the transition temperature is mainly due to desorption of adsorbates and condensates from the substrate surface when preheated above T_t. The contact between the splat and the substrate represents up to about 60% of the splat surface on substrates preheated above T_t, compared to possibly less than 20% on cold substrates. Accordingly, when spraying on rough substrates preheated above T_t, the coating adhesion is higher by a factor of 3–4 compared to that obtained on a cold substrate.

In real thermal spray situations, however, the phenomena involved are much more complex since the process involves the piling up of consecutive layers of splats, with every layer deposited acting as a substrate for the next layer. The time between two successive impacts is typically in the range of ten to a few tens of µs. Thus, the next particle impacts on an already solidified splat. The splat layering is controlled by the powder flow rate, the process deposition efficiency, and finally the spray pattern including the relative torch–substrate velocity. Changes taking place in the morphology of the substrate as well as its temperature will accordingly affect the nature of the splats and subsequently the quality of the deposit. Examples of defects in the coating that can be initiated at this stage are illustrated in Fig. 2.41. These include "shadow effect" resulting from the partial overlap of one splat over another, "narrow holes and gas inclusions" resulting from the bridging of a splat over two previous one, inclusions due to "unmolten" particles, and particle fragments due "exploded" particles. Each of these sources of defects as well as others such as surface oxidation and entrapments of debris contribute to the overall structure of the coating as illustrated in Fig. 2.42.

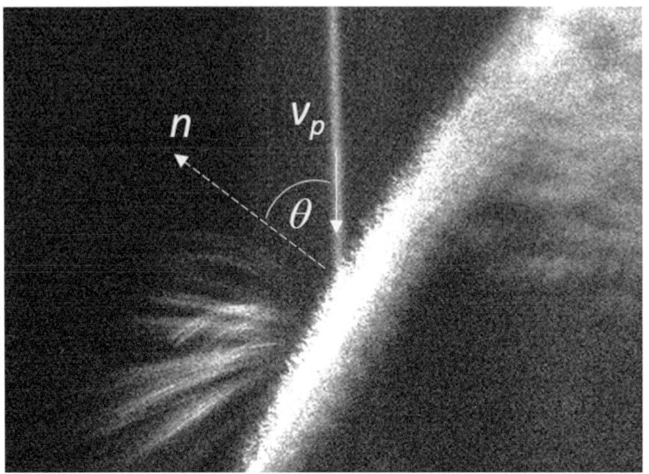

Fig. 2.39 Impact splashing on inclined substrates [Escure et al. (2003)]

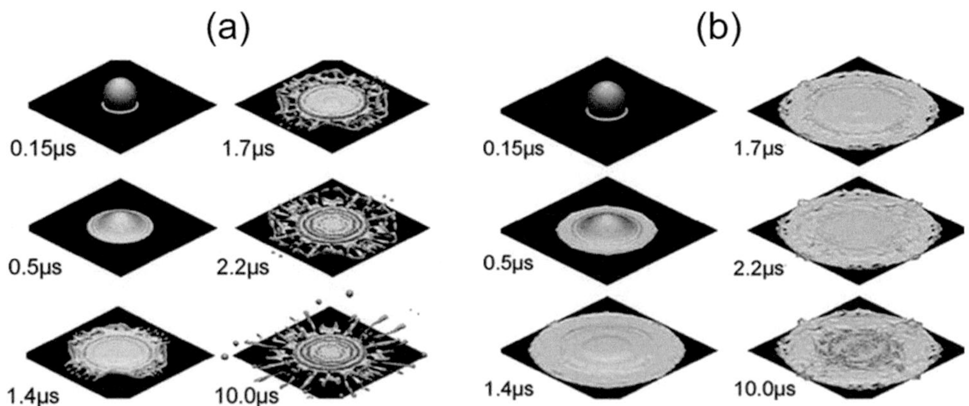

Fig. 2.40 Simulation results showing the different stages of splat formation for a 60 µm nickel droplet heated at 1600 K impacting at 73 m/s on a stainless-steel substrate at temperature of (**a**) $T_s = 563$ k and (**b**) $T_s = 673$ K [Pasandideh-Fard et al. (2002)]. (Reprinted with kind permission from Springer Business Media, Copy© ASM International)

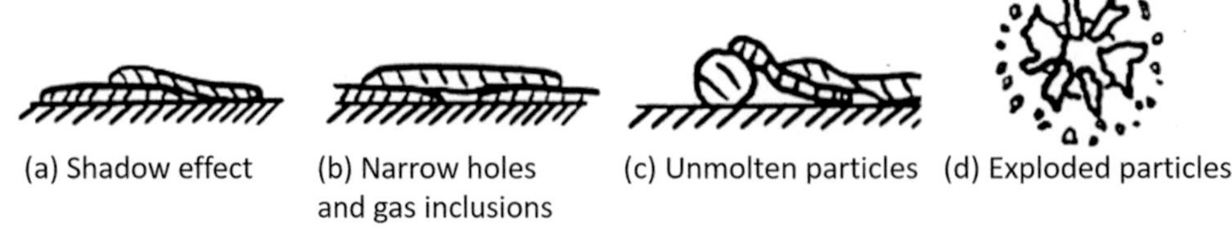

Fig. 2.41 Potential sources of coating defects in thermal spray operation

Coatings obtained by conventional spraying (sprayed particles with sizes of a few tens of micrometers) have thicknesses between about 50 μm and a few millimeters. Figure 2.43 presents the cross sections of two DC plasma-sprayed coatings of a metal (stainless-steel 304 L) and a ceramic (YSZ) on low-carbon steel. The microstructure of the SS coating given in Fig. 2.43a shows oxide inclusions, in spite of the fact that it was sprayed using Ar-H$_2$ as plasma gas, unmelted particles, and porosities. The microstructure of the YSZ coating shown in Fig. 2.43b reveals a lamellar structure, a limited number of unmelted particles, and visible porosity.

The microstructure of a cold-sprayed coating is significantly different from the plasma-sprayed one. As shown in Fig. 2.44, the microstructure of a cold-sprayed copper powder on a copper substrate shows the deformation of the sprayed spherical particles with a rather good contact between layered particles and few tiny pores less than 1%. No oxide inclusions are noticeable with an oxygen content of the coating of less than 0.1 wt%, which corresponds to that of the original particles [Stoltenhoff et al. (2002)].

In the case of coatings resulting from *suspension or solution plasma spraying*, the mechanism of coating formation is distinctly different from conventional plasma spraying processes. As mentioned in Sect. 2.3.3.3, these relatively new approaches in thermal spray coating were triggered by the need to deposit superior quality nanostructured coatings for which the precursor material would have to be in the micron or submicron particle size range [Fauchais et al. (2008)]. Such powders, because of their relatively small particle size and low inertia, would be impossible to feed into a combustion flame or a plasma flow using conventional pneumatic injection approach. The approach developed to overcome

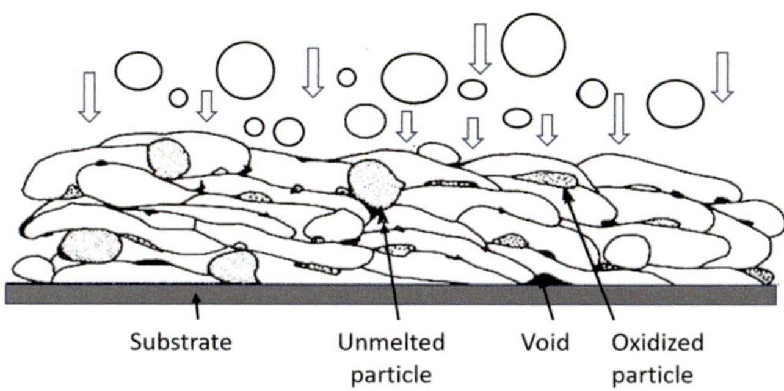

Fig. 2.42 Schematic of thermal spray coating formation and examples of possible defects (Reprinted with kind permission from Springer Science Business Media [45], copyright © ASM International)

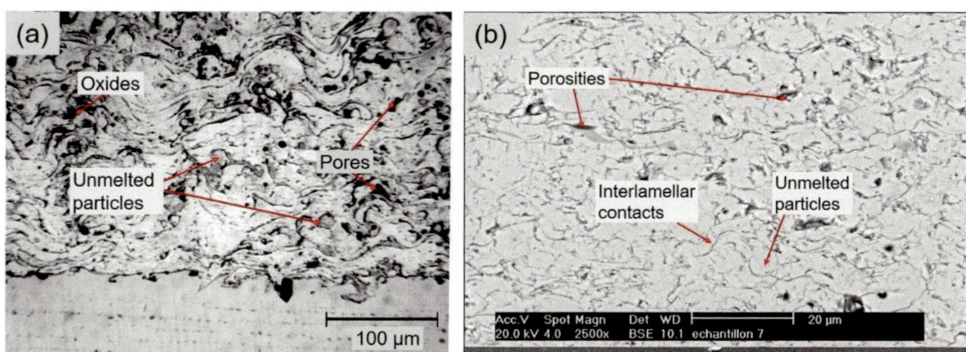

Fig. 2.43 (**a**) Stainless-steel coating 304 L deposited by air plasma spraying on a low-carbon steel 1040 substrate. (**b**) Plasma-sprayed coating of YSZ (8 wt%) on super alloy

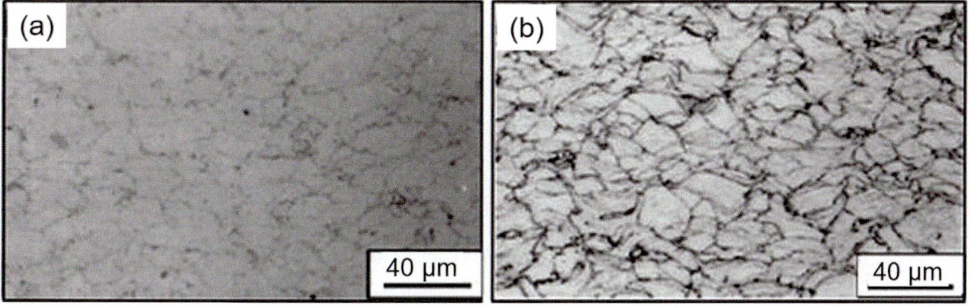

Fig. 2.44 Optical microscopy of the cross section of a cold-sprayed Cu coating: (**a**) polished, (**b**) etched. (Reprinted with kind permission from Springer Science Business Media [Stoltenhoff et al. (2002)], copyright © ASM International)

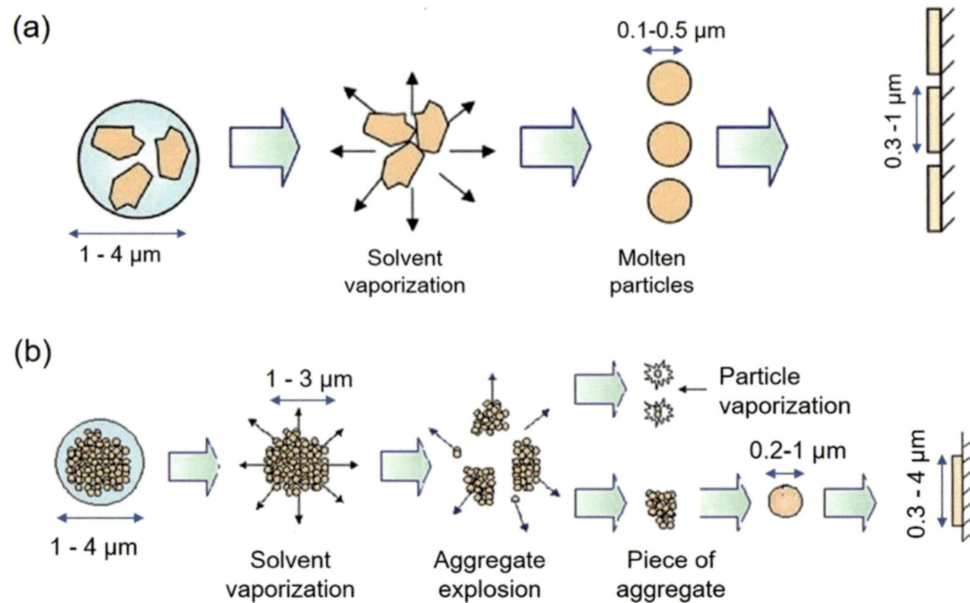

Fig. 2.45 Schematic representation of the different stages involved in suspension coating process when (**a**) containing attrition-milled YSZ particles and (**b**) containing Tosoh YSZ nanometric powders [Fauchais et al. (2008)]

such a problem is to either synthesize the particle in-situ from a solution, or to mix the fine/ultra-fine precursor powder with an appropriate liquid to form a suspension that is injected and atomized into the hot stream. As the formed droplets of the suspension are entrained by the hot gases, they are heated, and the liquid component evaporated releasing the individual fine/ultra-fine solid particles in the flow. The released particles will heat further, melt, and on impact with the substrate form splats or spherical microparticles. A schematic representation of the steps involved and an example of the formed micro splats are given in Figs. 2.45a and 2.46, respectively, for the case where the solid particles in the liquid suspensions were made by attrition-milled YSZ implying a precursor particle size in the 1–2 µm range. Alternately when using a nanometric size precursor powder in the suspension, with a submicron particle size in the range of 0.1–1 or 2 µm, the predominant mechanism, as illustrated in Fig. 2.45b, would deal with aggregates of the powder that could either melt in the hot stream giving rise to the formation of smaller molten droplets in the 0.2–1 µm size range or vaporize. An example of the YSZ splats obtained in this case is given in Fig. 2.47. The challenge in this area lies mostly in the improvement of the splat adhesion to the substrate and the improvement of the particle feed rate and deposition efficiency are greatly hindered by the disproportionately high energy load that is consumed for the evaporation of the liquid component in the suspension and the heating of the generated vapors to free stream temperatures.

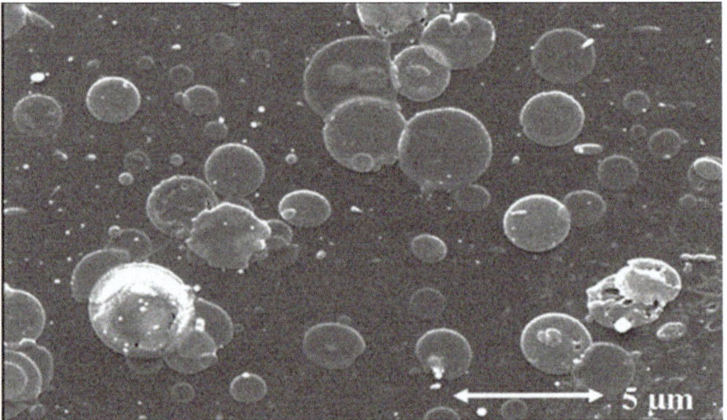

Fig. 2.46 Splats and particles (re-condensed or re-solidified) collected at 40 mm downstream of the nozzle exit on a smooth stainless-steel substrate when spraying attrition-milled YSZ particles (between 0.1 and 3 lm) with an Ar-H$_2$ (45–15 slm) plasma jet, hp. = 17.9 MJ/kg [Fauchais et al. (2008)]

In the case of Solution Precursor Plasma Spraying (SPPS), Gell et al. (2008) described the deposition mechanism, schematically illustrated in Fig. 2.48, as being one of solute evaporation including breakup, gelation, and precipitation followed by pyrolysis, sintering, melting, and crystallization. Upon impact on the substrate, fully melted particles will produce splats, which, depending on the spraying conditions, can have splat diameters almost twice that of the original granule or droplet. If the standoff distance increases, molten particles may solidify and crystallize before impact. For droplets traveling in the low-temperature regions of the jet,

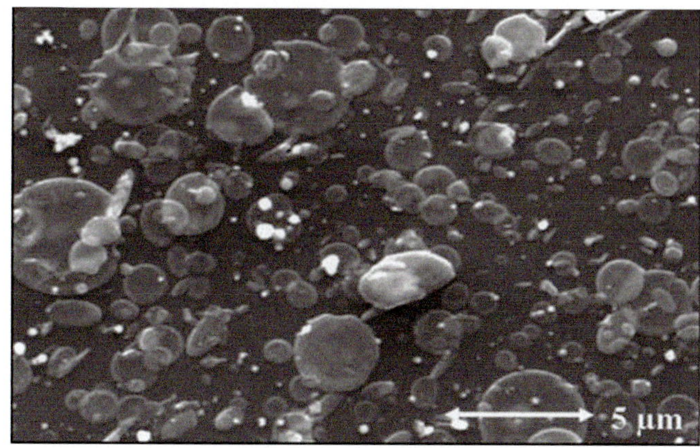

Fig. 2.47 Splats and particle collected at 40 mm downstream of the nozzle exit with a smooth stainless-steel substrate when spraying Tosoh-nanometric particles (between 0.1 and 3 lm) with an Ar-H$_2$ (45–15 slm) plasma jet, hp. = 17.9 MJ/kg [Fauchais et al. (2008)]

some precursor solution can reach the substrate in liquid form. Micrographs of typical splat morphologies, crystals, ruptured shell, and vapor-deposited films are shown in Fig. 2.49.

According to Gell et al. (2008), the SPPS produces a unique strain-tolerant, low-thermal-conductivity microstructure consisting of:

- Three-dimensional micrometer and nanometer pores
- Through-coating thickness (vertical) cracks
- Ultra-fine splats
- Inter-pass boundaries

Both thin (0.12 mm) and thick (4 mm) TBC coatings of YSZ were produced by the SPPS process. The volume fraction of porosity can be varied from 10% to 40% while

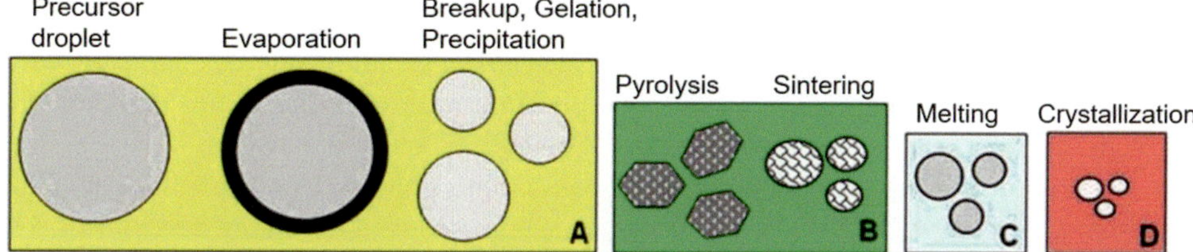

Fig. 2.48 Schematic representation of the different stages involved in SPPS coating process

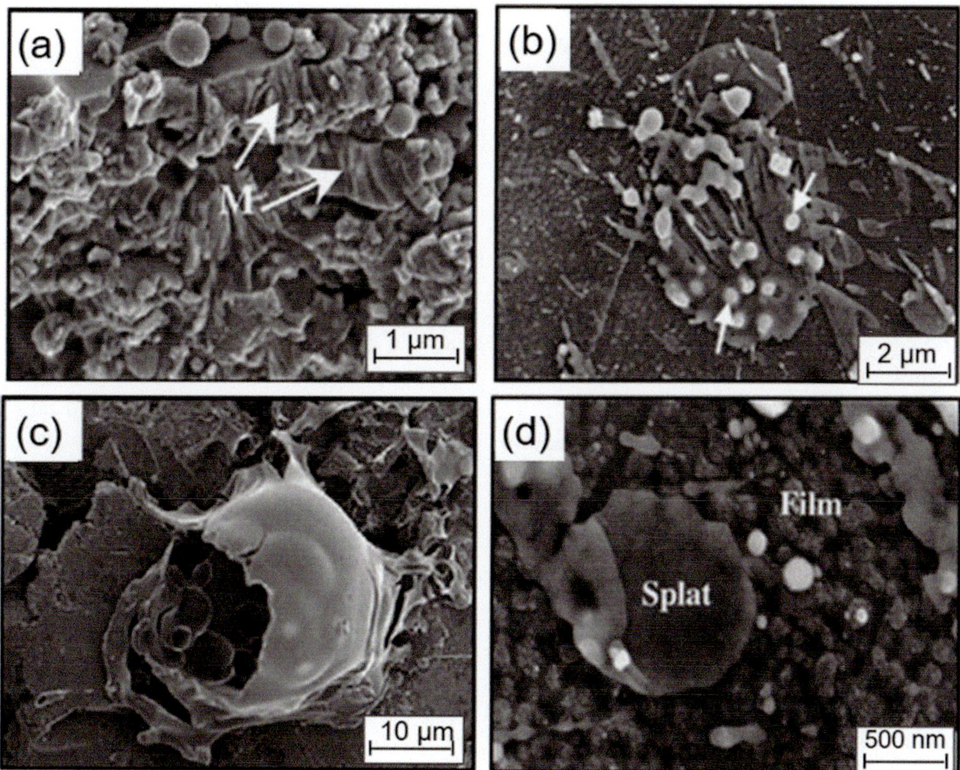

Fig. 2.49 Micrographs of thermal-sprayed coatings obtained using SPPS. (**a**) Fine splats, (**b**) crystallized spheres, (**c**) ruptured shell, and (**d**) vapor-deposited film

retaining the characteristic of ultra-fine splat microstructure with enhanced splat-to-splat bonding and strain relieving through vertical cracks. The substitution of liquid feed over the powder feed results in:

- Rapid and complete mixing of precursor material
- Precursor composition flexibility
- Rapid screening of compositions
- The creation of nanostructured coatings
- The creation of unique compositions, phases, and microstructures

Current challenges of the SPPS method include:

- Precursor solution chemistry
- Control over droplet entrainment
- Control of deposit homogeneity

2.3.6 Residual Stresses

The different steps in the thermal spray coating operation, whether for substrate preparation, coating formation, or post-deposition surface finishing, are responsible for the development of a broad range of stresses in the part being coated. According to Clyne et al. (1996), these include:

- *The splat quenching stress* (always tensile) due to cooling of each individual splat during its formation on the substrate. These stresses increase with the quality of the contact between the splat and the substrate or the splat and the previously deposited layer. This contact is improved by substrate preheating above the transition temperature.
- *The compressive stress* is due to particles impacting at high velocity during substrate preparation or coating formation (HVOF, D-Gun, cold spray). It is equivalent to shot peening, and can be affected by the coating temperature. These are often more important than the quenching stresses.
- *The temperature-gradient stress* develops especially within low-thermal-conductivity coatings. This stress must be avoided by suppressing or significantly reducing the temperature gradients within the coating during its formation.
- *The phase change stress* occurs as a result of the transformation into a new phase of the coating material following its deposition, which is smaller or larger than in the preceding phase. This is, for example, the case when sprayed coatings essentially composed of γ alumina transform into α alumina at about 950 °C. This situation is generally

encountered during service conditions, but it can also occur if too high temperature gradients are encountered during spraying of low-thermal-conductivity materials. Such transformations must be avoided because ceramic coatings are generally destroyed when this transformation occurs.

- *Stress due to mismatch of the coefficient of thermal expansion* occurs during substrate and coating cooling from their deposition temperature to room temperature. This stress increases with the increase in the mean deposition temperature and the importance of the difference between the expansion coefficients of coating and substrate. Here also, for low-thermal-conductivity coatings, a too fast cooling can induce damaging temperature gradients.

The final stress distribution within the substrate and the coating is the addition of all the previously described stresses, including those created in the substrate by its preparation such as grit blasting. Service conditions will bring new stress, generally due to temperature gradients developed between coating surface and substrate interface either during heating or cooling phases. This stress will be added to the residual one and can result in coating detachment. The control of these different stresses depends strongly on coating and substrate temperature during spraying and is of primary importance for coating service conditions. With a too high residual stress, coatings can peel off, especially thick ones [Clyne et al. (1996)].

2.3.7 Brief Descriptions of Thermal Spray Applications

Choosing a thermal spray material for an application and the thermal spray technology to use for its application is relatively challenging because coating properties are not as predictable as those of conventional materials. With the wider acceptance of thermal spray technology, a broader number of applications are well established, and new ones continue to be developed and qualified for industrial-scale applications. The optimal pairing of base material and surface coating properties is well recognized as an effective tool for improving performance compared to homogeneous materials. The aerospace industries have been a leader in this area by providing an ideal environment and technological incentives for testing and integrating a wide range of functional coatings. The technology has advanced to a point where it has increased the credibility and reliability of coatings. They have led to applications in other markets such medical, energy, oil and gas, chemicals, metal processing, textile, paper, printing, and more recently in the automotive industries. In choosing a specific thermal spray technology for a specific application, it must be kept in mind that different spray techniques are complementary and not competitive. The most common applications of thermal spray technology include the following.

Wear-Resistant Coatings These are used against abrasion, erosion, cavitation wear, galling, fretting, friction, etc., and often more than one aspect of wear can be addressed. For example, cermets combine hardness, ductility, and acceptable thermal conductivity. Nonstick materials with low-friction coefficients can be combined with hard coatings. Self-lubricating materials can be deposited (for example, materials containing free carbon, or MoS_2). Very hard materials can be obtained with ceramics, which can also be combined (for example, ZrO_2 and Al_2O_3). Corrosion and wear-resistant materials can be combined.

Corrosion-Resistant Coatings Many materials are used such as zinc, aluminum (which represents the broadest atmospheric protection), nickel-base alloys, copper–nickel alloys, chemically inert ceramics, plastics, and noble metals. One of the big concerns with thermally sprayed coatings used against corrosion is the interconnected porosity. To make the coatings impervious, high-energy thermal sprays are used, or sealants, of course, depending on service temperature. Coatings can achieve protections against high-temperature (<1500 °C) corrosion or oxidation. Cold spraying seems to offer interesting possibilities against corrosion due to their high density, especially for Al, Zn, Zn–Al coatings.

Thermal Insulation Coatings Thermal barrier coatings (TBCs) became a great success of thermal spray technology. They are made of stabilized or Partially Stabilized Zirconia (PSZ; a low-thermal-conductivity ceramic material: $K < 1.5$ W/m.K when sprayed) sprayed on a superalloy bond coat protecting the substrate from oxidation or corrosion. It is worth to note that corrosion by hot gases decreases through the temperature drop achieved within the ceramic coating. The latest TBCs are vertically segmented to have a better compliance to the stress generated when heating and cooling them. Much effort is also devoted to developing deposition processes for nanostructured TBCs, and new materials, for example, with pyrochlore structure, which has no phase transformation at high temperature as zirconia does.

Abradable and Abrasive Coatings These coatings are used in gas turbine engines for clearance control. Blade tips are designed or coated to make grooves in the relatively soft abradable coating face. The coating thus creates a gas-path seal that prevents gases from bypassing the blades, increasing the engine performance. Nickel/graphite, nickel/bentonite, nickel/polyester, and aluminum/polyester are used or more generally a metal matrix with a nonmetallic filler (graphite, polyester, polyimide, boron nitride, or a friable mineral). The abrasive coatings made of oxides or carbides, which can also be embedded in a metallic matrix, are applied to the blade tip to reduce wear as it rubs against the abradable coating.

Electrically Conductive Coatings Electrical contacts are made of silver, copper, aluminum, tin alloys, and bronze alloys. Thus electrical conductivity depends on the spray technique used and is generally between 40% and 90% of that of the bulk material.

Electromagnetic Shielding It can reduce electromagnetic or radio frequency interference, which can damage electronic components or disturb low-level signals. Such shielding is, for example, achieved by a metallic coating, often wire arc sprayed, on the inside surface of an instrumentation cabinet, which is usually made of plastic or a composite material. In another application, conductive material coatings are cold sprayed on aluminum surfaces. Microwave Integrated Circuits (MIC) are made by plasma-sprayed Mg–Mn ferrite. Al, Ta, and Nb capacitor electrodes are sprayed in air. Unusual conductors such as aluminum titanate, barium titanate, molybdenum di-silicide, and other ceramic compounds are also sprayed.

Electrically Resistive or Insulating Coatings Electrically insulating surfaces (mostly oxides and especially alumina) are used for dielectrics (for example, in ozonizers), high-temperature strain gages, oxygen sensors (ionic conduction of zirconia), and insulation of heating coils. Resistors are made of NiCr or super alloys doped with alumina particles to tailor their electrical conductivity.

Electrochemical Active Coatings These, along with Solid Oxide Fuel Cells (SOFCs) and High-Temperature Electrolyzer (HTE), are very efficient devices that electrochemically convert fuel energy into electricity and heat (SOFC) or Electricity into hydrogen (HTE). SOFCs are well suited as cogeneration units for providing electricity and heat to buildings. They could greatly reduce the consumption of fuels, lowering greenhouse gas and pollutant emissions, but their development is hindered by high cost of manufacturing through wet-ceramic techniques (tape casting and screen printing). The HTE, working according to the reverse reaction of SOFC, produces hydrogen from water vapor. Plasma spray processes instead of wet-ceramic techniques could significantly reduce manufacturing costs, and many works, with industrial applications, have been devoted to these techniques.

Heat Transfer Improvement   It is achieved with high-thermal-conductivity coatings such as Cu or Al, or BeO for ceramics when electrical insulation is mandatory.

Medical Coatings   These can be bioactive or biocompatible or bioinert. Bioactive coatings are made of hydroxyapatite, or tricalcium phosphate, that emulates the characteristics of bone material and induces the growth of new bone attached to the implant. Biocompatible materials are made of titanium alloys, mainly Ti-6Al-4 V, and Ti-6Al-7Nb, Ti-13Nb-13Zr, etc., as well as TiO_2. Titanium alloys have big pores at their surface allowing the growth of new tissue to secure the implant. SiO_2–CaO–P_2O_5-based bioactive glasses and glass ceramics are attractive materials for biomedical applications, because of the excellent levels of bioactivity. The main applications are dental, hip, and knee implants.

Dimensional Restoration Coatings   These are used for salvage of worn or over-machined parts. All spray processes are used for this salvage work and very often NiCr and NiAl materials are used. With plasma-transferred arc processes, coating thicknesses can reach a few mm. Cold spray has been successfully used to impart surface protection and restore dimensional tolerances to magnesium alloy components.

Free-Standing Shapes   Parts are fabricated from hard-to-machine materials by building a coating on a removal form. This process is often less costly than conventional processing methods such as sintering or hot isostatic pressing. Ceramic membranes, ceramic tubes (1 m in diameter, 10 m long, and with wall thickness of a few mm), rocket nozzles, and ceramic or refractory material crucibles are manufactured this way.

Nuclear Applications   In addition to the usual mechanical applications (however with short life time elements: no cobalt, for example), thermally sprayed coatings are used in Tokomak reactors and magnetic fusion devices.

Polymer Coatings   These are used as protection against chemical attack, corrosion, or abrasion. Unlike inorganic coatings, polymer coatings may have equal or better properties than their cast or molded counterpart.

To conclude, this list is far from being exhaustive, but it shows the wide variety of applications of thermal spray coatings. A more detailed discussion of thermal spray applications is given in Chaps. 19 and 20.

2.4   Summary and Conclusions

When considering "Thermal Spray Coating (TSC)" for a specific application, it is important to realize that it is only a specific niche area in the broad field of "surface modification technologies," which is historically split between "surface transformation" and "surface coating." Technologies in each of these two groups have their unique features, advantages, and limitations. In order to stay within the scope of this book, the focus is placed on the coating approach, introducing in this chapter a brief outline of the most widely used coating techniques. These are grouped into following three sub-categories based on the characteristics of the coating produced:

- *Coatings deposited at the atomic level*, which are essentially thin films of thickness typically in the few micrometers up to 20 or 30 μm. These include:
  - Plating, whether electrodeless or electroplating.
  - Physical Vapor Deposition (PVD), including evaporation by resistive heating, electron and ion beam plating, sputtering, plasma-assisted PVD, and laser-assisted PVD.
  - Chemical Vapor Deposition (CVD), including low-pressure CVD, plasma-enhanced CVD, and laser-enhanced CVD.
- *Thermal-sprayed coatings*, which include:
  - Combustion-based processes, such as flame spraying, detonation gun spraying, High-Velocity Oxy-Fuel (HVOF), and high-velocity air fuel (HVAF).
  - Plasma Spray Coating (PSC), including Wire Arc Spraying (WAS), DC-Plasma Spraying (DC-PS), RF induction plasma spraying (RF-IPS), Atmospheric Plasma Spraying (APS), and controlled atmosphere plasma spraying (CAPS).
- *Plasma-Transferred Arc (PTA) deposition*, which is distinguished from other plasma-based processes by the fact that the coating is generally quite thick and is fused into the substrate.

A comparison of the principal characteristics of these different coating techniques is given in Table 2.1 based on the work of T. Bernecki [Cartier M. (2003)]. The semiquantitative data provided can be useful in building-up a comparative image of the different technologies, their principal advantages and limitations, and the areas in which they would be best adapted for different coating needs.

Table 2.1 General characteristics of main coating methods [Cartier M. (2003)]

Characteristics	Plating	PVD	CVD	Thermal spray			
				Combustion	APS	CAPS	PTA
Equipment cost	Low	Moderate to high	Moderate	Low	Moderate	High	Low
Operating cost	low	Moderate to high	Low to moderate	Low	Moderate	High	Low
Process environment	Aqueous solution	High vacuum	Medium to vacuum	Atmospheric	Atmospheric	Soft vacuum	Atmospheric
Coating geometry	Omnidirectional	Line of sight	Omnidirectional	Line of sight	Line of sight	Line of sight	Line of sight
Coating thickness (μm)	Moderate 10-1000 μm	Very thin to Moderate 1–30 μm	Moderate 10-1000 μm	Thick 50 μm–10 mm	Thick 50 μm–10 mm	Thick 50 μm–10 mm	Thick 50 μm–10 mm
Substrate temperature	Low	Low	Moderate to high	Moderate to high	Moderate to high	Moderate to high	High
Adhesion	Moderat	Moderate	Very good to excellent	Good	Good to very good	Good to very good	Excellent
Surface finish	Coarse to glossy	Smooth to high gloss	Smooth to glossy	Coarse to smooth	Coarse to smooth	Coarse to smooth	Coarse
Coating material	Metals	Metals, ceramics & polymers	Metals, ceramics & polymers	Metals, ceramics & polymers	Metals, ceramics & polymers	Metals, ceramics & polymers	Metals, composites

Nomenclature

Units are indicated in parentheses; when no units are indicated, the parameter is dimensionless.

Latin Alphabet

a — Particle surface area (m^2)

c_p — Specific heat (J/kg K)

d_i — Nozzle internal diameter (mm)

d_p — Particle diameter (μm)

e_p — Effusivity $e_p = \sqrt{\rho_p \cdot \kappa_p \cdot c_p}$ $\left(J/m^2 \ K \ s^{0.5} \right)$

E_A — Activation energy (J)

h — Transfer coefficient (W/m^2 K)

k — Specific reaction rate constant

k_B — Boltzmann constant (1.38×10^{-23} J/K)

L_m — Latent heat of melting (J/kg)

m_p — Particle mass (kg)

q_{cv} — Heat-transferred rate (W)

Q_F — Energy necessary to melt the particle (J)

v_g — Gas velocity (m/s)

v_p — Particle velocity (m/s)

T_s — Gas temperature (K)

Greek Alphabet

ε — Emission coefficient

μ_g — Viscosity (Pa.s)

K — Thermal conductivity (W/m K)

ρ_p — Particle density (kg/m^3)

σ_S — Stefan–Boltzmann constant (5.671×10^{-8} W/m^2 K^4)

τ — Residence time of the particle in the hot jet (s)

References

Argarwala, R., and V. Argarwala. 2003. Electroless alloy/composite coatings. *A Review Sädhand* 28 (3&4): 475–493.

Bach, F.W., A. Laarmann, and T. Wenz, eds. 2006. *Modern surface technology*. Weinheim: J Wiley-VCH Verlag GmbH & Co.

Baptista, A., F.J.G. Silva, J. Porteiro, J.L. Míguez, G. Pinto, and L. Fernandes. 2018. On the physical vapour deposition (PVD): Evolution of magnetron sputtering processes for industrial applications. *Procedia Manufacturing* 17: 746–757.

Bhattacharya, S., and M. Meyer. 2003. *Large firms and the science-technology interface, patents, patent citations, and scientific output of multinational corporations in thin films*. Jointly published by Akadèmiai Kiado, Budapest Scientometrics, and Kluwer Academic Publishers, Dordrecht, 58 (2): 265–279.

Bianchi, L., A. Denoirjean, F. Blein, and P. Fauchais. 1977. Microstructural investigation of plasma sprayed ceramic splats. *Thin Solid Films* 299: 125–135.

Bobzin, K., T. Brögelmann, and N.C. Kruppe. 2018. Enhanced PVD process control by online substrate temperature measurement. *Surface and Coatings Technology* 354: 383–389.

Boulos, M. 1992. RF induction plasma spraying: State-of-the-art review. *Journal of Thermal Spray Technology* 1: 33–40.

———. 1990. Particle slip velocities in air jets under atmospheric pressure and soft vacuum conditions. *Canadian Journal of Chemical Engineering* 68: 353–359.

Bunshah, R.F., ed. 1982. *Deposition technologies for films and coatings*. Park Ridge: Noyes Publications.

Cartier, M. 2003. *Handbook of surface treatments and coatings*. New York: ASME Press.

Champagne, V.K., ed. 2007. *The cold spray materials deposition process*. Cambridge: Whoodhead Publishing Ltd.

Chan, S.-H., S.-H. Chen, W.-T. Lin, M.-C. Li, Y.-C. Lin, and C.-C. Kuo. 2013. Low-temperature synthesis of graphene on Cu using plasma-assisted thermal chemical vapor deposition. *Nanoscale Research Letters* 8: 285–290.

Clyne, T.W., and S.C. Gill. 1996. Residual stresses in thermal spray coatings and their effect on interfacial adhesion: A review of recent work. *Journal of Thermal Spray Technology* 5 (4): 401–418.

Costa, M. 1997. Toxicity and carcinogenicity of Cr(VI) in animal models and humans. *Critical Reviews in Toxicology* 27: 431–442.

Cruz-Barba, L.E., S. Manolache, and F. Denes. 2003. Generation of Teflon-like layers on cellophane surfaces under atmospheric pressure non-equilibrium SF6-plasma environments. *Polymer Bulletin* 50: 381–387.

Dayan, A.D., and A.J. Paine. 2001. Mechanisms of chromium toxicity, carcinogenicity and allergenicity: Review of the literature from 1985 to 2000. *Human & Experimental Toxicology* 20: 439–451.

Dhiman, R., A.G. McDonald, and S. Chandra. 2007. Predicting splat morphology in a thermal spray process. *Surface & Coatings Technology* 201: 7789–7801.

Directive 2002/95/EC. 2003. of the European Parliament and of the Council on the restriction of the use of certain hazardous substances in electrical and electronic equipment. Off. J. Eur. Union 1, 27, pp. 19–37.

Djokic, S.S., and P. Cavallotti. 2010. Electroless deposition: Theory and applications, in "Modern aspects of electrochemistry" No. 48, Chapter 6, pp 251–289 (pub.) Springer, New York, 2010.

Dobkin and Zuraw. 2003. *Principles of chemical vapor deposition.* Dordrecht: Springer-Science+ Business Media B.V.

Doll, G.L., B.A. Mensah, H. Mohseni, and T.W. Scharf. 2010. Chemical vapor deposition and atomic layer deposition of coatings for mechanical applications. *Journal of Thermal Spray Technology* 19 (1–2): 510–516.

Dykhuizen, R.C. 1994. Review of Impact and Solidification of Molten Thermal Spray Droplets. *Journal of Thermal Spray Technology* 3 (4): 351–361.

Escure, C., M. Vardelle, and P. Fauchais. 2003. Experimental and theoretical study of the impact of alumina droplet on cold and hot substrates. *Plasma Chemistry and Plasma Processing* 3: 291–309.

Fanelli, F., and F. Fracassi. 2014. Aerosol-assisted atmospheric pressure cold plasma deposition of organic–inorganic nanocomposite coatings. *Plasma Chemistry Plasma Processes* 34: 473–487.

Fauchais, P. 2004. Understanding plasma spraying, an invited review. *Journal of Physics D: Applied Physics* 37: 2232–2246.

Fauchais, P., and M. Vardelle. 2010. Sensors in spray processes. *Journal of Thermal Spray Technology* 19 (4): 668–694.

Fauchais, P., M. Fukumoto, A. Vardelle, and M. Vardelle. 2004. Knowledge concerning splat formation: An invited review. *Journal of Thermal Spray Technology* 13 (3): 337–360.

Fauchais, P., R. Etchart-Salas, V. Rat, J.-F. Coudert, N. Caron, and K. Wittmann-Ténèze. 2008. Parameters controlling liquid plasma spraying: Solutions, sols or suspensions. *Journal of Thermal Spray Technology* 17 (1): 31–59.

Fayyad, E.M., A.M. Abdullah, M.K. Hassan, A.M. Mohamed, G. Jarjoura, and Z. Farhat. 2018. Recent advances in electroless-plated Ni-P and its composites for erosion and corrosion applications: A review. *Emergent Materials* 1: 3–24.

Filkova, I., and P. Cedik. 1984. In *Nozzle atomization in spray drying, advances drying*, ed. A.S. Mujumdar, vol. 3, 181–215. New York: Hemisphere Pub. Corp.

Frey, H., and H. R. Khan. 2013. *Handbook of thin film technology*, (pub.) Springer 550 pages.

Fukumoto, M., and Y. Haang. 1999. Flattening mechanism in thermal sprayed Ni particles impinging on flat substrate. *Journal of Thermal Spray Technology* 8 (2): 427–432.

Fukumoto, M., S. Katoh, and I. Okane. 1995a. Splat Behavior of Plasma Sprayed Particles on Flat Substrate Surface. In Proc. of the 14th Int. Thermal Spray Conference, Vol. 1, A. Ohmori, ed., High Temp. Soc. of Japan, Osaka, Japan, pp. 353–59.

Fukumoto, M., H. Hayashi, and T. Yokoyama. 1995b. Relationship between particle's splat pattern and coating adhesive strength of HVOF sprayed Cu-alloy. *J. Japan Thermal Spray Soc* 32-33: 149–156. (in Japanese).

Fukumoto, M., E. Nishioka, and T. Matsubara. 1999. Flattening and solidification behavior of a metal droplet on a flat substrate surface held at various temperatures. *Surface and Coating Technology* 120-121: 131–137.

Fukumoto, M., K. Yang, K. Tanaka, T. Usami, T. Yasui, and M. Yamada. 2011. Effect of substrate temperature and ambient pressure on heat transfer at interface between molten droplet and substrate surface. *Journal of Thermal Spray Technology* 20 (1–2): 48–58.

Gärtner, F., T. Stoltenhoff, T. Schmidt, and H. Kreye. 2006. The cold spray process and its potential for industrial applications. *Journal of Thermal Spray Technology* 15 (2): 223–232.

Gell, M., E. Jordan, M. Teicholz, B.M. Cetegem, N.P. Padture, L. Xie, D. Chen, X. Ma, and J. Roth. 2008. Thermal barrier coatings made by the solution precursor plasma spray process. *Journal of Thermal Spray Technology* 17 (1): 124–135.

Gladush, G. G., and I. Smurov. 2011. Physics of laser materials processing: Theory and experiment. Springer series in Materials Science.

Glassmann. 1977. *Combustion.* New York: Academic Press.

Glocker, D.A., and S.I. Shah. 2002. *Handbook of thin film process technology* (2 vol. set). Bristol: Institute of Physics Pub.

———. 2011. Flattening and cooling of millimeter- and micrometer-sized alumina drops. *Journal of Thermal Spray Technology* 20 (1–2): 59–67.

Hatakeyama, R. 2017. Nanocarbon materials fabricated using plasmas. *Reviews of Modern Plasma Physics* 1: 1–7.

Hermanek, F.J. 2001. *Thermal spray terminology and company origins.* Materials Park: ASM International.

Hussary, N.A., and J.V.R. Heberlein. 2001 Atomization and particle jet interactions in the wire-arc spraying process. *Journal of Thermal Spray Technology* 10(4).604–610.

———, N.A., and J.V.R. Heberlein. 2007. Effect of system parameters on metal breakup and particle formation in the wire arc spray process. *Journal of Thermal Spray Technology* 16(1): 140–152.

Jiang, B.-B., M. Pan, C. Wang, H.-F. Li, N. Xie, H.-Y. Hu, F. Wu, X.-L. Yan, M.H. Wu, K. Vinodgopal, and G-Pi Dai. 2018. Morphology engineering and etching of graphene domain by low-pressure chemical vapor deposition. *Journal of Saudi Chemical Society* 2018.

Kadyrov, E. 1996. Gas-particle interaction in detonation spraying systems. *Journal of Thermal Spray Technology* 5 (2): 185–195.

Kadyrov, E., and V. Kadyrov. 1995. Gas dynamical parameters of detonation powder spraying. *Journal of Thermal Spray Technology* 4 (3): 280–286.

Kanani, N. 2004. *Electroplating, basic principles, processes and practice.* Oxford: Elsevier Advanced Technology.

Kashirin, A. I., O.F. Klynev, and T.V. Buzdygar. 2002a. Apparatus dynamic coating, US patent 6, 402, 050, June 11.

———. 2002b. Apparatus dynamic coating, US patent 6, 402, 050, June 11.

Kashirin, A., O. Klynev, T. Buzdygar, and A. Shkodin. 2007. DYMET technology evolution and application. In *Thermal Spray global coating solutions*, ed. B.R. Marple et al. Materials Park: ASM Int. e-proceedings.

Katsui, H., and T. Goto. 2017. Bio-ceramic coating of Ca–Ti–O system compound by laser chemical vapor deposition, Chapter 4, 47–62 Pub. Springer.

Lang, E., ed. 1983. *Coatings for high temperature applications.* London/New York: Applied Science Publishers.

Li, C.-J., J. Zou, H.-B. Huo, J.-T. Yao, and G.-J. Yang. 2016. Microstructure and properties of porous abradable alumina coatings flame-sprayed with semi-molten particles. *Journal of Thermal Spray Technology* 25 (1–2): 264–272.

Luscher, P.E., and D.M. Collins. 1979. Design considerations for molecular beam epitaxy systems. In *Crystal growth and characterization*, vol. 2, 15–32. Elsevier.

Maaß, P., and P. Peißker. 2011. *Handbook of hot-dip galvanization,* 494 p. New York: Wiley.

Mahan, J.E. 2000. *Physical vapor deposition of thin films.* New York: John Wiley & Sons.

Mauer, G., A. Hospach, N. Zotov, and R. Vaßen. 2013a. Process conditions and microstructures of ceramic coatings by gas phase deposition based on plasma spraying. *Journal of Thermal Spray Technology* 22: 83–89.

Mauer, G., A. Hospach, and R. Vaßen. 2013b. Process development and coating characteristics of plasma spray-PVD. *Surface and Coatings Technology* 220: 219–224.

Miller, R.A. 1987. Current status of thermal barrier coatings – An overview. *Surface and Coating Technology* 30: 1–11.

Mauer, G., N. Schlegel, A. Guignard, M.O. Jarligo, S. Rezanka, A. Hospach, and R. Vaßen. 2015. Plasma spraying of ceramics with particular difficulties in processing. *Journal of Thermal Spray Technology* 24 (1-2): 30–37.

———. 1997. Thermal barrier coatings for aircraft engines: History and directions. *Journal of Thermal Spray Technology* 6 (1): 35–42.

Morishita, T. 1991. In *Plasma technik 2nd symposium*, ed. S. Blum-Sandmeier et al., vol. 1, 137–145. Wohlen: Plasma Technik.

Neiser, R.A., M.F. Smith, and R.C. Dykhuisen. 1998. Oxidation in wire HVOF-sprayed steel. *Journal of Thermal Spray Technology* 7 (4): 537–545.

Nelea, V., H. Pelletier, M. Iliescu, J. Werckmann, V. Cracium, I.N. Mihailescu, C. Ristoscu, and C. Ghica. 2002. Calcium phosphate thin film processing by pulsed laser deposition and in situ assisted ultraviolet pulsed laser deposition. *Journal of Materials Science: Materials in Medicine* 13: 1167–1173.

von Niessen, K., M. Gindrat, and A. Refke. 2010. Vapor phase deposition using plasma spray-PVD. *Journal of Thermal Spray Technology* 19: 502–509

von Niessen, K., and M. Gindrat. 2011. Plasma spray-PVD: A new thermal spray process to deposit out of the vapor phase. *Journal of Thermal Spray Technology* 20: 736–743.

Pasandideh-Fard, M., V. Pershin, S. Chandra, and J. Mostaghimi. 2002. Splat shapes in a thermal spray coating process: Simulations and experiments. *Journal of Thermal Spray Technology* 11 (2): 206–217.

———. 2003. Dépôts physiques, (pub.) Presses Polytechnique et Univrsitaires Romandes (In French).

Pelletier, V.H., M. Iliescu, J. Werckmann, V. Cracium, I.N. Mihailescu, C. Ristoscu, and C. Ghica. 2002. Calcium phosphate thin film processing by pulsed laser deposition and in situ assisted ultraviolet pulsed laser deposition. *Journal of Materials Science. Materials in Medicine* 13: 1167–1173.

Rockett, A. 2008. *The material science of semiconductors,* Chapter

Sabiruddina, K., P.P. Bandyopadhyay, G. Bolelli, and L. Lusvarghi. 2011. Variation of splat shape with processing conditions in plasma sprayed alumina coatings. *Journal of Materials Processing Technology* 211: 450–462.

Schwartz, M. 1982. Deposition from aqueous solutions: An overview, Chapter 10. In *Bunshah RF et al 'deposition technologies for films and coatings'*, 385–453. Park Ridge: Noyes Publication.

Shkodkin, A., A. Kashirin, O. Klynev, and T. Buzdygar. 2006. Metal particle deposition simulation by surface abrasive treatment in gas dynamic spraying. *Journal of Thermal Spray Technology* 15 (3): 382–386.

Singh, J., D.E. Wolfe, and J. Singh. 2002. Architecture of thermal barrier coatings produced by electron beam-physical vapor deposition (EB-PVD). *Journal of Materials Science* 37: 3261–3267.

Singh, J., D.E. Wolfe, R.A. Miller, J.I. Eldridge, and D.M. Zhu. 2004. Tailored microstructure of zirconia and hafnia-based thermal barrier coatings with low thermal conductivity and high hemispherical reflectance by EB-PVD. *Journal of Materials Science* 39: 1975–1985.

Smith M. F., R. A. Neiser, and R.L. Dykhuizen. 1994. An investigation of the effects of droplet impact angle in thermal spray deposition, in *Thermal spray: Industrial applications*, eds. C.C. Berndt and S. Sampath (pub.) ASM Int., Materials Park, 603–608.

Smith, M.F., R.C. Dykhuisen, and R.A. Neiser. 1997. Oxidation in HVOF sprayed steels. In *Thermal spray: A united forum for scientific and technological advances*, ed. C.C. Berndt, 885–892. Materials Park: ASM Int.

Song, B., M. Bail, K.T. Voisey, and T. Hussain. 2017. Role of oxides and porosity on high-temperature oxidation of liquid-fueled HVOF thermal-sprayed Ni50Cr coatings. *Journal of Thermal Spray Technology* 26: 554–568.

Stoltenhoff, T., H. Kreye, and H.J. Richter. 1994. An analysis of the cold spray process and its coatings. *Journal of Thermal Spray Technology* 11 (4): 542–550.

———. 2002. An analysis of the cold spray process and its coatings. *Journal of Thermal Spray Technology* 11 (4): 542–550.

Syed, A.A., A. Denoirjean, P. Denoirjean, J.C. Labbe, and P. Fauchais. 2005. In-flight oxidation of stainless-steel particles in plasma spraying. *Journal of Thermal Spray Technology* 14 (1): 117–124.

Taki, Y., T. Kitagawa, and O. Takai. 1997. Preparation of diamond-like carbon thin films by shielded arc ion plating. *Journal of Materials Science Letters* 16: 553–556.

Thorpe, M.L., and H.J. Richter. 1992. A pragmatic analysis and comparison of HVOF processes. *Journal of Thermal Spray Technology* 1 (2): 161–170.

Tillmann, W., and E. Vogli. 2006. *Selecting surface treatment technologies', modern surface technologies* Bach et al. (eds). J Wily-VCH, 1–10.

Vardelle, M., A. Vardelle, P. Fauchais, K.-I. Li, B. Dussoubs, and N.J. Themelis. 2001. Controlling particle injection in plasma spraying. *Journal of Thermal Spray Technology* 10 (2001): 267–286.

Vidal-Setif, M.H., N. Chellah, C. Rio, C. Sanchez, and O. Lavigne. 2012. Calcium-magnesium alumino-silicate (CMAS) degradation of EB-PVD thermal barrier coatings: Characterization of CMAS damage on ex-service high-pressure blade TBCs. *Surface and Coatings Technology* 208: 39–45.

Vikharev, A.L., A.M. Gorbachev, A.B. Muchnikov, and D.B. Radishchev. 2007. Study of microwave plasma-assisted chemical vapor deposition of poly- and single-crystalline diamond films. *Radiophysics and Quantum Electronics* 50: 10–11. Pub. Springer.

Von Burg, R., and D. Liu. 1993. Chromium and hexavalent chromium. *Journal of Applied Toxicology* 13: 225–230.

Wilden, J., J.P. Bergmann, and H. Frank. 2006. Plasma transferred arc welding-modeling and experimental optimization. *Journal of Thermal Spray Technology* 15 (4): 779–784.

Wu, L.-P., Y.-Z. Li, B.-J. Wang, Z.-P. Mao, H. Xu, Y. Zhong, L.-P. Zhang, and X.-F. Su. 2018. Electroless Ag-plated sponges by tunable deposition onto cellulose-derived templates for ultra-high electromagnetic interference shielding. *Materials and Design* 159: 47–56.

Abbreviations

1-D	One Dimension
2-D	Two Dimensions
3-D	Three dimensions
CFD	Computational Fluid Dynamics
DC	Direct Current
D-Gun	Detonation Gun
HVAF	High-velocity Air Fuel
HVOF	High-Velocity Oxy Fuel
ICP	Inductively Coupled Plasma
LES	Large Eddy Simulation
LHS	Left-Hand Side
LTE	Local Thermodynamic Equilibrium
NS	Navier–Stokes
RF	Radio Frequency
RHS	Right-Hand Side

3.1 Introduction

Except for Cold spray, thermal spray processes are based on the use of either combustion or an electric discharge as an energy source for the spraying process. The general concept involved as illustrated in Fig. 3.1 makes use of a "Spray Torch" for the generation of a stream of hot gases at high velocity, either by combustion or by an electric discharge, in which the material to be sprayed is introduced in the form of a stream of powder, a solution or liquids suspension, wire or rod [Fauchais et al. (2014)]. As the spray material encounters the flow, it is heated, melted, atomized, and accelerated toward the substrate on which, on impact, forms splats which are the building blocks of the coating. With the accumulation of successive layers of splats, the coating is formed with a thickness that can vary between a few hundred micrometers to one or two millimeters. Unmolten particles would normally bounce off the substrate, though if ductile enough, and depending on their

velocity which needs to be of the order of 800–900 m/s, they will undergo plastic deformation giving rise to the formation of relatively dense coating with excellent adhesion to the substrate.

In the great majority of thermal spray applications, the chemical composition of the coating reflects closely the composition of the material feed into the spray torch which would have gone through a physical change of its morphology during its inflight melting, splat formation, and rapid solidification on impact with the substrate. In a few cases partial material decomposition can take place depending on the chemistry of the material being sprayed. In the case of spraying metals and alloys, or certain ceramics oxidation of the material during the spraying process is possible, either inflight or because of post-impact oxidation of the deposited splats exposed to the boundary layer flow. In both situations the source of oxygen would be due to excess oxygen in the combustion mixture or air entrained into the spray jet from the ambient atmosphere. The mechanism at play is illustrated in Fig. 3.2 representing the flow pattern in a combustion flame or a thermal plasma jet. Gan and Berndt (2013) identify three regions of the flow in which particle/deposit oxidation is possible. The first is at the exit level of the jet where inflight particle oxidation is only possible in combustion spraying where particles are oxidized through their contact with excess oxygen in the combustion gases. The second region is further downstream in flow in which the oxygen responsible of particle oxidation can be present either in combustion gases or in the plasma jet. In the latter case, the oxygen would originate from ambient air entrained by the plasma jet through a turbulent engulfment process as described by Pfender et al. (1990). The third region where particle oxidation is possible is in the deposited coating where oxidation takes place at the surface of the coating or in its internal pore structure because of their contact with the oxygen-laden boundary layer flow.

According to Gan and Berndt (2013) inflight oxidation does not imply simply the formation of a thin oxide layer on

M. I. Boulos et al. (ed.), *Thermal Spray Fundamentals*, https://doi.org/10.1007/978-3-030-70672-2_3

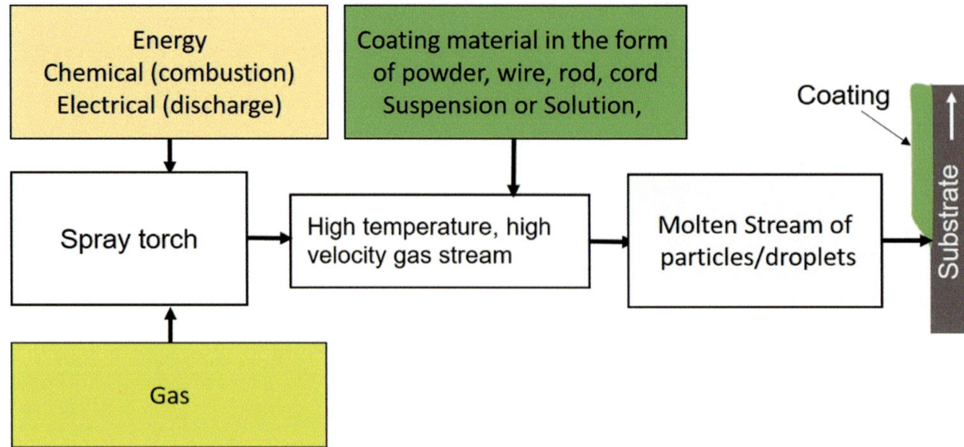

Fig. 3.1 Schematic of the thermal spray concept except Cold Spray and Weld coating

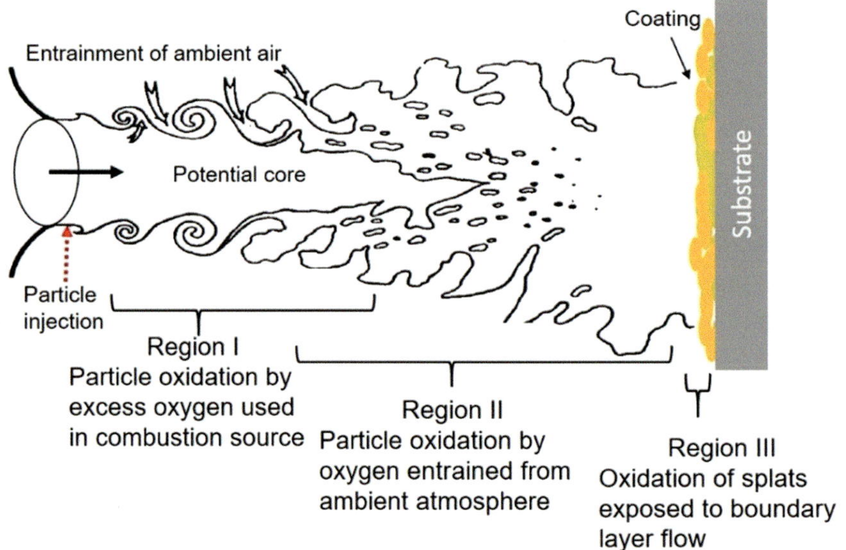

Fig. 3.2 Schematic of the potential oxidation regions within a thermal spray process. [Pfender et al. (1990), and Gan and Berndt (2013)]

the surface of the particle or droplet. Internal convection movements are possible in the molten droplets depending on the difference of kinematic viscosities between the liquid droplet and the surrounding media. This convective movement is mainly developed in plasma spraying where the gas temperatures can be as high as 10,000–14,000 K and particles velocities of 200–400 m/s. For example, Syed et al. (2004) have sprayed 316L austenitic stainless-steel powders, named 41C ($-106 + 45$ μm) from Sulzer Metco using a DC plasma gun (PTF-4 type) in an ambient air atmosphere. The plasma torch had an anode nozzle internal diameter i.d. of 7 mm, operating using an Ar/H_2 as plasma gas, with a flow rate of 45 slm (Ar) +15 slm (H_2). The torch current was 550 A, and the corresponding discharge power 33 kW. SEM sectional micrographs of the treated powders

are given in Fig. 3.3. These reveal the presence of oxide nodules inside the particle which can only result from surface oxidation of the particle followed by convective movements of fragments of the formed oxide layer into the molten liquid droplets. Such internal convection implies that the ratio of the plasma to molten particle kinematic viscosities was larger than 50, and that the particle Reynolds number (Re) was higher than 20.

This chapter recalls the basic phenomena involved allowing for the understanding of the generation of high temperature jets used for thermal spray operations, their principal aerodynamic and chemical properties, and the mechanism by which they could affect the quality of the coating obtained. General concepts and equations presented in this chapter are then used in subsequent chapters related to the various combustion and plasma spray processes.

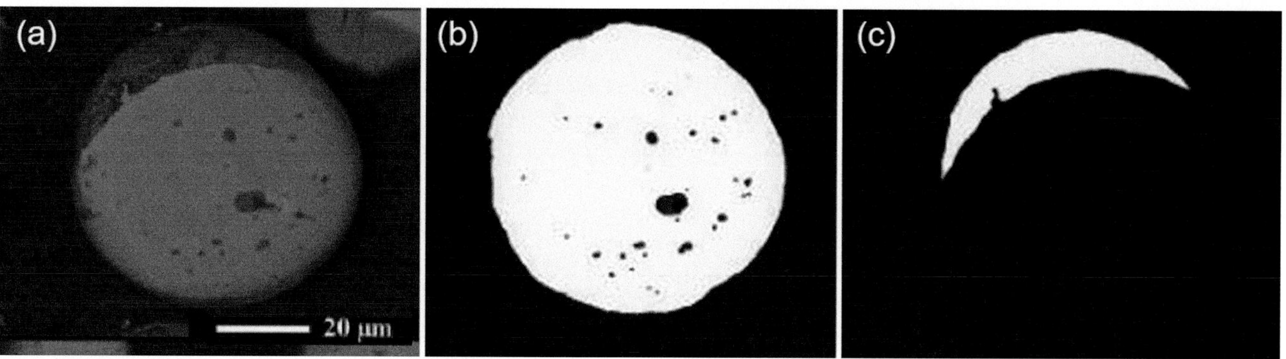

Fig. 3.3 Image treatment analysis of (**a**) the polished cross sections of a collected particle, (**b**) measurements of oxide nodules and (**c**) oxide cap. [Syed et al. (2004)]

3.2 Combustion

3.2.1 Description of Combustion Processes

Combustion-based thermal spray technologies make use of liquid or gaseous fuels for the generation of a high-temperature gaseous jet used for the inflight heating and melting of the material to be sprayed. Except for hydrogen, most of the fuel used is light hydrocarbon molecules of the nominal composition C_xH_y, including, though not limited to, methane CH_4, acetylene C_2H_2, methyl-acetylene C_3H_4, propene or propylene C_3H_6, propane C_3H_8, and liquid hydrocarbons such as kerosene. Air is commonly used as the oxidant though several systems also use pure oxygen O_2 to achieve high energy densities and temperatures. The combustion can be studied either at equilibrium, characterized by its temperature, pressure, and composition, or out of equilibrium, corresponding to fast reactions, where the chemical kinetics defines the flame propagation within the combustible mixture, its possible explosive behavior, and its flammability limits (explosive limits are quite different from flammability limits!)

3.2.2 Combustion at Equilibrium

The combustion reaction involves a significant heat release. The stoichiometry of the reaction corresponds to the case where all carbon and hydrogen atoms are reacted into carbon dioxide and water as in Eq. (3.1):

$$C_xH_y + \left(x + \frac{y}{4}\right).O_2 \;\rightarrow\; x.CO_2 + \left(\frac{y}{2}\right).H_2O \qquad (3.1)$$

In general, the combustion is characterized by the molar ratio of fuel to oxidizer (*F/A*):

$$(F/A) = nC_xH_y/mO_2 \qquad (3.2)$$

where n and m are mole numbers of hydrocarbon and oxygen molecules, respectively.

When the mixture is stoichiometric $(F/A) = 1/(x + y/4)$, with x and y being the coefficients defined in Eq. (3.1). R', called either equivalence, or fuel richness ratio, is defined as

$$R' = (F/A)/(F/A)_{\text{stoichiometry}} \qquad (3.3)$$

A fuel-rich system $(R' > 1)$ corresponds to a combustion mixture with more than the stoichiometric amount of fuel. An overoxidized or fuel-lean system, on the other hand, $(R' < 1)$ corresponds to a system with above stoichiometric concentration of oxidizing agent, while $(R' = 1)$ corresponds to a stoichiometry balanced system. As illustrated in Fig. 3.4, after Glassman (1977), in both fuel-rich and fuel-lean systems, the combustion temperature will be lower than that under stochiometric conditions. In the former case of a fuel-rich mixture, the excess of fuel will have to be heated by the reaction, thus lowering its temperature below that of a stoichiometric mixture, while in the lean-fuel case, the excess of oxidizer will cool the combustion mixture below its stochiometric combustion value.

The same argument also applies if air is used as oxidizer instead of oxygen, as 1 mole of air contains only 0.2 moles of O_2, the balance of 0.8 moles being mostly N_2. Accordingly, five times more air by volume will be needed to obtain the required stoichiometric quantity of oxygen. The combustion temperature with air will consequently be lower by 600–800 K compared to that obtained with pure oxygen as illustrated for different gaseous fuels in Table 3.1. It is to be noted that the use of air as oxidizer also results in the formation of NO_x, which is an air pollutant with serious health hazards. The concentration of formed NO_x increases with the increase of the combustion temperature.

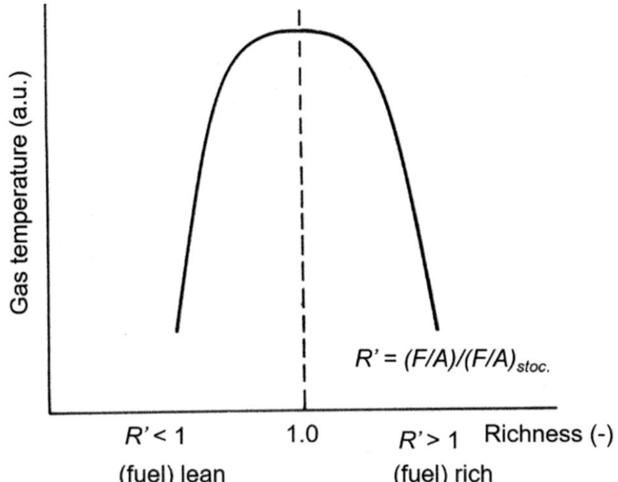

Fig. 3.4 Variation of combustion temperature with richness ratio. ([Glassman (1977)]. Reprinted with the kind permission of Elsevier)

Table 3.1 Approximate flame temperatures of various stoichiometric mixtures (initial temperature 298 K, atmospheric pressure) [Glassman (1977)]

	Flame temperature (K)	Flame temperature (K)
Fuel	Air as oxidizer	Oxygen as oxidizer
Hydrogen	2400	3080
Methane	2210	3030
Heptane	2290	3100
Acetylene	2600[a]	3410[b]
Carbon monoxide	2400	3220

[a]This maximum exists at $R' = 1.3$
[b]This maximum exists at $R' = 1.7$
Reprinted with the kind permission of Elsevier

It is to be noted, however, that mixing and mass transfer phenomena in the flame can have a significant influence on the optimal fuel richness ratio, R', that is needed to obtain a maximum flame temperature. As shown in Fig. 3.5, maximum flame temperature and maximum flame velocity can be obtained at values of R' varying between 1.05 for hydrogen and 1.45 for propane depending on the chemical nature of the fuel [Linde AG Gas group, Acetylene].

3.2.3 Combustion Kinetics

3.2.3.1 One-Step Reactions

A one-step chemical reaction can generally be represented by

$$\sum_{j=1}^{n^l} \nu'_j M_j \rightarrow \sum_{j=1}^{n^l} \nu''_j M_j \qquad (3.4)$$

where ν'_j is the stoichiometric coefficient of the reactants, ν''_j the stoichiometric coefficient of the products, M the arbitrary

specification of all chemical species and n' the total number of compounds involved. If a species involved does not occur as a reactant or product, its ν_j equals zero [Glassman (1977)]. In a reacting system, the rate of change of the concentration of a given species is given by

$$\mathrm{d}(M_i)/\mathrm{d}t = \left(\nu''_i - \nu'_i\right) \cdot k_R \cdot \left(M_j\right)^{\nu_j} \qquad (3.5)$$

where (M_i) is the concentration of species i and k_R the specific reaction rate constant.

This expression corresponds to the law of mass action stating that the disappearance of a chemical species is proportional to the product of the concentrations of the reacting chemical species, each concentration raised to a power equal to the corresponding stoichiometric coefficient.

For example, the formation of molecular hydrogen through recombination of hydrogen atoms in the presence of a third body M is written as

$$H + H + M \rightarrow H_2 + M \qquad (3.6)$$

and Eq. (3.5) becomes $d(H_2)/dt = k_R \cdot (H)^2 \cdot (M)$.

The specific reaction rate constants k_R are expressed by

$$k_R = c(T) \cdot \exp\left(-E_A/k_B \cdot T\right) \qquad (3.7)$$

where $c(T)$, the collision terms of reactants, have a mild temperature dependence, and E_A is the critical energy for the reaction to occur (Arrhenius law). Values of k_R can be found in papers devoted to specific reactions, such as Mackie J.C. and J.C. Smith (1990) for the mixture C_2H_2–O_2, or in tables such as those of Kee et al. (1998).

3.2.3.2 Simultaneous Interdependent and Chain Reactions

In combustion processes a simple one-step reaction is the exception rather than the rule. Generally, they involve simultaneous, interdependent reactions, or chain reactions Glassman (1977) which often involve the dissociation of molecules such as O_2 and H_2, to form radicals O* and H*, that initiate a chain of steps. This is illustrated below for the reaction $H_2 + Br_2 \rightarrow 2HBr$:

(i)	$Br_2 \rightarrow 2Br^* (k_1)$	initiation
(ii)	$Br^* + H_2 \rightarrow HBr + H^* (k_2)$	chaincarrying step
(iii)	$H^* + Br_2 \rightarrow HBr + Br^* (k_3)$	chain − carrying step
(iv)	$H^* + HBr \rightarrow H_2 + Br^* (k_4)$	chain − carrying step
(v)	$2Br^* \rightarrow Br_2 (k_5)$	chain breaking

In this case, the reaction is initiated by the dissociation of Br_2 molecules rather than the H_2 molecule, because the energy required at room temperature is about half that of H_2

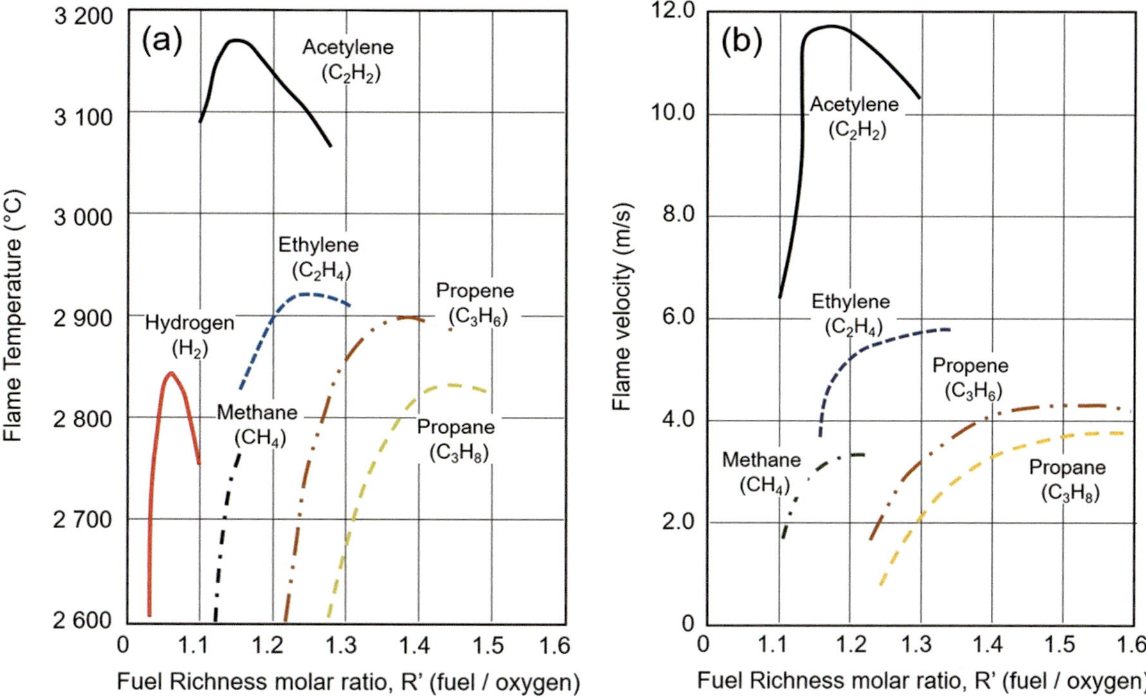

Fig. 3.5 (**a**) Flame temperatures and (**b**) flame velocity as function of the fuel/oxygen molar ratio for different hydrocarbon gaseous fuels. ([Linde AG, Gas group, Acetylene]. Reprinted with the kind permission)

dissociation. This is, however, not the case when the reaction takes place in gases at high temperatures since H_2 dissociation starts at about 3000 K. The HBr formation rate $d(HBr)/dt$ is calculated by considering reactions (ii), (iii) (HBr formation), and (iv) (HBr destruction). Assuming that radicals react very fast it can be postulated that their concentration reaches a steady state and thus the formation rates of both radicals dH^*/dt and dBr^*/dt can be set to zero. It allows calculating the HBr formation rate as a function of the five reaction rate constants of reaction $i \rightarrow v$ [Glassman (1977)].

3.2.3.3 Criterion for Explosion

For combustion, whether deflagrations (also called flames) or detonation, propagation of the combustion front depends on the reaction kinetics which must be fast enough for the mixture to be explosive. Explosions in fuel–oxidizer mixtures occur under certain conditions depending on temperature and/or pressure. They can occur only if in the chain

system the multiplication factor of radicals, often called α', is larger than unity (more radicals R created than destroyed). A simple calculation by Glassman (1977) illustrates this point: Assuming in a straight chain reaction there are 10^8 collisions/s, 1 chain particle/cm^3, and 10^{19} molecules/cm^3, it is estimated that the molecules will be consumed in 10^{11} s, corresponding to 30 years! Under the same conditions, a branched chain reaction is assumed with a multiplication factor $\alpha' = 2$ giving rise to

$$2^N = 10^{19} \rightarrow N = 62 \ \text{generations}$$

All molecules will be consumed in 62 generations corresponding to a time $62 \times 10^{-8} \approx 10^{-6}$ s. For $\alpha' = 1.01$ the time is 10^{-4} s, which is still very fast! In the following the branched chain reaction system is illustrated:

$$(i) \qquad M \rightarrow R^*(k_1) \qquad \text{initiation(radical production)}$$
$$(ii) \qquad R^* + M \rightarrow M + \alpha' R^*(k_2) \qquad \text{chain branching : increased } R \text{ production}$$
$$(iii) \qquad R^* + M \rightarrow P(k_3) \qquad \text{chain branching : increased } R \text{ production}$$
$$(iv) \qquad R^* + \text{wall} \rightarrow \text{destruction } (k_4) \qquad \text{chain termination (very efficient)}$$
$$(v) \qquad R^* + \text{gas} \rightarrow \text{destruction } (k_5) \qquad \text{chain termination}$$

At steady state, $d(R^*)/dt = 0$, the rate of product formation, $d(P)/dt$, becomes infinite, or the system is explosive, when the denominator becomes zero, corresponding to a critical value of $\alpha' = \alpha'_{crit}$:

$$\alpha'_{crit} = (1 + k_3/k_2) + (k_4 + k_5)/(k_2 \cdot (M)) \qquad (3.8)$$

The dependence of the critical value α_{crit} on the concentration (M), as given in Eq. (3.8), as well as the other multiplication factor of radicals, partially explains why α'_{crit} depends on pressure and temperature. In fact, the mixture is explosive if $\alpha' > \alpha'_{crit}$, while for $\alpha' < \alpha'_{crit}$ the product-forming reactions are too slow. Explosion limits for combustion mixtures are well defined and represented as the temperature–pressure boundary separating the regions of slow and fast reactions.

One must have fast reactions for a flame to propagate. *The explosive limits are not flammability limits, corresponding to the lean and rich fuel mixture ratios beyond which no flame will propagate.* These limits are calculated or measured according to the range of a concentration (by volume) of a gas or vapor that will burn (or explode) if an ignition source is introduced. For example, according to Glassman (1977) with a mixture of H_2–air, the flammability and explosive limits are respectively 4 and 18 for the lean part, and 74 and 59 for the rich part. Generally, detonation limits are narrower than flammability limits. A general rule of thumb is that with oxygen the upper limit is about three times the stoichiometric mixture ratio and the lower limit is about half the stoichiometry [Glassman (1977)]. The higher limit is much higher in oxygen than in air, while the lower limit is the same for oxygen or air.

3.2.4 Combustion (Deflagrations) or Detonations

It is first important to define the terms used according to Glassman (1977):

- Explosion refers to rapid heat release or pressure increase.
- Subsonic combustion or deflagration are terms used interchangeably.
- Detonation is a shock wave sustained by the energy of the chemical reaction in the highly compressed explosive medium.
- A pure explosion does not necessarily require the passage of a combustion wave through the explosive medium, whereas an explosive gas mixture must exist to have either combustion (deflagration) or detonation. It can also be stated that deflagrations (subsonic waves) or detonations (supersonic waves) require a rapid energy release, while

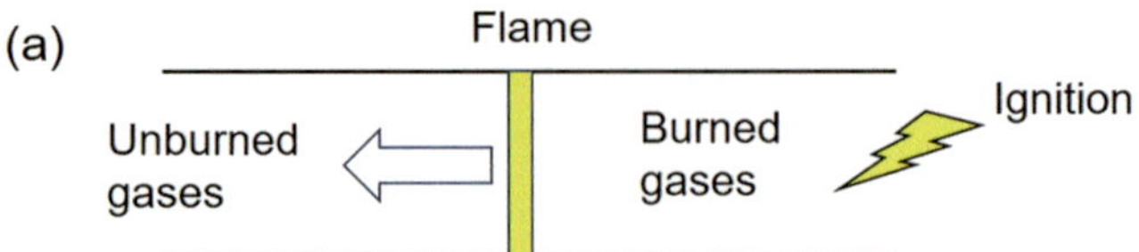

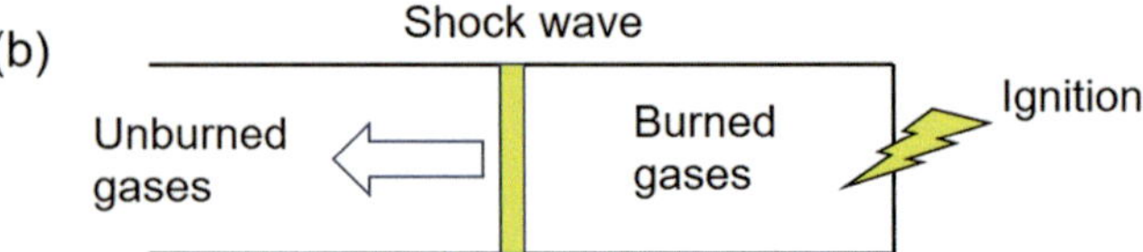

Fig. 3.6 Combustion in a tube initially filled with a premixed mixture of fuel and oxidizer: (**a**) Tube opened at both ends, with ignition at one end resulting in flame wave propagation. (**b**) Tube opened at one end only, with ignition at that location resulting in detonation wave propagation. ([Glassman (1977)]. Reprinted with the kind permission of Elsevier)

explosions do not require the presence of a waveform. Of course, explosion limits depend on the mixture composition (particularly the mixture ratio), its pressure, and temperature.

3.2.4.1 Combustion (Deflagration)

To explain simply the difference between combustion (deflagration) and detonation one can consider a tube filled with a premixed mixture of fuel and oxidizer (Fig. 3.6) [Glassman (1977)]. If both ends of the tube are closed to keep the mixture enclosed, then both ends are opened simultaneously, and an ignition source applied at one end (Fig. 3.6a), a flame appears in the tube. It is characterized by a very luminous zone less than 1 mm thick, which is composed of a preheat zone, a reaction zone and a recombination zone. In front of the flame the unburned gases are at room temperature, and at the end of the luminous zone the temperature is the highest and corresponds to the adiabatic flame temperature as shown in Fig. 3.5a. The reactions are very fast in the reaction zone, while being much slower in the burned gases. The flame can be considered as a subsonic wave sustained by combustion which propagates at velocities between 1 and 11.5 m/s with oxygen as shown in Fig. 3.5b. The flame velocity S_L, in laminar conditions, is given by Eq. (3.11)

$$S_L \approx (a.R_o)^{\frac{1}{2}} \qquad (3.9)$$

where a is the diffusivity in the combustion wave and R_o the reaction rate.

In a burner the combustion wave can be maintained at a fixed position as illustrated in Fig. 3.6a. This is possible because a competition exists between the wave velocity S_u° directing the flame toward the unburned gases flowing in the cylindrical tube of the Bunsen burner and the velocity of

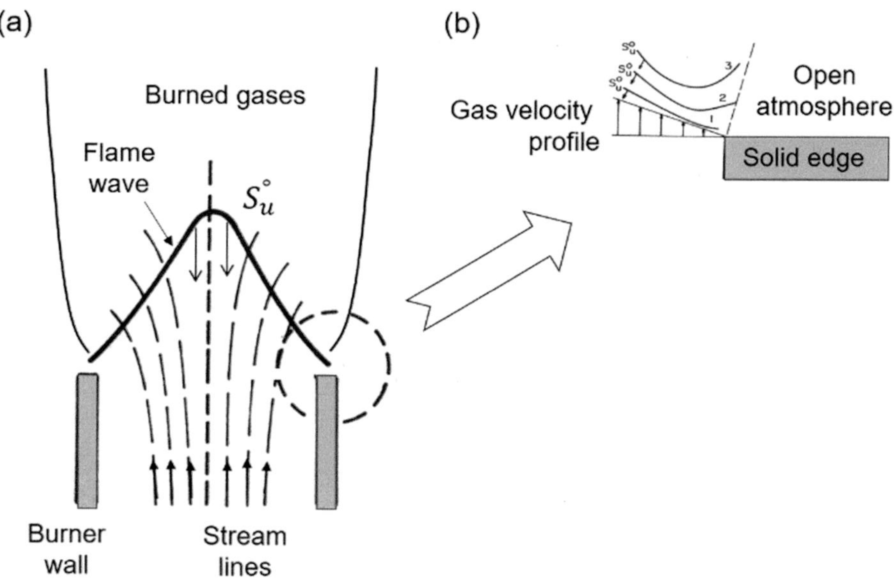

Fig. 3.7 (**a**) Schematic diagram of a Bunsen burner with the combustible gas flow streams, the wave flame and the burned gases. (**b**) Detail closeup to the burner wall: flame stabilization. [Glassman (1977)]

these gases flowing in the tube and directed toward the tube exit

Assuming the flow is laminar, the gas velocity profile at the tube exit, close to the wall, has been represented as being zero at the wall and varying linearly with the tube radius (see Fig. 3.7). The wave velocity far away from the rim attains its maximum S_u°, while when it approaches the burner rim it decreases as heat, and chain carriers are lost to the rim. If the distance of the wave from the rim increases, losses diminish and the wave velocity increases. This is illustrated in Fig. 3.7b for different positions of the wave, numbered 1–3. Finally, a position can be found for a cold gas mean velocity where equilibrium exists.

If the unburned gas velocities diminish too much, the flame will penetrate into the tube, a phenomenon called flash back. Close to the wall, the unburned gas velocity tends to zero, but quenching of radicals increases; thus, chain carriers are lost resulting in no possible flash back in this area. To avoid this unwanted phenomenon in the central part of the tube, one has to reduce its internal diameter to increase drastically the radical destruction and reduce the flame velocity. If the burner hole is small enough (smaller than the fluid boundary layer thickness), no flash back is possible. It is also possible on the other hand to blow out the flame if the hole is too small, and the velocity of unburned gases is increased too much, a phenomenon called blow-off. Flash arrestors which are commonly introduced in the flow of combustible gases to avoid "flash backs" consists essentially of a fine metal mesh which acts as a heat sink, thus cooling the flame and impairing its propagation across the metal wire mesh.

The color of the luminous zone of the flame depends on the fuel/oxidizer ratio [Glassman (1977)]; the radiation is deep violet for fuel-lean mixture (due to CH radicals), while it is green in fuel-rich mixtures (due to C_2 molecules). In the burned high-temperature gases, a reddish glow arises from CO_2 and H_2O vapor radiation. When the mixture is adjusted to be fuel rich, free carbon soot particles are formed. These are responsible for the intense yellow emission observed in such flames. In thermal spray processes, flame spraying uses flames at atmospheric pressure, while HVOF or HVAF spraying uses flames but at pressures generally below 1 MPa.

3.2.4.2 Detonation

As in the case of deflagration or flame, one considers a tube filled with a premixed mixture of fuel and oxidizer (*see* Fig. 3.6) [Glassman (1977)]. First, both ends of the tube are closed to keep the mixture enclosed, and then one end (instead of both) is opened and an ignition source is applied simultaneously at the closed end (see Fig. 3.6b). Again a flame or a wave propagates, except its velocity is not between a few tens of centimeters per second and about 10 m per second, but reaches thousands of meters per second. Such velocities are supersonic with respect to the unburned gases. Chapman performed the first studies of this phenomenon at the end of the nineteenth century followed by Jouget at the beginning of the twentieth century. It was demonstrated that the only possible velocity for a detonation wave was the velocity of sound in gases behind the detonation wave (burned gases), plus the velocity of the burned gases relatively to the tube, also called mass velocity [Glassman (1977)], The velocity of sound is depending on the square

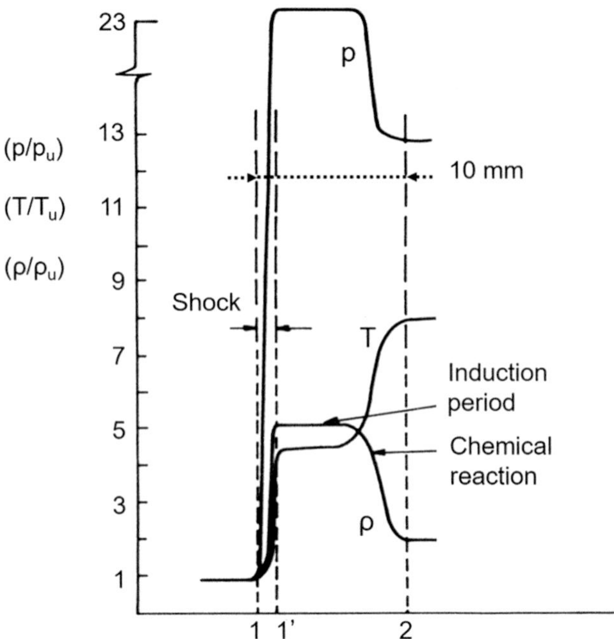

Fig. 3.8 Structure of a detonation wave: variation of physical parameters through it. ([Glassman (1977)]. Reprinted with the kind permission of Elsevier)

Table 3.2 Calculated values of the physical parameters at the positions 1, 1', and 2 indicated in Fig. 3.8 in a 20 vol.% H_2–air detonation [Glassman (1977)]

	1	1'	2
$M(-)$	4.5	0.42	1.0
u (m/s)	1524	305	549
p (MPa)	0.1	2.3	1.3
T (K)	298	1350	2425
ρ/ρ_u	1.0	5.0	1.8

Reprinted with the kind permission of Elsevier

root of temperature, which results in an adiabatic shock front for the propagating wave.

The Theoretical studies in the early forties and fifties, have clarified the structure of the detonation wave: it consists of a shock moving at the detonation velocity, leaving heated and compressed gases behind it. The chemical reaction then starts and as it progresses, the temperature rises, and the specific mass and pressure fall until they reach the Chapman–Jouget (C–J) values and the reaction attains equilibrium. Behind the C–J shock, energy is generated by thermal reactions [Glassman (1977)]. Figure 3.8 is a representation of the theory of Zeldovich–von Neumann showing the variation of the different physical parameters. Plane 1 corresponds to the shock front, plane 1' is that immediately after the shock, and plane 2 is the Chapman–Jouget plane. Table 3.2 presents the calculated values of the physical parameters related to a 20 vol.% H2–air detonation.

As it can be seen, when the gas passes from the shock front to the C–J state, its pressure drops by a factor of almost two, while its temperature is almost doubled, and its specific mass is reduced by about a factor of three. Compared to the flame wave, the ratio ρ_b/ρ_u is higher than ten (against 0.8–0.9), while the ratio ρ_b/ρ_u has values between 1.5 and 2 (against 0.06–0.25 for flames) with, of course, much higher velocities (about two orders of magnitude).

It must be emphasized that, as for flames, detonation limits exist depending on the mixture used (lean or rich), and which are different from deflagration limits. The difference comes from the fact that the shock provides the detonation mechanism whereby the combustion process is continually sustained. The energy to drive the shock is not the same as that for the combustion wave. In thermal spray processes detonation guns consist in a barrel close at one of its ends as shown in Fig. 3.6b.

3.3 Thermal Plasmas for Spraying

3.3.1 Comparison of Thermal Plasma and Combustion Spraying

As mentioned in the introduction of this chapter, "thermal plasma spraying" and "combustion spraying" are essentially comparable technologies which differ in the characteristics of the "spray torch" which as shown in Fig. 3.1 is the central part of the process. Both approaches give rise to the generation of a stream of gases at high temperatures and velocities in which the coating material in powder form, wire or rod, is introduced into the flow where it is heated, melted, atomized, and accelerated toward the substrate on which the molten particles deform forming splats that pile up on the substrate building the coating. While in combustion, the thermal energy is generated through a chemical reaction between the fuel and an oxidant, in plasmas the thermal energy is generated through a direct conversion of electrical energy to thermal energy. The basic mechanism involved is an electrical arc discharge between two electrodes as in atmospheric or vacuum DC plasma spray torches, wire-arc torches, or plasma transferred arc torches (PTA). Alternately, plasmas can also be generated through radio frequency (RF) inductively or capacitively coupled discharges. In each of these cases the thermophysical properties of the plasma can differ significantly from combustion-based sources as well as between the different plasma torch devices. A comparison between the gas temperature and velocity of different combustion and plasma sources is presented in Fig. 3.9. This shows that most combustion-based sources are limited to mean gas temperatures below 4000 K while offering relatively high gas velocities that can reach up to 2000 m/s in the case of HVOF, D-gun, and Cold spray. Plasma torches, on

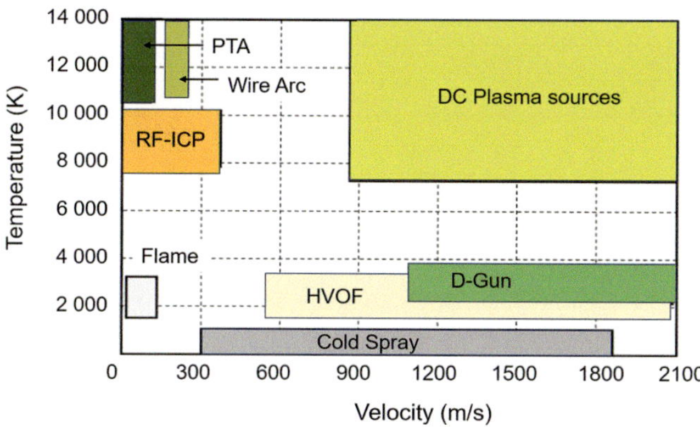

Fig. 3.9 Gas temperature-velocity diagram for different combustion and plasma-based thermal spray technologies. [Fauchais et al. (2014)]

the other hand, can have gas/plasma temperatures as high as 14,000 K or more, with relatively high velocities up to 2000 m/s in the case of DC plasmas. As shown in Fig. 3.9, RF induction plasma torches, wire-arc spray torches, and PTA system, operating over the same temperature range of DC torches, are limited in terms of gas velocity which is generally below 300 m/s. Exceptionally, RF induction plasma torches can generate supersonic plasma jets with velocities up to 2000 m/s or more when equipped with a Laval-type nozzle operating under low-pressure conditions [Hollenstein et al. (1999)].

3.3.2 Definition

Plasmas, considered in this book, consist of a mixture of electrons, neutral molecules, atoms and ions in the fundamental or excited states. The oscillation of molecules, atoms, and ions between their excited state and lower or ground state results in the emission of photons that constitute the luminous signature of the plasma. Such a mixture, however, is qualified as plasma only if the negative and positive charges balance each other, that is, overall the plasma is electrically neutral. Thermal plasmas exist only if their electrical conductivity exceeds a given threshold, which for the mixtures used in plasma spraying at atmospheric pressure corresponds to temperatures roughly between 7000 and 8000 K [Boulos et al. (1994)].

Commonly used thermal plasmas are optically thin, that is, all the radiation escapes and does not participate in establishing an equilibrium. However, it is no more the case when plasma contains metallic vapors resulting from sprayed particle evaporation. In plasmas the ions and electrons are accelerated, and thus gain energy, under the influence of the applied electric field. The exchange of energy between elementary particles either of the same species or of different species is a direct result of particle collisions. The energy

exchanged in any such collision between two species with masses m_1 and m_2 is proportional to [2 $m_1.m_2/(m_1 + m_2)^2$]. Thus, the energy exchange between two heavy species is rather fast (50% if they have the same mass). It is not the case with electrons that have a very low mass compared to that of other species except photons (m_H, the lightest atom, has a mass 1836 times larger than that of electrons!). Correspondingly the energy exchange between an electron and a heavy species is rather small (proportional to 2 m_e/m_h, that is, for an argon atom 1/37,120!). Accordingly, a very large number of collisions are required for the electrons and heavy particles to reach thermal equilibrium. A unique temperature can characterize thermal plasmas if collisions are numerous enough to thermalize all species including electrons. This condition is fulfilled in thermal plasmas where temperatures are between 8000 and 15,000 K with electron densities ranging from 10^{21} to 10^{24} m^{-3}. Of course, deviations from equilibrium must be expected in some areas, especially close to electrodes or where a cold gas or a liquid jet is injected with a high-energy momentum density or pressure ($\rho \cdot v^2$) compared to that of the plasma, and in the presence of steep temperature gradients. In the following this situation will not be discussed.

In combustion, the composition of the fuel–oxidizer mixture and its pressure determine the energy dissipated. In contrast, in thermal plasmas, external energy (electrical energy through a direct current arc or inductively coupled radio frequency discharge) is supplied to the plasma forming gas. If this energy is sufficiently high, it results first in the dissociation of molecules (presenting the case of molecular gases) and then in ionization of atoms. Table 3.3 summarizes the ionization ($X \rightarrow X^+ + e$) and dissociation energies ($X_2 \rightarrow 2 X$) of the main plasma forming gases and air (X represents the species considered). The first observation is that the ionization energies of all plasma gases, except He, are rather close together (less than 2.1 eV difference). This means that ionization for all gases, except for helium, will be completed at about the same temperature (about 15,000 K at atmospheric pressure). However, it does not mean that the energy required for achieving a number density of electrons within the plasma forming gas (which will result in an electrical conductivity sufficiently high to sustain the plasma with an arc or a r.f. discharge) will be the same whatever the plasma gas may be. Dissociation energies are lower than ionization energies, which means that dissociation will be completed at temperatures lower than those of ionization (at atmospheric pressure about 3500 K for hydrogen and oxygen, and 7500 K for nitrogen). The ionization of molecular gases is far more energy demanding than most monatomic gases because of the need to supply both dissociation and ionization energies (4.588 + 13.481 eV) for H$^+$ ion compared with (15.755 eV) for Ar$^+$. It is important to note that in all plasma sources used in thermal spray applications, the degree of ionization of the

Table 3.3 Ionization and dissociation energies of the main spray plasma gases and surrounding atmosphere [Boulos et al. (1994)]

Species	Ar	He	H	N	O	H_2	N_2	O_2
Ionization energy (eV)	15.755	24.481	13.659	14.534	13.614	15.426	15.58	12.06
Dissociation energy (eV)	–	–	–	–	–	4.588	9.756	5.08

Reprinted with the kind permission of Springer

gas is relatively small (less than 1–3%) which explains why it is possible to have a plasma with an average bulk gas temperature (specific energy level) that is almost one order of magnitude smaller than that needed for full ionization of the plasma gas.

3.3.3 Plasma Composition

Plasma composition is calculated by minimizing the Gibbs free enthalpy, considering the van't Hoff's laws for dissociation and ionization, the conservation of different elements, the electrical neutrality and Dalton's law. For details the interested reader is referred to Boulos et al. (1994). Results are traditionally represented as the evolution of the species number densities with temperature at a given pressure. In the following only atmospheric pressure conditions, at which most plasma torches work, will be considered.

Figure 3.10 represents the dependence of the argon number density on temperature at atmospheric pressure. As the temperature increases, n_{Ar}, the density of argon atoms, decreases monotonically first due to the temperature increase ($n_{Ar} = p/k_B \cdot T$) and then due to ionization, with the progressive increase of n_{Ar}^+ and correspondingly of n_e (electrical neutrality). Ionization is completed at about 15,000 K. It must be noted that the second ionization (Ar^{++}) represents a density of 10^{15} m^{-3} at about 10,700 K, that is, 10^{-9} less than Ar atoms. At 8000 K, electrons and ions represent about 2% of the total species.

With helium results are quite similar, except of course that ionization is completed at about 25,000 K (against 15,000 K for argon). The percentage of electrons reaches about 1% at temperatures higher than 15,000 K. Thus, in any Ar–He mixtures (except with less than 5 vol.% Ar), electrons and ions are all due to argon ionization.

With nitrogen, as shown in Fig. 3.11, the evolution is different from that obtained with argon. As a first step nitrogen molecules must be dissociated and, at the beginning, n_{N2} diminishes fast with temperature ($n_{N2} = p/k_B \cdot T$) and then faster with N_2 dissociation. The latter is completed at about 10,000 K. As the ionization energies of N and N_2 are rather close (see Table 3.3), the ionization of these species starts at about the same temperature.

However, despite the slightly lower ionization energy of atomic Nitrogen compared to molecular nitrogen, N_2^+ ions appear before N^+ ions because the number density of N_2 is

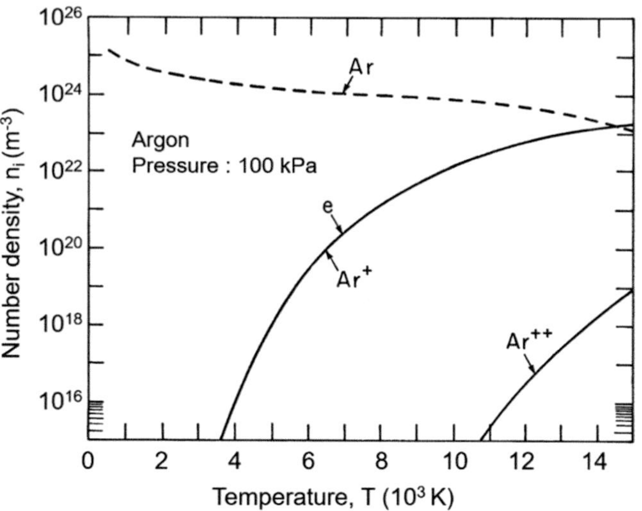

Fig. 3.10 Temperature dependence of number densities of argon plasma species at atmospheric pressure. ([Boulos et al. (1994)]. Reprinted with the kind permission of Springer)

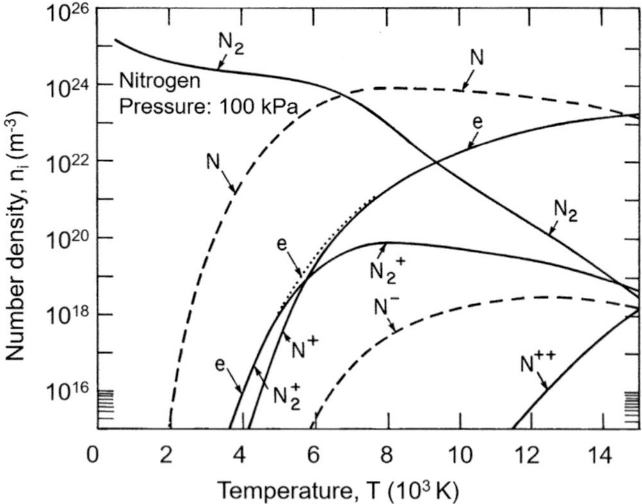

Fig. 3.11 Temperature dependence of number densities of nitrogen plasma species at atmospheric pressure. ([Boulos et al. (1994)]. Reprinted with the kind permission of Springer)

higher than that of N at this temperature. The density of N_2^+ ions decreases above about 7000 K because of their dissociation. The presence of negative ions N^- must also be noticed. However, under spraying conditions N_2^+ and N^- ions have

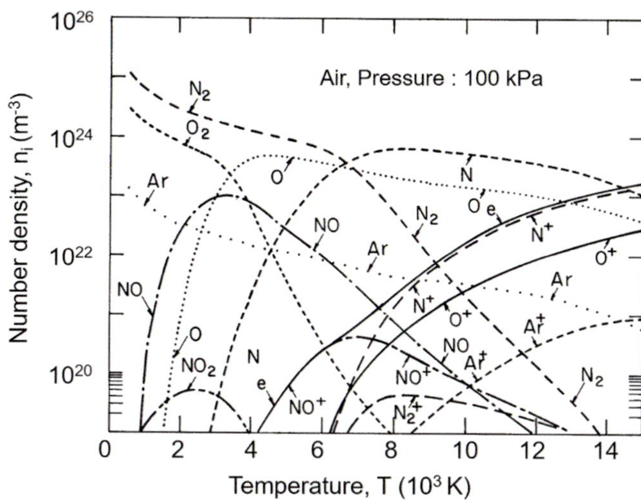

Fig. 3.12 Temperature dependence of number densities of air plasma species at atmospheric pressure. ([Boulos et al. (1994)]. Reprinted with the kind permission of Springer)

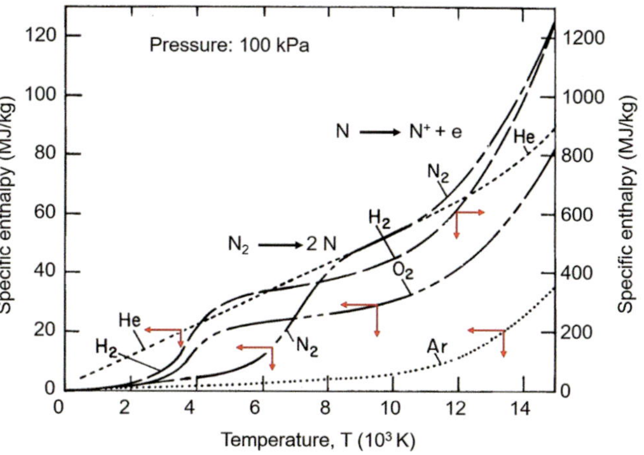

Fig. 3.13 Temperature dependence of the specific enthalpy (MJ/kg) of Ar, He, N_2, O_2, and H_2 at atmospheric pressure. ([Boulos et al. (1994)]. Reprinted with the kind permission of Springer)

no practical influence. At the end of dissociation, nitrogen atoms start to ionize, and ionization is completed above 15,000 K.

Dissociation and ionization steps comparable to those of nitrogen are also observed with other diatomic molecules, used in plasma spraying or existing in the surrounding atmosphere. This is the case of O_2, but considering that the dissociation energy for oxygen is about half that of nitrogen (see Table 3.3), dissociation is completed above about 3500 K and ionization around 15,000 K. Molecular ions, O_2^+ and O^- ions, also exist at levels slightly different from those of nitrogen. With hydrogen, dissociation is completed above 3500 K, and if H_2^+ does not exist, H^- is formed above 4000 K. As for nitrogen, these positive molecular and negative atomic ions have no practical influence in spraying conditions. Figure 3.12 presents the composition of dry air plasma, with no water vapor, at atmospheric pressure.

In this calculation dry air is considered to be a mixture of nitrogen (78.09 vol.%), oxygen (20.95 vol.%) and argon (0.93 vol.%), with no water vapor. The following species are important: e, N, O, Ar, N^+, O^+, Ar^+, N_2, N_2^+, O_2, O_2^+, NO, NO^+, NO_2, and N_2O. below 6000 K; the main ions present are NO^+, while N^+ and O^+ are dominant above 9000 K. In practice, the surrounding air cools plasma jets faster than flames because the air entrained into the plasma jet reaches rapidly temperatures above 3000–3500 K, and the induced molecular oxygen dissociation reaction reduces the energy in the plasma.

It is to be noted that the specific mass of all gases decreases almost linearly with the increase of the temperature at temperatures below dissociation and ionization. Between room temperature and 15,000 K, for heavy gases which entrain particles in plasma spraying, $\rho_{300}/\rho_{15\,000} = 46$ for argon and 145 for nitrogen. This implies that the spray particle acceleration will be the smallest for the highest plasma temperature at the same flow velocity.

3.3.4 Thermodynamic Properties

The plasma-specific enthalpy depends strongly upon the dissociation and ionization phenomena. Figure 3.13 represents the variation with temperature of the specific enthalpy (MJ/kg) of the different plasma gases (Ar, He, N_2, O_2 and H_2). Details about the calculations of the such data can be found in Boulos et al. (1994). The steep variations of enthalpy are due to the heats of reaction (dissociation and ionization). The very high enthalpy of pure hydrogen is due to its low mass. At the maximum temperature of this figure, the ionization for helium has not yet started; thus, in spite of its low mass, helium enthalpy is lower than that of hydrogen but higher than that of argon. The figure illustrates the important economics of using plasmas, in which the energy supply is independent of the gas, and the temperature is not determined by the chemical reactions (as in flames). Specifically, if an oxygen-fuel flame at 3000 K is used to heat a particle up to 2500 K, only 20% of its energy is used. If nitrogen plasma at 10000 K is used for the same purpose, it is possible to recover almost 95% of the available energy in the gas. Adding secondary gases such as helium or hydrogen to the primary ones, argon or nitrogen, increases the specific enthalpy of the mixture.

The corresponding values of specific heat at constant pressure (100 kPa) for Ar, He, N_2, and H_2 as function of temperature are given in Fig. 3.14. All curves show peaks at dissociation and ionization temperatures. As it could be

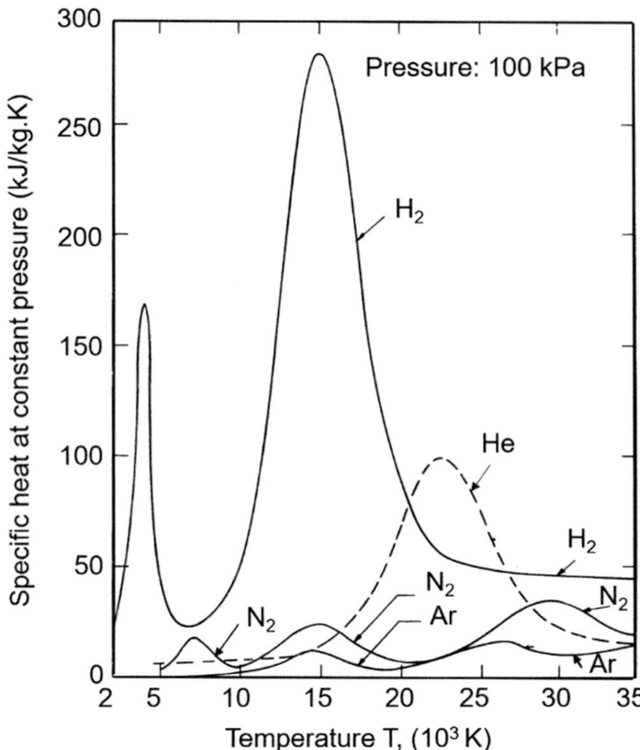
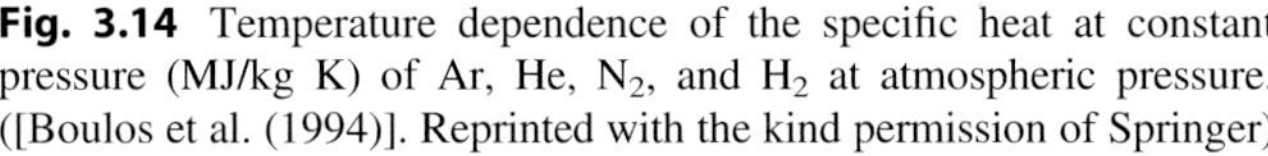

Fig. 3.14 Temperature dependence of the specific heat at constant pressure (MJ/kg K) of Ar, He, N_2, and H_2 at atmospheric pressure. ([Boulos et al. (1994)]. Reprinted with the kind permission of Springer)

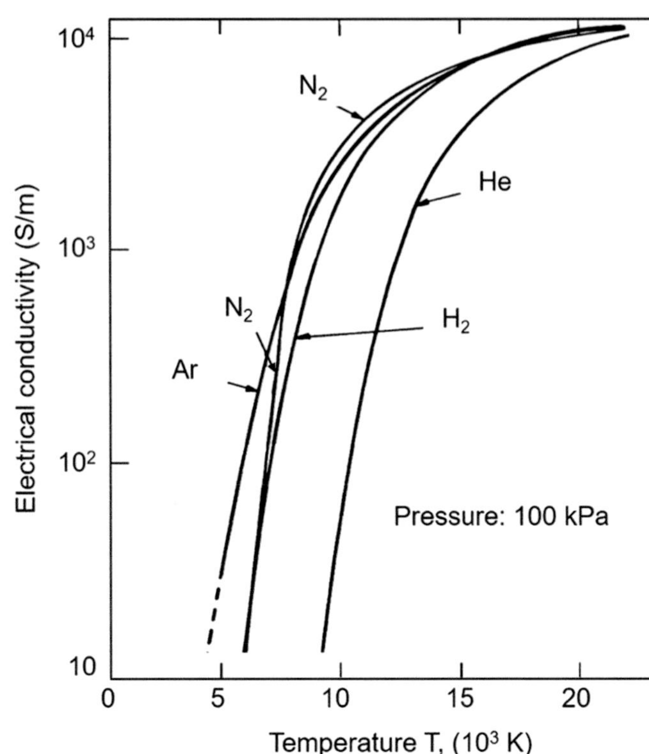

Fig. 3.15 Temperature dependence of the electrical conductivity (decimal logarithmic scale) (S/m) of Ar, He, N_2, and H_2 at atmospheric pressure. ([Boulos et al. (1994)]. Reprinted with the kind permission of Springer)

expected, these peaks are the highest for the light gases: helium and hydrogen.

3.3.5 Transport Properties

3.3.5.1 Electrical Conductivity

The electrical conductivity σ_e is proportional to the electron number density. Because of the close ionization energies (see Table 3.3) of Ar, N_2, H_2, and O_2, it is not surprising that the corresponding values of σ_e are rather close (within a temperature difference of 1000 K) as given in Fig. 3.15. This is not the case for helium with significant difference, about 5000 K in its electrical conductivity over the same temperature range. It is important to note that Fig. 3.15 has a decimal logarithmic scale for σ_e and that for electrical conductivities below a few hundreds of Siemens per meter, no arc or RF discharge at atmospheric pressure can be sustained.

It can be noted from Fig. 3.16 that, on a linear σ_e scale, the electrical conductivities of Ar, Ar–H_2 (25 vol.%), and N_2 are very close together. This observation explains why in plasma spraying the temperature values T_c, where current flow can be established, are in general above 7000–8000 K. For example, for an electric arc, one can define an isothermal envelope at $T = T_c$, outside of which no

electrical conduction is possible, and which appears as a boundary between the arc column and the "cold" sheath. If the electrical conductivity is expressed as function of the gas-specific enthalpy, one can see large differences between the properties of different gases. The electric conduction requires much more energy for nitrogen (about ten times more) than for argon, as shown in Fig. 3.17, with the argon–hydrogen mixture being in between (about five times less energy than for nitrogen).

3.3.5.2 Molecular Viscosity

The molecular viscosity, μ, establishes the proportionality between the friction force in the direction of the flow and the velocity gradient in the orthogonal direction. Figure 3.18 presents the dependence on temperature of μ for Ar, H_2, and He at atmospheric pressure.

The molecular viscosity of thermal plasmas reaches its maximum when the volume fraction of electrons in this plasma reaches about 3%. For higher percentages of charged particles, the long-distance interactions between such particles diminish significantly the interactions between the gas particles and the viscosity decreases. At its maximum, the plasma molecular viscosity is about ten times that of the cold gas. This difference partly explains the difficulty of mixing cold gases with plasmas. However, the difference also

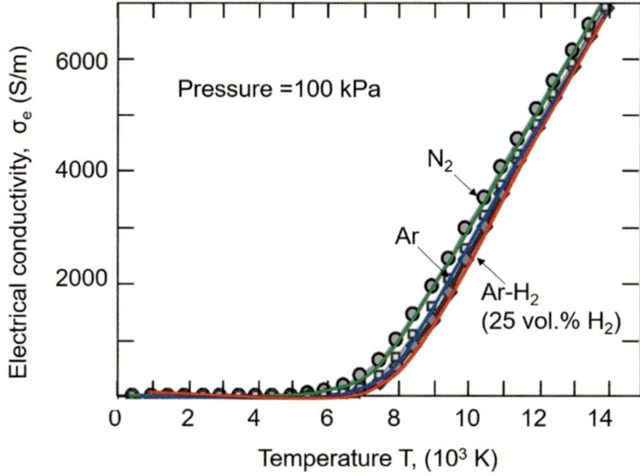

Fig. 3.16 Electrical conductivity (linear scale) (S/m) as a function of temperature for Ar, N_2, and Ar–H_2 (25 vol.% H2) at atmospheric pressure. ([Boulos et al. (1994)]. Reprinted with the kind permission of Springer)

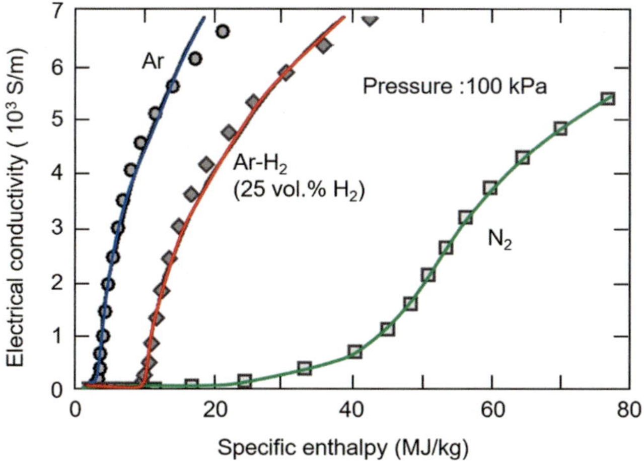

Fig. 3.17 Electrical conductivity (linear scale) (S/m) as a function of the specific enthalpy for Ar, N_2, and Ar–H_2 (25 vol.% H2) at atmospheric pressure. ([Boulos et al. (1994)]. Reprinted with the kind permission of Springer)

protects the plasma core from rapid cooling by mixing with the air entrained by the high velocity plasma jet.

When mixing hydrogen with argon (less than 30 vol.% H_2 in plasma spraying), the viscosity of argon is only slightly reduced because the mixture mass is mainly controlled by that of argon. The viscosity of nitrogen is about the same as that of argon, and hydrogen addition does not modify it appreciably. When adding helium (up to 60 vol.% He) to argon, the decrease of the viscosity argon normally observed above 10,000 K is not so fast, and the viscosity of the mixture continues to increase up to about 15,000 K where it starts to decrease due to the beginning of helium ionization. This

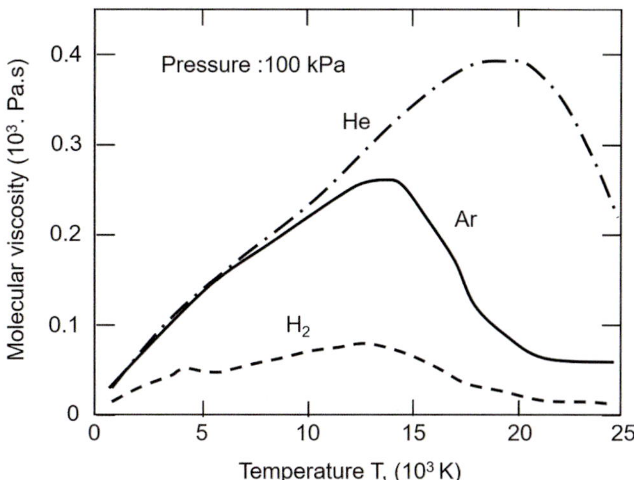

Fig. 3.18 Molecular viscosity (Pa s) as a function of temperature for Ar, He, and H_2 at atmospheric pressure. ([Boulos et al. (1994)]. Reprinted with the kind permission of Springer)

effect allows running slightly longer Ar–He jets than those obtained with Ar–H_2 mixtures in an ambient air atmosphere.

3.3.5.3 Thermal Conductivity

The thermal conductivity of thermal plasmas is of primary importance because it controls the energy losses in the arc or RF discharge fringes, and thus the discharge behavior as well as the heat transfer to solid materials or sprayed particles.

Thermal conductivity κ can be split into three important terms as presented in

$$\kappa_{\text{total}} = \kappa_{\text{tr}}^{\text{h}} + \kappa_{\text{tr}}^{\text{e}} + \kappa_{\text{R}} \qquad (3.10)$$

- $\kappa_{\text{tr}}^{\text{h}}$ corresponding to the translational thermal conductivity of heavy species and thus decreasing when the electron percentage is about 3% and tending to zero for temperatures above 15,000 K for argon where ionization is almost completed
- $\kappa_{\text{tr}}^{\text{e}}$ corresponding to the translational thermal conductivity of electrons, which becomes important above 8000 K (electron percentage over 1%)
- κ_{R} corresponding to reactional thermal conductivity, which can be very important as soon as dissociation and ionization occur

The temperature dependence of these three terms and their sum, κ_{total}, for argon at atmospheric pressure is illustrated in Fig. 3.19. The reactional thermal conductivity appears when ionization starts and goes toward zero when ionization is fully completed.

The dominance of the reactional thermal conductivity when dissociation occurs explains why hydrogen is added

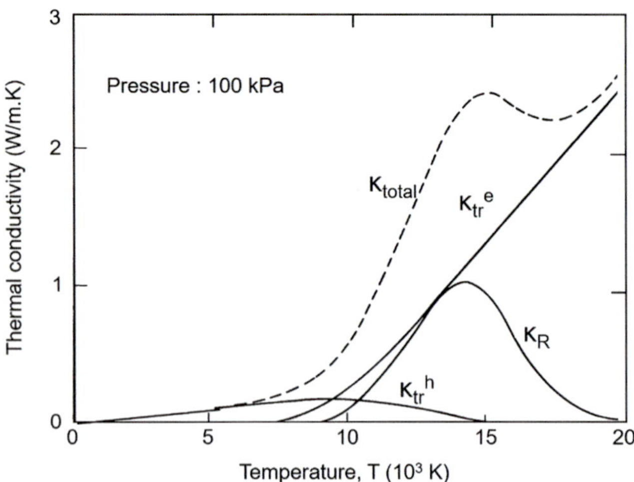

Fig. 3.19 Temperature dependence of the total, κ_{total}, electron translational, κ_{tr}^{e}, heavy species translational, κ_{tr}^{h}, reactional, κ_{R}, and total thermal conductivities (W/m K) of Ar at atmospheric pressure. ([Boulos et al. (1994)]. Reprinted with the kind permission of Springer)

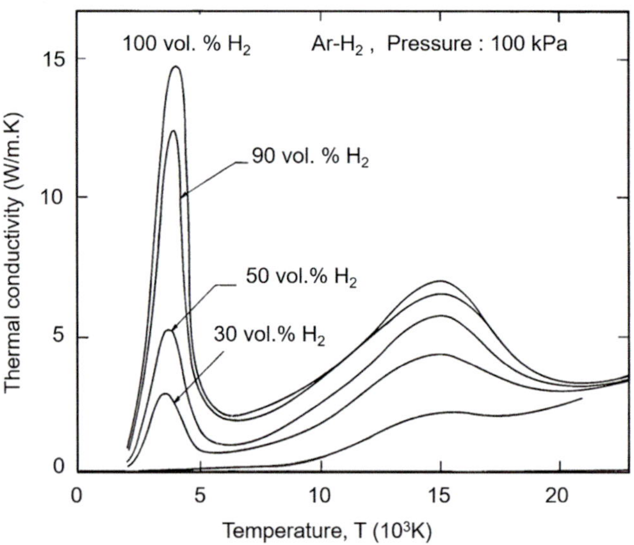

Fig. 3.20 Temperature dependence of the total, κ_{total}, thermal conductivity (W/m K) of the mixture Ar–H$_2$, with volume percentages varying between 0% and 100%, at atmospheric pressure. ([Boulos et al. (1994)]. Reprinted with the kind permission of Springer)

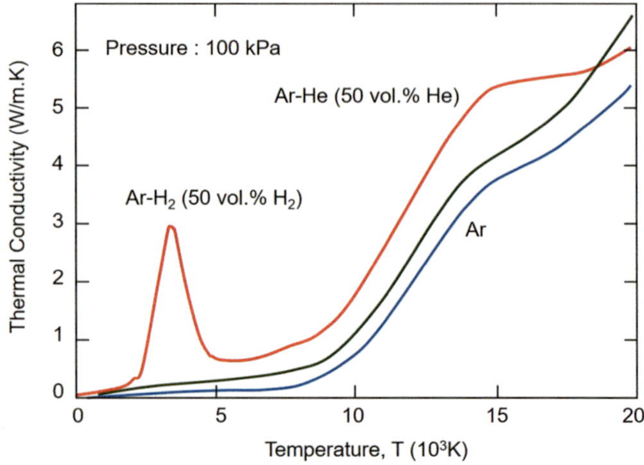

Fig. 3.21 Temperature dependence of the total, κ_{total}, thermal conductivity (W/m K) of mixtures Ar–H$_2$ (25 vol.% H2), Ar–He (50 vol.% He), and pure Ar at atmospheric pressure. ([Boulos et al. (1994)]. Reprinted with the kind permission of Springer)

The influence of the addition of light gases to argon on the thermal conductivity of the mixture is illustrated in Fig. 3.21 presenting its dependence on temperature for three conventional spray mixtures: Ar, Ar–He (50 vol.% He), and Ar–H$_2$ (25 vol.% H$_2$). As can be seen, the best result is obtained with the Ar–H$_2$ mixture with the strong dissociation peak between about 2500 and 5000 K. The addition of helium increases the thermal conductivity more regularly, resulting in a value between that of pure argon and the argon–hydrogen mixture. Of course, at temperatures above 15,000 K, the ionization of helium boosts the mixture thermal conductivity.

In the presence of steep temperature gradients, such as those observed in the thermal boundary layer surrounding a particle in flight in a plasma jet, careful attention must be given to the way the heat transfer is calculated. Bourdin et al. (1983) proposed using the integrated mean thermal conductivity κ' to calculate the heat transfer coefficient with κ' defined as

$$\kappa' = \frac{1}{(T_p - T_s)} \int_{T_s}^{T_p} \kappa(T) \cdot dT = \frac{I_p - I_s}{(T_p - T_s)} \qquad (3.11)$$

with

$$I_x = \int_{T_0}^{T_x} \kappa(T) \cdot dT \qquad (3.12)$$

to argon: it improves the heat transfer at temperatures above 3000 K where dissociation starts. Figure 3.20 shows how the thermal conductivity increases with the volume percentage of hydrogen in an argon–hydrogen mixture, especially when dissociation and ionization occur. With 30 vol.% H$_2$, which is generally the maximum used in plasma spraying, the thermal conductivity of the mixture is increased by a factor of 12.

where T_x corresponds to the temperature limits across which the energy transfer is taking place, T_p the plasma temperature outside the boundary layer, and T_s the particle surface temperature. Figure 3.22 presents κ' for three different plasma spray mixtures: Ar–H$_2$ (26 vol.% H$_2$), Ar–H$_2$ (11 vol.% H$_2$), and Ar–N$_2$ (11 vol.% N$_2$). The figure highlights the

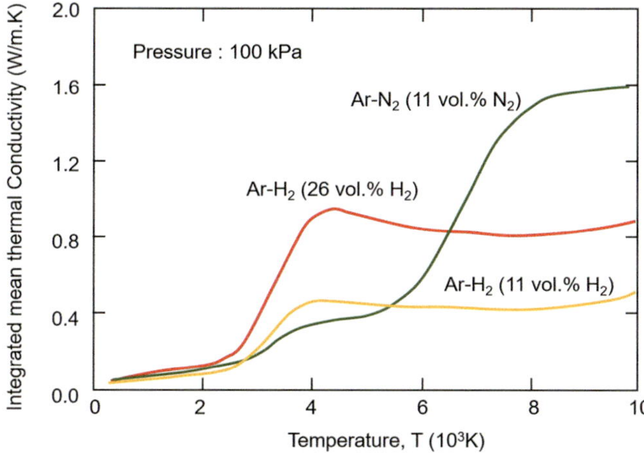

Fig. 3.22 Temperature dependence of the mean effective or integrated thermal conductivity κ' (W/m K) of mixtures Ar–H$_2$ (26 vol.% H2), Ar–H$_2$ (11 vol.% H2), and N$_2$–H$_2$ (11 vol.%) at atmospheric pressure. ([Boulos et al. (1994)]. Reprinted with the kind permission of Springer)

advantage of using diatomic gases to improve the integrated thermal conductivity of the mixture. When adding hydrogen to argon, κ' increases over 3000 K and becomes almost constant over 4000 K, but larger by a factor of about four for the 11 vol.% of H$_2$ mixture and of about ten for the 26 vol.% H$_2$ mixture. In nitrogen–hydrogen plasmas, where the hydrogen percentage is generally below 12 vol.% H$_2$, first κ' increases between 3000 and 4000 K, but then the nitrogen dissociation, between about 6000 and 8500 K, raises it significantly, with κ' being multiplied by a factor of about 16 above temperatures of 8500 K compared to its value at 2000 K. This sharp increase of the heat transfer at 8500 K, which is also the temperature at which the plasma can sustain an arc, explains the constriction of nitrogen arc column compared to argon–hydrogen plasma in which the heat transfer between the arc column and its cold boundary layer depends strongly on the hydrogen concentration (Nogues et al. 2007).

3.4   Basic Concepts in Modeling of Plasma Spraying Processes

3.4.1   Introduction

Mathematical modeling of plasma spray processes is a very powerful tool that has been used extensively for the developing of our fundamental understanding of the basic phenomena involved, optimization of process performance, and the development of new advanced processes. Being a heterogeneous process involving the injection of the material to be

sprayed in the form of a powder, solution, suspension, or other forms into a hot gas stream, whether a combustion flame or an electrical discharge, in order to heat, melt, and project the molten droplets on the surface of the substrate to be coated, it is necessary to deal with a wide range of physical phenomena including:

- Combustion and flame dynamics.
- Electric discharges and thermofluidic dynamics of plasma flows.
- Pneumatic transport of powders, liquids, and suspensions for their controlled feeding and injection into the combustion or plasma flow.
- Liquid atomization, whether a solution or suspension, as injected into a combustion or plasma flow.
- In-flight interaction between the particles and the combustion gases or the plasma stream resulting in particle heating, melting and possible particle evaporation.
- Plasma-particle interactions under dense loading conditions. That includes the effect of the plasma on the particle inflight, and the effect of the particles on the plasma flow and its temperature fields.
- Molten droplet/particle impact and deposition on the substrate. Splat formation and rapid cooling and solidification of the molten droplets.
- In-flight interaction between the particles and the combustion gases or the plasma stream results in particle heating, melting, and possible particle evaporation.
- Health and safety issues associated with the spraying operation.

Each one of these phenomena is a field by itself involving basic science, mathematical modeling, and diagnostics to be fully understood and carefully managed. Fundamental issues related to the first two areas of research, that is, combustion and electric discharge physics and fluid dynamics, have been briefly reviewed in this chapter. A discussion and examples of modeling of plasma flows are presented. Further information is also presented in Chaps. 4, 6, 7, 8, 9, and 10 dealing with combustion or plasma source design and operation. A review of topics related to powder injection into the flow and combustion gas/plasma-particle interaction is presented in Chap. 3. Particle deposition, splat, and coating formation are discussed in Chaps. 13, 14, and 15. A more detailed discussion and complementary material on the subject can also be found in "Handbook of thermal plasmas," by Boulos, Fauchais, and Pfender, to be published by Springer (2020).

Such developments have been possible, thanks to the improvement of our knowledge base of thermodynamic and transport properties that are a prerequisite for modeling of thermal plasmas. Advanced modeling of plasma flows offers several important challenges such as deviations from local thermodynamic equilibrium (LTE) which can be encountered when operating the plasma sources under very low pressures (below 10 kPa). A detailed discussion of such phenomena is beyond the aim of this book and will not be discussed.

3.4.2 Conservation Equations for the Modeling of Plasma Flows

Serious modeling efforts must be based on first principles, that is, on solutions of the conservation equations with appropriate boundary conditions [Oran and Boris (2001), Hirsch (1988), Hoffman and Chiang (1993), Kundu Pijushand and I. Cohen (2008)]. The conservation equations represent a set of coupled, nonlinear differential equations, which must be solved simultaneously to obtain the desired results that include temperature and velocity fields, species concentrations, heat and enthalpy fluxes and electrical characteristics of the plasma. The general equations used for the modeling of combustion and thermal plasma flows are presented below. These have to be adapted on a case-by-case basis to best represent the geometry of the combustion or plasma source and its operating conditions. In each of these cases the reader will be redirected to the corresponding references.

3.4.2.1 Continuity Equations

The continuity equations comprise a set of equations describing conservation of mass, conservation of species, and conservation of electrical charges (electrical current in plasmas). Conservation of electrical current applies only to "active," current-carrying plasmas (e.g., to electric arcs), not to "passive," non-current-carrying plasmas such as plasma jets. Considering a gas comprised of k different species (e.g., electrons, ionic and neutral species for plasmas), the species conservation equation for species of type k may be written as

$$\frac{\partial \rho_k}{\partial t} + \nabla \cdot \left(\rho_k \vec{v}_k \right) = \Gamma_k \tag{3.13}$$

where ρ_k represents the mass density, v_k the velocity, and Γ_k the mass generation rate of species of type k. By introducing the concentration

$$c_k = \frac{\rho_k}{\rho} \text{ with } \rho = \sum_k \rho_k \tag{3.14}$$

and the mass flux

$$\vec{I}_k = \rho_k \left(\vec{v}_k - \vec{v}_g \right) \tag{3.15}$$

where v_g is the velocity of the center of mass, Eq. (3.13) may be transformed into

$$\rho \frac{Dc_k}{dt} + \nabla \cdot I_k = \Gamma_k \tag{3.16}$$

with the substantive derivative defined as

$$\frac{D}{dt} = \frac{\partial}{\partial t} + \left(\vec{v}_g . \nabla \right) \tag{3.17}$$

Equation (3.16) is an alternate equation describing species conservation. Summation over all k species, and making use of the fact that $\sum_k \Gamma_k = 0$, Eq. (3.16) transforms into the following overall mass conservation equation:

$$\frac{\partial \rho}{\partial t} + \nabla . \left(\rho \vec{v}_g \right) = 0 \tag{3.18}$$

An alternate form of the continuity equation may be obtained by introducing Eq. (3.15) into Eq. (3.16):

$$\frac{d\rho}{dt} + \rho . \nabla . v_g = 0 \tag{3.19}$$

For plasmas, the space charge density is given by

$$\rho_{el} = e \left(n_i - n_e \right) \tag{3.20}$$

The charge or current conservation equation may be expressed by

$$\frac{\partial \rho_{el}}{\partial t} + \nabla . \vec{j} = 0 \tag{3.21}$$

where e is the elementary charge, n_i and n_e are ion and electron density, respectively, and $\vec{j}$ is the current density vector.

By specifying a coordinate system (one-, two-, or three-dimensional), these continuity equations can be expressed in terms of the adopted coordinate system. For example, a rotationally symmetric steady-state plasma flow, $\left(\frac{\partial}{\partial t} = 0 \right)$, can be represented in cylindrical coordinates (r, z, φ) as follows:

$$\frac{\partial}{\partial z} (\rho_k u_k) + \frac{1}{r} \frac{\partial}{\partial r} (\rho_k r v_k) = \Gamma_k \tag{3.22}$$

where the index k indicates the different species present in the plasma (electrons, ions, neutrals), Γ_k accounts for the mass generation rate of species of type k due to chemical reactions, and u_k and v_k are respectively the velocity components in the axial and radial directions. In a similar way, Eq. (3.19) translates into

$$\frac{\partial}{\partial z}(\rho u) + \frac{1}{r}\frac{\partial}{\partial r}(\rho r v) = 0 \qquad (3.23)$$

which is the overall mass conservation equation for this particular situation.

Assuming the electric current has only a z- and r-component, Eq. (3.21) translates into

$$\frac{\partial}{\partial z}j_z + \frac{1}{r}\frac{\partial}{\partial r}(r.j_r) = 0 \qquad (3.24)$$

Besides this simple example, the previously discussed continuity equations may be written for other coordinate systems and for steady- as well as for non-steady-state situations.

3.4.2.2 Momentum Equations

The overall momentum equation, which is identical to the basic equation of hydrodynamics, may be written as

$$\rho \, \frac{D\vec{v}_g}{dt} = -\nabla p + \sum_k \rho_k \vec{F}_k \qquad (3.25)$$

where p is the total pressure considering contributions from all species present in the plasma and $\vec{F}_k$ represents external forces. Possible external forces are important electric forces due to applied electric fields, magnetic forces due to external or self-induced magnetic fields for plasmas, and pressure forces provided there are significant pressure gradients, pressure jump when the detonation wave front passes. Forces due to gravity are generally negligible for plasma spray torches.

Internal forces as friction and thermal diffusion may appear in the momentum equation of individual species. For species of type k the momentum equation assumes the form

$$\frac{\partial \rho_k \vec{v}_k}{\partial t} + \nabla.\left(\rho_k \vec{v}_k \, \vec{v}_k\right) = -\nabla p_k - \nabla.\vec{\tau}_k + \rho_k \, \vec{F}_k + \sum_j \vec{F}_{k_J} \qquad (3.26)$$

where v_k is the velocity vector of species k, $\vec{\tau}_k$ is the stress tensor of species k, and p_k the partial pressure of species k given by

$$p_k = n_k.k_B.T \qquad (3.27)$$

$\vec{F}_k$ represents the rate of momentum transfer between species k and another species j. For example, in DC plasmas, assuming that only electric and magnetic forces are present, the term describing external forces can be expressed by

$$\rho_k \vec{F}_k = e_k n_k \vec{E} + e_k n_k \left(\vec{v}_k \times \vec{B}\right) \qquad (3.28)$$

where e_k is the electric charge, n_k is the number density of species of type k, $\vec{E}$ is the applied electric field, and $\vec{B}$ represents the magnetic induction.

Assuming charge neutrality

$$\sum_k e_k n_k = 0 \qquad (3.29)$$

the mass-averaged momentum equation assumes the form

$$\frac{\partial \rho \vec{v}_g}{\partial t} + \nabla.\left(\rho \vec{v}_g \, \vec{v}_g\right) = -\nabla p - \nabla.\vec{\tau}_k + \vec{J} \times \vec{B} \qquad (3.30)$$

This equation may be written for any desired coordinate system. As an example, the same two-dimensional situation will be considered specified in the previous section

(cylindrical coordinates r, z, φ). In this case, Eq. (3.30) transforms into two equations for the z- and r-direction:

z-direction:

$$u\frac{\partial u}{\partial z} + \rho v\frac{\partial u}{\partial r} = -\frac{\partial p}{\partial z} + 2\frac{\partial}{\partial z}\left(\mu\frac{\partial u}{\partial z}\right) + \frac{1}{r}\frac{\partial}{\partial r}\left(\mu r\frac{\partial u}{\partial r}\right) + \frac{1}{r}\frac{\partial}{\partial r}\left(\mu r\frac{\partial v}{\partial r}\right) + j_r B_\varphi \qquad (3.31)$$

r-direction:

$$\rho u \frac{\partial v}{\partial z} + \rho v \frac{\partial v}{\partial r} = -\frac{\partial p}{\partial z} + \frac{\partial}{\partial z}\left(\mu \frac{\partial v}{\partial z}\right) + \frac{2}{r}\frac{\partial}{\partial r}\left(\mu r \frac{\partial v}{\partial r}\right) + \frac{\partial}{\partial z}\left(\frac{\partial v}{\partial r}\right) - \frac{2\mu v}{r^2} - j_z B_\varphi \tag{3.32}$$

3.4.2.3 Energy Equations

The energy equation may be written in the form (for derivation see Chap. 5 in [Boulos et al. (1994)]

$$\rho \frac{d}{dt}\left(\frac{v_g^2}{2} + u_g\right) = -\nabla \cdot \left(p\vec{v_g} + \vec{W}\right) + \sum \vec{F}_k \cdot \rho_k \, \vec{v_k} \tag{3.33}$$

In this equation, u_g is the internal energy of the system and $\vec{W}$ is the heat flux vector. This equation represents a typical balance equation, that is, (rate of change of the total energy in the system) = (net flux of work and heat into the system) + (energy dissipation in the system).

By introducing the expression for the substantive derivative, Eq. (3.31) transforms into

$$\rho \frac{\partial}{\partial t}\left(\frac{v_g^2}{2} + u_g\right) + \rho \vec{v}_g \nabla\left(\frac{v_g^2}{2} + u_g\right) = -\nabla \cdot \left(p\vec{v}_g + \vec{w}\right) + \sum \vec{F}_k \cdot \rho_k \vec{v}_k \tag{3.34}$$

For example, in the case of steady-state plasma in LTE with negligible viscous dissipation and under optically thin conditions, this equation reduces to

$$\rho \vec{v_g} \nabla\left(\frac{v_g^2}{2} + h\right) + \nabla \cdot \vec{W} = \vec{j} \cdot \vec{E} + \frac{5}{2}\,\vec{j} \cdot \nabla T - S_R(\rho, T) \tag{3.35}$$

where h is the enthalpy. The first two terms on the RHS of Eq. (3.35) represent the Ohmic heating term and an energy dissipation term which becomes important if $\vec{j}$ and ∇T are parallel which is, for example, the case in the electrode boundary layer of electric arcs. In the case of a fully developed arc column, $\vec{j}$ and ∇T are perpendicular to each other, that is, $\vec{j} \cdot \nabla T = 0$.

Again, the energy equations [Eq. (3.33) or (3.34)] may be written for any desired coordinate system. As an example, the same two-dimensional situation as discussed above will give rise to the following equation:

$$\rho u \frac{\partial h}{\partial z} + \rho v \frac{\partial h}{\partial r} = \frac{\partial}{\partial z}\left(\frac{\kappa}{c_p}\frac{\partial h}{\partial z}\right) + \frac{1}{r}\frac{\partial}{\partial r}\left(\frac{\kappa}{c_p}\frac{\partial h}{\partial r}\right) + \frac{j_z^2 + j_r^2}{\sigma_e} + \frac{5}{2}\frac{k}{e}\left(\frac{j_z}{c_p}\frac{\partial h}{\partial z} + \frac{j_r}{c_p}\frac{\partial h}{\partial r}\right) - S_R(\rho, T) \tag{3.36}$$

where c_p is the specific heat at constant pressure and $S_R\,(\rho,T)$ the optically thin radiation term. The LHS of this equation represents two convection terms, followed by two heat conduction terms on the RHS. The next term accounts for Joule heat dissipation, followed by another heat dissipation term which is not negligible for situations in which $\frac{\partial h}{\partial z} \neq 0$ and $j_r \neq 0$. The last term in Eq. (3.35) accounts for optically thin radiation.

3.4.2.4 Electromagnetic Field Equations

Electromagnetic field equations required for modeling of arcs and plasma jets include:

- Charge or current conservation equation
- Poisson equation
- Ohm's law
- Magnetic field equation

For $\frac{\partial \vec{B}}{\partial t} = \frac{\partial \vec{E}}{\partial t} \approx 0$ modeling of arcs, changes of electric or magnetic fields may be assumed to be relatively slow.

Conservation of charge has been already discussed in part (a) and will not be reiterated here. The Poisson equation reads as

$$\nabla . \vec{E} = \frac{\rho_{el}}{\varepsilon_0} = \frac{e.\,(n_i - n_e)}{\varepsilon_0} \tag{3.37}$$

where ρ_{el} is the space charge density, n_i and n_e the ion and electron density, respectively, and ε_0 is the dielectric constant. With

$$\vec{E} = -\nabla \Phi \tag{3.38}$$

Equation (3.37) becomes

$$\nabla^2 \Phi = -\frac{\rho_{el}}{\varepsilon_0} \tag{3.39}$$

Ohm's law in the simplest form may be written as

$$\vec{j} = \sigma_e\,\vec{E} \tag{3.40}$$

which implies that there are no other significant driving forces for the electric current besides the applied electric field. As shown in Chap. 5 of the book of Boulos et al. (1994), there may be additional driving forces such as temperature gradients, pressure gradients, $\vec{v_e} \times \vec{B}$ force induced by a magnetic field on electrons, which contribute to the current flow. The magnetic induction, which appears in the momentum equations, may be determined from one of Maxwell's equations, that is,

$$\nabla \times \vec{B} = \mu_0\,\vec{j} \tag{3.41}$$

where μ_0 is the magnetic permeability constant. The electromagnetic field equations may be written in terms of any desirable coordinate system. As an example, the transformation of Eq. (3.41) into cylindrical coordinates will be considered assuming that $\vec{B}$ has only a φ-component, that is,

$$B_\varphi = \frac{\mu_0}{r} \int_0^r j_z\,\zeta\,\mathrm{d}\zeta \tag{3.42}$$

where ζ is a dummy variable.

3.4.2.5 Laminar or Turbulent Flows

The assumption of laminar flow entails relatively low mass flow rates. For example, arcs with little or no superimposed flow are mainly used in laboratories. In contrast, arcs exposed to substantial flows are of great practical interest, as, for example, in the development of arc gas heaters and plasma torches. The flow in this case is usually turbulent, except that in the plasma core it is laminar, which requires models to be able to treat the presence of laminar and turbulent conditions in different regions of the flow.

Turbulence is one of the most puzzling phenomena in fluid dynamics and even more so in thermal plasma technology, which is, to a large degree, governed by turbulent flow situations. In spite of many successful commercial developments over the years, the underlying physics of turbulence in plasma flows is still poorly understood. The phenomena are particularly challenging in thermal plasma flows due to the inherent difficulty of making dynamic, time-dependent, velocity measurements of plasmas, and the occasional presence of both laminar and turbulent flow regimes in the same flow field depending on the local temperature distribution. It is not surprising that the presence of turbulence adds another dimension of complexity to the already complex situation in thermal spray flow systems.

Turbulence is characterized by highly random, non-steady, and three-dimensional effects in the flow. Turbulent flow, in principle, can be described by the Navier–Stokes (NS) equations. Unfortunately, these equations cannot be solved for most practical problems, because such turbulent flows encompass a wide range of turbulent eddy sizes from the size of the domain of interest, down to sizes below those at which viscous dissipation and chemical reactions occur. This fact translates into immense requirements in terms of the number of spatial grid cells and the sheer volume of the data that has to be manipulated for an adequate modeling of the process.

Turbulent transport properties also require closure assumptions, because information is lost in the averaging process. Due to the nonlinearity of the governing equations, the averaging process introduces unknown correlations, for example, between fluctuating velocities and between velocities and scalar fluctuations. This implies that the governing equations can only be solved if the turbulence correlations are specified. Determining such correlations is one of the main problems in calculating turbulent flow. The choice of the closure assumptions requires specification of the proper turbulence model. A number of such models have been proposed to solve this closure problem, including:

- Zero-equation models
- One-equation models
- Two-equation models
- Stress/flux-equation models
- Scalar-fluctuation models

These models have been classified according to the number of turbulence parameters that appear as dependent variables in the differential equations. All these models listed above have been reviewed in detailed surveys [Launder and Spalding (1974), Rodi (1980), Favre et al. (1977), Shih et al. (1995), Pope (2004), Bazilevs et al. (2007) and Shigeta M (2016)]. The simplest models relate the turbulence correlations directly to the local mean flow quantities, while the most complex models solve differential transport equations for the individual turbulence correlations. The choice of the proper turbulence model depends on the requirements of the problem and thus on the compromise between the complexity of the model and the availability of computational resources, and the applicability and accuracy of the information needed.

A good turbulence model should generally provide sufficient universality without being too complex to use. The mixing length model is a good example because of its simplicity in assuming the turbulent viscosity μ_t directly proportional to the local velocity gradient as given by

$$\mu_t = L_m \frac{du}{dy} \tag{3.43}$$

The main problem is to formulate an expression giving the variation of the mixing length L_m in the flow field. The k–ε turbulence model is a widely used alternate approach in which the three turbulent fluctuation components, u', v', and w', are integrated in the expression of the turbulent kinetic energy through

$$k = \frac{1}{2} \left(\overline{u'^2} + \overline{v'^2} + \overline{w'^2} \right) \tag{3.44}$$

This formulation allows for the modeling of *isotropic* turbulence by solving one equation related to k representing turbulence generation, and another one to ε the turbulent energy dissipation. It is possible to combine a k–ε turbulence model with a mass-weighted averaging concept in his model.

A multiple time scale turbulence model, using one-time scale for velocity fields and another for scalar fields, was first proposed for parabolic-type flows. The ability to predict both unweighted and mass-weighted mean temperatures can be achieved by adopting mass-weighted averaging for transport equations and a probability density function. This model offers a linkage for correlating different measurement methods, for example, spectroscopy and enthalpy probes for plasma temperatures. The aforementioned parabolic model has been extended to more general situations for two-dimensional plasma flows, allowing, for example, the presence of a swirl component, a cross-stream pressure gradient, or even recirculation. However, the coefficients involved in the closure model must be reoptimized to achieve better overall agreement with available experimental data.

Caution has to be exercised in applying the various turbulence models. For example, the traditional high-Reynolds-number k–ε modeling method, which has been successfully used in fully developed flow regions, may not be appropriate for modeling of plasma jets. k–ε models are not well adapted to describe flows where the turbulence intensity is low or comprising a laminar core followed by a turbulent plume as in plasma jets. The turbulence intensity in k–ε low Reynolds models is addressing this issue by defining a turbulent Reynolds number Re_t representing the ratio between turbulent and laminar viscosities

$$Re_t = \frac{\overline{\rho}\, k^2}{\mu \varepsilon} \tag{3.45}$$

and by modifying the turbulence constants C_μ and C_z of the standard k–ε model by expressions depending on Re_t. Another point where the k–ε models are poor (even the low Reynolds ones) is when the jet impacts on a surface (e.g., the substrate in various spray processes). The model cannot be applied for boundary layers where the flow is laminar and also where laws are used to describe the friction and the thermal exchange with the wall. To solve this problem, Yap (1987) proposed to add a source term in the equation of turbulent energy to modify the characteristic turbulence length toward a local equilibrium value close to the wall. Chen and Kim (1987) proposed on the other hand to add a source term in the equation of turbulent energy which represents the transition of the turbulence energy transfer at large scale to that at small scale within the boundary layer.

Other models have also been used for the turbulence. The R_{ij}–ε model solves transport equations of Reynolds stress $R_{ij} = \overline{u'_i u'_j}$, and the heat fluxes of turbulence. This approach allows solving the weakness of the k–ε model linked to the turbulent viscosity and the assumption of isotropic

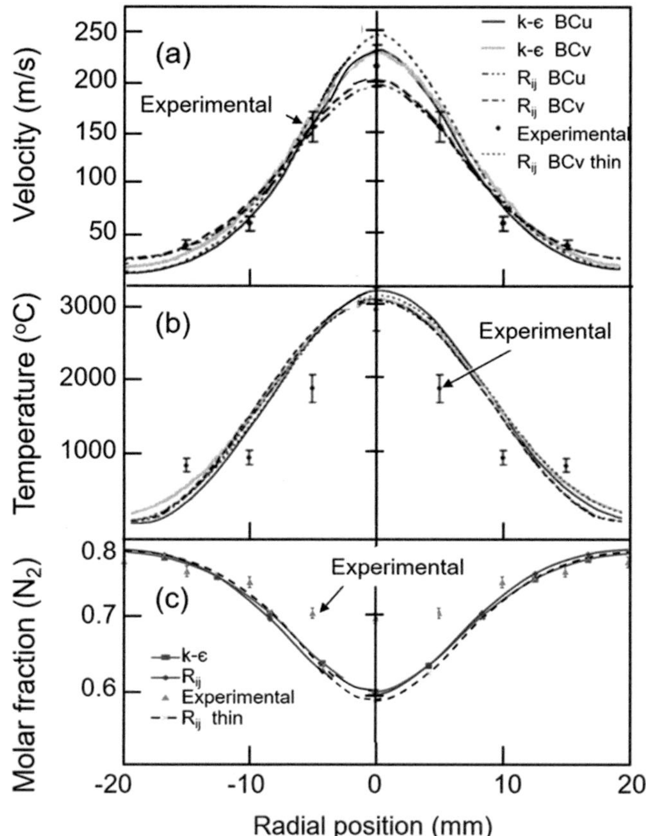

Fig. 3.23 (a) Velocity, (b) temperature, and (c) nitrogen concentration profiles in the radial direction for an Ar/H$_2$ DC plasma jet at 80 mm from torch nozzle exit. ([Fauchais and Vardelle (2004)]. Reproduced by permission of IOP Publishing)

turbulence. This model is however time expensive. In 3D cases it is necessary to solve eight equations instead of two for the standard $k - \varepsilon$ model and the use of a finer grid. It also requires more empirical constants than the standard $k - \varepsilon$.

The "Reynolds Normalization Group" (RNG) $|k - \varepsilon|$ model is derived from the use of a rigorous statistics applied to Navier–Stokes equations. It represents better than the $k-\varepsilon$ low Reynolds model and flows with low Reynolds numbers, especially close to walls.

To illustrate the differences between the turbulence models, Legros (2003) calculated a 2D-stationary DC plasma jet using the $k-\varepsilon$ and two versions of the $R_{ij}-\varepsilon$ models and compared results with experimental values. The plasma jet was produced by a DC plasma torch PTF4 type with a 7 mm i.d. anode nozzle working with Ar/H$_2$ mixture as plasma gas (45 slm Ar + 15 slm H$_2$), arc current 600 A, voltage 65 V, and a thermal efficiency of 55%. Figure 3.23a, b represents respectively the velocities and temperatures profiles at an axial position of 80 mm downstream of the torch exit nozzle. As can be seen, the differences between the different turbulence models are relatively small while the computation time between $k-\varepsilon$ and $R_{ij}-\varepsilon$ is more than one order of magnitude.

As noted in Fig. 3.23b the cooling down of the jet predicted by the model is less than that given by the experimental results. The air entrainment as indicated by the nitrogen concentration profiles given in Fig. 3.23c is also underestimated compared to experiments.

3.4.3 Gas Composition, Thermodynamic, and Transport Properties

3.4.3.1 Gas Composition

The gas composition depends on reaction rates linked to local temperature and pressure. Different assumptions can be assumed regarding the reaction rates, including the following:

Infinitely fast reaction rate (equilibrium calculation): Equilibrium compositions are obtained by minimizing the Gibbs free energy [Storey and van Zeggeren (1970)]. The data are given in different tables such as those of Thermodata, Data Bank, Janaf Thermochemical Tables (1985), and Gurvich et al. (1990). The data in each table are coherent among themselves, but not necessarily between tables, especially for plasmas over 6000 K. In most cases plasma flows are calculated at equilibrium.

Finite reaction rate in Arrhenius (see Eq.3.7): The multi-reaction simulations use a number of intermediate chemical species the number of which increases with temperature when dissociation and of course ionization (at higher temperatures) occur. For example, in combustion a few tens of reactions should be considered and a higher number of reversible reactions. Then the time rate of change of the concentration of a given species is given by Eq. (3.5). This results in a set of nonlinear equations, which can be solved numerically, knowing the initial values of n_i to determine the composition at time t [Benson (1976)]. The data for calculations have to be checked carefully and they must satisfy the following equation:

$$K_x = \frac{k_f}{k_r} \tag{3.46}$$

where k_f is the forward reaction rate, k_r the reverse reaction rate, and K_x the molar fraction equilibrium constant related to the considered reaction. Many of the data are gathered in the database. However, these rates can be computed from the Arrhenius rate expressions (see Eq. 3.7) when turbulence is not the dominating mechanism, which is not the case in flames and HVOF. For plasmas, kinetic calculations are used when a cold gas, reacting or not, is injected into the plasma jet and the induced reactions are of interest. In most cases in spray processes the plasma jet perturbation by the cold carrier gas is neglected as soon as the carrier gas mass flow rate does not exceed 10% of the plasma gas mass flow rate. In flame or HVOF spraying, such kinetic calculations have been used in most cases to demonstrate that equilibrium calculation is sufficient to characterize the flame.

Finite reaction rate is limited by turbulent mixing, which assumes that the reaction rate is limited by the turbulent mixing rate of components, or the reaction is instantaneous as the reactants are mixed together. This approach is currently used in combustion processes; see, for example, Bandyopadhyay and Bandyopadhyay and Nylén (2003), but not in plasmas where the jet characteristics mainly depend on phenomena occurring in the laminar core.

3.4.3.2 Thermodynamic Properties

Once the composition is determined, the different thermodynamic properties are relatively easy to calculate, even in plasmas [Boulos et al. (1994)].

3.4.3.3 Transport Properties

They are rather complex to calculate in plasmas, because, up to now, they have been calculated only for a few species and mixtures containing only two different species. The calculation of complex mixtures is rather long even if all interaction potentials are known, which is not necessarily the case. Thus, simplified mixing rules are used [Boulos et al. (1994)]. However, the diffusion coefficients, which must be used to account for species diffusion within flows, are often unknown.

3.4.4 Examples of DC Torch Modeling Results

To illustrate the basic concepts in plasma modeling described previously, results reported by Alaya et al. (2015, 2016) of models applied to a conventional DC plasma torch are presented. These were carried out using a standard stick-type cathode/anode arrangement as shown in Fig. 3.24. The objective of the first study [Alaya et al. (2015)] is the investigation of the effect of the torch geometry and operating parameters on the electric arc characteristics. Emphasis is placed on the description of the cathode arc attachment

which is an important part of a realistic model of the plasma torch operation as the properties of electric arcs at atmospheric pressure depend not only on the arc plasma medium, but also on the electrodes.

The model is based on a magnetohydrodynamic approach (MHD) coupling the Navier–Stokes equations for non-isothermal fluid with the Maxwell equations of electromagnetic field. The model which provides a 3D, time-dependent, numerical simulation of a plasma arc including the cathode boundary region assumes laminar, incompressible flow for an optically thin plasma in local thermodynamic equilibrium (LTE) and quasi-steady electromagnetic phenomena. At this stage of model development, the cathode was supposed not to evaporate.

The study focused on the cathode-plasma interaction of the flow using a two-step approach. The first step using a Gaussian-like current density profile imposed at the cathode surface represented by

$$j = j_0 \left(\frac{-r}{R} \right)^n \tag{3.47}$$

where j is the current density over the cathode surface and r the radial coordinate measured from the torch axis. The adjustable parameters j_0, R, and n control the shape of the profile. This served to determine the effect of the "profile parameters" on the plasma gas temperature, velocity, and electromagnetic fields.

The second step included the cathode in the computation domain and calculated the corresponding temperature and electromagnetic fields within the cathode and the fluid. In both steps a given electrical conductivity was imposed at the anode wall in order not to implement any arc attachment models on the anode side.

The proposed approach has the advantage of allowing for:

- To get rid of the need for boundary condition at the cathode tip which is difficult to establish
- To better predict the arc dynamics, thanks to more realistic predictions of the current density distribution, magnetic field, arc temperature, and velocity field, and finally, arc voltage
- To investigate the effect of cathode shape and position on the arc behavior.
- To predict, in the long term, the wear of the cathode.

Calculations were performed for an arc current of 600 A, using a mixture of argon-hydrogen (45 slm Ar + 15 slm H$_2$) as plasma forming gas. The calculation domain as illustrated in Fig. 3.24 is 42 mm long with a numerical grid consisting of 350,000 nodes, of which 50,000 are in the cathode. The minimum grid cell size is about 40 μm close to the cathode tip. The time step is 0.1 μs. For more details about the

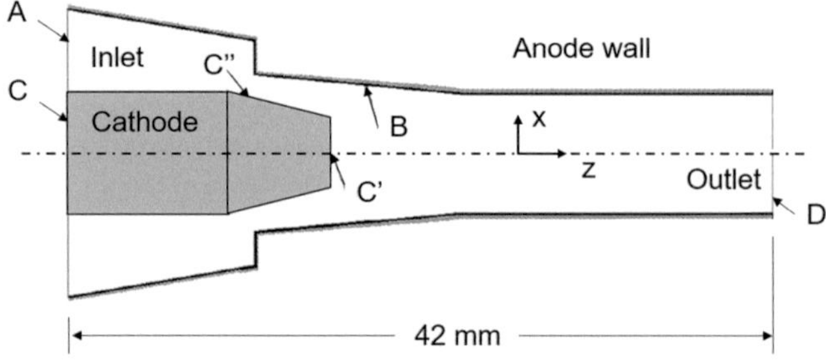

Fig. 3.24 Nozzle geometry, computational domain, and boundaries for DC plasma torch. [Alaya et al. (2015)]

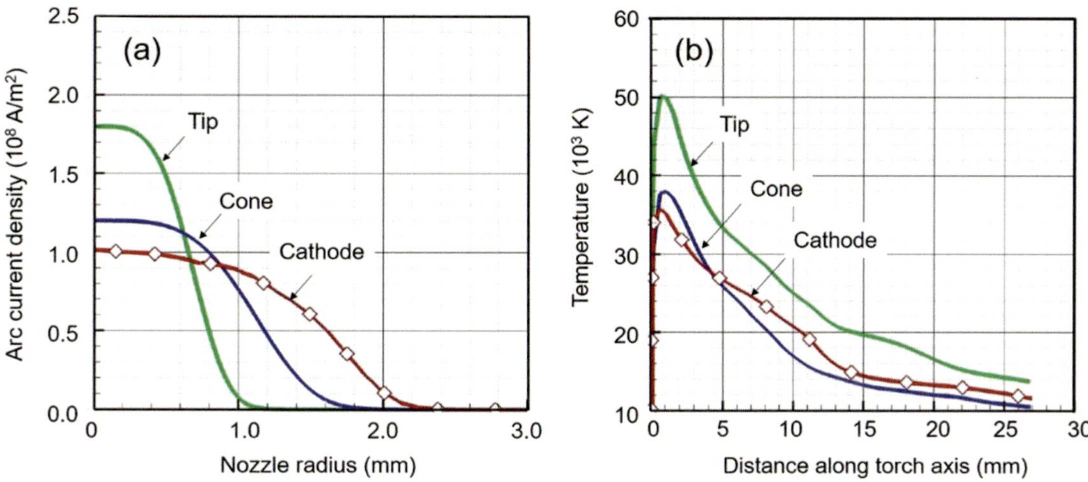

Fig. 3.25 (**a**) Current density profiles imposed at the cathode tip and calculated with the arc-cathode coupling and (**b**) corresponding temperature profile along the centerline of the torch. [Alaya et al. (2015)]

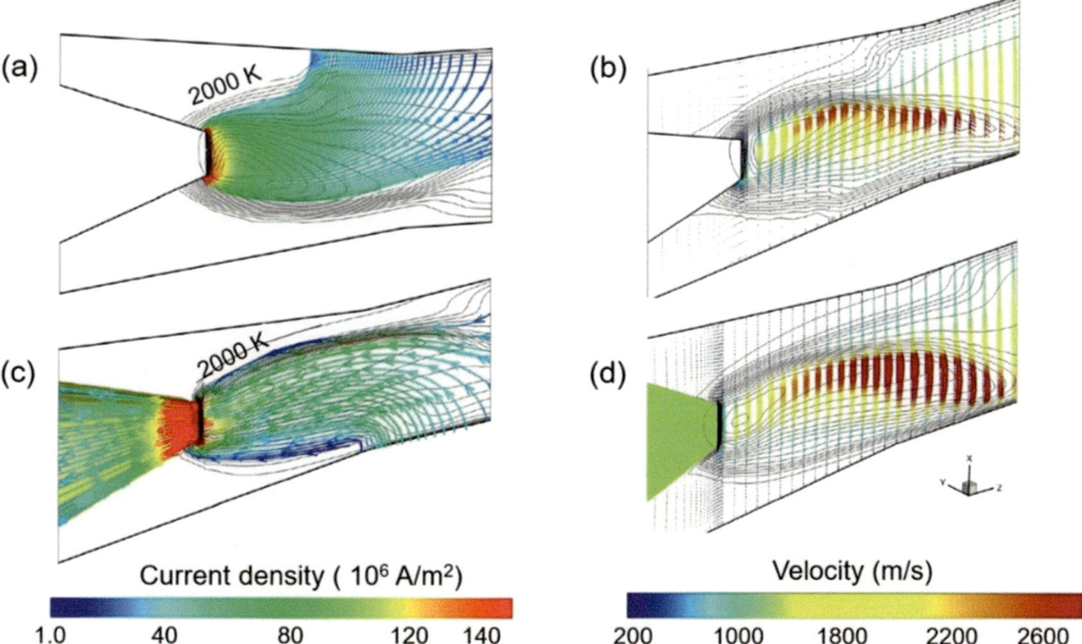

Fig. 3.26 (**a**) Current density distribution (color scale) and (**b**) corresponding velocity vectors (color scale) with imposed current density profiles at cathode surface, and (**c** and **d**) corresponding values with arc-cathode coupling. Corresponding temperature isotherms are indicated in all figures in black. [Alaya et al. (2015)]

boundary conditions and calculations procedure, see Alaya et al. (2015).

Results obtained using the first-step approach are given in Fig. 3.25. These show the calculated current density profiles imposed at the cathode tip, cathode cone, and cathode and the corresponding axial temperature profile along the centerline of the torch.

Thermal and magnetic coupling between the cathode and arc made it possible to predict the temperature distribution within the cathode, and the distribution of the magnetic field close to the cathode. Figure 3.26a, b shows the current streamlines and the corresponding velocity vectors when a current density profile is imposed on the tip of the cathode. The corresponding values of the current streamlines and velocity vectors obtained when the cathode is coupled with the arc are given in Fig. 3.26c, d. The predicted value of the current density in the cathode is in good agreement with experimental value (10^8 A/m^2) and as expected increases in

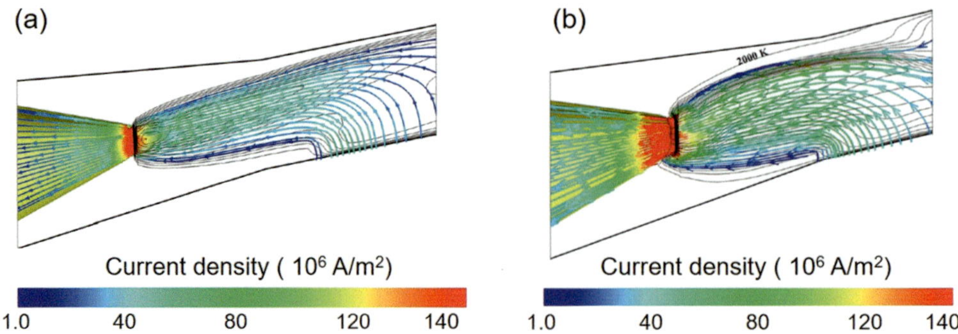

Fig. 3.27 Current density distribution (vectors) in cathode and electric arc close to the cathode tip for arc currents of (**a**) 300 and (**b**) 600 A. [Alaya et al. (2016)]

the conical section of the cathode. Comparing Fig. 3.26b, d reveals that the arc velocity is higher when the magnetic coupling of the arc and the cathode is considered which shows the effect of a better description of the magnetic fields around the cathode as it directly affects the Lorentz $(\vec{j} \times \vec{B})$ forces.

To conclude the coupling of the electric arc and electrodes seems to be the first step toward a consistent and realistic model that should use as input data torch operating parameters (torch internal geometry, arc current, gas type, and flow rate). In a subsequent report, Alaya et al. (2016) coupled the electromagnetic and heat transfer phenomena in a non-transferred arc plasma torch based on the observation of the current density profile experimentally and used in the computations an estimation of current distribution at cathode tip. The cathode and nozzle geometries were similar to those used in their earlier study as given in Fig. 3.24, except for its principal dimensions which were 10.0 mm inlet diameter, 6.0 mm outlet diameter, and 60 mm overall length. The cathode was included in the computational domain with the arc current imposed on the rear surface of the cathode. The electromagnetism and energy conservation equations for the fluid and the electrode were coupled. The solution of this system of equations was implemented in a CFD computer code to model various plasma torch operating conditions. Typical results are given in Fig. 3.27 in terms of current streamlines in the cathode and electric arc for arc currents of 300 and 600 A. These show a significant widening of the arc column with the increase of the arc current which results in the reduction of the thickness of the boundary layer at the anode surface and an increase of the electromagnetic forces exerted on the attachment of column at the anode. Such effect which is consistent with the experimental observations of Maecker (1955) favors the attachment of the arc closer to the cathode tip and thus the shortening of the arc column.

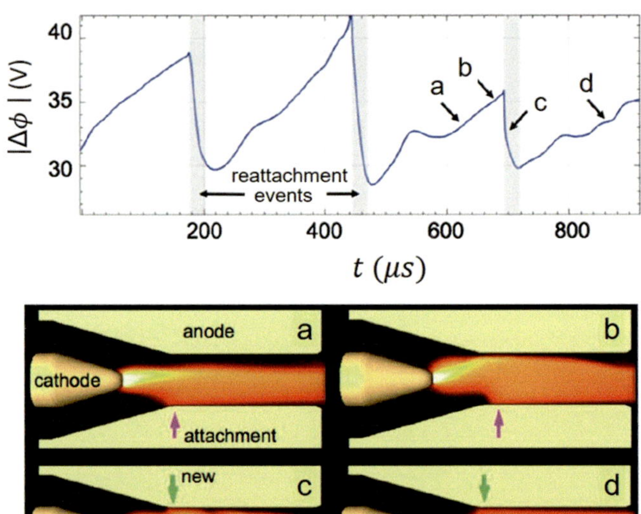

Fig. 3.28 Arc dynamics events for a DC plasma torch, nozzle diameter 8 mm, arc current 400 A, plasma gas flow rate of 60 slm (Ar): (top) time evolution of the voltage difference (work function) across the torch; (bottom) temperature field snapshots at representative instants. [Chazelas et al. (2017)]

Chazelas et al. (2017) point out that the operation mode of the electric arc in plasma torches is controlled by dynamic, thermal, electromagnetic, acoustic, and chemical phenomena that take place at different scales and whose interactions are not completely understood. Even though no single model of plasma torch operation fully addresses these phenomena, most of these models are useful tools for parametric studies, provided that their use is reinforced by knowledge of torch operation and that model predictions are validated against experimental data. A specific issue in plasma spray torch modeling that is not addressed in transferred arc models is the movement of the arc attachment at the anode wall that makes the implementation of an anode sheath model much more complex. Figure 3.28 depicts the displacement of arc-anode root attachment and arc reattachment process. The displacement of the anode attachment is a direct consequence of the imbalance between the drag force due to the interaction of the working gas flow on the plasma and the electromagnetic (Lorentz) force caused by the local curvature of the arc. The dominance of the drag force causes the displacement of the arc-anode attachment along the direction of the flow, linearly in the case of straight injection or following a corkscrew trajectory for swirl injection. The displacement of the arc-anode attachment is depicted in Fig. 3.28 (top) by the variation in arc voltage (work function) between the points a and b. It can also be identified in the temperature distribution snapshots given in Fig. 3.28 (bottom) by comparing the locations of the anode attachment (indicated by the arrows) in frame a with respect to that in frame b. The snapshots provided in this figure were obtained for a DC torch with a discharge nozzle id of 8 mm, operated using pure argon as plasma gas at a flow rate of 60 slm (Ar) and an arc current of 400 A. The snapshots given in Fig. 3.28 (bottom) reveal in (a) single established anode attachment; (b) dragging of the attachment along the flow direction; (c) formation of a new attachment, which is more thermodynamically favorable due to its proximity to the cathode; and (d) predominance of the new attachment and extinguishment of the old one.

3.5 Summary and Conclusions

Many spray processes are based on combustion flames at atmospheric or higher pressures, or on thermal plasmas produced by blown or transferred arcs, or radio frequency inductively coupled discharges, mainly at atmospheric pressure but also at lower pressures and in some cases above atmospheric pressure. The principal objective of this chapter is to present a comparative overview of the different available technologies for thermal spraying, without going at this stage into details of the properties of flames and plasmas to which books have been devoted.

This chapter is devoted to the description of the thermodynamic and transport properties of these high-energy jets: composition of the gas, enthalpy and specific heats, transport properties such as electrical conductivity, thermal conductivity, and viscosity, which are of critical importance for the understanding of the basic phenomena in the transfer heat between hot jets and particles and particles entrainment. For flames and plasmas, the differences between kinetic and equilibrium conditions are also discussed because of their importance close to walls, controlling the flame, or plasma development.

Of course for flames, besides hot gas properties, flammability and explosive conditions are very important for working and safety conditions. The design principle of the gun to achieve flame or detonation is discussed to introduce D-guns.

At last, the conservation equations are represented by a set of coupled, nonlinear differential equations, which have to be solved simultaneously to obtain the desired results that include temperature and velocity fields, species concentrations, heat and enthalpy fluxes, and electrical characteristics for plasmas. The heat, mass, and momentum transfers' hot gas particles are discussed in this chapter, while the description of spray torches is presented in Chap. 4 for flames and Chaps. 6, 7, 8, 9 and 10 for plasmas.

Nomenclature

Units are indicated in parentheses; when no units are indicated, the parameter is dimensionless.

Latin Alphabet

a	Thermal diffusivity of the combustion wave ($\kappa/\rho\, c_\mathrm{p}$) (m^2/s)
B	Magnetic induction (T)
c_k	Concentration of species of type k
c_p	Specific heat at constant pressure (J/K kg)
D	Detonation velocity (m/s)
e	Elementary charge (C)
E	Electric field strength (V/m) or energy (J)
E_I	Ionization energy (J or eV)
F	External forces per unit mass (N/kg)
F_k	Represents external forces (N)
F_{kj}	Rate of momentum transfer between species k and other species j (kg/m^2 s^2)
F/A	Molar ratio of fuel to oxidizer
g	Statistical weight
h	Planck's constant (6.6×10^{-34} J s)
h_m	Enthalpy per unit mass (J/kg)
k_B	Boltzmann constant (1.13×10 J/kg particle)
k_R	Specific reaction rate constant
I_k	Mass flux of particles of type k (kg/m^2 s)
I_p	Heat flux potential: $I_\mathrm{p} = \int_{T_0}^{T_\mathrm{p}} \kappa(T)\mathrm{d}T$ (W/m)
I_s	Heat flux potential: $I_\mathrm{s} = \int_{T_0}^{T_\mathrm{s}} \kappa(T)\mathrm{d}T$ (W/m)
j	Current density (A/m^2)
k_f	Forward reaction rate, $k_\mathrm{f} = A \times T^B \times \exp.(-E/kT)$ (m^3/part s)
k_r	Reverse reaction rate, for a binary reaction (m^3/part s)
K_x	Molar equilibrium constant ($K_\mathrm{x} = k_\mathrm{f}/k_\mathrm{r}$)
M	Arbitrary specification of all chemical species
M_e	Mach number
(M_i)	Concentration of species i (m^{-3})
m	Mass (kg)
N_i	Number species i
n'	Total number of compounds involved in a chemical reaction
n	Number density of particles (m^{-3})
n_i	Number density of species i: $n_i = N_i/V$ (m^{-3})
p_i	Pressure of species i (Pa)
p	Total pressure (Pa)
q	Chemical energy release at constant pressure (J/kg)

Q	Partition function
R	Perfect gas law constant (J/K kg)
R'	Equivalence ratio of richness, Eq. (3.3)
R^*	Radical
R_o	Reaction rate (units depend on the species number involved)
R	Radial coordinate (m)
S	Entropy (J/K)
S_L	Flame velocity (m/s)
S_R	Optically thin radiation losses (W/m^3)
S_u^0	Combustion wave velocity (m/s)
T	Temperature (K)
t	Time (s)
u	Axial velocity component (m/s)
V	Volume (m^3)
v	Radial velocity component (m/s)
v_g	Velocity of the center of mass (m/s)
v_k	Velocity of particles of type k (m/s)
W	Heat flux (W/m^2)
x	Coordinate (m)
z	Axial coordinate (m); ion charge number

Mathematical Symbols

| ∇ | Spatial derivative (m^{-1}) |

Greek Alphabet

α'	Multiplication factor of radicals
A	Thermal diffusion factor (A/V m)
ΔE_I	Lowering of the ionization energy (J or eV)
ε_o	Dielectric constant (A s/V m)
Φ	Electric potential (V)
Φ	Dummy variable
Φ	Time-averaged dummy variable
Φ'	Time fluctuating dummy variable
Γ_k	Mass generation rate of species of type k (kg/m^3 s)
Γ	Ratio of specific heats
K	Thermal conductivity (W/m K)
κ'	Integrated mean thermal conductivity, Eq. (3.11) (W/m K)
κ_total	Thermal conductivity also noted κ (W/m K)
$\kappa_\mathrm{tr}^\mathrm{h}$	Translational thermal conductivity of heavy species (W/m K)
$\kappa_\mathrm{tr}^\mathrm{e}$	Translational thermal conductivity of electrons (W/m K)
κ_R	Reactional thermal conductivity (W/m K)
μ	Molecular viscosity (Pa s)
ν'_j	Stoichiometric coefficient of the reactants
ν''_j	Stoichiometric coefficient of the products
ρ_i	Specific mass (kg/m^3)
ρ	Total mass density (kg/m^3)
ρ_el	Space charge density (C/m^3)
σ	Electrical conductivity (S/m)
$\bar{\bar{\tau}}_k$	Stress tensor of species k (Pa)
ζ	Dummy variable (m)

References

Alaya, M., C. Chazelas, G. Mariaux, and A. Vardelle. 2015. Arc-cathode coupling in the modeling of a conventional DC plasma spray torch. *Journal of Thermal and Spray Technology* 24 (1–2): 3–10.

Alaya, M., C. Chazelas, and A. Vardelle. 2016. Parametric study of plasma torch operation using a MHD model coupling the arc and electrodes. *Journal of Thermal and Spray Technology* 25 (1–2): 36–43.

Bandyopadhyay, R., and P. Nylén. 2003. A computational fluid dynamic analysis of gas and particle flow in flame spraying. *Journal of Thermal Spray Technology* 12 (4): 492–503.

Bazilevs, Y., V.M. Calo, J.A. Cottrell, T.J.R. Hughes, A. Reali, and G. Scovazzi. 2007. Variational multiscale residual-based turbulence modeling for large eddy simulation of compressible flows. *Computer Methods in Applied Mechanics and Engineering* 197 (1–4): 173–201.

Benson, S.W. 1976. *Thermochemical kinetics*. 2nd ed. New York: Wiley.

Boulos, M., P. Fauchais, and E. Pfender. 1994. *Thermal plasmas, fundamentals and applications*. New York\London: Plenum Press.

Bourdin, E., M. Boulos, and P. Fauchais. 1983. Transient conduction to a single sphere under plasma conditions. *International Journal of Heat and Mass Transfer* 26: 567–579.

Chazelas, C., J.P. Trelles, and A. Vardelle. 2017. The main issues to address in modeling plasma spray torch operation. *Journal of Thermal and Spray Technology* 26: 3–11.

Chen, Y.S., and S.W. Kim. 1987. *Computation of turbulent flows, turbulence closure models using and extended k-ε*. NASA, CR-179204.

Fauchais, P., and M. Vardelle. 2004. Understanding plasma spraying, Topical review. *Journal of Physics D: Applied Physics.* 37: R86–R108.

Fauchais, P., J. Heberlein, and M. Boulos. 2014. *Thermal spray fundamentals, from powder to part*. New York: Springer. 1550 pages.

Favre, A., L.S.G. Kovasznay, R. Dumas, J. Gaviglio, and M. Coantic. 1977. *The turbulence in fluid mechanics*. Paris: Gauthier-Villars Editors.

Fluid Mechanics. 2011. *A short course for physists*. Cambridge: Cambridge University Press.

Gan, J.A., and C.C. Berndt. 2013. Review on the oxidation of metallic thermal sprayed coatings: A case study with reference to rare-earth permanent magnetic coatings. *J. of Thermal Spray Technology* 22 (7): 1069–1091.

Glassman, I. 1977. *Combustion*. New York: Academic.

Gurvich, L.V., I.V. Veyts, and C.B. Alcock. 1990. *Thermodynamic properties of individual substances*. 4th ed. Washington, DC/Philadelphia/London: Hemisphere Pub. Corp., A member of the Taylor and Francis Group N.Y.

Hirsch, C. 1988. *Numerical computation of internal and external flows; Vol. I: Fundamentals of numerical discretization and numerical computation of internal and external flows and Vol. II: Computational methods for inviscid and viscous flows*. New York: Wiley.

Hirsch, C. 1988. *Numerical Computation of Internal and External Flows; Vol. I: Fundamentals of Numerical Discretization and Numerical Computation of Internal and External Flows and Vol. II: Computational Methods for Inviscid and Viscous Flows*. (New York: John Wiley & Sons).

Hoffman, K.A., and S.T. Chiang. 1993. *Computational fluid dynamics for engineers*. Vol. 1. Kansas: Engineering Educational Systems.

Hollenstein M., Rahmane M., and Boulos M.I. (1999) "Aerodynamic Study of the Supersonic Induction Plasma Jet', 14th. International Symposium on Plasma Chemistry – ISPC-14, Prague (Czech Republic), I, pp. 257–262

Janaf Thermochemical Tables. 1985. Part I and II, Published by the American Chemical Society and the American Institute of Physics for the National Bureau of Standards Michigan, USA.

Kee R.J., F.M. Rupley and J.A. Miller (1998) *A Fortran chemical kinetics package for the analysis of gas-phase chemical kinetics*, Sandia National Laboratories Report, SAND89–8009.

Kundu, Pijush K., and M. Ira Cohen. 2008. *Fluid mechanics*, 4th revised ed. New York: Academic.

Launder, B.E., and D.B. Spalding. 1974. The numerical computation of turbulent flow. *Computer Methods in Applied Mechanics and Engineering* 35: 269–289.

Legros, E. 2003, November. *Contribution to 3D modelling of the plasma spray process and application to a two-torch process*. PhD Thesis, University of Limoges, France (in French).

Launder, B.E., and D.B. Spalding. 1974. The numerical computation of turbulent flow. *Computer Methods in Applied Mechanics and Engineering* 35: 269–289.

———. 2009. Modeling and control of high-velocity oxygen-fuel (HVOF) thermal spray: A tutorial review. *Journal of Thermal and Spray Technology* 18 (5–6): 753–768.

Linde catalogue, Acetylene… there is no better fuel gas for Oxy-fuel gas processes.

Lyphout, C., and S. Björklund. 2015. Internal diameter HVAF spraying for wear and corrosion applications. *Journal of Thermal and Spray Technology* 24 (1–2): 235–243.

Mackie, J.C., and J.C. Smith. 1990. Inhibition of C_2 oxidation by methane under oxidative coupling conditions. *Energy & Fuels* 4 (3): 277–285.

Maecker, N.Z. 1955. Plasmastromungen in lichtbogen infloge eigenmagnetischer compression (in German). *Zeitschrift für Physik* 141 (1–2): 198–216.

Nogues, E., P. Fauchais, M. Vardelle, and P. Granger. 2007. Relation between the arc-root fluctuations, the cold boundary layer thickness and the particle thermal treatment. *Journal of Thermal Spray Technology* 16 (5-6): 919–926.

Oran, E.S., and J.P. Boris. 2001. *Numerical simulation of reactive flow*. 2nd ed. Cambridge: Cambridge University Press.

Pfender, E., W.L.T. Chen, and R. Spores. 1990. A new look at the thermal and gas dynamic characteristics of a Plasma Jet. In *Proceedings of the 3rd. NTSC-1990, Long Beach, California*, ed. C.C. Berndt, 1–10. Materials Park: ASM International.

Pope, S.B. 2004. Ten questions concerning the large-eddy simulation of turbulent flows. *New Journal of Physics* 6 (35): 1–24.

Rodi, W. 1980. Turbulence models for environmental problems. In *Prediction methods for turbulent flows*, ed. W. Kollmann, 260. Washington, DC: Hemisphere Publishing Co.

Sagaut, P. 1998. *Introduction to large eddy scale simulations for the modeling of uncompresible flows* (in French). Mathématiques et Applications, 30. Springer.

Shigeta, M. 2016. Turbulence modelling of thermal plasma flows. *Journal of Physics D: Applied Physics* 49: 493001.

Shih, T.H., W.W. Liou, A. Shabbir, Z. Yang, and J. Zhu. 1995. A new eddy viscosity model for high Reynolds number turbulent flows model development and validation. *Computers and Fluids* 24: 227.

Storey, S.H., and F. van Zeggeren. 1970. *The computation of chemical equilibria*. Cambridge University: Press.

Syed, A.A., A. Denoirjean, P. Denoirjean, J.C. Labbe, and P. Fauchais. 2004. In-flight oxidation of stainless-steel particles in plasma spraying. *Journal of Thermal Spray Technology* 14 (1): 117–124.

Thermodata, Data Bank, BP 22, 38402 St Martin d'Heres, France.

Yap, C. 1987. *Turbulent heat and momentum transfer in recirculating 1798 and impinging flows*. Ph. D, University of Manchester, G.B.

Abbreviations

DC	Direct Current
D-Gun	Detonation Gun
FTIR	Fourier Transform Infra-Red
GLR	Gas-to-Liquid mass Ratio
HVAF	High Velocity Air Fuel
HVOF	High Velocity Oxygen Fuel
i.d.	internal diameter
ICP	Inductively Coupled Plasma
LTE	Local Thermodynamic Equilibrium
PIV	Particle Image Velocimetry
RF	Radio Frequency
SHS	Self-propagation High temperature Synthesis
XRD	X-Ray Diffraction
YSZ	Yittria Stabilized Zirconia

4.1   Introduction

Thermal spray coatings technology is essentially based on the in-flight heating, melting and atomization of the material to be used for the coating followed by the deposition of the formed molten droplets on the substrates as successive splats, which pile up forming the coating. A combustion flame or an electrical discharge is used as the heating medium in which the coating material is injected in the form of powder, wire, solution, or liquid suspension. Since the coating material is in contact with the heating medium for less than a few milliseconds, the proper control of the trajectories of the particles or droplets and their temperature history is of critical importance for the overall success of the operation. A slight deviation from near-optimal conditions can easily lead to poor results due to either the lack of melting of particles, insufficient impact velocities, or the modification of their chemical composition due to inflight particle evaporation or unwanted chemical reactions.

This chapter is mainly devoted to a review of the different fundamental phenomena governing the inflight interaction between the coating material and the heating medium during their short contact time. Comprehensive studies on the subject were reported in the eighties and nineties by Boulos (1976)(1978)(1985), Mostaghimi J et al. (1984)(1987) (1989), Proulx P et al. (1985)(1987)(1990)(1991)a,b, Vardelle M et al. (1983)(1988), Bourdin et al. (1993), Fauchais et al. (1997)a,b, Chen and Pfender (1982a, b), Pfender E. (1985) (1989) (1999), Pfender et al. (1985, 1998), Boulos et al (2004)(2016) and Xue S et al. (2001) (2019). Studies and developments were also published for cold spray, HVOF, and D-gun and more recently for suspension and solution plasma spraying. Following a brief introduction of the properties of particles and powders, the basic phenomena governing the momentum and heat transfer between a single spherical particle and a high-energy gas flow is discussed. This includes such phenomena as plasma-particle momentum and heat transfer, transient heating, melting and evaporation of particles under plasma conditions, mass transfer and chemical reactions for liquid or gaseous phases. Particle trajectory and temperature history calculations are discussed next for ensemble of particles/ powder injected into the plasma flow under dilute loading condition, which is limited to situations where the mass feed rate of particles/powder in the flow is sufficiently low so as not to have any impact on the plasma flow and temperature fields. This is followed by a discussion of the treatment of ensembles of particles/powder in gas flows including particle injection techniques and thermal loading effects resulting from plasma–particle interactions under dense loading conditions, which is of critical importance in any realistic discussion of the inflight transient heating and melting of the particles.

M. I. Boulos et al. (ed.), *Thermal Spray Fundamentals*, https://doi.org/10.1007/978-3-030-70672-2_4

4.2 Overview of Powder Characteristics

Powders are by definition composed of individual particles, which can be of different sizes, shapes and compositions. Their properties are closely dependent on their manufacturing technique and the thermophysical properties of the material. Particles are generally characterized by the following:

- Individual particle diameter and morphology, including their shape, sphericity, and porosity
- Particle chemical composition and crystal structure

Powders on the other hand are characterized by the following:

- Particle size distribution, mean particle diameter, and standard deviation
- Apparent density and tap-density of the powder, the latter corresponding to the highest packing density of the powder
- Powder flowability (e.g., Hall-flow) reflecting its ability to free flowing
- Angle of internal friction and angle of repose, both of which relate to properties of the packed powder

4.2.1 Individual Particle Size and Morphology

Powders used in thermal spray operating are composed of ensembles of particles, which can have a wide range of shapes and properties depending on their manufacturing technique. Typical examples of thermal spray powders are given in Fig. 4.1. These include the following:

(a) Water-atomized iron powder (Fe-C-Si) with a bulky rounded edged form (Tsunekawa et al. 2006)
(b) Sintered and crushed tungsten carbide powder (WC-Co-Cr) which gives rise to bulky, porous, round edged particles [*Starck Amperit 553.065*].
(c) Fused and crushed alumina powder (Al_2O_3) with its sharp-edged fully dense particles (Laha et al. (2005).
(d) Bulk molybdenum disulfide powder (MoS_2) with their natural flaky structure
(e) Plasma spheroidized molybdenum powder (Mo) with its generally perfect spherical dense particles (Boulos 2016)
(f) Plasma atomized titanium alloy powder (Ti-6Al-4V) showing a dense spherical morphology (Polak et al 2017)

It may be noted that the production of dense spherical metallic powders is generally limited to plasma spheroidized

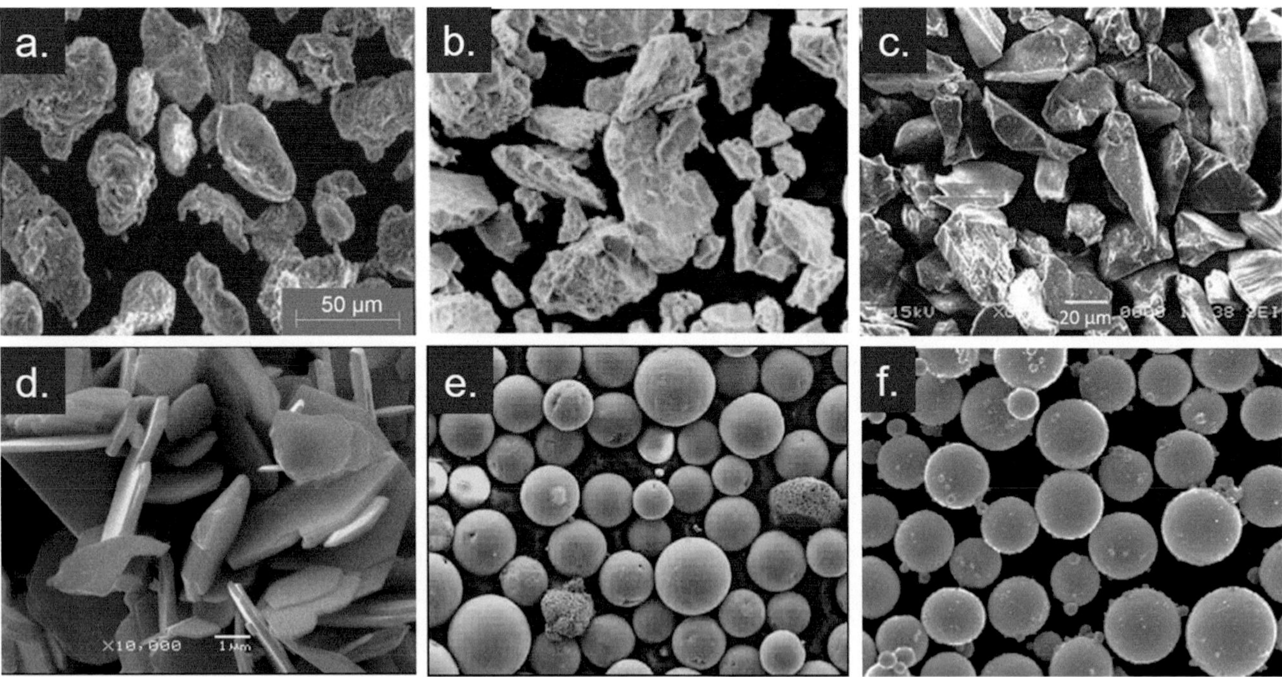

Fig. 4.1 Examples of plasma-spray powder particles reflecting different manufacturing routes

Powder type Manufacturer	Fused and milled	Sintered and milled	Agglomerated and sintered	Spheroidized	Atomized
Particle shape	Blocky - angular	Blocky - angular	Spherical	Spherical	Spherical - irregular
Porosity	Dense	Dense - porous	Porous	Dense - hollow	Porous - hollow
Crystalline size	Course- fine	Course - fine	Medium - fine	Medium - fine	Fine
Homogeneity	Alloyed	Alloyed	Alloyed - heterogeneous	Alloyed - heterogeneous	Alloyed

Fig. 4.2 Specific YSZ (ZrO_2–8 wt.% Y_2O_3) powder characteristics depending on their manufacturing route, Schwier (1986)

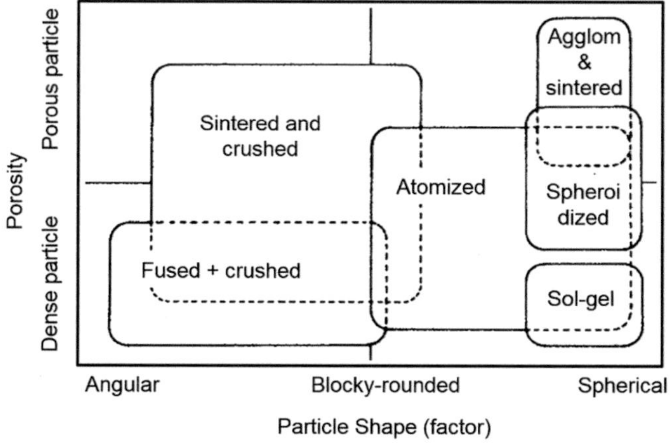

Fig. 4.3 Powder characteristics, porosity, and particle shape, as a function of their manufacturing process, Schwier (1986)

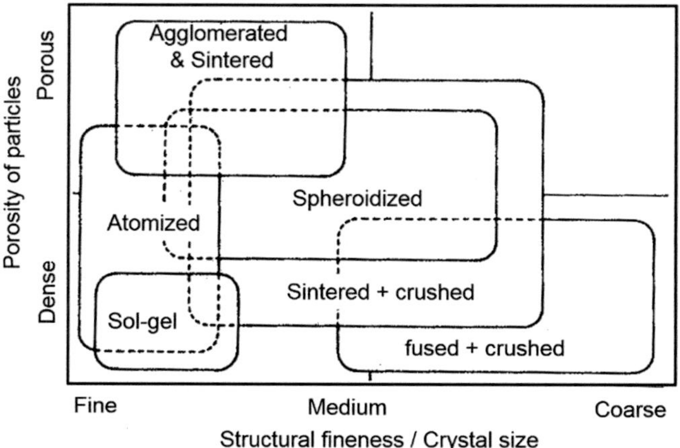

Fig. 4.4 Powder characteristics—porosity and structural fineness—as a function of their manufacturing process, Schwier (1986)

or plasma atomized powders, Fig. 4.1e, f, which are formed in the molten state and acquire their spherical shape through surface tension effect. A study of the link between the powder manufacturing technique and the morphology of the particles obtained for Yttria-stabilized Zirconia (8 wt.% Y_2O_3) was reported by Schwier (1986). The results given in Figs. 4.2, 4.3 and 4.4 show agglomerated and sintered powders commonly used in plasma spray coating tend to be composed of porous particles with a rounded shape and a rather fine crystal structure. Fused and crushed powders, are on the other hand, are composed of blocky and dense particles with sharp edges, and medium to coarse gain structure.

A systematic effort has been devoted in literature in order to develop appropriate parameters to characterize the different particle morphologies. According to Reist (1993), particles with a bulky shape, whether rounded or sharp edged, can be described by an equivalent particle diameter and a shape factor, such as "sphericity." The Feret's diameter and the Martin's diameter, illustrated in Fig. 4.5, are commonly used based on the image analysis of projected particle micrographs. The *Feret's* diameter is defined as the maximum distance from edge to edge of each particle in the image along an arbitrary but fixed axis applicable to all particle. The *Martin's* diameter, on the other hand, is the length of the line that separates each particle into two equal surfaces. Reist (1993) points out that since these measurements can vary depending on the orientation of the particle, they are only valid if averaged over a number of particles and if all measurements are made parallel to one another.

Alternately the following definitions of an equivalent particle diameter are commonly used, such as:

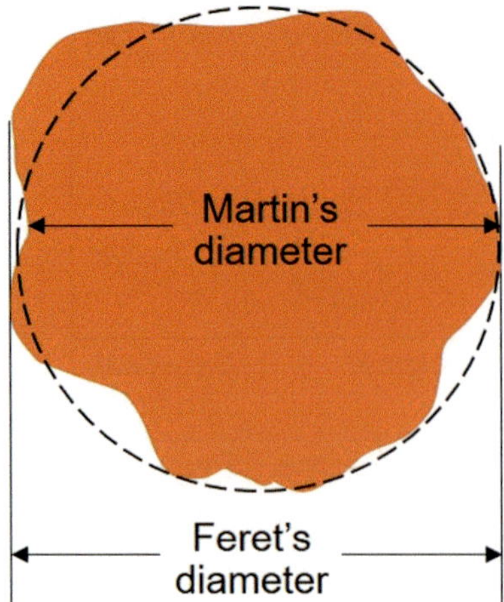

Fig. 4.5 Illustration of Feret's and Martin's diameters, Reist (1993)

Equivalent perimeter diameter, d_{pp}, is defined as the diameter of the circle, which has the same perimeter as that of the projected particle image:

$$d_{pp} = \frac{P_p}{\pi} \tag{4.1}$$

where P_p is the perimeter of the projected image of the particle.

Equivalent projected area diameter, d_{pa}, is defined as the diameter of the circle having the same projected surface area as that of the particle.

$$d_{pa} = \sqrt{\frac{4a_p}{\pi}} \tag{4.2}$$

where a_p is the projected surface area of the particle.

Equivalent volume diameter, d_{pv}, is defined as the diameter of the sphere having the same volume as that of the particle.

$$d_{pv} = \left(\frac{6V_p}{\pi}\right)^{\frac{1}{3}} \tag{4.3}$$

where V_p is the volume of the particle.

The projected area diameter, and the equivalent volume diameter, are often combined with secondary parameter, which reflects how close are the particle's from being a sphere. These are defined as "***Circularity***" and "***Sphericity***." The circularity "**C**" is defined as the ratio of the perimeter of the circle with the same projected surface area of the particle, $\pi\, d_{pa}$, to that of the actual projected image of the particle, P_p.

$$C = \frac{\pi\, d_{pa}}{P_p} \tag{4.4}$$

The ***sphericity***, ψ on the other hand is defined as the ratio of the surface area of the sphere of equal volume as the particle to the surface area of the particle.

$$\psi = \frac{\pi d_{pv}^2}{A_p} \tag{4.5}$$

Obviously both these parameters, C and ψ would tend to unity for spherical particles. Finally, the "**Stokes diameter, d_{st},**" is defined as the diameter of the sphere of the same density and the same terminal settling velocity, u_t as that of the particle.

$$d_{st} = \sqrt{\frac{18\, \mu\, u_t}{(\rho_p - \rho_o)g}} \tag{4.6}$$

where ρ_p and ρ_o are the respective densities of the particle and the fluid, μ the dynamic viscosity of the fluid and g the gravitational acceleration.

In the present chapter, unless indicated differently, the equivalent sphere diameter, d_{pv} will be used for the characterization of non-spherical bulky particles, the subscript "$_v$" will be dropped for simplicity.

4.2.2 Particle Size-Distribution

When dealing with powders, it is most unlikely to have all of the particles of the same diameter. Powders are generally polydisperse when formed, some more than others depending on their manufacturing process. Proper representation of the particle size requires a statistical analysis of the particle size distribution (PSD) of a representative sample of the powder that could be carried out using optical or electron microscopy coupled with a detailed image analysis, light diffusion or instruments based on the particle terminal settling velocity, or mechanical screening using a set of calibrated sieves. The results could then be presented in graphical or in terms of key statistical parameters. Most random phenomena in nature can be represented by the standard normal distribution shown in right-hand side of Fig. 4.6 represented by the relationship:

$$F(x) = \frac{1}{\sigma_x\sqrt{2\pi}} \exp\left(-(x - \bar{x})^2/2\sigma_x^2\right) \tag{4.7}$$

where $\bar{x}$ and σ_x are, respectively, the mean value and standard deviation defined as:

$$\bar{x} = \frac{\sum\limits_{i=1}^{\infty} n_i x_i}{\sum\limits_{i=1}^{\infty} n_i} \tag{4.8}$$

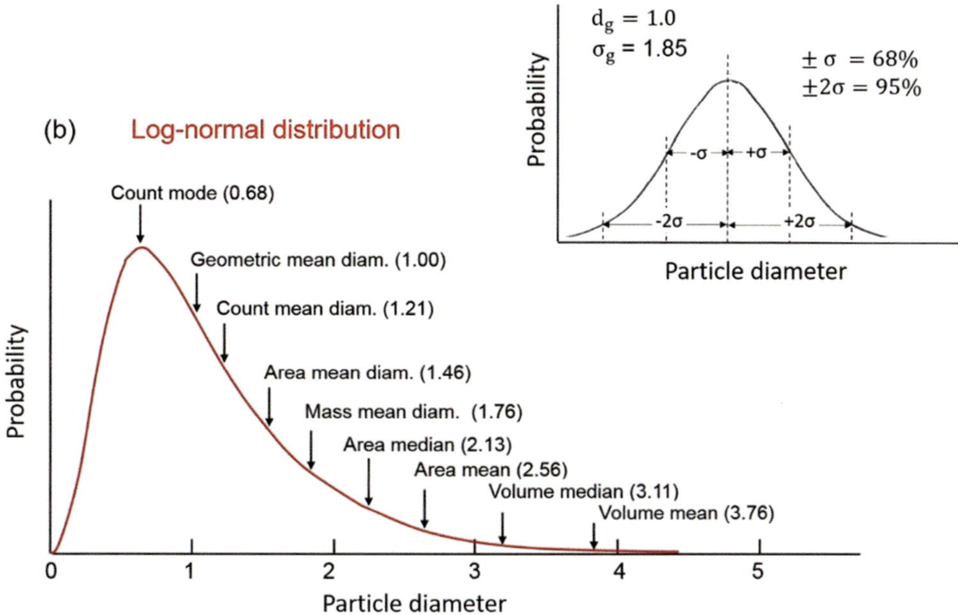

Fig. 4.6 Normal and Log-normal particle size distribution, after Reist (1993)

and

$$\sigma_x = \left[\frac{\sum\limits_{i=1}^{\infty} n_i(x_i - \bar{x})^2}{\left(\sum\limits_{i=1}^{\infty} n_i\right) - 1} \right]^{\frac{1}{2}} \tag{4.9}$$

Statistically, a variable with a normal distribution will have a 68% probability to be within $\pm\sigma_x$, and 95% probability to be within $\pm 2\sigma_x$.

For a variable to have a normal statistical distribution, it has to have equal probability to have a higher or lower value relative to its mean value (Fig. 4.6a). This condition is rarely satisfied for particle size distributions of powders that tend generally to be biased toward the finer size fractions. These distributions, however, would fit a normal distribution if the probability is plotted against the logarithm of the particle diameter giving rise to what is known as "lognormal distributions" as shown in lower LHS of Fig. 4.6b. The corresponding parameters for such a distribution are indicated on the curve.

These can be calculated as follows:

Geometric mean diameter, $\overline{d_g}$ and geometric standard deviation, σ_g are defined for a number particle size distribution as:

$$log\,\overline{d_g} = \frac{\sum n_i \, log\, d_i}{\sum n_i} \tag{4.10}$$

and

$$log\,\sigma_g = \left[\frac{\sum n_i \left(log\, d_i - log\, d_g \right)^2}{\sum n_i - 1} \right]^{\frac{1}{2}} \tag{4.11}$$

The corresponding values for a particle size distribution in terms of mass fraction rather than number fraction are given as:

$$log\,\overline{d_g} = \frac{\sum n_i d_i^3 \, log\, d_i}{\sum n_i d_i^3} \tag{4.12}$$

and

$$log\,\sigma_g = \left[\frac{\sum n_i d_i^3 \left(log\, d_i - log\, d_g \right)^2}{\sum n_i d_i^3 - 1} \right]^{\frac{1}{2}} \tag{4.13}$$

It is important to note the fundamental difference between these two types of particle size distributions since in a number distribution the fine particles are given the same weight as any large particles in the powder and the distribution is accordingly biased toward fine particle presentation. The effect is clearly demonstrated in Fig. 4.7, which shows the same particle size histogram or frequency distribution function, represented both in terms of number fraction and mass fraction of the particles. For most engineering applications, the mass fraction distribution is the most pertinent since it reflects the mass distribution of the powder between different particle sizes.

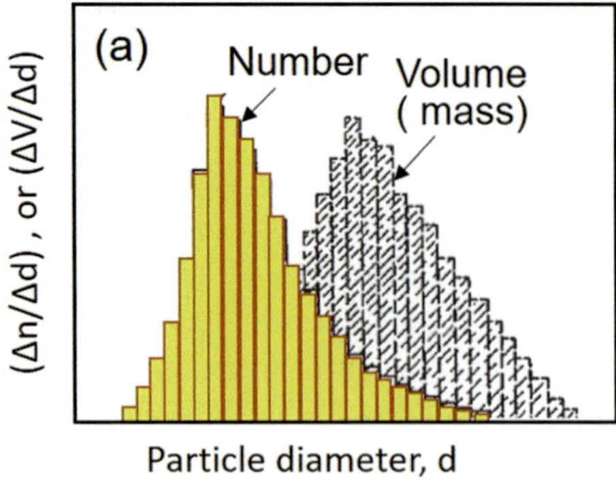

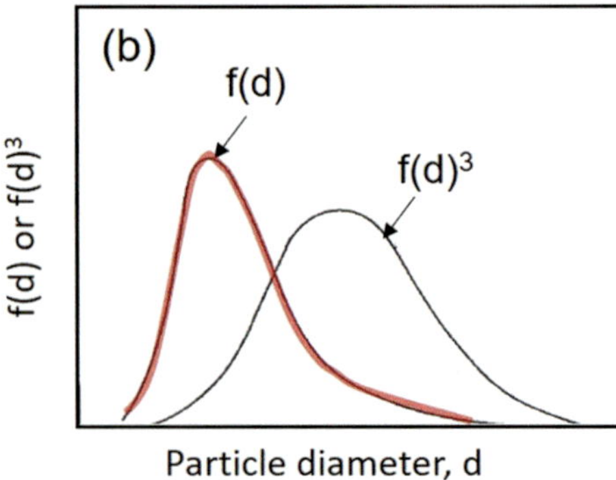

Fig. 4.7 Particle size histogram and frequency distribution function presented in terms of number and mass fraction distributions, after Lefebvre (1989)

4.3 Plasma–Particle Momentum Transfer

While particles used in thermal spray tend to be of wide range of sizes and shapes, analysis of the plasma–particle interactions is developed for the simple geometry of a spherical particle in an infinite plasma stream. The justification for such an approach is based on the notion that most particles used tend to be of a considerably smaller dimension compared with the characteristic dimensions of the plasma and that once melted the particles acquire a spherical shape due to surface tension effects. Only during the initial part of the trajectory of a particle in a plasma its original shape is maintained. Non-spherical bulky particles could then be treated using an equivalent particle volume diameter, Eq. 4.3 combined with a corresponding shape factor if necessary such as the sphericity, given by Eq. 4.5.

4.3.1 Flow around Single Sphere and Drag Coefficient

The flow around a single sphere or any blunt body has been the subject of intensive study for many decades mostly with regard to aerospace, environmental, and chemical engineering applications. Numerous textbooks on the subject are available in literature, such as Clift et al. (1978) and Rudinger (1980). According to these studies, the flow field around a single sphere is governed by the standard Navier–Stokes (N–S) equations, which represents a balance between the inertia and viscous forces characterized by the Reynolds number, Re, in defined as:

$$Re = \frac{\rho\, u_R\, d_p}{\mu} \quad (4.14)$$

where

ρ mass density of the fluid (kg/m³)
μ dynamic viscosity of the fluid (Pa.s)

d_p diameter or characteristic dimension of the particle (m),
u_R relative velocity between the fluid and the particle defined as

$$u_R = \sqrt{(u - u_p) + (v - v_p)} \quad (4.15)$$

With, u and u_p are, respectively, the fluid and particle velocities in the axial direction (m/s), while v and v_p represent the corresponding fluid and particle velocities in the radial direction (m/s).

For a single sphere at very low Reynolds numbers (RE <0.01), the inertia forces are negligible compared with the viscous forces, the N–S equation can then be solved analytically giving rise to a symmetrical streamline upstream and downstream of the particle, as shown on the LHS of Fig. 4.8a. With the gradual increase of the Reynolds number in the range (0.01< RE <1.0) the flow remains dominated by viscous effects though with a loss of the upstream to downstream symmetry of the flow as illustrated by the Oseen solution given in the RHS of Fig. 4.8b.

With the further increase of the Reynolds number, the flow on the downstream side of the particle starts to show the beginning of flow separation at values of RE = 10 with full wake formation behind the sphere taking place at higher Reynolds number as shown in Fig. 4.9 after Taneda (1956). Beyond a Reynolds number of 130–200 vortex shedding downstream of the sphere starts reaching an intense level at a Reynolds number above 1000 as shown in Fig. 4.10.

Such important changes in the flow pattern around the particle have a direct impact on the drag coefficient, C_D between the particle and its surrounding defined as the ratio of the drag force per unit projected surface area of the particle (F_D/A_p) and the fluid inertia forces $(\frac{1}{2}\rho U_r^2)$.

$$C_D = \frac{(F_D/a_p)}{(\frac{1}{2}\rho u_R^2)} \quad (4.16)$$

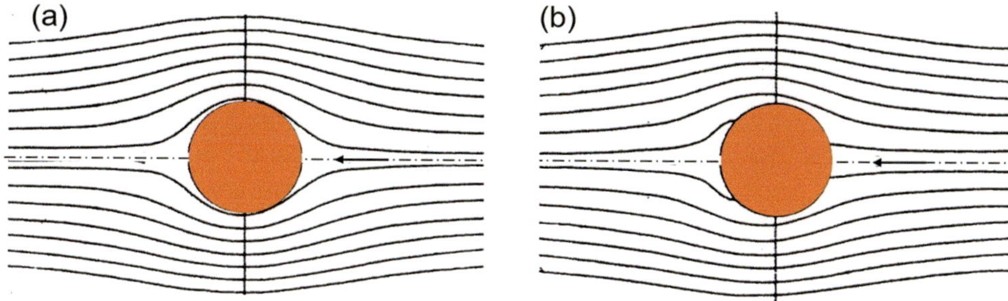

Fig. 4.8 Streamlines over a spherical particle at low Reynold numbers (**a**) Re < 0.01 Stoke's flow (**b**) 0.01 < Re < 1.0 Oseen approximation, after Clift et al. (1978)

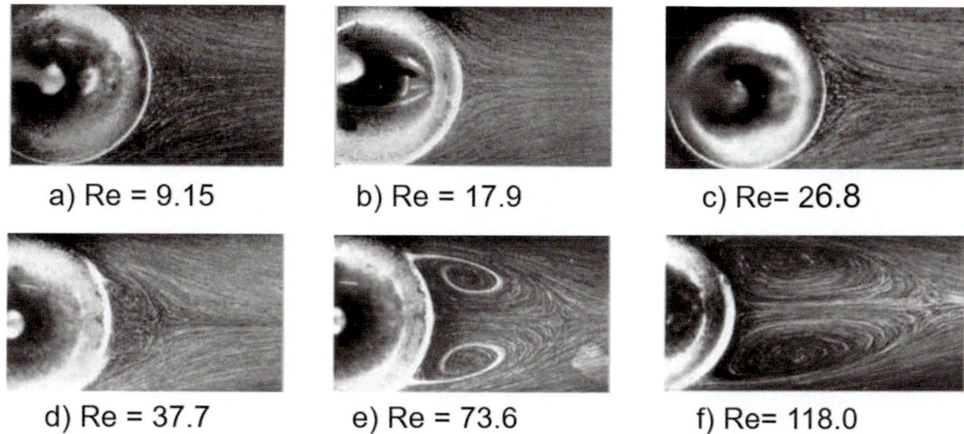

Fig. 4.9 Flow visualization around a spherical solid particle at Reynolds numbers between 9 and 120. Flow from left to right, after Taneda (1956)

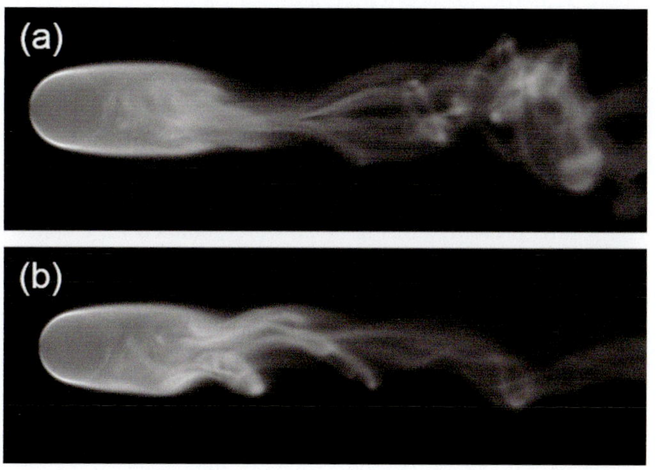

Fig. 4.10 Flow visualization of vortex shedding in the wake of a sphere at ambient temperature

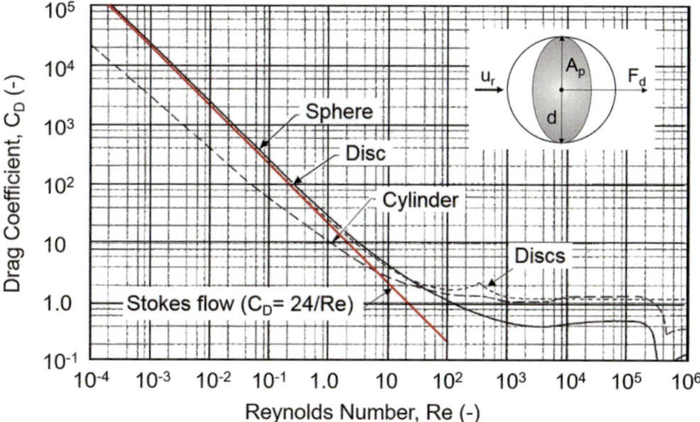

Fig. 4.11 Drag coefficient as function of the Reynolds number for a single sphere, a cylindrical body, and a disc, after Lapple and Shepherd 1940

where

F_D is total drag force exerted by the fluid on the particle (N), a_p is projected surface area of the particle perpendicular to the flow $(a_p = \pi d_p^2 /4)$ (m^2).

The evolution of the drag coefficient, C_D as function of the Reynolds number, Re, for a sphere, a circular disc and a cylindrical object are given in Fig. 4.11. These are mostly based on flow modeling studies and experimental data under normal ambient temperatures and pressures. For the case of a

Table 4.1 Drag coefficient for a single sphere as function of the Reynolds number

$C_D = \frac{24}{Re}$	$Re \leq 0.2$
$C_D = \frac{24}{Re}\left(1 + 0.1\ Re^{0.99}\right)$	$0.2 < Re \leq 2$
$C_D = \frac{24}{Re}\left(1 + 0.11\ Re^{0.81}\right)$	$2 < Re \leq 20$
$C_D = \frac{24}{Re}\left(1 + 0.189\ Re^{0.62}\right)$	$20 < Re \leq 500$
$C_D = 0.44$	$Re > 500$

Table 4.2 Correlations for the drag coefficient for a single sphere as function of the Reynolds number, after Oberkampf and Talpallikar (1994)

$C_D = \frac{24}{Re}$	$Re < 1$
$C_D = \frac{24}{Re}\left(1 + 0.15\ Re^{0.687}\right)$	$1 < re < 1000$
$C_D = 0.44$	$Re > 1000$

sphere moving in a fluid at Reynolds number less than unity, known as the Stokes flow regime, the drag on the sphere is governed mainly by viscous forces, with the drag coefficient inversely proportional to the Reynolds number ($C_D = 24/Re$). At higher Reynolds number ($Re > 1$), the increase of the inertia forces gives rise to what is known as "form drag," which is responsible for the deviations of the drag coefficient from the Stokes flow relationship. Beyond $Re = 10^3$, up to 10^5, the drag coefficient remains essentially constant at around 0.44, which is predominantly governed by form drag. During the plasma spraying of micrometer sized powders, the range of Reynolds number is rarely above 100 because of the relatively small particle diameter ($10 < d_p < 100$ μm) normally used. The drag confident over this range is best evaluated using the curve-fitted correlations given in Table 4.1. Alternate correlations proposed by Oberkampf and Talpallikar (1994) for Reynolds numbers in the range 1–1000 are given in Table 4.2. The correlation given by Eq. 3.17 was proposed by White (1974) for the Reynolds number range $Re < 100$.

$$C_D = \frac{24}{Re} + \frac{6}{1 + Re^{0.5}} + 0.4 \quad Re < 100 \qquad (4.17)$$

4.3.2 Corrections to the Drag Coefficient

Under thermal plasma conditions whether for plasma spraying, powder spheroidization, or chemical synthesis, the presence of steep temperature gradients across the boundary layer surrounding the particle has a significant effect on the flow field around the particle and consequently on momentum and heat exchange between the plasma and the particle. Special attention has, therefore, to be given to the

correction of the drag and heat transfer coefficients predicted using standard correlations obtained under normal temperatures and pressures. Other corrections may also be necessary under rarified flow conditions mostly encountered during low-pressure plasma spraying. Under these conditions, the mean free path of the gas molecules at high temperature can reach a few μm, which starts to be comparable with the characteristic dimension of the particles. Non-continuum effects, often referred to as Knudsen effect, can be important in this case and require special corrections to the drag coefficient predicted using continuum fluid mechanics.

4.3.2.1 Effect of the Temperature Gradients

As discussed by Boulos et al. (1994), the plasma temperature outside the boundary layer surrounding a particle can exceed 10,000 K, while the particle surface temperature may not exceed 3000 K. Such a condition will give rise to steep temperature gradients across a boundary layer over a few hundreds of micrometers thick, causing strong non-linear variations of transport properties Lee et al. (1981). As a first approximation, Lewis and Gauvin (1973) proposed to evaluate the thermodynamic and transport properties used for the calculation of the drag and heat transfer coefficients based on the arithmetic mean film temperature, T_f, across the boundary layer surrounding the sphere:

$$T_f = \frac{(T_s + T_\infty)}{2} \qquad (4.18)$$

where T_s is the surface temperature of the sphere and T_∞ is the free stream plasma temperature. In a subsequent study, Sayegh and Gauvin (1979) carried out computations of the flow and temperature fields surrounding a spherical particle moving into an argon plasma at a Reynolds number of 50, assuming in one case constant fluid properties across the boundary layer, and in another case variable properties. The results presented in Fig. 4.12 were obtained for a sphere surface temperature, $T_s = 0.25 \times T_\infty$. These are presented in terms of stream lines, and temperature isocontours, T^* is defined as ($T^* = T/T_\infty$) with the constant fluid property case presented on the lower part of the figure, while the variable property results are presented on the upper part of the figures, respectively.

Based on these results, Sayegh and Gauvin (1979) recommended the estimation of the fluid properties (kinematic viscosity, $\nu = \mu/\rho$, and thermal conductivity, κ, at a reference temperature defined as $T_{0.19}$, where

$$T_{0.19} = T_s + 0.19\ (T_\infty - T_s) \qquad (4.19)$$

Numerous other corrections for the effect of temperature gradients across the boundary layer on the drag and the heat

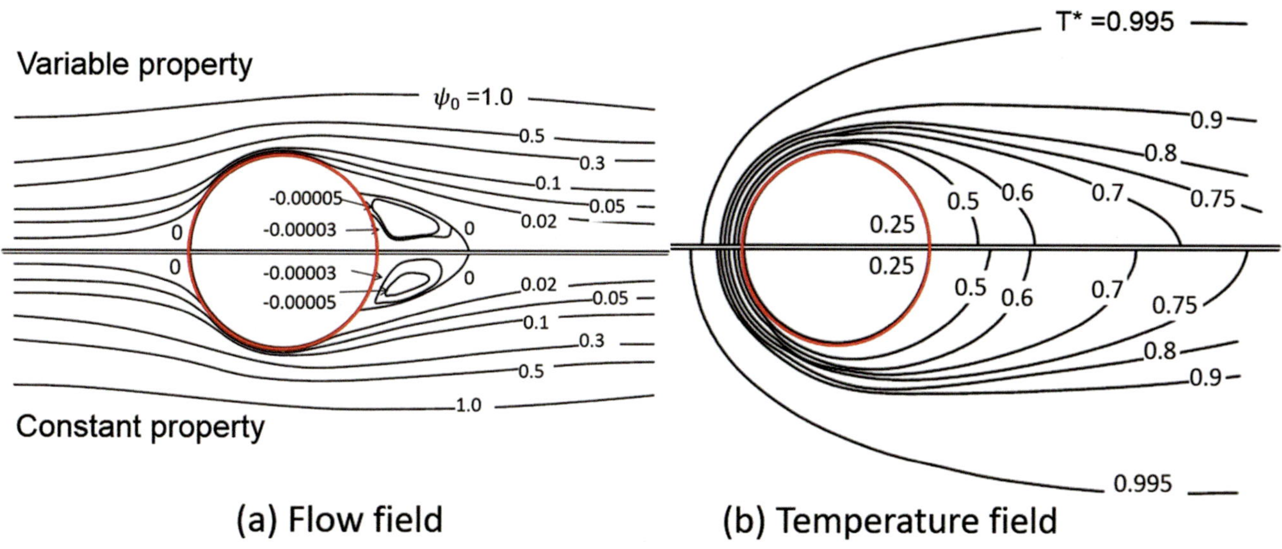

Fig. 4.12 (**a**) Streamlines and (**b**) temperature isocontours for constant and variable fluid property flows, $N_{Re} = 50$, $T_s = 2500$ K, $T_\infty = 10{,}000$ K, $T^* = \frac{T}{T_\infty}$, after Sayegh and Gauvin (1979)

Table 4.3 Drag coefficients for an argon plasma calculated using different correction factors Pfender and Lee (1985) and Pfender (1989)

			Reynolds number				
T_s(K)	T_∞(K)	Reference	0.1	1.0	10	20	50
1000	4000	Simulation by Lee et al. (1982)	100.8	15.4	2.9	1.8	1.0
		Properties evaluated at film temperature	100.6	12.3	2.4	1.6	1.1
		Lee et al. (1981)	115.6	14.2	2.7	1.9	1.2
		Lewis and Gauvin (1973)	146.1	17.9	3.4	2.4	1.6
2500	10,000	Simulation by Lee et al. (1982)	151.9	17.6	3.3	2.1	1.2
		Properties evaluated at film temperature	112.9	13.7	2.5	1.7	1.1
		Lee et al. (1981)	136.9	16.6	3.1	2.1	1.4
		Lewis and Gauvin (1973)	164.5	19.9	3.7	2.5	1.7
3000	12,000	Simulation by Lee et al. (1982)	199.9	22.4	4.2	2.6	1.4
		Properties evaluated at film temperature	137.6	16.2	2.9	1.9	1.2
		Lee et al. (1981)	202.3	23.9	4.2	2.8	1.8
		Lewis and Gauvin (1973)	226.8	26.8	4.7	3.2	2.0

transfer coefficients have also been proposed and widely accepted in the literature (Pfender 1985). These involved the use of the mean film temperature defined by Eq. 4.18, for the estimation of the thermodynamic and transport properties of the plasma, followed by a further correction of the drag or heat transfer coefficient obtained. The ratio of kinematic viscosity estimated at the mean film temperature, ν_f, to that estimated at the free stream temperature, ν_∞, to the power 0.15 was used giving rise to:

$$C_D = C_{Df} \left(\nu_f / \nu_\infty \right)^{0.15} \tag{4.20}$$

where C_{Df} is the drag coefficient evaluated at the mean film temperature T_f. Lee et al. (1981) proposed a different correction factor based on their computation's studies (performed mostly for argon):

$$C_D = C_{Df} \left(\rho_\infty \mu_\infty / \rho_s \mu_s \right)^{0.45} \tag{4.21}$$

The index "s" means that the plasma properties are evaluated at the particle surface temperature T_s and "∞," meaning properties are evaluated at the free-stream plasma temperature.

The difference between these two corrections, Eqs. 4.20 and 4.21, is well within the experimental error of the available data ($\pm$ 25%). Results obtained with different expressions used for the correction of the drag coefficient in a DC plasma jet are summarized in Table 4.3. These show that, for a given Reynolds number, the drag coefficient calculated by different methods can vary by up to 50% between different correlations and simulation methods. They also show a strong dependence on the particle and free stream plasma temperatures.

4.3.2.2 Effect of Particle Shape

In all preceding discussion, the particles have been assumed to be spherical. While this is the case for agglomerated, spray-dried, atomized or spheroidized powders, it is not true for fused and crushed particles which, as shown earlier, can have angular shapes with a low shape factor. The shape of the particle modifies its drag coefficient (Ganser 1993), which, for example, can be correlated to the particle sphericity factor in a limited range of shape effects (Fukanuma et al. 2006). It must be emphasized, however, that particles when injected into the plasma are rapidly (~ a few tens of µs) heated to their melting temperature, thus attaining a spherical shape, which limits the effect of non-sphericity to the early part of their trajectory in the plasma close to their point of injection. The increase of drag force for spheroids compared with those for spheres with the same volume, may appreciably affect the particle trajectory, thus its temperature history in the plasma (Fukanuma et al. 2006; Xu et al. 2002). It is to be noted that the non-sphericity of the particles may also have a significant effect on the behavior of particles within the injector. More work is needed to understand and quantify this effect. Fukanuma et al. (2006) proposed that the influence of the shape of the particle on its drag coefficient can be correlated with the particle sphericity factor for a limited range of shapes. Equations have also been proposed to calculate the acceleration of non-spherical particles.

4.3.2.3 Non-continuum Effect

Other corrections that might be necessary when dealing with the transport and heating of fine particles ($d_p < 10$ µm) under plasma conditions, whether at atmospheric pressure or under soft vacuum conditions, are due to the non-continuum effect. These can be particularly important when the mean free path of the plasma, λ, is of the same order of magnitude as the diameter of the particles, d_p. Chen and Pfender (1983a, b) proposed the following correction to the drag coefficient in order to take into account the Knudsen effect, in the Knudsen number range $(0.01 < Kn < 1.0)$,

$$C_D = (C_D)_{cont.} \left[\frac{1}{1 + \left(\frac{2-a}{a}\right)\left(\frac{\gamma}{1+\gamma}\right)\frac{4}{Pr_s}(Kn)} \right]^{0.45} \tag{4.22}$$

where

$(C_D)_{cont.}$ drag coefficient evaluated using standard continuum fluid mechanics,

Kn Knudsen number ($Kn = \lambda / d_p$)
a thermal accommodation coefficient $(-)$
d_p particle diameter (m)
λ mean free path (m)

Table 4.4 Knudsen numbers (Kn) for a 1 µm particles in plasmas at different absolute pressures and temperatures

Pressure (kPa)	Temperature (K)				
	1000	3000	5000	7500	10,000
5	5.26	15.79	26.32	39.48	52.64
20	1.32	3.95	6.59	9.87	13.16
40	0.66	1.97	3.30	4.94	6.58
70	0.38	1.13	1.88	2.82	3.76
100	0.26	0.79	1.32	1.97	2.63

Table 4.5 Correction factor to the Drag coefficient due to Knudsen effects for small zirconia particles immersed in an infinite Ar-H_2 (25 vol. % H_2) plasma at 10,000 K Fazilleau (2003)

d_p (µm)	5.0	1.0	0.1
$T_p = 1000$ (K)	0.33	0.17	0.06
2000	0.28	0.14	0.05
3000	0.26	0.13	0.046

γ specific heat ratio ($\gamma = c_p/c_v$)
Pr_s Prandtl number of the gas at the surface temperature of the particle

Typical values of the Knudsen number for a 1 µm particle in an argon plasma at different absolute pressures and temperatures are given in Table 4.4. Numerical values of the correction factor, in square brackets, vary from 1.0 to 0.4, with the increase of the Knudsen number from 0.01 to 1.0. The correction to the drag coefficient can be important when considering particles in the size range 0.1 to 5 µm as used in suspension plasma spraying. This is illustrated in Table 4.5 from Fazilleau (2003), related to zirconia particles immersed in an infinite Ar-H_2 (25 vol. % H_2) plasma at 10,000 K.

4.3.2.4 Effect of Particle Charging

Particle charging can also have a limited effect on its trajectory in plasma jets, since a particle injected into thermal plasma will always assume a negative charge due to the difference between the thermal velocities and mobilities of electrons and ions. Whether or not this affects particles drag in thermal plasmas has never been explored. Chen and Ping (1986), Pfender (1989) and Chyou and Pfender (1989) have shown that under LTE conditions, particle charging will be of minor importance, because of the low charge concentration that can exist in the region near the particle surface. This may, however, be different for frozen chemistry, under non-continuum conditions in the case of low-pressure plasma spraying.

Deviations from LTE conditions in the plasma can also be important, particularly close to the particle surface. Whether or not such deviations will substantially affect heat and momentum transfer to particles in the condensed phase, remains a matter for further study. Studies of momentum and heat transfer between low-pressure plasma and particles are generally based on molecular dynamics and they are beyond the scope of this book. The interested reader can find information in references Chen and Chen (1989), Uglov and Gnedovets (1991), Soo (1967).

4.4 Plasma–Particle Heat Transfer

As illustrated in Fig. 4.13, a single spherical particle immersed in a plasma will exchange heat with its surrounding by conduction, convection, and radiation. The net energy received by the particle, Q_n, is given by Eq. 4.23, which represents a simple energy balance between heat received by the particle from the plasma by conduction and convection, Q_{cv}, and heat lost by radiation from the surface of the particle to the surrounding, Q_{sr}.

$$Q_n = Q_{cv} - Q_{sr}$$
$$= h\, a_p\, (T_\infty - T_s) - A_p\, \varepsilon\, \sigma_s\, \left(T_s^4 - T_a^4\right) \qquad (4.23)$$

where

h heat transfer coefficient (W/m^2. K)
a_p surface area of the particle, for a sphere $a_p = \pi\, d_p^2$ (m^2)
T_∞ the free-stream plasma temperature (K)
T_s the surface temperature of the particle (K)
ε particle emissivity (varies from 0 to 1.0)
σ_s Stephan–Boltzmann constant (5.67×10^{-8} W/m^2. K^4)

It is to be noted that any heat received by the particle as a result of radiation from the hot plasma is neglected since the plasma is generally considered as optically thin and the radiation incident angle is small.

The key parameter that needs to be specified in Eq. 4.23 is the heat transfer coefficient h, which is function of the relative velocity between the plasma and the particle, the composition of the plasma and its thermodynamic and transport properties.

$$Q_{CV} = h\, a_p\, (T_\infty - T_s) \qquad Q_{sr} = a_p\, \varepsilon\, \sigma_s\, (T_s^4 - T_a^4)$$

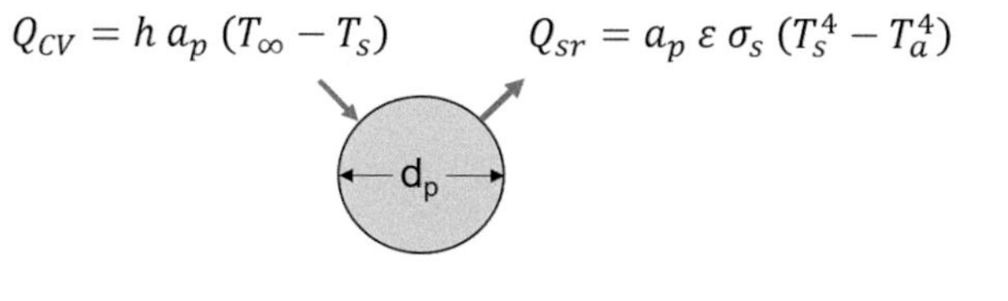

Fig. 4.13 Schematic representation of the net heat exchanged between a particle and its surrounding

4.4.1 Heat Transfer Coefficient

The heat transfer coefficient, h, between the plasma and the particle may be expressed in terms of the Nusselt number, Nu, defined as:

$$Nu = \left(\frac{h d_p}{\kappa}\right) \qquad (4.24)$$

where d_p is the particle diameter and κ is the thermal conductivity of the plasma in the boundary layer surrounding the particle. For a spherical particle immersed in an infinite plasma of uniform temperature, in the absence of any convention effects, it can be demonstrated that $Nu = 2.0$. In the presence of a relative motion between the particle and the plasma, a boundary layer develops around the sphere as shown earlier in Figs. 4.8, 4.9 and 4.10 giving rise to the gradual increase of the Nusselt number as a function of the Reynolds number, Re, and the Prandtl number, Pr. The data given in Fig. 4.14. are for heat transfer between a single sphere and air under normal temperature and pressure conditions. At low Reynolds numbers, the data can be correlated by the Ranz–Marshall correlation, Eq. 4.25 which is valid for $Re < 200$ and $0.5 < Pr < 1.0$,

$$Nu = 2.0 + 0.6\, Re^{0.5} Pr^{0.33} \qquad (4.25)$$

For gases other than air and in the presence of steep temperature gradients in the boundary layer surrounding the particle, a number of corrections have been proposed to the Ranz–Marshall equation as discussed in the next sections.

4.4.2 Corrections to the Heat Transfer Coefficient

4.4.2.1 Effect of the Temperature Gradients

As with the case of corrections to the drag coefficient under plasma conditions, corresponding efforts were devoted to the identification of the best approach to account for the significant variation of the thermodynamic and transport properties in the boundary layer surrounding the particle in a plasma flow and its influence on the heat transfer coefficient. As a first approximation, Lewis and Gauvin (1973) suggested the use of the mean film temperature, T_f, defined as, $T_f = (T_s + T_\infty)/2$, for the evaluation of the fluid properties used in the calculation of the Reynolds, Prandtl, and Nusselt numbers. A correction factor in the form of the kinematic viscosity ratio to the power 0.15 was also included in a modified form of the Ranz–Marshall equation for pure argon as follows:

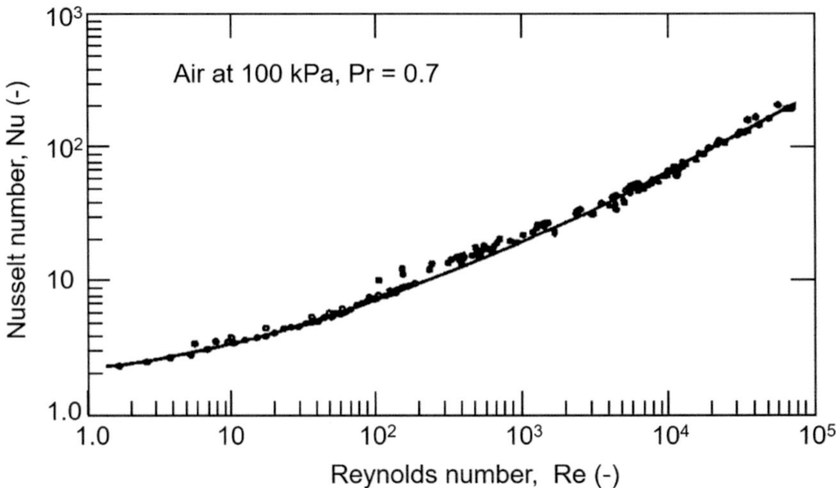

Fig. 4.14 Nusselt number as function of the Reynolds number for a single sphere in air, Pr = 0.7, after Clift et al. (1978)

$$Nu = \left(2.0 + 0.515\ Re^{0.5}\right)\left(v_f/v_\infty\right)^{0.15} \qquad (4.26)$$

Based on their modeling work for an argon plasma, Fig. 4.12, Sayegh and Gauvin (1979) proposed a similar approach as that used for the correction of the drag coefficient correction. This involved the calculation of the fluid properties at the reference temperature $T_{0.19}$ defined as $T_{0.19} = T_s + 0.19\ (T_\infty - T_s)$ and to use the following rather complex empirical correlation for the calculation of the Nusselt number:

$$Nu = 2.0\ f_0 + 0.473\ Re^{0.552} Pr^{m} \qquad (4.27)$$

where

$$m = 0.78\ Re^{0.552}\ Pr^{0.36} \qquad (4.28)$$

and

$$f_0 = \frac{\left(1 - T_0^{1+x}\right)}{\left[(1+x)(1 - T_0)T_0^{x}\right]} \qquad (4.29)$$

where x, the value of the exponent of T in Eq. 4.29 relating the viscosity and the thermal conductivity to the absolute temperature ($x = 0.8$ for Ar at $T < 10{,}000$ K and $p = 100$ kPa).

Fizdon (1979), on the other hand, suggested using the following simpler correction:

$$Nu = \left(2.0 + 0.6\ Re^{1/2}\ Pr^{1/3}\right)\left(\frac{\rho_\infty \mu_\infty}{\rho_s \mu_s}\right)^{0.16} \qquad (4.30)$$

where Re and Pr are calculated at the film temperature, T_f.

Lee et al. (1981) proposed the following slightly modified correlation:

$$Nu = \left(2.0 + 0.6\ Re^{1/2}\ Pr^{1/3}\right)\left(\frac{\rho_\infty \mu_\infty}{\rho_s \mu_s}\right)^{0.6}\left(\frac{c_{p\infty}}{c_{p\,s}}\right) \qquad (4.31)$$

Chen and Pfender (1982a, b), Chen and Pfender (1983a, b) and Chen (1988) gave in a subsequent study an even more complex relationship:

$$Nu = 2.0\left[1.0 + 0.63\ Re_\infty\ Pr_\infty^{0.8}\left(\frac{Pr_s}{Pr_\infty}\right)^{0.42}\left(\frac{\rho_\infty \mu_\infty}{\rho_s \mu_s}\right)^{0.52} C^2\right] \qquad (4.32)$$

in which,

$$C = \frac{1 - (h_s/h_\infty)^{1.1}}{1 - (h_s/h_\infty)^2} \qquad (4.33)$$

where h_s and h_∞, in Eq. 4.33, are the specific enthalpy of the plasma calculated, respectively, at the particle surface temperature T_s, and the free-stream plasma temperature T_∞.

The values of the Nusselt number obtained using these different correlations and computation schemes are given in Fig. 4.15 for an atmospheric argon plasma at 4000 K, as function of the Reynolds number (Pfender 1985). While these are reasonably consistent at moderate plasma temperatures, significant difference is observed at plasma temperatures above 10,000 K due to ionization and recombination effects.

These differences are particularly important when considering diatomic gases such as Ar-H$_2$ (17 vol % H$_2$) mixture. This is

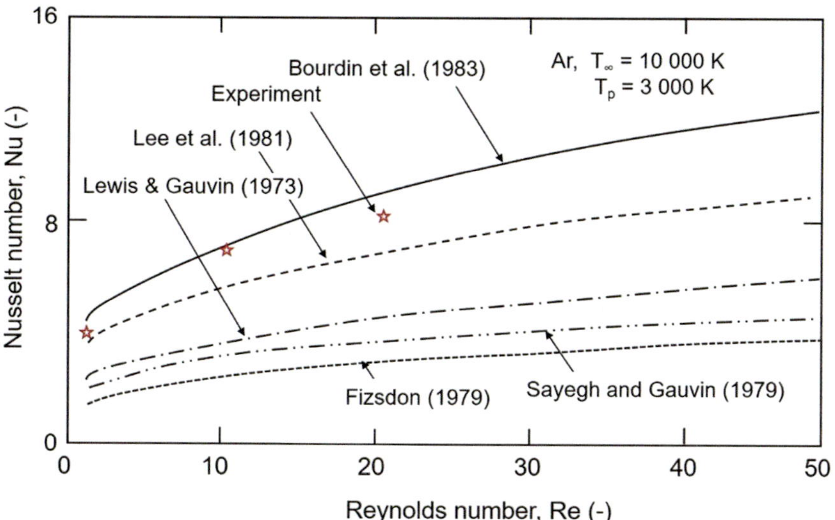

Fig. 4.15 Nusselt number derived by different authors and by computer simulation, after Pfender (1985)

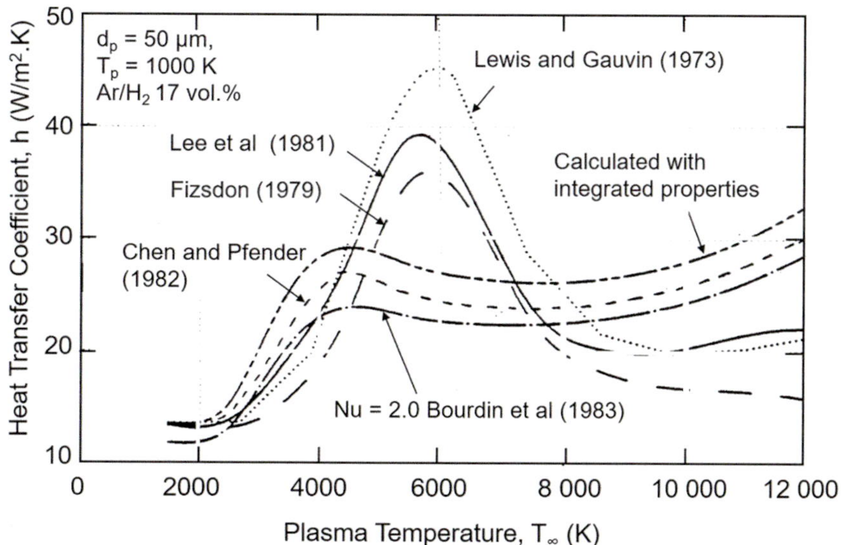

Fig. 4.16 Heat transfer coefficient for a 50 µm diameter particle travelling at 50 m/s with T_s = 1000 K in an Ar-H₂ (17vol.%) plasma calculated by Vardelle (1988)

illustrated in Fig. 4.16, representing the heat transfer coefficients calculated by Vardelle (1988) versus temperature for a 50 µm particle, with a surface temperature, T_s = 1000 K, and relative velocity between the plasma and the particle of 50 m/s.

The results by Lewis and Gauvin (1988), Fizdon (1979), and Lee et al. (1981) in Fig. 4.16 were evaluated at T_f and they exhibit similar evolution with a peak close to 6000 K corresponding to T_f = (6000 + 1000)/2 = 3500 K, that is, the dissociation temperature of hydrogen where κ, as well as c_p, exhibit high values. It can be seen that the difference between the values of the heat transfer coefficient, h, for pure conduction (Nu = 2.0, Bourdin et al. (1983) in Fig. 4.16) and that including convection effects (calculated with integrated properties in Fig. 4.16) is less than 20%.

Because of high temperature gradients and strong non-linear variations of the thermal conductivity, κ, with temperature, essentially for plasmas, one of the key problems for heat transfer coefficient calculations using the Nusselt number (Eq. 4.25) is to determine at which temperature the thermal conductivity has to be evaluated. It can be the film temperature T_f (Eq. 4.26), $T_{0.19}$ (Eq. 4.27) or any other temperature. This problem is important when considering diatomic gases for which a strong peak of κ is observed at the dissociation temperature of the gas.

A detailed theoretical analysis of the problem of pure conduction by Bourdin et al. (1983) showed that the standard heat transfer equation, Eq. 4.25 holds even under plasma

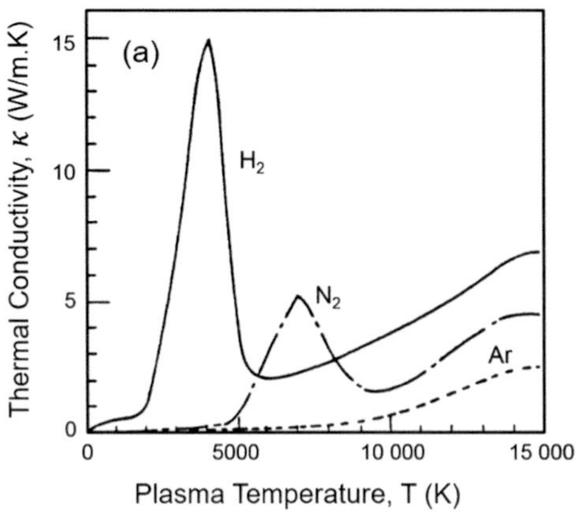
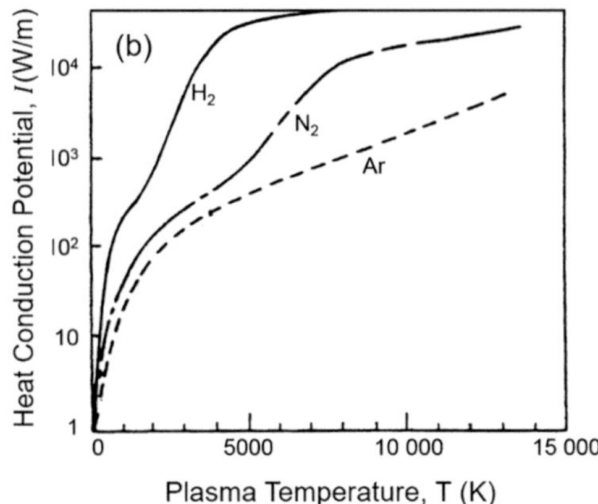

Fig. 4.17 Thermal conductivity (**a**) and heat conduction potential (**b**) for different gases at atmospheric pressure as function of temperature, after Bourdin et al. reprinted with kind permission from Elsevier for Journal of Heat and Mass Transfer Bourdin et al. (1983)

conditions, provided that the properties are evaluated as an integrated mean value defined as:

$$\bar{\kappa} = \frac{1}{(T_\infty - T_p)} \left[\int_{T_p}^{T_\infty} \kappa(T)dT \right] \quad (4.34)$$

It is interesting to note that if the thermal conductivity, κ is a linear function of temperature, Eq. (4.34) reduces to the commonly used practice of evaluating the property values at the arithmetic mean, that is, $\bar{\kappa}(T) = \kappa\left(T_f\right)$, or film temperature.

The application of Eq. (4.34) can be considerably simplified by splitting the integral with respect to some reference temperature, $T_o = 300$ K, as follows:

$$\bar{\kappa} = \frac{1}{(T_\infty - T_p)} \left[\int_{300}^{T_\infty} \kappa(T)dT - \int_{300}^{T_p} \kappa(T)dT \right] \quad (4.35)$$

$$\bar{\kappa} = \frac{1}{(T_\infty - T_p)} \left[S(T_\infty) - S(T_p) \right] \quad (4.36)$$

where $S(T)$ is the heat conduction potential. Values of κ and $S(T)$ are given in Fig. 4.17 for different plasma gases at atmospheric pressure as a function of temperature.

It should be noted that the analysis of Bourdin et al. (1983) is limited to pure conduction, which provides a reasonable estimate of the heat transfer to the particles considering that in most practical cases for plasma spraying ($Re < 20$ and $Pr < 1$) the convection term contributes less than 25% to the total heat transfer to the particles. Examples of the mean integrated thermal conductivity of Ar/H$_2$ and Ar/He plasmas at atmospheric pressure as function of temperature are given in Figs. 4.18 and 4.19, respectively. A comparison of the data for Ar/H$_2$–30 vol.% and Ar/He-60 vol.% is given in

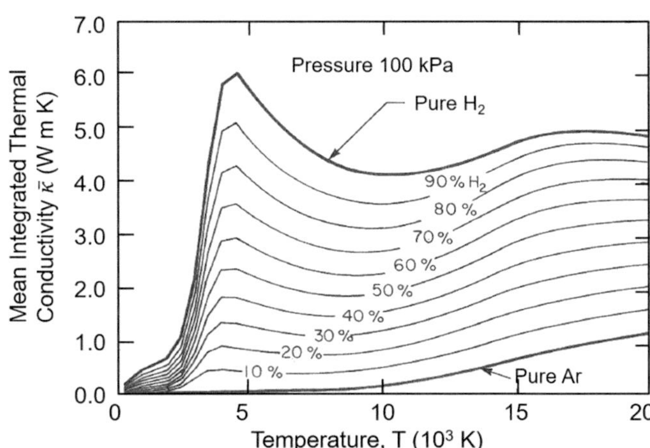

Fig. 4.18 Mean integrated thermal conductivity of Ar/H$_2$ plasmas at atmospheric pressure as function of temperature

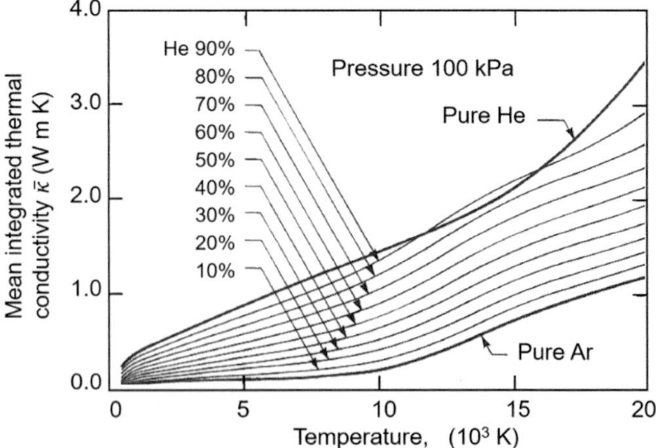

Fig. 4.19 Mean integrated thermal conductivity of Ar/He plasmas at atmospheric pressure as function of temperature

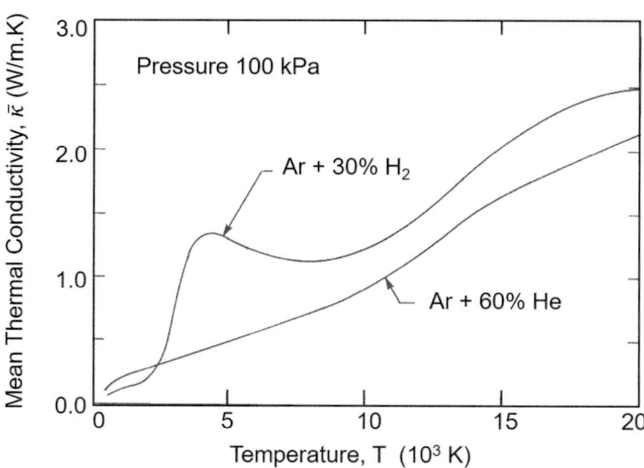

Fig. 4.20 Mean integrated thermal conductivity of Ar/H₂ and A/He plasmas at atmospheric pressure as function of temperature

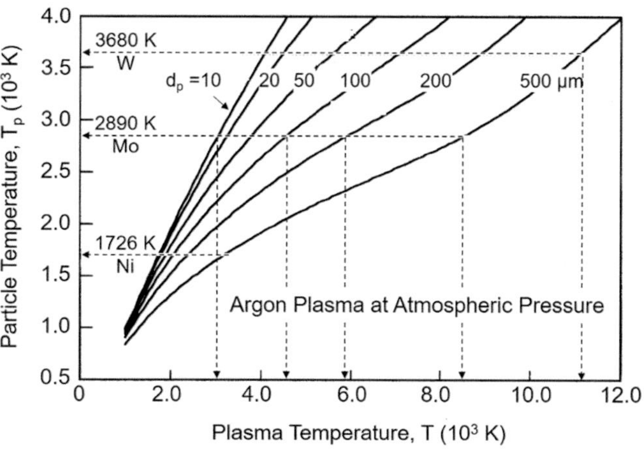

Fig. 4.21 Limiting temperature for metallic particles immersed in atmospheric pressure Ar plasma, particle emissivity assumed to be unity, after Boulos (2004)

Fig. 4.20, showing a high mean thermal conductivity for the Ar/H₂ mixture (30 vol.%) compared with Ar/He (60 vol.%), in the 3000–6000 K temperatures range.

4.4.2.2 Non-continuum Effect

The effect is quite similar to that observed for the drag coefficient using Eq. 4.22 with a thermal accommodation coefficient of 0.8. A corresponding correction to account for non-continuum effects in plasma–particle heat transfer over the same Knudsen number range *(0.01 < Kn < 1.0)* has also been proposed by Chen and Pfender (1983b). In this case, they made use of the concept of temperature jump, resulting in the following correction:

$$q = q_{cont.} \left[\frac{1}{1 + (2\,Z^*/d_p)} \right] \qquad (4.37)$$

where $(q)_{cont}$ is the heat flux calculated using an available correlation for continuum flow and Z^* is the jump distance. Typical numerical values of the correction factor in square brackets in this case could be as low as 0.4 to 0.5 for a Knudsen number of about 0.1. It is important to note that the Knudsen effect is stronger for plasmas with higher enthalpies Chen and Pfender (1983b). For small particles (e.g., 0.1–5 μm), corrections with a magnitude similar to those already presented for C_D have to be made (Fazilleau 2003).

4.4.3 Radiation Energy Losses from the Surface of the Particle

As stated earlier, Eq. 4.23, the net energy that contributes to the transient heating of the particle, is the difference between the heat received by the conduction and convention from the plasma, Q_{cv}, and heat lost by radiation from the surface of the particle to its surrounding, Q_{sr}.

$$\begin{aligned} Q_n &= Q_{cv} - Q_{sr} \\ &= h\,a_p\,(T_\infty - T_s) - a_p\,\varepsilon\,\sigma_s\,(T_s^4 - T_a^4) \end{aligned} \qquad (4.23')$$

Because of the dependence of the radiative energy losses on $(T_s^4 - T_a^4)$, Q_n decreases rapidly with the increase of the particle surface temperature T_s and eventually can be equal to zero or even acquire a negative value *even when the particle temperature is lower than that of the surrounding plasma*. At this point the particle temperature will remain constant (for $Q_n = 0$) or start cooling down ($Q_n < 0$). A conservative estimate of the limiting temperate that a particle can reach is obtained by equating the conduction heat transfer between the plasma and the particle ($Nu = 2.0$) with radiation losses as function of the surface temperature of the particle. Assuming the particle emissivity to be equal to unity allows for the calculation of limiting particle temperature that could be achieved as function of the plasma composition and temperature for particles of different diameters. Typical results obtained for an atmospheric pressure argon plasma are given in Fig. 4.21 Boulos (2004). These show for each particle diameter the gradual increase of the limiting particle temperature as function of the plasma temperature. The relationship is almost linear for particles of diameters less than 10 μm. With the increase of the particle diameter, radiation losses increase, requiring a higher plasma temperature to compensate for the increase of radiation losses. In specific terms if we consider, for example, a molybdenum particle with a diameter of 10 μm, immersed into an argon plasma, there would be no difficulty to melt the particle with a plasma temperature of about the same temperature as the melting

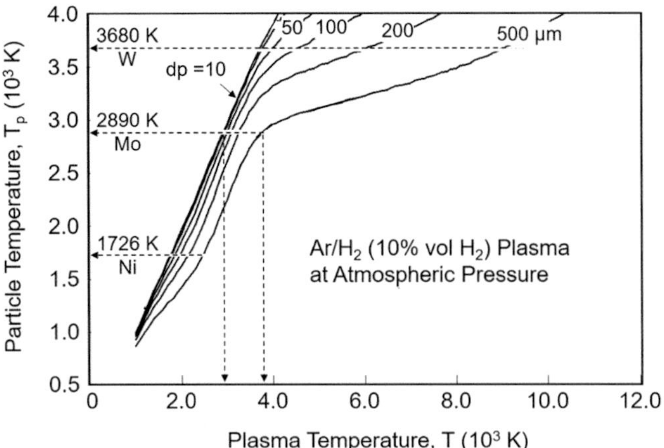

Fig. 4.22 Limiting temperature for metallic particles immersed in atmospheric pressure Ar/H₂ 10 vol.% plasma, particle emissivity assumed to be unity, after Boulos (2004)

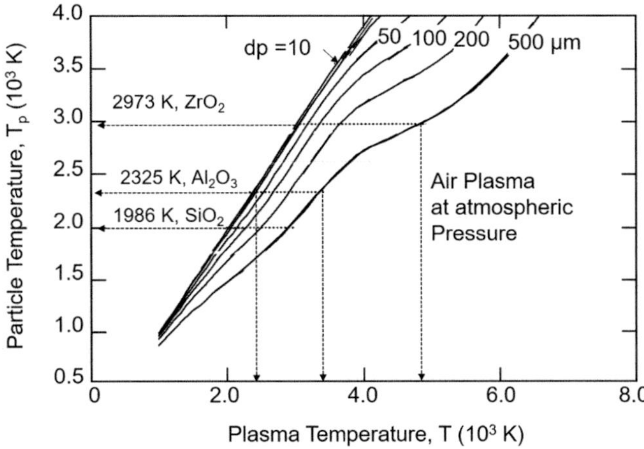

Fig. 4.23 Limiting temperature for ceramic particles immersed in atmospheric pressure Air plasma, particle emissivity assumed to be unity, after Boulos (2004)

temperature of molybdenum (2890 K). In contrast, for a molybdenum particle with a diameter of 100 µm, the plasma temperature necessary for its melting would be 4500 K. Corresponding values for particles of diameters of 200 µm and 500 µm would be 6000 K and 8300 K, respectively. It is to be noted that these are asymptotic limiting temperatures in which the time of contact between the particles with the plasma is not limiting.

The limiting role of radiation losses from the surface of the particle becomes increasingly more important with the increase of the melting temperature of the material. A 500 µm tungsten particle (melting temperature 3680 K) would accordingly require an argon plasma temperature of more than 11,000 K to reach its melting temperature (Fig. 4.22). Corresponding data for an Ar/H₂ 10 vol.% H₂ plasma given in Fig. 4.22 reveal essentially the same trends,

though the effect is significantly reduced for materials of low to medium melting temperature such as Ni and Mo due to the high thermal conductivity of the Ar/H₂ plasma requiring a smaller temperature difference for compensating for radiation heat losses from the surface of the particle.

Similar effects are also observed in Fig. 4.23 for ceramic particles in an air plasma with the extended range of the required plasma temperature for melting of the ceramic increasing significantly for high temperature ceramic materials such as zirconia (ZrO₂) compared with Alumina (Al₂O₃) or Silica (SiO₂).

4.5 Transient Heating and Melting of a Particle

This section deals exclusively with the transient heat conduction inside a single spherical particle immersed in a thermal plasma. The analysis is largely based on the work of Chen and Pfender (1982a, b) and Bourdin et al. (1983), who were the first to address this relatively complex problem. For simplicity, the analysis is limited to conduction heat transfer between the plasma and the particle surface, which correspond to the case of a stationary particle inside an infinite plasma volume. The model formulation used is based on the following assumptions:

- Spherical particle immersed in an infinite plasma volume
- Negligible radiation losses and vaporization from the surface of the particle
- Conduction heat flux to the surface of the particle ($Nu = 2$)

The assumption that $Nu = 2$, implies the absence of convective heat transfer between the plasma and the particle and that the boundary layer relaxation time around the particle is short in comparison to the heating required for melting the particle.

Chen and Pfender (1982a, b) and Bourdin et al. (1983) independently further extended their analysis to the moving boundary problem during the melting phase of the particle. While the proposed model was solved for a dense spherical particle, it is equally applicable to porous particles through the use of an equivalent effective thermal conductivity of the particle material.

4.5.1 Spherical Particle with Infinite Thermal Conductivity

For a metallic particle with a high thermal conductivity, it is reasonable to assume the internal temperature gradients in the particle to be negligible and the particle is essentially at a uniform temperature. The evolution of the temperature of a particle in this case, when suddenly immersed in a plasma environment, can be calculated by a simple energy balance on the particle as follows:

$$\frac{dT_p}{dt} = \frac{-12\,\bar{\kappa}}{\rho_p c_{ps} d_p^2}\,(T_p - T_\infty)$$ (4.38)

where

T_p particle temperature
T_∞ free-stream plasma temperature
$\bar{\kappa}$ integral mean thermal conductivity of the plasma
ρ_p density of the particle
c_{ps} specific heat of the particle
d_p particle diameter

Assuming constant thermodynamic properties of the material (ρ_p, c_{ps}) and transport properties of the plasma ($\bar{\kappa}$), it is possible to obtain an exact solution to Eq. 3.38, as follows:

$$\frac{T_p - T_\infty}{T_0 - T_\infty} = \exp\left(\frac{12\,\bar{\kappa}\,t}{\rho_p c_{ps} d_p^2}\right)$$ (4.39)

As expected, Eq. 4.39 gives rise to as faster increase of the particle temperature with the increase of the thermal conductivity of the plasma, and the decrease of the specific heat of the particle material. The effect of the different simplifying assumption on the predicted temperature history of a 100 µm alumina particle immersed in nitrogen plasma at 10,000 K can be identified in Fig. 4.24 where the transient temperature

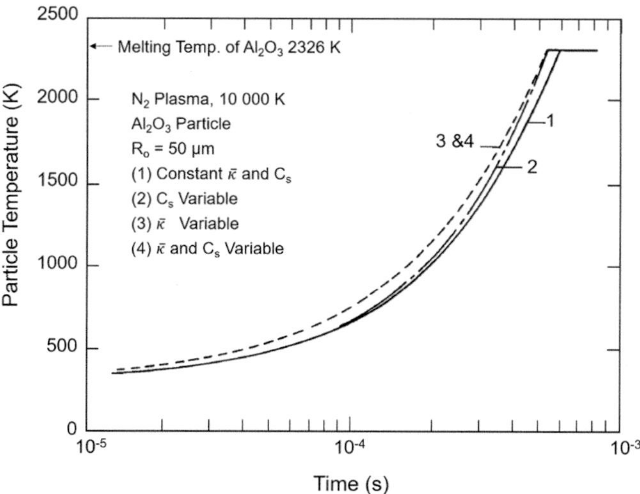

Fig. 4.24 Temperature history for a 100 µm alumina particle immersed in a nitrogen plasma at 10,000 K, assuming infinite thermal conductivity of the particle, Bourdin et al. (1983)

rise of the particle is plotted as function of time using assumptions Bourdin et al. (1983).

4.5.2 Effect of Internal Heat Conduction

The following assumptions are made:

- The particle is spherical with finite but constant specific heat and thermal conductivity.
- The particle has a uniform initial temperature, T_0 before being suddenly immersed in a plasma of a temperature T_∞.
- Negligible radiation losses from the particle surface.
- The particle temperature is followed only to the point where its surface reaches the melting point of the material.

Bases on the above assumptions, the equation governing the transient heat transfer in the particle can be written as:

$$\frac{1}{\alpha_p}\frac{\partial T}{\partial t} = \frac{1}{r^2}\frac{\partial}{\partial r}\left(r^2\frac{\partial T}{\partial r}\right)$$ (4.40)

where

α_p thermal diffusivity of the particle material ($\alpha_p = \kappa_p/\rho_p c_{ps}$)

Equation 4.40 can be solved with the following boundary conditions:

$$t = 0, \quad 0 < r < R_0, \quad T(r,t) = 300\ K$$

$$t > 0 \quad r = 0, \quad \left.\frac{\partial T}{\partial r}\right|_{r=0} = 0$$

$$r = R_0 \quad \kappa\left.\frac{\partial T}{\partial r}\right|_{r=R_0} = q$$

where q is the external heat flux to the surface of the particle given for $Nu = 2$, as follows:

$$q = \frac{2\bar{\kappa}}{d_0}\,(T_\infty - T_s)$$ (4.41)

The results obtained for the same conditions as those used in Fig. 4.24, that is, 100 µm diameter alumina particle suddenly immersed in a 10,000 K nitrogen plasma are given in Fig. 4.25. The computation is carried out in this case, included internal heat conduction into the particle. For comparison, the results obtained assuming a "uniform temperature of the particle" are superposed on the figure identified as the "simple case."

It is to be noted that when the surface temperature of the particle reaches its melting point 2326 K, the temperature of

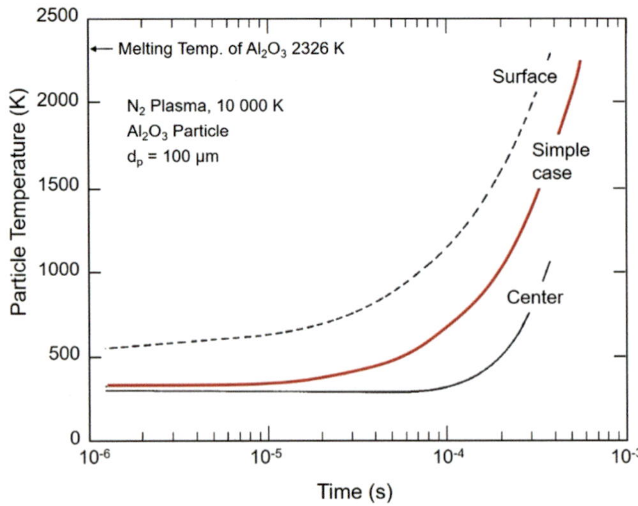

Fig. 4.25 Temperature history for a 100 μm alumina particle immersed in a nitrogen plasma at 10,000 K, including internal heat conduction, Bourdin et al. (1983)

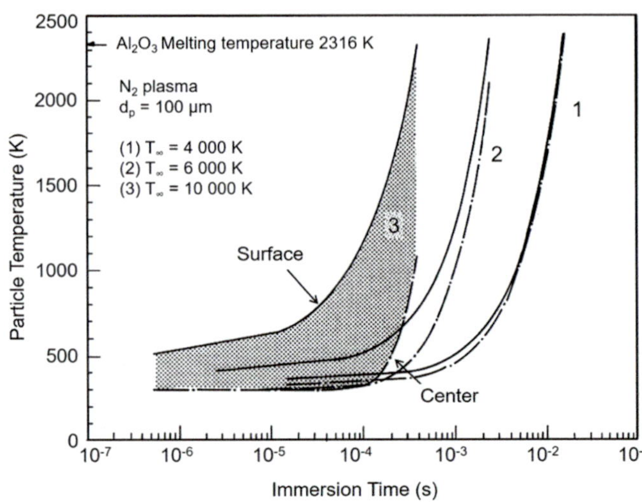

Fig. 4.27 Temperature history of 100 μm diameter alumina particles immersed in nitrogen plasma at different temperatures, after Bourdin et al. (1983)

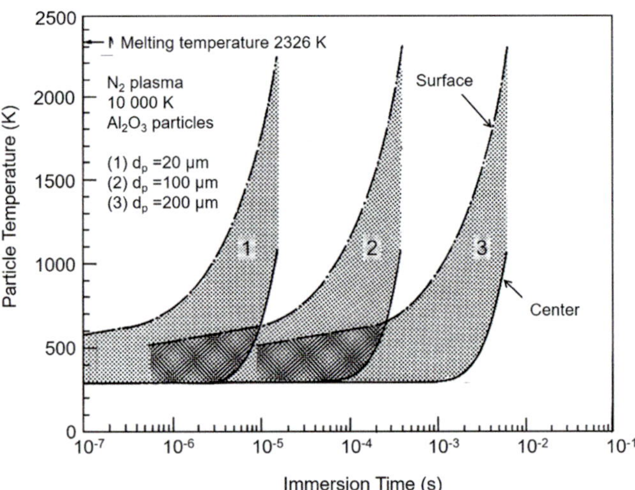

Fig. 4.26 Temperature history of alumina particles of different diameters suddenly immersed in a nitrogen plasma at 10,000 K, after Bourdin et al. (1983)

the center of the particle will be lagging behind at 1075 K, which is 1251 K lower than that of the surface temperature. Obviously the so-called simple case, assuming an infinite thermal conductivity of the particle, while representing reasonably well the mean temperature of the particle, would be seriously an error in terms of internal temperature distribution in the particle.

Bourdin et al. (1983) carried out a systematic study of the transient heating of spherical particles of different materials (Ni, Si, Al_2O_3, W, SiO_2) and different particle sizes ($d_p = 20$–400 μm) as they are suddenly immersed in different plasmas (Ar, N_2, H_2) at atmospheric pressure and temperatures varying between $T_\infty = 4000$ and 10,000 K.

Typical results obtained for alumina particles of different diameters immersed in a nitrogen plasma at 10,000 K are given in Fig. 4.26. These show that the heating time of the particles depends strongly on the particle diameter with the surface temperature of the particles increasing at a much faster rate than its center. This gives rise to the development of significant internal temperature gradients in the particle with temperature differences between the surface of the particle and its center of the order of 1000 K independent of the particle diameter. As shown in Figs. 4.27 and 4.28, the effect is strongly dependent on the external heat flux to the particle, the lower the plasma temperature or the thermal conductivity of the plasma gas, the more uniform will be the temperature distribution in the particle. As noticed in Fig. 4.29, the effect is also dependent on the thermal conductivity of the particle material. The higher is the thermal conductivity of the particle the more uniform its internal temperature distribution.

Based on the above results Bourdin et al. (1983) concluded that a good estimate of the relative importance of the phenomena of internal heat conduction in the particle can be made based on the value of the *Biot number*, *Bi*, which represents the ratio of the mean integral thermal conductivity of the plasma across its boundary layer surrounding the particle, $\bar{\kappa}$ and the thermal conductivity of the particle material, κ_p defined as:

$$Bi = \bar{\kappa}/\kappa_p \qquad (4.42)$$

As a general rule it was observed that important differences can exist between the surface temperature of the particle and that at its center, when the Biot number is larger than 0.03.

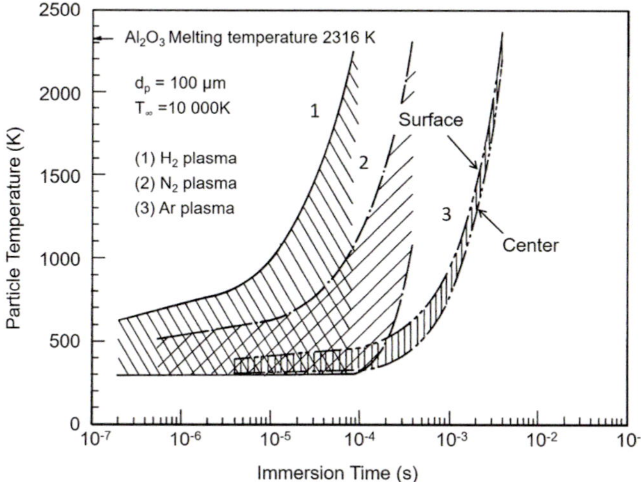

Fig. 4.28 Temperature history of 100 μm diameter alumina particles immersed in different plasmas at 10,000 K, after Bourdin et al. (1983)

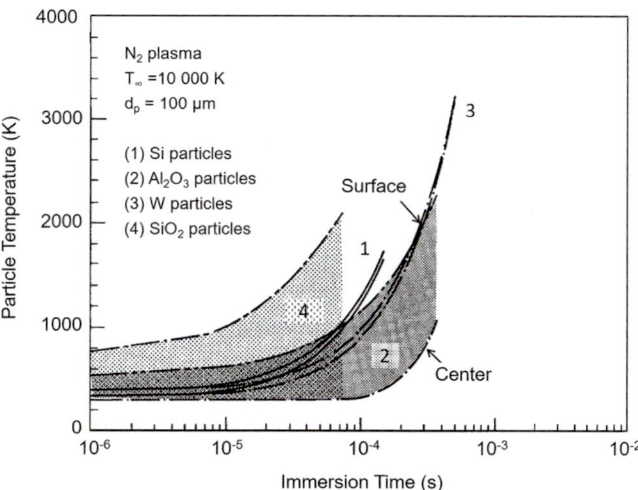

Fig. 4.29 Temperature history of 100 μm diameter particles of different materials immersed in a nitrogen plasma at 10,000 K, after Bourdin et al. (1983)

4.5.3   The Moving Boundary Problem

In the previous section dealing with internal heat conduction in the particles, results were given in which computations were limited to particle surface temperature below the melting point of the material. To take the computation beyond this point to include particle melting, the model formulation has to be modified including the propagation of the melting front as semantically represented in Fig. 4.30. The radius of the melting front, r_m defines the boundary between the solid core of the particle $(0 < r < r_m)$ and molten region $(r_m < r < R_0)$. The governing equations, after Murray and Landis (1959), can be written for both regions of the particle as follows:

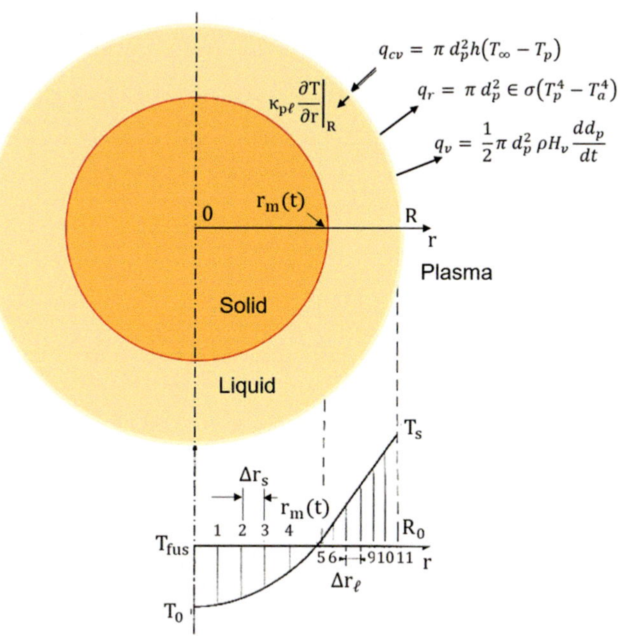

Fig. 4.30 Boundary conditions for the propagation of a melting front within a spherical particle Bourdin et al. (1983)

$$\frac{1}{\alpha_p} \frac{\partial T}{\partial r} = \frac{1}{r^2}\left[\frac{\partial}{\partial r}\left(r^2 \frac{\partial T}{\partial r}\right)\right] \tag{4.43}$$

$$\frac{1}{\alpha_\ell} \frac{\partial T}{\partial r} = \frac{1}{r^2}\left[\frac{\partial}{\partial r}\left(r^2 \frac{\partial T}{\partial r}\right)\right] \tag{4.44}$$

With α_p, and α_ℓ are, respectively, the thermal diffusivities of the particle material in the solid and liquids states, defined as $(\alpha_p = \kappa_p/\rho_p c_{ps})$ and $(\alpha_\ell = \kappa_\ell/\rho_\ell c_{p\ell})$.

The boundary conditions for Eqs. 4.43 and 4.44 are:

$$\text{at } r = 0 \quad \left.\frac{\partial T}{\partial r}\right|_{r=0} = 0 \tag{4.45}$$

and

$$\text{at } r = R_0 \quad \left.\frac{\partial T}{\partial r}\right|_{r=R_0} = q \tag{4.46}$$

at the solid–liquid interface, $r = r_m$, the velocity of propagation of the melting front is given by the equation:

$$\frac{\partial r_m}{\partial t} = \frac{1}{\rho H_m}\left[\kappa_\ell \left.\frac{\partial T_\ell}{\partial r}\right|_{r=r_m} - \kappa_p \left.\frac{\partial T_s}{\partial r}\right|_{r=r_m}\right] \tag{4.47}$$

where

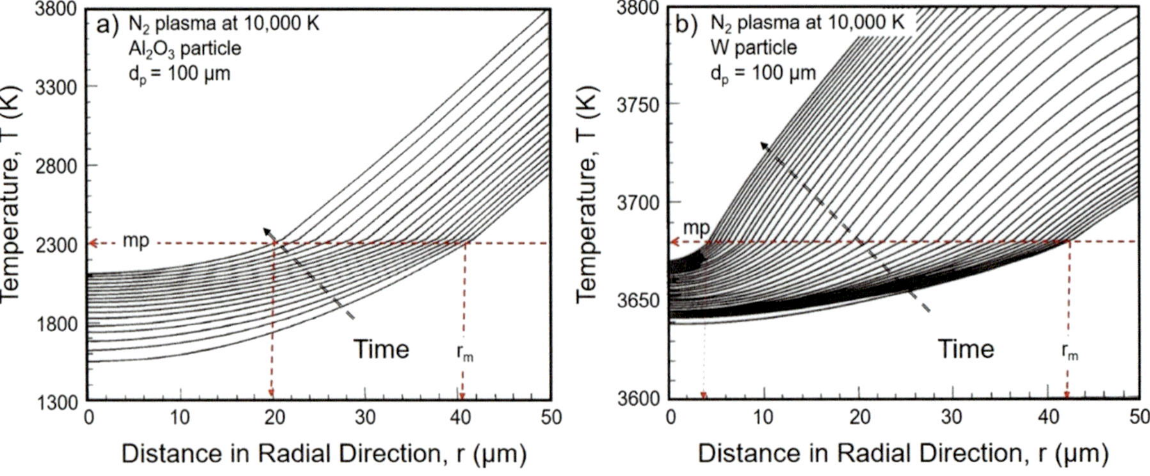

Fig. 4.31 Transient heat propagation and melting of a solid spherical particle. (**a**) Alumina, (**b**) tungsten Bourdin et al. (1983)

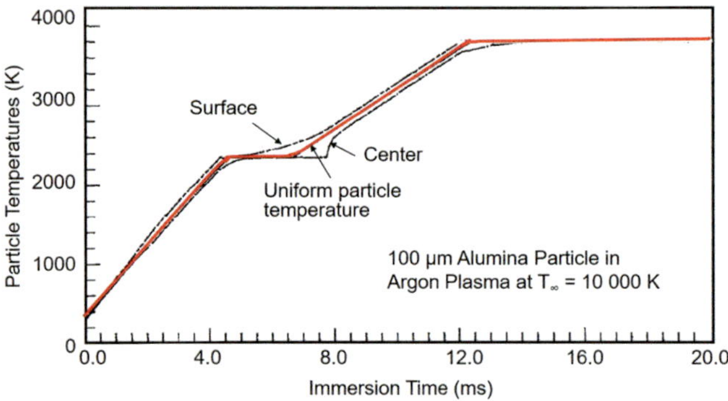

Fig. 4.32 Temperature history of a 100 µm alumina particle exposed to an argon plasma at 10,000 K, after Chen and Pfender (1982a, b)

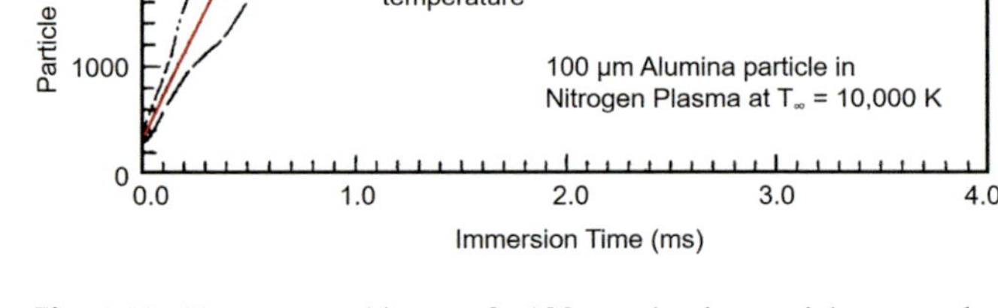

Fig. 4.33 Temperature history of a 100 µm alumina particle exposed to a nitrogen plasma at 10,000 K, after Chen and Pfender (1982a, b)

H_m latent heat of fusion of the particle material
κ_p thermal conductivity of the particle material in solid state
κ_ℓ thermal conductivity of the particle material in liquid state

Typical results reported by Bourdin et al. (1983) given in Fig. 4.31 show the evolution with time of the temperature profile across an alumina and a tungsten particle suddenly immersed into a nitrogen plasma at 10,000 K. Figure 4.31a shows that during the transient heating and melting of an alumina particle with a diameter of 100 µm, the surface temperature of the particle can reach the boiling temperature of alumina (3800 K) before the solid core of the particle is molten. In this particular case the radius of the solid core is 20 µm, at the time its surface reaches 3800 K, corresponding to 6.4% of the mass of the particle still not molten. The effect is a direct consequence of the high heat flux to the surface of the particle and the relatively low thermal conductivity of alumina. Figure 4.31b shows that the situation is quite different for a metallic particle, such as tungsten, where internal temperature gradients during the transient healing and melting of the particle are considerably lower than those observed for alumina.

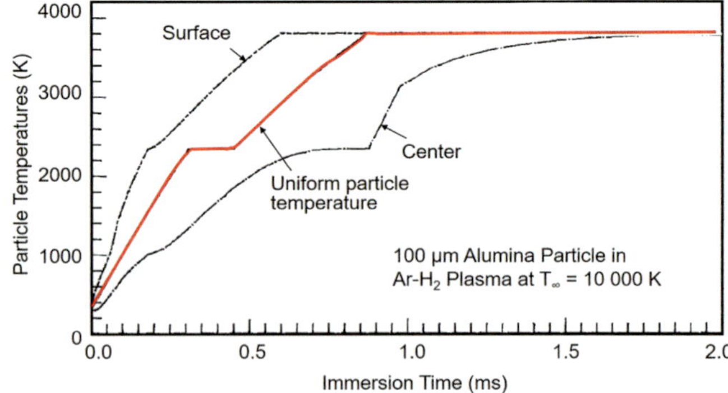

Fig. 4.34 Temperature history of a 100 µm alumina particle exposed to an Ar/H₂ plasma at 10,000 K, after Chen and Pfender (1982a, b)

Chen and Pfender (1982a, b) reached a similar conclusion in their study of the unsteady heating of small particles in thermal plasmas. Typical results given in Figs. 4.32, 4.33 and 4.34 show the evolution with time of the temperature of the surface and center of a 100 µm diameter. Alumina particle

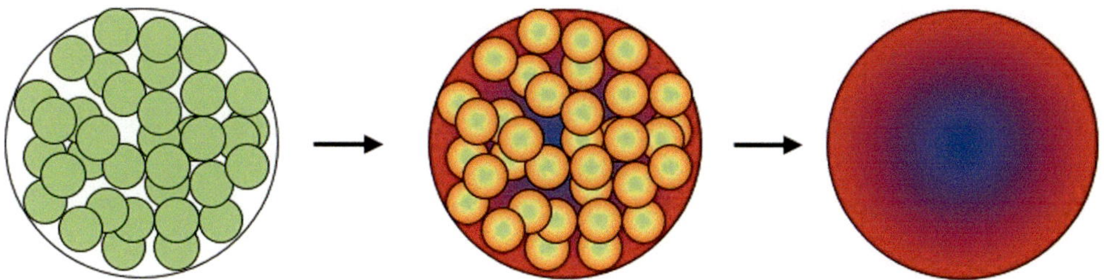

Fig. 4.35 Schematic representation of the progression of the heating and melting of a porous agglomerated particle

suddenly immersed in a plasma of different compositions at 10,000 K. Superposed on the figures is the corresponding temperature evolution of the same alumina particle assuming a "uniform particle temperature," which is equivalent to assuming an "infinite thermal conductivity" of the particles. The results show for an argon plasma (Fig. 4.32), which is characterized as being a "low-thermal conductivity plasma," the uniform particle temperature assumption is quite acceptable with relatively small internal temperature gradients observed during the melting phase of the particles. The situation is different, however, with a nitrogen or an Ar/H$_2$ plasma (25 vol. % H$_2$) (Figs. 4.33 and 4.34) with significant differences observed between the temperatures of the surface and that of the center of the particles. While the assumption of uniform particle temperature is not acceptable in this case, the approach can still be used as a first approximation for the estimation of the time required for the heating and melting of the particle. It should be pointed out that in the case of the treatment of an alumina particle in an argon plasma, the Biot number, defined by Eq. 4.42, was 0.019, while it was equal to 0.16 and 0.28 for the nitrogen and Ar/H$_2$ cases, respectively. It should be stressed that the phenomenon of internal heat propagation in the particle should be taken into account mostly in the case of low thermal conductivity ceramic particles injected in high thermal conductivity plasmas for which the Biot number is larger than 0.03.

4.5.4   Transient Heating and Melting of Porous Spherical Particle

With porous particles such as those obtained by the agglomeration of finer particles, or the atomization of a melt, as shown in Figs. 4.1 and 4.2, the particle can have a void fraction as high as 45% or more with a significant impact on its effective thermal conductivity compared with that of a dense material. Heat propagation phenomenon in such particles during heating and melting steps can be particularly important (Hurevich et al. 2002; Vardelle et al. 1990; Diez et al. 1993). The heat transfer process as schematically represented in Fig. 4.35 involves the transfer of heat through contact points between individual particles in the early stages

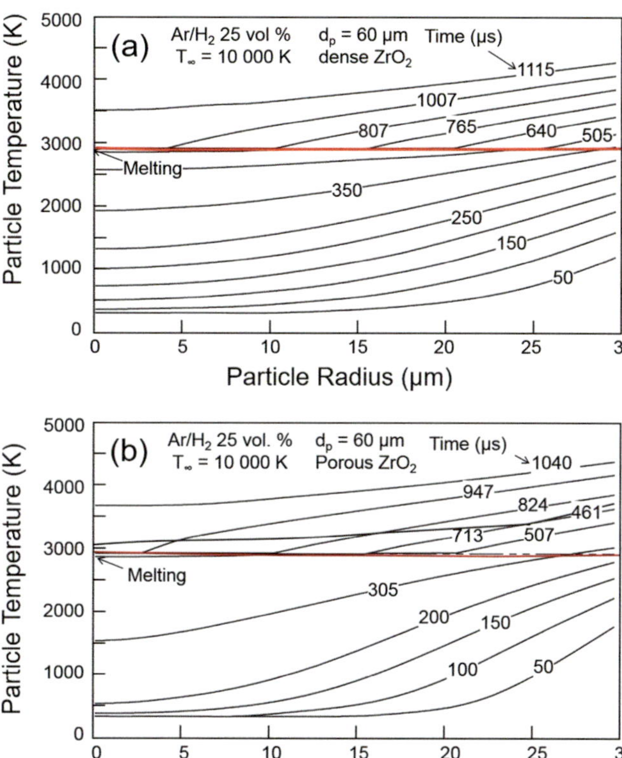

Fig. 4.36 Transient temperature profiles for a 60 μm diameter ZrO$_2$ particle in an Ar/H$_2$ (17 vol.%) plasma at 10,000 K, (**a**) dense particle (**b**) agglomerated particle with 45% void fraction and a mean pore diameter 50 nm Vardelle et al. (1990)

of the heating followed by a gradually improved transfer through the partially melted external shell of the particle. Typical results showing the temperature profile in a 60 μm diameter zirconia (ZrO$_2$) particle immersed in Ar/H$_2$ (17 vol. % H$_2$) plasma at 10,000 K are given in Fig. 4.36. The profiles given in Fig. 4.36a were computed for a fully dense particle, while those in Fig. 4.36b correspond to a porous particle (45% vol.) with a mean pore diameter of 50 nm.

As shown by Hurevich et al. (2002), the effect of porosity is taken into account by reducing the effective thermal conductivity of the bulk material. Experiments have shown, however, that the phenomenon is more complex since the gas in the pores, or that created by the evaporation of an organic binder in agglomerated particles, must find its way from the particle. This is relatively easy when spraying small particles ($d_p \sim 30$ μm) where the molten shell reaches temperatures beyond 3500 to 4500 K, thus allowing the gas to escape through the low viscosity liquid meniscus (Vardelle et al. 1990) giving rise to the formation of dense spherical droplets, which can be recovered as dense particles once they cool down and solidify. In contrast, larger porous particles ($d \geq 60$ μm) might not have the time to be sufficiently heated to allow the entrapped gas to escape through the molten shell. The collected particles in this case would have residual porosity and could constitute hollow spheres containing unmolten grains of the finer unmolten particles Diez and Smith (1993) and Chang and Khor (1996) reported similar results in a study of heating and melting of agglomerated hydroxyapatite powders. However, in most of these cases, dealing with agglomerated large particles, the powders are treated in a furnace to get rid of the binder material before being used in a plasma spraying operation. It should be pointed out that full densification and melting of powders is not required for the spraying of nanometer-sized agglomerated particles where it is important to limit the heating in order to achieve only partial melting of the powder to protect the nanostructure of the material in the coating (Fauchais et al. 2011).

4.6 Particle Vaporization Under Plasma Conditions

4.6.1 Basic Mechanism of Particle Vaporization

Particle evaporation under plasma conditions can be either a result of an undesirable side effect during the treatment of powders in a thermal plasma due to the overheating and vaporization of the finer size fractions or an integral part of the process for the synthesis of nanopowders. In the first case, especially for powders with a broad particle size distribution, the treatment will result in the overheating of the fine powder fractions before the large[-size fractions can receive the energy required for their heating and complete melting. Moreover, when processing alloys composed of elements with a wide range of thermophysical properties, the treatment can result in the partial loss of the lighter elements through in-flight vaporization. Processes developed for the synthesis of nanopowders of metals and alloys rely, on the other hand, on the complete vaporization of the precursor in the form of fine powder followed by the condensation of the formed vapor as a fine nanopowder with the required particle size distribution. In either case, it is essential for optimal process control to have a fundamental understanding of the basic

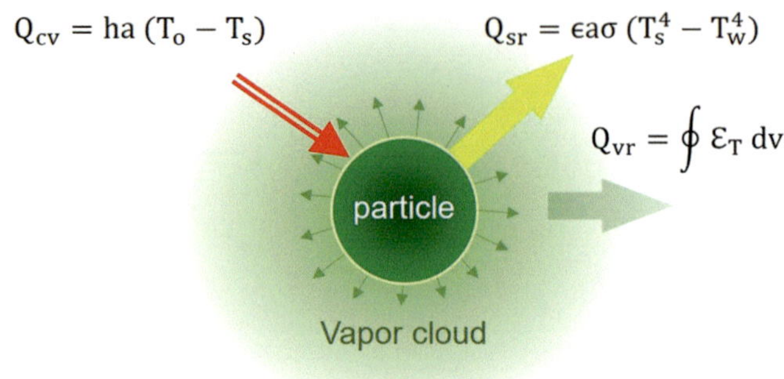

Fig. 4.37 Schematic representation of the basic phenomena involved during in-flight particle heating, melting and vaporization under plasma conditions

phenomena involved and their impact on process efficiencies and product quality.

A schematic representation of the basic phenomena involved in-flight particle heating, melting, and vaporization of a spherical particle is given in Fig. 4.37. This shows that the particle receives heat from the plasma by conduction and convection, Q_{cv}. The particle, in turn, losses heat to the surrounding by radiation from its surface Q_{sr}. The net energy received by the particle, given by Eq. 4.23, $Q_n = (Q_{cv} - Q_{sr})$, is responsible for the heating, melting, and eventually the vaporization of the particle material. Since conduction is generally the predominant heat transfer mechanism for particle heating, the analysis given in this section will also be limited to pure conduction for which $Nu = 2.0$. It is important to note, however, that the value of Q_{cv} evaluated on the basis of pure conduction to the particle surface, will be reduced in the presence of particle evaporation due to the local cooling of the boundary layer surrounding the particle by the vapor species liberated at the surface. In the case of metallic particles, the heat transfer to the particle surface is further significantly reduced as a result of the intense volumetric radiation losses emitted from the hot vapor cloud surrounding the particle, Q_{vr}.

At equilibrium, depending on the thermophysical properties of the particle and the nature of the evaporated elements, the process can be either mass-transfer or heat-transfer controlled. *Mass-transfer controlled*, in which the evaporation rate is controlled by the diffusion rate of the vapors across the boundary layer, will be favored for non-metallic, low boiling point materials, and materials with a low latent heat of vaporization. *Heat transferred controlled*, on the other hand, in which the particle evaporation rate is controlled by the heat transfer rate, will be the dominating mechanism for metallic materials with a high melting and vaporization temperatures. In the latter case, the particle vaporization rate is given by the following energy balance equation:

$$\frac{dm_p}{dt} = \frac{Q_n}{H_v} \qquad (4.48)$$

where m_p is the mass of the particle and Q_n is the net heat exchange between the plasma and the particle, and H_v is the latent heat of vaporization of the particle material.

4.6.2   Effect of Vaporization on Heat Transfer

For an evaporating particle, the vapors liberated at the surface will diffuse through the boundary layer surrounding the particle resulting in the significant cooling of the boundary layer and the corresponding increase of its thickness with the combined effect of reducing the heat flux to the particle. With the further increase of the evaporation rate, the heat transfer coefficients continue to decrease giving rise to a strong non-linearity in the transport equations. According to Chen and Pfender (1982a, b and Chen et al. (1985), the heat received by a spherical particle immersed in a uniform temperature plasma in the absence of evaporation, neglecting radiation exchange from and to the particle, can be written as:

$$Q_o = 2\pi d_p (S_{T\infty} - S_{Ts}) \qquad (4.49)$$

With

$$S_T = \int_{T_0}^{T} \kappa \, dT \qquad (4.50)$$

T_0 is a reference temperature, S_{Ts} and $S_{T\infty}$ are the values of the heat transfer potential at the surface of the particle and far away from the particle. The corresponding value of the heat flux to the particle per unit surface area, q_0, is inversely proportional to the particle diameter, as given by Eq. 4.51:

$$q_o = \frac{2(S_\infty - S_s)}{d_p} \qquad (4.51)$$

In the presence of evaporation from the surface of the particle, an additional assumption has to be made that the density of the gas phase is considerably lower than that of the condensed phase, and consequently that mass transfer can be treated as quasi-steady processes. The total heat transfer rate, Q_e, and the corresponding heat flux values, q_e, to the particle surface are given by:

$$Q_e = 2\pi d_p H_v \int_{T_s}^{T\infty} \frac{\kappa_\infty \, dT}{h - h_s + H_v} \qquad (4.52)$$

$$q_e = \frac{2\,H_v}{d_p} \int_{T_s}^{T\infty} \frac{\kappa_\infty \, dT}{h - h_s + H_v} \qquad (4.53)$$

Dividing Eq. 4.53 by Eq. 4.51 gives:

$$\frac{q_e}{q_o} = \frac{H_v}{(S_\infty - S_s)} \int_{T_s}^{T\infty} \frac{\kappa_\infty \, dT}{h - h_s + H_v} \qquad (4.54)$$

With the integral being only a function of plasma temperature for a given plasma composition and particle material. Calculations of the heat transfer ratio (Q_e/Q_0) for water droplets, alumina, graphite, and tungsten in a pure argon plasma were reported by Chen and Pfender (1982a) using the thermophysical material properties listed in Table 4.6. The results given in Fig. 4.38 show the strongest effect of the evaporation on the heat flux to the particles at high plasma temperatures, and for materials with low latent heat of evaporation such as water.

Chen and Pfender (1982a) developed further their analysis of the quasi-steady evaporation of particles under plasma conditions in order to calculate the evaporation time, t_e, of a particle with an initial diameter of d_{po}, which they gave as:

$$t_e = \frac{d_{po}^2}{4K} \qquad (4.55)$$

With

$$K = \frac{2 \int_{T_s}^{T\infty} \frac{\kappa \, dT}{(h - h_s + H_v)}}{\rho_p} \qquad (4.56)$$

Table 4.6   Thermophysical properties, after Chen and Pfender (1982a)

Material	Water	Alumina	Graphite	Tungsten
T_s (K)	370	3800	4100	5950
H_v (MJ/kg)	2.26	24.7	59.7	4.62

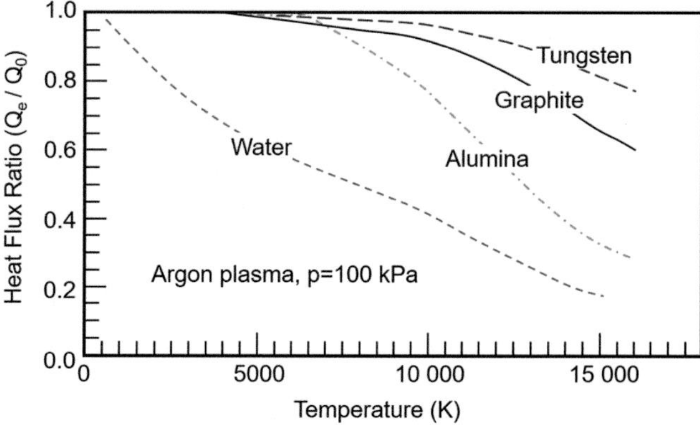

Fig. 4.38   Effect of evaporation (or sublimation) on the heat flux to a spherical particle for Re = 0, Q_e: heat flux in the presence of evaporation, Q_o heat flux in the absence of evaporation, after Chen and Pfender (1982a)

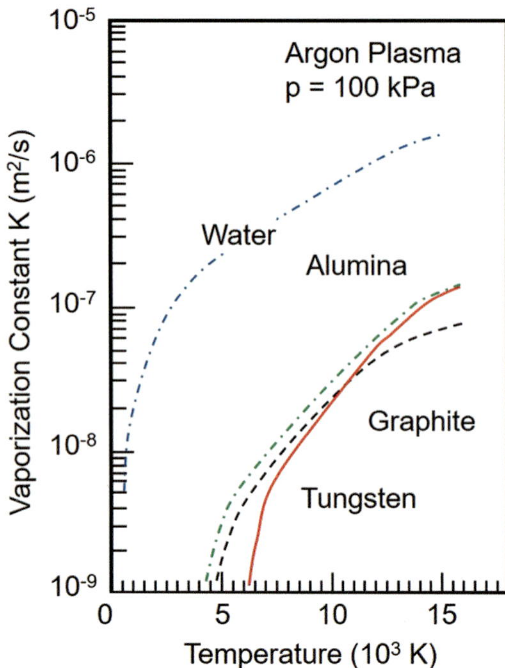

Fig. 4.39 Vaporization constant for different materials in an atmospheric pressure argon plasma, After Chen and Pfender (1982a)

It is to be noted that the evaporation time calculated using Eq. 4.55 does not include the time required for the initial heating and melting of the particle and the subsequent heating of the liquid droplet up to its vaporization (boiling) temperature. Eq. 4.55 also shows that, K, is a function of the thermophysical properties of the plasma and the particle, and the vaporization time for a given material and plasma system is directly proportional to the square of the initial particle diameter. Variations of the value of K with temperature for atmospheric pressure argon plasma are given in Fig. 4.39 for water, alumina, graphite, and tungsten. According to Eq. 4.55 and using the corresponding values of the vaporization constant, K, the vaporization time for a 100 μm water droplet in a 12,000 K argon plasma would be 2.43 ms. Corresponding values for an alumina, graphite or tungsten particles of the same diameter would be 41.5, 60.1, and 50.2 ms, respectively. As mentioned earlier, these values do not include the initial particle heating, melting, and subsequent liquid superheating time to bring the particle temperature to its evaporation (boiling) temperature. Moreover, the heat transfer mechanism involved in these calculations was limited to pure conduction between the plasma and the particle surface.

The effect of the composition of the plasma gas on the evaporation rate of different materials, and subsequently on the ratio (Q_e/Q_0) as function of the plasma temperature, was

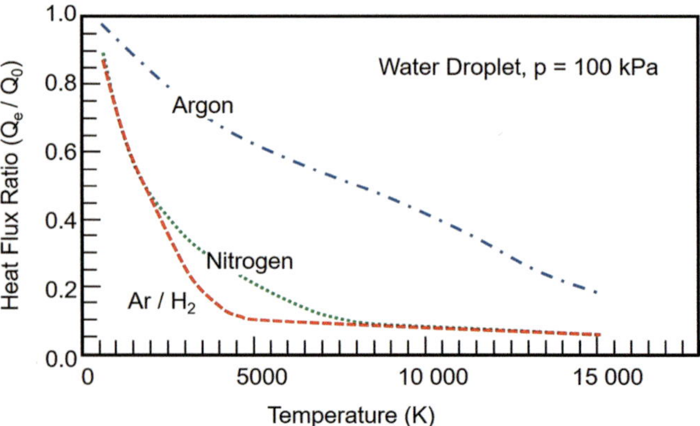

Fig. 4.40 Effect of evaporation on heat transfer to a water for Ar, N_2, and Ar/H_2 20 vol.% H_2 plasmas at atmospheric pressure, after Chen and Pfender (1982a)

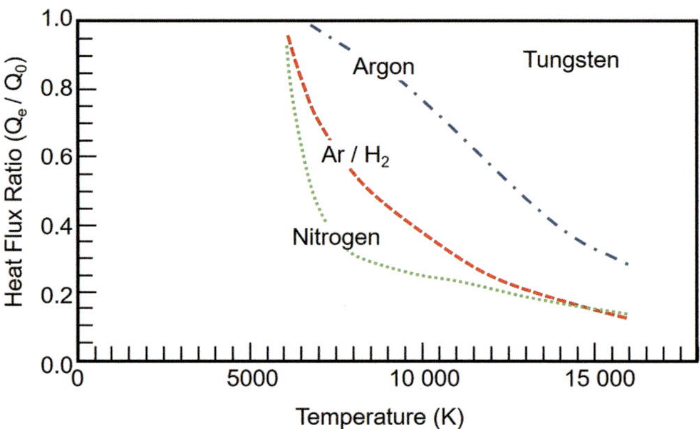

Fig. 4.41 Effect of evaporation on heat transfer to a tungsten particle for Ar, N_2, and Ar/H_2 20 vol.% H_2 plasmas at atmospheric pressure, after Chen and Pfender (1982a)

further investigated by Chen and Pfender (1982a). Typical results obtained for water droplets and tungsten particles immersed in an Ar, A/H_2 (1:4 molar ratio) and N_2 are presented in Figs. 4.40 and 4.41. These show a significant reduction of the ratio of the heat rate to the particle with the increase of the vaporization rate. The effect is more pronounced for an Ar/H_2 and N_2 plasma compared with that for a pure Ar plasma.

The corresponding vaporization constant, K, for these two materials in Ar, Ar/H_2 (1:4 molar ratio) and N_2 plasmas are given in Fig. 4.42. In contrast to the effect of vaporization on heat transfer, which was observed in Figs. 4.40 and 4.41 to be stronger for materials with a lower latent heat of evaporation, the vaporization constant seems more dependent on the density of the condensed phase (Figs. 4.43 and 4.44).

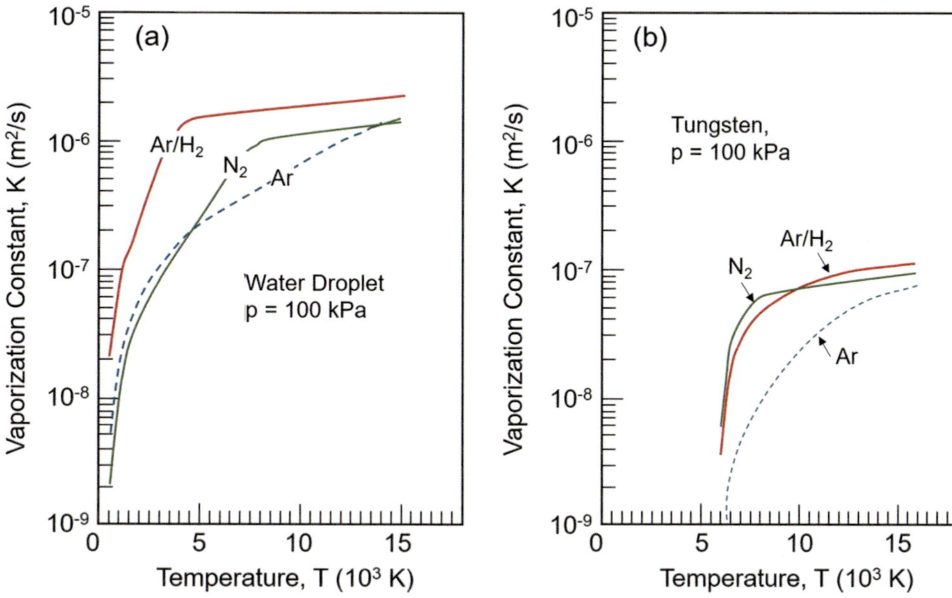

Fig. 4.42 Effect of the plasma composition on the vaporization constant as function of the plasma temperature for water and tungsten Chen and Pfender (1982a)

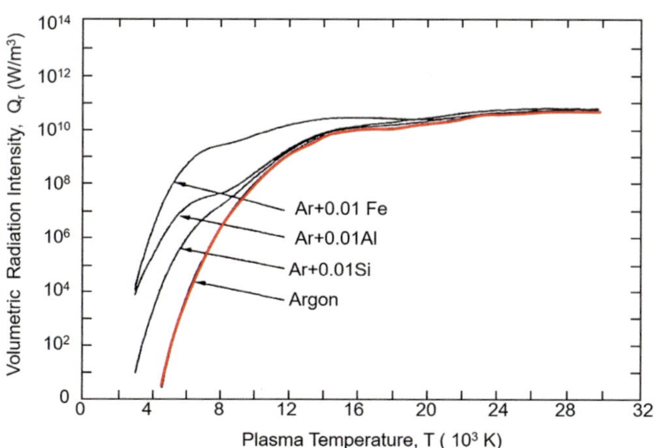

Fig. 4.43 Volumetric emission for an argon plasma in the presence of 1% molar fraction of different metallic vapors, plasma radius = 1 mm, after Essoltani et al. (1994)

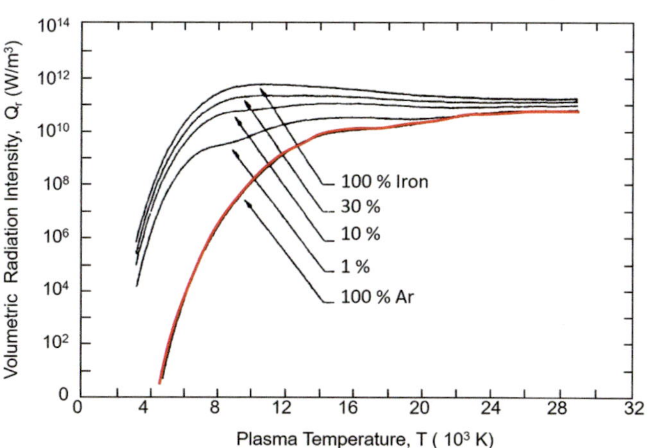

Fig. 4.44 Volumetric radiation losses for an Ar/Fe plasma as function of temperature and the molar fraction of iron vapor, plasma radius R = 1 mm, after Essoltani et al. (1990)

4.6.3 Effect of Radiation on Particle Vaporization

As illustrated in Fig. 4.37, radiation effects can have an impact on particle heating and vaporization through:

- Radiation exchange between the surface of the particle and its surrounding
- Volumetric radiation losses from the vapor cloud surrounding the particle

As mentioned earlier, the heating of the particle through radiation received from the surrounding can generally be negligible. On the other hand, heat losses by radiation from the surface of the particle can be significant, limiting the ability of heating the particle to reach its melting or vaporization temperature, as discussed earlier in Sect. 4.4.3. Chen and Pfender (1982b) integrated in Eqs. 4.52 and 4.53 radiation heat losses from the surface of the particle, giving rise to Eq. 4.57:

$$Q_{sr} = 2\pi d_p\, H_v \int_{T_s}^{T_\infty} \frac{\kappa_\infty\, dT}{h - h_s + c} \qquad (4.57)$$

with

$$c = H_v + \frac{2\pi\, d_p^2\, \varepsilon\, \sigma_s\, T_s^4}{\frac{dm}{dt}} \qquad (4.58)$$

where h is the plasma specific enthalpy at T, while h_s is the plasma specific enthalpy at T_s. This expression has been deduced assuming that the plasma thermodynamic and transport properties (h, κ) are not modified by the vapor diffusion.

Assuming $\left(\kappa/c_p = \overline{\kappa}/\overline{c_p}\right)$ in the boundary layer, Eq. 4.57 applies:

$$Q_{sr} = \frac{H_v}{h - h_s} \, ln \left(1 + \frac{h - h_s}{H_v}\right) 2\pi \, d_p \left(\overline{\kappa}/\overline{c_p}\right) (h - h_s)$$

$$(4.59)$$

Equation 4.57 is equivalent to the introduction of a correction factor λ_c to the thermal flux due to the plasma, Borgianni et al. (1969):

$$\lambda_c = \frac{H_v}{h - h_s} \ln \left[1 + \frac{h - h_s}{H_v}\right] \qquad (4.60)$$

Volumetric radiation losses from the plasma surrounding a particle can also result in a significant cooling of the boundary layer and the reduction of the heat transfer rate to the particle. The effect is particularly important for metallic particles since radiative energy losses from the metal vapors under plasma conditions can be important as reported by Cram (1985), Essoltani et al. (1990), Essoltani et al. (1991), Essoltani et al. (1994), and Vardelle et al. (1996). The effect is illustrated in Fig. 4.46, after Essoltani et al. (1990) shows that in the temperature range from 3000 to 8000 K the radiation losses of the vapor in the presence of only 1% molar of concentration of metallic vapors (Si, Al or Fe) can be several orders of magnitude higher than that of pure argon plasma. In this figure the effective radiation (including self-absorption) has been calculated for volume element of 1 mm in radius. The effect increases rapidly with the increase of the metal vapor concentration as illustrated in Fig. 4.47 for the Fe/Ar system (Essoltani et al. 1994). A detailed discussion of the effect of metal vapors on volumetric radiation losses under plasma conditions can be found in Boulos et al. (2021), Handbook of thermal plasmas, Part I, "Fundamentals of thermal plasmas," Chap. 8, "Plasma radiation transport."

A study of the effect of the radiative energy losses from the vapor surrounding an evaporating particle on the heat transfer rate to the particle was reported by Essoltani et al. (1993). The proposed transient model follows the development of the thermal boundary layer surrounding a single, 100 µm diameter, spherical iron particle suddenly immersed in a pure argon plasma at atmospheric pressure and a temperature of 10,000 K. The particle was assumed to be at the boiling temperature of iron, 3000 K and the relative velocity between the plasma and the particle was 10 m/s. The model is based on the solution of the corresponding 2-D continuity, momentum, energy, and mass transfer equations. The

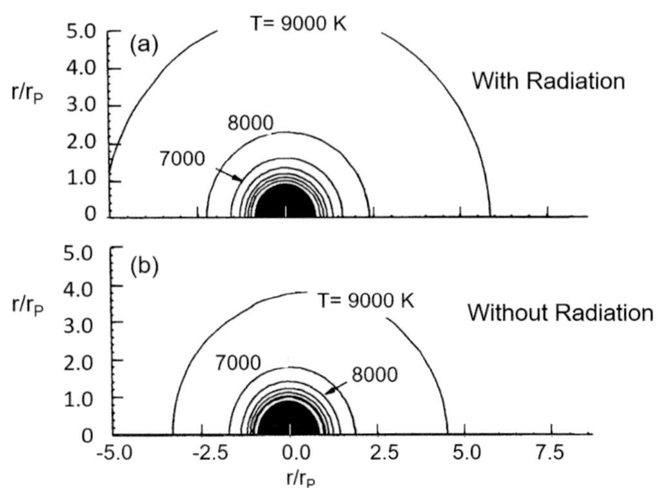

Fig. 4.45 Temperature contours around a 100 µm iron particle immersed in argon plasma at 10,000 K with a relative velocity of 10 m/s, (**a**) Considering radiation from the iron vapor and (**b**) neglecting radiation effects, after Essoltani et all (1993)

computation was limited to a short time period of the order of 100 µs, over which the particle diameter was assumed to be constant.

Typical results showing the temperature contours around the particle are given in Fig. 4.45. The results given in the upper part of the figure (Fig. 4.45a) were obtained taking into account radiation losses from the iron vapor surrounding the particle. The corresponding isocontours obtained neglecting radiation effects are given in the lower part of the figure (Fig. 4.45b). The importance of the radiative cooling of the plasma close to the surface of the particle is clearly demonstrated by the displacement of the high temperature isotherms away from the surface of the particle giving rise to smaller temperature gradients and lower heat transfer rates to the particle. Due to convective effects, the vapor concentration and temperature profiles in the boundary region surrounding the particle on the upstream and downstream side of the particle are different as shown in Fig. 4.46. The time-dependent variation of the temperature profile at 90° from the stagnation point of the flow ($\theta = 90$°) as function of the reduced radius (r/r_p) is given in Fig. 4.47a. The corresponding variation with time of the average Nusselt number over the entire surface of the particle is given in Fig. 4.47b. This shows a rapid drop of the Nu, from its pure conduction value of 2.0 to almost 0.6 in less than 10 µs due to the establishment of an iron vapor-rich boundary layer surrounding the particle responsible for the significant local cooling of the plasma due to radiative effects.

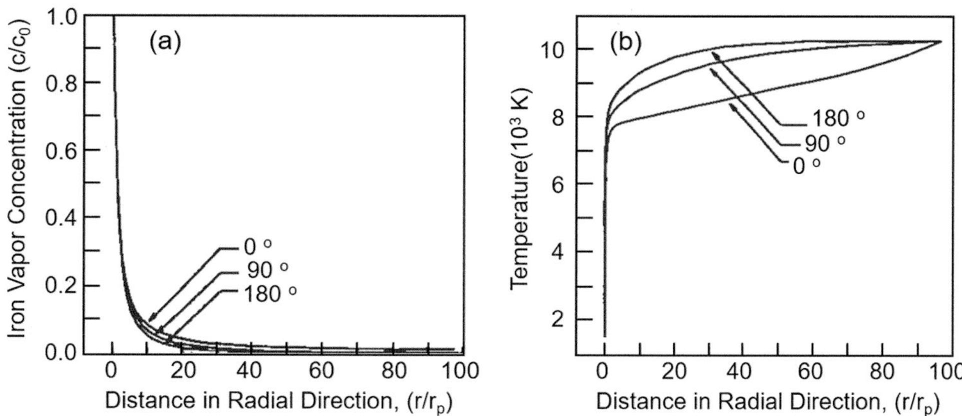

Fig. 4.46 (**a**) Local iron vapor concentration and (**b**) temperature radial profiles at different angles from the stagnation point of the flow ($\theta = 0\,°$) at $t = 1$ μs, after Essoltani et al. (1993)

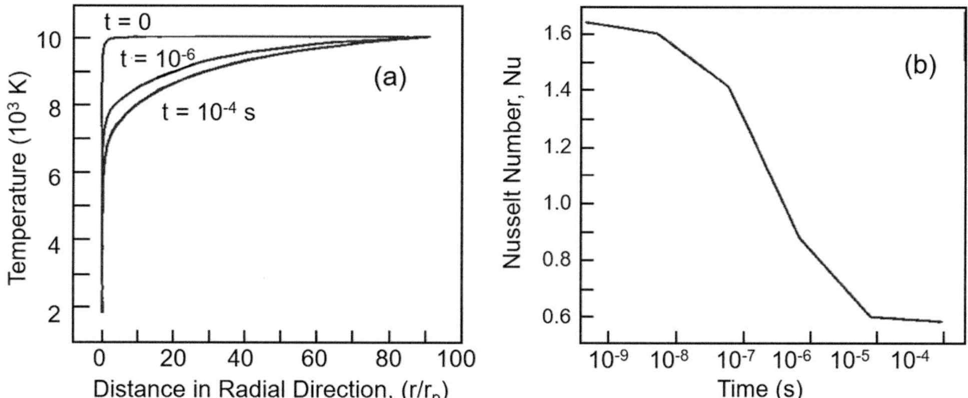

Fig. 4.47 (**a**) Transient temperature profiles at $\theta = 90\,°$, and (**b**) Corresponding variation with time of the Nusselt number, after Essoltani et al. (1993)

4.6.4   Effect of Mass Transfer and Chemical Reactions

In the case of metal particles injected in a plasma flow in an open-air atmosphere, such as in atmospheric plasma spraying (APS) operations, particle vaporization can be controlled by transport process involving the diffusion of oxygen from the plasma to the particle and chemical process leading to homogeneous oxidation reaction, which consumes the metal vapor and thereby increases the evaporation rate from the liquid surface. Above a critical value of oxygen partial pressure, the flux of oxygen molecules toward the particle is greater than the counter flux of metal vapor away from the particle. A solid or liquid oxide layer may then form on the particle surface reducing significantly the rate of vaporization.

The effect of chemical reaction on the vaporization of metallic particles was demonstrated by Vardelle et al. (1996), comparing the evaporation rates of iron particles injected into an Ar/H_2 plasma jet in an inert atmosphere, with that in the presence of ambient air. When operating in

an inert atmosphere, the vapor atoms/molecules produced by particle vaporization diffuse without reacting through the boundary layer surrounding the molten metal droplet, reducing the heat transfer rate to the droplet, as discussed earlier in Sect. 4.6.2, "Effect of vaporization on the heat transfer to a spherical particle," and Sect. 4.6.3, "Effect of radiation on particle vaporization." In the presence of oxygen in the ambient atmosphere, the counter diffusion of oxygen towards the surface of the evaporating droplet and its reaction with metal vapor results in a decrease of the metal vapor concentration at the surface of the particle, which helps to sustain the higher rate of the evaporation. The proposed mechanism was supported by measurements of the radial density profile of iron atoms at a distance of 80 mm from the plasma jet exit nozzle, for an Ar/H_2 (25 vol.% H_2) plasma jet in which iron particles (15–45 μm) are injected at a rate of 50 g/h. The results given in Fig. 4.48. show that with air as ambient gas the density of iron atoms in vapor phase is almost two orders of magnitude higher than that measured in an argon atmosphere.

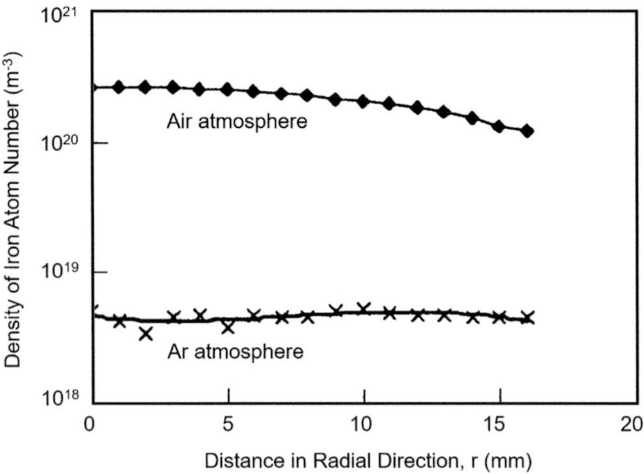

Fig. 4.48 Radial profiles of iron atom distribution in an Ar-H$_2$ plasma jet in which iron particles are injected, issuing into air or argon atmosphere, after Vardelle et al. (1996)

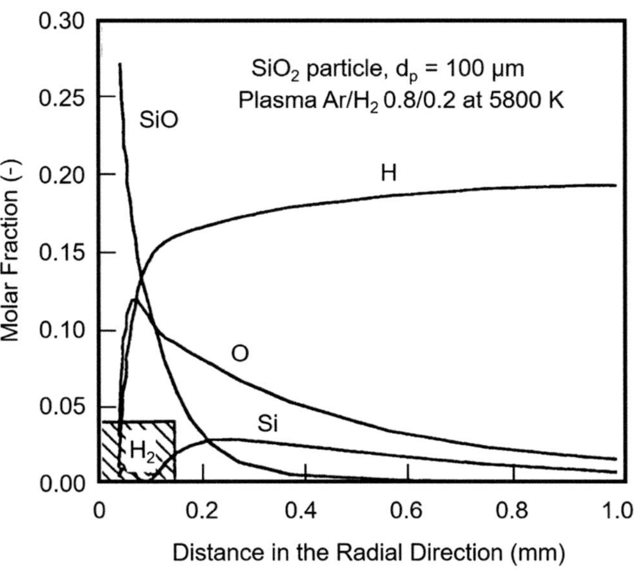

Fig. 4.49 Evolution of the molar fraction of the different chemical species around a 100 μm diameter SiO$_2$ particle in an Ar/H$_2$ (20 vol. %) plasma at 5800 K, calculated according to data of Humbert (1991)

In the case of nonmetallic particles, particle evaporation can also be associated with chemical decomposition of the particle material, giving rise to more complex mechanisms controlling the evaporation rate. A typical result is illustrated in Fig. 4.49 after Humbert (1991) for the evaporation of SiO$_2$ particles, 100 μm in diameter, immersed in Ar/H$_2$ plasma (20 vol. % H$_2$) at 5800 K. The results given in this figure show that close to the particle surface, the vapor comprises mostly of SiO, which is further reduced by H atoms as the vapor diffuses away from the particle surface.

4.7 Chemical Reactions and Melt Circulation

4.7.1 Diffusion Controlled Reaction

When the partial pressure of the reacting gas in the bulk of the hot gas surrounding a particle reaches a specific value, defined as the critical pressure, the flux of reacting species toward the surface of the droplet exceeds the counter flux of metal vapor, and a liquid or solid oxide, nitride or carbide layer (depending on the hot gas forming gas composition and ambient atmosphere) forms on the surface of the droplet. Typical values of the critical pressure of oxygen for an oxidation reaction can be calculated from:

$$p_{O_2\,max} = \frac{p_i}{\alpha_s\,h_{O_2}}\sqrt{\frac{RT}{2\pi M_i}} \qquad (4.61)$$

where α_s is the number of g-atoms of metal vapor required to combine with 1 mole of oxygen at the surface of the droplet, h_{O_2} is the mass transfer coefficient of oxygen (m/s), M_i is the atomic mass of the metal (kg) and p_i is the partial pressure of the vapor.

Under typical plasma spraying conditions, oxidation of the sprayed metal can take place either in the vapor phase surrounding the droplet, or at the surface of the droplet. In the latter case the formed oxide layer can result in a considerable reduction of the evaporation rate from the droplet surface. The effect is strongly dependent on the particle diameter, as illustrated in Fig. 4.50, for iron particles of diameters 40 and 80 μm, injected at 10 m/s into an Ar/H$_2$ DC plasma jet (conditions same as that of Fig. 4.48). The results are given in terms of molar ratio of the evaporated to the oxidized fraction of the iron particle as function of its axial position on its trajectory. As expected the smaller particles, $d_p = 40\,\mu m$, evaporate at a faster rate compared to the larger 80 μm particles, with comparable in-flight oxidation. This implies that if the trajectory of the 40 μm particle is optimum, the 80 μm particle will cross earlier the jet and thus is less heated. This results in a much lower evaporation in the latter case with a strong oxidation, while the opposite is true for the smaller particle.

At the end of the trajectory of the particle in the plasma, the composition of its surface depends on its temperature history, and the reactive species present in the plasma, provided that they can reach the particle surface. Thermodynamic equilibrium calculation can serve for the calculation of the chemical composition of the particles prior to their impact on the substrate. While the plasma conditions often deviate from equilibrium, the knowledge of phases which are thermodynamically stable under given conditions, and how they compare with those present in real systems, is very

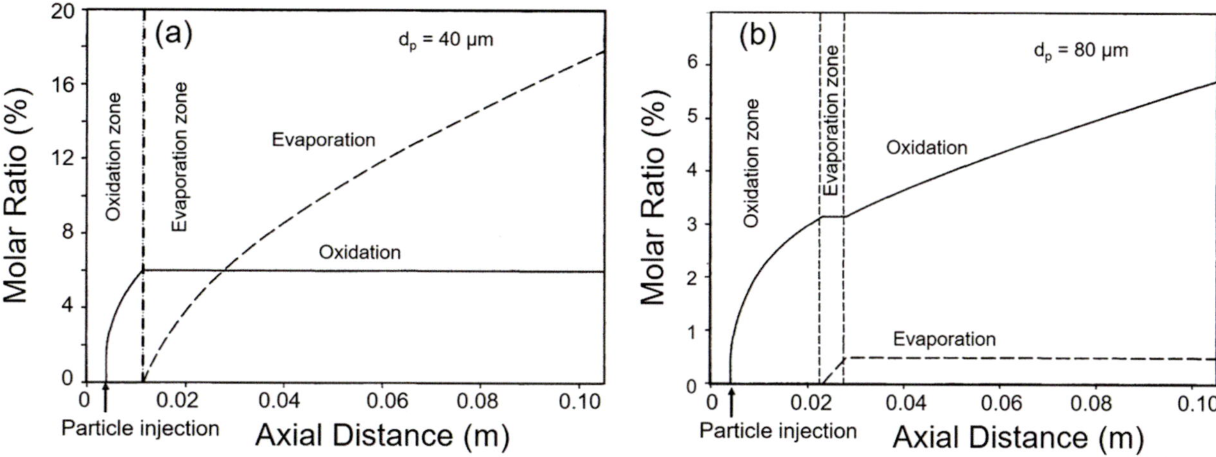

Fig. 4.50 Molar ratio of evaporated and oxidized fractions of an iron particle along its trajectory in an Ar/H$_2$ plasma jet. (**a**) d$_p$ = 40 μm, (**b**) d$_p$ = 80 μm, after Vardelle et al. (1996)

informative. This has been demonstrated, for example, for the oxides in plasma-sprayed chromium steel (Volenik 1997).

The next step is to calculate the diffusion phenomena occurring at the interfaces, diffusion of the reactant through the shell formed at the particle surface (in liquid or solid state), diffusion of gaseous reaction products formed (if any) through the shell and then through the boundary layer (see the shrinking core model described Asaki et al. (1974), Arnauld et al. (1985), and Amouroux et al. (1985)). The diffusion coefficients can be strongly affected by the expansion mismatch between the non-reacted particle core and the reacted shell formed around the core, which can be easily broken if it is in the solid phase.

A few examples of oxidation reactions can be found in the following references: Essoltani et al. (1990), − Vardelle et al. (1996), Li et al. (1995), Vardelle et al. (2002), Volenik et al. (1997), Vardelle et al. (1998), Espié et al. (2001), Volenik et al. (2003), Espié et al. (2005), Seyed et al. (2005). Pertinent references for other reactive systems, such as TiC + Ti, SiC + Si, W + W$_x$C$_y$, Mo + MoC$_2$, NiCr/Ti + TiC + Cr$_x$C$_y$, FeCrAlY + Cr$_x$Fe$_y$ + Fe$_x$C$_y$, Mo + MoSi$_2$, and Ti + TiB$_2$ can be found in Smith and Matasin (1992), Dallaire (1992), Jiang et al. (1994), Eckardt et al. (1994), Fauchais et al. 1997a, b), Dai et al. (1998), Fan and Ishigaki (1998), Denoirjean et al. (2003)).

4.7.2   Reactions Taking Place Between Condensed Phases

In this case, the basic reactants are in the solid phase either as cladded or agglomerated particles (Dallaire 1992; Borisov and Borisova 1993; Shaw et al. 1994; Deevi et al. 1997; Haller et al. 2004; Bach et al. 2001; Haller 2006) (see Fig. 4.51). The most common cladded particle is the Ni–Al where an Al core is surrounded by a Ni shell. Upon melting, Al reacts with the Ni creating intermetallic species such as

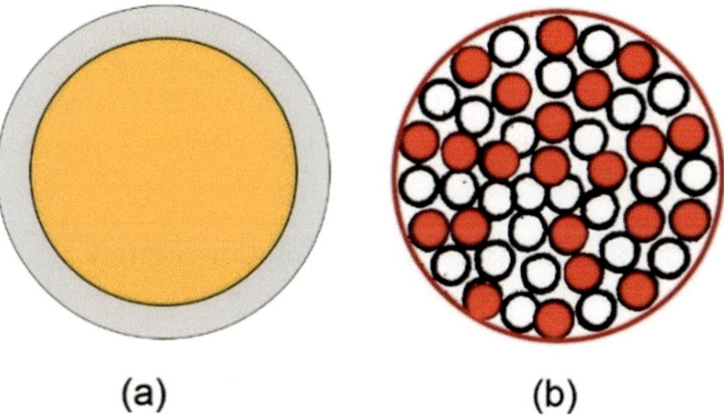

Fig. 4.51 (**a**) Cladded or (**b**) agglomerated, composite particles, which can be used for SHS plasma spraying Borisov and Borisova (1993)

Ni$_3$Al, NiAl. The same holds for agglomerated particles of Ti and C, for example, (Dallaire 1992). Hot gases heating of the particle triggers the reaction and initiates the self-propagation high temperature synthesis (SHS). The reaction depends strongly on the size of agglomerated particles and the possibility to heat the agglomerated particles without destroying the agglomerates by the produced gas expansion. As the speed of SHS is typically between 1 and 150 mm/s, the SHS reaction propagation is not necessarily completed during the flight time of the particles and may proceed after their impact resulting in very dense and hard coatings (Haller 2006).

Numerous coatings have been produced using the following approach:

- Transition metal/non-metal refractory compounds with mixtures of Cr and SiC or B$_4$C, Ti and SiC or B$_4$C or Si$_3$N$_4$ producing coatings with silicides, carbides, borides, which show excellent resistance to wear (Cliche and Dallaire 1991; Dallaire 1992; Dallaire and Cliche 1992; Borisov et al. 1986; Legoux and Dallaire 1993; Shaw et al. 1994)

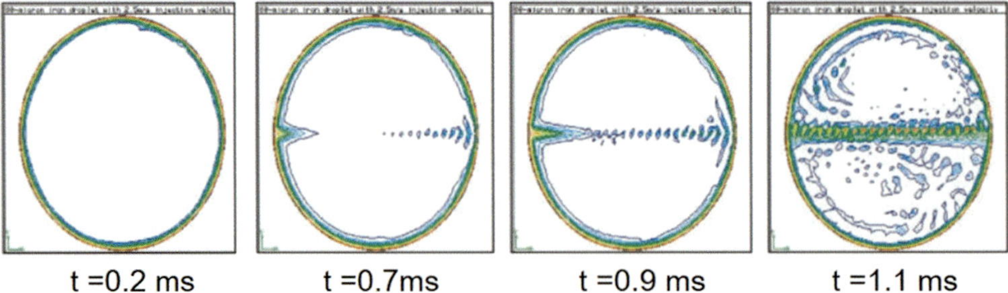

Fig. 4.52 Time-resolved modeling of internal convection and oxide penetration by a spherical vortex of Hill in a molten iron particle Espié et al. (1999)

- Copper-TiB_2 coatings starting from agglomerated particles of Ti-bronze and boron, the TiB_2 particles in coatings increase considerably the hardness of copper (Legoux and Dallaire 1993)
- TiB_2 or TiC particles in ferrous matrices to increase their hardness and wear resistance (Dallaire and Champagne 1984; Cliche and Dallaire 1991; Dallaire and Cliche 1992)
- Al and metal oxides to produce intermetallic compounds with aluminum particles (Bach et al. 2001)

4.7.3   Reactions Controlled by Convection Within Liquid Phase

Such reactions within droplets occur only in DC plasma jets as well as in wire-arc spraying where the shear stress at the surface of the molten droplet created by the large velocity difference between the molten particle in flight and the plasma flow induces a convective motion within the droplet. The phenomena is observed when the ratio of the kinematics viscosities of the plasma and the droplet is greater than 50 and the Reynolds number of the flow relative to droplet greater than 20 (Neiser et al. 1998). The motion inside the droplet can be represented by a Hill vortex that is an inviscid axisymmetric vortex (Espié et al. 1999). For a low carbon 60 µm iron droplet in a DC Ar–H_2 plasma jet, time-dependent computations performed by using a commercial code, FIDAP7-62, dedicated to the materials processing field [FIDAP code], indicated the formation of internal convection motion in the molten iron particle with oxide penetration inside the droplet caused by the spherical Hill's vortex as shown in Fig. 4.52.

The internal circulation in the liquid metal droplet continuously sweeps fresh liquid to its surface and makes it available for oxidation. As a result of this circulation, portions of the outer layer or of the dissolved oxygen at the droplet surface are entrained and transported to the core of the droplet. For a low-carbon iron particle, the oxide formed at the surface is liquid Fe_xO. Liquid iron and liquid Fe_xO are not miscible due to the significant difference in their surface tension (1778 mJ.m^2 for pure iron and 585 mJ. m^2 for the wüstite Volenik et al. (1997)). Upon cooling,

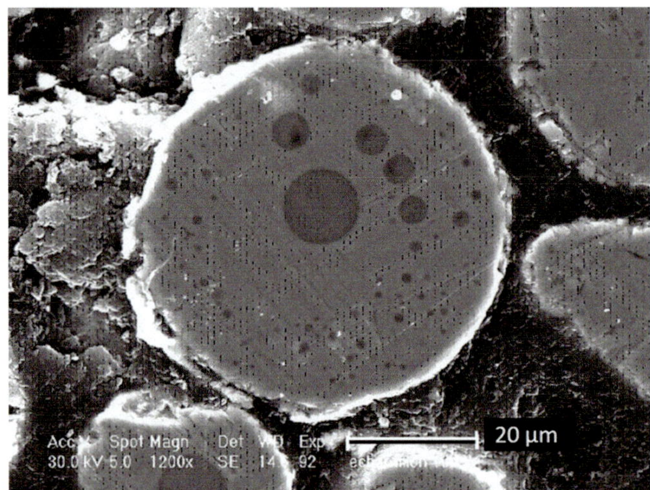

Fig. 4.53 Cross sections of a low carbon steel particle collected at z = 100 mm after its flight in a DC plasma jet: Ar 50 slm, H_2 10 slm, I = 500 A, nozzle i.d. 7 mm. SEM analysis at 30 keV. Scale 20 µm Espié et al. (1999)

both phases separate and spherical Fe_xO nodules are formed within the solidified particle, as shown in Fig. 4.53. Such recirculating flow within the droplet gives rise to an increase of the oxygen content in the solidified particle compared with that estimated by assuming a pure diffusion model Espié et al. (1999). For the iron particles shown in Fig. 4.53, the captured FeO corresponding to a mass percentage is 15 wt. %, while diffusion calculations estimate the Fe_2O_3 and Fe_3O_4 concentration to be less than 3 wt. %. The oxide shell formed at the surface of particles, with no inside convective movement, is made of Fe_2O_3 and Fe_3O_4. Such oxides are observed in the thin shell surrounding the particle formed at the end of the particle trajectory when the particle velocity is about the same as that of the surrounding plasma plume (no more convective effect). The same effect was observed with stainless steel particles plasma sprayed with Ar–H_2 plasma forming gas (Seyed et al. 2005).

Similar results have been observed when spraying Ti-6Al-4V particles in a DC nitrogen plasma under controlled atmosphere or in ambient air. TiN and TiO_2 contaminants are formed, in the particles with significantly less TiO_2 in controlled atmosphere spraying (Ponticaud et al. 2001).

4.7.4 Nano- and Micrometer-Sized Particles and Coating Structures

Over the past three decades, the interest for developing and studying nanostructured coatings has grown from about 300 peer-reviewed published in the 1980s to more than 2000 in the 1990s and about 28,000 in the last decade (Fauchais et al. 2011). This was mostly motivated by the superior properties of nanostructured coatings compared with those where the structure is micrometer-sized. Reducing the scale of the coating structure down to the nanometer level gives rise to an increased strength, improved toughness, and increased coefficient of thermal expansion while reducing apparent density and elastic modulus and lowering apparent thermal conductivity, among other numerous potential improvements (Gell 1995). Among the thermal routes, plasma spraying micrometer-sized particles, essentially ceramics, made of agglomerated nanometer sized particles is a possible way. Figure 4.54a (Lima and Marple 2008) shows a spray-dried "nanostructured" agglomerated Al_2O_3–13 wt.% TiO_2 (alumina–titania) powder (Nanox S2613S, Inframat Corp.,

Farmington, CT, USA). Examining the microstructure of these particles at higher magnification (Fig. 4.54b), it is possible to observe that the agglomerate is composed of individual particles varying from about 15 to 300 nm.

Complete melting of such particles during conventional plasma spraying operation will result in a complete loss of their initial nano-structure. Obviously, the spraying parameters have to be optimized to produce conditions (particle temperatures and velocities) that are limited to the partial melting of the agglomerates (to avoid the complete loss of the nanostructure) while maintaining a sufficiently high degree of melting to ensure effective deposition on the substrate and the formation of so-called nanozones (Lima et al. 2007). One of the critical goals of parameter optimization is to control the density of nanozones, that is, the density of the semi-molten nanostructured agglomerates embedded in the coating microstructure. This is achieved by finding the conditions needed to adjust the amount of the molten part of each semi-molten particle that penetrates into the capillaries (i.e., the non-molten particle core) of the agglomerates (Fig. 4.55) during flight within the thermal spray jet and/or at impact on the substrate surface and subsequent

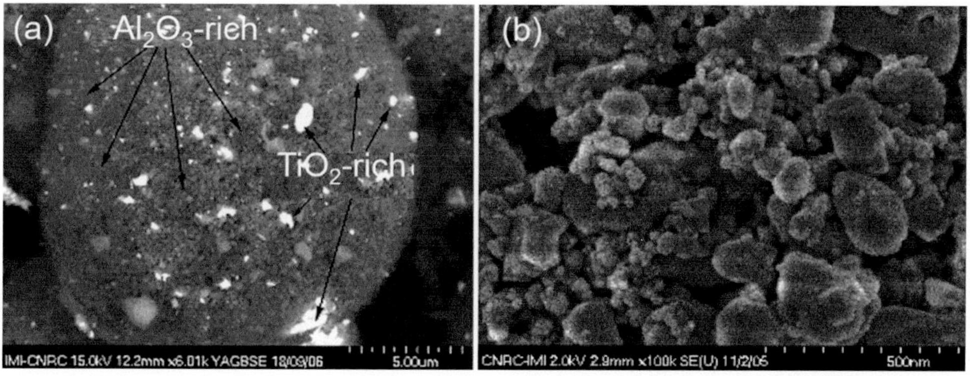

Fig. 4.54 (a) Spray-dried agglomerated Al_2O_3–13 wt. %TiO_2 particle. (b) Higher magnification view showing the "ultra-fine" character of the agglomerate Lima and Marple (2008)

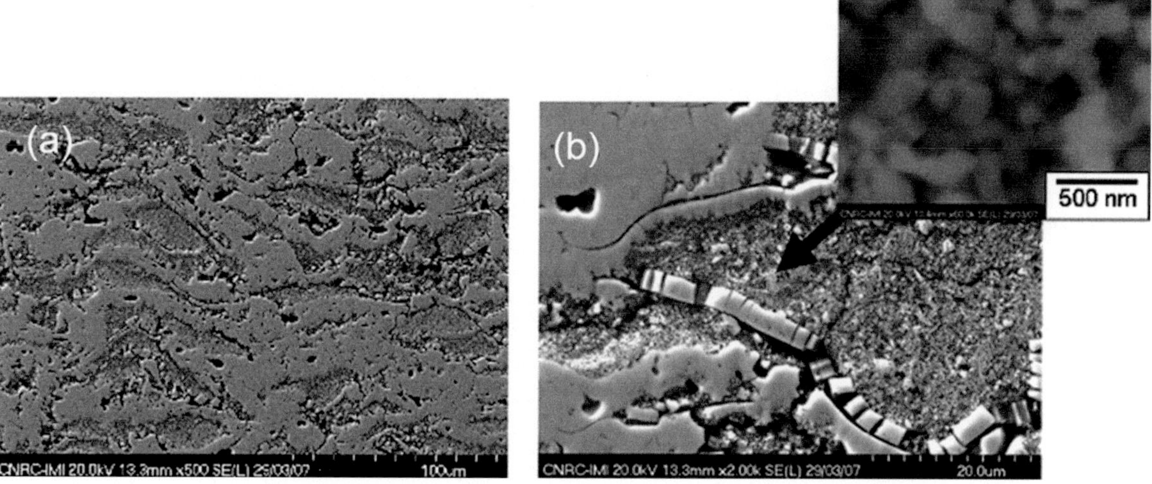

Fig. 4.55 (a) Plasma-sprayed coating (cross-section) engineered from the nanostructured-agglomerated YSZ particles. (b) High magnification view showing a semi-molten nanostructured agglomerate (porous nanozones) embedded in the coating microstructure Lima and Marple (2008)

re-solidification. It is important to point out Fauchais et al. (2011) that by introducing porous or dense nanozones throughout the coating microstructure it is possible to engineer coatings with very different and even opposite properties for a variety of purposes. For example, for thermal barrier and abradable seal applications, the presence of porous nanozones is paramount. On the other hand, for anti-wear applications, dense nanozones are absolutely required to induce high levels of wear resistance. The main technique employed for optimizing the spray parameters to engineer coatings produced from nanostructured-agglomerated particles is in-flight particle temperature and velocity monitoring. Moreover, the use of large agglomerates is paramount to engineer architectures that exhibit porous nanozones embedded in the coating micro-structure. For more details, see Lima and Marple (2008), Fauchais et al. (2011).

4.8   Summary and Conclusions

Three-dimensional simulations with full coupling between a single particle and gas flow, employing a Lagrangian particle-tracking frame coupled with a steady-state gas flow, have been developed to examine particle motion and heat transfer. These models, when necessary, are compressible and consider the effect on particles of compression and expansion waves. Since the 1990s' models have shown the importance of the different corrections adopted to the process for flames, HVOF or HVAF, D-Gun, and plasma spraying processes: high temperature gradients, particle evaporation or vaporization, rarefaction effect (Knudsen effect), shock waves, supersonic flow effect, such models have shown the drastic influence of the particle injection (position, velocity vector) on its velocity and temperature at impact. Different effects such as loading effect, particle morphology, and shape have been considered. However, the way the particle velocity vector distribution at the injector exit is calculated is still in its infancy in spite of its drastic influence on the particle treatment within the hot or cold gas flow. The first studies of the interaction between hot gas flows and a liquid (solution or suspension spraying) have also been presented; this technique is very promising for the deposition of finely or nanostructured coatings with thicknesses between a few micrometers and hundreds of them.

Nomenclature

Units are indicated in parentheses; when no units are indicated, the parameter is dimensionless.

Latin Alphabet

a	accommodation coefficient, Eq. 4.22
a_i	sound velocity (m/s)
a_p	surface area of particle $\left(a_p = \pi\, d_p^2\right)$ (m^2)
A_p	particle projected surface area perpendicular to the flow ($A_p = \pi d_p^2/4$) (m^2)
Bi	Biot number ($Bi = \kappa/\kappa_p$)
c	molar density of the bulk gas (mol/m^3)
C	circularity of a particle, Eq. 4.4 (-)
C_D	drag coefficient, Eq.4.16 (F_D/a_p)/($0.5\rho v^2$)
c_i	mass fraction of metal vapor at the location i
c_{ps}	specific heat of the particle
c_{pi}	specific heat at constant pressure in the state i (J/kg.K)
c_{vi}	specific heat at constant volume in the state i (J/kg.K)
d_d	drop diameter (m)
d_l	drop or liquid jet diameter (m)
d_p	particle diameter (m)
d_{pp}	equivalent particle perimeter diameter, Eq. 4.1 (m)
d_{pa}	equivalent particle projected area diameter, Eq. 4.2 (m)
d_{pv}	equivalent particle volume diameter, Eq. 4.3 (m)
d_{st}	Stokes diameter, Eq. 4.6 (m)
D_{vg}	diffusion coefficient of metal vapor through the surrounding gas (m^2/s)
F_B	Basset history term (N)
F_c	Coriolis force (N)
F_b	body force per unit particle mass (N)
F_d	force related to the pressure jump in the detonation wave $\left(F_d = \frac{\pi\, d_p^2}{4}\Delta p\right)$ (N)
F_D	drag force exerted by the fluid on the particle (N)
F_g	gravity force (N)
F_i	inertia force ($m_p \cdot \gamma_p$) (N)
F_p	pressure gradient force (N)
F_S	surface tension force (N)
F_T	thermophoresis force (N)
g	gravitational acceleration (g = 9.81 m/s^2)
h	heat transfer coefficient (W/m^2·K)
h_g	specific enthalpy (J/kg)
h'_i	specific enthalpy of the plasma calculated at i (J/kg)
I	arc current (A)
Kn	Knudsen number ($Kn = \lambda/dp$)
k_d	mass transfer coefficient (m/s)
L	mixing length in turbulent model (m)
m^o_{cg}	carrier gas mass flow rate (kg/s)
m_p	particle mass (kg)
M	molecular weight (kg/mol)
Ma	mach number ($Ma = v_g/a_g$)
N_{max}	maximum molar flux of vaporization from a droplet (m^2/s)
Nu	Nusselt number ($Nu = hd/k$)
N_v	molar flux of vapor (mol/m^2·s)
Oh	Ohnesorge number $\left(Oh = \mu_l/\sqrt{\rho_l \times d_l \times \sigma_l}\right)$
p	total pressure (Pa)
p_i	partial pressure of species i (Pa)
p_o	saturation vapor pressure of a liquid (Pa)
P	torch power (kW)
P_l	splat parameter (m)
P_{eff}	torch effective power (kW)
Pr	Prandtl number ($Pr = \mu \cdot c_p/\kappa$)
P_p	Perimeter of the projected image of the particle (m)
q	heat flux (W/m^2)
Q	heat transferred to a particle (W)
Q_{cv}	heat transferred to a particle by conduction and convection (W)
Q_{sr}	heat lost from the surface of the particle by radiation to surrounding (W)
Q_{vr}	volumetric radiation losses emitted from the hot vapor cloud (W)
Q_n	Net energy received by the particle, Eq. 4.23 (W)
r	plasma or particle radius (m)
r_d	liquid droplet radius (m)
r_s	initial liquid drop radius (m)
R	plasma torch internal or particle radius (m)

R'	universal ideal gas constant, ($R' = 8.32$ J/K $\cdot$ mol)
Re	Reynold's number, ($Re = \rho\, u_R d_p/\mu$)
R_{inj}	internal radius of injection tube (m)
S	cross section of injection tube (m^2)
S_T	heat conduction potential, Eq. 4.50 (W/m)
Sc	Schmidt's number ($Sc = \nu/D_{v,g}$)
Sh	Sherwood's number, ($Sh = k_d.d_p/D_{v,g}$)
St	Stokes number ($St = \rho_p\, d_p^2\, v_p/\mu_g l_{BL}$)
Ste	Stephan's number; ($Ste = c_{pi}(T_m - T_s)/\Delta H_m$)
t_d	drop fragmentation time (s)
t_r	time of flight of particles (s)
t_s	vaporization time (s)
T_a	ambient temperature (K)
T_f	mean film temperature, Eq. 4.18 ($T_f = (T_s + T_\infty)/2$) (K)
T_s	particle surface temperature (K)
T_∞	free stream plasma temperature (K)
U	mean arc voltage (V)
$U(t)$	transient arc voltage (V)
u	gas velocity component in the axial direction (m/s)
u_p	particle velocity component in the axial direction (m/s)
u_R	relative velocity between the particle and its surrounding (m/s)
u_t	terminal settling velocity (m/s)
v	gas velocity component in the radial direction (m/s)
v_p	particle velocity component in the radial direction (m/s)
V	volume (m^3)
V_p	particle volume (m^3)
V_s	liquid drop volume (m^3)
w_s	work resulting from the drag force (J)
We	Weber number $\left(We = \rho_g u_r^2 d_l/\sigma_l\right)$

Greek Alphabet

α_p	thermal diffusivity of the particle material ($\alpha_p = \kappa_p/\rho_p c_{ps}$)
α_l	thermal diffusivity of the molten particle material ($\alpha_l = \kappa_l/\rho_l c_{ls}$)
α_s	atoms of metal vapor required to combine with 1 mole of oxygen
δ	boundary layer thickness (m)
ΔE_s	variation of surface energy (J)
ΔH_m	latent heat of fusion (J/kg)
ΔH_v	latent heat of vaporization (J/kg)
ε	particle emissivity (integrated over all wave lengths)
$\phi(r)$	function representing temperature, enthalpy, velocity
γ	specific heat ratio ($\gamma = c_p/c_v$)
γ_p	particle acceleration (m/s^2)
η_{th}	torch thermal efficiency (%)
κ	thermal conductivity of the fluid (W/m K)
κ_p	thermal conductivity of the particle (W/m K)
$\bar{\kappa}$	mean integrated thermal conductivity, Eq. 4.34 (W/m K)
λ	plasma mean free path (m)
μ	dynamic viscosity (Pa·s)
ν_{cg}	carrier gas velocity (m/s)
ν	kinematic viscosity ($\nu = \mu/\rho$) (m^2/s)
ρ_g	gas mass density (kg/m^3)
ρ_o	fluid density or specific mass (kg/m^3)
ρ_p	particle density or specific mass (kg/m^3)
σ	droplet surface tension (N/m)
σ_s	Stephan–Boltzmann constant ($\sigma_s = 5.670 \times 10^{-8}$ W/m^2·K^4)
σ_x	standard deviation
ξ	ratio of splat to droplet diameter ($\xi = D_p/d_p$)
ψ	sphericity of a particle, Eq. 4.5 (m)
ψ	particle vaporization constant, Eq. 4.56 (m^2/s)

References

Amouroux, J., A. Gicquel, S. Cavadias, D. Morvan, and F. Arefi. 1985. Progress in the applications of plasma surface modifications and correlations with the chemical properties of the plasma phase. *Pure and Applied Chemistry* 57 (9): 1207–1222.

Arnauld, Ph., S. Cavadias, and J. Amouroux. 1985. *The interaction of a fluidized bed with a thermal plasma. Application to limestone decomposition, ISPC-7*, ed. Timmermans, 1195–1200. Eindhoven: University of Technology of Eindhoven.

Asaki, Z., Y. Fakunaka, T. Nagasi, and Y. Kondo. 1974. Thermal decomposition of limestone in a fluidized bed. *Metallurgical Transactions* B-5: 381–390.

Bach, Fr.W., Z. Babiak, T. Duda, T. Rothardt, and G. Tegeder. 2001. *Impact of self propagating high temperature synthesis of spraying materials on coatings based on aluminum and metal-oxides, ITSC-2001: New surfaces for a new millennium*, ed. C.C. Berndt, K.A. Khor and E.F. Lugsheider, 497–502. Materials Park: ASM International.

Borgianni, C., M. Capitelli, F. Cramarossa, L. Triolo, and L. Molinari. 1969. The behavior of metal oxides injected into an argon induction plasma. *Combustion and Flame* 13: 181–194.

Borisov, Y., and A. Borisova. 1993. *Application of self-propagating high-temperature synthesis in thermal spraying technology, ITSC-1993: Research, design and applications*, ed. T.F. Bernicki (pub.), 139–144. Materials Park: ASM International.

Borisov, Yu.S., A.L. Borisova and L.K. Shvedova. 1986. Transition metal-nonmetallic refractory compound composite powders for thermal spraying. *Advances in Thermal Spraying*: 323–329. Pergamum Press.

Boulos, M.I. 1976. Flow temperature field in the fire ball of an inductively coupled plasma. *IEEE Transactions on Plasma Science* PS-4: 28–39.

———. 1978. Heating of powders in the fire ball of an induction plasma. *IEEE Transactions on Plasma Science* PS-6 (2): 93–106.

———. 1985. The inductively coupled R.F. Plasma. *Journal of Pure and Applied Chemistry* 57: 1321–1352.

———. 1992. RF induction plasma spraying: State-of-the-art review. *Journal of Thermal Spray Technology* 1 (1): 33–40.

———. 2003. Spheroidization and densification of powders. In *Continuing education course on Thermal Plasmas, held in conjunction with ISPC-16*, ed. P. Fauchais. University of Limoges.

———. 2004. Plasma interaction with a dispersed medium Ch 6.2. In *Continuing education course in conjunction with ITSC-2004*, ed. J. Heberlein. Minneapolis: University of Minnesota.

———. 2016. The role of transport phenomena and modelling in the development of thermal plasma technology. *Journal of Plasma Chemistry Plasma Processing* 36: 3–29.

Boulos, M.I., and D.C.T. Pei. 1969. Simultaneous heat and mass transfer from a single sphere to a turbulent air stream. *Canadian Journal of Chemical Engineering* 47: 30–34.

Boulos, M.I., P. Fauchais, A. Vardelle, and E. Pfender. 1993. Fundamentals of plasma particle momentum and heat transfer. In *Plasma spraying theory and applications*, ed. R. Suryanarayanan. World Scientific Singapore.

Boulos, M., P. Fauchais, and E. Pfender. 1994. *Thermal plasmas fundamental and applications*. Plenum Press, NY, 452 pages.

Bouneder, M. 2006. *Modelling of heat and mass transfer within composite metal/ceramic particles in DC plasma spraying*. PhD thesis University of Limoges, France.

Bouneder, M., M. El Ganaoui, B. Pateyron, and P. Fauchais. 2003. Thermal modeling of composite iron/alumina particles sprayed under plasma conditions Part I: Pure conduction. *High Temperature Material Processes* 7 (4): 547–555.

Bourdin, E., M.I. Boulos, and P. Fauchais. 1983. Transient conduction to a single sphere under plasma conditions. *International Journal of Heat and Mass Transfer* 26: 567–579.

Bouyer, E., F. Gitzhoffer, and M.I. Boulos. 1997a. Experimental study of suspension plasma spraying of hydroxyapatite. In *Progress in plasma processing of materials*, ed. P. Fauchais, 735–750. Begell House.

Bouyer, E., M. Müller, N. Dard, F. Gitzhofer, and M.I. Boulos. 1997b. Suspension plasma spraying for powder preparation. In *Progress in plasma processing of materials*, ed. P. Fauchais, 751–759. Begell House.

Bouyer, E., F. Gitzhofer, and M.I. Boulos. 1997c. The suspension plasma spraying of bioceramics by induction plasma. *JOM* 49 (2): 58–68.

Bouyer, E., D.W. Branston, G. Lins, M. Müller, J. Verleger, and M. Von Bradke. 2001. Deposition of yttria-stabilized zirconia coatings using liquid precursors. In *Progress in plasma processing of materials*, ed. P. Fauchais, 501–506. Begell House.

Bouyer, L., X. Ma, A. Ozturk, E.H. Jordan, N.P. Padture, B.M. Cetegen, D.T. Xiao, and M. Gell. 2004. Processing parameter effects on solution precursor plasma spray process spray patterns. *Surface Coating Technology* 183 (1): 51–61.

Chang, P., and K.A. Khor. 1996. Influence of powder characteristics on plasma sprayed hydroxyapatite coatings. *Journal of Thermal Spray Technology* 5 (3): 310–316.

Chen, Xi. 1988. Particle heating in a thermal plasma. *Pure and Applied Chemistry* 60: 651–662.

Chen, Xi, and Xioming Chen. 1989. Drag on a metallic or non-metallic particle exposed to a rarefied plasma flow. *Plasma Chemistry and Plasma Processing* 9 (3): 387–408.

Chen, Xi, and He Ping. 1986. Heat transfer from a rarefied plasma flow to a metallic or nonmetallic particle. *Plasma Chemistry and Plasma Processing* 6 (4): 313–333.

Chen, Xi, and E. Pfender. 1982a. Unsteady heating and radiation effects of small particles in a thermal plasma. *Plasma Chemistry and Plasma Processing* 2: 293–316.

———. 1982b. Heat transfer to a single particle exposed to a thermal plasma. *Plasma Chemistry and Plasma Processing* 2 (2): 185–212.

———. 1983a. Effect of the Knudsen number on heat transfer to a particle immersed into a thermal plasma. *Plasma Chemistry and Plasma Processing* 39: 7–113.

———. 1983b. Behavior of small particles in a thermal plasma flow. *Plasma Chemistry and Plasma Processing* 3: 351–366.

Chen, Xi, Y.P. Chyou, Y.C. Lee, and E. Pfender. 1985. Heat transfer to a particle under plasma conditions with vapor contamination from the particle. *Plasma Chemistry and Plasma Processing* 5 (2): 119–141.

Chen, K., and M.I. Boulos. 1994. Turbulence in induction plasma modelling. *Journal of Physics D: Applied Physics* 27: 946–952.

Chyou, Y.P., and E. Pfender. 1989. Behavior of particulates in thermal plasma flows. *Plasma Chemistry and Plasma Processing* 9 (1): 45–71.

Cliche, G., and S. Dallaire. 1991. Synthesis and deposition of TiC-Fe coatings by plasma spraying. *Surface and Coating Technology* 46: 199–206.

Clift, R., J.R. Grace, and J.E. Weber. 1978. *Bubbles, drops and particles*. Academic.

Cram, L. 1985. Statistical evaluation of radiative power losses from thermal plasmas due to spectral lines. *Journal of Physics D: Applied Physics* 18: 401–411.

Dai S., J.-P. Delplanque, R.H. Rangel, and E.J. Lavernia. 1998. *Modeling of reactive spray atomization and deposition, NTSC-1998: Meeting the challenges of the 21st century*, ed. C. Coddet (pub.), Vol. 1, 341–346. Materials Park: ASM International.

Dallaire, S. 1992. Thermal spraying of reactive materials to form wear-resistant composite coatings. *Journal of Thermal Spray Technology* 1 (1): 41–47.

Dallaire, S., and B. Champagne. 1984. Plasma spray synthesis of TiB2-Fe coatings. *Thin Solid Films* 118: 477–483.

Dallaire, S., and G. Cliche. 1992. The influence of composition and process parameters on the microstructure of TiC-Fe multiphase and multilayer coatings. *Surface and Coating Technology* 50: 233–239.

Deevi, S.C., V.K. Sikka, C.J. Swindeman, and R.D. Seals. 1997. Reactive spraying of nickel-aluminide coatings. *Journal of Thermal Spray Technology* 6 (3): 335–344.

Denoirjean, A., P. Lefort, and P. Fauchais. 2003. Nitridation process and mechanism of Ti-6Al-4V particles by plasma spraying. *Physical Chemistry Chemical Physics* 5: 5133–5138.

Diez, P., and R.W. Smith. 1993. The influence of powder agglomeration methods on plasma sprayed yttria coatings. *Journal of Thermal Spray Technology* 2 (2): 165–172.

Ducos, M. 1988. Rechargement par plasma a arc transferé (in French), In Arc plasma processes, Union Internationale de l'Electrothermie, DOPEE diffusion, Av. F. Roosevelt, 77210, Avon, France, 251–263 (in French).

Eckardt, T., W. Malleaer, and D. Stove. 1994. *Reactive plasma spraying of silicon in controlled nitrogen atmosphere, NTSC-1994: Industrial applications*, ed. C.C. Berndt and S. Sampath, 515–520. Materials Park: ASM International.

Espié, G., P. Fauchais, B. Hannoyer, J.C. Labbe, and A. Vardelle. 1999. Effect of metal particles oxidation during the APS on the wettability. In *Heat and mass transfer under plasma condition*, ed. P. Fauchais, J. Van der Mullen, and J. Heberlein, vol. 891, 143–151. Annals of NY Academy of Sciences.

Espié, G., P. Fauchais, J.C. Labbe, A. Vardelle, and B. Hannoyer. 2001. *Oxidation of iron particles during APS, ITSC-2001: New surface for a new Millenium*, ed. C.C. Berndt, K.A. Khor and E. Lugscheider, 821–828. Materials Park: ASM International.

Espie, G., A. Denoirjean, P. Fauchais, J.C. Labbe, J. Dudsky, O. Scheeweiss, and K. Volenik. 2005. In flight oxidation of iron particles sprayed using gas and water stabilized plasma torches. *Surface and Coatings Technology* 195: 17–28.

Essoltani, A., P. Proulx, M.I. Boulos, and A. Gleizes. 1990. Radiation and self-absorption in argon - Iron plasmas at atmospheric pressure. *Journal of Analytical Atomic spectrometry* 5: 543–547.

———. 1991. *Radiative effects on plasma-particle heat transfer in the presence of metallic vapors, ISPC-10*, ed. U. Ehlemann et al. University of BOCHUM, Germany.

———. 1993. A combined convective, conductive and radiative heat transfer study for an evaporating metallic particle under plasma conditions. *High Temperature Materials and Processes* 2: 37–46.

———. 1994. Effect of the presence of iron vapors on the volumetric emission of Ar/Fe and Ar/Fe/H_2 plasmas. *Plasma Chemistry Plasma processing* 14 (3): 301–315.

Fan, X., and T. Ishigaki 1998. *Fabrication of composite SiC-MoSi2 powders through plasma reaction process, NTSC-1998: Meeting the challenges of the 21st century*, ed. C. Coddet, Vol. 2, 1161–1166. Materials Park: ASM International.

Fauchais, P. 2004. Understanding plasma spraying. *Journal of Physics D: Applied Physics* 37: 86–108.

Fauchais, P., and G. Montavon. 2010. Latest developments in suspension and liquid precursor thermal spraying. *Journal of Thermal Spray Technology* 19: 226–239.

Fauchais, P., J.F. Coudert, and M. Vardelle. 1989. *Diagnostics in thermal plasma processing, plasma diagnostics*, ed. O. Ociello and D.L. Flamm, Vol. 1, 349–446. Academic.

Fauchais, P., A.C. Léger, M. Vardelle, and A. Vardelle. 1997a. Formation of plasma sprayed oxide coatings. In *Proceedings of the Julian Szekely Memorial symposium on materials processing*, ed. H.Y. Sohn, J.W. Evans, and D. Apelian, 571–582. T.M.S.

Fauchais, P., A. Vardelle, and A. Denoirjean. 1997b. Reactive thermal plasmas: Ultrafine particle synthesis and coating deposition. *Surface and Coatings Technology* 979: 66–78.

Fauchais, P., V. Rat, C. Delbos, J.F. Coudert, T. Chartier, and L. Bianchi. 2005a. Understanding of suspension dc plasma spraying of finely structured coating for SOFC. *IEEE Transactions on Plasma Science* 33: 920–930.

Fauchais, P., M. Vardelle, J.F. Coudert, A. Vardelle, C. Delbos, and J. Fazilleau. 2005b. Plasma spraying from thick to thin coatings and micro to nano structured coatings. *Pure and Applied Chemistry* 77: 475–485.

Fauchais, P., R. Etchart-Salas, C. Delbos, M. Tognovi, V. Rat, J.F. Coudert, and T. Chartier. 2007. Suspension and solution plasma spraying of finely structured coatings. *Journal of Physics D: Applied Physics* 40: 2394–2406.

Fauchais, P., R. Etchart-Salas, V. Rat, J.F. Coudert, N. Caron, and K. Wittmann. 2008. Parameters controlling liquid plasma spraying: Solutions, sols or suspensions. *Journal of Thermal Spray Technology* 17 (1): 31–59.

Fauchais, P., G. Montavon, R. Lima, and B. Marple 2011. Engineering a new class of thermal spray nano-based microstructures from agglomerated nanostructured particles, suspensions and solutions: An invited review. *Journal of Physics D: Applied Physics* 44: 093001 (53p).

Fauchais, P., A. Joulia, S. Goutier C. Chazelas, M. Vardelle, A. Vardelle, and S. Rossignol. 2013. Suspension and solution plasma spraying. *Journal of Physics D: Applied Physics* 46: 224015 (14pp).

Fauchais, P., J.V. Heberlein, and M.I. Boulos. 2014. *Thermal spray fundamentals, From powder to part.* Springer, 1550 pages.

Fauchais, P., M. Vardelle, A. Vardelle, and S. Goutier. 2015. What do we know, what are the current limitations of suspension plasma spraying? *Journal of Thermal Spray Technology* 24 (7): 1120–1129.

Fazilleau, J. 2003. *Contribution to the understanding of the phenomena implied in the achievement of finely structured oxide coatings by suspension plasma spraying.* PhD. Thesis, In French, University of Limoges, France.

FIDAP Code distributed by Fluent Inc., Lebanon, NH, USA.

Fizdon, J.K. 1979. Melting of powder grains in a plasma flame. *International Journal of Heat and Mass Transfer* 22: 749–761.

Fukanuma, H., N. Ohno, Bo Sun, and R. Huang. 2006. *The influence of particle morphology on in-flight particle velocity in cold spray, ITSC-2006.* Materials Park: ASM International.

Ganser, G.H. 1993. A rational approach to drag prediction of spherical and non-spherical particles. *Powder Technology* 77: 143–152.

Gell, M. 1995. Application opportunities for nanostructured materials and coatings. *Materials Science and Engineering* 204: 246–251.

Gell, M., E.H. Jordan, Y.H. Sohn, D. Goberman, L. Shaw, and T.D. Xiao. 2001. Development and implementation of plasma sprayed nanostructured ceramic coatings. *Surface and Coatings Technology* 146–147: 48–54.

El Hage, M., J. Mostaghimi, and M.I. Boulos. 1989. A turbulent flow model for the R.F. inductively coupled plasma. *Journal of Applied Physics* 65: 4178–4185.

Haller, B. 2006. *Study of a process combining plasma spraying and SHS: application to Ti-graphite mixtures.* PhD Thesis, University of Limoges, France (in French).

Haller, B., J.P. Bonnet, P. Fauchais, A. Grimaud, and J.C. Labbe. 2004. *TiC based coatings prepared by combining SHS and plasma spraying, ITSC-2004 Innovative Equipment and Process Technology* V (pub.). Düsseldorf: DVS.

He, P.-J., S. Yin, C. Song, F. Lapostolle, and H.-L. Liao. 2016. Characterization of Yttria-stabilized zirconia coatings deposited by low-pressure plasma spraying. *Journal of Thermal Spray Technology* 25 (3): 558–566.

Heberlein, J., J. Mentel, and E. Pfender. 2007. The anode region of electric arcs - A survey. *Journal of Physics D: Applied Physics* 43: 023001.

Humbert, P. 1991. *Development of pilot set-up for silicon purification by induction thermal plasma and modelling of mass and heat transfers plasma-particles.* Ph.D. Thesis, Univ. of Paris VI, ENSCP, Paris, France.

Hurevich, V., I. Smurnov, and L. Pawlowski. 2002. Theoretical study of the powder behavior of porous particles in a flame during plasma spraying. *Surface and Coatings Technology* 151–152: 370–376.

Jiang, X.L., F. Gitzhoffer, M.I. Boulos, and R. Tiwari. 1994. *Induction plasma reactive deposition of tungsten and titanium carbides, NTSC-1994: Industrial applications,* ed. C.C. Berndt and S. Sampath (pub.), 451–456. Materials Park: ASM International.

Laha, T., K. Balani, A. Agarwal, S. Patil, and S. Seal. 2005. Synthesis of nanostructured spherical aluminum oxide powders by plasma engineering. *Metallurgical and Materials Transactions A* 36A (2005): 301–309.

Lapple, C.E., and C.B. Shepherd. 1940. Calculation of particle trajectories. *Industrial and Engineering Chemistry* 32 (5): 605–617.

Lee, C.S., and R.D. Reitz. 2001. Effect of liquid properties on the break-up mechanism of high-speed liquid drops. *Atomization and Sprays* 11: 1–18.

Lee, Y.C., C. Hsu, and E. Pfender. 1981. *Modelling of particle injection into a DC plasma jet ISPC-5.* Vol. 2, 795–801. Edinburgh.

Lee, Y.C., Y.P. Chyou, and E. Pfender. 1985. Particle dynamics and particle heat and mass transfer in thermal plasmas. Part II. Particle heat and mass transfer in thermal plasmas. *Plasma Chemistry and Plasma Processing* 5 (4): 391–414.

———. 1997. Particle dynamics and particle heat and mass transfer in thermal plasmas. Part III. Thermal plasma jet reactors and multi-particle injection. *Plasma Chemistry and Plasma Processing* 7 (1): 1–27.

Lefebvre, A.H. 1989. *Atomization and sprays.* Hemisphere Publishing Corp.

Legoux, J.G., and S. Dallaire. 1993. Copper-TiB$_2$ coatings by plasma spraying reactive micropellets. In *Proc. NTSC-1993,* ed. C.C. Berndt, 429–432. Materials Park: ASM International.

Lewis, J.W., and W.H. Gauvin. 1973. Motion of particles entrained in a plasma jet. *AICHE Journal* 19 (6): 982–990.

Li, K.-I., M. Vardelle, A. Vardelle, P. Fauchais, and C. Trassy. 1995. Comparisons between single and double flow injectors in the plasma spraying process, NTSC-1005.

Lima, R.S., and B.R. Marple. 2007. B thermal spray coatings engineered from nano-structured ceramic agglomerated powders for structural, thermal barrier and biomedical applications: A review. *Journal of Thermal Spray Technology* 16: 40–63.

———. 2008. Nanostructured YSZ thermal barrier coatings engineered to counteract sintering effects. *Materials Science and Engineering A* 485: 182–193.

Marchand, C., C. Chazelas, G. Mariaux, and A. Vardelle. 2007. *Liquid precursor plasma spraying: Modelling the interaction between the transient plasma jet and the droplets, ITSC-2007: Global coating solutions,* ed. B R Marple et al., 196–201. Materials Park: ASM International.

Marchand, C., A. Vardelle, G. Mariaux, and P. Lefort. 2008. Modelling of the plasma spray process with liquid feedstock injection. *Surface and Coating Technology* 202: 4458–4464.

Marchand, O., L. Girardot, M.P. Planche, P. Bertrand, Y. Bailly, and G. Bertrand. 2011. An insight into suspension plasma spray: Injection of the suspension and its interaction with the plasma flow. *Journal of Thermal Spray Technology* 20 (6): 1310–1320.

Mostaghimi, J., and E. Pfender. 1984. Effects of metallic vapor on the properties of an argon arc plasma. *Plasma Chemistry Plasma Processing* 4 (2): 129–139.

Mostaghimi, J., P. Proulx, and M.I.A. Boulos. 1987. Two-temperature model of the inductively coupled R.F. Plasma. *Journal of Applied Physics* 61 (5): 1753–1760.

Mostaghimi, J., and M.I. Boulos. 1989. Two-dimensional electromagnetic field effects in R.F. inductively coupled plasmas. *Journal of Plasma Chemistry and Plasma Processing* 9: 23–42.

Mostaghimi, J., P. Proulx, M.I. Boulos, and R.M. Barnes. 1985. Computer modeling of the emission patterns for a spectrochemical ICP. *Spectro-chemical Acta* 40B: 153–166.

Murray, W.D., and F. Landis. 1959. Numerical and machine solutions of transient heat-conduction problems involving melting or freezing-Part I. Method of analysis and sample solutions. *Journal of Heat Transfer* 81: 106112.

Neiser, R.A., M.F. Smith, and R.C. Dykhuisen. 1998. Oxidation in wire HVOF-sprayed steel. *Journal of Thermal Spray Technology* 7 (4): 537–545.

Oberkampf, W.L., and M. Talpallikar. 1994. Analysis of a high velocity oxygen-fuel (HVOF) thermal spray torch, part 1: Numerical formulation. In *Thermal spray industrial applications*, ed. C.C. Berndt and S. Sampath, 381–386. ASM International, Materials Park.

Pfender, E. 1985. Heat and momentum transfer to particles in thermal plasma flows. *Pure and Applied Chemistry* 57: 1179–1196.

———. 1989. Particle behavior in thermal plasmas. *Plasma Chemistry and Plasma Processing* 9 (1): 167S–194S.

———. 1999. Thermal plasma technology: Where do we stand and where are we going? *Plasma Chemistry and Plasma Processing* 19 (1): 1–31.

Pfender, E., and C.H. Chang. 1998. *Plasma spray jets and plasma-particulate interaction: Modeling and experiments, ITSC-1998: Meeting the challenges of 21st century*, ed. C. Coddet, Vol. 1, 315–328. Materials Park: ASM International.

Pfender, E., and Y.C. Lee. 1985. Particle dynamics and particle heat and mass transfer in thermal plasmas. Part I. The motion of a single particle without thermal effects. *Plasma Chemistry Plasma Processing* 5 (3): 211–237.

Ponticaud, C., A. Grimaud, A. Denoirjean, P. Lefort, and P. Fauchais. 2001. Titanium powder nitridation by reactive plasma spraying. In *Progress in plasma processing of materials*, ed. P. Fauchais, J. Amouroux, and M.F. Elchinger, 527–536. Begell House Inc.

Proulx, P., J. Mostaghimi, and M.I. Boulos. 1985a. Plasma-particle interaction effects in induction plasma modelling under dense loading conditions. *International Journal of Heat and Mass Transfer* 28: 1327–1336.

———. 1987a. Heating of powders in an R.F. inductively coupled plasma under dense loading conditions. *Plasma Chemistry and Plasma Processing* 7 (1): 29–53.

———. 1990a. Loading and radiation effects in plasma jet modelling. *Journal de Physique Colloques* 51 (C5): 263–270.

———. 1991a. Modelling of the vaporization of small metallic particles in a D.C. plasma jet. *Heat Transfer in Thermal Plasma Processing* 161: 155–160.

———. 1991b. Radiative effects in ICP modelling. *International Journal of Heat and Mass Transfer* 31 (10): 2571–2579.

Reist, P.C. 1993. *Aerosol science and technology.* New York: McGraw Hill.

Rudinger, G. 1980. Fundamentals of gas-particle flow. In *Handbook of powder technology*, ed. J.C. Williams and T. Allen, vol. 2. Amsterdam: Elsevier.

Sayegh, N.N., and W.H. Gauvin. 1979. Analysis of variable property heat transfer to a single sphere in high temperature surrounding. *AICHE Journal* 25: 522–534.

Schwier, C. 1986. *Plasma spray powders for thermal barrier coating. Advance thermal spray*, 277–286. Pergamon Press.

———. 2005. In-flight oxidation of stainless steel in plasma spraying. *Journal of Thermal Spray Technology* 14 (1): 177–124.

Shaw, K.G., K.P. McCoy, and J.A. Trogolo. 1994. *Fabrication of composite spray powders using reaction systems, NTSC-1994: Industrial applications*, ed. C.C. Berndt and S. Sampath, 509–514. Materials Park: ASM International.

Smith, R.W., and Z.Z. Matasin. 1992. Reactive plasma spraying of wear-resistant coatings. *Journal of Thermal Spray Technology* 1 (1): 57–63.

Soo, S.L. 1967. *Fluid dynamics of multiphase systems.* New York: Blaisdell.

Taneda, S. 1956. Flow past a sphere. *Journal of the Physical Society of Japan* 11 (10): 1104.

Tsunekawa, Y., I. Ozdemir, and M. Okumiya. 2006. Plasma Sprayed Cast Iron Coatings Containing Solid Lubricant Graphite and h-BN Structure. *Journal of Thermal Spray Technology* 15 (2): 239–245.

Uglov, A.A., and A.G. Gnedovets. 1991. Effect of particle charging on momentum and heat transfer from rarefied plasma flow. *Plasma Chemistry Plasma processing* 11 (2): 251–267.

Vardelle, M., A. Vardelle, P. Fauchais, and M.I. Boulos. 1983. Plasma-particle momentum and heat transfer, modelling and measurements. *AIChE Journal* 29: 236–243.

———. 1988a. Particle dynamics and heat transfer under plasma conditions. *AIChE Journal* 34 (4): 567–573.

Vardelle, M., A. Vardelle, A. Denoirjean, and P. Fauchais. 1990. Heat treatment of zirconia powders with different morphologies under thermal spray conditions. In *MRS Spring meeting proceedings*, ed. D. Apelian and J. Szekely, vol. 190, 175–183. MRS.

Vardelle, A., M. Vardelle, P. Fauchais, P. Proulx, and M.I. Boulos. 1992 *Loading effect by oxide powders in DC plasma jets, NTSC-1992: International advances in coatings technology*, ed. C.C. Berndt, 543–548. Materials Park: ASM International.

Vardelle, M., A. Vardelle, K.-I. Li, P. Fauchais, and N.J. Themelis. 1996. Coating generation: Vaporization of particles in plasma spraying and splat formation. *Pure and Applied Chemistry* 68 (5): 1093–1099.

Vardelle, A., N.J. Themelis, B. Dussoubs, M. Vardelle, and P. Fauchais. 1997a. Transport phenomena in thermal plasmas. *Journal of High Temperature Material Processes* 1 (3): 295–317.

Vardelle, A., P. Fauchais, and P. Fauchais. 1997b. *Vaporization and ultra-fine particle generation during the plasma spraying processNTSC-1997: A United Forum for scientific and technological advances*, ed. C.C. Berndt, 543–548. Materials Park: ASM International.

Vardelle, A., P. Fauchais, B. Dussoubs, and N.J. Themelis. 1998. Heat generation and particle injection in a thermal plasma torch. *Plasma Chemistry and Plasma Processing* 18 (4): 551–574.

Vardelle, M., A. Vardelle, P. Fauchais, K.-I. Li, B. Dussoubs, and N.J. Themelis. 2001. Controlling particle injection in plasma spraying. *Journal of Thermal Spray Technology* 10 (2): 267–284.

Vardelle, A., M. Vardelle, H. Zhang, N.J. Themelis, and K. Gross. 2002. Controlling particle injection in plasma spraying. *Journal of Thermal Spray Technology* 11 (2): 244–284.

Volenik, K., J. Leitner, F. Hanousek, J. Dubsky, and B. Kolman. 1997. Oxides in plasma-sprayed chromium steel. *Journal of Thermal Spray Technology* 6 (3): 327–334.

Volenik, K., P. Chraska, J. Dubsky, J. Had, J. Leitner, and O. Schneewein. 2003. Oxidation of Ni-based alloys sprayed by a water-stabilized plasma gun (WSP). In *Thermal Spray-2003; Advancing the science and applying the technology*, ed. C. Moreau and B. Marple, 1033–1041. Materials Park: ASM Int.

White, F.M. 1974. *Viscous fluid flow.* McGraw Hill.

Xi, Chen, and E. Pfender. 1982a. Unsteady heating and radiation effects of small particles in a thermal plasma. *Plasma Chemistry and Plasma Processing* 2: 293–316.

———. 1982b. Heat transfer to a single particle exposed to a thermal plasma. *Plasma Chemistry and Plasma Processing* 2 (2): 185–212.

Xue, S., P. Proulx, M.I. Boulos. 2001. Extended − field electromagnetic model for the inductively coupled plasma. *Journal of Physics D: Applied Physics* 34: 1897–1906.

Xue, S., and M.I. Boulos. 2019a. Transient heating and evaporation of metallic particles under plasma conditions. *Journal of Physics D: Applied Physics* 52: 454002. (12 pages).

Xu, D.-Y., X.-C. Wu, and Chen Xi. 2002. Motion and heating of non-spherical particles in a plasma jet. *Surface and Coatings Technology* 171 (1–3): 149–156.

Ye, R., P. Proulx, and M.I. Boulos. 2000. Particle turbulent dispersion and loading effects in an inductively coupled radio frequency plasma. *Journal of Physics D: Applied Physics* 33: 2154–2162.

Abbreviations

CFD	Computational Fluid Dynamic
DC	Direct Current
D-gun	Detonation Gun
DSD	Droplet Size Distribution
FTIR	Fourier transform infra-red
GLR	Gas-to-Liquid mass Ratio
HA	Hydroxyapatite
HVAF	High Velocity Air Fuel
HVOF	High Velocity Oxygen Fuel
i.d.	internal diameter
LDA	Laser Doppler Anemometry
LES	Large Eddy Simulation
LHS	Left-Hand Side
LTE	Local Thermodynamic Equilibrium
PIV	Particle Image Velocimetry
PSD	Particle Size Distribution
PSI-Cell	Particle-Source-In-Cell
PTA	Plasma Transferred Arc
PVD	Physical Vapor Deposited
RF	Radio Frequency
RHS	Right-Hand Side
slm	standard liter per minute
SMD	Sauter Mean Diameter
SPPS	Solution Precursor Plasma Spraying
SPS	Suspension Plasma Spraying
TS	Thermal Spray
XRD	X-Ray Diffraction
YSZ	Yttria Stabilized Zirconia

5.1 Introduction

As discussed in Chaps. 1 and 2 of this book, thermal spray coatings are formed through the injection of powders or liquids as solutions or suspensions into a high energy source where the individual powder particles are entrained by the high energy fluid, heated and melted prior to their impact on the substrate forming the coating. In the case of using liquids and suspensions as precursor to the spraying operation, the process involves the added steps of liquid or suspension atomization and solute/liquid evaporation prior to the heating and melting of the solid particle grain produced in the process. When using wires, rods or cords, as in "wire arc spraying" process, the initial wire or cords are heated and melted by the arc followed by the atomization of the molten material and the entrainment of the formed molten droplets by the flow where they are projected against the substrate forming the coating. In all of these cases the basic phenomena controlling the process is the inflight momentum and energy exchange between the free stream combustion or plasma gases and the entrained particles or droplets. In Chap. 4, "Plasma–particle momentum and heat transfer" a review was presented of fundamental phenomena governing the inflight interaction between a spherical droplet/particle and the heating medium during their short contact time. The topics covered included, momentum and heat transfer, transient heating and melting of the particles, particle evaporation and possible chemical reaction between the molten particle material and the ambient gas.

In the present chapter, attention is given to particle/powder injection into the plasma or combustion flow as illustrated in Fig. 5.1 for the case of powder injection in a DC plasma torch. Particle trajectory and temperature history calculations are discussed together with the inflight transformations taking place in the particles as they melt, and possibly partially or fully vaporize and/or react with the ambient gases. These are treated in the first place assuming dilute loading conditions that represents the limiting situations where the mass feed rate of the precursor particles/powder in the flow is sufficiently low so as not to have any impact on the flow and temperature fields. Attention is given, however, to the local cooling effect of the carrier gas used for the pneumatic transport and injection of the powder into the hot combustion or plasma stream. This is followed by a discussion of the

© Springer Nature Switzerland AG 2021
M. I. Boulos et al. (ed.), *Thermal Spray Fundamentals*, https://doi.org/10.1007/978-3-030-70672-2_5

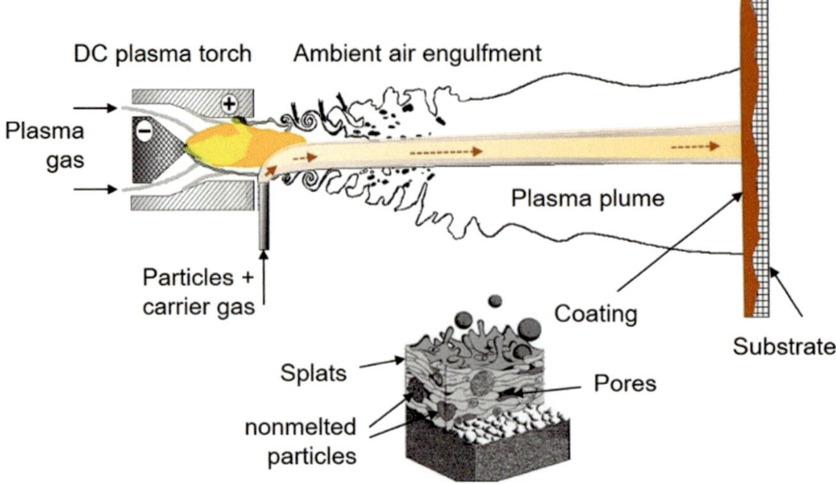

Fig. 5.1 Schematic representation of DC plasma spraying process through the particle injection into the base of a plasma jet

treatment of ensembles of particles/powder in gas flows introducing the formulation of stochastic models for a more realistic representation of a thermal spray process. Finally, in the last section of this chapter, the analysis is further expanded to the case of powder treatment under dense loading conditions in which the powder feed rate into the combustion or plasma source is sufficiently high in order to have a significant impact on the flow and temperature fields and consequently on the particle trajectories and their corresponding particle temperature history. Since such phenomena have been mostly observed in RF induction plasma spraying and/or powder spheroidization processes, the analysis will be focused on such systems though the phenomena and the proposed approach are equally valid in all combustion and thermal plasma spray operations under high precursor feed rate conditions.

5.2   Particle Injection in Plasma Spray

5.2.1   Design Considerations of Particle Injection Systems

Considering that most of the thermal spray applications use precursors in powder form, the discussion in this section will be focused on the transport and injection of pounders into the combustion or plasma source. A separate analysis will be given in Sect. 5.2.2 for liquid and suspension injection and atomization in thermal spray operations. Considering that the use of wires, rods, or cords is only applicable to "Wire Arc Spraying" technology, the feeding of wires and atomization of wires and cords will be treated separately in Chap. 10 of this book.

While the pneumatic transport of powders is a well-known technology that is widely used in the mining and chemical process industries, its adaptation to thermal spraying is rather delicate. This is due to the need to exercise a close control on the injection conditions of the powders into the hot plasma or combustion stream, while keeping the flow rate of the required transport gas at a strict minimum to avoid the excessive local cooling of the plasma or combustion gases at the point of powder injection. A common solution to this problem is the use of relatively small diameter cylindrical powder injection tubes as shown in Figs. 5.1 and 5.2a. The internal diameter of such powder injectors can vary between 1.5 and 3 mm, with powder carrier gas flow rates typically in the range of 2–10 slm. An example of such a powder injector configuration is given in Fig. 5.3, for the Oerlikon-Metco F4MB-XL plasma spraying torch. Two options are illustrated in this case, involving in Fig. 5.3a, the radial injection of the powder at an angle of 90 ° to the torch axis, or as illustrated in Fig. 5.3b the tilting of the injector by a small angle toward the upstream direction of the flow. Coaxial powder injection of the powder into the plasma flow is also possible in DC plasma sources such as the Mettech Axial III torch shown in Fig. 5.4 and is standard practice in inductively coupled RF induction plasma torches such, Tekna PL-50 torch shown in Fig. 5.5.

In certain cases, such as with the radial injection of the powder shown in Figs. 5.2 and 5.3, it might be desirable, in order to facilitate the handling of the torch by the operator, to bend the injector line toward the upstream end of the torch using a curved injector tube shown in Fig. 5.2b. In such cases it is necessary to maintain a minimum length, L_2 of the injector tube in the radial direction in order to insure the uniform distribution of the powder into the injector. It is to be noted that because of the centrifugal forces, the powder will be projected in the bend region toward the outer region of the bend and would normally require a distance equivalent to around 20 tube diameters or more for the complete redistribution of the powder in the flow.

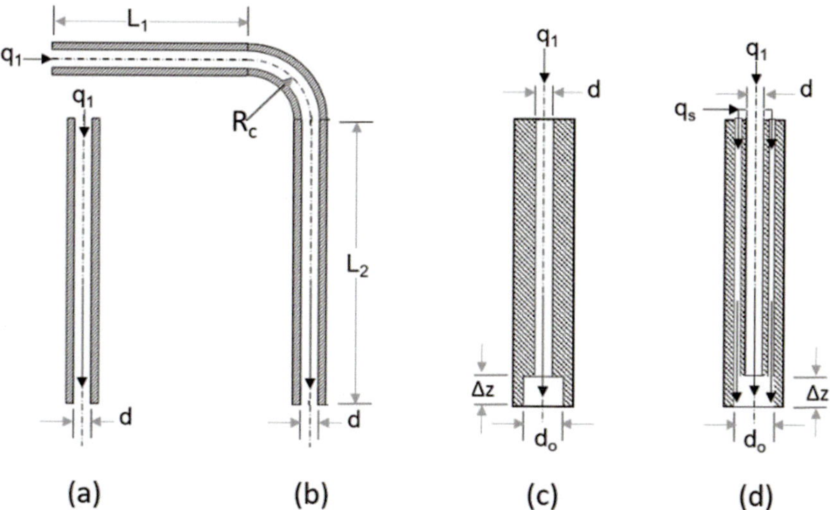

Fig. 5.2 Injector geometries typically used for radial injection (**a**) straight injector, (**b**) curved injector, (**c**) double diameter injector, and (**d**) double-flux injector [Vardelle et al. (2002) and Bronet et al. (1990)]

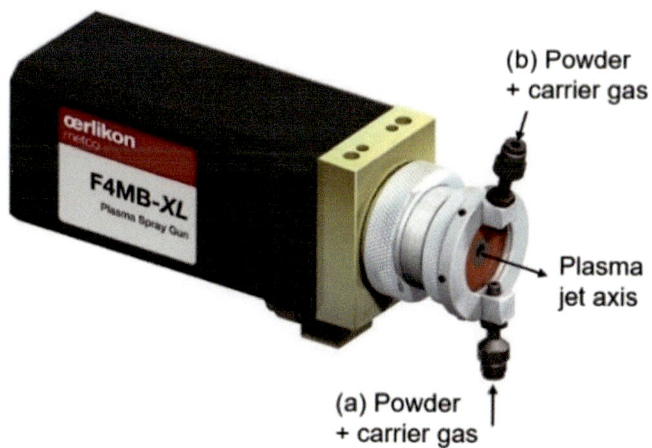

Fig. 5.3 Powder injector arrangements in the Oerlikon Metco F4MB-XL torch showing two injector orientations (**a**) radial injector at 90° to the torch axis (**b**) radial injector tilted to the upstream direction of the flow. [Reproduced with kind permission of Oerlikon-Metco Corp.]

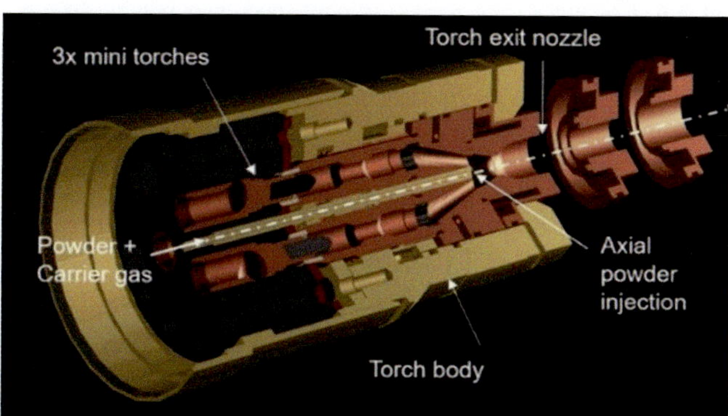

Fig. 5.4 Coaxial powder injector arrangements in the Mettech Axial III plasma torch. [Reproduced with kind permission of Mettech Corp.]

In all cases, however, the individual powder particles will be bouncing continuously against the inner surface of the tube against which they lose a significant portion of their inertia with the direct consequence of the slowing down of the particles transport velocity compared to the mean velocity of the carrier gas. The phenomenon was investigated by Vardelle et al. (2001) who computed the velocity distribution of ZrO_2 powder with a particle size of 5–60 μm, in a straight and curved powder injection tube with an internal diameter of 1.75 mm and argon as carrier gas at a flow rate of 4 slm. The mean transport gas velocity in the tube was 27.7 m/s. The results given in Fig. 5.6 show that in a straight tube the mean particle transport velocity was around 18 m/s that corresponds to about 65% that of the mean carrier gas. In the presence of a bend in the transport line, Fig. 5.2b, the particle velocity at the exit of the bend was considerably lower than that in a straight tube. The results given in the lower part of Fig. 5.6 show mean particle transport velocities in the range of 7–11 m/s corresponding to only 25–40% of the mean carrier gas velocity. The results also indicate a larger drop of the particle velocity for the case of a large radius of curvature of the bend ($R_c = 50.8$ mm) comparted to that obtained for a radius of curvature of $R_c = 12.7$ mm. The effect was attributed to the longer path of the particles along the inner wall of the tube in the case of the bend with a larger radius of curvature.

The slip velocity between the powder and carrier gas at the exit of a 2.2 mm i.d. straight powder injector, Fig. 5.2a, was investigated by Bronet et al. (1990). Measurements made using Laser Doppler Anemometry (LDA) were reported for the velocity of spherical glass and Nickel alloy powders with a particle size distribution in the range of 53 to 74 μm, a carrier gas flow rate of 12 slm (Air), and a powder feed rate of 3 and

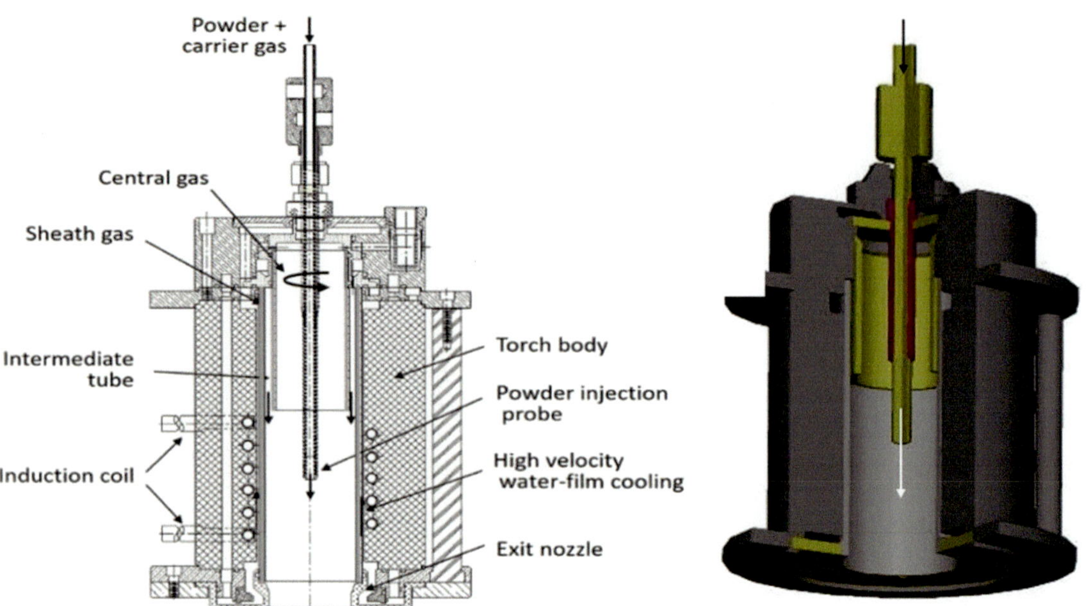

Fig. 5.5 Coaxial powder injector arrangements in the Tekna PL-50 Inductively coupled RF plasma torch [reproduced with kind permission of Tekna Plasma Systems Inc.]

7 g/min for the glass and nickel alloy powders respectively. The velocity profiles of the carrier gas at the exit of the injector were also measured by the same technique by seeding the carrier gas with a finely dispersed tracer talcum powder with a particle diameter less than 1 μm. The results given in Fig. 5.7a, b show respectively the radial profiles of the carrier gas and particle velocities at a location 5 mm downstream of the exit level of the probe for a free jet at atmospheric pressure (101 kPa) and under soft vacuum (20 kPa). It may be noted from Fig. 5.7a that the carrier gas velocity is considerably higher than that of the transported particles with the slip between the gas and particle velocity smaller for the glass powder compared to that of the Nickel alloy powder that reflect the significant difference in the solid particle densities of these two materials (2500 kg/m^3 for the glass, compared to 8200 kg/m^3 for the Nickel). The mean carrier gas velocity calculated as 52.6 m/s corresponds in this case to 83% of centerline jet velocity that is typical for turbulent flow in a circular conduit. On the other hand, the maximum particle velocity at the center line of the flow corresponds only to 48% and 32% of the maximum carrier gas velocities (62 m/s) for the glass and nickel alloy powders respectively. When operating under soft vacuum conditions ($p = 20$ kPa), the slip velocity between the particles and the carrier gas is accentuated with the maximum particle velocities being 44% and 29.4% of the maximum carrier gas velocity (170 m/s) for the glass and nickel alloy powders.

In an attempt to reduce the slip velocity between the particles and the carrier gas at the point of powder injection into the plasma, Bronet et al. (1990) proposed a modified

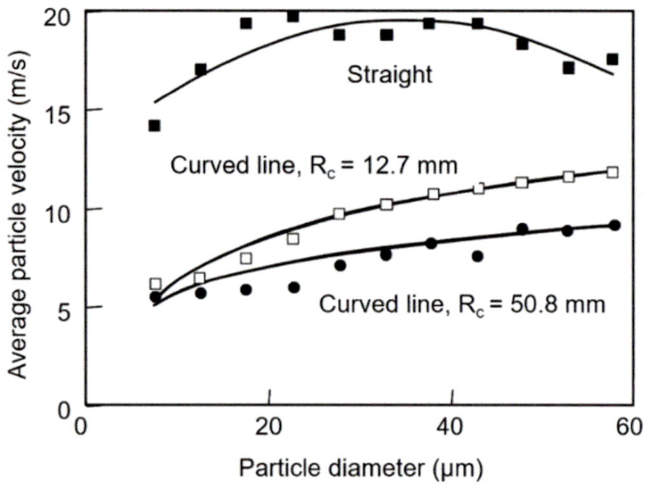

Fig. 5.6 Computed mean particle velocity function of particle diameter for ZrO$_2$ powder injected into a 1.75 mm i.d. powder injection tube with a straight and curved configuration; carrier gas flow rate: 4 slm, and L$_2$ = 35 mm [Vardelle et al. (2001)]

design of the tip of the injection probe as illustrated in Fig. 5.2c, in which the internal diameter of the injector is increased near the exit level of the probe for a short distance Δz that was varied between 0 and 40 mm. The results given in Fig. 5.8 were obtained for the same system dimensions and operating conditions under soft vacuum with a carrier gas flow rate of 12 slm (Air). The measured carrier gas velocity profiles are given in Fig. 5.8a, for $\Delta z = 20$ and 40 mm, and compared with those obtained in absence of this expansion ($\Delta z = 0$). The results reveal a significant drop of the carrier

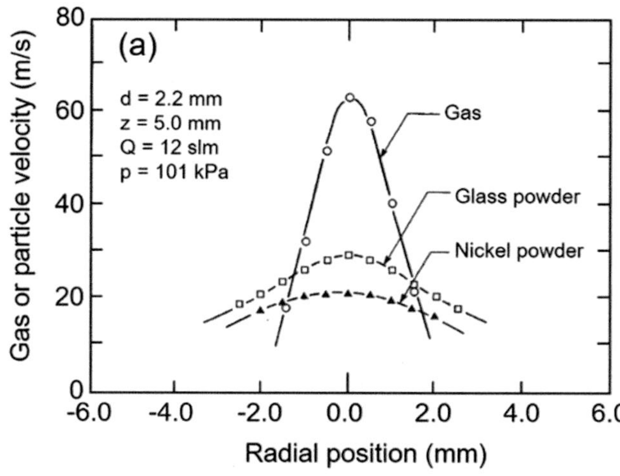
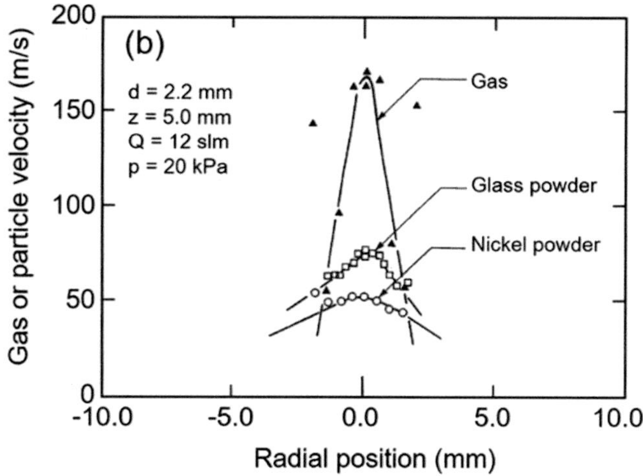

Fig. 5.7 Axial velocity profiles in the radial direction for the carrier gas, glass, and Nickel alloy powders at the exit level ($z = 5$ mm) of a standard 2.2 mm i.d. powder injection probe, carrier gas flow rate of 12 slm (Air) (**a**) atmospheric pressure, $p = 101$ kPa, and (**b**) soft vacuum, $p = 20$ kPa [Bronet et al. (1990)]

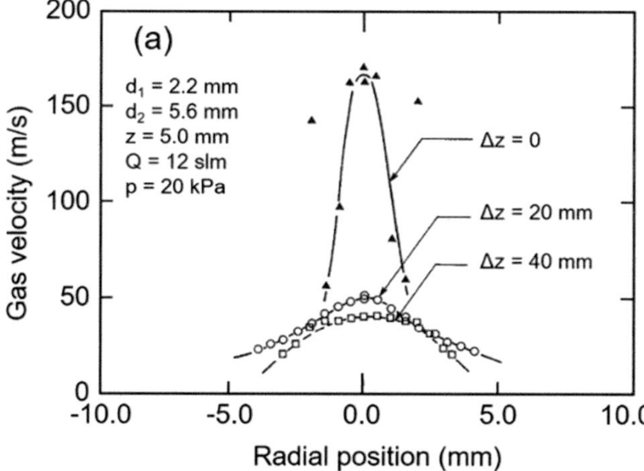
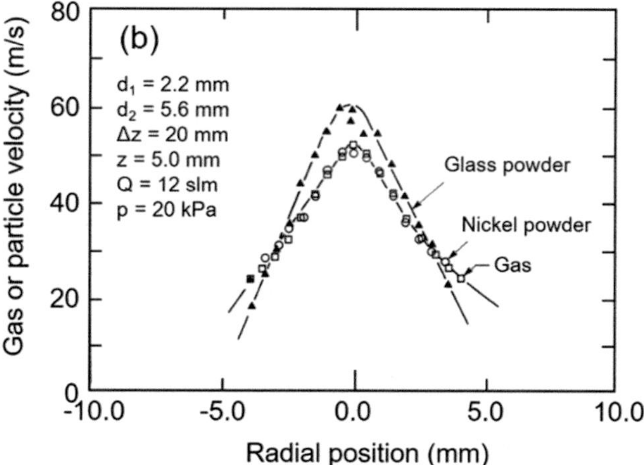

Fig. 5.8 Axial velocity profiles for powder injection under soft vacuum conditions using a double diameter probe design (Fig. 5.2c) $d_1 = 2.2$ mm and $d_2 = 5.6$ mm (**a**) the carrier gas velocity profiles for different length of diameter expansion zone $\Delta z = 20$ and 40 mm, and (**b**) carrier gas, glass and Nickel alloy powders velocity profiles at the exit level ($z = 5$ mm) for $\Delta z = 20$ mm. [Bronet et al. (1990)]

gas velocity by a factor of three in the presence of this short expansion zone at the end of the probe, with little difference observed between the profiles obtained for $\Delta z = 20$ mm compared to those for $\Delta z = 40$ mm. The presence of this expansion zone at the end of the injection probe does not seem to affect the particle velocity profiles as can be observed in Fig. 5.8b, where the velocity profiles of the carrier gas and those of glass and Nickel alloy powders are given for a probe with $\Delta z = 20$ mm. The three velocity profiles seem to almost overlap with the particle velocity slightly higher than that of the carrier gas.

Special attention was devoted to the prediction and control of the dispersion angle of the particle-laden jet immerging from the powder injector that can have a significant impact on the deposition efficiency of the plasma spraying operation

(Vardelle et al. (2001). The "double flux injector" shown in Fig. 5.2d represents a typical injector design composed of two concentric tubes allowing for the injection of a shroud gas in the annular space surrounding the central powder-carrying injector tube. As shown in Figs. 5.9 and 5.10 this type of injector can be efficiently used for the culmination of zirconia particles (45–5 µm) with shroud gas flow rate as low as 1 to 5 slm depending on the density and particle size distribution of the injected powder. It may be noted that the powder jet is almost twice less dispersed with double injector. The usefulness of such injectors is particularly important for small ($d_p < 20$ µm) light particles, which are difficult to collimate [Renouard-Vallet (2003)].

Alternately in the treatment of powders in an inductively coupled RF plasma torch, whether for powder

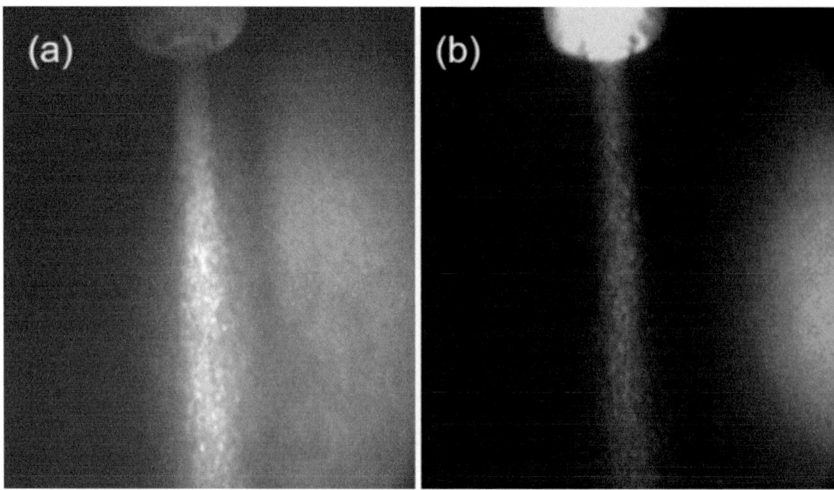

Fig. 5.9 Laser flashes photograph of zirconia particles (45–5 µm) (the exit of conventional and double flux injectors (**a**) Conventional injector (**b**) in the presence of shroud gas. [Reprinted with kind permission from ASM International, Renouard-Vallet (2003)]

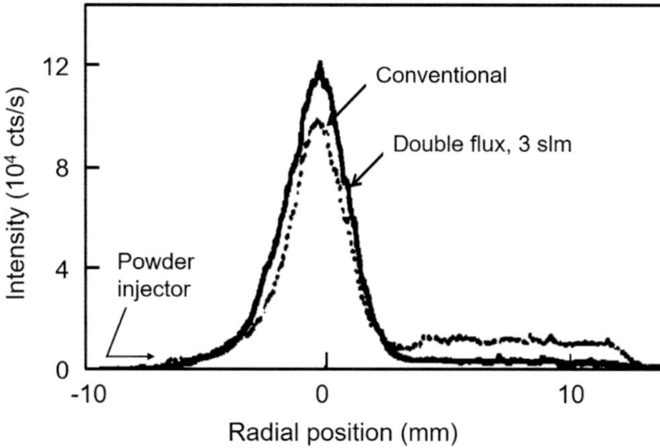

Fig. 5.10 Penetration of particles (zirconia 45–5 µm) into a DC plasma jet. Measurements performed at 20 mm downstream of the nozzle exit with conventional and double flux injectors. (Reprinted with kind permission from ASM International [Li et al. (1995)])

spheroidization of plasma spray operation, it might be advantageous to disperse the powder uniformly in the discharge cavity. For such applications, dispersion powder injection probes were developed through the introduction of spiral generator in the powder injection tube that depending on operating parameters can result in an increase of the dispersion of the powder jet as shown in Fig. 5.11.

5.2.2 Effect of Carrier Gas

Studies of the effect of powder carrier gas on the flow and temperature fields in a DC plasma jet were reported by [Vardelle et al. (2001)]. Calculations were performed for using the ESTET commercial fluid dynamic code in which the mass diffusion of species is assumed to behave similarly

to heat diffusion, that is, the effective diffusivity of each species is equal to its thermal diffusivity. In addition, the Schmidt turbulence number, Sc, (the ratio of momentum diffusivity to molecular mass diffusivity), is fixed at 0.7. The thermodynamic and transport properties of the gas mixture, consisting of the plasma-forming gas, the carrier gas, and the ambient atmosphere, are determined on the basis of the mixture laws using the properties of pure argon, air, and argon-hydrogen mixtures. A rectangular mesh is used, subdivided into a non-uniform $49 \times 49 \times 43$ grid. The length of the calculation domain is 100 mm and its radius of 21 mm is about 7 times the radius of the torch nozzle. At the two inlets of the domain, that is, at the plasma torch and injector exits, the values of all variables are defined.

Results given in Fig. 5.12a, b corresponding respectively to external and internal injection of the powders in the plasma torch. It may be noted that the effect of the powder carrier gas is more noticeable for internal injection Fig. 5.12b, compared to that for external injection Fig. 5.12a. The plasma forming gas in these two cases is a mixture of 27 slm (Ar) + 7 slm (H$_2$) and the carrier gas flow consists of 8 slm (Ar). The effective power dissipated in the gas is 13 kW, the plasma torch nozzle i.d. is 6 mm, and the injector i.d. is 1.6 mm. Under these conditions, the angles of deflection of the axis of the plasma jet with respect to the torch axis were calculated to be 6° and 2° for internal and external injection, respectively. For internal injection, the deflection of the jet is noticeable when the carrier gas mass flow rate exceeds 10% of the plasma gas mass flow rate.

In the case of plasma torches or combustion torches with axial injection such as the HVOF (Sulzer-Metco Diamond Jet gun) the carrier gas flow rate (nitrogen), while being very low compared to that of the fuel (propylene)-oxygen-air flow rate, 4.45×10^{-5} kg/s (2.14 slm) against 6.9×10^{-3} kg/s (331.2

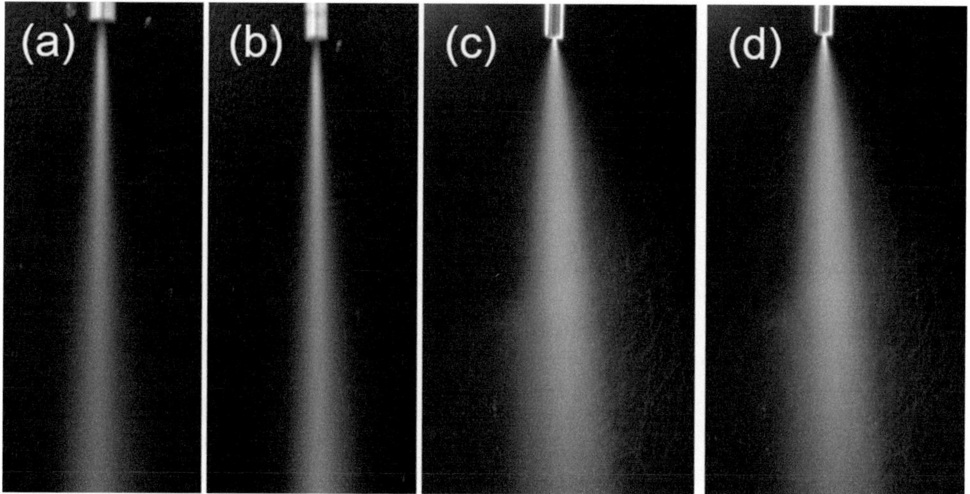

Fig. 5.11 Photographs of a powder-laden gas jet in a free discharge using different levels of powder dispersion conditions. [Photograph courtesy of Tekna Plasma Systems Inc., Sherbrooke, Québec]

slm), can perturb the flow close to the axis [Chen et al. (1995)]. With the increase of the nitrogen flow rate the flame temperature and velocity, especially the temperature, near the centerline of the nozzle, are decreased. With the flow rate values given above, the maximum gas temperature along the centerline is about 3000 K. When doubling the nitrogen flow rate, the maximum temperature decreases significantly to 1700 K and the velocity at the nozzle exit decreases from 1950 to 1800 m/s. Since nitrogen gas does not take part in the chemical reaction, it absorbs the thermal energy generated by the oxy-fuel reaction. As a result, the flame temperature and velocity are decreased, which, in turn, decrease the efficiency of the thermal spray device. Nevertheless, the calculated results show that it is wise to keep a minimum nitrogen flow rate to ensure smooth feeding of particles [Chen et al. (1995)] and get a maximum flame velocity and temperature over the entire particle trajectory.

The local cooling effect of the powder carrier gas is even more evident in the case of RF induction plasma sources that are characterized by their relatively low plasma gas velocity and laminar flow in the center of the discharge. Typical results given Figs. 5.13 and 5.14 show the temperature and flow fields in the discharge region of PL-35 Tekna Induction plasma torch, at a discharge power of 13 kW, for powder gas flow rates, $Q1 = 0$. 2 and 4 slm (Ar) in the absence of powder feed into the discharge at this stage. These are obtained for a torch with an internal diameter of the plasma confinement tube of 35 mm, in which powder would be centrally injected using SI-635 water-cooled injection probe with an internal channel diameter of 1.67 mm. The torch was operated at atmospheric pressure, with an Ar/H$_2$ mixture as sheath gas, $Q_3 = 34$ slm Ar +6 slm H$_2$. Central gas, $Q_2 = 12$ slm (Ar) and a powder carrier gas, $Q_1 = 0$–5 slm (Ar). The oscillator frequency was 5 MHz and the plate power of 20 kW (65%

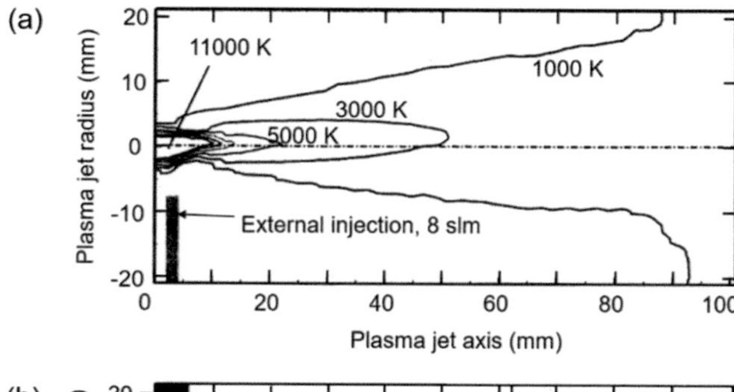

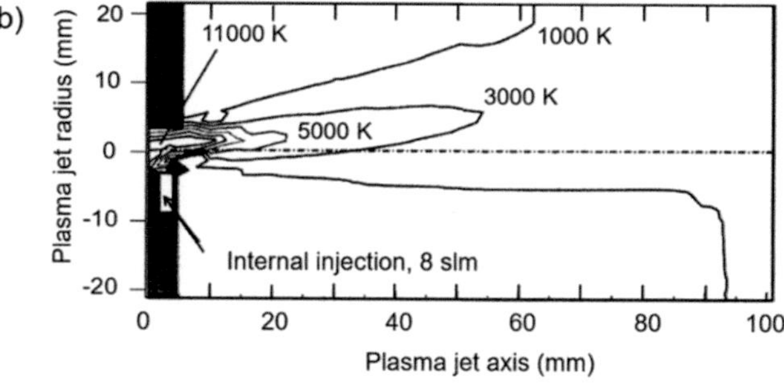

Fig. 5.12 Isotherms at intervals of 2000 K for (**a**) external powder injection (injector i.d. 1.8 mm, plasma forming gas: 27 slm Ar + 7 slm H$_2$, carrier gas flow rate 8 slm, $I = 500$ A, V = 65 V, thermal efficiency 56% and plasma anode-nozzle i.d. 6 mm), (**b**) internal injection. [Vardelle et al. (2001)]. Reprinted with kind permission from Springer Science Business Media, copyright © ASM International

coupling, 13 kW coupled into the discharge). The results given in Fig. 5.13 are presented in terms of 2-D temperature isocontours (Figs. 5.13a, c, e) with the corresponding streamlines of the flow in (Figs. 5.13b, d, f) for powder carrier gas flow rates of $Q_1 = 0$, 2, and 4 slm (Ar). It should be stressed, however, that these results were obtained with only the carrier gas injected into the flow in the absence of any powder feed. The zone of local cooling at the tip of the

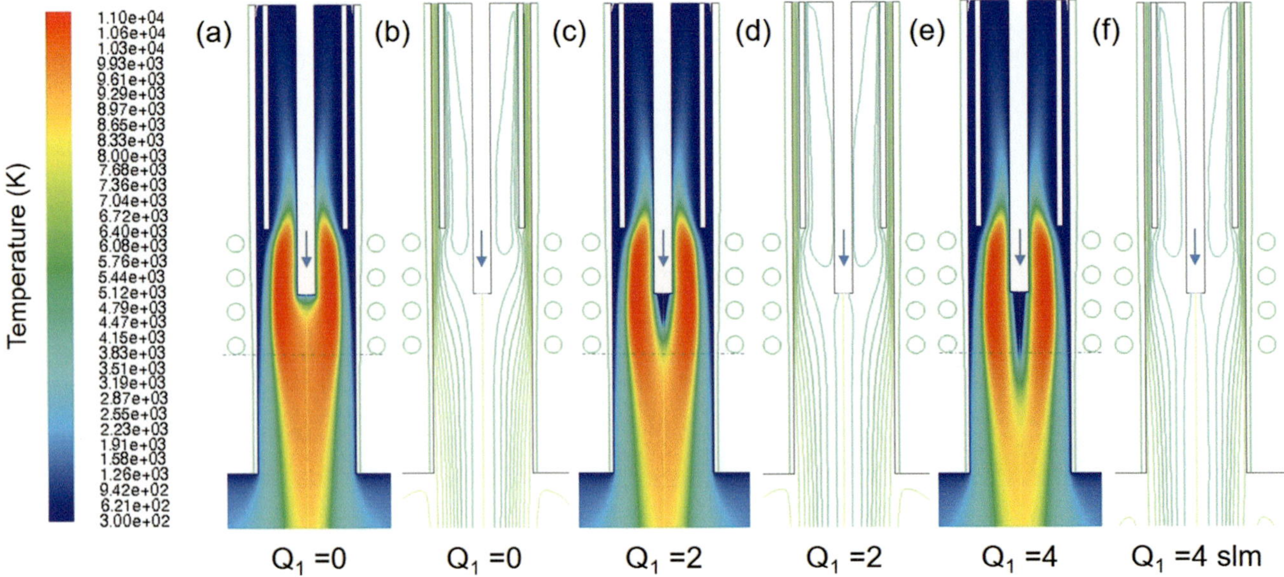

Fig. 5.13 Temperature and flow fields in the discharge region of a PL-35 Induction plasma torch in the presence of increasing levels of axial powder injection carrier gas flow rate $Q_1 = 0–5$ slm, Coupled power $= 13$ kW [Computations courtesy of Dr. Siwen Xu, Tekna Plasma System Inc.]

injection probe is noted to increase gradually with the increase of the gas flow rate.

The corresponding temperature profiles along the centerline of the torch given in Fig. 5.14 are shown the displacement of the temperature rise in front of the probe tip with the increase of the carrier gas flow rate by a distance of more than 40 mm, together with a drop of the maximum centerline temperature by almost 2000 K. The corresponding radial temperature profiles at the downstream level of the induction coil are given in Fig. 5.15. These show that the local cooling effect of the carrier gas, to be relatively negligible at low carrier gas flow rates of 1–2 slm, increasing rapidly with the increase of the carrier gas flow rates with its effect felt over a 10–12 mm diameter region in the center of the discharge that is significant considering that the diameter of the discharge is about 25 mm. To overcome this cooling effect, dispersion probes such as those shown in Fig. 5.11 have been used in order to spread the trajectory of the injected particles over a larger volume of the discharge for superior powder processing efficiencies.

5.3   Suspension or Solution Injection into Plasma Flows

Over the past two decades, there has been a growing interest in the plasma spraying of nano-structured coatings that were identified for their superior thermomechanical properties compared to standard micron-sized structured coatings. Being generally thinner than conventional thermal spray coatings with a finer grain size nano-structured coatings

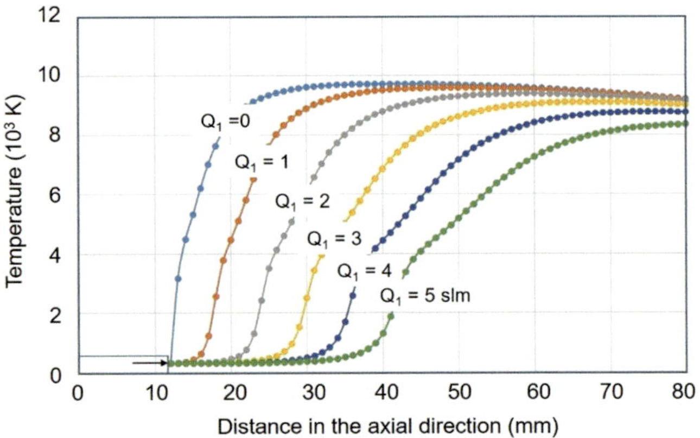

Fig. 5.14 Axial temperature profiles along the centerline of the discharge for a PL-35 Induction plasma torch in the presence of increasing levels of axial powder injection carrier gas flow rate $Q_1 = 0–5$ slm, coupled power $= 13$ kW [Computations courtesy of Dr. Siwen Xu, Tekna Plasma System Inc.]

bridge the gap between Thermal-Sprayed (TS) coatings and Physical Vapor Deposited (PVD) coatings [Fauchais et al. (2008)]. The principal challenge in producing such coatings by injecting nanopowders with a carrier gas into a plasma flow is that nanoparticles have a considerably lower mass than micron-sized particles, and accordingly have a significantly lower inertia for the same particle injection velocity. Moreover, as soon as the nano-sized particles come in direct contact with a plasma flow, they almost instantly melt and vaporize. Alternately, pre-agglomeration of nanoparticles in the form of micron-sized particles are a viable option that overcomes the challenge of nanoparticle injection into the

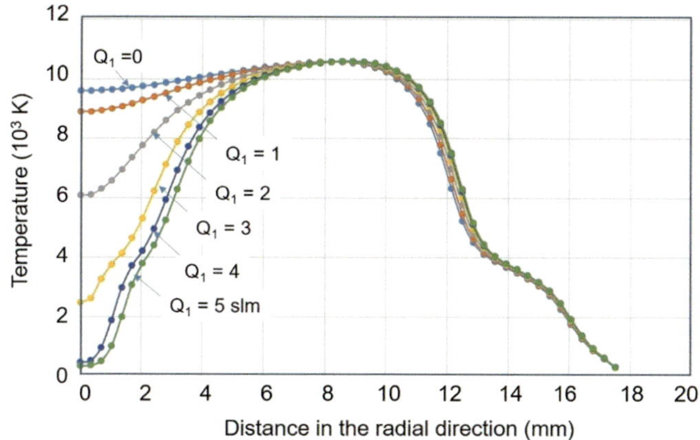

Fig. 5.15 Radial temperature profiles at the downstream end of the induction coil, dotted line on Fig. 5.13, for PL-35 Induction plasma torch in the presence of increasing levels of axial powder injection carrier gas flow rate $Q_1 = 0\text{–}5$ slm, coupled power $= 13$ kW [Computations courtesy of Dr. Siwen Xu, Tekna Plasma System Inc.]

combustion stream or plasma flow. They require, however, avoiding their complete in-flight melting in order to maintain their nanostructure.

Suspension and solution spraying offer an alternate route that has gained wide acceptance over the past decade. In the case of Suspension Plasma Spraying (SPS) the precursor material is fed into the combustion flame or plasma in the form of a suspension of a fine or ultrafine powder mixed with an appropriate liquid. As the suspension is injected into the flow it will be atomized or fragmented forming fine droplets that are entrained by the hot gases. On further heating of the formed droplets, the liquid in the droplets is vaporized leaving behind a dispersed or partially agglomerated pack of fine or ultrafine particles that subsequently melt on further heating. On impact of the molten droplets with the substrate they form of micro-splats, a few micrometers in diameter, which are the building blocks of the coating. Solution Precursor Plasma Spraying (SPPS), on the other hand, is based on the complete dissolution of the coating precursor in an appropriate liquid that is injected/atomized into the high temperature combustion or plasma stream. As the liquid is vaporized the dissolved solid precipitates as granules in the partially vaporized droplets. On complete vaporization of the droplets the residual solid grain melts/crystalizes forming on impact with the substrate a coating composed of a blend of crystals and splats.

In either cases of suspension or solution plasma spraying the controlled injection and of the liquid or suspension into the high temperature combustion or plasma flow represents a critical step on which the quality of the formed coating strongly depends. Basically, two approaches have been widely used:

- **_Spray Atomization_** through the use of an atomizing medium such as a compressed gas that supplies the energy

required for the atomization of the liquid or suspension. Also referred to as pneumatic or gas or two-fluid atomization.
- **_Mechanical Atomization_** where the liquid or suspension is pressurized and pushed through an appropriate nozzle resulting its fragmentation into small droplets. This approach is also used to provide the liquid/suspension with the necessary momentum energy in order to penetrate the combustion or plasma flow. On reaching to core high velocity region of the flow the liquid is fragmented before vaporization using the knetic energy of the plasma. Mechanical atomization can also be accomplishhed using a vibrating device or ultrasound source.

Whatever the liquid/suspension injection mode used, the main questions that need to be addressed are: how to control the atomization or fragmentation process? and what happens when the injected liquid droplets interact with the plasma flow?

5.3.1 Gas Atomization

Gas Atomization has been used for a long time for suspensions and solutions for a wide range of industrial applications including powder manufacturing with many textbooks written on the field such as Lefebvre (1989). The handbook by Neikov et al. (2019) is of interest since it devotes one chapter to the atomization and granulation of molten metals using a wide range of technologies such as: water or oil atomization, gas atomization, and centrifugal atomization. While none of these techniques are directly applicable for plasma spraying of liquids or suspensions, the basic concepts of gas or high-pressure mechanical atomization are relevant to the design of the liquid suspension injection device that could be used in combustion and plasma spray systems.

Early studies in this field have identified the following principal parameters that control the sizes and velocities of drops exiting the atomization nozzle [Filkova and Cedik (1984), Bouyer et al. (1996), Wang et al. (1998), Bouyer et al. (2004), Rampon et al. (2006), Toma et al. (2006a, b), Gell et al. (2008), Karthikeyan et al. (1997), Kassner et al. (2008), Marchand et al. (2011)].

- Atomization nozzle design.
- Thermophysical properties of the liquid/suspension (soild to fluid ratio, density, surface tension, dynamic viscosity).
- Gas-to-Liquid mass Ratio (GLR).

Basically, for liquids with a viscosity between a few tenths to a few tens of mPa s, their atomization, or break-up, into drops depends on the Weber number, which represents the ratio of the force exerted by the atomizing gas on the liquid, to the surface tension of the liquid:

$$We = \frac{\rho_g . U_r^2 . d_l}{\sigma_l} \qquad (5.1)$$

where ρ_g is the gas mass density (kg/m^3), U_r the relative velocity gas-liquid (m/s), d_l the drop or jet diameter (m), and σ_l (N/m) the surface tension of the liquid. It is generally accepted that fragmentation of the liquid occurs when We >12–14. This means that, for a liquid with a given surface tension, atomization will strongly depend on the relative gas velocity and specific mass of the atomizing gas.

According to Rampon et al. (2008) and Marchand et al. (2011) atomization also depend, but to lesser extent, on the Ohnesorge number, Oh, which is defined as:

$$Oh = \frac{\mu_l}{\sqrt{\rho_l . d_l . \sigma_l}} \qquad (5.2)$$

where μ_l is the viscosity of the liquid (Pa s), ρ_l the liquid mass density (kg/m^3), d_l the drop or jet diameter (m), and σ_l (N/m) the surface tension of the liquid.

In spray or gas atomization, the liquid or suspension is fragmented into small droplets through the action of high-pressure gas stream. As show in Fig. 5.16, a broad range of nozzle designs have been developed using either internal or external mixing of the gas with the liquid stream. In the internal-mixing design, Fig. 5.16a, the contact of the gas with the liquid/suspension takes place inside the nozzle with the shearing of the liquid/suspension taking place at the nozzle exit where the gas and liquid co-exit under high shear conditions. On the other hand, in the external mixing design, illustrated in Fig. 5.16 b&c, both liquid/suspension and gas streams follow separate routes until their point of contact that can be either in a co-flowing mode, or at right angle (90°). While each of these approaches have their individual advantages and limitations, they generally offer, compared to mechanical atomization, greater flexibility for the independent control of the liquid feed rate and droplet size distribution.

An example of the range of droplet size distribution that can be obtained with right-angle external mixing atomizers is given in Fig. 5.17 after [Jordan et al. (2007)]. Three different types of atomizers were used: (a) a home-made capillary atomizer, (b) a narrow angle fan nozzle atomizer, (c) a transverse air blast atomizing nozzle with a relatively large spray angle, and (d) for comparison the size distribution with a nebulizer is also shown. It is apparent that the broadest droplet size distribution is for the air cap atomizer and the narrowest is for the capillary atomizer with the finest mean droplet sizes (d_{50} < 20 μm) Fig. 5.16c.

A comparative study of the characteristics of nozzles with internal and external mixing was reported by Marchand et al. (2011) for the atomization of Yttria Stabilized Zirconia (YSZ, 8 wt.% yttria) suspensions in water and alcohol. The mean particle diameter of the powder used was $d_{50} = 0.6$ μm. The suspension (20 wt.% solid) feed rate was varied between 25 and 50 ml/min, with an atomizing gas flow rate of 2–4 slm (Ar). The droplet size distribution was recorded at a distance of 6 mm from the exit of the atomizing nozzle. Typical results are given in Fig. 5.18 in terms of high-speed images and Droplet Size Distributions (DSD) of the atomized

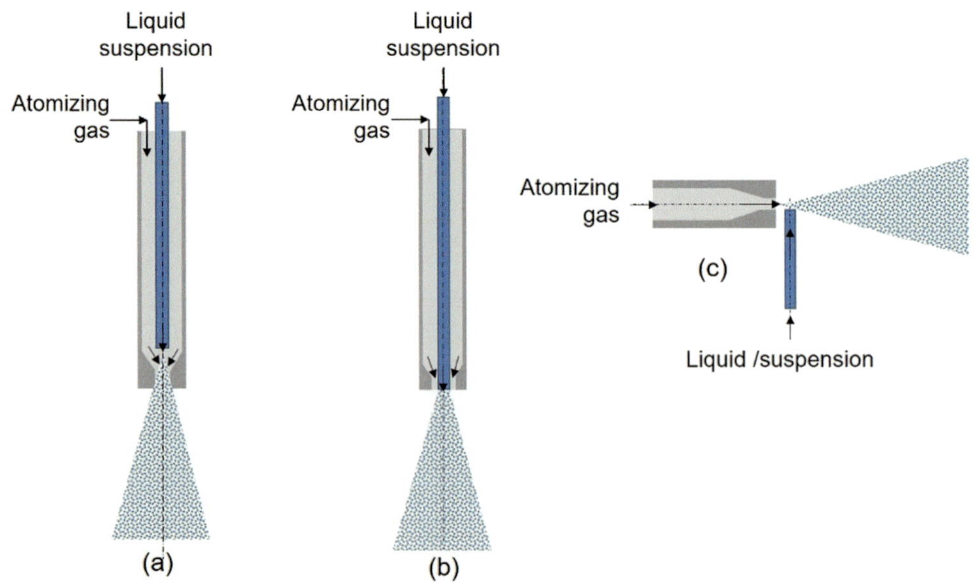

Fig. 5.16 Schematics of different gas atomization nozzle designs (**a**) internal mixing (**b** and **c**) external mixing

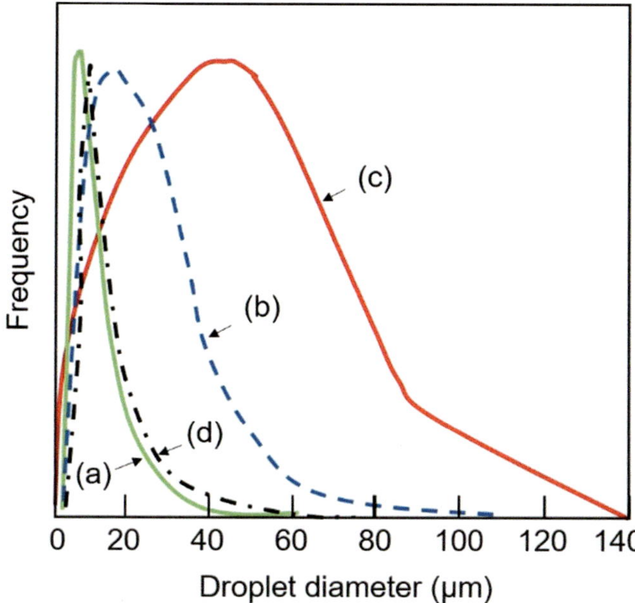

Fig. 5.17 Droplet size distributions for different atomizers. (**a**) Capillary atomizer, (**b**) fan nozzle, (**c**) transverse air blast atomizer, (**d**) nebulizer [Jordan et al. (2007)]

suspensions as function of the nozzle design and operating conditions. These show, for both external and internal mixing, a strong dependence of the level of atomization on the Gas to Liquid mass Ratio (GLR). The images given on the LHS of the figure, corresponding to external mixing, show for (Fig. 5.18a) GLR = 0.06, a stable jet that breaks up according to Rayleigh break-up mode into large drops at 20 mm from the nozzle exit. The corresponding DSD is given on the RHS of the figure, showing a mean particle diameter d_{50} = 350 μm. With the increase of the atomizing gas flow rate to GLR = 0.3, membrane-type ligaments are formed which on breaking up give rise to a finer spray with a broader size distribution and a d_{50} = 150 μm. By switching the atomizing liquid from water to alcohol, with a significant decrease in the surface tension of the liquid from 38 to 25 mN/m, a more efficient atomization is observed with the further decrease of the mean particle diameter to d_{50} = 90 μm. As observed in the corresponding DSD given on the RHS of the figure the droplet size distributions in this case is multi-modal. The corresponding results obtained using the nozzles with internal mixing, Fig. 5.18d–g, show essentially the same trends though the mead droplet diameter in the case of (g), being much finer with a d_{50} = 20 μm.

For more details about gas atomization see the book of Lefebvre (1989) where one can find, for example, the relationship between the Sauter Mean Diameter (SMD), the diameter of the discharge orifice of the atomizing probe, the GLR, and the liquid properties σ_l, ρ, ρ_l, μ [Fauchais et al. (2011)].

It must be stressed, however, that the atomization of the liquid or suspension is only an intermediate step in the thermal plasma spraying process since the challenge is to obtain the best uniform liquid atomization with a minimum amount of atomizing gas and the narrowest possible spray cone in order to facilitate the injection of the spray droplets into the combustion or plasma stream that is a requirement for the building of a uniform adherent coating on the substrate. An example of a gas-atomized ethanol stream and its interaction with a DC plasma jet are shown in Fig. 5.19a, b after [Wittmann et al. (2002)]. It is obvious in this case that none of the above-mentioned objectives are completely meet in terms of maximizing the percentage of the droplets that are entrained into the combustion or plasma stream and minimizing the impact on the local temperature field in the plasma jet.

With combustion or plasma sources allowing axial injection of the powder, liquid, or suspension into the discharge region, the problem of injection and dispersion of the atomized droplets is less acute since the configuration of the flow allows for easy access to the discharge zone. As an example of the technology developed in the nineties for the induction plasma spraying of Hydroxyapatite (HA) [Ca$_{10}$ (PO$_4$)$_6$ (OH)$_2$] coatings for biomedical applications is illustrated in Fig. 5.20 [Gitzhofer et al. (1997) and Bouyer et al. (1997a, b)]. The precursor in this case is composed of nano-sized, needle-shaped crystals of HA that was maintained in a stable aqueous suspension using an appropriate polymeric surface-active agent such as methyl polymetacrylate Darvan 7 (<1% by volume). Through dilution, the apparent viscosity of the suspension needed to be maintained below 0.8 Pa s in order to allow for its easy injection and atomization into the plasma. As the suspension is locally gas atomized at the exit of the water-cooled injection probe Fig. 5.20b, the generated droplets are entrained by the plasma flow in which they go through a series of transformations as shown in Fig. 5.20c. These involve in the first place the evaporation of the water-content of the droplets followed by flash sintering and, fusion of the formed granules. On impact of the molten, or partially molten, particles on the substrate, they form splats that are the building blocks of the coating. Obviously, the microstructure of the coating will depend on the degree of melting of the particles prior to their impact on the substrate and the temperature history of the coating during the spraying process.

The interaction of the atomizing probe gas with the central discharge can be observed in Fig. 5.21a, b that shows photographs of a laboratory scale induction plasma torch, 50 mm i.d., operated using argon as plasma gas at a frequency of 3 MHz, and nominal power of 10 kW in which a probe gas seeded with a nano-sized zirconia powder in injected at flow rates of Q_l = 1 and 5 slm (Ar). It can be noted with the increase of the probe gas a central channel is formed that is

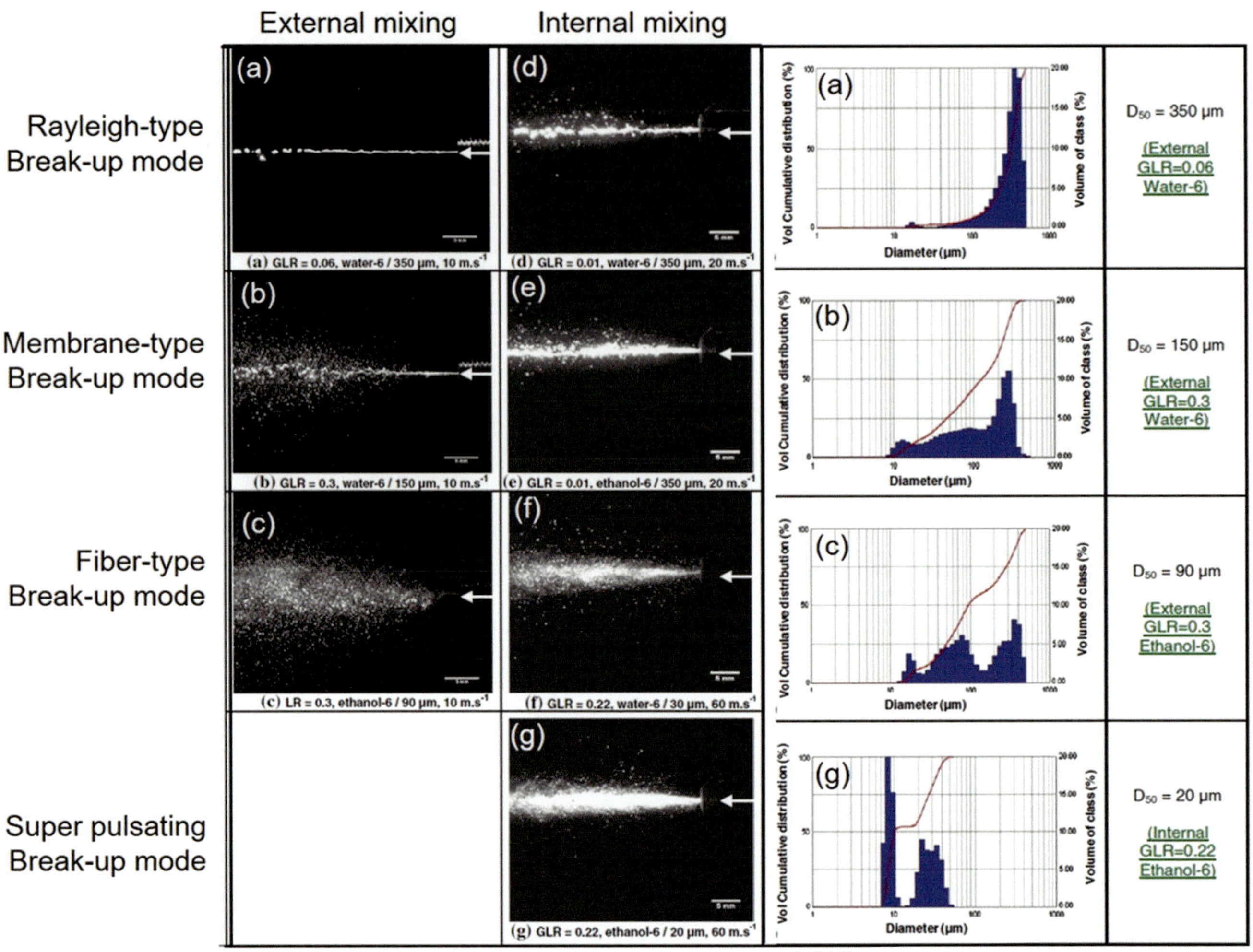

Fig. 5.18 Images and droplet size distributions (DSD) of atomized YSZ suspension in water and alcohol for nozzles with external mixing (**a, b,** and **c**) and internal mixing (**d, e, f,** and **g**). DSD given only for the external mixing nozzles [Marchand et al. (2011)]

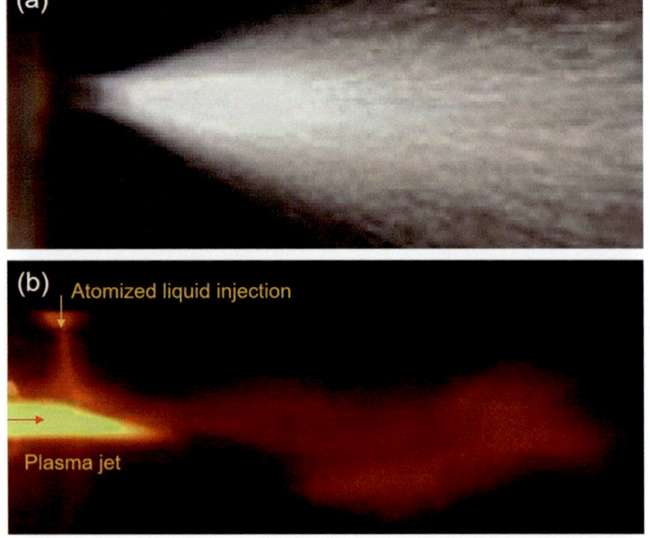

Fig. 5.19 (**a**) Atomized ethanol jet (**b**) Interaction of atomized ethanol flow with an Ar-H$_2$ (45–15 slm) DC plasma jet (I = 600 A, anode-nozzle i.d. 7 mm [Wittmann et al. (2002)]

defined by the bright emitting seed powder flowing through the central region of the discharge bending completely with the plasma gas at the downstream discharge end of the torch.

5.3.2 Mechanical Atomization

Mechanical injection is achieved by using a pressurized container in which a liquid or suspension is stored and forced through a nozzle of internal diameter, $d_n = 50$–300 μm. At the injector exit, a liquid jet is generated with a diameter is about 1.2 to 1.5 time the nozzle diameter, d_n depending on the tank pressure and nozzle shape [Delbos et al. (2006), Etchart-Salas (2007) and Fauchais et al. (2008)] (Fig. 5.22a). After a length of about 100–150 times d_n Rayleigh–Taylor instabilities lead to fragmentation of the jet into droplets with a diameter of about 1.3–1.6 times that of the jet, which are entrained by the plasma flow as illustrated in Fig. 5.22 b, c. m_l^0 is the mass flow rate of the liquid injected into the plasma flow, d_n is expressed as

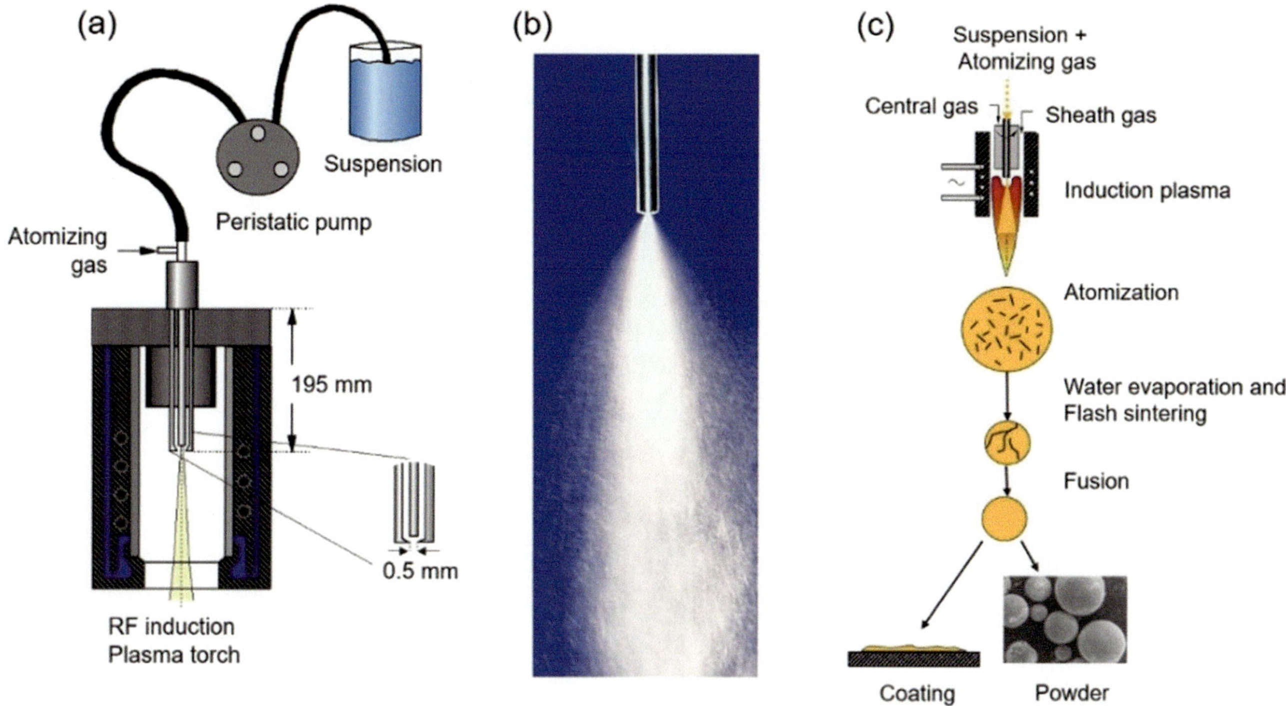

Fig. 5.20 (**a**) Basic concept of suspension plasma spraying for an induction plasma torch (**b**) Image of a gas atomization suspension and (**c**) Schematic representation of the transformation steps of a suspension droplet through its trajectory in the plasma [Bouyer et al. (1997a, b)]

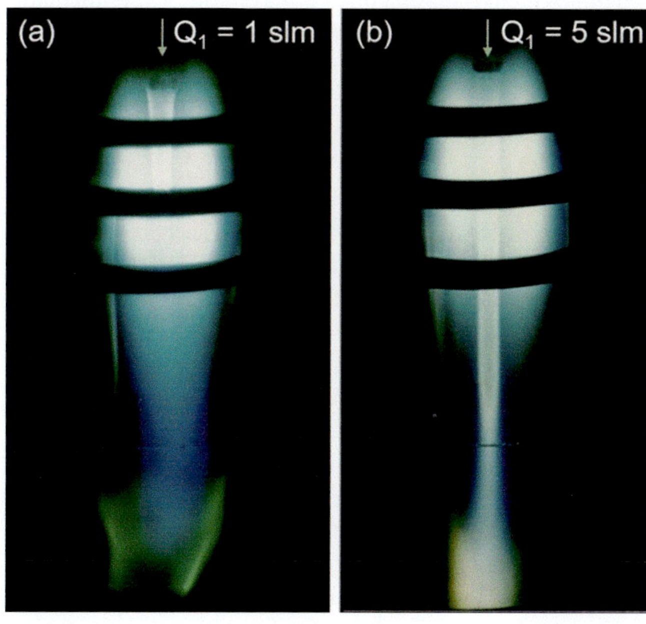

Fig. 5.21 Photographs of 10 kW induction plasma in the presence of two levels of probe gas injection seeded with fine zirconia powder (**a**) $Q_1 = 1$ slm (**b**) $Q_1 = 5$ slm [Courtesy of the Plasma Technology Research Center, University of Sherbrooke]

function of the nozzle diameter, ρ_l is liquid density, and v_l is the average velocity of the liquid in the nozzle [Fauchais et al. (2008)];

$$m_l^o = \frac{\pi d_n^2}{4}\, v_l \rho_l \tag{5.3}$$

The differential pressure, Δp across the nozzle is depicted as follows:

$$\Delta p = \frac{1}{2} f\, \rho_l v_l^2 \tag{5.4}$$

where f is the friction coefficient of the liquid in the nozzle that depends on the rheological properties of the liquid and the nozzle design. Typically for an aqueous solution at room temperature, flowing at a rate of 0.47 cm³/s, the pressure-drop across a 150 μm diameter nozzle will be expected to be in the range of 0.5 MPa. For a smaller nozzle diameter, $d_n = 50$ μm, the corresponding value of $\Delta p = 41$ MPa.

5.3.2.1 Liquid Penetration into the Plasma Flow

In conventional particle spraying the optimum particle trajectory is achieved when the injection force of the particle is about that imparted to it by the plasma jet, F_{ip}.

$$F_{ip} = s_p \rho v^2 \tag{5.5}$$

where s_p is the projected surface of the particle, $s_p = \pi d_p^2/4$ and ρv^2 the local momentum density of the plasma jet,

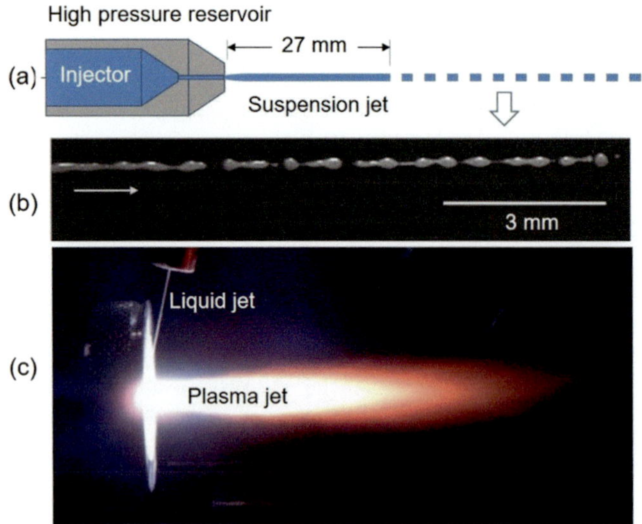

Fig. 5.22 (**a**) Schematic of a mechanical injector showing the transition distance after which the Rayleigh–Taylor instability develops forming the liquid droplets (**b**) Photo of the formed droplets (**c**) Plasma jet in the presence of a suspension jet injected towards the plasma at exit level of the torch [Fauchais et al. (2008)]

corresponding to a pressure (of course varying along the particle trajectory within the jet). For liquid jet injection the momentum density of the liquid $\rho_l v_l^2$ has to be compared to that of the plasma ρv^2. As shown in the next section, when the liquid jet or liquid droplets penetrate the jet, they are progressively fragmented and thus their volumes and apparent surfaces become smaller. Accordingly, the injection force of the droplets is reduced as well as the force imparted to them by the plasma jet. Thus, rapidly their penetration ceases, and the condition good penetration implies that:

$$\rho_l v_l^2 \gg \rho v^2 \tag{5.6}$$

High-speed photographic images of the penetration and atomization of zirconia suspension in ethanol, injected into Ar-He DC plasma jet with velocities of 27 and 33,5 m/s are given in Fig. 5.23. These show, as expected, an increase of the penetrates of the suspension into the hot core of the jet with the increase of its injection velocity. The observation is not surprising since at the high velocity condition, the momentum density of the liquid jet is 0.96 MPa which is significantly higher than the momentum density of the plasma flow that was 0.02 MPa.

Upon penetration within the plasma jet the liquid drops are exposed to a strong shear stress due to the plasma flow, which further fragments the liquid drops into smaller droplets that are exposed to a very high heat flux resulting in the rapid vaporization of the liquid.

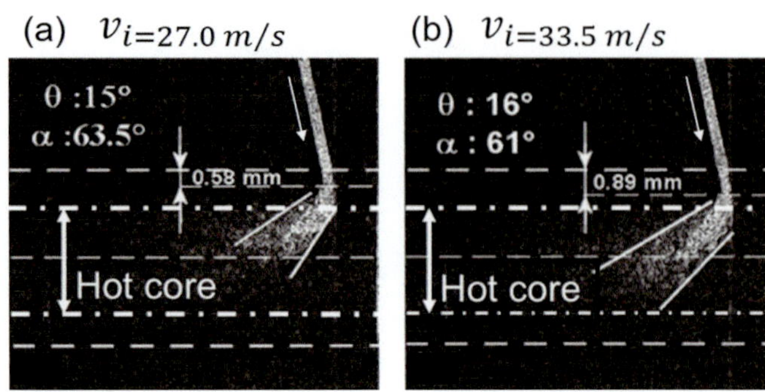

Fig. 5.23 Penetration of a zirconia suspension in ethanol injected with two different injection velocities (27 and 33.5 m/s) into an Ar-He DC plasma jet (60 slm vol. 50% He), 700 A, anode-nozzle i.d. 6.0 mm [Fauchais et al. (2008)]

5.3.2.2 Liquid Fragmentation

In itself droplet fragmentation is a very complex phenomenon, which has been extensively studied for fuel injection. Different break-up regimes exist depending on the Weber number and break-up should happen according to the catastrophic regime where Rayleigh–Taylor and Kelvin–Helmholtz waves develop at different stages of process [Hwang et al. (1996)]. The problem is even more complex when the drop encounters steep temperature and velocity gradients as it penetrates the DC plasma flow. For example, a 150 μm diameter drop crossing a DC plasma flow will be exposed, between its front and rear ends, by a temperature difference of 500 to 1000 K and a velocity difference of about 50 m/s. This will result in a strong deformation of the drop with a significant impact on its drag coefficient. With very high momenta of the atomizing gas flow as those obtained, for example, in wire arc spraying, molten metal drops are torn into sheets of metals, as shown by [Hussary and Heberlein (2001)]. The eddy structure in the flow leads to the sheet disintegration in a shower of droplets with varying trajectories.

To determine an order of magnitude, estimate of what happens to a suspension drop entering a hot gas flow, only a simple force balance can be considered. The drop will be fragmented as soon as the aerodynamic force acting on it will exceed the surface tension force. This practically never occurs in RF induction plasmas where flow velocities are only a few tens of m/s but can occurs in DC plasmas and HVOF jets where velocities are much higher (up to 2200 m/s). To determine the diameter of the resulting droplet, d_d, the equality of the aerodynamic force F_{ip} acting on it and the surface tension force F_s will be considered giving rise to:

$$\frac{\pi}{8} C_D d_d^2 \rho U_r^2 = \pi d_d \sigma_l \tag{5.7}$$

where C_D is the flow drag coefficient, ρ is its specific mass of the gas (kg/m^3), u_r, the relative velocity between flow and

Table 5.1 Calculation of droplet diameters d_d (μm) for different plasma jet conditions

Plasma velocity (m/s)	500 m/s	1000 m/s	2000 m/s
Water $\sigma = 72 \times 10^{-3}$ N/m	40.4 μm	10.1 μm	2.5 μm
Ethanol $\sigma = 22 \times 10^{-3}$ N/m	12.4 μm	3.1 μm	0.8 μm

Table 5.2 Calculation of fragmentation times t_d (μs) for different plasma jet velocities

Plasma velocity (m/s)	500 m/s	1000 m/s	2000 m/s
Water $\sigma = 72 \times 10^{-3}$ N/m	0.52 μs	0.29 μs	0.15 μs
Ethanol $\sigma = 22 \times 10^{-3}$ N/m	0.57 μs	0.30 μs	0.14 μs

drop; $u_r = u - u_d$ (m/s) where u is the flow velocity and u_d the drop velocity, and σ_l is the surface tension of the drop (N/m). If it is assumed that $u_r \gg u_d$ the formed droplet diameter is then given by:

$$d_d = \frac{8\,\sigma_l}{C_D \rho u_r^2} \tag{5.8}$$

of course, d_d depends on the properties of the liquid used and the flow velocity as shown in Table 5.1 for an Ar-H$_2$ DC plasma jet assumed to be at 10000 K. It should be noted, however, that the calculations were made for pure solvents. Considering that the viscosity of a suspension or solution will be higher than that of the solvent, and that adjustment need to be made to account for such an increase in viscosity, the proposed approach provides a reasonable estimate of trends.

A preliminary estimate of fragmentation time can be obtained by an energy balance on the fragmented particle. For that it can be assumed that an initial drop of radius r_s is fragmented into n droplets of radius r_d where $(n\,r_d^3 = r_s^3)$. The variation of the surface energy for a liquid of surface tension σ, during the fragmentation step, results in work:

$$\Delta E_s = \sigma\,\Delta s = 4\pi\sigma\left(nr_d^2 - r_s^2\right) = 4\pi\sigma r_s^2\left(\frac{nr_d^2}{r_s^2} - 1\right) \tag{5.9}$$

The work, W_F, of the drag force F_D during the fragmentation time, t_d, can be calculated from:

$$W_F = F_D\,u_r\,t_d \tag{5.10}$$

resulting in:

$$t_d = \frac{\Delta E_s}{F_D\,U_r} = \frac{8\sigma\left(\frac{nr_d^2}{r_s^2} - 1\right)}{C_D \rho u_r^3} \tag{5.11}$$

The estimation of the fragmentation time for an ethanol drop, 300 μm in diameter, entering an Ar/H$_2$ plasma jet at 10000 K with different velocities is given in Table 5.2. The results show that the fragmentation time of the initial drops takes place in about 1 μs.

The study of [Hwang et al. (1996)] describes fragmentation of fuel drops 189 μm in diameter injected at 16 m/s into an air jet with velocities between 70 and 200 m/s at 450 K.

The values correspond to typical sizes of droplets used in plasmas with similar gas momentum densities. The different mechanisms depend strongly on the Weber number Eq. 5.1 and the different regimes for break-up are defined in [Gelfand (1996)]. For practical purposes, We ≈ 14 has been chosen as the critical value over which particles break up [Watanabe and Ebihara (2003)]. Other authors propose We ≈ 12. The former value is the criterion adopted by [Ozturk and Cetegen (2004, 2005)] and [Basu et al. (2006)]; with Ar-H$_2$ and Ar-He conventional DC plasma jets, in the jet core and its fringes ($T > 3000$ K), ρv^2 values are between 15 and 120 kPa. Thus, ethanol droplets down to 15 μm can be broken up in the jet fringes, but if $\rho v^2 = 50$ kPa, 4 μm droplets will be broken up. In good agreement with the simplified calculations presented in the previous section, this particle fragmentation will be more the rule than the exception. Of course, increasing σ_l, for example, by replacing ethanol with water modifies the minimum size of droplet break-up that is multiplied by a factor of 3.27 (ratio of the surface tensions) and thus small droplets (below 40–50 μm) will be mainly fragmented deeper in the plasma jet and not in its fringes.

Another criterion has to be taken into account when considering suspensions: what changes in the fragmentation mechanism that can be expected when the weight percentage of solid particles in the suspension is increased? The answer lies in the Ohnesorge number, Oh (Eq. 5.2). With the increase of the particle loading in the suspension, its viscosity increases but not its surface tension, thus the Ohnesorge number will increase. The critical value of the Weber number over which fragmentation occurs also increases and fragmentation is thus delayed. This is illustrated in Fig. 5.24 representing the penetration of mechanically injected (a) pure ethanol and (b) suspension with 7 wt.% zirconia particles in ethanol. The addition of solid particles and dispersant is noted to reinforce the cohesion of the drops that is responsible for the observed delay fragmentation. The fragmentation of the ethanol (left) starts before that of the suspension (right), the white dotted line corresponding to the nozzle diameter.

Droplet trajectories are determined using the momentum equations, in the axial and radial directions, as used when spraying micrometric particles [Boulos et al. (1993a, b)]. Along the trajectory, when they penetrate into the plasma jet, their break-up can be calculated by simple expressions such as those from which Eq. 5.7 is derived. More sophisticated models such the Taylor Analogy Break-up model can

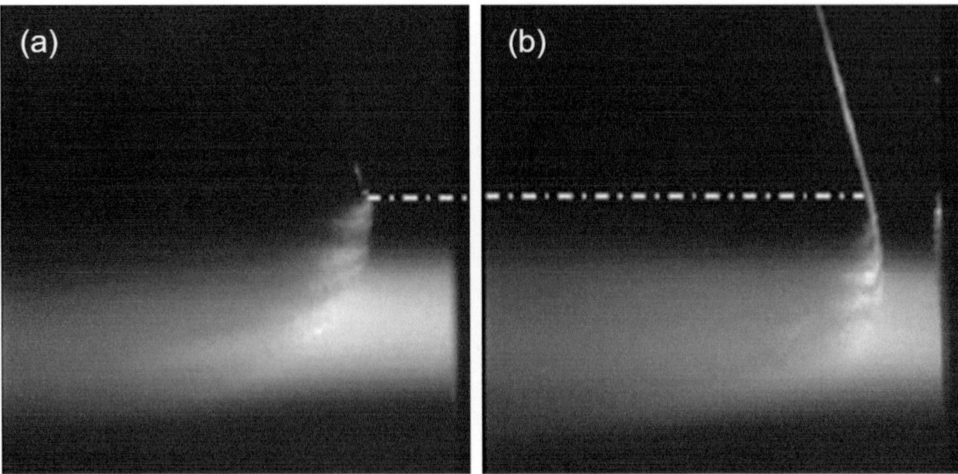

Fig. 5.24 Influence of the liquid jet composition upon its break-up by Ar-H$_2$ DC plasma jet. (**a**) Pure ethanol and (**b**) Zirconia suspension in ethanol. Exposure time 50 µs, five laser flashes of 1 µs duration with 5 µs between each [Fauchais et al. (2008)]

be used (see [Basu et al. (2006)]). The new trajectories of the new droplets are again calculated through the momentum equations, taking into account their vaporization. In Eqs. 5.7 and 5.1, the surface tension σ_l of drops and droplets plays a key role. It diminishes with liquid heating:

$$\sigma_l = \sigma_l^0 - aT \tag{5.12}$$

and in principle, the drop or droplet heating along its trajectory should be calculated. However, taking into account the considerations presented in the next section, vaporization is a very slow process compared to fragmentation, and droplet heating can be calculated on the basis of a lumped capacity model.

A complete model of the liquid–plasma interaction is still a challenge and would require as far as possible the independent study and validation of the various sub-processes that govern the droplet behavior in the flow and at impact. Sophisticated numerical models have also been developed for fragmentation and vaporization of drops [Lee and Reitz (2001), Marchand et al. (2008), Meillot et al. (2009), Carruyer et al. (2011), Vincent et al. (2011), Basu et al. (2008), Jordan et al. (2007)]. However, the effect of vaporization on the plasma flow is generally neglected, which is far from the reality. The break-up models that have been used for plasma conditions are essentially those developed for the interaction of a cold flow with a liquid. The Taylor-analogy break-up (TAB) model [Lee and Reitz (2001), Marchand et al. (2008), Meillot et al. (2009), Carruyer et al. (2011)] is based on the analogy between a spring-mass system and an oscillating and distorting droplet and the wave model that supposes that the break-up time and the resulting droplet size are related to the fastest growing Kelvin–Helmholtz instability and applies better for larger Weber numbers. However, in both models, the

We and *Oh* numbers as well as the constants and parameters of the models have been established for conditions rather different of that prevailing in liquid plasma spraying and their extension to these conditions should be validated.

5.3.2.3 Droplets Fragmentation and Vaporization

It is interesting to evaluate the vaporization time of drops and droplets. As already discussed in Sect. 4.3, the heat flux balance for heating and vaporizing a drop is given by:

$$4\pi r_s^2\, h\, (T - T_s) = \left(H_v + c_{pl}(T_s - T)\right).\rho_l \frac{dV_s}{dt} \tag{5.13}$$

where $\left(\frac{dV_s}{dt}\right)$ is the variation of the volume of the drop $\left(\frac{dV_s}{dt} = 4\pi r_s^2\, \frac{dr_s}{dt}\right)$ and c_{pl} the specific heat at constant pressure of the liquid (J/kg K).

Assuming that the vaporization time, t_s, is that necessary for the drop to shift from radius r_s to radius zero one finds:

$$t_s = \frac{\left(H_v + c_{ps}(T_s - T)\right)\rho_s.2r_s^2}{(T - T_s)\,\kappa.Nu} \tag{5.14}$$

As shown in Fig. 5.25, the characteristic time for the fragmentation and vaporization of droplets of different sizes along the radius of a DC plasma jet (Ar-H$_2$, 45–15 slm, 500 A, anode nozzle i.d. 7 mm) differ by at least 3 to 4 orders of magnitude for diameters over 1 µm. The vaporization time of a 300 µm ethanol drop is 500 µs in a 10,000 K plasma against 1 µs for its fragmentation. For a fragmented droplet of 1.5 µm diameter the evaporation time is only 0.05 µs. Corrections of these time estimates for the Knudsen effect, temperature gradient and the evaporation effects should be made. The net result of these corrections is the lengthening of the time, t_s, by a factor of up to 60, depending on particle sizes.

It is important to note that as the plasma jet is cooled by the liquid injection and with the very low inertia of particles below 1 μm, the spray distances with a conventional DC plasma torch with a stick type cathode are very short compared to conventional spraying. With stand-off distances as short as 25–35 mm, instead of 100–120 mm in conventional spraying, the transient heat fluxes of the plasma jet on the substrate will be multiplied by about one order of magnitude: in the 20–40 MW/m^2 instead of 2 MW/m^2!

Fauchais et al. (2013) reported a study of the suspension plasma spraying of YSZ using successively ethanol or water as solvent; both being injected in the plasma jet with the same injector and the same injection parameters. The liquid was injected at an injector pressure corresponding to 300 μm drops directed toward the nozzle exit axis of the plasma torch. Shadowgraph images of the atomization of the alcohol and water-based suspensions in the plasma stream given in Fig. 5.26 show a significant impact of the Weber number (We) on the nature of the atomization. For the ethanol-based suspension (Fig. 5.26a), with Weber number $We = 563$, the suspension seems to go through a catastrophic breakdown with an almost instantaneous evaporation of the formed droplets. In contrast, with the water-based suspension, Fig. 5.26b with $We = 170$, the suspension seems to go more through a stripping break-up. Measurements of corresponding droplet size distributions are presented in Fig. 5.27 confirm the behaviors presented in Fig. 5.26a, b. Water droplets measured 15 mm downstream of the nozzle exit shown in Fig 5.27b had a mean size of drops of 28 μm while at 40 mm downstream of the nozzle exit, that the corresponding mean droplet size is 22 μm. In contrast to water, ethanol at 15 mm downstream of the nozzle exit had a mean size in the range of 30 μm and 40 mm with nothing is observable further downstream at distances above 15 mm from the torch nozzle.

5.3.2.4 Influence of Arc Root Fluctuations

The voltage fluctuations (in the restrike mode with a PTF-4 torch operating with an Ar-H$_2$ mixture) correspond to a voltage varying between 40 and 80 V for the same arc current of 503 A. The torch velocity follows the corresponding enthalpy variations while the temperature is almost constant (<1000 K difference when the enthalpy is doubled) due to the "inertia wheel" effect of ionization. Thus, the momentum density of the plasma varies drastically, especially in the jet fringes. [Marchand et al. (2007)] have calculated the transient momentum density values. They have used a three-dimensional (3-D) time-dependent model of the plasma jet issuing into air in conjunction with a Large Eddy Simulation (LES) turbulence model to examine the influence of the arc voltage fluctuation amplitude and frequency on the time evolution of the plasma jet flow fields and droplet injection. Typical results of the values of the momentum density of the plasma, v^2, in a plane orthogonal to the jet axis and located at 8 mm from the nozzle exit, are presented in Fig. 5.28. These

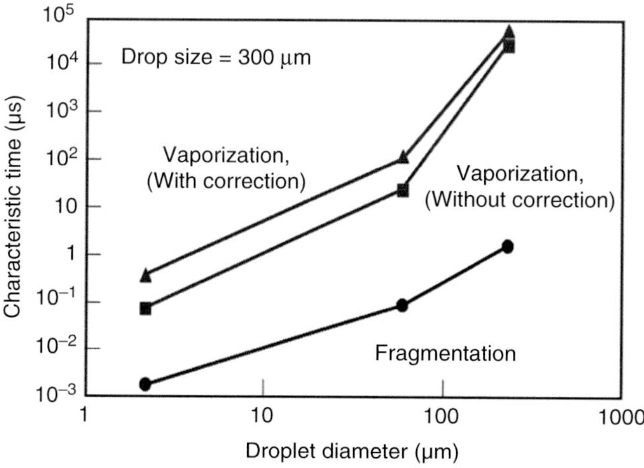

Fig. 5.25 Evolution along the Ar-H$_2$ plasma jet (45-15slm, 500 A, anode-nozzle i.d. 7 mm) of the vaporization and fragmentation times for different droplet radii [Fauchais and Montavon (2010)]

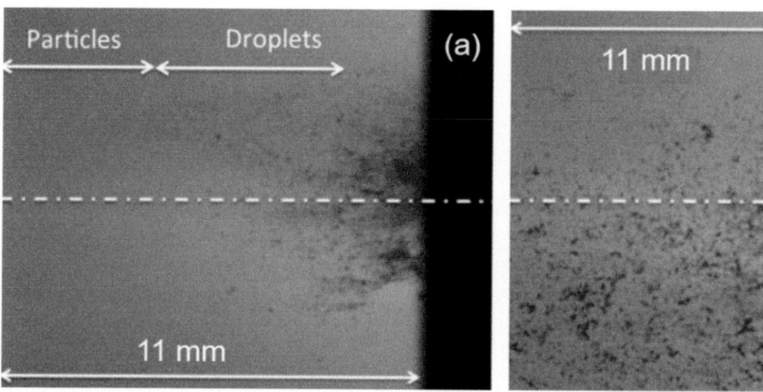

Fig. 5.26 Shadowgraph measurements showing the penetration (high pressure injector) of (**a**) ethanol suspension, (**b**) water suspension (Ar-He 40–20 slm, anode-nozzle i.d. 6 mm, enthalpy 14 MJ/kg). The black part at the right of figures corresponds to injection system [Fauchais et al. (2013)]

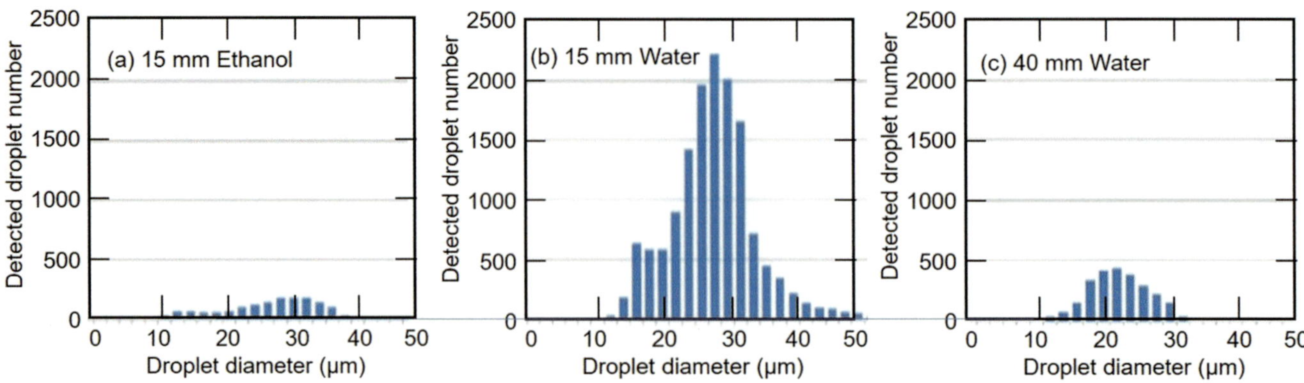

Fig. 5.27 Distribution by number of ethanol and water-based droplets (>5 μm) 15 and 40 mm downstream of the injection point in the Ar-He-H$_2$ plasma jet (45–45–3 slm, anode-nozzle i.d. 6 mm, enthalpy 169 MJ/kg [Fauchais et al. (2013)]

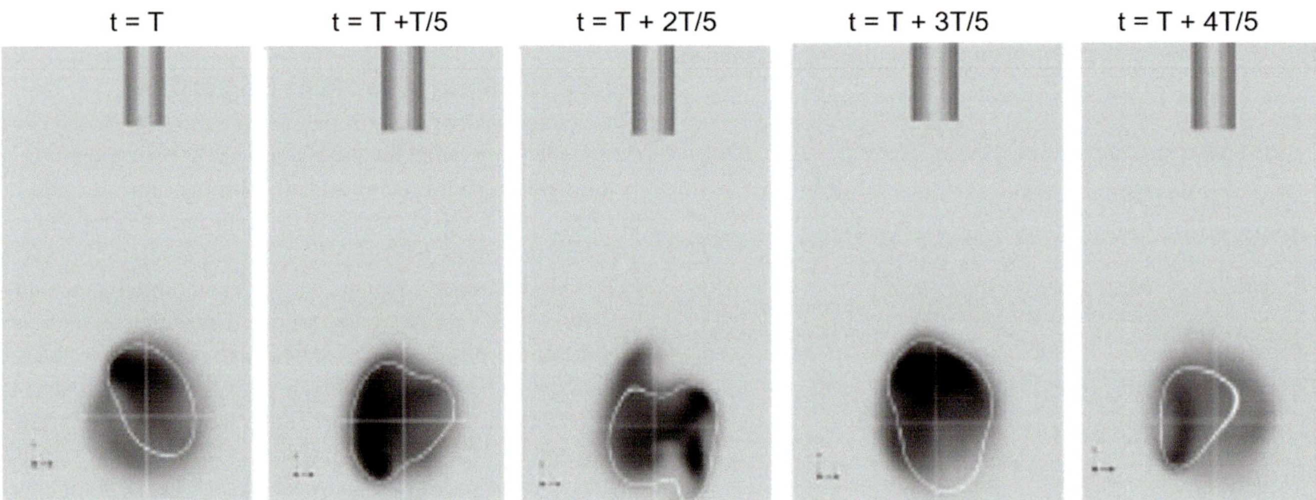

Fig. 5.28 Calculated momentum density (Pa) of a DC plasma jet in an orthogonal plane to the jet axis at 8 mm from the nozzle. Ar-H$_2$ (45–15 slm), I = 600 A. Reprinted with kind permission from ASM International [Marchand et al. (2007)]

results were obtained for an Ar-H$_2$ plasma jet at 600 A. With T as the period of the restrike mode, Fig. 5.28 represents five images of ρv^2 calculated at times differing by T/5; in this figure the isotherm 8000 K is represented as a white line. This figure shows that the momentum density varies drastically with time for the same position as well as the position of the hot zone of the plasma (8000 K isotherm).

As already emphasized, this variation is due to the variation in length of the electric arc, and the location of the anodic arc root. Such calculations well underline the very inhomogeneous dynamic treatment that droplets could undergo. To get more information from images as those presented in Fig. 5.24, up to 10 images have been superposed and their treatment has allowed determining their envelope and thus measuring the liquid droplet dispersion angle θ and mean deviation α, as shown in Fig. 5.23 for an Ar-He stable plasma jet. Of course, the results obtained depend on the torch voltage level at which images are taken, as illustrated in

Fig. 4.100 for zirconia suspension in ethanol. It can be readily seen that the deviation angle is not very sensitive to the voltage fluctuation (58° at 40 V against 60.5° at 80 V). However, this is not the case of the dispersion angle that varies from 33° at 80 V against 64° at 40 V (Fig. 5.29).

In contrast, with an Ar-He plasma that is more stable, such as that presented in Fig. 5.23 (ΔV/V $= 0.25$ instead of 1 for the Ar-H$_2$ plasma), the liquid penetration is better, and especially the dispersion angle, θ is much lower (more uniform treatment of droplets) and less fragmentation occurs in the jet fringes.

5.3.3 Cooling of Plasma Flow by the Liquid

When injecting water drops 300 μm in diameter into a DC plasma jet (45 slm (Ar) + 15 slm (H$_2$), P$_{eff}$ = 25 kW, nozzle i.d. 7 mm) the plasma jet is cut into two parts almost instantly [Wittmann et al. (2002)]. This is illustrated in Fig. 5.30

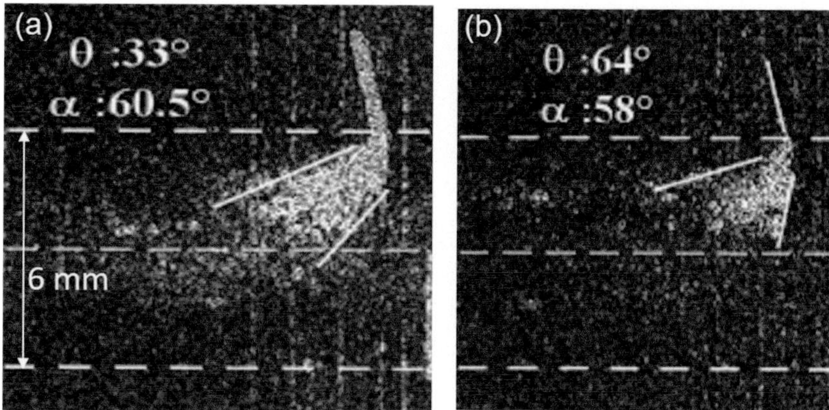

Fig. 5.29 Dependence of the dispersion angle θ and deviation angle α of the liquid droplets cloud (Ar-H$_2$ DC plasma jet) on arc voltage for two transient voltages corresponding to their (a) maximum 80 V and (b) minimum 40 V values [Etchart-Salas et al. (2007)]

showing two orthogonal pictures of the jet: one taken orthogonally to the injection, while the other is parallel to it. This local cooling is caused by drop vaporization, dissociation of water, and ionization of H and O (only about 3% of the jet energy is consumed). But 15 mm downstream from the injection point, the plasma jet has recovered its axial symmetry and is a uniform mixture of Ar, H and O in its hottest zone (T > 8000 K), as shown in Fig. 5.31 [Wittmann et al. (2002)]. Almost no more liquid is present after 15 mm, and the solid particles contained initially within a droplet are then accelerated and heated by the plasma.

[Fazilleau (2003)] has compared DC plasma jets working in the same conditions with Ar/H$_2$ (0.984/0.16 mass fractions) and Ar/H$_2$/H$_2$O (0.750/0.040/0.210 mass fractions). Calculations of jet temperatures were performed assuming that Ar, H$_2$, and O$_2$ are pure gases mixed and they do not react with electrodes. It is of course over-simplified, but the trends are significant as presented in Fig. 5.32. They show clearly that the plasma jet core length (T > 8000 K) is shifted to shorter axial distances. In the reality the cooling is more important because in this calculation the water vaporization has completely been neglected. The latter is not at all negligible because the vaporization of 0.47 cm^3/s requires 3 MJ/kg when the plasma enthalpy in these conditions is 12 MJ/kg. When comparing images of plasma jets (Ar + H$_2$ 25 vol.%) where nothing was injected and then successively where water and ethanol were injected (with the same flow rates) the longer plasma core (about 45 mm) was naturally with no injection, followed by ethanol injection (about 35 mm) and water injection (about 27 mm) (Fig. 5.32).

5.4   Particles and Droplets in Combustion and Thermal Plasmas

As individual particles whether solid, liquids, or suspensions are injected into a combustion or plasma stream and entrained by the high temperature flow they are heated and melted in-flight before being projected against the substrate on which they are deposited in the form of splats that are the building blocks of the coating. The quality of the splats and subsequently of the formed coating depends directly on the interaction between the particles and the hot stream during their relatively short contact time in-flight. A good knowledge of the characteristics of the combustion or plasma stream combined with the proper control of the particle trajectories and their temperature histories is, therefore, essential to achieve the necessary particle properties prior to their impact on the surface of the substrate. It is not surprising that because of the importance of this critical step and its impact on the quality of the coating, that an intense research effort was devoted over the past four decades to the study of the flow and temperature fields in the plasma and of particle injection these flows, plasma–particle interactions and splat formation on impact with the substrate. In this section an example highlights of some of the early diagnostic studies carried out with the objective of learning as much as possible about the plasma flows and particle trajectories, particle velocity and temperature, and particle flux density distribution in the plasma stream.

5.4.1   Flow and Temperature Fields in DC Plasma Jets

Two examples are given for at atmospheric pressure DC plasma jet that were reported by Vardelle et al. (1982). The reader is referred to the original papers as well as specialized references son the subject for details of the experimental technique used. The results given are for standard DC plasma torch with a stick-type cathode and a water-cooled copper anode. Velocity field measurements in the plasma jet were carried out using Lase Doppler Anemometry (LDA) whose temperature field measurements were carried out using mission spectroscopy. Typical results given in Figs. 5.33 and 5.34 were obtained for Ar-H$_2$ and N$_2$-H$_2$ plasmas, respectively. The measurement on the Ar-H$_2$ discharge was carried out using 8 mm i.d. anode discharge nozzle, with a plasma

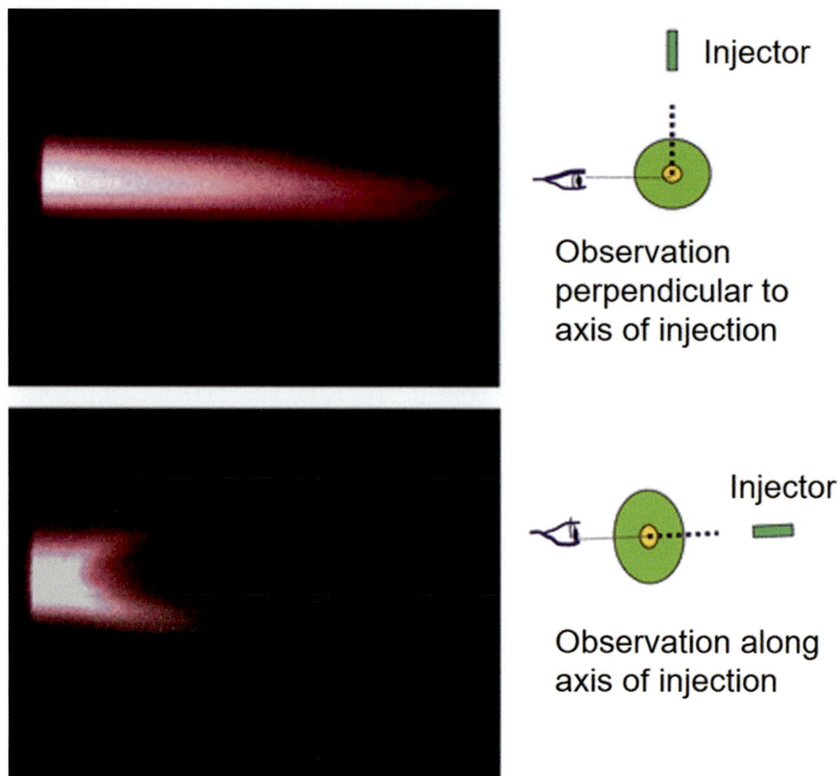

Fig. 5.30 Two orthogonal pictures of the plasma jet, one taken orthogonally to the injection, while the other is parallel to it (45 slm (Ar) + 15 slm (H_2), 600 A, 65 V, 6 mm i.d. nozzle): Reprinted with kind permission from ASM International [Wittmann et al. (2002)]

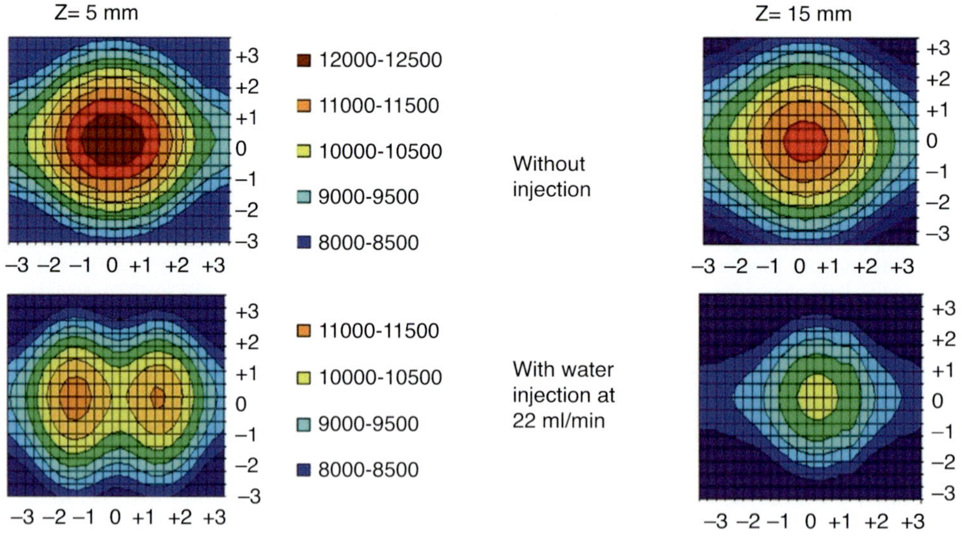

Fig. 5.31 Spectroscopic measurements at 5 and 15 mm downstream of the nozzle exit of the plasma jet presented in Fig. 5.30 without and with water injection [Wittmann et al. (2002)]

gas flow rate of 75 slm (Ar) + 37 slm (H_2) and a power of 29 kW. The measurements with the N_2-H_2 case made use of a smaller i.d. discharge nozzle, 6 mm i.d. anode discharge nozzle, with a plasma gas flow rate of 37 slm (N_2) + 27 slm (H_2) at the same power rating of 29 kW. In none of these cases a powder gas was injected in the flow. As mentioned earlier in Sect. 5.2 powders used for plasma spraying are typically pneumatically injected into the plasma flow using an appropriate carrier gas which unfortunately is responsible for the local cooling of the plasma jet at the point of powder

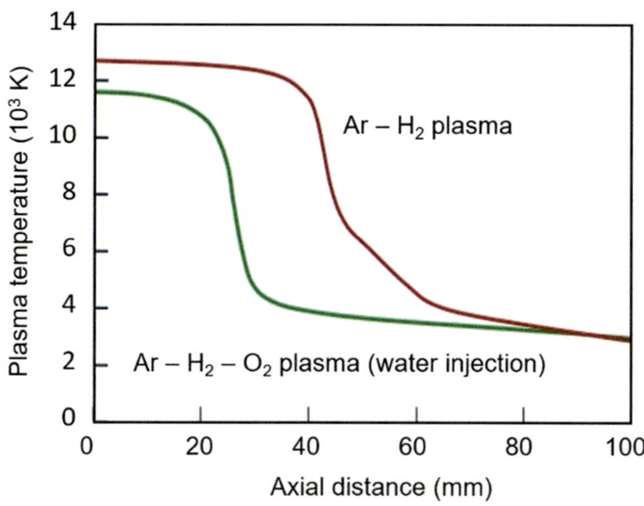

Fig. 5.32 Temperature profile along the centerline of a DC plasma jet for (a) Ar-H$_2$ plasma (0.984/0.016 mass fractions) and (b) Ar-H$_2$-O$_2$ plasma (0.75/0.04/0.21 mass fractions) [Fazilleau (2003)]

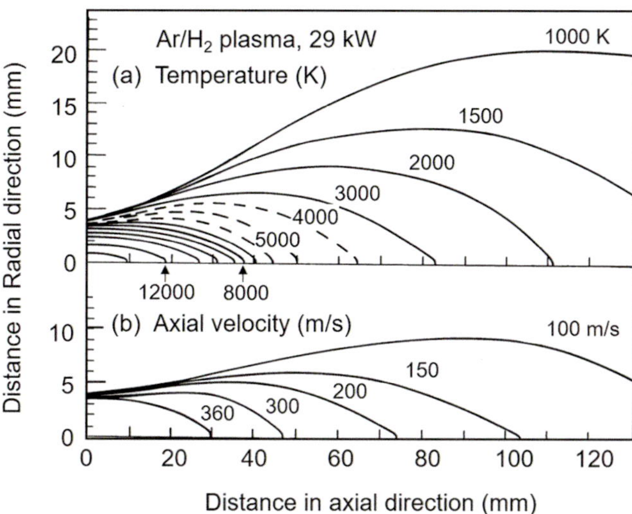

Fig. 5.33 (a) Temperature and (b) velocity fields for an atmospheric pressure DC plasma jet with a plasma gas flow rate of 37 slm (Ar) + 27 slm (H$_2$), and power level of 29 kW [Vardelle et al. (1982)]

injection as shown in Fig. 5.12. The effect is also associated with a loss of the axisymmetric of the flow that becomes in these cases fully 3-dimensional.

Temperature and flow fields in the discharge for the Ar-H$_2$ case are given respectively in Fig. 5.33a, b. These show a very high concentration of energy in a relatively small core of the plasma jet about 5 mm in diameter and 30 mm long in which plasma temperature can reach 12,000 K or more. The temperature fields are also characterized by steep temperature gradients in the fringes of the jet that corresponds to close to 5000–6000 K/mm. The corresponding velocity field given in Fig. 5.33b shows centerline exit jet velocities above 360 m/s, dropping rapidly in about 100 mm to below 150 m/s. As with the temperature fields, the velocity isocontours also show very steep gradients in the fringes of the jet of hundreds of m/s/mm. The corresponding results for the N$_2$-H$_2$ case are given in Fig. 5.34a, b show essentially the same features as the Ar-H$_2$ case with a higher concentration of energy in the nozzle discharge zone due to the predominantly molecular nature of the gases used. Maximum observed temperatures are in the 12,000–13,000 K range and jet centerline exit velocities above 400 m/s. It should be pointed out that these are typical values for an atmospheric pressure DC discharge in this powder range. Subsequent studies show considerably higher velocities for low pressure or higher power operations though the plasma temperature remains essentially in the same range.

5.4.2   Particle Trajectories in DC Plasma Spraying

Considerable research effort has also been devoted to the development of new tools for in-flight particle velocity and surface temperature measurements. These were also extensively used for the determination of the in-flight particle conditions prior to their impact on the substrate. Both plasma and particle diagnostics also played a key role, as will be seen later for the validation of model predations that is an essential requirement for the development of appropriated mathematical models with reliable predations.

Early studies by Vardelle et al. (1982) also provided particle velocities and surface temperatures measurements for alumina particles, with different particle sizes in the range of 3 to 46 µm, injected into the above given atmospheric pressure, N$_2$-H$_2$ plasma DC plasma jet, 29 kW. Typical results of the evolution of the axial particle velocities along the centerline of the jet are given in Fig. 5.35. These were injected with a near optimal radial injection velocity in the range of 14–25 m/s, depending on the particle diameter, in order to maintain a trajectory a close as possible to the axis of the jet. It may be noted from Fig. 5.35 that the first part of the trajectory the plasma velocity (200–600 m/s) is considerably higher than the axial particle velocities of about 20 m/s. The important slip velocity between the gas and the particles is responsible for the development of a large drag force giving rise to particle accelerations that can be as high as 10^5 gravitational acceleration (g). This force, however, drops rapidly as the particle velocity increases and that of the gas drops, until a point is reached where the particle velocity is equal to that of the gas. From this point onward, the relative velocity between the gas and the particle becomes negative (particle velocity higher than that of the gas) giving rise to a breaking effect that is responsible for the gradual slowing down of the particle through the latter part of its trajectory.

The same phenomena are also observed for the particle surface temperature, though for not the same reason. Measurements of the temperature histories of 18 $\pm$ 2 µm

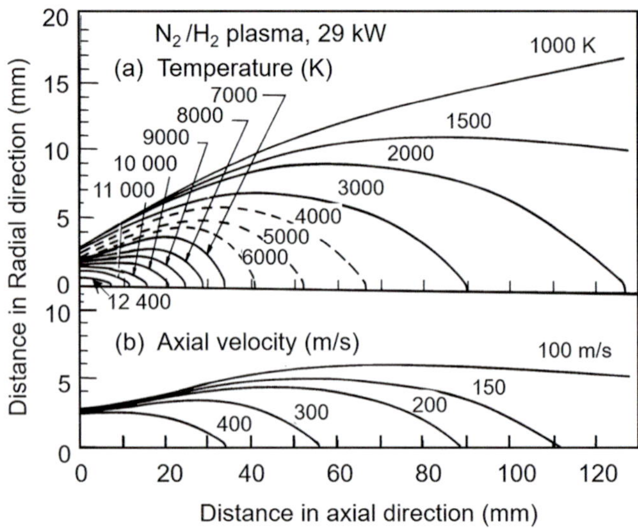

Fig. 5.34 (a) Temperature and (b) velocity fields for an atmospheric pressure DC plasma jet with a plasma gas flow rate of 37 slm (N_2) + 27 slm (H_2) and power level of 29 kW [Vardelle et al. (1982)]

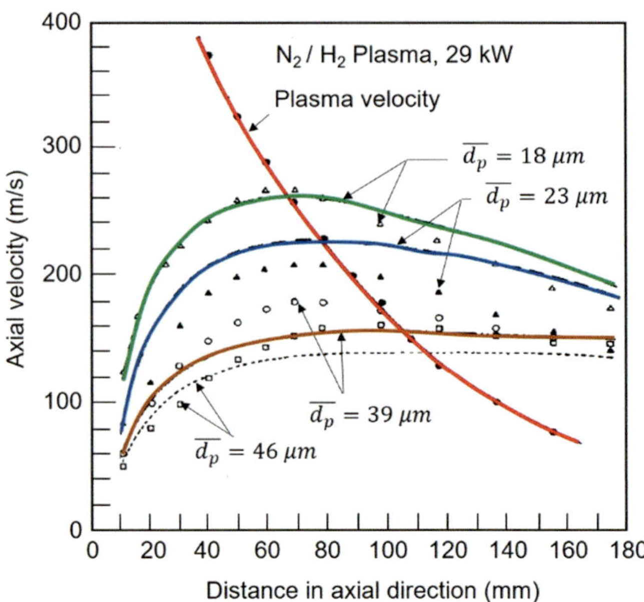

Fig. 5.35 Particle velocity measurements along the centerline of a N_2-H_2 plasma at 29 kW, for alumina particles of different diameters [Vardelle et al. (1982)]

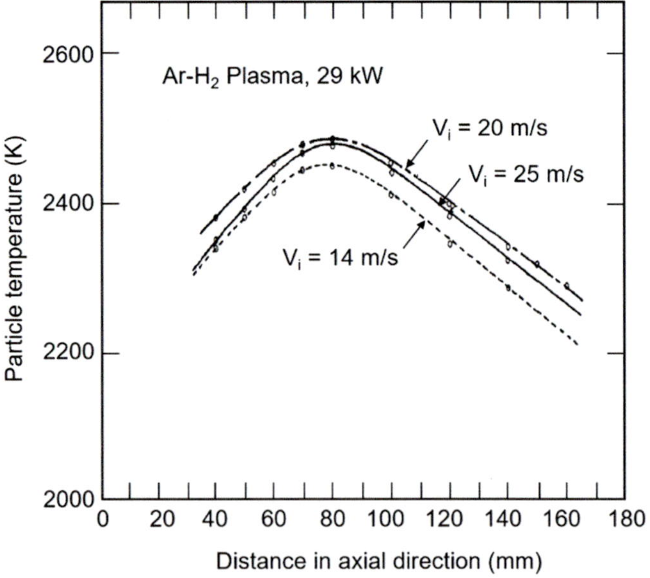

Fig. 5.36 Particle surface temperature measurements along the center-line of an Ar-H_2 plasma at 29 kW, for 18 $\pm$ 2 µm alumina particles injected at injection velocities v_i of 14 to 25 m/s into the plasma jet [Vardelle et al. (1982)]

alumina particles injected into Ar-H_2 plasmas, with a power rating of 29 kW at radial velocities varying between 14 and 25 m/s is given in Fig. 5.36. These show a rapid increase of the surface temperature of the particles within the first 50 mm of their trajectories. With the normal drop of the temperature of the plasma with the distance from the torch exit nozzle a point is reached where the heat transfer rate to the particle is equal to the heat losses by radiation from the surface to the particle to its surroundings. From this point onward, the temperature of the particle reaches a maximum beyond that the particle temperature will drop gradually depending on it the thermophysical properties of its material. It is to be noted that the point of maximum temperature of the alumina particle is essentially the same as that for the maximum particle velocity, with similar results observed for different diameters as well.

An extensive diagnostic study of plasma–particle interactions under DC plasma spray conditions was reported by Vardelle et al. (1988). Measurements were carried out for a standard DC plasma torch, 8 mm i.d. anode exit nozzle, with an external powder injection, 2 mm i.d. as schematically illustrated in Fig. 5.37a. The tip of the powder injector had to be pulled back a distance of 4 mm from the edge of the plasma discharge nozzle in order to avoid its damage by the intense heat flux from the plasma. The plasma gas was a mixture of 75 slm (Ar) + 15 slm (H_2), with a torch power varying between 17.1 and 29.2 kW. Two types of alumina powder were used in the study. The Particle Size Distribution (PSD) of the fine powder, given in Fig. 5.37b, had a mean particle diameter of 18 $\pm$ 3 µm, while the course powder had a particle size in the range of 45–90 µm and a mean value of 60 $\pm$ 15 µm. The powder feed rate was 4 g/min. The powder carrier gas flow rate was varied between 4.5 and 8.5 slm with a statistical distribution of the powder injection velocities, v_{pi}, at the exit of the injector tube for the fine powder is given in Fig. 5.38. It is interesting to note that the statistical mean particle injection velocities are 25.0, 30.0, and 35.0 m/s for carrier gas flow rates of 5.5, 7.0, and 8.5 slm, respectively. Based on these flow rates, with an internal diameter of the

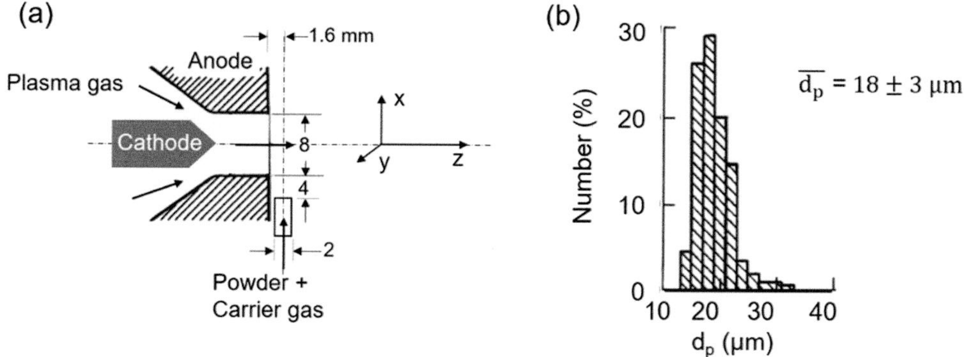

Fig. 5.37 (**a**) Schematic of the plasma torch nozzle design showing the position of the external powder injector and system of coordinates (**b**) Particle Size Distribution (PSD) of the fine alumina powder used in the study [Vardelle et al. (1988)]

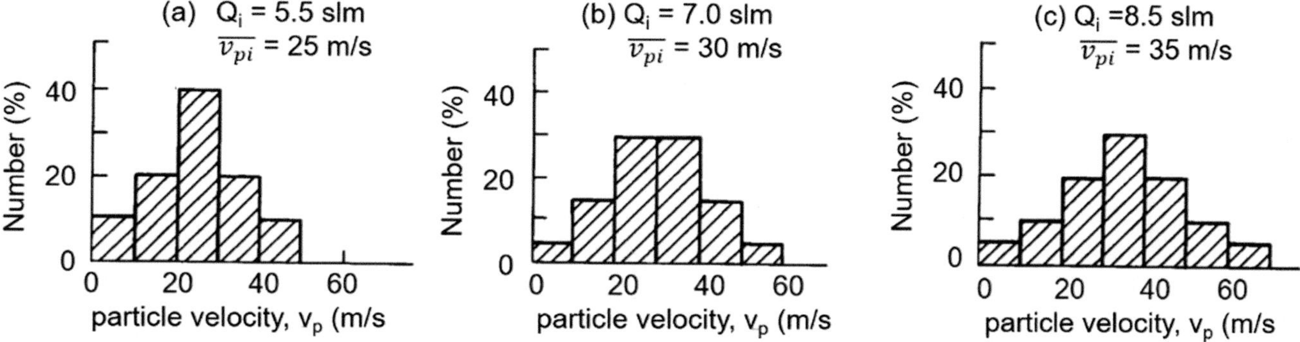

Fig. 5.38 Variation of the statistical particle injection velocity distributions of fine alumina powder at the exit of a 2 mm i.d. injector, as function of powder carrier gas flow rates, $Q_i = 5.5$, 7.0 and 8.5 slm. $d_p = 18 \pm 3\mu$m, PSD is given in Fig. 5.37 [Vardelle et al. (1988)]

feed tube of 2.0 mm, the corresponding mean carrier gas velocity can be calculated as 29.2, 37.2, and 45.0 m/s, respectively. These show the statistical mean particle velocity to be about 80% of the mean carrier gas velocity that is higher than the corresponding values of around 60% given in Sect. 5.2.1. The difference could well be particle size and density of the particle material in this case. The authors in the same paper give for the injection of the course powders, $d_p = 60 \pm 15\,\mu$m an injection velocity of 15 m/s for a carrier gas flow rate of 4.5 slm that would correspond to about 60% of the mean carrier gas velocity of 23.9 m/s.

Typical trajectory data and particle flux number density distributions are given in Fig. 5.39. These were obtained with the fine alumina powder, $d_p = 18 \pm 3\,\mu$m with a carrier gas flow rate of 5.5 slm for a corresponding mean injection velocity of the powder of 25 m/s. It may be noted that because of the relatively small and light powder particles, they spread over a wide range in the plasmas as indicated by the particle flux isocontours given on the LHS of the figure. The corresponding profiles of the particle number flux densities in two orthogonal plans given in the same figure show particle fluxes as high as 10^5 particles/mm^2 s at the

center of the particle stream. It also shows important asymmetry in the plan of particle injection (X) but good symmetry in the orthogonal plane (Y).

In order to determine the effect of particle injection conditions on the particle flux number density, particle velocity and particle temperature distributions, measurements were carried out at a single level a distance of 75 mm downstream of the torch nozzle exit, in the plane of particle injection, that is, X-direction. The plasma operating conditions were the same as described earlier Ar-H$_2$, 29.2 kW, using the fine alumina powder $d_p = 18 \pm 3\,\mu$m with carrier gas flow rates of 5.5, 7.0, and 8.5 slm (Ar). The results given in Fig. 5.40a show the particle velocity at a maximum at the centerline of the flow with a maximum value in the 210–220 m/s range. The particle velocity it drops rapidly with distance from the center to around 180 m/s at ± 5 mm from the center. Figure 5.40b shows the corresponding particle surface temperature distributions with essentially the same observation, maximum temperature being for particles around the axis of the flow, with particle surface temperatures in the 3000–3200 K temperature range that indicates the alumina to particles to be fully molten. The last graph in Fig. 5.40c

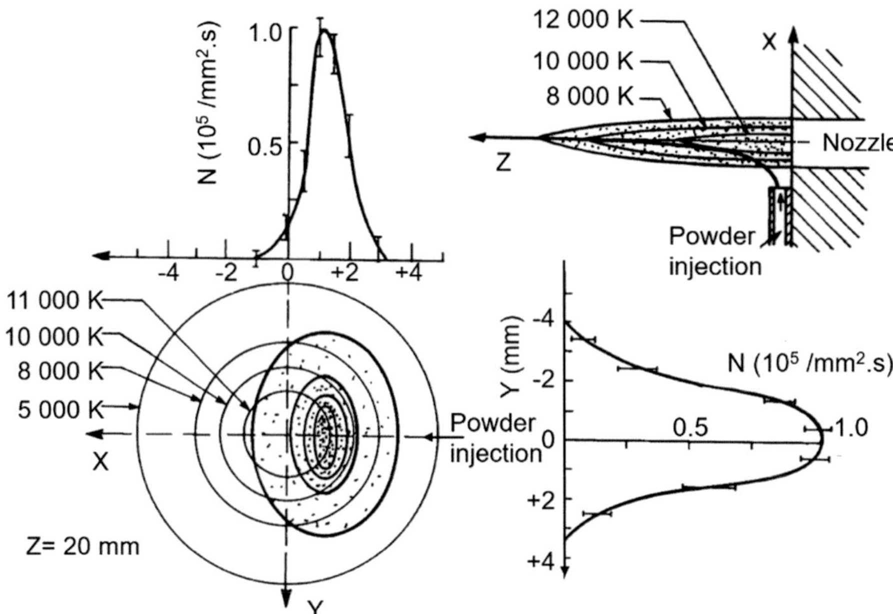

Fig. 5.39 Particle trajectories and number flux density distribution for the injection of the fine alumina powder, 18 3 μm in a 29.2 kW Ar-H2 plasma, 75 slm (Ar) + 15 slm (H2), with a carrier gas flow rate of 5.5 slm (Ar) [Vardelle et al. (1988)]

gives the corresponding particle number flux distributions that show a strong dependence on the powder injection velocity. Knowing that the injection position is on the negative side of the X-direction, it is consistent to observe the position of maximum flux density shifting in the opposite direction with the increase of the powder gas flow rate.

While having maximum particle velocities and temperatures seems consistent with the fact that the plasma jet also has its maximum velocity and temperature in the same location, it is disruptive to see that only at a carrier gas flow rate of 5.5 slm that the maximum particle flux density also coincides with the maximum velocity and temperature, meaning that the great majority of the particles would have acquired the necessary energy level for full melting of the particles and their acceleration that are necessary conditions obtaining a good quality coating with as satisfactory deposition efficiency, η_D. The latter being defined as the fraction of the powder feed into the plasma that are deposited on the substrate and contribute to the formation of the coating. For higher carrier gas flow rates, $Q_i = 7$ or 8.5 slm, it is obvious that an increasing fraction of the powder will not acquire the same conditions in terms of particle velocity, temperature, and degree of melting giving rise to a poorer deposition efficiency. Moreover, since non-molten particles tend to be at the source of defects and porosities in the coating, the quality of the coating would also suffer. It is therefore important to stress that when measuring parameters such as the velocity or temperature of particles in a spray environment, it is important to integrate the mass flux density in the analysis in order to evaluate the full picture of the spraying operation.

5.4.3 Flow and Temperature Fields in RF Induction Plasmas

The flow and temperature fields in an RF inductively coupled discharge are rather complex due to the interaction between the applied magnetic field and the induced electric fields in the discharge. This results in the creation of a high-pressure zone in the middle of the coil region due to electromagnetic pumping effect as illustrated in Fig. 5.41b. As noted, a circulation zone is thus created with a reversed flow stream along the axis of the discharge directed toward the upstream end of the plasma torch. By comparing the LHS "1" and the RHS "2" of this figure that corresponds to two different discharges at the same power level except for the use of different plasma gas flow rates, 40 slm (Ar) for side "1," and 5 slm (Ar) for side "2" it is noted that the position of the discharge with respect to the induction coil depends strongly on the plasma gas flow rate.

Any axial injection of powders into the discharge using the upstream probe indicated "in blue" on the figure, would require use of a sufficiently high powder carrier gas flow rate in order to overcome the reverse flow that the particles would be facing that could disperse them to the fringes of the discharge and the wall of the plasma confinement tube. One approach commonly used to avoid such an effect and the need for a high powder carrier gas flow rate is to displace the tip of the powder injection probe further downstream until the

Figure 5.41a shows temperature isocontours in the discharge for the same two plasma gas flow rates described earlier with the LHS "1" of the figure corresponding to the high gas flow rate case and those on the RHS "2" corresponding to the low plasma gas flow rate. As with the gas recirculation zone, the position of the high temperature region of the discharge is noted to depend on the total plasma gas flow rate. The higher is that flow, the further downstream is the discharge zone pushed, until eventually it is completely blown off and extinguished. In both low and high plasma gas flow rate cases, the discharge zone has a rather uniform temperature filed with gas temperatures ranging from a maximum of 10,500 K down to 8000 K. The gas temperature continues to drop rapidly close to the walls of the plasma confinement tube to below 1000 K. Obviously with the increase of the power level and depending on the composition of the plasma gas, the heat flux to the walls of the plasma confinement tubes will increase requiring efficient cooling for an adequate protection.

Another example of the temperature fields in an atmospheric pressure inductively coupled open discharge is given in Fig. 5.42 for two different types of plasma gases. The results on the LHS of the figure are for operating using pure argon (30 slm) at a power level of 12 kW, while that on the RHS of the figure is for operation with pure oxygen (30 slm) at a power level of 8 kW. The oscillator frequency in both cases was 10 MHz. The uniform temperature fields in the discharge region are noticeable, though the range of plasma temperatures for each car is very much dependent on the nature of the plasma gas.

5.4.4 Particle Velocity Distributions in RF Plasma Spraying

As mentioned earlier, inductively coupled discharges have the main advantage that they are electrodeless with easy access for axial injection of the spraying powder in the center of the discharge. In Sect. 5.4.3, the important issue of internal gas recirculation in the coil region, and the presence of a back flow along axis of the torch, have been discussed stressing the need for an increase of the powder carrier gas flow rate in order provide the injected particles with the necessary momentum to overcome the resistance caused by the back flow. Alternately, the tip of the powder injection probe should be pushed further into the discharge to the high-pressure region at the center of the coil thus avoiding the back flow in the torch.

Measurements were reported by [Lesinski and Boulos (1988)] of the plasma and particle velocities in an inductively coupled discharge using Laser Doppler Anemometry (LDA). A schematic of the principal torch dimensions is given in Fig. 5.43c showing the 50 mm i.d., air-cooled, quartz plasma

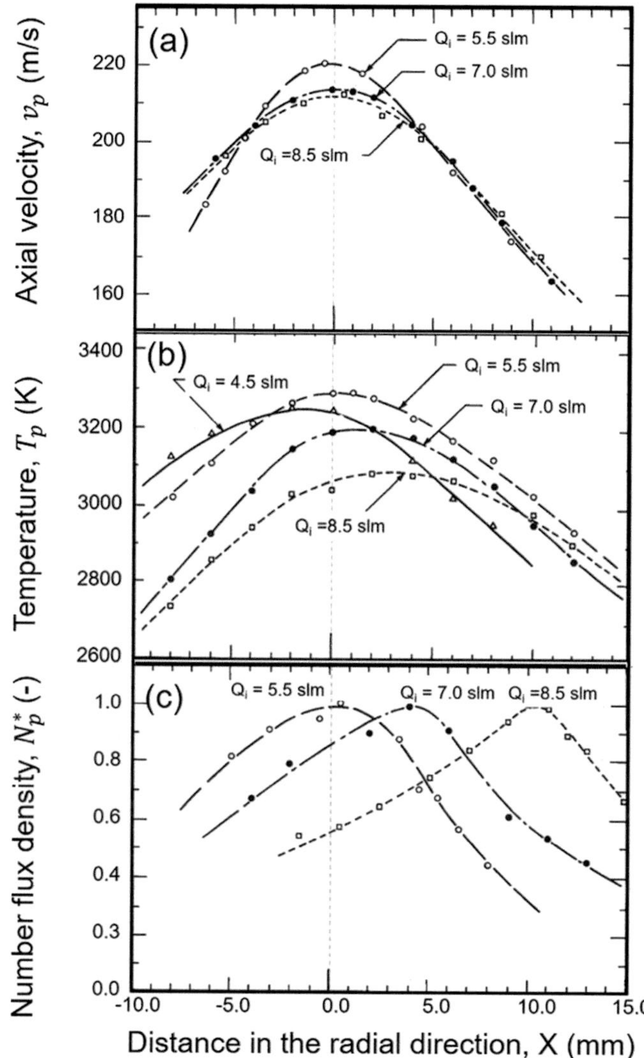

Fig. 5.40 (a) Axial velocity, (b) Temperature, and (c) Number flux density of the particles at 75 mm from the exit of the DC plasma torch operating at atmospheric pressure using 75 slm (Ar) +15 slm (H_2) as plasma gas and power of 29.2 kW. Alumina powder $d_p = 18 \pm 3$ μm injection at different carrier gas flow rate, Q_i (Ar), [Vardelle et al. (1988)]

middle level of the induction coil, around 26 cm from the upstream end of the torch as illustrated in Fig. 5.41a, b. In such a position, powder carrier gas will flow co-current with the plasma gases toward the downstream end of the torch allowing to the particles for a good contact time with the plasma of the order of 5–10 ms depending on the operating conditions. The photograph on the RHS of the figure (Fig. 5.41c) provides a clear picture of such a discharge in an air-cooled, 70 mm i.d. quartz tube, operating with pure argon as plasma gas at a power of 12 kW. In this particular case, the powder injection probe is absent. The lower section of the photograph is a composite picture showing an upward view of the discharge in the presence of a probe (shown in blue) surrounded by bright discharge current-carrying ring.

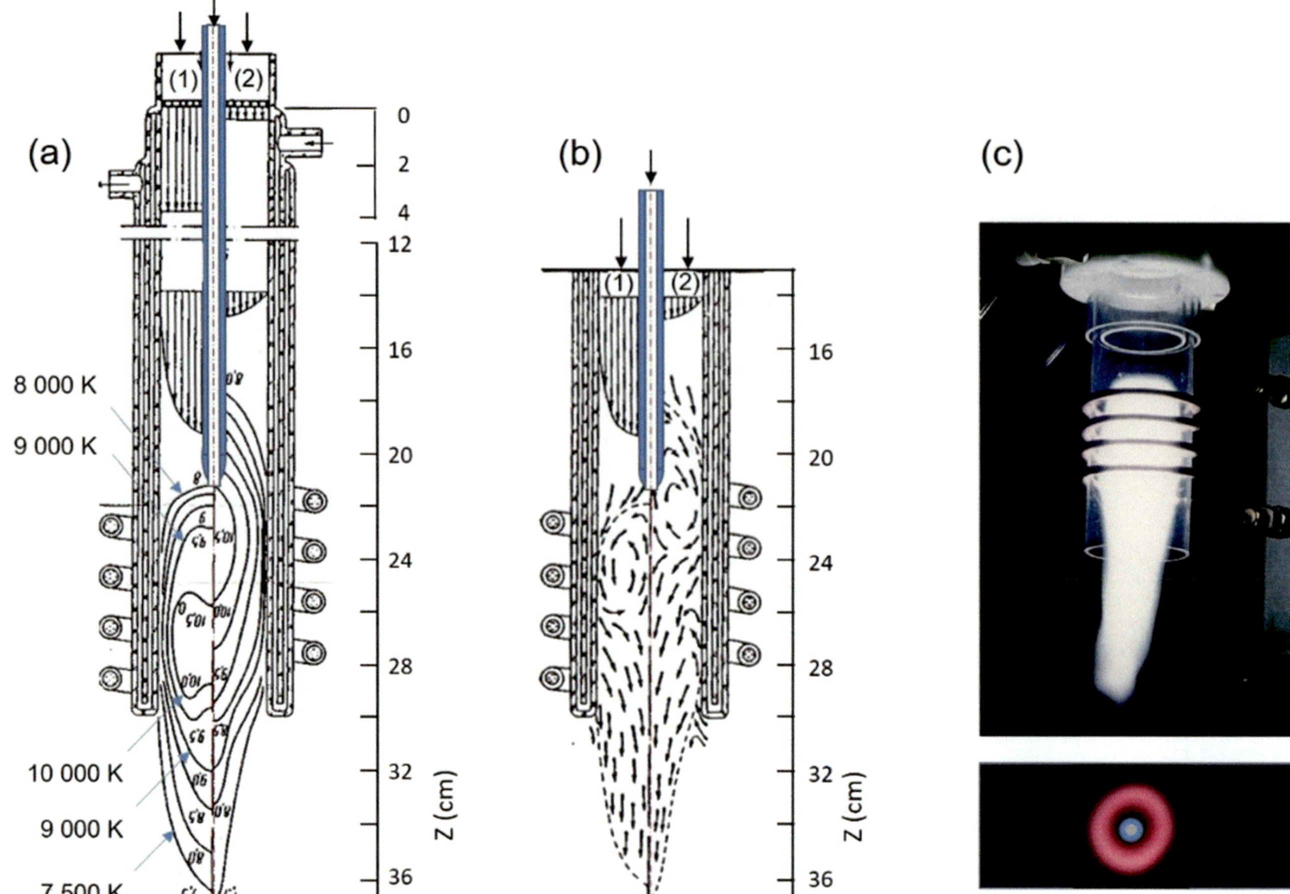

Fig. 5.41 Schematics of (**a**) Temperature fields (**b**) flow fields for an RF induction plasma discharge, side "1" of each of these two figures is for high plasma gas flow rate, Q = 40 slm (Ar), side "2" is for low gas flow rate, Q = 5 slm (Ar), torch i.d. 40 mm, f = 6 MHz, P = 7.6 kW after, Klubnikin (1975). (**c**) Photograph of an induction plasma discharge. (Courtesy of the University of Sherbrooke, Québec Canada)

confinement tube surrounded by a four-turn induction coil. The plasma gas was introduced in three independently controlled streams. The powder carrier gas was introduced through a water-cooled central injection probe, 2 mm i.d. with its tip place 10 mm above the induction coil level. The powder gas flow rate was varied between $Q_1 = 4.8$ and 7.4 slm (Ar). The other two gas streams, also of pure argon, were introduced as central gas $Q_2 = 11.7$ slm (Ar) introduced with a tangential velocity component, and sheath gas $Q_3 = 63$ slm (Ar) introduced along the inner walls of the plasma confinement tube. The oscillator frequency of the power supply was f = 3 MHz, the plate power Po was varied between 4.6 and 10.5 kW with a coupling efficiency $\eta_c = 65\%$ defined as the fraction of the plate power that is coupled into the discharge.

Two sets of measurements were carried out, the first in which a seed refractory powder with a particle size in the 1–2 μm rage was introduced via the central probe a sufficient carrier gas in order to disperse in powder in the flow. Because of the low inertia of such a powder the particle velocities were closed if not identical to the plasma velocity. The results given in Fig. 5.43a are presented in terms of the radial distribution of the axial velocity of the plasma at Z level between 10 and 82 mm from the probe tip as illustrated in Fig. 5.43c. A summary of the experimental variables is given in Table 5.3. The results show axial velocities in the range of 5–20 m/s with a maximum coinciding with the central axis of the plasma torch. Increasing the probe gas flow rate, Q_1, or the discharge powder results in a systematic increase of the ventral velocity of the plasma. It is noted that the radial profiles of the velocity distribution become increasingly flatter with the increase of the distance from probe tip level. Subsequent measurements were carried out with the injection of silicon particles with a diameter of 33 ± 13 μm through the central probe. The results given in Fig. 5.43b show mean particle velocities close to those reported for the plasma with a close concentration of the particle flux over the central 10 mm diameter around the torch axis, at a Z = 10 mm. With the increase of distance from the point of injection, the particle flux spreads over a larger surface of the flow field covering a spot diameter close to 20 mm at $Z = 84$ mm, an increase of fourfold compared to their distribution at $Z = 10$ mm.

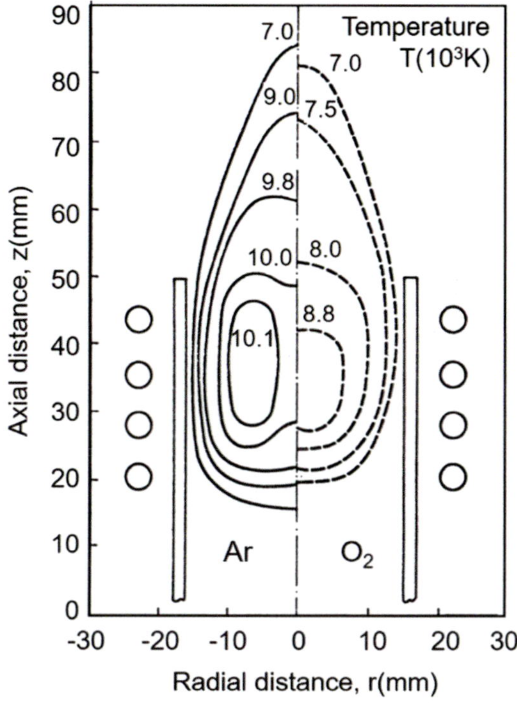

Fig. 5.42 Two-dimensional temperature mapping of, inductively coupled, atmospheric pressures, discharges of Argon (left) or oxygen (right), under comparable conditions. Gas flow rate = 30 slm (Ar or O_2), f = 10 MHz, power 12 kW for Ar and 8 kW for O_2 [Dresvin (1977)]

It should be pointed that this example provides a good idea about the particle density distribution in the flow of an RF inductively coupled discharge, compared to that of DC plasma jets that as noted earlier are characterized by considerably smaller plasma volumes, high particle flux densities, and particle velocities. Progress on both fronts for the DC and RF induction plasma sources has resulted in considerable increase in torch power ratings, powder throughputs, and the conditions of the treatment of these powders. The fundamentals behind these two approaches remain solid and valid.

5.5 Particle Trajectory and Temperature History

5.5.1 Model Formulation

In the present section, attention is focused on computation of the trajectories and temperature history of individual particles during their short residence time in the combustion or plasma stream. Fundamentals of momentum, heat, and mass transfer between an individual particle and a

Table 5.3 Experimental variables used for the LDA measurements in the induction plasma Lesinski and Boulos (1988)

Q_1 slm (Ar)	4.8	7.4	7.4
Q_2 slm (Ar)	11.7	11.7	11.7
Q_3 slm (Ar)	63.0	63.0	63.0
Plate power (kW)	4.6	4.6	10.3

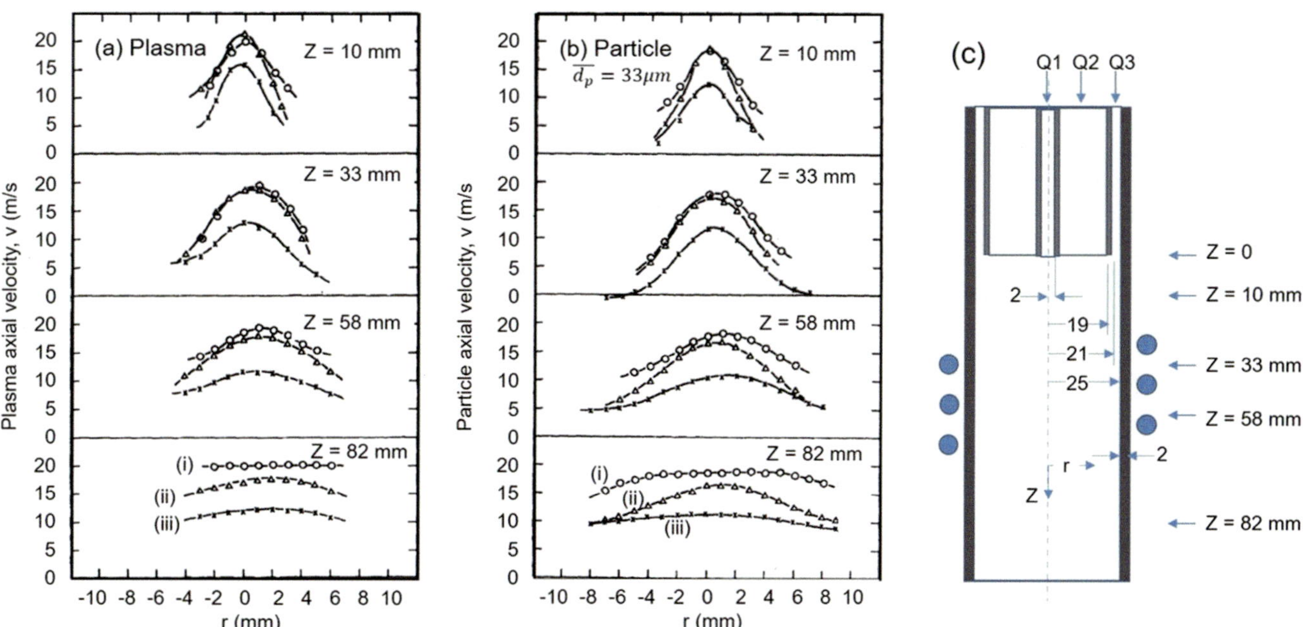

Fig. 5.43 (a) Plasma and (b) silicon particle, $d_p = 33 \pm 13$ μm, axial velocity distributions in an induction plasma flow and (c) schematic and principal dimensions of the plasma torch. Measurements carried out by LDA, see Table 5.3 for experimental variables [Lesinski and Boulos (1988)]

combustion or plasma stream were separately discussed in Chap. 4 *"Plasma–particle momentum and heat transfer."* The modeling of such systems requires the step-by-step computation of the motion of the individual particles in the combustion or high temperature plasma flow, coupled with the computation of plasma–particle heat exchange and thermophysical transformations in the particle such as particle melting, evaporation, and in some cases transformation in the particle chemical structure due to reactions between the particle and its environment. A typical example in this case is particle surface oxidation when carrying out the plasma spraying operation in an open atmosphere.

Considering that a powder, or a liquid spray, is normally composed of multiple particles, or drops, with different particle diameters and morphologies, a stochastic approach was proposed Boulos and Gauvin (1974), Vardelle et al. (1988) based on the division of the particle size distribution into a number of discrete subgroups, n_d, of particles as discussed in Chap. 4, Sect. 4.2, see Figs. 4.7 and 5.37. Each such subgroup is attributed a nominal diameter, d_{pi}, and an associated probability, f_{di}. Moreover, for a given powder carrier gas flow rate, the injection velocity of the individual particles will vary depending on their respective position and/or interactions in the powder transport line. Figure 5.38 shows a typical example of such a statistical distribution of the particle injection velocities for the transport of a fine alumina powder with three different powder carrier gas flow rates. To model such a situation, the respective distribution of the particle injection velocities can be divided into a number, n_v discrete injection velocities, v_{pi} with an associated probabilities, f_{vi}. The system model could then be carried out by repeating the individual particle trajectory computation for each size fraction, d_{pi} at the different injection velocities, v_{pi} giving rise to a total $(n_d \times n_v)$ trajectories with associates probabilities equal to $(f_{di} \times f_{vi})$. The final step of the model involving the processing of these results will depend on the objective of the model whether it is in terms of estimation of the molten fraction of the powder, or the temperature and velocity distributions of the particles prior to their impact on the substrate.

An important point to note at this stage is identify that values of the velocities and temperature fields of the plasma flow to use in these computations. Two approaches have been used:

The dilute flow approach involves the assumption that the powder feed rate in the process is sufficiently low that the energy transferred, momentum or thermal, from the plasma to the particles is negligible and accordingly the temperature and flow filed in the plasma is not affected by the presence of the particles. In this case, the model can be developed using experimentally available flow and temperature fields. Alternately, and more frequently, by running an appropriated

Computational Fluid Dynamic (CFD) model with the applicable boundary conditions and thermophysical properties, to calculate in a first step the flow and temperature fields in the plasma field in the absence of the particles, and use the results for the computation of the corresponding trajectories and temperature histories of the particles.

The way the flow is calculated of course plays a key role on the particle trajectories and, accordingly, its temperature and velocity. The temperature and velocity distributions of the gaseous flow have to be checked carefully, if possible, by comparison with measurements. When using, for example, a simple turbulence model with a mixing length as defined in the Genmix 2-D axisymmetric algorithm [Delluc et al. (2004)], different results can be obtained according to the choice of the mixing length L (m). This length can be taken from the Genmix program (calculated for flows below 1500 K) with the value corrected by:

- factor of 0.5 to achieve an under-expanded jet,
- by a factor of 1.5 (over-expanded jet),
- corrected by a factor related to the radial temperature distribution $[(T(r)/300)^{1/9}]$ to account for the laminar behavior of the jet at $T > 8000$ K and resulting in a slightly under-expanded jet.

The influence of the choice of the turbulence model has also been emphasized by Legros (2003) using 3D models: K–ε and different R_{ij}–ε models resulting, for the same spraying conditions, in different temperature and velocity distributions

The dense loading approach is applicable in the case of high powder feed rates into the plasma flow with the momentum and heat, transferred from the plasma to the particles sufficiently high in order to affect significantly the original plasma temperature and flow fields prior to the introduction of the powder. In this case an iterative procedure is necessary to account for the plasma–particle interactions as described by Proulx et al. (1985, 1987b) based on the Particle-Source-In-Cell approach by Crowe et al. (1977). Detailed description of this model is provided later in this chapter, Sect. 5.6 Plasma–particle interaction under dense loading conditions.

5.5.2 Single Particles Motion in Combustion or Plasma Stream

5.5.2.1 Equations of Motion

During the displacement of a particle in a hot gas or high energy plasma stream it is subjected to a number of forces, which act simultaneously on it and have varying influence on its trajectory and residence time in the plasma. These include:

- Inertia force, F_i function of the particel mass and velocity.
- Drag force, F_D function of the relative velocity between the particle and free stream.
- Gravity force, F_g function of the particlee mass.
- Added mass, F_{add} and Basset history term that are significant in the case of intense particle acceleration.
- Bronian force, F_B due to molecular collisions between the particle and elemental atomic or molecular particels. It is important for very fine submicron particles.
- Pressure gradient force, F_p in the presence of steep pressure gradients in the flow filed.
- Coriolis forces, F_C resulting from the rotation of the particles about an axis parallel to the direction of relative motion.
- Thermophorises forces, F_{Th} that is important for very fine particels $d_p < 1$ µm in the presence of steep temerpature gradients. It drives the particles toward the low tremperature end of the flow.
- Electrostatic forces, F_{elec} in the case of charged particles.

The most important among these forces are the ***inertia, F_i*** and ***drag F_D*** forces that need to be taken into account. To these, the ***gravity F_g*** force is added in situation when dealing with relatively large particles in low velocity conditions [Boulos et al. (1993a, b), Chyou and Pfender (1989)]:

$$F_i = \frac{\pi d_p^3}{6} \cdot \rho_p \cdot \frac{du_p}{dt} \tag{5.15}$$

$$F_D = \frac{\pi d_p^2}{4} \cdot C_D \cdot \frac{1}{2}\rho \cdot U_r^2 \tag{5.16}$$

$$F_g = \frac{\pi d_p^3}{6} \cdot \rho_p \cdot g \tag{5.17}$$

with d_p particle diameter (m), ρ_p particle density (kg/m^3), ρ plasma density (kg/m^3), C_D drag coefficient (−), u_p particle velocity (m/s), U_r relative velocity between the plasma and the particle (m/s), t time (s).

Considering that in most cases encountered in melting and processing of powders under plasma conditions, F_P (except for D-gun, cold spray, and to some extent HVOF) F_B, F_T, and F_C are negligible compared to F_i, F_D, and F_g. Therefore, a force balance around a single particle in motion in a plasma flow can be written simply as:

$$F_i = F_D + F_g \tag{5.18}$$

Substituting Eqs. (5.15), (5.16), and (5.17) in Eq. (5.18), which can be written in axisymmetric cylindrical coordinates as follows:

$$\frac{du_p}{dt} = -\frac{3}{4} C_D (u_p - u) u_R \left(\frac{\rho}{\rho_p d_p}\right) \pm g \tag{5.19}$$

$$\frac{dv_p}{dt} = -\frac{3}{4} C_D (v_p - v) v_R \left(\frac{\rho}{\rho_p d_p}\right) \pm g \tag{5.20}$$

u_p is the particle velocity (m/s) in the z direction of the plasma along the jet axis, u is the corresponding plasma velocity (m/s), and u_R is the relative velocity between the plasma and the particle in the z direction, v_p is the particle velocity (m/s) in the r-direction, v is the corresponding plasma velocity (m/s), and v_R is the relative velocity in the r direction. The acceleration due to gravity force ($\pm g$) has to be added or subtracted from either the axial or radial velocity components equations depending on flow orientation.

In the case of RF plasmas, generally oriented vertically, the (+) or (−) sign in front of the gravitational acceleration, g, corresponds to a flow in axial direction with the positive direction for the velocity of the particles being downwards or upwards, respectively.

For a DC plasmas, HVOF, or flames, where the axis is generally horizontal, (+) corresponds to a vertical injection downward and (−) to the upward injection.

If cylindrical symmetry does not exist anymore, which is the case when injecting particles radially, Eq. (5.19) or (5.20) holds but with the velocity vector and its derivative, as well as the gravity acceleration vector, must be described in a specific equation. When turbulent effects are taken into account, which is particularly important for small particles ($d_p < 20$ µm) with a low specific mass such as alumina, the turbulent component u' and v' must be added to the velocity components u and v [Rudinger (1980)], Chang (1993), Chyou and Pfender (1989)].

As discussed in Chap. 4 Plasma–particle momentum and heat transfer, Sect. 4.3, the drag coefficient, C_D, to be used in Eqs. (5.19) and (5.20), is a function of the Reynolds number of the flow around the particle defined as:

$$Re = \frac{\rho\, U_r d_p}{\mu} \tag{5.21}$$

where U_r is the relative velocity between the particle and the free stream, defined as

$$U_r = \sqrt{(u_p - u) + (v_p - v)} \tag{5.22}$$

Different correlations were given for the calculation of the drag coefficient of a spherical particle as function of the Reynolds number, the most widely used are given in Table 5.4.

As mentioned in Sect. 4.3.2, the drag coefficient calculated using these correlations needs to be corrected for

Table 5.4 Drag coefficient for a spherical particle as function of the Reynolds number

$C_D = \frac{24}{Re}$	$Re \leq 0.2$
$C_D = \frac{24}{Re}\left(1 + 0.1\,Re^{0.99}\right)$	$0.2 < Re \leq 2$
$C_D = \frac{24}{Re}\left(1 + 0.11\,Re^{0.81}\right)$	$2 < Re \leq 20$
$C_D = \frac{24}{Re}\left(1 + 0.189\,Re^{0.62}\right)$	$20 < Re \leq 500$
$C_D = 0.44$	$Re > 500$

thermal plasma conditions integrating the effect of temperature gradients as:

$$C_D = C_{Df}\left(\nu_f/\nu_\infty\right)^{0.15} \tag{5.23}$$

where C_{Df} is the drag coefficient evaluated at the mean film temperature, $T_f = \frac{(T_s + T_\infty)}{2}$ and ν_f, and ν_∞ are the kinematic viscosity ($\nu = \mu/\rho$) evaluated respectively at the mean film temperature, T_f and free stream plasma temperature, T_∞ for operation under low pressures, further corrections of the drag coefficient might be necessary for continuum effects as discussed in Sect. 4.4.2.3.

5.5.2.2 In-Flight Particle Heating, Melting, and Evaporation

The particle temperature calculations during its trajectory in a combustion flow or plasma are carried out by an energy balance on the particle (see Chap. 4 Plasma–particle momentum and heat transfer, Sect. 4.4). The net energy contributing to the heating or cooling of the particle, Q_n, is the difference between the heat received by the particle from the free stream by conduction and convection, Q_{cv}, and that lost from the surface of the particle by radiation to its surrounding, Q_{sr}.

$$\begin{aligned}
Q_n &= Q_{cv} - Q_{sr} \\
&= h\,a_p\left(T_\infty - T_s\right) - a_p\,\varepsilon\,\sigma_s\left(T_s^4 - T_a^4\right)
\end{aligned} \tag{5.24}$$

where

h heat transfer coefficient (W/m^2 K),
a_p surface area of the particle, for a sphere $a_p = \pi\,d_p^2$ (m^2),
T_∞ free-stream plasma temperature (K)
T_s surface temperature of the particle (K).
ε particle emissivity (varies from 0 to 1.0)
σ_s Stephan–Boltzmann constant (5.67×10^{-8} W/m^2 K^4).

Based on available literature data, the heat transfer coefficient can be calculated using the following correlation function of the Reynolds $\left(Re = \frac{\rho\,U_r d_p}{\mu}\right)$ and the Prandtl numbers $\left(Pr = \frac{C_p \mu}{k}\right)$.

$$Nu = 2.0 + 0.6\,Re^{0.5}Pr^{0.33} \tag{5.25}$$

where

Nu the Nusselt number $\left(Nu = \frac{h d_p}{\kappa}\right)$.
κ Thermal conductivity of the plasma boundary layer

Under plasma conditions, the thermal conductivity of the gas is evaluated at the mean film temperature $T_f = \frac{(T_s + T_\infty)}{2}$. Different corrections have been proposed such as due to Fizsdon (1979);

$$Nu = \left(2.0 + 0.6\,Re^{1/2}\,Pr^{1/3}\right)\left(\frac{\rho_\infty \mu_\infty}{\rho_s \mu_s}\right)^{0.16} \tag{5.26}$$

Or, Lee et al. (1981)

$$Nu = \left(2.0 + 0.6\,Re^{1/2}\,Pr^{1/3}\right)\left(\frac{\rho_\infty \mu_\infty}{\rho_s \mu_s}\right)^{0.6}\left(\frac{c_{p\infty}}{c_{p\,s}}\right) \tag{5.27}$$

It has also been proposed by Bourdin et al. (1983), to use an integral-mean thermal conductivity, $\bar{\kappa}$ defined as;

$$\bar{\kappa} = \frac{1}{\left(T_\infty - T_p\right)}\left[\int_{T_p}^{T_\infty}\kappa(T)dT\right] \tag{5.28}$$

$$\bar{\kappa} = \frac{1}{\left(T_\infty - T_p\right)}\left[\int_{300}^{T_\infty}\kappa(T)dT - \int_{300}^{T_p}\kappa(T)dT\right] \tag{5.29}$$

$$\bar{\kappa} = \frac{1}{\left(T_\infty - T_p\right)}\left[S(T_\infty) - S(T_p)\right] \tag{5.30}$$

where $S(T)$ is the heat conduction potential

Based on the value of Q_n the thermophysical transformations in the particles can be calculated assuming either, a uniform particle temperature, which is generally valid for metallic particles. Alternately calculations have to be made of the heat conduction in the particle as discussed in great details in Chap. 4 Plasma–particle momentum and heat transfer, Sect. 4.5. It might be useful to note that the Bio number $\left(Bi = \bar{\kappa}/\kappa_p\right)$, which represents the ratio integral mean thermal conductivity of the gas in the boundary layer surrounding the particle, $\bar{\kappa}$ to the thermal conductivity of the particle, κ_p is a good indicator whether the particle can be assumed to have a uniform temperature or not. According to Bourdin et al. (1983) important differences can exist between the surface temperature of the particle and its center, when the Biot number is larger than 0.03. In other words, the particle can be assumed to have a uniform temperature if $Bi < 0.03$.

Further corrections of the calculated heat transfer rate to the particles need to be applied in the case the particle temperature approaches its boiling point of its material giving rise to vaporization of the particle. As discussed in Chap. 4 Plasma–particle momentum and heat transfer, Sect. 4.6, a significant drop in the heat transfer rate to the particle can be expected in this case due to the thickening of the boundary layer surrounding the particle and its cooling by the evolved vapors. In the case of evaporation of metallic particles, the cooling of the boundary layer surrounding the particle is significantly enhanced resulting from intense radiation losses from the generated metal vapor cloud at plasma surrounding the particle Chap. 4 Plasma–particle momentum and heat transfer, Sect. 4.6, and Xue et al. (2019).

The governing equations for particle heating, melting, and evaporating throughout its trajectory in the combustion or high temperature plasma flows can be summarized as follows:

For $T_p < T_m$ or $T_m < T_p < T_v$

$$\frac{dT_p}{dt} = \frac{6Q_n}{\pi d_p^3 \rho_p c_{ps}} \tag{5.31}$$

where T_m and T_v are respectively the melting and boiling temperatures of the particle material, and c_{ps} is the specific heat of the particle.

For $T_p = T_m$, the particle undergoes a phase change with the mass fraction of the molten material, x, given by;

$$\frac{dx}{dt} = \frac{6Q_n}{\pi d_p^3 \rho_p H_m} \tag{5.32}$$

where H_m is the latent heat of fusion of the particle material.

For $T_p = T_v$, the particle evaporates with the rate of change of the particle diameter given by;

$$\frac{dd_p}{dt} = \frac{6Q_n}{\pi d_p^3 \rho_p H_v} \tag{5.33}$$

where H_v is the latent heat of vaporization of the particle material.

It should be pointed out that Eq. (5.33) is only an approximation and that evaporation is likely to take place well below the boiling point of the material (see section "Radiation of the Vapor"). As shown by Chen et al. (1985) the shielding effect of the particle by the evolving vapor could lead to considerable slowing down of the heating rate of the liquid droplet with the possibility of complete evaporation of the particle before it ever reaches its boiling temperature. Interestingly, Chen et al. (1985) also show that these phenomena do not seem to affect, to any considerable extent, the estimate of the time required for the complete evaporation of the particle.

5.5.3 Particle Trajectory in Combustion and DC Plasmas

5.5.3.1 Influence of the Injection Conditions

The importance of the injection conditions in DC plasma jets has been emphasized by many authors. This is illustrated below with the results of Delluc et al. (2004) for fully dense zirconia particles of different diameters injected into a DC plasma torch with an exit nozzle 7.0 mm i.d., operating using Ar-H$_2$ mixture as plasma gas (45 slm (Ar) + 15 slm H$_2$) working at 43 kW with a thermal efficiency of 60%. The particles are externally injected 4.5 mm downstream of the nozzle exit and at a distance of 8.5 mm from the torch axis. The injection is orthogonal to the jet axis. The trajectories followed by 30 μm diameter zirconia particle injected with velocities between 4.71 and 47.15 m/s are given in Fig. 5.44a, b. The radial distance = 0 corresponds to the torch axis. Obviously, with an injection velocity of 4.71 m/s the particle inertia is too small, and the particle crosses the torch axis only 150 mm downstream of the nozzle exit, essentially in the plasma plume where the temperature is below 2000 K. A particle injection velocity of 47.15 m/s is, on the other hand, excessive since the particle crosses the core of the plasma jet immerging from the other side. Optimal injection velocity seems to be in the range 10–20 m/s giving the particle a trajectory with a maximum residence time in the core of the plasma jet for maximum heating and acceleration by the plasma flow. Such a trajectory would be at an angle of 3.5–4° with respect to the axis of the torch that is recognize as being optimal in plasma spraying.

The effect of particle injection angle on the particle trajectory was also investigated by Delluc et al. (2004) for the same zirconia powder and operating conditions of the DC plasma torch. The results given in Fig. 5.44c, d for injection angles of 70° and 110° with respect to the axis of the torch show only a marginal influence of the injection angle on the particle trajectories over this range.

The effect of torch power on particle trajectories and the ease by which 30 μm zirconia particles can penetrate an Ar-H$_2$ plasma (45 slm (Ar) + 15 slm H$_2$) generated at power levels varying between 22 and 60 kW with a thermal efficiency of 60% was investigated by Delluc et al. (2004). The particle injection velocity was set at 20 m/s that was noted, based on earlier results (Fig. 5.44a, b), to be optimal for a plasma power of 43 kW. The results given in Fig. 5.45a, b show that as the plasma momentum density increases with increasing power level, the particle penetration of the plasma is significantly reduced. On the other hand, the particles are observed to have less difficulty to penetrate the plasma at lower power levels due to the decrease of the plasma momentum density. It is also noted that the particle velocity increases with the increase of the power level provided the

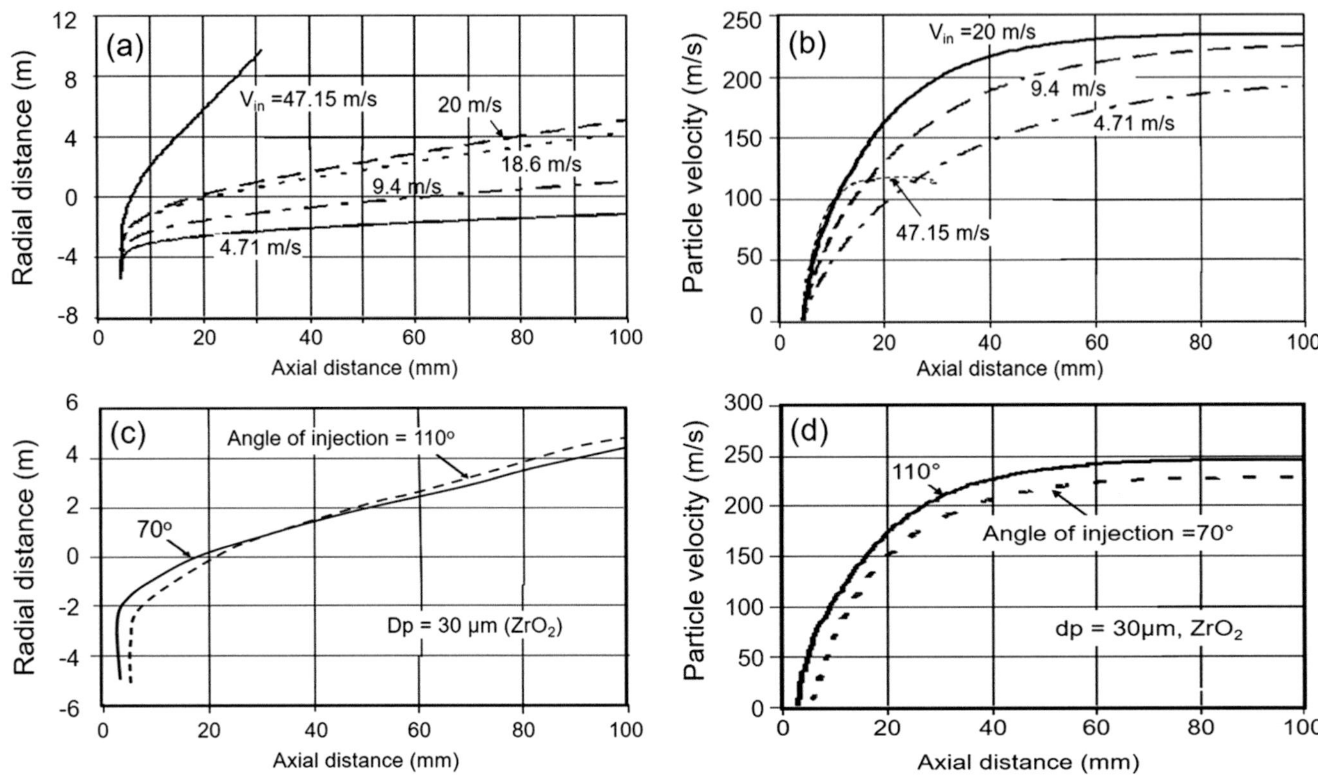

Fig. 5.44 (**a** and **c**) Particle trajectory (**b** and **d**) particle velocity, for 30 μm zirconia powder in externally injected into a DC. Plasma jet (45 slm Ar + 15 slm H$_2$) 43 kW, with different injection velocities and injection angles, radial distance = 0 corresponds to torch axis [Delluc et al. (2004)]

injection velocity is adapted to the each power level to achieve the optimum trajectory.

Similar results are obtained with a combustion source such as HVOF, when the particle injection is performed radially downstream of the nozzle throat [Kamnis et al. (2008)]. In the calculation presented in Fig. 5.46, 20 μm diameter Inconel 718 particles are injected at the axial distance of 120 mm. The injection velocities vary in the range 0–40 m/s. The particle trajectories show that at low injection velocities, close to zero, the particle is entrained by the gas flow and travels along the edge of the barrel. The increase of injection velocity to 8 and 10 m/s enables the particle to penetrate the gas flow and move toward the center of the jet. At an injection velocity of 40 m/s the particle penetrates the hot gas flow, crosses it immerging from the opposite side hitting the internal surface of the barrel. At this point the particle trajectory is changed to the opposite direction as a result of the elastic collision. Nozzle wear is one of the most frequently encountered problems for operating HVOF guns and the nozzle needs to be replaced after about 10 h spraying. Thus, the injection velocity of particles has to be carefully chosen.

It should be pointed out that, in real world, where a powder is injected and not a particle, particles have different trajectories, due to their collisions between themselves and with the injector wall acquiring a divergence angle relative to the injector axis that can reach a total value of 40°, especially for small (d$_p$ < 20 μm) and specially for light particles such as alumina. Such trajectory distributions make calculations more difficult necessitating the use of stochastic models with probability distributions linked to different particle trajectories. As a first approximation, the trajectory of a single particle with a mean diameter (d_{50}) injected with the proper injection velocity can be reasonably representative of the mean trajectory of an ensemble of particles.

It is also worth noting that when the injection distance increases, due to the jet expansion, the jet momentum density decreases and, accordingly, the injection velocity for a trajectory close to the torch axis decreases. This is illustrated by the work of Hanson et al. (2002) who have studied an HVOF thermal spray torch that has a conical supersonic nozzle with several different particle injection locations. Particle injection location was found to determine the residence time of particles within the nozzle, thus providing a means to control particle temperature. Finally, injection location had a negligible effect on particle velocity according to the numerical model.

5.5.3.2 Optimization of the Injection

Vardelle et al. (1998) calculated the rate of change of momentum of particles due to the drag force in their direction

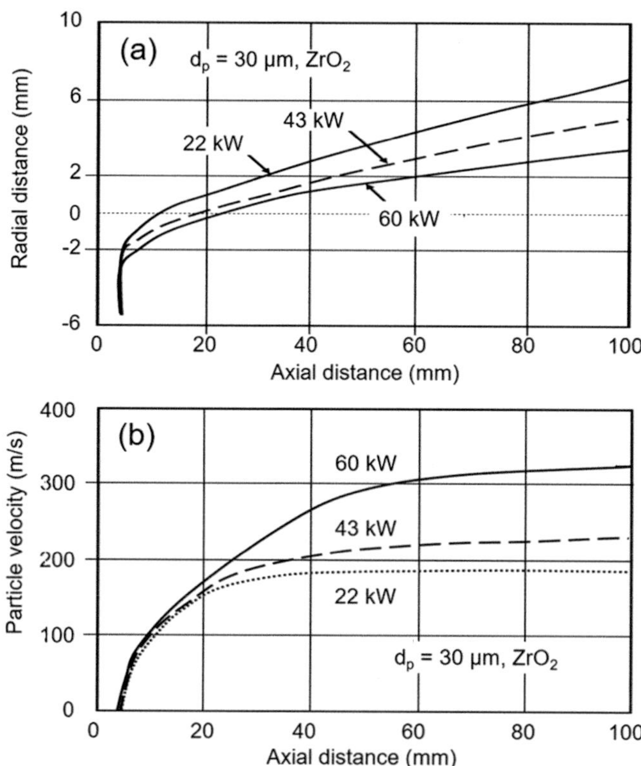

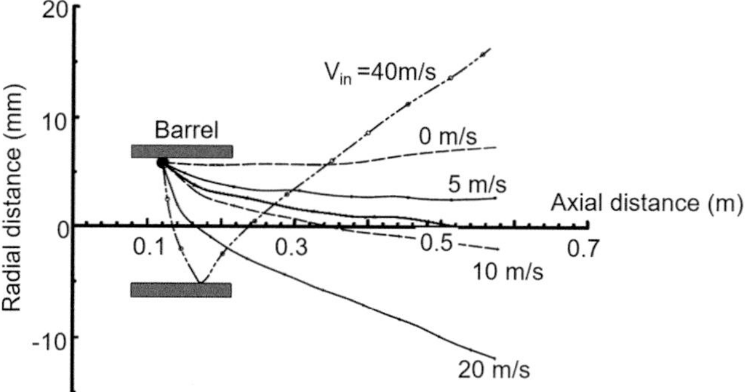

Fig. 5.46 Inconel 718 20 µm particle trajectories at different injection velocities, V$_{in}$: radial injection downstream the nozzle throat in HVOF gun working with kerosene and oxygen. Reprinted with kind permission from Elsevier [Kamnis et al. (2008)]

Fig. 5.45 (**a**) Particle trajectory (**b**) Particle velocity, for 30 µm zirconia powder in externally injected into a DC. Plasma jet (45 slm Ar + 15 slm H$_2$) operated, with power levels varying between 22 and 60 kW, and injection velocity of 20 m/s in an orthogonal direction to the flow, radial distance = 0 corresponds to torch axis [Delluc et al. (2004)]

of motion for a DC plasma jet. Assuming that the drag force corresponds to one in a Stokes regime with a constant plasma viscosity μ_g, the total time of flight of the particle depends on a reference time t_r:

$$t_r = \frac{\rho_p \cdot d_p^2}{\mu_g} \qquad (5.34)$$

The results given in Fig. 5.47 show the residence time for various particles as a function of their specific mass and diameter as they are injected in an Ar-H$_2$ (45 slm–15 slm) DC plasma jet with an effective power level of 21.5 kW and a nozzle i.d. of 7 mm. The required injection velocity for the different particles to "land" on nearly the same location on the substrate was calculated and the results integrated on the right-hand y-coordinate of Fig. 5.47. For a given plasma jet, the injection velocity varies drastically with the particle size and specific mass.

5.5.3.3 Influence of Plasma Jet Fluctuations

As will be described later in Chap. 8 on plasma spraying using DC plasma sources, the restrike mode is observed with plasmas containing diatomic gases, giving rise to voltage

fluctuations with the highest voltage more than twice the lowest one. As the power supply is generally a current source, it results in power fluctuations proportional to those of the voltage with subsequent plasma jet fluctuations as shown in Fig. 5.48. The frequency of these fluctuations being a few 1000 Hz. Since the carrier gas flow rate cannot be made to follow such intense power changes, it is usually set to achieve the optimum trajectory with the time averaged value of the power dissipated in the torch. This is illustrated in Fig. 5.49 [Moreau et al. (2006)] showing the trajectories, in the y–z plane, of 25 µm alumina particles injected into the Ar-H$_2$ plasma jet at different instants of the arc fluctuation period but with the same injection conditions: particles are located on the axis of the injector, and they leave the injector with an axial velocity of 20 m/s. Under the conditions of the study, the way the trajectory is affected by flow fluctuations depends essentially on the particle momentum with respect to the instantaneous momentum of the gas at the point where the particle penetrates the jet flow. The width of the particle spray jet, on impact at the substrate, is 3.5 mm for 25 µm particles, but is as large as 6.6 mm for 10 µm particles. Meillot and Balmigere (2008) obtained similar results.

5.5.4 Trajectory Corrections Due to Various Effects

As already emphasized in Chap. 4, Plasma–particle momentum and heat transfer, the unique features of the flow field and temperature fields under plasma conditions can have a significant impact on plasma–particle momentum and heat transfer phenomena. A number of approaches were proposed for the correction of the drag and heat transfer coefficients for the effect of steep temperature gradients, turbulence, rarefication of the flow under low pressure conditions as well as particle-related phenomena such as particle, shape, particle porosity, particle charging, and particle vaporization. In this section, a detailed analysis is given of the major phenomena involved and the impact such corrections could have on the predicted particle trajectory and temperature history in combustion and DC plasma sources.

5.5.4.1 Effect of Temperature Gradient

A number of approaches were proposed in Chap. 4, Sect. 4.3.2 for the correction of the drag coefficient for thermal plasma conditions. These were tested by [Chyou and Pfender (1989)] for the case of a 20 μm alumina particle injected internally into an Ar-H$_2$ DC plasma jet using the isotherms

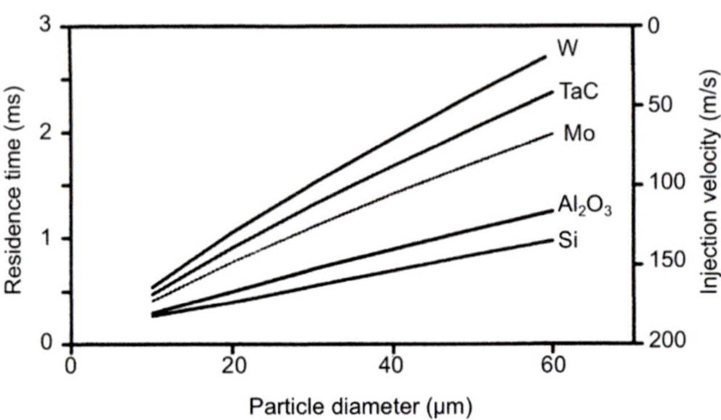

Fig. 5.47 Particle residence time as a function of particle diameter, for different materials with a wide range of specific mass (plasma forming gas: 45 slm Ar + 15 slm H$_2$, effective power: 21.5 kW, nozzle i.d.: 7 mm) [Vardelle et al. (1998)]. Reprinted with kind permission from Springer Science Business Media

and velocities measured by Fauchais et al. (1989). The results given in Fig. 5.50 show clearly a strong dependence of the predicted trajectories on the choice of the correction used for the drag coefficient.

5.5.4.2 Effect of Rarefaction and Vaporization

Once the correction for the steep gradients has been chosen, the corrections for non-continuum effects and vaporization have also to be determined. This is illustrated in Fig. 5.51 for 20 μm diameter alumina particles (this size is chosen because it shows a significant non-continuum effect) injected at 25 m/s (to obtain about the same trajectory as that of the 60 μm particles). Here again different types of corrections are used to show the impact they have on the calculated particle trajectory. The corresponding effect on the particle velocity along each of these trajectories is shown in Fig. 5.52.

5.5.4.3 Effect of Turbulence

The effect of turbulence on the particle motion has also been shown to have an important influence on the small particle trajectory by Pfender (1989). Typical results for 10–50 μm alumina particles injected into an argon DC turbulent plasma jet are given in Fig. 5.53.

The ranges of dispersed particle trajectories are obviously larger for the smaller particles (10 μm). This is due to the small mass of such particles that follow easier the motion within eddies. It is also observed that the dispersion becomes larger downstream. This is caused by the

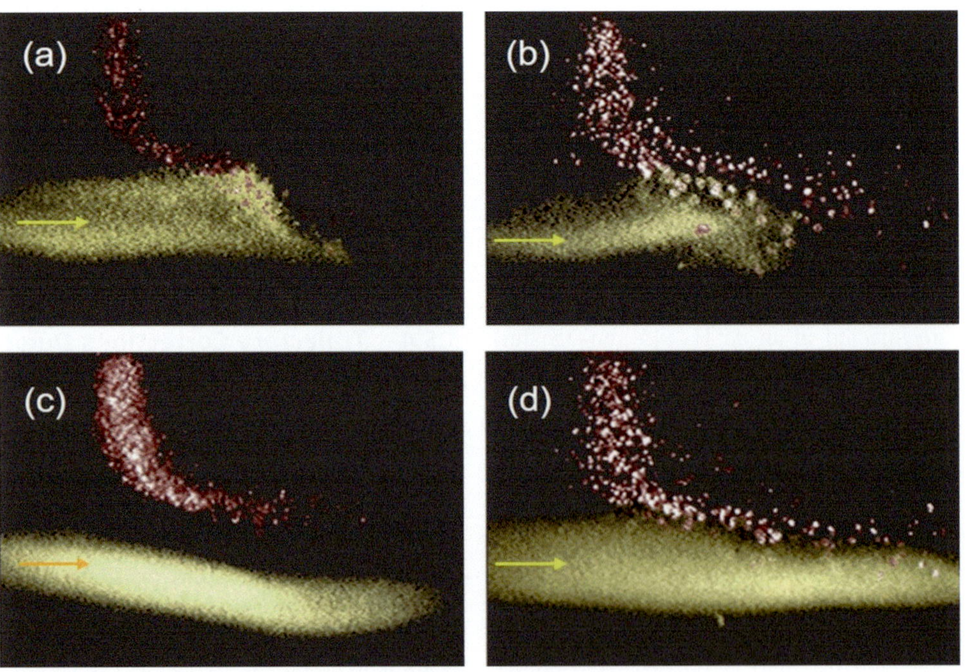

Fig. 5.48 High-speed images showing the fluctuations of the plasma jet and the spray particle fluxes being exposed to varying environments [Fauchais (2004)] (Copyright IOP Publishing. Reproduced with permission. All rights reserved)

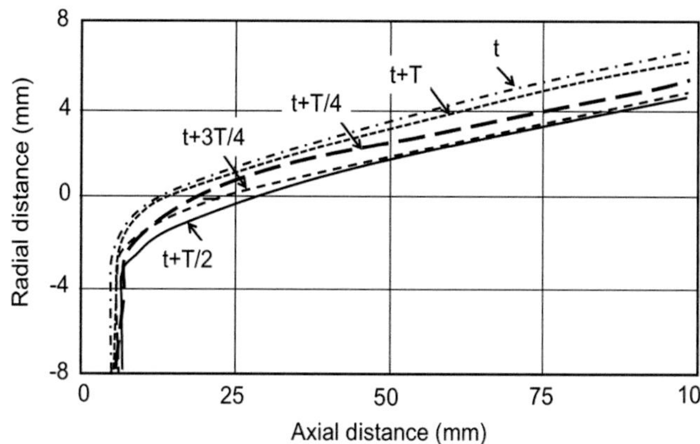

Fig. 5.49 Trajectories of 25 μm alumina particles injected into an Ar-H$_2$ DC plasma jet (45–15 slm, 600 A, anode-nozzle i.d. 7 mm) at different instants of the fluctuation period with an injection velocity of 20 m/s (y–z plane) [Moreau et al. (2006)]. Reprinted with kind permission from Springer Science Business Media, copyright © ASM International

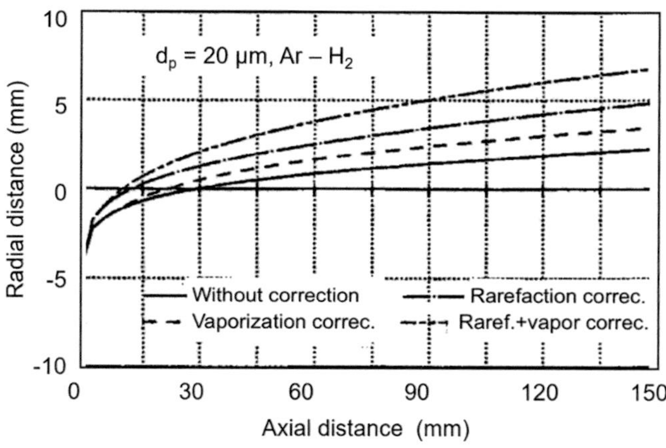

Fig. 5.51 Trajectories of 20 μm in diameter alumina particles injected at 25 m/s into the plasma jet depicted in Fig. 5.50. The temperature gradient correction was calculated using the mean integrated properties, Vardelle et al. (1997a, b)

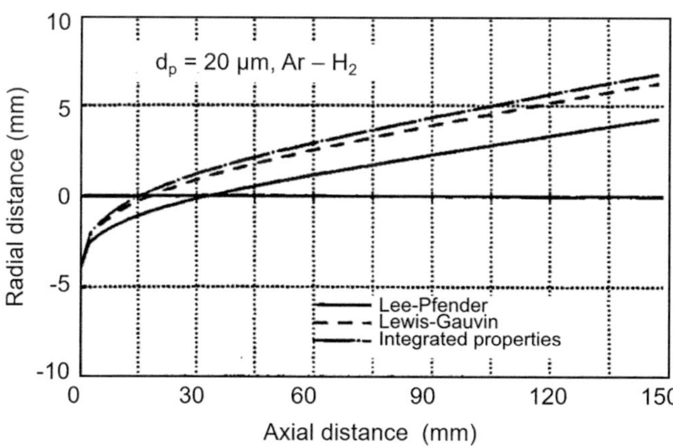

Fig. 5.50 Effect of the correction used for the drag coefficient on the trajectories of 20 μm diameter alumina particle injected at 8 m/s into the plasma jet (75 slm (Ar) +15 slm H$_2$), P = 29 kW, ρ_{th} = 63%, nozzle i.d. = 8 mm [Chyou and Pfender (1989), Lee et al. (1981), Lewis and Gauvin (1973), Bourdin et al. (1983)]

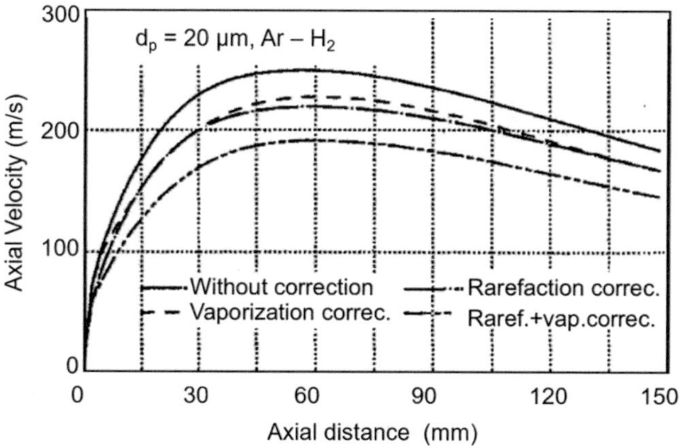

Fig. 5.52 Particle velocity along the trajectories shown in Fig. 5.51 with different drag coefficient corrections for non-continuum effects and particle Vardelle et al. (1997a, b)

$$F_{th} = \frac{p.\lambda.d_p^2.\Delta T}{T_p} \tag{5.35}$$

accumulated "random walk" influence. Comparing Fig. 5.53 with Figs. 5.50, 5.51 and 5.52 it seems that the different corrections of the equations of motion may still exceed the influence of turbulence, for particles larger than 15 μm in diameter.

5.5.4.4 Thermophoresis Effect

Thermophoresis is a phenomena that can affect the trajectory of relatively fine submicron particles ($d_p < 1$ μm) in the presence of steep temperature gradients in the flow. Under such condition the developed thermophoretic force, F_{th} given by Eq. 5.35, drives the particles in the negative direction of the temperature gradient towards the colder regions of the flow [Pfender E (1989)].

where p is the gas pressure, λ the mean free path of the gas, d_p the particle diameter, ΔT the temperature gradient between the front and rear faces of the particle, and T_p the particle temperature. This effect is important for particle sizes below 1 μm and becomes negligible for particles over 5 μm. Figure 5.54 shows the influence of the thermophoresis effect on a 1 μm particle trajectory injected at a radial distance of 2 mm from the jet axis and 15 mm from the nozzle exit. The velocities and temperature fields are for an Ar-H$_2$ plasma jet with a 6 mm nozzle diameter and a net power of 32 kW supplied to the gas. The bright part of the jet core is surrounded by the 11,000 K isotherm. The particle trajectory is almost parallel to the torch axis (an angle of 3–4°) and once it arrives in strong gradient

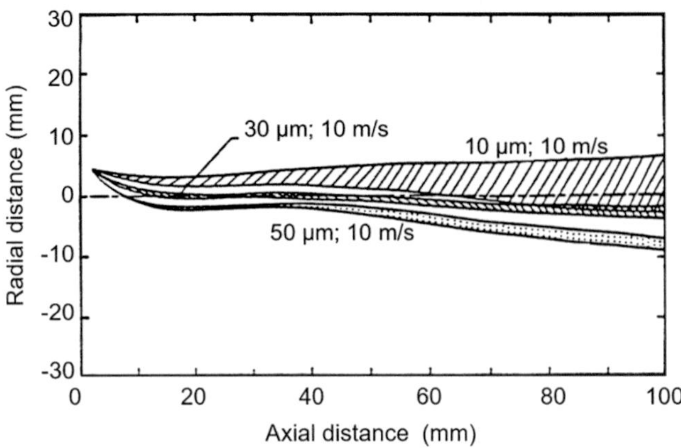

Fig. 5.53 Dispersed 10–50 μm alumina particle trajectories in an argon, DC turbulent plasma jet. Reprinted with kind permission from Springer Science Business Media Pfender (1989)

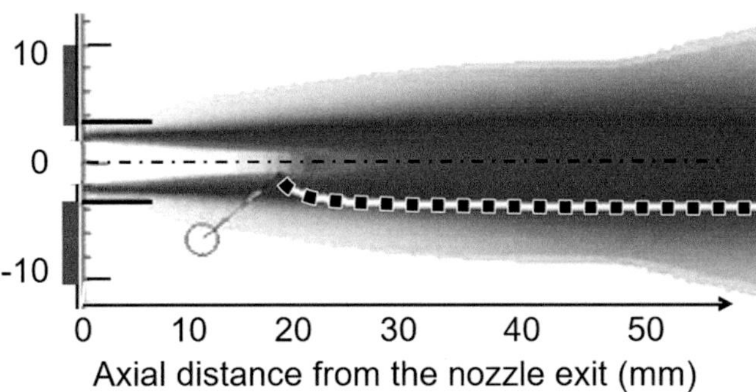

Fig. 5.54 Thermophoresis effect on the trajectory of a 1-μm zirconia particle [Chen et al. (1985)]. Reprinted with kind permission from Springer Science Business Media

region, it is ejected toward the jet periphery and once the temperature gradient is reduced, the particle tends to come back parallel to the jet axis. The phenomenon is enhanced as the particle size decreases. It means that small particles, even correctly entering the plasma, can escape from the core and travel in the fringes.

5.5.4.5 Other Effects
(a) *Particle Shape*

In all preceding calculations, the particles have been assumed to be spherical. If this is the case for agglomerated, spray dried, atomized powders, or those produced by sol–gel techniques, it is not at all true for fused and crushed particles, which can have angular shapes with a low shape factor, see Chap. 4, Fig. 4.1. The shape of the particle modifies its drag coefficient [Ganser (1993)], which, for example, can be correlated to the particle sphericity factor in a limited range of shape effects [Fukanuma et al. (2006a, b)]. However, it must be emphasized that particles when injected into the plasma or HVOF or flame jets are rapidly (approximately a few tens of μs) heated above or close to their melting point, thus attaining a spherical shape. For example, in DC plasma spraying, the non-sphericity effect plays a role mainly close to the injector. The increase in drag force of oblate spheroids compared to those of equal volume spheres may appreciably affect the particle trajectory, thus its temperature history [Fukanuma et al. (2006a, b) and Xu et al. (2002)]. This non-sphericity modifies much more the behavior of particles within the injector.

(b) *Particle Charging*

Particle charging can also have a limited effect on particle trajectory in plasma jets, since a particle injected into thermal plasma will always assume a negative charge due to the difference between the thermal velocities and mobilities of electrons and ions. Whether or not this affects particles drag in thermal plasmas has never been explored. Under LTE conditions [Pfender (1989), Chyou and Pfender (1989), Chen and Xiaoming Chen (1986)] in the boundary layer, particle charging will be of minor importance, because of the low charge concentration that can exist in the region near the particle surface. This may, however, be different for frozen chemistry, under non-continuum conditions in the case of low-pressure plasma spraying.

Deviations from LTE conditions in the plasma can also be important, particularly close to the particle surface. Whether or not such deviations from LTE will substantially affect heat and momentum transfer to particles in the condensed phase, remains still a matter for further study. Studies of momentum and heat transfer between low-pressure plasma and particles are generally based on molecular dynamics and they are beyond the scope of this book. However, the interested reader can find information in refs. [Chen and Chen (1989), Uglov and Gnedovets (1991), Gnedovets and Uglov (1992), Chen et al. (1995) and Soo (1967)].

(c) *Problem of Particle Inertia*

When spraying fine particles in the micrometer or nanometer size range they tend to follow the flow over the surface where they are to be deposited. The following question must be considered: under which conditions will such small particles reach the surface of the substrate and not continue to follow the gas flow, as expected according to their low inertia? The answer is given by the Stokes number [Fauchais et al. (2008)]:

$$St = \frac{\rho_p.d_p^2.v_p}{\mu_g.l_{BL}} \qquad (5.36)$$

where the index p relates to the liquid particle, g to the gas, and BL is the fluid boundary layer thickness. St must be larger than 1 for the particle to cross the boundary layer and reach the substrate. Considering a zirconia particle with an impact velocity of 300 m/s, a boundary layer of 1 mm, and an Ar-H$_2$ plasma containing 30% air (50 mm spray distance), with a Stokes number of 10, a 1-μm particle impacts onto the substrate, but $St = 1$ for a 0.3 μm particle and with sizes below none of them will impact the substrate. The limiting size depends strongly on the particle impact velocity at the edge of the boundary layer, and on the flow velocity close to the substrate, controlling the boundary layer thickness. The latter is inversely proportional to the square root of the flow velocity. Particle impact thus depends on the spray gun used, its working conditions, especially its nozzle design, the gas flow rate, the power dissipated, and the spray distance, especially when it is below 50 mm as in suspension plasma spraying. The longer the spray distance, the more the hot jet expands and mixes with the surrounding atmosphere, thus increasing the flow boundary layer thickness and reducing the Stokes number.

5.5.5 Particle Trajectory in Induction Plasmas

The modeling of RF inductively coupled discharges is particularly complex because of the interaction between the applied magnetic field and the induced electromagnetic fields in the discharge. This gives rise to recirculation flows in the induction coil region. The flow field is also characterized by its relatively low velocity except for the entrance region of the sheath gas where the high flow gives rise to intense shear and localized turbulence. A number of studies have been reported since the early seventies on the modeling of induction plasmas [Boulos (1976, 1985, 1997), Mostaghimi et al. (1985a, b) and Colombo et al. (2010)].

A study of particle trajectories and temperature history calculations in an inductively coupled plasma was reported by Boulos (1978). A schematic of the inducting plasma torch is given in Fig. 5.55a. Details of the principal dimensions and operating conditions are listed in Table 5.5. The modeling of the particle dynamics was based on initial results of the flow and temperature fields for a pure argon plasma at atmospheric pressure in the absence of the particle given in Fig. 5.55b, c [Boulos (1976)]. The 2-D particle trajectory was computed using Eqs. 5.19, 5.20, 5.21 and 5.22 and drag coefficient corrections for presence of steep temperature gradients. Particle temperature calculations were based on an energy balance using Eqs. 5.24 and 5.25 combined with Eqs. 5.31, 5.32 and 5.33. The individual particles were assumed to be at all time with a uniform temperature with negligible internal temperature gradients.

Computations were carried out for alumina particles with a particle size distribution in the range from 10 to 200 μm, injected axially into the discharge upstream of the induction coil with a distance of 42 mm between the tip of the powder injection probe and the first turn of the induction coil. Typical results are given in Fig. 5.56. These show the particle trajectory for different particle sizes and initial injection velocities, V_i. The open circles are representing the solid particles, while the dark circles represent a fully molten particle. Figure 5.56a shows the trajectory of a 10 μm, particle injected at a location $(r_i/R_o) = 0.05$ (i.e., $r_i = 0.7$ mm from the axis of the torch) at an injection velocity $V_i = 1.5$ m/s. It is noted that this particle has too small inertia to overcome the back flow to which it is exposed on the upstream end of the discharge, as shown in Fig. 5.55b. It is therefore deflected outward in the radial direction only to be entrained by the recirculating flow to the center of the recirculation in the discharge. As the particle is heated along its trajectory, it is melted and completely vaporized during its total residence time in the discharge of $t_s = 23.5$ ms. The result given in Fig. 5.56b is for a 100 μm particle injected at the same location, $(r_i/R_o) = 0.05$, and with the same injection velocity, $V_i = 1.5$ m/s as the previous particle. Having still a too small initial inertia, the particle is unable to overcome the back flow on the upstream end of the discharge and is projected outward toward the wall of the plasma confinement tube. Being still in the solid state as indicate by the open circle, it bounces off the torch wall and is entrained by the plasma flow where it is further heated and completely melted prior to immerging from computation domain after a residence time $t_s = 105.4$ ms in the discharge. It is to be noted that particle bouncing above an inductively coupled discharge was experimentally observed as reported by Chase (1971). Figure 5.56c shows another example of a 100 μm particle injected closer to the axis of the torch, $(r_i/R_o) = 0.01$, at an injection velocity of $V_i = 5.5$ m/s. The increase of the injection velocity in this case enhanced its initial inertia level and allowed the particle to penetrate the discharge against the reversed flow. By slowing down in the discharge region, the particle is completely molten following a liner trajectory in the flow with a residence time in the discharge of $t_s = 32.4$ ms.

In practice, due to the flow pattern in the discharge region, the position of the tip of the powder injection probe is of critical importance [Boulos (1992)]. If located above the first turn of the induction coil, it could result in excessive deposition of powder on the walls of the confinement tube, which may lead to its eventual failure. A low position of the powder-injection probe at the downstream end of the induction coil will result in excessive cooling of the central part of the plasma (heated mainly by conduction–convection) and a reduction of the particle residence time in the discharge. This will cause a reduction of the ability of the plasma to completely melt the particles in-flight. Practically, in industrial RF plasma torches by Tekna Plasma Systems Inc., the

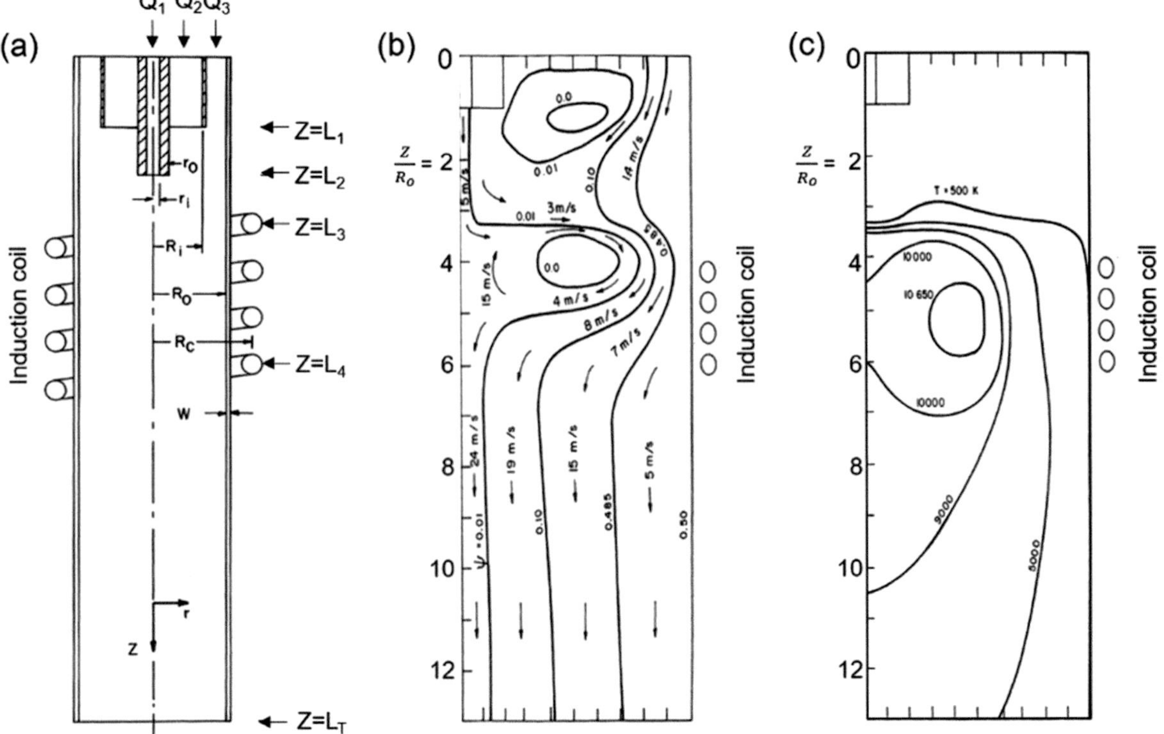

Fig. 5.55 (**a**) Schematic of the induction plasma torch showing a definition of its principal dimension and operating conditions given in Table 5.2. (**b**) Flow field, (**c**) Temperature field for a pure argon plasma discharge at atmospheric pressure after [Boulos (1978)]. Reprinted with kind permission from Springer Science Business Media, copyright © ASM International

Table 5.5 Principal dimensions and operating conditions used for the particle trajectory computations in an RF Induction plasma [Boulos (1978)]

r_i (mm)	1.68	L_0 (mm)	14.0	Q_1 (slm)	0–0.6
r_0 (mm)	2.80	L_1 (mm)	56.0	Q_2 (slm)	0.2
R_i (mm)	10.0	L_2 (mm)	87.0	Q_3 (slm)	18.0
R_0 (mm)	14.0	L_T (mm)	182.0	f (MHz)	3.0
R_c (mm)	24.0	w (mm)	1.0	Po (kW)	3.77

double-walled and water-cooled injector tip is disposed about at the level of the upper third of the coil (see Chap. 9). With a Mettech DC torches with axial injection the maximum temperature of particles injected was at the same location as that of the maximum number of particles, which is not the case with radial injection [Abukawa et al. (2006)] This demonstrates that axial injection is the most efficient way of heating particles.

5.6 Plasma–Particle Interactions Under Dense Loading Conditions

While the assumption of dilute systems has generally been accepted for the calculation of individual particle trajectories and temperature histories under plasma conditions, the interpretation of the results obtained is greatly hindered by the simple fact that any application of plasma technology for the in-flight processing of powders will have to be carried out under sufficiently high loading conditions, in order to make efficient use of the energy available in the hot or cold flow. With hot flows the main effect is the local modification of the flow energy due to the presence of the particles, and model predictions using the dilute-loading assumptions can be misleading when the currently used powder loadings in spray processes are considered. It is generally recognized that the loading effect has to be considered as soon as the particle mass flow rate is higher than 4% of the mass flow rate of the gases. Such a situation occurs in flame spraying, plasma spraying (DC and RF) but is generally negligible in PTA and has almost no possibility to occur in HVOF or HVAF spraying as well as in cold spray.

In an attempt to take into account the plasma–particle interaction effects, Boulos and his collaborators [Proulx et al. (1983, 1985, 1987)] developed a mathematical model, which, through the interactive procedure illustrated in Fig. 5.57. This procedure updates continuously the computed plasma flow, temperature, and concentration fields to take into account the plasma–particle interaction effects. The interaction between the stochastic single particle trajectory calculations and those of the continuum flow, temperature,

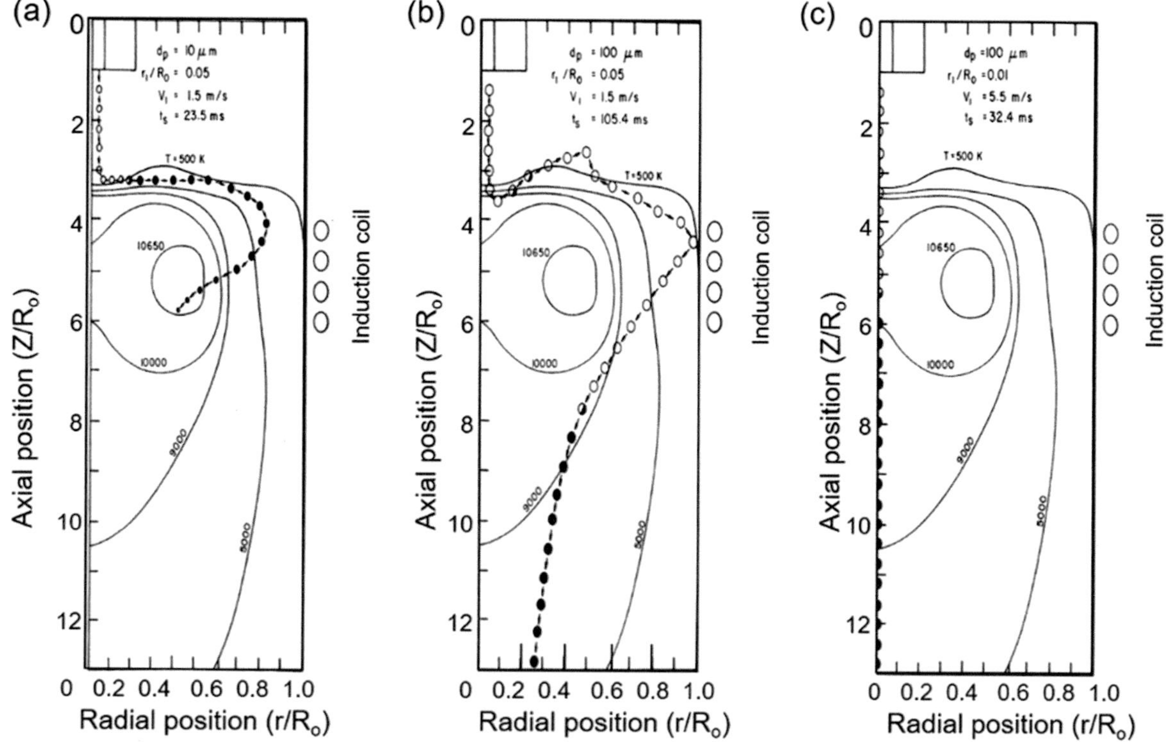

Fig. 5.56 Particle trajectories for alumina powder injected into an RF induction plasma: f = 3 MHz, P_o = 3.77 kW, powder gas Q_1 = 0.4 slm (Ar), central gas, Q_2 = 2.0 slm (Ar), sheath gas Q_3 = 16.0 slm (Ar) (**a**) 10 μm particle, (**b** and **c**) 100 μm particle after [Boulos (1978)]. Reprinted with kind permission from Springer Science Business Media, copyright © ASM International

and concentration fields is incorporated through the use of appropriate source–sink terms in the respective continuity, momentum, energy, and mass transfer equations. These are estimated using the so-called particle-source-in-cell model (PSI-Cell) after Crowe et al. (1977).

As an example of possible plasma–particle interaction effects in induction plasma spraying under heavy loading conditions, results reported by Proulx et al. (1985), will be given in this section. These were obtained for a 50 mm i.d. inductively coupled plasma torch operated with argon at atmospheric pressure, 3 kW power level, working at frequency of 3 MHz. Schematic of the torch design, its principal dimensions and operating conditions are given in Fig. 5.58. The operating frequency was f = 3.0 MHz and power coupled into the discharge P_o = 3 kW.

Copper powder with a mean particle diameter of 70 μm and a standard deviation of 30 μm was injected through the central powder injection probe into the coil region of the discharge at the upstream level of the induction coil. The thermophysical properties of copper used in the calculations are given in Table 5.6. The availability of data on the electrical conductivity and transport properties of Ar/Cu vapor mixtures as function of concentration and temperature by Mostaghimi and Pfender (1984) allowed for the integration in the model the effect of particle vaporization on the local

flow and temperature field calculation in the plasma torch. The Gaussian particle size distribution of copper powder was discretized, and the injection velocity of the particles was assumed to be equal to that of the mean carrier gas velocity at the point of injection.

Computations were carried out of the flow, temperature, and copper vapor concentration fields for copper powder injection feed rates varying between 0.1 and 20 g/min (0.006–1.2 kg/h). Typical results given in terms of the 2-D flow, temperature, and copper vapor isocontours for copper powder feed rate of 5 g/min are given in Fig. 5.59. These show on the LHS of Fig. 5.59a, the temperature isotherms, while on its RHS, the streamlines of the flow. As expected, the circulation pattern of the flow is noticeable in the induction coil region. The relatively high carrier gas flow rate, Q_1 = 3 slm, seems to be sufficient to overcome the revers flow along the axis of the discharge, with the majority of the injected copper particles assuming an axial trajectories. Figure 5.59b shows the 2-D isocontours of the copper vapor mapping in the flow. It may be noted that the copper vapor on the centerline of the discharge reaches a maximum concentration of 10 wt.% dropping outward to 1–5 wt.% in the fringes of the discharge. The corresponding temperature profiles in the axial and radial directions are given in Figs. 5.60 and 5.61, respectively. As noted in Fig. 5.60,

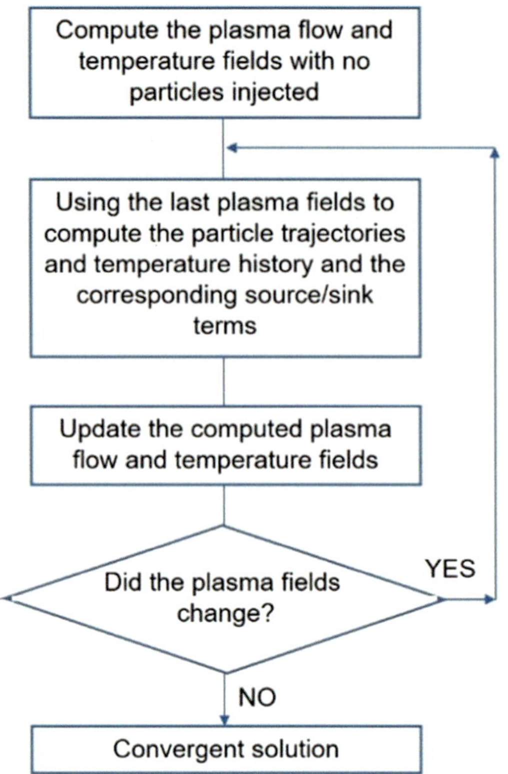

Fig. 5.57 Flow chart for computational approach integrating plasma–particle interactions under dense loading conditions in particle trajectory and temperature history modeling after [Proulx et al. (1985)]

the increase of the copper feed rate from 0 to 20 g/min is responsible for a major drop of the centerline discharge temperature from an initial maximum value above 9000 K in the absence of copper power injection, down to just below 4000 K at a copper powder feed rate of 20 g/min. As noted, however, from the radial temperature profiles given in Fig. 5.61, obtained at $Z = 71$ mm, corresponding to the lower end of the induction coil, the cooling effect powder injection is a rather localized and limited to the central 5–6 mm radius of the discharge where most of the copper powder trajectories are concentrated. Beyond this point the plasma temperature is not affected by the injected powder or by the contamination with the copper vapor. As expected, the localized cooling of the plasma with the increase powder feed rate gives rise to a significant reduction of the heating rate of the copper particles and their rate of evaporation. The results clearly demonstrate that although the overall loading ratio of the copper powder to plasma gas might be small (0.19 g copper/3 g argon), the local cooling effects are significant. This is a clear indication that the plasma–particle interaction effects could be locally very important under loading conditions for which it is frequently assumed to neglect the changes in plasma temperature due to the presence of the powder.

It should be noted that momentum transfer between the plasma and the particles is generally negligibly, and the flow field is affected only through changes of the local plasma temperature. As the particles pass through the plasma, a

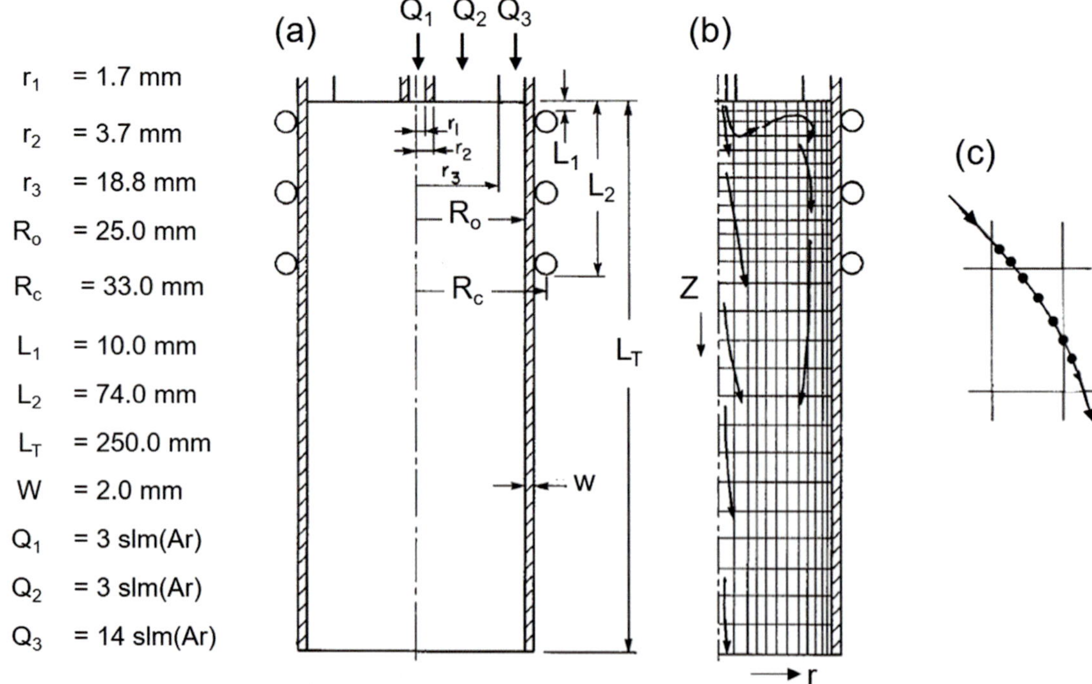

Fig. 5.58 (**a**) Schematic of the induction torch design, its principal dimensions and operating conditions (**b**) Grid structure used in the computation and (**c**) basic concept of the source-in-cell model [Proulx et al. (1985)]

Table 5.6 Thermophysical properties of copper after Mostaghimi and Pfender (1984)

$T_m = 1350$ K	$H_m = 204.7$ kJ/kg	$C_{ps} = 0.425$ kJ/kg K
$T_v = 2840$ K	$H_v = 4794.0$ kJ/kg	$C_{pv} = 0.480$ kJ/kg K
$\rho_p = 8900$ kg/m^3		

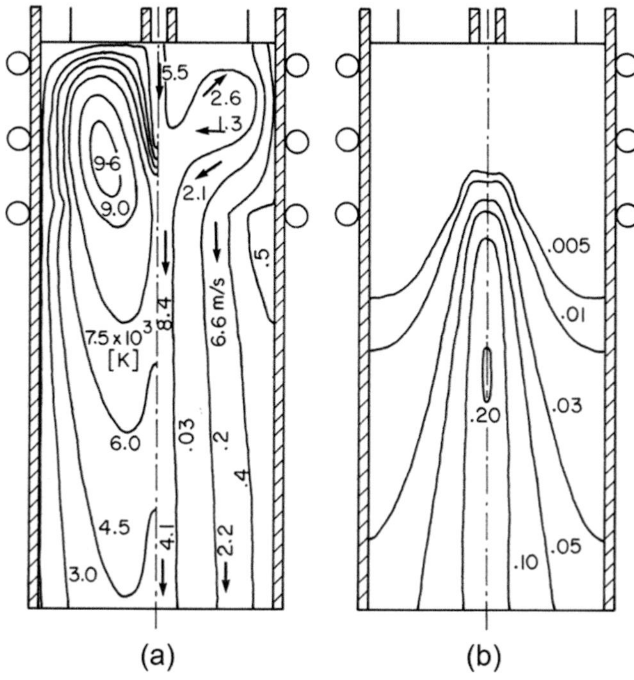

Fig. 5.59 (**a**) 2-D temperature and flow fields in the discharge (**b**) copper vapor concentration mapping in wt.%, for the injection of 5 g/min of 70 μm mean copper powder with a carrier gas flow rate of 3 slm (Ar) in a pure Argon plasma at a discharge power of 3 kW [Proulx et al. (1985)]

portion of the powder evaporates, and the vapor diffuses into the plasma medium and the total energy absorbed by the powder is between 3.1% and 17.8% of the input plasma power. Thus, the injection conditions will have a very important effect on the loading effect in RF plasma torches.

The study of plasma–particle interactions under dense loading conditions was further extended to other materials and powder feed rates. The results reported by Proulx et al. (1987) was developed for a RF induction plasma torch of the same dimensions as that given in Fig. 5.58, operating using pure argon as plasma gas, $Q_1 = 3$ slm, $Q_2 = 3$ slm, and $Q_3 = 33$ slm, a discharge power of 5 kW and an oscillator frequency f = 3 MHz. Three powders were tested in such as system, Nickel, Alumina, and Tungsten, with thermophysical properties as listed in Table 5.7. All powders had a mean particle diameter of 60 μm, injected into the center of the discharge at feed rates varying between 1 and 50 g/min (0.06–3.0 kg/h).

Typical particle temperature history along the centerline of the discharge for each of these three powders Ni, Al$_2$O$_3$,

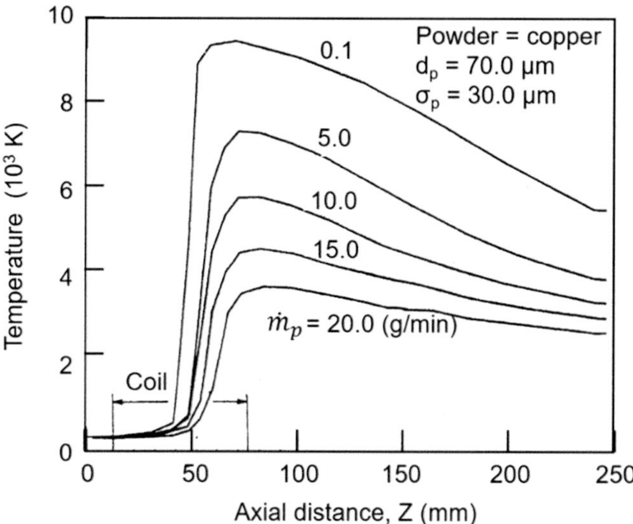

Fig. 5.60 Effect of the powder feed rate on the axial temperature profile at r = 0.0 mm for an induction plasma torch operating with pure argon at discharge power of 3 kW, see Fig. 5.58 for detailed dimensions and operating conditions after Proulx et al. (1985). Reprinted with kind permission from Elsevier for Journal of Heat and Mass Transfer

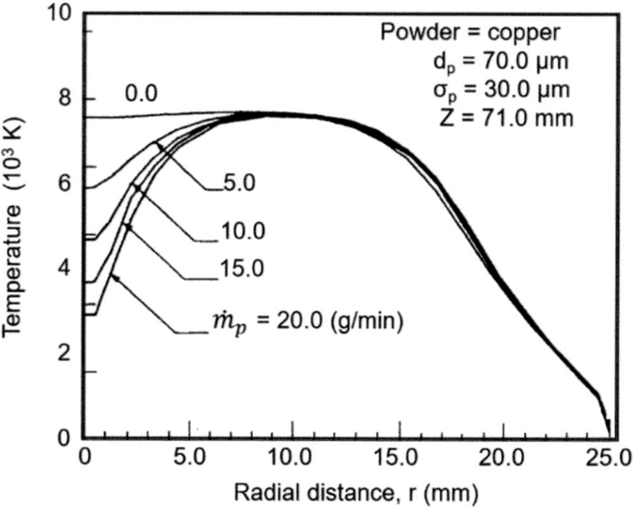

Fig. 5.61 Effect of the powder feed rate on the radial temperature profile at z = 71 mm corresponding to the lower end of the induction coil, for an induction plasma torch operating with pure argon at discharge power of 3 kW, see Fig. 5.58 for detailed dimensions and operating conditions after Proulx et al. (1985). Reprinted with kind permission from Elsevier for Journal of Heat and Mass Transfer

and W is given in Figs. 5.62, 5.63, and 5.64, respectively. Two radial positions in the powder injector (r = 0.25 and r = 1.75 mm) were considered for initial particle injections for each of these powders.

The thermal loading effect due to particle injection in each of these cases is significant. The effect is best demonstrated by comparing the plasma temperature along the centerline of the torch in the presence of the powders at their maximum

Table 5.7 Thermophysical properties of Ni, Al$_2$O$_3$ and W powders, Proulx et al. (1987)

Material	Ni	Al$_2$O$_3$	W
Melting temperature (K)	1727	2323	3680
Boiling temperature (K)	3005	3800	6200
Density (kg/m^3)	8900	3900	19,350
Specific heat, solid (kJ/kg K)	0.54	1.20	0.17
Specific heat, liquid (kJ/kg K)	0.54	1.20	0.17
Latent heat of fusion (MJ/kg)	0.31	1.00	0.19
Latent heat of vaporization (MJ/kg)	0.62	25.0	2.26

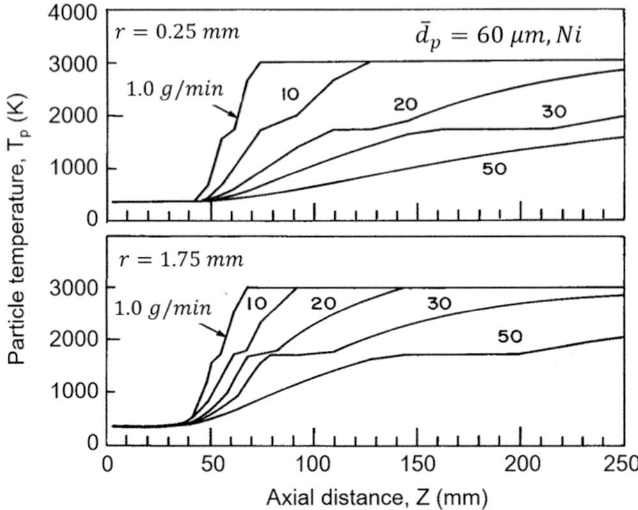

Fig. 5.62 Temperature history for a 60 μm Nickel powder along the centerline of the discharge for different powder feed rates for initial radial positions in the injector of 0.25 and 1.75 mm Proulx et al. (1987)

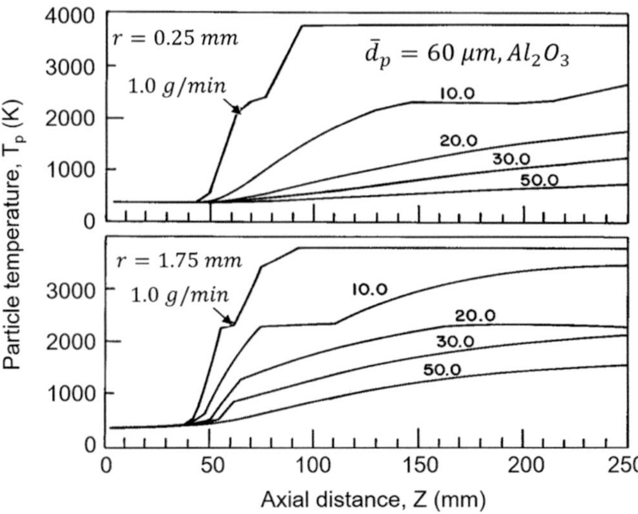

Fig. 5.63 Temperature history for a 60 μm Alumina powder along the centerline of the discharge for different powder feed rates for initial radial positions in the injector of 0.25 and 1.75 mm Proulx et al. (1987)

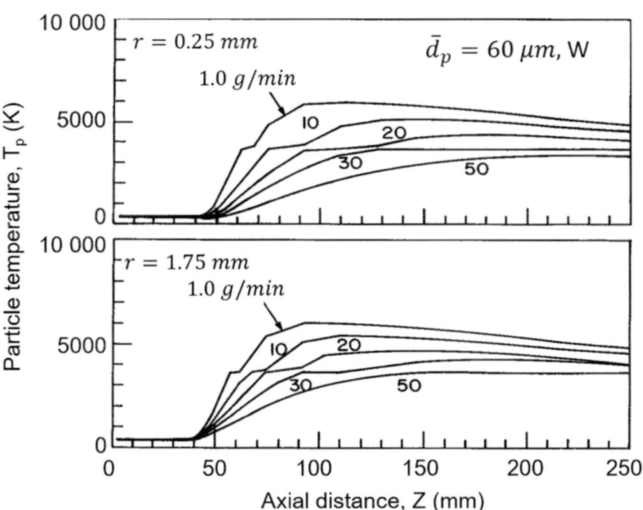

Fig. 5.64 Temperature history for a 60 μm Tungsten powder along the centerline of the discharge for different powder feed rates for initial radial positions in the injector of 0.25 and 1.75 mm Proulx et al. (1987)

feed rates of 50 g/min with that in the absence of particle injection. The results given in Fig. 5.65 show a significant cooling of the plasma in the presence of the different powders. The effect is strongest for alumina where the plasma temperature drops from a maximum of 9000 K down to less than 1000 K.

Integrating the total energy received by the particles during their trajectory in the plasma as function of the powder feed rate, Fig. 5.66 reveals a distinctly nonlinear behaviors with the total energy received by the particles reaching a plateau independent of the powder feed rate. In the best case of Alumina, the plateau is at around 900 W, which represents about 18% of the energy in the discharge (5 kW). Nickel and tungsten seem to reach a plateau at a lower value of 800 and 680 W, respectively, corresponding to 16% and 14% of the discharge energy.

As would be expected, the thermal loading effect has a major impact on the thermophysical transformations taking place in the powder as a result of its passage through the plasma. The results are summarized in Table 5.8 that gives for these three materials the fraction of the powder that would

have remained unmelted, the fraction that melted and that which would have vaporized. The results show that for the Nickel powder, the plasma treatment will result in the near complete melting of the powder for powder feed rates up to 30 g/min, beyond which at the fraction of powder melted will drop to 81% at 50 g/min. It may also be noted that at low feed rates, a significant amount of the powder would be vaporized, 73% at 1 g/min. Corresponding results given for alumina show near complete melting limited to feed rates up to 10 g/min, with powder vaporization limited to 26% at 1 g/min. With tungsten on the other hand, near complete melting can be achieved at feed rates up to 30 g/min, with a negligible

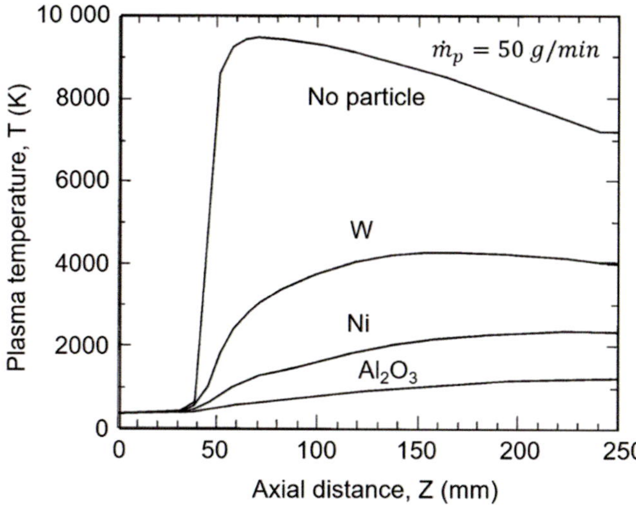

Fig. 5.65 Plasma temperature profiles along the centerline of the discharge in the presents of different powders injected at a powder feed rate of 50 g/min, Proulx et al. (1987)

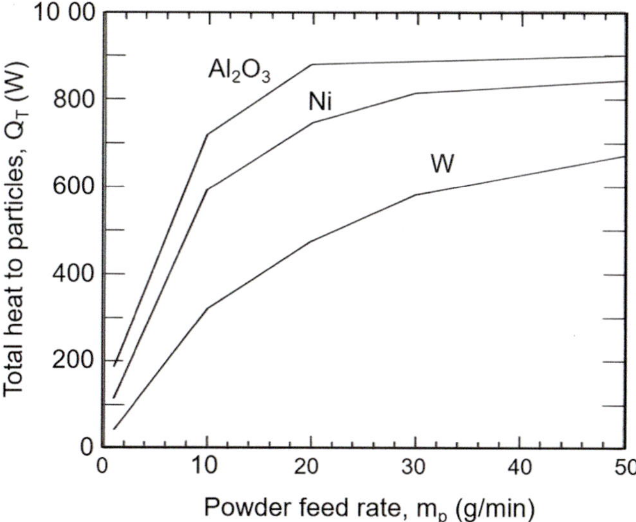

Fig. 5.66 Integral energy received by a 60 μm particles of different powders during their trajectory along the centerline of the RF induction plasma discharge as function of the powder feed rate Proulx et al. (1987)

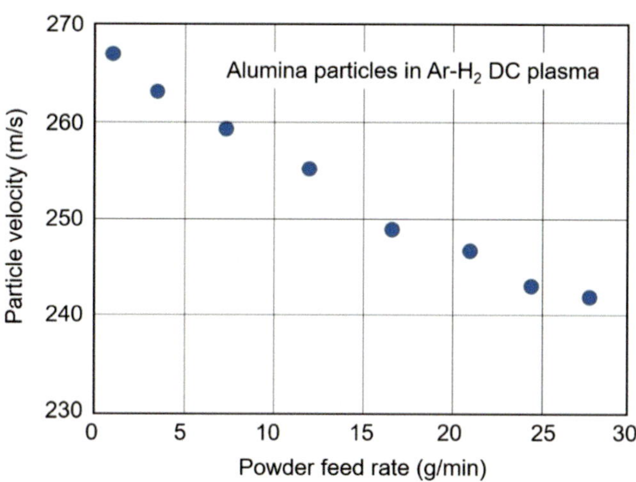

Fig. 5.67 Measured Al_2O_3 particle velocity as function of the powder feed rate at location 20 mm from the nozzle exit, after Vardelle et al. Reprinted with kind permission from ASM International [Vardelle et al. (1992)]

Table 5.8 Thermophysical composition of the powder as it immerges from the plasma torch as function of the powder feed rate. Values given in mass fraction of the injected powder [Proulx et al. (1987)]

Material	State	Feed rate (g/min)				
		1.0	**10.0**	**20.0**	**30.0**	**50.0**
Nickel	Solid	0.00	0.00	0.00	0.02	0.19
	Liquid	0.27	0.69	0.88	0.94	0.81
	Vapor	0.73	0.31	0.12	0.04	0.00
Alumina	Solid	0.00	0.02	0.43	0.84	1.00
	Liquid	0.74	0.98	0.57	0.16	0.00
	Vapor	0.26	0.00	0.00	0.00	0.00
Tungsten	Solid	0.00	0.00	0.00	0.05	0.70
	Liquid	0.90	1.00	1.00	0.95	0.30
	Vapor	0.10	0.00	0.00	0.00	0.00

level of vaporization (10%), even at 1 g/min. Proulx et al. (1987) underline the importance of the loading effect is clearly linked to thermophysical properties of particles: alumina, for example, has the highest latent heat of melting and boiling as well as high specific heats of the solid and liquid phase. The lowest values are those of tungsten, nickel being between alumina and tungsten.

Corresponding calculations and measurements were performed [Vardelle et al. (1992)] for alumina particles (fused and crushed – Alumina particles 62 + 18 μm) injected into a DC plasma flows. The spraying parameters in this case were: nozzle i.d. 7 mm, 32 slm Ar + 12 slm H_2, $I = 600$ A, $V = 58$ V, $\eta = 52\%$, injector i.d. 1.8 mm, external injection $r = 9$ mm, $z = 6$ mm. The distributions of particle velocities and surface temperatures were measured at a point situated on the centerline of the plasma jet at a location 120 mm from the nozzle exit for different powder mass flow rates. Results given in Fig. 5.67 show that the velocity of the particles is reduced by 11% when the powder mass flow rate is increased from 0.5 to 33.3 g/min (0.03–2 kg/h) while their surface temperature is reduced by about 14%. When calculating the particle velocities, the mean value is within 10% of the experimental values for a low mass flow rate of 3.3 g/min (0.2 kg/h). However, significant differences are observed for a powder flow rate of 33.3 g/min (2 kg/h). The effect is the same, but more pronounced for the particle temperature. This discrepancy underlines the importance of representing the particle injection more accurately.

5.7 Summary and Conclusions

Three-dimensional simulations with full coupling between a single particle and gas flow, employing a Lagrangian particle-source-in-cell particle-tracking frame coupled with a steady-state gas flow, have been developed to examine particle motion and heat transfer. These models, when necessary, are compressible and consider the effect on particles of compression and expansion waves. Since the ninety's models have shown the importance of the different corrections adopted to the process for flames, HVOF or HVAF, D-Gun, and plasma spraying processes: high temperature gradients, particle evaporation or vaporization, rarefaction effect (Knudsen effect), shock waves, and supersonic flow effect. Such models have played a key role in the study of particle trajectory and temperature history in a wide range of thermal spray devices whether combustion based or plasmas. In particular they allowed the for the determination of the effect of particle injection conditions (position and velocity vector) on the particle parameters prior to their impact on the substrate. Different effects such as loading effect, particle morphology, and shape have been considered. However, the way the particle velocity vector distribution at the injector exit is calculated is still in its infancy in spite of its influence on the particle treatment within the hot or cold gas flow. The first studies of the interaction between hot gas flows and a liquid (solution or suspension spraying) have also been presented, this technique being very promising for the deposition of finely or nanostructured coatings with thicknesses between a few micrometers and hundreds of them.

Nomenclature

Units are indicated in parentheses; when no units are indicated, the parameter is dimensionless.

Latin Alphabet

a	Accommodation coefficient
a_i	Sound velocity (m/s)
Bi	Biot number, $Bi = \frac{\kappa}{\kappa_p}$
c	Molar density of the bulk gas (mol/m^3)
C_D	Drag coefficient, $C_D = (F_D/s_p)/\frac{1}{2}\rho v^2$
c_i	Mass fraction of metal vapor
c_p	Specific heat at constant pressure (J/kg K)
c_v	Specific heat at constant volume (J/kg K)
d_d	Drop/droplet diameter (m)
d_p	Particle diameter (m)
D_{vg}	Diffusion coefficient of metal vapor through the surrounding gas (m^2/s)
f	Friction coefficient
F_B	Basset history term (N)
F_c	Coriolis force (N)
F_b	Body force per unit particle mass (N)
F_d	Force resulting from detonation wave, $F_d = \frac{\pi d_p^2}{4}\Delta p$ (N)
F_D	Drag force exerted by the fluid on the particle (N)
F_g	Gravity force (N)
F_i	Inertia force, $F_i = m_p \cdot \gamma_p$ (N)
F_{ip}	injection force of the particle (N)
F_p	Pressure gradient force (N)
F_S	Surface tension force (N)
F_T	Thermophoresis force (N)
g	Gravity acceleration, g = 9.81 (m/s^2)
h	Heat transfer coefficient (W/m$^2 \cdot$K)
h_g	Specific enthalpy (J/m^3, or J/kg)
H_m	Latent heat of fusion (J/kg)
H_v	Latent heat of vaporization (J/kg)
I	Arc current (A)
$I(T)$	Heat conduction potential $I(T) = \int_{T_{300}}^{T}\kappa(s)ds$ (W/m)
Kn	Knudsen number, $Kn = \lambda/d_p$
k_d	Mass transfer coefficient (m/s)
L	Mixing length in turbulent model (m)
m_{cg}^o	Carrier gas mass flow rate (kg/s)
m_p	Particle mass (kg)
M	Molecular weight (kg/mol)
Ma	Mach number ($Ma = v_g/a_i$)
N_{max}	Maximum molar flux of vaporization from a droplet (m^2/s)
Nu	Nusselt number, $Nu = (hd/\kappa)$
N_v	Molar flux of vapor (mol/m$^2 \cdot$s)
Oh	Ohnesorge number, $Oh = \mu_l/\sqrt{\rho_l d_l \sigma_l}$
p	Total pressure (Pa)
p_i	Partial pressure of species, i (Pa)
p_o	Saturation vapor pressure of a liquid (Pa)
P_o	Torch power (kW)
P_1	Splat parameter (m)
P_{eff}	Torch effective power (kW)
Pr	Prandtl number, $Pr = c_p\,\mu/\kappa$
q	Heat flux (W/m^2)
Q	Heat transferred to a particle (W)
Q	Gas flow rate (slm)
r	Plasma or particle radius (m)
r_d	Liquid droplet radius (m)
r_s	Initial liquid drop radius (m)
R	Plasma torch internal or particle radius (m)
R'	Universal ideal gas constant $R' = 8.32$ (J/K mol)
Re	Reynolds' number, $Re = \rho\,U_R\,d_p/\mu$
r_{inj}	Internal radius of injection tube (m)
s_p	Projected surface area of particle normal to flow, $s_p = \frac{\pi d_p^2}{4}$ (m^2)
S	Cross section of injection tube (m^2)
Sc	Schmidt number, $Sc = \nu/D_{v,g}$
Sh	Sherwood number, $Sh = k_d \cdot d_p/D_{v,g}$
St	Stokes number, $St = \rho_p \cdot d_p^2 \cdot v_p/\mu_g l_{BL}$
Ste	Stephan number, $Ste = c_{pi} \cdot (T_m - T_s)/H_m$
t_d	Drop fragmentation time (s)
t_r	Time of flight of particles (s)
t_s	Vaporization time (s)
t_s	Residence time (s)
T_a	Ambient temperature (K)
T_f	Mean film temperature, $T_f = \frac{(T_s+T_\infty)}{2}$ (K)
T_s	Particle surface temperature (K)
T_∞	Free stream temperature (K)
U	Mean arc voltage (V)
$U(t)$	Transient arc voltage (V)
u	Axial gas velocity (m/s)
u_p	Axial particle velocity, (m/s)

u_R	Gas-particle relative velocity, (m/s)
u_r	Gas-liquid relative velocity, (m/s)
v	Radial gas velocity (m/s)
v_l	velocity of the liquid (m/s)
v_p	Particle velocity component in the radial direction (m/s)
V	Volume (m^3)
V_p	Particle volume (m^3)
V_s	Liquid drop volume (m^3)
W_s	Work resulting from the drag force (J)
We	Weber number, $We = \frac{\rho_g . U_r^2 . d_l}{\sigma_l}$

Greek Alphabet

δ	Boundary layer thickness
ΔE_s	Variation of surface energy (J)
Δp	Differential pressure (kPs)
ΔT	Temperature difference between front and rear ends of particle (K)
ε	Particle emissivity (integrated over all wave lengths)
$\phi(r)$	Function representing temperature, enthalpy, velocity
γ	Isentropic coefficient, $\gamma = c_p/c_v$
γ_p	Particle acceleration (m/s^2)
η_{th}	Torch thermal efficiency (%)
κ	Thermal conductivity of the fluid (W/m K)
κ_p	Thermal conductivity of the particle (W/m K)
$\bar{\kappa}$	Integrated mean thermal conductivity, Eq. (5.28), (W/mK)
λ	Plasma mean free path (m)
μ_i	Fluid viscosity (Pa s)
ν_{cg}	Carrier gas velocity (m/s)
ν_i	Kinematic viscosity (μ/ρ) (m^2/s)
ρ_g	Gas mass density (kg/m^3)
ρ_i	Fluid density or specific mass (kg/m^3)
ρ_l	Liquid density
σ	Droplet surface tension (N/m)
σ_s	Stephan-Boltzmann constant, $\sigma_s = 5.67 \times 10^{-8}$ (W/m^{2}K^4)
ξ	Ratio of splat to droplet diameters, $\xi = D_p/d_p$

References

Abukawa, S., T. Takabate, and K. Tani. 2006. Effect of powder injection of deposit efficiency in plasma spraying. In *Proceedings of the 2006 International Thermal Spray Conference*, ed. B. Marple et al. Materials Park: ASM International. e-processing.

Arnauld, Ph., S. Cavadias, and J. Amouroux. 1985. The interaction of a fluidized bed with a thermal plasma. Application to limestone decomposition. In *Proceedings of International Symposium on Plasma Chemistry*, ed. Timmermans (Pub.), 195–200. University of Technology of Eindhoven NL1.

Basu, S., E.H. Jordan, and B.M. Cetegen. 2006. Fluid mechanics and heat transfer of liquid precursor droplets injected into high temperature plasmas. *Journal of Thermal Spray Technology* 15 (4): 576–581.

———. 2008. Fluid mechanics and heat transfer of liquid precursor droplets injected into high-temperature plasmas. *Journal of Thermal Spray Technology* 17: 60–72.

Boulos, M.I. 1976. Flow temperature field in the fire ball of an inductively coupled plasma. *IEEE Transactions on Plasma Science* PS-4: 28–39.

———. 1978. Heating of powders in the fire ball of an induction plasma. *IEEE Transactions on Plasma Science* PS-6 (2): 93–106.

———. 1985. The inductively coupled R.F. (radio frequency) plasma. *Pure and Applied Chemistry* 57 (9): 1321–1352.

———. 1992. RF induction plasma spraying: State-of-the-art review. *Journal of Thermal Spray Technology* 1 (1): 33–40.

———. 1997. The inductively coupled radio frequency plasma. *Journal of High Temperature Materials Processes* 1: 17–39.

———. 2003. Spheroidization and densification of powders, Continuing education course on Thermal Plasmas, held in conjunction with 16th International Symposium on Plasma Chemistry (ed.) P. Fauchais, University of Limoges.

———. 2004. Chapter 6.2: Plasma interaction with a dispersed medium. In *Continuing Education Course in Conjunction with ITSC-2004*, ed. J. Heberlein. Minnesota: University of Minnesota.

Boulos, M.I., and W.H. Gauvin. 1974. Powder processing in a plasma jet, a proposed model. *Canadian Journal of Chemical Engineering* 52: 355–363.

Boulos, M.I., P. Fauchais, A. Vardelle, and E. Pfender. 1993a. Fundamentals of plasma particle momentum and heat transfer. In *Plasma Spraying Theory and Applications*, ed. R. Suryanarayanan. Singapore: World Scientific.

———. 1993b. Fundamentals of plasma particle momentum and heat transfer. In *Plasma Spraying Theory and Applications*, ed. R. Suryanarayanan. Singapore: World Scientific.

Boulos, M., P. Fauchais, and E. Pfender. 1994. *Thermal Plasmas Fundamental and Applications*. Vol. 1, 452. London: Plenum Press.

Bourdin, E., P. Fauchais, and M.I. Boulos. 1983. Transient heat conduction under plasma conditions. *International Journal of Heat and Mass Transfer* 26: 567–582.

Bouyer, E., F. Gitzhofer, and M. Boulos. 1996. Parametric study of suspension plasma sprayed hydroxyapatite. In *Thermal Spray: Practical Solutions for Engineering Problems*, ed. C.C. Berndt, 683–691. Materials Park: ASM International.

Bouyer, E., F. Gitzhofer, and M.I. Boulos. 1997a. Suspension plasma spraying for hydroxyapatite powder preparation by RF plasma. *IEEE Transactions on Plasma Science* 25 (5): 1066–1072.

———. 1997b. The suspension plasma spraying of bioceramics by induction plasma. *JOM: The Journal of The Minerals, Metals & Materials Society* 49 (2): 58–68.

Bouyer, E., F. Gitzhofer, and M. Boulos. 2000. Morphological study of hydroxyapatite nanocrsystal suspension. *Journal of Materials Science* 11: 523–531.

Bouyer, L., X. Ma, A. Ozturk, E.H. Jordan, N.P. Padture, B.M. Cetegen, D.T. Xiao, and M. Gell. 2004. Processing parameter effects on solution precursor plasma spray process spray patterns. *Surface and Coating Technology* 183 (1): 51–61.

Bronet, M., and M.I. Boulos. 1989. Particle velocity measurements in induction plasma spraying. *Journal of Plasma Chemistry and Plasma Processing* 9: 343–353.

———. 1990. Particle slip velocities in air jets under atmospheric and soft vacuum conditions. *The Canadian Journal of Chemical Engineering* 68: 353–359.

Carruyer, C., S. Vincent, E. Meillot, Caltagirone. 2011. Modelling of the fragmentation of a water jet in the liquid precursor plasma spraying. In 5th *International of Workshop on Suspension and Solution Plasma Spraying*, Tours September 4–6 (2011) (pub.) CEA Le Ripault.

Chase, J.D. 1971. Theoretical and experimental investigation of pressure and flow in induction plasmas. *Journal of Applied Physics* 42: 4870.

Chen, X.I., Y.P. Chyou, Y.C. Lee, and E. Pfender. 1985. Heat transfer to a particle under plasma conditions with vapor contamination from the particle. *Plasma Chemistry Plasma Processing* 5 (2): 119–141.

Chen, Xi, and E. Pfender. 1982a. Heat transfer to a single particle exposed to a thermal plasma. *Journal of Plasma Chemistry Plasma Processing* 2 (2): 185–212.

———. 1982b. Unsteady heating and radiation effects of small particles in a thermal plasma. Journal of *Plasma Chemistry Plasma Processing* 2 (3): 293–316.

Chen, Xi, and Ping He. 1986. Heat transfer from a rarefied plasma flow to a metallic or non-metallic particle. *Journal of Plasma Chemistry Plasma Processing* 6 (4): 314–333

Chen, Xi, Ji Chen, and Yandan Wang. 1995. Unsteady heating of metallic particles in a rarefied plasma. *Plasma Chemistry Plasma Processing* 15 (2): 199–219.

Chen, Xi, and Xiaoming Chen. 1989. Drag on metallic or non-metallic particle exposed to a rarefied plasma flow. *Journal of Plasma Chemistry Plasma Processing* 9 (3): 387–408.

Chyou, Y.P., and E. Pfender. 1989. Behavior of particulates in thermal plasma flows. *Journal of Plasma Chemistry Plasma Processing* 9 (1): 45–71.

Colombo, V., E. Ghedini, and P. Sanibondi. 2010. A three-dimensional investigation of the effects of excitation frequency and sheath gas mixing in an atmospheric-pressure inductively coupled plasma system. *Journal of Physics D: Applied Physics* 43: 105202. (13pp).

Crowe, C.T., M.P. Sharma, and D.E. Stock. 1977. The particle-source-in-cell (PSI-cell) model for gas-droplet flows. *Journal of Fluids Engineering* 99: 325–332.

Delbos, C., J. Fazilleau, J.-F. Coudert, P. Fauchais, L. Bianchi, and K. Wittmann-Ténèze. 2003. Plasma spray elaboration of fine nanostructured YSZ coatings by liquid suspension injection. In *Thermal Spray 2003: Advancing the Science and Applying the Technology*, ed. C. Moreau and B. Murple, 661–667. Materials Park: ASM International.

Delbos, C., J. Fazilleau, V. Rat, J.F. Coudert, P. Fauchais, and B. Pateyron. 2006. Phenomena involved in suspension plasma spraying, part 2: Zirconia particle treatment and coating formation. *Plasma Chemistry and Plasma Processing* 26 (4): 393–414.

Delluc G., L. Perrin, H. Ageorges, P. Fauchais, and B. Pateyron. 2004. Modelling of plasma jet and particle behavior in spraying conditions, in ITSC-2004, in *Modeling and Simulation* V (pub.) DVS, Düsseldorf, Germany, C.D. Rom.

Dresvin, S.V. 1977. *Physics and technology of low-temperature plasmas*, Moscow Atomizat (1972), English Edition, Iowa State University press, Ames.

———. 2002. Arc instabilities in a plasma spray torch. *Journal of Thermal Spray Technology* 11 (1): 44–51.

Etchart-Salas, R., V. Rat, J.F. Coudert, P. Fauchais, N. Caron, K. Wittman, and S. Alexandre. 2007. Influence of plasma instabilities in ceramic suspension plasma spraying. *Journal of Thermal Spray Technology* 16 (5–6): 857–865.

Fauchais, P. 2004. Understanding plasma spraying, topical review. *Journal of Physics D: Applied Physics* 37: R86–R108.

Fauchais, P., and G. Montavon. 2010. Latest developments in suspension and liquid precursor thermal spraying. *Journal of Thermal Spray Technology* 19: 226–239.

Fauchais, P., J.F. Coudert, and M. Vardelle. 1989. Diagnostics in thermal plasma processing. In *Plasma Diagnostics*, ed. O. Ociello and D.L. Flamm, vol. 1, 349–360. New York: Academic Press.

Fauchais, P., R. Etchart-Salas, C. Delbos, M. Tognovi, V. Rat, J.F. Coudert, and T. Chartier. 2007. Suspension and solution plasma spraying of finely structured coatings. *Journal of Physics D: Applied Physics* 40: 2394–2406.

Fauchais, P., R. Etchart-Salas, V. Rat, J.F. Coudert, N. Caron, and K. Wittmann. 2008. Parameters controlling liquid plasma spraying: Solutions, sols or suspensions. *Journal of Thermal Spray Technology* 17 (1): 31–59.

Fauchais, P., G. Montavon, R. Lima, and B. Marple. 2011. Engineering a new class of thermal spray nano-based microstructures from agglomerated nanostructured particles, suspensions and solutions: An invited review. *Journal of Physics D: Applied Physics* 44: 093001. (53p).

Fauchais, P., A. Joulia, S. Goutier, C. Chazelas, M. Vardelle, A. Vardelle, and S. Rossignol. 2013. Suspension and solution plasma spraying. *Journal of Physics D: Applied Physics* 46: 224016.

Fazilleau, J. 2003. *Contribution to the Understanding of the Phenomena Implied in the Achievement of Finely Structured Oxide Coatings by Suspension Plasma Spraying*. PhD. thesis, In French, University of Limoges France.

Fazilleau, J., C. Delbos, V. Rat, J.F. Coudert, P. Fauchais, and B. Pateyron. 2006. Phenomena involved in suspension plasma spraying, part 1: Suspension injection and behavior. *Plasma Chemistry and Plasma Processing* 26 (4): 371–391.

Filkova, I., and P. Cedik. 1984. Nozzle Atomization. In *Spray Drying, Advances Drying*, ed. A.S. Mujumdar, vol. 3, 181–215. Hemisphere Publishing Corporation.

Fizdon, J.K. 1979. Melting of powder grains in a plasma flame. *International Journal of Heat and Mass Transfer* 22: 749–761.

Fukanuma, H., N. Ohno, Bo Sun, and R. Huang. 2006a. The influence of particle morphology on in-flight particle velocity in cold spray. In *Proceedings of the 2006 International Conference of Thermal Spray*. ASM International, Materials Park. e-proceedings.

———. 2006b. In-flight particle velocity measurements with DPV-2000 in cold spray. *Surface & Coatings Technology* 201: 1935–1941.

Ganser, G.H. 1993. A rational approach to drag prediction of spherical and non-spherical particles. *Powder Technology* 77: 143–152.

Gelfand, B.E. 1996. Droplet break-up phenomena in flows with velocity lag. *Progress in Energy and Combustion Science* 22: 201–265.

Gell, M., E.H. Jordan, M. Teicholz, B.M. Cetegen, N. Padture, L. Xie, D. Chen, X. Ma, and J. Roth. 2008. Thermal barrier coatings made by the solution precursor plasma spray process. *Journal of Thermal Spray Technology* 17 (1): 124–135.

Gitzhofer, F., E. Bouyer, M.I. Boulos. 1997. *Suspension Plasma Spray Deposition*, US Patent, 5,609,921 (March, 11th, 1997).

Gnedovets, A.G., and A.A. Uglov. 1992. Heat transfer to non-spherical particles in a rarefied plasma flow. *Plasma Chemistry Plasma Processing* 12 (4): 383–402.

Gross, K.A., P. Fauchais, M. Vardelle, J. Tikkanen, and J. Keskinen. 1997. Vaporization and ultrafine particle generation during the plasma spraying process. In *Thermal Spray: a United Forum for Scientific and Technological Advances*, ed. C.C. Berndt, 543–548. Materials Park: ASM International.

Gu, S., D.G. McCartney, C.N. Eastwick, and K. Simmons. 2004. Numerical modeling of in-flight characteristics of Inconel 625 particles during high-velocity oxy-fuel thermal spraying. *Journal of Thermal Spray Technology* 13 (2): 200–211.

Hanson, T.C., C.M. Hackett, and G.S. Settles. 2002. Independent control of HVOF particle velocity and temperature. *Journal of Thermal Spray Technology* 11 (1): 75–85.

Hussary, N.A., and J.V.R. Heberlein. 2001. Atomization and particle-jet interactions in the wire-arc spraying process. *Journal of Thermal Spray Technology* 10 (4): 604–610.

Hwang, S.S., Z. Liu, and R.D. Reitz. 1996. Breakup mechanisms and drag coefficients of high-speed vaporizing liquid drops. *Atomization and Sprays* 6: 353–376.

Iinoya, K., K. Gotoh, and K. Higashitani, eds. 1991. *Powder Technology Handbook*. New York/Basel/Hong Kong: M. Dekker, Inc.

Janisson, S., E. Meillot, A. Vardelle, J.F. Coudert, B. Pateyron, and P. Fauchais. 1999. Plasma spraying using Ar-He-H2 gas mixtures. *Journal of Thermal Spray Technology* 8 (4): 545–552.

Jordan, E.H., L. Xie, X. Ma, M. Gell, N.P. Padture, B. Cetegen, A. Ozturk, J. Roth, T.D. Xiao, and P.E.C. Bryant. 2004. Superior thermal barrier coatings using solution precursor plasma spray. *Journal of Thermal Spray Technology* 13 (1): 57–65.

Jordan, E.H., M. Gell, P. Bonzani, D. Chen, S. Basu, B. Cetegen, F. Wu, and X. Ma. 2007. Making dense coatings with the solution precursor plasma spray process. In *Thermal Spray 2007: Global Coating Solution*, ed. B.R. Marple, M.M. Hyland, Y.-C. Lau, C.-J. Li, R.S. Lima, and G. Montavon, 463–470. Materials Park: ASM International, e-proceedings.

Kamnis, S., S. Gu, and N. Zeoli. 2008. Mathematical modelling of Inconel 718 particles in HVOF thermal spraying. *Surface & Coatings Technology* 202: 2715–2724.

Karthikeyan, J., C.C. Berndt, J. Tikkanen, S. Reddy, and H. Herman. 1997. Plasma spray synthesis of nanomaterial powders and deposits. *Surface and Coating Technology* 238 (2): 275–286.

Kassner, H., R. Siegert, D. Hathiramani, R. Vassen, and D. Stöver. 2008. Application of the suspension plasma spraying (SPS) for the manufacture of ceramic coatings. *Journal of Thermal Spray Technology* 17 (1): 115–123.

Klubnikin, V.S. 1975. Thermal and gas dynamic characteristics of an argon induction discharge. *High Temperature* 13: 439–446.

Lee, C.S., and R.D. Reitz. 2001. Effect of liquid properties on the break-up mechanism of high-speed liquid drops. *Atomization and Sprays* 11: 1–18.

Lee Y.C., C. Hsu, and E. Pfender 1981. *Modelling of Particle Injection into a D.C Plasma Jet 5th International Symposium on Plasma Chemistry*, Edinburgh, Scotland, 2, 795–801.

Lefebvre, A.H. 1989. *Atomizations and Sprays*. Hemisphere Pub Corp.

Legros, E. 2003. *3D Modeling of the Spray Process*. Ph.D. Thesis, University of Limoges, France, November.

Lesinski, J., and M.I. Boulos. 1988. Laser Doppler anemometry under plasma conditions, part II measurements in an inductively coupled rfr plasma. *Plasma Chemistry Plasma Processing* 8: 133–144.

Lewis, J.W., and W.H. Gauvin. 1973. Motion of particles entrained in a plasma jet. *AICHE Journal* 19 (6): 982–990.

Li, K.-I., M. Vardelle, A. Vardelle, P. Fauchais, and C. Trassy. 1995. Comparisons between single and double flow injectors in the plasma spraying process. In *Thermal Spray: Practical Solutions for Engineering Problems*, ed. C.C. Berndt, 45–50. Materials Park: ASM International.

Marchand, C., C. Chazelas, G. Mariaux, and A. Vardelle. 2007. Liquid precursor plasma spraying: Modelling the interaction between the transient plasma jet and the droplets. In *Thermal Spray 2007: Global Coating Solutions*, ed. B.R. Marple et al., 196–201. Materials Park: ASM International.

Marchand, C., A. Vardelle, G. Mariaux, and P. Lefort. 2008. Modelling of the plasma spray process with liquid feedstock injection. *Surface and Coatings Technology* 202: 4458–4464.

Marchand, O., L. Girardot, M.P. Planche, P. Bertrand, Y. Bailly, and G. Bertrand. 2011. An insight into suspension plasma spray: Injection of the suspension and its interaction with the plasma flow. *Journal of Thermal Spray Technology* 20 (6): 1311–1320.

Meillot, E., and G. Balmigere. 2008. Plasma spraying modelling: Particle injection in a time-fluctuating plasma jet. *Surface & Coatings Technology* 202: 4465–4469.

Meillot, E., S. Vincent, C. Caruyer, J.-P. Caltagirone, and D. Damiani. 2009. From DC time-dependent thermal plasma generation to suspension plasma-spraying interactions. *Journal of Thermal Spray Technology* 18: 875–886.

Metz, R., C. Machado, M. Houabes, M. Elkhatib, and M. Hassanzadeh. 2008. Nitrogen spray atomization of molten tin metal: Powder morphology characteristics. *Journal of Materials Processing Technology* 195: 248–254.

Moreau, E., C. Chazelas, G. Mariaux, and A. Vardelle. 2006. Modeling the restrike mode operation of a D.C. Plasma spray torch. *Journal of Thermal Spray Technology* 15 (4): 524–530.

Mostaghimi, J., and E. Pfender. 1984. Effects of metallic vapor on the properties of an argon arc plasma. *Plasma Chemistry and Plasma Processing* 4 (2): 129–139.

Mostaghimi, J., P. Proulx, and M.I. Boulos. 1985a. Computer modelling of the emission patterns for a spectrochemical ICP. *International Journal of Heat and Mass Transfer* 28: 1327–1336.

Mostaghimi, J., P. Proulx, M.I. Boulos, and R.M. Barnes. 1985b. Computer modelling of the emission patterns for a spectrochemical. *ICP Spectro-Chemical Acta* 40B: 153–166.

Neikov, OD, Naboychenko SS and Yefimov NA (2019) *Handbook of non-ferrous metal powders, Technology, and applications,* Elsvier, Cambridge MA.

Ozturk, A., and M. Cetegen. 2004. Modeling of plasma assisted formation of precipitates in zirconia containing liquid precursor droplets. *Materials Science and Engineering A* 384: 331–351.

———. 2005. Experiments on ceramic formation from liquid precursor spray axially injected into an oxy-acetylene flame. *Acta Materialia* 2005 (53): 5203–5211.

Pfender, E. 1989. Particle behavior in thermal plasmas'. *Journal of Plasma Chemistry Plasma Processing* 9 (1): 168S–194S.

Pfender, E., and C.H. Chang. 1998. Plasma spray jets and plasma-particulate interaction: Modeling and experiments. In *Thermal Spray: Meeting the Challenges of 21st Century*, ed. C. Coddet, 1315–1328. Materials Park: ASM International.

Proulx, P., J. Mostaghimi, and M.I. Boulos. 1983. Plasma-particle interaction effects in induction plasma modelling under dense loading conditions, ISPC-6, Montréal. In , ed. M.I. Boulos and R.J. Munz, vol. 1, 59–68. University of Sherbrooke, Canada

———. 1985. Plasma-particle interaction effects in induction plasma modelling under dense loading conditions. *International Journal of Heat and Mass Transfer* 28: 1327–1336.

———. 1987. Heating of powders in an R.F. inductively coupled plasma under dense loading conditions. *Journal of Plasma Chemistry and Plasma Processing* 7 (1): 29–53.

———. 1990. Loading and radiation effects in plasma jet modelling. *Journal de Physique* 51 (C5): 263–270.

Rampon, R., G. Bertrand, F.L. Toma, and C. Coddet. 2006. *Liquid Plasma Sprayed Coatings of Yttria stabilized for SOFC Electrolyte, ITSC-2006*. Materials Park: ASM International, e-proceedings.

Rampon, P., C. Filiatre, and G. Bertrand. 2008. Suspension plasma spraying of YPSZ coatings for SOFC: Suspension atomization and injection. *Journal of Thermal Spray Technology* 17 (1): 105–114.

Renouard-Vallet, G. 2003. *Plasma Spraying of Dense YSZ Electrolyte for SOFCs*. PhD Thesis, University of Limoges, France and University of Sherbrooke, CN, November. (in French).

Rudinger, G. 1980. *Fundamentals of Gas-Solid Particle Flow*. Amsterdam: Elsevier.

Soo, S.L. 1967. *Fluid Dynamics of Multiphase Systems*. New York: Blaisdell.

Toma, F.L., G. Bertrand, R. Rampon, D. Klein, and C. Coddet. 2006a. *Relationship Between the Suspension Properties and Liquid Plasma Sprayed Coating Characteristics, ITSC-2006*. Materials Park: ASM International, e-proceedings.

Toma, F.L., G. Bertrand, D. Klein, C. Coddet, and C. Meunier. 2006b. Nanostructured photocatalytic Titania coatings formed by suspension plasma spraying. *Journal of Thermal Spray Technology* 15 (4): 587–592.

Tsunekawa, Y., I. Ozdemir, and M. Okumiya. 2006. Plasma sprayed cast Iron coatings containing solid lubricant graphite and h-BN structure. *Journal of Thermal Spray Technology* 15 (2): 239–245.

Uglov, A.A., and A.G. Gnedovets. 1991. Effect of particle charging on momentum and heat transfer from rarefied plasma flow. *Plasma Chemistry Plasma Processing* 11 (2): 251–267.

Vardelle, A., M. Vardelle, and P. Fauchais. 1982. Influence of velocity and surface temperature of alumina particles on the properties of plasma sprayed coating. *Plasma Chemistry Plasma Processing* 2: 255–291.

Vardelle, A., M. Vardelle, P. Fauchais, and M.I. Boulos. 1988. Particle dynamics and heat transfer under plasma conditions. *AICHE Journal* 34: 567–573.

Vardelle, A., M. Vardelle, P. Fauchais, P. Proulx, and M.I. Boulos. 1992. Loading effect by oxide powders in DC plasma jets. In *Thermal Spray: International Advances in Coatings Technology*, ed. C.C. Berndt, 543–548. Materials Park: ASM International.

Vardelle, M., A. Vardelle, and P. Fauchais. 1993. Spray parameters and particle behavior relationships during plasma spraying. *Journal of Thermal Spray Technology* 2 (1): 79–92.

Vardelle, M., A. Vardelle, K.-I. Li, P. Fauchais, and N.J. Themelis. 1996. Coating generation: Vaporization of particles in plasma spraying and splat formation. *Pure and Applied Chemistry* 68 (5): 1093–1099.

Vardelle, A., M. Vardelle, H. Zhang, N.J. Themelis, and K. Gross. 2002. Controlling particle injection in plasma spraying. *Journal of Thermal Spray Technology* 11 (2): 244–284.

Vardelle, A., N.J. Themelis, B. Dussoubs, M. Vardelle, and P. Fauchais. 1997a. Transport phenomena in thermal plasmas. *High Temperature Materials and Processes* 1 (3): s295–s317.

Vardelle, A., P. Fauchais, and P. Fauchais. 1997b. Vaporization and ultra-fine particle generation during the plasma spraying process. In *Thermal Spray: A United Forum for Scientific and Technological Advances*, ed. C.C. Berndt, 543–548. Materials Park: ASM International.

Vardelle, A., P. Fauchais, B. Dussoubs, and N.J. Themelis. 1998. Heat generation and particle injection in a thermal plasma torch. *Plasma Chemistry Plasma Processing* 18 (4): 551–574.

Vardelle, M., A. Vardelle, P. Fauchais, K.-I. Li, B. Dussoubs, and N.J. Themelis. 2001. Controlling particle injection in plasma spraying. *Journal of Thermal Spray Technology* 10 (2): 244–284.

Vardelle, A., M. Vardelle, H. Zhang, N.J. Themelis, and K. Gross. 2002. Controlling particle injection in plasma spraying. *Journal of Thermal Spray Technology* 11 (2): 244–284.

Vaßen, R., H. Kaßner, G. Mauer, and D. Stöver. 2010. Suspension plasma spraying: Process characteristics and applications. *Journal of Thermal Spray Technology* 19 (1–2): 219–225.

Vincent, S., E. Meillot, C. Carruyer, and J.-P. Caltagirone. 2011. Analysis by modelling of the interaction between a thermal flow and a liquid. In *5th International Workshop on Suspension and Solution Plasma Spraying, Tours* September 4–6, CEA Le Ripault.

Volenik, K., P. Chraska, J. Dubsky, J. Had, J. Leitner, and O. Schneewein. 2003. Oxidation of Ni-based alloys sprayed by a water-stabilized plasma gun (WSP). In *Thermal Spray 2003: Advancing the Science and Applying the Technology*, ed. C. Moreau and B. Marple, 1033–1041. Materials Park: ASM International.

Wang, J.Y., A.H. King, and H. Herman. 1998. Nanomaterial deposits formed by dc plasma spraying of liquid feedstocks. *Journal of the American Ceramic Society* 81 (1): 121–128.

Watanabe, T., and K. Ebihara. 2003. Numerical simulation of coalescence and break-up of rising droplets. *Computers and Fluids* 32: 823–834.

White, F.M. 1974. *Viscous Fluid Flow*. New York: McGraw Hill.

Wilden, J., H. Frank, and J.-P. Bergmann. 2006. Process and microstructure simulation in thermal spraying. *Surface & Coatings Technology* 201: 1962–1968.

Wittmann, K., F. Blein, J.-F. Coudert, and P. Fauchais. 2001. Control of the injection of an alumina suspension containing nanograins in a D.C. plasma. In *Thermal Spray 2001: New Surface for a New Millennium*, ed. C.C. Berndt, K.A. Khor, and E. Lugscheider, 375–382. Materials Park: ASM International.

Wittmann, K., J. Fazilleau, J.-F. Coudert, P. Fauchais, and F. Blein. 2002. A new process to deposit thin coatings by injecting nanoparticles suspensions in a D.C. Plasma Jet. In *Proceedings of ITSC-2002*, ed. E. Lugscheider, 519–522. Düsseldorf: DVS.

Xue, S., and M.I. Boulos. 2019b. Transient heating and evaporation of metallic particles under plasma conditions. *Journal of Physics D: Applied Physics* 52: 454002. (12 pages).

Xu, D.-Y., X.-C. Wu, and Xi Chen. 2002. Motion and heating of non-spherical particles in a plasma jet. *Surface and Coatings Technology* 171 (1–3): 149–156.

Part II

Thermal Spray Technologies

Cold Spray

Abbreviations

ABS	Acrylonitrile Butadiene Styrene
ASB	Adiabatic Shear Band
CFD	Computational Fluid Dynamics
CFRP	Carbon Fibber Reinforced Polymer
CGDS	Cold Gas Dynamic Spray process
CS	Cold Spray
CW	Compression Wave
EBSD	Electron Back Scattered Diffraction
EDS	Energy-Dispersive X-ray Spectroscopy
FEA	Finite Element Analysis
FEM	Finite-Element Modeling
GN	Graphene-Nanosheet
HA	Hydroxyapatite
HBN	Hexagonal-Boron-Nitride
HVOF	High Velocity Oxygen Fuel
ITO	Indium Tin Oxide
KS	Kinetic Spray
LACS	Laser-Assisted Cold-Sprayed
LHS	Left-Hand Side
LPCS	Low Pressure Cold Spray
PEEK	Polyether Ether Ketone
P-GDS	Pulsed-Gas Dynamic Spray
PIC	Particle Impact Conditions
PSD	Particle Size Distribution
PVA	Poly Vinyl Alcohol
RAS	Russian Academy of Sciences
RHS	Right-Hand Side
SEM	Scanning Electron Microscopy
SIMAT	Supersonically Induced Mechanical Alloy
SLD	Supersonic Laser Deposition
TEM	Transmission Electron Microscopy
TST	Thermal Spray Toolkit
VCS	Vacuum Cold Spray
WS	Window of Spray ability
XRD	X-ray diffraction

6.1　Introduction

Cold Spray (CS), a technology discovered in the early nineteenth century, has made a huge leap forward in the last two decades, leading to a wide range of industrial applications [Raoelison et al. (2018)]. Its recognition in the mid-1980s is credited to a group of scientists from the Institute of Theoretical and Applied Mechanics of the Siberian Branch of the Russian Academy of Sciences (ITAM of RAS) in Novosibirsk. Irissou et al. (2008), who developed the Cold-Spray (CS) process as we know it today, successfully deposited a wide range of pure metals, alloys, and composites on a variety of substrate materials and demonstrated the feasibility of cold spray for various applications. A US patent on the cold-spray technology was issued in 1994 [Davis et al. (2004)] and a European patent was granted in 1995 [Dykhuizen and Smith (1998)]. Since then, interest in the field of cold spray has grown exponentially and it is probably one of the spray processes to which the largest numbers of publications and patents have been devoted over the past decades [Irissou et al. (2008)]. In this chapter, an overview is presented of the cold-spray technology and its evolution over the past decade. This is followed by a detailed analysis of gas dynamics in the cold-spray process and coating formation. Deposition parameters and process controls are discussed next, followed by a brief review of cold-spray coating applications. Extensive references are given for the interested reader in this rapidly developing field of surface modification technologies.

M. I. Boulos et al. (ed.), *Thermal Spray Fundamentals*, https://doi.org/10.1007/978-3-030-70672-2_6

6.2 Overview of Cold Spray Technologies

6.2.1 Conventional Cold Spray

Cold Spray (CS) is a kinetic-spray process that uses supersonic jets of compressed gas to accelerate, near-room temperature, powder particles to ultra-high velocities [Champagne (2007), Papyrin (2001)]. The solid ductile particles, travelling at velocities between 300 and 1500 m/s, plastically deform on impact with the substrate and consolidate to create the coating [Papyrin (2001), Handbook of Thermal Spray (2004), Champagne (2007)]. The process requires particle velocities on impact that are higher than the so-called *critical velocity*, which depends on the nature of the particle material, its Particle Size Distribution (PSD) and particle morphology.

The fundamental phenomenon in the cold-spray process is the gas-dynamic acceleration of particles to supersonic velocities and hence to high kinetic energies. This is achieved using convergent-divergent de Laval nozzles, as shown in Fig. 6.1 [Dykhuizen and Smith (1998)]. The upstream gas pressure in a cold-spray system varies between 2 and 4 MPa (300–600 psig). Gases used include N_2, He, or their mixtures, at relatively high flow rates (up to 4 m^3/min). The gas must be preheated up to 700–800 °C to avoid condensation once it undergoes sonic expansion. This heating also increases the particle temperature and velocity, and consequently assists in particle deformation upon impact. When spraying with helium, the system needs to be operated in an enclosure in order to allow for the recycling of the gas, which is essential for the economic viability of the process.

Since the particles are injected upstream of the nozzle, as shown in Fig. 6.2, the powder feeder design should allow for operation at relatively high pressures, slightly above the upstream pressure of the process gas supply. Particles adhere onto the substrate only if their impact velocity is above a critical value, typically in the range of 500 to 900 m/s, depending on the sprayed material. The cold-spray process works best with fine feedstock powders with a mean particle diameter in the range of 1 to 50 μm. Deposition efficiencies can be as high as 70 wt.% with spray rates of 3–5 kg/h, with a deposition "sweet spot" below 25 mm^2. It is important to emphasize that throughout the cold-spray process, the substrate is not heated. Since the feed powder remains in the solid state [Champagne (2007), Papyrin (2001)], the deleterious effects of oxidation, evaporation, melting, crystallization, residual stresses (except peening effect), debonding, gas release, and other common problems in traditional thermal spray processes can all be avoided. Owing to the impact-fusion coating process, only ductile metals such as Zn, Ag, Cu, Al, Ti, Nb, Mo, or alloys such as Ni-Cr, Cu-Al, MCrAlYs, and polymers can be cold sprayed.

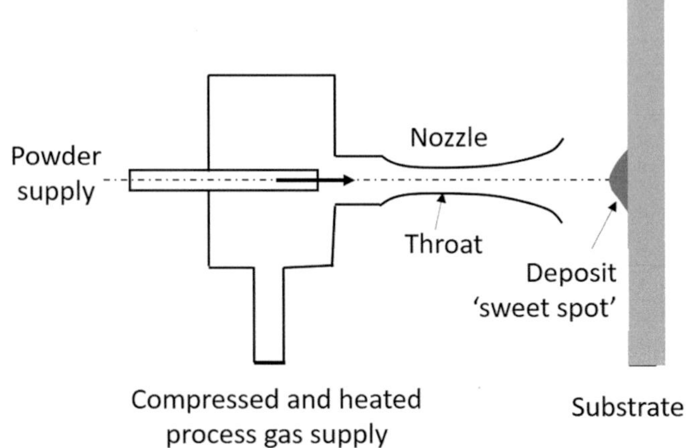

Fig. 6.1 Typical Cold Spray system geometry [Dykhuizen and Smith (1998)] Reprinted with kind permission from Springer Science Business Media copyright © ASM International

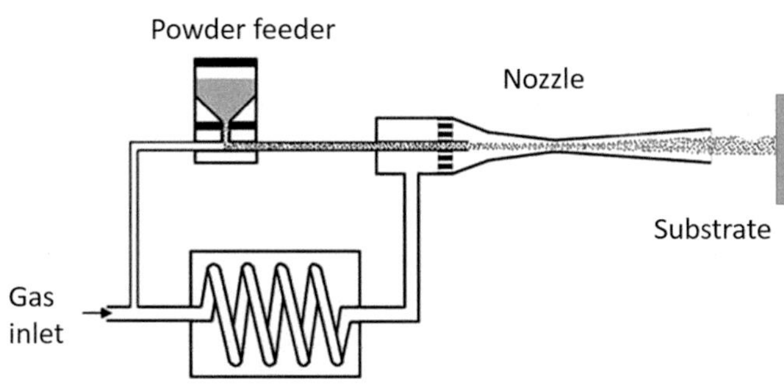

Fig. 6.2 Schematic of the conventional cold-spray process [Champagne et al. (2005)] Reprinted with kind permission from Springer Science Business Media copyright © ASM International

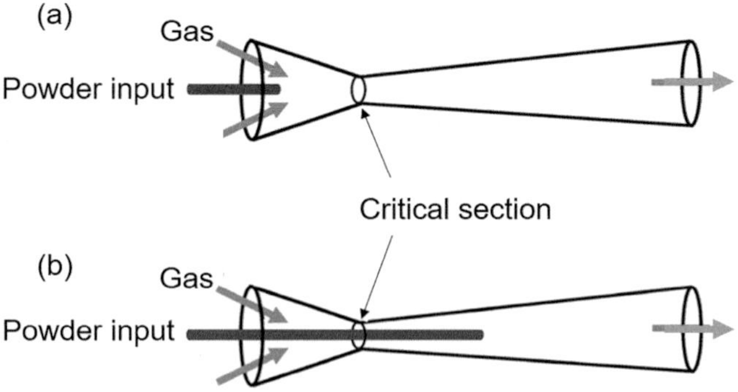

Fig. 6.3 Cold Spray nozzle designs with the different location of the tip of the powder injection probe (**a**) subsonic flow region and (**b**) supersonic flow [Klinkov et al. (2008)]

According to Klinkov et al. (2008), when using cold spray for the deposition of multicomponent coatings, the spraying conditions should be set in order that the particle velocities be above the "critical velocity" of all of the components in the

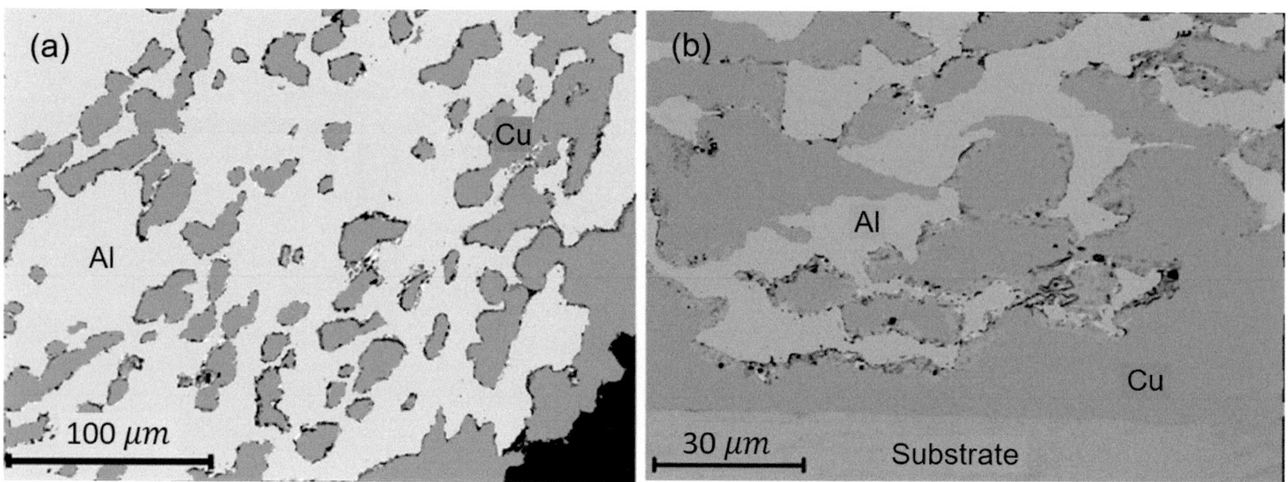

Fig. 6.4 Cross-sections of the two-component copper–aluminum coating (**a**) at the coating surface, (**b**) at the substrate–coating interface [Klinkov et al. (2008)]

mixture, which might not always be attainable. The most-effective method to spray multicomponent coatings is to treat each component under spraying conditions appropriate for this particular component. [Klinkov et al. (2008)] recommend using nozzles with different locations of powder feeders, as illustrated in Fig. 6.3. The feasibility of the proposed approach was experimentally proven by forming a composite coating of copper and aluminum, which are two significantly different materials. The microstructures of the coatings obtained are given in Fig. 6.4.

The overall success of conventional cold spray is probably due to the following process characteristics [Handbook of Thermal Spray (2004), Champagne (2007), Papyrin (2001)]:

- Oxide level in the coatings of metals or alloys is limited to that of the feedstock.
- Coatings can be deposited on temperature-sensitive materials such as polymers without using sophisticated cooling devices.
- Coatings are very dense and exhibit microstructures identical to those of the feed-stock materials.
- The spray trace is well defined and relatively small (between 1 and 25 mm^2), allowing for a precise control of the deposition area.
- The stress generated is compressive and allows for spraying thick coatings (up to 50 mm) without adhesion failure.
- The deposition rates are relatively high, which makes it attractive for industries.

However, while only ductile particles can be sprayed, the high impact velocities and cold deformation of the sprayed metal give rise to a coating with low ductility. [Wu et al. (2012)] underline that with its low temperature and high velocity compared with thermal spraying, the cold-spray process offers competitive advantages for a wide range of industrial applications. The first users were the automotive and aerospace industries, and a strong interest seems to be developing in the electronic industry [Champagne (2007), Champagne et al. (2005)].

6.2.2   Kinetic Spray

"Kinetic Spray" (KS) is an equivalent "cold spray" process developed by [Van Steenkiste et al. (2000)a,b, Van Steenkiste et al. (2002), Van Steenkiste et al. (2004)]. As illustrated in Fig. 6.5, instead of using nitrogen or helium as working gas, the "Kinetic Spray" process works with preheated air (2 MPa, 18 g/s), with the powder premixed with the working gas in a chamber. The powder carrier gas, nitrogen, can be also preheated to increase particle velocity. The ratio of the cross-sectional surface area of the working gas to that of the powder injector is between 126 and 388, which is 2.5–7.5 less than in conventional cold spray. The mixture of gas and particles, with particle sizes in the range of 50 and 200 μm, flows through a de Laval-type nozzle, with the particles accelerated by the high-velocity gas stream. Mean critical velocities of the larger particles (d > 50 μm) are significantly smaller than those of the smaller particles in the cold-spray process (d < 50 μm). The smaller diameters of the powder-injection tubes in the kinetic-spray process appear to be the most important difference with the cold-spray process. According to [Van Steenkiste et al. (2004)], the higher gas temperatures and possibly lower turbulence upstream of the throat are among the principal features that distinguish the kinetic-spray process compared to conventional cold-spray processes.

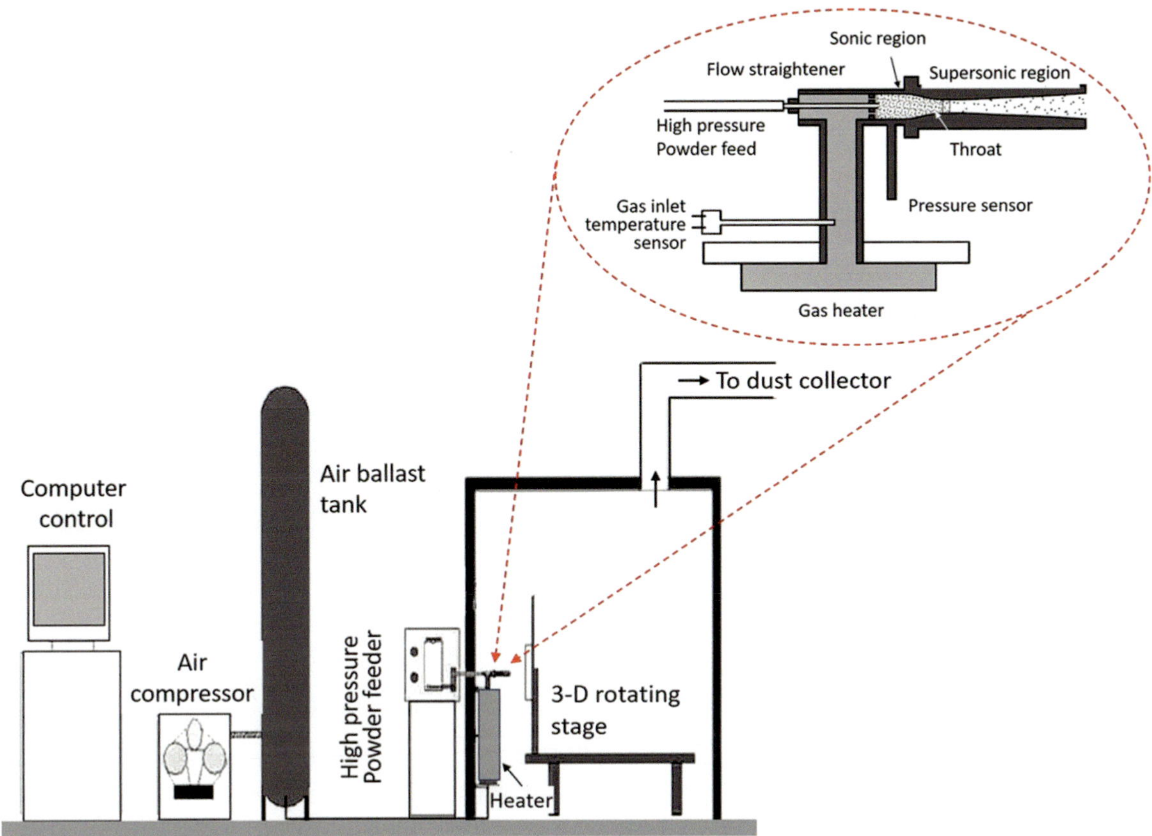

Fig. 6.5 Schematic diagram of the Kinetic Spray equipment with Close-up of the spray gun region [Van Steenkiste et al. (2002)] Reprinted with kind permission from Elsevier

6.2.3 Pulsed-Gas Dynamic Spray

The low particle-impact temperature in the conventional cold-spray process is problematic when spraying hard particles, for which it is difficult to achieve sufficiently high velocities to reach the critical value. A new coating process identified as "Pulsed-Gas Dynamic Spraying" (P-GDS) was developed by Jodoin et al. (2007), Yandouzi et al. (2007), Yandouzi et al. (2008). This process allows for accelerating in an inert gas the feedstock particles to high impact velocities at intermediate temperatures. The higher particle temperature prior to its impact on the substrate enhanced plastic deformation, thus requiring a lower particle impact velocity "critical velocity" compared to the conventional cold-spray process. A schematic of the PGDS process is given in Fig. 6.6. It comprises a long smooth wall pipe, of circular cross-section, which is divided into two compartments separated by a valve ($V_{H/L}$). The first compartment, called the shock generator, connected to a high-pressure gas supply, is kept at high-pressure between each pulse p_{int} (3.0 MPa), controlled by a pressure regulator. The second compartment, the spraying gun, is opened at the end and kept at atmospheric pressure (p_{atm}). When the valve ($V_{H/L}$) separating the two sections is opened rapidly for a short period, a shock wave, depending on the initial pressure ratio

(p_{int}/p_{atm}), is generated and propagates into the spraying gun, accelerating and heating the powder present in the gun. This operation is repeated at a set frequency. During each cycle, or pulse, a controlled quantity of the feedstock powder is dropped down from the feeder to the tube by opening the feeder valve (V_p) and closing it prior to the passage of the shock wave. Electrical heaters allow for preheating both the gas and the powder. The aim of this process is to exploit the combined benefits of HVOF and cold-spray processes. It is expected to achieve high impact velocities, similar to those obtained in conventional cold spray at intermediate temperatures (below the melting point of the powder material) in a non-reacting gas.

A detailed description of the process is presented in Fig. 6.7 [Jodoin et al. (2007)]. The process uses Compression Waves (CW) produced at constant frequency and injected into a quiescent inert gas in a spray gun containing the powder feedstock material (Fig. 6.7a). The CW coalesces to form shock waves and generates behind their passage in the gun a high-speed intermediate temperature flow (Fig. 6.7b). The induced flow heats and accelerates the powder feedstock material toward the substrate (Fig 6.7b' & b"). Modeling of this process has shown that helium gives the highest velocities and temperatures [Jodoin et al. (2007)].

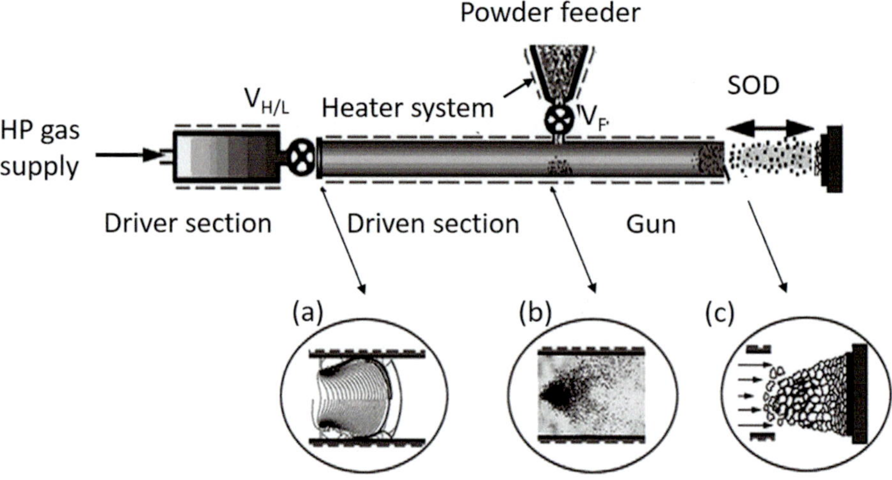

Fig. 6.6 Principle of the Pulsed Gas Dynamic Spray process (PGDS) [Yandouzi et al. (2008)]. Reprinted with kind permission from Elsevier

6.2.4 Low Pressure Cold Spray

Blends of ductile materials (> 50 vol.%) with brittle metals or ceramics can also be used in a modified version of cold spray called the "Low Pressure Cold Spray" (LPCS) process. In this case, the upstream pressure is below 1 MPa, typically 0.5 MPa, with a much lower (about one order of magnitude) gas flow rate of about 0.4 m³/min. A lower power is accordingly required to preheat the spray gas, which is normally air and, therefore, cheaper than N_2 or He [Boro Djordjevic and Maev (2006), Kashirin et al. (2007), Kosarev et al. (2003), Alkimov et al. (1990)]. Particles are injected downstream of the nozzle throat (see Fig. 6.8, which is an illustration of the ("Supersonically Induced Mechanical Alloy Technology" SIMAT process). With gas velocities in the 300–400 m/s range, the critical velocity of particles is not attained. To achieve coatings, metal particles are mixed with ceramic ones, which mostly rebound upon impact, as shown in Fig. 6.9. Their role is to apply compressive stress to the metal particles on the substrate and the previously deposited layers [Alkimov et al. (1995)]. As expected, the deposition efficiency is lower compared to the more standard high-pressure cold-spray process. The LP process is attractive for short run-production [Boro Djordjevic and Maev (2006), Kashirin et al. (2007), Kosarev et al. (2003), Alkimov et al. (1990), Alkimov et al. (1994), Alkimov et al. (1995)].

[Koivuluoto and Vuoristo (2010b)] report that powder type and composition have a very important role in the production of metallic and metallic-ceramic coatings using the Low-Pressure Cold-Spray (LPCS) process. Furthermore, structure and mechanical properties of Cu and Cu + Al_2O_3 coatings are strongly influenced by the powder characteristics of Cu particles. The aim of this study was to evaluate the

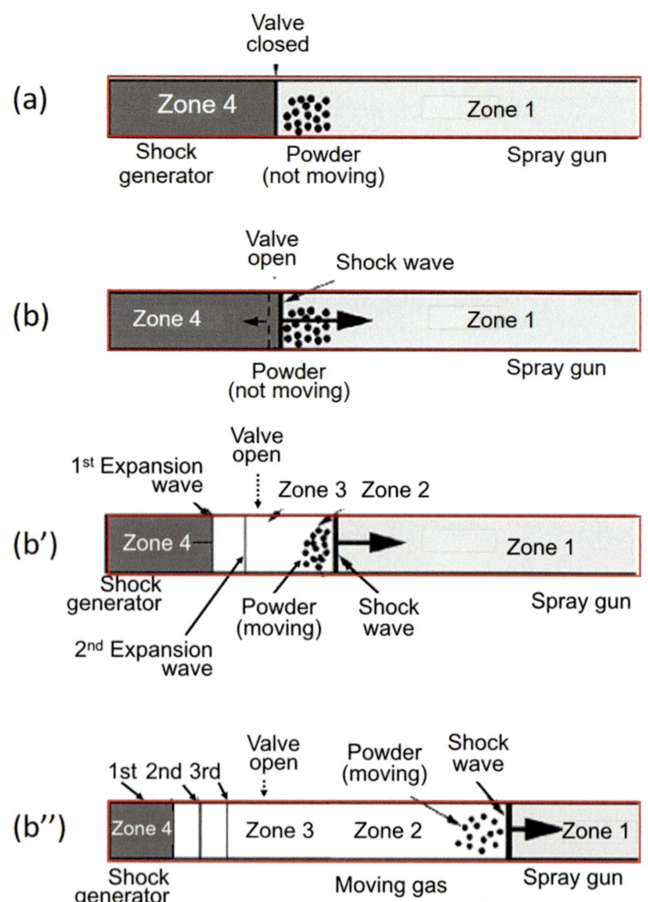

Fig. 6.7 Schematic of one pulse spray cycle. The opening and closure of the valve located between the shock generator and the spraying gun result in a shock wave moving toward the spray gun exit entraining the powder in a high-velocity intermediate temperature flow [Jodoin et al. (2007)] Reprinted with kind permission from Elsevier

effect of different particle types of Cu powder and different compositions of added Al_2O_3 particles on the microstructure, fracture behavior, density, and mechanical properties such as

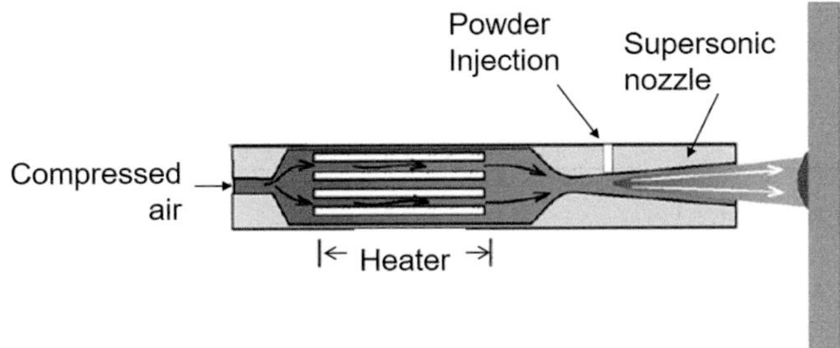

Fig. 6.8 Schematic of the Supersonically Induced Mechanical Alloy Technology (SIMAT®), also known as Gas-Dynamic Spraying, low-pressure cold-spray process principle [Boro Djordjevic and Maev (2006)]. Reprinted with kind permission of ASM

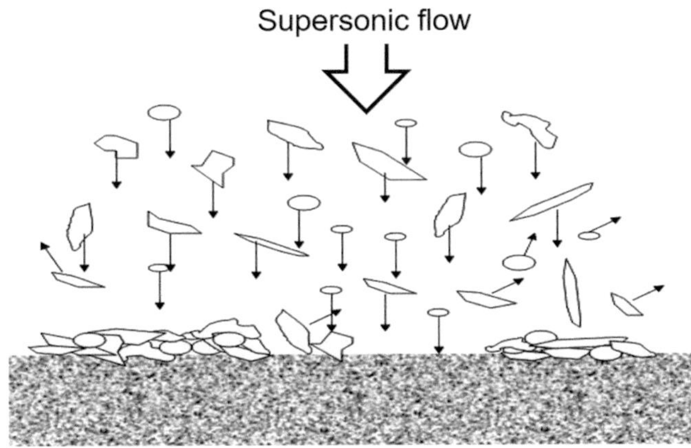

Fig. 6.9 Schematic of the particle deposition by SIMAT ® low-pressure cold-spray process. [Boro Djordjevic and Maev (2006)]. Reprinted with kind permission of ASM

hardness and bond strength. Spherical and dendritic Cu particles were tested together with 0, 10, 30, and 50 vol.% Al_2O_3 additions. Coating density and particle deformation level increased with the addition of hard particles. Furthermore, hardness and bond strength increased with increase in Al_2O_3 fractions. In the comparison between different types of powders, spherical Cu particles led to the denser and least oxide-containing coating structure due to the highly level of particle deformation.

[Ning et al. (2010)] established, using Computational Fluid Dynamics (CFD), a 2-D model of the low-pressure cold spray with a radial powder feeding software. The flow field was simulated for both propellant gases, nitrogen and helium. To predict the in-flight particle velocity and temperature, a discrete phase model was introduced to simulate the interaction of the particles and the supersonic gas jet. The experimental velocity of copper powder with different sizes was used to validate the calculated one for the low-pressure cold-spray process. The results show that the computational model can provide a satisfactory prediction of the supersonic gas flow, which is consistent with the experimental Schlieren photos. As the shape factor was estimated, the reasonable prediction of velocity for non-spherical particle, can be obtained, to compare with the experimental results.

6.2.5   Vacuum Cold Spray

A very simple system, shown schematically in Fig. 6.10, was used to deposit TiO_2 nanometer-sized powders (~ 200 nm) using Vacuum Cold Spraying (VCS). It consists essentially of a vacuum chamber and a vacuum pump, an aerosol chamber, an accelerating gas feeding unit, a particle-accelerating nozzle, a two-dimensional worktable, and a

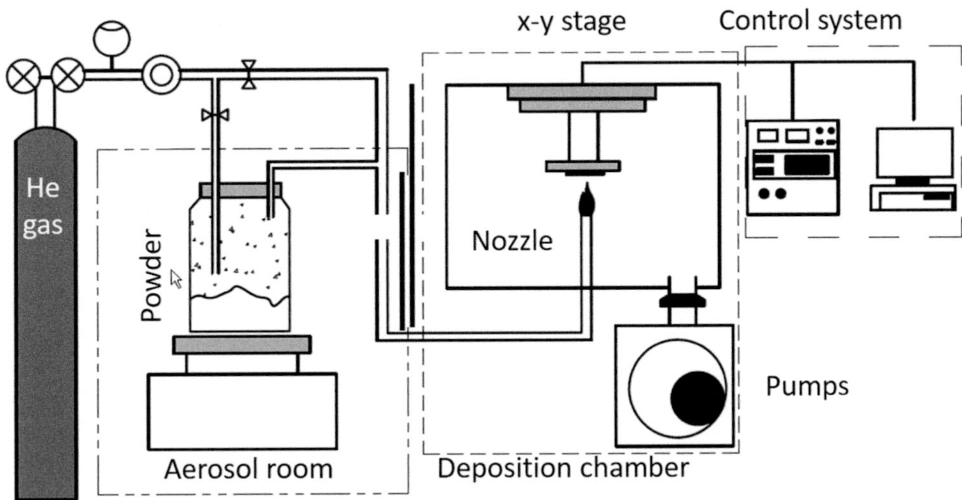

Fig. 6.10   Schematic diagram of Vacuum Cold Spray (VCS) system [Fan et al. (2006)]. Reprinted with kind permission of ASM

control unit. Helium was used as accelerating gas (3 slm), the chamber pressure was 2×10^{-3} Pa, and the orifice of the square nozzle was 2.5×0.2 mm^2. The system deposited TiO$_2$ coatings on ITO (Indium Tin Oxide) conducting glass [Cinca et al. (2010)]. The influence of annealing of the coating between 450 and 500 °C gave rise to both better adhesion and higher photo-catalytic activity [Fan et al. (2006), Yang et al. (2007), Fan et al. (2006)]. Other vacuum cold-spray systems are described in patents [Irissou et al. (2008)].

6.2.6   Laser-Assisted Cold-Spray

A review of Laser-Assisted Cold-Sprayed (LACS) Corrosion- and Wear-Resistant Coatings was reported by Olakanmi and Doyoyo (2014). They underline that this technique presents some advantages: solid-state deposition of dense, homogeneous, and pore-free coatings onto a range of substrates; and high build rate at reduced operating costs without the use of expensive heating and process inert gases. As illustrated in Fig. 6.11, the laser heats both the particles and the substrate to between 30 and 80% of particle melting point. By varying the LACS process parameters and their interactions, the functional properties of coatings can be manipulated. Moreover, thermal effects due to laser irradiation and microstructural evolution complicate the interpretation of LACS mechanical deformation mechanism, which is essential for elucidating its physical phenomena. The investigations of all process sequences, from laser irradiation of the powder-laden gas stream and the substrate, to the

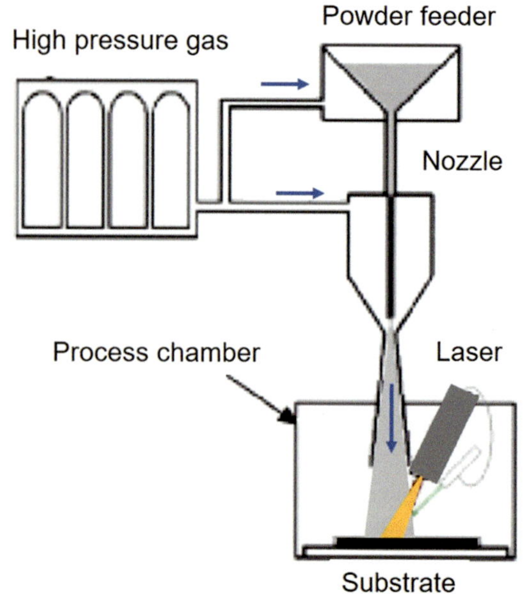

Fig. 6.11 Schematic of Laser-Assisted Cold-Sprayed system (LACS) [Olakanmi and Doyoyo (2014)]

impingement of thermally softened particles on the deposition site, and subsequent further processes, are described.

According to Li et al. (2015), the coating microstructure and properties are greatly influenced by particle preheating and substrate softening. That is why they studied the Cold-Spray (CS) process of copper powder, with assistance of synchronous laser irradiation. The results show that the coating surface with laser irradiation is smoother than that without laser irradiation. The peak coating thickness increases by about 70% as synchronous laser irradiation is employed, indicating an improvement in deposition efficiency. It is also found that with synchronous laser irradiation, the coating is denser, and the coating-substrate interfacial bonding is better compared with that without laser irradiation. Moreover, the EDS and XRD analyses find Cu oxidation occurrence in the SLD (Supersonic Laser Deposition) coating, but the oxide is trivial. The aforementioned improvements on the coating largely arise from particle preheating and substrate softening by synchronous laser irradiation in the CS process.

6.3   Gas Dynamics in Cold Spray Process

As with other spray processes, the gas stream, whether combustion or plasma, is the prime source of kinetic energy for particle acceleration and projection toward the substrate. The cold-spray process is no exception, with the maximum achievable particle velocity depending directly on the maximum gas velocity. Considering that in cold spray, the particle velocity needs to be above a critical value for deposit formation, the control of the gas flow pattern in the spray gun has been one of the key aspects of process development [Champagne (2007), Papyrin (2001)]. Numerous studies have been devoted to the optimization of the process of gas dynamics through improved nozzle designs [Dykhuizen and Smith (1998), Champagne et al. (2005), Kosarev et al. (2003), Stoltenhoff et al. (2002a, b), Alkhimov et al. (2001), Sakaki and Shimizu (2001), Sakaki and Shimizu (2001)].

6.3.1   Isentropic Expansion of the Flow

Historically, a one-dimensional (1-D) isentropic flow analysis has been used to show how to generate a supersonic flow. This analysis is based on the assumption that under steady-state conditions, the flow inside a variable area duct is one dimensional with no chemical reaction, no external work done on the flow, no change in potential energy, and no heat transfer.

The simplest models consider a 1-D geometry of the de Laval nozzle, as shown in Fig. 6.12, with an isentropic expansion of the flow from the nozzle throat toward its exit [Dykhuizen and Smith (1998)].

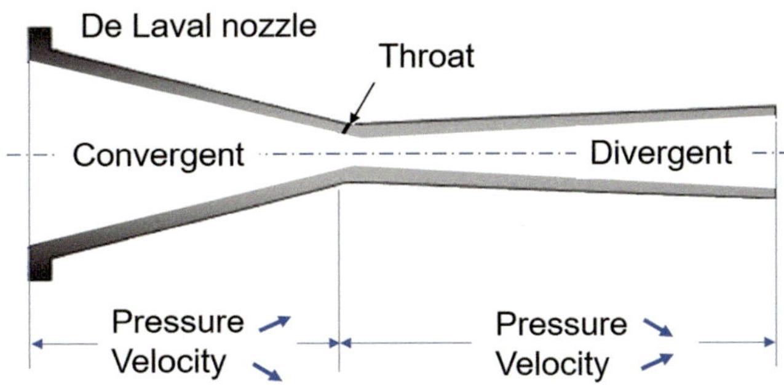

Fig. 6.12 Schematic diagram of de Laval nozzle [Dykhuizen and Smith (1998)]

The process gas flow is assumed to originate from a large chamber or duct where the pressure is equal to the stagnation pressure (p_o), at a temperature (T_o), where the gas velocity is zero. For practical purposes, this matches the conditions upstream of the nozzle throat, provided that the flow area is at least three times larger than the throat area [Dykhuizen and Smith (1998)]. At the throat, the Mach number is unity. The local velocity in the divergent section of the nozzle can be calculated as follows:

$$u_g = M \sqrt{T \gamma R/M_g} \qquad (6.1)$$

Where, u_g is the gas velocity, M the Mach number, R the universal gas constant, γ the specific heats ratio ($\gamma = c_p/c_v$), M_g the molar mass of the gas, and T the temperature at the location considered, which is linked to the initial gas temperature T_o by the relationship:

$$T_o = T \left(1 + \frac{(\gamma - 1)}{2}\right) \qquad (6.2)$$

For monatomic gases, $\gamma = 1.66$, while for diatomic gases, $\gamma = 1.4$. Equation (6.1) demonstrates well the interest of using He instead of N_2: the quantity R/M_g is 296.78 (J/kg.K) for nitrogen compared to 2007.15 (J/kg.K) for helium. The mass balance yields the gas density at the throat:

$$\rho^* = \frac{\dot{m}}{(v^* . A^*)} \qquad (6.3)$$

where m is the gas mass flow rate, v^* and A^* are respectively the throat velocity and area. Using the perfect gas law [Dykhuizen and Smith (1998)], the pressure at the throat is obtained:

$$p^* = \rho^* R_u T^* \qquad (6.4)$$

with $R_u = R/M_g$. From the throat pressure, the stagnation pressure can be calculated. If sonic conditions are to be maintained at the throat, the throat pressure must be above the ambient pressure, or the pressure in the spray chamber. Current cold-spray systems have an exit pressure well below the ambient pressure to obtain maximum spray particle velocities. Thus, the gas flow is said to be over-expanded. This results in a flow outside of the nozzle that cannot be solved by simple one-dimensional gas dynamic equations. However, the one-dimensional results [Dykhuizen and Smith (1998)] still apply inside the nozzle as long as the overexpansion is not so important as to cause a normal shock inside the nozzle. The exit Mach number can then be calculated when the exit area is specified. With the exit Mach number known, the other gas conditions can be obtained from conventional isentropic flow relationships.

The acceleration of one particle is a function of the drag force exerted on it (see Chap. 4, Sect. 4.2.2.1). Assuming that the particle velocity is low compared to that of the gas, the particle velocity becomes:

$$u_p = u_g \sqrt{\frac{C_D \, a_s \, \rho_g \, x}{m_p}} \qquad (6.5)$$

where u_p is the particle velocity, u_g that of the gas, C_D the drag coefficient, a_s the particle cross-sectional area, m_p the particle mass, ρ_g the specific mass of the gas, and x the distance travelled by the particle. Equation (6.5) implies that for a constant drag coefficient, C_D, the particle velocity is proportional to the square root of the ratio of the distance travelled, x, divided by the particle diameter. It also shows the important influence of the specific mass of the gas (ρ_g). The gas dynamic equations reveal, on the other hand, that the gas mass density decreases as the gas velocity increases through the supersonic nozzle. Thus, there exists an optimal condition that yields a maximum acceleration, which corresponds to conditions where a maximum drag force is exerted on the particle. The latter can be estimated by differentiating the RHS of the drag force equation with respect to pressure (or temperature for they are related via the isentropic flow assumption). The maximum drag force can then be estimated using the perfect gas relations. Note that both the gas mass density and the gas velocity are dependent on the pressure level to which the gas is expanded [Dykhuizen and Smith (1998)]. Finally, one finds that a gas velocity corresponding to a Mach number of 1.4 yields near-optimal conditions that maximize the acceleration of the sprayed particles.

With the convergent-divergent nozzle presented in Fig. 6.12, the fluid is continuously accelerated from low subsonic to supersonic velocity. Another requirement must also be fulfilled to create a supersonic flow: the pressure ratio between the stagnation pressure p_o of the flow and the backpressure p_b outside the nozzle must be sufficiently

high. The design pressure ratio $(p_b/p_o)_d$ for a shockless supersonic nozzle flow is found with the one-dimensional isentropic relation for an adiabatic perfect gas. Under the stated assumptions, the back pressure and the stagnation pressure are the only parameters controlling the design exit Mach number (M) [Alkhimov et al. (2001), Shapiro (1953)]:

$$\left(\frac{p_o}{p_b}\right)_d = \left(\frac{\gamma + 1}{2 + (\gamma - 1)\,M_e^2}\right)^{\frac{\gamma}{\gamma-1}} \quad (6.6)$$

Once the desired exit Mach number and the backpressure are known, the required stagnation pressure is found from Eq. (6.6). However, even if the design pressure ratio is used $(p_b/p_o)_d$, not all convergent-divergent nozzles will produce the desired supersonic flow. The right specific nozzle geometry must be used. The design requirement from the one-dimensional isentropic flow analysis is entirely fulfilled if the design pressure ratio is used with the proper throat (A^*) to exit area (A_e) ratio given by [Shapiro (1953), Alkhimov et al. (2001)]:

$$\frac{A_e}{A^*} = \frac{1}{M_e}\left[\left(\frac{2}{\gamma+1}\right)\left(1 + \left(\frac{\gamma-1}{2}\right)M_e^2\right)\right]^{\frac{\gamma+1}{2(\gamma-1)}} \quad (6.7)$$

According to Eq. (6.6), the stagnation pressure, but not the stagnation temperature, plays a fundamental role in the process of generating a supersonic flow at a specific exit Mach number. The pressure ratio and the specific geometry (through its exit to throat area ratio, A_e/A^*) determine the exit Mach number. However, the stagnation temperature has a direct influence on the exit velocity of the fluid [Alkhimov et al. (2001)]. For a perfect gas, the speed of sound, c, is given by:

$$c = \sqrt{\frac{\gamma\,R\,T}{M_g}} \quad (6.8)$$

where R is the perfect gas constant (8.32 J/mole.K) and M_g the molar mass of the gas (kg/mol). Since the velocity of the fluid is given by the relation $v = c.M_e$, the temperature plays a direct role in the exit velocity of the fluid. From the energy conservation law, the perfect gas law and the expression of sound velocity, one obtains:

$$\frac{T_o}{T_e} = \left[1 + \left(\frac{\gamma-1}{2}\right)M_e^2\right] \quad (6.9)$$

The results of calculations [Alkhimov et al. (2001)] show that the required stagnation gas temperature drops with the increase in the gas exit Mach number, resulting in a smaller heater needed. The solid particles injected in the jet will encounter temperatures ranging from the exit jet temperature

T_e to the stagnation temperature T_o. The latter is found not only inside the tank at the stagnation section but also at the stagnation point, on the substrate, as the flow is heated during its deceleration. Setting the maximum allowable temperature, T_o 1000 K, based on the current practice in cold spray shows that the minimum exit Mach number should generally be kept above 1.5. This limit increases with the required exit velocity [Alkhimov et al. (2001)]. For more details, the interested reader is referred to the following papers [Dykhuizen and Smith (1998), Alkhimov et al. (2001), Alkhimov et al. (2001), Samareh and Dolatabadi (2007)].

A three-dimensional, Computational Fluid Dynamics (CFD) model of the cold-spray gas conditions process was developed by [Zahiri et al. (2014)]. The proposed holistic model is important to track the state of the gas and particles from the injection point to the substrate surface with significant benefits for optimization of very rapid "nanoseconds" cold-spray deposition. Careful attention was given to computation time to benefit the broader cold-spray industry, which has limited access to supercomputers. The k-ε type CFD model is evaluated using measured temperature for a titanium substrate exposed to cold-spray nitrogen at 800 °C and 3 MPa. The principal model parameters are detailed including domain meshing method with turbulence, and dissipation coefficients during spraying.

It is important to point out that particle acceleration in the nozzle is dependent on the residence time of the particle in the nozzle and consequently on the nozzle length L. On the other hand, as the nozzle length increases, the boundary layer thickness of the flow in the nozzle section increases, leading to a decrease in the effective nozzle cross-sectional area in comparison to its geometrical value, and the corresponding decrease in gas velocity at the nozzle exit decreases in comparison to the ideal gas flow velocity [Alkhimov et al. (2001)]. This effect should be considered especially in the case of the cold-spray nozzle design for operation with small particles that require high critical impact velocities for coating formation.

On the substrate side, as illustrated in Fig. 6.13, when the supersonic jet impinges on the surface of the substrate positioned perpendicularly to the flow, a shock wave appears at some distance Z_w from the substrate surface. The effect is accompanied by a gradual drop in the axial velocity of the supersonic jet and the deviation of the gas stream in the radial direction outward over the surface of the substrate. A high-pressure and high-density gas layer is thus created between the substrate surface and the shock, which offers significant resistance to the spray particles to cross it, especially for fine particles with diameters less than 5 μm. The sprayed powder coming through this compressed layer will lose some of its velocity while crossing the shock front. The greater this loss, the greater the thickness of the compressed layer [Sakaki and Shimizu (2001)], which is another good reason to limit the Mach number.

6.3.2 Compressible Flow Models

The equations governing the steady-state flow are the mass, momentum, and energy conservation equations (see Chap. 3 and especially Sect. 3.3.7.2). According to [Jodoin (2002)], due to the transonic nature of the flow under study, this set of equations is classified as elliptic in the subsonic part of the flow, parabolic in the sonic part of the flow, and hyperbolic in the supersonic part of the flow [Jodoin (2002)]. This difference in the classification of the equations is linked to the way the perturbations are transmitted in the flow. To properly represent the physical system, the numerical discretization scheme must reproduce this physical feature. This feature increases the level of complexity involved in solving the governing equations since the numerical discretization scheme used to solve each set of equations is different. A k-ε model represents the turbulence in the flow. As the flow is compressible, the gas mass density is not constant. Using, as a first approximation, the ideal gas law to calculate mass density, in order to have the compressibility effects considered, closes the system of equations [Jodoin (2002)].

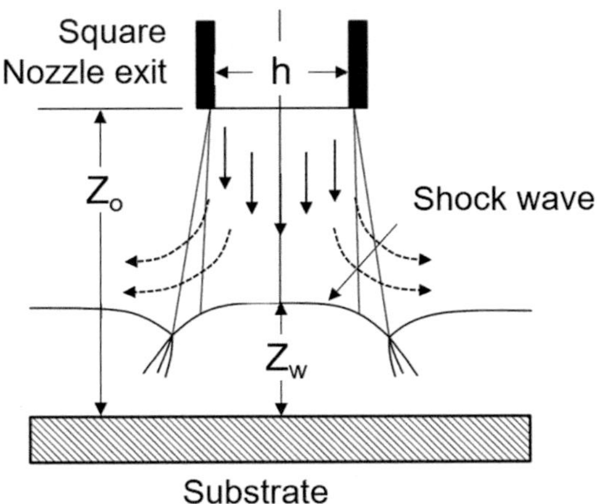

Fig. 6.13 Schematic of the supersonic gas jet impinging on a flat unrestricted obstacle. [Alkhimov et al. (2001)]. Reprinted with kind permission from Springer Science Business Media, copyright © ASM International

Models can be 2-D [Jodoin (2002), Li Wen-Ya and Chang-Jiu Li (2005)] or 3-D [Samareh and Dolatabadi (2007)]. In the latter case, 3-D modelling allowed for the study of the effect nozzle exit-to-substrate distance and substrate shape on the shock wave in front of the substrate. Results reported by Samareh and Dolatabadi (2007), for the cold spray torch configuration given in Fig. 6.14, showed the variation of the local gas pressure and Mach number profiles along the axis of the torch for a substrate placed orthogonally to the torch axis.

Calculations of [Samareh and Dolatabadi (2007)] show that placing the substrate at a distance of 10 mm from the nozzle exit will result in the lowest peak pressure (see Fig. 6.15a). Indeed, the substrate location is of primary importance as it determines the landing situation for particles. Because of the strong bow shock formed on the substrate, a high stagnation pressure zone is created near the plate, resulting, as explained before, in a deceleration of the particles as well as off-normal impact on the substrate. The deviation of particle trajectories from the centerline and those particles that hit the substrate with a low-normal impact velocity, reducing their bond to it, are directly responsible for the reduction in the deposition efficiency. The Mach number also varies with the substrate position, as illustrated in Fig. 6.12b.

For the gas-particle interactions, the conventional equations are used with the drag coefficient, C_D, for particle acceleration, and with the Nusselt number, Nu, for the heat transfer. Of course, these coefficients must be corrected, especially to account for the compressibility effect. For example, for C_D, the approximation of Henderson is used (see [Samareh and Dolatabadi (2007)]) and also Chap. 4 Sect 4.2.2.1.b).

[Voyer et al. (2003)] have used FLUENT to calculate the compressible flow of N_2 gas with an initial temperature of 593 K and a pressure of 2.5 MPa in a standard nozzle. Results are presented in Fig. 6.16 for the gas temperature and velocity, and for the temperature and the velocity of a Cu particle 15 µm in diameter. As can be seen, the gas acceleration starts before the throat and is almost finished in the first third of the divergent section. The gas temperature starts to decrease before the throat and, when the jet expands, falls far below room temperature. In the free jet, after the nozzle exit,

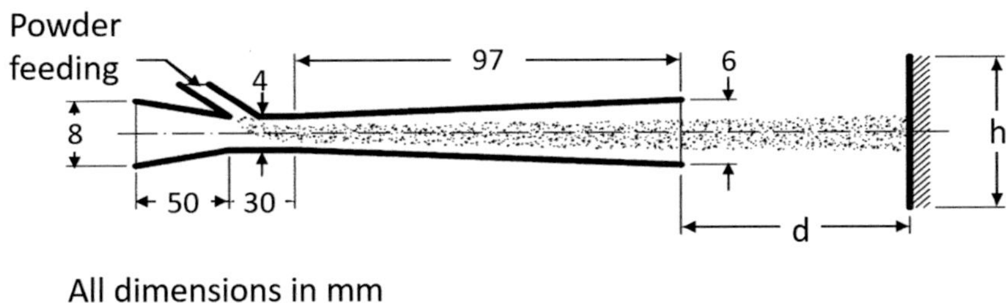

All dimensions in mm

Fig. 6.14 Schematic of the cold spray torch nozzle and substrate used by Samareh and Dolatabadi (2007). Reprinted with kind permission from Springer Science Business Media, copyright © ASM International

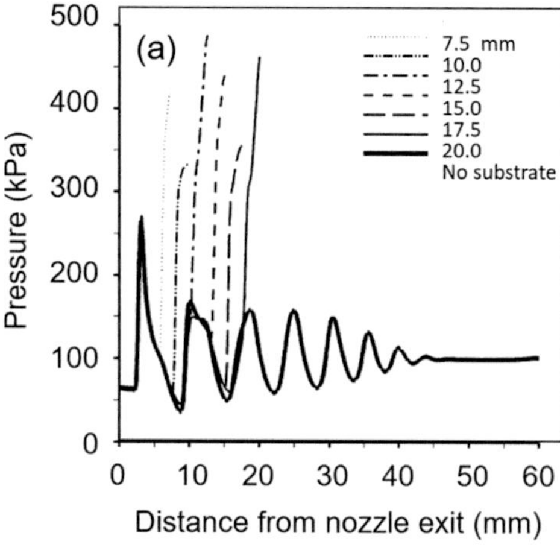
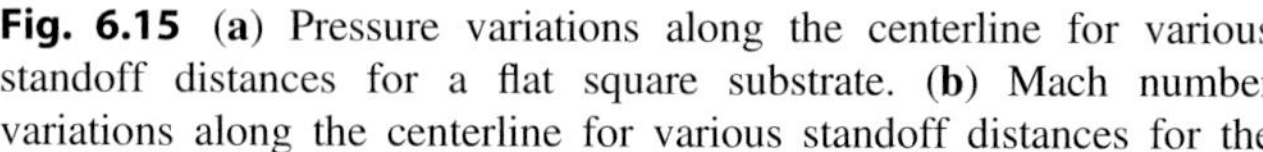
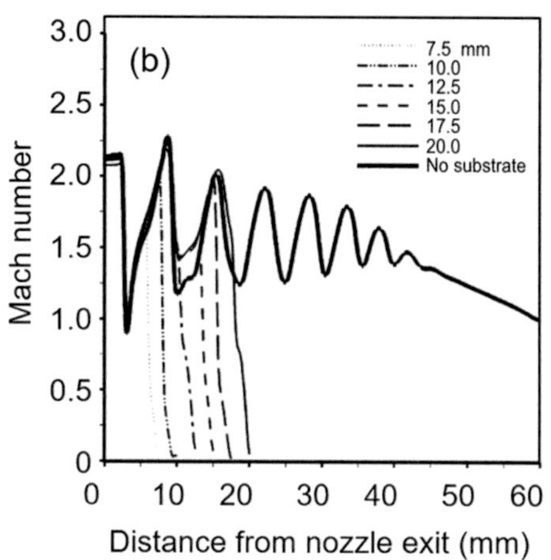

Fig. 6.15 (a) Pressure variations along the centerline for various standoff distances for a flat square substrate. (b) Mach number variations along the centerline for various standoff distances for the same substrate [Samareh and Dolatabadi (2007)]. Reprinted with kind permission from Springer Science Business Media, copyright © ASM International

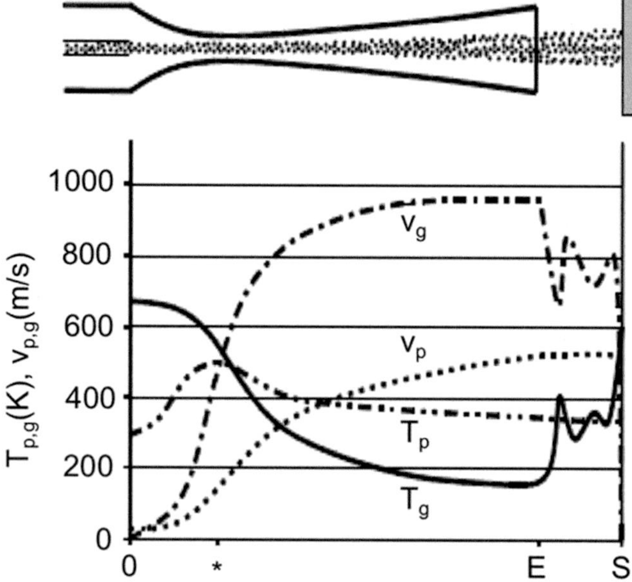

Fig. 6.16 Gas temperature, T_g, and velocity, v_g, evolutions along the nozzle axis for a N_2 flow ($T_o = 573$ K, $p_o = 2.5$ MPa, standard nozzle), and Cu particle (15 μm) temperature, T_p and velocity, v_p, along their trajectory (* nozzle throat, E nozzle exit, S substrate) [Voyer et al. (2003)]. Reprinted with kind permission of ASM

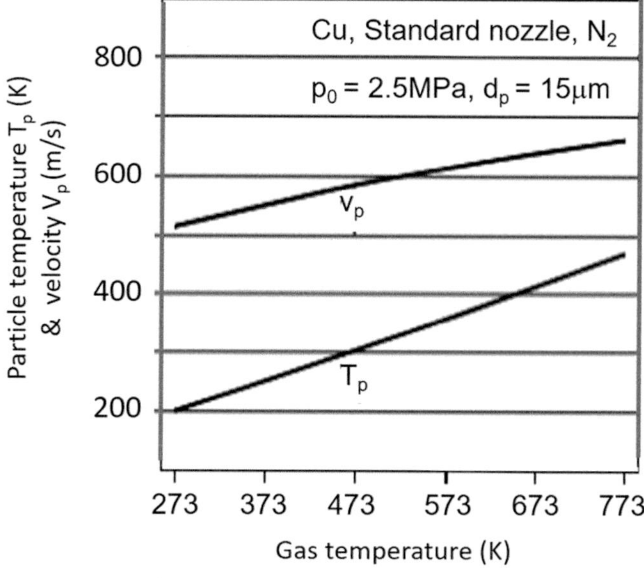

Fig. 6.17 Dependence of the particle temperature and velocity at the nozzle exit on the initial gas temperature for a standard nozzle, N_2 gas, initial pressure 2.5 MPa [Voyer et al. (2003)]. Reprinted with kind permission of ASM

irregular changes of gas properties (T_g, v_g) are due to compression shocks. Particle velocity, v_p, rises sharply in the first half of the divergent section of the nozzle. As the gas velocity is larger than that of the particle, its acceleration is still important even when the gas velocity has reached a plateau. Particle heating occurs in the convergent part of the nozzle, and then the particle temperature decreases but much less than that of the gas. The shock waves do not affect particle temperature, T_p, and velocity.

[Voyer et al. (2003)] have also calculated the influence of the initial pressure and temperature on the temperature and velocity of 15 μm Cu particles at the nozzle exit, using a 1-D model, which is good for showing trends. When the initial temperature increases from 373 to 873 K for a fixed inlet gas pressure of 2.5 MPa, the particle temperature increases linearly from 200 K to about 440 K, while the velocity increases almost linearly from about 520 to 660 m/s (see Fig. 6.17), that is, 25% against the more than 200% increase in temperature.

Figure 6.18 represents the variation with inlet pressure of the temperature and velocity of the 15 μm Cu particle at the

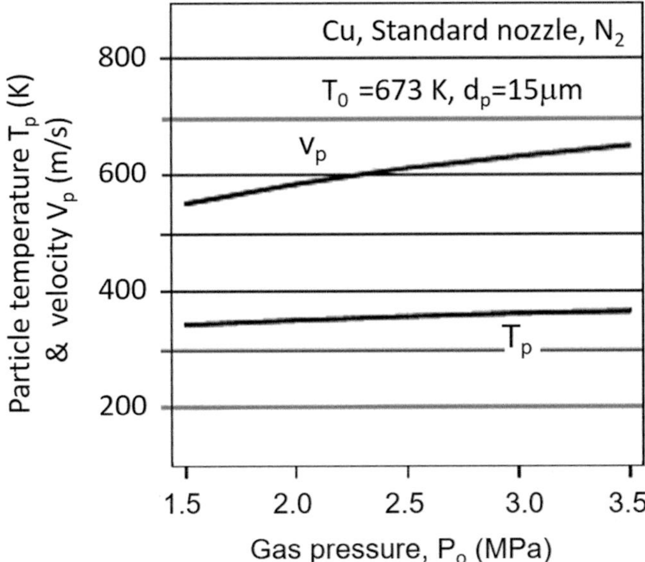

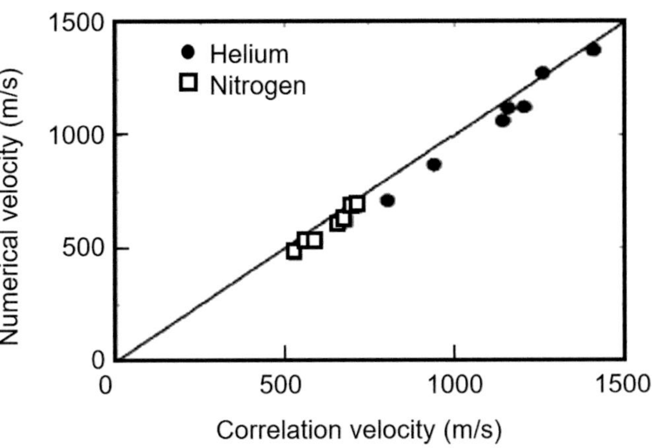

Fig. 6.20 Comparisons of the Alkhimov correlation to numerical results from Sandia's one-dimensional computer code for aluminum and copper powders [Dykhuizen and Neiser (2003)]. Reprinted with kind permission of ASM

Fig. 6.18 Dependence of the particle temperature and velocity at the nozzle exit on the inlet gas pressure for a standard nozzle, N_2 gas, initial temperature 673 K [Voyer et al. (2003)]. Reprinted with kind permission of ASM

As expected, smaller particles (about <25 µm) possess a much higher velocity and, accordingly, their residence time is smaller as well as their temperature.

[Dykhuizen and Neiser (2003)] have studied the influence of the process gas: nitrogen versus helium with the Sandia nozzle. They have compared copper and aluminum particle (about 15 µm in diameter) velocities measured at the nozzle exit, with the values calculated from the empirical expression proposed by [Alkimov et al. (2000)]. Figure 6.20 presents the results of these measurements and calculations. It is clear that higher velocities are obtained with helium.

Such results are confirmed by the work of [Li Chang-Jiu et al. (2005)], who have studied the influence of the spray gas (N_2 and He) and copper particle sizes on their velocities at impact. Results, calculated with a conventional model, are presented in Fig. 6.21. It is interesting to note that the velocity evolution with particle diameters, d_p, is well represented by the empirical expression:

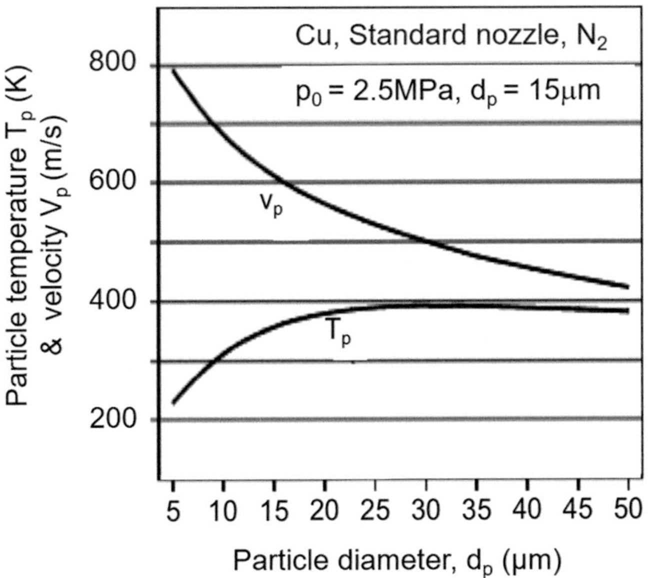

Fig. 6.19 Dependence of the Cu particle temperature and velocity on its size at the nozzle exit: standard nozzle, N_2 gas, 2.5 MPa, 673 K [Voyer et al. (2003)]. Reprinted with kind permission of ASM

$$v_p = \frac{k}{d_p^n} \qquad (6.10)$$

nozzle exit for a standard nozzle and the N_2 gas at an initial temperature of 673 K. The results show that the doubling of the inlet gas pressure to the torch can result in an increase of the particle velocities from 550 to 650 m/s (almost + 18%) with, as expected, minor changes of the particle temperature. Of course, the particle size also plays an important role. Figure 6.19, from [Voyer et al. (2003)] shows the influence of the Cu particle size on its temperature and velocity at the nozzle exit.

One of the main advantages of 2- or 3-D compressible models is that they allow representation of shock waves and their interactions with small particles. As emphasized by [Jodoin (2002)], the discontinuity has no effect on the particles by itself, but since the shock wave considerably changes the gas flow properties, and in particular the gas mass density and velocity, this effect is referred to as the shock-particle interaction. The particles injected in the jet will be exposed to temperatures ranging from the nozzle exit temperature (the coldest temperature in the flow) to the stagnation temperature. The latter is experienced by the particles near the stagnation point on the substrate. For example, in the calculation (two-dimensional axisymmetric

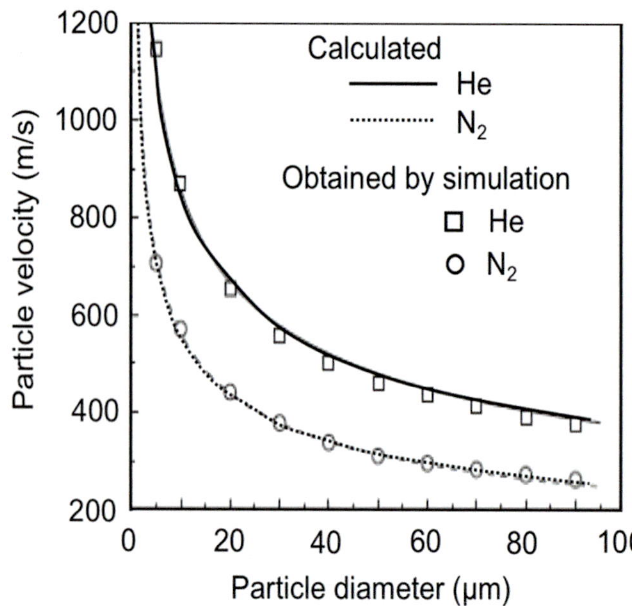

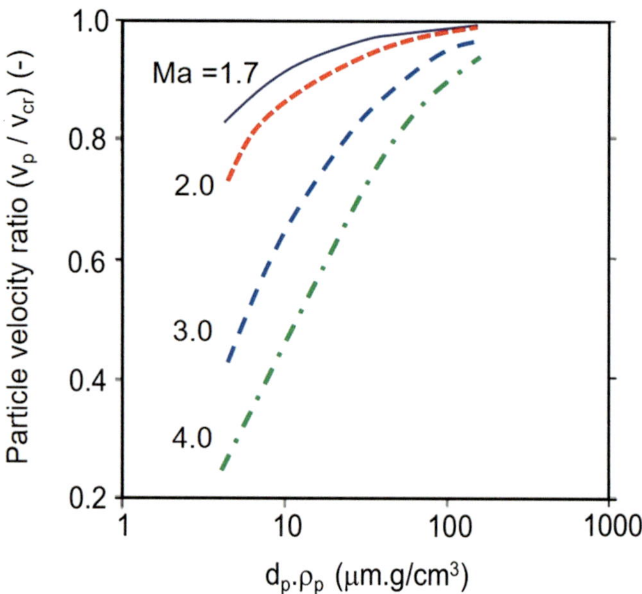

Fig. 6.21 Effect of particle diameter on particle velocity with He and N$_2$ gases operated at the inlet pressure of 2 MPa and a temperature of 340 °C [Li Chang-Jiu et al. (2005)]. Reprinted with kind permission from Elsevier

Fig. 6.23 Particle velocity ratio (v_p/v_{cr}) after the shock wave for particles with different $r_p.\rho_p$ based on the two-dimensional flow field. Four pre-shock Mach numbers are shown: 1.7, 2, 3, and 4 [Jodoin (2002)]. Reprinted with kind permission from Springer Science Business Media copyright © ASM International

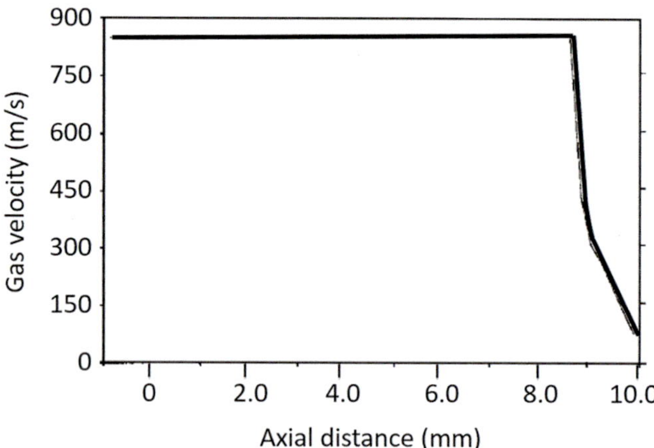

Fig. 6.22 Gas velocity along the axis of the Mach = 1.7 jet. Axial positions = 0 at nozzle exit, 10 mm substrate. [Jodoin (2002)]. Reprinted with kind permission from Springer Science Business Media, copyright © ASM International

analysis) of [Jodoin (2002)] for a $Ma = 1.7$ impinging jet, the stagnation pressure and the stagnation temperature are, respectively, 145 kPa and 22 °C. The exhaust backpressure is fixed by the atmospheric pressure at 100 kPa. The non-slip and adiabatic conditions are used at the nozzle walls. With the presence of a shock wave inside the nozzle, the flow decelerates to subsonic speed. The substrate standoff distance is set at 10 mm from the exit of the nozzle. The no-slip velocity condition is used at the substrate surfaces and these surfaces are assumed to be at a uniform and constant

temperature of 22 °C. The stagnation temperature after the shock wave is not affected significantly and remains nearly constant. In contrast, Fig. 6.22 shows the axial gas velocity profile along the centerline of the jet for a Mach number of 1.7 [Jodoin (2002)]. The velocity is noted to decrease sharply at the shock wave close to the substrate [Jodoin (2002)].

To predict the particle position and velocity along the axis, the one-dimensional particle model has been used with the 2-D flow calculation results. From a kinematic point of view, the equations allowing for the calculation of the velocity of a particle show that the important physical parameter controlling the response of the particles is the product $r_p.\rho_p$ [Jodoin (2002)], where r_p is the particle radius and ρ_p its specific mass. This factor takes into consideration the type and size of the particle. Particles of two different materials will experience the same trajectory and velocity if they have the same $r_p.\rho_p$. Four different supersonic flows were studied with different pre-shock Mach numbers: respectively 1.7, 2, 3, and 4. The exit gas velocity and pressure were set constant at 850 m/s and 100 kPa [Jodoin (2002)]. Figure 6.23 shows the effects of the shock wave on the velocity ratio v_p/v_{crit}. For particles having different $r_p.\rho_p$. The effect of the shock wave is rather important on particles having low values of $r_p.\rho_p$, see Fig. 6.23. For example, even for an exit Mach number of 1.7, the velocity ratio of the particles at the substrate surface for $r_p.\rho_p = 8$ µm·g/cm³ is reduced by a factor of 0.90 over the short shock-substrate distance and to as low as 0.40 for the case of an exit Mach number of 4.

The conclusions of [Jodoin (2002)] are that in order to achieve a sufficiently high exit jet velocity to ensure that the particle velocities are over their critical value, different nozzle designs need to be used. Nozzles with a low design Mach number will require a higher stagnation temperature to produce the required velocity. This may lead to temperatures in the vicinity of the substrate that are high enough to reduce the benefits of the cold-spray process since the substrate will be exposed to these temperatures. Calculations showed that for the range of velocity used in cold spray the design Mach number of an air nozzle should be limited to 1.5 and above to avoid too high temperatures in the vicinity of the substrate. Analyses showed that for large/heavy particles ($r_p.\rho_p > 200$ μm.g/cm^3), the shock wave has a limited but noticeable effect on the impact velocity. This effect increases with the Mach number and becomes important at a Mach number around 3. For small/light particles ($r_{part}.\rho_{part} < 50$ μm·g/cm^3), the effect of the shock-particle interactions is very strong, even at a short standoff distance and at a Mach number as low as 2 [Jodoin (2002)]. Of course, other authors, for example [Katanoda et al. (2007a, b)], have used the two-dimensional axisymmetric, time-dependent Navier-Stokes equations along with the k-ε turbulence model [Samareh and Dolatabadi (2007)]. The governing equations are solved sequentially in an implicit, iterative manner using a finite difference formulation.

The best description of 3-D models used is that of [Samareh and Dolatabadi (2007)]. The ideal gas law is used to calculate mass density, in order to have the compressibility effects considered. Since for a high-speed, compressible flow, viscosity changes with temperature can be important. These need to be taken into account using the three-coefficient Sutherland viscosity law which is best suited for such high-speed flow configuration [Samareh and Dolatabadi 2007)]. Due to the presence of diamond shocks at the nozzle exit and a bow shock on the substrate, the flow will experience sharp gradients and steep changes in pressure and velocity. Therefore, the RSM (Reynolds Stress Model) turbulence model is used in the simulation of [Samareh and Dolatabadi (2007)] as it takes into account the compressibility effect, as well as streamline curvature, swirl, rotation, and rapid changes in strain rate, in a more rigorous manner compared to other models and will result in higher accuracy. However, as this method creates a high degree of coupling between the momentum equation and the turbulence stresses in the flow, calculations can be more susceptible to stability and convergence difficulties compared to the k-ε model. In order to overcome this problem, for each case, the calculation begins with the k-ε model with low under-relaxation factors and is then switched to the RSM scheme.

[Samareh et al. (2009)] have compared calculations with measurements (cutting-edge flow visualization and particle velocimetry) for the torch from CGT company equipped with nozzles: A "v27", called 1 in the following, and "v24," called 2. The torch was run with nitrogen with a stagnation pressure of about 3 MPa. The copper particles (1–25 μm) were injected with nitrogen at a rate of 3 g/s. They found that using RSM, a drag law covering all speed flows, and two-way coupled Lagrangian particle tracking result in capturing shocks and expansion waves virtually identical to those observed in the experiments. Injecting the particles to the gas flow is accompanied with a momentum exchange between the gas and solid particle phases. Particles are accelerated and the gas flow decelerates accordingly. The higher the particle loading, the larger the momentum deficit of the gas flows. Typical results are illustrated in Fig. 6.24 and 6.25.

A schematic of nozzle 1 used in the calculations is given at the top of Fig. 6.24. Results presented include the calculated and measured pressure gradient (gas flow was mapped using a non-commercial optical flow imaging system developed at Siemens). The uncertainties of the measured pressures and mass flow are estimated to be 0.1 MPa and 10%, respectively. Direction is from left to right. Flow separation occurs inside the nozzle. No substrate was present. A visible difference between both flows in Fig. 6.24 (a-1, and a-2), is that the point of separation from the interior nozzle wall that occurs earlier without the powder injection. The free jet shock and expansion waves are altered when coating particles are injected (even with 10% of the mass flow rate of the gas). The Mach number at the nozzle exit decreases from 2.4 for the free jet case to 2.1 when particles are injected (3 g/s), while the gas velocity at the nozzle exit falls from 820 to 770 m/s and the temperature rises by 40 K.

Particle velocities where measured with the SprayWatch® system. Figure 6.25 shows the decreasing gas velocity with increasing feed rate. Particles 25 μm in diameter lose 6% of their velocity when the injection rate is increased from 1 to 3 g/s. At 3 g/s, the calculated particle velocity for 3 g/s powder injection is in good agreement with measurements.

[Rahmati and Ghaei (2014)] recall that many numerical studies have been carried out in the literature in order to study this process in more depth. Despite the inability of the Johnson-Cook plasticity model in predicting material behavior at high strain rates, it is the model that has been frequently used in the simulation of cold spray. They compared the performance of different material models in the simulation of the cold-spray process. Six different material models, appropriate for high strain-rate plasticity, were employed in a finite element simulation of cold-spray process for copper. The results showed that the material model had a considerable effect on the predicted deformed shapes.

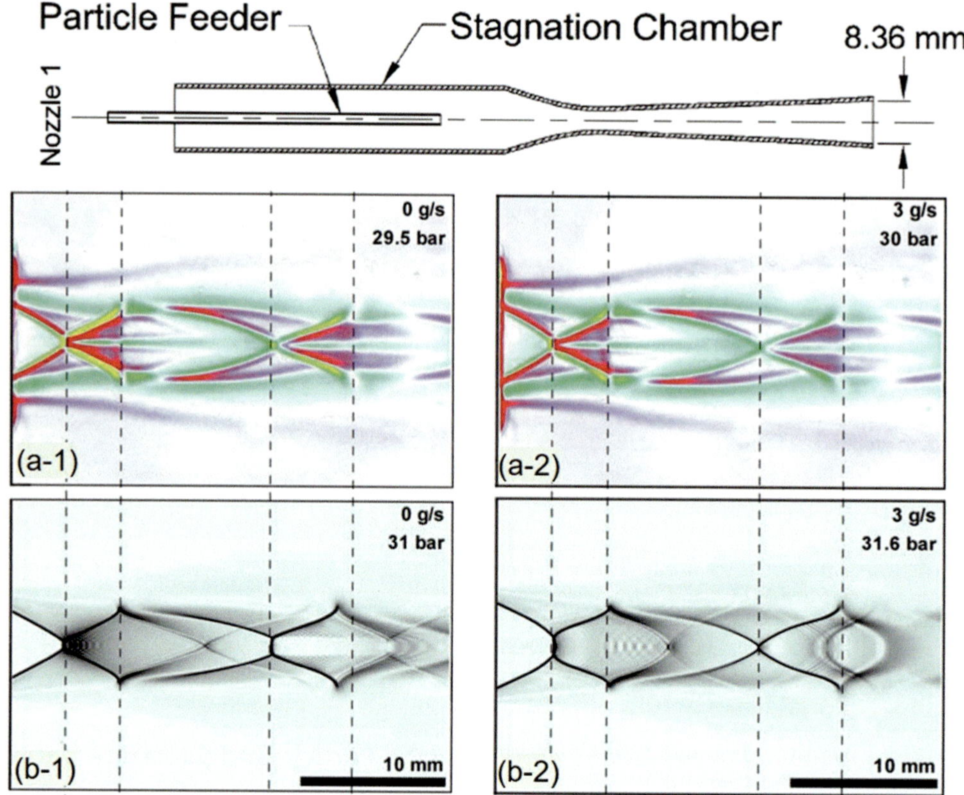

Fig. 6.24 (**a**) Measured and (**b**) calculated pressure gradient contours of the flow behind nozzle 1(top of figure) at constant gas flow rates of 25.4 and 27.8 g/s, respectively, (1) without and (2) with particle injection, 3 g/s. [Samareh et al. (2009)]. Reprinted with kind permission from Springer Science Business Media, copyright © ASM International

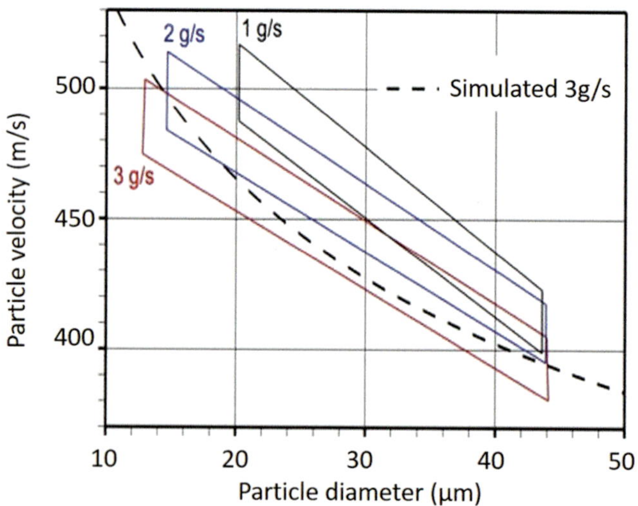

Fig. 6.25 Particle velocities measured in the 2-mm radial distance from the centerline of the flow behind nozzle 1, as function of particle diameter for varying powder injection rates. The trapezoidal regions indicate the 2σ confidence bands for the mean velocity v [Samareh et al. (2009)]. Reprinted with kind permission from Springer Science Business Media, copyright © ASM International

6.3.3   Nozzle Design

As mentioned earlier, particle acceleration in cold spray depends on the nature of the particle material and its particle size distribution. For example, for copper particles, using air as driving gas at 25 °C and a pressure of 2.1 MPa, a 5-µm diameter particle was accelerated to a velocity of 600 m/s, while a 22 µm particle was limited to velocities of around 400 m/s [Gilmore et al. (1999)]. Different nozzle designs were developed with the objective of achieving the most uniform particle velocities possible [Heinrich et al. (2005)]. Results reported by [Blose et al. (2003)] using two rectangular nozzles (2x10 mm), and one circular nozzle (7.1 mm i. d.) are given in Fig. 6.26. These show that for Cu particles (5–20 µm), sprayed with helium (2.1 MPa, 325 °C), the velocity distributions obtained at a stand-off distance of 25 mm were considerably more uniform with the rectangular-shaped nozzles compared to the circular ones. [Gärtner et al. (2006)] made calculations for two nozzle configurations, a standard trumpet-shaped nozzle and a bell-shaped (MOC-designed) nozzle. Results show significant differences in particle acceleration for 20 µm copper

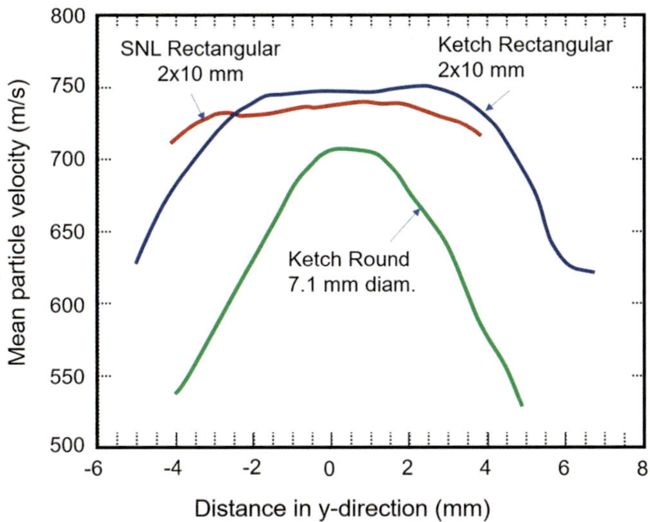

Fig. 6.26 Comparison of mean particle velocity distributions, at 25 mm standoff distance, for circular (d = 7.1 mm) and rectangular (2 × 10 mm) spray nozzles: He, 2.1 MPa, 325 °C, Cu particles 20–5 μm [Blose et al. (2003)]. Reprinted with kind permission of ASM

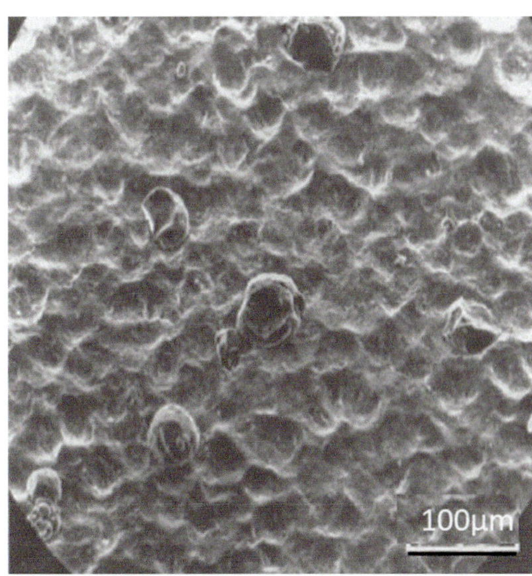

Fig. 6.27 Copper substrate surface after exposure for 25 s to a two-phase aluminum laden jet, particle diameter (30 μm) [Klinkov and Kosarev (2006)]. Reprinted with kind permission from Springer Science Business Media, copyright © ASM International

particles sprayed with nitrogen (3 MPa and 320 °C). With the standard trumpet-shaped nozzle, the particle is accelerated to 500 m/s, compared to 580 m/s for the bell-shaped nozzle.

6.4 Coating Formation

6.4.1 Induction Time

When starting the cold-spray process with a smooth surface, it is commonly observed that during the first tens of seconds, particles bounce off the surface and do not get attached to the substrate. During this short period, generally referred to as *induction time, t_i*, the substrate surface is being cleaned from any oxide layer and is deformed through "shot pinning," as seen in Fig. 6.27 after [Klinkov and Kosarev (2006)]. At a time > t_i (~ few tens of seconds), particles begin to attach to the substrate, provided their velocity is higher than the critical velocity, v_{cr}. Beyond this initial induction period, which depends on the nature of the substrate, the particle concentration in the flow and their impact velocity on the substrate, the attachment rate increases rapidly, in an avalanche-like manner, forming the coating [Klinkov and Kosarev (2006)]. The example given in Fig. 6.27. is for the spraying of aluminum on copper substrate using helium at 2 MPa, and 300 K as process gas. The induction time in this case is $t_i = 25$ seconds.

6.4.2 Particle and Substrate Deformation

As mentioned earlier, powders deposited by cold spray do not undergo melting before or upon impacting the substrate [Rokni et al. (2017)]. This feature makes CS suitable for deposition of a wide variety of materials, most commonly pure metals and metallic alloys, as well as some ceramics and composites. During processing, the particles undergo severe plastic deformation, creating a more mechanical and less metallurgical bond with the substrate material. The deformation behavior of an individual particle depends on multiple material and process parameters that are classified into three major groups:

- Powder characteristics.
- Geometric parameters.
- Processing parameters.

each with their own sub-categories. Changing any of these parameters leads to the evolution of a different microstructure and consequently changes the mechanical properties of the deposit.

While the cold-spray technology has matured during the last decade, the process being inherently complex, our understanding of the effects of deposition parameters on particle deformation, deposit microstructure, and mechanical properties remains far from complete. Upon impact, when the energy is sufficient, the stress applied to the particle is

generally higher than its yield stress, leading to its plastic deformation. The pressure at impact, depending on the particle size, specific mass of the material, and impact velocity, can easily reach 40 to 50 MPa (for about a few tens of nanoseconds), diminishing rapidly afterward with the increase of the contact area. The resulting plastic deformation wave corresponds to the propagation of structure defects. Upon impact, high plastic strain rates occur in the immediate vicinity of the contact zone, resulting in adiabatic heating (during the short time, the high-pressure period). More than 90% of the impact energy is converted into heat, softening locally the material [Grujicic et al. (2004)]. The impact of the particles onto the substrate takes place in two consecutive steps [Schmidt et al. (2006)]:

- Pressure buildup and *elastic deformation* of the particles up to the dynamic flow limit,
- *Plastic deformation* with significant deformation of the material structure and warming due to deformation.

According to the work of [Assadi et al. (2003)], the strong and fast heating of the particle induces locally a transition of the deformation mode from a plastic behavior to a more viscous phenomenon. The mechanical resistance of the material in this transient period is much reduced and a structural instability appears, easing the adiabatic shear phenomenon. This is illustrated in Fig. 6.28 after [Grujicic et al. (2004)] showed the relationships existing between stress and strain during the dynamic deformation of a solid. The stress–strain curve called *"isotherm"* in Fig. 6.28 shows a monotonic increase in the flow stress with plastic strain. Under *"adiabatic"* conditions, the temperature increases with the heat generated by the plastic deformation giving rise to material softening. The *Adiabatic* curve in Fig. 6.28 reveals that the flow stress reaches a maximum value because of the decrease in the rate of strain hardening. However, as emphasized by [Grujicic et al. (2004)] "fluctuations in stress, strain, temperature or microstructure, and the instability of strain softening can give rise to plastic flow (shear) localization." While "shearing and heating" (and consequently softening) become highly localized, the straining and heating in the surrounding material regions practically stops. This, in turn, causes the flow stress to quickly drop to zero, as illustrated with the *Localized* curve in Fig. 6.28.

Numerical simulations of the particle impact were performed [Grujicic et al. (2004), Schmidt et al. (2006) and Schmidt et al. (2009a, b)] to obtain information about the impact loading of the particles: local distribution and temporal evolution of pressure, stress, strain, and temperature. A common feature in all simulations is that there is markedly inhomogeneous deformation of particles and localized heating of the interacting surfaces as a result of the impact. An analysis of the first instant of impact of a Cu particle onto a Cu substrate was reported by Schmidt et al. (2009a, b) using a thermally coupled axisymmetric model. The results show that at the beginning of the impact, a strong pressure field propagates spherically (elastic deformation) into the particle and substrate from the point of first contact, Fig. 6.29(a). The pressure gradient at the gap between the colliding interfaces generates a shear load, which accelerates the material laterally and thereby causes localized shear straining. Figure 6.29(b). When the impact pressure and the respective deformation are high enough, this shear straining leads to adiabatic shear instabilities. This means that thermal softening is locally dominant over strain and strain-rate hardening, which leads to a discontinuous jump in strain and temperature, and an immediate breakdown of stress. This is illustrated in Fig. 6.29 c and d giving the corresponding temperature fields at the beginning of the impact (0.05 μs) and later after the impact (0.5 μs). An out-flowing material jetting with material temperatures close to the melting temperature is observed in Fig. 6.26(b). This phenomenon of jetting is also a known feature in explosive cladding [Schmidt et al. (2006)].

The shear instabilities result in solid-state jets of material extruded from the interface between the impinging particle and substrate. The oxide layer is partially removed allowing true metal-to metal contact [Schmidt et al. (2006), (2009a, b), Assadi et al. (2003)]. Such contacts have been observed, for example, by following the formation of non-uniform intermetallic layers at the particle-substrate boundaries on Al-Cu [Price et al. (2007)] or by using high resolution microstructure investigations at interfaces [Wank et al. (2006)].

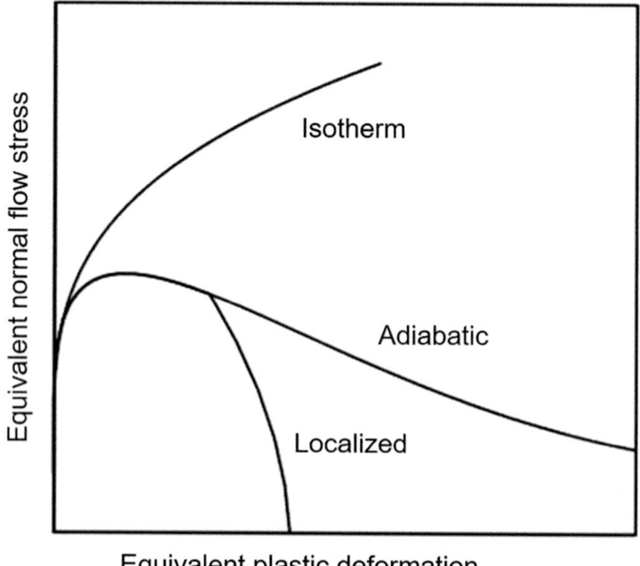

Fig. 6.28 Curves representing the stress-strain in a solid for isothermal, adiabatic, and localized deformation [Grujicic et al. (2004)]. Reprinted with kind permission from Elsevier

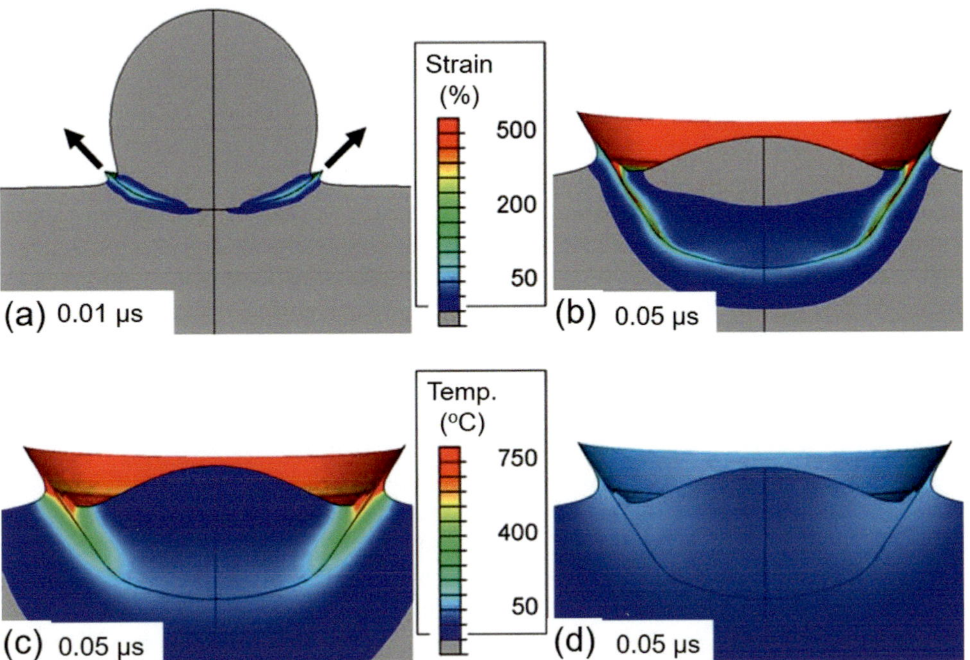

Fig. 6.29 Single sequences of an impact of a 25 µm Cu particle on a Cu substrate at a velocity of 500 m/s and an initial temperature of 20 °C. (**a, b**) Strain field (**c, d**) temperature field. [Schmidt et al. (2009a, b)]. Reprinted with kind permission from Springer Science Business Media, copyright © ASM International

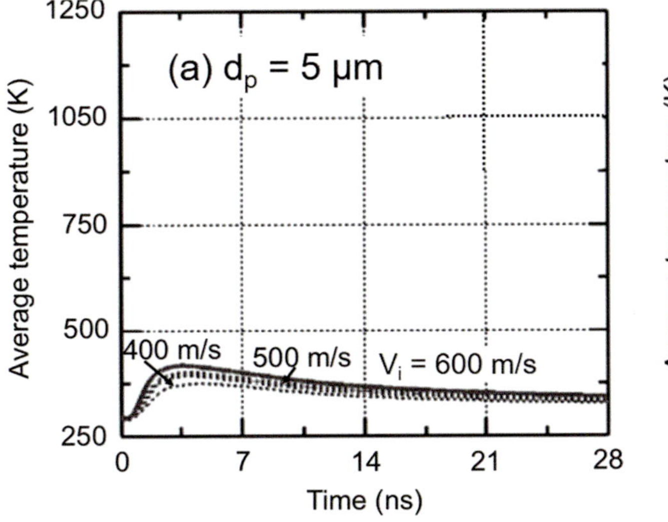

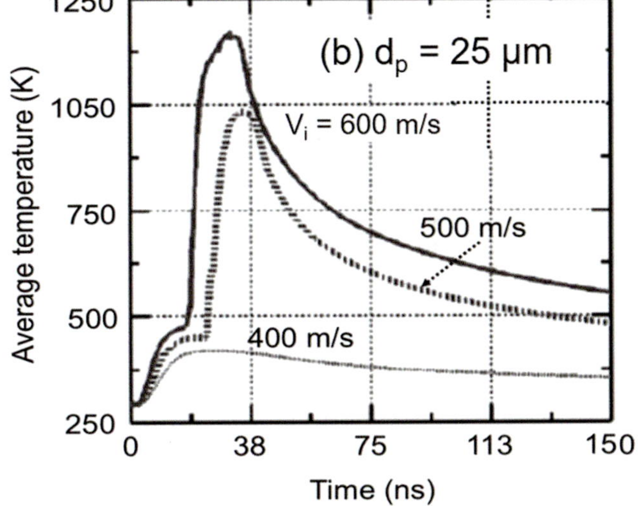

Fig. 6.30 Temporal evolution of the temperature at the monitored volume (sheared interface) of copper particles impacting a copper substrate with different velocities, V_i for two particle sizes: (**a**) 5 µm and (**b**). 25 µm [Schmidt et al. (2009a, b)]. Reprinted with kind permission from Springer Science Business Media copyright © ASM International

According to [Grujicic et al. (2004)], adhesion in cold spray is a nano-length-scale phenomenon involving atomic interactions between the contacting surfaces and not atomic diffusion. It requires clean surfaces and relatively high contact pressures to make the surfaces mutually conforming. The adiabatic softening of the material in the particle/substrate interfacial region, combined with relatively high contact pressures and the partial removal of oxide layers, promotes the formation of mutually conforming contacting surfaces via plastic deformation of the contacting surfaces.

To illustrate the importance of particle size and velocity on the adiabatic softening, [Schmidt et al. (2006), (2009a, b)] calculated the temporal evolution of the temperature of copper particle on impact on a copper substrate. This temperature is the average of the elements in the highly strained contact zone. Typical results are given in Fig. 6.30

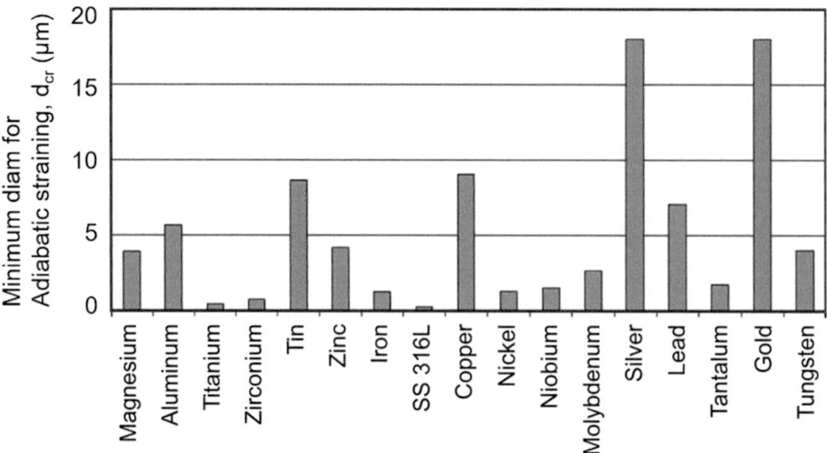

Fig. 6.31 Critical particle diameter for localized adiabatic straining during impact calculated for different materials with empirical Eq. (6.11) [Schmidt et al. (2006)]. Reprinted with kind permission from Elsevier

for 5 and 25 µm Cu particles impacting at impact velocities of $V_i = 400$, 500 and 600 m/s. The average temperature at the selected part of the interface for 5 µm particles is only slightly increased (Fig. 6.30a), and even higher impact velocities do not lead to shear instabilities. The 25 µm particles, on the other hand, show sudden temperature rise, indicating shear instabilities, which occur at velocities of and above 500 m/s.

The possibility for the particle to melt upon impact was also studied by Li W.-Y. et al. (2007a). who showed that for most spray materials, it is possible to experience a local melting at the contact interfaces between the deposited particles. Low melting point and reactions with atmosphere are the two main factors contributing to impact fusion. Poor thermal diffusivity can also play a role. For example, impact fusion has been observed when spraying zinc [Li Chang-Jiu et al. (2004)].

As noted in Fig. 6.30a, the temperature of copper particles 5 µm in diameter impacting at different velocities remains systematically rather low, which limits the chances for shear instabilities and accordingly for particle bonding to the substrate. Schmidt et al. (2006), (2009a, b) identified the presence of a critical particle diameter below which it is not possible to form a coating with a reasonable adhesion to the substrate. These authors have proposed the following empirical equation, Eq. 6.11, giving the critical particle diameter, above which thermal diffusion is slow enough to allow localized shear instability to occur at the surface of an impacting spherical particle. Smaller particles would not be able to reach conditions for bonding.

$$d_{cr} = \frac{36\,\kappa_p}{c_p\,\rho_p\,v_p} \qquad (6.11)$$

where κ_p, ρ_p, c_p are respectively the thermal conductivity, specific mass, and specific heat of the particle, while v_p is its impact velocity. Figure 6.31 represents d_{cr} for different materials. Empirical values of d_{cr} presented in Fig. 6.31 indicate that for tin, copper, silver, and gold, thermal diffusion limits bonding of small particles, whereas spraying of titanium and steel 316 L is less restricted by this effect.

6.4.3 Critical Impact Velocity

As illustrated schematically in Fig. 6.32a from [Gärtner et al. (2006)], particles impacting at low velocities will abrade the substrate material, whereas at higher velocities, deposition and coating formation can occur. The transition velocity between these two regimes defines the "critical velocity" v_{crit}, which, as expected, depends on the material properties. As shown in Fig. 6.32b, based on experimental and modelling data reported by [Schmidt et al. (2009a, b)], the critical particle velocity is also reported to be a function of the particle diameter especially in the particle size range below 50 µm.

As the particle transport velocity in a cold-spray jet varies with the particle size (Fig. 6.25) (smaller particles are significantly faster), the amount of adhering material is correlated to the distribution of a given powder. The critical velocity is, therefore, that of the largest particle that bonds to the substrate. [Papyrin et al. (2003)] have developed a simple model based on the comparison of adhesion energy and energy of elastic deformation generated under particle impact. They have studied the influence of particle size and velocity for which the adhesion energy exceeds that of elastic deformation, thus resulting in the formation of the first coating layer.

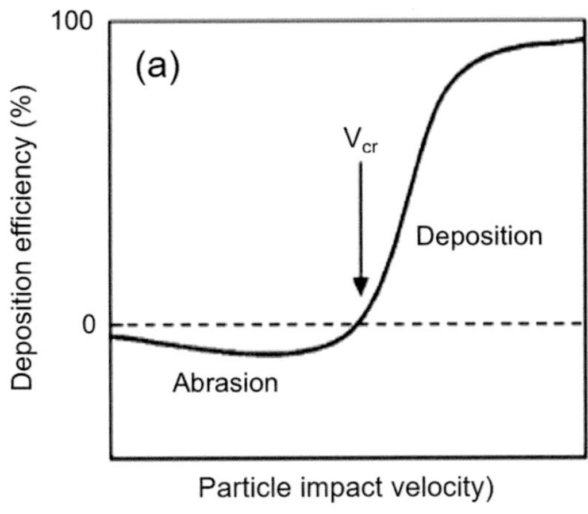

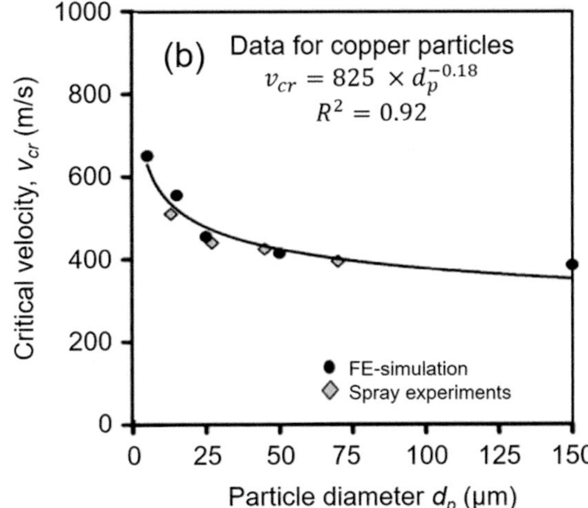

Fig. 6.32 (**a**) schematic of the correlation between particle velocity and deposition efficiency. [Gärtner et al. (2006)] (**b**) critical velocities of copper particles as function of particle diameter [Schmidt et al. (2006)].

Reprinted with kind permission from Springer Science Business Media, copyright © ASM International

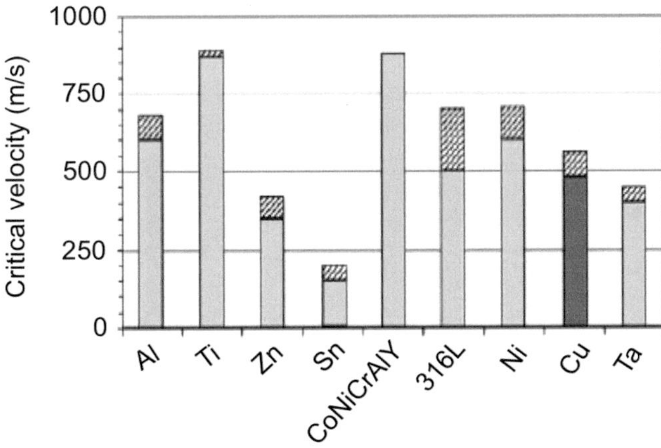

Fig. 6.33 Experimentally determined critical velocities of various spray materials onto a copper substrate. The error bar accounts for differences caused by the range of available powder sizes [Gärtner et al. (2006)]. Reprinted with kind permission from Springer Science Business Media, copyright © ASM International

Based on this concept of bonding by shear instabilities and combining the results from modeling and experimental investigations, analytical expressions have been established to predict the ranges of optimum spray conditions. Those have been determined according to properties of sprayed materials and substrates, spray particle sizes, and temperatures [Schmidt et al. (2006), Schmidt et al. (2009a, b)].

Critical velocities have been determined for various spray materials and substrates. For example, [Gärtner et al. (2006)] have studied the critical velocities of different sprayed materials onto copper substrate. If the size cut of the powder contains only negligible amounts of fine particles, the cumulative size distribution can be correlated directly to the deposition efficiency in cold spray. Results are summarized in Fig. 6.33.

In materials of similar crystallographic structures (i.e., face-centered cubic) and stiffness (such as aluminum and copper), mass density plays a role in bonding. When comparing Ti and Zn (hexagonal close-packed metals), the melting temperature or mechanical strength plays a role. The very high values of v_{crit} for Ti and CoNiCrAlY require He for spraying them. For more details, see [Schmidt et al. (2006), Schmidt et al. (2009a, b)].

With the kinetic-spray process [Van Steenkiste et al. (2002), Van Steenkiste et al. (2004)] have sprayed aluminum and copper particles with sizes between 50 and 200 µm and they found this coating process fundamentally different from that for smaller particles. It seems that mean critical velocities for the larger particles ($d_p > 50$ µm) are substantially less than those reported for particles of average diameter 10 or 20 µm [Van Steenkiste et al. (2002), Van Steenkiste et al. (2004)]. This is perhaps due to the fact that the particles are not perfectly spherical (the contact point has a smaller radius of curvature than $d_p/2$) and have a larger momentum that can cause an impact stress exceeding the yield stress. Perhaps also the fracture of the particle oxide layer before plastic deformation is more important here. Moreover, the smaller injector tube (compared to conventional cold spray) increases deposition efficiencies for larger particles, but there is no definitive answer to the physics behind this observation [Van Steenkiste et al. (2004)].

Of course, this section deals with conventional cold spray. [Assadi et al. (2003)] have modeled the deformation of particles upon impact by using the finite element program ABAQUS/Explicit version 6.2–1. The analysis accounted for strain hardening, strain-rate hardening, and thermal softening and heating due to frictional, plastic, and viscous dissipation. The heating was assumed to be adiabatic. The model was developed for copper and results were in relatively good

agreement with experimental results. To extend the results of the modelling to other materials and spraying conditions, the effect of various process parameters and material properties on the calculated critical velocity was estimated by using a specific method [Assadi et al. (2003)]. In this method, the effect of small changes in material and process parameters on the critical velocity has been estimated. The overall effect of these parameters can be summarized in the following simple formula:

$$V_{cr} = 667 - 14\,\rho_p + 0.08\,T_m + 0.1\,\sigma_u - 0.4\,T_{po} \quad (6.12)$$

where ρ_p is the mass density in g/cm^3, T_m is the melting temperature in °C, σ_u is the ultimate strength in MPa, and T_{po} is the initial particle temperature in °C.

[Schmidt et al. (2009a, b)] used a three-step method to calculate the critical velocity:

- First, they considered the impact dynamic effects through the interplay between material strength and dynamic load, resulting in a *critical mechanical velocity*.
- Second, they considered the energy balance between thermal dissipation and provided kinetic energy, resulting in a *critical thermal velocity*.
- Third, they combined both velocities by using a weighting factor of 0.5, which matches better with experimental results than each of these two velocities.

Of course, both equations contain calibration factors (respectively F_1 and F_2), which were generated by correlating calculated critical velocities on the basis of material properties with experimentally determined critical velocities, resulting in $F_1 = 1.2$, $F_2 = 0.3$.

Finally, the expression obtained is as follows:

$$v_{cr}^{th,mech} = \sqrt{\frac{4\,F_1\,\sigma_{TS}\,(1 - (T_i - T_R)/(T_m - T_R))}{\rho}} + F_2\,c_p\,(T_m - T_i)$$

$$(6.13)$$

Where, c_p is the specific heat (J/K.kg), T_i the impact temperature (K), T_m the melting temperature (K), T_R the reference temperature (293 K), σ_{TS} the tensile strength (MPa), and ρ the specific mass (kg/m^3).

Figure 6.34, from [Schmidt et al. (2006)], represents the results of the calculation of critical impact velocities using Eq. (6.13) for 25 µm particles of different metals.

A comparison of the calculated and experimentally determined critical impact velocities for cold spraying and for 20 mm ball impacts is presented in Fig. 6.35 [Schmidt et al. (2006)]. The correlation is much better than the results obtained using Eq. (6.7) of [Assadi et al. (2003)]. This is especially the case with tin and tantalum, which have significantly different properties from copper. The difference between V_{crit} of copper calculated with Eq. (6.12) and (6.13) is based on differences in calibration data, in terms of particle size. With experimentally determined hardness values of spray powders, the tensile strength in Eq. (6.13) can also be replaced by the Vickers hardness to estimate critical velocity (here with $F_1 = 3.8$ and the same thermal correlation factor $F_2 = 0.3$).

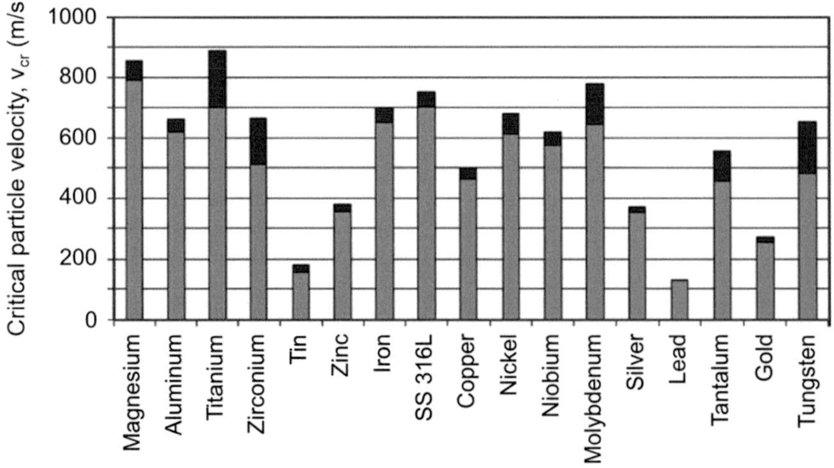

Fig. 6.34 Critical impact velocity for a 25 µm particle calculated for different materials with Eq. (6.13). The dark grey levels indicate a range of uncertainties with respect to available materials data [Schmidt et al. (2006)]. Reprinted with kind permission from Elsevier

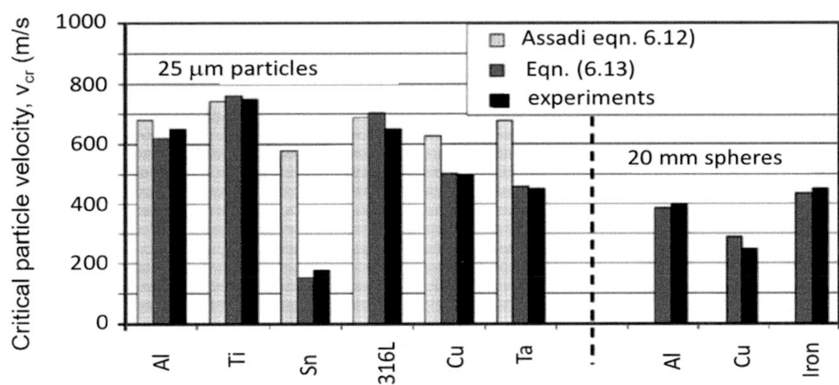

Fig. 6.35 Comparison of calculated V_{cr}, using Eq. (6.12) and 6.13) with experimental results of spray experiments ($d_p = 25$ µm) and impact tests ($d = 20$ mm) [Schmidt et al. (2006)]. Reprinted with kind permission from Elsevier

6.4.4 Material and Substrate Compatibility

A systematic approach of material suitability analysis for the cold-spray process can be found in [Papyrin et al. (2003), Vlcek et al. (2005)]. The authors point out that materials' suitability depends mainly on its deformation properties and what actually happens inside the metals during deformation. The deformation process is principally determined by the mobility of dislocations in the crystal lattice and their interactions. In particular, the crystal structure and the type of bonding, as well as further parameters such as structure, grain size, and foreign atoms or phases, determine the resistance to deformation. Vlcek et al. (2005) identifies the most important isomechanical groups of metal as [Frost and Ashby (1982)]:

- Face-Centered Cubic (fcc) with large number of sliding planes results in good deformability (Al, Cu, Ag, Au, Pt, Ni, and γ-Fe).
- Close-packed hexagonal (cph), where the number of sliding planes is greatly reduced, corresponds to poorer deformability (Cd, Zn, Co, Mg, and Ti).
- Body-Centered Cubic (bcc) (also transitional states) represents the lowest deformability of the three structures (W, Ta, Mo, Nb, V, Cr, α-Fe, and β-Ti).

On the other hand, groups of tetragonal or trigonal crystal systems including oxides, which exhibit low plasticity, are not suitable for the cold-spray process.

In the cold-spray process, the substrate plays an equally critical role in the deposit formation. For example, adhesion has been studied for tin, ABS, copper, Al alloy, brass, different tool steels, carbon steel, glass, and alumina substrates (hardness varying from 0.08 GPa to 10.7 GPa) on which aluminum powder (15–75 µm) was sprayed [Zhang et al. (2003), Zhang et al. (2005)]. The results confirmed that initiation onto soft metallic substrates is difficult due to the lack of deformation of the aluminum particles. The initiation of aluminum deposition increases with the substrate hardness. Clear evidence of metal jetting was observed upon deposition onto steel substrates, where the most rapid initiation is observed. Deposition onto aluminum was difficult due to the aluminum oxide layer on the substrate. Deposition onto non-metallic substrates was difficult due to lack of a metallic bond that can be formed [Zhang et al. (2005)]. For particles with a low hardness sprayed onto a hard substrate, the adhesion can occur only on rough substrates (mechanical adhesion) [Zhang et al. (2003), Dykhuizen et al. (1999)].

As discussed earlier, the deformation of the substrate and the impacting particles is strongly linked to the particle velocity above the critical velocity [Zhang et al. (2003), Dykhuizen et al. (1999)] and Fig. 6.36 from [Raletz (2005)] presents the deformation of Ni particles with different impact velocities on a stainless steel (304 L) substrate. With increase in velocity, images show better penetration of particles into the substrate and a lip angled away from the splat, which is composed mostly of nickel.

[Fukumoto et al. (2007)] investigated fundamentally the deposition behavior of sprayed individual metallic particles on the substrate surface in the cold-spray process. As a preliminary experiment, pure copper (Cu) particles were sprayed on mirror-polished stainless steel and aluminum (Al) alloy substrate surfaces. Process parameters that changed systematically were particle diameter, working gas, gas pressure, gas temperature, and substrate temperature, and the effect of these parameters on the flattening or adhesive behavior of an individual particle was precisely investigated. The deposition ratio on the substrate surface was also evaluated using these parameters. From the results obtained, it was quite noticeable that the higher substrate temperature brought about a higher deposition rate of Cu particles, even under the condition where particles were kept at room temperature. This tendency was promoted more effectively using helium instead of air or nitrogen as a working gas. Both

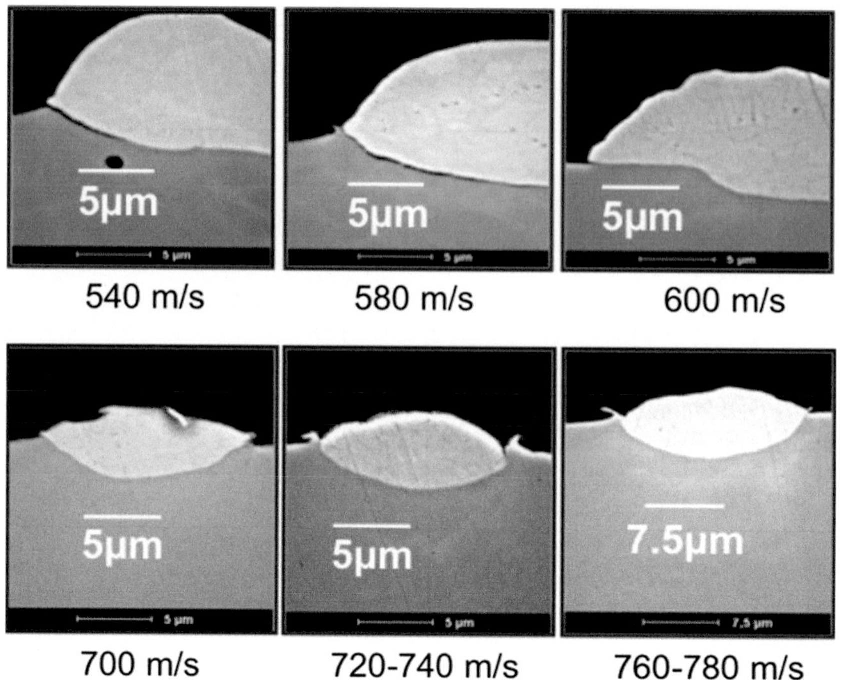

Fig. 6.36 Deformation of Ni particles impacting with different velocities onto stainless-steel substrate (304 L) [Raletz (2005)]. Reprinted with kind permission of Dr. F. Raletz

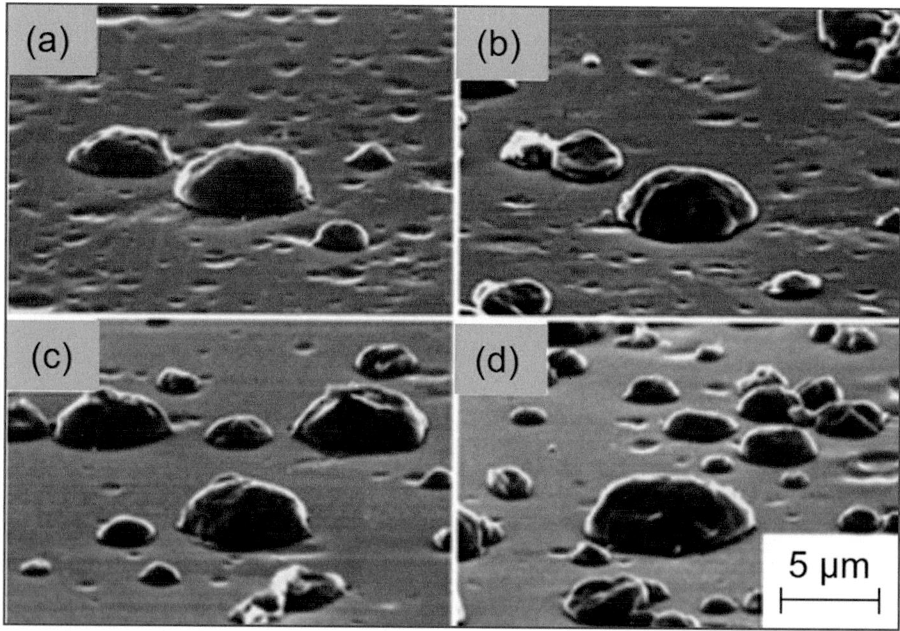

Fig. 6.37 Morphologies of collected particles, gas pressure: 0.3 MPa, gas temperature 300 K, particle size: 5 µm at different substrate temperatures. (**a**) Substrate temperature: 373 K. (**b**) 473 K. (**c**) 573 K. (**d**) 673 K [Fukumoto et al. (2007)]

higher velocity and temperature of the particles sprayed are the necessary conditions for the higher deposition ratio in cold spraying. However, instead of particle heating, substrate heating may bring about the equivalent effect for particle deposition, see Fig. 6.37.

6.4.5   Particle Shock Consolidation

As mentioned earlier in Sect. 6.2.4, Gas Dynamic Spray (GDS) is a cold spray process developed using low-pressure ($p_g < 1$ MPa) spray guns [Kashirin et al. (2007); Kosarev et

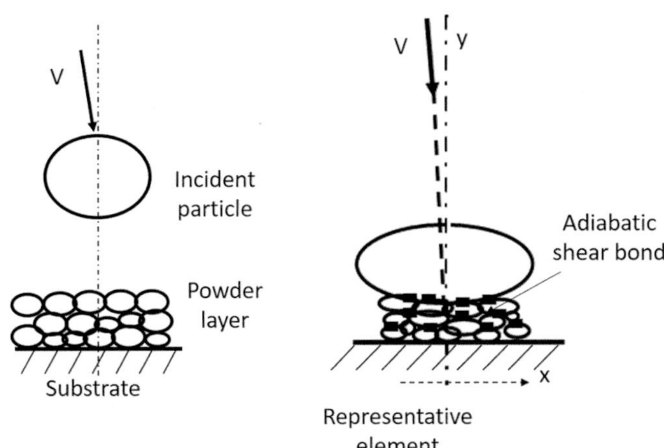

Fig. 6.38 Bimodal powder mixture: large particle impinges upon the layer of smaller particles with the formation of adiabatic shear stress. Reprinted with kind permission from Springer Science Business Media [Champagne et al. (2005)], [Maev and Leshchinsky (2006)], copyright © ASM International

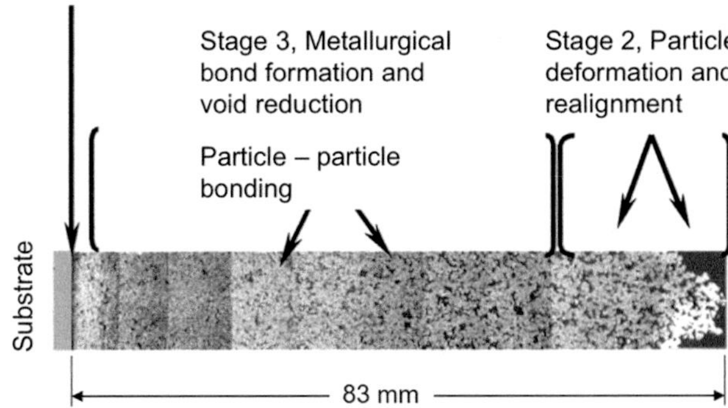

Fig. 6.39 Optical micrograph of an etched cross-section of a thick Al sample prepared at 315 °C, showing stage 1-type, stage 2-type, and stage 3-type behavior of the bonding process [Van Steenkiste et al. (2002)]. Reprinted with kind permission from Elsevier

al. (2003); Alkimov et al. (1990); Alkimov et al. (1994)]. These include a heated compressed air or nitrogen or helium supply, a supersonic nozzle with radial injection of the powder downstream of the throat and a powder feeder (see Fig. 6.6). With these guns, special high-pressure pumps are not necessary. The gas flow rates are in the range 0.2–0.5 m^3/min compared to 1.3–5 m^3/min, and thus the heating power is much less, about 10 kW, compared to the 40–70 kW required in conventional cold spray. However, particles generally do not reach the critical velocity and the compaction of the particles must be optimized by the selection of powder types and powder particle size distributions [Kashirin et al. (2007), Kosarev et al. (2003), Alkimov et al. (1990), Alkimov et al. (1994), Alkimov et al. (1995)]. Thus, one drawback of the system is that powder recuperation is hardly possible for multi-component mixtures because of the change in composition after spraying. The addition of very dissimilar powder components, such as ceramic and metal, can enhance processing efficiency by improving the self-cleaning of the substrate or, in the case of ceramic particles, by contributing to the consolidation of the coating via mechanical peening during its formation and consolidation [Alkimov et al. (1994)]. In these systems (DYMET, SIMAT...), the dominant mechanism is believed to be analogous to impact compaction, wherein localized pressure waves develop in the solid material upon impact of the incoming particles producing deformation both of the impacting particle and of the substrate [Alkimov et al. (1995)]. [Maev and Leshchinsky (2006)] have studied this particle shock consolidation. Their work is based on the "adiabatic shear band" ABS [Wright (2002)] and is applied to mixtures consisting of particles of various sizes and specific mass. The principle is illustrated in Fig. 6.38.

6.4.6 Coating Buildup

Once the first layer has been deposited, the next particles impact the previously deposited layer, which is of the same material as that of the spay particles. A decrease or increase in the growth rate should be considered according to the changes in the interaction conditions with the surface [Klinkov and Kosarev (2006)]. [Van Steenkiste et al. (2002)] sprayed aluminum particles (mostly over 50 μm) with the kinetic-spray process onto brass substrates with a de-Laval rectangular nozzle using air at 315 °C and a pressure of 2 MPa. They deposited, at above critical velocity $v_{cr.}$, with a stationary torch, 83 mm thick coating, the etched cross-section of which is presented in Fig. 6.39. It can be observed that the coating drops with increase in the coating thickness. The effect was attributed to the fact that successive impacts have a peening effect and create further densification and work hardening of the coating. That is why the top layers that have received much less impacts than the deeper ones are less dense than the bottom layers of the coating.

The coating adhesion is due to the importance of plastic deformation (adiabatic shear instabilities), which occurs at the particle-particle interfaces. [Gärtner et al. (2006)] sprayed copper powder (−25 + 5 μm) onto copper substrate with N$_2$ as process gas (3 MPa and 320 °C) using standard trumpet-shaped and bell-shaped nozzles. The micrographs given in Figure 6.40a, obtained with the trumpet-shaped nozzle, show a porosity below 1%. A higher magnification and etched cross-section of the same coating is given in (Fig. 6.40b), which reveals the internal grain boundaries of former spray particles. Authors point out that such a chemical attack would be too strong for more open defects. Corresponding micrograph given in Fig. 6.40c, for a coating obtained using copper powder (−38 + 10 μm), with bell-shaped gun nozzle, using

Fig. 6.40 (**a**) and (**b**) Cu on Cu substrate cold-sprayed with a trumpet-shaped standard nozzle (**c**). Cu on Cu substrate cold-sprayed with a bell-shaped MOC-designed nozzle [Gärtner et al. (2006)]. Reprinted with kind permission from Springer Science Business Media, copyright © ASM International

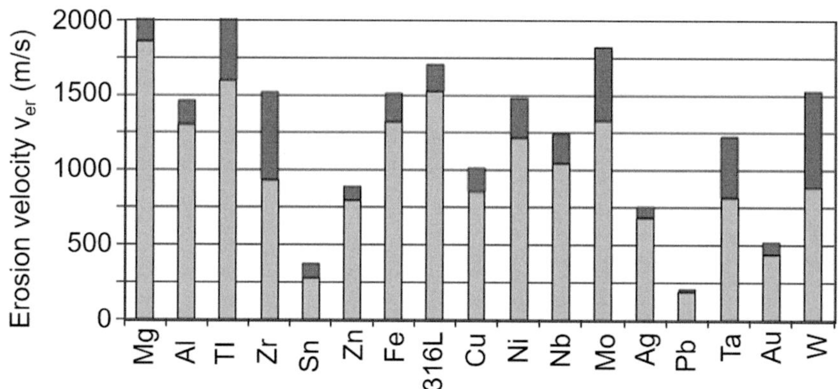

Fig. 6.41 Erosion velocities of different metals for 25 µm particles calculated by [Schmidt et al. (2006)]. Reprinted with kind permission of Elsevier

nitrogen as process gas (3 MPa, 500°C), reveals the grain boundaries inside the former coating.

The analysis of Gärtner et al. [Gärtner et al. (2003)] suggests that the specific mass and temperature of particles have significant effects on critical velocities. If the particle velocity becomes too high and reaches the velocity range of hydrodynamic penetration, the impacting particle will cause strong erosion. Schmidt et al. have calculated [Schmidt et al. (2006), (2009a, b)] such velocities, called erosion velocity, v_{er}. It can be observed that they are higher than the critical velocities (see Fig. 6.41).

Such calculations have allowed defining the variation of v_{cr} and v_{er} with particle temperature at impact. The area between v_{cr} and v_{er} describes the window of spray ability, WS, which can additionally be limited by a region of brittle material or limited ductility at lower temperatures. In the WS region, impacting particles cause deposition. The Particle Impact Conditions (PIC) represent the optimal condition in Fig. 6.42.

The erosion velocity also depends on the particle size, as illustrated in Fig. 6.43 from J. [Wu et al. (2006)]. v_c is almost independent of the particle size, as already emphasized by numerous authors, while the v_{ero} diminishes drastically with

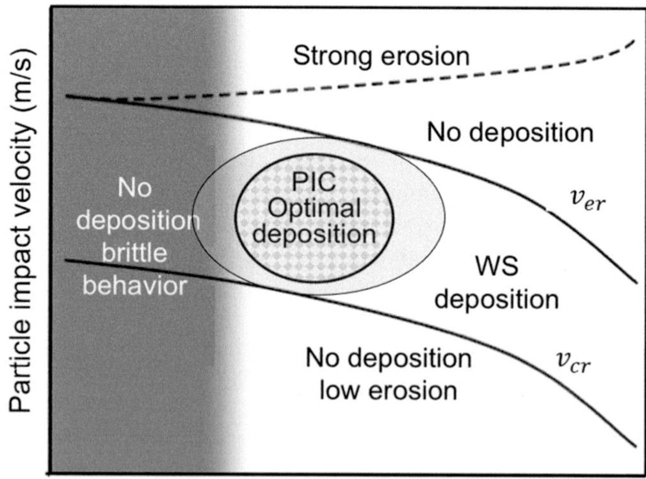

Fig. 6.42 Optimum Particle Impact Conditions (PIC) in the particle velocity vs particle temperature diagram. [Schmidt et al. (2006)]. Reprinted with kind permission from Elsevier

the particle size. This erosion velocity is obtained by comparing the adhesion energy to the rebound energy.

Special attention was also given to heat generation during the cold-spray process as a result of extreme plastic

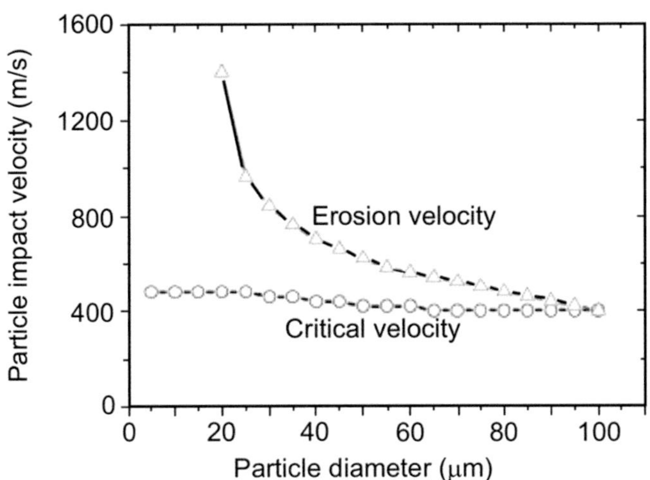

Fig. 6.43 Calculated Critical particle impact deposition and Erosion velocities for Al-Si powder on a mild steel substrate as function of particle diameter [Wu et al. (2006)]. Reprinted with kind permission of ASM

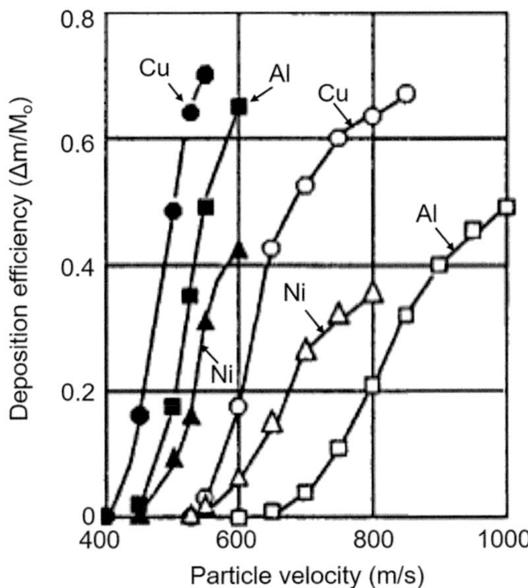

Fig. 6.44 Dependence of the coefficient of particle deposition $\Delta m/M_o$ on the particle impact velocity for aluminum, copper, and nickel. Open symbols are for air-He at 300 K, closed symbols obtained with heated air [Alkimov et al. (1998)]. Reprinted with kind permission from Woodhead Publishing

deformation of the impacting particles. [Ozdemir et al. (2018)] point out that the heating of the coating and the substrate during the spraying process can lead to residual stress, which may cause coating–substrate delamination. In this study, heat generation due to gas impingement and particle plastic deformation has been predicted from Computational Fluid Dynamics (CFD) and Finite Element Analysis (FEA) simulations, respectively. Furthermore, a finite volume method has been presented for transiently simulating the coating buildup and bulk heat generation in the coating and the substrate. The model is intended to assist researchers in understanding thermal affects in the coating process and help design more informed coating patterns to reduce negative thermal effects.

It should also be emphasized that after deposition, the heat treatment of the coating can favorably affect the adhesion strengths of the cold-sprayed Cu and Ni-20Cr coatings. However, according to Mäkinen et al. (2007), while the effect increases with the increase in the annealing temperatures, the annealed coatings are softer than the as-sprayed coatings due to recrystallization of the coating resulting from the heat treatment (Mäkinen et al. 2007).

6.4.7   Deposition Efficiency

[Papyrin (2001) Papyrin et al. (2006)] studied the coating formation using an analytical approach. These authors considered the influence of the coating erosion by the reflected particles. Qualitative dependences of the increase in the coating mass were calculated for different values of probabilities of adhesion and erosion processes. In case of significant

erosion (q > 0.5), the thickness of the first layer does not increase during the coating process. The parameter q is defined as:

$$q = \frac{p_{er}}{(p_c + p_{er})} \qquad (6.14)$$

where p_c is the probability of deposition of the sprayed material, while p_{er} is the probability of its loss from the coating by erosion. When adhesion prevails over erosion (q < 0.5), the mass of the coating increases linearly with time, with its growth rate equal to the difference between adhesion and erosion rates [Papyrin et al. (2006)]. [Alkimov et al. (1998)] reported experimental results on the dependencies of the coefficient of particle deposition $\Delta m/M_o$, defined as the ratio of the substrate mass increase Δm to the total mass of the powder consumed M_o, on the particle impact velocity. The coatings were applied to stationary substrates under strictly controlled conditions of various metallic powders with a particle size of 1–50 µm. The experimental values of the deposition coefficient for aluminum, copper, and nickel powders versus the particle velocity, which were obtained by accelerating the particles with a mixture of air and helium, are shown in Fig. 6.44 (curves 1–3). It can be seen that, for the considered metallic particles (d_p < 50 µm), there exist critical velocities $v_{cr} \sim 500$–600 m/sec for their interaction with the substrate. The conventional process of relatively low erosion is observed for $v_p < v_{cr}$ for $v_p > v_{cr}$, a dense metallic layer is formed on the substrate surface, and the character of coating

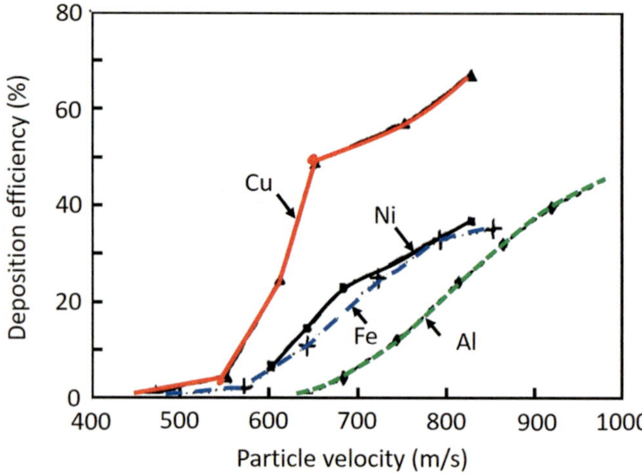

Fig. 6.45 Deposition efficiency as function of particle impact velocity for Cu, Ni, Fe, and Al [Dykhuizen and Smith (1998)]. Reprinted with kind permission from Springer Science Business Media copyright © ASM International

2.1). For example, the maximum velocity for Cu particles (19 μm in diameter) is 62% that of the nitrogen gas at the nozzle exit compared to 42% of helium gas velocity [Gilmore et al. (1999)].

If the influence of the gas pressure, p_g, is significant for low-pressure cold spray, it is much less at higher pressures: roughly, the particle velocity varies as the logarithm of the pressure. For example [Gilmore et al. (1999)], on increasing the gas pressure from 1.5 to 3 MPa, v_p increases by 15%, but the increase is only 5% for a pressure increase from 2.4 to 2.9 MPa. [Stoltenhoff et al. (2002a, b)] obtained similar results. The gas pressure has little influence on the particle temperature (see Fig. 6.15). When increasing the gas temperature, T_g, the particle velocity increases (see Fig. 6.14 from [Voyer et al. (2003)]) as does, almost linearly, the particle temperature. If the particle velocity is kept constant while the gas temperature is varying, the deposition efficiency is not modified, the coating density is similar, but adhesion improves [Han et al. (2004)].

formation changes abruptly as the velocity is increased further. In particular, the value of the deposition coefficient for the examined powders increases from zero to 0.4–0.8 for $v_p \sim 1000$ m/s. For Al, Cu, and Ni, the curves 4–6 in Fig. 6.44 show the same data, but with the particle velocity calculated for the temperature to which the air is heated. Compared to results obtained with the air and helium mixture at $T_o = 300$ K, it can be concluded that the higher particle temperatures significantly affect the spray process [Champagne (2007), Papyrin (2001), Alkimov et al. (1998)]. Similar results, given in Fig. 6.45, were reported by the same authors cited by Dykhuizen and Smith (1998) for Cu, Fe, Ni, and Al coatings.

6.5 Deposition Parameters

6.5.1 Spray Gases

As previously indicated, helium is the primary gas that can be used for cold spray because of its low density, which gives rise to the highest gas and particle velocities [Stoltenhoff et al. (2002a, b), Gärtner et al. (2006), Raletz (2005)]. Using He instead of N_2 for the same gun nozzle, gas pressure and temperature results in an increase of aluminum particle velocities by about 60% [Neiser and Neiser (2003)]. If 5 vol. % air is introduced into the He gas, the gas velocity, v_g, drops by 14%, with 10 vol.% air, the drop is 22%, and with 15 vol. % air, the reduction is further accentuated to 29%. If air is used instead of Helium, the gas velocity drops by 66%. On the other hand, helium, due to its low specific mass, requires more time for particles to reach their maximum velocity (see the expressions of the drag coefficient in Chap. 4 Sect. 4.2.

6.5.2 Spray Powders

6.5.2.1 General Remarks

A wide range of powders have been used with the cold-spray process. These have generally been on the fine side, with particle diameters less than 50 μm. Because of the requirement that the individual particles used for the spray be ductile, the large majority of powders used are either of pure metals or alloys. Metal matrix composites have also been used including up to 50 wt.% of a ceramic. In this section, the different types of powders used for CS are reviewed with examples of powder compositions and performance.

[Koivuluoto and Vuoristo (2010b)] showed that powder type and composition have a very important role in the production of metallic and metallic-ceramic coatings using the low-pressure cold-spray process. Furthermore, the structure and mechanical properties of Cu and Cu + Al_2O_3 coatings are strongly influenced by the powder characteristics of Cu particles. In the comparison between different powder types, pure spherical Cu powders led to the denser coating structure due to the highly deformed particles. The aim of their study was to evaluate the effect of addition of different percentages of Al_2O_3 particles on the microstructure of the coating, its density, and mechanical properties including fracture behavior, hardness, and bond strength. Spherical and dendritic Cu particles were tested together with 0, 10, 30, and 50 vol.% Al_2O_3 additions. Coating density and particle deformation level, hardness, and bond strength increased with increasing Al_2O_3 fractions.

When pure aluminum particles are cold-sprayed, particle surface oxidation can have an important impact on the necessary coating parameters and the properties of the coating

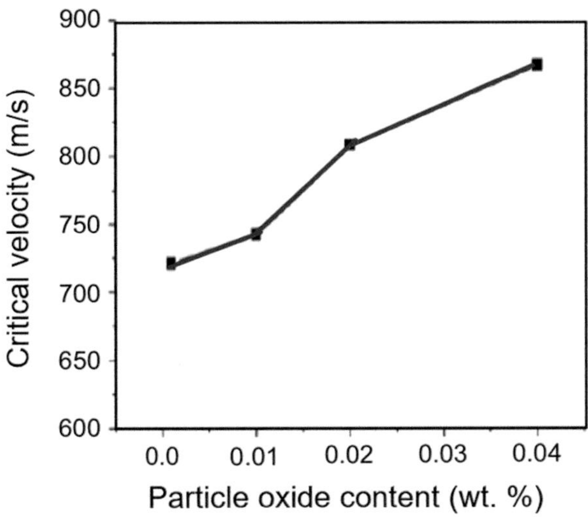

Fig. 6.46 Critical velocity of four kinds of Aluminum powder feedstock with different oxygen contents [Kang et al. (2007)]. Reprinted with kind permission from Springer Science Business Media, copyright © ASM International

obtained, as shown in Fig. 6.46 [after Kang et al. (2007)]. This was demonstrated through the preoxidation of as-received aluminum powder, by etching, oxidizing in a furnace at 150 °C for 15 min, followed by exposure of the powder to atmospheric conditions at room temperature for 7 days. The use of the preoxidized powder, with an oxygen content of close to 0.04 wt.%, for cold spraying on aluminum substrates, the corresponding critical velocity for coating formation was reported to be more than 850 m/s which is considerably higher than the standard value of 750 m/s normally used with nonoxidized powders.

6.5.2.2 Influence of Particle Diameter, Density, and Specific Heat

Numerous systematic studies were reported on the influence of particle diameter, its density, and specific heat on the critical velocity and deposition efficiency in the CS process [Stoltenhoff et al. (2002a, b), Jodoin (2002), Li Wen-Ya and Chang-Jiu Li (2005), Klinkov and Kosarev (2006), Li Chang-Jiu et al. (2005), Lee et al. (2006), Katanoda et al. (2007), Mäkinen et al. (2007), Helfritch and Champagne (2006), Alkhimov et al. (2006), Katanoda et al. (2007a, b)].

When the particle diameter increases, the particle velocity decreases for the same spray conditions, as illustrated in Fig. 6.19, for copper particles with N_2. Of course, it must be kept in mind that particles have size distributions: the bigger particles will be slower, and the smaller particles will be faster. Thus, not all of the particles in the powder would contribute to the coating since particles over a certain size may have velocities below the critical one and will bounce off the surface of the substrate. Based on numerical

simulation of the cold spray process, Katanoda et al. (2007) have identified two parameter combinations that are independent of the material type and that affects the particle velocity and temperature. Using nitrogen and helium as the process gases with stagnation pressures of 3 MPa and gas temperatures upstream of the nozzle throat of 576 K, for the spraying of Cu, Ti, and WC-12Co, they have shown that (d_p. ρ_p), called particle-velocity-parameter (PVP), can be used as material-independent parameter, which affects the particle velocity. They found that the maximum particle velocity is obtained, independent of the particle material, was (d_p. ρ_p=0.020 kg/m^2) when spraying with nitrogen, and (d_p. ρ_p=0.0065 kg/m^2) for Helium as spray gas. The corresponding group ($d_p^2.\rho_p.c_p$) called particle-temperature-parameter (PTP) was found to be material-independent affecting the particle temperature.

[Schmidt et al. (2006)] underline that in cold spraying, bonding is associated with shear instabilities caused by high strain rate deformation during the impact. It is well known that bonding occurs when the impact velocity of an impacting particle exceeds a critical value. This critical velocity depends not only on the type of spray material, but also on the powder quality, the particle size, and the particle impact temperature. Up to now, optimization of cold spraying mainly focused on increasing the particle velocity. The new approach presented in this contribution demonstrates capabilities to reduce critical velocities by well-tuned powder sizes and particle impact temperatures. A newly designed temperature control unit was implemented in a conventional cold-spray system, and various spray experiments with different powder size cuts were performed to verify the results from calculations. Microstructures and mechanical strength of coatings demonstrate that the coating quality can be improved significantly by using well-tuned powder sizes and higher process gas temperatures. The presented optimization strategy, using copper as an example, can be transferred to a variety of spray materials and thus, should boost the development of the cold-spray technology with respect to the coating quality.

6.5.2.3 Particle Temperature

The deposition efficiency increases with the particle temperature at impact for the same velocity (for all sprayed materials, temperature increase implies the Young's modulus reduction). For example, [Lee et al. (2006)] cold-sprayed Cu-20 wt.% Sn alloy (bronze) particles with a mean size of 20 μm onto aluminum with nitrogen, different process gas pressures and temperatures. For particles with velocities of 550 m/s, the deposition efficiency was about 35% with the gas at 300 °C, 62% at 400 °C, and 75% at 500 °C. The critical velocity diminishes with increasing process gas temperature.

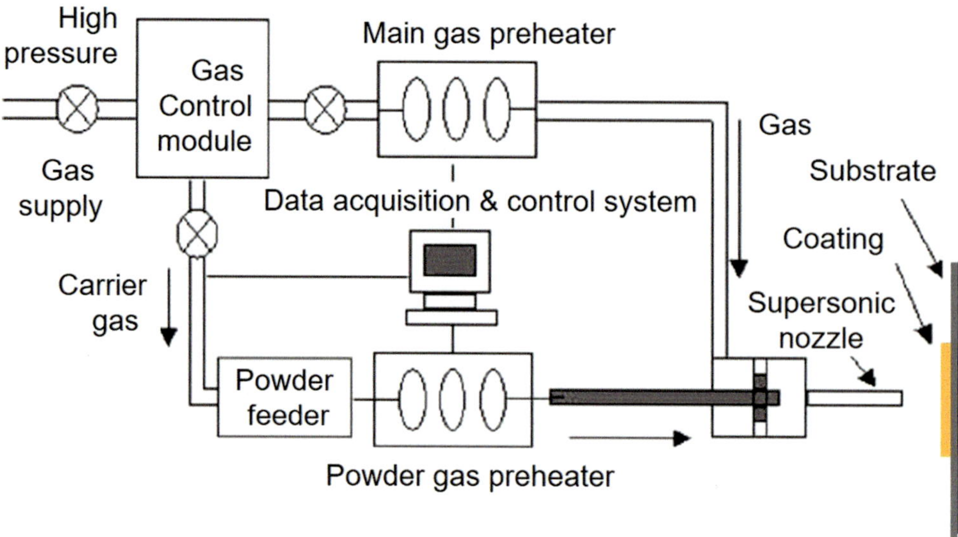

Fig. 6.47 Diagram of the cold-spray equipment with a powder heater [Kim et al. (2005)]. Reprinted with kind permission from Elsevier

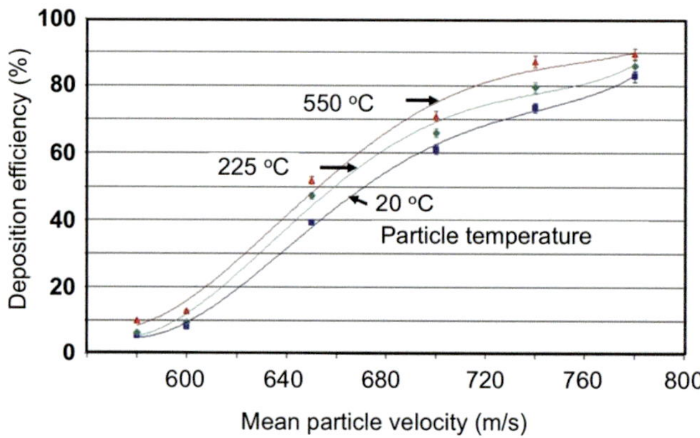

Fig. 6.48 Deposition of Ni particles on a stainless-steel substrate, dependence of deposition efficiency on the mean particle velocity for different particle preheating temperatures [Raletz (2005)]. Reprinted with kind permission from Dr. F. Raletz

One way to achieve higher temperatures of the particles, independently of that of the spray gas, consists in heating the carrier gas separately. New cold-spray devices are equipped with a particle heating system before their injection into the nozzle by using a heated carrier gas as proposed by [Papyrin et al. 1993)]. The principle of the device is shown in Fig. 6.47 from [Kim et al. (2005)]. A gas heater is added downstream of the powder feeder and as close as possible to the injector. For example, the heater consists of a stainless-steel tube (about 3 mm in internal diameter and 300 mm in length), thermally insulated and heated by induction [Dykhuizen et al. (1999)]. Depending on the carrier gas flow rate, the carrier gas can reach up to 680 °C. However, if particles are heated too much, they become sticky and the control of their mass flow is difficult. Nozzle clogging and sticking to interior walls can also start. Surface oxidation of particles can also

be increased especially if air is used as spray gas (kinetic process).

Figure 6.48 shows a strong dependence of the deposition efficiency of Ni particles on stainless steel substrate, on the mean particle velocity, and on the particle preheating temperature [Dykhuizen et al. (1999)]. It can be seen that increasing the particle temperature by 530 °C increases the Ni deposition efficiency by about 13% for the different velocities considered, which is less than that obtained with the lower-melting-point Cu-Sn alloy.

6.5.2.4 Composite Materials

The easiest way to spray composite coatings made of different metals or alloys is to use preliminarily prepared powder either as mixtures of both powders or as composite particles made by agglomeration or cladding or mechanical alloying. While each of these methods offers an important potential for the CS deposition of a wide range of materials, they also have their limitations such as [Kosarev et al. (2003)]:

- Impossibility to change the ratio of the different components in the powder mixture during the spraying process, which in certain cases can be of interest for the spraying of graded coatings with varying compositions across the coating thickness.
- Considering that particles of different materials and sizes can require different spraying conditions such as particle temperature and critical velocity; it could be challenging to optimize the spraying conditions for the different constituents of the powder.
- The presence in the powder mix of brittle particles such as ceramics can lead to substrate and spray nozzle erosion, especially around the throat region.

The last two points can be addressed through the use of specialized nozzle designs that allow for the injection of different powders in different locations in the spay gun. This option will be discussed further in details in the section dealing with nozzle design and powder injection conditions (Sect. 6.5.4).

A typical example of composite deposition is the CS of cermets containing carbides, which have the tendency to decompose when sprayed by conventional thermal spray processes such as plasma or HVOF. This is especially the case when the carbide particles are small or nanometer sized. The cold spray of such particles should be possible if the impacting particles and the previously deposited underlying particles have sufficient deformation. [Siao Ming Ang et al. (2011)] cold-sprayed, on stainless steel substrates, micron- and nano-structured powders with similar Co content (17 wt.%). The working gas was high-pressure helium at 1.2–1.5 MPa, with the gas preheated to ~873 K. The mean particle size of the powder was 31 µm. Micro-structured ones were sintered and crushed with WC grains of 2–3 µm, while nano-structured ones were wet chemically synthesized and agglomerated with WC grain sizes of 40–700 nm and 80–800 nm, respectively. The micrometer-structured WC–Co feedstock did not permit coating buildup due to simultaneous erosion and deposition effects on the surface. It seems that the substrate deformation allows particle penetration, resulting in the formation of the first layer of the coating via interlocking with the metallic Co, WC particles, and steel substrate. Fine WC particles (< 1 µm) were observed near the substrate interface and larger WC particles (1–2 µm) in the vicinity of the first layer surface. It seems that particles have not enough metallic Co binder and critical velocities and thus erode and reduce WC particle sizes of the first WC–Co layer. With the nano-structured particles, the minimum conditions for deformation and deposition seem to be satisfied. [Siao Ming Ang et al. (2011)] have introduced the mean free path of the WC particles, measuring the thickness of the binder layers between the WC particles and related to the plastic deformation. The effect of this mean free path is illustrated in Fig. 6.49, showing how the coating hardness is expected to increase as the particle size of the ceramic decreases and with a greater number of fragmented WC particles.

[Gao et al. (2008)] studied deposition behavior of nano-structured WC-Co spray particles with different hardness impacting the substrate during the cold-spray process. The porosities of the three powders were respectively 43.8 ± 1.3%, 29.8 ± 1.1%, and 4.9 ± 1.2%, and the corresponding micro-hardness of the three powders were 0.78 ± 0.02, 5.48 ± 0.46, and 13.17 ± 2.47 GPa. The process used helium for both accelerating and feeding the powder, the He gas being at a pressure of either 2.0 MPa or 2.5 MPa and a temperature 650 °C. The standoff distance was 20 mm and

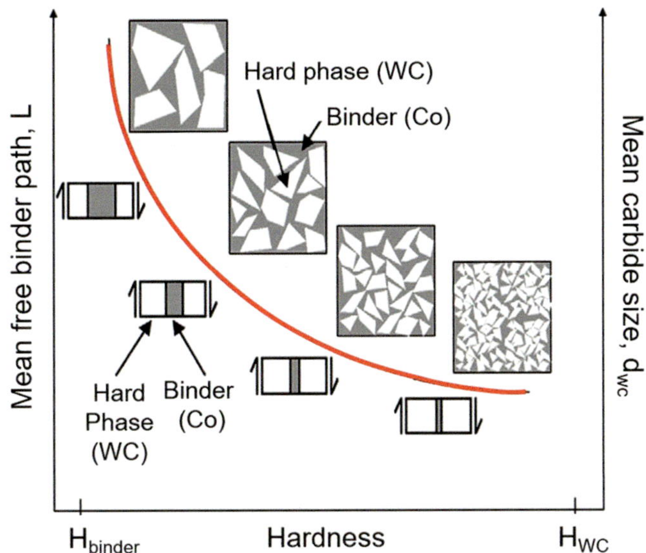

Fig. 6.49 Effect of mean free binder path (L) and mean carbide size d_{WC} on coating hardness [Siao Ming Ang et al. (2011)]. Reprinted with kind permission from Elsevier

different substrates were used with hardness from 2 to 8 GPa. With the two porous powders, coatings were formed on all substrates, while with the dense one (porosity around 5%), sprayed particles can only be embedded into soft stainless-steel surfaces. Kim et al. [6.85] obtained similar results with nano-structured WC–12%Co particles sprayed with helium at 3.1 MPa and preheated to 600 °C. They showed that there was no detrimental phase transformation and/or decarburization of WC by cold-spray deposition. The coating showed a very high hardness (~2050 HV). They emphasized that powder preheating (500 °C for the powder) was one of the most crucial parameters.

[Yandouzi et al. (2007)] compared nanostructured WC-Co coatings sprayed by Conventional Cold Spray (CGDS process) and the pulsed gas Dynamic Spray Process (PGDS), the advantage of the latter being the high impact velocities and intermediate temperatures. They found that there was no degradation of the phase with both processes, but dense coatings without major defects were difficult to obtain when using the CGDS process. With the PGDS process, it was possible to produce coatings with low porosity if the feedstock powder was preheated above 573 K. The same authors [Yandouzi et al. (2008)] sprayed six commercial types of WC–Co powders with the PGDS process. The PGDS optimum process parameters were material dependent. It was found that sintered and cast and crushed powders with similar chemical composition produced coatings with similar porosity level and hardness but the latter type produced coatings with more cracks, attributed to the bulky dense aspect of the powder. The study also revealed that larger agglomerated and sintered powder particles deformed better on impact with the

substrate and formed better coatings (reduced porosity and higher hardness) due to the larger volume of softer material available for deformation.

[Sova et al. (2009), Sova et al. (2010), Sova et al. (2011)] have sprayed soft metal particles (Al with $d_{50} = 25$ μm, Cu $d_{50} = 39$ μm) with fine (around 20 μm) or large (around 130 μm) ceramic powders (Al_2O_3 and SiC) onto grit-blasted aluminum substrates. Ceramic powders were injected either downstream of the nozzle (configuration A) or with metal particles in the subsonic part of it (configuration B). The ceramic particle velocity before impact was measured in both spray configurations. The difference in velocity of the ceramic particles of the same size injected in configurations A and B does not exceed 100 m/s. The velocity of ceramic particles larger than 100 μm is less than 200 m/s. Ceramic powders of fine fractions enhance the coating growth, while ceramic powders of large fractions considerably hinder coating growth. Only a small part of fine ceramic particles is embedded in the coating. Metal particles form a metal matrix with uniformly distributed separate ceramic particles interlocked within it. No large ceramic particles >100 μm were found in the coating when spraying coarse powders. [Sova et al. (2009)] conclude by underlying the importance of the ceramic particle velocity that must have a minimum velocity to penetrate the coating. When soft metals (Al, Cu) are sprayed with fine ceramic powders, the "critical" temperature of the metals decreases compared to pure metals.

Cold-spray coatings are essentially achieved by spraying mixtures of the components, as illustrated by the few following examples. [Lee et al. (2007a)] have cold-sprayed (air at 0.8 MPa and 280 °C) pure Al and Ni powder mixtures (respectively 75:25 and 90:10 wt.%). The Ni particles were finely dispersed and embedded in the Al matrix. Above 450 °C of post-annealing temperature, the Al_3Ni and Al_3Ni_2 phases were observed in the cold-sprayed Al–Ni coatings, with the Ni particles being fully consumed via compounding reaction with Al at 550 °C. In another paper, [Lee et al. (2010)] investigated the effects of gas pressure on the morphological evolution of the cold-sprayed Al-Ni system. Under low pressure (0.7 MPa), a large number of defects were created at the interfaces. These effects gradually decreased as pressure increased. They supposed that the peening effects, depending on gas pressure, could alter the main route of Al consumption during annealing. [Novoselova et al. (2007)] cold-sprayed titanium and aluminum powder mixtures, which were then converted to titanium aluminide with suitable heat treatments. The desirable set of intermetallic phases ($TiAl_3$, r-$TiAl_2$, TiAl, and Ti_3Al) was obtained by varying heat treatment conditions. While the porosity in the as-sprayed precursor deposits was about zero, significant porosity developed during heat treatment.

6.5.2.5 Metal-Ceramic Blends

In this case, the blend is injected in the subsonic zone. Besides those already cited [Klinkov et al. (2009); Sova et al. (2011)], numerous studies have been reported using this technique for a wide range of applications such as the cold spraying of SiC-reinforced aluminum alloy [Sansoucy et al. (2008)], Al with Al_2O_3 or SiC [Tao et al. (2009)], Al reinforced with TiN [Li et al. (2008)], rare earth $Nd_2Fe_{14}B$ permanent magnet with Al [King et al. (2007)], and Ni–Al_2O_3 coating [Hu et al. (2011)]. Only two typical examples are presented.

To produce $Nd_2Fe_{14}B$ permanent magnet/aluminum-composite coatings, King et al. (2007) blended $Nd_2Fe_{14}B$ powder with aluminum powder (20–80 vol.%). They used nitrogen at 2.6 MPa heated to 200–480 °C. The hard $Nd_2Fe_{14}B$ particles tended to fracture and fragment upon impact, while aluminum underwent severe plastic deformation, eliminating pores and trapping $Nd_2Fe_{14}B$ within the coating. Higher spray temperatures and finer $Nd_2Fe_{14}B$ particle sizes improved their retention rate within the composite structure. Of course, the magnetic properties of $Nd_2Fe_{14}B$ remained unaffected by the cold spray process. According to King et al. (2007), with the spray conditions used, only a small proportion of the $Nd_2Fe_{14}B$ particles were able to penetrate deeply into aluminum. But among elastically rebounding particles, those with a small enough rebound momentum to be overcome by new arriving particles were imbedded into the coating.

The cold-sprayed Ni–Al_2O_3 coating on Inconel 600 substrate with the ratio of Ni-60 wt.% and α-Al_2O_3–40 wt.% [Hu et al. (2011)] is another example presenting a compact microstructure and low porosity. The bonding mechanism was dominated by deformation and interlocking of the inter-particles accompanied with local metallurgical bonding.

6.5.2.6 Metal-Cladded Composite Particles

According to [Li et al. (2008)], the use of metal cladding of ceramic powders is an effective way to increase the volume fraction of hard particles in the composite. For that, they used the hydrothermal hydrogen reduction method commonly used to prepare Ni-cladded composite materials with graphite, diamond, and alumina filling. The cold-spray process gives rise, however, to a loss of some of the filling materials through the possible tearing of the cladding envelope and the rebounding of some of the fine-particle ceramic filling, as shown in Fig. 6.50. The volume fraction of Al_2O_3 in the deposited composite in this case was about 29 ± 6 vol.% against 40–45 vol.% in the feedstock. The microhardness of the composite coating was beyond the strain-hardening effect, which was about $173 \pm 33 HV_{0.2}$. With this technique, fine Al_2O_3 particles are clad by a non-uniform Ni layer about 20 μm thick.

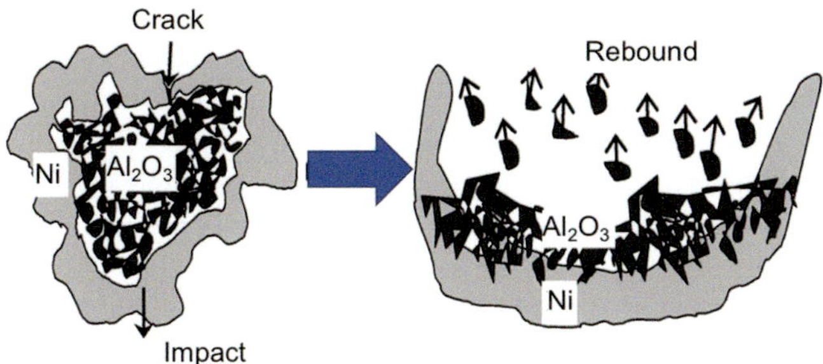

Fig. 6.50 Schematic diagram of the impacting process of a Ni-cladded Al₂O₃ composite particle resulting in the rebound and embedding of Al₂O₃ particles [Li et al. (2008)]. Reprinted with kind permission from Elsevier

[Wang et al. (2009)] have cold-sprayed (N₂ at 2 MPa and 500 °C) a ball-milled mixture Fe₆₀Al₄₀–10 vol.% Al₂O₃, the starting alumina particles' mean size being 5 μm. The metastable microstructure of the milled feedstock was completely retained in the coating. The Fe/Al alloy in the coating was completely transformed to an FeAl intermetallic phase by the annealing treatment of the as-sprayed coatings at a temperature higher than 600 C. [Luo et al. (2011)] have deposited (He at 2.2 MPa and 600 °C) the mechanically alloyed composite powder cBNp/NiCrAl. The coating presented a dense microstructure, and cBN particles were uniformly dispersed in the NiCrAl matrix although a limited fraction of NiCrAl lamellae were present in the coating.

6.5.2.7 Nano Composites

As with other sprayed coatings, especially those plasma sprayed, the interest in using nano-meter sized particles is developing. [Liu et al. (2010)] sprayed three different contents of nano-sized TiN powers of 20 nm in size added to nano-sized SiC powders of 40 nm, see Fig. 6.51.

The Vacuum Cold-Spray (VCS) process was used to deposit SiC-TiN composite coatings on Al₂O₃ substrates. Figure 6.52 shows the typical fracture morphologies of the SiC-50%TiN coatings deposited on Al₂O₃ substrates by VCS. The phenomenon of coating spallation and peeling from substrate does not occur during the course of fracturing. It indicates that the composition coating adheres well to the substrate.

A micro-structure and phase structure analysis of the samples was performed by Scanning Electron Microscopy (SEM) and X-ray diffraction (XRD). Sheet resistance of the VCS coatings was measured using a four-point probe method. The influences of TiN additions on the electrical resistivity of SiC-TiN composite coatings and the conductive mechanisms were investigated. The electrical resistivity of

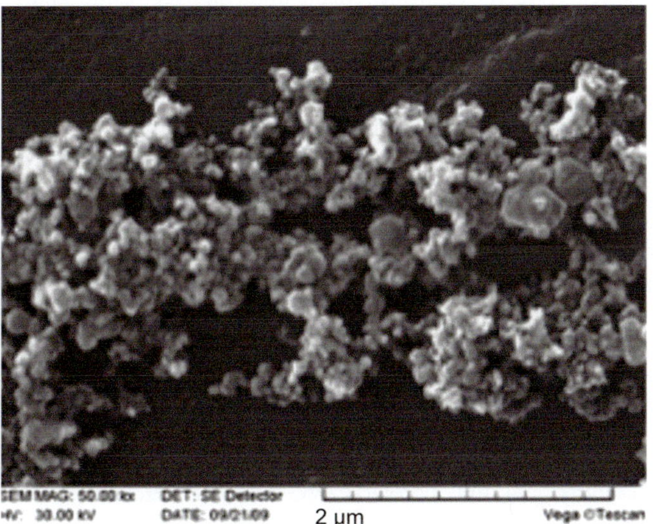

Fig. 6.51 Morphologies of SiC-TiN powers (50 mol% TiN) [Liu et al. (2010)]

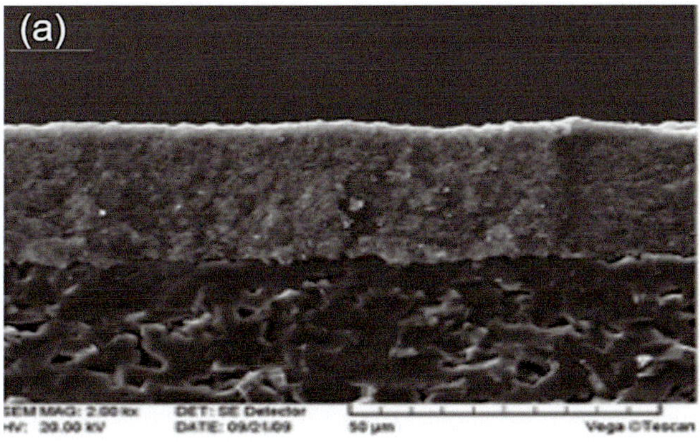

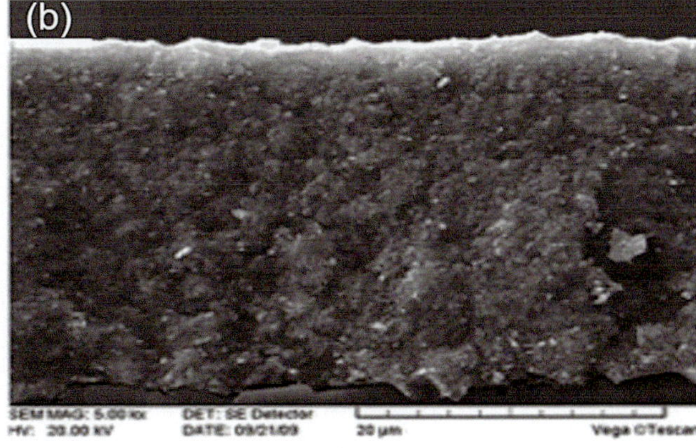

Fig. 6.52 Fractured morphologies of VCS SiC-50%TiN composite coating ((**a**) low resolution; (**b**) high resolution) [Liu et al. (2010)]

SiC-TiN coatings decreases with increase in TiN contents, reaching a minimum of 1.82 Ω.m with 50 mol.% TiN.

[Liu et al. (2014)] recall that development of novel biocompatible nanomaterials has provided insights into their potential biomedical applications. Unfortunately, bulk fabrication of the nanomaterials in the form of coatings remains challenging. They report hydroxyapatite (HA)/Graphene-Nanosheet (GN) composite coatings deposited by vacuum cold spray (VCS). Significant shape changes of HA nanograins during the coating deposition were revealed. The nano-structural features of HA together with curvature alternation of GN gave rise to dense structures. Based on the microstructural characterization, a structure model was proposed to elucidate the nano-structural characteristics of the HA-GN nanocomposites. Results also showed that addition of GN significantly enhanced the fracture toughness and elastic modulus of the HA-based coatings, which is presumably accounted for by the crack bridging offered by GN in the composites. The VCS HA-GN coatings show potential for biomedical applications for the repair or replacement of hard tissues.

[Yang et al. (2012)] recall that the pore structure in nanoporous TiO_2 coatings influences the ion diffusion property and the photo-voltaic performance of dye-sensitized solar cells. In their paper, TiO_2 coatings were deposited by vacuum cold spray (VCS) using a strengthened nano-structured powder. The pore structure, ion diffusion, and dye infiltration properties were examined to understand the coating deposition mechanism. Results showed that the pores in the VCS TiO2 coatings presented a bimodal size distribution with two peaks at ~15 and ~ 50 nm. Based on the impact behavior of spray powder particles, a deposition model was proposed to explain the formation mechanism of the pores in the VCS coating using strengthened nanostructured powder. It was found that, compared to the conventional unimodal-sized nano-pores in TiO_2 coatings, the bimodal-sized nano-pores contributed to a higher ion diffusion coefficient of the coatings and thereby a higher photovoltage of the solar cells. Neshastehriz et al. (2014)] have developed for cold-spray coating of aluminum components nano-engineered self-lubricating particles comprising hexagonal-boron-nitride powder (hBN) encapsulated in nickel. The nickel encapsulant consists of several nano-sized layers, which are deposited on the hBN particles by electroless plating. In the cold-spray deposition, the nickel becomes the matrix in which hBN acts as the lubricant. The coating demonstrated a very promising performance by reducing the coefficient of friction by almost 50% and increasing the wear resistance more than tenfold. The coatings also exhibited higher bond strength, which was directly related to the hardenability of the particles. During the encapsulation process, the hBN particles agglomerate and form large clusters. De-agglomeration has been studied through low- and high-energy ball milling to create more uniform and consistent particle sizes and to improve the cold-spray deposition efficiency. The un-milled and milled particles were characterized with Scanning Electron Microscopy, Energy-Dispersive X-Ray Spectroscopy, BET, and hardness tests. It was found that in low-energy ball milling, the clusters were compacted to a noticeable extent. However, the high-energy ball milling resulted in the breakup of agglomerations and destroyed the nickel encapsulant.

[Cavaliere et al. (2014)] in the present paper describe the main processing parameters affecting the microstructural and mechanical behavior of metal-metal cold-spray deposits. The effect of process parameters on grain refinement and mechanical properties are analyzed for composite particles of Al-Al_2O_3, Ni-BN, Cu-Al_2O_3, and Co-SiC. The properties of the formed nanocomposites were compared with those of the parent materials sprayed under the same conditions. The process conditions, leading to a strong grain refinement with an acceptable level of the deposit mechanical properties such as porosity and adhesion strength, are discussed.

[Ravi et al. (2016)] deposited by cold-spray Ultra-High Molecular Weight Polyethylene (UHMWPE) powder mixed with nano-alumina, fumed nano-alumina, and fumed nano-silica. They used two different substrates, namely, polypropylene and aluminum. The coatings with UHMWPE mixed with nano-alumina, fumed nano-alumina, and fumed nano-silica were highly contrasting in terms of coating thickness. Nano-ceramic particles played an important role as a bridge bond between the UHMWPE particles. Gas temperature and pressure played an important role in the deposition. The differential scanning calorimetry results of the coatings showed that UHMWPE was melt-crystallized after the coating.

The experiment of [Hall et al. (2008)] clearly demonstrates that cold spray can be used to refine the microstructure of an ultra-fine-grained (100 s of nm grain size) powder and consolidate it to create a homogenous nanocrystalline (20–40 nm) micro-structure. When compared to the work of Ajdelsztajn, which shows that the cold-spray process can be used to consolidate nanocrystalline aluminum without causing recrystallization, it illustrates the flexibility of the cold-spray process. This flexibility is typical of spray processes and highlights the need to further understand the process-microstructure-property relationships in nanocrystalline cold-spray coatings. The cold-spray process is likely an extremely controllable method for preparing metal coatings and bulk metal shapes with homogenous nanocrystalline microstructures. The cold-spray process can cause the refinement of ultra-fine-grained powdered aluminum feed stocks, creating a homogenous nanocrystalline microstructure. Grain refinement observed in Al due to cold spray occurs because of the severe plastic deformation associated with the cold-spray process.

6.5.3 Substrate

As already emphasized previously, the substrate hardness relative to that of the impacting particles plays a key role in the cold spray process [Zhang et al. (2003); Vlcek et al. (2005); Zhang et al. (2005)] (see the Sect. 6.1.2.a. devoted to the deposition of the first layer). For example, Vlcek et al. (2005) have demonstrated that the initiation of deposition of aluminum particles, onto soft metallic substrates, is difficult due to the lack of deformation of the aluminum particles. In fact, deposition onto a tin substrate could even result in melting of the tin.

The ease of initiation of aluminum deposition increases as the hardness of the metallic substrates increases. The substrate temperature also is important since the deformation process leading to particle bonding and coating formation takes place both on the particle and the substrate side [Legoux et al. (2007)]. Of course, the deposition efficiency changes as a function of the process gas temperature, which influences both the particle velocity and the substrate surface temperature. Legoux et al. (2007) have heated the carbon-steel substrate with a resistor at constant gun temperature. For a low-pressure gun (N_2 below 0.62 MPa), the substrate temperature increase has raised the deposition efficiency of Al, reduced that of Zn (the surface temperature might have been too high for this low-melting-temperature material), and has had no effect on Sn.

Fukumoto et al. (2007), with a rather low-pressure gun (air, 0.4–1 MPa), have shown for Cu particles cold-sprayed on mirror polished stainless steel and aluminum alloy, that the deposition efficiency was increased when the substrate temperature was raised. It can be seen in Fig. 6.53 that the deposition efficiency on AISI 304 substrates can be improved significantly by increasing the substrate temperature, even under conditions without any heating of the working gas and consequently of the particles.

[Sakaki et al. (2004)] obtained similar results with a conventional cold-spray gun (N_2 at 3 MPa) spraying Cu and Ti onto a mild steel substrate. They also have shown that an increase in the gun traverse speed (from 20 to 100 mm/s) has a negative effect on the Deposition Efficiency (DE). With the increase in the traverse speed in both cases, the substrate temperature diminished from about 390 K to 360 K, but DE was reduced from 68% to 55% for copper, while that for Ti DE dropped from 19% to a few%.

[Goyal et al. (2012)] used Taguchi L18 orthogonal array for depositing the electro-conductive coatings by varying different process parameters, that is, substrate material, type of powder feeding arrangement, stagnation gas temperature, stagnation gas pressure, and stand-off distance. The response parameter of the coatings so produced is measured in terms of surface roughness. The significant process parameters in the order of their decreasing percentage contribution are:

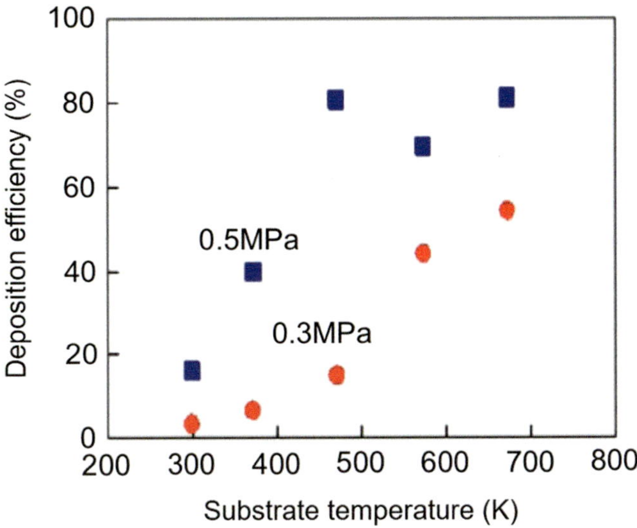

Fig. 6.53 Relation between deposition efficiency and both substrate temperature and gas pressure (He gas temperature: RT, Cu particle size: 5 μm). Reprinted with kind permission from Springer Science Business Media [Fukumoto et al. (2007)], copyright © ASM International

- Stagnation pressure.
- Stand-off distance.
- Substrate material.
- Stagnation temperature of the carrier gas.
- Feed arrangement of the powder particles.

The average surface roughness of the low-pressure cold-spray process was measured as 4.92 μm, which was within the confidence interval of the predicted optima of surface roughness for the coatings. The carrier gas stagnation pressure helps in imparting the high velocity to the impacting particles and thus aids in the formation of the coating and a better surface roughness value of the deposited coating. The stand-off distance inversely affects the surface roughness. The surface roughness obtained depends on the type of substrate material used, the minimum being for the ASTM B 435 (Ni) alloy, followed by ASTM B36 (Brass) alloy and the ASTM B 221 (Al) alloy.

6.5.3.1 Substrate Roughness

[Nastic et al. (2017)] studied the effect of substrate surface topography on the creation of metallurgical bonds, and mechanical anchoring points have been studied for the cold-spray deposition of pure aluminum on 300 M steel substrate material. The coating's adhesion strength showed a significant decrease from 31.0 ± 5.7 MPa on polished substrates to 6.9 ± 2.0 MPa for substrates with roughness of 2.2 ± 0.5 μm. Strengths in the vicinity of 45 MPa were reached for coatings deposited onto forced pulsed-waterjet-treated surfaces with roughness larger than 33.8 μm. A finite element analysis has confirmed the sole presence of mechanical anchoring in

coating adhesion strength for all surface treatment except polished surfaces. Grit embedment has been shown to be non-detrimental to coating adhesion for the current deposited material combination. The particle deformation process during impacts has been studied through a finite element analysis using the Preston–Tonks–Wallace (PTW) constitutive model. The obtained equivalent plastic strain, temperature, contact pressure, and velocity vector were correlated to the particle ability to form metallurgical bonds. Favorable conditions for metallurgical bonding were found to be highest for particles deposited on polished substrates, as confirmed by fracture surface analysis. The findings showed in this study not only explain the causes of the drop in cold-spray coating adhesion strength for low substrate roughnesses compared to polished surfaces but also prove the irrelevance of grit embedment when mechanical anchoring is the only bonding mechanism responsible for adhesion.

6.5.3.2 Spray Distance

[Cai et al. (2014)] point out that in the process of cold-spray applications, robot kinematic parameters have significant influences on the coating quality. Those parameters include spray distance, spray angle, gun relative velocity to substrate, scanning step, and cycle numbers. The combined effects, which are caused by their interactions, determine the coating thickness. Thus, it becomes necessary to optimize the robot trajectory for the spraying process in order to obtain the desired coating thickness. Their study aims at investigating the relationship between the coating profile and the spray distance, the scanning step, and introducing the basic principle of a software toolkit named Thermal Spray Toolkit (TST), developed in their laboratory to generate the optimized robot trajectories in spray processes including thermal spray and cold spray. Experiments have been carried out to check the reliability of the simulated coating profile and the calculated coating thickness by TST.

6.5.3.3 Spray Angle

For optimal spray deposition efficiencies, the particle velocity must be above the critical velocity, v_{cr}, and the impact angle, θ, must be as close as possible to 90°. With increasing deviation from orthogonal impact on the substrate surface, the particle velocity component normal to the substrate varies as ($v_p \sin\theta$); thus with decreasing angle, v_p can become lower than the critical velocity, v_{cr}, leading to a significant drop of deposition rate. Li Chang-Jiu et al. (2005) and Li et al. (2006) calculated the deposition efficiencies for various cold spray conditions of copper particles, with particle sizes between 22.6 and 142.6 μm. Results given in Fig. 6.54 [Li Chang-Jiu et al. (2005)] show good agreement between the calculated values and experimental deposition efficiencies measure using helium gas (2 MPa, 340 °C). These show a rapid drop

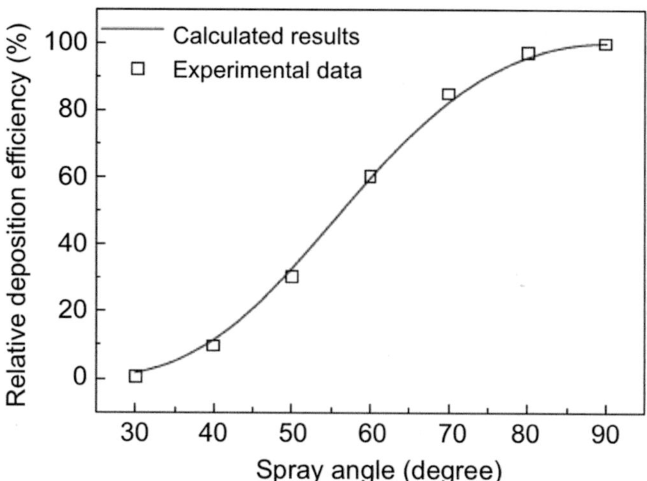

Fig. 6.54 Computation and experimental data of the deposition efficiency as function of the spray angle for the CS of copper powder (5–25 μm) on copper substrate. Helium was used as process gas at 2 MPa and 340 °C [Li Chang-Jiu et al. (2005)]. Reprinted with kind permission from Elsevier

of the relative deposition efficiency with the reduction of the impact angle below 70°.

6.5.3.4 Substrate Oxidation

[Ichikawa and Ogawa (2015)] reported that the impinging velocity and deformation behavior of the particles are believed to be critical deposition parameters. Other factors such as oxide-film thickness and mechanical properties may also affect cold-spray deposition. In their study, the effect of substrate conditions, especially oxide film thickness, on the deposition process was investigated. The efficiency of deposition and the relationship between the impinging position of the particles and the number of particles deposited were investigated for two substrates coated with oxide films of different thicknesses; the particles were deposited sparsely on the substrates. Results showed that efficiency of deposition was lower for the oxide film with the greater thickness. Deposition tendency was also associated with impinging velocity, that is, the critical velocity on which oxide film thickness depends; this is the reason for the different efficiency of deposition. An oxide film may inhibit deposition on a new surface. These results indicate that oxide film thickness has a substantial effect on the efficiency of deposition on the substrate. This is illustrated in Fig. 6.55, where the morphology of the substrates at the center of the sprayed region was observed by SEM. Particles of different sizes, from less than 5 μm to 50 μm, were found to be deposited on the mirror-polished substrate, as shown in Fig. 6.55(a). In contrast, only a few particles were found to be deposited on the heat-treated substrate, as shown in Fig. 6.55(b). The region observed was the same for both substrates; however, their deposition efficiencies were clearly different.

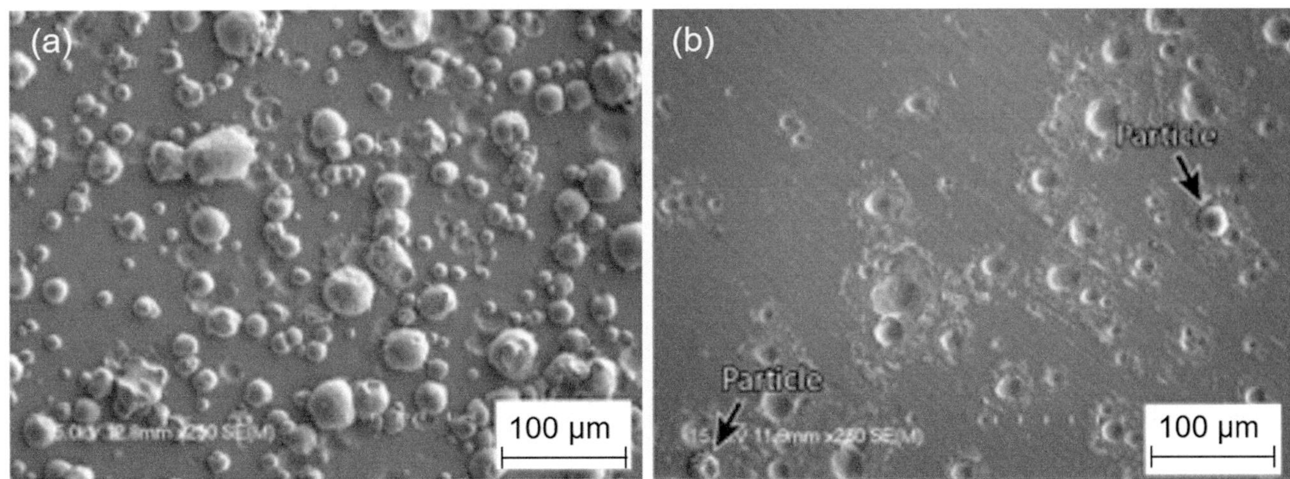

Fig. 6.55 Morphology of the substrates observed at the center of the sprayed region, (**a**) particles deposited on mirror-polished substrate, (**b**) particles deposited on oxide-film-coated substrate

6.5.3.5 Laser Preheating of the Substrate

When spraying harder materials than Cu, Al, Ag, and Zn, such as Ti alloys or Mo, it becomes necessary to heat the spray gas more and to use helium instead of nitrogen with pressures up to 4 MPa, corresponding to He flow rates in the order of 3 m^3/min. However, heating the spray gas too much increases the risk of nozzle fouling with low-melting-temperature particles. That is why it has been proposed to preheat the substrate with a laser just prior to particle deposition [Zieris et al. (2003, Kulmala and Vuoristo (2008), Bray et al. (2009)]. Impact temperatures are then increased, through the local heating of the substrate by the laser (of course below the melting temperatures of both the substrate and the particles). Thus, coatings can be obtained with less spray gas preheating and lower particle velocities (no need to use helium, which costs about 80 times more than nitrogen with no helium recycling).

Compared to conventional cold spray, the amount of heat and the rate at which it is applied and removed from the impact zone affect both the rate of deposition and the properties of the material [Bray et al. (2009)]. It is also possible to see an annealing effect if the deposited particles are cooled at a suitable rate to allow recovery. [Bray et al. (2009)] have demonstrated that this method was viable for the deposition of metallic coatings, especially titanium. Oxide-free titanium coatings have been deposited without the use of gas heating at velocities around half of those required in cold spray. For impact velocities of approximately 400 m/s, dense coatings (porosity below 1%) were produced between 650 and 900 °C, well below the melting point of titanium (1668 °C). The oxygen content of cold-sprayed titanium coatings, including that laser treated, was below 0.6 wt.% to be compared with the 5 wt.% obtained with shrouded HVOF-sprayed ones.

6.5.4 Nozzle Design and Powder Injection

As already mentioned, the nozzle geometry plays a very important role for particle axial and radial velocities [Alkhimov et al. (2001), Sakaki and Shimizu (2001), Li Wen-Ya and Chang-Jiu Li (2005), Heinrich et al. (2005), Blose et al. (2003), Gärtner et al. (2006), Schmidt et al. (2006), Li et al. (2006), Stoltenhoff et al. (2002a, b)]. For example, [Stoltenhoff et al. (2002a, b)] have shown that with four different nozzles characterized by different expansion ratios (6 or 9), shapes (conical and bell), and lengths of the diverging sections, the deposition efficiency varies from 38% to 72%! The highest deposition efficiency is obtained with the longest nozzle, an expansion factor of 9, and a bell-shaped nozzle. However, there is an optimum in the gas velocity. [Dykhuizen and Smith (1998)] were the firsts to show that when the gas velocity or its Mach number, M, was increased, a maximum velocity of the particles was obtained for a value $M = \sqrt{2}$. [Jodoin (2002)] has shown that the design of the Mach number of an air nozzle should not be higher than 1.5 to avoid the use of too high temperatures.

Nozzle clogging is sometimes a concern when spraying metallic powders. A patent [Papyrin (2001)] has been granted to a method describing a mixture with two particle populations. One of the populations is selected such that its velocity is below v_{cr} and it differs from the other in terms of size, yield strength and material nature. It thus maintains the supersonic nozzle in a non-obstructed condition by cleaning it. It also allows for, raising the main gas operating temperature, thereby increasing the deposition efficiency of the powder the velocity of which is above the critical value. Besides the size of the particles, their morphology also plays a key role [Fukanama et al. (2006)]. For example, angular particles

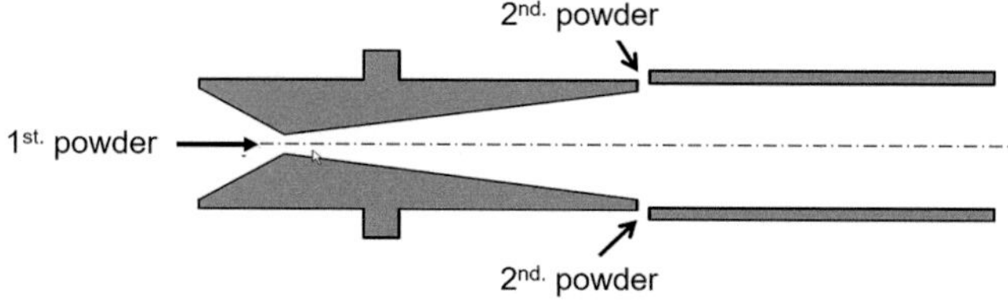

Fig. 6.56 Nozzle design with two points of powder injection into the stream [Klinkov et al. (2008)]. Reprinted with kind permission from Elsevier

of both stainless steel and bronze powders were significantly faster (~ 80 m/s) than spherical particles.

[Kosarev et al. (2003)], [Shkodkin et al. (2006)], [Klinkov and Kosarev (2006)], [Sova et al. (2009)], [Sova et al. (2010)], [Sova et al. (2011)], [Spencer et al. (2009)], among others, have used this technique. According to [Klinkov et al. (2008)], the most efficient method to spray multicomponent coatings would be heating and accelerating each component under spraying conditions appropriate for this particular component. For that, they have designed an axisymmetric nozzle with two points of powder injection, schematically presented in Fig. 6.56. The metal powder is injected in the subsonic part for effective heating of the particles. As for the ceramic particles, their temperature is not important; they are injected in the supersonic part, preventing erosion of the nozzle throat. Of course, this process requires two powder feeders.

It is important to point out, however, that the interaction between high-velocity gas jet and the substrate is not simple for supersonic nozzles with rectangular sections and a Mach number at the nozzle exit in the range of 2–3.5 [Kosarevt al. (2003)]. The problem has been recognized by Samareh and Dolatabadi (2007) who have shown that various modes of interaction can take place of which they identified the classic mode, or the mode with the jet oscillations. The main parameters influencing the transition from one mode to another are the jet pressure ratio, the stand-off distance, and jet thickness. The classic mode occurs at near isobaric exhausting and small stand-off distances. According to Jodoin (2002), optimum choice of the stand-off distance has to be made based on careful examination of the interactions between the shock diamonds formed in the jet and the bow shock formed at the substrate where significant drop in the axial gas velocity takes place, associated with the increased gas temperature which interacts mostly with the smallest particles with sizes below 25 μm.

Reducing the intensity of this bow shock implies also diminishing the Mach number to $M = 1.5$ [Jodoin (2002)]. The problem is then to accelerate the particles sufficiently to a value above the critical velocity value, which sometimes requires using a higher inlet gas temperature. However, to keep the benefits of the cold spray, depending on the material sprayed, the temperature to which the substrate is subjected must also be limited. The nozzle roughness also plays a role in determining the gas velocity [Alkhimov et al. (2001)]. A roughness, Ra, of 1 μm has a negligible influence but when particles stick on the nozzle wall, the roughness can reach 100 μm, and comparing a new nozzle and a nozzle coated with particles, the velocity loss is about 80 m/s. Thus, this phenomenon must be considered in the nozzle design and the choice of materials (for example, by use of hard ceramic nozzles).

[Li and Li (2005)], using a computational fluid dynamics program, FLUENT, conducted a numerical analysis for the accelerating behavior of spray particles in cold spraying. The optimal design of the spray gun nozzle was achieved based on simulation results to solve the problem of coating for the limited inner wall of a small cylinder or pipe. It was found that the nozzle expansion ratio, the particle size, the nature of the process gas, the operating pressure, and temperature were the main factors influencing the accelerating behavior of spray particles in a limited space. The experimental results using the designed short nozzle with a whole gun length of <70 mm confirmed the feasibility of optimal design for a spray gun nozzle used in a limited space. Based on the results presented above, a short cold-spray gun was designed and manufactured for use in a limited space, especially for the inner wall of a cylinder or pipe with a total length of no longer than 70 mm in the nozzle-axial direction. Figure 6.57 shows the typical microstructure of a cold-sprayed Cu coating using the designed spray gun. It was shown that a dense Cu coating could be deposited using the short spray gun in comparison with the conventional cold-spray gun. The thickness of the sprayed Cu coating was about 1 mm. The spray particles have experienced intensive deformation to form splats during deposition.

[Han et al. (2009)] point out that one of the requirements for cold spray is to inject particles to be sprayed into a pre-nozzle chamber where both the particles and the powder feed gas are entrained into the primary gas stream. In their study, they investigated the effects of powder injection on coating formation through both experimental studies and computational simulations. Several issues related to powder injection are examined, including the size of powder injector, the differential pressure, powder gas flow, and injector clogging. It is shown that an improved powder injector design not only enables the use of a reduced amount of powder carrier gas flow but also maintains steady, clogging-free spraying conditions. Combined with properly selected injection conditions, it can also lead to enhanced coating deposition by the kinetic-spray process.

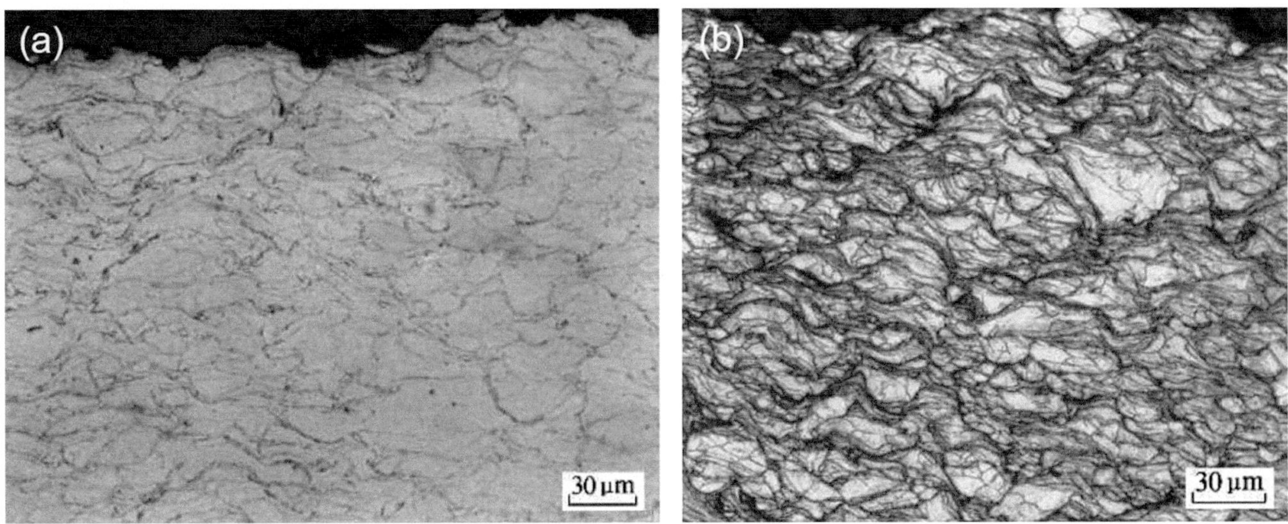

Fig. 6.57 (**a**) Typical microstructure of a Cold-Sprayed Cu coating obtained using a short spray gun, N_2 at 2 MPa and $\sim$300 °C. (**b**) Etched microstructure revealing interparticle boundaries [Li and Li (2005)]

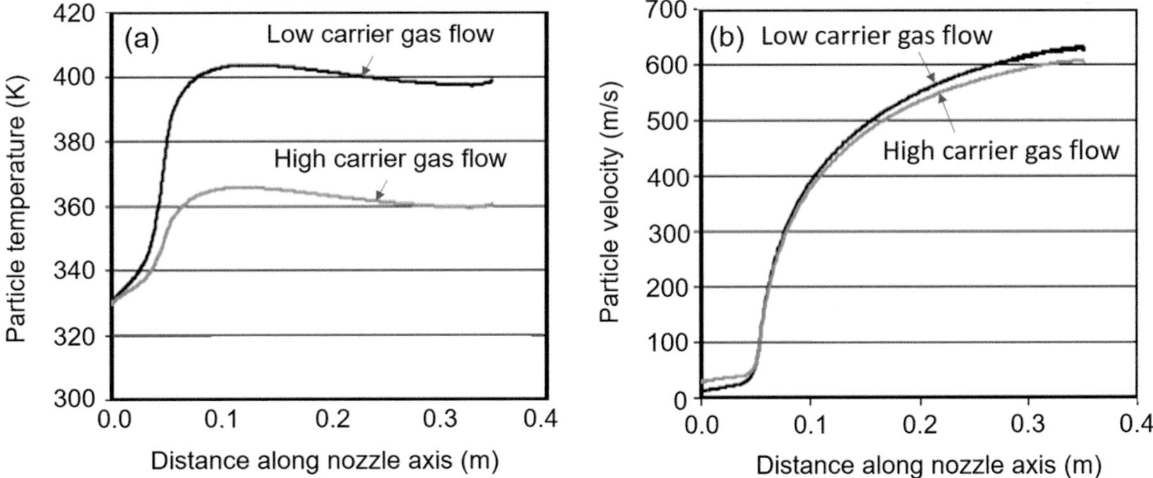

Fig. 6.58 Computational results showing the axial aluminum particle (**a**) temperature and (**b**) velocity distributions for two-carrier gas flow rates. Reprinted with kind permission from Springer Science Business Media [Han et al. (2004)], copyright © ASM International

6.5.4.1 Carrier Gas

A minimum gas flow rate is necessary to avoid injector clogging, but too high flow rates accelerate particles and reduce their residence time, and also modify the gas temperature at the nozzle throat. A compromise must be found between spraying hotter particles (thus more ductile) and faster particles [Han et al. (2004)]. Both computational simulations and experiments demonstrated that a lower carrier gas flow rate produced a higher mean gas temperature at the throat. This increased mean gas temperature tends to increase the particle temperature, Fig.6.58a, but not the particle velocity, as shown in Fig 6.58b.

Moreover, in most cases, even when He is used as the process gas, the carrier gas is nitrogen, which mixes with the helium in the nozzle and reduces the particle velocities. Balani et al. [Balani et al. (2005)] have shown that aluminum coatings were denser with pure He as carrier gas when compared to a mixture He–20 vol.% N_2. [Goldbaum et al. (2012)] recall that cold spray is a complex process where many parameters have to be considered in order to achieve optimized material deposition and properties. In the cold-spray process, the deposition velocity influences the degree of material deformation and material adhesion, as shown in Fig. 6.59, illustrating three commonly known deposition mechanisms in cold-spray splats. However, not all the splats in the jet undergo deposition. When the particle velocity is lower than the critical velocity, some particles do not deposit and instead rebound from the substrate.

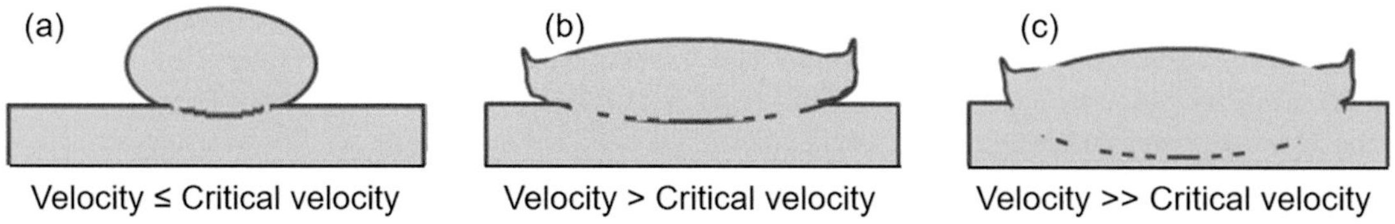

Fig. 6.59 General appearance of cold-sprayed splats as a function of particle velocity v_p with respect to the critical velocity v_{cr} (**a**) $v_p \leq v_{cr}$, (**b**) $v_{cr} \leq v_p$. (**c**) $v_p >> v_{cr}$ [Goldbaum et al. (2012)]

At medium particle velocities (b), there is significant deformation of the splat and formation of adiabatic shear and metallurgical bonding in these local regions. However, the substrate remains undeformed. At the highest particle velocities (c), both the splat and the substrate deform extensively, and a combined adiabatic shear process leads to an increase in the length of the conformal interface and an increase in the density of regions that have become metallurgically bonded. While most materials can be easily deposited at relatively low deposition velocity (<700 m/s), this is not the case for high-yield-strength materials like Ti and its alloys. Authors evaluate the effects of deposition velocity, powder size, particle position in the gas jet, gas temperature, and substrate temperature on the adhesion strength of cold-sprayed Ti and Ti6Al4V splats. The splats were deposited onto Ti or Ti6Al4V substrates over a range of deposition conditions with either nitrogen or helium as the propelling gas. The results demonstrated that optimization of spray conditions makes it possible to obtain splats with continuous bonding along the splat/substrate interface and measured adhesion strengths approaching the shear strength of bulk material. The parameters shown to improve the splat adhesion included an increase in the splat deposition velocity well above the critical deposition velocity of the tested material, an increase in the temperature of both the powder and the substrate material, a decrease in the powder size, and optimization of the flow dynamics for the cold-spray gun nozzle.

[Huang and Fukanuma (2012)] studied the adhesion mechanism of the deposit/substrate interface. It seems that the adhesion strength is mainly determined by the mechanical (including the plastic deformation of particle and substrate) and thermal interaction between the particle and the substrate when the particles impact onto the substrate at a high velocity. In order to understand the adhesion mechanism, a novel adhesive strength test was developed to measure the higher bonding strength of cold-sprayed coatings in this study. The method breaks through the limits imposed by glue strength in the conventional adhesive strength test, and it can be used to measure the coatings with a higher adhesive strength. The particle velocity was obtained with DPV-2000 measurement and CFD simulation. The relationships between the adhesion strength of the deposits/substrate interface and the particle velocity were discussed. The results show that a stronger adhesion strength can be obtained with an increase in particle velocity. There are two available ways to improve adhesion strength. One is to increase the temperature of the working gas, and another is to employ helium gas as the working gas instead of nitrogen gas.

[Ozdemir et al. (2016)] point out that cold spray is a developing technology that is increasingly finding applications in the coating of similar and dissimilar metals, repairing geometric tolerance defects to extend expensive parts' life, and additive manufacturing across a variety of industries. Expensive helium is used to accelerate the particles to higher velocities in order to achieve the highest deposit strengths and to spray hard-to-deposit materials. Minimal information is available in the literature studying the effects of He-N_2 mixing on coating deposition efficiency, and how He can potentially be conserved by gas mixing. In this study, a one-dimensional simulation method is presented for estimating the deposition efficiency of aluminum coatings, where the He-N_2 mixture ratios are varied. The simulation estimations are experimentally validated through velocity measurements and single-particle-impact tests for Al6061.

6.5.4.2 Critical Velocity

[Li et al. (2006)] estimated both experimentally and theoretically the critical velocity of copper (Cu) particles for deposition in cold spraying. An experimental method was proposed to measure the critical velocity based on the theoretical relationship between deposition efficiency and critical velocity at different spray angles. A numerical simulation of particle-impact deformation was used to estimate the critical velocity. The theoretical estimation was based on the critical velocity corresponding to the particle velocity at which impact begins to cause adiabatic shear instability. The experimental deposition was conducted using Cu particles of different sizes, velocities, oxygen contents, and temperatures. The dependency of the critical velocity on particle temperature was examined. Results show that the critical velocity can be reasonably measured by the proposed test method, which detects the change in critical velocity with particle temperature and oxygen content. The Cu particles of oxygen content 0.01 wt.% yielded a critical velocity of about 327 m/s. Experiments show that

the oxygen content of the powder significantly influences the critical velocity. Variations in oxygen content can explain the large discrepancies in critical velocity that have been reported by different investigators. Critical velocity is also found to be influenced by particle temperature, as well as the types of materials. High particle temperature causes a decrease in critical velocity. This effect is attributed to the thermal softening at elevated temperatures.

[Li et al. (2016)] conducted a numerical analysis for the accelerating behavior of spray particles in cold spraying using a computational fluid dynamics program, FLUENT. They found that the nozzle expansion ratio, particle size, accelerating gas type, operating pressure, and temperature were the main factors influencing the accelerating behavior of spray particles in a limited space. The particle impacting deformation behavior during CS has been simulated by the use of both the Lagrangian and Eulerian methods, while the Eulerian method seemed more feasible considering extreme element distortion. The simulations offered abundant information, such as particle/substrate deformation morphology, metal jet, temperature, stress and strain fields, and even energy evolution during the impact process.

6.5.4.3 Particle Loading Effect

Varying the mass flow rate at which the feed stock particles are fed can change the thickness of the coating [Taylor et al. (2006)]. Is it a loading effect? Once a maximum powder flow rate is reached, too many particles impact the surface, resulting in excessive residual stress causing the coating to peel. This phenomenon is linked to excessive particle bombardment per unit area of the substrate. This is illustrated in Fig. 6.60, representing the coating thickness evolution, as determined from the SEM

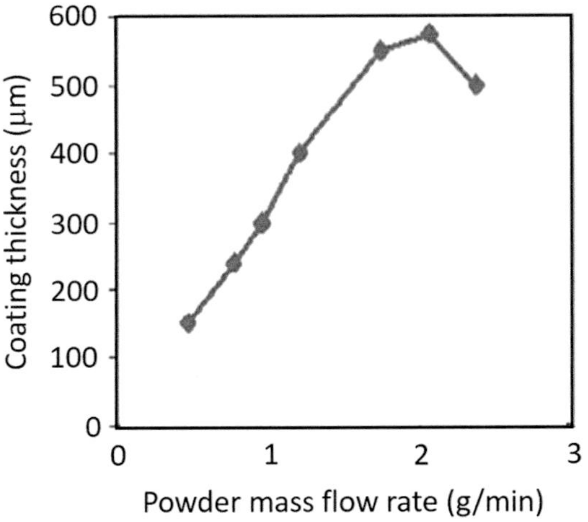

Fig. 6.60 Cold spray copper coating thicknesses on aluminum substrates as a function of the powder mass flow rate and a propellant gas temperature of 5 °C [Taylor et al. (2006)]. Reprinted with kind permission from Springer Science Business Media, copyright © ASM International

images of the various coatings, on aluminum substrates as a function of the copper powder mass flow rate used.

6.6 Coating Materials and Process Applications

6.6.1 General Remarks

In his editorial of a special issue of Journal of Thermal Spray Technology, [Lugsheider (2006)] has emphasized the broad variety of feedstock materials used in cold spray. As with other thermal spray coating technologies, cold spray have been successfully used in a wide range of applications:

– Corrosion and wear resistance applications [Marx et al. (2006)],
– Repair and restoration of parts, especially those made of magnesium alloys [Marx et al. (2006)],
– Shielding for electromagnetic interference [Marx et al. (2006)],
– Demanding electric, electronic, or thermal applications.
– Deposition of soldering or brazing alloys and generation of solderable surfaces on materials with poor wettability, such as heat sinks with copper on aluminum [Li et al. (2007b)].
– Conducting structures on non-metal composite layers (the cold-sprayed metal-matrix composite coating is dense, with a strong bonding between the metal matrix and the dispersant [Li et al. (2007b)].

[Gärtner et al. (2006)] point out that cold spraying has attracted serious attention since unique coating properties can be obtained by the process that are not achievable by conventional thermal spraying. This uniqueness is due to the fact that coating deposition takes place without exposing the spray or substrate material to high temperatures and, in particular, without melting the sprayed particles. Thus, oxidation and other undesired reactions can be avoided. Spray particles adhere to the substrate only because of their high kinetic energy on impact. For successful bonding, powder particles have to exceed a critical velocity on impact, which is dependent on the properties of the particular spray material. This requires new concepts for the description of coating formation but also indicates applications beyond the market for typical thermal spray coatings.

It should be pointed out, however, that compared to conventional thermal spray coating, cold-spray processing typically employs finer, smaller-diameter metal powders. Furthermore, cold-sprayed particles exhibit fewer surface oxides than thermally sprayed particles due to the absence of particle melting during spraying. For these reasons, it is

important to consider the potential for dust explosions or fires during cold-spray processing, for both industrial and R&D applications. [Sakata et al. (2014)] examined the dust explosion characteristics of metal powders typically used in cold-spray coating for the purpose of preventing dust explosions and fires, and thus protecting the health and safety of workers and guarding against property damage. In order to safely make use of the new cold-spray technology in industrial settings, it is necessary to manage the risks based on an appropriate assessment of the hazards. However, there have been few research reports focused on such risk management. Therefore, in this study, the dust explosion characteristics of aluminum, titanium, zinc, carbonyl iron, and eutectoid steel containing chromium at 4 wt.% (4 wt.% Cr-eutectoid steel) powders were evaluated according to the standard protocols JIS Z 8818, IEC61241–2-3(1994–09) section 3, and JIS Z 8817. This chapter reports the results of [Sakata et al. (2014)] concerning the dust explosion properties of the above-mentioned metal powders.

6.6.2 Metals

A variety of coatings have been applied using the cold-spray process. These include pure metals (Ag, Al, Cu, Fe, Mo, Ta, Ti, Zn...), low-alloy steels, Ni-Cr alloys, Ni base superalloy, stainless steel, Zn alloys, Al alloys, Cu alloys, MCrAlY, among others. A correlation between deformation properties and bonding type for the three iso-mechanical groups (fcc: face-centered cubic, bcc: body-centered cubic, and hcp: hexagonal close-packed) is supplied by plotting the product of the shear modulus and the modulus of compression as a function of the normalized temperature $T_0/(T_0 + T_m)$ with $T_0 = 273$ K and T_m the melting temperature of the material [Papyrin et al. (2003)] (Fig. 6.61). The product of the shear modulus and modulus of compression takes into consideration two important material parameters and clarifies the distribution of the data points. The representation permits a rough classification of suitability for the cold-spray process. Empirically, the plotted line corresponds to a T_m of ~1600 °C, at which temperature difficulties must be expected with regard to compacting.

In addition, the bonding type can be characterized in a simplified manner by the melting temperature, which can be used as an evaluation criterion. At approximately 1600 °C, a significantly more difficult compaction ability is to be defined in the cold-spray process on the basis of the experiments and the data published in the literature. Excellent results have been obtained with materials that have low melting point and mechanical strength, with Zn, Al, and Cu being ideal. Good results are obtained with Fe and Ni base materials, provided higher temperatures and velocities are used to spray. Recent application development efforts are based on several advantages of the cold-spray process or cold-sprayed layers. Table 6.1 from [Marx et al. (2006)] contains the decisive advantages for some applications.

Cold spray produces ultra-pure coatings with unique characteristics that can be used in the following applications:

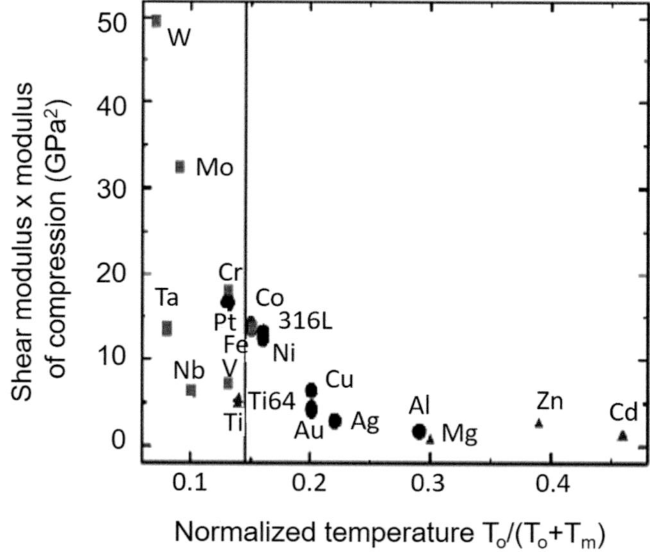

Fig. 6.61 Variation of the plastic properties of various materials represented by the product of (shear modulus x compression modulus) as a function of the normalized temperature $T_0/(T_0 + T_m)$ with $T_0 = 273$ K and T_m the melting temperature of the material. [Vlcek et al. (2005)]. Reprinted with kind permission from Springer Science Business Media, copyright © ASM International

Table 6.1 Properties, materials, and applications of cold spray. [Marx et al. (2006)], Reprinted with kind permission from Springer Science Business Media copyright © ASM International

Application	Typical materials	Advantages
Corrosion protection	Zn, Ni, brass	Little porosity
Conductor, thermal management	Cu, Al, steel, Ni	Little porosity, low oxygen
Repair, structural coatings	Alloys, soldering/brazing	Little porosity, no phase change
Solder/braze deposition	Alloys	Strong bonding, no phase change, Low oxygen content
Solderability	Cu	Low oxygen content

6.6.2.1 Aluminum

When cold-sprayed, the oxygen content increases at the maximum by 0.02 wt.% and porosity is close to zero. Thus, the process can be used for the spray forming of aluminum components with elastic modulus, ultimate tensile strength, and elongation similar to those of wrought aluminum. The high bond strength of cold-sprayed aluminum layers on glass or ceramic substrates seems also promising, and such layers can be used as intermediate or bond layers for other coatings. Aluminum alloy coatings can be deposited for brazing. They comprise Al-12 wt.% Si (for the brazing filler metal), Zn (for corrosion protection), and $KAlF_4$ (flux powder), and they are increasingly being used in brazed aluminum heat exchangers, for example, in air conditioning equipment. Al-13Co-26Ce alloys give rise to a substantial reduction in fatigue effects in comparison with uncoated and Al-clad-coated substrates. It was proposed that any crack propagation in the coating was arrested by the residual compressive stresses contained in the coating [Li et al. (2007b), Hall et al. (2006), Yoon et al. (2007a, b), Makinen et al. (2006), Sansoucy et al. (2007a), Lee et al. (2007b)]. [Lee et al. (2007b)] have repaired aluminum molds by cold-spraying aluminum. To check the recovered molding capability, polymer (Polystyrol) products were fabricated by injection molding using the repaired mold. Experimental results showed that the mold repair process using cold-spray deposition and machining can be applied in the mold repair industry [Bobzin et al. (2008)] have cold-sprayed AlSi12 and AlSi10Cu4 onto Mg-containing aluminum alloys 6063 and 5754. Some coated samples were brazed under argon atmosphere without any fluxes. The results show that AlSi12 had much better deposition behavior than AlSi10Cu4. Due to the rupture of oxide scales, Cu and Si diffused into the substrate and a metallurgical bond formed between the brazing alloys and the substrates during heat-treatment. The coated samples could be brazed without any fluxes.

As magnesium alloys have poor corrosion resistance, [DeForce et al. (2011)] have applied to a magnesium alloy (ZE41A-T5 Mg) a corrosion-resistant barrier coating made of high-purity Al-5 wt.% Mg that has galvanic compatibility with magnesium and a hardness value as high as that of magnesium. The coating resulted in the best overall performance, including a high hardness, 125 HV_{1N}, and adhesion strength, over 60 MPa, when treated for over 1000 h in a salt spray chamber and with a low galvanic current.

Nano crystalline coatings have also been sprayed by cold spray:

[Richer et al. (2006)] have cold-sprayed with helium Al-Mg feedstock powder, characterized by a broad particle-size distribution. Hardness measurements and TEM inspection demonstrated that the microstructure of the original feedstock powder was conserved throughout the cold-spray process, and that no increase in hardness due to cold working of particles was observed.

[Zhang et al. (2011)] have cold-sprayed with nitrogen aluminum-alloy 2009 using a powder mixture of nano-crystalline and as-atomized powders, the latter being bigger than the former. The density of the coating was much higher than that of a coating using pure nano-crystalline powder. [Ajdelsztajn et al. (2006a)] have successfully cold-sprayed conventional and nano-crystalline Al-Cu-Mg-Fe-Ni-Sc coatings. The conventional cold-sprayed coating showed negligible porosity and an excellent interface with the substrate material. This was not the case for the nano-crystalline one, in which the porosity level was in the range of 5–10%. The microstructure of the feedstock powders (atomized and cryo-milled) was retained after the cold-spray process. The difference in porosity between both coatings was explained by the hardness and microstructure of the corresponding feedstock powder: the softer and coarse-grain conventional powder experienced generalized adiabatic shear instability upon impact, resulting in a dense coating, whereas the harder and fine-grain cryo-milled powder did not exhibit the same extent of deformation, resulting in a less dense coating. The conventional powder showed work hardening during impact, thus increasing the hardness of the conventional coating compared with the feedstock powder. On the other hand, the hardness of the nano crystalline coating was similar to that of the feedstock powder, suggesting that no work hardening took place in this case. The Vickers microhardness of the nano crystalline coating was, despite the porosity, much higher than that of the conventional coating.

[Bérubé et al. (2012)] studied by Cold Gas Dynamic Spray Process (CGDS) aluminum alloy Al-5Fe-V-Si (in wt.%) obtained with feedstock powder produced by Rapid Solidification (RS) using the gas atomization process. This technic was selected to produce high-temperature resistant Al-alloy coatings using the Cold Gas Dynamic Spraying Process (CGDS). The alloy composition was chosen for its mechanical properties at elevated temperature for potential applications in Internal-Combustion (IC) engines. The CGDS spray process was selected due to its relatively low operating temperature, thus preventing significant heating of the particles during spraying and as such allowing the original phases of the feedstock powder to be preserved within the coatings. The study revealed the conservation of the complex microstructure of the rapid solidified powder during the spray process. Four distinct micro--structures were observed as well as two different phases, namely an $Al_{13}(Fe,V)_3Si$ silicide phase and a metastable $(Al,Si)_x(Fe,V)$ micro-quasi-crystalline icosahedral (MI) phase. Aging of the coating samples was performed and it was confirmed that the transformation of the metastable phases and coarsening of the nanosized precipitates occurs at around 400 °C. The metastable MI phase was determined to be thermally stable up to 390 °C,

after which a phase transformation to silicide starts. Compared to the feedstock powder, the as-sprayed Al-5Fe-V-Si coatings seem to preserve their microstructure during the cold-spray process. Fine precipitates' micro-structure and metastable (Al,Si)x(Fe,V) MI phase show to be unaffected during spraying.

[Cavaliere and Silvello (2017)] recall that the cold-spray process has recently been recognized as a very useful tool for repairing metallic sheets, achieving desired adhesion strengths when employing optimal combinations of material process parameters. They present the possibility of repairing cracks in aluminum sheets by cold spray. A 2099 aluminum alloy panel with a surface 30° V notch was repaired by cold-spraying 2198 and 7075 aluminum alloy powders. The crack behavior of V-notched sheets subjected to bending loading was studied by Finite-Element Modeling (FEM) and mechanical experiments. The simulations and mechanical results showed good agreement, revealing a remarkable K factor reduction, and a consequent reduction in crack nucleation and growth velocity. The results enable the prediction of the failure initiation locus in the case of repaired panels subjected to bending loading and deformation. The stress concentration was quantified to show how the residual stress field and failure are affected by the mechanical properties of the sprayed materials and by the geometrical and mechanical properties of the interface. It was demonstrated that crack resistance increases more than sevenfold in the case of repair using AA2198 and that cold-spray repair can contribute to increased global fatigue life of cracked structures.

[Qiu et al. (2017)] investigated the microstructure and mechanical properties of cold-sprayed bulk A380 alloy after heat treatment at various conditions, using optical and electron microscopy, and tensile and hardness tests, respectively. The results revealed that heat treatment increased the strength and ductility of the cold-sprayed A380 alloy deposits compared with the as-sprayed state. Heat treatment showed two different effects on the mechanical properties of the deposits. On the one hand, it resulted in effective diffusion at interparticle boundaries that altered the particle-bonding mechanism from pure mechanical interlocking to metallurgical bonding. Thus, the strength and ductility of the material were greatly enhanced. On the other hand, interparticle diffusion during high-temperature heat treatment resulted in growth of the Si phase and pores, which ultimately reduced the strength and elongation of the alloy. This observation was consistent with the hardness results.

[Sova et al. (2017)] point out that the spatial resolution of cold spray depends on the exit dimension of the converging-diverging nozzle used for particle acceleration. The typical diameter of a cold-spray nozzle lies in the range between 4 and 10 mm. In this study, a micronozzle with a 0.5 mm throat diameter and a 1 mm exit diameter is applied for cold-spray deposition of aluminum powder. Numerical simulation and experimental velocity measurements demonstrate that aluminum particles could be accelerated to the velocity sufficient for coating deposition if room temperature helium is applied as working gas. The diameter of the spraying spot in this case does not exceed 1.7–1.8 mm, which is significantly lower than that for a standard cold-spray nozzle. However, the density and the bond strength of the obtained aluminum coating are significantly lower than those for standard cold-spray aluminum deposits.

6.6.2.2 Copper

Metallization: the combination of the low porosity and the low O_2 content contributes to excellent electrical properties of coatings (close to those of bulk material). Cu coatings also display excellent tensile properties. They are used extensively in electrical metallization including plastic parts, current-carrying thick films in the automotive industry: for example, on a heat sink of aluminum to solder power transistors. The cold-sprayed copper layer removes the natural oxide film and provides a solder-able surface for the tin- or copper-plated area of the electronic part. Moreover, copper conducts very rapidly the heat dissipated to the heat sink. Papers about cold-sprayed copper can be found, for example, in references [Kairet et al. (2007), Sudharshan Phani et al. (2007a, b), Kairet et al. (2006), Calla et al. (2006), Zahiri et al. (2006a, b), McCune et al. (2000), Koivuluoto et al. (2007), Fukumoto et al. (2010), Eason et al. (2011), Donner et al. (2011), Jin et al. (2011)]. [Fukumoto et al. (2010)] have systematically studied copper splat formation on smooth substrates, with particularly the metal jetting behavior of the particles related to some process factors. [Koivuluoto et al. (2007)] showed that cold-sprayed Cu coatings (coarser particle size) were dense according to SEM analysis and open cell potential measurements. [Eason et al. (2011)] have produced dense bulk material obtained by continuous spray deposits 25 mm thick, the elevated microhardness in the as-sprayed coatings resulted from the cold work hardening. Post-annealed microhardness values indicated the potential for strengthened bulk materials through cold-spray processing, exceeding the performance of typical P/M material. [Donner et al. (2011)] have shown that copper coatings with high electrical conductivity (98% of that of bulk material) were successfully deposited onto insulating alumina thermal spray ceramic coatings by cold gas spraying. Adhesion was good when achieved via an activation of the ceramic surface. Activation consisted of preheating the substrates to remove adsorbed water and of using chemical activation by interlayers to improve coating buildup. [Jin et al. (2011)] attempted to manufacture a Cu-In coating layer via the cold-spray process and investigated the applicability of the layer as a sputtering target material. They demonstrated that the Cu-In coating layer manufactured via the cold-spray

process with an annealing heat treatment can be applicable as a sputtering target in the future. The Cu coatings have also been used as magnetic shielding coatings on electronic components [Handbook of Thermal Spray (2004), Marx et al. (2006)].

[Goyal et al. (2014)] used a low-pressure cold-spray process to deposit copper coatings, for optimization. Two potential response parameters, that is, coating thickness and coating density, have been studied. Utility values based on these response parameters have been analyzed for optimization using the Taguchi approach. The selected input parameters of powder feeding arrangement, substrate material, air stagnation pressure, air stagnation temperature, and stand-off distance significantly improve the utility function (raw data) comprising quality characteristics (coating thickness and coating density). The percentage contribution of the parameters to achieve a higher value of the utility function is substrate material (50.03%), stand-off distance (28.87%), air stagnation pressure (6.41%), powder feeding arrangement (4.68%), and air stagnation temperature (2.64%).

6.6.2.3 Nickel

In general, nickel coatings are used for corrosion and wear protection layers. Cold-sprayed nickel layers are used as base or intermediate layers and they are less expensive than electroplated coatings. They are also used in coatings for induction heating on cooking appliances or cooking pots. However, [Koivuluoto and Vuoristo (2010a)] have shown that if cold-sprayed, Ni and Ni-30%Cu coatings seemed to be dense according to the SEM studies, corrosion tests showed some through-porosity in the cold-sprayed Ni and Ni-30% Cu coatings. The weak points were on the particle boundaries where the bonding between the particles is not strong and uniform enough. [Koivuluoto and Vuoristo (2010a)] also studied the denseness improvement of cold-sprayed Ni, Ni-20Cu, Ni-20Cr, Ni-20Cr + Al_2O_3, and Ni-20Cr + WC-10Co-4Cr coatings. The denseness, evaluated by corrosion tests, of the Ni coating was improved with optimized spraying parameters whereas the denseness of the Ni-20Cr coatings was increased by adding hard particles in the powder mixture. In addition, the denseness of the Ni-20Cu coatings was improved by heat treatments. With cold spray, very dense and well-adhering thick induction heating layers are deposited by a single processing step onto metallic or even non-metallic cooking utensils [Blose et al. (2006)]. [Zhang et al. (2008)] have synthesized by ball milling of nickel-aluminum powder a mixture with a Ni/Al atomic ratio of 1:1. The powder was deposited by cold spraying using N_2 as accelerating gas. The annealing of the resulting Ni/Al alloy coating at a temperature higher than 850 °C leads to the complete transformation from the Ni/Al

alloy to a NiAl intermetallic compound. By cold-spraying with helium Ni-50Cr on two boiler steels SA-213-T22 and SA 516 (Grade 70), [Bala et al. (2010)] demonstrated that coatings were very useful in developing hot corrosion resistance in the aggressive environment of Na_2SO_4–60% V_2O_5 at 900 °C.

[Ortega et al. (2015)] recall that in the case of fabrication of rotational molds, the multilayer concept with the combination of nickel and copper as primary materials for shells seems to be promising. In their work, a hybrid approach combining the advantages of the electroforming process (accuracy, reproducibility, and quality of material) and cold spray (small amount of heat transferred to the substrate, retention of microstructure and properties of feedstock, and fast speed of process) for producing a multi-material mold is discussed. Experimental results of the feasibility study presented in this paper show that the cold-spray deposition of thick (2–4 mm) copper layers on nickel substrate produced by an electroforming technique is possible if an intermediate copper layer produced by electroplating is applied. Addition of ceramic particles to the copper feedstock powder permits to perform cold-spray deposition directly on an electroformed nickel surface without an intermediate layer, but bonding strength in this case is lower. Thermal shock tests did not reveal any evidence of layers delaminating in both cases.

[Xiong et al. (2010)] studied the impact and deposition behavior of nickel particles onto relatively soft 6061-T6 aluminum alloy and copper substrates in a kinetic-spray process, comparing individual particle impact with full-coating deposition. The results indicated that the deposition onset of nickel coatings on the two substrates follows different deposition mechanisms depending on the corresponding deformability of the impact couples (substrate and particle). Nickel particles were hardly attached onto the relatively soft 6061-T6 substrate in case of individual impact, but the deposition onset of full coating took place depending on embedding, tamping of successive impact, and metallurgical "cold welds" of the viscous metal at impact interface when the impinging particle velocity was relatively low. In case of Ni-Cu impact, the bonding formed at the peripheral impact interface dominated the deposition onset of nickel coating due to the comparable deformability of the impact couple (Ni and Cu).

6.6.2.4 Selective Galvanizing

Corrosion protection for automotive body and chassis structures relies primarily on the use of percolated (galvanized) steel and post-manufacturing processes. Cold spraying has been found to be a useful approach for selective protection of a localized area. Zn and Al-Zn coatings are used [Handbook of Thermal Spray (2004), Blose et al. (2003), Makinen et al. (2006)].

6.6.2.5 Superalloys

Typical applications studied are oxidation-resistant MCrAlY bond coats [Ichikawa et al. (2007)] and IN718 coatings for localized repair [Marrocco et al. (2006a)]. It is possible to deposit the CoNiCrAlY coating on an Inconel 625 substrate by the cold-spray technique. From the results of a laser shock impact test, the adhesion requires higher velocity and specific phase combination. From STEM-EDX results of as-sprayed coatings and SEM, EPMA results, the bonding between CoNiCrAlY coating and an Inconel 625 substrate occurred to only the γ/γ'-(Ni/Ni$_3$Al) phase of CoNiCrAlY within the substrate [Richer et al. (2008)]. IN718 coatings were produced on IN718 plates with very good bonding and very low oxide content [Marrocco et al. (2006a)]. Such coatings have been studied as bond coats of TBCs [Richer et al. (2008), Chen et al. (2011)]. [Richer et al. (2008)] demonstrated the occurrence of microstructural changes during the deposition process of CoNiCrAlY coatings. Evidence of grain refinement of the initially large γ-matrix grains, as well as dissolution of the fine β-phase precipitates, was observed. Such transformations of the deposited material's microstructure were attributed to the intense plastic deformation. [Chen et al. (2011)] showed that TBCs with HVOF- and cold-spray-CoNiCrAlY bond coats had extended durability, compared to TBCs with APS-CoNiCrAlY bond coats. However, the durability of cold-spray-CoNiCrAlY was not as good as that of the HVOF-CoNiCrAlY.

6.6.2.6 Titanium and TiO$_2$ and TiN

Cold-sprayed coatings are very promising against fatigue and for wear resistance, for example, on internal surfaces of cylinder blocks of internal combustion engines, pumps, and hydraulic cylinders for automotive, heavy machinery, forgings, and extrusions. Numerous studies are devoted to Ti or TiAl coatings to achieve layers as dense as possible with limited residual stress [Li Chang-Jiu and Wen-Ya Li (2003), Novoselova et al. (2006), Price et al. (2006a, b), Marrocco et al. (2006b), Binder et al. (2011), Christoulis et al. (2011), Lima et al. (2002), Zahiri et al. (2009a, b), Li et al. (2009), Cinca et al. (2010)]. [Binder et al. (2011)] have shown that using nitrogen as process gas at a pressure of 4 MPa and a temperature of 1000 °C resulted in ultimate strengths of about 450 MPa, similar to those of pure bulk material. Bonding is most prominently caused by shear instabilities under plastic deformation and only to a minor extent by the heat of friction. [Christoulis et al. (2011)] studied the deposition of cold-sprayed titanium onto various substrates.

The use of helium as the processing gas led to a significant improvement in coating properties: deposition efficiency, mean thickness, and porosity. Despite the use of a coarse powder, the formation of titanium cold-sprayed coatings of high density at high deposition efficiency was achieved. [Zahiri et al. (2009a, b)] also showed that coatings densification was improved by using helium. Densification (2% porosity) was achieved by heating particles up to 700 °C, and the metallurgical bond with the substrate was obtained by vacuum heat treatment at 955 °C for 8 hours. [Li et al. (2009)] showed that porous coatings were obtained when using large particles (45–160 µm). After annealing at 850 °C for 4 h under vacuum condition, the Ti coating also presented a porous structure with more uniformly distributed small pores. A metallurgical bonding between the deposited particles was formed through the annealing treatment. The adhesive strength of the coating was significantly improved after annealing, as well as the microhardness.

[Cinca et al. (2010)] cold-sprayed Ti onto aluminum alloy 7075-T6, which is highly used in aeronautical engineering, to protect it from corrosion. They have studied it as an alternative to other deposition techniques such as electro-deposition, chemical vapor deposition, and vacuum plasma spray, which are more expensive. They obtained dense Ti coatings with thicknesses larger than 300 µm and no microstructural changes. [Binder et al. (2011)] recall that titanium coatings have a high potential for various applications and can be produced in high quality by cold spraying. In their contribution, the two major challenges are addressed: (i) optimizing mechanical properties by systematic variation of process parameters, and (ii) evaluating the influence of the spray angle with respect to complex geometries. High deposition efficiencies of more than 95% can be obtained and the coatings show very low porosities, as well as high tensile strength of over 450 MPa, by using nitrogen as process gas. The influence of process conditions on the mechanical properties is discussed on the basis of single-impact morphologies, coating microstructures, tubular coating tensile, as well as shear tests. The influence of process conditions on the mechanical properties is discussed on the basis of single-impact morphologies, coating microstructures, tubular coating tensile, as well as shear tests.

[Gardon and Guilemany (2014)] recall that titanium dioxide has been the most investigated metal oxide due to its outstanding performance in a wide range of applications, chemical stability, and low cost. Recent findings on titanium dioxide coatings deposited by cold gas spray and the capacity of this technology to prevent loss of the nanostructured anatase metastable phase are reviewed.

[Price et al. (2006a, b)] have examined the deposition of titanium on a titanium alloy substrate for potential use as a surface treatment for medical prostheses. A Ti6Al4V alloy was coated with pure titanium by cold gas dynamic spraying. Coatings were deposited onto samples with two different surface preparation methods (as-received and grit-blasted). The fatigue life of the as-received and grit-blasted materials, both before and after coating, was measured with a rotating-bend fatigue rig. A 15% reduction in the fatigue endurance limit was observed after an application of the coating to the

as-received substrate, but no significant reduction was observed on its application to the grit-blasted substrate. The reduction in the fatigue endurance limit has been related to the substrate-coating interface properties, the elastic modulus, and the residual stress states.

6.6.2.7 Iron and Steel

Cold spraying of high-purity iron as an intermediate layer was found to allow for the use of more conventional welding processes for the repair and material buildup on dies formed by the thermal spray process [Sun et al. (2006)]. Fe-based amorphous alloy (Fe-Cr-Mo-W-C-Mn-Si-B) coatings on Al-6061 were successfully deposited with negligible porosity and good values of hardness and elastic modulus [Ajdelsztajn et al. (2006b)]. Metastable alloys (Al-Fe-V-Si), which have good weld ability and a low thermal expansion coefficient, were successfully cold-sprayed and the complex microstructure of the powder was retained after spraying [Jodoin et al. (2006)]. A nanostructured Fe/Al alloy powder was prepared by a ball-milling process in the study by [Wang et al. (2007)]. The cold-sprayed Fe/Al alloy coating evolved in-situ into an intermetallic compound coating through a post-heat treatment. Results showed that the milled Fe-40Al powder exhibits a lamellar microstructure. The microstructure of the as-sprayed Fe(Al) coating depended significantly on that of the as-milled powder. The heat-treatment temperature significantly influenced the in-situ evolution of the intermetallic compound. The heat treatment at a temperature of 500 °C resulted in the complete transformation of Fe(Al) solid solution to a FeAl intermetallic compound.

Stainless steel was sprayed on aluminum against wear (for example on a cylinder, with a specially designed gun [Alexandre et al. (2007)]). [Spencer and Zhang (2011)] studied the porosity of 316 L stainless steel cold-spray coatings. They applied the powder metallurgy practice of mixing particle-size distributions to improve coating density. Results showed that coatings sprayed using mixed-particle-size distributions can have properties similar to those sprayed using fine particles alone. Beyond a critical coating thickness, the substrate is no longer attacked in polarization tests.

[AL-Mangour et al. (2014)] studied the effects of heat treatment on the microstructure and mechanical properties of cold-sprayed stainless steel 316 L coatings using N_2 and He as propellant gases. Both propellant gases have been successfully employed to produce dense 316 L coatings, with no phase change obtained during deposition. Low porosity ($\leq 0.3\%$) was obtained for He-sprayed coatings in the as-sprayed condition and N_2-sprayed coatings annealed at temperatures >800 °C. Annealing of N_2-sprayed coatings also increased tensile strength and ductility, although tensile properties were lower than that of the bulk material due to defects remaining in the annealed coating (e.g., porosity,

incomplete bonding). The Electron Back Scattered Diffraction (EBSD) analysis revealed that annealing at a temperature > 1000 °C produced a uniform, fully recrystallized microstructure with grains more refined than those in the bulk material.

[Sova et al. (2017)] studied experimentally the influence of machining by turning and ball burnishing on the surface morphology, structure, and residual stress distribution of cold-sprayed 17–4 PH stainless steel deposits. It is shown that cold-spray deposits could be machined by turning under parameters closed to turning of bulk 17–4 PH stainless steel. The ball-burnishing process permits to decrease surface roughness. A cross-sectional observation revealed that the turning and ball-burnishing process allowed microstructural changes in the coating near-surface zone. In particular, significant particle deformation and particle boundary fragmentation areobserved. Measurements of residual stresses showed that residual stresses in the as-spray deposit are compressive. After machining by turning, tensile residual stresses in the near-surface zone were induced. Further surface finishing of turned coating by ball burnishing allowed for the establishment of the compressive residual stresses in the coating.

6.6.2.8 Tantalum

Cold spray seems to be an alternative technique to the currently used vacuum plasma spraying. The main advantage of cold spray is the very low level of oxidation during processing of this metal, which is very sensitive to oxidation. It has been demonstrated that Ta coatings can be produced in air with excellent adhesion, low porosity, and increased hardness. Of course, better coatings were obtained when He was used [Van Steenkiste and Gorkiewicz (2004)]. [Koivuluoto et al. (2009)] have investigated the microstructural details, denseness, and corrosion resistance (in 3.5 wt.% NaCl and 40 wt.% H_2SO_4 solutions at room temperature and 80 °C) of two cold-sprayed tantalum coatings. Standard and improved tantalum powders were tested with different spraying conditions. The cold-sprayed tantalum coating prepared with improved tantalum powder showed excellent corrosion resistance.

6.6.2.9 Pure Silicon

It has been deposited as anode for lithium secondary batteries. The average thickness of the as-deposited silicon anodes is between 0.4 and 0.7 μm. Silicon anodes (20 mm by 20 mm) have a capacity of about 1.5 to 2.5 mA.h [Sakaki et al. (2007)].

6.6.2.10 Pure Silver

[Manhar Chavan et al. (2011)] have investigated the process parameters and heat treatment on the thermal and electrical properties of silver powders (water atomized) cold sprayed. The optimized Ag coating, in the as-coated condition,

exhibits unusually high electrical conductivity of the order of 70% of the bulk Ag value. The influence of heat treatment of the "as-coated" cold-sprayed Ag coating on its electrical conductivity is complex and depends also strongly upon the atmosphere where it is performed.

6.6.2.11 Metallic Coatings on Polymers

[Zhou et al. (2011)] have cold-sprayed (with nitrogen as spray and carrier gas) Al and Al/Cu coatings on the surface of the carbon fiber-reinforced Polymer Matrix Composite (PMC). Results have shown the deposition of metallic coatings directly onto the PMC surface was possible with reasonable bonding of feedstock and substrate material. [Lupoi and O'Neil (2010)] have studied the potential of the cold-spray process to produce metallic coatings on non-metallic surfaces such as polymers and composites for engineering applications. They analyzed copper, aluminum, and tin powders on a range of substrates such as PC/ABS, polyamide-6, polypropylene, polystyrene, and a glass-fiber composite material.

[Lupoi and O'Neil (2010)] points out that current coating technologies such as plasma spray, High Velocity Oxygen Fuel (HVOF) and laser cladding involve the delivery of molten materials during the deposition process. However, such techniques are not well suited to the deposition of metallic coatings on polymers and composites, especially when particles are very hot or molten. Cold Spray (CS) has attracted much industrial interest over the past two decades. In this method, a material in powder form is accelerated on passage through a converging–diverging nozzle to high speeds via a high-pressure coaxial carrier gas jet and the particle temperatures are rather low. The high-impact kinetic energy deforms the particles, which creates effective bonding to the substrate. In their paper experimental and Computational Fluid Dynamics (CFD) results when spraying copper, aluminum and tin powder on a range of substrates such as PC/ABS, polyamide-6, polypropylene, polystyrene and a glass-fiber composite material are presented and analyzed. Based on these results, a process characterization chart has been initialized showing the relationship between properties of feedstock powders against deposition behavior. Results have demonstrated that materials in the range of copper travelling at deposition velocities can generate single-particle impact energies in the order of 0.02 mJ. This leads to severe contact stresses; therefore, the predominant effect is erosion of the polymer. Aluminum, due to its low specific weight, does not bring any considerable damage to the surface; however, its critical velocity could not be achieved with the cold-spray system used in the experiments. When adjusting the spray parameters and selecting a suitable nozzle type, coating of tin on a variety of plastic substrate materials was achieved.

[Che et al. (2018)] studies the cold-spray-ability of various metal powders on different polymeric substrates. Five different substrates were used, including Carbon Fiber Reinforced Polymer (CFRP), Acrylonitrile Butadiene Styrene (ABS), Polyether Ether Ketone (PEEK), Poly-Ethylenimine (PEI); and mild steel as a benchmark substrate. The CFRP used in this work has a thermosetting matrix, and the ABS, PEEK, and PEI are all thermoplastic polymers, with different glass transition temperatures, as well as a number of distinct mechanical properties. Three metal powders, tin, copper, and iron, were cold-sprayed with both a low-pressure system and a high-pressure system at various conditions. In general, cold spray on the thermoplastic polymers rendered more positive results than the thermosetting polymers, due to the local thermal softening mechanism in the thermoplastics. Thick copper coatings were successfully deposited on PEEK and PEI. Based on the results, a method is proposed to determine the feasibility and deposition window of cold-spraying specific metal powder/polymeric substrate combinations.

[Gonzalez et al. (2016)] presented a literature review on the thermal spray deposition of metals onto polymer-based structures. The deposition of metals onto polymer-based structures has been developed to enhance the thermal and electrical properties of the resulting metal-polymer material system. First, the description of the thermal spray metallization processes and technologies for polymer-based materials are outlined. Then, polymer surface preparation methods and the deposition of metal bond coats are explored. Moreover, the thermal spray process parameters that affect the properties of metal deposits on polymers are described, followed by studies on the temperature distribution within the polymer substrate during the thermal spray process. The objective of this review is devoted to testing and potential applications of thermal-sprayed metal coatings deposited onto polymer-based substrates.

6.6.2.12 Complex Alloys

[Ajdelsztajn et al. (2006a)] describe recent progress in cold-spray processing of conventional and nanocrystalline 2618 (Al- Cu-Mg-Fe-Ni) aluminum-alloy containing scandium (Sc). As-atomized and cryo-milled 2618 + Sc aluminum powders were sprayed onto aluminum substrates. The mechanical behavior of the powders and the coatings were studied using micro- and nanoindentation techniques, and the microstructure was analyzed using scanning and transmission electron microscopy (SEM and TEM). The influence of powder microstructure, morphology, and behavior during

deposition on the coating properties was analyzed. This work shows that Al-Cu-Mg-Fe-Ni-Sc coatings with a nanocrystalline grain structure can be successfully produced by the cold-spray process. Inspection of the scientific literature suggests that this is the first time a hardness value of 181 HV has been reported for this specific alloy. Cold-spray deposition of conventional and nanocrystalline Al-Cu-Mg-Fe-Ni-Sc coatings was successfully achieved. The conventional cold-sprayed coating showed negligible porosity and an excellent interface with the substrate material. This was not the case for the nanocrystalline coating, in which the porosity level was in the range of 5–10%. The microstructure of the feedstock powders (atomized and cryo-milled) was retained after the cold-spray process. Similarly, the conventional powder showed work hardening during impact, thus increasing the hardness of the conventional coating compared with the feedstock powder. On the other hand, the hardness of the nanocrystalline coating was similar to that of the feedstock powder, suggesting that no work hardening took place in this case.

[Bush et al. (2017)] recall that when a solid, ductile particle impacts a substrate at sufficient velocity, the resulting heat, pressure, and plastic deformation can produce bonding between the particle and the substrate. The cold-spray process has been commercialized for some metallic materials, but further research is required to unlock the exciting potential material properties possible with polymeric particles. In this work, a combined computational and experimental study was employed to study the cold-spray deposition of high-density polyethylene powders over a wide range of particle temperatures and impact velocities. Cold-spray deposition of polyethylene powders was demonstrated across a broad range of substrate materials including several different polymer substrates with different moduli, glass, and aluminum. A material-dependent window of successful deposition was determined for each substrate as a function of particle temperature and impact velocity. Additionally, a study of deposition efficiency revealed the optimal process parameters for high-density polyethylene powder deposition, which yielded a deposition efficiency close to 10% and provided insights into the physical mechanics responsible for bonding while highlighting paths toward future process improvements.

6.6.2.13 Submicronic Ceramic Powders

[Moreau et al. (2017)] report an approach using cold-spray technology to coat polyvinyl alcohol (PVA) in polymer and hydrogel states with hydroxyapatite (HA). Using porous aggregated HA powder, they hypothesized that fragmentation of the powder upon cold spray could lead to the formation of a ceramic coating on the surface of the PVA substrate. However, direct spraying of this powder led to the complete destruction of the swollen PVA hydrogel substrate. As an alternative, HA coatings were successfully produced by spraying onto dry PVA substrates prior to swelling in water. Dense homogeneous HA coatings composed of submicron particles were obtained using rather low-energy spraying parameters (temperature 200–250 °C, pressure 1–3 MPa). Coated PVA substrates could swell in water without the removal of the ceramic layer to form HA-coated hydrogels. Microscopic observations and in situ measurements were used to explain how local heating and the impact of sprayed aggregates induced surface roughening and strong binding of HA particles to the molten PVA substrate. Such an approach could lead to the design of ceramic coatings whose roughness and crystallinity can be finely adjusted to improve interfacing with biological tissues.

6.6.3 Composites

Such coatings are mainly used for wear protection. They have been developed, for example, to spray the following materials:

6.6.3.1 Pure Iron (99.5%) or Stainless Steel (304 L) Reinforced by Diamond

The composites Fe (5–25 μm) and SUS304 stainless steel (5–45 μm) contained, respectively, 46 and 12% diamond fraction [Kim et al. (2006)]. The coating adhesion on an Al alloy (ASTM 6061) was over 100 MPa. Some of the diamond particles within coatings were fractured. Some of these coatings were coated with Ti.

6.6.3.2 Aluminum and Copper

Aluminum has been reinforced by different ceramic particles. Aluminum (6061-T6) was reinforced by B_4C (up to 20 vol.%) [Haynes et al. (2006)]. Post-deposition heat treatments and mechanical property tests were performed, and results indicated that adequate strength and ductility were obtained. However, the elastic modulus was lower than expected, probably due to B_4C particles fracturing upon impact. Aluminum coatings have also been reinforced with SiC [Sansoucy et al. (2008), (2007b)]. The composite coating microstructure, thickness, porosity, and composition were compared to an Al-12Si coating. The results show that 45% of the SiC mixed with the aluminum matrix was retained in the composite coatings. These SiC particles exhibit a reasonably uniform distribution within the aluminum matrix. The coating thickness and adhesion strength were not affected by the SiC content. However, the coatings became more porous on increasing the SiC volume fraction of the blended powders [Sansoucy et al. (2007b)]. [Yong et al. (2004a, b)] have successfully cold-sprayed SiC and Al_2O_3 with Al. The adhesion and compactness

depended on the quantities and sizes of the starting materials.

It was experimentally shown that when soft metals (Al, Cu) were sprayed with fine ceramic powders, the "critical" temperature of the metals decreased compared to pure metals. Ceramic powders of fine fractions produce a strong activation effect on the spraying process and enhance the coating growth, while ceramic powders of large fractions considerably hinder coating growth [Klinkov et al. (2008), Sova et al. (2010)]. The experiments demonstrated the importance of the ceramic particle velocity for the cermet coating formation. Stable cermet coating formation is possible if the ceramic particles are accelerated to an optimal level: ceramic particles accelerated to a high velocity penetrate the coating, while low-velocity ceramic particles rebound from its surface. [Irissou et al. (2007)] showed that the addition of Al_2O_3 to Al powder increased the adhesion of the coating on the substrate (up to 60 MPa for samples produced with more than 30 wt.% Al_2O_3) due to the creation of micro-asperities by the hard-ceramic particles. Moreover, the inclusion of alumina particles in the aluminum coatings had no detrimental effect on the corrosion protection of the substrate. [Poirier et al. (2011)] used cold spray to consolidate milled Al and Al_2O_3/Al nano-composite powders, as well as the initial un-milled and un-reinforced Al powder. Results showed that the large increase in the hardness of the Al powder after mechanical milling is preserved after cold spraying. A good-quality coating with low porosity is obtained from milled Al. However, the addition of Al_2O_3 to the Al powder during milling decreased the powder and coating nano-hardness. The coating produced from the milled Al_2O_3/Al mixture also showed lower particle cohesion and higher amount of porosity.

Aluminum and Al6061 aluminum-based alloys Al_2O_3 particle-reinforced composite coatings were sprayed by [Spencer et al. (2009)] on AZ91E Mg substrates using cold spray. The strength of the coating/substrate interface in tension was found to be higher than that of the coating itself. The coatings have corrosion resistance similar to that of bulk pure aluminum in both salt spray and electrochemical tests. The wear resistance of the coatings is significantly better than that of the AZ91 Mg substrate, but the significant result is that the wear rate of the coatings is several decades lower than that of various bulks Al alloys tested for comparison. [Kong et al. (2010)] prepared by cold-spraying a novel $TiAl_3$-Al coating onto Ti-22Al-26Nb (atm.%) substrate for high temperature protection. The coating was sprayed with pre-mixed Al and Ti powders. The as-sprayed specimens were subjected to heat treatment (630 °C for 5 h) under Argon gas flow of 40 mL/min. The composite coating was mainly composed of $TiAl_3$ embedded in the matrix of residual aluminum. The oxidation test indicated that this composite coating was very effective in improving the high-temperature oxidation resistance of the substrate alloy at 950 °C in the 150 tested cycles without any sign of degradation.

[Van Steenkiste (2006)] produced composite coatings with different rare-earth iron alloys (REFe2) and several ductile matrices. Composite coatings of Terfenol-D [(Tb$_{0.3}$Dy$_{0.7}$)Fe$_{1.9}$] and SmFe$_2$ were combined with ductile matrices of aluminum, copper, iron, and molybdenum. Evidence of an induced magnetic coercivity was measured for the REFe2-Mo (H_{ci} = 3.7 k Oe) and Fe composite coatings. Coatings were produced on flat substrates and shafts. $Nd_2Fe_{14}B$ permanent magnet/aluminum composite coatings were produced by cold-spray deposition by [King et al. (2007)]. Isotropic $Nd_2Fe_{14}B$ powder was blended with aluminum powder to make mixtures of 20–80 vol.% $Nd_2Fe_{14}B$, and these mixtures were sprayed at temperatures of 200–480 °C. The hard $Nd_2Fe_{14}B$ particles tended to fracture and fragment upon impact, while aluminum underwent strong plastic deformation, eliminating pores and trapping $Nd_2Fe_{14}B$ within the coating. It was found that higher spray temperatures and finer $Nd_2Fe_{14}B$ particle sizes improved their retention rate. The magnetic properties of $Nd_2Fe_{14}B$ remained unaffected by the cold-spray process.

6.6.3.3 Aluminum and Silicon

[Hao et al. (2016)] used the cold-spray technique to deposit Al-12Si coatings on AZ31 magnesium alloy. The influence of gas pressure and gas temperature on the microstructure of coatings was investigated so as to optimize the process parameters. OM, SEM, and XRD were used to characterize the as-sprayed coatings. Mechanical properties including Vickers microhardness and adhesion strength were measured in order to evaluate coating quality. Test results indicate that the Al-12Si coatings possess the same crystal structure with powders, sufficient thickness, low porosity, high hardness, and excellent adhesion strength under optimal cold-spray process parameters.

6.6.3.4 Fabrication of Cermet Coatings

WC–Co with nano- and micro-structured feedstock powders by cold-spray deposition [Kim et al. (2005)]. There was no detrimental transformation and/or decarburization of WC by cold-spray deposition. It was observed that nano-sized WC in the feedstock powder is maintained in the cold-spray deposits. It seems that nano-sized WC is advantageous over micron-sized WC in cold-spray deposition because higher particle velocities can be obtained with the same gas velocity. [Kim et al. (2005)] deposited by the cold-spray process using helium with reasonable powder preheating WC–12%Co powders with nano-sized WC. The results showed that there were no detrimental phase transformation and/or decarburization of WC, and the nano-sized nature of the WC in the feedstock powder was maintained in the coatings. The porosity was low and the hardness high ($\sim$2050 HV$_5$). [Kim et al. (2005)] cold-sprayed the same particles using nitrogen and helium gases. Best results were obtained with nanometer-sized WC particles. Powder preheating was one of the most

crucial parameters for the cold-spray deposition from the commercial production point of view.

The cold-spray process was successful in applying Cr_3C_2-25wt.%NiCr and Cr_3C_2-25wt.%Ni coatings on 4140 alloys for wear-resistant applications [Wolfe et al. (2006)]. The average Vickers hardness of the Cr_3C_2-based coatings ranges from 300 to 900 HV_3. This strongly suggests that the hardness (and thus wear resistance) can be tailored over a wide range of values to meet a variety of hardness specifications.

In order to improve the wear and corrosion resistance of commonly used magnesium alloys, Chen et al. (2017) cold-sprayed 316 L stainless steel and 316 L-SiC composite coatings on commercial AZ80 magnesium alloy. The microstructure, hardness, and bonding strength of as-sprayed coatings were studied, and their tribological properties such as sliding against Si_3N_4 and GCr_{15} steel under unlubricated conditions were evaluated by a ball-on-disk tribometer. Corrosion behaviors of coated samples were also evaluated and compared to that of an uncoated magnesium alloy substrate in 3.5 wt.% NaCl solution by electrochemical measurements. Scanning electron microscopy was used to characterize the corresponding wear tracks and corroded surfaces to determine wear and corrosion mechanisms. The results showed that the as-sprayed coatings possessed higher microhardness and more excellent wear resistance than the magnesium alloy substrate. Meanwhile, the 316 L and 316 L-SiC coating also reduced the corrosion current density of the magnesium alloy, and the galvanic corrosion of the substrates was not observed after 200-h neutral salt spray exposure, which demonstrated that the corrosion resistance of a magnesium alloy substrate could be greatly improved by cold-sprayed stainless steel-based coatings.

6.6.3.5 Fe-Al Inter-Metallic Compounds

At last, Fe-Al inter-metallic compounds have been produced using a ball-milled metastable Fe-Al-WC powder, spraying it by cold spray and then annealing the coatings at a temperature above 500 °C to form Fe-Al intermetallic compounds. The as-sprayed coating hardness was 648 HV, and it decreased upon annealing only if the corresponding temperature was above 550 °C [Wang et al. (2006)].

6.6.3.6 Cermets

[Sova et al. (2009)] experimentally studied the influence of the ceramic particle size on the process of formation of cermet coatings by cold spray. An especially developed nozzle with separate injection of ceramic and metal powders into the gas stream was used in the experiments. The results obtained demonstrate that fine ceramic powders (Al_2O_3, SiC) produce a strong activation effect on the process of spraying soft metal (Al, Cu) and increase the deposition efficiency of the metal component of the mixture compared to the pure metal spraying. It has also been experimentally shown that the addition of fine hard powder to soft metals such as Al and Cu results in a significant reduction of the "critical" temperature for spraying these metals (the minimum gas stagnation temperature at which a nonzero particle deposition is observed). It should be noted, on the other hand, that the addition of coarse ceramic powder can produces a strong erosion effect that considerably reduces the coating mass growth and deposition efficiency of the metal component.

6.6.3.7 Ceramics

To the best of our knowledge, TiO_2 is the only ceramic material that has been cold-sprayed. In spite of its non-ductile behavior, it can be successfully sprayed if a powder is prepared by agglomerating ultra-fine (between 10 to 100 mm) primary particles, with such agglomerated particles presenting a "ductile" behavior. The interest in TiO_2 lies in its photo catalytic properties, and a pure anatase crystalline structure is used as feedstock. The advantage of the cold-spray process is the avoidance of transformation of anatase by avoiding the heating present in HVOF-sprayed titanium or plasma spray processes. [Yang et al. (2008)] successfully deposited through cold-sprayed TiO_2 photo catalytic films up to 15 μm thick with a rough surface and porous structure, as it could be expected. The photocatalytic performance was examined through acetaldehyde degradation under ultraviolet illumination, for which the film was very active. [Yamada et al. (2010)] also cold-sprayed TiO_2. They showed that the process gas used was not an important factor in fabricating anatase TiO_2 coatings, and the use of inexpensive nitrogen instead of helium was an important economic factor. The thickness of coatings increased with increase in process gas temperature and the adhesion strength on the Cu substrate reached 25 MPa at nitrogen pressure of 1 MPa. The powder preparation was a key factor in achieving good coatings. They used a simple hydrolysis method of titanyl sulfate ($TiOSO_4$) in distilled water with a small addition of inorganic salt [Tjitra Salim et al. (2011)]. The solution was stirred on a hot plate to hold the temperature at 120 °C for 8 h after which a white precipitate was formed. The precipitate was then dried in an oven for 10 h at 120 °C to obtain a powder of pure anatase [Tjitra Salim et al. (2011)]. The powder was then agglomerated with fine nano primary particles, with different post-synthesis treatments leading to different TiO_2 nanostructures but without alteration of the anatase phase. Annealing at 600 °C for 1 h caused a significant growth in primary particles with the existence of internal pores within a particle. On the other hand, hydrothermal treatment, performed by soaking the TiO_2 powder with distilled water in an autoclave and heat-treating it at 150 °C for 5 h produced a unique oriented agglomeration structure where the primary particles were agglomerated in one single crystal axis. Figure 6.62 shows the powder morphologies obtained by the FE-SEM.

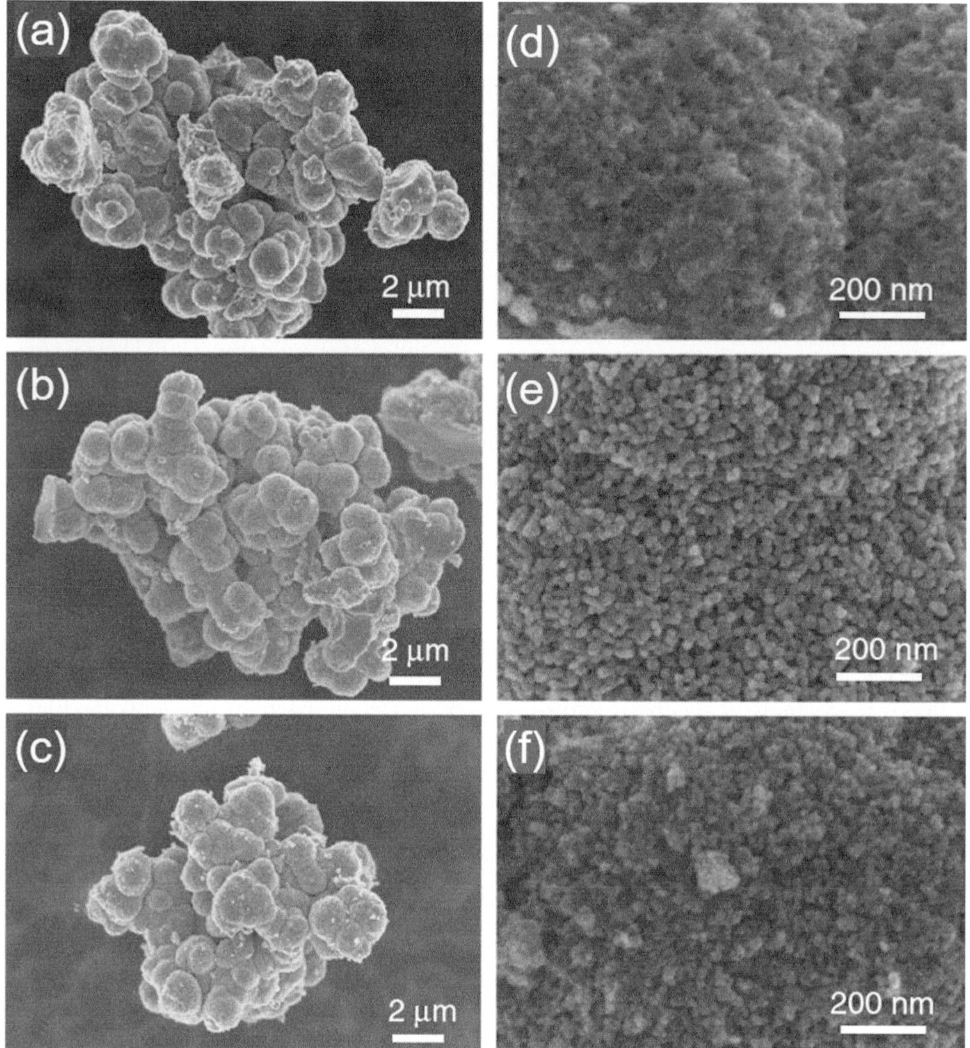

Fig. 6.62 FE-SEM images taken at low magnification of (**a**) as-synthesized, (**b**) annealed, and (**c**) hydrothermal treated TiO$_2$ powders; and at high magnification of (**d**) as-synthesized, (**e**) annealed, and (**f**) hydrothermal treated TiO$_2$ powders [Tjitra Salim et al. (2011)]. Reprinted with kind permission from Elsevier

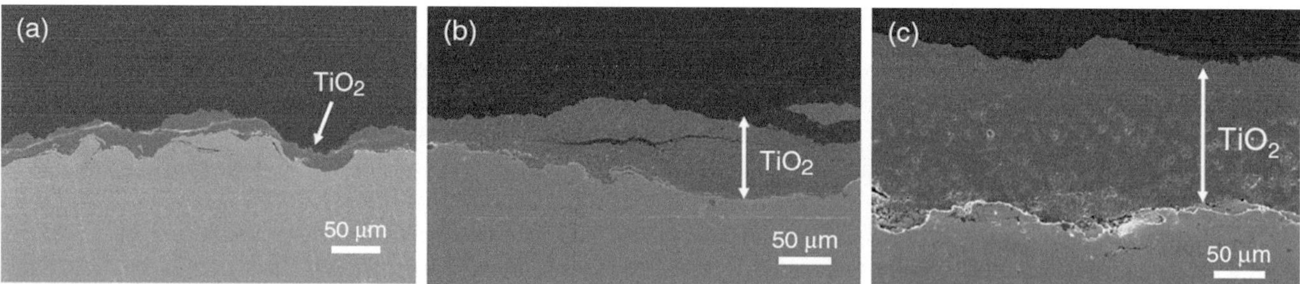

Fig. 6.63 Cold-sprayed coatings (same spray conditions) of (**a**) as synthesized, (**b**) annealed, and (**c**) hydrothermally treated TiO$_2$ powders deposited on copper [Tjitra Salim et al. (2011)]. Reprinted with kind permission from Elsevier

From the TEM images [Tjitra Salim et al. (2011)], the particle sizes of as-synthesized, hydrothermally treated, and annealed powders were determined to be about 3, 5 and 25 nm, respectively. Coatings were achieved with the three powders and results are presented in Fig. 6.63. With as-synthesized particles, only embedment was observed (Fig. 6.63a), while the other two powders formed coatings (Fig. 6.63b and c), with the hydrothermally treated TiO$_2$ being the thicker one with about 150 μm (Fig. 6.63c).

The photo-catalytic activity of the coatings was similar to or better than that of the feedstock powder due to the formation of a large reaction area. To conclude this work, Yamada et al. [Yamada et al. (2010), Tjitra Salim et al. (2011)] concluded that the three major factors in achieving TiO_2 coatings on different substrates (with hardness between 45 and 700 HV) are the following:

- Porous agglomerates with sizes between 10 and 20 µm.
- Nano-scaled primary particles agglomerated with a porous structure.
- Primary particles agglomerated and oriented with a single crystal axis.

6.7 Immerging Technologies and Applications of Cold Spray

6.7.1 Low Pressure Cold Spray (LPCS)

A few coatings, such as Cu-based W-reinforced and Ni-based TiC reinforced composite coatings, have been produced [Beneteau et al. (2006)] to increase the load-bearing capacity, and hence the load and sliding speed range within which the dry sliding wear is low.

Near-net shape parts have been produced with Al particles (150–45 µm) and small Al_2O_3 particles (5–10 µm). Coatings presented very low residual porosity (more than 97% of theoretical density) [Weinert et al. (2006)]. However, if coating properties were close to those of wrought aluminum, they were less ductile. The hardness was in the range 90–100 $HV_{0.1N}$.

The work of [Irissou et al. (2007)] was related to the influence of the Al particle size and Al_2O_3 mass fraction in Al-Al_2O_3 powder mixtures on the coating deposition and properties. For the two Al powders used, d_{50} was either 91 or 36 µm. Because the mean velocity of the particles was a function of their size, the Al powder with the larger particle size distribution had a significantly smaller volume of particles with velocities higher than the critical one. The deposition efficiency was consequently lower. However, coatings with the starting powder based on the larger Al particles were harder than the coatings made with the smaller size Al powder mixtures. This was likely due to the more important peening effect of the large particles due to their higher kinetic energy. The addition of Al_2O_3 ($d_{50} = 25.5$ µm) to the Al powders helped improving the coating deposition. Optimal deposition efficiency was found for a mass fraction of about 30% of Al_2O_3 in the starting powder. Abrasion resistance was found to be independent of the alumina mass

fraction in the coatings. Al_2O_3 particles played only the role of peening and roughening the layers during deposition. However, their inclusion in the coatings was limited to about 25 wt.%. The bonding between Al and Al_2O_3 was therefore believed to be weak and resulted in poor cohesion between the two powder types. This limited any possible improvement of the abrasion resistance of the composite coatings. The adhesion was higher than 60 MPa for samples produced with more than 30 wt.% Al_2O_3. The Al-Al_2O_3 coatings proved to be as efficient as pure Al coatings in providing corrosion protection against alternating immersion in saltwater and exposure to salt spray. The inclusion of alumina particles in the aluminum coatings had no detrimental effect on corrosion protection [Irissou et al. (2007)].

[Ogawa et al. (2008)] have studied the possibility of using the low-pressure cold-spray technique of aluminum-on-aluminum alloys instead of welding for repairing cracks. Their work was focused on the active newly formed surface after the aluminum particle deposition. They found that for efficient particle deposition, it was necessary to activate the newly formed surface by several impingements because the existence of inactive native oxide films had an adverse effect on the deposition. Furthermore, the strength of a cold-sprayed specimen was found to be higher than that of a cold-rolled specimen under compressive loading.

[Koivuluoto et al. (2008)] have used the low-pressure cold-spray process (0.9 MPa and preheating to 650 °C) to spray metallic powders with alumina particles. Soft metallic coatings (Cu, Ni, and Zn) with a ceramic hard phase can be used for different application areas, for example, thick coatings and coatings for electrical and thermal conduction and corrosion protection applications. LPCS coatings seemed to be dense according to (SEM), but through-porosity was observed in structures through corrosion tests. Bond strengths of LPCS Cu and Zn coatings were found to be 20–30 MPa, and hardness was high, indicating reinforcement and work hardening.

[Koivuluoto and Vuoristo (2010b)] have shown that the powder type and composition have a very important role in the production of metallic and metallic-ceramic coatings by using the low-pressure cold-spray process. They have evaluated the effect of different particle types of Cu powder and different compositions of added Al_2O_3 particles on the microstructure, fracture behavior, denseness, and mechanical properties. Spherical and dendritic Cu particles were tested together with 0, 10, 30, and 50 vol.% Al_2O_3 additions. Coating denseness and particle deformation level increased with the hard particle addition. Furthermore, hardness and bond strength increased with increasing Al_2O_3 fractions. In the comparison between different powder

types, spherical Cu particles led to denser and less oxide-containing coatings structure due to highly deformed particles.

[Kulmala and Vuoristo (2008)] have applied the laser-assisted cold spraying (see Sect. 6.2.3.2) to low-pressure cold spraying and called it Laser Assisted Low Pressure Cold Spray (LALPCS). The purpose of the additional energy from the laser beam is to create denser and more adherent coatings, enhance deposition efficiency, and increase the variety of coating materials. They studied copper and nickel powders with additions of alumina powder on carbon steel. Coatings were sprayed using air as process gas. A 6 kW continuous-wave high-power diode laser and a low-pressure cold-spray unit were used in the experiments. The coating density was tested with open cell potential measurements. Results showed that laser irradiation improved the coating density and also enhanced deposition efficiency.

Aluminum components are difficult to repair by welding due to its high specific thermal conductivity and high coefficient of thermal expansion. The low-pressure cold-spray technique can be used instead of welding for repairing cracks, as shown by [Ogawa et al. (2008)]. These authors have focused their study to the effect of surface conditions on aluminum particle deposition and mechanical properties of resulting coatings. They found that for efficient particle deposition, it was necessary to obtain an active newly formed surface of the substrate and particle surfaces by several impingements because of the existence of inactive native oxide films. Furthermore, the strength of a cold-sprayed specimen was found to be higher than that of a cold-rolled specimen under compressive loading.

Al-10Sn and Al-20Sn binary alloy coatings have been successfully deposited onto Al6061, copper, and SUS304 substrates with the low-pressure cold-spray process with He as propellant gas. However, with particle sizes between 5 and 40 μm, the deposition efficiency was 11.2%.

6.7.2 Additive Manufacturing

[Sova et al. (2013)] discussed in this paper the application of Cold-Spray (CS) coating deposition technology as an additive manufacturing technique. Absence of material melting during CS deposition permits to obtain deposits with low value of residual stresses and to preserve the phase composition of the source material, which is a very important advantage. In this paper, the latest developments in the field of cold spray, such as a micronozzle device and a new multi-material deposition approach permitting to significantly enlarge the potential of cold spray as an additive manufacturing technology is discussed.

6.8 Summary and Conclusions

Cold spray processes are rapidly developing technologies that address significant shortcomings of the other thermal spray processes. The technology is based on the concept of accelerating the sprayed material in the form of fine powder to relatively high velocities using a "cold" high pressure gas. On impact on the substrate, the individual powder particle goes through ductile deformation forming a dense coating with good to excellent adhesion to the substrate. Cold spray technology is consequently applicable only to the spraying of ductile materials such as metals or alloys and a limited number of cermets. The technology can be used with high pressure nitrogen or helium as process gas, with best results obtained with helium at gas flow rates as high as a few m^3/min. For economic reasons, the need for high flow rates of helium as process gas imposes on the technology the need of operation in a closed vessel with 90–95% recycling of the process gas. To avoid liquefaction of the spray gas when accelerated in a Laval type nozzle as well as for the improvement of the ductility of impacting particles, the process gas is often heated to temperatures up to 650 $^{\circ}$C depending on the sprayed material (for example, 120 $^{\circ}$C for Al and 650 $^{\circ}$C for MCrAlY). Further improvements of coating qualities can also be achieved through the preheating of the sprayed powders, and/or the substrate or previously deposited layers with a laser just prior to particle deposition. In parallel to the relatively high-pressure (1-4 MPa) cold spray guns, equipment has been developed to spray with air (or nitrogen) as gas, at a working pressure below 1 MPa and a rather low air flow rate (< 0.4 m^3/min) requiring less heating energy (<10 kW). These low-pressure guns are well suited for short-run production. As most particles do not reach the critical velocity, their compaction must be optimized through the selection of powder types and powder particle size distributions, multicomponent mixtures (metal and ceramic) being generally used.

The main advantage cold spray technologies lie in their ability to spray high-purity metals on oxidation-sensitive substrates, with no modification of their oxide content. Coating properties are also close to those of wrought materials with increasingly high-definition spraying capabilities that makes them ideal for high end applications in a wide range of industries. The first and very successful application of cold spray is for copper metallization. Other industrial-scale applications include its use for metal restoration and sealing, engine blocks, castings, molds and dies, refrigeration equipment, heat exchangers, aluminum piston heads, manifolds, disk brakes, heat sinks for microelectronics (Al and Cu), solid lubricant matrix with base metals, electronically conductive coatings (Cu or Al on ceramic or polymeric components), electromagnetic shielding, as well as localized corrosion protection (Zn or Al coatings).

Nomenclature

Units are indicated in parentheses; when no units are indicated, the parameter is dimensionless.

Latin Alphabet

$A*$	throat area (m^2)
A_e	exit area (m^2)
a_s	particle cross-section (m^2)
c	speed of sound (m/s)
C_D	drag coefficient ($-$)
c_p	specific heat (J/K.kg)
d_p	particle diameter (m)
$\dot{m}_g$	gas mass flow rate (kg/s)
m_p	particle mass (kg)
M	Mach number
Me	nozzle exit Mach number
M_o	mass of the consumed powder (kg)
M_g	molar mass of the gas (kg/mol)
p_b	back pressure
p_o	stagnation pressure (kPa)
$p*$	pressure at the nozzle throat (Pa)
p_{er}	probability of erosion
p_c	probability of deposition
r_p	particle radius (m)
R	universal gas constant $R = 8.32$ (J/K.mol)
R_u	universal gas constant, $R_u = R/M_g$ (J/K.kg)
s_p	particle correctional area (m^2)
T	Temperature (K)
T_o	initial gas temperature (K)
T_o	stagnation temperature (K
$T*$	temperature at the nozzle throat (K)
T_e	exit jet température (K)
T_{pi}	impact temperature (K),
T_m	melting temperature (K),
T_p	particle temperature (K)
T_{po}	initial particle temperature (K)
T_R	reference temperature (293 K),
u_g	gas velocity (m/s)
u_p	particle velocity (m/s)
$v*$	gas velocity at nozzle throat (m/s)
v_{cr}	critical particle impact velocity (m/s)
v_{er}	particle erosion velocity (m/s)
v_{pi}	particle impact velocity (m/s)
x	distance travelled by the particle (m)

Greek Alphabet

Δm	increment of mass of deposited coating (kg)
γ	specific heat ratio ($\gamma = c_p/c_v$)
ρ_g	specific mass of the gas (kg/m^3)
$\rho*$	specific mass of gas at nozzle throat (m/s)
ρ_p	specific mass of the particle (kg/m^3)
σ_{TS}	tensile strength (MPa)

References

Ajdelsztajn, L., A. Zúñiga, B. Jodoin, and E.J. Lavernia. 2006a. Cold-spray processing of a Nanocrystalline Al-Cu-Mg-Fe-Ni alloy with Sc. *Journal of Thermal Spray Technology* 15 (2): 184–190.

Ajdelsztajn, L., B. Jodoin, P. Richer, E. Sansoucy, and E.J. Lavernia. 2006b. Cold gas dynamic spraying of Iron-based amorphous alloy. *Journal of Thermal Spray Technology* 15 (4): 495–500.

Alexandre, S., T. Laguionie, and B. Baccaud. 2007. Realization of an internal cold spray coating of stainless steel in an aluminum cylinder. In *Thermal spray 2007: Global coating solutions*, ed. B.R. Marple et al. Materials Park, Ohio: (Pub.) ASM International®. e-proc.

Alkhimov, A.P., V.F. Kosarev, and A.N. Papyrin. 1998. Gas dynamic spraying. An experimental study of the spraying process. *Journal of Applied Mechanics and Technical Physics* 39 (2): 318–323.

Alkhimov, A.P., S.V. Klinkov, and V.F. Kosarev. 2000. Study of heat exchange of supersonic plane jet with obstacle at gas-dynamic spraying. *Thermophysics and Aeromechanics* 7 (3): 375–382.

Alkhimov, A.P., V.F. Kosarev, and S.V. Klinkov. 2001. The features of cold spray nozzle design. *Journal of Thermal Spray Technology* 10 (2): 375–381.

Alkhimov, A.P., S.V. Klinkov, S.M. Shin, C.H. Lee, and H.J. Kim. 2006. Effect of particle temperature on the critical velocity for particle deposition by kinetic spraying. In *Thermal spray 2006*, ed. B. Marple. Materials Park, Ohio: (pub.) ASM International. e-proc.

Alkimov A.P., V.F. Kosarev, N.I. Nesterovich, and A.N. Papyrin (1990) "Method of Applying Coatings," Russian Patent No. 1618778, Sept 8, 1990 (Priority of the Invention: Jun 6, 1986).

Alkimov A.P., A.N. Papyrin, V.F. Kosarev, N.I. Nesterovich, and M.M. Shushpanov (1994) "Gas Dynamic Spraying Method for Applying a Coating," U.S. Patent No. 5,302,414, Apr 12, 1994, Re-examination Certificate, Feb 25, 1997.

——— (1995) "Method and Device for Coating," European Patent No. 0484533, Jan 25, 1995.

——— (1997) "Gas Dynamic Spraying Method for Applying a Coating," U.S. Patent No. 5,302,414, Apr 12, 1994, Re-examination Certificate, Feb 25, 1997.

Al-Mangour, B., P. Vo, R. Mongrain, E. Irissou, and S. Yue. 2014. Effect of heat treatment on the microstructure and mechanical properties of stainless steel 316L coatings produced by cold spray for biomedical applications. *Journal of Thermal Spray Technology* 23 (4): 641–652.

Assadi, H., F. Gärtner, T. Stoltenhoff, and H. Kreye. 2003. Bonding mechanism in cold gas spraying. *Acta Materialia* 51: 4379–4394.

Bala, N., H. Singh, and S. Prakash. 2010. Accelerated hot corrosion studies of cold spray Ni–50Cr coating on boiler steels. *Materials and Design* 31: 244–253.

Balani, K., T. Laha, A. Agarwal, J. Karthikeyan, and N. Munroe. 2005. Effect of carrier gases on microstructural and electrochemical behavior of cold-sprayed 1100 aluminum coating. *Surface & Coatings Technology* 195: 272–279.

Beneteau, M., W. Birtch, J. Villafuerte, J. Paille, M. Petrocik, R.Gr. Maev, E. Strumban, and V. Leshchinsky. 2006. Gas dynamic spray composite coatings for Iron and steel castings. In *Proceedings of the 2006 thermal spray conference*, ed. B. Marple et al. Materials Park, OH: (pub.) ASM Int. e-proceedings.

Bérubé, G., M. Yandouzi, A. Zuniga, L. Ajdelsztajn, J. Villafuerte, and B. Jodoin. 2012. Phase stability of Al-5Fe-V-Si coatings produced by cold gas dynamic spray process using rapidly solidified feedstock materials. *Journal of Thermal Spray Technology* 21 (2): 240–254.

Binder, K., J. Gottschalk, M. Kollenda, F. Gärtner, and T. Klassen. 2011. Influence of impact angle and gas temperature on mechanical

properties of titanium cold spray deposits. *Journal of Thermal Spray Technology* 20 (1–2): 234–242.

Blose, R.E., T.J. Roemer, A.J. Mayer, D.E. Beatty, and A.N. Papyrin. 2003. Automated cold spray system: Description of equipment and performance data. In *Thermal spray 2003: Advancing the Science & Applying the technology*, ed. C. Moreau and B. Marple, 103–111. Materials Park, Ohio: (Pub) ASM International.

Blose, R.E., B.H. Walker, and S.H. Froes. 2006. Depositing titanium alloy additive features to forgings and extrusions using the cold spray process. In *Proceedings of the 2006 thermal spray conference*, ed. B. Marple et al. Materials Park, OH: (pub.) ASM Int. e-proceedings.

Bobzin, K., L. Zhao, F. Ernst, and K. Richardt. 2008. Flux-free brazing of mg-containing aluminumalloys by means of cold spraying. *Frontiers of Mechanical Engineering in China* 3 (4): 355–359.

Boro, Djordjevic B., and R.Gr. Maev. 2006. SIMAT application for aerospace corrosion protection and structural repair. In *ITSC 2006*, ed. B. Marple. Materials Park OH: (pub.) ASM Int. e-proceedings.

Bray, M., A. Cockburn, and W. O'Neill. 2009. The laser-assisted cold spray process and deposit characterization. *Surface & Coatings Technology* 203: 2851–2857.

Bush, T.B., Z. Khalkhali, V. Champagne, D.P. Schmidt, and J.P. Rothstein. 2017. Optimization of cold spray deposition of high-density polyethylene powders. *Journal of Thermal Spray Technology* 26: 1548–1564.

Cai, Z., S. Deng, H. Liao, C. Zeng, and G. Montavon. 2014. The effect of spray distance and scanning step on the coating thickness uniformity in cold spray process. *Journal of Thermal Spray Technology* 23 (3): 354–362.

Calla, E., D.G. McCartney, and P.H. Shipway. 2006. Effect of deposition conditions on the properties and annealing behavior of cold-sprayed copper. *Journal of Thermal Spray Technology* 15 (2): 255–262.

Cavaliere, P., and A. Silvello. 2017. Crack repair in aerospace aluminum alloy panels by cold spray. *Journal of Thermal Spray Technology* 26: 661–670.

Cavaliere, P., A. Perrone, and A. Silvello. 2014. Processing conditions affecting grain size and mechanical properties in nanocomposites produced via cold spray. *Journal of Thermal Spray Technology* 23 (7): 1089–1096.

Champagne, V., ed. 2007. *The cold spray materials deposition process: Fundamentals and applications*. Cambridge: Woodhead pub.

Champagne, V.K., Jr., D. Helfritch, P. Leyman, S. Grendahl, and B. Klotz. 2005. Interface material mixing formed by the deposition of copper on aluminum by means of the cold spray process. *Journal of Thermal Spray Technology* 14 (3): 330–335.

Chang-Jiu, Li, and Wen-Ya Li. 2003. Deposition characteristics of titanium coating in cold spraying. *Surface and Coatings Technology* 167: 278–283.

Li Chang-Jiu, Wen-Ya Li and H. Fukanuma (2004) Impact fusion phenomenon during cold spraying of zinc, in ITSC 2004 (pub.) DVS, Dusseldorf, Germany, e-proceedings.

Chang-Jiu, Li, Wen-Ya Li, Yu-Yue Wang, Guan-Jun Yang, and H. Fukanuma. 2005. A theoretical model for prediction of deposition efficiency in cold spraying. *Thin Solid Films* 489: 79–85.

Chang-Jiu, Li, Wen-Ya Li, and Hanlin Liao. 2006a. Examination of the critical velocity for deposition of particles in cold spraying. *Journal of Thermal Spray Technology* 15 (2): 212–222.

———. 2006b. Examination of the critical velocity for deposition of particles in cold spraying. *Journal of Thermal Spray Technology* 15 (2): 212–222.

Che, H., X. Chu, P. Vo, and S. Yue. 2018. Metallization of various polymers by cold spray. *Journal of Thermal Spray Technology* 27: 169–178.

Chen, W.R., E. Irissou, X. Wu, J.-G. Legoux, and B.R. Marple. 2011. The oxidation behaviour of TBC with cold spray CoNiCrAlY bond coat. *Journal of Thermal Spray Technology* 20 (1–2): 132–138.

Chen, J., B. Ma, G. Liu, H. Song, J. Wu, L. Cui, and Z. Zheng. 2017. Wear and corrosion properties of 316L-SiC composite coating deposited by cold spray on magnesium alloy. *Journal of Thermal Spray Technology* 26: 1381–1392.

Christoulis, D.K., S. Guetta, V. Guipont, and M. Jeandin. 2011. The influence of the substrate on the deposition of cold-sprayed titanium: An experimental and numerical study. *Journal of Thermal Spray Technology* 20 (3): 523–533.

Cinca, N., M. Barbosa, S. Dosta, and J.M. Guilemany. 2010. Study of Ti deposition onto Al alloy by cold gas spraying. *Surface & Coatings Technology* 205: 1096–1102.

Davis J.R. 2004. *Handbook of thermal spray technology*. ASM International, Materials Park (pub.).

DeForce, B.S., T.J. Eden, and J.K. Potter. 2011. Cold spray Al-5% mg coatings for the corrosion protection of magnesium alloys. *Journal of Thermal Spray Technology* 20 (6): 1352–1358.

Delloro, F., M. Jeandin, D. Jeulin, H. Proudhon, M. Faessel, L. Bianchi, E. Meillot, and L. Helfen. 2017. A morphological approach to the modeling of the cold spray process. *Journal of Thermal Spray Technology* 26: 1838–1850.

Donner, K.-R., F. Gaertner, and T. Klassen. 2011. Metallization of thin Al_2O_3 layers in power electronics using cold gas spraying. *Journal of Thermal Spray Technology* 20 (1–2): 299–306.

Dykhuizen, R.C., and R.A. Neiser. 2003. Optimizing the cold spray process. In *ITSC 2003*, ed. C. Moreau and B. Marple, 19–26. Materials Park, OH: (pub.) ASM Int.

Dykhuizen, R.C., and M.F. Smith. 1998. Gas dynamic principles of cold spray. *Journal of Thermal Spray Technology* 7 (2): 205–212.

Dykhuizen, R.C., M.F. Smith, D.L. Gilmore, R.A. Neiser, X. Jiang, and S. Sampath. 1999. Impact of high velocity cold spray particles. *Journal of Thermal Spray Technology* 8 (4): 559–564.

Eason, P.D., J.A. Fewkes, S.C. Kennett, T.J. Eden, K. Tello, M.J. Kaufman, and M. Tiryakioglu. 2011. On the characterization of bulk copper produced by cold gas dynamic spray processing in as fabricated and annealed conditions. *Materials Science and Engineering A* 528: 8174–8178.

Fan, S.-Q., G.-J. Yang, G.-J. Liu, C.-X. Li, C.-J. Li, and L.Z. Zhang. 2006. Characterization of microstructure of TiO_2 coating deposited by vacuum cold spraying. In *Proceedings of the 2006 thermal spray conference*, ed. B. Marple et al. Materials Park, OH: (pub.) ASM Int. e-proceedings.

Frost, H.J., and M.F. Ashby. 1982. *Deformation mechanism maps: The plasticity and creep of metals and ceramics*. Oxford: Pergamon Press.

Fukanama, H., N. Ohno, B. Sun, and R. Huang. 2006. The influence of particle morphology on in-flight particle velocity in cold spray. In *Thermal spray 2006*, ed. B. Marple. Materials Park, Ohio: (pub.) ASM International. e-proc.

Fukumoto, M., H. Wada, K. Tanabe, M. Yamada, E. Yamaguchi, A. Niwa, M. Sugimoto, and M. Izawa. 2007. Effect of substrate temperature on deposition behavior of copper particles on substrate surfaces in the cold spray process. *Journal of Thermal Spray Technology* 16 (5–6): 643–650.

Fukumoto, M., M. Mashiko, M. Yamada, and E. Yamaguchi. 2010. Deposition behavior of copper fine particles onto flat substrate surface in cold spraying. *Journal of Thermal Spray Technology* 19 (1–2): 89–94.

Gao, P.-H., C.-J. Li, G.-J. Yang, Y.-G. Li, and C.-X. Li. 2008. Influence of substrate hardness on deposition behavior of single porous WC-12Co particle in cold spraying. *Surface & Coatings Technology* 203: 384–390.

Gardon, M., and J.M. Guilemany. 2014. Milestones in functional titanium dioxide thermal spray coatings: A review. *Journal of Thermal Spray Technology* 23 (4): 577–595.

Gärtner, F., C. Borchers, T. Stoltenhoff, and H. Kreye. 2003. Numerical and microstructural investigations of the bonding mechanisms in cold spraying. In *Thermal spray 2003: Advancing the Science & Applying the technology*, ed. C. Moreau and B. Marple, 1–8. Materials Park, Ohio: (Pub.) ASM International.

Gärtner, F., T. Stoltenhoff, T. Schmidt, and H. Kreye. 2006. The cold spray process and its potential for industrial applications. *Journal of Thermal Spray Technology* 15 (2): 223–232.

Gilmore, D.L., R.C. Dykhuizen, R.A. Neiser, T.J. Roemer, and M.F. Smith. 1999. Particle velocity and deposition efficiency in the cold spray process. *Journal of Thermal Spray Technology* 8 (4): 576–582.

Goldbaum, D., J.M. Shockley, R.R. Chromik, A. Rezaeian, S. Yue, J.-G. Legoux, and E. Irissou. 2012. The effect of deposition conditions on adhesion strength of Ti and Ti6Al4V cold spray splats. *Journal of Thermal Spray Technology* 21 (2): 288–303.

Gonzalez, R., H. Ashrafizadeh, A. Lopera, P. Mertiny, and A. McDonald. 2016. A review of thermal spray metallization of polymer-based structures. *Journal of Thermal Spray Technology* 25 (5): 897–919.

Goyal, T., R. Singh Walia, and T.S. Sidhu. 2012. Surface roughness optimization of cold-sprayed coatings using Taguchi method. *International Journal of Advance Manufacturing Technology* 60: 611–623.

Goyal, T., T.S. Sidhu, and R.S. Wa. 2014. Multi-response optimization of process parameters for low-pressure cold spray coating process using Taguchi and utility concept. *Journal of Thermal Spray Technology* 23 (1–2): 114–122.

Grujicic, M., C.L. Zhao, W.S. DeRosset, and D. Helfritch. 2004. Adiabatic shear instability based mechanism for particles/substrate bonding in the cold-gas dynamic-spray process, mater. *Design* 25: 681–688.

Ha, Yong Lee, Young Ho Yu, Young Cheol Lee, Young Pyo Hong, and Kyung Hyun Ko. 2004. Cold spray of SiC and Al$_2$O$_3$ with soft metal incorporation: A technical contribution. *Journal of Thermal Spray Technology* 13 (2): 184–189.

———. 2005a. Thin film coatings of WO3 by cold gas dynamic spray: A technical note. *Journal of Thermal Spray Technology* 14 (2): 183–186.

———. 2005b. Thin film coatings of WO$_3$ by cold gas dynamic spray: A technical note. *Journal of Thermal Spray Technology* 14 (2): 183–186.

Hall, A.C., D.J. Cook, R.A. Neiser, T.J. Roemer, and D.A. Hirschfeld. 2006. The effect of a simple annealing heat treatment on the mechanical properties of cold-sprayed Aluminium. *Journal of Thermal Spray Technology* 15 (2): 233–238.

Hall, A.C., L.N. Brewer, and T.J. Roemer. 2008. Preparation of aluminum coatings containing homogenous Nanocrystalline microstructures using the cold spray process. *Journal of Thermal Spray Technology* 17 (3): 352–359.

Han, T., Z. Zhao, B.A. Gillispie, and J.R. Smith. 2004. Effect of spray conditions on coating formation by the kinetic spray process. *Journal of Thermal Spray Technology* 14 (3): 373–383.

Han, T., B.A. Gillispie, and Z.B. Zhao. 2009. An investigation on powder injection in the high-pressure cold spray process. *Journal of Thermal Spray Technology* 18 (3): 320–330.

Yi, Hao, Ji-qiang Wang, Xin-yu Cui, Jie Wu, Tie-fan Li, and Tian-ying Xiong. 2016. Microstructure characteristics and mechanical properties of Al-12Si coatings on AZ31 magnesium alloy produced by cold spray technique. *Journal of Thermal Spray Technology* 25 (5): 1020–1028.

Haynes J., A. Pandey, J. Karthikeyan, A. Kay (2006) Cold Sprayed Discontinuously Reinforced Aluminium, Proc. of the 2006 Thermal Spray Conf. (ed.) B. Marple et al (pub.) ASM Int. Materials Park, OH, USA, e-proceedings.

Heinrich, P., H. Kreye, and T. Stoltenhoff (2005) Laval Nozzle for Thermal and Kinetic Spraying, U.S. Patent 2005/0001075 A1, January 6, 2005.

Helfritch D. and V. Champagne (2006) Optimal Particle Size for the Cold Spray Process, Thermal Spray 2006 (ed.) B. Marple (pub.) ASM International, Materials Park, Ohio, USA e-proc.

Hu, H.X., S.L. Jiang, Y.S. Tao, T.Y. Xiong, and Y.G. Zheng. 2011. Cavitation erosion and jet impingement erosion mechanism of cold sprayed Ni–Al$_2$O$_3$ coating. *Nuclear Engineering and Design* 241: 4929–4937.

Huang, R., and H. Fukanuma. 2012. Study of the influence of particle velocity on adhesive strength of cold spray deposits. *Journal of Thermal Spray Technology* 21 (3–4): 541–549.

Hyung-Jun, Kim, Chang-Hee Lee, and Soon-Young Hwang. 2005. Fabrication of WC–co coatings by cold spray deposition. *Surface & Coatings Technology* 191: 335–340.

Ichikawa, Y., and K. Ogawa. 2015. Effect of substrate surface oxide film thickness on deposition behavior and deposition efficiency in the cold spray process. *Journal of Thermal Spray Technology* 24 (7): 1269–1276.

Ichikawa Y., K. Sakaguchi, K. Ogawa, T. Shoji, S. Barradas, M. Jeandin, M. Boustie (2007) Deposition Mechanisms of Cold Gas Dynamic Sprayed MCrAlY Coatings, Proc. of the 2007 Thermal Spray Conf. (ed.) B. Marple et al (pub.) ASM Int. Materials Park, OH, USA, e-proceedings.

Irissou, E., J.-G. Legoux, B. Arsenault, and C. Moreau. 2007. Investigation of Al-Al$_2$O$_3$ cold spray coating formation and properties. *Journal of Thermal Spray Technology* 16 (5–6): 661–668.

Irissou, E., J.-G. Legoux, A.N. Ryabinin, B. Jodoin, and C. Moreau. 2008. Review on cold spray process and technology: Part I – Intellectual property. *Journal of Thermal Spray Technology* 17 (4): 495–516.

Jin, Y.-M., J.-H. Cho, D.-Y. Park, J.-H. Kim, and K.-A. Lee. 2011. Manufacturing and macroscopic properties of cold sprayed cu-in coating material for sputtering target. *Journal of Thermal Spray Technology* 20 (3): 497–507.

Jodoin, B. 2002. Cold spray nozzle Mach number limitation. *Journal of Thermal Spray Technology* 11 (4): 496–507.

Jodoin B., L. Ajdelsztajn, G. Bérubé (2006) Cold Spray Deposition of Metastable Alloys, Proc. of the 2006 Thermal Spray Conf. (ed.) B. Marple et al (pub.) ASM Int. Materials Park, OH, USA, e-proceedings.

Jodoin, B., P. Richer, G. Bérubé, L. Ajdelsztajn, A. Erdi-Betchi, and M. Yandouzi. 2007. Pulsed-gas dynamic spraying: Process analysis, development and selected coating examples. *Surface & Coatings Technology* 201: 7544–7551.

Kairet T., G. Di Stephano, M. Degrez, F. Campana, J-P. Janssen (2006) Comparison Between Coatings from two Different Copper Powders: Mechanical Properties, Hardness and Bond Strength, Proc. of the 2006 Thermal Spray Conf. (ed.) B. Marple et al (pub.) ASM Int. Materials Park, OH, USA, e-proceedings.

Kairet T., M. Degrez, F. Campana, J-P Janssen (2007) Influence of the Powder Size Distribution on the Microstructure of Cold Sprayed Copper Coatings Studied by X-ray Diffraction, Thermal Spray 2007: Global Coating Solutions, (Ed.) B.R. Marple et al (Pub.) ASM International®, Materials Park, Ohio, USA e-proc.

Kang K. C., S. H. Yoon, Y. G. Ji, C. Lee (2007) Oxidation Effects on the Critical Velocity of Pure Al Feedstock Deposition in the Kinetic Spraying Process, Thermal Spray 2007: Global Coating Solutions (Ed.) B.R. Marple et al (Pub.) ASM International®, Materials Park, Ohio, USA e-proc.

Kashirin A., O. Klynev, T. Buzdygar and A. Shkodin (2007) DYMET technology evolution and application, in Thermal Spray 2007: global coating solutions, (ed.) B.R. Marple et al, (pub.) ASM Int. Materials Park OH, USA, e-proceedings.

Katanoda, H., M. Fukuhara, and N. Iino. 2007. Numerical study of combination parameters for particle impact velocity and temperature in cold spray. *Journal of Thermal Spray Technology* 16 (5-6): 627–633.

Katanoda, H., M. Fukuhara, and N. Iino. 2007a. Numerical study of combination parameters for particle impact velocity and temperature in cold spray. *Journal of Thermal Spray Technology* 16 (5–6): 627–633.

Katanoda H., M. Fukuhara and N. Iino (2007b) Numerical Study of Combination Parameters for Particle Impact Velocity and Temperature in Cold Spray, in Thermal Spray 2007: Global Coating Solutions (Ed.) B.R. Marple et al (Pub.) ASM International®, Materials Park, Ohio, USA, Copyright© e-proceedings.

Kim, H.-J., C.-H. Lee, and S.-Y. Hwang. 2005. Superhard nano WC–12%co coating by cold spray deposition. *Materials Science and Engineering A* 391: 243–248.

Kim H.J., D.H. Jung, J.H. Jang and C.H. Lee (2006) Assessment of Metal/Diamond Composite Coatings by Cold Spray Deposition, Proc. of the 2006 Thermal Spray Conf. (ed.) B. Marple et al (pub.) ASM Int. Materials Park, OH, USA, e-proceedings.

King, P.C., S.H. Zahiri, and M.Z. Jahedi. 2007. Rare earth/metal composite formation by cold spray. *Journal of Thermal Spray Technology* 17 (2): 221–227.

Klinkov, S.V., and V.F. Kosarev. 2006. Measurements of cold spray deposition efficiency. *Journal of Thermal Spray Technology* 15 (3): 365–371.

Klinkov, S.V., V.F. Kosarev, A.A. Sova, and I. Smurov. 2008. Deposition of multicomponent coatings by cold spray. *Surface & Coatings Technology* 202: 5858–5862.

———. 2009. Calculation of particle parameters spraying of metal-ceramic mixtures. *Journal of Thermal Spray Technology* 18 (5–6): 944–956.

Koivuluoto, H., and P. Vuoristo. 2010a. Structural analysis of cold-sprayed nickel-based metallic and metallic-ceramic coatings. *Journal of Thermal Spray Technology* 19 (5): 975–998.

———. 2010b. Effect of powder type and composition on structure and mechanical properties of cu + Al$_2$O$_3$ coatings prepared by using low-pressure cold spray process. *Journal of Thermal Spray Technology* 19 (5): 1081–1092.

Koivuluoto, H., J. Lagerbom, and P. Vuoristo. 2007. Microstructural studies of cold sprayed copper, nickel, and Nickel-30% copper coatings. *Journal of Thermal Spray Technology* 16 (4): 488–497.

Koivuluoto, H., J. Lagerbom, M. Kylmälahti, and P. Vuoristo. 2008. Microstructure and mechanical properties of low-pressure cold-sprayed (LPCS) coatings. *Journal of Thermal Spray Technology* 17 (5–6): 721–727.

Koivuluoto, H., J. Nakki, and P. Vuoristo. 2009. Corrosion properties of cold-sprayed tantalum coatings. *Journal of Thermal Spray Technology* 18 (1): 75–82.

Kong, L.Y., L. Shen, B. Lu, R. Yang, X.Y. Cui, T.F. Li, and T.Y. Xiong. 2010. Preparation of TiAl$_3$-Al composite coating by cold spray and its high temperature oxidation behavior. *Journal of Thermal Spray Technology* 19 (6): 1206–1210.

Kosarev, V.F., S.V. Klinkov, A.P. Alkhimov, and A.N. Papyrin. 2003. On some aspects of gas dynamics of the cold spray process. *Journal of Thermal Spray Technology* 12 (2): 265–281.

Kulmala, M., and P. Vuoristo. 2008. Influence of process conditions in laser-assisted low-pressure cold spraying. *Surface & Coatings Technology* 202: 4503–4508.

Lee J.H., J.S. Kim, S.M. Shin, C.H. Lee and H.J. Kim (2006) Effect of Particle Temperature on the Critical Velocity for Particle Deposition by Kinetic Spraying, Thermal Spray 2006 (ed.) B. Marple (pub.) ASM International, Materials Park, Ohio, USA e-proc.

Lee, H.Y., S. Hun Jung, S. Yong Lee, and K. Hyun Ko. 2007a. Alloying of cold-sprayed Al–Ni composite coatings by post-annealing. *Applied Surface Science* 253: 3496–3502.

Lee, J.C., H.J. Kang, W.S. Chu, and S.H. Ahn. 2007b. Repair of damaged Mold surface by cold-spray method. *Annals of the CIRP* 56 (1): 577–580.

Lee, H., H. Shin, and K. Ko. 2010. Effects of gas pressure of cold spray on the formation of Al-based intermetallic compound. *Journal of Thermal Spray Technology* 19 (1–2): 102–109.

Legoux, J.G., E. Irissou, and C. Moreau. 2007. Effect of substrate temperature on the formation mechanism of cold-sprayed aluminum, zinc and tin coatings. *Journal of Thermal Spray Technology* 16 (5–6): 643–650.

Chang-Jiu, Li, and Wen-Ya Li. 2003. Deposition characteristics of titanium coating in cold spraying. *Surface & Coatings Technology* 167: 278–283.

Li, Wen-Ya, and Chang-Jiu Li. 2005. Optimal Design of a Novel Cold Spray gun Nozzle at a limited space. *Journal of Thermal Spray Technology* 14 (3): 391–396.

Li W.-Y., H.L. Liao, G. Zang, C. Coddet, H.-T. Wang, C.-J. Li (2006) Optimal Design of a Convergent-Barrel Cold Spray Nozzle by Numerical Method, Thermal Spray 2006 (ed.) B. Marple (pub.) ASM International, Materials Park, Ohio, USA e-proc.

Li W.-Y., C. Zhang, X.P. Guo, L. Dembinski, H. Liao, C. Coddet (2007a) Impact Fusion of Particle Interfaces in Cold Spraying and Its Effect on Coating Microstructure, Thermal Spray 2007: Global Coating Solutions (Ed.) B.R. Marple, M.M. Hyland, Y.-C. Lau, C.-J. Li, R.S. Lima, and G. Montavon, Published by ASM International®, Materials Park, Ohio, USA, Copyright©.

Li W.-Y., C. Zhang, X.P. Guo, H.L. Liao, C. Coddet (2007b) Deposition Characteristics of Al-12Si Alloy Coating Fabricated by Cold Spraying with Relatively Large Powder Particles, Thermal Spray 2007: Global Coating Solutions, (Ed.) B.R. Marple et al (Pub.) ASM International®, Materials Park, Ohio, USA, e-proc.

Li, W.-Y., G. Zhang, H.L. Liao, and C. Coddet. 2008. Characterizations of cold sprayed TiN particle reinforced Al2319 composite coating. *Journal of Materials Processing Technology* 202: 508–513.

Li, W.-Y., C. Zhang, H. Liao, and C. Coddet. 2009. Effect of heat treatment on microstructure and mechanical properties of cold sprayed Ti coatings with relatively large powder particles. *Journal of Coating Technology and Research* 6 (3): 401–406.

Li, B., L. Yang, Z. Li, J. Yao, Q. Zhang, Z. Chen, G. Dong, and L. Wang. 2015. Beneficial effects of synchronous laser irradiation on the characteristics of cold-sprayed copper coatings. *Journal of Thermal Spray Technology* 24 (5): 836–847.

Li, W.Y., K. Yang, S. Yin, and X.P. Guo. 2016. Numerical analysis of cold spray particles impacting behavior by the Eulerian method: A review. *Journal of Thermal Spray Technology* 25 (8): 1441–1460.

Lima, R.S., A. Kucuk, and C.C. Berndt. 2002. Deposition efficiency, mechanical properties and coating roughness in cold-sprayed titanium. *Journal of Materials Science Letters* 21: 1687–1689.

Liu, Y., Y.-Y. Wang, G.-J. Yang, J.-J. Feng, and K. Kusumoto. 2010. Effect of Nano-sized TiN additions on the electrical properties of vacuum cold sprayed SiC coatings. *Journal of Thermal Spray Technology* 19 (6): 1238–1243.

Liu, Y., J. Huang, and H. Li. 2014. Nanostructural characteristics of vacuum cold-sprayed hydroxyapatite/graphene-Nanosheet coatings for biomedical applications. *Journal of Thermal Spray Technology* 23 (7): 1149–1156.

Lugsheider, E. 2006. High kinetic process developments in thermal spray technology. *Journal of Thermal Spray Technology* 15 (2): 155–156.

Luo, X.-T., G.-J. Yang, and C.-J. Li. 2011. Multiple strengthening mechanisms of cold-sprayed cBNp/NiCrAl composite coating. *Surface & Coatings Technology* 205: 4808–4813.

Lupoi, R., and W. O'Neil. 2010. Deposition of metallic coatings on polymer surfaces using cold spray. *Surface & Coatings Technology* 205: 2167–2173.

Maev R. Gr., V. Leshchinsky (2006) Low Pressure Gas Dynamic Spray: Shear Localization during Particle Shock Consolidation, Proc. of the 2006 Thermal Spray Conf. (ed.) B. Marple et al (pub.) ASM Int. Materials Park, OH, USA, e-proceedings.

Makinen H., J. Lagerbom, P. Vuoristo (2006) Mechanical Properties and Corrosion Resistance of Cold Sprayed Coatings, Proc. of the 2006 Thermal Spray Conf. (ed.) B. Marple et al (pub.) ASM Int. Materials Park, OH, USA, e-proceedings.

Mäkinen H., J. Lagerbom, P. Vuoristo (2007) Adhesion of Cold Sprayed Coatings: Effect of Powder, Substrate, and Heat Treatment, in Thermal Spray 2007: Global Coating Solutions (Ed.) B.R. Marple et al (Pub.) ASM International®, Materials Park, Ohio, USA, Copyright© e-proceedings.

Manhar, Chavan N., M. Ramakrishna, P. Sudharshan Phani, D. Srinivasa Rao, and G. Sundararajan. 2011. The influence of process parameters and heat treatment on the properties of cold sprayed silver coatings. *Surface & Coatings Technology* 205: 4798–4807.

Marrocco T., D.G. McCartney, P.H. Shipway and A.J. Sturgeon (2006a) Comparison of Microstructure of Cold Sprayed and Thermally Sprayed IN718 Coatings, Proc. of the 2006 Thermal Spray Conf. (ed.) B. Marple et al (pub.) ASM Int. Materials Park, OH, USA, e-proceedings.

Marrocco T., D.G. McCartney, P.H. Shipway, and A.J. Sturgeon (2006b) Production of Titanium Deposits by Cold-Gas Dynamic Spray: Numerical Modeling and Experimental Characterization, Proc. of the 2006 Thermal Spray Conf. (ed.) B. Marple et al (pub.) ASM Int. Materials Park, OH, USA, e-proceedings.

Marx, S., A. Paul, A. Köhler, and G. Hüttl. 2006. Cold spraying: Innovative layers for new applications. *Journal of Thermal Spray Technology* 15 (2): 177–183.

McCune, R.C., W.T. Donlon, O.O. Popoola, and E.L. Cartwright. 2000. Characterization of copper layers produced by cold gas-dynamic spraying. *Journal of Thermal Spray Technology* 9 (1): 73–82.

Moreau, D., F. Borit, L. Corté, and V. Guipon. 2017. Cold spray coating of submicronic ceramic particles on polyvinyl alcohol in dry and hydrogel states. *Journal of Thermal Spray Technology* 26: 958–969.

Nastic, A., M. Vijay, A. Tieu, S. Rahmati, and B. Jodoin. 2017. Experimental and numerical study of the influence of substrate surface preparation on adhesion mechanisms of aluminum cold spray coatings on 300M steel substrates. *Journal of Thermal Spray Technology* 26: 1461–1483.

Neiser, R.C., and R.A. Neiser. 2003. Optimizing the cold spray process. In *ITSC 2003*, ed. C. Moreau and B. Marple, 19–26. Materials Park, OH: (pub.) ASM Int.

Neshastehriz, M., I. Smid, and A.E. Segall. 2014. In-situ agglomeration and De-agglomeration by milling of Nano-engineered lubricant particulate composites for cold spray deposition. *Journal of Thermal Spray Technology* 23 (7): 1191–1198.

Ning, X.-J., Q.-S. Wang, Z. Ma, and H.-J. Kim. 2010. Numerical study of in-flight particle parameters in low-pressure cold spray process. *Journal of Thermal Spray Technology.* 19 (6): 1211–1217.

Novoselova, T., P. Fox, R. Morgan, and W. O'Neill. 2006. Experimental study of titanium/aluminumdeposits produced by cold gas dynamic spray. *Surface & Coatings Technology* 200: 2775–2783.

Novoselova, T., S. Celotto, R. Morgan, P. Foxa, and W. O'Neill. 2007. Formation of TiAl intermetallics by heat treatment of cold-sprayed precursor deposits. *Journal of Alloys and Compounds* 436: 69–77.

Ogawa, K., K. Ito, K. Ichimura, Y. Ichikawa, S. Ohno, and N. Onda. 2008. Characterization of low-pressure cold-sprayed aluminum coatings. *Journal of Thermal Spray Technology* 17 (5–6): 728–735.

Olakanmi, E.O., and M. Doyoyo. 2014. Laser-assisted cold-sprayed corrosion- and Wear-resistant coatings: A review. *Journal of Thermal Spray Technology* 23 (5): 765–778.

Ortega, F., A. Sova, M.D. Monzón, M.D. Marrero, A.N. Benítez, and P. Bertrand. 2015. Combination of electroforming and cold gas dynamic spray for fabrication of rotational moulds: Feasibility study. *Journal of Thermal Spray Technology* 76: 1243–1251.

Ozdemir, O.C., C.A. Widener, D. Helfritch, and F. Delfanian. 2016. Estimating the effect of helium and nitrogen mixing on deposition efficiency in cold spray. *Journal of Thermal Spray Technology* 25 (4): 660–671.

Ozdemir O. C., Q. Chen, S. Muftu, V. K. Champagne Jr. (2018) Modeling the Continuous Heat Generation in the Cold Spray Coating Process, Journal of Thermal Spray Technology, Published online 15 November.

Papyrin A. (2001) Cold spray technology, Adv. Materials Process. Sept. 49–51.

Papyrin A.N., A.P. Alkhimov, and V.F. Kosarev (1993) Device for Applying Coatings by Deposition, SU1674585, year of priority (issued): 1989.

Papyrin A. N., S. V. Klinkov, V. F. Kosarev (2003) Modeling of Particle-Substrate Adhesive Interaction Under the Cold Spray Process, Thermal Spray 2003: Advancing the Science & Applying the Technology, (Ed.) C. Moreau and B. Marple,Published by ASM International, Materials Park, Ohio, USA 27–35.

Papyrin A.N., V.F. Konsarev, S.V. Kinkov (2006) Effect of coating erosion on the cold spray process, Thermal Spray 2006 (ed.) B. Marple (pub.) ASM International, Materials Park, Ohio, USA e-proc.

Poirier, D., J.-G. Legoux, R.A.L. Drew, and R. Gauvin. 2011. Consolidation of Al$_2$O$_3$/Al nanocomposite powder by cold spray. *Journal of Thermal Spray Technology* 20 (1–2): 275–284.

Price, T.S., P.H. Shipway, and D.G. McCartney. 2006. Effect of cold spray deposition of a titanium coating on fatigue behavior of a titanium alloy. *Journal of Thermal Spray Technology* 15 (4): 507–512.

Price T.S., P.H. Shipway, D.-G. McCartney (2006a) The Effect of Cold Spray Deposition of a Titanium Coating on the Fatigue Behaviour of a Titanium Alloy, Proc. of the 2006 Thermal Spray Conf. (ed.) B. Marple et al (pub.) ASM Int. Materials Park, OH, USA, e-proceedings.

Price, T.S., P.H. Shipway, and D.G. McCartney. 2006b. Effect of cold spray deposition of a titanium coating on fatigue behavior of a titanium alloy. *Journal of Thermal Spray Technology* 15 (4): 507–512.

Price, T.S., P.H. Shipway, D.G. McCartney, E. Calla, and D. Zhang. 2007. A method for characterizing the degree of inter-particle bond formation in cold sprayed coatings. *Journal of Thermal Spray Technology 2007* 16 (4): 566–570.

Qiu, X., Ji-Q. Wang, N.U.H. Tariq, L. Gyansah, J.-X. Zhang, and T.-Y. Xiong. 2017. Effect of heat treatment on microstructure and mechanical properties of A380 aluminum alloy deposited by cold spray. *Journal of Thermal Spray Technology* 26: 1898–1907.

Rahmati, S., and A. Ghaei. 2014. The use of particle/substrate material models in simulation of cold-gas dynamic-spray process. *Journal of Thermal Spray Technology* 23 (3): 530–540.

Raletz F. (2005) Contribution to the development of a Cold Gas Dynamic Spray system (C.G.D.S.) for the realization of nickel coatings, Ph. D, University of Limoges, France 17 Oct. (in French).

Raoelison, R.N., Y. Xie, T. Sapanathan, M.P. Planche, R. Kromer, S. Costil, and C. Langlade. 2018. Cold gas dynamic spray technology: A comprehensive review of processing conditions for various technological developments till to date. *Additive Manufacturing* 19 (2018): 134–159.

Ravi, K., Y. Ichikawa, K. Ogawa, T. Deplancke, O. Lame, and J.-Y. Cavaille. 2016. Mechanistic study and characterization of cold-sprayed ultra-high molecular weight polyethylene-Nano-ceramic

composite coating. *Journal of Thermal Spray Technology* 25 (1–2): 160–169.

Richer, P., B. Jodoin, L. Ajdelsztajn, and E.J. Lavernia. 2006. Substrate roughness and thickness effects on cold spray Nanocrystalline Al-mg coatings. *Journal of Thermal Spray Technology* 15 (2): 246–254.

Richer, P., A. Zúñiga, M. Yandouzi, and B. Jodoin. 2008. CoNiCrAlY microstructural changes induced during cold gas dynamic spraying. *Surface & Coatings Technology* 203: 364–371.

Rokni, M.R., S.R. Nutt, C.A. Widener, V.K. Champagne, and R.H. Hrabe. 2017. Review of relationship between particle deformation, coating microstructure, and properties in high-pressure cold spray. *Journal of Thermal Spray Technology* 26: 1308–1355.

Sakaki, K., and Y. Shimizu. 2001. Effect of the increase in the entrance convergent section length of the gun nozzle on the high-velocity oxygen fuel and cold spray process. *Journal of Thermal Spray Technology* 10 (3): 487–496.

Sakaki K., T.Tajima, H. Li, S. Shinkai and Y.Shimizu (2004) Influence of Substrate Conditions and Traverse Speed on Cold Sprayed Coatings, ITSC 2004, DVS, Düsseldorf, Germany e.proceedings.

Sakaki K., S. Shinkai, Y. Shimizu (2007) Investigation of Spray Conditions and Performances of Cold-sprayed Pure Silicon Anodes for Lithium Secondary Batteries, Thermal Spray 2007: Global Coating Solutions, (Ed.) B.R. Marple et al (Pub.) ASM International®, Materials Park, Ohio, USA e-proc.

Sakata, K., K. Tagomori, N. Sugiyama, S. Sasaki, Y. Shinya, T. Nanbu, Y. Kawashita, I. Narita, K. Kuwatori, T. Ikeda, R. Hara, and H. Miyahara. 2014. Dust explosion characteristics of aluminum, titanium, zinc, and Iron-based alloy powders used in cold spray processing. *Journal of Thermal Spray Technology* 23 (1–2): 123–130.

Samareh, B., and A. Dolatabadi. 2007. A three-dimensional analysis of the cold spray process: The effects of substrate location and shape. *Journal of Thermal Spray Technology* 16 (5–6): 634–642.

Samareh, B., O. Stier, V. Lüthen, and A. Dolatabadi. 2009. Assessment of CFD modeling via flow visualization in cold spray process. *Journal of Thermal Spray Technology* 18 (5–6): 934–943.

Sansoucy, E., G.E. Kim, A.L. Moran, and B. Jodoin. 2007a. Mechanical characteristics of Al-co-Ce coatings produced by the cold spray process. *Journal of Thermal Spray Technology* 16 (5–6): 651–660.

Sansoucy E., L. Ajdelsztajn, B. Jodoin, P. Marcoux (2007b) Properties of SiC-Reinforced AluminumAlloy Coatings Produced by the Cold Spray Deposition Process, Thermal Spray 2007: Global Coating Solutions (Ed.) B.R. Marple et al (Pub.) ASM International®, Materials Park, Ohio, USA.

Sansoucy, E., P. Marcoux, L. Ajdelsztajn, and B. Jodoin. 2008. Properties of SiC-reinforced aluminum alloy coatings produced by the cold gas dynamic spraying process. *Surface & Coatings Technology* 202: 3988–3996.

Schmidt, T., F. Gärtner, H. Assadi, and H. Kreye. 2006. Development of a generalized parameter window for cold spray deposition. *Acta Materialia* 54: 729–742.

Schmidt, T., H. Assadi, F. Gärtner, H. Richter, H. Kreye, and T. Klassen. 2009a. From particle acceleration to impact and bonding in cold spraying. *Journal of Thermal Spray Technology* 18 (5–6): 794–808.

———. 2009b. Erratum to: From particle acceleration to impact and bonding in cold spraying. *Journal of Thermal Spray Technology* 18 (5–6): 1038.

Shapiro, A.H. 1953. *The dynamics and thermodynamics of compressible fluid flow*. New York: Ronald Press.

Sheng-Qiang, Fan, Chang-Jiu Li, Guan-Jun Yang, Ling-Zi Zhang, Jin-Cheng Gao, and Ying-Xin Xi. 2007. Fabrication of Nano-TiO$_2$ coating for dye-sensitized solar cell by vacuum cold spraying at room temperature. *Journal of Thermal Spray Technology* 16 (5–6): 893–897.

Shkodkin, A., A. Kashirin, and O. Klyuev. 2006. T. Buzdygar, metal particle deposition stimulation by surface abrasive treatment in gas dynamic spraying. *Journal of Thermal Spray Technology* 15 (3): 382–386.

Siao Ming Ang, A., C.C. Berndt, and P. Cheang. 2011. Deposition effects of WC particle size on cold-sprayed WC–co coatings. *Surface & Coatings Technology* 205: 3260–3267.

Sova, A., A. Papyrin, and I. Smurov. 2009. Influence of ceramic powder size on process of cermet coating formation by cold spray. *Journal of Thermal Spray Technology* 18 (4): 633–641.

Sova, A., D. Pervushin, and I. Smurov. 2010. Development of multi-material coatings by cold spray and gas detonation spraying. *Surface & Coatings Technology* 205: 1108–1114.

Sova, A., V.F. Kosarev, A. Papyrin, and I. Smurov. 2011. Effect of ceramic particle velocity on cold spray deposition of metal-ceramic coatings. *Journal of Thermal Spray Technology* 20 (1–2): 285–291.

Sova, A., S. Grigoriev, A. Okunkova, and I. Smurov. 2013. Potential of cold gas dynamic spray as additive manufacturing technology. *International Journal of Advanced Manufacturing Technology* 69: 2269–2278.

Sova, A., C. Courbon, F. Valiorgue, J. Rech, and Ph. Bertrand. 2017. Effect of turning and ball burnishing on the microstructure and residual stress distribution in stainless steel cold spray deposits. *Journal of Thermal Spray Technology* 26: 1922–1934.

Spencer, K., and M.-X. Zhang. 2011. Optimization of stainless-steel cold spray coatings using mixed particle size distributions. *Surface & Coatings Technology* 205: 5135–5140.

Spencer, K., D.M. Fabijanic, and M.-X. Zhang. 2009. The use of Al–Al$_2$O$_3$ cold spray coatings to improve the surface properties of magnesium alloys. *Surface & Coatings Technology* 204: 336–344.

Stoltenhoff T., J. Voyer and H. Kreye (2002a) Cold Spraying: State of the Art and Applicability, ITSC 2002 (ed.) E Lugsheider, (pub.) DVS, Dusseldorf, Germany, e-proceedings.

Stoltenhoff, T., H. Kreye, and H.J. Richter. 2002b. An analysis of the cold spray process and its coatings. *Journal of Thermal Spray Technology* 11 (4): 542–550.

Sudharshan, Phani P., D. Srinivasa Rao, S.V. Joshi, and G. Sundararajan. 2007a. Effect of process parameters and heat treatments on properties of cold sprayed copper coatings. *Journal of Thermal Spray Technology* 16 (3): 424–434.

Sudharshan Phani P., D. Srinivasa Rao, K. Tanabe, M. Yamada, E. Yamaguchi, A. Niwa, M. Sugimoto, M. Izawa (2007b) Deposition Behavior of Sprayed Metallic Particle on Substrate Surface in Cold Spray Process, Thermal Spray 2007: Global Coating Solutions, (Ed.) B.R. Marple et al (Pub.) ASM International®, Materials Park, Ohio, USA e-proc.

Sun B., R.-Z. Huang, N. Ohno and H. Fukanuma (2006) Effect of Spraying Parameters on Stainless Steel Particle Velocity and Deposition Efficiency in Cold Spraying, Proc. of the 2006 Thermal Spray Conf. (ed.) B. Marple et al (pub.) ASM Int. Materials Park, OH, USA, e-proceedings.

Tao, Y., T. Xiong, C. Sun, H. Jin, H. Du, and T. Li. 2009. Effect of α-Al$_2$O$_3$ on the properties of cold sprayed Al/α-Al$_2$O$_3$ composite coatings on AZ91D magnesium alloy. *Applied Surface Science* 256: 261–266.

Taylor, K., B. Jodoin, and J. Karov. 2006. Particle loading effect in cold spray. *Journal of Thermal Spray Technology* 15 (2): 273–279.

Tjitra Salim, N., M. Yamada, H. Nakano, K. Shima, H. Isago, and M. Fukumoto. 2011. The effect of post-treatments on the powder morphology of titanium dioxide (TiO2) powders synthesized for cold spray. *Surface & Coatings Technology* 206: 366–371.

Trinchi, A., Y.S. Yang, A. Tulloh, S.H. Zahiri, and M. Jahedi. 2011. Copper surface coatings formed by the cold spray process: Simulations based on empirical and phenomenological data. *Journal of Thermal Spray Technology* 20 (5): 986–991.

Van Steenkiste, T. 2006. Kinetic sprayed rare earth Iron alloy composite coatings. *Journal of Thermal Spray Technology* 15 (4): 501–506.

Van Steenkiste, T., and D.W. Gorkiewicz. 2004. Analysis of tantalum coatings produced by the kinetic spray process. *Journal of Thermal Spray Technology* 13 (2): 265–273.

Van Steenkiste, T., and J.R. Smith. 2004. Evaluation of coatings produced via kinetic and cold spray processes. *Journal of Thermal Spray Technology* 13 (2): 274–282.

Van Steenkiste T.H., J.R. Smith, R.E. Teets, J.J. Moleski, and D.W. Gorkiewicz (2000) Kinetic Spray Coating Method and Apparatus U.S. Patent 6 139 913, (Oct. 31, 2000).

——— (2001) Kinetic Spray Coating Apparatus U.S. Patent 6 283 386B1, (Sept. 4, 2001).

Van Steenkiste, T.H., J.R. Smith, and R.E. Teets. 2002. Aluminum coatings via kinetic spray with relatively large powder particles. *Surface & Coating Technology* 54: 237–252.

Vilardell, A.M., N. Cincal, A. Concustell, S. Dosta, I.G. Cano, and J.M. Guilemany. 2015. Cold spray as an emerging technology for biocompatible and antibacterial coatings: State of art. *Journal of Materials Science* 50: 4441–4462.

Vlcek, J., L. Gimeno, H. Huber, and E. Lugscheider. 2005. A systematic approach to material eligibility for the cold-spray process. *Journal of Thermal Spray Technology* 14 (1): 125–133.

Voyer J., T. Stoltenhoff, H. Kreye (2003) Development of cold gas sprayed coatings, in ITSC 2003 (ed.) C. Moreau and B. Marple (pub.) ASM Int. Materials Park, OH, USA 71–78.

Wang, F. 2017. Deposition characteristic of Al particles on mg alloy micro-channel substrate by cold spray. *International Journal of Advanced Manufacturing Technology* 91: 791–802.

Wang H.-T., C.-J. Li, G.-J. Yang, C.-X. Li (2006) Formation of Fe-Al Intermetallic Compound Coating Through Cold Spraying, Proc. of the 2006 Thermal Spray Conf. (ed.) B. Marple et al (pub.) ASM Int. Materials Park, OH, USA, e-proceedings.

Wang, H.-T., C.-J. Li, G.-J. Yang, C.-X. Li, Q. Zhang, and W.-Y. Li. 2007. FeAl intermetallic compound coating and its ball-milled feedstock powders. *Journal of Thermal Spray Technology* 16 (5–6): 669–676.

Wang, H.-T., C.-J. Li, G.-J. Yang, and C.-X. Li. 2009. Effect of heat treatment on the microstructure and property of cold-sprayed nanostructured FeAl/Al2O3 intermetallic composite coating. *Vacuum* 83: 146–152.

Wang, Y.-Y., Y. Liu, G.-J. Yang, J.-J. Feng, and K. Kusumoto. 2010. Effect of microstructure on the electrical properties of Nano-structured TiN coatings deposited by vacuum cold spray. *Journal of Thermal Spray Technology* 19 (6): 1231–1237.

Wank, A., B. Wielage, H. Podlesak, and T. Grund. 2006. High-resolution microstructural investigations of interfaces between light metal alloy substrates and cold gas-sprayed coatings. *Journal of Thermal Spray Technology* 15 (2): 280–283.

Weinert H., E. Maeva, V. Leshchinsky (2006) Low Pressure GAS Dynamic Spray Forming Near-net Shape Parts, Proc. of the 2006 Thermal Spray Conf. (ed.) B. Marple et al (pub.) ASM Int. Materials Park, OH, USA, e-proceedings.

Wen-Ya, Li, and Chang-Jiu Li. 2005. Optimal Design of a Novel Cold Spray gun Nozzle at a limited space. *Journal of Thermal Spray Technology* 14 (3): 391–396.

Widener, C.A., M.J. Carter, O.C. Ozdemir, R.H. Hrabe, B. Hoiland, T.E. Stamey, V.K. Champagne, and T.J. Eden. 2016. Application of high-pressure cold spray for an internal bore repair of a navy valve actuator. *Journal of Thermal Spray Technology* 25 (1–2): 193–201.

Wolfe, D.E., T.J. Eden, J.K. Potter, and A.P. Jaroh. 2006. Investigation and characterization of Cr3C2-based Wear-resistant coatings applied by the cold spray process. *Journal of Thermal Spray Technology* 15 (3): 400–412.

Wright, J.T. 2002. *The physics and mathematics of adiabatic shear bands*. Cambridge: Cambridge University Press.

Wu J.W., H.Y. Fang, C.H. Lee, S.H. Yoon, H.J. Kim (2006) Critical and maximum velocities in kinetic spraying, Thermal Spray 2006 (Ed.) B.R. Marple et al (pub.) ASM International®, Materials Park, Ohio, USA e-proc.

Wu, X.K., J.S. Zhang, X.L. Zhou, H. Cui, and J.C. Liu. 2012. Advanced cold spray technology: Deposition characteristics and potential applications. *SCIENCE CHINA Technological Sciences* 55 (2): 357–368.

Xian-Jin, Ning, Jae-Hoon Jang, Hyung-Jun Kim, Chang-Jiu Li, and Changhee Lee. 2008. Cold spraying of Al–Sn binary alloy: Coating characteristics and particle bonding features. *Surface & Coatings Technology* 202 (9): 1681–1687.

Xiong, Y., G. Bae, X.G. Xiong, and C. Lee. 2010. The effects of successive impacts and cold welds on the deposition onset of cold spray coatings. *Journal of Thermal Spray Technology* 19 (3): 575–585.

Yamada, M., H. Isago, H. Nakano, and M. Fukumoto. 2010. Cold spraying of TiO2 Photocatalyst coating with nitrogen process gas. *Journal of Thermal Spray Technology* 19 (6): 1218–1223.

Yandouzi, M., E. Sansoucy, L. Ajdelsztajn, and B. Jodoin. 2007. WC-based cermet coatings produced by cold gas dynamic and pulsed gas dynamic spraying processes. *Surface & Coatings Technology* 202: 382–339.

Yandouzi, M., L. Ajdelsztajn, and B. Jodoin. 2008. WC-based composite coatings prepared by the pulsed gas dynamic spraying process: Effect of the feedstock powders. *Surface & Coatings Technology* 202: 3866–3877.

Yang, Guan-Jun, Chang-Jiu Li, Sheng-Qiang Fan, Yu-Yue Wang, and Cheng-Xin Li. 2007. Influence of annealing on photocatalytic performance and adhesion of vacuum cold-sprayed nanostructured TiO2 coating. *Journal of Thermal Spray Technology* 16 (5–6): 873–880.

Yang, G.-J., C.-J. Li, F. Han, W.-Y. Li, and A. Ohmori. 2008. Low temperature deposition and characterization of TiO2 photo catalytic film through cold spray. *Applied Surface Science* 254: 3979–3982.

Yang, G.-J., K.-X. Liao, C.-J. Li, S.-Q. Fan, C.-X. Li, and S. Li. 2012. Formation of pore structure and its influence on the mass transport property of vacuum cold sprayed TiO2 coatings using strengthened nanostructured powder. *Journal of Thermal Spray Technology* 21 (3–4): 505–513.

Yang, G.-J., S.-N. Zhao, C.-X. Li, and C.-J. Li. 2013. Effect of phase transformation mechanism on the microstructure of cold-sprayed Ni/Al-Al2O3 composite coatings during post-spray annealing treatment. *Journal of Thermal Spray Technology* 22 (2–3): 398–405.

Yi, Hao, Ji-qiang Wang, Xin-yu Cui, Jie Wu, Tie-fan Li, and Tian-ying Xiong. 2016. Microstructure characteristics and mechanical properties of Al-12Si coatings on AZ31 magnesium alloy produced by cold spray technique. *Journal of Thermal Spray Technology* 25 (5): 1020–1028.

Yin, S., X. Suo, J. Su, Z. Guo, H. Liao, and X. Wang. 2014. Effects of substrate hardness and spray angle on the deposition behavior of cold-sprayed Ti particles. *Journal of Thermal Spray Technology* 23 (1–2): 76–83.

Yin, S., M. Meyer, W. Li, H. Liao, and R. Lupoi. 2016a. Gas flow, particle acceleration, and heat transfer in cold spray: A review. *Journal of Thermal Spray Technology* 25 (5): 874–896.

———. 2016b. Gas flow, particle acceleration, and heat transfer in cold spray: A review. *Journal of Thermal Spray Technology* 25 (5): 874–896.

Yong, Lee Ha, Young Ho Yu, Young Cheol Lee, Young Pyo Hong, and Kyung Hyun Ko. 2004a. Cold spray of SiC and Al2O3 with soft metal incorporation: A technical contribution. *Journal of Thermal Spray Technology* 13 (2): 184–189.

———. 2004b. Cold spray of SiC and Al2O3 with soft metal incorporation: A technical contribution. *Journal of Thermal Spray Technology* 13 (2): 184–189.

———. 2005. Thin film coatings of WO_3 by cold gas dynamic spray: A technical note. *Journal of Thermal Spray Technology* 14 (2): 183–186.

Yoon, S.H., C. Lee, and H.J. Kim. 2007a. Process development of brazed AluminumHeat exchanger using a kinetic spraying process. In *Thermal spray 2007: Global coating solutions*, ed. B.R. Marple et al. Materials Park, Ohio: (Pub.) ASM International®. e-proc.

———. 2007b. Process development of brazed AluminumHeat exchanger using a kinetic spraying process. In *Thermal spray 2007: Global coating solutions*, ed. B.R. Marple et al. Materials Park, Ohio: (Pub.) ASM International®. e-proc.

Zahiri, S.H., D. Fraser, S. Gulizia, and M. Jahedi. 2006a. Effect of processing conditions on porosity formation in cold gas dynamic spraying of copper. *Journal of Thermal Spray Technology* 15 (3): 422–430.

———. 2006b. Effect of processing conditions on porosity formation in cold gas dynamic spraying of copper. *Journal of Thermal Spray Technology* 15 (3): 422–430.

Zahiri, S.H., C.I. Antonio, and M. Jahedi. 2009a. Elimination of porosity in directly fabricated titanium via cold gas dynamic spraying. *Journal of Materials Processing Technology* 209: 922–929.

———. 2009b. Elimination of porosity in directly fabricated titanium via cold gas dynamic spraying. *Journal of Materials Processing Technology* 209: 922–929.

Zahiri, S.H., T.D. Phan, S.H. Masood, and M. Jahedi. 2014. Development of holistic three-dimensional models for cold spray supersonic jet. *Journal of Thermal Spray Technology* 23 (6): 919–933.

Zhang, D., P.H. Shipway, and D.G. McCartney. 2003. Particle-substrate interactions in cold gas dynamic spraying. In *Thermal spray 2003: Advancing the Science & Applying the technology*, ed. C. Moreau and B. Marple, 45–52. Materials Park, Ohio: (Pub.) ASM International.

———. 2005. Cold gas dynamic spraying of aluminum: The role of substrate characteristics in deposit formation. *Journal of Thermal Spray Technology* 14 (1): 109–116.

Zhang, Q., C.-J. Li, X.-R. Wang, Z.-L. Ren, C.-X. Li, and G.-J. Yang. 2008. Formation of NiAl intermetallic compound by cold spraying of ball-milled Ni/Al alloy powder through post-annealing treatment. *Journal of Thermal Spray Technology* 17 (5–6): 715–720.

Zhang, Y.Y., X.K. Wu, H. Cui, and J.S. Zhang. 2011. Cold-spray processing of a high-density Nano crystalline aluminum alloy coating using a mixture of as-atomized and as-Cryomilled powders. *Journal of Thermal Spray Technology* 20 (5): 1125–1132.

Zhou, X.L., A.F. Chen, J.C. Liu, X.K. Wu, and J.S. Zhang. 2011. Preparation of metallic coatings on polymer matrix composites by cold spray. *Surface & Coatings Technology* 206: 132–136.

Zhu, L., T.-C. Jen, Y.-T. Pan, and H.-S. Chen. 2017. Particle bonding mechanism in cold gas dynamic spray: A three-dimensional approach. *Journal of Thermal Spray Technology* 26: 1859–1873.

Zieris, R., S. Nowotny, L.-M. Berger, L. Haubold, and E. Beyer. 2003. Characterization of coatings deposited by laser-assisted atmospheric plasma spraying. In *CD-ROM proceeding of the international thermal spray conference*. Materials Park, Oh: (Pub.) ASM International.

Abbreviations

APS	Atmospheric Plasma Spraying
BC	Bond Coat
CFD	Computational Fluid Dynamic
DC	Direct Current
D-gun	Detonation Gun
EDS	Energy Dispersive Spectroscopy
FS	Flame Spraying
GS-HVOF	Gas Shroud High-Velocity Oxy Fuel
HVAF	High-Velocity Air Fuel
HVOF	High-Velocity Oxy Fuel
HV-SPS	High velocity Suspension Plasma Spraying
LHS	Left-Hand Side
LT-HVOF	Low-Temperature High-Velocity Oxy Fuel
MPD	Maximum Power Density
PDTS	Pulse Detonation Thermal Spray
PFS	Powder Flame Spraying
PTS	Polymer Thermal Spray
RHS	Right-Hand Side
RT	Room Temperature
SBF	Simulated Body Fluid
SEM	Scanning Electron Microscopy
SFS	Solution Flame Spraying
SHS	Self-propagating High Temperature Synthesis
slm	standard liters per minute
SoD	Standoff Distance
WFS	Wire Flame Spraying
WS-HVAF	Warm Spray High-Velocity Air Fuel
XRD	X-ray Diffraction
YSZ	Yttria-stabilized zirconia

7.1 Overview of Combustion-Based Spray Technologies

"Combustion spraying," often referred to as "Flame spraying (FS)," is at the base of the wide field of thermal spray technologies which was developed in the early twentieth century and remains one of the most widely used spray coating processes because of their favorable economics. Its most common application is in the corrosion protection of infrastructure buildings such as bridges and steel structures. Over the years, the growth of combustion processes was driven by both scientific and technical developments and market requirements. The development of High-Velocity Oxy Fuel (HVOF) spraying by Browning in 1983 is a typical example which was pushed forward by the need to produce WC-Co cermet coatings with superior properties as a replacement for the environmentally polluting chrome plating technologies. Table 7.1 lists the milestones in the development of some of the most important combustion-based spray processes.

In this chapter, a review is made of three combustion-based spray technologies covering for each, the basic principal behind the process, their main characteristics, as well as potential applications and limitations. These are:

- *Flames spraying (FS)* using mostly oxy-acetylene torches achieving premixed combustion temperatures up to about 3000 K. Sprayed materials are introduced generally axially into the spray gun. Flame velocities below 100 m/s characterize this process. This technique was the first to be developed in 1909 by Maximilian Ulrich Schoop.

© Springer Nature Switzerland AG 2021
M. I. Boulos et al. (ed.), *Thermal Spray Fundamentals*, https://doi.org/10.1007/978-3-030-70672-2_7

Table 7.1 Milestones in the development of key combustion-based spray processes

Year	Process	Energy generation	Inventor(s)	Reference
1909	Flame spraying (FS)	Combustion (diffusive)	Schoop (Switzerland)	Anon. (1913)
1955	Detonation gun spraying (D-gun)	Combustion (detonation)	Gfeller and Baiker[a] (USA)	Smith (1991)
1983	High velocity oxy fuel (HVOF)	Combustion (deflagration)	Browning (USA)	Browning (1983)

[a]Union Carbide, today Praxair Surface Technology (USA)

- ***High-Velocity Oxy Fuel flame (HVOF)*** spraying (or supersonic flame spraying) where the flame (or deflagration) of a gaseous or liquid hydrocarbon molecule (C_xH_y for example) is achieved at an adjustable stoichiometry with an oxidizer, which is either oxygen or air in a chamber at pressures between 0.24 and 0.82 MPa. The chamber is connected to a convergent-divergent (C-D) Laval nozzle achieving very high gas velocities (up to 2200 m/s), which would be in the range of Mach 2 when considering the local pressure and temperature. Recently, new spray guns using kerosene as fuel have been developed with chamber pressures up to 1 MPa. The trend is also to reduce the combustion temperature and increase the velocity of the combustion products by injecting non-combustible gases such as nitrogen in the combustion chamber. *High-velocity air-fuel flame (HVAF)* is a variant of this technology in which oxygen is replaced by air, which reduces the cost of the spraying process without compromising the quality of the coating. The principal difference is in the flame temperature, which is typically 500 to 1000 °C lower for air/fuel compared to oxygen/fuel combustion. For example, propane/air combustion has an adiabatic flame temperature of 1967 °C, while propane/oxygen combustion temperature is 2526 °C. This results in a lower particle temperature for HVAF processes, which can have either positive or negative effects on the properties of the coating depending on the materials being sprayed. Cermet-type coatings and lower-melting-point materials may benefit from less thermal degradation and oxidation with HVAF.
- ***Detonation gun spraying (D-gun)*** generally using acetylene-oxygen or hydrogen-oxygen mixture contained in a tube closed at one of its ends. The shock wave generated by the reaction in the highly compressed explosive medium (about 2 MPa) heats and accelerates the particles to be sprayed, which are projected toward the substrate. Gas velocities above 2000 m/s are commonly achieved in detonation spray guns. Contrary to the two previous processes where combustible gases and powders are continuously fed within the gun, combustible gas and powder are fed in D-guns in cycles repeated at a frequency ranging from 3 to 100 Hz.

In the following, each of these three techniques will be described in detail. References are given for expanded exposure to recent developments and immerging technologies and their applications.

7.2 Flame Spraying

7.2.1 Basic Concepts

Flame spraying requires the melting of the sprayed material, generally metals or alloys. The process relies on the combustion of a gaseous fuel for the generation of a high-temperature flue gas stream which is used to heat and melt, in-flight, the precursor powder and accelerates the formed droplets toward the substrate on which they are deposited forming the coating. Oxy-acetylene torches are most commonly used for flame spraying. The mixture of acetylene (C_2H_2) and oxygen (O_2) at a stoichiometric ratio gives rise to a combustion temperature of 3410 K at atmospheric pressure, which is the highest temperature of all hydrocarbon gaseous fuels (C_nH_m) at this pressure (see Chap. 3, Table 3.1). Stoichiometry corresponds to the full transformation of C into CO_2 and H into H_2O, according to the global reaction:

$$2\,C_2H_2 + 5\,O_2 \rightarrow 4\,CO_2 + 2\,H_2O \qquad (7.1)$$

The parameter R', defined as the fuel to oxidant *(F/A)* ratio used, compared to that of their stoichiometric value $(F/A)_{St}$, is a good indicator of the combustion conditions.

$$R' = (F/A)/(F/A)_{St.} \qquad (7.2)$$

where F is the volume or mass feed rate of the fuel and A the corresponding value for the oxidizer (see Chap. 3 for more details). The mixture is called fuel rich if $R' > 1$ and fuel lean if $R' < 1$. It should be noted that combustion reactions may also be defined by the Stoichiometric factor, S_t, defined as:

$$S_t = Q_{O_2}/(Q_{O_2})_{St.} \qquad (7.3)$$

where Q_{O_2} is the oxygen flow rate (slm), and $(Q_{O_2})_{St.}$ the stoichiometric value of oxygen flow rate necessary for the

complete combustion of the fuel. Obviously R' and S_t are interrelated with $S_t = 1/R'$.

As the flame length is limited and the particle velocity is only a few tens of m/s, the maximum temperatures that the particles can achieve after their flight is about $T_p = 0.7xT_g$ to $0.8xT_g$, with T_g being the gas temperature, that is, $T_p \approx$ 2300 K for a flame temperature of about 3000 K, which is too low to melt most ceramic materials.

Flame spray has also been used with precursors in the form of wires, cords, and rods, which are introduced axially through the rear of the nozzle into the flame burning at the nozzle exit. Cords being formed of a continuous thin metallic foil folded on itself after being filled with the material to be sprayed as fine powder, mostly ceramics. The resultant, powder-packed, continuous tube has a diameter compatible with most spray wires. Rods, on the other hand, are made of a sintered ceramic with a diameter in the range of 3.17 to 6.35 mm, though generally limited in length to less than 1.0 m. As the material fed is melted, it is atomized, forming droplets which are accelerated toward the substrate surface by the expanding hot gas flow and air jets. The main difference between the delivery of the feedstock as either powder or wire, cord, or rod is the melting capability. In contrast, powders, the metal wire, cords, or rods can be advanced in the flame in such a way as to allow their tip to be reached high temperatures for melting and atomization. The formed droplets of the molten material can reach in this case temperatures as high as $0.95T_g$ ($T_p \approx 2850\ K$ for a flame temperature of about 3000 K), which would allow even zirconia to be melted [Thermal Spraying (1985)].

Numerous specialized torch designs have been developed for the "Flame Spraying" of different materials, in powder, wires, cords, or rods, for different applications. Examples are given in the following highlighting design features and typical operating conditions and applications.

7.2.2 Powder Flame Spraying

7.2.2.1 Spray Gun Design and Process Characteristics

A schematic of a typical Sulzer-Metco powder flame spraying (PFS) torch is given in Fig. 7.1. The acetylene and oxygen premixed streams are fed to the nozzle extremity through a ring of 16 to 18 orifices each with an i.d. ≤ 1 mm to avoid the flashback of the flame (see Chap. 3, section 3.1.4.1). Proper control of the gas feeding velocity v_u compared to the flame velocity, V_f, is critical for flame stability. As shown in Fig. 7.2, if $v_u \gg V_f$, the flame is blown out, while if $v_u \ll V_f$, flash-back occurs, except if the feeding orifice diameters are small enough to quench the flame and destroy all radicals. In this case, the flame extinguishes.

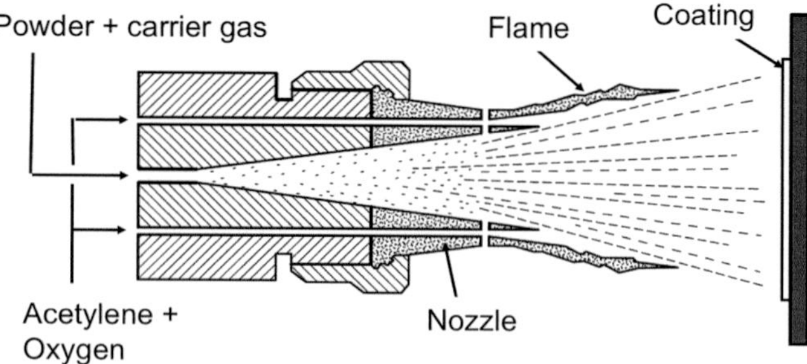

Fig. 7.1 Principal of the Flame Powder Spraying torch. (Courtesy of Sultzer Metco)

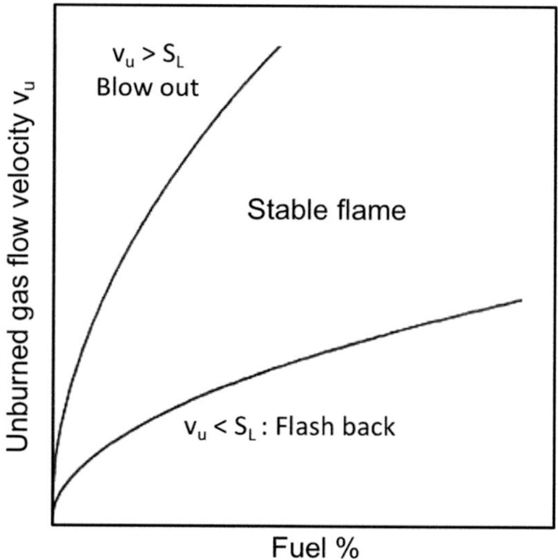

Fig. 7.2 Pre-mixed flame stability diagram [Wigren et al. (1996)]. (Reprinted with kind permission from ASM International)

It is important to emphasize that the power generated in the flame, P_F, depends on the nature of the fuel and oxidizer chosen (highest values being obtained with O_2 and C_2H_2 mixture as shown in Table 3.1 in Chap. 3). P_F depends also on the fuel/oxidizer ratio (about 1/1.6 for the O_2-C_2H_2 mixture) and on the combustible gas flow rate $\dot{m}_u$. Practically, for the oxy-acetylene mixture, P_F is proportional to the oxygen gas flow rate Q_{O_2} (slm), which is the limiting factor.

$$P_F \approx 1.354\ Q_{O_2} \tag{7.4}$$

For $Q_{O_2} = 27\ slm$, $P_F = 36.6\ kW$. It is important to note that in flame spraying, the generated power P_F is typically in the 10 to 50 kW range, which is comparable to that of most plasma spray torches [Wigren et al. (1996)].

In a more recent design of the powder flame spray torch, by Oerlikon-Metco, given in Fig. 7.3, the spray powder is introduced laterally into the central axial channel of the torch, from which it is transported by the gaseous fuel into the

combustion region of the torch. The oxidizer, which is oxygen in this case, is introduced in a similar way as in earlier designs through a series of orifices surrounding the fuel-powder mixture. A photograph of the 5P-II handheld torch is given in Fig. 7.4. According to Oerlikon-Metco, such a torch can operate using either acetylene or hydrogen as fuel with a nominal operating power of 35 kW.

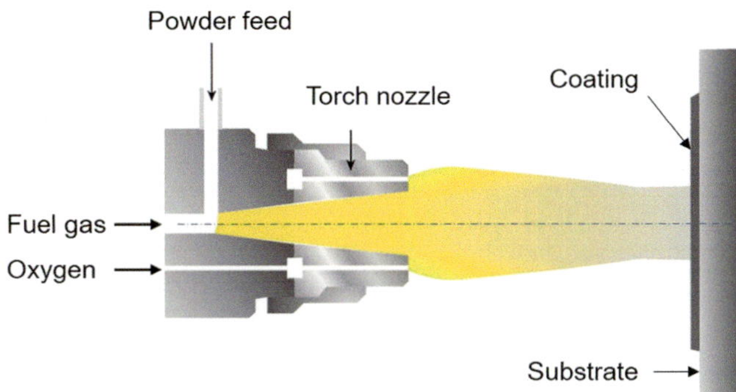

Fig. 7.3 Schematic of the Oerlikon-Metco Flame Powder Spraying torch. (Courtesy of Oerlikon-Metco)

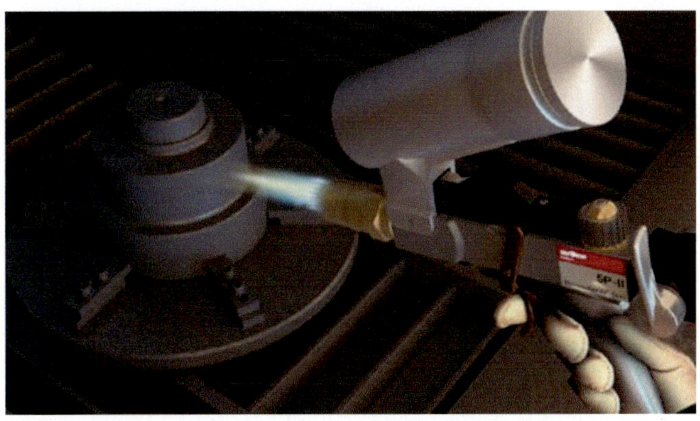

Fig. 7.4 Photograph of the hand-held Oerlikon-Metco Flame Powder Spraying torch. (Photograph Courtesy of Oerlikon-Metco)

Fundamental studies of the flow and temperature fields in low-velocity powder flame spraying (FPS) systems were reported by Nylén and Bandyopadhyay (2000), Bandyopadhyay and Nylén (2003)]. A simplified, and not to scale, schematic representation of the torch configuration used in their study is shown in Fig. 7.5.

The cooling air is injected through an outer ring of orifices. The pre-mixed acetylene-oxygen mixture (22 slm C_2H_2 + 34 slm O_2) was fed through a ring of 16 orifices (0.8 mm in internal diameter) around the nozzle rim, yielding a symmetric flame geometry, while the powder is transported by argon through a central axial channel. The diffusion combustion of the fuel extends in a free jet projecting the particles on the substrate located at 20 cm from the nozzle exit.

The mixture of the reacting gas flow was modeled using the standard k-ε turbulence gas model. In the derivation of this model, it is assumed that the flow is fully turbulent, and the effects of molecular viscosity are negligible. To accurately model the combustion process with reasonable computational effort, single- or multi-step reduced-reaction chemistry models could be used. The choice of this turbulence model also enables the use of the eddy-dissipation model as a competing mechanism for species creation/destruction, together with the Arrhenius finite-rate mechanism. As the process involves chemical reactions, and fast-moving and turbulent gas flow, a generalized finite rate formulation was applied. This approach is based on the solution of the species transport equations for reactants and product concentrations, based on the general chemical reaction given in Eq. 7.1. The gas temperature and velocity fields obtained by this relatively simple, single general chemical reaction simulation are closely comparable with those predicted by significantly more complex multireaction models [Nylén and Bandyopadhyay (2000)]. The convergence time for the simulations without the reversible reactions was significantly shorter, taking only a number of hours instead of a number of days.

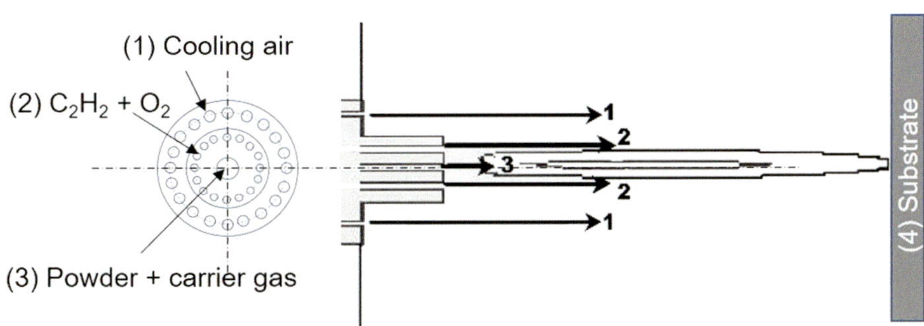

Fig. 7.5 Schematic of the Flame Powder Spraying torch (1) cooling air injected through outer ring of nine orifices at 200 m/s, (2) Fuel-oxygen mix, C2H2/O2 injected through inner ring of 16 orifices at 117 m/s, (3) argon as powder carrier gas at 12 m/s, (4) Substrate [Bandyopadhyay and Nylén (2003)]

Typical results in terms of the gas temperatures and flow fields in the flame are given in Fig. 7.6a and b, respectively. These show an average flame temperature above 2600 °C, while gas velocities are typically below 100 m/s. Bandyopadhyay and Nylén (2003) also calculated and measured the velocities and temperatures of nickel-covered bentonite particles (5–120 μm) *(Metco 312 NS)* injected axially into the flame. The results given in Fig. 7.7a show the probability distributions of particle velocities in the flame prior to their impact on the substrate. The dark bar chart provides data measured using a DPV-2000 system [Bandyopadhyay and Nylén (2003)], while the dotted lines refer to the results of the simulation study. The two sets of results are in reasonable agreement with the experimental distribution, showing a pronounced peak at a particle velocity of 22 m/s. The simulated distribution has a somewhat less pronounced peak, but this also occurs around 22 m/s. The measured distribution has a wider range than the results of the simulation with a small number of particles having velocities exceeding 30 m/s. The corresponding particle temperature distributions are given in Fig. 7.7b. Both measurement results and simulations are also in reasonable agreement with the distribution obtained by the modeling work showing a particle temperature in the range of 1000 to 2600 K, with a peak at 2300 K. The measured data show a broader particle temperature distribution in the range of 1000 to 3500 K, with a peak value around 2400 K.

By varying the flow rates of acetylene and oxygen, the flame can be made either oxidizing (fuel lean) or reducing (fuel rich). A reducing-flame limits metal oxidation. The powder is fed either by gravity or preferably with a powder feeder. The carrier gas flow rates, generally argon or nitrogen, have limited influence (a few m/s) on particle velocities, but they increase the spray spot diameter. Reducing the fuel (C_2H_2) and oxygen (O_2) flow rates from 22/34 slm to 18/24 slm results in a reduction of the power dissipated by 21% and the gas velocity by 22%. The corresponding drop of the particle mean velocity is only 13% and that of the mean temperature is 4% [Wigren et al. (1996)]. The highest temperature of NiCrAl/bentonite particles is obtained with a ratio of O_2/C_2H_2 lower than 1.4 (1.6 at stoichiometry) [Wigren et al. (1996)]. As in all spray processes, increasing the powder feed rate has a significant local cooling effect, reducing the flame velocity and temperature (loading effect). The

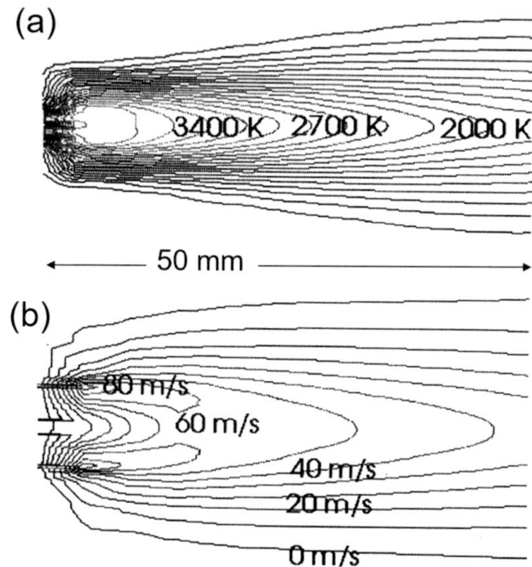

Fig. 7.6 (a) Gas temperature and (b) flow fields, for an oxy-acetylene flame [Nylén and Bandyopadhyay (2000)]. (Reprinted with kind permission from Springer Science Business Media, copyright © ASM International)

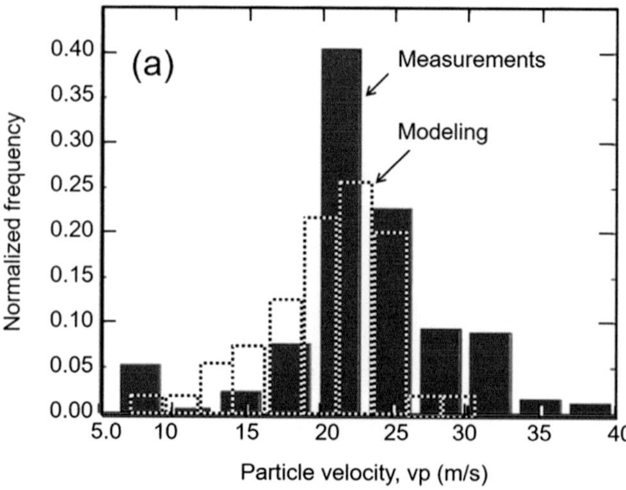

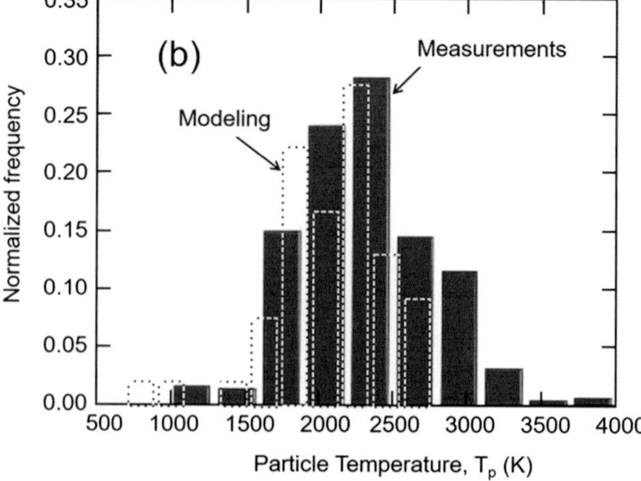

Fig. 7.7 (a) Particle velocity and (b) particle temperature distributions for Nickel covered bentonite particles (5-120 μm) injected axially into the Oxy-acetylene combustion flame. Dark bar-chart correspond to measurements, while the dotted lines correspond to the results of the modeling work [Bandyopadhyay and Nylén (2003)]. (Reprinted with kind permission from ASM International)

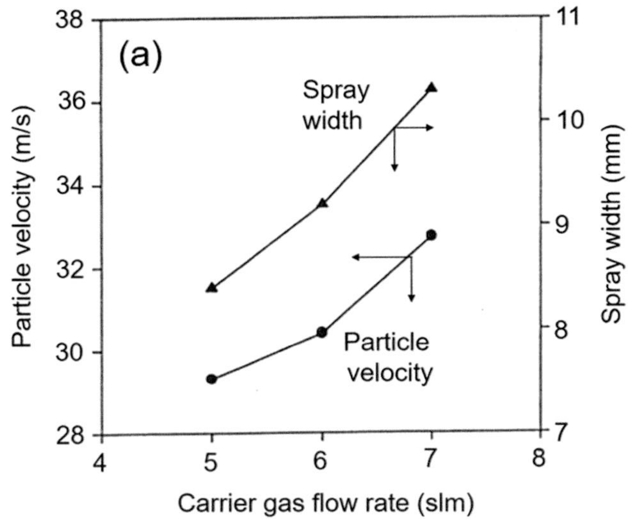

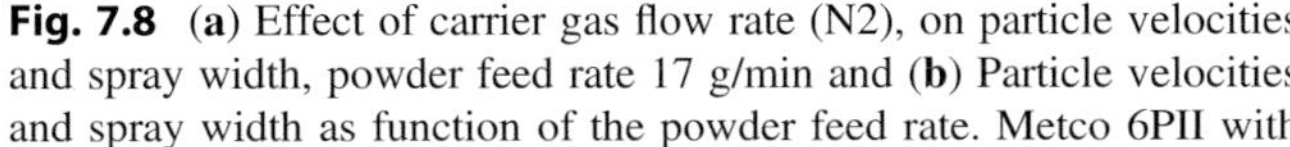

Fig. 7.8 (**a**) Effect of carrier gas flow rate (N2), on particle velocities and spray width, powder feed rate 17 g/min and (**b**) Particle velocities and spray width as function of the powder feed rate. Metco 6PII with 7C-M nozzle (18 slm C2H2 + 28 slm O2) [Wigren et al. (1996)]. (Reprinted with kind permission from ASM International)

results given in Fig. 7.8a [Wigren et al. (1996)] show that increasing the powder carrier gas flow rate gives rise to a slight increase of the particle velocity and an increase of width of the spray spot. Increasing the powder feed rate, on the other hand, Fig. 7.8b, results in a significant drop of the particle velocities, mostly due to the thermal loading effect, and an increase of the width of the spray spot.

[Hall et al. (2010)] studied the heating and acceleration of particles feed stock with an oxy-fuel flame. As for other spray processes, coating microstructure and properties strongly depend on particle velocities and temperatures at impact. They reviewed the different process factors affecting particle temperature and velocity. They showed that multiple process inputs, especially torch hardware, can significantly affect particle temperature and velocity distributions in the powder flame spray process.

7.2.2.2 Applications

As mentioned earlier, powder flame spraying (PFS) has been widely used for a large number of applications. Examples are given in this section of some of the most important and developing applications. These are presented based on the nature of the sprayed material.

(a) *Metals:*

Mainly those that are easy to melt (Pb, Ag, Zn, Al, Al-12 wt % Si, Al - 5 wt % Mg, Al-Zn, Cu and Cu alloys: Cu-10 wt % Al, Cu-10 wt % Zn, Cu-40 wt % Zn, Cu-30 wt % Zn, Cu-5 wt % Sn, Ni, Ni-20 wt % Cr, steel (0.1 wt % C or 0.1 wt % Cr, 0.25 wt % C, 0.4 wt % C, 0.8 wt % C) and Mo

[Rodriguez et al. (2007), Panossian et al. (2005), Paredes et al. (2006), Vijaya Babu et al. (1996), Brantner et al. (2003), Laribi et al. (2006)].

[Babu et al. (1996)] tested flame-sprayed molybdenum to enhance the performance of engineering components such as pistons, piston rings, and shafts. Microstructural studies were conducted on the sprayed coatings and the changes in the microstructure with different spraying conditions were correlated with the strength variations of the coatings.

[Planche et al. (2005)] sprayed NiCrBSi powder successively with plasma, flame spraying and HVOF process. The coatings were characterized in term of microstructure and composition, hardness, Young's modulus, porosity, and oxide level. The results revealed significant modifications in coating properties depending on the spray process used. Comparing the particle characteristics obtained with the different processes allows to better understand the properties of the formed coatings and to investigate the causes of coating quality changes. Differences were interpreted based on in-flight diagnostic results.

[Sakata et al. (2007)] described microstructure control aimed at improving the wear-resistance properties of Co-based (Co-Cr-W-B-Si) self-fluxing alloy coating by diffusion treatment. The treatment of thermally sprayed Co-based self-fluxing alloy coating on steel substrate was carried out at 1370–1450 K for 10–100 min under an Ar gas atmosphere. This diffusion treatment temperature substantially improved the abrasion resistance of the coating.

[Bergant and Grum (2009)] studied the properties of NiCrBSi coatings, produced by a two-step process of flame deposition followed by a furnace post-treatment. Based on

the evaluation of coating properties, the best overall quality was obtained after remitting at a peak temperature 1080 °C with 5 minutes of holding time, followed by slow air cooling.

[Huang et al. (2014)] studied aluminum-alumina composite coatings obtained by flame spraying for potential marine applications against corrosion and wear. Microstructure examination suggested dense coating structures and the evenly distributed alumina splats formed a hard skeleton connecting individual Al splats. The anti-corrosion and wear performance of the coatings were enhanced significantly by the presence of alumina. Failure analyses of the coatings after accelerated corrosion testing disclosed the intact alumina skeleton, which prevented further advancement of the corrosion.

[Navidpour et al. (2017)] considered the potential photoactivity of zinc ferrite, extensively used as a photocatalytic material. They synthesized zinc ferrite powder using mechanical alloying of hematite and zinc oxide at a molar ratio of 1:1, followed by sintering and crushing the alloy for deposition by flame spraying on 316 stainless-steel plate. They concluded that the zinc ferrite coating deposited by flame spraying possesses high photo-absorption properties as well as sufficient photoactivity under visible-light irradiation.

(b) *Self-Fluxing Alloys:*

These contained Si and B (e.g., CrBFeSiCNi), which are used as deoxidizers with a reaction of the type:

$$(FeCr)_x O_{x+y} + 2\,B + 2\,Si \rightarrow x\,Fe + x\,Cr + B_2 O_x.SiO_y$$

$$(7.5)$$

$B_2 O_x.SiO_y$ is a borosilicate. This process requires a thermal post treatment of the coating involving a "fusing" step carried out over 1040 °C using an oxy-acetylene torch which is well suited to reheat the coating. Other means can be used such as a furnace, induction (if the part diameter is almost constant), and lasers. This "fusing" treatment excludes treating coatings on aluminum alloys! During reheating, oxide diffuses toward the coating surface where it is mechanically removed (turning, milling...). The process results in rather dense and hard coatings [Lin and Han (1998), Navas et al. (2006), Sakata et al. (2007)]. These self-fluxing alloys can be reinforced with hard ceramic particles such as CrC [Harsha et al. (2007)].

[Lin and Han (1998)] investigated the process of boriding of flame-sprayed coatings. The combination of flame-spraying and boriding has been shown to improve the mechanical properties, especially the surface properties, of steels. The flame-spray process offered good workability, whereas boriding enhanced the hardness, wear resistance, and corrosion resistance of the surface, thus optimizing the surface properties. Different boriding media, boriding temperatures, and times were investigated.

(c) *Composites:*

[*Torres* **et al.** (2009)] fabricated, using oxyacetylene, thermal spraying aluminum matrix composites reinforced with more than 50 vol.% of SiC particles. The sprayed material consisted of mixtures of aluminum powder with 60–85 vol.% of SiC particles. To favor the processing of the composite, in some cases, the SiC particles were coated with silica following a sol-gel route. This allowed for obtaining as-sprayed samples with thickness above 2 mm and with porosity values below 2%. Post-processing of the samples by hot pressing allowed to reduce further the porosity of the composites and to enhance their microstructural homogeneity.

[Wang et al. (2009)] made a TiC-reinforced Ni-based composite coating by flame spraying a mixture of titanium, nickel, and sucrose (to carbonize the sucrose source of carbon). The carbon obtained by pyrolysis of sucrose was a reactive constituent as well as the binder in the composite powder. The titanium and nickel particles were bound by the carbon to form granules of the composite powder. This powder feedstock was used to prepare in situ TiC-reinforced Ni-based composite coating by oxyacetylene flame spraying. The TiC-Ni composite coating was made of TiC, Ni, and some Ni_3Ti. In the coating, a mass of fine TiC particles was uniformly distributed within the metallic matrix. The microhardness and surface hardness of the coating were, respectively, 1433 $HV_{0.2kg}$ and 62 ± 6 (HR30N). The wear resistance was much better for the TiC-Ni composite coating than for the substrate and Ni60 coating.

[Zhang et al. (2014)] deposited on 1045 carbon steel by a flame spraying and melting processing NiCr alloy coatings with 0.5, 1.0, 1.5 and 2.0 wt.% of La_2O_3. The effect of La_2O_3 addition on the tribological properties of the coatings was investigated under dry sliding wear conditions. The result showed that the microstructure of the NiCr alloy coatings is refined with proper amounts of La_2O_3, and the microhardness and wear resistance of the coatings show best enhancement with 1.0% La_2O_3.

(d) *Reactive Spraying:*

TiC cermets were obtained using asphalt as a carbonaceous precursor. The powder is made by mixing ferrotitanium powder and asphalt at 300 °C, which is heated for 2 h at 600 °C under argon atmosphere (asphalt being completely carbonized). The resulting porous mass is then crushed to form a flame sprayable powder. In situ reactions form TiC particles, which are fine, spherical, and uniformly

dispersed in the coating [Liu et al. (2002), Liu and Huang (2005)].

TiC-Fe fine-grained multiphase and multilayer coatings, composed of alternate TiC-rich and TiC-poor lamellae with different microhardness, can be obtained by reactive flame spray technology using ferrotitanium and graphite as the starting materials. Liu et al. (2002) found that parameters related to the melting and reaction of the reactive micro-pellets are the factors influencing the hardness of TiC-rich layers. A high hardness can be favored by more titanium reacted with graphite (that is, less Fe content and more C/Ti atomic ratio within the reactive micro-pellets), smaller micro-pellets and properly longer spray distance. Fe content (or the expected TiC content) and spray distance are the main parameters to affect the microstructure of the coatings.

[Hui Yuan Liu and Jihua Huang (2006)] prepared a kind of Ti–Fe–C compound powder for Reactive Flame Spray (RFS), using ferrotitanium and asphalt, which was used as the origin of carbon in the mix. The agglomerated Ti-Fe-C compound spraying powder was prepared by heating a mixture of ferrotitanium and asphalt to pyrolyze the asphalt. The TiC/Fe cermet coating prepared by RFS showed high hardness and wear resistance properties. The surface hardness of the TiC/Fe cermet coating was 65 ± 6 (HR30N). In the same fretting conditions, the wear area of the Ni60 coating is much more than that of the TiC/Fe cermet coating.

[Wang et al. (2009)] made a Ni-Ti-C composite powder for reactive thermal spraying by heating a mixture of titanium, nickel, and sucrose to carbonize the sucrose, which is used as the source of carbon. The carbon obtained by pyrolysis of sucrose was a reactive constituent as well as the binder in the composite powder. The titanium and nickel particles were bound by the carbon to form granules of the composite powder. This powder was used to prepare in situ TiC-reinforced Ni-based composite coating using oxy-acetylene powder flame spraying. The TiC-Ni composite coating is made of TiC, Ni, and some Ni_3Ti. In the coating, a mass of fine TiC particles was uniformly distributed within the metallic matrix. The microhardness and surface hardness of the coating were, respectively, 1433 HV 0.2 kg and 62 ± 6 (HR30N). The wear resistance was much better for the TiC-Ni composite coating than for the substrate and Ni60 coating.

(e) *Photocatalytic Zinc Ferrite:*

Considering its potential photoactivity, zinc ferrite has been extensively used as a photocatalytic material. Navidpour et al. (2017) presented investigation with zinc ferrite powder synthesized using mechanical alloying of hematite and zinc oxide powder at a molar ratio of 1:1 sintered and crushed for deposition by flame spraying on a 316 stainless-steel plate. Solutions were employed to study the photocatalytic activity of the coating. They concluded that the zinc ferrite coating deposited by flame spraying not only possessed high photo-absorption ability, but also exhibited sufficient photoactivity under visible-light irradiation.

(f) *Polymers*:

To reduce the risk of overheating or burning of polymer particles, flame spray guns were modified by equipping them with an air or nitrogen distributor that could be used to inject cold air or nitrogen inside the combustion zone (see Fig. 7.9a). The cold gas flow acts as a buffer zone between the flame and the particle protecting them against overheating.

The heating and fusing operation of polymer particles using flame spray guns remains; however, a challenge due to the inherently low thermal conductivity of polymers (< 0.5 W/m.K), resulting in high temperature gradients within coating [Gawne et al. (2001)], and the transient heat flux imparted to the coating surface must be carefully controlled.

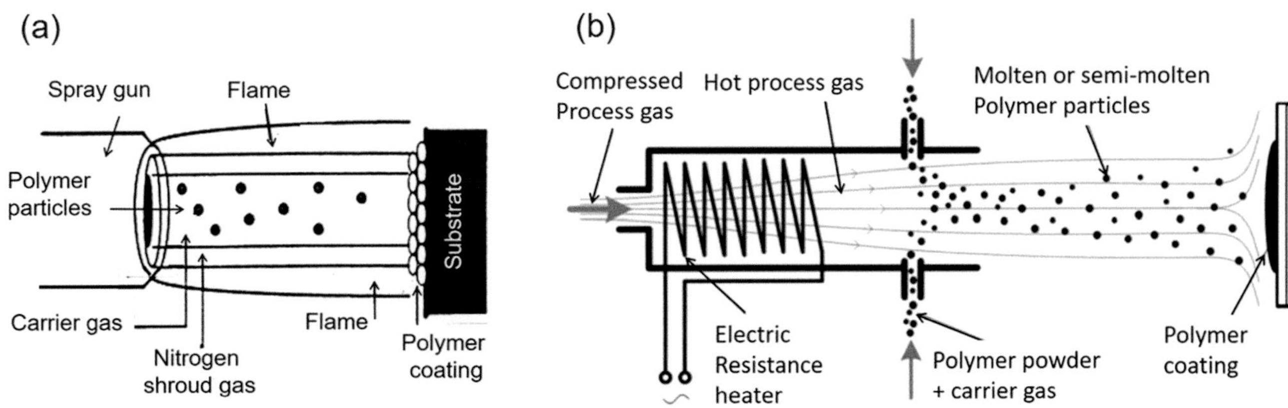

Fig. 7.9 (a) Modified flame spray gun for polymer particles spraying [Ivosevic et al. (2009)] (b) Resistively heated gas spray gun used for polymer spraying [Ivosevic et al. (2007)]. (Reprinted with kind permission from ASM International)

The limitations of the combustion-based processes for thermal spray of polymers were the motivation of Ivosevic et al. (2009) for the development of a low-temperature spray process called polymer thermal spray (PTS) based on the resistive heating of the process gas in the spray gun. A schematic of such a PTS torch is given in Fig. 7.9.*b*. An electro-resistive heating element is used to heat the main process gas, either pure air, nitrogen, an inert gas, or a mixture. The electric heating element can be set to the required temperature depending on polymer properties. The velocity of the main process gas at the nozzle exit can be adjusted in the range of 20 to 60 m/s. It controls particle in-flight residence time as well as the intensity of forced convection heat over the substrate and/or previously deposited coating layers. Polymer powder is injected downstream of the main process gas heating module using an appropriate carrier gas (see Fig. 7.9.*b*). PTS guns were developed for operation at power levels of 6 kW and 18 kW.

On the other hand, deposition of metallic coatings on elastomeric polymers is a challenging task due to the heat sensitivity and soft nature of these materials and the high temperatures in thermal spraying processes [Ashrafizadeh et al. (2016)]. They employed a flame spraying process to deposit conductive coatings of aluminum-12 silicon on polyurethane elastomers. They found that the coating porosity and electrical resistance decreased by increasing the pressure of the air injected into the flame spray torch during deposition. The latter also allowed the reduction of the stand-off distance of the flame spray torch. They found that the spray process did not significantly change the storage modulus of the polyurethane substrate material.

[Chebbi and Stokes (2012)] studied flame-sprayed biocompatible polymer coatings, made of biodegradable and non-biodegradable polymers as single coatings on titanium and as top coatings on plasma-sprayed hydroxyapatite. Biocompatible polymers can act as drug carriers for localized drug release following implantation. The polymer matrix consisted of a biodegradable polymer, poly-hydroxybutyrate 98%/poly-hydroxy valerate 2% (PHBV), and a non-biodegradable polymer, poly-methyl-methacrylate (PMMA). They studied the effects of spray parameters on coating characteristics (thickness, roughness, adhesion, wettability) and optimized the coating properties accordingly. They showed that optimized flame-sprayed biocompatible polymers underwent little chemical degradation and did not produce acidic by-products in vitro and that cells proliferated well on their surface.

[Gonzalez et al. (2016)] made a review of the deposition of metals onto polymer-based structures to enhance the thermal and electrical properties of the resulting metal-polymer material system. They explored the polymer surface preparation methods and the deposition of metal bond-coats. The objective of their review was devoted to the potential applications of thermal-sprayed metal coatings deposited onto polymer-based substrates. Their review aimed to summarize the state-of-the-art contributions on the thermal spray metallization of polymer-based materials, which has gained recent attention for potential and novel applications.

(d) *Glass:*

Refractory glass such as (Al_2O_3-Y_2O_3-BiO_2) has been flame sprayed onto stainless steel where it formed hard, uniform and well-adherent coatings [Sainz et al (2008)]. The properties of these coatings were comparable to the bulk properties of glasses of the same composition. Patents, especially in Japan, were also granted for the flame spraying of a wide range of glasses and enamels [Masataka and Kazumi (1989); Koji et al (1990); Hideki et al (1991); Tatsuya et al (1988)]. Bio-glass has been sprayed by Lugsheider et al. (1995), and the spraying of enamels on substrates prone to decompose has been reported by Arcondéguy et al (2007). According to Arcondéguy et al. (2007) and Zhang et al. (2000), the success of these applications requires a close control of the heat flux to the coating/substrate during the spraying operation, and significant reduction of residual stresses in the coating during subsequent cooling down stage.

According to Arcondéguy et al. (2007), mechanisms occurring during glaze-coating manufacturing by flame spraying may differ from those usually encountered when considering more traditional materials such as oxides or even carbides. Indeed, the coating results from the coalescence of molten particles and is manufactured in one pass rather than the stacking of individual lamellae in multiple passes. It is, therefore, important to estimate the thermal flux transmitted from the torch to the substrate, as it has to be high enough to improve the particles' spreading and wettability without leading to thermal decomposition of the substrate nor to the development of unacceptable levels of residual stresses.

7.2.3 Solution Flame Spraying (SFS)

Solution flame spraying (SFS) is a variant of powder flame spraying which has been introduced in the mid-1990s. Its novelty is in the use of a liquid feedstock, whether pure liquid, solution, or suspension, which is atomized and injected in an oxygen-hydrogen or oxy-acetylene flame where the liquid phase is evaporated, and thermo-chemical reactions produce fine or nanometer-sized particles [Tikkanen et al. (1997)]. According to these authors, atomization is optimized with an organic solvent, such as isopropanol, nebulized with hydrogen gas at a high flow rate. Liquid droplets injected into the flame are subjected to a maximum temperature of 2600 °C and are accelerated to

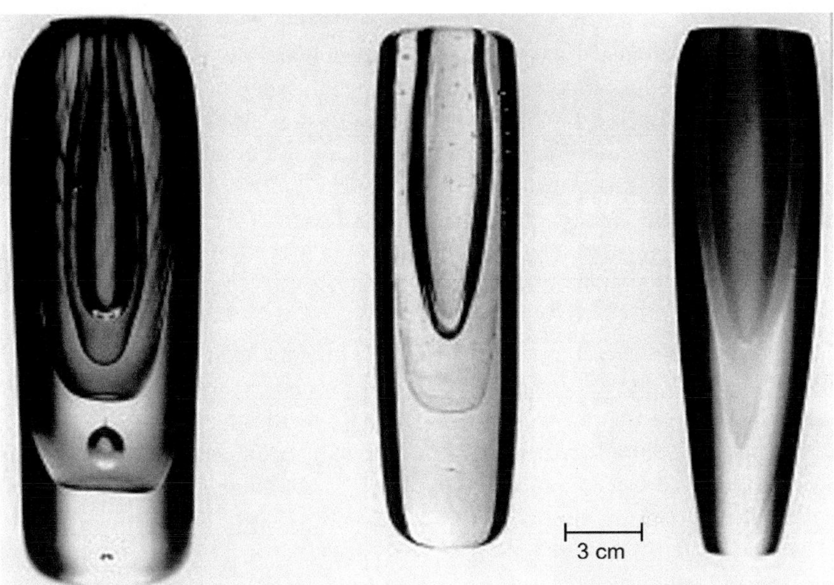

Fig. 7.10 Blue, blue-green, and yellow colored glassware produced by glass blowing. The light blue-green color indicates the weak but eloquent coloring effect of copper ions [Gross et al. (1999)]

about 160 m/s. The flame length can be controlled by flame velocity and the type of solvent. Water produces a shorter flame, whereas isopropanol extends the flame. Injection of the aerosol produces a "pencil-like" region that does not experience turbulence for most of the flame length. Experimentation with manganese nitrate and aluminum iso-propoxide or aluminum nitrate showed conversion to a manganese oxide and alumina, respectively. The process was used for glass coloring by flame spraying Co, Cu, and Ag nitrates dissolved in alcohol or water. After evaporation, precursors are transformed into oxides and sprayed onto soda-lime silica glass at 900 °C to 1000 °C. This process has produced blue, blue-green, and yellow colors [Gross et al. (1999)]. Figure 7.10, unfortunately in black and white, shows colored glass vases produced by glass blowing. The process was also used to deposit finely structured or nanostructured coatings: zirconia, starting from zirconium oxy-acetate precursor gas atomized in the flame [Poirier et al. (2003)] and nanostructured TiO_2 coatings.

[Poirier et al. (2003)] combined the atomization of a colloidal suspension with the lateral injection of the aerosol in a flame, aerosol flame spraying (AFS). The aerosol droplets were partially dried when crossing the flame and then deposited as a coating onto a substrate. Afterward, the coating was consolidated by heat treatment without extensive grain growth. They also modeled the trajectories, acceleration, and vaporization of the droplets to predict the impact conditions of the in-flight dried particles, their size, and water content when they impinge onto the substrate.

The reactive flame-sprayed deposition of photocatalytic nanostructured TiO_2 coatings was reported by Yang et al (2005) using a 30 wt.% solution of butyl titanate in pure ethanol. The formed nanostructured TiO_2 deposit, which was mostly in the anatase phase, could be transformed to rutile by annealing at temperatures above 400–500 °C. The grain size of rutile phase obtained was, however, larger than that of anatase. Moreover, it was noted that while the deposits annealed at temperatures below 450 °C were photocatalytically active, those annealed at 500 °C became photocatalytically inactive in spite of the fact that they contained 95% anatase crystal structure.

7.2.4 Wire, Rod, and Cord Spraying

7.2.4.1 Spray Gun Design and Process Characteristics

Wire, cord, and rod fed devices, as illustrated in Fig. 7.11, use air turbines or electrical motors, built into the torch, that power drive rolls, which pull feedstock from the source and push it through the nozzle at a controlled velocity. If the velocity is too high, the tip of the wire cannot melt because its residence time in the flame is too short. The uniform motion of the wire, adapted to the gas composition, flow rate, and the wire diameter and composition, is the key issue. Typical gas flow rates are as follows: 10 to 30 slm acetylene, 20 to 100 slm oxygen, and up to 17 slm (1 m^3/h) air for atomization of the molten tip.

Typical wire velocities, depending on the wire material, are between 0.6 and 14 m/min. Higher speeds are used to spray lower-melting-point metals such as Babbitt, tin, and zinc. Figure 7.12 shows the details of the nozzle of the gun.

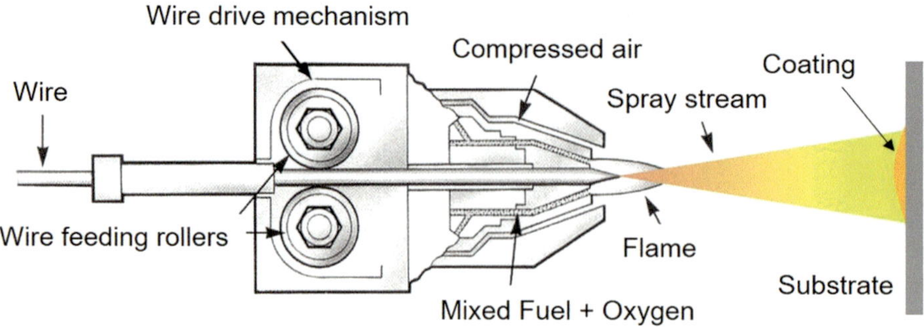

Fig. 7.11 Schematic of a Wire, Rod, or Cord Flame Spray gun [Thermal Spraying (1985)]. (Courtesy of Metallisation Flamespray Dudley, Pear Tree Lane, West Millands, DY2 OXH, England)

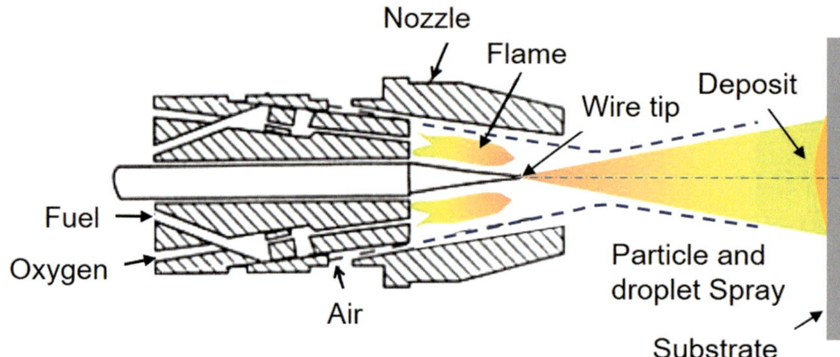

Fig. 7.12 Details of the nozzle design for a Wire, Rod, or Cord Flame Spray gun [Thermal Spraying (1985)]. (Reprinted with kind permission from American Welding Society)

Fig. 7.13 Photograph of the Oerlikon-Metco 16E Wire Flame Spray gun. (Photograph Courtesy of Oerlikon-Metco)

The nozzle produces and shapes the flame and guides the wire along its axis. This nozzle can be changed to adapt to wires of different diameters. In Fig. 7.12, the conical shape of the tip can be observed. Assuming that the liquid metal is withdrawn by the atomization gas as soon as it forms, a constant gas temperature and heat transfer coefficient result in a conical wire tip (cone angle of 25 to 30°), as observed experimentally. A photograph of an Oerlikon-Metco Model 16E Wire Flame Spray gun is shown in Fig. 7.13. Different models of this gun are offered for operation with different types of wires (Hard or soft), standard and high wire speeds, using acetylene, propylene, or propane as gaseous fuel.

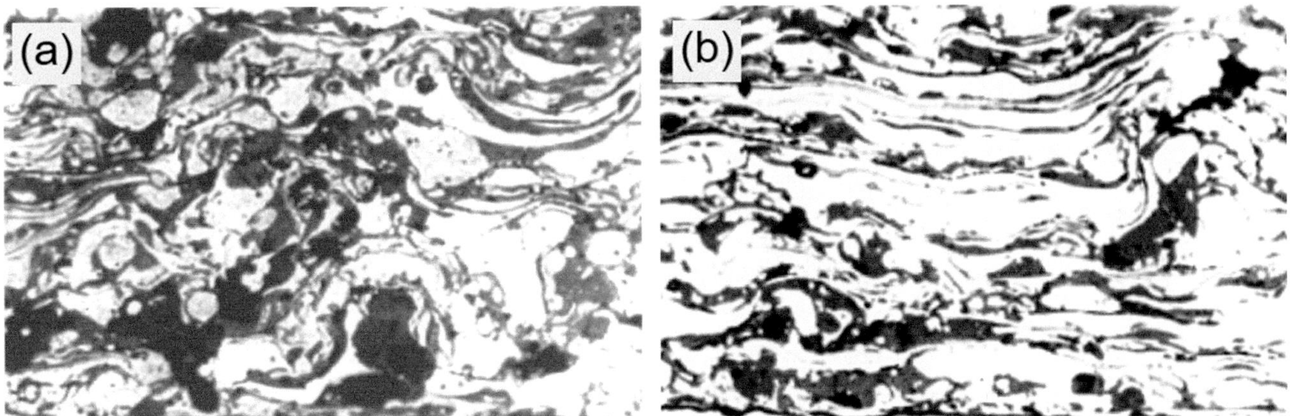

Fig. 7.14 Cross-sections of Wire Flame Sprayed Aluminum Coating (**a**) air atomized (**b**) nitrogen atomized

Typical wire diameters used are between 1.2 and 4.76 mm. They must be clean and smooth with a precise dimensional uniformity (between 0.01 and 0.05 mm). They are stored in coils, spools, or barrels. Wire feed rates depend on sprayed materials and wire diameters, for example: Al: 2–8 kg/h, Zn: 8–30 kg/h, Steel: 2–4.5 kg/h and Mo: 0.7 to 2.5 kg/h.

Particles are produced by the flow of compressed atomizing gas (generally air or nitrogen), which creates liquid metal sheets that become self-aligned in the flow (see *Fig. 11.5* of Chap. 11, pictures obtained for wire arc spraying). Waves are created at the sheet surface due to hydrodynamic instabilities, forming protuberances and inducing sheet disintegration. The flapping motion of the sheets creates showers of drops, which increases the divergence angle. The large eddy structures formed in the flow also modify particle trajectories, as shown by Hussary et al. (2007) for wire arc spraying (Chap. 11).

Coatings obtained by this process generally have high porosity (~ 10%) and can be highly oxidized, for example, to more than 25% for air atomized Al. As can be observed in Fig. 7.14, this oxidation can be reduced by a factor 2 when using nitrogen instead of air. However, considering the flow rates of the atomization gas needed, this solution is rather expensive.

Wire flame sprayed (WFS) coatings are extensively used for corrosion protection, in the automotive industry (Mo on piston rings and for "rain drop erosion" of piston heads), against aqueous corrosion [Ishikawa et al. (1999)] with Al and 80 Ni-20 Cr coatings. However, because of the coating porosity, coatings to be used for corrosion resistance must be sealed using, for example, epoxy or silicone if the service temperature is relatively low (< 200 °C) [Ishikawa et al. (2001)]. Finally, the process is rather economical and simple; it has high deposition rates (10–40 kg/h) and very good thermal efficiency to melt metals (60–70%). Moreover, substrate heating by the flame is limited.

Alternately, WFS can also be used with rods or cords (referred to Rod FS or Cord FS). The process developed by Rokide®, now Saint Gobain®, aimed at the adaptation of the flame spraying technology to the spraying of ceramics. The ceramic particles to be used were sintered to form a rigid rod (diameters between 3.17 and 6.35 mm) and continuously feed into the flame in the same way as the WFS technology, at feed rates between 10 and 20 cm/min (1.5 to 3 mm/s). Unfortunately, due to the limitation of 608 mm, on the length of the rod, the spray cycle is limited to an operation between 3 and 6 min, after which the operation has to be stopped for rod replacement. The problem was partially overcome through the use of cords which are made of ceramic particles agglomerated with either an organic binder (the evaporation or decomposition of which starts at 250 °C and is completed at 400 °C), or a mineral binder [Al(OH)$_3$ that remains up to 1500 °C]. The diameters are the same as those of rods, but the cord length can reach 120 m, which extends the cycle time to 10 to 20 h. The sprayed ceramics are mainly alumina, chromia, titania, zirconia (with calcia, magnesia, yttria as stabilizer) zircon, and alumina-titania. Spray conditions are similar to those of wires.

7.2.4.2 Applications

The materials sprayed are:

- ***Pure metals***: Al 99 wt %, Cu (99.5 and 99 wt %), Mo (99.5, 99 and 98 wt %), Sn, Zn (99.95 wt %).
- ***Alloys***: Cobalt, Copper, Nickel base, MCrAlY (cored wires), NiAl.
- ***Carbide Cermets (cored wires)***: Cr$_3$C$_2$ with Fe and FeC, WC/W2C + Fe, WC/TiC + Fe, Cr, Ni *Steels*: low carbon and stainless steel (ductile), Bronze + Al or NiAl.
- High-velocity flame spraying.

7.3   High-Velocity Flame Spraying

High-velocity flame spraying encompasses HVOF, which stands for "High Velocity Oxy-Fuel," in which the spray gun is fed with combustible gas or liquid fuel and oxygen, while HVAF stands for "High Velocity Air-Fuel," where oxygen is replaced by air, resulting in lower temperatures and higher gas velocities.

7.3.1   Basic Concepts

7.3.1.1   Spray Gun Design and Process Characteristics

Union Carbide (now Praxair Surface Technology) introduced the HVOF process in 1958 though it was not really commercialized until the early 1980s, when the Jet Kote (Deloro Stellite) system was introduced by Browning J.A. (1983). The principle of the Jet Kote system, shown schematically in Fig. 7.15a, consists in feeding a high volume of combustible gases into a water-cooled pressurized combustion chamber. The exit of this chamber is at right angle to the exit nozzle. The powder is introduced through a water-cooled central injector on the axis of the gas stream in the throat region [Thorpe and Richter (1992)]. The design of the axial flow HVOF torch shown in Fig. 7.15b was developed further through the introduction of water cooling, which improved the robustness of the torch and its thermal efficiency while maintaining axial powder injection [Ishikawa et al. (2001)]. As with the Jet Kote type gun, axial injection of

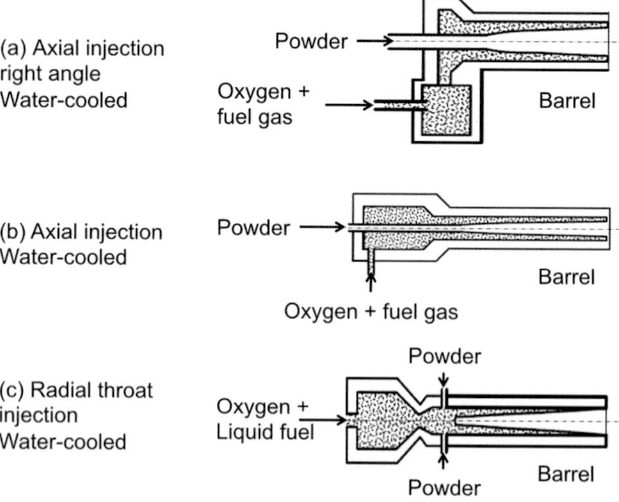

Fig. 7.15 Typical design evolution of HVOF guns (**a**) Principle of the Jet Kote, (**b**) Axial injection in the combustion chamber, (**c**) Axial chamber with radial powder injection [Thorpe and Richter (1992)]. (Reprinted with kind permission from Springer Science Business Media, copyright © ASM International)

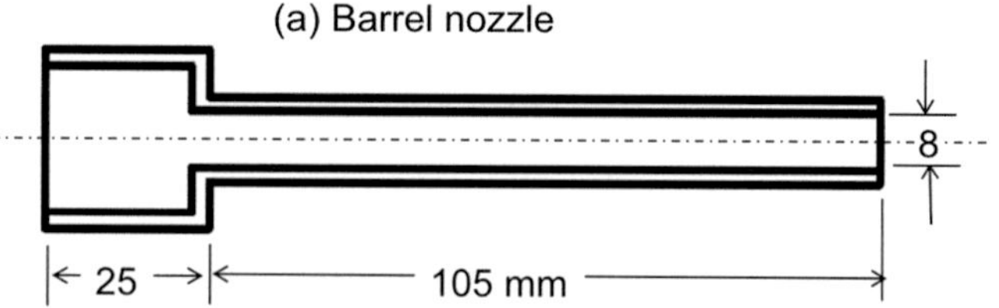

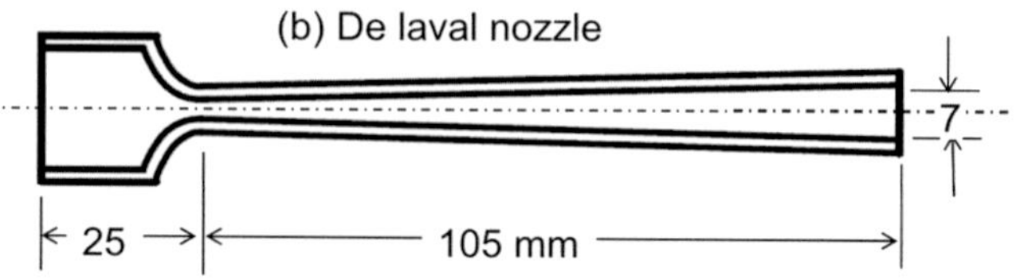

Fig. 7.16 (**a**) Barrel and (**b**) de Laval nozzles. Dimensions given in millimeters [Korpiola et al. (1997)]. (Reprinted with kind permission from Springer Science Business Media], copyright © ASM International)

the powder implies using a pressurized powder feeder. In the torch design presented in Fig. 7.15c, the combustion chamber is still co-axial with the nozzle, but the powder is injected radially beyond the throat at the beginning of the barrel or in the divergent part of the nozzle. Downstream of the throat, the pressure is much lower than in the combustion chamber upstream of the nozzle throat. Injecting powders at this point simplifies the process allowing, if necessary, multiple powder injection ports for a more uniform loading of the exit stream and efficient use of the available energy. Generally, operating data show that at least twice the spray rate per unit of energy can be achieved with radial injection versus axial injection. Moreover, this design has allowed increasing the combustion gas pressure (up to 0.8–0.9 MPa).

As in all spray processes, when a hot gas exits in the ambient air, the hot jet cools down rather fast due to its expansion and the entrainment of the surrounding air. To impede this phenomenon, it has been proposed to extend the nozzle by a water-cooled barrel (up to 30 cm long), where some energy of the jet is lost to the walls of the barrel but much less than through mixing with the ambient air. Of course, this also requires that particles are not over-heated, or remain below the melting temperature, to avoid deposition on the barrel wall. Korpiola et al. (1997) have shown, based on the 1-D compressible flow model, that with the replacement of a simple straight barrel by a de Laval nozzle, with a divergent section of the same length as the barrel, Fig. 7.16, the gas velocities increased by about 300 m/s, Fig. 7.17. They have also shown that the gas velocity at the exit plane depends on the fuel/oxygen ratio, while the combustion velocity is almost independent of the total gas flow at a constant fuel/oxygen ratio. Typical working pressures are in the range of 0.3–0.6 MPa.

7.3.1.2 High-Power HVOF

Almost simultaneously, guns with the general designs of Fig. 7.15b and b were developed using a liquid fuel such as kerosene instead of combustible gas and oxygen, allowing very high dissipated powers (almost 300 kW). With these guns, powders can also be injected radially downstream of the torch throat. Such high-power guns result in large thermal stresses and oxidation of gun components, particularly the combustion chamber and nozzle, for which the water-cooling must be carefully designed. However, the use of a liquid fuel allowed for a significant increase of the torch power, simplifies the spraying process, improves operational safety, and decreases costs without degrading operational parameters. Examples of commercially available gaseous and liquid-fueled HVOF torches are given in Fig. 7.18, provided courtesy of Oerlikon Metco. Both torches are water-cooled, with axial powder injection in the case of the torch for gaseous fuel, Fig. 7.18a, and radial injection of the powder in the case of the liquid-fueled torch given in Fig. 7.18b. A photograph of the Woka-star 610 liquid-fueled torch is given in Fig. 7.19. According to Oerlikon-Metco, this type of torch is rated for operation at power levels of 293 kW.

7.3.1.3 Evolution of the HVOF Gun Design

The spraying of many metallic or cermet materials (mostly with carbides) using HVOF is sensitive to particle oxidation and partial decarburization, which must be eliminated or at least significantly reduced in order to avoid compromising the quality of the coating. Dedicated efforts deployed over the past two decades toward this objective have led to the development of numerous innovative solutions and design modifications of the standard HVOF gun. In the following, a brief review is given highlighting some of these design proposals that have been widely accepted and used for a wide range of applications. These mostly aim at reducing the gas temperature, increasing its velocity, and protecting the particles as much as possible from exposure to the oxidizing atmosphere.

(a) *High-Velocity Air Fuel (HVAF) gun.*

The HVOF design given in Fig. 7.15c has allowed J. W. Browning [Browning (1992)] to develop the high-velocity air

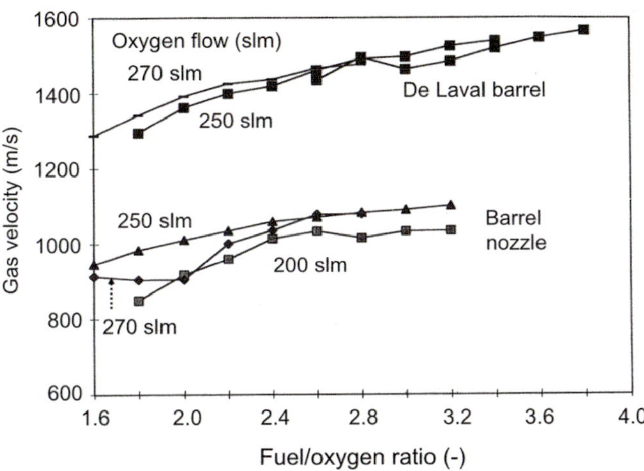

Fig. 7.17 Gas velocity at the exit plane of the HVOF spray gun (H2/02), for different fuel/oxygen ratios) for the two types of gun design: combustion chamber followed by (a.) barrel, (b.) convergent-divergent (de Laval) nozzle (see Fig. 7.16) [Korpiola et al. (1997]. (Reprinted with kind permission from Springer Science Business Media, copyright © ASM International)

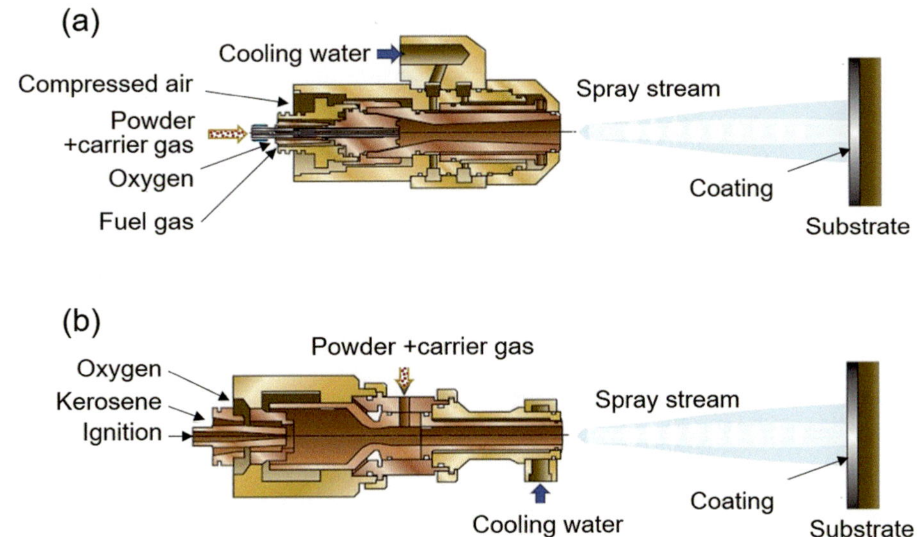

Fig. 7.18 Typical design of the Oerlikon-Metco, HVOF spray guns (**a**) water-cooled, gaseous fuel gun, (**b**) water-cooled, liquid-fueled gun (Schematics Courtesy of Oerlikon-Metco)

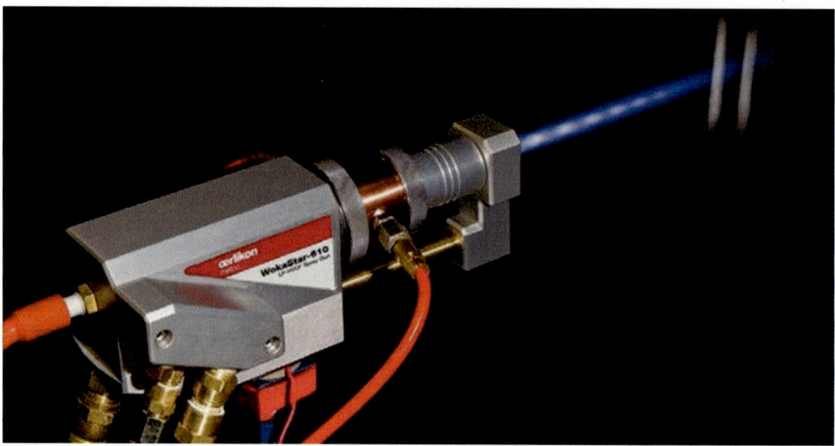

Fig. 7.19 Photograph of the Oerlikon-Metco Woka-star 610 Liquid Fuel HVOF gun, Power rating 293 kW (Schematics Courtesy of Oerlikon-Metco

fuel (HVAF) process in which he replaced the oxygen by air as oxidant. This required an increase of the oxidant gas flow rate by a factor of five for the same fuel/oxygen ratio. Nitrogen in the air, which does not participate in the combustion, must be heated, resulting in lower flame temperatures and higher gas velocities. The pressure in the combustion chamber in this case is over 0.8 MPa with airflows in the 8–10 m^3/min range (for an HVOF gun of the first generation, gas flow rates are below 1 m^3/min), giving rise to very high gas velocities (up to 2000 m/s) [Browning (1999)]. Lower temperatures and higher velocities are beneficial to limit particle oxidation. Moreover, the gun can be cooled by the airflow, and its efficiency can reach 90%. Flame ignition is generally made with pure oxygen. For example, [Evdokimenko et al. (2001)] calculated the temperature in the chamber of an HVAF gun working at 1 MPa with three different gases at stoichiometric composition: with hydrogen T = 2383 K, with methane T = 2267 K and finally with kerosene T = 2321 K.

According to Matikainen et al. (2018), particle heating and acceleration in the HVAF spray process can be efficiently controlled by changing the nozzle geometry. They sprayed fine WC-10Co-4Cr and Cr$_3$C$_2$–25NiCr powders with three different nozzle geometries (cylindrical exit diameter d = 19 mm, convergent-divergent with d = 22.5 mm, convergent-divergent with d = 26 mm) to investigate their effect on the particle temperature, velocity, and coating microstructure. Their results showed that:

- Changing the HVAF nozzle geometry from cylindrical to convergent–divergent increased the particle velocities while maintaining the particle temperature.
- Heating, melting degree, and the resulting carbide dissolution and de-carburization were effectively controlled by different nozzle geometries.

- Higher particle velocity increased the HVAF-sprayed coating hardness and cavitation erosion resistance with WC-10Co$_4$Cr feedstock.
- HVAF-sprayed WC-10Co$_4$Cr and Cr$_3$C$_2$–25NiCr coatings demonstrated significantly improved cavitation erosion resistance compared to other HVOF spray processes and bulk materials.

(b) *Gas Shrouded (GS-HVOF) and Warm Spray (WS-HVOF) guns.*

Gas shrouding is an efficient solution for the protection of the sprayed powder from contact with oxygen in the ambient air, which is engulfed into the HVOF high-velocity jet at the exit of the spray gun. It does not protect, however, the particles from oxidation by any excess oxygen in the fuel-oxygen mixture, or from partial decarburization due to excessive heating. Dilution of the combustion products by an inert gas allows, on the other hand, for the independent control of the temperature and velocity of the spray jet. An interesting review of HVOF process development is this area was reported by Kuroda et al. (2011), in which they identified shrouding and inert gas injection, proposed by Ishikawa et al. (2005), and Kawakita et al. (2006), as a viable option for the reduction of the contact of the sprayed particle with oxygen, reduction of the flame temperate, and increase of the flame velocity.

A schematic of the three HVOF torch designs considered are given in Fig. 7.20. These include, in Fig. 7.20a, a standard high-pressure HVOF (JP5000) gun design as a basis for comparison. An example of a gas-shroud attachment developed by Ishikawa et al. (2005) is given in Fig. 7.20b, which is identified as GS-HVOF. The attachment consists of a cylindrical tube added at the exit nozzle of the spray gun, which offers provision for the injection of the shroud gas through a

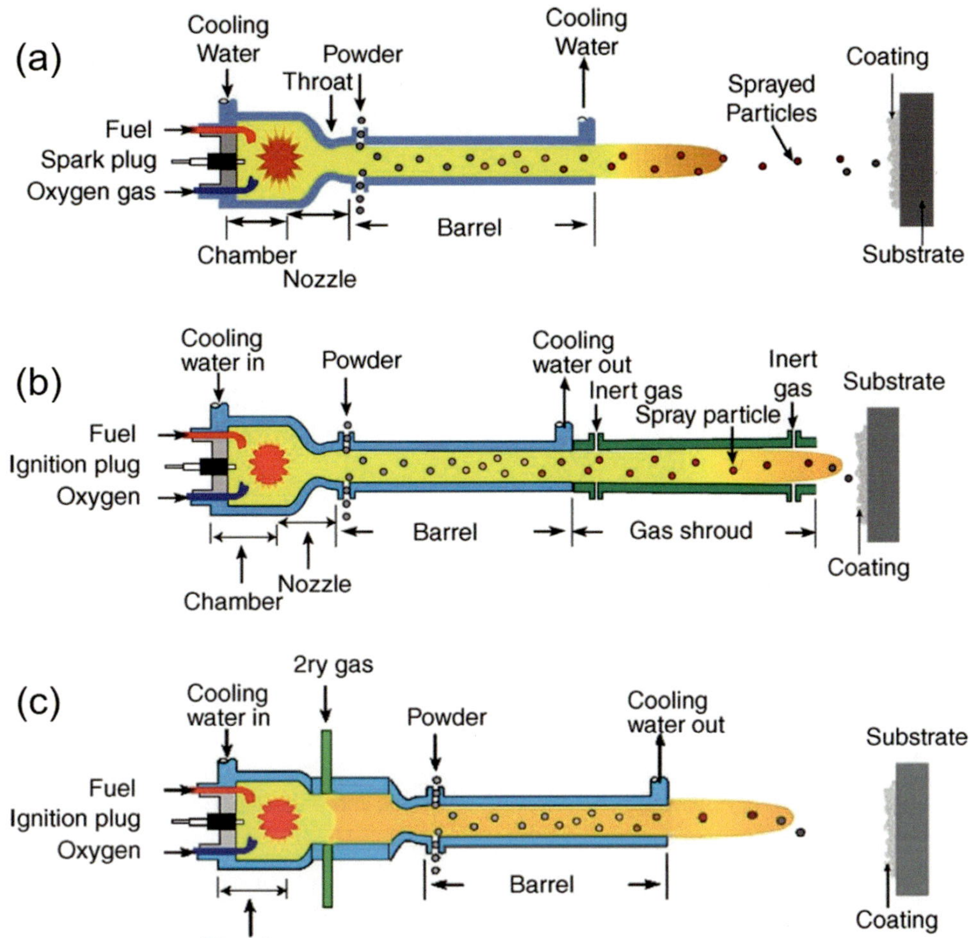

Fig. 7.20 Schematic of (**a**) commercially available high-pressure HVOF (JP-5000) gun, (**b**) gas shrouded (GS-HVOF) gun, (**c**) two-stage HVOF gun with inert gas injection identified as warm spray (WS-HVOF) gun. Guns [Kuroda et al. (2011))]. (Reprinted with kind permission from Springer Science Business Media, copyright © ASM International)

set of annular holes in its wall. The gas shroud set-up has two important effects on (WC-Co) particles injected into the flow; mean particle velocities are increased (with a reducing flame at a pressure of 0.72 MPa) from 760 m/s to 850 m/s with the shroud, while their mean temperatures drop from 1950 °C to 1830 °C. Accordingly, the density of the sprayed WC-Co coating was improved, the degree of decomposition of WC dropped from 6% to 2.5%, and coatings showed superior corrosion and wear resistance properties compared to those obtained in the absence of a shroud.

An alternative to shrouding proposed by Kawakita et al. (2006) referred to as warm spray (WS-HVOF) is based on the addition of a mixing chamber between the combustion chamber and the nozzle, as shown in Fig. 7.20c. The combustion gas generated is mixed with an inert gas such as nitrogen to lower its temperature. The objective is to limit the gas temperature between 1000 and 2300 K. Compared to the HVAF process, dilution of the combustion gases with nitrogen gives more flexibility to tailor temperatures and velocities by adjusting the nitrogen flow rate. Typical working conditions for a modified JP-5000 gun are the following: kerosene from 0.29 to 0.35 L/min (17.4–21 L/h), oxygen between 0.55 and 0.73 m³/min, nitrogen from 0.5 to 2 m³/min. When spraying titanium on a steel substrate using the modified torch design, dense coatings were obtained, which provide excellent corrosion protection in seawater in a laboratory test for over one month [Kawakita et al. (2006)]. The results clearly reveal the effectiveness of controlling the temperature and the composition of the gas environment where the titanium or titanium-alloy powder is being sprayed. Cooling of the jet flame prior to injection of the powder allowed a reduction of the oxygen content of the resulting coating to the level of the titanium feedstock, suggesting this process might compete with cold spraying [Kuroda et al. (2011)].

Figure 7.21 shows the variation in the oxygen content in the titanium coating and coating porosity with the nitrogen flow rate [Wu et al. (2006)]. The gas cooling and the resulting lower particle temperatures led to a decrease of the oxidation

level from 5.4 wt.% at 500 slm N_2 to 0.22 at 2000 slm (spray distance 280 mm). Over 1000 slm (N_2), the porosity of the coating increases with large average pore sizes because particle temperatures become too low, and particles lose their plasticity and deformability at impact. Thus, as usual, a compromise must be found.

(c) *Low-Temperature (LT-HVOF) gun.*

The low-temperature (LT-HVOF) gun proposed by *Lin* et al. *(2014)* was developed from the conventional kerosene-fueled HVOF spraying system GTV-K2 by reducing the critical diameter of the nozzle at the exit of the combustion chamber in order to increase the pressure of the combustion chamber and consequently the velocity of the gases at the exit of the spray gun. A schematic of the LT-HVOF gun is given in Fig. 7.22. In this process, the flow rate of the hot gas in the combustion chamber was controlled to be relatively low through the restriction of the exit nozzle diameter. The

plume of the modified process has a higher velocity compared to the conventional HVOF spray process. With the increase of the plume velocity, the heating time of particles decreased, resulting in a decrease of the heat transferred to the spray powder. The kerosene feed rate was reduced from 0.43 L/min (26 L/h) for a conventional HVOF gun to 0.18 L/min (11 L/h) for the LT-HVOF gun. The corresponding oxygen feed rate to the LT-HVOF gun was 88 m^3/h, compared to 90 m^3/h for a conventional HVOF gun. The combustion chamber pressure increased from 0.85 MPa for conventional HVOF to 1.55 MPa for the LT-HVOF. Results from this study *[Lin* et al. *(2014)]* demonstrate that the deposition mechanisms of the LT-HVOF process are closely related to the particle size of the feedstock materials and that the LT-HVOF process is a feasible way to spray dense TiNi composite coatings, on stainless steel 316 L substrate, with low oxygen content combined with the utilization of large-sized Ni-clad Ti powders.

(d) *Liquid Injection (LT-HVAF) gun.*

To reduce the HVAF process temperature and particle oxidation further, the injection of a liquid such as water, downstream of the nozzle, but upstream of the powder injection port, as shown in Fig. 7.23, was proposed [Yuan et al. (2006)]. The evolution of the gas temperature and velocity in such a process, referred to as LT-HVAF, as a function of the water injection rate, is given in Fig. 7.24. The results are based on computer modeling for operating using 5.47 kg/h (6.8 L/h) kerosene and water injection rate varying between 20 and 30 kg/h (20−30 L/h). Considering the high latent heat of evaporation of water, it is not surprising under these conditions to observe the rapid decrease of the gas temperature and velocity to the 600–700 K and 700–800 m/s range, respectively. It is to be noted that as the gas temperature and velocity continues to drop, the ability to use such flame for the spraying of metals and ceramics diminishes. With a water injection flow rate of 20 kg/h, the characteristics of the flame approaches essentially those of the cold spray process.

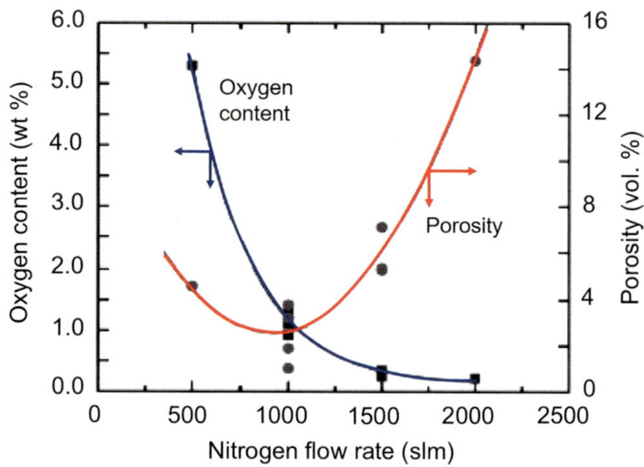

Fig. 7.21 Dependence of titanium (average particles size 28 μm) coating oxygen content and porosity on nitrogen flow rate, sprayed with the gun depicted in Fig. 7.20c [Wu et al. (2006)]. (Reprinted with kind permission from ASM International)

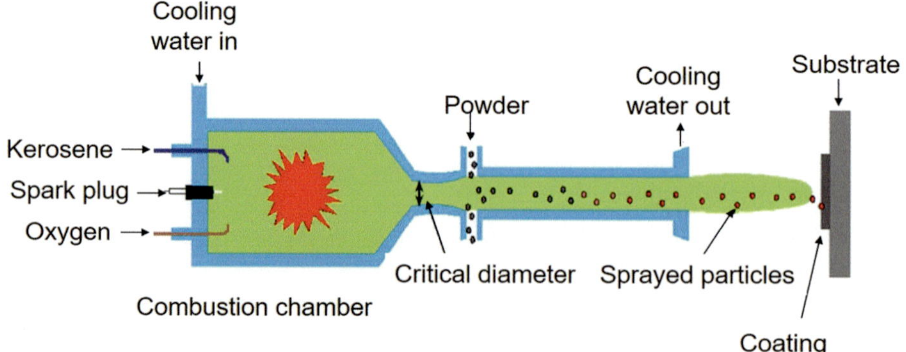

Fig. 7.22 Schematic of LT-HVOF gun by reducing the critical diameter of the combustion chamber [Lin et al. (2014)

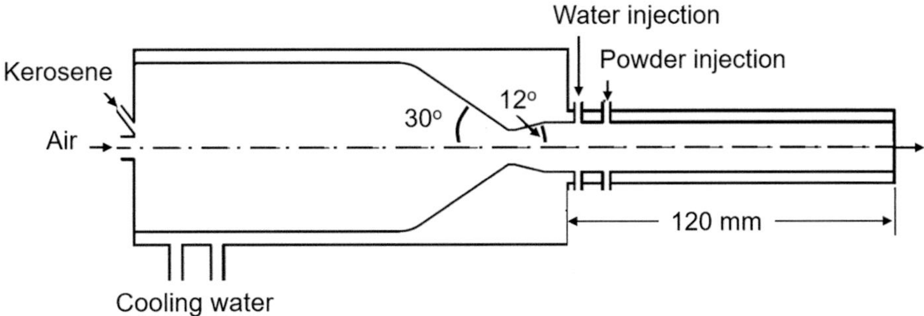

Fig. 7.23 Schematic diagram of low-temperature (LT-HVAF) spraying gun and nozzle with water injection downstream of the nozzle throat [Yuan et al. (2006)]. (Reprinted with kind permission from Springer Science Business Media, copyright © ASM International)

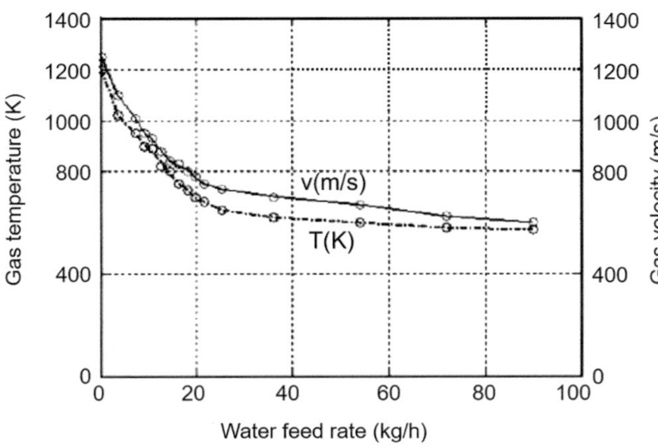

Fig. 7.24 Variation, on a logarithmic scale, of gas the velocity and temperature with water injection rate into the LT-HVAF-gun illustrated in Fig. 7.20 [Yuan et al. (2006)]. (Reprinted with kind permission from Springer Science Business Media, copyright © ASM International)

Fig. 7.25 Copper and nickel particle behavior at LT-HVAF gun exit with water injection [Yuan et al. (2006)]. (Reprinted with kind permission from Springer Science Business Media, copyright © ASM International)

In their study, Yuan et al. (2006) also computed individual particle trajectories and temperature history for copper and nickel particulate with diameter in the range of 2 to 50 µm for the same operating conditions in the presence 20 kg/h of water atomized into the flow. The particle velocities and temperatures at the gun exit are given in Fig. 7.25 as a function of the particle diameter. These show maximum copper particle temperatures of 1200 K for particles less than 10 µm in diameter, dropping down to 650 K for 50 µm particles. The corresponding velocities for copper particles are 800 m/s for less than 10 µm particles down to 400 m/s for 50 µm particles. Slightly lower particle temperatures are observed for nickel with essentially the same velocities as for that of copper particles. It is to be noted, however, that in none of these two cases, the particle temperatures remain below the corresponding melting temperature of the particles, T_m for copper being 1357 K and that for Nickel 1728 K.

7.3.1.4 Gas and Particle Dynamics in HVOF Systems

As discussed earlier, HVOF/HVAF systems can be operated using either gaseous or liquid fuels. The typical gaseous fuels used include Hydrogen, acetylene, methane, propane, and propylene. Liquid fuels include kerosene and possibly liquid propane. As an oxidizer, oxygen or air is commonly used. The typical velocities at the exit of the nozzle can be as high as 1900 m/s. Such flows generate oblique shock waves and shock diamonds, as seen in Fig. 7.19 and Fig. 7.26. The diamonds are brighter because of higher local pressure and the temperature in those regions.

Much effort has been devoted to the study and modeling of the flow and temperature fields in HVOF/HVAF jets. These varied from simple isentropic 1-D models [Thorpe and Richter (1992)] and [Korpiola et al. (1997) to 3-D supersonic compressible flow models [Tawfik and Zimmerman (1997)]. The chemical reaction involved,

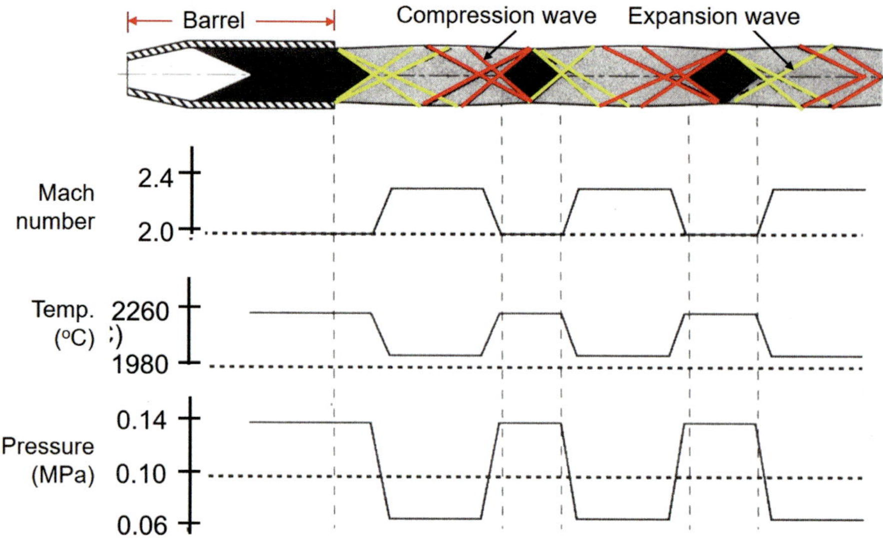

Fig. 7.26 Schematic of oblique shock waves at the barrel exit of an HVOF gun together with axial velocity (Mach number), temperature, and pressure distributions [Thorpe and Richter (1992)]. (Reprinted with kind permission from Springer Science Business Media, copyright © ASM International)

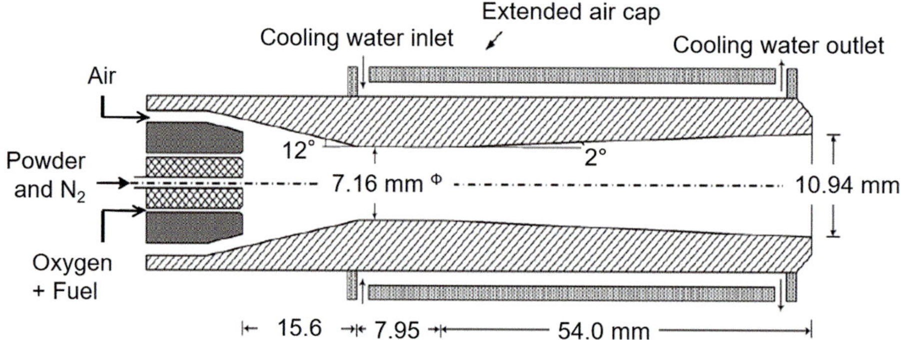

Fig. 7.27 Schematic diagram of the diamond jet hybrid thermal spray gun [Li M. and P. D. Christofide (2006)]. (Reprinted with kind permission from Elsevier)

which is a rather complex problem, is generally treated by an approximate, single-step, general formulation as follows [Oberkampf and Talpallikar (1996a)], [Oberkampf and Talpallikar (1996b)]:

$$C_xH_y + z\,O_2 \rightarrow y_1\,CO_2 + y_2\,CO + y_3\,H + y_4\,H_2 + y_5\,H_2O + y_6\,O + y_7\,OH \qquad (7.6)$$

where x, y, z, and y_i are stoichiometric coefficients, which depend on the chemical composition of the hydrocarbon fuel (C_xH_y) used. Oberkampf and Talpallikar (1996a), Oberkampf and Talpallikar (1996b) were the firsts who suggested this approach. In general, modeling results match rather well with experimental results, as for example the pressure variation along the centerline of the nozzle. It should be pointed out that in a typical de Laval nozzle configuration, with convergent/divergent expansion sections, the flow is ideally expanded if the pressure at the nozzle exit, p_e, is equal to the ambient pressure in the surrounding atmosphere, p_a (i.e., $p_e = p_a$). It is, however, known to be underexpanded if $p_e > p_a$, and overexpanded if $p_e < p_a$.

Typical results from [Li and Christofides (2005)] are presented to illustrate possibilities of 2-D models developed for the diamond jet hybrid gun. A schematic diagram of the torch is given Fig. 7.27, which is of the type shown in Fig. 7.15b.

The propylene fuel gas used in this case is mixed with oxygen through a siphon system, and fed to the air cap, where they react to produce high-temperature combustion gases. The exhaust gases, together with the air injected from the annular inlet orifice, expand through the nozzle to reach supersonic velocity. The air cap is cooled by both water and air ("hybrid") to prevent it from melting. It is assumed that the walls of the torch are maintained at a constant temperature of 400 K. The powder particles are injected at the axial powder feed channel in the inlet nozzle using nitrogen as the carrier gas. The different gas flow rates are the following: propylene 83 slm, oxygen 273 slm, air 404 slm, nitrogen carrier gas 13.45 slm; the total mass flow rate of the gaseous feed is 18.10 g/s. The flame richness ratio, R', see Eq. 7.2 $[R' = (fuel/oxygen)/(fuel/oxygen)_{St}$, is 1.045, reflecting a slight excess of propylene.

First, a 1-D simplified model formulation was used by the authors [Li and Christofides (2005)] to calculate the chamber pressure. A chemical equilibrium code developed by Gordon and McBride (1994) was next used to define the reaction formula (see Eq. 5.5) for the given partial pressures of oxygen and propylene. Subsequently, the CFD 2-D simulation was run and the final pressure compared with the one used for deriving the reaction. The trial and error approach shows that the chamber pressure is about 0.6 MPa and the partial pressures of oxygen and propylene are about 0.34 MPa. The results obtained are presented in Fig. 7.28 in terms of the isocontours and profiles, along the centerline of the torch, for the static pressure, axial velocity, Mach number, and temperature in the internal and external fields, which are given in Li and Christofides (2005). The reaction of the pre-mixed oxygen and propylene results in an increase of gas temperature above 3000 K, and a pressure of 0.6 MPa is maintained. At the exhaust end of the torch, gases expand through the convergent–divergent nozzle, pressure (Fig. 7.28a) decreases, and the gas velocity (Fig. 7.28b) increases continuously. At the throat of the nozzle, the Mach number is close to 1. The gas is accelerated to supersonic velocity in the divergent section of the nozzle and reaches a Mach number of 2 at the exit of the nozzle (Fig. 7.28c). Simulation shows that the pressure at the exit of the air cap is 0.06 MPa, which implies that the flow is over-expanded. The over-expanded flow condition gives a slightly higher gas velocity, and more kinetic energy can be transferred to the powders. It is important to note that although the gas temperature inside of the torch is very high, its value at the centerline is less than that outside of the torch (Fig. 7.28d). This also implies that the external thermal field plays a very important role in particle heating.

The over-expanded flow pattern involved in the HVOF thermal spray process is illustrated in Fig. 7.29 [Li and Christofides (2005)]. At the exit of the nozzle, the shock front begins obliquely as a conical surface and is cut off by a "Mach shock disc" perpendicular to the axis of the flow.

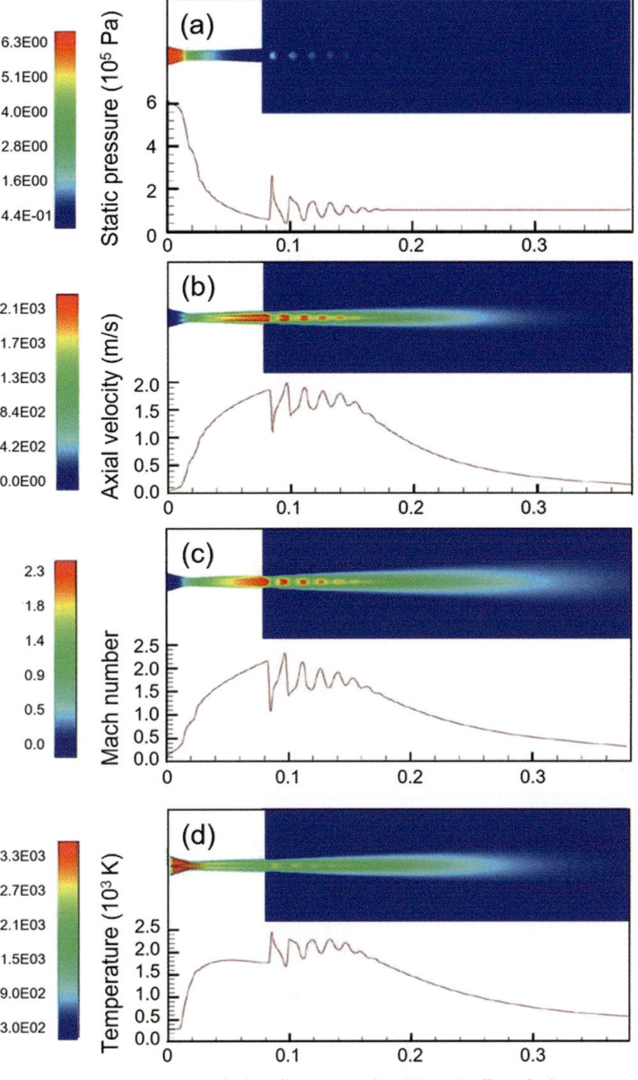

Fig. 7.28 Contours of gas properties (upper plot) and centerline profiles of gas properties (lower plot): (**a**) static pressure, (**b**) axial velocity, (**c**) Mach number and (**d**) static temperature [Li and Christofides (2005)]. (Reprinted with kind permission from Elsevier)

Behind the incident and Mach shock front, a reflected shock front and a jet boundary develop. As the reflected shock front meets the jet boundary, reflected expansion waves develop. These reflected expansion waves converge before reaching the opposite boundaries and give rise to shock fronts, which meet the jet boundary again and the whole process repeats. This periodic jet pattern is eventually blurred and dies out due to the action of viscosity at the jet boundary. It is shown that the gas temperature is relatively low at the exit of the torch (approximately 1800 K). However, passing through the first shock leads to a sharp increase in the gas temperature to around 2500 K. The location of the first shock is 7 mm from the nozzle exit.

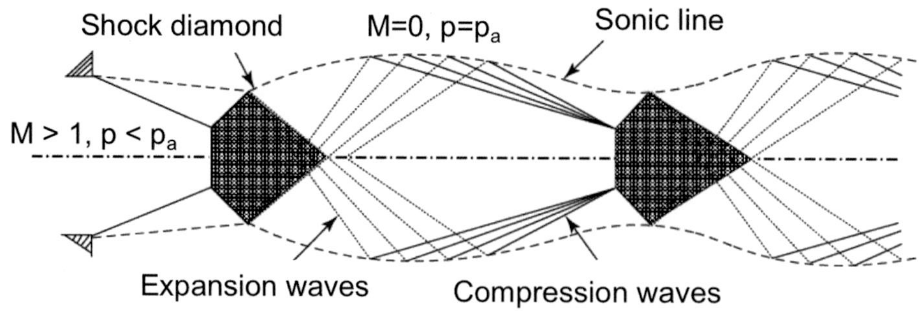

Fig. 7.29 Schematic of wave structure in the over-expanded jet [Li and Christofides (2005)]. (Reprinted with kind permission from Elsevier)

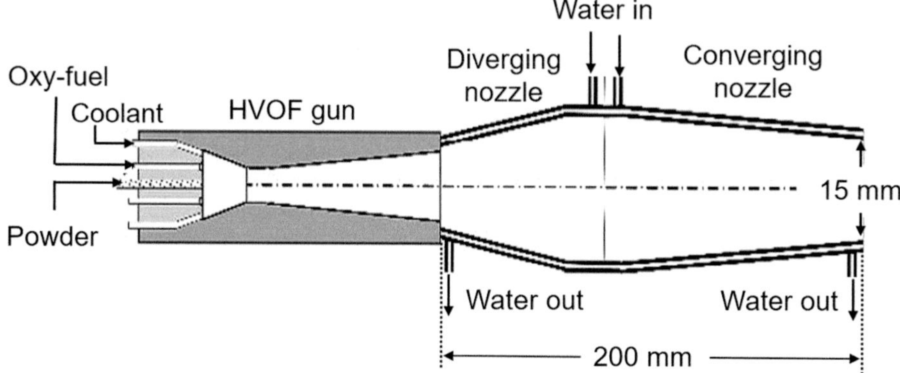

Fig. 7.30 New attachment configurations: Diverging–converging [Dolatabadi et al. (2005)]. (Reprinted with kind permission from Springer Science Business Media, copyright © ASM International)

The main parameters controlling the flow are the nozzle and barrel designs, the choice of gases, the flame richness of the mixture, and the total gas flow rate (often linked to the gas pressure in the combustion chamber).

Extensive modeling studies were dedicated to the nozzle design [Tawfik and Zimmerman (1997), Oberkampf and Talpallikar (1996a, b), Gu et al. (2001), Sakaki and Shimizu (2001), Katanoda et al. (2005), Dolatabadi et al. (2005), Yuan et al. (2006), Li and Christofides (2006)]. Improvements were made in the areas of deposition efficiency, control of in-flight particle oxidation, and flexibility to allow deposition of ceramic coatings. Based on a numerical analysis, a new attachment to a standard HVOF torch, illustrated in Fig. 7.30, was developed, tested, and used to produce thermal spray coatings [Dolatabadi et al. (2005)]. Its performance was evaluated by spraying several coating materials including metal and ceramic powders. Particle characteristics and spatial distribution, as well as gas phase composition, were compared for the new attachment and standard HVOF guns. The attachment provides better particle spatial distribution, combined with higher particle temperature and velocity.

The choice of the fuel/oxygen (F/A) ratio is important for the operation of an HVOF/HVAF system because it controls the gas temperature and the particle oxidation level. For example, calculations by Gu et al. (2001), with a gun, similar to that presented in Fig. 7.15b, working with the mixture C_3H_6-O_2, show that the temperatures that are reached within the combustion chamber vary with the (F/A) ratio. A dependence on the total gas flow rate was also observed with the higher flow rates giving marginally higher combustion temperatures. Compared to the combustion temperature calculated by another author for the same propylene-oxygen mixture at 0.1 MPa, the temperature difference is rather small. Yang et al. (2005) calculated the adiabatic combustion temperature of natural gas (CH_4) as a function of the stoichiometric factor S_t, using different O_2 + air mixtures. The stoichiometric factor S_t, as defined in Eq. 7.3, is the ratio of the oxygen flow rate used to that necessary for the complete combustion of the fuel, that is, its stoichiometric value. The results given in Fig. 7.31 show a maximum adiabatic flame temperature of 3250 K for a (CH_4 + O_2) mixture at Stoichiometric factor $S_t = 0.9$–1.0. With the addition of air (between 50 and 145 slm) to the combustion mixture, with a corresponding decrease of the oxygen flow rate in order to control the S_t values, the maximum flame temperature drops in steps down to 3150 K, while the optimum value of S_t approaches 1.0.

An extensive CFD modeling study of the HVOF process was reported by Cheng et al. (2001) with the objective of providing fundamental understanding of the effect's

controllable experimental parameters on the gas and particle dynamics in the thermal spray process. The study was carried out on the Sulzer-Metco diamond jet gun equipped with a convergent–divergent nozzle, as shown in Fig. 7.27. Since particle loading in the HVOF process is very low, it was assumed that the presence of the particles had a negligible effect on the gas velocity and temperature fields. The model is based on the solution of the corresponding mass, momentum, and energy conservation equations for turbulent flow using the $k - \varepsilon$ formulation, combined with a simplified combustion model as proposed by Oberkampf and Talpallikar (1994, 1996a, b). The study was carried out for a propylene-oxygen mixture with the chemical reaction represented by a simplified single-step equilibrium formulation given by:

$$C_3H_6 + 13.552\,O_2 \rightarrow 1.036\,CO_2 + 1.964\,CO + 0.441\,H + 0.476\,H_2 + 1.937\,H_2O + 0.398\,O + 0.734\,OH \tag{7.7}$$

The fuel gas (propylene) is premixed with oxygen in the front portion of the gun prior to its injection in the annular gap to the convergent–divergent nozzle where it is ignited. The exhaust gases, along with the compressed air injected from the annular inlet orifice, form a circular flame configuration that surrounds the powder particle input from the

central inlet hole. The parameters investigated were essentially the total gas flow rates fed into the HVOF gun and the barrel length. The results are compared to a reference case "REF" corresponding to operation under the recommended manufacturer standard conditions. The composition of the gases used in the four cases studied are given in Table 7.2. These are given in (g/s) for each component with the reference case for a total gas flow rate of 6.94 g/s. The corresponding mass fraction for each component is given as "*REF, mass fraction*." The total gas flow rate for cases *TF1*, *TF2*, and *TF3* varied between 2.28 and 9.21 g/s while keeping the ratio of fuel to oxygen, nitrogen, and air constant, as given in the *REF* case.

Results obtained for the reference case (REF) [Cheng et al (2001)], which corresponds to a relatively high total gas flow rate, are given in Fig. 7.32. The exit pressure in this case is higher than the atmospheric pressure and consequently an under-expanded jet is formed. Figure 7.32a shows the Mach number contours over the entire calculation domain. In the convergent portion of the nozzle, the velocity of the gas mixture is low, but is then accelerated to M = 1 at the nozzle throat. In the divergent portion, the gas is further accelerated to a supersonic velocity and reaches M = 2.1 (V = 2032 m/s) at the nozzle exit. It is also noted that as the gases enter the ambient environment, they expand to match the ambient pressure, accelerating to the highest velocity of 2530 m/s. The corresponding gas temperature contours and the flow streamlines are given in Fig. 7.32b and c. These show that the temperature of the gases increases rapidly in the convergent portion of the gun, where the fuel combustion takes

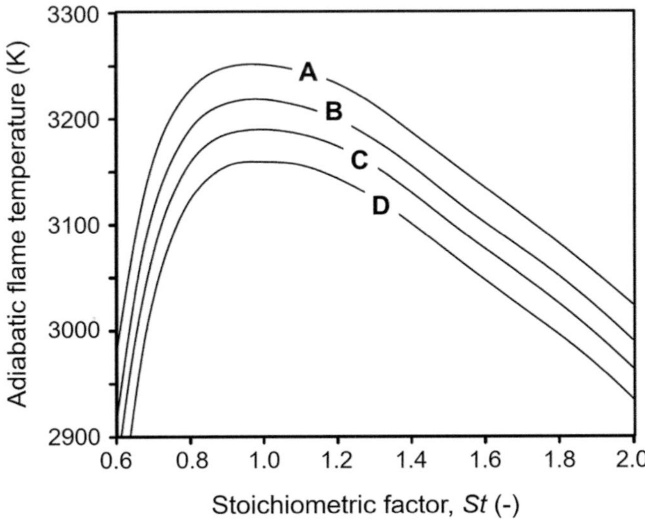

Fig. 7.31 Theoretical adiabatic flame temperatures for different combustion systems at 0.5 MPa (absolute pressure) (**a**) CH4 (200 slm) + O2 (**b**) CH4 (200 slm) + O2 + air (50 slm) (**c**) CH4 (200 slm) + O2 + air (95 slm), (**d**) CH4 (200 slm) + O2 + air (145 slm) [Yang et al. (2002)]. (Reprinted with kind permission from Springer Science Business Media, copyright © ASM)

Table 7.2 Relative mass flow rates of inlet gases for different cases compared to those of the standard case after [Cheng et al. (2001)]

Cases	Propylene (g/s)	Oxygen (g/s)	Nitrogen (g/s)	Air (g/s)	Total (g/s)
REF, mass flow rate	1.3	3.05	0.0445	2.55	6.9445
REF, mass fraction	18.72%	43.92%	0.64%	36.72%	100%
TF1	0.43	1.0	0.0101	0.84	2.2801
TF2	0.87	2.04	0.0204	1.71	4.6404
TF4	1.73	4.05	0.0405	3.39	9.2105

Reprinted with kind permission from Springer Science Business Media

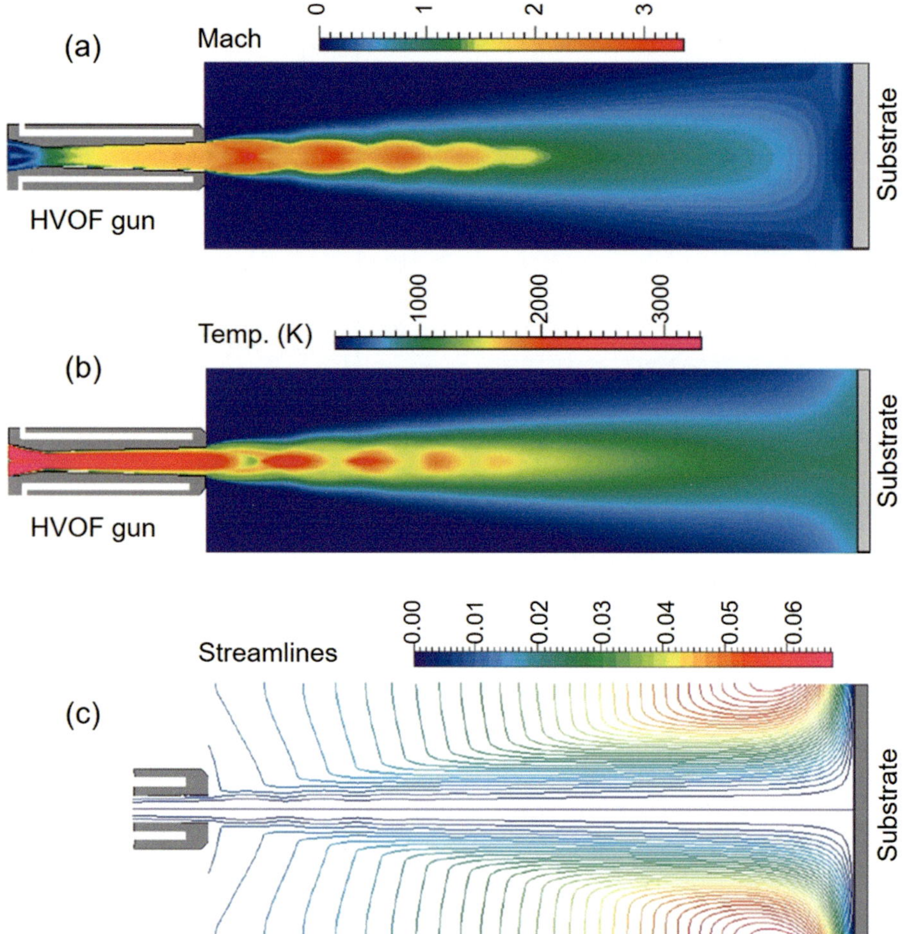

Fig. 7.32 (**a**) Mach number contours, (**b**) temperature and (**c**) streamlines for the HVOF gun with gas flow rates given in Table 7.2, REF case. (propylene-oxygen mixture, Sulzer-Metco Diamond Jet gun) [Cheng et al. (2001)]. (Reprinted with kind permission from Springer Science Business Media)

place, reaching its highest value (about 3200 K) at the throat. As the gases move in the divergent section and exit the torch, their temperature decreases as a result of energy transfer to kinetic energy. Beyond the exit of the torch, the temperature of the gases oscillates as they go through repetitive expansion and compression cycles associated with the creation of shock diamonds. The pattern of expansion and compression waves is repeated until mixing with the entrained ambient atmosphere dissipates the supersonic jet, as shown in Fig. 7.32c, giving the streamlines of the flow.

As reported by Cheng et al. (2001) (Fig. 7.30), increasing the total gas flow rate, cases *TF1*, *TF2*, and *TF3*, has little effect on the gas velocity and temperature inside the nozzle in the range investigated. This implies that the gas flow inside the nozzle is choked. However, the increase in the total gas flow rate leads to a significant increase in the pressure inside the nozzle, as shown in Fig. 7.30a. At the nozzle exit, the pressures on the centerlines reach either positive or negative values with respect to the surroundings. A positive value, obtained for high gas flow rates, indicates that the static

pressure is higher than the ambient pressure, as was in the REF case, and that the flow is under-expanded. A negative value indicates, on the other hand, that the static pressure is lower than the ambient pressure and that the flow is over-expanded [Cheng et al. (2001)]. Whatever the flow at the nozzle exit may be, under- or over-expanded, shock diamonds are observed, resulting in velocity and temperature variations along the jet axis, as illustrated, respectively, in Fig. 7.30b and c.

Such temperature and velocity variations can affect the heat and momentum transfers to small particles (below 30–20 μm depending on their specific mass). The shock diamonds and the bow shock created close to the substrate affect mainly the small particles, for example, those below 25 μm for MCrAlY particles [Yang et al. (2002)].

Mathematical modeling has also been a key tool for the study of the gas dynamics in novel HVOF/HVAF devices such as the warm spray (WS-HVAF) gun reported by Kuroda et al. (2011), as described in "*section 7.3.1.3 Evolution of the HVOF gun design*." The study made use of gas dynamic

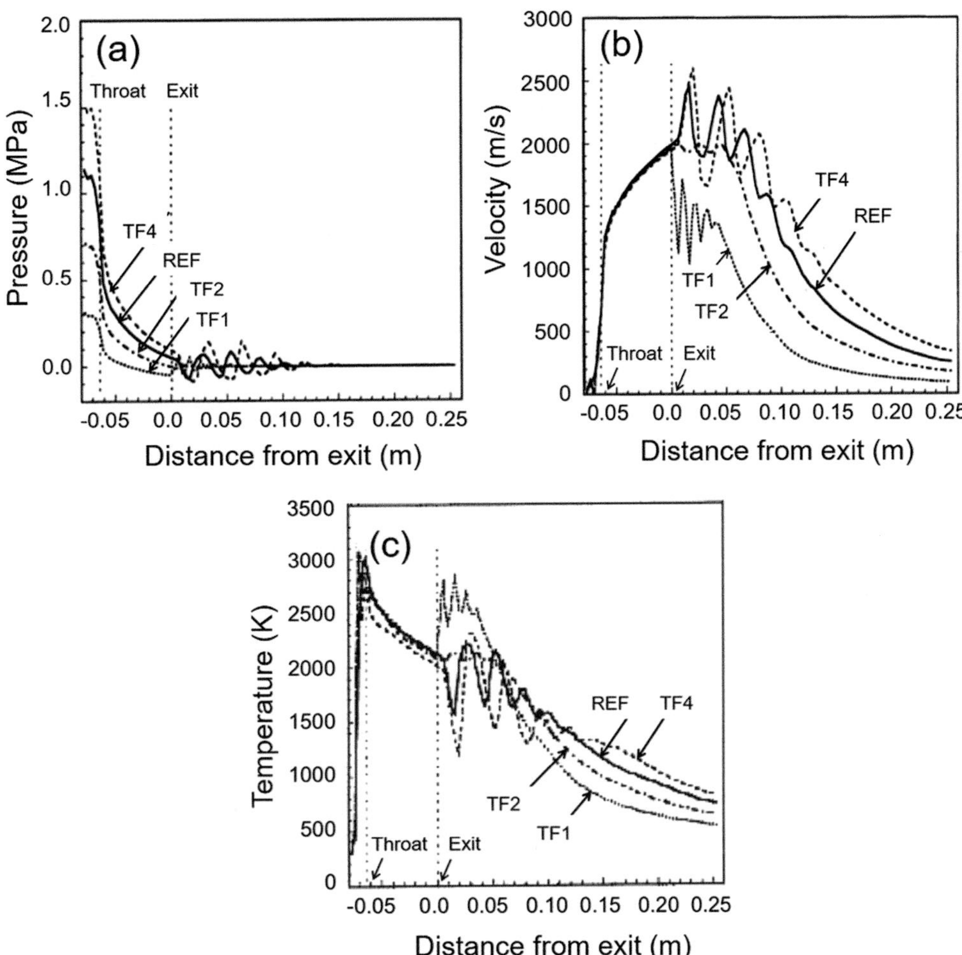

Fig. 7.33 Effect of total gas flow rate on the (**a**), pressure (**b**) velocity and (**c**) temperature, along the centerline (propylene-oxygen mixture, Sulzer-Metco Diamond Jet gun) REF and TF4 are under-expanded jets, while TF1 and TF2 are over-expanded [Cheng et al. (2001)]. (Reprinted with kind permission from Springer Science Business Media)

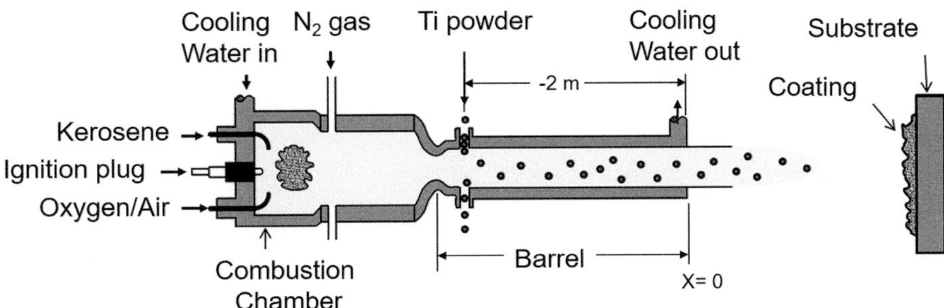

Fig. 7.34 Schematic of the WS-HVAF torch used for the numerical simulation [Kuroda et al. (2011)]

simulation for the calculation of the flow and temperature fields, particle velocity and temperatures in a WS-HVAF gun, and a standard HVAF gun for comparison. A schematic of the WS-HVAF gun used for the numerical simulation is given in Fig. 7.34. Four cases were considered for the operating conditions, given in Fig. 7.35a &b, of the WS-HVAF and the standard HVAF guns, respectively.

In the four cases considered, the fuel (kerosene) feed rate in the WS-HVAF gun varied linearly from 0.35 L/min in case 1 to 0.25 in case 4, with the mixing ratio of kerosene to oxygen set at the stoichiometric value for complete combustion. For the HVAF torch, the fuel feed rate was increased steadily from 0.17 L/min for case 1 to 0.22 L/min for case 4. The feed rate of nitrogen injected into the mixing chamber

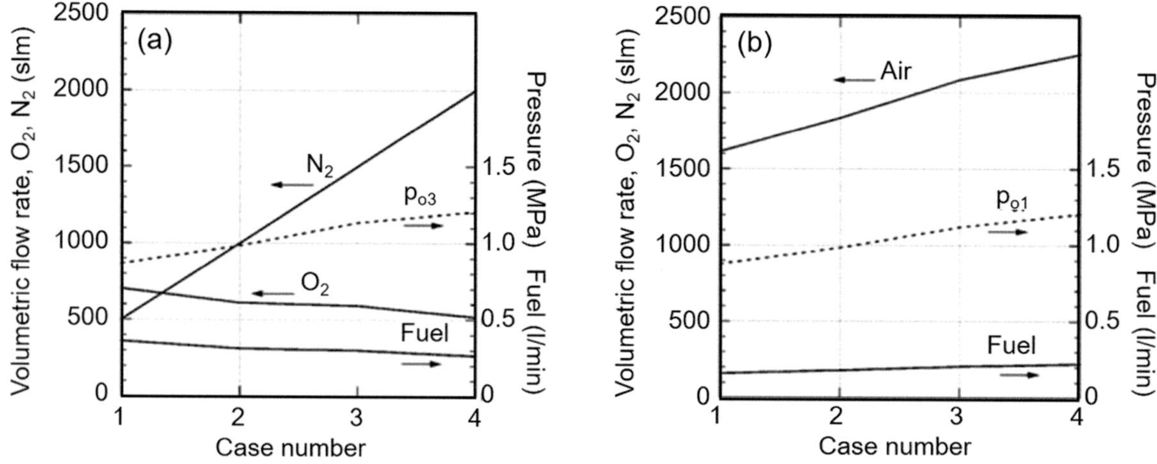

Fig. 7.35 Operating conditions of (**a**) WS-HVAF and (**b**) standard HVAF used for numerical simulation [Kuroda et al. (2011)]

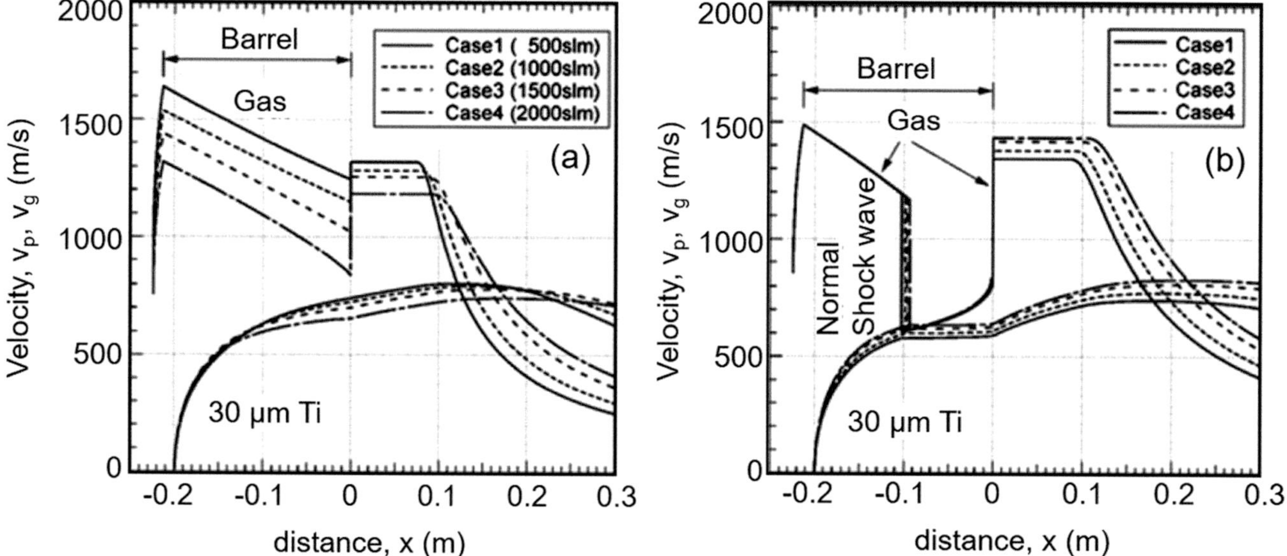

Fig. 7.36 Gas/particle velocities along the centerline (**a**) WS-HVAF gun and (**b**) standard HVAF gun based on numerical simulation [Kuroda et al. (2011)]

of the WS-HVAF torch was increased linearly from 500 slm (N_2) for case 1 up to 2000 slm (N_2) for case 4, with the pressure in the mixing chamber, p_{03}, varying from 0.88 to 1.2 MPa for case 1 and case 4, respectively. For the HVAF gun, the corresponding pressure in the combustion chamber, p_{01}, followed essentially the same curve as a result of the linear increase of the fuel and combustion air feed rate. Under these conditions, the temperature in the mixing chamber of the WS-HVAF gun was evaluated by chemical equilibrium calculations to be 2740 K for case 1 and 1780 K for case 4. The corresponding value for the combustion chamber of the HVAF gun was 2250 K for cases 1–4.

After obtaining the stagnant gas conditions just upstream of the convergent-divergent (C-D) nozzle for both the WS-HVAF and HVAF guns, a quasi 1-D gas dynamics model [Katanoda et al. (2006)] was used for the region inside the spray gun, which starts from the entrance of the C-D nozzle to the exit of the barrel. Outside the barrel, semi-empirical equations [Kleinstein et al. (1964)] were used to calculate the gas velocity and temperature of the gas jet along the centerline. The model calculations were complemented by an experimental determination of the total power generated by the combustion in the WS-HVAF gun to be around 200 kW with heat losses to the guns cooling water varying between 80 and 110 kW. Modeling results reported by Kuroda et al. (2011) are given, respectively, in Fig. 7.36 and Fig. 7.37 for the velocity and temperature of the gas and particles along the centerline of the WS-HVAF and the standard HVAF gun. Distance x = 0 corresponds to the barrel exit of the nozzle. The powder feed ports are situated about

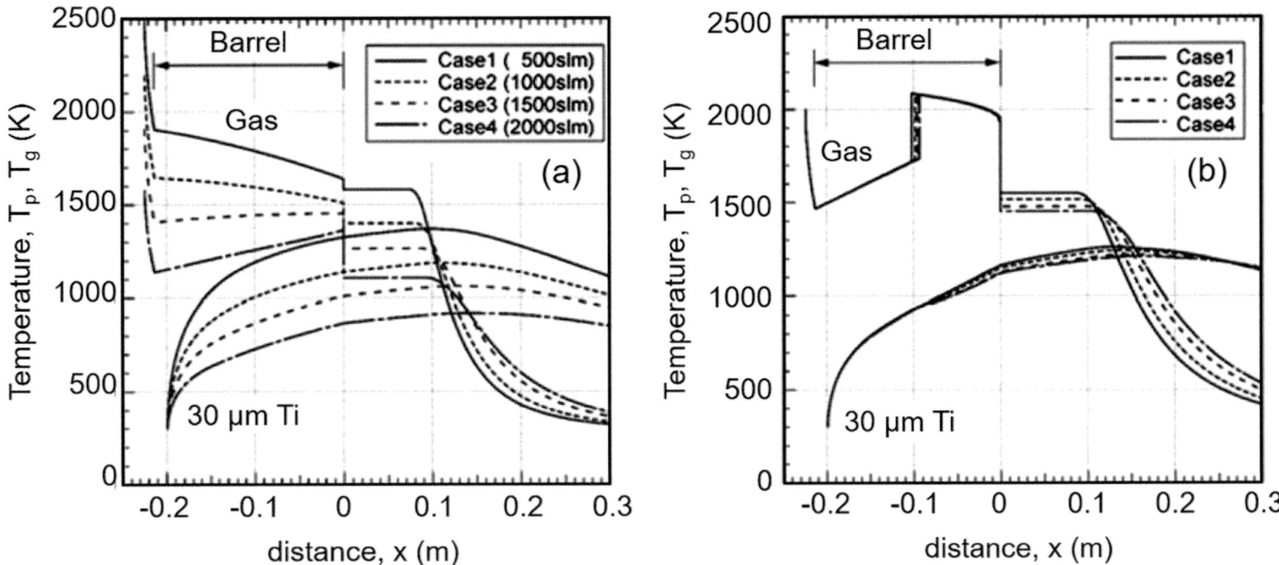

Fig. 7.37 Gas/particle temperatures along the centerline (**a**) WS-HVAF gun and (**b**) standard HVAF gun based on numerical simulation [Kuroda et al. (2011)]

−0.2 m upstream. It may be noted in Fig. 7.36a that as the gas expands through the C-D nozzle, its velocity increases rapidly to about 1300–1600 m/s depending on the nitrogen flow rate, corresponding to a Mach = 2, and then decreases gradually along the barrel length due to friction with the inner wall of the barrel. Outside the barrel, the gas velocity increases stepwise, which represents the under-expansion of the jet flow at the barrel exit, and it remains constant in the potential core of the jet outside the barrel, followed by a gradual decreases of the velocity due to mixing with the ambient air. In the case of the standard HVAF gun, Fig. 7.36b, the gas velocity at the inlet of the barrel is 1500 m/s (Mach = 2.0), which is almost the same as that for Case 2, of the WS gun. However, the gas decreases more rapidly after entering the barrel as compared to the WS gun, due to a larger decelerating effect by the pipe friction against the acceleration effect of cooling. The gas flow cannot maintain a supersonic flow until the end of the barrel, causing a normal shock wave in the barrel at around x = −0.1 m for all the four cases. In the lower part of both Figs. 7.36 *a&b,* the velocity of 30 μm diameter Ti particles is given along the centerline of the WS and standard HVAF gun. These show relatively little dependence on the flow rate of the injected nitrogen gas (cases 1–4) with an exit velocity from the gun barrel of about 600–700 m/s.

Figure 7.37a shows the significant effect of mixing with the cold nitrogen gas on the temperature of the gas and particles. After exiting the barrel, the gas temperature remains constant in the potential core region of the jet, and finally decreases again due to mixing with the ambient air. It is also noted that the injection of nitrogen gas has a significant impact on the in-flight maximum particle temperature in the

gun barrel, which drops from 1400 K to less than 1000 K with increase of the gas flow rates from 200 to 2000 slm (N$_2$). The effect translates in a drop of the particle temperature at the substrate position, x = 0.2 m, for the WS-HVAF process, from 1250 to 900 K while keeping its velocity relatively unchanged. The corresponding temperature profile for the standard HVAF gun given in *Fig. 7 37b* shows relatively higher temperatures of the gases in the barrel region, about 2000 K, with the particle temperatures impacting on the substrate placed at x = 0.2 m, around 1200 K.

7.3.2 Powder Spraying Using HVOF/HVAF

7.3.2.1 Particle Temperatures and Velocities

As has been well recognized, the quality of a coating depends directly, though not exclusively, on the particle velocity and temperature distributions prior to their impact on the substrate. It is, therefore, not surprising that much attention has been given to the calculation and control of the in-flight particle temperature and velocities within such hot flows. This is achieved using the conventional equations considering the drag force exerted on the particle [Tawfik and Zimmerman (1997), Oberkampf and Talpallikar (1996a), Oberkampf and Talpallikar (1996b), Gu et al. (2001), Sakaki and Shimizu (2001), Yuan et al. (2006), Gu et al. (2001), Yang et al. (2002), Srivatsan and Dolatabadi (2006), Cheng et al. (2001)1, Gu et al. (2004), Hanson et al. (2002), He et al. (2001)] (see also for details Chaps. 4 and 5). The drag coefficient is corrected for the particle morphology [Cheng et al. (2001b)] or, what is particularly important, for gas compressibility effects at supersonic velocities [Gu et al.

(2004)]. Such calculations show, as for other spray processes, the strong dependence of the particle temperatures and velocities at impact on their size and trajectories in the flow.

For example, in flow conditions similar to those presented in Figs. 7.32 and 7.33 [Cheng et al. (2001b)], the temperature evolution along the axial trajectory of WC-18Co particles (82 wt.% WC+ 18 wt.% Co) with six different diameters is presented in Fig. 7.38 [Cheng et al. (2001a)].

As can be seen in Fig. 7.38, the 2 and 5 µm particles follow closely the gas temperature. These small particles reach gas temperature before entering the divergent portion inside the gun, and continuously cool down with the decrease in gas temperature. As cold ambient air is entrained, it rapidly cools the gas jet, and the particle temperature drops as well. Beyond a distance of 0.24 m from the nozzle exit, the temperature of 2 µm particles equals the gas temperature. Particles with diameters of 10 and 20 µm reach the gas temperature inside the convergent portion of the nozzle, and after exiting the gun nozzle, their temperatures decrease monotonically. The temperatures for particles larger than 20 µm remain relatively unchanged after exiting the nozzle due to their high thermal inertia. The oblique shocks at the nozzle exit perturb strongly the temperature of particles smaller than 20 µm.

Similar results were obtained by Srivatsan and Dolatabadi (2006), who studied the influence of the oblique shock at the nozzle exit and the bow shock in front of the substrate on MCrAlY particles, 15 and 30 µm in diameter. It was found that 15 µm particles, being very light, are largely affected by the shock diamonds at the nozzle exit and bow shock near the substrate, whereas 30 µm particles are least affected by either shock diamonds or bow shocks. This is due to the larger Stokes numbers associated with 30 µm particles, which is the ratio of particle response time to a time characteristic of the fluid motion. They have also studied the choice of substrate configuration. A convex substrate is better when compared with flat and concave configurations. Although the shape of the concave substrate is favorable for capturing all the particles, the strength of the bow shock formed on a concave surface is very high, which leads to dispersion of most of the lighter particles.

The impact temperatures and velocities of particles thus depend strongly on their sizes, as illustrated in Fig. 7.39a and b from [Cheng et al. (2001a)], and on the distance between

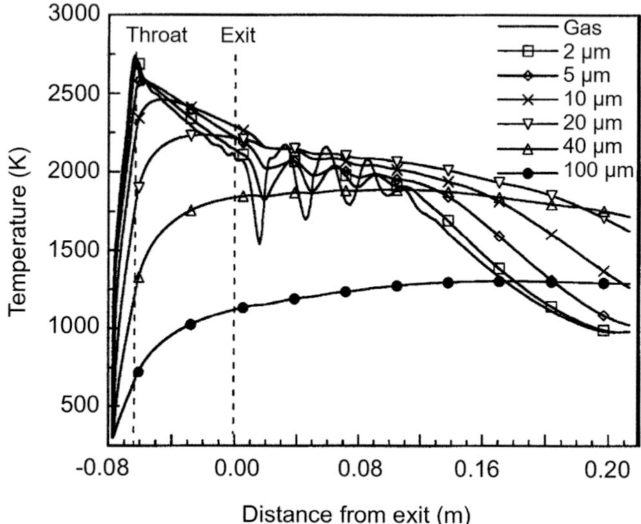

Fig. 7.38 Predicted temperatures of spherical WC-Co (82 wt.% WC) particles with different diameters as a function of distance from the gun exit for axial flow. Gas flow rates: propylene $= 6.2 \times 10^{-5}$ mol/s; oxygen $= 20.3 \times 10^{-5}$ mol/s; nitrogen $= 0.99 \times 10^{-5}$ mol/s, and cooling air $= 30 \times 10^{-5}$ mol/s [Cheng et al. (2001a)]. (Reprinted with kind permission from Springer Science Business Media)

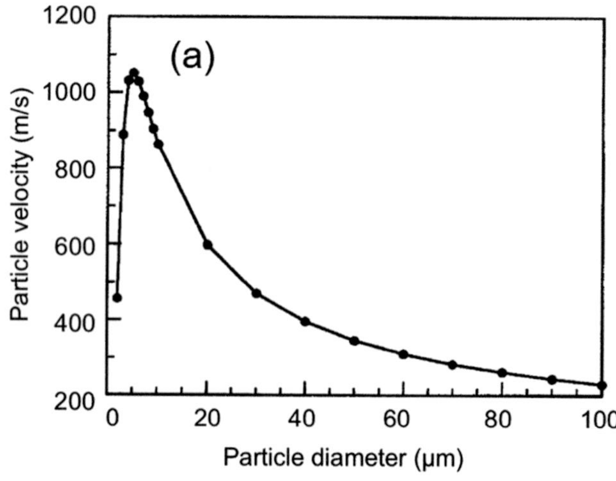

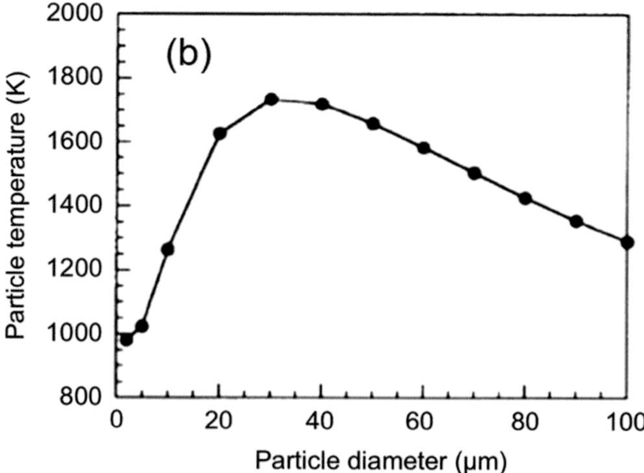

Fig. 7.39 Predicted effect of particle size on its (**a**) velocity and (**b**) temperature, at a spray distance of 0.254 m. Gas flow rates: propylene, 6.2×10^{-5} mol/s; oxygen, 20.3×10^{-5} mol/s; nitrogen, 0.99×10^{-5} mol/s; and cooling air 30×10^{-5} mol/s [Cheng et al. (2001a)]. (Reprinted with kind permission from Springer Science Business Media)

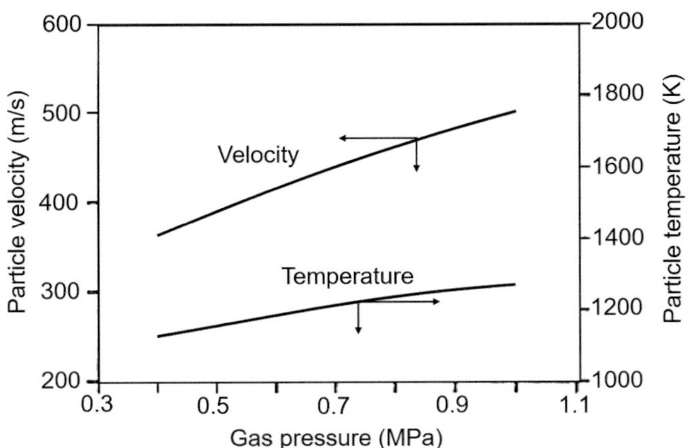

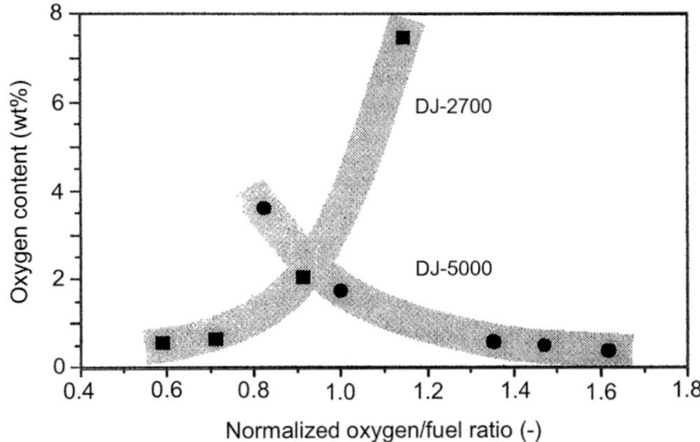

Fig. 7.40 Calculated velocity and temperature of a 38 μm stainless steel particles at the exit of the conical nozzle as function of chamber pressures. Modified TAFA JP-5000 system HVOF gun using kerosene/oxygen [Hanson et al. (2002)]. (Reprinted with kind permission from Springer Science Business Media, copyright © ASM International)

Fig. 7.41 Variation of the oxidation level of stainless steel 316 L particles with the oxygen/fuel ratio of combustible gases for the DJ-2700 and DJ-5000 HVOF gun [Dobler et al. (2000)]. (Reprinted with kind permission from Springer Science Business Media, copyright © ASM International)

nozzle exit and the substrate. These show that the most critical parameter to be concerned about for HVOF spraying is the particle velocity. As seen in Fig. 7.39a, for particles with a diameter between 10 and 60 μm, typically used in thermal spraying applications, particle velocities are in the range between 310 and 860 m/s at a spray distance of 0.254 m. This difference in velocities is equivalent to an order of magnitude difference in the kinetic energy per unit mass of the impinging particle. This will lead to different deformation conditions as splats are being formed, and the higher the velocity, the better is the coating quality. As shown in Fig. 7.39b, the effect of size distribution is less important on particle temperatures.

Calculations by Li et al. (2004b) have also shown that for a given spray gun (type of Fig. 7.15c) and a combustible gas mixture, the velocity of stainless-steel particles (28 μm in mean diameter) is almost independent of the position of the axial injector and varies almost linearly with the chamber pressure, as shown in Fig. 7.40. The observed temperature variations are significantly less than velocity. However, it was reported that particle temperature is very sensitive to the radial injector position, which has been varied from upstream of the nozzle throat to different positions downstream of it. The particle injection location was found to determine the residence time of particles within the nozzle, thus providing a means to control particle temperature. On the other hand, the injection location had a negligible effect on particle velocity according to the numerical model predictions. Experiments revealed that for 38 μm stainless steel particles, their impact velocity and temperature could be controlled within the ranges 340 to 660 m/s and 1630 and 2160 K, respectively.

7.3.2.2 Particle Oxidation

Oxidation is a key parameter for a process that is mainly devoted to the spraying of metals and metal-matrix composites. Moreover, composite particles often contain fine carbide particles, which are very sensitive to oxidation. Carbides also tend to dissolve into the metal matrix if the particle temperature is too high.

Oxidation is present inside the gun and in the flame core due to the excess oxygen (lean flame). However, the particle temperature plays a key role in the oxidation process, especially if it is above the melting temperature: the oxygen content within a steel coating varies from 0.25 wt. % for particle temperatures below 1700 K to 0.8 wt.% for particles at 1800 K! Particle oxidation does not vary in a systematic fashion with the oxygen/fuel ratio of the combustible gases [Dobler et al. (2000)] (Fig. 7.41). With the DJ-2700 gun with axial injection of particles (particle temperatures higher than when injected radially into a JP-5000 gun), the excess of oxygen increases the oxygen content of the coating (316 L). In contrast, with the JP-5000, where the particle temperature is almost 800 K lower, the excess of oxygen cools the particles, and the effect is reverse, that is, the more the oxygen in the flame, the less the oxide in the coating.

Oxidation, as shown by Dobler et al. (2000), occurs only slightly in the jet core but more in the mixing region with the surrounding air (see Fig. 7.42 [Hackett et al. (1995)]) and within the coating during its formation. To reduce oxidation, a shroud gas can be used [Hackett et al. (1995)]. However, shroud gas flow rates (nitrogen) must be very high. For an O_2 flow rate of 727 slm and kerosene flow rate of 0.42 L/min (gun type of Fig. 7.15c), the nitrogen flow rate reaches 0.45 kg/s (about 21.6 m^3/s!) to reduce the oxygen content of pure iron from 3 wt. % to 1 wt.%, as shown in Fig. 7.43 [Hackett et al. (1995)].

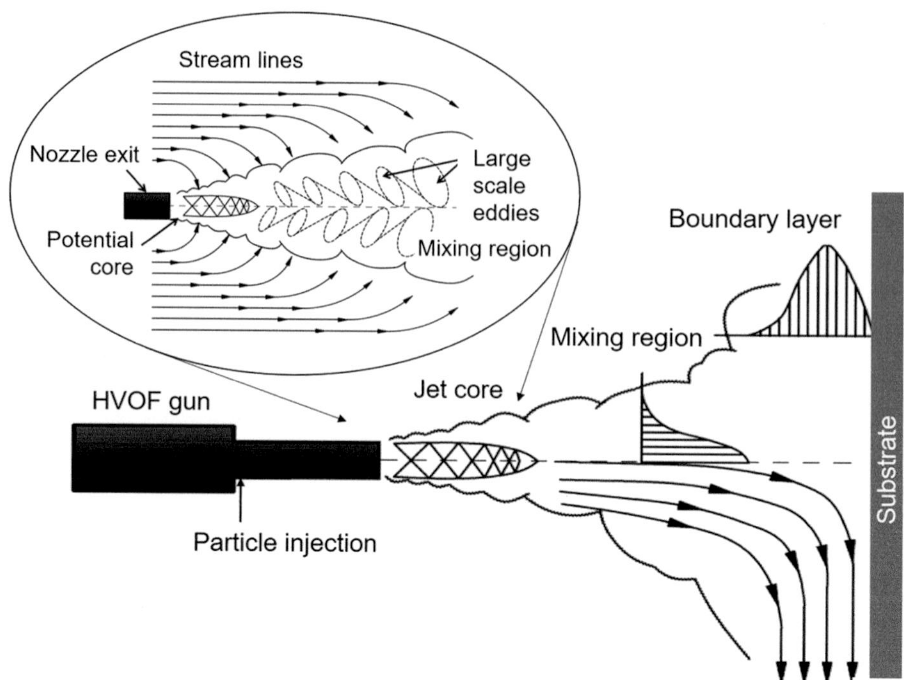

Fig. 7.42 Schematic drawing of HVOF jet mixing with the surrounding gas (**a**) details of the flow (**b**) different regions of the spray process [Hackett et al. (1995)]. (Reprinted with kind permission from ASM International)

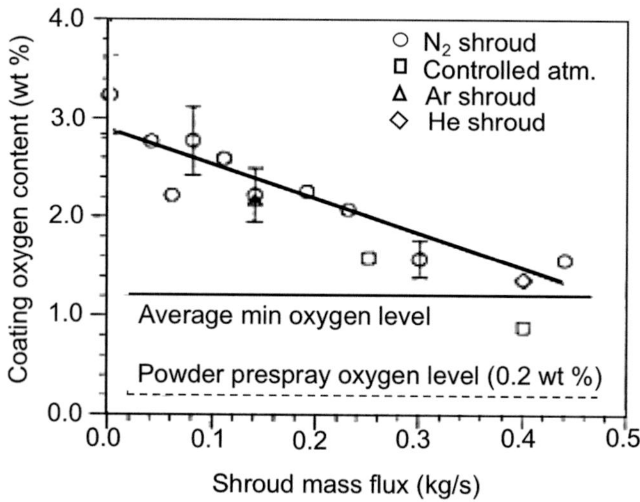

Fig. 7.43 Evolution of the coating oxygen content with the mass flow rate of shroud gas for different shroud gases [Hackett et al. (1995)]. (Reprinted with kind permission from ASM International)

The gas shroud set-up has two important effects: mean particle (WC-Co) velocities increase (with a reducing flame at a pressure of 0.72 MPa) from 760 m/s to 850 m/s with the shroud, while their mean temperatures drop from 1950 °C to 1830 °C. Accordingly, the density of the sprayed coating was improved, the degree of decomposition of WC dropped from 6% to 2.5%, and coatings were superior both in corrosion and wear resistance compared to those obtained with no shroud.

As discussed earlier in sect. 7.3.1.3, a number of HVOF/HVAF gun design modifications were made, leading to the development of the so-called gas shroud (GS-HVOF), the warm spray (WS-HVAF), and the low-temperature (LT-HVAF) guns, which offer distinct advantages in terms of reduction of in-flight particle temperature, without compromising, or at times with an increase of the corresponding particle velocity, which plays a key role in reducing particle oxidation and ceramic degradation in the spray coating process.

7.3.2.3 Coating Formation

[Trompetter et al. (2006)] made an interesting study of splat formation. They used an HVAF gun with a gas temperature of about 1300 °C, below the melting point of NiCr powder (T_m = 1400 °C). The particle velocity was about 670 m/s. When sprayed on soft substrates (compared to the NiCr hardness), predominantly solid splats with deep penetration were observed, while on hard substrates that resisted to particle penetration, a higher percentage of molten splats were observed. They attributed this behavior to increased conversion of particle kinetic energy into heat with increased particle plastic deformation with harder substrates.

The second effect of the high-velocity impacts is a peening action that results in compressive residual stresses in coatings. The first work on this topic was that of Kuroda et al. (2001). They found that the intensity of the peening action and the resultant compressive stress by HVOF-sprayed particles increase with the kinetic energy of the sprayed particles.

Measurements revealed that the residual stress at the surface of the HVOF coatings is low, often in tension, but the stress inside the coatings is at a high compression level. This low value at the surface reflects the lack of peening effect of the last deposited layer. Normally, the quenching stress due to fast cooling of the splat is always tensile, and only the peening effect can transform this tensile stress into a compressive one.

The link between the compressive peening stresses induced in the deposited coating layers and the impact velocity of the sprayed particles have also been confirmed by other researchers [Totemeier et al. (2002); Totemeier et al. (2004); Lima et al. (2006); Yilbas and Arif (2007)]. Residual stresses of +531 MPa were reported by Yilbas and Arif (2007) for the spraying of diamalloy alloy (Inconel 625) onTi-6Al-4V. Comparable values were reported by Lima et al. (2006) for the spraying of superalloy bond coats such as Amdry 9951, Sulzer Metco or Amdry 997, Sulzer Metco on UNS G41350 steel. Such values, strongly linked to the impact velocity, vary from about 100 MPa at 500 m/s to 400 MPa at 650 m/s for 316 L coatings [Totemeier et al. (2004)]. Of course, the substrate is also affected by this peening effect, which, for example for a stainless-steel substrate, affects a region of approximately 100 μm deep, producing an increased subsurface hardening. When these coatings are submitted to high temperature conditions, a recrystallization process takes place and the hardness decreases, as well as the residual compressive stress.

Diamond-Cu composites have been considered to be the next generation of electronic packing materials. One of the key stumbles for such an application is the joining problem between diamond-Cu composites and other materials due to the poor wettability of the diamond particles in the composites. In order to overcome this hurdle, [Liu et al. (2016)] thermally sprayed pure Cu powder onto diamond-Cu substrate by low-temperature high-velocity oxygen fuel spraying process. The spraying was conducted by the LT-HVOF process. As the critical diameter of the combustion chamber was reduced from 7.8 to 5.0 mm, the pressure of the combustion chamber was increased obviously from 0.85 to 1.55 MPa. The flame of the modified process was faster and more intensive than the conventional HVOF spraying process. As a result, the heating time of particles was shortened correspondingly. In order to obtain coatings with low oxygen content and porosity, the spraying parameters such as the flow rate of oxygen and kerosene, the powder feeding rate, and the critical diameter of combustion chamber were optimized. The microstructure and some fundamental properties of the coating obtained were systematically investigated, and morphologies of the single splat deposited on the diamond-Cu substrate were also observed. The splats obtained had good adhesion with the substrate as fine particles flattened sufficiently, while the coarse particles were significantly deformed. The coating was quite dense, with porosity lower than 1%, oxygen content under 0.5%,

and thermal conductivity about 266 W/m.K, and still remained on the diamond-Cu substrate after 50 thermal shock cycles between 300 °C and water bath at room temperature. Meanwhile, the solderability of the coating was significantly improved. Therefore, the Cu coating deposited on the diamond-Cu substrate by the low-temperature high-velocity oxygen fuel spraying process can be beneficial in the electronic industry, assisting with soldering and improving the wettability for joining of other materials.

Finally, the HVOF process provides high-density coatings (less than 3% porosity) due to the high impact energy of particles (between 400 and 650 m/s) [Sidhu et al. (2005)]. The short residence times of particles with rather low temperatures (by increasing the gas flow rate) allow for the deposition of particles slightly below the melting temperature, thus reducing the oxidation and decomposition of materials such as WC. The particles have better bonding due to the high impact velocity and low degradation of the sprayed materials. Thus, such coatings have better wear resistance and higher hardness than those sprayed by flames or plasmas, and now, with the *"third generation"* of spray guns, sufficiently low porosity is obtained that these coatings can be used for corrosion protection. Coatings have mostly compressive residual stress, while those sprayed by flames or plasma are often tensile.

Figure 7.44 shows the evolution of the HVOF/HVAF gun designs over the last few years tending toward a reduction of particle temperature in favor of an increase of particle velocity. HVOF spray systems of the *"third generation"* (DJ-2700, DJ-2800, JP-5000) using chamber pressures of up to 1 MPa accelerate spray particles to velocities of about 650 m/s. For HVAF guns, pressures can reach 2 MPa. Coating microstructures demonstrate that only small particles or fractions of larger ones are molten before the impact onto the substrate. A further reduction in particle temperatures below the melting temperatures of metals requires a substantial increase in velocity, which can only be realized by optimizing the expansion ratio in the diverging nozzle section and by using higher chamber pressures through the injection of non-combustible gases. The last developments correspond to power levels up to 300 kW (kerosene 31 L/h, O_2 965 slm, air 500 slm), with the flame being ignited in the combustion chamber with a hydrogen pilot flame (88 slm).

To check the influence of the gun design on coating quality, a series of experiments [Schwetzke and H. Kreye (1999)] were performed with the same powder, an agglomerated and sintered WC-Co 83–17 with a particle size range of −45 + 10 μm. The powder was sprayed with the HVOF systems Jet Kote (Stellite Coatings, Goshen, IN), Top Gun (UTP Schweißmaterial, Bad Krozingen, Germany), Diamond Jet (DJ) Standard, DJ -600, DJ-2700 (Sulzer Metco, Westbury, NY), JP-5000 (Tafa, Concord, NH), and Top Gun-K (GTV, Luckenbach, Germany). The degree of phase transformations depends on the heat

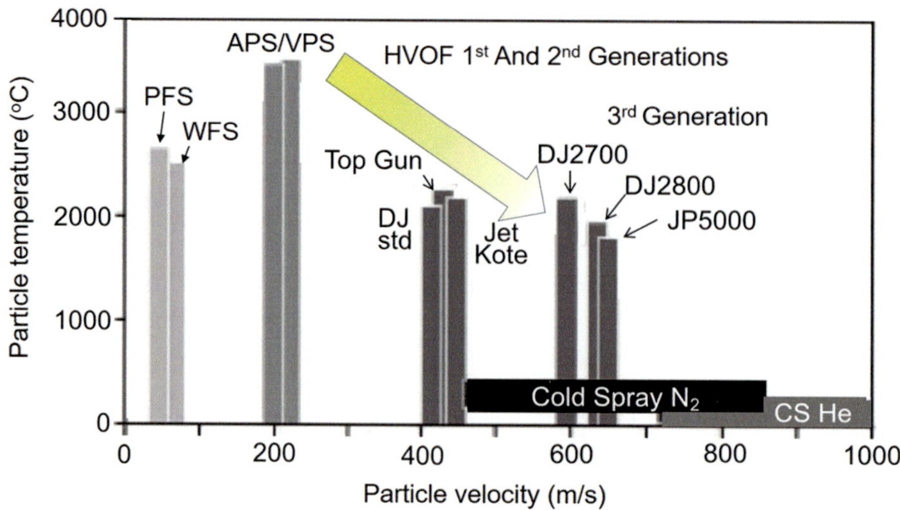

Fig. 7.44 Particle temperatures and velocities obtained in different thermal spray processes, as measured for high-density materials. (PFS) Powder Flame Spraying, (WFS) Wire Flame Spraying, (APS) Air Plasma Spraying, (VPS): Vacuum Plasma Spraying, (CS) Cold Spray) [Gärtner et al. (2006)]. (Reprinted with kind permission from Springer Science Business Media], copyright © ASM Int.)

transferred to the particles, which depends in turn on the spray system and fuel used, the flame temperature, and the type of spray powder [Schwetzke and H. Kreye (1999)]. Phase transformations increase when the injection of the powder occurs in a region where the flame temperature is highest, such as in the top gun system, where the powder is injected directly into the combustion chamber. Fewer phase transformations occur when the powder is injected behind the combustion chamber in a region where the flame temperature is low, as in the JP-5000 and Top Gun-K system, or when the flame temperature is lowered by cooling air, as in the DJ-2600 and-2700 systems [Hackett et al. (1995)]. The use of dense spray powders, which are heated up less in the spray process, or powders that already contain some amounts of h-phase, reduces phase transformations. Decarburization of an agglomerated and sintered WC-Co (83-17 wt %) powder ranges from 25 to 70% for the various spray systems and fuels. However, the properties of the coatings such as hardness and wear resistance are not influenced when the carbon loss remains below 60%. Hardness and bond strength of the coatings are mainly determined by the impact velocity of the particles, which are highest when systems with a converging-diverging Laval nozzle are used, due to the superior particle velocities.

Wielage et al. (2006) point out that new HVOF spraying guns operating at increased combustion chamber pressures have shown high potential for spraying of coatings consisting of metals that do not feature the outstanding ductility of pure copper or aluminum. High deposition efficiency, that is, up to 85%, at a considerable powder feed rate of 4.5 kg/h is already possible for spraying corrosion-protective iron or nickel-based coatings like AISI 446, AISI 316 L, or MCrAlYs. Also spraying of highly reactive materials like titanium under atmospheric conditions becomes feasible. Both dense coatings for corrosion-protection purposes and porous coatings for biomedical applications can be produced. [Wielage et al. (2006)].

According to Tillmann et al. (2018), current developments in different industrial sectors also showed an increasing demand of thermally sprayed internal diameter (ID) coatings. The most recent research and development was mainly focused on commercial applications such as arc spraying (AS), atmospheric plasma spraying (APS), and plasma transferred wire arc spraying, especially for cylinder liner surfaces. However, efficient HVOF torches are meanwhile available for ID applications as well, but in this field, there is still a lack of scientific research. Especially, the compact design of HVOF-ID and APS-ID spray guns, the need for finer powders, and the internal spray situation led to new process effects and challenges, which had to be understood in order to achieve high-quality coating properties comparable to outer diameter coatings. Thus, in the present work, the focus was on the ID spraying of bond coats (BC) and thermal barrier coatings (TBC) for high-temperature applications. An HVOF-ID gun with a N_2 injection was used to spray dense bond-coat (MCrAlY) coatings. The TBCs (YSZ) were sprayed by utilizing an APS-ID torch. Initially, flat steel samples were used as substrates. The morphology and properties of the sprayed ID coating systems were investigated with respect to the combination of different HVOF and APS spray parameter sets. The results of the conducted experiments show that the HVOF-ID spray process with N_2 injection allowed to adjust the particle temperatures and speeds within a wide range. CoNiCrAlY bond coats with porosity from 3.09 to 3.92% were produced. The spray distance was set to 53 mm, which led to the smallest coat-able ID of 133 mm, see *Fig. 5.44*. The porosity of the TBC ranged from 7.2 to 7.3%. The spray distance for the APS-ID process was set to 70 mm, which leads to the smallest coat-able ID of 118 mm.

7.3.3 Wire Spraying Using HVOF /HVAF

One of the drawbacks of wire flame spraying is the relatively low velocity of particles at impact, resulting in very porous coatings (over 10%), which are only tolerated because of the high deposition rates of the process (up to 10 kg/h), a key factor in process economics. HVOF wire spraying has been intensively studied for the automotive industry. A schematic of a typical wire HVOF torch (Metco DJRW Diamond Jet gun) is given in Fig. 7.45.

The nozzle exit diameter is 8.7 mm and a typical wire diameter is 3.2 mm. The gas flow rates are 47 slm propylene, 212 slm oxygen, and 519 slm atomizing air. With a steel wire, these conditions resulted in a droplet size distribution between 10 and 80 µm with velocities around 250 m/s. According to Neiser et al. (1995), Lopez et al. (1998), Neiser et al. (1998), stainless steel droplet velocities were measured to be 540 m/s for 10 µm and 395 m/s for 20 µm, while their temperature was close to 2200 K.

[Neiser et al. (1995), Neiser et al. (1998), Lopez et al. (1998)] have shown that the high gas velocity induces a convection phenomenon within liquid droplets removing continuously fresh metal at their surface and moving the oxygen or the oxides formed at the droplet surfaces to the inside. The formation of such as internal circulation pattern inside the droplets gives rise to coatings where FeO is detected in the resulting splats. The phenomenon is limited to the region in the close vicinity of the exit nozzle of the torch. Further downstream in the jet particle/droplet oxidation is limited to its surface. To spray steel inside aluminum-silicon cylinder bores for the automotive industry, a curved air cap has been adapted to the Metco rotating torch (DJRW Diamond jet) (Fig. 7.46). Calculations and measurements reported by Hassan et al. (1998) and Lopez et al. (1998) with the axial position of the wire give particle velocities at impact between 200 and 250 m/s.

Molybdenum has also been sprayed by wire flame and D.C. plasma jets [Modi and Calla (2001)]. Coatings obtained with Mo + 25 wt % NiCrBSi were compared to those sprayed by HVOF. The latter had a lower friction resistance than the plasma-sprayed coatings. They were also harder and more wear resistant than wire flame-sprayed coatings. Corrosion protection coatings of Ni-Cr wires [Sidhu et al. (2006)1] and NiCrBSi, Cr_3C_2-NiCr and Stellite-6 [Sidhu et al. (2006)2] sprayed with HVOF wire fuel-oxygen gun had porosities of less than 1%, and a good resistance to hot corrosion, especially the Ni-20Cr coatings.

It is interesting to note that the *"hybrid spray process"* has been developed by Stanisic et al. (2006) through a combination of arc spraying with HVOF/plasma jet gun. In this system, schematically illustrated in Fig. 7.47, the material to be sprayed is introduced either via arcing of wires only or a full-hybrid mode with both arcing of wires and a powder or wire through the HVOF gun. It was also possible to operate in

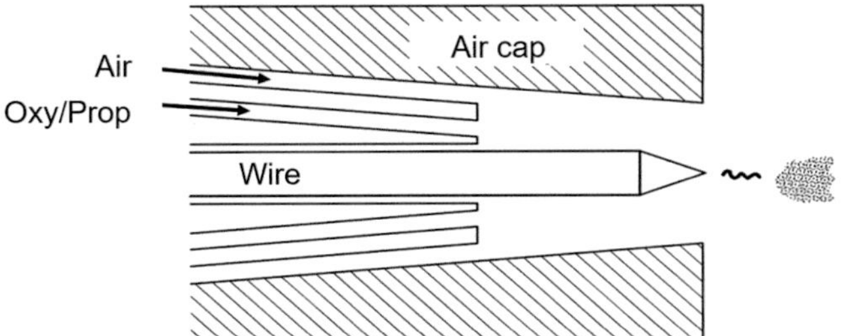

Fig. 7.45 Schematic diagram of the HVOF torch and the wire melting and atomization process [Neiser et al. (1995)]. (Reprinted with kind permission from ASM International)

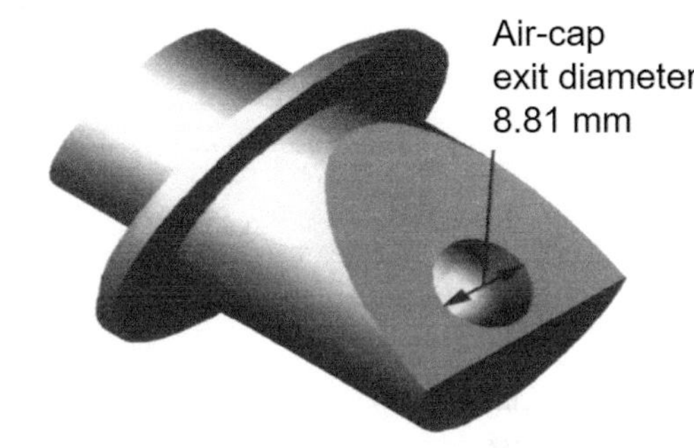

Fig. 7.46 Curved air cap adapted to the HVOF wire gun schemed in Fig. 5.37 [Hassan et al. (1998)]. (Reprinted with kind permission from Springer Science Business Media, copyright © ASM International)

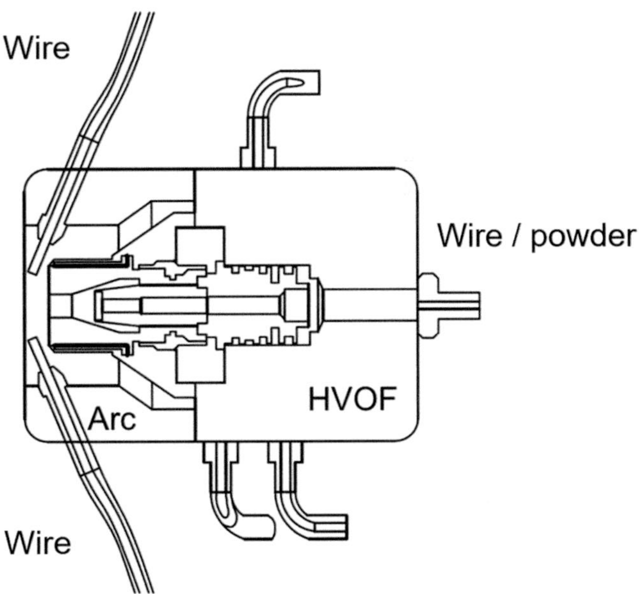

Fig. 7.47 Hybrid spray torch comprising wire arc and HVOF gun. [Stanisic et al. (2006)]. (Reprinted with kind permission from Springer Science Business Media, copyright © ASM International)

a pure HVOF mode with either wire or powder fed to the HVOF gun. According to the authors, this system provides very dense coatings compared to arc-sprayed ones [Stanisic et al. (2006)].

7.3.4   High-Velocity Suspension Flame Spraying (HV-SFS)

According to Killinger et al. (2006), the use of thermal spray technologies for the deposition of nanostructured coatings is a challenging approach with new promising applications. It requires the processing of very fine-grained powders with grain size in the nanoscale. As nano- and sub-micrometer powders cannot be handled using mechanical powder feeders, new concepts have to be developed. Among these, suspension spraying is one of the most promising. Standard HVOF techniques are not easily adapted to the injection either of nanoparticles in a liquid carrier (suspension) or in a solution that upon heating transforms in solid particles that can be very small. High-velocity suspension flame spraying (HV-SPS) is a new approach to spray micron, submicron, or nanoparticles with hypersonic speed to form thin and dense coatings. The process is based on the dispersion of the powder in an aqueous or organic solvent which is fed axially into the combustion chamber of a modified high-velocity oxyfuel (HVOF) spray torch. Several suspension feeder concepts were tested to ensure a constant flow of the suspension and, thus, a stable spray process. The study was carried out using a suspension containing sub-micrometer or nano-sized powders of alumina, titania, and yttrium-stabilized zirconia (YSZ).

[Dongmo et al. (2009a)] and [Dongmo et al. (2009b)] point out that the HV-SPS system is characterized by a two-step combustion process. The premixed combustion of propane and oxygen is run with an excess of oxygen (lean-burn combustion), leading to excess oxygen after propane combustion. Using ethanol as dispersant for the nano-structured feed powder, the ethanol vapors liberated as the suspension droplets evaporate are burned in a diffusion flame reaction using the excess oxygen in the system. Experimental studies show a jet stream, or open jet formation, of the suspension (in this case, ethanol solution containing nanoparticles [Dongmo et al. (2009b)]) for injection in a TopGun-G® combustion chamber at high pressure (0.3–0.4 MPa) with an injection nozzle with an internal diameter < 1 mm . The mechanism of droplet formation in the form of a dense cloud of droplets containing nanoparticles with uniform diameter at the nozzle outlet followed by droplet breakup is illustrated in Fig. 7.48. Open jet formation means that the fluid is injected as a coherent stream into the surrounding medium (combustion chamber with high pressure), while, for example, for automotive injection systems, a spray injection is applied with a diverging jet of atomized fluid. In case of the HV-SPS injection, the suspension jet breaks up only downstream under the influence of fluid dynamic forces. By calculating the suspension jet's Weber number, it can be evaluated if breakup takes place under the injection conditions and which breakup model is appropriate for the modeling of breakup and droplet.

Bolelli et al. (2009) manufactured Al_2O_3 coatings using the high-velocity suspension flame spraying (HV-SPS) technique with a nanopowder suspension. Their structural and microstructural characteristics, micromechanical behavior, and tribological properties were studied and compared to conventional atmospheric plasma-sprayed and high-velocity oxygen-fuel-sprayed Al_2O_3 coatings manufactured using commercially available feedstock. The HV-SPS process enabled near-full melting of the nanopowder particles, resulting in very small and well-flattened lamellae (thickness range 100 nm to 1 μm), almost free of transverse microcracking, with very few unmelted inclusions. Thus, porosity was much lower, and pores were smaller than in conventional coatings. Moreover, few interlamellar or interlamellar cracks existed, resulting in reduced pore interconnectivity (evaluated by electrochemical impedance spectroscopy). Such strong interlamellar cohesion favored dry sliding wear resistance at room temperature and at 400 °C.

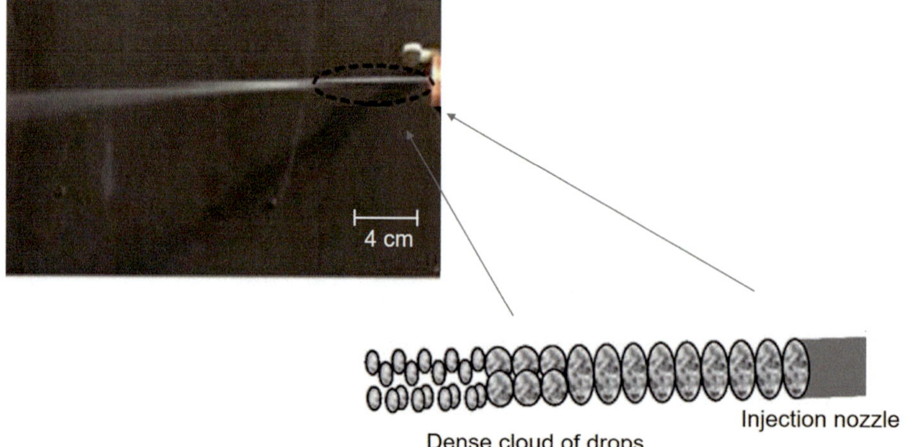

Fig. 7.48 Suspension droplets emerging from the injection nozzle and their subsequent breakup in the HVSFS jet stream [Dongmo et al. (2009b)]

[Altomare et al. (2011)] used the high-velocity suspension flame spraying technique (HV-SPS) to deposit 45S5 bioactive glass coatings onto titanium substrates, using a suspension of micron-sized glass powders dispersed in a water + iso-propanol mixture as feedstock. By modifying the process parameters, five coatings with different thicknesses and porosities were obtained. The coatings were entirely glassy but exhibited a through-thickness micro-structural gradient, as the deposition mechanisms of the glass droplets changed at every torch cycle because of the increase in the system temperature during spraying. After soaking in simulated body fluid, all of the coatings were soon covered by a layer of hydroxyapatite; furthermore, the coatings exhibited no cytotoxicity, and human osteosar-coma cells could adhere and proliferate well onto their surfaces. HV-SPS-deposited 45S5 bioglass coatings are therefore highly bioactive and have potentials as replacement of conventional hydroxyapatite in order to favor osseointegration of dental and prosthetic implants.

[Killinger et al. (2015)] point out that suspension spraying has evolved during the past decades and now is at the threshold of commercial utilization. Compared to standard powder spray methods, mainly DC plasma spraying and (high velocity) flame spraying, it is quite clear that suspension spraying will not replace these well-established technologies but can extend them by adding new coating properties. Still there remain many issues to be resolved. Suspension interaction with the hot gas stream is much more complex than in ordinary powder spray processes. In case of HVOF, when axial injection into the combustion chamber is used, a direct observation of the liquid flame interaction is not possible. This paper discusses the present status of suspension HVOF spraying (high-velocity suspension flame spraying) including torch concepts, torch configuration in case of a top gun system as well as different injector concepts and their influence on suspension atomization. The role of suspensions is discussed regarding their rheological and thermo-dynamical properties, mainly given by the solvent type and the solid content. An overview of different available diagnostic methods and systems, and the respective applicability is given. Coating properties are shown and discussed for several oxide ceramics with respect to their possible applications.

Müller et al. (2012) compared two different spray processes: suspension plasma spraying (SPS) and high-velocity suspension flame spraying (HV-SPS), which are under focus in the field of suspension spraying. Both techniques are suitable for manufacturing finely structured coatings. The differences in particle velocity and temperature of these two processes cause varying coating characteristics. The high particle velocity of the HV-SPS process leads to more dense coatings with low porosity values. Coatings with higher and also homogeneous porosity, which can be generated by SPS, also have high potential, for example, for thermal barrier coatings. In this study, both the processes, SPS and HV-SPS, were compared using alumina as feedstock material mixed with different solvents. Besides the characterization of the microstructure and phase composition of the applied coatings, the focus of this study was the investigation of the melting behavior of the particles in-flight and of single-splat characteristics.

Using the SPS process, high porosities up to 40% were achieved, with the applied solvent strongly influencing the melting behavior. Much better melting can be achieved by means of organic solvents of the additional energy that is delivered during their combustion. Further improvements concerning suspension injection will lead to better melting of the particles. Comparison of HV-SPS and SPS coatings revealed differences in phase composition. In SPS coatings, a high amount of alpha alumina was detected, whereas in HV-SPS coatings, the gamma phase was dominant.

The use of suspension plasma spraying (SPS) was also used by Oberste Berghaus et al (2008) for the manufacturing of thin and low-porosity electrolytes in an effort to develop a cost-effective and scalable fabrication technique for high-performance, metal-supported SOFCs. Three substrates metal-supported solid oxide fuel cells (SOFCs) were thermally sprayed on Hastelloy. These were composed of a samarium-doped ceria (SDC) ($Ce_{0.8}Sm_{0.2}O_{2-\delta}$) electrolyte layer and Ni-$Ce_{0.8}Sm_{0.2}O_{2-\delta}$ (Ni-SDC) cermet anode. The cathode, a $Sm_{0.5}Sr_{0.5}CoO_3$ (SSCo)-SDC composite, was screen-printed and fired in situ. The anode was produced by suspension plasma spraying (SPS) using an axial injection plasma torch. The SDC electrolyte was produced by high-velocity oxy-fuel (HVOF) spraying of liquid suspension feedstock, using propylene fuel (DJ-2700).

Saeidi et al. (2011) sprayed using a Praxair (CO-210-24) high-velocity oxy-fuel (HVOF) and vacuum plasma spraying (VPS), coatings made of CoNiCrAlY powder. Free-standing coatings underwent vacuum annealing at different temperatures for times of up to 840 h. Feedstock powder, and as-sprayed and annealed coatings, were characterized by scanning electron microscopy (SEM), energy dispersive spectroscopy (EDS), and X-ray diffraction (XRD). The hardness and Young's modulus of the as-sprayed and the annealed HVOF and VPS coatings were measured, including the determination of Young's moduli of the individual phases via nanoindentation and measurements of Young's moduli of coatings at temperatures up to 500 °C. The Eshelby inclusion model was employed to investigate the effect of microstructures on the coatings' mechanical properties. The sensitivity of the mechanical properties to microstructural details was confirmed. Young's modulus was constant up to ~200 °C, and then decreased with increase in measurement temperature. The annealing process increased Young's modulus because of a combination of decreased porosity and β volume fraction. Oxide stringers in the HVOF coating

maintained its higher hardness than the VPS coating, even after annealing.

Stiegler et al. (2012) deposited hydroxyapatite (HA) coatings on Ti plates by the high-velocity suspension flame spraying (HV-SPS) technique. The process characteristic, the microstructure, and phase composition of the coatings were significantly influenced by the solvent and by the design of the combustion chamber of the HV-SPS torch. Water-based suspensions always led to fairly low surface temperatures (350 °C) and deposition efficiencies <40% and produced coatings with low amount of crystalline, which tends to dissolve very rapidly in simulated body fluid (SBF) solutions. DEG-based suspensions, when sprayed with a properly designed combustion chamber, produce deposition efficiencies of 45–55% and high surface temperatures (550–600 °C). In these coatings, the degree of crystallinity increased from the bottom layer to the top layer, probably because the increasingly large surface temperature can eventually favor re-crystallization of individual lamellae during cooling. These coatings were much more stable in simulated body fluid solutions.

7.3.5 Industrial Applications

HVOF/HVAF and associated technologies have been widely used for the spraying of metals, cermets, and a few ceramic protective coatings, which are typically 100–300 μm thick, onto surfaces of engineering components to allow for their use under demanding conditions. Initially, the focus was on wear-resistant coatings, but now the coatings are extensively studied for their corrosion and oxidation resistance, which can be better than those produced by other thermal spraying technologies. These coatings have found wide applications in marine, aircraft, automotive, and other industries. They are also used for reclaiming a wide range of petrochemical process components.

7.3.5.1 Metals

Pure aluminum coatings with low porosity were obtained with the lowest oxygen/fuel ratio with HVOF [Evdokimenko et al. (2001)], and porosity values down to 1% were obtained with HVAF [Evdokimenko et al. (2001)]. Iron-aluminide was sprayed with HVOF with 7 to 15 wt.% oxide inclusion [Totemeier et al. (2003)]. Among the alloys, the most popular is Ni-Cr, for which adherence is very high on stainless steel substrates (more than 100 MPa) and is not affected by thermal fatigue [Higuera et al. (2002)]. When HVAF sprayed onto aluminum substrates, NiCr particles exhibit a strong bond with the substrate with an important penetration into it but with no evidence of chemical bonding [Trompetter et al. (2005)]. NiCr or NiCrSiB coatings are used against the severe corrosive environment in waste-to-energy boilers

[Li and Wang (2002) and Kawahara (2007)]. NiCrMoNb is also used against corrosion [Yilbas et al. (2003)].

In thermal barrier coatings, upon exposure to high-temperature gases, a thin oxide scale, the thermally grown oxide (TGO), forms at the bond-coat/topcoat interface and continues to grow in thickness during thermal cycling. The TGO's uneven and rapid growth results in localized stress concentrations where cracks can nucleate and initiate the failure dynamics. To achieve a dense and uniform α-Al_2O_3 TGO scale with a slow growth rate, HVOF coatings of superalloys have been used [Yang et al. (2002), Rajasekaran et al. (2011), Lima et al. (2007), Yang et al. (2002), Lima et al. (2007), Ni et al. (2011), Chen et al. (2008), Brandl et al. (1997), Scrivani et al. (2003), Rajasekaran et al. (2011), Ni et al. (2011), Lima and Guilemany (1997), Richer et al. (2010), Fossati et al. (2010), Jang et al. (2006), Yuan et al. (2008)]. Coatings have been studied for the influence of various parameters such as bond-coat thickness and roughness, composition, mechanical properties evolution through the topcoat, bond coat and substrate, spray process. HVOF-sprayed bond-coat coatings have been compared to those obtained by plasma spraying (APS and VPS), cold spraying, and D-gun. They have higher aluminum content, and the thermally grown oxide (TGO) developed during cyclic oxidation contains fewer mixed oxide clusters and a stable Al_2O_3 layer. Thus, the TGO has a low growth rate and a low tendency for crack propagation, which leads to improved TBC durability. A typical HVOF bond coat of NiCrAlY (Cr: 24–26, Al: 4–6, Y: 0.4–0.7, Ni: bal.) with a narrow size distribution (47–57 μm) sprayed with JP-5000 HP/HVOF (kerosene-oxygen) is presented in Fig. 7.49a, representing substrate, bond coat, and topcoat, the bond coat microstructure being presented in Fig. 7.49b. It has a lamellar structure containing dispersed inclusions of aluminum oxide, as well as pores of varying sizes for a total porosity of 3.2%. Its average roughness Ra is 5.38 μm. The bond coat presents broader and lower diffraction peaks of the γ-Ni_3Al phase than the starting NiCrAlY powder.

HVOF metal coatings (Stellite®6) have excellent resistance to cavitation erosion [Kumar et al. (2005)]. Bulk amorphous coatings (NiTiZrSiSn) have also been successfully sprayed when reducing in-flight oxidation by reducing the oxygen to fuel gas ratio [Yuan et al. (2008)].

[Sidhu et al. (2006)] analyzed and compared the mechanical properties and microstructure details at the interface of HVOF-sprayed NiCr-coated boiler tube steels, namely ASTM- SA-210 grade A1, ASTM-SA213-T-11, and ASTM-SA213-T-22. Coatings were developed by two different techniques, and in these techniques, liquefied petroleum gas was used as the fuel gas. First, the coatings were characterized by metallographic, scanning electron microscopy/energy-dispersive X-ray analysis, X-ray diffraction, surface roughness, and microhardness, and then

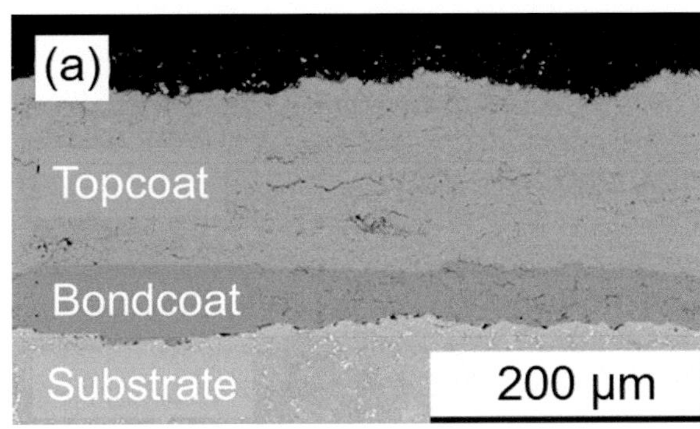

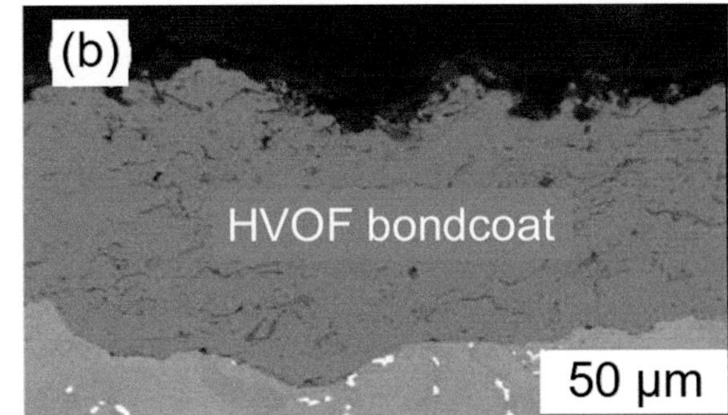

Fig. 7.49 (**a**) Cross-section of detonation-sprayed TBC using hollow sphere YSZ powder D-gun sprayed with the HVOF bond coat, (**b**) Microstructure of the HVOF NiCrAlY coating in the as-sprayed state. (Reprinted with kind permission from Elsevier [Ahmed et al. (2002)])

were subjected to erosion testing. An attempt has been made to describe the transformations taking place during thermal spraying. It was concluded that the HVOF wire-spraying process offered a technically viable and cost-effective alternative to the HVOF powder spraying process for applications in an energy generation power plant with a point view of life enhancement and to minimize the tube failures because it gives a coating having better resistance to erosion.

Trompetter et al. (2010) investigated the effect of substrate surface oxides on splat-substrate bonding by thermally spraying NiCr particles onto aluminum substrates with surface oxide layers grown hydrothermally and electro-chemically. Cross-sections of bonded solid and molten splats revealed substantial deformation of both the substrate and the surface oxide. In spite of the substantial substrate deformation, there was no significant loss of the surface oxide material and there was no observed diffusion of the substrate oxide into the NiCr particle or vice versa. For solid splats, the substrate oxide was still present over the entire splat-substrate interface; however, for molten splats, the oxide had been penetrated in several locations, allowing close proximity of the splat metal to the substrate metal. These results strengthen the theory that oxide layers impede bonding and that successful bonding occurs only when the surface oxide is substantially deformed or disrupted to produce mechanically interlocking features at the interface.

Trompetter et al. (2016) studied Ni-Cr alloy particles thermally sprayed onto a variety of substrate materials using the high-velocity air fuel (HVAF) technique. Although the various substrate materials were sprayed using identical powder material and thermal spray conditions, the type and variation of splat morphologies were strongly dependent on the substrate material. Predominantly solid splats were observed penetrating deeply into softer substrates, such as aluminum, whereas molten splats were observed on harder substrates, which resisted particle penetration. The observed correlation between molten splats and substrate hardness

could be due a dependency of deposition efficiencies of solid and molten splats on the substrate material. However, it was found that the conversion of particle kinetic energy into plastic deformation and heat, depending on substrate hardness, can make a significant contribution toward explaining the observed behavior.

[Han et al. (2017)] deposited the chemical composition of NiCoCrAlHfYSi with a suitable particle size, using an activated combustion-high velocity air fuel (AC-HVAF) spray. AC-HVAF is a potentially promising process because dense, continuous, and pure alumina can be formed on the surface of the MCrAlY metallic coatings after isothermal oxidation exposure. The NiCoCrAlHfYSi (Amdry386) and NiCo-CrAlTaY (Amdry-997) coatings were produced using AC- HVAF and APS, respectively. Isothermal oxidation was subsequently conducted at 1050 °C in air for 200 h. This paper compared the characteristics of four coated samples, including the surface roughness, elastic modulus, hardness, oxide content, microstructural characteristics, and phase evolution of thermally grown oxides (TGO). The growth of both the TGO and alumina scales in the TGO of the HVAF386 coating was relatively rapid. The θ- to α-alumina phase transformation was strongly determined by the Hf and Si dopants in the HVAF-386 coating. Finally, the extent of grain refinement and deformation storage energy in the HVAF997 coatings were determined to be significantly crucial for the θ- to α-alumina phase transformation.

[Sadeghimeresht et al. (2016)] considered that the selection of the thermal spray process is the most important step toward a proper coating solution for a given application, as important coating characteristics such as adhesion and microstructure are highly dependent on it. In their work, they performed a process-microstructure-properties-performance correlation in order to figure out the main characteristics and corrosion performance of the coatings produced by different thermal spray techniques such as high-velocity air fuel (HVAF), high-velocity oxy fuel (HVOF), and atmospheric

plasma spraying (APS). Previously optimized HVOF and APS process parameters were used to deposit Ni, NiCr, and NiAl coatings and compare with HVAF-sprayed coatings with randomly selected process parameters. As the HVAF process presented the best coating characteristics and corrosion behavior, a few process parameters such as feed rate and standoff distance (SoD) were investigated to systematically optimize the HVAF coatings in terms of low porosity and high corrosion resistance. Ni and NiAl coatings with lower porosity and better corrosion behavior were obtained at an average SoD of 300 mm and feed rate of 150 g/min. The NiCr coating sprayed at a standoff distance of 250 mm and feed rate of 75 g/min showed the highest corrosion resistance among all the investigated samples.

[Milanti et al. (2015)] studied novel Fe-based coatings to evaluate the microstructural details and corrosion properties using two different generations of HVAF spray guns. These two generations of HVAF guns are activated combustion HVAF (AC-HVAF, second generation) M2 gun and supersonic air fuel HVAF (SAF, third generation) M3 gun. Structural details were analyzed using X-ray diffractometry and a field-emission scanning electron microscope. Higher denseness with homogeneous microstructures was achieved for the Fe-based coating deposited by the M3 process. Such coatings exhibited higher particle deformation and lower oxide content compared to coatings manufactured with M2 gun. Corrosion properties were studied by open-cell potential measurements and electrochemical impedance spectroscopy. The lower porosity and higher interlamellar cohesion of the coating manufactured with the M3 gun prevent the electrolyte from penetrating through the coating and arriving to the substrate, enhancing the overall corrosion resistance. This can be explained by the improved microstructures and coating performance.

Gupta et al. (2018) pointed out that the development of thermal barrier coatings (TBCs) manufactured by suspension plasma spraying (SPS) is of high commercial interest as SPS has been shown capable of producing highly porous columnar microstructures similar to the conventionally used electron beam–physical vapor deposition. However, the lifetime of SPS coatings needs to be improved further to be used in commercial applications. The bond-coat microstructure, as well as top-coat–bond-coat interface topography, affects the TBC lifetime significantly. The objective of this work was to investigate the influence of different bond coat deposition processes on SPS top-coats. In this work, a NiCoCrAlY bond coat deposited by high-velocity air fuel (HVAF) was compared to commercial vacuum plasma-sprayed NiCoCrAlY and PtAl diffusion bond coats. All bond-coat variations were prepared with and without grit blasting the bond-coat surface. SPS was used to deposit the top coats on all samples using the same spray parameters.

Warm spraying has been developed, in which powder particles are accelerated and simultaneously heated, and deposited onto a suitable substrate in thermally softened solid state. Accordingly, Kim et al. (2010) sprayed commercially available titanium powder onto a steel substrate by the spraying process. Microstructural developments and deposition behaviors from a deposited single particle to a thick coating layer were observed by high-resolution electron microscopes. A single titanium particle sprayed onto the substrate was severely deformed and grain-refined mainly along the interfacial boundary of the particle/substrate by the impact of the sprayed particle. A successive impact by another particle further deformed the previously deposited particle and induced additional grain refinement of the remaining part. In a thick coating layer, severe deformation and grain refinement were also observed. The results have demonstrated the complex deposition behavior of the sprayed particles in warm spraying using thermally softened metallic powder particles.

When comparing Ni-20 wt % Cr sprayed with gaseous or liquid fuel, the latter results in less oxidation, less porosity, and less evidence of melting. Accordingly, such coatings are able to withstand more than 3000 h exposure to neutral salt spray tests without substrate corrosion, which is two orders of magnitude more than what is obtained with gaseous fuels [Aalamialeagha et al. (2003)]. Materials sprayed are almost the same as those sprayed with gaseous fuel. The main differences are the higher velocities and the lower temperatures.

With copper particles sprayed at 500 m/s, very dense coatings were obtained when the particle temperatures were such that the particles were highly deformable just below the melting temperature [Kawakita et al. (2006)]. The same results were obtained with stainless steel, which, with addition of molybdenum, resulted in good corrosion-resistance coatings [Kawakita et al. (2005)] that were also harder than wrought material [Totemeier et al. (2005)]. Stellite®-6 was also excellent for repairing cavitation damage in turbines and pumps [Kumar et al. (2005)]. Similar results were obtained with Ni-Cr coatings, which had adhesion values of over 70 MPa on 9Cr-1Mo steel, and which presented excellent steam-oxidation resistance [Sundararajan et al. (2004a, b)]. When spraying MCrAlY coatings with such HVOF guns with an inert gas shroud, the coating oxygen levels were almost as low as those obtained in vacuum plasma spraying (VPS) [Pant et al. (2007)]. It was also possible to achieve amorphous phase formation of Zr-based alloy coatings (up to 62%) due to low oxide formation. Of course, if the coating is overheated by spraying it with too low stand-off distance [Kim et al. (2001)], no amorphization is possible. When heat treating Co-Mo-Cr-Si coatings at 600 °C, the hardness increases and the friction between the coating and an alumina pin diminishes [Bolelli and Lusvarghi (2006)].

Liquid-fuel HVOF is also used to spray materials in which solid lubricants are incorporated such as WC-Co and fluorinated ethylene-propylene (FEP) copolymer-based material [Marple and Voyer (2001)] or iron sulfide [Wang (2004)].

As with flame or gaseous-fuel HVOF, SHS (self-propagating high-temperature synthesis) reactions can be produced upon spraying, for example, with SiO_2/Ni/Al-Si-Mg powder, resulting in composite materials of $MgAl_2O_4$, Mg_2Si in an Al-Si matrix [Ozdemir et al. (2005)]. A rapid formation of aluminide ($NiAl_3$) can be produced by an exothermic reaction of plated nickel with Al-Si-Mg core powder [Ozdemir et al. (2005)]. In both cases, reactions start in-flight and finish during splat layering. Such composites are rather hard.

Cetegen and Basu (2009) presented a review of the current state-of-the-art in the modeling of liquid chemical precursor droplets and particles injected into high-temperature jets in the form of DC-arc plasmas and high-velocity oxy-fuel flames to form coatings. Conventional thermal-spray processes have typically utilized powders that are melted and deposited as a coating on hardware surfaces. However, production of coatings utilizing liquid precursors has emerged in the last decade as a viable alternative to powder deposition. Use of liquid precursors has advantages over powder in terms of their relative ease of feeding and tailoring of chemical compositions. In this, the authors reviewed the modeling approaches to injection of liquid precursors and particles into plasmas and high-velocity oxy-fuel flames. Modeling approaches for the high-temperature DC-arc plasma and oxy-fuel flame jets were first reviewed. This was followed by the liquid spray and droplet-level models of the liquid precursors injected into these high-temperature jets. The various knowledge gaps in detailed modeling are identified and possible research directions are suggested in certain areas.

7.3.5.2 Cermets

These coatings have been intensively studied from the beginning, and the success of HVOF or HVAF processes relies on them [Ahmed and Hadfield (2002), Parco et al. (2006), Perry et al. (2002), Deng et al. (2007), Jacobs et al. (1998), Jacobs et al. (1999), Marple and Lima (2005), Moskowitz and Trelewicz (1997), Guilemany et al. (2005), Ji et al. (2007), Matthews et al. (2004), Mizuno and Kitamura (2007)]. The cermets, which are mostly sprayed, are WC-Co and Cr_3C_2-NiCr. One of the key issues for coating quality is to avoid carbide decomposition, or at least reducing it as much as possible. Spraying with fuel-rich mixtures [5.122], and using HVAF preferentially to HVOF [Parco et al. (2006), Jacobs et al. (1999)], allows for reducing drastically or even avoiding carbide decomposition. These coatings are used for their wear resistance (friction and erosion), and corrosion resistance, especially for those containing NiCr. Other

composites are also sprayed, such as MoB/CoCr [Mizuno and Kitamura (2007)], against erosion by molten Al-Zn alloys, Ni-Ti-C [Horlock et al. (2005)], which, by SHS reaction, after spraying, contains TiC in a Ni-rich solution together with NiTi, TiO_2, and $NiTiO_3$, and lastly silicon nitride-based coatings [Thiele et al. (2002)] where the silicon nitride is imbedded in a complex oxide binder matrix. Such coatings exhibit satisfactory corrosion and thermal-shock resistance.

With cermets such as WC-Co and Cr_3C_2-NiCr, rather dense coatings are obtained with Cr_3C_2-NiCr coatings being more protective against corrosion than WC-Co ones [de Villiers Lovelock (1998); Otsubo et al. (2000); Sidhu et al. (2007); Bolelli et al. (2006); Ishikawa et al. (2007); Maiti et al. (2007); Sidhu et al. (2006); Sidhu et al. (2007)]. This fact is attributed to the formation of Cr_2O_3, NiO, and $NiCr_2O_4$ [Bolelli et al. (2006)]. WC-20 wt% Cr_3C_2-7 wt% Ni cermet coatings have good resistance to sliding wear [Bolelli et al. (2006)]. Adding WC to the powder [Maiti et al (2007)] improved the hardness of WC-Co-Cr. Cr_3C_2-NiCr coatings on Ni base super alloys provided a good resistance to hot corrosion [Sidhu et al. (2007)]. At last, WC-Co + CoMoCrSi coatings are harder but less tough than electrolytic hard chrome (EHC) coatings, and their two-body sliding resistance definitively overcomes that of EHC coatings, but they experience a comparable or even higher mass loss when subjected to three-body abrasion conditions [Ishikawa et al. (2007)]. WC-NiCrFeSiB coatings on Ni- and Fe-based super alloys show excellent oxidation and hot corrosion resistance at 800°C [Sidhu et al. (2007b)].

Optimized processing windows for spraying high-quality metal carbide-based coatings were developed using particle diagnostic technology by Ang et al. (2016). The cermet coatings were produced via the high-velocity oxygen-fuel (HVOF) spray process and were proposed for service applications such as marine hydraulics. The traditional "trial and error" method for developing coating process parameters is not technically robust, as well as being costly and time consuming. Instead, this contribution investigated the use of real-time monitoring of parameters associated with the HVOF flame jets and particles using in-flight particle diagnostics. Subsequently, coatings can be produced with knowledge concerning the molten particle size, temperature, and velocity profile. The analytical results allowed identification of optimized coating process windows, which translate into coatings of lower porosity and improved mechanical performance.

Bartuli et al. (2005) performed the parametric study of nanocrystalline WC-Co coatings deposited by high-velocity oxyfuel from commercial nanostructured composite powders. Processing parameters were optimized for maximal retention of the nanocrystalline size and for minimal de-carburation of the ceramic reinforcement. Thermochemical and gas-dynamical properties of gas and particle flows

within the combustion flame were identified in various operating conditions by computational fluid-dynamics (CFD) simulation. Significant improvements in the mechanical properties of the coatings were obtained: a decrease of the friction coefficient was measured for the nanostructured coatings, together with an increase of microhardness and fracture toughness.

According to Bobzin et al. (2016), thermally sprayed Fe-based coatings reinforced by TiC particles are a cost-effective alternative to carbide coatings such as WC/CoCr, Cr_3C_2/NiCr, and hard chrome coatings. They feature good wear resistance and—with sufficient number of alloying elements like Cr and Ni—high corrosion resistance. In hydraulic systems, the piston is usually coated with hard chrome coatings for protection against corrosion and wear. New water-based hydraulic fluids require an adaption of the coating system. In order to investigate the wear and corrosion resistance of Fe/TiC, a novel powder consisting of a $FeCr_{27}Ni_{18}Mo_3$ matrix and 34 wt.% TiC was applied by HVOF and compared to reference samples made of WC/CoCr (HVAF) and hard chrome. Besides an in-depth coating characterization (metallographic analyses, electron microprobe analyzer–EMPA), wear resistance was tested under reverse sliding in a water-based hydraulic fluid. The novel Fe/TiC coatings showed good wear protection properties, which are comparable to conventional coatings like WC/CoCr (HVAF) and electroplated hard chrome coatings. Corrosion resistance was determined by polarization in application-oriented electrolytes (hydraulic fluid at 60 °C, artificial sea water at RT). The corrosion resistance of the investigated iron-based coatings at 60 °C was superior to that of the reference coatings for both hydraulic fluids. Select coatings were tested in an application-oriented hydraulic test bench with HFC hydraulic fluid (water polymer solutions), showing comparably good wear and corrosion resistance as the hard chrome-coated reference.

7.3.5.3 Ceramics

Ceramic coatings are not the strongest point of HVOF or HVAF processes because the process is more prone to achieve high velocities than high temperatures. Titania is rather well melted in processes working with propylene [Lima and Marple (2003a) and (2003b)]. High Weibull modulus values are obtained compared to other spray techniques and coatings are very dense and uniform, with rutile being the major phase. Alumina is mostly sprayed with chromia, which allows for stabilizing the α phase [Stahr et al. (2006)]. At last, zirconia stabilized with yttria, with particles below 10 μm, can be sprayed, and it is suggested that sintering can play a role with good adhesive and cohesive coatings [Dobbins et al. (2003)].

7.3.5.4 Polymers

Among the sprayed polymers, Nylon 11 is the one most frequently used, with or without ceramic doping [Petrovicova et al. (2000), Ivosevic et al. (2005), Jackson et al. (2007)]. The residence time in the HVOF is generally short enough (~ 1 ms) to limit its degradation and a change of its color. Such coatings, when used for their dry sliding wear performance, are more wear resistant without doping powders because those additions may lead to abrasive and fatigue wear.

Thermal spray of polymers has had limited investigation due to the narrow processing windows that are inherent to polymer powders, especially their low temperatures of thermal degradation [Withy et al. (2008)]. The polymer poly-aryl-ether-ether-ketone (PEEK) has a continuous use temperature of 260 °C, does not suffer significant thermal degradation below 500 °C [Lu et al. (1996)], and has high resistance to alkaline and acidic attack. These properties led [Withy et al. (2008)] to select PEEK for investigation. To minimize thermal degradation of the particles, the high-velocity air-fuel technique was used. To investigate the effect of substrate pretreatment on single-splat properties, single splats were collected on aluminum 5052 substrates with six different pretreatments. The single splats collected were imaged by scanning electron microscopy and image analysis was performed with ImageJ, an open source scientific graphics package. On substrates held at 323 °C, it was found that substrate pretreatment had a significant effect on the circularity and area of single splats, and also on the number of splats deposited on the substrates. Increases in splat circularity, area, and the number of splats deposited on the surface were linked to the decrease in chemisorbed water on the substrate surface and the decrease of surface roughness. This proved that surface chemistry and roughness are crucial to forming single splats with good properties, which will lead to coatings of good properties.

7.4 Pulse Detonation Thermal Spray

The pulse detonation thermal spray (PDTS), commonly known as detonation gun (D-gun), was developed in the 1950s in the USA by Union Carbide Corporation (Now Praxair) in 1955, Poorman et al. (1955), and independently in 1969 at the Institute of Materials Science (Kiev, Ukraine) in the former Soviet Union, Nevgod et al. (1987a, b), and Barisov et al. (1990).The technology is closer to cold spray or HVOF/HVAF than conventional thermal spray coating since it relies heavily on the acceleration of particles to be sprayed to high velocities, projecting them toward the substrate in

essentially a softened solid-state where they deform on impact, forming the coating. In contrast to other combustion processes, where the flame is a subsonic wave sustained by a chemical reaction, detonation wave is a shock wave sustained by the energy of chemical reactions in a compressed explosive gas mixture. The pressure ratio of burned to unburned gases, $p_b/p_u \approx$ 13–25 and $\rho_b/\rho_u \approx$ 1.4–2, where ρ_b is the specific mass of burned gases and ρ_u that of unburned ones [Glassman (1977)]. These ratios are considerably higher than those of atmospheric pressure flames where the density ratio of the hot gases is typically in the range ($\rho_b/\rho_u \approx$ 0.06–0.25), at pressures slightly below atmospheric pressure ($p_b \approx$ 80 to 90 kPa). The ideal detonation wave travels in gases at constant speed close to the Chapman–Jouguet velocity (v_{CJ}), which is between 1500 and 3000 m/s, depending on the type and composition of the fuel-oxidizer mixture. The pressure just behind the detonation wave can be as high as 20 ~ 30 times the ambient pressure. In this section, a brief review of the basic concepts behind the technology is presented, followed by typical D-gun spray coating applications.

7.4.1 Basic Concepts

As illustrated in Fig. 7.50, the basic concept behind a detonation gun (D-gun) is to generate a detonation wave in a confined space (Barrel) open at one end, through the combustion of a well-defined mixture of oxygen and an appropriate fuel such as hydrogen, acetylene, or propane. As the fuel/oxygen mixture is ignited, a high-pressure shock wave (detonation wave) is generated which, depending on the composition of the fuel/oxygen mixture, can have a gas temperature as high as 4000 K and a velocity of the shock wave up to 3500 m/s. As the wave propagates toward the open end of the barrel, it entrains, heats, and accelerate a small charge of powder pre-placed in the combustion chamber to a plasticizing stage (skin melting) and relatively high velocities of the order of 1200 m/s. On immerging from the open end of the barrel, the powder is projected towards the substrate (work piece), placed a distance typically between 100 and 200 mm from the end of the barrel, on which the coating is formed on impact through plastic deformation. The process is cyclic, involving the repetitive spraying of the powder in successive "shots" driven by a repetitive explosive combustion at a frequency in the 1–15 Hz range. At the end of each cycle, the chamber is flushed with nitrogen to remove any remaining "hot" powder particles from the chamber as these can otherwise detonate the explosive mixture in an irregular fashion and render the whole process uncontrollable. Recent studies by Endo et al. (2016) reported experiments using a high-frequency pulse detonation combustor (HFPDC)

operated in the liquid-purge mode at frequencies up to 150 Hz.

In a typical D-gun, the barrel is made of a heavy wall, water-cooled cylindrical tube closed at one end with an internal dimeter, $d_i = $ 10 to 40 mm, and length, L_T, of up to 1 or 1.5 m. As a rule of thumb, $L_T = $ 40 to 80 x d_i up to 100 d_i. The geometry of the detonation chamber must allow for steady detonation, which also depends on the mixture composition, temperature, and pressure. The fuel and oxidizer feed are either separately injected into the closed end of the barrel tube or premixed prior to injection. Charge ignition is made using a sparkplug. The detonation wave tends to propagate in all directions, including into the gas supply, creating an explosion hazard. The effect is called "backfiring," which must be considered in the design of the gas detonation equipment. The problem is usually solved through the use of an inert gas, such as nitrogen, in order to separate different portions of the fuel from one another and to prevent propagation of the detonation wave into the gas distributor. Numerous other safety features are introduced in the gun design to avoid faulty ignition of the fuel; these generally involve the periodic purging of the barrel tube with nitrogen at the end of every ignition and the nitrogen purging the fuel and oxidant mechanical or electromagnetic valves. The material to be sprayed is introduced into the gun barrel in the form of small charges, typically of less than 1 g each, of fine powder with a particle size in the range of 10 to 40 μm. The position of the powder feed in the barrel is a critical parameter in the design and operation of the gun. The powder charge is either axially introduced from the closed end of the barrel using a coaxial probe that can deliver the powder to different locations on the barrel axis, as shown in Fig. 7.50a, or laterally (radially) further downstream at any desired locations, as shown in Fig. 7.50b.

According to [Kadyrov E. and V. Kadyrov (1995)], the process consists essentially of the following four principal steps, schematically represented in Fig. 7.51:

(a) Injection of an oxygen and fuel mixture into the combustion chamber

(b) Injection of the powder "charge" followed by a nitrogen purge between the ignition point and the combustion gas mixture to separate the gas supply from the explosive gases and prevent backfiring

(c) Ignition of the combustion gas mixture giving rise to detonation and the creation of a detonation wave, which entrains, heats, and accelerates the powder toward the open end of the barrel and projects it on the substrate, forming the coating

(d) Nitrogen purge of the barrel at the end of the cycle [Kadyrov E. and V. Kadyrov (1995)]

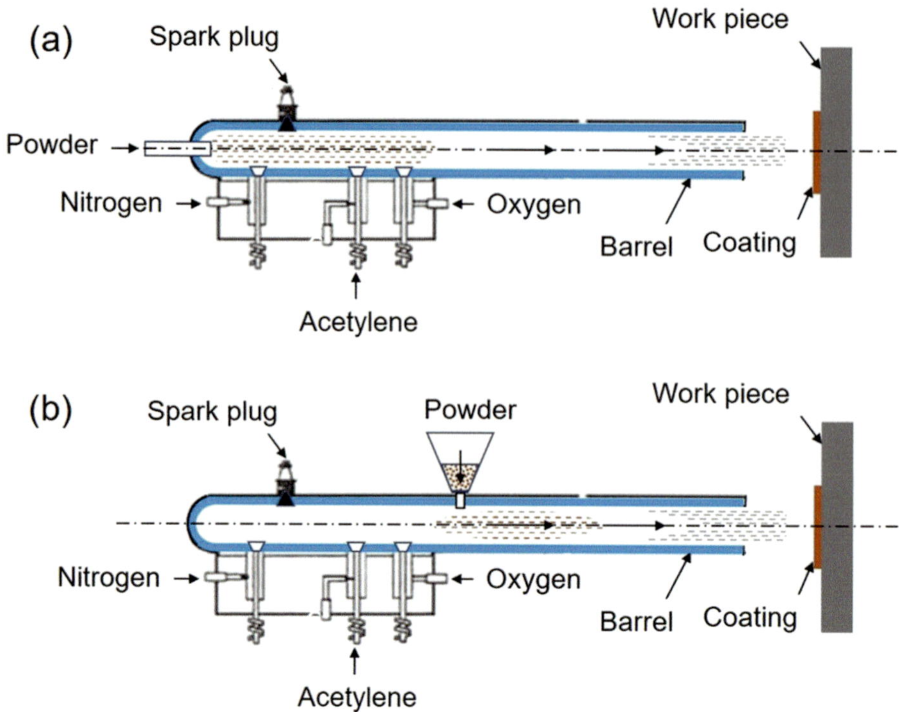

Fig. 7.50 Schematic of detonation gun spray coating process (**a**) axial powder feed (**b**) lateral (radial) powder feed

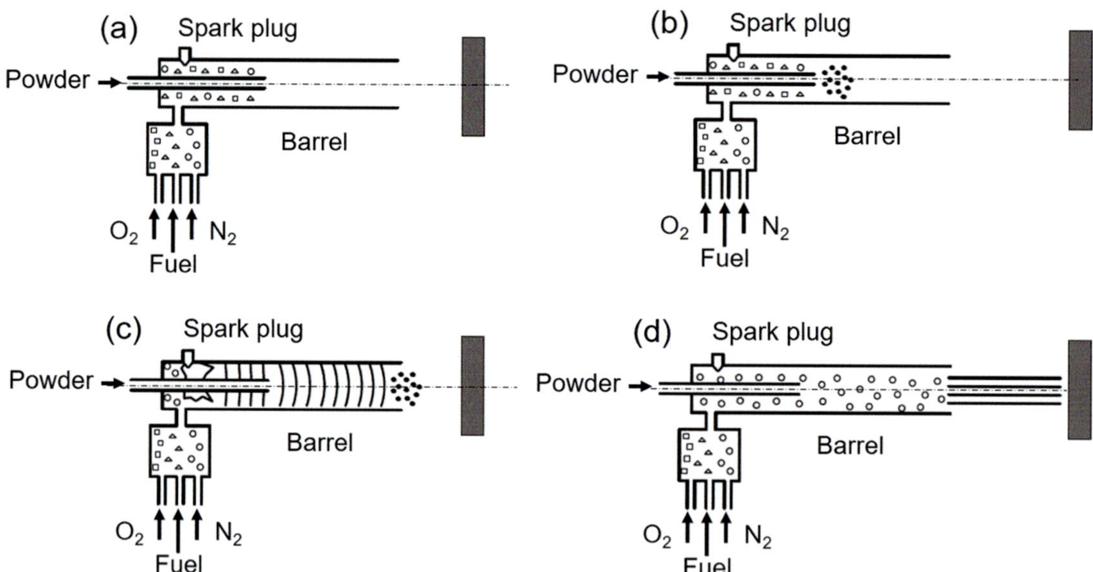

Fig. 7.51 Schematic of the detonation process cycle (**a**) injection of fuel and oxygen into combustion chamber, (**b**) injection of powder charge and nitrogen gas, (**c**) detonation of the combustion mixture, powder acceleration, and projection on the substrate, (**d**) nitrogen purge of the chamber in preparation for the next cycle [Kadyrov and Kadyrov (1995)]. (Reprinted with kind permission from Springer Science Business Media, copyright © ASM International)

Steps (a) and (b) are the preparatory ones, while steps (**c**) and (**d**) are self-governing ones independent of the process control system. The injection of fuel into the D-gun and the mixing of the fuel with the incoming oxidizer or injection of the fuel-oxidizer mixture start the new operation cycle. A mechanical or electromagnetic valve is used in industrial

guns to prevent detonations or shocks into the feeding system and provide sufficient time for the mixing of the fuel with the oxidizer. It is important to note that the detonation coating process is, by definition, an intermittent process that allows the temperature of the substrate to be kept low, generally below 100 °C, avoiding undesirable thermal deformation of the substrate.

The deposition rate of the coating material using the D-gun is directly dependent on the mass of the *charge* that is projected at every cycle, the *frequency of ignition* cycle, and the *deposition efficiency*. The latter is the fraction of the projected powder that is deposited on the substrate. The ignition frequency, f, is inversely proportionate to the duration of the cycle, t_c, which is composed of six characteristic time intervals [Kharlamov (2004)]:

$$f = 1/t_c \qquad (7.8)$$

with

$$t_c = \Delta t_{gfl} + \Delta t_{pfl} + \Delta t_{pr} + \Delta t_{in} + \Delta t_{tr} + \Delta t_{exp} \qquad (7.9)$$

where

Δt_{gfl} barrel filling time by fresh gas mixture
Δt_{pfl} barrel filling time with powder charge
Δt_{pr} time for nitrogen purging
Δt_{in} time for detonation initiation
Δt_{tr} time for detonation traversing the barrel Δt_{exp} exhaust of detonation products and powder particles and nitrogen purge

The dynamic filling, Δt_{gfl}, and exhaust/purging, Δt_{exp}, processes have the longest durations, which can be shortened by operating at higher dynamic pressures, with an upper limit due to filling losses. The length and the volume of the barrel, together with the pre-detonation chamber and the filling velocity, determine the filling time, Δt_{gfl}, since the mass flow into the combustor has to traverse the combustor length [Kharlamov (2004)]. The compressibility effects occurring when injected gas velocity is over Mach 0.5 limit the fill rate. A full cycle of the detonation gun, t_c, can then be calculated by summing all of characteristic times. An optimum exists because of the coupling of the length, dynamic pressure, and operating frequency of a barrel. Practical values are up to 10 Hz for a 1 m long combustor operating at an initial pressure of 0.1 MPa [Kharlamov (2004)]. When using multiple injection locations, this frequency limit could be overcome [Roy et al. (2004)]. While the length of the barrel can reach 100xd_i [Kharlamov (2004)], consideration should be

given to the fact that half of the gas thermal energy is lost after 40xd_i.

The minimum barrel diameter depends on heat losses to the tube wall and is typically 15–20 mm for most fuels and particle materials. A barrel 15 mm in diameter is sufficient for spraying cermets of tungsten carbide-metal when using acetylene, and a barrel 40 mm in diameter is necessary when using propane [Nikolaev et al. (2003)]. According to [Kharlamov (2004)], varying the barrel cross-sectional area along its length allows for a more uniform filling of the barrel. It is possible to use a nozzle at the barrel end to improve the performance [Roy et al. (2004)]. Contrary to nozzles of steady-flow devices, D-gun nozzles operate at essentially unsteady conditions and their design and optimization require considering the whole operation process. Attaching a nozzle to the end of the detonation tube makes it possible to gradually expand gases and decrease the rate of pressure drop in the tube. However, the nozzle increases the length of the barrel and thereby decreases the operation frequency.

Turbulence, caused by the roughness of the internal surfaces of the D-gun barrel, while promoting a better mixing of the powder with the combustion products, leads to increased heat losses to the wall. Roughness can be achieved by threaded grooves of various profiles, made along the entire length of the barrel, or in individual sections [Kharlamov (2004)]. When the grooves are located in the barrel inlet section, the pre-detonation distance is shortened, whereas their location at its outlet improves powder mixing and intensifies heat exchange.

Typical supply gas pressures to D-guns are 140–200 kPa (20 to 30 psig). The consumption of gas per unit mass of sprayed matter is four to eight times less than that for HVOF spraying [Kadyrov E. and V. Kadyrov (1995)]. The frequency and noise levels (about 145 dBA) require the detonation to be confined in acoustical enclosures. According to Kadyrov E. and V. Kadyrov (1995), one can image propagation of the detonation wave as a stationary distribution of density, temperature, pressure, and velocity moving with some speed distribution. Figure 7.52 gives a qualitative image of the detonation parameters [Kadyrov E. and V. Kadyrov (1995)]. It is assumed that the thickness of the reaction zone is very thin and that the zone of the reaction products is separated from the initial gas mixture by the detonation wave front. At the leading edge of the detonation wave, the temperature increases rapidly, and then behind the front rises gradually in the reaction zone before decreasing gradually with increasing distance from the wave front (Fig. 7.52b). There are also step-like increases in pressure and specific mass density at the detonation wave front followed by a gradual decrease behind (Fig. 7.52a and c).

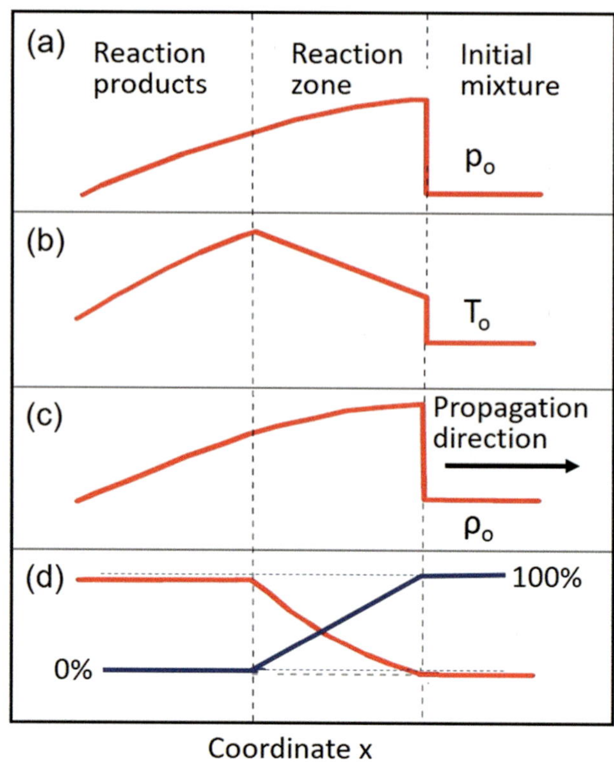

Fig. 7.52 Qualitative picture of detonation parameter variations in the gas (**a**) pressure, (**b**) temperature, (**c**) specific mass, (**d**) distribution of the detonation velocity, v_D, and un-burned gases [Kadyrov and Kadyrov (1995)]. (Reprinted with kind permission from Springer Science Business Media, copyright © ASM International)

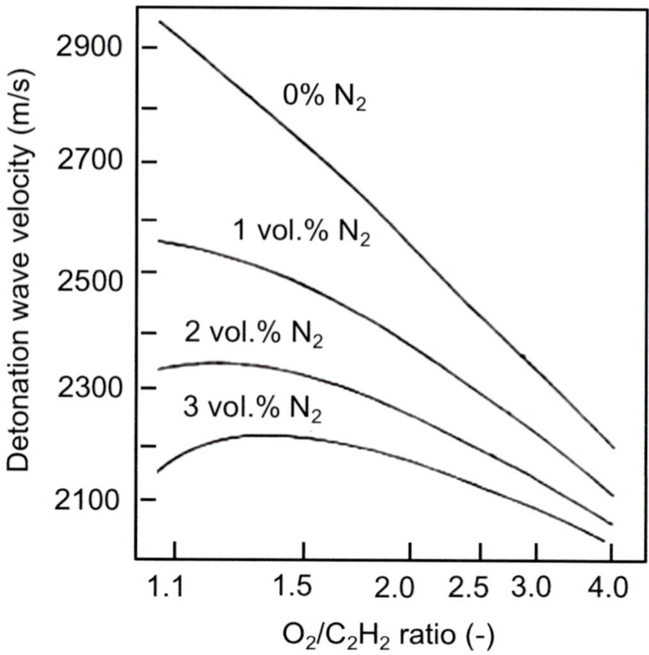

Fig. 7.53 Effect of nitrogen gas on the detonation velocity for different O_2/C_2H_2 mixtures in the presence of different vol. % of nitrogen addition [Kadyrov et al. (1995)]. (Reprinted with kind permission from Springer Science Business Media, copyright © ASM International)

7.4.2 Gas and Particle Dynamics

Numerous studies have been devoted to the understanding of the gas and particle dynamics under pulse detonation thermal spray (PDTS) conditions due to their critical impact on the spraying process and the quality of the coating obtained. The work of [Kadyrov et al. (1995) and Kadyrov (1996)] is pioneering in this area by studying gas dynamic parameters of the detonation powder spray coating process and the mechanism of particle acceleration by the shock wave inside the coating apparatus. Velocities of gas detonation in different gas mixtures are analyzed by applying conventional fluid dynamic theory of detonation for oxygen/hydrogen and oxygen/acetylene mixtures. Their results show that the velocity of the detonation wave varies between 1000 and 3000 m/s depending on the composition of the gas mixture. It is independent of the barrel diameter, provided it exceeds a critical diameter. Once the detonation occurs, its theoretical velocity v_D (m/s) can be calculated by:

$$v_D = \sqrt{2(\gamma^2 - 1)\, Q_o} \qquad (7.10)$$

$$T_1 = \frac{2\gamma\, Q_o}{(\gamma + 1)\, c_v} \qquad (7.11)$$

$$p_1 = 2(\gamma - 1)\rho_o Q_o \qquad (7.12)$$

Where;

γ specific heats ratio of the burned gases ($\gamma = c_p/c_v$)
c_p specific heat at constant pressure
c_v specific heat at constant volume
Q_o specific energy of chemical reaction during detonation,
T_1 temperature of detonation products
p_1 pressure of detonation products

They also point out that detonation parameters are sensitive to the composition of the combustible mixture. The addition of light gases (He, H_2) increases the detonation velocity, while the addition of heavier gases (such as Ar, N_2) decreases the velocity. This is illustrated in Fig. 7.53, representing the velocity of the detonation wave calculated using Eq. (7.10) for different O_2/C_2H_2 mixtures in the presence of nitrogen at different concentrations. The detonation velocity decreases with the increase of the oxygen/acetylene ratio. The introduction of a small percentage of

nitrogen decreases rapidly the velocity of the detonation wave especially for sub-stoichiometric combustion mixtures ($O_2/C_2H_2 < 2.5$). For example, 3 vol.% of nitrogen reduces the detonation velocity by 600 to 800 m/s, with a corresponding drop in particle velocity, leading to a degradation of the quality of the coating. The results of similar calculations using Eqs. (7.10 to 7.12) for different hydrogen/oxygen fuel mixtures in the presence of different additives such as Ar and N_2 are summarized in Table 7.3.

Heat and momentum transfer to particles entrained by the detonation wave are calculated using the conventional equations of motion and heat transfer with or without corrections for supersonic velocities [Kadyrov et al. (1995)], see also Chaps. 4 and 5. Computations were carried out for 20 μm alumina particles injected at three different

Table 7.3 Parameters of the detonation wave in hydrogen/oxygen mixtures with the addition of different gases [Kadyrov et al. (1995)]

Fuel mixture	p_2/p_1	$T_2(K)$	$v_D(m/s)$
$2H_2 + O_2$	18.0	3283	2806
$(2H_2 + O_2) + O_2$	17.4	3390	2302
$(2H_2 + O_2) + 4H_2$	16.0	2976	3627
$(2H_2 + O_2) + N_2$	17.4	3367	2378
$(2H_2 + O_2) + 3 N_2$	15.6	3003	2033
$(2H_2 + O_2) + 1.5Ar$	17.6	3412	2117

Reprinted with kind permission from Springer Science Business Media, copyright © ASM International

axial locations in the gun's barrel, $x_i = 0.2$, 0.6 and 1.5 meters, with x_i being the "loading distance" from the point of initiation of the detonation ($x = 0$) to the point of powder injection into the gun. The result given in Fig. 7.54a shows the position of the particle in a 2 m barrel as a function of time for oxygen/acetylene and oxygen/hydrogen stoichiometric mixtures, with initial particle injection at $x_i = 0.2$ m. These show a time-of-flight of the particles until it reaches the exit level of the gun barrel of 2.0 ms, for the oxygen/acetylene mixture, which is shorter than that for the oxygen/hydrogen mixture, 2.7 ms.

The results given in Fig. 7.54b, for the same combustion mixtures and particle system, show that the acceleration and the flight history of the particles in the gun barrel, until they reach its exit level, strongly depend on the injection location of the powder. For example, with a loading distance of 0. 2 m in an O_2/C_2H_2 mixture, the maximum velocity of the particles of 837 m/s is reached against 1152 m/s for a loading distance of 60 cm in the same combustion mixture. Delaying the injection of the powder to the downstream end of the gun barrel, $x_i = 1.5$ m in a 2.0 m gun, gives rise to a very short period of particle entrainment and acceleration by the detonation wave, not allowing the particle to reach its maximum value. These results exhibit several interesting features. The acceleration of the particles from zero to the maximum velocity occurs in a relatively short distance, 0.2 to 0.4 m, in less than 0.6 ms. According to Kadyrov et al. (1995), this is due to the fact that the initial particle

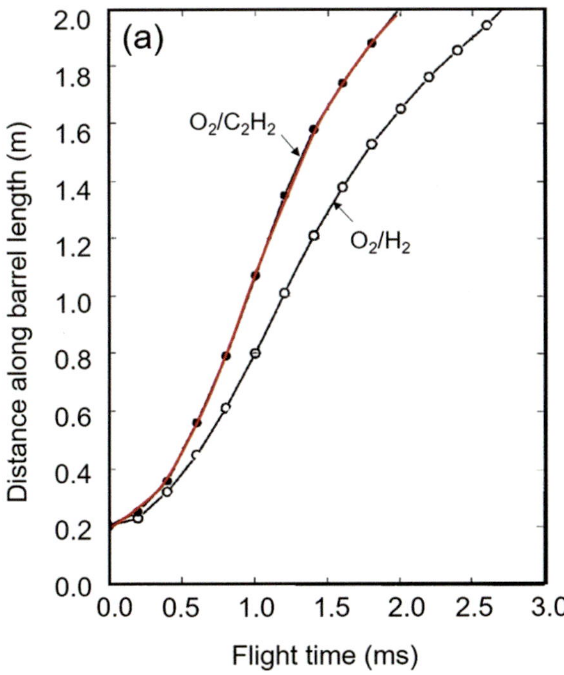

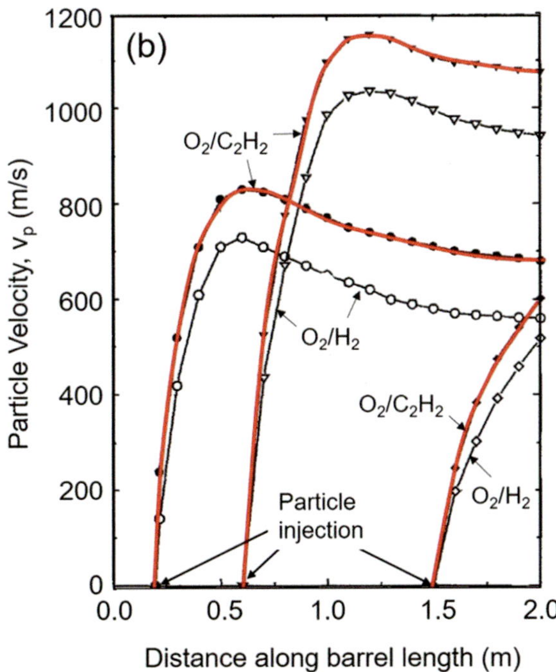

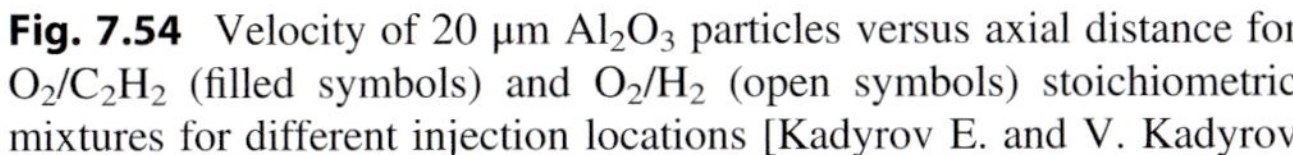

Fig. 7.54 Velocity of 20 μm Al_2O_3 particles versus axial distance for O_2/C_2H_2 (filled symbols) and O_2/H_2 (open symbols) stoichiometric mixtures for different injection locations [Kadyrov E. and V. Kadyrov (1995)]. (Reprinted with kind permission from Springer Science Business Media, copyright © ASM International)

acceleration is provoked by the high-speed gas flow following the detonation wave front. As the detonation wave propagates forward and the particle separates from the wave front, the velocity of the detonation products decreases, and at some point, becomes smaller than that of the particle. With the particle moving faster than the detonation products, the drag force reverses direction and the particle starts to slow down. As can be observed in Fig. 7.54b, the particle exit velocity at the limit of the D-gun barrel, v_{po}, remains strongly dependent on the composition of the detonation mixture and the position of the powder injection in the barrel. For an oxygen/acetylene detonation mix, $v_{po} = 690$, 1080 and 600 m/s for values of $x_i = 0.2$, 0.5 and 1.5 m. The corresponding particle exit velocities for an oxygen/hydrogen mix are 560, 950, and 510 m/s, respectively. As pointed out by Kadyrov et al. (1995), such variations of the particle velocity along its trajectory inside the D-gun are distinctly different from those observed in alternate thermal spray technologies such as HVOF and plasma spraying.

A comprehensive analysis of particle dynamics and heat transfer in pulse detonation thermal spray (PDTS) was reported by Ramadan et al. (20042004). The proposed axisymmetric, 2-D transient compressible gaseous detonation model is based on a three-stage representation of the process, as illustrated in Fig. 7.55. These consist of:

(a) Simultaneous feeding of the reaction mixture (oxygen and acetylene) through a mixing chamber, and of the powder into the tubular D-gun barrel.

(b) Ignition of the combustion mixture using a spark plug near the closed end of the barrel.

(c) Transition from deflagration where the combustion wave propagates at subsonic speed to detonation with the generated high-pressure reactive shock wave (detonation wave) propagates at supersonic speed toward the open end of the barrel entraining and accelerating the supplied powder charge. The collision of the high-velocity, high-temperature powder particles with the substrate forms the coating.

Ramadan et al. (2004) point out that following the detonation front is an expansion wave that propagates rearward toward the breech end of the tube (Fig. 7.55b) and that once the shock front exits the barrel, it starts decaying as the gas expands outside the barrel. Depending on the stand-off distance (SoD), defined as the distance between the barrel exit and the substrate, the shock wave can reach the substrate with a large pressure ratio reflecting back into the path of the incoming particles, as shown in Fig. 7.55c. The interaction between the reflected wave and the particles depends on the particle properties and their size distribution. As the particles impact on the surface of the substrate, they form splats adhering to the substrate depending on their temperature and impact velocity. Following a purge of the barrel to remove any residual powders and combustion products, a new cycle is initiated through the feeding of new fuel-oxidant gas mixture-powder charge into the barrel.

The basic assumptions made in their model can be summarized as follows:

- Axisymmetric 2-D system of coordinates.
- Transient inviscid compressible flow model with the viscous effects confined to the interaction between the gas and the particles.
- Combustion chemical reactions take place in a one-step irreversible process.
- Negligible particle loading effects on the gas flow, that is, one-way coupling between the gas and particulate phases. This is justified by the fact that the mass of powder injected into the barrel is usually smaller than 10–20% of the gas weight.
- Eulerian and Lagrangian representation of the gas and particulate phase governing equations, respectively.

Results were given for a cylindrical detonation gun, 25 mm in internal diameter, with different lengths, L. The reactive gas mixture used was $C_2H_2/O_2/N_2$, with $O/C = 1$, $N_2(\%) = 40$. The resulting detonation wave speed for this mixture was 2464 m/s. The sprayed particles were alumina with three different diameters: 10, 20 and 30 μm. The stand-off distance (SoD) taken is six tube diameters, that is, the substrate was far enough from the barrel exit plane so that the

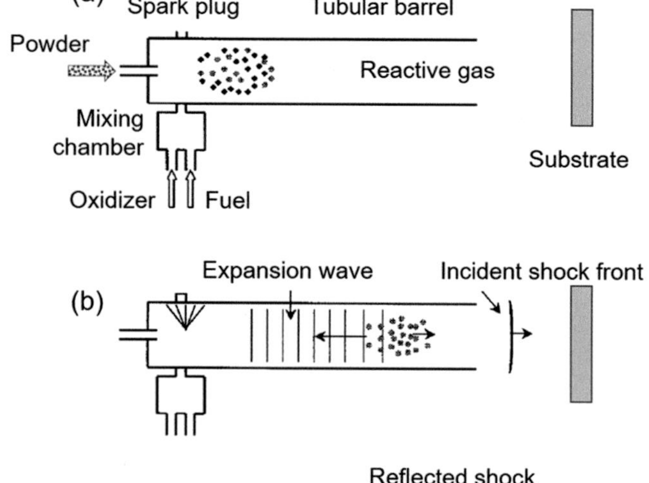

Fig. 7.55 Three-stage representation of the PDTS process [Ramadan et al. (2004)]

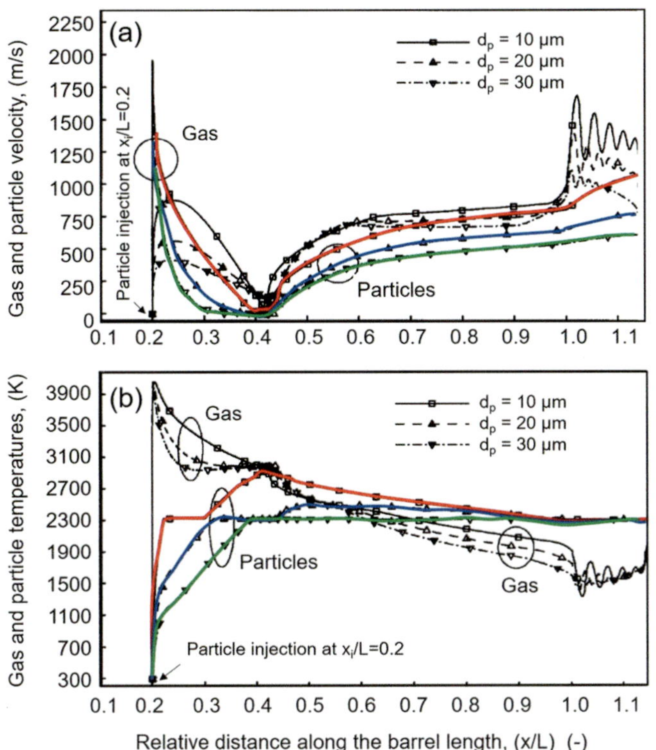

Fig. 7.56 (a) Particle velocity and gas velocity in its location, as function of distance along the barrel length, (b) Particle temperature and gas temperature at its location as function of distance long the barrel length. Alumina particles in an $O_2/C_2H_2/N_2$ (40 vol.%) axial loading distance $x_i/L = 0.2$ and radial loading distance $y_i/r = 0.05$, L = 1.0 m, $d_i = 25$ mm and SoD $= 6d_i$ [Ramadan et al. (2004)]. (Reprinted with kind permission from Springer Science Business Media, copyright © ASM International)

gas expanded freely with negligible shock reflection effects. Figures 7.56a and b present the dependencies of solid particle velocities and temperatures on their axial location, x, as well as the gas velocities and temperatures at the particle location. In this calculation, the initial loading location is $x_i/L = 0.2$ and L = 1 m (injection position far from optimal). Hot gases also heat particles to high temperatures that can reach the melting point, depending on their sizes (Fig. 7.51b) and injection location (not represented here).

Ramadan et al. (2004) pointed out that the geometrical configuration of the detonation gun, the particle size of the powder feed, the loading locations, and the SoD were important parameters in determining the properties of the end product. In general, increasing the standoff distance resulted in an increase in velocity and a decrease in temperature of the particles at impact on the target surface. Increasing the stand-off distance also resulted in less uniform properties of the coating layer, where the properties of the coating layer were dependent on the particle characteristics upon impact on the target surface. As shown in Fig. 7.57, the particle velocity on impact on the substrate is strongly dependent on the particle axial loading location and the SoD. The effect is less

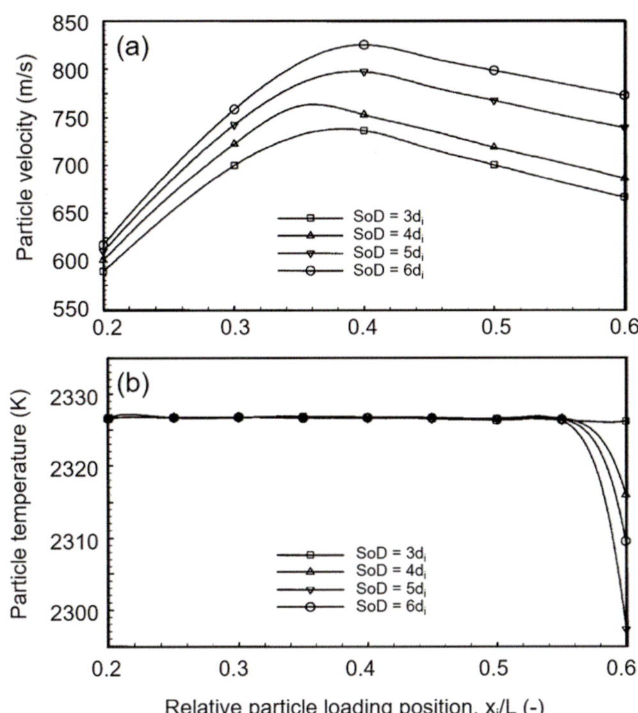

Fig. 7.57 (a) Alumina particle velocity and (b) temperature prior to their impact on the substrate as function of the relative loading position, for different Stand-off Distance (SoD). $d_i = 25$ mm, L = 1.0 m [Ramadan et al. (2004)]. (Reprinted with kind permission from Springer Science Business Media, copyright © ASM International)

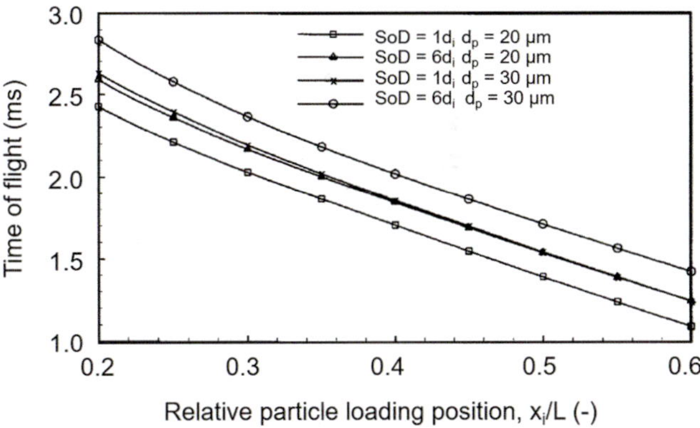

Fig. 7.58 Time of flight for alumina particles as a function of the relative particle loading position, particle diameter, and Stand-off Distance (SoD). $d_i = 25$ mm, L = 1.0 m [Ramadan et al. (2004)]. (Reprinted with kind permission from Springer Science Business Media, copyright © ASM International)

pronounced on the particle temperature prior to its impact on the substrate. The time of flight of the particles from imitation of the detonation to their impact on the substrate is given in Fig. 7.58. As expected, the results are relatively sensitive to the particle diameter and the stand-off distance (SD), and inversely proportional to the relative particle loading position. It is to be noted that these results were obtained

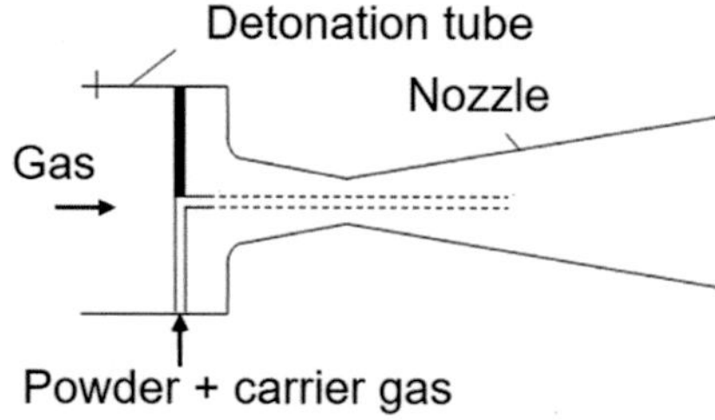

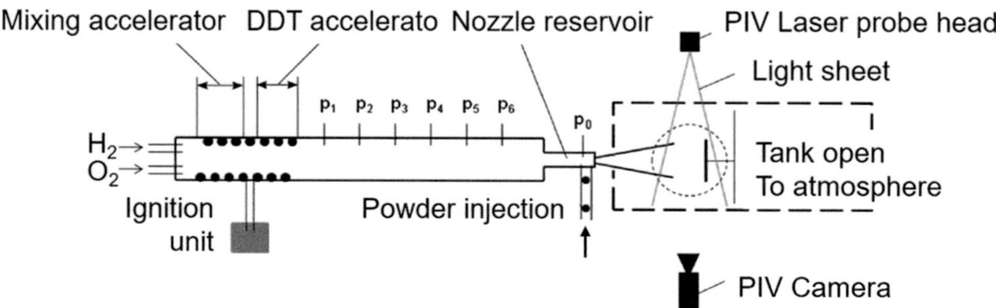

Fig. 7.59 Modified D-Gun test rig with convergent-divergent nozzle extension. [Henkes et al. (2014)]

for a barrel length L = 1 m. Operation with a longer or shorter barrel length has a significant influence on the reported data.

An interesting development reported by Henkes et al. (2014) describes a modified detonation gun through the addition of a convergent-divergent nozzle that allowed for a significant increase of the particle projection velocity over 1200 m/s, leading to improved coating qualities. The device, schematically represented in Fig. 7.59, could be operated either as a conventional D-gun in which the powder is accelerated in a blast wave, or as an intermittent shock tunnel process in which the particles are accelerated in a high-enthalpy nozzle flow with high reservoir conditions. Experimental data and modeling results are presented for the spraying of WC-Co (88/12), NiCr (80/20), Al_2O_3, and Cu using different substrate/powder combinations.

7.4.3 Coating Formation

As mentioned earlier, thermal spray coating using the pulse detonation thermal spray (PDTS) or D-gun technology is based on the cyclic projection of the material to be deposited in the form of small charges of fine powders each of a mass less than 1 g at frequencies up to 10 to 15 Hz. The technology relies heavily on the in-flight heating and acceleration of particles to be sprayed to temperatures near their melting point and velocities of the order of 1000 m/s, projecting them toward the substrate in essentially a softened solid state where they deform on impact forming a high-density well-bonded coating. The rate of coating deposition, $\dot{m}_d$, can be simply calculated as:

$$\dot{m}_d = \eta_d f \, m_c \tag{7.13}$$

where;

η_d deposition efficiency
f detonation frequency
m_c powder charge per detonation cycle

It is to be noted that the mass of powder charge used per shot has to be carefully controlled. In principle, it should be less than 10–20% of the mass of the combustion gas products, which provides the particles with the necessary heat and mechanical energy to raise their temperature and accelerates and projects the individual powder particles onto the substrate. The coating produced by each individual shot should mainly consist of individual particles distributed uniformly over the surface of the substrate. If the powder charge is too high, a continuous film is formed in a single shot, which on cooling will develop cracks in the film, which weakens it and induces a certain level of porosity [Nikolaev et al. (2003)].

With a typical charge of 0.2 g of powder per shot, operating at a frequency of 5 Hz, a deposition rate of around 60 g/min (3.6 kg/h) can be expected depending on the deposition efficiency, η_d. The latter depends, in turn, on the properties of the material deposition, the substrate preparation, and the operating parameters including the stand-off distance (SoD) between the end of the spray gun and the substrate. Normally with D-gun technology, deposition efficiencies close to 100% can be attained with the diameter

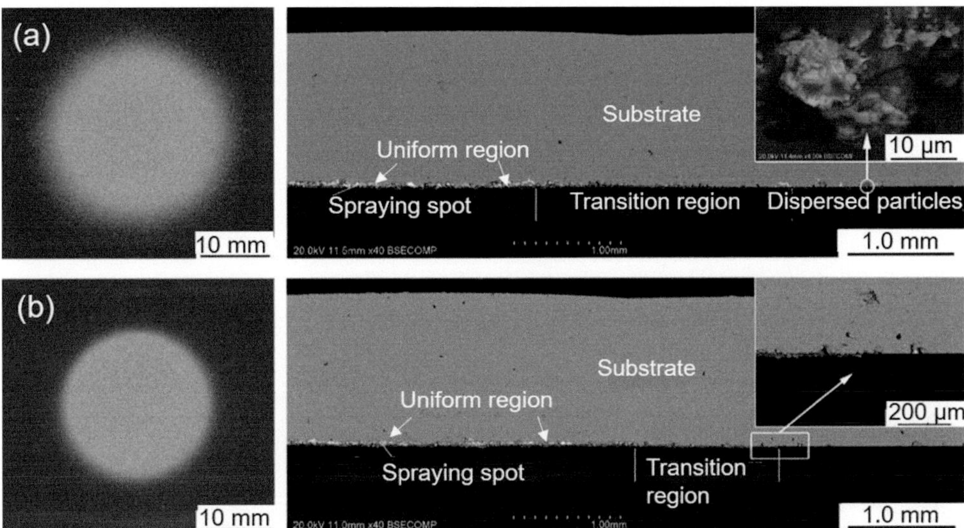

Fig. 7.60 Image of a "sweet spot" and cross-section microstructure for a single shot of WC-Co powder on polished stainless-steel substrate using modified D-gun (**a**) without separator mask (**b**) with separator mask [Wang et al. (2009)]

of the deposit "sweet spot" created by a single shot equal to the inner diameter of the gun barrel.

Wang et al. (2009) reported, however, that in certain cases, the sprayed powder from a D-gun can show an undesirable lateral dispersion as it exit the gun barrel at high speed, giving rise to a lower-quality deposit in the outer fringes of the deposited sweet spot. The study was carried out for the spraying of WC-Co (12wt.%Co) powder with a particle diameter in the 30–60 µm range on a stainless-steel substrate, which was grit blasted to a roughness, R_a, close to 7 µm. The spray gun had a 1.25 m long barrel, with a powder loading distance of 0.5 m from the discharge end of the barrel and an SoD of 110 mm (corresponding to 5.5 d_i, for a barrel i.d. of 20 mm). The combustion mixture was oxygen/acetylene with an O_2/C_2H_2 volume ratio of 1.06. The gun was operated at a detonation frequency of 4 shots/s, and a powder feed rate of 1.0–1.4 g/s (corresponding to a single powder charge of 0.25–0.35 g/shot). Typical results are given in the upper part of Fig. 7.60a, which shows on its LHS an image of the deposited "sweet spot" from a single shot, which has a central core with a diameter of about 23 mm and a halo of scattered particles with an outer diameter of 27 mm. Details of the microstructure of the deposit are given in the RHS of the same figure. This shows the central region with a uniform distribution of the deposited particles and a gradual transition to a region covered with dispersed particles in its fringes.

In an attempt to get rid of this particle-dispersed region surrounding the core of the deposit, Wang et al. (2009) designed and executed a separator mask, which they attached to the original D-gun, as shown in Fig. 7.61. The mask, which was placed at a distance of about 100 mm from the end of the gun barrel, had a circular opening of the same diameter as the barrel internal diameter (estimated 20 mm) coaxial with the

gun barrel. The outer diameter of the separator mask was 5x the diameter of the opening (i.e., 100 mm). Spraying experiments carried out with the separator mask using the same powder and gun operating parameters gave rise to a significantly better-defined deposition spot, as shown on LHS of Fig. *7.60b*. The halo around the deposited spot formed by the dispersed particles observed earlier in Fig. 7.60a was completely eliminated, giving rise to a superior-quality coating.

Comparing WC-Co coatings with the separation mask to those without mask, Wang et al. (2009) reported that the surface roughness S_a and R_a, the porosity, and the wear rate of the coating decreased respectively by 77%, 41%, 40%, and 20%, while the Vickers microhardness HV, the Knoop microhardness in both directions (parallel and orthogonal to the coating) $HKcs_\perp$, $HKcs_\parallel$, the elastic modulus $Ecs_\perp$, $Ecs_\parallel$, and the adhesive strength increased respectively by 12%, 13%, 13%, 32%, 36%, and 16%. They also pointed out that the weight of the coating received on the substrate in the presence of the separator mask was 12% less than that of the coating without the mask. Such a drop of deposition efficiency to 88% of the powder consumed in the process gave rise to a corresponding increase of the cost of the coating, which has to be evaluated against the improved properties and performance of the coating. Considering that the fringe powders would be intercepted by the separator mask, attention has to be given in such a setup to the periodic replacement of the mask for cleaning and recycling if possible.

Considering that operating parameters can be modified between successive shots of the detonation gun, the "shot control method" offers an excellent advantage for co-spraying of metal and ceramic powders with their optimum spraying parameters. Basically, only one type of

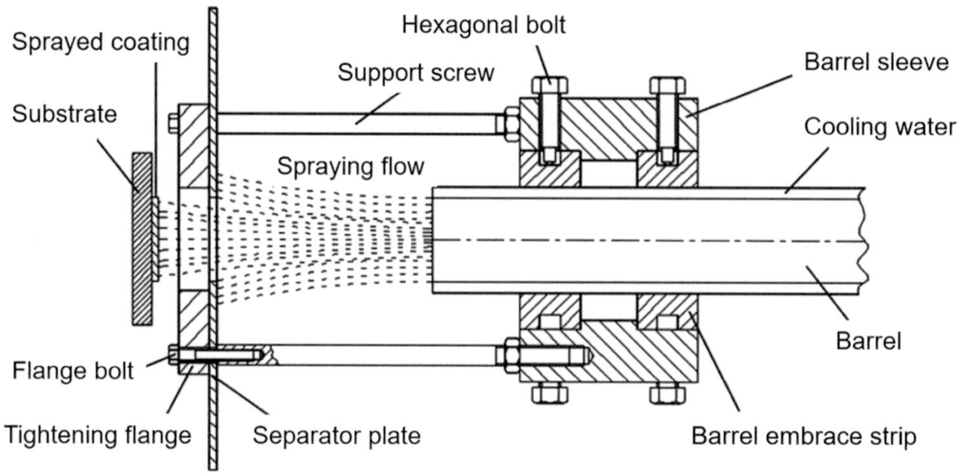

Fig. 7.61 Schematic of separation mask attachment used in D-gun spraying system [Wang et al. (2009)]

powder is deposited per individual shot to produce a coating mixture of ceramics and metals [Kim et al. (2003)]. Graded coatings with a compositional gradient through thickness could thus be produced by spraying several coating layers with increasing ratios of ceramic to metal shots in a sequence, as shown in Fig. 7.62. Other methods for the spraying of graded coatings, such as the mixed powder methods, are difficult to execute and optimize due to the important differences in the respective spraying parameters used for ceramics and metals.

According to Cannon et al. (2008), the particle size distribution of the feed powder has a significant impact on the coating properties. A decrease in powder size reduces roughness and porosity in the case of NiCr and Al_2O_3 but has a marginal effect for WC-Co due to the flow of the softer binder (Co) phases, especially when a coarser particle size is utilized. The toughness of the Al_2O_3 and WC-Co coatings are dependent on the presence respectively of W_2C in WC-Co, and α –Al_2O_3 in alumina ones. The wear resistance is only marginally affected by the powder size for WC-Co coatings, in contrast to Al_2O_3 coatings where it is very sensitive to the particle size distribution. Finally, the authors [Cannon et al. (2008)] conclude that commercial powders with size distributions of 10–44 μm can be used.

To summarize, the design of a D-gun with all its devices is not simple. Kharlamov (2004) summarized all the issues to ensure rapid development of a detonation wave within a short cycle time. With high impact velocities, and temperatures, which can be controlled in the range 0.9 T_m to T_m, T_m being the melting temperature, very dense coatings can be obtained. D-gun spraying usually produces denser coatings than any other thermal spray technique.

7.4.4 Coating Properties

The most important and well-developed applications of the D-gun are the deposition of wear-resistant, thermo-barrier, electro-insulating, and high-temperature oxidation-resistant coatings.

7.4.4.1 Plain Fatigue and Fretting Fatigue

Cu-Ni-In coatings were used to test the improvement in plain fatigue and fretting fatigue of an Al-Mg-Si alloy substrate [Rajasekaran et al. (2011)]. Coatings 40 μm thick were found to give the best result. The same Cu-Ni-In coatings deposited on Ti alloys were beneficial. The detrimental effect of life reduction due to fretting was relatively larger in the Al-alloy substrate compared to the Ti-alloy one. This result was explained in terms of differences in the values of surface hardness, surface roughness, surface residual stress, and friction stress [Ganesh Sundara Raman et al. (2007)]. Cu–Ni–In powder was deposited on Ti–6Al–4 V fatigue test samples using plasma spray and detonation gun (D-gun) spray processes [Rajasekaran et al. (2008)]. The D-gun sprayed coating was dense with lower porosity compared with the plasma-sprayed coating. The hardness value of the D-gun sprayed coating was higher than that of the plasma-sprayed coating and of the substrate because of its higher density and cohesive strength. Coating surfaces were very rough in both coatings. While the D-gun sprayed coating had higher compressive residual stresses, the plasma-sprayed coating exhibited lower values of compressive residual stress and even tensile residual stress. The higher surface hardness and higher compressive residual stress of the D-gun sprayed specimens were responsible for their superior fretting fatigue lives compared with the plasma-sprayed specimens [Rajasekaran et al. (2008)].

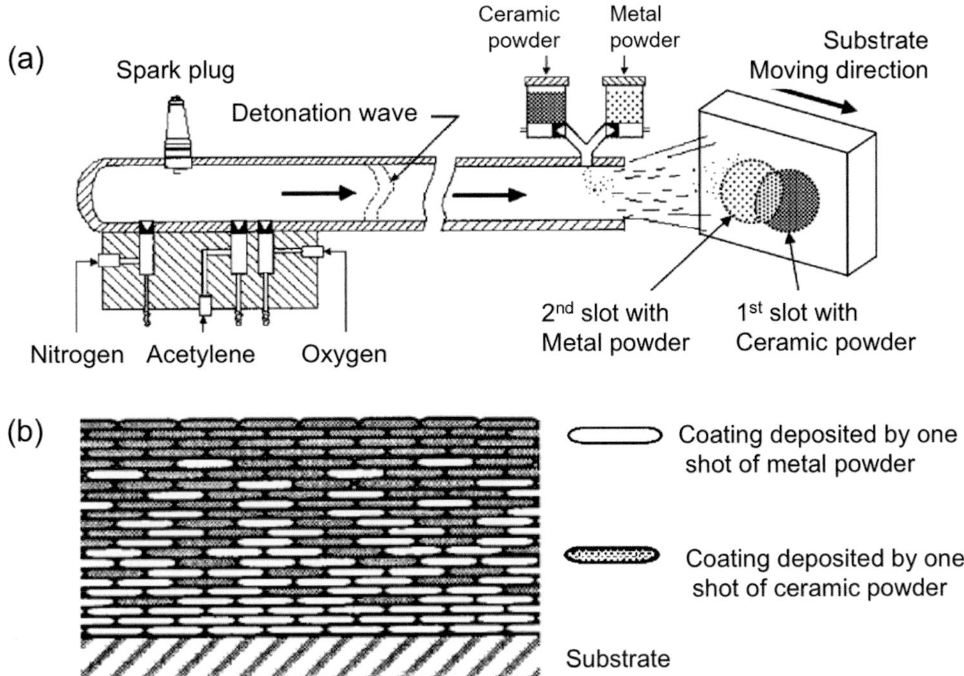

Fig. 7.62 Schematic showing the deposition scheme to produce, graded coatings using the "shot control method" (**a**) D-gun with one radial injection port and two powder feeders, (**b**) typical microstructure of a graded coating [Kim et al. (2003)]. (Reprinted with kind permission from Elsevier)

7.4.4.2 Oxidation Resistant Coatings

In the initial oxidation process of eutectic (β - NiAl + γ-Re) coatings, rhenium partially evaporates in the form of volatile oxides. Instead of rhenium oxides, a protective alumina layer forms on the coating surface. The oxidation process is reduced as Al_2O_3 is formed. It has been established that the air oxidation resistance of eutectic (β-NiAl + γ-Re) alloy coatings substantially increases at 800 to 1100 °C as compared with single-phase NiAl intermetallic coatings. The ratio of alumina and nickel formation rates is the key factor for the oxidation of these coatings. The scale formed on the single-phase NiAl intermetallic coating consists of a Al_2O_3 monolayer with spinel inclusions, while a two-layer protective scale forms on the eutectic (β-NiAl + γ-Re) alloy coating: spinel external layer and alumina internal layer, which has a better resistance to oxidation up to 1100 °C [Oliker et al. (2007)1 and 2].

7.4.4.3 Thermal Barrier Coatings (TBCs)

(i) TBC's bond coat

For thermal barrier bond coats, Ti-Al-Cr [Oliker et al. (2007)2] and MCrAlY [Yuan et al. (2008), Belzunce et al. (2002), Zhang et al. (2002), Taylor and Knapp (1995), Zhang et al. (2003)] coatings were tested. The behavior of TiAlCr depends on the substrate structure, which determines the nature of diffusion. NiCrAlY coatings displayed very favorable oxidation resistance; the aluminum layer formed was intact for over 300 h at 1050 °C. NiCrAlY coatings sprayed using a D-gun and LPPS showed a splat layer structure in the as-sprayed state [Zhang et al. (2003)2]. The difference is that an oxidation reaction took place during D-gun spraying and Al_2O_3 distributed on the splat boundaries of as-sprayed D-gun NiCrAlY coatings. Isothermally oxidized at 1050 °C for 300 h, both coatings showed favorable oxidation resistance and an Al_2O_3 scale was formed on their surface, although the oxidation resistance of D-gun NiCrAlY coatings was inferior to that of LPPS NiCrAlY coatings. D-gun NiCrAlY coatings have higher surface hardness. D-gun NiCrAlY coatings should be considered at first when the working conditions are in the high dynamic load or low oxidizing air range [Zhang et al. (2002)]. HVOF and D-gun techniques were employed to deposit the Ni–25Cr–5Al–0.5Y bond coat (BC) of a thermal barrier coating with the YSZ topcoat detonation-sprayed using hollow spherical YSZ powder [Yuan et al. (2008)]. The oxidation rate of the TBC system at 1100 °C is two times lower with the HVOF-sprayed NiCrAlY (see Fig. 7.41b) than with the D-gun sprayed one (see Fig. 7.63), with both processes using the same NiCrAlY powder. A significant number of Cr_2O_3 and $NiCr_2O_4$ oxide nodules can be formed on top of the D-gun NiCrAlY coating after 10 h oxidation at 1100 °C, while more α-Al_2O_3 is present for the HVOF sprayed coating. The coarse surface

Fig. 7.63 Cross-sections of detonation-sprayed TBCs using YSZ fused and crushed powder for the topcoat and NiCrAlY for the bond-coat [Yuan et al. (2008)] (Reprinted with kind permission from Elsevier)

roughness in the case of the D-gun bond coat ($R_a = 6.61$ µm against 5.38 µm for the HVOF bond coat) and the higher porosity (7.8% for the D-gun BC against 3.2% for the HVOF BC) have a detrimental effect on the oxidation behavior of the TBC system. The fine α-Al_2O_3 dispersion formed during HVOF spraying exerts a beneficial effect on the oxidation resistance of the HVOF NiCrAlY coatings by serving as initial alumina nuclei and promoting the development of a protective TGO scale [Yuan et al. (2008)].

(ii) TBC's ceramic layer

Yttria partially stabilized zirconia was D-gun sprayed on a Ni-base superalloy, M38G [Ke et al. (2005)]. The D-gun-sprayed TBC had a very low thermal conductivity of 1.0–1.4 W/m K, close to that of the plasma-sprayed YSZ coatings and much lower than their EB-PVD counterparts. The TBCs exhibited excellent resistance to thermal shock up to 400 cycles of 1050 °C to room temperature (forced air cooling) and 200 cycles of 1100 °C to room temperature (forced water quenching). As cycles proceeded, the development of Ni/Co rich TGO and t-to-m transformation within the ceramic coat could account for eventual spallation [Zhang et al. (2002)]. Similar results were obtained through optimizing the spray parameters (especially the ratio of C_2H_2 to O_2) [Wu et al. (2003)]. The oxidation behaviors at 1000 and 1100 °C were studied. TBCs detonation sprayed were uniform and dense, with a few micro-cracks in the ceramic coats and a rough surface of bond coats. At the high temperature, the dense detonation-sprayed ceramic coats with low porosity could obviously decrease the diffusive channels for oxygen and reduce the oxygen pressure (P_{O_2}) at the ceramic-bond coat interface [Wu et al. (2004)]. NiCrAlY/YPSZ and NiCrAlY/NiAl/YPSZ thermal barrier coatings (TBCs) were also successfully deposited by detonation spraying [Ke et al. (2005)]. TBCs included a uniform

ceramic coat containing a few microcracks and a bond coat with a rough surface. Results were similar to those previously discussed. *Figure* 7.64 from [Yuan et al. (2008)] presents a fractured YSZ TBC's topcoat obtained by D-gun (acetylene-oxygen [Suresh Babu et al. (2008)]) sprayed hollow sphere particles 10–75 µm in diameter. The microstructure of the topcoat is similar to APS ones with regard to the lamellar microstructure (Fig. 7.64-*A*). It means that it was built up by layering individual splats resulting from impacting melted or semi-molten particles. The large, equiaxed voids are probably due to particles that did not fully melt (Fig. 7.64-*B*). Splat are approximately 3–6 µm thick and hundreds of micrometers in diameter. The thermal conductivity is similar to that obtained with APS coatings.

(iii) Graded coatings

Functionally graded thermal barrier coatings (FGM TBC) have been deposited by the D-gun spray process according to the "shot-control method" [Ramadan and Butler (2004)]. The gradient ranged from 100% NiCrAlY metal on the substrate to 100% ZrO2–8 wt.% Y_2O_3 ceramic for the topcoat and consisted of a finely mixed microstructure of metals and ceramics with no obvious interfaces between the layers.

7.4.4.4 Wear Resistant Coatings

The hardness of coatings is conventionally used as the primary correlating parameter for evaluating wear resistance [Wu et al. (2003)]. Authors have studied large varieties of coatings (WC-Co, TiMo(CN), TiN-CoN, Al_2O_3, Al_2O_3-TiO_2 with different size distributions and obtained by different manufacturing processes deposited with the D-gun process with a range of process parameters for each coating. Coatings were characterized in terms of phase content and distribution, porosity, micro hardness, and evaluated for erosion, abrasion, and sliding wear resistance. Results [Sundararajan et al. (2005)] show that tribological properties of the coatings are more strongly influenced by the coating process parameters themselves rather than microstructural parameters like phase content and distribution, and porosity.

(i) WC-Co

Suresh et al. (2008) have shown that the microstructure, in terms of nature/extent of decomposition of WC, as well as properties of WC–12 wt.% Co coatings, is critically dependent on the variations of the oxygen–fuel ratio. Murthy et al. [5.196] have studied WC–10Co–4Cr and Cr_3C_2–20(NiCr) coatings deposited by HVOF and D-gun processes. The abrasion tests were done using a three-body solid particle rubber wheel test rig using silica grits as the abrasive medium. The D-gun coating performed slightly better than the HVOF one possibly due to the higher residual

compressive stresses of the former, and the WC-based coating had higher wear resistance compared to the Cr_3C_2-based coating. Also, the thermally sprayed carbide-based coatings have excellent wear resistance with respect to the hard chrome coatings [Murthy et al. (2006)]. Wang et al. (2010) have emphasized that the compressive stress of WC-Co coatings (peening effect of high spraying velocity and kinetic energy during the D-gun process) could significantly improve the coating properties, whereas the tensile stress impaired the coating properties. Park et al. (2007) have sprayed WC–12 wt. % Co, commercial feedstock, with carbide grain size of approximately 100–200 nm. Post-heat treatment of WC-Co coatings modified their resistance because microhardness increased, while fracture toughness and wear resistance increased by increasing the temperature to 800 °C, but decreased after heat treatment at 900 °C. The amorphous phase disappeared and other carbide phases such as W_3Co_3C

and W_6Co_6C formed during heat treatment above 700 °C. The improved properties were attributed to microstructural changes. Du et al. (2005) have studied the influence of process variables on the qualities of D-gun-sprayed WC–Co coatings. Suresh Babu et al. (2007) have shown that the feedstock size influences the phase composition and porosity of WC-Co coatings, imparting variations in the coating hardness, fracture toughness, and wear properties. The fine and narrow size range WC-Co coating exhibited superior wear resistance. Oliker et al. (2004) have also shown that powder structure is important. The use of powders with highly porous particles results in the formation of porous coatings with low wear resistance. Du et al. (2006) have studied the fabrication and evaluation of D-gun-sprayed WC–Co coating with self-lubricating property.

Murthy et al. (2006) have studied WC-10Co-4Cr and Cr3C2–20(NiCr) coatings, deposited by HVOF and D-gun processes, and low stress abrasion wear resistance of these coatings have been compared. *Figure* 7.64 from Murthy et al. (2001) presents SEM micrographs of transverse section of WC-10Co-4Cr coatings HVOF and D-gun sprayed. The DJ-2600 HVOF gun was working with H_2 (3.6 m^3/h), O_2 (1.8 m^3/h), and air (1.2 m^3/h), while the D-gun used O_2/C_2H_2 (1/1.23 volume ratio) and the powder size was between 11 and 53 μm. Figure 7.65 shows that the D-gun coating is slightly denser than the HVOF one, as confirmed by a porosity of 1.3% against 1.6%. Correlatively, the microhardness ($HV_{0.3}$) is 792 for the HVOL coating against 849 for the D-gun one. Micrographs of traverse sections of as-sprayed WC-10Co-4Cr HVOF and D-gun coatings are given in Fig. 7.65a and b, respectively, showing in both cases damage indued by surface grinding. Coatings were ground with a diamond resin bonded wheel to achieve a final surface roughness of $R_a < 0.2$ μm. The grinding induced surface damage in both the cases, as shown in Fig. 7.65a and b. The

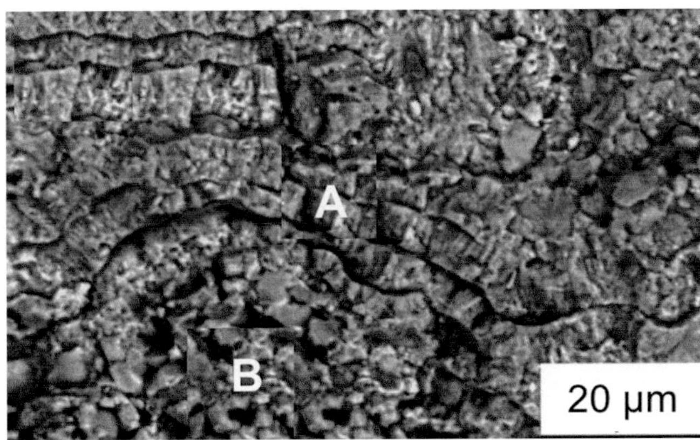

Fig. 7.64 Fractured YSZ topcoat obtained by D-gun spraying (acetylene/oxygen mixture) hollow sphere particles 10–75 μm in diameter (as-sprayed state) [Yuan et al. (2008)]. (Reprinted with kind permission from Elsevier)

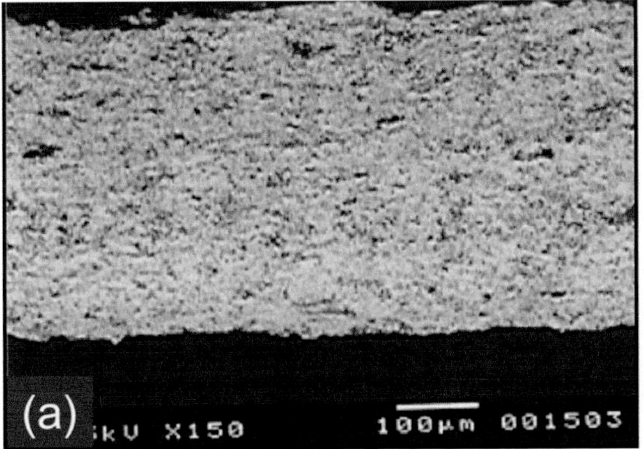

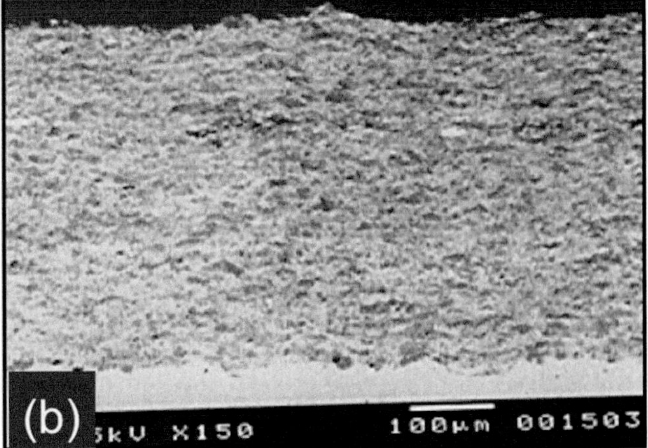

Fig. 7.65 SEM micrographs of transverse section of WC-10Co-4Cr coating in as-sprayed condition: (**a**) HVOF coating; (**b**) D-gun sprayed coating [Murthy et al. (2001)]. (Reprinted with kind permission from Elsevier)

microstructure of the HVOF coating developed numerous cracks beneath the surface up to a depth of 150–200 μm. However, such cracks were minimum in the case of D-gun-sprayed samples (Fig. 7.65b). The erosion resistance of the D-gun coating was found to be higher compared to that of the HVOF coating in the as-coated condition. This is possibly due to the slightly higher microhardness, lower porosity, and possibly higher residual compressive stresses of the DS coating [Murthy et al. (2001)].

Usually results show that the D-gun coatings perform slightly better than the HVOF coatings possibly due to the higher residual compressive stresses induced by the former process, and WC-based coatings have higher wear resistance in comparison to Cr_3C_2-based coating. Also, the thermally sprayed carbide-based coatings have excellent wear resistance with respect to the hard chrome coatings.

Du et al. (2006) deposited by the detonation gun (D-gun) process a WC–Co coating with self-lubricating property, using a commercial WC–Co powder doped with a MoS_2–Ni powder, under a proper spray condition. The results indicate that the MoS_2 content was maintained, if not slightly increased, in the coating which was attributed to the protection offered by the nickel coating of the MoS_2 doping material in the feed powder. Evaluation on sliding wear property indicates that the MoS_2 composition plays an important role in lowering both coefficient of friction and wear rate for the resulting coating, which is confirmed by observations on wear track, as well as X-ray photoelectron spectroscope (XPS) results on worn surface. It suggests that the deposition of a WC–Co coating with self-lubricating property by D-gun spray is feasible by controlling lubricant powder and spray conditions, which can exhibit higher sliding wear resistance. To analyze the improved sliding wear property of the WC–Co coating containing MoS_2 composition, worn surface after tests were observed and compared with that of the pure one, as shown in Fig. 7.66.

It is apparent that the groove is shallower for the WC–Co coating with MoS_2 composition. It is supposed that the wear particles produced in the sliding process are comprised of WC, Co, MoS_2, and Ni or some of them, which underwent crushing, compaction, and smearing at the contact area between tested material/counter body, leading to the diminution of their size, due to the developed shear stresses.

(ii) Alumina

The structure of D-gun-sprayed Al_2O_3 coatings was investigated by Li and Ohmori (1996) using a copper electroplating technique with a typical layer structure with poor bonding at interfaces between flattened particles. The mean bonding ratio of the bonded interface to total apparent bonding interface area was about 10%. However, a high particle velocity may result in a rough surface for individual flattened particles, which is effective for better interlocking of flattened particles. The good wear resistance might be mainly due to the mechanical interlocking effect between flattened particles. As for WC-Co, the feedstock size was also found to influence the phase composition of Al_2O_3. The course ($d_{50} = 18$ μm) and narrow ($d_{50} = 10$ μm) size distribution Al_2O_3 coating exhibited better performance under abrasion and sliding wear modes; however, under erosion wear mode, the as-received Al_2O_3 ($d_{50} = 14$ μm) coating exhibited better performance. Saravanan et al. (2000) have conducted a Taguchi-full factorial (L16) design parametric study to optimize the D-gun spray process parameters. For hardness, the significant factors of influence were spray distance, carrier gas flow rate, and fuel-to-O_2 ratio selected, in that order. They showed that alumina coatings of the highest hardness 1363 H and the lowest porosity 1.45% could be obtained for specific spray parameters. Other authors have also sprayed alumina coatings [Niemi et al. (1994), Saravanan et al. (2000), Pogrebnyak et al. (2000), Sobiecki et al. (2004)].

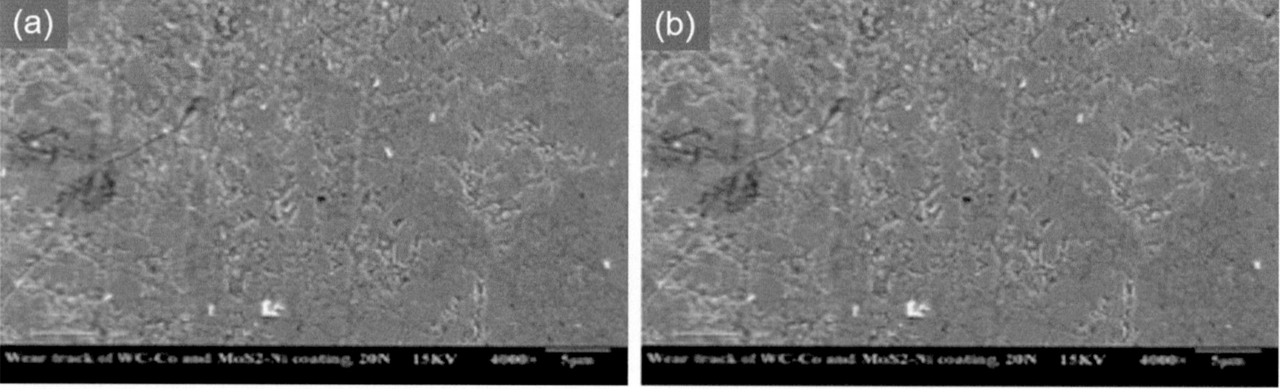

Fig. 7.66 SEM morphologies of wear surface of the WC–Co coatings against WC–Co ball under normal load of 20 N; **(a)** pure WC–Co coating, **(b)** WC–Co coating containing MoS_2 composition [Du et al. (2006)]

(iii) Alumina-titania

D-gun-sprayed coatings of Al_2O_3-TiO_2 [Venkataraman et al. (2006)] (13 wt.%), with a particle size of 22–45 μm, have low porosity, high density, hardness above 1000 $HV_{0.5}$, and, at the same time, considerably improved frictional wear resistance compared to that of the AISI-1045 steel substrate. A preliminary preparation of the substrate surface by compressed air-enhanced grit blasting is necessary for the coating to adhere well to the substrate. Venkataraman et al. (2006) have studied the phases present in D-gun-sprayed Al_2O_3-TiO_2 (13 wt%) and compared them to those obtained with plasma spraying. Semenov and Cetegen (2002) have studied D-gun spraying of nano-structured alumina–titania (12 wt.%) coatings, starting from commercial agglomerated nanostructured particles. XRD spectrum peaks of α-Al_2O_3 (113) and γ-Al_2O_3 (400) for the feedstock powder allowed for determining the fraction of melted powder. Coating obtained by D-gus spraying contained a smaller concentration of α-Al_2O_3 phase compared to that of γ-Al_2O_3 which is an indication of limited melting of the powder prior to their deposition on the substrate. It also suggests that the technology could be successfully employed for the deposition of nanostructured alumina–titania coatings. These findings combined suggest that the detonation deposition process could be successfully employed in the deposition of nano-structured alumina–titania coatings.

Coatings exhibited lower erosion wear resistance than bulk alumina but better resistance to abrasion wear, especially Al_2O_3 + 3 wt % TiO_2 and Al_2O_3 + 40 wt % ZrO_2. Compared to alumina coatings that are plasma sprayed and contain up to 98% of γ-alumina, D-gun coatings have γ-alumina phase content of less than 60%, and about 30% of α-alumina phase content, the balance being β- and δ-alumina phases [Semenov and Cetegen (2002)].

Bhandari et al. (2012) investigated the slurry erosion performance of detonation gun (D-gun) spraying ceramic coatings (Al_2O_3 and Al_2O_3-$13TiO_2$) on CF8M steel. The slurry collected from an actual hydro power plant was used as the abrasive media in a high-speed erosion test rig. An attempt has been made to study the effect of concentration (ppm), average particle size, and rotational speed on the slurry erosion behavior of these ceramic-coated steels under different experimental conditions. The analysis of eroded samples was done using SEM, XRD, and stylus profilometry. The slurry erosion performance of the D-gun spray Al_2O_3-$13TiO_2$-coated steel has been found to be superior to that of Al_2O_3-coated steel. Both the coatings showed brittle fracture mechanism y of material removal during the slurry erosion exposure. During the slurry erosion of Al_2O_3-coated steels, slurry concentration and average particle size were found to be relatively more dominant factors in comparison with rotational speed. On the other hand, in the case of Al_2O_3-$13TiO_2$-coated steel y, rotational speed was found to be more dominant in comparison with slurry concentration and average particle size. Thus, D-gun-sprayed Al_2O_3-$13TiO_2$ coatings can be used in low-speed hydro turbines with high concentration slurries of large particles. Fatigue and brittle failure were found to be dominating material removal mechanism in the Al_2O_3-$13TiO_2$ coating whereas the Al_2O_3 coating showed brittle behavior mechanism of material removal. Under stated experimental conditions, there was no chemical effect of the used slurry on the coatings.

(iv) Tribological coatings

Sundararajan et al. (2010) have studied the tribological performance of 200 μm thick TiMo(CN)-28Co and TiMo (CN)–36NiCo coatings obtained using a D-gun. Powders were prepared by SHS and had the following characteristics: TiMo(CN)-36NiCo, d_{50} = 36.4 μm irregular and blocky shaped, phases being $TiC_{0.7}N_{0.3}$, Co, Ni, and TiMo(CN)-28Co, d_{50} = 23.3 μm, irregular shaped, phases being $TiC_{0.7}N_{0.3}$, Co. In both coatings, the best tribological performance and the lowest porosity were obtained at intermediate oxygen/acetylene ratios. When comparing the tribological performance of the optimized TiMo(CN) type coatings with that of optimized WC-Co ones, the abrasion resistance is comparable. However, the erosion and sliding wear resistance of TiMo(CN) type coatings were considerably lower than those of the WC-Co coatings. Du et al. (2007) have deposited WC-Co/MoS_2–Ni coatings by D-gun, using a commercial WC-Co powder and a MoS_2-Ni powder, with the spray condition being adapted to both powders. The MoS_2 composition was kept and distributed homogeneously. Coatings had self-lubricating properties. MoS_2 lowered the wear rate under dry sliding conditions when its content was lower than 4.9 wt.%.

(v) Corrosion and wear-resistant coatings

Cr_3C_2 (75 wt %)-NiCr D-gun-sprayed coatings on high-temperature structure steel DIN 12CrMo44 [Wang et al. (2000)] were dense with good bonding to substrate as well as high resistance to high-temperature oxidation and wear. Reports from the in-service test at Bao Shan Steel Company indicate that such coatings have at least doubled the roll life. Cr_3C_2–NiCr cermet coatings were deposited on two Ni-based superalloys, namely Superni 75 and Superni 718, and one Fe-based superalloy Superfer 800H by the D-gun thermal spray process [Sundararajan et al. (2010)]. The cyclical hot-corrosion studies were conducted on uncoated as well as D-gun-coated superalloys in the presence of a mixture of 75 wt.% Na_2SO_4 + 25 wt.% K_2SO_4 film at 900 °C for 100 cycles. It was observed that the Cr_3C_2–NiCr-coated superalloy showed better hot-corrosion resistance than the

uncoated superalloys, as a result of the formation of continuous and protective oxides of chromium, nickel, and their spinel. Cr_3C_2–NiCr coatings were D-gun sprayed on Superni 75, Superni 718, and Superfer 800H superalloys [Kamal et al. (2008)]. Coatings exhibited nearly uniform, adherent, and dense microstructures with porosity less than 0.8%. These coatings were found to be very effective in decreasing the corrosion rate in a molten salt environment (Na_2SO_4–60% V_2O_5) at a high temperature of 900 °C for 100 cycles. Particularly, the coating deposited on Superfer 800H showed better hot corrosion protection as compared to Superni 75 and Superni 718. The cyclical oxidation behavior of the same coatings on the same superalloy at 900 °C for 100 cycles in air under cyclic heating and cooling conditions has been investigated [Kamal et al. (2009)]. Among all the coated superalloys, chromium, iron, silicon, and titanium were oxidized in the inter-splat region, whereas splats that consisted mainly of Ni remained un-oxidized. The parabolic rate constants of Cr_3C_2-NiCr-coated alloys were lower than those of the bare superalloy. The cyclical oxidation behavior of the same coatings on the same superalloy at 900 °C for 100 cycles in air under cyclic heating and cooling conditions has been investigated [Kamal et al. (2009)]. Among all the coated superalloys, chromium, iron, silicon, and titanium were oxidized in the inter-splat region, whereas splats, which consisted mainly of Ni remained un-oxidized. The parabolic oxidation rate of Cr_3C_2-NiCr-coated alloys were lower than those of the bare superalloys. To prolong the life of steel slab continuous casting rolls, Cr_3C_2-NiCr D-gun spray coatings were processed on the roll surface (DIN 12CrMo44) in a steelmaking plant in China [Wang et al. (2002)]. The wear resistance of the coated samples reduced the risk of seizure compared to uncoated samples, at room and elevated temperatures with any load and sliding velocity. It must also be emphasized that Cr_3C_2–20(NiCr) coatings have excellent wear resistance [Murthy et al. (2006), Wang et al. (2002)], however lower than those of the WC-Co ones. When they are exposed to high temperatures (over 600 °C), their corrosion resistance is improved [Wang et al. (2002b), Murthy et al. (2007)].

Kamal et al. (2011) investigated rare earth oxide (CeO_2) incorporated in the NiCrAlY alloy and the hot corrosion resistance of detonation-gun-sprayed NiCrAlY +0.4 wt.% CeO_2 coatings on superalloys, namely, superni 75, superni 718, and superfer 800H in molten 40% Na_2SO_4–60% V_2O_5 salt environment at 900 °C for 100 cycles. The coatings exhibited a characteristic splat globular dendritic structure with diameters similar to those of the original powder particles. The weight change technique was used to establish corrosion kinetics. X-ray diffraction (XRD), field emission scanning electron microscopy/energy-dispersive analysis (FE-SEM/EDAX), and X-ray mapping techniques were used to analyze the corrosion products. The coated superfer

800H alloy showed the highest corrosion resistance among the examined superalloys. CeO_2 was found to be distributed in the coating along the splat boundaries, whereas Al streaks distributed non-uniformly. The main phases observed for the coated superalloys are the oxides of Ni, Cr, Al, and spinels, which are suggested to be responsible for developing corrosion resistance. The surface morphology of the as-sprayed coating depicts the formation of un-melted particles in the form of a globular dendritic structure. The diameter of the dendritic structure is equal to that of original powder particles. Cerium oxide is uniformly distributed along the Ni-rich splat boundaries. Traces of Al are distributed along the Ni-rich splat boundaries. The bare and coated Fe-based superfer 800H superalloy showed least and highest resistance to hot corrosion, respectively. D-gun-sprayed NiCrAlY +0.4 wt.% CeO2 coating was found to be effective in imparting hot corrosion resistance to superfer 800H in the molten salt environment. The formation of oxides along the splat boundaries and within the open pores of the coatings might have acted as a diffusion barrier to the inward diffusion of molten salt. A dense oxide scale formed on the coated superalloys and the hot corrosion resistance of the coating might be due to the formation of protective phases like NiO, Cr_2O_3, Al_2O_3, $NiCr_2O_4$, and Ni Al_2O_4.

Ma et al. (2015) prepared a pure Al coating by a detonation gun (D-gun) spraying process to protect sintered NdFeB magnets. The detonation gun sprayed coating was very uniform and had a low porosity of 0.77%. The thickness of the Al coating was approximately 16 μm. The corrosion current density for the coated sample was 1.30 $\times 10^{-5}$ A/cm^2 immediately after immersion in 3.5% NaCl solution, compared to 6.54 x 10^{-5} A/cm^2 for the uncoated sample. X-ray photoelectron spectrometry results indicated that the formation of Al_2O_3 film contributed to the increased corrosion resistance of Al coating. Meanwhile, electrochemical impedance spectroscopy with an equivalent electrical circuit was used to ascertain the corrosion process of the Al coatings. The micrographs of the surface and the cross-section of the Al coating on sintered NdFeB permanent magnets are revealed in Fig. 7.67. Results showed that the corrosion procedure consisted of two stages, which agreed with the potential-dynamic polarization test. It was concluded that the Al coating deposited by the D-gun spray process can improve the corrosion resistance of the sintered NdFeB.

Corrosion-resistant Cr_2O_3- Al_2O_3 (50 wt.% Al_2O_3) coatings were sprayed using D-gun process by Rani et al (2017)a on ASTM-SA213-T-22 boiler steel, and Fe-based superalloy Superfer 800 H. The high-temperature corrosion performance of the coated as well as the bare alloys was evaluated in Na_2SO_4–60%V_2O_5 molten salt with an aggressive environment at 900 °C under cyclic conditions. The kinetics of the corrosion were analyzed by the change in

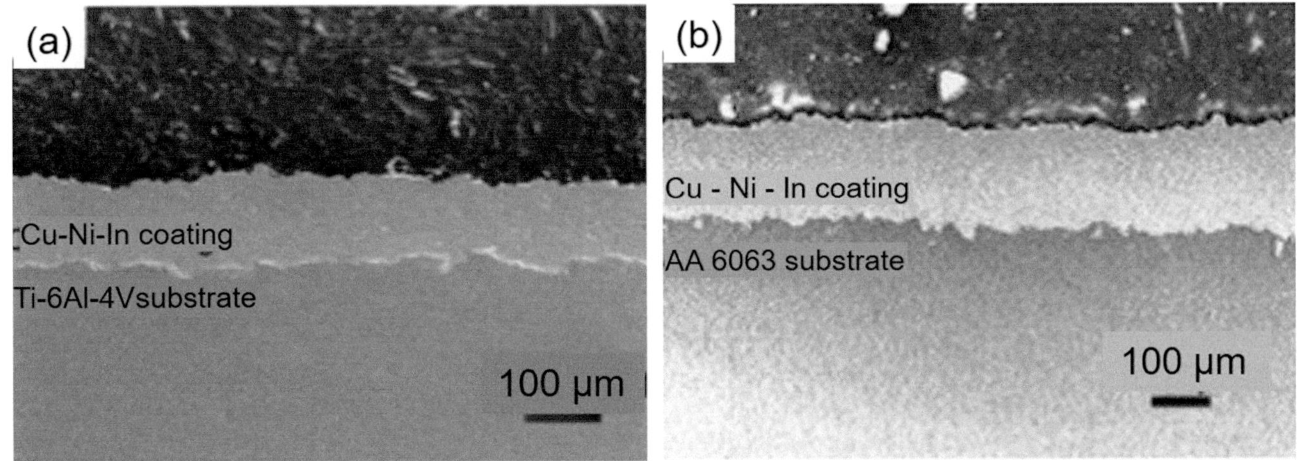

Fig. 7.67 SEM morphologies of the Al coating: (**a**) surface, (**b**) cross-section [Ma et al. (2015)]

weight measurements, which were taken after each cycle (i.e., 1-h heating in a tube furnace followed by 20-min cooling in ambient air) for a total period of 50 cycles. The X-ray diffraction and scanning electron microscopy/energy-dispersive X-ray analysis techniques were used for the analysis of the corrosion products. During investigations, it was found that both the selected bare alloys have suffered intensive spallation in the form of removal of their oxide scales, which may be attributed to the formation of non-protective Fe_2O_3-dominated oxide scales, whereas the coated alloys have shown lesser weight gains along with better adhesiveness of the oxide scales with the substrate till the end of the experiment. The oxides of chromium and aluminum were the main phases revealed in the oxide scales of the coated specimens, which are reported to be protective against the hot corrosion. The Cr_2O_3–50% Al_2O_3 coating was successful in reducing the corrosion rate of the T-22 steel by 97%, in terms of overall weight gains, whereas the reduction was 19% for the Superfer 800H alloy. In case of the Cr_2O_3–50% Al_2O_3-coated T-22 alloy, the weight change was almost negligible throughout the study, whereas in case of the Cr_2O_3–50% Al_2O_3-coated Superfer 800H superalloy, the weight gain abruptly increased after the 25th cycle, but remained lesser than the bare counterpart.

Rani et al. (2017)2 comparatively discussed in the present study the high-temperature hot corrosion behavior of bare and detonation-gun-sprayed Ni-5Al coatings on Ni-based superalloy Inconel-718. Hot corrosion studies were carried out at 900 °C for 100 cycles in the Na_2SO_4–60% V_2O_5 molten salt environment under cyclic heating and cooling conditions. The thermo-gravimetric technique was used to establish the kinetics of hot corrosion. X-ray diffraction, SEM/EDAX, and X-ray mapping techniques were used to analyze the hot corrosion products of bare and coated superalloys. The results indicated that the Ni-5Al-coated superalloy showed very good hot corrosion resistance. The overall weight gain and parabolic rate constant of the Ni-5Al-coated superalloy were less in comparison with the bare superalloy. The D-gun-sprayed Ni-5Al coating was found to be uniform, adherent, and dense in a hot corrosion environment. The formation of nickel- and aluminum-rich oxide scales might have contributed to the better hot corrosion resistance of the coated superalloy. The Ni-5Al coating exhibited dense and uniform lamellar structure with cross-sectional porosity values of around 0.9%, micro-hardness values in the range of 253–306 HV, and surface roughness values in the range of 4.5–5.6 µm.

Mudgal et al. (2015) pointed out that Cr_3C_2–25(NiCr) coatings were widely used in wear, erosion, and corrosion applications. In the present study, D-gun-sprayed Cr_3C_2–25 (NiCr) coatings with and without 0.4 wt.% ceria incorporated were deposited on Superni 718, Superni 600, and Superco 605 substrates. Hot-corrosion runs were conducted in 40%Na_2SO_4–40%K_2SO_4–10%NaCl-10%KCl environment at 900 °C for 100 cycles. Corrosion kinetics were monitored using weight-gain measurements. Characterization of corrosion products was carried out by field-emission scanning electron microscopy (FESEM)/energy-dispersive spectroscopy (EDS) and X-ray diffraction (XRD) techniques. It was observed that Cr_3C_2–25(NiCr) coating with and without added ceria deposited on both of the Ni-based alloys showed resistance to corrosion under the given environment. Addition of ceria enhanced the adherence of the oxide to the coating during the corrosion run and reduced the overall weight gain. However, Cr_3C_2–25 (NiCr)-coated Superco 605 did not perform satisfactorily under this environment. In the case of Superco 605, Cr_3C_2-NiCr with and without ceria did not perform very well in the hot-corrosion run, with a lot of delamination and spallation being observed from a very early stage of exposure. SEM

analysis of all the coatings clearly showed the variation in surface morphology of the coatings with and without ceria doping. In the case of coating without ceria doping, massive oxide along with needle-like morphology could be seen, while in ceria-doped coating some dense patches were observed throughout the surface, with clusters containing ceria in substantial amounts. In the case of corroded coated Superni 718 and Superni 600, Cr_2O_3 and $NiCr_2O_3$ were the major oxides providing protection.

[Kaushal et al. (2014)] recalled that to protect materials from surface degradations such as wear, corrosion, and thermal flux, a wide variety of materials can be deposited on the materials by several spraying processes. Their paper examined and compared the microstructure and high-temperature corrosion of Ni-20Cr coatings deposited on T22 boiler steel by high-velocity oxy-fuel (HVOF), detonation gun spray, and cold spraying techniques. The coatings' microstructural features were characterized by means of XRD and FE-SEM/EDS analyses. Based on the results of mass gain, XRD, and FE-SEM/EDS analyses, it may be concluded that the Ni-20Cr coating sprayed by all the three techniques was effective in reducing the corrosion rate of the steel. Among the three coatings, D-gun spray coating proved to be better than HVOF-spray and cold-spray coatings.

(vi) Erosion resistance

It is known that the addition of Cr to WC–Co improves the binding of the metallic matrix with the WC grains and provides better wear-resistant coatings. Thus, WC–Co–Cr is considered to be a potentially better wear-resistant coating material as compared to WC–Co. The thermally sprayed carbide coatings are in general surface finished by machining or grinding after the coating process. A WC-10Co-4Cr powder has been sprayed [Murthy et al. (2001)] on medium-carbon steel using D-gun and HVOF processes. Some of the coated specimens were further ground by a diamond wheel with controlled parameters, and both "as-coated" and "as-ground" conditions have been tested for solid particle erosion behavior. It has been found that surface grinding improved the erosion resistance. A detailed analysis indicates that the increase in residual stress in the ground specimen is a possible cause for the improvement in erosion resistance. Sand erosion studies of D-gun sprayed WC-Co-Cr have been undertaken [Wood et al. (1997)] using a sand/water jet impingement rig. The erosion rate of coatings compared well with sintered tungsten carbide-cobalt-chrome for low-energy impacts, but the sintered material outperforms the sprayed material for high-energy impacts by a factor of four. This fact is the result from the anisotropic microstructure of coatings with preferred crack propagation parallel to the coating surface, followed by crack interlinking and spalling. The influence of the angle of impingement of the sand/water slurry jet with respect to the surface of the coated part was found to be most pronounced for low energy slurries compared to that under high energy conditions. Maximum erosion rates in this case were obtained with the slurry jet impinging the surface of the part at 90°, while minimal erosion rates were observed at angles below 30°. In contrast, the erosion rate using high energy slurries were essentially independent of the jet angel.

WC-12%Co and WC-17%Co coatings were deposited [Manish (2002)] by detonation spraying on three different substrate materials—mild steel, commercially pure (CP) aluminum, and CP titanium—to test their dynamic hardness by a drop weight system. WC-Co coatings exhibit higher hardness under impact at a high strain rate. The dynamic hardness of a WC-Co coating is maximum on aluminum substrate and minimum on mild steel substrate. The identical coating exhibits intermediate hardness on titanium substrate.

Harpreet Singh Grewal et al. (2012) deposited WC-Co-Cr coatings on some hydro-turbine 13Cr4Ni and 16Cr5Ni steels by the detonation-gun spray process. An in-depth characterization of the as-sprayed coating was done using X-ray diffraction (XRD) and scanning electron microscopy (SEM)/energy-dispersive X-ray spectroscopy (EDS) techniques. Microhardness and porosity measurements were also made. The coating was found to have a typical splat-like morphology with some indications of un-melted carbide particles. The XRD results showed the presence of WC as the primary phase along with W_2C and Co_6W_6C as secondary phases. Furthermore, the slurry erosion behavior of the coatings was investigated to ascertain the usefulness of the coatings to reduce the slurry erosion of the steels. The effect of four operating factors viz. the velocity, impact angle, concentration, and particle size on the slurry erosion of coated and bare steels have been studied using a high-speed jet-type test rig. The sand used as an erodent was collected from a power plant to replicate the actual turbine conditions. It has been observed that the given cermet coating can enhance the erosion resistance of the steel. Velocity was found to be the most significant factor affecting the erosion behavior of the coating, whereas it was the erodent particle size in the case of uncoated steel. As evidenced from the SEM images, the platelet mechanism of erosion seemed to be the prominent one, causing the removal of material from the surface of the steel, whereas for the coating, the formation and interlinking of cracks resulted in the removal of material.

(vii) Fatigue

Ganesh Sundara Raman et al. (2007) sprayed, using the detonation gun (D-gun) spray process, Cu-Ni-In coating on two substrate materials: Ti-alloy (Ti-6Al-4 V) and Al-alloy (AA 6063) fatigue test specimens. Coatings on both substrates were dense with low porosity, high hardness, and high surface roughness, see Fig. 7.68. Relatively higher surface compressive residual stress was present at the coating on

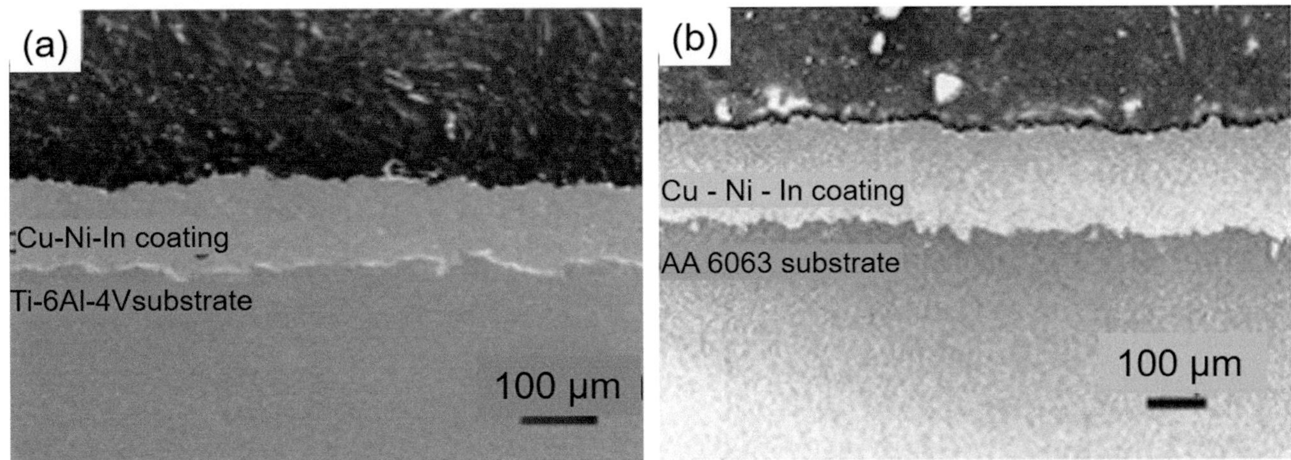

Fig. 7.68 SEM micrographs showing the cross section of Cu-Ni-In coating on two different substrates: (**a**) Ti-6Al-4 V; (**b**) AA 6063 [Ganesh Sundara Raman et al. (2007)]

Ti-alloy specimens. In case of the coating on Al-alloy samples, tensile residual stress was also present in some places. Uniaxial plain fatigue and fretting fatigue experiments were conducted on uncoated and coated specimens. The detrimental effect of life reduction due to fretting was relatively larger in the Al-alloy compared to the Ti-alloy. While Cu-Ni-In coating was found to be beneficial on the Ti-alloy, it was deleterious on the Al-alloy substrate under both plain fatigue and fretting fatigue loading. The results were explained in terms of differences in the values of surface hardness, surface roughness, surface residual stress, and friction stress.

The mismatch in the properties of coating and substrate was less in case of the Cu-Ni-In coating on Ti- 6Al-4 V alloy than the Cu-Ni-In coating on AA 6063. The detrimental effect of life reduction due to fretting was relatively larger in the Al-alloy compared with that in the Ti-alloy. While the Cu-Ni-In coating was found to be beneficial on the Ti-alloy, it was deleterious on the Al-alloy substrate under both plain fatigue and fretting fatigue loading. The results were explained in terms of differences in the values of surface hardness, surface roughness, surface residual stress, and friction stress.

(viii) Boiler environment

Kaushal et al. (2011) used two spray techniques—high-velocity-oxy-fuel (HVOF) spray and detonation-gun (D-gun) spray—to deposit Ni-20Cr coatings on a commonly used boiler steel ASTM-SAE 213-T22. The specimens, with and without coating, were subjected to molten salt (Na_2SO_4–60% V_2O_5) deposition in a laboratory furnace at 900 °C to determine hot-corrosion resistance. Specimens were also exposed to the superheater zone of a thermal power plant boiler at an average temperature of 700 °C under cyclic conditions to ascertain their erosion-corrosion (E-C) behavior. Mass-change measurements were taken to approximate the kinetics

of corrosion and erosion-corrosion. In the case of E-C, the thickness lost data were also taken at the end of the exposure. The exposed specimens were characterized by X-ray diffraction (XRD) and field-emission scanning electron microscopy/ energy dispersive spectroscopy (FE-SEM/EDS). The HVOF-sprayed coating was found to be intact during exposure to both given environments, whereas the D-gun coating showed spallation of its oxide scale during exposure to the molten salt environments. An overall analysis of the results indicated that the HVOF-sprayed Ni-20Cr coating should be a better choice for the given.

7.4.4.5 Other Cermets

Metal-matrix composites (MMCs), containing large ceramic particles as super-abrasives, are typically used for grinding stone, minerals, and concrete. Sintering and brazing are the key manufacturing technologies for grinding tool production. However, restricted geometry flexibility and the absence of repair possibilities for damaged tool surfaces, as well as difficulties of controlling material interfaces, are the main weaknesses of these production processes. Thermal spraying offers the possibility to avoid these restrictions [Oliker et al. (2006)]. Cu-based coatings containing large ceramic particles (Al_2O_3 and SiC particles larger than 150 μm) gave high abrasiveness for grinding stones and concrete [Tillmann et al. (2007)].

FeTi-SiC coatings gave very high hardness (7900 MPa) when spraying agglomerated powders with fine SiC. The coatings contained $TiSi_2$, TiSi, Ti_5Si_3, and FeTiSi, and the porosity was below 2% [Oliker et al. (2005)]. Detonation coatings of mechanically Ti-Al alloyed powders [5.231] contained titanium aluminide with the inclusion of nitrides, depending on the working gas environment. With the mechanically alloyed powder Ti-50Al, coatings can be consolidated by the formation of the compound Al_2TiO_5 obtained when the working gas is oxidizing [Oliker et al.

(2005)]. Filimonov et al. (2008) have considered the mechanisms of structure formation during gas detonation spraying of coatings of $TiAl_3$ and Ni_3Al intermetallic compounds produced under equilibrium and non-equilibrium synthesis conditions.

Podchernyaeva et al. (2003) have investigated the composition, structure, and wear rate of detonation coatings on steel 30KhGSNA deposited from composite powders based on $TiC_{0.5}N_{0.5}$ with refractory additions of SiC, AlN, and a Ni-Cr metallic binder. It was shown that at a load of 10 MPa, these coatings exhibited substantially less wear and a larger range of sliding velocities with a stable value of wear than coatings of the hard alloy WC-15% Co.

A kind of Ti-Fe-Ni-C compound powder was prepared by a novel precursor pyrolysis process using ferro-titanium, carbonyl nickel powder, and sucrose as raw materials [Zhu et al. (2008)]. The powder had a very compact structure and was uniform in particle size. The TiC-Fe36Ni composite coatings were simultaneously in-situ synthesized by reactive D-gun spraying using this powder. The coatings presented the typical morphology of thermally sprayed coatings with two different areas: one was the area of TiC distribution where the round fine TiC particles (from 300 nm to 1 μm) were dispersed in the Fe36Ni alloy matrix; the other was the area of TiC accumulation (from 2 to 4 μm). The surface hardness of the composite coating reached about 94 ± 2 (HV_{15N}).

The effects of the powder particle size and the acetylene/oxygen gas flow ratio during the D-gun spray process on the amount of molybdenum phase, porosity, and hardness of the coatings using MoB powder were investigated by Yang et al. (2001). The results show that the presence of metallic molybdenum in the coating results from decomposition of MoB powder during thermal spray. The compositions of the coatings are metallic Mo, MoB, and Mo_2B, which are different from the phases of the original powder. Similar results were obtained by Gao et al. (2001).

7.4.4.6 Alloys

[Senderowski and Bojar (2008), Senderowski and Bojar (2009)] have studied D-gun sprayed NiAl and NiCr intermediate layers underneath the intermetallic Fe–Al type coatings on a plain carbon steel substrate. The interface layers are responsible for the hardness, bond strength, thermal stability, and adhesive strength of the whole coating structure. The physical–chemical properties of the intermediate layers, combined with a unique, very dense, and pore-free intermetallic Fe–Al coating obtained from self-decomposing powders resulted in a more complex structure. It enabled independent control of its functional properties and considerably reduced negative gradients of stress and temperature influencing the substrate and increasing adhesion strength.

7.5   Summary and Conclusions

This chapter was devoted to spray processes where thermal and kinetic energies are generated chemically through the combustion (or detonation) of fuels (gaseous or liquid) with oxygen or air. The oldest process is flame spraying, which appeared in 1910 and is still in common use. Oxyacetylene torches are the most common, and undoubtedly this process is the cheapest one. The use of rods or cords has allowed spraying ceramics, which otherwise as powders would not have been melted. The main limitation of the flame process is the relatively low velocity of sprayed particles (< 60–80 m/s) resulting in porous coatings. Oxide inclusions are also present due to the high degree of interaction between fully molten droplets and the surrounding atmosphere.

The detonation gun (D-gun) appeared in Western countries in the 1950s. In this process, very high thermal and kinetic energies were achieved by producing a detonation in an explosive mixture (mostly acetylene-oxygen) confined in a tube closed at one end into which the powder is introduced. Of course, the process is cyclic (6–100 Hz) with the introduction of the explosive gases and powder dose, the detonation, ignited by a spark plug followed by the purging of burned gases. The process generates high particle velocities (500–900 m/s) and fully molten particles (even ceramic ones, provided their size is small <30 μm). Thus, coatings are dense (less than 2% porosity), well bonded, with compressive stress, and their oxidation is limited.

In the 1960s, to compete with the D-gun deposition, the HVOF process was introduced where the combustion flame is produced in a pressurized chamber followed by a convergent-divergent nozzle and a barrel to limit the surrounding atmosphere entrainment. First developed with gaseous combustible gases and oxygen, it was extended by replacing oxygen by air (HVAF process) to reduce the gas temperature and increase its velocity. The next step was the high-power HVOF torches working with kerosene followed by HVOF processes where non-combustible gas (nitrogen) was introduced in the combustion chamber or water injected downstream of the nozzle throat. These new torches were aimed at reducing the gas temperature and increasing the velocity. Coatings obtained with HVOF torches are rather dense and their oxidation level is rather low. Of course, the density increases with particle impact velocity that varies with the combustible gas or liquid used with either oxygen or air or nitrogen addition in the combustion chamber. Velocities between 400 and 900 m/s can be achieved for the same particles depending on the spray conditions and gun used. With high velocities (> 700 m/s), it becomes possible to spray particles below their melting point and thus with a low oxidation level. However, it must be emphasized that these processes are not suited to spraying ceramics.

Nomenclature

Units are indicated in parentheses; when no units are indicated, the parameter is dimensionless.

Latin Alphabet

a_i local sound velocity, $a_i = \sqrt{\gamma \rho / \rho_g}$ (m/s)

A oxidizer volume or mass (m^3 or kg)

A_i cross-sectional area perpendicular to the direction of the flow (m^2)

A_t throat area (m^2)

D detonation wave velocity (m/s)

F fuel volume or mass (m^3 or kg)

F/A fuel to oxidizer molar ratio

h_i enthalpy (J/kg)

$\dot{m}_{O_2}$ oxygen gas feed rate (kg/s)

M_a Mach number, $Ma = vg/ai$

p pressure (Pa)

p_t gas pressure at the throat (Pa)

P_F power dissipated in the flame (kW)

q chemical energy release at constant pressure (J/kg)

Q Gas flow rate (slm)

E_D specific energy of detonation (J)

R perfect gas constant (J/K.kg)

S_t stoichiometric factor $S_t = Q_{O_2} / \left(Q_{O_2} \right)_{St.}$

R_u universal gas constant (8.32 J/K.mole)

R flame richness ratio $R' = (F/A)/(F/A)_{St.}$

t_c detonation cycle duration time (s)

T_g gas temperature (K)

T_m melting temperature (K)

T_p particle temperature (K)

T_t temperature at the throat (K)

u_b burned gas velocity (m/s)

u_u unburned gas velocity (m/s)

v_f flame velocity (m/s)

v_g g velocity (m/s)

Greek Alphabet

b burned gases (in the immediate vicinity behind the front)

g gas

p particle

t throat

u unburned gas (before the detonation wave front)

Subscripts

Δt_i characteristic time intervals of a detonation (s)

ϕ dummy variable

ϕ' time- (or Reynolds) averaged dumb variable

ϕ'' density-averaged dumb variable

γ specific heats ratio $\gamma = c_p/c_v$

ρ_g gas mass density (kg/m^3)

ρ_p particle mass density (kg/m^3)

ρ_t gas mass density at the throat (kg/m^3)

References

Aalamialeagha, M.E., S.J. Harris, and M. Emamighomi. 2003. Influence of the HVOF spraying process on the microstructure and corrosion behavior of Ni-20%Cr coatings. *Journal of Materials Science* 38: 4587–4596.

Ahmed, R., and M. Hadfield. 2002. Mechanisms of fatigue failure in thermal spray coatings. *Journal of Thermal Spray Technology* 11 (4): 551–558.

Altomare, L., D. Bellucci, G. Bolelli, B. Bonferroni, V. Cannillo, L. De Nardo, R. Gadow, A. Killinger, L. Lusvarghi, A. Sola, and N. Stiegler. 2011. Microstructure and in vitro behaviour of 45S5 bioglass coatings deposited by high velocity suspension flame spraying (HVSFS). *Journal of Materials Science: Materials in Medicine* 22: 1303–1319.

American Welding Society. 1985. *Thermal spraying, practice*. Miami: Theory and Application.

Ang, A.S.M., H. Howse, S.A. Wade, and C.C. Berndt. 2016. Development of processing windows for HVOF carbide-based coatings. *Journal of Thermal Spray Technology* 25 (1–2): 28–35.

Anon. 1913. Metal plating with the air brush. *Scientific American* 1: 346–352.

Arcondéguy, A., A. Grimaud, A. Denoirjean, G. Gasnier, C. Huguet, B. Pateyron, and G. Montavon. 2007. Flame-sprayed glaze coatings: Effect of operating parameters and feedstock characteristics onto coating structures. *Journal of Thermal Spray Technology* 16 (5–6): 978–990.

Ashrafizadeh, H., A. McDonald, and P. Mertiny. 2016. Deposition of electrically conductive coatings on Castable polyurethane elastomers by the flame spraying process. *Journal of Thermal Spray Technology* 25 (3): 419–430.

Astakhov, E.A. 2008. Controlling the properties of detonation-sprayed coatings: Major aspects. *Powder Metallurgy and Metal Ceramics* 47 (1–2): 70–79.

Bandyopadhyay, R., and Per Nylén. 2003. A computational fluid dynamic analysis of gas and particle flow in flame spraying. 12 (4): 494–503.

Barisov, Yui S., E.A. Asstachov, and V.S. Klimenko. 1990. *Detonation Spraying equipment, materials and Applications*, 26–32. Essen, Germany: Thermische Spritzkonferenz.

Barthel, K., S. Rambert, and St. Siegmann. 2000. Microstructure and polarization resistance of thermally sprayed composite cathodes for solid oxide fuel cell use. *Journal of Thermal Spray Technology* 9 (3): 343–347.

Bartuli, C., T. Valente, F. Cipri, E. Bemporad, and M. Tului. 2005. Parametric study of an HVOF process for the deposition of nanostructured WC-co coatings. *Journal of Thermal Spray Technology* 14 (2): 187–195.

Belzunce, F.J., V. Higuera, S. Poveda, and A. Carriles. 2002. High temperature of HFPD thermal-sprayed MCrAlY coatings in simulated gas turbine environments. *Journal of Thermal Spray Technology* 11 (4): 461–467.

Bergant, Z., and J. Grum. 2009. Quality improvement of flame sprayed, heat treated, and Remelted NiCrBSi coatings. *Journal of Thermal Spray Technology* 18 (3): 380–391.

Berghaus, Oberste J., J.-G. Legoux, C. Moreau, R. Hui, C. Decès-Petit, W. Qu, S. Yick, Z. Wang, R. Maric, and D. Ghosh. 2008. Suspension HVOF spraying of reduced temperature solid oxide fuel cell electrolytes. *Journal of Thermal Spray Technology* 17 (5–6): 700–707.

Bhandari, S., H. Singh, H.K. Kansal, and V. Rastogi. 2012. Slurry Erosion behaviour of detonation gun spray Al_2O_3 and Al_2O_3–$13TiO_2$-coated CF8M steel under hydro accelerated conditions. *Tribology Letters* 45: 319–331.

Bobzin, K., M. Ote, T.F. Linke, and K.M. Malik. 2016. Wear and corrosion resistance of Fe-based coatings reinforced by TiC particles for application in hydraulic systems. *Journal of Thermal Spray Technology* 25 (1–2): 365–374.

Bolelli, G., and L. Lusvarghi. 2006. Heat treatment effects on the Tribological performance of HVOF sprayed co-Mo-Cr-Si coatings. *Journal of Thermal Spray Technology* 15 (4): 802–810.

Bolelli, G., V. Cannillo, L. Lusvarghi, and S. Ricco. 2006. Mechanical and tribological properties of electrolytic hard chrome and HVOF-sprayed coatings. *Surface & Coatings Technology* 200: 2995–3009.

Bolelli, G., J. Rauch, V. Cannillo, A. Killinger, L. Lusvarghi, and R. Gadow. 2009. Microstructural and Tribological Investigation of High-Velocity Suspension Flame Sprayed (HVSFS) Al_2O_3 Coatings. *Journal of Thermal Spray Technology* 18 (1): 35–49.

Brandl, W., D. Toma, J. Kruger, H.J. Grabke, and G. Matthäus. 1997. The oxidation behavior of HVOF thermal-sprayed MCrAlY coatings. *Surface and Coatings Technology* 93-95: 21–26.

Brantner, H.P., R. Pippan, and W. Prantl. 2003. Local and global fracture toughness of a flame sprayed molybdenum coating. *Journal of Thermal Spray Technology* 12 (4): 560–571.

Browning, J.A. 1983. *Highly concentrated supersonic liquefied material flame spray method and apparatus*. US patent # 4,416,421, November.

———. 1992. Hypervelocity impact fusion-a technical note. *Journal of Thermal Spray Technology* 1 (4): 289–229.

———. 1999. Viewing the future of HVOF and HVAF thermal spraying. *Journal of Thermal Spray Technology* 8 (3): 351–356.

Cannon, J.E., M. Alkam, and P.B. Butler. 2008. Efficiency of pulsed detonation thermal spraying. *Journal of Thermal Spray Technology* 17 (4): 456–464.

Cetegen, B.M., and S. Basu. 2009. Review of modeling of liquid precursor droplets and particles injected into plasmas and high-velocity oxy-fuel (HVOF) flame jets for thermal spray deposition applications. *Journal of Thermal Spray Technology* 18 (5–6): 769–793.

Chang-Jiu, Li, and Yu-Yue Wang. 2002. Effect of particle state on the adhesive strength of HVOF sprayed metallic coating. *Journal of Thermal Spray Technology* 11 (4): 523–552.

Chebbi, A., and J. Stokes. 2012. Thermal spraying of bioactive polymer coatings for orthopedic applications. *Journal of Thermal Spray Technology* 21 (3–4): 719–730.

Chen, W.R., X. Wua, B.R. Marple, D.R. Nagy, and P.C. Patnaik. 2008. TGO growth behavior in TBCs with APS and HVOF bond coats. *Surface & Coatings Technology* 202 (12): 2677–2683.

Cheng, D., Q. Xu, G. Trapaga, and E.J. Lavernia. 2001. The effect of particle size and morphology on the in-flight behavior of particles during high-velocity oxyfuel thermal spraying. *Metallurgical and Materials Transactions B* 32 (3): 525–535.

———. 2001b. A numerical study of high-velocity oxygen fuel thermal spraying process. Part I: gas phase dynamics. *Metallurgical and Materials Transactions A* 32A: 1609–1620.

Chow, R., T.A. Decker, R.V. Gansert, D. Gansert, and D. Lee. 2003. Properties of aluminium deposited by a HVOF process. *Journal of Thermal Spray Technology* 12 (2): 208–213.

Deng, C., M. Liu, C. Wu, K. Zhou, and J. Song. 2007. Impingement resistance of HVAF WC-based coatings. *Journal of Thermal Spray Technology* 16 (5–6): 604–609.

Dent, A.H., S. DePalo, and S. Sampath. 2002. Examination of the wear properties of HVOF sprayed nanostructured and conventional WC-co Cermets with different binder phase contents. *Journal of Thermal Spray Technology* 11 (4): 551–558.

Dobbins, T.A., R. Knight, and M.J. Mayo. 2003. HVOF thermal spray deposited Y_2O_3- ZrO_2 coatings for thermal barrier applications. *Journal of Thermal Spray Technology* 12 (2): 214–225.

Dobler, K., H. Kreye, and R. Schwetzke. 2000. Oxidation of stainless steel in the high velocity oxy-fuel process. *Journal of Thermal Spray Technology* 9 (3): 407–413.

Dolatabadi, A., V. Pershin, and J. Mostaghimi. 2005. New attachment for controlling gas flow in the HVOF process. *Journal of Thermal Spray Technology* 14 (1): 91–99.

Dongmo, E., R. Gadow, A. Killinger, and M. Wenzelburger. 2009a. Modeling of combustion as well as heat, mass, and momentum transfer during thermal spraying by HVOF and HVSFS. *Journal of Thermal Spray Technology* 18 (5–6): 896–908.

Dongmo, E., A. Killinger, M. Wenzelburger, and R. Gadow. 2009b. Numerical approach and optimization of the combustion and gas dynamics in High Velocity Suspension Flame Spraying (HVSFS). *Surface & Coatings Technology* 203: 2139–2145.

Du, H., W. Hua, J. Liu, J. Gong, C. Sun, and L. Wen. 2005. Influence of process variables on the qualities of detonation gun sprayed WC–Co coatings. *Materials Science and Engineering A* 408: 202–210.

Du, H., C. Sun, W.G. Hua, Y.S. Zhang, Z. Han, T.G. Wang, J. Gong, and S.W. Lee. 2006. Fabrication and evaluation of D-gun sprayed WC–Co coating with self-lubricating property. *Tribology Letters* 23 (3): 261–266.

Du, H., C. Sun, W. Hua, T. Wang, J. Gong, X. Jiang, and S. Wohn Lee. 2007. Structure, mechanical and sliding wear properties of WC-Co/MoS2-Ni coatings by detonation gun spray. *Materials Science and Engineering A* 445-446: 122–134.

Edrisy, A., T. Perry, and A.T. Alpas. 2005. Wear mechanism maps for thermal-spray steel coatings. *Metallurgical and Materials Transactions* 36A: 2737–2750.

Endo, T., R. Obayashi, T. Tajiri, K. Kimura, Y. Morohashi, T. Johzaki, K. Matsuoka, T. Hanafusa, and S. Mizunari. 2016. Thermal spray using a high-frequency pulse detonation combustor operated in the liquid-purge mode. *Journal of Thermal Spray Technology* 25 (3): 494–508.

Evdokimenko, Y.I., V.M. Kisel, V. Kh, A. Kadyrov, A. Korol, and O.I. Get'man. 2001. High-velocity flame spraying of powder Aluminium protective coatings. *Powder Metallurgy and Metal Ceramics* 40 (3–4): 121–126.

Filimonov, V. Yu, V.I. Yakovlev, M.A. Korchagin, M.V. Loginova, A.S. Semenchina, and A.V. Afanas'ev. 2008. Structure formation during gas-detonation spraying of coatings from composite powders $TiAl_3$ and Ni_3Al. *Combustion, Explosion, and Shock Waves* 44 (5): 591–596.

Fossati, A., M. Di Ferdinando, A. Lavacchi, U. Bardi, C. Giolli, and A. Scrivani. 2010. Improvement of the isothermal oxidation resistance of CoNiCrAlY coating sprayed by high velocity oxygen fuel. *Surface & Coatings Technology* 204: 3723–3728.

Furuhata, T., S. Tanno, T. Miura, Y. Ikeda, and T. Nakajima. 1997. Performance of numerical spray combustion simulation, energy convers. *Manage* 38 (10–13): 1111–1122.

Ganesh Sundara Raman, S., B. Rajasekaran, S.V. Joshi, and G. Sundararajan. 2007. Influence of substrate material on plain fatigue and fretting fatigue behavior of detonation gun sprayed cu-Ni-in coating. *Journal of Thermal Spray Technology* 16 (4): 571–579.

Gao, Y., H. Zu-kun, X. Xiaolei, and X. Gang. 2001. Formation of molybdenum boride cermet coating by the detonation spray process. *Journal of Thermal Spray Technology* 10 (3): 456–460.

Gärtner, F., T. Stoltenhoff, T. Schmidt, and H. Kreye. 2006. The Cold Spray Process and Its Potential for Industrial Applications. *Journal of Thermal Spray Technology* 15 (2): 223–232.

Gavrilenko, T.P., and Yu.A. Nikolaev. 2006. Limits of gaseous detonation spraying. *Combustion, Explosion, and Shock Waves* 42 (5): 594–597.

———. 2007. Calculation of detonation gas spraying. *Combustion, Explosion, and Shock Waves* 43 (6): 724–731.

Gawne, D.T., T. Zhang, and Y. Bao. 2001. Heating effect of flame impingement on polymer coatings, thermal spray 2001. In *New surfaces for a new millennium*, ed. C. Berndt, K. Khor, and E. Lugscheider. Materials: ASM International.

Glassman, I. 1977. *Combustion*. New York: Academic.

Gonzalez, R., H. Ashrafizadeh, A. Lopera, P. Mertiny, and A. McDonald. 2016. A review of thermal spray metallization of polymer-based structures. *Journal of Thermal Spray Technology* 25 (5): 897–919.

Gordon, S., and B.J. McBride. 1994. *Computer program for calculation of complex chemical equilibrium compositions and applications*, NASA reference publication 1311. Cleveland: Lewis Research Center.

Grewal, Harpreet Singh, Sanjeev Bhandari, and Harpreet Singh. 2012. Parametric study of slurry-Erosion of hydro-turbine steels with and without detonation gun spray coatings using Taguchi technique. *Metallurgical and Materials Transactions A* 43A: 3387–3401.

Gross, K.A., J. Tikkanen, J. Keskinen, V. Pitkänen, M. Eerola, R. Siikamaki, and M. Rajala. 1999. Liquid flame spraying for glass coloring. *Journal of Thermal Spray Technology* 8 (4): 583–589.

Gu, S., C.N. Eastwick, K.A. Simmons, and D.G. McCartney. 2001. Computational fluid dynamic modelling of gas flow characteristics in a high-velocity oxy-fuel thermal spray system. *Journal of Thermal Spray Technology* 10 (3): 461–469.

Gu, S., D.G. McCartney, C.N. Eastwick, and K. Simmons. 2004. Numerical modeling of in-flight characteristics of Inconel 625 particles during high-velocity oxy-fuel thermal spraying. *Journal of Thermal Spray Technology* 13 (2): 200–213.

Guilemany, J.M., N. Espallargas, P.H. Suegama, A.V. Benedetti, and J. Fernández. 2005. High-velocity oxyfuel Cr_3C_2-NiCr replacing hard chromium coatings. *Journal of Thermal Spray Technology* 14 (3): 335–341.

Gupta, M., N. Markocsan, X.-H. Li, and L. Östergren. 2018. Influence of bond-coat spray process on lifetime of suspension plasma-sprayed thermal barrier coatings. *Journal of Thermal Spray Technology* 27: 84–97.

Hackett, C.M., and G.S. Settles. 1995. Research on HVOF gas shrouding for coating oxidation control. In *Thermal spray: Science and technology*, ed. C.C. Berndt and S. Sampath, 21–29. Geauga County: ASM International, Materials Park.

Hall, A., D. Urrea, J. Mccloskey, D. Beatty, T. Roemer, and D. Hirschfeld. 2010. The effect of torch hardware on particle temperature and particle velocity distributions in the powder flame spray process. *Journal of Thermal Spray Technology* 19 (4): 824–827.

Han, Y., H. Chen, D. Gao, G. Yang, B. Liu, Y. Chu, J. Fan, and Y. Gao. 2017. Microstructural evolution of NiCoCrAlHfYSi and NiCoCrAlTaY coatings deposited by AC-HVAF and APS. *Journal of Thermal Spray Technology* 26: 1758–1775.

Hanshin, C., S. Lee, B. Kim, H. Jo, and C. Lee. 2005. Effect of in-flight particle oxidation on the phase evolution of HVOF NiTiZrSiSn bulk amorphous coating. *Journal of Materials Science* 40: 6121–6126.

Hanson, T.C., C.M. Hackett, and G.S. Settles. 2002. Independent control of HVOF particle velocity and temperature. *Journal of Thermal Spray Technology* 11: 75–85.

Harsha, S., D.K. Dwivedi, and A. Agarwal. 2007. Influence of CrC addition in Ni-Cr-Si-B flame sprayed coatings on microstructure, microhardness and wear behaviour. *Surface & Coatings Technology* 201: 5766–5775.

Hassan, B., A.R. Lopez, and W.L. Oberkampf. 1998. Computational analysis of a three-dimensional High-Velocity Oxygen Fuel (HVOF) thermal spray torch. *Journal of Thermal Spray Technology* 7 (1): 71–77.

He, J., M. Ice, and E. Lavernia. 2001. Particle melting behavior during high-velocity oxygen fuel thermal spraying. *Journal of Thermal Spray Technology* 10 (1): 83–93.

Henkes and H. Olivier, J (2014) Particle Acceleration in a High Enthalpy Nozzle Flow with a Modified Detonation Gun. *J Thermal Spray Technology* 23(4) 626-640

Hideki, I., U. Shibakumaran, and G. Kazumasa. 1991. *Method for thermally spraying and grazing cement martial*. Japanese patent JP3033084, 13-02-1991.

Higuera, V., F.J. Belzunce, A. Carriles, and S. Poveda. 2002. Influence of the thermal-spray procedure on the properties of a nickel-chromium coating. *Journal of Materials Science* 37: 649–654.

Horlock, A.J., Z. Sadeghian, D.G. McCartney, and P.H. Shipway. 2005. High-velocity Oxyfuel reactive spraying of mechanically alloyed Ni-Ti-C powders. *Journal of Thermal Spray Technology* 14 (1): 77–84.

Huang, J., Y. Liu, J. Yuan, and H. Li. 2014. Al/Al_2O_3 composite coating deposited by flame spraying for marine applications: Alumina skeleton enhances anti-corrosion and Wear performances. *Journal of Thermal Spray Technology* 23 (4): 676–683.

Hussary, N.A., and J.V.R. Heberlein. 2007. Effect of system parameters on metal breakup and particle formation in the wire arc spray process. *Journal of Thermal Spray Technology* 16 (1): 140–152.

Ishikawa, K., T. Suzuki, Y. Kitamura, and S. Tobe. 1999. Corrosion resistance of thermal sprayed titanium coatings in chloride solution. *Journal of Thermal Spray Technology* 8 (2): 273–278.

Ishikawa, K., T. Suzuki, S. Tobe, Y. Kitamura, and K. Ishikawa. 2001. Alloy against aqueous corrosion. *Journal of Thermal Spray Technology* 10 (3): 520–525.

Ishikawa, Y., J. Kawakita, S. Osawa, T. Itsukaichi, Y. Sakamoto, M. Takaya, and S. Kuroda. 2005. Evaluation of corrosion and Wear resistance of hard cermet coatings sprayed by using an improved HVOF process. *Journal of Thermal Spray Technology* 14 (3): 384–390.

Ishikawa, Y., S. Kuroda, J. Kawakita, Y. Sakamoto, and T. Matsufumi. 2007. Sliding wear properties of HVOF sprayed $WC–20\%Cr_3C_2–7\%Ni$ cermet coatings. *Surface & Coatings Technology* 201: 4718–4727.

Ivosevic, M., R. Knight, S.R. Kalidindi, G.R. Palmese, and J.K. Sutter. 2005. Adhesive/cohesive properties of thermally sprayed functionally graded coatings for polymer matrix composites. *Journal of Thermal Spray Technology* 14 (1): 45–51.

Ivosevic, M., R.A. Cairncross, and R. Knight. 2007. Melting and degradation of Nylon-11 particles during HVOF combustion spraying. *Journal of App. Pol. Science* 105 (2): 827–837.

Ivosevic, M., S.L. Coguill, and S.L. Galbraith. 2009. Polymer thermal spraying: A novel coating process. In *Thermal spray 2009: Proceedings of the international thermal spray conference*, ed. B.R. Marple, M.M. Hyland, Y.-C. Lau, C.-J. Li, R.S. Lima, and G. Montavon, 1078–1083. Materials Park, OH, USA: ASM International.

Jackson, L., M. Ivosevic, R. Knight, and R.A. Cairncross. 2007. Sliding Wear properties of HVOF thermally sprayed Nylon-11 and Nylon-11/ceramic composites on steel. *Journal of Thermal Spray Technology* 16 (5–6): 927–932.

Jacobs, L., M.M. Hyland, and M. De Bonte. 1998. Comparative study of WC-cermet coatings sprayed via the HVOF and the HVAF process. *Journal of Thermal Spray Technology* 7 (2): 213–218.

———. 1999. Study of the influence of microstructural properties on the sliding-Wear behavior of HVOF and HVAF sprayed WC-cermet coatings. *Journal of Thermal Spray Technology* 8 (1): 125–132.

Jang, H.-J., D.-H. Park, Y.-G. Junga, J.-C. Jang, S.-C. Choi, and U. Paik. 2006. Mechanical characterization and thermal behavior of HVOF-sprayed bond coat in thermal barrier coatings (TBCs). *Surface & Coatings Technology* 200: 4355–4362.

Jayaganthan, R., S. Prakash, and Sanjay Kumar. 2008. Hot corrosion behavior of detonation gun sprayed Cr_3C_2–NiCr coatings on Ni and Fe-based superalloys in Na_2SO_4–60% V_2O_5 environment at 900 °C. *Journal of Alloys and Compounds* 463: 358–372.

Ji, G.-C., C.-J. Li, Y.-Y. Wang, and W.-Y. Li. 2007. Erosion performance of HVOF-sprayed Cr_3C_2-NiCr coatings. *Journal of Thermal Spray Technology* 16 (4): 557–565.

Jin, Kawakita, Seiji Kuroda, Takeshi Fukushima, and Toshiaki Kodama. 2005. Improvement of corrosion resistance of high-velocity oxyfuel-sprayed stainless-steel coatings by addition of molybdenum. *Journal of Thermal Spray Technology* 14 (2): 224–230.

Jorge, Lino F., Teresa P. Duarte, and Ricardo Maia. 2003. Development of coated ceramic components for the aluminum industry. *Journal of Thermal Spray Technology* 12 (2): 250–257.

Kadyrov, E. 1996. Gas-particle interaction in detonation spraying systems. *Journal of Thermal Spray Technology* 5 (2): 185–195.

Kadyrov, E., and V. Kadyrov. 1995. Gas dynamical parameters of detonation powder spraying. *Journal of Thermal Spray Technology* 4 (3): 280–286.

Kamal, S., R. Jayaganthan, S. Prakash, and Sanjay Kumar. 2008. Hot corrosion behavior of detonation gun sprayed Cr_3C_2–NiCr coatings on Ni and Fe-based superalloys in Na_2SO_4–60% V_2O_5 environment at 900 °C. *Journal of Alloys and Compounds* 463: 358–372.

Kamal, S., R. Jayaganthan, and S. Prakash. 2009. High temperature oxidation studies of detonation-gun-sprayed Cr_3C_2–NiCr coating on Fe- and Ni-based superalloys in air under cyclic condition at 900 °C. *Journal of Alloys and Compounds* 472: 378–389.

———. 2011. Hot corrosion studies of detonation-gun-sprayed NiCrAlY + 0.4 wt.% CeO_2 coated Superalloys in molten salt environment, Journal of. *Materials Engineering and Performance* 20 (6): 1068–1077.

Kamnis, S., and S. Gu. 2006. Numerical modelling of propane combustion in a high velocity oxygen–fuel thermal spray gun. *Chemical Engineering and Processing* 45: 246–253.

Katanoda, H., H. Yamamoto, and K. Matsuo. 2006. Numerical simulation on supersonic flow in high velocity oxy fuel thermal spray gun. *Journal of Thermal Science* 15 (1): 65–70.

Katanoda, H., T. Matsuoka, S. Kuroda, J. Kawakita, H. Fukanuma, and K. Matsuo. 2005. Aerodynamic study on supersonic flows in high-velocity oxy-fuel thermal spray process. *Journal of Thermal Science* 14: 126–129.

Katanoda, H., H. Yamamoto, and K. Matsuo. 2006. Numerical simulation on supersonic flow in high velocity oxy fuel thermal spray gun. *Journal of Thermal Science* 15 (1): 65–70.

Kaushal, G., H. Singh, and S. Prakash. 2011. Comparative high temperature analysis of HVOF- sprayed and detonation gun sprayed Ni–20Cr coating in laboratory and actual boiler environments. *Oxidation of Metals* 76: 169–191.

Kaushal, G., N. Bala, H. Singh, N. Kaur, and S. Prakash. 2014. Comparative high-temperature corrosion behavior of Ni-20Cr coatings on T22 boiler steel produced by HVOF, D-Gun, and Cold spraying. *Metallurgical and Materials Transactions A* 45A: 395–410.

Kawahara, Y. 2007. Application of high temperature corrosion-resistant materials and coatings under severe corrosive environment in waste-to-energy boilers. *Journal of Thermal Spray Technology* 16 (2): 202–213.

Jin, Kawakita, Seiji Kuroda, Takeshi Fukushima, and Toshiaki Kodama. 2005. Improvement of corrosion resistance of high-velocity oxyfuel-sprayed stainless steel coatings by addition of molybdenum. *Journal of Thermal Spray Technology* 14 (2): 224–230.

Kawakita, J., S. Kuroda, T. Fukushima, H. Katanoda, K. Matsuo, and H. Fukanuma. 2006. Dense titanium coatings by modified HVOF spraying. *Surface & Coatings Technology* 201: 1250–1255.

Ke, P.L., Y.N. Wu, Q.M. Wang, J. Gong, C. Sun, and L.S. Wen. 2005. Study on thermal barrier coatings deposited by detonation gun. *Surface & Coatings Technology* 200: 2271–2276.

Kharlamov, Y.A. 2004. Gaseous pulse detonation spraying: Current status, challenges, and future perspective. In *ITSC-2004: Thermal spray crossing borders*, ed. E. Lugsheider. Düsseldorf, Germany: DVS. E-proceedings.

Killinger, A., M. Kuhn, and R. Gadow. 2006. High-velocity suspension flame spraying (HVSFS), a new approach for spraying nanoparticles with hypersonic speed. *Surface & Coatings Technology* 201 (2006): 1922–1929.

Killinger, A., P. Müller, and R. Gadow. 2015. What do we know, what are the current limitations of suspension HVOF spraying? *Journal of Thermal Spray Technology* 24 (7): 1130–1142.

Kim, J.H., K.M. Lim, B.G. Seong, and C.G. Park. 2001. Amorphous phase formation of Zr-based alloy coating by HVOF spraying process. *Journal of Materials Science* 36: 49–54.

Kim, J.H., M.C. Kim, and C.G. Park. 2003. Evaluation of functionally graded thermal barrier coatings fabricated by detonation gun spray technique. *Surface and Coatings Technology* 168: 275–280.

Kim, K.H., S. Kuroda, and M. Watanabe. 2010. Microstructural development and deposition behavior of titanium powder particles in warm spraying process: From single splat to coating. *Journal of Thermal Spray Technology* 19 (6): 1244–1254.

Kleinstein, G. 1964. Mixing in turbulent axially symmetric free jets. *Journal of Spacecraft* 1 (4): 403–408.

Koji, N., H. Kunio, and Y. Eiichi. 1990. *Glass-coated metallic work piece*. Japanese patent JP2011749, 16-01-1990.

Korpiola, K., J.P. Hirvonen, L. Laas, and F. Rossi. 1997. The influence of the nozzle design on HVOF exit gas velocity and coating microstructure. *Journal of Thermal Spray Technology* 6 (4): 469–474.

Kumar, Ashok, J. Boy, Ray Zatorski, and L.D. Stephenson. 2005. Thermal spray and weld repair alloys for the repair of cavitation damage in turbines and pumps: A technical note. *Journal of Thermal Spray Technology* 14 (2): 177–182.

Kuroda, S., Y. Tashiro, H. Yumoto, S. Taira, H. Fukanuma, and S. Tobe. 2001. Peening action and residual stresses in high-velocity oxygen fuel thermal spraying of 316L stainless steel. *Journal of Thermal Spray Technology* 10 (2): 367–374.

Kuroda, S., M. Watanabe, K.H. Kim, and H. Katanoda. 2011. Current status and future prospects of warm spray technology. *Journal of Thermal Spray Technology* 20 (4): 653–676.

Kwon, J.-Y., J.-H. Lee, Y.-G. Jung, and U. Paik. 2006. Effect of bond coat nature and thickness on mechanical characteristic and contact damage of zirconia-based thermal barrier coatings. *Surface & Coatings Technology* 201: 3483–3490.

Laribi, M., A.B. Vannes, and D. Treheux. 2006. On a determination of wear resistance and adhesion of molybdenum, Cr–Ni and Cr–Mn steel coatings thermally sprayed on a 35CrMo4 steel. *Surface & Coatings Technology* 200: 2704–2710.

Li, M., and P.D. Christofides. 2005. Multi-scale modelling and analysis of an industrial HVOF thermal spray process. *Chemical Engineering Science* 60: 3649–3669.

Li, S., and Panagiotis D. Christofides. 2006. Computational study of particle in-flight behavior in the HVOF thermal spray process. *Chemical Engineering Science* 61: 6540–6552.

Li, C.-J., and A. Ohmori. 1996. The lamellar structure of a detonation gun sprayed Al_2O_3 coating. *Surface and Coatings Technology* 82: 254–258.

Chang-Jiu, Li, and Yu-Yue Wang. 2002. Effect of particle state on the adhesive strength of HVOF sprayed metallic coating. *Journal of Thermal Spray Technology* 11 (4): 523–552.

Li, M., P. Shi, and P.D. Christofides. 2004a. Diamond jet hybrid HVOF thermal spray: Gas-phase and particle behavior modelling and feedback control design. *Industrial and Engineering Chemistry Research* 43: 3632–3652.

Li, J.F., L. Li, and F.H. Stott. 2004b. Multi-layered surface coatings of refractory ceramics prepared by combined laser and flame spraying. *Surface and Coatings Technology* 180–181: 500–505.

Lima, C.R.C., and J.M. Guilemany. 1997. The oxidation behavior of HVOF thermal-sprayed MCrAlY coatings. *Surface and Coatings Technology* 93-95: 21–26.

Lima, C.R.C., and J.M. Guilemany. 2007. Adhesion improvements of thermal barrier coatings with HVOF thermally sprayed bond coats. *Surface & Coatings Technology* 201: 4694–4701.

Lima, R.S., and B.R. Marple. 2003a. High Weibull modulus HVOF titania coatings. *Journal of Thermal Spray Technology* 12 (2): 240–249.

———. 2003b. Optimized HVOF titania coatings. *Journal of Thermal Spray Technology* 12 (3): 360–369.

Lima, C.R.C., J. Nin, and J.M. Guilemany. 2006. Evaluation of residual stresses of thermal barrier coatings with HVOF thermally sprayed bond coats using the modified layer removal method (MLRM). *Surface & Coatings Technology* 200: 5963–5972.

Lin, L., and K. Han. 1998. Optimization of surface properties by flame spray coating and boriding. *Surface and Coatings Technology* 106: 100–105.

Lin, Q.S., K.S. Zhou, C.M. Deng, M. Liu, L.P. Xu, and C.G. Deng. 2014. Deposition mechanisms and oxidation behaviors of Ti-Ni coatings deposited in low-temperature HVOF spraying process. *Journal of Thermal Spray Technology* 23 (6): 892–902.

Liu, Hui, and Jihua Huang. 2005. Reactive thermal spraying of TiC-Fe composite coating by using asphalt as carbonaceous precursor. *Journal of Materials Science* 40: 4149–4151.

Liu, Hui Yuan, and Ji Hua Huang. 2006. Reactive flame spraying of TiC-Fe cermet coating using asphalt as a carbonaceous precursor. *Surface & Coatings Technology* 200: 5328–5333.

Liu, C.S., J.H. Huang, and S. Yin. 2002. The influence of composition and process parameters on the microstructure of TiC-Fe coatings obtained by reactive flame spray process. *Journal of Materials Science* 37 (2002): 5241–5245.

Liu, M., K. Yang, C.-M. Deng, C.-G. Deng, and K.-S. Zhou. 2016. Microstructure and properties of Cu coating fabricated onto diamond-Cu substrate by low-temperature HVOF process. *Journal of Thermal Spray Technology* 25 (8): 1516–1525.

Lopez, A.R., B. Hassan, W.L. Oberkampf, R.A. Neiser, and T.J. Roemer. 1998. Computational fluid dynamics analysis of a wire-feed, high-velocity oxygen fuel (HVOF) thermal spray torch. *Journal of Thermal Spray Technology* 7 (3): 374–382.

Lugsheider, E., P. Remer, A. Nyland, and R. Siking. 1995. Thermal spraying of Bio-active glass ceramics. In *Thermal Spray: Science and Technology*, ed. C.C. Berndt and S. Sampath, 583–587. Materials Park, OH, USA: ASM International.

Lyphout, C., and S. Björklund. 2015. Internal diameter HVAF spraying for Wear and corrosion applications. *Journal of Thermal Spray Technology* 24 (1–2): 235–243.

Lyphout, C., K. Sato, S. Houdkova, E. Smazalova, L. Lusvarghi, G. Bolelli, and P. Sassatelli. 2016. Tribological properties of hard metal coatings sprayed by high-velocity air fuel process. *Journal of Thermal Spray Technology* 25 (1–2): 331–345.

Ma, J., X. Liu, W.O. Qu, and C. Zhou. 2015. Corrosion behavior of detonation gun sprayed Al coating on sintered NFeB. *Journal of Thermal Spray Technology* 24 (3): 394–400.

Maiti, A.K., N. Mukhopadhyay, and R. Raman. 2007. Effect of adding WC powder to the feedstock of WC-Co-Cr based HVOF coating and its impact on erosion and abrasion resistance. *Surface & Coatings Technology* 201: 7781–7788.

Manish, R. 2002. Dynamic hardness detonation sprayed WC-co coatings. *Journal of Thermal Spray Technology* 11 (3): 393–399.

Marple, B.R., and R.S. Lima. 2005. Process temperature/velocity-hardness-Wear relationships for high-velocity Oxyfuel sprayed nanostructured and conventional cermet coatings. *Journal of Thermal Spray Technology* 14 (1): 67–76.

Marple, B.R., and J. Voyer. 2001. Improved Wear performance by the incorporation of solid lubricants during thermal spraying. *Journal of Thermal Spray Technology* 10 (4): 626–636.

Masataka, M., and K. Kazumi. 1989. *Substrate for flame-sprayed tile.* Japanese patent JP1192777, 02-08-1989.

Matikainen, V., H. Koivuluoto, P. Vuoristo, J. Schubert, and S. Houdkova. 2018. Effect of nozzle geometry on the microstructure and properties of HVAF-sprayed WC-10Co4Cr and Cr3C2-25NiCr Coatings. *Journal of Thermal Spray Technology* 27: 680–694.

Matthews, S., M. Hyland, and B. James. 2004. Long-term carbide development in high-velocity oxygen fuel/high-velocity air fuel Cr_3C_2-NiCr coatings heat treated at 900 °C. *Journal of Thermal Spray Technology* 13 (4): 526–536.

Mesrati, N., H. Ajhrourh, Du Nguyen, and D. Treheux. 2000. Thermal spraying and adhesion of oxides onto graphite. *Journal of Thermal Spray Technology* 9 (1): 95–99.

Milanti, A., H. Koivuluoto, and P. Vuoristo. 2015. Influence of the spray gun type on microstructure and properties of HVAF sprayed Fe-based corrosion resistant coatings. *Journal of Thermal Spray Technology* 24 (7): 1312–1322.

Min'kov, D.V., V.Yu. Lakunin, S.T. Kartashov, O.M. Bashkirov, A.S. Ivanov, A.V. Kasatkin, and M.D. Min'kov. 2008. Restoration of drying cylinders on spinning machines by gas detonation spraying. *Fibre Chemistry* 40 (6): 545–547.

Mingheng, Li, and Panagiotis D. Christofides. 2006. Computational study of particle in-flight behavior in the HVOF thermal spray process. *Chemical Engineering Science* 61: 6540–6552.

Mizuno, H., and J. Kitamura. 2007. MoB/CoCr cermet coatings by HVOF spraying against Erosion by molten Al-Zn alloy. *Journal of Thermal Spray Technology* 16 (3): 404–413.

Modi, S.C., and E. Calla. 2001. A study of high-velocity combustion wire molybdenum coatings. *Journal of Thermal Spray Technology* 10 (3): 480–486.

Moskowitz, L., and K. Trelewicz. 1997. HVOF coatings for heavy-Wear, high-impact applications. *Journal of Thermal Spray Technology* 6 (3): 294–299.

Mudgal, D., S. Singh, and S. Prakash. 2015. Evaluation of ceria-added Cr_3C_2-25(NiCr) coating on three Superalloys under simulated incinerator environment. *Journal of Thermal Spray Technology* 24 (3): 496–514.

Müller, P., A. Killinger, and R. Gadow. 2012. Comparison between high-velocity suspension flame spraying and suspension plasma spraying of alumina. *Journal of Thermal Spray Technology* 21 (6): 1120–1127.

Murthy, J.K.N., and B. Venkataraman. 2006. Abrasive wear behaviour of WC–CoCr and Cr_3C_2–20(NiCr) deposited by HVOF and detonation spray processes. *Surface & Coatings Technology* 200: 2642–2652.

Murthy, J.K.N., D.S. Rao, and B. Venkataraman. 2001. Effect of grinding on the erosion behaviour of a WC–Co–Cr coating deposited by HVOF and detonation gun spray processes. *Wear* 249: 592–600.

Murthy, J.K.N., S. Bysakh, K. Gopinath, and B. Venkataraman. 2007. Microstructure dependent erosion in Cr_3C_2–20(NiCr) coating deposited by a detonation gun. *Surface & Coatings Technology* 202: 1–12.

Navas, C., R. Colaço, J. de Damborenea, and R. Vilar. 2006. Abrasive wear behaviour of laser clad and flame sprayed-melted NiCrBSi coatings. *Surface & Coatings Technology* 200: 6854–6862.

Navidpour, A.H., M. Salehi, H.R. Salimijazi Y. Kalantari, and M. Azarpour Siahkali. 2017. Photocatalytic activity of flame-sprayed coating of zinc ferrite powder. *Journal of Thermal Spray Technology* 26: 2030–2039.

Neiser, R.A., J.E. Brockmann, T.J. O'Hern, R.C. Dyhkuizen, M.F. Smith, T.J. Roemer, and R.E. Teets. 1995. Wire melting and droplet atomization in a HVOF jet. In *Thermal Spray Science and Technology*, ed. C.C. Berndt and S. Sampath, 99–104. Materials Park OH, USA: ASM International.

Neiser, R.A., M.F. Smith, and R.C. Dykhuizen. 1998. Oxidation in wire HVOF-sprayed steel. *Journal of Thermal Spray Technology* 7 (4): 537–545.

Nevgod, V.A., V.H. Kadyrov, and A. Khairutdinov. 1987a. *Gas detonation apparatus.* U.S. Patent 4 669 658.

———. 1987b. *Gas detonation apparatus*. U.S. Patent 4 687 135.

Ni, L.Y., C. Liu, H. Huang, and C.G. Zhou. 2011. Thermal Cycling behavior of thermal barrier coatings with HVOF NiCrAlY bond coat. *Journal of Thermal Spray Technology* 20 (5): 1133–1138.

Niemi, K., P. Vuoristo, and T. Mantyla. 1994. Properties of alumina-based coatings deposited by plasma spray and detonation gun spray processes. *Journal of Thermal Spray Technology* 3 (2): 199–203.

Nikolaev, Y.A., A.A. Vasil'ev, and B.Yu. Ul'yanitskii. 2003. Gas detonation and its application in engineering and technologies (review). *Combustion, Explosion, and Shock Waves* 39 (4): 382–410.

Nylén, P., and R. Bandyopadhyay. 2000. A computational fluid dynamic analysis of gas and particle flow in flame spraying. In *Thermal Spray: Surface Engineering via Applied Research*, ed. C.C. Berndt, 237–244. Materials Park, Ohio, USA: ASM International®.

Oberkampf, W.L., and M. Talpallikar. 1994. Analysis of a high velocity oxygen-fuel (HVOF). In *Proceedings of 7th. National Thermal Spray Conf., Boston, MA*, 381–392.

———. 1996a. Analysis of a high velocity oxygen-fuel (HVOF) thermal spray torch part1: Numerical simulation. *Journal of Thermal Spray Technology* 5 (1): 53–61.

———. 1996b. Analysis of a high velocity oxygen-fuel (HVOF) thermal spray torch part 2: Computational results. *Journal of Thermal Spray Technology* 5 (1): 62–68.

Oliker, V.E., E.F. Grechishkin, V.V. Polotai, M.G. Loskutov, and I.I. Timofeeva. 2004. Influence of the structure and properties of WC-Co alloy powders on the structure and wear resistance of detonation coatings. *Powder Metallurgy and Metal Ceramics* 43 (5–6): 258–264.

Oliker, V.E., V.L. Sirovatka, I.I. Timofeeva, E.F. Grechishkin, and T. Ya Gridasova. 2005. Effects of properties of titanium aluminide powders and detonation spraying conditions on phase and structure formation in coatings. *Powder Metallurgy and Metal Ceramics* 44 (9–10): 472–480.

Oliker, V.E., V.L. Sirovatka, I.I. Timofeeva, T. Ya Gridasova, and Ye F. Hrechyshkin. 2006. Formation of detonation coatings based on titanium aluminide alloys and aluminium titanate ceramic sprayed from mechanically alloyed powders Ti—Al. *Surface & Coatings Technology* 200: 3573–3581.

Oliker, V.E., M.Yu. Barabash, E.F. Grechishkin, I.I. Timofeeva, and T. Ya Gridasova. 2007a. High-temperature air oxidation based on eutectic (β-NiAl + γ- and NiAl) Intermetallide. *Powder Metallurgy and Metal Ceramics* 46 (3–4): 175–181.

Oliker, V.E., A.A. Pritulyak, V.L. Syrovatka, E.F. Grechishkin, and T. Ya Gridasova. 2007b. Formation and high-temperature oxidation of thermal-barrier coatings with Ti–Al–Cr binding layer. *Powder Metallurgy and Metal Ceramics* 46 (9–10): 483–491.

Otsubo, F., H. Era, T. Uchida, and K. Kishitake. 2000. Properties of Cr_3C_2-NiCr cermet coating sprayed by high power plasma and high velocity oxy-fuel processes. *Journal of Thermal Spray Technology* 9 (4): 499–504.

Ozdemir, I., I. Hamanaka, Y. Tsunekawa, and M. Okumiya. 2005. In-process exothermic reaction in high-velocity oxyfuel and plasma spraying with SiO_2/Ni/Al-Si-mg composite powder. *Journal of Thermal Spray Technology* 14 (3): 321–329.

Panossian, Z., L. Mariaca, M. Morcillo, S. Flores, J. Rocha, J.J. Pen, F. Herrera, F. Corvo, M. Sanchez, O.T. Rincon, G. Pridybailo, and J. Simancas. 2005. Steel cathodic protection afforded by zinc, aluminium and zinc/aluminium alloy coatings in the atmosphere. *Surface & Coatings Technology* 190: 244–248.

Pant, Bharat K., Vivek Arya, and B.S. Mann. 2007. Development of low-oxide MCrAlY coatings for gas turbine applications. *Journal of Thermal Spray Technology* 16 (2): 275–280.

Parco, M., L. Zhao, J. Zwick, K. Bobzin, and E. Lugscheider. 2006. Investigation of HVOF spraying on magnesium alloys. *Surface & Coatings Technology* 201: 3269–3274.

Paredes, R.S.C., S.C. Amico, and A.S.C.M. d'Oliveira. 2006. The effect of roughness and pre-heating of the substrate on the morphology of aluminium coatings deposited by thermal spraying. *Surface & Coatings Technology* 200: 3049–3055.

Park, S.Y., M.C. Kim, and C.G. Park. 2007. Mechanical properties and microstructure evolution of the nano WC–Co coatings fabricated by detonation gun spraying with post heat treatment. *Materials Science and Engineering A* 449–451: 894–897.

Perry, J.M., A. Neville, and T. Hodgkiess. 2002. A comparison of the corrosion behaviour of WC-Co-Cr and WC-Co HVOF thermally sprayed coatings by in situ atomic force microscopy (AFM). *Journal of Thermal Spray Technology* 11 (4): 536–541.

Petrovicova, E., R. Knight, L.S. Schadler, and T.E. Twardowski. 2000. Nylon 11/Silica nanocomposite coatings applied by the HVOF process. II. mechanical and barrier properties. *Journal of Applied Polymer Science* 78: 2272–2289.

Planche, M.P., H. Liao, B. Normand, and C. Coddet. 2005. Relationships between NiCrBSi particle characteristics and corresponding coating properties using different thermal spraying processes. *Surface & Coatings Technology* 200: 2465–2473.

Podchernyaeva, I.A., V.V. Shchepetov, A.D. Panasyuk, V.Yu. Gromenko, D.V. Yurechko, and V.P. Katashinskii. 2003. Refractory and ceramic materials structure and properties of wear-resistant detonation coatings based on titanium carbonitride. *Powder Metallurgy and Metal Ceramics* 42 (9–10): 497–502.

Pogrebnyak, A.D., Yu.N. Tyurin, Yu.F. Ivanov, A.P. Kobzev, O.P. Kul'ment'eva, and M.I. Il'yashenko. 2000. Preparation and investigation of the structure and properties of Al_2O_3 plasma-detonation coatings. *Technical Physics Letters* 26 (11): 960–963.

Poirier, T., A. Vardelle, M.F. Elchinger, M. Vardelle, A. Grimaud, and H. Vesteghem. 2003. Deposition of nanoparticle suspensions by aerosol flame spraying: Model of the spray and impact processes. *Journal of Thermal Spray Technology* 12 (3): 393–402.

Poorman, R.M., H.B. Sargent, and R. Lamprey. 1955. *Method and apparatus utilizing detonation waves for spraying and other purposes'*. US patent 2 714 563.

Rajasekaran, B.S., S. Ganesh Sundara Raman, V. Joshi, and G. Sundararajan. 2006. Effect of detonation gun sprayed Cu–Ni–In coating on plain fatigue and fretting fatigue behavior of Al–Mg–Si alloy. *Surface & Coatings Technology* 201: 1548–1558.

Rajasekaran, B.S., S. Ganesh Sundara Ramana, V. Joshi, and G. Sundararajan. 2008. Performance of plasma sprayed and detonation gun sprayed Cu–Ni–In coatings on Ti–6Al–4V under plain fatigue and fretting fatigue loading. *Materials Science and Engineering A* 479: 83–92.

Rajasekaran, B., G. Mauer, and R. Vaßen. 2011. Enhanced characteristics of HVOF-sprayed MCrAlY bond coats for TBC applications. *Journal of Thermal Spray Technology* 20 (6): 1209–1216.

Ramadan, K., and P. Barry Butler. 2004. Analysis of particle dynamics and heat transfer in detonation thermal spraying systems. *Journal of Thermal Spray Technology* 13 (2): 248–264.

Rani, A., N. Bala, and C.M. Gupta. 2017. Accelerated hot corrosion studies of D-gun-sprayed Cr_2O_3–50% Al_2O_3 coating on boiler steel and Fe- based Superalloy. *Oxidation of Metals* 88: 621–648.

Richer, P., M. Yandouzi, L. Beauvais, and B. Jodoin. 2010. Oxidation behaviour of CoNiCrAlY bond coats produced by plasma. *HVOF and cold gas dynamic spraying, Surface & Coatings Technology* 204: 3962–3974.

Roberson, J.A., and C.T. Crowe. 1997. *Engineering fluid dynamics*. 6th ed. New York: Wiley.

Rodriguez, R.M.H.P., R.S.C. Paredes, S.H. Wido, and A. Calixto. 2007. Comparison of aluminium coatings deposited by flame spray and by electric arc spray. *Surface & Coatings Technology* 202: 172–179.

Roy, G.D., S.M. Frolov, A.A. Borisov, and D.W. Netzer. 2004. Pulse detonation propulsion: Challenges, current status, and future perspective. *Progress in Energy and Combustion Science* 30: 545–672.

Sadeghimeresht, E., N. Markocsan, and P. Nylen. 2016. A comparative study on Ni-based coatings prepared by HVAF, HVOF, and APS methods for corrosion protection applications. *Journal of Thermal Spray Technology* 25 (8): 1604–1616.

Saeidi, S., K.T. Voisey, and D.G. McCartney. 2011. Mechanical properties and microstructure of VPS and HVOF CoNiCrAlY coatings. *Journal of Thermal Spray Technology* 20 (6): 1231–1243.

Sainz, M.A., M.I. Osendi, and P. Miranzo. 2008. Protective Si-Al-O-Y glass coatings on stainless steel in situ prepared by combustion flame spraying. *Surface & Coatings Technology* 202: 1712–1717.

Sakaki, K., and Y. Shimizu. 2001. Effect of the increase in the entrance convergent section length of the gun nozzle on the high-velocity oxygen fuel and cold spray process. *Journal of Thermal Spray Technology* 10 (3): 487.

Sakata, K., K. Nakano, H. Miyahara, Y. Matsubara, and K. Ogi. 2007. Microstructure control of thermally sprayed co-based self-fluxing alloy coatings by diffusion treatment. *Journal of Thermal Spray Technology* 16 (5–6): 991–997.

Saladi, S., J.V. Menghani, and S. Prakash. 2015. Characterization and evaluation of cyclic hot corrosion resistance of detonation-gun sprayed Ni-5Al coatings on Inconel-718. *Journal of Thermal Spray Technology* 24 (5): 778–788.

Saravanan, P., V. Selvarajan, D.S. Rao, S.V. Joshi, and G. Sundararajan. 2000. Influence of process variables on the quality of detonation gun sprayed alumina coatings. *Surface and Coatings Technology* 123: 44–54.

Schwetzke, R., and H. Kreye. 1999. Microstructure and properties of tungsten carbide coatings sprayed with various high-velocity oxygen fuel spray systems. *Journal of Thermal Spray Technology* 8 (3): 433–439.

Scrivani, A., U. Bardi, L. Carrafiello, A. Lavacchi, F. Niccolai, et al. 2003. A comparative study of high velocity oxygen fuel, vacuum plasma spray, and axial plasma spray for the deposition of CoNiCrAlY bond coat alloy. *Journal of Thermal Spray Technology* 12 (4): 504–507.

Semenov, S.Y., and B.M. Cetegen. 2002. Experiments and modeling of the deposition of nano-structured alumina–titania coatings by detonation waves. *Materials Science and Engineering* A335: 67–81.

Senderowski, C., and Z. Bojar. 2008. Gas detonation spray forming of Fe–Al coatings in the presence of interlayer. *Surface & Coatings Technology* 202: 3538–3548.

———. 2009. Influence of detonation gun spraying conditions on the quality of Fe-Al intermetallic protective coatings in the presence of NiAl and NiCr interlayers. *Journal of Thermal Spray Technology* 18 (3): 435–447.

Sidhu, T.S., S. Prakash, and R.D. Agrawal. 2005. Studies on the properties of high-velocity oxy-fuel thermal spray coatings for higher temperature applications. *Materials Science* 41 (6): 805–823.

Sidhu, H.S., B.S. Sidhu, and S. Prakash. 2006. Comparative characteristic and erosion behavior of NiCr coatings deposited by various high-velocity Oxyfuel spray processes Journal of. *Materials Engineering and Performance* 15 (6): 699–704.

Sidhu, T.S., S. Prakash, and R.D. Agrawal. 2006a. Characterisation of NiCr wire coatings on Ni- and Fe-based superalloys by the HVOF process. *Surface & Coatings Technology* 200: 5542–5549.

———. 2006b. Characterizations and hot corrosion resistance of Cr_3C_2-NiCr coating on Ni-base superalloys in an aggressive environment. *Journal of Thermal Spray Technology* 15 (4): 811–816.

———. 2006c. Hot corrosion resistance of high-velocity oxyfuel sprayed coatings on a nickel-base superalloy in molten salt environment. *Journal of Thermal Spray Technology* 15 (3): 387–399.

Sidhu, H.S., B.S. Sidhu, and S. Prakash. 2007a. Hot corrosion behavior of HVOF sprayed coatings on ASTM SA213-T11 steel. *Journal of Thermal Spray Technology* 16 (3): 349–354.

Sidhu, T.S., A. Malik, S. Prakash, and R.D. Agrawal. 2007b. Oxidation and hot corrosion resistance of HVOF WC-NiCrFeSiB coating on Ni- and Fe-based superalloys. *Journal of Thermal Spray Technology* 16 (5-6): 844–849.

Smith, R.W. 1991. Plasma spray processing... The state of the art and future. From a surface to a materials processing technology. In *Proceedings of the 2nd Plasma Technik symposium*, vol. 1, 13–38. Wohlen, Switzerland: Plasma Technik.

Sobiecki, J.R., J. Ewertowski, T. Babul, and T. Wierzchon. 2004. Properties of alumina coatings produced by gas-detonation method. *Surface and Coatings Technology* 180–181: 556–555.

Srivatsan, V.R., and A. Dolatabadi. 2006. Simulation of particle-shock interaction in a high velocity oxygen fuel process. *Journal of Thermal Spray Technology* 15 (4): 481–487.

Stahr, C.C., S. Saaro, L.-M. Berger, J. Dubsky, K. Neufuss, and M. Hermann. 2006. Dependence of the stabilization of α-alumina on the spray process. *Journal of Thermal Spray Technology* 16 (5–6): 822–830.

Stanisic, J., D. Kosikowski, and P.S. Mohanty. 2006. High-speed visualization and plume characterization of the hybrid spray process. *Journal of Thermal Spray Technology* 15 (4): 750–758.

Stiegler, N., D. Bellucci, G. Bolelli, V. Cannillo, R. Gadow, A. Killinger, L. Lusvarghi, and A. Sola. 2012. High-velocity suspension flame sprayed (HVSFS) hydroxyapatite coatings for biomedical applications. *Journal of Thermal Spray Technology* 21 (2): 275–287.

Sundararajan, T., S. Kuroda, and F. Abe. 2004a. Steam oxidation of 80Ni-20Cr high-velocity oxyfuel coatings on 9Cr-1Mo steel: Diffusion-induced phase transformations in the substrate adjacent to the coating. *Metallurgical and Materials Transactions A* 36A: 2165–2174.

———. 2004b. Effect of thermal spray on the microstructure and adhesive strength of high-velocity oxy-fuel–sprayed Ni-Cr coatings on 9Cr-1Mo steel. *Metallurgical and Materials Transactions A* 35A: 3187–3199.

Sundararajan, G., D. Sen, and G. Sivakumar. 2005. The tribological behavior of detonation sprayed coatings: The importance of coating process parameters. *Wear* 258: 377–391.

Sundararajan, G., G. Sivakumar, D. Sen, D. Srinivasa Rao, and G. Ravichandra. 2010. The tribological behaviour of detonation sprayed TiMo(CN) based cermet coatings. *International Journal of Refractory Metals & Hard Materials* 28: 71–81.

Suresh, Babu P., D.S. Rao, G.V.N. Rao, and G. Sundararajan. 2007. Effect of feedstock size and its distribution on the properties of detonation sprayed coatings. *Journal of Thermal Spray Technology* 16 (2): 281–290.

Suresh, Babu P., B. Basu, and G. Sundararajan. 2008. Processing–structure–property correlation and decarburization phenomenon in detonation sprayed WC–12Co coatings. *Acta Materialia* 56: 5012–5026.

Tatsuya, N., K. Izozou, M. Haruyuki, and S. Yasuhi. 1988. *Improved glaze application for coating and flame spraying and glaze*. Japanese patent JP63277583, 15-11-1988.

Tawfik, H.H., and F. Zimmerman. 1997. Mathematical modelling of the gas and powder flow in HVOF systems. *Journal of Thermal Spray Technology* 6 (3): 345–352.

Taylor, T.A., and J.K. Knapp. 1995. Dispersion-strengthened modified MCrAlY coatings produced by reactive deposition. *Surface and Coatings Technology* 76-77: 34–40.

Thermal Spraying, Practice, Theory and Application. 1985. *American welding society*. Miami.

Thiele, S., R.B. Heimann, L.-M. Berger, M. Herrmann, M. Nebelung, T. Schnick, B. Wielage, and P. Vuoristo. 2002. Microstructure and

properties of thermally sprayed silicon nitride-based coatings. *Journal of Thermal Spray Technology* 11 (2): 218–225.

Thorpe, M.L., and H.J. Richter. 1992. A pragmatic analysis and comparison of HVOF processes. *Journal of Thermal Spray Technology* 1 (2): 161–170.

Tikkanen, J., K.A. Gross, C.C. Berndt, V. Pitkatnen, J. Keskinen, S. Raghu, M. Rajala, and J. Karthikeyan. 1997. Characteristics of the liquid flame spray process. *Surface and Coatings Technology* 90: 210–216.

Tillmann, W., E. Vogli, and J. Nebel. 2007. Development of detonation flame sprayed Cu-Base coatings containing large ceramic particles. *Journal of Thermal Spray Technology* 16 (5–6): 751–758.

Tillmann, W., C. Schaak, L. Hagen, G. Mauer, and G. Matthäus. 2018. Internal diameter coating processes for bond coat (HVOF) and thermal barrier coating (APS) systems. *Journal of Thermal Spray Technology*. Published online: 30-10-2018.

Torres, B., P. Rodrigo, M. Campo, A. Urena, and J. Rams. 2009. Oxy-acetylene flame thermal spray of Al/SiCp composites with high fraction of re-enforcements. *Journal of Thermal Spray Technology* 18 (4): 642–651.

Totemeier, T.C. 2005. Effect of high-velocity oxygen-fuel thermal spraying on the physical and mechanical properties of type 316 stainless steel. *Journal of Thermal Spray Technology* 14 (3): 369–372.

Totemeier, T.C., R.N. Wright, and W.D. Swank. 2002. Microstructure and stresses in HVOF sprayed Iron aluminide coatings. *Journal of Thermal Spray Technology* 11 (3): 400–408.

———. 2003. Mechanical and physical properties of high-velocity oxy-fuel–sprayed Iron aluminide coatings. *Metallurgical and Materials Transactions A* 34A: 2223–2231.

———. 2004. Residual stresses in high-velocity oxy-fuel metallic coatings. *Metallurgical and Materials Transactions A* 35A: 1807–1814.

Trompetter, W.J., M. Hyland, P. Munroe, and A. Markwitz. 2005. Evidence of mechanical interlocking of NiCr particles thermally sprayed onto Al substrates. *Journal of Thermal Spray Technology* 14 (4): 524–529.

Trompetter, W., M. Hyland, D. McGrouther, P. Munroe, and A. Markwitz. 2006. Effect of substrate hardness on splat morphology in high-velocity thermal spray coatings. *Journal of Thermal Spray Technology* 15 (4): 663–669.

———. 2010. The effect of substrate surface oxides on the bonding of NiCr alloy particles HVAF thermally sprayed onto aluminum substrates. *Journal of Thermal Spray Technology* 19 (5): 1024–1031.

Venkataraman, R., B. Ravikumar, R. Krishnamurthy, and D.K. Das. 2006. A study on phase stability observed in as sprayed Alumina-13 wt.% Titania coatings grown by detonation gun and plasma spraying on low alloy steel substrates. *Surface & Coatings Technology* 201: 3087–3095.

Vijaya, Babu M., R. Krishna Kumar, O. Prabhakar, and N. Gowri Shank. 1996a. Simultaneous optimization of flame spraying process parameters for high quality molybdenum coatings using Taguchi methods. *Surface & Coatings Technology* 79: 276–288.

Vijaya, Babu M., R. Krishna Kumar, O. Prabhakar, and N. Gowri Shankar. 1996b. Simultaneous optimization of flame spraying process parameters for high quality molybdenum coatings using Taguchi methods. *Surface and Coatings Technology* 79: 276–288.

Villiers Lovelock, H.L., P.W. Richter, J.M. Benson, and P.M. Young. 1998. Parameter study of HP/HVOF deposited WC-Co coatings. *Journal of Thermal Spray Technology* 7 (1): 97–107.

Voyer, Joel. 2013. Flexible and conducting metal-fabric composites using the flame spray process for the production of Li-ion batteries. *Journal of Thermal Spray Technology* 22 (5): 699–709.

Wang, Y. 2004. Nano- and submicron-structured sulfide self-lubricating coatings produced by thermal spraying. *Tribology Letters* 17 (2): 165–168.

Wang, J., Zhang Li, B. Sun, and Y. Zhou. 2000. Study of the Cr_3C_2-NiCr detonation spray coating. *Surface and Coatings Technology* 130: 69–73.

Wang, Jun, Baode Sun, Qixin Guo, Mitsuhiro Nishio, and Hiroshi Ogawa. 2002. Wear resistance of a Cr_3C_2-NiCr detonation spray coating. *Journal of Thermal Spray Technology* 11 (2): 261–265.

Wang, J., Baode Sun, Qixin Guo, Mitsuhiro Nishio, and Hiroshi Ogawa. 2002b. Wear resistance of a Cr_3C_2-NiCr detonation spray coating. *Journal of Thermal Spray Technology* 11 (2): 261–265.

Wang, T.-G., S.-S. Zhao, W.-G. Hua, J. Gong, and C. Sun. 2009. Design of a separation device used in detonation gun spraying system and its effects on the performance of WC–Co coatings. *Surface & Coatings Technology* 203: 1637–1644.

Wang, T.-G., S.-S. Zhao, W.-G. Hua, J.-B. Li, J. Gong, and C. Sun. 2010. Estimation of residual stress and its effects on the mechanical properties of detonation gun sprayed WC–Co coatings. *Materials Science and Engineering A* 527 (3): 454–461.

Wang, X., Q. Song, and Z. Yu. 2016. Numerical investigation of combustion and flow dynamics in a high velocity oxygen-fuel thermal spray gun. *Journal of Thermal Spray Technology* 25 (3): 441–450.

Wielage, B., A. Wank, H. Pokhmurska, T. Grund, Ch. Rupprecht, G. Reisel, and E. Friesen. 2006. Development and trends in HVOF spraying technology. *Surface & Coatings Technology* 201: 2032–2037.

Wigren, J., et al. 1996. On-line diagnostics of traditional flame spraying as a tool to increase reproducibility. In *Proceedings of National Thermal Spray Conference*, ed. C.C. Berndt, 675–681. Materials Park, OH, USA: ASM International.

Withy, B.P., M.M. Hyland, and B.J. James. 2008. The effect of surface chemistry and morphology on the properties of HVAF PEEK single splats. *Journal of Thermal Spray Technology* 17 (5–6): 631–636.

Wood, R.J.K., B.G. Mellor, and M.L. Binfield. 1997. Sand erosion performance of detonation gun applied tungsten carbide/cobalt-chromium coatings. *Wear* 211: 70–83.

Wu, Y.N., F.H. Wang, W.G. Hua, J. Gong, C. Sun, and L.S. Wen. 2003. Oxidation behavior of thermal barrier coatings obtained by detonation spraying. *Surface and Coatings Technology* 166: 189–194.

Wu, Y.N., P.L. Ke, Q.M. Wang, C. Sun, and F.H. Wang. 2004. High temperature properties of thermal barrier coatings obtained by detonation spraying. *Corrosion Science* 46: 2925–2935.

Wu, T., S. Kuroda, J. Kawakita, K. Katanoga, and R. Reed. 2006. Processing and properties of titanium coatings produced by warm spraying. In *Thermal spray: Building on 100 years of success*, ed. B. Marple et al. ASM International, Materials Park, OH, USA. e-proceedings.

Yang, G., H. Zu-kun, X. Xiaolei, and X. Gang. 2001. Formation of molybdenum boride cermet coating by the detonation spray process. *Journal of Thermal Spray Technology* 10 (3): 456–460.

Yang, Y.-M., H. Liao, and C. Coddet. 2002. Simulation and application of a HVOF process for MCrAlY thermal spraying. *Journal of Thermal Spray Technology* 11 (1): 36–43.

Yang, G.-J., C.-J. Li, and Y.-Y. Wang. 2005. Phase formation of Nano-TiO_2 particles during flame spraying with liquid feedstock. *Journal of Thermal Spray Technology* 14 (4): 480–486.

Yasunari, Ishikawa, Seiji Kuroda, and Jin Kawakita. 2007. Yukihiro Sakamoto, Matsufumi Takaya, sliding wear properties of HVOF sprayed WC-20%Cr_3C_2-7%Ni cermet coatings. *Surface & Coatings Technology* 201: 4718–4727.

Yilbas, B.S., and A.F.M. Arif. 2007. Residual stress analysis for HVOF diamalloy 1005 coating on Ti–6Al–4V alloy. *Surface & Coatings Technology* 202: 559–568.

Yilbas, B.S., M. Khalid, and B.J. Abdul-Aleem. 2003. Corrosion behavior of HVOF coated sheets. *Journal of Thermal Spray Technology* 12 (4): 572–575.

Yuan, X., H. Wang, G. Hou, and B. Zha. 2006. Numerical modelling of a low temperature high velocity air fuel spraying process with injection of liquid and metal particles. *Journal of Thermal Spray Technology* 15 (3): 413–421.

Yuan, F.H., Z.X. Chen, Z.W. Huang, Z.G. Wang, and S.J. Zhu. 2008. Oxidation behavior of thermal barrier coatings with HVOF and detonation-sprayed NiCrAlY bondcoats. *Corrosion Science* 50: 1608–1617.

Yuuzou, Kawahara. 2007. Application of high temperature corrosion-resistant materials and coatings under severe corrosive environment in waste-to-energy boilers. *Journal of Thermal Spray Technology* 16 (2): 202–213.

Zhang, T., Z. Qiu, Y. Bao, G.T. Gawne, and K. Zhang. 2000. Temperature profile and thermal stress analysis of plasma sprayed glass-composite coatings. In *Thermal spray: Surface engineering via applied research*, ed. C.C. Berndt, 355–361. ASM International, Materials Park, OH, USA.

Zhang, Yu-Juan, Xiao-Feng Sun, Heng-Rong Guana, and Zhuang-Qi Hua. 2002. 1050°C isothermal oxidation behavior of detonation gun sprayed NiCrAlY coating. *Surface and Coatings Technology* 161: 302–305.

Zhang, Y.J., X.F. Sun, Y.C. Zhang, T. Jin, C.G. Deng, H.R. Guan, and Z.Q. Hu. 2003. A comparative study of DS NiCrAlY coating and LPPS NiCrAlY coating. *Materials Science and Engineering A* 360: 65–69.

Zhang, G., H. Liao, H. Yu, S. Costil, S.G. Mhaisalkar, J.-M. Bordes, and C. Coddet. 2006. Deposition of PEEK coatings using a combined flame spraying–laser remelting process. *Surface & Coatings Technology* 201: 243–249.

Zhang, Z., B. Liang, and H. Guo. 2014. Interface microstructure and Tribological properties of flame spraying NiCr/La$_2$O$_3$ coatings. *Journal of Thermal Spray Technology* 23 (8): 1404–1412.

Zhu, J.L., J.H. Huang, H.T. Wang, J.L. Xu, X.K. Zhao, and H. Zhang. 2008. In-situ synthesis and microstructure of TiC–Fe36Ni composite coatings by reactive detonation-gun spraying. *Materials Letters* 62: 2009–2012.

Abbreviations

APS	Atmospheric Plasma Spraying
CAPS	Controlled Atmosphere Plasma Spraying
CARS	Coherent Anti-Stokes Raman spectroscopy
CBL	Cold Boundary Layer
CFD	Computational Fluid Dynamics
CVD	Chemical Vapor Deposition
DC	Direct Current
D-Gun	Detonation Gun
eV	Electron Volt
HVOF	High-Velocity Oxygen Flame
i.d.	internal diameter
ISPC	International Symposium on Plasma Chemistry
ITSC	International Thermal Spray Conference
LDA	Laser Doppler Anemometry
LTE	Local Thermodynamic Equilibrium
NLTE	Non-equilibrium
NTSC	National Thermal Spray Conference
PSD	Particle Size Distribution
PTA	Plasma Transferred Arc
PVD	Physical Vapor Deposition
RF-ICP	Radio Frequency Inductively Coupled Plasma
slm	standard liter per minute
TS	Thermal Spray
TTPR	Triple Torch Plasma Reactor
ULPPS	Ultra-Low-Pressure Plasma Spraying
UTSC	United Thermal Spray Conference
VPS	Vacuum Plasma Spraying

8.1 Introduction

Materials have been developed for exceptional functional performance in specific applications, such as a number of specialty steels or super alloys. The increasing demand for combined functional requirements, such as high temperature resistance to corrosivec atmospheres in addition to abrasive wear resistance, and the ease of machining to the final form at a competitive cost, has led to the ever-increasing demand for coatings. According to [Fauchais et al. (2014)], the motivation for coating structural parts can be summarized by the following needs:

(a) improve functional performance by, for example, allowing higher-temperature exposure through the use of thermal-barrier coatings,

(b) improve component life by reducing wear due to abrasion, erosion, and corrosion,

(c) extend functional use by rebuilding the worn part to its original dimensions, thus avoiding the need for replacing the entire component, for example, a shaft or axle, and,

(d) reduce component cost by improving the functionality of a low-cost material with an expensive coating. In each of these uses of coating technologies, there should be no, or minimal, machining required of the coated part.

As discussed in Chap. 2, coating technologies can be roughly divided into deposition of thin films (thickness e <20 μm) and deposition of thick films (30 μm < e < several mm). Thin films offer excellent enhancement of surface properties. Physical Vapor Deposition (PVD) or Chemical Vapor Deposition (CVD) can provide surfaces with

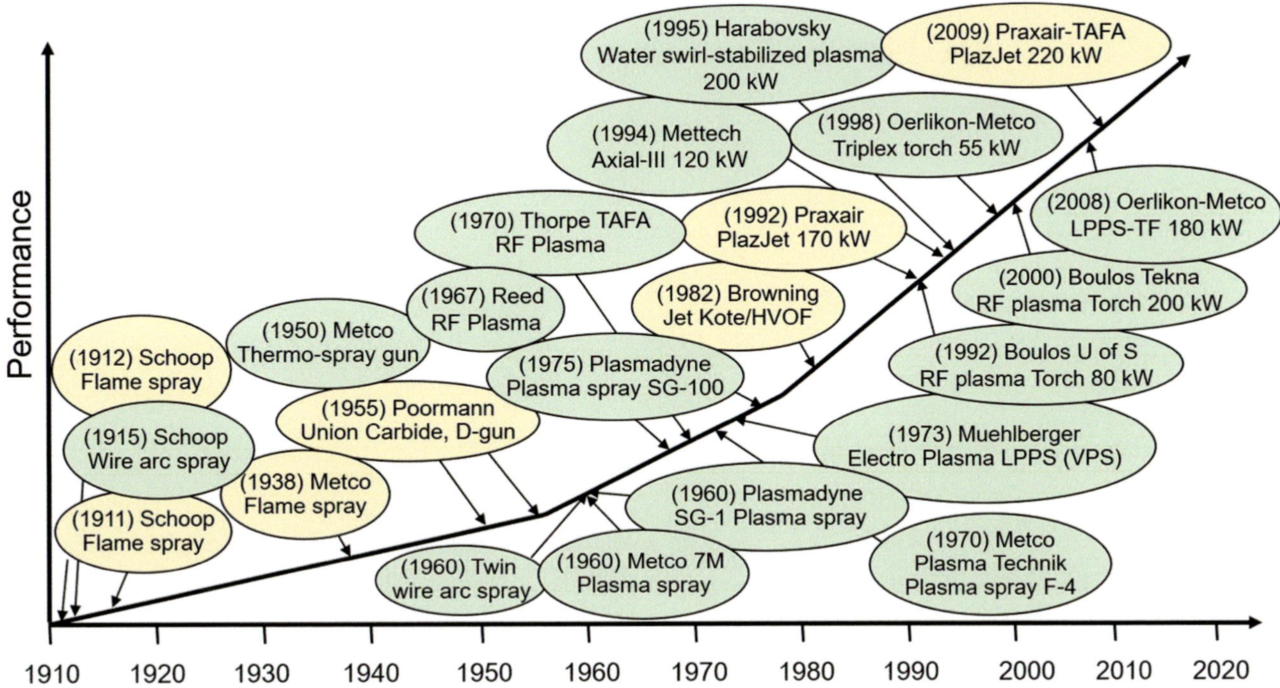

Fig. 8.1 Milestones in the development of the thermal spray industry, Yellow-combustion; Green-Plasma. [(Ang et al. (2013)]

unparalleled hardness or corrosion resistance. As most thin-film technologies require a reduced pressure environment, they are more expensive and impose a limit on the size and shape of the part to be coated. Thick films with e >30 μm are required when the functional performance depends on the layer thickness, for example, in thermal-barrier coatings, when strong erosion and corrosion conditions result in wear and the component life depends on the layer thickness, or when the original dimensions of the worn parts are being restored. Thick-film deposition methods include chemical/electro-chemical plating, brazing, weld overlays, and thermal spray. Each of these methods offers certain advantages and has certain limitations. In the following, only plasma spraying will be considered. It belongs to thermal-spray processes for which the definition is "Thermal spraying comprises a group of coating processes in which finely divided metallic or non-metallic materials are deposited in a molten or semi-molten condition to form a coating. The coating material may be in the form of powder, ceramic rod, wire or molten materials." Weld overlays could be included in this definition; however, from all the weld-coating processes, only Plasma Transferred Arc (PTA) deposition is usually included as a thermal-spray process. The coating material may be in the form of powders, ceramic rods, wires, or molten materials. In this chapter, only coatings resulting from powders will be discussed.

While thermal spray (TS) was developed in the beginning of the twentieth century, the technology was limited, for a long period, to essentially flame spraying. In the mid-1950s, the first plasma-spray torch was developed by Thermal Dynamics Corporation, followed by Metco and Plasmadyne, which developed the basis of the torch designs currently used in this field. Figure 8.1, based on data from [Ang et al. (2013)], illustrates the evolution of thermal-spray technology in terms of the turning points in plasma torch designs used in the thermal spray industry. In this figure, the yellow entries refer to combustion-based technologies, while the green ones refer to plasma-based technologies. Intense activities in this field over the past three to four decades has led to the development of a number of novel spray-coating devices/processes, such as the Water-stabilized plasma torch by [Hrabovsky (1997)], the RF Induction Coupled Plasma torch by [Boulos MI (1992)], the Axial III torch central powder injection DC plasma torch by [Northwest Mettech corp (1994)], and the triplex Torch and the Ultra Low-Pressure Plasma Spraying process (ULPPS) by [LPPS-TF, Oerlikon–Metco Cop. (2008)].

The driving force behind many of these developments has been the need to increase the temperature and gas velocity of the thermal-spray source beyond that offered by combustion-spraying technologies. They also aimed at better control on the chemical nature of the gases used, increased process

productivity, and improved economics. Compared to other thermal-spray technologies, the use of DC plasma sources has been central to such developments since they are at the high end of the temperature vs velocity diagram, as shown in Fig. 8.2.

The principal components of a DC plasma-spray system/process are illustrated in Fig. 8.3. Central to this system is the DC plasma torch generating a high-temperature, high-velocity plasma stream in which the coating material is injected in the form of either a fine powder using an appropriate carrier gas or a liquid as in the case of solution or suspension plasma spraying. The system also consists of an electrical, current-controlled, power supply, gas supply, powder, or liquid/suspension feeder and a closed-loop cooling water circuit. The auxiliary equipment necessary for the operation of the system include a manual or robotic torch controller, substrate support, environmental control of the ambient air, and general instrumentation and control systems.

This chapter deals essentially with the fundamentals of the DC plasma-spraying technology, highlighting the principal concepts of DC plasma torch design. This is followed by a discussion on gas and particle dynamics under plasma-spraying conditions including a brief review of mathematical models developed for the simulation of the plasma-spray process. In order to keep a balance between the size of the different chapters of this book and for ease of referencing, a review of the different DC plasma-spray processes is presented separately, in Chap. 9. These include Atmospheric Plasma Spraying (APS), Controlled Atmosphere Plasma Spraying (CAPS), Vacuum Plasma Spraying (VPS), and the relatively novel process of Ultra-Low-Pressure Plasma Spraying (ULPPS). A detailed discussion of powder/wire or cord preparation, substrate preparation, coating formation, and characterization are covered in Part III of this book, while process integration, industrial applications, and process economics are covered in Part IV.

8.2   Basic Concepts

Electric arcs generate high-temperature plasma through resistive energy dissipation when the current flows through the gas at sufficiently high temperatures to have an appreciable degree of ionization and high electrical conductivities. For most plasma-forming gases, the required temperatures at atmospheric pressure are above 8000 K [Boulos et al. (1994)]. By definition, in a thermal arc, the thermodynamic state of the plasma generally approaches Local Thermodynamic Equilibrium (LTE), which includes kinetic and chemical equilibrium, except for the arc fringes or close to a cooled wall. Such arcs, known as high-intensity arcs, typically require currents above 100 A and pressures in excess of 10 kPa. At lower pressures, electron and heavy particle temperatures are significantly different, as shown in Fig. 8.4. [Pfender (1978)].

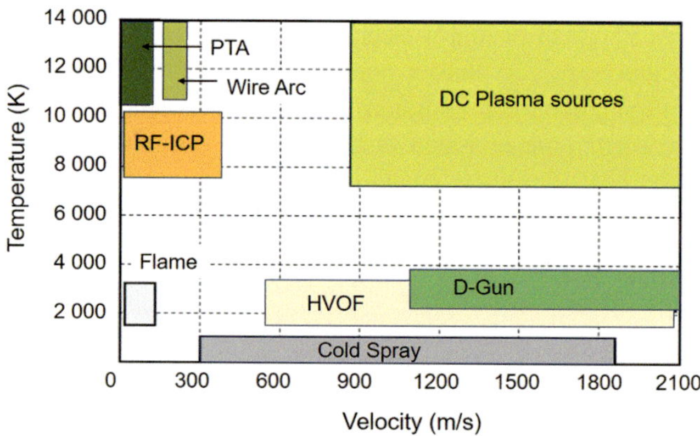

Fig. 8.2 Gas temperatures vs velocity mapping associated with different thermal-spray processes

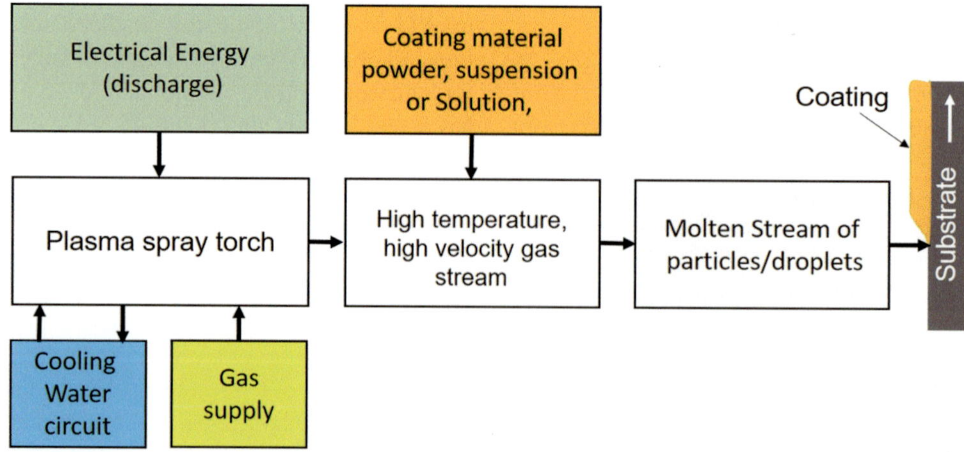

Fig. 8.3 Block diagram of a conventional DC plasma spray system

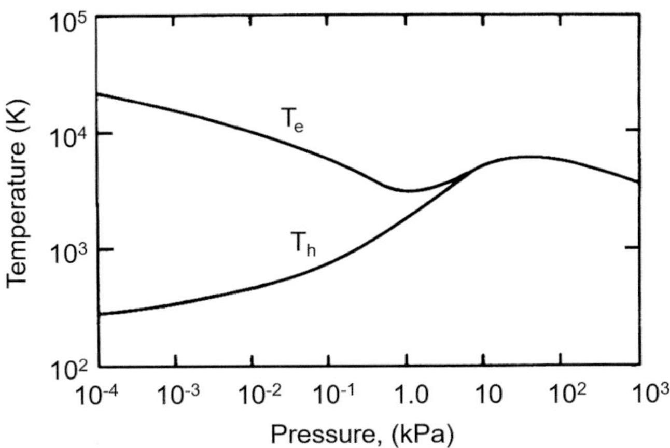

Fig. 8.4 Behavior of electron temperature (T_e) and heavy particles temperature (T_h) in an arc plasma. [(Pfender 1978)]

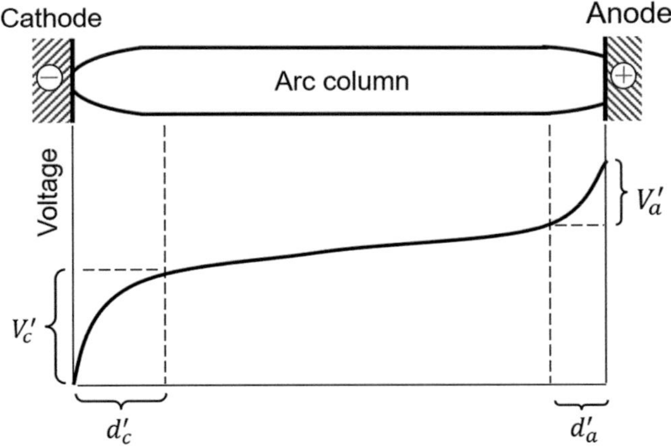

Fig. 8.5 Typical potential distributions along an arc

An electric arc struck between two electrodes is schematically represented in Fig. 8.5. There are three principal features that distinguish an arc from other discharge modes. For the sake of simplicity, the following discussion will be restricted to steady-state (DC) arcs

- *Relatively high current densities*, in excess of 10^6 A/m². The situation is even more pronounced at the electrodes. Particular to the cathode, arcs may attach to the electrodes in the form of tiny spots in which current densities can be as high as 10^{10} A/m². The associated heat-flux densities are of the order of 10^9 to 10^{11} W/m², values requiring special precautions to protect the mechanical integrity of the electrodes.

- *Low cathode fall*. As indicated in Fig. 8.5, the potential distribution in an electric arc changes rapidly in front of the electrodes, forming the so-called cathode and anode fall. The cathode fall is of particular interest; it assumes values of around 10 V in contrast to the typical cathode

falls in glow discharges, which usually exceed 100 V. This relatively low cathode fall is a consequence of the more efficient electron-emission mechanisms at the cathode compared with those prevailing in glow discharges. It should be pointed out that the thicknesses d_c' and d_a' of the electrode regions in Fig. 8.5 are enlarged, particularly if high-pressure arcs are considered. Also, V_c' does not, in general, represent the true cathode fall. The actual cathode fall, V_c, corresponding to the electron space charge region may be several volts less than V_c'.

- *High luminosity of the arc column*. This criterion provides a useful distinction between an arc and other discharge modes, provided the pressure is sufficiently high (p ≥ 1 kPa).

An electric arc can be divided into three regions, the cathode region, the arc column, and the anode region.

8.2.1 Cathode

The cathode must supply electrons to the arc. In plasma torch designs used for plasma spraying, electrons are supplied through thermionic emission from a hot cathode material. The thermionic emission current can be described by the Richardson–Dushman equation relating the emission to the cathode temperature T_c and work function ϕ_w.

$$j = AT_c^2\, exp\left(-e\,\phi_w/k\,T_c\right) \tag{8.1}$$

Where, A is an empirical constant, which is about 6×10^5 A/m² K² for the majority of the cathode materials, e is the electronic charge, and k is the Boltzmann constant. The work function for most materials is around or above 4.0 eV; however, some oxides have work function values of between 2.5 and 3 eV. To obtain a current density of approximately 10^8 A/m², as required by the arc, the cathode temperature will have to be about 4500 K for a cathode material with a work function of 4.5 eV (e.g., for tungsten), but only about 2700 K for a material with a work function of 2.5 eV. Since tungsten has a very high melting point (4500 K) and, therefore structural stability, it is widely used as torch cathode. It is commonly used, however, alloyed with 1 or 2% by weight of a material with a lower work function. The consequence is a significantly reduced cathode operating temperature and longer cathode life. Examples of such materials are ThO_2, La_2O_3, or LaB_6. Cathodes based on tungsten cannot be used with oxidizing gases because the formation of volatile tungsten oxides, at temperatures already below 2000 K, will result in rapid cathode erosion. For operation with oxidizing gases, button-type electrodes are used where the thermionic emission material is inserted in the form of a button into a water-cooled copper holder. The button typically consists of

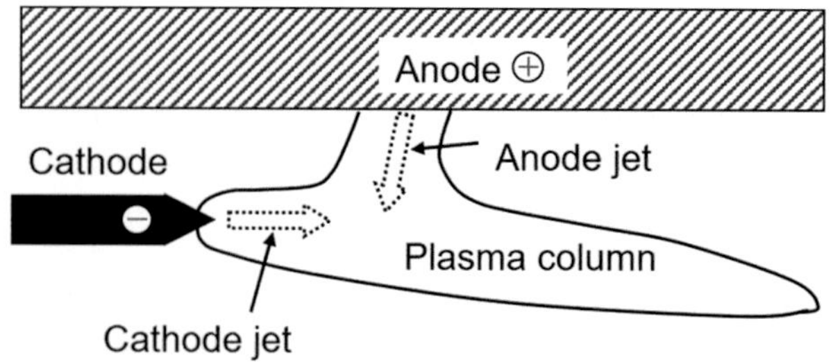

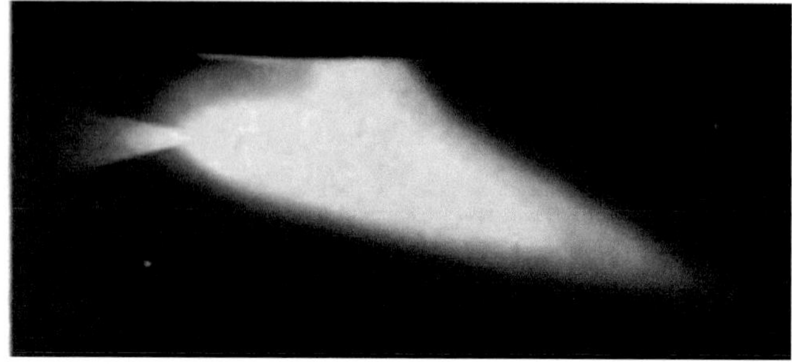

Fig. 8.6 Interaction of cathode and anode plasma jets. [Pfender (1978)]

hafnium or zirconium, and the surface of the material is molten. However, the pressure exerted by the arc due to the ion flux to the cathode and due to the magnetic forces resulting from the self-magnetic field and the change in current density in front of the cathode will keep the molten material in place [Peters et al. (2005)].

The self-magnetic field generates a higher pressure inside the arc than the pressure surrounding the arc, and this pressure differential is proportional to the arc current density. The arc is usually constricted in front of the cathode because of the heat loss from the plasma to the solid material. As a consequence, the pressure close to the cathode is higher than it is farther downstream, resulting in a flow away from the cathode, the cathode jet [Maecker (1955)], as illustrated in Fig. 8.6. The superposition of the cathode jet to the axial gas flow used in a plasma-spray torch results in the increased convective cooling of the arc in the vicinity of the cathode, leading to even stronger constriction. The resulting flow velocities can reach several 100 m/s. The highest temperatures in the arc are observed in front of the cathode, reaching values between 20,000 and 30,000 K depending on the current density and the fluid dynamics of the flow in the cathode region.

8.2.2 Arc Column

The arc column characteristics are determined by the energy dissipation per unit length, that is, by the arc current, the plasma gas flow and composition, and the arc channel diameter. The arc column can be described by the conservation equations for mass, momentum, and energy. In principle, Joule heat dissipation, expressed by the product of current

density and electrical field strength, is balanced by the heat addition to the plasma gas and the heat losses to the surroundings. If the heat losses to the surroundings increase, the power dissipation has to increase. Since the plasma torches operate with a constant current, higher electric-field strengths and arc voltages will compensate for the increased heat loss. Increased heat losses are the result of:

- reduced nozzle diameter,
- higher total plasma gas flow rates, and,
- increased thermal conductivities and specific heats of the plasma gases.

As shown in Chap. 3, the thermal conductivities and specific heats increase from argon to nitrogen and helium (depending on the temperature range) to hydrogen. Accordingly, the column-field strengths and arc voltages are increasing from argon arcs to helium and nitrogen arcs to hydrogen arcs, with pure hydrogen arcs typically having four times the arc voltage of an argon arc for similar conditions. However, it must once again be emphasized that to operate the hydrogen arc in a plasma torch, significantly higher volumetric flow rates will be required, with considerably higher voltages as the result. The arc diameter will decrease with increasing energy loss rates.

The species present in the arc column are molecules, atoms, ions, and electrons, and for gas mixtures, there will be multiple types of atoms and ions. The strong radial temperature and density gradients result in strong diffusion fluxes, and the diffusion coefficients are quite different for the different species. The consequence is a "demixing" effect where lighter species have higher concentrations in the arc fringes than predicted according to an equilibrium temperature [Murphy (1996) and Snyder et al. (1995)]. These effects must be taken into account in particular when heat transfer rates are estimated from plasmas consisting of gas mixtures. The result is that heat transfer rates in the fringes of the arc or arc jet can be higher than expected if diffusion is not taken into account. While electrons would diffuse at very high rates, their diffusion rate is limited by the electric fields generated when electron densities are different from ion densities. Therefore, electrons diffuse only in conjunction with an ion with the ambipolar diffusion rate, which is approximately twice the ion diffusion rate [Fincke et al. (1993a, b)]. Furthermore, any mixture containing helium is unlikely to contain any helium ions because of the high ionization energy of helium compared to all other gases. The consequence is that there is no ambipolar diffusion of He ions and lower He concentrations are found in the arc fringes [Fincke et al. (1993a, b)].

While immense progress has been made with the modeling of the arc column in plasma torches [Mariaux et al. (2001), Baudry et al. (2005), Dussoubs (1998), Trelles

et al. (2006), Trelles (2007)], it still requires significant effort and computational time to use CFD codes as a design tool. However, an integral energy balance (integrated over the radial coordinate) can provide some information on the arc-heating process inside the torch with information, which can be easily obtained:

$$\dot{m}\,\frac{\Delta \bar{h}}{\Delta z} + Q_{cond} + Q_{rad} = I \times E \tag{8.2}$$

with $\dot{m}$ being the mass flow rate of the plasma gas, $\Delta \bar{h}$ the increase in enthalpy averaged over the cross-section per unit length Δz, and Q_{cond} and Q_{rad} the losses to the torch wall by heat conduction and radiation, and $I \times E$ the electric energy dissipated per unit length of the arc. Further integration over the entire arc length will give the energy balance for the torch

$$\dot{m}\,h_{ave} = \eta P_{el} = P_{el} - Q_{loss} \tag{8.3}$$

with h_{ave} the average enthalpy of the gas leaving the torch, η the torch gas heating efficiency, P_{el} the electric power input into the torch (current × voltage), and Q_{loss} the heat lost to the torch cooling water. The torch average enthalpy is defined as

$$h_{av} = 2\pi\,\frac{\int_{o}^{R} \rho v h r\,dr}{\dot{m}} \tag{8.4}$$

with ρ, v, and h the plasma mass density, velocity, and enthalpy at a given radial location at the torch exit, and R the channel radius. Sometimes the term "average temperature" is used, related to the average enthalpy

$$h_{av} = \int_{T_{ref}}^{T_{ave}} c_p\,dT \tag{8.5}$$

with T_{ref} the reference temperature for a zero value of the enthalpy and c_p the specific heat at constant pressure of the plasma gas. The average velocity is defined accordingly

$$v_{ave} = \frac{\dot{m}}{\rho(T_{ave})\,A} \tag{8.6}$$

with $\rho(T_{ave})$ the mass density of the plasma gas at the average temperature, and A the channel cross-sectional area. A torch energy balance easily allows determination of the average values of temperature and velocity, but one needs to keep in mind that the peak values for temperature and velocity can easily be twice the average value.

8.2.3 Anode

The anode usually consists of a water-cooled copper channel, sometimes lined with a tungsten or tungsten–copper sleeve, and numerous designs exist for tailoring the fluid dynamics

within this channel. Nozzle designs can emphasize high gas velocities, high gas temperatures, narrower or wider temperature profiles, and different average arc lengths and consequently arc voltages and torch powers. Typical nozzle diameters are between 6 and 10 mm for arc currents between 300 and 1000 A. The effect of the diameter on the plasma temperature depends on the operating conditions, but in general, a temperature increase is noticed for smaller channel diameters, because the increased electric field strength resulting from the increased heat loss leads to increased local heating of the plasma gas. Because of the nonlinearly varying dependence of the enthalpy on the temperature (e.g., a strong increase with temperature in the ionization region 12,000–15,000 K), a strong increase in power dissipation will increase the enthalpy but only change the temperature by 1000–2000 K. The effect of the nozzle diameter on the plasma velocity is stronger, because the flow velocity is inversely proportional to the channel cross-section area, in addition to showing increases due to higher temperatures and lower plasma densities.

It should be noted, however, that the anode is a passive component, just collecting electrons to allow the current to flow from the solid conductors of the electrical circuit to the plasma. The difficulty is that a cold gas boundary layer exists between the plasma and the anode and that nonequilibrium conditions exist in this boundary layer, that is, higher electron temperatures than heavy particle temperatures, necessary for the current to flow across this boundary layer [Paik et al. (1993), Trelles et al. (2007a, b, c), Dimulescu et al. (1980) and Leveroni et al. (1987)]. The current flow is additionally strongly dependent on the electron-density gradients and can be expressed by [Dimulescu et al. (1980)].

$$j = \sigma E_x - \frac{\sigma}{e n_e}\frac{dp_e}{dx} \tag{8.7}$$

where j is the electron current density, σ the electrical conductivity, E_x the electric field normal to the anode surface, e the electronic charge, and n_e and p_e the electron number density and partial pressure, respectively. The strong partial pressure gradient in front of the cooled anode surface results in strong electron diffusion fluxes toward the surface, constituting the major component of the current flow.

The heat flux to the anode is strongly tied to the current flow [Pfender (1994), Jenista et al. (1997a, b)]:

$$q = j\phi_w + \left(\frac{5k}{2e}\right) jT_e - \kappa_h\frac{dT_h}{dx} - \kappa_e\frac{dT_e}{dx} + j_i(\varepsilon_i - \phi_w) \tag{8.8}$$

with ϕ_w the work function of the anode material, k the Boltzmann constant, κ_h and κ_e the heavy particle and electron

thermal conductivities, respectively, T_e and T_h the electron and heavy particle temperatures, respectively, j_i the ion current toward the anode, ε_i the ionization energy of the plasma gas atoms. The first term on the right-hand side is associated with energy release when the electrons carrying the current to the anode are incorporated into the crystal lattice (condensation energy); the second term is the thermal energy transported by the electrons; the third and fourth terms are the heavy particle and electron conduction terms, respectively; and the fifth term represents the energy released when an ion recombines with an electron at the anode surface. The first three terms are usually the dominant ones [Jenista et al. (1997a, b)].

The attachment of the arc at the anode surface may occur diffusely, as well as severely constricted (spot), and it is still not entirely clear under which conditions constriction will occur. Chemical reactions on the anode surface as, for example, encountered in arcs operated in atmospheric air or in other oxidizing fluids seem to favor a constricted anode root which, at the same time, may move more or less randomly over the anode surface with appreciable velocities. Anode evaporation is another mechanism, which leads, in general, to spot formation. In high-current arcs with relatively small electrode separation (a few centimeters), the diffuse anode attachment is directly associated with the cathode jet. The well-known bell shape of a free-burning high-intensity arc is a typical example. The intense cathode jet impinging on the anode surface pushes hot plasma against the anode, eliminating the need for ionization in the anode fall zone. By increasing the electrode gap under otherwise identical conditions, the influence of the cathode jet at the anode is diminished and, finally, at sufficiently large gaps, the arc forms a single or several spots at the anode surface. Further evidence that anode spot formation and cathode jet are intimately related is illustrated in Fig. 8.6., in which the axis of the cathode is parallel to the surface of a plane anode so that the cathode jet does not impinge on the anode. The deflected cathode jet provides evidence that there must be an appreciable constriction of the anode attachment. Any constriction of the current path leads to the previously described pumping action, which results in this case in an anode jet that causes the observed deflection of the cathode jet from the anode surface. The relative strength of the two jets determines the angle of deflection.

While most commercial plasma torches use anodes with cylindrical nozzle bores, anodes with Laval-type diverging nozzles exits have been used. This type of nozzle provides a somewhat more uniform velocity and temperature distribution at the nozzle exit and somewhat reduced turbulent cold gas entrainment, resulting in more uniform particle heating and acceleration [Roumilhac et al. (1990a, b), Rahman et al. (1998)].

8.2.4 Arc Stability

Arc stability has to be of central concern in plasma torch design. The arc plasma is easily influenced by asymmetric cooling or by effects of magnetic fields that can lead to arc extinction. The principal counter-acting effect promoting arc stability is an increase in heat loss due to steeper temperature gradients. However, arc voltage and power dissipation will increase. Steenbeck's minimum principle states that the arc will adjust its position such that the overall voltage drop will be minimum. That means the preferred arc position is that where the energy loss is minimized [Pfender (1978)]. An excellent review on the subject of electric arc fluctuations in DC plasma spray torches is given by [Rat et al. (2017)].

Arcs can be stabilized by a cylindrical wall (e.g., water-cooled metal tube surrounding the arc) forcing the arc to remain on the axis of the cylinder (wall stabilization), or it can be stabilized by a superimposed axial flow of a cold sheath gas, again increasing cooling when the arc moves off the axis (gas stabilization). Flow stabilization is particularly effective when a swirl (or vortex) flow is used because the vortex creates a low-pressure zone on the axis of the flow in which the low-density, high-temperature arc gases are confined, while the heavier cold gases are flowing along the external regions of the vortex. In plasma torches, one usually has a combination of wall and flow stabilization. The preferred position of the arc would be the one where the cathode-anode distance is the smallest. However, at this location, the cold gas velocity is usually the highest, resulting in strong cooling of the arc and higher arc voltages. Consequently, the arc attachment is usually found where the fluid dynamic condition brings the plasma close to the anode surface, either through recirculation eddies or through heating of the boundary layer flow [Sun and Heberlein (2005), Yang and Heberlein (2007)].

In plasma spraying, the most important instability is the anode attachment instability. As illustrated in Fig. 8.7, the arc root struck between the column and the anode surface is exposed to several forces, the dominant being the drag force, F_D, exerted by the cold gas flow in the boundary layer over the anode surface, and the electromagnetic force, F_m, which is a Lorentz force due to the self-induced magnetic field [Rat et al. (2017)]. Because of the stiffness of the arc root the cold boundary layer flow will go around, it is exerting a drag force on the arc root in the downstream direction. As a consequence, the arc root attachment is pushed downstream until the voltage between the column and the anode surface becomes sufficiently high for a local breakdown to occur at an upstream location (restrike) [Wutzke et al. (1968)]. Fig. 8.8 shows a schematic representation of a restriking arc and the associated voltage oscilloscope trace, indicating strong voltage oscillations giving rise

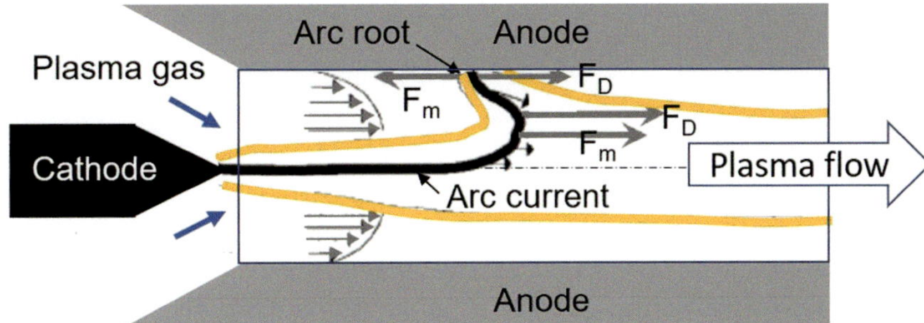

Fig. 8.7   Schematic representation of the main forces acting on the arc column and its root attachment. [Rat et al. (2017)]

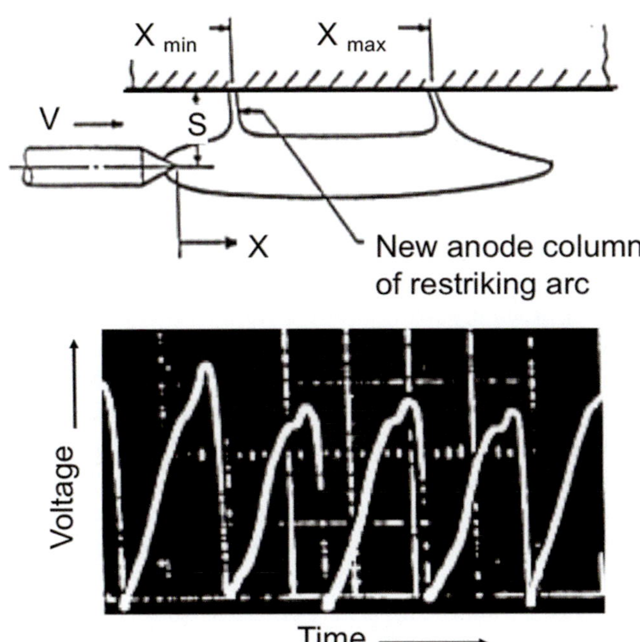

Fig. 8.8   Illustration of the restrike mode of operation of a plasma torch with an upstream connection forming between the arc column and the anode nozzle. The oscilloscope image shows the associated voltage trace. (Courtesy of Emil Pfender and Steven Wutzke)

to important fluctuations of the arc power. The effect of such fluctuations on particle heating and acceleration will be discussed later. Recent flow visualization experiments with a similar set-up of an anode with cold gas flow between the arc and parallel to the anode surface have demonstrated that the location of the restrike is determined by fluid dynamic instabilities.

[Paik et al. (1993)] demonstrated that the Steenbeck minimum principle is a useful criterion for predicting the position of the anode arc root in a plasma torch. Therefore, this criterion should provide useful guidelines for the design of plasma torches. Although their analysis is based on a two-dimensional model, the results are in qualitative agreement with experimental findings, which show that:

- the axial distance from the cathode tip to the stable arc-root position, x, increases with increase in the gas flow rate due to the stronger gas dynamic drag force,
- the axial distance x, decreases with increase in the arc current, due to the stronger magnetic body force,
- the arc attachment on the anode surface is closer to the cathode tip with reduction in the diameter of the arc channel,
- a more constricted anode root attachment is observed with hydrogen plasma compared to argon plasma.

The movement of the point of anode attachment, as indicated by the arc voltage fluctuations given in Fig. 8.9, reveals three different modes of operation [Coudert et al. (1994), Coudert et al. (1995), Planche et al. (1997), Duan et al. (1999a), Dorier et al. (2001), Duan et al. (2000, 2002)].

- the "*steady*" mode corresponding to a fixed location of the arc root attachment on the anode with very low fluctuations of the voltage trace,
- the "*take-over*" mode with the arc root moving back and forth along the main flow direction and with a sinusoidal wave form of the arc voltage,
- the "*restrike*" mode where the arc attachment moves downstream inside the anode nozzle until an upstream restrike occurs with a new connection closer to the cathode between the arc column and the anode nozzle. The arc voltage has a saw-tooth-shaped form.

The *restrike* mode is mostly observed when using molecular plasma forming gases. It was observed in both upstream and downstream restrike configurations with high-voltage amplitude fluctuations, which can reach values of 30 to 70% of the average arc voltage, and frequency typically between 2 and 6 kHz [Duan and Heberlein (2002)]. The *take-over* mode is mostly observed with pure noble gases such as Ar and He; the magnitude of the fluctuations is directly connected with the arc current and usually of the order of 20 to 50% of the mean arc voltage. It has a much wider Fourier spectrum of the fluctuation frequencies. For the *steady* mode, only minimal fluctuations are observed at low

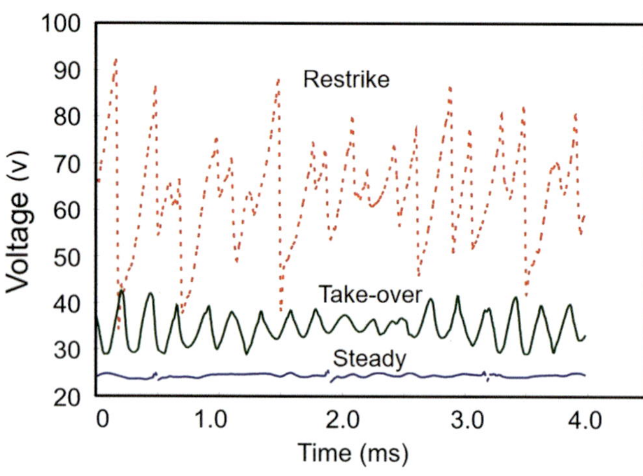

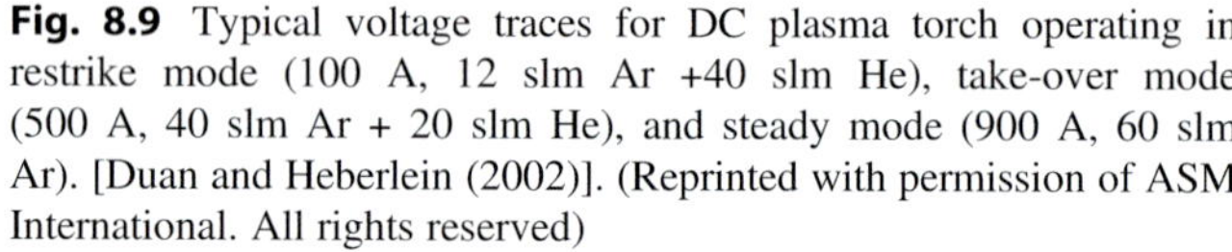

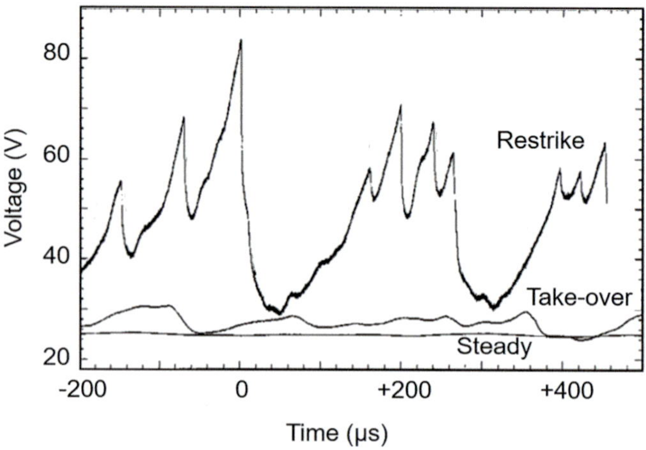

Fig. 8.9 Typical voltage traces for DC plasma torch operating in restrike mode (100 A, 12 slm Ar +40 slm He), take-over mode (500 A, 40 slm Ar + 20 slm He), and steady mode (900 A, 60 slm Ar). [Duan and Heberlein (2002)]. (Reprinted with permission of ASM International. All rights reserved)

Fig. 8.10 Typical voltage traces for an F4 plasma torch, in the restrike, take-over, and steady modes of operation obtained with warn electrodes, and a straight gas injector. (a) restrike mode, 500 A, 40 slm Ar + 4 slm H_2, (b) take-over mode 500 A, 40 slm Ar, (c) steady mode, 600 A, 30 slm Ar. [Dorier et al. (2001)]

frequencies, and these fluctuations are not associated with the fluid dynamic behavior of the arc.

A study of the performance of the F4 torch by Oerlikon-Metco was reported by [Dorier et al. (2001)]. Typical voltage traces under different operating conditions are given in Fig. 8.10. These were obtained using worn electrodes with a straight gas injector. The "*restrike*" mode trace was obtained for a current of 500A, and gas composition of (40 slm Ar + 4 slm H_2) show high voltage fluctuations formed of large voltage drops with smaller irregular voltage drops in between. For the "*take-over*" mode, obtained with a current of 500 A and plasma gas of pure argon (40 slm Ar), the voltage signal exhibits smooth fluctuations of weak amplitude (less than 10 V). Operating at high current with a reduced gas flow of pure argon (600 A, and 30 slm Ar), the take-over mode switches to the so called "*steady*" mode for which the voltage is nearly constant. The typical power spectra of the voltage fluctuations in the *restrike* and *take-over* modes are given in Fig. 8.11. These were obtained with new electrodes and Ar/H_2 plasma gas with a swirl injector. The spectra given in Fig. 8.11a were in the *take-over* mode with currents of 200 and 600 A. These show that the fluctuations of the arc voltage are reduced especially at high currents. The corresponding power spectrum for operation in the *restrike* mode, with an arc current of 400 A and plasma gas of (40 slm Ar + 4 slm H_2), is given in Fig. 8.11b. This spectrum is characterized by a strong peak at 3.6 kHz, corresponding to the frequency of the major restrikes.

According to [Dorier et al. (2000)], the gas injection mode (straight or vortex) can also modify the shape of the restrike signal, as shown in Fig. 8.12a, for N_2-H_2 plasma [Dorier et al. (2000)]. The corresponding power spectrum, given in

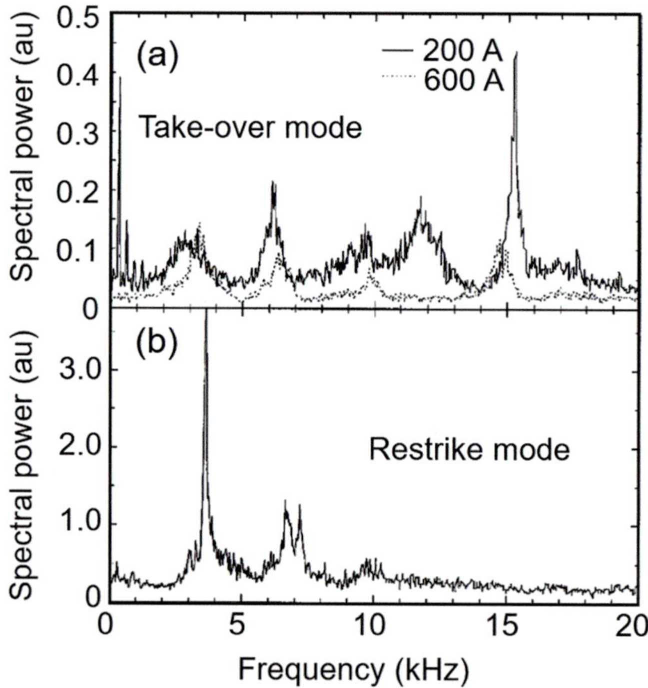

Fig. 8.11 Power spectra of the voltage fluctuations for an F4 plasma torch using new electrodes and a swirl gas injector (a) take-over mode with arc currents of 200 and 600 A, 40 slm Ar, (b) restrike mode with arc current of 400 A, plasma gas (40 slm Ar +4 slm H_2). [Dorier et al. (2001)]

Fig. 8.12b, shows significant differences for these two types of injection modes of the plasma forming gas.

By assigning a "mode value" of "zero" to the *steady* mode, of "one" to the *take-over* mode, and of "two" to the *restrike* mode, and using an algorithm based on the ratio of

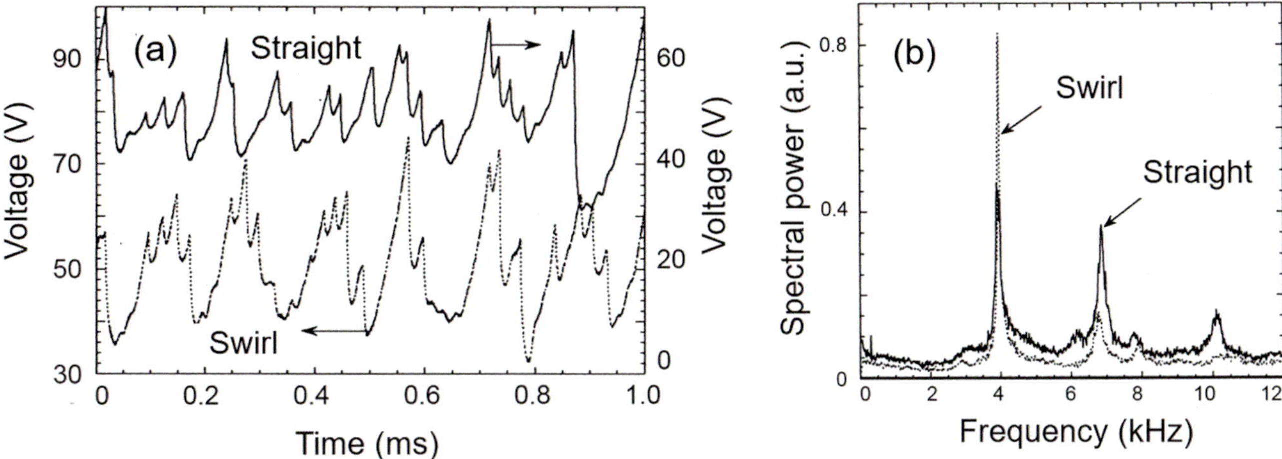

Fig. 8.12 (**a**) typical voltage traces and (**b**) corresponding power spectra for an F4 torch operating with N_2-H_2 mixture with vortex and straight injection of the plasma forming gas. [Dorier et al. (2000)]

the rate of voltage increase vs. the rate of voltage decrease, and of the absolute value of the voltage fluctuation, any arc appearance between the extremes can be quantified by assigning a specific mode value between zero and two to the voltage trace [Duan and Heberlein (2000)]. The mode value for distinguishing the *restrike* mode from the *take-over* mode is characterized by the shape factor "S," defined as:

$$S = Time\ of\ voltage\ increase/Time\ of\ voltage\ decrease,$$

while the mode value for distinguishing the *take-over* mode from the *steady* mode is based on the amplitude factor "A" of the voltage trace:

$$A = (\Delta V/V)\ 100\%$$

$A \geq 10\%$ and S 5	the arc is in a *restrike* mode with a mode value of 2
$A \geq 10\%$ and $S < 1.1$	the arc is in a *take-over* mode with a mode value of 1
$A \geq 10\%$ and $S < 5$	the arc is in a *restrike – take-over* mixed mode, and the mode value is determined as [1 + (S-1.1)/ 3.9]
$A < 2\%$,	the arc is in a *steady* mode with a mode value of 0
$2\% < A < 10\%$,	the arc is in a *take-over – steady* mixed mode with a mode value of [(A – 2%)/8%]

The physical significance of this mode value can be explained by the end-on picture of the arc inside the nozzle shown in Fig. 8.13a, together with an intensity scan giving the arc cross-section and the thickness of the boundary layer between the arc and the anode nozzle wall. Most of the data given in Fig. 8.13b show a clear correlation between the mode value and the thickness of the boundary layer. A

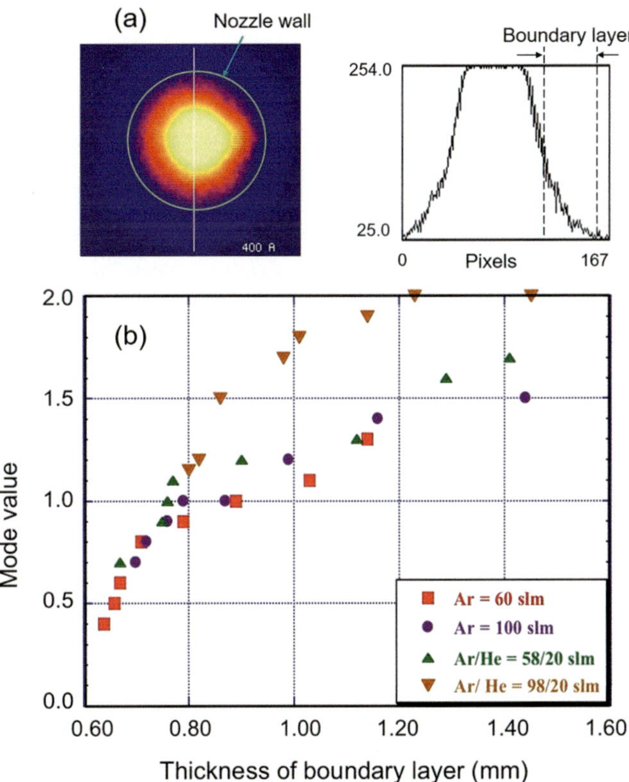

Fig. 8.13 (**a**) End-on image of the arc inside the anode nozzle and the associated intensity scan used for boundary layer thickness measurements. (**b**) The relation between torch operating mode value and boundary layer thickness. [Duan and Heberlein (2002)]. (Copyright © ASM International. Reprinted with permission)

thicker boundary layer generated by higher gas flow rates, higher hydrogen or helium concentrations in the plasma gas, or lower currents, results in a more *restrike*-like mode. In

general, a *restrike* mode does not occur with a pure argon arc, but addition of hydrogen or nitrogen beyond a certain amount will constrict the arc and increase the boundary layer thickness, favoring a *restrike-like* behavior. A study of the effect of operating parameters on the boundary layer thickness and the voltage fluctuations for different nozzle designs show that the arc current has the strongest effect on reducing the boundary layer thickness and the arc root fluctuations, but that the effect of increasing the total mass flow rate and the concentration of the secondary gas such as hydrogen can be different for different nozzle designs, and will depend on the nature of the primary gas, being argon or nitrogen [Nogues et al. (2007)]. For plasma spraying, well controlled arc motion to distribute the heat input to the anode surface is necessary to reduce anode erosion and extend its life. The preferred mode should not give rise to excessive variations of the arc voltage as experienced in the *restrike* mode since it

would result in significant fluctuations of the arc power, that is, a mode close to the *take-over* mode value of "one."

According to [Duan and Heberlein (2002)], the control over coating quality in plasma spraying is partly dependent on the arc and jet instabilities of the plasma torch. Different forms of instabilities have been observed with different effects on the coating quality. Authors report on an investigation of these instabilities based on high-speed end-on observations of the arc. The framing rate of 40,500 frames per second has allowed for the visualization of the anode attachment movement and the determination of the thickness of the cold-gas boundary layer surrounding the arc. The images have been synchronized with voltage traces. The arc instabilities can be seen to enhance the plasma jet instabilities and the cold-gas entrainment. These results are particularly useful for guiding plasma torch design and operation in minimizing the influence of plasma jet instabilities on coating properties. The investigation has been carried out with a Praxair SG-100 plasma-spray torch operating in the subsonic mode. A #2083–129 cathode and a #2083–720 anode have been used. The anode has a straight bore nozzle with an 8 mm diameter and 25 mm in length. The torch is operated with argon as the primary gas and helium as the secondary gas. The operating parameters include arc current, mass flow rate of the plasma gases, gas injection method, and anode condition. The spray parameters are summarized in Table 8.1.

Figure 8.14 [Duan and Heberlein (2002)] shows the mode value distribution for different currents and mass flow rates of

Table 8.1 Plasma torch operating conditions

Parameter	Setting
Arc current	200 to 900 (A)
Primary gas	40 to 120 slm (Ar)
Secondary gas	0 to 40 slm (He)
Gas flow	20% vortex +80% straight
Anode internal diameter	8 and 10 (mm)
Anode condition	New, used, burnt

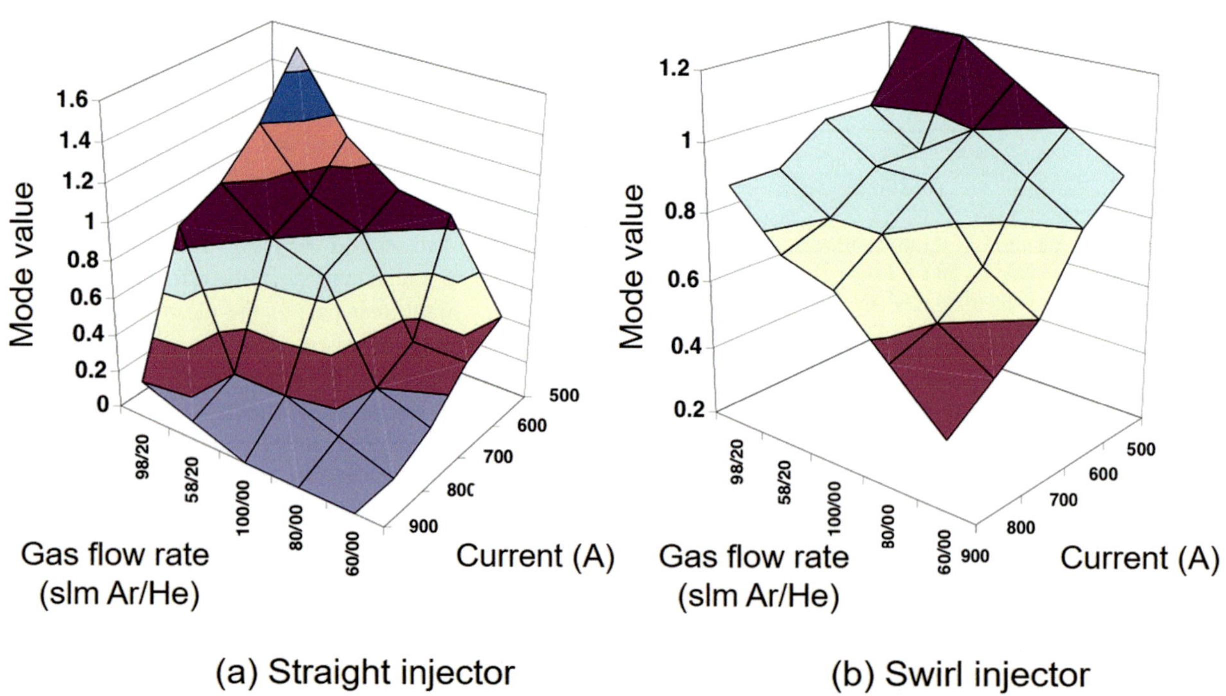

Fig. 8.14 Mode value contours for different arc currents and plasma gas flow rates for a Praxair SG 100 torch with (**a**) straight injection and (**b**) swirl injection (right). [Duan and Heberlein (2002)]. (Copyright © ASM International. Reprinted with permission)

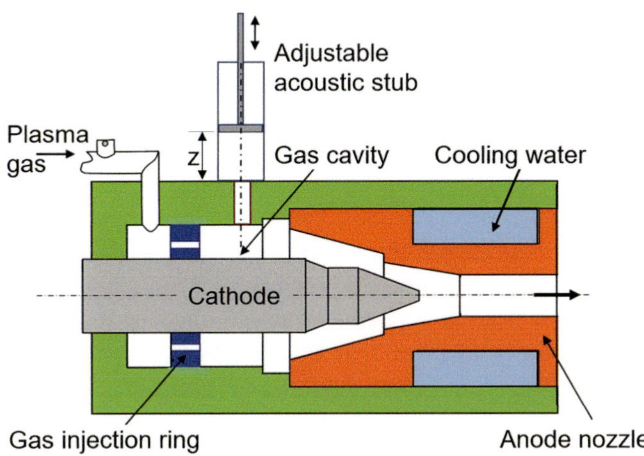

Fig. 8.15 Schematic of (**a**) the PT F4 torch with the cathode cavity volume V, (**b**) the acoustic stub fixed on the pressure probe position and acting as resonator on the cathode cavity [Rat and Coudert (2011)]. (Copyright © ASM International. Reprinted with permission)

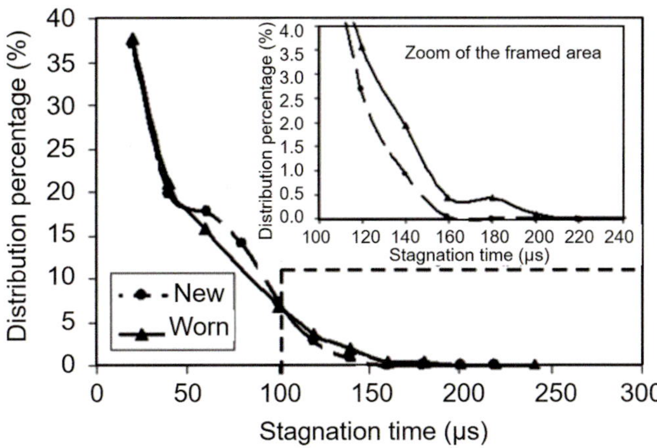

Fig. 8.17 Fraction of occurrences that an arc has a specified residence time at a given location, for "New" and "Worn" anode. The inset shows an increase in residence time above 120 µs for the used anode, indicating accelerating erosion. [Fauchais (2004)]. (Copyright IOP Publishing. Reproduced with permission. All rights reserved)

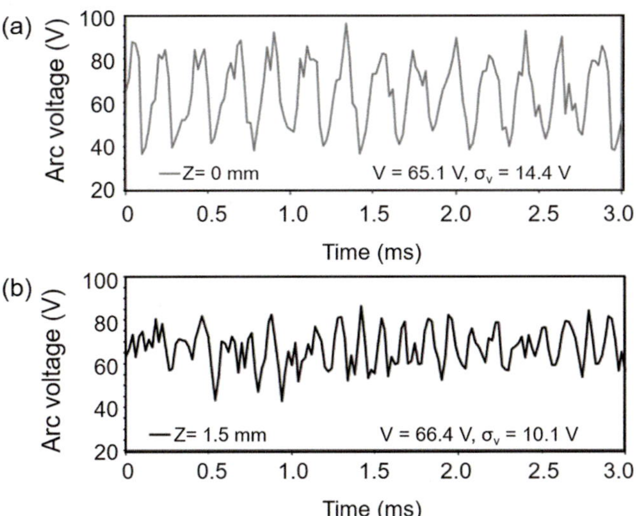

Fig. 8.16 Arc voltage fluctuations for different piston positions of the acoustic stub, Ar-H$_2$ (45–10 slpm), 500 A. (**a**) z = 0 mm, (**b**) z = 1.5 mm [Rat and Coudert (2011)]. (Copyright © ASM International. Reprinted with permission)

a Praxair SG-100 plasma torch (see Sect. 8.2.3.1 for torch description) for two different plasma gas injection methods, straight injection and tangential (swirl) injection. It is clear that the distributions for both injection methods have a plateau with a mode value between 0.8 and 1 (*take-over* mode), which indicates a stable mode of operation where small variations in the operating parameters do not change the mode value very much. Lower currents, higher mass flow rates, and higher concentrations of the auxiliary gas such as helium in the plasma gas lead to higher mode values, that is, a more *restrike-like* behavior.

[Coudert et al. (2007) and Rat et al. (2008)] studied the arc voltage fluctuations for a commercial plasma torch, Sulzer Metco F4, with a 6 mm i.d. anode-nozzle, arc current of 600 A, working with Ar-H$_2$ mixtures as plasma gas (45 slm (Ar) + different flow rates of H$_2$). Power spectra of the arc voltage fluctuations revield main fluctuation frequencies between 4 and 5 kHz, with a secondary peak observed at 7.5 kHz for low hydrogen flow rates. Attributing the mean arc spot lifetime exclusively to the Cold Boundary Layer (CBL) is not fully justified since the fluctuation frequency increases as the hydrogen flow rate increases. Moreover, according to the voltage signal, the duration of arc restrikes is about 30–100 µs, that is, frequencies higher than 10 kHz and the evolution of the main fluctuation frequency (4–5 kHz) cannot be described in terms of CBL thickness. [Coudert and Rat (2008) and Rat and Coudert (2011)] showed that the restrike mode is superimposed on the main fluctuation of arc voltage, observed around 4–5 kHz. This main oscillation is mainly due to the compressibility effects of the plasma-forming gas in the rear part of the plasma torch, that is, in the cathode cavity represented in Fig. 8.11 [Coudert et al. 2007; Coudert and Rat 2008)].

The plasma torch behaves like a Helmholtz resonator where pressure inside the cathode cavity is coupled with the arc voltage. The authors showed that pressure inside this cavity is correlated with the arc voltage. They have correlated the intermittent behavior of the torch, and the oscillation modes exchange energy. They also showed that the Helmholtz mode can be damped by the use of an acoustic stub (see Fig. 8.15b) mounted on the plasma torch body at the pressure probe position (see Fig. 8.15a). By adjusting the position, z, of the piston (see Fig. 8.15b), the amplitude of the voltage fluctuations can be reduced. This is illustrated in Fig. 8.16 showing the change of the time dependence of the

arc voltage with a varying z, generated with an Ar-H$_2$ (45–10 slm), a 500 A arc, and a nozzle with an inner channel diameter of 6 mm. In this figure, arc voltages are raw signals and contain all contributions (modes) described previously. The position z = 0 mm corresponds to the situation (Fig. 8.16a) where the acoustic stub is inactive and z = 1.5 ± 0.5 mm to the most effective position to damp the Helmholtz mode. Of course, such studies are preliminary ones and additional work is still necessary.

The electrical stability of the arc is determined by its volt-ampere characteristic. The free-burning arc has over a range of conditions a falling characteristic, meaning that the voltage decreases with increasing current. Stable operation requires current control by the power supply, which is offered by practically all modern DC power supplies. For most plasma torches, the average arc voltage does not change strongly with changing current. However, increasing the gas flow rate will increase the arc voltage. The strongest effect has a change in plasma gas composition. A pure argon arc will have the lowest arc voltage, addition of helium or nitrogen will increase the voltage, and hydrogen in the arc will provide the highest voltage, that is, for a constant current also the highest power dissipation and strongest spray particle heating and acceleration. A variation in pressure will also change the voltage, with lower voltage at lower pressures (reduced heat losses) and higher pressures resulting in increased radiation losses and higher voltages. This holds, of course, only for the same arc current and the same arc length.

8.2.5 Electrode Erosion

Cathode erosion occurs mainly when the current density at the cathode attachment is high enough to result in larger molten volumes. This occurs when

- the attachment area is constricted, for example, by having a pointed conical cathode tip,
- there is a lack of low work-function material within the cathode,
- there is strong arc constriction due to the use of a molecular plasma gas such as hydrogen,
- the use of high currents exceeds the cathode design value.

The lack of low work-function material can occur when the material addition to the tungsten matrix evaporates faster than it can diffuse to the surface. This effect can be mitigated by the redeposition of the material when it is ionized after evaporation and driven back to the cathode surface by the cathode voltage drop [Zhou and Heberlein (1998)]. In general, cathode erosion affects torch operation during the first few hours, in particular during the first hour, resulting in a decrease in the arc voltage.

Anode erosion is caused by the strong heat fluxes of the constricted arc attachment between the arc column and the anode surface. For a constricted attachment through a cold gas boundary layer, it has been shown that an instability exists, leading to an electron temperature run-away and to localized evaporation of anode material [Yang and Heberlein (2007)]. Anodes of plasma-spray torches exhibit this type of erosion following the first few hours of operation. The erosion patterns can result in preferred arc attachment spots, leading to the establishment of preferred tracks for the arc attachment movements. A study of the erosion patterns and the associated arc voltage traces was reported. [Rigot et al. (2003)] have shown that the preferred anode attachment slightly moves upstream and becomes slightly more constricted with arc operation. The amount of erosion has been found to correlate with the residence time of the arc (time of a constant voltage value) at a specific location. Under normal conditions, this residence time varies from 30 μs to 150 μs, with the lower values occurring much more frequently. The relative distribution of the stationary arc attachment times for "New" and "Worn" anodes are given in Fig. 8.17 [Fauchais (2004)]. While for the new anode the attachment times remain smaller than 150 μs, the worn anode shows an increasing fraction of residence time at a specific attachment location exceeding this value. A thermal analysis indicates that with residence times of 150 μs, a strong increase in erosion can be expected, that is, once a slightly eroded spot has formed, the arc prefers to attach at this spot, rapidly increasing the erosion rate [Rigot et al. (2003)].

Anode erosion can also be strongly influenced by the anode design and the gas injection method. The results of a fluid dynamics study for a commercial plasma torch are given in Fig. 8.18 [Sun and Heberlein (2005)]. To observe the fluid flow in the torch, a glass torch was constructed; the arc heated gases were represented by a flow of heated helium emanating from the cathode tip, and cold argon gas was injected at the base of the cathode as in a conventional torch with three different gas injectors:

- commercial radial gas injector,
- commercial swirl gas injector with four injection holes,
- modified commercial gas injector with eight injection holes.

The flow was visualized by introducing fine particles with the gas flow. Flow simulations performed using a commercial CFD code indicated regions of flow recirculation, and the flow visualization experiments indicated powder deposits on the nozzle wall where the recirculation was predicted. Comparison with the erosion patterns of an actual plasma torch with an identical internal geometry indicated that erosion started at the

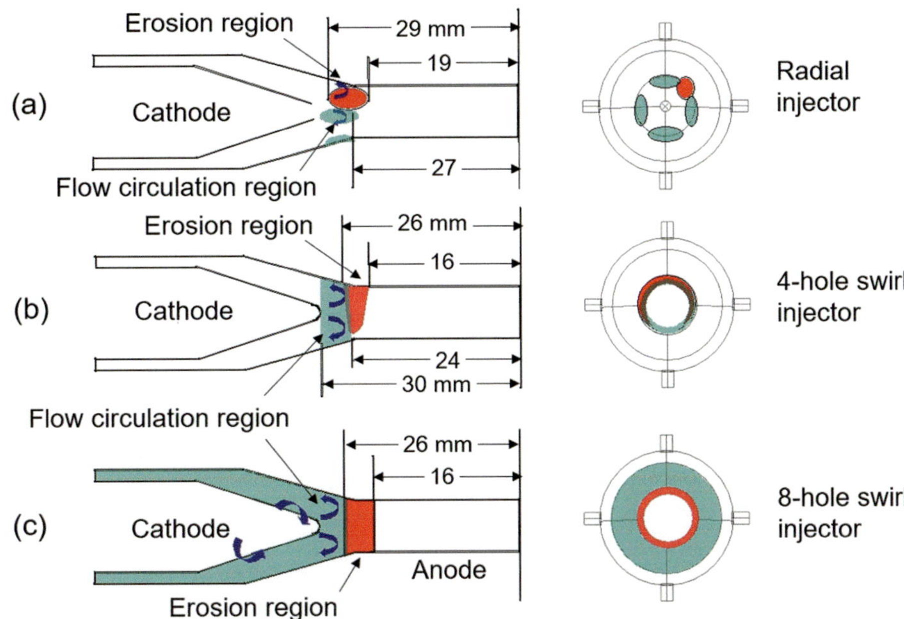

Fig. 8.18 Schematic indication of locations of the flow recirculation region, powder deposition, and observed erosion pattern for (**a**) radial gas injector, (**b**) 4-hole swirl gas injector, (**c**) 8-hole swirl gas injector. [Sun and Heberlein (2005)]. (Copyright © ASM International. Reprinted with permission)

spots where recirculation was predicted and powder deposits were observed. Clearly, radial gas injection results in rapid erosion in a region where recirculation can occur, in a circumferential location between two injection holes. The flow pattern of the four-hole swirl injector indicates that the gas is not swirling uniformly, but that stratified flow zones exist with stronger and weaker swirl velocities. Accordingly, there is circumferentially non-uniform erosion of the nozzle wall. The eight-hole swirl gas injector has the least amount of erosion, most uniformly distributed around the circumference, but predominantly in the region of recirculation [Sun and Heberlein (2005)].

Typically, plasma spray torch anodes can be used for 30 to 60 hours before they are eroded to a degree that the coating quality is strongly compromised. The number of arc-starts has also a strong impact on anode erosion due to the high currents associated with the high-frequency breakdown resulting in a pitted anode surface. Such an erosion mode can be limited by the use of argon as plasma gas for torch start-up and by limiting the start-up current to 100–150 A. A study of the long-term stability of the spray torch has shown that over the period between 10 and 40 hours of operation, the voltage and torch power decrease steadily, and the average particle temperatures proportionally [Leblanc and Moreau (2002)]. Since coating porosity is a strong function of particle temperature, maintaining a constant coating quality requires adjustment of torch power over this time period.

It is of interest to point out that while anode erosion reduces the average value of the arc voltage, it is responsible for increased voltage fluctuations. Both the absolute value of the voltage, and the amplitude and frequency of the voltage fluctuations can be used as a measure for erosion and process stability. Since the acoustic spectrum of the noise emitted by the arc is directly linked to such voltage fluctuations, recording this spectrum with a microphone can also be used as a means of detecting the erosion level of the electrodes and check if it has reached unacceptable levels [Duan et al. (2000)].

8.3   Plasma Torch Design

The basic DC plasma spray torch design has evolved considerably over the past two to three decades. The main objectives have been

- Improve plasma jet stability.
- Extend electrode life.
- Extend the range of plasma gases that could be used.
- Facilitate access to the plasma stream for injection of the spray material.
- Reduce ambient air entrainment into the plasma jet.
- Increase torch power for higher productivity.

Schematic representations of some of the basic concepts commonly used in the design of DC plasma spray torches are illustrated in Fig. 8.15. These differ mostly in the cathode design, which ranges from "stick-type" hot cathode

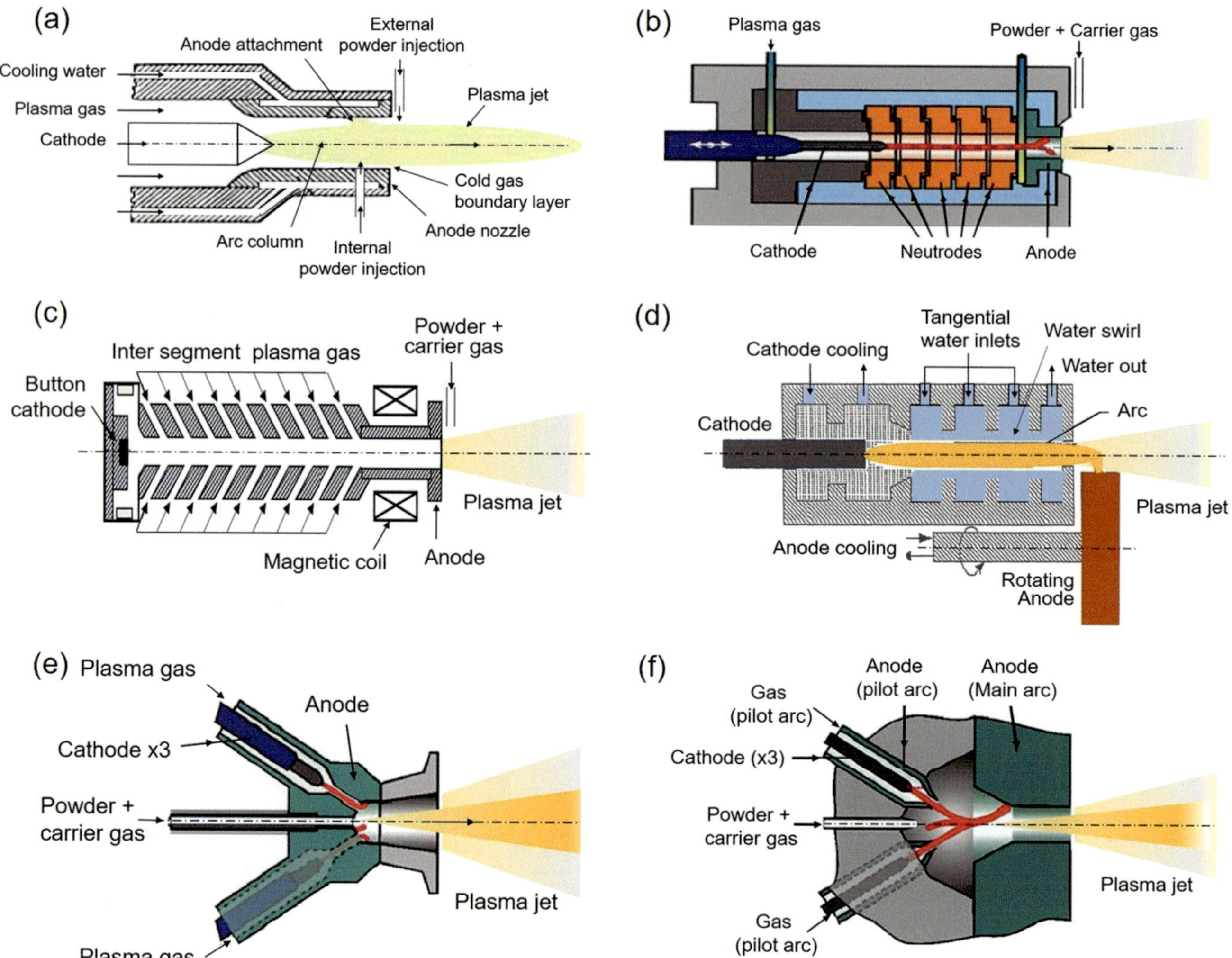

Fig. 8.19 Schematics of basic design concepts used for DC plasma spray torches (**a**) Gas-stabilized DC torch design, (**b**) Wall-stabilized torch design [Marqués et al. (2009)], (**c**) Segmented, Cascade plasma torch with inter segment gas injection [Zhukov (1979)], (**d**) Water-stabilized arc torch [Hrabovsky (1992)], (**e**) Converging plasma jet configuration [Marqués et al. (2009)], (**f**) Multi-cathode plasma torch. [Marqués et al. (2009)]

(Fig. 8.19a, b, e and f) to "button" cathode (Fig. 8.19c), consumable cathode (Fig. 8.19d), and multiple cathode arrangement (Fig. 8.19e, f). They also differ in the arc stabilization technique used, which ranges from "gas stabilized" (Fig. 8.19a, e, and f) to wall stabilized (Fig. 8.19b), segmented wall with inter-segment gas injection (Fig. 8.19c), and water-stabilized (Fig. 8.19e). In the following, a few examples the torch designs developed and used in plasma spray coating operations have been given.

8.3.1 Gas-Stabilized DC Plasma Torch

The design illustrated in Fig. 8.19arepresents the configuration used in the majority of "conventional" DC plasma spray torches in which an arc is struck between a water-cooled, "stick" type thoriated tungsten cathode, and a cylindrical,

water-cooled copper anode coaxial with the cathode. Most of the torches manufacture the anode using oxygen-free copper for its higher thermal conductivity in order to ensure effective cooling of the anode walls and reduce anode erosion. The anode nozzle is typically between 5 and 8 mm in diameter, with its length between 1 and 5x its diameter. The plasma gas is introduced into the discharge cavity through a ceramic annular diffuser ring with a tangential velocity component, in most cases, in order to induce into the flow a swirling velocity component in order to ensure a uniform distribution of the plasma gas in the annular space between the cathode and the anode. The swirling motion of the plasma gas is also necessary for maintaining the arc root in constant movement over the surface of the anode in order to distribute the thermal load of the arc root heat losses on as large an area of the anode as possible and thus reduce its erosion. Because of the use of a tungsten cathode, this type of torches could

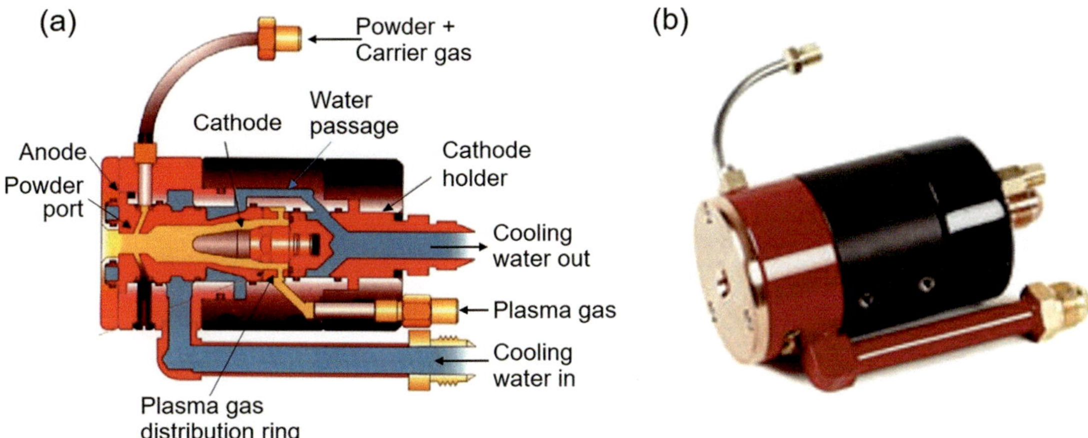

Fig. 8.20 (**a**) Schematic representation and (**b**) photograph of the SG-100, DC plasma spraying torch with internal powder injection by Praxair-TAFA cop Fig. 8.20 (a) Sectional

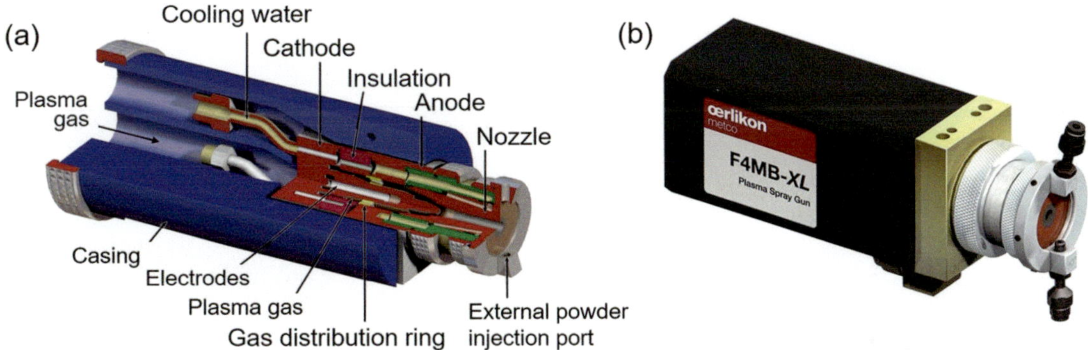

Fig. 8.21 (**a**) Sectional view and (**b**) photograph of the Oerlikon-Metco F4 MB-XL DC plasma gun with external powder injection. [Reproduced with kind permission of Oerlikon-Metco Corp]

only be used with inert or reducing plasma-forming gas. The presence of any oxygen in the plasma gas would result in the rapid oxidation of the tungsten cathode, forming a volatile tungsten oxide, leading rapidly to its destruction. Powder injection into the plasma stream is made radially, either internally through an appropriate channel into the anode wall, or externally at the exit level of the plasma jet from the anode nozzle.

Schematics and photographs of two of the leading torches used for plasma spraying are illustrated in Figs. 8.20 and 8.21, which are rated for operation at power levels up to 80 kW with anode nozzle diameters in the range of 5 to 8 mm i.d. Fig. 8.20 shows a schematic of the SG-100 by Praxair-TAFA DC gun with its high-pressure cooling water circuit (typically between 1 and 1.18 MPa) entering first on the anode side and exiting from the cathode side.

The reverse sequence is sometime used depending on pressure drop across each of these two components. In many of these designs, the cooling-water hoses also carry the electric power cables for each of the two electrodes. As shown in the figures, the plasma gas is injected separately into the discharge cavity through a ceramic gas distribution ring, which serves to control the flow pattern in the discharge. The powder to be sprayed is injected into the plasma stream using a carrier gas internally into the anode nozzle. Fig. 8.21 shows a sectional view and a photograph of the F4 MB-XL DC plasma gun by Oerlikon-Metco Corp. It features a compact, relatively light-weight design (below 3 kg without the connecting cables) that which allows for easy mounting on a robotic arm. The torch is equipped with a double-port powder-injecting ring externally attached to the anode nozzle. One of the ports can be used for 90°

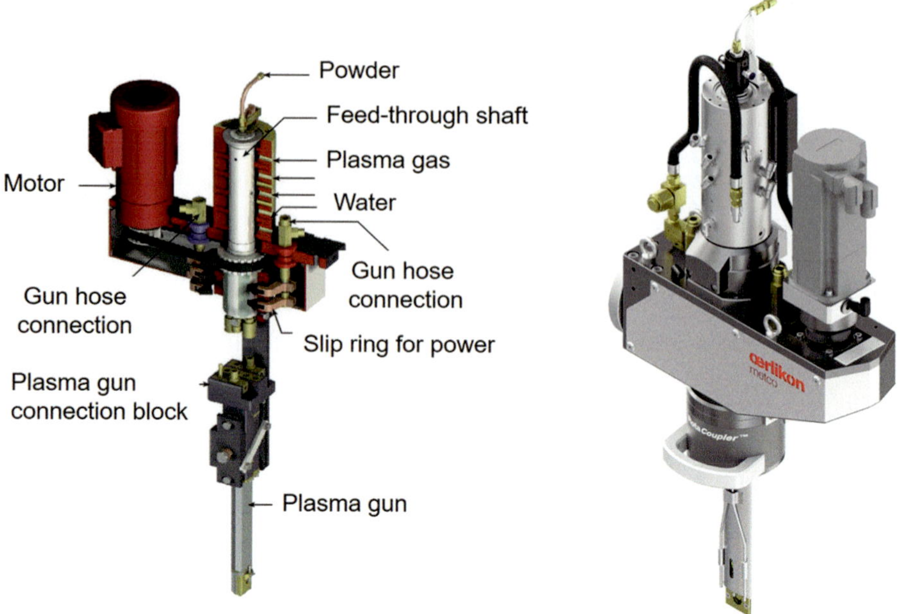

Fig. 8.22 Schematic of Oerlikon-Metco Rota-plasma torch for coating inside surfaces of cylinders. (Reproduced with kind permission of Oerlikon-Metco AG, Switzerland)

injection of the powder into the plasma stream, while the second port allows for a slight upstream inclination of the powder injection port.

With increasing demand for the use of plasma spraying for the coating of the interior surfaces of cylinder bores for aluminum cylinder blocks in the automobile industry, Sulzer-Metco developed a compact design of a DC plasma torch, commercialized under the name "Rota-Plasma," which is capable of rotation on its axis [Barbezat (2001)]. A schematic representation and photograph of an updated version of this torch is given in Fig. 8.22 (courtesy of Oerlikon-Metco Corp). The most notable of these developments is the compact torch design with the axis of the spray nozzle perpendicular to the axis of rotation of the torch body combined with its ability to operate under a harsh environment inside the cylinder bore during the coating process. The gas and cooling-water connections are made through a rotating feed-through shaft. The torch can be used for the coating of cylinder bores with diameters of as small as 60 mm, rotating at speeds up to 800 rpm, and power levels of 16 kW. Modular plasma-extension guns have also been developed for operating at higher powers, up to 90 kW (for the iPro-90 model), capable of coating internal bores as small as 150 mm and 600 mm deep.

8.3.2 Wall-Stabilized DC Plasma Torch

One of the problems with the arc-root fluctuations in conventional plasma spray torch presented in Fig. 8.15a are the development of large arc voltage fluctuations, which can be of the same order of magnitude as the mean arc voltage (typically a few tens of volts). These would result in a strong variation of the power dissipated in the torch and serious degradation of its powder melting capabilities. One way of overcoming such a difficulty is by significantly increasing the mean arc voltage through an increase in arc length and by improving its stability in the discharge cavity. The principle of wall stabilization of arcs has been known for more than 60 years; it was introduced in connection with arc lamps. The basic concept behind it is to confine the arc within a water-cooled circular channel in which the arc is to assume a rotationally symmetric, coaxial position. In the event of an accidental displacement of the arc toward the wall of the channel, the increase in heat conduction to the wall will result in the local cooling of the plasma in this location and, consequently a drop in its electrical conductivity, which will force the arc to return to its coaxial equilibrium position at the center of the channel. In this situation, increased thermal conduction to the wall of the arc confinement tube and the associated secondary effects represent the stabilizing mechanism.

Plasma torch designs providing stable arc operation by fixing the anode attachment location have long been used for plasma arc property studies [Maecker (1955), Zhukov (1994)]. Two methods can be used to elongate the arc: either by using a nozzle with a step diameter change in it or by having the channel consisting of a number of water-cooled segments, electrically insulated from each other and from the anode, called a segmented or cascaded arc. One problem with these designs is an enhancement of the anode erosion that

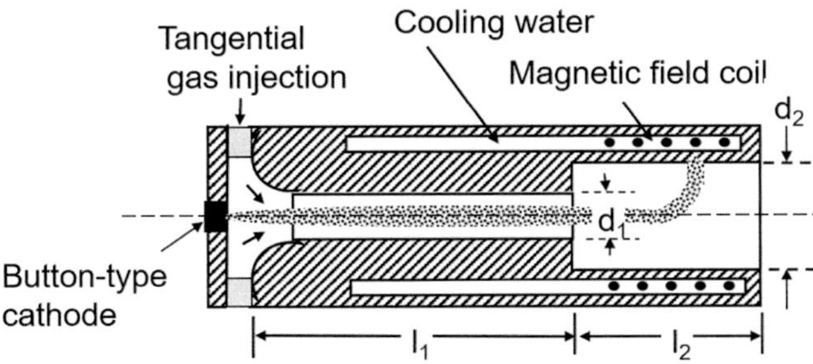

Fig. 8.23 Fixed anode attachment with a button-type cathode, injection of the plasma-forming gas as a vortex and an anode with a step. [Zhukov (1989a, b)]. (Reproduced with permission of VSP)

$$\int_0^{\delta_z} E \, dz \le (V_c + V_a) \tag{8.9}$$

occurs with one fixed anode attachment. Limiting it requires reducing the arc current, but the higher arc voltage compensates for the lower power levels (about 3 to 6 times that of conventional spray torches.

The anode-nozzle with a step change in diameter is represented schematically in Fig. 8.23 [Zhukov (1989a, b)]. The length l_1 of the smaller diameter (d_1) part has to be shorter than the length l_{ca} that the arc would assume without the step change in the anode diameter d_2. Under such conditions due to the recirculation, which occurs downstream of the step change, the arc strikes close to this step change. To achieve a long free attachment in the nozzle of diameter d_1, a strong vortex must be used with a high swirl, which can be achieved when the ratio d_v/d_1 is high, d_v being the diameter of the gas injection tangentially to the arc chamber. This high vortex requires using a button type cathode. For low arc currents, the arc characteristic is first slightly decreasing and then slowly increasing [Zhukov (1994)].

In order to cope with the extremely high wall heat fluxes experienced with high-intensity arcs enclosed in small-diameter channels, Maecker introduced the concept of using a stack of electrically insulated, water-cooled disks (usually made of Cu), often referred to as "neutrodes," as shown in Fig. 8.24, leading to what is commonly known as the wall-stabilized, cascaded arc, which has been extensively used as a basic research tool. The same principle found widespread application in the design of high-power arc gas heaters with considerably higher arc voltages. Due to the much higher electrical conductivity of metals compared with that of the arc column, segmentation of the channel enclosing the arc is necessary because a continuous metal tube would cause a double arc through the metal wall (arcing from the cathode to the metal tube and from the metal tube to the anode) seeking the path of least resistance. If E is the field strength in the arc column and δ_x the thickness (in field direction) of an individual segment, the following condition has to be met for avoiding double arcing:

The minimum voltage required for establishing and maintaining an arc is the sum of the cathode fall, V_c, and the anode fall, V_a. A reduction in the inside diameter of the segments (arc constriction) leads to a marked increase of the field strength, keeping the other arc parameters the same, and accordingly to higher energy losses by heat conduction. In this situation, the maximum thickness of a segment may be limited to $\delta_z = 1.2$ to 2.0 mm. It is essential for retaining the wall-stabilizing effect that the diameter of the channel containing the arc be smaller than the diameter of a free-burning arc operated under the same conditions. If the channel diameter is too large, the stabilizing effect due to heat conduction is lost and, at the same time, free convection effects may cause serious distortions to the arc column.

With these torches, a much higher voltage between the cathode and the anode is obtained. Limiting its axial movement to the anode length stabilizes the long arc. If at the anode, voltage fluctuations ΔV can reach ± 40 to 50 V with molecular gases and especially hydrogen, as the mean arc voltage V_m can be higher than 150 V, effects of fluctuations are significantly reduced. As the erosion of electrodes depends essentially on the arc current and fairly little on the voltage, power levels 3 to 4 times higher than with conventional torches without compromising anode erosion and electrode life.

8.3.3 Laminar Plasma Torch

Numerous studies, especially in China, have been devoted to the development of laminar plasma jets because of their inherent stability and long contact time between the particle and the plasma flow, which could represent a significant improvement in system controllability and material-processing technology [Pan et al. (2001), Liu et al. (2018), Liu et al. (2019)]. The approach used has mostly been centered on improving arc stability through modification of the internal geometry of the arc confinement channel and operating conditions. A schematic of the DC plasma torch used by [Pan et al. (2001)] is given in Fig. 8.25. This features the introduction of an interelectrode insert between the cathode and anode of the torch which serves mainly to confine the arc root movement of the surface of the anode to a limited circular range. The anode nozzle diameter was varied over the range of 4 to 10 mm, separate main gas inlets were incorporated in the design, allowing for the introduction of the gas in the torch either tangentially to the wall of the discharge cavity or in a linear flow mode in the axial direction

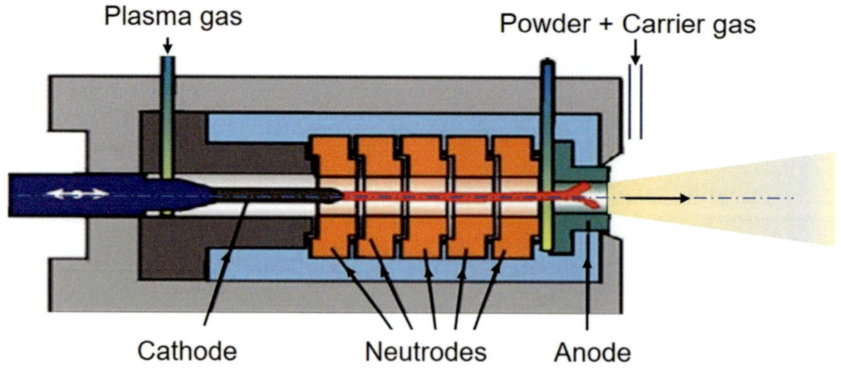

Fig. 8.24 Schematic of a plasma torch with neutrodes: Advanced Plasma Gun (APG). [Marqués et al. (2009)]

of the torch. Provision was also made for the tangential introduction an auxiliary gas stream downstream of the cathode region for a closer control of the flow pattern and the composition distribution in the gases while operating with mixed plasma gases. Results are given operating with a plasma gas composed of [4.8 slm (N_2) +8.4 slm (Ar)] with an 8 mm i.d. anode nozzle at an arc current of 160 A. A photograph of the plasma jet obtained, given in Fig. 8.22, shows a long, stable plasma jet with a length-to-diameter ratio of 45. The effect of the total gas flow rate and the mode of injection of the gas into the discharge cavity were investigated for a pure argon plasma generated by the same torch with an anode diameter of 8 mm, operated at essentially the same conditions as those used earlier.

Photographs given in Fig. 8.27a show a rather turbulent jet generated with a plasma gas flow rate of 27 slm (Ar) and an arc current of 160 A. Fig. 8.27b and c, obtained with a lower gas flow rate of 13.2 slm (Ar), shows, on the other hand, a considerably longer and more stable plasma jet due to the lower gas flow rate for the same arc current. The authors point out that the apparent difference in arc length of the two jets was caused by the different inlet modes of introduction of the plasma working gas. The gas was fed along the torch axis for the jet shown in Fig. 8.27b and tangentially with the channel wall for the jet shown in Fig. 8.27c. The results indicate that a vortex motion of the gas in the discharge cavity would be favorable for lengthening the plasma jet compared to the axial injection mode. The authors attributed the effect to arc stabilization by the vortex motion in the discharge cavity. Throughout this study, the authors point out to a direct correlation between the operation of the plasma torch in the laminar mode with a stable plasma jet and the significant reduction in the noise level associated with running the plasma torch.

A wall-stabilized design of a DC plasma torch was reported by [Liu et al. (2019)] for the generation of silent, stable, and super-long laminar plasma jets in atmospheric air. A schematic of the DC, non-transferred laminar plasma torch (ZH-30, Zhenhuo Plasma Technology Co. Ltd., Chengdu,

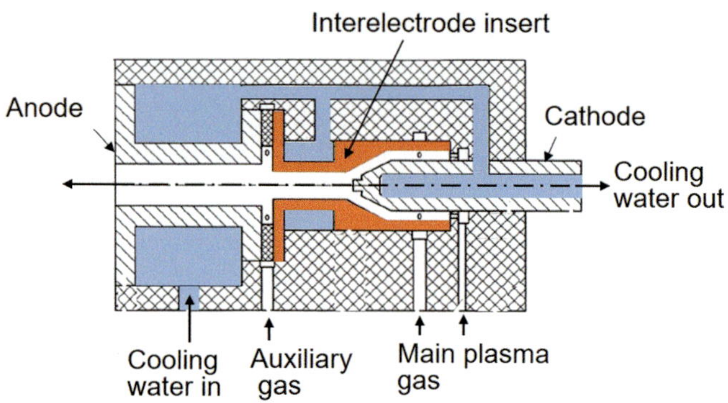

Fig. 8.25 Schematic of the modified DC plasma torch used for operation in the laminar plasma mode. [Pan et al. (2001)]

Fig. 8.26 Argon/Nitrogen laminar plasma jet generated using 4.8 slm (N_2) +8.4 slm (Ar) as plasma gas and arc current of 160 A. [Pan et al. (2001)]

China) used in this study is given in Fig. 8.28. Basically, it is comparable with the design proposed earlier by [Marqués et al. (2009)], see Fig. 8.24, composed of a cathode, insulated inserts (neutrodes), an anode, and a nozzle outlet. The diameter of the nozzle exit is 5 mm, with the neutrodes assembled using intersegment insulating rings to extend and constrict the arc column in a cylindrical channel. The length of the insulated inserts can be changed depending on the desired applications. The trumpet-like structure is innovative for laminar jets and beneficial for their stable generation.

The results given in Fig. 8.29 were obtained at atmospheric pressure, with a mixture of Ar/N_2 as plasma gas (30 vol.%

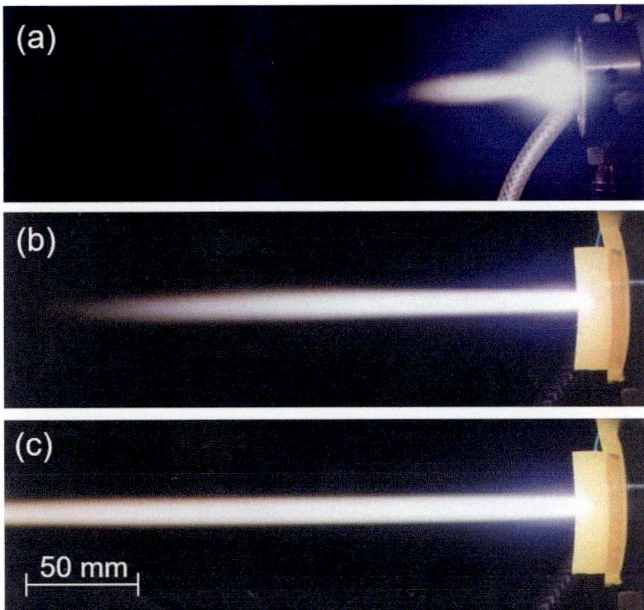

Fig. 8.27 Argon plasma jets generated with an arc current of 160 A, and gas flow rate of (**a**) 27 slm, (**b**) 13.2 slm, axial gas injection, (**c**) 13.2 slm, tangential gas injection. [Pan et al. (2001)]

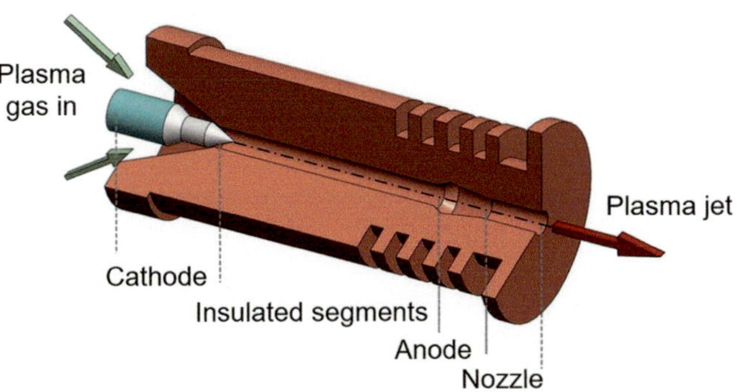

Fig. 8.28 Schematic diagram of internal structure of the laminar plasma torch. [Liu et al. (2019)]

Ar + 70 vol.% N_2) at gas flow rates of 8.5 to 15.00 slm, arc currents varying from 60 to 165 A, and torch power of 8.5 to 28 kW. The length of the corresponding laminar plasma jets varied from 100 to 720 mm with an ambient generated noise level of <80 *dB*. When the inlet gas rate was outside this range, lower than 8.5 slm or higher than 15 slm, the plasma jets become uniformly shorter (about 80 mm) and extremely noisy (about 110 dB) because of intense entrainment of ambient air, just as in the case of conventional direct-current non-transferred arc plasma jets. The photographs given in Fig. 8.29a, obtained for a gas flow rate of 8.5 slm and arc currents varying between 60 and 165 A, show a steady s of the

length of the plasma jet with increase in the arc current. The arc stability, however, seems to be very sensitive to the plasma gas flow rate since, as shown in Fig. 8.29b, the increase in the plasma gas flow rate from 8.5 to 15 slm results in a significant decrease in the length of the plasma jet for the same current rating. Observations of temporal evolution of the plasma jet appearance and the voltage demonstrated, however, that the jet is highly stable in the atmospheric environment over the range investigated.

8.3.4   Cascade DC Plasma Torch with Intersegment Gas Injection

As demonstrated by [Zhukov (1994)] and his coworkers, the need for additional gas injection arises from the behavior of long arcs stabilized by a vortex flow as shown schematically in Figs. 8.19c and 8.30. In these conditions, the length of the segments surpasses the limit at which double arcing would occur. The cold gas injection, which creates a new cold boundary layer simultaneously, reduces the heat losses to the wall. This principle has been successfully used in high-enthalpy, high-power torches for re-entry simulation. The arc voltage in these conditions is much higher than those obtained with conventional plasma spray torches.

The fraction of the total mass flow rate of the sheath gas, which is injected in the ith segment, is defined as $x_i = (\rho_i v_i / \rho_o v_o)$, whose value can vary between 0.1 and 1. The index "o" corresponds to the mean gas conditions in the torch. v_i is the gas injection velocity, which can be in the range of 0.5–0.7 the velocity of sound in the cold plasma forming. [Zhukov (1994)] proposed a semi-empirical relationship between the mean ratio, $\overline{R_z}$, of the segment length $\overline{\delta_z}$ to its diameter d_c by

$$\overline{R_z} = \frac{\overline{\delta_z}}{d_c} = \frac{1.3 \times R_e^{0.25}}{\left(1 + 1.85 \times 10^{-3}\left(I / \left(d_c \times \sqrt{\mu_0 h_0 \sigma_0}\right)\right)\right)} \quad (8.10)$$

With, I, μ_o, h_o, and σ_o being respectively the current (A), the viscosity (kg/m. s), the specific enthalpy (J/kg), and the electrical conductivity (A/V. m) of the plasma at the temperature T_o corresponding to 1 vol.% of electrons. The Reynolds number, Re, is calculated using the following expression:

$$Re = \frac{\dot{m}_g}{\mu_o d_o} \quad (8.11)$$

Where, $\dot{m}_g$ is the total gas flow rate (kg/s) and d_o the internal diameter (i.d.) of the cylindrical nozzle, and μ_0 is the viscosity of the plasma at the temperature T_o corresponding to 1%

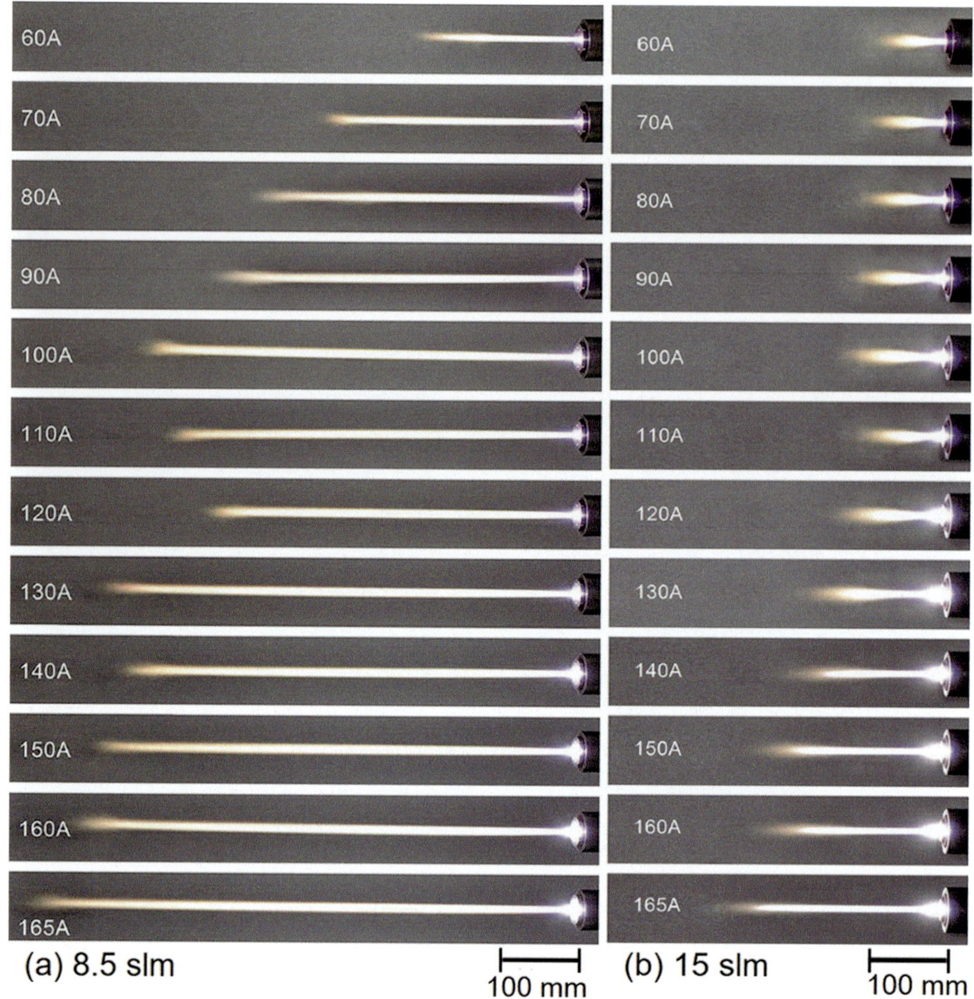

Fig. 8.29 Variation in the length of an atmospheric pressure laminar plasma jet (Ar 30 vol% + 70 vol.% N_2) for different plasma gas flow rates and arc currents (**a**) 8.5 slm, (**b**) 15 slm. [Liu et al. (2019)]

of electrons. For (I/d_c) values between $1-10^4$ A/m and $Re = 10^4$ to 10^5, Eq. 8.10 predicts values of ($\overline{R_z}.Re^{-0.25}$) between 1.3 and 0.8.

High electric fields and correspondingly high voltages (up to 6–8 kV) are feasible with such torches, which are commercially available, for example [Acurex, Aero Therm Mountain View, CA 94043, SKF, Santen et al. (1986) and Thörnblom (1989), and Aerospatiale, Van den Brook et al. (1987)]. They have also been extensively developed in the former USSR [Pustogarov (1994), Zhukov (1989a, b)]. This principle has been successfully used in high-enthalpy, high-power plasma torches for re-entry simulation, with power ratings in the range from 50 kW to 5 MW. The EDP-119 Plasma torch (referred to in the Russian literature as Plasmatron), illustrated in Fig. 8.31, is an example rated for 500 kW [Zhukov and Zasypkin (2007)]. A scaled-down version of this design to the 100 kW power level was recently adapted by [Belashchenko (2010)] for plasma spray coating applications.

8.3.5 Water-Stabilized DC Plasma Torch

Water was used as a stabilizing medium for an electric arc in the early 1920s [Gerdien and Lotz (1922)], mainly to achieve higher plasma temperatures. The high specific heat of water, steam, and hydrogen/oxygen was used in this case to constrict the arc, increase its voltage and local energy generation, and consequently increase the arc temperature. These basic phenomena were used at the Institute for Plasma Physics of the Czech Academy of Sciences [Hrabovsky (1992), Chraska and Hrabovsky (1992)] for the development of the water-stabilized plasma torch, as schematically illustrated in Fig. 8.19d. The first version of the torch was based on the use of a consumable carbon electrode as cathode and a water-cooled external, rotating copper disc as anode. The electric arc crossing the plasma torch was confined by a narrow circular channel created by a high-velocity vortex flow of water in the discharge cavity. The water is introduced into the arcing chamber with a strong swirl component in several sections,

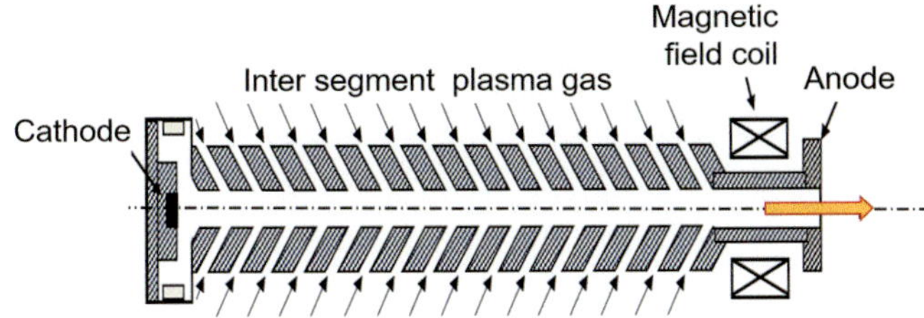

Fig. 8.30 Principle of an arc constrictor with insulated segments and intersegment gas injection. [Zhukov (1979)]

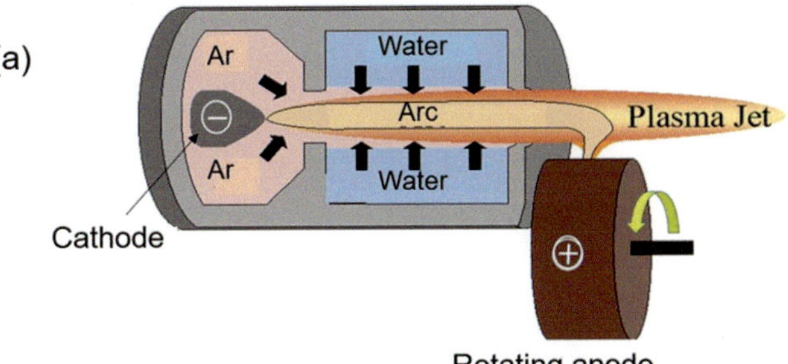

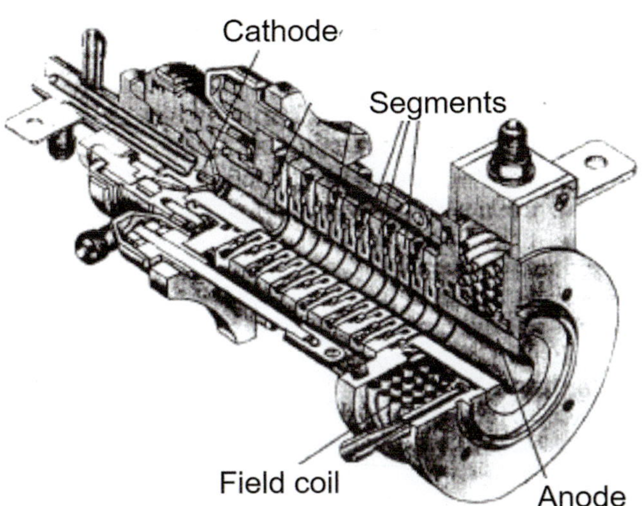

Fig. 8.31 Schematic of the EDP-119 plasma torch. [Zhukov (1977)]

Fig. 8.32 (**a**) Schematic and (**b**) photograph of a water-stabilized arc plasma torch with an external rotating anode. (With kind permission of Milan Hrabovsky and the Institute of Plasma Physics of the Czech Academy of Sciences)

and an outlet right before the torch nozzle allows a return of the unused water. The arc is established through a contact electrode and transferred from the cathode to the external anode. The plasma gas is provided by the evaporation of the water that is heated by the electric arc before exiting the torch through a nozzle, as a high-velocity plasma jet. The anode is rotating at a high speed to distribute the high heat load received from the plasma and the anode-root attachment.

As the torch design evolved, a water-cooled cathode with an argon gas shield was used to replace the consumable graphite electrode, which was limiting the ease at which the torch could be integrated in normal plasma-spraying operations. Environment in the upstream part of the torch (hybrid torch) providing significantly increased operation times. A schematic of the torch using the shrouded cathode, in combination with the rotating water-cooled anode, is given in Fig.8.32a; a photograph of the torch, courtesy of Dr. Milan Hrabovsky, is given in Fig. 8.32b. The strong cooling of the arc results in very high arc voltages (260 to 300 V) and high torch powers (80 to 200 kW) for moderate currents (300 to 600 A). Temperatures at the nozzle exit can reach 28,000 K, and peak plasma velocities of 7000 m/s have been reported [Hrabovsky et al. (1995)]. A comparison between the performance of water-stabilized torches and that of gas-stabilized

torches in terms of power per unit mass of plasma medium is given in Fig. 8.33. As can be noted, for an equivalent power rating, water-stabilized torches operate at a considerably lower mass flow rate of the plasma medium (water + Ar as shield gas) than gas-stabilized torches (Ar/H_2 or Ar/N_2). Only a small fraction of the mass is used, with the torch achieving the high power and enthalpy levels in the plasma jet. These values allow for spray rates in the range of 25 to 45 kg/h for ceramics and 80 to 100 kg/h for metals, which are an order of magnitude higher than that for regular plasma torches operating at 40 to 50 kW, and still significantly higher than that of high-power plasma torches operating at 200 kW with a nitrogen-hydrogen mixture. The high deposition rates make this torch particularly useful for the manufacturing of free-standing shapes or for coating large areas. However, the external rotating anode makes torch movement somewhat more difficult.

8.3.6 Multi-Electrode DC Plasma Torch

As was discussed earlier in Chap. 5, "Plasma and particle dynamics in thermal spray," special attention has been devoted to powder injection in plasma streams in order to optimize their trajectories and extend their residence time in the plasma stream to achieve the best thermal processing conditions for the precursor material. When using

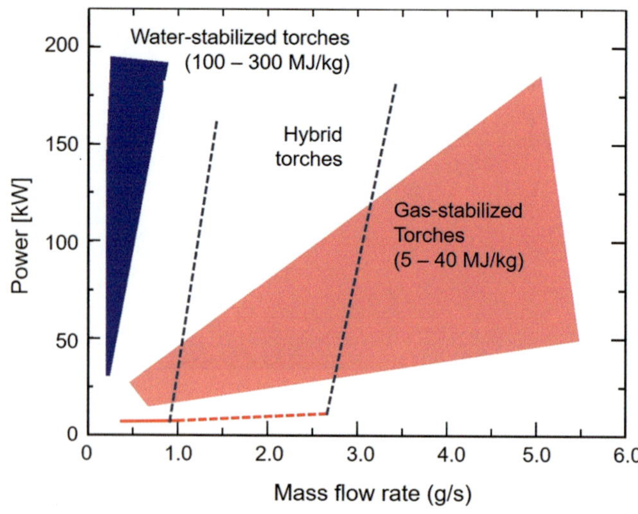

Fig. 8.33 Illustration of torch power variation with plasma mass flow rate for water swirl stabilized and for gas-stabilized torches. (With kind permission of Milan Hrabovsky)

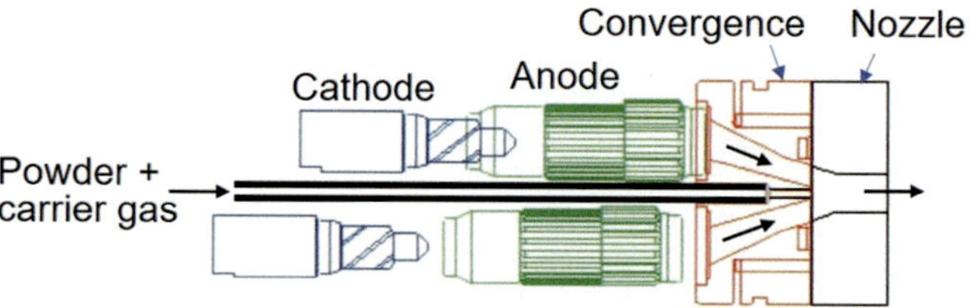

Fig. 8.34 Schematic of Northwest Mettech Axial III central injection torch. (Courtesy of Northwest Mettech)

conventional DC plasma sources for spraying operation, considerable challenges are met due to the high energy density, high velocity, and relatively small diameter of DC plasma jets, which are dependent on the torch nozzle diameter, typically in the range of 4 to 8 mm i.d. Particle injection from one side of the plasma flow also leads to the classification of particle trajectories according to their initial momentum. Particles with higher mass and injection velocity can penetrate the plasma jet and traverse it, having the majority of their trajectories in the low temperature fringes of the jet on the opposite side of their injection location. In contrast, particles with too low a momentum (lower mass and/or lower velocity) will not be able to penetrate the plasma jet and will remain in the fringes on the injection side. Neither of these groups of particles will have the required temperature and velocity for the formation of a dense coating with good adhesion to the substrate.

The use of multiple plasma sources converging to a single point has long been considered a viable option for increasing the volume of the discharge, allowing for an easier access for central injection of the precursor into the plasma. Specific design options were proposed by [Fukanuma (1988), Marantz and Herman (1991, 1992), Muehlberger et al. (1994), Chen et al. (2000) and Marqués et al. (2009)] for the use of a multiple converging plasma torch configuration or multiple cathodes, with a single-anode configuration, with the arc transferred to either an internal or an external common anode, as illustrated respectively in Fig. 8.19e and f. These would allow for the stabilization of the arc length, a reduction in the fluctuation of the arc voltage and torch power, and an increase in plasma volume. As pointed out by [Marqués et al. (2009)], one of the inherent problems in all of these systems is the need to ensure that the reactant stream is uniformly

contacted and distributed in the plasma stream formed from the converging discrete plasma jets. Nozzle blocking or "spitting," defined as the periodic burst of released precursor material built-up on the system walls, or exit nozzle, is another common problem that has been observed. None of these systems finally had a significant technological or commercial success.

The commercial torch Axial III was developed at the University of British Colombia by Doug Ross in 1990 and exclusively licensed to Northwest Mettech (Canada) [Moreau et al. (1995), Burgess (2002)]. The basic design of the Axial III torch, as illustrated in Fig. 8.34, consists essentially of three cathode-three anode spray torches, arranged such that their axes are parallel, surrounding the powder feed tube placed along the central axis of the torch, followed by a convergent section. The powder is injected with an axial trajectory at the point where the three plasma streams converge. As the powder is entrained by the plasma gas, the individual particles are accelerated and heated to their melting temperature before exiting the torch through the torch nozzle (8–25 mm i.d.). A sectional 3-D view of the internal torch construction and a photograph are given in Fig. 8.35 (courtesy of Northwest Mettech). The torch operates using either argon or nitrogen as the primary plasma forming gas to which hydrogen or helium can be added depending on the required spraying conditions. The maximum gas flow rates are factory set at 400 slm (Ar), 200 slm (N_2), 100 slm (H_2), and 140 slm (He). The individual arc current can be set at 3×30–250 A, with electrode voltages 60–200 V, for a maximum power rating between 50 and 150 kW. Cooling-water requirement is 50 l/min at a pressure of 1.4 MPa (200 psig). The principal advantage of the Axial III torch is its high powder-feed rate, up to 120 g/min (7.2 kg/h) and high deposition efficiency (80%). The torch has been successfully demonstrated for use with powders, liquids, or suspensions as feed material with coating quality less sensitive to the powder material, that is, size distribution and morphology, compared to standard DC plasma spray torches. However, the problem of plasma/arc fluctuations remains.

A development by the Universität der Bundeswehr, Munich, and Sulzer Metco, has succeeded in solving the problem of plasma arc instability and increased anode erosion by dividing the arc current into three separate arcs with three

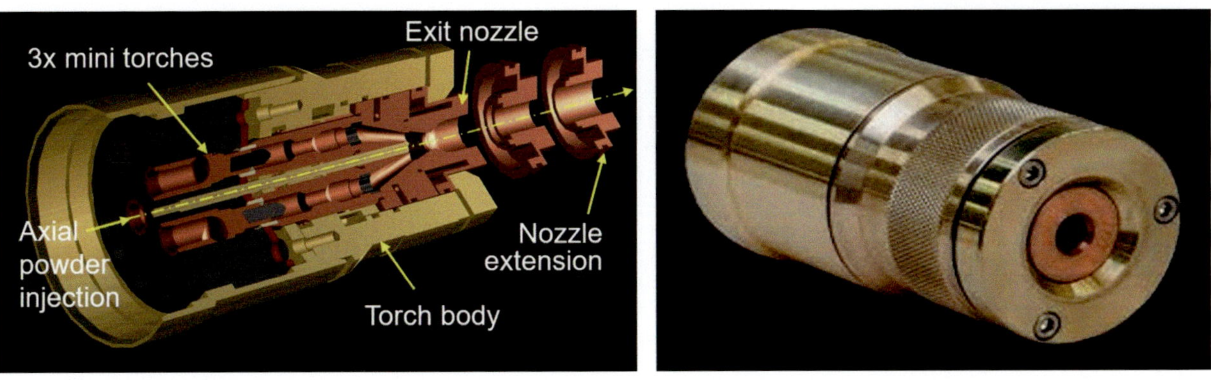

Fig. 8.35 (**a**) Sectional view and (**b**) Photograph of the Northwest Mettech Axial III central injection torch. (Courtesy of Northwest Mettech)

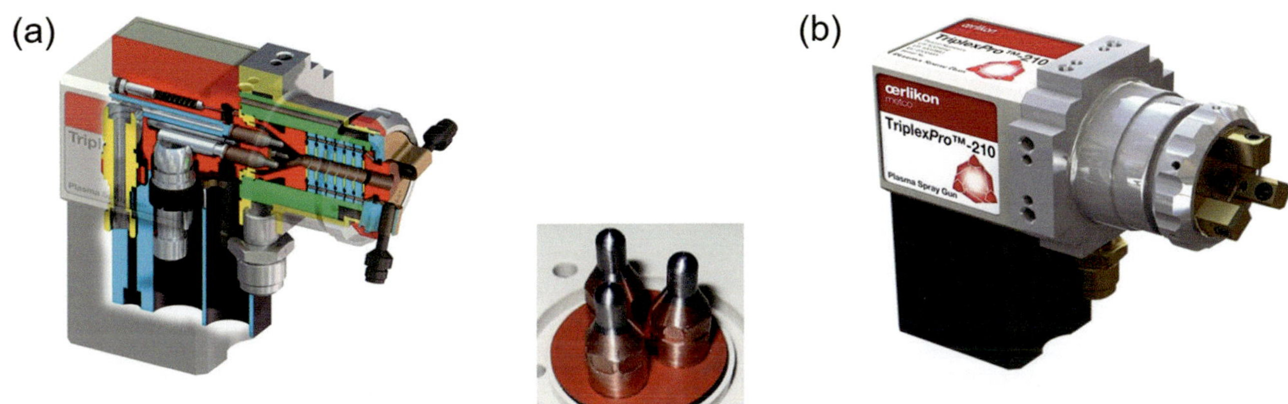

Fig. 8.36 Sectional view of Orelikon-Metco Triplex Torch. (Reproduced with kind permission of Orelikon-Metco AG, Switzerland)

cathodes and three separate anode attachments ("Triplex" torch, see Fig. 8.36, courtesy of Orelikon-Metco) [Zierhut et al. (1998), Barbezat and Landes (2000a, b)]. As shown in Figs. 8.36a and 8.37a, the arc length is extended through the use of several insulated segments, which constitute the wall of the arc channel, giving rise to a higher operating voltage of the torch (around 100 V). While the individual arc root fluctuations at the anode are the same as those for a conventional DC plasma spray torch with a stick-type cathode (for example, about $\Delta V = \pm 10 - 15$ V), their relative value to the mean arc voltage for Ar-He plasma is considerably lower for the Triplex torch compared to a conventional torch, Fig. 8.38. Considering that for a conventional torch operating with Ar/He mixture as plasma gas, the mean arc voltage V_m can be around 35 V and the ratio $\Delta V/V_m$ would be at 15/35 (43%). In contrast, the corresponding value of the $\Delta V/V_m$ ratio for the Triplex torch would be 15/100 (15%).

This torch also offers a very stable operation with argon or argon-helium mixtures, with a significant increase in the powder feed rate. However, powder injection is external to the plasma torch immediately downstream of the anode, using single or multiple three injectors adjusted 120° around the periphery of the exit nozzle for the torch, as shown in Fig. 8.36b. The improved arc stability, lower turbulence in the jet, and the use of three injection systems allow for higher deposition rates in addition to better process control and longer operating time. The anode-nozzle i.d. is between 6 and 9 mm. When observing the plasma jet for small anode-nozzle diameters, it can be observed that they consist of three lobes, as shown in Fig. 8.37b (courtesy of Oerlikon-Metco). Thus, particles can be injected either into the lobes or between the lobes, the latter injection being called the cage effect.

Emission tomography studies of the temperature field of the generated plasma jet, using three CCD cameras rotating around the jet axis, were reported by [Mauer et al. (2011a, b)]. The results allowed for the tomographic reconstruction of the gas temperature distribution and a better positioning of the powder injectors in the regions of high gas temperature and thus of high gas velocity and viscosity. In all investigated cases, the most efficient particle heating corresponded to an injection direction between two high temperature lobes, thus confirming the so-called "cage effect" [Mauer et al. (2011b)]. As can be noted in Fig. 8.37b, the lobe opposite to the injector kept in particular the larger particles properly in the jet center and prevented them from passing through the plume due to their too high momentum. As particle velocities were found to vary only moderately, the particle temperatures were more meaningful

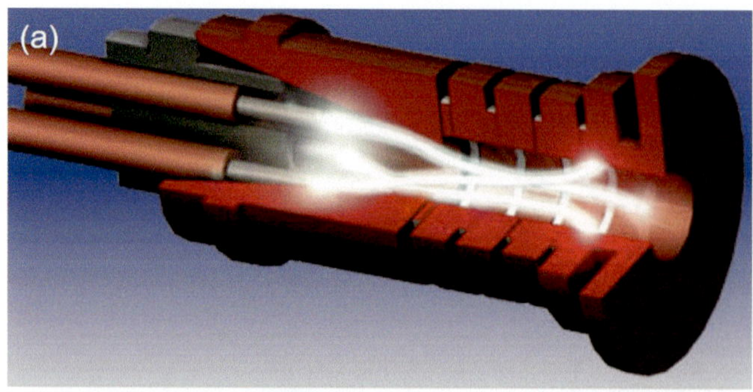

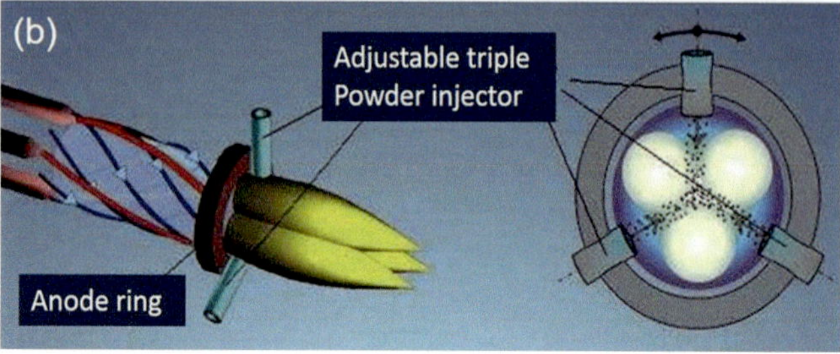

Fig. 8.37 (**a**) Sectional view, (**b**) 3-D representation of the arc lobs inside the of Sulzer Metco Triplex Torch. [Mauer et al. (2011b)]. (Reproduced with kind permission of Orelikon-Metco AG, Switzerland)

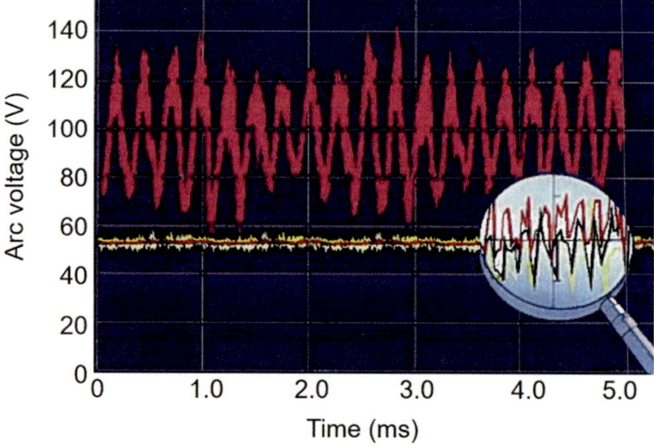

Fig. 8.38 Comparison of a typical voltage fluctuation signal for a standard DC plasma torch (toptrace) compared to the a Triplex torch (bottom trace [Barbezat G and K Landers])

when the injection conditions were to be optimized. The results showed that particle injection between the hot lobes, combined with an appropriate carrier gas mass flow rate, allowed for the optimization of the process effectiveness as a function of the size of the injected particles.

The Triple Torch Plasma Reactor (TTPR) developed by Pfender and his collaborators at the University of Minnesota represents another use of multiple electrode arrangement for powder synthesis and spraying experiments [Lu and Pfender (1989)], [Young and Pfender 1989)]. As shown schematically in Fig. 8.39 [Chen et al. (2000)], the system

is composed of three torches arranged inside a controlled atmosphere chamber with an adjustable angle to the chamber axis and distance to a central substrate holder. The three plasma jets join to form an enlarged plasma region. The powder is injected through a central water-cooled injection tube on the axis of the reactor, close to the location where the three jets join to form the combined jet. Three power supplies are necessary to operate the reactor. Operating currents can be varied between 250 and 320 A with Ar or Ar-H$_2$ mixtures as the plasma gas. Enthalpy probe measurements of the plasma flow show a rather uniform temperature and velocity field over a relatively large surface area of the substrate [Asmann et al. (2001)]. The temperature and velocity in the central portion of the plasma jet can be adjusted by changing the carrier gas flow rate, that is, higher carrier gas flow rates will result in cooling of the central portion of the jet and reduced particle heating. Clearly, coating of complicated shapes will be difficult because the torches cannot be moved independently, but the substrate can be moved. This torch has been used for specialty coatings, such as the three layers of a solid oxide fuel cell (SOFC), where in a single process, the porous Ni-YSZ anode layer, the high-density YSZ electrolyte layer (with 99.5% density), and the porous lanthanum manganite perovskite cathode layer were deposited [Chen et al. (2000)].

8.4   Gas and Particle Dynamics in DC Plasma Spraying

This section is devoted to a discussion on the gas and particle dynamics in DC plasma torches and their dependence on the torch design and its operating conditions. The subject is treated in three district subsections dealing first with the characteristics of arc and gas dynamics, which build on Sect. 8.2.4 arc instabilities, and their impact on the gas dynamics in the plasma jet. This is followed by a discussion on the time-averaged, steady-state characteristics of the plasma jet and its interactions with ambient atmosphere. The important topic of particle injection into the plasm jet is treated next, which builds up on the fundamental concepts developed earlier in Chap. 5, Gas and Particle Dynamics in Thermal Spray.

8.4.1   Arc and Plasma Jet Dynamics

As mentioned earlier in Sect. 8.2.4, arc instabilities are inherent to the basic configuration of the design of DC plasma spray torches with a central hot cathode surrounded by a water-cooled, coaxial annular anode forming the torch nozzle. The arc is stuck between the cathode tip, which is usually in the form of a sharp cone, "stick-type cathode," and the anode walls. The arc-root bridging the anode boundary layer,

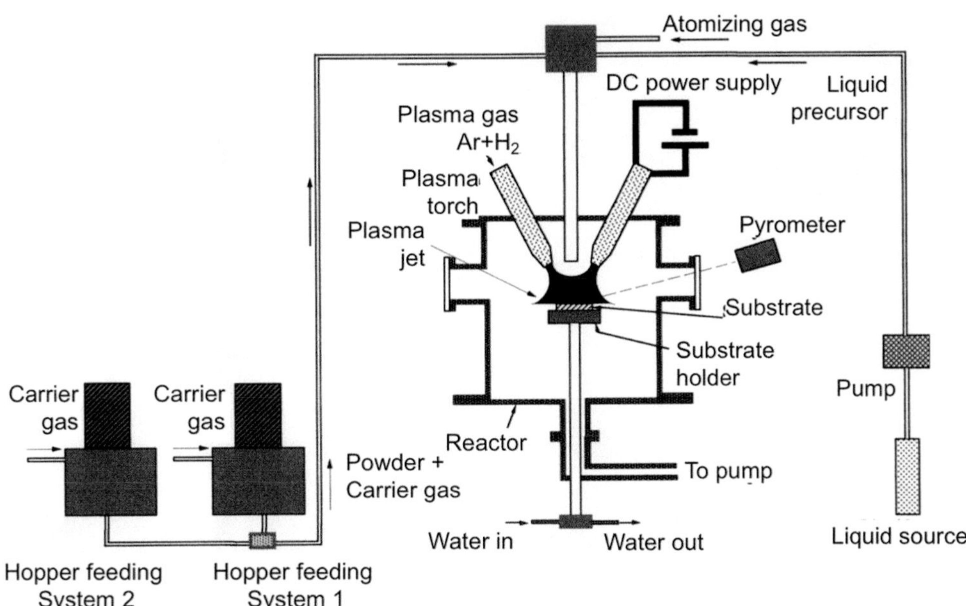

Fig. 8.39 Triple Torch Plasma Reactor with three torches in controlled atmosphere chamber and central injection and multiple powder/liquid feeders. [Chen et al. (2000)]. (Copyright © ASM International. Reprinted with permission)

connecting the arc column to the point of attachment on the anode surface, is exposed to multiple forces, the most important of which are the drag force exerted by the gas flow in the direction of the gas stream and the electromagnetic force due to the self-induced magnetic field. In the case of torch designs incorporating a magnetic field coil such as those illustrated in Figs. 8.23, 8.30 and 8.31, the applied field will also exert an additional electromagnetic force on the arc root forcing its continuous movement on the surface of the anode. The resulting arc movement illustrated on the LHS of Fig. 8.40a and supported by the highspeed images given on the RHS Fig. 8.40a, after [Wutzke et al. (1967)], gives rise to corresponding arc-voltage fluctuations typical of the restrike mode of operation, as illustrated in Fig. 8.40b, after [Coudert et al. (1966)].

Such strong arc voltage fluctuations encountered in the restrike mode of operation are promoted by thick boundary layers between the arc and the anode nozzle wall, which mostly exist under conditions of low currents, high plasma gas flow rates, and high percentages of molecular gases such as hydrogen or helium. However, restrike-like behavior can also be observed when the anode is eroded, and a preferred track exists for the motion of the anode root attachment. The restrike motion is usually characterized by a specific frequency in the low kHz range (2–6 kHz), and increased erosion will give rise to a voltage fluctuation with this frequency. Fig. 8.41 [Duan et al. (1997)] shows the Fourier transform of arc voltage fluctuations progressing from a random distribution, at the top of the figure for a new anode, to a spectrum dominated by a specific frequency as the anode becomes increasingly eroded. Even more sensitive

than the Fourier transform of the voltage trace is the spectrum of the sound pressure level as obtained with a microphone. For a new anode, two frequency peaks are observed, one coinciding with the frequency of the voltage peak (in the 2–6 kHz range) and another in the 8–10 kHz range, as shown in Fig. 8.42 [Duan et al. (2000)]. As the anode becomes increasingly eroded, the lower frequency peak will shift to lower frequencies and its height increases with respect to the high frequency peak. While such changes in the acoustic spectrum with the erosion of the anode would be somewhat different for other torch nozzle designs, the acoustic spectrum still offers an interesting diagnostic indicator for the anode state.

Such stochastic arc movement gives rise to a complex flow pattern in the discharge cavity as well as in the emerging plasma jet, as illustrated in Fig. 8.43a, for a conventional DC plasma spray torch. The intense and close interaction between the arc column and the plasma forming gas in the discharge cavity results in the rapid heating of the plasma-forming gas and its acceleration through the exit nozzle forming the plasma jet with corresponding plasma jet oscillations, as shown in the high-speed images given in Fig. 8.43b.

Such plasma-jet oscillations are also sensitive to the anode conditions, as can be observed in Fig. 8.44 [Duan et al. (1997)], which shows two high-speed images of a plasma jet obtained with a Laser-Strobe camera (Control Vision) at 100 nm exposure time. Fig. 8.44a shows a typical image of the jet with a new anode, while Fig. 8.44b shows the image of a strongly eroded anode. The jet length is strongly decreased.

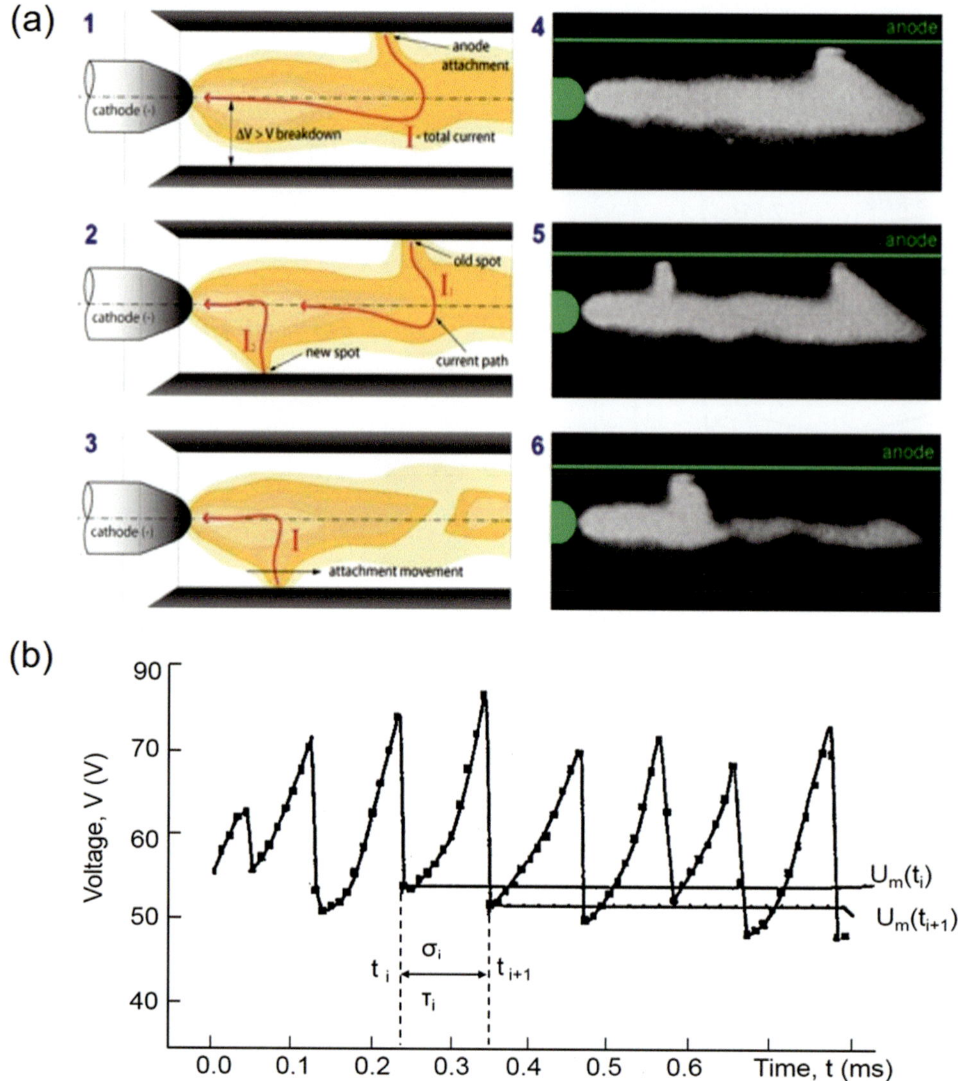

Fig. 8.40 Arc attachment dynamics and associated arc voltage for a DC plasma spray torch of conventional design. (**a**) Schematic of the arc attachment process (left) and supporting experimental high-speed images (right) [Wutzke et al. (1967)], (**b**) reconstruction of the temporal evolution of the arc voltage. [Coudert et al. (1966)].

Since the turn of the century, a systematic effort has been devoted to the dynamic modeling of the arc in the discharge cavity using 3-D, time-dependent, laminar flow simulation involving the numerical solution of the transient continuity, momentum, and energy-conservation equations simultaneously with the electric current and electromagnetic equations [Moreau et al. (2006), Trelles et al. (2006a, b, 2007a, b, c, 2008, 2009, 2011, 2013]. Additional assumption made are:

- gas is treated as incompressible, that is, its Mach number is <0.3,
- gas properties are temperature dependent,
- optically thin plasma in local thermodynamic equilibrium (LTE).

While the assumption of LTE is valid in the arc column, it is questionable in the electrode boundary layers where noticeable deviations from thermal and chemical equilibrium may occur. This was overcome by [Moreau et al. (2006)] assuming an initial high temperature "appendage" between the arc column and the anode wall, often referred to as "arc-root." Upstream restrike was obtained by re-establishing such an arc-root between the main arc column and the anode surface at a location where the electric field at the anode surface exceeds a critical value. This allowed for the reproduction of the different voltage fluctuation modes of steady, takeover, and restrike for an argon-hydrogen arc with the LTE assumption.

A comparable three-dimensional, transient, equilibrium (LTE) model was developed by [Trelles et al. (2006) and Trelles (2006c)] through the solution of the fluid and

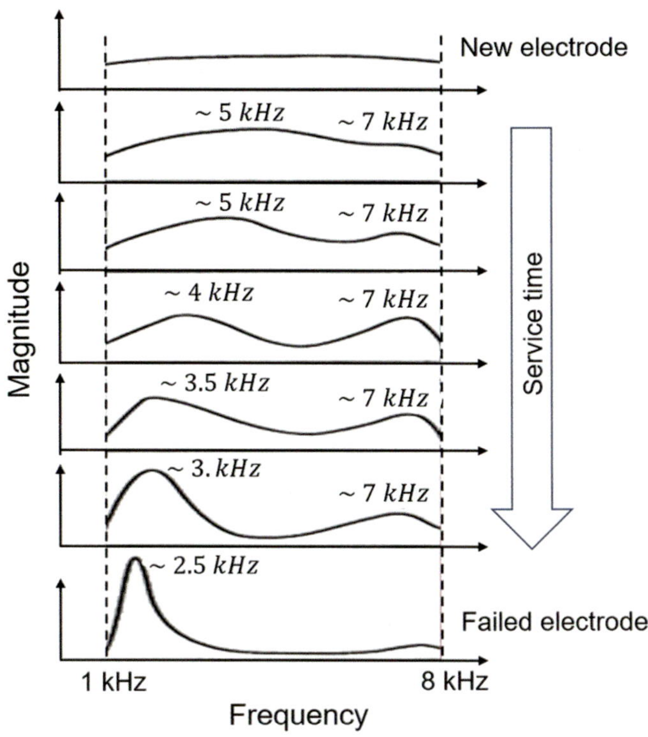

Fig. 8.41 Schematic representation of the Fourier transform power spectrum of the voltage trace of an arc with a new anode (top) to a progressively eroded anode (bottom) showing the increase in the power in the 2–4 kHz range and the shift to a somewhat lower frequency. [Duan et al. (1997)]. (Reprinted with permission of ASM International. All rights reserved)

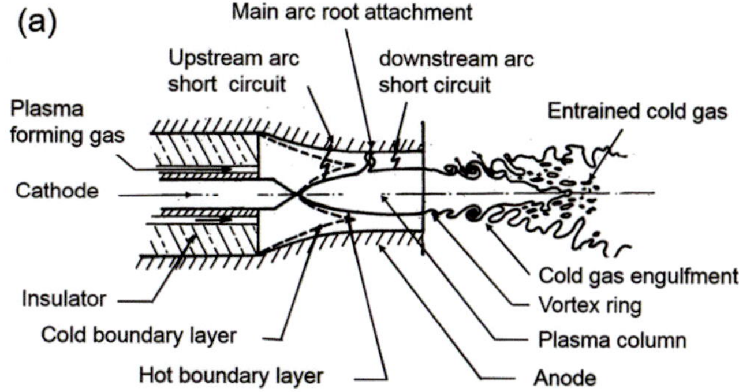

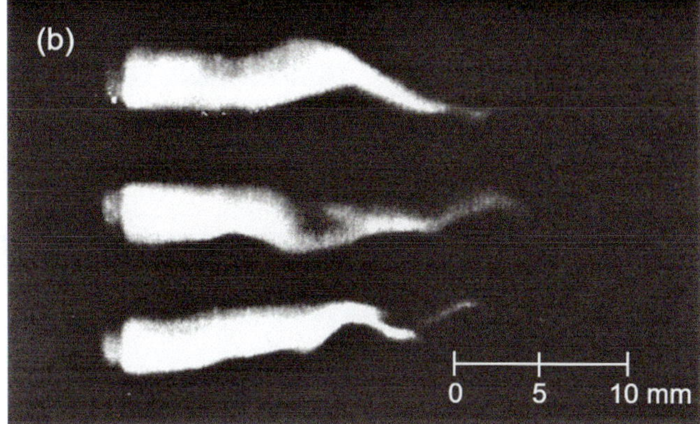

Fig. 8.43 Schematic representation of a conventional DC plasma-spraying torch (**a**) Flow structure [Fauchais and Vardelle (1997)], (**b**) supporting high-speed images (50 ns exposure) of plasma-jet oscillations [Pfender (1999)]

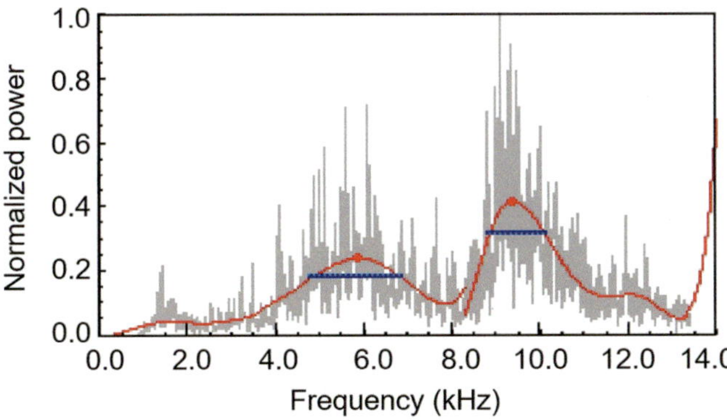

Fig. 8.42 Acoustic spectrum for the Praxair SG 100 torch showing two distinct peaks. The lower frequency peak increases with increasing erosion. [Duan et al. (2000)]. Copyright © ASM International. (Reprinted with permission)

electromagnetic equations in a fully coupled approach by a multiscale finite element method. The model allowed for the prediction of the operation of the torch in steady and take-over modes without any further assumption on the reattachment process, except for the use of an artificially high electrical conductivity region near the electrodes needed because of the equilibrium assumption. Typical results obtained for an

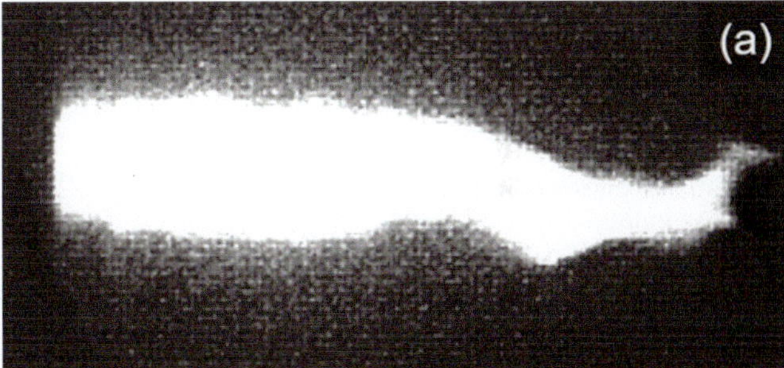

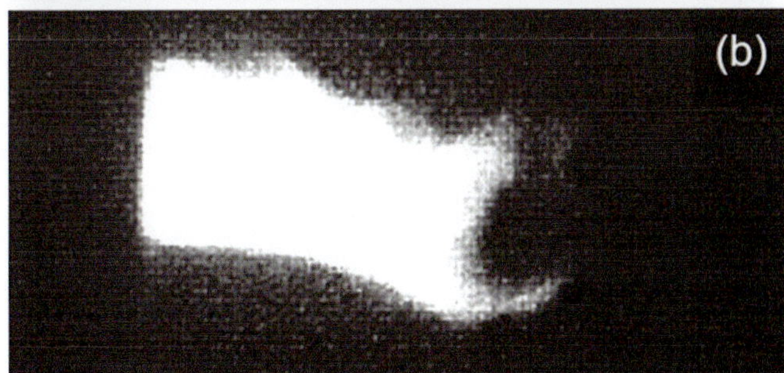

Fig. 8.44 High-speed images (100 ns) of a plasma jet with a new anode (top) and an eroded anode (bottom). [Duan et al. (1997)]. (Reprinted with permission of ASM International. All rights reserved)

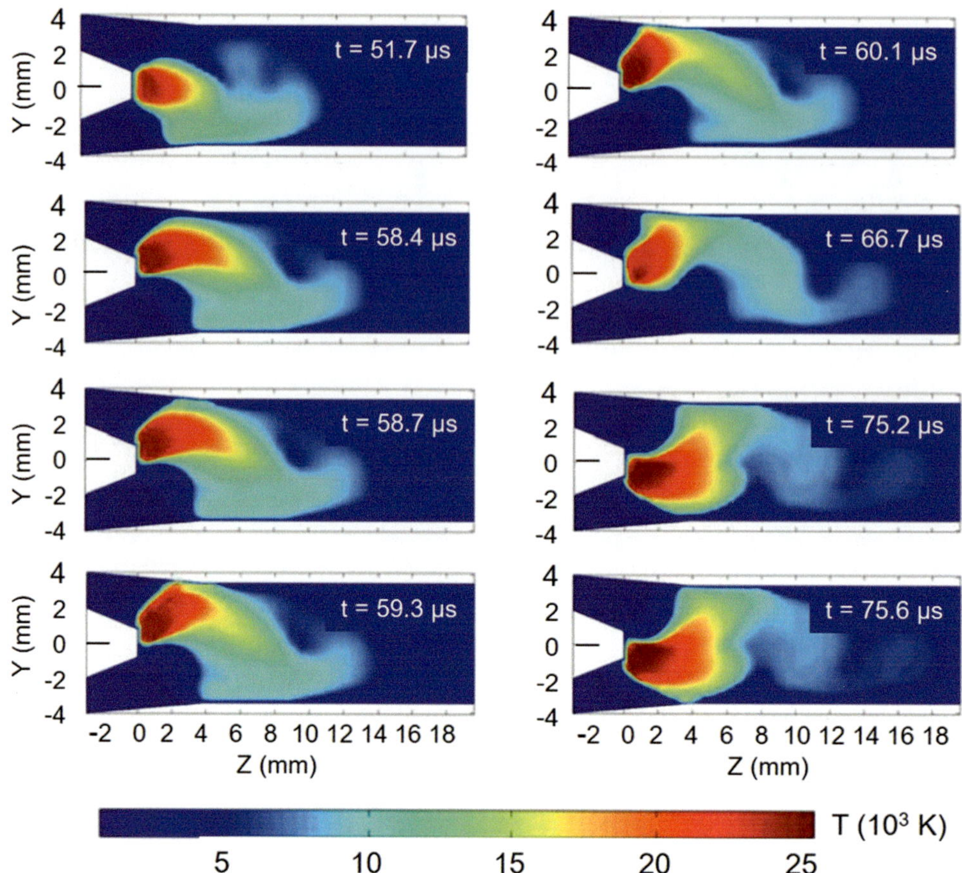

Fig. 8.45 Temperature distribution through a vertical plane during the different time steps of formation of a new arc-root attachment, Ar/H$_2$ plasma (45 slm Ar +15 slm H$_2$), 600 A. [Trelles et al. (2006)]

F4-MB torch geometry operating with an arc current of 400A using Ar/H$_2$ as plasma gas at a flow rate of (45 slm Ar + 15 slm H$_2$) are given in Fig. 8.45. This shows a time sequence of the reattachment process formed whenever the arc gets "close enough" to the anode. The required proximity being a function of the thickness of the region in front of the anode specified with an artificially high electrical conductivity ($\sim$0.1 mm). The results obtained indicate that the reattachment process in these operating modes may be driven by the movement of the arc rather than by a breakdown-like process used by [Moreau et al. (2006)]. Trelles et al. (2006) reports that for a torch operating in these modes using straight gas injection, the arc reattachment tends to be on the opposite side of its original attachment point due to the imbalance between electromagnetic and fluid drag forces.

The results show that, when straight or swirl injection is used, the arc is dragged by the flow and then jumps to form a new attachment, preferably at the opposite side of the original attachment, as has been observed experimentally. The predicted restrike frequencies, arc length, and anode spot size were in fairly good agreement with experimental values. Arc voltages obtained were, however, too high compared to experimentally observed voltages.

This model was also used to compare the arc movement inside anode nozzles for different torch designs operating with argon-hydrogen or argon-helium mixtures as plasma gas [Trelles et al. (2006), Trelles et al. (2007a)]. Simulations of an F4-MB torch from Sulzer-Metco (presently Oerlikon-Metco) and two configurations of the SG-100 torch from Praxair. A schematic of the attachment scheme proposed by [Trelles et al. (2007a)] for the LTE arc modeling is given in Fig. 8.46. It consists of the following four steps:

(a) arc deflected by incoming flow and location where the breakdown condition is satisfied,
(b) insertion of an electrically conductive channel,
(c) formation of a new attachment,
(d) predominance of the new attachment over the old one.

Non-equilibrium, two-temperature modeling of the DC arc plasma torches was reported by [Trelles et al. (2007b)]. Because of the significant increase in computation load, the model was limited to thermal non-equilibrium, thus neglecting chemical non-equilibrium effects. Computations were carried out for a conventional torch configuration

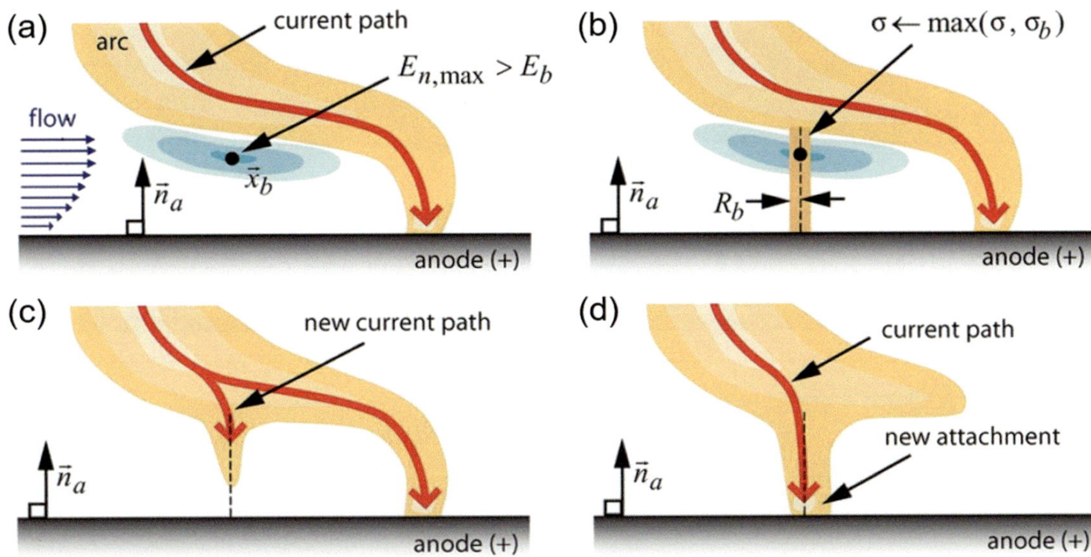

Fig. 8.46 Schematic of the different steps of the reattachment model used in LTE simulation. [Trelles et al. (2007a)]

operating with pure argon as plasma gas (60 slm) with 45° swirl injected, and arc currents of 400 and 800 A. Typical results given in Fig. 8.47 present snapshots in time of the reattachment process for the 400 A arc current case. The temperature distributions given in Fig. 8.47a, b, obtained using the non-equilibrium model (NLTE), are presented, respectively, in terms of the heavy particle temperature, T_h, and electron temperatures, T_e. For comparison, the equivalent results obtained at the same time steps using the equilibrium model (LTE) are given in Fig. 8.46c. The NLTE results show the cathode jet, the movement of the anode column, the formation of a new arc attachment, the small fluctuation of the arc column, and the undulating behavior of the flow as it leaves the torch. It can also be observed how the anode attachment is driven by the net angular momentum over the arc toward the opposite side of the original attachment point. For the conditions studied, a reattachment model was not needed for the NLTE simulation since the formation of a new attachment, as observed by the appearance of a small high temperature "appendage" (see arrows in Fig. 8.47a, b), seems to be driven by high values of the local effective electric field and high values of electron temperature. The reattachment process occurs in this case in a natural manner mimicking the *take-over* and/or *steady* modes of operation. This is not necessarily the case for the *restrike* mode where the use of a reattachment model such as the one used in the LTE simulations (see Fig. 8.46) would be required even in non-equilibrium simulations.

Figure 8.48 shows a time sequence of the dynamics of the arc inside the torch and the plasma jet immerging from the torch nozzle as obtained numerically with an NLTE model, represented by iso-contours of heavy-particle temperature, as

well as high-speed images of the plasma jet for the same torch, conventional design with an 8.0 mm nozzle i.d. and closely similar, though not identical, operating conditions. Modeling being done for operation at 400 A, pure argon plasma (60 slm (Ar), with 45° swirl injection with an open discharge in an argon ambient atmosphere, with images given at ~0.1 *ms* interval. The experimental images were taken also at 400 A arc current, for Ar/He plasma (48 slm Ar + 14 slm He) 45° swirl injection, with an open discharge in ambient air with images given at an interval of ~2 *ms*. Both modeling and experimental observations are consistent and in general agreement showing large-scale structures of the plasma jet oscillation that can be traced back to the dynamics of the arc inside the torch, while the fine-scale structures are a consequence of the interaction of the jet with the cold surrounding gas. No turbulence model has been employed in those simulation results.

Further development of DC non-transferred plasma torch modeling was reported by [Trelles (2001, 2013)], linking for the first time the dynamic plasma jet characteristics with the transient arc motion in the discharge and anode nozzle regions. The flow of a plasma jet produced by a non-transferred arc DC plasma torch impinging on a flat substrate is simulated using a fully coupled thermodynamic non-equilibrium (two-temperature) model. For simplicity, the chemical equilibrium assumption was maintained. The modeling results indicate that the dynamics of the arc inside the torch cause the quasi-periodic movement of the jet, which leads to the formation of complex patterns in the heavy-species and electron temperature distributions, together with significant deviations from thermodynamic equilibrium, near the substrate. It is observed that the evolution of the degree of non-equilibrium near the substrate is strongly correlated with the direction of movement of the jet and, hence, with the arc

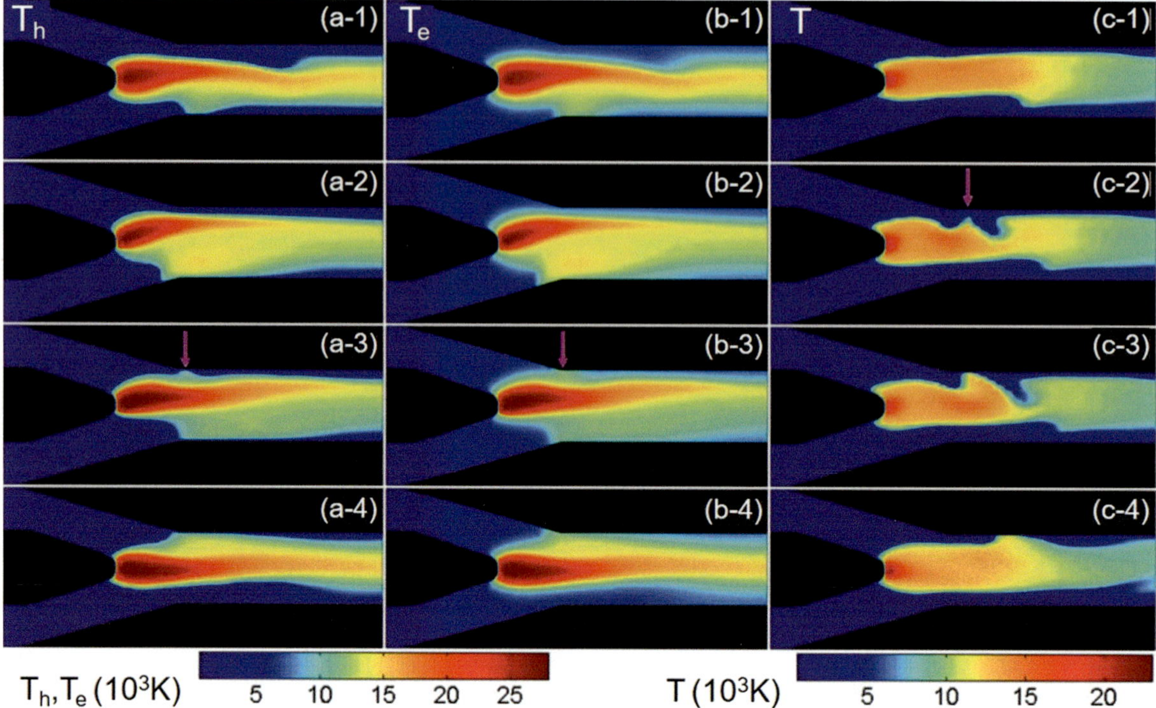

Fig. 8.47 Temperature fields inside the torch during the reattachment process for a DC arc plasma torch, 8 mm nozzle i.d., 60 slm (Ar) 45° swirl injection, and 400 A obtained using NLTE model (**a**) heavy-particle, (**b**) electron temperatures and LTE model, and (**c**) equilibrium temperature. The arrows indicate the location where the new attachment starts forming. [Trelles et al. (2007b]

dynamics inside the torch. Simulation results also uncover distinct aspects of the flow dynamics, including the jet forcing due to the movement of the electric arc, the prevalence of deviations between heavy-species and electron temperatures in the plasma fringes, the development of shear-flow instabilities around the jet, the occurrence of localized regions with high electric fields far from the arc, and the formation and evolution of coherent flow structures.

Typical results given in Fig. 8.49 depict the difference between the heavy-species and the electron temperature fields during an arc attachment event. The time instants corresponding to the snapshots are indicated. According to Trelles (2013), comparing the location of the anode attachment and the curvature of the jet in Fig. 8.49a, b-1 with those of Fig. 8.49a, b-4, it is possible to notice a correlation between the dynamics of the arc inside the torch and the large-scale structure of the plasma jet. Trelles (2013) also proposes that the turbulent nature of the plasma jet could be linked to shear flow instabilities (also known as Kelvin–Helmholtz) near the torch exit, as shown by the sequence of snapshots given in Fig. 8.50, which were obtained at time sequels indicated in the figure. It is underlined, however, that the inherently sensitive nature of instability development phenomena makes imperative the use of high-accuracy numerical simulations, which was not the case in this study. By contrasting the results in Fig.8.49 with those in

Fig. 8.50, it is observed that the shear-flow instabilities are correlated with the distribution of T_h and not with the distribution of T_e.

8.4.2   Ambient Gas Entrainment by the Plasma Jet

The unsteady nature of the plasma jet has been the subject of numerous studies [Spores and Pfender (1989), Spores et al. (1990), Pfender et al. (1990), Spores et al. (1990), Fincke and Swank (1991), Coudert et al. (1995), Planche et al. (1997), Duan et al. (1999a, b), Duan and Heberlein (2000, 2002), Gregori et al. (2003), Bisson et al. (2003), Bisson and Moreau (2003), Nogues et al. (2007), Coudert et al. (2007), Coudert and Rat (2008)]. There are two primary causes for this unsteady behavior:

- the constant movement of the arc root, resulting in voltage and power fluctuations and strong temporal variations of the jet temperature and velocity distributions, and,
- the extreme density and viscosity gradients between the low-density, high-viscosity plasma jet and the cold gas surroundings [Pfender et al. (1990)].

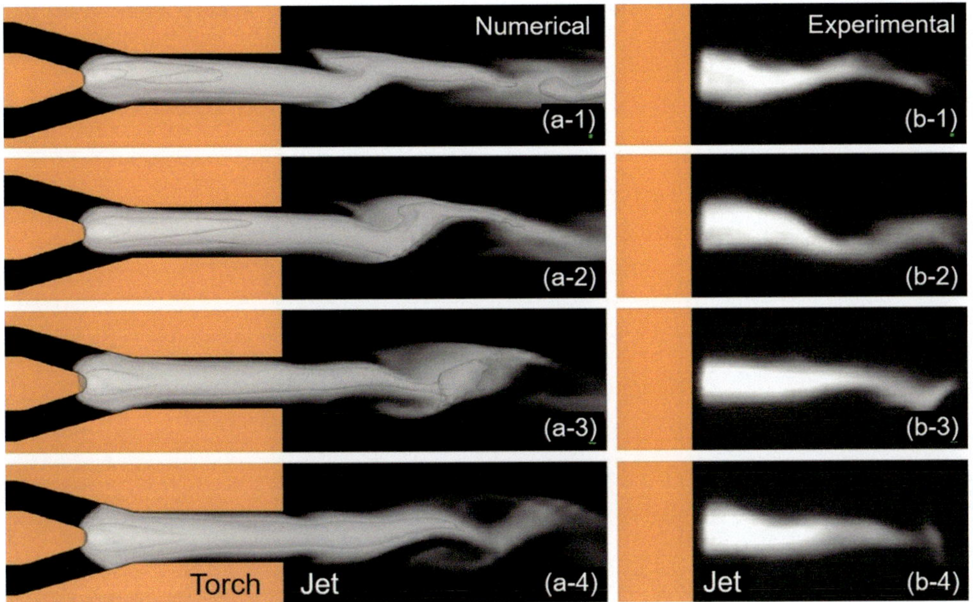

Fig. 8.48 Arc and plasma jet dynamics for an 8 mm i.d. torch nozzle operated at 400 A (**a**) sequence of heavy-particle temperature distributions obtained using an NLTE model, pure argon plasma gas 60 slm (Ar) 45° swirl injection, with ∼100 µs time interval between frames (Numerical) and (**b**) experimental highspeed images, plasma gas (48 slm Ar + 14 slm He) 45° swirl injection, ∼2 ms time interval between frames (Experimental). [Trelles et al. (2008)]

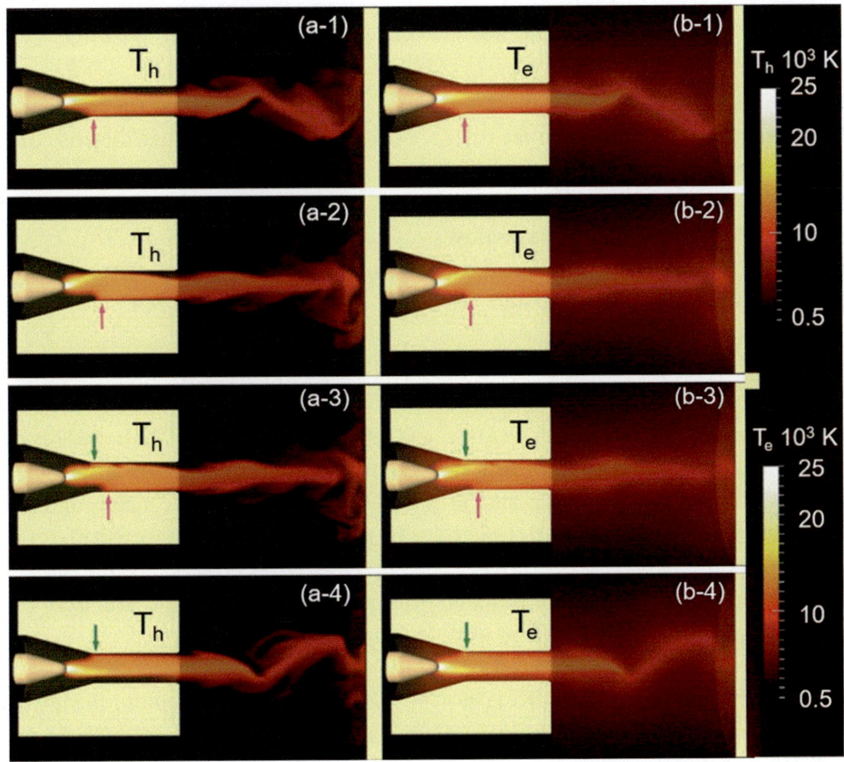

Fig. 8.49 Temperature dynamics during an arc reattachment event (**a**) T_h and (**b**) T_e; sequence 1 to 4, respectively, at 1.16, 1.29, 1.31, and 1.41 ms. [Trelles (2013)]

These give rise to intense plasma jet fluctuations shown in the series of high-speed images given in Fig. 8.43 [Pfender (1999)], Fig. 8.44 [Duan et al. (1997)], and Fig. 8.48 [Trelles et al. (2008)]. As schematically illustrated in Fig. 8.51a [Pfender et al. (1990)], a rather complex flow pattern develops rapidly around the plasma jet as it exits the plasma

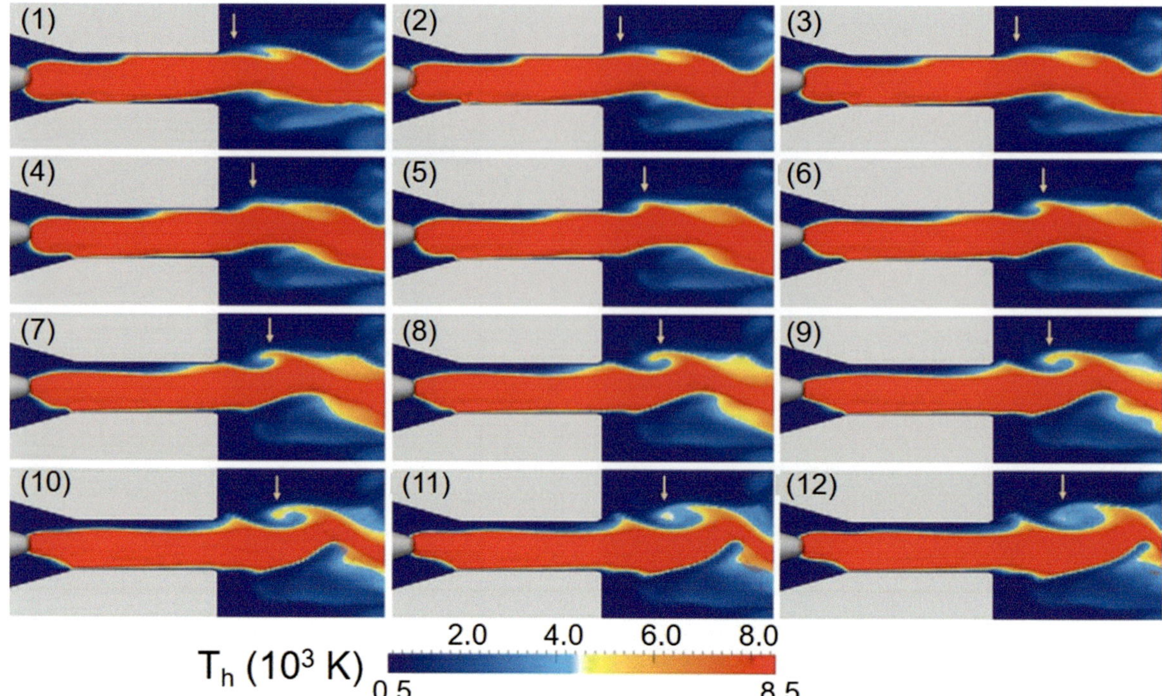

Fig. 9.50 Formation and development of shear-flow instabilities: sequence of snapshots of T_h distribution in the region encompassing the cathode tip and the torch discharge. The arrows indicate the location where an instability develops. The color scale used emphasizes the development of the instabilities. [Trelles (2013)]

torch nozzle and penetrates the ambient gas. The large velocity difference between the plasma jet and the ambient gas causes the rolling up of the flow around the nozzle exit into a ring vortex, which is pulled downstream by the flow, allowing the process to repeat itself again at the nozzle exit. Adjacently formed vortex rings at the outer edge of these vortices then lead to wave instabilities growing around the entire ring. Next, the distorted vortex rings start entangling themselves with adjacent rings, finally resulting in a total breakdown of the vortex structure into large-scale eddies and the onset of turbulent flow. This entanglement process of adjacent unstable vortices results in the first large-scale engulfment of the ambient gas, although some entrainment also takes place during the roll-up process of the jet's shear layer. These large eddies of cold gas that are entrained have a much higher density and thus greater inertia than their high-temperature counterparts. The eddies of cold gas travel in the axial direction at much lower velocity, while the hot plasma gas essentially accelerates around and stagnates on the cold gas eddies with little initial mixing. All eddies in the flow are continually breaking down into smaller and smaller eddies, while diffusion is taking place on the molecular level at all eddy boundaries. The mixing and diffusion process eventually reaches the centerline of the jet, foretelling the end of the laminar core. The jet now undergoes transition and eventually becomes fully turbulent, while eddies of external gas continue to be engulfed and then absorbed into the main

jet along its entire length, further reducing both the mean temperature and velocity.

This qualitative description of turbulence development and cold gas entrainment is corroborated by the pulsed-laser Schlieren photograph (exposure time 10 ns) of argon DC plasma jet given in Fig. 8.51b after [Finke et al. (2003)], obtained for an SG-100 plasma torch with a nozzle i.d. of 8.0 mm operated using pure argon as plasma gas (35.4 slm), 900 A and 15.4 V arc current and voltage, respectively. Based on measurements of temperature and velocity, the centerline Mach number at the torch axis is ~0.49.

Further corroboration of the intense mixing between the plasma jet and the ambient atmosphere is given in Fig. 8.52, after [Pfender et al. (1991)], showing 2-D isocontours of relative nitrogen concentration in DC plasma jet, 12.7 mm nozzle id, operated using pure argon as plasma gas (9.8 slm), and arc currents of 250 and 500 A, with a corresponding torch power of 5 and 10 kW, respectively. The data given in this figure were obtained using CARS, which is a nonintrusive diagnostic technique capable of measuring nitrogen molecular concentrations which are given in this figure normalized to the number density of N_2 at ambient conditions (No $\approx$ 1.7 10^{19} cm^{-3}). The rapid entrainment of air into the core flow of the jet, coincident with the onset of turbulence and jet breakup, is evident in the concentration contours. It is interesting to note that even though the onset of turbulence occurs earlier in the 5 kW case, a laminar, all-argon core

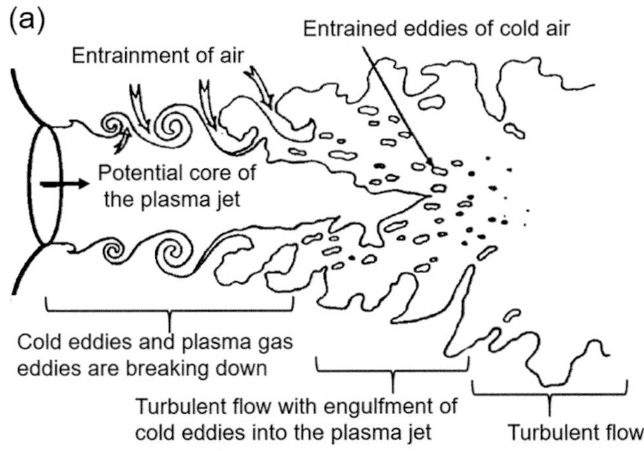

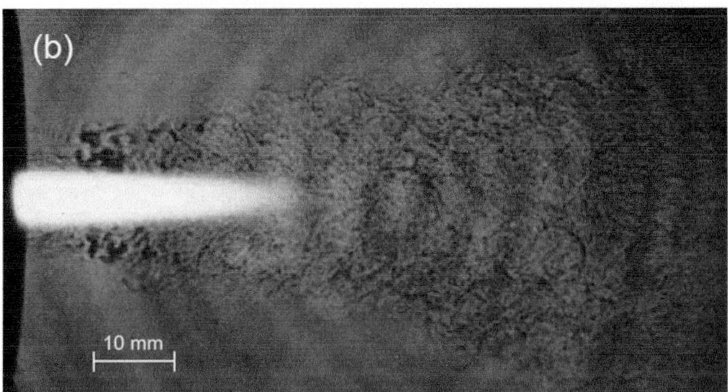

Fig. 8.51 Schematic of the large-scale turbulence and cold gas entrainment in a DC plasma jet issuing into a cold gas environment [Pfender et al. (1990)] and pulsed-laser Schlieren images of a turbulent Ar plasma jet, SG-100 torch, 8 mm nozzle id, 35.4slm, 900 A. 15.4 V. [Finke et al. (2003)] (Reprinted with permission of ASM International. All rights reserved)

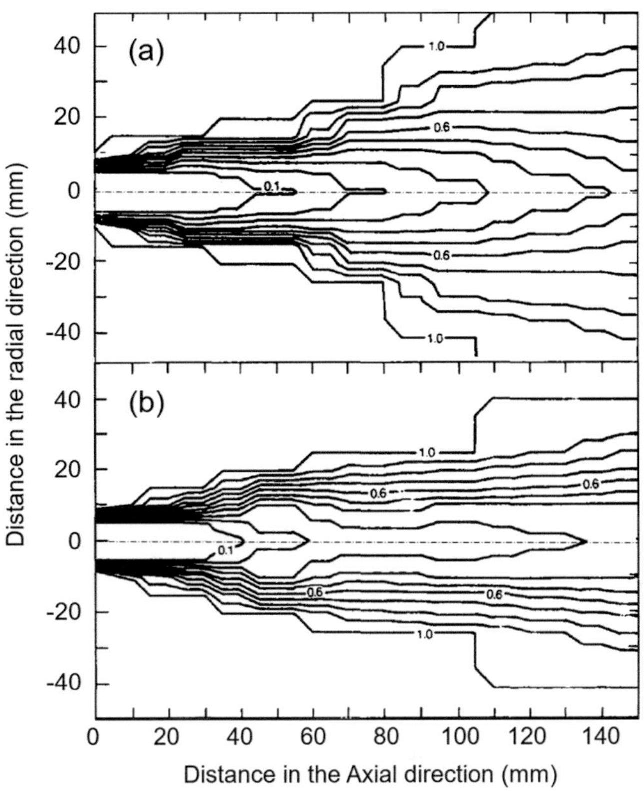

Fig. 8.52 Contours of nitrogen entrained from an ambient air atmosphere into a pure-argon DC plasma jet 9.8 slm (Ar). Values give are normalized to the number density of N_2 at ambient conditions, (**a**) 250 A, 5 kW, (**b**) 500 A, 10 kW. [Pfender et al. (1991)]

appears to persist for approximately the same distance in both cases, and initially the mixing rates are comparable. Farther downstream, the lower power 5 kW case spreads more rapidly and entrains air at a greater rate.

Measurement of the amount of air entrained in an argon plasma jet issuing into an air environment, using gas sampling with enthalpy probes, showed that at an axial location 20 mm downstream from the nozzle exit, 50% of the gas consisted of air [Pfender et al. (1990)]. Furthermore, in a set of definitive experiments using enthalpy probe measurements, two-photon Laser-Induced Fluorescence and CARS, the entrainment of air into an argon plasma jet was demonstrated, and the CARS measurements showed that the entrained oxygen remains at a low temperature of about 2000 K [Finke et al. (2003)]. Fig. 8.53 shows for the plasma torch and operating conditions used for Fig. 8.51b, the axial temperature decay measured with an enthalpy probe, the fraction of the air in the gas sample obtained with the enthalpy probe, and the temperature of the entrained oxygen derived from CARS measurements of the rotational–vibrational level populations of molecular oxygen. The results clearly indicate the significant difference in temperatures

between the plasma jet and the entrained air, as confirmed by numerical simulations results [Williamson et al. (2003)]. Corresponding radial profiles of the measured plasma axial velocity, temperature, and entrained air concentration profiles at different distances downstream from the nozzle exit level of the torch are given in Fig. 8.54.

Figure 8.55a [Gregori et al. (2003)] shows 100 measured values of peak temperatures in an atmospheric-pressure argon plasma jet (35 slm) at an axial location 11 mm from the nozzle exit, derived from spectroscopic measurements of an absolute line intensity with a time resolution of 50 μs. The strong variation in this peak temperature ($\pm$ 2000 K) is evident. However, what is even more interesting is the radial position at which this peak temperature has been measured (see Fig. 8.55b), ranging from the axis to a position 3 mm from the axis for a nozzle radius, $r_o = 4$ mm [Gregori et al. (2003)]. Clearly, the nice axisymmetric distributions obtained with measurements over a time scale of seconds cannot be assumed when processes are considered at or below the millisecond time scale.

As described earlier, the mixing of the ambient air with the plasma jet flow cools it down rather rapidly, mainly due to oxygen dissociation starting at 3000 K. To delay this entrainment effect, the mixing with the ambient gas, and the cooling of the plasma, it is possible to use an anode

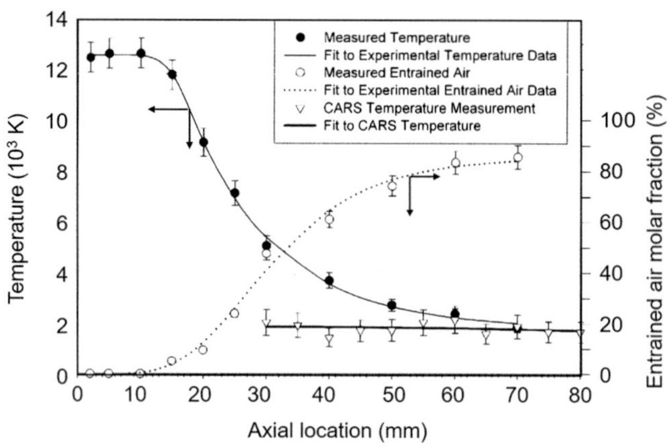

Fig. 8.53 Axial distributions of enthalpy-probe measurements of plasma temperatures, volume fraction of entrained air in the plasma jet, and CARS-temperatures of the entrained oxygen showing the low temperatures of the entrained oxygen. [Finke et al. (2003)]. (Reproduced with kind permission of Elsevier)

nozzle extension with the same diameter as that of the nozzle. Such an approach, generally called "solid shroud," has been proven to be effective in delaying the contact of the plasma gas with the ambient atmosphere with limited loss in terms of plasma energy (generally of the order of 10 to 15%) to the wall of the extension tube. According to [Roumilhac (1990)], the plasma jet core exiting the anode nozzle extension had almost the same length as that of the jet core without the anode nozzle extension. Another important feature of these extensions is that, even with the power lost to cool the extension, the temperatures and velocities of the jet are not significantly reduced at the extension exit. [Betoule (1994)] reported that a DC plasma torch with an anode nozzle i.d. of 7 mm working with an Ar–H_2 mixture (45–15 slm) and an arc current of 600 A corresponding to 40 kW, has measured 2 mm downstream of the nozzle exit on the torch axis a gas temperature of 13,000 K and a velocity of 2000 m/s. 52 mm downstream on the torch axis, the temperature was 6200 K and the velocity was about 500 m/s. With an insertion of a 50 mm long nozzle extension of the same i.d., the temperature was 11,000 K and the velocity 1700 m/s due to 5.7 kW lost in the extension cooling. Unfortunately, these nozzle extensions are difficult to use with powder radially injected in the anode nozzle downstream of the arc root, due to their divergent trajectories. However, such extensions can be used with axial injection torches, where the particle jet divergence is much less than for radial powder injection. In this case, at the exit of the extension, particles reach high velocities, especially for submicron- or nano-sized particles which, in spite of the Knudsen effect, reach velocities between 500 and 700 m/s. Extensions that have a divergent shape and are rather long (about 100 mm) have also been used for spraying. The main problem is again the tendency of the

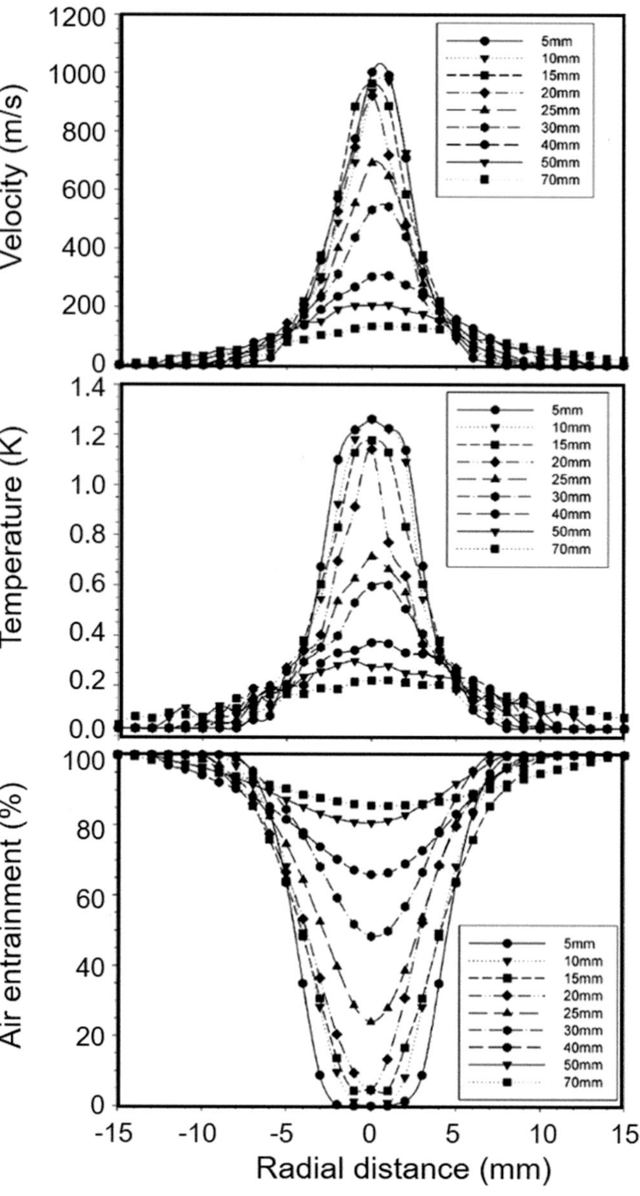

Fig. 8.54 Enthalpy probe measurements of the radial profiles (**a**) axial velocity, (**b**) temperature, and (**c**) entrained air mole fraction, for a pure-argon DC plasma jet SG-100, 8.0 mm nozzle id, 35.4 slm (Ar), 900 A, 15.4 V in an open discharge in ambient air. Distance in legend are axial location measured from the face of the torch. [Finke et al. (2003)]. (Reproduced with kind permission of Elsevier)

fully molten particles to stick to the inner wall of the solid shroud. To avoid that, the total angle of the divergent shape must have a value between 40 and 50°. While this configuration reduces plasma losses to the shroud wall, boundary layer separation gives rise to recirculating flow and the entrainment of ambient air in the shroud space, which defeats its purpose.

Fluid dynamic shrouds, without nozzle extension, provide a curtain of an inert gas between the plasma jet and the surrounding atmosphere. The problem with such shrouds is

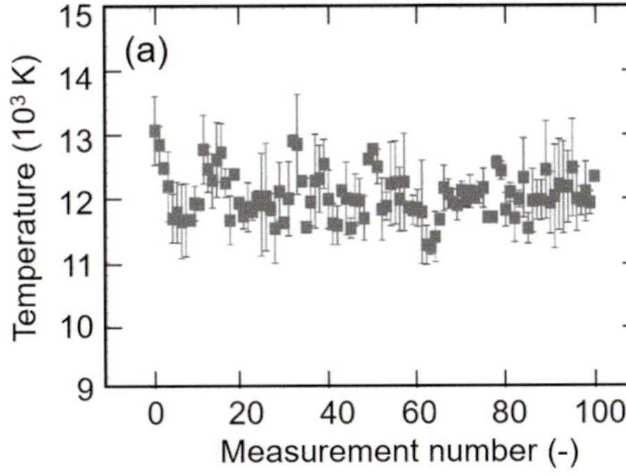
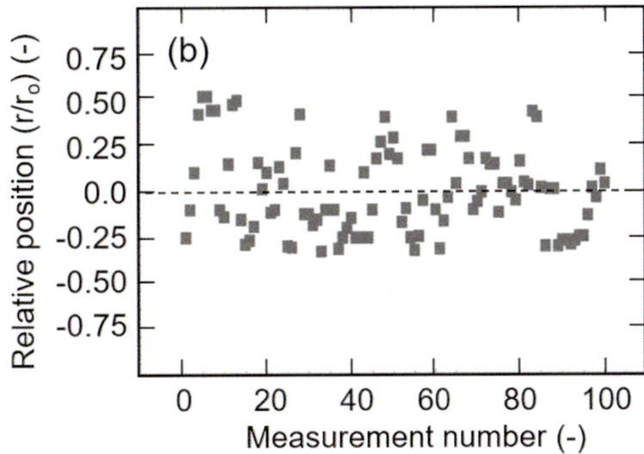
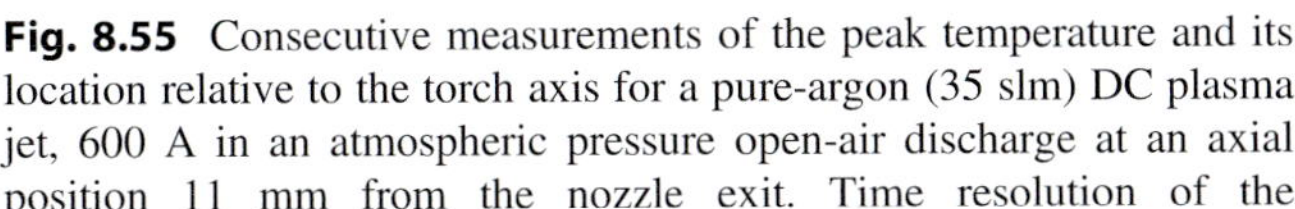

Fig. 8.55 Consecutive measurements of the peak temperature and its location relative to the torch axis for a pure-argon (35 slm) DC plasma jet, 600 A in an atmospheric pressure open-air discharge at an axial position 11 mm from the nozzle exit. Time resolution of the measurement 50 μs, 5 Hz repetition rate (**a**) peak temperature (**b**) relative distance to the torch axis (r/Ro), where Ro is the nozzle radius. [Gregori et al. (2003)]. (Reproduced with kind permission of IPCS)

that the amount of gas in the shroud has to be rather large to match the boundary layer velocity because of the relatively large cross-sectional area. With a shroud where argon gas was injected parallel to the flow of the plasma jet through a ring-shaped slit surrounding the jet, it was possible to spray reactive metal particles in an air environment with minimal oxidation. The amount of shroud gas was varied from 100% to 300% of the plasma gas flow rate, and at 300% shroud mass gas flow, the oxygen fraction at the center of the jet was less than 0.5% at a standoff distance of 85 mm. The appropriate selection of the distance between the nozzle wall and the ring-shaped slot was crucial for the success [Mohanty et al. (1996)].

A similar approach has been taken with an axial injection torch (Axial III from Northwest Mettech, Canada) in a spray process depositing a Hastelloy bond coat [Burgess (2002)]. The oxidation of the particles was minimized by using a double shroud with a total gas flow rate of 300 slm for a plasma gas flow rate of 240 slm (75% Ar, 15% H_2, 10% N_2). An inner argon-gas shroud surrounded the plasma jet as it exited the nozzle, and an outer shroud of nitrogen was introduced further downstream at the end of a cylindrical guide tube for the inner shroud.

In a somewhat modified approach, an anode nozzle was modified at its downstream end by fitting a porous ring surrounding the original copper nozzle wall [Malmberg et al. (1995), Chen et al. (1996)]. One of the two powder feed channels of an SG-100 torch was used to provide the shroud gas (see Fig. 8.56) [Chen et al. (1996)]. In some experiments, this shroud was complemented by injecting part of the gas tangentially between the point of the arc attachment and the powder injection point to provide a swirl component opposing the swirl of the plasma gas. The results showed that the particle fluxes were more confined to the central hot core part of the jet. The coatings showed lower porosity and higher deposition efficiencies. The amount of shroud gas in this case did not exceed the plasma gas flow.

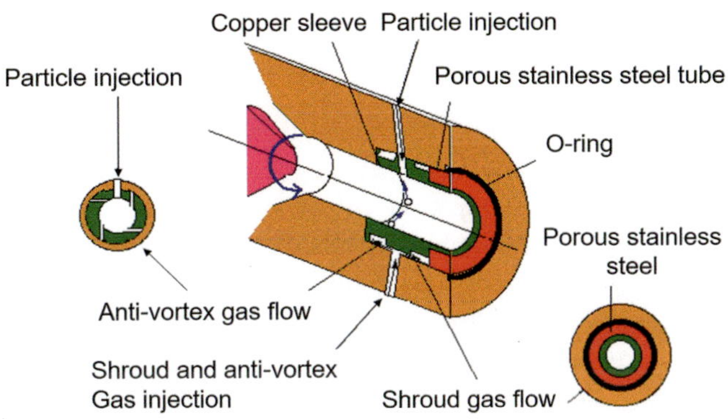

Fig. 8.56 Schematic of modification to SG-100 anode with porous ring shroud and counter swirl. [Chen et al. (1996)]. (Reprinted with permission of ASM International. All rights reserved)

8.4.3 Plasma Jet Characteristics

8.4.3.1 Plasma Torch Operating Parameters

Most DC plasma torches operate in a current control mode, that is, the current is kept constant through a feedback control loop in the power supply which compensates by changing the voltage in order to maintain the current at the required set value. The current set point and the plasma gas flow rate and composition are the independent parameters and are chosen by the operator. A specific plasma torch is then characterized by the time-averaged arc voltage, its standard deviation, and the cooling water temperature rise, and these two quantities allow for the determination of the torch operation, its energy efficiency, and the power carried into the plasma jet, as well as the average specific enthalpy of the plasma gas, its average temperature, and velocity. The typical voltage–current diagram for a DC plasma torch of conventional design operating at atmospheric pressure using Ar/H_2 (45 slm Ar + 15 slm H_2) as plasma gas is given in Fig. 8.57. This shows a strong dependence of the arc voltage on the mode of plasma gas injection into the discharge cavity, with axial injection giving rise to the

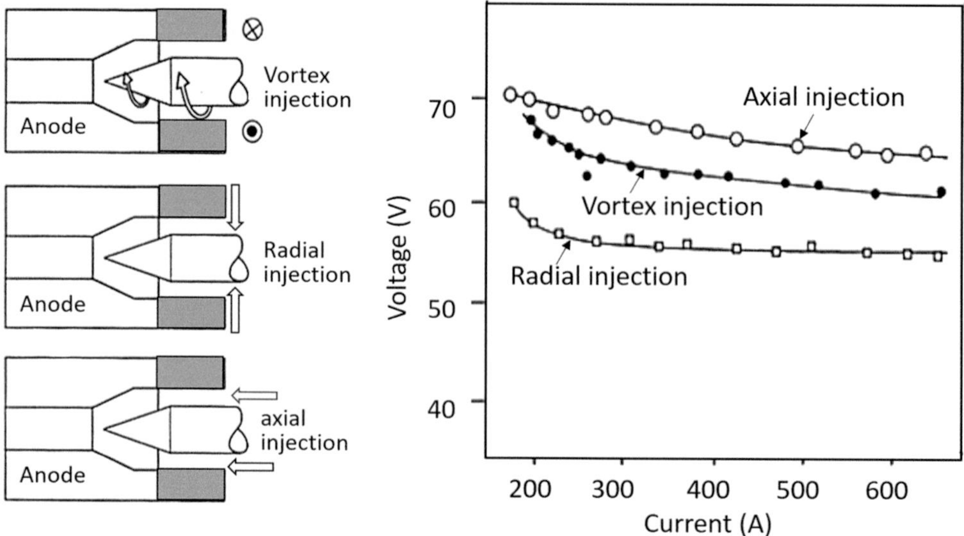

Fig. 8.57 Arc voltages as a function of arc current for Ar/H$_2$ plasma using three different plasma gas injection modes. [Roumilhac (1990]. (Courtesy of P. Roumilhac)

highest voltages, corresponding to the longest arc length, with vortex and radial injection modes giving rise to shorter arc lengths and successively lower voltages. In all three cases, the arc voltage drops gradually with the increase in the arc current.

The corresponding variation in the arc voltage [Roumilhac et al. (1990a, b)] and energy efficiency [Roumilhac (1990a, b)] plasma torch with increase in the molar fraction of the secondary gas (H2 or He) are given in Fig. 8.58. These were obtained for a DC torch of conventional design using an axial injection of the plasma gas and arc current of 600 A. The addition of a secondary gas such as hydrogen or helium to the argon plasma gas has visibly a strong influence on the arc voltage and torch efficiency. A small addition of hydrogen in concentrations of the order of 10 to 20 vol.%, leads to a significant increase in the arc voltage and torch efficiency. This is mostly caused by the increased thermal conductivity of the arc column due to the fast diffusion of the smaller hydrogen atoms to its fringes. The resultant local cooling effect in the fringes of the arc column will cause a severe constriction of the arc and an increase in the arc voltage. The higher efficiency can be explained by the fact that for a given arc current, the anode root heat losses, which are primarily determined by the arc current, remain constant, while the arc voltage and arc power increase significantly with the introduction of the secondary gas. Energy losses to the anode will accordingly represent a smaller fraction of the total arc power, which combined with the reduced radiation losses from the arc smaller arc column diameter will give rise to an increase of the overall energy efficiency of the torch.

8.4.3.2 Velocity and Temperature Fields

The plasma jet can be characterized by several diagnostic methods. Emission spectroscopy, enthalpy probes, and laser scattering can yield the distribution of temperature and electron density in the jet, and laser Doppler anemometry will allow for the determination of the velocity distributions [Vardelle et al. (1982)].

Figure 8.59 [Coudert et al. (1995)] shows the radial velocity distribution of an argon jet 5 mm from the nozzle exit and that of an argon arc with 20 vol.% hydrogen addition, for approximately the same current but considerably higher gas flow rate. A small addition of hydrogen causes the jet to be severely constricted and more than doubles the peak velocity. It should be noted that the effect on temperature is much less because the high specific heat of hydrogen reduces the temperature.

Figures 60 and 8.61 [Vardelle et al. (1982)] give, respectively, examples of the temperature and velocity distributions in the jet of a plasma spray torch in an open discharge in ambient air operating with a mixture of N$_2$/H$_2$ or Ar/H$_2$ as plasma gases. These were obtained for a DC plasma torch of conventional design with the following nozzle dimensions and operating conditions:

- for the N$_2$/H$_2$ case, Fig. 8.60 shows an exit nozzle i.d. of 6.0 mm with a length of the cylindrical channel 26 mm, plasma gas flow rate (37 slm N$_2$ + 27 slm H$_2$), DC power 29 kW,
- for the Ar/H$_2$ case, Fig. 8.61 shows an exit nozzle i.d. of 8.0 mm with a length of the cylindrical channel 22 mm, plasma gas flow rate (75 slm Ar +37 slm H$_2$), DC power 29 kW.

Measurements were carried out using emission spectroscopy for the 2-D temperature fields, and laser Doppler anemometry for the velocity fields which required that the flow be seeded by small particles, which are assumed to

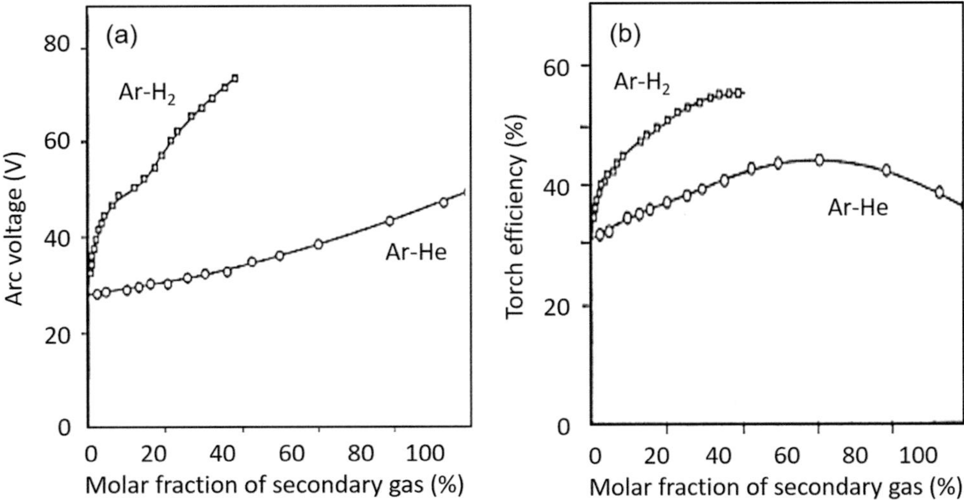

Fig. 8.58 Effect of secondary gas concentration (vol%) at constant arc current, 600 A, on the (**a**) Arc voltage and (**b**) torch energy efficiency, for a conventional DC plasma torch. [Roumilhac et al. (1990a, b)]. (Reprinted with permission of ASM International. Courtesy of P. Roumilhac)

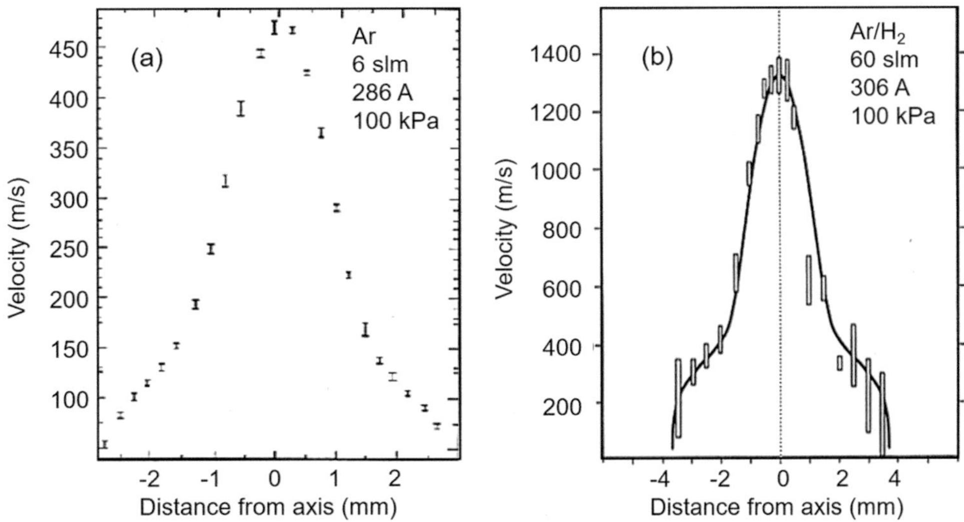

Fig. 8.59 Effect of plasma gas composition on the radial profiles of the axial plasma velocity for (**a**) a pure argon jet with 6 slm at 286 A and (**b**) an Ar/H$_2$ (20 vol.% H$_2$) jet with 60 slm at 306 A. [Coudert et al. (1995)]. (With kind permission from Springer Science+Business Media B.V)

follow the flow field. In both cases, Fig. 8.60a and Fig. 8.61a, the plasma jet can be divided into three regions. The core region is the one in which the plasma temperature is relatively constant, 12,000–12,500 K, extending to about 10 to 12 mm from the torch nozzle. This is followed by a transition region in which the plasma temperature falls rapidly to less than 3000 K at 100 mm from the nozzle exit. In the last region, the gas temperature drops gradually as the gas is mixed with the entrained ambient air. The radial temperature profiles are particularly steep over the core region. The presence of such steep temperature gradients explains the problem of the thermal treatment of the

particles in a DC plasma jet, which requires very close control of the particle injection conditions into the plasma.

In Fig. 8.60b, the axial velocity isocontours are shown for the N$_2$/H$_2$ case. Beyond the core region, the plasma velocity drops rapidly with increase in distance from the torch nozzle. Similar results at the same power level of 29 kW are represented in Fig. 8.61b for an Ar/H$_2$ mixture. It can be seen that the diameter of this jet is greater than the one of N$_2$/H$_2$ (due to the increase in the nozzle diameter) and that the core and plume of the jet are also longer (for the core, about 30 mm for N$_2$/H$_2$ and 50 mm for Ar/H$_2$ on the axis). The velocities in the N$_2$/H$_2$ jet are higher

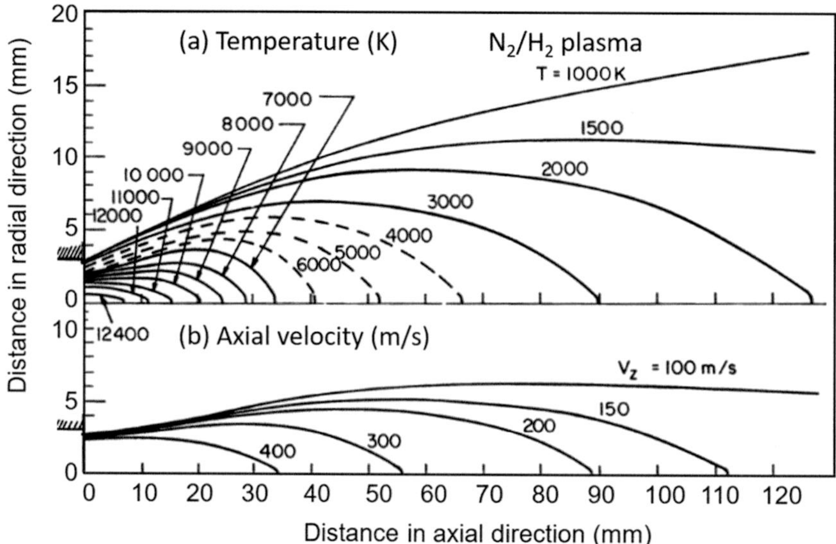

Fig. 8.60 (**a**) Temperature (**b**) axial velocity isocontours for an N$_2$/H$_2$ DC plasma jet (37 slm N$_2$ + 27 slm H$_2$), power 29 kW, [Vardelle et al (1982)]. (With kind permission from Springer Science Business Media B.V.)

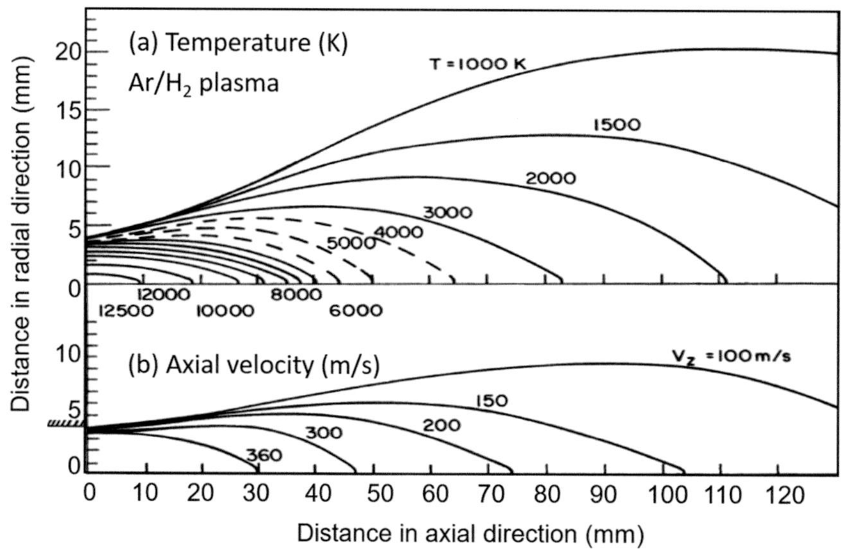

Fig. 8.61 (**a**)Temperature (**b**) axial velocity isocontours for an Ar/H$_2$ DC plasma jet (75 slm Ar + 37 slm H$_2$), power 29 kW. [Vardelle et al (1982)]. (With kind permission from Springer Science Business Media B.V.)

than the ones in the Ar/H$_2$ jet, a phenomenon due probably to the lower viscosity of the N$_2$/H$_2$ mixture and to the smallest nozzle diameter used in the N$_2$/H$_2$.

The design of the anode nozzle and the mode of gas injection of the plasma gas into the discharge cavity affect the length of the arc and consequently arc stability, power loss to the cooling water, and the temperature and velocity distributions in the plasma jet. Numerous nozzle designs have been proposed and tested, allowing for the control of the plasma jet temperature and velocity distributions, as function of the operating parameters [Rahman et al. (1998)]. An example of the impact of small changes in the nozzle design,

such as a slight divergence of the nozzle downstream of the arc attachment, on the temperature distributions in the plasma jet is given in Fig. 8.62 [Coudert et al. (1993)]. There are few general rules on how the nozzle influences the jet because the primary influence is provided by the location of the arc attachment. A smaller nozzle diameter or a constriction of the nozzle downstream of an arcing chamber will result in shorter arcs, lower temperatures at the nozzle exit, though possibly higher velocities for the same current and mass flow rate. The effect of a divergent anode nozzle or a Laval-type anode nozzle has been shown to result in less cold gas

entrainment by the plasma jet [Henne et al. (2001)] and higher deposition efficiencies [Schwenk et al. (2004)].

The mode of gas injection into the discharge cavity, whether radial, axial, or tangential (swirl), offers an important design variable for the control of the fluid dynamics of the flow in the discharge cavity, the arc stabilization, and the characteristics of the plasma jet. An example of the temperature contours for two Ar/H$_2$ plasma jets, with radial and axial gas injection modes, is given in Fig. 8.63 [Romilhac et al. (1990a, b)]. In both cases, plasma gas composition, flow rates, and arc currents are comparable. Significantly lower temperatures are observed for the plasma jet, with radial gas injection compared to that with axial injection. The effect is attributed to the longer arc length, with axial injection giving

rise to a higher arc voltage, and higher power for the same arc current. It should be mentioned that the change from swirl injection to radial injection can be achieved in steps by simply changing the orientation of the gas injection orifices in the gas distribution ring from pointing toward the torch axis (radial injection) to having a purely tangential orientation. Intermediate positions have shown to provide for a stable plasma operating and plasma jet structure [Outcalt et al. (2007)].

The effect of the surrounding atmosphere is frequently overlooked in plasma spraying. Both the atmospheric pressure and the composition (e.g., humidity) have a strong influence on the jet appearance, and the heat and momentum transfer to the spray particles because the surrounding gas is mixed with the plasma gas in the turbulent jet as described in the previous "Section 8.4.2 ambient gas entrainment by the plasma jet." As an example of this influence, Fig. 8.64 [Romilhac et al. (1990a, b)] shows the constant temperature contours in an argon–hydrogen plasma jet issuing into an air, nitrogen, or argon ambient atmosphere. It is clear that mixing with air results in the strongest quenching of the jet. The quenching will be even stronger when there is high humidity in the air. Lower atmospheric pressures will not only result in less quenching and longer jets but also reduce the heat and momentum transfer to the particles, as discussed in "Chapter 5, Plasma and particle dynamics."

Close attention has also been given to the impact of the electrode design and the operating conditions on the radial profiles of the plasma axial velocity at the torch exit. A study by [Planche et al. (1998)] reported time-of-flight image velocimetry measurements of the plasma axial velocity for a DC plasma torch of conventional design with a cylindrical anode nozzle with an i.d. varying between 6 and 10 mm. The plasma gas was a mixture of Ar/H$_2$ (25 vol% H$_2$) or pure N$_2$, at flow rates varying between 30 and 80 slm. Typical results given in Fig. 8.65 show the radial profiles of the axial plasma velocity for the 10 mm i.d. anode, at a distance of 4 mm from

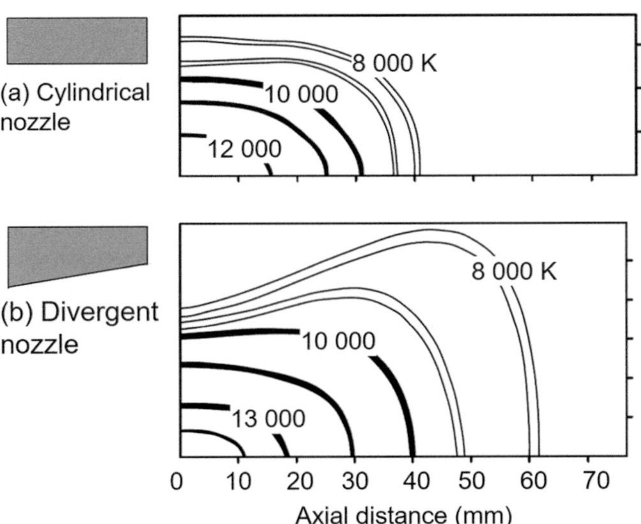

Fig. 8.62 Plasma jet temperature contours for two different nozzle shapes showing the longer and broader jet with a divergent nozzle, (**a**) cylindrical nozzle, (**b**) divergent nozzle. [Coudert et al. (1993)]. (With kind permission from Springer Science+Business Media B.V)

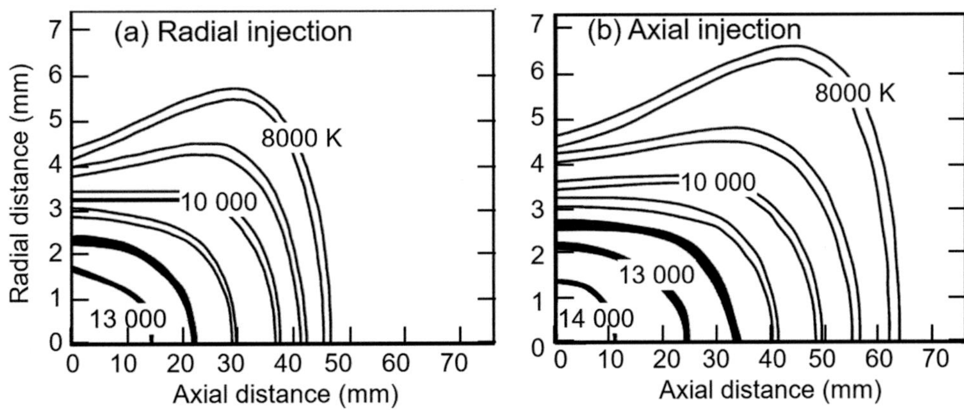

Fig. 8.63 Plasma jet temperature contours for two different plasma gas injection modes indicating the longer high-temperature zone with the axial injection, (**a**) radial injection, (**b**) axial injection (derived from data published in [Romilhac et al. (1990b]

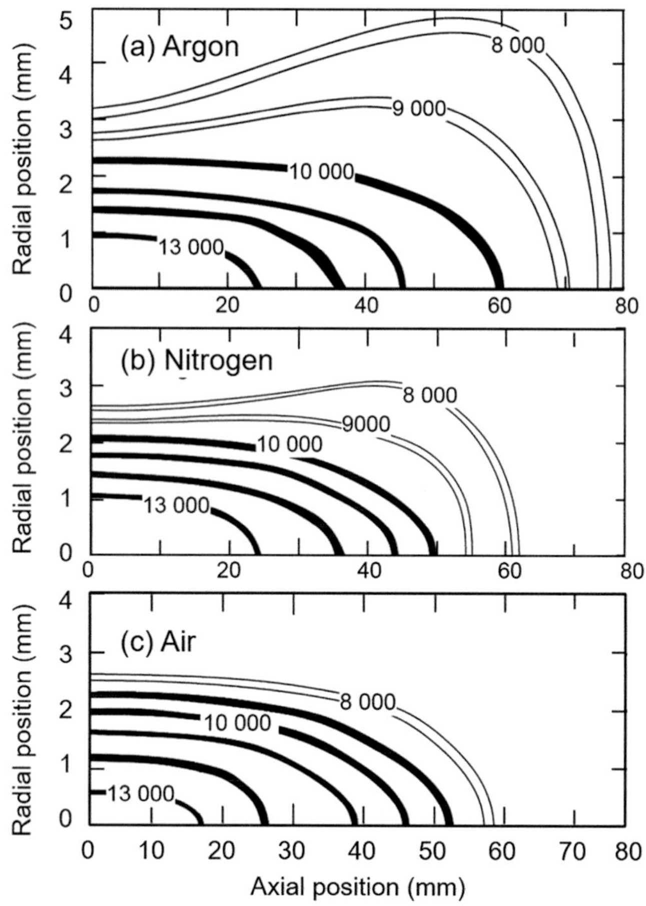

Fig. 8.64 Effect of the ambient gas entrainment on the temperature contours of an argon plasma jet, with air showing the strongest quench effect. (**a**) Argon ambient gas. (**b**) Nitrogen. (c) Air, derived from data published by [Romilhac et al. (1990a, b)]

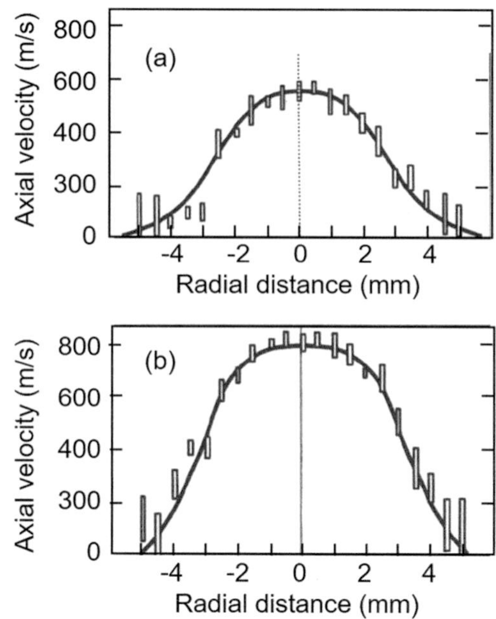

Fig. 8.65 Effect of arc current on the radial profiles of the axial plasma velocity for Ar/H$_2$ plasma gas flow rate of 60 slm (25 vol.% H$_2$), *nozzle i.d.* = *1*0 mm, z = 4 mm from the nozzle exit (**a**) I = 299 A, (**b**) I = 632 A. [Planche et al. (1998)]

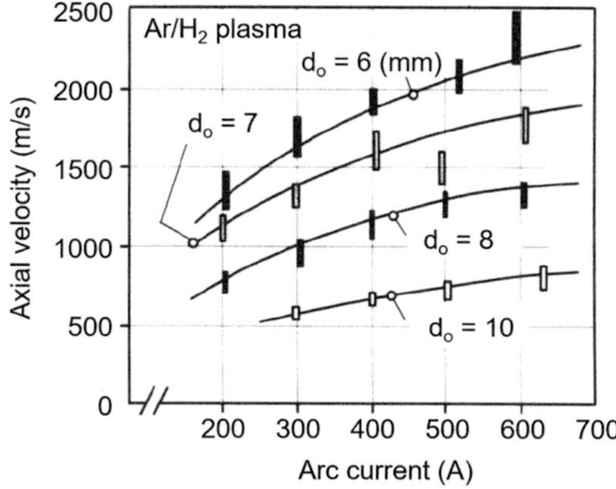

Fig. 8.66 Evolution of the maximum axial plasma velocity on the torch axis, with the arc current for different nozzle i.d.s varying from 6–10 mm, z = 4 mm, Ar/H$_2$ plasma gas flow rate of 60 slm (25 vol.% H$_2$). [Planche et al. (1998)]

the nozzle exit, for a single plasma gas flow rate of (45 slm Ar +15 slm H$_2$) and arc currents of 299 and 632 A. The corresponding arc voltages were respectively 64.5 and 48 V (arc power of 19.3 and 30.3 kW). The results indicate that the increase in the arc current from 299 to 632 A gives rise to an increase in the maximum plasma velocity on the axis of the torch, from 550 to 780 m/s, together with the flattening of the profile. These results are summarized in Fig. 8.66, where the center line velocities at the nozzle exit (4 mm) are plotted as functions of arc current for a 60 slm flow rate and nozzle internal diameters ranging from 6 to 10 mm. As can be expected, the tendencies shown in this figure indicated the strong influences of both the diameter and the arc current intensity on the velocity values.

The effect of the gas flow rate on velocity distributions is illustrated in Fig. 8.67 for the same plasma torch using an 8 mm i.d. anode nozzle and the same plasma gas composition of Ar/H$_2$ (25 vol.% H$_2$). For gas flow rate value of 32, 48, and 68 slm. The arc current was fixed in all three cases at 600 A. The corresponding arc voltage increased steadily

with the increase in the plasma gas flow rate to 45, 55, and 58 V, respectively (torch power 27.0, 33.0, and 24.8 kW). It may be noted that the maximum plasma velocity did not follow the flow rate changes and seemed to be almost constant (1200 ± 100 m/s), despite the fact that the flow rate was increased by a factor higher than 2. As suggested by the solid curves plotted in these figures, by reducing the velocity profile width when it was increased, the flow rate seemed to participate efficiently to the thermal pinch of the arc column.

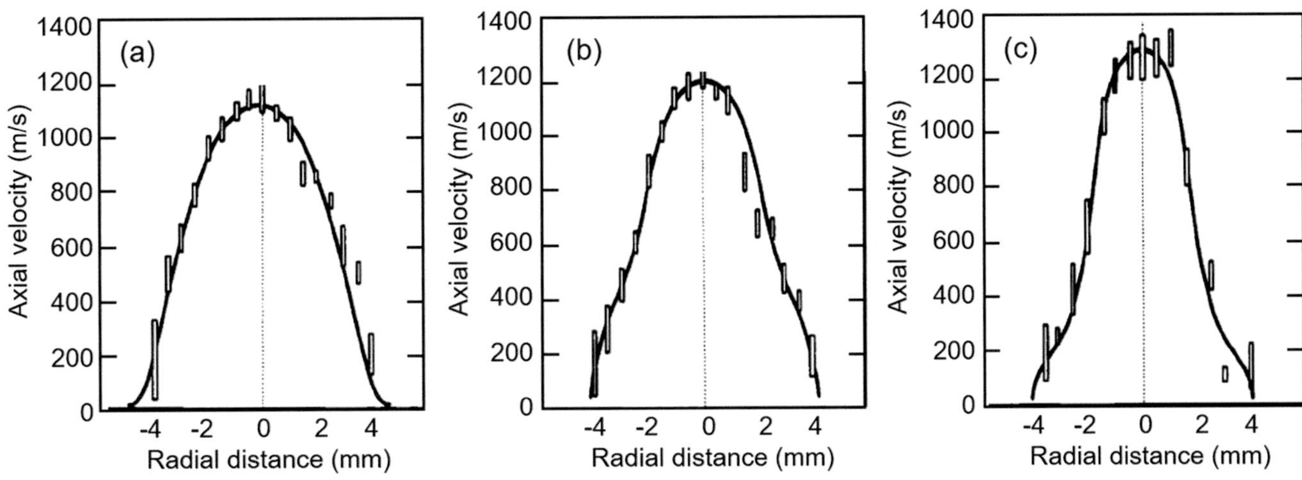

Fig. 8.67 Effect of plasma gas flow rate on the radial profiles of the axial plasma velocity for DC plasma torch, do = 8 mm, z = 4 mm. Plasma gas Ar/H$_2$ (25 vol.% H$_2$) I = 595 ± 5 A, (**a**) 32 slm (**b**) 48 slm; (**c**) 68 slm. [Planche et al. (1998)]

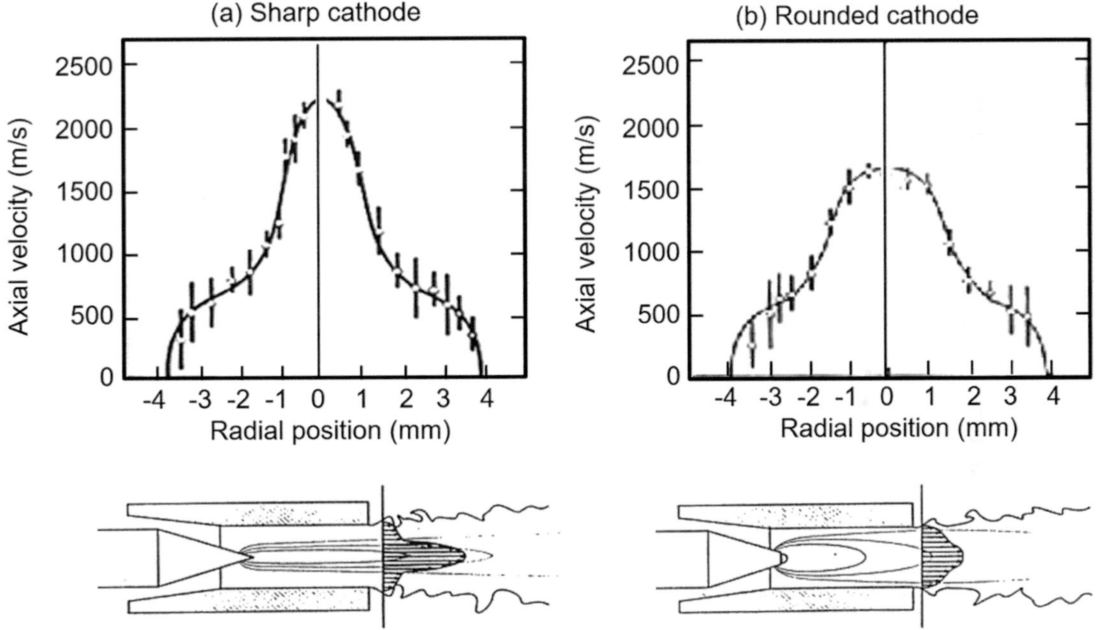

Fig. 8.68 Effect of cathode tip melting and erosion on plasma jet axial velocity profiles, Ar/H$_2$ plasma (45 slm Ar + 15 slm H$_2$), 604 A, 7 mm nozzle i.d. [Planche et al. (1995)]. (Reproduced with kind permission of IPCS)

An interesting observation reported by [Planche et al. (1995)] is the dependence of plasma jet characteristics on the state of erosion of the torch electrodes. Fig. 8.68 shows the velocity profiles obtained under essentially the same conditions (d$_o$ = 7 mm, z = 4 mm, Ar/H$_2$ (25 vol.% H$_2$) flow rate of 60 slm, and arc current 604 A), with a cathode with a sharp conical tip (Fig. 8.68a) and with one that has a rounded tip due to tip melting and erosion wear (Fig. 8.68b). The higher peak velocities and steeper radial profiles with the sharp-tipped cathode are evident. A narrow conical cathode tip will be molten during operation and therefore eroding

quickly due to the ejection of small molten droplets, resulting in a change of cathode shape within the first hour of operation. This change will be accompanied by a change in the plasma jet characteristics, and in the particle heating and acceleration, and consequently in the coating characteristics. From a cathode design point of view, a "burn-in" period will be required before reproducible coatings can be obtained.

Measurements [Hrabovsly et al. (1997)] and modeling studies [Jenista (2017)] were reported of the plasma velocity and temperature fields for a steam plasma torch operating arc currents varying between 300 and 600 A, with corresponding

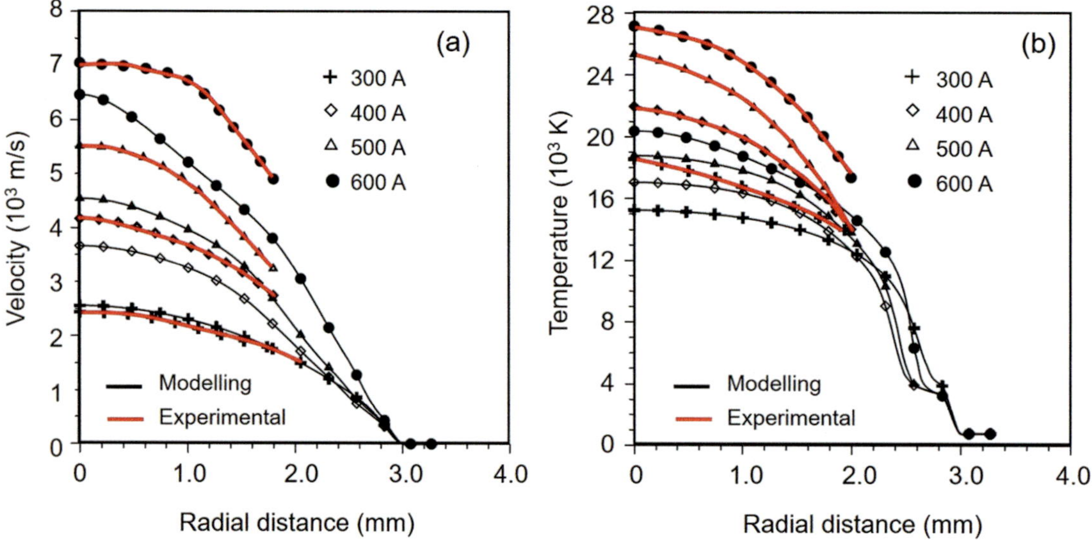

Fig. 8.69 Calculated and experimental radial profiles of the plasma (**a**) velocity and (**b**) temperature for a steam plasma torch as function of the arc current (300–600 A), exit torch nozzle i.d. = 6 mm. [Jenista (2017)]

DC torch power of 55.5 to 123.7 kW; water evaporation rate was estimated to vary between 0.204 and 0.325 g/ s (0.734–1.17 kg/h). The torch nozzle diameter was 6.0 mm, with measurement carried out at a distance of 2 mm downstream of the nozzle exit. The results given in Fig. 8.69 show centerline plasma velocity increasing from 2500 to close to 7000 m/s and centerline maximum plasma temperatures of 18,000 to 27,000 K, with an increase in the arc current from 300 to 600 A. While the measured and model-predicted velocity profiles seem to be consistent in reasonable agreement, important differences are observed in terms of the temperature fields, which vary sometimes by a few thousands of degree K. The data provide, however, some indication of upper velocity and temperature limits meet in spray plasma torches which have been successfully used for the spraying of ceramic coatings, such as alumina, on an industrial production level.

8.4.4   Particle Dynamics in Plasma Flows

As mentioned in "Chapter 5 Plasma and particle dynamics in thermal spray," and earlier in this chapter, the injection of the coating precursor, whether it is as a dry powder, solution, or suspension, is the most critical step in the plasma-spraying process, on which depends, to a large extent, the quality of the coating produced. A slight deviation from near-optimal conditions can lead to widely varying changes in the quality of the coating, whether in terms of coating density, adhesion, and microstructure, or the process efficiency in terms of the fraction of the powder deposited on the substrate. The instabilities and continuous oscillation of the generated plasma jet, as described earlier in Sect. "8.2.4 Arc stability," and "8.4.1 Arc and plasma jet dynamics," and demonstrated by the high-speed photographs given in Fig. 8.70, offer a significant challenge that requires a thorough understanding of particle dynamics under plasma conditions for process optimization. In this figure, the spray particles are injected from the top, and the plasma jet exits the nozzle at the right-hand side of the images. It can be seen that the arc fluctuations lead to uneven particle heating and the entrainment of cold surrounding gas into the jet, rapidly reducing its average temperature and velocity. In this section, the focus will be on the feeding of dry powder using pneumatic transport and injection into the plasma stream. Current DC torches used in plasma spray operations, as described in Sect. "8.3 Plasma Torch Design," mostly allow for the radial injection of the powder into the plasma either internally, in the anode nozzle such as in the SG-100 TAFA-Praxair torch, or externally at the exit level of the nozzle, such as in the F4 MB-XL or Triplex torches by Oerlikon-Metco. The Northwest Mettech Axial III is one of the few DC plasma spray torches that allow for axial injection of the powder into the plasma stream. For simplicity, the discussion in this section will be limited to the radial injection of the powder into the plasma stream. A detailed discussion on particle injection in plasma torches, and particle trajectory and temperature history calculations under plasma conditions can be found in "Chapter 5 Plasma and particle dynamics in thermal spray."

In a conventional plasma-spraying torch, the particles are injected into the plasma jet either inside the anode nozzle or right outside of the nozzle exit, with an injector tube of diameter typically between 1.2 and 2 mm. The direction of injection can be perpendicular to the jet axis or at a positive or

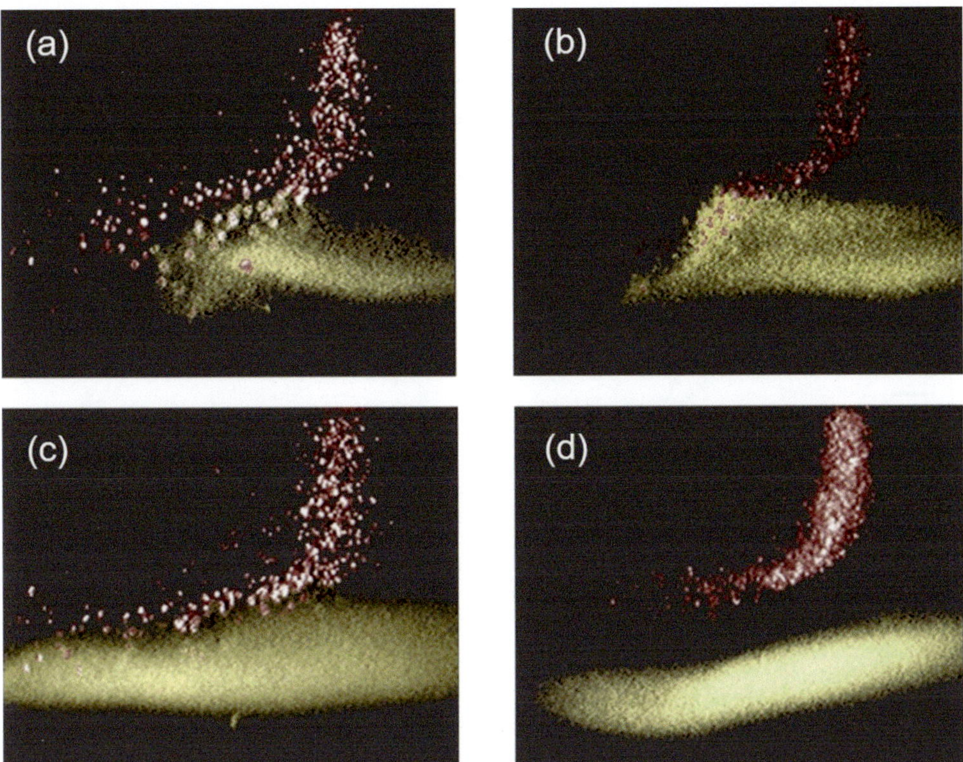

Fig. 8.70 High-speed images showing the fluctuations of the plasma jet and the spray particle fluxes being exposed to varying environments. [Fauchais (2004)]. (Copyright IOP Publishing. Reproduced with permission. All rights reserved)

negative angle with respect to the perpendicular direction. Injection with a negative angle with respect to the injector axis (backward injection) can result in higher particle temperatures but somewhat lower particle velocities [Bisson et al. (2005)]. Internal injection requires careful adjustment of the carrier gas flow; too high flow rates will result in the particles impinging on the opposite side of the anode nozzle. It would also create strong deflection of the plasma jet. As a guideline, the carrier gas mass flow rate should be below 10% of the mass flow rate of the plasma gas.

Among the different independent parameters that control the particle trajectory in a DC plasma jet, the size distribution of the spray powder is one of the most important since the momentum of the particle is proportional to the third power of the particle diameter. For example, if the spray powder has a nominal size distribution of 10–50 μm, the momentum of the particles can vary by a factor of 125 for the same velocity, resulting in vastly different trajectories. In general, small particles remain in the fringes of the jet, never reaching the hot core. However, their residence times in the lower-velocity regions of the jet are longer and the time needed for their melting is shorter because of their lower-mass particles. The large particles will traverse the jet and remain in the fringes on the opposite side, and it is likely that these particles are not fully molten when they hit the substrate. The intermediate-size particles will have the optimal trajectories for particle

heating and melting. In general, it is recommended to use powders with a particle size distribution (PSD) as narrow as possible and adjust the carrier gas flow rate such that the mean particle momentum is equal to the plasma jet momentum flux. It is possible also to reduce the carrier gas flow rate to allow a more favorable trajectory of larger particles. There are several instruments that allow online observation of the particle fluxes/trajectories and that can be used to adjust the carrier gas flow rate to an optimal value for a given particle mass flow rate and torch-operating conditions.

A schematic of a typical arrangement for internal powder injection in a DC plasma torch is given in Fig. 8.71a [Vardelle et al. (2001)], together with radial profile powder distribution at a distance of 70 mm from the exit level of the torch nozzle, Fig. 8.71b [Vardelle et al. (2001)]. The plasma jet in this case was an Ar/H_2 plasma (45 slm Ar + 15 slm H_2), an arc current of 600 A, a torch nozzle i.d. of 7 mm, an injector i.d. of 1.75 mm, and the powder injection point = 3 mm upstream of nozzle exit. Injected alumina powder had a particle diameter (−45 + 22 μm). The result shows a gaussian powder distribution, with its maximum further away from the torch axis with increase in the powder carrier gas flow rate.

The dependence of the particle trajectories on their mass and injection velocities is best demonstrated by the experimental and numerical simulation studies reported by

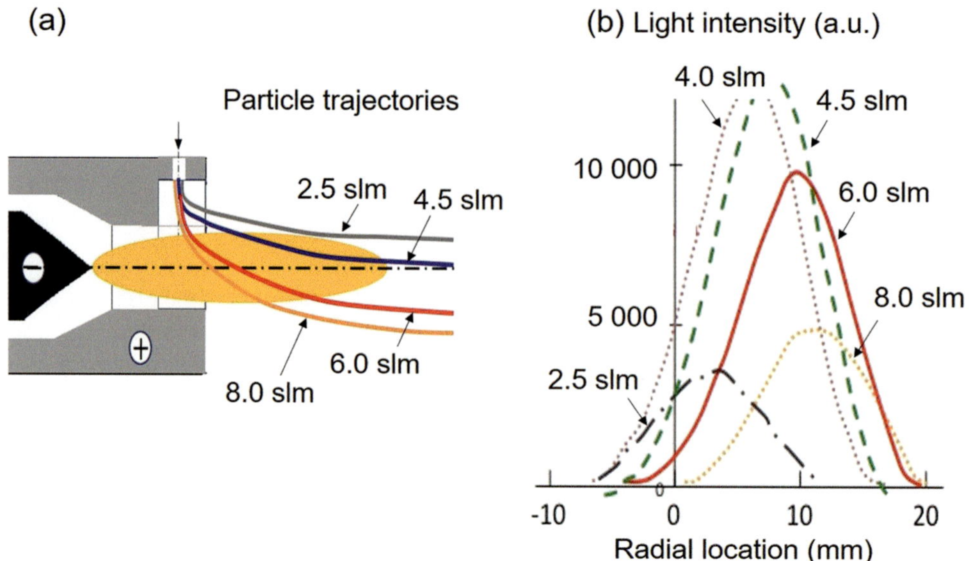

Fig. 8.71 (a) schematic of internal powder injection in a DC plasma torch (b) Effect of carrier gas flow rate on radial distribution of alumina powder (22 to 45 μm) in an Ar/H$_2$ plasma jet (45 slm, Ar + 15 slm H$_2$) P_o = 20 kW, Nozzle i.d. = 7 mm, injector i.d. = 1.8 mm, powder injection = 3 mm upstream of nozzle exit, Measurement = 70 mm downstream of nozzle exit. [Vardelle et al (2001)]

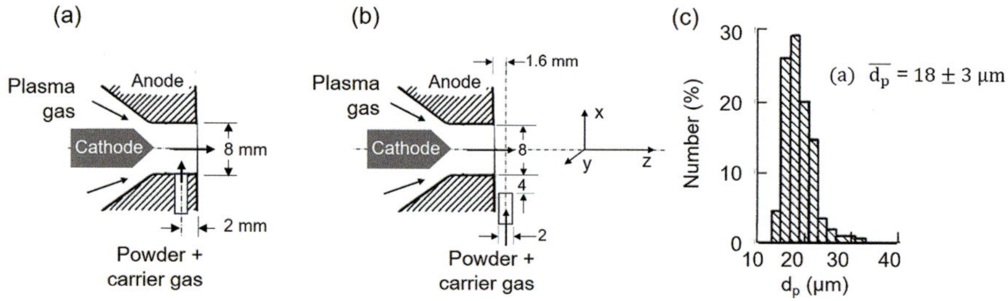

Fig. 8.72 Schematics of the plasma torch nozzle designs used with (a) internal powder injection, (b) external powder injection, and (c) particle size distribution (PSD) of the fine alumina powder used in the study. [Vardelle et al. (1983, 1988)]

[Vardelle et al. (1983, 1988)] for alumina powder injection into Ar/H$_2$ DC plasma jet generated by a torch of conventional design with both internal and external powder injection configurations. In the first study [Vardelle et al. (1983)], the anode nozzle of the plasma torch had an 8 mm i.d., with an internal powder injection port, as illustrated in Fig. 8.72a, with a diameter of 2 mm and its axis 2.0 mm upstream of the nozzle exit. The torch was operated with a mixture of Ar/H$_2$ (75 slm Ar + 15 slm H$_2$), or N$_2$/H$_2$ (37 slm N$_2$ + 11 slm H$_2$) as plasma gas, arc current of 290 A, and voltage 100 V, giving rise to a DC torch power of 29 kW in both cases, and energy efficiency, η, of 63% for the Ar/H$_2$ operation and 72% for the N$_2$/H$_2$ operation. Where, η is defined as the fraction of the DC power that is coupled into the plasma jet at the exit of the torch. Measurements were made of the temperature field in the jet using a combination of emission spectroscopy and probing using thermocouples. Plasma velocity measurements were carried out using Laser Doppler Anemometry (LDA)

after seeding the gas with ultrafine tracer particles. With the injection of course alumina powders into the plasma jet, particle velocities and surface temperature were measured, respectively, using LDA and two-wavelength pyrometry. A summary of the alumina powder characteristics and their corresponding radial injection velocities is given in Table 8.2. Typical results for torch operation with Ar/H$_2$ mixture are given in Fig. 8.73 in terms of the measured axial particle velocity profiles along the centerline of the torch for different powders. Superposed on the graph is the measured axial velocity of the plasma flow along the centerline of the torch, and the results modeling work, as discussed in "Chapter 5 Plasma and particle dynamics in plasma spraying," of the axial velocity of individual particles with the mean particle diameter of the different powders and their corresponding injection velocities in the jet. As expected, the smaller particles, because of their lower inertia, are accelerated much faster by the flow compared to the heavier

Table 8.2 Summary of alumina powder characteristics and their corresponding injection velocities in the plasma flow

Powder	$\overline{d}_p$ (µm)	σ_g	v_{pi}(m/s) (N$_2$/H$_2$) plasma	v_{pi}(m/s) (Ar/H$_2$) plasma
A	18	1.18	25.0	32.0
B	23	1.19	22.0	29.0
C	39	1.20	18.0	24.0
D	46	1.20	14.0	19.0

[Vardelle et al. (1983)]

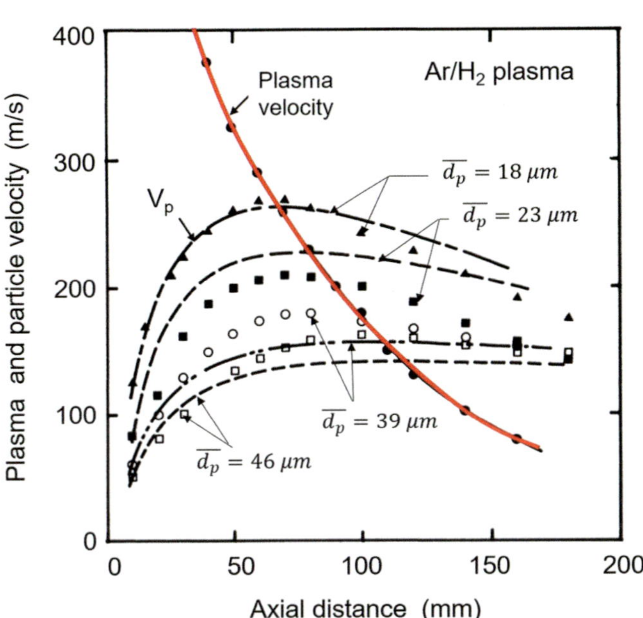

Fig. 8.73 Experimental (symbols) and modeling (solid lines) results of plasma and particle axial velocity along the centerline of Ar/H$_2$ plasma jet. [Vardelle et al. (1983)]. (With kind permission from Springer Science+Business Media B.V)

of 4 mm from the edge of the plasma jet to avoid overheating and melting. The plasma gas flow rate was (75 slm Ar + 15 slm H$_2$) and DC power 29.0 kW. Close-up of the particle trajectories and the number flux powder distribution along two orthogonal axes, at 20 mm from the exit level of the plasma jet, are given in Fig. 8.74. The results given for a carrier gas feed rate of 5.5 slm (Ar) show a skewed distribution in the powder injection plane while a centered symmetrical one in the orthogonal plan. The corresponding values of the distribution of the particle parameters (particle velocity, surface temperature, and number flux density) in the injection plane (x-axis) are given in Fig. 8.75. The measurements carried out using laser anemometry and two-wavelength pyrometry were carried out in a plane at 75 mm downstream from the exit level of the nozzle of the plasma torch for different powder carrier gas flow rates over the range of 5.5 to 8.5 slm (Ar).

Results given in Fig. 8.75a show maximum particle velocities at the centerline of the flow of 210–220 m/s range. The particle velocity drops rapidly with distance from the center to around 180 m/s at ±5 mm. Fig. 8.75b shows the corresponding particle surface temperature distributions, with essentially the same observation of maximum temperature for particles around the axis of the flow, with particle surface temperatures in the 3000–3200 K temperature range, which indicates the alumina particles to be fully molten. The last graph Fig. 8.75c gives the corresponding particle number flux distributions, which shows a strong dependence on the powder injection velocity. Knowing that the injection position is on the negative side of the x-direction, it is consistent to observe the position of maximum flux density shifting in the opposite direction with increase in the powder gas flow rate. While having maximum particle velocities and temperatures seems consistent with the fact that the plasma jet also has its maximum velocity and temperature in the same location, it is disruptive to see that only at a carrier gas flow rate of 5.5 slm that the maximum particle flux density also coincides with the maximum velocity and temperature, meaning that the great majority of the particles would have acquired the necessary energy level for full melting of the particles and their acceleration, which are necessary conditions for obtaining a good-quality coating with a satisfactory deposition efficiency, η_D. The latter is defined as the fraction of the powder feed into the

particles. The latter, however, once they attain their maximum velocity at a distance of about 60 to 80 mm from the nozzle exit, maintain their momentum and eventually move at a higher velocity than that of the plasma at a distance of 140 to 160 mm from the nozzle. It may also be noted that the measurement particle velocity profiles (symbols) are consistent with the modeling results (solid lines), though at time significant differences in their numerical values can be observed. This could be largely due to the significant simplifying assumptions that have been addressed in subsequent modeling studies.

Subsequent studies [Vardelle et al. (1988)] were carried out for essentially the same type of plasma torch, with the exception of using external powder injection, as shown in Fig. 8.72b, and limiting the work to a single alumina powder with the particle size distribution given in Fig. 8.72c, with a mean particle diameter of 18.0 ± 3 µm. The powder is injected at the base of the plasma jet using a 2.0 mm i.d. injector tube, with its axis at 1.6 mm downstream from the surface of the nozzle and its tip withdrawn by a distance

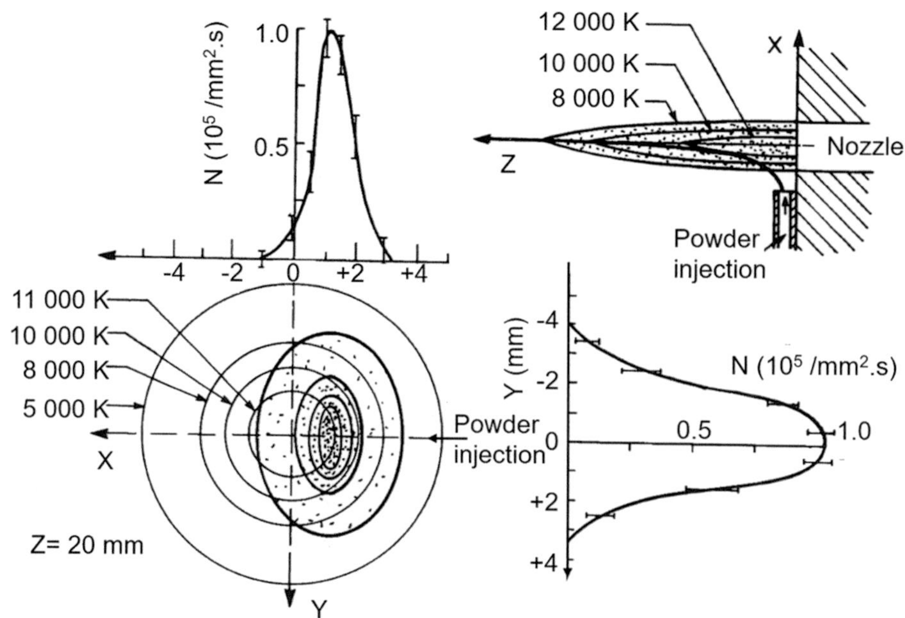

Fig. 8.74 Particle trajectories and number flux density distribution for the injection of the fine powder, 18 ± 3 µm in a 29.2 kW Ar-H$_2$ plasma, 75 slm (Ar) + 15 slm (H$_2$), with a carrier gas flow rate of 5.5 slm (Ar). [Vardelle et al. (1988)]

plasma that are deposited on the substrate and contribute to the formation of the coating. For higher carrier gas flow rates, $Q_i = 7$ or 8.5 slm, it is obvious that an increasing fraction of the powder will not acquire the same conditions in terms of particle velocity, temperature, and degree of melting giving rise to a poorer deposition efficiency. Moreover, since non-molten particles tend to be at the source of defects and porosities in the coating, the quality of the coating would also suffer. It is, therefore, important to stress that when measuring parameters such as the velocity or temperature of particles in a spray environment, it is important to integrate the mass flux density in the analysis in order to evaluate the full picture of the spraying operation.

Special attention has also been given to the anode design and its impact on the plasma flow and consequently on the particle trajectories and their temperature history. While as seen earlier, most commercial plasma torches use anodes with cylindrical nozzle bores, Laval-type nozzles with a convergent/divergent profile have also been used. As observed in Fig. 8.62 [Coudert et al. (1995)], a small change in the nozzle design by having a slight divergence of the nozzle downstream of the arc attachment can significantly change the jet temperature distributions. This type of nozzle provides a somewhat more uniform velocity and temperature distribution at the nozzle exit and a reduced turbulent entrainment of cold gas [Henne et al. (2001)], resulting in more uniform particle heating and acceleration [Roumilhac et al. (1990b), Rahmane et al. (1998)]. There are few general rules on how the nozzle influences the jet because the primary through its impact on the location of the arc attachment. A

smaller nozzle diameter or a constriction of the nozzle downstream of an arcing chamber will always result in shorter arcs and lower temperatures at the nozzle exit, but higher velocities for the same current and mass flow rate of the plasma-forming gas. The effect of a divergent anode nozzle or a Laval-type anode nozzle has been shown to result in higher deposition efficiencies [Schwenk et al. (2004)].

Figure 8.76 [Coudert et al. (1993)] shows that for Al$_2$O$_3$ and YSZ particles, the maximum particle velocity, measured using LDA at a location 100 mm downstream of the nozzle exit, increases linearly with current for a nozzle diameter of 7 mm, with a considerably smaller increase with the current observed for a 10 mm diameter nozzle.

A comparison of plasma jet and particle characteristics obtained with a supersonic nozzle (the authors derived from measurements of the pressure drop across the nozzle a Mach number value of $M = 1.6$ at the nozzle exit) and a commercial $M = 1$ nozzle shows that the cold air entrainment is lower by a factor of 2 in the supersonic jet [Fincke et al. (1993a, b)]. At a distance of 100 mm from the torch, WC:Co particle velocities at the axis are 380 m/s for the supersonic nozzle vs. 200 m/s for the sonic nozzle. Particle temperatures are given as 3200 K for the supersonic nozzle vs. 2300 K for the sonic nozzle. It is interesting to note that the plasma temperatures at the nozzle exit are higher with the sonic nozzle, but the stronger entrainment of ambient air leads to a faster decrease in temperatures, and at an axial position 100 mm from the nozzle, the gas temperatures are about 1600 K for the supersonic jet and about 1100 K for the sonic jet.

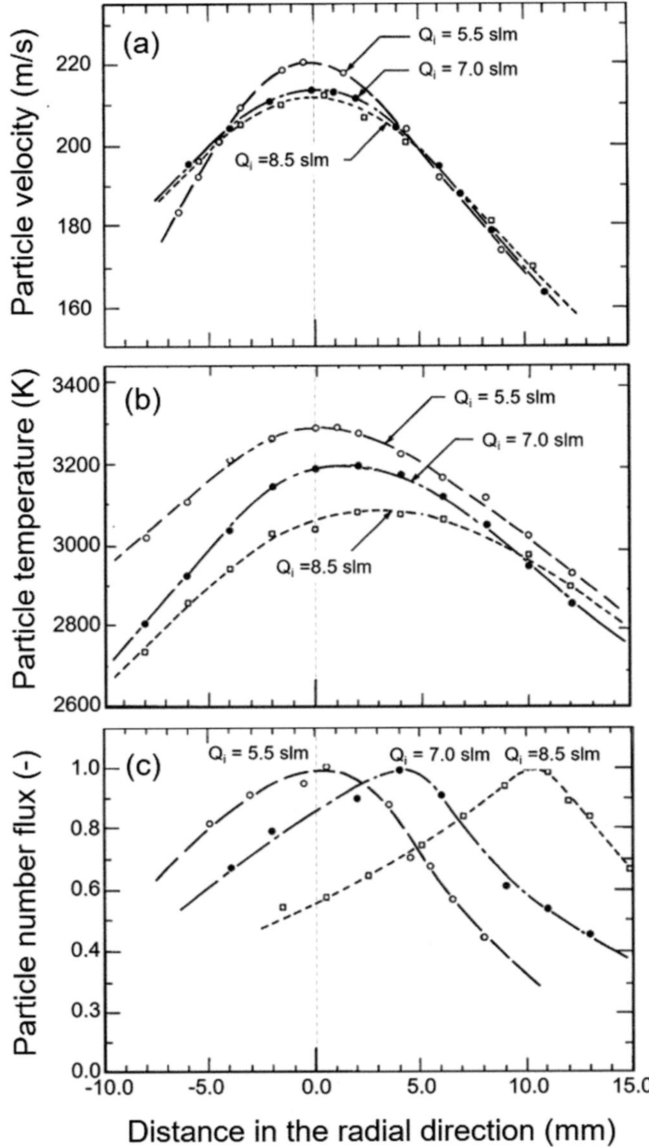

Fig. 8.75 (a) Axial velocity, (b) Temperature, and (c) Number flux density of the particles at 75 mm from the exit of the DC plasma torch operating at atmospheric pressure using 75 slm (Ar) +15 slm (H$_2$) as plasma gas and power of 29.2 kW. Alumina powder d$_p$ = 18 ± 3 μm injection at different carrier gas flow rate, Q$_i$ (Ar). [Vardelle et al. (1988)]

In Fig. 8.77, two different commercial torch nozzle designs (SG-100/175 and SG-100/324) are shown for the SG-100 torch (Praxair-TAFA). The corresponding particle velocity, temperature, particle diameter, and flux distributions are given in Fig. 8.78. Yttria-stabilized Zirconia (YSZ) (20%) particles are injected from the positive y-direction into the plasma jet. The figure shows the particle characteristics in a plane of particle injection (y-direction) perpendicular to the torch axis, as a function of the distance from the torch axis, at an axial location 80 mm from the nozzle exit. Particle velocities, temperatures, diameters, and fluxes are all averaged values

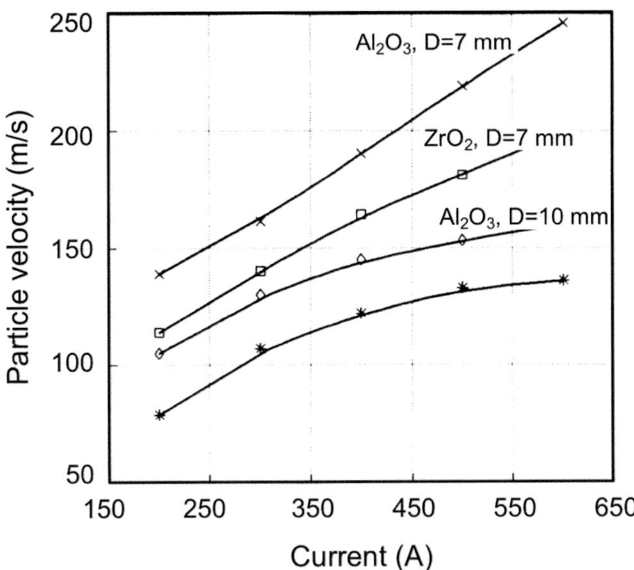

Fig. 8.76 Effect of arc current on particle velocity for two nozzle diameters (7 mm and 10 mm), and two powder materials (Al$_2$O$_3$ and YSZ). [Coudert et al. (1993)]. (Reprinted with permission of ASM International. All rights reserved)

over 1000 particles. The normalized flux shown is the ratio of particles detected at a specific y-location in particles per second, divided by the particle flux at all y-locations. All the data have been obtained with the DPV 2000 instrument (Tecnar Automation, Montreal, Canada), at 800 A arc current and plasma gas flow rate of (48 slm Ar +12 slm He). The powder feed rate was 5.8 g/min with a carrier gas of 2.5 slm (Ar). It may be noted that:

(a) the highest velocities and temperatures are for the particles on the axis;

(b) the highest particle fluxes are below the axis for particle injection above the axis;

(c) the largest particles are farthest from the jet axis, and have lower temperatures and velocities;

(d) the supersonic anode provides higher velocities but lower temperatures.

Comparing these results with those presented by [Fincke and Swank (1992)], one must conclude that the specific design of the supersonic nozzle can affect entrainment and the temperature decay, and that the observed results may be different for different materials.

8.4.5 Plasma Torch and Spray Process Modeling

[Watson and Pegot (1962)] have reported one of the earliest models in this field for the entrance, as well as for the

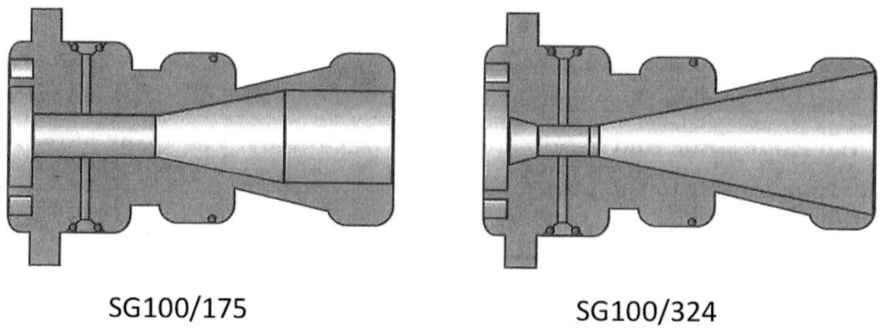

Fig. 8.77 Anode nozzle profiles for the Praxair-TAFA SG 100 torch, No. 175 (left) and No. 324 (right) used in the study. (Reproduced with kind permission of Praxair-TAFA)

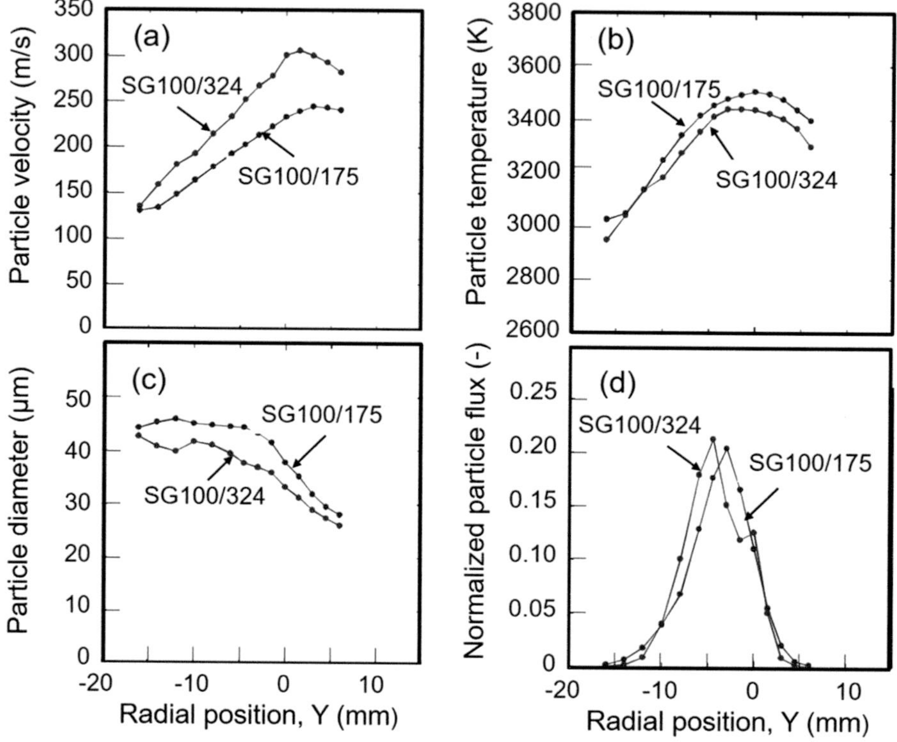

Fig. 8.78 Measurements of the distribution of average values of (**a**) particle velocity, (**b**) particle temperature, (**c**) particle diameter, and (**d**) relative particle flux for two Praxair-TAFA SG-100 anode nozzles. Measurements made at 80 mm from the nozzle exit. 800 A, Ar/He plasma gas (48 slm Ar + 12 slm He), carrier gas flow 2.5 slm (Ar), YSZ powder feed rate 5.8 g/min

asymptotic region of arcs in laminar flow. Fig. 8.79 shows select results for a nitrogen arc, for a current of 580 A, and ambient pressure slightly above atmospheric. The plasma enthalpy (Fig. 8.79a) increases rapidly in the entrance region, reaching a peak before it levels off toward the fully developed (asymptotic) region of the arc. In spite of the high axial velocities (see Fig. 8.79d), the mass flow rate within the constrictor (Fig. 8.79b) is essentially confined to a relatively cold boundary layer close to the wall of the arc confinement tube, especially in the vicinity of the entrance. This effect is due to the low mass density of the plasma in the arc core (see Fig. 8.79c representing $1/\rho$ for N_2), which is a consequence of

the high enthalpies (temperatures) in the arc axis. With increasing distance from the entrance, more and more of the cold gas permeates into the arc.

Numerous models exist for both plasma spray torches and plasma spray processes including the particle trajectories [Pfender et al. (1985), Pfender (1989), Chyou et al. (1989a, b), Li and Pfender (2005)]. Only a brief overview will be given here; a more detailed discussion has been presented in "Chapter 4 Plasma-particle momentum and heat transfer" and "Chapter 5 Plasma and particle dynamics in thermal spray." The reader is also encouraged to consult the relevant literature for details.

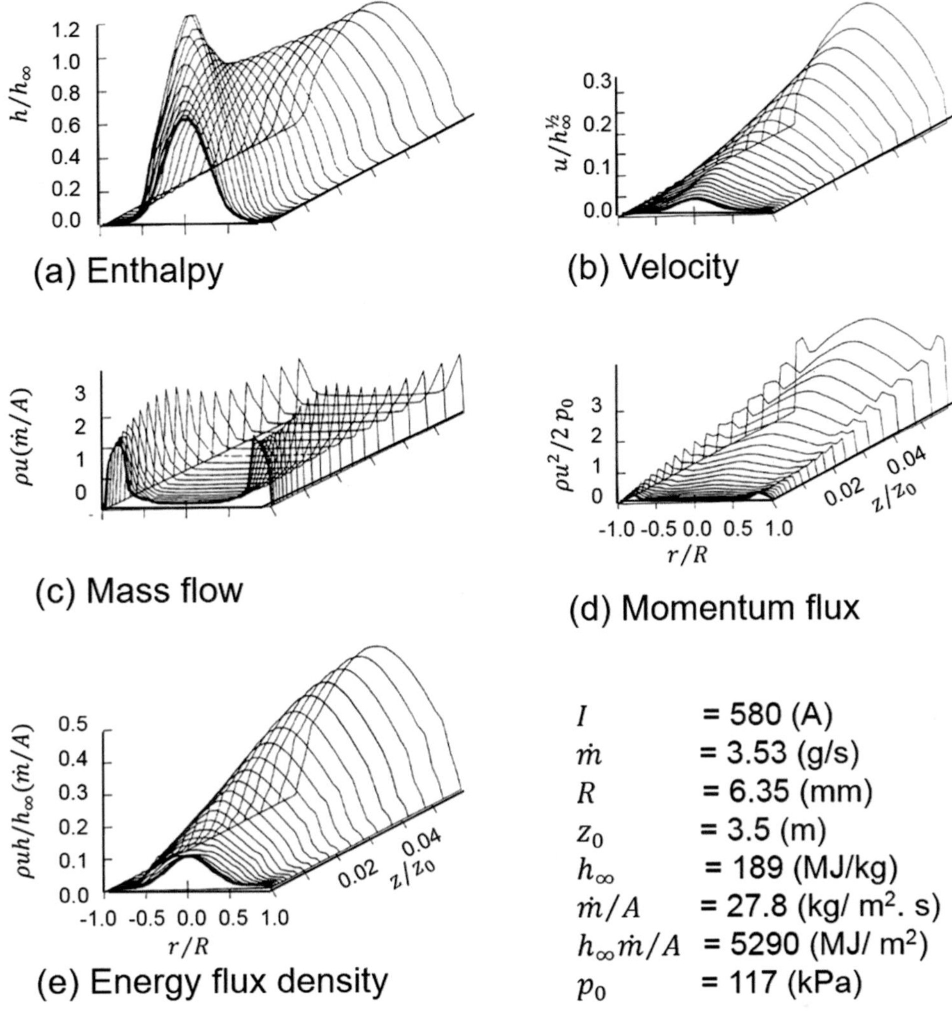

Fig. 8.79 Calculated performance of a wall-stabilized nitrogen arc in axial flow. [Watson and Pegot (1962)]

The traditional models assumed axisymmetric, steady-state plasma flow, allowing a two-dimensional treatment of the torch fluid dynamics. The conservation equations are solved numerically together with the relevant Maxwell equations and thermodynamic equations. Typical boundary conditions are an assumed current density profile at the cathode and an artificially high electrical conductivity in the anode boundary layer to allow current transfer through the low temperature layer while keeping the assumption of LTE. Turbulence is treated using either a K–ε or a Reynolds Stress approach, but usually for low Reynolds numbers. Two-dimensional particle trajectories are calculated in a plane formed by the direction of the particle injection and the torch axis, using the local plasma properties for determining the particle heating and acceleration. Steady-state three-dimensional models have demonstrated the effect of the carrier gas on the temperature and velocity distributions of the plasma jet and the particle trajectories [Li and Pfender (2005)]. The results show good agreement with time-averaged experimental measurements. However, one has to keep in mind that very strong fluctuations exist of the plasma properties, and averages can reproduce only a part of the process. Nevertheless, such models have been used to evaluate operational characteristics of new torch designs, for example, for the Sulzer Metco Triplex torch [Muggli et al. (2007, Molz et al. (2007)], and the potential for expanding their range of operating conditions.

Among the approaches that attempt description of the large-scale turbulence in the jet are the assumption of a time-varying radial profile for the plasma temperatures and velocities at the nozzle exit [Dussoubs et al. (1999), Park et al. (1997, 1999), and the use of a two-fluid model [Huang et al. (1995a, b, c), where one of the fluids is the cold gas moving toward the jet axis, while the other fluid is the hot plasma moving in the outward direction, in addition to the movement in the axial direction (see Fig.8.80) [Huang et al. (1995c]. Calculations have been performed only for argon jets into an argon environment with the assumption of LTE. Both

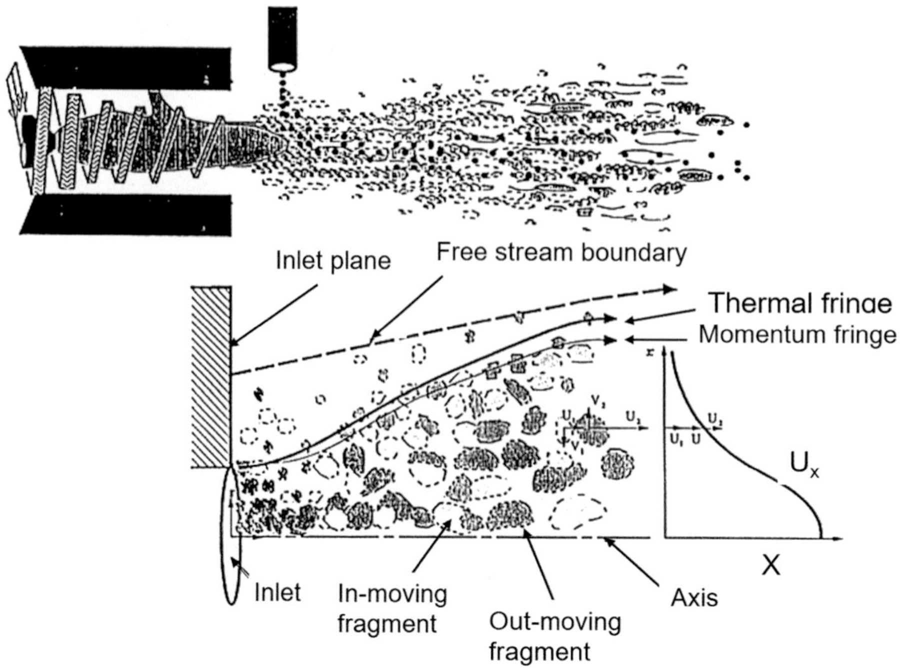

Fig. 8.80 Illustration of two-fluid approach for simulating cold gas entrainment by the plasma jet. [Huang et al. (1995)]. (With kind permission from Springer Science+Business Media B.V)

approaches have demonstrated that particles injected with uniform size and velocity experience different heating and acceleration conditions due to the plasma property fluctuations, resulting in different particle trajectories and temperatures at the substrate location. This is shown in Fig. 8.81 [Huang et al. (1995a), in which the results of the two-fluid model show that because of the different radial velocities in the different gaseous environments, there is a divergence in the particle trajectories (Fig. 8.81a); particles with the same mass and injection velocities can be exposed to plasma velocities of 80 to 400 m/s (Fig. 8.81b) and plasma temperatures of 4000 K to about 11,000 K (Fig. 8.81c) depending on whether the particle sees the hot plasma or the entrained cold gas bubble. This fact leads to a wide range of particle temperatures and velocities at the axial position where the substrate would be located. The two-fluid model has demonstrated further the experimentally confirmed observation that the entrained cold gas bubbles require a significant amount of time to break up and mix with the hot plasma gas, and only at a location typically four to five nozzle diameters from the nozzle exit is the jet composition relatively uniform and less entrainment of larger cold gas bubbles occurs (see Sect. 8.4.2 Ambient gas entrainment by the plasma jet). A simulation of the plasma jet using the LAVA code [Williamson et al. (2003)] showed that even with the assumption of a steady arc and turbulence as modeled using a K–ε approach, a significant amount of entrainment can be predicted being responsible for the rapid axial temperature

drop three to four nozzle diameters downstream of the nozzle exit, results corroborated by experiments [Finke et al. (2003)].

8.5 Summary and Conclusions

The increasing demand for combined functional requirements such as high temperature resistance to corrosive atmospheres, in addition to abrasive wear resistance, and the ease of machining to the final form, at competitive costs has led to the ever-increasing demand for coatings. DC plasma spray coating has emerged as one of the leading plasma spray processes that is presently widely used on an industrial scale. According to [Fauchais et al. (2014)], the motivation for coating structural parts can be summarized by the following needs:

(a) improve functional performance by, for example, allowing higher temperature exposure through the use of thermal barrier coatings,

(b) improve component life by reducing wear due to abrasion, erosion, and corrosion,

(c) extend functional use by rebuilding the worn part to its original dimensions, avoiding the need for replacing the entire component, for example, a shaft or an axle, and,

(d) reduce component cost by improving the functionality of a low-cost material with an expensive coating.

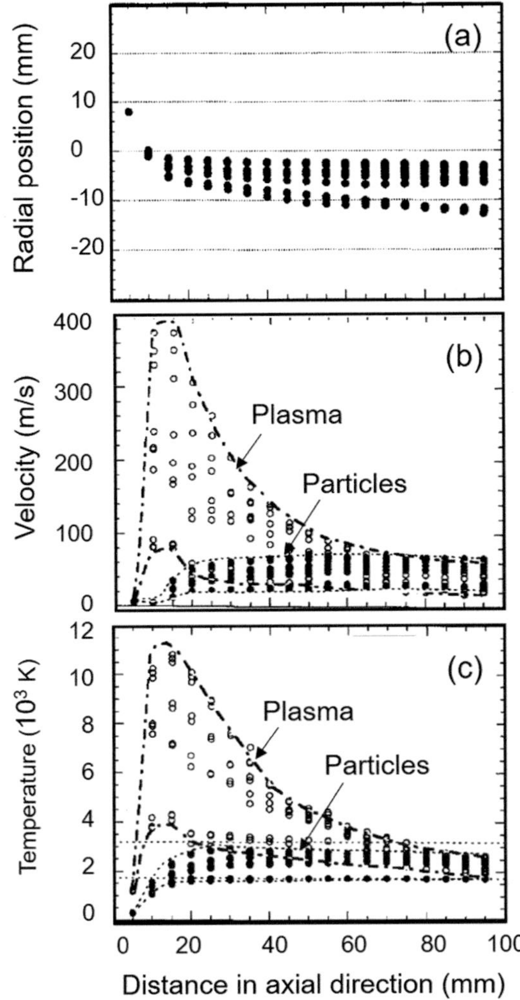

Fig. 8.81 Results of stochastic particle injection model into two-fluid plasma jet; 30 μm diam. Ni particles injected 5 mm downstream of nozzle exit at 5 m/s, anode nozzle i.d. 8.5 mm. Torch operation at 400 A, 23.9 V, thermal efficiency 42.2%, gas flow rate of 35.4 slm (Ar), (**a**) particle trajectories (**b**) axial particle velocities, and (**c**) plasma temperatures; (open symbols) representing plasma values and (closed symbols) representing particle values. [Huang et al. (1995)]. (Reproduced with kind permission of Elsevier)

In each of these uses of coating technologies, there should be no, or minimal, machining required of the coated part.

Because of the intensive R&D efforts in this field over the past few decades, the presentation of this technology has been split into two complementary chapters, with the present chapter dedicated to a discussion on the basic concepts behind the technology, including the properties of electric arcs, arc stability, and electrode erosion. This is followed by a description of the principal design features of DC plasma sources used in plasma spray operations. The important topic of gas and particle dynamics in DC plasma spraying is discussed next, with an emphasis on plasma jet dynamics, ambient gas entrainment mechanism into the plasma jet, and its effect on the flow and temperature fields in the plasma jet. The discussion on particle dynamics presented in this chapter

builds on "Chapters 4 Plasma-particle momentum and heat transfer" and "Chapter 5 Gas and Particle Dynamics in Thermal Spray" covering such topics as particle trajectory and thermal history calculations and plasma-particle interactions.

The different DC plasma spray technologies are presented in the following: "Chapter 9 DC plasma Spray Processes," which includes Atmospheric Plasma Spraying (APS), Controlled Atmosphere Plasma Spraying (CAPS), Vacuum Plasma Spraying (VPS), and the relatively novel process of Ultra-Low-Pressure Plasma Spraying (ULPPS). Detailed discussions on substrate preparation, coating formation, and characterization are covered in Part III, while process integration including powder/wire or cord preparation, instrumentation, industrial applications, and process economics are covered in Part IV of this book.

Nomenclature

Units are indicated in parentheses; when no units are indicated, the parameter is dimensionless.

Latin Alphabet

A	empirical constant (6×10^5 A/m^2 K^2)
A	channel cross-sectional area (m^2)
c_p	specific heat at constant pressure (J/kg.K)
d_o	anode nozzle internal diameter (mm)
d_a'	anode boundary layer thickness (m)
d_c'	cathode boundary layer thickness (m)
d_p	particle diameter (μm)
$\overline{d_p}$	mean particle diameter (μm)
e	electric charge (Coulomb)
E	electric field strength (V/m)
E_x	electric field strength normal to the anode surface (V/m)
h	specific enthalpy (J/kg)
h_{av}	average enthalpy (J/kg)
I	arc current (A)
j_e	electrons current density (A/m^2)
j_i	ions current density (A/m^2)
k	Boltzman constant (1.38×10^{-23} J/K)
$\dot{m}_g$	mass flow rate of plasma gas (kg/s)
$\dot{m}_p$	powder mass flowrate (kg/s)
M	Mach number [$M = v/v_c$]
n_e	electron number density (m^{-3})
p	pressure (Pa)
p_e	partial pressure (Pa)
P_{el}	electric power input into the torch [$P_{el} = I \times V$] (W)
Q_{cond}	losses to the torch wall by heat conduction (W)
Q_{rad}	losses to the torch wall by radiation (W)
Q_{loss}	heat lost to the torch cooling water (W)
Q_i	Carrier gas flow rate (slm)
r	distance in the r-direction (m)
r_o	nozzle radius (mm)
Re	Reynolds number $Re = \rho v d/\mu$
$\overline{R_z}$	segment length to diameter ratio, Eq. 8.10 (−)
T_c	cathode temperature (K)

T_h	heavy species temperature (K)
T_m	Particle melting temperature (K)
T_e	electrons temperature (K)
T_{ref}	reference temperature (K)
V'_a	Anode voltage fall (V)
V'_c	Cathode voltage fall (V
v	gas velocity (m/s)
v_c	Velocity of sound (m/s)
v_p	particle velocity (m/s)
v_{pi}	particle injection velocity (m/s)
x	distance in the x-direction (m)
y	distance in the y-direction (m)
z	distance in the z-direction (m)

Greek Alphabet

δ_z maximum thickness of a segment (mm)

$\Delta\bar{h}$	increase in enthalpy averaged over the cross section per unit length
Δz	distance (m)
η	thermal efficiency of the torch (%)
η_D	deposition efficiency (%)
κ_e	thermal conductivity of electron (W/m.K)
κ_h	thermal conductivity of heavy particles (W/m.K)
μ	dynamic viscosity
ρ	mass density of the gas (kg/m^3)
ρ_p	mass density of the particle or droplet (kg/m^3)
σ	electrical conductivity (A/V.m)
σ_g	geometric standard deviation
σ_p	liquid drop surface tension (N/m)
ϕ_w	work function (eV)

References

Acurex, Aero Therm. Mountain View, CA, 94043.

Ang, Andrew Siao Ming, Noppakun Sanpo, Mitchell L. Sesso, Sun Yung Kim, and C.C. Berndt. 2013. Thermal spray maps: Material genomics of processing technologies. *Journal of Thermal Spray Technology* 22 (7): 1170–1183.

Asmann, M., A. Wank, H. Kim, J. Heberlein, and E. Pfender. 2001. Characterization of the converging jet region in a triple torch plasma reactor. *Plasma Chemistry Plasma Process* 21: 37–63.

Barbezat, G., and K. Landes. 2000a. Plasma torch system triplex, increased production and a more stable process. *Sulzer-Metco Report.*

———. 2000b. Plasma technology Triplex for the deposition of ceramic coatings in the industry. In *Proc. 1st; ITSC-2000 Montreal, Canada*, ed. C.C. Berndt, 881–885. Materials Park: ASM International.

Barbezat, G. 2001. The internal plasma spraying on powerful technology for the aerospace and automotive industries. In *Proc. ITSC, Singapore, 2001*, ed. C.C. Berndt, K.A. Khor, and E.F. Lugscheider, 135–140. Materials Park: ASM International.

Baudry, C., A. Vardelle, G. Mariaux, M. Abbaoui, and A. Lefort. 2005. Numerical modeling of a D.C. non-transferred plasma torch: movement of the arc anode attachment and resulting anode erosion. *High Temperature Material Processes* 9 (1): 1–18.

Belashchenko, V. 2010. *Multi-electrode plasma system and method for thermal spraying.* US Patent 7 750 265 B2 July 6th. 2010.

Betoule, O. 1994. *Influence of velocity and temperature distributions of D.C. plasma jets and alumina particles in flight onto coating properties.* Limoges: University of Limoges.

Bisson, J.F., and C. Moreau. 2003. Effect of direct current plasma fluctuations on in-flight particle parameters: Part II. *Journal of Thermal Spray Technology* 12 (2): 258–264.

Bisson, J.F., B. Gauthier, and C. Moreau. 2003. Effect of plasma fluctuations on in-flight particle parameters. *Journal of Thermal Spray Technology* 12 (1): 38–43.

Bisson, J.F., C. Moreau, M. Dorfman, C. Dambra, and J. Mailon. 2005. Influence of hydrogen on the microstructure of plasma-sprayed Yttria-stabilized zirconia coatings. *Journal of Thermal Spray Technology* 14 (1): 85–90.

Borisov, Y. 1993. *Structure and parameters of heterogeneous plasma jets shielded by water flow.* Proc. 2nd Plasma Technik Symposium, Wohlen, Switzerland, 1–8.

Boulos, M.I. 1992. Radio-frequency plasma development, scale-up and industrial applications. *Journal of High Temperature Chemical Processes* 1: 141–148.

Boulos, M., P. Fauchais, and E. Pfender. 1994. *Thermal plasma fundamentals and applications.* Vol. 1. New York: Plenum.

Boulos, M. I., P. Fauchais, and E. Pfender (2016) Thermal Arcs, Handbook of Thermal Plasmas, (eds.) M. Boulos et al. (eds.) 1–45.

Burgess, A. 2002. Hastelloy C-276 parameter study using the axial III plasma spray system. In *Proc. ITSC-2002, Essen, Germany*, ed. E. Lugsheider, 516–518. Materials Park: ASM International.

Chen, H.C., Z. Duan, J. Heberlein, and E. Pfender. 1996. Influence of shroud gas flow and swirl magnitude on arc jet stability and coating quality in plasma spray. In *Proc. thermal spray: Particle solutions for engineering problems, Cincinnati, Ohio, 1996*, ed. C.C. Berndt, 533–561. Ohio: AMS International.

Chen, H.C., J. Heberlein, and R. Henne. 2000. Integrated fabrication process for solid oxide fuel cells in a triple torch plasma reactor. *Journal of Thermal Spray Technology* 9 (3): 348–353.

Chraska, P., and M. Hrabovsky. 1992. An overview of water stabilized plasma guns and their applications. In *Proc. ITSC-1992, Orlando, Florida*, ed. C.C. Berndt, 81–85. Materials Park: ASM International.

Chyou, Y.P., and E. Pfender. 1989a. Behavior of particulates in thermal plasma flows. *Plasma Chemistry and Plasma processing* 9 (1): 45–71.

———. 1989b. Modeling of plasma jets with superimposed vortex flow. *Plasma Chemistry and Plasma Process* 9 (2): 291–328.

Coudert, J.-F., M.-P. Planche, and P. Fauchais. 1994. Anode-arc attachment instabilities in a spray plasma torch. *Journal of High Temp. Mater. Process* 3: 639–651.

———. 1995. Characterization of D.C. plasma torch voltage fluctuations. *Plasma Chem Plasma Process* 16 (1): 211S–227S.

Coudert, J.F., M.P. Planche, and P.L. Fauchais. 1966. Characterization of DC plasma torch voltage fluctuations. *Journal Plasma Chemistry Plasma Processing* 16 (1): 211S–227S.

Coudert, J.-F., V. Rat, and D. Rigot. 2007. Influence of Helmholtz oscillations on arc voltage fluctuations in a dc plasma spraying torch. *Journal of Physics D: Applied Physics* 40: 7357–7366.

Coudert, J.-F., and V. Rat. 2008. Influence of configuration and operating conditions on the electric arc instabilities of a plasma spray torch: Role of acoustic resonance. *c* 41: 205–208.

Dinulescu, H.A., and E. Pfender. 1980. Analysis of the anode boundary layer of high intensity arcs. *Journal of Applied Physics* 51 (6): 3149–3157

Dorier, J.L., C. Hollenstein, A. Salito, M. Loch, and G. Barbezat. 2000. Influence of external parameters on arc fluctuations in a F4 DC plasma torch used for thermal spraying. In *Proceedings of the 1st ITSC-2000, Montreal, Quebec, Canada*, ed. C.C. Berndt, 37–43. Materials Park: ASM International.

Dorier, J.L., M. Gindrat, C. Hollenstein, A. Salito, M. Loch, and G. Barbezat. 2001. Time-Resolved imaging of Anodic Arc root behavior during fluctuations of a DC plasma spraying torch. *IEEE Transactions on Plasma Science* 29 (3): 494–501.

Duan, Z., S. Janisson, K. Wittmann, J.-F. Coudert, J. Heberlein, and P. Fauchais. 1999a. Effects of nozzle fluid dynamics on the dynamic characterisitcs of a plasma spray torch. In *Proc. UTSC-(1999), Düsseldorf, Germany*, ed. E. Lugscheider and P.A. Kammer, 247–252. Materials Park: ASM International.

Duan, Z., K. Wittmann, J.-F. Coudert, J. Heberlein, and P. Fauchais. 1999b. Effects of the cold gas boundary layer on arc fluctuations. In *Proc. ISPC-14, Prague, Czech Republic*, ed. M. Hrabovsk'y, M. Konrád, and V. Kopeck'y, 233–238. Prague: Institute of Plasma Physics, CAS.

Duan, Z., L. Beall, M.P. Planche, J. Heberlein, E. Pfender, and M. Stachowicz. 1997. Arc voltage fluctuations as an indication of spray torch anode condition. In *Proceedings of thermal spray: A united forum for scientific and technological advances*, ed. C.C. Berndt, 407. Material Park, ASM International.

Duan, Z., and J. Heberlein. 2000. Anode boundary layer effects in plasma spray torches. In *Proc. 1st ITSC-2000, Montreal, Quebec*, ed. C.C. Berndt, 1–7. Materials Park: ASM International.

Duan, Z., L. Beall, J. Schein, J. Heberlein, and M. Stachowicz. 2000. Diagnostics and modeling of an Argon/Helium plasma spray process. *Journal of Thermal Spray Technology* 9 (2): 225–234.

Duan, Z., and J. Heberlein. 2002. Arc instabilities in a plasma spray torch. *Journal of Thermal Spray Technology* 11 (1): 44–51 ex 32.

Dussoubs, B. 1998 *3D modeling of the plasma spray process*. PhD thesis, University of Limoges, Limoges, France (in French).

Dussoubs, B., A. Vardelle, G. Mariaux, P. Fauchais, and N.J. Themelis. 1999. Modeling of simultaneous plasma spraying of two powders. In *Proceedings united thermal spray society, Düsseldorf, Germany, 1999*, ed. E. Lugsheider and P.A. Kammer, 793–798. Materials Park: ASM International.

Dussoubs, B., A. Vardelle, G. Mariaux, N.J. Themelis, and P. Fauchais. 2001. Modeling of plasma spraying of two powders. *Journal of Thermal Spray Technology* 10 (1): 105–110.

Fauchais, P., and A. Vardelle. 1997. Thermal plasmas. *IEEE Trans. on Plasma Science* 25 (6): 1258–1280.

Fauchais, P. 2004. Understanding plasma spraying. *Journal of Physics D: Applied Physics* 37: 86–108.

Fauchais, P., J. Heberlein, M. Boulos. 2014. Thermal spray fundamentals from powder to part, Chapter 5 Combustion spraying systems.

Fincke, J., and W.D. Swank. 1991. The effect of Plasma jet fluctuations on particle time-temperature histories. *In Proceedings of 4th NTSC, Pittsburgh, Pennsylvania, 1991*, ed. T.F. Bernecki, 193–198. Materials Park: ASM International

Fincke, J.R., D.M. Crawford, S.C. Snyder, W.D. Swank, D.C. Haggard, and R.L. Williamson. 2003. Entrainment in high-velocity, high temperature plasma jets, Part I: Experimental results. *International Journal of Heat Mass Transfer* 46: 4201–4213.

Fukanuma, H. 1988. Japanese Patent 230300 JP. 04/24/1988,

Fukumoto, M., I. Ohgitani, H. Nagai, and T. Yasni. 2005. Effect of substrate surface change by heating on flattening behavior of thermal sprayed particles. In *ITSC-2005*, ed. E. Lugscheider. Düsseldorf: DVS. e-proceedings.

Fukumoto, M., H. Nagui, and T. Yasui. 2006. Influence of surface character change of substrate due to heating on flattening behavior of thermal sprayed particles. In *ITSC-2006*, ed. B. Marple et al. Materials Park: ASM International.

Fukumoto, M., T. Yamaguchi, M. Yamada, and T. Yasui. 2007. Splash splat to disk splat transition behavior in plasma-sprayed metallic materials. *Journal of Thermal Spray Technology* 16 (5–6): 905–912.

Fukumoto, M., K. Yang, K. Tanaka, T. Usami, T. Yasui, and M. Yamada. 2011. Effect of substrate temperature and ambient pressure on heat transfer at interface between molten droplet and substrate surface. *Journal of Thermal Spray Technology* 20 (1–2): 48–58.

Gerdien, H., and A. Lotz. 1922. Wiss. Veröff. Siemens-Konz. 489–4.

Gregori, G., U. Kortshagen, E. Pfender, and J. Heberlein. 2003. Time-resolved temperature measurements in a turbulent argon plasma jet. In *Proc. 16th Intern. Symp. Plasma Chemistry, Taormina, Italy, 2003*, ed. R. d'Agostino, P. Favia, F. Fracassi, and F. Palumbo. University of Bari. unpaginated CD.

Guest, C.J.S., and K.G. Ford. 1975. US Patent 3892882,

Huang, P.C., J. Heberlein, and E. Pfender. 1995. Particle behavior in a two-fluid turbulent plasma jet. *Surface and Coatings Technology* 73 (3): 142–151.

Hardwicke, C.U., and Y.-C. Lau. 2013. Advances in thermal spray coatings for gas turbines and energy generation: A review. *Journal of Thermal Spray Technology* 22 (5): 564–576.

Harry, J.E., and L. Hobson. 1979. Production of a large discharge using a multiple arc system. *IEEE Transactions on Plasma Science* 7: 157–162.

Harry, J.E., and R. Knight. 1981. Simultaneous operation of electric arcs from the same supply. *IEEE Transactions on Plasma Science* 9: 248–254.

Haure T., *Multifunctional layers obtained by a multi-technique process*. Ph.D University of Limoges, France, Nov. 2003 (In French).

Henne, R., E. Bouyer, V. Borck, and G. Schiller. 2001. Influence of anode nozzle and external torch contour on teh quality of the atmospheric DC plasma spray process. In *Proceedings of the ITSC-2001, Singapore*, ed. C.C. Berndt, K.A. Khor, and E. Lugsheider, 471–478. Materials Park: ASM International.

Hrabovsky, M. 1992. Plasma generator with water stabilized DC arc. In *Proc. industrial applications of plasma physics, Varenna, Italy, 1992*, ed. W. Hooke and E. Sindoni, 557–562. Italian Physical Society.

Hrabovsky, M., V. Kopecky, and V. Sember. 1995. Water stabilized arc as a source of thermal plasma. In *Proc. Intern. Symp. Heat and Mass Transfer under Plasma Conditions, Cesme, Turkey, 1995*, ed. P. Fauchais, 91–98. New York: Begell House, Inc.

Hrabovsky, M., M. Konrad, V. Kopecky, and V. Sember. 1997. Processes and properties of electric arc stabilized by water vortex. *IEEE Transactions on Plasma Science* 25: 833–839.

Jenista, J. 2017. Steam torch plasma modelling. *Plasma Chemistry Plasma Process* 37: 653–687.

Jenista, J., J. Heberlein, and E. Pfender. 1997a. Model for anode heat transfer from an electric arc. In *Proc. fourth international thermal plasma processes conference, Athens, Greece*, ed. P. Fauchais, 805–815. New York: Begell House, Inc.

———. 1997b. Numerical model of the anode region of high-current electric arcs. *IEEE Transactions on Plasma Science* 25 (5): 883–890.

Leblanc, L., and C. Moreau. 2002. The long-term stability of plasma spraying. *Journal of Thermal Spray Technolgoy* 11 (3): 380–386.

Leveroni, E., A.M. Rahal, and E. Pfender. 1987. Electron temperature measurements in the near-wall region of wall-stabilized arcs. In *Proc. ISPC-8, Tokyo, Japan, 1987*, ed. K. Akashi and A. Kinbara, 346–351. Tokyo: University of Tokyo.

Li, H.P., and E. Pfender. 2005. Three-dimensional effects inside a dc arc plasma torch. *IEEE Transactions on Plasma Science* 33 (2): 400–401.

Liu, S.-H., C.-X. Li, L. Li, J.-H. Huang, P. Xu, Y.-Z. Hu, G.-J. Yang, and C.-J. Li. 2018. Development of long laminar plasma jet on thermal spraying process: Microstructures of zirconia coatings. *Surface & Coatings Technology* 337: 241–249.

Liu, S.-H., S.-L. Zhang, C.-X. Li, L. Li, J.-H. Huang, J.P. Trelles, A.B. Murphy, and C.-J. Li. 2019. Generation of long laminar plasma jets: Experimental and numerical analyses. *Plasma Chemistry and Plasma Processing* 39: 377–394.

Lu, Z.P., and E. Pfender. 1989. Synthesis of AIN powder in a triple torch plasma reactor. In *Proc. ISPC-9, Pugnochiuso, Italy*, ed. R. d'Agostino, 675–680. Bari: University of Bari.

Malmberg, S., K. Leung, J. Heberlein, and E. Pfender. 1995. Particle Trajectory Control for DC Plasma Spraying. In *ITSC-1995 - Current*

status and future trends, Kobe, Japan, ed. A. Ohmori, 371–375. High Temperature Society of Japan.

Huang, P.C., J. Heberlein, and E. Pfender. 1995. Particle behavior in a two-fluid turbulent plasma jet. *Surface and Coatings Technology* 73 (3): 142–151.

Marantz, D.R., and H. Herman. 1991. Plasma generating apparatus and method, US Patent 4982067.

———. 1992. Plasma spray gun and method of use, US Patent 5144110.

Mariaux, G., P. Fauchais, A. Vardelle, and B. Pateyron. 2001. Modeling of the plasma spray process: From powder injection to coating formation. *Jouurnal of High Temperature Materials and Processes* 5: 61–85.

Mariaux, G., and A. Vardelle. 2005. 3-D time-dependent modelling of the plasma spray process. Part 1: flow modelling. *International Journal of Thermal Sciences* 44: 357–366.

Marqués, J.-L., G. Forster, and J. Schein. 2009. Multi-electrode plasma torches: Motivation for development and current state-of-the-art. *The Open Plasma Physics Journal* 2: 89–98.

Mauer, G., R. Vaßen, D. Stöver, S. Kirner, J.L. Marque ́s, S. Zimmermann, G. Forster, and J. Schein. 2011a. Improving power injection in plasma spraying by optical diagnostics of the plasma and particle characterization. *Journal of Spray Technology* 20 (1–2): 3–11.

Mauer, G., R. Vaßen, and D. Stöver. 2011b. Plasma and particle temperature measurements. *Journal of Thermal Spray Technology* 20 (3): 391–406.

Mohanty, M., R.W. Smith, R. Knight, W.L.T. Chen, and J. Heberlein. 1996. Shrouded air plasma processing of lightweight coatings. In *Proceedings of 9th NTSC, Cincinnati, OH, 1996*, ed. C.C. Berndt. Materials Park: ASM International.

Molz, R., R. McCullough, D. Hawley, and F. Muggli. 2007. Improvement of plasma gun performance using comprehensive fluid element modeling II. *Journal of Thermal Spray Technology* 16 (5–6): 684–689.

Moreau, C., P. Gougeon, A. Burgess, and D. Ross. 1995. Characterization of particle flows in an axial injection plasma torch. In *Proc. NTSC-1995, Houston, Texas*, ed. C.C. Berndt and S. Sampath, 141–147. Materials Park: ASM International.

Moreau, E., C. Chazelas, G. Mariaux, and A. Vardelle. 2006. Modeling the restrike mode operation of a DC plasma spray torch. *Journal of Thermal Spray Technology* 15 (4): 524–530.

Muehlberger, E., S.E. Muehlberger, and A. Sickinger. *Modular segmented cathode plasma generator.* US Patent 5298835, 1994.

Muggli, F., R. Molz, R. McCullough, and D. Hawley. 2007. Improvement of plasma gun performance using comprehensive fluid element modeling: Part I. *Journal of Thermal Spray Technology* 16 (5–6): 677–683.

Murphy, A.B. 1996. The influence of demixing on the properties of a free-burning arc. *Applied Physics Letters* 69: 328–330.

Outcalt, D., M. Hallberg, G. Yang, J. Heberlein, P. Strykowski, and E. Pfender. 2007. Diagnostics and control of instabilities in a plasma spray torch. In *Proc. ITSC-18, Kyoto, Japan, 2007*, ed. K. Tachibana. Kyoto University. p unpaginated CD.

Paik, S., P.C. Huang, J. Heberlein, and E. Pfender. 1993. Determination of the arc-root position in a DC plasma torch. *Plasma Chemistry and Plasma Processing* 13 (3): 379–397.

Pan, W., W. Zhang, W. Zhang, and C. Wu. 2001. Generation of long, laminar plasma jets at atmospheric pressure and effects of flow turbulence. *Plasma Chemistry and Plasma Processing* 21 (1): 23–35.

Park, J.H., Z. Duan, J. Heberlein, E. Pfender, Y.C. Lau, and H.P. Wang. 1997. Modeling of fluctuations experienced in N2 and N2H2 plasma jets issuing into atmospheric air. In *Proceedings of 13th international symposium on plasma chemistry, Beijing, P.R. China, 1997,* ed. C.K. Wu, 326–331. International Union of Pure and Applied Chemistry.

Park, J.H., J. Heberlein, E. Pfender, Y.C. Lan, J. Rund, and H.P. Wang. 1999. Particle behavior in a fluctuating plasma jet. In *Proceedings of 2nd international symposium on heat and mass transfer under plasma conditions, Antalya, Turkey, 1999,* ed. P. Fauchais, J. Van der Mullen, and J. Heberlein, 417–424. New York: Academy of Sciences.

Peters, J., F. Yin, C. Borges, J. Heberlein, and C. Hackett. 2005. Erosion mechanisms of hafnium cathodes at high current. *Journal of Physics D: Applied Physics* 38: 1781–1794.

Pfender, E., and Y.C. Lee. 1985. Particle dynamics and particle heat and mass transfer in thermal plasmas. Part I. The motion of a single particle without thermal effects. *Journal of Plasma Chemistry Plasma Processing* 5 (3): 211–237.

Pfender, E. 1989. Multiple arc plasma device with continuous gas jet. US patent 4818837, 1989.

Pfender, E. 1978. Electric arcs and arc gas heaters. In *Gaseous electronics: Electrical discharges,* ed. M.N. Hirsh and H.J. Oskam, vol. 1, 291–398. Academic Press, Inc.

———. 1998. *Multiple arc plasma device with continuous gas jet.* US patent 4818837.

Pfender, E., W.L.T. Chen, and R. Spores. 1990. A new look at the thermal and gas dynamic characteristics of a plasma jet. In *Proc. NTSC-1990, Long Beach, California,* ed. C.C. Berndt, 1–10. Materials Park: ASM International.

Pfender, E., J. Fincke, and R. Spores. 1991. Entrainment of cold gas into thermal plasma jets. *Plasma Chemistry and Plasma Processing* 11 (4): 529–543.

Pfender, E. 1994. Thermal plasma-wall boundary layers. In *Proc. international symposium on heat and mass transfer under plasma conditions, Cesme, Turkey,* ed. P. Fauchais, 223–235. New York: Begell House.

———. 1999. Thermal plasma technology: Where Do we stand and where are we going? *Plasma Chemistry and Plasma Processing* 19 (1): 1–31.

Planche, M.-P., J.-F. Coudert, and P. Fauchais. 1995. Evolution of the plasma flow characteristics, during the first hours, after replacing the set of electrodes. In *ISPC-12, Minneapolis, MN,* ed. J.V. Heberlein, D.W. Ernie, and J.T. Roberts, 1475–1480. Minneapolis: University of Minnesota.

Planche, M.-P., Z. Duan, O. Lagnox, J. Heberlein, J.-F. Coudert, and E. Pfender. 1997. Study of arc fluctuations with different plasma spray torch configurations. In *ISPC-13, Beijing, P.R. China,* ed. C.K. Wu, 1460–1465. International Union of Pure and Applied Chemistry.

Planche, M.P., J.F. Coudert, and P. Fauchais. 1998. Velocity measurements for arc jets produced by a DC plasma spray torch. *Plasma Chemistry and Plasma Processing* 18 (2): 263–283.

Prystay, M., P. Gougeon, and C. Moreau. 1996. Correlation between particle temperature and velocity and the structure of plasma sprayed zirconia coatings. In *Proc. NTSC-1996, Cincinnati, Ohio,* ed. C.C. Berndt, 511–516. Materials Park: ASM International.

Rahmane, M., G. Soucy, M. Boulos, and R. Henne. 1998. Fluid dynamic study of direct current plasma jets for plasma spraying applications. *Journal of Thermal Spray Technology* 7 (3): 349–356.

Ramesh, C.S., D.S. Devaraja, R. Keshavamurthya, and B.R. Sridharb. 2011. Slurry erosive wear behavior of thermally sprayed Inconel-718 coatings by APS process. *Wear* 271 (9–10): 1365–1371.

Rat, V., and J.-F. Coudert. 2011. Improvement of plasma spray torch stability by controlling pressure and voltage dynamic coupling. *Journal of Thermal Spray Technology* 20 (1-2): 28–38

Rat, V., F. Mavier, and J.F. Coudert. 2017. Electric arc fluctuations in DC plasma spray torch. *Plasma Chemistry Plasma Process* 37: 549–580.

Rigot, D., G. Delluc, B. Pateyron, J. Coudert, P. Fauchais, and J. Wigren. 2003. Transient evolution and shift of signals emitted by a DC plasma gun (type PTF4). *Journal of High Temperature Material Processes* 7: 175–186.

Roumilhac, P., J.-F. Coudert, and P. Fauchais. 1990b. Influence of the arc chamber design and the surrounding atmosphere on the characteristics and temperature distribution of Ar-H2 and Ar-He spraying plasma jets. In *Plasma processing and synthesis of materials*, ed. D. Apelian and J. Szekely, vol. 190, 227–333. Pittsburgh: MRS.

Roumilhac, P., J.-F. Coudert, and P. Fauchais. 1990a. Designing parameters of spraying plasma torches. In *Proc. NTSC-1990., Long Beach, CA*, ed. T. Bernecki, 11–19. Materials Park: ASM International.

———. 1990b. Influence of the arc chamber design and the surrounding atmosphere on the characteristics and temperature distribution of Ar-H$_2$ and Ar-He spraying plasma jets. In *Plasma processing and synthesis of materials*, ed. D. Apelian and J. Szekely, vol. 190, 227–333. Pittsburgh: MRS.

Roumilhac, P. 1990. *Contribution to the metrology and understanding of the DC plasma spray torches and transferred arcs for reclamation at atmospheric pressure*. PhD thesis, University of Limoges, Limoges, France (in French).

Santen, S., L. Bentell, B. Johansson, and P. Westerland. 1986. Chapter 9: Applications of plasma technology in ironmaking. In *Plasma technology in metallurgical processing*, ed. J. Feinmann. Warrendale: AIME Iron and Steel Soc.

Schmidt, T., H. Assadi, F. Gärtner, H. Richter, H. Kreye, and T. Klassen. 2009. From particle acceleration to impact and bonding in cold spraying. *Journal of Thermal Spray Technology* 18 (5–6): 794–808.

Schwenk, A., H. Gruner, X. Zimmermann, K. Landes, and G. Nutsch. 2004. Improved Nozzle Design of de-Laval-type Nozzles of the Atmospheric Plamsa Spraying. In *Proc. Proceedings of the International Thermal Spray Conference, Osaka, Japan, 2004*, ed. A. Ohmori, 600–605. Materials Park: ASM International.

Snyder, S.C., A.B. Murphy, D.L. Hofeldt, and L.D. Reynolds. 1995. Diffusion of atomic hydrogen in an atmospheric pressure free-burning arc discharge. *Physical Review E* 52: 2999–3009.

Spores, R., and E. Pfender. 1989. Flow structure of a turbulent thermal plasma jet. *Surface and Coatings Technology* 37 (3): 251–270.

Spores, R., E. Pfender, and J. Heberlein. 1990. Fluid-dynamic characteristics of a plasma spray jet. In *Proc. plasma jets in the development of new material technology, Frunze, USSR*, ed. O.P. Solonenko and A.I. Fedorchenko, 699–716. Utrecht: V.S.P.

Sun, X., and J. Heberlein. 2005. Fluid dynamic effects on plasma torch anode erosion. *Journal of Thermal Spray Technology* 14 (1): 39–44.

Thörnblom, J. 1989. Industrial plasma applications. In *Proc. ISPC-9, workshop on industrial applications*, 25. University of Bari, Department of Di Chemica, Italy.

Trelles, J.P., E. Pfender, and J.V.R. Heberlein. 2006. Multiscale finite element modeling of arc dynamics in a DC plasma torch. *Journal of Plasma Chemistry Plasma Processing* 26: 557–575.

Trelles, J.P., and J.V.R. Heberlein. 2006b. Simulation results of arc behavior in different plasma spray torches. *Journal of Thermal Spray Technology* 15 (4): 563–569.

Trelles, J.P., and J. Heberlein. 2006c. Simulation results of arc behavior in different plasma spray torches. *Journal of Thermal Spray Technology* 15 (4): 563–569.

Trelles, J.P. 2007. *Finite element modeling of flow instabilities in arc plasma torches*. Ph.D. Thesis, University of Minnesota, Minneapolis.

Trelles, J.P., E. Pfender, and J.V.R. Heberlein. 2007a. Modelling of the arc reattachment process in plasma torches. *Journal of Physics D: Applied Physics* 40: 5635–5648.

Trelles, J.P., J. Heberlein, and E. Pfender. 2007b. Non-equilibrium modeling of arc plasma torches. *Journal of Physics D: Applied Physics* 40: 5937–5952.

Trelles, J.P., J. Heberlein, and E. Pfender. 2008. The reattachment process in nonequilibrium arc simulation. *IEEE Transactions on Plasma Science* 36 (4): 1024–1025.

Trelles, J.P., E. Pfender, and J. Heberlein. 2007c. Thermal nonequilibrium simulation of an arc plasma jet. *IEEE Transactions on Plasma Science* 36 (4): 1026–1027.

Trelles, J.P., C. Chazelas, A. Vardelle, and J.V.R. Heberlein. 2009. Arc plasma torch modeling. *Journal of Thermal Spray Technology* 18 (5–6): 728–751.

Trelles, J.P. 2011. Nonequilibrium thermal plasma jet impinging on a substrate. *IEEE Transactions on Plasma Science* 39 (11): 2870–2871.

———. 2013. Computational study of flow dynamics from a dc arc plasma jet. *Journal of Physics D: Applied Physics* 46: 255201. (17pp).

Tucker, R.C., Jr., ed. 2013. *Thermal spray technology Vol. 5A, ASM handbook*. Materials Park: ASM International.

Van den Brook, J., M. Labrot, and D. Pineau. 1987. *Revue Générale de Thermique* 310: 527–541. (in French).

Vardelle, A., M. Vardelle, and P. Fauchais. 1982. Influence of velocity and surface temperature of alumina particles on the properties of plasma sprayed coatings. *Journal of Plasma Chemistry Plasma Processing* 2 (3): 255–291.

Vardelle, M., A. Vardelle, P. Fauchais, and M.I. Boulos. 1983. Plasma-particle momentum and heat transfer: Modelling and measurements. *AIChE Journal* 29 (2): 236–243.

———. 1988. Particle dynamics and heat transfer under plasma conditions. *AIChE Journal* 34 (4): 568–573.

Vardelle, M., A. Vardelle, P. Fauchais, K.-I. Li, B. Dussoubs, and N.J. Themelis. 2001. Controlling particle injection in plasma spraying. *Journal of Thermal Spray Technology* 10 (2): 267–284.

Vardelle A., C. Moreau, J. Akedo, H. Ashrafizadeh, C. C. Berndt, J. Oberste Berghaus, M. Boulos, J. Brogan, A. C. Bourtsalas, A. Dolatabadi, M.l Dorfman, T. J. Eden, P. Fauchais, G. Fisher, F. Gaertner, M. Gindrat, R. Henne, M. Hyland, E. Irissou, E. H. Jordan, K. A. Khor, A. Killinger, Y.-C. Lau, C.-J. Li, Li Li, J. Longtin, N. Markocsan, P. J. Masset, J. Matejicek, G. Mauer, A. McDonald, J. Mostaghimi, S. Sampath, G. Schiller, K. Shinoda, M. F. Smith, A. A. Syed, N. J. Themelis, F.-L. Toma, J. P. Trelles, R. Vassen, and P. Vuoristo. 2016. The 2016 thermal spray roadmap. *Journal of Thermal Spray Technology* 25(8): 1376–1440.

Watson V.R., and E.B. Pegot. 1967. *Numerical calculation for the characteristics of a gas flowing axially through a constricted arc*, NASA, TD, D-4042.

Wigren, J., and K. Täng. 2007. Quality considerations for the evaluation of thermal spray coatings. *Journal of Thermal Spray Technology* 16 (4): 533–540.

Williamson, R.L., J.R. Finke, D.M. Crawford, S.C. Snyder, W.D. Swank, and D.C. Haggard. 2003. Entrainment in high-velocity, high temperature plasma jets, Part II: computational results and comparison to experiment. *International Journal Heat and Mass Transfer* 46: 4215–4228.

Wutzke, S.A., E. Pfender, and E.R.G. Eckert. 1968. Symptomatic behavior of an electric arc with a superimposed flow. *AIAA Journal* 6 (8): 1474–1482.

Yang, G., and J. Heberlein. 2007. The anode region of high intensity arcs with cold cross flows. *Journal of Physics D: Applied Physics* 40: 5649–5662.

Yang, E.-J., G.-J. Yang, X.-T. Luo, C.-J. Li, and M. Takahashi. 2012. Epitaxial grain growth during splat cooling of alumina droplets produced by atmospheric plasma spraying. In *ITSC-2012*. Materials Park: ASM International. e-proc.

Young, R.M., and E. Pfender. 1989. A novel approach for introducing particulate matter into thermal plasmas: The triple-cathode arc. *Plasma Chemistry and Plasma Processing* 9 (4): 465–481.

Zhukov M.F. 1989. *Electric arc generators of low temperature plasma.* High Temperature Dust Laden Jets. VPS, NL.

Zhukov, M.F. 1989a. *Electric arc generators of low temperature plasma. High temperature dust laden jets.* Utrecht: VPS.

———. 1998. Linear direct current plasma torches. In *Proc. thermal plasma and new materials technology,* ed. O.P. Solonenko and M.F. Zhukov, 9–21. Cambridge, UK: Cambridge Interscience.

Zhukov, M.F. 1994. Linear direct current plasma torches. In *Proceedings of thermal plasma and new materials technology,* ed. O.P. Solonenko and M.F. Zhukov, 9–21. Cambridge: Cambridge Interscience.

Zierhut, J., P. Haslbeck, K.D. Landes, B. G, M. Muller, and M. Schutz. 1998. TRIPLEX - an innovative three-cathode plasma torch. In *Proceeding of the international thermal thermal spray conference, Nice, France, 1998,* ed. C. Coddet, 1374–1380. Materials Park: ASM International.

Zhukov, M.F., ed. 1977. *Electric arc plasmatrons.* The USSR Academy of Science, Siberian Chapter, Institute of Thermal Physics, Novosibirsk/USSR.

———. 1989b. *Electric arc generators of low temperature plasma. High temperature dust laden jets.* Utrecht: VPS.

Zhou, X., and J. Heberlein. 1998. An experimental investigation of factors affecting arc-cathode erosion. *Journal of Physics D: Applied Physics* 31 (19): 2577–2590.

Zhukov, M.F. 1994. Linear direct current plasma torches. In *Proc. thermal plasma and new materials technology,* ed. O.P. Solonenko and M.F. Zhukov, 9–21. Cambridge, UK: Cambridge Interscience.

Zhukov, M.F., and I.M. Zasypkin, 2007. *Thermal plasma torches: Design, characteristics, applications,* 596 pages. Cambridge Int Science.

Abbreviations

APS	Atmospheric Plasma Spraying
BET	Brunauer, Emmett and Teller
BL	Boundary Layer
CAPS	Controlled Atmosphere Plasma Spraying
CCD	Charge Coupled Device
CFD	Computational Fluid Dynamics
CNT	Carbon NanoTubes
CVD	Chemical Vapor Deposition
DC	Direct Current
EB-PVD	Electron Beam-Physical Vapor Deposition
ITSC	International Thermal Spray Conference
KH	Knoop Hardness
LPPS	Low Pressure Plasma Spraying
LPPS-TF	Low Pressure Plasma Spraying-Thin Film
LTE	Local Thermodynamic Equilibrium
PS-PVD	Plasma Spraying-Physical Vapor Deposition
PTA	Plasma Transferred Arc
PVD	Physical Vapor Deposition
SOFC	Solid Oxide Fuel Cells
SPS	Suspension Plasma Spraying
TBCs	Thermal Barrier Coatings
TEM	Transmission Electron Microscopy
ULPPS	Ultra Low-Pressure Plasma Spraying
VLPPS	Very Low-Pressure Plasma Spraying
VPS	Vacuum Plasma Spraying
VPSF	Vacuum Plasma Spray Forming
YSZ	Yttria Stabilized Zirconia

9.1 Introduction

As discussed in Chap. 8, DC plasma spraying stands out as one of the most widely used thermal spray technologies that is broadly applicable for tribological and wear resistance, corrosion and/or oxidation resistance, thermal protection, biomedical applications, and the deposition of free-standing spray-formed parts. The technology is based on a simple concept of in-flight melting of the material to be sprayed in the form of fine dispersed powder, solution or suspension, followed by the deposition of the molten droplets on the substrate to be sprayed on which they form splats freezing on impact. The coating is thus formed through the accumulation of successive layer of splats which can grow to hundreds of micrometers thick and more.

The motivation for coating process can be summarized by the following needs:

(a) Improve functional performance, for example, by allowing higher temperature exposure through the use of thermal barrier coatings.

(b) Improve component life by reducing wear due to abrasion, erosion, and corrosion.

(c) Extend functional use by rebuilding the worn part to its original dimensions avoiding the need for replacing the entire component, for example, a shaft or axle.

(d) Reduce component cost by improving functionality of a low-cost material with an expensive coating. In each of these uses of coating technologies, there should be no, or minimal, machining required of the coated part.

Because of the intensive R&D efforts in this field over the past few decades, the presentation of this technology has been split into two complementary chapters, with "Chapter 8 DC Plasma Spraying –Fundamentals" dedicated to a discussion of basic concepts behind the technology including properties of electric arcs, arc stability, and electrode erosion. This is followed by a description of the principal design features of DC plasma sources used in plasma spray operations. The important topic of gas and particle dynamics in DC plasma spraying is discussed next with emphasis on plasma jet dynamics, ambient gas entrainment mechanism into the plasma jet, and its effect on the flow and temperature fields

© Springer Nature Switzerland AG 2021
M. I. Boulos et al. (ed.), *Thermal Spray Fundamentals*, https://doi.org/10.1007/978-3-030-70672-2_9

in the plasma jet. The discussion on particle dynamics builds on "Chapters 4 Plasma-particle momentum and heat transfer" and "Chapter 5 Gas and Particle Dynamics in Thermal Spray" covering such topics as particle trajectory and thermal history calculations and plasma–particle interactions.

The present chapter follows up with a description of DC Plasma Spray technology under a wide range of conditions including:

- Atmospheric Plasma Spraying (APS)
- Controlled Atmosphere Plasma Spraying (CAPS)
- Vacuum Plasma Spraying (VPS)
- Ultra-Low-Pressure Plasma Spraying (ULPPS)

Detailed discussions of substrate preparation, coating formation, and coating characterization are covered in Part III, while process integration including powder/wire or cord preparation, instrumentation, industrial applications, and process economics are covered in Part IV of this book.

9.2 Atmospheric Plasma Spraying

Atmospheric Plasma Spraying (APS) is the simplest form of plasma spraying operations in which the plasma torch is operated in an open discharge mode generating a high temperature, high velocity plasma jet in which the coating material is injected in the form of fine powder, liquid solution, or suspension. As the individual particles are entrained by the plasma gas, they are heated, melted, and accelerated by the flow before being projected at high speeds toward the substrate on which they impact forming splats which are the building blocks of the coating. When feeding solutions or suspension, the liquid component is first evaporated leaving behind small gains of solid material that on further heating are partially or completely melted before being projected toward the substrate forming the coating. As will be discussed further in this section, the quality of the coating depends on a large number of parameters reflecting the plasma jet characteristics, the powder/precursor properties and its injection conditions, the substrate preparation, and stand-off distance between the plasma torch and the substrate. By operating a TS system in an open discharge mode, the entrainment and mixing of ambient air with the plasma jet offers an added complexity to the process due to the quenching of the jet by the entrained air, resulting in a modification of the jet structure and of the temperature field to which the sprayed particles are exposed. While the effect is of no serious consequences when spraying oxide ceramics, it can be responsible for the partial oxidation of metallic particles and oxidation-sensitive ceramics. However, the economic advantage of the APS process and the lack of any limitation regarding the shape and dimensions of the part to

be coated in the open discharge mode of operation make it one of the most competitive and commonly used configurations of the spray of ceramics and high melting temperature metals and alloys.

9.2.1 Atmospheric DC Plasma Spraying Equipments

The principal components of a DC plasma spray system/ process are illustrated in Fig. 9.1. Central to this system is the DC plasma torch which was described in great detail in Sect. 8.3 "Plasma Torch Design." The system also constitutes an electrical power supply, high frequency starter, gas supply, powder or liquid/suspension feeder and a closed loop high pressure cooling water circuit. A Process Control Console allows adjustment of the operating parameters, that is, arc initiation, the control of arc current, plasma gas, and powder and carrier gas flow rates. The Control Console also houses safety interlocks to avoid starting the arc without cooling water flow or primary plasma gas flow. Auxiliary equipments are necessary for system operation, which include a manual or robotic torch controller, substrate support, spray booth, and general instrumentation and control subsystems.

The power supply is usually a current controlled rectifier. While thyristor-controlled rectification is still being frequently used, modern rectifiers are designed using high frequency switching to reduce the size of the inductor necessary for obtaining a smooth DC current. The problem with thyristor-controlled rectifiers is that there is an inherent harmonic of the DC current which is noticeable in the jet as a 300/360 or 600/720 Hz power ripple. Inverter type power supplies rely on high frequency (>15 kHz) switching to deliver smooth current. The open circuit voltages are typically in the 100 V to 200 V range. As a rule of thumb, it is necessary to have an open circuit voltage about twice ($2\times$) the working voltage. To establish the arc initially, a starting circuit is employed, consisting of a high voltage transformer and capacitor, which allows the breakdown of a spark gap and the initiation of the current flow. However, this type of starter generates a high frequency noise signal detrimental to electronic devices, which must be disconnected automatically during arc start-up stage.

The plasma gas supply consists of two or more (up to 12 or 24) high pressure gas cylinders, with the gas flow rates controlled separately at the control console, frequently by means of sonic orifices or mass flow controllers. The gases are mixed and introduced into the plasma torch. The primary gas should be heavy and must efficiently push the arc root downstream. The most frequently used primary gas is argon because of its low breakdown voltage for arc ignition and low energy density. It is often necessary to add a secondary gas

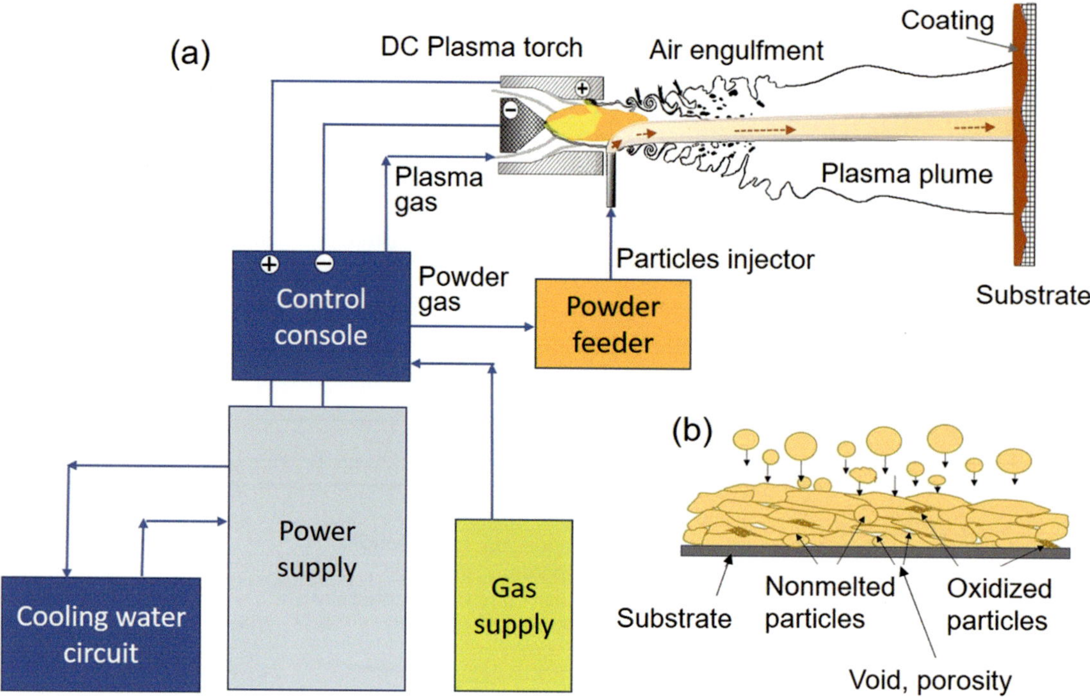

Fig. 9.1 Schematic of a plasma spray system showing plasma torch, cooling water supply, gas supply, power supply, spray powder feeder, and control unit

such as hydrogen or helium to the basic argon flow, to increase the plasma power density, gas velocity, and the heat transfer rates to powders. Nitrogen, carbon dioxide, or water vapor are also used as secondary gas, or even as primary plasma gases in specially designed torches. For applications requiring high deposition rates and where high-power levels are desirable, mixtures of nitrogen and hydrogen are used. It must be emphasized that it is the mass flow rate that is important for controlling the arc inside the torch and not the volumetric flow rate. Increasingly, ternary gas mixtures are being also used as plasma gases, such as Ar-H_2-N_2 or Ar-H_2-He, to tailor the plasma properties for optimal heat and momentum transfer to the particles while keeping electrode erosion at acceptable levels [Janisson et al. (1998); Fauchais and Vardelle (2003); Fauchais (2004)].

The cooling water supply consists of a closed-loop, de-ionized or distilled, water circuit. Depending on the plasma torch design, the water pressure should be maintained in excess of 1 MPa (146 psig), preferably between 1.2 and 1.7 MPa (175–247 psig) and cooled by a heat exchanger placed on the return line of the cooling water circuit. The water-flow control valve should also be placed on the return line of the cooling circuit in order to maintain a high backup pressure in the torch circuit to avoid film boiling under the high heat flux conditions in the anode region. The necessary cooling water flow rate is determined depending on the torch power rating, the ambient cooling water supply temperature,

and the allowable temperature rise. For safety reasons, cooling water temperature at the exit of the torch should not be allowed to exceed 50–60°C.

The spray powder supply system consists of a powder hopper, which is heated and vibrating to avoid powder agglomeration and moisture pick-up. The powder feeding rate is controlled using either a rotating screw system or a rotating wheel or disc with a scraper or a slot, mounted at the bottom of the powder hopper. From the feeder, the powder is transported by the carrier gas to the torch using a pneumatic line. Depending on the powder particle size distribution, the material density, and the i.d. of the powder injector probe, the required powder injection velocity should be set for optimal particle trajectory in the plasma stream. Based on the required powder injection velocity and the i.d. of the powder injector, the carrier gas flow rate is calculated using a mean velocity of the carrier gas in the injector that is double (2×) that of the particle injection velocity. Once the carrier gas flow rate was set, the i.d. of the powder transport line is calculated with the objective of maintaining a carrier gas velocity in the transport line above (10 m/s) to avoid particle deposition and partial blockage of the transport line and/or pulsating slug flow. Assuring good and steady powder flow by avoiding any accumulation of powder in a section of the supply line where the flow could slow down is of utmost importance because next to the torch operational stability, the stability of the powder injection has the strongest influence on coating quality.

Additional ancillary equipments consist of robots moving the torch and the part to be sprayed such that the particle-laden plasma jet impinges as much as possible perpendicularly to the surface of the part being sprayed. A spray booth with an efficient filtration system is necessary to avoid contamination of the environment with hazardous fumes generated during the process through partial vaporization of the powder and its subsequent condensation in the form of nanometric-sized fumes. A spray booth also shields the operator from the ultraviolet radiation from the plasma jet and the relatively high level of acoustic noise generally associated with the process. A detailed analysis on plasma spray system integration is given in Chap. 18 "System Integration."

9.2.2 Plasma Spraying Parameters

The coating properties, as for most thermal spray processes, depend on a large array of process parameters. These can be grouped under the following five primary sub-systems as presented in Table 9.1.

Table 9.1 Factors influencing the coating properties in plasma spray process

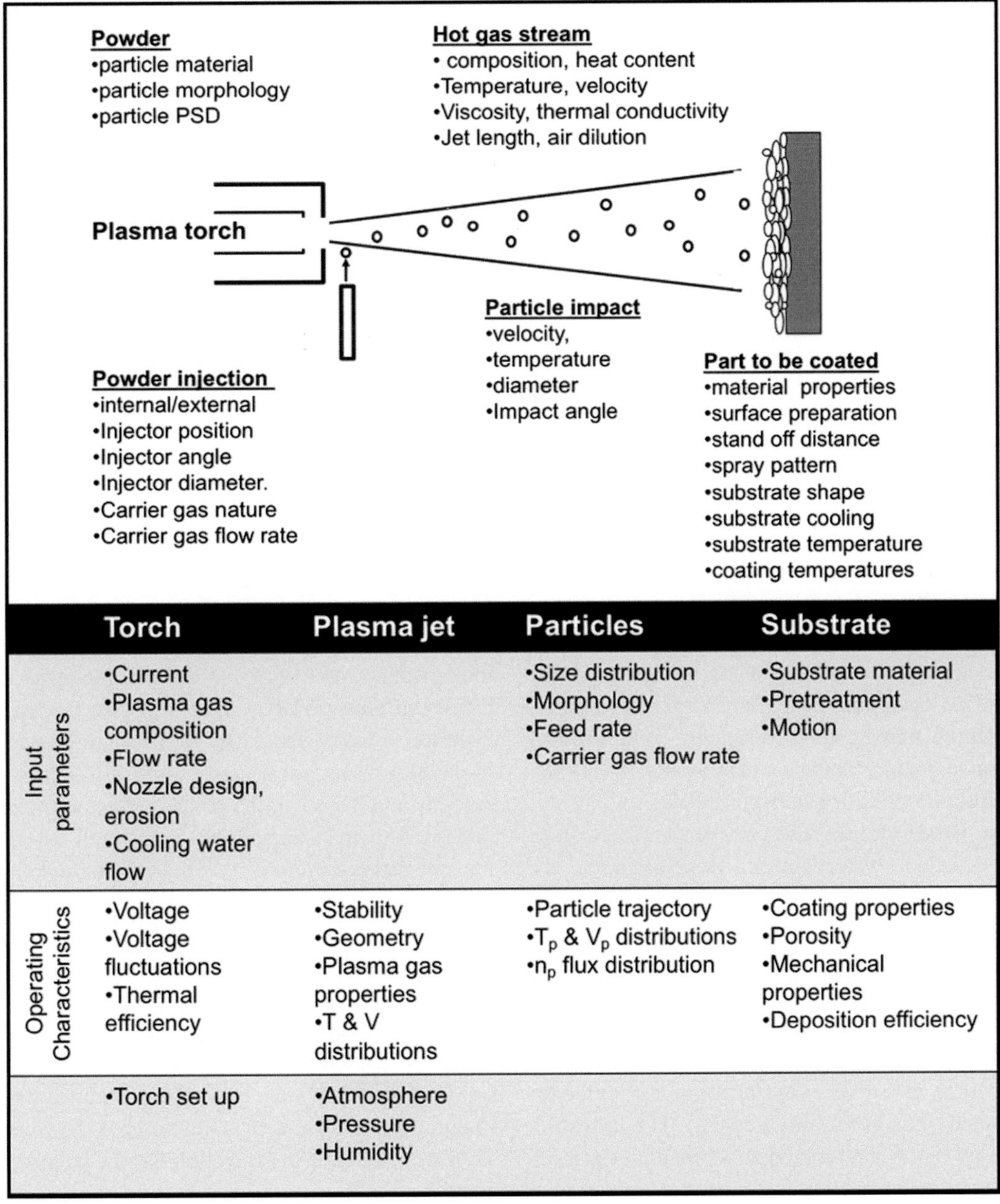

	Torch	Plasma jet	Particles	Substrate
Input parameters	•Current •Plasma gas composition •Flow rate •Nozzle design, erosion •Cooling water flow		•Size distribution •Morphology •Feed rate •Carrier gas flow rate	•Substrate material •Pretreatment •Motion
Operating Characteristics	•Voltage •Voltage fluctuations •Thermal efficiency	•Stability •Geometry •Plasma gas properties •T & V distributions	•Particle trajectory •T_p & V_p distributions •n_p flux distribution	•Coating properties •Porosity •Mechanical properties •Deposition efficiency
	•Torch set up	•Atmosphere •Pressure •Humidity		

- *Plasma torch parameters* are centered around their specific design and independent operating conditions, including torch current, plasma gas composition and flow rate, cooling water pressure, and flow rate. The corresponding torch-dependent parameters are the mean torch voltage and standard deviation which is an indication of the level of arc fluctuations, and the overall torch energy efficiency.
- *Plasma jet characteristics,* which are dependent on the torch/nozzle design and torch operating conditions such as temperature and velocity distributions and jet stability, thermal conductivity, and viscosity of the plasma gas.
- *Spray particle parameters* include material properties such as density, thermal conductivity, specific heat, melting and vaporization temperatures, latent heat of melting and of vaporization, surface tensions, and emissivity of the molten material, molten material. To this, the collective powder properties need to be added which includes particle size distribution (PSD), its mean and standard deviation, particle porosity, morphology, and flowability.
- *Particle injection parameters*, which include injector dimensions, position and orientation with respect to the plasma flow, as well as carrier gas flow rate. These would translate into dependent variables of particle injection such as particle injection velocity and temperature. Powders being composed of particles of different sizes and shapes will be injected into the plasma with a broad range of particle velocities and spray patterns for a given carrier gas flow rate.
- *Substrate parameters*, such as substrate material, its thermophysical properties, substrate shape, and its principal dimensions as well as surface preparation in cleanliness and roughness. It is also necessary to avoid excessive gas adsorption which can interfere with splat formation and reduce coating adhesion. Substrate temperature is also another important substrate parameter which has to be closely controlled during the spray coating operation due to its impact on splat formation and the properties of the coating obtained, as discussed later in Sect. 9.2.4. shape.

9.2.3 Substrate Preparation

Thermal spraying begins with proper surface preparation, which is absolutely essential. Steps must be undertaken correctly in order for the coating to perform the design expectation [Davis (2004)]. Without surface preparation failure of the coating becomes highly probable because coating adhesion quality is directly related to the cleanliness, the roughness, and sometimes the proper machining for optimal coating performance.

The coating material and the nature of the substrate are the major factors in determining what kind of surface preparation is necessary to achieve a resistant bonding. It is also

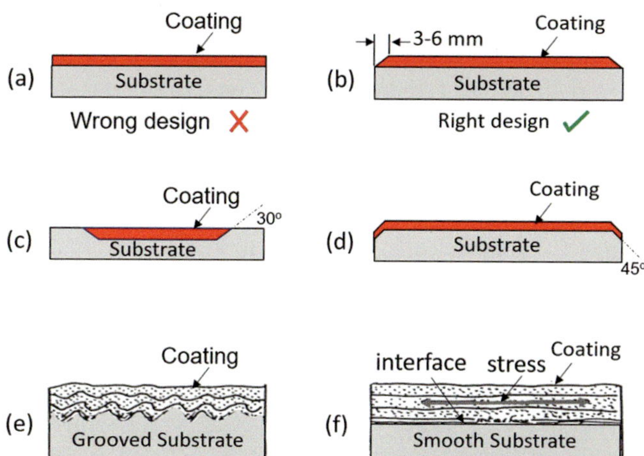

Fig. 9.2 Coating and substrate preparation: (**a**) wrong design with the coating abruptly ending at the extremity, (**b**) right design with a narrow-feathered band, (**c**) recommended undercut machining, (**d**) edge machining, (**e**) grooved substrate, (**f**) smooth substrate. Reprinted with kind permission from ASM International [Davis (2004)] and American Welding Society [Thermal Spraying (1985)]

important to keep in mind that the coating must never end abruptly at the part extremity (see Fig. 9.2a), where it may act as a stress raiser and where cracks will develop, especially when loads are applied. Coating at the extremity of a part should present a narrow-feathered band, rather than a sharp edge (see Fig. 9.2b).

Conventional machining: turning or milling allows undercutting the substrate target area to accept the coating. As without undercutting, the coating deposited must never end abruptly. In the undercuts the corners must be chamfered, or its cutting edge removed before spraying. The chamfer angle should be about 30° (see Fig. 9.2c) or up to 45° (see Fig. 9.2d). Sharp corners capture loose spray particles, dusts, and debris resulting in porous areas.

Machining is also used for creating grooves or threads into the surface to be sprayed (see Fig. 9.2e) to restrict shrinkage stresses and to disrupt the lamellar pattern of particle deposition in order to break up the shear stresses parallel to the surface (see Fig. 9.2f) [Thermal Spraying (1985)]. This technique is mainly used for spraying thick coatings or coatings on a substrate with a short radius. The surface is generally roughened after grooving. Two types of grooves are used: V-shaped ones when the V angle relatively to the part surface is 70° with the root rounded or U-shaped grooves, between 1.1 and 1.4 mm in width.

As discussed in the review paper by [Chandra and Fauchais (2009)], the vast majority of substrates used in thermal spray applications are metallic. Metals (except gold and platinum) or some alloys have oxidized surfaces even without preheating them (native oxide). Unless oxide layers are removed prior to spraying, particles impact on the oxide layer existing on the substrate surface. Then for the following deposited layers, particles also mostly impact on oxides because particles have been oxidized either in-flight or during deposition in air atmosphere. Most studies were performed on smooth surfaces where the observation at the nanometer scale is possible and where the particle flattening can be followed. Several authors have pointed out that mean surface roughness Ra of smooth surfaces of metals (except Au and Pt) or alloys increases at the nanometer scale with substrate preheating. [Fukumoto et al. (2005)] have noted that when a surface was polished to a particular average roughness Ra, coating adhesion was not as high as it was when preheating produced the same roughness. To more accurately characterize the substrate surface, [Fukumoto et al. (2005)] introduced another index of surface topology: skewness (S_k), which is a measure of the symmetry of the peaks and valleys of the surface roughness of the substrate. The skewness increases from negative values to positive ones upon preheating a polished surface and splat shape changes correspondingly from fingered to disk shaped. [Cedelle et al. (2006)] have shown for a 304 L stainless steel substrate that changes in surface topography in the nanometer range have a large effect on the static wetting behavior of molten metal droplets placed on its surface. Hence substrate preheating does not modify just the chemical composition and thickness of the oxide layer on the metallic substrate, but also its physical aspect, which plays a key role in the behavior of impacting particles. For $S_k > 0$ (more peaks than undercuts) obtained for 304 L stainless steel, preheated at 673 K by a D.C. plasma jet [Fukumoto et al. (2005)], [Cedelle et al. (2006)], [McDonald et al. (2007)] or a laser (with a power density over 50 MW/ m^2) [Fukumoto et al. (2006)], the resulting zirconia splats were disk shaped and not extensively fingered as when sprayed on the cold substrate ($S_k \sim 0$). For example, Fig. 9.3 represents the surface topography of 304 L stainless steel after polishing at room temperature (Fig. 9.3a) and preheated at 673 K (Fig. 9.3b). The preheated surface with $S_k \sim 1$ presents more peaks than valleys compared to the not-preheated polished one where $S_k \sim 0$. When the substrate

surface is treated with a laser, average roughness Ra and skewness S_k vary with energy density and number of laser pulses.

Any surface, especially in the presence of an oxide layer, is polluted by adsorbates and condensates. Upon impact of a molten droplet on the substrate, it flattens trapping the adsorbates and any vapors between the flattened droplet surface and the substrate giving rise to the development of a sharp pressure wave which can reach up to thousands of MPa for a few hundreds of nanoseconds. More generally, the presence of an evaporable substance on the surface affects significantly the flattening process, through the droplet-substrate wettability and the contact resistance created between the flattening particle and substrate [Li et al. (1998), (1999), (2007); Fukumoto et al. (2006)].

A detailed study of adsorbates and condensates at the substrate surface was reported by [Li and Li (2004)]. Often, water is the main component at the surface. Its adsorption on an oxide surface is usually the result of one of three possible mechanisms, or combinations of them, depending on the temperature, the intrinsic reactivity of the surface, and the number of defect sites at the surface [Henrich and Cox (1994)]:

* Physisorption of molecular water
* Chemisorption of molecular water
* Chemisorption of molecular water followed by dissociation

Physisorption corresponds to very weak interaction between the substrate and adsorbates, while chemisorption is much stronger and may involve partial charge transfer, which more readily leads to dissociation [Henrich and Cox (1994). Stronger adsorption occurs at steps and defects and new OHy radicals, resulting from dissociation, bond with a surface metal ion.

When the surface is heated over a critical temperature, adsorbed water species, as well as other adsorbed liquids, will

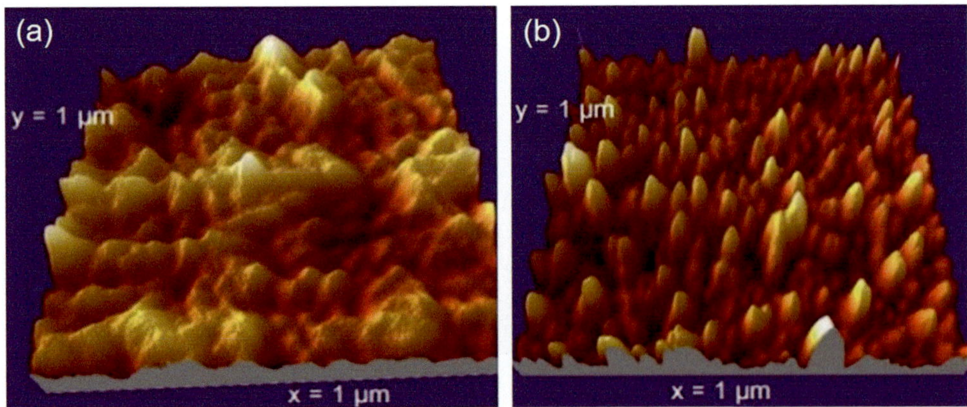

Fig. 9.3 Surface texture of 304 L stainless steel 1×1 µm2 (**a**) kept at room temperature after polishing, Ra =0.6 nm and Sk = ± 0.1 (b) preheated by a DC plasma torch (Ar-H2) at 250 °C during 120 s, Ra = 3.5 nm and Sk = 0.9. Reprinted with kind permission from Elsevier [Cedelle et al. (2006)]

be desorbed. This desorption process is linked to assorted products, surface structural features, and surface materials. Physisorbed molecular water is generally completely removed by preheating above 150 °C. However, with chemisorbed water, the thermal desorption temperature depends on adsorbed features and the subsequent desorbed product. For most oxide-covered metal surfaces, chemisorbed occurs at temperatures ranging from 127 to 320 °C as summarized in the table given by [Li and Li (2004)] for several oxides and nominal metal surfaces. The authors [Li and Li (2004)] underline that this desorption temperature range is in good agreement with the substrate preheating temperatures reported as transition temperature over which splats are disk shaped [Chandra and Fauchais (2009); Fukumoto et al. (2005); Cedelle et al. (2006); McDonald et al. (2007); Fukumoto et al. (2006)].

In an attempt to identify the effect of the nature of the adsorbent on the splat formation, Li and li (2004) brushed a stainless-steel substrate with either xylene, glycol, or glycerol prior to thermal spraying. The results showed that regular disc-shaped splats were obtained in all cases provided that the substrate preheating temperature was 50 °C above the corresponding boiling temperature of the organic adsorbent.

An interesting question was how long it takes after preheating and evaporating of all adsorbates and condensates, for the surface to be recovered again by adsorbents from the surrounding atmosphere. To answer the question, [Fukumoto et al. (2006)] exposed stainless steel 304 L substrates, which were heated to a temperature where disk-shaped splats were obtained, to ambient air at room temperature for different period of time (24, 48, 72, 96, and 120 h). They investigated the effect of elapsed time after heating on the splat morphology change. The result given in Fig. 9.4 shows that for up to 24 h following the heating and removal of the adsorbates, a high fraction of disk-shaped splats could still be observed. For longer periods, beyond 24 h, the fraction of disk-shaped splats began to decrease. Finally, the fraction of disk splat decreased to less than 50% over 72 h and more from the initial surface cleaning from adsorbates. Of course, both Ra and S_k of a once-heated substrate kept in an air atmosphere at room temperature for a long time were unchanged.

A study of the particle flattening behavior under reduced pressure was reported by [Fukumoto et al. (2007)] with the objective of determining the effect of pressure on desorption of adsorbates and condensates (plasma spraying under soft vacuum). Measurements made of the pressure below which splats became disk-shaped for different materials were used as indication of gas desorption. These have shown that desorption probably occurs at pressures between 92 and 26 kPa, depending on substrate material. A laser can also achieve evaporation of adsorbates and condensates provided its energy density is over a certain threshold [Coddet et al. q1999; Costil et al. (2005)].

According to [Dong et al. (2013)], the elimination of adsorbates and condensates can also be achieved by blasting the substrate surface with dry-ice (CO_2 at -79 °C). The process as schematically represented in Fig. 9.5 involves three steps. The first (Fig. 9.5a) involves the cooling and

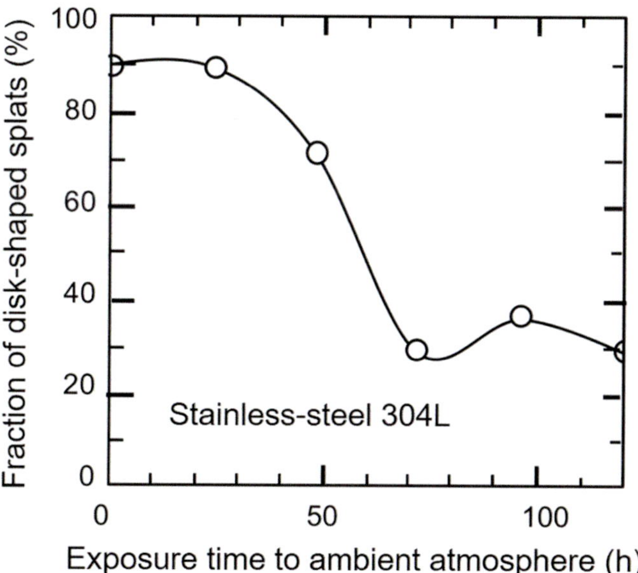

Fig. 9.4 Relationship between elapsed time after heating to 723 K and fraction of disk-shaped splats [Fukumoto et al. (2006)]. (Reprinted with kind permission from Springer Science Business Media copyright © ASM International)

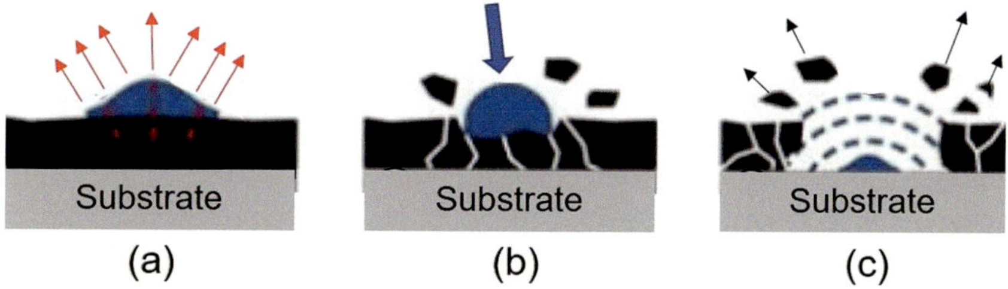

Fig. 9.5 Schematic representation of the elimination of adsorbates and condensates by dry ice pellet blasting (**a**) cooling of the substrate by impacting ice pellets, (**b**) solidification of adsorbates and condensates, (**c**) ejection of broken solid adsorbates and condensates [Dong et al. (2013)]. (Reprinted with kind permission from Elsevier)

solidification of the adsorbates and condensates. This is followed by the breakage (Fig. 9.5b) and the ejection from the surface (Fig. 9.5c) of the solidified adsorbates and condensates as a result of their bombardment with dry-ice granules.

Preheating the substrate over the transition temperature (most probably the temperature above which adsorbates and condensates are removed) modifies the oxide layer thickness, composition, and roughness.

When the substrate is preheated in a furnace with a controlled atmosphere and well-defined conditions (heating rate, maximum temperature, preheating time, oxygen partial pressure, or controlled atmosphere), oxide layers are reproducible. On the other hand, if the surface is heated with an electrical resistance while being exposed to ambient air, the oxide formation is less well controlled since the atmosphere in contact with the surface contains oxygen molecules and different amounts of water vapor. Moreover, contact between the substrate and the resistance heater used to raise its temperature is not perfect, resulting in uneven heating.

When a spray torch is used to preheat the surface, even if the mean heating rate and final temperature can be well defined, the local temperature varies with the movement of the torch relative to the substrate. The surrounding atmosphere is mainly air at atmospheric pressure. With combustion flames (flame spraying, HVOF, HVAF) gas temperatures are below 3000 K and the entrained air contains mainly molecular oxygen. With plasma jets, gas temperatures of up to 5000 K can be achieved at spray distances of 80 mm, so the oxygen in contact with substrate surface may be partially or totally atomic oxygen. These devices produce large heating rates (between 1 and 5 K/s). The formed oxide layers generally have compositions and morphologies different from those obtained by more conventional heating [Haure T (2003)], Pech et al. (1997), Pech et al. (2000), Syed et al. (2006)].

Preheating and/or modifying the substrate-surface with a laser may also result in oxide layers different from those obtained with a plasma or conventional heating. Very different results can be obtained, depending on the laser used and the substrate treated. [Fukumoto et al. (2006)] have shown that for laser power densities below 50 MW/m^2 the surface roughness does not change, and its skewness remains negative or close to zero. It is also possible that the laser irradiation removes part of the oxide layer at the surface [Li et al. (2006a, b)] if the laser energy density is sufficiently high or if the number of pulses at the same location is increased. It is worth noting that oxide layer elimination requires a much higher energy density than that necessary to remove adsorbates and condensates. Oxide removal is due to mismatch between the coefficients of expansion of the oxide layer and the underlying metal, which is heated by laser irradiation through the transparent or semi-transparent thin oxide layer. The oxide layer formed at the metal or alloy or cermet substrates plays a key role in coating adhesion. The latter, besides the spray conditions, depends on the oxide layer thickness, its structure and composition, and, sometimes, the corresponding composition gradients between the oxide layer top surface and its interface with the substrate. Many techniques exist to characterize the oxide layer, and the selection of any particular one depends upon the information required.

9.2.4 Splats and Coating Formation

Coatings are obtained with a given spray pattern dependent on the relative movement of the torch with respect to the substrate, and for a given standoff distance. Two modes of powder injection into the torch are used: radial, which is the case of most spray torches; and axial, as in the case of the Mettech Axial III torch. In the radial mode, the particle injection momentum has to be sufficiently high to allow the particle to penetrate into the high velocity plasma stream. In contrast, with axial particle injection, the powder carrier gas flow rate, and consequently the particle injection momentum, has much less effect on the particle trajectories.

Coating properties, whatever be the spray process used, depend on splat formation and layering which, in turn for micrometer sized particles, depend on their degree of superheat (above the melting temperature), their impact velocity (key parameter), and substrate temperature. For a substrate preheated above the splat transition temperature T_t, the ratio of the splat diameters, D_s, to the initial droplet diameter, d_p, is typically in the range $2 < D_s/dp < 6$. For most sprayed materials on different substrates, the transition temperature $T_t < 0.3 T_m$ (T_m being the particle melting temperature of the particle material). The contact between the splat and the substrate represents up to about 60% of the splat surface on substrates preheated above T_t, compared to possibly less than 20% on cold substrates. When spraying on rough substrates, preheated above T_t the coating adhesion is higher by a factor of 3 to 4× compared to that obtained on a cold substrate! The time between two successive impacts, typically in the order of a few tens µs, is long enough compared to droplet solidification time, to ensure that the next droplet impacts on an already solidified splat. The splat layering is controlled by the powder flow rate, the process deposition efficiency, and finally the spray pattern including the relative torch-substrate velocity. Coatings contain layered splats, porosities (often due to the poor ability of the flattening particle to follow the cavities present in the previously deposited layer), un-melted or partially melted particles, and oxides for metals and alloys. Splashing of the melted particles during flattening upon impact on substrate below transition temperature can also significantly affect the coating properties [Newbery and Grant (2000)]. Resulting coating is schematically illustrated in Fig. 9.6. Typical final coating thicknesses are between about 50 μm and a few millimeters.

Microstructures of typical metallic (Stainless steel 304) and ceramic (YSZ-8%Y_2O_3) coatings are given in Fig. 9.7. These show in the metallic coating (Fig. 9.7a) the typical features and coating defects such as pores, unmolten particles, and oxide inclusions to which reference is made in Fig. 9.6. The YSZ coating shown in Fig. 9.7b shows, on the other hand, a more pronounced lamellar structure. In these coatings, cracks can also be observed within splats or between layers due to stresses generated during the spray process such as:

- Quenching stress (always tensile) due to cooling of each individual splat.
- Temperature-gradient stress, which develops especially within low thermal conductivity coatings.
- Phase change stress occurring when the crystal in a new phase after deposition is smaller or larger than in the preceding phase.
- Expansion mismatch stress, which occurs during the cooling phase of the deposition process. This stress increases with the increase of mean deposition temperature (substrate and coating) and mainly generated due to the mismatch between the expansion coefficients of coating and substrate. For low thermal conductivity coatings, giving rise to steep temperature gradients within deposited layers, a too fast cooling of the substrate/coating, can induce damaging temperature gradients.
- Additional stresses can also be generated within the coating provoked by substrate-induced stresses generated during the surface preparation/grid blasting step.

9.2.4.1 General Remark

First it must be recalled that the impacting particle/molten droplet is not necessarily that injected in the plasma jet. It can be modified during its flight in the high-energy jet by its heating, melting, partial vaporization, and possible oxidation during the heating stage. Oxidation is controlled by diffusion and convection. Oxidation controlled by diffusion (oxide shell around the particle) occurs mainly in the plasma jet plume. According to the large difference between oxide and metal melting temperatures a solid crust can be formed (in the colder zones of the jet) around the particle before its impact on the substrate. This can be observed, for example, in the case of NiAl particles with the formation of an alumina crust. The other oxidation possibility is convection occurring in the plasma jet core where temperatures are over 8000 K but also surrounding atmosphere oxygen has been entrained. Convective oxidation can be the dominating oxidization process if the plasma to liquid particle kinematic viscosities ratio is higher than 55 and where the relative Reynolds number Re attains values higher than 20 [Syed et al. (2006)]. It results in oxide grains inside the particle prior to its impact.

These phenomena also vary with the temperature distribution within the sprayed particle. A uniform droplet temperature depends on its heat conduction, or more precisely its Biot number, Bi, which must be lower than 0.01 for a particle uniform temperature. $Bi = \kappa/\kappa_p$ where κ is the plasma thermal conductivity and κ_p that of the particle. If the condition Bi

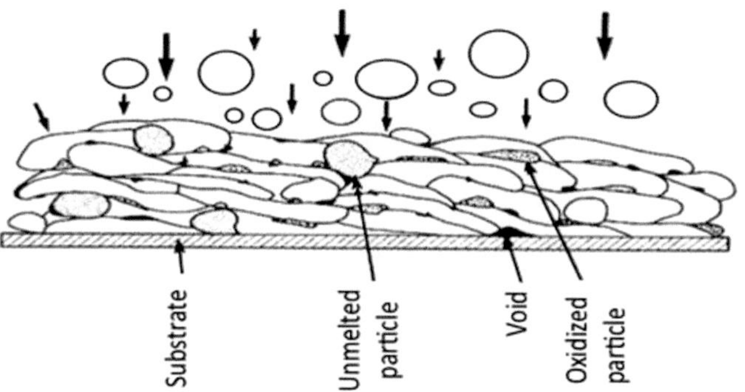

Fig. 9.6 Schematic cross section of thermally sprayed coating

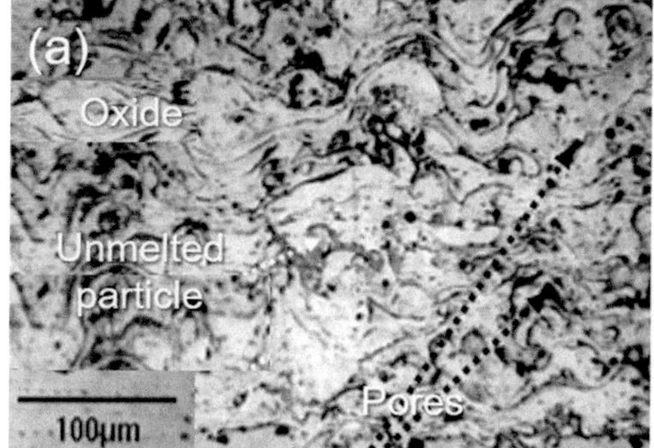

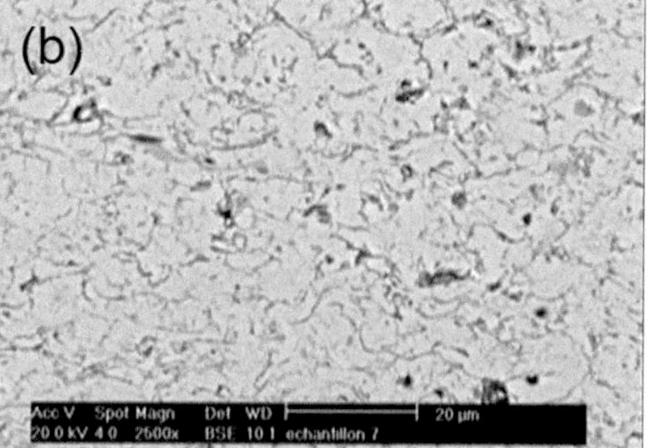

Fig. 9.7 (**a**) Stainless steel coating (304 L) deposited by air plasma spraying (APS) on a low carbon (1040) steel substrate, (**b**) YSZ (8 wt%) coating plasma sprayed on a super alloy (APS)

<0.01 is not fulfilled the core of the particle can be still solid at impact, modifying deeply its flattening. This is, for example, the case of polymer particles. Figure 9.8 summarizes where the different oxides can be formed. The way the splat will be formed on the substrate surface depends on the particle wettability and its thermal contact resistance.

9.2.4.2 Splat Formation

In principle particle/droplet impact on the substrate should be orthogonal to the substrate surface for optimum splat formation and coating generation. Unfortunately, even in the ideal situation where this condition is met for particles moving along the axis of the jet, the natural divergence of the plasma jet would result in the impact of the particles in its fringes on the surface of the substrate with an angle generally comprised between 70 and 110°. The orthogonal impact condition for optimal splat formation is even more complex and harder to maintain when the part to be sprayed has steep angular area, convex or concave zones, with small radii. [Kudinov et al. (1981)] at the beginning of the eighties were among the first to make some sort of classification of splat shapes on a smooth surface. They found 29 different shapes of splats depending on spray conditions. Splat shapes depend on a large number of factors, including the degree of superheating of the particle, partial solidification of the particle in-flight resulting in formation of a solid shell surrounding a molten core (this occurs especially for low thermal conductivity materials), degree of in-flight oxidation, and the presence of contaminants at the substrate surface. According to recent studies by [Dhiman et al. (2007); Goutier et al. (2011); Fukumoto et al. (2011); Sabiruddina et al. (2011)] the splat shapes given in Fig. 9.9 are representative of the different forms of splats observes. These are identified as:

* ***Extensively fragmented splats,*** Fig. 9.9a Alumina of ASI-304 SS (cold) *[Goutier et al. (2011)]* and Fig. 9.9b Nickel on ASI-304 SS pre-oxidized at 150 °C *[Dhiman et al. (2007)].*

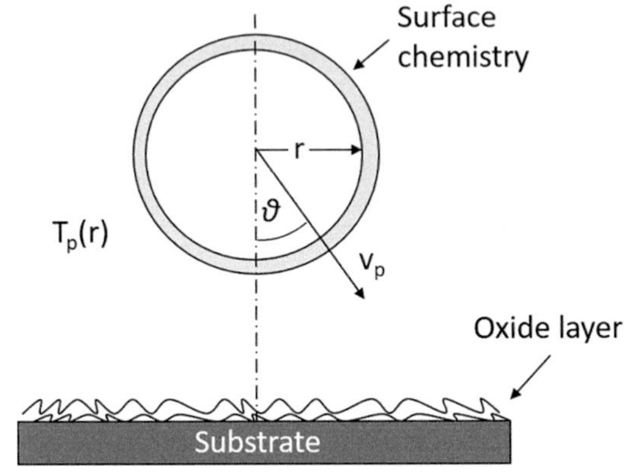

Fig. 9.8 Schematic of the parameters controlling the fully melted particle flattening

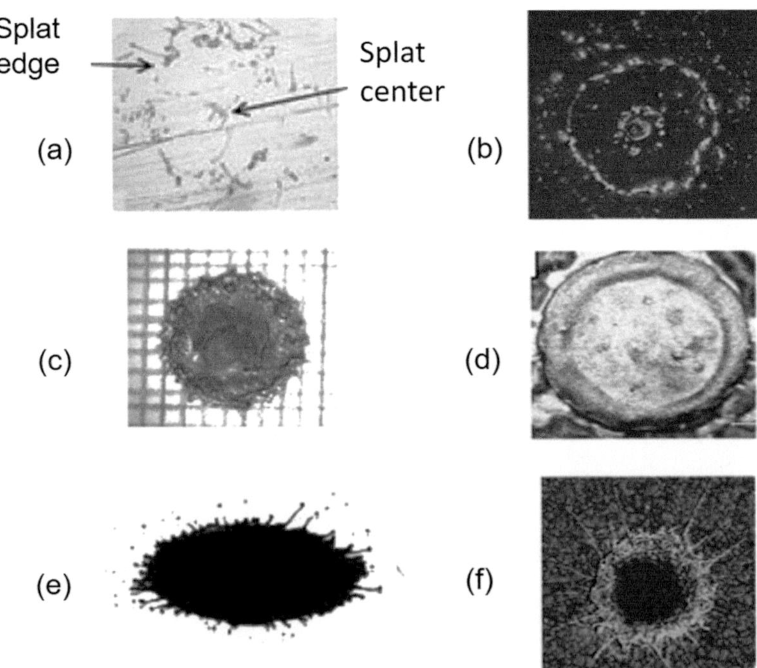

Fig. 9.9 Examples of plasma sprayed splats of different materials on Stainless Steel (SS-304). (**a**) Alumina on cold SS [Goutier et al. (2011)], (**b**) Ni on SS pre-oxidized at 150 °C [Dhiman et al. (2007)], (**c**) Cu on SS [Fukumoto et al. (2011)], (**d**) alumina on low-carbon steel preheated at 200 °C [Sabiruddina et al. (2011)], (**e**) alumina on SS preheated at 400 °C [Goutier et al. (2011)], (**f**) Ni on pre-oxidized SS at 650 °C [Dhiman et al. (2007)]

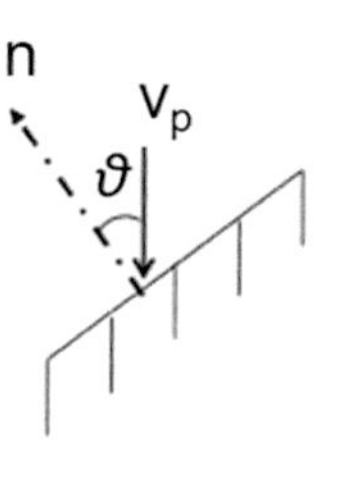

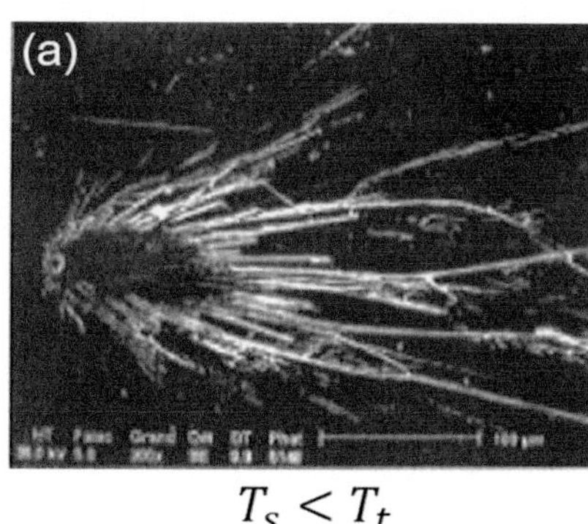

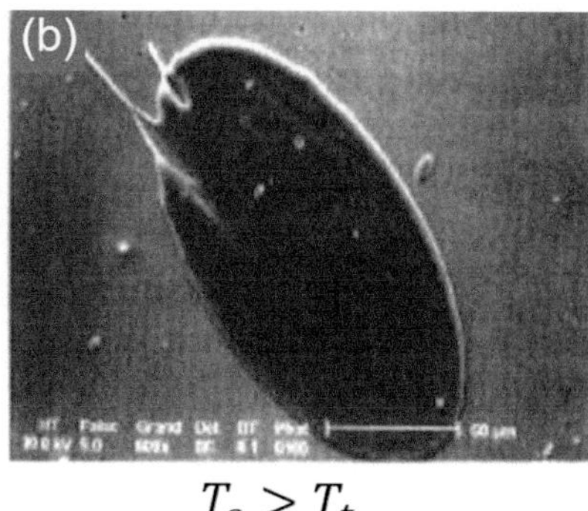

Fig. 9.10 Impact of alumina plasma sprayed particles (dp = 58 µm, vp = 138 m/s, Tp = 2400 K) onto an inclined ASI-304 SS (impact angle θ =30°) (**a**) room temperature substrate, (**b**) substrate preheated over transition temperature [Bianchi et al. (1977)]

- ***Disk-shaped splats***, Fig. 9.9c Copper on ASI-304 SS *[Fukumoto et al. (2011)]* and Fig. 9.9d alumina on low-carbon steel preheated at 200 °C*[Sabiruddina et al. (2011)]*.
- ***Fingered splats***, Fig. 9.9e Alumina on ASI-304 SS preheated at 400 °C*[Goutier et al. (2011)]*, and Fig, 9.9f, Nickel on pre-oxidized ASI-304 SS at 650 °C *[Dhiman et al. (2007)]*.

Roughly splats of the type a and b have a very low contact with the substrate, while those presented in c to e have a much better contact essentially due to substrate preheating over the transition temperature. In these conditions, the molten splat, at least in its central part, has a good contact with the substrate and its cooling depends strongly on its thermal contact resistance with the substrate. For more details, see [Fauchais 2004], [Chandra and Fauchais 2009]. The adhesion of these splats on smooth surfaces preheated over the transition temperature depends also on other parameters and when the conditions c to d presented in Fig. 9.9 are not fulfilled the adhesion on smooth substrates is close to zero.

In the event that the impact angle on the substrate is less, or significantly higher, than the range of normal impact (between 70 and 110°) splats on cold substrates are extensively fingered, especially in the direction of substrate inclination as shown in Fig. 9.10a.[Bianchi et al. (1977), Kang CW, H.W. Ng, and Kang et al (2006), Salimijazi et al. (2007)] have shown, on the other hand, that for stainless steel, copper, and a few other polished substrates (Ra < 0.1 µm), preheated over the transition temperature, splats have an elliptical shape, as shown in Fig. 9.10b when the spray angle, θ, increases from 0° to 75° with the ratio of the major to the minor axes increasing as the spray angle increases.

For different materials (alumina, zirconia, titania, Al, Ni, Astroloy, and Cu), the relationship between the major and minor axes shows a strong linearity over a wide range of splat sizes. This observation implies that the Elongation Factor EF (ratio between the major and minor axes) does not depend on particle diameter and impact velocity but only on spray angle. The splat thickness increases slightly along the inclined surface and it becomes progressively thicker in the direction of the inclined surface when the spray angle increases. This probably explains why, when the spray angle exceeds a critical value, which depends on the sprayed material and the substrate material, splashing occurs in the inclination direction even on substrates preheated above the transition temperature.

Attention was also given to the impact of the substrate on the crystallographic structure of the coating. Alumina particles were plasma sprayed onto polished (Ra ≈ 0.4 µm) plasma sprayed alumina coatings [Denoirjean et al. (1998)], which were kept at 250 °C before spraying to get rid of adsorbates and condensates. The latter were either as-sprayed (> 99 wt.% γ phase) or preheated at 1373 K at a rate of 5 °C/min, annealed for 6 h and cooled at a rate of 5 °C/min resulting in a 100% α-columnar structure. Some were also preheated to 1873 K at a temperature ramp of 5 °C/min, annealed for 3 h, and cooled at a rate of 5 °C/min resulting in an α-granular structure with grains between 3 and 5 µm. As shown in Fig. 9.11a, on the γ-alumina substrate, alumina splats (γ phase) exhibit a columnar and regular structure ~100–150 nm; the adhesion of the alumina coating (300 µm thick) obtained on this smooth substrate was 35 ± 3 MPa! On the columnar α-alumina substrate, splats (γ phase) form a columnar and irregular structure ~150–300 nm (see Fig. 9.11b); the adhesion of the coating is only 3 ± 1 MPa. On the α-alumina substrate with a granular structure, splats (γ phase) achieve a very irregular structure ~100–400 nm (see Fig. 9.11c), and splats have the tendency to peel off and it is impossible to achieve any coating.

[Yang E.-J. et al. (2012)] plasma sprayed alumina particles onto single crystalline alumina substrates (α phase) kept at 900 °C. The growth orientation of alumina splat crystal during rapid solidification was exactly the same as the orientation facets of [001] or [110] of the single crystalline alumina substrate, epitaxial solidification taking place during splat cooling.

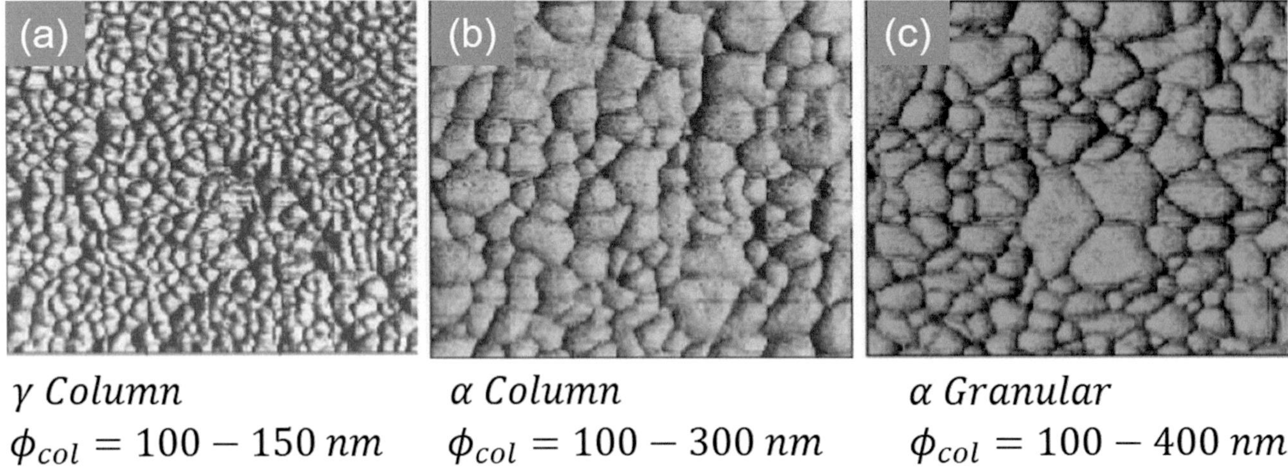

γ *Column*   α *Column*   α *Granular*

$\phi_{col} = 100 - 150\ nm$ $\phi_{col} = 100 - 300\ nm$ $\phi_{col} = 100 - 400\ nm$

Fig. 9.11 Alumina splats plasma sprayed onto smooth alumina substrates with different microstructures, (**a**) γ columnar, (**b**) α columnar, and (**c**) α granular [Denoirjean et al. (1998)]

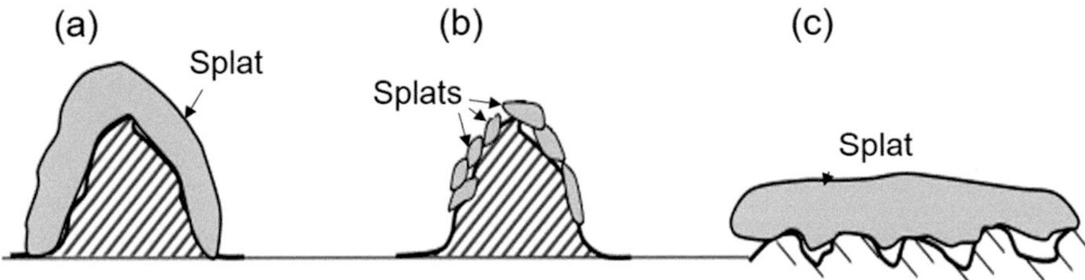

Fig. 9.12 Schematic representation of splat size relative to grit blasted surface peak sizes defined by Rt. (**a**) Splat size adapted to peak size, (**b**) too small splat sizes relatively to peak size, and (**c**) too large splat size relatively to peak sizes

9.2.4.3 Coating Adhesion

(a) ***Mechanical adhesion*** takes place when the substrate is roughened, for example, by grit blasting. For the best mechanical adhesion, intuitively the peak heights, characterized by the distance between the highest peak and the deepest undercut (R_t), must be adapted to the mean size of the splats, as illustrated in Fig. 9.12a. The spacing between peaks also plays a key role on the flattening of particles impacting on rough surfaces. The shrinkage of the resulting splats while cooling contributes to the adhesion of the splat to the surface because of the frictional force that develops [Fauchais et al. (1996), Morks et al. (2002), Bahbou and Nylen (2005), Bahbou and Nylen (2007)]. This adhesion depends both on the amplitude of roughness, and on the mean spacing, w, between peaks and valleys. A good mechanical adhesion is achieved when the mean splat diameter, d_s, which would be obtained on the same smooth substrate, is about 2 to 3 $\times$ R_t. If $d_s << R_t$, as illustrated in Fig. 9.12b, the adhesion is very poor. It is the same if $d_s >> R_t$, see Fig. 9.12c. However, the flattening liquid drop must also penetrate within the undercuts [Mehdizadeh et al. (2002)]. For that the

impact pressure ($\rho_p.v_p^2$) must be larger than the tension force. It results in the following expression, where σ_p is the liquid drop surface tension.

$$w > 4\sigma_p/\rho_p v_p^2 \tag{9.1}$$

[Bahbou and Nylén (2007)] established that a good correlation exists between the adhesion strength of a NiAl (5 wt%) coatings, on Ti-6Al-4 V substrate, and the root mean square value $R_{\Delta q}$ of the substrate roughness, which considers both the amplitude and spacing of peaks. The correlation is rather poor when variables such as R_a, R_t or the peak spacing are considered individually. For more details, see [Fauchais et al. (2014) Chap. 12 Surface preparation] and [Fauchais et al. (2014) Chap. 13 Conventional coating formation].

(b) ***Diffusive adhesion of metals and alloys*** occurs when the temperature during coating formation is sufficiently high and no oxide layer exists at the substrate surface. In a simplified approach the thermal diffusion distance, d_d, is given by:

$$d_d = \sqrt{D_{th}t} \tag{9.2}$$

where t is the time during which the contact occurs at temperature T, and D_{th} the diffusion coefficient varying as $e^{-E_A/kT}$ where E_A is an activation energy, T the absolute temperature, and k the Boltzmann constant. Essentially, good adhesion of a splat/or coating on a substrate can only be assured at sufficiently high substrate temperatures and sufficient contact time the coating material and the substrate. This is conditional, however, to the absence of an oxide layer on the surface of the substrate which in turn depend on the chemical nature of the substrate and that of the ambient atmosphere. Controlled atmosphere or soft vacuum plasma spraying offers means of guaranteeing operation in an inert or reducing atmosphere. Moreover, by operating the plasma torch at reduced ambient pressure in a reverse polarity transferred arc mode for a short period prior to the spray coating operation, it is possible to remove any oxide layer initially present on the surface of the substrate.

(c) **Chemical adhesion** occurs when the surface atoms of two separate surfaces form ionic, covalent, or hydrogen bonds. It requires that the impacting droplet melts the substrate and a chemical compound of both liquids exists. For example, when Mo particles impact on steel, the melting temperature and effusivity (defined as $(\rho_p . c_p . \kappa_p)^{0.5}$) of the Mo droplet are higher than those of the steel substrate, which melts and reacts to form a chemical compound, $MoFe_2$. Mo particle flattening on a stainless-steel substrate results in splats formed of different pieces as shown in Fig. 9.13a after [Li and Li (2006)]. This type of splashing is due to a localized melting of the substrate surface by the impacting droplet. The melted crater on the substrate surface alters the flow direction of the droplet fluid and tends to form a free liquid jet that detaches from the substrate surface (Fig. 9.13b). Cracking initiated within the Mo splat Fig. 9.13c occurs within it (Fig. 9.13d) due to the restraining of the splat contraction during cooling and also to the low plasticity of the solidified splat material [Li and Li (2006)]. The liquid film observed in Fig. 9.13c and d, mainly resulting from the steel substrate material, promotes the floating of Mo pieces initiated from cracks.

9.2.4.4 Spray Pattern

The movement of the torch has a significant influence on the heat and mass transfer during the coating process. The structure and final properties of the coating can be improved by adjusting the movement direction and speed profiles of the torch. Thus, the manufacturing of coatings on complex geometries requires the generation of an optimized movement with a defined torch velocity. A bead is obtained (Fig. 9.14a) when translating the torch relatively to the substrate (1-D movement) at a given relative velocity torch-substrate, v_r, for example, in the x direction. The bead parameters, its height h_b and its width w_b at mid-height, depend on the spray process, the torch working conditions, the relative velocity torch-substrate v_r, the powder mass flow rate $\dot{m}_p$, the particle size distribution, the deposition efficiency η_d, the torch nozzle internal diameter d_n, the injection conditions, and the spray distance.

In fact, a Gaussian shape is in good agreement with the distribution of hot particles. In the bead wings, particles

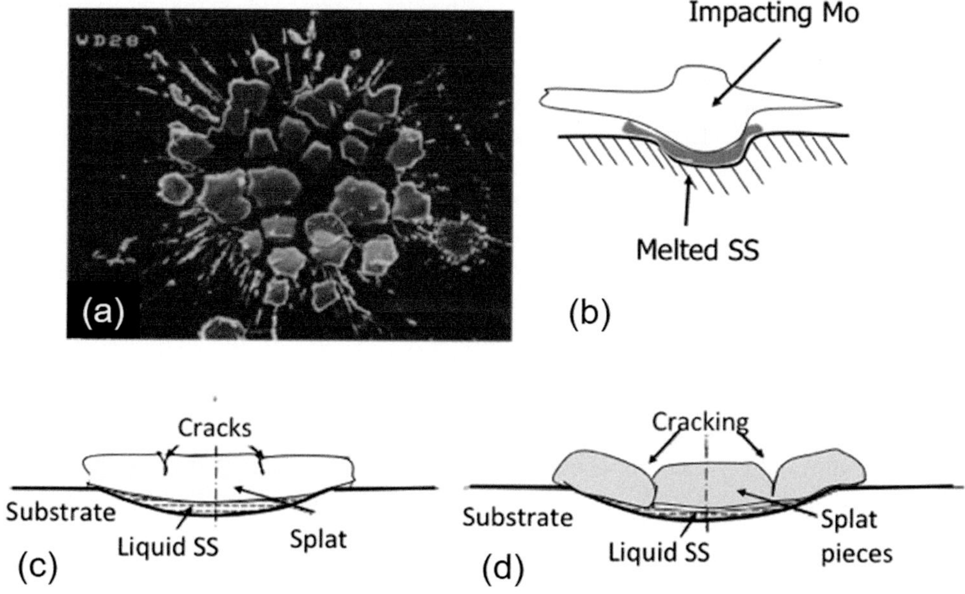

Fig. 9.13 (a) Mo splat on stainless steel (SS) substrate, (b) Schematic diagram of droplet impact inducing substrate melting, (c) Crack initiation in the Mo splat, (d) Displacement by floating of the cracked pieces on low melting point liquid [Li et al. (2006)]

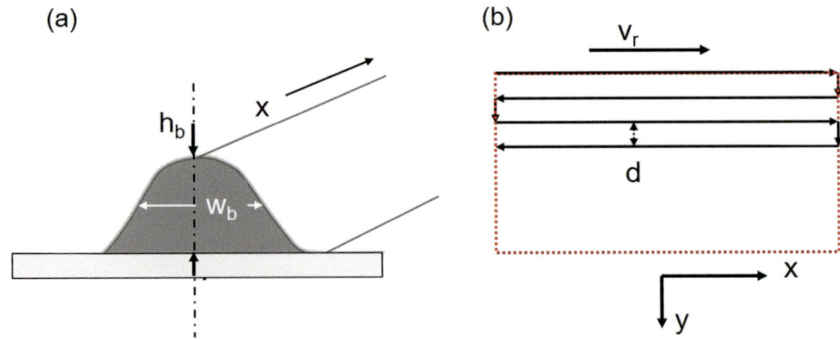

Fig. 9.14 Schematic representation of the coating formation on a flat substrate (**a**) spray-bead, (**b**) typical spray pattern

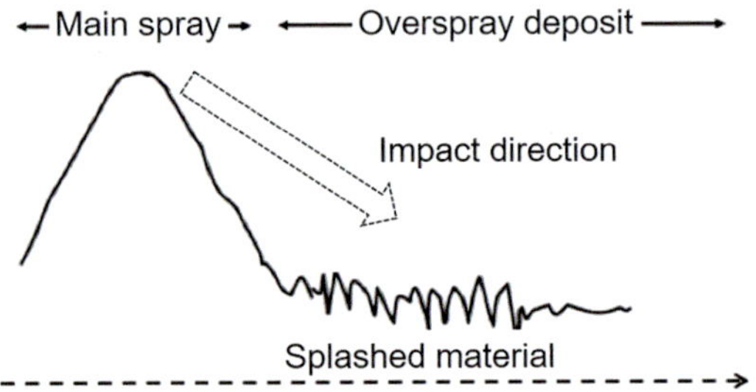

Fig. 9.15 Splashed material from inclined spraying where the bead intercepts part of the impacting particles

impact with an angle and lower velocities and temperatures resulting in a more porous area with a poorer contact between splats. Also, in the wings of the bead, frequently particles appear which are not well exposed to the hot jet but are sufficiently heated to stick to the bead wing. Moreover, these particles, at least the metallic ones, are easily oxidized in-flight and they may form the major source of oxide inclusions in sprayed coatings, and thus create defects. The coating is achieved for each successive layer by moving the torch as shown in Fig. 9.14b. Once the first layer is achieved the next one is built following the same process with every pass overlapping previous beads, deposited in the x direction (Fig. 9.11a), the next bead being deposited at a distance d in the y direction and its thickness depends on the overlapping ratio. For a plane substrate, the distance d between successive beads in the y direction is calculated as the bead width at mid-height w_b generally multiplied by a given number, generally called overlapping ratio, f, (typically 0.33, 0.5, 1.0, 2.0, 3.0). The choice of the beads overlapping ratio depends on their thickness and the acceptable level of surface undulation.

When spraying off-normal angles, there is always a risk of having particles intercepted by the bead rebounding and splashing on the other side of it, as shown in Fig. 9.15. The adhesion of the splashed material being very poor, it will be the same with the next bead deposited on top of it.

9.3 Controlled Atmosphere Plasma Spraying

9.3.1 Ambient Gas Entrainment in DC Plasma Jets

As discussed in "Chapter 8, DC Plasma Spraying – Fundamentals" DC plasma jets working in air are known for their high-energy density and high plasma temperature (above 12,000 K) and velocities (in the hundreds of m/s) with steep temperature and velocity gradients. As the plasma jets immerge into a stagnant ambient atmosphere, a strong shear layer develops in the interface between the jet and the ambient gas. The structure of such a flow has been the subject of intense studies in the 1980s and 1990s due to its impact on the overall jet behavior and the resulting flow and temperature fields [Pfender et al. (1991)], [Russ et al. (1994)]. Based on Schlieren photographs, emission spectroscopy, and enthalpy probe concentration [Pfender et al. (1991)] concluded that the large velocity difference between the jet and the ambient atmosphere causes rolling up of the flow around the nozzle exit into the ring vortex which is pulled downstream by the flow, allowing the process to repeat itself again at the nozzle exit. Adjacently formed vortex rings at the outer edge of the jet have the tendency to coalesce, forming large vortices. As schematically represented of the flow pattern given in Fig. 9.16a, and supported by the Schlieren image of a DC plasma jet in ambient air, Fig. 9.16b, the distorted vortex rings start entangling themselves with the adjacent rings, finally resulting in total breakdown of the vortex structure into large-scale eddies and the onset of turbulent flow. This process results in the first large-scale engulfment of the ambient gas into the jet flow. Some entrainment also takes place during the roll-up process of the jet shear layer. The eddies of cold gas traveling in the axial direction at lower velocities than the flow continuously break down into smaller eddies, while diffusion takes place on the molecular level at all eddy boundaries. The extent of the impact of ambient gas entrainment on the temperature fields of a DC plasma jet is demonstrated in Fig. 9.17 after [Roumilhac et al (1990)] showing the temperature isocontours for a pure argon DC plasma jet in an open discharge in an ambient argon atmosphere (Fig. 9.17a), in a nitrogen atmosphere (Fig. 9.17b) and in air (Fig. 9.17c). These show a clear quenching effect resulting from the engulfment of nitrogen or air into the argon plasma flow. The effect is quantitatively noticeable in Fig. 9.18 giving the

profile of the plasma temperature and the molar fraction of entrained air in the axial direction along the centerline of a DC pure argon plasma jet in an open discharge into ambient air. The results show that the mole fraction of air on the centerline of the jet reaches 50% only 34 mm downstream of the nozzle exit of the jet.

Other than the above discussed impact of ambient air entrainment on the temperature field in the plasma jet, the abundance of oxygen mixed with the plasma flow can have a detrimental effect of the chemistry of the material being sprayed especially when dealing with pure metals, alloys, or non-oxide ceramics (carbides, nitrides...). The air entrainment could result in these cases in the in-flight oxidation of sprayed material. [Syed et al. (2006)] investigated the in-flight oxidation of stainless-steel particles in the plasma spray jets in an open discharge mode. Two types of 316 L austenitic stainless-steel particles were sprayed by a DC plasma gun with a 7 mm anode nozzle i.d. varying gun parameters and surrounding gases composition. Electron micrographs given in Fig. 9.19 show the surface morphology and cross-sections of the powder used. The first (Fig. 9.19a, b) was a gas-atomized TY316L with a particle size in the range (63–50 μm) from Techphy H.T.M., France. The second, coarser powder (Fig. 9.19c, d), was a water-atomized feedstock referred to as 41C (106–45 μm) from Sulzer Metco, USA. Following the injection of each of these two powders in the plasma jet in an open discharge mode in air, the treated powder was collected and characterized to establish relationship between spray parameters and particle oxidation. The results given in Fig. 9.20 show that, besides diffusion-based oxidation, convective oxidation in the particles can also occur within the plasma jet core if plasma to molten particle kinematic viscosities ratio is superior to 55 and the Reynolds number (Re) of the flow relative to the particle/droplet is more than 20. In these conditions, the oxide formed, or oxygen dissolved at the surface of the liquid droplet, can be swept into its interior forming isolated oxide nodules. Fresh liquid metal is transported from interior toward particle surface.

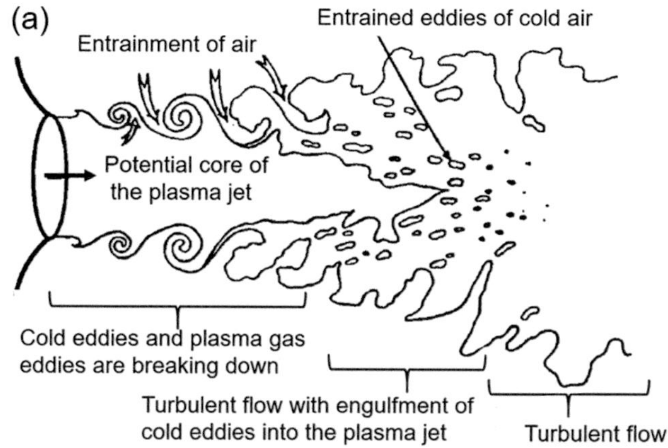

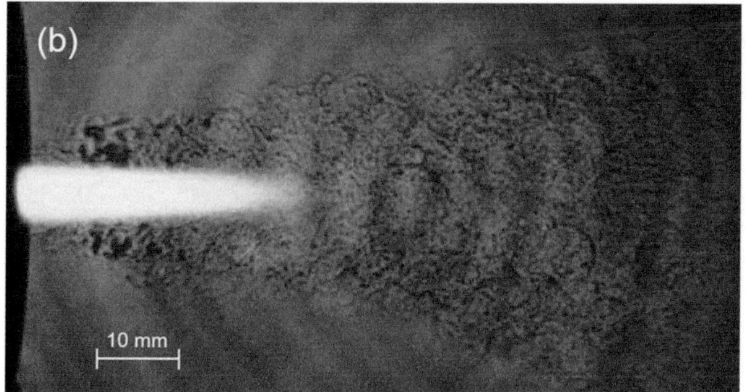

Fig. 9.16 (**a**) Schematic representation of the main regions of DC plasma jet showing the rolling of the flow around the nozzle exit, cold eddy engulfment, and breakdown followed by turbulent flow generation, (**b**) Schlieren image of a DC plasma jet in ambient air [Pfender et al. (1991)]

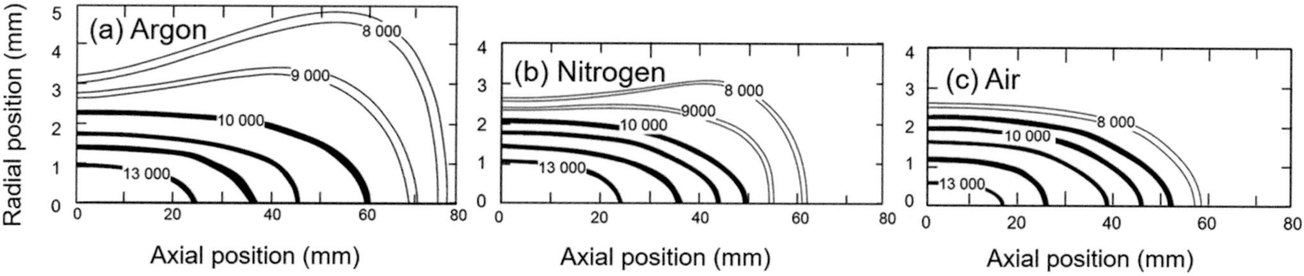

Fig. 9.17 Effect of the ambient gas entrainment on the temperature contours of an argon plasma jet, with air showing the strongest quench effect. (**a**) Argon ambient gas. (**b**) Nitrogen. (**c**) Air, derived from data published by [Roumihac et al. (1990)]

The oxidation rates were estimated to be higher compared to diffusion-based oxidation which is the dominant phenomenon in the plasma jet plume in the absence of convective oxidation. Spray parameters leading to higher kinematic viscosities ratio and Re, such as increasing arc current, hydrogen content in the plasma forming gases, or decreasing sprayed particle size range, resulted in enhanced convective oxidation in the plasma core. The diffusion-based oxidation of particles in the plasma jet plume can be principally controlled by their size (specific surface area), temperature and velocity (dwell time), and the molar fraction of oxidizing and reducing species in the plasma jet. While investigating the influence of the atmosphere of plasma jet on the in-flight particle oxidation, it was found that the surface area of the oxide nodules and the mass percentage of total oxygen in collected particles followed a parabolic and linear relationship with the partial pressure of oxygen, p_{O2} in the surrounding atmosphere. Keeping surrounding p_{O2} at 0.1 and altering N_2 and Ar content resulted in higher oxygen content in particles sprayed in Ar-rich surrounding, whereas no distinct difference in oxide nodules surface area was reported. [Syed et al. (2006)] studied systematically in-flight reactions especially those resulting in metastable oxides. Oxide nodules with mixed oxide of Cr, Fe, Mn, and Si as cations were detected. Spinel type $Fe_{3-x}Cr_xO_4$ form was the observed phase with x being between 1.5 and 1.9. No corundum type oxide phase was detected.

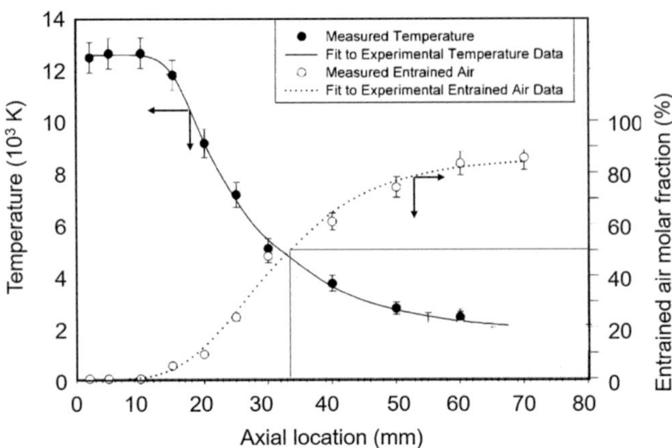

Fig. 9.18 Axial distributions of enthalpy-probe measurements of plasma temperatures, volume fraction of entrained air in the plasma jet, and CARS-temperatures of the entrained oxygen showing the low temperatures of the entrained oxygen [Finke et al. (2003)] (reproduced with kind permission of Elsevier)

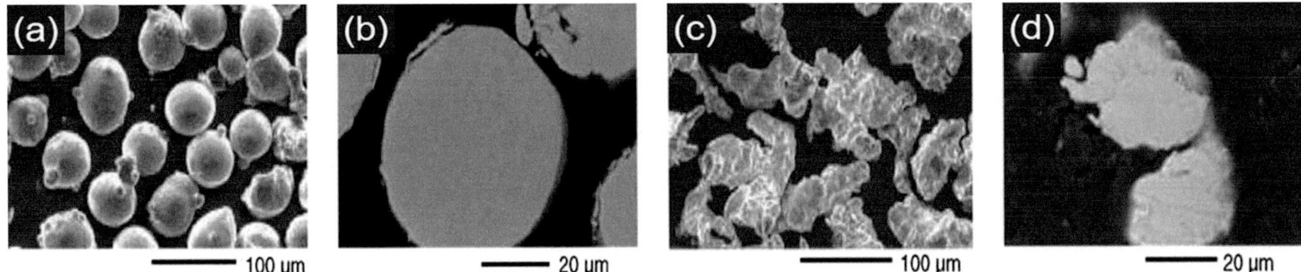

Fig. 9.19 (**a**) Electron micrographs of feedstock 316 L stainless steel particles, (**b**) cross section of sprayed particles, (**c**) Sulzer-Metco powder feedstock, (**d**) cross section [Syed et al. (2006)]

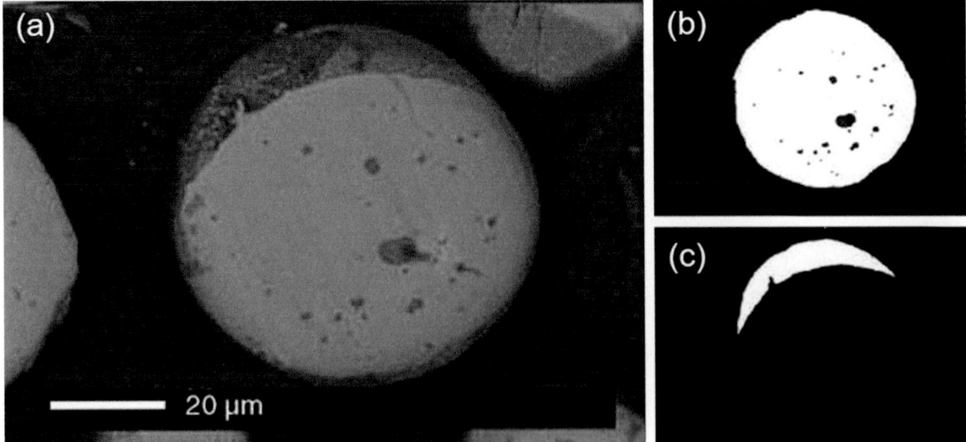

Fig. 9.20 Electron micrographs images of the plasma treated particles (**a, b**) of polished cross sections showing oxide nodules (**c**) oxide cap [Syed et al. (2006)]

It should be pointed out that other than the in-flight particle oxidation during the spray process, the presence of entrained air in the plasma plume would also result in the oxidation of the substrate surface and coating, especially between successive passes which would lead to the decrease of the adhesion/cohesion of coatings and hinder diffusion phenomena at the coating-substrate interface.

9.3.2 Systems and Operating Conditions

While the use of a gas shroud as described earlier in Chap. 8 "DC Plasma Spraying, Fundamentals" offers means of reducing the entrainment of ambient air/oxygen into the plasma jet and consequently the chances of in-flight oxidation of the sprayed particles, the spraying of oxidation sensitive materials in a controlled atmosphere plasma spraying (CAPS) chamber has been widely accepted as more effective and secure way to protect the contamination of the coating with ambient gas species. (CAPS) developed in the 1960s, uses essentially vacuum-tight chambers with a vacuum pump, for the initial evaluation of the chamber, before backfilling it with an inert gas, usually argon, to atmospheric pressure or higher. The oxygen content in an argon atmosphere can be lower than 7 ppm if a liquid argon source is used [Freslon A. (1995)]. As shown earlier plasma jets in argon atmosphere are broader and longer by a factor 1.5 to 2 compared to the plasma jet in air [Roumilhac (1990)]. Nitrogen atmosphere results in an intermediate plasma jet (Fig. 9.17). If the spray chamber is big enough ≈ 10 m^3 there is no need to water-cool its walls.

CAPS is normally operated at pressures in the range of 70 to 300 kPa for the spraying of materials which are very sensitive to oxidation such as carbides, borides, and refractory metals. For example, to spray carbides the jet in argon atmosphere must be longer than the jet obtained in air. Few tests have been performed using chambers at pressures over 300 kPa [Jäger et al. (1992)]. The plasma jets are shorter in this case compared to that at atmospheric pressure, and electrode erosion becomes more important due to the increased radiation from the plasma column. The increase of the ambient pressure also results in the increase of arc voltage since the arc column constricts and its temperature rises, resulting in higher losses to which the arc responds by an increase of the electric field strength (arc voltage). As the heat losses increase, the specific enthalpy of plasma jet shows a slight drop. At chamber pressure below 300 kPa specific mass, viscosity and thermal conductivity of Ar plasma jet are higher, which is necessary for the spraying of carbides. It should be pointed out that even when using CAPS, the chemistry of the coating remains sensitive to that of the composition of the plasma gas and/or the ambient atmosphere. For example, when spraying reactive metals such as

titanium with Ar-N$_2$ plasma-forming gas, in controlled argon or nitrogen atmosphere chamber, at different pressures, the nitrogen content of the coating in the form of TiN can be significant: the coating content (wt.%) is 2.8 with a chamber pressure of 20 kPa against 12.4 at 119 kPa [Jäger et al. (1992)]. [Guipont et al. (2002), Espanol et al. 2002, Guipont et al. (2010)] have sprayed hydroxyapatite with argon plasma working in argon atmosphere at pressure up to 0.3 MPa. They have shown that the decomposition level can be tailored without using N$_2$or H$_2$. Coatings present highly soluble as well as less soluble (crystalline) characteristics. The nature of the ceramic composite with multiphases can be adjusted through the chamber pressure. Alumina coatings sprayed in Ar atmosphere at different pressures exceeding atmospheric pressure have also been reported [Ma et al. (2002)]. Another advantage is the temperature regulation of both substrate and coating (up to 1000 °C if the treated material can sustain it) resulting in better adhesion. The promoted diffusion results in better densification of coatings. The absence of oxides and the better adhesion can result in higher coating thickness, superior hardness.

[Leylavergne et al. (1998)] compared plasma-sprayed coatings produced in argon or nitrogen atmosphere. The choice between argon and nitrogen atmosphere also depends on cost considerations, nitrogen being 25 to 30% cheaper than argon! They showed that in controlled atmosphere plasma spray process, the use of a nitrogen instead of an argon atmosphere results in coatings with comparable properties. The nitrogen content of the niobium and titanium carbide coatings analyzed in this paper was at most 1%. This result was confirmed with other ceramic and metal coatings. The diffusion of nitrogen seemed to occur in-flight in the molten droplets, with the cryogenic cooling of the substrate by liquid nitrogen jets preventing the reactions after impact of the droplets onto the substrate. Gas flow fields were computed in nitrogen and argon atmospheres. The predictions confirm the small differences observed in coating properties. Under the conditions of the present study, the replacement of argon by nitrogen as an inert atmosphere seems feasible, thus reducing process operating costs by approximately 30%.

In a study of the tribological behavior of Al$_2$O$_3$ coatings on AISI 316 stainless steel, [Sarafoglou et al. (2007)] operated a CAPS process at higher pressure in the 100 and 400 kPa range, referred to as High Pressure Plasma Spraying (HPPS). The results indicate that plasma composition, through its higher heat capacity, does influence the heat transfer to particles, and, consequently, their flattening and densification process, which govern coating properties. It was revealed that tribological behavior of coatings was also influenced by the applied spraying method. However, HPPS coatings led to worst wear behavior. In general, properties, such as microstructure, microhardness, coefficient of friction, and wear resistance depended on the processing conditions such as pressure and composition of the spraying chamber atmosphere. The dominant wear mechanism was that of abrasion for all coatings and conditions of wear tests, and that of adhesion for the antagonistic material.

9.4 Vacuum Plasma Spraying

9.4.1 Basic Concepts

Vacuum Plasma Spraying (VPS) also referred to as Low Pressure Plasma Spraying (LPPS) was developed in 1974 by Müehlberger (Electro-Plasma Inc., now Öerlikon Metco Inc.) [Muehlberger E. (1988), Meyer and Hawley (1991)]. It is a process carried out under soft vacuum conditions in a controlled atmosphere chamber at absolute pressures in the range of 5 to 70 kPa allowing for the coating of small to medium-sized parts of complex shapes with high-density coatings of metals and alloys including refractory metals. Compared to APS, its principal advantages are:

- Prevents in-flight and post-deposition oxidation of the spray powders insuring an oxide-free deposit.
- Substrate can be maintained at high temperatures (up to 950 °C for superalloys) without oxidation. Such temperatures promote inter-diffusion, thus enhancing the coating adhesion,
- High velocity plasma jet (2000 to 3000 m/s) giving rise to higher particle velocities compared to APS (800 to 900 m/ s for alumina and of 450 to 650 m/s for YSZ).
- Lower density gradients between the plasma jet and the surroundings atmosphere result in more stable jets and with less cold gas entrainment providing a more uniform

particle heating and acceleration as well as less divergence of the particle trajectories.

- Reduced heat transfer rates to particles largely compensated by longer jets and increased residence time in the plasma; reduced heating rates also result in less superheating of the particle surface and lower temperature differences between particle surface and its center.

The principal limitation of VPS is the size of the parts that can be coated and the melting temperature of the material to be sprayed which excludes ceramic materials particularly non-oxide ceramics with melting temperatures above 3000 K. The high capital investment needed for such installations also imposes an important limitation on the wider used of the technology.

As shown in Fig. 9.21, VPS is carried out in large vacuum chamber with volumes of a few m^3 up to 10–20 m^3, typically 1.5 to 2.5 m in diameter by 3 to 4 m long, which houses the substrate, often on a carousel holding more than one part to be coated, a DC plasma torch, and a robotic torch manipulator. The chamber is normally water-cooled with a large access door for ease of servicing, placing the parts to be coated on the carousel substrate holder and retrieving them at the end of the coating cycle. The gas exit port of the chamber is connected to an efficient dust/fume collection filter followed by a high-volume pumping station capable of the initial evacuation of the chamber below 1 kPa in a reasonable time (a few minutes) and to maintain the chamber pressure during the spraying cycle under soft vacuum (5 > p > 70 kPa). It is to be noted that under such pressure range of operation, the torch nozzle must be adapted for supersonic jet velocities to reduce or avoid the formation of diamond shock waves in the jet.

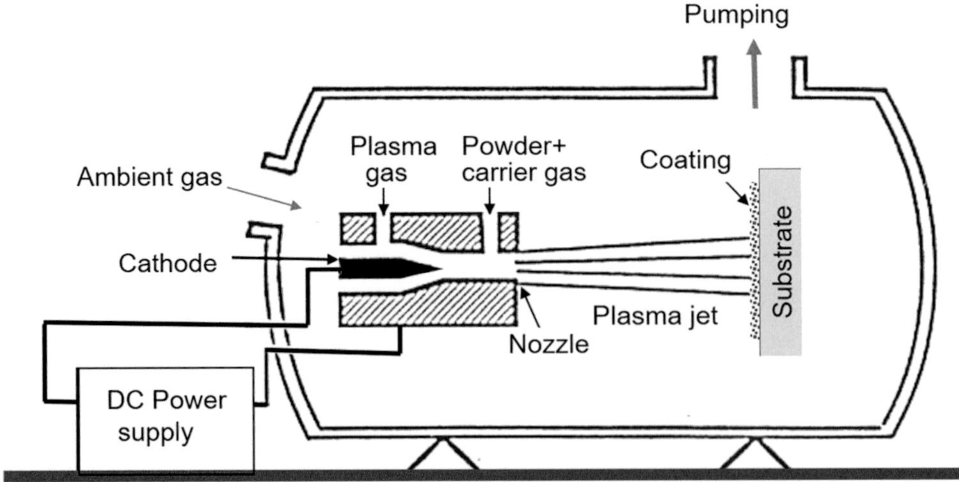

Fig.9.21 Schematic representation of a low-pressure plasma spray chamber. (Courtesy of Emil Pfender)

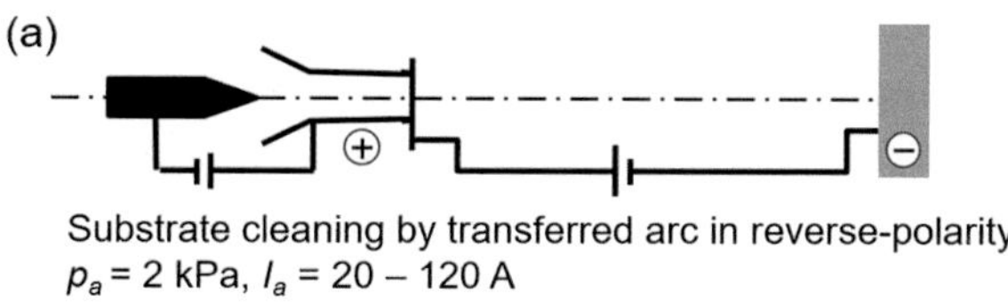

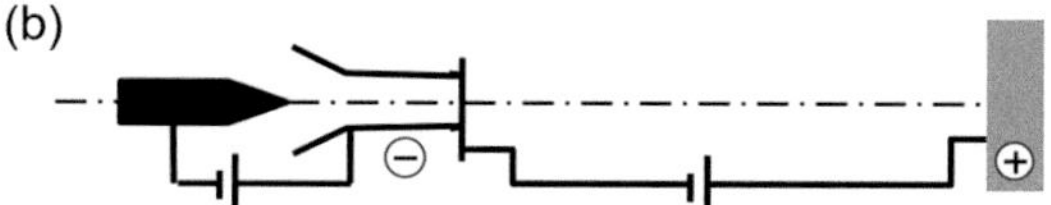

Fig. 9.22 Principle of the transferred arc in VPS (**a**) reverse polarity, substrate as cathode for the surface cleaning step (**b**) straight polarity, substrate as anode for the substrate preheating step

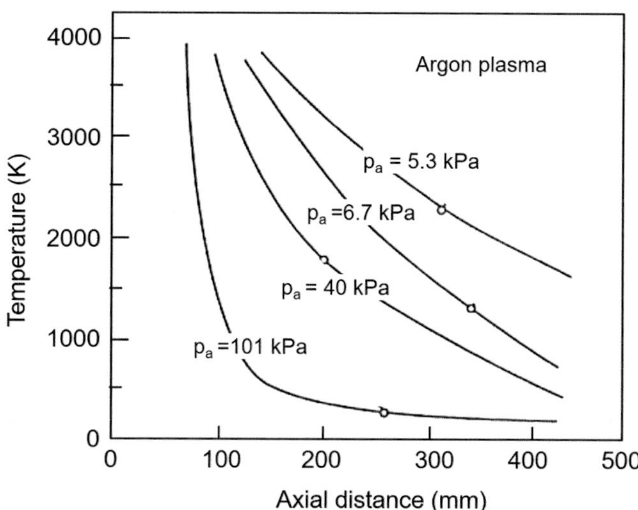

Fig. 9.23 Effect of chamber pressure on the axial temperature profile along the centerline of DC pure argon plasma jet. (Courtesy of Emil Pfender)

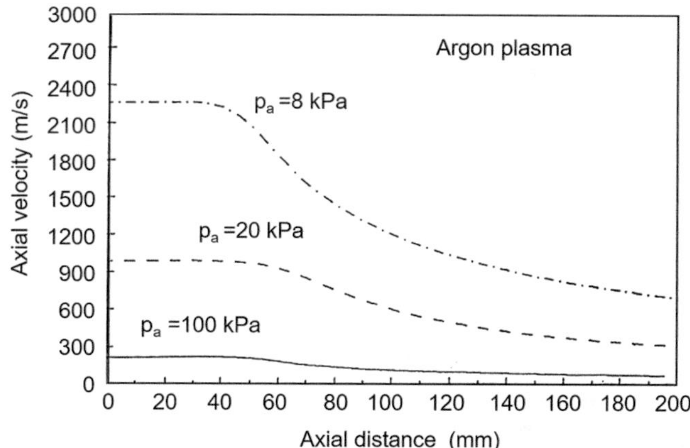

Fig. 9.24 Axial velocity profiles along the centerline of DC pure argon plasma jet at different chamber pressure. (Courtesy of Emil Pfender)

Torch operation is similar to that of APS torches, that is, with similar arc currents, plasma gas selections and flow rates, power levels, and powder flow rates. The torch-to-substrate distance is longer than normally used in APS, typically in the range of 250 and 500 mm. At the start of each operation cycle, the chamber is pumped down to a pressure of a few Pascal, and then backfilled with high purity argon to the desired operating pressure. Depending on the lowest absolute pressure that can be achieved by the pumping system, and the sensitivity to oxidation of the material to be sprayed, the cycle of chamber evacuation and its refilling with pure argon may need to be repeated more than once in order to bring down the residual oxygen level in the chamber to an acceptable level, typically in the range of a few ppm O_2.

Prior to the initiation of the spraying cycle and depending on the material to be sprayed and the substrate properties, a substrate surface cleaning and preheating step might be necessary. This is carried out using the plasma torch in a reversed-polarity transferred-arc mode as shown in Fig. 9.22a, in which a low-current arc is established between the substrate and the plasma torch with the substrate acting as cathode and the torch as anode. The substrate surface cleaning process, typically carried out with a chamber pressure of 2 kPa and an arc current between 20 and 120 A, results in the evaporation of impurities, in particular of oxides, in a timeframe of a few minutes [Itoh et al. (1990)]. Following the substrate surface cleaning step, substrate preheating to sufficiently high temperature to promote coating adhesion is possible by switching the transferred arc to a straight-polarity mode, as shown in Fig. 9.22b, with the substrate acting as anode and the torch as cathode. The chamber pressure is increased to 6 kPa, which is typically the working spraying pressure, and the arc current is kept between 50 and 150 A. In VPS, the substrate can be maintained at high temperatures (up to 950 °C for superalloys) without oxidation. Such temperatures promote inter-diffusion, thus enhancing significantly the coating adhesion.

With the ignition of the plasma torch for spraying operation, the plasma jet becomes increasingly longer and wider as the pressure decreases because the cooling through the entrained air is avoided [Roumilhac et al. (1990)]. Modeling results of axial temperature distributions along the centerline of an argon DC plasma jet issuing into an argon environment at different pressures are given in Fig. 9.23. These show an increasingly longer plasma jet, and the much slower temperature decay with the decrease of the chamber pressures. Figure 9.24 shows corresponding axial velocity distributions for a low power argon plasma jet (8 kW) with 11.6 g/min argon flow in an 8 mm diameter anode-nozzle and three different chamber pressures. The strong increase in the plasma velocity is noticeable compared to APS operation, even with pure argon. The lower axial temperature and velocity gradients are due to the lower large-scale turbulence and less cold gas entrainment, a consequence of the smaller density gradients resulting in less turbulent shear layers. The lower densities result, however, in lower electron-ion recombination rates, and deviations from composition and Local Thermodynamic Equilibrium (LTE). The effect of going to very low pressures on the plasma jet and particle properties has been shown by [Refke et al. (2003)] using an experimental vacuum chamber and a Sulzer-Metco O3CP torch, a set-up that has the capability to be operated down to 0.1 kPa (1.0 mbar) with up to 150 slm of total gas flow rate. The process, identified as Ultra Low-Pressure Plasma Spraying (ULPPS), is discussed in the next section.

9.4.2 Torch Nozzle Design and Gas Dynamics

For compressible nozzle flows reaching supersonic speeds inside the nozzle, shock diamonds are usually observed in the plasma jet (see schematic in Fig. 9.25). These occur when the pressure inside the jet is different from the surrounding atmosphere leading to sudden velocity drop and the creation of shock waves. The jet is identified as "over-expanded" when the pressure inside the jet is lower than the ambient pressure in the chamber, and "under-expanded" with a jet pressure higher than the ambient pressure. As a consequence, through the interaction with the environment the jet is compressed or expanded, but an over-compensation may occur resulting in another under-expansion/over-expansion of the flow. A complex flow structure evolves with shock waves starting at the

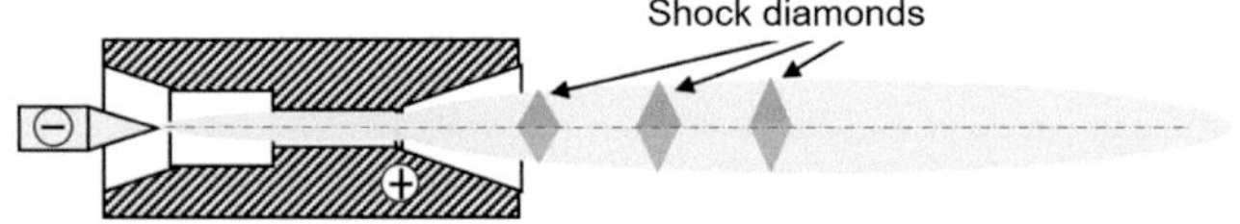

Fig. 9.25 Schematic illustration of shock diamond as consequence of inadequate equilibration between pressures in the jet and in the surroundings

nozzle exit being reflected at the discontinuity between jet and environment, and these reflected shockwaves create the shock diamonds. The more shock diamonds and the clearer they are visible, the worse the fluid dynamic condition. The use of a properly designed Laval nozzle either as part of the anode or as an attachment to the anode can significantly improve the results by avoiding the shock structures immediately downstream of the anode nozzle exit [Meyer and Hawley (1988)]. A schematic of a torch with a Laval nozzle attachment is shown in Fig. 9.26 to improve the problem cited above [Meyer and Hawley (1988)]. The Laval nozzle attachment offers an increase of particle velocities by 30% to 50% and particle trajectories parallel to the torch axis [Meyer and Hawley (1988), Henne et al. (1993)], resulting in a higher deposition efficiency [Rahmane et al. (1998)].

With a Laval-type nozzle design (Fig. 9.26) the jet will lengthen and enlarges in soft vacuum, as shown in Fig. 9.27. At 95 kPa the jet is only close to 60 mm long, at 5 kPa its length reaches 400 mm, and if the pressure is reduced down to 0.1 kPa the jet length can be more than 1200 mm and up to 200 mm in diameter.

9.4.3 Coating Microstructure

In the following paragraph, a few examples of coatings achieved with VPS (also LPPS) are presented:

NiCoCrAlY coating obtained by [Scrivani et al. (2003)] with a Central Injection APS torch, HVOF torch, and VPS system showed the VPS coatings to be significantly superior with lowest porosity, lower unmelt density, and practically no oxide formation. This is illustrated in Fig. 9.28 showing significant variations in the coating microstructures before thermal treatment, the denser structure being that of VPS (Fig. 9.28c). Axial plasma sprayed coatings show higher amounts of un-melted particles with respect to other traditional technologies. HVOF coating shows some un-melted particles and this could be expected because the HVOF flame

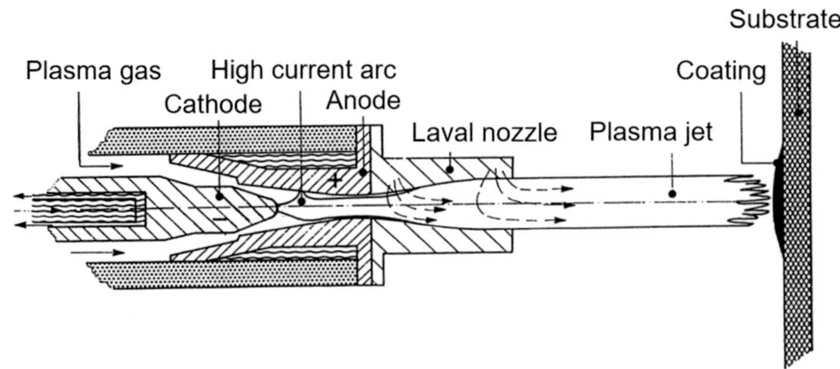

Fig.9.26 Schematic of torch nozzle with Laval attachment for improved supersonic flow structure [Mayr and Henne (1988)]. (Reproduced with kind permission of Sulzer Metco AG, Switzerland)

is colder than the plasma jet. Un-melted particles could not be observed in the coatings produced by VPS. VPS produces coatings that are quite completely oxide-free.

[McKechnie et al. (1994)] reported that VPS forming has been developed and characterized for near-net-shape fabrication of aerospace components. Applications require VPS forming of structural materials in both monolithic form (i.e., free-standing shapes) and as integral parts of complex components (e.g., liner for a rocket engine combustor). In these applications the material deposited on both the inside and the outside of large and small components must meet strict quality requirements. They used NARIoy-Z, which is a copper-base alloy with a nominal composition of Cu-3.0Ag-0.5Zr and high thermal conductivity. It is used in the combustion chamber liner of the Space Shuttle main engine and is the baseline material for current VPS forming development. Authors discussed metallurgical and processing comparisons between depositing material on inside and outside surfaces of symmetrical shapes. Specific examples of material properties (e.g., grain structure, hardness, and tensile properties) and process parameters (e.g., standoff distance and gun design) are presented in terms of the fabrication of large rocket engine combustion chambers (Fig. 9.29).

Microstructure of silicon coating prepared by [Niu et al. (2009)] using VPS technology is given in Fig. 9.30. This shows compact dense silicon coating with no identifiable areas of silicon oxide. Small ball-like particles of sizes of less than 1 μm was found on the surface and in the coatings, which seems to result from splash debris. The porosity was low and composed of pores with spherical and linear shapes. The coating exhibited typical two-layer microstructure in flattened splats which had equiaxed nanometer grains and overlying columnar grains in the longitudinal section.

[Niu et al. (2013)] reported a study of $MoSi_2$ coating prepared by the VPS technology. These exhibited a lamellar and dense microstructure, which was composed of grains with irregular shapes and different sizes of 0.1–0.2 μm. It was composed mainly of tetragonal and hexagonal $MoSi_2$ phases and a small amount of tetragonal Mo_5Si_3 phase. TEM observation confirmed that Mo_5Si_3 was randomly distributed in the matrix of $MoSi_2$. After being exposed to air at 1300 and 1500 °C for 6 h, a dense and thin film (<10 μm) of silicon oxide was formed on the surface of the $MoSi_2$ coating, providing excellent protection for the coating. Applying the heat treatment at 1700 °C for 6 h resulted, however, in the formation of a relatively thick layer (20–30 μm) of Mo_5Si_3 formed under the silicon oxide film indicating the oxidation of the $MoSi_2$ coating.

[Salimijazi et al. (2005a, b)] reported studies of the Vacuum Plasma Spraying of Ti-6Al-4V alloys and B_4C in an attempt to produce near net-shaped structural parts of layered composite materials for light weight ballistic protection applications. The work was undertaken in two steps, the first dealing with the deposition of Ti-6Al-4V alloys

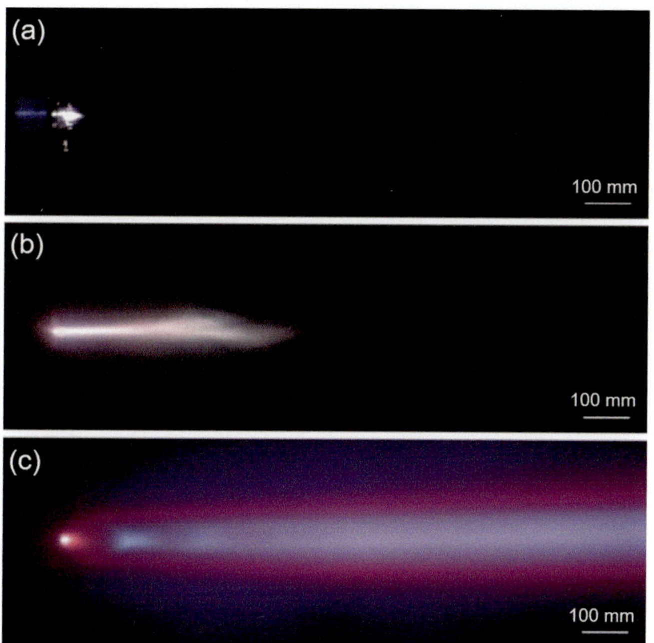

Fig. 9.27 Images of plasma jets in a controlled-atmosphere chamber expanding at different pressures (**a**) 95 kPa (APS), (**b**) 5 kPa (VPS), (**c**) 0.1 kPa (ULPPS) and deposition process [von Niessen and Gindrat (2011)]

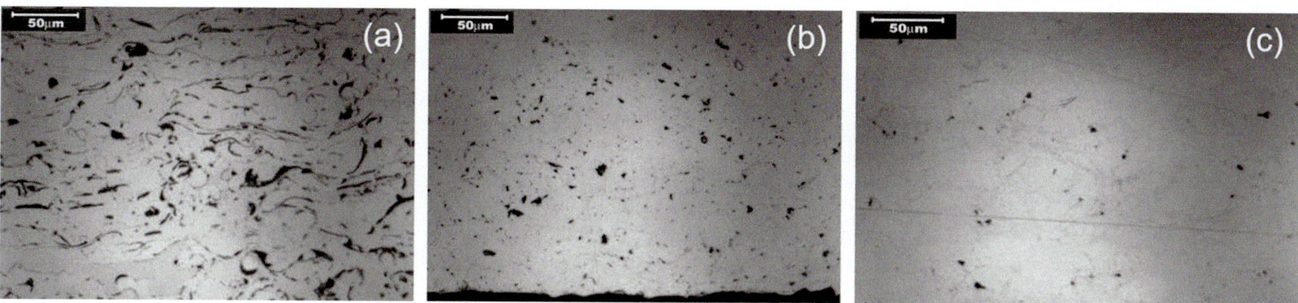

Fig. 9.28 Microstructure of CoNiCrAlY alloy coatings obtained by different thermal spray technologies (**a**) Central injection APS (**b**) HVOF (**c**). VPS [Scrivani et al. (2003)]

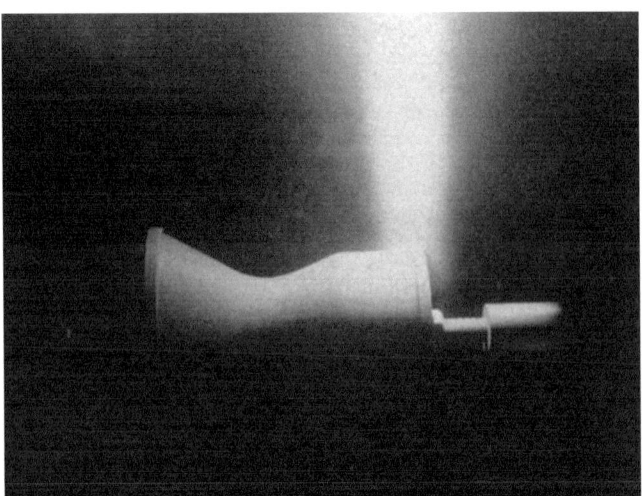

Fig. 9.29 Vacuum plasma spray forming of NARIoy-Z on the outside of a combustion chamber mandrel [McKechnie et al. (1994)]

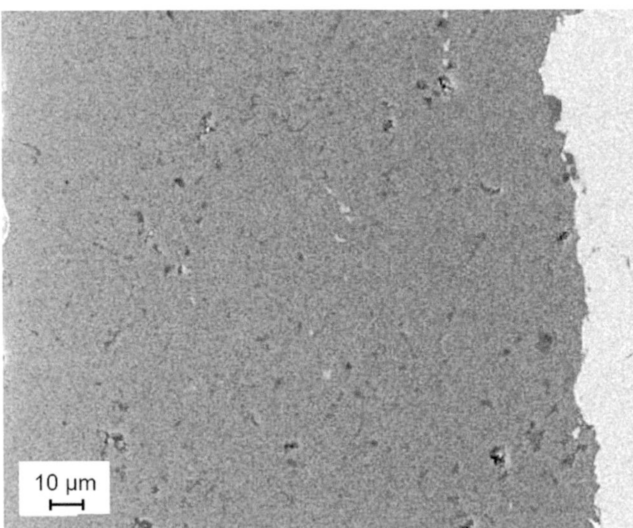

Fig. 9.30 Cross-sectional micrograph of VPS-Si coatings [Niu et al. (2009)]

[Salimijazi et al. (2005a)] on copper, as a dummy substrate. Detailed microstructural, chemical composition analysis of sprayed material was made. The observed chemistry and microstructure of the coating reflected the conditions under which the coating was formed and the phase equilibria in the Ti-alloy system. The porosity of the deposit was in the range of 3 to 5%. Slight decrease in the Al content and an increase in the amount of oxygen and hydrogen were found relative to the starting powder. The latter was attributed respectively to oxygen leakage into the vacuum deposition chamber and the use of hydrogen in the plasma forming gas. As with any structural material, attention was also given to metallurgical considerations such as grain size, phase composition, and phase distributions. The as-deposited material is $\geq$90 % α'martensite that is present in the form of fine lathes on the order of 500 nm in width surrounded by residual $\beta-$phase. SEM micrographs of an etched Ti-6Al-4V deposit given in Fig 9.31a show a columnar grain structure within the central core of the splat which are oriented in the direction of heat transfer to the substrate, or prior splat. Optical micrographs of the etched cross section of the deposit given in Fig 9.31a show the presence of some unmolten and partially molten particles, with isolated larger porosities detectable in the unetched micrograph given in Fig 9.31c.

The second part of the study [Salimijazi et al. (2005b)] was dedicated to the VPS deposition of B_4C on Ti-6Al-4V substrate with a coating thickness of 300 to 600 μm. The microstructure of a polished section of the B_4C deposit is given in Fig. 9.32a, b which shows the splat morphology and the pore structure in the deposit. According to [Salimijazi et al. (2005b)] the rounded pores could be caused by gas entrapment during the deposition while angular edged pores were attributed to loss of solid particles or fragments. The B_4C deposit was also subjected to surface hardness measurements using both Knoop and Vickers indenters. Knoop hardness tests were performed with the long axis of the Knoop indenter both parallel and perpendicular to the splat spread direction. The applied load for Knoop hardness

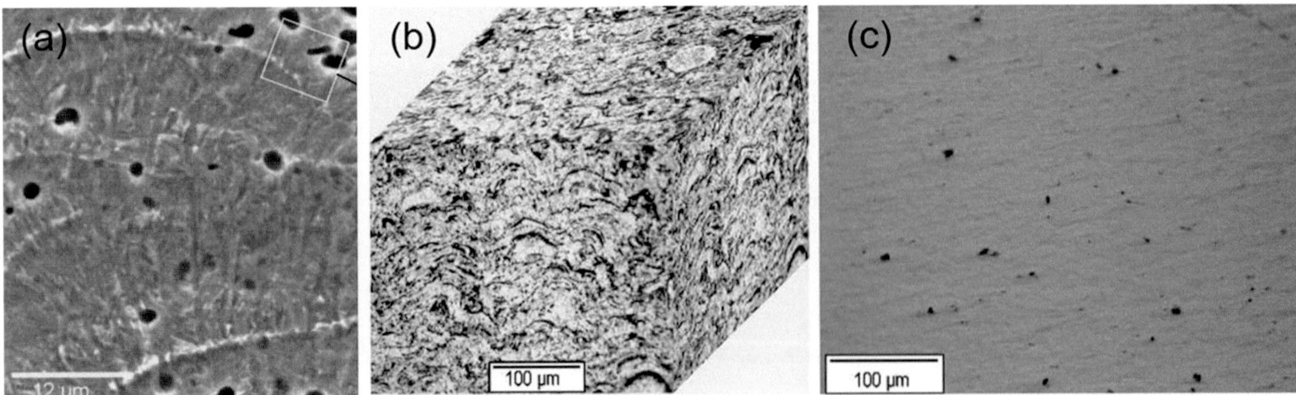

Fig. 9.31 Cross section micrographs of VPS deposited Ti-6Al-4 V alloy (**a**) SEM micrograph (**b**) etched optical micrograph (**c**) unetched optical micrograph [Salimijazi et al. (2005a)]

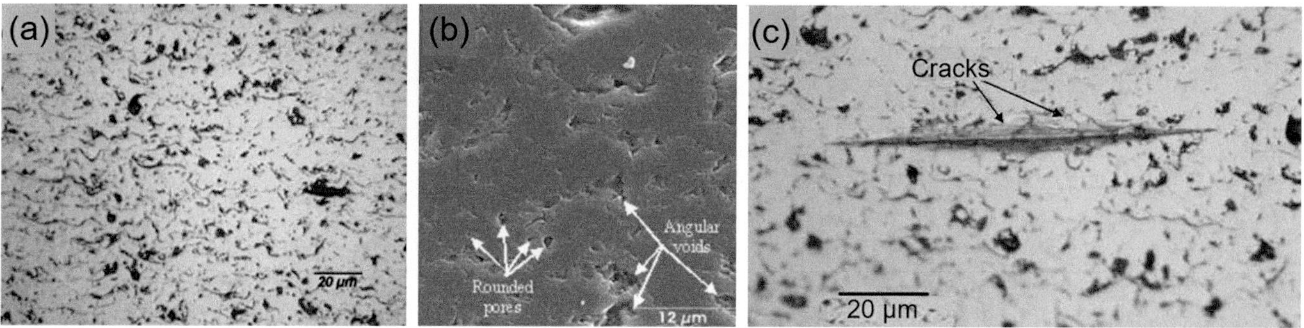

Fig. 9.32 Micrograph of polished cross section of the VPS B4C deposit (**a**) optical micrograph (**b**) SEM micrograph (**c**) micrograph showing cracks propagation around the microhardness indentation area [Salimijazi et al. (2005b)]

measurements was 500 g. The as-sprayed B_4C has an average hardness of 1236 $\pm$ 140 KH parallel to the splats and 1583 $\pm$ 103 KH in the perpendicular orientation. Some cracks along the splat boundaries oriented in the direction of the large diagonal of the Knoop indenter can be observed (Fig. 9.32c).

A study for the optimization of the operating conditions for the VPS deposition of boron carbide (B_4C) was reported by [Lin Chun-Ming (2012)]. According to results obtained by a Taguchi design method a deposit porosity of 8.5% was obtained using the following near optimal operating conditions: chamber pressure of 65 kPa, current 680 A, plasma gas flow rate [45 slm (Ar) +13.5 slm (H_2)], and spraying distance of 150 mm. Further optimization of the spraying parameters by [Lin Chun-Ming (2012)] involved the combination of Taguchi design method with a neutral network / Generic Algorithm, (GA), allowed for the reduction of the porosity level of the deposit to 5.6% using revised operating conditions: Chamber pressure of 66 kPa, current 680 A, plasma gas flow rate [39.7 slm (Ar) +12.5 slm (H_2)], and spraying distance of 236 mm.

[Liu et al. (2017)] showed that coating quality is affected by arc and plume instabilities during plasma spraying. In closed chamber plasma spraying, gradual drift is one of the intermediate instabilities, which is mainly due to the electrode erosion. Their work focuses on the source of the gradual drift of the plasma jet and the influence on coating quality. The variation in the plasma jet was observed by a particle flux image device based on a CCD camera. The optical spectrum of the plasma plume was measured and analyzed through an optical spectrometer. The results indicated that the addition of hydrogen to plasma gas induced the change in the plasma jet length and width with changing rates depending on the chamber state and the ventilation power. With poor ventilation, the intensity of H_α emission was found to become gradually stronger while H_β and $H\gamma$ were found to become weaker. On closing the chamber and retaining enough ventilation power, it was observed that the ambient gas slowly turned red. Simultaneously, the coating weight and thickness were slightly decreased. Meanwhile the porosity ratio was obviously increased. The red ambient gas has been proved to be able to acidify the city water with pH value decreased from 7 to 1–3. Without hydrogen, the plasma jet was found to be stable without reddening and variation, but the plasma enthalpy was unfortunately low.

An interesting study reported by [Zhang et al. (2006)] compared the performance of APS and VPS technologies for the manufacture of the electrolyte materials for Solid Oxide Fuel Cells (SOFC). The study was carried out using two commercially available yttria stabilized ZrO_2 (YSZ) 8 mol% Y_2O_3 powders:

- The first called "YSZ1" was a fused-crushed powder (Plasmatex, Sulzer-Metco, Winterthur, Switzerland) with particle size range 5 to 26 µm.
- The second called "YSZ2" was an agglomerated powder containing nanoscale substructure of 25 nm (TZ8Y, TOSOH, Tokyo, Japan) with particle size range 20 to 116 µm.

Coating layers with these two powders were deposited by Vacuum Plasma Spraying (VPS) and atmospheric plasma spraying (APS) to compare the impact of spray methods and particle morphology-size/manufacturer on the electrical conductivity and gas permeability of the coating obtained. The microstructure of YSZ coating was characterized using scanning electron microscopy and X-ray diffraction analysis. The results showed that for each of the two powders, the gas permeability of the coating made using VPS is lower than that obtained using APS. Different results were obtained, however, with different powders, with the gas permeability of the coating made by fused-crushed powder (YSZ1) generally one order of magnitude lower than those produced by the agglomerated powder (YSZ2). The electrical conductivity of the coatings, measured by potentiostat/galvanostat based on three-electrode assembly approach, using fused-crushed powders of small particle size, showed that coating deposited by VPS was 0.043 S/cm, at 1000 °C, which is about 20% higher than that of APS with the same powder.

9.5   Ultra Low-Pressure Plasma Spraying

9.5.1   Basic Concept and Gas Dynamics

Ultra-Low-Pressure Plasma Spray (ULPPS), also identified as Very Low Pressure Plasma Spraying (VLPPS), Plasma Spray-Physical Vapor Deposition (PS-PVD) and referred to by Sulzer-Metco as Low Pressure Plasma Spraying-Thin Film (LPPS®-TF) is a relatively new technology that bridges the gap between conventional VPS spraying processes in which the coating is formed through the building up of successive splats on the surface of the substrate, and Physical Vapor Deposition (PVD) processes in which the coating is formed through the deposition of the coating material from the vapor phase at lower temperatures at relatively low deposition rates. In fact, ULPPS is a process developed in direct competition to EB-PVD, for the production of ceramic TBC coatings with equivalent quality as EB-PVD technology, at considerably higher deposition rates and lower cost. The two technologies, ULPPS and EB-PVD, differs essentially in the way the precursor is transformed into the vapor phase with ULPPS based on the in-flight heating and evaporation of the precursor in powder form using the plasma jet, while EB-PVD makes use of an Electron Beam for the precursor evaporation form a solid target or powder. This difference has a significant impact on their respective operating pressure and the deposition rates associated with each of the two technologies, with the ULPPS process operating at pressures (< 200 Pa) with deposition rates in the range of μm/s, in contrast to the Electron Beam technology which required considerably lower operating pressures (<5 Pa), has lower deposition rates in the 0.1 to 100 μm/min range, and significantly higher investment cost.

[Niessen and [von Niessen and Gindrat (2011)], Hospach et al. (2011), Mauer et al. (2010, 2013, 2015), and Mauer (2014), 2019] were among the first who worked on the Ultra-Low-Pressure Plasma Spray (ULPPS), which they achieved using a modified single cathode vacuum plasma gun, the Sulzer Metco O3CP, which allows for the use of a high gas flow, up to 200 slm, and torch current up to 3000 A (180 kW). The gun can be run with either a two- or fourfold internal powder injection. Pre-heating and cleaning of the substrate can be performed by the plasma itself or by means of an integrated transferred arc processing to optimize the coating adhesion. The combination of a high-energy plasma gun and a low-pressure environment enables achieving a predefined evaporation of the injected powder material and the controlled formation of dense thin film coatings by vapor phase deposition. As shown in Fig. 9.33, [Sulzer-Metco bulletin (2008)] the plasma jet is typically more than 2 m long, with a jet diameter of 200 to 400 mm for a plasma torch power of 180 kW. A

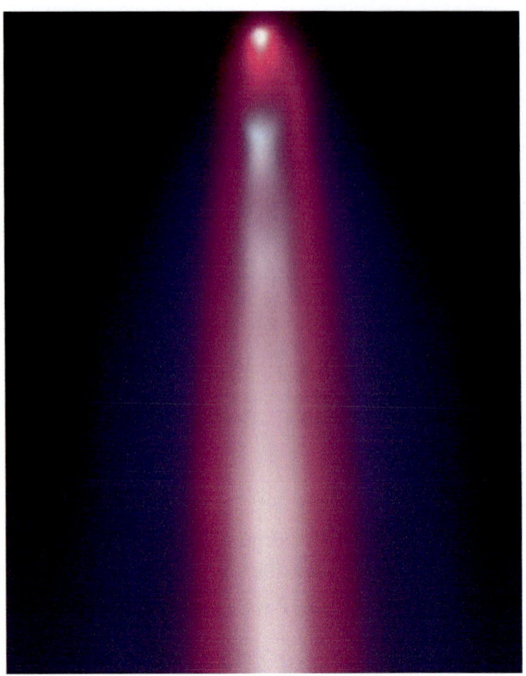

Fig. 9.33 Images of the plasma jets in the LPPS®-TF process, chamber pressure 200 Pa (2 mbar), plasma Jet length 2 m, Jet diameter 200–400 mm, torch power up to 180 kW with 4 powder injectors. [image and info courtesy of *Ö* relikon-Metco corporation]

photograph of such a ULPPS installation at the Forschungszentrum J *ü* lich GmbH, Institut f *ü* r Energieforschung J *ü* lich, Germany, which is an upgraded Sulzer Metco LPPS system, is given in Fig. 9.34. [Mauer et al. 2010].

At this pressure, the plasma jet interaction with the surrounding atmosphere is very weak. The plasma velocity is almost constant over a large distance from the nozzle exit, the collision frequency is distinctly reduced, and the mean free path is significantly increased. As a consequence, the specific enthalpy of the plasma is substantially high, though at low density. These particular plasma characteristics offer enhanced possibilities to spray dense, thin film, ceramics coatings compared to conventional processes which operate in the pressure range between 5 and 20 kPa.

An early diagnostic study was reported by [Refke et al. (2003)] for an Ar/H$_2$ [100 slm (Ar) + 3 slm (H$_2$)] plasma generated using a Sulzer-Metco-O3P torch operated at a current of 1500 A and absolute pressure of 150 to 200 Pa and 1 kPa in which fin YSZ powder was injected. Measurements were made, using enthalpy probe techniques, of the plasma velocity and temperature profiles in the radial direction at axial location 775 mm from the torch nozzle exit. DPV-2000 system was used for the measurement of the corresponding in-flight particle velocity and temperature at

Fig. 9.34 New LPPS-TF facility at the Institute of Energy Research Germany (IEF-1), Forschungszentrum Jülich [Mauer et al. (2010)]

an axial location of 975 mm from the torch nozzle exit. Typical results given in Fig. 9.34a, b are for the plasma and particle velocity profiles in the radial direction at absolute chamber pressures of 200 Pa and 2.0 kPa for the plasma velocity, 150 Pa and 1 kPa for the particle velocities. As expected, the reduction of the pressure from 1 kPa to 200 Pa gives rise to a significant increase of the maximum plasma velocity on the centerline of the flow from 1100 m/ s to 2800 m/s. The effect of pressure on the corresponding particle velocities is relatively small. The corresponding radial profiles for the plasma and in-flight particle temperature are given in Fig. 9.35c, d. The effect of the pressure on the temperature of the plasma is less pronounced with the maximum flow temperature increasing from 6500 K to almost 9000 K. The central part of the plasma jet flow is relatively uniform over a radius of about 40 mm beyond which the temperature of the plasma drops rather steeply. The decrease in chamber pressure from 1 kPa to 150 Pa has only a small impact on the particle temperature. The results also show relatively uniform particle velocity and temperature over a large central region of the flow with a radius of more than 100 mm.

Details of the O3CP torch nozzle design used by Refke et al. *(2003)* is given in Fig. 9.36. *[Mauer (2014)]* points out

that while the plasma velocity at the critical nozzle cross section cannot exceed the local speed of sound (Mach number Ma = 1), the jet becomes supersonic (Ma > 1) when subsequently passing through the nozzle expansion section. As the flow is faster than the pressure waves traveling in the plasma at the local speed of sound, no information on the chamber pressure is carried inside the nozzle. This means that the plasma gas can exit the nozzle at a pressure which is different from the chamber pressure. Since the chamber pressure under PS-PVD conditions is lower than the pressure in the exiting jet, it becomes under-expanded, that is, it tends to expand immediately after the nozzle to accommodate to the chamber pressure. Having reached pressure equilibrium, the free-flowing jet is highly laminar, and no turbulent transition occurs along the spray distance to the substrate.

In his study of the plasma characteristics and the interaction between the plasma and the coating powder precursor, [Mauer (2014); He et al. (2017)] conclude that:

- Plasma characteristics and torch gas dynamics in the ULPPS process can be approximated by minimizing the Gibbs energy assuming chemical equilibrium conditions.
- Calculation approaches based on the kinetic theory of gases are able to analyze the plasma–particle interaction

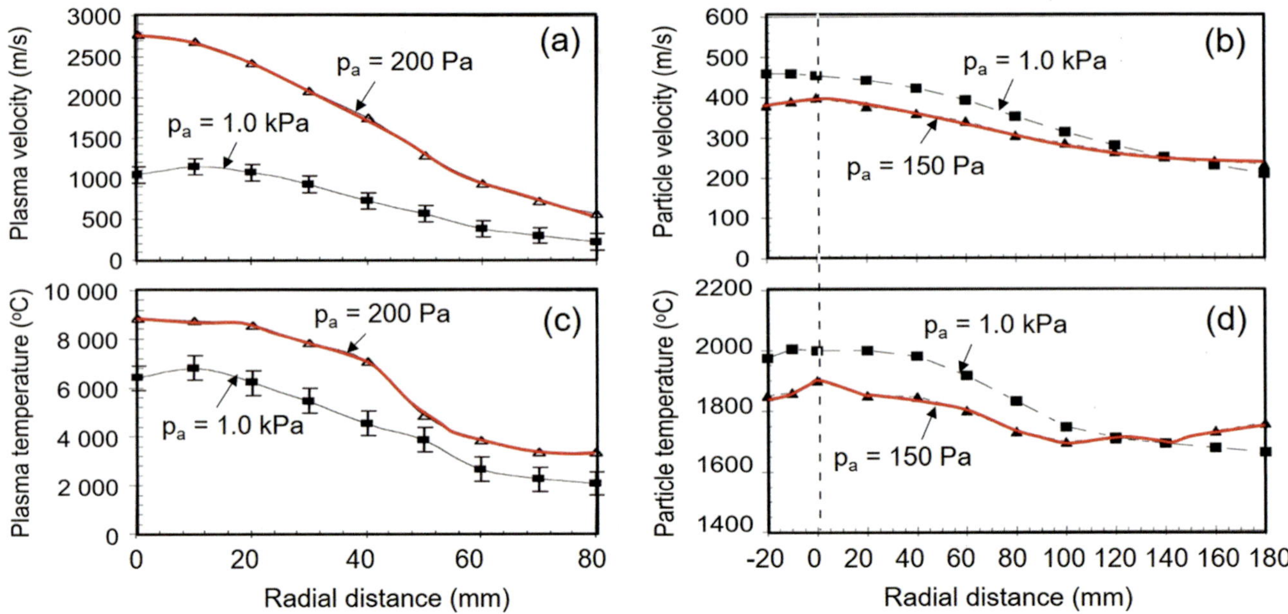

Fig. 9.35 Radial profiles of the velocity and temperature fields in an Ar/H2 [100 slm (Ar) +3 slm(H2)] plasma jet at 1500 A in the presence of YSZ particle injection (**a**) plasma velocity (**b**) particle velocity (**c**) plasma temperature and (**d**) in-flight particle temperature. Results given in (a&c) z = 775 mm (b&d) z = 975 mm from nozzle exit [Refke et al. (2003)]. (Reprinted with permission of ASM International. All rights reserved)

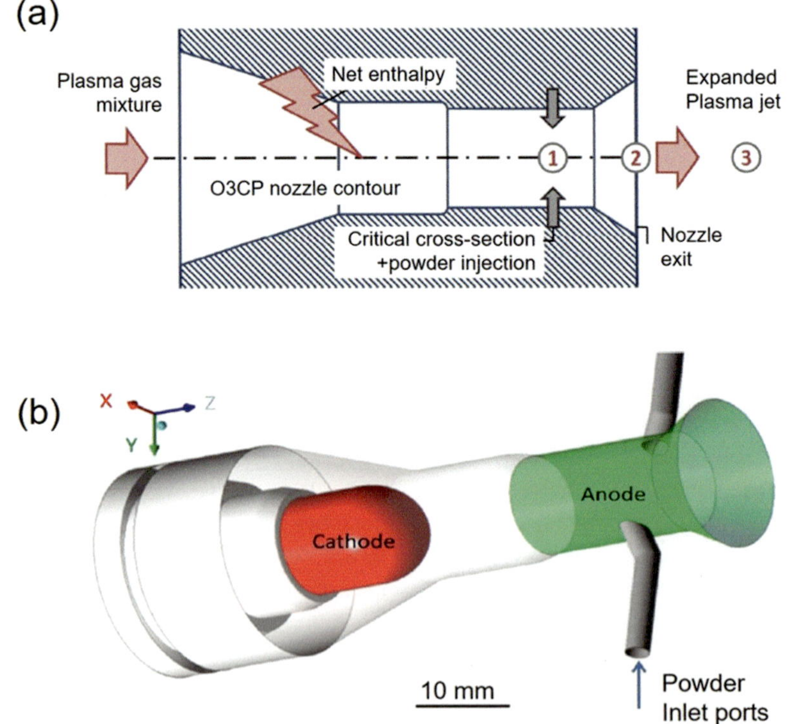

Fig. 9.36 (**a**) Cross-section through the O3CP torch nozzle [Mauer (2014)], (**b**) simulated representation of the O3CP torch nozzle [He et al. (2017)]

and to predict the degree of feedstock vaporization under ULPPS conditions. These show that the feedstock treatment along the very first trajectory segment between injector and jet expansion plays a key role.

- Ceramic thermal barrier coatings manufactured under three different process conditions confirm the modeling results and demonstrate the diversity of microstructural features achievable by ULPPS.

9.5.2   Coating Formation

According to [Niessen and Gindrat (2011)] the possibility to vaporize feedstock material and to produce layered coating out of the vapor phase results also in new and unique coating microstructures. The properties of such coatings are superior to conventional thermal spray (APS or VPS) coatings and electron beam-physical vapor deposition (EB-PVD) coatings. For example, in Fig. 9.37a–d the microstructure of four TBC ceramic coatings made with different spray parameters is given. The layered-type structure coating shown in Fig. 9.37a can be produced using the ULPPS technology by increasing the powder feed rate, thus limiting the process to partial evaporation of the powder precursor. Turning it from a splat-type layered to a porous coating, and finally columnar structures as shown in Fig. 9.37b, c with a layer thickness of approximately 250 μm, could be achieved by a combination of lowering the powder feed rate and the use of specific mixture of argon and secondary plasma gases and a large spray distance. The ULPPS produces columnar coatings consisting mainly of fine needles with a high defect density

and a high amount of internal porosity compared to those obtained in EB-PVD (Fig. 9.37d). With an adequate parameter optimization, the density of the coating structure could be improved reducing the inter-columnar gaps to a minimum as shown by the micrographs of 7YSZ, given in Fig. 9.38 compared to that of Fig. 9.37b and c.

[Niessen and Gindrat (2011)] point out that owing to the high velocity and the large dimension of the plasma jet, the gas stream is able to flow around complex geometries and is "forced" to go through the shadowed areas. Since the coating material is in vapor phase and transported inside this stream, the coating deposit is made whenever the plasma is in contact with the surface, thereby making the ULPPS a non-line-of-sight coating process. The entire surface of a double-nozzle-guided vane shown in Fig. 9.39 could thus be coated with a TBC, with a simultaneous coverage of both sides of the airfoils and the platforms during the same spray run.

[Mauer et al. (2013); Mauer (2014)] point out that by applying powder feedstock, it is possible to fragment the particles into very small nano-sized clusters or even to evaporate the material. As a consequence, the deposition

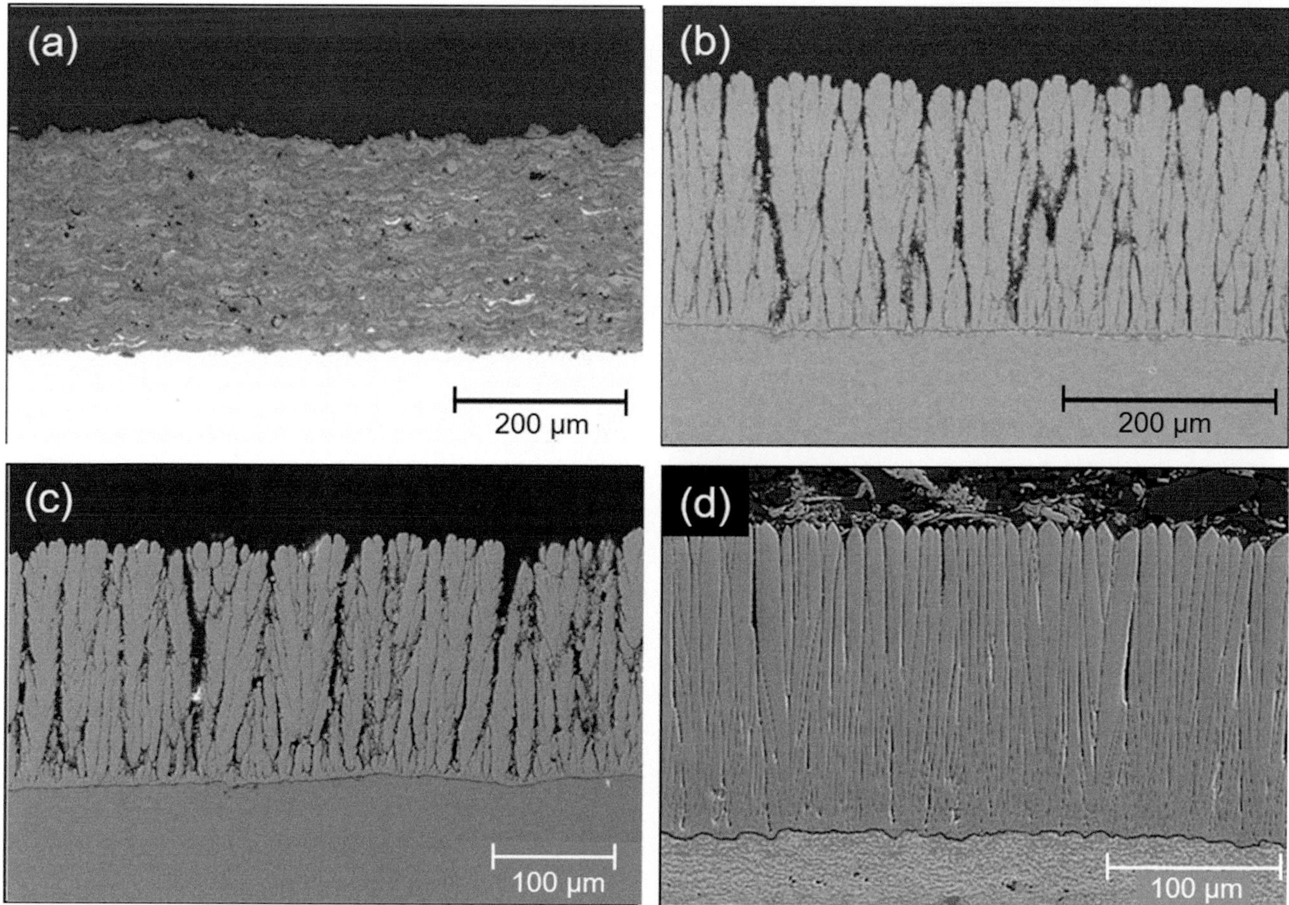

Fig. 9.37 TBC coating of 7YSZ using different deposition techniques (**a**) ULPPS-layered coating (**b & c**) ULPPS-columnar structure (**d**) EB-PVD columnar coating [Niessen and Gindrat (2011)] (EB-PVD coating microstructure, courtesy of U. Schulz, DLR, Inst. of Materials Research, Cologne, Germany)

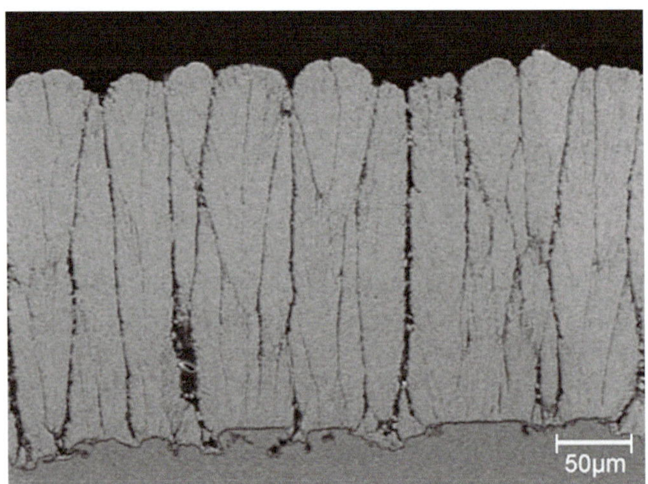

Fig. 9.38 Optimized ULPPS columnar TBC structure coating of 7YSZ deposited on MCrAlY bond coat [Niessen and Gindrat (2011)]

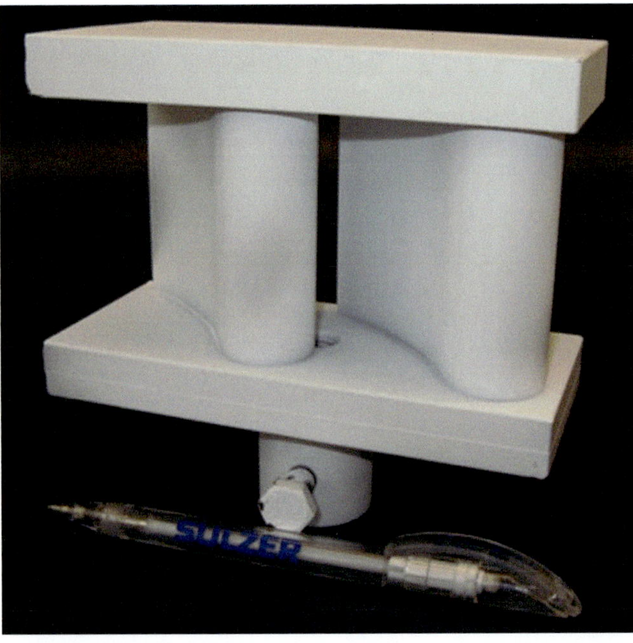

Fig. 9.39 Double vane dummy with a YSZ TBC coating, deposited with PS-PVD [Niessen and Gindrat (2011)]

mechanisms and the resulting coating microstructures can be quite different compared to conventional liquid splat deposition. Thin and dense ceramic coatings as well as columnar-structured strain-tolerant coatings with low thermal conductivity can be achieved offering new possibilities for application in energy systems. To exploit the potential of such a gas phase deposition from plasma spray-based processes, the deposition mechanisms and their dependency on process conditions must be better understood. Thus, plasma conditions were investigated by optical emission spectroscopy. Coating experiments were performed, partially at

extreme conditions. Based on the observed microstructures, a phenomenological model is developed to identify basic growth mechanisms. Typical spray parameters were Ar 35 slm and He 60 slm with a power of 60 kW for a current of 2600 A, chamber pressure 200 Pa, and a spray distance between 300 and 1400 mm. In APS or VPS, liquid droplets form splats piling up to build the coating. At ULPPS conditions, the deposits are formed predominantly from clusters and/or vaporized atomic species. Shadowing, adsorption, nucleation, and growth, as well as bulk recrystallization are the basic mechanisms characterizing the coating growth. In the case of cluster deposition, shadowing mainly occurs. This is the interaction between the roughness of the growing surface and the angular directions of the arriving particles. After sticking, the clusters are hardly mobile so that the surface roughness cannot be smoothed. Thus, the shadowing intensifies as the coating grows. The consequence is a microstructure consisting of tapered columns with domed tops and separated by voids. Figure 9.40 gives an example of typical fracture surface and a coating surface.

[Mauer et al. (2013)] conclude that by ULPPS, it is possible to evaporate the powder feedstock at appropriate parameters providing advanced microstructures and non-line of sight deposition. To ensure evaporation of the feedstock, the conditions inside the nozzle in proximity to the location of injection are crucial. High-power density and accommodated feedstock characteristics are required. Besides evaporation, the formation of nano-sized clusters is observed. Having excited the nozzle, further particle heating is reduced compared to atmospheric pressure due to the low density (Nusselt numbers are smaller). Typical results presented in Fig. 9.41 illustrate the importance of the plasma gases used. These show the microstructure of two YSZ TBC produced using Ar/He and Ar/He/ H_2 plasmas under conditions summarized in Table 9.2.

As pointed out by [Mauer et al. (2015)] complex compositions are often employed in ceramics to achieve advanced material properties, for example, high thermal stability, low thermal conductivity, high electronic and ionic conductivity as well as specific thermo-mechanical properties and microstructures. For example, pyrochlore lanthanum zirconate $La_2Zr_2O_7$ (LZO) is a promising candidate for thermal barrier coatings (TBCs) but its processing (APS) is challenging because La_2O_3 is prone to evaporate in the plasma plume. It shows a considerably higher vapor pressure, than ZrO_2. The same is the case for gadolinium zirconate Gd2Zr2O7 (GZO) and the inhomogeneous evaporation of Gd2O3 and ZrO2. In this work, examples of such challenging materials are investigated, namely pyrochlores applied for thermal barrier coatings as well as perovskites for gas separation membranes. In particular, new plasma spray processes like suspension plasma spraying and plasma spray-physical vapor deposition are considered. In some cases, plasma

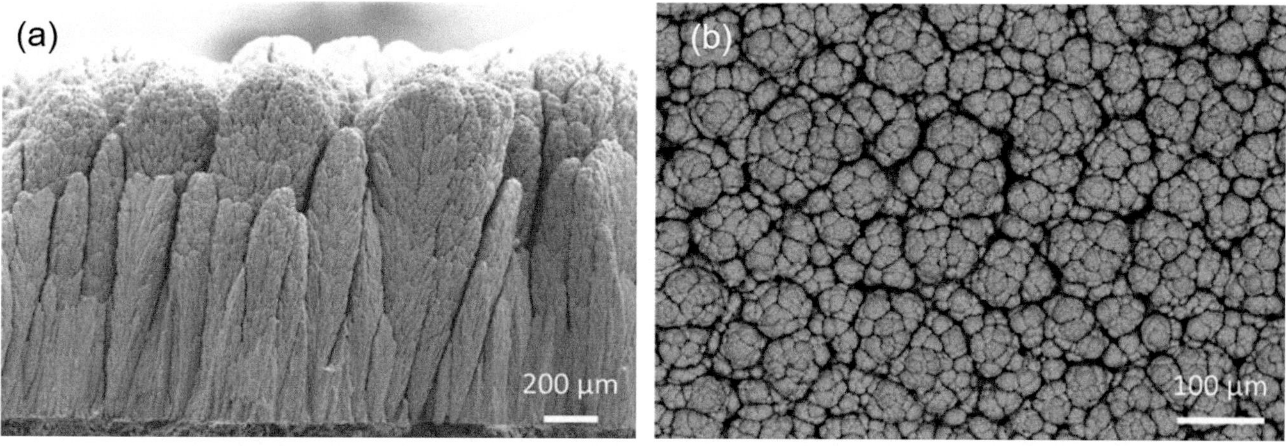

Fig. 9.40 Micrographs of columnar YSZ-TBC structures generated by shadowing (**a**) fracture surface (**b**) coating surface (back-scattered electron image) (20 g/min, 300 mm) [Mauer et al. (2013)]

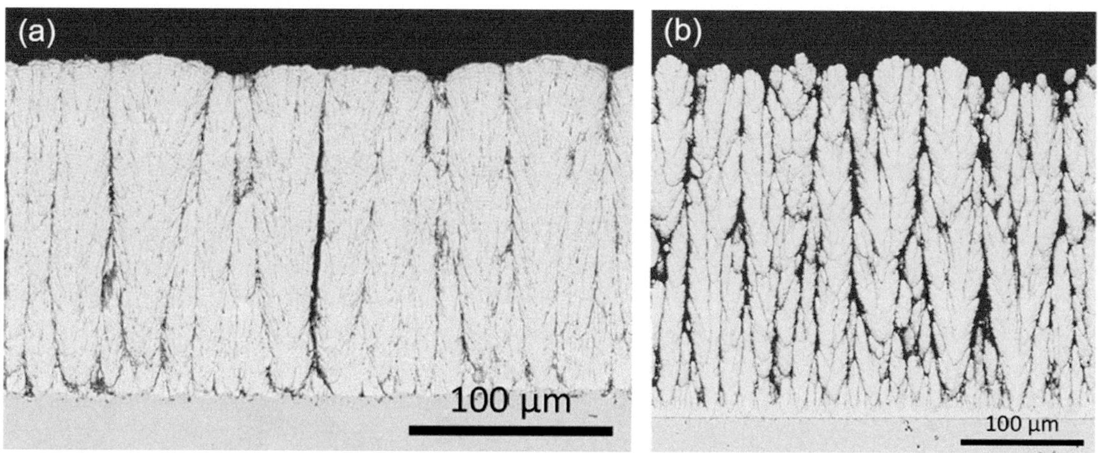

Fig. 9.41 SEM image of a cross-section of a YSZ TBC manufactured by ULPPS and deposited with (**a**) Ar/He plasma, condition "A" (**b**) Ar/He/H2 plasma, condition "B" as per Table 9.1 [Mauer (2014)]

Table 9.2 Summary of ULPPS operating conditions [Mauer (2014)]

Parameter	Units	Condition A	Condition B	Condition A
Plasma gas flow rate	slm (Ar)	35	35	100
	slm (He)	60	60	–
	slm (H_2)	–	10	10
current	(A)	2600	2200	2200
Power	(kW)	60	60	60
Pressure	(Pa)	200	200	200
Powder feed rate	g/min	2 × 8	2 × 8	2 × 20
Carrier gas flow rate	slm (Ar)	2 × 16	2 × 16	2 × 10
Spraying distance	(mm)	1000	1000	1000

diagnostics are applied to analyze the processing conditions. Such materials, however, often involve particular difficulties in processing by plasma spraying. The inhomogeneous dissociation and evaporation behavior of individual constituents can lead to changes of the chemical composition and the formation of secondary phases in the deposited coatings. At PS-PVD, the stoichiometry of the LZO coatings is strongly related to the homogeneity of the feedstock evaporation and of the deposition mode (vapor, cluster, or splat deposition). For GZO, it can be concluded that the trends are in principle the same as for LZO. If the rare earth constituent evaporates more intensively than others it is deposited in excess at PS-PVD. However, for GZO the inhomogeneity of evaporation is significantly lower than for LZO. Thus, the effects on stoichiometry and phase composition of the deposits are less pronounced.

At last [Mauer (2019)] studied the interaction between the gas jet and the particles feedstock, which can affect the coating build-up mechanisms considerably. In particular under high-kinetic and low-pressure conditions, small particles are subjected to rapid deflection and velocity changes close to the substrate. In his work, numerical studies were carried out to investigate the interaction between gas and particles in the substrate boundary layers (BL). Typical conditions for suspension plasma spraying (SPS), plasma spray-physical vapor deposition (PS-PVD), and aerosol deposition (AD) were taken as a basis. Particular importance was attached to the consideration of rarefaction and compressibility effects on the drag force. Typical Stokes numbers for the different thermal spray processes were calculated and compared. Possible effects on the resulting coating build-up mechanisms and microstructure formation are discussed. The results show that just for larger particles in the SPS process the laminar flow attached to the particles begins to separate so that the drag coefficients have to be corrected. Furthermore, slip effects occur in all the investigated processes and must be considered. The comparison of calculated Stokes numbers with critical values shows that there is a disposition to form columnar microstructures or stacking effects depending on the particle size for PS-PVD and SPS.

9.6 Plasma Sprayed Materials and Coatings

9.6.1 General Remarks

With plasma jet core temperatures over 8000 K any material whether metallic or ceramic can be melted. Using conventional DC plasma spray torches, with a spray distance of 100 to 120 mm, the heat flux to the substrate during a normal plasma spraying can reach 2 MW/m^2 requiring that substrate and the coating must be cooled. To avoid oxidation, spraying must be performed under controlled atmosphere or soft vacuum conditions. However, it is important to keep in mind that while the plasma torch and particle injection mode used have an important impact on the quality of the coating, the final properties of the coating obtained depends equally on the material sprayed and the operating conditions.

In general plasma sprayed coating porosities are between 2 and 8 vol. %, the oxygen content of metal or alloy coatings being between 1 and 5 wt. % and their adhesion is good (> 40–50 MPa). Such torches are mainly used to spray oxides ceramics with good resistance to abrasive, adhesive, fretting, or sliding wear. They also produce thermally conductive or

resistant surfaces and are used for salvage and restoration of worn parts.

It is not surprising that the great success of APS lies in the spraying of oxide materials. For non-oxide ceramics and refractory metals, used mostly in very high temperature and harsh conditions, CAPS is the most adopted process, even if VPS is sometimes used. For materials requiring very low oxide levels, VPS is the best suited process, and its great success is the spraying of super alloys for aerospace, aeronautic, and land-based turbines. APS is also used to spray metals, alloys, and cermets when oxidation is not a problem. In the following section, the previous remarks will be illustrated by a few examples, which represent only a very small part of plasma sprayed coatings.

9.6.2 Oxide Ceramics

The most frequently sprayed oxides are: Al_2O_2, TiO_2, ZrO_2, $ZrSiO_4$, Cr_2O_3, Y_2O_3, and their mixtures.

9.6.2.1 Alumina

When plasma sprayed, for all fully molten particles the initial α phase is transformed into γ phase upon splat cooling. Unfortunately, when the coating is preheated over 950 °C the γ phase is transformed back into the α phase with about 4 vol. % change resulting in coating spallation. Thus plasma-sprayed alumina coatings cannot be used over 800 °C. However, spray conditions have been sought, which would keep the α phase. [Stahr et al. (2007)] have studied the stabilization of α alumina by spraying, with APS, WPS, and HVOF, mechanical mixtures of alumina and chromia, as well as pre-alloyed powders consisting of solid solution α-Al_2O_3 with additions of Cr_2O_3. Stabilization of the α phase using the WPS process starting from mechanical mixtures was confirmed. For the WPS process, the positive effect of an increase in the α phase with increasing Cr_2O_3 content in the coating was clearly shown. Contrary to this, in the APS process starting from mechanical mixtures, no stabilization effect was found when mechanically mixed powders were sprayed. However, additional work is necessary to better understand the involved phenomena.

Alumina coatings are mainly used for their good resistance to abrasive wear, sliding wear, and oxidation. Alumina coatings are also used for their dielectric strength, suitable for insulating coatings extensively used in ozonizers [Kogelschatz (2003)]. [Kitamura et al. (2008)] have shown that coating density strongly affects dielectric strength for both Al_2O_3 and Y_2O_3 coatings, while the influence of purity is found to be much less important in their study. [Toma et al. (2011)] have used APS and HVOF processes to prepare alumina (Al_2O_3) and magnesium spinel ($MgAl_2O_4$) coatings designed for insulating applications. Electrical insulating properties (dielectric strength and electrical resistance/

resistivity) were very dependent on relative air humidity level and water vapor sorption. Alumina coatings obtained by APS are denser than those flame sprayed [Costil et al. (2007)] and are also used against wear, as well as Cr_2O_3, TiO_2. Wear resistance of sprayed oxide ceramics is not dependent only on the hardness of the coatings but also strongly on their toughness, which is not very high for alumina coatings. [Darut et al. (2010)] have studied the tribological properties of alumina coatings with two different structural scales, a micrometer-sized one manufactured by atmospheric plasma spraying and a sub-micrometer-sized one manufactured by suspension plasma spraying, SPS. Coating architectures were analyzed, and their friction coefficients were measured using a dry sliding mode. The friction coefficient was decreased by a factor of two when the characteristic scale was reduced by two orders of magnitude. Accordingly, the wear resistance of SPS coatings was significantly higher than that of APS coatings, thanks to a better toughness of the layer.

Because of their porosity, plasma sprayed Al_2O_3 and Cr_2O_3 coatings must be sealed to improve their corrosion resistance. [Leivo et al. (1997)] used aluminum phosphates to seal them. Phosphates were formed throughout the coating, down to the substrate. The sealing increased the hardness of the coatings by 200–300 Vickers hardness (HV) units. Abrasion and erosion wear resistances were increased by the sealing treatment. Sealing also substantially closed the open porosity, as shown in electrochemical corrosion tests. The sealed structures had good resistance against corrosion during 30 days of immersion in both acidic and alkaline solutions with pH values from 0 to 10. No decrease in abrasion wear resistance was observed after immersion. [Berard et al. (2008)] tested alumina plasma-sprayed coatings, manufactured with feedstock of different particle size distributions, graded alumina–titania coatings, and phosphate-sealed alumina coatings to improve the properties of metallic substrates operating in extreme environments encountered in the processing of nuclear wastes. Such conditions correspond, for example, to aging at moderate temperatures (several months at about 300 °C) coupled to thermal shocks (numerous cycles up to 850 °C for a few seconds, and a few treatments up to 1500 °C), under a reactive environment made of a complex mixture of acid vapors in the presence of an electric field of a few hundred volts and of radioactivity. Aluminum phosphate impregnation appeared to be an efficient post-treatment to fill connected porosity of these coatings. [Mesrati et al. (2000)] proposed to reduce graphite reactivity and permeability toward oxygen by plasma spraying of Al_2O_3 or ZrO_2. The process required bond coats of Cr_3C_2 or SiC. However, the protection was good only after a post-treatment of the coating based on an impregnation with enamel of the porous oxide.

[Keshri and Agarwal (2011)] have proposed to improve the wear resistance of plasma sprayed Al_2O_3 using Carbon NanoTubes (CNT). Composite coatings have been investigated at room temperature (298 K), elevated temperature (873 K), and in sea water. Lowest wear volume loss has been observed in the sea water as compared to dry sliding at 298 and 873 K. Relative improvement in the wear resistance of Al_2O_3–8 wt% CNT coatings compared to Al_2O_3 was 72% at 298 K, 76% at 873 K, and 66% in sea water. This was attributed to:

- Larger area coverage by the protective film on the wear surface at room temperature and in sea water
- Higher fracture toughness of Al_2O_3-CNT coatings due to CNT bridging between splats
- Antifriction effect of sea water

Another important application is the coating of rolls that are used to transport a polymer material past a corona discharge to modify the surface, for example, to allow printing on it.

9.6.2.2 Titania and Alumina–Titania Coatings

Titania can be used for applications similar to those of alumina but coating hardness is lower as well as the dielectric strength and the resistance to chemical attack.

Titania loses oxygen when plasma sprayed (its color tends to shift from white to gray or even black), the stoichiometry variation of the coatings resulting in a large variation of their electrical properties [Branland et al. (2006)]. [Ctibor et al. (2006)] have sprayed, with the high throughput Water-Stabilized Plasma (WSP), agglomerated nanometer-sized titania powder and a fused and crushed titania. Results indicate that the "nano"-coatings in general exhibit finer pores than coatings of the "conventional" micron-sized powders. The "nano"-coating properties (resistance to slurry abrasion, for example) were better only for carefully selected sets of spraying parameters, which seemed to have a very important impact. Titania coatings could be promising for their photocatalytic activity if the anatase structure is preserved after spraying, which unfortunately is not the case with plasma spraying. [Ye et al. (2007)] have studied the addition of hydroxyapatite, HAP, $(Ca_{10}(PO_4)6(OH)_2)$ to inhibit the phase transformation of anatase TiO_2 to rutile. The photocatalytic activity of TiO_2–10%HAP coating was highest among TiO_2, TiO_2–20%HAp, and TiO_2–30%HAP coatings for the optimum adsorption property and anatase content.

Al_2O_3–TiO_2 particles, generally fused and crushed, with different wt% of TiO_2 are extensively used against wear and for their resistance to acids and alkalis. Three different compositions are mainly used:

- **Al_2O_3–3TiO_2,** which is less brittle, but with lower dielectric strength, than pure Al_2O_3 coatings.
- **Al_2O_3–13TiO_2,** which has lower hardness, but higher toughness, and lower dielectric properties than Al_2O_3–

$3TiO_2$. This hypoeutectic composition results in coatings where γ-alumina is the main phase [Ilavsky et al. (1997)]. The wear resistance is better than that of alumina coatings, thanks to the increased toughness, in spite of a hardness reduction [Fervel et al. (1999), Yilmaz et al. (2007), Fervel et al. (1999), Normand et al. (2000)].

- **Al_2O_3–45TiO_2** coatings (hyper-eutectic composition) present lower hardness and less wear resistance than the preceding ones. This is probably due to the lower hardness and toughness of Al_2TiO_5 (and/or $Al_6Ti_2O_{13}$) aluminum–titanates contained in these coatings.

9.6.2.3 Chromium Oxide

Plasma-sprayed chromium oxide coatings are dense, corrosion, and wear resistant and used on pump seal areas, rolls and wear rings, ceramic anilox rolls used in printing machines working with flexography systems, and they are insoluble in acids, alkalis, and alcohol. In the printing industry the rolls plasma coated with chromium oxide are used to transport the precisely determined quantity of ink in the flexographic printing machines. The powder choice and spray conditions influence coating properties [Pawlowski L (1996)]. These coatings are then ground and polished before being laser engraved. Chromium oxide coatings are often used for their tribological properties, as reported, for example, by [Ahn and Kwon (1999)]. [Usmani and Tandon (1992)] have simulated the reciprocating sliding wear under lubrication encountered in internal combustion engines with plasma-sprayed chromium oxide, metal arc-sprayed martensitic stainless steel, and electroplated chromium coatings applied to a steel base material. The chromium oxide coating was found to perform equally well compared to the hard chrome coating that is conventionally used.

9.6.2.4 Zirconia

This is probably one of the most studied plasma-sprayed ceramic for its use in thermal barrier coatings because of its low thermal conductivity (<1.5 W/m K) and excellent thermal shock resistance. It must be stabilized to avoid phase transformation upon heating and cooling. This is achieved with Y_2O_3, CeO_2, or MgO, the higher stabilization temperatures being obtained with Y_2O_3 and CeO_2. When the plasma-sprayed thermal barrier coatings were first used in the late 1980s both for aero-engines [Miller (1997)] and land-based turbines [Parks et al. (1997)], the thermal insulation was achieved with zirconia stabilized by 8 wt% of yttria, PYSZ, which has been established as a standard for the last 20–30 years. Many studies have been devoted to these coatings and their qualities dependent on the powders used and the spray conditions. Ar–H_2 mixtures have been mostly used, although recently the advantages of N_2–H_2 mixtures have been reported [Marple et al. (2007)]. A recent

review by [Vaßen et al. (2010)] reports that during the last decade a number of ceramic materials, mostly oxides, have been suggested as new thermal barrier coating (TBC) materials. These new compositions (pyrochlores, zirconia doped by different rare-earth cations, hexaaluminates, and perovskites) have been developed to compensate the limited temperature capability of PYSZ, restricted to about 1200 °C, a value which is insufficient for advanced gas turbines.

It is also important to note that PYSZ are also used in solid oxide fuel cells, SOFC. Fuel cells directly convert chemical energy into electrical energy with the potential of very high-efficiency values, because they are not subject to the Carnot limitation. They typically operate at a high temperature around and above 800 °C and their operation is determined by the electrochemically active ceramic layer [Henne (2007)]. According to the review by [Henne (2007)] different manufacturing methods are in use or under development for the production of the cell components. Among these methods are also plasma spray technologies, as described by [Henne (2007)] and others [Chen et al. (2000), Tsukuda et al. (2000), Carayon and Lacout (2003)]. In these SOFCs both the cathode [generally strontium-doped lanthanum manganite (LSM)], the electrolyte (PYSZ or YSZ), and the anode (often PYSZ/Ni) are plasma sprayed. The main difficulties are to make the electrolyte as thin as possible (<10 μm, which seems possible with suspension plasma spraying) and impervious to gases, while cathode and anode must be porous. Other ceramics are currently under study for these plasma-sprayed SOFC components.

Zirconia has also good wear resistance and it is sprayed together with other oxides. [Abdel-Samad et al. (2000)] showed that by the addition of zirconia to an alumina matrix, the porosity, hardness, and coefficient of friction of the coating were decreased. At the same time, the wear resistance was increased due to the higher toughness of this ceramic.

9.6.2.5 Other Oxides

Besides oxides previously cited, which are the most frequently used for plasma spraying, other oxide materials consist of mixtures of different oxides such as Al_2O_3–SiO_2, Al_2O_3–$xTiO_2$–$ySiO_2$, Cr_2O_3–$xTiO_2$, Cr_2O_3–$xTiO_2$–$ySiO_2$, $ZrO2$–$xTiO_2$–yY_2O_3, ZrO_2–$xCaO$–yAl_2O_3–$zSiO_2$, but also Y_2O_3, HAP, TaO_2, and mullite, and only a few examples will be presented.

Because of the excellent biocompatibility of hydroxyapatite (HAP), plasma-sprayed HAP is widely used to coat orthopedic protheses. During the plasma spraying process, the thermal decomposition of HAP produces [Carayon and Lacout (2003)] tricalcium phosphate (TCP), tetracalcium phosphate (TeCP), calcium oxide (CaO), oxyhydroxyapatite (Oxy-HAP), and a molten phase. [Khor et al. (1998)] have shown that subsequent post-spray heat treatment at 600, 700, and 800 °C for 1 h restored most of the crystalline HA in the coatings. There was also a corresponding increase in the crystalline HAP phase when increasing particle size range. Bond coats based on bioinert ceramic materials such as titania and zirconia were developed to increase the adhesion strength of the coating system hydroxyapatite-bond coat to Ti–6Al–4V alloy surfaces used for hip enoprostheses and dental root implants [Heimann (1999), Besov and Batrak (2005)]. To summarize plasma spraying technology is more and more used in fabrication of articles for medical devices.

Plasma treatment is effectively used in the fabrication of microelectronic components especially in dry etching. The halogen-contained plasma at high power levels erodes the film protecting the chamber wall at a high rate, generating large quantities of particles/fume requiring frequent maintenance of the production equipment. Plasma-sprayed alumina or yttria coatings [Kitamura et al. (2006)] have technical and commercial advantages to overcome these problems. [Kitamura et al. (2011)] have shown very recently that yttria coatings obtained by suspension spraying with an axial torch (Mettech Axial III) presented high density, uniform structure, high hardness, high plasma erosion resistance, and retention of a smoother surface after plasma erosion. Although the properties of the best suspension coating are still inferior to bulk Y_2O_3, they are much better than those of the conventional plasma sprayed coatings.

[Moldovan et al. (2004)] have sprayed Ta_2O_5 coatings that can be considered for environmental barrier applications. The as-sprayed coatings were approximately 97–98% dense and were composed of the low-temperature polymorph, β-Ta_2O_5 and some α-Ta_2O_5. Upon thermal treatment, the α-phase was converted to the low-temperature β-Ta_2O_5 phase.

Among the different ceramic mixtures, plasma sprayed, YSZ-mullite multilayer architectures with compositional grading between the bond coat and YSZ topcoat are potential solutions to ease their coefficient of thermal expansion mismatch induced stress. [Cojocaru et al. (2011)] have identified a better match between the elastic modulus, E, and hardness, H, values of the bond coat, and subsequent layers obtained from powders having similar type and size distribution. [Das et al. (2005)] have plasma sprayed a mixture of zircon and alumina on a mild steel substrate. The coating appeared to be sound and continuous along the interfaces and in the bulk as well. During spraying, zircon sand dissociated into zirconia and silica and a fraction of this silica in turn combined with a fraction of the alumina to form the mullite phase. The presence of the mullite phase enhanced the thermal fatigue characteristics of the coating.

9.6.3 Non-Oxide Ceramics

Carbides, borides, and nitrides can be plasma sprayed when they have a melting point that is separated by at least by 300 K from evaporation or decomposition temperatures. These ceramics are also very sensitive to oxidation and must be sprayed in controlled atmosphere or under soft vacuum. They are used for their high temperature resistance in nuclear, military, and aerospace industries, and thus only a few papers are published about them. For example, [Zeng et al. (2002)] have sprayed in air B_4C particles. Besides B_xC, B_4C, and carbon, other impurity phases exist consisting of B, O, and Fe elements (i.e., Fe_3BO_6). As the carbon phase, the Fe_3BO_6 phase also has a much lower microhardness than the major phase, contributing to the decrease of microhardness of boron carbide coatings. B_4C is an important material for fusion reactors and it has been extensively studied [Matejicek et al. (2007)]. Its advantages include low atomic number, Z, a relatively high melting point, thermal shock resistance, low vapor pressure, oxygen gettering, chemical inertness, and easy hydrogen isotope release. [Greuner et al. (2004)] demonstrated that a VPS-based coating technique was suitable for manufacturing the B_4C protection layers on stainless steel wall panels of the W7-X fusion reactor. The other sprayed boride is ZrB_2 for its high hardness, associated with high electrical and thermal conductivities, together with a very high melting point ($T_m = 3313$ K). [Tului et al. (2002)] showed that ZrB_2 coatings, deposited by CPS, were good electrical conductors and compared favorably with the reference coatings, in terms of tribological performance, which was not the case of those sprayed by APS. studied the high temperature mechanical properties of ZrB_2–SiC (30% vol.)–$MoSi_2$ (10% vol.) plasma-sprayed coatings at different temperatures (up to 1500 °C) in air. Results showed an improvement in mechanical properties as the temperature increased up to 1000 °C. However, at 1500 °C, coatings maintained good mechanical properties but exhibited a plastic behavior.

Carbides are also used for high temperatures applications. [Li et al. (2011)] prepared a dense, thick ZrC coating on the surface of SiC-coated C/C composites by supersonic plasma spraying. The ZrC coating had good adhesion to the SiC inner layer. It greatly improved the ablation resistance of SiC-coated C/C composites.

Another way to prepare non-oxide ceramic coatings is reactive plasma spraying. For example, [Hu et al. (2012)] have prepared SiC coatings for carbon/carbon (C/C) composites by a combination of vacuum plasma spraying technology and heat treatment. The SiC coatings were formed by the reaction of C/C substrates with as-sprayed silicon coatings deposited by vacuum plasma spraying. The preparation temperature and the thickness of the original silicon coatings were shown to have a great influence on the microstructure and the thickness of the synthesized SiC coatings. [Siegmann et al. (2004)] have sprayed by VPS an optimized iron-based alloy consisting of $Fe_{17}Cr_{10}Mn_6VMo$. Particles were nitrogen-atomized to create nano-structured hard phases of vanadium nitrides, VN, in each droplet, their size and distribution being adjusted as a function of the N content during the formation process. Using a VPS facility, coatings were sprayed without any oxidation during coating formation and with excellent coating properties. Besides the preceding nitridation of the starting powders, the reactive alloying with N can be achieved during vacuum plasma spraying by the exposure of the particles to a N_2-rich plasma and by creating a N_2-rich environment in the VPS chamber. [Borisova and Borisov (2008)] have presented many carbide coatings that can be prepared by APS by self-propagating high-temperature synthesis, SHS. SHS is accompanied by heat generation in composite powder particles. Moreover, this technique allows extending the range of possible compositions of composite powders, compared with the conventional methods. The combination of heating of particles (by hot gas jet) and the additional heat source through the exothermic synthesis of coating material formation from components of composite powder particles increases the thermal energy of sprayed particles, which also can improve the quality of the resultant coatings.

9.6.4 Cermets

A cermet is a composite material composed of ceramic and metallic materials. A cermet is ideally designed to have the optimal properties of both the ceramic and those of the metal especially plastic deformation. The metal is used as a binder for oxides, borides, or carbides. Cermets are used not only for their superior wear and corrosion properties but also in the manufacture of resistors. It must be kept in mind that with the APS process, the metal part of the cermet will be oxidized, but when the ceramic is non-oxide it can be partially or totally decomposed and even dissolved in the molten metal. However, partially decomposed carbide can be useful because free carbon is produced resulting in coatings with a low friction coefficient against certain materials. Among the numerous carbide cermets, WC–Co is the most widely applied one, but when plasma sprayed WC is prone to oxidize and/or decompose, while Cr_3C_2-NiCr coatings are more resistant to decomposition at higher temperatures, where corrosion and oxidation can occur simultaneous. Naturally VPS or CPS allows avoiding these problems, but their costs cannot compete with those of an HVOF processes producing the same quality of coatings.

For example, [Basak et al. (2008)] have plasma sprayed agglomerated nano-sized powders of alumina (Al_2O_3) and cermet coatings containing alumina dispersed in a FeCu or a FeCuAl matrix. Nanostructured cermet coatings appeared to offer better wear resistance under sliding and abrasion tests than nanostructured Al_2O_3 coatings. The cermet coatings contained up to four different phases after plasma spraying. It was shown that an appropriate balance between hard and soft phases resulted in optimal tribological properties of the nanostructured cermet coatings. [Fervel et al. (1999)] have studied the tribological behavior of plasma-sprayed coatings Al_2O_3, Al_2O_3–TiO_2, Al_2O_3–TiO_2–Cu. in dry conditions. It was found that TiO_2 improved the fracture toughness of coatings and copper reduced friction and wear. [Hwang et al. (2006)] have investigated plasma spray-coated piston rings and cylinder liners for marine diesel engines to find the optimum combination of coating materials. Coating materials studied were Cr_2O_3–NiCr, Cr_2O_3–NiCr–Mo, and Cr_3C_2–NiCr–Mo. They found that a dissimilar coating combination of Cr_2O_3–NiCr–Mo and Cr_3C_2–NiCr–Mo provided the best anti-wear performance. The addition of molybdenum was found to be beneficial to improve the wear resistance of the coating.

Instead of using cermet powders, metal and ceramic powders can be sprayed separately as demonstrated by [Ageorges et al. (2006)] with Cr_2O_3 and stainless-steel powders. Increasing the stainless-steel percentage increased the coating cohesion, while more Cr_2O_3 in coatings resulted in higher hardness and lower weight losses during wear tests in dry abrasion. The study has also shown that the optimum stainless-steel percentages in coatings were not identical to reach their maximum resistance to slurry or dry abrasion. [Hashemi et al. (2009)] sprayed Ni–Al–SiC composite powder containing 12 wt% SiC. The powder was prepared by ball milling in a conventional tumbler mill. The results showed that spray parameters have a significant influence on the phase composition and mechanical properties of coatings.

[Economou et al. (2000)] have tested at room temperature and at 550 °C vacuum plasma sprayed NiCrFeAlTi–TiC, NiCr–(Ti, Ta)C, and NiCrMo–(Ti, Ta)C coatings. Air plasma-sprayed NiCr and NiCr–TiC coatings, as well as the high velocity oxy-fuel WC–Co coating, were investigated for comparison. The wear resistance of all coatings decreased at high temperature. The NiCrFeAlTi–TiC coating showed the best wear resistance among the TiC-based coatings. At room temperature and at 550 °C, the WC–Co coating was more wear resistant than the TiC-based coatings, even though the coating cracked in the high temperature tests. The coefficients of friction of the WC–Co coating, tested against sapphire, were lower than those of the TiC-based coatings at room temperature, but not lower than those obtained at 550 °C. These examples illustrate the diversity of plasma-sprayed cermets and of the results obtained. The latter can be very different for the same cermet composition depending not only on the spray parameters, but also on the spray process used and the powder morphologies.

9.6.5   Metals or Alloys

9.6.5.1   Vacuum Plasma-Sprayed Metal Coatings

The VPS development is linked to that of bond coats (MCrAlY coatings) for TBCs in aero-engines. Bond coats [Brindley (1997), Feuerstein et al. (2008)] are applied to enhance the adherence of the TBC coating to the substrate alloy and to increase the high-temperature oxidation and corrosion resistance of the underlying structural alloy. The protection against oxidizing environments is achieved by forming a thin oxide layer commonly termed as thermally grown oxide, TGO. This layer acts as a barrier between the alloy and the atmosphere and impedes further oxidation. Growth of the TGO occurs either by outward diffusion of metal ions or by inward diffusion of oxygen ions through the oxide layer. The most commonly used TGO is alumina Al_2O_3, which has a slow growth rate and is easily formed on metallic alloys [Feuerstein (2008)]. However, to reduce the costs, superalloy coatings are also produced using APS [Koomparkping et al. (2005)], sometimes with shrouded torches [Sidhu and Prakash (2007)].

VPS is also used to spray very specific metals such as beryllium [Castro et al. (1997)] for in-situ repair of damaged beryllium surfaces, which are facing the plasma in fusion reactors such as the international thermonuclear experimental reactor (ITER). The effective bond strength and failure characteristics of plasma-sprayed beryllium on beryllium surfaces were determined by mechanical interlocking at low substrate temperatures and increased metallurgical bonding at higher substrate temperatures. VPS can also be used to spray cast iron coatings [Morks et al. (2003)] to hinder graphite formation. [Schiller et al. (1995)] have produced by a VPS process electrode coating for advanced alkaline water electrolysis. Molybdenum-containing Raney nickel coatings were applied for cathodic hydrogen evolution. For the preparation of Raney nickel coatings, a precursor alloy Ni–Al–Mo was sprayed and leached subsequently in a caustic solution to remove the aluminum content, forming a porous, high-surface area nickel–molybdenum layer.

9.6.5.2   Air Plasma-Sprayed Metal Coatings

Most plasma-sprayed alloys are nickel or cobalt, or iron based, with, as mentioned previously, superalloys and molybdenum as the metals. Some abradable coatings are also sprayed. [Tani and Harada (2007)] have plasma sprayed, onto the combustion facing surface of steam generating tubes in a heavy oil-fired boiler, Ni–50 wt.% Cr alloy coatings for

their corrosion resistance. The Ni–50 wt.% Cr alloy coating was successful for the suppression of hot corrosion failure of the steam generating tubes of the boiler. Cr-Ni-2.5Mo-Si-0.5B alloys containing 55 to 58 wt.% Cr were also deposited by [Longa-Nava et al (1995)]. CO_2-1aser glazing subsequently modified the thermally sprayed coatings. Oxidation and corrosion resistance of the as-sprayed and laser-processed coatings were tested in the presence of $85V_2O_5$–Na_2SO_4 fused salt at 900 °C in air. Hot corrosion resistance of the CrNiMoSiB coatings increased in the following order: VPS, APS, laser-glazed VPS, and laser-glazed APS. Laser glazing enriches the silicon content in the top layers of the coating. Laser-glazed APS CrNiMoSiB coatings exhibited excellent corrosion resistance due to the continuous oxide film of chromia and silica formed on the top surface of the coating. Cyclic oxidation behavior of plasma-sprayed NiCrAlY, Ni–20Cr, Ni_3Al, and Stellite-6 coatings was investigated by [Singh et al. (2006)] in an aggressive environment of Na_2SO_4–$60\%V_2O_5$. These coatings were deposited on a Ni-base superalloy, namely Superni 600: 10Fe–15.5Cr–0.5Mn–0.2C–Bal Ni (wt%). All of the coatings were found to be useful in reducing the spallation of the substrate superalloy. Moreover, the coatings were successful in maintaining continuous surface contact with the base superalloy during the cyclic oxidation. [Miguel et al. (2011)] sprayed bronze coatings that showed good friction properties (friction coefficient of about 0.3) that remained constant during the entire test. However, bronze–alumina composite coatings obtained by spraying both powders with conditions allowing both powders to be melted produced dense composite coatings with good adhesion and with a general increase in abrasion resistance, even if accompanied by an increase of the friction coefficient.

Coatings of molybdenum, pure or blended with other elements, are often used for their wear resistance, for example, on synchronizer rings or piston rings. [Ahn et al. (2005)] have plasma sprayed four different powders, one of pure molybdenum, the others being blended powders of bronze and aluminum–silicon alloys mixed with molybdenum powders. The wear test results revealed that the wear rate of all coatings increased with increasing wear load and that the blended coatings exhibited better wear resistance than the pure molybdenum coating, although the hardness was lower. The molybdenum coating blended with bronze and aluminum–silicon alloy powders exhibited excellent wear resistance because hard phases such as $CuAl_2$ and Cu_9Al_4 formed inside the coating. [Niranatlumpong and Koipraser (2010)] studied coatings plasma sprayed with various amounts of Mo mixed with NiCrBSi at 0, 25, 50, 75, and 100 wt%. They found that as the Mo/NiCrBSi ratio increases, the wear mechanism changed. Coatings containing 75%Mo and 25%NiCrBSi exhibit the highest wear depths corresponding to the cracking of the thin

NiCrBSi splats. On the other hand, coatings containing 25% Mo and 75%NiCrBSi had the lowest wear depths with no surface cracks.

Plasma-sprayed coatings are also used to produce resistors. For example, [Prudenziati and Lassinantti Gualtieri (2008)] have deposited Ni and Ni20Cr powders to obtain resistors. A clear correlation was found between the oxygen content in the resistors and the temperature dependence of resistance values. The long-term stability of APS Ni-based resistors rendered them excellent candidates for temperature sensors and self-controlled high temperature heaters. [Floristán et al. (2010)] have sprayed by APS electrically conductive coatings on a glass ceramic substrate, using TiO_2 and mixed phases of TiO_2/NiCrAlY. Direct spraying on glass substrates is interesting for industrial applications but the expansion coefficient mismatch is a problem. TiO_2/NiCrAlY mixed phase coatings presented a change in the metal content depending on the spray distance. The most stable system was the TiO_2 coating, with the lowest tensile stresses and the best adhesion. However, the electrical conductivity of the coating restricted its usability to applications, where the resistivity values were not lower than 0.05 Ω cm.

The different coatings presented above illustrate the high versatility of the plasma spray processes. Almost any material with a melting temperature, T_m, that is at least 300 K lower than its decomposition or vaporization temperatures can be sprayed even if T_m is over 4000 K! However, the oxidation of materials sprayed in air can be significant. The oxidation can be controlled when spraying in soft vacuum or a controlled atmosphere, but at a much higher cost.

9.6.6 DC Plasma Spray Applications

Plasma spray coating is one of the most rapidly developing surface modification technologies which found numerous applications in a wide range of industrial sectors such as the aerospace and automobile industries, electrical and electronic industries, chemical process industries, printing and textile industries, and biomedical industries. The specific role played by the coating in each of these industries varies widely between wear resistance, component rebuilding and restoration, thermal and corrosion protection, electrical insulation, electromagnetic shielding, and functional surface modification. A broad review of all such applications is discussed in great detail in this book, Part IV "Process Integration and Industrial Applications," Chap. 20 "Industrial Applications of Thermal Spray Technology."

In the present chapter, we will limit our discussion to one of the most important and typical application, "*Wear*" which occurs in almost all industries and in all cases, involving the progressive loss of material at the active surface due to the relative movement of another part or particles on this surface. The lifetime of the coating used against wear depends on its resistance to wear and its thickness. The different types of wear, for which plasma sprayed coatings are used, are described below, according to [Davis (2004) and Dorfman (2013)].

9.6.6.1 Abrasive Wear

It represents more than 50% of wear and corresponds to weight loss with grooves, pits, and scores at the surface due to cutting or deformation. The phenomena are promoted when temperature, humidity, aggressiveness of the environment (corrosion) increase. The three-body abrasion depends also on the shape, grain size, and hardness of the abrasive particles and the relative speed of the two bodies. As a general rule abrasion increases drastically as soon as the hardness of one metal or alloy becomes equal to that of the abrasive particles. In most cases wear-resistant coatings are hard with a good resistance to heat and chemical attack. [Kulu et Pihl (2002)] have presented coating selection diagrams of wear resistance versus hardness. They also discussed the cost-effectiveness of coatings in the application areas that are more sensitive to cost, and composite coatings based on recycled materials are offered. Information can be also found in [Thermal spraying (1985)]. In the following we present a few coating materials:

Self-Fluxing Alloys (SFA) If their corrosion resistance is generally excellent, their hardness is not particularly high and they cannot compete with cermets for sliding wear resistance. Thus one can either spray them with hard particles or heat-treat them. Increasing the surface hardness and toughness of coatings was an effective way to improve the erosion resistance. For example, [Zheng et al. (2009)] have studied the oxidation behavior of Ni–20Cr alloy and Ni-base self-fluxing (NiCrSiBC) alloy sprayed by APS. The oxygen content in the NiCrSiBC coating was lower by a factor of over 10 than that in the Ni20Cr coating. The mechanism of protecting NiCrSiBC alloy particles from oxidation is preferential oxidation of C, Si, and B and simultaneous vaporization of the formed oxides. [Natarajan et al. (2016)] APS sprayed NiCrBSi-graphite composite coatings. With addition of graphite, phases such as CrB, Cr_7C_3 emerge in composite coating. Using an abrasive wear test rig according to ASTM G65 at room temperature, addition of graphite in the coatings resulted in reduction of volume loss and wear rate significantly. The NiCrBSi-8 wt %C composite coating presented excellent abrasion resistance. [Wen et al. (2016)] subjected to vacuum re-melting Ni60-NiCrMoY composite coatings prepared by supersonic atmospheric plasma spraying. Vacuum re-melting greatly improved the bonding strength and erosion resistance of the as-sprayed coatings. [Hou et al. (2015)] used APS to deposit $Mo(Si, Al)_2$ coating on substrate of niobium alloy. The anti-ablation property of the coating was investigated under supersonic flame at 1400 °C for 600 s. The composition and structure of as-sprayed Mo(Si, Al)$_2$ coating were maintained under the Al-depletion layer. The epitaxial mixed oxide scale protected the rigid surface of Mo (Si, Al)$_2$coating and played a key role under the dynamic ablation.

Cermet Coatings [Gawne et al. (2001)] have plasma sprayed ball-milled mixture of glass and alumina powders to produce alumina-glass composite coatings. The alumina raised the mean hardness from 300 HV for pure glass coatings to 900 HV for a 60 wt. % alumina-glass composite coating. The scratch resistance increased by a factor of 3, and the wear resistance by a factor of 5, maximum value obtained with 40 to 50 vol. % alumina. This alumina content corresponded to the change over from a glass matrix to an alumina matrix. [Cipri et al. (2007)] produced thick alumino-silicate coatings by plasma-spray technique, powders being agglomerated with a mean size of 38 μm. Spray conditions were such that in-flight reactions between alumina and silica formed low-melting compounds. Coatings, efficiently coupled to metallic substrates even after plastic deformation, exhibited very interesting performance in terms of mechanical properties and fracture toughness (elastic modulus of 43 GPa and K1c of about 2 MPa.m$^{1/2}$ for 850 μm thick coatings) and compliance.

These coatings showed an excellent refractory behavior allowing a wide use as wear-resistant thermal barrier coatings, in metallurgical and glass plants and in high temperature heat exchangers. [Zhu et al. (2013)] deposited TiB$_2$–Ni cermets coating by APS using two kinds of feedstock: agglomerated powder and clad powder. The micro hardness (Hv) and the fracture toughness (K1$_c$) of the coatings were also evaluated. TiB$_2$ particle's rebounding in deposition and their dissolution were closely related to the feedstock's TiB$_2$ particle size, binder content, and its morphology, which played an important role on the coating microstructure and its mechanical properties. The coating deposited from agglomerated and sintered TiB$_2$–40 wt.% Ni powder presented better tribological properties, that is, low friction coefficient and wear rate. [Wang et al. (2015)] compared mechanical property and sliding wear performance of plasma sprayed boride based composite coating (TiB$_2$–NiCr) to traditional carbide based composite coating (Cr$_3$C$_2$–NiCr). The results indicated that the boride-based composite coating presented higher micro hardness and improved wear resistance under dry sliding condition. Moreover, the boride composite coating displayed less wear damage to the iron counterpart.

9.6.6.2 Erosive Wear

It occurs when hard particles carried by a fluid hit the surface. The impact angle plays an important role on the wear rate depending on the surface material: ductile, hard metal, ceramic… [Chattopadhyay (2001)]. To reduce the wear, one must choose a coating material harder than the abrasive particles, with toughness high enough, especially for inclined impact angle between 30 and 45°, which must be avoided if possible, and the coating roughness must be reduced to the minimum. It is also important to avoid turbulences in service conditions because they increase erosion. It is usually considered that the wear and erosion resistance of cermet coatings are predominating influenced by their microstructures including the splat size, carbide particle size, carbide content, and carbide distribution within a splat, and the cohesion between the splats [Kim et al. (1994)] showed that inter-splat cohesive strength of coatings, measured by a simple bonding test, was the most significant factor relating to the wear rate of plasma sprayed WC-12 wt. % Co coatings. [Krishnamurthy et al. (2012)] studied the erosion resistance of plasma sprayed alumina and calcia-stabilized zirconia coatings on Al-6061 substrate. It was found that erosion of coating systems occurred through spalling of lamella exposed on coating surface resulting from cracking along the lamellar interface. Erosion wear was more at 45° angle of impact. Pores seemed to act as stress concentrators and decreased the load-bearing surface. [Ramachandran et al. (2013)] studied the erosive wear of Yttria Stabilized Zirconia (YSZ) coatings and Lanthanum Zirconate (LZ) coatings on Inconel 738 base material (BM). It was found that in solid particle erosion, the wear resistances of YSZ and LZ coatings were the best at their lowest porosity volume and it decreased with the increase in the percentage volume of porosity. There was a linear increase in the wear resistance with the increase in hardness. [Li et al. (2006)] worked on the erosion rate of plasma-sprayed Al$_2$O$_3$ coatings at impact angle of 90°. The results clearly revealed that the erosion of plasma-sprayed ceramic coating was inversely proportional to the mean lamellar bonding ratio. Coating fracture toughness controlled the erosion resistance of a thermally sprayed ceramic coating.

9.6.6.3 Friction and Adhesive Wears

It occurs when particles are transferred from one interacting surface to the other. When different materials are in contact, particles are mainly transferred from the softer or weaker material onto the harder one. This wear is promoted when increasing the load and/or the temperature, and under dry friction, or poor lubrication. It depends on the structure, composition, hardness, and melting temperature of the material. To reduce it, dry friction must be avoided and, if it is not possible, coatings containing solid lubricants or retaining lubricants must be used. The compatibility of materials is also important: the friction coefficient, f (f = T_f/N, T_f = tangential force, N = Normal force), depends on the couple of materials rubbing against each other. It is also linked to the load, P, (or more precisely the pressure, p,) and the relative velocity, v, between both parts (in principle the product, p.v, must be below 1 MPa.m/s [Cartier (2003)]) The roughness of surfaces into contact must be as low as possible and from a rough surface (Ra of a few μm) to a smooth one Ra < 0.1 μm), f, can be reduced by 60%. Dry friction usually results in high local heating and, even with a very low relative velocity, the friction coefficient, f, increases with temperature. Low friction coefficients can be achieved with solid lubricants (f = 0.001 to 0.05), but the relative velocity must be low. Using materials with a high thermal conductivity (both coating and substrate) reduces the heating due to friction. As wear increases with the energy of adhesion, it is also important to avoid micro welding, that is, to use materials with a low solubility and a narrow range of solid solution (ex: Fe/Al, Fe/Cu). The work of adhesion decreases with the following couples: metal–metal, metal–ceramic, ceramic–ceramic, metal–polymer, ceramic–polymer, and polymer–polymer.

9.7   Summary and Conclusions

Following the presentation in Chap. 8. "DC plasma spraying-Fundamentals" which covered the scientific foundation of DC plasma spraying technology including plasma torch designs and gas-particle dynamics in DC plasma spraying, the present chapter is devoted to a review of the principal DC plasma spray technologies. Details are given in each case of process equipments and hardware, spraying parameters, and coating characteristics. These are presented in a distinct sequence based on the underlying technology.

Atmospheric Plasma Spraying (APS) in which the spraying process is carried out in the open atmosphere, involving the entrainments into the plasma jet affecting particle heating and melting and can be responsible for the in-flight, and post-deposition oxidation of the coating material and degradation of its properties. APS remains, however, one of the most widely used DC plasma spraying technologies widely applicable for the spraying of oxide ceramics and a limited number of metals and alloys. A discussion was also presented of substrate surface preparation and coating formation which will also be covered in more details in subsequent chapters of this book, Part III, Coating Formation and Characterization.

Controlled Atmosphere Plasma Spraying (CAPS) was discussed next in which the spraying process is carried out in a controlled/inert atmosphere chamber with pressures roughly between 100 and 300 kPa thus avoiding the problem of in-flight and post-deposition oxidation of the coating material at the expense of physical limitation of the size and shape of the components to be sprayed and a significant increase of the required capital investment. As the atmosphere inside the chamber is rather dusty with ambient temperatures between 150 and 200 °C the arms of the robotic plasma torch manipulator have to be protected against heat and dust. Filters must also be disposed between chamber and pumps to trap the dust.

Vacuum Plasma Spraying (VPS) is a natural extension of the CAPS technology in which the plasma spraying operation takes place in the controlled/inert atmosphere at soft-vacuum conditions with absolute pressures typically between 10 and 30 kPa. The process gives rise to superior quality coatings which are fully integrated in the most demanding aerospace and biomedical applications.

Ultra-Low-Pressure Plasma Spraying (ULPPS) was the next step pushing further the limit of the plasma spray technology to bridge the gap between conventional coatings built by splat-like building blocks, to coatings formed directly by vapor deposition much like the Physical Vapor Deposition (PVD) technology, or a mixture of both. The technology operates at pressures as low as 100–200 Pa making it possible to evaporate in-flight the powder feedstock material providing advanced microstructures of the deposited coatings. ULPPS offers new possibilities for the manufacturing of thermal barrier coatings (TBCs), with reduce thermal conductivity and strain tolerance columnar micro-structures, comparable with EB-PVD techniques at a considerably higher deposition rates and lower cost.

Nomenclature

Units are indicated in parentheses; when no units are indicated, the parameter is dimensionless.

Latin Alphabet

A	empirical constant (6×10^5 A/m^2 K^2)
A	channel cross-sectional area (m^2)
c_p	specific heat at constant pressure (J/kg.K)
d_o	anode nozzle internal diameter (mm)
d_a'	anode boundary layer thickness (m)
d_c'	cathode boundary layer thickness (m)
d_p	particle diameter (μm)
$\overline{d_p}$	mean particle diameter (μm)
e	electric charge (Coulomb)
E	electric field strength (V/m)
E_x	electric field strength normal to the anode surface (V/m)
E_I	ionization potential of the plasma gas (eV)
h	specific enthalpy (J/kg)
h_{av}	average enthalpy (J/kg)
I	arc current (A)
j_e	electrons current density (A/m^2).
j_i	ions current density (A/m^2)
k	Boltzman constant (1.38×10^{-23} J/K)
$\dot{m}_g$	mass flow rate of plasma gas (kg/s)
$\dot{m}_p$	powder mass flowrate (kg/s)

M Mach number $M = v/v_c$
n_e electron number density (m^{-3})
p pressure (Pa)
p_e partial pressure (Pa)
P_{el} electric power input into the torch $P_{el} = I \times V$ (W)
Q_{cond} losses to the torch wall by heat conduction (W)
Q_{rad} losses to the torch wall by radiation (W)
Q_{loss} heat lost to the torch cooling water (W)
Q_i Carrier gas flow rate (slm)
r distance in the r-direction (m)
r_o nozzle radius (mm)
Re Reynolds number, $Re = G/\mu_o\rho_o$
t   time during which the contact occurs at temperature T (s)
T_c cathode temperature (K)
T_h heavy species temperature (K)
T_m Particle melting temperature (K)
T_e electrons temperature (K)
T_{ref} reference temperature (K)
V'_a Anode voltage fall (V)
V'_c Cathode voltage fall (V
v gas velocity (m/s)
v_c Velocity of sound (m/s)
v_p particle velocity (m/s)
v_{pi} particle injection velocity (m/s)
x distance in the x-direction (m)
y distance in the y-direction (m)
z distance in the z-direction (m)

Greek Alphabet

δ_z maximum thickness of a segment (mm)
$\Delta\overline{h}$ increase in enthalpy
Δz distance (m)
η thermal efficiency of the torch (%)
η_D deposition efficiency (%)
κ_e thermal conductivity of electron (W/m.K)
κ_h thermal conductivity of heavy particles (W/m.K)
μ dynamic viscosity
ρ mass density of the gas (kg/m^3)
ρ_p mass density of the particle or droplet (kg/m^3)
σ electrical conductivity (A/V.m)
σ_g geometric standard deviation
σ_p liquid drop surface tension (N/m)
ϕ_w work function (eV)

References

Abdel-Samad, A.A., A.M.M. El-Bahloul, E. Lugscheider, and S.A. Rassoul. 2000. A comparative study on thermally sprayed alumina based ceramic coatings. *Journal of Materials Science* 35: 3127–3130.

Ageorges, H., P. Ctibor, Z. Medarhri, S. Touimi, and P. Fauchais. 2006. Influence of the metallic matrix ratio on the wear resistance (dry and slurry abrasion) of plasma sprayed cermet (chromia/stainless steel) coatings. *Surface and Coatings Technology* 201: 2006–2011.

Ahn, J., B. Hwang, and S. Lee. 2005. Improvement of wear resistance of plasma-sprayed molybdenum blend coatings. *Journal of Thermal Spray Technology* 14 (2): 251–257.

Bahbou, M.F., and P. Nylén. 2007. On-line measurement of plasma-sprayed Ni-particles during impact on a Ti-surface: Influence of surface oxidation. *Journal of Thermal Spray Technology* 16 (4): 506–511.

Basak, A.K., S. Achanta, J.P. Celis, M. Vardavoulias, and P. Matteazzi. 2008. Structure and mechanical properties of plasma sprayed nanostructured alumina and FeCuAl-alumina cermet coatings. *Surface and Coatings Technology* 202: 2368–2373.

Berard, G., P. Brun, J. Lacombe, G. Montavon, A. Denoirjean, and G. Antou. 2008. Influence of a sealing treatment on the behavior of plasma-sprayed alumina coatings operating in extreme environments. *Journal of Thermal Spray Technology* 17 (3): 410–419.

Besov, A.V., and I.K. Batrak. 2005. Application of technology of plasma spraying in the fabrication of articles for medical devices. *Powder Metallurgy and Metal Ceramics* 44 (9-10): 511–516.

Borisova, A.L., and Yu S. Borisov. 2008. Self-propagating high-temperature synthesis for the deposition of thermally-sprayed coatings. *Powder Metallurgy and Metal Ceramics* 47 (1-2): 80–94.

Branland, N., E. Meillot, P. Fauchais, A. Vardelle, F. Gitzhofer, and M. Boulos. 2006. Relationships between microstructure and electrical properties of RF and DC plasma-sprayed Titania coatings. *Journal of Thermal Spray Technology* 15 (1): 53–62.

Carayon, M.T., and J.L. Lacout. 2003. Study of the Ca/P atomic ratio of the amorphous phase in plasma-sprayed hydroxyapatite coatings. *Journal of Solid-State Chemistry* 172: 339–350.

Cartier, M., ed. 2003. *Handbook of surface treatments and coatings.* New York: ASME Press, 412 pages.

Castro, R.G., A.H. Bartlett, K.J. Hollis, and R.D. Fields. 1997. The effect of substrate temperature on the thermally diffusivity and bonding characteristics of plasma sprayed beryllium. *Fusion Engineering and Design* 37: 243–252.

Cedelle, J., M. Vardelle, and P. Fauchais. 2006. Influence of stainless-steel substrate preheating on surface topography and on millimeter- and micrometer-sized splat formation. *Surface and Coatings Technology* 201 (3–4): 1373–1382.

Chandra, S., and P. Fauchais. 2009. Formation of solid splats during thermal spray deposition. *Journal of Thermal Spray Technology* 18 (2): 148–180.

Chattopadhyay, R. 2001. *Surface wear: Analysis, treatment, and prevention.* Materials Park: ASM International, 307 pages.

Chen, H.C., J. Heberlein, and R. Henne. 2000. Integrated fabrication process for solid oxide fuel cells in a triple torch plasma reactor. *Journal of Thermal Spray Technology* 9 (3): 348–353.

Coddet, C., G. Montavon, S. Ayrault-Costil, O. Freneaux, F. Rigolet, G. Barbezat, F. Foliot, A. Diard, and P. Wazen. 1999. Surface preparation and thermal spray in a single step: The PROTAL process – Example of application for an aluminium-base substrate. *Journal of Thermal Spray Technology* 8 (2): 235–242.

Cojocaru, C.V., Y. Wang, C. Moreau, R.S. Lima, J. Mesquita-Guimaraes, E. Garcia, P. Miranzo, and M.I. Osendi. 2011. Mechanical behavior of air plasma-sprayed YSZ functionally graded mullite coatings investigated via instrumented indentation. *Journal of Thermal Spray Technology* 20 (1-2): 100–107.

Costil, S., H. Liao, A. Gammondi, and C. Coddet. 2005. Influence of surface laser cleaning combined with substrate preheating on splat morphology. *Journal of Thermal Spray Technology* 14 (1): 31–38.

Costil, S., C. Verdy, R. Bolot, and C. Coddet. 2007. On the role of spraying process on microstructural, mechanical, and thermal response of alumina coatings. *Journal of Thermal Spray Technology* 16 (5-6): 839–843.

Das, S., S. Ghosh, A. Pandit, K. Bandyopadhyay, A.B. Chattopadhyay, and K. Das. 2005. Processing and characterization of plasma sprayed zirconia-alumina-mullite composite coating on a mild-steel substrate. *Journal of Materials Science Letters* 40: 5087–5089.

Darut, G., F. Ben-Ettouil, A. Denoirjean, G. Montavon, H. Ageorges, and P. Fauchais. 2010. Dry sliding behavior of sub-micrometer-sized suspension plasma sprayed ceramic oxide coatings. *Journal of Thermal Spray Technology* 19 (1–2): 275–285.

Davis, J.R., ed. 2004. *Handbook of thermal spray technology*. Materials Park: ASM International.

Denoirjean, A., A. Grimaud, P. Fauchais, P. Tristant, C. Tixier, and J. Desmaison. 1998. Splat formation, first step for multi-technique deposition of plasma spraying and microwave plasma enhanced CVD. In *Thermal spray: Meeting in challenges of the 21st. century*, ed. C. Coddet, vol. 2, 1369–1374. Materials Park: ASM International.

Dhiman, R., A.G. McDonald, and S. Chandra. 2007. Predicting splat morphology in a thermal spray process. *Surface & Coatings Technology* 201: 7789–7801.

Dong, S., B. Song, B. Hansz, H. Liao, and C. Coddet. 2013. Improvement in the microstructure and property of plasma sprayed metallic, alloy and ceramic coatings by using dry ice blasting. *Surface and Coatings Technology* 220: 199–203.

Dorfman, M.R. 2013. Challenges and Strategies for growth of thermal spray markets: The six-pillar plan. *Journal Thermal Spray Technology* 22 (5): 559–563.

Economou, S., M. De Bonte, J.P. Celis, R.W. Smith, and E. Lugscheider. 2000. Tribological behaviour at room temperature and at 550°C of TiC-based plasma sprayed coatings in fretting gross slip conditions. *Wear* 244: 165–179.

Espanol, M., V. Guipont, K.A. Khor, M. Jeandin, and N. Llorca Isern. 2002. Effect of heat treatment on high pressure plasma sprayed hydroxyapatite coatings. *Surface Engineering* 18 (3): 213–218.

Fauchais, P. 2004. Understanding plasma spraying. *Journal of Physics D: Applied Physics* 37: 86–108.

Fauchais, P., and M. Vardelle. 2003. How to improve the reliability and reproducibility of plasma sprayed coatings. In *Proceedings of Thermal Spray: Advancing the Science and Applying the Technology, Orlando, FL*, ed. C. Moreau and B. Marple, 1165–1173. Materials Park, Ohio, USA: ASM International.

Fauchais, P., J. Heberlein, and M. Boulos. 2014. *Thermal spray fundamentals, from powder to part*. Springer. NY1550 pages.

Fauchais, P., M. Vardelle, and S. Goutier. 2016. Latest researches advances of plasma spraying: From splat to coating formation. *Journal of Thermal Spray Technology* 25 (8): 1534–1553

Fervel, V., B. Normand, and C. Coddet. 1999. Tribological behavior of plasma sprayed Ai$_2$O$_3$-based cermet coatings. *Wear* 230: 70–77.

Feuerstein, A., J. Knapp, T. Taylor, A. Ashary, A. Bolcavage, and N. Hitchman. 2008. Technical and economical aspects of current thermal barrier coating systems for gas turbine engines by thermal spray and EBPVD: A review. *Journal of Thermal Spray Technology* 17 (2): 199–213.

Fincke, J.R., D.M. Crawford, S.C. Snyder, W.D. Swank, D.C. Haggard, and R.L. Williamson. 2003. Entrainment in high-velocity, high temperature plasma jets, Part I: experimental results. *International Journal of Heat and Mass Transfer* 46: 4201–4213.

Floristán, M., R. Fontarnau, A. Killinger, and R. Gadow. 2010. Development of electrically conductive plasma sprayed coatings on glass ceramic substrates. *Surface and Coatings Technology* 205: 1021–1028.

Foster, T., and G. Liu. 1980. *Plasma arc heater design, Techniques for commercial process applications*. Acurex Corp./Aerotherm Division, Mountain view Calif., Internal report.

Freslon, A. 1995. Plasma spraying at controlled temperature and atmosphere. In *Thermal spray: Science and technology*, ed. C.C. Berndt and S. Sampath, 57–63. OH, USA: ASM International.

Fukanuma, H. 1988. Japanese Patent 230300 JP. 04/24/1988

Fukumoto, M., I. Ohgitani, H. Nagai, and T. Yasni. 2005. Effect of substrate surface change by heating on flattening behavior of thermal sprayed particles. In *ITSC-2005*, ed. E. Lugscheider. Düsseldorf, Germany, e-proceedings: DVS.

Fukumoto, M., H. Nagui, and T. Yasui. 2006. Influence of surface character change of substrate due to heating on flattening behavior of thermal sprayed particles. In *ITSC-2006*, ed. B. Marple et al. Materials Park, OH, USA: ASM International.

Fukumoto, M., T. Yamaguchi, M. Yamada, and T. Yasui. 2007. Splash splat to disk splat transition behavior in plasma-sprayed metallic materials. *Journal of Thermal Spray Technology* 16 (5–6): 905–912.

Fukumoto, M., K. Yang, K. Tanaka, T. Usami, T. Yasui, and M. Yamada. 2011. Effect of substrate temperature and ambient pressure on heat transfer at Interface between molten droplet and substrate surface. *Journal of Thermal Spray Technology* 20 (1–2): 48–58.

Goutier, S., M. Vardelle, J.C. Labbe, and P. Fauchais. 2011. Flattening and cooling of millimeter- and micrometer-sized alumina drops. *Journal of Thermal Spray Technology* 20 (1–2): 59–67.

Greuner, H., M. Balden, et al. 2004. Evaluation of vacuum plasma-sprayed boron carbide protection for the stainless steel first wall. *Journal of Nuclear Materials* 329–333: 849–854.

Guipont, V., M. Espanol, F. Borit, N. Llorca-Isern, M. Jeandin, K.A. Khor, and P. Cheang. 2002. High- pressure plasma spraying of hydroxyapatite powders. *Materials Science and Engineering A* 325 (1–2): 9–18.

Guipont, V., S. Bansard, M. Jeandin, K.A. Khor, M. Nivard, L. Berthe, J.P. Cuq-Lelandais, and M. Boustie. 2010. Bond strength determination of hydroxyapatite coatings on Ti-6Al-4V substrates using the LAser Shock Adhesion Test (LASAT). *Journal of Biomedical Materials Research. Part A* 95A (4): 1096–1104.

He, W., G. Mauer, M. Gindrat, R. Wäger, and R. Vaßen. 2017. Investigations on the nature of ceramic deposits in plasma spray–physical vapor deposition. *Journal of Thermal Spray Technology* 26: 83–92.

Heimann, R.B. 1999. Design of novel plasma sprayed hydroxyapatite-bond coat bioceramic systems. *Journal of Thermal Spray Technology* 8 (4): 597–603.

Henne, R., W. Mayr, and A. Reusch. 1993. Influence of nozzle geometry on particle behavior and coating quality in high velocity VPS. In *Proceedings of Thermal Spray Conference TS93, Aachen*, 7–11. DVS, Germany.

Henrich, V.E., and P.A. Cox. 1994. *The surface science of metal oxides*. Cambridge/New York: Cambridge University Press.

Hospach, A., G. Mauer, R. Vaßen, and D. Stöver. 2011. Columnar-structured thermal barrier coatings (TBCs) by thin film low-pressure plasma spraying (LPPS-TF). *Journal of Thermal Spray Technology* 20 (1–2): 116–120.

Hou, H., X. Ning, Q. Wang, Y. Liu, and Ying Liu. 2015. Anti-ablation behavior of air plasma-sprayed Mo(Si, Al)2 coating. *Surface and Coating Technology* 274: 60–67.

Hu, C., Y. Niu, H.J. Li, M. Ren, X. Zheng, and J. Sun. 2012. SiC coatings for carbon/carbon composites fabricated by vacuum plasma spraying technology. *Journal Thermal Spray Technology* 21 (1): 16–22.

Ilavsky, J., C. C. Berndt, H. Herman, P. Chraska and J. Dubsky. 1997. Alumina-base plasma-sprayed materials – Part II: Phase transformations in aluminas. *Journal of Thermal Spray Technology* 6 (4):439-444.

Itoh, A., K. Takeda, M. Itoh, and M. Koga. 1990. Pretreatments of substrates by using reversed transferred arc in low pressure plasma spray. In *Thermal spray research and application*, ed. T. Bernecki, 245–252. Materials Park: ASM International.

Jäger, D.A., D. Stöver, and W. Schlump. 1992. High pressure plasma spraying in controlled atmosphere up to 2 bars. In *Proceedings of ITSC-1992., Orlando, FL*, ed. C.C. Berndt, 69–74. Material Park: ASM International.

Janisson, J.S., E. Meillot, A. Vardelle, J.-F. Coudert, B. Pateyron, and P. Fauchais. 1998. Plasma spraying using Ar-He-H2 gas mixtures. In *Proceedings of the 15th International Thermal Spray Conference, Nice, France*, ed. C. Coddet, 803–814. ASM International.

Kang C.W., H.W. Ng, and S.C.M. Yu (2006) Imaging Diagnostics Study on Obliquely Impacting Plasma-SprayedParticles Near to the Substrate, *J Thermal Spray Technology* 15(1) 118–130

Keshri, A.K., and A. Agarwal. 2011. Wear behavior of plasma-sprayed carbon nanotube-reinforced aluminum oxide coating in marine and high-temperature environments. *Journal Thermal Spray Technology* 20 (6): 1217–1230.

Khor, K.A., P. Cheang, and Y. Wang. 1998. Plasma spraying of combustion flame spheroidized Hydroxyaptite (HA) Powders. *Journal of Thermal Spray Technology* 7 (2): 254–260.

Kim, H.J., Y.G. Kweon, and R.W. Chang. 1994. Wear and erosion behavior of plasma-sprayed WC-Co coatings. *Journal of Thermal Spray Technology* 3 (2): 169–178.

Kitamura, J., H. Ibe, F. Yuasa, and H. Mizuno. 2008. Plasma sprayed coatings of high-purity ceramics for semiconductor and flat-panel-display production equipment. *Journal of Thermal Spray Technology* 17 (5-6): 878–886.

Kitamura, J., Z. Tang, H. Mizuno, K. Sato, and A. Burgess. 2011. Structural, mechanical and erosion properties of Yttrium oxide coatings by axial suspension plasma spraying for electronics applications. *Journal of Thermal Spray Technology* 20 (1-2): 170–185.

Koomparkping, T., S. Damrongrat, and P. Niranatlumpong. 2005. Al-rich precipitation in CoNiCrAlY bondcoat at high temperature. *Journal of Thermal Spray Technology* 14 (2): 264–267.

Krishnamurthy, N., M.S. Murali, B. Venkataraman, and P.G. Mukunda. 2012. Characterization and solid particle erosion behavior of plasma sprayed alumina and calcia-stabilized zirconia coatings on Al-6061 substrate. *Wear* 274–275: 15–27.

Kudinov, V.V., et al. 1981. *High temp. dust laden jets*, 381–392. VSL, NL.

Kulu, P., and T. Pihl. 2002. Selection criteria for wear resistant powder coatings under extreme erosive Wear conditions. *Journal of Thermal Spray Technology* 11 (4): 517–522.

Leivo, E.M., M.S. Vippola, P.P.A. Sorsa, P.M. Vuoristo, and T.A. Mantyla. 1997. Wear and corrosion properties of plasma sprayed Al$_2$O$_3$ and Cr$_2$O$_3$ coatings sealed by aluminum phosphates. *Journal of Thermal Spray Technology* 6 (2): 205–210.

Leylavergne, M., B. Dussoubs, A. Vardelle, and N. Gobot. 1998. Comparison of plasma-sprayed coatings produced in argon or nitrogen atmosphere. *Journal of Thermal Spray Technology* 7 (4): 527–536.

Li, C.-J., and J.-L. Li. 2004. Evaporated-gas-induced splashing model for splat formation during plasma spraying. *Surface and Coatings Technology* 184: 13–23.

Li, C.-J., J.-L. Li, and W.B. Wang. 1998. The effect of substrate preheating and surface organic covering on splat formation. In *Thermal Spray: Meeting the challenges of the 21st century*, ed. C. Coddet, 473–480. Materials Park OH, USA: ASM International.

Li, C.-J., J.-L. Li, W.B. Wang, A.-J. Fu, and A. Ohmori. 1999. A mechanism of the splashing during droplet splatting. In *Thermal Spraying, UTSC-1999. Düsseldorf*, ed. E. Lugscheider and P.A. Kammer, 530–535. DVS Düsseldorf Germany.

Li, H., S. Costil, S.-H. Deng, H.-L. Liao, C. Coddet, V. Ji, and W.-J. Huang. 2006c. Benefit of surface oxide removal on thermal spray coating adhesion using the PROTAL process. In *ITSC-2006*, ed. B. Marple et al. ASM International, Materials Park, OH, USA. e-proc.

Li, H., Y. Danlos, S. Costil, and C. Coddet. 2007. Influence of laser induced surface topography on the surface static wettability in Protal®. In *ITSC-2007 Global Coating Solutions*, ed. B. Marple et al., 1070–1074. Materials Park, OH, USA., e-proceedings: ASM International.

Lin, Chun-Ming. 2012. Parameter optimization of a vacuum plasma spraying process using boron carbide. *Journal of Thermal Spray Technology* 21 (5): 873–881.

Liu, T., A. Ansar, and J. Arnold. 2017. A study of the influence of the surrounding gas on the plasma jet and coating quality during plasma spraying. *Plasma Chemistry Plasma Processes* 37: 1009–1032.

Longa-Nava, E., M. Takemoto, and K. Hidaka. 1995. High-temperature corrosion performance of plasma-sprayed CrNiMoSiB coatings. *Journal of Thermal Spray Technology* 4 (2): 169–174.

Ma, X.-Q., F. Borit, V. Guipont, and M. Jeandin. 2002. Thin alumina coating deposition by using controlled atmosphere plasma spray system. *Journal of Advanced Materials* 34 (4): 52–57.

Matejicek, J., P. Chraska, and J. Linke. 2007. Thermal spray coatings for fusion applications-review. *Journal of Thermal Spray Technology* 16 (1): 64–83.

Mauer, G. 2014. Plasma characteristics and plasma-feedstock interaction under PS-PVD process conditions. *Plasma Chemistry and Plasma Processing* 34: 1171–1186.

———. 2019. Numerical study on particle–gas interaction close to the substrates in thermal spray processes with high-kinetic and low-pressure conditions. *Journal of Thermal Spray Technology* 28: 27–39.

Mauer, G., R. Vaßen, and D. Stöver. 2010. Thin and dense ceramic coatings by plasma spraying at very low pressure. *Journal of Thermal Spray Technology* 19 (1–2): 495–501.

Mauer, G., A. Hospach, N. Zotov, and R. Vaßen. 2013. Process conditions and microstructures of ceramic coatings by gas phase deposition based on plasma spraying. *Journal of Thermal Spray Technology* 22 (2–3): 83–89.

Mauer, G., N. Schlegel, A. Guignard, M.O. Jarligo, S. Rezanka, A. Hospach, and R. Vaßen. 2015. Plasma spraying of ceramics with particular difficulties in processing. *Journal of Thermal Spray Technology* 24 (1–2): 30–37.

Mayr, G., and Henne. 1988. Investigation of a VPS Burner with Laval Nuzzle using an automated laser doppler measuring system. In *Proceedings of 1st. Plasma-Technik Symposium, Lucerne, Switzerland*, ed. H. Eschnauer, P. Huber, A.R. Nicoll, and S. Sandmeier, 87–97. Wohlen, Switzerland: Plasma Technik.

McDonald, A., C. Moreau, and S. Chandra. 2007. Effect of substrate oxidation on spreading of plasma-sprayed nickel on stainless steel. *Surface & Coatings Technology* 202 (2007): 23–33.

McKechnie, T.N., Y.K. Liaw, E.R. Zimmerman, and R.M. Poorman. 1994. Metallurgical and process comparison of vacuum plasma spray forming on internal and external surfaces--atechnical note. *Journal of Thermal Spray Technology* 3 (3): 270–274.

Mesrati, N., H. Ajhrourh, N. Du, and D. Treheux. 2000. Thermal spraying and adhesion of oxides onto graphite. *Journal of Thermal Spray Technology* 9 (1): 95–99.

Meyer, P.J., and D. Hawley. 1991. LPPS production systems. In *Proceedings of 4th. National Thermal Spray Conf., Pittsburgh, PA*, ed. T.F. Bernecki, 29–38. Materials Park, OH: ASM International.

Miguel, J.M., S. Vizcaino, C. Lorenzana, N. Cinca, and J.M. Guilemany. 2011. Tribological behavior of bronze composite coatings obtained by plasma thermal spraying. *Tribology Letters* 42: 263–273.

Moldovan, M., C.M. Weyant, D. Lynn Johnson, and K.T. Faber. 2004. Tantalum oxide coatings as candidate environmental barriers. *Journal of Thermal Spray Technology* 13 (1): 51–56.

Moreau, C., P. Gougeon, A. Burgess, and D. Ross. 1995. Characterization of Particle Flows in an Axial Injection Plasma Torch. In *Proceedings of NTSC-1995, Houston, Texas*, ed. C.C. Berndt and S. Sampath, 141–147. Materials Park, OH: ASM International.

Natarajan, S., E. Edward Anand, K.S. Akhilesh, Ananya Rajagopal, and Preeti P. Nambiar. 2016. Effect of graphite addition on the microstructure, hardness and abrasive wear behavior of plasma sprayed NiCrBSi coatings. *Materials Chemistry and Physics* 175: 100–106.

Niu, Y., X. Liu, X. Zheng, H. Ji, and C. Ding. 2009. Microstructure and properties characterization of silicon coatings prepared by vacuum plasma spraying technology. *Journal of Thermal Spray Technology* 18 (3): 427–434.

Niu, Y., X. Fei, H. Wang, X. Zheng, and C. Ding. 2013. Microstructure characteristics and oxidation behavior of molybdenum disilicide coatings prepared by vacuum plasma spraying. *Journal of Thermal Spray Technology* 22 (2–3): 96–103.

Normand, B., V. Fervel, C. Coddet, and V. Nikitine. 2000. Tribological properties of plasma sprayed alumina-titania coatings: Role and control of the microstructure. *Surface and Coatings Technology* 123: 278–287.

Pawlowski, L. 1996. Technology of thermally sprayed anilox rolls: State of art, problems, and perspectives. *Journal of Thermal Spray Technology* 5 (3): 317–334.

Pech, J., B. Hannoyer, L. Bianchi, A. Denoirjean, and P. Fauchais. 1997. Study of oxide layers obtained onto 304L substrate heated by a D.C. plasma jet. In *Thermal spray: A united forum for scientific and technological advances*, ed. C.C. Berndt, 775–782. Materials Park, Oh, USA: ASM International.

Pech, J., B. Hannoyer, A. Denoirjean, and P. Fauchais. 2000. Influence of substrate preheating monitoring on alumina splat formation in DC plasma process. In *Thermal spray: Surface engineering via applied research*, ed. C.C. Berndt, 759–765. Materials Park, Oh, USA: ASM International.

Ramachandran, C.S., V. Balasubramanian, P.V. Ananthapadmanabhan, and V. Viswabaskaran. 2012. Understanding the dry sliding wear behavior of atmospheric plasma-sprayed rare earth oxide coatings. *Materials and Design* 39: 234–252.

Rahmane, M., G. Soucy, M. Boulos, and R. Henne. 1998. Fluid dynamic study of direct current plasma jets for plasma spraying applications. *Journal of Thermal Spray Technolgoy* 7 (3): 349–356.

Refke, A., G. Barbezat, J.L. Dorier, M. Gindrat, and C. Hollenstein. 2003. Characterization of LPPS processes under various spray conditions for potential applications. In *Proceedings of the International Thermal Spray Conference, Orlando, Florida, 2003*, ed. B. Marple and C. Moreau, 581–588. Materials Park, OH: ASM International.

Rigot, D., G. Delluc, B. Pateyron, J. Coudert, P. Fauchais, and J. Wigren. 2003. Transient evolution and shift of signals emitted by a DC plasma gun (type PTF4). *Journal High Temperature Materials and Processes* 7: 175–186.

Roumilhac, P. 1990. *Contribution to the metrology and understanding of the DC plasma spray torches and transferred arcs for reclamation at atmospheric pressure*. Ph.D. thesis, Université de Limoges, Limoges, France.

Roumilhac, P., J.-F. Coudert, and P. Fauchais. 1990. Designing parameters of spraying plasma torches. In *Proceedings of National Thermal Spray Conference, Long Beach, CA, 1990*, 19, ed. T. Bernecki, 11. Materials Park, OH: ASM International.

Russ, S., P.J. Strykowski, and E. Pfender. 1994. Mixing in plasma and low-density jets. *Experiments in Fluids* 16: 297–3.07.

Sabiruddina, K., P.P. Bandyopadhyay, G. Bolelli, and L. Lusvarghi. 2011. Variation of splat shape with processing conditions in plasma sprayed alumina coatings. *Journal of Materials Processing Technology* 211 (2011): 450–462.

Salimijazi, H.R., T.W. Coyle, J. Mostaghimi, and L. Leblanc. 2005a. Microstructure and failure mechanism in As-deposited, vacuum plasma-sprayed Ti-6Al-4V alloy. *Journal of Thermal Spray Technology* 14 (2): 215–223.

Salimijazi, H.R., T.W. Coyle, and J. Mostaghimi. 2005b. Vacuum plasma spraying: A new concept for manufacturing Ti-6Al-4V structures. *JOM*: 50–56. September 2006.

Schiller, G., R. Henne, and V. Borck. 1995. Vacuum plasma spraying of high-performance electrodes for alkaline water electrolysis. *Journal of Thermal Spray Technology* 4 (2): 185–194

Scrivani, A., U. Bardi, L. Carrafiello, A. Lavacchi, F. Niccolai, and G. Rizzi. 2003. A comparative study of high velocity oxygen fuel, vacuum plasma spray, and axial plasma spray for the deposition of CoNiCrAlY bond coat alloy. *Journal of Thermal Spray Technology* 12 (4): 504–507.

Sidhu, B.S., and S. Prakash. 2007. Analytical studies on the behavior of Nickel and Cobalt base shrouded plasma spray coatings at elevated temperature in air. *Oxidation of Metals* 67: 279–298.

Siegmann, S., O. Brandt, and M. Dvorak. 2004. Thermally sprayed wear resistant coatings with nanostructured hard phases. *Journal of Thermal Spray Technology* 31 (1): 37–43.

Stahr, C., S. Saaro, L.-M. Berger, J. Dubsky, K. Neufuss, and M. Hermann. 2007. Dependence of the stabilization of alpha-alumina on the spray process. *Journal of Thermal Spray Technology* 16 (5–6): 822–830.

Syed, A.A., A. Denoirjean, P. Fauchais, and J.C. Labbe. 2006. On the oxidation of stainless steel particles in the plasma jet. *Surface & Coatings Technology* 200 (2006): 4368–4382.

Thermal Spraying. 1985. *Practice, theory and applications*. Miami: American Welding Soc.

Toma, F.L., S. Scheitz, L.-M. Berger, V. Sauchuk, M. Kusnezoff, and S. Thiele. 2011. Comparative study of the electrical properties and characteristics of thermally sprayed alumina and spinal coatings. *Journal of Thermal Spray Technology* 20 (1-2): 195–204.

Tsukuda, H., A. Notomi, and N. Hisatome. 2000. Application of plasma spraying to tubular-type solid oxide fuel cells production. *Journal of Thermal Spray Technology* 9 (3): 364–368.

Tului, M., F. Ruffini, F. Arezzo, S. Lasisz, Z. Znamirowski, and L. Pawlowski. 2002. Some properties of atmospheric air and inert gas high-pressure plasma sprayed ZrB2 coatings. *Surface and Coatings Technology* 151–152: 483–489.

Usmani, S., and K.N. Tandon. 1992. Evaluation of thermally sprayed coatings under reciprocating lubricated wear conditions. *Journal of Thermal Spray Technology* 1 (3): 249–255.

Vaßen, R., M.O. Jarligo, T. Steinke, D.E. Mack, and D. Stöver. 2010. Overview on advanced thermal barrier coatings. *Surface and Coatings Technology* 205: 938–942.

von Niessen, K., and M. Gindrat. 2011. Plasma spray-PVD: A new thermal spray process to deposit out of the vapor phase. *Journal of Thermal Spray Technology* 20 (4): 736–743.

Wang, H., Hui Li, Hongbin Zhu, Fangjie Cheng, Dongpo Wang, and Zhuoxin Li. 2015. A comparative study of plasma sprayed TiB_2–NiCr and Cr_3C2–NiCr composite coatings. *Materials Letters* 153: 110–113.

Wen, Z.H., Y. Bai, J.F. Yang, and J. Huang. 2016. Effect of vacuum remelting on the solid particle's erosion behavior of Ni60-NiCrMoY composite coatings prepared by plasma spraying. *Vacuum* 134: 73–82.

Ye, F.X., A. Ohmori, T. Tsumura, K. Nakata, and C.J. Li. 2007. Microstructural analysis and photocatalytic activity of plasma-sprayed titania-hydroxyapatite coatings. *Journal of Thermal Spray Technology* 16 (5-6): 776–782.

Yilmaz, R., A.O. Kurt, A. Demir, and Z. Tath. 2007. Effects of TiO_2 on the mechanical properties of the Al_2O_3-TiO_2 plasma sprayed coating. *Journal of the European Ceramic Society* 27: 1319–1323.

Zeng, Y., S.W. Lee, and C. Ding. 2002. Study on plasma sprayed boron carbide coating. *Journal of Thermal Spray Technology* 11 (1): 129–133.

Zeng, Z., S. Kuroda, and H. Era. 2009. Comparison of oxidation behavior of Ni–20Cr alloy and Ni-base self-fluxing alloy during air plasma spraying. Surface, and Coatings Technology 204: 69–77.

Zhang, C., H.-L. Liao, W.-Y. Li, G. Zhang, C. Coddet, C.-J. Li, C.-X. Li, and X.-J. Ning. 2006. Characterization of YSZ solid oxide fuel cells electrolyte deposited by atmospheric plasma spraying and low-pressure plasma spraying. *Journal of Thermal Spray Technology* 15 (4): 598–603.

Zhu, H.B., Hui Li, and Zhuo Xin Li. 2013. Plasma sprayed TiB_2–Ni cermet coatings: Effect of feedstock characteristics on the microstructure and tribological performance. *Surface, and Coatings Technology* 235: 620–627.

Induction Plasma Spraying **10**

Abbreviations

AC	Alternating Current
APS	Atmospheric Plasma Spraying
CCD	Charge Coupled Device
CFD	Computational Fluid Dynamics
CVD	Chemical Vapor Deposition
DC	Direct Current
HA	Hydroxyapatite
HF	High Frequency
i.d.	Internal Diameter
ICP	Inductively Coupled Plasma
IPS	Induction Plasma Spraying
ITSC	International Thermal Spray Conference
LHS	Left-Hand Side
LTE	Local Thermodynamic Equilibrium
MCVD	Modified Chemical Vapor Deposition
MS	Mass Spectrometry
OES	Optical Emission Spectroscopy
PVD	Physical Vapor Deposition
RC	Radiation Cooled
RF	Radio Frequency
RF-ICP	Radio Frequency Inductively Coupled Plasma
RF-IPS	Radio Frequency Induction Plasma Spraying
RHS	Right-Hand Side
SOFC	Solid Oxide Fuel Cells
SPS	Suspension Plasma Spraying
VIPS	Vacuum Induction Plasma Spraying
WC	Water Cooled
YSZ	Yttria-Stabilized Zirconia

10.1 Introduction

Induction plasma spraying (IPS), commonly identified as radio frequency induction plasma spraying (RF-IPS), inductively coupled plasma spraying (ICPS), or vacuum induction plasma spraying (VIPS), is a plasma technology that has immerged in the early sixties as the first electrodeless open discharge for the generation of a high purity atmospheric pressure plasma that could be used for crystal growing and the deposition of high purity ceramics [Reed's (1961a, b)]. As illustrated in Fig. 10.1, interest in the technology grew over the years pioneered by M Thopre of TAFA corpo, who developed it further to the hundreds of kW power level funded by NASA for the US aerospace program. Equivalent development was carried out over the same period in the former Soviet Union by S. Dresvin and his collaborators at the St. Petersburg Polytechnic Institute in the USSR [Dresvin S. Ed. (1977)]. The early development of the induction plasma technology was mostly funded by NASA in the USA in the early sixties for its aerospace program [TAFA (1966, 1968), Thorpe (1968, 1970a, b), Thorpe et al. (1968), Thorpe and Harrington (1970a, b), Thorpe and Scammon (1968, 1969), Thorpe and Suncook (1970), Dundas and Thorpe (1969), Dundas (1970), Dundas et al. (1975), Pool and Vogel (1972), Pool et al. (1973), and Vogel (1970, 1971)]. Similar development programs were also on their way in the former Soviet Union over the same period for essentially similar objectives. In both cases, the aim was to develop an appropriate source of high purity, high temperature, partially ionized, air stream for the testing of thermal shields needed to protect space vehicles from the high temperature environment encountered during their re-entry into the earth atmosphere.

Further development of the technology was motivated by a wide range of potential applications varying from small-scale systems for spectrochemical elemental analysis to large power installations for crystal growing, preform cladding in the fiber optics industry, and more recently for the synthesis and processing of advanced materials including nano- and micron-sized high purity spherical powders and the deposition of coatings and near net-shaped parts.

In this chapter, the basic concepts used for the generation of the RF inductively coupled plasmas are presented followed by a detailed discussion of the energy coupling mechanism and induction plasma torch design. The results

© Springer Nature Switzerland AG 2021

M. I. Boulos et al. (ed.), *Thermal Spray Fundamentals*, https://doi.org/10.1007/978-3-030-70672-2_10

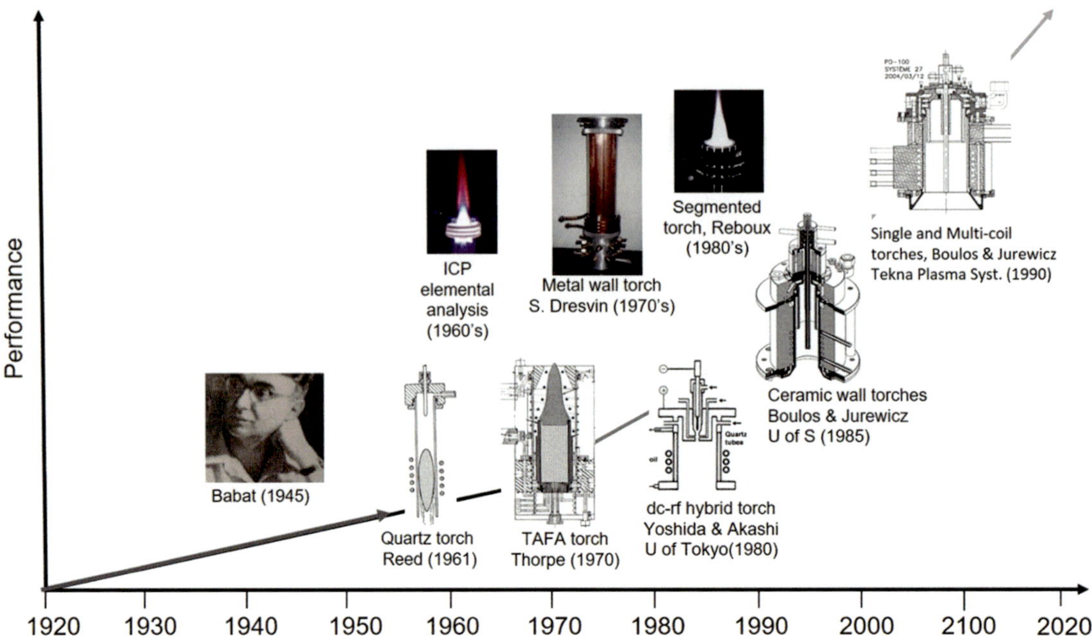

Fig. 10.1 Evolution of the inductively coupled plasma (ICP) technology over the period 1920–2020

of a wide range of diagnostic and modeling studies are reviewed next providing a description of the main features of the electromagnetic, temperature, flow and concentration fields in the discharge and of plasma-particle dynamics. A description of integrated RF induction plasma deposition systems and typical operating conditions is described followed by typical present and potential examples of industrial applications of the technology.

10.2 Basic Concepts

10.2.1 General Remarks

The first experiments with the atmospheric pressure electrodeless discharges can be traced back to Babat (1947). In his extensive study of this type of discharges, Babat identified two distinct electrodeless energy-coupling mechanisms that are represented on Fig. 10.2. "Capacitive coupling" or E-type discharges are excited by the electrical field with the conductance current not closed. "Inductive coupling" or H-type discharges are excited by alternating magnetic fields with the conductance current in the form of a closed loop or a short circuit secondary turn of a transformer.

Such a division of electrodeless discharges between capacitive and inductive is valid only when the wavelength (λ), corresponding to the excitation frequency, exceeds the characteristic dimension of the discharge space (l). Where, $\lambda = c/f$, with, c, the velocity of light ($c = 3 \times 10^8$ m/s) and, f,

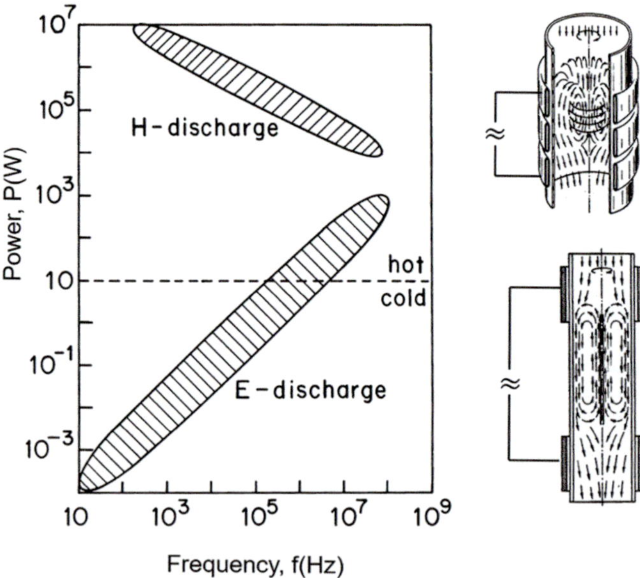

Fig. 10.2 Frequency dependence of the power density in E- and H-type discharges [Babat (1947)]

the oscillator frequency. Generally, this condition is satisfied at frequencies lower than 100 MHz ($\lambda > 3$ m). Babat (1947) proposed that the ratio (λ/l) be used as a basis for the classification of electrodeless discharges into the following three groups:

Low frequency $(\lambda / l) > 100$
High frequency $100 \geq (\lambda / l) \geq 10$
Ultra-high frequency $10 > (\lambda / l)$

Based on these criteria, for a 100 mm diameter discharge chamber, the definition of a high frequency discharges will be limited to operation at frequencies above 30 MHz, while ultra-high frequency discharges will refer to operation at frequencies above 300 MHz. While such a terminology was generally adhered to the former Soviet Union, it has not been universally adopted, with numerous studies in literature referring to high frequency (HF) or radio frequency (RF) discharges operating at frequencies in the range of 1–5 MHz.

At frequencies below 1 MHz, E-discharges are characterized by their low power densities (of the order of a few W/cm^3), and are generally referred to as "cold" or "nonequilibrium" discharges. With the increase of the excitation frequency and the corresponding decrease of the (λ / l) ratio, E-discharges approach a thermal discharge at frequencies above 10 MHz. H-discharges, on the other hand, can be generated at frequencies as low as 10–100 kHz with corresponding power densities in the hundreds of W/cm^3. The higher the frequency, the lower is the current in the inductor needed to sustain the plasma. A representation of the minimum power versus frequency relationship for these two types of discharges is shown on the LHS of Fig. 10.2 after Babat (1947). It may be noted that when the dimensions of the discharge space are one or two orders of magnitude less than the wavelength of the excitation frequency ($f \cong 100$ MHz) the difference between these two types of discharges, as regard to power, becomes insignificant. In the present chapter, we will deal exclusively with H-type inductively coupled discharges. The development of this type of discharge, or of the induction plasma torch as we know it today, is mainly due to Reed (1961a, b and 1963a, b) who demonstrated that inductively coupled plasma can be maintained in an open tube in the presence of a gas flow. A schematic of Reed's induction plasma torch is shown in Fig. 10.3. This was built using a 26 mm o.d. air-cooled, quartz plasma confinement tube surrounded by a 5-turn, water-cooled, copper coil. Electric power was supplied to the coil by a Lepel generator (Model T-10 RF) having a maximum power output of 10 kW and a nominal operating frequency of 4 MHz. Pure Argon, Ar/Air, Ar/O$_2$, Ar/He and Ar/H$_2$ were used as plasma gases in Reed's experiments.

A considerable research effort was devoted to the development of the induction plasma torch design for such applications as testing of materials for re-entry simulation for aerospace applications and the processing of high added-value advanced materials. A vast volume of literature is available in the open literature mostly reporting diagnostic and mathematical modeling studies of the induction plasma discharge as well as studies of materials synthesis and processing [Eckert (1974), Dresvin et al. (1977, 1993), Boulos (1985, 1992a, b, 1997)].

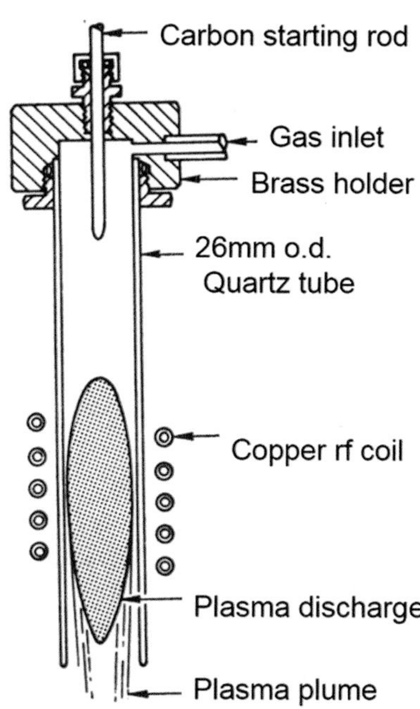

Fig. 10.3 Schematic of Reed's induction plasma torch [Reed (1963a, b)]

10.2.2 Energy Coupling Mechanism

The basic phenomena governing the operation of the RF inductively coupled plasma is similar to that of the induction heating of metals, which found numerous large-scale industrial applications [Davis and Simpson (1979), Zinn and Semiatin (1988)]. A photograph of a typical, low power (3 MHz, 12 kW), quartz tube, air-cooled, ICP discharge is shown on the RHS of Fig. 10.4. A schematic representation of the energy coupling mechanism is given on the LHS of the same figure. This shows the electromagnetic field lines crossing the coaxial plasma load within the induction coil region. The discharge is maintained through ohmic heating by the induced circular currents in the outer regions of the discharge as a result of the intersection of the applied oscillating magnetic field with the conducting load. As will be seen later, the fact that the "load" is a conducting gas with a lower electrical conductivity than most metals has a direct influence on the range of oscillator frequencies required to sustain an inductively coupled plasma discharges.

In an analogous way to the induction heating of a cylindrical metallic load, Freeman and Chase (1968) proposed the "channel model" for the representation of the induction heating of the plasma. According to this model (Fig. 10.5a), the electrical discharge in the coil region can be represented by a cylindrical load of radius, r_n, and uniform electrical conductivity, σ_o. Outside this region, ($r_n < r < r_o$), where r_o is the internal radius of the plasma confining tube, the gas is assumed to be nonconducting and spans the temperature difference between that of the plasma and the wall. The

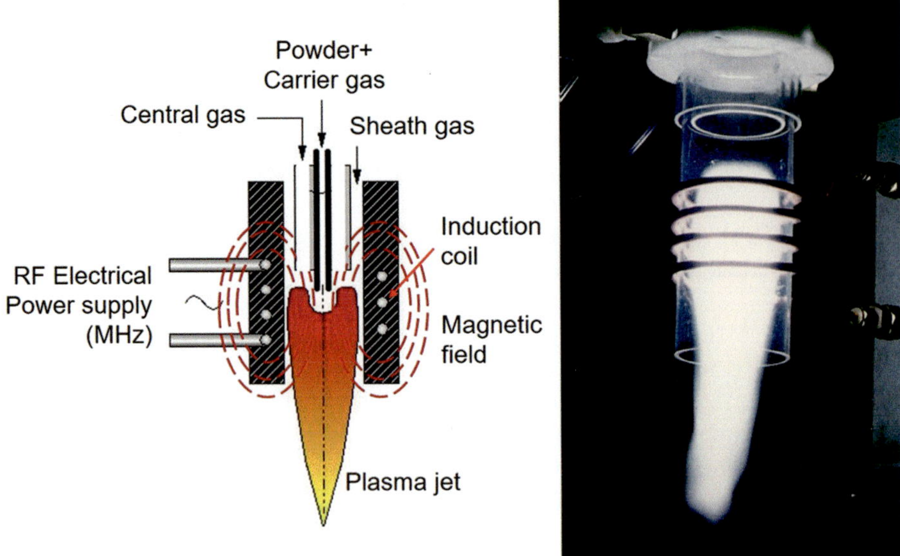

Fig. 10.4 Photograph of atmospheric pressure pure argon ICP discharge (right), and schematic representation of the electromagnetic energy coupling mechanism (left) [Photograph courtesy of the University of Sherbrooke]

internal radius of the induction coil is designated as r_c, the coil length, L_c, and its number of turns, n_c. The alternating electromagnetic field generated by the applied AC current through the induction coil will accordingly couple only into the cylindrical load in a similar fashion as that of induction heating of metals. Through the solution of the standard electromagnetic induction heating problem, Freeman and Chase (1968) demonstrated that the energy will be coupled into the outer shell of the cylindrical load over a thickness, δ_t, known as the "skin depth," which is a function of the average electrical conductivity of the plasma, σ_o, the oscillator frequency, f, and the magnetic permeability of the medium, μ_o. Based on standard electromagnetic formulation, the skin depth can be calculated using the following Eq. 10.1:

$$\delta_t = \frac{1}{\sqrt{\pi\,\sigma_o\,\mu_o\,f}} \qquad (10.1)$$

It is generally accepted that the magnetic permeability of the medium can be taken as that of free space ($\mu_o = 4\,\pi\,10^{-7}$ Hy/m). For a pure argon plasma at atmospheric pressure with an average temperature of 8,000 K, its corresponding average electrical conductivity, σ_o, is equal to 990 A/Vm. The skin depth in this case will be equal to approximately 8 mm for an oscillator frequency of 4 MHz. An end-view photograph of induction plasma is given in Fig. 10.5b showing clearly the annular nature of the discharge ring.

The results of computations carried out using this model for the case of atmospheric pressure N_2 plasma, and Ar/H_2 plasma at 8,000 K and pure Ar plasma at 10,000 K, are given in Fig. 10.6 together with corresponding values of the "skin depth" for cylindrical metallic loads of aluminum, copper,

mild steel, and stainless steel for comparison. These show "skin depth" values for the case of metallic loads that are almost two orders of magnitude smaller than those predicted for plasma, which explains the need for considerably higher frequencies, in the MHz range, for the excitation of ICP discharges, compared to power supplies traditionally used for the induction heating of metals that operate typically in the kHz range.

The energy coupling efficiency, η_c, is defined as the ratio of the power coupled into the discharge, P_o, divided by the reactive power in the oscillator circuit, P_{ac} has been shown by Mensingen and Boedecker (1968) to be a function of the ratio of the radius of the discharge, r_n to that of the skin depth, δ_t. Mensingen and Boedecker defined the coupling parameter, κ_c.

$$\eta_c = \frac{P_o}{P_{ac}} \qquad (10.2)$$

$$\kappa_c = \sqrt{2}\,(r_n/\delta_t) \qquad (10.3)$$

Where r_n is the radius of the discharge, and δ_t is the skin depth. The results given in Fig. 10.7 show a strong dependence of the energy coupling efficiency on the coupling parameter, κ_c and the ratio (r_n/r_c), where r_c is the radius of the induction coil. Maximum energy coupling achieved at κ_c values between 2.5 and 4 corresponding to values of $r_n/\delta_t = 1.8 - 2.5$. Because of the strong dependence of the skin depth on the oscillator frequency, and the electrical conductivity of the plasma, the discharge diameter to be used, has to be closely matched to the operating frequency and the nature of the plasma gas. For example, for an Ar/H_2

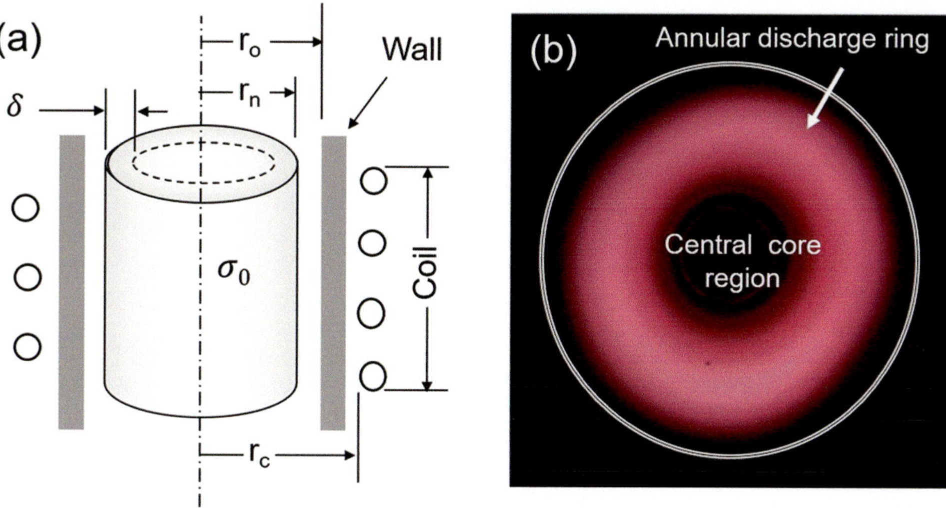

Fig. 10.5 (**a**) Equivalent channel model representation [after Freeman and Chase (1968)] and (**b**) end-view photograph showing the annular ICP discharge [photograph courtesy of M. Thorpe-TAFA Corp.]

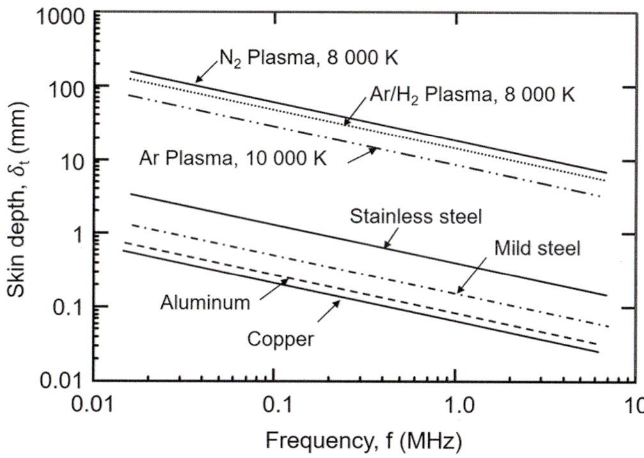

Fig. 10.6 Skin depth as a function of the oscillator frequency for different metals and plasma gases at atmospheric pressure

discharge operating at an average discharge temperature of 8000 K, with an oscillator frequency of 1 MHz, the skin depth is equal to 20 mm. The optimal discharge radius in this case will have to be around 36–50 mm, requiring a minimum of 100 mm internal diameter of the plasma confinement tube.

It may also be noted from Fig. 10.7 that the closer the ratio of (r_n/r_c) is to unity, the higher is the energy coupling efficiency. For most induction plasma torches operating at power levels less than 50 kW, geometric considerations limit this parameter to values in the range of 0.6–0.7. The reason is largely due to the need for adequate cooling of the plasma confinement tube and avoiding a direct electrical contact between the induction coil and the plasma. At higher power levels, requiring larger plasma confinement tubes, the ratio of (r_n/r_c) approaches a value of 0.8 depending on the details of

the induction plasma torch design, resulting in an improvement of the energy coupling efficiency.

Besides the above-indicated geometric coupling parameters, the efficiency of energy transfer between the electrical circuit and the plasma load depends on the proper tuning of the oscillating circuit. This is achieved in low power systems, up to 10 kW, through the introduction of an appropriate impedance matching circuit that protects the power supply from impedance fluctuations of the plasma load and allows the power supply to be tuned for a fixed load impedance (generally, 50 Ohm). In larger power systems, above 10 kW, power losses in the matching network are excessive making it more appropriate to operate with a free-oscillating circuit in which the plasma load is an integral part of the oscillating circuit. This results in a higher overall energy coupling efficiency, at the expense of tolerating small variations of the oscillating frequency depending on the plasma characteristics.

10.2.3 Minimum Sustaining Power

Sustaining a plasma discharge, whether DC or RF inductively coupled depends on a simple equilibrium between local energy dissipation through ohmic heating of the plasma, and energy losses from the discharge through wall conduction and convection, and radiative energy transfer. RF inductively coupled plasma requires accordingly a sufficiently high-energy generation in the plasma in order to compensate for energy losses and maintain the discharge at the lower limit of its electrical conductivity. Because of the high specific enthalpy required for the dissociation and ionization of molecular gases, it is not surprising that they would require

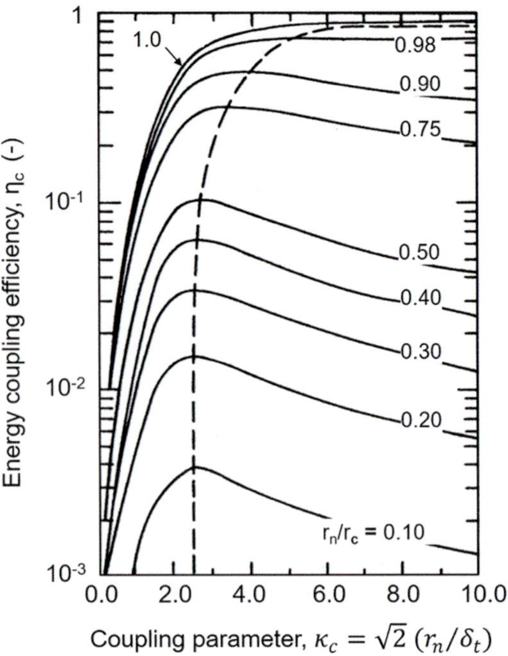

Fig. 10.7 Energy coupling efficiency as a function of the coupling parameter, κ_c and the ratio of the discharge diameter to induction coil diameter (r_n/r_c) [Mensingen and Boedecker (1968)]

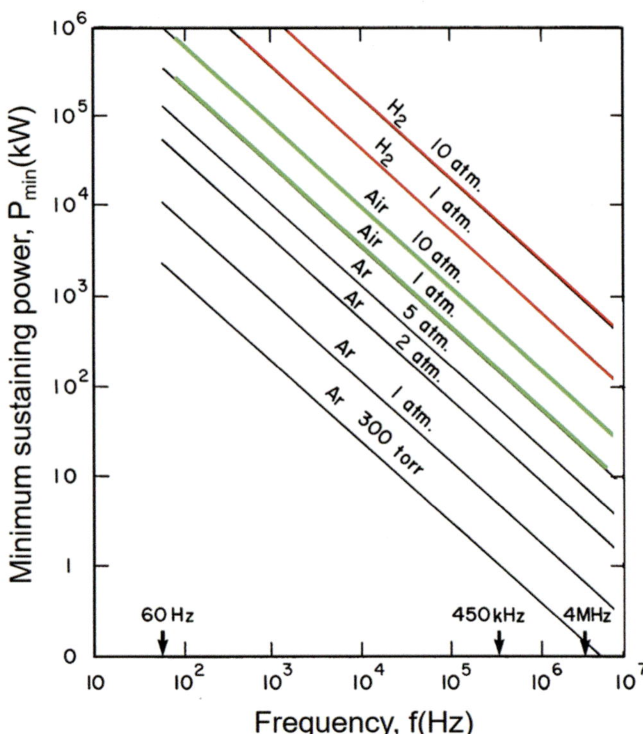

Fig. 10.8 Minimum discharge sustaining power as function of the oscillator frequency for different plasma gases and operating pressures [Thorpe and Scammon (1969)]

a higher energy density than monatomic gases to be maintained in the plasma state. It is also obvious that maintaining an electrical discharge at high pressure would require proportionally higher energy densities than that at atmospheric or lower pressures.

Based on the analyses presented earlier, Sect. 10.2.2, the energy is coupled into an annular region around the periphery of the discharge with a thickness equal to the skin depth, the energy density in the discharge, for a given plasma power, will be inversely proportional to the skin depth. Since the latter is inversely proportional to the square root of the oscillator frequency and the electrical conductivity of the plasma, there is a direct relationship between the energy density in the discharge and the oscillator frequency. The results presented in Fig. 10.8, based on a simplified mathematical model by [Thorpe (1968), Thorpe and Scammon (1968, 1969) and Pool et al. (1973)], clearly indicate that for a given plasma gas and a discharge pressure, the minimum power required to sustain a stable discharge drops exponentially with the increase in the oscillator frequency. While the absolute values given in this example depend on the particular size of the discharge cavity, the trend is clear with a drop of the minimum sustaining power of the discharge, by many orders of magnitude, with the increase in the frequency from the kHz to the MHz range. The lowest values of the minimum sustaining power are obtained for argon plasmas at low pressure and oscillator frequency in the MHz range. With the increase in pressure, the minimum

sustaining power requirement increases. The same effect is noted with the shift from argon plasma to a molecular gas, such as nitrogen or air, with the highest values of the minimum sustaining power required for hydrogen above atmospheric pressure.

These results are consistent with experimental observations made and provide a simple explanation for the commonly used practice of igniting an inductively coupled plasma discharge at low pressure with argon as the plasma gas. Once a discharge has been established, it is possible to increase the pressure with a corresponding increase in the plasma power. The addition of molecular gases into the discharge could also be made at this stage accompanied by an increase of the discharge power in order to avoid the extinction of the plasma.

10.3   Induction Plasma Torch Design

10.3.1   Flow Stabilization Mechanism

The stability of operation of the induction plasma torch depends to a large extent on the ability to control the aerodynamics of the flow in the discharge. Numerous plasma stabilization flow configurations were developed over the years. These are summarized in Fig. 10.9 after Dresvin (1977).

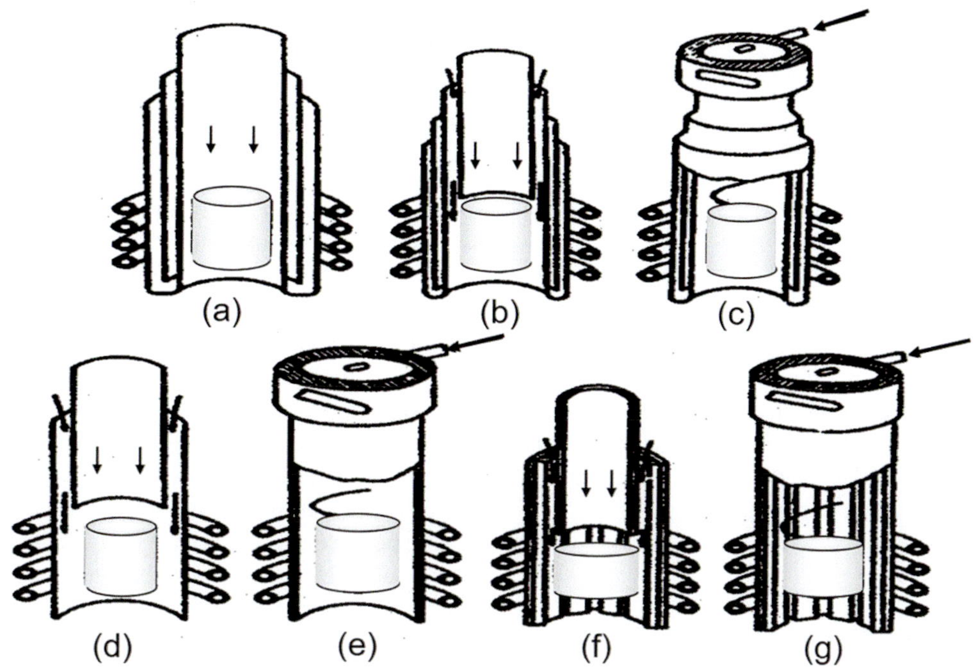

Fig. 10.9 Gas stabilization of induction plasma discharges in the single flux and double flux modes [Dresvin et al. (1977)]

They can be grouped into two broad categories: single flux and double flux torches.

In the "single flux torches" (Fig. 10.9a, c, e and g), the plasma gas is introduced into the discharge cavity as a single stream, either axially (Fig. 10.9a) or tangentially (Fig. 10.9c, e, g) oriented at different angles with respect to the axis of the plasma confinement tube. The role of the plasma gas in these cases is to keep the discharge in a central location in the coil region and avoid a too-close contact between the discharge and the inner surface of the plasma confinement tube. Protecting the walls of the gas-cooled or water-cooled plasma confinement tube from thermal damage is a major concern in all of these designs.

In the "double flux torches," the plasma gas is introduced into the discharge chamber as two different gas streams separated by a concentric quartz tube as shown in Fig. 10.9b, d, and f. These configurations are widely used because of their ability to properly protect the plasma confinement tube through the use of two separate gas streams injected into the discharge region, a high "sheath gas" flow stream evenly distributed along the inner perimeter of the plasma confinement tube, and a lower flow "central gas" stream directed toward the center of the discharge. It also offers the added advantage of the possible use of different gas compositions for each of the sheath gas and the central gas streams. In general, when the plasma is formed of a mixture of argon with a molecular gas, such as hydrogen or nitrogen, the molecular gas is introduced premixed with the sheath gas. It may be noted that the sheath gas is commonly introduced

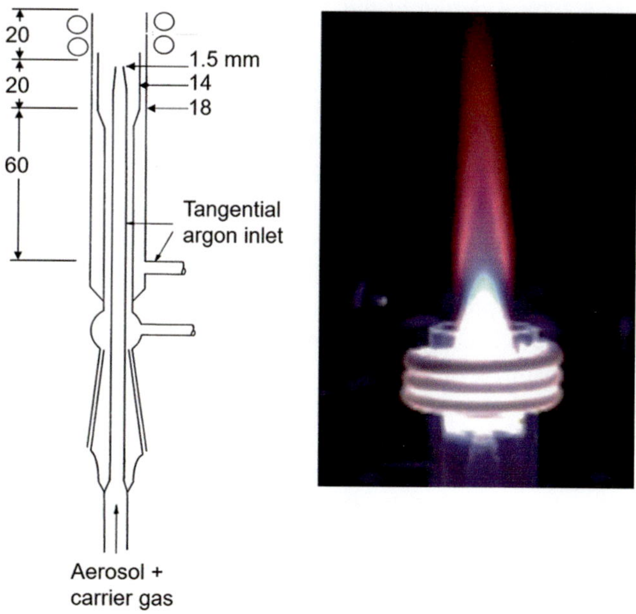

Fig. 10.10 Air-cooled, quartz, induction plasma torch used for spectrochemical elemental analysis

into the plasma torch in an axial flow mode, while the central gas is introduced in a tangential flow configuration.

At low pressures and low power levels, a simple air cooling of the external plasma confinement tube is adequate for its protection from thermal damage with either modes of plasma stabilization (Fig. 10.9d, e). At higher power levels, external air cooling is not sufficient to insure a proper protection of the plasma confinement tube. Water cooling is the next step up in these cases as shown in Fig. 10.9a, b and c. The use of a double or triple wall, water-cooled construction results in a significant increase of the outer diameter of the tube and the corresponding increase of the diameter of the induction coil for the same inner diameter of the plasma confinement tube. This results in a decrease of the plasma-to-coil diameter ratio, and a corresponding decrease in the energy coupling efficiency. At yet higher power levels, simple water cooling of the walls of the plasma confinement tube is not enough to ensure its adequate protection from the discharge, a water-cooled, segmented metal cage is inserted in this case into the discharge cavity of the torch as shown in Fig. 10.9f and g. Details of induction plasma torch designs are discussed in more details in the following sections.

10.3.2 Quartz-Wall Induction Plasma Torches

One of the first and most widely used induction plasma torch designs is that of the low-power (2–3 kW), air-cooled, quartz wall torch widely used for spectrochemical elemental analysis. A typical design of this type of torch and its principal dimensions are given in Fig. 10.10. The torch is basically constructed of 18-mm i.d., 100 mm long, precision-ground, quartz tube, in which a second smaller quartz tube, 14 mm i. d., 16 mm o.d. is placed concentric with respect to the outer tube. The 1-mm annular gap between the two tubes serves to introduce the sheath gas that flows tangentially along the inner wall of the plasma confinement tube in order to keep the discharge centered in the discharge cavity away from its wall and thus protecting it from thermal damage. A third, and smaller, precision ground, concentric quartz tube, with an orifice diameter of 1.5 mm, is placed axially into the center of the torch as shown in Fig. 6.10 with its tip 1– 2 mm upstream of the end of the intermediate quartz tube. This central tube serves to introduce the aerosol stream of the material to be analyzed with an appropriate carrier gas.

The torch is normally operated in a laminar flow mode with its discharge oriented vertically upward. The outer plasma confinement tube is surrounded by a 2 or 3 turn, water-cooled, copper induction coil, which is connected to the high frequency power supply though an appropriate impedance matching box. Because of the relatively small dimensions of the torch and the need to respect the rule of keeping the skin depth less than half of the radius of the plasma load, this type of induction plasma torches is typically excited at a frequency of 13.7 or 26.3 MHz. Operation at higher frequencies of 40 MHz or more has also been reported in the literature. When operating with pure argon at atmospheric pressure in the open discharge mode, with an oscillator frequency of 26.3 MHz, the skin depth can be estimated as being 2.5 mm for an average discharge temperature of 8000 K. For a radius of the discharge in the coil region of around 6 mm, the corresponding ratio (r_n/δ_t) is around 2.4.

A photograph of this type of torch in operation with argon as the plasma gas is shown on the RHS of Fig. 10.10. This shows the position of the discharge region in relation to the coil and the wall of the plasma confinement tube. Also noticeable on this photograph is the central channel created through the injection of the aerosol carrier gas stream into the base of the discharge. As the aerosol droplets enter the discharge region, they evaporate and the released aerosol vapors are dissociated and excited giving rise to the typical emission pattern, which is at the base of the elemental analysis techniques using optical emission spectroscopy (OES) or mass spectrometry (MS). A vast literature exists on the design of such inductively coupled plasma (ICP) torches and their use for elemental analysis.

A number of other quartz-tube torch designs have also been reported in the literature for operation at considerably higher power levels. These consisted essentially of a cylindrical quartz tube, with diameters varying between 30 and 300 mm, air or water cooled, surrounded by an induction coil made of water-cooled copper tubing wound in the form of a helical coil. The number of coil turns used varied from 3 to 7, depending on the operating frequency of the RF power supply, the dimensions of the coil, and the nature of the plasma gas. One of the first commercially used induction plasma torches, developed in the sixties and early seventies by TAFA Corporation of Concord NH, is shown in Fig. 10.11 [TAFA (1966, 1968), Thorpe (1966, 19681970a, b), Dundas and Thorpe (1969)]. This torch is of a water-cooled, quartz tube construction. It was rated for operation with a wide range of plasma gases at power levels up to 100 kW. The internal diameter of the plasma confinement tube was in this case in the range of 45–65 mm depending on the torch model and its power rating. These torches operated with air, or pure oxygen, as the plasma gas, at power level in the 40–50 kW range, with an oscillator frequency of 2–3 MHz.

10.3.3 Segmented Metal Wall Torches

Segmented metal wall induction plasma torches were mostly developed in the early seventies in the former Soviet Union by Dresvin and his collaborators Dresvin (1977) at the St Petersburg Technical University. These torches are based on the use of standard, single or double flux, gas flow configuration in conjunction with single wall quartz, plasma-confinement tube. A segmented, water-cooled, metal shield, often referred in literature as a "metal cage," is introduced

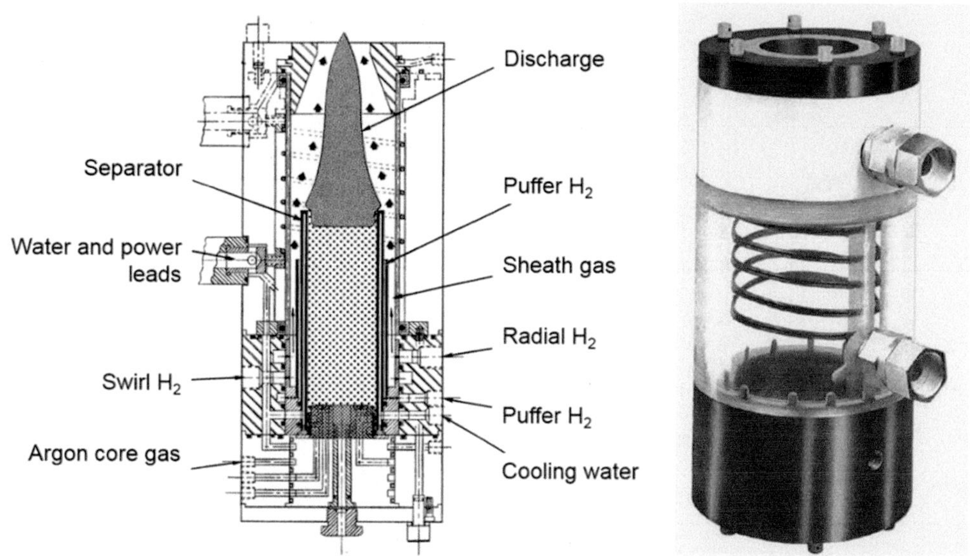

Fig. 10.11 Schematic and photograph of water-cooled, quartz tube, induction plasma torch developed in the seventies by M Thorpe, TAFA corp., [TAFA Bulletin (1968)]

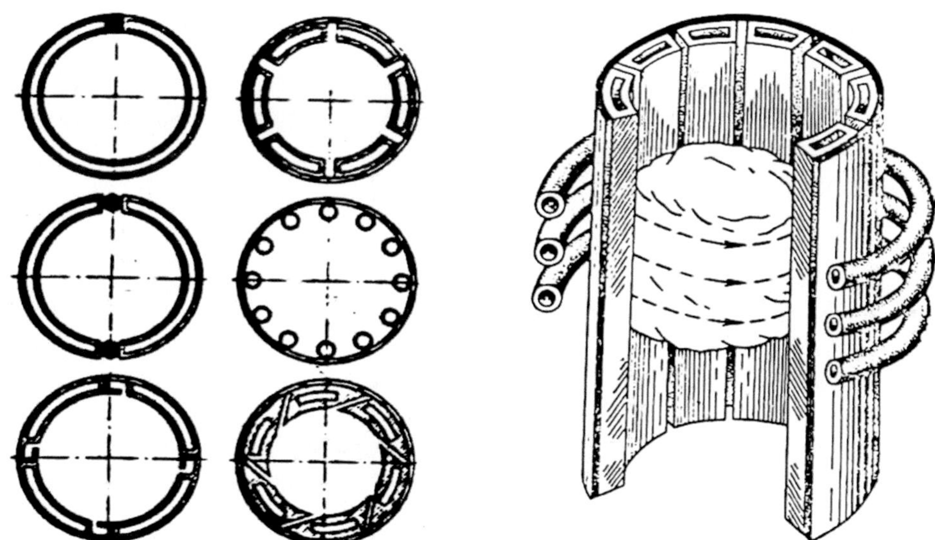

Fig. 10.12 Segmented metal wall induction plasma torches developed in the sixties in the former Soviet Union [Dresvin et al. (1977)]

into the plasma confinement tube in order to provide an adequate protection to the external quartz tube from the plasma discharge. Different designs were adopted with varying shapes and number of segments used as shown in Fig. 10.12, [Dresvin (1977)]. The principal feature of these segmented separators lies in the cooling water circuit design, which needs a continuous flow of coolant through all of the segments while avoiding the formation of a closed-loop electromagnetic coupling into the metal wall. The larger the number of segments, the more effective is the suppression of the induced currents in each segment and consequently the lower will be the energy loss in each segment. The way the

segments work is essentially through the suppression of the induced currents in any given segment through the action of the magnetic fields developed by the currents induced in the neighboring segments. The best results are obtained through the reduction of the gap between successive segments without compromising the electrical insulation between the segments. Figure 10.13 shows a schematic and a photograph of one of the largest segmented induction plasma torches built in the seventies by Dresvin and his collaborators for operation at nominal power levels up to 1 MW.

Variations of this design were developed by TAFA, and Los Alamos National Labs in the USA [Hollabaugh et al.

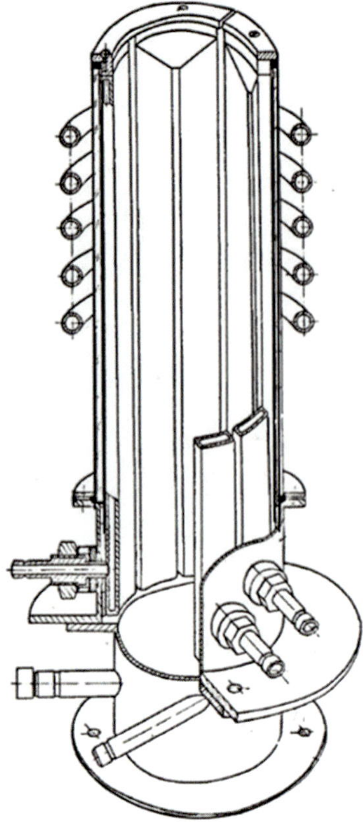

Fig. 10.13 Schematic and photograph of a 1 MW, segmented metallic wall, induction plasma torch [drawing and Photograph courtesy of S. Dresvin]

(1983)]. A number of segmented metal wall torch designs, shown in Fig. 10.14, were also developed in France by [Reboux (1971) and Galtier et al.(1973)], and were the subject of numerous investigations by Devely and his collaborators [Fouladgar et al. (1993)] at St Nazaire in France. These included both experimental and mathematical modeling of the discharge and of the induced currents in the segmented metal separator.

Globally the segmented metal wall design, while quite effective in reducing thermal stresses on the quartz plasma confinement tube, is more difficult to ignite and has a lower overall energy efficiency. The use of metallic components in contact with the discharge can also be a source of contamination of the plasma. The effect is particularly detrimental with the use of oxidizing plasma gases such as air or oxygen, or in the event of introducing corrosive vapors, such as halide-based chemical precursors, into the discharge.

10.3.4 Ceramic Wall Torches

Ceramic wall induction plasma torches make use of the availability of advanced ceramics for the replacement of the quartz plasma confinement tube. The principal properties to

look for in the ceramic material to be used are their inertness and high thermal conductivity combined with a high thermal shock resistance [Boulos and Jurewicz, (1992, 1993, 1996a, b, 1997, 1999, 2001)]. Tekna induction plasma torch design, schematically given in Fig. 10.15a with photograph of a standard Laboratory-scale (PS-50) torch in Fig, 10.15b, have the following distinct features:

- Use of a ceramic plasma confinement tube
- High velocity, (above 10 m/s) water-film cooling flowing in a narrow annular chamber surrounding the ceramic tube
- Induction coil cast in a polymer-matrix composite torch body for ease of assembly and precision alignment of all the torch component
- Water-cooled gas distributer head for precise control of the aerodynamics of the double flux gas flow into the discharge chamber with axial sheath gas and tangential central gas
- Ease of access for axial central introduction of the powder injection probe into the discharge cavity

This torch design allows for the internal axial injection of reactants or materials to be processed in the form of powder, liquid, or suspension into the center of the discharge cavity

Fig. 10.14 Segmented metal wall induction plasma torch [Reboux (1971)]

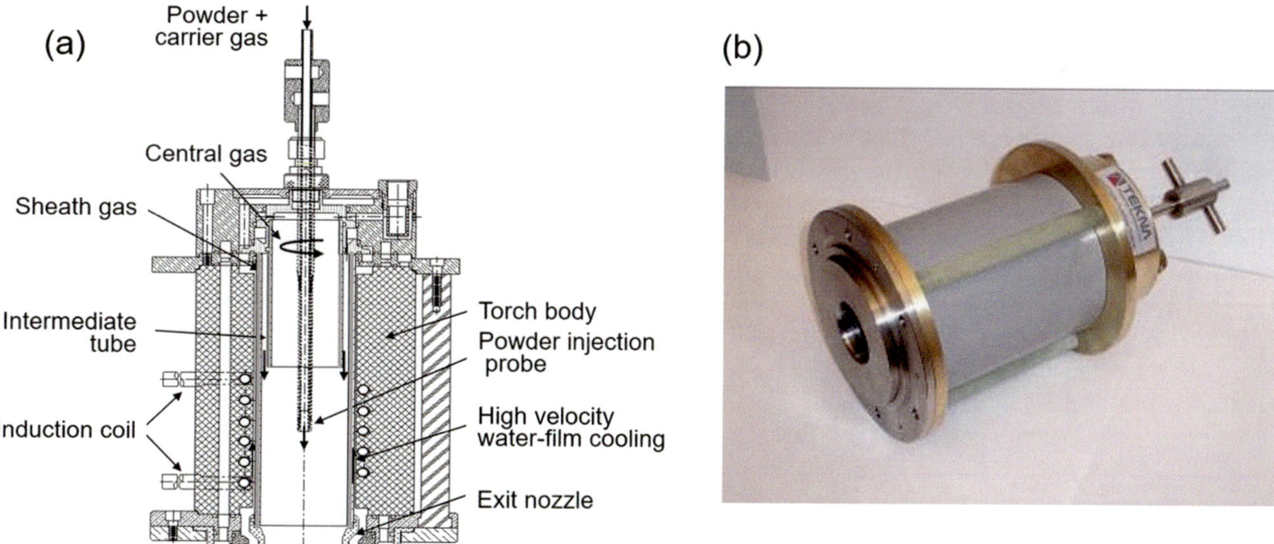

Fig. 10.15 (**a**) Schematic drawing and (**b**) photograph of the ceramic wall induction plasma torch [Boulos and Jurewicz Canadian patent 2,085,133 (1992) and US. Patent 5,200,595 (1993)] [courtesy Tekna Plasma Systems Inc.]

through the use of a water-cooled metallic injection probe. This insures for the proper dispersion and the intimate contact between the injected material and the plasma stream.

Because of the close interdependence between the physical dimension of the induction plasma torch, its nominal design power rating and operating frequency, a single

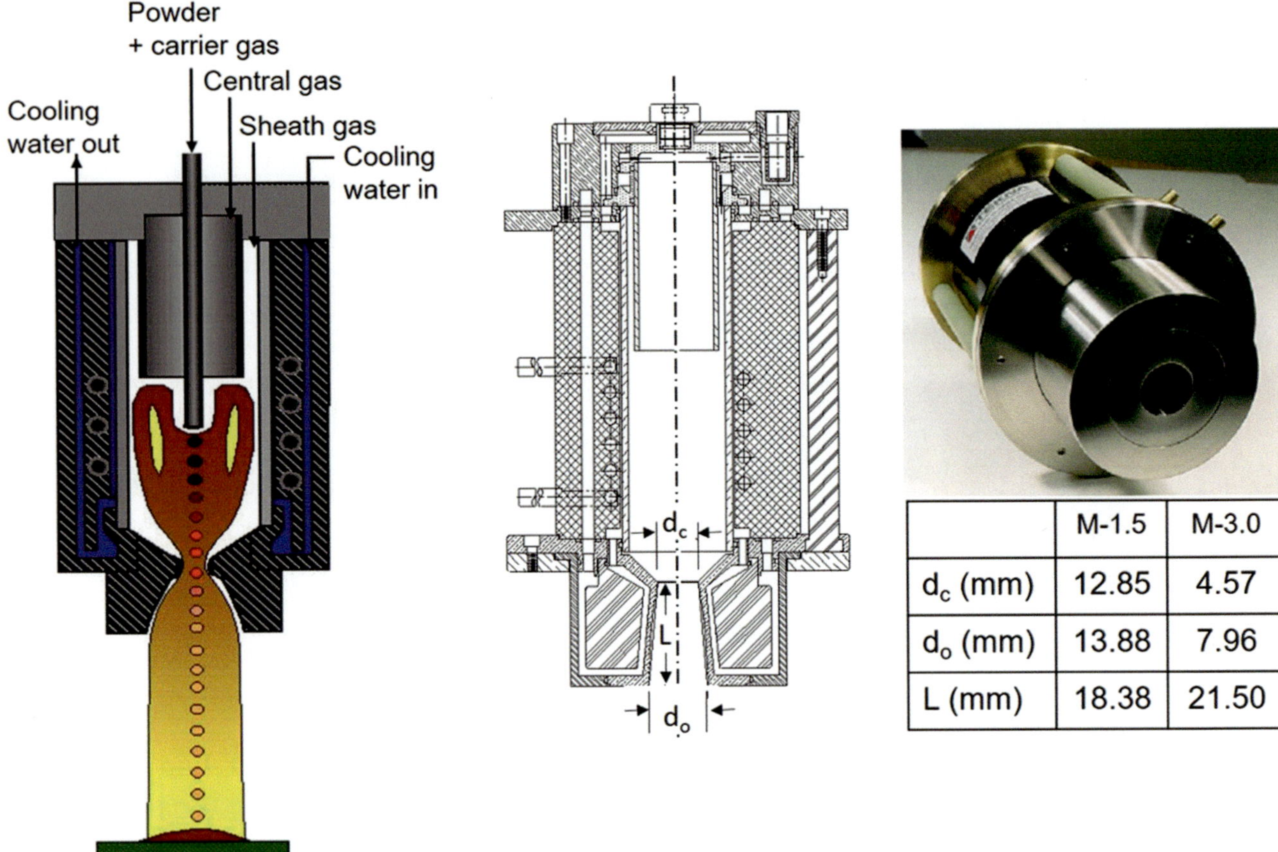

	M-1.5	M-3.0
d_c (mm)	12.85	4.57
d_o (mm)	13.88	7.96
L (mm)	18.38	21.50

Fig. 10.16 (**a, b**) Schematic drawings and (**c**) photographs of Tekna PS-50 induction plasma torch with Laval-type nozzle for operation at supersonic flow conditions over the Mach number range of 1.5–3.0 [courtesy Tekna Plasma Systems Inc.]

induction plasma torch design tends to have a relatively narrow range of operating power. Tekna Plasma Systems Inc. developed a complete line of RF-ICP torches identified by the internal diameter of their plasma confinement tube. For example, Model PS-35 would correspond to a torch with a 35 mm i.d. ceramic tube, rated for operation over the power range of 5–45 kW and a frequency of 3–4 MHz. The next size up would be the PS-50 for power rating of 20–70 kW, PS-70, for 30–120 kW and PS-100 for 50–200 kW. Obviously, these power ratings are nominal, subject to change as the technology evolves, and function of the torch operating conditions in terms of gas composition, gas flow rates and discharge pressure.

The integration of a front-mounting flange into the torch design, with an exchangeable discharge nozzle, allows for the mounting of the torch on the top of plasma spraying chamber, and the control of the plasma flow characteristics at the exit of the torch. An example of a Laval-type nozzle, with its principal dimensions for operation at supersonic velocities in the Mach number range of 1.5–3, is shown in Fig. 10.16.

10.3.5 Multi-coil Plasma Torches

Multi-coil induction plasma torches were developed with the specific objective of allowing for the scaling-up of the induction plasma technology to the hundreds of kW power range. The critical issue in this case was to allow for the use of high power, solid state, RF power supplies. Such power supplies, which are currently available for operation at power ratings up to the MW range, are characterized by their low frequency range of operation (100–500 kHz). They are also low voltage and high current devices, which require massive induction coil structure to allow for high power operation. The multi-coil induction plasma torch design [Boulos and Jurewicz (2004, 2005)] has allowed for the optimal matching of the coil impedance with the electrical characteristic of the solid-state power supply. This was achieved through the use of multiple coils (up to three or four independent coils) interconnected in series–parallel combination in order to achieve any required equivalent coil impedance, including fractional coil-turns, which could not be achieved otherwise. A schematic representation and a photograph of a Tekna PD-100 multi-coil induction plasmas torch are shown in Fig. 10.17.

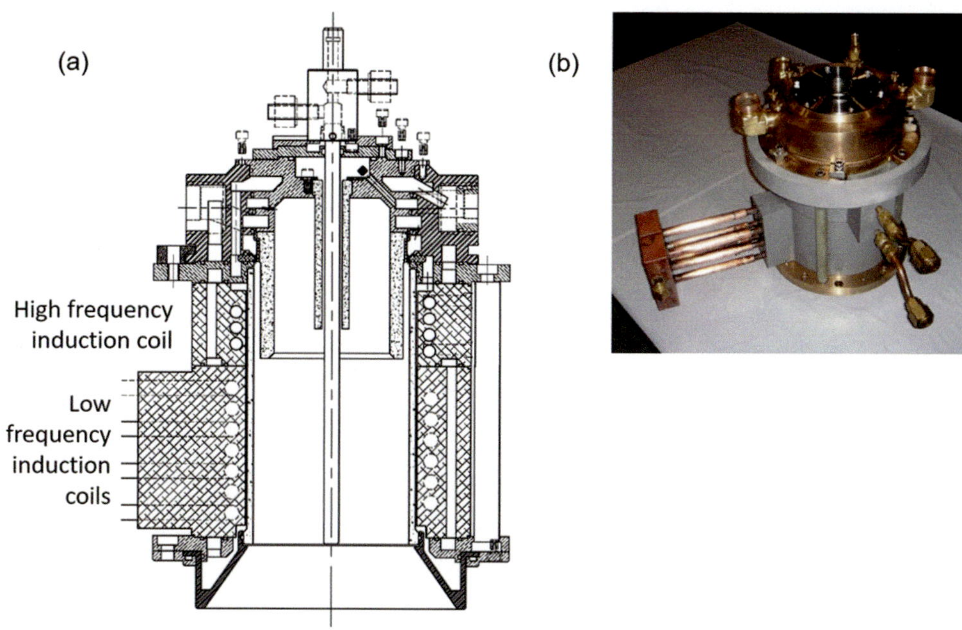

Fig. 10.17 (**a**) Schematic drawing and (**b**) photograph of a PD-100 multi-coil induction plasma torch [Boulos and Jurewicz (2004, 2005)] [Photograph courtesy Tekna Plasma Systems Inc.]

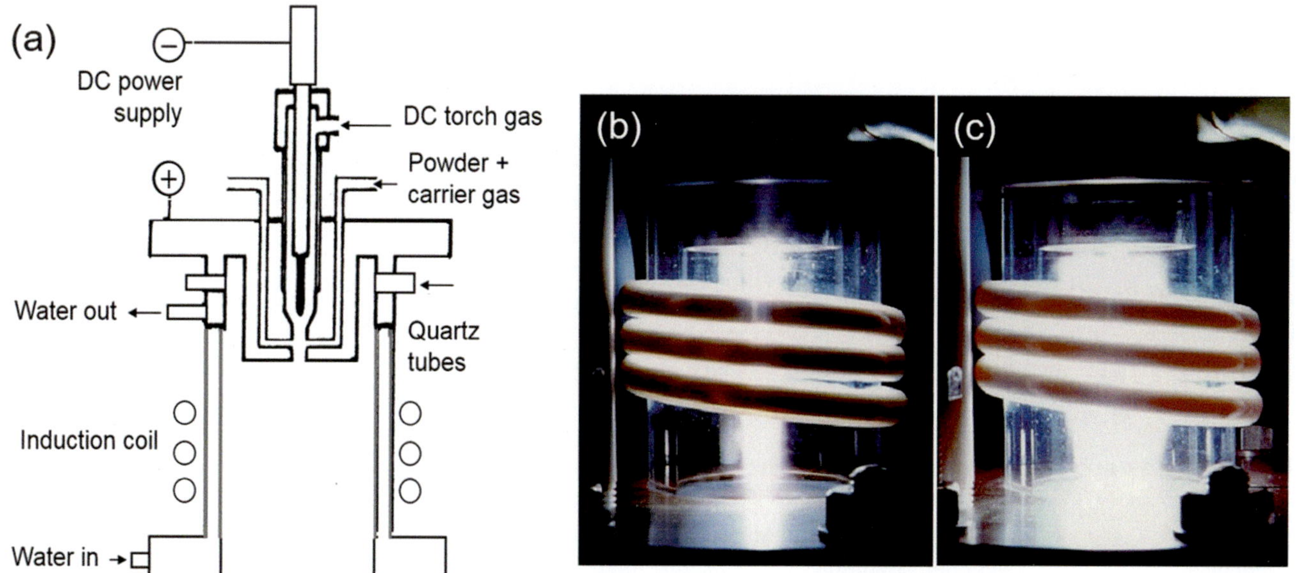

Fig. 10.18 Schematic representation and photographs of DC/RF hybrid plasma torch, [Yoshida et al. (1981)] [Photographs courtesy of the University of Tokyo]

10.3.6 DC-RF Hybrid Plasma Torches

The concept of a hybrid induction plasma torch was developed in the early eighties by Yoshida and Akashi at the University of Tokyo in Japan [Yoshida et al. (1981, 1983) and Uesugi et al. (1988)]. Schematic and photographs of a DC/RF hybrid plasma torch given in Fig. 10.18 show a typical hybrid torch design, in which a DC plasma torch is co-axially mounted at the top of an RF induction plasma torch. The plasma jet from the DC torch flows axially into the center of the discharge cavity of the induction plasma torch. The applied RF electromagnetic fields generated through the induction coil could then couple into the partially conducting gas and provide the necessary energy to increase the volume of the discharge that filled the discharge cavity defined by the water-cooled, quartz or ceramic plasma confinement tube. Photographs provided by Professor Yoshida of Tokyo University, Fig. 10.18b and c, show,

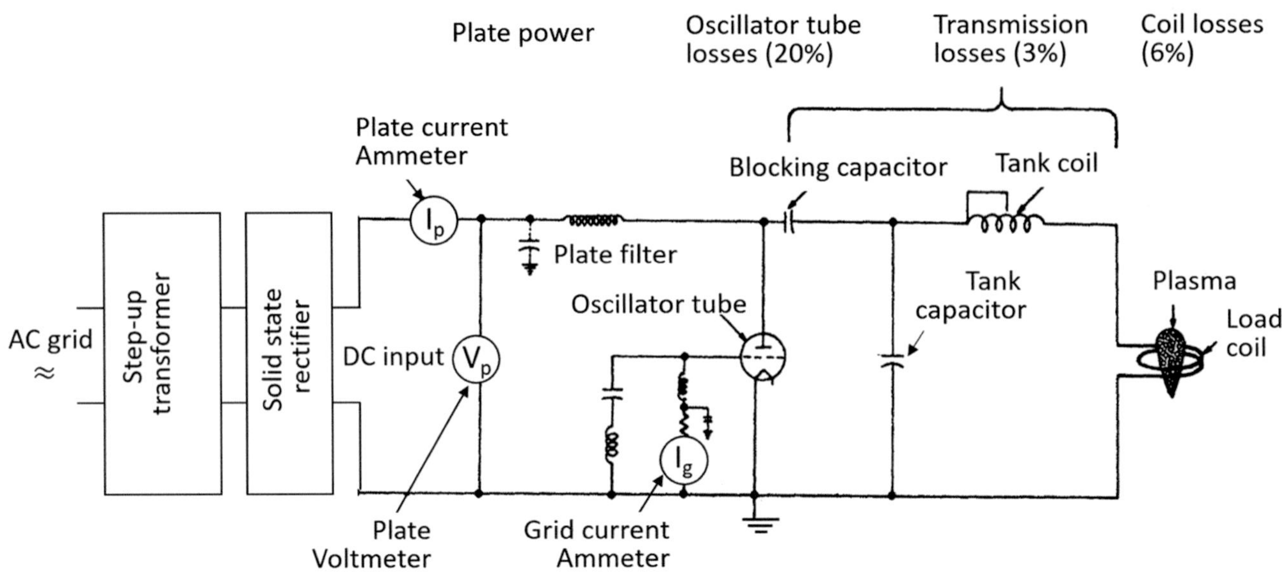

Fig. 10.19 Basic components of a "vacuum oscillator tube" RF powder supply

respectively, the discharge in the presence of the DC plasma plume only, and that in the presence of the combined effects of the DC and RF inductively coupled discharges.

Numerous studies reported by [Yoshida et al. (1983)] were devoted to the study of the characteristics of this type of discharge that was successfully used in a large number of applications in the area of material processing, plasma spraying, and the synthesis of ultrafine powders. Systematic efforts were devoted to the modeling of these discharges in order to identify their principal characteristics and fluid dynamic conditions. Generally, these were limited to operation in the power range of 30–100 kW with the DC power typically around 10–15% of the total power applied to the discharge.

10.3.7 RF Power Supply and Energy Balance

One of the first steps in the operation of a thermal plasma system is to optimize its operating conditions and system tuning in order to insure the maximum conversion of the electrical energy from the grid into plasma energy. Figure 10.19 shows a block diagram of a typical RF induction plasma power supply using a "vacuum tube oscillator" technology.

The overall electrical circuit consists essentially of the following five major components:

- Step-up transformer that serves to convert the line voltage of the supply grid into a high voltage, AC current at the same frequency as that of the supply grid (50 or 60 Hz)
- Solid-state rectifier that converts the current from AC to DC

- Oscillator circuit built around a "vacuum-tube" or "solid-state" oscillator to convert the current from high voltage DC to high voltage AC at the resonant frequency
- Tank circuit that serves to transmit the power from the oscillator circuit to the load coil in the plasma torch
- The induction plasma torch, an integral part of the oscillator circuit, in which the energy is transferred from the electric circuit to the plasma through electromagnetic coupling

Each of these transformations is associated with different levels of energy losses. The first two transformations of the energy form the supply grid to high voltage DC at the exit of the solid-state rectifier are generally achieved with a relatively high efficiency and less than 5% of energy losses in the circuit. For this reason, and in consideration that the DC plate voltage, V_p, and plate current, I_p, are commonly used as an indication of the power input to the oscillator circuit. The plate power, P_{pt}, being defined as:

$$P_{pt} = I_p \times V_p \tag{10.4}$$

Oscillator tube energy losses, on the other hand, represent the most important single energy loss factor in the circuit, which can account to close to 20–25% of the DC plate power. While this percentage changes slightly with the proper tuning of the oscillator circuit, it is very much inherent to the basic design of vacuum tube oscillators. Recent developments in RF power supply design has evolved toward the replacement of the traditional "vacuum tube oscillator" by a solid-state circuit. This is generally accompanied by a significant

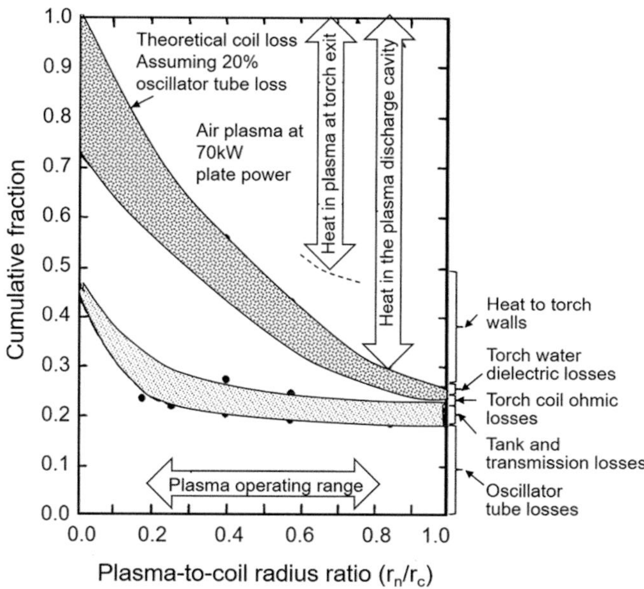

Fig. 10.20 Energy distribution in a typical RF plasma torch [TAFA (1966)]

reduction of the oscillator circuit losses down to possibly 5–10%, which would have a significant impact on overall energy efficiency of the system. Unfortunately, at the moment, the response time of some of the major semiconductor components in such circuits limits their oscillator frequency to the hundreds of kHz range up to 1 MHz. For this reason, standard RF induction plasma systems still rely on the traditional "vacuum-tube-oscillator," which is robust and has been successfully in operation for many decades.

On the RF part of the power supply circuit, on the RHS of Fig 10.19, energy losses are due to transmission line and coil losses, which can account respectively to 3 and 6% of the plate power. To this of course one has to add electromagnetic coupling losses between the induction coil and the plasma and finally thermal losses in the plasma torch. These depend on the particular design of the plasma torch and the overall tuning of the oscillator and transmission circuits.

An energy distribution diagram for a typical TAFA induction plasma system including a TAFA torch, and Lepel RF power supply, is given in Fig. 10.20. [TAFA (1966, 1968)]. This gives the different energy losses in the circuit as function of the ratio of the plasma-to-coil radius ratio. The data are consistent with the analysis given earlier by Mensingen and Boedecker (1968) in Sect. 10.2.2 (Fig. 10.7), which shows a strong dependence of the energy coupling efficiency on the coupling parameter, $\kappa_c = \sqrt{2}\,(r_n/\delta_t)$ and the ratio of the plasma load to coil radii (r_n/r_c). The results given in Fig. 10.20 show that at near optimal coupling conditions with an (r_n/r_c) ratio between 0.7 and 0.9 about 65% of the plate power is coupled into the discharge. The overall energy efficiency of the system drops considerably as the plasma-to-

coil radius ratio decreases below 0.7 with a considerable increase in RF transmission losses. For this reason, it has generally been accepted, when using the induction plasma torch for material processing, that it is important to have the proper match between the torch dimensions, its power rating and plasma gas composition in order to maintain the (r_n/r_c) ratio as close as possible to 0.7 or 0.8. By going the extra step of injecting the material to be processed into the center of the discharge cavity would ensure better use of the available energy coupled into the discharge. Under optimal conditions, this could account to almost 65% of the plate power, with a significant gain of processing efficiency compared to downstream injection of the material to be processed into the plasma jet at the torch exit as is commonly done with DC plasma sources.

In the following energy balance, data are given as "typical" for the operation of induction plasma torches under different conditions. All the data are for double flux, ceramic tube torches mounted on a material spraying chamber in the presence of a central, water-cooled, powder injection probe. Since no powder was being processed in most of these cases, no carrier gas was introduced through the central probe. The data comprise three principal sets of information:

- The plasma operating conditions as identified by the flow and composition of the different gas streams (Sheath Q_{sh}, central Q_{ce}, and probe Q_{pb}) and the chamber pressure, p_a.
- The power supply setting as identified by the oscillator frequency, f, plate current, I_p, plate voltage, V_p, and grid current, I_g.
- Energy recovery system as defined by power recovered in the water-cooling circuit of the power supply, P_{gen}, the torch, P_{tor}, the probe P_{pb}, and the reactor P_{reac}. Each of these parameters is calculated by the standard energy balance Eq. (10.5):

$$P_i = m_i c_p \left(T_{out} - T_{in}\right) \qquad (10.5)$$

Where,

P_i power recovered in cooling-water stream i
i dummy variable standing for the cooling water streams
m_i cooling water flow rate (kg/s)
c_p specific heat of water ($c_p = 4.18$ kJ/kg °C)
T_{out} exit temperature of cooling water
T_{in} inlet temperature of cooling water
The energy coupled in the plasma, P_c is given by:

$$P_c = P_{tor} + P_{pb} + P_{reac} \qquad (10.6)$$

The total energy recovered in the cooling circuit of the system, P_w is given by:

Table 10.1 Energy balance for a PS-70 Tekna induction plasma torch operating at near atmospheric pressure using Ar/H_2 mixture as plasma gas (8.3 vol% H_2) and a frequency of 3 MHz [Data courtesy of Tekna Plasma Systems Inc.]

Gas supply					Reactor pressure	Power supply				Energy recovered					η_c	η_o
Q_3		Q_2	Q_1	Total		V_p	I_p	I_g	P_{pt}	P_{gen}	P_{tor}	P_{pb}	P_{reac}	P_w		
Slm Ar	Slm H_2	Slm Ar	Slm Ar	Slm Ar/H_2	kPa	kV	A	A	kW	kW	kW	kW	kW	kW	%	%
162	18	35	0	215	82.4	9.2	7.5	1.2	**68.5**	38.3	23.4	2.9	21.0	100.3	**69%**	**55%**
162	18	35	0	215	82.4	10.0	8.5	1.3	**84.9**	46.8	30.7	3.3	27.2	102.2	**72%**	**57%**
162	18	35	0	215	106.5	10.3	8.7	1.2	**89.3**	49.2	28.2	3.3	30.1	126.7	**69%**	**56%**
162	18	35	0	215	106.5	10.7	9.2	1.2	**98.5**	56.0	31.3	3.3	32.9	127.6	**69%**	**55%**

$$P_w = P_{gen} + P_c \tag{10.7}$$

The energy coupling efficiency is calculated as the ratio of the energy coupled into the plasma, P_c, divided by the plate power P_{pt}:

$$\eta_c = P_c/P_{pt} \tag{10.8}$$

The overall energy efficiency of the system is given as the ratio of the energy coupled into the plasma, P_c, divided by the total energy recovered in the cooling circuit, P_w, which is equivalent to the energy drawn from the electrical grid.

$$\eta_o = P_c/P_w \tag{10.9}$$

The first set of data is given for a Tekna PS-70 induction plasma torch with a 70 mm internal diameter of the plasma confinement tube. The oscillator frequency was around 3 MHz, and the gas flow rates through the torch were set as follows:

Sheath gas $Q_{sh} = 162$ slm (Ar) $+18$slm (H_2)
Central gas $Q_{ce} = 35$ slm (Ar)
Probe gas $Q_{pb} = 0$
Hydrogen concentration in sheath gas $= 10.0$ vol%
Overall hydrogen concentration in the plasma gas $= 8.3$ vol%

Results are given in Table 10.1 for reactor pressures of 82.4 and 106.5 kPa. These show negligible effects of the reactor pressure, or the plate power, on the system performance. The energy coupling efficiency is around 70% of the plate power, while the overall energy efficiency of the system is around 55%. The latter represents the ratio of the net energy coupled into the plasma to the total energy taken from the supply grid. The value given is on the low side, due to relatively high-energy losses in the power supply in this case. A typical value of the overall energy efficiency for this type of systems is normally around 65%. It may also be noted that of the energy coupled into the discharge, about 50% is lost to the walls of the plasma confinement tube and the central water-cooled probe. Overall, only 30% of the plate power is delivered to the plasma jet at the exit of the torch. As mentioned earlier, this justifies the internal axial injection approach, where the material to be processed is injected in the discharge cavity, usually around the center of the induction coil.

10.4 Gas and Particle Dynamics

In this section, a review is presented of the principal characteristics of RF inductively coupled plasma (RF-ICP). The topics discussed are the electrical and magnetic field characteristics, temperature, flow, and concentration fields under typical operating conditions. The analysis presented here is based on both measurements and mathematical modeling studies. No details of the measurement, or computational, techniques used will be discussed since these are covered in corresponding chapters of Part I of this book. Emphasis is placed on the analysis of typical results obtained and their interpretation as a means of developing a fundamental understanding of the basic phenomena involved.

10.4.1 Electrical and Magnetic Fields

The first measurements of the electrical and magnetic fields in an RF-ICP were reported in the seventies by Eckert (1971, 1972). These were carried out using a 3 mm diameter water-cooled probe inserted into a relatively large-volume, low-power density discharge. The discharge was confined in this case by a water-cooled quartz tube, 155 mm in i.d. The oscillator frequency was 2.6 MHz, and the applied RF power was estimated at about 25 kW. The discharge was of pure argon at atmospheric pressure with a total argon flow rate into the torch of 120 slm. Figure 10.21 shows variation of the RF magnetic field across the discharge at the middle level of the inductor coil. Curve (a) corresponds to the axial magnetic field in the coil region in the absence of the discharge, while curve (b) corresponds to the case after ignition of the discharge. In the absence of the plasma, the axial magnetic field is relatively uniform with the field decreasing slightly around the axis of the coil, as one would expect for a "short" inductor coil. On the other

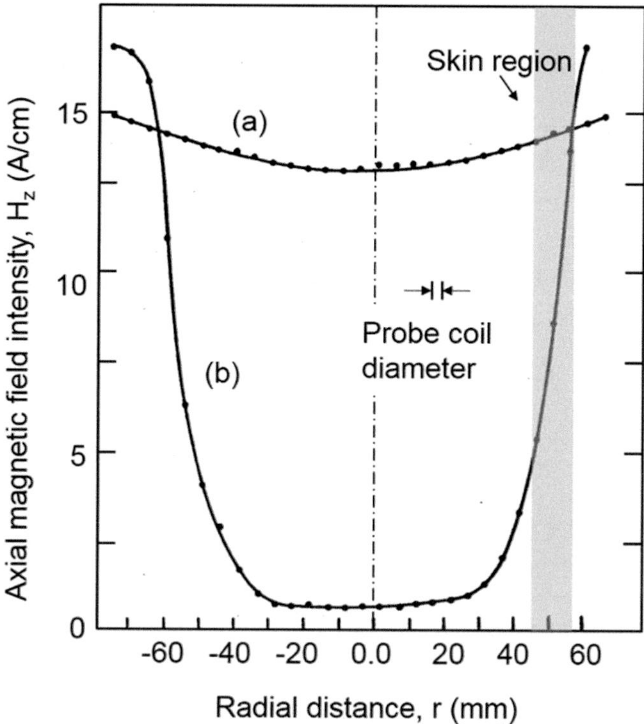

Fig. 10.21 Radial profiles of the magnetic field intensity at the mid-level of the induction coil (**a**) in the absence (**b**) and presence of the discharge [Eckert (1971, 1972)]

hand, in the presence of the plasma, the field drops steeply in the central region of the discharge to around 5% of its value outside the discharge region. The skin depth, δ_t, as identified in Fig. 10.21, is estimated to be of the order of 10 mm, which is consistent with its estimate using Eq.10.1 with μ_o the permeability of vacuum ($\mu_o = 4\pi\ 10^{-7}$ Hy/m), and $\sigma_o = 10$ mho/cm, corresponding to the electrical conductivity of argon at atmospheric pressure at 8000 K. Outside the discharge region, in the annular space between the plasma column and the wall of the plasma confinement tube, the field increases by about 10% over its value in the absence of the discharge (curve a).

In a subsequent publication, [Eckert (1972)] extended his magnetic field measurement technique, using a dual magnetic probe system, to phase angle measurements of the magnetic field profile. By keeping one of the probes in a fixed position near the plasma boundary and traversing the plasma with the second probe, it was possible to obtain information on the magnetic and electrical fields, and the electrical conductivity profiles in the discharge. Results reported for pure argon plasma at atmospheric pressure, in a 140 mm, i.d., plasma-confining quart tube are presented in Fig. 10.22. The measurements were carried out for an oscillator frequency of 2.6 MHz and an RF power of 25 kW. The data are expressed in terms of the electric field $|E|$ (V/cm), and current density $|j|$ (A/cm^2), and electrical conductivity

profiles σ (mho/cm) in the middle section of the inductor. In contrast to the monotonic increase of $|E|$ with r, $|j|$ passes through a sharp maximum at $r = 58$ mm, which amounts to about 0.8 of the radius of the plasma confinement tube ($r_o = 70$ mm). These two curves could be used to calculate the radial profile of the local energy dissipation, $P(r)$, into the discharge as:

$$P(r) = |j| \times |E| \qquad (10.10)$$

Superimposed on this figure are two vertical dotted lines, which identify the region of effective energy dissipation into the discharge corresponding to the "skin region" or "skin depth" defined in Sect. 10.2.2 and which is estimated to be about 10 mm.

From the ratio of the ($|j|$ / $|E|$), Eckert evaluated the radial variation of the electrical conductivity of the plasma, σ, which is also plotted in Fig. 10.22. The results show that σ increases from a value of about 600 mho/m on the axis of the discharge to a maximum of 1150 mho/m at $r = 55$ mm, beyond which the electrical conductivity drops sharply in the boundary layer between the plasma and the plasma confinement tube. This kind of variation reflects the circular doughnut shape of the energy dissipation zone in the discharge region and is qualitatively consistent with

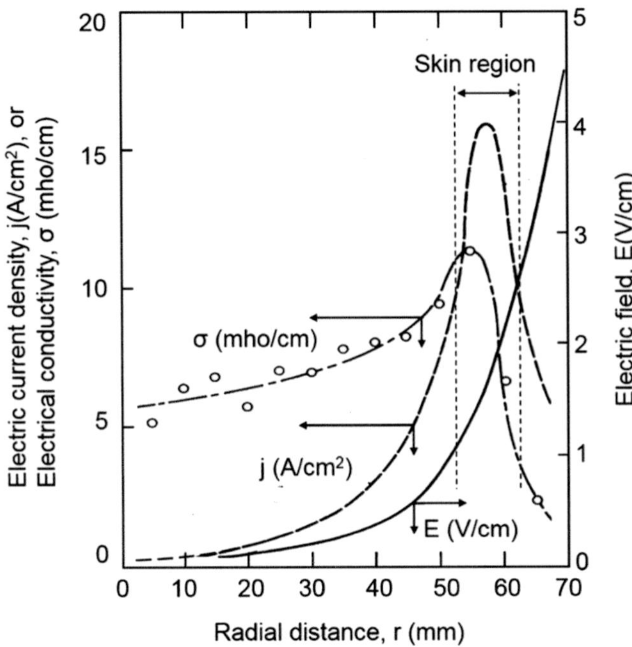

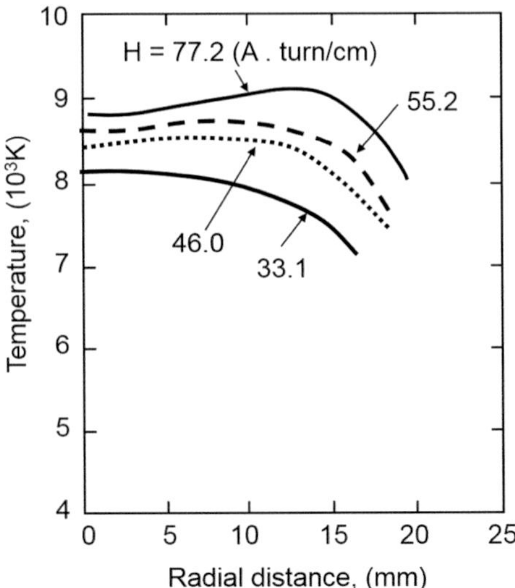

Fig. 10.23 Radial temperature profiles at the center of the induction coil for an argon discharge at atmospheric pressure for different axial magnetic field intensities [Douglas (1974)]

Fig. 10.22 Radial distribution of the magnitude of the induced electric field, E, current density, j, and electrical conductivity, σ, in an argon induction discharge, $f = 2.6$ MHz, $P_o = 25$ kW [Eckert (1972)]

observation revealed by the end view photographs of the discharge shown in Fig. 10.5b.

It should be pointed out that the decrease of the magnetic flux density in the center of the discharge compared to that in the absence of the discharge, as shown in Fig. 10.21, is a direct consequence of the interaction between the applied magnetic field, and the magnetic field generated by the current loop induced in the discharge. Throughout each oscillator half cycle, these two fields, the applied field and the induced field, are opposing each other in the center of the discharge while they act in the same direction outside the discharge giving rise to the observed suppression of the magnetic field in the center of the discharge, and the increase of the field strength outside the discharge region compared to its value in the absence of the discharge.

10.4.2 Temperature Fields

Numerous studies have been reported dealing with the measurement of the temperature field in an RF-ICP. One of the first of such reports by Reed (1961a) made an error in the interpretation of the emission spectroscopic data that gave rise to temperature mapping with a maximum temperature of 25,000 K on the axis of the discharge. The error was rapidly recognized and attributed to the annular shape of the energy dissipation zone in the discharge giving rise to an off-axis maximum in the temperature profiles that was later confirmed

by all subsequently reported data and modeling results. In this section, a few examples of typical temperature field measurements are reported with the specific objective of revealing the variation of the temperature field in the discharge with the intensity of the magnetic field and the nature of the plasma gas.

Douglas (1974) in the early 1970s reported measurements of the radial temperature profiles in the middle section of the induction coil. These were made using emission spectroscopy on a 45 mm i.d. quartz-wall, inductively coupled discharge operated at atmospheric pressure, with pure argon as the plasma gas at an oscillator frequency of 2 MHz. Typical results obtained at different intensities of the applied magnetic fields are given in Fig. 10.23. It may be noted that they all show the typical off-axis maximum, which is a direct consequence of the annular mode of energy dissipation into the discharge region. It is important to note that a significant increase in the intensity of the applied magnetic field results only in a slight increase in the temperature of the plasma. The radius of the discharge as estimated by the outer limit of the temperature profiles, however, increases from about 16 to 20 mm. This would correspond to an increase in volume of the discharge by more than 50% for a doubling of the applied magnetic field. The maximum reported temperature in the discharge in this case remains in the 8000–9000 K range.

Dresvin et al. (1977) reported a compilation of numerous measurements of the temperature field in RF-ICP. Typical results for pure argon induction plasma jets emerging from a 14 mm i.d. segmented, water-cooled, plasma torch are presented in Fig. 10.24. These were obtained for a generator frequency of 17 MHz and an RF power in the range of 6–8 kW at different total argon gas flow rates. The temperature

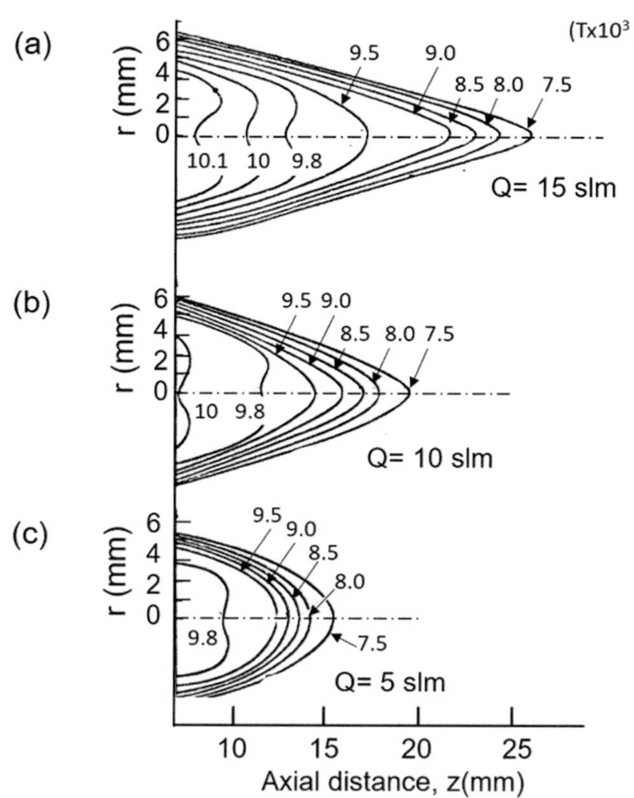

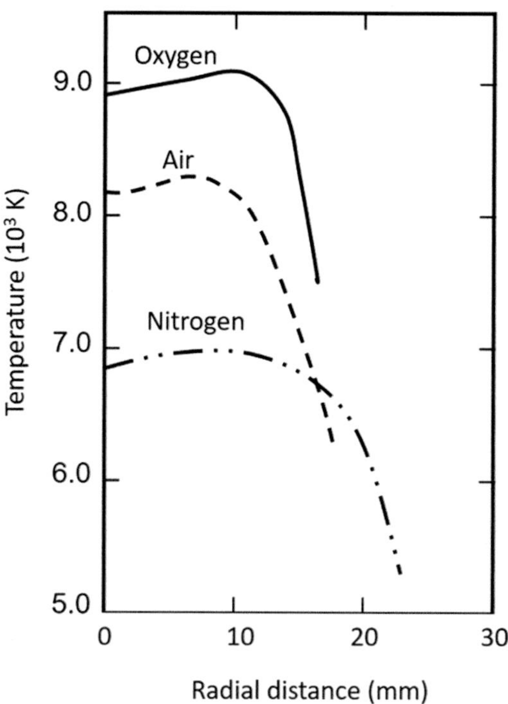

Fig. 10.24 Two-dimensional temperature mapping of the plasma jet emerging from an atmospheric pressure RF-ICP. Plasma confinement tube diameter = 14 mm, oscillator frequency = 17 MHz, RF power = 6–8 kW. Total Argon gas flow rate (a) 15 slm, (b) 10 slm, and (c) 5 slm [Dresvin et al. (1977)]

Fig. 10.25 Radial temperature profiles, at the middle section of the induction coil, for atmospheric pressure inductively coupled discharge with different molecular gases [Dresvin et al. (1977)]

mappings are presented as 2D iso-contours with the temperature levels indicated in 1000 K. The data show an increase in the length of the plasma jet with the increase of the total plasma gas flow rate. No significant influence of the plasma gas flow rate on the temperature field in the center of the discharge is observed.

Radial temperature profiles in the center of the discharge, and 2D temperature mappings for plasma of molecular gases generated by an inductively coupled discharge are given in Figs. 10.25 and 10.26, respectively. The radial profiles given in Fig. 10.25 show that, for a given power input, there is a strong dependence on the maximum temperature, and the size of the discharge region, on the nature of the plasma gas. Higher centerline temperatures ($T \approx 9000$ K) are noted for oxygen plasma compared to that for an air discharge ($T \approx 8200$ K). The lowest centerline temperature reported in this case is for nitrogen plasma ($T \approx 7000$ K).

Two-dimensional temperature mappings for atmospheric pressure argon and oxygen plasma under comparable operating conditions are given in Fig. 10.26. The total gas flow rate was set in each of these two cases at 30 slm (Ar) or (O_2). The oscillator frequency was 12 MHz and the RF power input 12 kW for the case of the Argon plasma, and 8 kW for

the oxygen plasma. These graphs show relatively large volume discharges with a relatively uniform temperature field in both the radial and axial directions. The power density in the discharge can be estimated in this case by dividing the applied plate power by the volume of the discharge within the coil region. The latter is calculated assuming the radius of the discharge as being about 80% of the radius of the plasma confinement tube, and a length of the discharge as being equal to the length of the induction coil. The power densities for these two cases are noted to be 0.102 and 0.068 kW/cm^3, respectively. For comparison, the corresponding power densities for the Tekna PS-35 and PS-70 torches are, respectively, 0.144 and 0.072 kW/cm^3. It should be underlined that these estimates are only approximate since they depend on the identification of the radius of the discharge, which can vary with the composition of the torch sheath gas, the operating pressure, the oscillator frequency, and the total applied plate power.

10.4.3 Flow Fields

The flow field in an inductively coupled RF discharge is rather complex due to the interaction between the applied magnetic field and the induced electric fields in the discharge. Chase (1969, 1971) was among the first to recognize that the resulting $\left(\vec{J} \times \vec{B} \right)$ electromagnetic forces, which act in the

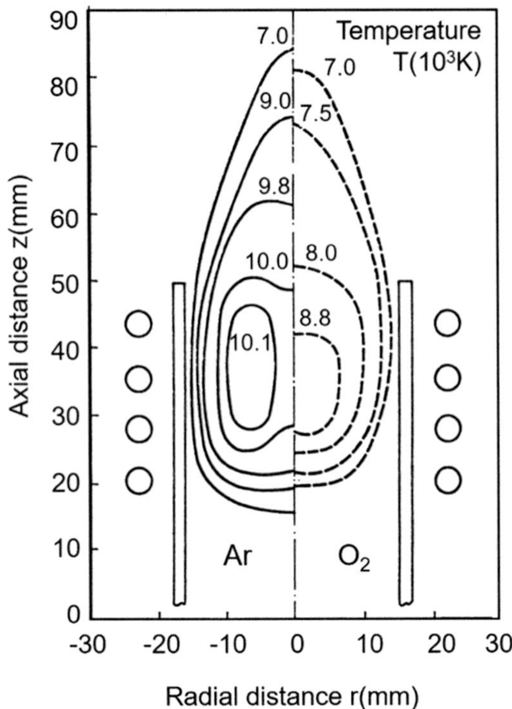

Fig. 10.26 Two-dimensional temperature mapping of inductively coupled, atmospheric pressures, discharges of argon (left) and oxygen (right) under comparable conditions. Total gas flow rate = 30 slm, oscillator frequency = 10 MHz, power = 12 kW for argon plasma and 8 kW for oxygen plasma. Temperature values are given in 1000 K [Dresvin et al. (1977)]

radial direction at the middle section of the coil, are responsible for the creation of a high-pressure zone in the center of the discharge. The magnetically produced pressure, p_{mo}, was calculated from:

$$p_{mo} = \int_{r=r_o}^{r=0} j_{\vartheta}.B_z \cos(\phi_E - \phi_H)dr \qquad (10.11)$$

Where, B_z is the magnetic flux intensity in the axial direction, j is the azimuthal current density, $cos(\phi_E - \phi_H)$ is the cosine of the difference in phase angle between the electric and magnetic fields. p_{mo}, j_{ϑ}, B_z are the time-averaged values (rms) of the parameters.

Calculation and probe measurements of the magnetically induced pressure inside the discharge indicated that it was of the order of 100 Pa for a 28 mm i.d. quartz tube plasma torch (Fig. 10.27a) operating in the plate power range of 2–8 kW. This gave rise to back flow velocities on the upstream end of the discharge of the order of 5 m/s. Further observation of the reversed flow caused by the electromagnetic pressure in the discharge was facilitated using a similar air-cooled, quartz tube torch except for the use of a flat, pancake-type induction coil (Fig. 10.27b), which allowed for the visual observation and the high speed photography of the motion of alumina

tracer particles of a diameter of 10 μm injected on the upstream end of the discharge. The photographs, shown in Fig. 10.27c and d, obtained using pure argon as plasma gas (16 slm) and a plate power of 6.9 kW, present tangible evidence of this observation that has been subsequently systematically observed in numerous experimental studies and supported by 2- and 3-dimensional modeling of the flow and temperature fields in the discharge as further discussed in the next section of this chapter.

Further study of the flow and temperature fields in an inductively coupled RF discharge was reported by Klubnikin (1975), who proposed the flow pattern given in Fig. 10.28a for two different flow conditions. The corresponding temperature fields in the discharge are shown in Fig. 10.28b. In each of these two figures, the LHS identified as (a-1) or (b-1) corresponds to operation with a low flow rate of the plasma gas into the discharge (5 slm, Ar), while the right-hand side, of each of the figures, identified as (a-2) and (b-2) corresponds to a high plasma gas flow rate operation (40 slm, Ar).

It may be noted that the flow pattern in the center of the discharge is characterized by an electromagnetically induced radial flow toward the center at the middle section of the induction coil. This effect is observed in both the low and high flow cases. It may also be observed in the low flow case (a-1), the pumping effect is strong enough to entrain gases from the downstream region of the torch, which move upstream along the inner wall of the plasma confinement tube. With the increase of the total gas flow rate into the torch, case (b-1), the discharge is slightly shifted downstream and the radial pumping action in the center of the coil is no longer capable of entraining gas from the downstream end of the torch. The only reverse flow that is maintained in this case is that along the centerline of the discharge on the upstream end of the induction coil. The corresponding temperature contours given on the RHS of Fig. 10.28b reveal a shift in the position of the discharge with the increase of the total plasma gas flow rate. This schematic is consistent with the observation that the position of the discharge in the torch cavity is a result of a delicate balance between the drag forces on the discharge by the plasma and sheath gas flow, and the stabilizing forces created by the electromagnetic pumping. The stability of the discharge is also affected by the swirl velocity component, which is often induced into the flow pattern of the gases introduced into the plasma torch (central gas). It should be mentioned that increasing the flow rate of the plasma gas into the torch can and will eventually lead to complete extinction of the discharge. The critical total plasma gas flow rate, which can be tolerated by an induction plasma torch, will depend on the dimension of the torch, the oscillator frequency, and the energy coupled into the discharge.

It should be pointed out that the observation of the reversed flow on the upstream side of the discharge is a

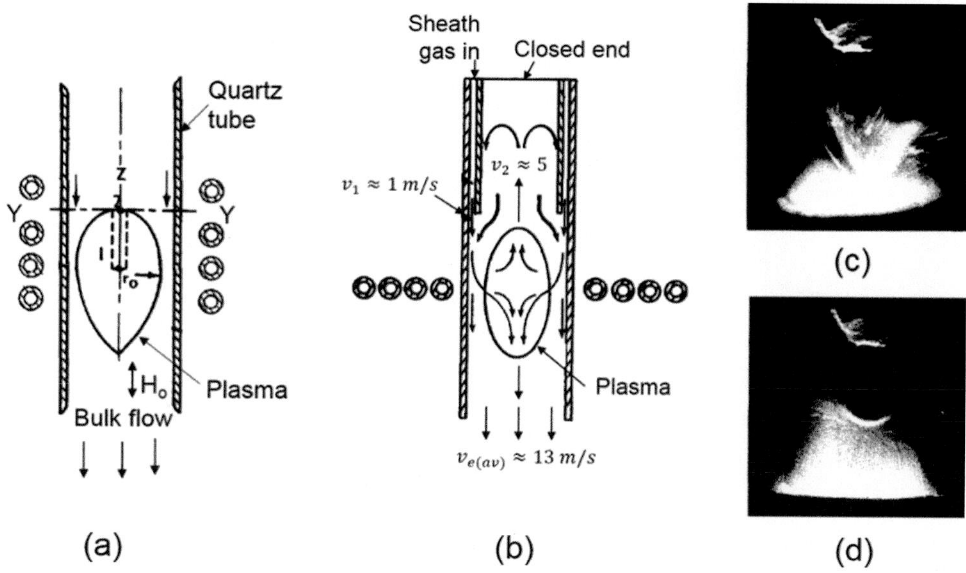

Fig. 10.27 (**a**) and (**b**) Air-cooled, quartz wall torches used by Chase in diagnostic study of the magnetic pressure in the center of an induction plasma discharge. (**c**) and (**d**) Photograph showing the magnetically induced backflow deflecting 10 μm alumina particles injected on the upstream end of the discharge [Chase (1971)]

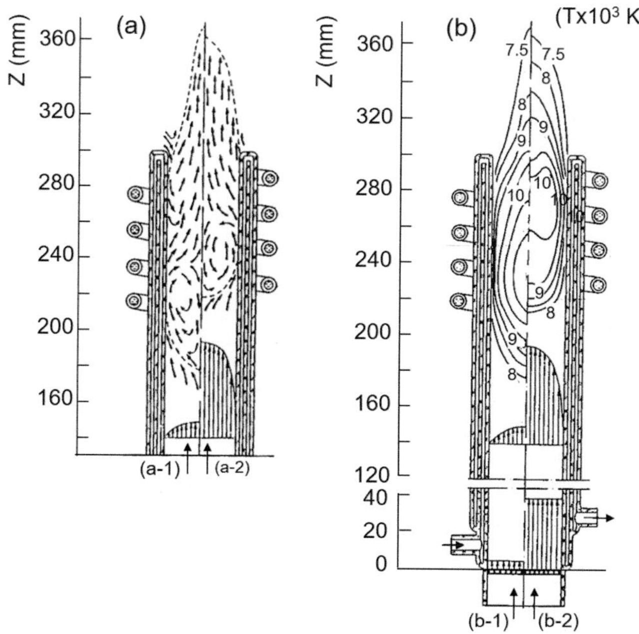

Fig. 10.28 (**a**) Flow field and (**b**) temperature field for Argon RF-ICP at atmospheric pressure. Torch diameter $= 40$ mm, $f = 6$ MHz, $P = 7.6$ kW, (a-1 and b-1) $Q_o = 5$ slm (Ar), (a-2 and b-2) $Q_o = 40$ slm (Ar) [Klubnikin (1975)]

distinct characteristic of the induction plasma discharge and requires special attention when using induction plasma torches for the treatment of solid, liquid, or gaseous precursors. As was shown earlier in Fig. 10.15, 10.16 and 10.17, induction plasma torches are generally equipped with a central port at the upstream end of the torch that allows for

the introduction of a water-cooled powder-feeding probe, which serves to inject the material to be treated into the center of the discharge. The exact position of the probe tip can vary depending on the particle size distribution and density of the material to be treated. In general, injecting the material from a position too high above the center of the coil can lead to internal recirculation of the powder in the discharge cavity and its partial deposition on the inner wall of the plasma confinement tube. On the other hand, injection of the material from a probe tip position that is too low beyond the center of the coil will result in a reduction of the trajectory of the powder in the hottest zone of the plasma and overall loss of the powder treatment efficiency.

At the exit of the plasma torch, the plasma flow emerges in the form of a low velocity laminar or turbulent jet, depending on the overall gas flow rate and operating conditions. Under low-pressure conditions, 33 kPa (250 Torr), power 60 kW, the length of the core region of the emerging plasma jet can be more than ten to fifteen times the diameter of the exit nozzle of the plasma torch. It is not unexpected to have such as long jet exhibiting low frequency fluctuations at its tail end as can be observed in the sequence of photographs given in Fig. 10.29 [Boulos MI (2001)]. These were taken at different exposure times varying from 1/1000th s for Fig. 10.29a to 1/60th s for Fig. 10.29e. It may be noted that while the jet shown in Fig. 10.29e could be interpreted as being a symmetrical stable flow, the same conclusion could not be made of the photograph shown in Fig. 10.29a, which reveals an interesting swirl flow component that can be traced back to the central gas injection mode into the discharge cavity. However, such fluctuations being limited to the tail part of

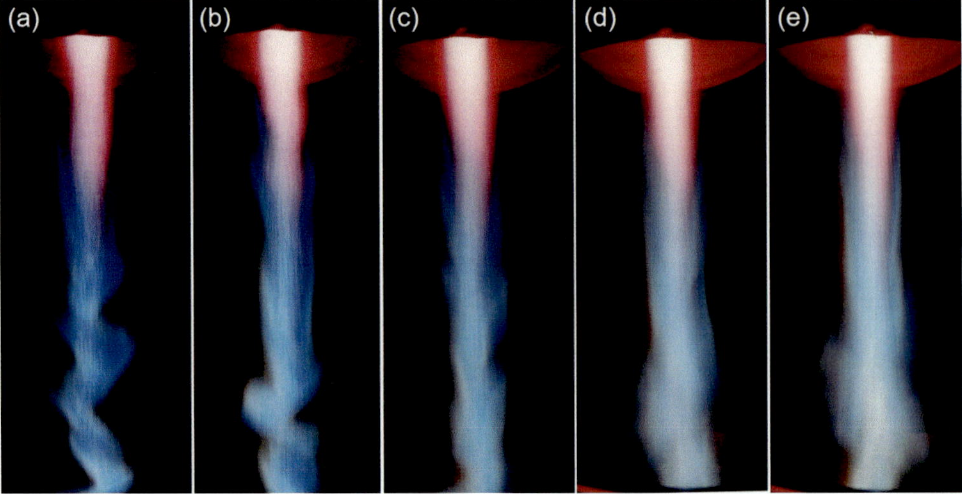

Fig. 10.29 Sequence of photographs taken at different exposure time of low-pressure RF-ICP Ar/H$_2$ plasma jet, immerging from a PS-70 plasma torch, nozzle i.d. = 64 mm, power = 60 kW, chamber pressure = 33 kPa, (**a**) 1/1000 s, (**b**)1/500 s, (**c**) 1/250 s, (**d**) 1/120 s, and (**e**) 1/60 s [Boulos MI (2001)]

the plasma jet with a total length of 700 mm would not affect the inflight heat transfer to the particles in the flow due to their relative low velocity and long residence, and the stability of the upper core region of the jet that is close to 300 mm long.

Typical axial velocity, v_z, and mass flux, (ρv_z), profiles at the exit of a 30 mm i.d. induction plasma torch are given in Fig. 10.30, after Dresvin et al. (1977). These were obtained for the same conditions as those for which temperature mapping were given in Fig. 10.28. The measurements were made using Pitot tube probes for argon (solid lines) and oxygen (dotted lines) plasma. The total gas flow rate was 30 slm, oscillator frequency, $f = 10$ MHz, with plate power, $P = 12$ kW for the argon plasma, and 8 kW for the oxygen plasma. The results show maximum centerline velocities in the range from 40 to 60 m/s depending on the nature of the plasma gas, which is considerably lower than that obtained in standard DC plasma jets. Important to note that, contrary to observations made with DC plasma jets, the maximum in the mass flux profile for an induction plasma jet is located on the centerline of the discharge rather than off axis. The effect is a direct consequence of the radial pumping and mixing of the sheath gas toward the center of the discharge at the middle section of the induction coil.

Developments of special Laval nozzle adapted to induction plasma torches, as shown earlier in Fig. 10.16, allowed for a significant increase of the plasma jet velocity at the exit of the torch to supersonic speed levels. The corresponding radial profiles of the temperature, axial velocity, local Mach number, and differential pressure are shown in Figs. 10.31 after [Hollenstein et al. (1999)]. These were obtained for an Ar/H$_2$ (2 vol. % H$_2$) plasma flow with water-cooled Laval nozzles at Mach number 3. Details of the nozzle dimensions

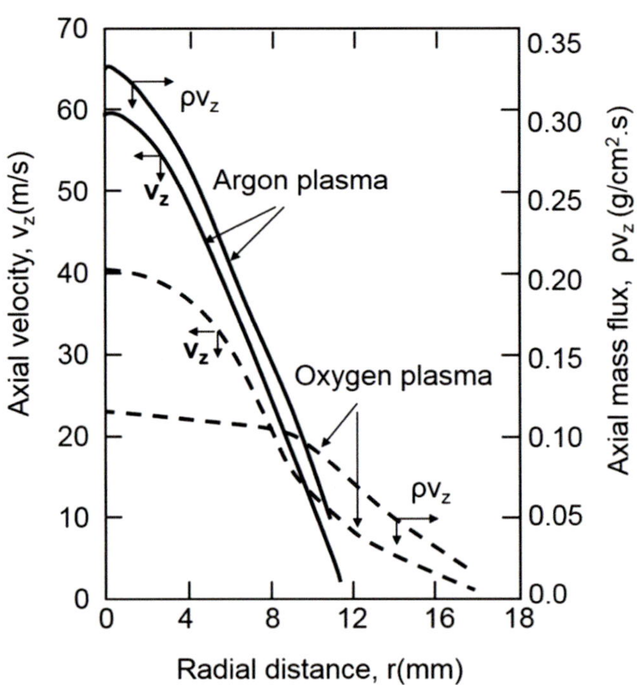

Fig. 10.30 Axial velocity (v_z) and mass flux (ρv_z), profiles at the exit of a 30 mm i.d. induction plasma torch, $Q_o = 30$ slm., $f = 10$ MHz, $P = 12$ kW for argon plasma (solid lines) and $P = 8$ kW for oxygen plasma (dotted lines) [Dresvin (1977)]

are given in Fig. 10.16. The plate power for the discharge was 25 kW with downstream chamber pressure at 17.1 kPa and back-pressure in the discharge cavity in the plasma torch 212 kPa. The results reported at a distance downstream of the nozzle exit, $z = 17$ mm, showing peak velocities of 2000 m/s reached on the centerline of the plasma jet. The

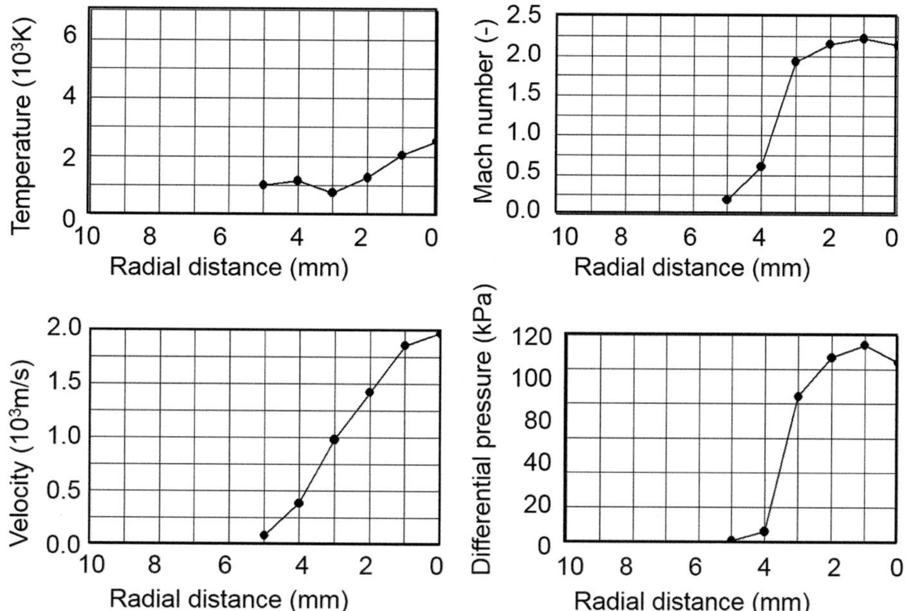

Fig. 10.31 Radial profiles of the temperature, velocity, Mach number, and differential pressure at axial distance $z = 17$ mm from the exit of M-3 supersonic plasma nozzle attached to Tekna PS-35 torch, operating with Ar/H$_2$ (2 vol% H$_2$), power $= 25$ kW, chamber pressure, $p_a = 17.1$ kPa, torch back pressure, $p_b = 212$ kPa, [Hollenstein et al. (1999)]

local Mach number in this case was close to 2.2, and the local temperature of the plasma flow was below 3000 K.

Further development along these lines aimed at achieving yet higher velocities and uniformity of the supersonic plasma flow. Léveillé (2002) and Léveillé et al. (2003) reported a systematic study of enthalpy probe measurements of the temperature and velocity profiles for an Ar/H$_2$ plasma jet using both water-cooled (WC) and radiation-cooled (RC) Laval nozzles under a wide range of operating conditions. The same plasma gas flow rates were maintained throughout these measurements. These were set at 40 slm (Ar) +1.8 slm (H$_2$) for the sheath gas and 20 slm (Ar) for the central gas of the plasma torch. The torch power was set at 20 kW, and the absolute pressure in the chamber was 1.5 kPa. Typical results are presented in Fig. 10.32. These show the axial profiles of the temperature and axial velocity along the centerline of the plasma jet. The maximum axial velocity achieved in these cases was between 2000 and 2300 m/s, which was obtained with the radiation-cooled (RC) M-2.45 nozzle. The corresponding velocities obtained with the water-cooled (WC) nozzles were lower by about 500 m/s. As expected, the corresponding temperature profiles show significant axial variations of the velocity corresponding with the observed shockwaves in the plasma jet. Photographs of different supersonic plasma jets obtained with different gas are shown in Fig. 10.33. These were all obtained at essentially the same power level in the 20–25 kW range with chamber pressures of 7.9–9.3 kPa (60–70 Torr)

10.4.4 Concentration Fields

The ability to axially inject reactants into the center of RF-ICP without disturbing the discharge has long been recognized as being of particular value for the use of this plasma source for plasma spraying and chemical synthesis. Special attention has therefore been given to the study of the mass transfer and mixing pattern in the discharge. Dundas (1970) reported in the early seventies one of the first studies of mass transfer in an inductively coupled discharge. His measurements were carried out on a TAFA Model 66 induction plasma torch with a 76.2 mm i.d. plasma confinement tube and a 50.8 mm i.d., 3.18 mm wall water-cooled metallic separator. Pure air was injected axially as a sheath gas in the annular space between the plasma confinement tube and the metallic separator, while pure argon was injected as central gas in the separator cavity with tangential and axial velocity components. The torch was connected to an RF power supply operating at an oscillator frequency in the range of 2.5–4.5 MHz and a maximum plate power of 90 kW. Concentration mapping of the air–argon mixture in the torch cavity was carried out using a water-cooled microsampling probe continuously withdrawing a sample of the gas, which was analyzed using a thermal conductivity cell of a gas chromatograph. Typical results are given in Fig. 10.34 for two cases with identical gas flow rates, with a sheath gas flow $= 484.6$ slm (air), central gas flow $= 63.2$ slm (argon). The mass ratio of the sheath gas to central gas $= 5.55$.

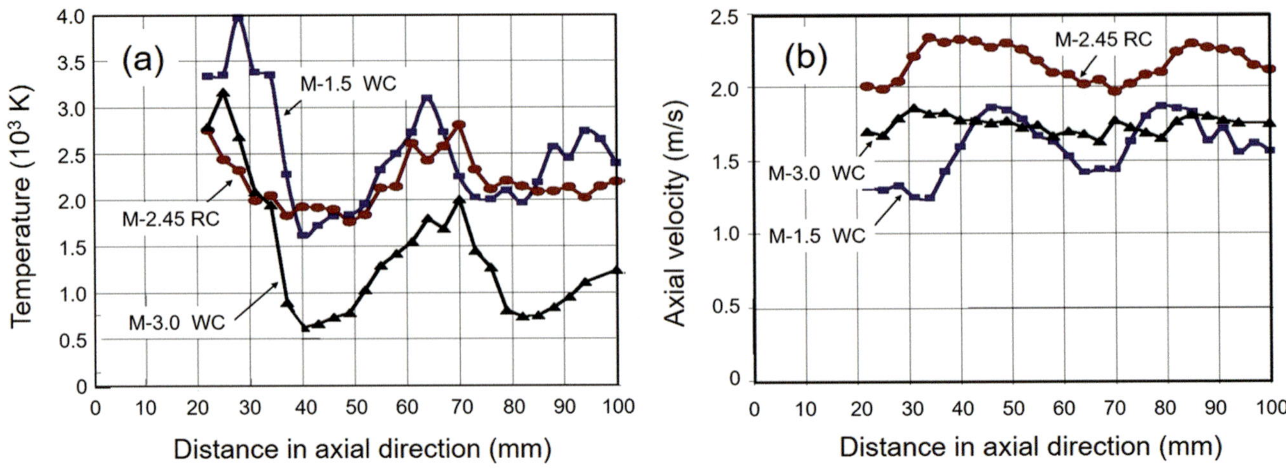

Fig. 10.32 Axial variation of (**a**) the temperature and (**b**) velocity along the centerline of a supersonic induction plasma jet for different Laval nozzle designs (Mach-1.5, Water-Cooled (WC), Mach-3, Water-Cooled (WC), and Mach-3 Radiation-Cooled (RC). Plasma gas Ar/H$_2$, power = 20 kW [Léveillé et al. (2002)]

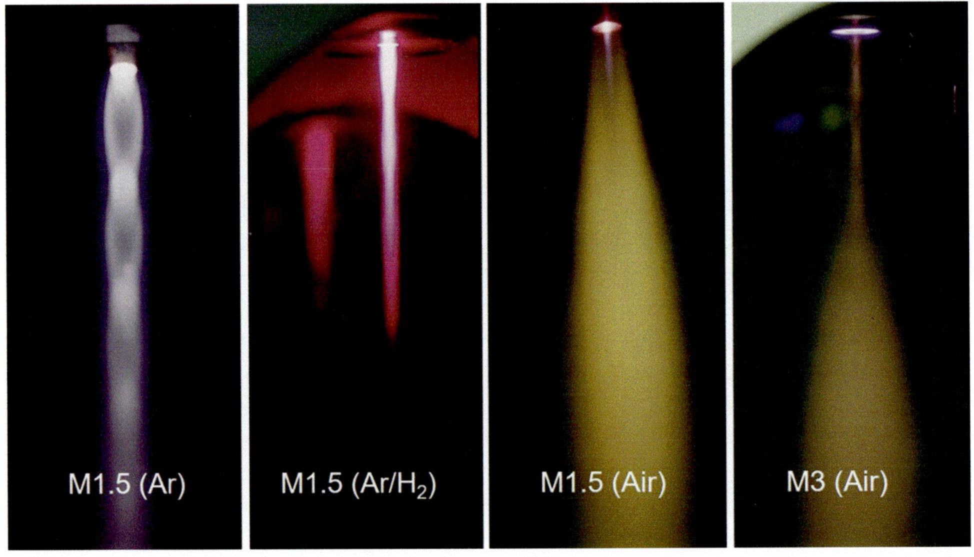

Fig. 10.33 Photographs of supersonic plasma jets of different gases Ar, Ar/H$_2$, and Air, generated using PS-50 induction plasma torch with a Laval nozzle [photographs courtesy of the University of Sherbrooke]

The 2D concentration mappings given in Fig. 10.34 are in presented in terms of molar fraction of air in the (air/argon) mixture. The results given in Fig. 10.34a were obtained at room temperature in the absence of the discharge. These show a very intense mixing between the two streams due to the high level of turbulence in the flow. The concentration of air in the mixture at the centerline of the torch at the gas entry level was as high as 70–75 vol%. In contrast to the room temperature case, measurements under plasma conditions, with a plate power of 35 kW, given in Fig. 10.34b, show significantly less mixing between the sheath gas and central gas streams. The air concentrations on the axis of the torch at the gas entry level being less than 1 vol%. The effect is

primarily due to the high viscosity of the plasma and complete suppression of turbulence and gas recirculation effects in this case. As will be discussed later, the nature of the flow in the discharge cavity is significantly more complex than initially thought and involves in most cases a combination of laminar and turbulent flow in the same flow field with laminar flow being predominant in the core, high temperature discharge region, and turbulence phenomena controlling the flow of the colder boundary layer close to the wall of the plasma confinement tube.

Subsequent studies by Boulos and his collaborators (2001) documented these phenomena using photographic and mass spectrometric measurements of the mixing pattern

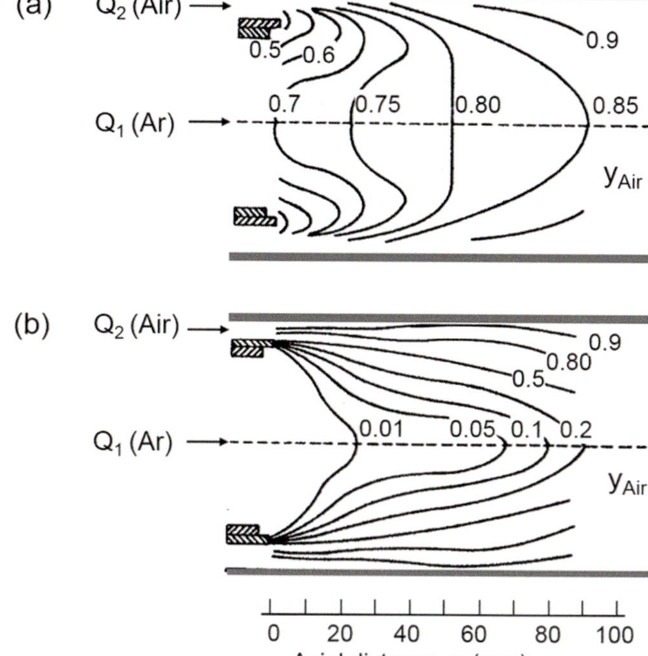

Fig. 10.34 Concentration mapping in induction plasma in the absence (top) and presence (bottom) of the discharge. Air/Argon mass flow ratio $= 5.55$ [Dundas (1970)]

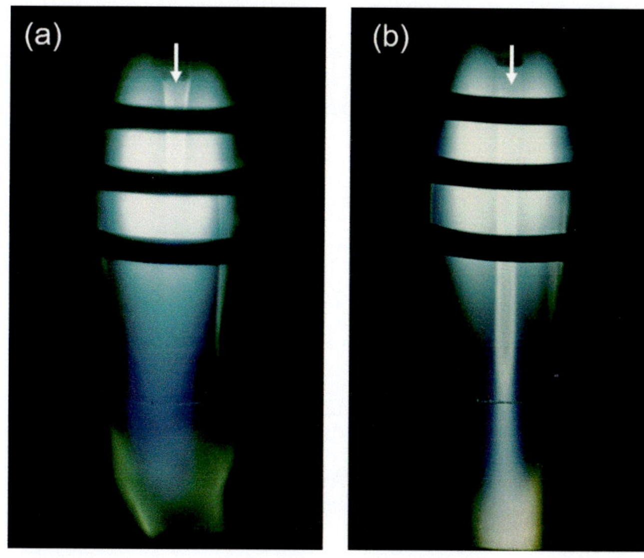

Fig. 10.35 Photographs of an RF inductively coupled discharge in the presence of axial injection of ultrafine zirconia particles with the probe gas along the centerline of the discharge. Pure argon plasma with an oscillator frequency of 3 MHz and a plate power of 7 kW, powder gas flow rate (**a**) 1 slm and (**b**) 5 slm [Boulos (2001)]

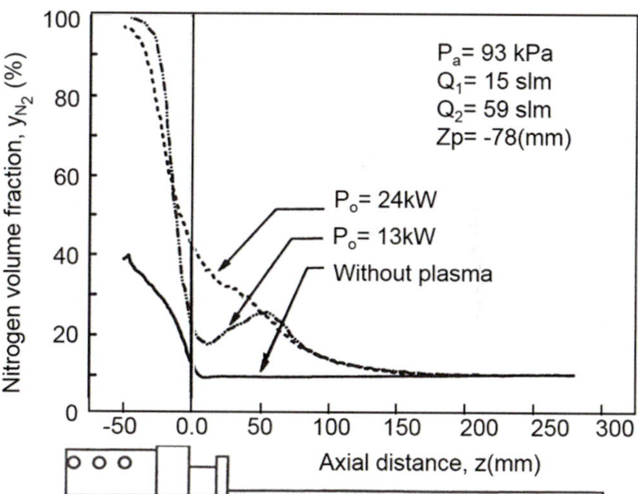

Fig. 10.36 Concentration profiles along the axis of the plasma torch/reactor system for the axial injection of nitrogen in an Ar/H$_2$ induction plasma, probe gas, $Q_{pb} = 15$ slm (N$_2$), $p_a = 93$ kPa [Soucy et al. (1994a)]

in an inductively coupled RF discharge. The photographs given in Fig. 10.35 show a 50 mm i.d. inductively coupled RF discharge, in the presence of tracer ultrafine zirconia particles mixed with the probe gas and axially introduced along the centerline of the discharge. Figure 10.35a is for a low powder gas flow rate (less than 1 slm of argon) while that of the Fig. 10.35b is for the same discharge in the presence of a higher probe powder-gas injection (above 5 slm). The plasma was of pure argon in this case, operated at atmospheric pressure, with an oscillator frequency of 3 MHz and a plate power of 7 kW.

The photographs clearly support the laminar flow hypothesis in this case with little mixing between the probe powder gas with the tracer particles and the plasma core. It is also noted that at low probe gas flow rates, the tracer particles do not penetrate the plasma and are only observed on the upstream side of the discharge. With the increase of the probe powder gas flow rate, the tracer particles are able to overcome the magnetically induced back-pressure in the center of the discharge and create a well-defined central channel in the plasma stream. It is important to underline the critical role of mixing and mass transfer in the design and scaling up of induction plasma reactors for the plasma chemical synthesis of materials.

The significant difference between the mixing patterns in the presence and absence of the discharge was studied by [Soucy et al. (1994a, b)] using mass spectrometric probing.

Typical results obtained using a Tekna PS-50 torch in the presence of axial and radial injection of a nitrogen tracer gas in an Ar/H$_2$ plasma stream are given in Figs. 10.36 and 10.37. The sheath gas flow rate was 89 slm Ar + 9.6 slm H$_2$, and central gas 59 slm (Ar). The chamber pressure was varied between 35 and 93 kPa with the plate power set in the range of 13–24 kW. Nitrogen was injected into the flow as tracer, either axially through the water-cooled probe at the middle of the induction coil or radially through multiple radial holes in

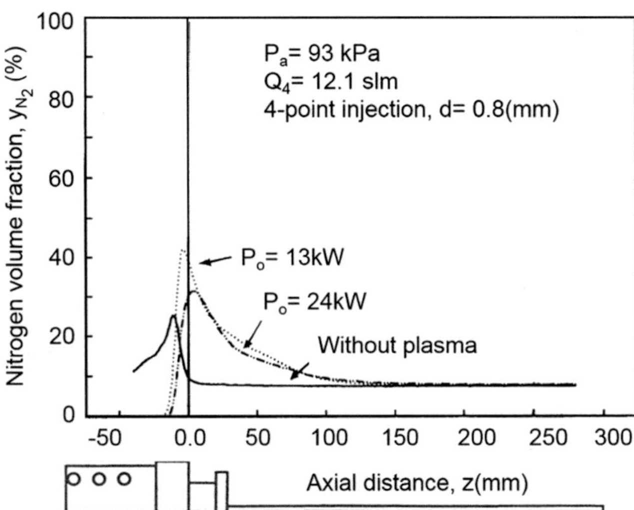

Fig. 10.37 Concentration profiles along the axis of the plasma torch/reactor system for the radial, four ports injection of nitrogen in an Ar/N_2 induction plasma, $Q_{in} = 15$ L/min; (N_2) STP, $p_a = 93$ kPa [Soucy et al. (1994b)]

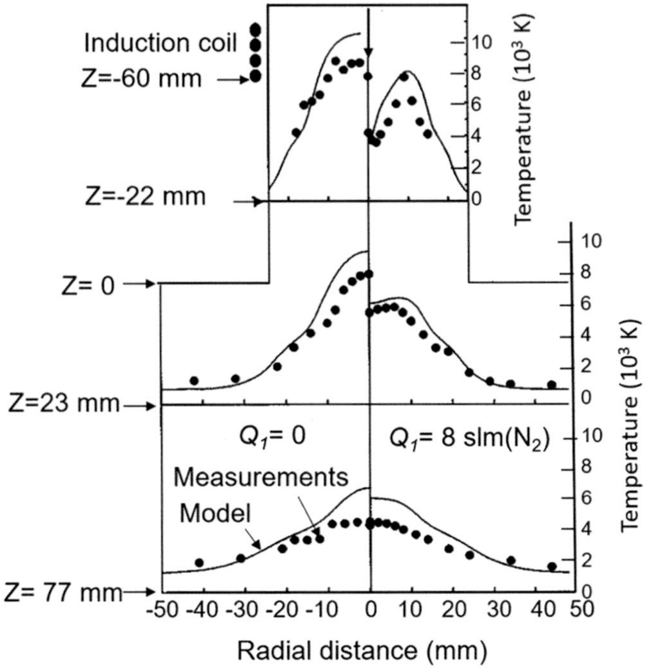

Fig. 10.38 Radial temperature profiles measured and calculated at different levels in an inductively coupled plasma jet in the presence and absence of "powder gas" injected through a central probe in the middle level of the induction coil. Ar/H_2 plasma at 20 kW, 250 Torr, (left) powder gas, $Q_1 = 0$, (right) powder gas $Q_1 = 8$ slm (N_2) [Rahmane et al. (1996)]

the exit nozzle of the plasma torch. Measurements were made of the nitrogen volume fraction in the flow through the sampling of the gases along the centerline of the reactor using a water-cooled sampling probe connected to a mass spectrometer for gas analysis.

The nitrogen gas concentration profiles of along the centerline of the plasma reactor are given in Fig. 10.36, with the tracer gas axially injected into the discharge at a probe gas flow rate of 15 slm (N_2). The probe tip was located at a distance of 78 mm upstream of the exit level of the plasma torch. The solid line, which refers to the no plasma case, that is, cold flow, shows that the nitrogen concentration drops rapidly as the gas emerges from the probe tip reaching almost its asymptotic value of 10 vol% at the exit of the torch. In contrast, in the presence of the discharge, dotted lines, whether at a power level of 13 or 24 kW, the nitrogen concentration along the axis remains essentially unchanged within the torch, dropping only to 30–40 vol% at the exit level of the torch. The asymptotic value of 10 vol% is reached in this case about 150 mm downstream of the exit of the torch. The observed increase in the nitrogen concentration between 0 and 60 mm from the exit of the torch for a plasma power of 13 kW is believed to be a result of a 3D effect on the flow pattern in the reactor causing the deviation of the flow from the centerline of the discharge over this region.

Corresponding axial profiles of nitrogen concentration are given in Fig. 10.37 for a radial injection of the tracer gas into the flow at the level of the torch nozzle. Here again one observes that in the cold flow case, "without plasma," the injected nitrogen flow mixes instantaneously with the mainstream with nitrogen detected even upstream of the point of injection. The asymptotic nitrogen concentration of 10 vol% is reached in this case almost instantaneously at the point of injection. In the presence of the discharge, whether at 13 or 24 kW, the mixing process is considerably slowed down with high concentrations (30–40 vol% N_2) observed at the centerline of the discharge at the level of tracer gas injection. The mixing process takes some time before the asymptotic value of 10 vol% is reached around 120 mm downstream of the point of injection. The results obtained in both axial and radial injection cases clearly demonstrate the considerable difficulty in the mixing under plasma conditions, which should be carefully considered when designing a plasma reactor for chemical synthesis or reactive plasma spraying applications

The effect of nitrogen gas injection through the central probe, often referred to as powder carrier gas injection, was also investigated by [Rahmane et al. (1994, 1996)] using an enthalpy probe system. Torch operating parameters were plate power = 20 kW, chamber pressure = 33 kPa, sheath gas flow of 71 slm Ar + 4 slm H_2, central gas flow of 33 slm of (Ar), and probe position 93 mm above the torch exit. Figures 10.38 and 10.39 show the radial profiles of the temperature and the velocity, respectively, at different axial locations, for the case without powder gas injection (LHS) and with nitrogen injection at a rate of 8 slm through the

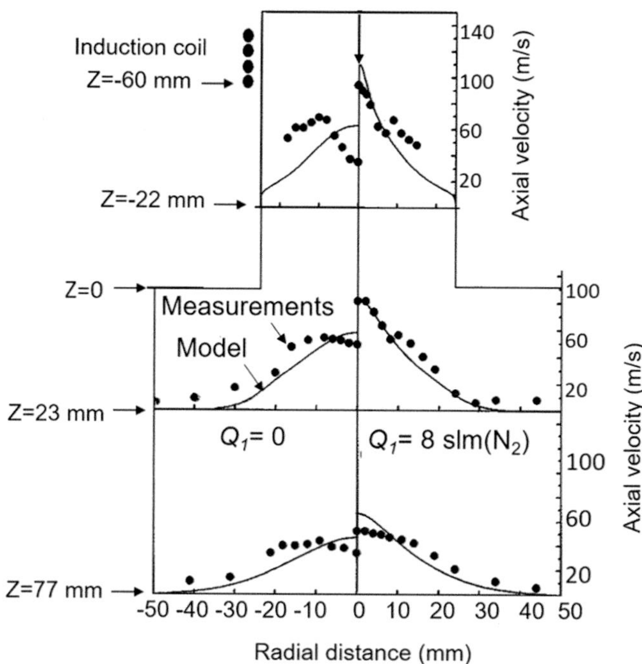

Fig. 10.39 Radial profiles of the axial velocity at different levels in an inductively coupled plasma in the presence and absence of "powder gas" injected through a central probe in the middle level of the induction coil. Ar/H$_2$ plasma at 20 kW, 250 Torr, (left) powder gas, $Q_1 = 0$, (right) powder gas $Q_1 = 8$ slm (N_2) [Rahmane et al. (1996)]

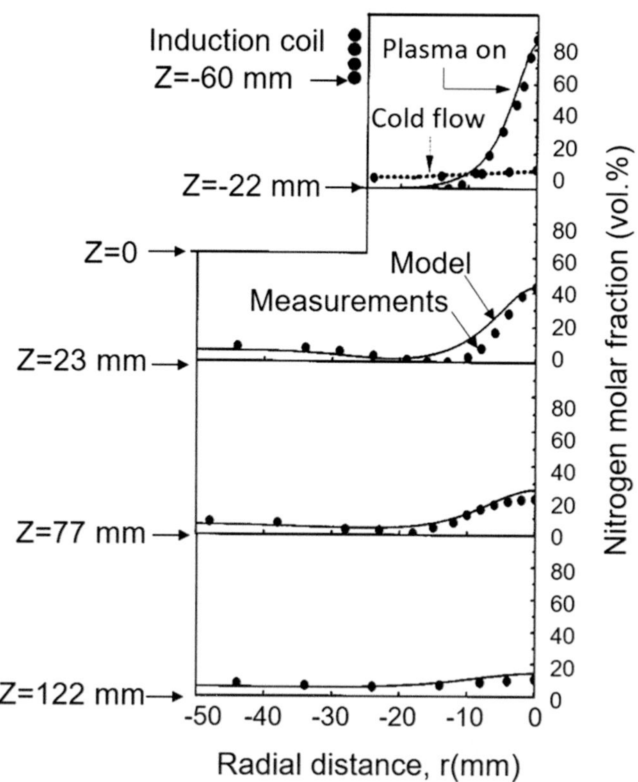

Fig. 10.40 Radial nitrogen concentration profiles at different levels in inductively coupled plasma in the presence of "powder gas" injection through a central probe in the middle level of the induction coil. Ar/H$_2$ plasma at 20 kW, 250 Torr, powder gas, $Q_{pb} = 8$ slm (N_2) [Rahmane et al. (1996)]

central probe (RHS). The peak temperature along the centerline of the torch is noted to be reduced, and the corresponding gas velocity increased, as a result of the probe gas injection. Superimposed on these figures are the results of flow modeling (solid lines). These show good agreement with the experimental measurements. Figure 10.40 shows the corresponding profiles of the volume fraction of nitrogen in the flow, which are in good agreement with the modeling results. For the position of 22 mm above the torch exit ($z = -22$ mm), the measured nitrogen volume fraction under "cold flow" conditions are also indicated. These were obtained in the absence of the discharge. They clearly show evidence of the rapid mixing of the flow under the "cold flow" conditions. In contrast, in the presence of the discharge, the mixing of the different streams is considerably hindered by the high viscosity and low turbulence level in the discharge.

10.4.5 Plasma and Particle Dynamics Modeling

Modeling is an effective tool for the fundamental understanding of the basic phenomena governing materials processing under plasma conditions. Through modeling it is possible to calculate the flow, temperature and concentration fields in plasma under a wide range of conditions. A considerable research effort was devoted over the past four decades to the modeling of the inductively coupled plasma and plasma–particle interactions. The fundamental phenomena involved have been discussed in Chap. 4, Plasma–Particle Momentum and Heat Transfer, and Chap. 5, Gas and particle Dynamics in Thermal Spray. In this section, no attempt is made to cover this topic in an exhaustive way. Modeling of the RF inductively coupled plasma spraying (RF-IPS) is presented simply as a research tool that contributes to our understanding of the basic phenomena that governs the processing conditions and plasma–particle interactions during typical plasma spraying operations. Extensive references are given for a more detailed documentation on the subject.

Most of the modeling results reviewed in this section are based on the following principle assumptions:

- Steady state, laminar and/or turbulent flow conditions
- Axisymmetric, 2D system of coordinates with a few examples presented of 3D modeling studies
- Optically thin plasma
- Plasma in local thermodynamic equilibrium (LTE)
- Negligible displacement current

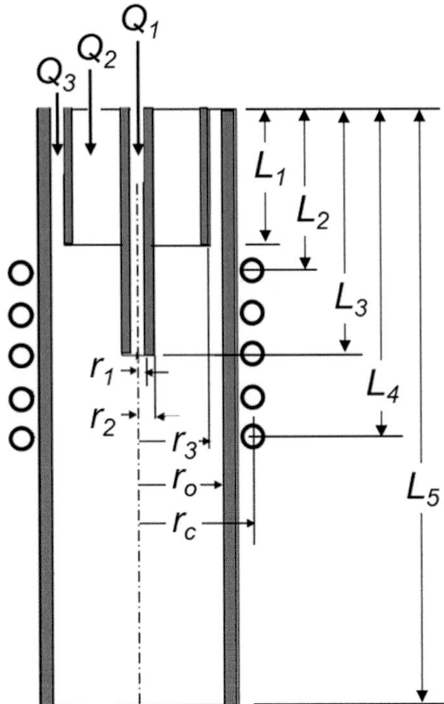

Fig. 10.41 Typical induction plasma torch configuration and principal dimensional parameters used in induction plasma modeling. After Rahmane et al. (1996), with kind permission from Springer Science + Business Media B.V

10.4.5.1 Flow, Temperature and Concentration Fields Modelling

A standard schematic representation of the induction plasma torch as mostly used in all modeling studies is shown in Fig. 10.41. This is based on the double flux torch design with three main gas streams introduced into the torch. The external sheath gas stream, Q_3, is introduced in the annular space between the plasma confinement tube and the intermediate separation tube. The plasma or central gas stream, Q_2, is introduced in the central region of the separation tube and mostly incorporates a tangential velocity component. The third gas stream, generally referred to as the powder or carrier gas stream, Q_1 is usually used for the introduction of the powder or precursor gas stream into the torch. All other dimensions given in Fig. 10.41 are self-explanatory. The most important of which are, r_o, inner radius of the plasma confinement tube; r_c, the inner radius of the induction coil; and r_1, the inner radius of the powder injection probe. It is important to point out that L_3 represents the position of the tip of the powder injection probe that is an important operational parameter that can be changed from one experiment to the other depending on the properties of the powder being treated/plasma sprayed. The relative position of L_1, the tip of the intermediate separator tube, compared to L_2, the top

level of the induction coil is also an important parameter that can have an influence on the flow and temperature fields and consequently on the stability of the discharge and the efficiency of the treatment.

Modeling and experimental results obtained by Rahmane et al. (1994, 1996) for a standard RF induction plasma torch model PS-50, with a 50 mm i.d. plasma confinement tube is presented in Figs. 10.38, 10.39 and 10.40. The tip of the injection probe is placed at 93 mm upstream of the torch exit level. Computations were carried for a plate power of 20 kW plate power, with an energy coupling efficiency of 60% (12 kW coupled into the discharge), a central gas flow rate of 33 slm (Ar) and Ar/H$_2$ sheath gas mixture (71 slm Ar + 4 slm H$_2$). Nitrogen is injected as the powder carrier gas, Q_1. The plasma is operated at near atmospheric, 82.4 kPa (620 Torr) and low pressure, 33.2 kPa (250 Torr) conditions. The 2D flow streamlines and temperature isocontours are shown on the LHS of Fig. 10.42. The corresponding concentration field for H$_2$ and N$_2$ are shown on the RHS of the same figure. It may be noted that since the hydrogen is injected with the cold and turbulent sheath gas, it mixes at a much faster rate than the nitrogen, which is injected in the center of the high temperature, laminar, and discharge zone. Particularly relevant to this study are the effective transport coefficients for the momentum, heat, and mass presented in Fig. 10.43. In each of these figures, the molecular properties are identified with the subscript "m" while the turbulent properties are identified with the subscript "t". It is significant to note that molecular properties of the gas are the controlling transport parameter in the high temperature core region of the flow ($-200 < z < +200$ mm), while turbulent transport mechanisms are predominant in the downstream end of the flow at axial distances beyond 300 mm ($z > 300$ mm).

Chen and Boulos (1994) reported similar observations by investigating turbulence effects in a RF-ICP using the k–ε turbulence model. The results reviled a high level of turbulence in the entrance region of the sheath gas near the wall of the plasma confinement tube, while the high temperature plasma region was always laminar. The discharge zone is pushed downstream for higher flow rates if turbulence is considered. Ye et al. (1999) further improved the turbulence model by including temperature fluctuations and the resulting density fluctuations in the model. No significant effect of these fluctuations on the turbulence level or the temperature distributions was observed.

10.4.5.2 Integrated Electrodynamic System Modelling

As pointed out earlier, since the plasma is an active part of the power supply circuit, the oscillator frequency and RF power coupled into the plasma changes with the change of the electrical properties of the load presented by the plasma. Merkhouf and Boulos (1998) addressed this issue in a

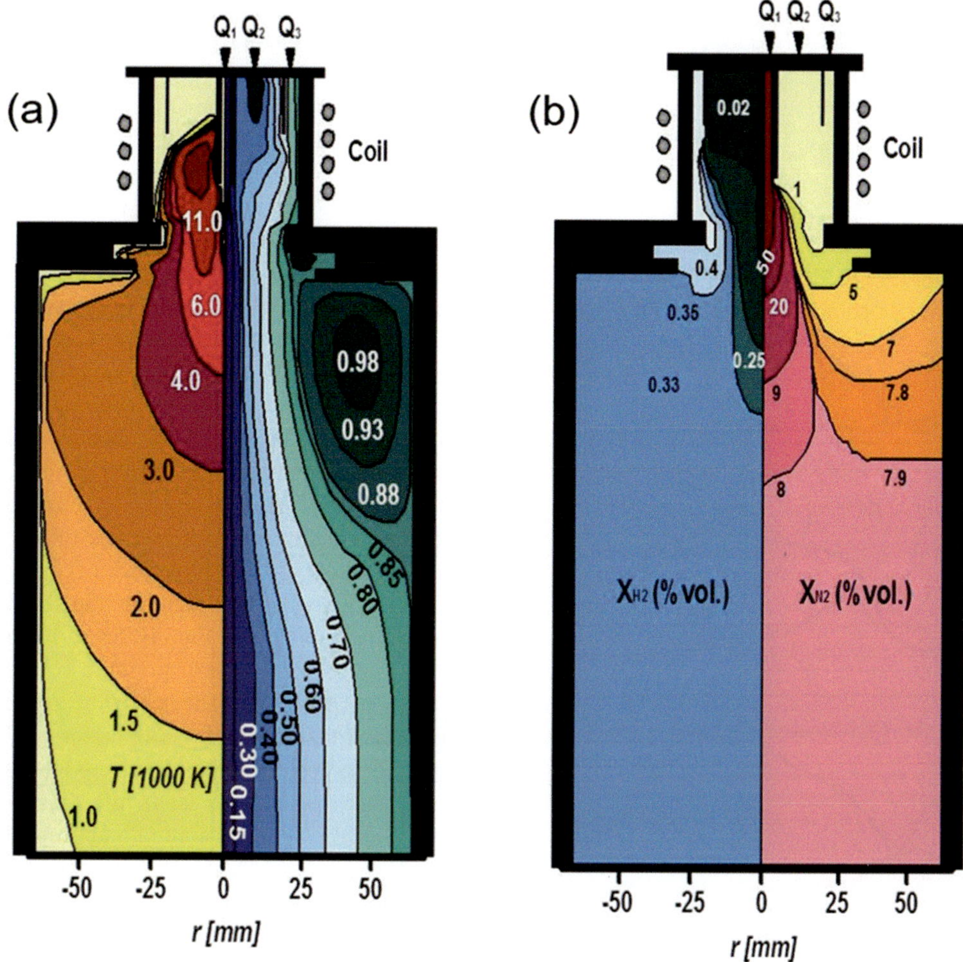

Fig. 10.42 (**a**) Flow and temperature fields, and (**b**) concentration fields for an Ar/H$_2$ plasma (71 slm Ar + 4 slm H$_2$) at 20 kW with a nitrogen gas injected through the probe with its tip at the center of the induction coil, pressure = 33 kPa (250 Torr). After Rahmane et al. (1994) with kind permission

model that includes the power supply circuit according to the circuit block diagram shown in Fig. 10.44. Results given in Fig. 10.45 show of the relationship between the plate current and the coil current for different plate voltages. The corresponding temperature fields in the RF discharge are shown in Fig. 10.46 with the radial temperature profile in the middle section of the coil shown in Fig. 10.47, indicating the strong widening of the plasma region for an increase of the plate voltage from 6 to 9 kV. The total input power and the corresponding coupling efficiency are shown in Fig. 10.48. The plasma and sheath gas flow rates were 30 slm (Ar) and 150 slm (Ar), respectively. In a series of electrical and calorimetric measurements, an energy balance was performed for plate voltages of 6.5, 7.0, and 7.5 kV. The power losses can be expressed as the sum of the power loss to the cooling water at the oscillator tube P_{os}, the RF coil P_{co}, the central injection tube P_i, and the torch wall P_w.

$$P_{lo} = P_{os} + P_{co} + P_i + P_w = P_{el} + P_{te} \qquad (10.12)$$

The energy in the plasma jet at the torch exit, P_{te} is measured by the energy transferred to the cooling water for the chamber into which the plasma jet is exhausted, P_{cw} and for the heat exchanger at the outlet of the chamber P_{ex}.

$$P_{te} = P_{cw} + P_{ex} \qquad (10.13)$$

The results are given in Table 10.2.

10.4.5.3 Induction Plasma Spray Modelling

CFD modeling studies by Xue et al. (2001) reported temperature, flow, and concentration fields in an inductively coupled plasma discharge under typical plasma spraying conditions. The results obtained for two chamber pressures of 101 kPa (14.7 psia) and 34.3 kPa (5 psia) are given in Figs. 10.49, 10.50, and 10.51. These were obtained for a Tekna PS-50 induction plasma torch (50 mm i.d.), operated using Ar/H$_2$ as the sheath gas (Q_3 = 100 slm Ar + 20 slm H$_2$), central gas (Q_2 = 30 slm Ar), and powder carrier gas (Q_1 = 5 slm Ar).

The plate power was 70 kW of which 65% is coupled into the discharge (45 kW). The oscillator frequency was 3 MHz. The temperature field shown in Fig. 10.49 reveals, as expected, a longer plasma jet under low pressure conditions compared to the atmospheric pressure case. The effect is a direct consequence of the proportional increase of the plasma velocity at the exit of the plasma torch. In both cases, we note that the powder gas allows for the axial injection of the "powder" into the center of the discharge. The corresponding streamlines in the plasma jet and upper section of the deposition chamber are shown in Fig. 10.50. The corresponding hydrogen concentration, introduced with Q_3, is given in Fig. 10.51. These show relatively little effect of the pressure on mass transfer phenomena under these conditions. The temperature and axial velocity profiles along the centerline of the plasma torch are given in Figs. 10.52 and 10.53. These show a maximum temperature along the centerline of the discharge of about 10,000 K, with a slower temperature drop with distance in the plume of the jet under low-pressure conditions compared to the atmospheric pressure case. The effect is compensated under atmospheric pressure conditions by a slight increase of the discharge power from 70 to 100 kW. The effect of pressure is, however, significantly more important on the axial velocity of the plasma that is noted in Fig. 10.53 to increase from around 70 m/s, under atmospheric pressure conditions, to close to 200 m/s at 34.3 kPa. The effect of power on the plasma velocity is considerably less important.

Further computations were carried out using this model for an induction plasma jet in the presence of a substrate under typical plasma spraying conditions. The torch geometry and its principal dimensions are given in Fig. 10.54. The plasma torch used in this case was the same at that used in earlier calculations, Tekna PS-50 torch. The substrate was in the shape of a disk of 100 mm diameter, placed at a distance of 100 mm downstream of the torch exit. All other operating conditions were identical to those used earlier in the computations given in Figs. 10.49, 10.50, 10.51, 10.52, and 10.53 with the exception of pressure, which was fixed at 34.3 kPa. The flow and temperature fields in the discharge are given in Fig. 10.55. These show a strong interaction

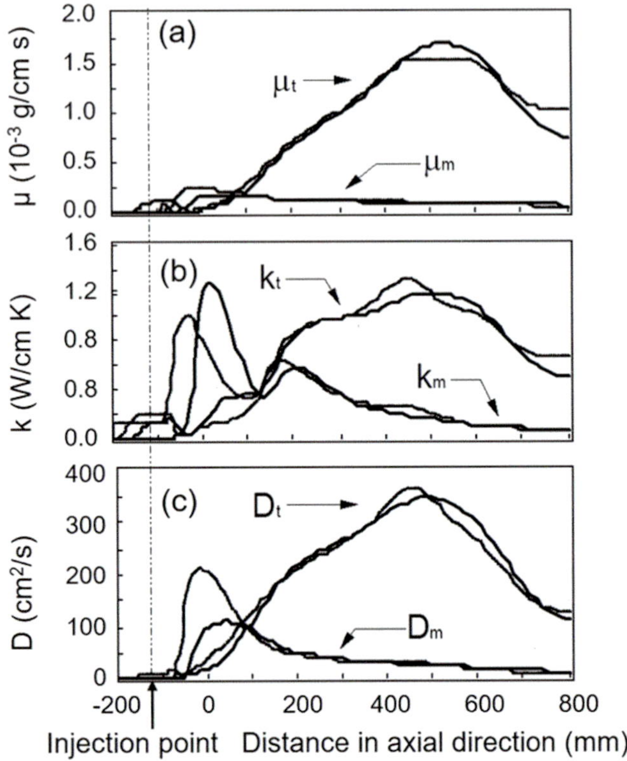

Fig. 10.43 Axial profile of the molecular and turbulent components of the (**a**) effective viscosity, (**b**) thermal conductivity, and (**c**) mass diffusivity along the centerline of the torch and reactor system. Similar plasma operating conditions as Fig. 10.42 After Rahmane et al. (1996) with kind permission from Springer

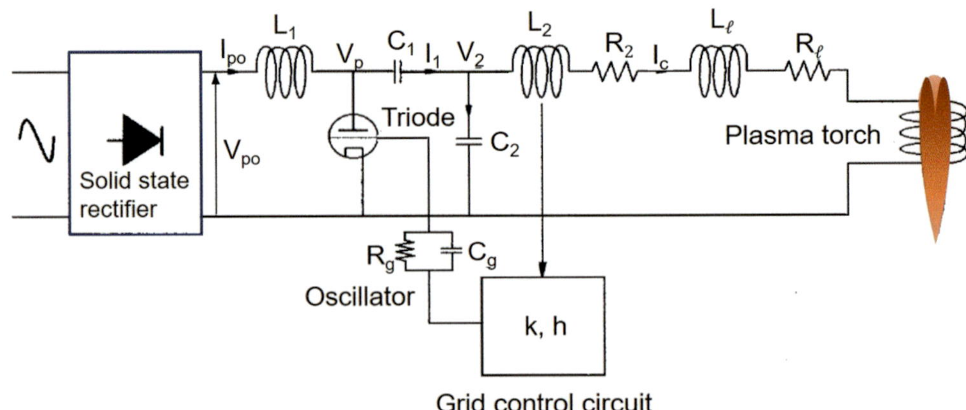

Fig. 10.44 Block diagram for the electrical circuit of the power supply integrating the RF plasma load. After [Merkhouf and Boulos (1998)]. Copyright IOP Publishing. Reproduced with permission. All rights reserved

between the plasma jet immerging from the torch and the substrate on which the flow is directly impinging. Side jets are also being generated in this case as a result of the impaction on the jet on the substrate, with the corresponding formation of strong recirculation eddies in the deposition chamber. Special attention is to be given in the design of the deposition chamber to avoid such recirculation patterns since they could cause the re-entrainment of powders back into the plasma jet and their entrapment into the coating without proper melting causing defects in the coating.

Further computations were carried out using essentially the same operating conditions as those used in Fig. 10.55, except for particles injection into the flow with the carrier gas through the central powder injection probe. Details of the model formulation including loading effects due to plasma–

particle interactions were discussed Chap. 5 Gas and Particle Dynamics in Thermal Spray, Sect. 5.6 Plasma–Particle Interactions Under Dense Loading Conditions, [Proulx et al. (1985, 1987)]. Molybdenum powder with a normal particle size distribution in the range of 40–80 µm and a mean particle diameter, $d_{50} = 60$ µm. The thermophysical properties p_f and the powder used are summarized in Table 10.3. The sheath gas was a mixture of Ar/H$_2$ (120 slm) with a hydrogen concentration of 16 vol%, while the central and powder carrier gas flow rates were, respectively, 30 slm (Ar) and 5 slm (Ar). The powder feed rate was varied among 4, 8, and 16 kg/h for the same plasma power of 70 kW and chamber pressure of 34.3 kPa (5 psia).

The effect of powder injection on the temperature and flow field in the discharge is shown in Fig. 10.56. The upper part of the figure shows effect of increasing the powder feed rate, from 4 to 8 kg/h and up to 16 kg/h. It is noted that the temperature of the central part of the discharge drops with the increase of the powder feed rate. The effect is less noticeable on the corresponding flow fields on the lower part of the figure. The effect of powder loading on the local radial temperature profiles at the downstream end of the coil ($z = 230$ mm, Fig. 10.54), is shown in Fig. 10.57. It is noted that the effect is localized around the centerline of the plasma jet with a drop of almost 6000 K of the axis temperature with a particle feed rate of only 4 kg/h. With the increase of the powder feed rate, the extent of the loading effect is noticeable further away from the flow axis. The effect of powder loading on the plasma temperature along the centerline of the flow is shown in Fig. 10.58 for powder feed rates of 4.0, 8.0, and 16.0 kg/h. The plasma temperature profile in the absence of any powder feeding is shown for

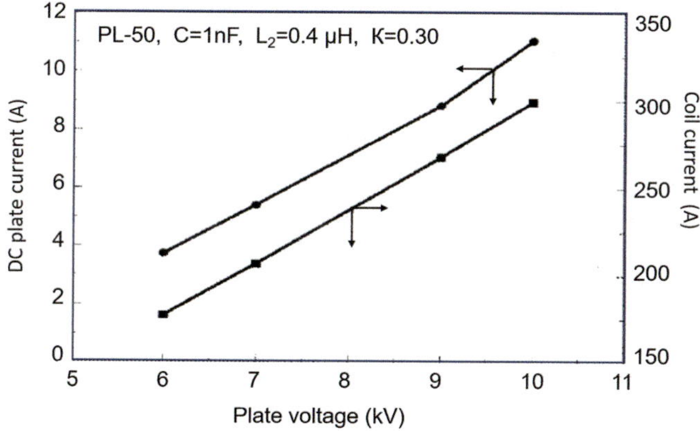

Fig. 10.45 Plate current and coil current as function of plate voltage. After [Merkhouf and Boulos (1998)]. Copyright IOP Publishing. Reproduced with permission. All rights reserved

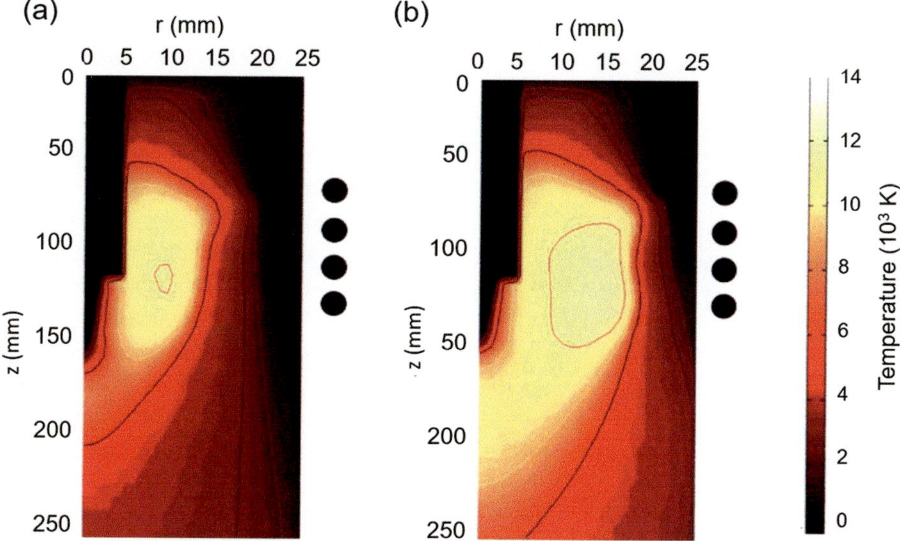

Fig. 10.46 Temperature fields in the RF discharge for plate voltages of (a) 6 kV and (b) 9 kV [Merkhouf and Boulos (1998)] Copyright IOP Publishing. Reproduced with permission. All rights reserved

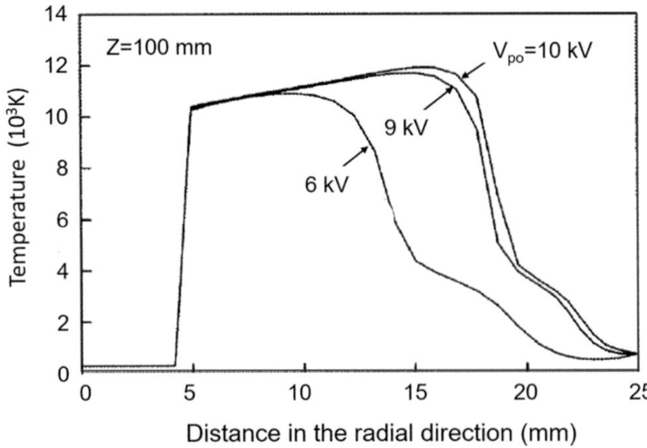

Fig. 10.47 Effect of the plate voltage on the radial temperature profiles at the middle of the induction coil, $z = 100$ mm. After [Merkhouf and Boulos (1998)]]. Copyright IOP Publishing. Reproduced with permission. All rights reserved

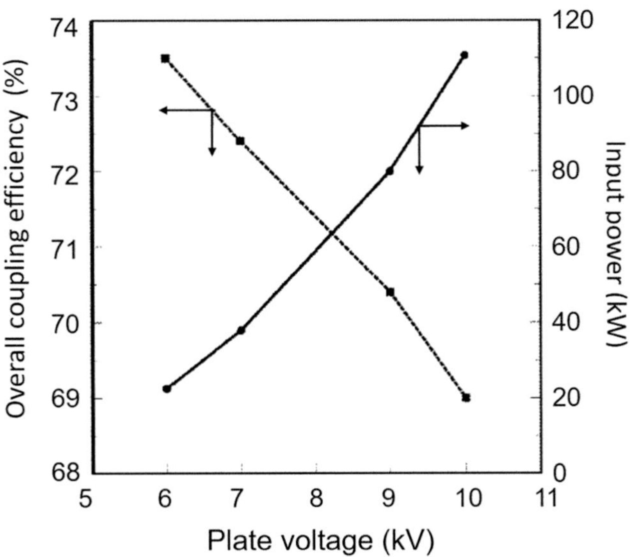

Fig. 10.48 Input power and overall coupling efficiency as a function of plate voltage, [Merkhouf and Boulos (1998)]. Copyright IOP Publishing. Reproduced with permission. All rights reserved

Table 10.2 Measured power distribution for the Tekna PL-35 plasma torch operated with 30 slm Ar central gas flow and 70 slm Ar sheath gas flow, after [Merkhouf and Boulos (2000)]

Plate voltage, kV	6.5	7.0	7.5
Plate current, A	2.6	3.0	3.5
Plate power, kW	16.9	21.0	26.25
High frequency current, A	152.5	163.5	181.5
High frequency voltage, V	571.2	600	641
Frequency, MHz	4.86	4.87	4.88
P_{OS} (%)	32.72	33.33	35.08
P_o (%)	64.08	63.14	59.81
P_e (%)	36.5	35.6	33.6

comparison. It is noted that the effect is significant and goes all the way until the plasma jet intercepts with the surface of the substrate on which the molten powder droplets are deposited.

The importance of the plasma–particle interactions phenomena and the resulting loading effects presented in the above discussion, underlined the need for effective means of proper powder dispersion in the plasma flow in order to achieve maximum treatment and/or deposition efficiency. Figure 10.59 shows photographs of a powder dispersion probe developed by Tekna Plasma Systems Inc. for the axial injection of powders in an inductively coupled plasma stream. The photograph on Fig. 10.59a shows the powder stream immerging from a standard probe, while the photographs at the center, Fig. 10.59b and on the RHS; Fig. 10.59c shows the corresponding powder stream in the presence of increasing rates of a dispersion gas using dispersion probes.

In order to quantitatively evaluate the impact of proper powder dispersion, computations were carried out for a standard Ar/H$_2$ plasma flow generated using a PS-70 torch at above atmospheric pressures ($p = 112$ kPa) and 100 kW plate power. A standard molybdenum powder, similar to that used in earlier modeling work, was injected axially in the center of the discharge at different feed rates and the fraction of the molten powder evaluated. The results show that with a dispersion angle of 15° the powder is fully molten (100% melted fraction) at a feed rate of 25.7 kg/h. A melting fraction of 100% is even obtained at a higher powder feed rate of 30 kg/h using such a dispersion powder injection probe.

Significant improvements in the modeling of plasma–particle interaction effects during induction plasma treatment of powders, whether for simple powder spheroidization or plasma spraying, were made over the past decade by Colombo and his collaborators at the University of Bologna [Bernardi et al. (2003a, b, 2004a, b, 2005a, b), Colombo et al. (2007, 2010a, b)]. Their main contribution is in taking the modeling one step further using 3D modeling codes, which significantly increases model complexity and required computation time. The results, however, reveal for the first time the 3D nature of the flow and the effect of geometrical asymmetries on the 3D plasma temperature field and particle trajectories and dispersion into the discharge. The results reported by Bernardi et al. (2004) were essentially for the same torch geometry and operating conditions as those used in an earlier publication by Proulx et al. (1987). The torch geometry used in this case is schematically represented in Fig. 10.60. Special attention was given in this case to the 3D effect of the coil design, as shown on the left-hand side of this figure, and its effect on the temperature fields in the discharge. The corresponding temperature fields for a pure argon plasma operated at a frequency of 3 MHz and a net power coupled into the plasma of 25 kW are shown in

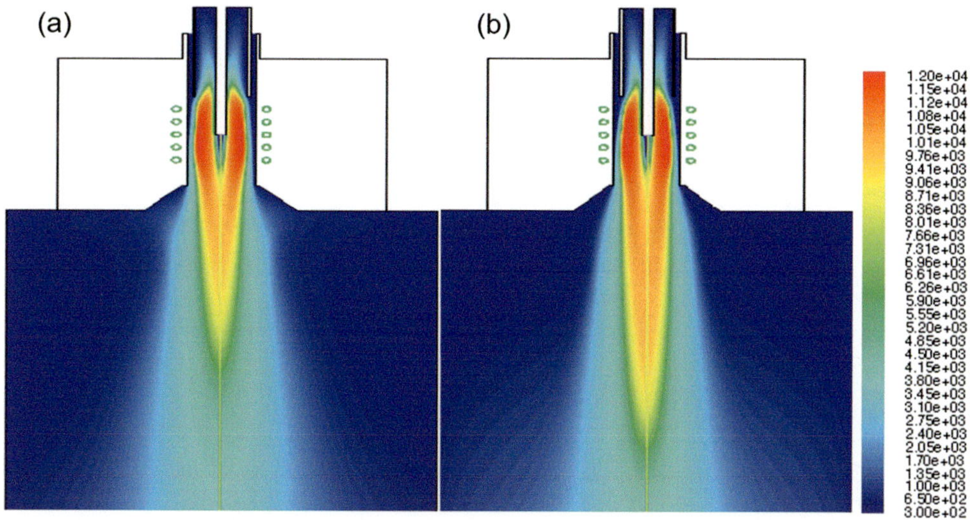

Fig. 10.49 Temperature field for Tekna PS-50 induction plasma flow Ar/H$_2$, plate power = 70 kW free jet at (**a**) pressure = 101 k Pa and (**b**) pressure = 34.4 kPa. Results supplied by Tekna Plasma Systems Inc. (2012)

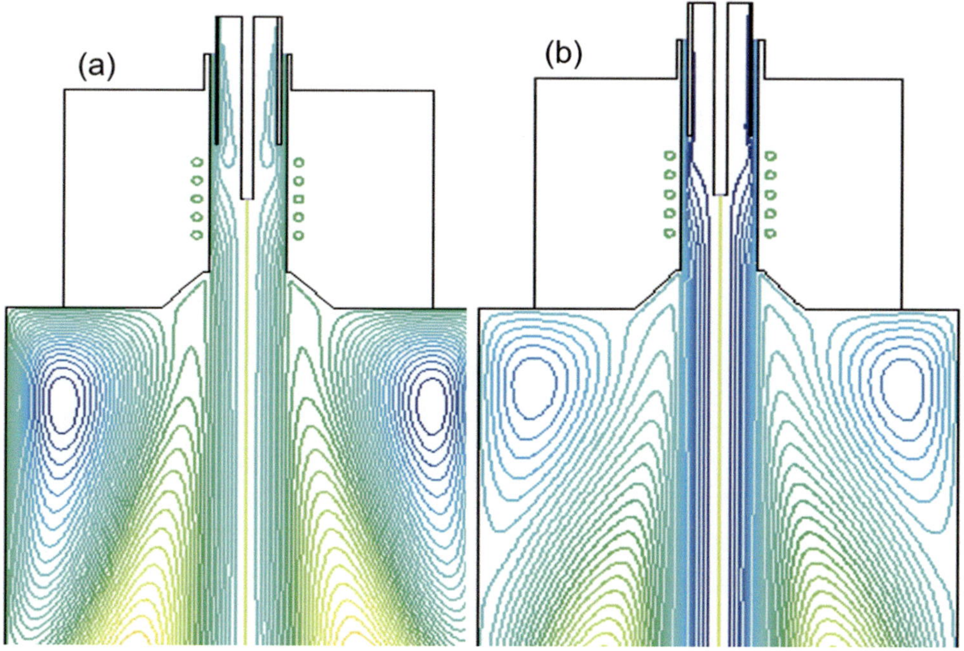

Fig. 10.50 Streamlines for Tekna PS-50 induction plasma flow Ar/H$_2$, plate power = 70 kW free jet at (**a**) pressure = 101 k Pa and (**b**) pressure = 34.4 kPa. Results supplied by Tekna Plasma Systems Inc. (2012)

Fig. 10.61. The results are presented for two cases in the absence (Fig. 10.61a, b) and the presence (Fig. 10.61c, d) of particles injected into the center of the discharge. For each of these two cases, the corresponding plasma temperature fields are given along two orthogonal planes passing through the axis of the discharge. The results reveal in both cases an important 3D asymmetry of the temperature fields that is due to 3D effect of the coil geometry. As expected, the results also reveal the local cooling of the plasma near the axis of the discharge as a result of Tungsten powder injection, $d_p = 30$ μm, at a feed rate of 20 g/min.

The subsequent study reported by [Colombo et al. (2010)] was focused on the particle loading effects using 3D representation of the flow, temperature, and particle trajectories. A schematic representation of the torch geometry (Tekna Plasma Systems Inc., Model PS-35 torch) and reactor design used for the computations are shown in Fig. 10.62. Computations were carried out for torch operation at atmospheric pressure, with a mixture of Ar/H$_2$ as sheath gas, and pure Ar as central gas, and a net power dissipated in the discharge of 15 kW. Two oscillator frequencies were used in the computation. These were three and 13.56 MHz. The

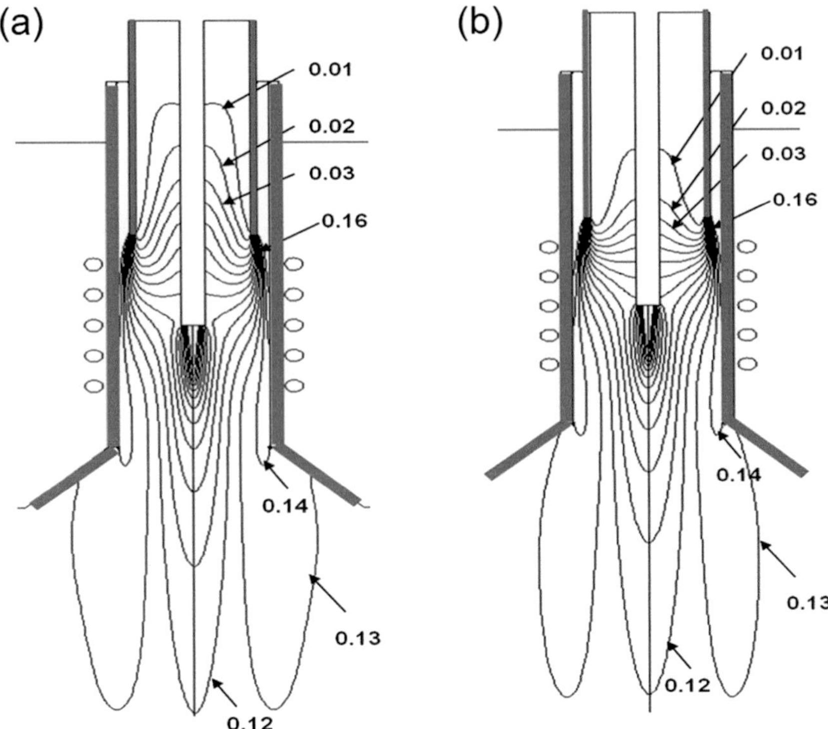

Fig. 10.51 Hydrogen concentration mapping, in vol% H_2, for Tekna PS-50 induction plasma Ar/H_2, plate power $=$ 70 kW free jet at (**a**) pressure $=$ 101 k Pa and (**b**) pressure $=$ 34.4 kPa. Results supplied by Tekna Plasma Systems Inc. (2012)

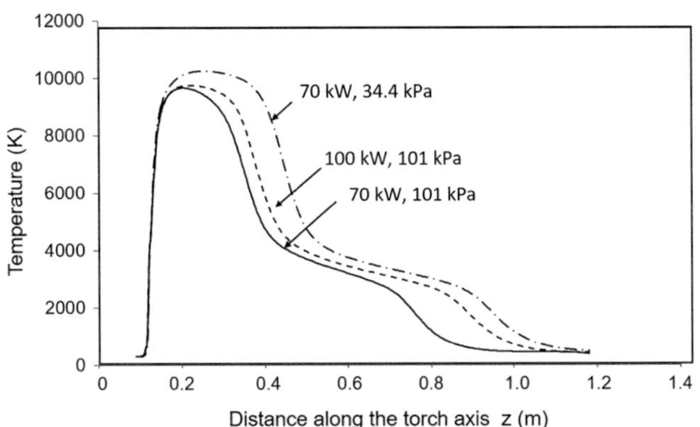

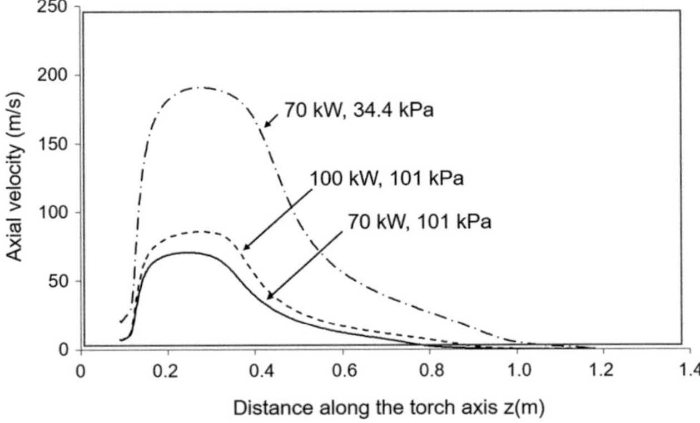

Fig. 10.52 Axial temperature profile along the centerline of a Tekna PS-50 induction plasma free jet at atmospheric (101 k Pa) and low pressure (34.3 k Pa). Ar/H_2 Plasma, plate power $=$ 70 kW. Results supplied by Tekna Plasma Systems Inc. (2012)

Fig. 10.53 Axial velocity profile along the centerline of a Tekna PL-50 induction plasma free jet at atmospheric (101 kPa) and low pressure (34.4 kPa) conditions. Ar/H_2 Plasma, plate power $=$ 70 kW. Results supplied by Tekna Plasma Systems Inc. (2012)

results, given in Fig. 10.63, show the effect of frequency on the temperature fields in the discharge. These are presented as temperature isocontours on two orthogonal planes passing through the axis of the torch, with the results obtained at an oscillator frequency of 3 MHz presented on the left-hand side of the figure (Fig. 10.63a, b), while the results obtained at a frequency of 13.56 MHz are represented on the right-hand side of the figure (Fig. 10.63c, d).

A 3D representation of alumina particle trajectories as they are injected into the center of the discharge is given in

Fig. 10.64. The particle size range used in these cases was varied between 25 and 125 μm, and the powder feed rate was 10 g/min. It is noted that the model predicts a widely dispersed and an asymmetric particle distribution, which can have a major impact on particle deposition efficiency during induction plasma spraying operations. The effect of the powder feed rate on the particle temperature history along their trajectories is shown in Fig. 10.65 for alumina powder feed rates of 2, 5, 10, and 20 g/min. These show an increasing loading effect with the increase of the powder feed rate resulting in a significant reduction of the ability to fully melt the powder especially the large article size fractions (75 and 125 μm). A compilation of the results obtained at the two oscillator frequencies considered and for different injection points of the powder along the centerline of the discharge are presented in Figs. 10.66 and 10.67. The location of the probe tip in this case is defined by $z_p = 45$ or 55 mm, representing the distance between the probe tip and the exit level of the torch nozzle. The results are presented in this case as the mean plasma temperature (K) along the particle trajectory (Fig. 10.65) or the mean plasma temperature as a function of the powder feeding rate (Fig. 10.66), or the integral of the heat transferred to the participle stream (kJ/kg) (Fig. 10.67), as function of the powder feed rate. The results are consistent with earlier observations made by [Proulx et al. (1987)] and given in Fig. 10.56, which shows that the total energy

transferred from the plasma to the particles reaching a plateau with the increase of the powder feed rate resulting in a linear drop of the specific energy transferred to the powder expressed in terms of kJ/kg of powder.

10.5   Induction Plasma Spraying

In this section, a review is presented on the basic elements that enter into the design and operation of an inductively coupled plasma spraying (IPS) system. This is followed by examples of applications of the technology under different system configurations and operating conditions.

10.5.1   Basic System Design

The basic mechanisms involved in IPS is essentially similar to that at play in DC plasma spraying that involves the in-flight heating and melting of the feed material in powder form followed by the projection of the formed molten droplets on the surface of the substrate forming splats that are the building blocks of the coating. As in the case of DC plasma spraying, the shape of the formed splats depends on the inflight particle parameters, velocity and temperature, prior to their impact on the substrate, and the substrate properties such as cleanliness and surface roughness and most of all the substrate temperature that was shown to play a key role in the splat formation. Coating formation is also possible through the vapor phase deposition route as in the case of physical, or chemical vacuum deposition (PVD) or (CVD), as in the case of DC ultra low-pressure plasma deposition.

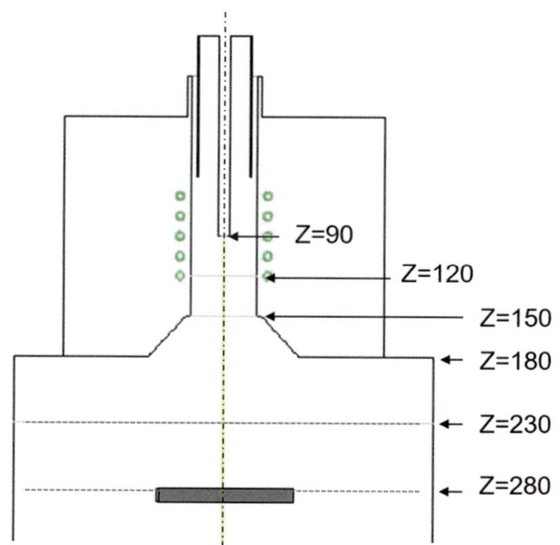

Fig. 10.54 Torch geometry and system of coordinates for an induction plasma vacuum spraying setup. Tekna Plasma Systems Inc. (2012)

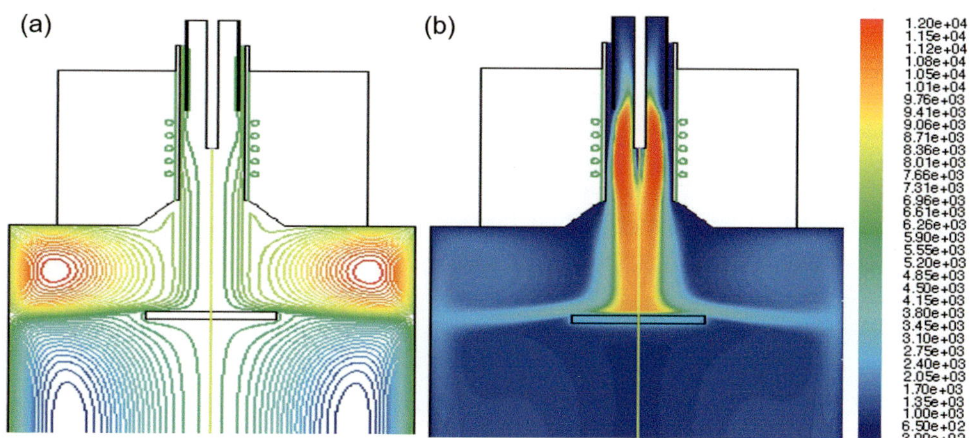

Fig. 10.55 Temperature and flow fields for an impinging induction plasma jet under vacuum plasma spraying conditions. Ar/H$_2$ plasma, pressure = 34.3 kPa, plate power = 70 kW. Results supplied by Tekna Plasma Systems Inc. (2012)

DC and induction plasma spraying differs, however, in a number of key features the most obvious of which is in the access to the plasma for powder injection. As shown in Fig. 10.68, while the large majority of DC plasma spray torches use radial powder injection mode into the plasma jet, whether internal to the torch nozzle or external at the exit level of the plasma jet, RF-IPS allows for axial injection of the powder into the center of the discharge. Exceptions do exist in both technologies with axial powder injection possible when using Axial III DC plasma torches, and radial

injection of the powder in the plasma jet at the exit of the ICP troch has also been successfully used in certain application such as preform cladding on the fiber optics industry. The second most significant difference between the two technologies, DC and IPS, lies in the fact that the use of RF power for the energy transfer to the plasma does not allow for the use of flexible leads as is the case with DC plasma torches. In contrast to DC plasma torches, which can generally be easily manipulated by a robotic arm, RF-IPS torches are fixed in rigid position connected to the power supply and it is the substrate that needs to be manipulated for the control of the spray pattern and the coating distribution on the part to be sprayed. Such a constrain offers important limitations on the shape, dimensions and weight of the part to be sprayed depending on the capacity of the robotic arm used for part manipulation and on the volume of the spray chamber, in the case of vacuum induction plasma spraying (VIPS). The net impact of such a constraint is in the limitation of the use of the IPS technology to coating of relatively small parts with high added value that would benefit from the fact that, being an

Table 10.3 Thermophysical properties of the powder used in the plasma pray modeling study

Material	Molybdenum
Mass density	10,200 kg/m^3
Specific heat	0.25 kJ/kg K
Melting temperature	2890 K
Latent heat of fusion	333.54 kJ/kg
Boiling temperature	4912 K
Latent heat of vaporization	6.23 MJ/kg

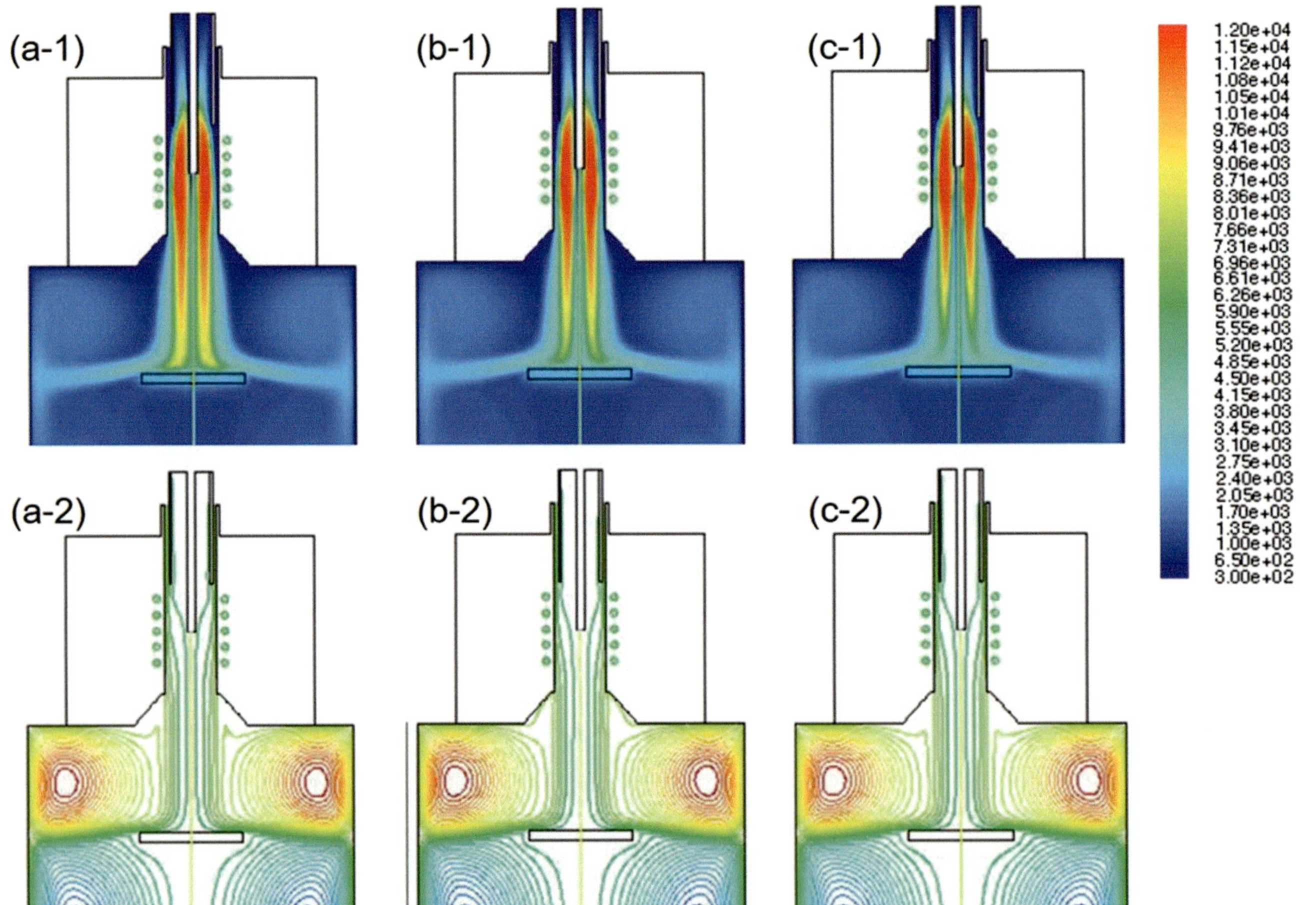

Fig. 10.56 Effect of powder feed rater on the (**a**) temperature and (**b**) flow fields for an impinging induction plasma jet under vacuum plasma spraying conditions. Ar/H$_2$, pressure = 34.3 k Pa, Plasma, plate power = 70 kW condition (1) 4.0 kg/h, (2) 8.0 kg/h, and (3) 16.0 kg/h. Results supplied by Tekna Plasma Systems Inc. (2012)

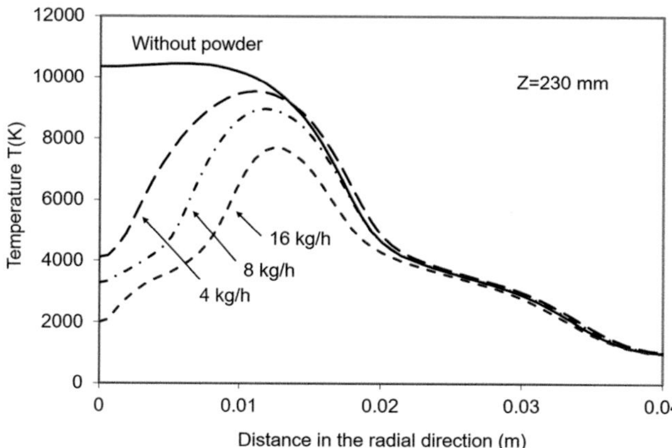

Fig. 10.57 Radial temperature profiles, at $z = 230$ mm, see Fig. 8.50, for an impinging induction plasma jet under vacuum plasma spraying conditions, with molybdenum powder feed rates varying between 4 and 16.0 kg/h. Ar/H$_2$ Plasma, plate power = 70 kW. Results supplied by Tekna Plasma Systems Inc. (2012)

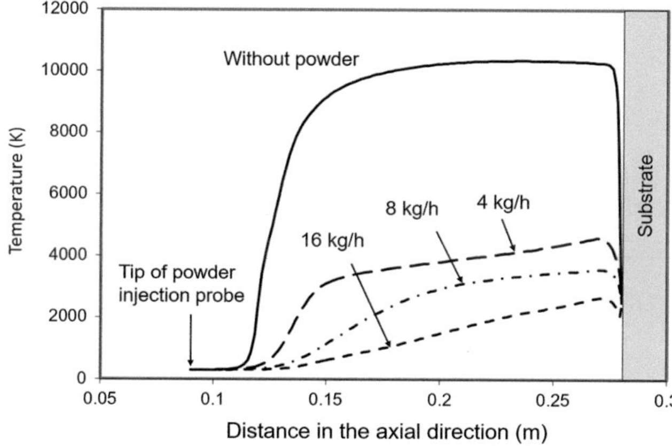

Fig. 10.58 Plasma temperature profiles along the center line of the discharge for an impinging induction plasma jet under vacuum plasma spraying conditions, with molybdenum powder feed rates of 4, 8.0, and 16.0 kg/h. Ar/H$_2$ Plasma, plate power = 70 kW. Results supplied by Tekna Plasma Systems Inc. (2012)

electrodeless plasma technology, it offers means of operating under ultrahigh purity conditions that are not always compatible with DC plasma technologies due to electrode erosion. The ultrahigh purity environment in the VIPS deposition chamber also allows for greater flexibility in the selection of the coating precursor, which is not limited to powders, and could be a gas, liquid, solution, or suspension.

10.5.2 Atmospheric Induction Plasma Spraying

One of the first applications of induction plasma spraying (IPS) was for the growth of refractory crystals using induction plasma torch [Reed TB (1961a, b)]. The study carried out

using a 25 mm i.d. plasma confinement quartz tube, Fig. 10.69, described the growth of crystals in a Verneuil-type geometry that could be carried out at higher temperatures than normally possible with flames in an inert or reactive atmosphere. The plasma was ignited at atmospheric pressure using argon as plasma gas, followed by the gradual additions of helium, or a molecular gas such as oxygen, nitrogen, or hydrogen depending on the crystal being grown. Total gas flow rate was 10–20 slm and power 1–10 kW. The material used for the crystal growing was supplied in powder form injected into the center discharge with a carrier gas. Sapphire crystals were grown in an argon atmosphere with a diameter of 5–15 mm and length up to 90 mm at a rate of 20–50 mm/h. Results were also reported for the growth of Calcia-stabilized zirconia crystals.

The same approach has been subsequently used in the fiber optics industry for the preform overcladding with high purity silica grain for the protection of the fiber core. As reported by [Sudarsan Neogi, private communication] in the process of manufacturing optical fiber, the primary preform is manufactured by modified chemical vapor deposition (MCVD) process in which a tube made of high purity silica is coated on its inner surface with multiple layers of doped synthetic silica. With the proper control of the nature of the dopant and its concentration profile between successive layers, it is possible to obtain, following the subsequent heating and sintering the deposited soot, a clear glass with the required refractive index profile necessary for the optimal performance of the optical fiber. Considering that after drawing the fiber, its optically active core is reduced to about 8 μm in diameter, it is necessary for the mechanical reinforcement of the core of the fiber to grow the preform to a diameter that would result in a final fiber diameter following its drawing that is about 15× that of the active core (i.e., external diameter of the final fiber of 125 μm). Atmospheric IPS is one of the techniques that gained wide acceptance for such a task because of its ability to maintain the highest level of purity in the deposited glass on the outer surface of the core. The process, as illustrated in Fig. 10.70, involves the heating of the outer surface of the primary preform using an ICP jet of pure oxygen or air (3 MHz frequency and 100 kW) directed at an angle toward the surface which is in continuous rotation and translation. High purity silica grain is injected into the plasma stream by an appropriate feeding tube. As the individual silica grain particles are entrained by the plasma stream, they are converted to a molten or semi-molten state before being projected toward the surface of the preform on which they adhere and further heated to sufficiently high temperature in order to form a uniform clear glass layer. The preform substrate is kept in rotation and back and forth translation in front of the plasma torch to build successive layers of high-quality glass until the required final diameter of the preform is reached. The process includes the torch

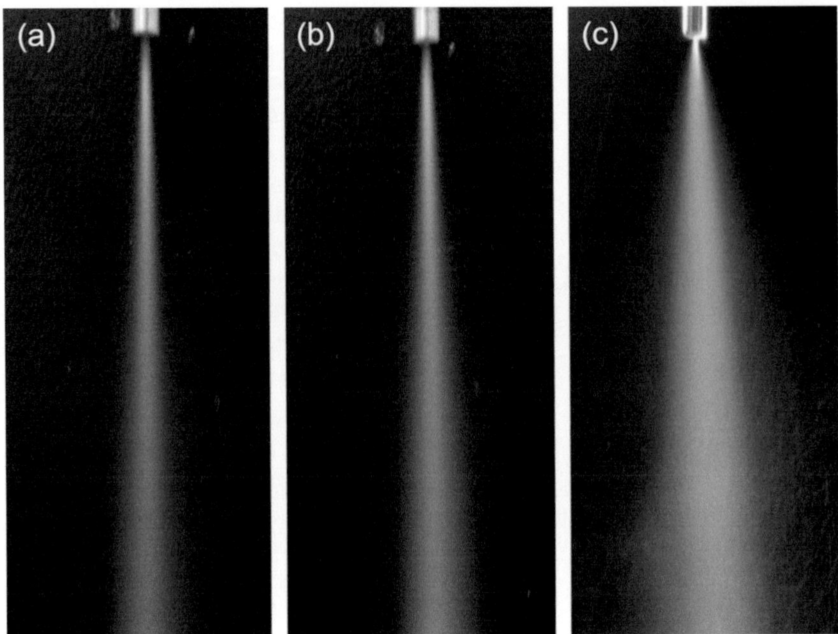

Fig. 10.59 Photographs of a particle jet illustrating the effect of probe design and dispersion gas flow rate on the powder dispersion. Results supplied by Tekna Plasma Systems Inc. (2012)

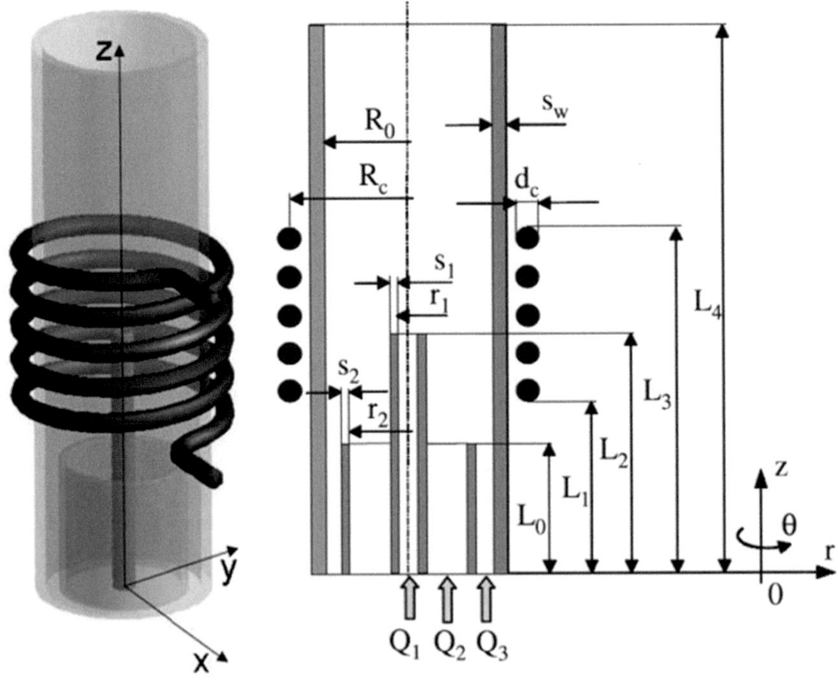

Fig. 10.60 3D schematic of the induction plasma torch. After Bernardi et al. (2004a)] with kind permission

power, plasma gas composition, spraying distance, grain size, and feed rate as well as the speed of rotation and translation of the substrate. Typical grain deposition rates are of the order to 15–20 g/min.

Variations on this, using ICP technology for the deposition of high purity silica has been reported in the patent literature such as the Alcatel process, US Patent application 2002/0011083 A1, illustrated in Fig. 10.71.

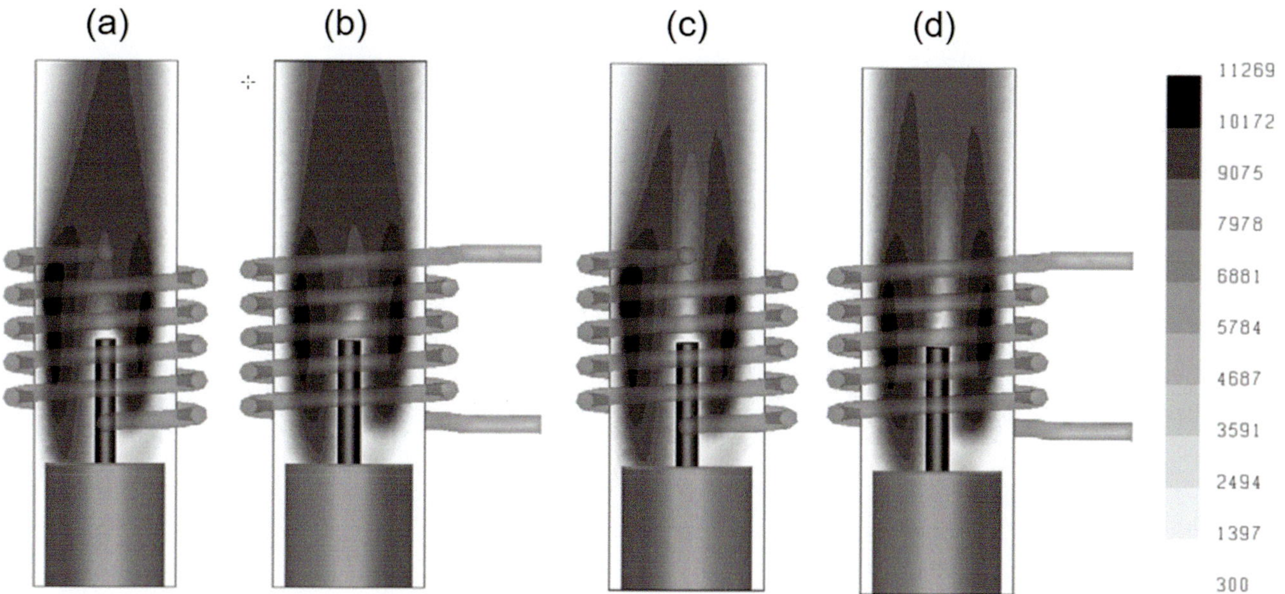

Fig. 10.61 Plasma temperature fields (K) on two orthogonal planes through the axis of the torch. Cases (**a**) and (**b**) without powder injection. Cases (**c**) and (**d**) with 30 μm W powder injection at 20 g/min. After Bernardi et al. (2004a)] with kind permission

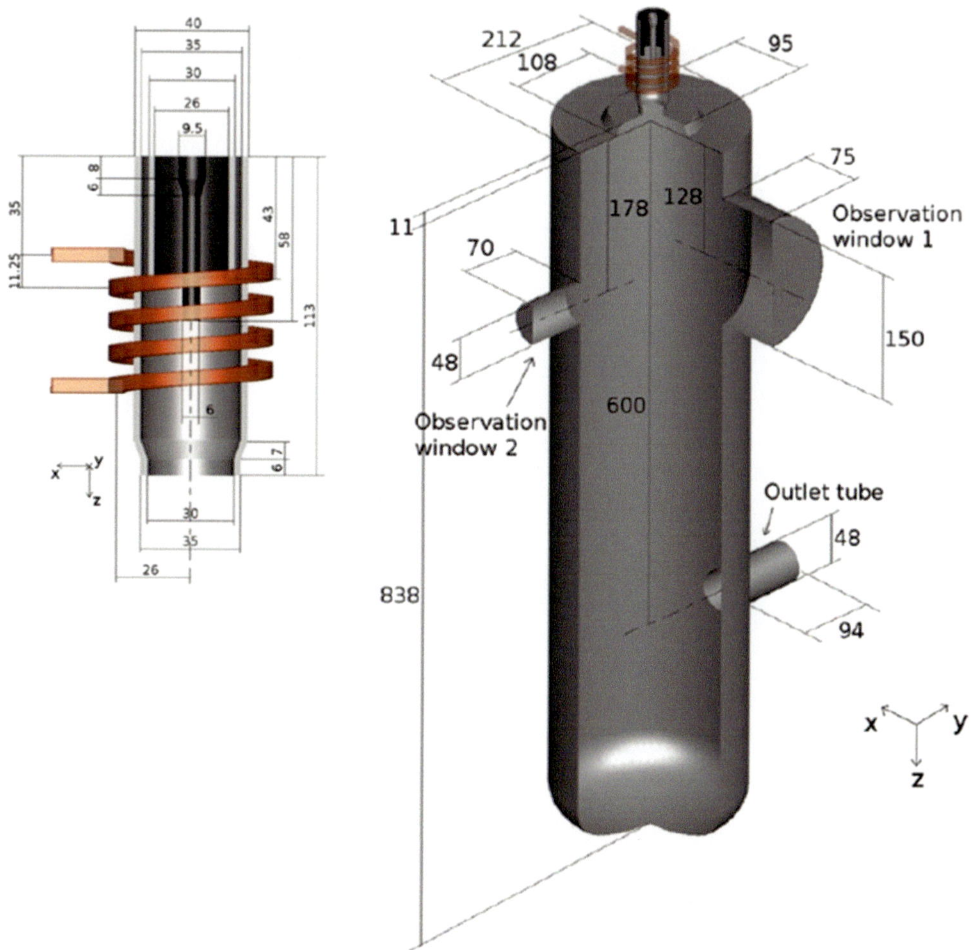

Fig. 10.62 Details of the PL-35 torch geometry and reaction chamber, dimensions in mm. After Colombo et al. (2010)]. Copyright [2010], American Institute of Physics, Reproduced with permission. All rights reserved

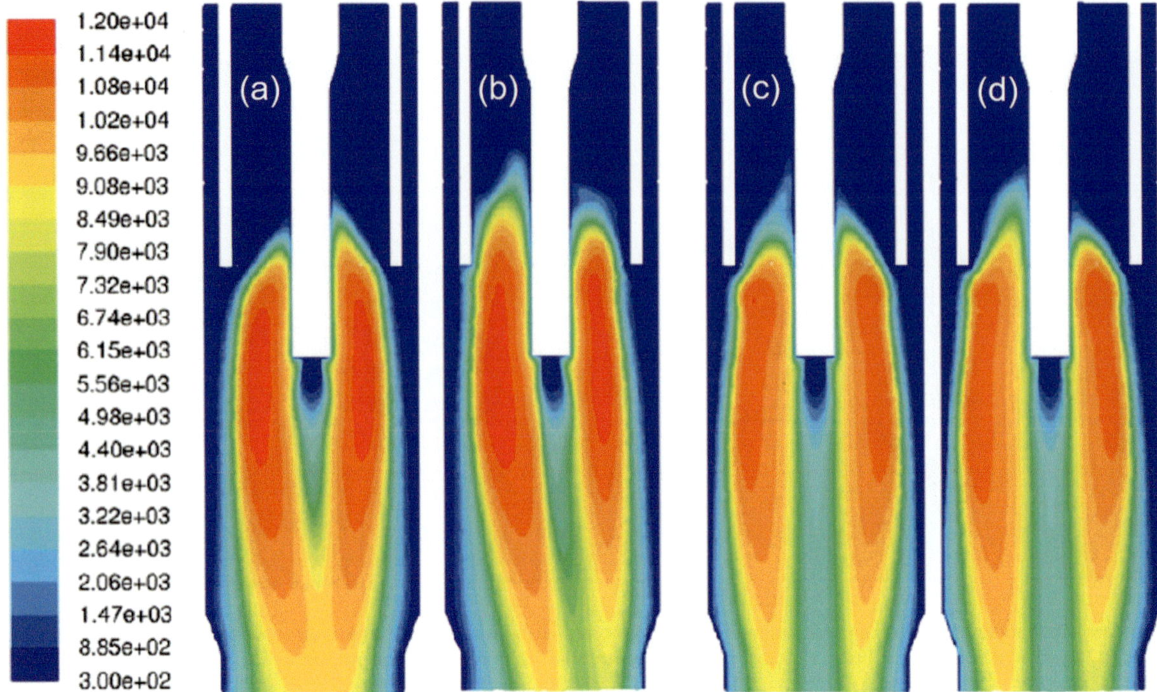

Fig. 10.63 Plasma temperature distribution (K) on two orthogonal planes through the axis of the torch. Cases (**a**) and (**b**) for 3 MHz operation. Cases (**c**) and (**d**) for 13.56 MHz operation, alumina powder feed rate at 10 g/min, with a probe tip at 55 mm upstream of the torch exit nozzle level. After [Colombo et al. (2010)]. Copyright [2010], American Institute of Physics, Reproduced with permission. All rights reserved

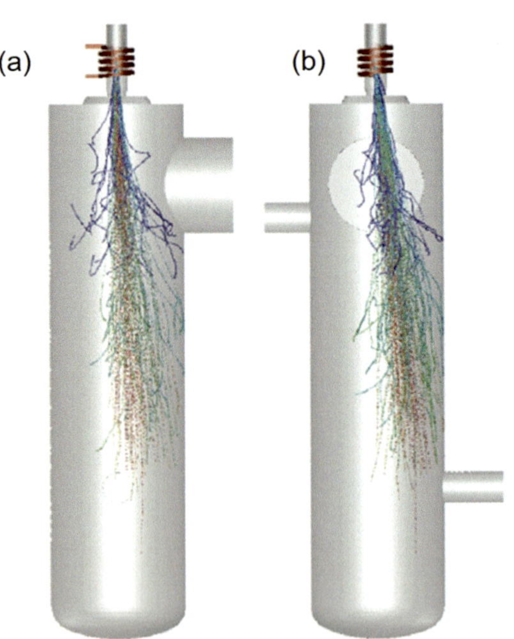

Fig. 10.64 Plasma trajectories on two orthogonal planes passing through the torch and reactor axis, for the case at 3 MHz, powder feed rate of 10 g/min, and probe tip position of 55 mm upstream of the torch exit nozzle. After Colombo et al. (2010)]. Copyright [2010], American Institute of Physics, Reproduced with permission. All rights reserved

An alternate approach proposed by Berthou et al. (1993) is based on the deposition of the high purity silica over-clade through an ICP thermal plasma CVD process in which the silica is synthesized through oxidation $SiCl_4$ vapor with an oxygen plasma:

$$Si\,Cl_4 + O_2 \rightarrow SiO_2 + 2Cl_2 \qquad (10.14)$$

A schematic of the setup used is illustrated in Fig. 10.72

The same approach was also used for the production of synthetic high purity silica boule through using an ICP torch for the heating of oxygen in which at the $SiCl_4$ is injected in vapor phase as shown in Fig. 10.73. The formed silica soot is deposited on the surface of the target that is vitrified to a clear glass through its further heating under controlled conditions.

10.5.3 Vacuum Induction Plasma Spraying

The principal components of an RF-IPS system areillustrated in Fig. 10.74. The RF power supply is the centraland largest single components of the system, whichaccording to current technology would mostly be of thevacuum tube design with a simplified circuit diagram givenearlier in Fig. 10.44. The power supply will have 3-phase ACpower input at nominal regional frequency and a high voltageRF output accessible

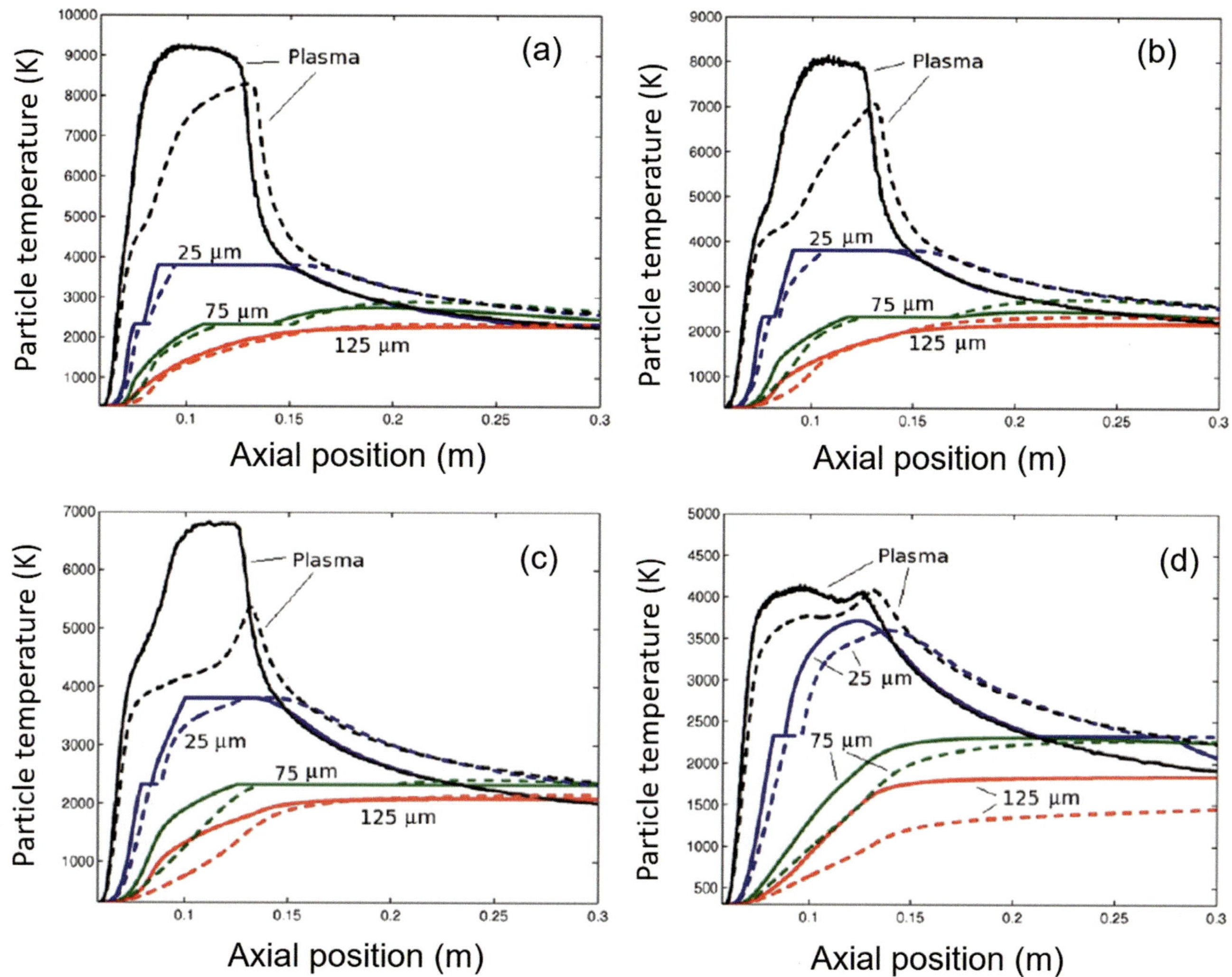

Fig. 10.65 Plasma and mean particle temperature computed along their trajectories for the 3 MHz (solid line) and 13.56 MHz (dashed line) for different particle diameters and powder feed rates. Probe tip at 55 mm upstream of the torch exit nozzle level. (**a**) mp = 2 g/min, (**b**) mp = 5 g/min, (**c**) mp = 10, and (**d**) mp = 20 g/min. After Colombo et al. (2010)]. Copyright [2010], American Institute of Physics, Reproduced with permission. All rights reserved

through a tank circuit for fine tuning theimpedance of the plasma torch load with that of the powersupply. Normally, the tank circuit is integrated in the mainpower supply cabinet with leads accessible from the side ofthe cabinet "auxiliary tank circuit RF outlet" (Fig. 10.69).Such an option is considered for small power R&D systemsor in the event the system is used in an open atmosphericpressure discharge mode in order to have the plasma torch/spraying system at a lower easily accessible level. Alternately,the tank circuit can be moved to a remote location atthe top of the cabinet as shown in Fig. 10.74 for easy connectionto the RF ICP torch when places at the top of a reactor orlarge vacuum deposition chamber.As mentioned earlier in Sect. 10.3 Induction Plasma TorchDesign, the choice of the type and dimension of the plasmatorch to be used has to be carefully made in order to match theoperating conditions, atmospheric or vacuum plasmaspraying, and power supply frequency and plate powerrange to be used. In vacuum inductively coupled plasmaspraying (VIPS) systems the ICP torch is mounted on thetop of a water-cooled, vacuum deposition chamber equippedwith the centrally located robotic arm with three or fourdegrees of freedom for the manipulation of the substrate inthe plasma/molten particle stream emerging from the plasmatorch. Because of the high ambient temperature in the chamber,the mechanical components of the robotic arm shouldalso be water-cooled and adequately protected against radiantheating and the dusty environment. Exhaust gases from thechamber loaded with residual powders and fumes exit thespray chamber from an appropriate outlet at the bottom of thechamber to the gas cleaning and vacuum pumping

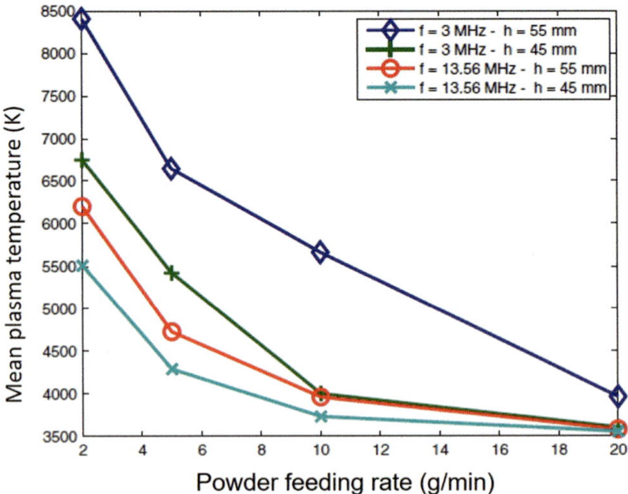

Fig. 10.66 Mean particle temperature computed along their trajectories for the 3 and 13.56 MHz cases. And for two probe tip positions, $h = 45$ and 55 mm, as function of the powder feed rate. After [Colombo et al. (2010)]. Copyright, American Institute of Physics, Reproduced with permission.

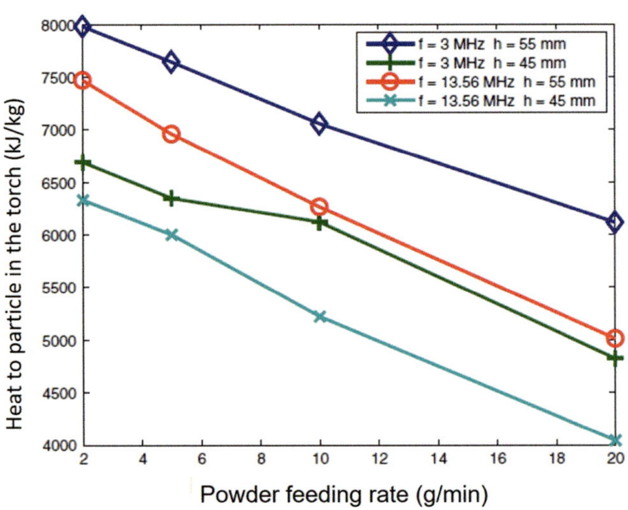

Fig. 10.67 Integral energy transferred from the plasma to the particles along their trajectories for the 3 and 13.56 MHz cases. And for two probe tip positions, $h = 45$ and 55 mm, as function of the powder feed rate. After Colombo et al. (2010)]. Copyright, American Institute of Physics, Reproduced with permission.

station.Other components of the system include a cooling waterschiller used with a closed loop cooling-water circuit for thecooling of the power supply components including the oscillatorvacuum tube, plasma torch, reactor or vacuum depositionchamber, and system accessories. A control system,either integrated in the power supply or in a separate cabinet,serves to monitor vital system parameters including torchignition, power control, current, voltage, gas and water flowrates, cooling-water temperature and safety interlocks. Anappropriate gas and powder feeder supply form an integralpart of the system. The latter is normally placed at a

relativelyhigh level above the torch to shorten as much as possible thelength of the powder transport line to the entrance of thewater-cooled powder injection probe centrally placed in the torch as shown in Fig. 10.74.

10.5.3.1 System Design

A typical medium size R&D VIPS system, as represented schematically in Fig. 10.75, is composed of a horizontally oriented, water-cooled, stainless steel cylindrical chamber, 0.5 m in internal diameter and 0.64 m long, on which the ICP torch is mounted on the top in a fixed position. A Faraday cage (not shown in the drawing) surrounds the torch to protect the surrounding equipment, instrumentation, and computer-based control modules from EMI originated from the induction coil. Coaxial with the torch is the exit port from the chamber through which the plasma gases and powder overspray are evacuated. This port is located directly underneath the torch on the opposite side of the chamber. It is of conical shape ending by a water-cooled cylindrical container in which any over-sprayed powder collects by gravity prior to the exit of the plasma gases from a side port. From this port, the gases are directed toward a filter to recover any entrained fumes and fine powders before being sent to the vacuum pumping station. An access door to the deposition chamber is located on the right-hand side of the figure. It is also water cooled and equipped with appropriate seal for maintaining the vacuum tightness of the chamber while allowing for quick opening and closing of the chamber for installation of the substrates to be coated and recovery of the coated components. Opposite to the access door is the mechanical module in which the motors and motion control components are located for a 3-axis robotic arm, which supports the substrate. The water-cooled arm has to allow for the precise control of the exposure of the substrate to the stream of plasma and molten particles immerging from exit nozzle of the plasma torch. In standard, pilot installations of the robotic arm is designed for the control of the linear motion of the arm in the horizontal direction, the vertical spraying distance between the substrate and the torch exit, and the speed of rotation of the substrate around a vertical or horizontal axis. In more elaborate and larger units, a fully equipped five-axis robot can be located in the deposition chamber in order to allow for the articulation of more complex parts underneath the plasma torch, at the expense of considerably increased level of complexity and the size of the spraying chamber.

A photograph of an induction plasma vacuum spraying chamber at Tekna Plasma Systems Inc. in Sherbrooke, Québec Canada, is shown in Fig. 10.76. One notices the powder feeder located on the top of the chamber above the induction plasma torch. The latter is not visible since the Faraday cage used for protection from electromagnetic emission covers the torch. On the bottom, right-hand side of the photograph, the RF power supply, and control panel for the operation of the system are noticeable.

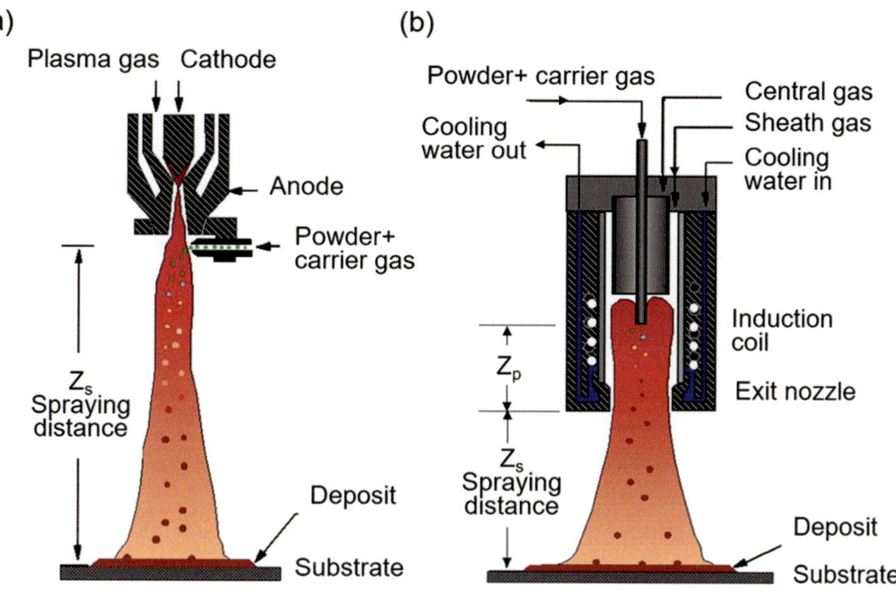

Fig. 10.68 Schematic representation of commonly used powder injecting techniques in (**a**) DC plasma spray torch and (**b**) RF-IPS

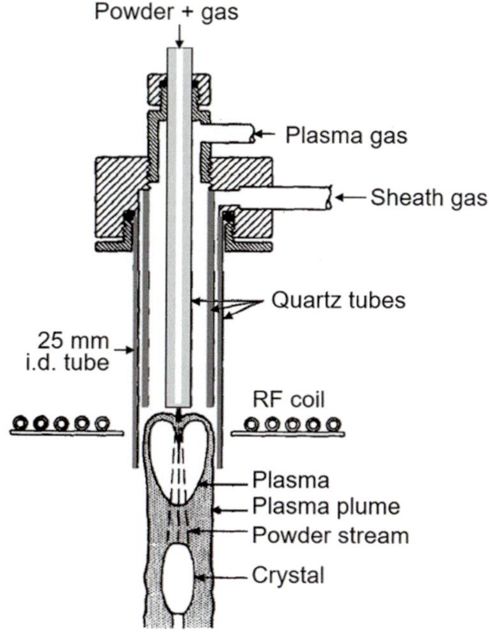

Fig. 10.69 Laboratory scale ICP torch used for crystal growing [Reed TB (1961a, b)]

10.5.3.2 Splats and Coating Formation

As with other thermal spray operations, VIPS depends largely on the ability to heat and melt, inflight, the precursor particles prior to their impact on the surface of the substrate. As discussed in Chaps. 8 and 9 dealing with DC plasma spraying, the splat formation on the substrate is a critical step in the building up of the coating that required complete melting of the particles prior to their impact on the substrate as well as substrate surface preparation and its temperature. The same elements apply to IPS with the exception that because of the relatively large volume and low velocity of the RF-ICP discharge, the residence time of the particles in the plasma stream is typically of the order of 10–20 ms, which is one order of magnitude larger than the 1–2 ms residence time typical of DC vacuum plasma spraying. The ability to melt and deposit with RF-IPS powders with a considerably larger diameter than with DC plasma sources is demonstrated by powder melting tests that are generally a prerequisite for the evaluation of the

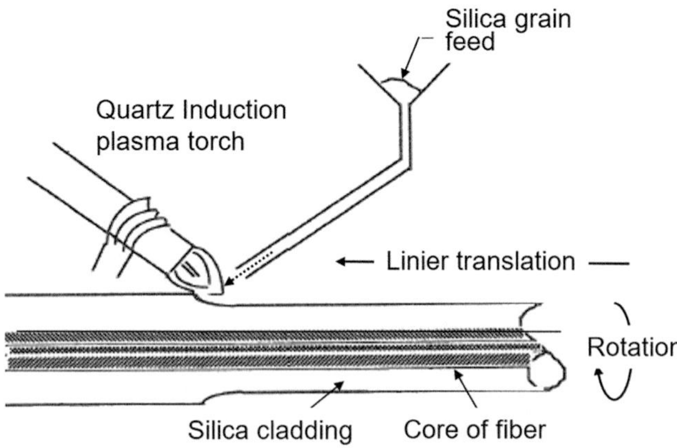

Fig. 10.70 Schematic of the preform plasma cladding process in the fiber optics industry [Sudarsan Neogi, private communication]

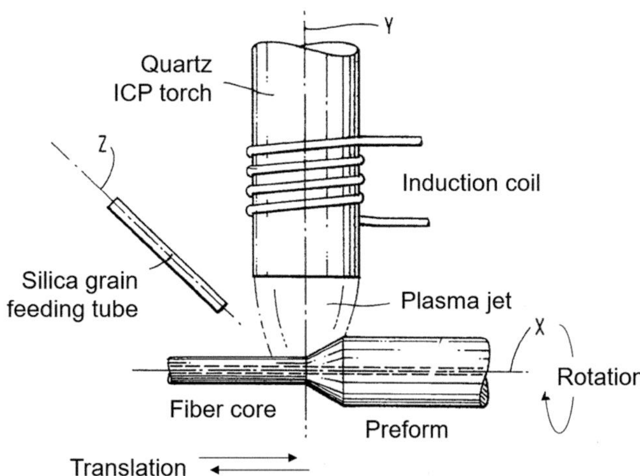

Fig. 10.71 Alcatel method and apparatus for manufacturing an optical fiber preform, US Patent application 2002/0011083 A1

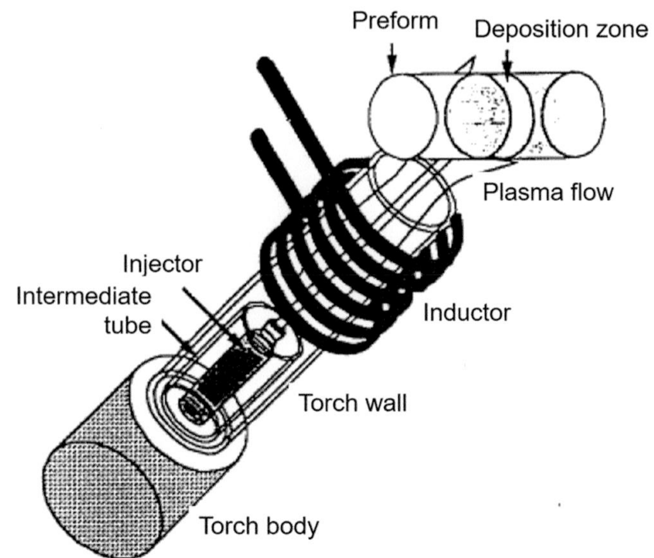

Fig. 10.72 Schematic of the system used for the thermal CVD of synthetic SiO_2 on tubular quartz substrate preform [Berthou et al. (1993)]

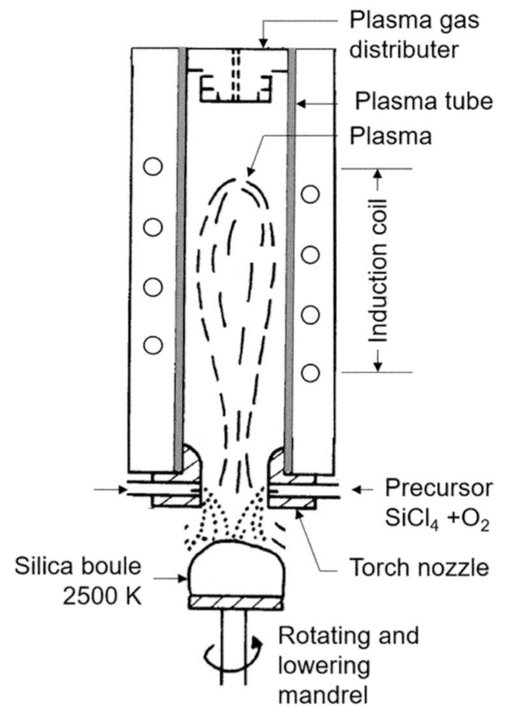

Fig. 10.73 ICP synthesis of high purity silica boule/ingot

possible use of a plasma source for the spraying of a given powder. A typical example is given in Fig. 10.77 for the melting of a molybdenum powder with a melting temperature ($T_m = 2896$ K), particle size in the range of 40–250 μm ($-60 + 400$ Mesh) with a relatively porous morphology as shown in Fig. 10.77a. Micrographs of the powder treated in a 100 kW, Ar/H$_2$, induction plasma at near atmospheric pressure are given in Fig. 10.77b, c, and d for powder feed rates of 7.0, 14.0, and 20 kg/h, respectively. These show complete melting of the powder forming perfectly spherical particles at feed rates up to 14.0 kg/h. At higher feed rates, an increasing fraction of the powder is partially melted that would lead to poor splatting formation and lower deposit density. Similar results obtained with other refractory metals and ceramics such as Rhenium ($T_m = 3453$ K) and Tungsten ($T_m = 3683$ K) are shown Figs. 10.78 and 10.79, respectively.

Corresponding splats obtained with finer molybdenum under VIPS conditions are given in Fig. 10.80. The powder

used in this case had a narrower particle size distribution (PSD) in the range of 53–63 μm, melted in an Ar/H$_2$ plasma at a plate power of 35 k with a powder feed rate of 26 g/min (1.56 kg/h), chamber pressure = 26.6 kPa (200 Torr), and a total spraying distance, $z_t = 350$ mm. Where z_t is the distance between the tip of the powder injection probe in the torch and the substrate, ($z_t = z_p + z_s$) as indicated on Fig. 10.68. These

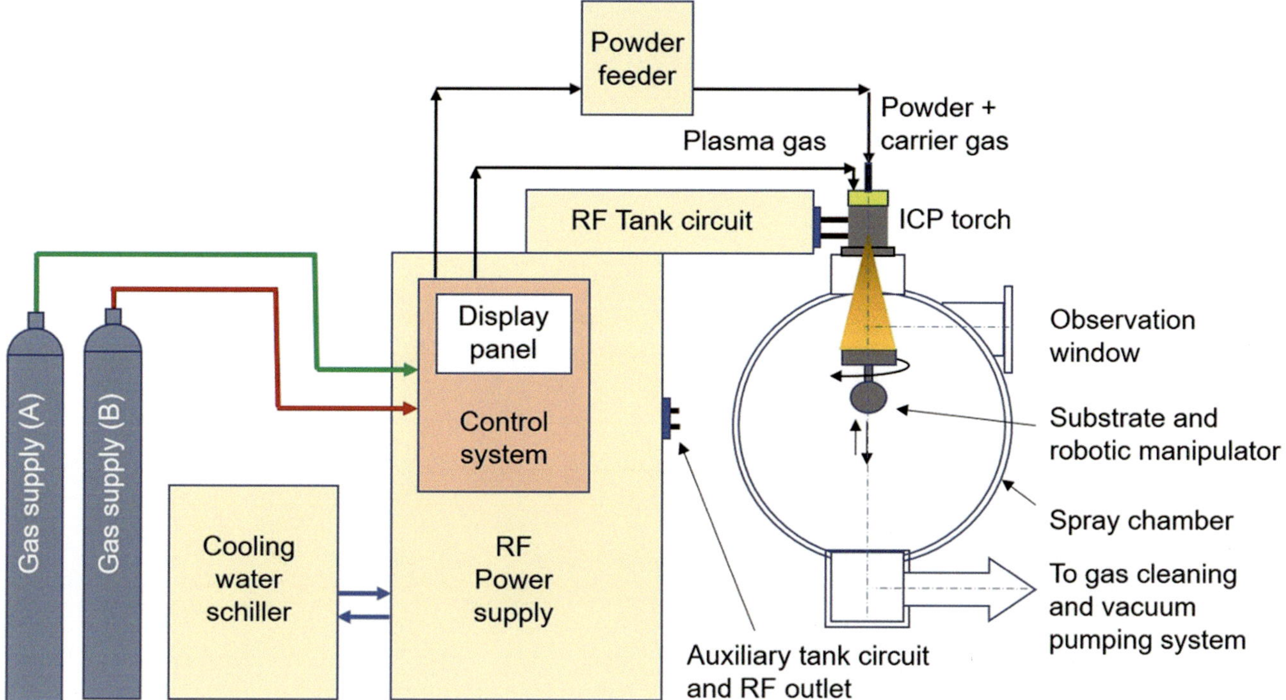

Fig. 10.74 Schematic representations of principal components of an integrated RF-VIPS system

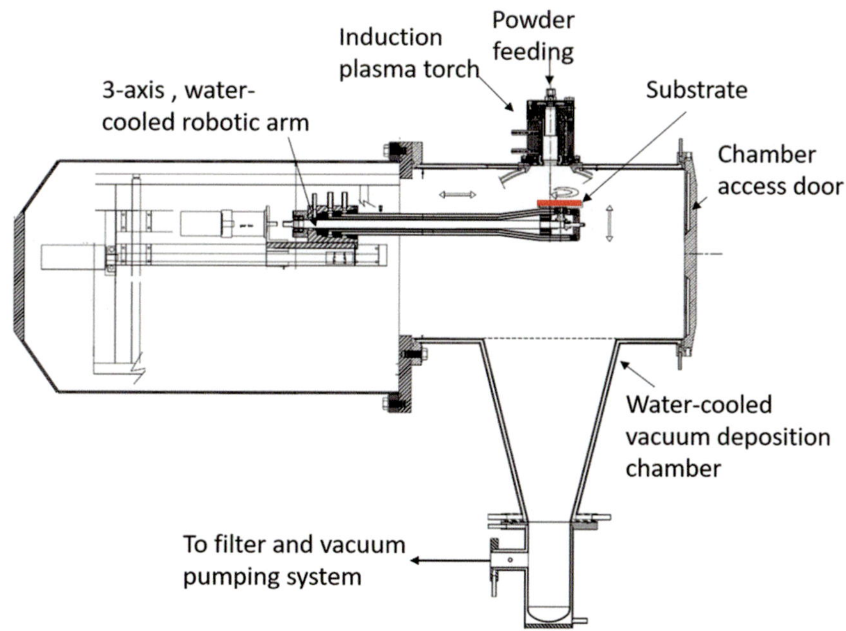

Fig. 10.75 Schematic of a pilot scale RF-VIPS chamber [Drawing courtesy of Tekna Plasma Systems Inc. (2012)]

show an excellent flattening of the molten Mo droplets with a minimal level of splashing at the outer rim of the splats.

Micrographs of splats obtained with a tungsten power of the same PSD, between 53 and 63 μm, and powder feed rate of 26 g/min (1.56 kg/h), with an Ar/H$_2$ plasma at slightly higher power of 40 kW with a chamber pressure = 16 kPa (120 Torr) are given in Fig. 10.81. The results obtained at a total spraying distance, z_t = 350 mm show essentially the same degree of droplet flattening as observed with the molybdenum powder (Fig. 10.80) except for the occasional

Fig. 10.76 Photograph of a 100 kW RF-VIPS system [Courtesy of Tekna Plasma Systems Inc. (2012)]

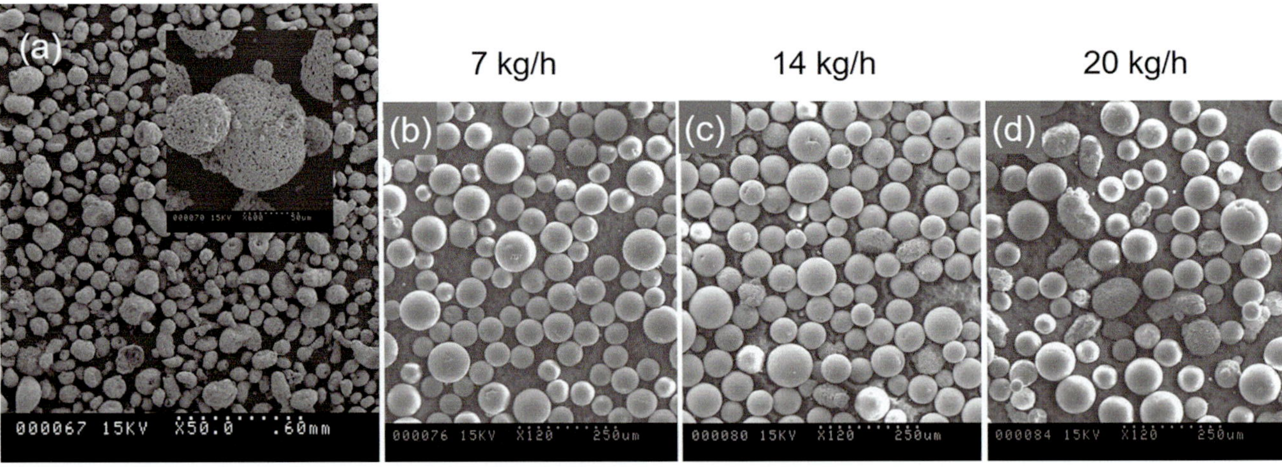

Fig. 10.77 Electron micrographs of molybdenum powder melted in an atmospheric pressure, 100 kW, Ar/H$_2$ IPS process at different powder feed rates: (**a**) raw powder and (**b–d**) processed powder. Please note change of magnification in (a) compared to other microgrpahs

presence of a small number of partially molten particles. In a rerun of these conditions with a slightly higher chamber pressure = 26.6 kPa (200 Torr) that is identical to that used with the molybdenum powder, it is observed that, at a spraying distance of $z_t = 320$ mm, Fig. 10.82a, a superior level of splat formation is obtained. This is not, however, the case if the spraying distance is increased to $z_t = 400$ mm, where visible signs of the presence of solid, or partially molten, particles are observed in Fig. 10.82b. Considering that all the particle flux was completely molten as identified by the splats obtained at $z_t = 320$ mm, the only possible explanation that could justify the reappearance of partially molten particles among the splats at $z_t = 400$ mm is the inflight solidification of molten droplets prior to their impact

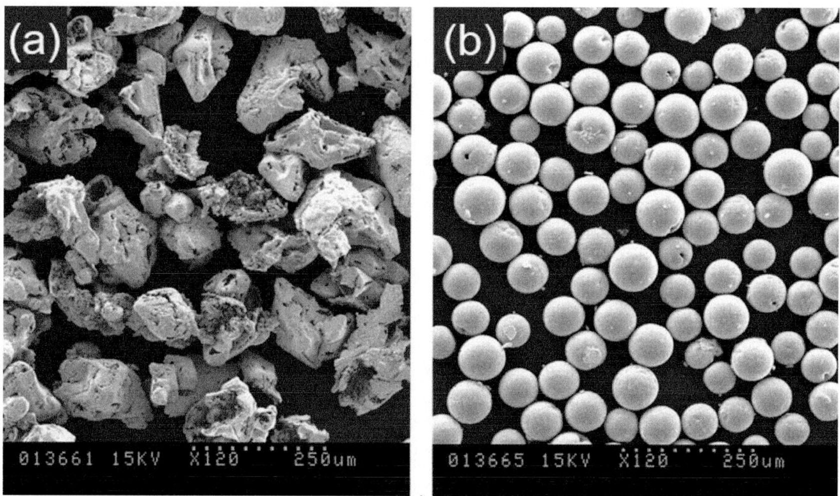

Fig. 10.78 Electron micrographs of rhenium powder melted in an atmospheric pressure, 100 kW, Ar/H$_2$ IPS process: (**a**) raw powder and (**b**) processed powder

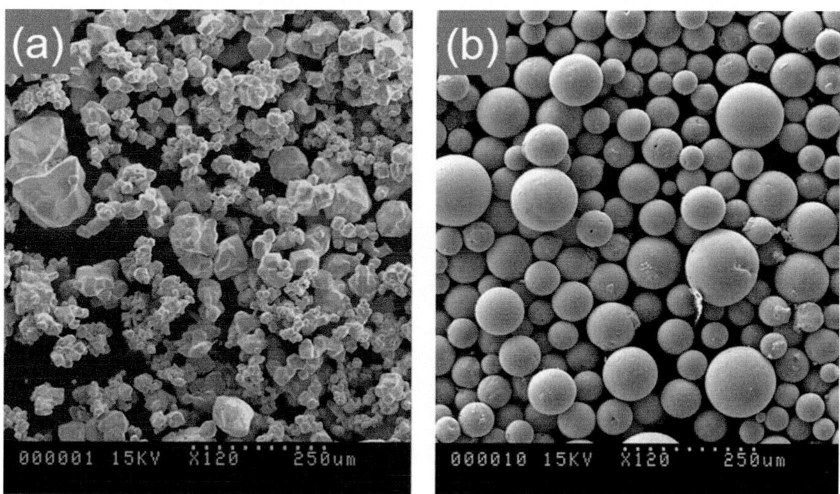

Fig. 10.79 Electron micrographs of tungsten powder melted in an atmospheric pressure, 100 kW, Ar/H$_2$ IPS process: (**a**) raw powder and (**b**) processed powder

on the substrate. The phenomena, while to some extent unexpected considering the high temperature of the plasma in the plume of the jet, could be a direct consequence of intense radiation energy losses from the surface of the particles at these high temperatures $T > 3683$ K, as discussed in details in Chap. 4, Plasma–Particle Momentum and Heat Transfer, Sect. 4.4.3 Radiation Energy Losses From the Surface of the Particle. The direct consequence of this observation is that when dealing with the spraying of refractory metals with melting temperature above 2500–3000 K, the proper control of the spraying distance is rather critical since a too short of a distance would not allow the particles the necessary time for complete melting, while a too long of a distance would risk in-flight particle

solidification due to radiation energy loses from the surface of the particle.

Comparable results were obtained with ceramic powders such as alumina with a melting temperature of $T_m = 2345$ K. The splats shown in Fig. 10.83 were obtained with an alumina powder with a PSD in the 53–63 µm range, Ar/H$_2$ plasma at 25 kW, powder feed rate of 15 g/min (0.9 kg/h), a spraying distance, $z_t = 270$ mm, at two chamber pressure of 33.2 kPa (250 Torr) and 66.5 kPa (500 Torr). These show at the higher pressure perfect rounded splats in contrast to the "splashed" splats observed when operating a lower pressure. A rather unexpected form of alumina splat is observed in Fig. 10.84, which could only originate from a partially molten particle as supported by the inflight particle

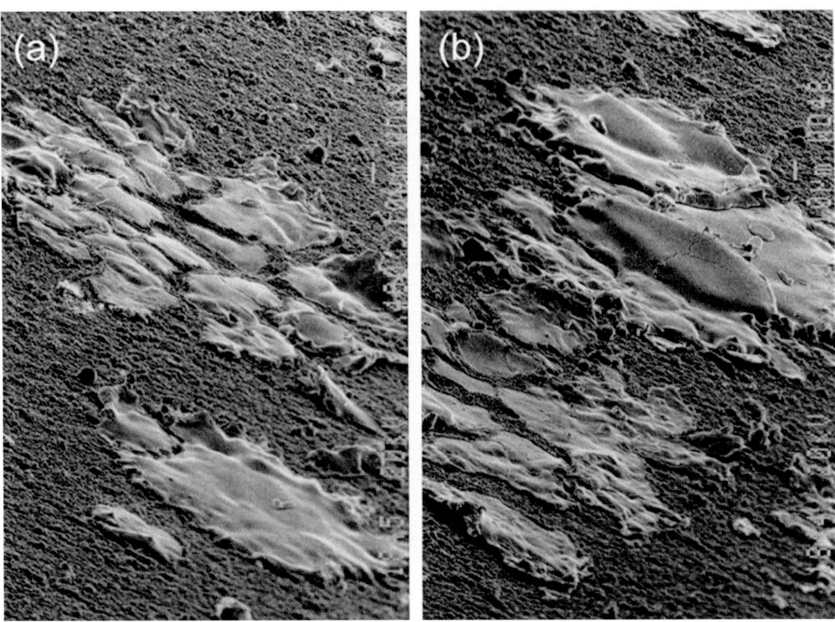

Fig. 10.80 Molybdenum splats obtained using VIPS Ar/H$_2$ plasma at plate power of 35 kW, chamber pressure 26.6 kPa (200 Torr), PSD of the powder 53-63 μm, powder feed rate, 26 g/min (1.56 kg/h), and total spraying distance, $z_t = 350$ mm

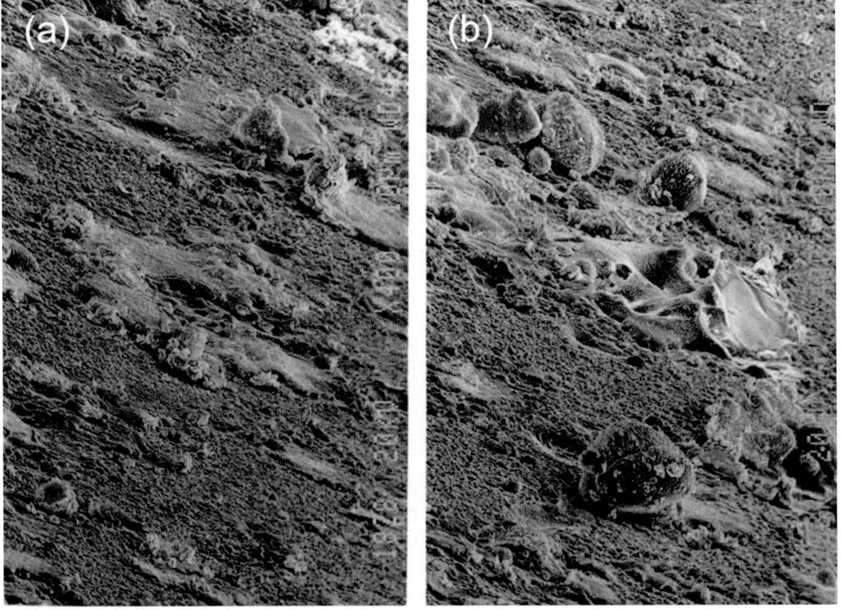

Fig. 10.81 Tungsten splats obtained using VIPS Ar/H$_2$ plasma at plate power of 40 kW, PSD of the powder 53–63 μm, powder feed rate, 26 g/min (1.56 kg/h), chamber pressure 16 kPa (120 Torr), and spraying distance $z_t = 350$ mm

temperature and velocity distributions prior to their impact on the substrate given, respectively, in Fig. 10.84b, c. The alumina powder used in this case was slightly courser, 33.2 kPa (PSD in the 63–75 μm range, 25 kW plasma, powder feed rate of 26 g/min (1.56 kg/h), spraying distance, $z_t = 210$ mm, and chamber pressure of 33.2 kPa (250 Torr).

10.5.4 Coatings and Near Net-Shaped Parts

Since the early nineties systematic studies on the induction plasma deposition of refractory metals were reported by Boulos and his collaborators [Jiang et al. (1992, 1993a, b, 1994, 2001, 2004), Kovarik et al. (2006, 2007), Fan et al. (1999)]. The study by Jiang et al. (1993b) focused on the

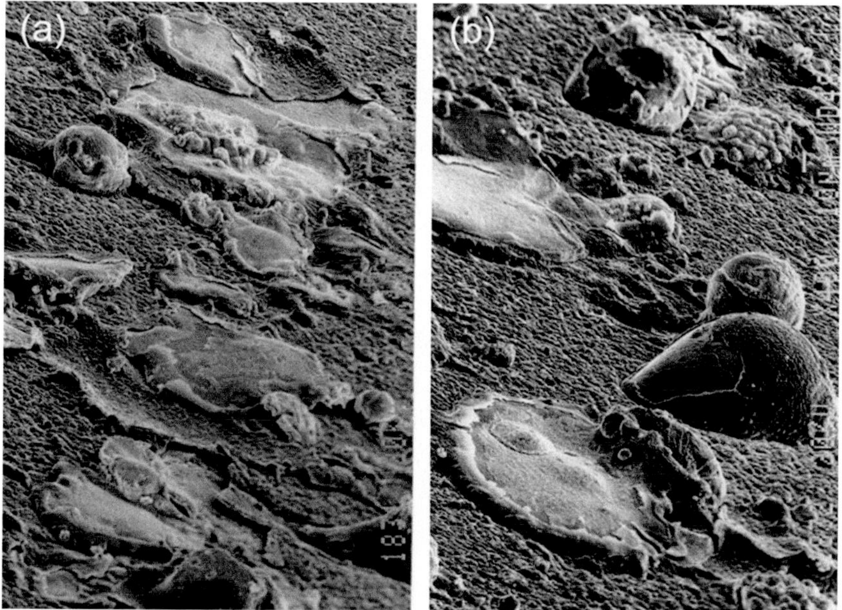

Fig. 10.82 Tungsten splats obtained using VIPS operated using Ar/H$_2$ plasma at plate power of 40 kW, PSD of the powder 53–63 µm, powder feed rate, 26 g/min (1.56 kg/h), chamber pressure 26.6 kPa (200 Torr), and spraying distance (**a**) $z_t = 320$ mm, (**b**) $z_t = 400$ mm

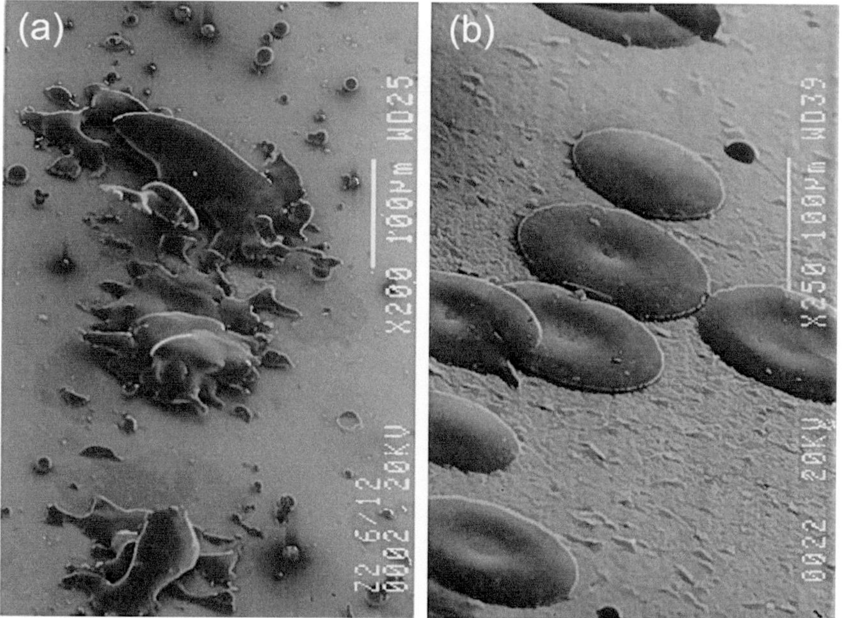

Fig. 10.83 Alumina splats obtained using VIPS Ar/H$_2$ plasma at plate power of 25 kW, PSD of the powder 53–63 µm, powder feed rate, 15 g/min (0.9 kg/h), spraying distance $z_t = 270$ mm, chamber pressure (**a**) 33.2 kPa (250 Torr) and (**b**) 66.5 kPa (500 Torr)

impact of the primary deposition parameters on the deposition efficiency and the density of the deposit. The deposition of tungsten on a stationary graphite substrate was investigated using a Tekna PS-50 torch operated at power levels between 25 and 45 kW using an Ar/H$_2$ (9% vol. H$_2$) as sheath gas and pure Ar as central gas. The chamber pressure was varied between 26.6 and 53.2 kPa (200–400 Torr), the

spraying distance was 175–300 mm, and the powder feed rates were varied between 12 and 40 g/min (0.72–2.4 kg/h). A photograph of the deposit during its formation at plate power level of 40 kW, a chamber pressure of 26.6 kPa (200 Torr), and a spraying distance of 200 mm is shown in Fig. 10.85a. A photograph of the deposit obtained and a cross-sectional mapping of its apparent density are given in

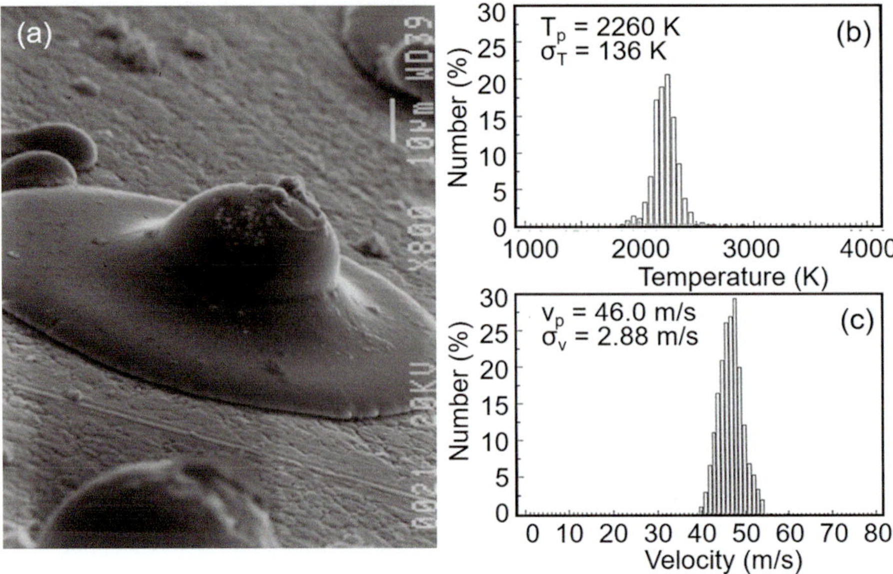

Fig. 10.84 (a) Alumina splats obtained using VIPS Ar/H$_2$ plasma at plate power of 25 kW, PSD of the powder 63-75 μm, powder feed rate, 26 g/min (1.56 kg/h), spraying distance $z_t = 270$ mm, chamber pressure 33.2 kPa (250 Torr) and corresponding distributions of (b) particle temperature and (c) particle velocity prior to their impact on the substrate

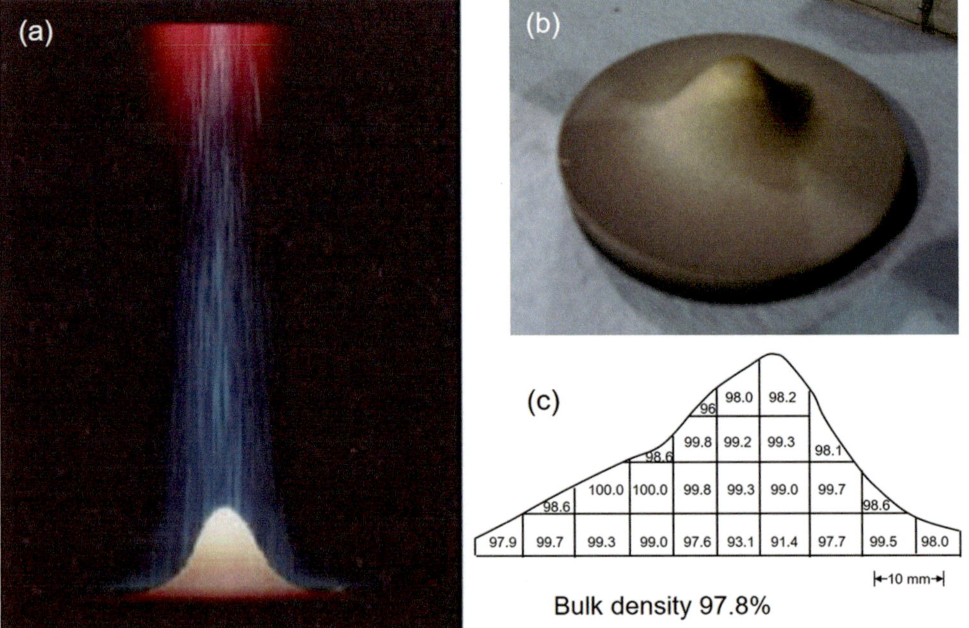

Fig. 10.85 (a) Photograph of VIPS of tungsten on graphite substrate, (b) photograph of the deposit obtained, and (c) cross-sectional mapping of the relative density of the deposit. Ar/H$_2$ plasma at plate power 40 kW, chamber pressure = 26.6 kPa (200 Torr), spraying distance = 200 mm. [Jiang et al. (1993b)]. Reprinted with permission of ASM International. All rights reserved

Fig. 10.85b and c. It may be noted that the deposit had a diameter of 72 mm and a maximum thickness at its center of 30 mm. The density varies between 91.4 and 100%, of the theoretical density of tungsten with the lowest reported densities mostly observed in bottom layer of the deposit and its fringes. The bulk relative density of the deposit is 97.8%.

Figures 10.86 and 10.87 show respectively the variation of the relative density of the deposit, and the deposition efficiency, with the increase of the spraying distance between 180 and 300 mm, and the chamber pressure from 26.6 kPa (200 Torr) to 53.2 kPa (400 Torr). The deposition efficiency is defined as the ratio of the powder deposited on the substrate to that injected into the plasma torch. It may be noted

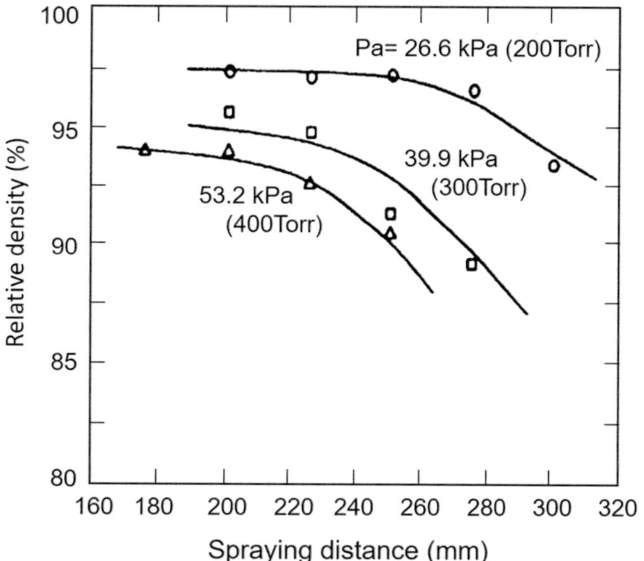

Fig. 10.86 Effect of spraying distance and chamber pressure on the relative density of tungsten deposit, plasma plate power 40 kW [Jiang et al. (1993b)]. Reprinted with permission of ASM International. All rights reserved

from Fig. 10.86 that the relative density of the deposit decreases rapidly with the increase of the spraying distance. The effect reflects the in-flight radiative cooling and eventual solidification of the molten tungsten droplets prior to their impact on the substrate. This results in the formation of porosity and defects in the coating and a drop of the deposition efficiency, which is observed in Fig. 10.87. It is important to stress that in the plasma spraying of refractory metals, the in-flight cooling of the particles is largely due the intense radiative energy losses from the molten liquid droplet at the melting/solidification temperature of such metals and can have a predominant effect on the temperature history of the particles, even during its flight in the plasma with a higher plasma temperature than that of the particle. With the drop of the chamber pressure, the flight velocity of the particles increases resulting in a drop of the time of flight of the particles between the exit of the plasma torch and the substrate. The effect results in a reduction of the cooling time of the particle and a corresponding increase of the relative density of the deposit and of the deposition efficiency even with the substrate locate at 250 or even 300 mm from the probe tip. The effect of the plasma power on the relative density of the deposit obtained and the associated deposition efficiency is shown in Fig. 10.88. It is noted that, as expected, an increase of the plasma power results in a corresponding increase of the deposition efficiency and of the apparent density of the deposit obtained. It is significant to note that at high power levels, deposits of refractory metals could be obtained with apparent densities above 98% of the theoretical density of the metal and deposition efficiencies as high as 90%. Such levels of deposition efficiencies are very

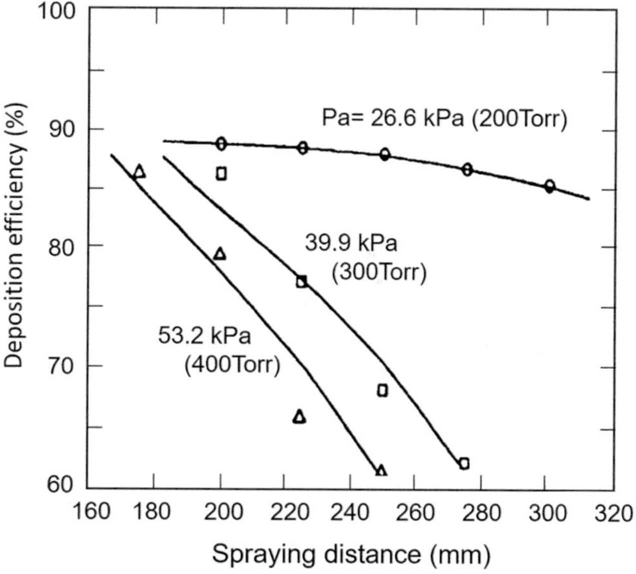

Fig. 10.87 Effect of spraying distance and the chamber pressure on the deposition efficiency of tungsten, Plasma plate power 40 kW. [Jiang et al. (1993b)]. Reprinted with permission of ASM International. All rights reserved

exceptional compared to those normally achieved using DC plasma spraying techniques.

Photographs of VIPS operations for different applications are given in Figs. 10.89, 10.90, and 10.91. The example given in Fig. 10.89 is for the deposition of a refractory metal such as molybdenum or tungsten on a cylindrical graphite mandrel in order to form a near net-shaped part, which is a standalone component once the graphite substrate is machined out following the deposition step. Examples of such parts of cylindrical, conical, or other more complex shapes are given in Fig. 10.92. Similar setup in operation for the coating of X-ray or a sputtering target using VIPS technology is shown in Fig. 10.93. The sheath gas used in this case is composed of 120 slm Ar + 20 slm H_2. The plasma gas is 30 slm Ar, while the powder gas is 8 slm He. The chamber pressure was 40 kPa (300 Torr), Plate power, 80 kW, and spraying distance 100 mm.

An extended discussion of the mechanism of coating formation and its strong dependence on the individual splat formation and solidification during the induction plasma spraying has been reported by [Jiang et al. (2001)] and [Kovarik et al. (2006, 2007)]. These underline the very important role of the substrate temperature on the splat formation mechanism and consequently on the properties of the coating obtained. The spatial variation of the properties of refractory metal deposits has also been the subject of study at Tekna Plasma Systems Inc. by [Fan and Boulos (2004)]. A typical cross section and apparent density profile for a 200-mm diameter thick tungsten disc made by VIPS on a graphite substrate is shown in Fig. 10.92. The plasma power was in this case 80 kW, the chamber pressure 40 kPa (300 Torr), and the spraying distance 100 mm. During the spraying operation, the substrate was translated underneath the plasma jet while being simultaneously rotated around a vertical axis. It is important to stress the need for a detailed simulation of the deposition pattern in such cases for proper matching between the speed of rotation and liner translation depending on the location of the disc in the stream of the molten metal spray. The results show slight variation of the thickness of the coating across the radius of the disc, which can be improved upon by fine tuning the control of the speed of rotation and linear translation. The corresponding density variations across the disc are noted to be in the range of 96–98% of theoretical density of tungsten. Details of metallographic microstructure of the deposit at different locations on the disc are shown in Figs. 10.92 and 10.93. The micrographs shown in Fig. 10.94 were etched in order to reveal details of the grain structure of the deposit (Fig. 10.95).

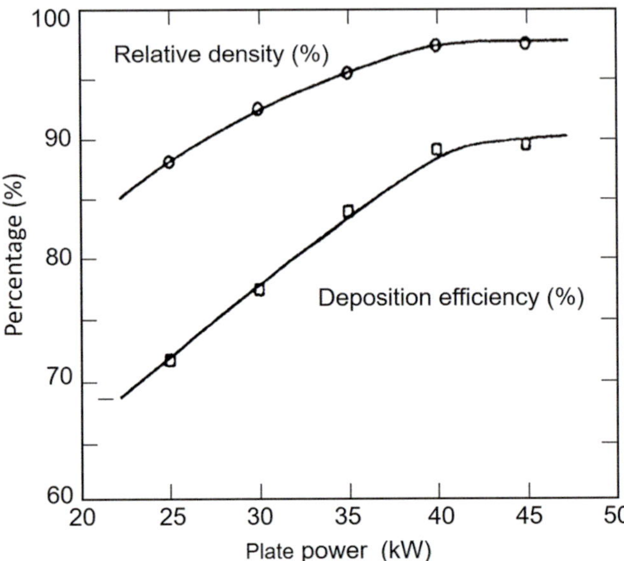

Fig. 10.88 Effect torch plate power on the relative density and the deposition efficiency of tungsten, Chamber pressure = 26.6 kPa (200 Torr), spraying distance = 200 mm. [Jiang et al. (1993b)]. Reprinted with permission of ASM International. All rights reserved

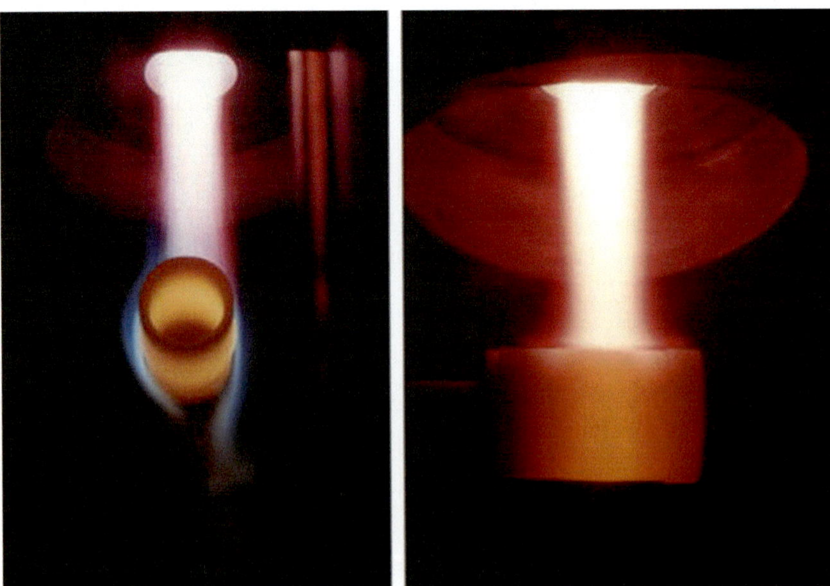

Fig. 10.89 Photograph of VIPS of tungsten on a graphite substrate. Courtesy of the CRTP, University of Sherbrooke (1995)

Fig. 10.90 Photograph of VIPS deposited tungsten and molybdenum near net-shaped parts. Courtesy of Tekna Plasma Systems Inc. (2005)

Fig. 10.91 Photograph of VIPS of tungsten coating on a graphite substrate. Courtesy of the CRTP, University of Sherbrooke (1995)

The ability to use VIPS technology for the deposition of relatively thick structural discs or cylinders attracted attention for its potential use for the manufacture of near net-shaped parts for a wide range of applications. The first of such applications was in the area of solid oxide fuel cells (SOFC) where VIPS was considered to manufacture the bipolar metallic plates that separated the stacks of the different cell components as shown in Fig. 10.96 after Henne et al. (1997, 1999). The results revealed, however, a strong dependence of the porosity of the coating on the spray angle. As shown in Figs. 10.96, 10.97 and 10.98, the best results in terms of deposit density are obtained when the spraying angle is close to 90° to the surface of the substrate. An increasing level of porosity is observed with the decrease in the spray angle below 45°. The effect is not unique to induction plasma spraying since such dependence of the microstructure of coatings has equally been observed in flame and DC plasma spraying as well.

Numerous studies dealing with induction plasma spraying of thick coatings or near net-shaped parts have also dealt with the deposition of sputtering targets as reported by Müller et al. (1995, 2000). Vacuum induction plasma spraying has also been used for the deposition of a wide range of ceramic materials such as Alumina, YSZ, and Titania as reported by [Gitzhofer et al. (1994)], Fan et al. (1998)], and Branland (2003a, b, 2008)]. These generally show comparable results as with refractory metals in terms of deposition efficiencies, but not in terms of porosity that is generally higher than the above reported values. Thick ceramic coatings are rather difficult to achieve, however, due to important internal stresses, which develop during the cooling step of the process following the deposition of the coating.

10.5.5 Supersonic Induction Plasma Spraying

Supersonic induction plasma spraying (SIPS) can be achieved through the addition of a converging–diverging Laval nozzle to the plasma torch as previously described in Sects. 10.3.4 Ceramic wall torches and 10.4.3 Flow field. Typical temperature and velocity profiles in the jet are given in Figs. 10.31, 10.32, and 10.33. A photograph of a

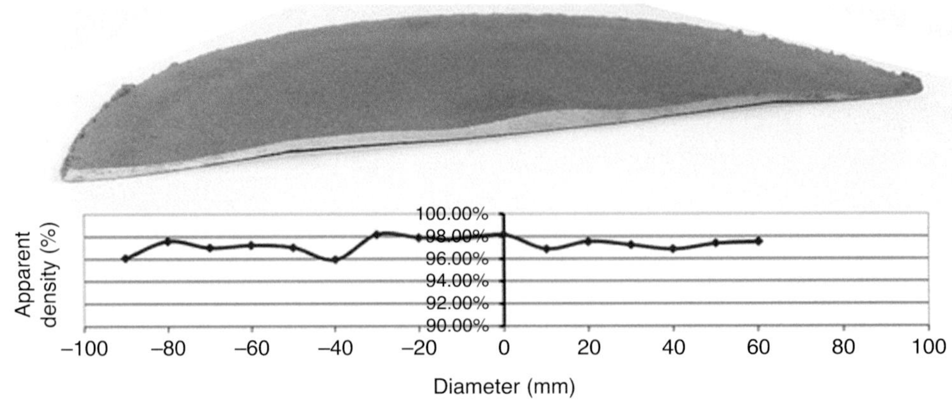

Fig. 10.92 Cross-section of a 200-mm diameter tungsten disc made by VIPS on a graphite substrate and the corresponding apparent density profile. Plasma power 80 kW, Chamber pressure, 40 kPa (300 Torr), and spraying distance 100 mm. After Fan et al. Tekna plasma systems (2012)

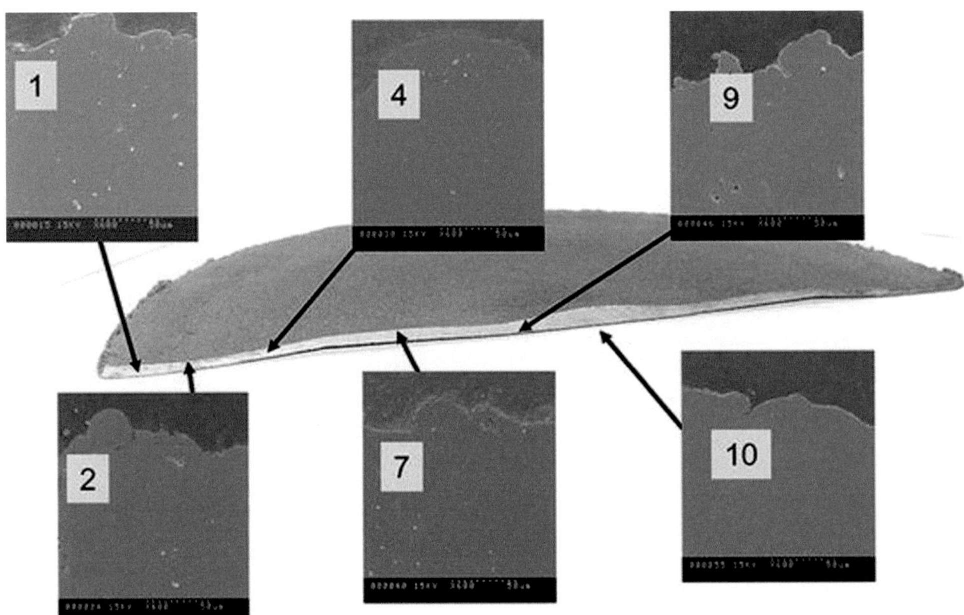

Fig. 10.93 Microstructure of a 200 mm diameter tungsten disc made by VIPS on a graphite substrate. Plasma power 80 kW, Chamber pressure, 40 kPa (300 Torr), and spraying distance 100 mm. After Fan et al. Tekna plasma systems (2012)

supersonic induction plasma spray jet in the presence of fine Yttria-Stabilized Zirconia (YSZ) powder is shown in Fig. 10.99. Numerous publications on the subject were reported by [Mailhot et al. (1997a, b, 1998a, 1999)] and [Theophile et al. (1996)] in connection with a study for the deposition of high density YSZ thin membrane that could be used as an electrolyte for solid oxide fuel cells (SOFC). The plasma gas in this case was a mixture of Ar/Air [Sheath = 68 slm (Air), Central = 21.9 slm (Ar), and powder = 0.83 slm (Ar)]. The YSZ powder feed rate was limited to a few g/min. The chamber pressure was 6.6 kPa (50 Torr) and spraying distance of 250 mm. In order to avoid the development of excessive residual stresses in the coating and risk its cracking, the torch power was modulated starting at 30 kW during a 30s preheating step, followed by powder deposition for 10 s at 40 kW, before dropping the torch power to 16 kW for 30 s during the final deposit cooling step. Microscopic images of the coating obtained using optical microscopy are given in Fig. 10.100, with the "a" series for cross-sectional view while the "b" series are for a longitudinal section in the coating. Apparent porosity measurement of the samples obtained in different reproducibility runs is given in

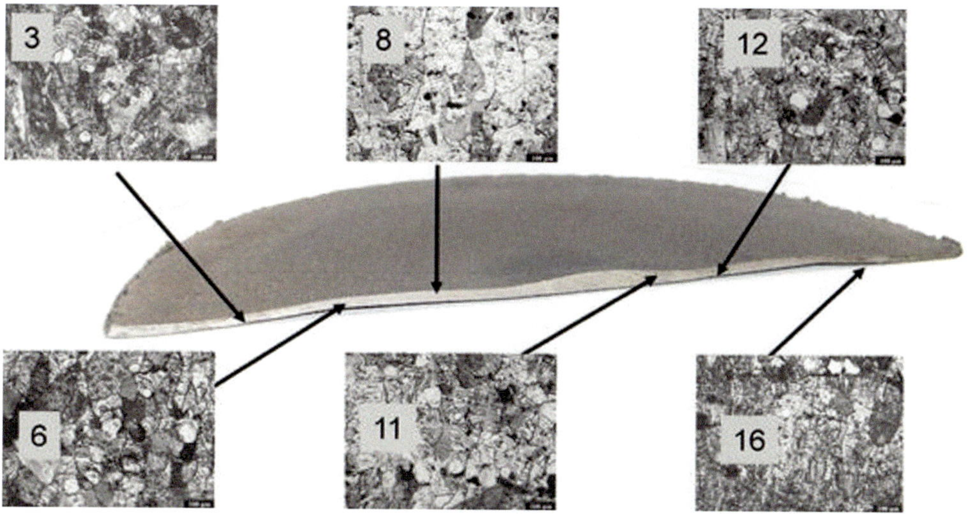

Fig. 10.94 Etched microstructure of a 200-mm diameter tungsten disc made by VIPS on a graphite substrate. Plasma power 80 kW, Chamber pressure, 40 kPa (300 Torr), and spraying distance 100 mm. After Fan et al. Tekna plasma systems (2012)

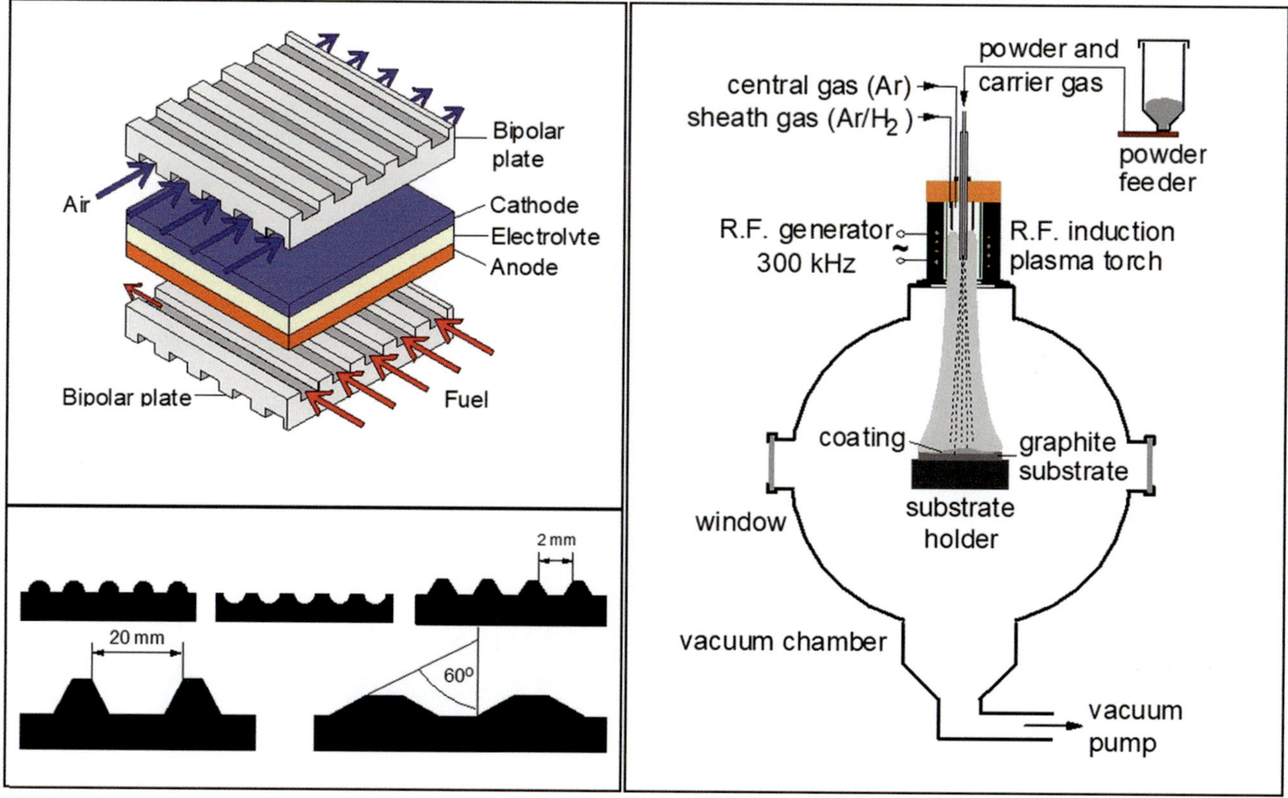

Fig. 10.95 Typical structure of a planar SOFC showing the profile of the bipolar plate (lower left-hand corner) and VIPS system [Henne et al. (1999)]. Reprinted with permission of ASM International. All rights reserved

Table 10.4. These were obtained using image analysis of the optical micrographs and buoyancy measurements. The micrographs given in Fig. 10.100 were those for the deposits obtained in run # 211. His results show apparent porosity of the coating in the range of 1.15–1.75%, which are consistent with apparent density measurements of the bulk coating using buoyancy measurement techniques. The corresponding apparent porosity of the coating based on bulk density measurement was in the range of 1.31–1.9.

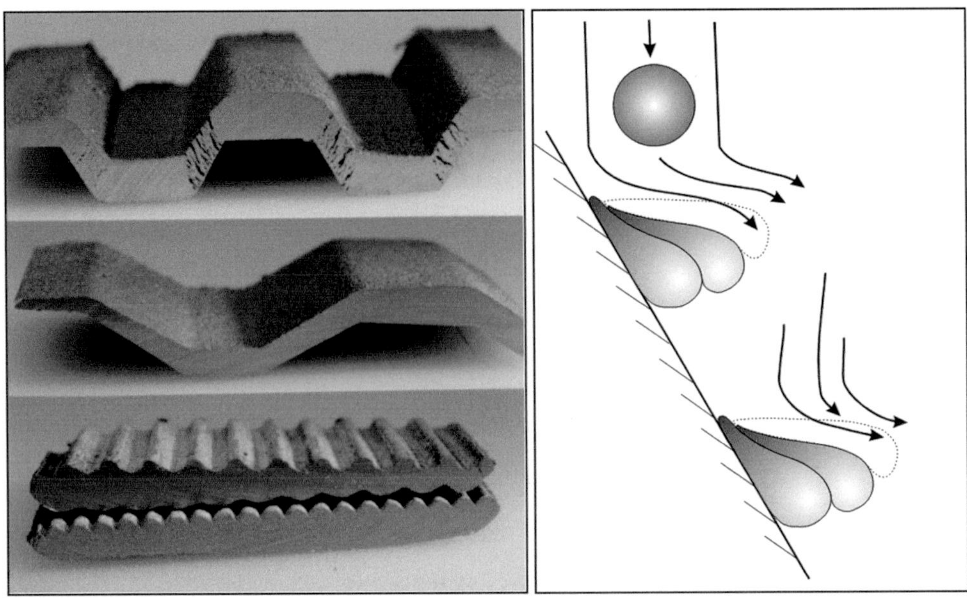

Fig. 10.96 Effect of angle of incidence on deposit formation and its properties [Henne et al. (1999)]. Reprinted with permission of ASM International. All rights reserved

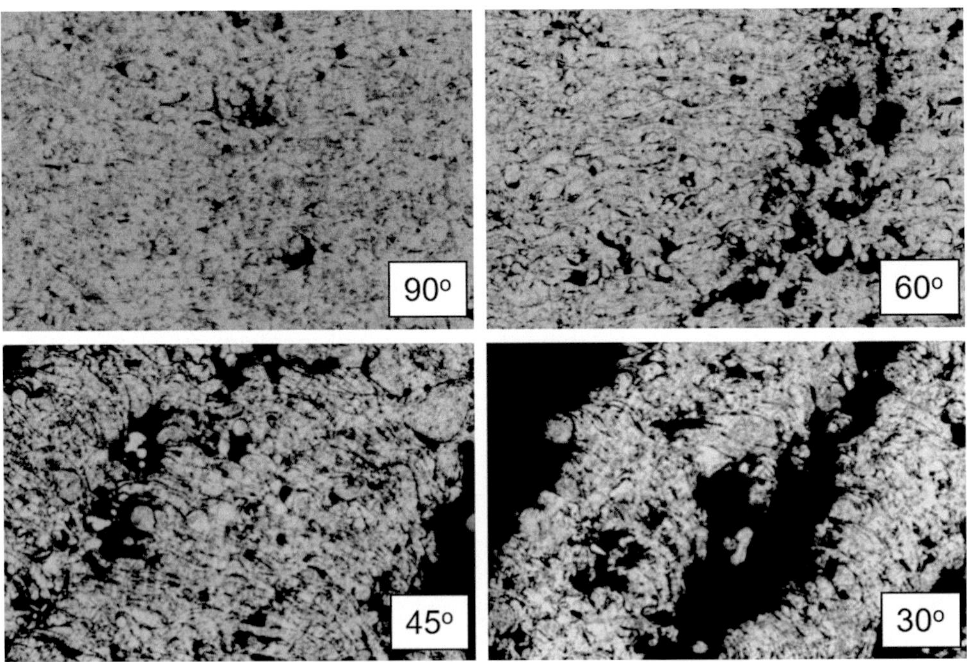

Fig. 10.97 Microstructure of coating made at different angles of incidence of the spray with respect to the substrate [Henne et al. (1999)]. Reprinted with permission of ASM International. All rights reserved

10.5.6 Reactive Induction Plasma Spraying

Reactive induction plasma spraying (RIPS) is a process developed in the 1990s by which the chemical composition of the coating is modified during the coating process. This results in a coating or deposit with a different chemical composition compared to that of the feed material. The chemical transformation is known to take place, either in-flight, through the reaction of the injected particles with the plasma gas, or after the deposition through the reaction of the plasma gas with the formed deposit. In most cases, it is believed that both mechanisms contribute to the process.

The studies reported by Jiang et al. (1194a, b, 1995, 2001, 2004) and by Pu and Boulos (2004) give typical examples of the formation of carbides of tungsten, titanium, and of boron through reactive induction plasma spraying. In the study by Jiang and Boulos (1995), the carbides coatings were obtained through the reaction of the corresponding metal, pure W or Ti, with a source of carbon such as CH_4 or C_2H_2. In the case of tungsten carbide deposition, the chemical route proposed goes through the formation free carbon as a result of methane thermal decomposition in the plasma, followed by the reaction between the free carbon and the metal in-flight according to the following chemical transformations.

$$CH_4(g) \rightarrow C(s) + 2\,H_2(g) \tag{10.15}$$

$$C(a) + 2W(\ell) \rightarrow W_2C\,(\ell) \tag{10.16}$$

This reaction route is supported by the fact that X-ray diffraction (XRD) pattern of the powder collected, in-flight, Fig. 10.101, shows the powder constituted principally of a mixture of W and W_2C. XRD pattern of the deposit, shown in Fig. 10.102, reveals on the other hand a continuation of the carburization reaction even after deposit formation giving rise to the transformation of the deposit to a mixture of exclusively WC and W_2C through the following chemical transformations.

$$2W(s) + C(s) \rightarrow W_2C(s) \tag{10.17}$$

$$W_2C(s) + C(s) \rightarrow 2WC\,(s) \tag{10.18}$$

A similar reaction route is proposed for the reactive deposition of titanium carbide through the reaction between titanium metal and methane according to the following chemical transformations.

In-flight carburization:

$$CH_4(g) \rightarrow C(s) + 2\,H_2(g) \tag{10.19}$$

$$Ti\,(\ell) + (1-x)C(s) \rightarrow TiC_{(1-x)}\,(\ell) \tag{10.20}$$

Secondary carburization reactions of the deposit:

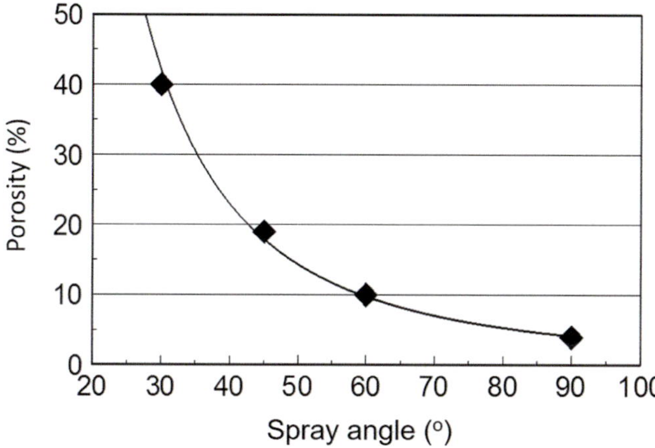

Fig. 10.98 Effect of angle of incidence (spray angle) on the porosity of the coating [Henne et al. (1999)]. Reprinted with permission of ASM International. All rights reserved

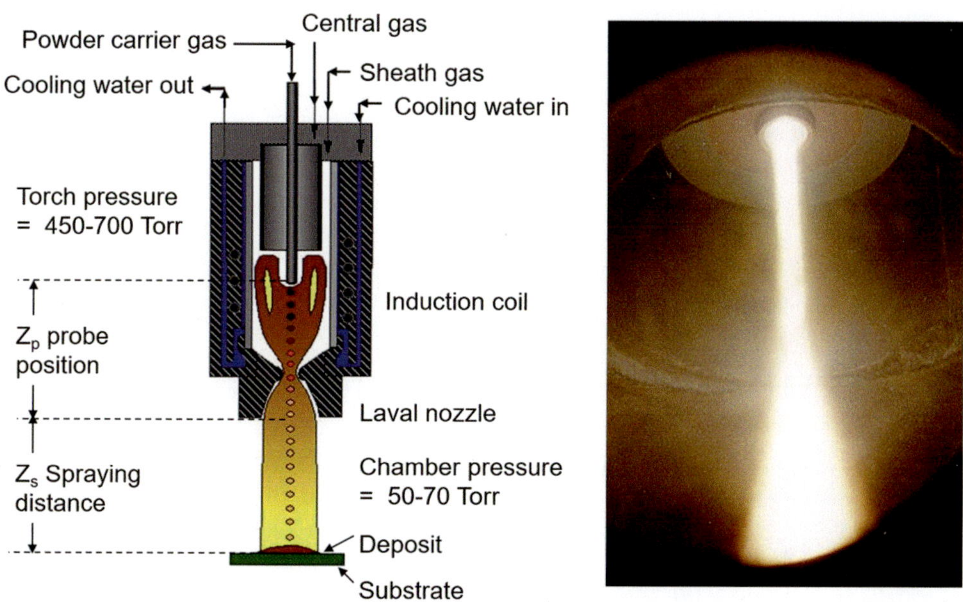

Fig. 10.99 Supersonic induction plasma spraying setup and photograph of the plasma jet in the presence of YSZ powder during a plasma spraying operation. After Mailhot et al. (1997). Reprinted with permission of ASM International. All rights reserved

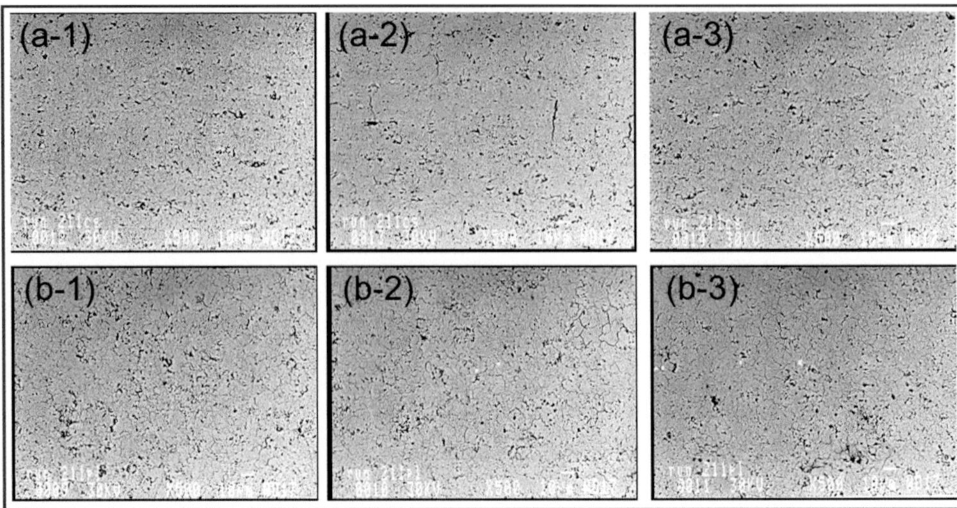

Fig. 10.100 Microstructure of YSZ coating obtained using supersonic IPS. (**a**) Cross-sectional view and (**b**) longitudinal section. After Mailhot et al. (1997). Reprinted with permission of ASM International. All rights reserved

Table 10.4 Apparent porosity measurement for the YSZ samples obtained using supersonic VIPS, runs # 210–213, after Mailhot et al. (1997)

Run #	Image analysis (IA)		Average IA	Buoyancy	Precision
	Cross-sectional	Longitudinal	Average	bulk	± %
210	1.76	1.68	1.72	1.90	9.5
211	1.96	1.65	1.81	1.83	1.1
212	1.15	1.14	1.15	1.31	12.2
213	1.45	1.37	1.41	1.49	5.7

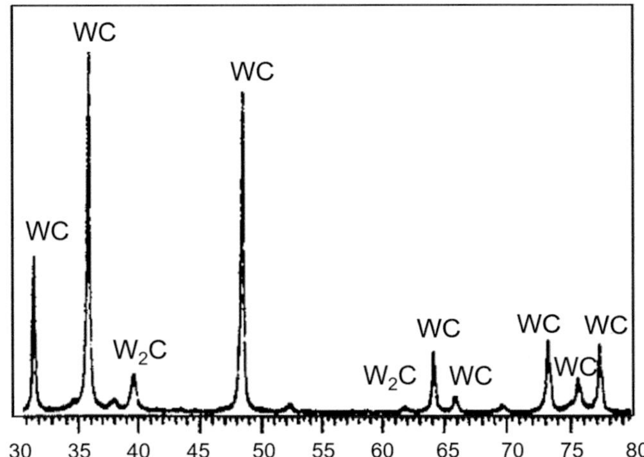

Fig. 10.102 XRD pattern of WC coating formed by in-flight reactive deposition of tungsten powder, after [Jiang and Boulos (1995)] with kind permission

$$TiC_{(1-x)}\,(\ell) + xC(s) \rightarrow TiC(s) \qquad (10.22)$$

These reaction steps are also supported by XRD patterns of the top and bottom of the deposit shown in Figs. 8.84 and 8.85, respectively. The observed difference in composition between the top and bottom layers of the deposit is a clear indication of the continuation of the carburization reaction in the solid state even after the deposit formation (Figs. 10.103 and 10.104)

The induction plasma reactive deposition of boron carbide, B_4C was also reported by [Pu and Boulos (2004)]. The reaction route used in this case is based on the, in-flight, carburization of boron using methane as the carburizing agent

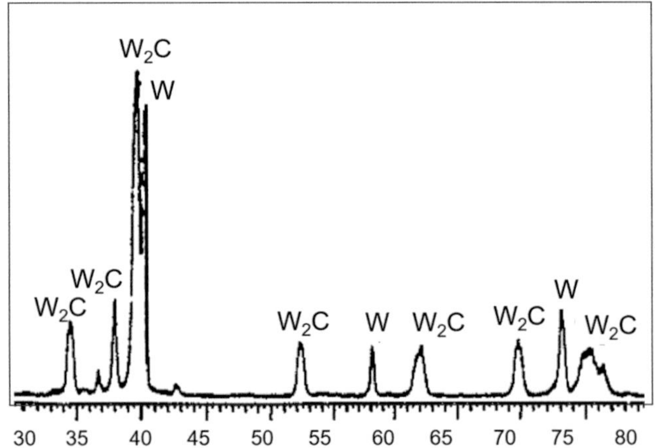

Fig. 10.101 XRD pattern of the in-flight carburized tungsten powder, after [Jiang and Boulos (1995)] with kind permission

$$Ti\,(s) + (1-x)C(s) \rightarrow TiC_{(1-x)}\,(s) \qquad (10.21)$$

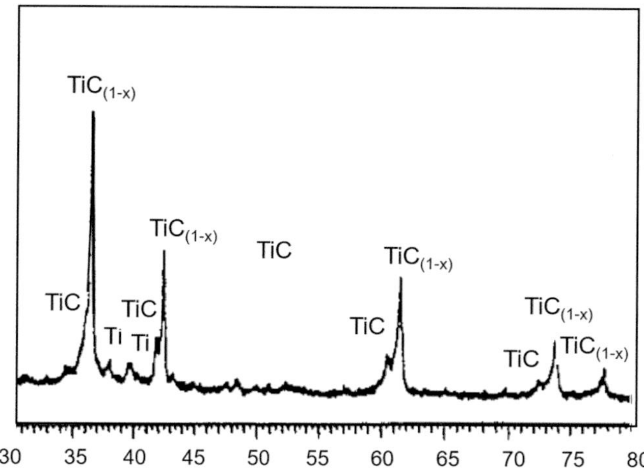

Fig. 10.103 XRD pattern of top layer of Ti/TiC(1–x)/TiC coating formed by in-flight reactive deposition of titanium powder, after [Jiang and Boulos (1995)] with kind permission

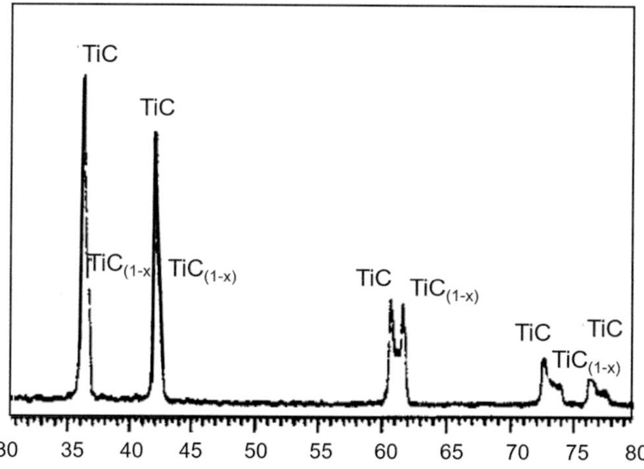

Fig. 10.104 XRD pattern of bottom layer of TiC(1–x)/TiC coating formed by in-flight reactive deposition of titanium powder. After [Jiang and Boulos (1995)] with kind permission

10.5.7 Suspension Induction Plasma Spraying

Plasma spraying with injection of a liquid with fine powders suspended in it into the plasma, known as suspension plasma spraying (SPS), was developed in the early 1990s [Gitzhofer et al. (1997)]. Over the past two decades, it has been increasingly used with both induction and DC plasmas spraying operations for the deposition of heat sensitive and nanostructured coatings and materials. Bouyer et al. (1995, 1997a, b) has reported numerous publications on the subject. The approach used for suspension plasma spraying of hydroxyapatite (HA) is based on the formulation of a stable suspension of HA in water that is the transported to the injection probe via a peristaltic pump as shown in Fig. 10.105. An atomizing gas is guided to the probe tip separately from the liquid, and atomization of the suspension takes place at the probe tip. The atomized droplets have typically diameters in the range of 80–150 μm. In the plasma, the water evaporates and the agglomerated HA crystals fuse into spherical molten droplets with typical diameters of 30 μm. These molten HA droplets can be allowed to freeze in-flight, forming a dense spherical powder or alternately they can be directed toward a substrate where they form a coating through the splat formation of the individual droplets on the substrate (Fig. 10.106)

Compared to DC suspension plasma spraying, it is easy in induction plasma spraying to bring the suspension into the discharge with the central injection probe, and the atomization is carried out locally through the use of an independently controlled atomizing gas stream. In contrast, the atomization of the suspension in DC plasma spraying is carried out by the intense shear of the plasma flow at the point of injection of the suspension into the jet.

Typical results of HA coatings obtained using SPS techniques in an induction plasma torch are given in Fig. 10.107. On the top left-hand side of the figure, micrographs of the HA crystals in the suspension are shown, while micrographs of the coating formed are shown on the bottom left-hand side of the figure. On the right-hand side of the same figure, XRD pattern of the, as synthesized, HA crystals is given, with the XTD corresponding pattern of the HA coating shown on the bottom right-hand side of the figure. It may be noted that the HA, which is known to be a rather heat sensitive materials, suffers from minor decomposition as identified by the tri-calcium phosphate (αTCP) and calcium oxide (CaO) peaks observed in the XRD pattern.

Other examples of applications of SPS using induction plasmas have been reported by Müller et al. (1997a, b, 1998) for the deposition of heat sensitive perovskites for solid oxide fuel cells and spinels for alkaline water electrolysis.

10.6 Summary and Conclusions

Induction plasmas have emerged over the past three or four decades as a potentially powerful tool for materials processing. Vacuum induction plasma spraying (VIPS) has also found niche applications for the deposition of protective coating and near net-shaped parts. The main limitation of the technology stems from the fact that the plasma torch cannot be easily translated with respect to the substrate. Rather, the substrate has to be translated with respect to the torch. This resulted in limiting the use of induction plasma technology on an industrial scale to powder processing and the coating of relatively small high added value parts. In this chapter, a review is presented of the main design features of induction plasma torches and vacuum induction plasma deposition systems. A description is

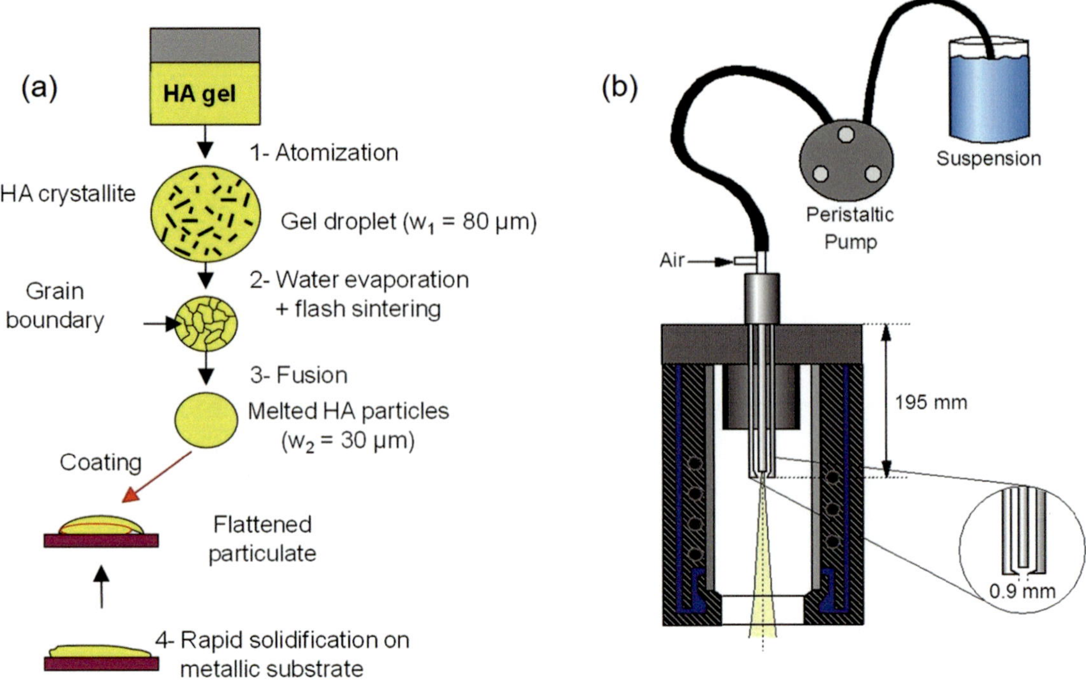

Fig. 10.105 Suspension plasma spraying of hydroxyapatite (HA). Left, illustration of the process scheme; right, schematic of an experimental setup. After Bouyer et al. (1997) with kind permission

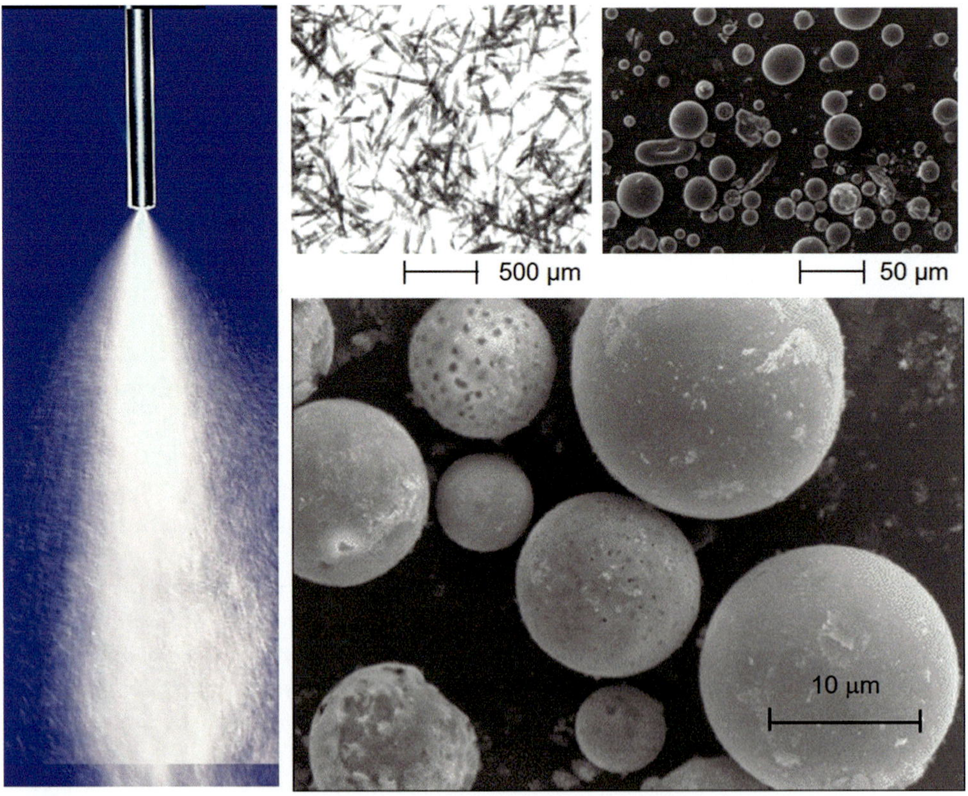

Fig. 10.106 Suspension plasma spraying of hydroxyapatite (HA). Left, photograph of atomized jet; right, electron micrographs of the HA crystals and SPS formed dense solid HA powders. After Bouyer et al. (1997) with kind permission

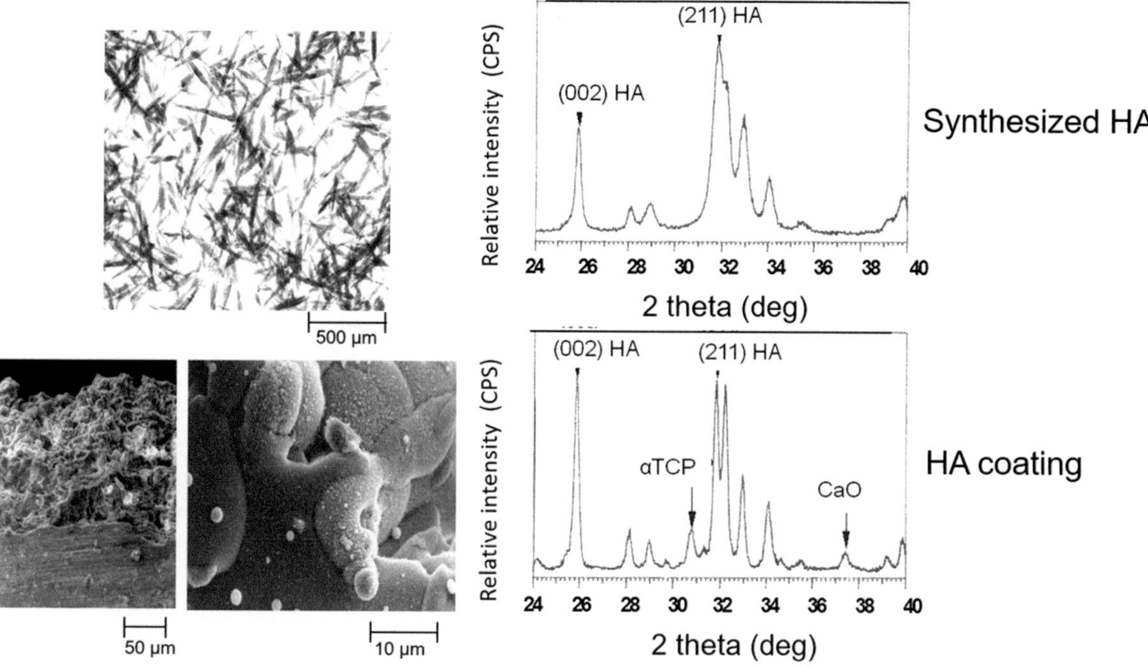

Fig. 10.107 Left, electron micrographs of HA crystals and of deposits made using induction plasma SPS. Right, XRD of as-synthesized HA crystals and XRD of HA coating. After [Bouyer et al. (1997)] with kind permission

provided of the main characteristics of the flow and temperature fields in the discharge and of plasma--particle interactions. These reviled the predominantly large volume, low velocity nature of the discharge, which allows for a long contact time between the plasma and the particles which can be used for the melting of relatively larger particles compared to DC plasma spraying. The absence of electrodes offered the added advantage of operation under high purity conditions that can be of critical importance in biomedical and electron applications. It also allows for powder treatment in a wide range of chemistries under inert, oxidizing, or reducing atmosphere. Typical examples are given for the use of induction plasma technology for the VIPS of refractory metals and ceramics as well as reactive and suspension plasma spraying, at atmospheric pressure, vacuum and ultravacuum supersonic flow conditions.

Nomenclature

Units are indicated in parentheses; when no units are indicated, the parameter is dimensionless.

Latin Alphabet

B_z	Magnetic flux intensity in the axial direction
c	Speed of light in space, $c = 299{,}792{,}458$ (m/s)
c_p	Specific heat (J/K kg)
d_p	Particle diameter (µm)
D_m	Molecular diffusion coefficient (m^2/s)
D_t	Turbulent diffusion coefficient (m^2/s)
E	Electric field intensity (V/m)
f	Oscillator frequency (Hz)
H	Magnetic field intensity (T)
i	Dummy variable
j	Current density (A/m^2)
I_c	Coil current (A)
I_p	Plate current (A)
I_g	Grid current (A)
I_{im}	Applied coil current (A)
I_{in}	Induced current in the coil due to plasma electromagnetic fields (A)
k_m	Molecular thermal conductivity (W/m K)
k_e	Effective thermal conductivity (W/m K)
k_t	Turbulent thermal conductivity (W/m K)
l	Characteristic dimension of the discharge space (m)
L_c	Coil length (m)
m_i	Cooling water flow rate (kg/s)
n_c	Number of coil turns
p	Pressure (Pa)
p_{mo}	Magnetically produced pressure (Pa)
P_c	Energy coupled in the plasma (W)
P_{gen}	Energy recovered in the generator cooling water (W)
P_o	Plasma power (W)
P_i	Power recovered in cooling-water stream I, (W)
P_{tor}	Energy losses to the torch (W)
P_{pb}	Energy losses to the probe (W)
P_{reac}	Energy losses to the reactor/chamber (W)
P_{pt}	Plate power, $P_{pt} = I_p \times V_p$ (W)
P_w	Total energy recovered in the cooling circuit of the system (W)
Q_{sh}	Sheath gas flow rate, also Q_3 (slm)
Q_{ce}	Central gas flow rate, also Q_2 (slm)
Q_{pb}	Probe gas flow rate, also Q_1 (slm)
r_c	Radius of the induction coil (m)

r_n	Radius of cylindrical load of discharge (m)
r_o	Radius of the plasma confining tube (m)
T_{out}	Exit temperature of cooling water (°C)
T_{in}	Inlet temperature of cooling water (°C)
v_z	Axial velocity component (m/s)
v_r	Radial velocity component (m/s)
V_p	Plate voltage (V)
w	Tangential velocity component (m/s)

Greek Alphabet

δ_t	Thickness of skin depth (m)
ϕ_E	Phase angle of the electric field
ϕ_H	Phase angle of the magnetic field
η_c	Energy coupling efficiency, $\eta_c = P_o/P_{ac}$
κ	Coupling parameter, $\kappa_c = \sqrt{2}\,(r_n/\delta_t)$
κ_m	Molecular thermal conductivity (W/mK)
κ_t	Turbulent thermal conductivity (W/mK)
λ	Wavelength (m)
μ_o	Magnetic permeability of space (Hy/m)
μ_o	Magnetic permeability of vacuum $[\mu_0 = 4\pi \times 10^7$ (H/m)$]$
μ_m	Molecular viscosity (Pa s)
μ_t	Turbulent viscosity (Pa s)
σ_o	Electrical conductivity (A/V.m)
ρ	Mass density (kg/m^3)

References

Babat, G.I. 1947. Electrodeless discharges and some allied problems. *Institute of Electrical Engineering Journal* 94: 27–37.

Bernardi, D., V. Colombo, E. Ghedini, and A. Mentrelli. 2003. Comparison of different techniques for the FLUENT-based treatment of the electromagnetic field in inductively-coupled plasma torches. *European Physical Journal D Atomic Molecular and Optical Physics* 27: 55–72.

Bernardi, D., V. Colombo, E. Ghedini, A. Mentrelli, and T. Trombetti. 2004. 3-D numerical simulation of fully-coupled particle heating in ICPTs. *European Physical Journal D Atomic Molecular and Optical Physics* 28: 423–433.

Boulos, M.I. 1992. Radio frequency plasma development, scaleup and industrial applications. *High Temperature Chemical Processes* 1: 401–411.

Bouyer, E., F. Gitzhofer, and M.I. Boulos. 1997a. The suspension plasma spraying of bioceramics by induction plasma. *Journal of Materials Science Letters* 49 (2): 58–62.

———. 1997b. Suspension plasma spraying for for hydroxyapatite powder preparation by R.F. Plasma. *IEEE Transactions on Plasma Science* 25 (5): 1066–1072.

Boulos, M.I., P. Fauchais, and R. Pfender. 1994. *Thermal plasmas fundamentals and applications*. Vol. 1, 452. New York: Plenum Press.

Boulos, M.I. 1985. The inductively coupled r.f. plasma. *Pure and Applied Chemistry* 57: 1321–1352.

———. 1992a. R.F. induction plasma spraying, state-of-the-art review. *Journal of Thermal Spray Technology* 1: 33–40.

———. 1992b. Radio-frequency plasma developments, scale-up and industrial applications. *Journal of High Temperature Chemical Processes* 1: 401–411.

———. 1997. The inductively coupled radio frequency plasma. *High Temperature Materials Processing* 1: 17–39.

———. 2001. Visualization and diagnostics of thermal plasma flows. *Journal of Visualization* 4: 19–28.

Boulos, M.I., and J. Jurewicz. 1992. High performance induction plasma torch with a water-cooled ceramic confinement tube, Canadian Patent, 2,085,133 (April 10th.)

———. 1993. High performance induction plasma torch with a water-cooled ceramic confinement tube, US Patent, 5,200,595.(April 6th)

———. 1996a. US Patent, 5,560,844, (October 1st).

———. 1996b. Chinese patent, ZL92103380.X (April 11th)

———. 1997. European Patent, 533884 (January 2nd.)

———. 1999. Korean patent, 203,994, (March 25th)

———. 2001. Japanese Patent, 3,169,962, (March 16th).

Boulos, M., and J. Jurewicz. 2004. Multi-coil induction plasma torch for solid-state power supply, US patent 6,693,253, Feb 17th.

———. 2005. Multi-coil induction plasma torch for solid-state power supply, US patent 6,919,527 July 19th.

Branland, N. 2002. *Plasma spraying of Titania coatings: Contribution to the study of their electrical properties and microstructures*, PhD thesis, University of Limoges, France/University of Sherbrooke, Canada. (in French).

Branland, N., E. Meillot, P. Fauchais, A. Vardelle, F. Gitzhofer, and M. I. Boulos. 2006. Relationships between microstructure and electrical properties of RF and DC plasma-sprayed Titania coatings. *Journal of Thermal Spray Technology* 15: 53–62.

Chase, J.D. 1969. Magnetic pinch effect in the thermal r.f. induction plasma. *Journal of Applied Physiology* 40: 318–325.

———. 1971. Theoretical and experimental investigation of pressure and flow in induction plasmas. *Journal of Applied Physics* 42: 4870–4879.

Chen, K., and M.I. Boulos. 1994. Turbulence in induction plasma modelling. *Journal of Physics D: Applied Physics* 27: 946–952.

Colombo, V., E. Ghedini, and P. Sanibondi. 2010a. Three-dimensional investigation of particle treatment in an RF thermal plasma with reaction chamber. *Plasma Sources Science and Technology* 19: 065024.

———. 2010b. A three-dimensional investigation of the effects of excitation frequency and sheath gas mixing in an atmospheric-pressure inductively coupled plasma system. *Journal of Physics D: Applied Physics* 43: 105202.

Colombo, V., C. Deschenaux, E. Ghedini, M. Gherardi, C. Jaeggi, M. Leparoux, V. Mani, and P. Sanibondi. 2012. Fluid-dynamic characterization of a radio-frequency induction thermal plasma system for nanoparticle synthesis. *Plasma Sources and Science Technology* 21: 045010.

Davies, J., and P. Simpson. 1979. *Induction heating handbook*. London: McGraw Hill.

Douglas, T.S. 1974. Radial temperature profiles in an R.F. plasma over a wide range of applied magnetic flux intensities: Theory and experiments, Ph.D. Thesis, Georgia Institute of Technology.

Dresvin, S.V., ed. 1977. *Physics and technology of low temperature plasmas*. Ames: Iowa State Univ. Press.

———. 1993. The fundamentals of theory and design of HF plasma generators, translated from Russian.

Dundas, P.H. 1970. Induction plasma heating, measurement of gas concentrations, temperatures and stagnation heads in a binary plasma system. *NASA-CR* 1527.

Dundas, and M. Thorpe. 1969. Economics and technology of chemical processing with electric-field plasmas', Chemical Eng., June 30th., 123–127.

Dundas, P. H., J. W. Pool, and C.E. Vogel. 1975 Low frequency induction plasma system, US Patent, 3 862 393, Jan 21st.

Fan, X., F. Gitzhofer, and M.I. Boulos. 1998. Investigation of alumina splats formed in the induction plasma process. *Journal of Thermal Spray Technology* 7 (2): 197–204.

Eckert, H.U. 1971. Measurement of the r.f. magnetic field distribution in a thermal induction plasma. *Journal of Applied Physiology* 42: 3108–3113.

———. 1972. Dual magnetic probe system for phase measurements in thermal induction plasmas. *Journal of Applied Physiology* 43: 2707–2713.

———. 1974. The induction arc: A state-of-the-art review. *High Temperature* 6: 99–134.

Fouladgar, J., A. Chentouf, and G. Develey. 1993. The calculation of the impedance of an induction plasma installation by a hybrid finite-element boundary-element method. *IEEE Trans* 29: 2479–2481.

Freeman, M.P., and J.D. Chase. 1968. Energy-transfer mechanism and typical operating characteristics for the thermal RF plasma generator. *Journal of Applied Physiology* 39: 180–190.

Galtier, F., Collongues, R., and J. Reboux. 1973. Les fours à plasma haute fréquence', Chaudron and trombe Eds., Les Hautes Températures, Masson et CI, 82–121.

Gitzhofer F., K. Mailhot, M.I. Boulos, I.H. Jung, J.S. Lee, and H.S. Park. 1998. *Fabrication of simulated nuclear fuel pellets by induction plasma deposition*. Proceedings (ITSC-1998), Nice (France), vol. 2, 1283–1288.

Gitzhofer F., M.I. Boulos, G. Schiller, and R. Henne. 1999. Materials synthesis using induction plasma. In *Proceedings of UTSC-1999 – 2nd United Thermal Spray Conference & Exposition – Coating in Practice*. Düsseldorf.

Henne, R., M. Müller, E. Pross, G. Schiller, F. Gitzhofer, and M.I. Boulos. 1999. Near-net-shape forming of metallic bipolar plates for planar solid oxide fuel cells by induction plasma spraying. *Journal of Thermal Spray Technology* 8 (1): 110–116.

Hollabaugh, C.M., D.E. Hull, L.R. Newkirk, and J.J. Petrovic. 1983. R.-R. plasma system for the production of ultrafine, ultrapure silicon carbide powder. *Journal of Materials Science* 18: 3190–3194.

Hollenstein, M., M. Rahman, and M.I. Boulos 1999 Aerodynamic study of the supersonic induction plasma jet, ISPC-14, Czech Republic, August 2–6.

Jiang, X.L., R. Tiwari, F. Gitzhofer, and M.I. Boulos. 1993. On the induction plasma deposition of tungsten metal. *Journal of Thermal Spray Technology* 2: 265–270.

———. 1995. Reactive deposition of tungsten and titanium carbides by induction plasma. *Journal of Materials Science Letters* 30: 2325–2328.

Klubnikin, V.S. 1975. Thermal and gas dynamic characteristics of an argon induction discharge. *High Temperature* 13: 439–446.

Kovarik, O., S. Xue, and M. Boulos. 2006. *RF plasma deposition of refractory metals – Case study for W*. Proceedings (ITSC-2006) Seattle, USA.

Kovarik, O., X. Fan, and M. Boulos. 2007. In-flight properties of W particles in an Ar-H$_2$ plasma. *Journal of Thermal Spray Technology* 16 (2): 229–237.

Léveillé, V. 2002. Diagnostic du jet de plasma hf supersonique, Université de Sherbrooke, M.Sc.A. Thesis.

Léveillé, V., D.V. Gravelle, and M.I. Boulos. 2003. Diagnostic study of supersonic plasma flows using enthalpy probe, Schlieren and high-speed photography, Int. Thermal Spray Conference (ITSC-2003), Orlando (Florida).

Mailhot K., F. Gitzhofer, and M.I. Boulos. 1999. Plasma deposition of dense ceramic coatings using supersonic induction plasma. In *UTSC-1099 – 2nd United Thermal Spray Conference & Exposition – Coating in Practice*. Düsseldorf.

Mailhot, K., F. Gitzhofer, and M.I. Boulos. 1997. Supersonic induction plasma spraying applied to dense Yttria stabilized zirconia electrolyte coatings. *Proceedings ISPC'13, Beijing (China)* 3: 1445–1450.

Mailhot, K., F. Gitzhofer, and M.I. Boulos. 1998a. Supersonic induction plasma spraying of Yttria stabilized Zirconia films. *International Thermal Spray Conference (ITSC'98), Nice (France)* 2: 1419–1424.

Mailhot K., F. Gitzhofer, and M.I. Boulos. 1998b. *Supersonic induction plasma spraying of Yttria stabilized Zirconia films*. Proceedings (ITSC-1998), Nice (France), vol. 2, 1419–1424.

Mensingen, A.E., and L.R. Boedecker. 1968. Theoretical investigations on RF induction heated plasmas. *NASA CR-1312*: 1–75.

Merkhouf, A., and M.I. Boulos. 1998. Integrated model for the radio frequency induction plasma torch and power supply system. *Plasma Sources Science Technology* 7 (4): 599–606.

———. 2000. Distributed energy analysis for an integrated radio frequency induction plasma system. *Journal of Physics D: Applied Physics* 33: 1581–1587.

Müller M., G. Schiller, F. Gitzhofer, and M.I. Boulos. 1997a. *Suspension plasma spraying for the preparation of Perovskite powders and coatings*. NTSC-1997, Indianapolis (IN) (USA).

Müller M., G. Schiller, F. Gitzhofer, M.I. Boulos, and R.B. Heinmann. 1997b. *Suspension plasma spraying of cobalt spinel*. Proceedings NTSC-1997, Indianapolis (IN) (USA).

Müller, M., R.B. Heinmann, F. Gitzhofer, M.I. Boulos, and K. Schwarz. 2000. Radio frequency plasma processing to produce chromium sputter targets. *Journal of Thermal Spray Technology* 9 (4): 488–493.

Pool, J.W., and C.E. Vogel. 1972. Induction torches and low frequency tests. *NASA* CR-2053.

Pool, J.W., M. P. Freeman, K. W. Doak, and M. L. Thorpe. 1973. Simulator tests to study hot-flow problems related to a gas core reactor, NASA CR-2309.

Proulx, P., J. Mostaghimi, and M.I. Boulos. 1985. Plasma-particle interaction effects in induction plasma modelling under dense loading conditions. *International Journal of Heat and Mass Transfer* 28: 1327–1336.

———. 1987. Heating of powders in an R.F. inductively coupled plasma under dense loading conditions. *Journal of Plasma Chemistry and Plasma Processing* 7 (1): 29–53.

Pu W., and M.I. Boulos. 2004. Induction plasma spraying of boron carbide coating. In *International Thermal Symposium Conference (ITSC-2004)*. Osaka.

Rahmane, M., G. Soucy, and M.I. Boulos. 1994. Mass transfer in induction plasma reactors. *International Journal of Heat and Mass Transfer* 37: 2035–2046.

———. 1996. Diffusion phenomena of a cold gas in thermal plasma stream. *Journal of Plasma Chemistry and Plasma Processing* 16: 169S–189S.

Reboux, J. 1971. L'utilisation du fours à plasma inductif dans le traitement et la préparation des matériaux réfractaires', A.I.M. Liège, 30 mars.

Reed, T.B. 1961a. Induction coupled plasma torch. *Journal of Applied Physics* 32: 821–824.

———. 1961b. Growth of refractory crystals using the induction plasma torch. *Journal of Applied Physics* 32: 2534–2536.

———. 1963a. Heat transfer intensity from induction plasma flames and oxy-hydrogen flames. *Journal of Applied Physics* 34: 2266–2226.

———. 1963b. High-power low-density induction plasma. *Journal of Applied Physics* 34: 3146–3147.

Soucy, G., J. Jurewicz, and M.I. Boulos. 1994a. Mixing study of the induction plasma reactor: Part I – Axial injection mode. *J. Plasma Chem. and Plasma Process.* 14: 43–57.

———. 1994b. Mixing study of the induction plasma reactor: Part II – Radial injection mode. *Journal Plasma Chemistry and Plasma Processing* 14: 59–71.

TAFA. 1966. Operating limits model 56, Engineering Data - Bulletin 52 E3.

——— 1968 Induction plasma heat source 200-1000kW, Bulletin P6 4/68.

Theophile E., F. Gitzhofer, and M.I. Boulos. 1999a. Manufacturing of solid oxide fuel cells components by induction plasma spraying. In *UTSC-1999 – 2nd United Thermal Spray Conference & Exposition – Coating in Practice*. Düsseldorf.

Theophile, E., F. Gitzhofer, and M.I. Boulos. 1999b. Supersonic induction plasma spraying of Yttria stabilized Zirconia coatings. *Proceedings ISPC-14, Prague (Czech Republic)* IV: 2145–2148.

Thorpe, M. L. 1966. High temperature heat with induction plasma', R&D Magazine, Jan.66 Thompson publications.

———. 1968. Induction plasma heating: System performance, hydrogen operation and gas core reactor simulation development, TAFA internal report.

———. 1970a. Process push for plasma, chemical week, June 17th., 119–120.

———. 1970b. Plasmas: The lab toy that grew up, business week, August 12th

Thorpe, M. L., and K.W. Harrington. 1970a. induction plasma generation having improved chamber structure and control, US Patent 3 530 334. Sept. 22nd.

———. 1970b. induction plasma generation with high velocity sheath, US Patent 3 530 335. Sept. 22nd.

Thorpe, M.L., and L.W. Scammon. 1968. *Induction plasma heating: High power, low frequency operation and pure hydrogen heating.* Cleveland: NASA 3-9375.

———. 1969. *Induction plasma heating: High power, low frequency operation and pure hydrogen heating.* Cleveland: NASA CR-1343.

Thorpe, M. L., K.W. Harrington, and N. H. Suncook. 1968. Induction plasma generator including cooling means, gas flow means and operating means therefor, US Patent 3 401 302. Sept. 10th.

Uesugi, T., O. Nakamura, T. Yoshida, and K. Akashi. 1988. A tandem radio-frequency plasma torch. *Journal of Applied Physics* 64: 3874–3879.

Vogel, C.E. 1970. Experimental plasma studies simulating a gas-core nuclear rocket, AIAA, San Diego, Paper #70–691.

———. 1971. Curved permeable wall induction torch tests. *NASA* CR-1764.

Xiaobao, F., and M. Boulos. 2005. *Near-net shape forming of tungsten material by induction plasma deposition, proceedings (ITSC-2005).* Basel.

Xue, S., P. Proulx, and M.I. Boulos. 2001. Extended-field electromagnetic model for the inductively coupled plasma. *Journal of Physics D: Applied Physics* 34: 1897–1906.

Yoshida, T., K. Nakagawa, T. Harada, and K. Akashi. 1981. New design of a radio frequency plasma torch. *Plasma Chemistry and Plasma Processing* 1: 113–129.

Ye, R., P. Proulx, and M.I. Boulos. 1999. Turbulence phenomena in the radio frequency induction plasma torch. *International Journal of Heat and Mass Transfer* 42 (9): 1585–1595.

Yoshida, T., T. Tawi, H. Nishimura, and K. Akashi. 1983. Characterization of a hybrid plasma and its application to a chemical synthesis. *Journal of Applied Physics* 54: 640–646.

Ye, R., T. Ishigaki, J. Jurewicz, P. Proulx, and M.I. Boulos. 2004. In-flight spheroidization of alumina powders in Ar–H$_2$ and Ar–N$_2$ induction plasmas. *Journal of Plasma Chemistry and Plasma Process* 24: 555–571.

Zinn, S., and S.L. Semiatin. 1988. *Elements of induction heating.* Palo Alto: EPRI and ASM International.

Abbreviations

APS	Atmospheric Plasma Spraying
BSE	Backscattered Electron
CAS	Cord Arc Spraying
CCD	Charge-Coupled Device
CFD	Computational Fluid Dynamics
COF	Coefficient of Fiction
DC	Direct Current
DTA	Differential Thermogravimetric Analysis
EDS	Electron Diffraction Spectroscopy
EIS	Electrochemical Impedance Spectroscopy
EMI	Electromagnetic Interference
HD-WAS	High Definition-Wire Arc Spraying
HV	High Velocity
ISPC	International Symposium on Plasma Chemistry
ITSC	International Thermal Spray Conference
LHS	Left-Hand Side
LP-WAS	Low Pressure-Wire Arc Spraying
MHP	Machine Hammer Peening
PSD	Particle Size Distribution
PT-WAS	Plasma Transferred-Wire Arc Spraying
PVD	Physical Vapor Deposition
RF	Radiofrequency
RHS	Right-Hand Side
SEC	Saturated Calomel Electrode
SEM	Scanning Electron Microscopy
STS	Special Treatment Steel
SWAS	Single Wire Arc Spraying
SW-VAS	Single Wire-Vacuum Arc Spraying
TEM	Transmission Electron Microscopy
TWAS	Twin Wire Arc Spraying
UTSC	United Thermal Spray Conference
WAS	Wire Arc Spraying
XRD	X-Ray Diffraction

11.1 Introduction

Wire arc spraying (WAS) is the oldest of thermal spray processes, with its first patent issued in the USA in 1915 by Schoop (1915). It was only in the 1960s that the real potential of the technology was recognized, and its applications greatly expanded. This was mainly due to significant improvements in our understanding of the fundamentals governing the process technology through systematic studies involving high-time resolution diagnostics (Steffens 1966). By the late 1990s and the beginning of the twenty-first century, several significant improvements in the equipment design and processes automation were achieved (Steffens et al. 1990; Marantz and Marantz 1990). Compared to alternate surface modification technologies, WAS is one of the most competitive technologies because of its high deposition rates up to 25 kg/h, favorable economics, and potential use for a wide range of applications mostly for corrosion protection of infrastructure such as bridges and metallic constructions in general as well as in the marine and automotive industry. The main limitation of the technology is that it works best with metallic ductile wires, such as aluminum, zinc, or steels. The development of the "cord–wire" approach in which a wire is formed through the use of a ductile metallic foil as envelop filled with a ceramic or composite fine powder allowed the expansion of the technology to new industrial-scale wear resistance applications involving the spraying of carbide-based cermets such as Cr_3C_2 with Fe and FeC, WC/W_2C + Fe, WC/TiC + Fe, Cr, Ni for wear resistance application. The technology can be combined with epoxy or silicon polymer coating for the sealing open porosity in the coating provided that the service temperature is below < 200 °C. In this chapter, the basic concepts behind

M. I. Boulos et al. (ed.), *Thermal Spray Fundamentals*, https://doi.org/10.1007/978-3-030-70672-2_11

the technology are discussed highlighting the fundamental phenomena involved. This is followed by a review of industrial torch designs with emphasis on their relative advantages, limitation, and arc droplet dynamics and their impact on coating quality. Process technology is reviewed next with examples of some of the most extensively used WAS applications.

11.2 Basic Concepts

11.2.1 General Remarks

Wire arc spray (WAS) is a plasma spray coating process based on the concept of melting the material to be sprayed in wire form using an electric arc struck between their tips of two wires, or a wire and a non-consumable electrode, and atomizing the formed molten metal by a high-velocity gas stream which projects the droplets toward the substrate. As with conventional thermal spraying, the molten droplets form splats that rapidly solidify on impact with the substrate surface building up the coating in successive layers (Davis 2004;

Tucker 2013). As schematically represented in Fig. 11.1, the process can be maintained in a continuous mode by electrically connecting the two wires to a DC power supply and continuously feeding the wires in a closely controlled speed to compensate for the melting of their respective tips such as to maintain a constant gap between them and consequently a constant arc voltage. A high-velocity gas flow injected between the two wires toward the arc removes constantly formed molten material from the wire tips, breaks down the larger droplets into smaller ones in a secondary atomization process, and propels them toward the substrate. The shearing of the liquid metal layer from the wire tips by the high-velocity gas flow can be perceived in the high-speed photographs given in Fig. 11.2.

Numerous wire and atomizing gas nozzle configurations have been used in the design of wire arc spraying torches. The setups schematically represented in Fig. 11.3. for the twin-wire arc spray (TWAS) torch can be characterized as follows: Fig 11.3a is the standard nozzle with a straight bore, Fig 11.3b is a converging–diverging nozzle allowing supersonic flow of the atomizing gas to extend farther into the atomizing region, Fig 11.3c is standard nozzle with secondary gas injection, and Fig 11.3d is the addition of a shroud to the nozzle with secondary gas injection. As will be discussed later in detail, the effects of each of these arrangements are mostly felt on the atomizing gas velocity, formed droplet size and velocity, droplet temperature, droplet trajectories, and coating characteristics.

Alternate torch designs also include the so-called single-wire arc spraying (SWAS) with a single consumable wire electrode and a non-consumable second electrode (Marantz and Marantz 1990; Marantz et al. 1991; Kowalsky et al. 1991, 1992; Steffens and Wewel 1991; Steffens and Nassenstein 1994; Carlson and Heberlein 2002). Such a design which has also been commonly referred to as the plasma transferred wire arc spraying (PT-WAS) has the advantages of potentially generating a narrower spray

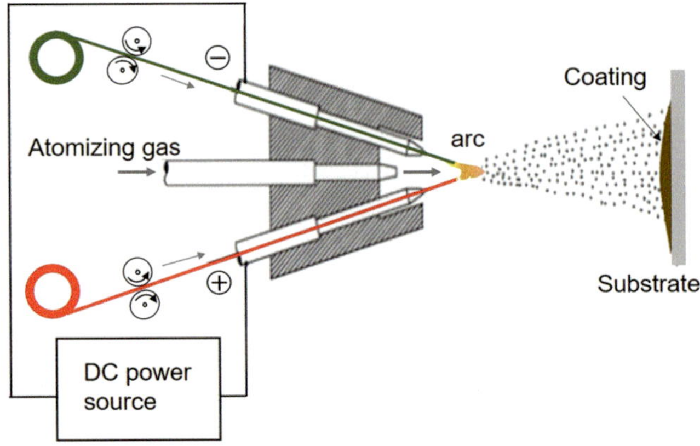

Fig. 11.1 Principle of wire arc spraying

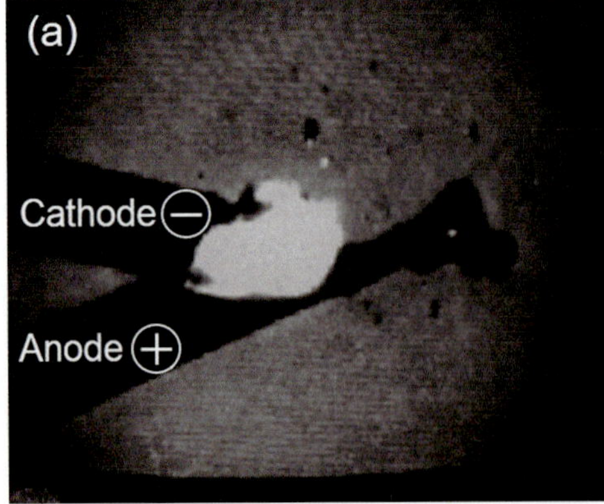

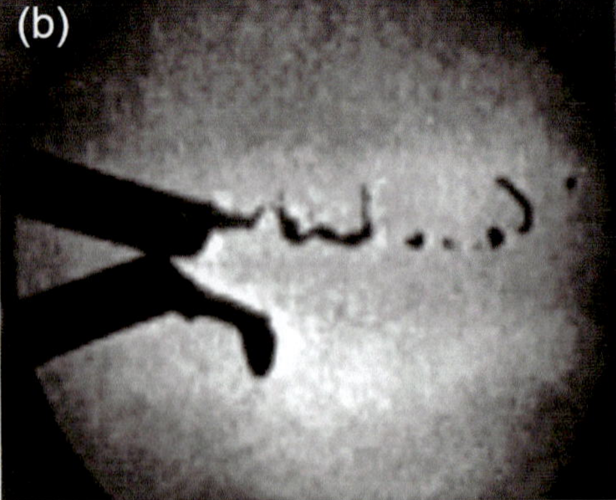

Fig. 11.2 High-speed images of liquid metal droplet formation with 100 ns exposure time

pattern, with a smaller and lighter-weight torch using significantly reduced atomization gas flow rates. All the materials normally used as wires for the twin-wire arc spray processes could be equally used with this process as well. Single-wire arc spraying torches are, however, still in its early stages of their development and less commonly used on an industrial scale compared to conventional twin-wire arc spraying devices.

Compared to plasma spraying, the heat transfer to wires is excellent, and the droplets are fully molten when they leave the wire tips. The atomizing gas plays a key role in droplets formation with the formed particles smaller than the wire diameters and their acceleration and impact velocity also strongly dependent on the atomizing gas and also on the design of different gas injection nozzles. The droplets are generally spherical in the size range from sub-micron (fume) up to 200 μm depending on spray conditions. Droplet sizes increase with the increase of the arc current, the atomizing gas flow, the wire size, and the decrease of the arc voltage. Atomizing gas flow rates are generally rather high up to 1.8 m³/min; air is mostly used for economic reasons at the expense of risking the potential oxidation of the material being sprayed either in-flight or after its deposition on the substrate. Considering that the atomizing gas is not heated, the heat flux to the substrate is limited to the sensible and latent heat of the molten droplets spray, which is significantly lower than that experienced in processes such as combustion DC or RF induction plasma spraying. Low melting point substrates can consequently be safely coated with this process without the need for additional cooling. On the other hand, preheating the substrate to improve the coating adhesion, when necessary, would require the use of an additional heat source. Droplet oxidation can be significantly reduced through the use of inert atomizing gas such as argon or nitrogen and reducing the spraying distance. Coatings with thicknesses over 1 mm are easily attainable. Typical operating ranges are 15–400 A, with open circuit voltage of 40 V. Scaled-up units operating with arc currents up to 1500 A are used with high throughput for the coating of large parts such as bridges. Typical material feed rates with standard WAS systems given in Table 11.1 (Davis 2004) are higher than comparable DC plasma spraying units.

To summarize, the principal advantages of the wire arc spray process are:

- Use of wire as the feed material, which is more economical than the use of powder.
- Efficient and complete wire melting resulting in high deposition efficiencies.
- High energy efficiency since the major portion of the electrical energy supplied is used for melting the wire tips and all material arriving at the substrate was initially molten.
- Minimal heat transferred to the torch, eliminating the need for water cooling and permitting smaller and simpler torch designs.
- High gas flow rates and the relatively low power result in low gas enthalpies and, consequently, minimal substrate heating.
- Small arcing gaps result in lower arc voltages,
- Arc initiation through touching of the wires eliminating the need for high frequency starting unit.

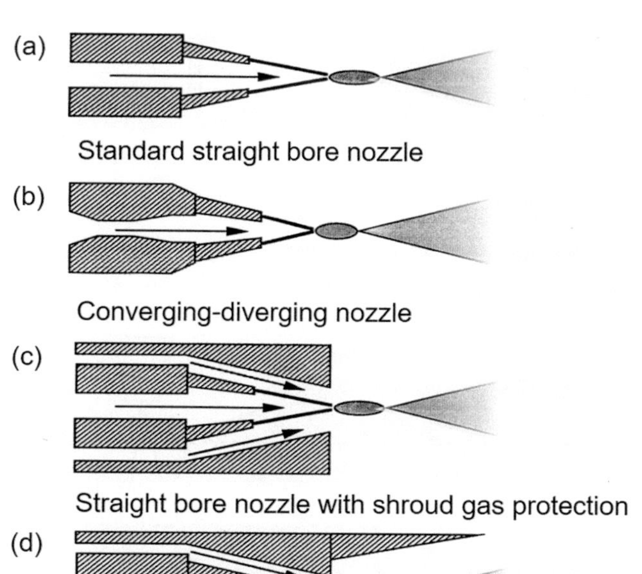

Fig. 11.3 Schematics of different wire and atomizing gas nozzle configurations used in the design of twin-wire arc spraying torches. (**a**) Standard straight bore nozzle. (**b**) Laval type nozzle. (**c**) Straight bore nozzle with secondary gas flow. (**d**) Straight bore nozzle with secondary gas flow and solid shield surrounding the spray jet (Wang et al. 1999)

11.2.2 Droplet Formation Mechanism

In this section, a discussion is presented regarding the basic mechanism involved in molten metal atomization and metal droplet formation. While the analysis is made for the

Table 11.1 Typical WAS deposition rates (kg/h) for various materials (Davis 2004)

Wire material	Al	Bi	Brass	Cu	Mo	Steel	Stainless steel	Tin	Ti	Zn
Material feed rate (kg/h)	2.7	22.7	5.0	5.0	4.6	4.6	4.6	20.5	1.4	10.9

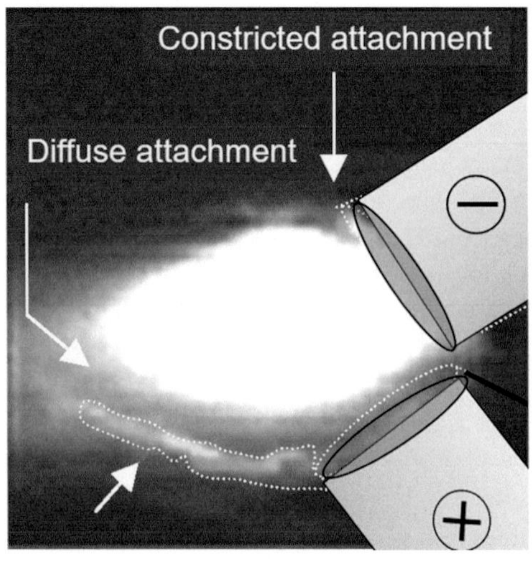

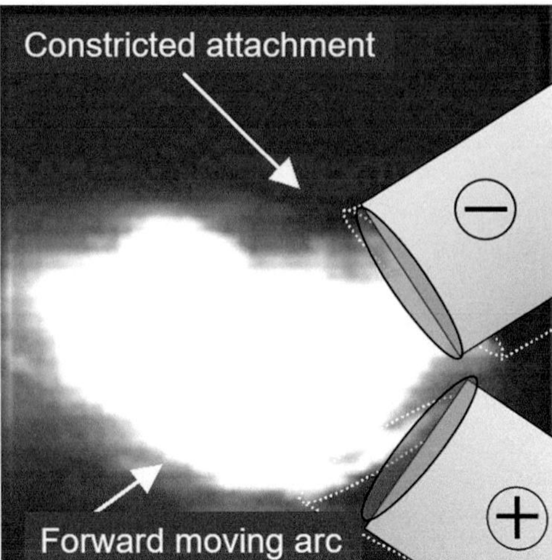

Fig. 11.4 High-speed images of arc attachment on wire tips showing a constricted attachment at the cathode (−) wire and a diffuse attachment at the anode wire (+) (Hussary and Heberlein 2001)

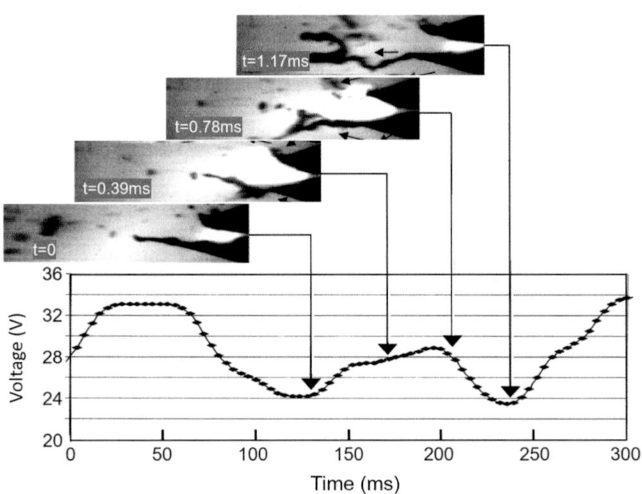

Fig. 11.5 High-speed images of metal droplet formation synchronized with arc voltage trace, showing the forward movement of the anode arc attachment with the associated increase in voltage. Operating parameters: current = 300 A, voltage = 33 V, pressure = 69 kPa (Hussary and Heberlein 2007)

conventional twin-wire arc spraying process, the basic concepts involved are equally valid for single-wire wire arc spraying torch design.

According to Lefebvre (1989), the atomization mechanisms, i.e., the way the molten metal layer is sheared from the wire tip and subsequently subjected to secondary atomization, are mainly governed by Raleigh breakup and membrane breakup mechanisms. While the velocity of the atomizing gas influences strongly the primary atomization, in many situations, the fluid dynamics of the torch has an even stronger influence on the secondary atomization downstream of the arc zone. The low-density, high viscosity, arc acts as a barrier for the cold gas cross flow, and a substantial portion of the atomizing gas flows around the arc column, forming a vortex street. Metal droplets are further atomized in this vortex sheet and can change direction, resulting in further dispersion of the spray pattern. A secondary nozzle close to the wire tip location strongly affects the gas flow, increases the gas velocity at the wire tips, and reduces the jet divergence.

An earlier study by Kawase et al. (1984a, b) in which the anode wire and cathode wire feed rates were varied independently contributed to the identification of:

- Stable operating region when the anode wire feed rate is approximately the same as that of the cathode wire
- Self-regulating operation with the cathode wire feed rate slightly higher than the anode wire feed rate
- Unstable operation when either feed rates (and the current) are too low to melt the wires

Having approximately the same feed rates for the anode and cathode wire results in the highest droplet temperatures and optimal coating adhesion (Kawase and Kureishi 1985a, b), with higher voltages improving the adhesion. The differences between the arc–anode interface compared to that of the arc–cathode results in differences in the droplet formation at the anode and cathode wires (Steffens et al. 1990; Wang et al. 1999). This is illustrated in Fig. 11.4 showing a constricted arc attachment at the cathode wire, and a more diffuse attachment at the anode wire, leading to the formation of a large molten metal sheet from the anode wire with the arc attachment traveling toward the downstream end of this sheet (Hussary and Heberlein 2001).

High-speed photographs cross-referenced to voltage values obtained through synchronization of an oscilloscope with the high-speed CCD camera are given in Fig. 11.5 (Hussary and Heberlein 2007). These show the forward movement of the anode arc attachment with the

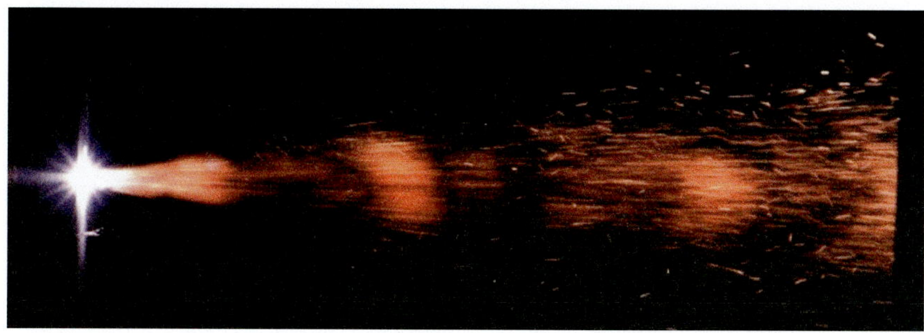

Fig. 11.6 Photograph of a wire arc spray jet showing the periodic variation of droplet flux (Hussary and Heberlein 2001)

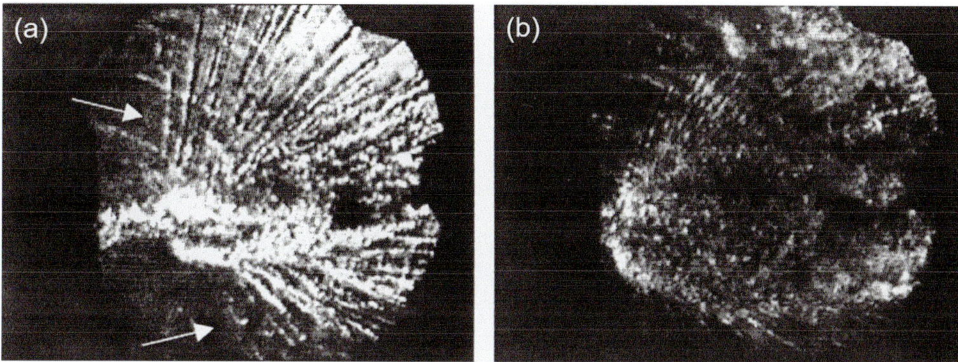

Fig. 11.7 High-speed Schlieren images of the droplet jet from the wire arc spray torch, positioned at the left of the figure, showing the divergence of the droplet beam during a large droplet formation event and secondary atomization event with droplet breaking up into a range of smaller droplets (Hussary 1999)

associated increase in voltage and formation of droplet and their detachment from the wire tip. Operating parameters were arc current = 300 A, voltage = 33 V, pressure = 69 kPa. A photograph of the entire jet of metal droplets given in Fig. 11.6 (Hussary and Heberlein 2001) shows periodic variations in particle fluxes. Typically, a burst of particles is emitted after a dip in the voltage, indicating a momentary peak in the arc current and melt rate. This is usually followed by a steady stream of particles for a good part of a millisecond, until the particle stream subsides, the voltage drops to a minimum followed by another burst of particles (Sheard et al. 1997). The voltage dip can reach a zero value, indicating a direct contact between the tip of the two wires. Such a situation results in an explosive emission of large metal droplets, which are broken up in the secondary atomization zone. Figure 11.7 shows Schlieren images of droplet dynamics in the arc. In particular, the LHS image reveals a secondary breakup of a large metal droplet (Hussary 1999). Clearly, this occurrence results in the broadening of the spray pattern and of the particle size distribution.

The primary causes for such voltage fluctuations are:

- Improper match of the voltage and wire feed rate settings, or too low open circuit voltage. Arc extinction may be responsible for a voltage spike before the dip to zero, caused by the arc blown toward the end of a liquid metal sheet followed by the detachment of the liquid metal.
- Uncontrolled movement of the wire tips due to the "cast and kink" in the wires or due to wear of the wire guide and contact tips.
- Differences in the melt rates of the anode and cathode wires.

The size distribution of the formed metal droplets depends on the atomization mechanisms, i.e., the way the liquid metal layer is sheared initially from the wire tips and on secondary atomization. Classifications of the various disintegration mechanisms in jets and liquid sheets have been the subject of numerous studies (Chigier 1981, Mansour and Chigier 1990, and Lin and Reitz 1998). These show that aerodynamic atomization of the liquid jets (liquid jet co-flowing with high-velocity gas stream) are governed by two important non-dimensional numbers, the Reynolds number, Re, and the Weber number, We, defined as:

$$Re = \frac{\rho_\ell u_\ell d_\ell}{\mu_\ell} \tag{11.1}$$

$$We = \frac{u_r^2 L_\ell \rho_g}{\sigma_\ell} \tag{11.2}$$

where:

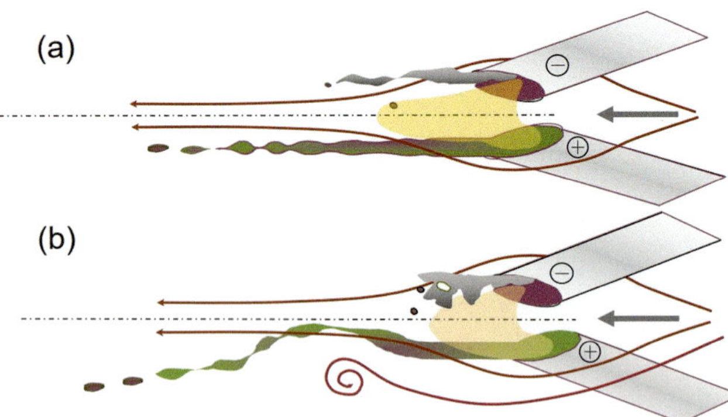

Fig. 11.8 Schematic illustration of (a) axisymmetric breakup of anode sheet and (b) non-axisymmetric breakup of the anode sheet and membrane-type breakup of the cathode sheet (Hussary and Heberlein 2007)

d_ℓ diameter of the liquid jet (m)

L_ℓ characteristic dimension such as the liquid sheet thickness (m)

u_ℓ velocity of the liquid (m/s)

u_r relative velocity of the liquid to gas stream (m/s)

ρ_ℓ liquid density (kg/m^3)

ρ_g gas density (kg/m^3)

μ_ℓ dynamic viscosity of the liquid (kg/m.s)

σ_ℓ surface tension of the liquid (N/m) (kg/s^2)

Liquid jet breakup is generally recognized to involve the following principal mechanisms:

- Axisymmetric Rayleigh breakup ($We < 15$),
- Non-axisymmetric Rayleigh breakup ($15 < We < 25$),
- Membrane breakup ($25 < We < 70$)
- Fiber-type breakup ($100 < We < 500$).

The first three types of primary atomization mechanisms are most frequently observed in wire arc spraying corresponding to relatively low Weber numbers (< 100) (Chigier 1981; Lefebvre 1989; Mansour and Chigier 1990). A schematic representation of the axisymmetric and non-axisymmetric Raleigh breakup mechanisms is presented in Fig. 11.8 (Hussary and Heberlein 2007). Primarily, the thermophysical properties of the liquid (molten metal) such as viscosity, surface tension and density, and the relative velocity between the atomizing gas and the liquid determine the predominant breakup mechanism. In the WAS process, the constant transient variation of the properties of the molten metal gives rise to corresponding variation in the breakup behavior. The axisymmetric Raleigh breakup of the anode liquid sheet is schematically illustrated in Fig. 11.8a, while Fig. 11.8b shows the non-axisymmetric Rayleigh breakup of the anode liquid metal sheet leading to droplet formation from extended liquid metal ligaments and the membrane-

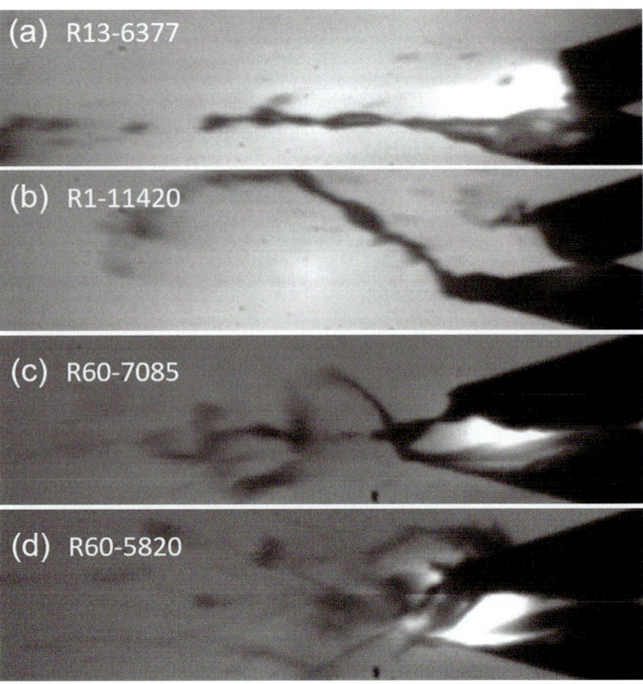

Fig. 11.9 High-speed images of liquid metal atomization of the carbon steel wire tips showing (a) disintegration of the anode sheet in an axisymmetric Raleigh breakup, (b) a non-axisymmetric Raleigh breakup of a curved anode sheet, (c) anode sheet curvature due to large eddies, and (d) a membrane-type breakup of the cathode sheet with holes forming in the membrane (Hussary and Heberlein 2007)

type breakup due to hole formation in a cathode liquid metal sheet.

High-speed images of such breakups at the wire tips during spraying using carbon steel wires are shown in Fig. 11.9 (Hussary and Heberlein 2007). The non-axisymmetric breakup is the most frequently observed mechanism at the anode for almost all operating conditions, while the cathode frequently shows a membrane-type breakup together with the non-axisymmetric breakup, especially at higher voltages. The high-speed videos (18,000–40,500 frames/s) also show that increasing the wire feed rate/current setting will increase the wire melting rate. Increasing the voltage, on the other hand, will in general not result in larger gaps but rather in arcs that are strongly bowed in the direction of the flow or more likely to move toward the end of a liquid metal ligament. Consequently, an increase in droplet temperatures is to be expected with higher-voltage settings.

It should also be noted that, while the atomizing gas velocity influences strongly the primary atomization, in many situations, the fluid dynamics of the torch has an even stronger influence on the secondary atomization downstream of the arcing zone. The low-density arc acts as a barrier for the cold gas cross flow, and a substantial portion of the atomizing gas flows around the arc, forming a vortex street behind the obstacle. Metal droplets are further atomized in this vortex street and can change direction, resulting in further dispersion of the spray pattern. A secondary nozzle close to the wire tip location will strongly affect the gas flow, increase the velocity at the wire tips, and reduce the jet divergence.

Droplet formation in the wire arc spray process is not, however, a closed issue. While higher velocity gas flows at the location of the wire tips will result in smaller droplet sizes, it is obvious that the entire fluid dynamics of torch and surroundings affect significantly this process. The role of secondary atomization is presently not fully understood. Furthermore, there appears to be some contradicting evidence reported in the open literature on the effect of the other operating parameters on droplet size distributions and the role of wire polarity. Relatively few studies have been performed on the electrical characteristics of the power supply and the rectifier's responds to voltage fluctuations. The variations in the droplet's formation are recognized to be linked to the instant voltage fluctuations which are linked in turn to:

- Improper match of voltage and wire feed rate settings, too low open circuit voltage
- Movement of the wire tips due to the cast and kink in the wires or wear of the contact tips
- Differences in the melt rates of the anode and cathode wires

11.2.3 Particle Size Distribution

The effect of the atomizing gas velocity on particle size distributions (PSD) is demonstrated in Fig. 11.10 (Wang et al. 1999). It is to be noted that the pressure values given in this figure were measured at the control panel. These would correspond to almost double the gas pressure at the torch. The PSD values are given in number, and not in the more conventional volume or mass fractions, which would move them toward the smaller particle sizes. The results show that an increase of the torch pressure will result in a finer average droplet diameter and narrower PSD due to the increase of the atomizing gas velocity. The data given in Fig. 11.10 are for aluminum particles obtained by spraying aluminum wire into ice and determining the particle size distribution of the powder obtained using SEM and image analysis.

An interesting observation was reported by Planche et al. (2003) who measured droplet size distributions at different distances from the torch axis for a TAFA 9000 torch used with converging cap nozzle, operating at 100, 150, and 200 A and air flow rates between 1567 and 2167 slm. The size distributions were different at different

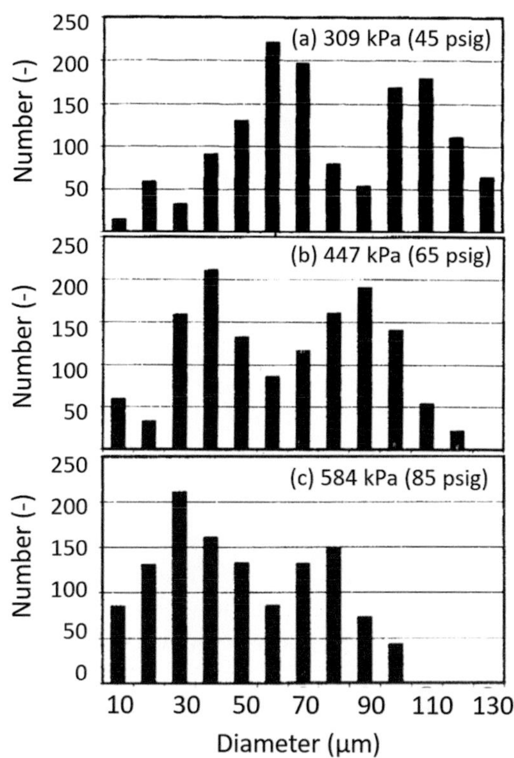

Fig. 11.10 Aluminum particles size distributions obtained for different atomizing gas pressures (**a**) 309 kPa (45 psig), (**b**) 447 kPa (65 psig), and (**c**) 584 kPa (85 psig) (Wang et al. 1999)

distances from the torch axis, ranging from relatively narrow and mono-modal on the axis to wider multimodal distribution with larger average particle diameters at off-axis locations.

In an attempt to explain the source of the bimodal nature of the PSD reported by different authors, and to determine the effect of nozzle design and operating parameters on the PSD for the droplets produced, Liao et al. (2005) carried out a systematic study using the three different nozzle configurations given in Fig. 11.11. These were adapted to a TAFA 9000 torch with nozzle (Fig. 11.11a) being a standard closed spray nozzle with a convergent orifice referred to as (C/CL nozzle). Figure 11.11b illustrates a second nozzle design which is of open configuration and incorporating an upstream convergent–divergent atomization nozzle without a cap (CD/OP nozzle). Figure 11.11c shows a modified closed nozzle with a converging–diverging orifice and secondary gas flow (Liao et al. 2005). To trace the source of a particle in the spray, i.e., whether it is resulting from the anode or cathode wires, wire materials with different magnetic properties were used for the anode and cathode, such as copper and steel. After spraying into water, the steel particles were separated from the copper ones using a magnetic field. Typical results presented in terms of volume fractions which are equivalent to the mass distribution of the powder are

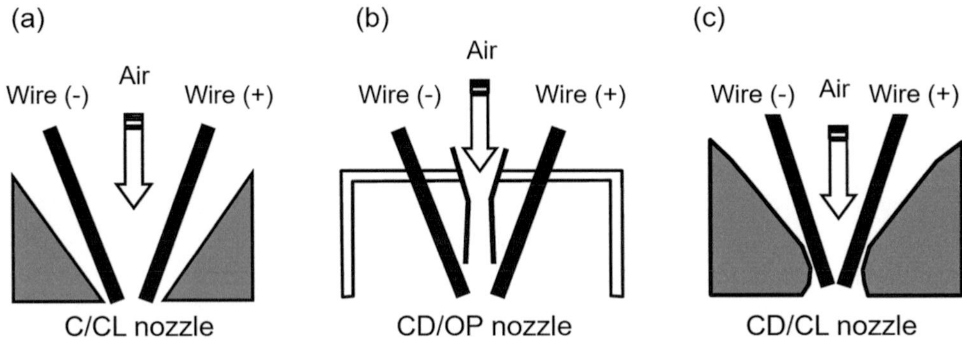

Fig. 11.11 Schematic of different nozzle configurations considered: (**a**) C/CL, standard TAFA 9000 spray nozzle, closed nozzle and convergent orifice. (**b**) CD/OP Open nozzle with convergent–divergent Laval orifice. (**c**) CD/CL closed nozzle with convergent-divergent Laval-type secondary nozzle (Liao et al. (2005)

given in Fig. 11.12 (Liao et al. 2005). These were obtained using a stand torch with a C/CL nozzle, with the particles sprayed into water with a standoff distance of 300 mm, a torch operating at 30 V and 200 A. Copper and steel wires were used in these experiments with the copper wire being alternately as anode (Fig. 11.12a & b) or cathode (Fig. 11.12c & d). Two different atomizing air supply pressures, 0.28/0.28 MPa (Fig. 11.12a & c) and 0.46/0.49 MPa (Fig. 11.12b & d), were used for primary and secondary gas flows, respectively.

The following observations are made (Liao et al. 2005):

- Both the anode and the cathode wire deliver droplets with binary size distributions, i.e., the binary size distributions observed in other experiments, are not necessarily due to the different droplet sizes from the anode and the cathode wires.
- The size distribution becomes close to mono-modal at high supply pressure of the atomizing gas.

The average particle diameter obtained from the copper and steel wires used in different polarity combination and operating conditions are given in Fig. 11.13 (Liao et al. 2005). These are presented under the following set of conditions:

- *Group 1:* regular atomizing gas nozzle with secondary nozzle, 0.28/0.28 MPa supply pressures, straight polarity
- *Group 2:* regular atomizing gas nozzle with secondary nozzle, 0.28/0.28 MPa supply pressures, reversed polarity
- *Group 3:* regular atomizing gas nozzle with secondary nozzle 0.46/0.49 MPa supply pressure, straight polarity
- *Group 4:* regular atomizing gas nozzle with secondary nozzle, 0.46/0.49 MPa supply pressure, reverse polarity
- *Group 5:* regular atomizing gas nozzle with a Laval-type secondary nozzle, 0.32/0.32 MPa supply pressure, reversed polarity

- *Group 6:* Laval-type atomizing gas nozzle without secondary nozzle, 0.32/0.32 MPa supply pressures, reversed polarity

The results show that, for copper, the average size of the droplets is not significantly different whether the wire is used as anode or as cathode. This is not the case for steel wires which show about 20% larger mean droplet diameter when the steel wire is used as anode compared to that when the wire is used as cathode. Moreover, considering the same polarity, the average size of the steel droplets is larger than that for the copper for droplets originating from the anode, while the reverse is observed for the droplets originating from the cathode. That implies that the wire material has a strong influence on the relative importance of the polarity.

A similar study on the of the effect of wire polarity on the PSD of copper and steel powders was reported by Pourmousa et al. (2005) using with a Sulzer Metco ValuArc 200 operated with a wire feed rate of 7 m/min, arc voltage of 32 V, and gas supply pressure of 0.208 MPa resulting in an air flow rate of approximately 1032 slm. The PSD of the formed droplets was interpreted as representing the superposition of two log-normal distributions which they interpreted as belonging respectively to particles from the cathode and the anode. Applying this assumption to the spraying with aluminum wires, and further assuming the feeding rate of the cathode and the anode wires were equal, they investigated the effect of operating parameters on the PSD of the powder obtained. As reported by earlier studies, the mass–mean diameter of the formed droplets was observed to decrease with the increase of the gas supply pressure. Variation of wire feed rate, arc current, and voltage had, on the other hand, little effect on the mass–mean diameter. They further mentioned that secondary atomization was unlikely since as the droplets leave the wire tip, the relative velocity between the droplets and the gas, and accordingly the Weber number, decreased rapidly. This observation, however, does not consider the large-scale

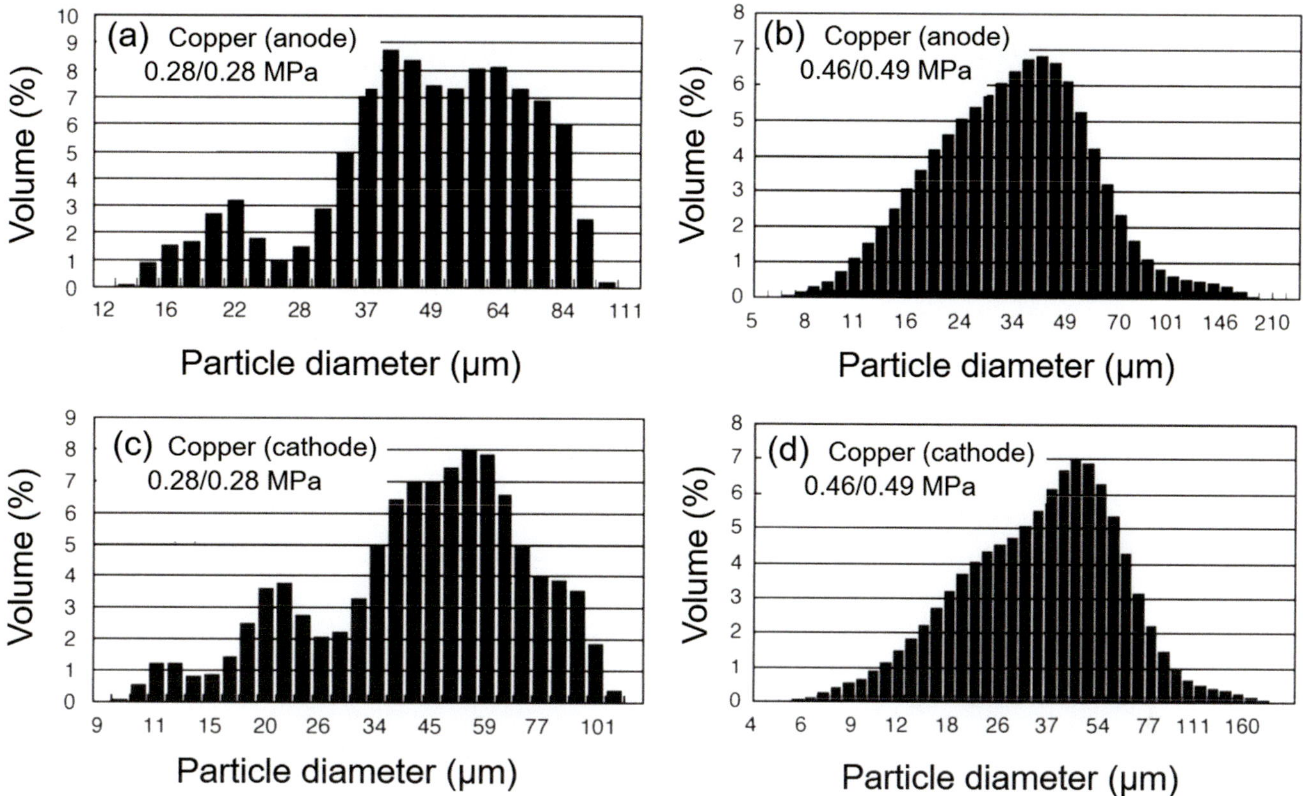

Fig. 11.12 Particle size distribution of droplets generated using a standard C/CL TAFA 9000 nozzle, from an anode copper wire (**a** & **b**) and a cathode copper wire (c &d) with supply pressure of the atomization and secondary gases of 0.28/0.28 MPa (**a** & **c**) and 0.46/0.49 MPa (b &d) (Liao et al. 2005)

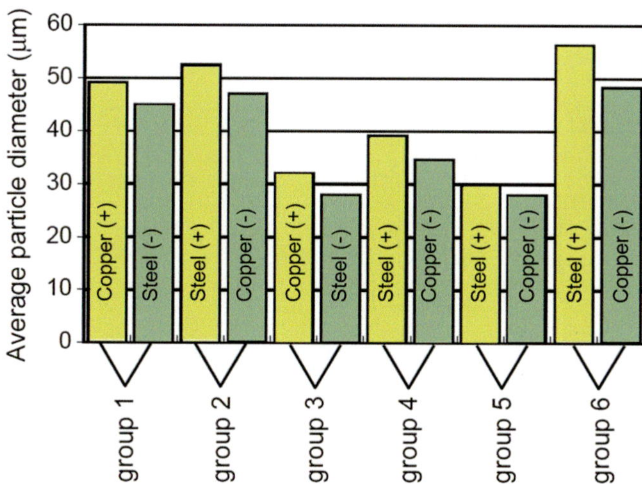

Fig. 11.13 Average droplet/particle sizes for anode wires and cathode wires for different configurations and operating conditions (Liao et al. 2005)

turbulence and the entrainment of slower moving gas from the surroundings that can give rise to secondary atomization, which is observed under some conditions in the wake of the arc using high-speed photography.

In an experiment to generate a Ti-Al intermetallic compound in a coating, the co-spraying of titanium and an aluminum wires was investigated (Watanabe et al. 2002). The work was carried out using a Sulzer Metco 4R arc spraying system using 1.6 mm diameter titanium and aluminum wires, with arc currents of 100 to 200A, arc voltages of 24 to 33 V, wire feed rates of 1.8 g/s, and atomizing gas pressure of 0.30 MPa. It was generally observed that operation with the Al wire as the anode, the arc was unstable with large voltage fluctuations, attributed to the different melt rates of the wires. With the Ti wire as the anode, a more stable operation was observed. Instabilities were also observed with two Al wires, while the voltage showed little fluctuation with two Ti wires. The conclusion reached from these experiments was that with low melting point materials as anode wires, it may be difficult to match the wire feed rate with the melt and removal rates, leading to the voltage fluctuations. Larger voltage fluctuations resulted in broader size distributions, and increasing the operating voltage resulted in increased average droplet diameters.

Figure 11.14a & b shows typical SEM micrographs of droplets produced with Al electrodes and Ti electrodes, Fig. 11.14c corresponding to the case with Ti anode and Al cathode, and Fig. 11.14d for the case with Al anode and Ti

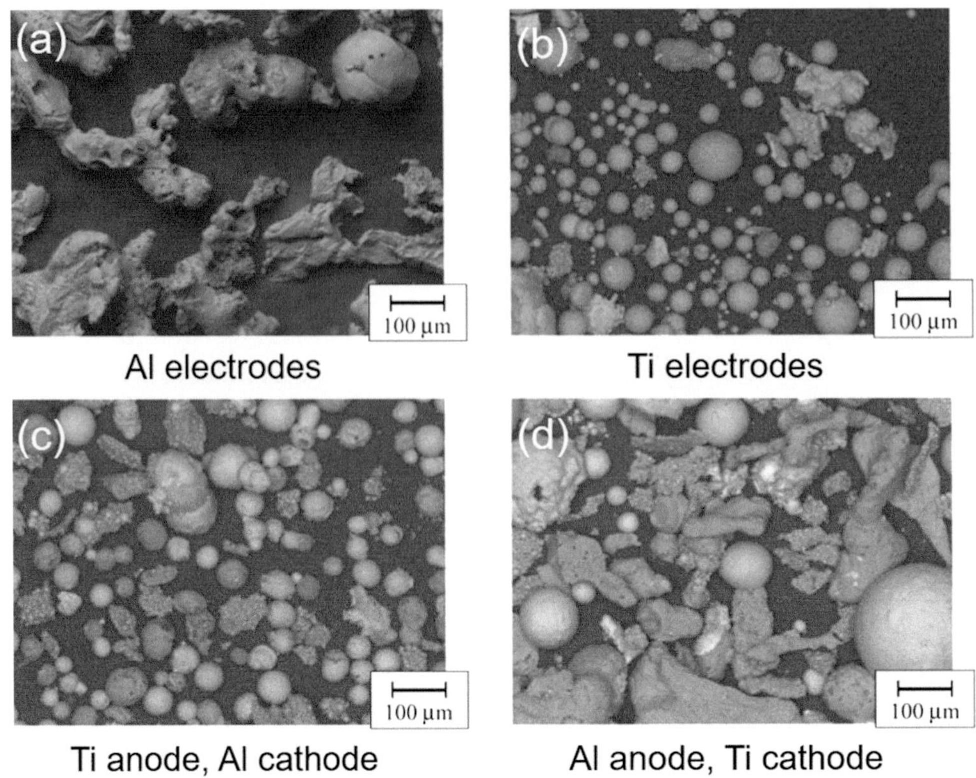

Fig. 11.14 Micrographs of droplets/particles produced with (**a**) Al electrodes, (**b**) Ti electrodes, (**c**) Ti anode and Al cathode, and (**d**) Al anode and Ti cathode (Watanabe et al. 2002)

cathode. The arc current in this case was 150 A, arc voltage of 30 V, and supply gas pressure of 0.3 MPa. The droplets produced from dual aluminum electrodes are strongly deformed, while the titanium droplets are almost spherical with a wide particle size distribution. The droplets produced from the combination of Ti and Al electrodes include relatively large particles which can be attributed to intense voltage fluctuations. It is not surprising to have aluminum particles as least spherical since they have a relatively low melting temperature and lower surface tension than titanium (0.91 N/m for Al, and 1.65 N/m for Ti). It has also to be considered that in wire arc spraying, the atomizing gas is at relatively low temperature with the molten droplets having a minimum degree of superheat which does not allow the necessary time for the molten droplets to acquire a spherical shape prior to freezing.

A systematic study of the effect of the operating torch parameters on the PSD for carbon steel particles obtained by spraying steel wires into a layer of ice using a Miller Thermal (now Praxair- TAFA) PB-400 torch was reported by Hussary and Heberlein (2007). The results given in Fig. 11.15 are presented in terms of mass fraction for different particle diameter ranges, obtained with an arc current of 100 A and 36 V setting. The atomizing gas was nitrogen, at a flow rate of approximately 1100 slm.

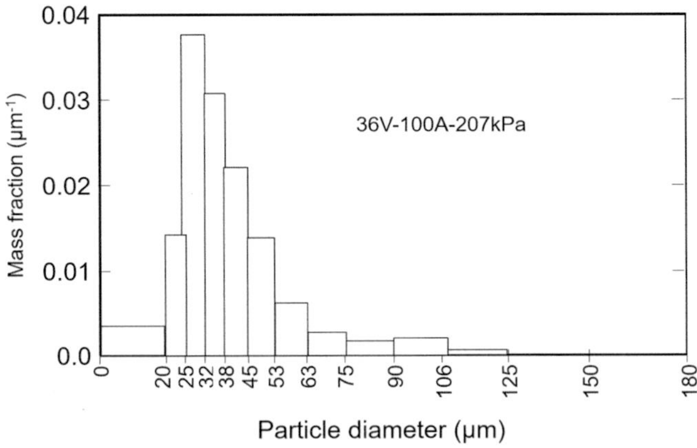

Fig. 11.15 Size distribution of carbon steel particles collected for the process parameters: I = 100 A, V = 36 V and p = 207 kPa (Hussary and Heberlein 2007)

Figure 11.16 (Hussary and Heberlein 2007) shows the evolution of the mass–mean diameter for steel particles as a function of the pressure for different arc currents and voltage settings. In this torch configuration, a pressure of 207 kPa would give rise to an atomizing gas flow rate of approximately 1100 slm. The results show a systematic decrease of the mass–mean droplet/particle diameter with the increase of the pressure and the corresponding increase of the atomizing gas flow rate. Increasing the current or the arc voltage, on the other hand, gives rise to the increase of the mean droplet diameter, due to the increase of the wire melting rate. The parametric dependence of the droplet size distribution on the atomizing gas flow rate, arc current, and arc voltage is also reflected on the microstructure of the deposit given in Fig. 11.17. Micrographs in Fig. 11.17a & b, obtained with a torch pressure of 90 kPa and arc currents of 100 and 300 A, respectively, show that the increasing the arc current gives rise to a significant increase of the average size of the formed droplets which results in turn in the increase of the splat thickness and the development of coarser-grain microstructure of the coating. The reverse effect is observed from a comparison of micrographs in Fig. 11.17b & c, where a significant reduction of the deposit grain size is noted with the increase of the torch pressure from 90 to 365 kPa for an arc current of 300A which is an indication of a finer mean droplet size.

11.2.4 Splat and Coating Formation

Splat formation and layered building of the coating is the final step of all thermal spray processes on which the quality of the coating and its adhesion to the substrate depends. As discussed earlier in combustion, DC plasma, and RF induction plasma spraying, a wide range of splat shapes and configurates depend on:

- In-flight particle/droplet parameters prior to their impact on the substrate. This includes particle/droplet temperature and velocity and thermophysical properties of the molten material such as density, viscosity, and surface temperature.
- Substrate surface preparation including cleanliness from pollutants and adsorbents, surface roughness, substrate temperature, and thermophysical properties such as thermal conductivity and heat capacity,
- Incidence angle of the particle trajectory on the substrate. An increasing level of porosity appearing in the deposit with deviation from orthogonal impact on the substrate.

Splat formation in wire arc spraying process is no exception to the above-listed parametric dependence. Fang et al. (2005) sprayed stainless-steel wires (3Cr13) using CMD-AS1620 WAS unit working with an arc current of 220 A, voltage of 35 V, and air as the atomizing gas. The in-flight particles' temperature measured using two-wave pyrometry prior to droplet impact on the substrate was 2400 °C. The splats were collected on 30×30 mm polished stainless-steel substrates, 4 mm thick, which was preheated to either 60 °C or 160 °C. Typical splats obtained for each of these two substrate temperatures are given in Fig. 11.18 with particle impacting the substrate at 90° to its surface (Fig. 11.18 a,c) or at 30° to the axis orthogonal to the substrate (Fig. 11.18 b,d). The results show that independent of the particle impact

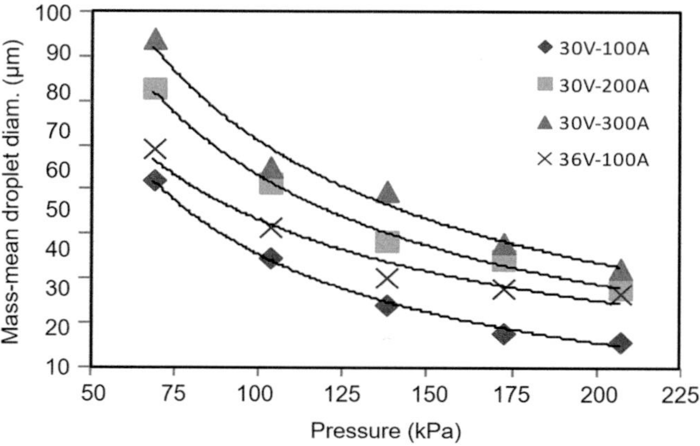

Fig. 11.16 Dependence, for carbon steel, of the mass–mean droplet diameter on atomizing gas pressure, arc current, and voltage (Hussary and Heberlein 2007)

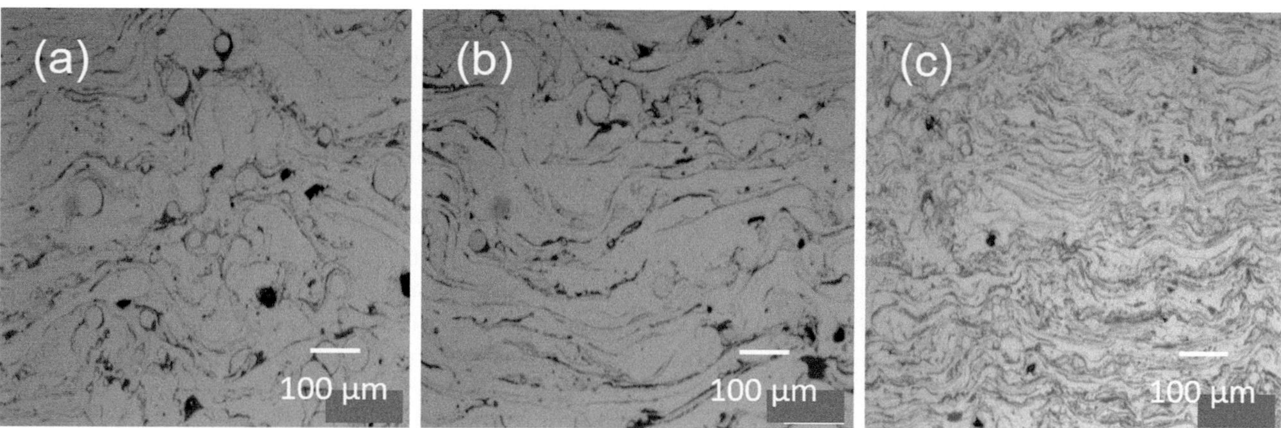

Fig. 11.17 Cross sections of carbon steel coatings deposited at (**a**) 90 kPa, 30 V, 100 A, (**b**) 90 kPa, 30 V, 300 A, and (**c**) 365 kPa, 30 V, 300 A (Hussary and Heberlein 2007)

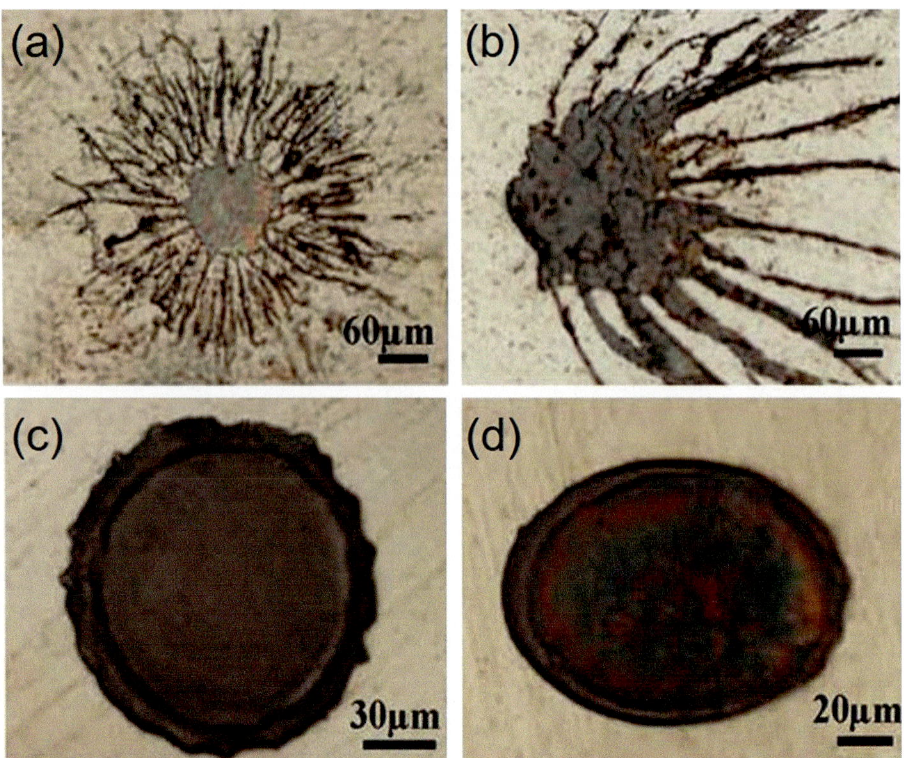

Fig. 11.18 Typical morphology of splats at different substrate temperatures and angle of impact on the substrate: (**a** and **b**) Substrate temperature at 60 °C, with (**a**) at 90° impact and (**b**) at 30° to the axis orthogonal to substrate surface. (**c** and **d**) Substrate temperature at 140 °C, with (**c**) at 90° impact and (**d**) at 30° to the axis orthogonal to substrate surface (Fang et al. 2005)

angle, the particle splats obtained with low substrate temperatures, 60 °C (Fig. 11.18 a, b), tend to be mostly of flat structure in the center with significant finger-splashing. On the other hand, with a substrate temperature of, 140 °C (Fig. 11.18 c,d), the particle splats tend to be of a flat circular shape with a distinct rim. It is to be noted, however, that in both the low- and high-temperature substrate cases, the angle of impact of the droplet on the substrate has a significant influence on the shape of the splats which tend to be elongated in the direction of the slope of the substrate.

These observations were further confirmed by Abedini et al. (2006) who sprayed aluminum wires onto polished AISI-304 L coupons maintained at temperatures ranging from 25 °C to 450 °C. In-flight droplet parameters (diameter, velocity, temperature) were measured using a DPV-2000 system and resulting splats were photographed. At low substrate temperature, droplets splashed, forming irregular splats, while at higher temperatures, there was no splashing and splats formed circular disks. The temperature at which the transition occurred between these two splatting modes decreased with increasing impact velocity. Raising substrate temperature increased both deposition efficiency and adhesion strength. Predictions from analytical models to calculate splat diameter and transition temperature were in good agreement with experimental data.

11.2.5 Coating Formation

A study of splats layering and coating formation was reported by Steffens and Nassenstein (1999) for the WAS of steel wire, 1.6 mm diameter, with 13 wt. % of chromium on steel substrate in the form of a rotating cylinder 90 mm o.d. and 3.2 mm wall thickness. The spray gun-type LD/U2, manufactured by OSU, Germany, was used operating with an arc current of 200 A, voltage 25 V, with air as atomizing gas at pressures varying between 0.3 and 0.5 MPa. The spray distance was kept constant at 150 mm. The study examined the effect of substrate surface velocity on the coating microstructure. This was achieved by rotating the substrate at different speeds providing linear surface velocities in the range of 5 to 100 m/min. In all test, the substrate translation velocity with respect to the spray jet was kept constant at 1 m/min. Figure 11.19 shows the microstructure structure of X46Cr13 coating obtained at surface velocities, $v_s = 5$ m/min, 35 m/min, and 100 m/s. Globally all the coatings obtained had typical lamellae character. The coatings prepared at surface velocities of 5 and 35 m/min are denser compared to that at 100 m/min. The higher surface velocity coatings exhibit an increasing level of lenticular-shaped pores, which can be partly due to the reduced level of densification by the subsequent droplets (less peening or self-

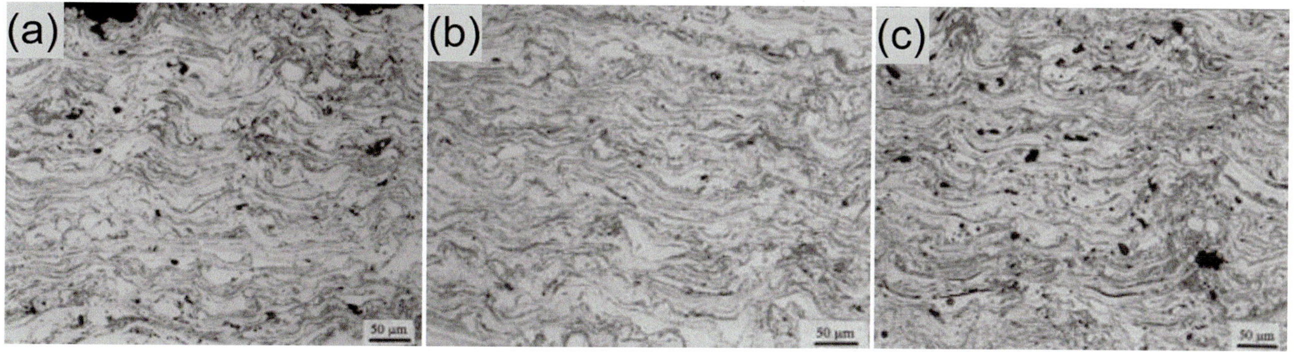

Fig. 11.19 Microstructure of cross section of steel X46Cr13 coating sprayed using closed nozzle system with a wire diameter 1.6 mm, gas pressure 0.45 MPa, I = 200 A, V = 25 V, substrate surface velocity (**a**) 5 m/min, (**b**) 35 m/min, (**c**) 100 m/min (Steffens 1999)

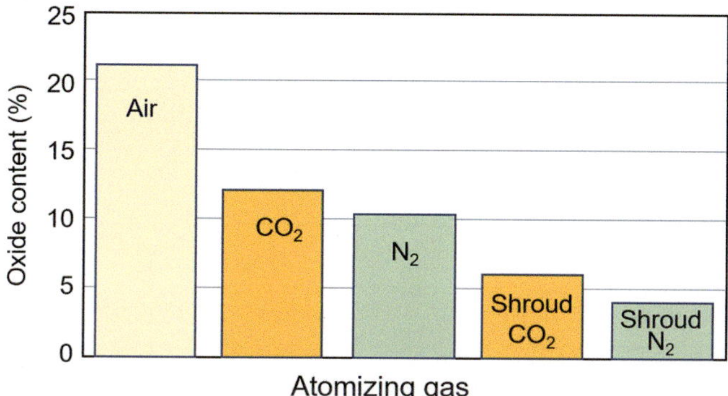

Fig. 11.20 Oxide content in chrome steel WAS coating deposited using different atomizing gases combined with additional use of shroud after (Wang et al. 1999)

densification effect). Moreover, the coating layers appear to have poor cohesion. This effect is supported by sickle-shaped inhomogeneities observed in Fig. 11.19b. Basically, two main physical parameters influenced the porosity, *impact particle velocity and temperature*. Increasing the particle temperature decreases its viscosity on impact with the substrate. Thus, the coating roughness is reduced. The closed nozzle system used in this study supports a finer atomization of the melt by a radial jet stream focused on the arc. Consequently smaller, hotter, and faster spray droplets are produced, which form a denser structure. Unfortunately, with the air as atomizing gas, oxide content increases for the hotter particles, as shown in Fig. 11.19c.

The principal characteristics of the coating are evaluated in terms of:

- *Porosity*
- *Oxide content*
- *Elemental composition*

Further characteristics are *surface roughness*, *adhesion*, and *functional values* like hardness. Coating porosity in general

decreases with decreasing droplet size; however, oxide content with air sprayed coatings increases with decreasing particle size because of the larger surface area-to-mass ratio of smaller particles and because of the increase in internal convection in the individual droplets. For example, for the case of WAS of aluminum using a Sulzer Metco 4RG torch with a 150 A, 30 V setting, and a standoff distance of 150 mm, an increase of the atomizing gas pressure from 450 kPa to 590 kPa has resulted in a decrease in the porosity from 18% to 12% and in an increase in the oxide content from 19.5% to 25% (Wang et al. 1999). Significant reduction of oxide content can be achieved with the use of nitrogen or CO_2 as atomizing gas and the addition of shroud gas, as shown in Fig. 11.20. Wang et al. (1999) reported that the oxide content of the coating could be reduced for the same operating conditions given above from 21% using air as atomizing gas down to 12% using CO_2, 10% using N_2. It could be even further reduced to 6% using CO_2 and 4% using N_2 when combining them with a secondary shroud gas.

The effect of the atomizing gas flow rate on the coating porosity and oxide content was investigated by Watanabe et al. (1996). Results given in Fig. 11.21 were obtained using Sulzer Metco 4RG torch operated at 200 A, 34 V, with air as atomizing gas using the following three torch nozzle configurations. Lowest porosity and highest oxide content were obtained with the high velocity cap nozzle (HV).

- Standard nozzle (std)
- High velocity cap offering a secondary nozzle configuration around the wire tips (HV)
- Modified version of the standard nozzle with a "Laval-type" profile (Laval)

It should be pointed out, however, that inflight oxidation of the droplets formed during WAS process is not always reflected in the increase of the oxygen content in the coating

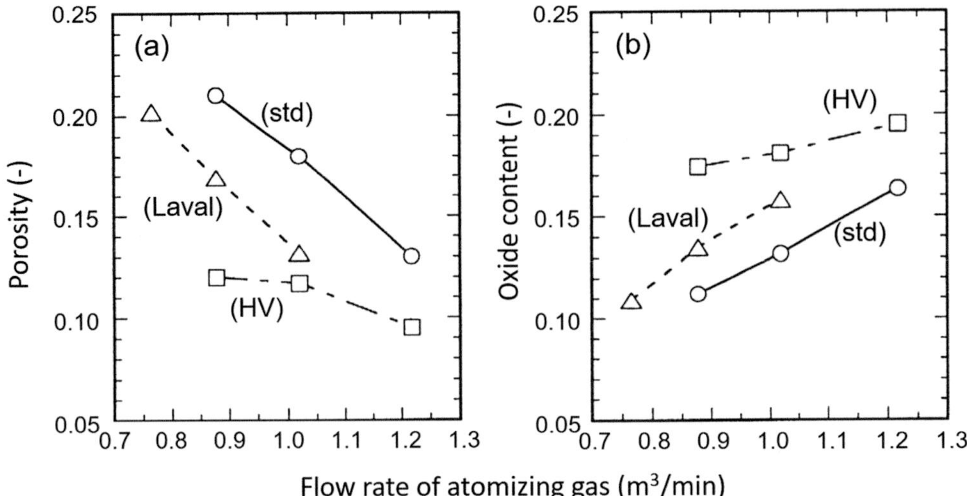

Fig. 11.21 (**a**) Porosity and (**b**) oxide content in aluminum coating as function of the atomizing gas flow rate for three different nozzle designs: standard (std), high-velocity cap (HV), and Laval-type nozzle (Laval), arc current = 200 A, voltage = 34 V (Watanabe et al. 1996)

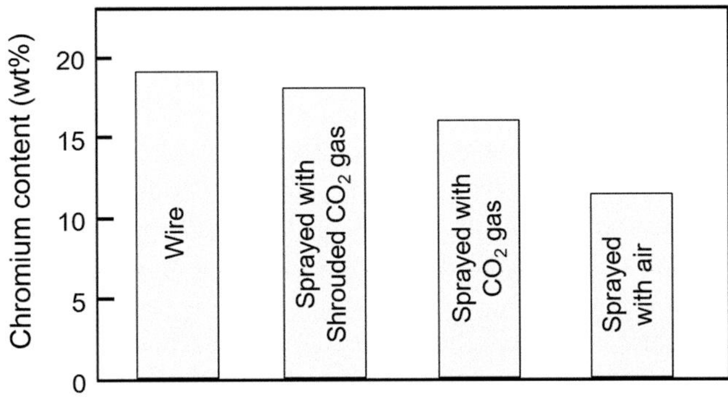

Fig. 11.22 Chromium content in wire and in coatings deposited using CO_2 atomizing gas with shroud, CO_2 atomizing gas without shroud, and air atomization showing the loss of Cr due to volatile oxide formation (Wang et al. 1995b)

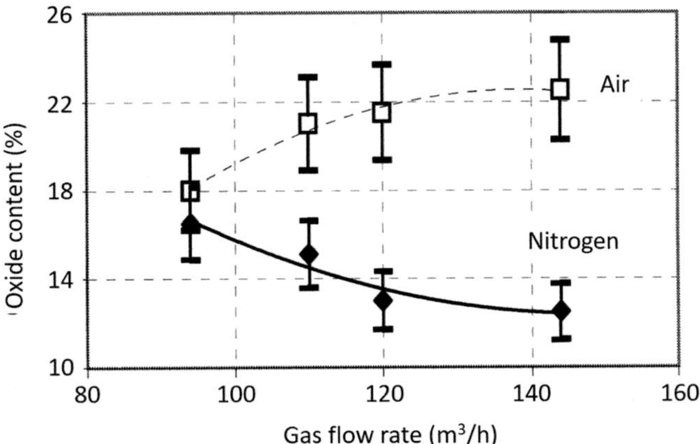

Fig. 11.23 Oxide content in carbon steel coating as function of atomizing gas flow rate for air and for nitrogen as atomizing gas. (Jandin et al. 2002)

since, depending on the alloy composition, in-flight droplet oxidation can result in the loss of some of the alloying elements through volatilization. For example, in chrome steel spraying, the formation of a volatile CrO_3 can reduce the Cr content from 19% in the wire to 11% in the coating (Wang et al. 1995b). The use of an inert atomizing gas and an inert secondary gas significantly reduces the Cr loss as shown in Fig. 11.22.

In general, the coating microstructure is more uniform with regular lamellae at the higher gas flow rates, and less surface roughness is observed (Wang et al. 1995a; Planche et al. 2003; Jandin et al. 2002). A study of the coating formation through investigation of individual splats (Planche et al. 2004) required preheating of the substrates with an auxiliary heat source, because the splats obtained with cold substrates were extensively fingered and made further analysis difficult.

Jandin et al. (2002) proposed that the coating chemistry can also depend on the atomization gas flow rate. The proposed mechanism is based on the observation that increasing the atomization gas flow rate gives rise to a decrease of the mean size of the atomized droplets with a proportional increase of the specific surface area per unit mass. In-flight oxidation of the droplets being essentially a surface phenomenon will consequently increase with the decrease of the droplet diameter. The results given in Fig. 11.23 offer strong support to this mechanism showing a steady increase of the oxide content of the coating with the increase of the atomization gas (air) flow rate. The reverse trend is observed, when using nitrogen as atomizing gas, where the oxide content of the coating is noted to drop steady with the increase of the atomizing gas flow rate. The authors also note that spraying

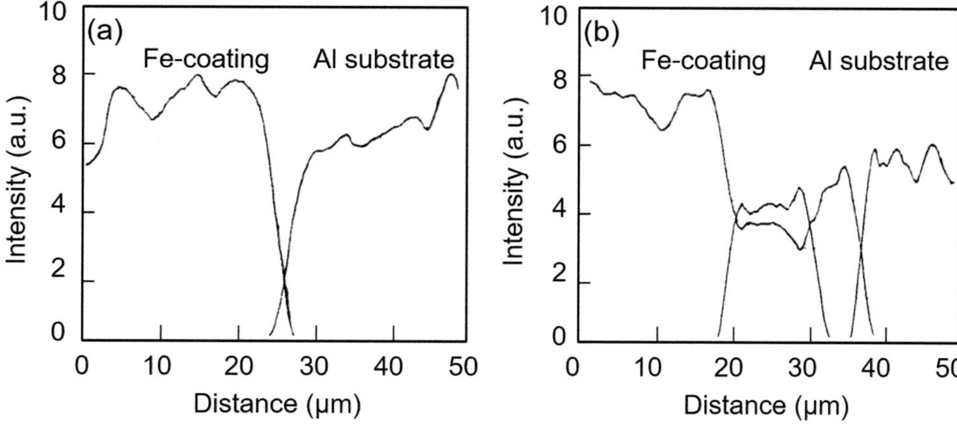

Fig. 11.24 Auger spectrum across the interface between steel coating and aluminum substrate (**a**) WAS using a commercial torch with primary gas atomization and (**b**) WAS with primary and secondary gas atomization (Wang et al. 1999)

carbon steel with nitrogen can, on the other hand, result in a reduction in the amount of carbon in the steel due to reaction with nitrogen and an associated reduction in hardness.

Increased oxide content usually translates into increased hardness while smaller particle sizes and associated higher droplet velocities result in lower surface roughness and usually better adhesion. An example of the improvement of adhesion is illustrated in Fig. 11.24 giving the elemental composition across the interface between a steel coating and aluminum substrate (Wang et al. 1999). As shown in Fig. 11.24a, the use of a regular nozzle shows a clear difference in the Auger spectrum between the Fe signal of the coating and that of Al signal of the substrate. In contrast, the coating obtained with secondary gas atomization (Fig. 11.24b) shows an intermetallic region, of a few tens of μm thick, between the coating and the substrate where both Al and Fe are present. XRD analysis of this region shows indeed the formation of intermetallic at the interface which is probably due to the higher velocity and higher temperature of the droplet on their impact with the substrate, resulting in local melting of the substrate material.

Matthews and Schweizer (2013) developed a full factorial experimental design to investigate the effect of four process variables *(current, voltage, spray distance,* and *atomizing air pressure)* on coating quality. Each of these variables had a high and low settings, on the performance of an industrial arc-spray system using 1.6 mm diameter wires of nominal composition Ni-43Cr-0.3Ti with the substrate placed orthogonal to the axis of the torch. The study focused on the impact of these variables on the *splat shape* and *thickness, coating porosity, oxide content,* and *microhardness.* The results showed that:

- The *splat thickness* increased with increasing voltage but decreased with increasing gas atomizing pressure

- High *coating thicknesses* were generated at high current due to the higher wire feed rate at this setting. Higher gas atomizing pressure and lower voltage also contributed to thicker coatings due to their effect on forming smaller particles which were assumed to have higher deposit efficiency.
- The *coating porosity* was reduced at high gas atomizing pressures due to the high-velocity achieved by the small particles generated under these conditions. A complex interaction between the arc current and voltage played a secondary role in the formation of coating porosity.
- All the variables considered contributed to the amount of *oxide formation* in the coating. High oxide contents were generated at high gas atomizing pressure, spray distance, and current and low voltage settings. These effects were related to the high surface area of the molten particles generated under these conditions and their residence time in-flight.
- High coating *microhardness* was also observed at high gas atomizing pressure and high current settings. This combination of parameters generated high oxide contents and low coating porosities, both of which contributed to the increase in coating microhardness.

Reactive twin-wire arc spraying (TWAS) where each of the two wires is of different composition has as objective the depositing an intermetallic compound of the two wires. According to Chang et al. (2011) when arc spraying Ti and Ni wires fed synchronously into the TWAS gun, some intermetallic compounds such as $TiNi_3$ and Ni-Ti alloy are synthesized in the Ni-Ti coating. The wear resistance of the Ni-Ti composite coating is superior to that of pure Ni-sprayed coating but slightly inferior to that of the titanium. The corrosion resistance of the

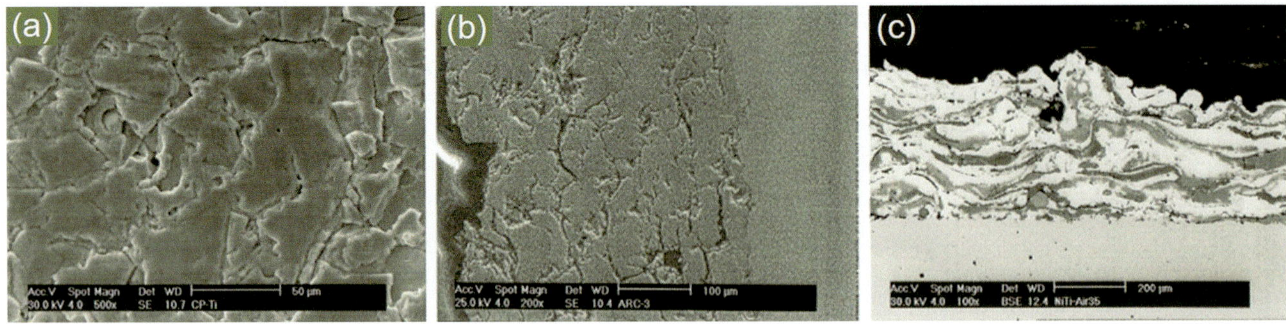

Fig. 11.25 SEM micrograph: (**a**) Ti coating, surface 5009; (**b**) Ti coating, cross section 2009; and (**c**) Ni-Ti composite coating, cross. (Chang et al. 2011)

arc-sprayed Ni-Ti coating is superior to that of Ti but inferior to that of Ni. This is due to numerous microcracks observed on both the surface and the cross section of the Ti coating; see Fig. 11.25a and b. On the other hand, the coating of Ni-Ti is a mixture of white and gray platelets that were identified as Ni and Ti, respectively, by EDS analysis. Cross-sectional SEM micrograph of the arc sprayed Ni-Ti coating is shown in Fig. 11.25c with no visible cracks detectable.

(Laik et al. 2005) studied metal–ceramic bonding produced by the technique of wire arc–plasma spraying of Ni on Al_2O_3 substrate, Ni being atomized with argon. The plasma deposited Ni layer shows a uniform lamellar microstructure throughout the cross section. The metal–ceramic interface was well bonded with no pores, flaws, or cracks. An annealing treatment at 1273 K for 24 h of the plasma-coated samples did not result in formation of any intermetallic compound or spinel at the Ni/Al_2O_3 interface.

11.2.6 Fume Formation

The high arc temperatures at the wire tips result in some evaporation of the metal or alloying elements. The evaporated metal oxidizes in the surrounding air, and the oxide is quenched to form ultra-fine nano-sized particles (fumes). The lower the boiling point of the metal, the larger the fume formation. These particles represent serious health hazard and require protective measures for the operator and the environment by operating in well aerated booth units equipped with adequate heppa filters and the use of personal protective equipment's such as respirators. Figure 11.26 (Watanabe et al. 1995) shows the aluminum vapor as observed with a CCD camera equipped with a narrow band filter having transmission at the 313 nm atomic aluminum line, with a 100 ns exposure time. The wire tips are on the lefthand side (LHS), the cathode wire in the top position.

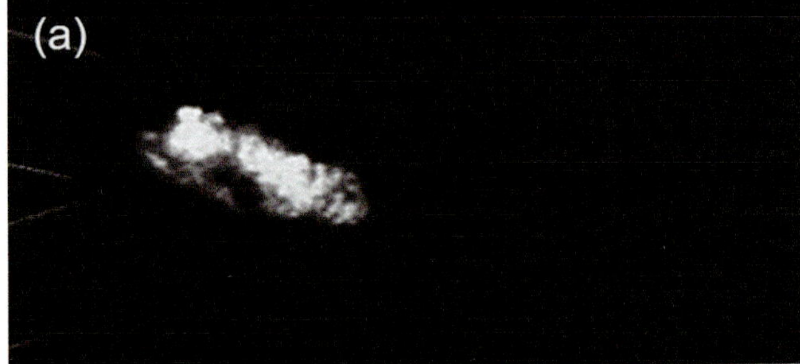

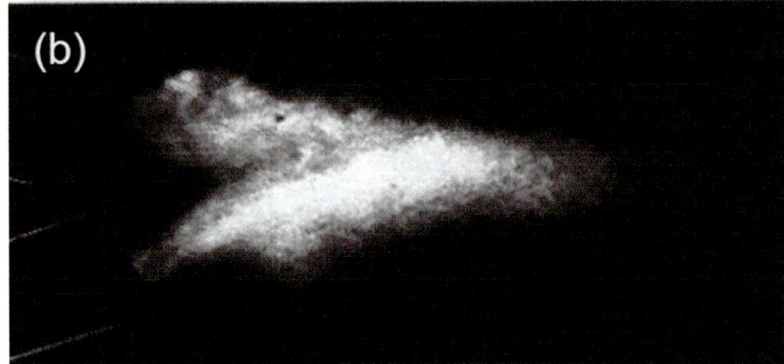

Fig. 11.26 Photograph of aluminum vapor in front of the wire tips (not seen at the LHS). The vapor was observed using a line filter for Al line and a high-speed CCD camera. (Watanabe et al. 1995)

There is more evaporation observed from the cathode wire, and the evaporation strongly increases with the increase of the arc current.

Figure 11.27 (Watanabe et al. 1995) shows the relative increase in fume formation with increasing arc current and the relative reduction of fume generation when using a Laval-type nozzle and nitrogen as atomizing gas. The stronger evaporation from the cathode is due to the higher current and heat flux densities at the cathode resulting in higher droplet temperatures. Use of nitrogen as atomizing gas and combined with a nozzle design that increases the gas velocity at the wire tips reduces the fume formation, as shown in Fig. 11.27 (Watanabe et al. 1995)

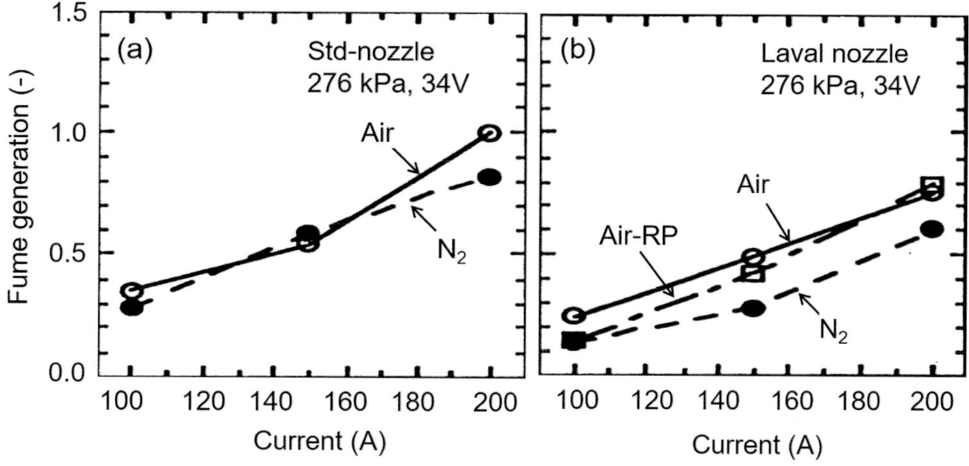

Fig. 11.27 Intensity of Al vapor spectral light emission as function of arc current atomization with air and N_2 using (**a**) standard and (**b**) Laval type nozzle with air, in straight and reverse polarity (RP),and N_2, pressure $=$ 276 kPa. (Watanabe et al. 1995)

11.3 Equipment Design and Operating Parameters

11.3.1 Conventional Twin-Wire Arc Spraying

Compared to other spray processes, arc spraying presents a higher deposition rate with values ranging, dependent on current (100–360A), from 3 to 15 kg/h for Al, 4.5 to 17 kg/h for stainless steel, to 10 to 33 kg/h for Zn. Considerably higher deposition rates, up to 200 kg/h, can be achieved with higher power installations (Steffens et al. 1990).

The main requirements for the material to be sprayed by the wire arc spray process are that the material must be electrically conducting and that it can be obtained in form of a wire without being too brittle or too stiff. Many metals or alloys fulfill this requirement. Therefore, the material to be sprayed is selected according to the application, for which the drawbacks of wire arc spraying, such as relatively high porosity and oxide contents, are less important than its advantages in terms of high deposition rates, low substrate heating, and overall favorable process economics.

A typical twin-wire arc spray setup is shown in Fig. 11.28. It is composed essentially of four main components:

- The DC power supply
- Dual wire rolls or spools and wire feeding mechanism
- Control unit for power supply, wire feeding, and atomizing gas (air)
- Plasma torch and its traverse moving mechanism

The power supply is usually a thyristor-controlled DC rectifier which must be able to withstand occasional shorting of the output when the wires touch. Since the electrodes are

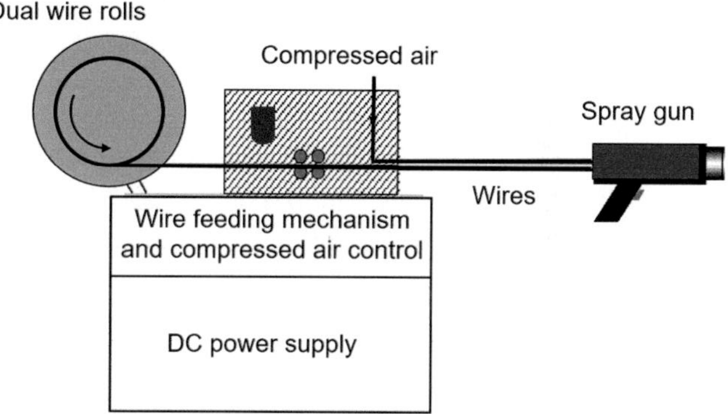

Fig. 11.28 Schematic of twin-wire arc spray (TWAS) system with power supply, control unit, wire supply rolls, compressed air line, and torch

consumable resulting in a varying arc length, the rectifier uses *voltage control* rather than *current control* as in conventional DC plasma spraying systems. The desired voltage is set, and when the melt rate of wires is higher than the feed rate, the arc voltage will increase surpassing the set point; the power supply control will respond by reducing the arc current, and therefore the melt rate until the set voltage value is again reached. Alternately, when the arcing gap gets too small resulting in too low voltage values, or if the wires even touch resulting in a momentary short, the current will increase to increase the melt rate. However, since the loss of a metal droplet from the wire tips usually leads to a stepwise change in the arc voltage, one has continuous fluctuations of the arc voltage and arc current as can be observed in Fig. 11.29 with the principal frequencies varying between 500 and 2000 Hz. These fluctuations can be minimized by judiciously adjusting the arc voltage, the wire feed rate, and the atomizing gas flow. Misalignment of the wire guides such that the two wires do not extend to a single

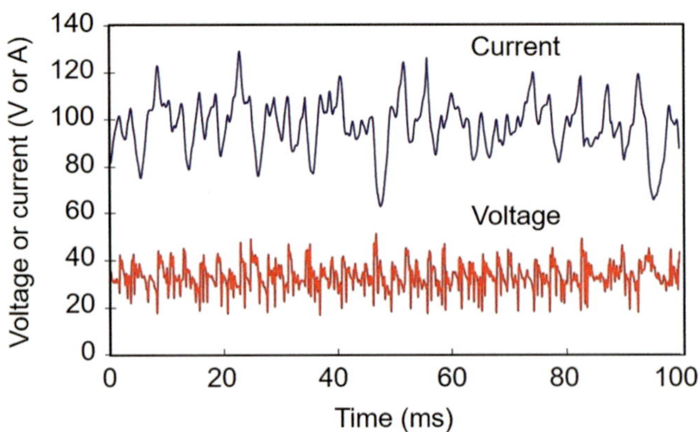

Fig. 11.29 Illustration of voltage (bottom) and current (top) signal in WAS system

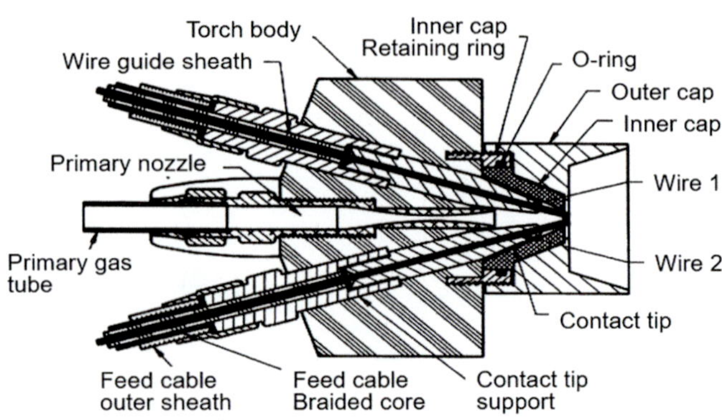

Fig. 11.30 Schematic of a twin-wire arc spray (TWAS) torch head (reproduced with kind permission of Carlson RR 2001)

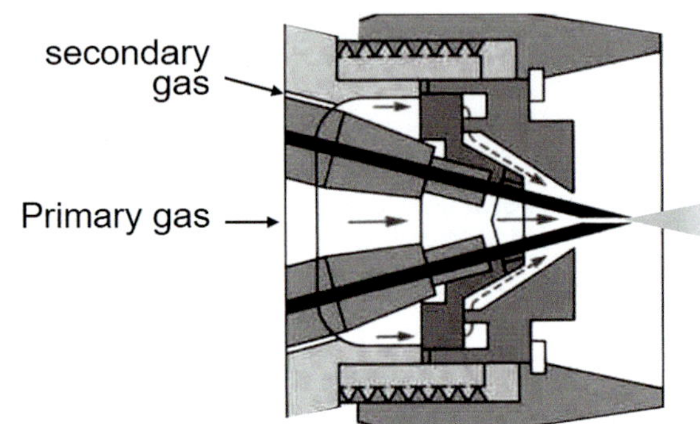

Fig. 11.31 Schematic of twin-wire arc spray (TWAS) torch head with primary and secondary gas flow. (Reproduced with kind permission of Praxair TAFA)

point can also lead to voltage fluctuations, as can a movement of the wire tips due to a remaining kinks or cuts in the wires. Most wire arc spray operations are performed with voltage settings below 40 V, with an open circuit voltage requirement between 60 and 80 V. It is to be noted that the response of the power supply to the constantly varying load can influence the coating quality (Marantz and Marantz 1990). The "current control" knob on the power supply controls the wire feed rate, because for a given voltage, higher wire feed rates require higher arc currents.

The wire spools are necessarily made of ductile and electrically conductive material taken from a large spool. In order to reduce the wire curvature (cast and kink), the wire is frequently run through a wire straightener, consisting of an assembly of rolls forcing the wire through a straight path. Without the straightener, the wire tips will have a periodic movement resulting in uneven melt rates. The wire transport rollers are either in the control unit, pushing the two wires to the torch head (push wire feed), or they are mounted on the torch head (pull wire feed). It is also possible to have both a push and pull wire feed for optimal feed speed control and allowing longer distances between the control unit and the torch head (up to 15 m) provided that exact synchronization of the drive motors are assured.

The control unit allows adjustment of the operating voltage, the wire feed rate via the feed motor speed, and the pressure of the atomizing gas in the line to the torch. The gas is fed from a compressor or high-pressure gas tanks at pressures of at least 0.6 MPa. Air is the most commonly used atomizing gas, although nitrogen and CO_2 are being used as well.

The wires arc torch is of a relatively simple air-cooled design in which the wires are guided inside coaxial hoses to the spray torch, with the metallic layer surrounding the plastic wire guide serving as the power conduit (see

Fig. 11.30). The gas supply tube connects the control unit with the torch. The nozzle providing the high-velocity gas flow directed toward the wire tips. Alignment of the contact tips such that the two wires meet is important for obtaining steady melt rates.

The primary atomizing gas nozzle consists usually of a straight bore operating under choked flow conditions, meaning that the flow velocity in the tube reaches sonic velocity. Alternate torch designs such as that of Praxair TAFA, Fig. 11.31, have a secondary gas flow, i.e., a portion of the atomizing gas flow is diverted to flow around the wire guides to the wire tips. A second nozzle (air cap) is mounted surrounding the wire tips or right upstream of the tips This secondary nozzle is reported not only to improve the atomization providing higher gas velocities at the wire tips but can also be used to reduce the jet divergence. The liquid metal atomization process is strongly dependent on the gas velocity and the fluid dynamic design of the torch which affect significantly the coating quality.

An example of a typical, general purpose WAS unit by Oerlikon-Metco, Model Flexi ArcTM300, is show in

Fig. 11.32. It is commercialized for applications of standard coatings with high output rates. It features maximum current rating of 300A and use of handheld spray gun (LD/U2) with a pneumatic push/pull wire feed system.

Typical operating parameter ranges of WAS units in general are listed in Table 11.2. It should be noted that high-power twin-wire arc coating installations exist operating at arc currents up to 1500 A. These installations which require water-cooling of the contact tips are mainly used for anti-corrosion coatings on large surfaces.

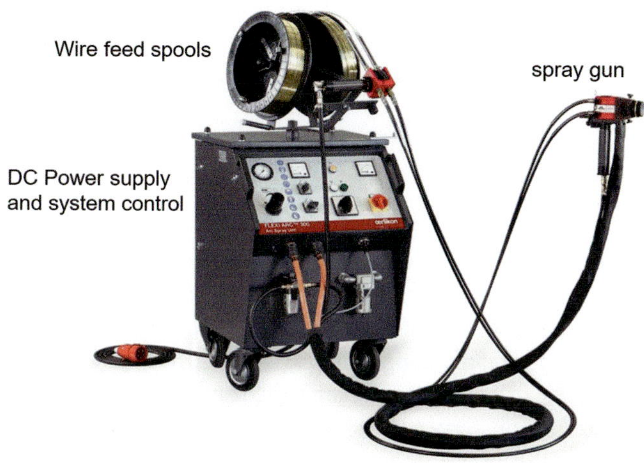

Fig. 11.32 General view of an Oerlikon Metco Flexi Arc™ 300 wire arc spray unit. (Reproduced with kind permission of Oerlikon Metco)

Table 11.2 Typical wire arc operating ranges

Arc voltage	20–40 V
Arc current	100–400 A
Wire feed speed	7–10 m /min
Standoff distance	100–200 mm
Plasma gases	Air, nitrogen, CO_2
Plasma gas flow rates	800–2400 slm
Gas supply pressure	0.27–0.6 MPa

11.3.2 High-Velocity Twin-Wire Arc Spraying

A different nozzle design aiming at significantly increasing the droplet velocities has been proposed by (Hussary et al. 1999). The study involved the modification of commercial Miller Thermal (now Praxair TAFA) BP 400 by replacing its upstream nozzle by a Laval-type nozzle with the wire tips at the nozzle exit plane (CDCP) as shown in Fig. 11.33a. The nozzle design and dimensions were based on compressible flow computations for the specific operating conditions. A ring of small orifices was also added surrounding the nozzle and providing a shroud gas flow, as shown in Fig. 11.33b, with the orifices angled toward the jet axis. Two different angles were proposed (SCDCP1 and SCDCP2) (Hussary 1999).

Further development of the standard twin-wire arc spraying torch design involved the use of a high-velocity gas stream to atomize the arc-melted material at the tip of the wires and propel the droplets toward the substrate surface. The gas stream velocity conditioning the droplets impacts velocity and thus the splats diameters and thicknesses and consequently the coatings properties. As demonstrated by different authors (Kelkar and Heberlein 2002; Liao et al. 2005; Chen et al. 2012; Wu et al. 2014), high impact droplet velocities will result in better coating properties. The principle used to achieve high-velocity particles is presented in Fig. 11.34 where the gas injection is achieved with a convergent–divergent nozzle.

A novel design of the torch head configuration was put forward by Chen et al. (2012) based on 3-D modeling of the flow field around the wire tip, the arc, and molten metal atomization region. A schematic representation of the proposed geometry for the wire guides, and the profile of the gas atomization nozzle is shown in Fig. 11.35. When comparing the velocity developments along with spray distance for the droplets with different sizes, it was predicted that higher velocities and a shorter acceleration distance would be obtained for the smaller droplets. For most moderate size droplets, their velocity increased rapidly in the early stages

(a)   (b)

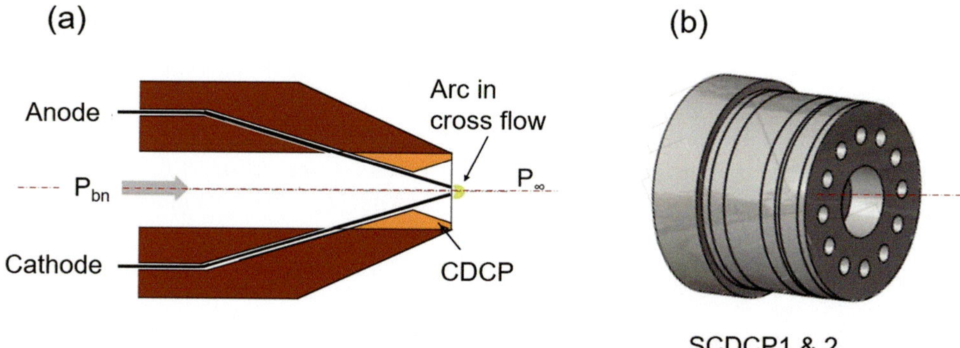

Fig. 11.33 Schematic of (**a**) Laval-type nozzle surrounding wire tips; (**b**) gas shroud arrangement. (Hussary 1999)

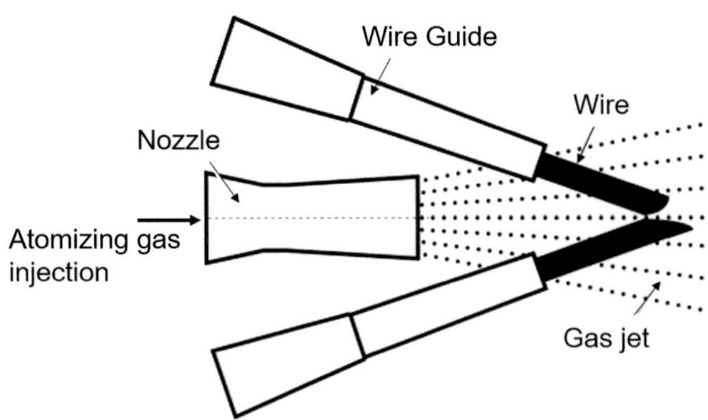

Fig. 11.34 Schematic of the original HAS-01 type torch configuration. (Chen et al. 2012)

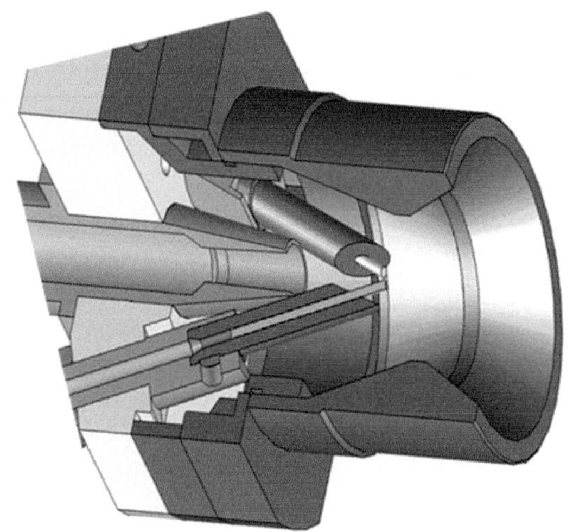

Fig. 11.35 Schematic of the newly designed twin-wire arc spray (TWAS) torch head configuration. (Chen et al. 2012)

of their trajectory reaching a rather stable value for the remaining of their trajectory until their impact on the substrate. Experiment measurements of the droplet velocities along their trajectories were in good agreement with the modeling results.

11.3.3 Single-Wire Arc Spraying

A typical design of single-wire arc spraying (SWAS) torch is given in Fig. 11.36 *after* Marantz et al. (1991). This designed, often referred to as plasma transferred wire arc spraying (PT-WAS), is based on the use of a non-consumable thoriated tungsten cathode coaxially placed at the at the center of an air-cooled copper nozzle, with consumable-wire anode placed at 90° to the torch axis downstream of the nozzle exit. The arc is initiated by a high-frequency ignition between the cathode and the nozzle, followed by the transfer of the arc to the consumable-wire

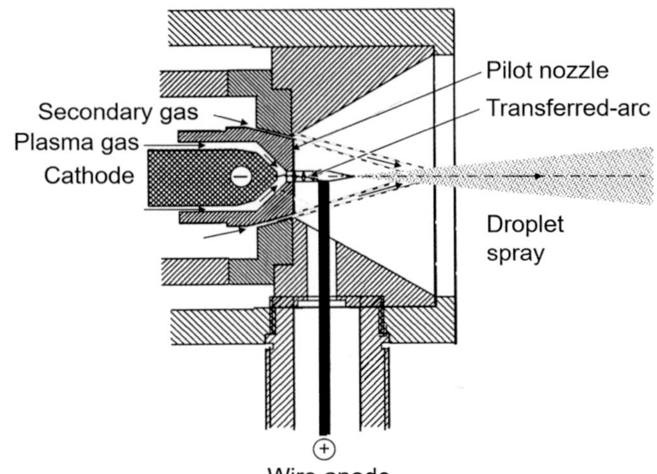

Fig. 11.36 Schematic of plasma transferred wire arc spraying (PTWAS) torch. (Marantz et al. 1991)

anode. The plasma gas is preferably argon–hydrogen mixture, though other gases have also been used including air and nitrogen, which is injected in the annular space between the cathode and the copper nozzle. A secondary gas injected as a series of high-velocity micro-jets surrounding the central arc nozzle serves to atomizing molten metal formed at the tip of the wire (Cook et al. 2003). It also serves as shroud gas for narrowing the molten droplet spray pattern (Marantz et al. 1991; Kowalsky et al. 1992).

A constant current power supply is used, with currents ranging from 20 to 100 A, with typical operating voltages of around 120 V. Compared to twin-wire arc spraying, smaller droplet sizes are generated resulting in a finer grain structure of the coating comparable to that of plasma-sprayed coatings. The position of the wire tip can be adjusted by adjusting the operating parameters (current and wire feed rate) as an additional control of coating quality. Typical plasma gas flow rates are in the range of 51 to 71 slm, while considerably higher flow rates are needed for the secondary/shroud gas in the range of 621 to 793 slm (Kowalsky et al. 1992). Droplet velocities, measured using laser Doppler anemometry, are in the range of 80 to 125 m/s varying depending on the wire material and nature of the atomizing gas. The arc current has little effect on the droplet velocity.

A modified version of the PT-WAS torch, schematically illustrated in Fig. 11.37, was developed for the coating of inner walls of the cylinders in automobile aluminum engine blocks which can have an inner diameter as small as 75 mm (Marantz et al. 1991; McCune Jr. et al. 1993; Cook et al. 2003). The small-sized, rotating atomization module allowed for the uniform coating of the wall of relatively small engine cylinders. This development which targeted automotive applications had to comply also with the need of good coating adhesion as a prime concern, which could be achieved

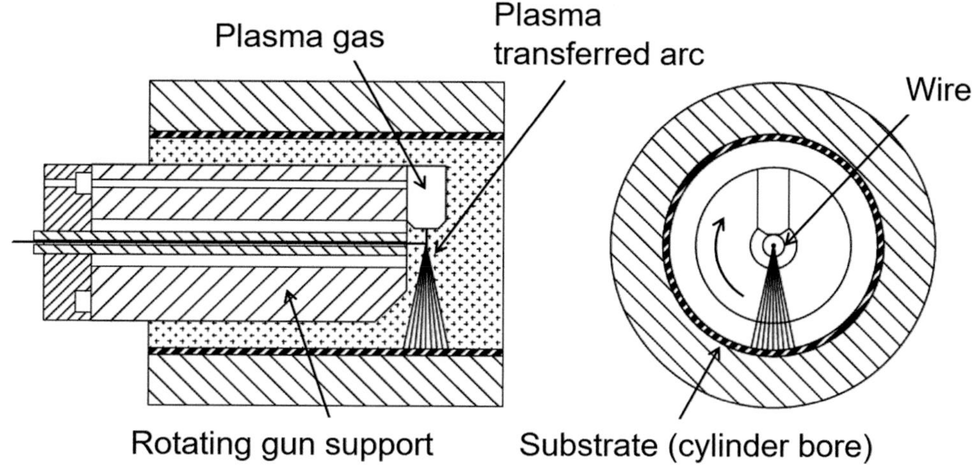

Fig. 11.37 Schematic of plasma transferred wire arc spraying (PTWAS) torch developed for the coating the wall of cylinder bore. (Marantz et al. 1991)

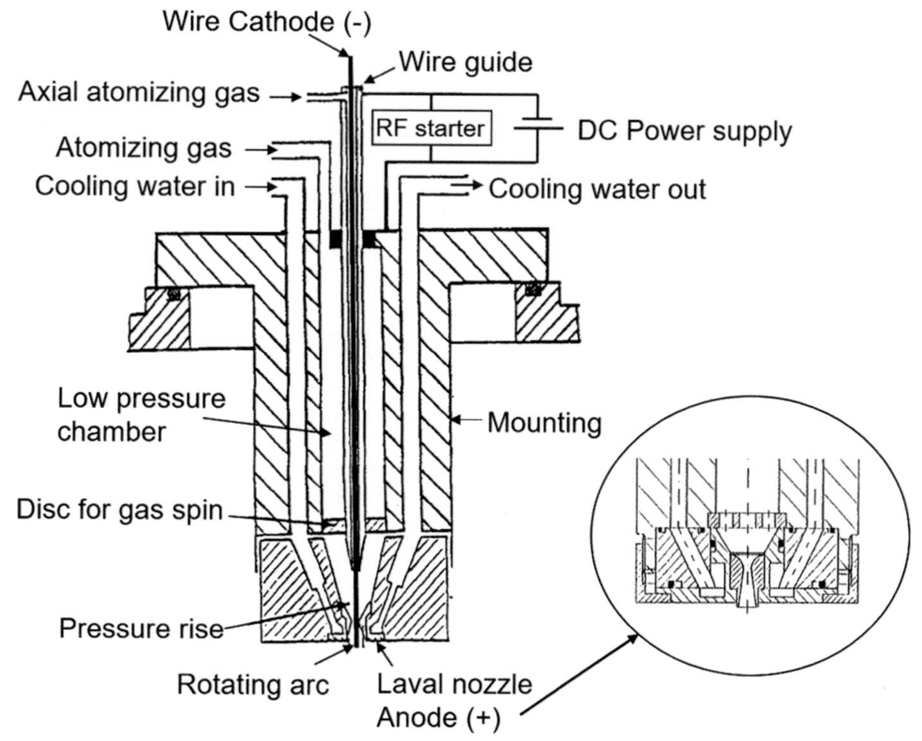

Fig. 11.38 Schematic of a single-wire vacuum arc spray (SW-VAS) torch. (Steffens and Nassenstein 1994)

using a pretreatment with a fluoride-based flux and a NiAl bond coat before the final low carbon steel coat is applied (Cook et al. 2003). Coatings with less than 2% porosities were reported with excellent adhesion to the substrate.

A different approach for the design of the PT-WAS torch was proposed by Steffens and his research group at the University of Dortmund, (Steffens and Wewel 1991; Steffens and Nassenstein 1994). As schematically illustrated in Fig. 11.38, the proposed design was closer to a conventional DC plasma torch with the feed wire (1.6 mm diam.) acting as cathode, centrally located in a water-cooled anode nozzle 2.4 mm i.d. surrounding the tip of the wire. The arc is initiated by an RF discharge and operated with a constant current power supply at arc currents between 100 A (for stainless steel) and 140 A (for Ti). The arc voltage ranged from 30 to 45 V. The process known as single-wire vacuum arc spraying (SW-VAS) was mostly developed for the spraying of high purity coatings of Ti or Ta, since a single wire as coating precursor material offers the lowest chances of surface contamination. The spraying was performed in a controlled atmosphere mostly under reduced pressure (10–80 kPa) with Argon as atomizing gas at flow rates of 158 slm. Maximum wire feed had to be dropped sharply from 3.5 m/min to less than 2.0 m/min with the increase of the chamber pressure from 10 to 80 kPa to maintain the quality of the coating. The corresponding metal spraying rate in this case would be around 1 kg/h (Ti) and 1.8 kg/h (steel). Coating porosity, on the other hand, was observed to increase from less than 3% to above 10% with the increase of the camber pressure over the same pressure range.

A similar approach was used by Heberlein and his research group at the University of Minnesota for the development of the high-definition single wire arc spray (HD-WAS) torch (Carlson et al. 2000; Carlson and Heberlein 2001, 2002). As schematically illustrated in Fig. 11.39a, the wire, which in this case acts as anode, is surrounded by a nozzle at a floating voltage. A tungsten rod attached to the downstream face of the nozzle serves as cathode. A constant current welding power supply was used with typical arc currents between 35 and 125 A and voltages are between 19 and 22 V. Argon was used as the atomizing gas with backpressures between 100 and 400 kPa above the atmosphere. This development was mainly pursued for obtaining a narrow particle stream for high-definition coating of small areas such as valve seats. Figure 11.39b (Carlson 2005) shows the particle stream exiting the device. Measured divergence angles are between 2 and 3 degrees, slightly increasing with higher currents and higher back pressures. The wire is fed through a wire straightener before entering the torch to assure minimal lateral movement of the wire tips.

Results have been reported using mild steel wires, 0.584 and 0.762 mm, diameter at feed rates between 3 and 8.4 m/min with matching arc currents of 35 and 125 A. The gas nozzle diameter is typically 2.5 to 3 times the wire diameter. Droplet sizes are significantly larger than in any other thermal spray process, with typical mass mean diameters in the order of 300 μm. Figure 11.40 shows corresponding droplet/particle size distributions obtained by spraying into water for these two wire sizes.

Droplet size decreases with increasing backpressure and decreases slightly with increasing current. Figure 11.41 (Carlson and Heberlein 2001) shows examples of traces sprayed with this torch, and it is seen that with the smaller wire diameter and the lower back pressure, the deposition foot print is about 3 mm wide. The coating cross sections show dense equiaxed material without the layer structure typical of any other spray coating (see Fig. 11.42) (Carlson and Heberlein 2001). Deposition efficiencies are between

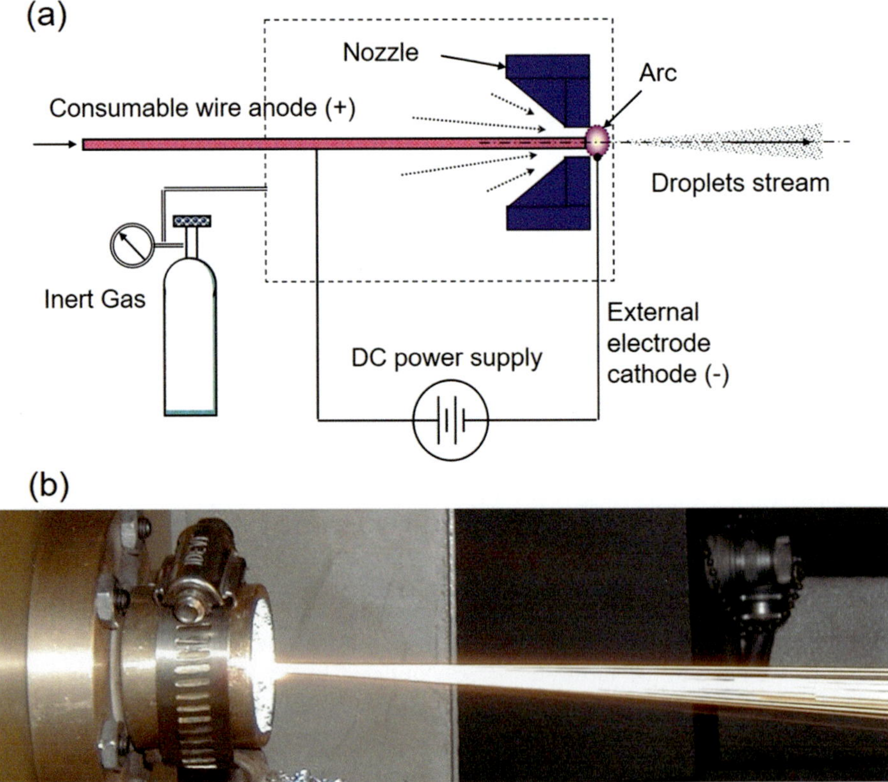

Fig. 11.39 (**a**) Schematic of high-definition single-wire arc spray (HD-WAS) torch and (**b**) photograph of droplet stream. (Carlson and Heberlein 2001; Carlson 2005)

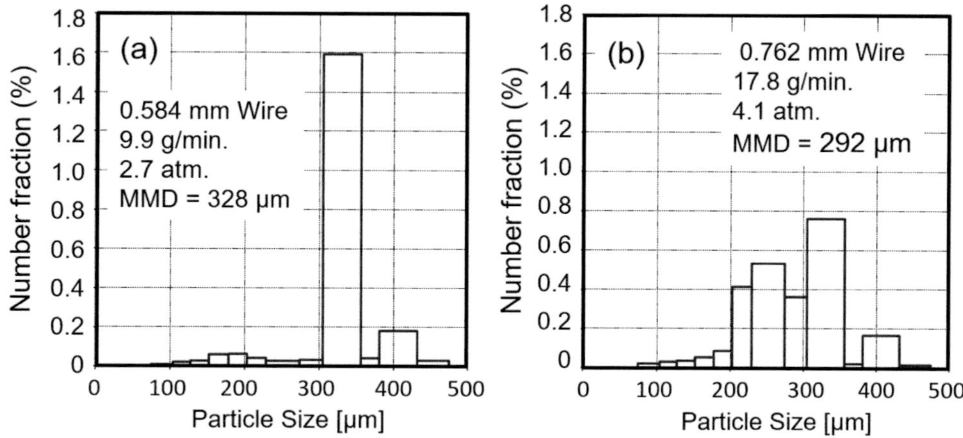

Fig. 11.40 Carbon steel droplet size distributions obtained with the HD-WAS torch for two different wire diameters, 0.584 and 0.762 mm diameter (Carlson and Heberlein 2001, Carlson 2005)

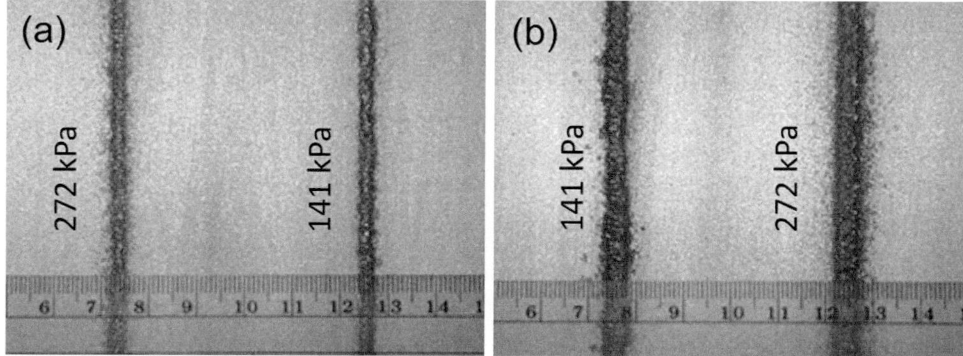

Fig. 11.41 Photograph of mild steel deposit traces on Al substrates for two different wire diameters (**a**) wire diam. =0.584 mm, feed rate = 4.9 m/min (**b**) wire diam. =0.762 mm, feed rate = 8.4 m/min, and two different and atomizing gas back pressures, 141 and 272 KPa..(Carlson and Heberlein 2001)

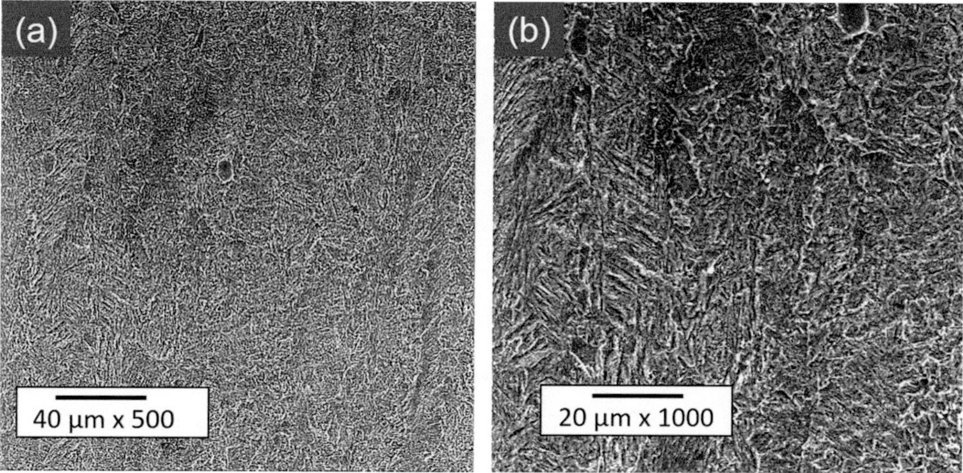

Fig. 11.42 Micrograph of coating cross sections obtained with the HD-WAS torch showing equiaxed large grain deposits, wire diam. = 0.762 mm,, feed rate = 23.3 g/min, and arc current = 108 A. (Carlson and Heberlein 2001, Carlson 2005)

75 and 86%, larger for the larger wire size and for higher currents. Deposition rates are up to 15.6 g/min for the smaller diameter wire and up to 25.8 g/min for the larger diameter wire.

11.4 Gas and Particle Dynamics in Wire Arc Spraying

11.4.1 Particle Velocity and Flux Distribution

The velocity of the droplets/particles at the exit of the atomization zone are closely dependent on the velocity of the atomizing gas in the arc region between the two-wire tips, or between the wire and auxiliary electrode tip depending on the torch configuration. The atomizing gas velocity reflects, in-turn:

- Upstream pressure of the atomizing gas
- The primary atomization nozzle shape, i.e., Laval-type nozzle vs. straight bore nozzle
- Secondary nozzle around the wire tips ("high-velocity cap")
- Secondary gas flow rate along wire guides
- Shroud gas flow rate from a secondary nozzle

Schlieren images of the region between the atomizing gas nozzle exit and the wire tips in a twin-wire arc spray torch are given in Fig. 11.43 (Hussary 1999) for air supply pressures of 482 kPa (70 psig), 550 kPa (80 psig), and 826 kPa (120 psig). These clearly reveal well-defined shock diamonds which are due to an under-expanded flow pattern exiting the atomizing gas nozzle at sonic velocity. Results of pitot tube velocity measurements in a cold compressed air flow, in the absence of arcing, at the location of the wire tips (50 mm from the nozzle exit) are given in Table 11.3 (Hussary 1999). The results obtained using a straight bore nozzle "regular, "and a converging–diverging "Laval-type" nozzle show that for an identical gas supply pressure of 450 kPa (65 psig), the exit gas velocity can be increased to supersonic values of 352 m/s (Mach number = 1.02) with the use of a "Laval-type" nozzle.

It should be noted that despite the choked nozzle flow conditions the mass flow rate of the atomizing gas will increase with the increase of the upstream pressure because the gas density increases and some of the gas is diverted to flow along the wire guides. Both effects will increase the shearing actions on the liquid metal film resulting in smaller droplet sizes.

Measurement of droplet velocities have been reported in wire arc spray conditions using streak photography (Wang et al. 1996) and the DPV-2000 (Hussary 1999) in-flight particle diagnostic techniques. Figure 11.44 shows the axial distributions of the centerline velocity of aluminum droplets atomized using a BP 400 twin-wire arc spray torch operating at 100 A and 30 V for different supply gas pressures of 276 kPa (40 psig), 344.5 kPa (50 psig), 413.5 kPa (60 psig), and 482.3 kPa (70 psig). The corresponding atomizing gas flow rates were respectively, 750, 935, 1133, and 1331 slm. The results further confirmed that the increase of the supply pressure is responsible for the increase of the droplets velocities which reaches its maximum value at 50 to 100 mm from the nozzle exit level. Velocities averaged over the cross section and weighted by the droplet number density at radial locations were, however, almost 10% lower than the peak values over the axis of the torch due to lower velocity particles in the fringes of the jet.

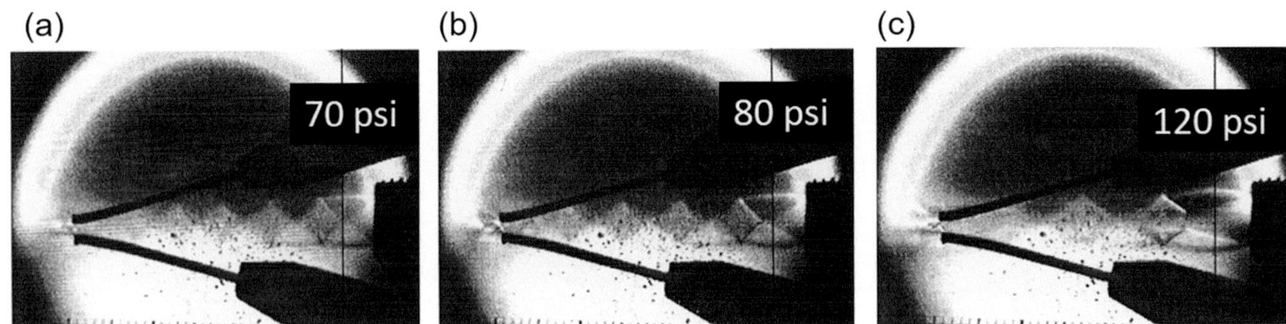

Fig. 11.43 Schlieren images of shock pattern outside atomizing gas nozzle showing under-expanded flow exiting straight bore nozzle. (Hussary 1999)

Table 11.3 Cold flow gas velocities in the absence of arcing, for two nozzle geometries at 50 mm from the nozzle exit (Hussary 1999)

Nozzle type	Supply pressure (kPa)	Torch inlet pressure (kPa)	Gas flow rate (slm)	Gas velocity (m/s)	Mach number
Regular	450	202	1232	334	0.97
Laval-type	450	220	1232	352	1.02

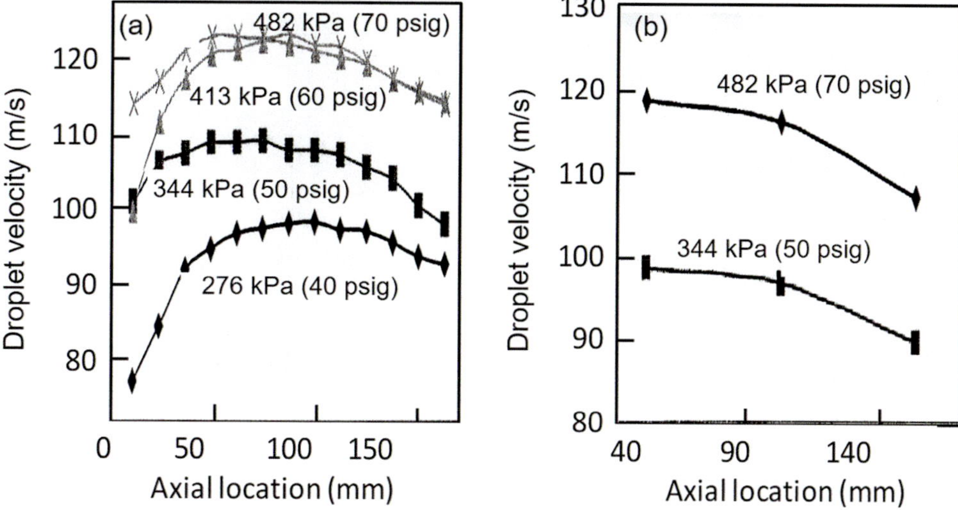

Fig. 11.44 Axial droplet velocity profiles for different atomizing gas pressures. (**a**) Axial velocity distribution along the centerline of the jet (**b**) Average velocity over the cross section weighted by local droplet flux (Hussary 1999). (Reproduced with kind permission)

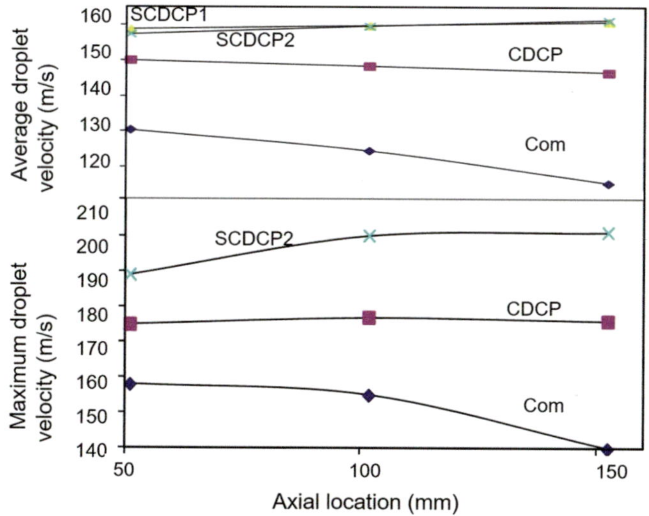

Fig. 11.45 Axial distributions of the maximum and weighted-average aluminum droplet velocities for different nozzle designs. (Hussary et al. 1999)

Special attention was also given to the effect of the atomizing gas nozzle design on the axial velocities of the generated droplets. Figure 11.44 (Hussary et al. 1999) shows the axial distributions of the droplet velocity along the center line of the jet and the flux-weighted average velocities for aluminum droplets generated using a modified BP 400 torch. For comparison, values obtained with the standard commercial torch/nozzle configuration are included identified as "com." A modified version of the torch illustrated earlier in Fig. 11.33a, was also tested. It involved the elimination of the upstream nozzle and replacing it by a Laval-type nozzle with the wire tips at the nozzle exit plane. The results obtained

with this nozzle are identified as "CDCP" nozzle on Fig. 11.45. A second modification tested, identified in Fig. 11.45 as "SCDP1 and SCDP2" involved the addition of a ring with small orifices surrounding the nozzle at two different angles providing a shroud gas flow as illustrates in Fig. 11.33b. The results show that the Laval-type nozzle (CDCP) provides a significant increase of the average and maximum droplet velocities compared to the standard torch design "com." Furthermore, the use of a shroud, whether of the "SCDP1" or "SCDP2" designs further increased the measured droplet/particle velocity. Hussary (1999) reported that the use of the modified nozzle and the shroud gas significantly reduces the divergence of the droplet jet as shown in the Schlieren images of the particle jet given in Fig. 11.46 and the $r_{0.5}$ values given in Fig. 11.47. The latter represent the radial location at which the droplet flux reached a value corresponding to 50% of its maximum value on the jet axis. The narrowing of the divergence of the droplet jet is illustrated in Fig. 11.48 by the profile of the "footprint" deposited on a stationary substrates, commonly referred to as the "sweet spot" (Hussary 1999).

Measurements of the velocities of carbon steel droplet sprayed by a TAFA-9000 system with a converging nozzle and secondary gas flow, operated at 100 A, 30 V for a range of atomizing gas flow rates varying from 90 to 150 m³/ h (1500–2500 slm) obtained using a DPV-2000 are given in Fig. 11.49 (Jandin et al. 2002). The corresponding droplet/ particle diameter distributions were obtained by spraying into oil and measuring the particle diameter using optical or electron microscopy.

This study was further extended by Planche et al. (2003, 2004) with the objective of formulating an empirical correlation between the droplet characteristics and various operating

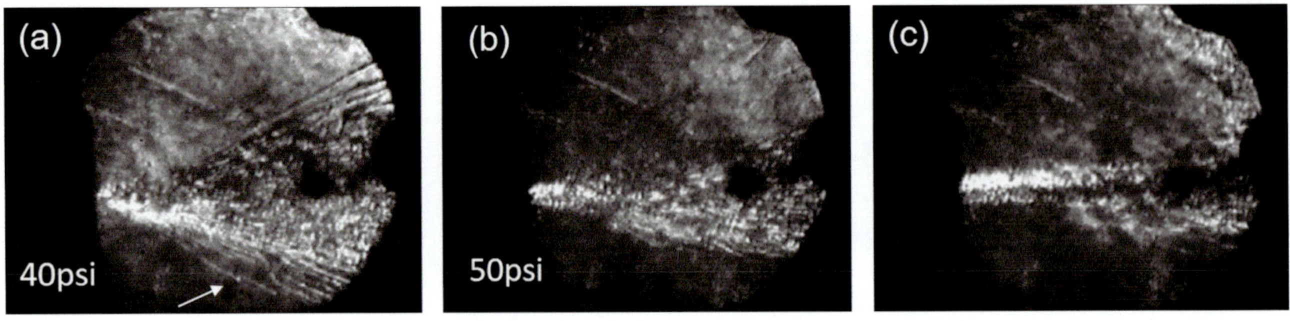

Fig. 11.46 Schlieren images of Al spray droplet jet for different nozzle designs: (**a**) commercial nozzle with cap "com," (**b**) Laval-type nozzle "CDCP," and (c) Laval-type nozzle with additional shroud gas "SCDP1" (Hussary 1999)

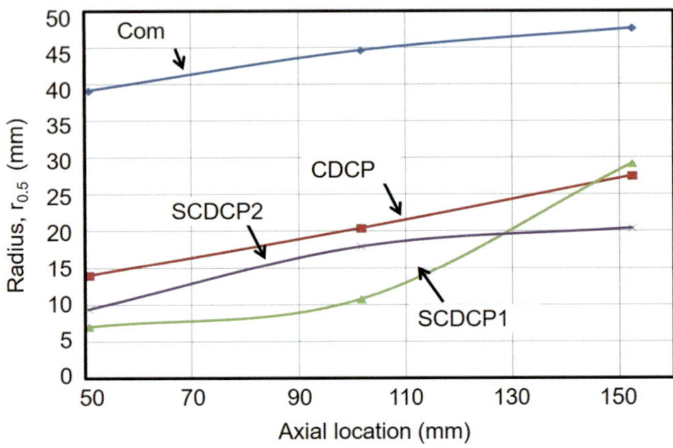

Fig. 11.47 Radial location of $r_{0.5}$ where the local mass flux is 50% of the maximum at three different axial locations and for four different nozzle designs. (Hussary et al. 1999)

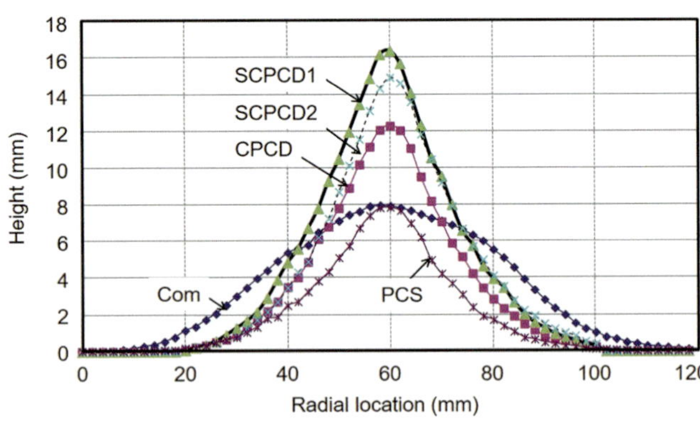

Fig. 11.48 Aluminum "footprint–sweet spot" deposit on stationary substrate with four different nozzle designs. (Hussary 1999)

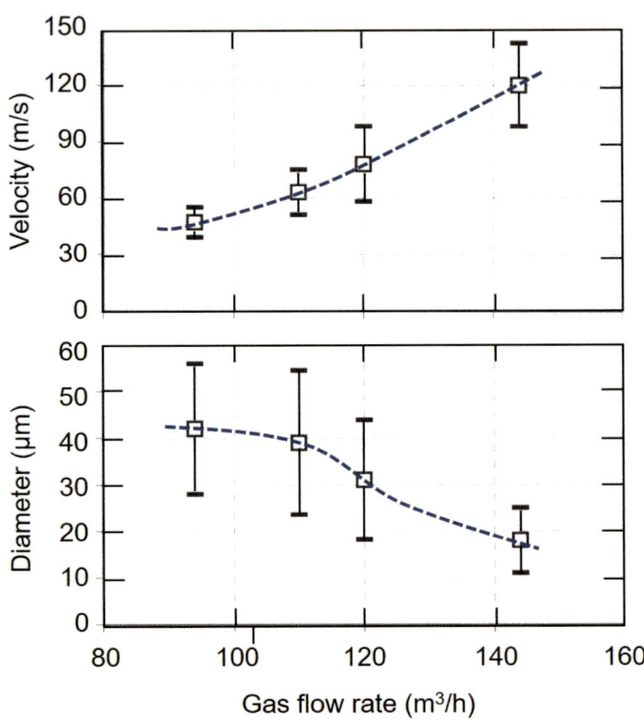

Fig. 11.49 Average carbon steel droplet velocity (top) and diameter (bottom) as function of atomizing gas flow rates with converging nozzle in cap and secondary gas flow. (Jandin et al. 2002)

parameters and spatial locations in the jet. The arc current was varied over the range 100 to 200 A, with a total atomizing air flow rates from varied from 1500 to 2167 slm, axial spray distances from z = 200 to 300 mm, and horizontal (x) and vertical (y) variations of up to 20 and

30 mm from the spray axis, respectively. The following relationships have been found for I in A and $\dot{Q}$ in m³/ s (Planche et al. 2004) relating the droplet properties temperature, velocity, and diameter with the operating parameters and the position in the jet, given by the coordinates x (perpendicular to the spray direction and perpendicular to the plane of the wires, zero on the torch axis), y (perpendicular to the spray direction and in the plane formed by the two wires, zero again being the torch axis), and z the axial coordinate with zero being the torch exit, all coordinates measured in m:

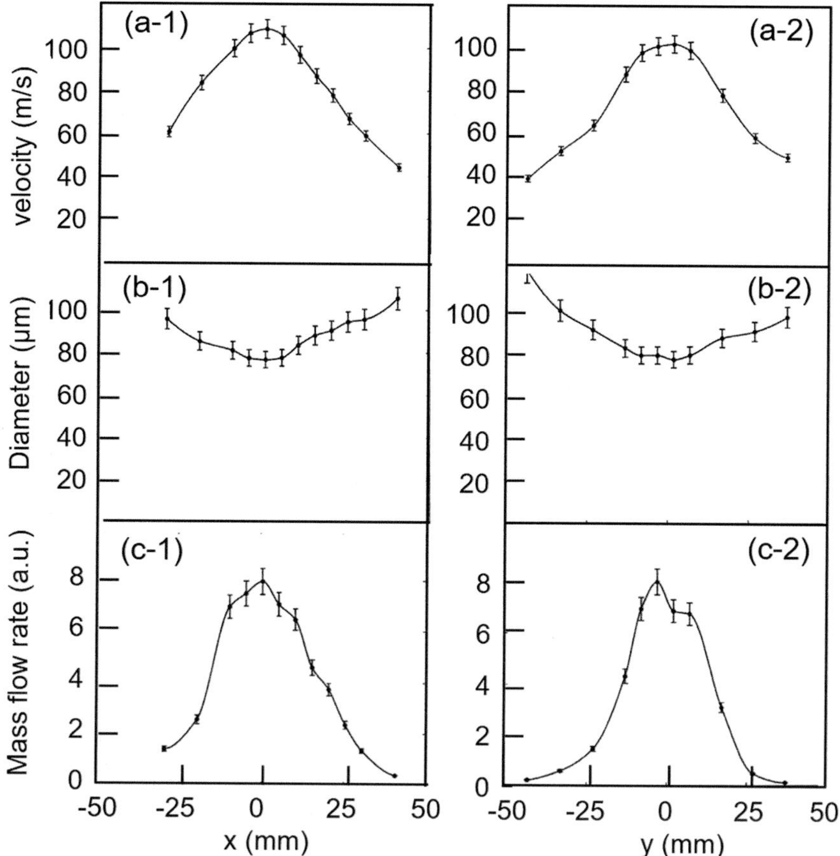

Fig. 11.50 Lateral profiles of (**a**) droplet velocity, (**b**) mean droplet diameter, and (**c**) normalized mass flux, of aluminum droplets, along x (left) and y (right) directions, at an axial location of 50 mm from the nozzle exit. The wires are in the y plane. (Pourmousa et al. 2005)

Velocity in (m/s)

$$v \propto \frac{\dot{Q}^{0.99}}{x^{0.021}\, y^{0.025}\, z^{0.744}\, I^{0.07}} \qquad (11.3)$$

Temperature in (K):

$$T \propto \frac{I^{0.004}}{x^{0.006}\, y^{0.002}\, z^{0.033}\, \dot{Q}^{0.068}} \qquad (11.4)$$

Droplet diameter in μm:

$$d \propto \frac{I^{0.022}\, x^{0.308}}{y^{0.006}\, z^{0.322}\, \dot{Q}^{2.569}} \qquad (11.5)$$

It is interesting to note the strong inverse correlation of droplet diameter with atomizing gas flow rate. The stronger acceleration due to lower mass appears to offset the lower drag due to a smaller cross section, resulting in an almost linear increase of the droplet velocity with atomizing gas flow rate.

Similar measurements were reported by Pourmousa et al. (2004, 2005) using a commercial twin-wire arc spray system (ValuArc from Sulzer Metco) operated at 32 V, aluminum wire feed rate of 7 m/min, and atomizing gas supply pressure of 208 kPa (30 psig) with a corresponding air flow rate of 1032 slm. The results given in Fig. 11.50, obtained with a DPV-2000 instrument at 50 mm from the wire tips, are presented in terms of droplet/particle velocity profiles, mean particle diameters distributions, and the normalized mass flux distributions along two orthogonal axis x and y, with the y plane being that of the two wires in the torch. The results show reasonable spatial symmetry of the distributions in the x and y directions with the maximum droplet/particle velocity of about 100 m/s on the axis of the jet. The mean droplet/particle diameter does not seem to change excessively in the measurement plane with the droplets/particles in the fringes having a slightly larger diameter compared to those moving along the centerline of the jet. Particle flux distributions drops significantly with distance from the axis of the jet covering a spraying area of about 50 mm in diameter.

The spraying of chrome steel wires using air as atomizing gas and secondary shroud gas was reported by (Wilden et al. 2007b). Operating parameters were varied from 24 to 36 V, 100 to 300 A, and gas supply pressures from 200 to 400 kPa. The droplet velocity was strongly affected by the gas supply pressure increasing from about 75 m/s to about 130 m/s for 32 V 100 A operation (not quite a proportional increase as predicted by Eq. 11.3). Increasing the arc current resulted in a slight decrease of the droplet/particle velocity (from about 130 m/s at 100 A to about 115 m/s at 300 A and 32 V, 400 kPa). This effect could be explained by a larger droplet diameter at higher currents in qualitative agreement with Eq. 11.3. The effect of voltage on droplet velocity was negligible, though considerably more pronounced on droplet temperatures.

A study of droplet velocities and temperatures for nine different wire materials and different nozzle sizes was reported by Mohanty et al. (2003). A TAFA 8830 torch was used with different caps, operated at 200 and A 36 V with air or nitrogen as atomizing gas at a flow rate of about 1420 slm. At 160 mm from the nozzle, the average velocities ranged from about 90 m/s for Ni to about 115 m/s for NiAl and about 121 m/s for Mo, with the droplet/particle velocities of different types of steel and copper essentially in the same velocity range. No direct relationship with material properties is reported. No significant difference in droplet velocity was found when nitrogen gas was used instead of air. It is to be noted that droplet velocities are strongly influenced by the torch design and optimizing droplet velocities will require evaluation of the fluid dynamics inside the torch.

11.4.2 Particle Temperature

In contrast to metal particle velocity and flux distributions which are strongly affected by the fluid dynamics of the flow, droplets/particles temperatures generally slightly above the melting temperature of the material. During particle flight in the atomizing gas jet, the particle surface temperature can increase due to surface oxidation resulting from their contact with the atomizing air. In-flight temperature measurements of aluminum droplets have often been reported as being as high as the melting point of Al_2O_3, 2045 °C, because the surface of the droplet is oxidized. It also must be mentioned that wire arc spray droplets are in a condition where convection-induced oxidation has been observed with oxide content more than 20% for aluminum coatings, as in plasma spraying. This effect of internal convection can explain the very high oxide content values more than 20% observed with aluminum coatings, and the increase of the oxide content

with atomizing gas flow rate is also expected because of the convection of the oxide inside the particle.

As seen from the empirical relation Eq. 11.4, the droplet temperature does not change very much with arc current, atomizing gas flow rate, and horizontal or vertical position, and only a gradual decrease with axial distance is seen. Jandin et al. (2002) report a wider spread of droplet temperatures for smaller particles (less than 20 µm diameter), i.e., at higher atomizing gas flow rates. Pourmousa et al. (2004, 2005) report temperatures of Al droplets decreasing from 2160 °C to 2130 °C with an increase of the supply pressure from 236 to 542 kPa with little variation in the horizontal or vertical direction. Wilden et al. (2007a) report an increase in the temperatures of chrome steel droplets from 2070 to 2180 °C when the arc voltage was increased from 24 to 36 V; however, little change of droplet temperatures with current is found, consistent with Eq. 11.3. All temperatures reported indicate a superheating of the droplets significantly above their melting points. For example, Mohanty et al. (2003) report temperatures of about 2275 °C for stainless steel, about 2370 °C for Ni (melting point 1455 °C), about 2610 °C for Cu (melting point 1085 °C), and about 2860 °C for Mo (melting point 2617 °C); all obtained for 32 V, 200 A operation. While no significant dependence on atomizing gas flow rate is seen, a significant effect of the atomizing gas is observed, with steel droplets showing temperatures of about 2600 °C when atomized with air compared to about 2430 °C when atomized with nitrogen. This effect is probably due to oxidation providing additional heat. The observation that droplet temperatures increase somewhat for higher gas flow velocities obtained through a different nozzle, i.e., with smaller droplets, supports this conclusion.

The lack of dependence on arc current is clear if one considers that the arc current is an indication of wire feed rate, and increased power by increasing the current is used for higher melting rates. The increase in droplet temperature for increasing arc voltage can be explained by the high-speed video observations that increased arc voltage allows the arc to bow more in the downstream direction and to provide more heat to the extended metal ligament. Operation at low arc voltages can result in solidification of particles, particularly when coating at larger standoff distances. There are no droplet temperature measurements that separate the droplets originating from the cathode and those from the anode. The higher current densities at the cathode would lead one to expect higher droplet temperatures. The higher droplet temperatures observed by Jandin et al. (2002) for small droplet sizes may be an indication for a temperature difference between the cathode and the anode droplets.

11.4.3 Process Modeling

Only a few models have been formulated for the wire arc spray process because of its complexity, requiring by nature three-dimensional, time-dependent treatment, including phase change of the metal electrode and atomization of the liquid metal. One modeling effort has split up the problem into several separate parts:

- Compressible flow model for the atomizing gas flow upstream of the arc
- 3-D model of an arc in cross flow (steady state), with different anode and cathode attachments, yielding temperature, and velocity fields
- 2-D turbulent plasma jet model
- Liquid metal atomization models for the cathode and for the anode and a secondary atomization model (Kelkar et al. 1968; Kelkar and Heberlein 2000, 2002)

The compressible flow model provided input for the arc in cross-flow model, and this model was verified with some previously published experimental data as well as some data obtained for a wire arc configuration but without consumable electrodes and using enthalpy probes to determine plasma temperatures and velocities. Figure 11.51 (Kelkar and Heberlein 2000) shows the temperature and velocity distribution in the center plane of the arc. The model captures the differences between the arc anode and cathode attachments, with higher temperatures and velocities at the cathode. But it shows that 5 cm downstream of the electrodes the flow is essentially axisymmetric, justifying a 2-D model for the jet region.

The cathode atomization model is essentially a model as developed for metal welding arcs, whereas the anode atomization model uses the boundary layer stripping of the liquid metal sheets and the breakup due to the Kelvin–Helmholtz instability. The secondary breakup model follows an empirical relationship developed by Hsiang and Faeth (1992).

Figure 11.52 (Kelkar and Heberlein (2002) shows calculated particle velocities as function from the distance from the wire tips for different particle sizes. Also shown in the figure are two measured particle velocities using the DPV-2000 and their respective sizes, showing a surprisingly good agreement considering the modeling simplifications.

Figure 11.53 (Kelkar and Heberlein (2002) show the calculated droplet size distributions (bars) compared to

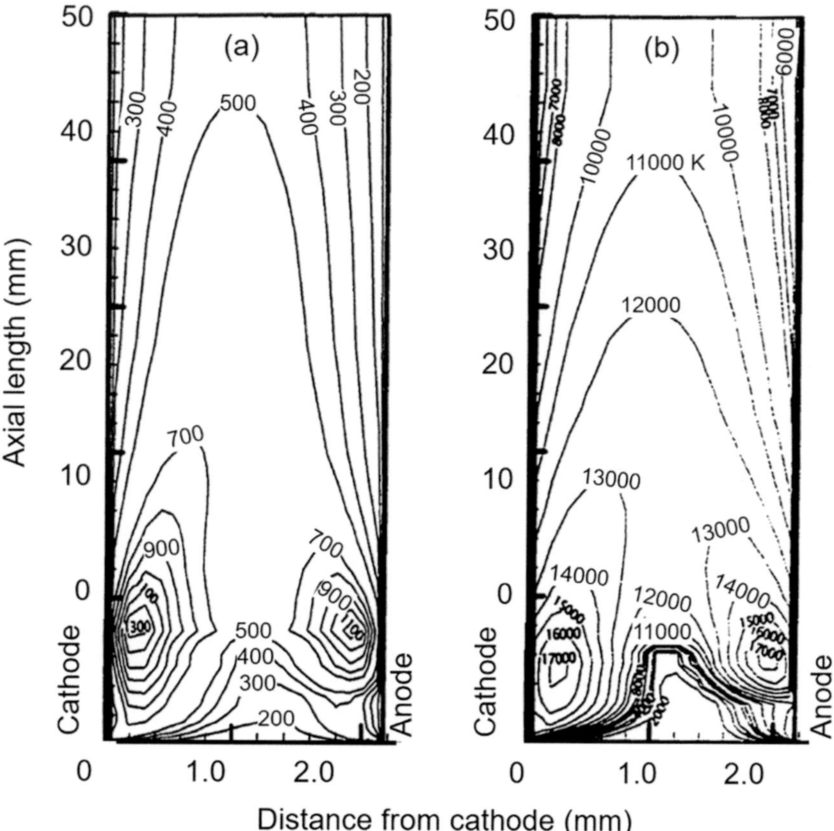

Fig. 11.51 Computed (**a**) velocity (m/s) (**b**) temperature (K) in the mid-plane of an arc with cross flow (from the bottom). (Kelkar and Heberlein 2000)

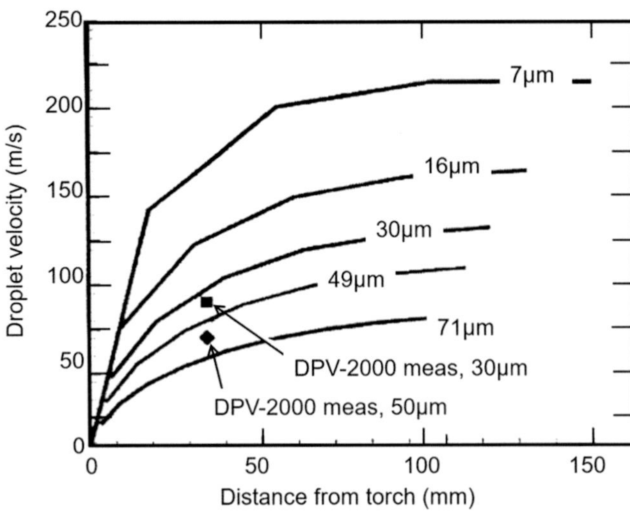

Fig. 11.52 Computed axial droplet velocity profiles for different droplet diameters. For comparison, two experimentally determined (DPV-2000) droplet velocities are included. (Kelkar and Heberlein 2002)

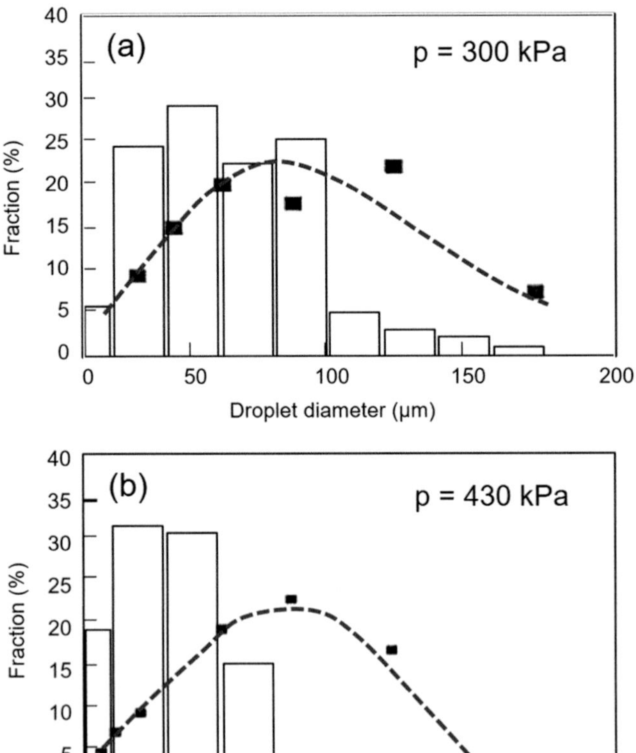

Fig. 11.53 Modeling results (bars) and measured (squares) of the droplet size distributions as function of atomization gas pressure (**a**) 300 kPa, (**b**) 430 kPa. (Kelkar and Heberlein 2002)

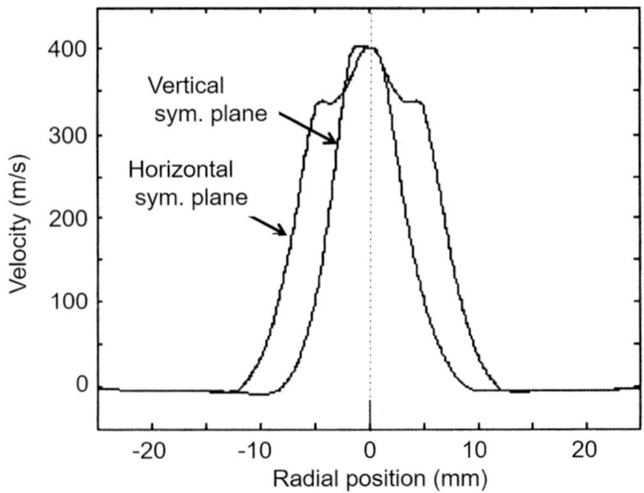

Fig. 11.54 Computed radial distributions of the axial gas velocity 25 mm from the nozzle exit for two symmetry planes, vertical (plane of the wires) and horizontal, showing the broader profile in the plane perpendicular to the wires. Gas flow rate 1833 slm (gas supply pressure 285 kPa). (Bolot et al. 2001)

experimentally measured ones (squares) for two different nozzle inlet pressures The shape of the size distribution, going from a bimodal to an almost mono-modal distribution when the nozzle inlet pressure is increased from 300 kPa to 430 kPa, is well replicated and shows the increased importance of secondary atomization. However, the calculated sizes are smaller than the measured ones. Clearly, more effort is required for a complete modeling description of the wire arc spray process.

Several CFD calculations were performed for simulating the performance of the TAFA 9000 torch and for optimizing the nozzle designs. Bolot et al. (2001)used the commercial PHOENICS code to simulate the exact configuration of the torch. The effect of the arc on the flow was neglected. Measured relationships between gas flow rate and upstream pressure were used. The results show the gas velocity distributions inside the torch and in the jet. Figure 11.54 (Bolot et al. 2001) shows as an example calculated horizontal and vertical distributions of the axial velocity, indicating the difference.

The power required to melt the wires at a given feed rate and to heat the droplets was estimated to be 65% of the electrical input power. A 2-D version of PHOENICS was used to identify nozzle geometries for the same torch that would yield increased droplet velocities (Gedzevicius et al. 2003). Several designs of the nozzle surrounding the wire tips were simulated, and a design was chosen that had an extension of the nozzle in the downstream direction with a constant cross-sectional area. Experiments performed with such a modified nozzle demonstrated an increase in the droplet velocity by about 20%. The velocity distributions for the three configurations of the TAFA 9000 torch shown in Fig. 11.11 and particle velocities were obtained considering different droplet sizes and acceleration by the gas with the calculated velocity (Planche et al. 2003), again using the PHOENICS code. The results are shown in Fig. 11.55

(Liao et al. 2005) where the gas velocities and the velocities of droplets with three different droplet sizes (10, 51.7 and 70 μm) are displayed for two different nozzle arrangements, the Laval-type nozzle without cap (open configuration) and the Laval-type nozzle in the cap. The advantage of having the peak of the gas velocity at the location of the wire tips in the case of the cap is shown clearly, leading to a better acceleration of the droplets.

A similar approach but based on the commercial code FLUENT with a k-ε turbulence 3-D model was again applied to the TAFA 9000 torch (Bolot et al. 2008). The wire material, TAFA 38 T 1.6 mm Mn-steel wire was assumed to be heated to the boiling point (3133 K) and use of experimentally determined droplet size distributions avoided the requirement of an atomization model. Gas and steel droplet velocity distributions were calculated for an atomization air pressure of 0.414 MPa (60 psig), air flow rate of 1680 slm, arc voltage of 31 V, and arc currents varying between 100 and 250 A. The corresponding wire atomization rate was estimated at 2.62 to 9.35 kg/h with mean droplet sizes of 18.8 to 20.5 μm, for arc currents of 100 and 250 A, respectively.

Computations were carried out of the gas flow and temperature fields including particle trajectories and gas particle loading effects. Typical results given in Fig. 11.56 and 11.57 obtained for an arc current of 100 A, 2.62 kg/h wire atomization rate show respectively the gas velocity fields and particle trajectories in two orthogonal planed over an extended region of interest of 200 mm downstream of the WAS torch. No particle loading was considered in this case. Figure 11.56 shows the velocity magnitude increases along the primary nozzle with air flowing through the secondary path around the primary nozzle. There is a very high-velocity region just

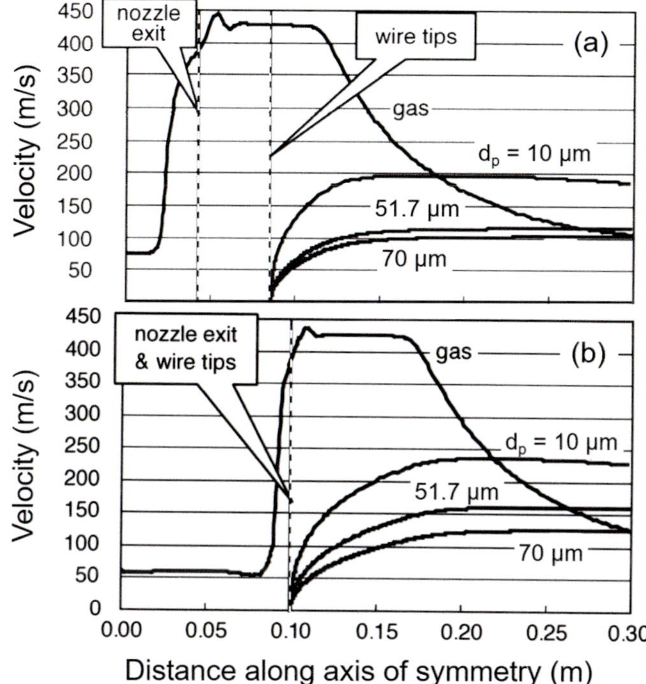

Fig. 11.55 Calculated centerline atomizing gas and droplet velocities for three droplet sizes of 10, 51.7, and 70 μm (**a**) the arrangement of a Laval type atomizing gas nozzle without cap, and (**b**) for a secondary nozzle (cap) with Laval type nozzle. (Liao et al. 2005)

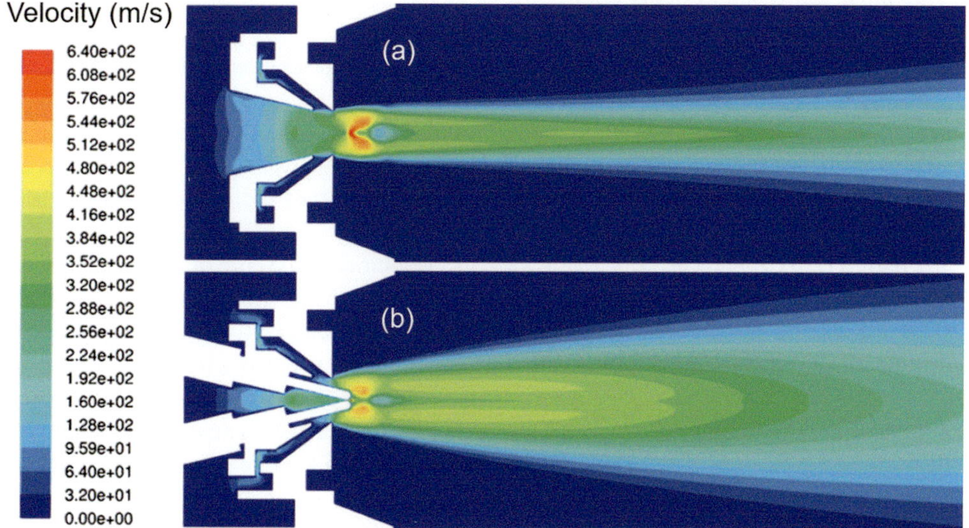

Fig. 11.56 Gas velocity mapping in two orthogonal plans for WAS TAFA 38 T steel wire atomization, air pressure = 0.414 MPa (60 psig), arc current = 100 A, (**a**) vertical plane, (**b**) horizontal plane. (Bolot et al. 2008)

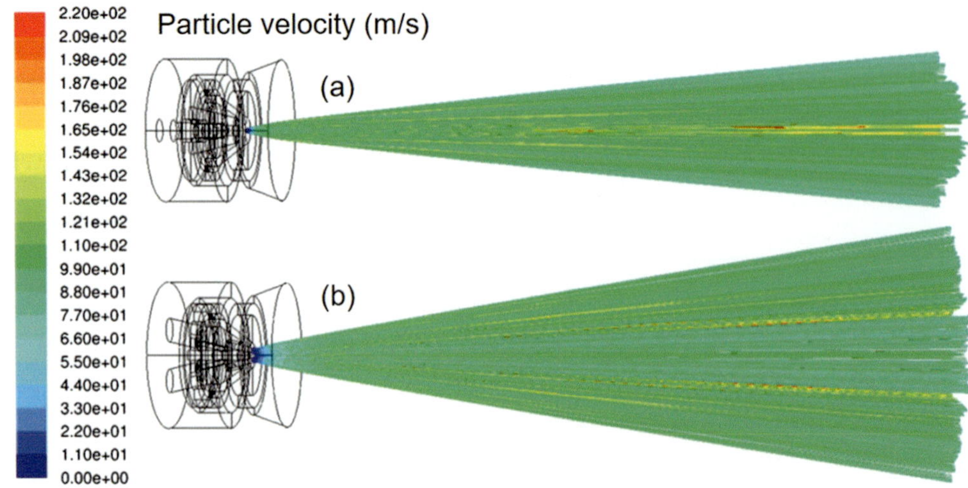

Fig. 11.57 Droplet trajectory mapping in two orthogonal plans for WAS TAFA 38 T steel wire atomization, air pressure = 0.414 MPa (60 psig), arc current =100 A, (**a**) vertical plane, (**b**) horizontal plane. (Bolot et al. 2008)

at the wires meeting point which forms a horizontal V shape in the vertical plan (i.e., perpendicular to the plan of the wires). A strong difference in the jet divergence in the vertical and horizontal plans is also noticeable. In particular, the jet divergence is much more pronounced in the plan of the wires than in the vertical plan. Figure 11.57 shows the droplet trajectories viewed in the two orthogonal plans. The particle jet divergence in the vertical plan (Fig. 11.57a) is lower in comparison with that observed in the plan of the wires (Fig. 11.57b). The range of particle velocity is comprised between 50 and 220 m/s at the distance of 200 mm. Computations were also carried out at higher arc currents and consequently higher atomized droplet loading in the gas jet (9.35 kg/h at 250 A). These show an increasing "loading effect" with the axial gas velocity suppressed by 50 m/s at a location 30 mm downstream from the nozzle exit when the interaction with droplets was considered. The effect became less noticeable at farther downstream locations. Again, a different divergence of gas velocities and particle trajectories is observed, with a larger divergence angle in the plane of the wires. This observation is in contrast to that of (Pourmousa et al. 2005) who observed a larger jet divergence in the plane perpendicular to the plane of the wires, but with a different spray torch. Figure 11.58 (Bolot et al. 2008) shows a comparison of the calculated droplet velocities with measured (DPV-2000) for an arc current of 100 A. The error bars represent standard deviations. Good agreement is seen for the smaller droplet sizes.

Chen et al. (2009b) presented a CFD model for a wire arc spray gun, to optimize the configuration parameters on the basis of numerical simulation. The FLUENT commercial code was used to study the gas flow inside and outside the spray gun. A comparison between the converging nozzles and converging/diverging nozzles was done to analyze the gas flow dynamics distributions. The compression wave and

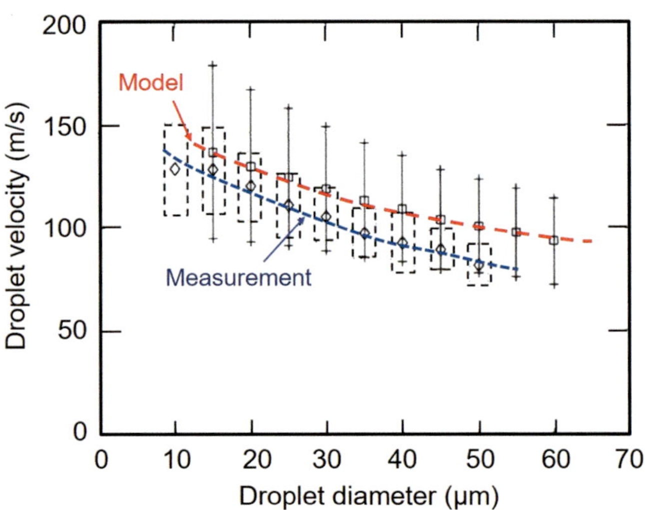

Fig. 11.58 Comparison of calculated and measured particle velocities for different particle sizes, arc current = 100 A. (Bolot et al. 2008)

expanding phenomenon of the converging–diverging nozzle were relatively weaker as compared with the converging nozzle. For the optimized converging–diverging nozzle with varying inlet pressures, from 0.4 MPa to 0.6 MPa, the gas velocities in the core of the spray jet were still kept at high values even if the spray distance was about 200 mm. They showed that other parameters were also important, such as the distance from the two wires' intersection point to the nozzle exit and the wires intersection angle, because they significantly affected the dynamics properties of the external gas flow. The 3-D model provided an analysis of the above configuration parameters. The results showed that moving the intersection point from the nozzle outside to the nozzle exit and that moderate intersection angle was better for the flow velocity distribution and droplets atomization. Results allowed modifying the gun design to upgrade the original

HAS-01 gun, and the modified design was experimentally compared with the original one by measuring the in-flight droplet size and velocity.

Toma (2013) has also discussed the effects of velocity and temperature of sprayed particles on physical and mechanical properties of coatings. These parameters are in a close relationship with the drive jet velocity and temperature. They made a comparative research of steel (C-0.14, Mn-0.43, Cr-0.3, Ni-0.3, Si-0.15, P0.04 and S-0.04 wt.%) deposits, carried out by two spraying procedures: the classic wire arc spraying procedure plus a new one, which represents a combination between the classic wire arc spraying and the flame spraying procedure. The use of a new procedure allowed the increase of the drive jet temperature and the study of its effect created on deposit properties. The modeling of the arc spraying and the analysis with finite elements in a coupled field allowed the determination of the drive jet temperature variation and the fuel flow, necessary for the temperature to be maintained over 1000 K at spraying distance of 150 mm. The investigations carried out on steel coatings, obtained by this procedure, demonstrate that the spraying jet temperature increase determines the average growth of the coating's adherence over 18% for cylindrical surface and over 5% for flat surface. Also, these investigations demonstrate the decrease of the average porosity by over 22% for cylindrical surface and by over 17% for flat surface. Because of the difficulty of obtaining reliable gas velocity measurements, these simulations are valuable for optimization of nozzle designs and understanding the atomization and droplet acceleration behavior.

Toma (2013) has compared the classic wire arc spraying procedure with a new one, which is a combination between the classic wire arc spraying and the flame spraying procedure. For steel spraying, it was possible to increase the temperature of the flame up to 2000 K and keep its value above 1000 K at the distance up to 150 mm. The increase in the temperature allowed better melting of the sprayed particles and improving the properties of deposits. The coating porosity decreased by over 22%, for deposits carried out on cylindrical surfaces, and by over 17%, for deposits carried out on flat surfaces. The modification of the spray technique also improved the coatings adherence: 18% for cylindrical surfaces and of 5% for flat surfaces.

11.5 Applications of Wire and Cord Arc Spraying

As mentioned earlier, wire arc spraying (WAS) is one of the oldest thermal spray technologies that has gained wide acceptance for a broad range of applications because of its flexibility, reliability, and favorable economics compared to alternate more technologies. It is, however, limited to a large extent to metallic coatings for which the feed material is available in the form of flexible ductile wires. The use of cords allowed to a limited degree the extension of the use of the technology to metal-ceramic composites for special applications. Cord arc spraying (CAS) is predominantly used for wear resistant coatings with the hard and brittle compounds used as filler material in low carbon steel sheath. It is also possible to make use of generating alloys in flight or on the surface, such as NiCr with Ni and Cr powders, or compounds like Ni_3Al with a Ni and Al powder mixture (Steffens et al. 1990). In this section, a brief review is presented of principal applications of wire arc spraying as tools for functional surface modifications and the rebuilding of worn surfaces or near net-shaped parts.

The large majority of materials used for WAS or CAS are available either as standard wire, 1.6–4.76 mm diameter (1/16–3/16 inch), of pure metals and alloys, or in the form of a cord in the same size range manufactured through the rolling of a thin metal foil in the form of a tube packed with fine powder mix of the required composition of the coating material. An example of some of the materials commonly used in this field and their recommended applications are given in Table 11.4 after (*Oerlikon Metco technical bulletin, issue 6*).

Table 11.4 Applications for the different families of metals and their alloys recommended for wire arc spraying (Oerlikon Metco technical bulletin, issue 6).(reproduced with permission)

Base metal	Applications
Aluminum	Corrosion and galvanic protection, oxidation resistance, dimensional restoration and repair, net shape parts and dimensional buildup, decorative, optical/reflective, electrical thermoconductance, EMI shielding
Copper	Corrosion/galvanic protection, biofouling control, dimensional restoration and repair, decorative, electrical and thermoconductance, low friction
Iron	Corrosion/galvanic protection, erosion/wear and cavitation control, net shape parts and dimensional buildup, dimensional restoration and repair, gripping/antiskid, low friction
Nickel	Corrosion/galvanic protection, erosion/wear and cavitation control, bond coat, oxidation resistance, dimensional restoration and repair, gripping/antiskid
Silver	Clearance control, electrical and thermoconductance
Tin	Corrosion/galvanic protection
WC	Erosion/wear and cavitation control
Zinc	Corrosion/galvanic protection, net shape parts and dimensional build up, electrical and thermoconductance, EMI shielding

It may be noted from this table that Corrosion and galvanic protection is one of the most important applications of WAS using mostly aluminum, copper, iron, and nickel and their alloys. Erosion/wear protection and cavitation control comes next using iron, nickel, and their alloys as well as tungsten carbine cords. Iron and its alloys are also used for gripping/antiskid and low friction while Nickel and its alloys for bond and oxidation-resistant coatings. Other applications of varying scope include net shape parts and dimensional restoration, clearance control, electro- and thermoconductance, EMI shielding, biofouling control, and decorative.

A comprehensive review of present and potential wire arc spray technology in China was reported by Liu (2001) who points out that it is the most efficient way for long life corrosion protection of steel structures. The arc spraying process is also used for a wide range of applications such as renovation and surface modification of machine components, mold making for plastic products, high-temperature corrosion resistance for boilers, anti-sliding coatings, and self-lubricating coatings. With the advances and developments in arc spray equipment and spray materials and the simplicity of operation, arc spray technology is used for demanding, high-quality coating applications, which are nearly equivalent to those previously served by alternate plasma technologies at significantly reduced capital and operating costs. Arc spraying is becoming a serious economic alternative to other thermal spraying processes while maintaining, and in many cases improving, the quality of the finished product.

11.5.1 Corrosion Protection

Corrosion protection is one of the most important industrial applications of wire arc spraying. Basically, two distinct approaches are used to achieve this objective, the galvanic protection or the sealed coating protection. Galvanic is mostly used when the part to be protected is completely or partially immersed in an electrolyte such as sea water, as in most marine applications (ship hulls, decks, or offshore oil drilling platforms). For structures made of mild steel or low alloy steel, the coating materials has to be more electronegative than the base material (-0.6 to -0.7 V) as given in Fig. 11.59 for the galvanic series of elements referenced to saturated calomel electrode (SCE). Zinc (-1.0 V) and aluminum alloys (-0.75 to -1.0 V) with a galvanic voltage lower than those of steel are good candidates for such coatings which is this case can be porous since the coating is sacrificial (cathodic protection). On the other hand, when the potential of the coating material is higher than that of steel, such as nickel (-0.1 to -0.2 V) (anodic protection), the coating needs to be impermeable with no open porosity in order to prevent the corrosive elements to reach the base surface.

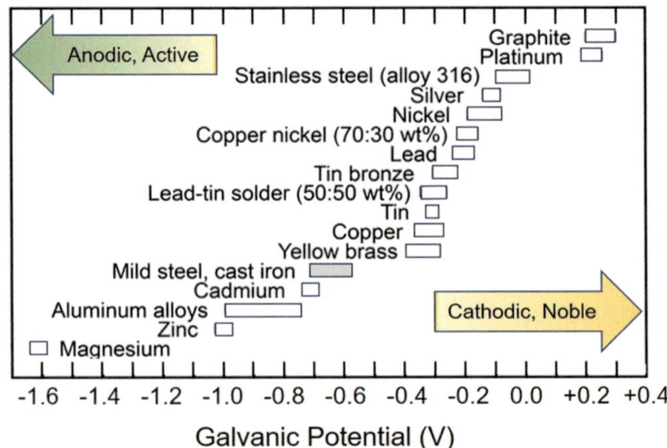

Fig. 11.59 Galvanic series of the elements referenced to saturated calomel electrode (SCE)

The wire arc spraying for anti-corrosion protection of the steel structure of ships is typical example. http://www.metallisation.com/ reports on a number of separate projects where a variety of ship were sprayed with pure zinc or a zinc-aluminum alloy (85/15%) including two trawlers, an oceanographic ship, and fish carrier which all external decks superstructures the bridge, mast, and chimney metal sprayed. An example is also given for the wire arc spraying of close to 5000 m^2 of its internal and external surfaces with zinc on a 3588-ton Norwegian fishing trawler, (Fig. 11.60). One of the advantages cited is the ease of operation and the long supply pack (20 m) between the spray unit power supply/wire spools and the spray torch which allowed for the spraying of hard-to-reach regions in the ship.

Another example cited by http://www.metallisation.com/ is the use of wire arc spraying for the cathodic protection of steel in concrete. The prime cause of corrosion in this case is the salt, chloride, and de-icing products often used in cold regions for road salting during the winter. As the water loaded with the chloride salts seeps into the concrete, the chloride ions attack the steel reinforcing bar transforming the ferric oxide film on their surface to red rust causing cracks and spalling of the concrete with potential failure of the structure. The long-term solution proposed was to wire arc spray the concrete with zinc or zinc alloys, Fig. 11.61, which offers cathodic protection operated in galvanic or sacrificial mode. Metallized zinc cathodic protection systems can also be operated in impressed current mode in which an external DC power supply is connected between the anode and the steel. The electrical circuit is completed between the steel bars and the zinc coating. The action of the corrosion cell causes the zinc to corrode in preference to the steel by the presence of moisture in the concrete. It is important to stress that the anode and the steel bars should not short out. According to Metallisation Ltd., if the coating is performed correctly and depending on the coating applied, the process

Fig. 11.60 Zinc arc spraying of Norwegian fishing trawler using Arcspray 140 system. http://www.metallisation.com/

Fig. 11.61 Al/Zn/In wire arc spraying of concrete beams and caps of the San Luis Pass Bridge. http://www.metallisation.com/

can offer corrosion protection for up to 20 years before the next significant maintenance is required.

A recently developed alloy of aluminum, zinc, and indium has been used in a small number of applications. This material is more active than zinc, and it is claimed to not require an impressed current to provide adequate levels of corrosion protection. The corrosion protection of precast concrete panels 2.6 × 2.8 m forming the underside of viaducts carrying heavy traffic to the Charles de Gaulle (Roissy) airport complex, France, is cited by Metallisation Ltd. as an example where a 300 μm coating of Al/Zn/In alloy was successfully adapted the corrosion protection of these panels. Another significant application of the Al/Zn/In alloy in the USA is the san Luis Pass Bridge near Galveston, Texas, which is also sited where more than 30,000 m^2 of concrete beams and caps are protected with this alloy Metallisation Lt. using ARC-700 units (Fig. 11.61).

According to Lester (2005) and Wang et al. (2010), aluminum is also used in high-temperature applications, such as flare booms on offshore oilrigs, as the aluminum effectively dissolves into the substrate when exposed to high temperatures (aluminizing). Zinc/aluminum alloy is used in environments where the corrosion resistance of zinc is borderline, for example, when sprayed on ductile iron pipes that are buried in the ground (Varis and Rajamake 2002; Lester 2005; Wang et al. 2010).

Sørensen et al. (2009) have presented a review of anti-corrosive coatings for marine structures in general such as containers, offshore constructions, wind turbines, ballast and storage tanks, bridges, rail cars, and petrochemical plants. Flame and wire arc spraying meet such requirements. However, coatings obtained by these methods are relatively porous (up to 20%). Sacrificial coatings (e.g., Zn or Al on steel) are mainly used. Referring to such coatings must have a cathodic behavior relatively to the ions of the metal to be protected, steels in almost all cases. The cathodic protection can be porous without any corrosion of the underneath metal. Zinc performs better than aluminum in alkaline conditions, while aluminum is better in acidic conditions. Zinc–aluminum (Zn-15wt%Al) seems to combine advantages of both materials.

Other treatments are possible such as plastic deformation. For example, the initial porosities of 4–14% (depending on spray conditions) of aluminum coatings, wire arc sprayed, was reduced to 0.16–0.83% after being shot peened with SiC glass beads of 0.21–0.3 mm (Pacheo da Silva et al. 1991). Sealers can also present anti-fouling properties. Chun-long et al. (2009) have arc sprayed aluminum on steel panels and then sealed coatings with nanocomposite epoxies especially developed for sealing them. In order to test their performance some panels were tested in the East China Sea. Test panels were mounted respectively for 3 years, in the marine atmosphere zone, seawater splash zone, tidal zone, and full-immersion zone. Tests included marine atmospheric outdoor exposure test, seawater corrosion test, and coating adhesion test. It was found that the appearance of coating panels was as fine as original but with a little sea species adhering to panels on tidal zone and full-immersion zone. Basically, no change in the morphology, bond strength, and any visible coating crack, blister, rust, and break off was observed.

Schmidt et al. (2006) investigated the corrosion protection performance of 17 different Zn and Al sacrificial coating system configurations during marine atmospheric exposure (20-month exposure time) at Kure Beach, NC, USA. The coating systems incorporated several conversion coating layers, primers, and organic topcoats. The sacrificial Zn coatings provided sacrificial protection at the defects. Of the two thermal spray Zn coatings, flame spray coatings were more protective than arc spray.

Han et al. (2009) used aluminum wire for the WAS of an aluminum coating on special treatment steel (STS) 304 base

metal. The electrochemical experiment was performed in natural seawater. The coating with the higher thickness represented good corrosion resistance in seawater.

Esfahani et al. (2012) deposited aluminum coating on mild steel by arc spraying. The corrosion behavior was evaluated by electrochemical impedance spectroscopy (EIS) and polarization tests in 3.5 vol. % NaCl solution. The as-coated samples were also subjected to a 1500-h salt spray assay. EIS measurements showed improved corrosion protection performance of the coating during long-time immersion and exposure to saline mist. This could be due to plugging of pores by corrosion products, which hinder further penetration of the electrolyte through the coating. To limit atmospheric or marine corrosion, if one uses an anodic protection, the coating should be as dense as possible (if necessary sealed). Cramer et al. (1999) as well as Holcomb et al. (1997) have proposed a cobalt-catalyzed, non-consumable, thermal sprayed titanium anode as an alternative to thermal sprayed zinc anodes for impressed current cathodic protection. Ti was wire arc sprayed (Holcomb et al. 1997), the atomization being performed with nitrogen to limit as much as possible the formation of γ-TiO$_2$. The titanium anode had a porous heterogeneous structure composed of α-titanium containing interstitial oxygen and nitrogen and a fcc (face cubic centered) phase thought to be Ti(O,N). The final titanium anode thickness was 50–150 μm.

Electrochemical aging was studied using catalyzed titanium anodes thermal sprayed on concrete slabs containing a steel mesh cathode. Cobalt catalyst, applied to the anode as an aqueous cobalt nitrate-amine complex, penetrated the anode and accumulated at the anode-concrete interface and in cracks within the concrete adjacent to the interface. Anodic polarization during and after application converted the cobalt to the active catalyst, Co$_3$O$_4$. With aging, cobalt catalyst dissolved in the increasingly acid environment of the interface and dispersed into the Ca-deficient, silica-rich reaction zone to re-precipitate and form a more diffuse site for the anode reactions. Stable operation of catalyzed anodes was maintained for a period equivalent to 23 years' service at Oregon DOT bridge ICCP conditions with no evidence that operation would degrade with further aging. Early results from the field experiment at the Depoe Bay Bridge suggested anodes may age more slowly at low current densities with lesser impact from acidification (Holcomb et al. 1997).

11.5.2 Wear Protection

The use of wire arc spraying (WAS) for wear protection is one of the fastest growing areas of application of the technology on an industrial scale. Special attention has been given over the past three decades to the potential use of the technology in the automobile industry where intense efforts are made for the improvement of the energy efficiency of the automobile through the increased use of aluminum and its alloys for the manufacture of key engine and structural components. The coatings of the cylinder wall of the engine and the valve lifters offer an attractive alternative to electroplating or the used of cast iron liners in aluminum alloy engine blocks which would allow for an important reduction of the weight of the engine. Several developments have been pursued for the coatings of cylinder walls, using either DC plasma spraying (Oerlikon-Metco Automotive solutions kit, BRO-0011.5), single wire arc spray systems or twin wire arc spray systems with a 90° nozzle diverting the particle stream (Nakagawa et al. 1990). These types of coatings are typically carbon steel based, deposited to thicknesses of 300 to 400 μm, with porosity required to be less than 5%, and must have excellent adhesion to the substrate (Marantz et al. 1991). Adhesion can be improved by pretreatment of the surface with a fluoride-based flux to reduce the oxide layer using a NiAl bond coat (Cook et al. 2003) model predicting the coating deposition as a function of the coating strategy.

Developments on diverging the droplet fluxes by 90° have been pursued for inside cylinder wall coating applications (Benary 2000; Dunkerley et al. 1999). Figure 11.62 (Bolot et al. 2007) illustrates the approaches used for a 90° bending of the droplet jet, as described in (Benary 2000; Dunkerley et al. 1999). Potential applications in the automobile industry for the coating of aluminum engine blocks have been the principal driver of such developments. For this application a coating of 300 to 400 μm is desired with very good adhesion and porosity of less than 5%. Benary et al. (1999) describe an extension of Oerlikon Metco Smartarc torch. Atomization occurs by blowing the gas through 90° nozzle consisting of several orifices immediately downstream of the initial droplet formation. This perpendicular gas flow is responsible for secondary atomization. The torch is operated with a 28 V, 200 A setting, and 448.2 kPa pressure

for the 90° nozzle. Velocities of NiAl droplets as measured with a DPV-2000 instrument are reduced from 88 m/s with straight flow to 66 m/s in the propagation direction with the cross flow. Little effect on the droplet temperature has been noted.

A dedicated effort by Bobzin et al. (2008) was devoted to the study of the best route for the development of a low friction cylinder liner technology to replace conventional cast iron liners. A novel and highly alloyed iron-based surface building materials is proposed for plasma transferred wire arc spraying (PT-WAS). Bobzin et al. (2008) describe the development of such materials used for the spraying of partially amorphous coatings with embedded boridic nanoscale precipitations. The coatings were applied onto the inner diameters of test liners made of aluminum EN AW 6060 using the PT-WAS system illustrated in Fig. 11.63a. Details of the torch designed for these applications, illustrated in Fig. 11.63b, was discussed in Sect. 11.3.3 Single Wire Arc Spraying. Prior to coating, all surfaces to be coated were pretreated by a novel fine boring process in order to create a surface topography which enables the adherence of the coatings. The nanocrystalline microstructure of the coating and its properties were analyzed by light optical microscopy, hardness measuring, and transmission electron microscopy. Furthermore, the oil storage capacities of the honed surfaces were determined.

Because of the critical importance of coating adhesion to the cylinder wall, Bobzin et al. (2008) experimented with different techniques which could be used for the surface preparation of the cylinder walls. This included the roughening of the substrate by grit blasting, high-pressure water jet with pressures up to 300 MPa or the use of a flux such as Nocolok™ to strip the oxide layer from the aluminum-based alloy. An alternate process developed at the Institute of Machine Tools and Production Technology of Braunschweig University (IWF) within the framework of the "NaCoLab" project, used a fine boring tool with one or two geometrically defined cutting edges to carve different topographies into the substrate. Figure 11.64 shows microstructure of coating and substrate cross-sections obtained using this technique which had an adhesion close to 60 MPa, the best among the different techniques tested.

Hahn and Fischer (2010) point out that demand for low-friction and wear-resistant materials coatings is also extended to diesel engines in order to increase the efficiency and achieve environmentally sound solutions. Thermally sprayed Fe-based coatings were investigated for application as cylinder liners in cast aluminum engine blocks. To understand the influence of the coating microstructures on its integrity and performance, coatings were examined in laboratory tests in terms of different loading situations: cavitation, tribological stability, and ability to resist high-frequency cyclic impact stresses.

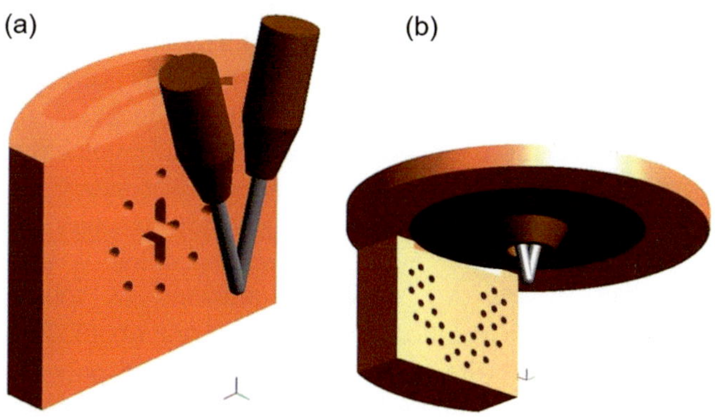

Fig. 11.62 Schematics of nozzle arrangements for 90° angle spraying (**a**) perpendicular atomization with cross-shaped orifice, (**b**) vertical atomization with cross flow. (Bolot et al. 2007)

(a)

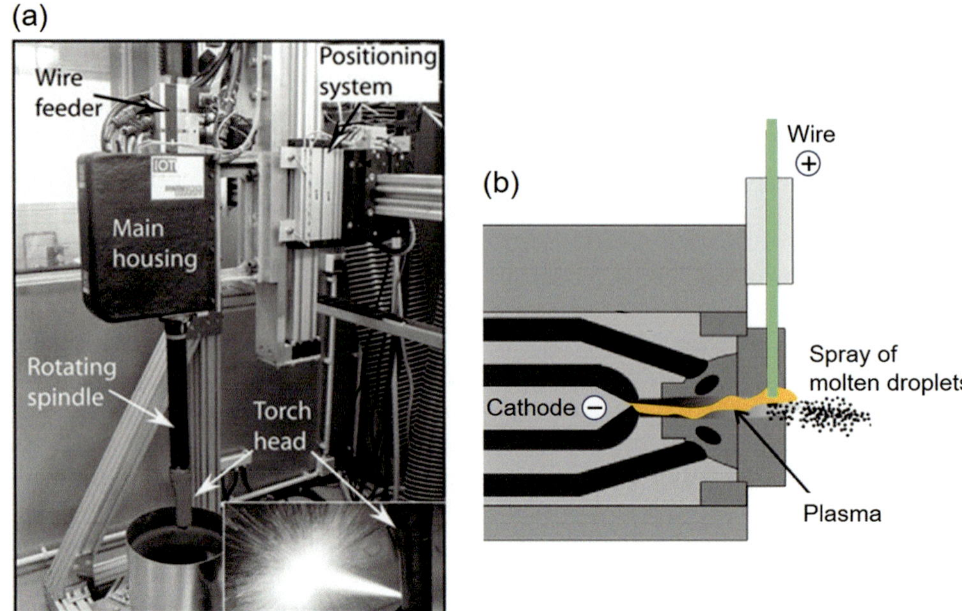

Fig. 11.63   (**a**) Photograph of the PTWAS system used and (**b**) schematic of the torch head. (Bobzin et al. 2008)

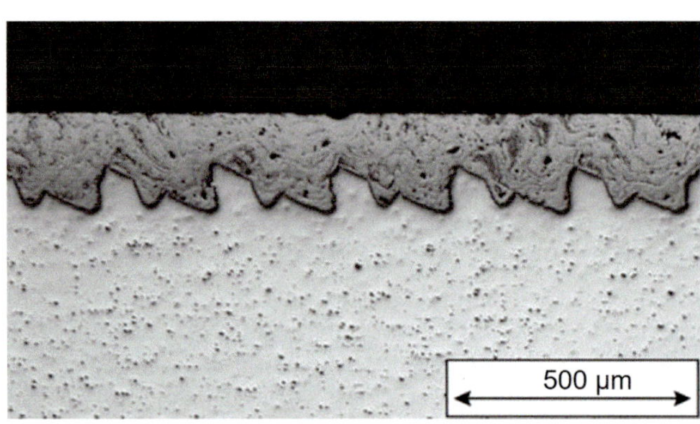

Fig. 11.64   Micrograph of coating on mechanically roughened/carved substrate surface. (Bobzin et al. 2008)

A recent study by Darut et al. (2015) examined the use of the plasma transferred wire arc spraying (PT-WAS) for the deposition of the cylinder liner in an engine block in a four-step process as illustrated in Fig. 11.65. These consisted of the following:

(a)   Machining of the cylinder in the engine block to its exact diameter
(b)   Surface activation by either sand blasting, high pressure water jet, or mechanical roughening
(c)   Coating application by PT-WAS
(d)   Honing to a smooth surface finish at the exact dimension

The coating is composed of iron in majority, oxides (around 20%), and pores (around 1%). The porosity in the coatings with 150 A torch is significantly higher in the middle and at the bottom of the bores, when compared to the 85 A torch. Very thick coatings of around 1 mm could be obtained without modification of the microstructure.

Due to the continuous development in automobile engines, the coating strategies needed to be adapted in parallel to achieve a quality-conformed coating result. The use of twin wire arc-spraying process for the coating of aluminum bore to create a thin, iron–carbon-alloyed coating which is surface-finished through honing was reported by König et al. (2015). The most important factors to this end are the controlled indemnification of a minimal coating thickness and a homogeneous coating deposition of the complete bore. They developed a specific system that enabled the measuring and adjusting of the part and the central plunging of the coating torch into the bore to achieve a homogeneous coating thickness. The coating thickness was determined by measurements of the bore diameter of the cylinder before and after coating. The information was then transferred by a specially developed software to a model that predicts the final coating deposition as function of the coating strategy.

(Bolot et al. 2007) report on a nozzle optimization effort for the TAFA torch, where a downstream secondary gas flow perpendicular to the original gas flow changes the droplet flow direction by 90°. Torch operation is with 30 to 35 V, 150 A, gas supply pressures of 400 to 600 kPa, and perpendicular gas flow rates of 1333 to 11,667 slm. The wire guides are stationary and only the secondary gas flow nozzle head rotates around the initially formed droplet stream.

He et al. (2007) deposited by Cord Arc Spraying two Fe-FeB-WC coatings on Q235 steel substrate using cord wires.

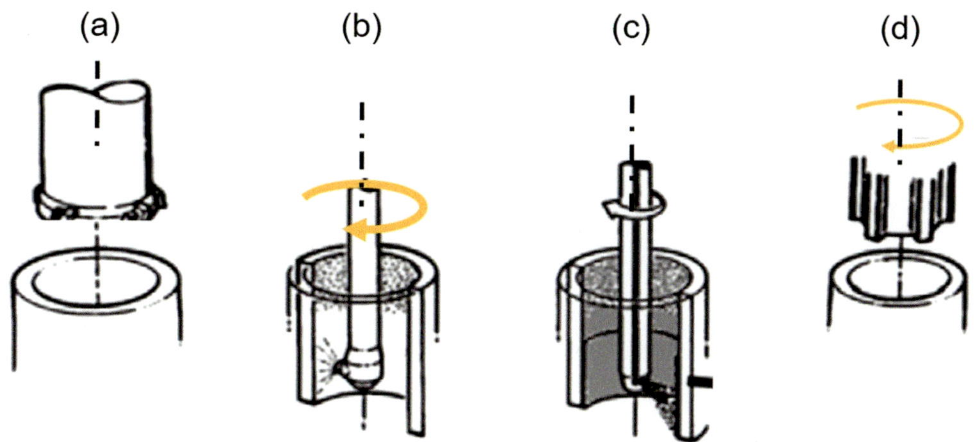

Fig. 11.65 Steps for coating application in an engine block cylinder: (**a**) initial machining, (**b**) surface activation/preparation, (**c**) WAS of the coating, (**d**) final finishing, honing. (Darut et al. 2015)

Table 11.5 Composition of the filler materials (Element, wt.%)

Coatings	Fe	Cr	Ni	B	W	Co	C
W1	70.12	12.67	7.14	4.05	5.07	0.72	0.23
W2	70.12	12.67	7.86	4.05	5.07	…	0.23

The sheath of the cored wire was 304 L stainless steel. The composition of both fillers considered is given in Table 11.5.

The coatings prepared by arc spraying using cored wires exhibited a typical lamellar structure, as seen in Fig. 11.66.

The majority of phases in both coatings were Fe-Cr and Fe_2B. A trace of WC and $a\text{-}W_2C$ was recognized from the XRD pattern of coating W1, while not recognized in coating W2. The results indicated that it was easier to deposit an alloy-based coating containing FeB than one containing WC by arc spraying using cored wire with hard phase fillers. The average microhardnesses of two arc sprayed Fe-FeB-WC coatings were 920 $HV_{0.1}$ and 872 $HV_{0.1}$, higher than that of an arc sprayed 3Cr13 coating. The results showed that a wear-resistant coating can be deposited by arc spraying using a cored wire of hard filler materials. The coatings exhibited excellent abrasive wear resistance, being 3.3–4.8 times higher than that of arc sprayed 3Cr13 coating.

Zhao et al. (2009) deposited a cored wire of 304 L stainless steel as sheath material and NiB and WC-12Co as filler materials to produce a new wear-resistant coating containing amorphous phase by arc spraying. The XRD and TEM analyses showed that there are high volume of amorphous phase and very fine crystalline grains in the coating. DTA measurements revealed that the crystallization of the amorphous phase occurred at 579.2 °C. Because metallurgical processes for single droplets were non-homogenous during spraying, the lamellae in the coating have different hardness values, which lie between about 700 and 1250 HV0.98 N. The abrasive wear test showed that the new Fe-based coating was very wear resistant.

Chen et al. (2012) synthesized Fe-CrB-Si-Nb-W coatings using robotically manipulating twin-wire arc spraying system with Fe-based corded wire consisting of mechanically mixed alloys powders such as ferrosilicon (75%Si), Ferroboron (17% B), ferrotungsten (55% W), and Niobite (50% Nb). The microstructure and mechanical properties of the coating were characterized. The coating had a laminated structure, and its porosity was 2.8%. The microstructure of the coating consisted of amorphous and α-(Fe,Cr) nanocrystalline. The nanocrystalline grains with a scale of 20–75 nm were homogenously dispersed in amorphous matrix. The results show that Fe-CrB-Si-Nb-W coating had excellent wear (about 4.6 times higher than that of 3Cr13) and corrosion resistance. In 3.5% NaCl aqueous solution, the amorphous/nanocrystalline composite coating had better corrosion resistance than that of the 0Cr18Ni9 stainless-steel coating.

Tian et al. (2014) studied how to improve the lamellar structure and wear resistance of arc-sprayed coatings of FeNiCrAl. As shown in Fig. 11.67a, the 150-μm thick coating presented typical thermally sprayed laminar structure which is responsible for its higher susceptibility to spalling and delamination. The coatings were remelted by the tungsten inert gas (TIG) welding method. The cross-section of the coating after TIG remelting is presented in Fig. 11.67b. The structure of the remelted coatings was characterized by its homogenous structure and the elimination of the lamination and porosity. In contrast to the interface between the remelted coating and the substrate, it was less noticeable compared to the original lamellar coating. Additionally, the consequence of remelting treatment was the transformation α-Fe phase into γ-Fe. Energy dispersive spectroscopic analysis confirm that oxide delamination is the dominant wear mechanism of the as-sprayed coating while for the TIG remelted coating, which was mostly pore- and crack-free, it was cutting and plowing. The wear resistance of coated specimens was far better than that of uncoated ones.

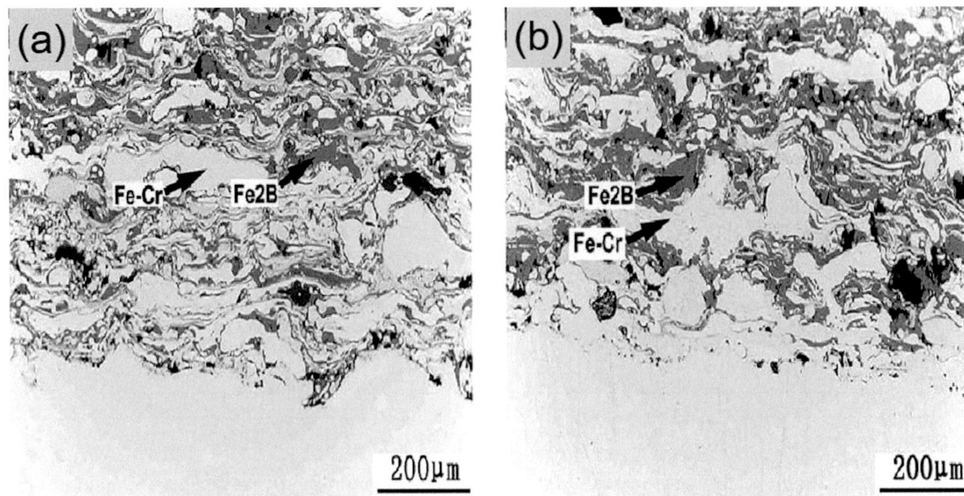

Fig. 11.66 Cross-sectional microstructure of the as-sprayed coatings. (**a**) Coating W1. (**b**) Coating W2. (He et al. 2007)

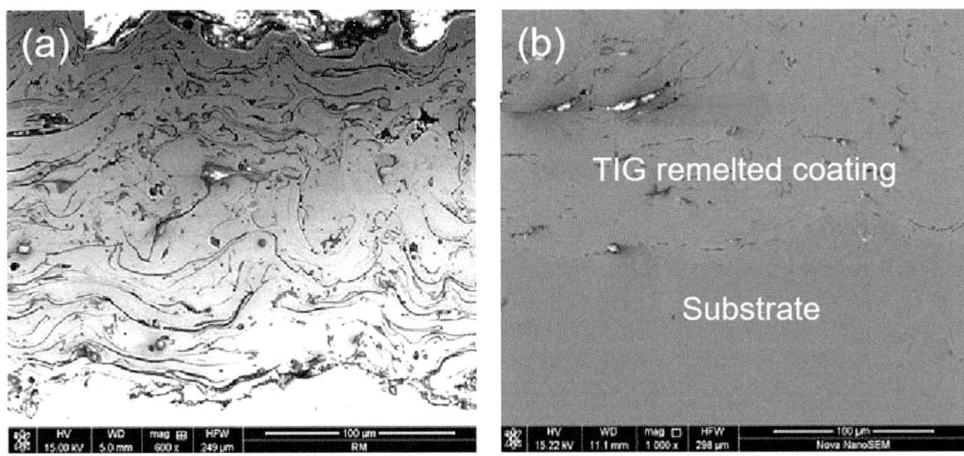

Fig. 11.67 SEM cross-sectional microstructure of (**a**) As-sprayed coating, (**b**) TIG remelted coating. (Tian et al. 2014)

Tian et al. (2014) evaluated the effect of the TIG remelting treatment of self-fluxing FeNiCrAlBRE alloy coatings, formed by means of high-velocity cord arc spraying on steel surfaces. The microstructure of the remelted coatings changed from lamellar to cellular structure with a visible reduction of coating defects. The shape of the Al_2O_3 phase changed from irregular block to slender strip depending on the heat input to the coating. Remelted coatings were dense with no visible cracks or pores. The worn volume and worn track width, length, and depth of the remelted coating were all smaller than that of the as-sprayed coating.

According to Tillmann et al. (2017), the outstanding properties of WC-W_2C iron-based cermet coatings, widely used in the field of wear protection, is mainly determined by the carbide grain size fraction. Coatings made using cored wires with tungsten carbides of different sizes as filling material were subjected to post-treatment processes such as Machine Hammer Peening (MHP). Figure 11.68 shows the microstructure of typical coating as sprayed and that following MHP treatment.

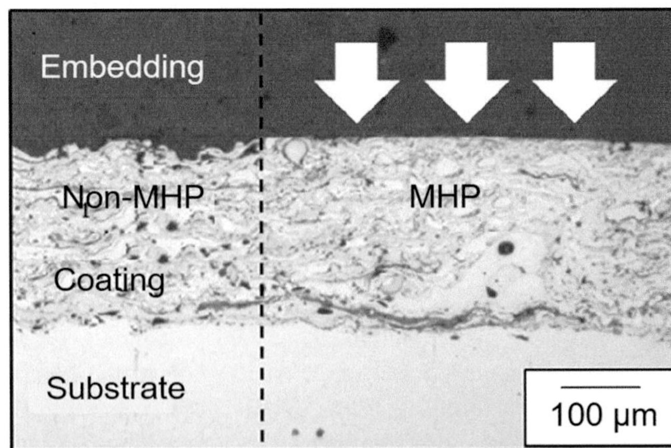

Fig. 11.68 Cross-sectional images taken by light microscopy showing the coating morphology across the non-MHP and MHP area. (Tillmann et al. 2017)

The results obtained indicated that finer carbide grain size fraction leads to:

- Decrease in wear coefficient for dry sliding experiments due to an enhanced distribution of loads to a larger amount of carbides
- Increase in the abrasive wear rate, shown by a higher abrasive wear coefficient of the stressed surface proven by dry rubber wheel tests
- Reduced coefficient of friction (COF) due to a fine lamellar microstructure

The post-treatment of surfaces with the MHP process leads to:

- Poorer wear behavior in dry sliding experiments due to the occurrence of cracks, which are previously induced by high loads of the MHP process
- Enhanced wear behavior against abrasion, which corresponds to strain hardening effects of the metallic matrix which was revealed by the mechanical response by means of nano-indentation
- Improved COF due to a reduced exposure of pores during ball-on-disk testing
- Higher ratio of the hard phases related to the metallic binder leads to
- Reduced wear coefficient in dry sliding experiments, which corresponds to the higher amount of W-rich hard phases
- Higher abrasive wear coefficient for deposits by means of dry rubber wheel tests, since both the MHP samples and the non-MHP feature a reduced hardness and Young's modulus
- Slight decrease in the COF for MHP coatings compared to non-MHP samples

In an attempt to improve heat transfer rates and wear resistance of the surface of drying cylinders in paper production machines, Yao et al. (2017) wire arc-sprayed coatings of different compositions using cord wires and compared the results to those obtained using conventional X30Cr13 solid wires. The generic composition of the cord wires used had a based composition of $(Fe_{87}\ Cr_{13})$ to which various concentrations of Boron were added giving compositions of $(Fe_{87-x}\ Cr_{13}\ B_x)$ with x = 1.0 wt.%, 1.5 wt.%, 2.0 wt.%, 2.5 wt.%, 3.0 wt.%, and 4.0 wt.%. Boron addition to the cored wires seemed to promote the formation of borides and amorphous phases in the FeCrB coatings, with decreased grain size, slightly increased porosity, and significantly decreased oxygen content with increasing boron content. As a result, the thermal conductivity of the coating changed significantly, reaching a maximum of $\kappa = 8.83$ W/m.K at 2 wt.% boron content, significantly higher than that of the X30Cr13 coating

($\kappa = 5.45$ W/m-K). In addition, the microhardness and relative wear resistance of the FeCrB coatings increased with increasing boron content in the cored wires, being much higher than for the X30Cr13 coating. These coatings are very promising with superior heat transfer ability and wear resistance for drying cylinders in the paper production.

11.5.3 High-Temperature Erosion and Oxidation Protections

The high-temperature protection of steels against erosion and oxidation are an industrial-scale problem that has been the subject of numerous investigations especially in the area of thermal power stations and boiler design and maintenance.

11.5.3.1 Erosion Protection

According to Chen et al. (2012) high-temperature erosive wear in boilers is one of the main causes of downtime and loss of productivity. The use of FeBSiNb amorphous coatings synthesized by wire arc spraying to improve elevated temperature erosion resistance for boiler applications has been the subject of numerous studies. The erosion resistance of the coating was reported to be sensitive to test temperature, decreasing with the increase of environment temperature. The relationship between microstructure and erosion resistance of the coating was also analyzed. The FeBSiNb coating had excellent elevated temperature erosion resistance up to 600 °C. SEM image of a 700 µm thick coating is presented in Fig. 11.69 showing a very dense and smooth coating well adhering to the substrate with no visible cracks.

The same coating cross-sections are represented in Fig. 11.70 following erosion tests at an impact angle of 30° (the worst for erosion) at different temperatures. Micrograph of test carried out at 25 °C, given in Fig. 11.70a, revile a number of inter-splat cracks near the coating surface which

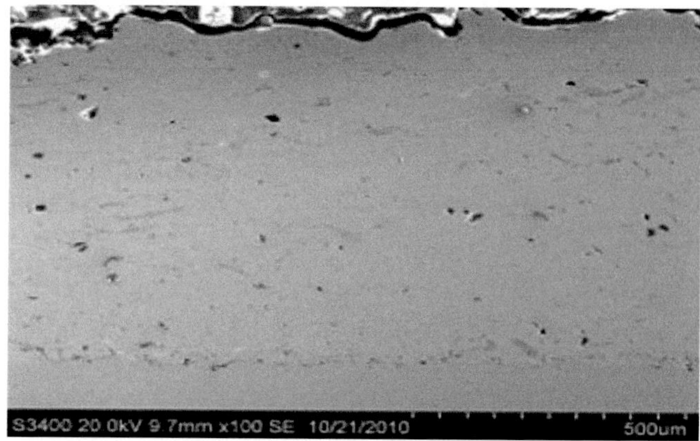

Fig. 11.69 SEM image of cross section of FeBSiNb amorphous coating. (Chen et al. 2012)

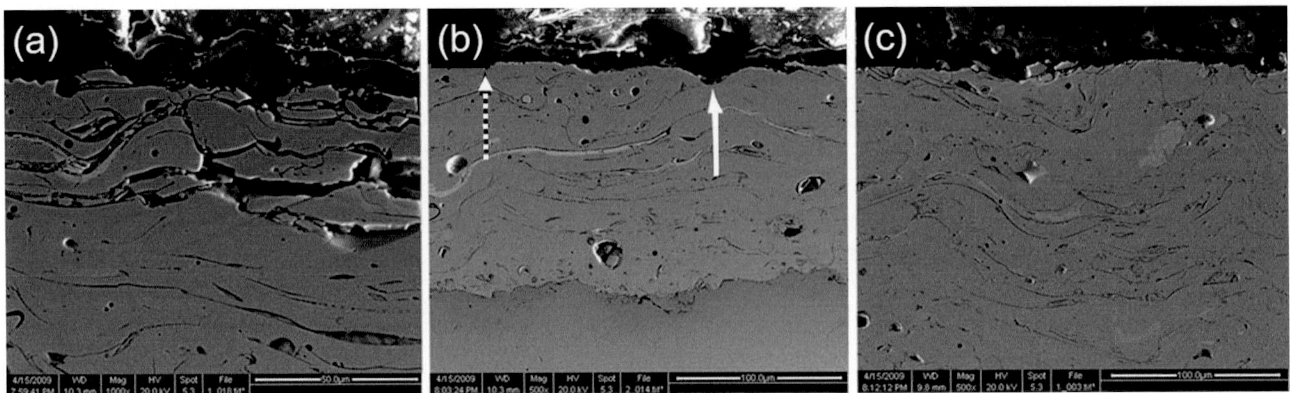

Fig. 11.70 Cross-section morphologies of the coatings after erosion test at impact angle of 30° and temperature of (**a**) 25 °C, (**b**) 450 °C, and (**c**) 600 °C. (Chen et al. 2012)

are attributed to poor bonding between splats. The micrograph given in Fig. 11.70b for a sample tested at 450 °C reveals a small groove in the eroded surface, marked with white dotted arrow, corresponding to the transition of erosion mechanism from a purely "brittle" to a relatively "ductile" behavior. In the same sample, a solid white arrow marks a visible pit in the coating. With the increase of the erosion testing temperature up to 600 °C, the surface of the coating became smoother as can be observed in Fig. 11.70c.

11.5.3.2 Oxidation Protection

The wire arc spray coating of aluminum bronze on mild steel substrates was investigated by Zhang et al. (2006) for the oxidation protection of steel exposures to air at 900 °C. The experiments showed that for exposures up to 24, the bond strength of the coatings on steel substrates remained stable and the substrate was well protected. The coatings withstood more than ten thermal shock tests without any coating separation. After exposure at high temperature, the coating maintained its excellent adhesion to the steel substrates.

Wielage et al. (2013) developed arc-sprayed iron-based coatings for high-temperature oxidation protection applications. Cored wires with a diameter of 1.8 mm with a fine powder mixture of chemical composition FeCr6B3Al14 was used with a filling coefficient of 20–27%. The sheath foil was made of low-carbon steel (0.08% C), with a thickness of 0.4 mm. The wire arc sprayed coating made demonstrated a 2.5 to 4× increase in gas abrasive wear resistance of 12CrMoV steel at temperatures in the range of 300–600 °C. At elevated temperatures, a transformation of tensile stresses, which are typical for arc-sprayed coatings, into compressive stresses (10 to 15 MPa) took place. This can be explained by the inner oxidation of the aluminum-rich coating. Additionally, the cohesion improves twofold due to reinforcement of coatings with interlamellar 100–150 nm thick oxide films. The formation of $(Fe,Al,Cr)_2O_3$ oxide film on the surface of

the coating was observed as well. Subsequent controlled interlamellar oxidation (oxidation rate 12–20 $g/m^2 \cdot h$ in the temperature range of 500–600 °C during 10–20 h) provided the best coating properties. Authors established that the presence on the surface of the coating of hematite oxide film alloyed with aluminum as well as the compressive stresses in them assures an improvement of the resistance of the investigated wire arc sprayed coating against hot gas abrasive wear.

Li et al. (2015) designed a new FeCrSiB-cored wire to produce protective coatings for the components used in high-temperature environment. They investigated microstructure, phase composition, microhardness, and high-temperature corrosion/erosion behavior of the new coating in comparison with FeCr coating and commercial NiCrTi coating. Results showed that the FeCrSiB coating was composed of dense lamella with much fewer oxide inclusions than the FeCr and NiCrTi coatings. The microhardness of the FeCrSiB coating was the highest of the three coatings. They studied the high-temperature corrosion behavior of these coatings in mixed salt Na_2SO_4–25% K_2SO_4 environment at 650 °C under cyclic condition. The high-temperature corrosion resistance of the FeCrSiB coating was significantly better than that of the FeCr coating and approaching to the commercial NiCrTi coating. Chen et al. (2012) synthesized by arc spraying FeBSiNb amorphous coatings to improve elevated temperature erosion resistance for boiler applications. The influence of test temperature, velocity, and impact angle on material wastage was revealed using air solid particle erosion rig. The experimental results showed that moderate degradation of the coating was predominant at lower impact velocity and impact angles, while severe damage arose for higher velocities and impact angles. The erosion behavior of the coating was sensitive to test temperature. The erosion rates of the coating decreased as a function of environment temperature: the FeBSiNb coating had excellent elevated-temperature erosion resistance at temperatures at least up to 600 °C during service.

11.5.3.3 Thermal Shock Resistance

Arc spraying with the cored wires was applied by Luo et al. (2010a) to deposit FeMnCrAl/Cr_3C_2 coatings on low carbon steel substrates, namely FM1, FM2, and FM3. Details of the composition of each of core material are given in Table 11.6. The study focused on the influence of Cr3C2 content on thermal shock resistance obtained.

The XRD analysis for FeMnCr/Cr_3C_2 coating revealed that the main phase was Fe-based solid solution and a spot of oxide and Cr_3C_2. The FM3 coating has darker agglomerate phase than FM1 and FM2 coatings. With increase of Cr_3C_2 content, hardness of the three coatings increased, bonding strength decreased slightly, and porosity rose. FM2 coating exhibited the best thermal shock resistance compared to that of FM1 and FM2.

In a follow-up publication, Luo et al. (2010b) extended their study to the determination of the effect of Al content on the high temperature erosion properties of arc-sprayed FeMnCrAl/Cr_3C_2. coatings. Three types of core materials were used with Al content of 0 wt.%, 8wt.%, and 15 wt. % as given in Table 11.7.

Low-carbon steel was used as a substrate with chemical composition (wt.%): 0.16–0.24 C, 0.35–0.65 Mn, 0.15–0.30 Si, $\leq$ 0.040 S, $\leq$0.035 P, and a balanced Fe. The FeMnCrAl/Cr_3C_2 coating consisted of slatted layers of mainly Fe solid solution phases mixed with oxide phases, un-melted

Table 11.6 Compositions of the core materials (wt. %) (Luo et al. 2010)

Samples	Mn	Cr	Al	FeSi	Cr_2C_3	Ni
FM1	30	10	10	2	20	Balance
FM2	30	10	10	2	30	Balance
FM3	30	10	10	2	45	Balance

Table 11.7 Compositions of the core material (wt %) (Luo et al. 2010)

Samples	Mn	Cr	Al	Cr3C2	FeB	FeSi
FeMnCrAl/Cr_3C_2	40	14	15	25	3	3

particles, and pores. Compared with the FeMnCr/Cr_3C_2 coating, the FeMnCrAl/Cr_3C_2–Ni_9Al coating exhibited higher bonding strength and better thermal shock resistance. The cracks in the coatings were mainly initiated and propagated along the oxide phases during the thermal shock test. The uniformly shaped pores in the coatings helped prevent crack initiation and propagation. Ni and Fe appeared in the inter-diffusion process, forming the diffusion layer that improved the thermal shock resistance of the coatings.

Zhang et al. (2015) studied the high-temperature oxidation behavior of FeCrNiNbBSiW and FeCrNiNbBSiMo coatings on a mild steel substrate in an open air atmosphere. The results indicated that the FeCrNiNbBSiW coating had better high-temperature oxidation resistance than did the FeCrNiNbBSiMo coating. This result is primarily due to its compact, lower porosity microstructure, combined with fattened and less splashed surface than FeCrNiNbBSiMo coating. The oxidation products of coatings depended on the oxidation temperature. When the temperature increased from 550 °C to 650 °C, the oxidation product of the coatings transformed from FeO·$(Fe,Cr)_2O_3$ and $(Fe,Cr)_2O_3$ to $(Fe, Cr)_2O_3$ and Cr_2O_3. During the oxidation, more chromium ions than iron ions diffuse to the surface layer of the coating and preferentially react with oxygen.

11.5.4 Rebuilding Worn Surfaces and Near Net-Shaped Parts

Another important application of wire arc spraying is rebuilding of worn surfaces using NiAl, NiCrAl, AlMg, or Mo, from jet engine parts to rolls, shafts, plungers, and crankshafts of major construction equipment (Steffens et al. 1990; Sampson 1993; Chang et al. 2011; Zajchowski and Crapo 1999). For these applications, low porosity and low levels of oxidation are desirable, as well as smooth surface finish. A comparison between Ni-Cr-Al coatings on steel substrate using convention atmospheric plasma spraying (APS) (Fig. 11.71a) and two different wire arc spraying

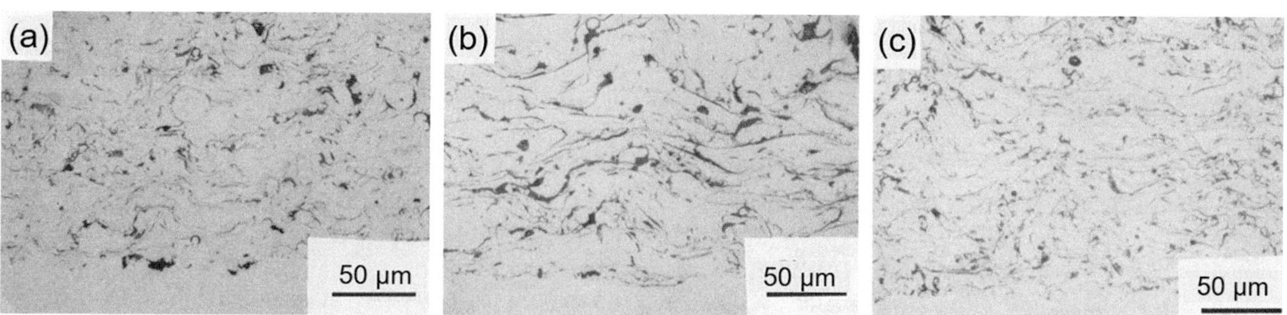

Fig. 11.71 Microstructures of Ni-Cr-Al coatings sprayed by three different processes. (**a**) Plasma spray process. (**b**) Standard wire arc configuration, (**c**) fine spray jet configuration. (Zajchowski and Crapo 1999)

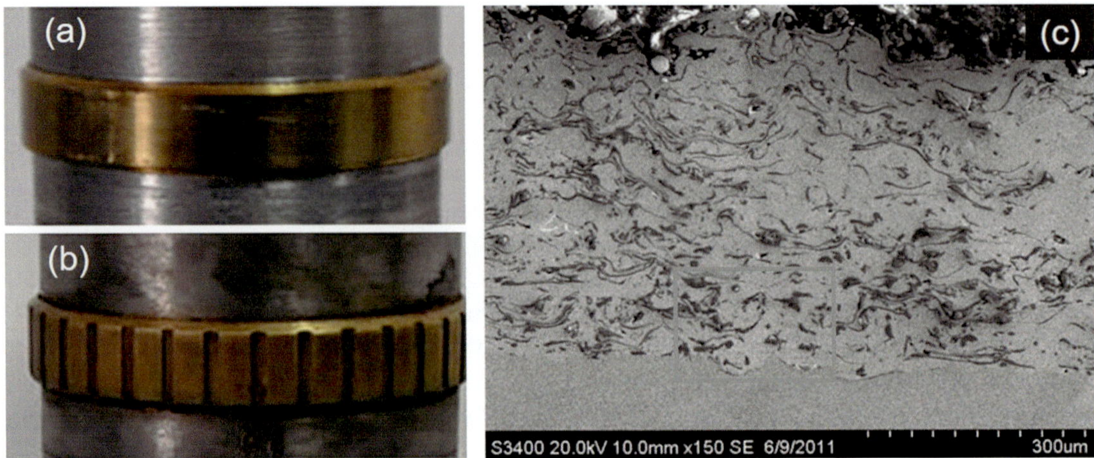

Fig. 11.72 Photograph of wire arc sprayed aluminum bronze rotating band (**a**) before engraving, (**b**) after engraving, (**c**) SEM micrographs of cross section. (Wu et al. 2014)

(WAS) torch configurations (Fig. 11.71b,c) was reported by (Zajchowski and Crapo 1999). The microstructure of each of these three coating displayed comparable levels of oxides and porosity. The microstructure of both wire arc coatings given in Fig. 11.71b,c shows typical lamellar structure with interlamellar oxides and voids. The wire arc coatings of Fig. 11.71c had a finer-sized microstructure with thinner oxide stringers than the nickel–chromium–aluminum dual-wire arc coating sprayed by the standard gun Fig. 11.71b. The latter had an average hardness and surface roughness that were slightly higher than those of the coating shown in Fig. 11.71c.

The study by Wu et al. (2014) for the manufacture of aluminum bronze rotating bands for high-load ballistic applications, using the high-velocity arc spraying (HVAS), is another example of the use of wire arc spray technology for the manufacture of near net-shaped parts. Photographs of a band before and after engraving are given in Fig. 11.72 a,b with SEM micrograph of the cross section of the coating given in Fig. 11.72c. These show that the coatings present a relatively dense structure with low porosity (1.6%) and good adhesion to the substrate. Its microhardness is better than that of pure copper coatings commonly used. The as-sprayed coatings present friction coefficient of 0.2 to 0.3 in steady state under dry sliding friction test conditions.

11.5.5 Metallic Membranes and Rapid Tooling

Madaeni et al. (2008) introduced a novel method for the preparation of stainless-steel porous metallic membranes for filtration applications using wire arc spraying. The distance between gun and substrate surface was selected as the variable of metal spraying. The effects of gun distance on coating properties and membrane performance were investigated with noticeable differences in various separation processes. The metallographic and performance data showed that the range of 350–400 mm is the optimum gun distance for spraying. Increase of gun distance leads to the increase of oxide content and membrane porosity. However, porosity was reduced at gun distances over 400–405 mm while the oxide content was stabilized at this range of spraying distance. Moreover, the filtration capability of the prepared membranes for blue indigo dye particles was investigated. The results indicate that the prepared stainless-steel membrane is able to efficiently remove particles from water. The low cost and rapid fabrication are probably the major advantages of wire arc spraying technique for the preparation of metallic membranes.

An application of increasing importance is that of spray forming of molds for rapid tooling (Fang et al. 2005; Grant et al. 2006). Typically, thick steel coatings are deposited on substrates like alumina. For obtaining a 10 mm thick coating of a 300 × 300 mm shell, multiple torches are used, and the deposition path of the torch movement has been devised such that internal stresses are minimized, i.e., thermally generated stresses are compensated by phase transition stresses. The structural performance of arc spraying rapid tooling (ASRT) directly affects its application in prototype automobile manufacturing (Wu et al. 2014). The conventional fabrication process of ASRT is shown in Fig. 11.73 a–d. The master pattern has the inverse shape of the product. Metal coating is sprayed on the master pattern to form the required shell. After framing and backing up, the shell is de-molded from the substrate.

Wu et al. (2014) sprayed low melting point materials such as ZnAl15 alloy, which has the same spraying performance

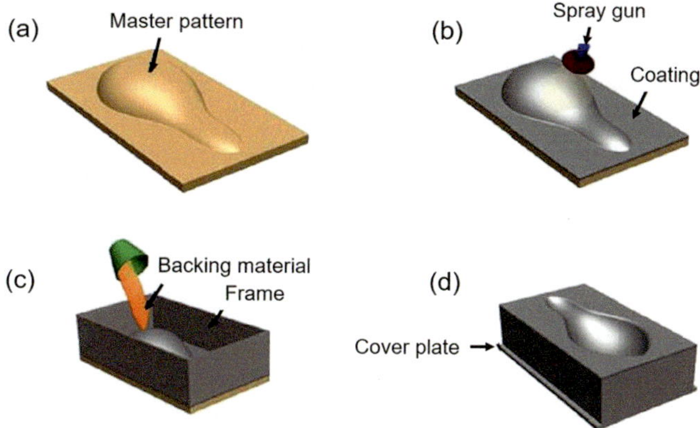

Fig. 11.73 The schematic drawing of ASRT process flow. (**a**) Master pattern preparation. (**b**) Arc spraying. (**c**) Framing and backing up. (**d**) Final tooling. (Wu et al. 2014)

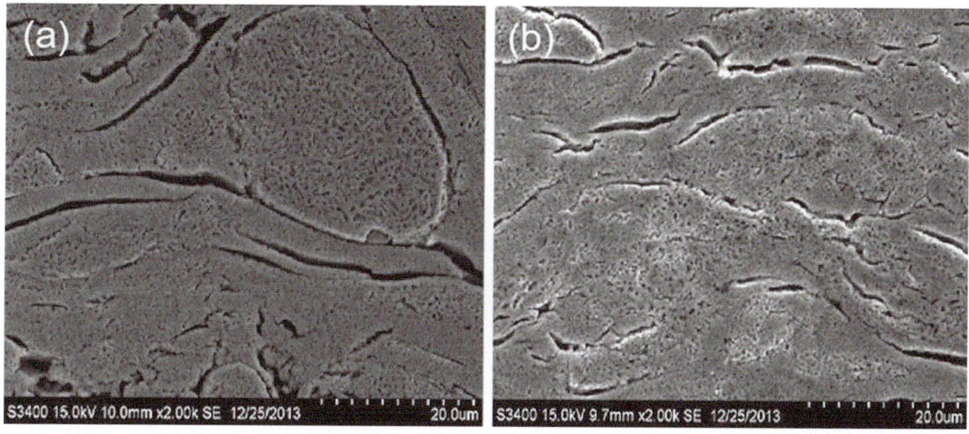

Fig. 11.74 Micrographs of coating perpendicular to the lamellar direction. (**a**) Before compression. (**b**) After compression. (Wu et al. 2014)

as zinc, with mechanical properties are close to those of aluminum. The microstructure of arc sprayed coating was composed of lamellar structures with the presence of some micro-defects such as pores and microcracks as can be observed in Fig 11.74a. The size and direction of microcracks and pores were reduced, Fig. 11.74b, and mechanical properties improved after compression.

11.5.6 Electrical and Electronic Industries

Wire arc spraying applications in the electrical and electronic industries include spraying of resistors, capacitors, conducting connections on insulators, magnetic, or radiofrequency shielding. Materials used include NiCr, Al, Cu, and Zn.

While thermal spray coatings cannot compete with PVD techniques to manufacture electronic devices, wire arc sprayed coatings (Osbond et al. 1992) are used as *electromagnetic shielding materials* to protect electronic circuits from radio frequency interference and dissipate static charges. The main advantage of WAS is the ability to relatively thin conductive coatings on heat sensitive substrates with a minimal heating of the substrate material. Zn and Al are commonly used to protect computers, electronic office equipment's, and medical monitoring devices, housed in heat sensitive plastics. The use of nitrogen for wire atomizing reduces possible oxide inclusions in the coating and helps keep the electrical conductivity of the coating as low as possible. For military applications, the whole room can be shielded using wire arc spraying of zinc or aluminum film.

Gonzalez et al. (2016) made a literature review of the thermal spray deposition of metals onto polymer-based structures. Attention has been given to limit the heat flux to the polymer substrate surface preparation of the substrate to insure good adhesion of the coating to the surface of the polymer.

Wilden et al. (2007a, b) have studied the influence of spray parameters, including the plume width, on wire arc coatings of martensitic stainless steel X46Cr13. They showed that for spraying this alloy, it is important to reduce the burn-off of elements. A low particle temperature process was important to maintain the alloy composition. By using low-voltage level at high atomizing gas pressure and high current, the particle temperature and surface roughness were reduced.

(Laik et al. 2005) wire arc sprayed Ni on Al_2O_3 substrate. The metal–ceramic interface was well bonded with no pores, flaws, or cracks in the as-sprayed condition. An annealing treatment at 1273 K for 24 h of the plasma-coated samples did not result in formation of any intermetallic compound or spinel at the Ni/Al_2O_3 interface.

11.6   Immerging Wire Arc Spraying Technologies

11.6.1   Antibacterial Coating

Infection of medical devices and health-care facilities can result in significant morbidity and mortality. Correspondingly antibacterial surfaces such as silver and copper coatings have been developed. Sharifahmadian et al. (2013a, b) used wire arc spraying technique to produce an ultra-fine microstructure copper coating on stainless steel 316 L substrates. The coating microstructure given in Fig. 11.75 showed stacked splats separated by thin oxide layers. Full validation of the technology and its biomedical advantages is far from being presently well established. A considerable R&D effort is needed to advance further this application.

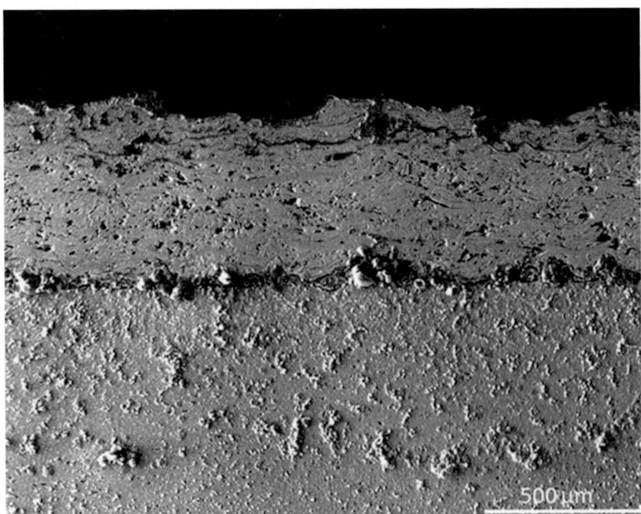

Fig. 11.75 SEM micrographs from cross section of wire arc sprayed copper coating. (Sharifahmadian et al. 2013b)

11.6.2   Low-Pressure Wire Arc Spraying

Low-pressure wire arc spraying (LP-WAS) is possible by placing the entire wire arc spraying system in a low pressure deposition chamber operated at 52.6 kPa (400 Torr) (Halter et al. 2003). The principal objective is the generation of high purity deposits, for example, of NiTi foils for shape memory applications. Wire arc spraying offers the potential of lower contamination of the feedstock because the wire has less surface area than the powders used for plasma spraying. Using argon as atomizing gas and operating the torch with 9.5 kW, 0.2 to 0.3 mm thick foils have been generated with about 5% porosity.

11.6.3   Nanostructured Coatings

According to Cheng et al. (2017), Fe-based amorphous and nanocrystalline coatings can potentially have extraordinary properties including high hardness, outstanding corrosion, and wear resistance. During conventional thermal spraying, grain growth is restricted because of rapid solidification rate, namely, 10^5–10^8 K/s (Davis 2004), thus permitting the formation of an amorphous structure. To achieve such coatings using WAS two similar and promising Fe-based powders compositions were used with compositions of $Fe_{48}Mo_{14}Cr_{15}Y_2C_{15}B_6$ and $Fe_{49.7}Cr_{17.7}Mn_{1.9}Mo_{7.4}W_{1.6}B_{15.2}C_{3.8}Si_{2.4}$, respectively. By using arc spraying process, Fe-based glassy coating in Fe-B-Si-Nb system could be formed on metal substrate. In their study, Cheng et al. (2017) developed a new type of FePSiBNb cord with the filler alloy powders made of ferro-boron (18 wt.% B), ferro-silicon (75 wt.% Si), ferro-phosphorous (25 wt.% P), and ferro-niobium (65 wt.% Nb), respectively. The range of the powders size distribution was 150–200 µm. At first, the reactant powders were mixed by using ball milling as the core. After ball milling, the powders size was about 17–70 µm. Figure 11.76 displayed the cross-sectional backscattered electron (BSE) images of the coating which exhibited a lamellar structure owing to the spraying process. The coating had an average thickness of 450 µm and porosity of less than 3%. In Fig. 11.76b, the as-sprayed coating revealed under high magnification a dense structure with well-flattened splats and partially un-melted particles. The crystalline phases were identified as bcc-Fe phase, without any compound phases. The TEM image showed a uniform nanoscale structure with grain size ranging from 12 to 50 nm. The average chemical composition of the nanostructured region with a few residual amorphous phases were $B_{3.45}$ $Si_{1.76}$ $P_{1.58}$ $Nb_{3.79}$ $Fe_{89.42}$ (wt.%). It was therefore concluded that the FePSiBNb coating exhibited nanostructural features.

The tribological properties of nanostructured FePSiBNb coating prepared by arc spraying sliding against WC ball at different loads and speeds were investigated. The coating consisted of nanoscale Fe structure with grain size ranging

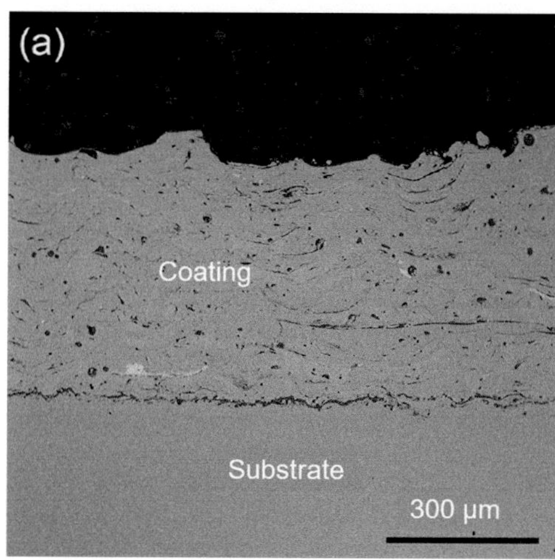

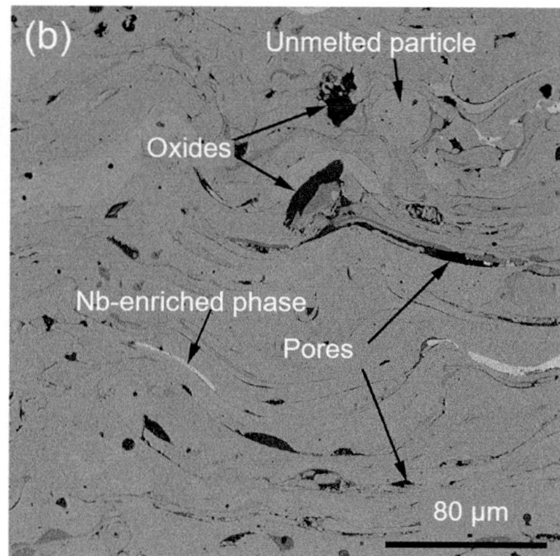

Fig. 11.76 Backscattered electron (BSE) images of the coating (**a**) low magnification, (**b**) high magnification (Cheng et al. 2017)

from 12 to 50 nm. The hardness and reduced elastic modulus of the coating were 12.3 and 204 GPa, respectively. The average porosity of the coating was less than 3%. The friction coefficient of the coating decreased gradually with increasing normal load. The wear rate of the coating was increasing linearly as a function of the normal load, whereas it showed an inverse trend with increasing sliding speed. The excellent wear resistance of the coating was attributed to its good mechanical properties and the formation of local tribo-oxide films on the worn surface. The dominating wear mechanism of the coating was a combination of oxidative wear coupled with delamination under dry sliding conditions. Shukla et al. (2013) by the high-velocity arc spraying process deposited on a 301S stainless steel substrate FeCr-based nanostructured coatings. The oxidation behavior of the coatings exposed to elevated temperatures (700 °C and 900 °C) under laboratory conditions as well as in an actual industrial environment of a coal-fired boiler (at 700 ± 10 °C) was investigated. The microhardness of the coating was found to be 520–1100 HV. The (FeCr)-based nanostructured coating showed good adherence to the 310S substrate and excellent oxidation resistance during the exposures with no tendency for spallation of its oxide scales in both environments. The nanometer-sized grain morphology of the coating facilitated the formation of protective scales, which is continuous, adherent, and nonporous due to the higher diffusivity of alloying elements in the coatings. It precludes high-temperature oxidation by acting as a diffusion barrier between the environment and the coating.

11.7 Summary and Conclusions

Compared to other thermal spray processes, wire arc spraying presents the following principal advantages:

- One of the simplest and most economical processes.
- Highest deposition rates.
- Thick coatings of up to a few millimeters can be easily obtained.
- Substrate heating is low and usually substrate cooling is not required,
- Compact and versatile systems are easy to use for on-site coating application such as bridges.

These advantages are countered by the following limitations:

The material to be sprayed must be ductile and electrically conducting easily obtainable in the form of wires. Cored wires containing brittle material such as refractories or ceramics in a ductile envelope offers a way around this limitation at a high wire cost. Coatings are rather porous, and their oxide content is higher than with standard thermal spray processes.

Wire arc spray coatings have found widespread use for corrosion protection and wear reduction in large-scale manufacturing or infrastructure maintenance applications where the advantages outweigh the shortcomings. Continuous improvement in process control and coating quality is leading to a penetration of a market where the coating represents a high added value to the component. A further diversification of process details and products must be expected.

Nomenclature

Units are indicated in parentheses; when no units are indicated, the parameter is dimensionless.

Latin Alphabet

d	Droplet/particle diameter (m)
d_ℓ	Liquid jet diameter (m)
I	Arc current (A)
L_ℓ	Characteristic dimension for the liquid (m)
L_s	Characteristic length (m)
$\dot{Q}$	Gas flow rate (m³/s)
$R_{0.5}$	Radius at which the droplet flux reached 50% of its maximum centerline value (m)
Re	Reynolds number ($Re = \rho_\ell u_\ell d_\ell / \mu_\ell$)
T	Droplet temperature (K)
u_ℓ	Liquid velocity (m/s)
u_r	Relative velocity of the liquid to gas stream (m/s)
v	Droplet velocity (m/s)
v_s	Substrate traverse velocity (m/s) or (m/min)
x, y, z	Lateral, perpendicular, and axial coordinates (m)
We	Weber number $\left(We = u_r^2 L_\ell \, \rho_g / \sigma_\ell \right)$

Greek Alphabet

ρ_g	Density of gas (kg/m³)
κ	Thermal conductivity (W/mK)
ρ_ℓ	Liquid density (kg/m³)
σ_ℓ	Surface tension (N/m) (kg/s²)
μ_ℓ	Dynamic viscosity of the liquid (kg/m.s)

References

Abedini, A., A. Pourmousa, S. Chandra, and J. Mostaghimi. 2006. Effect of substrate temperature on the properties of coatings and splats deposited by wire arc spraying. *Surface & Coatings Technology* 201 (2006): 3350–3358.

Benary, R. 2000. Thermal spray gun extension and gas jet member, therefore. US Patent US Patent No. 6,091,042.

Bobzin, K., F. Ernst, J. Zwick, T. Schlaefer, D. Cook, K. Nassenstein, A. Schwenk, F. Schreiber, T. Wenz, G. Flores, and M. Hahn. 2008. Coating bores of light metal engine blocks with a nanocomposite material using the plasma transferred wire arc thermal spray process. *Journal of Thermal Spray Technology* 17 (3): 344–351.

Bolot, R., R. Bonnet, G. Jandin, and C. Coddet. 2001. Application of CAD to CFD for the wire arc spray process. In *Proceedings of ITSC-2001 Singapore*, ed. C.C. Berndt, K.A. Khor, and E. Lugscheider, 889–894. Materials Park, OH: ASM International.

Bolot, R., H. Liao, C. Mateus, C. Coddet, and J.-M. Bordes. 2007. Optimization of a rotating twin wire-arc spray gun. In *Proceedings of ITSC-2007, Beijing, People's Republic of China*, ed. B. Marple, M.M. Hyland, Y.-C. Lau, C.-J. Li, R.S. Lima, and G. Montavon, vol. 5–6, 783–790. Materials Park, OH: ASM International.

Bolot, R., M.-P. Planche, H. Liao, and C. Coddet. 2008. A three-dimensional model of the wire-arc spray process and its experimental validation. *Journal of Materials Processing Technology*: 94–105.

Branagan, D.J., W.D. Swank, D.C. Haggard, and J.R. Fincke. 2001. Wear resistant amorphous and nanocomposite steel coatings. *Metallurgical and Materials Transactions A: Physical Metallurgy and Materials Science* 32A (10): 2615–2621.

Branagan, D.J., M. Breitsameter, B.E. Meacham, and V. Belashchenko. 2004. High-performance nanoscale composite coatings for boiler applications. *Journal of Thermal Spray Technology* 14 (2): 196–204.

Carlson R. R. 2005. *Single wire arc spray*. PhD Thesis, University of Minnesota.

Carlson, R., and J.V.R. Heberlein. 2001. Effects of operating parameters on high definition single wire arc spraying. In *Proceedings of ITSC-2001, Singapore*, ed. C.C. Berndt, K.A. Khor, and E. Lugscheider, 447–453. Materials Park, OH: ASM International.

Carlson R. and JVR Heberlein 2002. *Single-wire arc spray apparatus and methods for using same*. US Patent US Patent No. 6,610,959 B2,

Carlson, R., J.V.R. Heberlein, N.A. Hussary, and K. Shi. 2000. High-definition single-wire-arc-spray. In *Proceedings of 1ˢᵗ. ITSC-2000, Montreal, Quebec*, ed. C.C. Berndt, 709–716. Materials Park, OH: ASM International.

Chang, C.H., M.C. Jeng, C.Y. Su, and T.S. Huang. 2011. A study of Wear and corrosion resistance of arc-sprayed Ni-Ti composite coatings. *Journal of Thermal Spray Technology* 20 (6): 1278–1285.

Chen, Y., X. Liang, S. Wei, Y. Liu, and B. Xu. 2009a. Heat treatment induced intermetallic phase transition of arc-sprayed coating prepared by the wires combination of aluminum-cathode and steel-anode. *Applied Surface Science* 255: 8299–8304.

Chen, Y., X. Liang, Y. Liu, and B. Xu. 2009b. Numerical analysis of the effect of arc spray gun configuration parameters on the external gas flow. *Journal of Materials Processing Technology* 209 (2009): 5924–5931.

Chen, Y., X. Liang, S. Wei, X. Chen, and B. Xu. 2012. Numerical simulation of the twin-wire arc spraying process: Modeling the high velocity gas flow field distribution and droplets transport. *Journal of Thermal Spray Technology* 21 (2): 263–274.

Cheng, J., Q. Liu, B. Sun, X. Liang, and B. Zhang. 2017. Structural and tribological characteristics of nanoscale FePSiBNb coatings. *Journal of Thermal Spray Technology* 26: 530–538.

Chigier, N. 1981. *Energy, combustion and environment*. New York: McGraw Hill.

Chun-long, Y., A. Yun-qi, and S. Ya-tan. 2009. Three years corrosion tests of nanocomposite epoxy sealer for metalized coatings on the East China Sea, in thermal spray 2009. In *Proceedings of the international thermal spray conference*, ed. B.R. Marple, M.M. Hyland, Y.-C. Lau, C.-J. Li, R.S. Lima, and G. Montavon, 1090–1093. Materials Park: ASM International.

Cook, D., M. Zaluzec, and K.A. Kowalsky. 2003. Development of thermal spray for automotive cylinder bores. In *Proceedings of ITSC-2003, Orlando, Florida*, ed. B. Marple and C. Moreau, 143–147. Materials Park, OH: ASM International.

Cooke, K., G. Oliver, V. Buchanan, and N. Palmer. 2007. Optimization of the electric wire arc-spraying process for improved wear resistance of sugar mill roller shells. *Surface & Coating Technology* 202: 185–188.

Cramer, S.D., B.S. Covino Jr., G.R. Holcomb, S.J. Bullard, W.K. Collins, R.D. Govier, R.D. Wilson, and H.M. Laylor. 1999. Thermal sprayed titanium anode for cathodic protection of reinforced concrete bridges. *Journal of Thermal Spray Technology* 8 (1): 133–145.

Davis, J.R. (ed.). 2004. *Handbook of thermal spray technology* (Pub.) ASM international, Materials Park, OH 44073-0002, USA.

Darut, G., H. Liao, C. Coddet, J.M. Bordes, and M. Diaby. 2015. Steel coating application for engine block bores by plasma transferred wire arc spraying process. *Surface & Coatings Technology* 268: 115–122.

Dubourg, L., R.S. Lima, and C. Moreau. 2002. Properties of alumina titania coatings prepared by laser-assisted air plasma spraying. *Surface & Coatings Technology* 201 (2007): 6278–6284.

Dunkerley J. P., T. A. Friedrich and G. Irons 1999. *Apparatus for rotary spraying a metallic coating*. US Patent US Patent No. 5,908,670,

Esfahani, E.A., H. Salimijazi, M.A. Golozar, J. Mostaghimi, and L. Pershin. 2012. Study of corrosion behavior of arc sprayed aluminum coating on mild steel. *Journal of Thermal Spray Technology* 21 (6): 1195–1202.

Fang, J.C., W.J. Xu, and Z.Y. Zhao. 2005. Arc spray forming. *Journal of Materials Processing Technology* 164–165: 1032–1037.

Fauchais, P., J.V.R. Heberlein, and M. Boulos. 2014. *Thermal spray fundamentals*, 1566. New York: Springer.

Gedzevicius, I., R. Bolot, H. Liao, and C. Coddet. 2003. Application of CFD for Wire-Arc Nozzle Geometry Improvement. In *Proceedings of ITSC-2003, Orlando, Florida*, ed. B. Marple and C. Moreau, 977–980. Materials Park, OH: ASM International.

Gonzalez, R., H. Ashrafizadeh, A. Lopera, P. Mertiny, and A. McDonald. 2016. A review of thermal spray metallization of polymer-based structures. *Journal of Thermal Spray Technology* 25 (5): 897–919.

Grant, P.S., S.R. Duncan, A. Roche, and C.F. Johnson. 2006. Scientific, technological, and economic aspects of rapid tooling by electric arc spray forming. In *Proceedings of the ITSC-2006, Seattle, Washington*, ed. B. Marple, M.M. Hyland, Y.-C. Lau, R.S. Lima, and J. Voyer, vol. 4, 796–801. Materials Park, OH: ASM International.

Hahn, M., and A. Fischer. 2010. Characterization of thermal spray coatings for cylinder running surfaces of diesel engines. *Journal of Thermal Spray Technology* 19 (5): 866–887.

Han, M.-S., Y.-B. Woo, S.-C. Ko, Y.-J. Jeong, S.-K. Jang, and S.-J. Kim. 2009. Effects of thickness of Al thermal spray coating for STS 304. *Transactions of Nonferrous Metals Society of China* 19: 925–929.

Halter, K., A. Sickinger, L. Zysset, and S. Siegmann. 2003. Low pressure wire arc and vacuum plasma spraying of NiTi shape memory alloys. In *Proceedings of ITSC-2003, Orlando, Florida*, ed. B. Marple and C. Moreau, 289–295. Materials Park, OH: ASM International.

He, D., N. Dong, and J. Jiang. 2007. Corrosion behavior of arc sprayed nickel-based b coatings. *Journal of Thermal Spray Technology* 16 (5–6): 850–586.

Holcomb, G.R., S.D. Cramer, S.J. Bullard, B.S. Covino Jr., W.K. Collins, R.D. Govier, and G.E. McGill. 1997. Characterization of thermal-sprayed titanium anodes for cathodic protection. In *Thermal spray: A united forum for scientific and technological advances*, ed. C.C. Berndt, 141–150. Materials Park, OH: ASM International.

Hsiang, L.P., and G.M. Faeth. 1992. Near-limit drop deformation and secondary breakup. *International Journal of Multiphase Flow* 18: 635–652.

Hussary N.A. 1999. *Fluid dynamic investigations of wire arc spraying process*. MS Thesis, University of Minnesota, Minneapolis.

——— N.A. and JVR Heberlein. 2001. Atomization and particle jet interactions in the wire-arc spaying process. *Journal of Thermal Spray Technolgoy* 10(4) 604–610.

——— N.A. JVR Heberlein. 2007. Effect of system parameters on metal breakup and particle formation in the wire arc spray process, *Journal of Thermal Spray Technology* 16(1) 140–152.

——— N.A., J. Schein and JVR Heberlein 1999. Control of jet convergence in wire arc spray systems, *Proceedings of UTSC-1999, Düsseldorf, Germany* (E. Lugscheider and P. A. Kammer eds). ASM International, Materials Park, OH, pp 335–339.

Jandin, G., M.-P. Planche, H. Liao, and C. Coddet. 2002. Relationships between in-flight particle characteristics and coating microstructure for the twin wire arc spray process. In *Proceedings of ITSC-2002, Essen*, ed. E. Lugscheider, 954–959. Materials Park, OH: ASM International.

Kawase, R., and M. Kureishi. 1985a. Fused metal temperature in arc spraying. *Transactions Japan Welding Society* 16 (1): 82–88.

———. 1985b. Relation between adhesion strength of arc sprayed coating and fused metal temperature. *Transactions Japan Welding Society* 16 (2): 69–73.

Kawase, R., M. Kureishi, and S. Minehisa. 1984a. Relation between arc spraying condition and adhesive strength of sprayed coating. *Transactions Japan Welding Society* 15 (2): 27–33.

Kawase, R., M. Kureishi, and K. Maehara. 1984b. Arc phenomena and wire fusion in arc spraying. *Transactions Japan Welding Society* 15 (2): 34–39.

Kelkar, M., and J.V.R. Heberlein. 2000. Physics of an arc in cross flow. *Journal of Physics D: Applied Physics* 33: 2172–2182.

———. 2002. Wire-arc spray modeling. *Plasma Chemistry and Plasma Processing* 22 (1): 1–25.

Kelkar, M., N.A. Hussary, J. Schein, and J. Heberlein. 1968. Optical diagnostics and modeling of gas and droplet flow in wire arc spraying. In *Proceedings of 15th. ITSC-1998, Nice, France*, ed. C. Coddet, 329–324. Materials Park, OH: ASM International.

Kharlamov, M.Yu., I.V. Krivtsun, V.N. Korzhyk, Y.V. Ryabovolyk, and O.I. Demyanov. 2015. Simulation of motion, heating, and breakup of molten metal droplets in the plasma jet at plasma-arc spraying. *Journal of Thermal Spray Technology* 24 (4): 659–670.

König, J., M. Lahres, and O. Methner. 2015. Quality designed twin wire arc spraying of aluminum bores. *Journal of Thermal Spray Technology* 24 (1–2): 63–74.

Kowalsky, K.A., D.R. Marantz, and H. Herman. 1991. Characterization of coatings produced by the wire-arc-plasma spray process. In *Proceedings of 4th NTSC-1991, Pittsburgh, Pennsylvania*, ed. T.F. Bernecki, 389–394. Materials Park, OH: ASM International.

Kowalsky, K.A., R. Neiser, and M.F. Smith. 1992. Diagnostic Behavior of the Wire-Arc-Plasma Spray Process. In *Proceedings of Thermal Spray: NTSC-1992 Orlando, Florida*, ed. C.C. Berndt, 337–342. Materials Park, OH: ASM International.

Laik, A., D.P. Chakravarthy, and G.B. Kale. 2005. On characterisation of wire-arc-plasma-sprayed Ni on alumina substrate. *Materials Characterization* 55: 118–126.

Lefebvre, A.H. 1989a. *Atomization and sprays*, 421 pages. New York: Hemisphere Publishing Corporation.

Lester T. 2005. *Metallized coatings application, An overview of the flame-and arc-spraying processes used for surface finishing.* Organic Finishing July/August:35–38.

Li, R., Z. Zhou, D. He, Y. Wang, X. Wu, and X. Song. 2015. Microstructure and High Temperature Corrosion Behavior of Wire-Arc Sprayed FeCrSiB Coating. *J. of Thermal Spray Technology* 24 (5): 857–864.

Liao, H., Y.L. Zhu, R. Bolot, C. Coddet, and S.N. Ma. 2005. Size distribution of particles from individual wires and the effects of nozzle geometry in twin wire arc spraying. *Surface and Coatings Technology* 200: 2123–2130.

Liu, X. 2001. Arc spraying in China. *Journal of Thermal Spray Technology* 10 (1): 40–43.

Luo, L., J. Yu, S. Liu, and J. Li. 2010a, April. Thermal shock resistances of FeMnCr/Cr$_3$C$_2$ coatings deposited by arc spraying. *Journal of Wuhan University of Technology-Material Science Edition*.

Luo, L.M., S.G. Liu, J. Yu, J. Luo, and J. Li. 2010b. Effect of Al content on high temperature erosion properties of arc-sprayed FeMnCrAl/Cr$_3$C$_2$ coatings. *Transactions of Nonferrous Metals Society of China* 20: 201–206.

Madaeni, S.S., M.E. Aalami-Aleagha, and P. Daraei. 2008. Preparation and characterization of metallic membrane using wire arc spraying. *Journal of Membrane Science* 320: 541–548.

Mansour, A., and N. Chigier. 1990. Disintegration of liquid sheets. *Physics of Fluids A* 2 (5): 706–719.

Marantz, D., and D.R. Marantz. 1990. State of the art arc spray technology. In *Proceedings of thermal spray research and applications, Proceedings of the 3rd. NTSC-1990, Long Beach, California*, ed. T.F. Bernecki, 113–118. Materials Park, OH: ASM International.

Marantz, D.R., K.A. Kowalsky, and D. Marantz. 1991. Wire-Arc-Plasma Spray Process – Basic Principles and Its Versatility. In *Proceedings of the 4th. NTSC-1991, Pittsburgh, Pennsylvania*, ed. T.F. Bernecki, 381–387. Materials Park, OH: ASM International.

Metallization Ltd, see https://www.metallisation.com/

Matthews, S., and M. Schweizer. 2013. Optimization of arc-sprayed Ni-Cr-Ti coatings for high temperature corrosion applications. *Journal of Thermal Spray Technology* 22 (4): 538–550.

McCune R. C. Jr., L. V. Reatherford and M. Zaluzec 1993. *Thermally spraying metal/solid lubricant composites using wire feedstock.* US Patent No. 5,194,304,

Meng, F.J., B.S. Xu, S. Zhu, S.N. Ma, and W. Zhang. 2005. Oxidation performance of Fe-AI/WC composite coatings produced by high velocity arc spraying. *Journal of Central South University of Technology* 12 (2): 222–225.

Mohanty, P.S., R. Allor, P. Lechowicz, R.A. Parker, and J.E. Craig. 2003. Particle Temperature and velocity characterization in spray tooling process by thermal imaging technique. In *Proceedings of ITSC-2003, Orlando, Florida*, ed. B. Marple and C. Moreau, 1183–1190. Materials Park, OH: ASM International.

Nakagawa, M., K. Shimoda, T. Tomoda, M. Koyama, Y. Ishikawa, and T. Nakajima. 1990. Development of mass production technology of arc spraying for automotive engine aluminum alloy valve lifters. In *Proceedings of the 3rd NTSC*, 457–464.

Osbond, P. 1992. Plasma sprayed anti-reflection coatings for microwave optical components. *Advanced Materials* 4: 807–809.

Pacheo da Silva, C. et al. 1991. *2nd Plasma Technik Symposium,* vol. 1, 363–373 (Pub.) Plasma Technik Wohlen, CH.

Planche, M.P., A. Lakat, H. Liao, and C. Coddet. 2003. Investigations of in-flight particle characteristics through DPV measurements and correlation with impact analysis and coating properties. In *Proceedings of ITSC-2003, Orlando, FL*, ed. B. Marple and C. Moreau, 1175–1182. Materials Park, OH: ASM International.

Planche, M.-P., H. Liao, and C. Coddet. 2004. Relationship between in-flight particle characteristics and coating microstructure with a twin wire arc spray process and different working conditions. *Surface and Coatings Technology* 182 (2–3): 215–226.

Pokhmurskii, V., M. Student, V. Dovhuny, I. Sydorak, and H. Pokhmurska. 2002. Wear resistance arc-sprayed coating from power wires. In *Proceedings of ITSC, Essen, 2002*, ed. E. Lugscheider, 559–562. ASM Thermal Spray Society.

Pourmousa, A., A. Abedini, J. Mostaghimi, and S. Chandra. 2004. Particle Diagnostics in Wire-Arc Spraying System. In *Proceedings of the ITSC-2004, Osaka, Japan*, ed. A. Ohmori, 962–967. Material Park, OH: ASM International.

Pourmousa, A., J. Mostaghimi, A. Abedini, and S. Chandra. 2005. Particle size distribution in a wire-arc spraying system. *Journal of Thermal Spray Technology* 14 (4): 502–510.

Sakoda, N., M. Hida, Y. Takemoto, A. Sakakibara, and T. Tajiri. 2003. Influence of atomization gas on coating properties under Ti arc spraying. *Materials Science and Engineering* A342: 264–269.

Sampson, E.R. 1993. The economics of arc vs. plasma spray for aircraft components. In *Proceedings of 5th. NTSC-1993, Anaheim, CA*, ed. C.C. Berndt and T.F. Bernecki, 257–262. Material Park OH: ASM International.

Sampson, E.R., and M.P. Zwetsloot. 1997. Arc spray process for the aircraft and stationary gas turbine industry. *Journal of Thermal Spray Technology* 6 (2): 150–152.

Schmidt, D.P., B.A. Shaw, E. Sikora, W.W. Shaw, and L.H. Laliberte. 2006. Corrosion protection assessment of sacrificial coating systems as a function of exposure time in a marine environment. *Progress in Organic Coating* 57: 352–364.

Schoop M. U. 1915. *Apparatus for spraying molten metal and other fusible substances.* US Patent No. 1,133,507,

Sharifahmadian, O., H.R. Salimijazi, M.H. Fathi, J. Mostaghimi, and L. Pershin. 2013a. Relationship between surface properties and antibacterial behavior of wire arc spray copper coatings. *Surface & Coatings Technology* 233: 74–79.

———. 2013b. Study of the antibacterial behavior of wire arc sprayed copper coatings. *Journal of Thermal Spray Technology* 22 (2-3): 371–379.

Sheard, J., J.V.R. Heberlein, K. Stelson, and E. Pfender. 1997. Diagnostic development for control of wire-arc spraying. In *Proceedings of 1st UTSC-1997, Indianapolis, Indiana*, ed. C.C. Berndt, 613–618. Materials Park, OH: ASM International.

Shukla, V.N., R. Jayaganthan, and V.K. Tewari. 2013. Degradation behavior of nanostructured coatings deposited by high-velocity arc spraying process in an actual environment of a coal-fired boiler. *JOM* 65 (6): 784–791.

Skoblo, T.S., V.M. Vlasovets, and V.V. Moroz. 2001. Structure and distribution of components in the working layers upon reconditioning of parts by electric-arc metallization. *Metal Science and Heat Treatment* 43 (11–12): 497–500.

Sørensen, P.A., S. Kiil, K. Dam-Johansen, and C.E. Weinell. 2009. Anticorrosive coatings: A review. *Journal of Coatings Technology and Research* 6 (2): 135–176.

Steffens, H.D. 1966. Metallurgical changes in the arc spraying of steel. *British Welding Journal* 13 (10): 597–605.

Steffens, H.D., and K. Nassenstein. 1994. Recent developments in single-wire vacuum arc spraying. *Journal of Thermal Spray Technology* 3 (4): 412–417.

Steffens, H.-D., and K. Nassenstein. 1999. Influence of the spray velocity on arc-sprayed coating structures. *Journal of Thermal Spray Technology* 8 (3): 454–460.

Steffens, H.D., and M. Wewel. 1991. One wire vacuum arc spraying-A new modified process. In *Proceedings of 4th NTSC-1991, Pittsburgh, Pennsylvania*, ed. T.F. Bernecki, 395–398. Materials Park, OH: ASM International.

Steffens, H.D., Z. Babiak, and M. Wewel. 1990. Recent developments in arc spraying. *IEEE Transactions on Plasma Science* 18 (6): 974–979.

Tian, H.L., S.C. Wei, Y.X. Chen, H. Tong, Y. Liu, and B.S. Xu. 2014. Microstructure and wear resistance of an arc-sprayed Fe-based coating after surface remelting treatment. *Strength of Materials* 46 (2): 229–234.

Tillmann, W., and M. Abdulgader. 2012. Particle size distribution of the filling powder in cored wires: Its effect on arc behavior, in-flight particle behavior, and splat formation. *Journal of Thermal Spray Technology* 21 (3–4): 706–718.

Tillmann, W., E. Vogli, and M. Abdulgader. 2008a. Asymmetric melting behavior in twin wire arc spraying with cored wires. *Journal of Thermal Spray Technology* 17 (5–6): 974–982.

Tillmann, W., E. Vogli, M. Abdulgader, M. Gurris, D. Kuzmin, and S. Turek. 2008b. Particle behavior during the arc spraying process with cord wires. *Journal of Thermal Spray Technology* 17 (5–6): 966–973.

Tillmann, W., E. Vogli, and M. Abdulgader. 2010. The correlation between the coating quality and the moving direction of the twin wire arc spraying gun. *Journal of Thermal Spray Technology* 19 (1–2): 409–421.

Tillmann, W., L. Hagen, and P. Schröder. 2017. Investigation on the tribological behavior of arc-sprayed and hammer-peened coatings using tungsten carbide cored wires. *Journal of Thermal Spray Technology* 26: 229–242.

Toma, S.L. 2013. The influence of jet gas temperature on the characteristics of steel coating obtained by wire arc spraying. *Surface & Coatings Technology* 220: 261–265.

Tucker R.C. Jr. (ed.). 2013. *ASM handbook Vol. 5A Thermal Spray technology*, (Pub.) ASM international, Materials Park, OH 44073-0002, USA.

Varis, T., and E. Rajamake. 2002. Effect of the nozzle design and atomization gas on the properties of the electric arc sprayed Ni18Cr6A12Mn-coatings. In *Proceedings of ITSC-2002, Essen*, ed. E. Lugscheider, 550–552. Materials Park, OH: ASM Thermal Spray Society.

Wang, X., J.V.R. Heberlein, E. Pfender, and W. Gerberich. 1995a. Enhancement of coating uniformity with secondary gas atomization in wire arc spray. In *Proceedings of ISPC-12, Minneapolis, Minnesota*, ed. J. Heberlein, D.W. Ernie, and J.T. Roberts, 907–913. Minneapolis: International Union of Pure and Applied Chemistry.

———. 1995b. Effect of shrouded CO2 gas atomization on coating properties in wire arc spray. In *Proceedings of 8th NTSC Conference, Houston, Texas, 1995*, ed. C.C. Berndt and S. Sampath, 31–37. Materials Park, OH: ASM International.

———. 1996. Effect of gas velocity and particle velocity on coating adhesion in wire arc spraying. In *Proceedings of 9th NTSC, Cincinnati, OH, 1996*, ed. C.C. Berndt, 807–811. Materials Park, OH: ASM International.

———. 1999. Effect of nozzle configuration, gas pressure and gas type on coating properties in wire arc spray. *Journal of Thermal Spray Technolgoy* 8 (4): 565–575.

Wang, R., D. Song, W. Liu, and X. He. 2010. Effect of arc spraying power on the microstucture and mechanical properties of Zn-Al coating deposited onto carbon fiber reinforced epoxy composites. *Applied Surface Science* 257: 203–209.

Watanabe, T., X. Wang, J.V.R. Heberlein, and E. Pfender. 1995. Fume generation mechanism in wire arc spraying. In *Proceedings of ISPC-12*, ed. J. Heberlein, D.W. Ernie, and J.T. Roberts, 889–894. Minneapolis: International Union of Pure and Applied Chemistry.

Watanabe, T., X. Wang, J.V.R. Heberlein, E. Pfender, and W. Herwig. 1996. Voltage and current fluctuations in wire arc spraying as indications for coating properties. In *Proceedings of 9th NTSC-1996, Cincinnati, OH*, ed. C.C. Berndt, 577–583. Materials Park, OH: ASM International.

Watanabe, T., T. Sato, and A. Nezu. 2002. Electrode phenomena investigation of wire arc spraying for preparation of Ti-Al intermetallic compounds. *Thin Solid Films* 407: 98–103.

Wielage, B., H. Pokhmurska, M. Student, V. Gvozdeckii, T. Stupnyckyj, and V. Pokhmurskii. 2013. Iron-based coatings arc-sprayed with cored wires for applications at elevated temperatures. *Surface & Coatings Technology* 220: 27–35.

Wilden, J., J.P. Bergmann, S. Jahn, S. Knapp, F. van Rodijnen, and G. Fischer. 2007a. Investigation about the Chrome Steel Wire Arc Spray Process and the Resulting Coating Properties. In *Proceedings of ITSC, Beijing, People's Republic of China, 2007*, ed. B. Marple, M. Hyland, Y.C. Lau, C.J. Li, R.S. Lima, and G. Montavon, 759–767. Materials Park, OH: ASM International.

———. 2007b. Investigation about the chrome steel wire arc spray process and the resulting coating properties. *Journal of Thermal Spray Technology* 16 (5–6): 759–767.

Wu, B., L.-h. Fang, X.-l. Chen, Z.-q. Zou, X.-h. Yu, and G. Chen. 2014. Fabricating aluminum bronze rotating band for large-caliber projectiles by high velocity arc spraying. *Journal of Thermal Spray Technology* 23 (3): 447–455.

Yao, H.H., Z. Zhou, Y.M. Wang, D.Y. He, K. Bobzin, L. Zhao, M. Öte, and T. KÖnigstein. 2017. Microstructure and properties of FeCrB alloy coatings prepared by wire-arc spraying. *Journal of Thermal Spray Technology* 26: 483–491.

Zajchowski, P., and H.B. Crapo. 1999. Evaluation of three dual-wire electric arc-sprayed coatings: Industrial note. *Journal of Thermal Spray Technology* 5 (4): 457–462.

Zhang, Z.L., D.Y. Li, and S.Y. Wang. 2006. High temperature performance of arc-sprayed aluminum bronze coatings for steel. *Transactions of Nonferrous Metals Society* 16: 868–872.

Xin, Zhang, Jinran Lin ZehuaWang, and Zehua Zhou. 2015. A study on high temperature oxidation behavior of high-velocity arc sprayed Fe-based coatings. *Surface & Coatings Technology* 283: 255–261.

Zhao, L., B. Fu, D. He, and P. Kutschmann. 2009. Development of a new wear resistant coating by arc spraying of a steel-based cored wire. *Frontiers of Mechanical Engineering* 4 (1): 1–4.

Abbreviations

CA-TIG	Cap active tungsten inert gas
CFD	Computational fluid dynamics
DC	Direct current
DE-GMAW	Double Electrode-GMAW
FC	Friction coefficient
FWHM	Full width at half maximum
GMAW	Gas Metal Arc Welding
GTAW	Gas tungsten arc welding
HAZ	Heat-affected zone
HE-PTA	High energy plasma transferred arc
HVF	High-volume fraction
HVOF	High-velocity oxy fuel
i.d.	Internal diameter
ISPC	International Symposium on Plasma Chemistry
ITSC	International Thermal Spray Conference
LHS	Left-Hand Side
LRM	Laser rapid manufacturing
MIG	Metal inert gas
MMC	Metal matrix composite
NTSC	National Thermal Spray Conference
PAC	Plasma arc cutting
PAW	Plasma arc welding
PC-TIG	Pulsed current tungsten inert gas
PP-AM	Pulsed plasma additive manufacturing
PTA	Plasma transferred arc
PWHT	Post weld heat treatment
RHS	Right-Hand Side
RT	Room temperature
SAW	Submerged arc welding
slm	Standard liters per minute
TBC	Thermal barrier coating
TIG	Tungsten inert gas
UTSC	United Thermal Spray Conference
YSZ	Yttria stabilized zirconia

12.1 Introduction

The plasma transferred arc (PTA) coating process was developed in the 1960s for the rebuilding and repair of worn parts and the hard facing of parts exposed to extreme abrasion and corrosion. Regular steel parts with an appropriate PTA coating can exhibit superior corrosion and wear-resistant behavior even compared to specialty alloys. The process is significantly different from other coating processes, including plasma spraying, as the surface of substrate is locally molten in the process creating a strong metallurgical bond between the molten coating material and the substrate if they are metallurgical compatibles.

The substrate, in most of the cases, serves as the anode for the transferred arc and only exceptionally as cathode to significantly limit the heat flux to the substrate and clean its surface. The powder to be deposited is introduced into the arc plasma through two or more orifices located on a ring surrounding the exit of the plasma torch nozzle. A further annular slot surrounding the powder injection ring provides the shield gas flow, necessary to avoid reaction of the molten metal with the environmental air. Once the molten metal cools and freezes, it creates a metal matrix composite coating, the mechanical and metallurgical properties of which can differ from those of the substrate or the coating material. The PTA process combines good control over the process with high deposition rates (typically 3–10 kg/h up to 20 kg/h) and good coating quality with practically no porosity and dilution of less than 7% (better than 5% have been reported). Feeding of multiple powders allows for a range of coating compositions to be obtained. Compared to other thermal spray processes, PTA has the major advantage that alloy formation can take place in the metal pool during the process, allowing for a wider range of combinations of metal compounds or metal matrix composite compositions than with any other spray deposition process.

In a variance of PTA coating technology, tungsten inert gas (TIG) welding, which is a free burning, un-constricted arc

M. I. Boulos et al. (ed.), *Thermal Spray Fundamentals*, https://doi.org/10.1007/978-3-030-70672-2_12

welding process, has also been used for coating applications with the filler material introduced in form of a wire. Because of the less stable arc attachment, the coating obtained has higher dilution values and larger heat affected zones compared to the standard PTA process. Deposition rates are of the order of 3 kg/h or less. Metal inert gas (MIG) welding using a wire as anode and the workpiece as the cathode is also a possible alternative to PTA. This has the advantage that oxide layers on light metals can be cleaned off before deposition. Deposition rates are in the order of 6–9 kg/h [Gebert and Bouaifi (2005)], and the dilution is somewhat less than with the TIG welding-coating process. Compared to other weld overlay processes, the coating quality is less sensitive to the torch–substrate distance.

This chapter is devoted to PTA used for coatings and hard facing, describing their basic design features, typical performance characteristics, and applications.

12.2 Basic Concept

12.2.1 General Remarks

The PTA process is closer to plasma arc welding than other plasma spray coating processes since the coating is fused to the substrate which is part of the electrical circuit serving in most of the cases serves as the anode to the transferred arc and only in a few cases as the cathode. The technique is comparable to tungsten inert gas (TIG) and metal inert gas (MIG) welding except for the feeding of the material to be deposited in the form of powder. It is also generally operated at higher power levels than TIG and MIG welding which gives rise to higher deposition rates of the order of tens of

kg/h compared to 3 kg/h or less for TIG and 6 to 9 kg/h for MIG [Gebert and Bouaifi (2005)]. Laser cladding or laser weld deposition, on the other hand, involves the local melting of the substrate by a laser beam and the introduction of the cladding or welding material into the molten pool in form of powder or wire [Lewis and Schlienger (2000)]. As the cladding/coating material is injected into the molten pool, it either melts inflight or, as it enters the molten pool, bends with the pool material forming a metallurgical bond or remained is solid form, as in the case of ceramic powders, which disperses into the molten pool forming a metal matrix composite. The laser cladding/coating process, in spite of its high capital investment and low deposition rates comparable with TIG and MIG, is used advantageously, for high precision coating of small areas because of the ability to precisely control the movement of the laser beam. The same is true for electron beam overlaying, which is used for some special applications.

12.2.2 Plasma Transferred Arc Deposition

A schematic of the basic concept used in plasma transferred arc (PTA) deposition is illustrated in Fig. 12.1. The PTA torch consists of a hot, thoriated tungsten cathode surrounded by a water-cooled copper nozzle constricting the arc between the cathode and the workpiece. The plasma gas is introduced at the base of the cathode tip. The nozzle also acts as anode for the "pilot arc," i.e., there is a current supplied by a separate power supply flowing from the nozzle/pilot arc anode to the cathode. This pilot arc pre-ionizes the gap between the cathode and the workpiece and allows easy

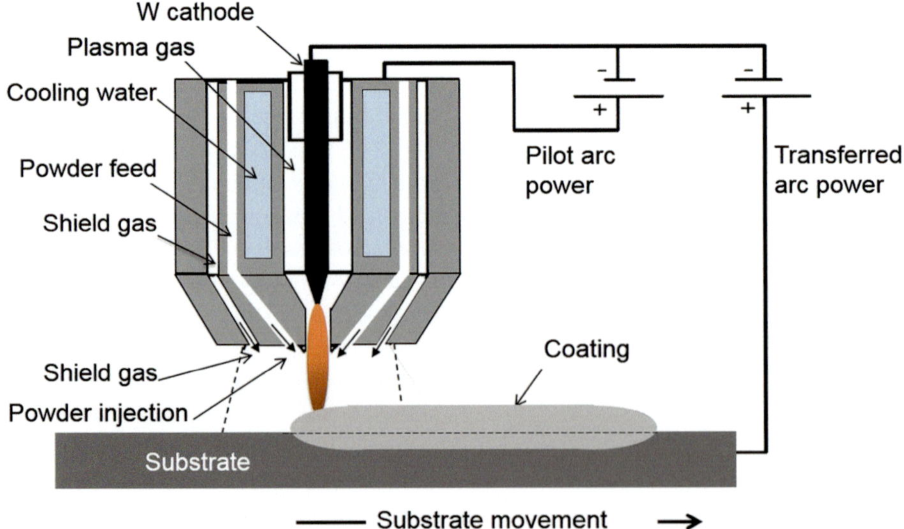

Fig. 12.1 Schematic of the basic concept of PTA deposition

establishment of the transferred arc and a good control of its arc current. The pilot arc further stabilizes the transferred arc when the plasma is cooled as a result of the injection of the feed material in the form of powder or wire and assures the correct positioning of the anode attachment and the area of powder deposition. The powder is introduced into the arc plasma through two or more orifices located on a ring surrounding the exit of the nozzle. A further annular slot surrounding the powder injection ring provides the shield gas flow, necessary to protect the molten metal from contact with the ambient air.

The coating is different from those obtained using other thermal spray coating processes because of the local melting of the substrate material at the anodic arc root attachment. The powder feed into the arc is also mostly melted, in-flight, and recovered into the molten pool of the substrate material with which it is mixed and alloyed. The consequence is that a good metallurgical bond or fusion bond is achieved between the coating and the substrate with a very low porosity. The heat penetration region and in particular the region where the coating material is mixed with the substrate material can change the properties of both the substrate and the coating. This mixing region is expressed in terms of the "dilution." As illustrated in Fig. 12.2, representing a cross section through a bead of deposited material mixed with the substrate material, the dilution (D) is defined as the ratio of the area where the substrate and coating materials are mixed (B), to the total area in which the coating material appears ($A + B$). That is:

$$D = B/(A + B)\% \tag{12.1}$$

The dilution is a major consideration when the functionality of the coating is evaluated. It is typically less than 5% for coatings of a thickness larger than 1 mm. It is also considerably lower than corresponding values obtained in TIG and MIG welding as well as submerged arc welding (SAW) in which dilution values of 15–40% are reached, requiring the deposition of several layers [Hallen et al. (1991)].

The PTA process combines good control over the process with high deposition rates (typically 3–10 kg/h, with up to 20 kg/h) and good coating quality with practically no porosity, and dilution of less than 7%. Compared to other thermal spray processes, PTA has the major advantage that alloy formation takes place in the molten metal pool during the process, provided the two metals/materials are compatible, allowing for a wider range of combinations of metal compounds or metal matrix composites compared to spray deposition process. According to DuMola and Heath (1997), PTA has the following advantages:

- High powder feed rates and deposition efficiencies (>95%).
- Tolerance to variations in feed powder quality, allowing for the use of low cost powders.
- Low porosity coatings (almost theoretical density),
- Thicker coatings in one pass can be achieved requiring fewer passes for part building.
- Metallurgical bond with strong adhesion to the substrate.
- Fine microstructure coating with smooth surfaces finish of the coating.
- Molten metal pool protected by a gas shield preventing oxidation during the deposition process.

PTA technology has, on the other hand, the following limitations:

- The requirement of a metallic substrate.
- Limitations on the substrate shape complexity.
- Dilution changing some of the coating and the substrate properties.
- High heat input into the substrate, requiring process modifications or active substrate cooling when smaller size substrates are coated.
- The process is not adapted to vertical surfaces due to the influence of gravity on the fluid flow and the coating geometry [Wilden et al. (2006)].
- Coatings are poorly adapted to complex substrate shapes including corners and grooves [Gatto et al. (2004)].
- Residual stresses can be rather high.
- Metallurgical compatibility between coating and substrate must be considered.

12.2.3 Arc Stabilization Mechanism

A wide range of plasma cutting/welding and coating torch designs and arc stabilization techniques have been developed and commercially used in industry. Most of these are covered by a large number of patents and proprietary information, which makes a detailed discussion of their design and performance rather limited [Colombo et al. (2011)]. Schematic representations of two typical PAC torch designs are given in Fig. 12.3. The design given in Fig. 12.3a shows a torch with a stick-type thoriated tungsten cathode with *axial* injection of the plasma forming gas. Because of the sensitivity of

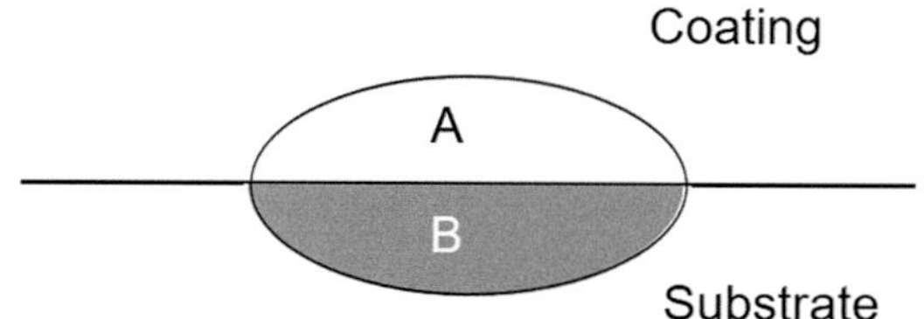

Fig. 12.2 Defining the dilution of a PTA coating

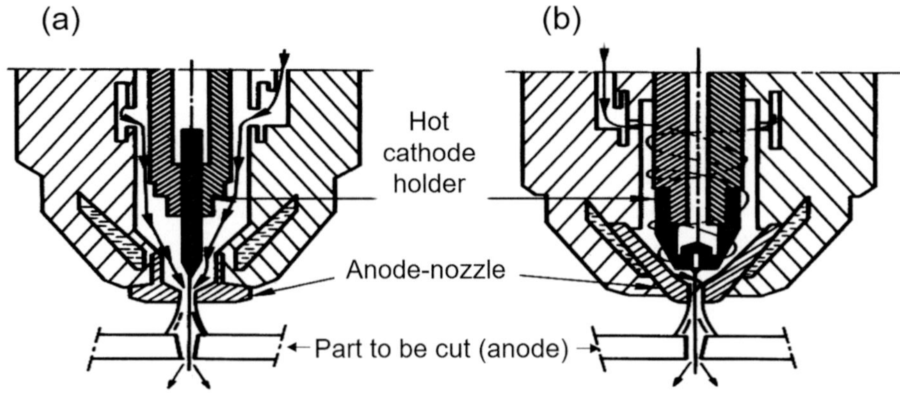

Fig. 12.3 Plasma cutting torches with (**a**) stick-type cathode, and axial plasma forming gas injection, and (**b**) button-type cathode, and vortex plasma forming gas. [Eliot (1991)]

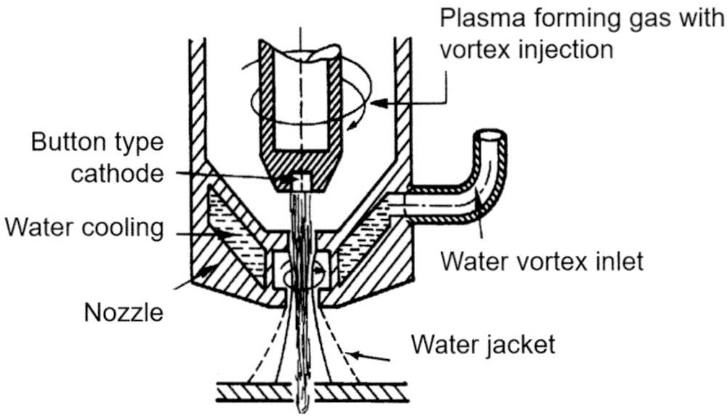

Fig. 12.4 Plasma cutting torch with button-type cathode, vortex plasma forming gas, and external water vortex [Eliot (1991)]

thoriated tungsten to the presence of oxygen in the plasma gas, these torches can only be used with non-oxidizing gases such as N_2 or Ar-H_2. The torch design given in Fig. 12.3b uses, on the other hand, a button-type cathode with *vortex* injection of the plasma-forming gas. The cathode is made in this case of either hafnium (Hf) or zirconium (Zr) for operation with air or oxygen or of thoriated tungsten for operation with non-oxidizing gases. In each of these torch designs, with axial or vortex arc stabilization, the arc is struck between the central electrode which acts as cathode, and the part to be cut which acts as anode, with currents up to 200 A or more.

In order to increase the arc constriction, decrease the fume emission, and reduce the thermally affected zone of the workpiece, a water vortex can also be added at the nozzle exit as shown schematically in Fig. 12.4. Due to additional energy losses in the water, this torch is operated at higher power levels. The arc current can vary between 15 and 640A but remains limited to 200A with Hf or Zr cathodes.

In order to keep the current density high, the nozzle orifice is kept as small as possible with its internal diameter directly dependent on the arc current. For example, with nitrogen as plasma gas, the i.d. of the cutting torch nozzle can be as small

as 0.9 mm for arc currents up to 30 A, 1.2 mm for currents of 60 A, and 1.5 mm for 120 A. The gas flow rates are adjusted with the arc chamber pressure reaching up to sonic velocities in some cases.

12.2.4 Choice of Plasma Gas

The choice of the proper plasma gas for the operation of plasma arc cutting/welding or PTA deposition torches has a critical influence on the torch performance and the process economics. Gases best adapted for plasma cutting must have a high specific enthalpy and a high thermal conductivity, which is generally the case with diatomic gases such as H_2, N_2, and O_2. They should also have a high molecular mass in order to compensate for the significant drop in the specific mass of the gas at high temperatures. Pure argon is only used to start the arc, but never for cutting due to its relatively low specific enthalpy and thermal conductivity. Ar-H_2 mixtures are typically used with 20 vol. % H_2. The presence of hydrogen gives rise to a significant increase of the specific enthalpy and thermal conductivity of the mixture compared to that of pure argon. Hydrogen also helps to trap, at least partially, oxygen and provides neat and clean cutting grooves. N_2 is a good alternative plasma gas due to its high specific enthalpy and high thermal conductivity, which allows fast cutting/ welding or coating without excessive oxide and slag formation. The thickness of the workpieces in this case is limited to 25–30 mm. The addition of H_2 to N_2 contributes to achieving better performance. It is to be noted, however, that nitrogen is susceptible to partial desolation in the molten metal in the cut or coating area, which could make it more difficult to weld the part in the same cut/deposited area. O_2 while commonly used in cutting operations is not acceptable in PTA deposition because of the formation of oxides which can blend into the coting structure.

The proper choice of the plasma gas has to be made taking also into account the electrode material since it will have an impact on the electrode lifetime. Zirconium or hafnium cathodes exhibit much higher wear, and consequently shorter lifetime, than that of the tungsten electrodes when working with non-oxidizing gases. Moreover, as the arc current is limited, by electrode design, to a 200–300A, the maximum power of the plasma torch will depend on the nature of the plasma gas used, which has a direct effect on the torch voltage.

12.3 Equipment Design and Operating Parameters

12.3.1 Basic Design Features

As illustrated in Fig. 12.5, the principal components of a typical PTA deposition installation are:

PTA torch is built around a central thoriated tungsten (2% ThO_2) cathode, typically 6–8 mm diameter (although smaller diameters have been used). The plasma gas is introduced into the torch at the base of the cathode and usually high purity argon or an argon-helium mixture. Downstream of the cathode tip is a water-cooled copper nozzle, typically 2–4 mm i. d., depending on the operating power of the torch. This nozzle serves as anode to the pilot arc and constricts/ stabilizes the arc. Alignment of the cathode with the nozzle axis is of critical importance for obtaining quality coatings because a slight misalignment can lead to unstable or strongly asymmetric arcs, uneven powder heating, uncontrolled substrate heating, and reduced electrode life. Surrounding this nozzle is a ring with powder injection orifices. The powder is injected, depending on the process and material, either into the arc where it is fully melted or close to be melted or into the fringes of the arc to keep it only partially melted. In the first case, there is a secondary nozzle (sometimes called focusing nozzle) with a diameter of about double that of the constricting nozzle surrounding the powder injection ring. Torches that do not have this secondary nozzle have the ring surrounding the powder injection orifices at the same level as the constriction nozzle exit, and this arrangement results in powder injection into the fringes of the arc with less heating of the powders. The powder can also be fed directly into the molten metal pool by an extension of the tube that transports the powder beyond the secondary nozzle to a location close to the substrate surface. Surrounding this secondary nozzle is the ring introducing the shield gas which consists typically of argon with 5–7% hydrogen added and provides a curtain that protects the molten metal from reacting with the surrounding air atmosphere. Shield gas flow rates are multiples of the plasma gas flow rates. For deposition of metals that will react with hydrogen such as Ti or Al, pure argon can be used as shield gas.

Two independent power supplies, one for the pilot arc and the second for the transferred arc. Both power supplies are controlled in the constant current mode with relatively low open circuit voltages (<100 V). Alternately, a single two-stage DC power supply can also be used or a single DC power supply with the pilot arc anode connected to the positive power supply terminal via a resistor [Wassermann et al. (1978)]. A high-frequency starter unit in the pilot arc circuit facilitates the ignition of the pilot arc.

Closed-loop high-pressure cooling water supply for the torch; typically 6–8 l/min (1.5–2 USGPM) are required at a pressure of 0.45–0.6 MPa (66–88 psig). The heat from the

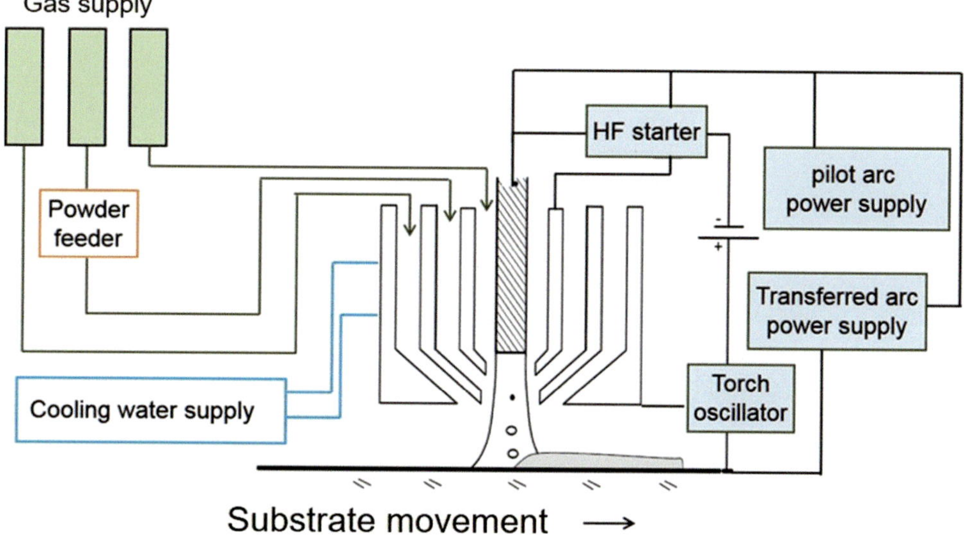

Fig. 12.5 Principal components of PTA deposition system

torch is removed by a heat exchanger which is connected to a chilled water circuit or city water circuit.

Gas supplies and flow controls for the plasma gas and the shield gas. Usually high purity argon is used as the plasma gas, helium being a possible addition. Ar/H$_2$ 5–7 vol.% H$_2$ is used as the shield gas.

Powder feeder is essentially the same as in other thermal spray processes, although the use of larger size powders (typically 50–150 μm) may require some adjustments; multiple powder feeders are used for composite coatings. The powder is transported by a carrier gas (usually argon or argon with a few percent hydrogen) to the torch.

Control console allows for the selection and control of the pilot and arc currents setting, the gas flow rates, and the powder feed rates. The control console also houses the safety interlocks that prevent arc initiation under unsafe conditions such as in the absence of plasma gas or cooling water flow.

Mechanical systems as schematically illustrated in Fig. 12.6 allows for the translation of the substrate at speeds between 1 and 2 mm/s with potential additions of an oscillatory movement of the torch, at frequencies of 0.5–1.0 Hz and amplitude of about 10–40 mm in a direction perpendicular to that of the substrate motionwith. These are necessary in order to limit the local heat flux to the substrate and to produce flatter beads. The oscillation width needs to be adjusted such that the solidification of the molten metal pool during one oscillation is controlled and that exposure of a part of the molten metal pool to air outside the shield gas envelope is avoided [Hallen et al. (1992)]. A multi-axis robot could be needed for that purpose depending on the complexity of the shapes that can be coated.

Examples of the as sprayed texture of the coating obtained on different substrate shapes and configurations for a wide range of wear resistance applications are given in Fig. 12.7 [courtesy of 2017 DURM Verschleiss-Schutz GMBH].

Depending on the size of the part to be coated and the associated power requirement, PTA deposition has been developed at the following three nominal production levels [DuMola and Heath (1997), Bouafi et al. (1996)]. The smallest level of operation of a PTA system is generally referred to as *micro-PTA* (μ-PTA). It is used at deposition rates of 0.1–2 kg/h for the coating of small parts or components with complex shapes, which can tolerate little heat [Shubert (1987), Shubert (1987), Lindland and Shubert (1988), Saltzmann et al. (1989), Saltzman and Sahoo (1991)]. The next size up is the *Regular PTA* used at deposition rates of 2–10 kg/h for the bulk of the PTA coating applications. The *High power PTA* is generally reserved for deposition rates between 10 and 20 kg/h (or even larger) for rapid coating of large surfaces or large components [Hallen et al. (1991), Hallen et al. (1992)]. A summary of the corresponding operating conditions for each of these operation levels are summarized in Table 12.1.

The properties of the coating obtained are dependent on the operating parameters and the material used. For regular PTA operation, the following represent typical values of system performance:

Bead thickness = 0.5–6 mm

Bead width = 2–40 mm, mostly around 10 mm

Coating thickness = 20 mm

Deposition efficiency = 95–98%

Dilution = 5–7%, possibly higher for coatings <1 mm thick

For micro-PTA (μ-PTA) bead widths can be as small as 0.5 mm, ranging up to 3 mm, with single pass bead thicknesses 0.5–3 mm, giving bead height to width ratios of 0.5–2 [Shubert et al. (1987)]. The dilution increases with the power. The fastest solidification and finest microstructure is obtained with the micro-PTA process.

While the majority of the equipment are automated, handheld PTA torches are also frequently used [Kammer et al. (1991), Schreiber and Krefeld (2002)]. These are primarily used for smaller parts with complex shapes. There are also numerous modifications of the standard PTA equipments and processes principally to improve control over the powder heating, the heat flux to the substrate, and to adjust the process to specific substrate sizes and shapes. The loss of deposition material can occur through evaporation which makes the deposition of low boiling point materials such as zinc rather challenging. Homogenous condensation of the metal vapor leads to the formation of ultra-fine fume, which can be deposited on parts of the torch or on the substrate outside the bead. Avoiding injection of the powder into the center of the arc as well as using larger size particles (>60 μm), and increasing the shield gas flow rate, and the

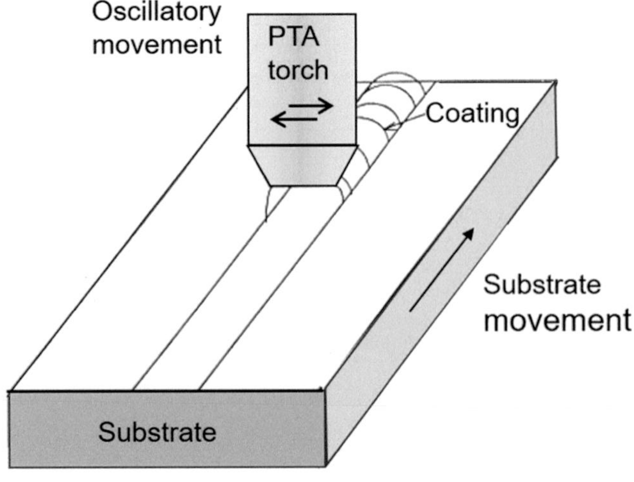

Fig. 12.6 Schematic of PTA deposition process

Fig. 12.7 Photographs of typical PTA process applications [2017 DURM Verschleiss-Schutz GMBH]

Table 12.1 Summary of operating conditions for different PTA scales of operation

Operating parameter	Micro-PTA (µ-PTA)	Regular PTA	High power PTA
Pilot arc current (A)	–	50–100	50–100
Pilot arc voltage (V)	–	40	40
Transferred arc current (A)	5 to 50	100–300	200–400
Transferred arc voltage (V)	20	40	40
Nozzle diameter (mm)	1–2	2–4	2–4
Torch-to-substrate dist. (mm)	5–10	10–20	15–20
Plasma gas flow rate (slm)	0.4–1.4 Ar	2–10 Ar or Ar/He	3–10 Ar or Ar/He
Shield gas flow rate (slm)	10–24 Ar or Ar + 5–7%H_2	10–40 Ar or Ar + 5–7%H_2	10–50 Ar + 5–7%H_2
Carrier gas flow rate Ar (slm)	1–3.5 Ar	1–20 Ar or Ar + 5%H_2	5–20 Ar or Ar + 5%H_2
Powder feed rate (kg/h)	0.1–2	2–10	10–20
Powder particle size (µm)	20–100	60–200	60–200
Translation speed (mm/s)	2–3.5	1–2	1–4
Oscillation frequency (Hz)	Non	0.5	0.3–1.0
Amplitude of oscillation (mm)	Non	10–20	20–40

torch-to-substrate distance can minimize the problem of the fumes.

Typical examples of commercially available PTA torches are given in Fig. 12.8 and Fig. 12.9. These vary in the details of the design configurations, power rating, and usable range of powder feed rate. These operate at pilot arc currents up to 30 A and maximum transferred arc currents up to 300 A and an open circuit voltage of 100 V DC. They vary in their duty cycle between 35% and 60% depending on the arc current. Examples of the torches shown in Fig. 12.9, by Durum GMBH, are given for a handheld torch (Fig. 12.9a) and machine-mounted, automatic or semi-automatic torches (Fig. 12.9b, c). Not included in the figure is the Durum PT-200i-80ID torch specially designed for deposition of internal surfaces with a minimum diameter of 80 mm with a maximum current rating of 200 A and powder feed rate up to 50 g/min. It should be noted that the given powder feed rate values are material and powder density defendant.

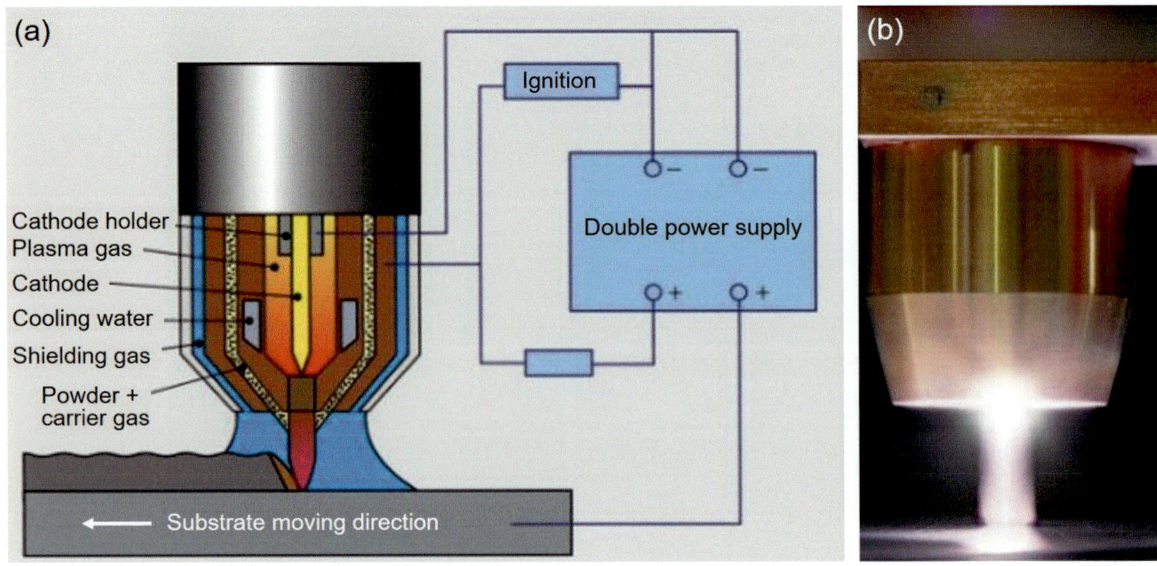

Fig. 12.8 (**a**) Schematic and (**b**) photograph of PTA welding/over laying plasma torch by Castelin Eutectic Corp. [www.eutectic.com]

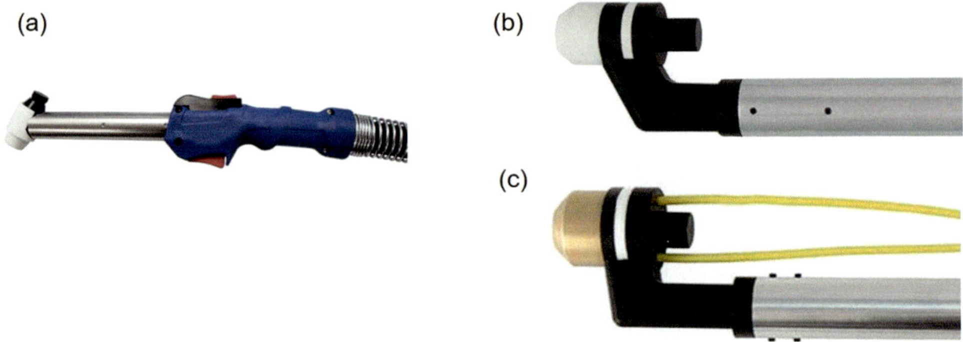

Fig. 12.9 Photograph of PTA Durmat Torches (**a**) PT-150 M manual, handheld torch, 150A, 30 g/min; (**b**) PT-300 AUT machine torch, 280 A, 75 g/min; (**c**) PT-400 machine torch 300 A, 100 g/min. [2017 DURM Verschleiss-Schutz GMBH]

12.3.2 Effect of Process Parameter Changes on Coating Properties

The primary qcuantities that are of concern are:

- Bead thickness and width
- Deposition rate, efficiency, and dilution
- Adhesion
- Functional performance such as hardness, porosity, defect density, and resistance to cracking

The functional performance parameters are usually addressed through the selection of the powder to be deposited, though they can also be influenced by the process parameters such as:

- Pilot arc and transferred arc currents
- Plasma gas, powder gas, and shield gas flow rates

- Powder feed rate
- Torch substrate distance
- Substrate translation and oscillation velocities

In the following, a brief discussion is presented on the influence of each of the principal process parameters.

12.3.2.1 Pilot Arc Current

The pilot current has little effect on the coating quality if all other parameters remain unchanged. However, it is mainly beneficial for the stability of the transferred arc and can in certain cases be used to provide added particle heating when the transferred arc current has to be reduced because of the sensitivity of the substrate to excessive heating. This condition is encountered when coating low melting point substrates or when dilution needs to be minimized, e.g., in thin coatings. The condition of high pilot arc current and low transferred arc current approaches the situation of a plasma spray process.

12.3.2.2 Transferred Arc Current

Increasing the transferred arc current has the strongest effect on the substrate heating, the arc diameter, the plasma velocity, the heating and melting of the powder, and the movement in the molten metal pool. It also increases dilution, adhesion, bead width, and deposition efficiency, and lowers porosity, but *decreases* bead thickness, and may also reduce hardness as a consequence of increased dilution. In the case of hard surfacing involving the deposition of carbide powders for enhanced wearer resistance, high currents may lead to the partial decomposition or dissolution of the carbide phase, reducing the wear resistance. On the other hand, too low currents may lead to incomplete melting of the powder.

Typical results reported by Wang et al (2002) on the use of a microplasma PTA deposition system for free form fabrication are given in Figs. 12.10, 12.11, and 12.12 in terms of deposit bead width and thickness as function of the arc current, powder feed rate, and substrate displacement speed, respectively. The torch-to-substrate distance in this case was 3 mm, plasma gas flow rate 0.4 slm, powder gas flow rate 3.5 slm, and shield gas flow rate 10 slm.

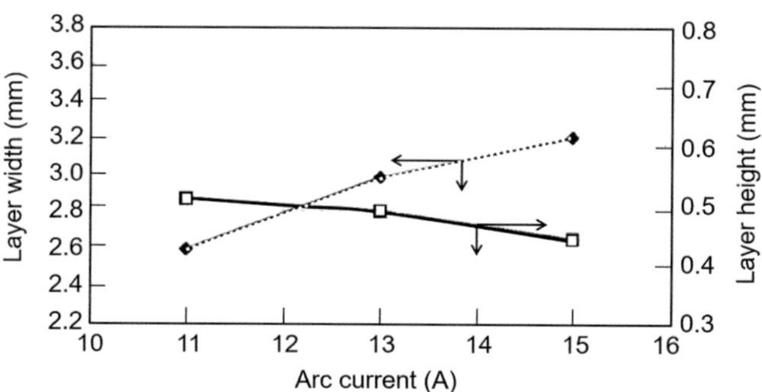

Fig. 12.10 Measured dependencies of bead width and thickness on arc current for powder feed rate of 2.3 g/min and substrate displacement speed of 0.52 mm/s [Wang et al. (2002)]

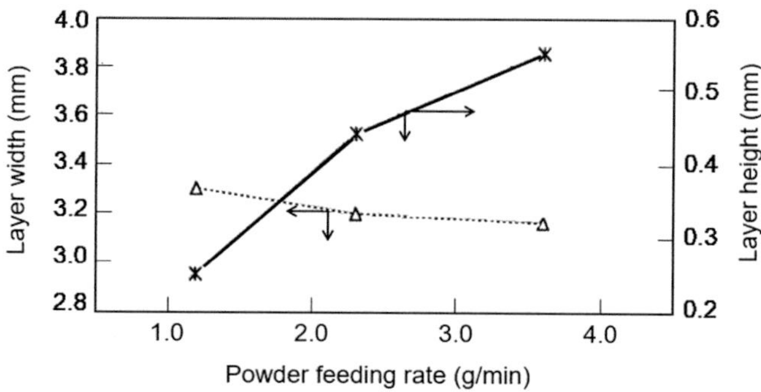

Fig. 12.11 Measured dependencies of bead width and thickness on powder feed rate for an arc current of 15 A and substrate displacement speed of 0.52 mm/s [Wang et al. (2002)]

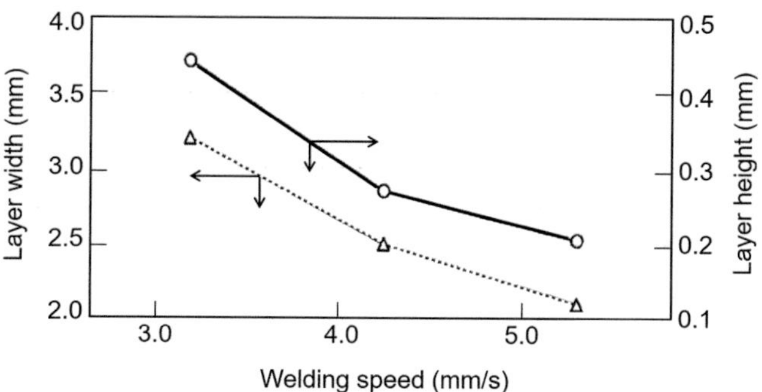

Fig. 12.12 Measured dependencies of bead width and thickness on substrate displacement speed for powder feed rate of 2.3 g/min and arc current of 15 A [Wang et al. (2002)]

12.3.2.3 Torch-to-Substrate Distance

Increasing the torch-to-substrate distance, gives rise to an increase of the arc voltage for the same current and consequently the increase of the total arc power. This would result in a corresponding, though not proportional, increase of energy transmitted to the substrate. The reverse effect is to be expected, however, on the gas velocity which decreases with the increase of the torch-to-substrate distance giving rise to a reduction of the movement of the molten metal pool and of pool dilution. The latter effect is stronger at higher arc currents. At a substrate distance of 20 mm, an increase in current from 150 A to 250 A can be balanced by an increase in powder feed rate from 6 kg/h to 10 kg/h, keeping the dilution below 10%. At a substrate distance of 15 mm, the dilution will increase with the increase of the arc current even with the increase in powder feed rate [Hallen et al. (1991)].

12.3.2.4 Plasma Gas Flow Rate

Increasing the plasma gas flow rate results in an increase of the arc width, and plasma velocity, as well as the movement in the molten metal pool. It can also result in increased heating of the substrate and increased dilution [Hallen et al. (1991), Deuis et al. (1998)]. The effect is dependent on the arc current and the traverse speed. At high heat transfer and substrate melting rates, the higher plasma gas flow rate can push the molten material toward the sides and cause turbulence and increase dilution in the molten metal pool.

12.3.2.5 Powder Feed Rate

Increasing the powder feed rate reduces the dilution, increases bead thickness, but reduces deposition efficiency. Too high powder feed rates for a given current will eventually lead to incomplete melting of the powder. Typical results by Wang et al (2002) on the effect of power feed rate on the width and height of the deposited bead is given in Fig. 12.11.

12.3.2.6 Substrate Material Properties

The nature of the substrate material has an influence on the arc voltage for a given arc current and consequently on the heat flux to the substrate. For cooled substrates, the substrate heat flux increases with the increase of the substrate thermal conductivity, while for an uncooled substrate, the heat flux increases with the increase of heat capacity of the substrate.

12.3.2.7 Substrate Motion

An increase in the translation velocity of the substrate will generally reduce dilution. The reverse effect has also been reported by Lakshminarayanan et al. (2008). It also can result in a finer microstructure in high power PTA deposition due to faster solidification of the molten metal [Hallen et al. (1992)]. At the optimal translation speed, the arc attachment is in front of the molten metal pool. Excessively high speeds will result in reduced fusion bonding, while too slow speeds will give rise to increased porosity [Hallen et al. (1992)]. The oscillation width needs to be adjusted such that the solidification of the molten metal pool during one oscillation is controlled or that exposure of a part of the molten metal pool to air outside the shield gas envelope is avoided. Typically, oscillation widths remain below 40 mm [Hallen et al. (1992)]. Experimental data by Wang et al (2002) obtained using a microplasma PTA system given in Fig. 12.12 show a steady decrease of both width and high of the deposited bead with the increase of the speed of the substrate displacement.

Special attention should be given to the dilution values in the bead because of their direct impact on the functional properties of the coating. While low dilution values improve the wear resistance of the coating, they are usually accompanied by low fusion values, i.e., small areas where a metallurgical bond is obtained. DuPont (1998) presents a correlation of the "fusion factor," i.e., the width of the dilution region over the width of the bead in a coating cross section, as function of the ratio of powder feed rate to arc power as parameter. They report a strong drop off of the fusion factor at a certain value of the ratio of the metal powder feed rate to the arc power, i.e., if the power is insufficient to melt all of the powder. A drop in the deposition efficiency usually accompanies the drop of the fusion factor. While these data were obtained for plasma arc weld overlays (no pilot arc), a similar effect can be expected for PTA coatings.

A study of the effect of the arc current, powder feed rate, and powder gas flow rate on dilution was reported by [Hallen et al. (1991)]. Typical results given in Fig. 12.13 were obtained for a workpiece St-37, 300 × 100 × 30 mm, coated using a torch-to-substrate distance of 20 mm, translation speed of 1.3 mm/s, amplitude of oscillation of 40 mm at an average oscillation speed of 1.8 mm/s, and plasma gas flow rate of 10 slm (Ar), with varying powder feed rates (6–12 kg/h) and transport gas flow rates (1–4 slm Ar). The results show

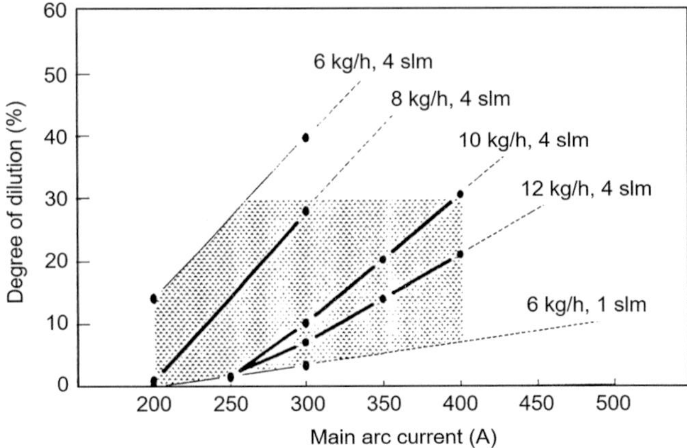

Fig. 12.13 Degree of dilution as function of arc current, powder feed rate, and powder gas flow rate. Workpiece St-37, 300 × 100 × 30 mm, plasma gas = 10 slm (Ar), amplitude of oscillation = 40 mm, oscillation speed = 1.8 mm/s, translation speed =1.38 mm/s, torch-to-substrate distance = 20 mm. [Hallen et al. (1991)]

Table 12.2 Effect of process parameters on coating properties, (+) for positive impact and (–) for negative impact [Ducos (1985)]

	I_{pilot}	I_{arc}	m_{powder}	m_{gas}	$v_{travers}$
Bead width	-	++	-		-
Thickness	+	-	++		--
Deposition efficiency		+	--		
Surface		+++	-		-
Dilution	-	+++	---	++/--	--
Hardness	+	---	+++	++	++

a steady increase of dilution with the increase of the arc current. As expected, however, the dilution drops with the increase of the powder feed rate and the powder transport gas flow rate. A discussion of operating parameter influences on dilution using a response surface analysis is presented by Lakshminarayanan et al. (2008).

[Just et al. (2010)] reported higher hardness at lower currents, faster cooling, lower dilution, higher impact resistance, but not necessarily better wear because of less bonding between the carbide and the matrix. A summary of the effect of process parameters compiled by Ducos (1985) is given in Table 12.2.

12.3.3 Process Modifications and Adaptations

The principal process modifications other than the adjustments of process parameters for high deposition rate, high power PTA, or low deposition rate micro PTA, or for adjustments for a specific coating/substrate combination, are:

- Limiting the use of the pilot arc to the arc ignition stage of operation.

- Special powder injection to avoid too overheating of the powder or substrate.
- Operating with nitrogen for surface nitriding.
- Operation of the transferred arc in reverse polarity for surface cleaning of the substrate, especially those with thin and impervious oxide layers such as Al or Mg.
- Modulating the plasma gas flow and/or the arc current, or torch design modification to affect the heat transfer to the substrate.
- Increasing the energy density at low total energies for high-velocity, small-area depositions (high energy PTA).
- Combining tape casting with PTA for complex shape coatings.
- Combining PTA with MIG or laser welding processes.

12.3.3.1 Limiting the Use of the Pilot Arc to Arc Ignition

An example of zero pilot arc current deposition (except for arc starting) in a micro-PTA deposition process is described by Wang et al. (2003) for the production of 3D free-standing parts. A transferred arc current of 10–15 A was used to deposit a 2–3 mm wide, 0.45–0.55 mm high bead with a "tool steel" powder feed rate of 2.3 g/min (0.138 kg/h). Deposition parameters were 3 mm standoff distance, plasma gas flow rate = 0.4 slm (Ar), carrier gas flow rate = 3.5 slm (Ar), shield gas flow rate = 10 slm (Ar), and substrate translation speed of 188 mm/min. Tungsten carbide powder was added with a second powder feeder, and the substrate was preheated to reduce the time for melting. Good control over the arc length was important for getting a uniform product.

12.3.3.2 Variation of Powder Feed Rate

To avoid overheating of the powders, e.g., carbides leading to their dissolution, powder injection can be limited to the fringes of the arc or directly into the molten metal pool through extended powder feeding tube [DuMola and Heath (1997), Bouaifi et al. (1996, 1997)]. Bouaifi (1997) describes the coating of industrial knife-edges with a chrome-vanadium steel, with the addition of vanadium carbide made using a separate powder feeder into the molten metal pool. The approach allowed for the increase of the VC particle loading fraction in the coating to 70 vol. %. Deposition parameters were arc current = 160–230 A, powder mass feed rate = 1.5–5.5 kg/h, and a translation velocity 0.67–2.5 mm/s. Some edges had to be ground after the coating deposition.

12.3.3.3 Nitriding of Coating

Efforts were made to increase the coating hardness through online nitriding resulting from carrying out the deposition process in a nitrogen atmosphere. Bouaifi et al. (2001) describe experiments for the deposition of Fe-Cr-V-C alloys with the addition of nitrogen to the plasma gas (Ar) and the use of pure nitrogen as powder and shield gases. Problems were encountered with nitrogen in the plasma gas since for volume concentrations above 50–70% of N_2 the arc became unstable. The stronger constriction of the arc also led to increased fume formation and shorter electrode lifetimes, as well as reduced coating adhesion at the bead edges. While the nitrogen content in the shield gas did not influence the coating properties notably, the use of 33 vol.% nitrogen in the plasma gas significantly increased the wear resistance of some the alloys deposited which contained up to 0.28 wt% nitrogen. A similar experiment is reported by Bach and Zühlsdorf (1999) using plasma weld deposition in a chamber with a nitrogen atmosphere. At 1 atm pressure in the chamber and 10 vol.% N_2 in the plasma gas, nitrogen incorporation into a Cr-Ni steel of up to 0.37 wt. % could be achieved. Increasing the chamber pressure reduced the nitrogen content and led to material evaporation and fume formation.

In an attempt to further increase the nitrogen content of the coating, Wang et al. (2009) pretreated the base metal alloy (7005 Al alloy) by depositing a TiN/Ni composite coatings on it using high-speed jet electroplating, followed by a PTA scanning process. The parts produced had a rapidly solidified microstructure consisting of uniformly distributed TiN phase and fine Al_2Ni_3 intermetallic phases. The microhardness was about HV 852 which is seven times higher than that of the Al alloy substrate. The friction coefficient for PTA scanning treated specimen was 0.25 which is significantly lower than 0.35 for TiN/Ni composite coating. The corrosion resistance of the composite coating was also significantly improved.

12.3.3.4 Modulation of Deposition Parameters

Modulation of the plasma gas flow rate or of the arc current has been the subject of numerous investigations for specialized PTA coating applications. Plasma gas modulation at frequencies between 1 and 5 Hz with an amplitude of 1–3 slm (Ar) was successfully used by Ebert et al. (2009) for the improvement of the dispersion of heavy WC particles in a Co-Alloy PTA coating containing 40 vol. % WC. The arc current in this case was 95 A, powder gas flow rate 2 slm (Ar), shield gas 10 slm (Ar), and traverse velocity of 1.67 mm/s.

Arc current modulation was used by Shubert (1987) in a micro-PTA application in which a bead is deposited as thin as 0.5 mm, with a bead height to width ratio that could be varied between 0.5 and 2 and with a ratio of bead width to substrate width of 0.5–1. An "auxiliary" current of 4–15 A is used for the arc on which high current pulses of 0.4 s duration, with a 25% duty cycle are superimposed to yield a time-averaged current between 12 and 40 A. The deposition rate was between 0.3 and 0.5 kg/h and the translation speed between 2.3 and 3.5 mm/s. This process has been applied in particular to the refurbishing of gas turbine parts [Shubert (1987), Lindland and Shubert (1988)]. Saltzmann et al. (1989) describe the refurbishing of super-alloy parts in gas turbines with average currents of 3–20 A and with 10–40 ms long high current pulses during which the weld pool is created, while it is allowed to solidify during the low current time. The quick solidification times lead to very fine grain structure in the deposits. Very short torch-to-substrate distances are used (5–10 mm for currents from 10–50 A). The powder is delivered into the weld pool through an external feed tube.

A similar approach of pulsing the arc current was also used for the reduction of workpiece distortion for a near net-shaped application in which a stellite-6 coating was deposited on steel with an intermediate Ni alloy coating using a pulsed plasma weld coating technique [Matthes and Alaluss (1996)]. The amount of transferred arc current was periodically varied between a base current value and a pulse current value, and the effect of the current values and the duty cycle of the pulsed current investigated. Three passes were used to reach the required thickness of 8 mm. The dilution was found to be increasing with the arithmetic mean of the arc current, and with the increase of the ratio of base current to pulse current, as well as with the increase in duty cycle. All these factors influenced the heat input into the substrate, which are generally responsible not only for the dilution of the coating but also the workpiece distortion and reduction of its hardness. It was generally reported that current modulation was effective for the preparation of good coatings on small near net-shaped parts with a minimum distortion of the substrate material. D'Oliveira et al. (2005) evaluated the effect of pulsed current on a high carbon cobalt alloy deposited by PTA on carbon steel and stainless steel.

Razal Rose et al. (2012) pointed out metallurgical advantages of pulsed current tungsten inert gas welding (PC-TIG) frequently reported in literature. These include reduced width of heat-affected zone and refinement of the grain size of fusion zone. In their paper, an attempt was made to establish an empirical relationship to predict tensile strength of the PC-TIG welded AZ61A magnesium alloy by incorporating process parameters such as pulse current, base current, pulse frequency, and pulse on time. Statistical tools such as design of experiments, analysis of variance, and regression analysis were used to develop the relationships. The developed empirical relationship can be effectively used to predict the tensile strength of PC-TIG welded AZ61A magnesium alloy joints at the 95% confidence level. The results indicated that the pulsed current had the greatest influence on tensile strength, followed by pulsed on time, pulse frequency, and base current.

12.3.3.5 Pulsed Plasma Additive Manufacturing

Recently, Lin et al. (2016) developed a novel application of current-pulsed micro-PTA technology called, pulsed plasma additive manufacturing (PP-AM), in which a micro-PTA system is used for the deposition of thin coatings or standalone parts for additive manufacturing using a wire as feed material introduced into the weld pool as illustrated in Fig. 12.14a. In this research, several Ti-6Al-4V thin wall coatings or standalone parts were deposited by an optimized PP-AM process, in which the heat input was gradually decreased layer by layer. The deposited thin wall consisted of various morphologies with different microstructure, such as epitaxial growth of prior β-grains, martensite, and horizontal layer bands of Widmanstätten, which depend on the heat input, multiple thermal cycles, and gradual cooling rate in the deposition process. Both of the size and growth direction of prior β grains are affected by the gradual reduced heat input of the pulsed plasma arc. The technology offers an important potential due to its efficiency, convenience, and cost-savings comparing with other AM process.

12.3.3.6 High Energy PTA

The use of high energy PTA (HE-PTA) developed by SNMI, Avignon, France [Proner et al. (1997a, b)], was proposed for

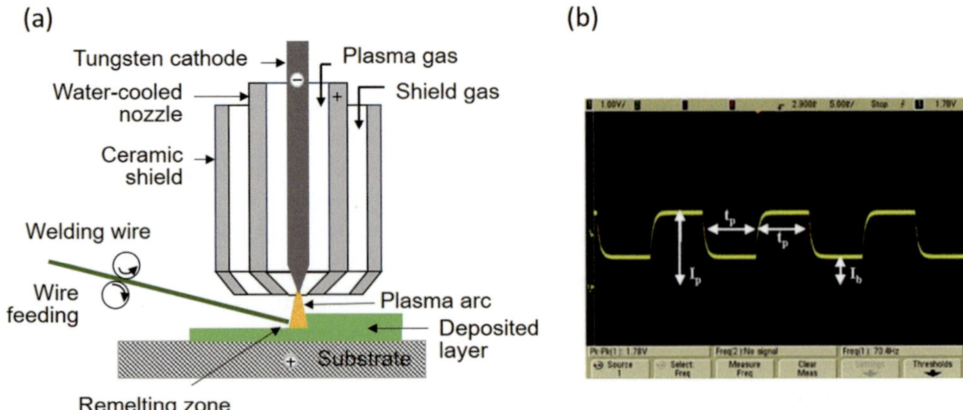

Fig. 12.14 (**a**) Schematic representation of the PP-AM process, (**b**) measured current [Lin J.J et al. (2016)]

the reduction of the heat input to small parts. The development aimed at the preparation of wear-resistant coatings of smaller parts such as internal combustion engine valves, with minimal needs for water-cooling. The arc in this case is highly constricted providing a small attachment spot diameter at the workpiece, with high current densities and high energy densities. The relative torch-to-substrate translation velocity can be very high giving rise to low heating of the substrate, low dilution values, and high cooling rates resulting in a coating with a fine microstructure. The required translation velocity, v_{tras}, is function of the attachment spot diameter D, melted layer thickness δ, and thermophysical properties of the substrate as given in Eq. 12.2:

$$v_{tras} = \left(\frac{D}{5\,\delta^2}\right)\left(\frac{\kappa}{\rho\,c_p}\right) \qquad (12.2)$$

with κ, ρ, and c_p being respectively the thermal conductivity, density, and specific heat of the substrate material. The condition for the required power input at a specific velocity and melt depth is given in Eq. 12.3:

$$\left(\frac{V\,I}{v\,D}\right) \geq 10^7 \left(\frac{J}{m^2}\right) \qquad (12.3)$$

High energy PTA (HE-PTA) coatings have been successfully used for the coating of steel plates with a thin layer of Co-alloy with minimal dilution [Proner et al. (1997a, b)]. Microhardness profiles of the coatings obtained are given in Fig. 12.15 which shows (HE-PTA) values significantly higher than those obtained by conventional PTA technology.

12.3.3.7 PTA Operating in Reverse Polarity Mode

A number of studies have been directed toward process development for coatings of light metals such as aluminum and magnesium alloys. These have a low melting points

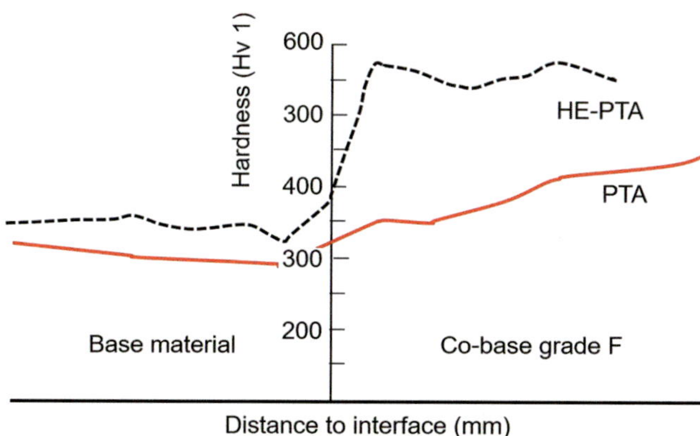

Fig. 12.15 Comparison of hardness profiles across a Co-alloy coating on steel for regular (PTA) and high energy PTA (HE-PTA) [Proner et al. (1997a, b)]

requiring reduced heat input to the substrate and an oxide surface layer that needs to be removed to obtain good adhesion of the coating [Gebert and Bouaifi (2005)]. They also have high thermal diffusivities resulting in rapid transfer of the heat from the arc attachment to the rest of the substrate. The consequence is that it will take a relatively long time to reach the melting point, but then a major part of the substrate may be molten. Reduction of the overall heat transfer is necessary accompanied by an increase in the local heat flux, as can be achieved by a reverse arc heating though small cathode spots. Removal of the oxide layer is also possible by thermal means, heating the substrate and melting the oxide with He as addition in the plasma gas. However, this process creates larger melt pools and movement of the molten material [Dilthey et al. (1999)]. Operating the arc in a reverse polarity mode, with the substrate serving as the cathode, will result in lower overall heat transfer to the substrate, and a high heat flux density in the continuously moving cathode spots. Cathodic arc deoxidation has been a standard practice in the metallurgical industry.

Dilthey et al. (1999) and Lugscheider et al. (1999) describe a PTA process with direct current combined polarity heating, in which the pilot arc is established between the tungsten cathode and the pilot arc anode nozzle, but the transferred arc is struck between the pilot arc anode nozzle and the workpiece as the cathode. Significant increases in wear resistance have been achieved by either dispersing a hard material such as WC/W_2C into the molten surface of a $AlMgSi_{0.5}$ alloy, depositing an aluminum alloy together with a hard material, or alloying the substrate material with a deposited material to generate hard phases. For this process, the transferred arc current was reduced and the pilot arc current increased. The pilot arc anode had been designed for improved water-cooling to reduce the erosion due to higher thermal fluxes from two arcs. The same arrangement of a DC combined polarity has been used for increasing the wear resistance of AlMg4.5Mn [Dilthey et al. (2008)]. The coating material was AlCuMg alloy reinforced by up to 70 wt % fused tungsten carbide (WC + W_2C) or up to 50 wt% TiC. Hard coatings were deposited using pilot arc current of 120 A, transferred arc current of 100 A, plasma gas flow rate of 0.5–1 slm, carrier gas flow rate of 4 slm, and shield gas flow rate of 10 slm, oscillation with an amplitude of 10 mm and frequency of 1.8 Hz. These had significantly improved wear characteristics compared to the base material, by a factor of 40 for the tungsten carbide and a factor 25 for the TiC-containing coatings [Dilthey et al. (2008)]. The carbide particles were uniformly distributed in the coating. Addition of Ni into the coating material further improved hardness through the formation of hard intermetallic phases which also resulted in more brittle coatings.

12.3.3.8 Hard Coatings on Magnesium

Magnesium and its alloys are attractive as structural material for a wide range of applications allowing for more weight reduction than aluminum for an equivalent strength. However, because of their poor corrosion and wear resistance properties, they require coatings as described by Gebert et al. (2002). Coatings have been obtained using a matrix material for the composite coating of either Al or $AlSi_{12}$ with the addition of up to 10 vol.% Mn or Si and reinforcing particles of TiC, SiC, or B_4C. The best coatings were obtained with the workpiece at a positive polarity and with about 30 vol.% He added to the plasma gas to increase penetration depth to 2 or 3 mm. While the TiC and SiC particles have been non-uniformly distributed in the coating, a uniform distribution with the lighter B_4C was achieved. The hardness increases, as well as the wear resistance increases were dependent on the amount of B_4C addition, with 7.5–10% carbide giving strong improvements particularly to the wear rate.

12.4   Gas and Particle Dynamics in Plasma Transferred Arc Coating

Compared to other plasma spray coating processes, relatively few process characterization studies have been devoted to PTA. The process, however, bears enough similarity to TIG and MIG welding, and plasma arc cutting, that some of the studies involving these processes can be used for the understanding of the basic phenomena involved in the PTA process.

12.4.1 Temperature Distributions in the Arc

The temperature distributions and heat fluxes in a TIG welding arc have been well characterized by spectroscopic temperature measurements and modeling [Hsu et al. (1983), Lowke et al. (1997), Lowke et al. (2005), Bini et al. (2007), Etemadi (1982)]. Figure 12.16 shows its LHS typical temperature distributions obtained for a 200 A argon arc at atmospheric pressure and corresponding modeling results on its RHS [Young et al. (1983)]. The constriction encountered in a PTA arc will increase the peak temperatures on the arc axis and result in higher temperature gradients. The temperature distributions in a plasma cutting arc at different axial locations between the nozzle and the workpiece (z) are given in Fig. 12.17 for an oxygen arc at a current of 200 A [Peters et al. (2007)]. Because of the smaller nozzle diameter (2 mm), the constriction in this configuration is somewhat higher than that in a typical PTA torch. The plasma gas being oxygen at a relatively higher plasma gas flow rate also gives rise to higher constriction.

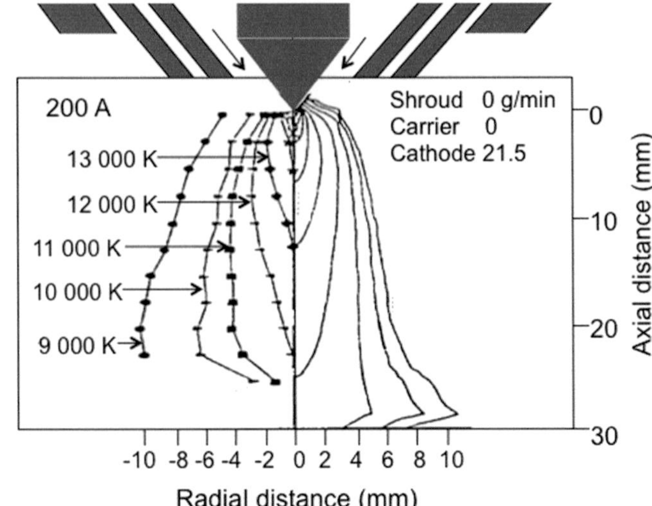

Fig. 12.16 Examples of temperature fields in an argon arc with I = 200 A, 1 atm, and torch-to-substrate distance = 30 mm, (LHS) measurements, (RHS) modeling results [Young et al. (1983)]

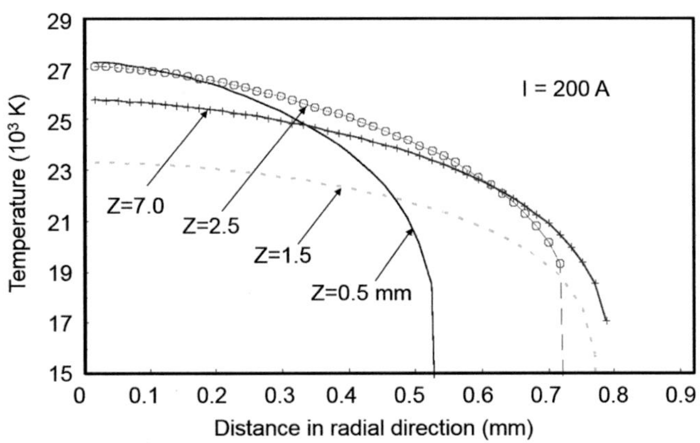

Fig. 12.17 Radial temperature distributions at different axial locations in a cutting arc derived from spectroscopic measurements, I = 200 A, nozzle i.d. = 2 mm, plasma gas flow rate = 39 slm (O2), torch-to-substrate distance = 13 mm. [Peters et al. (2007)]

Figure 12.18 shows a plasma welding torch with a nozzle i.d. = 3.45 mm, the effects of variation of the arc current (100–260 A), and plasma gas flow rate (2–8 slm Ar), on the temperature distributions between the nozzle exit and the substrate [Evrard and Blanchet (1970), Ducos (1985)]. It can be seen in Fig. 12.18a, b,c that an increase in arc current from 100 A to 260 A results in a corresponding increase of the arc diameter and of the anode attachment, giving rise to a higher temperature in front of the substrate, with hardly any increase of the peak temperature at the cathode attachment. The equivalent arc diameter, d_e (diameter of current path), can be estimated using Ohm's law according to the approximate relations:

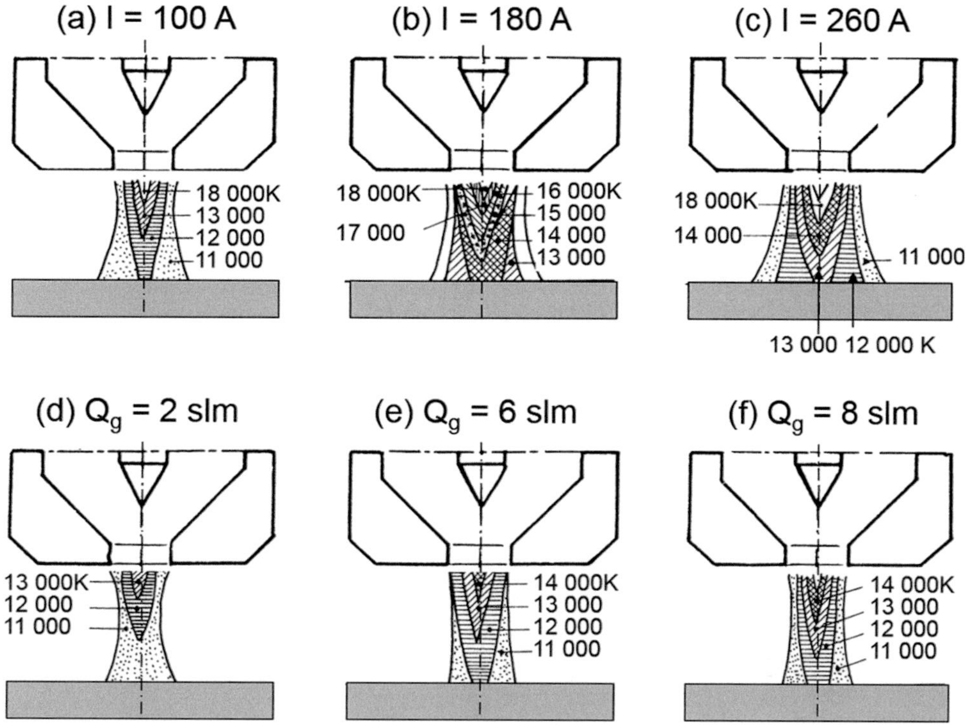

Fig. 12.18   Temperature fields in an arc struck between a plasma welding torch and the substrate for an anode nozzle i.d. = 3.45 mm (a, b, c) for different arc currents at a gas flow rate of 4 slm (Ar) and (d, e, f) for different argon gas flow rates at an arc current of 100 A [Evrard and Blanchet (1970)]

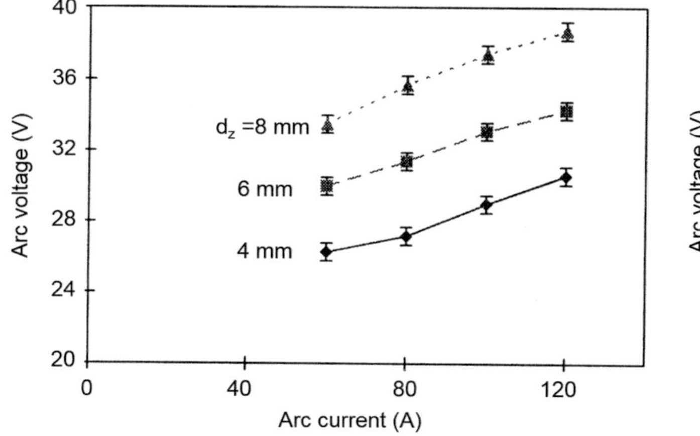

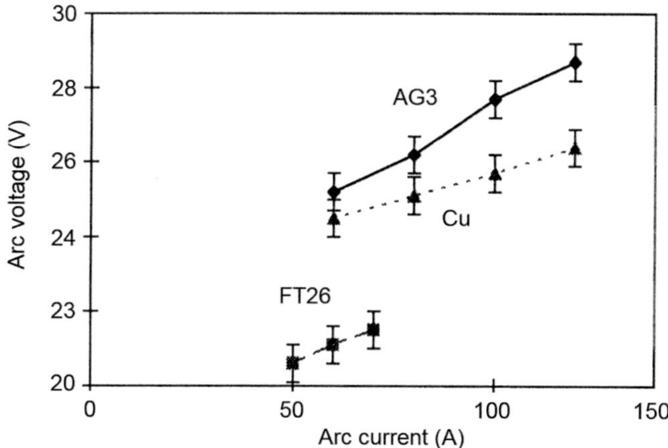

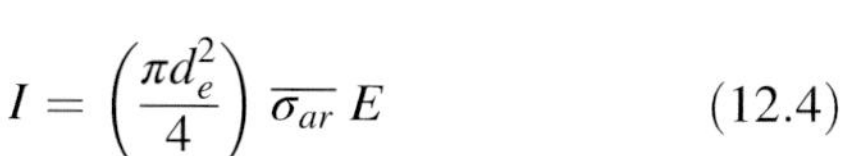

Fig. 12.19   Arc voltage as function of arc current for a PTA torch operating with 4 slm of (Ar) and Cu substrate for different torch-to-substrate distances(dz) [Leylavergne et al. (1998)]

Fig. 12.20   Arc voltage as function of arc current for different anode materials for a PTA torch operating at 4 slm of (Ar) and with a torch-to-substrate distance of 6 mm, substrate materials were Al alloy (AG3), cast iron (FT 26), and copper [Leylavergne et al. (1998)]

$$I = \left(\frac{\pi d_e^2}{4}\right) \overline{\sigma_{ar}}\, E \qquad (12.4)$$

$$d_e = 2 \times \sqrt{\frac{I}{\pi\, \overline{\sigma_{ar}}\, E}} \qquad (12.5)$$

with I the arc current, $\overline{\sigma_{ac}}$ the electrical conductivity of argon averaged over the arc cross section, and E the electric field intensity. Increase of the plasma gas flow rate (Figs. 12.18d, e, f) has essentially a similar constricting effect on the arc

giving rise to an increase of the plasma temperature in front the substrate and consequently an increase of the heat flux distribution to the substrate.

Because of the constricted nature of the arc in a PTA torch, an increase of the arc current gives rise to an almost linear increase of the arc voltage as shown in Fig. 12.19. The same trend is observed at all torch-to-substrate distances "standoff distances" with the arc voltage increasing, almost linearly, with the increase of the arc length [Leylavergne et al. (1998)]. Figure 12.20 shows the arc voltage Vs. Current relationship depends also on the nature of the substrate material (anode). The highest voltage was observed for Al alloy (25 V at 60 A and 6 mm) and the lowest voltage for cast iron (~21 V at 60 A and 6 mm).

12.4.2 Heat Flux to the Substrate

Analytical expressions for the heat fluxes to the substrate (anode) derived by Jenista et al. (1997a) and Heberlein et al. (2007) showed that the anode heat flux reflects the contribution of a wide range of different phenomena involved as given in Eq. 12.6:

$$q_a = j_e\phi_a + q_e - \kappa_e \frac{dT_e}{dz} - \kappa_h \frac{dT_h}{dz} + j_i(E_i - \phi_a) + Q_R \quad (12.6)$$

with q_a the anode heat flux; j_e the electron current density; ϕ_a the workfunction of the anode material; q_e the electron enthalpy flux; κ_e, T_e, and κ_h, T_h the thermal conductivities and temperatures of the electrons and the heavy particles, respectively; j_i the ion current density, E_i the ionization potential of the plasma gas; and Q_R the heat transfer by radiation [Heberlein et al. (2007)]. The first term on the right-hand side is due to the incorporation of the electrons into the metal lattice (electron condensation); the second term is the enthalpy flux carried by the electrons (electron enthalpy); the third and fourth terms are the (electron conduction) and (heavy particle conduction), respectively; and the fifth term is the energy release due to recombination of ions diffusing to the anode surface (ion recombination); and the last term expresses the radiation heat flux from the plasma to the anode.

Radial profiles of the local heat flux distributions for a 100 A argon arc are given in Fig. 12.21. These show that the

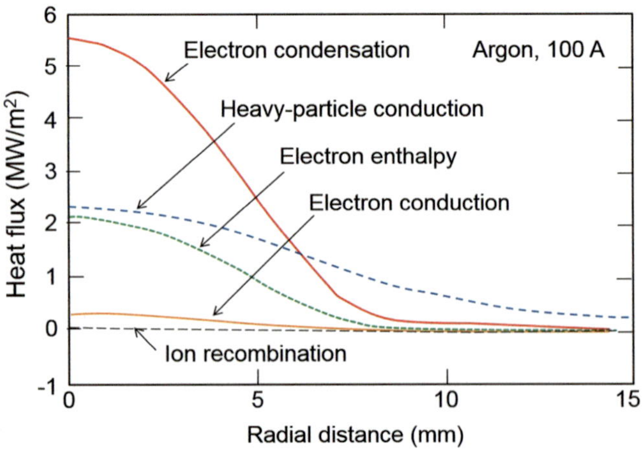

Fig. 12.21 Contributions to the heat flux to the substrate surface for a 100 A argon arc, at atmospheric pressure with a diffuse anode attachment [Jenista et al. (1997)]

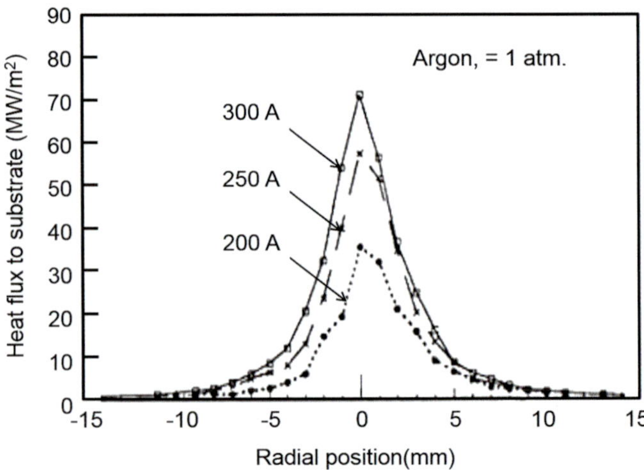

Fig. 12.22 Measured anode heat flux distributions for different arc currents in argon at 1 atm. [Menart (1996)]

electron condensation, electron enthalpy, and heavy particle conduction terms are the most important contributors to the heat transfer to the anode. [Jenista et al. (1997), Heberlein et al. (2007), Yang et al. (2008)]. The model predictions are in good agreement with experimental data obtained for plasma arc welding. Heat fluxes' distributions to the substrate measured using an especially designed heat flux probe for TIG type welding arcs for different arc currents in argon are given in Fig. 12.22 [Menart (1996)]. These show the heat flux to the substrate increases with the increase of the arc current. Higher heat fluxes in a PTA torch are expected due to the higher constriction of the arc the higher plasma gas flow rates.

Measurements of heat fluxes to the substrate were also reported by Leylavergne et al. (1998), for a PTA torch with the pilot arc off and without powder injection. These were obtained using a calorimetric method, for different arc currents (60–120 A), different torch-to-substrate distances (4, 6, 8 mm), different plasma gas flow rates (3, 4, 5 slm Ar), different shield gas flow rates (0, 20, 30, 40 slm He), and substrates consisting of Al alloy, cast iron, and copper. Typical results are given in Fig. 12.23, and Fig. 12.24 shows a practically linear increase of the heat flux to the substrate (anode) with the increase of the arc current and the torch-to-substrate distance. Assuming that only the heat transfer by convection and radiation is dependent on the anode distance, an extrapolation to a zero distance will give the heat flux due to the current transfer to the anode, which is independent of distance. The total arc power, torch efficiency, and different heat flux values are given in Fig. 12.24 and in Table 12.3 [Leylavergne et al. (1998)].

Considering that, as noted earlier, the arc voltage shows a dependence on the anode material, it is not surprising that the anode heat flux is also anode–material dependent with, for

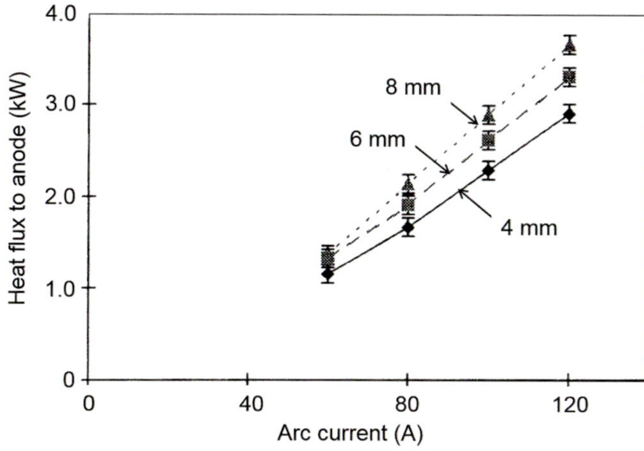

Fig. 12.23 Anode heat flux as function of arc current and arc length for a Cu anode without shield gas and 4 slm (Ar) plasma gas flow [Leylavergne et al. (1998)]

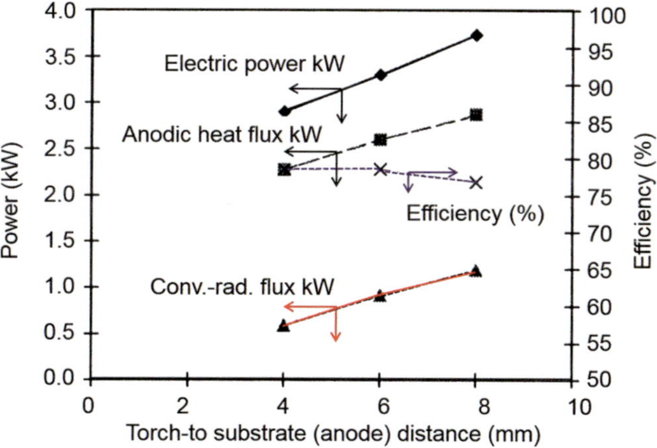

Fig. 12.24 Measured values of electric power input, anode heat flux, and derived values of convective–radiative power loss from the arc, and the arc heating efficiency for arc with plasma gas flow rate of 4 slm (Ar), arc current of 100 A and Cu substrate [Leylavergne et al. (1998)]

Table 12.3 Arc power and heat transfer to the anode for different torch-to-anode distances for Cu anode and 100 A arc current, data from Leylavergne et al. (1998)

Torch-to-anode distance (mm)	4	6	8
Arc power ($V \times I$) (kW)	2.9	3.3	3.7
Q_{an} (kW)	2.3	2.55	2.8
Q_{con} (kW)	0.6	0.9	1.2

example, the transfer to a copper anode being about 50% higher than that for a cast iron anode. While the cast iron substrate has the largest extent of evaporation and molten droplet ejection, the mechanisms could not account for the difference in the measured heat fluxes. Based on a dimensionless analysis, the following correlation between the operating parameters and the substrate heat transfer to the anode was reported by Leylavergne et al. (1998)

$$Q_{an} = 62.4\, I^{1.36}\dot{m}_g^{0.046} d_z^{0.382} \qquad (12.7)$$

with Q_{an} the anode heat flux in (W), I the arc current in (A), $\dot{m}_g$ the plasma gas flow rate in (kg/s), and d_z the torch-to-anode distance in (m).

Measurements of substrate heat flux distribution in PTA system were reported by Dilthey et al. (1993) using a split anode arrangement similar to one described first by Nestor (1962). The total heat flux to the substrate was measured thanks to two halves of a copper anode, thermally insulated from each other and separately cooled. The heat flux distribution was determined calorimetrically by moving the substrate across the arc attachment. Assuming a rotational symmetry, the heat flux distribution was described by as function of a "constriction parameter, k"[Dilthey et al. (1993)] defined as:

$$q(r) = q_m\, exp\left(-kr^2\right) \qquad (12.8)$$

with $q(r)$ the local heat flux to the substrate, q_m the maximum heat flux for a given condition, and r the distance in the radial coordinate. The authors reported that for a nozzle i.d. = 4 mm, the constriction parameter k varied for a nozzle diameter of 4 mm, plasma gas flow rate of 3.9 slm, without transferred arc, from 1.9 to 2.9 cm^{-2} when the pilot arc current was increased from 120 to 250 A. For a transferred arc current of 120 A, the corresponding values were 2.6 to 3.9 cm^{-2}. The associated heat fluxes are shown in Table 12.4 [Dilthey et al. (1993)]. Decreasing the nozzle i.d. to 2.5 mm, and the plasma gas flow rate to 1.5 slm, the constriction factor values ranged from 2.45 to 3.45 cm^{-2} without transferred arc for a pilot arc current variation of 100–300 A and from 4.15 to about 4.75 cm^{-2} for pilot arc current variation from 40 to about 250 A, with a transferred arc current of 120 A. The associated heat fluxes were 150–500 W and 2130 to about 2300 W, respectively.

It must be emphasized that Q_{sub} is the total heat transfer to the substrate and that these values include the contribution of the pilot arc power, i.e., are not only due to the anode heat transfer. Substrate melting and coating dilution are determined not only by the substrate heat transfer but

Table 12.4 Substrate heat transfer for different pilot arc and transferred arc currents, for nozzle diameter of 4 mm and a plasma gas flow rate of 3.9 slm(Ar) [Dilthey et al. (1993)]

I_{pilot} (A)	I_{arc} (A)	Q_{sub} (W)	$k\ (cm^{-2})$
120	0	275	1.9
200	0	500	
250	0	530	2.9
120	120	2220	2.6
200	120	2265	
250	120	2280	3.9

also by the distribution of the heat flux, expressed by the constriction parameter k. It can be clearly seen that substrate heating is due to the transferred arc, and the pilot arc contributes only in the order of 10–20%. About 32–48% of the electrical energy goes into heating of the substrate [Dilthey et al. (1993)].

The same authors also measured calorimetrically the heat content of the particles of the material to be deposited and derived mean particle temperatures from these values for CrNi steel particles with a diameter range from 100 to 160 μm. The results are shown in Table 12.5 [Dilthey et al. (1993)]. It is clear that most particle heating is due to the transferred arc. It can be seen from these results that the transferred arc current has the strongest effect on the process, both with respect to substrate heating/dilution and with respect to particle heating/deposition rates.

In the development of a process in which the plasma gas flow was modulated, Ebert et al. (2009) measured the stagnation pressure at the substrate surface with an orifice in the substrate connected to a pressure sensor. Moving the torch over the substrate gave radial pressure profiles which can be used to estimate the corresponding impact velocity profiles of the plasma on the substrate. Variation of the peak stagnation pressures with the plasma gas flow rate, for an arc current of 95 A, is given in Table 12.6.

The full width at half maximum (FWHM) of the stagnation pressure profile remained essentially constant between 3 and 4 mm with the increase of the gas flow rate from 1 to 4 slm. The pressure is also strongly influenced by the arc current increasing with the increase of the current and to a lesser extent by the nozzle design (decreasing with increasing diameter) and the substrate distance (decreasing with increasing distance). With higher stagnation pressure on the substrate, a displacement of the molten metal must be expected. Modulating the plasma gas flow or arc current will result in molten metal movement that will give a more uniform distribution of the particles feed into the molten metal pool.

Table 12.5 Mean temperatures for CrNi steel particles, 100–160 μm diameter with plasma gas flow rate of 1.3 slm (Ar) [Dilthey et al. (1993)]

I_{pilot} [A]	I_{arc} [A]	Mean particle temperature [K]
150	0	848
0	120	1641
150	120	1845

Table 12.6 Stagnation pressure on the substrate as function of the plasma gas flow rate for an arc current of 95 A [Ebert et al. (2009)]

Plasma gas flow rate (slm) Ar	1.0	2.0	3.0	4.0
Stagnation pressure (Pa)	330	430	525	620

12.4.3 Process Modeling

12.4.3.1 Arc Gas Dynamics and Heat Transfer

While numerous models and simulations exist for tungsten inert gas (TIG) and metal inert gas (MIG) welding arcs, very little modeling effort of the PTA process has been dedicated to the PTA deposition process. Because of the similarity between the PAC deposition process and arc welding processes, a few examples of modeling studies of arc constriction and weld pool profile in gas tungsten arc welding (GTAW) are given.

Particularly relevant is the study by Toropchin et al. (2014) of arc constriction due to the ceramic nozzle surrounding the tungsten electrode in TIG or GTAW. Schematics of some of the configurations studied are given in Fig. 12.25, showing a straight ceramic nozzle Fig. 12.25a, b, and a converging nozzle Fig. 12.25c. The corresponding shape of the arcs obtained with pure argon with each of these nozzle designs is superposed on the figure. It may be noted that the convergent nozzle design gives rise to a more constricted arc compared to that of the other two design options. While it is relatively easy to calculate the temperature and velocity distributions for the arc with pure argon as plasma gas, the problem is considerably more complex once the workpiece starts melting and vaporizing. The metal vapors evolving from the molten metal pool diffuse into the arc modifying significantly the thermodynamic, transport and radiative properties of the plasma. Moreover, in order to model the weld pool, one must also consider the different forces acting on the melt, which gives rise to internal recirculation of the molten metal. Among the forces that need to be considered are:

- *The Marangoni effect* by which the presence of a gradient in surface tension of the melt induces the liquid to flow away from regions of low surface tension
- *The drag force* acting opposite to the direction of the flow on the surface of the pool
- *The electromagnetic forces* due to the arc current
- *The buoyancy effects* which gives rise to an upward force exerted by the molten metal in an opposition to the weight of an immersed object

The conservation equations of fluid dynamics and heat transfer are used to derive the flow and temperature fields in the arc and weld profile of the molten metal pool. Typical results obtained for each of the geometries identified in Fig. 12.25 are given in Fig. 12.26 for an argon plasma with arc currents of 200 A (Fig. 12.26a) and 150 A (Fig. 12.26b, c). These show temperatures as high as 18,000 to 20,000 K near the cathode tip, with strong temperature gradients in the radial direction specially in the presence of the constricted

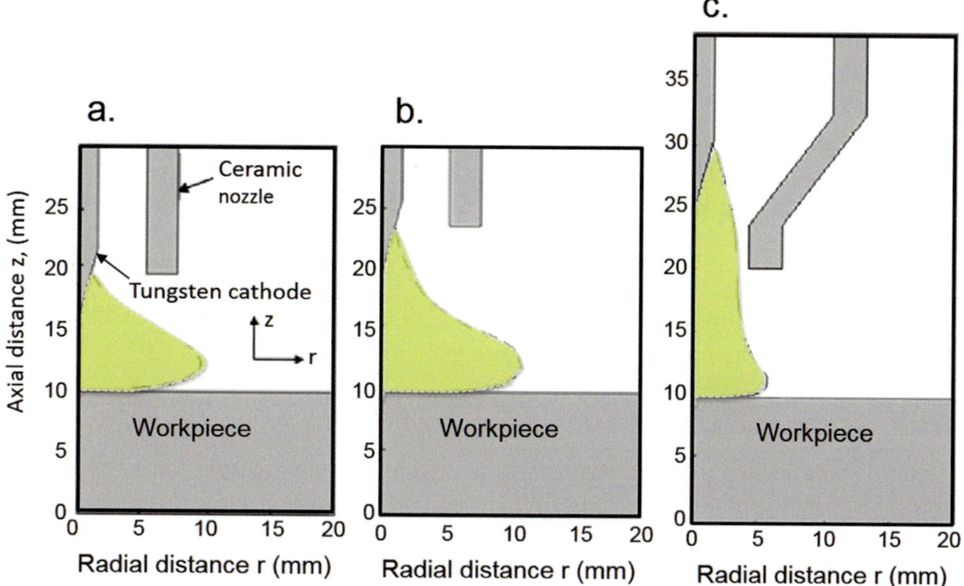

Fig. 12.25 Geometric configurations of the arc confinement nozzle with different internal diameters of the nozzle, locations of the cathode tip (Δz is the distances between the cathode tip and the anode) and the nozzle exit (**a**) i.d. $= 11$ mm, $\Delta z = 6$ mm (**b**) i.d. $= 11$ mm, $\Delta z = 10$ mm (**c**) i.d. $= 8.1$ mm, $\Delta z = 16$ mm [Toropchin et al. (2014)]

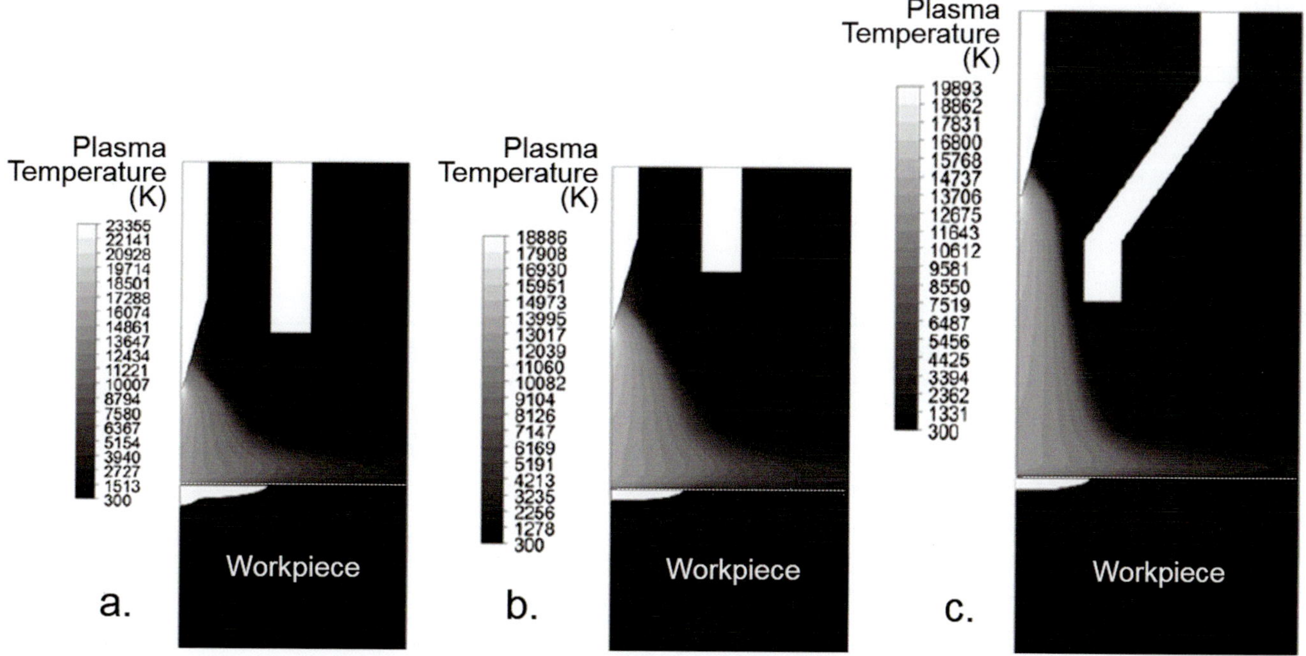

Fig. 12.26 Corresponding weld pool profiles and temperature fields in an argon arc plasma for the same geometrical parameters given in Fig. 12.24 for arc currents (**a**) I $= 200$A (**b**) and (**c**) I $= 150$A [Toropchin et al. (2014)]

ceramic nozzle shown in Fig 12.26c. The lower part of Fig. 12.26 shows the corresponding profile of the molten metal pool in the workpiece. The results are blown out for clarity in Fig. 12.27, for the case of 150 A arc using the ceramic nozzle (Fig. 12.25c). The results given in Fig. 12.27a and b also show respectively the corresponding flow vectors in the weld pool "with" and "without" inclusion of the Marangoni force in the calculation model. The markedly different weld depths in each of these two cases are related to basic material properties such as specific heat, electrical and thermal conductivity and model formulation. According to Toropchin et al. (2014), including the Marangoni effect gives rise to a dependence of the direction of melt recirculation in the molten metal pool on the variation of surface tension of the melt with temperature. Comparing a material with falling surface tension characteristic, versus a material, such as mild steel, with a rising characteristic, reviles a significantly different weld pool profiles and circulation patterns of in the molten metal pool in each of these two cases as shown in Fig. 12.28.

Fig. 12.27 Flow vectors in the weld pool for mild steel S235 at a current of 150A. (**a**) Computation including the Marangoni forces and (**b**) computation neglecting the Marangoni forces [Toropchin et al. (2014)]

It is important to point out that the weld pool profile formed in plasma arc welding depends on a large number of parameters, which includes the chemistry of the materials to be welded and that of the plasma and shield gases. The effect is clearly demonstrated experimentally by Tanaka and Lowke (2007) for the welding of stainless steel plates (SUS304) under essentially the same physical conditions with tungsten electrode 3.2 mm in diameter, distance of 5 mm between the electrode tip and the steel plate, and arc current of 150 A, at atmospheric pressure with argon as plasma gas. Photographs of the cross section of two stainless steel plates with sulfur contents of 40 and 220 ppm are given in Fig. 12.29. These show a significant difference in the corresponding weld pool profiles for each of these two plates depending on their sulfur content. The weld profile for the plates with 40 ppm sulfur being wide and shallow in sharp contrast to the profile obtained with the plates containing 220 ppm of sulfur in which the weld pool is narrow and deep.

Further support for the influence of the composition of the plasma and shield gases on the weld metal pool profile was provided by Tanaka and Lowke (2007). Micrographs given in Fig. 12.30 show typical cross sections of welds for stainless steel, conducted with an arc current of 150 A and arc gap of 3 mm, using Ar and He as plasma gases and a mixture of

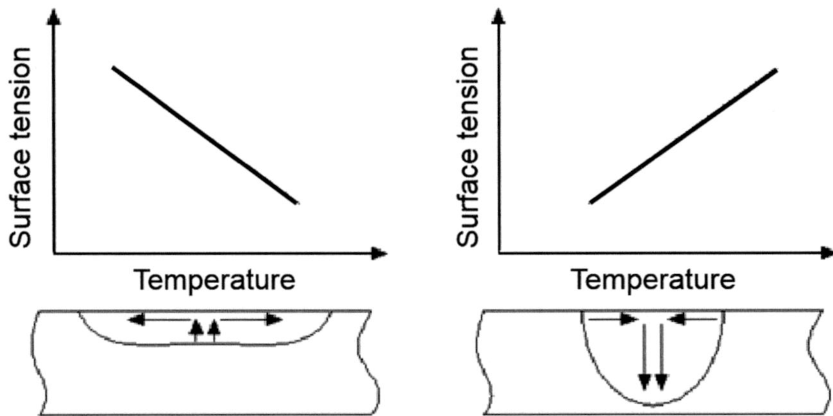

Fig. 12.28 Schematic illustration of convection in the weld pool driven by Marangoni force resulting from the temperature dependence of the surface tension [Toropchin et al. (2014)]

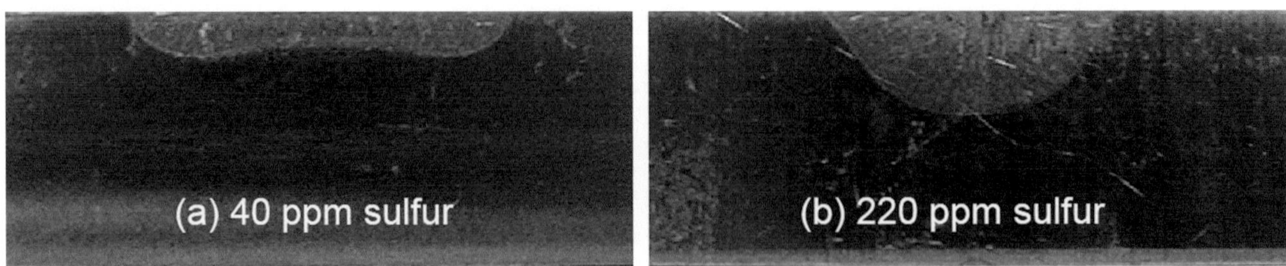

Fig. 12.29 Cross sections showing weld regions for stainless steel with different sulfur content, 150 A arcs with a 5 mm electrode separation at atmospheric pressure [Tanaka and Lowke (2007)]

Ar-CO$_2$ or Ar-10 vol.%H$_2$-CO$_2$ as sheath gas. The corresponding arc voltages varied with the plasma and shield gas composition being at 11.8 V for pure Ar while for an Ar arc shielded with CO$_2$ 15 V, and 18.9 V for Ar-10 vol. %H$_2$-CO$_2$ shielded. These show the penetration depth for the argon plasma gas CO$_2$-shielded arc to be 4.9 mm (Fig. 12.30b) and 6.9 mm for an Ar + 10 vol.% H2 mixture as the inner nozzle gas CO2-shielded (Fig. 12.30c) while the depth for the argon arc is only 2.0 mm (Fig. 12.30a). The effect of CO$_2$ shielding is due to the better arc constriction, as predicted by calculations.

Tanaka and Lowke (2007) point out that mathematical modeling is a powerful tool for the understanding and control of the arc welding process provided that the model incorporates both the arc plasma as the heat source and the workpieces. The need for such an integration of these two distinct fields is due to "the close interaction between the electrode, the arc plasma and the weld pool, which constitute the welding process, and must be considered as a unified system." Murphy et al. (2009, 2010) further confirmed the necessity to include the arc plasma in the computational domain taking into account the influence of the metal vapors released from the molten metal pool on the transport and radiative properties of the plasma. The most important effect of the metal vapor is the increased electrical conductivity at low temperatures and the significant increase of radiation heat losses from the arc column, which leads to lower heat flux density and current density at the weld pool, implying a shallower weld pool. They pointed out that the product of specific heat and mass density is particularly important in determining the arc constriction and the weld pool depth.

A modeling study of the influence of the nature of the plasma gas, Ar or He, on the arc and the weld pool was reported by Yamamoto (2008). Figure 12.31 shows the

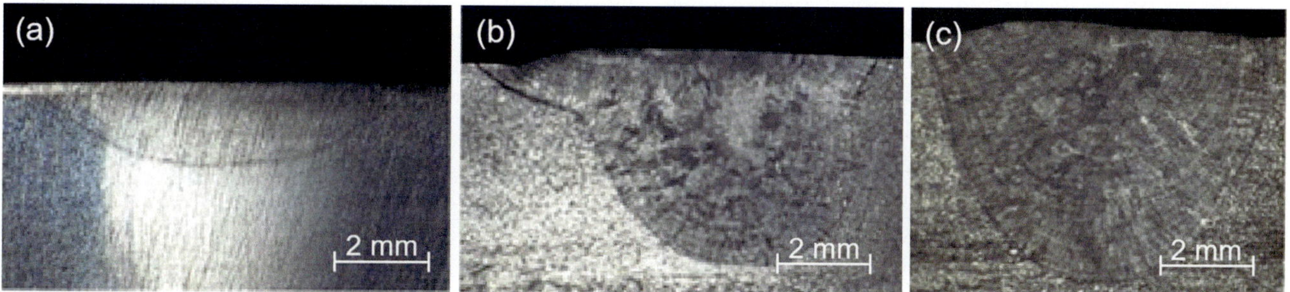

Fig. 12.30 Measured weld shape in stainless steel (SUS-304, 30 ppm sulfur) for an arc current = 120 A, arc gap =3 mm, welding speed =90 mm/min and different welding gases (**a**) Ar-TIG, (**b**) Ar + CO2, and (**c**) Ar + 10 vol.% H2 + CO2 [Tanaka and Lowke (2007)]

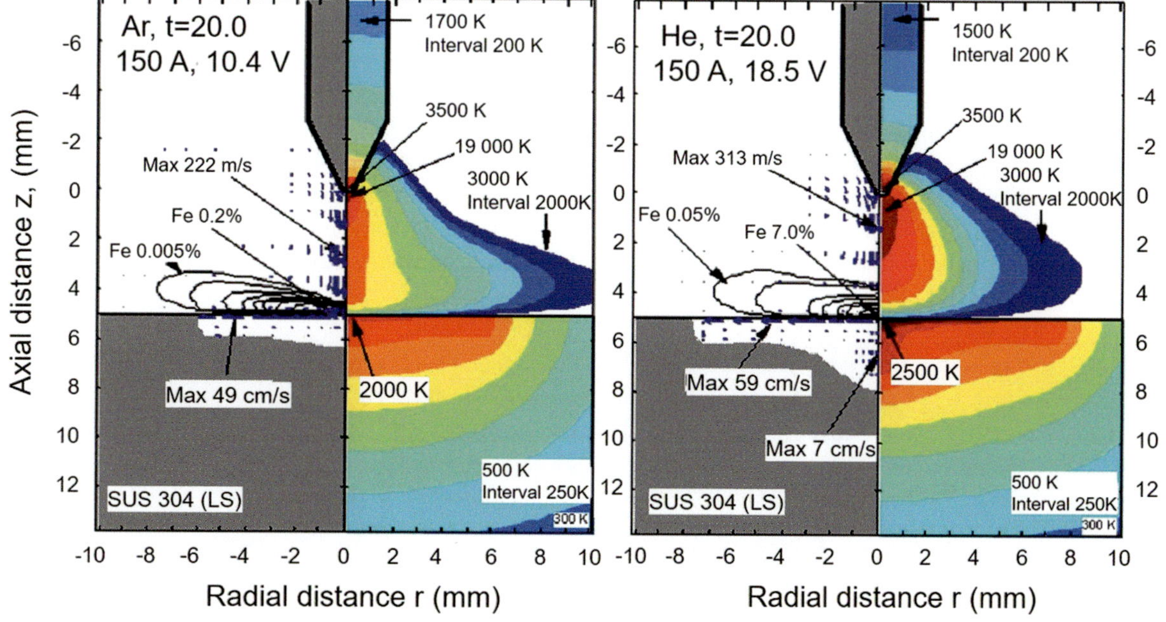

Fig. 12.31 Temperature fields (RHS) and iron vapor mole fraction fields and velocity vectors (LHS) in the arc and electrodes for 150 A arc in argon and helium at 100 kPa with a 304 stainless steel anode, after 20 s of operation. [Yamamoto (2008)]

temperatures, velocities, and iron vapor molar fractions (other components were not considered) in 150 A arc in argon and helium, with 304 stainless steel workpiece as anode. The temperature interval in the iso-contours for the plasma is 2000 K, while it is 200 K in the cathode, and 250 K in the anode. The concentration of the iron vapor released from the surface of the melt pool increases sharply with time following the initiation of the arc only to stabilize shortly after only 5 s. The maximum concentration of the iron vapor in the arc depends on the arc current and the nature of the plasma gas, reaching 0.2 vol. %, for an argon arc compared to 7.0 vol.% for the helium arc at the same arc current of 150 A. The effect is due to the higher melt pool temperature with a helium arc compared to that of an argon arc. Such high metallic vapor concentration levels can be detrimental to the welding process not only because of their influence on the plasma properties and the process energy efficiency but also because of the eventual homogenous condensation of the vapor in the colder regions of the flow forming a fine nanometer-sized fume which represent a serious health hazard to the operator [Tashiro et al. (2010)]. The use of active flux for deeper penetration can also modify the weld pool characteristics. Morisada et al. (2014) have proposed what they called "cap active" flux tungsten inert gas (CA-TIG) welding using atmospheric oxygen, to increase the penetration depth of a weld. Specially designed nozzle cap entrained air in the molten pool. The penetration depth for SUS304 stainless steel was increased by the reversal of the Marangoni convection due to the entrained oxygen, and it reached three times deeper than that of the conventional TIG welding. Authors point out that the effect of the entrained air on the mechanical properties of the joint formed by the CA-TIG welding is negligible, and the joint efficiency is high enough for conventional applications.

12.4.3.2 Melt Pool Modeling

According to Karanunakarani and Balasubramanian (2011) in conventional welding, fusion zones typically exhibit coarse columnar grains because of the prevailing thermal conditions during weld metal solidification. It often results in inferior weld mechanical properties and poor resistance to hot cracking. It is thus necessary to control solidification structure, temperatures, and thermal gradients in welds. This is especially the case when welding thin sheet materials with high thermal conductivity and diffusivity such as aluminum and magnesium alloys. Using pulsed current welding can solve the problem. When welding AA6351-T6 aluminum alloy joints pulsed current resulted in lower peak temperatures, lower magnitude of residual stresses and superior tensile properties is compared to constant current welding technique with the formation of finer grains. Razal Rose et al. (2012) studied the influence of pulse current, base current, pulse frequency, and pulse on time on tensile strength of the PC-TIG welded AZ61A magnesium. They observed that the effect of peak current and pulsing frequency initially increase the tensile properties which subsequently decreased irrespective of changes in base current and pulse on time. A maximum tensile strength of 199.5 MPa was obtained with the optimum welding parameters. Pulse current was more sensitive than the other parameters followed by pulsed on time, pulse frequency, and base current.

Modeling studies were also dedicated to the identification of the impact of substrate position on the final profile of the formed coating. The problem stems from the fact that PTA welding is presently restricted to substrates which are in a flat horizontal position. This means that damaged parts have to be dismounted to be processed. Furthermore, industry is interested in the development of strategies for PTA coating in constraint positions as complex three-dimensional (3-D) parts could be easily processed as well.

Photographs [Wilden et al. (2004)] of nickel coating on flat steel substrates obtained using PTA under "standard conditions" in horizontal and vertical orientations are given in Fig. 12.32. It can be noted that the weld obtained with the substrate in a horizontal position, Fig. 12.32a with an arc current of 80 A, powder feed rate of 20 g/min, and substrate displacement speed of 30 mm/s, is well defined with single weld beads clearly visible. In contrast, the weld obtained with the substrate in vertical position, Fig. 12.32b with the same conditions shows the molten metal flowing downward due to

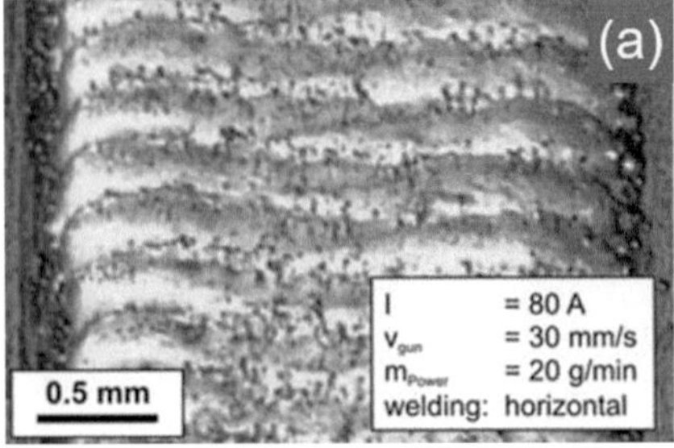

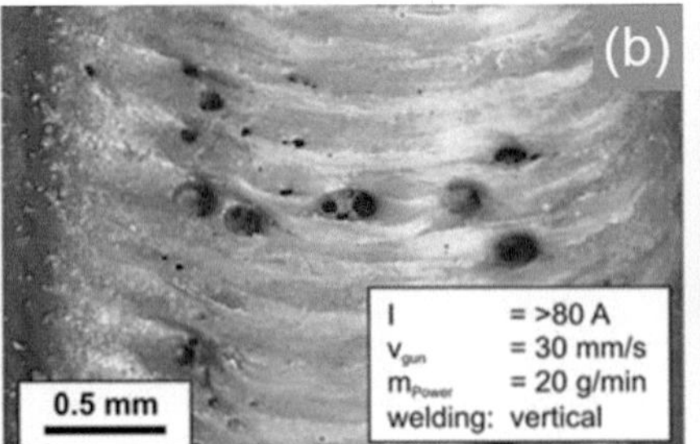

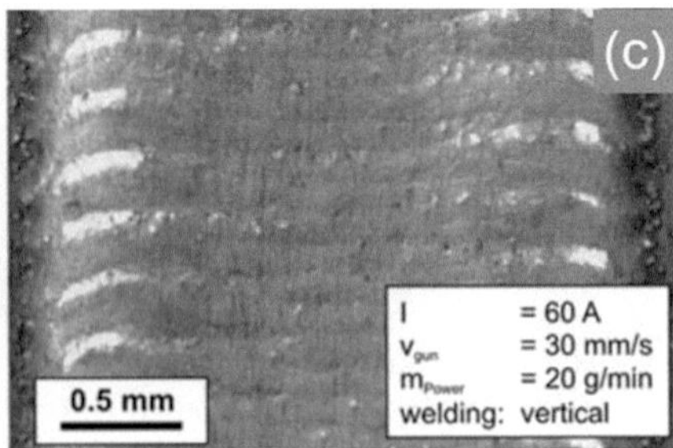

Fig. 12.32 Photographs of PTA welded nickel coating on steel substrate in different welding positions (**a**) horizontal, 80 A (**b**) vertical, 80 A (**c**) vertical, 60 A [Wilden et al. (2004)]

gravitation resulting in the formation of inhomogeneous coating with visible pores which was attributed by Wilden et al. (2004) to a too long solidification time. The reduction of the arc current from 80 A to 60 A seems to reduce such downward flow of the molten, leading to the formation of well-defined single overlapping weld beads as shown in Fig. 12.32c.

In an effort to better understand the problem and identify appropriate remedies that would allow for the use of the PTA technology in constraint positions, Wilden et al. (2006) developed a three-dimensional (3-D) model of the PTA process which they applied to the deposition of thin coatings ($\sim$500 µm) of Ni on steel with the substrate either in its conventional horizontal position or in a vertical orientation. Their model is based on the solution of the corresponding conservation equations, with the exception that the arc channel is simply represented by a 2–2.5 mm diameter channel with uniform heat dissipation, and no interaction with the pilot arc anode. The energy is transferred to the substrate and the feed material. The following operating parameters were used: pilot arc current, $I_{pilot} = 10$–50 A, plasma arc current, $I_{arc} = 70$–170 A, $V_{pilot} = 20$ V, arc voltage = 20 A, plasma gas flow rate = 1.5 slm (Ar), powder gas flow rate = 0.4 slm (Ar), powder feed rate 0.48–1.2 kg/h, and substrate velocity = 10–90 mm/s for 7 mm, with motion reversal following a step of 1 mm.

Typical simulation results obtained for a substrate in a horizontal position with a powder feed rate of 12 g/min, different arc powers and substrate speeds, are given in Fig. 12.33. These show that with a plasma arc power of 3.0 kW and a process velocity of 30 mm/s, Fig. 12.32a, a relatively flat surface is formed. If energy input is reduced to 2.4 kW and process speed is increased to 90 mm/s (reduced energy input per unit length), Fig. 12.33b, a rough surface with weld bead irregularities is observed.

Wilden et al. (2006) also investigated the effect of the substrate thickness on the molten pool formation during the PTA coating and its impact on the quality of the coating.

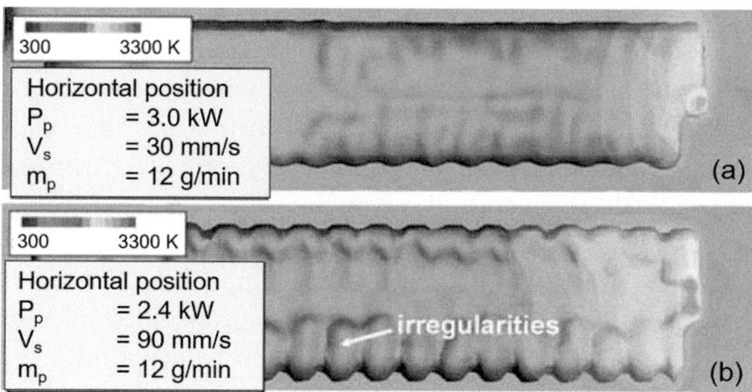

Fig. 12.33 Geometry of PTA welded Ni coating on a substrate. Temperature field (**a**) arc power 3.0 kW, process speed 30 mm/s; and (**b**) arc power 2.4 kW, process speed 90 mm/s [Wilden et al. (2006)]

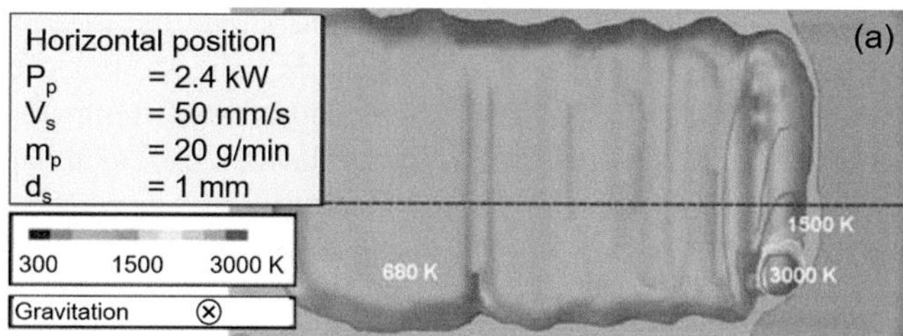

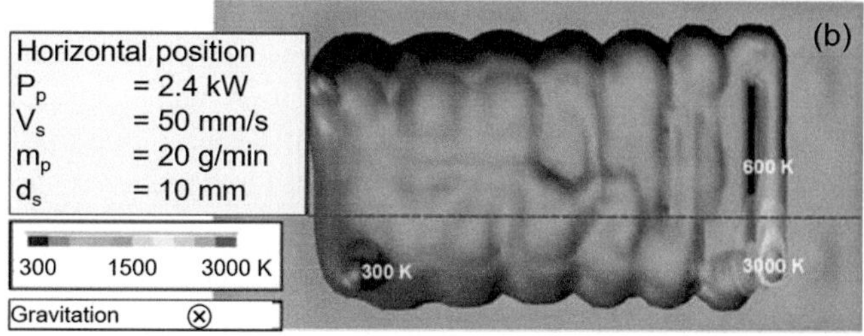

Fig. 12.34 Geometry of PTA welded Ni coating on a steel substrate in a horizontal position: (**a**) substrate thickness, 1 mm; (**b**) substrate thickness, 10 mm [Wilden et al. (2006)]

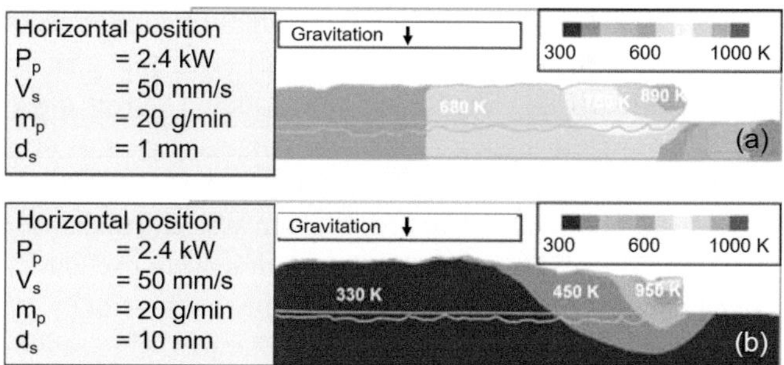

Fig. 12.35 Cross section of PTA welded Ni coating on a steel substrate in a horizontal position: (**a**) substrate thickness, 1 mm; (**b**) substrate thickness, 10 mm [Wilden et al. (2006)]

Results given in Figs. 12.34 and 12.35 were obtained for the deposition of a Ni coating on steel substrates of thickness $d_s = 1$ mm and 10 mm in a horizontal position, with Ni powder feed rates of 20 g/min, arc power of 2.4 kW, and substrate translation speed of 50 mm/s. The cross-sectional views given in Fig. 12.35 were made along the dotted line shown in Fig. 12.34 across the welded Ni coating. It may be noted that with a thin substrate, $d_s = 1.0$ mm, shown in Figs. 12 34a and 12.35a, significant heating of the substrate occurs during the coating process giving rise to the formation of a large melt pool which is difficult to handle. In contrast, the results obtained with the thicker substrate, $d_s = 10.0$ mm, shown in Figs. 12.34b and 12. 35b, the melt pool and the high temperature region are considerably less expanded than with the thin substrate which can be attributed to the considerably high heat dissipation capacity of the thick substrate.

The reduced coating temperature increases the viscosity and the surface tension and reduces the size of the melt pool as well. The corresponding cross section (along the dashed line shown in Fig. 12.34) is given in Fig. 12.35. For the thinner substrate, Fig. 12.35a, coating and substrate temperature is reduced from about 1000 K near the melt pool to about 600 K in the cooled area. This is about 300 K higher than for the thick substrate, Fig. 12.35b. Additionally, the higher heat capacity and the lower coating temperature of the thicker substrate slightly reduce the material dilution in the weld bead.

Vertical welded coatings show modified properties compared to welding in flat horizontal position with identical parameters. During welding in constraint position, one must consider that coating geometry is smoother and molten material can flow downward in the direction of gravitation [Wilden et al. (2006)]. However, the power level and the deposition velocity must be adapted, generally reduced, when coating a substrate in a vertical position. Due to the flow of molten material, a better surface quality can be reached in vertical deposition, as illustrated in Fig. 12.36 representing coatings obtained in vertical spray positions with two different welding velocities.

Further modeling studies of the coating formation process were reported by Matthes et al. (2002) with the goal to identify the stress distributions in free standing shapes made by PTA. The heat input into the substrate was given as heat flux across circular area with Gaussian distribution. The size of the weld pool and the bead formation was calculated for linear motion of the torch with superimposed oscillatory movement with an amplitude of 28 mm and velocity of 7 mm/s, while the traverse velocity was 2 mm/s. The oscillation resulted in lower molten pool temperature but with larger area covered by the molten metal. Calculated temperature distributions of the substrate/coating surface were in good agreement with experimentally determined values. The stresses after solidification were tensile close to the bead and compressive farther away from it. The calculated stress values also agreed with the experimentally determined ones.

It should be noted that while many of these modeling studies have been performed without powder injection, they allow estimates of how the fundamental properties change with the process parameters. Powder deposition would likely enhance the heat transfer because the deposition of the molten material would transfer some of the arc power that would otherwise be lost to the torch.

12.5 Coating Materials and Applications

The PTA process gives access to a large number of coating materials, various alloys, metal compounds and metal matrix composites, and there are continuously new coating materials being developed. Most applications of PTA coatings deposition technologies are for corrosion and wear resistant, taking full advantage of the superior adhesion and high density of the PTA coatings, and the possibility to select a wide range of materials and composite systems by mixing the appropriate powders. Wear is the problem, and it occurs in almost all industries where thermal sprayed coatings are used. In all cases, it is a progressive loss of material at the active surface due to the relative movement of another part or particles on this surface. The lifetime of the coating depends on its resistance to wear and its thickness, which can reach up to 10 mm with PTA coatings. Wear can also be linked to thermal fatigue initiating cracks generally propagating into the coating along carbide boundaries; these hard particles being generally used to reinforce the coating hardness and wear resistance. Such coatings are achieved using two types of powders: a metal powder close to the PTA nozzle exit and heated below their melting temperature and a ceramic one close to the molten bath to be included unmolten within the coating. Coatings have been developed for wear under different corrosive conditions, for abrasive wear, and for wear by erosion due to interaction with particulate matter.

It is important to point out that PTA coatings can be built to thicknesses of several millimeters with excellent high temperature stability, which is not always possible with other plasma spraying techniques. For example, d'Oliveira et al. (2002) characterized coatings of a cobalt-based alloy, deposited by PTA welding on an AISI 304 stainless steel substrate. Samples were subjected to temperature cycling at 1050 °C. As-deposited features and their stability to high temperature were studied; results were presented and compared with similar samples obtained by laser cladding. The performance of hard-faced material was evaluated by microstructure, chemical analysis, and microhardness measurements. For the processing conditions tested, results

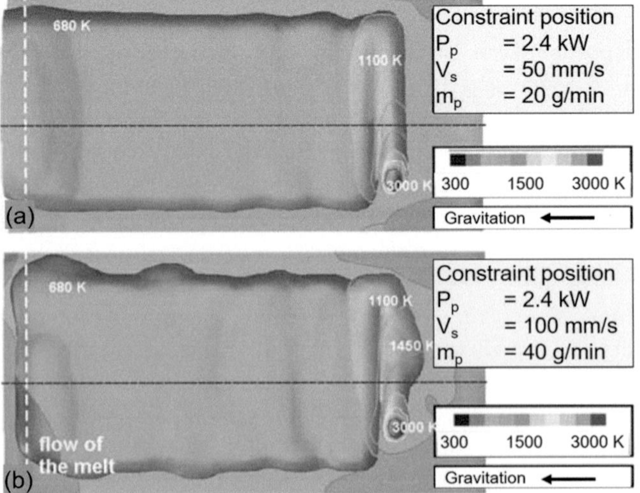

Fig. 12.36 Simulation of PTA welded Ni coating in vertical welding position with different process parameters. Temperature field (**a**) 2.4 kW, 50 mm/s; (**b**) 2.4 kW, 100 mm/s [Wilden et al. (2006)]

showed that, for operating conditions that do not require elevated temperature, laser hard-facing showed the bests results, due to the refined solidification structure produced. However, for the high temperature tested, PTA deposits exhibited superior microstructural stability, whereas laser deposits undergo significant changes. High-temperature treatment led to changes on carbides morphology and also to precipitation, in the case of laser deposits.

Another important application is the restoration of worn parts to original dimensions. Frequently the restored parts have better corrosion and wear characteristics than the original ones [Ducos (1985)]. The reclamation of large parts such as extruder screw, vanes, turbine blades, engine valves (especially big diesel engines), and rail train car wheels are commonly carried out using standard PTA technology. Micro-PTA which meet requirements of the meso-sized fabrication are used for the repair of small parts and high value components.

The majority of the substrate materials used are steels or other ferro-alloys, although with the increase in use of lighter metals for reducing component weight, coatings on aluminum or magnesium alloys are becoming more important. Light metals, on the other hand, tend to have considerably higher thermal diffusivities (9.8×10^{-2} m²/s for Al and 8.77×10^{-2} m²/s for Mg) compared to that of steel AISI 1010 (1.88×10^{-5} m²/s) resulting in a broad heating zone, which is not easy to control as the substrate temperature approaches its melting point.

PTA coatings are used in automotive industry, petroleum, chemical industries, and mining industries, where slurries are used. Micro-PTA coatings are also used with electrical and electronic industries, because very thick (up to 10 mm or more) copper coatings can be sprayed. When applied on powerful computer systems, they improve their cooling and achieve a good heat distribution.

12.5.1 Metals and Alloys

Plasma transferred arc (PTA) weld surfacing has attracted increasing attention for deposition of wear, corrosion, and heat resistance materials at high deposition rates and low heat input to the substrate [Hou et al. (2006)]. In general, coatings consist of a Fe-Cr or Fe-Cr-Ni alloys with the addition of either a refractory metal such as W, Mo, Nb, Ta, or B, Si, and C [DuMola and Heath (1997)]. The most widely used wear-resistant coatings are Fe-Cr-Mo-C (e.g., for equipment in the construction and mining industries) or Co-Cr-W-C for wear with better corrosion resistance (e.g., food processing knives, drills, valves). Stellite coatings of valve seats are also widely used. Alloys of Fe-Cr-V-C-B have excellent wear properties and are more economical than some of the cobalt or nickel-based alloys, with the hardness attributed to the formation of carbo-borides.

Other alloys that can form hard phases during spraying and solidification or after treatment are also used. For example, Fe-C-Cr-Cu coatings forming carbides upon spraying and ε-Cu phase upon treatment at 773 K for 35 h, or boride layers on AISI 1018 steel developed using the PTA alloying technique. Hou Q.Y. et al. (2006) have deposited by PTA (210–215 A) spherical particles (75–180 μm), with the chemical composition (C-1.4, Si-1.45, Cr-29.5, Fe-3, Ni-3, W-8.25, Mo-1, Co-Bal) on substrates $200 \times 35 \times 20$ mm annealed low carbon steel plates containing 0.12 wt.% carbon and which have not been preheated before PTA treatment. They also studied aging effects on the structure and wear resistance of the cobalt-based PTA layer. They found that the As-welded coating consisted of a cobalt-based solid solution with face-centered cubic crystal structure and hexagonal (Cr, Fe)7C$_3$. Unfortunately, there were lots of stacking faults existing in the cobalt-based solid solution. After aging at 600 °C for 60 h, precipitates of other carbide $M_{23}C_6$ with face-centered cubic crystal structure were formed. While the precipitation of this carbide can in principle increase the hardness, their course microstructure and the change in the morphology of the coating led to the increase of the wear weight loss.

The addition of molybdenum (6 wt. %) on nickel-based alloy coatings obtained by PTA was studied by Hou Q.Y. et al. (2007). The morphology of Cr-rich carbides observed in the inter-dendritic region of the nickel-based alloy coating changed from plate-like to net-like, as well as the refinement of Ni-rich dendrites because of this addition. The weight fraction of Cr-rich carbides increased from 36 wt. % to 45 wt. %. The wear resistance of the modified nickel-based alloy coating increased by 47.2%. The microstructural change and phase contents variation were responsible for the improvement of the wear resistance.

Hou Q.Y. et al. (2007) deposited FeCrBSi alloy powders without and with 2–6 wt. % Mo on steel using plasma transferred arc (PTA). The results showed that the Mo-free coating consisted of γ (Fe, Ni), $M_7(C, B)_3$, and $(Fe, Cr)_2B$ phases. Adding Mo leads to the formation of $M_{23}(C, B)_6$, $Mo_2(B, C)$, and $Fe_3Mo_3(C, B)$ phases, in addition to the phases existed in the Mo-free coating. A hypoeutectic microstructure was seen in the Mo-free and 2 wt. % Mo-added coatings. Increasing Mo addition to 4 wt.% or 6 wt.%, a hypereutectic microstructure was obtained. The microstructure of the Mo-free coating was refined after adding 2–6 wt.% Mo. The finest microstructure was obtained in the 4 wt. % Mo-added coating. The thermal shock resistance of the Mo-free coating was improved after adding 2–6 wt.% Mo. The best abrasive wear resistance and thermal shock resistance were obtained in the 4 wt. % Mo-added coating.

Liu Y.-F. et al. (2006) deposited wear-resistant Fe_2TiSi reinforced composite coating with a microstructure consisting of fine Fe_2TiSi primary dendrites uniformly distributed in the super fine γ-Fe/Ti_5Si_3 eutectic matrix was in situ fabricated on a substrate of 0.20 wt. %C plain low carbon steel substrate by the plasma transferred arc (PTA) cladding process using Fe-Ti-Si-Cr powders blend as the precursor material. The mixed powder blend can be directly fed into the plasma. Wear-resistant Fe_2TiSi/γ-Fe/Ti_5Si_3 multi-phase composite coating was fabricated on the carbon steel substrate. Microstructure of the coating consisted of fine primary Fe_2TiSi dendrites uniformly distributed in the super fine γ-Fe/Ti_5Si_3 eutectic matrix. A small amount of Ti_2N course particles was scattered in some local area of the coating. The Fe_2TiSi/γ-Fe/Ti_5Si_3 composite coating has excellent abrasive and adhesive wear resistance under dry sliding wear test conditions at room temperature.

Guoqing et al. (2013) studied microstructure and wear properties of nickel-based surfacing deposited by PTA. The chemical composition of the Ni-based particles was composed of Fe 3.0.5, Cr 10.47, Si 3.87, C 0.48, B 2.35, and Ni bal. (wt. %). The coating on AISI 304 L substrate was obtained by using the PTAW process. The As-deposited coating was composed of Ni-rich γ (Ni, Fe) phase, Cr_7C_3, CrB, Cr_3C_2, $M_{23}C_6$, Ni_3B, and Ni_3Si. At the applied load of 120 N, the wear rate was almost twice than the rates under loads of 30 N and 70 N. The main wear mechanism varied with the wearing time and the applied load. At the initial stage of the wear test, the wear mechanism was abrasive wear under low load and adhesive wear under high load. After long-time wearing, the wear mechanisms were adhesive wear and oxidation wear under low load, and fatigue wear appeared under high load.

Liyanage et al. (2010 studied the influence of alloy chemistry on microstructure and properties in NiCrBSi overlay coatings deposited by plasma transferred arc welding. Two alloys with slightly different percentages of species were used. It was found that the hardness of a typical NiCrBSi alloy deposit was controlled not only by the number of Cr-rich particles present but also by the volume fraction of the inter-dendritic phase. The inter-dendritic phase was shown to exhibit high hardness of about 860 HV relative to the primary dendritic phase, which had a hardness of 405 HV. Since the volume fraction of the inter-dendritic phases is higher for the harder alloy (~55% in Alloy B), it significantly contributes to the overall properties of the deposit. Although there was only a minor increase in Cr, C, and B content in Alloy B, this led to a large increase in the number of Cr-particles from 1.5 to about 15 vol.% between the two alloys examined. Ternary phase diagrams suggest that the addition of Fe or Si will decrease the solubility of Cr in Ni; however no significant change Cr segregation was observed in the alloy compositions examined. It was also found that boron segregated to the inter-dendritic regions in both alloys and promoted the formation of Ni_3B and CrB. Micro-hardness testing revealed that the primary Ni dendrite, inter-dendritic, and Cr-particle phases had average hardness values of 405, 860, and 1200 HV, respectively. An increase in the volume fraction of hard eutectics and Cr-particles led to a substantial increase in hardness and wear resistance.

Ozel et al. (2008) coated NiTi powder mixture on the surface of an austenitic stainless steel (AISI 304) by plasma transferred arc (PTA). Thickness of coating increased with current density. No cracks or pores were detected in the interface. Higher arc current densities caused a coating layer poor in NiTi alloy. Secondary phases such as Cr_2Fe_7Ni, Fe_7Cr_2Ni and Ni_3Ti are also formed in the coating layer.

Sigolo et al. (2016) tested the wear resistance of coatings of boron-modified stainless steels deposited by PTA. The deposition was carried out on AISI 4140 steel substrate. Microstructural characterization revealed dendritic growth in both cases; however, the number of borides formed was quite different, around 14% for super-martensitic with 1 wt. % B and 32% for super-duplex with 3 wt.% B. Thermodynamic calculations were in accordance with the experimental results. In dry sand/rubber wheel, the volume loss of quenched and tempered AISI 4140 was superior to super-martensitic steel with 1 wt.% B; nonetheless it was comparable to super-duplex steel with 3 wt.% B. The formations of hard borides increased both the hardness and overall wear resistance of boron-modified stainless steels

Lu et al. (2011) produced functionally graded coatings by plasma transferred arc centrifugal cladding. The torch (320 A-40 to 50 V) was moved axially inside the rotating cylinder to be coated (80 mm in internal diameter with wall 5 mm thick). Fe–Cr–M–C (M = Mo, W) powders were selected as the powder to fabricate functionally graded coating on the internal walls of cylinders. Wear resistance of the graded coating could be improved by around 18 times compared with the plain cylindrical substrate.

[Zhao and Liu (2006)] manufactured structural Ni-based super alloy (GH163) by PTA on a carbon steel (A3) substrate. The deposited samples exhibited a fine and homogenous cellular dendrite structure with a stronger crystallographic orientation along <001> direction. After ageing treatment, γ' phase and two types of metal carbides (i.e., MC and $M_{23}C_6$) with a size of 20–30 and 50–100 nm, respectively, precipitated. γ' phase was mainly located inside the grains and appeared perfectly coherent in relationship with γ matrix. On the other hand, metal carbides mainly precipitated at grain boundaries.

d'Oliveira et al. (2006) evaluated the effect of pulsed current on high carbon cobalt alloy deposited by PTA on carbon and stainless steel. Results showed that the use of pulsed current leads to a finer microstructure, higher hardness, and lower dilution. The role of the substrate steel depended on the set of processing parameters used, but for a same set of parameters, it determined microstructural features of the coatings. Pulsed current processing resulted on finer and more homogenous solidification structures and lower dilution levels and as a consequence on coatings exhibiting higher hardness.

Iakovou et al. (2002) boride the surface of a tool steel using boron powder and PTA. It was shown that this method is an easy and effective technique in producing uniform alloyed layers with a thickness of about 1.5 mm and a hardness between 1000 and 1300 HV. The microstructure of the borided surfaces consists of primary Fe_2B-type borides and a eutectic mixture of borides and martensite. Some cracks were observed in the eutectic regions but they did not seem to critically affect the behavior of the coatings in sliding wear. The wear rate of pin on disc tests is primarily affected by the applied load, and it lies between 10^{-5} mm^3/m for low loads and 10^{-2} mm^3/m for high loads. Two distinct regimes of mild and severe wear were obtained separated by a critical load. Mild wear was due to the load supporting effect of borides, and severe wear was due to their breakage above a critical load. The wear rate was not significantly affected by the sliding velocity and was consistent with the friction coefficient. The friction coefficient varied from 0.13 to 0.23 and depended strongly on the oxidation status of the wear track. The sliding velocity affected the sliding distance where the coefficient of friction reached equilibrium.

[Kim et al. (2003)] Pin-on-disc sliding wear tests were conducted in molten Zn–0.18%Al (wt. %) bath at 470 "C to evaluate the high temperature wear performance of cobalt-based alloys and their carbide composites processed by PTA weld-surfacing technique. The main results are as follows:

- Wear performances of Stellite 6 and Tribaloy 800 with carbide composite materials are not good when coupled with same materials although their hardnesses are higher and their microstructures are finer than those of Stellite 6 and Tribaloy 800.
- Wear test results from different materials combination show that as the hardness of the counterpart disk is increased, the wear rate of the Tribaloy 800 pin is usually increased.
- Although overall wear rates of both pin and disc are similar between Stellite 6 and Tribaloy 800, wear mechanism is quite different. While harder eutectic carbides are broken and detached in Stellite 6, softer matrix is preferentially damaged without pulling out of harder phases in Tribaloy 800 during the wear testing.

Pukasiewicz et al. (2014) studied cavitation wear induced by the collapse of vapor bubbles in a fluid (usually because of a localized pressure fluctuation) which leads to mass loss as a result of fatigue caused by the successive impacts generated by the collapse of bubbles near the surface of a material. They recalled that surface re-melting is an important technique for modifying the microstructure of thermally sprayed coatings as it reduces the porosity and promotes a metallurgical bond between substrate and coating. In their work, an Fe-Mn-Cr-Si alloy was deposited by arc spraying and then re-melted by a plasma-transferred arc process. The base metal was a soft martensitic stainless steel. The use of a lower mean current and a pulsed arc reduced the thickness of the heat-affected zone. In specimens re-melted with constant arc current, dendrites were aligned parallel to the path followed by the plasma torch; while in those re-melted with a pulsed plasma arc, the alignment of the microstructure was disrupted. The use of a higher peak current in pulsed-current plasma transferred arc re-melting reduced mass loss due to cavitation. Fe-Mn-Cr-Si coatings exhibited cavitation-induced hardening, with martensite formation during cavitation tests.

Bhargava et al. (2013) investigated the deposition of Inconel-625 using laser rapid manufacturing (LRM) and plasma transferred arc (PTA) deposition in individual and tandem mode. LRM has advantages in terms of dimensional accuracy, improved mechanical properties, finer process control, reduced heat input, and lower thermal distortion, while PTA scores more in terms of lower initial investment, lower running cost, and higher deposition rate. A number of samples were deposited with both processes at different parameters such as power, scan speed, and powder feed rate. They were subjected to tensile test, adhesion-cohesion test, impact test, and microhardness measurement. The results of individual tests showed the comparable mechanical properties with ±20% variation. The mixed dendritic-cellular and dendritic-columnar microstructures were respectively observed for LRM and PTA deposits with a distinct interface for the case of tandem deposition. The interface strength of tandem deposits was evaluated employing adhesion-cohesion test, and it was found to be (325 ± 35) MPa. The study confirmed the viability of LRM and PTA deposition in tandem for hybrid manufacturing. Figure 12.37a presents typical microstructure observed in LRM deposits of Inconel-625. The direction of dendrite growth was along the direction of deposition. Figure 12.37b presents the typical microstructure observed in PTA deposited Inconel-625 samples. For hybrid mode of material deposition, LRM and PTA processes were used to deposit material in successive tracks, and the interface was also studied. Figure 12.37c presents the microstructure of the deposit in tandem mode. Two distinct microstructures were observed in the regions deposited with different processes. The coarser columnar dendritic structure was observed in the region of PTA deposits, while mixed columnar and cellular dendrites were observed in the region of laser deposits. It was found that the maximum deposition rates

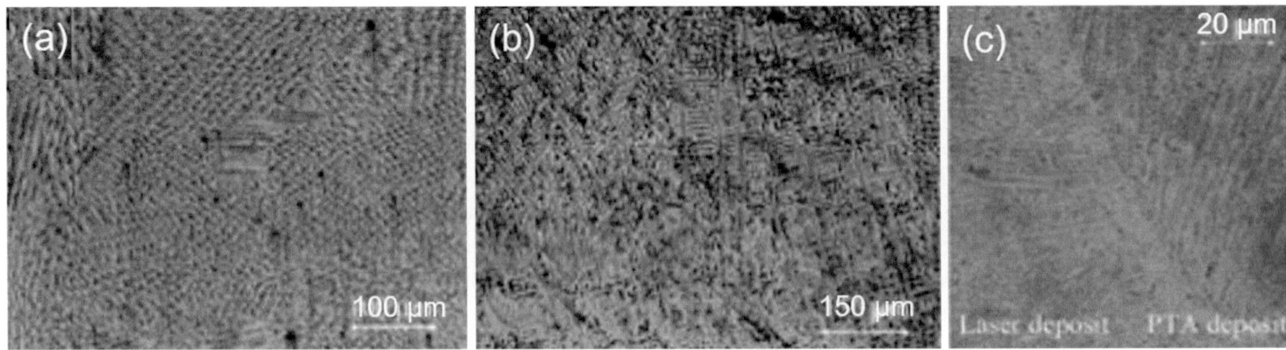

Fig. 12.37 Effect of the deposition technique on the microstructure of Inconel-625 deposited on steel substrate using (**a**) LRM, (**b**) PTA deposit, (**c**) LMR/PTA in tandem showing the interface between the two coatings. [Bhargava et al. (2013)]

were 12 g/min and 60 g/min for LRM and PTA process, respectively.

Tigrinho et al. (2007) studied the potential of PTA hardfacing beyond the surface welding of superalloys. This work evaluated low carbon steel surface modification by PTA deposition of fine WCoC carbides and mixtures of Fe powders and 5–35 wt% carbides. PTA processing allowed for the dissolution of carbides confirmed by X-ray diffraction, leading to homogeneous microstructures. Microstructures' morphology reaches typical dendritic solidification structure upon the WCoC content. Surface soundness depended on powder preparation and composition. Sound surfaces exhibiting hardness up to 700 HV were obtained for the 35 wt% WCoC powder mixture.

Reinaldo and D'Oliveira (2013) gathered microstructure data from the literature and analyzed Colmonoy 6 (NiCrSiB) which is a Ni-based alloy coating deposited by PTA hardfacing. This Ni-based alloy is recognized for its superior mechanical properties, attributed to the presence of a dispersion of hard carbides and borides, which is strongly dependent on processing technique. The aim of authors was to determine the influence of PTA deposition parameters and substrate chemical composition on NiCrSiB coating characteristics. Coatings were characterized in terms of their hardness, dilution, and microstructure, as well as mass loss during abrasive sliding wear tests. The results showed that coating performance is strongly dependent on the chemical composition of the substrate. Carbon steel substrate yielded coatings with greater wear resistance. Processing parameters also alter the performance of coatings with lower current and lower travel speeds resulting in reduced mass loss.

12.5.2 Composite Coatings Doped with Ceramics

There are fundamentally two approaches for obtaining a wear- and corrosion-resistant coating. In one approach, a metal alloy is deposited and during the cooling from the molten state hard phases or carbides precipitate. The other approach consists of adding carbides as hard phases into a suitable metal matrix. The addition of carbides will increase the abrasive and erosive wear resistance but may reduce the corrosion resistance. In most cases powder mixtures are used for the deposition, for example [Gallo et al. (2013)], blended powders of pure iron (~120 μm), carbon (~15 μm), titanium (~15 μm), and FeSi master alloy (~100 μm) and gray cast iron (FeSiMnC; ~125 μm). The powders were mixed thoroughly, dried at 150 °C for 1 h, and loaded into the hopper of a fluidized bed powder feeder. Powders with mean size around 150 μm can also be prepared and sprayed. Hou et al. (2015b) mixed the powders in a planetary ball mill, which were then made into a pasty mass in an agate mortar using a chemical binder. The obtained pasty mass was spread on a carbon steel plate to form prefabricated layers with thickness of about 3 mm, dried in an oven to vaporize the chemical binder before being processed by PTA surfacing to form thick coatings.

An example of such a coating on the tooth of an excavator is presented in Fig. 12.38a, with a cross section of the coating (Fig. 12.38b) showing a good dispersion of WC irregular particles (with no evidence of melting and spheroidizing).

Huang Z. et al. (2008) have deposited nickel-based alloy (C-0.6, Si-4.0, B-4.5, Cr-14, Fe-14.8, Ni-Bal) with 30 wt.% chromic carbide (Cr_3C_2) particles on Q235-carbon steel (including 0.12 wt.% C) substrate ($200 \times 35 \times 20$ mm^3) using PTA. They found that the γ(Ni, Fe), M7(C,B)3, Ni4B3, and (Cr,Fe)2B phases existed in the Cr_3C_2-free nickel-based alloy coating obtained by PTA process. The addition of 30 wt.% Cr_3C_2 particles led to the existing of Cr_3C_2 phase and the microstructure changing from hypoeutectic structure into hypereutectic structure. The average microhardness of the Cr_3C_2-free nickel-based alloy coating increased by 450–500 HV after the addition of 30 wt.% Cr_3C_2 particles. The partial dissolution of Cr_3C_2 particles led to the enrichment of carbon and chromium in the molten pool and hence caused the formation of more chromium-rich carbides after the solidification process.

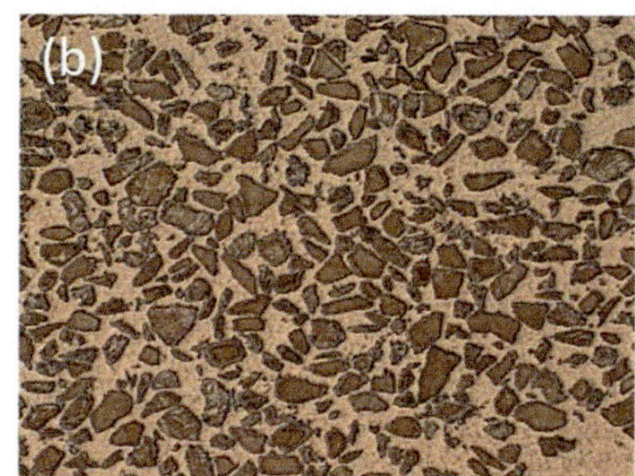

Fig. 12.38 (**a**) PTA coated tooth of excavator with Ni base coating + WC (25 kg/h). (**b**) cross section of the coating (courtesy of Castolin)

Hou et al. (2005) studied the microstructure and wear characteristics of Co-Cr-W series cobalt-based alloys deposited by PTA. A cobalt-based solid solution with face-centered cubic crystal structure and hexagonal $(Cr,Fe)_7C_3$ was obtained. There were lots of stacking defaults existing in the cobalt-based solid solution. After aging at 600 °C for 60 h, the precipitation of carbide $M_{23}C_6$ increased the hardness, but unfortunately the course of the microstructure and the change of the morphology led to the increase of the wear weight loss.

Wear resistance has also been improved by adding alumina particles in the PTA deposit. Hou et al. (2011) deposited nickel-based alloys powders with and without nano-α-Al_2O_3 addition on Q235A low carbon steel using PTA. Results show that the typical hypoeutectic microstructure and the component segregation exist in the Al_2O_3-free coating consisting of γ (Ni, Fe), $(Fe, Cr)_7C_3$, and $(Fe, Cr)_2B$ phases. The addition 0.8 wt. % nano-Al_2O_3 particles does not change the hypoeutectic microstructure characteristics of the Al_2O_3-free coating, though it refines its microstructure and decreases its component segregation. The nano-Al_2O_3 particles are mainly tetragonal lattice (γ-Al_2O_3). The substructure of the Al_2O_3-free coating is mainly dislocation. Adding nano-Al_2O_3 particles promotes to form stacking faults and dislocation cell wall. The sliding wear resistance of the Al_2O_3-added coating is higher than that of the Al_2O_3-free coating.

Hou et al. (2015a) prefabricated W-based composite powders containing NbC, TiC, and VC. PTA then processed these powders to form thick W-NbC, W-TiC, and W–VC coatings on a low carbon steel (0.454 wt. %) substrate. The best ability to resist elastic strain-to-failure and the highest toughness were obtained with the W-VC coating.

Guoqing et al. (2013) deposited nickel-based alloy coating on AISI 304 L stainless steel substrate using plasma transferred arc welding (PTAW). Microstructure and tribological characteristics of the coating were studied. The as deposited microstructure mainly consisted of Ni-rich γ (Ni, Fe) phase, Cr_7C_3, CrB, Cr_3C_2, $M_{23}C_6$, Ni_3B, and Ni_3Si. Compared with that of the 304 L stainless steel substrate, the wear resistance

of the coating was greatly improved. The wear test demonstrated that the wear mechanism depended on wearing time and applied load. When the wearing time was short, the wear mechanism was abrasive wear under low load and adhesive wear under high load. When the time was relatively long, the wear mechanisms were adhesive wear and oxidation wear under low load; fatigue wear appeared under high load.

Rokanopoulou et al. (2014) studied the use of the more thermodynamically stable ceramic, aluminum oxide, in order to reinforce the surface of the austenitic–ferritic steel SAF 2205 by PTA technique. The austenitic–ferritic microstructure was maintained, the Al_2O_3 reinforcement particles were finely distributed, and no other micro-constituents were found, leading to enhanced tribological properties, which is not the case when carbides are used for the reinforcement.

Rokanopoulou A. et al. (2016) deposited by PTA a powder mixture of Al_2O_3, TiS_2, and Fe. During solidification, Al_2O_3 forms and Al_2TiO_5 becomes thermodynamically stable and forms on the existing Al_2O_3 particles. As the melt cools further, the solubility of sulfur in molten steel decreases and the previously formed complex oxides spontaneously act as nucleation sites for sulfides, whereas Al_2TiO_5 is reduced to Al_2O_3. The Al_2O_3 reinforcement particles are surrounded by sulfides, and thus they have better wettability by the molten stainless steel and are less detachable compared without the addition of TiS_2. These sulfides also act as lubricants.

According to Deuis et al. (1997a) at the end of 1990s, PTA surfacing had already current commercial applications against wear with mostly metal coatings doped with ceramics, mainly TiC, WC, WC-W_2C, Cr_3C_2. The applications were for nuclear industry (fluid control), automobile (piston for tractor engine), power generation (steam turbine), chemical (ball valves), mining (slurry mixing paddles), polymers (extruder screw flights), sewage treatment (conveyors screws), cement (draft fan blades), steel making (basic oxygen furnace fume hood), and mining (liner plates, slurry paddles). Moreover, potential commercial applications are in automobile (engine components), mining (ore bucket lining), plastics (extruder screw), petroleum/chemical (protection of wear/corrosion areas), aerospace (high-temperature components, dry sliding couple system)), and heavy industry (wear components). If the hardness and wear resistance are usually significantly improved by the addition of carbides, the carbides may decompose when they are too strongly heated, e.g., in the transferred arc or in a high-temperature molten metal pool. Special injection techniques can remedy this situation. Some dissolved carbides will form mixed carbide hard phases in the coating.

As one of the most popular carbide addition giving the highest hardness is tungsten carbide [Lugscheider and Ait-Mekideche (1991)], which can be added (a) in form of macro-crystalline WC, large-enough particles that undergo minimal decomposition (although too large particles tend to settle at the bottom of the weld pool); (b) in form of fused tungsten carbide (FTC), an eutectic mixture of WC/W2C; and (c) in form of sintered WC in a cobalt. Other carbides include TiC, VC, NbC, or mixed carbides (W,Ti)C [Saltzmann (1986)]. Different carbides are used with different matrix materials, e.g., (V,W)C, NbC, TiC with a Fe-Cr-C matrix, (W,Ti)C, and W_2C with a Co-Cr-W-C matrix. For example TiC-W-Cr powders formed complex structure comprising primary austenite, martensite, an eutectic of (Fe, $Cr)_7C_3$ carbide and austenite as well as the uniformly distributed un-melted TiC particles. Un-dissolved Cr_3C_2 particles and chromium-rich carbides enhanced the hardness and wear resistance of the nickel-based alloy with 30 wt. % Cr_3C_2 within a coating deposited by the PTA process [Huang Z. et al. (2008)]. Nanometer-sized Al_2O_3 particles in nickel-based alloy powders promoted the formation of stacking faults and dislocation cell wall. The sliding wear resistance of the coating with Al_2O_3 added was higher than that of the Al_2O_3-free coating [Shi et al. (2011)].

[Kim et al. (2002)] PTA deposited two Fe-based gas atomized powders, proposed to as metamorphic alloys (Armacor M and Armacor C). The microstructures of both of them were mainly composed of Fe-based solid solution and chromium boride precipitates. However, the hardness and the boride size of the Armacor M were higher by about 200 HV and four times larger than those of the Armacor C, respectively. Armacor M coating exhibited better wear performance than Armacor C one.

Liu et al. (2007) deposited by PTA wear-resistant Fe_2TiSi reinforced composite coating by using a cladding process with Fe-Ti-Si-Cr powders blend, the substrate being plain low carbon steel. Under dry sliding wear conditions at an ambient temperature, the PTA clad Fe_2TiSi/γ-Fe/Ti_5Si_3 composite coating exhibited excellent abrasive and adhesive wear resistance.

[Skarvelis P. and G. D. Papadimitriou (2009a)] used PTA to produce composite coatings based on co-melting of MoS2, TiC, and iron ingredients, in an attempt to obtain wear-resistant layers with self-lubricating properties. The wear tests performed with a pin-on-disk apparatus have shown that the coatings which were produced using both TiC and MoS_2 ingredients exhibited improved friction coefficients and lower wear rates in comparison to the steel substrate as well as to the composite containing TiC, but exempted from MoS_2.

Demian C. et al. (2016) deposited by PTA yttria stabilized zirconia (YSZ) filler with two different particle sizes into NiCrAlY (respectively, 10 and 20 vol. %) and metal matrix composite (MMC) to develop cermet coatings on medium carbon steel type AISI 1035 25 mm thick. The YSZ filler in a concentration of 20 vol.% into the NiCrAlY powder was the optimal choice. The mixtures were mechanically blended and dried at a temperature of about 120 °C for 24 h. The substrate surface was prepared by grinding and preheating in an open-air furnace at temperatures over 365 °C, increasing with the YSZ content. When using high currents of transferred arc, good metallurgical bonding between the coating and substrate were obtained in both cases. Finally, it was found out that cermet coatings including small particles of YSZ filler in a concentration of 20 vol.% into a NiCrAlY matrix were the optimal choice among all tested in the presented conditions.

Fisher et al. (2013) recall that it is common to apply tungsten carbide-based composite overlays to improve the reliability and extend service lives of equipment and components, for example, to remove bitumen from oil sands deposits. The performance of the applied overlays is largely dependent on the selection of the carbide type and the wear environment. Evaluated overlays containing macro-crystalline, angular eutectic, and spherical eutectic tungsten carbides are deposited by PTA. They used angular WC, angular WC/W_2C, and spherical WC/W_2C. The three overlays contained relatively homogeneous distributions of primary tungsten carbide particles in the nickel-based matrix alloys. As an example, Fig. 12.39 shows a micrograph of the sprayed 6030 M overlay. The carbides are well distributed, and there is little or no evidence of primary carbide dissolution.

- Eutectic (WC/W_2C) primary carbides are more susceptible to dissolution due to heat input upon deposition. The levels of secondary carbide formation for the mono-crystalline (WC)-containing 6030 M overlay were significantly lower.
- Abrasion by sands sized less than 350 μm (50 mesh) preferentially removed the matrix alloy in the MMC.
- The abrasion rates recorded by the WC-MMC were related to the size of the abrasive and the associated applied load. Increases in these two factors produced higher levels of mass loss due to primary carbide fracturing.

Gallo et al. (2014) studied the in situ synthesis of titanium carbides in iron alloys using PTA. They showed that it was possible to convert synthetic rutile (TiO_2) into TiC by a reactive deposition process with a PTA technique. The TiC precursors, consisting of graphite (C) and synthetic rutile (TiO_2) in a 6:1 molar ratio, were easily introduced in the blend with iron powder up to a concentration of 30 wt.%. The addition of 1 wt.% of an alkali salt resulted in marked increases in the volume fraction of TiC precipitates in the overlays. They showed that silicon and manganese on the TiC precipitates had a significant effect on TiC size and morphology. However, an increase in the volume fraction of TiC required additional processing, more efficient ways to introduce the precursors and effective extraction of the gas produced in the reaction.

Gallo et al. (2014) evaluated the in situ synthesis of TiC-Fe composite overlays for low-cost TiO_2 precursors using PTA. Mild steel AISI 1020 was selected as a substrate with slabs 100 mm long, 32 mm wide, and 5 mm thick. Three different powder mixtures were used for the deposition: Fe + C + Ti, FeSi + C + Ti and FeSiMnC + Ti. Results showed that:

Fe-TiC composite overlays can be produced by in situ synthesis:

- The morphology of TiC precipitates was affected by the chemical composition of the iron matrix. Additions of silicon reduced the occurrence of script carbides and manganese hinders dendritic growth.
- The size and volume fraction of TiC precipitates in the FeSiMn matrix increased with increasing heat input and decreasing cooling rate. A coarsening and softening of the microstructure of the matrix accompanied this.
- The FeSiMn matrix promoted the precipitation of spheroidal or faceted TiC carbides, evenly distributed in the matrix, under a wide range of deposition conditions.

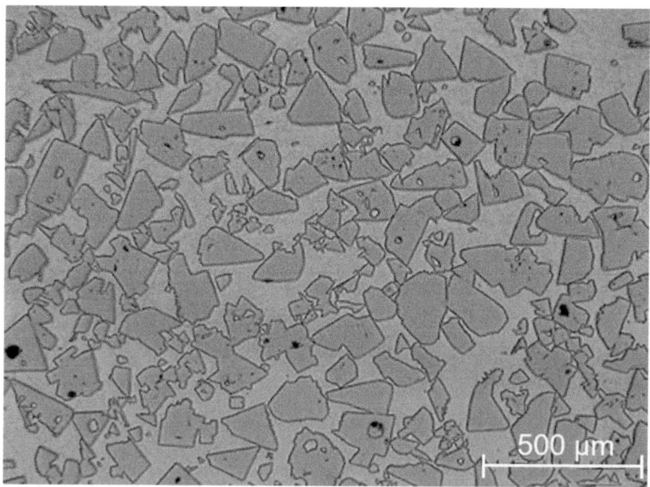

Fig. 12.39 Micrograph of the 6030 M overlay, showing the distribution of WC particles in the Ni-alloy matrix. Magnification 140x [Fisher et al. (2013)]

Figure 12.40 illustrates the typical microstructures observed in the overlays. Figure 12.40a corresponds to the pure Fe overlay, produced for reference purposes, and

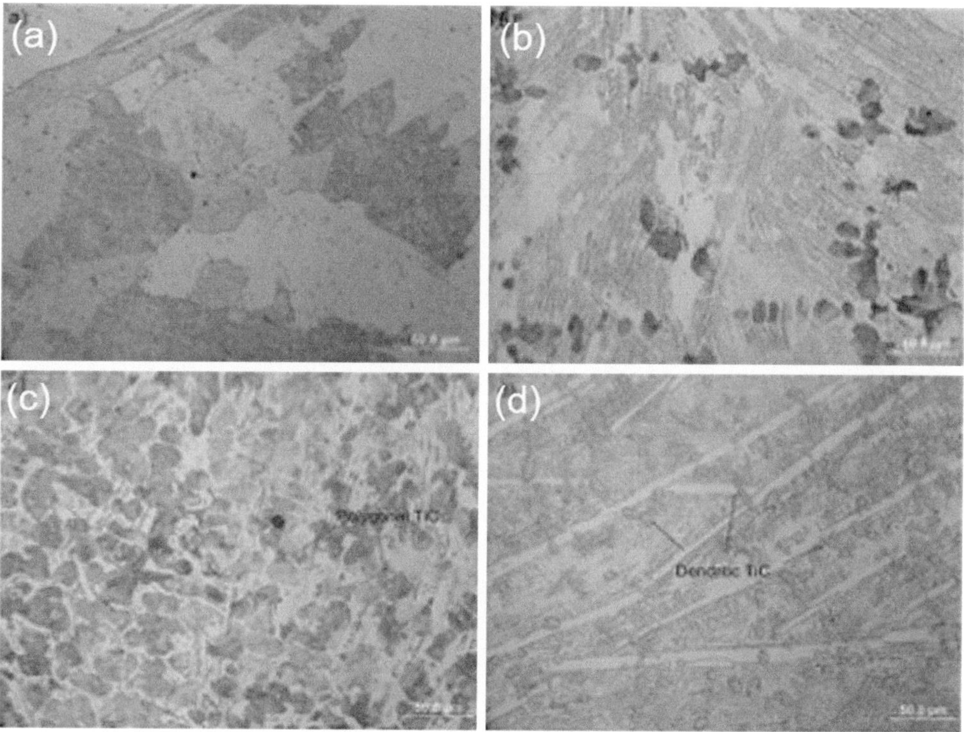

Fig. 12.40 Optical micrographs of overlays produced with (**a**) Fe, (**b**) Fe + C, (**c**) Fe + 20MM (TiO2 + C), and (**d**) Fe + 30MM (TiO2 + C) with alkali salt. MM mechanically mixed TiC precursors (TiO + C) [Gallo S. C. et al. (2014)]

exhibits the characteristic ferritic microstructure. Figure 12.40b shows the microstructure of the second overlay produced for reference purposes, which consisted of the pure Fe powder blended with 5 wt.% graphite. The carbon content of this specimen was equivalent to the carbon content of sample Fe + 10MM(TiO$_2$ + C). The microstructure of this overlay was typically martensitic-pearlitic, with a high-volume fraction of cementite (lighter phase in Fig. 12.40b). This reflects the high carbon content of the alloy resulting from the addition of graphite to the powder blend. The two reference samples provided a basis for the subsequent study. The typical microstructure of overlays produced with powder blends containing TiO$_2$ and graphite consisted of a pearlitic matrix with an even dispersion of precipitates. The addition of TiO$_2$ produced an apparent reduction in the carbon content of the iron matrix of the overlays, which becomes evident from the lower volume fraction of cementite in the typical microstructure of overlays produced with powder blends containing TiO$_2$ and graphite consisted of a pearlitic matrix with an even dispersion of precipitates. The addition of TiO$_2$ produced an apparent reduction in the carbon content of the iron matrix of the overlays, which becomes evident from the lower volume fraction of cementite in Fig. 12.40c and d.

Bourithis et al. (2002) produced coatings with boride, 1–1.5 mm thick, on AISI 1018 with different boron contents. The deposited powder contained C 4.1%, Cr 33.4%, Mo 12.5%, TiC 50% (wt %). The PTA alloying process was successful in producing MMC layers of 0.8–1 mm thickness on the surface of construction carbon steel. The microhardness of the alloyed surface was stable and attained values between 850 and 900 HV. The microstructure of the MMC layer consisted of a fine martensitic matrix with retained austenite and TiC carbides as reinforcing particles. The wear rate of the MMC is approximately 10^{-3} mm^3/m and that of the plain steel used as base metal is of the order of 10^{-2} to 10^{-1} mm^3/m, i.e., 10–100 times greater. The wear mechanism of the alloyed surface depends on the operating conditions. The dominant wear mechanism for low sliding speeds is plastic deformation. At higher sliding speeds an oxidation mechanism predominates. In particular a transition from mild to severe oxidation is observed when the applied load exceeds a certain value (between 29.4 and 39.2 N).

Liu et al. (2006) used Fe-Cr-C-Ni powder blends to deposit PTA coatings on steel substrate (0.45 wt.%.C). Results showed that the composite coating consisted of primary (Cr,Fe)$_7$C$_3$ as the reinforcing phase and interdendritic (CrFe)$_7$C$_3$/γFe eutectics as the matrix. The composite coating was metallurgically bonded to the substrate. The (CrFe)$_7$C$_3$ carbide reinforced composite coating has high hardness (around 440 HV) and excellent wear resistance under dry sliding wear test conditions, the worn surface of the carbide reinforced composite coating being smooth with only slight scratches.

Werry et al. (2016) studied multiscale-structured composite coatings by plasma-transferred arc for nuclear applications. PTA coatings of nickel-based alloys reinforced with alumina particles were deposited on 316 L stainless steel substrates. Metallic and alumina powders were oven dried at 80 °C for 5 h. Then, they were mixed into a stirrer with attrition balls (Ø 0.6 mm) for 12 h and sieved through 400 µm mesh sieve to remove the attrition balls and the largest alumina agglomerates. Under the conditions of this study, the addition of alumina particles resulted in a refinement of coating microstructure and the improvement of their resistance to abrasive wear. However, it does not bring about any change in coating micro-hardness.

Hung et al. (2006) used the PTA technique to overlay NbC reinforcing particles on the surface of commercially pure Ti in order to investigate the microstructural features of the over-layer and the interface between the over-layer and base metal by changing the overlaying current. The results indicated that the matrix phase of the over-layer was α-Ti containing about 10 at.% Nb and 1 at.% C. NbC and precipitated TiC produced by dissolved NbC reacted with Ti dispersed in the matrix. The microstructure of the cross-section of the over-layer (from surface to base metal), which was composed of α-Ti, can be separated into three layers: an upper over-layer with TiC, a middle over-layer with TiC and NbC, and a lower over-layer (interfacial layer and Heat Affected Zone, HAZ). Due to solidification beginning at the interface and the effect of dilution, the TiC in the interfacial layer was finer. Owing to faster solidification under low-current conditions, the TiC particles were finer than under high-current conditions. Also, dendritic TiC under a low-current in the upper over-layer was also finer than under a high current. Meanwhile, TiC precipitate that resulted from heterogeneous nucleation and Gibbs' free energy was also found around NbC. This NbC diffusion layer between TiC and NbC may have been βNb$_2$C phase.

Yuan and Li (2014) studied the fabrication of high-volume fraction (HVF) M$_7$C$_3$ (M = Cr, Fe) reinforced Fe-based composite coating on ASTM A36 steel plate using plasma transferred arc (PTA) welding. The results showed that the volume fraction of carbide M$_7$C$_3$ was more than 60%, and the relative wear resistance of the coating tested on a block-on-ring dry sliding tester at constant load (100 N) and variable loads (from 100 to 300 N) respectively was about 9 and 14 times higher than that of non-reinforced α-Fe coating. In addition, under constant load condition the friction coefficients (FCs) of two coatings increased first and then decreased with increasing sliding distance. However, under variable loads' condition, the FCs of non-reinforced α-Fe based coating increased gradually, while that of HVF M7C3 reinforced coating decreased as the load exceeded 220 N. The worn surface of non-reinforced α-Fe based coating was easily deformed and grooved, while that of the HVF

M7C3 reinforced coating was difficult to be deformed and grooved. Carbide Cr_3C_2 can act as substrate for the nucleation of M_7C_3. The mass loss and FCs of HVF M_7C_3 reinforced coating are lower than that of non-reinforced α-Fe coating under the conditions of constant and variable loads dry sliding wear. The main wear mechanism of HVF M_7C_3 reinforced coatings is micro-cutting, abrasive wear, and oxidation wear. Micro-cracks and spalling pits formed on the worn surfaces of carbide M_7C_3 particles as the loads increased from 100 to 300 N. The α-Fe matrix can prevent cracks extending in the coating and no M7C3 particle has been pulled out the matrix due to its long rod shape and high interface binding force derived from the in situ reaction.

Werry et al. (2016) studied the replacement in nuclear plants of hard-facing Stellite, a cobalt-based alloy, on parts of the piping system in connection with the reactor that has been investigated since the late 1960s. Various Fe-based or Ni-based alloys, Co-free or with a low content of Co, have been developed but with mechanical properties generally lower than that of Stellite. The fourth-generation nuclear plants impose additional or more stringent requirements for hard-facing materials. Plasma transferred arc (PTA) coatings of cobalt-free nickel-based alloys with the addition of sub-micrometric or micrometric alumina particles are thought to be a potential solution for tribological applications in the primary system of sodium-cooled fast reactors. In this study, PTA coatings of nickel-based alloys reinforced with alumina particles were deposited on 316 L stainless steel substrates. Under the conditions of this study, the addition of alumina particles resulted in a refinement of coating microstructure and the improvement of their resistance to abrasive wear. However, it does not bring about any change in coating microhardness.

Shi et al. (2012) subjected TiC-Cr overlay formed by plasma transferred arc to tempering treatments at 473, 673, and 873 K for 2 h, respectively. Microstructure investigation by SEM and TEM, microhardness, and fatigue test were performed on As-deposited condition and after tempering. Evolution of microstructure as well as its relationship with microhardness and fatigue property of coating was described. The results suggest that both microhardness and fatigue strength could be improved by tempering at 473 K; however, further elevation of the temperature directs degradation.

12.5.3 Slurry Erosive Wear

When comparing the influence of dry and slurry erosion on HVOF coatings, it was shown that erosion rates in dry particle impact were about three orders of magnitude higher than those in slurry systems. This difference probably reflects the real erodent target impact velocities, which are mitigated in the slurry test by the water medium [Hawthorne et al. (1999)].

The erosion occurred dominantly by spalling of splats from the lamellar interfaces, spalling resulting from the propagation of cracks parallel to the interfaces between the lamellae exposed at the surface and underlying coating. The carbide particle size and content in the coating influenced significantly the erosion performance of Cr_3C_2-NiCr coatings.

Santa et al. (2009) studied in laboratory the slurry and cavitation erosion resistance of six thermal spray coatings and compared to that of an uncoated martensitic stainless steel. Nickel, chromium oxide and tungsten carbide coatings were applied by oxy fuel powder (OFP) process and chromium and tungsten carbide coatings were obtained by high-velocity oxy fuel (HVOF) process. The results showed that the slurry erosion resistance of the steel can be improved up to 16 times by the application of the thermally sprayed coatings. On the other hand, none of the coated specimens showed better cavitation resistance than the uncoated steel in the experiments.

Flores et al. (2009a) studied the erosion–corrosion behavior of two tungsten carbide (WC)- metal matrix composites (MMCs) PTA deposited. Two variations of the NiCrBSi matrix with different concentration of chromium (Cr), carbon (C), and boron (B) were evaluated with and without reinforcing hard phase. The erosion–corrosion behavior was evaluated as a function of sand concentration and temperature. To determine the influence of the mechanical degradation, the study was conducted using 1 and 5 wt.% sand concentrations. Important microstructural differences were identified between the two MMCs. Elongated precipitates rich in W and Cr were identified near the WC grains and in the matrix phase of the WC-hard overlay. Needle-like precipitates with a lower concentration of W were identified in the WC-soft overlay. For both MMCs the α-Ni phase and silicide were present in the matrix phase. The addition of WC dramatically improved the erosion–corrosion resistance of the matrix-only overlays by almost 70% when 20 °C and 1 wt.% sand loading were used. At 20 °C and 1 wt. %, sand loading erosion had a predominant impact over corrosion in the erosion–corrosion degradation mechanisms. The effects of corrosion were more evident at 65 °C and 1 wt. % solid loading. A higher degradation of the matrix phase, due to corrosion effects, left the WC grains unprotected and their degradation was more pronounced. The matrix microhardness and precipitates distribution played an important role in the erosion–corrosion resistance of the MMCs. In the zone where low impact angles were predominant, the elongated precipitates provided a better protection against sand erosion to the matrix of the WC-hard overlay compared to the protection provided by the needle-like precipitates of the WC-soft overlay.

Flores et al. (2009b) investigated the microstructure and erosion–corrosion behavior of a Fe-Cr-C overlay (FeCrC–matrix) produced by plasma transferred arc welding (PTA) and its metal matrix composite (FeCrC–MMC). The latter was obtained by the addition of 65 wt.% of tungsten carbide. The erosion–corrosion tests (ECTs) were carried out using a submerged impinging jet. At 20 °C the FeCrC–matrix overlay showed micro-cutting and micro-plowing degradation processes followed by plastic flow. At 65 °C it suffered a higher degradation due to the selective dissolution of the dendritic structure by corrosion attack.

The removal of the matrix phase in the FeCrC–MMC was mainly by an extrusion process due to the erosive effects of the sand particles. Under the conditions of 10 g/l and 20 °C and at medium to low impact angles, the presence of inter-metallic phases distributed in the ductile matrix had a benefi-cial effect on the erosion–corrosion resistance of the FeCrC–MMC overlay. When the severity of the erosive conditions was increased, the mechanical degradation dominated the erosion–corrosion process. At 65 °C the FeCrC–MMC over-lay suffered severe corrosion attack; the effect of temperature in the corrosion behavior of the overlay under erosion–corro-sion conditions was significantly higher than the effect of the sand concentration.

[Gatto et al. (2004)] analyzed the influence of PTA as a thermal treatment on the base material. They showed that structural modifications, with the generation of three different zones (melted zone, HAZ, unaffected zone), depend most on the physical effect of heat transfer and almost not on the alloys. Sample observation showed the process to be ruled mostly by physical than by chemical phenomena (for the same geometrical configuration, identical results were obtained for different materials). Optimized process parameters were found to be the same for all the deposited alloys, varying only as a function of the geometrical configu-ration. It was proved that the quality of the deposition was strongly determined by an optimal reciprocal positioning and movement between the torch and the part to be coated, especially for critical geometries. Specimen characterization through liquid penetration inspection, optical and scanning electron microscopy, and microhardness tests proved that process parameter optimization depends only on the geomet-rical configuration and not on the deposited alloy.

12.5.4 Reclamation and Resurfacing

The effects of wear and corrosion can only be retarded, but not be stopped forever and wear parts must be replaced or resurfaced. It is the same for undersize parts due to manufacturing error. The ability to apply coatings by thermal spraying with a broad spectrum of thickness, finish and composition requirements makes the different thermal spray processes ideal solutions to resurface components, avoiding their costly replacement. Restoration is performed either with the same material as the base metal or with a more corrosion-

and/or wear-resistant material. However, it is better to choose coatings that can be machined. In a wide range of industrial applications, thermal coatings are used to renew components by restoring their specified dimensions and matching and sometimes improving their original performance. Very often it is possible to design a coating that meets the func-tional requirements of the component, and moreover offers better resistance to corrosion, oxidation, and mechanical wear than the original product. Of course, the cost of the resurfacing must be lower than that of a new part (a fraction of it). For example, this technique is particularly used for roller faces and journals, dryer drums for papermaking, pump seals (shafts and sleeves), pump housings, compressor rods, rotary airlocks and feeders, and conveyer screws [Davis J.R. Publisher (2004), Tucker R. C., Jr. Ed. (2013)].

Alberti et al. (2016) have defined the additive manufacturing, where PTA is also used. It is a process used to fabricate and repair parts that have a complex geometry or need to be functionally graded. The technique involves depositing multiple layers to produce the component. The success of the procedure depends on the deposition technique used, the parameters selected, and the alloy deposited. The deposition conditions, such as temperature and protective atmosphere, determine whether cracking and oxidation of the deposited layers occur. In their study the authors presented the potential of plasma transferred arc to produce thin walls by additive manufacturing. Two nickel-based alloys were used on an Ni-based substrate: a γ' precipitation-hardened alloy and a solid-solution-hardened alloy. During the study, the processing parameters required to produce a thin wall with each alloy were determined, and the use of preheating at 300 °C was analyzed. The results showed that the chemical composition of the alloy being processed and preheating influences the geometry of the wall. A fine dendritic solidification structure exhibiting epi-taxial growth between layers was observed. The precipitation-hardened alloy showed banding of a γ' precipitate-rich region that caused oscillations in the hard-ness. Dilution with the substrate was the main factor affecting the hardness profile of the wall processed with the Ni-based solid solution alloy, which did not change following post-deposition heat treatment. This study has shown that sound thin walls can be successfully processed by additive manufacturing using plasma transferred arc.

Lü et al. (2013) remanufactured by PTA the thrust of engine cylinder body. The Ni15 alloy was deposited on the thrust face of the body in order to recover its dimension. In addition, the remanufacturing forming with Fe-based, Inconel 625 alloy was studied. The microstructure and hard-ness of the As-deposited materials were investigated. In order to control the deforming of the formed parts, the PTA current must be controlled. When the PTA current was limited to 30 A, it was called micro-plasma arc. Ni_{15} alloy was

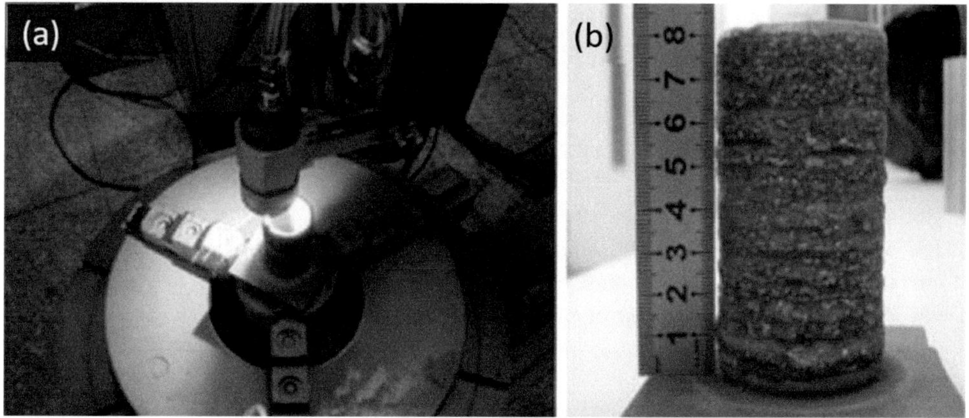

Fig. 12.41 Micro-plasma arc direct metal forming (**a**) process and (**b**) rough part [Lü Y.-H.et al. (2013)]

deposited on the thrust face of the body in order to recover its dimension. In addition, the remanufacturing

forming with Fe-based, Inconel 625 alloy was studied. The microstructure and hardness of the as-deposited materials were investigated. Figure 12.40 shows the micro-plasma arc direct metal forming process (see Fig. 12.41a) and the rough part (see Fig. 12.41b). Microhardness measurements were carried out on the transverse and longitudinal cross sections of the As-deposited Inconel 625 material. Results revealed microhardness on the transverse cross section in the range of 260–285 $HV_{0.2}$ with an obvious fluctuation.

Tigrinho et al. (2007) evaluated low carbon steel surface modification by PTA deposition of fine WCoC carbides and mixtures of Fe powders and 5–35 wt% carbides. PTA processing allowed for the dissolution of carbides confirmed by X-ray diffraction, leading to homogeneous microstructures. Sound surfaces exhibiting hardness up to 700 HV were obtained. Surfaces modified by melting Fe–35 wt% WCoC powder mixtures exhibited a typical solidification structure with a ferrite dendrites and an inter-dendritic Fe/Carbide eutectic. Increasing dilution with the steel substrate altered phase distribution and surface hardness. Surfaces modified with Fe–5 wt. % WCoC powder mixtures resulted in a slight increase hardness compared to the base steel.

According to Jhavar et al. (2014), micro-plasma transferred arc (μ-PTA) deposition process has potential to meet requirements of the meso-sized fabrication and repair of the high value components. The overall objective was to repair and/or remanufacture the defective dies and molds. An experimental setup was developed to deposit 300 μm diameter wire of AISI P20 tool steel on the substrate of the same material which is one of the most commonly used ones for making the dies and molds used for various applications. The μ-PTA deposition process was found to be capable of fabricating straight walls having total wall width of 2.45 mm and

effective wall width of 2.11 mm. The deposition efficiency was found to be 87% for the maximum deposition rate of 42 g/h. The deposited wall was free from cracks, porosity, and inclusions.

Jhavar et al. (2016) deployed an automatic micro-plasma deposition setup to deposit a wire of 300 μm of AISI P20 tool steel on the substrate of same material for the potential application in remanufacturing of the die and mold surface. Bead-on-plate trials were conducted to deposit single bead geometry at various processing parameters. A set of parameters leading to reproducible regular and smooth single bead geometry were identified and used to prepare a thin wall for mechanical testing. In their study, plasma power of 500 W, travel speed of 80 mm/min, and wire feed rate of 1275 mm/min producing optimum measurable values for deposition width, height, aspect ratio, percent dilution, and contact angle of 1.7 mm, 0.7 mm, 2.42, 3.4%, and 74, respectively, were used to produce multi-bead and multilayer depositions. The study opens the avenue for the deployment of PTA deposition for dies and molds repair/remanufacturing.

Zhao et al. (2016), to improve the tribological properties of automobile stamping dies, synthesized Cu-rich particulate composite coatings via PTA alloying of copper on ferritic nodular cast iron substrates. The plasma transferred arc heat input had an important effect on the synthesis of composite coatings. At a low heat input of 202 J/mm, a double-layer structure with both a Cu-rich layer and a Fe-rich layer was observed. As the heat input increased, the volume of the Cu-rich layer significantly decreased because the Cu is further mixed into Fe-rich layer. When the heat input was increased to 260 J/mm, a single Cu-rich particulate composite coating was synthesized in the molten pool. Microstructural analysis revealed that Fe-rich particles were embedded in the Cu-rich matrix, while Cu-rich particles were embedded in the Fe-rich matrix, which consisted predominantly of residual austenite, acicular martensite, and inter-dendritic eutectic carbides. The microhardness of the Cu-rich particulate composite coatings was measured, which was much higher than that of the untreated substrate material but lower than that of the remolten layer without the Cu additive. Sliding wear tests are carried out on the composite coatings at room temperature (RT) and 773 K. The results exhibited a much better wear resistance of the composite coating compared to the untreated material. The Cu-rich particles contained in the composite coatings played an important role in the antifriction properties, which was attributed to the formation of protective Cu films on the friction surface.

12.6 Summary and Conclusions

Among the global family of surface modification technologies, plasma transferred arc (PTA) coating is closer to plasma arc welding (PAW), metal inert gas (MIG), and tungsten inert gas (TIG) welding than conventional wire arc or DC plasma spraying. One of the unique features of PTA coating is that the substrate is actively serves as one of the arc electrodes, resulting in high heat transfer rates, the local melting of the substrate, and the fusion of the coating material to the substrate. The direct consequence of such a coating route is the formation of a good metallurgical bond of the coating to the substrate, very high coating density, and the possibility to generate thick coatings (up to 10 mm). The major drawback is the limited ability to coat complex shapes, the high level of residual stress, and the requirement of metallurgical compatibility between substrate and coating. There are numerous reports of successful uses of PTA deposited coatings for reducing wear and corrosion and rebuilding metallic components in almost every industry where thick coatings are of advantage. This process has led to a number of developments of new materials combinations in metallurgy and in composites, and it is expected that this trend will continue. In this chapter a review is presented of basic concepts at play in the technology, arc stabilization mechanism, and the importance of the choice of the proper plasma gas and its impact on the coating properties. Following a discussion of basic design features of PTA systems, a review is presented of the effect of process parameter changes on the coating properties. Attention is given to the concept of "dilution" which is a measure of the ratio of the area where the substrate and coating materials are mixed, to the total area in which the coating material appears as defined in Fig. 12.2. Typical rang of operating conditions for three scales of PTA operation are given: micro-PTA, regular PTA, and high-power PTA in Table 12.1 together with the impact of the effect of process parameter values on coating properties, Table 12.2.

Attention is also given to the important area of process modifications and adaptations. In this section a brief discussion is presented on different approaches undertaken in order to expand on the potential use of the technology including such topic as coating nitriding, modulation of deposition parameters, pulsed arc additive manufacturing, high energy PTA, and the operation of PTA in reverse polarity mode. Gas and particle dynamics in PTA coating operations is discussed next providing examples of typical temperatures fields in the arc column and the molten metal pool formed locally in the substrate where the coating material is injected and heat flux to the substrate. Results of process modeling studies are discussed including the relatively important area of the effect of substrate position, horizontal or vertical on the shape of the formed coating bead, and ns of controlling it which is of vital importance for industrial utilization of the technology for coating applications in constrained conditions. Finally, a brief review is presented of coating materials in the range of industrial applications of the technology which mostly in the areas of wear and corrosion protection as well as for the rebuilding of worn-up parts to their original dimensions.

Nomenclature

Units are indicated in parentheses; when no units are indicated, the parameter is dimensionless.

Latin Alphabet

A_{conv} – Area over which the convective heat transfer takes place (m^2)
A – Area where coating material is not mixed with the substrate (m^2)
B – Area where substrate and coating material are mixed (m^2)
c_p – Specific heat at constant pressure (J/kg.K)
d_z – Torch-to-substrate (anode) distance (m)
de – Equivalent diameter of current path (m)
D – Dilution
D – Arc attachment spot diameter (m)
E – Electric field (V/m)
E_i – Ionization potential of the plasma gas (eV)
I – Arc current (A)
I_{arc} – Arc current (A)
I_{pilot} – Pilot arc current (A)
j_e – Electrons current density (A/m^2)
j_i – Ions current density (A/m^2)
k – Constriction parameter (cm^{-2})
k_B – Boltzmann constant ($1.38\ 10^{-23}$ J/K)
$\dot{m}_p$ – Powder feed rate (kg/s)
$\dot{m}_{gas}$ – Plasma gas feed rate (kg/s)
q_a – Heat flux received by the anode (W/m^2)
q_e – The electron enthalpy flux (W/m^2)
q_m – Maximum heat flux (W/m^2)
$q(r)$ – Substrate heat flux (W/m^2)
Q_{an} – Heat transfer to the anode (W)
Q_{con} – Conduction heat transfer (W)
Q_{sub} – Total heat transfer to the substrate (W)
Q_{melt} – Power involved in melting steel (W)
Q_{oxy} – Power released from iron oxidation (W)
Q_R – The heat transfer by radiation (W/m^2)
r – Distance in the radial direction (m)
r_e – Arc radius (m)

T_e – Electrons temperature (K)
T_h – Heavy species temperature (K)
v – Specific velocity (m/s)
v_{tras} – Substrate translation speed (m/s)
V – Arc voltage (V)
V_{pilot} – *Pilot arc* voltage (V)
z – Distance in the axial direction (m)

Greek Alphabet

α – Fraction of the electrical power transferred to the anode by convection and radiation
δ – Melted layer thickness (m)
Φ_a – The work function of the anode material (eV)
κ – Thermal conductivity (W/m K)
ρ – Mass density (kg/m^3)
$\overline{\sigma_{ac}}$ – eEectrical conductivity of argon averaged over the arc cross section (A/V.m)
Δz – Distances between the cathode tip and the anode (m)

References

Alberti, E.A., B.M.P. Bueno, and A.S.C.M. D'Oliveira. 2016. Additive manufacturing using plasma transferred arc. *International Journal of Advanced Manufacturing Technology* 83: 1861–1871.

Bach, F.-W., and J. Zühlsdorf. 1999. Plasma powder welding under raised pressure environment, Proc. UTSC., Düsseldorf, Germany, 1999, eds. E. Lugscheider and P. Kammer (Pub.) DVS, Düsseldorf, Germany, pp 757–760.

Bini, R., M. Monno, and M.I. Boulos. 2007. Effect of cathode nozzle geometry and process parameters on the energy distribution for an argon transferred arc. *Plasma Chemistry and Plasma Processing* 27: 359–380.

Bhargava, P., C.P. Paul, C.H. Premsingh, S.K. Mishra, Atul Kumar, D. C. Nagpure, G. Singh, and L.M. Kukreja. 2013. Tandem rapid manufacturing of Inconel-625 using laser assisted and plasma transferred arc depositions. *Advanced Manufacturing* 1: 305–313.

Bouaifi, B., A. Gebert, and H. Heinze. 1993. Plasma-Pulver-Auftragsch-weißungen zum Verschleißschutz abrasiv beanspruchter Bauteile mit Kantenbelastung. *Schweissen und Schneiden* 45: 506–509.

Bouaifi, B., F. Schreiber, J. Göllner, and S. Schulze. 1996. Eigenschaften und Beständigkeit von Plasma-Pulver-Auftragsch-weißungen aus hartstoffverstärkten. *CrNiMoN-legierten Duplex-Stählen in DVS-Berichte* 175: 425–428.

Bouaifi, B., A. Ait-Mekideche, A. Gebert, and D. Wocilka. 2001. Nutzung von stickstoffhaltigen Hochtemperaturplasmen zum reaktiven beschichten mittels Plasmaauftragschweißen. *Schweissen und Schneiden* 53: 478–482.

Bourithis, L., S. Papaefthymiou, and G.D. Papadimitriou. 2002. Plasma transferred arc boriding of a low carbon steel: Microstructure and wear properties. *Applied Surface Science* 200: 203–218.

Camélia, Demian, Alain Denoirjean, Lech Pawłowski, Paule Denoirjean, and Rachida El Ouardi. 2016. Microstructural investigations of NiCrAlY + Y$_2$O$_3$ stabilized ZrO$_2$ cermet coatings deposited by plasma transferred arc (PTA). *Surface & Coatings Technology* 300: 104–109.

Colombo, V., A. Concetti, E. Ghedini, and S. Dallavalle. 2011. Design oriented simulation for plasma arc cutting consumables and experimental validation of results. *Plasma Sources Science and Technology* 20: 035010 (10pp).

Davis, J.R., ed. 2004. *Handbook of thermal spray technology.* Materials Park: ASM international.

Deuis, R.L., J.V. Bee, and C. Subramanian. 1997a. Investigation of interfacial structures of plasma transferred arc deposited Aluminum based composites by transmission Electron microscopy. *Scripta Materialia* 31 (6): 721–727.

Deuis, R.L., J.M. Yellup, and C. Subramanian. 1998. Metal-matrix composite coatings by PTA surfacing. *Composites Science and Technology* 58: 299–309.

Dilthey, U., J. Ellermeier, P. Gladkij, and A.V. Pavlenko. 1993. Kombiniertes Plasma-Pulver-Auftragschweissen. *Schweissen und Schneiden* 45 (5): 241–244.

Dilthey, U., L. Kabatnik, E. Lugscheider, K. Schlimbach, and G. Langer. 1999. Möglichkeiten zur Steigerung der Oberflächen-festigkeit bei Aluminiumlegierungen mit Plasma-Pulver-Schweißverfahren (Improving of the Wear Resistance of Aluminum Alloys by Plasma Transferred Arc Welding). *Mat.-wiss. u. Werkstofttech* 30 (11): 697–702.

Dilthey, U., S. Kondapalli, B. Balashow, and F. Riedel. 2008. Improving wear resistance of aluminium alloys by developing FTC and TiC based composite coatings using plasma powered arc welding process. *Surface Engineering* 24 (1): 75–80.

D'Oliveira, C.M., R.S. Paredes, and R.L. Santos. 2005. Pulsed current plasma transferred arc hardfacing. *Materials Processing Technology* 171: 167–174.

Ducos, M. 1985. Applications industrielles des plasmas d'arc de faible puissance et des plasmas inductifs, Rechargement par plasma à arc transferré, in Les Plasmas dans l'industrie, (Pub.). Dopée/85, EDF France (in French).

DuMola, R.J., and G. R. Heath. 1997. New Developments in the Plasma Transferred Arc Process, Proc. UTSC-1997, Indianapolis, IN, (ed.) C. C. Berndt (Pub.) ASM International, Materials Park, OH, pp 427–434.

DuPont, J.N. 1998. On optimization of the powder plasma arc surfacing process. *Metallugical and Materials Transactions, B* 29B: 932–934.

Ebert, L., S. Thurner, and S. Neyka. 2009. Beeinflussung der Hartstoff-verteilung beim Plasma-Pulver-Auftragschweißen, Influencing the distribution of reinforcing particles in plasma transfer arc welding. *Mat.-wiss u. Werkstofttech* 40 (12): 878–881.

Eddie, P. 2011. *Plasma cutting handbook,* (Pub.) HP books 1569 Penguin group, NY

Eliot, D. 1991. Technologie et spécificité du découpage par plasma dans "les plasmas dans l'industrie" (Pub.) Dopée Avon, France, 229–241 (in French).

Etemadi, K. 1982. Investigation of high-current arcs by computer-controlled plasma spectroscopy, PhD Thesis, University of Minnesota, Mn, USA

Evrard, M., and B. Blanchet. 1970. Etudes des plasmas d'arc du point de vue du soudage. *Soudages et Techniques Connexes* 7: 261–298. (in French).

Fisher, G., T. Wolfe, and K. Meszaros. 2013. The effects of carbide characteristics on the performance of tungsten carbide-based composite overlays, deposited by plasma-transferred arc welding. *Journal of Thermal Spray Technology* 22 (5): 764–771.

Flores, J.F., A. Neville, N. Kapur, and A. Gnanavelu. 2009a. An experimental study of the erosion–corrosion behavior of plasma transferred arc MMCs. *Wear* 267: 213–222.

———. 2009b. Erosion–corrosion degradation mechanisms of Fe–Cr–C and WC–Fe–Cr–C PTA overlays in concentrated slurries. *Wear* 267: 1811–1820.

Freton, P., J.J. Gonzalez, F. Camy Peyret, and A. Gleizes. 2003. Complementary experimental and theoretical approaches to the determination of the plasma characteristics in a cutting plasma torch. *Journal of Physics D: Applied Physics* 36: 1269–1283.

Gallo, S.C., N. Alam, and R. O'Donnell. 2013. In-situ synthesis of titanium carbides in iron alloys using plasma transferred arc welding. *Surface & Coatings Technology* 225: 79–84.

Gallo, S., C.N. Alam, and R.O. Donnell. 2014. In situ synthesis of TiC-Fe composite overlays from low cost TiO2 precursors using plasma transferred arc deposition. *Journal of Thermal Spray Technology* 23 (3): 551–556.

Gatto, A., E. Bassoli, and M. Fornari. 2004. Plasma transferred arc deposition of powdered high-performance alloys: Process parameters optimization as a function of alloy and geometrical configuration. *Surface & Coatings Technology* 187: 265–271.

Gebert, A., and B. Bouaifi. 2005. *Oberflächenschutz durch Auftragschweißen*, Moderne Beschichtungsverfahren. Wiley

Gebert, A., D. Wocilka, B. Bouaifi, and M. Schütz. 2002. Wear and corrosion prevention at light metals by means of welding methods, Proc. ITSC-2002, Essen, Germany, ed. E. Lugscheider (Pub.) DVS, Düsseldorf, Germany, pp 268–272.

Guoqing, R., M. Monno, and M.I. Boulos. 2007. Effect of cathode nozzle geometry and process parameters on the energy distribution for an argon transferred arc. *Plasma Chemistry and Plasma Processing* 27: 359–380

Guoqing, C., F. Xuesong, W. Yanhui, L. Shan, and Z. Wenlong. 2013. Microstructure and wear properties of nickel-based surfacing deposited by plasma transferred arc welding. *Surface & Coatings Technology* 228: S276–S282.

Hallen, H., E. Lugscheider, and A. Ait-Mekideche. 1991. Plasma Transferred Arc Surfacing with High Deposition Rates, Proc. of the 4th. NTSC-1991, Pittsburgh, PA, ed. T. Bernecki (Pub.) ASM International, Materials Park, OH, pp 537–539.

Hallen, H., H. Mathesius, A. Ait-Mekideche, F. Hettiger, U. Morkramer, and E. Lugscheider. 1992. New Applications for High Power PTA Surfacing in the Steel Industry, Proc. ITSC-1992 Orlando, Florida, ed. C. C. Berndt (Pub.) ASM International, Materials Park, OH, pp 899–902.

Hawthorne, H.M., B. Arsenault, J.P. Immarigeon, J.G. Legoux, and V. R. Parameswaran. 1999. Comparison of slurry and dry erosion behavior of some HVOF thermal sprayed coatings. *Wear* 225–229: 825–834.

Heberlein, J., J. Mentel, and E. Pfender. 2007. The anode region of electric arcs – A survey. *Journal of Physics D: Applied Physics* 43: 023001.

Heberlein, J. 2007. Observations on thermionic cathode erosion, 1st Inter. Round Table on Thermal Plasmas (Sharm El Sheikh, Egypt, January 2007).

Hou, Q.Y., J.S. Gao, and F. Zhou. 2005. Microstructure and wear characteristics of cobalt-based alloy deposited by plasma transferred arc weld surfacing. *Surface & Coatings Technology* 194: 238–243.

———. 2006. Microstructure and properties of Fe–C–Cr–Cu coating deposited by plasma transferred arc process. *Surface & Coatings Technology* 201: 3685–3690.

Hou, Q.Y., Y.Z. He, Q.A. Zhang, and J.S. Gao. 2007. Influence of molybdenum on the microstructure and wear resistance of nickel-based alloy coating obtained by plasma transferred arc process. *Materials and Design* 28: 1982–1987.

Hou, Q.Y., Z. Huang, and J.T. Wang. 2011. Influence of nano-Al2O3 particles on the microstructure and wear resistance of the nickel-based alloy coating deposited by plasma transferred arc overlay welding. *Surface & Coatings Technology* 205: 2806–2812.

Hou, Q.Y., L.M. Luo, Z.Y. Huang, P. Wang, T.T. Ding, and Y.C. Wu. 2015a. Comparison of three kinds of MC-type carbide modified thick W coatings fabricated by plasma transferred arc surfacing. *Surface & Coatings Technology* 283: 52–60.

Hou, Q.Y., T.T. Ding, Z.Y. Huang, P. Wang, L.M. Luo, and Y.C. Wu. 2015b. Microstructure and properties of mixed Cu–Sn and Fe-based alloys without or with molybdenum addition processed by plasma transferred arc. *Surface & Coatings Technology* 283: 184–193.

Hou, Q.Y. 2013a. Influence of molybdenum on the microstructure and properties of a FeCrBSi alloy coating deposited by plasma transferred arc hardfacing. *Surface & Coatings Technology* 225: 11–20.

———. 2013b. Microstructure and wear resistance of steel matrix composite coating reinforced by multiple ceramic particulates using SHS reaction of Al–TiO2–B2O3 system during plasma transferred arc overlay welding. *Surface & Coatings Technology* 226: 113–122.

Hsu, K.C., K. Etemadi, and E. Pfender. 1983. Study of the free-burning high-intensity argon arc. *Journal of Applied Physics* 54 (3): 1293–1301.

Hung, F.-Y., Z.-Y. Yan, L.-H. Chen, and T.-S. Lui. 2006. Microstructural characteristics of PTA-overlayed NbC on pure Ti. *Surface & Coatings Technology* 200: 6881–6887.

Huang Zhenyi , Qingyu Hou, Ping Wang (2008) Microstructure and properties of Cr3C2-modified nickel-based alloy coating deposited by plasma transferred arc process, Surface & Coatings Technology, 202, 2993-2999

Jenista, J., J. Heberlein, and E. Pfender. 1997. Numerical model of the anode region of high-current electric arcs. *IEEE Transactions on Plasma Science* 25 (5): 883–890.

Jenista, J., J. Heberlein, and E. Pfender. 1997a. Model for anode heat transfer from an electric arc, Proc. 4th Inter. Thermal Plasma Processes Conference, Athens, Greece, 1997, ed. P. Fauchais Begell House Inc., New York, NY, pp 805–815.

———. 1997b. Numerical model of the anode region of high-current electric arcs. *IEEE Transactions on Plasma Science* 25 (5): 883–890.

Jhavar, S., N.K. Jain, and C.P. Pau. 2014. Development of micro-plasma transferred arc (PTA) wire deposition process for additive layer manufacturing applications. *Journal of Materials Processing Technology* 214: 1102–1110.

Jhavar, S., C.P. Paul, and N.K. Jain. 2016. Micro-plasma transferred arc additive manufacturing for die and Mold surface remanufacturing. *JOM* 68 (7): 1801–1809.

Just, C., E. Badisch, and J. Wosik. 2010. Influence of welding current on carbide/matrix interface properties in MMCs. *Journal of Materials Processing Technology* 210: 408–414.

Kammer, P.A., M. Weinstein, and R. J. DuMola. 1991. Characteristics and applications for composite wear-resistant overlays, Proc. 4th. NTSC-1991, Pittsburgh, PA, ed. T. Bernecki (Pub.) ASM International, Materials Park, OH, pp 513–518.

Karanunakarani, N., and V. Balasubramanian. 2011. Effect of pulsed current on temperature distribution, weld bead profiles and characteristics of gas tungsten arc welded aluminum alloy joints. *Transactions of the Nonferrous Metals Society of China* 21: 278–286.

Kavka, T., O. Chumak, J. Sonsky, M. Heinrich, T. Stehrer, and H. Pauser. 2013. Experimental study of anode processes in plasma arc cutting. *Journal of Physics D: Applied Physics* 46 (2013): 065202 (11pp).

Kim, H.-J., B.-H. Yoon, and C.-H. Lee. 2002. Wear performance of the Fe-based alloy coatings produced by plasma transferred arc weld-surfacing process. *Wear* 249 (2002): 846–852.

———. 2003. Sliding wear performance in molten Zn–Al bath of cobalt-based over-layers produced by plasma-transferred arc weld-surfacing. *Wear* 254: 408–414.

Iakovou, R., L. Bourithis, and G. Papadimitriou. 2002. Synthesis of boride coatings on steel using plasma transferred arc (PTA) process and its wear performance. *Wear* 252: 1007–1015.

Lakshminarayanan, A.K., V. Balasubramanian, R. Varahamoorthy, and S. Babu. 2008. Predicting the dilution of plasma transferred arc hardfacing of stellite on carbon steel using response surface methodology. *Metals and Materials International* 14 (6): 779–789.

Lewis, G.K., and E. Schlienger. 2000. Practical considerations and capabilities for laser assisted direct metal deposition. *Materials & Design* 21 (4): 417–423.

Leylavergne, M., H. Valetoux, J. F. Coudert, P. Fauchais, and V. Leroux. 1998. Comparison of the behaviour of copper, cast iron and aluminum alloy substrates heated by a plasma transferred arc, in Proc. 15th ITSC-1998, Nice, France, ed. C. Coddet (Pub.) ASM International, Materials Park, OH, pp 489–495.

Li, G., C. Zhang, M. Gao, and X. Zeng. 2014. Role of arc mode in laser-metal active gas arc hybrid welding of mild steel. *Materials and Design* 61: 239–250.

Lin, J.J., Y.H. Lv, Y.X. Liu, B.S. Xu, Z. Sun, Z.G. Li, and Y.X. Wu. 2016. Microstructural evolution and mechanical properties of Ti-6Al-4V wall deposited by pulsed plasma arc additive manufacturing. *Materials and Design* 102: 30–40.

Liu, Y.-F., Z.-Y. Xia, J.-M. Han, G.-L. Zhang, and S.-Z. Yang. 2006. Microstructure and wear behavior of (Cr,Fe)7C3 reinforced composite coating produced by plasma transferred arc weld-surfacing process. *Surface & Coatings Technology* 201: 863–867.

Liu, Y.-F., X.-B. Liu, X.-Y. Xu, and S.-Z. Yang. 2007. Microstructure and dry sliding wear behavior of Fe2TiSi /γ-Fe/Ti5Si3 composite coating fabricated by plasma transferred arc cladding process. *Surface & Coatings Technology* 205: 814–819.

Lu, F., H. Li, Q. Ji, R. Zeng, S. Wang, J. Chi, M. Li, L. Chai, and H. Xu. 2011. Characteristics of the functionally graded coating fabricated by plasma transferred arc centrifugal cladding. *Surface & Coatings Technology* 205: 4441–4446.

Lü, Y.-H., Y.-X. Liu, F.-J. Xu, and B.-S. Xu. 2013. Plasma transferred arc forming technology for remanufacture. *Advanced Manufacturing* 1: 187–190.

Lindland, D., and G. Shubert. 1988. Method for applying a weld bead to a thin section of a substrate. *US Patent*: 4,739,146.

Lowke, J.J., R. Morrow, and J. Haidar. 1997. A simplified unified theory of arcs and their electrodes. *Journal of Physics D: Applied Physics* 30: 2033–2042.

Lowke, J.J., M. Tanaka, and M. Ushio. 2005. Mechanisms giving increased weld depth due to a flux. *Journal of Physics D: Applied Physics* 38 (18): 3438–3445.

Lugscheider, E., and A. Ait-Mekideche. 1991. Advances in PTA surfacing. In *Proceedings of the fourth thermal spray conference Pittsburgh, PA*, ed. C.C. Berndt, 529–535. Materials Park: ASM International.

Lugscheider, E., G. Langer, K. Schlimbach, U. Dilthey, and L. Kabatnik. 1999. Possibilites for improving wear-properties of Aluminum-alloys by plasma powder welding process. In *Proceedings of the UTSC, Düsseldorf, Germany*, ed. E. Lugscheider and P. Kammer, 410–413. Düsseldorf: DVS.

Matthes, K.-J., and K. Alaluss. 1996. Formgebendes Plasma-Pulverauftragschweissen mit Impulslichtbogen unter Beachtung minimaler Verformung. *Schweissen und Schneiden* 48 (9): 668–672.

Matthes, K.-J., K. Alaluss, and F. Riedel. 2002. Nutzung der Finite-Elemente-Methode zur Optimierung des formgebenden Pulver-Plasmaauftragschweissens für die Herstellung hoch beanspruchbarer Umformwerkzeuge. *Schweissen und Schneiden* 54 (4): 178–184.

Menart, J. A. 1996. Theoretical and experimental investigations of radiative and total heat transfer in thermal plasmas, Ph.D. Thesis, University of Minnesota, Mn, USA

Morisada, Y.I., H. Fujii, and N. Xukun. 2014. Development of simplified active flux tungsten inert gas welding for deep penetration. *Materials and Design* 54: 526–530.

Murphy, A.B., M. Tanaka, S. Tashiro, T. Sato, and J.J. Lowke. 2009. A computational investigation of the effectiveness of different shielding gas mixtures for arc welding. *Journal of Physics D: Applied Physics* 42: 115205(14pp).

Murphy, A.B. 2010. The effects of metal vapor in arc welding. *Journal of Physics D: Applied Physics* 43: 434001(31pp).

Murphy, A.B., M. Tanaka, K. Yamamoto, S. Tashiro, J.J. Lowke, and K. Ostrikov. 2010. Modelling of arc welding: The importance of including the arc plasma in the computational domain. *Vacuum* 85: 579–584.

Nemchinsky, V.A., and M.S. Showalter. 2003. Cathode erosion rate in high-pressure arcs influence of swirling gas flow. *Journal of Physics D: Applied Physics* 36: 704–712.

Nemchinsky, V.A. 2005. Anode layer in a high-current arc in atmospheric pressure nitrogen. *Journal of Physics D: Applied Physics* 38 (2005): 4082–4089.

Nemchinsky, V.A., and W.S. Severance. 2006. What we know and what we do not know about plasma arc cutting. *Journal of Physics D: Applied Physics* 39: R423–R438.

Nemchinsky, V. 2012. Cathode erosion in a high-pressure high-current arc: Calculations for tungsten cathode in a free-burning argon arc. *Journal of Physics D: Applied Physics* 45: 135201(8pp).

Nestor, O.H. 1962. Heat intensity and current density distributions at the anode of high current, inert gas arcs. *Journal of Applied Physics* 33 (5): 1638–1648.

d'Oliveira, A.S.C.M., R. Vilar, and C.G. Feder. 2002. High temperature behavior of plasma transferred arc and laser co-based alloy coatings. *Applied Surface Science* 201: 154–160.

d'Oliveira, A.S.C.M., R.S.C. Paredes, and R.L.C. Santos. 2006. Pulsed current plasma transferred arc hard-facing. *Journal of Materials Processing Technology* 171: 167–174.

Ozel, S., B. Kurt, I. Somunkiran, and N. Orhan. 2008. Microstructural characteristic of NiTi coating on stainless steel by plasma transferred arc process. *Surface & Coatings Technology* 202: 3633–3637.

Peters, J., J. Heberlein, and J. Lindsay. 2007. Spectroscopic diagnostics in a highly constricted oxygen arc. *Journal of Physics D: Applied Physics* 40: 3960–3971.

Proner, A., J.P. Dacquet, and R. Rouanet. 1997a. In *The plasma high energy: A new hardfacing technique*, ed. P. Fauchais, 787–794. New York: Begell House, Inc.

Proner, A., M. Ducos, and J.P. Dacquet. 1997b. Process for coating of hardfacing a part by means of a plasma tranferred arc. *US Patent US*: 5,624,717.

Pukasiewicz, A.G.M., P.R.C. Alcover Jr., A.R. Capra, and R.S.C. Paredes. 2014. Influence of plasma Remelting on the microstructure and cavitation resistance of arc-sprayed Fe-Mn-Cr-Si alloy. *Journal of Thermal Spray Technology* 23 (1–2): 51–59.

Razal, Rose A., K. Manisekar, V. Balasubramanian, and S. Rajakumar. 2012. Prediction and optimization of pulsed current tungsten inert gas welding parameters to attain maximum tensile strength in AZ61A magnesium alloy. *Materials and Design* 37 (2012): 334–348.

Reinaldo, P.R., and A.S.C.M. D'Oliveira. 2013. Coatings deposited by plasma transferred arc on different steel substrates. *Journal of Materials Engineering and Performance* 22 (2): 590–597.

Rokanopoulou, A., P. Skarvelis, and G.D. Papadimitriou. 2014. Microstructure and wear properties of the surface of 2205 duplex stainless steel reinforced with Al2O3 particles by the plasma transferred arc technique. *Surface & Coatings Technology* 254: 376–381.

———. 2016. Improvement of the tribological properties of Al_2O_3 reinforced duplex stainless steel MMC coating by the addition of TiS2 powder. *Surface & Coatings Technology* 289: 144–149.

Saltzmann G.A., T.A. Wertz, and I.L. Friedman. 1989. Method for refurbishing cast gas turbine engine components and refurbished component, US Patent 4,878,953.

Saltzman, G., and P. Sahoo. 1991. Applications of Plasma Arc Weld Surfacing in Turbine Engines, Proc. Proceedings of the Fourth National Thermal Spray Conference, Pittsburgh, PA, 1991 (ed.) C. C. Berndt (Pub.) ASM International, Materials Park, OH, pp 541–548

Santa, J.F., L.A. Espitia, J.A. Blanco, S.A. Romo, and A. Toro. 2009. Slurry and cavitation erosion resistance of thermal spray coatings. *Wear* 267: 160–167.

Schnick, M., U. Füssel, M. Hertel, A. Spille-Kohoff, and A.B. Murphy. 2010. Metal vapour causes a central minimum in arc temperature in gas–metal arc welding through increased radiative emission. *Journal of Physics D: Applied Physics* 43: 022001(5pp).

Schreiber, F., and D. Krefeld. 2002. Mobile plasma powder hand deposition welding: practice experience, Proc. Intern. Thermal Spray

Conference 2002, Essen, Germany, 4–6 March, 2002, ed. E. Lugscheider (Pub.) DVS, Düsseldorf, Germany, pp 273–277

Shi, K., S. Hu, and H. Zheng. 2011. Microstructure and fatigue properties of plasma transferred arc alloying TiC-W-Cr on gray cast iron. *Surface & Coatings Technology* 206: 1211–1217.

Shi, K., S. Hu, and L. Liang. 2012. Effect of tempering treatment on microstructure and fatigue life of TiC–Cr overlay, produced by plasma transferred arc alloying. *Journal of Materials Science* 47: 720–729.

Sigolo, E., J. Soyama, G. Zepon, C.S. Kiminami, W.J. Botta, and C. Bolfarini. 2016. Wear resistant coatings of boron-modified stainless steels deposited by plasma transferred arc. *Surface & Coatings Technology* 302: 255–264.

Shubert, G.C. 1987. Welding apparatus method for depositing wear surfacing material and a substrate having a weld bead thereon. *US Patent*: 4,689,463.

Skarvelis, P., and G.D. Papadimitriou. 2009. Plasma transferred arc composite coatings with self -lubricating properties, based on Fe and Ti sulfides: Microstructure and tribological behavior. *Surface & Coatings Technology* 203: 1384–11394.

Sung, J.K. 2009. Fluid dynamic instabilities in plasma arc cutting, (Pub.) University of Minnesota, 127 pages

Tanaka, M., and J.J. Lowke. 2007. Topical review, predictions of weld pool profiles using plasma physics. *Journal of Physics D: Applied Physics* 40: R1–R23.

Tashiro, S., T. Zeniya, K. Yamamoto, M. Tanaka, K. Nakata, A.B. Murphy, E. Yamamoto, K. Yamazaki, and K. Suzuki. 2010. Numerical analysis of fume formation mechanism in arc welding. *Journal of Physics D: Applied Physics* 43: 434012(12pp).

Tigrinho, J.J., A. Sofia, and C.M. d'Oliveira. 2007. Plasma transferred arc surface modification of a low carbon steel. *Journal of Materials Science* 42 (17): 7554–7557.

Toropchin, A., V. Frolov, A.V. Pipa, R. Kozakov, and D. Uhrlandt. 2014. Influence of the arc plasma parameters on the weld pool profile in TIG welding. *Journal of Physics: Conference Series* 550: 012004.

Tucker, R.C., Jr., ed. 2013. *ASM handbook Vol. 5A thermal spray technology*. Materials Park: ASM international.

Valetoux, H. 1998. Experimental approach of the phenomena implied in transferred arc plasma cutting. Contribution to the study of instabilities and thermal transfer, (in French) Ph.D. Thesis, University of Limoges, France.

Werry, A., C. Chazelas, A. Denoirjean, S. Valette, A. Vardelle, and E. Meillot. 2016. Multi-scale-structured composite coatings by plasma-transferred arc for nuclear applications. *Journal of Thermal Spray Technology* 25 (1–2): 375–383.

Wang, H., W. Jiang, M. Valant, and R. Kovacevic. 2003. Microplasma powder deposition as a new solid freeform fabrication process. *Proceedings of the Institution of Mechanical Engineers, Part B: Journal of Engineering Manufacture* 217: 1641–1650.

Wang, W., S.Q. Qian, and X.Y. Zhou. 2009. Microstructure and properties of TiN/Ni composite coating prepared by plasma transferred arc scanning process. *Transactions of the Nonferrous Metals Society of China* 19: 1180–1184.

Wassermann, R., J. Quaas, J.-C. Chalard, L. Noel, and H.-T. Steine. 1978. Installation for surfacing using plasma-arc welding. *US Patent*: 4,125,754.

Wilden, J., J. P. Bergmann, H. Frank, S. Pinzl, and F. Schrieber. 2004. Thin plasma-transferred-arc welded coatings – An alternative to thermally sprayed coatings? Proc. ITSC-2004, Osaka, Japan, ed. A. Ohmori (Pub.) ASM International, Materials Park, OH, pp 556–561.

Wilden, J., J.P. Bergmann, and H. Frank. 2006. Plasma transferred arc welding-modeling and experimental optimization. *Journal of Thermal Spray Technology* 15 (4): 779–784.

Wu, C.S., L. Wang, W.J. Ren, and X.Y. Zhang. 2014. Plasma arc welding: Process, sensing, control and modeling. *Journal of Manufacturing Processes* 16: 74–85.

Yamamoto, K., M. Tanaka, S. Tashiro, K. Nakata, K. Yamazaki, E. Yamamoto, K. Suzuki, and A.B. Murphy. 2008. Metal vapor behavior in thermal plasma of gas tungsten arcs during welding. *Science and Technology of Welding and Joining* 13: 566–572.

Yang, G., and J. Heberlein. 2008. Anode heat transfer by electron current and electron conduction in an atmospheric pressure argon arc. Proc. Intern. Conf. Electrical Contacts, Saint Malo, France, 2008, ed. N. B. Jemmaa (Pub.) Universite de Rennes, France pp 313–316.

Yang, G., and J. Heberlein. 2008. Anode heat transfer by electron current and electron conduction in an atmospheric pressure argon arc. In *Proceedings of the international conference on electrical contacts, Saint Malo, France*, ed. N.B. Jemmaa, 313–316. Rennes: Universite de Rennes.

Young, R.M, Y. P. Chyou, E. Fleck, and E. Pfender. 1983. An experimental arc plasma reactor for the synthesis of refractory materials, Proc. ISPC-6, Montreal, Quebec, 1983, eds. M. I. Boulos and R. J. Munz (Pub.) IUPAC, pp 211–218

Yuan, Y., and Z. Li. 2014. Microstructure and wear performance of high-volume fraction carbide M_7C_3 reinforced Fe-based composite coating fabricated by plasma transferred arc welding. *Journal of Wuhan University of Technology. Materials Science Edition* 29 (5): 1028–1035.

Zhao, W., and L. Liu. 2006. Structural characterization of Ni-based superalloy manufactured by plasma transferred arc-assisted deposition. *Surface & Coatings Technology* 201: 1783–1787.

Zhao, H., J. Li, Z. Zheng, A. Wang, D. Zeng, and Y. Miao. 2016. The microstructures and tribological properties of composite coatings formed via PTA surface alloying of copper on nodular cast iron. *Surface & Coatings Technology* 286: 303–312.

Zhenyi, Huang, Qingyu Hou, and Ping Wang. 2008. Microstructure and properties of Cr_3C_2-modified nickel-based alloy coating deposited by plasma transferred arc process. *Surface & Coatings Technology* 202: 2993–2999.

Part III

Coating Formation and Characterization

Abbreviations

AM	Additive Manufacturing
APD	Automatic powder diffraction
AS	Agglomerated and sintered
ASTM	American Society for Testing and Materials
BET	Brunauer, Emmett, and Teller
CET	Coefficient of thermal expansion
CNT	Carbon nano tubes
CS	Cold Spray
CVD	Chemical Vapor Deposition
DC	Direct Current
DC-TPAP	DC-Triple Plasma Atomization Process
D-gun	Detonation gun
DNA	Deoxyribo Nucleic Acid
ECTFE	Ethylene chlorotrifluoroethylene
EDS	Energy Dispersion Spectroscopy
EDXA	Energy dispersion X-ray absorption
EIGA	Electrode Induction-melting Gas Atomization
EMAA	Ethylene-methacrylic copolymer
EMPA	Electron Micro Probe Analysis
ETFE	Ethylene tetrafluoroethylene
EVA	Ethylene vinylene acetate
FC	Fused and crushed
FEP	Fluorinated ethylene propylene
GAP	Gas Atomization Process
GDC	Gadolinia-doped ceria
HA	Hydroxyapatite
HDH	Hydride-Dehydride
HIP	Hot Isostatic Pressing
HOSP	Hollow spherical particles
HVAF	High-velocity air fuel
HVOF	High-velocity oxy-fuel flame
ICP	Inductively Coupled Plasma
ICP-EA	ICP-Elemental Analysis
IMTA	Integrated Mechanical and Thermal Activation
JCPDS	Joint Committee on Powder Diffraction Standards, USA
LCP	Liquid-crystalline polymers
MA	Mechanical alloying
MIM	Metal Injection Moulding
MSR	Mechanical Size Reduction
NIST	National Institute of Standards and Technology
NS	Net Shape
OM	Optical Microscopy
OSD	Oxide-Dispersion Strengthened
PA	Plasma Atomization
PAEK	Polyaryletherketone
PAS	Polyarylene sulfide
PC	Polycarbonate
PC	Powder Consolidation
PE	Polyethylene
PEA	Polyether amide
PEEK	Poly-ether-ether-ketone
PES	Polyester
PET	Polyethylene Terephthalate
PFA	Perfluoroalkoxy
PMMA	Polymethylmethacrylate
PP	Polypropylene
PPS	Polyphenylene sulfide
PREP	Plasma Rotating Electrode Process
PSD	Particle Size Distribution
PTA	Plasma Transferred Arc
PTFE	Polytetrafluoroethylene
PVD	Physical Vapor Deposition
PVDF	Polyvinylidene fluoride
REP	Rotating Electrode Process
RF	Radio Frequency
RF-ICP	Radio Frequency Inductively Coupled Plasma
RF-IPAP	RF-Induction Plasma Atomization Process
RF-IPS	RF-Induction Plasma Spraying
RS	Ratio of Sedimentation height

© Springer Nature Switzerland AG 2021

M. I. Boulos et al. (ed.), *Thermal Spray Fundamentals*, https://doi.org/10.1007/978-3-030-70672-2_13

SD	Spray Drying
SEM	Scanning Electron Microscope
SHS	Self-propagating High-temperature Synthesis
SOFC	Solid Oxide Fuel Cell
SPPS	Solution Precursor Plasma Spraying
SPS	Suspension Plasma Spraying
SS	Stainless Steel
TEM	Transmission Electron Microscopy
TS	Thermal Spraying
UHMWPE	Ultra-high-molecular weight polyethylene
WAP	Water Atomization Process
WAS	Wire Arc Spray
XRD	X-Ray Diffraction
XRF	X-Ray Fluorescence spectroscopy
YSZ	Yttria-Stabilized Zirconia

13.1 Introduction

Powders, wires, or cords are the key components on which the quality of the coating obtained using Thermal Spraying strongly depends. As schematically illustrated in Fig. 13.1, thermal spray technology centers around the use of mechanical, chemical, or electrical energy to develop a high energy gas stream in which the material to be sprayed is injected in the form of a powder, wire, cord, suspension, or solution. On contact with the high energy spraying medium, the individual particles, liquid or suspension droplets, are transformed into a high energy spray of solid gains or molten droplets that are entrained by the process gas and projected at high velocity against the substrate forming splats which are the building blocks of the coating. The coating is formed through the successive layering of the splats with adequate adhesion between splats and between the coating and the surface of the substrate.

The properties of the precursor used in the thermal spray process depends essentially on two principal factors:

- Process requirements
- Coating requirements

Process requirements can vary widely depending on the spray process technology used. For example, combustion and plasma spraying require the use of powders of stable morphological and chemical properties that are free flowing with a relatively narrow Particle Size Distribution (PSD). The morphological stability of the powder implies that the particles are not friable, and their shaper and particle size distribution does not deteriorate through their transport and manipulation in the process. Chemical stability implies that the chemistry of the powder does not change in the process through chemical decomposition and/or loss of alloying elements through vaporization. The PSD of the powder used should be well adapted to the process needs. Generally, the use of a narrow PDF is recommended to avoid, on one hand, having excessive fractions of fine powders that would tend to overheat and possibly vaporize on contact with the hot spray medium, or, on the other hand of the distribution, having too large particles that would be difficult to melt in the short contact time available. The mean particle diameter should be large enough to good powder flowability and avoid particle overheating and possible vaporization, while being small enough to ensure their complete melting during their short residence time in the flame or plasma jet. Cold spray process is generally limited, on the other hand, to metallic powders with good ductility, while flame and wire arc spray process requires the precursor to be in the form of a wire with a not too large diameter (generally between 1.6 and 4.76 mm, 1/16 and 3/16 inch) that is not brittle with an acceptable ductility allowing for its continuous feed into the combustion or arc plasma torch. Cords offer means of expanding the range of materials that can be used with combustion and wire arc spray technologies. These are formed using a folded

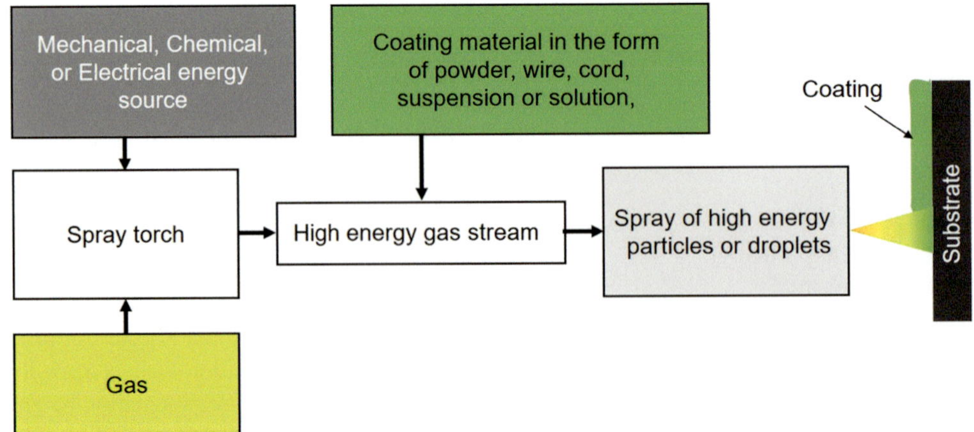

Fig. 13.1 Block diagram of generalized thermal spray process

metallic foil forming a tube that is filled with the material to be sprayed in the form of fine powder.

Coating requirements relate to the functional properties which are expected of the coating. Wear-resistant coating, for example, requires a coating material with favorable tribological properties, low friction coefficient, and in some cases toughness to stand mechanical shocks. Corrosion-resistant coatings should have a minimum or no open porosity, good adhesion to the substrate, and chemical stability against the corrosive elements to which it is exposed. Thermal barrier coatings should have, on the other hand, need to have a good thermal shock resistance, low thermal conductivity, and stable high temperature chemistry. Materials used for biomedical applications such as coating on prothesis implants should be biocompatible with stable chemical composition and excellent adhesion to the substrate.

In this chapter, a brief discussion of the principal powder parameters and characterization methods is presented, followed by a review of powder manufacturing techniques and the predominant properties of the powder obtained in each of these powder production routes. A comparative analysis of powder/precursor feeding techniques commonly used in thermal spray applications is presented next, including novel approaches such as liquid and suspension plasma spray techniques. It is important to underline the close link between this chapter and earlier topics presented in this book, in Part I, Chaps. 4 and 5 on plasma–particle interactions, and throughout Part II Chaps. 6, 7, 8, 9, 10, 11 and 12 discussing different spray techniques and mechanism by which the powder/pressor is injected into the high energy gas stream whether it is a high velocity gas flow as in "Cold Spray" (CS), a combustion product flow "Flame Spraying, HVOF or D-gun" or plasma flow "DC and RF Induction Plasma Spray" (RF-IPS) or electric arc as in "Wire Arc Spraying" (WAS).

13.2 Overview of Powder Characteristics

Thermally sprayed coatings can exhibit significant variations in coating quality, even though the starting powders appear to be equivalent with respect to its chemical composition and particle size distribution [Kubel EJ Jr. (1990)]. This has been demonstrated for plasma spraying [Vinayo et al. (1985), Denoirjean et al. (1992), Diez P and Smith RW (1993), Kolman et al. (1994), Wigren et al. (1996), Wang M and Shaw LL (2007)], HVOF [Beczkowiak J and Schwier G (1991), Jarosinski et al. (1993), Vuoristo et al. (1997), de Villiers Lovelock HL (1998)], D-gun [Kim et al. (1997), Oliker et al. (2004)], and pulsed gas dynamic spraying process [Yandouzi et al. (2008)].

The dependence of powder morphology on the manufacturing process is observed through variations in the shape and microstructure of the particles [Salman AD,

Ghadiri M, Hounslow MJ (eds) (2007)]. Irregular-shaped particles range from cubic-like structures to needles, while for blocky-shaped particles the largest and smallest dimensions of each individual particle are rather close. Moreover, while blocky and spheroidized particles are generally dense, irregular-shaped ones can have a wide range of porosities which impact the apparent density of the particle, its thermal conductivity, and diffusivity. These properties have, in turn, a significant impact on the in-flight heating and melting of the particle prior to its impact on the surface of the substrate and consequently on the quality of the coating formed [Davis JR (ed) (2004), AWS (1985), Pawlowski L (1995), Fauchais P et al. (2010)].

The flowability of powders is also another important parameter in all thermal spray processes. Powders with a poor flowability (generally measured by Hall Flow technique as described in Sect. 13.2.2.3) give rise to large variations in powder feed rates resulting in poor coatings. The powder mass flow rate also depends directly on its specific mass [Anderson et al. (1998)], which is, for a given material, highest with dense spherical particles such as those obtained using RF induction plasma spheroidization [Fan X et al. (1998)]. Particle morphology also plays an important role regarding the particle behavior upon their entrainment by the plasma flow [Vardelle M et al. (1993)]. The fragmentation of agglomerated particles upon penetration into the plasma jet results in the formation of large number of very fine particles, which tend to vaporize during their trajectory in the plasma, thus significantly reducing the ability of the plasma to melt the larger particles on which the coating formation is completely dependent.

The homogeneity of particles composition [Denoirjean A et al. (1992), and Kolman B et al. (1994)] is a key parameter for coating properties, especially at high temperatures. For example, fused and crushed yttria-stabilized zirconia (YSZ) (8 wt.%) keep their tetragonal non-transformable structure up to about 1300 °C, while those prepared by solgel process can be heated up to 1400 °C before transformation occurs [Boch P et al. (1985)]. This is due to the homogeneous distribution of yttria grains (at the nano-scale level) with the solgel technique. For complex coatings, such as quasi-crystalline [Sordelet D and Besser M (1996), Sordelet DJ et al. (1998)] or mullite [Wang Y et al. (2009)], there is also a close link between processing conditions, phase structure, crystallinity, and coating properties. The oxidation level of particles during their manufacturing process is also a very important issue because the end user of the coating must never forget that the oxide content in a coating will never be lower than that of powder used in the spraying process. Finally, it must be kept in mind that powder characterization is not necessarily straightforward; for example, the characterization of particle size distribution [Pei P et al. (1996)] may vary depending on the particle morphology. An exhaustive list of powders, wires, and cords offered by

Oerlikon Metco Corp. for thermal spray applications is given in [Oerlikon Metco (2018)].

Powders are by definition composed of individual particles which can be of different sizes, shapes, and compositions. Their properties are closely dependent on their manufacturing technique and the thermophysical properties of the material. *The properties of individual particles in a powder are generally characterized by:*

- Individual particle diameter and morphology, including their shape, sphericity, porosity, and friability
- Particle chemical composition and crystal structure

The aggregate properties of powders, on the other hand, are characterized by:

- Particle size distribution, mean particle diameter, and standard deviation
- Apparent density and tap density of the powder, the latter corresponding to the highest packing density of the powder
- Powder flowability (e.g., Hall flow) reflecting its ability to free flowing
- Angle of internal friction and angle of repose, both of which relate to properties of the packed powder

13.2.1 Individual Particle Properties

13.2.1.1 Particle Size and Morphology

Powders used in thermal spraying are composed of ensembles of particles which can have a wide range of sizes, shapes, and properties depending on their manufacturing technique and the thermal spray technology used. Typical examples of thermal spray powders manufactured using different production routes are given in Fig. 13.2. These include the following:

- *Water-atomized iron powder* (Fe-C-Si) Fig. 13.2a with a bulky round-edged form [Tsunekawa et al. (2006)]
- *Sintered and crushed Tungsten Carbide powder* (WC-Co-Cr) Fig. 13.2b, which gives rise to bulky, porous, round-edged particles [Starck Amperit 553.065]
- *Fused and crushed alumina powder* (Al_2O_3) Fig. 13.2c with its sharp-edged fully dense particles [Laha et al. (2005)]
- *Bulk Molybdenum disulfide powder* (MoS_2) Fig. 13.2d with their natural flaky structure
- *Plasma spheroidized Molybdenum powder* (Mo) Fig. 13.2e with its generally perfect spherical dense particles [Boulos (2016)]
- *Plasma atomized titanium alloy powder* (Ti-6Al-4V) Fig. 13.2f showing a dense spherical morphology [Polak et al. (2017)]

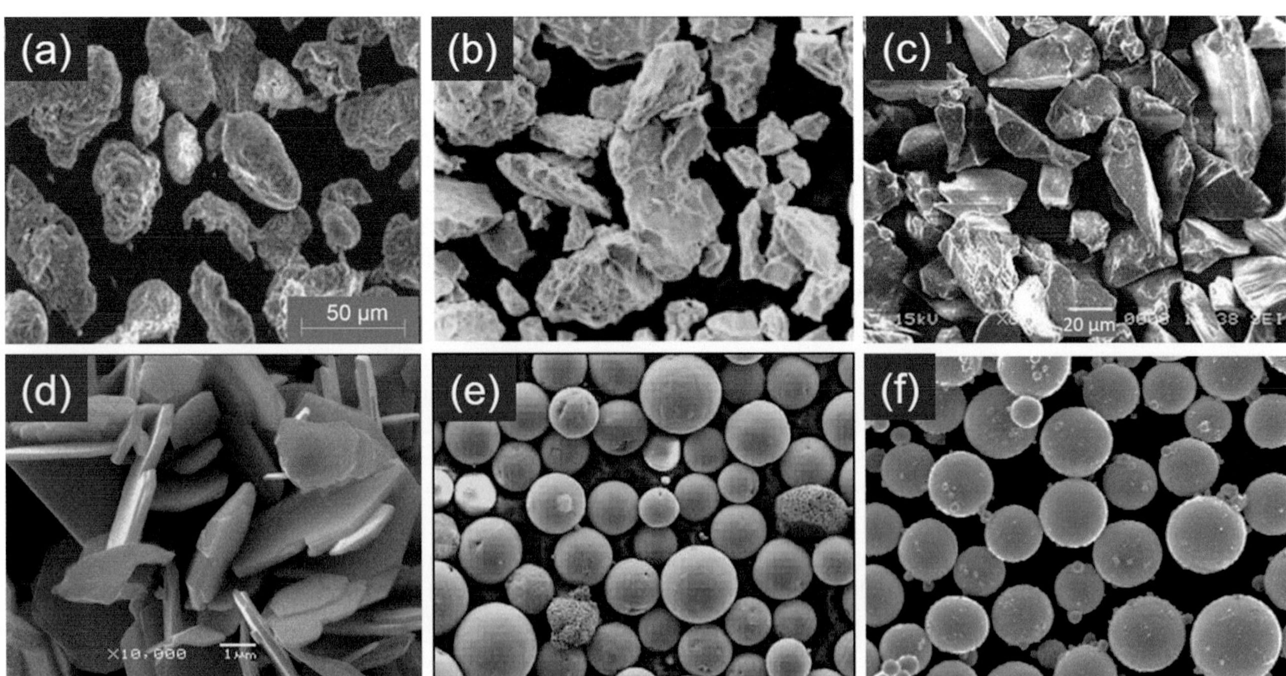

Fig. 13.2 Examples of metallic and ceramic plasma-spray powders reflecting different manufacturing routes

Powder type Manufacturer	Fused and milled	Sintered and milled	Agglomerated and sintered	Spheroidized	Atomized
		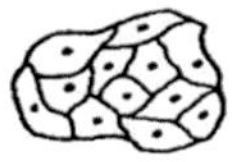	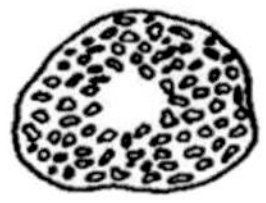		
Particle shape	Blocky - angular	Blocky - angular	Spherical	Spherical	Spherical - irregular
Porosity	Dense	Dense - porous	Porous	Dense - hollow	Porous - hollow
Crystalline size	Course- fine	Course - fine	Medium - fine	Medium - fine	Fine
Homogeneity	Alloyed	Alloyed	Alloyed - heterogeneous	Alloyed - heterogeneous	Alloyed

Fig. 13.3 Powder characteristics of YSZ (ZrO_2–8 wt.% Y_2O_3) obtained using different manufacturing routes, after [Schwier (1986)]

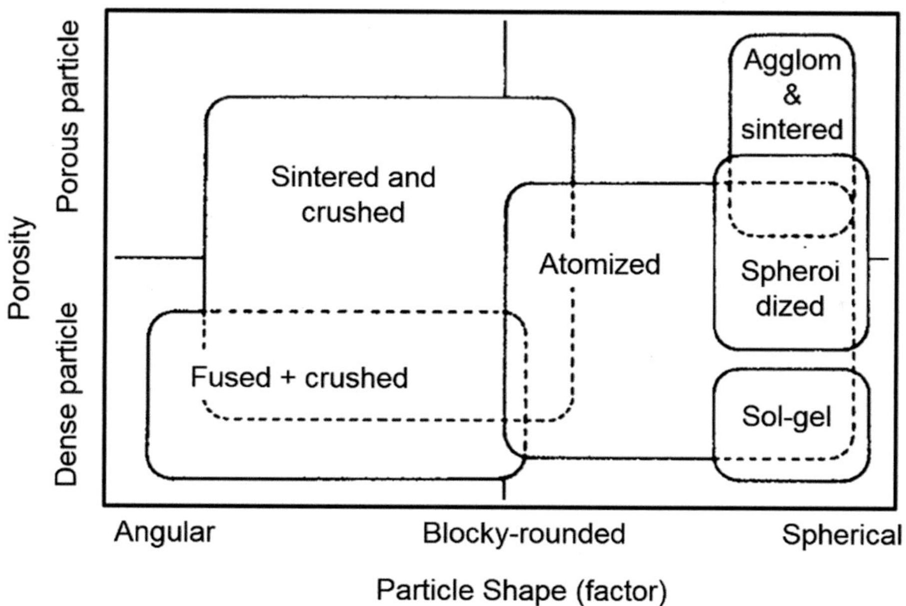

Fig. 13.4 Powder characteristics, porosity as function of particle shape, for different manufacturing routes, after [Schwier (1986)]

The production of dense spherical metallic powders is generally limited to gas, or plasma atomized or plasma spheroidized powders which are formed in the molten state and acquire their spherical shape through surface tension effect prior to solidification (Fig. 13.2e and f).

A study of the link between the powder manufacturing technique and the morphology of the particles obtained for yttria-stabilized zirconia (8 wt.% Y_2O_3) was reported by [Schwier (1986)]. The results given in Figs. 13.2, 13.3, 13.4, and 13.5 show agglomerated and sintered powders commonly used for the plasma spraying of Thermal Barrier Coatings (TBCs) tend to be composed of porous particles with a rounded shape and a rather fine microstructure. Fused and crushed powders are, on the other hand, composed of blocky and dense particles with sharp edges, which show medium to coarse gain structure.

A systematic effort has been devoted in literature in order to develop appropriate parameters to characterize the different

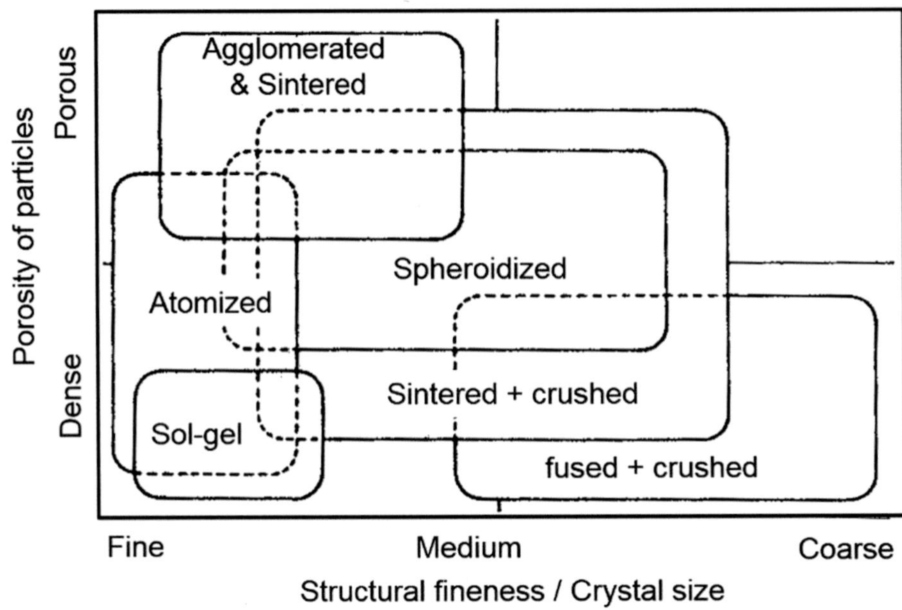

Fig. 13.5 Powder characteristics, porosity as function of structural fineness/crystal size for different manufacturing routes, after [Schwier (1986)]

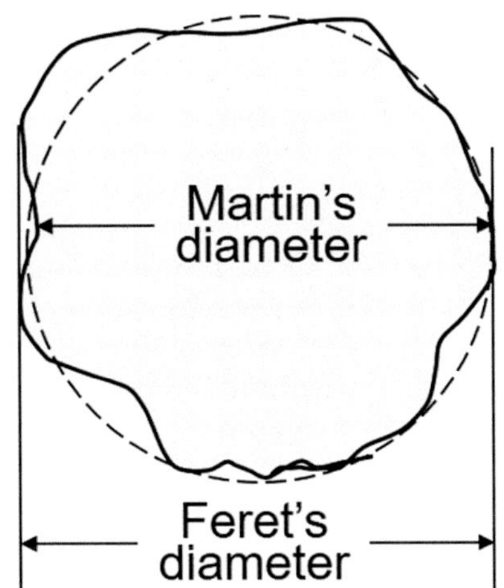

Fig. 13.6 Illustration of Feret's and Martin's diameters, after Reist (1993)

particle morphologies. According to [Reist (1993)] particles with a bulky shape, whether rounded or sharp-edged, can be described by an equivalent particle diameter and a shape factor, such as "sphericity." The Feret's diameter and the Martin's diameter, illustrated in Fig. 13.6, are commonly used based on the image analysis of projected particle micrographs. The *Feret's* diameter is defined as the maximum distance from edge to edge of each particle in the image along an arbitrary but fixed axis applicable to all particles in the photograph. The *Martin's* diameter, on the other hand, is the length of the line that separates each particle into two equal surfaces. [Reist (1993)] points out that since these measurements can vary depending on the orientation of the

particle, they are only valid if averaged over a reasonably large number of particles with all measurements along parallel lines.

Alternately the following definitions of an equivalent particle diameter are commonly used:

Equivalent perimeter diameter, d_{pp}, defined as the diameter of the circle which has the same perimeter as that of the projected particle image

$$d_{pp} = \frac{P_p}{\pi} \tag{13.1}$$

where P_p is the perimeter of the projected image of the particle.

Equivalent projected area diameter, d_{pa}, defined as the diameter of the circle having the same projected surface area as that of the particle

$$d_{pa} = \sqrt{\frac{4a_{po}}{\pi}} \tag{13.2}$$

where a_{po} is the projected surface area of the particle.

Equivalent volume diameter, d_{pv}, defined as the diameter of the sphere having the same volume as that of the particle

$$d_{pv} = \left(\frac{6V_p}{\pi}\right)^{\frac{1}{3}} \tag{13.3}$$

where V_p is the volume of the particle.

The above defined equivalent projected area diameter, and equivalent volume diameter need often be combined with a secondary parameter which reflects how close are the particles from a perfect spherical shape. These are defined as "*Circularity*" and "*Sphericity.*" The circularity "*C*" is defined as the ratio of the perimeter of the circle with the same projected surface area as that of the particle ($\pi\,d_{pa}$), to the perimeter of the projected image of the particle, P_p.

$$C = \frac{\pi d_{pa}}{P_p} \tag{13.4}$$

$C \leq 1.0$ approaching unity as the particle approaches a spherical shape. The *sphericity*, ψ, on the other hand is defined as the ratio of the surface area of the sphere of equal volume as the particle, to the surface area of the particle, A_p.

$$\psi = \frac{\pi d_{pv}^2}{A_p} \tag{13.5}$$

ψ would also be limited to values ≤ 1.0 tending to unity for spherical particles.

Finally, the "Stokes diameter," d_{st}, is defined as the diameter of the sphere of the same density and the same terminal settling velocity, u_t, as that of the particle, assuming to be in the Stokes flow regime

$$d_{st} = \sqrt{\frac{18\,\mu u_t}{(\rho_p - \rho_o)g}} \tag{13.6}$$

where ρ_p and ρ_o are respectively the densities of the particle and the fluid, μ the dynamic viscosity of the fluid, and g the gravitational acceleration.

In the present chapter, unless indicated differently, the equivalent sphere diameter, d_{pv}, will be used for the characterization of non-spherical bulky particles, with the subscript "$_v$" dropped for simplicity.

13.2.1.2 Particle Chemical Composition

Chemical composition of particles can be obtained by different techniques:

Inductively Coupled Plasma Elemental Analysis (ICP-EA) [Trassy and Mermet (1984)]. A suspension or solution of fine particles is injected in an RF induction plasma where the solvent is first vaporized. The sold material contained in the suspension/solution is further heated, melted, and vaporized. The formed vapors mix with the plasma gases and are heated to sufficiently high temperatures to be identified through their own emission spectrum which serves for the estimation of their initial concentration in the suspension/solution against a calibrated standard using elemental emission spectroscopy.

X-ray fluorescence spectroscopy. The principle is the same as that of EDS, except that the high-energy beam is a primary X-ray excitation source from an X-ray tube [Beckhoff et al. (2006)]. Atoms absorb the X-ray that transfers all of its energy to an innermost electron: "photoelectric effect." If the primary X-ray energy is sufficient, electrons are ejected from the inner shells, creating vacancies. As the atom returns to its stable condition, electrons from the outer shells are transferred to the inner shells giving off a characteristic emission whose energy is the difference between the two binding energies of the corresponding shells, allowing one to nondestructively measure the elemental composition of a sample. This process is called "X-ray Fluorescence" or XRF. A typical X-ray spectrum from an irradiated sample will display multiple peaks of different intensities. The spectral lines used for chemical analysis are selected on the basis of intensity, accessibility by the instrument, and lack of line overlaps.

13.2.1.3 Elements Distribution

Most powders contain different materials and it might be important to determine their distribution in order to see how they are homogenously distributed within each particle. For such analysis, the particles must be mounted in epoxy resin, using vacuum impregnation, and then the samples polished, using standard metallographic procedures. The polished sample is disposed in a Scanning Electron Microscope (SEM) equipped with Electron Micro Probe Analysis (EMPA) or Energy Dispersion Spectroscopy (EDS). These probes allow mapping of the elements (or chemical characterization) within the particles. The principle of the method is based on the interactions between electromagnetic radiation and matter, analyzing X-rays emitted by the matter when hit with charged particles [Goldstein et al. (2003)].

13.2.1.4 Crystallographic Structure

Phase identification and crystallinity analysis of the powders is carried out with the X-ray diffraction (XRD) technique [Goldstein et al. (2003)]. The most popular technique is the Debye–Scherer method, employing monochromatic X-radiation. The surface analyzed is typically 15×3 to 5 mm and the depth analyzed, depending on the material, is in the range of a few μm up to 50 μm. The method is based on the relationship existing between the diffraction angle θ and the d-spacing characterizing phase between the crystallographic planes according to the Braggs equation. The spectra obtained are compared to databases such as the Joint Committee on Powder Diffraction Standards (JCPDS) USA, using computer codes that permit to identify the crystallized phases. The relative phase ratios can be calculated by comparing the areas under peaks in the

X-ray diffraction (XRD) pattern. To determine the distribution of phases, the limit of detection corresponds to about 5 wt.%. Peak area under corresponding peaks of different phases can be profile fitted using an Automatic Powder Diffraction (APD) profile-fitting program. The size of the individual crystals or grains can be determined by measuring the width of X-ray peaks.

13.2.2 Aggregate Powder Properties

As already underlined, while coating properties depend on the spray process chosen, its physical and functional properties depend to a large extent on the powder properties. It is thus of primary importance to characterize the sprayed powder parameters such as its chemistry (composition and purity, elements distribution, phase analysis), size distribution, particles morphology and shape, apparent density, and flowability [Iinoya K et al. (1991), Allen T (1997), ASTM E2651-10, ASTM-B214-07 (2011)].

13.2.2.1 Powder Sampling

To characterize powders, generally a rather small quantity (a few grams to a few tens of grams) is necessary, compared to a batch of powder representing a few tens of kilograms. It is thus important work with a representative sample of the powder. A number of such powder sampling devices have been developed as described by ASTM standards [ASTM WK 19862]. It consists of two concentric cylinders that can be twisted relative to each other, capturing powder from specific location within a container. Samples, used in the different characterization technique [Davis JR (ed) (2004)] from different locations, are then combined, blended, and redevised until a small enough sample size is reached as required by the analytical technique to be used.

13.2.2.2 Particle Size Distribution

When dealing with powders, it is most unlikely to have all of the particles of the same diameter. Powders are generally polydisperse when formed, some more than others depending on their manufacturing process. Proper representation of the particle size requires a statistical analysis of the particle size distribution (PSD) of a representative sample of the powder that could be carried out using either optical or electron microscopy coupled with a detailed image analysis, light diffusion instruments, or measurements based on the particle terminal settling velocity. Mechanical separation of the particles into fractions of different particle sizes is also a relatively simple technique using appropriate sets of calibrated sieves or pneumatic powder separators. These are often used for cross calibration of the different measurement techniques. The results could then be presented in graphical terms such as the standard normal distribution shown in Fig. 13.7a represented by the relationship:

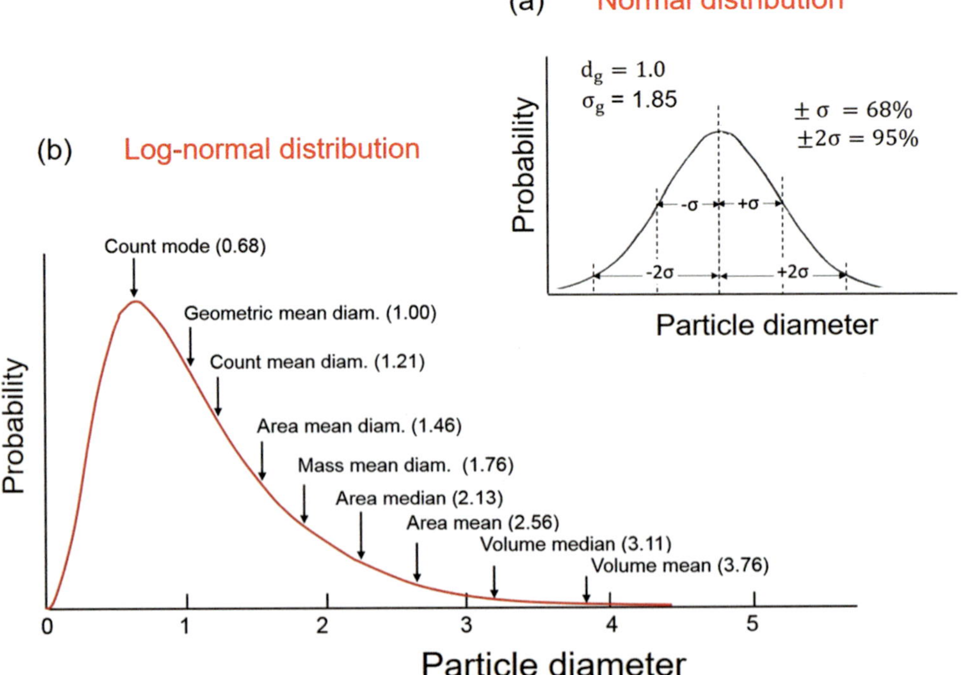

Fig. 13.7 (a) Normal and (b) Log-normal particle size distribution, after Reist (1993)

$$F(d) = \frac{1}{\sigma_d \sqrt{2\pi}} \exp\left(-(d - \overline{d})^2 / 2\sigma_p^2\right) \qquad (13.7)$$

where $\overline{d}$ and σ_d are respectively the mean value and standard deviation defined as

$$\overline{d_n} = \frac{\sum\limits_{i=1}^{\infty} n_i d_i}{\sum\limits_{i=1}^{\infty} n_i} \qquad (13.8)$$

and

$$\sigma_{p_n} = \left[\frac{\sum\limits_{i=1}^{\infty} n_i \left(d_i - \overline{d_n}\right)^2}{\left(\sum\limits_{i=1}^{\infty} n_i\right) - 1}\right]^{\frac{1}{2}} \qquad (13.9)$$

where n_i is the number of particles of diameter d_i, with $\overline{d_n}$ the number-mean value, and σ_{p_n} the standard deviation of the number particle size distribution. Statistically, a variable with a normal distribution will have a 68% probability to be within $\pm\sigma_p$, and 95% probability to be within $\pm 2\sigma_p$. The corresponding relationships in terms of the volume, or mass, fraction of the particles is,

$$\overline{d_x} = \frac{\sum\limits_{i=1}^{\infty} x_i d_i}{\sum\limits_{i=1}^{\infty} x_i} \qquad (13.10)$$

and

$$\sigma_{p_x} = \left[\frac{\sum\limits_{i=1}^{\infty} x_i \left(d_i - \overline{d_x}\right)^2}{\left(\sum\limits_{i=1}^{\infty} x_i\right) - 1}\right]^{\frac{1}{2}} \qquad (13.11)$$

where x_i is the volume, or mass-fraction of particles of diameter d_i, with $\overline{d_x}$ mass-mean value, and σ_{p_x} the standard deviation of the particle volume, or mass size distribution. The conversion from a number-based distribution to a mass- or volume-based distribution can be simple by weighing the number distribution to d_i^3 as given by Eq. 13.12.

$$x_i = \frac{n_i d_i^3}{\sum\limits_{i=1}^{\infty} n d_i^3} \qquad (13.12)$$

Statistically for a variable to have a normal distribution, Fig. 13.7a, it has to have equal probability for larger or smaller values relative to its mean. This condition is rarely satisfied for powders, since powder crushing, for example, is obviously biased toward the finer size fractions. These distributions, however, would fit a normal distribution if the probability is plotted against the logarithm of the particle diameter giving rise to what is known as "lognormal" distributions as shown in Fig. 13.7b. The corresponding geometric mean diameter, $\overline{d_{gn}}$, and geometric standard deviation, σ_{gn}, can be calculated using Eqs. 13.13 and 13.14, respectively.

$$\log \overline{d_{gn}} = \frac{\sum n_i \log d_i}{\sum n_i} \qquad (13.13)$$

and

$$\log \sigma_{p_{gn}} = \left[\frac{\sum n_i \left(\log d_i - \log d_g\right)^2}{\sum n_i - 1}\right]^{\frac{1}{2}} \qquad (13.14)$$

where n_i is the number of particles of diameter d_i, with $\overline{d_{gn}}$ mean value, and $\sigma_{p_{gn}}$ the standard deviation of the particle volume, or mass-based geometric size distribution. The corresponding values of the mass-mean particle diameter, d_{gx} and its standard deviation, $\sigma_{p_{gx}}$ are given by Eq. 13.15 and 13.16 respectively

$$\log \overline{d_{gx}} = \frac{\sum x_i d_i^3 \log d_i}{\sum x_i d_i^3} \qquad (13.15)$$

and

$$\log \sigma_{p_{gx}} = \left[\frac{\sum x_i d_i^3 \left(\log d_i - \log d_g\right)^2}{\sum x_i d_i^3 - 1}\right]^{\frac{1}{2}} \qquad (13.16)$$

It is important to note that in a number distribution the fine particles are given the same weight as any large particles in the powder and accordingly the distribution is biased toward fine particle presentation. The effect is clearly demonstrated in Fig. 13.8, which shows the same particle size histogram or frequency distribution function, represented both in terms of number fraction and volume/mass fraction of the particles. For most engineering applications, the mass fraction distribution is the most pertinent since it reflects distribution of the mass powder of different particle sizes.

Considering that most powders do not necessarily have a fully "normal" or "log-normal" size distribution, the average or mean particle diameter and standard deviation are not sufficient for a full description of the PSD of the powder. Besides having access to the PSD of the powder it is necessary to know its upper and lower limits as indicated by the d_{10} and d_{90} which represent the particle diameter limits between

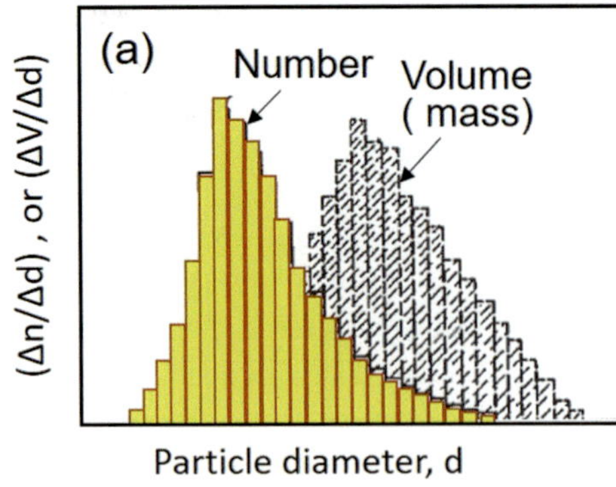

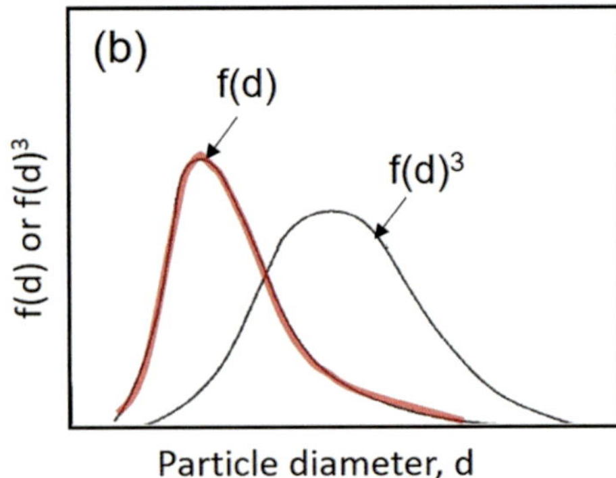

Fig. 13.8 Particle size histogram and frequency distribution function presented in terms of number and volume or mass fraction distributions, after Lefebvre (1989)

which 80% of the weight of the powder is situated. The d_{10} defines the particle diameter below which 10 wt.% of the powder is found, while d_{90} defines the particle diameter above which 10 wt.% of the powder is found. In exceptional cases, particle size distribution can also have a bimodal (two-peaks) size distribution, or a highly skewed distribution with the skewness coefficient defined as the difference between the mean particle size and the 50th percentile of the powder distribution. Skewness can be generally important with bimodal PSDs.

Measurements of particle size distributions are generally obtained using one or more of the following standard techniques.

(a) *Sieving, screen analysis*

Screening is one of the basic techniques used for the determination of the particle size distribution of a powder based on the physical separation of the powder into different size fractions identified by the screens used. A wide range of wove wire screens are available for analytical purpose according to either the "Tyler series" or the "ASTM-E11" US standard as shown in Table 13.1. The Mesh number is defined as the number of openings per linear inch of the screen material which is normally made of either woven brass or stainless-steel wire of tightly calibrated dimension. The standard "Tyler series" contains screens with Mesh number 10 to Mesh 400 with a corresponding screen opening of 1700 to 37 µm respectively. While the Tyler series shown in Table 13.1 is expanded down to Mesh 2500 corresponding to an opening of 5 µm, measurements with screen sizes in the range of Mesh 500 to Mesh 2500 are particularly challenging since the individual powder particles tend to blind the screen, requiring intense ultrasound vibration in order to achieve a reasonable and reliable separation of the powder in this size range. The ASTM-E11 standard shown in Table 13.1 covers essentially the same range of Mesh size from 12 to 400 with minor variations of the corresponding screen openings

Table 13.1 Standard screens giving the Tyler and ASTM-E11 sieves mesh sizes

Mesh size	Tyler	ASTM-E11	Mesh size	Tyler	ASTM-E11
µm	Mesh	No	Mm	Mesh	No
5	2500		180	80	80
10	1250		212	65	70
15	800		250	60	60
20	625		300	48	50
25	500		355	42	45
37	400	400	425	35	40
45	325	325	500	32	35
53	270	270	600	28	30
63	250	230	710	24	25
74	200	200	850	20	20
90	170	170	1000	16	18
106	150	140	1180	14	16
125	115	120	1400	12	14
150	100	100	1700	10	12

compared to the "Tyler series" caused by slight differences of the wire diameter used in their respective series.

(b) *Optical or electron microscopy coupled with image analysis*

The most direct way to determine the size and shape of a particle is to look at it [Barth H. Ed (1984), Iinoya et al. (1991)]. To get complete information, it is however necessary to observe particles both from the outside and inside. If the size of the particle is >1 µm, then optical microscopy (OM) working with visible light is fine. For more precision, Scanning Electron Microscopy (SEM) is used.

To observe particles from the outside it is necessary to give special attention to sample preparation to ensure regular distribution of particles on the surface of the sample holder without overlapping. Particle size measurements are made either directly at the microscope or from photographs taken using the microscope with appropriate magnification scale measurement integrated into the photograph. The main challenge is in determining the size and shape of three-dimensional particles on the basis of planar images. Several efficient image analysis software are available allowing for the measurement of a large number of particles in a relatively short time. The technique, however, becomes significantly more challenging to use with submicron particles.

(c) *Light scattering*

When a beam of light strikes a particle, part of it is transmitted, part is absorbed, and part is scattered. Light scattering is divided into three domains [van de Hulst HC (1981)] based on a dimensionless size parameter, α_d, defined as

$$\alpha_d = \pi d_p / \lambda \qquad (13.17)$$

where πd_p is the circumference of the particle and λ the wavelength of incident radiation.

$\alpha_d \ll 1$: Rayleigh scattering ($d_p \ll \lambda$)
$\alpha_d \approx 1$: Mie scattering ($d_p \cong \lambda$)
$\alpha_d \gg 1$: Geometric scattering ($d_p \gg \lambda$).

When $\alpha_d \gg 1$ particles cause Fraunhofer diffraction. The intensity of the forward-scattered light (i.e., light traveling in roughly the same direction as the incident light) is proportional to d_p^2. The relationship between scattering angle (θ_o) and d_p is

$$\sin \theta_o = 1.22 \, \lambda / d_p \qquad (13.18)$$

The light source commonly used is a He–Ne laser with $\lambda = 0.63$ μm which is compatible for the size measurement of particles in the range of 2 to 100 μm. Light-scattering methods can be used with particles suspended in either a gas or a liquid. They are rather fast and accurate and only a small sample is needed.

(d) *Coulter counter*

Particles suspended in an electrolyte are forced to flow through a narrow orifice on both sides of which an electrode is immersed. As a particle passes through the orifice, it displaces an equivalent volume of electrolyte and causes a change in resistance [Kinsman S (1979)]. The magnitude of this change is proportional to the particle size. Resistance changes are converted to voltage pulses that are amplified, sized, and counted to produce data for the size distribution of the suspended particles. The Coulter counter measures particles in a size range 0.5 to 100 μm.

13.2.2.3 Specific Surface Area

Specific surface area measurements are particularly useful for fine and ultrafine powders with a submicron PSD. While the PSD of such powders can be determined using Scanning Electron Microscopy (SEM) coupled for image analysis, or light scattering/diffusion methods, the most commonly used are based on the adsorption of gases onto a particle surface at low temperature and identified after its developers, Brunauer, Emmett, and Teller (BET). The method consists of measuring the surface area of the powder using N_2 adsorption isotherm at the boiling temperature of liquid nitrogen. The mass of gas adsorbed N_2 is measured as a function of gas pressure at a fixed temperature. Since one nitrogen molecule requires a surface area of 0.162 mm^2, the surface area of the particle can be easily estimated in m^2/g. The adsorbates volume, V_m, per unit mass of solid for monolayer coverage is determined by varying the relative pressure, p/p_0 pressure ratio in the range between 0.05 and 0.2. p, being the gas pressure and p_0 the saturation vapor pressure for the adsorbates at the adsorption temperature. If V_a is the volume of the adsorbates at relative pressure p/p_0, V_m is determined by plotting [$p/V_a \cdot (p-p_0)$] versus p/p_0 that gives a straight line from which V_m can be determined. The specific surface area, a_p, depends linearly on V_m. Once the particle mass mean specific surface of the particle is known, its equivalent mean particle diameter, d_{BET} can be calculate using Eq. 13.19

$$d_{BET} = 6 / \left(\rho_p a_p \right) \qquad (13.19)$$

With a_p the BET specific surface given in m^2/g, the particle density ρ_p in g/cm^3, the mean particle diameter, d_{BET} is obtained in μm.

A typical example of SEM electron micrograph for aluminum nanopowder, with its corresponding PSD in mass fraction obtained using a light scattering instrument, is given in Fig. 13.9. The indicated mean particle diameter $\overline{d_p} = 136$ nm with a standard deviation $\sigma_p = 18$ nm. The BET value for this powder was $a_p = 17.2$ m^2/g. With a density for aluminum $\rho_p = 2.7$ g/cm^3, the mean particle diameter is calculated using Eq. 13.19 $\overline{d_{BET}} = 130$ nm which is rather close to the value of 136 nm obtained using light scattering measurement.

13.2.2.4 Flowability

Flowability measurements are easy and inexpensive. The most used one is the Hall flow test presented in Fig. 13.10a. It consists essentially of a precisely machined brass or stainless-steel funnel with the indicated dimensions, tolerances, and surface finish. A photograph of a typical commercially available Hall flow setup is shown in Fig. 13.10b. The measurement consists essentially determining the time required for a 50 g sample of the powder to flow

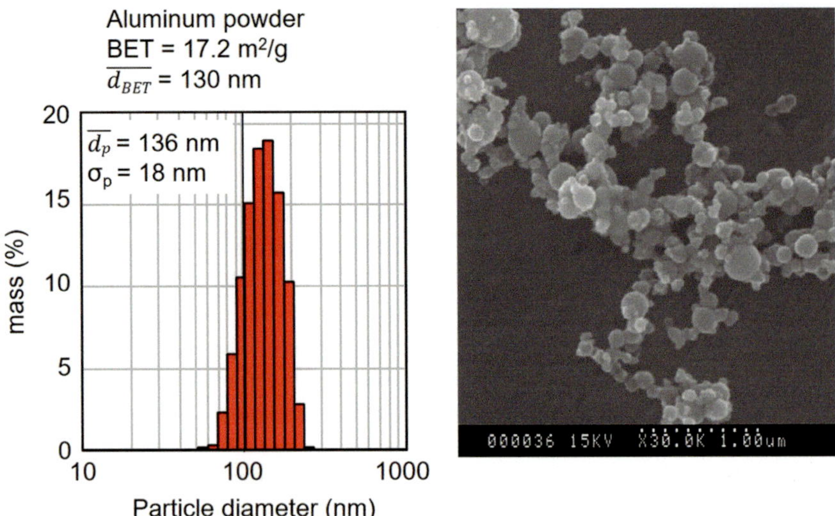

Fig. 13.9 Particle size distribution and high magnification electron micrograph of nano-sized aluminum powder. Mean particle diameter $\overline{d_p}$ = 136 nm, and standard deviation σ_p = 18 nm. [Data and image supplied by Tekna Plasma Systems Inc.] Reprinted with kind permission

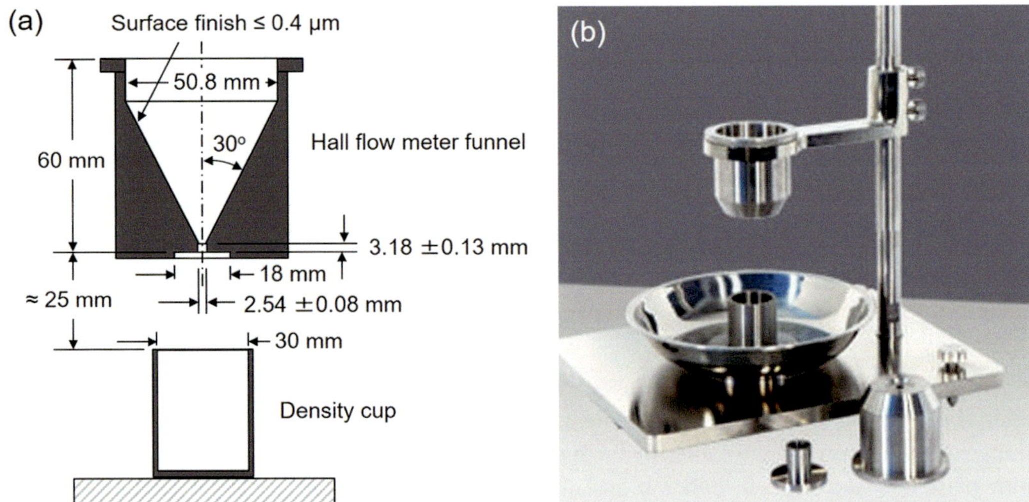

Fig. 13.10 (a) ASM Standard Hall flowability test funnel and (b) Photograph of a commercial unit by Qualtech Products Industry [Davis JR (ed) (2004)] Reprinted with kind permission from ASM Inter

through the funnel. The longer is the time for a given powder mass to flow, the poorer is its flowability [ASTM B213]. Best results are obtained for spherical and dense particles. Another test used for the quantification of the flowability of a given "free-flowing" powder is that of "angle of internal friction α" and "angle of repose θ" as illustrated in Fig. 13.11. These are defined as follows.

The angle of internal friction, α^o is defined as the angle with the horizontal assumed by a moving core of powder in a vessel provided with a central opening in its bottom through which the powder can flow in free fall. The line of demarcation between the stationary and flowing powder is usually well defined. This angle is usually critical for the design of hoppers for storage of powders in powder feeding systems. The smaller the value of α^o the easier will be to maintain an uninterrupted powder flow.

Angle of repose, θ^o is defined either as the angle with the horizontal formed by a pile of powder deposited on a flat plate as illustrated in the bottom part of Fig. 13.11, or the angle formed at the surface of the powder charge, in a reservoir drained by free flow through an orifice in its bottom as shown in the upper part of Fig. 13.11. The difference between the two definitions of the angle of repose, the one formed by piling and the other by draining a powder, is a function of the particle size distribution of the powder. The smaller is the angle of repose, the more the powder will be free flowing.

It is important to stress the need to have a close control on the humidity of the powder for both the "Hull flow" and "angle of repose" test due to its important potential impact on the results obtained. As powders have generally rather high specific surface area, they can retain or absorb moisture from the ambient air, especially in the case of hygroscopic materials (water absorbent) which can significantly affect the flowability measurement results [Stanford MK, Della Corte C (2006)]. This is achieved by placing the powder prior to the test in a drying oven at a temperature slightly over 100 °C for about 8 h. Obviously when dealing with polymer powders, the drying temperature has to be adjusted depending on the softening temperature of the polymer material. It is to be underlined that the requirement for powder drying is necessary prior to the use of the powder in any thermal spray operation. The drying time necessary depends on the mass of the powder to be dried and its PSD, allowing for more time the smaller is the mean particle diameter.

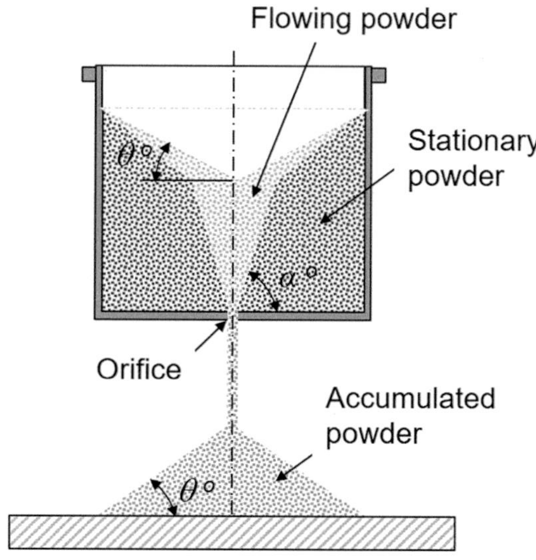

Fig. 13.11 Definition of the "angle of internal friction, α" and "angle of repose, θ" for a free-flowing powder

13.2.2.5 Apparent and Tap Density

Powders packed together will never reach the density of the material with which they are made. Their *apparent density*, under gravity conditions, depends on particles size distribution, shape, and morphology of the individual particles in the powder. It is defined as the weight the powder freely collected into a container of known volume. As volumetric feeders are based on the apparent density the feed rate can change for the same powder (shape and morphology) if the particles size distribution is modified. The *apparent density of the powder* is very simply measured as shown in Fig. 13.12. For powders with a broad PSD, changes in the PSD of the powder can change gradually in the course of operation through either attrition of the powder or segregation; the latter refers to the preferential accumulation of the finer fractions of the powder in the bottom of the container compared to the powder at the top levels in the same container. The *tap density of the powder*, on the other hand, is measured using essentially the same setup except for the measurement procedure which included the tapping on the container using a standard drop weight with the occasional topping of the powder content of the reservoir to ensure the best packing for the given standard volume of the reservoir. In general, powders with a wide PSD will have higher tap densities compared to their apparent density.

13.3 Powder Manufacturing Techniques

Powder manufacturing for thermal spray applications is of critical importance in the spray coating industry due to the strong dependence of the quality and performance of the coating quality on the material feed into the spray system whether in the form of powder, wire, or cord. Powders, which represent the majority of feed materials used in thermal spray, need to be manufactured in a reliable, efficient, economical, and environmentally responsible fashion that complies with the strict technical requirements of the thermal spray process. The wide range of approaches used can be grouped under the following generic approaches:

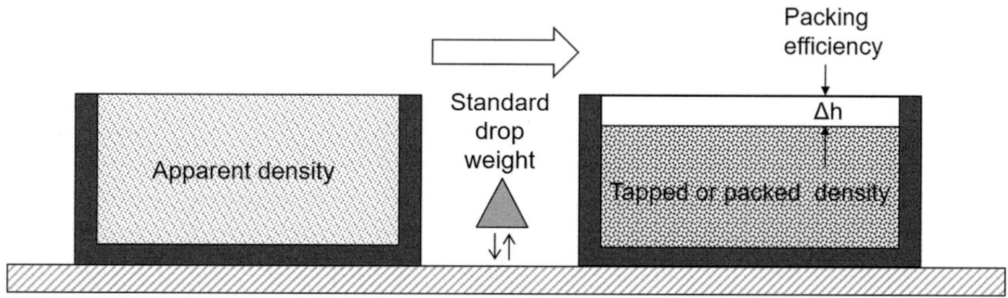

Fig. 13.12 Principle of apparent and tap/packing density measurement. Reprinted with kind permission from ASM International [Davis JR (ed) (2004)]

- Mechanical size reduction
- Powder consolidation
- Spheroidization
- Atomization
- Chemical synthesis

13.3.1 Mechanical Size Reduction

Mechanical Size Reduction (MSR) is the oldest and most widely used technique for the production of powders from natural materials such as rocks or minerals, fused or sintered ceramics, metals, and alloys. Depending on the nature of the raw material used, the powder obtained is mostly bulky with sharp-edged particles. A wide range of processes and equipments have been developed for this operation ranging from crushers used to produce relatively coarse materials, to mills which are used for the production of finer powders. A few examples of techniques commonly used are presented.

13.3.1.1 Fusing/Sintering and Crushing

This technique is devoted to brittle materials, mainly ceramics. The first part of the process is typically carried out in an arc furnace, which can require 10–20 h to melt the core of mixed components in the crucible. After cooling, the outer shell, made of non-melted, but partially sintered initial components, is removed and the frozen ceramic block is crushed using conventional crushers such as jaw-crusher, gyratory crushers, and attrition mills [Dunkley JJ (1998)]. Particles obtained are angular, dense, and blocky as illustrated in Fig. 13.13 for alumina particles [Laha T et al. (2005)].

13.3.1.2 Ball Milling

Milling is used when considering brittle particles such as ceramics, oxides, nitrides, or carbides and certain metals or metal alloys. Ball milling is one of the most common methods used to reduce ceramic particle sizes and eliminate aggregates [Davis JR (ed) (2004), Salman AD, Ghadiri M, Hounslow MJ (eds) (2007), Neikov OD (2009)]. A ball mill consists essentially of a barrel usually made of vulcanized rubber, high-density alumina, or steel that rotates on its axis. Laboratory scale units are often made of porcelain or metal covered with polyurethane or tungsten carbide cermet coating. The barrel is partially filled with a grinding material (called media) in the form of spheres, cylinders, or rods. Figure 13.14 shows a cross section of a ball mill. The quantity of the media is such that the rotation of the mill causes it to cascade, creating both shearing and crushing actions on the powder.

The media with a high specific mass provides greater impact during tumbling. Alumina, steel, zirconia, and mullite

Fig. 13.13 Low-magnification SEM micrograph of irregularly shaped alumina powder obtained by fusing and crushing [Laha T et al. (2005)]. Reprinted with kind permission from Springer Science Business Media

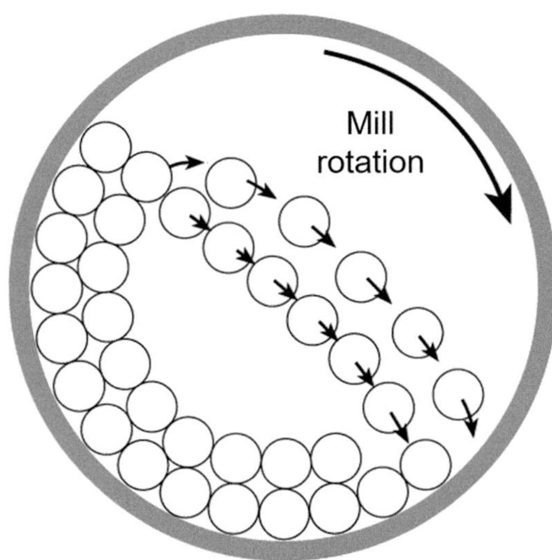

Fig. 13.14 Cross section of a ball mill showing the movement of the media as the mill rotates about its axis [Neikov OD (2009)]. Reprinted with kind permission from Elsevier NY

are mostly used media. The choice of media is also based on its cost, wear resistance, and the necessity to limit, as much as possible, contamination of the milled powder by fragments of the milling media. The media size depends on the size of the mill, quantity of powder to be milled, and the final particle size required. Typical diameter values range from a few millimeters to a few centimeters in diameter, the former being used for fine grinding. The main parameters controlling the PSD of the powder produced are:

- size of the media,
- speed of rotation of the mill,
- milling time.

The speed of rotation of the mill has to be sufficiently high to keep, by centrifugal force, the balls or rods close to the wall of the mill for about one third to half of a turn, dropping and tumbling from the top of the mill onto the material to be grounded, as shown in Fig. 13.14. At too low velocities the media rolls around the bottom of the mill, while at too high velocities the media maintains its position close to the wall for the full turn of the mill shell which would stop the grinding process.

13.3.1.3 Attrition Milling

Attrition milling differs from conventional ball milling in that it is carried out in the presence of a liquid with the slurry containing the particles to be milled and the milling media which is stirred continuously at frequencies of 1–10 Hz. The stationary grinding chamber, often water cooled, is aligned either vertically, as shown in Fig. 13.15, or horizontally, with the stirrer located in the center of the chamber. Chamber volume varies between 4 and 400 l (0.004 to 0.4 m^3). The stirrer rotates at about 250 rpm. The media consists of small spheres (0.2–5 mm) that make up between 60 and 90% of the available mill volume. Most attrition mills work on a continuous basis with the powder to be milled fed in at one end and the milled product collected at the other.

Attrition mills with typical product particle sizes between 0.1 and 5 µm can handle higher solid contents in the slurry and are more energy efficient with relatively short milling time, compared to standard ball milling. The use of small media combined with the lining of the chamber with a polymer or a ceramic and the use of ceramic stirrers and media help reduce potential product contamination. According to [Camerucci MA and Cavalieri AL (1998)] milling of

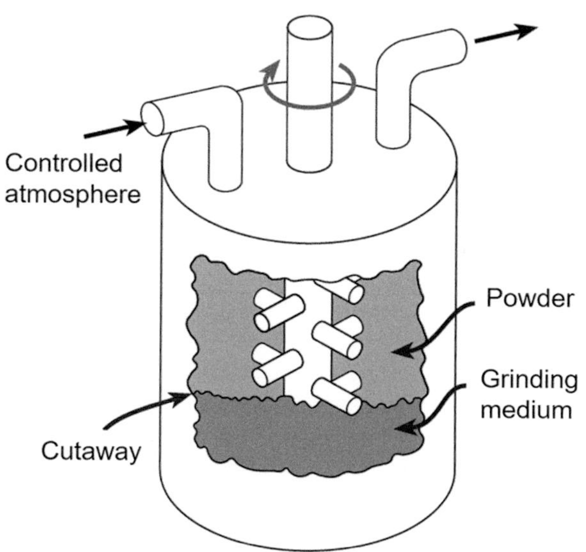

Fig. 13.15 Principle of attrition milling [Neikov OD (2009)]. Reprinted with kind permission from Elsevier NY

powders to submicron particles sizes is possible using intermediate rotor speeds and lower solid contents, and short milling times. To limit agglomeration, organic additives are added in the slurry. The separation of fine particles from the slurry represents one of the main limitations and challenges of attrition mills.

13.3.1.4 Alternate Crushing and Milling Routes

Globally crushing and grinding is very convenient, rather simple and inexpensive, allowing for size reduction of a wide range of materials down to 1 µm. It is, however, rather energetically inefficient with less than 2% of the energy consumed goes into the creation of new surfaces. It is also rather slow, with relatively long milling time required, and the risk of contamination of the final product by impurities originating from the grinding media. The main limitation of crushing and grinding is, however, related to the fact that it could only be used to brittle materials which exclude most metals which are mostly ductile tending to deform rather than break when subject to mechanical impact. A few techniques have been developed to overcome this limitation on a case-by-case basis such as

(a) *Cryo-milling,*

which involves the cooling of the material to be crushed and ground as well as the milling media to cryogenic temperatures, using for liquid argon or more often liquid nitrogen, which renders most metals brittle and easier to mill [Lavernia EJ et al (2008)]. Compared to standard milling at room temperature, cryo-milling has the following advantages:

* Powder agglomeration and welding to the milling media are suppressed, resulting in a more efficient milling outcome.
* Oxidation reactions during milling are reduced under the protection of a nitrogen environment.
* Milling time required to attain submicron sized powders is significantly reduced.

The main limitation of cryo-milling, however, is the added complexity and cost of the process for the cryogenic medium.

(b) *Reversible chemical transformations,*

such as the Hydride-Dehydride (HDH) process developed by Reading Alloys for the grinding of Titanium metal and its alloys down to particle size of 25 µm and lower [McCracken CG and D Barbis (2008), McCracken CG (2009), McCracken CG, and D.P. Barbis, (2010), McCracken CG, et al. (2010a, b), McCracken CG et al. (2012) and Froes, FH (2012a, b)]. The process is based on the concept of pure

Fig. 13.16 SEM images of HDH Titanium powders produced using different feedstocks (**a**) wrought Ti, (212–300 µm), (**b**) Na-reduced Ti, (212–300 µm), (**c**) Mg-reduced Ti, (212–300 µm), (**d**) wrought Ti, (< 25 µm), (**e**) Na-reduced Ti, (< 25 µm), (**f**) Mg-reduced Ti, (< 25 µm) [McCracken CG and D Barbis (2008)] Reprinted with kind permission

titanium or titanium alloy to a temperature above 350 °C and cooled in a hydrogen atmosphere as according to Eq. 13.20 forming a staple phase of titanium hydride (TiH$_2$) which contains about 4 wt.% H$_2$.

$$Ti + H_2 \rightleftarrows TiH_2 \tag{13.20}$$

The hydride having a brittle structure can be readily crushed and milled using conventional powder sizing equipment. Once the TiH$_2$ powder has been reduced to its powder-size objective, it can be reconverted to its initial composition (Dehydrated) to pure Ti or Ti- alloy, using the reverse reaction Eq. 13.20, by heating it under high vacuum at a temperature of 750 °C. Typical results given by McCracken CG and D Barbis (2008)] are given in Fig. 13.16, showing SEM images of coarse ($d_p = 212$–300 µm) and fine ($d_p < 25$ µm) powders of wrought titanium (Fig. 13.16a and d), Na-reduced Ti sponge (Fig. 13.16b and e), and Mg-reduced Ti sponge (Fig. 13.16c and f). The corresponding PSD of the three fine Ti powder fractions ($d_p < 25$ µm) are given in Fig. 13.17.

13.3.2 Powder Consolidation

Powder Consolidation (PC) is essentially the reverse process of Mechanical Size Reduction (MSR) by which particles of

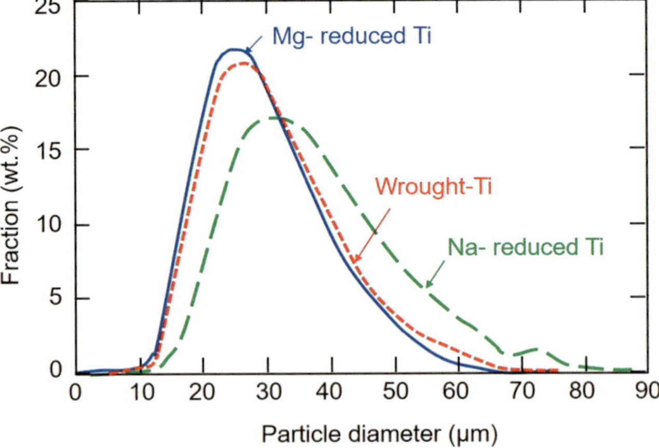

Fig. 13.17 PSD of HDH produced Ti powders using different feedstocks ($d_p < 25$ µm) McCracken CG and D Barbis (2008)] Reprinted with kind permission

smaller size, of pure or blended material, are consolidated together to give larger sized particles with a novel composition and morphology. Powder consolidation included such technologies as spray drying, sintering, mechanical alloying, and cladding. Occasionally, both PC and MER are used in tandem in order to achieve the optimal particle composition, morphology, and size distribution needed to meet the thermal spray process requirement.

13.3.2.1 Spray Drying

Spray drying is probably one of the most versatile techniques that can be used to consolidate fine particles into agglomerates with a well-controlled particle size distribution and uniform composition.

The first step in the process involves the formation of a stable suspension (often water-based) of the fine particles to be agglomerated, with a particle size typically below 5 μm. The formed slurry can be made of a single powder or a mixture of different constituents of compatible particle sizes with the addition of a suspension stabilizing and binding agents. The process is particularly suited to recycle off-spec fine powders obtained through other powder manufacturing routes such as crushing, grinding, milling, where many fine particles are produced which are not usable for conventional spraying. Moreover, the process can also be used to produce cermet particles formed by mixing rather fine ceramic and metal or alloy particles. The same approach is also used with nanometer-sized particles that can be sprayed either in suspensions injected directly into the hot gases or agglomerated as micrometer-sized particles for more conventional thermal spraying.

The process is not without its challenges since the formation of a stable suspension containing metal and ceramic particles is not a simple task. If suspensions of oxide particles are rather easy to stabilize, it is more difficult with a cermet such as WC-Co. This is due to the high specific gravity of WC-Co powders and the widely different acid/base properties of the two main particle constituents [Oberste Berghaus J et al. (2008)]. The formation of a stable suspension for spray drying requires often the addition of a stabilizing surface-active agent which helps to avoid the rapid sedimentation of the solids in suspension in order to

ensure a uniform composition of the powder produced. The addition of a binder may also be necessary in most cases in order to avoid the disintegration of the spray-dried powder prior to its final calcination or sintering step. The proper choice of the chemical composition of such suspension stabilizer and binder is of critical importance to avoid the addition of impurities and/or residual carbon into the powder.

Once a stable and rheological acceptable suspension (often referred to as slurry) has been produced, the next step is to inject the suspension into a large spray-drying chamber where the suspension is atomized into fine droplets dispersed into a hot air stream. As the droplets are entrained by the hot air stream, their liquid constituent vaporized in-flight, with the formed solid particle becoming completely dry before reaching the bottom of the spray chamber. The heating of the droplets by the hot air stream can also be enhanced, if necessary, by the radiation heating from the hot chamber walls. Different flow configurations are used, in co-current or counter current modes, to achieve the necessary contact between the hot air stream and the atomized droplets as shown in Fig. 13.18, with each having its advantages and limitations. In either case, however, it is essential to properly match the thermophysical properties of the liquid used, the rheological properties of the slurry, droplet size distribution, and residence time of the droplets in-flight, to ensure the complete vaporization of the liquid in the droplets and the drying of the formed granules before reaching any of the chamber wall or bottom. As can be noted in Fig. 13.18, the majority of the formed granules are collected by gravity in the bottom of the chamber, with the finer fractions recovered from the exhaust air/vapor stream at the exit of the spray-dryer using appropriate cyclone and filter collectors. The formed granules present large variety of shapes, varying

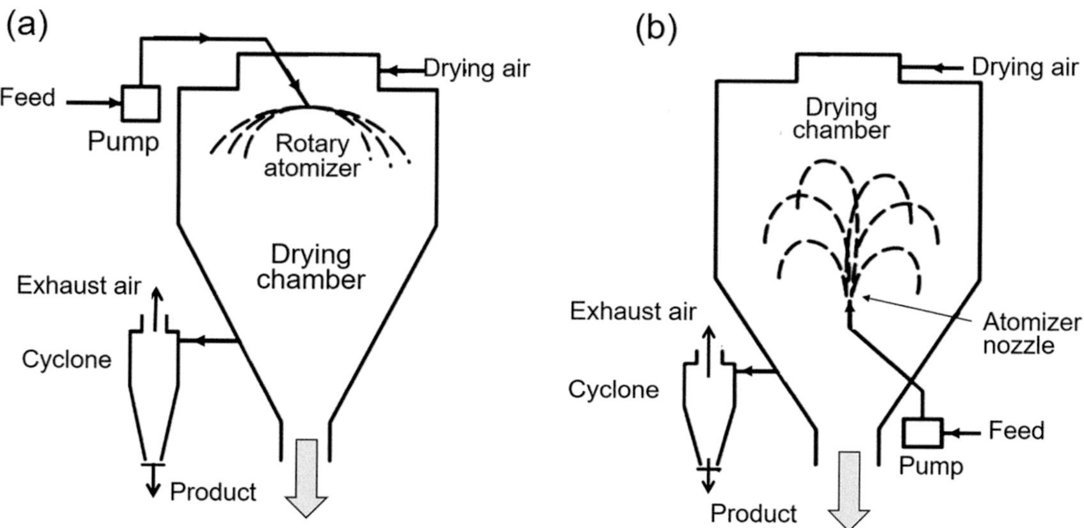

Fig. 13.18   Principle of a spray drier (**a**) Centrifugal atomizer with co-current air flow and (**b**) Nozzle atomizer using mixed-flow conditions

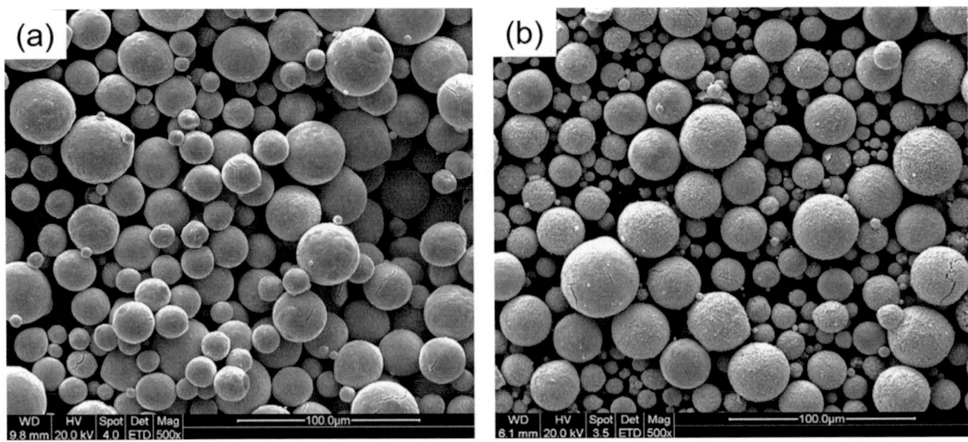

Fig. 13.19 Examples of SEM photos of spray-dried WC–Co (10 wt.% Co) particles (**a**) as spray dried and (**b**) calcined at 700 °C [Xiong Z et al. (2008)]. Reprinted with kind permission from Elsevier

from solid or elongated spheres, pancake shaped, donut shaped, needle-like or hollow granules, depending on the solid content of the slurry, its rheological properties, and the thermophysical properties of the liquid [Walton DE, Mumford CJ (1999)].

The next step of the process is the calcination of the powder obtained in order to eliminate the binder used and provide extra strength against attrition to the formed powder followed typically by a final screening step in order to classify the powder into fractions with different particle size distributions. SEM photographs of the as-spray-dried WC–Co (10 wt.%) particles and following their calcination at 700 °C are given in Fig. 13.19a and b, respectively. The spray-dried powder exhibits smooth surface and spherical shape without agglomeration. After calcination at 700 °C, surfaces of the particles became rough and porous due to the escape of the gases (NH_3, CO_2, and H_2O) resulting from the decomposition of the spray-dried powder.

The spray drying process depends strongly on the way the suspension is prepared. [Bertrand G. et al. (2005)] have determined the quantitative relation existing between slurry behavior and ceramic granule characteristics. They have shown that the shape of the spray-dried ceramic granules can be predicted by slurry studies, their work being performed with two ceramic oxides: alumina and zirconia. In order to produce spray-dried ceramic powders, the slurry must contain high solid content (from 15 vol.% for zirconia slurry to 30 vol.% for alumina slurry, which means that the solid fraction is 50 wt.% for both concentrated ceramic slurries) and be stable during the process. Changing the formulation of the slurry (by varying the pH, the amount of dispersant and the binder) modifies the interparticle interactions and, consequently the state of dispersion of the suspension. The sedimentation test (following the change in sediment heights over a long period as well as sedimentation rates) appears to be appropriate to assess the particles behavior. The Ratio of

Sedimentation height (*RS*) defined as the ratio of the sediment height after about 60 days of sedimentation, H_s compared to the sediment initial height H_i

$$RS = H_s/H_i \qquad (13.21)$$

Generally, it was demonstrated that low *RS* values (below, respectively, 0.6 and 0.7 for zirconia and alumina) attributed to disperse slurries provide hollow dried spheres, whereas high SR values, achieved with flocculated slurries, lead to solid dried granules, as illustrated in Fig. 13.20. Besides a linear relation was established for hollow spheres between the sedimentation rate and the shell thickness [Bertrand G. et al. (2005)].

Spray-drying processes can be used with either micrometer or nanometer-sized particles, as illustrated in Fig. 13.21 for titania nanoparticles [Lima RS and Marple BR (2008)]. It can be noted in this figure that the spray-dried and sintered titania particles are hollow and quasi-spherical.

Spray-drying also allows for the manufacturing of powders made of very dissimilar elements. For example, agglomerates of particles of aluminum–silicon (Al-Si) eutectic alloy ($d_{50} = 2.4 \pm 1.2$ µm, Al-11.6 wt.% Si-0.14 wt.% Fe) in which multiwalled carbon nanotubes (CNT) (purity higher than 95%, diameter = 40–70 nm, length = 1–3 µm) are dispersed by spray-drying [Baksh SR et al. (2008)]. Resulting micrometer-sized particles displayed in Fig. 13.22 had a CNT content of 5 wt.% within the spray-dried powders.

For high added-value applications such as in the biomedical field, where a reduction of powder production steps needs to be minimized to ensure the high purity of the powder produced and reduce the chances of contamination, the option of combining the spray-drying and powder consolidation/spheroidization steps was proposed by [Bouyer E et al. (1997a, b) (2000)]. The process identified as "Suspension Plasma Spraying" (SPS) [Gitzhofer et al. (1997)] illustrated

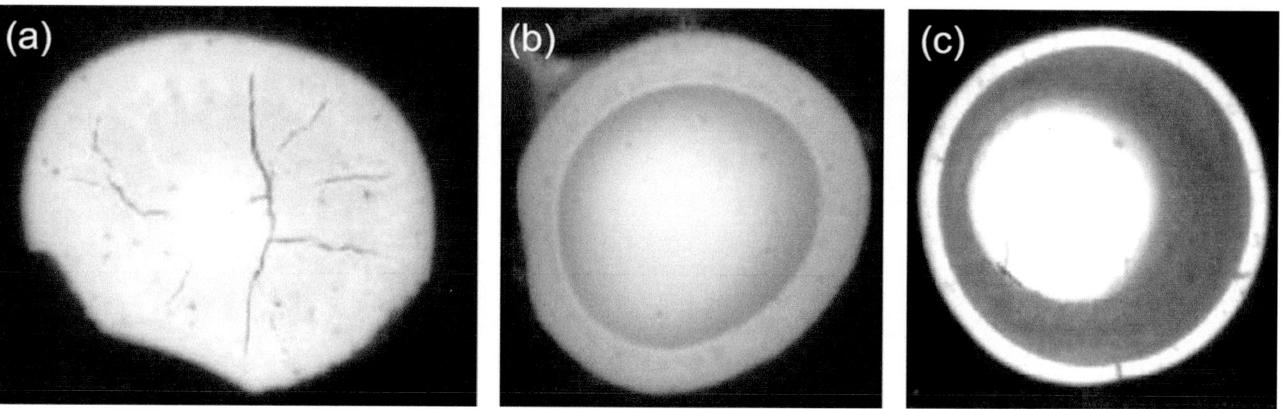

Fig. 13.20 Polished cross sections of dried granules realized from different zirconia-based slurry formulations with various sedimentation ratios (**a**) RS = 0.8, (**b**) 0.5, and (**c**) 0.3 [Bertrand G. et al. (2005)]. Reprinted with kind permission from Elsevier

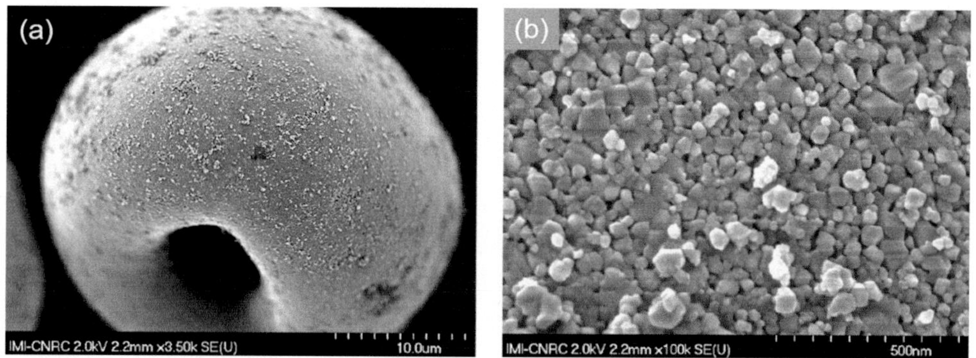

Fig. 13.21 (**a**) Agglomerate and spray-dried nano-sized titania powder. (**b**) Higher magnification view showing the nanostructure of the agglomerate [Lima RS and Marple BR (2008)]. Reprinted with kind permission from Elsevier

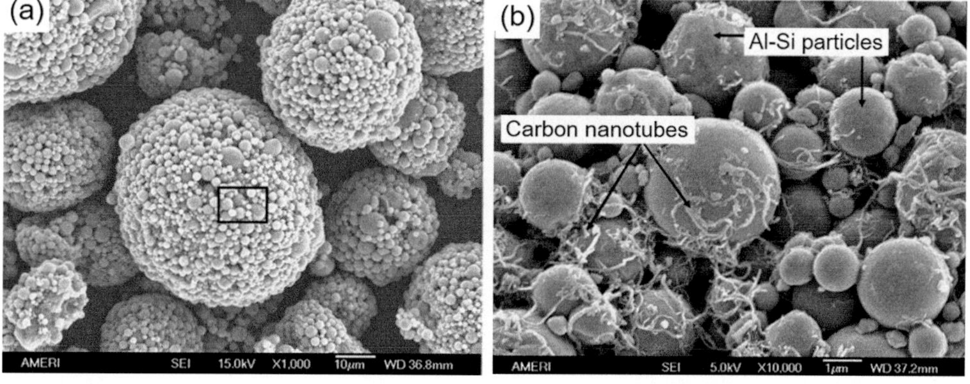

Fig. 13.22 (**a**) Agglomerated Spray-dried Al–Si powder, (**b**) Magnified region within the rectangle in (**a**) showing CNTs within the agglomerate [Baksh SR et al. (2008)]. Reprinted with kind permission from Elsevier

in Fig. 13.23a reduces considerably the processing steps compared to alternate routes involving either raw material drying, calcination, crushing, screening and spheroidization, or the atomization, calcination screening, and spheroidization routes. The cases considered in this example applies to the synthesis of ceramic material such as Hydroxyapatite (HA) for which the details of the transformations taking place in the SPS step are illustrated in Fig. 13.23b. The aqueous HA suspension used as the process starting material is synthesized by wet chemistry according to [Tagai and Aoki (1980)] involving the chemical reaction

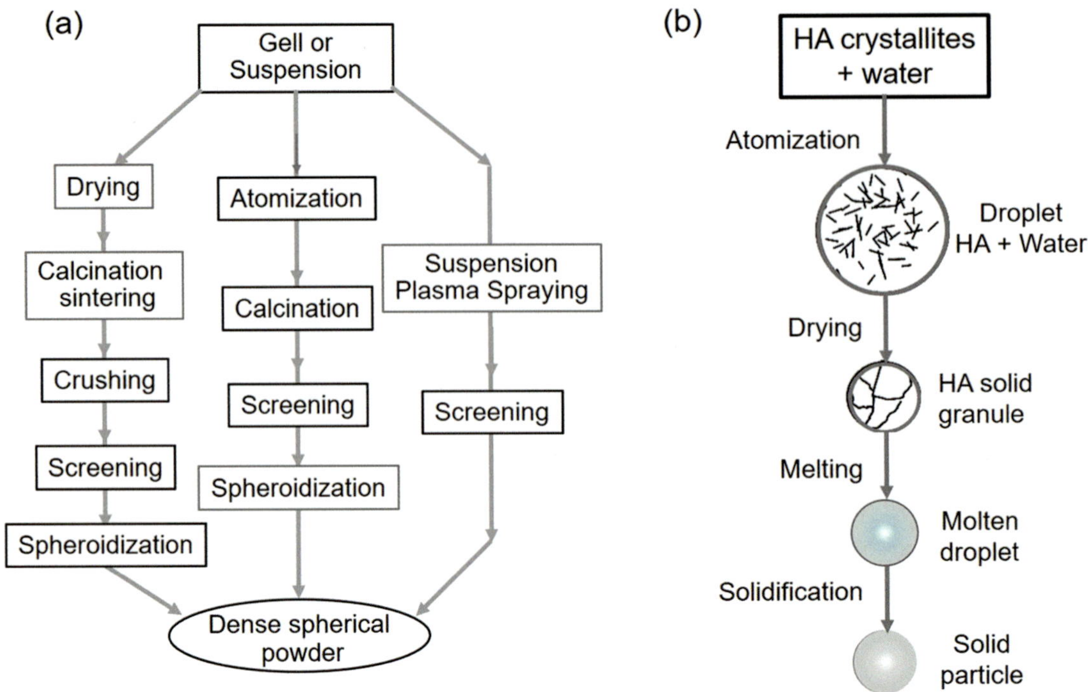

Fig. 13.23 Comparison of the process steps for the synthesis of HA spherical particles using conventional crushing and spheroidizations or spray-draying routes, with the SPS approach [Bouyer E et al. (1997a)]. Reprinted with the kind permission of IEEE Trans Plasma Sci

$$10\,\mathrm{Ca(OH)_2} + 6\,\mathrm{H_3PO_4} \rightarrow \mathrm{Ca_{10}(PO_4)_6(OH)_2} + 18\,\mathrm{H_2O}$$

$$(13.22)$$

The temperature is kept constant at 42° C during the reaction step. The resulting aqueous precipitate is concentrated by centrifugation, and heated until reaching a high percentage of solid in the suspension: around 40 wt.%. Without further processing, the viscosity of the suspension is too high to be fed into the plasma. Consequently, Darvan 7, a polymeric surface-active agent (1% by volume), is added to the suspension to decrease the viscosity and improve flowability. The viscosity of the suspension measured at room temperature is around 0.4 Pa s. The suspension is fed via a peristaltic pump and is gas-atomized into the plasma flow to produce droplets with an average diameter in the range of 50–100 μm. Because of the low relative velocity between the droplets and the plasma, the Weber number is below 14 and the droplets are not fragmented during their flight. As the droplets are entrained by the plasma, they are heated and their water content is evaporated giving rise to the formation of solid particles which upon further heating, flash sinter, melt, and finely solidify resulting in the formation of dense spherical particles with an average diameter considerably smaller than the original droplet size in the range of 20–50 μm. TEM photographs of the as synthesized nanosized HA crystals and the consolidated, dense spherical particles

formed by the SPS process are shown in Fig. 13.24a and b respectively. A close-up SEM photographs at two different magnifications of the formed particles are given Fig. 13.25. X-ray analysis of the formed HA powder had essentially the same chemical composition of the as synthesized powder with the exception of partial decomposition of the HA to lime (CaO) and tricalcium phosphate ($\mathrm{Ca_3(PO_4)_2}$, $\propto\,-\,$TCP) with the degree of decomposition suppressed by the presence of water vapor generated during the droplet evaporation step.

LaMnO$_3$ particles were also produced by SPS-ICP technology [Schiller G, et al. (1999)]. However, they were partially decomposed in spite of the use of an oxygen sheath gas. La$_{0.8}$Sr$_{0.2}$Mo$_{3-\delta}$ particles for SOFC's cathode material were also successfully produced [Bouchard D et al. (2006)]. This technology has also an important potential for the preparation of functional materials such as yttrium aluminums garnet (YAG) [Ravi BG et al. (2008)].

13.3.2.2 Sintering

During the grinding operation of a powder, fine to ultrafine particles with an average diameter less than 5 μm are also produced in the process which have to be removed from the product in order to "clean" the powder from contaminants which are not suitable for spraying. These small particles can, however, be agglomerated, spray-dried or pressed and sintered in order to recover their value and/or use them for new powder formulations. In the latter case, they are first

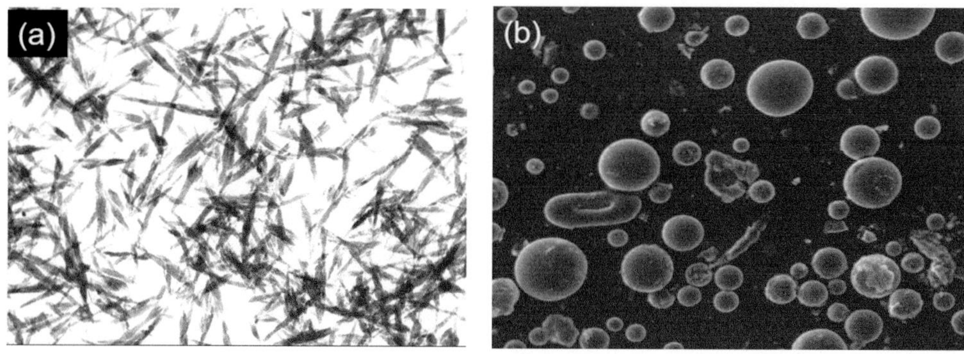

Fig. 13.24 (**a**) TEM photographs of the as synthesized HA nanocrystals. (**b**) SEM photographs of particles obtained the SPS process [Bouyer E et al. (1997a)]. Reprinted with the kind permission of IEEE Trans Plasma Sci

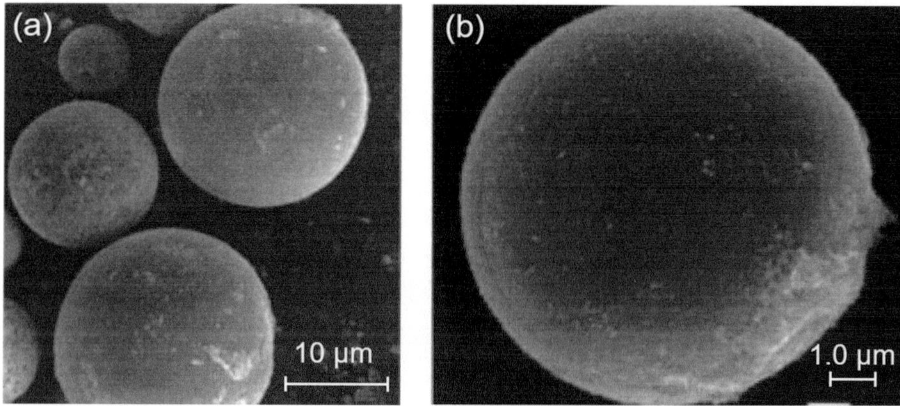

Fig. 13.25 (**a**) SEM photograph of SPS atomized HA particles, (**b**) view at larger magnification [Bouyer E et al. (1997a)]. Reprinted with the kind permission of IEEE Trans Plasma Sci

compacted (one-direction compaction), with or without a binder, and then sintered by heating them in a furnace to a temperature below the melting temperature, but sufficiently high to allow their binding by chemical diffusion between the individual particles [Davis JR (ed) (2004), and Salman AD, et al. (eds) (2007)]. The density of sintered material depends on their compaction and the size distribution of starting powders [Lee HM et al. (2009)]. When increasing the compaction pressure, sintering time, and temperature, the porosity of the particles obtained is reduced. The sintering and densification steps, on the other hand, are accompanying by grain growth which increases with the increase of the sintering temperature, with detrimental impact on the mechanical properties of the particles and coating, especially when dealing with nano-sized grains for the purpose of achieving nanostructure coatings.

Following the sintering step, the cake obtained is crushed to obtain the size distribution adapted to thermal spraying. Crushing is easier in this case compared to that of fused block as described in the previous section. As schematically illustrated in Fig. 13.26 the particles generated through this process are blocky and angular with varying level of porosity.

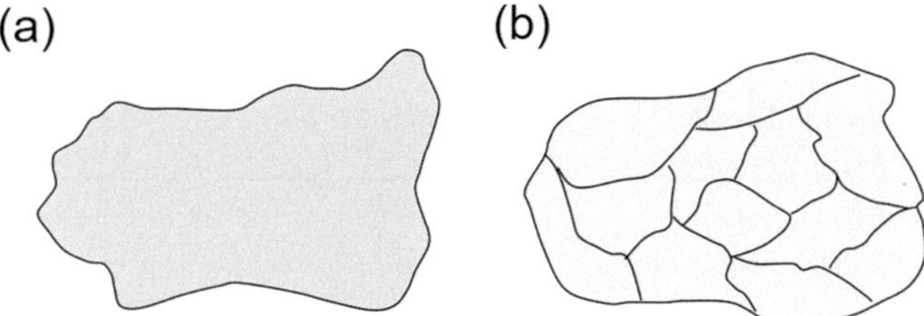

Fig. 13.26 Schematic representation of the microstructure of (**a**) fused and crushed and (**b**) sintered and crushed particles [Laha T et al. (2005)] Reprinted with kind permission from Pergamon press now Elsevier

Many carbide-metal or alloy particles are produced by this method. Micrographs of fused and crushed WC-Co (17 wt.% Co) from two different suppliers are shown in Fig. 13.27.

13.3.2.3 Mechanical Alloying

Mechanical Alloying (MA) is a solid-state powder processing technique involving repeated welding, fracturing, and re-welding of powder particles in a high-energy ball mill as detailed in the review paper of Suryanarayana [Suryanarayana C (2001)]. Originally developed to produce Oxide-Dispersion Strengthened (ODS) nickel- and iron-base

Fig. 13.27 SEM micrographs of sintered and crushed WC-Co-Cr (10 wt.% Co + 4 wt.% Cr) particles, (**a**) Oerlikon Metco Amdry 5843 and (**b**) Starck Amperit 553.065. [Berger L-M et al. (2001)] Reprinted with kind permission from Springer Science Business Media, copyright © ASM International

superalloys for applications in the aerospace industry, MA has now been shown to be capable of synthesizing a variety of equilibrium and nonequilibrium alloy phases starting from blended elemental or pre-alloyed powders. It is a complex process and hence involves optimization of a number of variables to achieve the desired product phase and/or microstructure. Commercially significant quantities of nanostructured materials with grain size in the range of 10–200 nm can be processed using mechanical alloying, which involves the transformation of plasticity-induced dislocation structures into high-angle grain boundaries in metallic powders.

Most important parameters that have an effect on the final constitution of the powder are:

- Type of mill
- Milling container
- Milling speed
- Milling time

The type and size distribution of the grinding medium depend on

- Ball-to-powder weight ratio
- Extent of filling the vial
- Milling atmosphere
- Process control agent
- Milling temperature

Mechanical alloying is mostly carried out using high-energy attritors (see Fig. 13.15), working either with liquid medium such as ethanol, methanol, inert gas atmosphere such as argon, or liquid nitrogen as in cryo-milling. The extremely low milling temperature in cryo-milling suppresses the recovery and recrystallization and leads to finer grain structures and superior rapid grain refinement [Lavernia EJ et al. (2008)]. Experimental and theoretical studies suggest that for each material, there is a minimum grain size that is obtainable by milling, and that its value is related to the intrinsic properties of the material, such as crystal structure. The results reviewed by [Lavernia EJ et al. (2008)] also demonstrate that cryo-milling can be effectively used to manufacture ceramic reinforced metals, in which the size of the reinforcement phase can be modified, depending on the milling parameters. In the following, five typical examples are discussed.

Figure 13.28a and b shows the morphological differences between the NiCrAlY powders before and after the cryo-milling [Picas JA et al. (2004)]. Mechanical cryo-milling of the as-received spherical NiCrAlY powders leads to the formation of irregular and flake-shaped agglomerates, as shown in the SEM image (Fig. 13.28b). This morphology is attributed to the continuous welding and fracturing of the powder particles during the milling process. The average particle size of the powder increases from its as-received value of 28.7 μm, to an average size of the agglomerate and cryo-milled powders of 106.0 μm. The cold welding seems to overcome the fracture process with milling time, resulting in an increase in agglomerate size. The formation of the nanocrystalline structure during cryo-milling is considered to be a consequence of plastic deformation at high strain rates [Suryanarayana C (2001)]. The grain size is in the nanometric range (lower than 50 nm). The increase in oxygen (0.027–0.17 wt.%) and nitrogen (0.0048–0.26 wt.%) concentration is believed to be a consequence of the incorporation of these elements from the air and liquid nitrogen environment, respectively.

A metastable composite $Fe_{60}Al_{40}$ powder was produced in an attritor filled with argon in which Fe (99.8 wt.%) particles below 54 μm and Al (99.5 wt.%) with a mean particle size

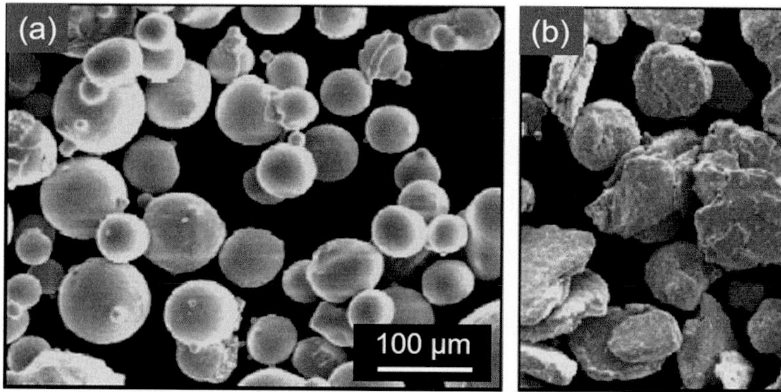

Fig. 13.28 Morphological characteristics of the NiCrAlY powders (**a**) before cryo-milling (as received) and (**b**) after cryo-milling process [Suryanarayana C (2001)]. Reprinted with kind permission from Elsevier

Fig. 13.29 Morphology of (**a**) as-milled $Fe_{60}Al_{40}$ powder and (**b**) cross-sectional microstructure [Wang H-T et al. (2007)]. Reprinted with kind permission from Springer Science Business Media, copyright © ASM International

below 74 µm were introduced (with the required Fe/Al wt.% ratio). The milling media being stainless steel balls 5 or 10 mm in diameter with a ball-to-powder weight ratio of 10:1 [Wang H-T et al. (2007)]. Figure13.29 shows typical SEM images of the morphologies and cross-sectional microstructures of the powder milled for 36 h. The as-milled powder, presented in the aggregated condition, shows an irregular angular morphology (Fig. 13.29a).The cross-sectional microstructure of the as-milled powder, shown in Fig. 13.29b, presents a fine lamellar structure within each particle. Meanwhile, a few relatively thicker layers are still present in the powder. Two distinguishable regions with different microstructural characteristics can be clearly seen which mainly differ in the lamellar thickness. EDXA analysis suggests that the thicker layer is a Fe-rich phase, while the fine lamella is a Fe–Al solid solution with high-Al content [Wang H-T et al. (2007)].

In contrast to the previous two cases where the particles were formed by cold welding, the milling of mixtures of brittle and ductile particles forms binder-induced agglomerates, which are primarily bonded by milling environment and can be decomposed back to their original powder form [He J and Schoenung JM (2002)]. By varying the milling conditions, the size of the hard phase obtained [He J and Schoenung JM (2002)] can be smaller than the starting ones. For example, in the course of milling of Cr_3C_2-25 (Ni_{20}–Cr) powder in hexane, the pure carbide phases detectable in the original powder mixture, shown in Fig. 13.30a, is completely undetectable after 16 h of milling, as shown in Fig. 13.30b. The ductile (soft) Ni-rich phase is distributed among the homogeneous finer particles and no longer seen as spherical particles in the milled powder. With an increase in milling time from 16 to 20 h the surface of the powder becomes rough with complete self-agglomeration in the powder as shown in Fig. 13.30c and d. X-ray mapping results reveal the particles in Fig. 13.30d to be neither pure carbide nor pure NiCr phase [He J and Schoenung JM (2002)].

Mechanical Alloying was also used to produce TiC–Ni-based cermet powders [Eigen N, et al. (2002)] with fine nano-sized TiC hard phases homogeneously distributed in the particles. The process was carried out using high-energy ball milling with an industrial-like vibration mill as well as with an attrition mill. It should be noted that at least 95% of the hard phase particles are smaller than 300 nm. Milling of the composite material leads to a uniform particle size of 20 µm.

13.3.2.4 Cladding

Cladding allows producing composite particles with their core of one material, coated with another dense and continuous coating or a heterogeneous layer. An excellent review of the different powder cladding processes reported by [Lugscheider E et al. (1991)] identifies the following commonly used techniques.

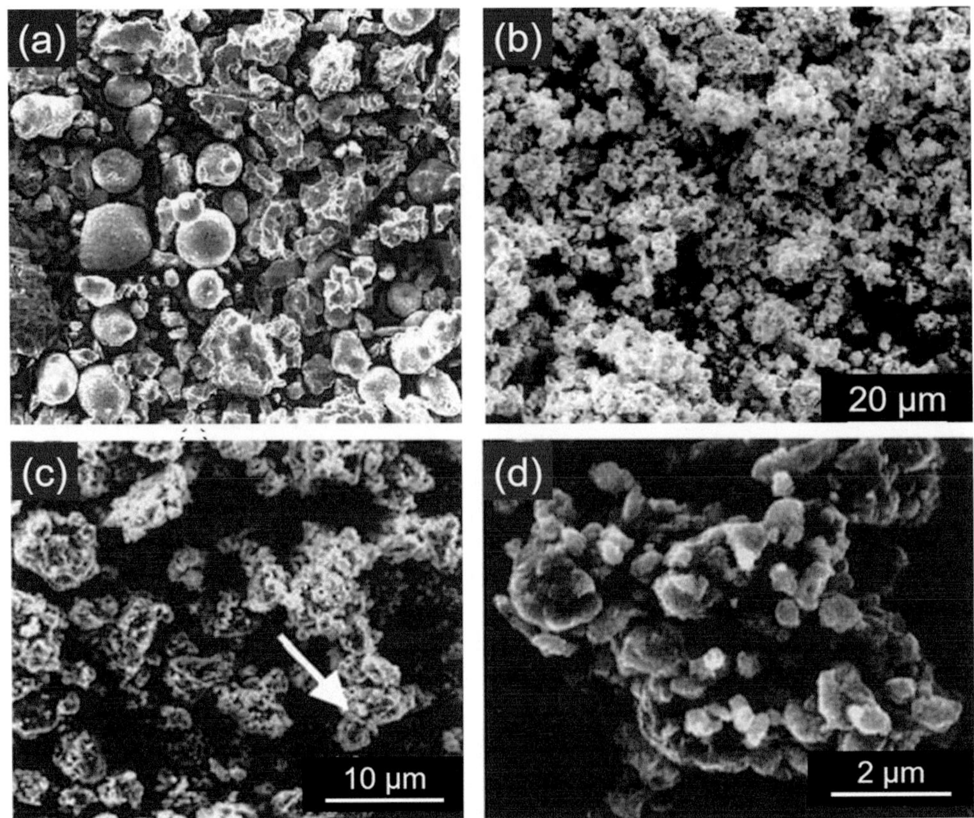

Fig. 13.30 SEM micrographs showing the morphology of the Cr_3C_2-NiCr powder (**a**) As-received powder and (**b**) powder after 16 h milling (**c**) and (**d**) powder after 20 h of milling with (**d**) higher magnification [He J and Schoenung JM (2002)]. Reprinted with kind permission from Elsevier

(a). *Mechanical alloying with attritors*

The core particles must be mechanically stable during the attrition milling and the ratio of core particles and coating material must be accurately calculated. If the quantity of the coating material is too low, the core material will be crushed, while if it is too high, the coating process will not take place

(b). *Sintering*

Coated powders can also be obtained using either impact-process sintering or by mixing large particles of one material with submicron-sized particles of the other one before sintering them. The result of the latter process is illustrated in Fig. 13.31 for alumina–titania particles in which Al_2O_3 core powder (10–45 µm) are mixed with submicron-sized TiO_2 particles before sintering [Wang M, Shaw LL (2007)].

The different types of coatings that can be obtained by cladding onto the core material are summarized in Fig. 13.32. Electron micrographs of Ni/Al- and Ni/graphite-cladded particles are shown in Fig. 13.33.

(c). *Mechanofusion*

This process is used to manufacture composite powders starting from two or more raw powdered materials having different grain sizes [Ito H et al. (1991)]. Its principle is illustrated in Fig. 13.34 where a rotating cylinder entrains the powder mixture below two inner compression pieces with adjustable clearance between the compression piece and the inner wall of the cylindrical housing. Two scrapers clean the cylinder wall and help the entrainment of the particle's mixture below the compression pieces. The raw particles are thoroughly mixed and subjected to powerful compression, resulting in strong mechanical forces generating thermal energy. Thereby, the material with the lowest melting temperature is heated to a plastic state or even melts and becomes the core of a composite particle to which the material with the highest melting temperature is welded as a surrounding shell. Figure 13.35 shows 316 L stainless-steel particles with a size distribution of 50–63 µm coated with pure α-alumina (99.99%) with a mean particles diameter of 0.6 µm [Ageorges H, Fauchais P (2000)]. The composite particles

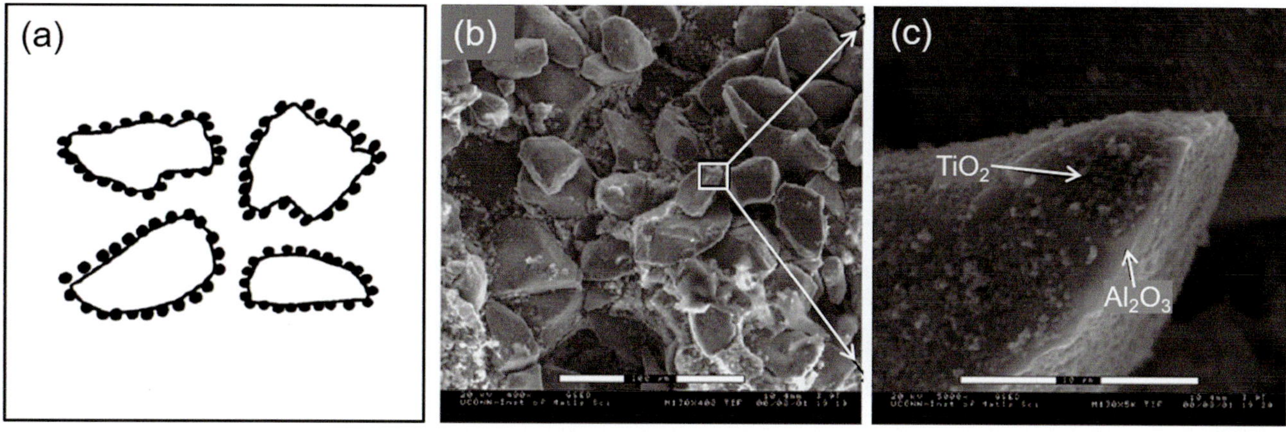

Fig. 13.31 (**a**) Schematic representation of core particles coated with a fine particle (**b**) and (**c**) SEM images of Metco 130 powder at two different magnifications. [Wang M, Shaw LL (2007)]. Reprinted with kind permission from Elsevier

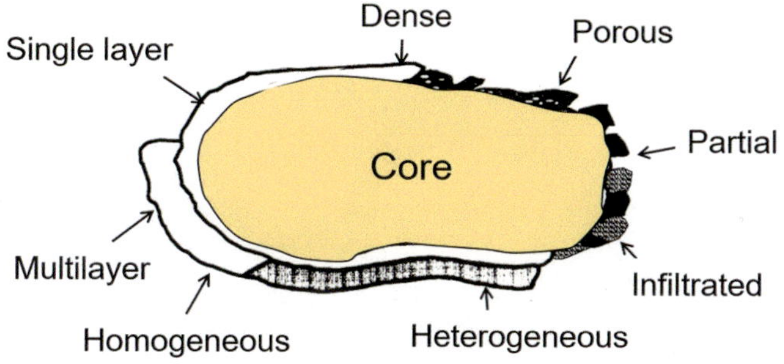

Fig. 13.32 Different types of particle cladding [Lugscheider E et al. (1991)]. (Courtesy of Sulzer Metco for Plasma Technik)

have a near spherical shape and their mean diameter, after sieving, is 65 µm, the alumina shell being 2 ± 4 µm thick.

(d). *Deposition from gaseous phase*

Chemical Vapor Deposition (CVD) is one of the most widely used techniques for powder cladding. It basically consists of coating the particles in a fluidized bed reactor through the thermal decomposition of the vapor precursor. The process, which is mostly a pyrolysis or chemical reduction, requires a close control of the temperature of the fluidized bed and the uniform distribution of the vapor precursor throughout the bed. The core of the particle to be coated must be stable at the reaction temperature. Coating with thickness up to 20 µm can be deposited using CVD on metals, alloys, carbides, nitrides, borides, silicides, and oxides. The purity of the coating, which is not the highest, and the stability of the fluidized bed requiring large quantities of fluidizing gas represent some of the challenger with this approach [Lugscheider E et al. (1991)]. In principle PVD based on evaporation, sputtering, or ion plating can also be used for particle cladding [Lugscheider E et al. (1991)]. Compared to CVD the cost is much higher and the particles must withstand temperatures up to 800 °C and pressures between 1 and 10^{-4} Pa. The list of

the deposited materials is the same as that of CVD. For more information see [Lugscheider E et al. (1991)].

(e). *Electrolytic coatings (also called galvanic plating)*

Electrolytic coating is applicable to electrically conductive particles which are at the anode potential, while the material to be deposited is in its metal salt solution. Coating thickness can vary over a wide range 0.5–2000 µm [Lugscheider E et al. (1991)]. The coatings are, however, rather porous and spongy and often irregular.

(f). *Chemical precipitation*

It is a very important process based on the electroless reduction of a metal salt solution, also called "hyperpoc" process (hydrogen-pressure-reducing-powder-coating-process). The core powder to be coated is suspended by mechanical agitation in an aqueous metal salt solution. Introducing hydrogen at elevated temperature and pressure in the autoclave containing the solution and the particle suspension reduces the metal. It requires that the hydrogen potential of the system exceeds the electrode potential of the metal ions. When this condition is fulfilled, the hydrogen reacts with the solution and the metal precipitates. The hydrogen potential increases when the pH of the solution and the partial pressure of hydrogen are raised. Typical pressures are in the 2–3 MPa, with the solution is heated to 180–200 °C. The core particles to be coated must be nonreactive to the solution. The metal salts are basically sulfates, nitrates, acetates, carbonates, and also organometallic salts [Lugscheider E et al. (1991)]. The deposited layer (Pt, Pd, Au, Ag, CU, Ni, Co, Mo) on the core particles are either homogenous or heterogeneous. Sintering can further consolidate composite particles. For example, this process is used to make NiCrAl/bentonite (silica-based clay) particles. Bentonite core particles are coated with layer (about

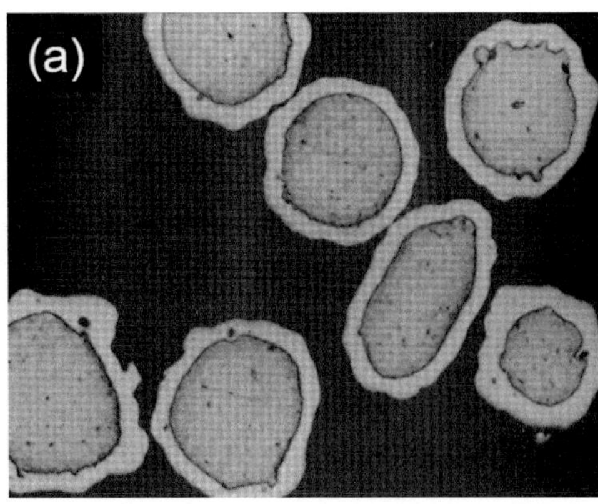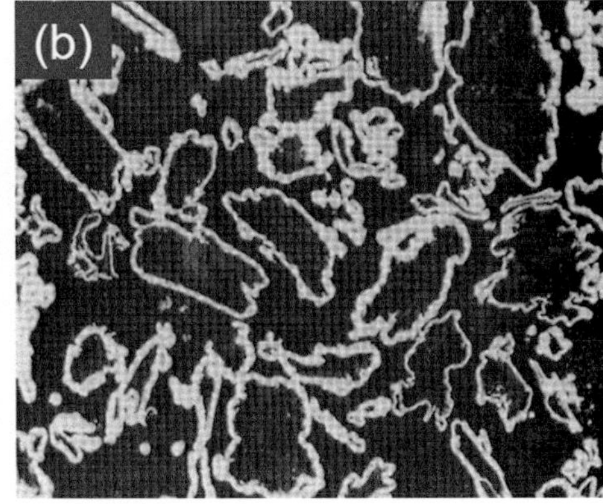

Fig. 13.33 Particles obtained by cladding (**a**) NiAl (20 wt.% Al) and (**b**) Ni–graphite (15 wt.% graphite) [Lugscheider E et al. (1991)]. (Courtesy of Sulzer Metco for Plasma Technik)

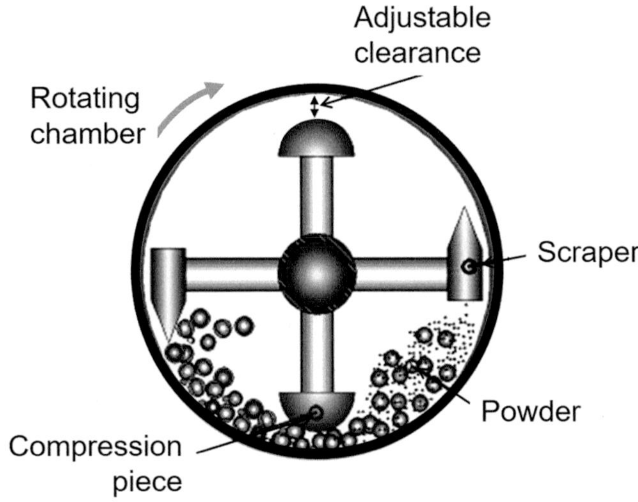

Fig. 13.34 Principle of a mechanofusion process [Ageorges H, Fauchais P (2000)]. Reprinted with the kind permission of Dr. R. Cuenca-Alvarez

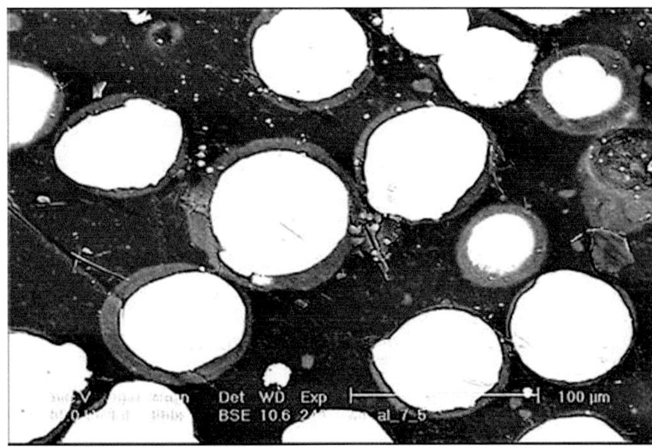

Fig. 13.35 SEM observation of cross section of mechanofused stainless steel/alumina particles [Ageorges H, Fauchais P (2000)]. Reprinted with kind permission from Elsevier

13.3.3 Plasma Spheroidization

13.3.3.1 Basic Concept

Using dense and spherical particles for spraying results in:

- improved powder flowability allowing for smooth particles flow with stable powder feed rates.
- higher particles density allows for uniform heating of the particles which is critically important low thermal conductivity materials [Ben Ettouil F et al. (2008)].
- Less friable powders ensuring the stability of its PSD and reducing its contamination with fines which tend to vaporize when injected into the plasma.

Plasma spheroidization of powders [Dignard N and M Boulos (2000)] is a relatively simple process which is similar to plasma spraying involving the in-flight melting of the particles injected into a plasma stream. The only difference in comparison to plasma spraying is that the formed molten

2 µm thick) of nickel (starting from a nickel amine sulfate). The nickel is further alloyed with chromium and/or aluminum by pack diffusion [Davis JR (ed) (2004)]. Typically, the average core diameter is 85 µm and the deposited layer is about 6 µm, the composition in wt.% being: 20% bentonite, 5% chromium, 3% aluminum, with the balance 72% nickel. This process can also be used to prepare materials such as graphite or silicate for abradable coatings. MoS_2 powder can be coated with Ni by chemical method. The coated powder has a nominal composition of 22 wt.% MoS_2 and 78 wt.% Ni, with an average particle size of 50 µm, and an average thickness of the Ni coating of 5 µm [Lugscheider E et al. (1991)].

droplets are allowed to cool down and solidify, in-flight, prior to hitting the wall or the bottom of the spheroidization chamber [Dzur B. (2008)]. The process gives rise to the formation of dense, free-flowing solid particles, which can be handled easily by most thermal or cold gas spray processes. In a number of cases, the in-flight, plasma melting of the particles also results in the purification of the particle material through the loss of more volatile impurities such as alkali metals and adsorbed gases. While both DC and RF Inductively Coupled Plasma (ICP) sources have been successfully used for powder spheroidization, Induction Coupled Plasmas offer in most cases a distinct advantage due to the following reasons:

- Low plasma flow velocity (about 50 m/s) resulting in residence times almost one order of magnitude longer than in DC jets
- Long residence times allowing for the complete heating and melting of most low thermal conductivity materials for which internal propagation of the thermal front is the limiting step in the particle heating and melting process
- Possibility to use almost any gas chemistry, whether reducing, inert, or oxidizing, as required by the chemical nature of the processed material. A typical example is the need to use an oxygen-rich plasma gas for the spheroidization of oxide ceramics such as silica, to avoid the loss of oxygen in the melting process.

Different RF-ICP systems were developed by Tekna Plasma Systems Inc. [Boulos MI (2004)] for powder

spheroidization with power levels up to 400 kW working at frequencies between 2 and 4 MHz with a wide range of gases at absolute pressures varying between 50 and 150 kPa. A schematic of 100 kW pilot-scale system is shown in Fig. 13.36. It comprises essentially of an induction plasma torch, PS-50 or PS-70 mounted at the top of a large-volume, water-cooled, stainless steel spheroidization reactor. The powder to be spheroidized is injected axially at the center of the discharge using a water-cooled powder injection probe. The tip of the probe is usually placed at the mid-height of the induction coil. With relatively coarse and high specific grav-ity particles, the tip of the probe can be placed slightly above the coil center, while with smaller, lighter particles, the probe tip is placed below the coil center in order to avoid the deposition of the molten particles on the inner surface of the plasma confinement tube. As the plasma gas entrains the injected particles, the individual particles are heated and melted in-flight. The particle free-fall velocity essentially governs the remaining part of the particle trajectories in the spheroidization reactor. Over this period, the particle cools down and freezes into a solid dense sphere. Large dense spherical particles are recovered at the reactor bottom of the reactor, while smaller finer powder fractions are entrained with the plasma gasses exiting from the side of the spheroidization reactor. The plasma gases with any entrained fine powders are directed to a water-cooled cyclone where the fine particles are recovered. Any remaining submicron particles are finally collected in a filter prior to the exhaust

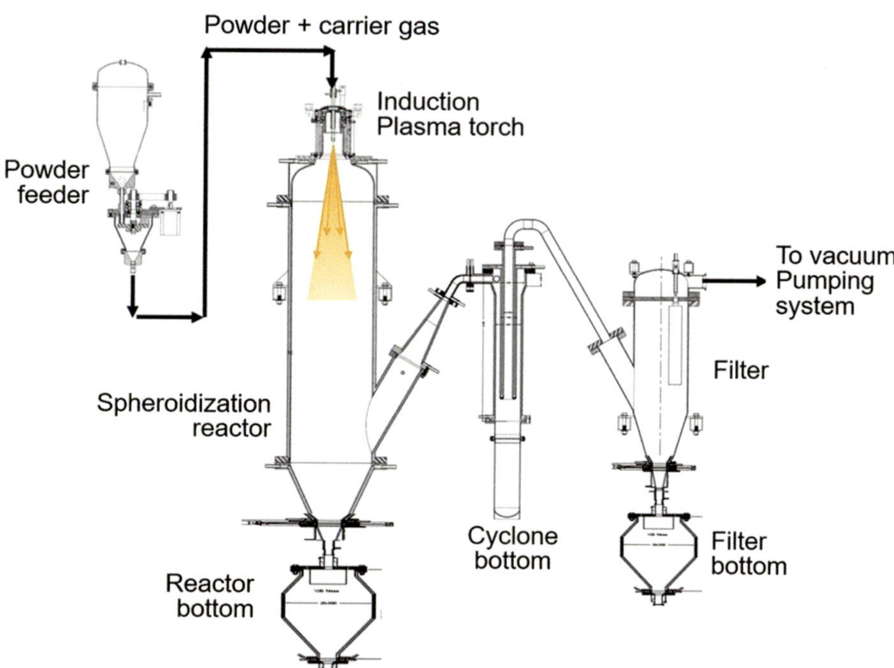

Fig. 13.36 RF-ICP pilot scale system developed by Tekna Plasma Systems Inc., for powder spheroidization [Boulos (2004) Boulos MI et al. (2006a)] (Courtesy of Tekna Plasma Systems Inc)

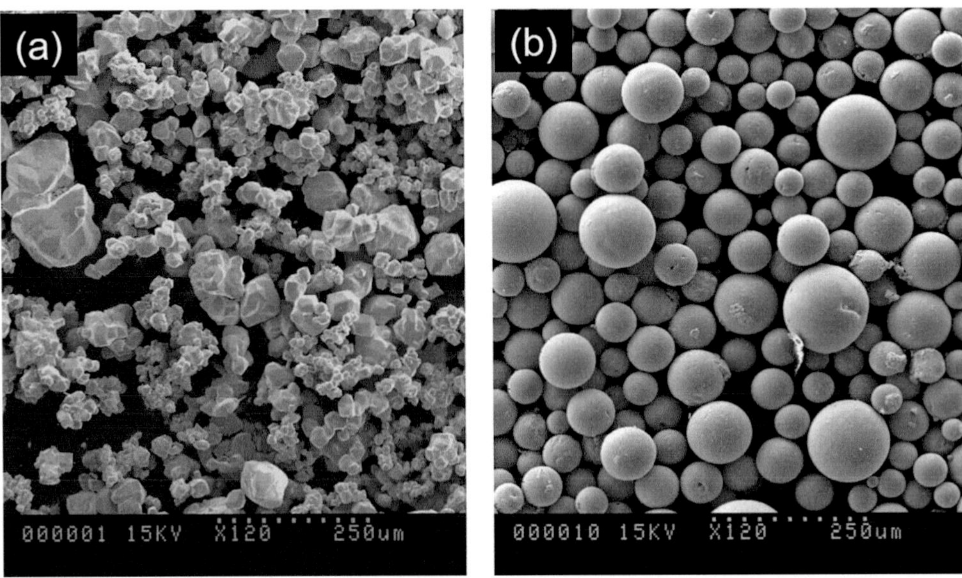

Fig. 13.37 Electron micrographs of tungsten powder (**a**) before and (**b**) after RF-ICP spheroidization [Boulos M (2004)] (Images courtesy of Tekna Plasma Systems Inc.)

of the plasma gases to the environment via the gas pumping system. The cyclone cut-off particle diameter varies between 1 and 5 μm depending on the density of the material of the powder being spheroidized.

Most powder spheroidization operations are carried out under near atmospheric pressure, though depending on the nature of the powder being treated, the process can also be carried out under soft vacuum or pressures above atmospheric. Typical results obtained for the spheroidization of tungsten and tungsten carbide powders are given in Figs. 13.37 and 13.38 respectively. These show electron micrographs of the powders before and after RF-ICP spheroidization. It should be pointed out that the melting temperature of tungsten = 3410 °C and its specific gravity = 19.3 g/cm^3. The tap density of the spheroidized powder increased from 7500 to 11,700 kg/m^3 as a result of the plasma spheroidization treatment. The powder flowability also improved considerably from no-flow to Hall flow index of 7 s/50 g. The corresponding values for tungsten carbide are: melting temperature = 2800 °C, specific gravity = 15.7 g/cm^3, Hall flow index before and after spheroidization 19 s/50 g, and 7.0 s/50 g respectively.

As discussed in great details in Part I, Chap. 5 "Gas and Particle Dynamics in Thermal Spray," Sect. 5.6. With the increase in the powder feed rate in a plasma spheroidization operation, the thermal loading effect caused by the powder will locally cool down the plasma flow reducing its ability to completely melt the powder with the effect being most noticeable on the larger particles. The effect can be clearly observed in Fig. 13.39, which shows electron micrographs of molybdenum powder, with a particle size in the range of (−60 + 400 Mesh/40–250 μm), spheroidized in a 100 kW,

PS-70 RF-ICP torch, using a mixture of Ar/H$_2$ as the plasma sheath gas, at different powder feeding rates.

The micrograph of the internal microstructure of the feed material which was prepared by spray-drying and sintering of a fine ($d_p \leq$ 1–5 μm) molybdenum powder is shown in Fig. 13.39a. For comparison, the feed material had a tap density of 3.2 g/cm^3 and a Hall flow index of 36 s/50 g. SEM micrographs of the plasma-treated powders at feed rates of 7 to 14, 20, and 25 kg/h are given on Fig. 13.39b–e. These show full powder spheroidization at powder feed rates of 7 and 14 kg/h. With the further increase of the powder fed rate to 20 and 25 kg/h, the presence of partially molten particles becomes increasingly apparent with some particles having their sharp corners rounded off without being completely melted. The effect is also reflected on the tap density and flowability, Hall flow index, of the powder given in the lower part of the figure. It may be noted that the spheroidization effect is most important at the low powder feed rate of 7 kg/h, where the tap density of the powder increased from its initial value of the raw powder of 3.2 g/cm^3 to 6.35 g/cm^3, with the Hall flow index decreasing from 36 s/50 g to 13.0 s/50 g. At a powder feed rate of 14 kg/h, the efficiency of the spheroidization process did not seem to suffer too much with the tap density and Hall flow index of the treated powder reaming essentially stable. With the further increase of the powder feed rate to 20 and 25 kg/h, a significant effect of the thermal loading of the plasma is observed, as demonstrated by the lower degree of melting of the powder, the drop in the tap density of the powder to 5.8 and 5.0 g/cm^3, and the increase in its Hall flow index to 16 and 19 s/50 g respectively. It is important to note that the treatment is accompanied by a drop in the level of

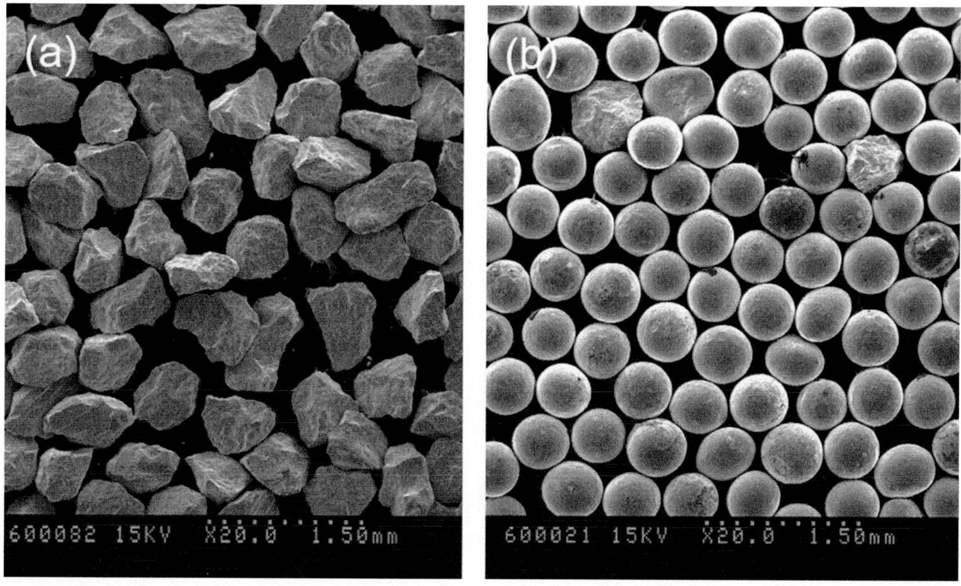

Fig. 13.38 Electron micrographs of tungsten carbide powder (**a**) before and (**b**) after RF-ICP spheroidization [Boulos M (2004)] (Images courtesy of Tekna Plasma Systems Inc.)

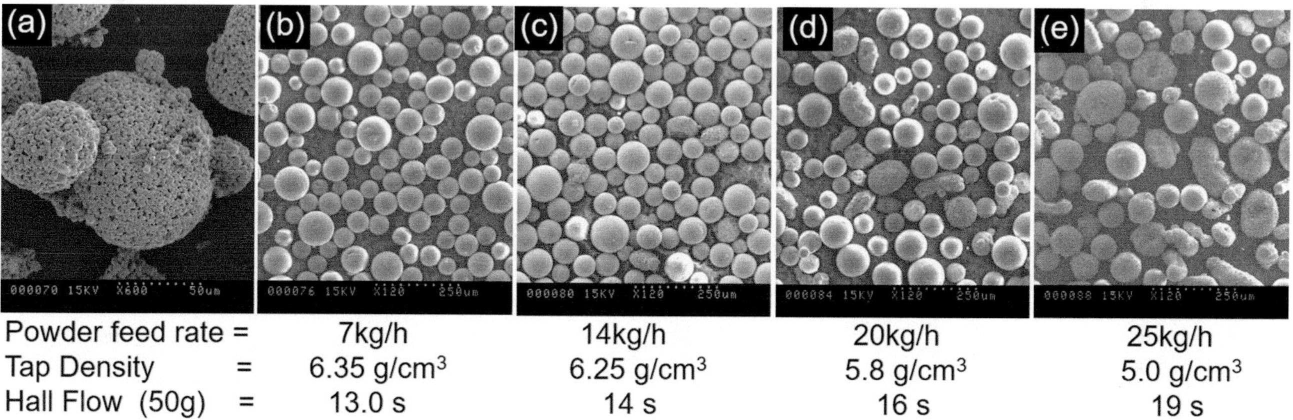

Powder feed rate =	7kg/h	14kg/h	20kg/h	25kg/h
Tap Density =	6.35 g/cm³	6.25 g/cm³	5.8 g/cm³	5.0 g/cm³
Hall Flow (50g) =	13.0 s	14 s	16 s	19 s

Fig. 13.39 SEM of molybdenum powder spheroidization in an RF-ICP system operated using Ar/H$_2$ as the plasma sheath gas, at a power level of 100 kW, and powder feed rates varying between 7 and 25 kg/h [Boulos M (2016)] Reprinted with kind permission

impurities in the treated powder, which, at a feed rate of 10 kg/h, presents a reduction of Cr content of the powder from 100 to 30 ppm, Fe from 70 to 25 ppm, and Ni from 100 to 37. The oxygen content of the treated powder was also observed to drop from 1000 down to 100 ppm.

13.3.3.2 Modeling Results

Supporting results obtained using mathematical modeling based on modeling formulation including the plasma-particle interaction and loading effects are given in Figs. 13.40 and 13.41. The background of the model development and assumptions involved have been discussed in detail in Chap. 5 "Gas and Particle Dynamics in Thermal Spray," Sect. 5.6. In these calculations, radiative energy losses from the surface of the particles during its transient heating and melting time is taken into account as well as radiation losses from the metallic vapor cloud generated through partial/complete particle evaporation particle evaporation which was recently shown to contribute significantly to the local cooling of the plasma [Boulos M (2016) and Siwen Xue and M Boulos (2019)]. The results shown in Fig. 13.40 are consistent with the experimental observation by showing a significant thermal loading effect resulting due to the molybdenum powder injection into the discharge, responsible for cooling of the central region of the plasma stream. A corresponding drop of the in-flight particle temperature and the melted fraction of the powder is observed in Fig. 13.41 with the increase of the powder feed rate.

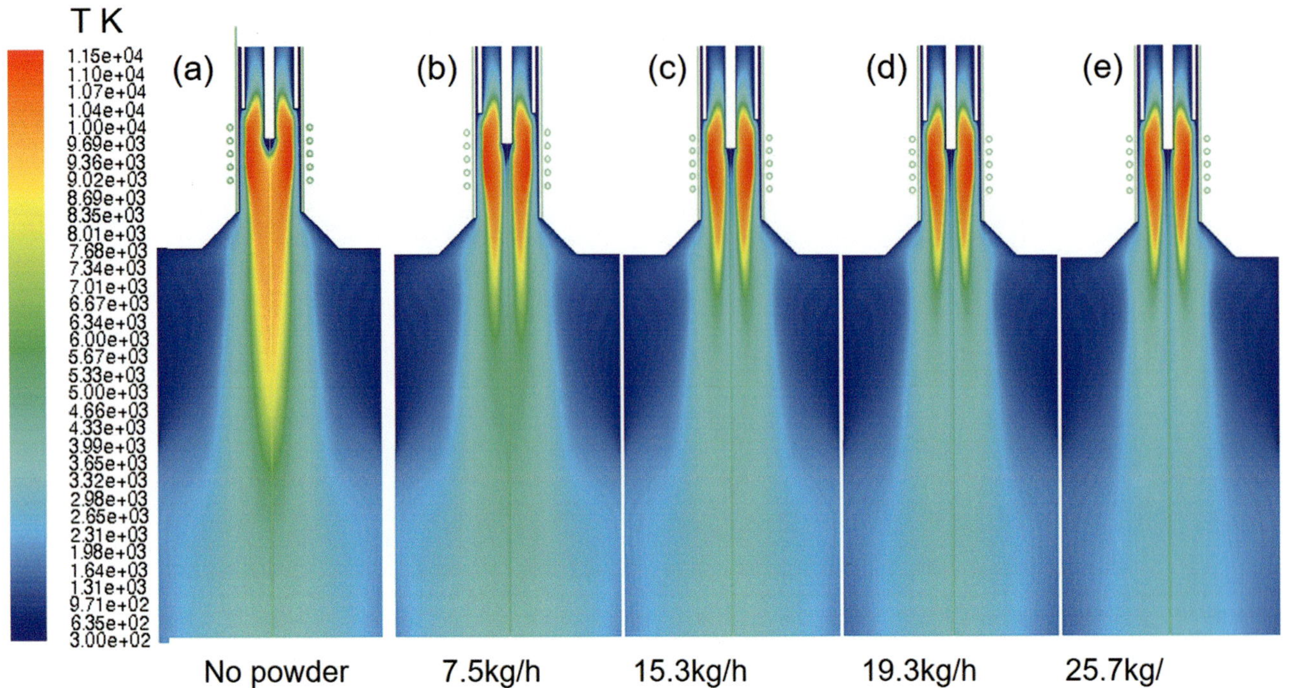

Fig. 13.40 Temperature field in an RF-ICP, PS-70, operated at plat power = 100 kW, pressure = 108 kPa, using Ar/H$_2$ as plasma gas, in the presence of increasing feed rate of molybdenum powder into the center of the discharge (**a**) no powder injection (**b**) 7.5 (**c**) 15.3 (**d**) 19.3 and (**e**) 25.0 kg/h [Boulos M (2016)] Reprinted with kind permission

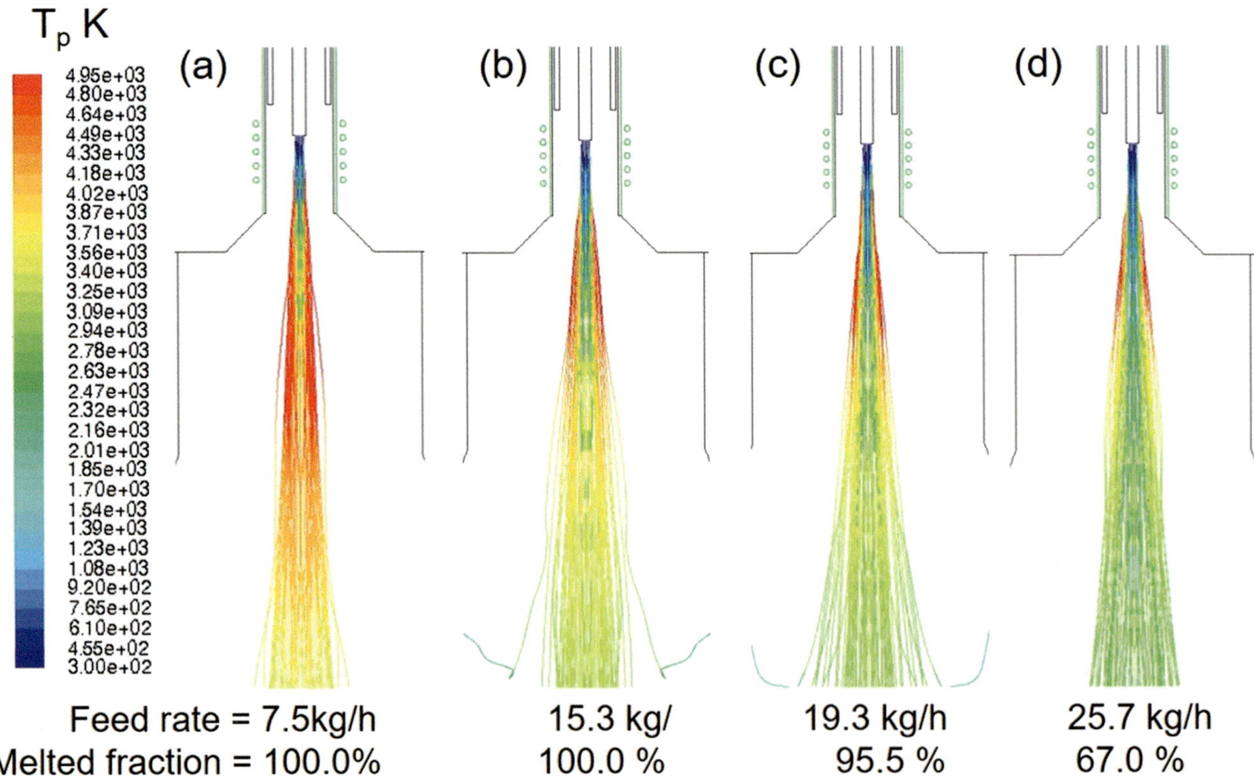

Fig. 13.41 Molybdenum particle trajectories and temperature history in an RF-ICP, PS-70, operated at plat power = 100 kW, pressure = 108 kPa, using Ar/H$_2$ as plasma gas, in the presence of increasing feed rate of molybdenum powder into the center of the discharge (**a**) 7.5 (**b**) 15.3 (**c**) 19.3 and (**d**) 25.7 kg/h. [Boulos M (2016)] Reprinted with kind permission

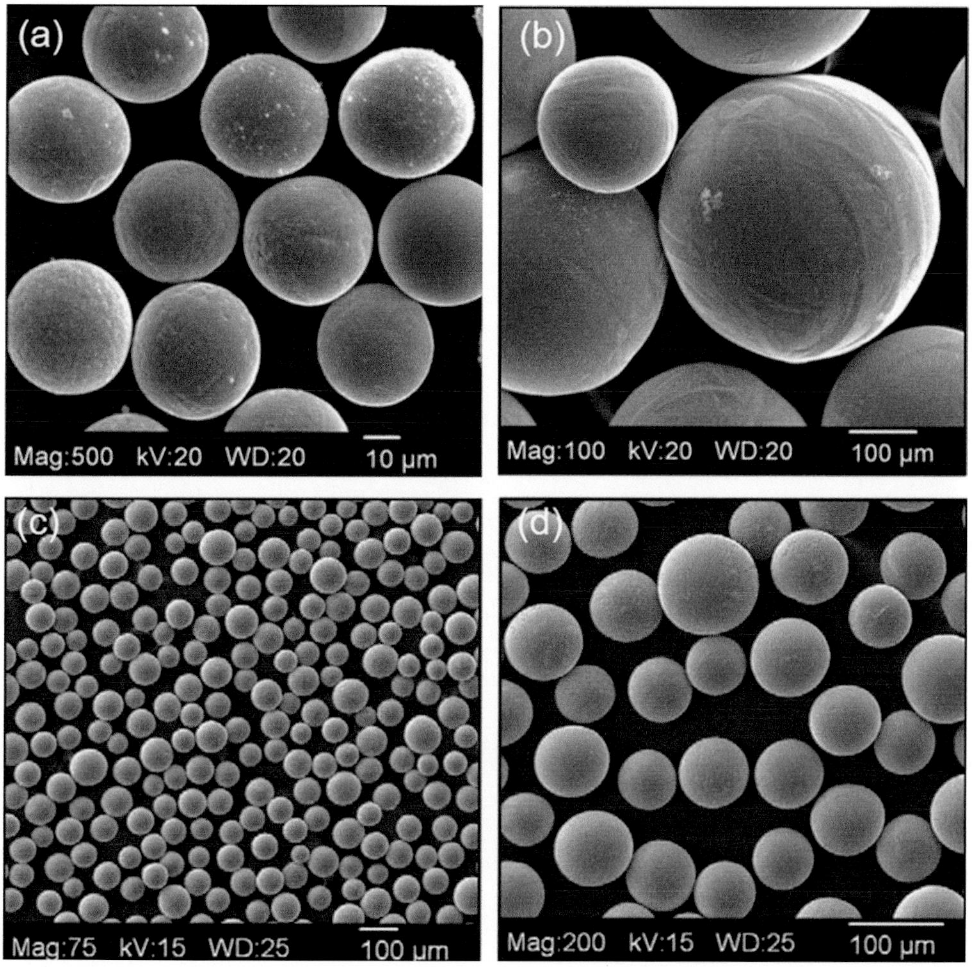

Fig. 13.42 SEM images of PF-ICP treated Ti-6Al-4V powders for different particle size ranges (**a**) 25–45 µm $\times$ 500 (**b**) 350–500 µm $\times$ 200 (**c**) 45–75 µm $\times$ 75 (**d**) 45–75 µm $\times$ 200 [McCracken CG et al. (2012)] Reprinted with kind permission

RF Inductively Coupled Plasmas (RF-ICP) was also used for the spheroidization of Titanium alloy powder (Ti-6Al-4V) obtained using Hydride-Dehydride process (HDH) as described by [McCracken CG et al. (2012) Froes, FH (2012a)]. SEM images, at different magnifications, of spherical powders obtained using different powder size fractions are given in Fig. 13.42. The images given in Fig. 13.42a, brepresent the extremes of the size of the different sizes fractions processed, with (a) corresponding to the smallest size fraction 25–45 µm with a magnification of $\times$500, while (b) are for the largest of the size fractions processed 350–500 µm with a magnification of $\times$100. The images given in Fig. 13.42c, dare for the same size fraction 45–5 µm with different magnifications of $\times$75 and $\times$200. All the images show excellent roundness and flowability. As the powder was pre-screened prior to plasma treatment the PSD of the powder was relatively narrow with little oversize, or undersize fractions.

DC plasma jets have also been used to spheroidize particles; however, it is more complex than RF plasmas, especially for ceramics with low thermal conductivities. DC plasma jets have higher velocities (more than one order of magnitude) than RF plasmas which reduces significantly the residence time of particles within plasma stream and requires using plasma-forming gases with higher thermal conductivity than pure argon. Unfortunately, higher thermal conductivities promote the heat propagation phenomenon and the nonuniform treatment of particles. The melting in a plasma jet (RF or DC) of nanostructured particles result in the destruction of their nanostructure. This is illustrated in Fig. 13.43 representing the same particle as that shown in Fig. 13.21 after they have been melted in an Ar-H_2 plasma jet produced with the APS-3 MB torch of Oerlikon-Metco [Lima RS and Marple BR (2008)].

Attempts made to produce nanostructure alumina particles by melting them in an Ar-He DC plasma jet and collecting the treated powder in water or liquid nitrogen [Dignard NM and Boulos MI (1997)]. Particles were spheroidized indicating their complete in-flight melting with those collected in liquid nitrogen (cooling rates calculated in the range of 3.33×10^6

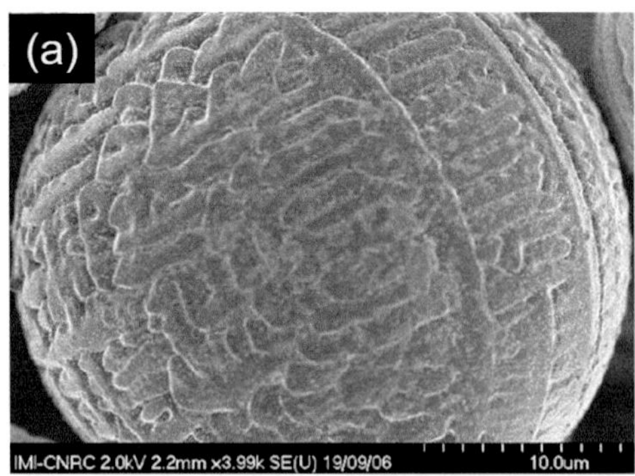

Fig. 13.43 (**a**) DC plasma spheroidized Spray-Dried titania powder (**b**) Higher magnification view showing the absence of the ultrafine character after the plasma-fusing process [Lima RS and Marple BR (2008)]. Reprinted with kind permission from Elsevier

to 2.27×10^8 K/s) showing a distinct nanostructured, as confirmed by the diffraction peaks. The TEM studies corroborated the presence of nanostructure, metastable phase, and amorphous phase formation in the powders. Al_2O_3-TiO_2 (87-13 wt.%) particles were also spheroidized by flame spraying.

13.3.4 Atomization

Atomization is a well-established process that is used commercially to produce the largest tonnage of metal and ceramic powders. In this process a molten metal or ceramic is broken into small droplets that are rapidly solidified before their impact onto the walls or bottom of the atomization chamber. The starting material, in the form of pure metal, multielement metallic alloy, or ceramic, is melted in an induction furnace, arc furnace, or other type of furnace. Following the complete melting and homogenization of the charge, the melt is transferred to a melt reservoir from which a constant controlled flow of molten material is supplied to the atomizing chamber. The molten material stream, exiting the melt reservoir ladle, is intercepted by either a disc rotating at high velocity or the high-velocity stream of the atomizing medium such as inert gas, air, or water disintegrating it into fine droplets that solidify in-flight during their fall through the atomization chamber. The general physics of conventional atomization processes have been reviewed by [Lefebvre (1989) and Fritsching U and Uhlenwinkel V (2012)]. Comprehensive reviews on melt atomization can be found in the "Handbook of non-ferrous metal powders" by [Neikov et al. Eds (2019)] and [Antony LVM and RG Reddy (2003) and Salman AD, Ghadiri M, Hounslow MJ (eds) (2007) Lawley (2003), Yule & Dunkley (1994) and Özbilen S (1999), Dunkley JJ (1998), and Mates SP and Settles GS (2005)].

In this section the focus is placed mainly on metal powder production using inert gases atomization such as argon,

helium, or nitrogen which generally gives rise to the formation of particles with good degree of sphericity and low levels of oxidation. Gas Atomization Process (GAP) requires, however, large water-cooled vessels, with controlled atmosphere which makes them rather expensive. Gas flow rates, depending on the metal mass treated, are in the range 0.5–2 m³/kg of atomized metal powder. Water Atomization Process (WAP), on the other hand, while being a relatively cheaper process, gives rise to particles with irregular shapes and relatively high degree of oxidation. The water/metal ratio is important since there must be enough water to break the metal apart, typical values are 4–10 ℓ/kg of atomized metal. Plasma Atomization (PA) has emerged over the past decades as a novel approach for the production of powders of high purity metals, alloys, and certain ceramics. The approach avoids the need for the initial melting of the material to be atomized which is fed into the plasma atomizing chamber in the form of a rod, wire, or cord and makes use of the plasma energy for the in-situ melting of the feed material. In the Plasma Rotating Electrode Process (PREP) the feed material in the form of a rod serves anode to an electric arc used to heat and melt the tip of the rod with the formed molten metal atomized as it is formed through the high-speed rotation of the rod. Alternately the feed material in the form of wire or cord is melted and atomized using high velocity (supersonic) plasma jets such as in the AP&C DC-Triple Plasma Atomization Process (DC-TPAP) or the Tekna RF-Induction Plasma Atomization Process (RF-IPAP). In the following sections, a brief description is given of each of these powder manufacturing processes.

13.3.4.1 Gas Atomization

Gas Atomization (GA), also referred to as "twin-fluid" atomization, is used for the production of a wide range of powders. As underlined by [Fritsching and Uhlenwinkel (2012)] since solid state materials such as metals, alloys, minerals, ceramics, and glass may be converted into liquid through melting, powders may be produced from all of these materials via atomization of the melt by a second fluid (gas) which provides the mechanical energy for the breakdown of the melt into small droplets that, on solidification, forms the powder. The main advantages of gas atomization are:

- Possibility of high throughputs
- Generally spherical powders with good flowability
- Reduced risk of oxidation when using inert gases

Gas atomization of melts, on the other hand, differs in several ways from conventional atomization techniques by the fact that it is a thigh temperature process dealing with fluids (melts) with a high surface tension and generally high viscosity. The process is also susceptible to premature melt freezing problems which can have a major disruptive impact on the atomization process.

As schematically represented in Fig. 13.44, two principal configurations are commonly used for gas atomization of melts. In the *close-coupled* atomizer configuration illustrated in Fig. 13.44a, the atomizing gas introduced at the base of the melt-delivery nozzle provides for an efficient atomization of the melt in terms of powder PSD with smaller particle sizes obtained at identical energy consumption, due to the short distance between the gas injection point and the melt exit. It is, however, for the same reason, that the *close-coupled* configuration is more susceptible to premature melt freezing problems. On the other hand, in the *free-fall* atomizer configuration, shown in Fig. 13.44b, the melt is allowed to move a certain vertical distance, in the direction of gravity, before being intercepted by the high velocity atomizing gas jets. As the melt stream and gas streams are well separated, premature freezing of the melt is considerably less problematic compared to *close-coupled* atomizer. However, the PSD of the powder obtained tends to be coarser than that obtained with the *close-coupled* configuration. Because of its simplicity, robustness, and reliability the *free-fall* atomizer configuration is widely used in spray drying, powder production, and spray forming processes [Yule and Dunkley (1994) and Fritsching (2008)].

A schematic view of a typical melt-gas atomization process system using the *close-coupled* approach is given in Fig. 13.45 [Mates SP and Settles GS (2005)]. It consists essentially of five main components, namely gas supply system, metal/alloy melting crucible and melt delivery nozzle, atomization gas nozzles, atomization chamber, powder collection and gas extraction system. System performance depends largely on the detailed design of the melt, and atomizing gas delivery systems, melt temperature and its feed rate, as well as atomizing gas pressure and flow rate.

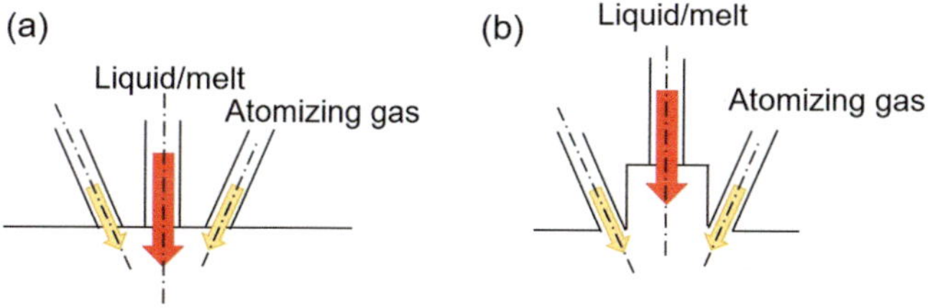

Fig. 13.44 Schematic representation of (**a**) close-coupled atomizer and (**b**) free-fall atomizer configurations

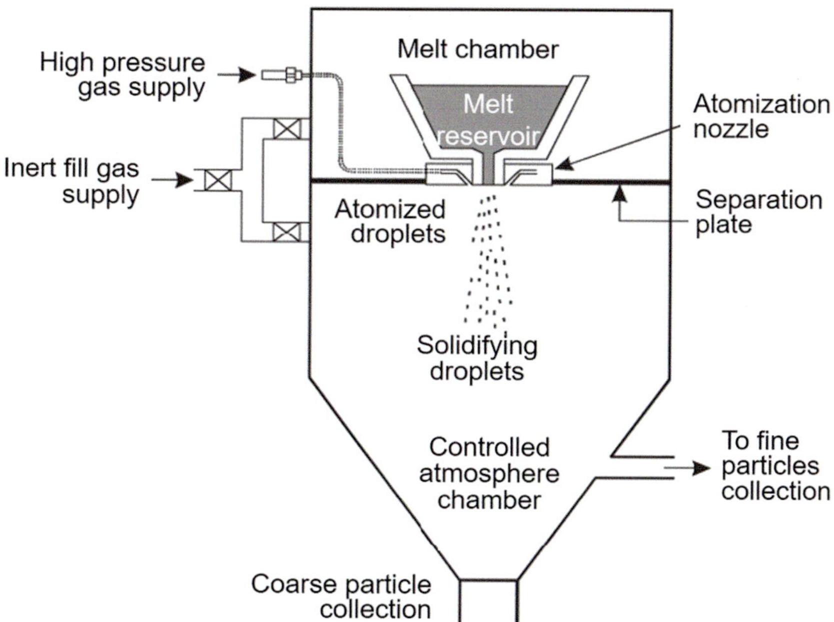

Fig. 13.45 Schematic of the molten metal-gas, close-coupled, atomization process [Davis JR (ed) (2004)] reproduced with kind permission from ASM International

Fig. 13.46 Photograph of the Dumlupinar University (Kütahya, Turkey) gas atomization unit. Reprinted with kind permission from Elsevier [Ünal R (2006)]

The atomizing chamber design is a function of the dispersion angle of the atomizing nozzles and the size distribution of the formed molten droplets. The latter has a direct influence on the height of the atomization chamber which should be sufficient in order to allow for the in-flight solidification of the largest formed melt droplets before reaching the bottom of the chamber. For inert gas-atomized powders, the particle shape depends on the nature of the atomizing gas [Dunkley JJ (1998)], the way the gas is directed to the molten metal stream [Mates SP and Settles GS (2005)] and on its pressure [Ünal R (2006), Antipas GSE (2006) and Allimant A et al. (2009)]. The secondary droplet breakup in the atomization process [Zambon A (2004), Özbilen S (1999)].

A photograph of a typical pilot-scale experimental molten metal gas atomization system is shown in Fig. 13.46 [Ünal R (2006)]. Such a unit, which can process molten metal batches up to 300 kg, has an atomizing chamber, 1.25 m in diameter by 4.5 m high [Ünal R (2006)]. The atomizing nozzle used is shown in Figure 13.47a with the liquid metal jet exiting along its centerline, in the absence of atomizing gas. A close-up photograph of the atomizing nozzle in operation is given in Fig. 13.47b showing the atomized spray pattern in which 20 discrete supersonic nitrogen jets located concentrically around the metal delivery tube are directed toward the molten metal stream. Lateral spreading occurs at the tip of the melt delivery tube forming a conical-shaped spray pattern as illustrated in Fig. 13.47c. The gas recirculation zone created by the atomizing gas jets forces the molten metal to flow outwards to the rim of the melt delivery nozzle where it is atomized by the high velocity atomizing gas jets. Details of the mechanism of the metal flow breakdown and droplet formation is discussed by [Antipas GSE (2006)].

The main design and operating parameters of gas atomization systems are:

- The atomizing jet geometry, pressure, velocity, and distance from the liquid jet
- The liquid nozzle geometry and metal flow velocity
- The melt superheat above the metal melting point
- The gas nature, especially its cooling capacity and its purity (to achieve low metal oxidation, purity levels of at least 99.99% are used)

When comparing the use of helium versus nitrogen as the atomization gas, it is observed that droplets are cooled faster with helium, giving rise to higher content of finer particles in the powder obtained with helium compared to that obtained using nitrogen as atomizing gas. Moreover, with helium, formation of nitrides is avoided and, due to larger extents of undercooling a higher amount of amorphous, or at least meta-stable, material can be retained [Zambon A (2004)]. Particle cooling rates (forced convection) in gas atomization depend on particle sizes and the nature of the atomizing gas, the fastest rates are obtained with helium. Roughly they range between 10^3 and 10^5 K/s using argon or nitrogen as atomizing gas increasing to 10^7 K/s with helium for 10 µm particles. Typical

Fig. 13.47 The formation of the atomization process using a modified close-coupled atomizer configuration. Images obtained from the recorded video with 40 ms time intervals. (**a**) Melt flow in the absence of atomizing gas (**b**) with gas atomization pattern and (**c**) atomization nozzle configuration [Ünal R (2006)] Reproduced with kind permission from Elsevier

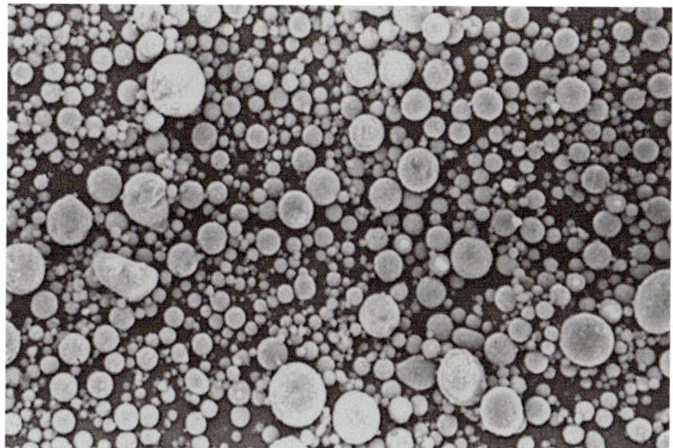

Fig. 13.48 SEM micrographs of Mg powder ($d_p < 32$ µm) produced using low-pressure argon gas atomization [Özbilen S (1999)] reproduced with kind permission from Elsevier

Fig. 13.49 SEM micrograph of Diamalloy 1005 (21.5 Cr, 2.5 Fe, 9.0 Mo, 3.7 Nb + Ta, Bal. Ni) showing the gas-atomized morphology typical of these materials. (Courtesy of Sulzer Metco)

gas flow rates are between 0.5 and 2 m^3/kg, for gas pressures varying from 1.4 to 4.2 MPa [Salman AD, Ghadiri M, Hounslow MJ (eds) (2007) and Davis JR (ed) (2004)].

Depending on the atomization parameters, especially the atomization gas pressure, different particle distributions are obtained: mono-modal fine particles distribution to coarse and bimodal one. Typical particle sizes can vary between 10 and 100 µm. The fine size range of the powder particles is generally spherical with smooth surface texture, as illustrated in Fig. 13.48 for Mg particles produced with low-pressure argon as atomizing gas [Özbilen S (1999)].

The powders treated by this technique are Co–Cr-base alloys, Ni–Cr-base alloys, Ni-base alloys, Cu-base alloys, Zn, stainless steels, Ni-base superalloys (see, for example, Fig. 13.49 and MCrAlYs [Davis JR (ed) (2004)].

Design modifications aiming at improving the atomization level of "free-fall" atomizer nozzle design were proposed by [Fritsching and Uhlenwinkel (2012)]. The design as illustrated in Fig. 13.50 is based on the combination of two gas nozzle systems, namely a primary and a secondary gas nozzle. The secondary nozzle is the main atomization unit. The concentric gas jets from the secondary nozzle impinge onto the central liquid jet that is disintegrated due to instabilities from the shearing action of the secondary gas flow and its relative velocity. The disintegration process of the liquid jet in the atomization zone is located underneath the secondary gas nozzle. In order to avoid the formation of the recirculation zone in the secondary atomization area and possible deposition of melt splashes on the atomizer body, [Fritsching and Uhlenwinkel (2012)] reported on the addition of a deflector ring named "Coanda-flow device" which seems to improve the quality of the atomization as shown in Fig. 13.51 after [Czisch & Fritsching, (2008)].

[Fritsching and Uhlenwinkel (2012)] proposes further the use of a "Hybrid" atomization system which combines the concept mechanical and gas atomization as illustrated in Fig. 13.52. The aim of the development is an atomizer design for highly viscous liquids and melts at high throughputs that produce a considerably low droplet size. The hybrid atomizer introduced here is a combination of a single-fluid rotary atomizer and an external mixing twin-fluid atomizer. The advantage of this specific design is that the feed material is first spread out by a spinning disc due to centrifugal forces, thereby increasing the initial liquid surface prior to the gas atomization process. The rotary atomizer is operated in sheet formation mode so that in the vicinity of the rotary disc a thin liquid film is formed which is subsequently guided into the atomization zone of the external mixing atomizer by the gas flow field. To ensure stable atomization, aerodynamic forces are used to protect the atomizer body from coming into

contact and being blocked by parts of the liquid (melt). A sufficiently small gas mass flow rate from the inside of the atomizer protects the atomizer body.

One of the most common problems with conventional gas atomization approaches when used for the atomization of high purity reactive metals is the potential contamination of the powder as a result of its reaction between the melt and the refractory material used for the lining of the melting crucible or the tundish feeder of the molten metal. Moreover, melting of refractory alloys in a ceramic-lined crucible is not possible due to their high melting temperature. In order to overcome this "ceramic problem" ALD Vacuum Technologies GmbH developed and is marketing a "ceramic-free" metal powder production system for reactive and refractory metals based on the concept of "Drip melting gas atomization" illustrated in Fig. 13.53a [Hohmann et al. (1989), and Hohmann M and N Ludwig (1991)]. The Electrode Induction-Melting Gas Atomization (EIGA) system 50/100 applies to a crucible-free melting and atomization design in which the material to be atomized is used in the form of a pre-alloyed electrode (cast, remelted or sintered) with diameter up to 50 or 100 mm and 800 to 1000 mm long. As the electrode, in slow rotation, is introduced into a conical-shaped induction coil, it is heated, and its tip melted with the stream of molten metal flowing directly into the center of the atomization nozzle where it is atomized by high velocity inert gas jets. A photograph of molten metal flow for a 100 mm diameter titanium rod is shown in Fig. 13.53b [courtesy of Helmholtz-Zentrum Geesthacht].

A schematic of EIGA system is illustrated in Fig. 13.54a shows on the top section of the powder recovery tower, an electrode atomization module including the electrode rotating and feeding system, the conical-shaped induction heating coil, and the gas atomization unit. A close-up view is provided of the tip of the electrode showing the melting pattern on its conical surface which reflects the tuning of

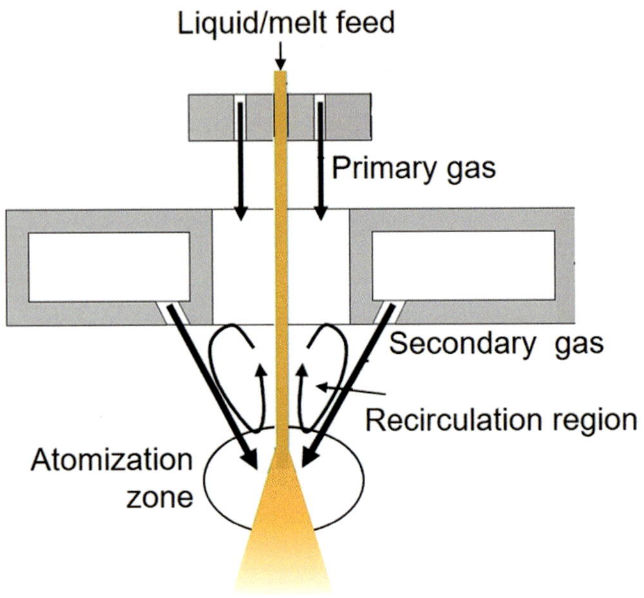

Fig. 13.50 Main components of a conventional "free-fall" gas atomizer with primary and secondary atomizing gas streams, after [Czisch & Fritsching, (2008)]

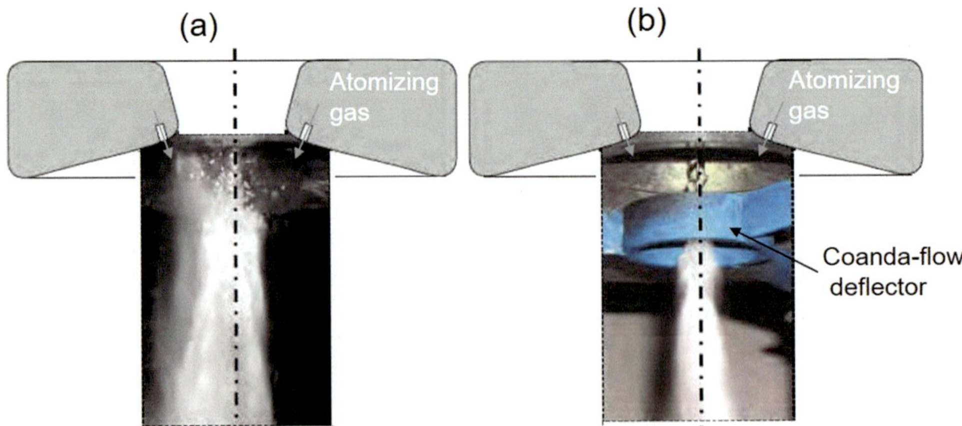

Fig. 13.51 Effect of the use of a "Coanda-flow" deflector in a model experiment on the atomization pattern of water using air as atomizing gas, after [Czisch & Fritsching, (2008a)]

the process conditions in terms of the speed of rotation, feed rate, and induction power supplied. According to [Hohmann et al. (1989)] typical EIGA operating conditions are: electrode diameter 25–70 mm, electrode melting temperature 1200–2100 °C, electrode feed rate 40–60 mm/min, and power requirement 50–100 kW. The powder cooling tower and recovery systems are of conventional design. Photograph of EIGA-50 production installations is shown in Fig. 13.54b.

Results of an experimental program carried out with Stainless steel (d_o = 45–65 mm, T_m = 1450 °C), Titanium (d_o = 45–60 mm, T_m = 1650 °C), and Niobium alloy (1 wt % Zr, d_o = 35 mm, T_m = 2100 °C) electrodes were very promising. These were obtained using nitrogen as atomization gas for the stainless test at a pressure of 2.5 MPa (362 psig). Argon was used for the atomization of Ti and Nb alloys at the same pressure. Electron micrograph of the Ti-powder given in Fig. 13.55a shows spherical powder with a smooth surface texture and relatively few satellites.

Test with 45 mm diameter, stainless steel electrodes could be run, after process optimization, at feed rates of 100 mm/min (75 kg/h). The corresponding PSD of the powder obtained given in Fig. 13.55b shows a mean particle diameter, d_{50} = 40–45 µm. The atomization of the 65 mm electrode was more challenging and needed further parameter optimization and operation at a lower electrode feed rates. Equally satisfactory results were obtained with Titanium electrodes, d_o = 45 mm, with electrode feed rate of 70 mm/min (30 kg/h). The PSD of the powder obtained had a d_{50} = 60 µm with a yield of around 25 wt.% for the size fraction d_p < 45 µm. The drip melting and atomization of the Nb alloy rod was most challenging because of its high melting temperature, high viscosity, and high surface temperature of the melt. Powders obtained with the 35 mm diameter Nb alloy electrode had a mean particle size d_{50} = 80 µm.

13.3.4.2 Water Atomization

Because of its favorable economics, water atomization is widely used for large-scale powder production. As illustrated in Fig. 13.56 [Dunkley JJ (1998)] the process is based on the use of water jets to hit a stream of the molten metal. Droplets are formed by "splash" or "scrape," according to disintegration models, and then freeze, in-flight, before being collected in the water pool at the bottom of the atomizer. To break the molten metal jet apart, the water quantity to molten metal ratios have to be in the range of 4–10 ℓ/min per kg of molten metal. Water pressures vary between 30 kPa and 60 MPa. Water atomization results in particle cooling rates in the range of 10^4–10^6 K/s. The high production rates are balanced by the necessity to dewater and dry the particles [Salman AD, Ghadiri M, Hounslow MJ (eds) (2007)]. The powder produced using water atomization has a typical particle size in the range of 150–400 µm. Finer particles can be obtained

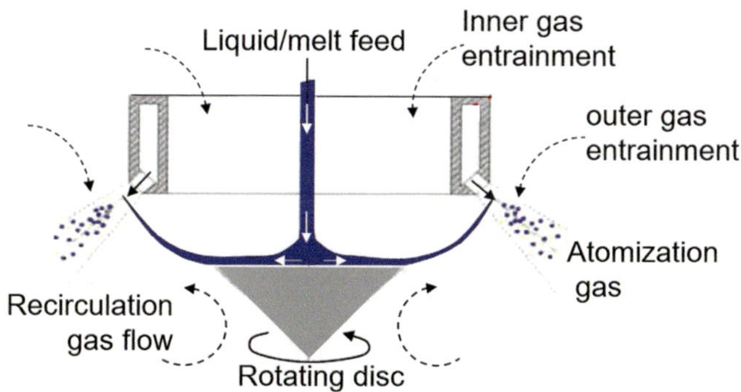

Fig. 13.52 Schematic representation of a "Hybrid" atomizer combing rotating disc approach with gas atomization [Czisch & Fritsching (2008a)]

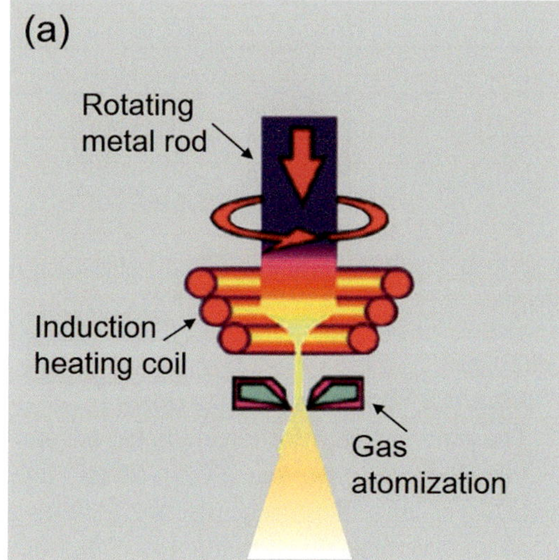

Fig. 13.53 Basic concept used by ALD in its Ceramic-free EIGA metal powder production systems [courtesy of Helmholtz-Zentrum Geesthacht]

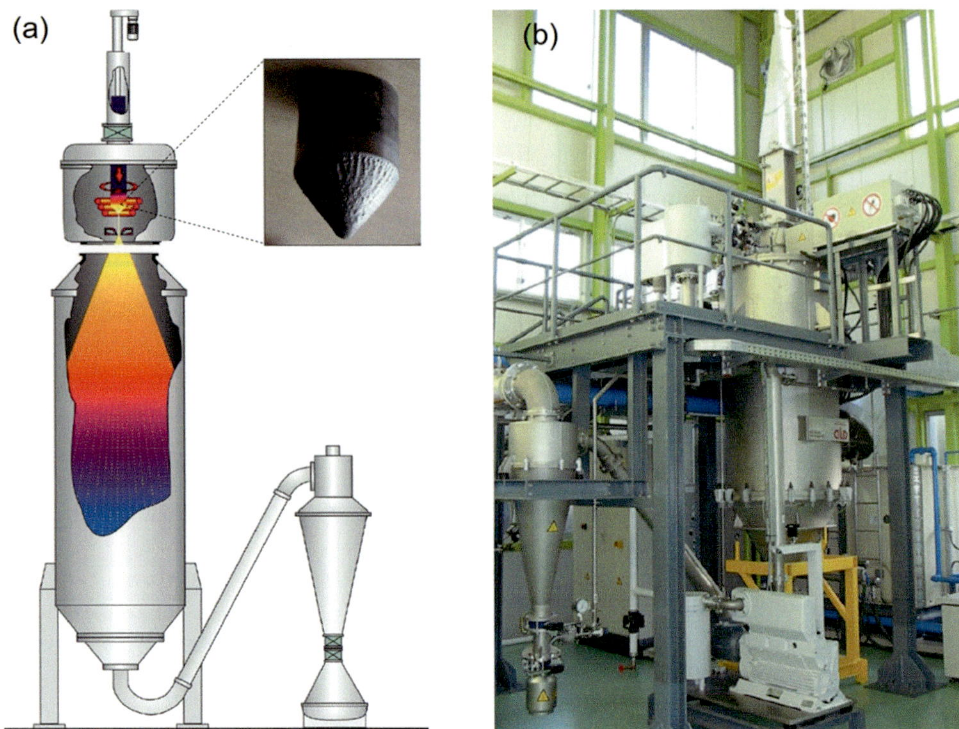

Fig. 13.54 (a) Schematic of EIGA metal powder production system with photograph of "correctly tune" drip melt rod end, and (b) photograph of EIGA-50 installation for reactive and heavy metals powder production [courtesy of Helmholtz-Zentrum Geesthacht]

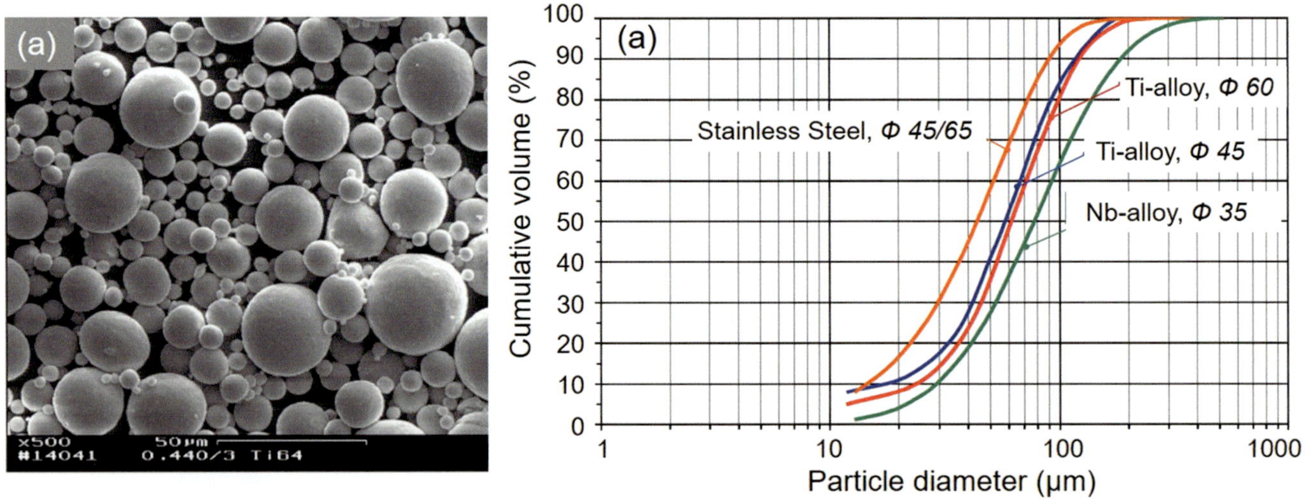

Fig. 13.55 (a) Electron micrographs of Ti grade EIGA powder and (b) Particle Size Distribution of various alloys obtained with different electrode diameters, values given in (mm). [courtesy of Helmholtz-Zentrum Geesthacht]

using high-water jet velocities. The irregular shape of particles (see, for example, Fig. 13.57) may necessitate a further spheroidization step which can be partially avoided through the superheating of the molten metal prior to atomization. Water atomization can result in relatively high-oxygen content of the powder obtained which can require a separate deoxidation step as a post-treatment of the powder.

13.3.4.3 Plasma Atomization

The growing need for a wide range of high-quality powders for Thermal Spraying (TS), Metal Injection Moulding (MIM), and more recently for Additive Manufacturing (AM) has been the driving force for the development of innovative technologies for the one-step Plasma Atomization (PA) of wires or rods to spherical, dense powders with good flowability. Plasma atomization of metals has emerged over the past two to three decades as the ultimate "ceramic -free" metal powder production technology in which the metal melting and atomization steps are combined in a single operation with the raw material supplied in the form of rod or wire. The operation is carried out under inert atmosphere which guarantees the purity of the powder, avoids any oxygen pickup or ceramic contamination in the process. By using a wire, cord, or rod as starting material, the process had little production inertia allowing for the production of relatively small batches of powders for R&D material development and the rapid switchover from one material to another without the risk of cross contamination from one powder production to the other. Three leading technologies have been developed in this area.

(a). *Plasma Rotating Electrode Process*

The technology was developed in the early Eighties (US patents 3,099,041, July 1963 and 3,802,816 April 1974) for the production reactive metal powders such as beryllium, molybdenum, titanium, and zirconium [Roberts et al. (1989)]. Early versions of the technology, Rotating Electrode Process (REP), was based on the use of an electric arc struck between a stationary, tungsten electrode (cathode), and a metal bar of the material to be atomized which acts as (anode), rotating at high speed around its axis. The molten metal pool generated at the anode arc attachment on the bar is ejected by the centrifugal forces in the form of spherical droplets which on cooling, in-flight, freezes forming a dense spherical particle of diameter which depends on the thermophysical properties of the material, the arc current, and the speed of rotation of the metal bar. The technology evolved rapidly toward the replacement of the tungsten electrode by a plasma torch operating in the transferred arc mode, Plasma Rotating Arc Process (PREP), as illustrated in Fig. 13.58. By protecting the water-cooled tungsten cathode of the DC plasma torch using an argon, or helium gas flow, the arc stability was enhanced, cathode life extended, and the risk tungsten-contamination of the produced powder significantly reduced.

A schematic representation of a typical PREP installation given in Fig. 13.59 [Wosch et al. (1997)] shows the controlled atmosphere deposition chamber, 2–3 m in diameter, the plasma torch, the feed bar with a diameter in the range of 50 to 75 mm diameter by 250 to 1500 mm long, the variable speed motor drive (5000–15,000 rpm) and linear translator for the continuous feeding of the rotating rod in order to

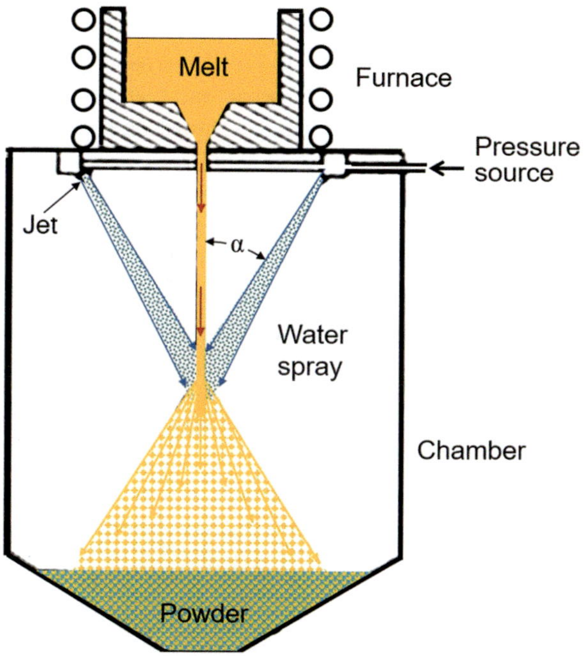

Fig. 13.56 Schematic of a water atomization process. Reprinted with kind permission from ASM International Dunkley JJ (1998)]

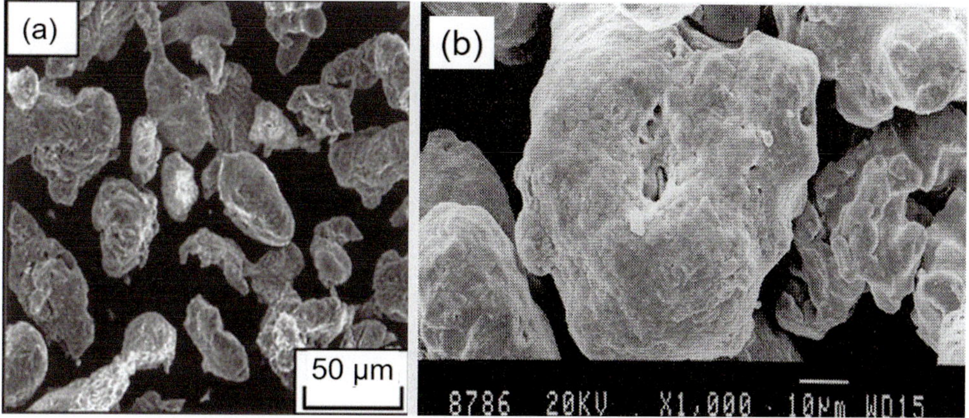

Fig. 13.57 SEM micrographs showing water-atomized cast iron powder: Fe-2.17 atm.% C-9.93 atm.% Si-3.75 atm.% Al. Reprinted with kind permission from Springer Science Business Media [Tsunekawa Y et al. (2006)], copyright © ASM International

maintain a constant arc length. Most of the powder is collected in the "catch-pot" with the fin fractions collected from the exhaust gas stream using appropriate powder collection system. Experimental results reported by [Wosch et al. (1997)], for the atomization of an austenitic steel using a nitrogen plasma, showed perfectly spherical powders with a strong dependence of the mean particle diameter on the speed of the rotation of the rod. Data provided show a steady decrease of the mean particle diameter from 379 μm to 137 μm with the increase of the rod rotation speed between 5000 and 15,000 rpm. Nitrogen mass pick up ranged between 0.48 and 0.962 wt./$N_2$with the negligible variation of the nitrogen mass content of the

powder with decreasing particle size for a given set of experimental conditions.

As reviewed by [McCraken et al. (2012)] the PREP method is used mainly for the production of the high strength Ti-6Al-4V alloy powders which are characterized as highly spherical powders with a high packing (Tap and Apparent) density. A typical PSD range would be 45–450 μm. As these spherical droplets are produced and radially dispersed away from each other, there is little or no opportunity for inter-particle collisions between solidifying droplets that may produce irregularly shaped powders, attached satellites, and powder clusters or agglomerates and therefore exhibit very good powder flow characteristics. Figure 13.60 shows a PREP Ti-6Al-4V powder with a PSD of <180 um (−80 mesh) and a Gas Atomized (GA) Ti-6Al-4V powder, 45–106 μm (−140 + 325 Mesh) for comparison.

[McCraken et al. (2012)] point out that the use of a mechanically rotating electrode does, however, result in practical limitations both in the maximum rotational speed and maximum anode diameter. The PREP process is also a batch process in which the bar needs to be repeatedly replaced. These limitations, in turn, restrict the particle size distribution (PSD) particularly in the production of very fine powder fractions and the overall throughput. All of these powder characteristics make PREP powders ideally suited for Hot Isostatic Pressing (HIP) applications of Net Shape (NS) complex components which require a free-flowing powder with a high tap density. Bulk Laser and Plasma Transfer Arc deposition processes also require free-flowing powders which can utilize PREP powders. By contrast spherical

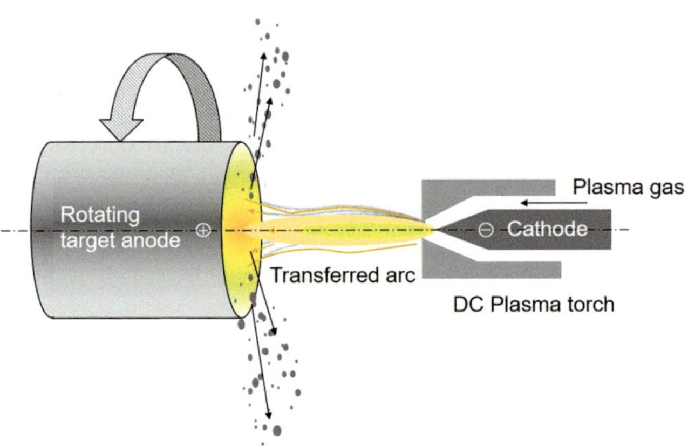

Fig. 13.58 Basic concept of Plasma Rotating Electrode Process for powder production

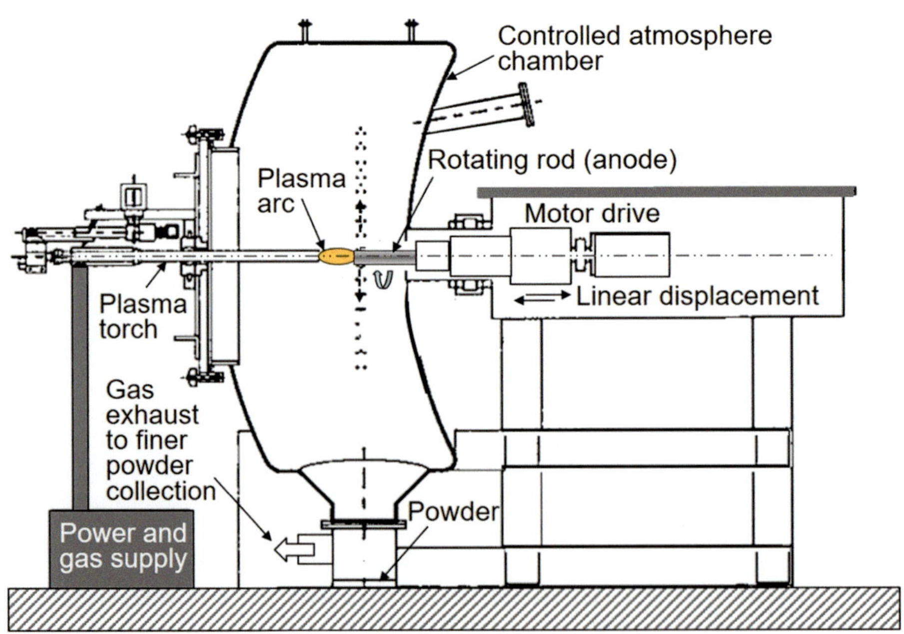

Fig. 13.59 Schematic of a PREP powder production setup [Wosch et al. (1997)]

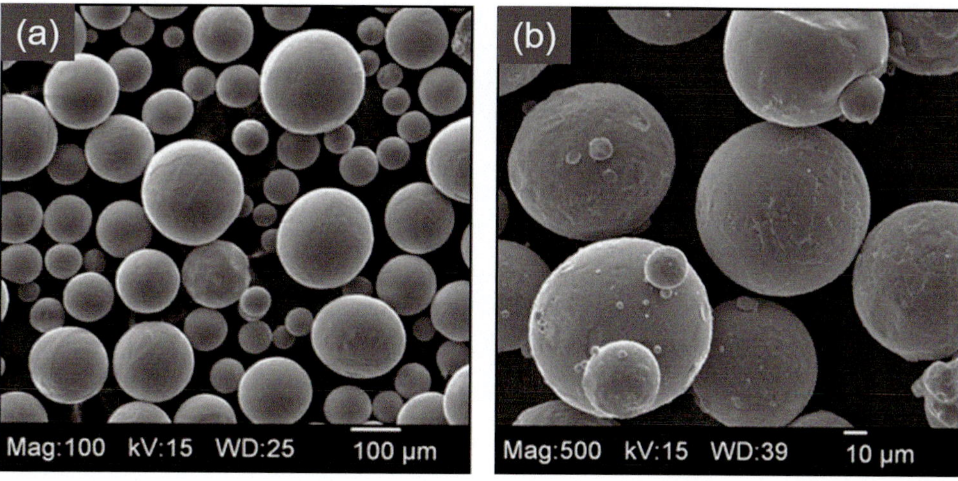

Fig. 13.60 Representative SEM images of spherical Ti-6Al-4V powder produced by (**a**) PREP, < 180 μm, (− 80 Mesh) (**b**) GA, 45–106 μm (−140 + 325 Mesh) [McCraken et al. (2012)] Reprinted with kind permission

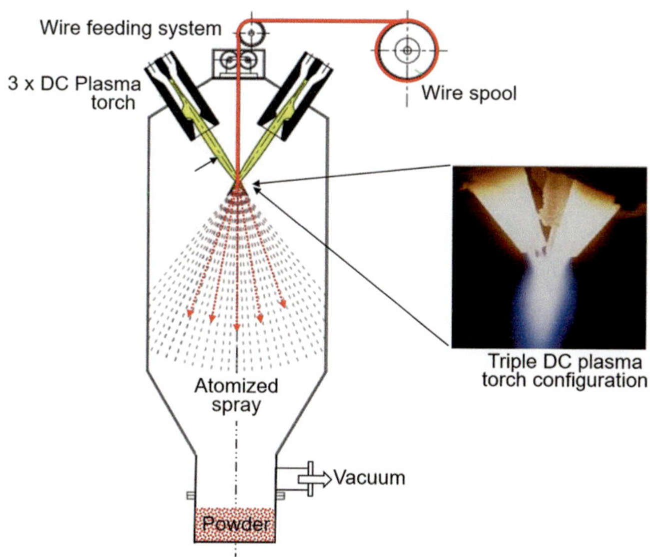

Fig. 13.61 Schematic representation of the DC Triple Plasma Atomization Process (DC-TPAP) including photograph of the 3-torch arrangement [Kroeger J and F Marion (2011)]

powders exhibit very low green strength and so are not compatible with the HDH powder consolidation processes

(b). *DC-Triple Plasma Atomization Process*

The DC-Triple Plasma Atomization Process (DC-TPAP) was developed in the late 90s jointly by Raymor AP&P and Hydro-Quebec, Montreal, Québec, Canada [Entezarian et al. (1996), Tsantrizos et al. (1998)]. As illustrated in Fig. 13.61, the process is essentially centered around the feeding of the material to be atomized in the form of wire, with a diameter in the range of 1.5–3.2 mm, into the apex of three DC plasma torches operating at nominal power level in

the 20 to 40 kW each, using Aron as plasma gas with a flow rate of 100–120 slm per torch. As shown in the photograph given in Fig. 13.61, the three torches are mounted with an inclination of 30° to the vertical axis with the ability to move them in/out individually with respect to their apex point thus creating an intense high temperature plasma zone at the point where the three plasma jets converge. As the wire is fed into this apex point, its tip is rapidly heated and melted and the formed melt atomized into a spray of small molten droplet by the high velocity, high temperature plasma jets. According to [Entezarian et al. (1996)] provision was also made to equip the plasmas a torch with a Laval-type convergent divergent nozzle in order to increase the plasma velocity at the torch exit and accordingly enhance the atomization process. As the molten metal droplets are formed, they drop in a free fall toward the bottom of the atomization chamber, cooling down in-flight and freezing into a solid particle. The largest-size particles settle by gravity into the "catch pot" in the bottom of the chamber, while the finer-size fractions are entrained by the gases exiting the chamber and recovered in the cyclone/filter collector. The final separation of the powder in distinct size fractions is carried out by a series of sieving operations carried out under controlled conditions. Micrographs of different fractions of (Ti-6Al-4V) plasma atomized powders are given in Fig. 13.62, with the corresponding PSD distribution given in Fig. 13.63. [Kroeger J and F Marion (2011)].

(c). *RF-Induction Plasma Atomization Process*

An alternative innovative approach for the production of high purity, spherical powders thought he atomization of the feed stock in the form of wires, cords or rods was developed by Tekna Plasma Systems Inc., in Sherbrooke, Que. Canada, taking full advantage of the unique characteristics of Inductively Coupled Plasma (ICP) technology as discussed in great

Fig. 13.62 SEM images of Plasma-atomized Ti-6Al-4V powders using AP&C DC-TPAP. [Kroeger J and F Marion (2011)]

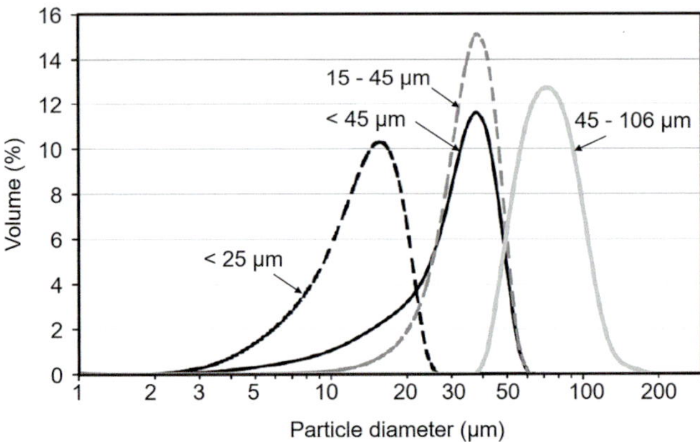

Fig. 13.63 Laser diffraction PSD of Plasma-atomized Ti-6Al-4V powders using AP&C-DC-TPAP. [Kroeger J and F Marion (2011)]

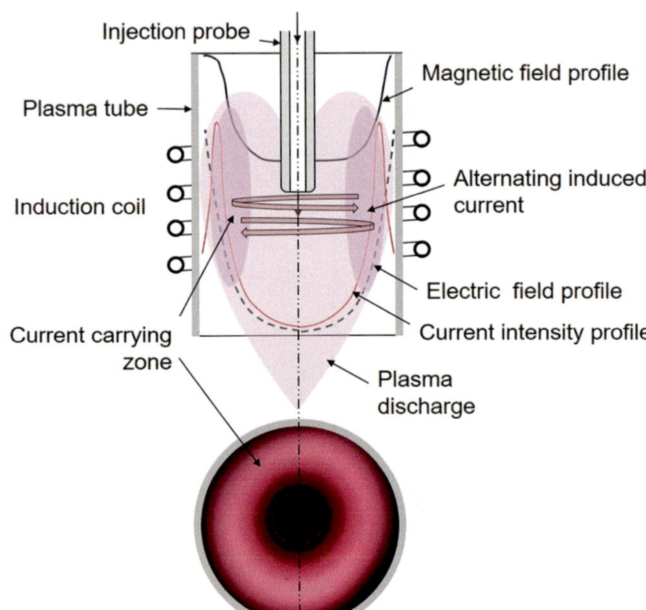

Fig. 13.64 Schematic representation of the core region of an induction plasma torch (top) and end-view photograph of the annular current-carrying region (bottom)

details in Part II, Chap. 10 "Induction Plasma Spraying." The basic concept involved, which is unique to RF-inductively coupled plasmas, lies in the ability to feed the material to be processed, whether in the form of powder, liquid, suspension, wire, or cord, axially into the center of the plasma torch as illustrated in Fig. 13.64, without perturbing the energy coupling into the discharge. As discussed in Chap. 10, for the case of RF-induction plasma spraying, the approach takes full advantage of the fact that energy coupling into the ICP is in the form of a donut shape as can be seen in the photograph at the bottom of Fig. 13.64, with the central darker region of the discharge not current-carrying. For more details on the electromagnetic field's distribution into the discharge cavity, see Chap. 10 Induction Plasma Spraying, Sect. 10.4.1 Electric and Magnetic Fields.

In the RF Induction Plasma Atomization Process (RF-IPAP) [Boulos et al. (2017a, b)], the same concept is used as in RF-IPS with the exception that flow of pneumatically transported powders introduced into the powder injection probe is replaced by the material to be atomized in the form of a wire, cord, or rod as schematically shown in Fig. 13.65. By properly adjusting the position of the probe tip

with respect to the exit nozzle level of the torch, it is possible to control the transit heating of the feed material and accordingly the tip temperature of the wire once it reaches the exit nozzle level of the torch. In the atomization nozzle the preheated wire, cord, or rod are exposed to further intense heating by an annular high-velocity plasma flow combined with a ring of high-speed (supersonic) atomization plasma jets which completes the melting of the wire tip and its atomization into a fin spray of molten metal droplets.

Mathematical modeling of the process was carried out for 50 mm i.d. induction plasma torch, operated at atmospheric pressure using Ar/H$_2$ (15% vol%H$_2$) as sheath gas, an RF frequency of 3 MHz, and plasma plate power of 60 kW. Computations were carried out for a 3.2 mm diameter stainless steel (SS) wire introduced through the central injection probe at different wire speeds and for different probe tip positions as indicated by the z distance, on Fig. 13.66a which also shows the temperature field in the discharge zone surrounding the stainless steel wire. Results given in Fig. 13.66b show the wire tip temperature systematically decreasing with the increase of the wire speed, while increasing significantly with the increase of the distance, z, over which the wire is being directly heated by the plasma. It is to be noted that for exposure distances below 50 mm, the wire temperature stays well below the melting temperature of stainless steel ($T_m = 1670$ K), while for values of $z > 100$ mm and wire speeds below 40–50 mm/s the tip temperature of the wire may exceed its melting point. Considering that the model did not take into account the wire melting process, wire temperature values above the melting point of stainless steel are only indicative of the increased heat flux to the wire under these conditions.

By mounting the RF-IP atomization torch on the top of the atomization chamber as shown in Fig. 13.67, the atomized molten metal droplets are allowed to cool down and freeze in free-fall under controlled conditions and recovered in the atomized powder catch pot at the bottom of the atomization chamber as well as from the gas stream by appropriate cyclone/filter collection. Final step of the process would involve the screening of the powder to required size fractions. Photograph of the plasma jet emerging from the nozzle of the RF-IPAP in the absence of the wire feed given in Fig. 13.68a shows the standard shock nodes characteristics of supersonic flow conditions. With the introduction of the wire into the torch, a dispersed spray of atomized droplets emerges from the torch nozzle exit as shown in Fig. 13.68b. SEM micrographs of (Ti-6Al-4V) powders obtained using this process are given in Fig. 13.69. These show excellent sphericity with limited number of satellite or particle defects. Details of the corresponding properties of the different powder fractions are summarized in Table 13.2.

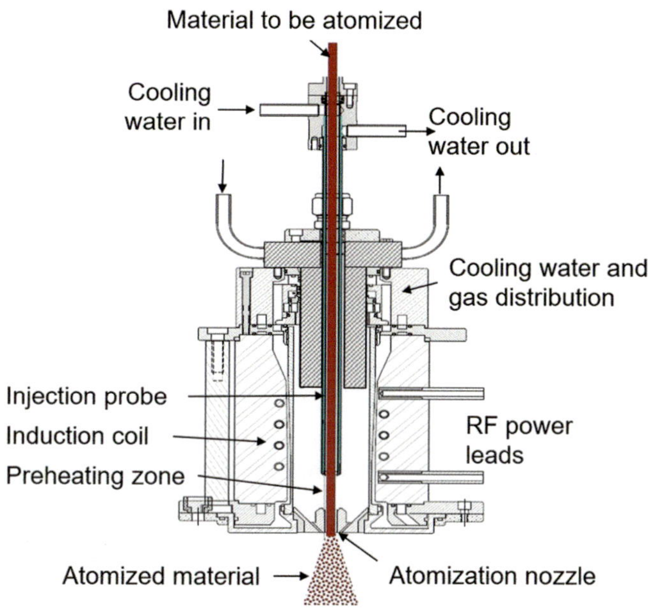

Fig. 13.65 Schematic of an RF-Induction Plasma Atomization Torch [Boulos M et al. (2017a, b)]

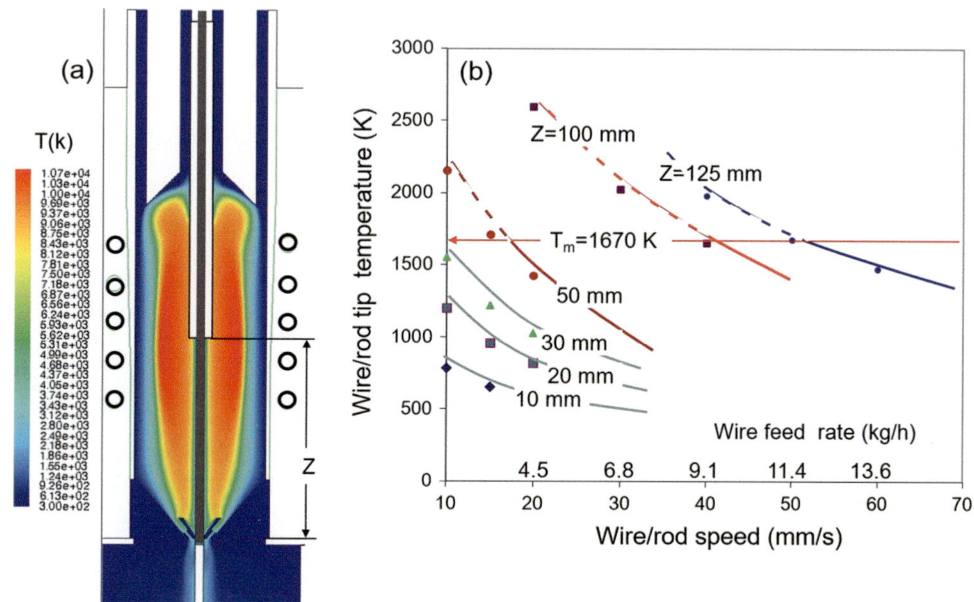

Fig. 13.66 (**a**) Temperature fields in an RF Induction Plasma Atomization torch, (**b**) Wire/rod tip temperature as function of injection probe position and wire/rod advancement speed [Boulos M et al. (2017a, b)]

13.3.5 Powder Chemical Synthesis

A broad range of technologies are used for the synthesis of powders used in by the thermal spray industries. No attempt is made to cover the subject in an exhaustive way. Examples are given of some of the key processing routes used, the majority of which are combined with some of the above-described powder manufacturing processes such as crushing, grinding, spray drying, sintering, spheroidization, and atomization in order to produce a final powder with the required chemical composition and powder morphological and rheological properties.

13.3.5.1 Sol–Gel Process

Sol–gel processing is commonly used to produce high purity fine oxide ceramic powders (alumina, chromia, partially or totally stabilized zirconia), with a high specific surface area which allows sintering to nearly full density at lower temperatures than normally required when the particles are made by alternate techniques [Davis JR (ed) (2004)].

The process comprises essentially of hydrolyzing the reactants in the form of metal alkoxy compounds (resulting from the reaction of metals with alcohols) by water in the presence of a catalyst to accelerate the reaction. The condensation and gel formation steps depend on the type of alkoxide used, the reaction temperature, the amount of water added, and the pH of the solution. A close control of the pH is very important to avoid or promote the formation of agglomerates. Gelation times vary from seconds to several days. As the gel forms, containing up to 5 vol.% of the oxide, it is dried and calcined to completely convert it to oxide. Powders produced by the sol-gel method are amorphous. A crystallization step is required to produce crystalline bodies, which is often performed after sintering.

[Wang et al. (2010)] synthesize nanostructured composite powder of MgO-Y$_2$O$_3$, using magnesium acetate (CH$_3$COO)$_2$ Mg·4H$_2$O and yttrium nitrate Y(NO$_3$)$_3$·6H$_2$O as starting chemicals. These products were first dissolved in

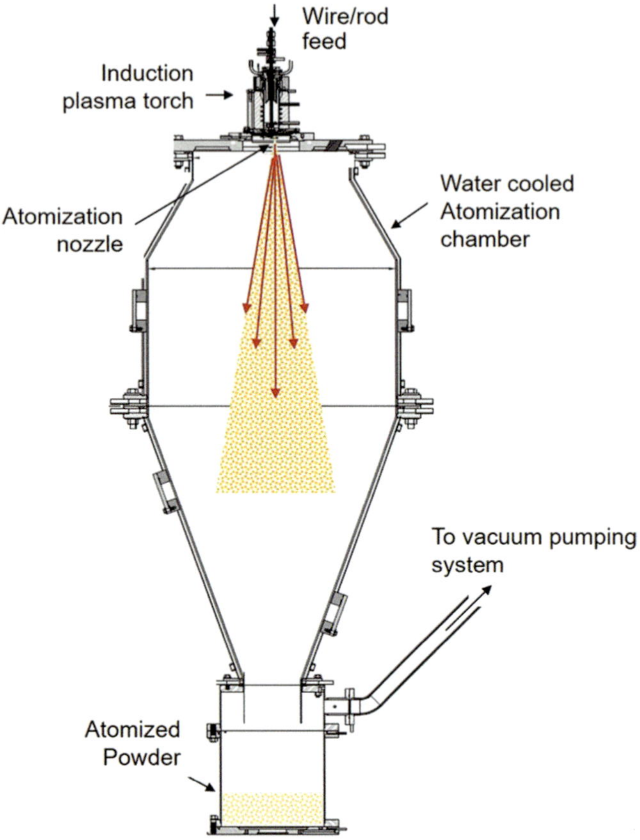

Fig. 13.67 Schematic of an RF Induction Plasma Atomization System [Boulos M et al. (2017a, b)]

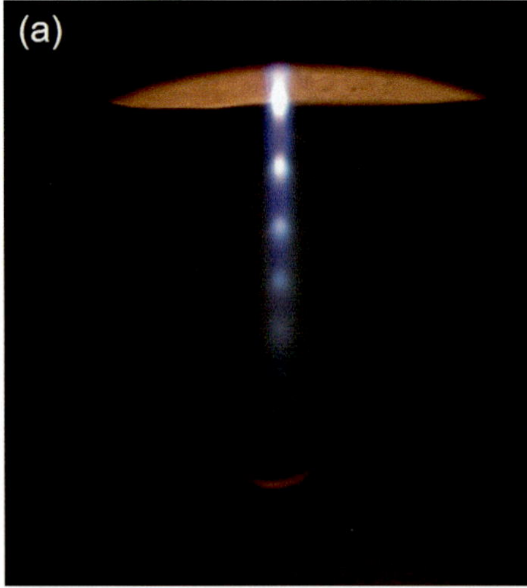

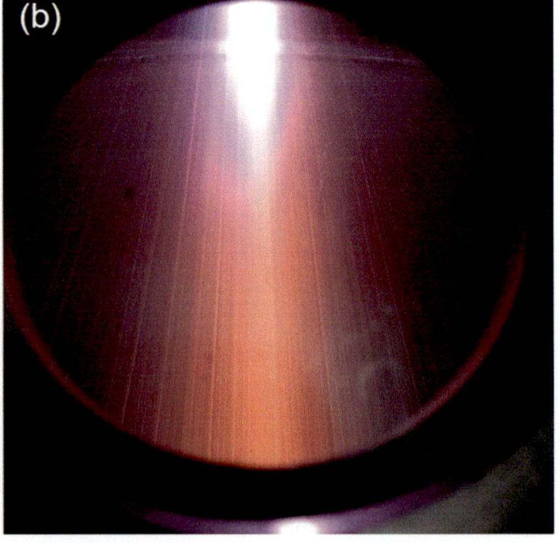

Fig. 13.68 Photographs of the plasma jet immerging from the induction plasma atomization torch nozzle (**a**) in the absence of the wire/rod feed and (**b**) in the presence of the wire/rod feed. [Photographs Courtesy of Tekna Plasma Systems Inc.] Reprinted with permission

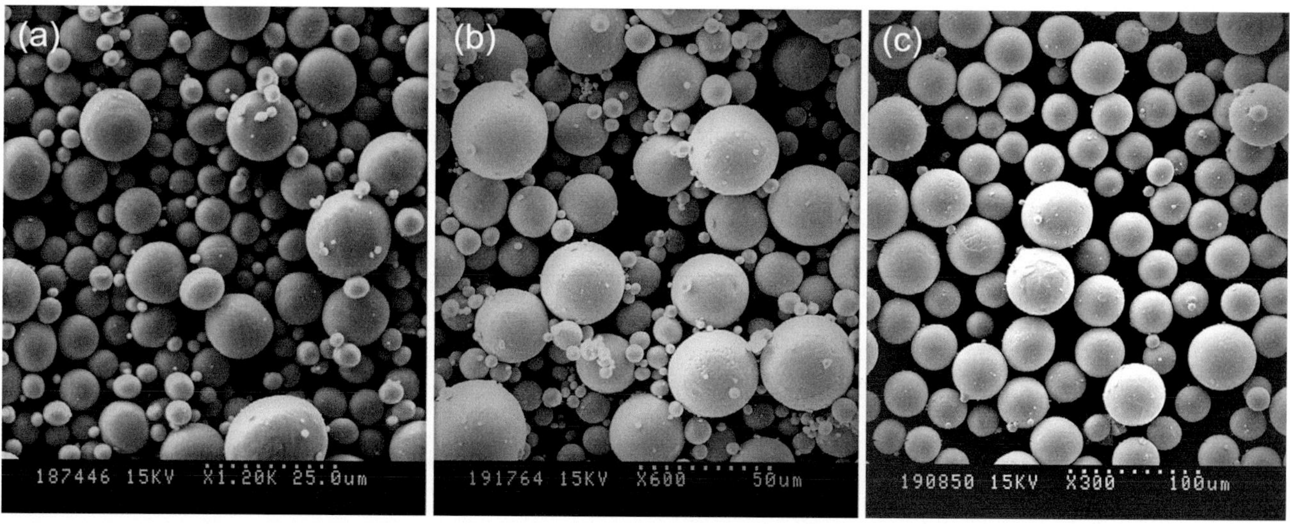

Fig. 13.69 SEM images of Plasma-atomized Ti-6Al-4V powders using Tekna's RF-IPAP (**a**) +5–25 μm, (**b**) +20–53 μm, and (**c**) +45–105 μm. [Images courtesy of Tekna Plasma Systems Inc] Reprinted with permission

Table 13.2 Technical characteristics of different size fractions of (Ti-6Al-4V) powders produced using the RF-IPAP [Data courtesy of Tekna Plasma Systems Inc.]

	d_{10} (μm)	d_{50} (μm)	d_{90} (μm)	App. Density g/cm^3	Tap. Density g/cm^3	Hall flow s/50 g	Oxygen ppm
+5–25 μm	9	15	24	≥ 2.10	≥ 2.70	No flow	1300
+20–53 μm	29	39	56	≥ 2.42	≥ 2.80	≤ 34	1000
+45–105 μm	50	68	96	≥ 2.45	≥ 2.80	≤ 29	500

deionized water, based on the volume fraction ratio of 50:50 of the oxides. After mixing for 10 h, the solution was dried in an oven at 200 °C to form a white gel, then heated in an oven to 600 °C with a heating rate of 20 °C/min. During the heating process, rapid combustion occurred, and voluminous fluffy powder was formed. After decarbonization at 600 °C for 2 h, white powder was produced and collected. A ball-milling process with ZrO_2 balls allowed breaking down large agglomerates into micrometer-sized particles. MgO and Y_2O_3 crystals (10–20 nm) were created. The $MgO–Y_2O_3$ nano-powder feedstock was successfully deposited on substrates by APS without the use of spray dry agglomeration.

[Chen D et al. (2009)] prepared $Al_2O_3–ZrO_2$ amorphous powders by a chemical reaction of aluminum nitrate Al$(NO_3)_3 \cdot 9H_2O$, and zirconium acetate $ZrO(OOCCH_3)_3$, in deionized water-based solution to produce a ceramic composition of Al_2O_3–40 wt.% ZrO_2. The gel is formed as the solution is heated to 80 °C with continuous stirring. The dried gel powder is heated at a rate of 10 °C/min up to 750 °C and maintained at this temperature for 2 h. The powder obtained was amorphous with no crystalline peaks detectable by XRD analysis.

[Prakash S et al. (2012)] explored the possibility of synthesizing thick film of gadolinia-doped ceria (GDC) using solution combustion synthesis approach. Nanoparticle-sized GDC powder prepared by using oxalyl dihydrazide fuel exhibited higher sintering activity giving dense pellet with high conductivity (3×10^{-4} S/cm at 400 °C). These powders exhibited necessary characteristics suitable for wet powder spraying. Due to the high reaction temperature agglomerated particles were formed on using glycine as fuel. Use of ammonium acetate fuel along with reduced amount of glycine yielded micron-sized, plasma spray-able grade GDC powder. These powders resulted in a dense atmosphere plasma-sprayed coating with superior inter-splat adhesion.

13.3.5.2 Self-Propagating High-Temperature Synthesis (SHS)

[Borisova AL and Borisov Yu S (2008)] published recently an excellent review of the Self-propagating High temperature Synthesis (SHS) process in which the reaction is locally initiated in a thin layer of a reactive mixture of reagents. Once initiated, it is self-propagating, in a wave of gasless combustion, over the entire system through heat transfer from hot reaction products to the next layer of the initial mixture. The thermal activity of the reaction is sufficient to maintain the process until completion. The approach was first used to obtain powders of refractory inorganic compounds such as carbides, borides, nitrides, silicides, and intermetallic. The approach is essentially a two-step process involving the heating of the SHS reactants-mixture in a furnace over the necessary reaction-triggering temperature, followed by the crushing/grinding of the formed cake to produce a powder with the required particle size distribution. Numerous experiments on metal–metal systems in which inter-metallics form (Ni-Al, CuAl, Ni-Ti, etc.) and on more complex metal-oxygen-free refractory compound composites (Ti-SiC, Ti-Si$_3$N$_4$, Cr-SiC, etc.) have shown that the abrupt peak of exothermal interaction in differential thermal curves mainly coincides with the moment when the system reaches the melting temperature of the low-melting component or reaction product (e.g., low-melting eutectic) [Borisova AL and Borisov Y (2008)].

It is also possible to combine the *SHS synthesis* with the coating deposition step [Borisova AL and Borisov Y (2008), Borisov Y et al. (1986), Borisov Y and Borisova A (1992), Deevi SC et al. (1997), Haller B et al. (2005), Dallaire S (1982), Legoux JG and Dallaire S (1993), Dallaire S and Cliche G (1992)]. In this case, the basic reactants are either as agglomerated or cladded to the required particle size distribution. As the powder is injected into the plasma and the individual particle entrained by the flow, they are heated attaining rapidly the reaction-triggering temperature, casing the reaction to start in the periphery of the particle. From this point on the reaction propagates toward the particle center. Depending on the particle size and its residence time in hot gases, the reaction may not be completed by the time the particle impacts on the surface of the substrate. In such a case, the reaction continues until completion within the deposited particle.

The spraying and synthesis of the material sprayed, which are accompanied by an exothermic effect, were first combined by the Metco Company that used nickel–aluminum composite in the form of aluminum particles coated with nickel [Borisov Y et al. (1986)]. Owing to the clad particle structure, components can interact through dissolution when one of them turns into melt. However, the SHS reaction is more efficient when the particle consists of uniformly mixed exothermically reacting components, as illustrated in Fig. 13.70.[Borisova AL and Borisov Y (2008)] calculated the evolution with time of the extent of interaction, called α, for the 80Ni–20Al-cladded particles and the composite one (for two speeds of the progression of the reaction front 23.5 and 47.0 mm/s). Results, presented in Fig. 13.71, show that in the composite particles, the SHS interaction substantially intensifies the reaction and heat releases in a shorter period of time, which increases the absolute temperature rise due to the internal heat source. Of course, to heat particles without fragmenting them implies that the produced gas expansion during the reaction can escape the particle where SHS reaction occurs.

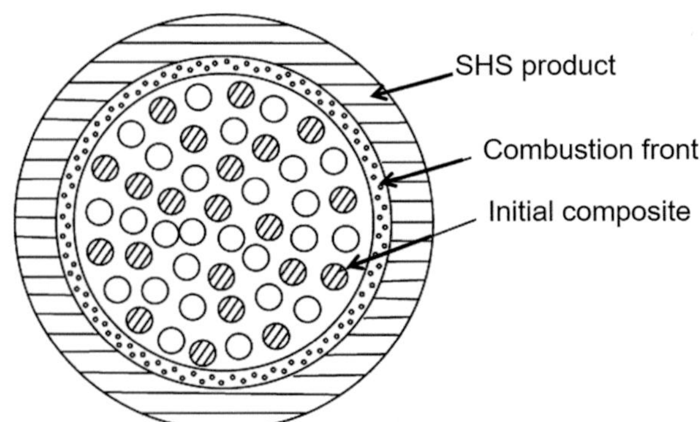

Fig. 13.70 SHS in a spherical powder particle [Borisova AL and Borisov Y (2008)]. Reprinted with kind permission from Springer Science Business Media

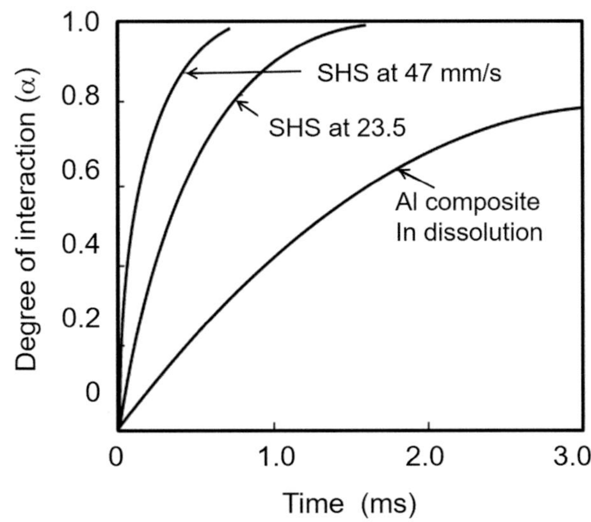

Fig. 13.71 Kinetics of the interaction of components in 80 Ni-20 Al composite in dissolution, and combustion (SHS) at 23.5, and 47.0 mm/s [Borisova AL and Borisov Y (2008)]. Reprinted with kind permission from Springer Science Business Media

An example of structure of Cr-SiC composite powder particle and the typical structure of the plasma-sprayed coating are shown in Fig. 13.72. The coating consists of Cr$_7$C$_3$, Cr$_{23}$C$_6$, Cr$_5$Si$_3$–xC$_x$. High hardness and strength of adhesion to steel (σ_{adh} = 26–29 MPa) and wear resistance surpass that of molybdenum coating in oil-abrasive medium and twice that of galvanic chromium coating in dry friction.

An extensive review of such composite powers and coatings is presented by [Borisova AL and Borisov Y (2008)] including transition metal-nonmetal refractory compounds with mixtures of Cr and SiC or B$_4$C, Ti, and SiC or B$_4$C or Si$_3$N$_4$. Silicide, carbide, and boride coating produced exhibited excellent resistance to wear [Borisov Y et al. (1986), Borisov Y and Borisova A (1992), Haller B et al. (2005), Deevi SC et al. (1997), Dallaire S (1982), Legoux JG and Dallaire S (1993)]; copper-TiB$_2$ coatings

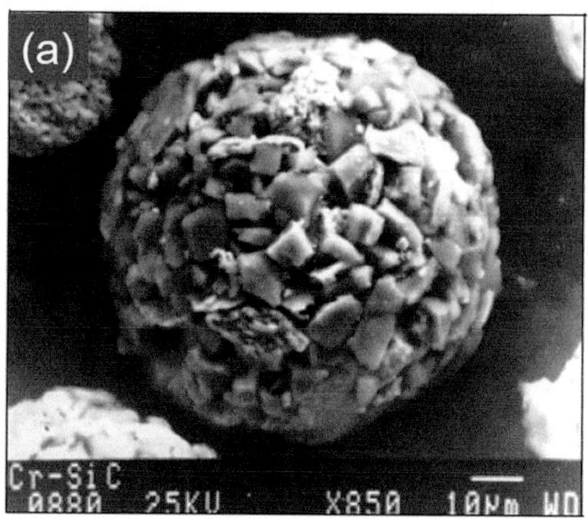

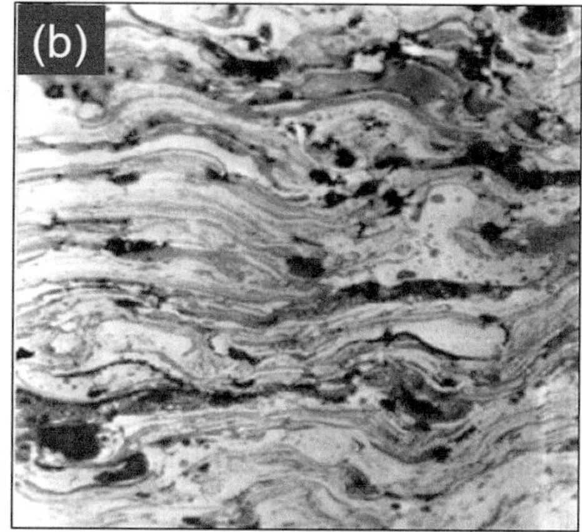

Fig. 13.72 External view of particles in Cr + SiC composite powders conglomerated with organic binder (**a**) and microstructure of a plasma coating of this powder (**b**) × 800 [Borisova AL and Borisov Y (2008)]. Reprinted with kind permission from Springer Science Business Media

Table 13.3 Techniques commonly used for the synthesis of nanopowders [Boulos MI (2002)]

Technique	Physical concept	Productivity	Capital investment	Material
Sol-gel	Wet chemistry	High	Low	Oxides
Combustion	Vapor phase react.	High	Medium	Oxides
Laser techniques	Vapor/cond.	Low	High	Metals/ceramics
Exploding wire	Vapor/cond.	Low	Low	Metals
Induction plasma	Vapor/cond.	High	High	Metals/ceramics
Transferred arc	Vapor/cond.	High	High	Metals/ceramic

starting from agglomerated particles of Ti–bronze and boron, the TiB_2 particles in coatings increasing considerably the hardness of copper matrix [Legoux JG and Dallaire S (1993)]; TiB_2 or TiC particles in ferrous matrices to increase their hardness and wear resistance [Dallaire S and Cliche G (1992)].

13.3.6 Nanosized Powder Synthesis

13.3.6.1 Basic Concept

With the growing acceptance of Suspension Plasma Spraying (SPS) as a means of spraying high performance nanostructured coatings [Fauchais et al. (2008a, b) (2010) (2011) (2013) (2015a, b)], a growing interest has immerged for the synthesis of ultrafine, nanosized powders of metals and ceramics [Boulos MI (2002), Guo et al. (2010), Boulos (2006b) (2009) (2011) (2015)]. The production of such powders of pure metal, alloys, or ceramic such as oxides, carbides, nitrides, or borides with a particle size in the range of a few hundreds nm is mostly carried out using a "bottom-up" approach rather than the more conventional "top-down" approach.

The principal limitation of the "top-down" approach involving the grinding and milling of the starting material from relatively large chuck to progressively finer powders is its intensive energy requirement and the significant risk of product contamination in the process. The bottom-up approach, on the other hand, involves the use of wet chemistry through which the reactants are activated in solution leading the formation of the reaction product as a fine precipitate or gel as described earlier in Sect. 13.3.5.1 "Sol-Gel processes." By adjusting the concentration of the formed nanosized powder/precipitate suspension, and the addition of an appropriate suspension stabilizing agents, the suspension could then be used directly in a thermal spray operation.

An alternate, bottom-up approach which has been gaining increasing attention is the use of thermal plasma technology for the complete vaporization of the precursor material, whether in the form of micron sized powder or organometallic compound, followed by the condensation of the formed metallic vapors under controlled conditions to the appropriate particle size distribution. The formed nanopowder, being reactive due to its high specific surface area, needs to be kept under inert atmosphere, or recovered in the liquid limitation forming a suspension with the required solid content to which surface active agents need to be added in order to stabilize the suspension which could then be used directly in a thermal spray operation.

A comparative analysis of alternate nanopowder synthesis techniques is presented in Table 13.3 highlighting the different approaches used and principal process characteristics. It may be noted that Sol-gel and combustion route processes are most adapted for the synthesis of oxide-type nanopowders. In the latter case, the precursor can be in the form of organo-metallic or a chloride. The synthesis of high purity nanopowder of silicon oxide (synthetic silica), or titanium dioxide, through the oxidation of Silicon tetrachloride or titanium tetrachloride respectively represent typical examples of the technology. Laser and exploding wire approaches are alternate routes which allow for better precision in terms of the PSD of the nanopowder obtained at the expenses of a significantly reduced productivity which make them mostly suitable for small-scale experimental studies. Thermal plasma technology offers on the other hand a viable approach which is scalable to industrial production levels for high quality powders. While the capital investment is relatively high, the technologies offer the advantage of high level of flexibility in terms of the precursor material and the chemical nature of the powder produced.

13.3.6.2 Plasma Synthesis of Nanopowders

As schematically illustrated in Fig. 13.73, the synthesis of nanopowders using thermal plasma technology is mostly carried out through the vaporization of the precursor material followed by the rapid quench of the formed vapors under controlled conditions in order to condense the vapors in the form of an ultrafine aerosol with a controlled morphology, crystal structure, and particle size distribution. Two techniques have been mostly used for feedstock vaporization. These are essentially base on either the in-flight vaporization of the feed material in the form of fine powder in a DC or RF inductively coupled discharge, or the vaporization of the precursor from a molten metal pool in a transferred arc plasma furnace.

The use of in-flight particle heating and vaporization of the feedstock for the synthesis of nanopowders has the main advantage of being essentially applicable to all materials without distinction to their physical, chemical, or electrical properties. The high temperatures and energy densities characteristic of thermal plasmas allow for the vaporization of even the most refractory materials under inert, oxidizing, or reducing atmosphere depending on the materials need. Refractory metals such as tungsten and tantalum can be successfully vaporized using these techniques in an argon-hydrogen plasma. Ceramic oxides on the other hand are best vaporized using an air plasma, an argon–oxygen plasma, or in some cases a pure oxygen discharge. As mentioned earlier, the plasma source to be used with this approach can be either a DC plasma jet or an inductively coupled RF discharge. The absence of electrodes in the case of the induction plasmas allows for more flexibility in the choice of the composition of the plasma gas used in comparison to DC plasma jets. Because of the generally lower plasma velocities associated with RF-ICP induction, plasmas also

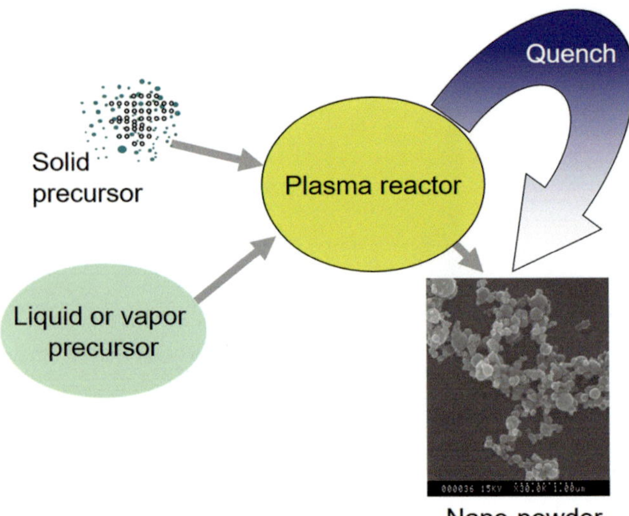

Fig. 13.73 Schematic representation of the plasma synthesis of ultra-fine nanostructured powders [Boulos MI (2002)]

allow for a longer contact time between the plasma and the feed powder resulting in a more complete in-flight vaporization of the feed material. A typical example of aluminum nanopowder produced through the ICP route is given in Fig. 13.74. The PSD of the powder as measured using light scattering technique given in Fig. 13.74a reveals a narrow particle size distribution, based on volume/mass fraction, with a mean particle diameter of $\overline{dp} = 136$ nm and a standard deviation of $\sigma_p = 18$ nm. As mentioned earlier in this chapter, Sect. 13.2.2.3 nanopowders can have a relatively high specific surface area which can be measured using the gas adsorption BET technique. The BET specific surface area value measured for the aluminum nanopowder is 17.2 m^2/g, with a corresponding mean parti-cle diameter of $\overline{d_{BET}} = 130$ nm, which is quite close to the value obtained by the light-scattering technique. SEM of the powder is given in Fig. 13.74b. The image magnification in this case is ×30,000 with the indicated scale bar corresponds to a length of 1 μm. The individual nanoparticles seem in this case reasonably spherical with a limited level of agglomeration.

The question of completion of feed material vaporization step is of critical importance in this approach for the synthesis of nanopowders due to the fact that any non-vaporized pre-cursor material will remain with the gas stream at the exit of the vaporization section of the reactor and will invariably find its way to the quench section where the nanosized aerosol is formed through condensation of the precursor vapors. This will result in the formation of an aerosol with a bimodal particle size distribution, as shown in Fig. 13.75a, obtained for tantalum nanopowder produced using the RF-ICP route ($T_m = 2996$ $^\circ$C, $T_v = 6100$ $^\circ$C) with SEM images of the powder given in Fig. 13.75b. It may be noted that in the

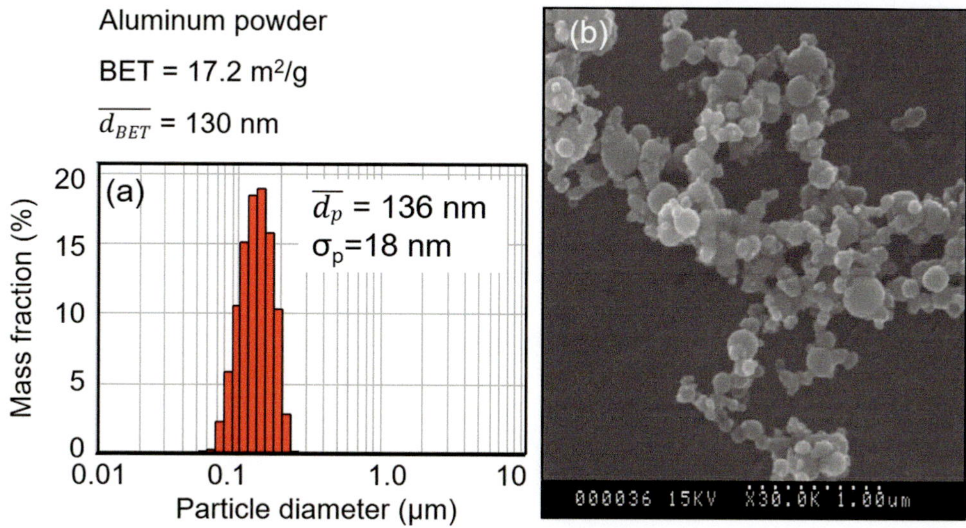

Fig. 13.74 (**a**) PSD of aluminum nanopowder obtained using light scattering and (**b**) SEM image of the powder [data and image supplied by Tekna Plasma Systems Inc.], reproduced with kind permission

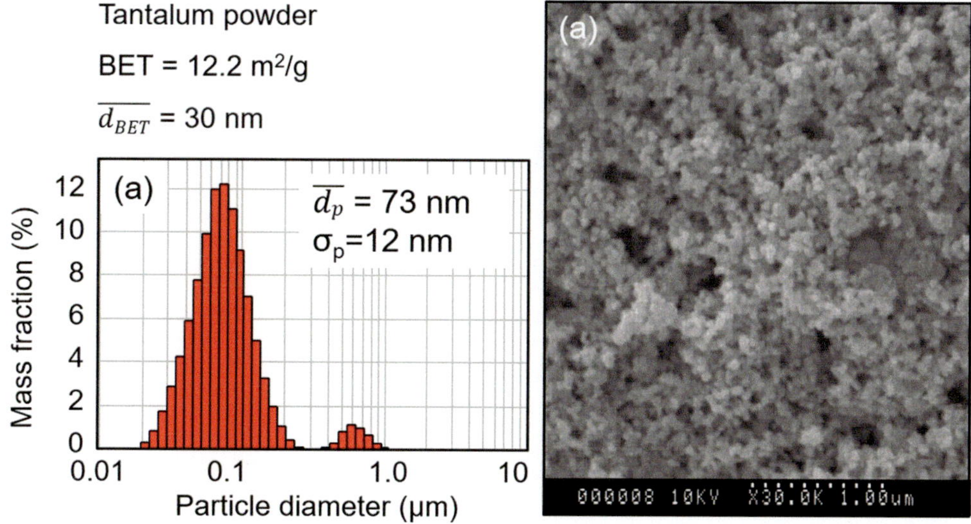

Fig. 13.75 (**a**) PSD of tantalum nanopowder obtained using light scattering and (**b**) SEM image of the powder [data and image supplied by Tekna Plasma Systems Inc.], reproduced with kind permission

bimodal PSD shown in Fig. 13.75a, the larger peak centered around 0.1 μm is representative of nanopowder formed through the vapor condensation process, while the second smaller around 0.6 μm is due to the residual partially vaporized precursor material. It is important to point out that the statistical mean particle diameter of $\overline{d_p} = 73\ nm$ is no more valid since it implies the assumption of a "normal" PSD, which is obviously not the case. A more realistic value would be in this case that obtained using the BET measurement, $\overline{d_{BET}} = 20\ nm$.

Examples of the effect of gas quench flow rate on the morphology and size distribution of the powder produced using the RF-ICP route can be noted in the SEM micrographs given in Fig. 13.76 for aluminum (Fig. 13.76a, b and c) and Germanium oxide (Ge) powders (Fig. 13.76d, e and f). A summary of BET specific surface area measurement for each of these two powders, and the corresponding mean particle diameter, $\overline{d_{BET}}$ for increasing levels of the quench gas flow rates are given in Table 13.4. These show mean particle diameters in the range of 20 to 100 nm with, as expected, decreasing mean diameter of the powder with the increase of the quench gas flow rate.

One of the principal concerns in the development of an appropriate technology for the synthesis of nanopowders is the ability to exercise a proper control on particle agglomeration and on the surface reactivity of the powder. The latter is

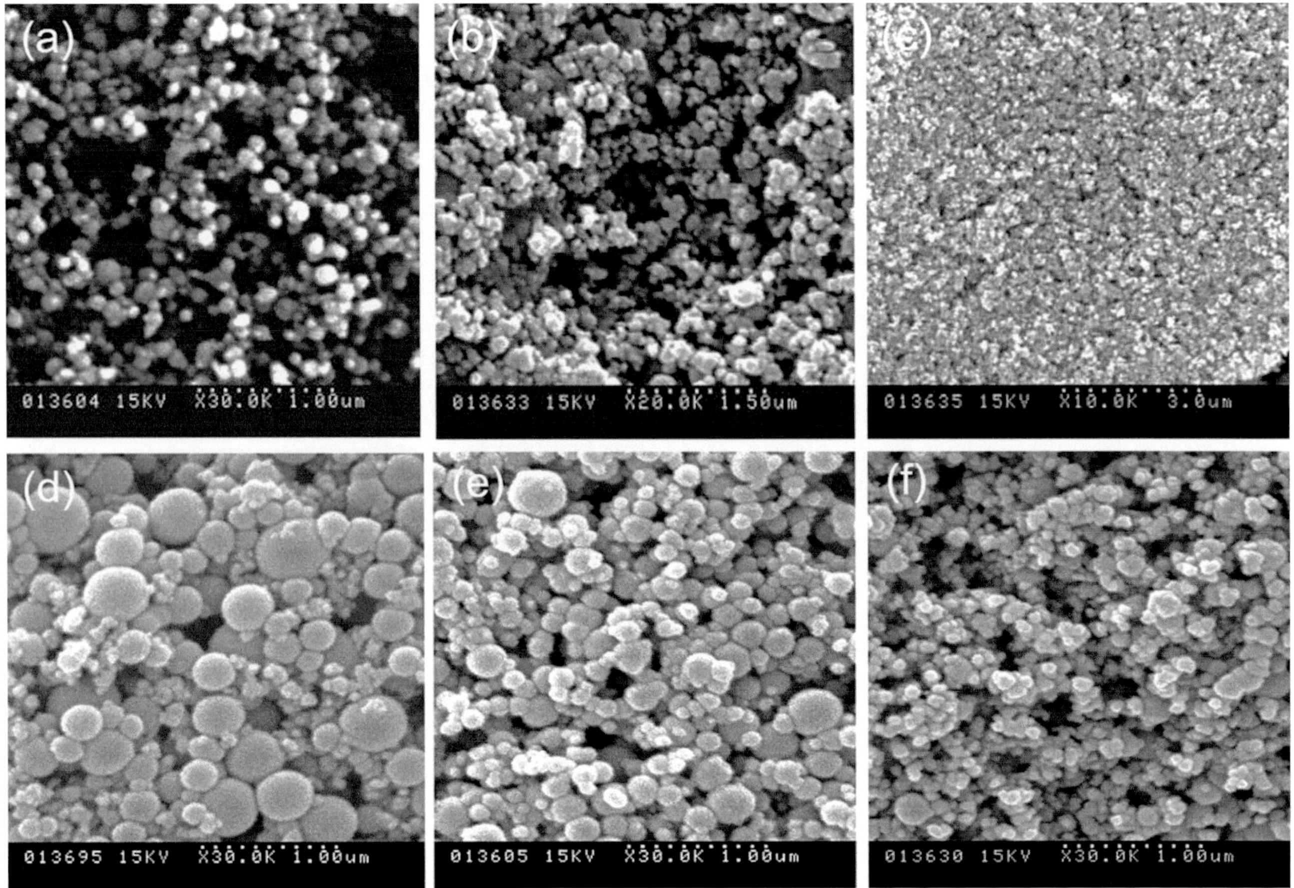

Fig. 13.76 (**a**, **b**, **c**) SEM of aluminum nanopowder obtained at increasing quench gas flow rates (**d**, **e**, **f**) corresponding SEM images for GeO nanopowder [data and image supplied by Tekna Plasma Systems Inc.] reproduced with kind permission

Table 13.4 Characteristics of aluminum and germanium oxide nanopowders function of the quench gas flow rates

Quench gas flow rate	Aluminum		Germanium Oxide	
	Sp. surface area (m^2/g)	$\overline{d_{BET}}$ (nm)	Sp. Surface area (m^2/g)	$\overline{d_{BET}}$ (nm)
I	21.5	100	9.1	105
II	36.5	60	11.0	87
III	96.5	25	23.0	41

important for metals which, because of the relatively high specific surface area of nanopowders, can be sufficiently reactive to ignite spontaneously when exposed to the air. Figure 13.77a, b shows an electron micrograph of a Fe$_{50}$-Co$_{50}$ nanopowder produced through the induction plasma vaporization of a mixture of iron and cobalt powder in an Ar/H$_2$ plasma. The upper part of the figure (Fig. 13.77a) shows the produced nanopowder without any surface treatment, while the lower part of the figure (Fig. 13.77b) shows the same powder with a carbon coating produced during the synthesis step through the addition of a carbon source, such as acetylene, into the quench section of the reactor. The protective action of the deposited carbon layer is evident from the associated X-ray diffraction patterns shown in Fig. 13.77c which reveals no oxidation of the carbon-coated

powder (top part of Fig. 13.77c) in contrast to the heavily oxidized nature of the non-coated powder (bottom part of Fig. 13.77c).

It is to be noted, however, that the problem of oxidation and the need of surface treatment of the powder is limited to metallic nanopowders. Moreover, the use of metallic or ceramic powders in suspension plasma spraying does not require special powder surface treatment since by keeping the powder in suspension in a liquid, its surface is protected from contact with the ambient air. Attention should be given, however, to the use of appropriate surface-active agents which are necessary to stabilize the suspension and avoid rapid sedimentation and segregation of the powder.

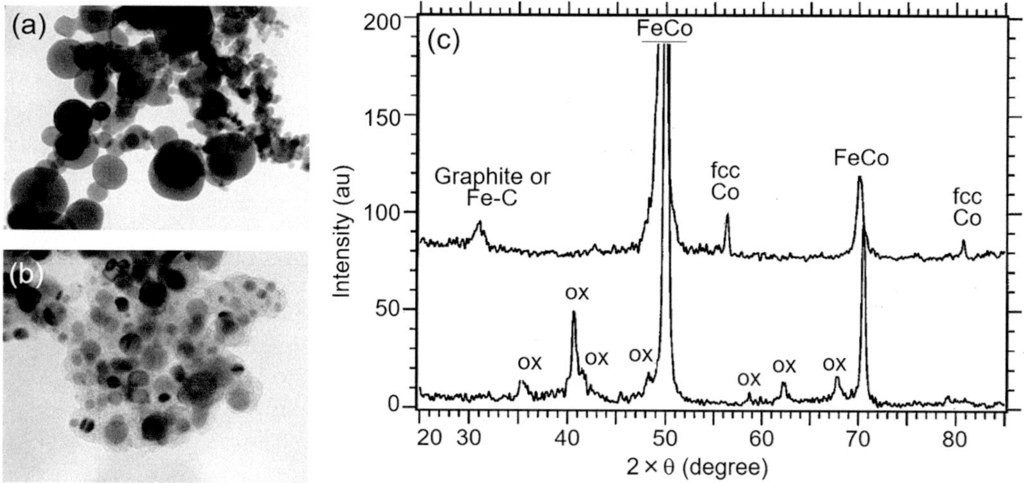

Fig. 13.77 (a) SEM of non-coated Fe$_{50}$-Co$_{50}$ nanopowder, (b) carbon-coated powder, and (c) X-ray diffraction of non-coated (top) and carbon-coated (bottom, Fe$_{50}$-Co$_{50}$ nanopowder [Scott et al. (1997)]

13.3.7 Polymer Powders

Polymers are often deposited directly from dry powders to avoid the use of organic solvents. Electrostatic spraying and fluidized-bed dip coating are considered to be the two primary methods for polymer powder deposition [Ivosevic M, et al. (2003), Åström BT (1997)]. Both methods are based on capillary fusion of powder particles on either preheated substrate surfaces or during baking. That is why thermal spraying has been successfully used to deposit polymer coatings [Petrovicova E et al. (2002)]. According to [Ivosevic M (2006)] the key advantages of using thermal spray are the following:

- Solvent-free deposition
- Possibility to coat large objects in the field under different environmental conditions
- Ability to process polymers with high melt viscosity ($>10^2$ Pa s)
- Ability to produce "ready-to-use" coatings without the need for post-deposition processing
- Possibility to spray polymer/ceramic composites with high reinforcement contents (>5 vol.%)

Unfortunately, these advantages are counterbalanced by:

- Lower deposition efficiency than conventional means, lower quality surface finish, higher process complexity, and jet/flame temperatures
- Potentially degrade polymeric feedstock or previously deposited material with hot gases of the spray process (plasma or flame)

- Polymers physical, thermal, and mechanical properties are very different from those of metals, alloys, ceramics, and cermets, thus requiring adapted spray conditions

Thermally sprayed polymer coatings are increasing in importance as protection against corrosion (gasoline, diesel oil, alcohol, salt spray, more or less acidic liquids) and wear (impact, abrasion, cavitation, sliding) due to their low friction coefficient.

Compared to conventional powders (metals, alloys, ceramics, cermets), polymer particles have very different properties. Polymer thermal conductivity is lower (a few tenths of W/m K) than that of other materials, even lower than that of zirconia (about 1 W/m K). Thus, during its flight in the hot gases, steep temperature gradients will develop within the particle with the core unmolten and the surface overheated. For relatively small particles, overheating can result in polymer degradation,

According to the rather low mass density of polymers (typically between 1000 and 2200 kg/m^3) particle velocities achieved prior to impact will be higher than those of conventional particles for the same particle size and trajectory in the same flame or plasma jet. Polymers typically have three to seven orders of magnitude higher melt viscosity and an order of magnitude lower surface tension than metals [Ivosevic M et al. (2006)]. The viscosity is strongly influenced by the polymer molecular weight, M_c, especially over a critical value of M_c. With metals, at the end of particle flattening the inertia force is balanced by the surface tension force (high flattening Weber number), while for polymer particle it is the viscous flow resistance (high flattening Reynolds number). Moreover, the viscosity of molten polymers is often non-Newtonian with shear thinning properties, implying that the apparent viscosity is a function of both the temperature and shear rate.

The kinetics of spreading during splat formation and solidification of polymer particles are also significantly different from those of conventional particles [Ivosevic M et al (2006a)]. For example, under typical flame spray conditions, the Peclet number, Pe, for a zinc droplet is in the range 1–10, whereas Pe for a Nylon-11 droplet is almost three orders of magnitude higher ($Pe \approx 3\,000$). Accordingly, on impact of a molten Nylon-11 droplet on a substrate, splat formation and cooling take place over two significantly different time scales. By definition, $Pe = t_c/t_s$ with the heat conduction time $t_c = d_p^2/\alpha_p$, d_p being the particle diameter (m) and α_p its thermal diffusivity (m²/s), while the droplet spreading time $t_s \approx d_p/v_p$ with v_p the particle impact velocity (m/s).

At last larger thermal expansion coefficients of polymers (values up to 150×10^{-6} K^{-1}), compared to those of metal substrates, require spraying rather thin passes. The other problem is that polymer powders sprayed by electrostatic spraying and fluidized-bed dip have size distributions that are not adapted to thermal spray, requiring sizes approximately between 50 and 200 μm [Soveja A et al. (2010)]. Thus, in the available powders those below 50 μm must be eliminated and the bigger ones cryo-milled in liquid nitrogen. The different thermal sprayed polymers are the following:

Nylon 11 has been flame sprayed to study its slurry erosion resistance [Ivosevic M et al. (2006)], HVOF sprayed [Ivosevic M (2006)] to characterize the resulting splats and coatings, PEEK (Poly-ether-ether-ketone) was flame sprayed and laser remelted [Soveja A et al. (2010), Zhang C et al. (2009b)],flame sprayed to study its crystallinity [Zhang C et al. (2009b)], HVAF sprayed to study its degradation and stability [Patel K et al. (2010)], plasma sprayed [Wu J et al. (2010)], and HVAF [Withy BP et al. (2008)] sprayed to study splats formation.
PTFE (Polytetrafluoroethylene), polyamides, and UHMWPE (ultra high molecular weight polyethylene) were sprayed with high-energy plasma [Niebuhr D and Scholl M

(2005)] to study their low friction for high contact pressure rolling/sliding systems.
EMAA (Ethylene-methacrylic copolymer) was flame sprayed to study the splats formation [Xie W et al. (2013)].
PVDF (fluoropolymer, polyvinylidene fluoride), ECTFE (ethylene chlorotrifluoroethylene), PFA (perfluoroalkoxy alkane), and FEP (fluorinated perfluoroethylene-propylene) were flame and plasma sprayed against corrosion: salt test and liquid immersion tests (pH values −0.7 and 14) [Leivo E et al. (2004)].

According to Ivosevic [Ivosevic M (2006)] up to 2006 many other polymers have been thermally sprayed: Ethyltetrafluoroethylene (ETFE), Liquid-crystalline polymers (LCP), Polyethylene (PE), Polyaryletherketone (PAEK), Polycarbonate (PC), Polyester (PES), Polypropylene (PP), Polyimide (PMR), Polyethylene terephthalate (PET), Polyphenylene sulfide (PPS), Polymethylmethacrylate (PMMA), Bismaleimide-phenolic resin, Cyanate ester thermosets, Ethylene vinylene acetate (EVA), Ethylene-acrylic acid (EAA) copolymer, Polyamides (PA), Phenolic thermosets, Polyarylene sulfide (PAS) Polyester-epoxy resin, Polyether block amide copolymer, Polyether-amide (PEA) copolymer, Polyethylene-polypropylene copolymer, Polysulfone Polyvinylidene fluoride (PVDF) Post-consumer commingled polymers (PCCP), and since 2006, some others, especially fluoro polymers: A few properties of some polymers are summarized in Tables 13.5 and 13.6.

13.3.8 Composite Powders

Blended powders, generally of two-powder feedstock, are mechanically mixed and then fed to the spray gun through the same port. However, if one wants to have similar trajectories for both types of particles, the particle distributions of both powders must be adjusted according to their specific mass. Otherwise spray processes are excellent classifying media, resulting in quite inhomogeneous

Table 13.5 Properties of different fluoropolymers used in the thermal spray industry [Leivo E et al. (2004)]

Property	PVDF Polyvinylidene fluoride	ECTFE Ethylene chloro trifluoro ethylene	PFA Perfluoro alkoxy alkane	FEP Fluorinated perfluoro ethylene propylene
Mass density (kg/m³)	1780	1680	2150	2150
Melting point (°C)	155–170	240	300–310	250–280
Max service T (°C)	150	166	260	205
Hardness (shore D)	D70–80	D75	D63–65	D55_66
Tensile strength (MPa)	35–52	31–48	28–30	21–28
Flexural strength (MPa)	59–75	48	–	21
Water absorption ASTM D-570%	0.04	<0.01	<0.03	<0.01

Reprinted with kind permission from Elsevier

Table 13.6 Thermophysical properties of selected polymer powders, data according to powder suppliers

Property	Nylon 11	PET	PEEK	PEK	PPS	PES
Mass density (kg/m^3)	1040	1340–1400	1320	1300	1400	1370
Melting point (°C)	175–191	>250	343			
Max service T (°C) Vicat[a]	140–160	170	160			
Hardness (MPa)	100 (R)		28 (HV)	34.4 (HV)	26.5 (HV)	50 (HV)
Tensile strength (MPa)	15–44	55–75	90–100	110	86	90
Flexural strength (MPa)	51.6–89.6		185	183	145	120
Glass trans. T (°C)	42	75	143–148	162	90	225
Water absorption ASTM D-570%	0.230–1.9	0.16	0.1			
Thermal conductivity (W/m K)	0.230–0.346	0.15–0.24	0.25			
CTE (μm/K m)	100–150	70				

[a]Vicat: determination of the softening point for materials that have no definite melting point, such as plastics. It is taken as the temperature at which the specimen is penetrated to a depth of 1 mm by a flat-ended needle with a 1 square mm circular or square cross section. For the Vicat A test, a load of 10 N is used. For the Vicat B test, the load is 50 N

coatings. The best way is to inject dissimilar particles at two different locations (of course it requires two powder feeders and two injection ports judiciously disposed).

One of the main difficulties when spraying particles made of a metal matrix reinforced by carbides is to avoid or at least limit the decarburization. A typical example is that of WC–Co with WC particles in the few micrometer size range, generally produced by spray drying followed by heat treatment. When replacing the micrometer-sized WC particles by nanometer-sized ones, the high surface area of the nanocomposite WC/Co particles facilitates all types of high-temperature reactions, which makes not only the spraying but also the manufacturing of such particles difficult. To overcome the problem, two techniques are used for synthesizing finely divided carbide powders.

13.3.8.1 Spray Drying Solutions

According to [Seegopaul P et al. (1997)], the combination of solution chemistry with gas–solid reaction processing imparts significant advantages to the spray conversion process (SCP). The process starts from an aqueous solution of soluble salts and any compatible combinations of salts can be spray dried to yield the desired products, such as WC–Co, WCCo–VC, WC–Co–Cr$_3$C$_2$, WC–Co–VC–Cr$_3$C$_2$ mixtures. Gas phase carburization is then conducted in a fluid bed reactor, with a good control of carbon variation, to yield the nanocomposite hard-metal powder. During the cool down cycle, the high-surface area powder is passivated with oxygen-rich species to avoid spontaneous combustion when exposed to air [Kear BH et al. (2000)]. The as-produced powder consists of solid agglomerates, which are screened to a relatively narrow size distribution, typically within the range of 15–40 μm [Kear BH et al. (2000)]. [Seegopaul P et al. (1997)] use milling to break the large particles and ensure uniform mixing of the additives, such as grain growth inhibitors [Seegopaul P et al. (1997)]. Sintering is then performed in hydrogen atmosphere to remove adsorbed oxygen for reduced porosity and optimized carbon balance. Sintering temperature must be well controlled to avoid the destruction of the nanostructured particles [Seegopaul P et al. (1997)]. Enhanced WC grain growth occurs early in the sintering cycle, even below the temperature at which the liquid phase is formed. This growth can be largely restricted by the addition of VC [Schubert WD, Bock A and Lux B (1995)].

13.3.8.2 Integrated Mechanical and Thermal Activation

Ban and Shaw [Ban Z-G and Shaw L (2002)] have developed a novel approach, termed as the integrated mechanical and thermal activation (IMTA) process, which was used to synthesize nanostructured WC–Co powder. The basis of the IMTA process in making nanostructured WC–Co powders is to mechanically activate reactants WO$_3$, CoO, and graphite at room temperature through high energy milling (the mechanical activation step), followed by completing the synthetic reaction at high temperatures (the thermal activation step) [Ban Z-G and Shaw LL (2003)]. The amount of free carbon present in the composite powder can be precisely tailored in the IMTA process to prevent decarburization during thermal spraying. IMTA process uses low-cost materials while producing nanostructured WC–Co powder that has better powder characteristics (e.g., solid particles rather than hollow ones). Particle sizes ranging from 0.3 to 0.5 μm and an average WC grain size around 30 nm were obtained by IMTA. As these particles were not suitable for thermal spraying (where a range of 5–45 μm, with a mean size around 20 μm is required), they were agglomerated using methylcellulose as the organic binder. These particles HVOF sprayed [Ban Z-G and Shaw LL (2003)] showed that the ratio of oxygen-to-hydrogen flow rate and the starting powder microstructures had strong effects on decarburization of the nano-coatings. IMTA process has also been used to produce pure nanostructured carbides (SiC and TiC) [Ren R-M et al. (1998)] and nitrides (Si_3N_4, CrN, and TiN) [Shaw LL (2000)].

[Kurlov AS et al. (2011)] have synthesized by the plasma-chemical method with thermochemical treatment WC powder with an average particle size of 35 nm. They produced microcrystalline Co powder with an average particle size of 2.2 µm and nanocrystalline vanadium carbide ($VC_{0.87}$) powder obtained by high-energy grinding. Then a powder mixture containing 93 wt% WC, 6 wt.% Co, and 1 wt% $VC_{0.87}$ was prepared by multistage mixing in a planetary ball mill with a complex mixing procedure. The nanopowder mixture contained the main components (WC and Co) and inhibiting additives (0.3–0.5 wt.% nanocrystalline carbides VC, Cr_3C_2, and TaC). Then powers must be densified.

13.4 Powder and Suspension Feeding

13.4.1 Conventional Powder Classification

Powder classification is the final and rather critical step in the powder manufacturing process. It is also essential for optimal matching of the feed material characteristics with the Thermal Spray Process requirement. It is not surprising that significant attention has been to this step in the thermal spray industry [Allen T (1997), ASTM-B214-07 (2011) ASTM E2651-10, Barth H. (ed.) (1984), Berndt C (2007), Davis JR (ed) (2004), and Iinoya K et al. (1991)]. Sieving and air/inert gas classification are among the most used techniques for coarse and medium-size particles (dp > 45 µm, corresponding to 325 Mesh). For powders with mean particle diameter below 45 µm classification becomes increasingly challenging requiring the use of screens with ultrasound vibrators. Below 5–10 µm, screens are no more of an option and alternate techniques continue to be developed mostly based on air classification using high performance cyclones. The morphology of the individual particles also plays a key role in the classification process with the screening of spherical particles being the easiest compared to that of angular or needle-shaped particles.

13.4.1.1 Sieving

Sieving is the oldest method to sort particles according to size. Typically, sieves with decreasing mesh size are stacked with the largest mesh at the top (Fig. 13.78). A representative weighed sample is poured into the top sieve that has the largest screen openings. Each lower sieve in the column has smaller openings than the one above. The column is typically placed in a mechanical shaker. After the shaking is complete the material on each sieve is weighed. The particle size

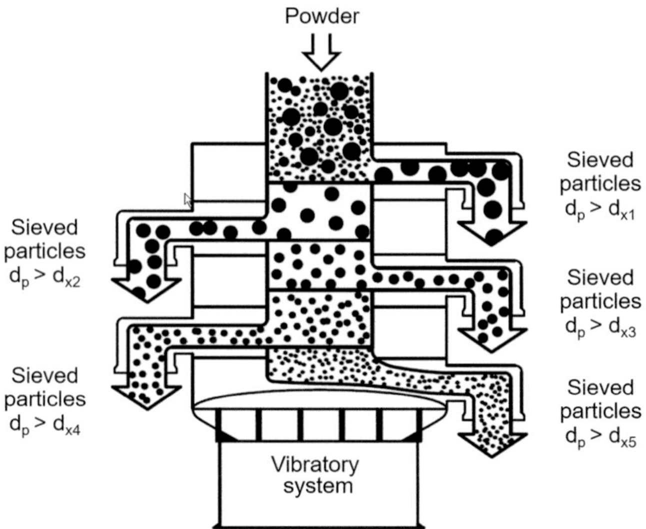

Fig. 13.78 Particle classification by vibratory sieving. Reprinted with kind permission from ASM International [Berndt C (2007)]

distribution corresponds to the weighing of amounts of powder in each sieve.

The term "mesh size" denotes the number of openings per linear length unit in the sieve screen. The National Institute of Standards and Technology (NIST) developed the sieve numbering system based on the "fourth root of two" ratio; this series is known as the ASTM E-11 standard (ASTM for American Society for Testing and Materials). This ratio (=1.189) means that the sieve openings are an exact geometric series. It can also be said that the mesh opening of two consecutive sieves has a 1.414 ($\sqrt{2}$) ratio. Table 13.1[Davis JR (ed) (2004)] gives the relationship between sieve (or mesh) number and mesh opening size, for the meshes used in thermal spray.

The normal wear (deformation, rupture, corrosion of the wires) must be checked regularly to keep the process reliable. However, one of the main problems is the embedding of individual particles in the sieve grids, especially for the small mesh opening, thus reducing the sieving efficiency and polluting the next batches. To limit this phenomenon each sieve is vibrated and recently a combination of ultrasonic and conventional vibration was developed which can be retrofitted onto any new or existing vibrating separator or screener. The acoustically developed transducer applies an ultrasonic frequency directly to the screener mesh to break down surface tension, effectively rendering the stainless-steel wires friction-free. Accurate separation down to 20 µm on even difficult powders can thus be achieved. The limit of such system is about 5 µm. However, in production, sieving is not used below 44 µm according to the too long times required and the risk of clogging. Instead of using dry sieving, wet sieving is interesting for materials that are not

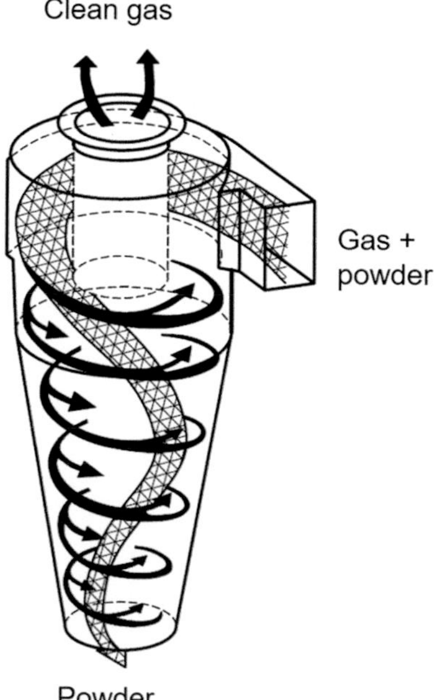

Fig. 13.79 Schematic of a cyclone used for air classification. Reprinted with kind permission from ASM International [Davis JR (ed) (2004), Berndt C (2007)]

sensitive to corrosion. Compared to dry sieving, wet sieving allows high throughputs under high-pressure conditions. It implies putting particles in a solution. It also allows separating particles down to a few μm.

13.4.1.2 Air Classification

The other method used employs cyclones and filters to separate particles according to their sizes [Davis JR (ed) (2004), Berndt C (2007)]. The principle of a cyclone is given in Fig. 13.79[Davis JR (ed) (2004), Berndt C (2007)]. The powder is introduced into a spiral-shaped air current flowing toward the center. The air vortex has suitable values of tangential and radial velocities so that a certain part of the powder is accelerated by the centrifugal force toward the periphery of the whirl, the other part being carried by the air current toward the center of the whirl by aerodynamic means. The shape, size, and specific mass of particles individually determine which direction they will take in the air current. The particle trajectories depend on their momentums when they follow a circular trajectory. The heavier particles will follow the gravity force and thus will be separated from the lighter ones following the swirl flow. The control parameters are the gas pressure drop, the cyclone dimensions, and the inlet particle concentration. It is possible to separate particles down to a few tenths of μm. This method is often used in production to classify particles with sizes below 44 μm, because screens below that value are impractical for production processes [Davis JR (ed) (2004)].

13.4.2 Powder Feeding

The powder feeder is as important as the spray gun to achieve reliable and reproducible coatings with required properties. It must produce a carrier gas stream in which the powder feedstock is entrained and delivered to the powder injector fixed on the spray gun. The powder issued from the powder feeder must have good flow ability, a regular and reproducible flow rate, whatever may be the particles size distribution, specific mass, and morphology. Of course, the carrier gas flow rate must be adjustable, independently of the particle mass flow rate, to transfer to particles the acceleration needed for optimum trajectories when radially injected or that adapted to optimum axial injection. Depending on the spray process used, the powder feeder must be capable of operation at low pressures, atmospheric pressures, or higher as needed for example in HVOF, HVAF, and cold spray processes.

Of course, all these requirements are difficult to satisfy, especially if the powder flow ability is poor and/or the particle size is below 10–15 μm. The conventional powder feeders (distributing particles with sizes larger than 10–15 μm) are generally categorized in three basic types: gravity-fed hoppers, volumetric powder feeders, and fluidized-bed powder feeders [Davis JR (ed) (2004)].

13.4.2.1 Gravity Fed Hoppers

This is the oldest type. It comprises a simple cylindrical container, often with a funnel-shaped bottom, in which the particles flow freely by gravity. The powder flow rate is controlled by a needle valve located in the funnel exit port. The powder falls in a pipe where the carrier gas flows. The feeder can be assisted by a vibration device to improve the powder flow. The system is of course very simple and inexpensive, but it works well only if the power flow ability is good, that is, with powders, with particles diameters above 10–15 μm, and with adapted morphologies (close to spherical). The powder flow rate is linked to the quantity of powder contained in the container and thus is not very accurate. Of course, such feeders cannot be pressurized. It is still used, mainly in powder flame spraying.

13.4.2.2 Volumetric Powder Feeders

They utilize a mechanism to deliver packets of powder in a carburetor, from where the carrier gas entrains them to the particle's injector. The different mechanisms involved are screws, slotted wheel, and disk with holes, as shown in Fig. 13.80. In these cases, the transport wheel rotates within the powder bed. Slots or holes are filled with powder from the container, scalpers scraping the excess of powder from the tops of slots or holes in order to feed the carburetor with a predetermined volume of material with each rotation of the wheel or disk. Holes or slots filling can be improved by

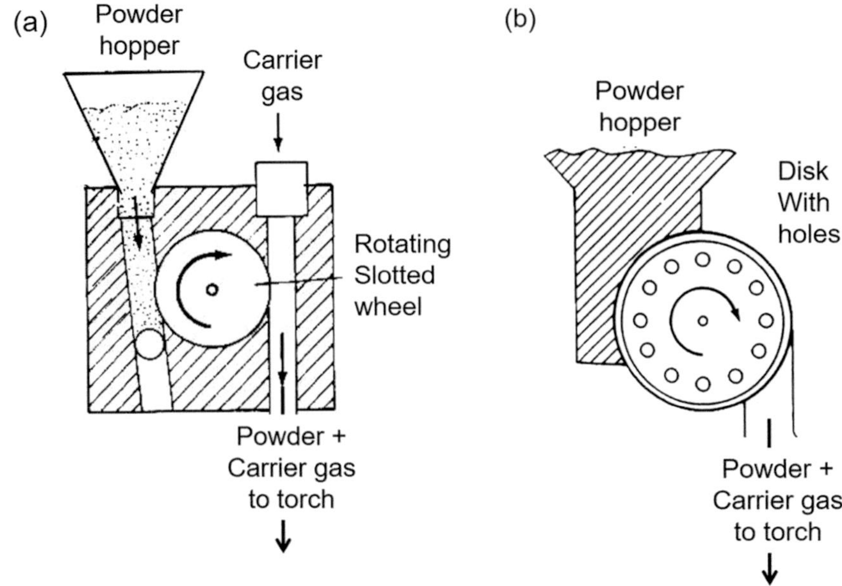

Fig. 13.80 Volumetric powder feeder (**a**) rotating wheel system and (**b**) rotating disk system. Reprinted with kind permission from ASM International [Berndt C (2007)]

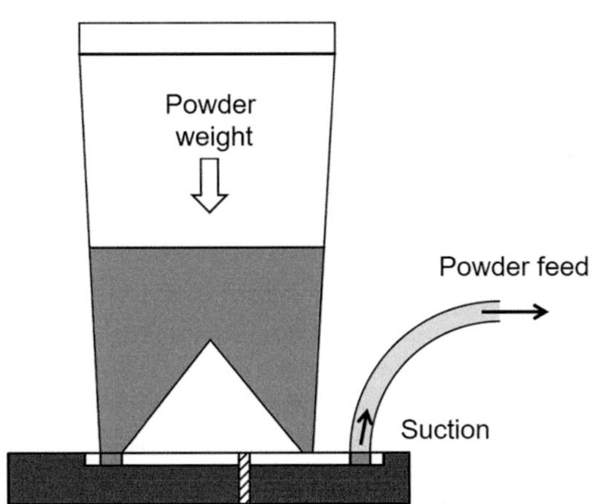

Fig. 13.81 Suction-type volumetric powder feeder. Reprinted with kind permission from ASM International [Davis JR (ed) (2004)]

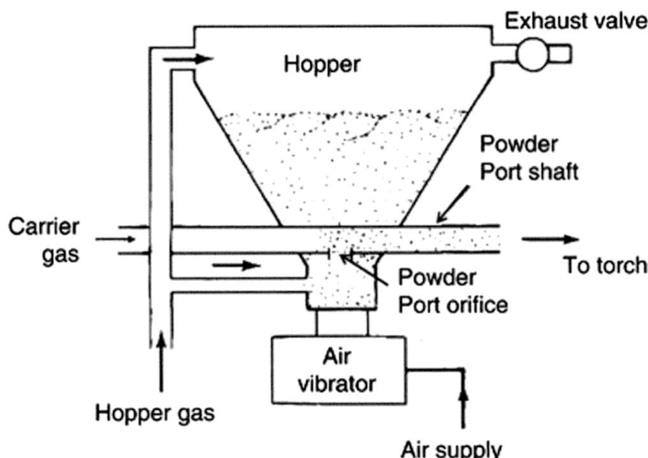

Fig. 13.82 Fluidized-bed powder feeder. Reprinted with kind permission from ASM International [Berndt C (2007)]

vibrating the powder feeder, especially for poor flow ability powder. For screw feeders, the end of the screw enters a tube that has the same function of the scalper to control the volume distributed by each rotation of the screw into the carburetor.

In most cases particles, fed by gravity into the carburetor, are entrained to the feed hose by a positive gas pressure, with a controlled flow rate. It is also possible to use suction-type volumetric feeder. They comprise a continuous circumferential groove in a powder-transport wheel, filled with powder by gravity [Davis JR (ed) (2004)], as shown in Fig. 13.81. A counter-rotating, inverter conical stirrer helps maintaining a more uniform powder bed density over the feed slots to pack powder into the groove and achieve a uniform stable filling. Powder is picked up by a suction Venturi, through which the

carrier gas flows, creating a vacuum that sucks the powder out of the transport wheel [Davis JR (ed) (2004)].

One of the main drawbacks of these powder feeders is the pulsation of the powder feed rate due to the periodic feeding of holes, slots, and screw pitches. Different solutions can be used to circumvent the problem: The use of a "cyclone mixer" where powder, exiting the carburetor, is fed tangentially and exit it axially, is an option which can help in damping feed rate pulsations under certain conditions. The use of multiple powder feeders operating out of phase feeding the same carburetor is an option to dampen powder feed rate pulsations, though at a considerable increase of complexity and cost.

13.4.2.3 Fluidized-Bed Feeders

In these systems the powder is continuously agitated in a fluidized bed by the fluidizing gas. A pipe with a hole, or port orifice, crosses the bed and the carrier gas entrains the powder at a uniform rate, as depicted schematically in Fig. 13.82 [Berndt C (2007)]. The bed must be continuously fluidized, and the volumes of gas and powder removed continuously replaced [Davis JR (ed) (2004)]. A gravity and/or vibrator feeder is used to maintain the flow of the powder from the hopper into the bed. While such powder feeders are less sensitive to particles morphology, they are more dependent of particle size distribution and mass density. This is due to the way fluidized beds work: particles are fluidized when the pressure drop across the bed is comparable with the weight of the bed. Moreover, for powders with a broad particle size distribution, the smaller and lighter particles will have a tendency to be transported outside the fluidized bed and entrained with the fluidizing gas. The separate control of fluidizing and carrier gases is also trickier.

13.4.2.4 Powder Feeders for Small Particles

To feed powders with particle sizes below 10–15 µm is quite a challenge due to their poor flow ability and their natural tendency to agglomerate and stick together forming bridging and "rate-holes" in the hoppers. Conventional powder feeders cannot be used in these cases and a wide range of novel approaches have to be developed for such powders on a case-by-case basis. The "powder pump" is a typical example developed for the transport of fine powders, $d_p < 10$–15 µm, based on the concept of feeding the power in small, well-defined successive doses. As schematically represented in Fig. 13.83 the powder pump contains a vacuum source, a pressure source/powder feeding gas, a powder suction line, the powder transport line to the torch, four valves made of special elastic tubes that can be opened and closed mechanically by pistons and a defined volume (dose). If the valves numbers 1 and 2 are closed and the valves 3 and 4 are open,

the powder will be sucked through the suction line out of the powder container into the dose due to the pressure drop produced by the vacuum pump. The dose can have a calibrated volume in the range of 10–1000 mm^3 as a function of application. As the valves 3 and 4 are closed and valves 1 and 2 opened, the powder will be pressed out of the dose into the powder feeding line by the pressurized transport gas. Under atmospheric conditions the working pressure will be less than 0.2 MPa depending on the powder to be fed and the length of the powder transport line. Alternately, the powder container can be as simple as an open box from which the powder is directly sucked out.

Vibration feeders have also been used for the feeding of fine powders either through a linear or angular vibration motion, which pushed the powder upstream of a lightly inclined linear or spiral ramp. A thin layer of powder is transported this way at a rate, which depends on the amplitude and frequency of the vibration. At the end of the ramp, the powder falls by gravity into the carburetor where it is mixed with the powder transport gas and conveyed into the powder transport line. For a close control on the transport gas flow rate, the ramp and the vibration mechanism have to be placed inside a gas tight container in which the carrier gas is introduced at a well-controlled gas flow rate. While this type of powder feeders works well with most powder at low feed rates (of the order of grams per minute), very fine submicron powders will have the tendency to agglomerate, as they are transported over the ramp, forming small "balls" at the carburetor discharge point. These will induce important fluctuations of the powder feed rate.

13.4.2.5 Powder Feed Rate Control

Closed-loop feed-rate controllers are necessary on powder feeders to keep the powder mass feed rate as constant as possible [Davis JR (ed) (2004)]. Most of them operate on a loss-in-weight principle in which the powder feeder is placed on a sensitive electronic balance that measures the rate of weight loss over time, as powder is fed to the transport line and injector. The main challenge with this approach is that, for a reasonable monitoring frequency of the powder feed rate, the changes in the weight of the powder feeder with its powder content can be rather small compared to the "tar" weigh of the powder feeder itself which results in a significant deterioration of the precision of the powder feed rate measurement. Moreover, any vibration introduced to ensure a reliable flowability of the powder into the powder transport line is usually counter-indicated for a reliable measurement of the weight change as perceived by the load cell of the electronic balance.

Alternate approach put forward by [Stanowski and Boulos (2014)] involves the monitoring of the opacity of the flow in the powder feed line which is a direct measurement of the particle loading of the transport gas and consequently the

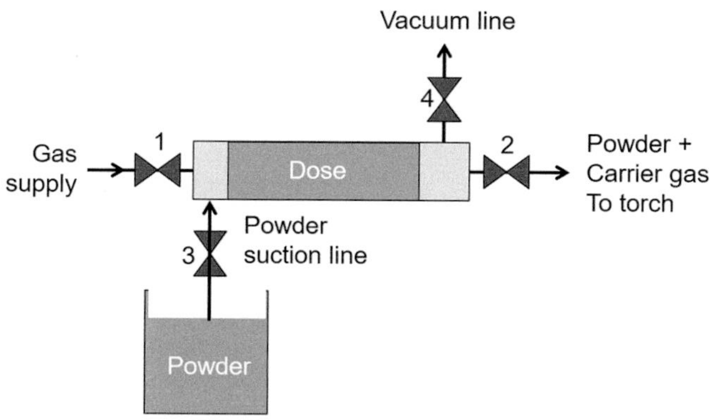

Fig. 13.83 Basic components of the "powder pump" arrangement for fine powder feeding

powder feed rate. The measurement can be made using a small module placed in the powder transport in which a light emitting diode (LED) or micro laser is used to illuminate the flow in a small transparent section of the powder transport line, with the light intensity on the opposite side of the tube monitored using a calibrated photodetector. As the powder loading of the flow increases, its opacity increases, with a corresponding drop of the transmitted light intensity received by the photodetector. The photodetector signal is consequently a direct measure of the powder loading and consequently the powder flow rate. The signal being in arbitrary units (au) needs to be calibrated as function of the powder used its particle size and particle density. The technique has an excellent dynamic range capable of detecting of pulsation of the powder feed rates which could be provoked by powder feeder irregularities which invariable would influence the thermal loading of the plasma and consequently the quality of the deposit.

With either approaches for the detection of the powder feed rate, the signal of the monitor device can be used for the appropriate corrective action changing the rotation speed of the powder transport wheel, disk, screw, or the gas pressure in order to keep the powder mass flow rate as constant as possible.

13.4.3 Liquid and Suspension Feeding

Over the past two decades, there has been a growing interest in the plasma spraying of nano-structured coatings which were identified for their superior thermomechanical properties compared to standard micron-sized structured coatings. Being generally thinner than conventional thermal spray coatings with a finer grain size, nano-structured coatings bridge the gap between Thermal-Sprayed (TS) coatings and Physical Vapor Deposited (PVD) coatings [Fauchais et al. (2008a, b)]. The principal challenge is that nanoparticles have a considerably lower mass than micron-sized particles, and accordingly have a significantly lower inertia for the same particle injection velocity. Moreover, as soon as the nanosized particles come in direct contact with a plasma flow, they almost instantly melt and vaporize. Alternately, pre-agglomeration of nanoparticles in the form of micron-sized particles is used as viable option that overcomes the challenge of nanoparticle injection into the combustion stream or plasma flow. They require, however, avoiding their complete in-flight melting in order to maintain their nanostructure.

Suspension or solution spraying, on the other hand [Gitzhofer, F. et al. (1997) Fauchais et al. (2008a, b), Fauchais and Montavon (2010), Fauchais et al. (2011)

(2013) (2015a, b)], offers an alternate route that has gained wide acceptance over the past decade. In the case of Suspension Plasma Spraying (SPS) the precursor material is fed into the combustion flame or plasma in the form of a suspension of a fine or ultrafine powder mixed with an appropriate liquid. As the suspension is injected into the flow it will be atomized or fragmented forming fine droplets which are entrained by the hot gases. On further heating of the formed droplets, the liquid in the droplets is vaporized leaving behind a dispersed or agglomerated pack of fine or ultrafine particles which subsequently melt on further heating. On the impact of the molten droplets with the substrate they form of micro-splats, a few micrometers in diameter, which are the building blocks of the coating.

Solution Precursor Plasma Spraying (SPPS) is based on the complete dissolution of the coating precursor in an appropriate liquid which is injected/atomized into the high temperature combustion or plasma stream. As the liquid is vaporized the dissolved solid precipitates as granules in the partially vaporized droplets. On complete vaporization of the droplets the residual solid grain melts/crystallizes forming on impact with the substrate a coating composed of a blend of crystals and splats.

In either case of solution or suspension plasma spraying, the stable and controlled feeding of the liquid or suspension into the high-temperature combustion or plasma stream represents a critical step on which the quality of the formed coating strongly depends. The following two approaches have been widely used:

- **Spray Atomization** through the use of an atomizing medium, such as a compressed gas, which supplies the energy required for the atomization of the liquid or suspension. Sometime also referred to as pneumatic or gas or two-fluid atomization.
- **Mechanical Atomization,** where the liquid or suspension is pressurized and pushed through appropriate nozzle, resulting its fragmentation into small droplets. This approach is also used to provide the liquid/suspension with the necessary momentum energy in order to penetrate the combustion or plasma flow. On reaching to core high velocity region of the flow the liquid is fragmented before vaporization using the kinetic energy of the plasma. Mechanical atomization can also be accomplished using a vibrating device or ultrasound source.

A detailed discussion on the atomization of suspensions or solutions injected into a DC or RF induction plasma stream is presented in Part I of this book, Chap.5 Gas and Particle Dynamics in thermal Spray.

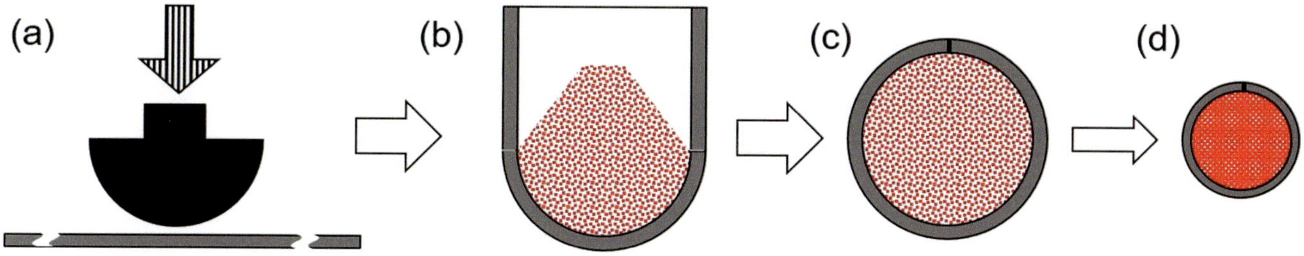

Fig. 13.84 Schematic representation of the manufacturing steps of a cord wire (**a**) ductile metal foil, (**b**) U-shaped metal foil with powder mixture deposited in its cavity, (**c**) folded and rolling foil to 3.8 mm diameter tube, and (**d**) rolling tube to its final 2 mm diameter cored wire [Liu S-G et al. (2007)]. Reprinted with kind permission from Elsevier

13.5 Wires and Cords

13.5.1 Manufacturing Routes for Wires and Cords

Contrarily to powder spraying, the use of wires or cords as feed material in any of the Thermal Spray technologies, whether Combustion Spraying (CS), Wire-Arc Spraying (WAS) or Plasma Transferred Arc (PTA) deposition requires that the wire or cord material be ductile and mailable available in diameters between 1.2 and 4.76 mm (1/16 to 3/16 in.) in coil wound on plastic spools, or production packs (drums). Larger, stiffer wires ($d > 2$ mm) are usually more difficult to feed as uncoiling requires more force and larger wire tension. The stiffer is the wire, the more complex is the wire feeding mechanism to be used and the larger is the wire guide erosion. For a regular and reproducible feeding, it is essential that wire diameter be well calibrated and its surface as smooth as possible. Materials used for wire arc spraying include aluminum, copper, iron, molybdenum, nickel, tin, titanium, and zinc base. For flame spraying, the same materials are commonly used with the exception of titanium but with the addition of lead and silver base.

Cords are the equivalent of wires, but they are devoted to feed into flame spray guns ceramic or non-ductile materials. They are typically made of a ductile metallic foil, cellulosic or plastic casing filled with hard metallic/alloy or ceramic particles with either an organic or a mineral binder [Liu S-G et al. (2007), Norouzi S (2004)]. As illustrated in Fig. 13.84, the casing of the cord in the form of a metallic foil 0.3 mm thick 12 mm wide, Fig. 13.84a, is rolled to give it a U-shaped cross section, as shown in Fig. 13.84b. The filling of the cord made of fine well-mixed powder of the required composition whether a mixture of different alloy/metals or ceramics is deposited in the cavity of the U-shaped metal foil which is then further rolled to form tube 3.8 mm in diameter, Fig. 13.84c. The formed tube with its charge of powder mixture is next pulled in sequence through a series of 14 dies densely compacting the powder charge and reducing

Fig. 13.85 Photograph of a Flexicord spool from Saint Gobain [courtesy of Saint Gobain]

gradually its diameter to the required final cord size of 2 mm as shown in Fig. 13.84d.

Alternately cords can also be manufactured using external sheath of plastic or cellulosic material allowing the flexibility of the cord. They are manufactured by co-extrusion of the sheath and internal material. Ceramic or non-ductile material particles are agglomerated either with an organic binder, decomposing around 250 °C, and that has completely disappeared at 400 °C, or a mineral binder (e.g., boehmite: $Al(OH)_3$) that keeps the cohesion of ceramic particles up to temperatures close to their melting point. These flexible cords have lengths up to 120 m. A photograph of a flexicord spool from Saint Gobain is given in Fig. 13.85.

13.5.2 Wire and Cord Feeders

Feeding the wires or cords into the TS unit is typically achieved either by air motor or electric motor and gearbox arrangement [Davis JR (ed) (2004), Lester T (2005)]. Three different configurations are used to drive wires.

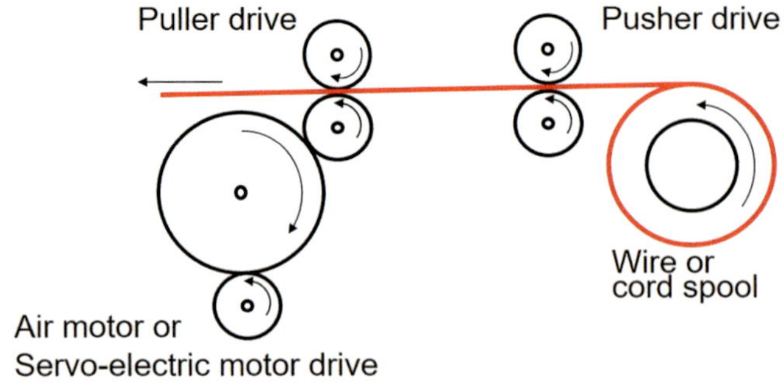

Fig. 13.86 Wire feeding principle. Reprinted with kind permission from ASM International [Davis JR (ed) (2004)]

- *Push only*: The wire is pushed from the drive unit to the gun. If the gun is lightweight, the wire position can fluctuate from the drive to the gun reducing the stability and limiting the pushing length to around 5 m. This system is convenient for hand spraying.
- *Pull only*: A drive unit mounted in the gun pulls the wire. The wire position is well defined, and in wire arc spraying they are as much more stable, thanks to the feeding rollers that are much closer to the last wire guide. Of course, the gun weight is increased, and the length of the wire supplied is limited to 5 m. This system is commonly used for automatic spraying.
- *Push/pull*: This is obviously a combination of the above two methods, the wire being driven to and pulled from the gun. The wire tip is very stable, and its length can reach 20 m. Synchronizing the two-drive units is more complex and the gun weight is higher. Figure 13.86 from [Davis JR (ed) (2004)] schematically represents the principle of wire entrainment device.

Compared to powders the main advantage of wires is the deposition rate. For example, Lester T (2005) indicates that Zn/Al (85/15) wires with a diameter of 4.76 mm can be sprayed using an oxy-propane flame at a throughput of 38 kg/h. With WAS, working with an arc current of 700 A, the throughput can be increased to 53 kg/h in continuous mode with 3.17 mm wires. The choice between both technologies depends strongly on the engineering situation. The following remarks [Lester T (2005)] can be made:

- Typically for Zn and Al, flame spray has higher deposition efficiency than arc spray, which is not necessarily the case for all materials.
- Flame spray has typically a finer coating finish, with less porosity and oxide than arc-spray coatings at equivalent spray rates.
- Flame-spray parameters need to be reset each time the gas supply is turned off or changed.

- Arc-spray parameters are set by simple switches and remain stable for the full operation.
- Arc spray is very simple to use; once parameters are set, the operator pushes a button, pulls the trigger, and is quickly spraying, with no consideration needed for the safe storage, handling, and use of flammable gases.
- Arc spray only requires the wire to be changed, whereas flame spraying also requires gas bottles to be replaced.
- Arc-sprayed aluminum has bond strength approximately 2.5 times higher than flame-sprayed aluminum, but its oxidation is almost twice higher.

13.6 Summary and Conclusions

Powders, wires, and cords are the feed stock for the great majority of the thermal spray industry on which its success heavily depends. The growing interest in the potential role of thermal spraying for a wide range of applications has stimulate the development of a most diverse industry with a wide range of technologies varying from cold spraying, combustion spraying, DC atmospheric and vacuum plasma straying, RF induction plasma spraying, wire arc spraying and Plasma transferred arc deposition. In each of these cases, the feed material is exposed to very different conditions which impose strict requirements on the range of materials and the form in which they could be used whether powders, dry or in suspension, wires, or cords. Especially for powders, other than their chemical composition, requirements are also imposed on their particle size distribution, individual particle morphology, powder flowability, and ruggedness while still being economically viable. Such requirements have triggered in turn a rush to the development of a wide range of technologies for powder, wire, and cord manufacturing.

In this chapter, a review is presented in the first place of powder characteristics in terms of individual particle and aggregate powder properties. Emphasis is placed on the definition of the most important powder parameters and the range of experimental techniques available for their measurements. No attempt is made to cover the subject in an exhaustive manner since numerous textbooks and archival references are available on the subject. The key point is to identify the material parameters that need to be measured to exercise a tight control on the processing conditions and the quality of the coatings obtained.

Powder manufacturing techniques are discussed next linking the properties of the powders obtained to their manufacturing routs whether through mechanical size reduction, powder consolidation, plasma spheroidization, atomization, or chemical synthesis. Considering the growing inters in the area of nanostructured coatings, a brief introduction is made to the synthesis of ultrafine nanopowders which when combined with such techniques as suspension plasma spraying have shown an important potential for the deposition of a novel coatings for high end applications. Attention is also given to the important area of powder and suspension feeding into the different thermal spray sources. A brief review is also made of wire and cord manufacturing and feeding in thermal spray applications.

Nomenclature

Units are indicated in parentheses; when no units are indicated, the parameter is dimensionless.

Latin Alphabet

a_p	Specific surface area of the particle, measured by BET technique (m^2/g)
a_{po}	Projected surface area of the particle (m^2)
A_p	Surface area of the particle (m^2)
C	Circularity Eq. 13.4
$\overline{d}$	Mean particle diameter of number or mass PSD (m) Eq. 13.8 and Eq. 13.9
d_{10}	Particle diameter below which 10 wt. % of the powder is found (m)
d_{50}	Particle diameter below which 50 wt. % of the powder is found (m)
d_{90}	Particle diameter below which 90 wt. % of the powder is found (m)
d_{BET}	Equivalent mean particle diameter evaluated using BET technique, (m) Eq. 13.19
$\overline{d_g}$	Geometric mean diameter (m) Eq. 13.13 and Eq. 13.15
d_p	Particle diameter (m)
d_{pa}	Equivalent projected area diameter, (m) Eq. 13.2
d_{pp}	Equivalent perimeter diameter (m) Eq. 13.1
d_{pv}	Equivalent volume diameter, (m) Eq. 13.3
d_{st}	Stokes diameter, (m) Eq. 13.6
g	gravitational acceleration (m/s^2)
m_p	Particle mass (kg)
p	Pressure (Pa)
p_0	Saturation vapor pressure (Pa)
P_p	Perimeter (m)
Pe	Peclet number ($Pe = t_c/t_s$)
r_p	Particle radius (m)
S	Specific surface area (m^2/g)
t_c	Heat conduction time ($t_c = d_p^2/\alpha_p$) (s)
t_s	Splash spreading time ($t_s = d_p/v_p$) (s)
T_m	Melting Temperature (^oC or K)
T_v	Vaporization temperature (^oC or K)
u_t	Terminal settling velocity (m/s)
v_p	Particle impact velocity on substrate (m/s)
V_m	Adsorbate volume per unit mass of solid for monolayer coverage (m^3/kg)
V_a	Adsorbate volume at relative pressure (m^3)
V_p	Particle volume (m^3)
x_i	Volume or mass fraction (−)

Greek Alphabet

α	Angle of internal friction (^{o})
α_d	Dimensionless size parameter (−) Eq. 13.17
α_p	Thermal diffusivity of the particle ($\alpha_p = \kappa /\rho_p c_p$)
θ	Angle of repose (^{o})
θ_o	Scattering angle ($^\circ$)
κ	Thermal conductivity (W/m K)
λ	Wavelength of incident radiation (nm)
μ	Dynamic viscosity (Pa.s)
ρ_o	Fluid density (kg/m^3)
ρ_p	Particle specific mass (kg/m^3)
σ_d	Standard deviation of number or mass PSD (m) Eq. 13.9 and Eq. 13.11
σ_g	Standard deviation of number or mass geometric PSD (m) Eq. 13.14 and Eq. 13.16
ψ	Sphericity, Eq. 13.5

References

Ageorges, H., and P. Fauchais. 2000. Plasma spraying of stainless-steel particles coated with an alumina shell. *Thin Solid Films* 370: 213–222.

ALD. n.d. *Ceramic -free metal powder production for reactive and refractory metals*, Technical Bulletin MetaCom/Eiga-e/05.11, www.ald-vt.de

Allen, T. 1997. *Particle size measurements*, Vol. 1: Powder sampling and particle size measurements. Vol. 2: Surface area and pore size determination. 5th ed. London: Chapman & Hall.

Allimant, A., M.P. Planche, Y. Bailly, L. Dembinski, and C. Coddet. 2009. Progress in gas atomization of liquid metals by means of a De Laval nozzle. *Powder Technology* 190: 79–83.

Anderson, I.E., D.J. Sordelet, M.F. Besser, and R.L. Terpstra. 1998. Effects of powder morphology on pneumatic feeding and plasma spray deposition. In *Thermal spray: Meeting the challenges of the 21st century*, ed. C. Coddet, 911–916. Materials Park: ASM International.

Antipas, G.S.E. 2006. Modelling of the breakup mechanism in gas atomization of liquid metals. Part I: The surface wave formation model. *Computational Materials Science* 35: 416–422.

Antony, L.V.M., and R.G. Reddy. 2003. Processes for production of high purity metal powders. *JOM NY* 55 (3): 14–18.

AP&C. n.d. *Unique plasma atomization process*, Technical Bulletin.

ASTM B213. n.d. *Flow rate of metal powders.*

ASTM E2651-10. n.d. *Standard guide for powder particle size analysis.*

ASTM WK19862. n.d. *Standard practices for sampling metal powders*, Revision of B215–04, Active Standard: B215–08.

ASTM-B214-07. 2011. *Standard test method for sieve analysis of metal powders.*

Åström, B.T. 1997. *Manufacturing of polymer composites*. London: Chapman & Hall.

AWS. 1985. *Thermal spraying, practice, theory and application*. Miami: American Welding Society.

Baksh, S.R., V. Singh, K. Balani, D.G. McCartney, S. Seal, and A. Agarwal. 2008. Carbon nanotube reinforced aluminum composite coating via cold spraying. *Surface Coating Technology* 202: 5162–5169.

Ban, Z.-G., and L. Shaw. 2002. Synthesis and processing of nanostructured WC-Co materials. *Journal of Materials Science* 37: 3397–3403.

Ban, Z.-G., and L.L. Shaw. 2003. Characterization of thermal sprayed nanostructured WC-Co coatings derived from nanocrystalline WC-18wt.% Co powders. *Journal of Thermal Spray Technology* 12 (1): 112–119.

Barth, H., ed. 1984. *Modern methods of particle size analysis*. New York: Wiley.

Beckhoff, B., B. Kanngießer, N. Langhoff, R. Wedell, and H. Wolff. 2006. *Handbook of practical X-ray fluorescence analysis*. Berlin: Springer.

Beczkowiak, J., and G. Schwier. 1991. Tailoring carbides and oxides for HVOF. In *Thermal spray: Coatings properties, processes and applications*, ed. C.C. Berndt and T.F. Bernecki, 121–126. Materials Park: ASM International.

Ben Ettouil, F., O. Mazhorova, B. Pateyron, H. Ageorges, M. El-Ganaoui, and P. Fauchais. 2008. Predicting dynamic and thermal histories of agglomerated particles injected within a DC plasma jet. *Surface Coating Technology* 202: 4491–4495.

Berger, L.-M., P. Ettmayer, P. Vuoristo, T. Mäntylä, and W. Kunert. 2001. Microstructure and properties of WC-10%Co-4% Cr spray powders and coatings: Part 1. Powder characterization. *Journal of Thermal Spray Technology* 10 (2): 311–325.

Berndt, C. 2007. *Thermal spray workshop/courses in conjunction with ITSC-2007 Beijing, May 2007*. Materials Park: ASM International.

Bertrand, G., P. Roya, C. Filiatre, and C. Coddet. 2005. Spray-dried ceramic powders: A quantitative correlation between slurry characteristics and shapes of the granules. *Chemical Engineering Science* 60: 95–102.

Boch, P., P. Fauchais, D. Lombard, B. Rogeaux, and M. Vardelle. 1985. Plasma sprayed zirconia coatings. *Journal of Advanced Ceramics* 12: 488–503.

Borisov, Y., and A. Borisova. 1992. Application of self-propagating high-temperature synthesis in thermal spraying technology. In *Thermal spray: Research, design and applications*, ed. T.F. Bernicki, 139–144. Materials Park: ASM International.

Borisov, Y., A. Borisova, and L. Shvedova. 1986. *Transition metal – Nonmetallic refractory compound composite powders for thermal spraying. Proceedings of ITSC-86, Montreal, Canada, 8–12 Sept 1986*, 323–330. Oxford: Pergamon Press.

Borisova, A.L., and S. Borisov Yu. 2008. Self-propagating high-temperature synthesis for the deposition of thermal-sprayed coatings. *Powder Metallurgy and Metal Ceramics* 47 (1–2): 80–94.

Bouchard, D., L. Sun, F. Gitzhofer, and G.M. Brisard. 2006. Synthesis and characterization of $La_{0.8}Sr_{0.2}MO_{3-\delta}$ (M = Mn, Fe, or Co) cathode materials by induction plasma technology. *Journal of Thermal Spray Technology* 15 (1): 37–45.

Boulos, M.I. 2002. *Induction plasma synthesis and processing f nanostructured materials*, IMP-2002, Miyagi Japan, Nov.27-29 (10 pp).

———. 2004. Plasma power can make better powders. *Metal Powder Report, Metal-powder.net* 59 (5): 16–21.

———. 2016. The role of transport phenomena and modelling in the development of thermal plasma technology. *Plasma Chemistry Plasma Processes* 36: 3–28.

Boulos, M., P. Fauchais, A. Vardelle, and E. Pfender. 1993. Injection of hot particles in a plasma flame. In *Plasma spraying theory and applications*, ed. R. Suryanarayanun, 3–61. Singapore: World Scientific.

Boulos, M.I., J.E. Jurewicz, and A.C.Nessim. 2006a. *Plasma synthesis of metal oxide nanopowder and apparatus therefor*, US patent 6,994,837B2, Feb. 6th.

Boulos, M., M. Ducos, P. Fauchais. 2006b. (in French) *Plasma technology and its potential for powder treatment*, Matériaux 2006, Dijon, France, 13–17 Nov 2006. SF2M, Paris, France.

Boulos, M.I., J.E. Jurewicz, and A.C.Nessim. 2009. *Apparatus for plasma synthesis of metal oxide nanopowder*, US patent 7,501,599 B2, Mar. 10th.

Boulos, M.I., J.E. Jurewicz, and J. Guo. 2011. *Induction plasma synthesis of nanopowders*, US Patent 8,013,269 B2, Sept 6th.

Boulos, M.I., J.E. Jurewicz, J. Guo, X. Fan, and N. Dignard. 2014. *Plasma synthesis of nanopowders*, US Patent.US 8,859,931 B2, Oct 14th.

Boulos, M.I., J.W. Jurewicz, and J. Gup. 2016. *Plasma reactor for the synthesis of nanopowders and materials processing*, US Patent 9,516,734 B2, Dec. 6th.

Boulos, M.I., A. Auger, and J.W. Jurewicz. 2017a. *Process and apparatus for producing powder particles by atomization of feed material in the form of an elongated member*, US patent 9,718,131 B2, August 1st.

———. 2017b. *Process and apparatus for producing powder particles by atomization of feed material in the form of an elongated member*, US patent 9,751,129 B2, Sept 5th.

Bouyer, E., F. Gitzhofer, and M. Boulos. 1997a. Suspension plasma spraying for hydroxyapatite powder preparation by RF plasma. *IEEE Transactions on Plasma Science* 25 (5): 1066–1072.

———. 1997b. The suspension plasma spraying of bioceramics by induction plasma. *Journal of Materials Science Letters* 49: 58–62.

———. 2000. Morphological study of hydroxyapatite nanocrystal suspension. *Journal of Materials Science: Materials in Medicine* 11: 523–531.

Camerucci, M.A., and A.L. Cavalieri. 1998. Process parameters in attrition milling of cordierite powders. *Journal of Materials Synthesis and Processing* 6 (2): 115–121.

Chen, D., E.H. Jordan, and M. Gell. 2009. Microstructure of suspension plasma spray and air plasma spray Al_2O_3-ZrO_2 composite coatings. *Journal of Thermal Spray Technology* 18 (3): 421–442.

Czisch, C., and U. Fritsching. 2008. Atomizer design for viscous-melt atomization. *Materials Science and Engineering A* 477 (1–2): 21–25.

Dallaire, S. 1982. Influence of temperature on the bonding mechanism of plasma-sprayed coatings. *Thin Solid Films* 95: 237–241.

Dallaire, S., and G. Cliche. 1992. The influence of composition and process parameters on the microstructure of TiC-Fe multiphase and multilayer coatings. *Surface Coating Technology* 50: 233–239.

Davis, J.R., ed. 2004. *Handbook of thermal spray technology*. Materials Park: ASM International.

de Villiers Lovelock, H.L., P.W. Richter, J.M. Benson, and P.M. Young. 1998. Parameter study of HP/HVOF deposited WC-Co coatings. *Journal of Thermal Spray Technology* 7 (1): 97–107.

Deevi, S.C., V.K. Sikka, C.J. Swindeman, and R.D. Seals. 1997. Reactive spraying of nickel-aluminide coatings. *Journal of Thermal Spray Technology* 6 (3): 335–344.

Denoirjean, A., A. Vardelle, A. Grimaud, P. Fauchais, E. Lugsheider, I. Rass, H.L. Heijen, P. Chandler, R. McIntyre, and T. Cosak. 1992. Comparison of the properties of plasma sprayed stabilized zirconia coatings for different powder morphologies. In *Thermal spray: International advances in coating technology*, ed. C.C. Berndt, 975–982. Materials Park, OH: ASM International.

Diez, P., and R.W. Smith. 1993. The influence of powder agglomeration methods on plasma sprayed yttria coatings. *Journal of Thermal Spray Technology* 2 (2): 165–172.

Dignard, N.M., and M.I. Boulos. 1997. Ceramic and metallic powder spheroidization using induction plasma technology. In *Thermal spray: A united forum for scientific and technological advances*, ed. C.C. Berndt, 1–7. Materials Park: ASM International.

———. 2000. *Powder Spheroidization using induction plasma technology*, 887–893. Montreal: ITSC.

Dunkley, J.J. 1998. *Atomization, powder metal technologies and applications, vol 7, ASM handbook*, 35–52. Materials Park: ASM International.

Dzur, B. 2008. Plasma puts heat into spherical powder production. *MPR* 2008 (Feb): 12–15.

Eigen, N., F. Gärtner, T. Klassen, E. Aus, R. Bormann, and H. Kreye. 2002. Microstructures and properties of nanostructured thermal sprayed coatings using high-energy milled cermet powders. *Surface Coating Technology* A336: 274–319.

Entezarian, M., F. Allaire, P. Tsantrizos, and R.A.L. Drew. 1996. Plasma atomization: A new process for the production of fin spherical powders. *JOM*: 53–55.

Fan, X., F. Gitzhofer, and M.I. Boulos. 1998. Statistical design of experiments for the spheroidization of powdered alumina by induction plasma processing. *Journal of Thermal Spray Technology* 7 (2): 247–253.

Fauchais, P., and G. Montavon. 2010. Latest developments in suspension and liquid precursor thermal spraying. *Journal of Thermal Spray Technology* 19 (1–2): 226–239.

Fauchais, P., V. Rat, J.F. Coudert, R. Etchart-Salas, and G. Montavon. 2008a. Operating parameters for suspension and solution plasma spray coatings. *Surface and Coating Technology* 202: 4309–4437.

Fauchais, P., R. Etchart-Salas, V. Rat, J.F. Coudert, N. Caron, and K. Wittmann-Ténèze. 2008b. Parameters controlling liquid plasma spraying: Solutions, sols or suspensions. *Journal of Thermal Spray Technology* 17 (1): 31–59.

Fauchais, P., G. Montavon, and G. Bertrand. 2010. From powders to thermally sprayed coatings. *Journal of Thermal Spray Technology* 19 (1–2): 56–80.

Fauchais, P., G. Montavon, S. Lima, and B.R. Maple. 2011. Engineering a new class of thermal spray nano-based microstructures from agglomerated nanostructured particles, suspension and solutions: an invited review. *Journal of Physics D: Applied Physics* 44. 093001 (53 pp).

Fauchais, P., A. Joulia, S. Goutier, C. Chazelas, M. Vardelle, A. Vardelle, and S. Rossignol. 2013. Suspension and solution plasma spraying. *Journal of Physics D: Applied Physics* 46. 224015 (14 pap).

Fauchais, P., M. Vardelle, S. Goutier, and A. Vardelle. 2015a. Key challenges and opportunities in suspension and solution plasma spraying. *Plasma Chemistry Plasma Process* 35: 511–525.

———. 2015b. Specific measurements of in-flight droplet and particle behavior and coating microstructure in suspension and solution plasma spraying. *Journal of Thermal Spray Technology* 24 (8): 1498–1505.

Fritsching, U., and V. Uhlenwinkel. 2012. Hybrid gas atomization for powder production. *Power Metallurgy* 99–124.

Froes, F.H. 2012a. Titanium powder metallurgy, a review part 1. *Advanced Materials & Processes* 2012 (September): 16–22.

———. 2012b. Titanium powder metallurgy, a review part 2. *Advanced Materials & Processes* 2012 (September): 26–29.

———. 2013. Titanium powder metallurgy: Development and opportunities in a sector poised for growth. *Powder Metallurgy Review* 2013 (Winter): 29–43.

Gitzhofer, F., E. Bouyer, and M.I. Boulos. 1997. *Suspension plasma spray deposition*, United States Patent, 5,609,921 (March 11th. 1997)

Goldstein, J.I., D.E. Newbury, P. Echlin, D.C. Joy, C.E. Lyman, E. Lifshin, L.C. Sawyer, and J.R. Michael. 2003. *Scanning electron microscopy and X-ray micro- analysis*. New York: Springer.

Guo, J., X. Fan, R. Dolbec, S. Xue, J.W. Jurewicz, and M. Boulos. 2010. Development of nanopowder synthesis using induction plasmas. *Plasma Science and Technology* 12 (2): 188–199.

Haller, B., J.-P. Bonnet, P. Fauchais, A. Grimaud, and J.-C. Labbe. 2005. TiC based coatings prepared by combining SHS and plasma spraying. In *Thermal spray connects: Explore its surfacing potential!* ed. E. Lugscheider. Düsseldorf: DVS-Verlag GmbH.

He, J., and J.M. Schoenung. 2002. Nanostructured coatings, a review. *Materials Science and Engineering* A336: 274–319.

Hohmann, M., and N. Ludwig. 1991. German patent DE 4102 101 A1

Hohmann, M., M. Ertl, and S. Jonsson. 1989. Experience and powder production by crucible free induction drip-melting combined with inert-gas atomizing. In *Advances in powder metallurgy*. New York: Metal Powder Industries Federation.

Iinoya, K., K. Gotoh, and K. Higashitani. 1991. *Powder technology handbook*. New York: Dekker.

Ito, H., M. Umakoshi, R. Nakamura, T. Yokoyama, K. Urayama, and M. Kato. 1991. Characterization of Ni-Al composite powders formed by mechanofusion. In *Thermal spray coatings: Properties, processes and applications*, ed. T.F. Bernecki, 405–411. Materials Park: ASM International.

Ivosevic, M. 2006. *Splat formation during thermal spraying of polymer particles: mathematical modeling and experimental analysis*. Ph.D. thesis, Faculty of Drexel University

Ivosevic, M., R. Knight, S.R. Kalidindi, G.R. Palmese, J.K. Sutter, and A. Tsurikov. 2003. Optimal substrate preheating model for thermal spray deposition of thermosets onto polymer matrix composite. In *Proceedings of the ITSC-2003, Orlando, FL*, ed. C. Moreau. Materials Park: ASM International.

Ivosevic, M., R.A. Cairncross, and R. Knight. 2006. 3D predictions of thermally sprayed polymer splats: Modeling particle acceleration, heating and deformation on impact with a flat substrate. *International Journal of Heat and Mass Transfer* 49: 3285–3297.

Jarosinski, W.J., M.F. Gruninger, and C.H. Londry. 1993. Characterization of tungsten carbide cobalt powders and HVOF coatings. In *Thermal spray coatings: Research, design and applications*, ed. C.C. Berndt and T.F. Bernecki, 855–861. Materials Park: ASM International.

Kear, B.H., R.K. Sadangi, M. Jain, R. Yao, Z. Kalman, G. Skandan, and W.E. Mayo. 2000. Thermal sprayed nanostructured WC/co hardcoatings. *Journal of Thermal Spray Technology* 9 (3): 399–406.

Kim, M.C., S.B. Kim, and J.W. Hong. 1997. Effect of powder types on mechanical properties of D-Gun coatings. In *Thermal spray: A united forum for scientific and technological advances*, ed. C.C. Berndt, 791–795. Materials Park: ASM International.

Kinsman, S. 1979. Aerosol particle size measurements using the electrical resistance principle. In *Aerosol measurements*, ed. D.L. Lundegren et al., 650–654. Gainesville: University of Florida Oppress.

Kolman, B., J. Forman, J. Dubsky, and P. Chraska. 1994. Homogeneity studies of powders and plasma sprayed deposits. *Mikrochimica Acta* 114–115: 335–342.

Kroeger, J., and F. Marion. 2011. Raymore AP&C leading the way with plasma atomized Ti spherical powder for MIM. *Powder Injection Moulding International* 5 (4): 55–57.

Kubel, E.J., Jr. 1990. Powders dictate thermal spray coating properties. *Advanced Materials and Processes*: 24–32.

Kurlov, A.S., A.A. Rempel, V. Blagoveshenskii Yu, A.V. Samokhin, and V. Tsvetkov Yu. 2011. Hard alloys WC–Co (6 wt %) and WC–Co (10 wt %) based on nanocrystalline powders. *Chemical Technology* 439 (1): 213–218.

Laha, T., K. Balani, A. Agarwal, S. Patil, and S. Seal. 2005. Synthesis of nanostructured spherical aluminum oxide powders by plasma engineering. *Metallurgical and Materials Transactions A: Physical Metallurgy and Materials Science* 36A: 301–309.

Lavernia, E.J., B.Q. Han, and J.M. Schoenung. 2008. Cryomilled nanostructured materials: Processing and properties. *Materials Science and Engineering A* 493: 207–214.

Lawley, A. 2003. *Atomization*. Princeton: Metal Powder Industries Federation Publishers, and cross -link to the citation.

Lee, H.M., C.Y. Huang, and C.J. Wang. 2009. Forming and sintering behaviors of commercial α-Al$_2$O$_3$ powders with different particle size distribution and agglomeration. *Journal of Mater Process Technology* 209: 714–722.

Lefebvre, A.H. 1989. *Atomization and sprays*, 421 pages. New York: Hemisphere Publishing Corporation.

Legoux, J.G., and S. Dallaire. 1993. Copper-TiB2 coatings by plasma spraying reactive micropellets. In *Thermal spray: Research, in design and applications*, ed. C.C. Berndt and T.F. Bernecki, 429–432. Materials Park: ASM International.

Leivo, E., T. Wilenius, T. Kinos, P. Vuoristo, and T. Mäntylä. 2004. Properties of thermally sprayed fluoropolymer PVDF, ECTFE, PFA and FEP coatings. *Progress in Organic Coating* 49 (1): 69–73.

Lester, T. 2005. Metallized coatings application: an overview of the flame- and arc-spraying processes used for surface finishing. *Organic Finishing* 103 (July/August): 35–39.

Lima, R.S., and B.R. Marple. 2008. Process–property–performance relationships for titanium dioxide coatings engineered from nanostructured and conventional powders. *Materials and Design* 29: 1845–1855.

Liu, S.-G., J.-M. Wu, S.-C. Zhang, S.-J. Rong, and Z.-Z. Li. 2007. High temperature erosion properties of arc-sprayed coatings using various cored wires containing Ti–Al intermetallic. *Wear* 262: 555–561.

Lugscheider, E., M. Loch, and H. Eschnauer. 1991. Coated powders-the status today. In *2nd. Plasma-Technik symposium, June 1991*, ed. P. Huber and H. Echnauer, vol. 2, 339–351. Wholen: Plasma Technik.

Mates, S.P., and G.S. Settles. 2005. A study of liquid metal atomization using close-coupled nozzles. Part1: Gas dynamic behavior. *Atomization and Sprays* 15: 19–40.

McCracken, C.G. 2009. *Manufacture of hydride-dehydride low oxygen Ti-6Al-4V (Ti-6-4) powder incorporating a novel powder de-oxidation step*, Euro PM conference.

McCracken, C.G., and D. Barbis. 2008. Production of fine titanium powders via the hydride-dehydride (HDH) process. *Powder Injection Moulding International* 2 (#2 June): 55–57.

McCracken, C.G., and D.P. Barbis. 2010. Review of current Hydride-Dehydride (HDH) powder production methods which impact the chemistry, morphology and particle size distribution of titanium based powders. In *Advances in powder metallurgy & particulate materials*, ed. M. Bulger and B. Stebick, vol. 1, 81–86. Princeton: Metal Powder Industries Federation.

McCracken, C.G., C.A. Motchenbacher, and D.P. Barbis. 2010a. Review of titanium powder production methods. *International Journal of Powder Metallurgy* 46 (5): 19–26.

McCracken, C.G., D.P. Barbis, and R.C. Deeter. 2010b. *Key titanium powder characteristics manufacture using the Hydride-Dehydride (HDH) process*, Powder characterisation, EURO PM2010, European Powder Metallurgy Association, Talbot House, Market Street, Shrewsbury, SY1 1 LG, UK.

McCracken, C.G., C. Motchenbacher, and D.P. Barbis. 2012. Production of a spherical Hydride-Dehydride Titanium powder. In *Proceedings on international conference on powder metallurgy & particulate materials*, Metal Powder Industries Federation (MPIF), Nashville, TN, USA

Neikov, O.D., S.S. Naboychenko, I. Murashova, V.G. Gopienko, I.V. Frishberg, and D.V. Lotsko, eds. 2009. *Handbook of nonferrous metals powders*. New York: Elsevier.

Neikov, O.D., S.S. Naboychenko, and N.A. Yefimov. 2019. *Handbook of non-ferrous metal powders, technology and applications*. 2nd ed. Cambridge, MA: Elsevier. 973 pages.

Niebuhr, D., and M. Scholl. 2005. Synthesis and performance of plasma-sprayed polymer/steel coating system. *Journal of Thermal Spray Technology* 14 (4): 487–494.

Norouzi, S. 2004. *Contribution to the development of hard metal coatings against abrasive wear sprayed by wire arc process*. Ph.D. thesis, University of Limoges, France, 30 Nov

Oberste Berghaus, J., J.-G. Legoux, C. Moreau, F. Tarasi, and T. Chraska. 2008. Mechanical and thermal transport properties of suspension thermal-sprayed alumina-zirconia composite coatings. *Journal of Thermal Spray Technology* 17 (1): 91–104.

Oerlikon Metco. 2018. *Thermal spray materials guide*, BRO-0001-17

Oliker, V.E., E.F. Grechishkin, V.V. Polotai, M.G. Loskutov, and I.I. Timofeeva. 2004. Refractory and ceramics materials influence of the structure and properties of WC-Co alloy powders on the structure and wear resistance of detonation coatings. Powder metallurgy. *Metal Ceram* 43 (5–6): 258–264.

Özbilen, S. 1999. Influence of atomizing gas pressure on particle shape of Al and Mg Powders. *Powder Technology* 102: 109–119.

Patel, K., C.S. Doyle, B.J. James, and M.M. Hyland. 2010. Valence band XPS and FT-IR evaluation of thermal degradation of HVAF thermally sprayed PEEK coatings. *Polymer Degradation and Stability* 95: 792–797.

Pawlowski, L. 1995. *The science and engineering of thermal spray coatings*. New York: Wiley.

Pei, P., J. Kelly, S. Malghan, and S. Dapkunas. 1996. Analysis of zirconia powder for thermal spray: Reference material for particle size distribution measurements. In *Thermal spray: Practical solutions for engineering problems*, ed. C.C. Berndt, 263–273. Materials Park: ASM International.

Petrovicova, E., R. Knight, L.S. Schadler, and T.E. Twardowski. 2002. Thermal spraying of polymers. *International Materials Review* 47 (4): 169–190.

Picas, J.A., A. Forn, L. Ajdelsztajn, and J. Schoenung. 2004. Nanocrystalline NiCrAlY powder synthesis by mechanical cryo-milling. *Powder Technology* 148: 20–23.

Pollak, J., O. Bailly, and R. Dolbec. 2017. Production of spherical metallic powders dedicated to additive manufacturing. *Powder Metallurgy*: 436–443.

Prakash, S., V.K.W. Grips, and S.T. Aruna. 2012. A single step solution combustion approach for preparing gadolinia doped ceria solid oxide fuel cell electrolyte material suitable for wet powder and plasma spraying processes. *Journal of Power Sources* 214: 358–364.

Ravi, B.G., A.S. Gandhi, X.Z. Guo, J. Margolies, and S. Sampath. 2008. Liquid precursor plasma spraying of functional materials: A case study for yttrium aluminium garnet (YAG). *Journal of Thermal Spray Technology* 17 (1): 82–90.

Reist, P.C. 1993. *Aerrosol science and technology*. 2nd ed, 379 pages. New York: McGraw-Hill Inc.

Ren, R.-M., Z.-G. Yang, and L.L. Shaw. 1998. A novel process for synthesizing nanostructured carbides: Mechanically activated synthesis. *Ceramic Engineering and Science Proceedings* 19 (4): 461–468.

Roberts, P.R., James J. Airey, Joseph J. Airey, and J.E. Blout. 1989 *Method and apparatus for producing fine metal powder*, US Patent No. 4,824,478, April 25.

Salman, A.D., M. Ghadiri, and M.J. Hounslow, eds. 2007. *Particle breakage, vol 12, Handbook of powder technology*, 1–1227. Amsterdam: Elsevier.

Schiller, G., M. Müller, and F. Gitzhofer. 1999. Preparation of perovskite powders and coatings by radio frequency suspension plasma spraying. *Journal of Thermal Spray Technology* 8 (3): 389–392.

Schubert, W.D., A. Bock, and B. Lux. 1995. General aspects and limits of conventional ultrafine WC powder manufacture and hard metal production. *International Journal of Refractory Metals and Hard Materials* 13: 281–296.

Schwier, C. 1986. Plasma spray powders for thermal barrier coating. In *Advances in thermal spraying*, 277–286. Oxford: Pergamon Press.

Scott, J.H.J., S.A. Majetich, Z. Turgut, M.E. McHenry, and M. Boulos. 1997. Carbon coated nanoparticle composites synthesized in an RF plasma torch. *Materials Research Society Symposia Proceedings* 457: 219–224.

Seegopaul, P., U.E. McCandlish, and F.M. Shinneman. 1997. Production capability and powder processing methods for nanostructured WC-Co powder. *International Journal of Refractory Metals and Hard Materials* 15: 133–138.

Siwen, Xue, and M.I. Boulos. 2019. Transient heating and evaporation of metallic particles under plasma conditions. *Journal of Physics D: Applied Physics* 52: 454002.

Sordelet, D., and M. Besser. 1996. Effect of starting powder particle size and composition on chemistry and structure of Al-Cu-Fe quasicrystalline plasma sprayed coatings. In *Thermal spray: Practical solutions for engineering problems*, ed. C.C. Berndt, 419–428. Materials Park: ASM International.

Sordelet, D.J., M.F. Besser, and M.J. Kramer. 1998. Thermal spray quasicrystalline coatings. Part 1: Relationships among processing phase structure and splat morphology. In *Thermal spray: Meeting the challenges of the 21st century*, ed. C. Coddet, 467–472. Materials Park: ASM International.

Soveja, A., S. Costil, H. Liao, P. Sallamand, and C. Coddet. 2010. Remelting of flame spraying PEEK coating using lasers. *Journal of Thermal Spray Technology* 19 (1–2): 439–447.

Stanford, M.K., and C. Della Corte. 2006. Effect of humidity on the flow characteristics of a composite plasma spray powder. *Journal of Thermal Spray Technology* 15 (1): 33–36.

Stanoeski, R., and M. Boulos. 2014. *Powder flow monitor and method for in-flight measurment of a flow of powder*, US Patent No: 2014/03456078 A1, Dec 4th.

Suryanarayana, C. 2001. Mechanical alloying and milling. *Progress in Materials Science* 46: 1–184.

Tagai, H., and H. Aoki. 1980. Chapter 39: Preparation of synthetic hydroxyapatite and sintering of apatite ceramics. In *Mechanical properties of biomaterials*, ed. G.W. Hastings and D.F. Williams. New York: Wiley.

Trassy, C., and J.M. Mermet. 1984. *Analytical applications of RF plasmas (in French)*. Paris: Lavoisier.

Tsantrizos, P., F. Allaire, and M. Entezarian. 1998. *Method and production of metal and ceramic powers by plasma atomization*, US patent 5,707,419 July 13th.

Tsunekawa, Y., I. Ozdemir, and M. Okumiya. 2006. Plasma sprayed cast iron coatings containing solid lubricant graphite and h-BN structure. *Journal of Thermal Spray Technology* 15 (2): 239–245.

Ünal, R. 2006. The influence of the pressure formation at the tip of the melt delivery tube on tin powder size and gas/melt ratio in gas atomization method. *Journal of Mater Process Technology* 180: 291–295.

van de Hulst, H.C. 1981. *Light scattering by small particles*. New York: Dover Publications. 470 p.

Vardelle, M., A. Vardelle, and P. Fauchais. 1993. Spray parameters and particle behavior relationships during plasma spraying. *Journal of Thermal Spray Technology* 2 (1): 79–92.

Vinayo, M.E., F. Kassabji, J. Guyonnet, and P. Fauchais. 1985. Plasma sprayed WC-co coatings influence of the spray conditions. *Journal of Vacuum Science and Technology* 3 (6): 2483–2489.

Vuoristo, P., T. Mantyla, L.M. Berger, and M. Nebelung. 1997. Sprayability and properties of TiC-Ni based powders in the detonation gun and HVOF processes. In *Thermal spray: A united forum for scientific and technological advances*, ed. C.C. Berndt, 855–861. Materials Park: ASM International.

Walton, D.E., and C.J. Mumford. 1999. Spray dried products-characterization of particle morphology. *Transactions. Institute of Chemical Engineers* 77: 21–38.

Wang, M., and L.L. Shaw. 2007. Effects of the powder manufacturing method on microstructure and wear performance of plasma sprayed alumina–titania coatings. *Surface Coating Technology* 202: 34–44.

Wang, H.-T., C.-J. Li, G.-J. Yang, C.-X. Li, Q. Zhang, and W.-Y. Li. 2007. Microstructural characterization cold-sprayed nanostructured FeAl intermetallic compound coating and its ball-milled feedstock powders. *Journal of Thermal Spray Technology* 16 (5–6): 669–676.

Wang, Y., R.S. Lima, C. Moreau, E. Garcia, J. Guimaraes, P. Miranzo, and M.I. Osendi. 2009. Mullite coatings produced by APS and SPS: Effect of powder morphology and spray processing on the microstructure, crystallinity and mechanical properties. In *Thermal spray 2009: Expanding thermal spray performance to new markets and applications*, ed. B.R. Marple, M.M. Hyland, Y.-C. Lau, C.-J. Li, R.S. Lima, and G. Montavon, 97–102. Materials Park: ASM International.

Wang, J., E.H. Jordan, and M. Gell. 2010. Plasma sprayed dense $MgO\text{-}Y_2O_3$ nanocomposite coatings using sol-gel combustion synthesized powder. *Journal of Thermal Spray Technology* 19 (5): 873–878.

Wigren, J., J.F. de Vries, and D. Greving. 1996. Effects of powder morphology, microstructure and residual stresses on thermal barrier coating thermal shock performance. In *Thermal spray: Practical solutions for engineering problems*, ed. C.C. Berndt, 855–861. Materials Park: ASM International.

Withy, B.P., M.M. Hyland, and B.J. James. 2008. The effect of surface chemistry and morphology on the properties of HVAF PEEK single splats. *Journal of Thermal Spray Technology* 17 (5–6): 631–636.

Wosch, E., A. Prikhodovski, S. Feldhaus, and T. El Gammal. 1997. Investigation on the rapid solidification of steel droplets in the Plasma Rotating Electrode Process. *Steel Research* 68 (6): 239–246. Investigations.

Wu, J., P.R. Munroe, B. Withy, and M.M. Hyland. 2010. Study of the splat-substrate interface for a PEEK coating plasma-sprayed onto aluminum substrates. *Journal of Thermal Spray Technology* 19 (1–2): 42–48.

Xie, W., J. Wang, and C.C. Berndt. 2013. Analysis of EMAA deposits on glass and mild steel substrates. *Journal of Thermal Spray Technology*. (published online).

Xiong, Z., G. Shao, X. Shi, X. Duan, and L. Yan. 2008. Ultrafine hard metals prepared by WC–10 wt.% Co composite powder. *International Journal of Refractory Metals and Hard Materials* 26: 242–250.

Yandouzi, M., L. Ajdelsztajn, and B. Jodoin. 2008. WC-based composite coatings prepared by the pulsed gas dynamic spraying process: Effect of the feedstock powders. *Surface coating Technology* 202: 3866–3877.

Yule, A.J., and J.J. Dunkley. 1994. *Atomization of melts*. Oxford: Oxford University Press Publishers.

Zambon, A. 2004. Nitrogen versus helium: Effects of the choice of the atomizing gas on the structures of $Fe_{50}Ni_{30}Si_{10}B_{10}$ and $Fe_{32}Ni_{36}Ta_7Si_8B_{17}$ powders. *Materials Science and Engineering A* 375–377: 630–637.

Zhang, C., G. Zhang, V. JI, H. Liao, S. Costil, and C. Coddet. 2009b. Microstructure and mechanical properties of flame-sprayed PEEK coating remelted by laser process. *Progress in Organic Coating* 66: 248–253.

Abbreviations

2-D	Two-Dimensional
3-D	Three-Dimensional
AWJ	Abrasive Water Jetting
BS	Bond Strength
CS	Cold Spray
DC	Direct Current
DC-APS	DC-Atmospheric Plasma Spraying
FS	Flame Spraying
HA	Hydroxyapatite
i.d.	internal diameter
LHS	Left-Hand Side
MEK	Methyl-ethyl-ketone
PSD	Particle Size Distribution
PTA	Plasma Transferred Arc
RF	Radio Frequency
RF-IPS	RF-Induction Plasma Spraying
RHS	Right-Hand Side
SMT	Surface Modification Technologies
WAS	Wire Arc Spraying

14.1 Introduction

As mentioned in Chap. 1, Introduction to Thermal Spray, of this book, the increasing demands for combined functional requirements of materials while keeping the final cost of the part at an acceptable level led to the ever-increasing demand which allows the decoupling of the surface properties of a part from its bulk and structural properties. Numerous thermal spray technologies have been reviewed in this context in Chaps. 6, 7, 8, 9, 10, 11 and 12, covering Cold Spray (CS), Flame Spraying (FS), DC and RF Induction Plasma Spraying (DC-APS) (RF-IPS), Wire Arc Spraying (WAS), and Plasma Transferred Arc (PTA) deposition. One common aspect of all

of these technologies is that they require proper surface preparation in order to ensure the adhesion of the coating to the substrate and achieving the functional modification of the surface properties in a reliable and economical fashion.

Surface preparation for thermal spraying is a relatively simple and intuitive task that starts at the design stage of the part in order to avoid sharp edges and corners, which can be the starting point of crack initiation in a coating. The following five principal steps of surface preparation include, *machining, cleaning, masking, roughening, and finishing.* This chapter is devoted to a brief description of each of these steps with emphasis on thermal spray process requirement and coating performance needs.

14.2 Basic Concepts

Thermal spraying begins with proper surface preparation, which is absolutely essential. Steps must be undertaken correctly in order for the coating to perform the design expectation [Davis J.R. ed. (2004)]. Without surface preparation, failure of the coating becomes highly probable because coating adhesion quality is directly related to the cleanliness, the roughness, and sometimes the proper machining for optimal coating performance. The coating material and the nature of the substrate are the major factors in determining what kind of surface preparation is necessary to achieve a resistant bonding.

14.2.1 Substrate Design

Good surface preparations start at the design stage. A number of important rules have to be respected in the design of the part to be coated in order to avoid creating weakness points in the coating where cracks would initiate once the coated part is exposed to mechanical or thermal stresses. The principal guidelines recommended by the Thermal Spraying,

© Springer Nature Switzerland AG 2021
M. I. Boulos et al. (ed.), *Thermal Spray Fundamentals*, https://doi.org/10.1007/978-3-030-70672-2_14

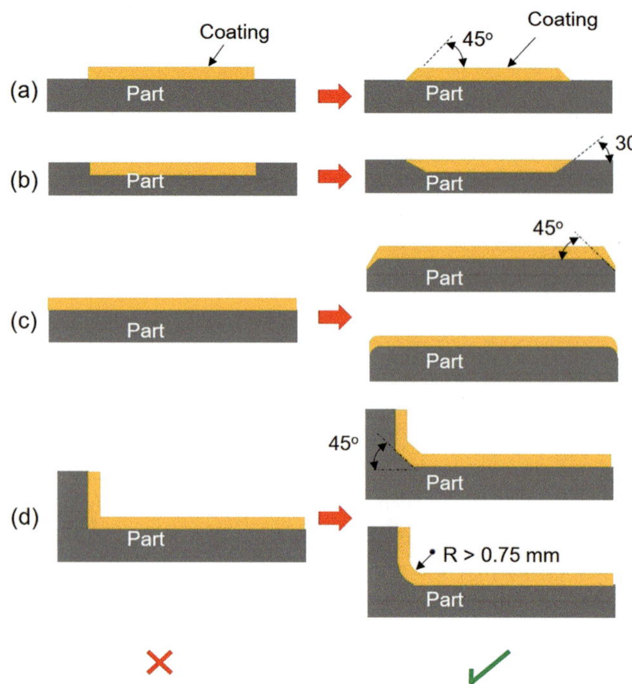

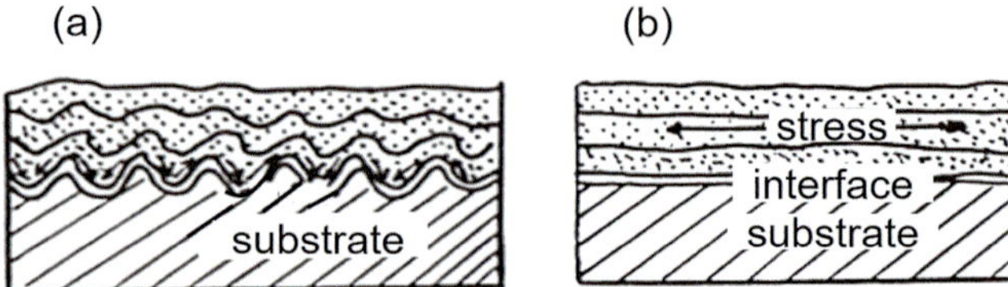

Fig. 14.2 (**a**) Grooved substrate and (**b**) smooth substrate. Reprinted with kind permission from American Welding Society [Davis J.R. ed. (2004)]

Fig. 14.1 Recommended machining for areas to be sprayed: (**a**) best design with a narrow-feathered band at the coating extremities, (**b**) undercut shape chamfered at about 30°, (**c**) corners must be chamfered at 45° or rounded with radius > 0.75 mm, (**d**) shoulders at 90° must be chamfered at 45° or rounded with radius of >0.75 mm. Reprinted with kind permission from ASM International [Davis J.R. ed. (2004) and Thermal Spraying, American Welding Soc. (1985)]

American Welding Soc. (1985) [Davis J.R. ed. (2004)], illustrated in Fig. 14.1, can be summarized as follows. The LHS of the figure representing what NOT to do while the RHS representing what is recommended. While these recommendations are illustrated for flat surfaces, they apply equally well to cylindrical or other surfaces:

- A coating must never end abruptly at the part extremity. A sharp edge, as illustrated on the LHS of Fig. 14.1a, may act as a stress raiser where cracks will develop, especially when loads are applied. Coating at the extremity of a part should accordingly present a narrow-feathered band, rather than a sharp edge.
- When undercutting a substrate area to accept the coating, the coating deposited must never end abruptly. In the undercuts the corners must be chamfered, or its cutting edge removed before spraying. The chamfer angle should be about 30° as illustrated in Fig. 14.1b. Sharp corners capture loose spray particles, dusts, and debris, resulting in porous areas.
- For a part with a sharp corner at 90°, the corner must be chamfered at 45°, as illustrated in Fig. 14.1c, or rounded with radius of >0.75 mm. Otherwise, with an abrupt edge, cracks will develop.

- For a part with a sharp shoulder at 90o the shoulder must be chamfered at 45°, as illustrated in Fig. 14.2d, or rounded with radius of >0.75 mm to avoid the development of cracks.

For the spraying of thick coatings or coatings on a substrate with rounded profile with a short radius of curvature, it is recommended to machine grooves or threads into the surface to be sprayed. As illustrated in Fig. 14.2a, such groves will reduce shear stresses parallel to the surface by restricting shrinkage stresses, and disrupting lamellar patters of splat deposition on the surface of the substrate. [Thermal Spraying American Welding Soc. (1985)]. The surface is generally roughened after grooving. Two types of grooves are used: V-shaped ones, when the V angle relatively to the part surface is 70° with the root rounded, or U-shaped grooves, between 1.1 and 1.4 mm in width.

14.2.2 Masking

Masking is necessary to prevent the deposition of the coating on areas where it is not wanted. It also improves the uniformity of the deposit when coating limited areas. A wide range of coating materials are used, including self-adhesive tapes or hard masks. In either case, the material use should be resistant to the conditions to which it is exposed in the roughening and thermal spray coating stages of the operation. Its attachment to the substrate should be reliable and strong, while being readily removal at the end of the coating step. Two types of masks are commonly used:

- *Contact masks*, using self-adhesive tapes, as illustrated in Fig. 14.3a, are applied to the substrate on the areas that are to be protected. The tape is withdrawn after grit blasting and thermal spraying operation. Any residual glue in the areas that were covered must be cleaned by an appropriate solvent. As shown in Fig. 14.3b, the edge of the coating in this case is sharp and can induce, in that area, debonding due to stresses when loads are applied. Contact masking is mostly used when the number of parts to be coated is small.

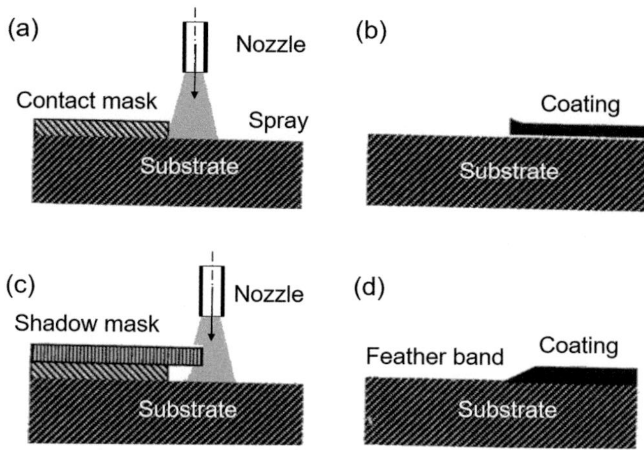

Fig. 14.3 Effect of masking technique on coating profile: (a) contact masking, (b) coating profile obtained with contact masking, (c) shadow masking, (d) coating profile obtained with shadow masking [Davis J.R. ed. (2004)]. Reprinted with kind permission from ASM International

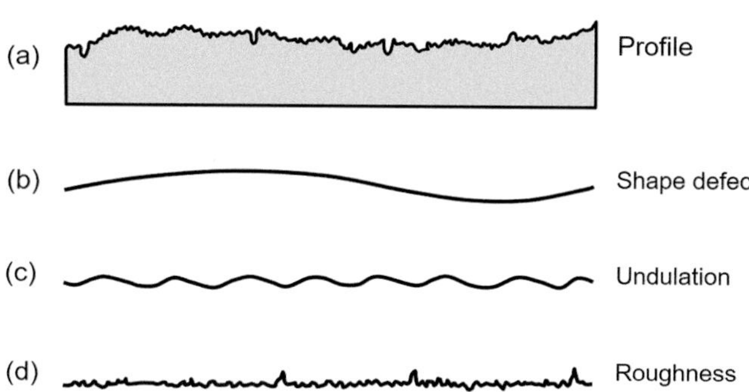

Fig. 14.4 Schematic representation of different types of surface defects as measured using contact-skidded instruments. (a) Combination of all three types of surface defects; (b) shape-type defects; (c) undulation-type defects and (d) roughness—type defects

- **Shadow masks**, consisting of placing a metal shield machined to the dimensions and profile of the areas to coated and firmly fixing it to the substrate, as shown in Fig. 14.3c, for both grid blasting and thermal spraying operations. Because of the small spacing between the shield and the substrate, a narrow-feathered band at the edges of the coating is formed as recommended for non-protected parts, Fig. 14.3d. Masks are generally recommended for the coating of a large number of parts. While they are reusable, they still need to be renewed on a regular basis due to normal wear and tear depending in its utilization conditions.

14.2.3 Surface Roughening

14.2.3.1 Definitions

Surface profile characterization is rather complex because of the different type of defects that can be simultaneously present in a surface. As illustrated in Fig. 14.4, these can generally be classifieds in the following three broad groups based on their linear dimension "wave-length," with Fig. 14.4a representing the combination of all of the three in a single part.

- Shape defects are observed at rather large scales of few centimeters, Fig. 14.4b.
- Undulation is generally present at the millimeters scale level, Fig. 14.4c.
- Roughness is observed at yet a smaller scale, Fig. 14.4d.

Roughness is typically considered to be the high frequency, short wavelength component of a measured surface characteristic that is the most pertinent to thermal spraying. This is due to the fact that the roughness scale is comparable with the size distribution of the sprayed powder/molten droplets impinging on the surface of the substrate. As will be discussed in Chap. 15, Coating Formation, the compatibility between the size of the spray droplets and surface roughness can have an important impact on the level of mechanical adhesion of the coating to the surface of the substrate.

14.2.3.2 Measurement Techniques

Surface roughness is a critical parameter on which the quality of the bonding between the coating and the substrate strongly depends. It can be measured using "contact-type" instrument [Dong S. et al. (2011), Montavon G. (2004)] in which a styles scans over the surface to detect the surface profile characteristics along a given trace. Alternately, a "non-contact approach" can be used based on optical, ultrasonic, and capacitance methods [Vorburger T.V. and E.C. Teague (1981), Tonshoff H.K. et al. (1988), Shin Y. C. et al. (1995), Bradley C. (2000)].

With a "contact-type" instrument, the profilometer diamond stylus scans repeatedly over the surface to be evaluated in different random orientations, measuring in each scan the surface profile characteristics. The collected data is then analysis, and statistically averaged over at least over 12–18 scans in order to have a reasonable assessment of the properties of the surface. A typical trace of a single traverse of length l_t is given in Fig. 14.5. A cut-off length is defined in order to discriminate the possible undulation from the roughness. If the cut-off length is below 1 mm the undulation is reduced. The assessment length l_n must be at least 5–6 times the cut-off length in order to define a length comprising undulation and roughness. The previous and post-running segments of the scan are discarded from the acquisition [Dong S. et al. (2011), Montavon G. (2004)]. The main limitation of the technique is the size of the stylus compared

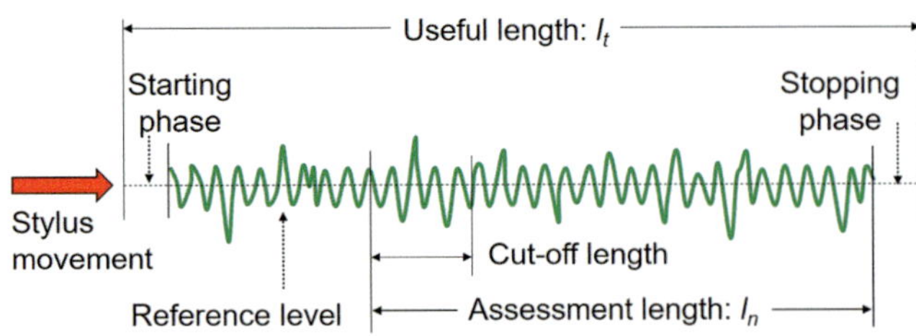

Fig. 14.5 Definitions of different length for roughness measurement: With a contact skidded instrument, such as the profilometer diamond stylus, the instrument probes the surface over a traverse length l_t [Reprinted with kind permission from Prof. G. Montavon]

to that of undercuts. The stylus may be either too blunt to reach the bottom of deep valleys or too sharp and scratch the surface of the substrate or round the tips of sharp peaks. In such cases the probe acts as a physical filter that limits the accuracy of the instrument.

Non-contact instruments, on the other hand, employing optical, ultrasonic, and capacitance [Vorburger T.V. and E.C. Teague (1981), Tonshoff H.K. et al. (1988), Shin Y. C. et al. (1995), Bradley C. (2000)], offer a better precision at the risk to have the surface profile modified through the sample preparation protocol. There are also limitations to non-contact instruments that rely on optical interference that cannot resolve surface features that are smaller than some fraction of their operating wavelength. This is especially true when measuring very smooth surfaces [Bradley C. (2000), and Cedelle J. et al. (2006)] for which there is a growing interest in conjunction with the plasma spraying of nanopowders using suspension or solution spraying. The splat sizes in these cases are in the range 0.1–1.0 μm that is in the same size range as that of the roughness levels of smooth surfaces.

An interesting comparison between the different techniques used for the measurement of surface roughness at the micrometer scale level is presented in the paper by [Conroy M. et al. (2005) and Hu Z. et al. (2008)]. [Al-Kindi GA and Shirinzadeh B (2007)] point out that accurate surface characterization would require three-dimensional (3-D) surface measurements for a better understanding of the surface properties. Such measurements are, however, rather complex and demanding in terms of data processing compared to traditional two-dimensional (2-D) surface measurements. In the following, the discussion of surface characterization parameters will be limited to 2-D measurements based on the amplitude and spacing criteria [Tomovich SJ and Peng Z (2005)]. The most frequently used parameters are:

The arithmetical average, AA;

$$AA = \frac{h_1 + h_2 + \dots h_N}{N} \tag{14.1}$$

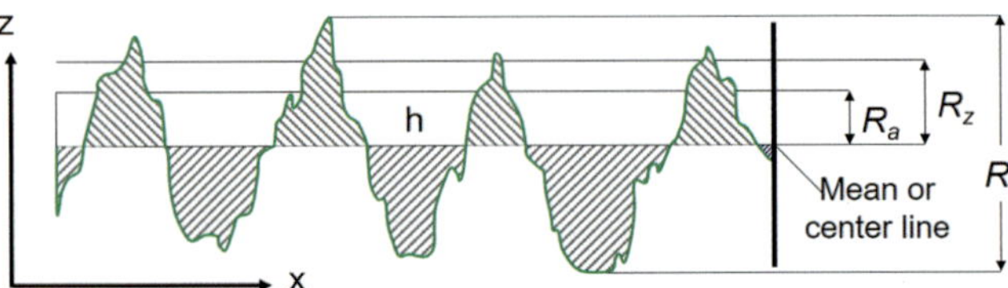

Fig. 14.6 Definitions of the average roughness R_a, root mean square roughness R_z, distance between highest peak and deepest undercut, R_t [Montavon G. (2004)]

where h_i *with* $i = 1$ *to* N, the height of all peaks and undercuts (without regard to the sign) from the centerline and, N the number of measurements.

As illustrated in Fig. 14.6, the following parameters are commonly used for the characterization of the value of the surface roughness over a scan of length, l.

The average roughness, R_a;

$$R_a = \frac{1}{l} \int_0^l z(x).dx \tag{14.2}$$

The root-mean square roughness, R_z

$$R_z = \sqrt{\frac{1}{l} \int_0^l z(x)^2 dx} \tag{14.3}$$

The R_t is the distance between the highest peak and the deepest undercut and which describes local singularities (Fig. 14.6). This value is very important for the mechanical adhesion of splats.

The standard deviation of the profile angle, $R_{\Delta q}$, represents the root mean-square roughness, is obtained from the local profile slope ($dz(x)/dx$) of the roughness profile as shown in Fig.14.7 given by Eq. 14.4.

$$R_{\Delta q} = \sqrt{\frac{1}{l} \int_0^l \left(\frac{dz(x)}{dx}\right)^2 dx} \tag{14.4}$$

The skewness parameter, S_k;

$$S_k = \frac{1}{\sigma^3} \int_{-\infty}^{+\infty} (z - m)^3 \varphi(x) d \tag{14.5}$$

where z is the surface height, m its mean value, x the sampling length and $\varphi(x)$ the distribution function of the surface heights. The influence of the skewness value on the surface morphology is illustrated in Fig. 14.8. The skewness has been shown to be particularly important at the nanometer-scale because it plays a key role on the liquid droplet wetting ability upon its flattening on the surface of the substrate and thus on splat formation [Thomas T.R. (1999), Bahbou F. and P. Nylen (2005), Fukumoto M. and Y. Huang (1999).

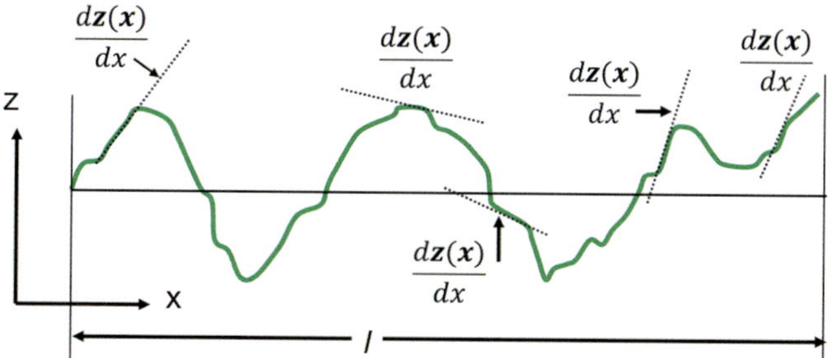

Fig. 14.7 Definition of the slopes used to define $R_{\Delta q}$ [Bahbou F. and P. Nylen (2005)]

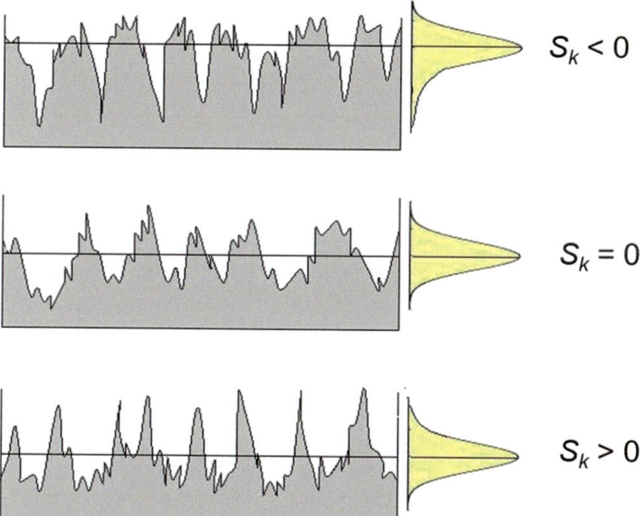

Fig. 14.8 Scheme of surfaces with different skewness. Reprinted with kind permission from [Montavon G. (2004)]

Figure 14.9 illustrates the skewness effect on a stainless steel 304 L kept at room temperature (Fig. 14.9a) with S_k close to 0 and preheated at 250 °C with a DC plasma torch (Ar–H$_2$ plasma) during 120 s, resulting in a skewness close to 1 due to surface oxidation (Fig. 14.9b).

The kurtosis Ku or third moment is used to quantify the anomalies in the heights, as illustrated in Fig. 14.10.

Parameters are also used to define the mean spacing R_{sm}, which is the arithmetic mean of the widths of the profile within the single measuring length that consecutively cross a lower threshold (C_2) and an upper threshold (C_1), as given by Eq. 14.6 and shown in Fig. 14.11.

$$R_{sm} = \frac{1}{m} \sum_{i=1}^{m} x_i \qquad (14.6)$$

The question is: which parameter is directly related to the coating adhesion? For example, Bahbou F. and P. Nylen (2005) have studied the influence of R_a, R_{sm}, and $R_{\Delta q}$ on the adhesion of Ni 5 wt.% Al coating plasma sprayed onto Ti-6Al-4 V substrates grit blasted with two brown alumina grit sizes (blasting angle 45°). Results show that the adhesion values measured are poorly correlated with R_a or R_{sm}. However, the hybrid parameter containing both amplitude and spacing properties of the surface $R_{\Delta q}$ (i.e., the average slope: see Fig. 14.7) is rather well correlated to the coating adhesion.

Of course, many other parameters can be used to characterize surfaces, including fractal dimensions [Amada A. and T. Hirose (2000), Fukumoto M. et al. (2004), Amada S., H. Yamada (1996), Amada S. and A. Satoh (2000), Guessasma S., et al. (2003)]. In the latter case; however, the methodology to accurately describe the surface state is very important and the fractal methodology is sensitive to the measurement scale and to the calculation protocol [Amada S. and A. Satoh (2000)].

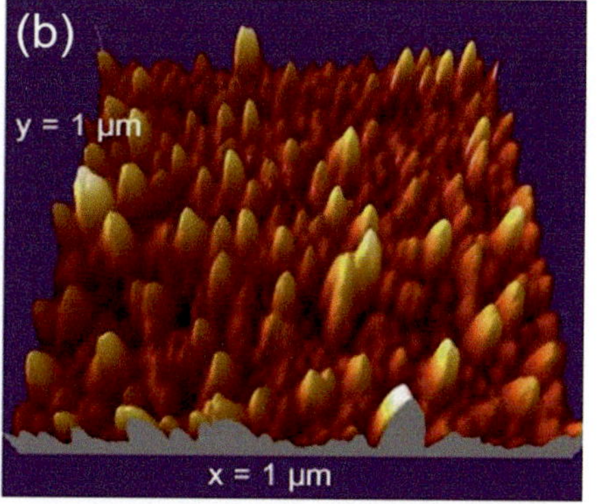

R_a = 0.6 nm; S_k = 0 – 0.1 R_a = 3.5 nm; S_k = 0.9

Fig. 14.9 304 L stainless steel surface 1×1 µm^2 (**a**) kept at room temperature after polishing, (**b**) preheated by a DC plasma torch (Ar-H$_2$) at 250 °C during 120 s. Reprinted with the permission of Elsevier [Cedelle J. et al. (2006)]

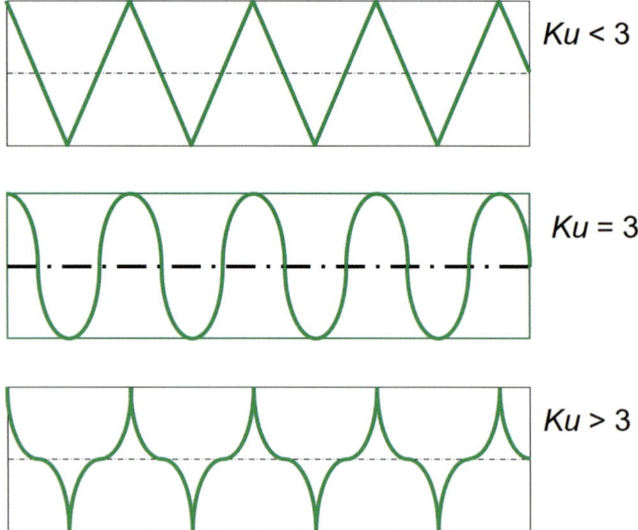

Fig. 14.10 Illustration of surface morphologies and resulting Kurtosis [Conroy M. et al. (2005)]. Reprinted with the permission of Prof. G. Montavon

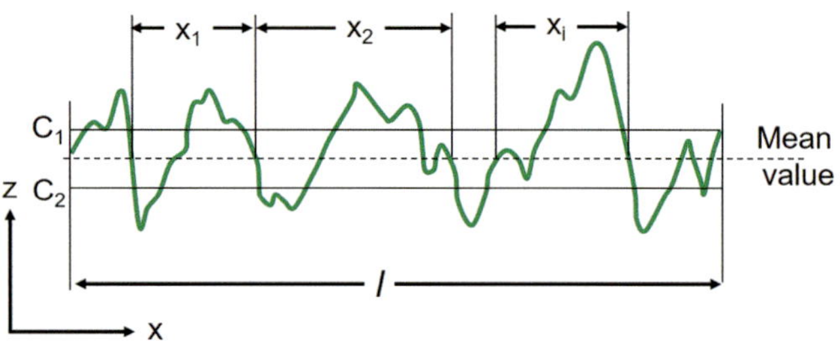

Fig. 14.11 Definition of the arithmetic mean spacing. Reprinted with kind permission from Springer Science Business Media

14.3   Cleaning

This is the first step for the substrate preparation. Before spraying, all contaminants, such as scale, oil, grease, and paint must be removed. After removing all contaminants, parts should be protected from airborne debris and fingerprints and should be handled with clean fixtures and materials. The main cleaning techniques are described in detail in ASM Handbook Volume 05: Surface Engineering (1994).

14.3.1 Solvent Degreasing

Solvent degreasing is a fast, efficient, and economical way commonly used to remove organic contaminants. Parts should be soaked 15–30 min to remove oil from interstices and surface pores. Porous materials must be soaked for

longer periods. Most solvents are hazardous, and manufacturer's instructions should be followed strictly (usage, location, aeration, and disposition). Recycling of solvents is strongly recommended. The solvents most often used include: Methyl-ethyl-ketone (MEK) and acetone. Aliphatic hydrocarbons are also used but in a blend of other solvents or diluents to be less aggressive and drying without forming a film. Isopropyl alcohols are used for precision cleaning, but they are not effective on heavy oils and some other contaminants. Large parts can be steamed or submerged into solvents, for example, aqueous washer solutions containing acetic acid. Chlorinated solvents, while very effective, are no more allowed due to environmental concerns.

14.3.2 Baking

The porous parts, such as those resulting from sand castings, that may absorb large quantities of oil are baked at 315 °C for 4 h to dry the oil and prevent bleeding.

14.3.3 Ultrasonic Cleaning

This process is used for relatively large pieces when contaminants are lodged in a restricted area. The solvent has to be adapted to the problem and the safety requirements. The equipment consists of a holding tank filled with a cleaning solution, mostly water with an appropriate detergent, in which the part is soaked while being exposed to strong ultrasonic vibration.

14.3.4 Wet or Dry Blasting

Wet abrasive blasting uses diluted slurry of abrasive media projected by an air jet onto the surface. Typical slurries are in the ratio 0.6 kg/L and rust inhibitors must be added. The abrasive size has to be adapted to the works to be done. Parts must be carefully rinsed after cleaning.

Dry abrasive is used to remove baked-on-deposits, scale, and oxides. A compressed air stream containing abrasive particles is directed, through a nozzle, toward the part to be treated [Lidong Wang (2004)]. The system is similar to grit blasting, but it should not be used for roughening because the contaminants introduced in the grit could pollute the surface.

14.3.5 Acid Pickling

Dilute acid etching is a very efficient method. The part is immersed totally in an acid solution for up to a few hours depending on the material and stock removal desired. After pickling, a hot water rinsing, an alkaline solution, and a careful hot water or steam blast cleaning are necessary.

14.3.6 Brushing

This process is used when only localized cleaning is needed. Small rotary wire brushes driven by a power tool are used in these cases to clean the surface.

14.3.7 Dry Ice Blasting

Dry ice blasting is a blasting method that uses small, compact dry ice pellets, which are not as hard as alternate blasting material. They are made of solid carbon dioxide at a temperature of about $-79\,^{\circ}\mathrm{C}$. Typical pellet sizes are in the few mm range (typically a diameter of 3 mm and 5–15 mm long). The dry ice pellets are accelerated in a jet of compressed air similar to that used in traditional blasting operations. Upon impact the energy transfer to the surface knocks contaminants as in grit blasting, while the dry ice pellet explodes and warms rapidly, the CO_2 gas produced expanding partly under the contaminant surface, thus participating in its removal. Moreover, they also create a micro-thermal shock between the surface contaminant and the substrate, inducing cracking and delamination of the contaminant. Dry ice blasting is a mild pretreatment method, since the dry ice pellets are relatively soft (hardness of 2–3 Mohs), thus causing only minimal damage to the surface and the fringe area. The system can be used on easily damaged surfaces like nickel, chromium, and soft aluminum for the removal of contaminants such as adhesives, varnish, oil, grease, and coal dust [Stratford S (2000), Spur G et al. (1999), Elbing F et al. (2003) and Bardi U et al. (2004)].

Dry ice blasting has occasionally been combined with the plasma spray coating operation. In this case, the dry ice blasting nozzle was located just before the plasma torch while the substrates were attached to a rotating holder with its axis orthogonal to those of the plasma torch and the dry ice blasting nozzle. With the simultaneous dry ice blasting of the substrate during the coating operation, the substrate and coating temperatures were reduced by a factor up to 2 compared to air jets cooling. [Dong S. et al. (2011)] have compared plasma sprayed Al_2O_3, CoNiCrAlY, and steel coatings, obtained with either conventional air cooling or dry ice blasting during plasma spraying. Denser and less oxidized steel and CoNiCrAlY coatings were obtained with dry ice blasting and the adhesive strength of the Al_2O_3 coating obtained was increased by about 30%. Authors attributed the improvement of the Al_2O_3 coating adhesion to the cleaning effect of the dry ice, the denser non-ceramic coating

properties, to the mechanical effect associated with the impact of the dry ice on the surface, and the oxide reduction due to the better cooling efficiency [Dong S. et al. (2011)].

14.4   Roughening by Grit Blasting

After cleaning and masking several methods are used to produce a surface to which the sprayed coating will adhere. Dry abrasive grit blasting is the most commonly used roughening technique by which dry abrasive particles are propelled toward the substrate at relatively high speeds. On impact, the sharp, angular particles act like chisels, cutting small irregularities into the surface creating the so called "roughness" depending on the nature and size of the grit used, the substrate material properties, as well as the grit blasting conditions, as described in the following section.

14.4.1 Grit Blasting Equipment

The type, size, and degree of automation of any blast system must be tailored to the part and to the scale of the operation. Blasting must always be performed in an enclosure designed for that purpose and equipped with exhaust and dust collection equipment's. As the grit particles fracture during the process and releases debris from the surface of the substrate in the form of fine dust, the used grit must be cleaned, re-screened, sized, and dust removed before recycling it in the process (generally it falls through a grated floor to a conveyor that transfers it to the cleaning system).

In thermal spraying, mainly two types of machines are used for the grit blasting operation:

- Blast cabinets with suction-type nozzles
- Pressure machines or blast generators

The principle of the suction-type gun is illustrated in Fig. 14.12.

The abrasive medium is dispensed from a hopper and air propelled to the substrate. These machines can be used with

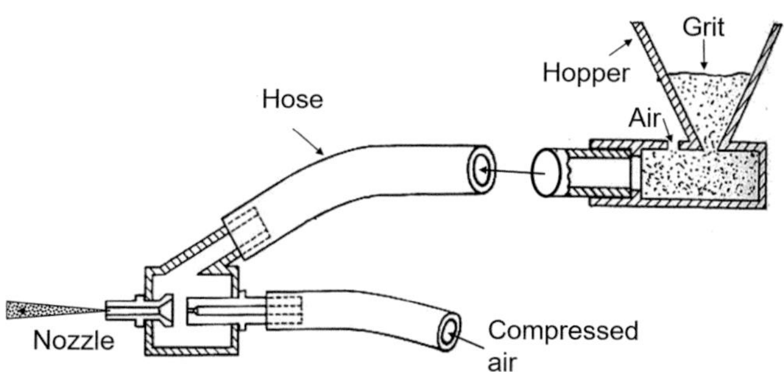

Fig. 14.12 Principle of the suction-type gun [Davis J.R. ed. (2004)]

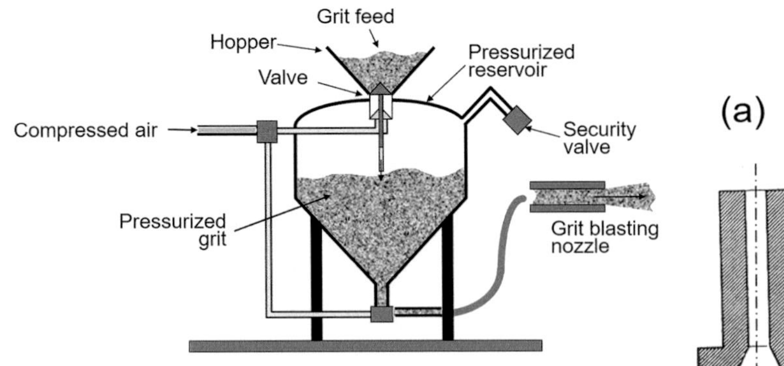

Fig. 14.13 Principle of the pressure-type grit-blasting machine

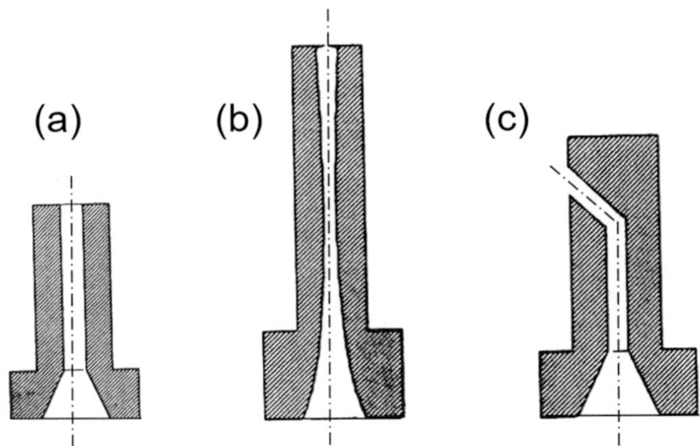

Fig. 14.14 Schematic of three typical grit-blasting nozzles: (**a**) cylindrical or conical (**b**) Venturi (**c**) with deviated jet

almost any type of abrasive with the exception of the larger sizes of crushed iron. This type of machine is less efficient than pressure-type machines, but the grit flow rate is lower (in a ratio of 3–6) and simpler. For more information about grit and air flow rates see the table given in [Davis J.R. ed. (2004)].

The pressure-type machine is shown schematically in Fig. 14.13. It is used for large work pieces requiring high production rates. They are more effective and efficient than suction-type machines, since higher velocities are imparted to the abrasive grains, thus allowing for greater cutting action. They use a closed reservoir for the abrasive, a carrier gas and a blasting nozzle. The reservoir is pressurized, and the grit carried, through the hose and nozzle to the part to be grit blasted. The reservoir must be periodically recharged with the abrasive, which is done automatically every time the compressed air supply is stopped. Different types of grit can be used with pressure-type machines with the grit velocity directly linked to the air pressure used.

14.4.2 Grit Blasting Nozzles

Straight nozzles are commonly used for commercial work. There are many sizes (diameters), lengths, and configurations as illustrated in Fig. 14.14 for the three most common ones:

Long wear nozzles are generally made of carbide (mainly WC) or a carbide-based composite [Bacova V. and D. Draganovska (2004) and Deng J., S. Junlong (2008)]. They are inserted into a steel or aluminum mounting to protect their hard, brittle, inner core.

The compressed air velocity v_g, which determines the grit particle velocity, depends strongly on the nozzle internal cross section area S_n (when compressibility effects are neglected, where the air mass flow rate $\left(\dot{m}_g = \rho_g\, v_g\, S_n\right)$ with ρ_g being its specific mass of the air. Because of the gradual wear of the nozzle inner surfaces during operation,

its principal dimensions need to be checked on a regular basis to avoid drifting of the grit blasting, which would obviously affect the surface properties of the grit-blasted parts. It is to be noted that, for example, for an 8 mm i.d. nozzle a 10% increase in its internal diameter results in a 27% increase of S_n with a corresponding drop of the grit velocity for the same compressed air flow rate.

14.4.3 Grit Material

The most commonly used grits are aluminum oxide, silicon carbide, and steel grit. Upon impact, the abrasive particles induce a plastic deformation of the substrate and the sharp angular particles act like chisels, cutting small irregularities in the surface. As shown by Griffiths B. J., D. T. Gawne et al. (1996), two main types of surface damage were observed.

- The first is produced by an impact from a sharp, angular grit particle that forms an indentation, often with a lip at its periphery. The crater formed is angular with steep sides and resembles the shape and form of the impinging grit. The crater is formed by plastic indentation of the surface region and elastic compression of the hinterland immediately beneath. It is designated indentation-deformation.
- The second type of damage in which an impinging particle strikes the surface and micro-machines a crater with a large lip or prow at its exit end. The sharp edge of the particle has cut into the metal, creating a fresh surface, pushing the metal and folding it over onto the undisturbed surface. The crater is more curved than the previous one but still has very angular features particularly in the region of the micro-chip. This is designated cutting-deformation since it involves fracture as well as elastic-plastic deformation.

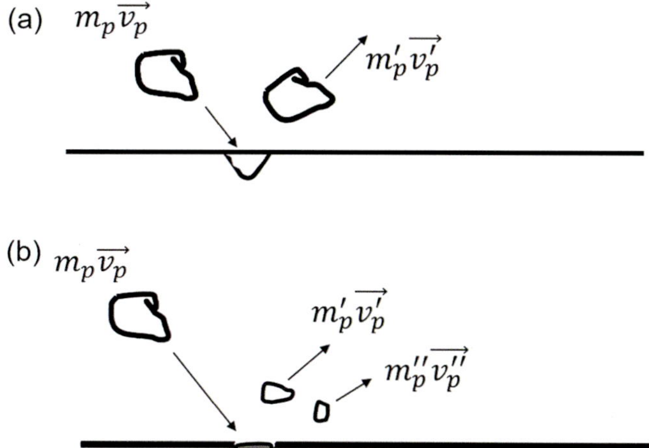

Fig. 14.15 Schematic of the grit particle impact creating an irregularity in the substrate before rebounding (**a**) without breaking, (**b**) with breaking

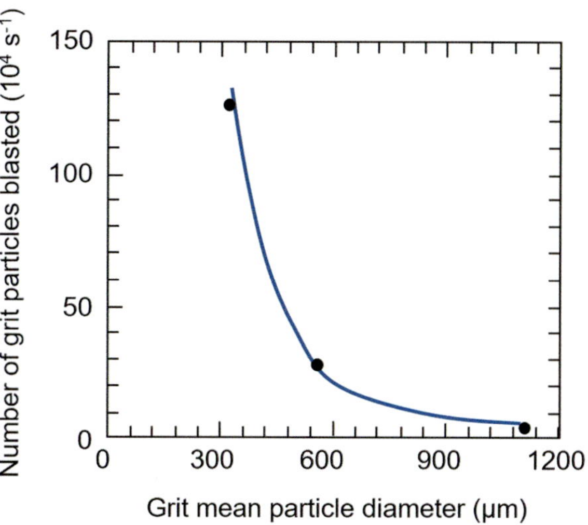

Fig. 14.16 Dependence of the number of grit particles per second on the mean grit diameter (suction-type machine, 6 mm i.d. nozzle, blasting pressure of 0.4 MPa [Bellmann R. and A. Levy (1981)], copyright ASM International, reproduced with kind permission

The amount of substrate deformation and irregularity created at the surface is a function of the size, specific mass, and hardness of the impacting particle and of its speed and angle at which it is directed toward the substrate (see Fig. 14.15a). Of course, the substrate deformation and the irregularity formed also depend on the substrate material (hardness, Young's modulus …). The surface roughness is obtained by brittle fracture and/or micro-cutting [Griffiths B. J., D. T. Gawne et al. (1996)]. On impact the grit particles rebound without breaking (Fig. 14.15a) or with breaking (Fig. 14.15b).

Again, it is very important to keep in mind that the particle momentum $m_p \overrightarrow{v_p}$ depends on its mass, proportional to the cube of its mean diameter. Thus, broken particles, if reused, will have a lower momentum, even if, for the same airflow rate, the particle velocity increases when its size decreases.

Since the roughness of the surface depends on the size of the grit, abrasives are supplied in different grades characterized by the screen or sieve used to determine powder size and/or its PSD. They are defined by standard mesh sizes (usually between 12 and 320). The smaller is the mesh number, the courser is the grit (see Chap. 13 Powders, wires and cords). Smaller particles allow for the preparation of a larger area per hour (more particles in 1 kg of powder) but the roughness is less. The bigger the particles are, the faster is the material removal from the substrate surface and thus the rougher is the finish. Of course, the larger the particle grit size is, the higher is the compressive peak close to the substrate surface.

It is also important to emphasize that the number of particles impacting on a given surface during one second varies with the grit size and decreases when the grit size increases. This is illustrated in Fig. 14.16 from [Maruyama T. et al. (2006)].

14.4.3.1 Aluminum Oxide

It is an angular and durable blasting abrasive that can be recycled (after sieving) many times. It is among the hardest grits that is used systematically for very hard surfaces (> 45 HRC). On soft surfaces (e.g., aluminum) it may be embedded. When properly crushed, aluminum oxide grits have sharp cutting edges. Its specific mass of 3900 kg/m^3 is about half that of chilled iron grit (about twice as many particles in a kg of grit). There are two types of aluminum oxide:

- Brown aluminum oxide, which contains less than 1.5% free silica, its composition is, for example, 93.3% Al_2O_3, 3.5% TiO_2, 1.8% SiO_2, 0.4% Fe_2O_3 (CaO and MgO < 1%), all percentages being in wt.%.
- White aluminum, which is 99.5% pure grade. It is very hard (> 9 on the Mohs scale).

14.4.3.2 Silicon Carbide Grit

It is the hardest blasting medium available. High quality silicon carbide is manufactured to a blocky grain shape that splinters and results in grit particles having rather sharp edges. It may react at high temperatures with various substrates. The concern is that damage to the base material may occur, and coating deterioration may be caused by the combination of silicon and carbon with oxygen as well as the elements present both in the base material and in the coating. On the other hand, SiC allows getting "near to zero" contamination at the interface coating/substrate because it dissolves during heat treatment [Maruyama T. et al. (2006)]. SiC has a much greater tendency to embed and it breaks down more rapidly than aluminum oxide. However, it results in a surface with higher

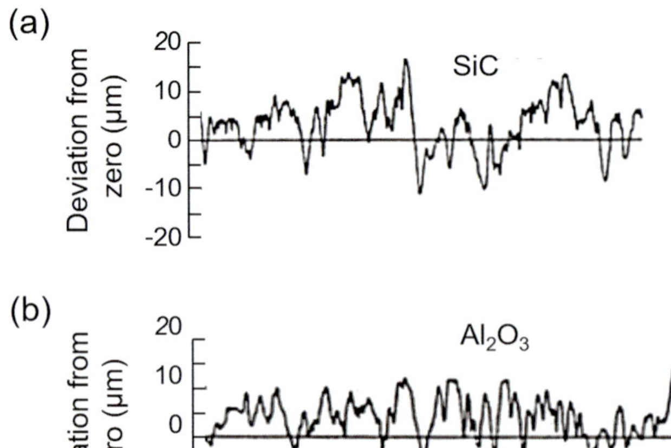

Fig. 14.17 Effect of grit material on surface roughness obtained (**a**) SiC grit (**b**) alumina grit. Same substrate, grit size distribution and blasting conditions [Scrivani A. et al. (2006)]

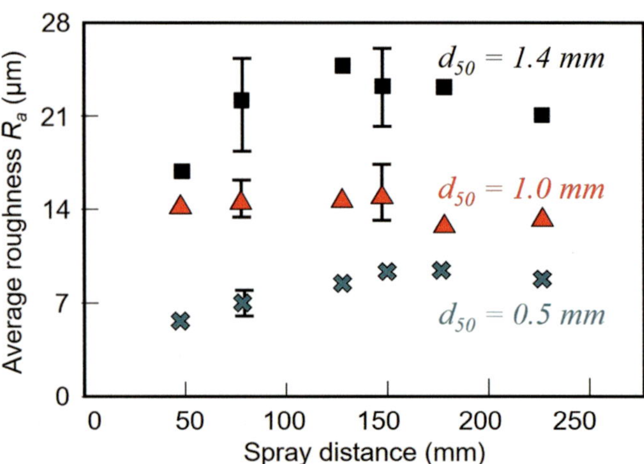

Fig. 14.18 Dependence of the average roughness R_a on the blasting distance for three different brown alumina grits, suction-type machine, nozzle i.d. 8 mm, aluminum substrate AU4G [Amada S. et al. (1999)]

peaks and deeper undercuts when considering the same substrate, the same grit size and the same blasting conditions. This is illustrated in Fig. 14.17 [Yankee S.J. et al. (1991)].

14.4.3.3 Angular Chilled Iron

The steel used contains more than 0.8 wt.% carbon (depending on its hardness) and less than 0.05 wt.% of sulfur and phosphorous. Its specific mass is about 7300 kg/m³. Different chemistries lead to different hardness:

- The softer one (40–50 HRC) rounds off rapidly and is mainly used for stripping oxides and cleaning,
- The harder one (55–65 HRC) maintains its angular shape and provides a good cutting action.

14.4.3.4 Other Grits

Silica is inexpensive and widely available. However, it presents an environmental hazard with the possibility of causing silicosis. In some countries, its use is either forbidden or submitted to strict regulations.

- Crushed garnet: its composition and blasting properties vary widely, depending on the origin of the ore.
- Crushed slag: it is made from the slag of certain types of furnaces. It is very angular, but its silica content must be checked.

14.4.4 Blasting Parameters

As mentioned earlier the quality of surface roughening by grit blasting depends on the blasting parameters, the properties of

the grit used, as well as the properties of the substrate. These can be summarized as follows:

- Blasting pressure/impact velocity
- Blasting distance
- Impact angle
- Blasting time
- Grit size and composition
- Substrate Young's modulus

14.4.4.1 Blasting Pressure

When increasing the blasting pressure, the particle impact velocity is increased. However, the effect is strongest with pressure-type machines, but rather small with suction-type machines [Amada S. et al. (1999) and Bahbou M.F. et al. (2004a)].

14.4.4.2 Blasting Distance

Results reported by [Mellali M. et al. (1997)] with three brown alumina grits, given in Fig. 14.18, shows the dependence of R_a on the blasting distance (suction-type machine, 8 mm i.d. nozzle). Whatever grit size is used, first R_a increases with the blasting distance reaching a plateau followed by a gradual decrease at distances above 150 mm.

Similar results were obtained with cast iron substrates FT-25 and hard steel 100C6. The results dispersion increases slightly with distance and the plateau is slightly broader when the Young's modulus of the substrates increases. When the blasting distance, d_b is too short, the rebounding particles reduce the number and/or the efficiency of impacting grit particles with which they collide in-flight. When d_b is too large, the particle velocity decreases, especially that of small particles, which have a lower inertia. It is worth noting that results are more variable when considering R_t instead of R_a. However, the general trend is the same.

14.4.4.3 Impact Angle

The influence of the impact angle is still subject to controversy. However, most authors agree on the fact that if a surface is blasted at an angle from one direction and then sprayed at an angle of the opposite direction the coating-substrate bond is very weak. This observation demonstrates that the sprayed particles must penetrate into the surface asperities in order to be mechanically bound to the substrate surface [Davis J.R. ed. (2004)]. It could also be said that blasting with an angle results in some sort of "shadowing" effect. The blasting angle is generally varied between 45° and 90°. Below 45° the coating adhesion becomes very poor. One observation is that R_a is not affected by the blasting angle variation (between 45° and 90°). However, the fractal dimension changes [Yankee S.J. et al. (1991)]. Similar results were obtained by Mellali M et al (1997), who showed that between 60° and 90° the R_a variation was within the error limit. The study of Bahbou F. and P. Nylen (2005) spraying Triballoy 800 on Ti-6Al-4 V substrate (grit blasted with alumina: 60 mesh and a suction-type machine) has allowed precisely determining the shadowing effect. The results presented in Fig. 14.19 show that adhesive strength increases with the increase of the blasting angle from about 22 MPa at 45° to 28 MPa at 90° (27% increase), and it increases even more when spraying with the same angle (see Fig. 14.19). This demonstrates the existence of the shadowing effect, which is maximal when blasting at 55° and spraying at 90°.

14.4.4.4 Blasting Time

Blasting time is a very important parameter. It is defined as the time which each surface unit will be exposed to grit particles. It depends on:

- The radius of the target area, r_t (mm), which increases with the blasting distance (almost linearly) [Amada S. et al. (1999)], and the nozzle i.d., and is almost insensitive to the grit size
- The displacement velocity of the blasting nozzle, v_m (m/s)

For example, Wigren J. (1988), using a sample support with a diameter ($Ds = 600$ mm), established that the blasting time t_b (s) can be estimated using Eq. 14.7 as follows:

$$t_b = \frac{3 \times 2 \times r_t^2 \times P}{\pi \times Ds \times v_m} \tag{14.7}$$

where P is the number of passes. The corresponding values of R_a for different alloys as a function of $\left(t_b/r_t^2, 10^{-3}\right)$ [according to Eq. 14.7] are given in Fig. 14.20.

The average roughness increases rapidly with the blasting time, reaching a plateau in a relatively short period ($\sim 0.003 \times r_t^2 \sim 2.5$–3 s). The time needed to reach this plateau (about 4 reduced times in Fig. 14.20) is the optimum blasting time [Celik E. et al. (1999)]. For longer blasting periods, the roughness decreases slowly, ut, as it will be shown later, the percentage of residual grit imbedded into

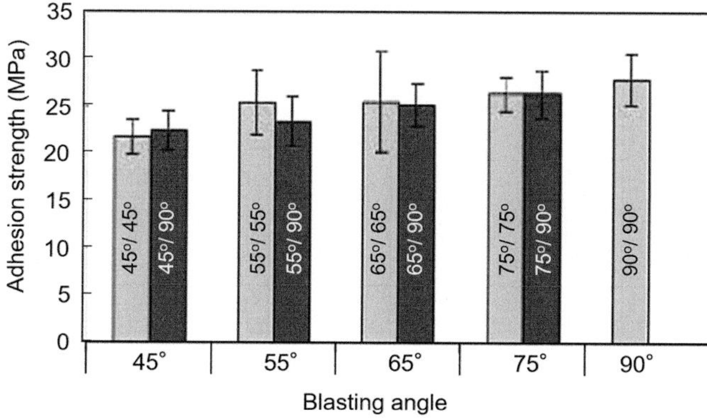

Fig. 14.19 Influence of the grit-blasting/spraying angles on adhesion (Triballoy 800 on Ti-6Al-4 V substrate). Numbers indicating blasting/spraying angles [Bahbou M.F. et al. (2004a)]

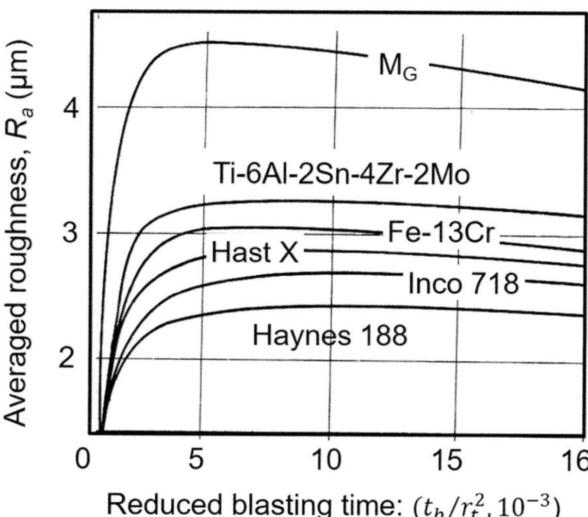

Fig. 14.20 Dependence of the average roughness R_a on the reduced blasting time $\left(t_b/r_t^2, 10^{-3}\right)$ for various alloys (alumina grit, suction-type machine, blasting distance 150 mm, grit flow 2 kg/mm) [Wigren J. (1988)]

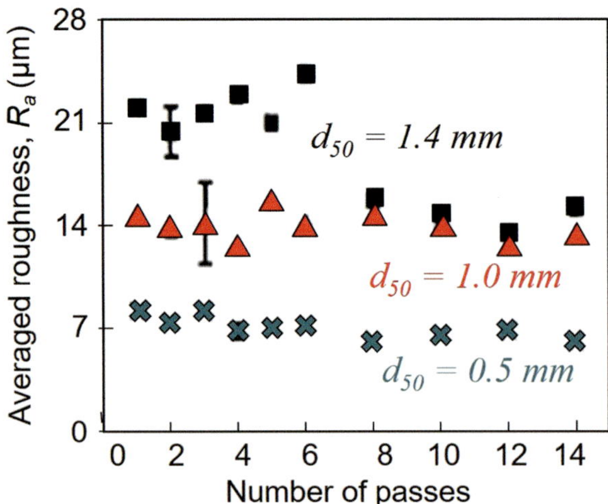

Fig. 14.21 Influence of the number of passes (each pass corresponding to the optimum blasting time) on the average roughness R_a of AU4G substrate for 3 brown alumina grit sizes (p = 0.3 MPa, d = 130 mm, suction-type machine, nozzle i.d. 8 mm) [Amada S. et al. (1999)], [Mellali M. et al. (1997)]

the surface increases (see Sect. 14.4.5). Moreover, when increasing the blasting time, the substrate hardness increases with the blasting resulting in some sort of cold work (shot peening effect [Wigren J. (1988) and Celik E. et al. (1999)]). Of course, the larger is the grit size, the faster is the substrate hardening. This illustrated in Fig. 14.21 from Amada S. et al. (1999), where an aluminum alloy AU4G substrate has been blasted with 3 grit sizes of brown alumina during different times. In this figure, the time is expressed in terms of number of passes, a pass corresponding to the optimum blasting time (i.e., about 4 reduced time see Fig. 14.21). After 6 passes with the largest grit, suddenly the roughness is reduced by about 39%. It corresponds to the crushing of the peaks when the underneath hardness increases. For smaller grit sizes, the phenomenon occurs later (for example, 21 passes with the 1 mm grit on AU4G). When considering a harder substrate, for example 100C6 instead of AU4G, the phenomenon occurs after 11 passes for the 1.4 mm grit.

14.4.4.5 Influence of the Grit

As it could be expected the R_a increases with the increase of the particle size of the grit. This illustrated in Fig. 14.22 as reported by [Mellali M. et al. (1997) and Al-Kindi GA, Shirinzadeh B (2007)].

A similar result has been obtained by Yang Y. et al. (2006), who have shown that for grit sizes over 1.4 mm the R_a increases more slowly.

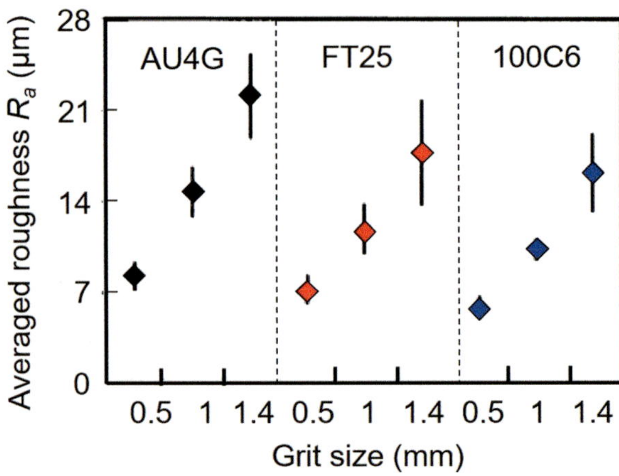

Fig. 14.22 Influence of the brown alumina grit size (0.5, 1, and 1.4 mm) on the average roughness R_a for three substrate materials: AU4G, FT25, and 100C6 (p = 0.3 MPa, d = 130 mm, optimum time, suction-type machine, nozzle i.d. 8 mm) [Mellali M. et al. (1997)]

The nature of the grit also plays a role. For example, [Tanaka Y. and M. Fukumoto (1999) and Mohammadi Z. et al (2007)] have blasted a Ti-6Al-4V substrate with alumina grit (300–425 µm) and silica grit (710–1400 µm) both with suction-type machine, and silica grit (355–710 µm) with a pressure type machine. The blasting conditions were adapted to obtain for the three grits the same $R_a = 3.51 \pm 0.15$ µm. When spraying hydroxyapatite in the same conditions, the following result was obtained for the bond strength (BS) of the coating:

- Alumina grit (300–425 µm) suction-type machine, BS = 24.1 ± 2 MPa
- Silica grit (710–1400 µm) suction-type machine, BS = 18.1 ± 3.4 MPa
- Silica grit (355–710 µm) pressure type machine, BS = 19.1 ± 3.4 MPa

No clear explanation is given to these results and authors do not precise the grit residues. The effect of the grit nature (steel grits and conventional grits) on the roughness of A36/1020 steel has also been studied by Varacalle D. J. et al. (2006).

14.4.4.6 Effect of Substrate Young's Modulus

For the same blasting conditions the Young's modulus of substrate controls the roughness obtained as shown in Fig. 12.23 from Wigren J. (1988) and Thomas T.R. (1999). For the same blasting conditions, the average roughness increases when the Young's modulus decreases (Fig. 14.23).

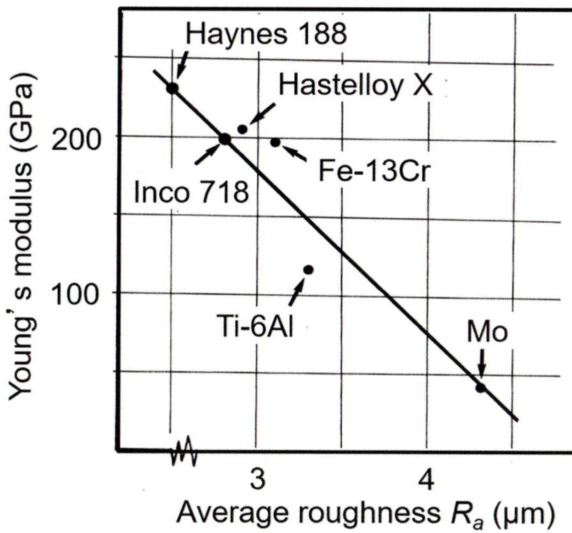

Fig. 14.23 Correlation between average roughness R_a and substrate Young's modulus [Wigren J. (1988)] (same conditions as in caption of Fig. 14.20)

14.4.5 Grit Residue

Unfortunately, grit blasting always leaves grit residues on the surface. In most cases the grit embedding happens as shown in Fig. 14.24. A grit particle is broken on impact with the substrate and one fragment of it remains at the surface while the others rebound. Then the next particle impacting on the remaining fragment imbeds it. These phenomena are controlled, on the one hand, by the number grit particles impacting the substrate at any given time and their momentum distribution and, on the other hand, the Young's modulus of the substrate material. The amount of grit residues increases with the substrate ductility, which becomes a problem for light weight alloys.

According to [Fukumoto M. and Y. Huang (1999) and Varacalle D. J. et al (2006)], the residues can have a significant impact on:

– Surface preparation
– Diffusion between substrate and coating
– Wetting properties of impacting molten spray droplets on the substrate
– Thermal stress due to thermal expansion coefficient mismatch between grit-matrix and first layers of the coating

In the following, a discussion is presented of the influence of the grit blasting parameters on the quantity of grit residues left at the surface and the available technique that could be used for their removal and then how to get rid of them. All measurements presented were performed at the optimum blasting distance for the different grit sizes and nozzle sizes used.

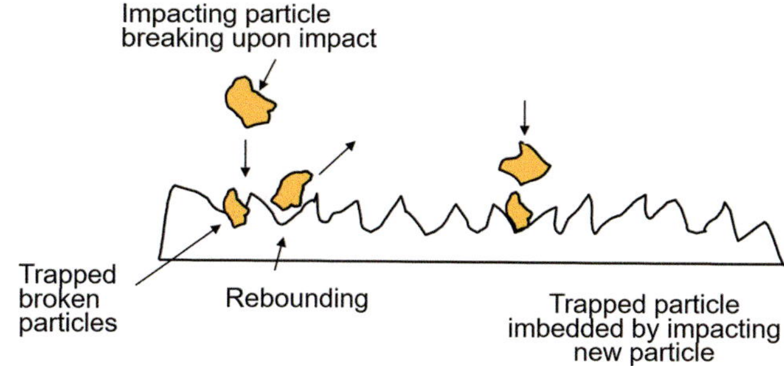

Fig. 14.24 Scheme of the embedding of grit particles at the substrate surface

The analytical techniques used for the evaluation of level of surface contamination by grit residue include:

- *X-ray fluorescence* [Mellali M. et al. (1997), Varacalle D. J. et al. (2006), Thomas T.R. (1999) and Tomovich S.J. and Z. Peng (2005)]: According to Wigren J. (1988), the analysis of 500 mm^2 area requires about 30 seconds with a precision better than $\pm 10\%$,

- *Optical microscopy coupled with image analysis*: When using white alumina grit the imbedded particles are clearly seen with an optical microscope and the ratio of their surface S' to the surface S of the substrate is easy to estimate [Amada S. et al. (1999)], [Al-Kindi GA, Shirinzadeh B (2007)],

- *Energy dispersive X-ray analysis*: For alumina grit containing mainly Al_2O_3 and ZrO_2, Amada S. et al. (1999), Maruyama T. et al. (2006) and Maruyama T. et al. (2007) measured Al and Zr at the substrate surface. The image indicating the content of one element was processed by image analysis using back-scattered mode to identify the alumina grit residue. The residual fraction of grits was determined as the ratio of the area covered by the residual grits to the total area of the substrate.

- *Dissolution of the blasted surface*: the method was used by Maruyama T. et al. (2006) and Maruyama T. et al. (2007), for carbon steel substrates grit blasted with alumina. After dissolution of the blasted surface in a mixed acid solution, the grit residue removed was weighted and the substrate thickness measured. The method also allows obtaining the grit particle penetration depth that can be compared with cross-sectional images of the blasted surface.

14.4.5.1 Influence of the Blasting Angle

As shown in Fig. 14.25 [Bahbou M.F. et al. (2004a)], grit residue increases monotonously with the increase of the blasting angle, reaching a maximum at 75°–90°. The effect is mainly due to the easier rebounding of grit particles

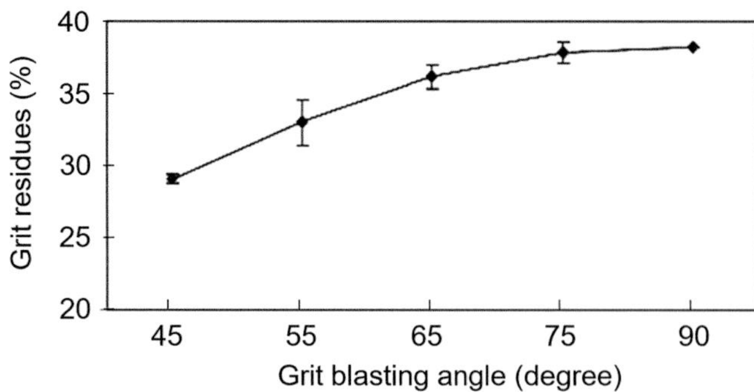

Fig. 14.25 Evolution with the reduced blasting time (see Fig. 12.21 caption) of the grit residue (surface %) Ti-6AL-4V substrate alumina grit [Bahbou M.F. et al. (2004a)]

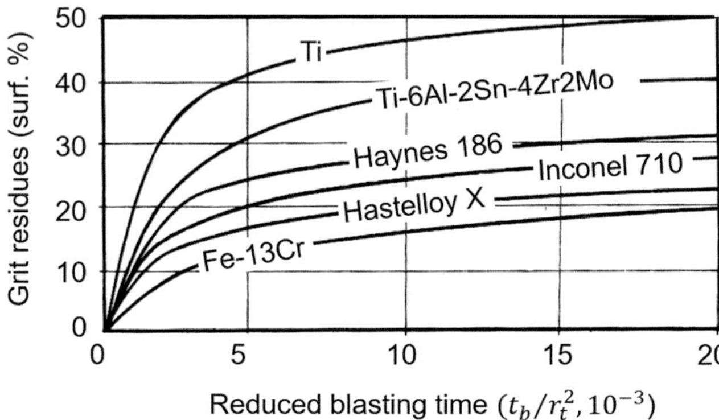

Fig. 14.26 Dependence of the grit residue (surface %) on the reduced blasting time (see Fig. 14.19 caption for different materials) [Wigren J. (1988)]. Reprinted with kind permission from ASM International

impacting the surface of the substrate at smaller angles between 45° and 75°.

14.4.5.2 Influence of the Blasting Time

The grit residue increases significantly with the blasting time. As with the substrate average roughness (see Fig. 14.20) the grit residue increases very fast up to a reduced time of about 4 (corresponding to 2.5 to 3.0 s) and after this knee increases more slowly but regularly as shown in Fig. 14.26 [Wigren J. (1988)]. Similar results were obtained by Mellali M. et al. (1997).

14.4.5.3 Influence of the Grit Size

The grit residue increases drastically with the blasting time. As for the substrate average roughness (see Fig. 14.20) the grit residue increases very fast up to a reduced time of about 4 (corresponding to 2.5–3 s) and after this knee increases more slowly but regularly (see Fig. 14.27 from [Maruyama T. et al. (2006)]. Similar results were obtained by [Mellali M. et al. (1997)].

The bigger are the grit particles, the deeper the grit residue is embedded in the substrate (particles mass increases as the

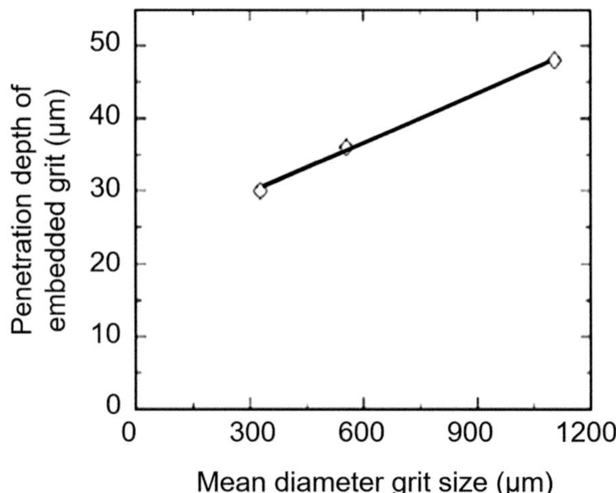

Fig. 14.27 Relationship between the mean diameter of the grit and the penetration depth of the imbedded grit (suction-type machine, blasting nozzle i.d. 6 mm, blasting pressure 0.4 MPa; white alumina grit, cold rolled carbon steel). Reprinted with kind permission from Springer Science Business Media [Maruyama T. et al. (2006)]

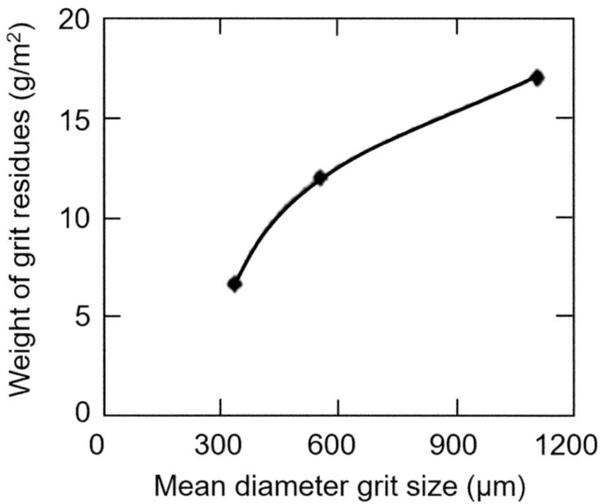

Fig. 14.28 Effect of the mean grit particle diameter on the weight of the residual grit (same conditions as those of (Fig. 14.26)). Reprinted with kind permission from Springer Science Business Media [Maruyama T. et al. (2006)]

cube of its diameter). This illustrated in Fig. 14.28 from [Maruyama T. et al. (2006)].

Figure 14.29 from [Wigren J. (1988)] represents the correlation between average roughness and grit residues for different blasting pressures, blasting times, and two blasting angles. It shows that an increased pressure always means increased R_a and increased amount of grit residue, due to the higher energy of the grit particle (the same effect is obtained for a given pressure by increasing the grit particle mean size), and the higher flow of grit particles. The high R_a and low-grit residue percentage are best obtained with high pressure and short time for a given grit size [Mellali M. et al. (1997)]. The blasting angle is not the most important parameter.

14.4.5.4 Grit Residue Removal

The cleaning procedure after grit blasting, to get rid of as much residue as possible, consists in blowing compressed air at a pressure of 0.4–0.5 MPa with a nozzle i.d. of 4 mm for a few tens of seconds and then immersing the substrate in an acetone bath solution ultrasonically agitated. As measured by Yankee S.J. et al. (1991) on Ti-6Al-4V substrates blasted with SiC or Al_2O_3 grit with a size of 36 mesh, the SiC

residues are easier to remove than the Al_2O_3 ones. This is illustrated in Fig. 14.30 where the percentage decreases in the original contamination ratio is plotted (the original ratio being normalized at 100%).

It can be seen that both methods work for SiC, but they are much less effective for Al_2O_3, which has the tendency to embed in the Ti-6Al-4V substrate surface. The SiC-roughened substrates were found to contain 7 area % of residue grit, while Al_2O_3-roughened substrate entrained 12 area %. However, in spite of that, the adhesion of hydroxyapatite (HA) coatings sprayed under the same conditions presented better adhesion on the alumina grit-blasted substrate and cleaned by air blast (compare Fig. 14.31a to Fig. 14.31b), in spite of the fact that the R_a was similar and the alumina grit residues more important with alumina [Yankee S.J. et al. (1991)].

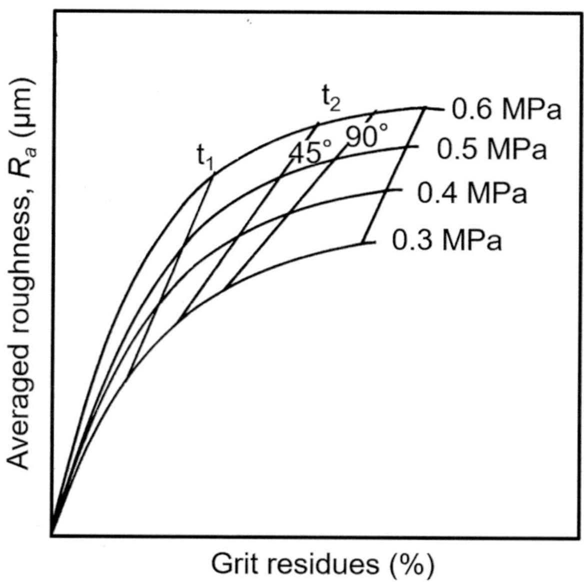

Fig. 14.29 Illustration of the correlation between surface roughness and grit residue, as a function of blasting pressure, angle and blasting time, see conditions in Fig. 12.20 caption: $t_1 = 10^{-3} r_t^2$ s, $t_2 = 3 \times 10^{-3} r_t^2$ s, $t_3 = 12 \times 10^{-3} r_t^2$ s, $t_4 = 12 \times 10^{-3} r_t^2$ s [Bellmann R. and A. Levy (1981)]

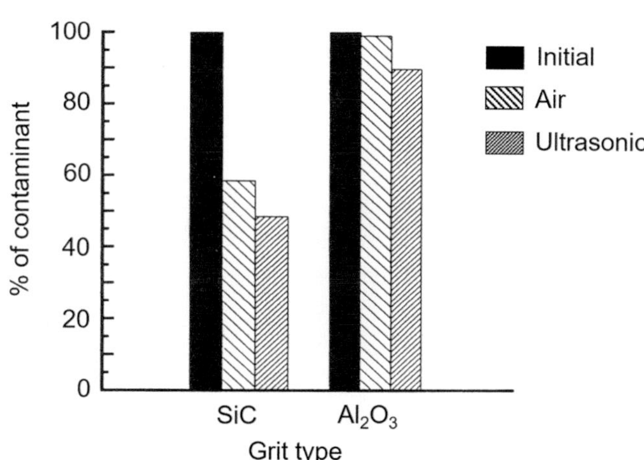

Fig. 14.30 Reduction in contamination, as determined via X-ray fluorescence, for two cleaning procedures (Ti-6Al-4V substrate, 36 mesh grit: SiC or Al2O, suction-type machine, 6.4 mm i.d. nozzle, 0.5 MPa). Reprinted with kind permission from ASM International [Yankee S.J. et al. (1991)]

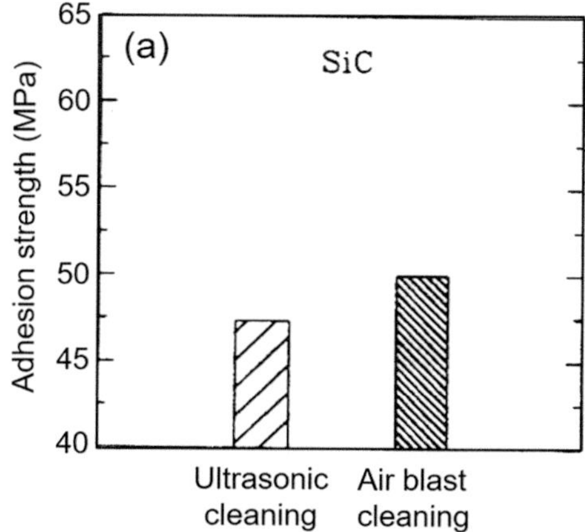

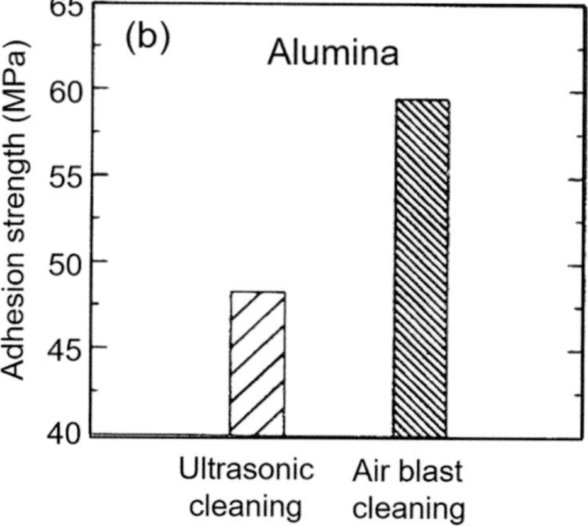

Fig. 14.31 Comparison of adhesion (breaking) strength of HA coatings plasma sprayed on Ti-6Al-4 V substrates roughened with (**a**) SiC and (**b**) Al_2O_3. Reprinted with kind permission from ASM International [Yankee S.J. et al. (1991)]

14.4.6 Grit Wear

The grit wear is strongly linked to the material used, the particle momentum at impact, that is, the blasting pressure and/or the grit size. For example, SiO_2 is more breakable than SiC, which is more breakable than alumina. For alumina, Mellali M. et al. (1997) have made a systematic study of the grit wear. Figure 14.32 shows three grit size distributions

after one cycle (corresponding to the optimum blasting time: the knee in the curves of (Fig. 14.21) for three blasting pressures and three grit sizes. The fragmented grit is important especially for the largest particles.

A single cycle of the grit significantly increases the quantity of small particles particularly when their velocity (blasting pressure) increases. The grit fragmentation increases with the number of cycles, as summarized in Table 14.1, for a suction-type machine and a XC38 low-carbon steel substrate.

The authors [Mellali M. et al. (1997)] also show that when using the same grit, without sieving it, to roughen surfaces, the R_a obtained decreased, as shown in Fig. 14.33, and the grit residue increases, especially for the larger grit for any substrate [Mellali M. et al. (1997)].

Table 14.1 Weight percentage of grit wear after one and five cycles for three different pressures and grit sizes [Mellali M. et al. (1997)]. Reprinted with kind permission from Springer Science Business Media, copyright © ASM International

Blasting pressure (MPa)	Grit size (mm)	Grit wear (wt.%)	
		1 cycle	5 cycles
0.2 MPa (30 psig)	0.5	–	–
	1.0	9	27
	1.4	17	39
0.3 MPa (45 psig)	0.5	13	31
	1.0	15	40
	1.4	28	66
0.5 MPa (75 psig)	0.5	20	44
	1.0	26	60
	1.4	31	77

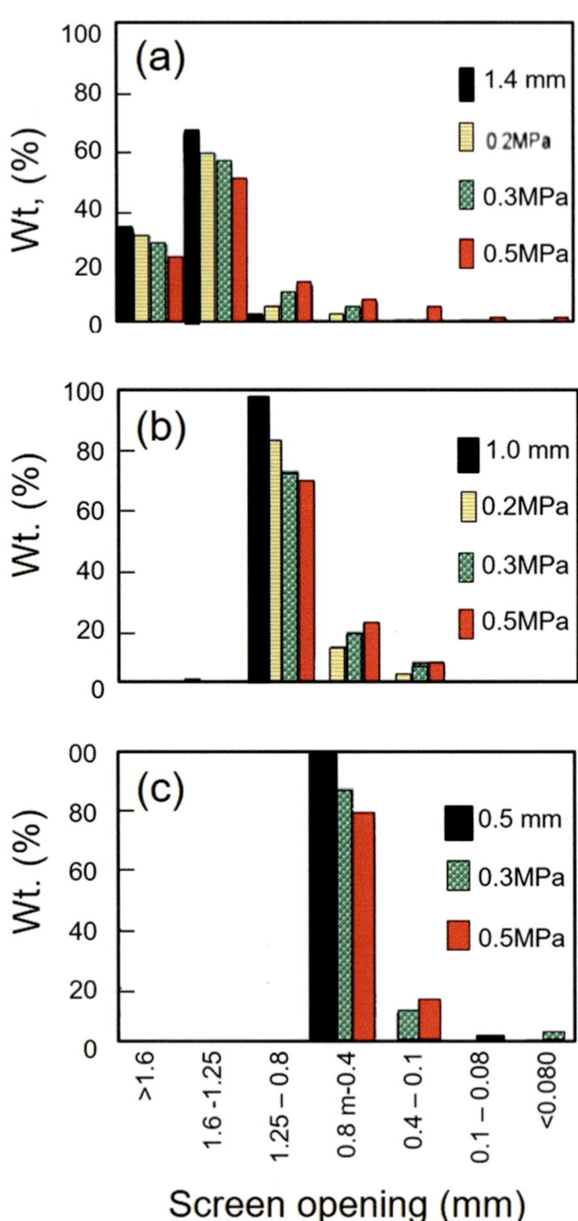

Fig. 14.32 Change of the alumina grit size distribution after one cycle (~3 s) for different blasting pressures and different initial grit sizes. Black bars correspond to the initial size distribution (suction-type machine, nozzle i.d. 8 mm) (**a**) 1.4 mm, (**b**) 1 mm, and (**c**) 0.5 mm. Reprinted with kind permission from Springer Science Business Media [Mellali M. et al. (1997)], copyright © ASM International

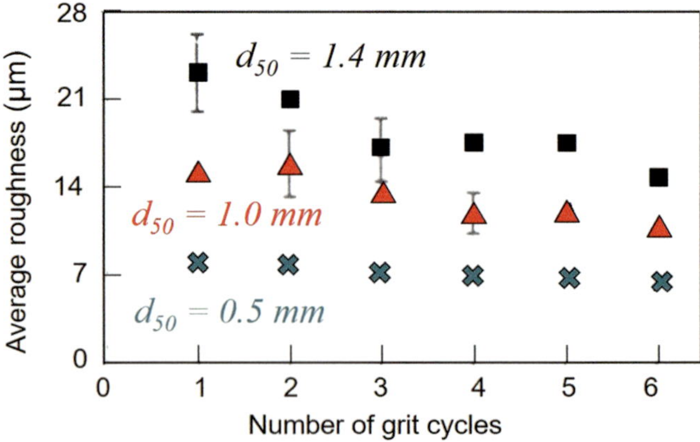

Fig. 14.33 Average roughness of AG4.5 substrates obtained with three different alumina grit sizes (1.4, 1, and 0.5 mm) used for up to six cycles without sieving it. Reprinted with kind permission from Springer Science Business Media [Mellali M. et al. (1997)], copyright © ASM International

14.4.7 Residual Stress Induced by Grit Blasting

Grit blasting induces compressive stress within metallic substrates [He Jianhong et al. (2008)] by plastic deformation of a layer of about 0.1–0.2 mm in thickness below its surface [Mellali M. et al. (1997)]. This is illustrated in Fig. 14.34 for a cast iron substrate blasted with 0.5 mm alumina grit. With 1.4 mm alumina grit, the maximum compressive residual stress can reach 1600–2000 MPa. This stress, of course, modifies the stress distribution between coating and substrate. When spraying materials with high heat content the blasting compressive stress can be partially released when the plastically deformed substrate layer is heated over the recovery temperature [Liao H. et al. (1999)]. This relaxation is only partial because the spray time is generally shorter than the time necessary for complete relaxation. To relax substrate compressive stress after grit blasting and before spraying, the substrates must be heated in a furnace, where the atmosphere is controlled to avoid oxidation, for example, according to the recommendation of ASM Metals Handbook (1964).

Mellali M. et al. (1997) have studied the influence of the grit-blasting parameters on the residual compressive stress induced by following the deflection of flat beams (2 × 10 × 100 mm) grit blasted under different conditions. Compared to the hole drilling method, the beam deflection gives only a mean value and no distribution along the substrate thickness, but the measurement is straightforward. Table 14.2a summarizes few results obtained with a low-carbon steel substrate (34CD4), grit blasted with three different alumina grits, and two blasting pressures. As could be expected, the deflection increases drastically with the grit size (roughly tripled between a grit of 1.4 mm against one of 0.5 mm) and the blasting pressure, especially for the 1.4 mm grit. The blasting time has also a non-negligible effect, especially for the largest grit (see Table 14.2b).

At last, when martensitic transformation occurs during blasting, for example, when grit blasting an austenitic stainless steel, large distortion can be observed [Abukawa S et al. (2006)].

14.4.8 Concluding Remarks on Grit Blasting

Grit blasting is a complex process, which has a significant impact on the adhesion of the sprayed coatings. For the best mechanical adhesion, intuitively the peak heights, characterized by R_t, must be adapted to the mean size of the splats, as illustrated in Fig. 14.35. Of course, the spacing between peaks also plays a key role [Mellali M. et al. (1997), Bahbou F. and P. Nylen (2005) and Fukumoto M. et al. (2004)].

To achieve good mechanical adhesion, it is not sufficient that the size of the peak is adapted to that of the splat size; the flattening liquid drop must also penetrate within the undercuts. For that the impact pressure ($\rho_p \cdot v_p^2$) must be larger than the surface tension force. It results in the following expression:

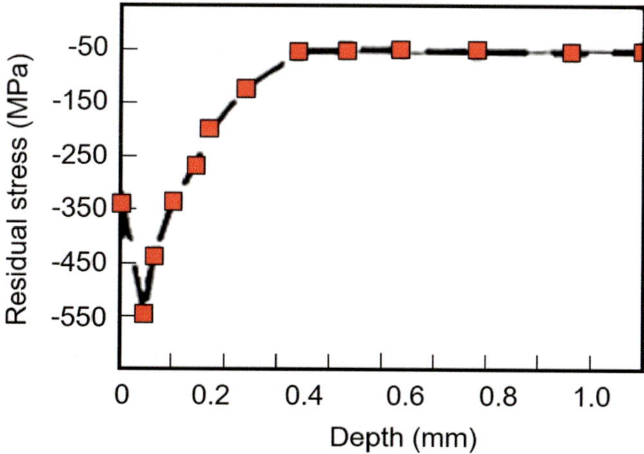

Fig. 14.34 Residual stress distribution, measured by hole drilling method, within a cast iron FT25-blasted substrate (suction-type machine, p = 0.3 MPa, nozzle i.d. 8 mm, alumina grit d_{50} = 0.5 mm). Reprinted with kind permission from Springer Science Business Media [Mellali M. et al. (1997)], copyright © ASM International

Table 14.2 Deflection (mm) due to grit blasting of 34CD4 steel flat beams (2 × 10 × 100 mm) (suction-type machine, nozzle i.d. 8 mm) [Mellali M. et al. (1997)]. Reprinted with kind permission from Springer Science Business Media, copyright © ASM International

(a) Effect of grit size and pressure

	Pressure (MPa)	Grit size (mm)		
		0.5	1.0	1.4
Deflection (mm)	0.3 (45 psig)	0.32 ± 0.07	0.66 ± 0.05	1.08 ± 0.15
	0.5 (75 psig)	0.43 ± 0.03	0.754 ± 0.04	1.89 ± 0.06

(b) Effect of grit size and blasting time (grit sieved between each cycle)

	Number of cycles	Grit size (mm)		
		0.5	1.0	1.4
Deflection (mm)	1	0.32 ± 0.07	0.66 ± 0.05	1.08 ± 0.15
	2	–	0.84 ± 0.06	1.12 ± 0.11
	4	0.37 ± 0.05	1.07 ± 0.05	1.56 ± 0.02

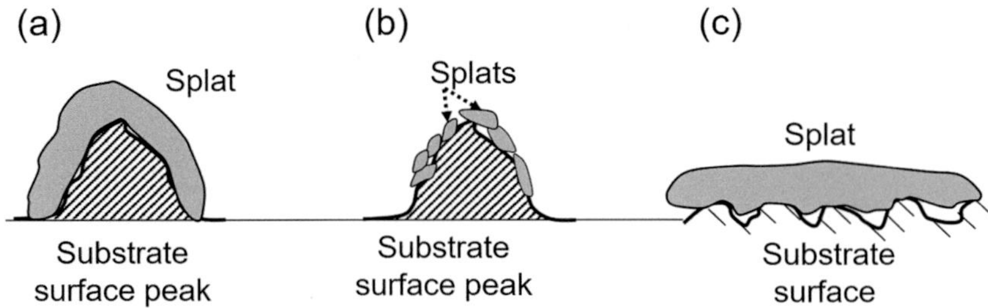

Fig. 14.35 Schematic of splat size relative to grit blasted surface peak sizes. (**a**) Splat size adapted to peak size. (**b**) Too small splat sizes relatively to peak size. (**c**) Too large splat size relatively to peak sizes

Table 14.3 Properties of the main grit materials

Grit material	Properties
Alumina	Hard, angular, durable, easily recycled, embedded in soft surfaces
Silicon carbide	Hardest, very angular, break down easily, produces sharp peaks
Angular chilled iron	Soft (40–50 HRC) rounds off rapidly, hard (40–50 HRC) maintains angular shape
Silica	Inexpensive, angular, breaks down very fast, may cause silicosis

Table 14.4 Influence of the different blasting parameters on the coating roughness, grit residues, and residual stress once blasting distance has been optimized

Grit-blasting parameters	Roughness	Grit residue	Grit wear	Residual stress
Grit size ↗	↗	↗	↗	↗
Impact velocity ↗ if p ↗ and/or nozzle i.d. ↘	↗	↗	↗	↗
Blasting time/optimal one t_b	$t < t_b$ ↗ cst. If $t > t_b$	$t > t_b$ ↗	$t > t_b$ ↗	$t > t_b$ ↗
Impact angle: Optimum: 90°	≈cst. "Shadowing" effect	45° ↗ 90		
Substrate Young's modulus ↗	↘	↘	↗	

$$\delta_{uc} > 4\sigma_p / \left(\rho_p v_p^2\right) \tag{14.8}$$

where σ_p is the liquid drop surface tension and δ_{uc} the characteristic dimension of the undercut. Finally, the good criteria must consider both phenomena. The parameter that seems the most suitable to characterize the surface roughness seems then to be the root mean square value $R_{\Delta q}$ related both to peak sizes and mean distances between peaks. $R_{\Delta q}$ has not yet been systematically studied, but the coating adhesion can be rather well linked to it [Bahbou F. and P. Nylen (2005)].

Once the peaks mean size and distance between them has been chosen for a mean splat dimension, the way to achieve the corresponding roughness depends on:

- Choice of the blasting machine (pressure or suction type)
- Blasting nozzle internal diameter and air pressure controlling the grit particle velocities
- Grit mean size and material controlling, with the particle mean velocity, the mean momentum of grit particles on impact with the substrate

It must be kept in mind, however, that the grit residues are also linked to the roughness and thus to particle size, velocity, and material, as well as the residual stress created within the top layer of the substrate. While residual stress can be released by heat treatment, removal of grit residue is more challenging. It may be simpler in this case to avoid, or reduce grit residue, by optimal selection of the blasting parameters (e.g., by choosing a smaller grit for obtaining the same roughness [Griffiths B. J., D. T. Gawne et al. (1996)].

Tables 14.3 and 14.4 summarize respectively the effects of the grit choice and of the grit-blasting parameters on the substrate modification (roughness, grit residue, residual stress induced) and the grit wear. While these offer good guidance in selecting the right grit and blasting conditions, it must be kept in mind that all these results depend strongly onto the substrate and especially its Young's modulus. For example, with the same grit blasting conditions, the substrate roughness decreases if the Young's modulus of the substrate increases.

14.5 High Pressure Water Jet Roughening

To avoid the embedded contaminants of the grit-blasting process, one of the alternative methods for surface preparation is the use of high-speed fluid jets. The advantage compared to grit blasting is no grit residues, less health hazard, and improved surface cleanness of the substrate. One of the first studies devoted to this technique using Inconel 718 substrates was reported by Taylor T. A. (1995a, b).

14.5.1 Equipment and Description of the Process

Pumps with pressures up to 360 MPa (51,000 psig) are used with the diameter of the water jet defined by a sapphire orifice with internal diameter of 0.3 or 0.4 mm. Water is the most widely used blasting fluid, often pretreated by reverse osmosis to fulfill the requirement of the pumping system. The velocity of the water jet at the exit of the sapphire orifice is a function of the water pressure, given by Eq. 14.9

$$v_j = \varphi \sqrt{\frac{p}{\rho_w}} \qquad (14.9)$$

where p is the water pressure, ρ_w the specific mass of water, and φ is the nozzle efficiency parameter [Momber A. W. et al. (2002)]. The energy of the jet is given by

$$E_j = \frac{m_w}{2} v_j^2 \qquad (14.10)$$

where $m_w = \dot{m}_w \times t_E$, with $\dot{m}_w$ the water jet mass flow rate and t_E is the local exposure time ($t_E = L_T/v_t$) with L_T being the length of the profiled section and v_t the transverse velocity of the jet with respect to the substrate, which is typically in the range of 50–2000 mm/min. The water jet mass flow rate is given by:

$$\dot{m}_w = \alpha \left(\pi d_o^2/4 \right) \rho_w v_j \qquad (14.11)$$

where d_o is the sapphire orifice internal diameter and α the outflow parameter. This results in the following expression for jet energy, E_j [Taylor T. A. (1995a)]:

$$E_j = \frac{\alpha \pi}{8} d_o^2 v_j^3 \frac{L_T}{v_t} \qquad (14.12)$$

For example [Abukawa S. et al. (2006)], with $d_o = 0.3$ mm, an exposure time of 600 ms and two water pressures, the following values are obtained:

$p = 200$ MPa (30,000 psig), $v_i = 600$ m/s, $E_i = 590$ J,
$p = 275$ MPa (41,000 psig), $v_i = 704$ m/s, $E_i = 960$ J.

The length of the water jet consists of a core zone, a transition zone, and a droplet-formation zone. The beginning of the droplet zone, L_c, is given by the following relationship [Taylor T. A. (1995a, b)]:

$$L_c \approx 300 \times d_o \qquad (14.13)$$

where d_o is the diameter of the sapphire orifice. For example, with $d_o = 0.3$ mm the droplet zone begins at $L_c \sim 90$ mm from the nozzle exit. Of course, the longer the standoff distance is, the larger is the water-spray region. The droplet velocity decreases with distance. A typical single erosion trace is between 1.5 and 2 mm. The pressure generated by an impacting water drop, p_o, is estimated as follows [Taylor T. A. (1995a)]:

$$p_o = \frac{v_j \rho_w c_s \rho_m c_m}{\rho_w c_w + \rho_m c_m} \qquad (14.14)$$

where c_s is the shock speed, c_m the sound velocity, and ρ_m the specific mass of the substrate and at last c_w the sound velocity in water. The shock speed is given by

$$c_s = c_w + 2v_j \qquad (14.15)$$

For example [Taylor T. A. (1995a)], for a jet velocity of 600 m/s, with $c_w = 1460$ m/s the impact pressure is $p_o = 1496$ MPa (217,700 psig)! Thus, the target, at a given standoff distance receives a number of high stress impacts resulting in a non-negligible residual stress below the substrate surface. The additional cyclic component (multiple droplet impact) gives rise to work hardens and embrittles of the surface. For example, after the water jet treatment, the surface hardness of hot-rolled low carbon steel (UH1) is significantly increased: from 24.3 HBR to 41.3 HBR [Taylor T. A. (1995a)].

The erosion of the substrate material by the water jet requires a minimum local exposure time. This is illustrated in Fig. 14.36 from Momber A. W. et al. (2002), representing the specific material removal in (mg/mm^2) as a function of the local exposure time. Below a critical exposure time t_{EC} no erosion occurs ($t_{EC} \sim 0.02$ s). When the exposure time increases, the specific material removal increases linearly in the first stage, followed by a gradual flattening of the curve at t_{Es}.

The successive phenomena are summarized in Fig. 14.37 from Momber A. W. et al. (2002): below t_{EC} probably a certain number of impacting water drops is required to erode material. However, the surface below is work harden and embrittle with micro-crack formation. The micro-cracks grow and over the time t_{EC} intersect with each other with material loss occurring until the saturation level is reached at which further materials loss is stopped.

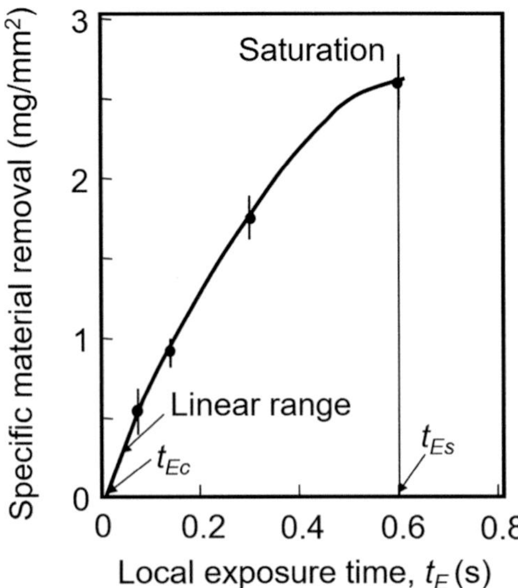

Fig. 14.36 Influence of the water jet time exposure on the specific mass removal of the substrate [Momber A. W. et al. (2002)]. Reprinted with kind permission from Elsevier

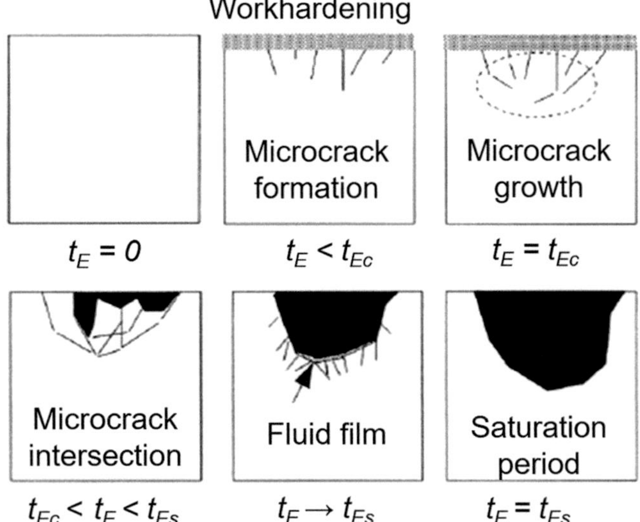

Fig. 14.37 Thresholds and material removal in water jet profiling. Reprinted with kind permission from Elsevier [Momber A. W. et al. (2002)]

14.5.2 Water Jet-Blasting Parameters

The main parameters are the water pressure, stand-off distance between the water jet nozzle and the substrate, blasting time as well as the material mechanical properties.

14.5.2.1 Water Pressure

As shown by Eq. 14.9 the water jet velocity depends on the square root of the pressure and which has a significant influence on the metal erosion. This is illustrated in Fig. 14.38

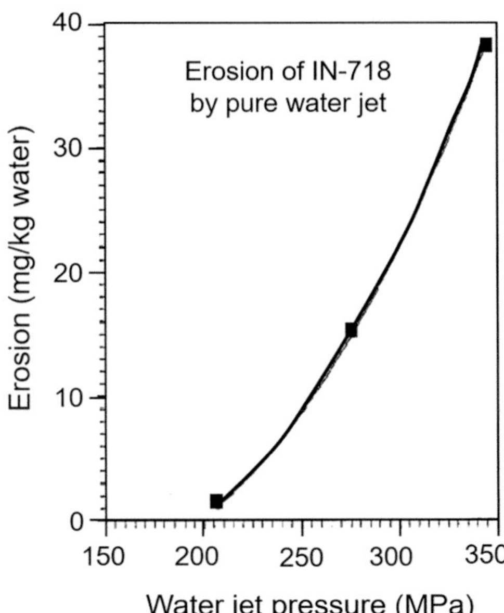

Fig. 14.38 Water jet erosion of IN-718, heat treated 1 h at 954 °C. Weight loss per kilogram of water delivered versus water pressure jet (nozzle i.d. 0.4 mm). Reprinted with kind permission from Elsevier [Taylor T. A. (1995b)]

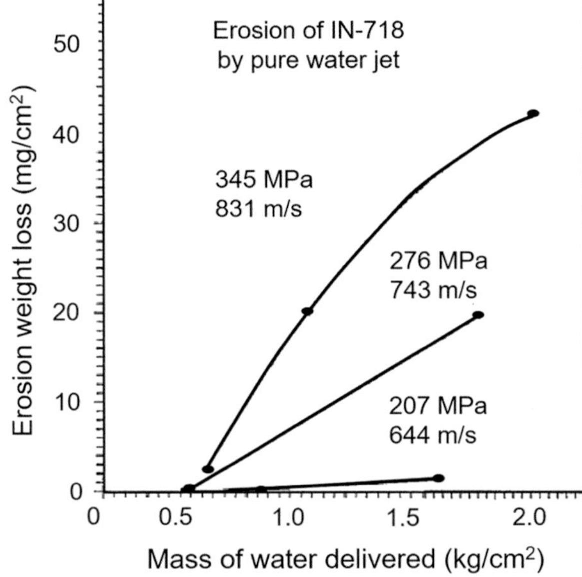

Fig. 14.39 Erosion (mg/cm^2) of IN-718 heat treated for 1 h at 954 °C, as a function of the mass of water delivered (kg/cm^2) for three blasting pressures (nozzle i.d. 0.4 mm). Reprinted with kind permission from Elsevier [Taylor T. A. (1995a)]

from Taylor T. A. (1995b) for IN-718 solution annealed (1 h at 954 °C).

This figure shows a parabolic dependence on pressure, which indicates the existence of a threshold pressure, which is here about 196 MPa (29,400 psig). This threshold pressure can also be observed in Fig. 14.39, where 207 MPa (31,050 psig) is probably very close to it.

14.5.2.2 Blasting Distance

As underlined by Taylor [Taylor T. A. (1995a) and Momber A. W. et al. (2002)], the optimum distance to achieve the best roughening is about 7.6 mm.

14.5.2.3 Blasting Time

The importance of a critical blasting time, t_{Ec} has been clearly shown in Fig. 14.36. [Taylor T. A. (1995a)] who emphasized that there is "a minimum mass of water required to impinge before there is measurable erosion that is an incubation period." The minimum mass of water per unit surface (kg/m^2) is calculated by dividing the water jet mass flow rate (kg/s) at a given pressure by the product of the water jet width (w) multiplied by its transverse velocity, v_t (m/s). In conventional grit blasting roughening occurs after a minimum blasting time, and for a water jet it can be assumed that it corresponds to a minimum mass of water per unit surface (kg/m^2). This is illustrated in Fig. 14.39 from Taylor T. A. (1995a), representing the metal eroded (mg/cm^2) as a function of the mass of water delivered by square centimeters (kg/cm^2) for three blasting pressures.

The effect of the blasting time is also shown in Fig. 14.40 from Taylor T. A. (1995a), where the logarithm of the erosion loss (mg/cm^2) is shown as a function of the logarithm of the traverse rate, vt (cm/min) for two blasting pressures: 207 and 345 MPa (31,050 and 51,750 psig). It can be seen that a small decrease of the blasting time induces a one order of magnitude decrease of the erosion.

The average surface roughness R_a is linked to the erosion loss (mg/cm^2), as shown in Fig. 14.41 from Taylor T. A. (1995b). Although R_a varies rapidly up to 15 μm with the erosion weight loss, this variation is strongly damped for larger values. The water jet pressure does not seem to have any effect.

The cleaning procedure after grit blasting, to get rid of as much residue as possible, consists in blowing compressed air at a pressure of 0.4–0.5 MPa with a nozzle i.d. of 4 mm during a few tens of seconds and then immersing the substrate in an acetone bath solution ultrasonically agitated. As measured by Yankee S.J. et al. (1991) on Ti-6Al-4V substrates blasted, with a 36 mesh SiC or Al$_2$O$_3$ grit, the SiC residues are easier to withdraw than the Al$_2$O$_3$ ones. It can be seen that both methods work for SiC but they are much less effective for Al$_2$O$_3$, which has the tendency to embed in the Ti-6Al-4V substrate surface. The SiC roughened substrates were found to contain 7 area % of residue grit, while Al$_2$O$_3$ roughened substrate entrained 12 area %.

14.5.2.4 Substrate Material

Figure 14.42 from Taylor T. A. (1995b) correlates the surface roughness of a number of superalloys with their erosion weight losses. There are substantial differences in the

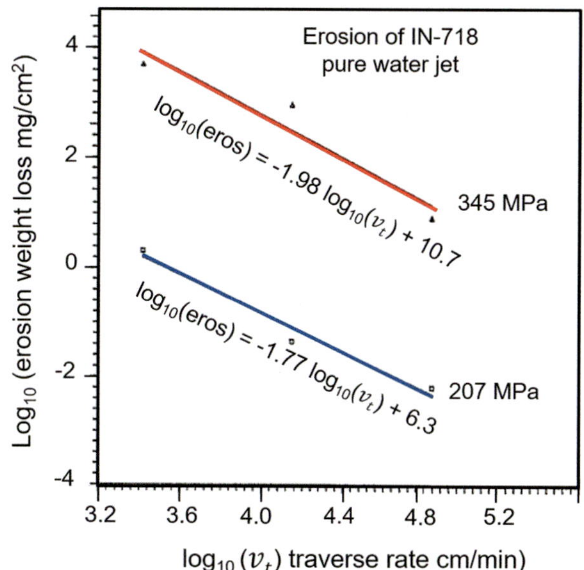

Fig. 14.40 Water jet erosion of IN-718, heat treated for 1 h at 954 °C, log10—log10 plot of weight loss versus jet traverse rate Tr (cm/min) showing an approximate slope of −2. Reprinted with kind permission from Elsevier [Taylor T. A. (1995a)]

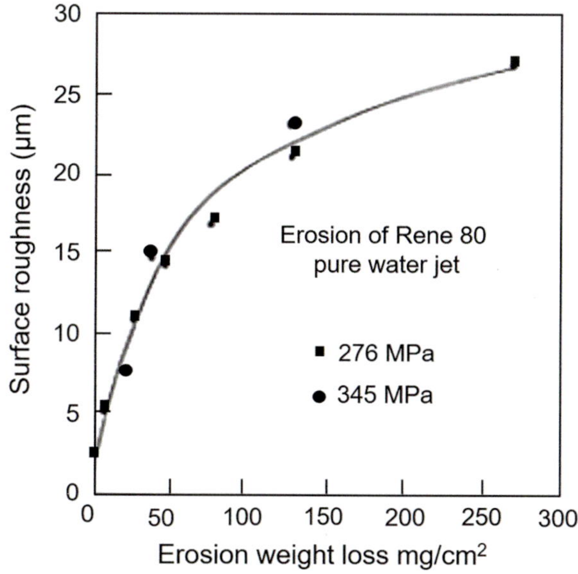

Fig. 14.41 Water jet erosion of Rene 80: dependence of the average roughness R_a on the erosion weight loss: two pressures: 276 and 345 MPa are considered, nozzle i.d. 0.4 mm. Reprinted with kind permission from ASM International [Taylor T. A. (1995b)]

roughness achieved with an Inconel sample depending on the heat treatment. The overexposure presents a substantially higher erosion loss and greater roughness. All alloys have a similar behavior for practical applications where R_a is in the range 3–8 μm.

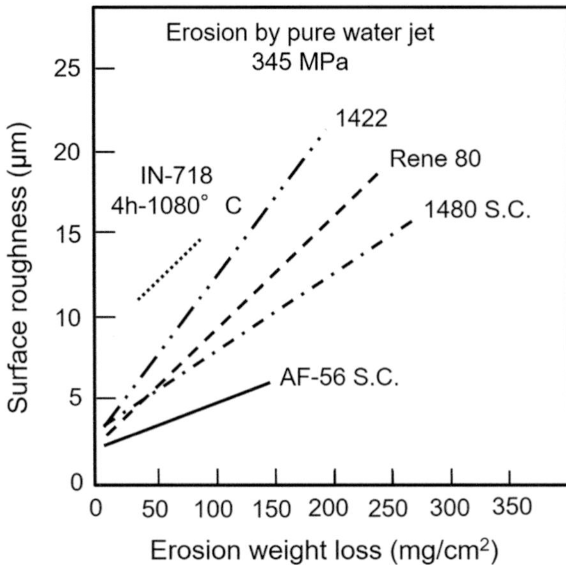

Fig. 14.42 Water jet erosion of various superalloys. Surface roughness versus weight loss. Reprinted with kind permission from Elsevier [Taylor T. A. (1995b)]

Fig. 14.43 Eroded surface sections (scale 100 μm) of hot-rolled low carbon steel (UH1) Left: grit-blasted surface (alumina grit $d_{50} = 165$ μm, nozzle i.d. 8 mm, p = 0.475 MPa, suction-type machine) Right: water jet blasted (nozzle i.d. 0.3 mm, p = 248 MPA). Reprinted with kind permission from Elsevier [Momber A. W. et al. (2002)]

14.5.2.5 Comparison Grit and Water Jet Blasting

Taylor T. A. (1995a) remarks, "the detail of eroded surface by water jet increases with increasing magnification, suggesting the water jet erosion produces a fractal surface. The highest magnification micrograph shows a multitude of granular fractures of about 2 μm in size and rather micro-faceted." When compared with conventional grit blasting at the same magnification and for similar average roughness, the feature size is at least an order of magnitude finer in the water jet surface. In contrast the grit-blasted surface appears smoother as the magnification is increased (see, e.g., Fig. 14.43 from Momber A. W. et al. (2002).

Knapp J.K. and J.A. Taylor (1996) have compared Inconel 718 and Mar-M 509 roughened by water jet and grit blasting and then plasma sprayed with the same superalloy. The have shown superior bounding capabilities of water jet roughened substrates compared to grit blasted substrates. For more details on water jet blasting, the interested reader can consult the reference by [Momber A.W. and R. Kovacevic (1998)].

Abrasive Water Jetting

It seems that Abrasive Water Jetting (AWJ) can improve pertinent surface characteristics with a higher degree of surface roughness due to the abrasion of the water jet and abrasive particles [Kovacevic R. et al. (1997)]. In this process the erosion of substrate material is primarily through the action of the abrasive particles, which are accelerated by a thin jet of high velocity water and directed through an AWJ nozzle [Wang L (2004)]. Abrasive particles are in the 100–150 mesh range and compared to conventional grit blasting, AWJ reduces grit embedding and grit residues. As for water blasting, the R_a increases with water pressure between (210 and 354 MPa) (31,500 and 53,100 psig), and with blasting time (up to a certain limit). The R_a increases also with the impingement angle, increasing progressively up to 60° and then slightly diminishing [Wang L (2004)]. This technique is mainly used for the abrasion and erosion of hard materials [Zeng J. and T.J. Kim (1996), and Ness E. and R. Zibbell (1996)].

14.6   Laser Treatment: Protal Process

The idea of laser treatment consists in coupling one or more Q-switched Nd:YAG lasers to the thermal spray torch. The PROTAL process has been developed by [Coddet C. and T. Marchione (1993), Coddet C. and T. Marchione (1997a), Coddet C. and T. Marchione (1997b), and Coddet C. and T. Marchione (1999)]. It combines in a single step the spraying operation and the surface preparation. The purpose of the laser irradiation is to eliminate the contamination films and oxide layers, to generate a surface state enhancing the deposit adhesion, and to limit the recontamination of the deposited layers by condensed vapors [Coddet, G. et al. (1999), and Coddet C. and T. Marchione (1997a)].

14.6.1 Laser Ablation

The interaction between a laser beam and a surface is a very complex problem that depends, among other parameters, on the nature of the substrate, its chemical and physical surface properties, the surface microgeometry (roughness), the beam energy density, the duration of irradiation, and the nature and pressure of the surrounding atmosphere. According to Coddet and Marchione (1997b), the incident laser energy is absorbed following two complementary mechanisms, namely the photonic absorption and the inverse Bremsstrahlung absorption. In both cases, the final result induces the excitation of electrons in the matter. The relaxation of these electrons follows three different mechanisms, depending on the electric properties of the substrate. For *insulating materials*, there is trapping of the excited electrons. For *semiconductor materials*, the relaxation is accomplished by heat radiation, while for *metallic materials*, which is the most common situation, the relaxation passes by the emission of a quantum of vibration energy (i.e., phonon). Basically, the phenomenon will consist of either thermal effects (the irradiation absorption increases the temperature locally until vaporization phenomena occur) or nonthermal effects such as photoablation. In every case, for each material, an intensity threshold needs to be observed.

The following parameters are to be considered in laser irradiation–material interaction:

- *Nature of the substrate*, its chemical and physical surface properties, its surface micro-geometry (roughness) and for semi-transparent materials (such as zirconia, for example, in the 1 μm wavelength range) pores and cracks which reduce the material transmissivity.
- *E04nergy density of the laser beam* (J/m^2) and its angle of incidence. The latter is very important because the absorption of the laser energy by the material diminishes rapidly with the increase of the angle beyond 25°, between the laser beam and the normal direction with respect to the substrate surface.
- *Nature and pressure of the surrounding atmosphere.* In most cases in thermal spraying, substrates are metallic. For a good adhesion of the sprayed particles to the substrate surface, it is necessary to clean the surface from adsorbates and condensates and, if possible, eliminate the oxide layer (when spraying a metal onto a metal).
- *The energy delivered by the laser irradiation* must be adapted to what is required and the nature of the contamination [Coddet and Marchione (1999)]. The lowest threshold laser beam energy is required to get rid of the adsorbates and condensates at the surface (the maximum energy of chemisorption is below a few eV). Higher threshold laser beam energy is required to ablate the oxide layer at the material surface (provided it is thin enough of the order of a few tens of nanometers). Depending on the properties of the oxide layer, three possibilities must be considered:

- If oxide layer is *transparent* to the irradiation wavelength, the energy will be absorbed by the metallic substrate underneath (in general a layer of a few tenths of μm). The resulting thermal effect abruptly expands the surface and induces the breaking and ejection of the oxide layer on top of it.
- If oxide layer is *semi-transparent*, absorption will take place by both the oxide layer and the substrate.
- If the oxide layer is *opaque,* the same phenomena occur as when it was transparent, but it is now the oxide layer which is heated.

14.6.2 Protal Experimental Setup

Depending on the thermal and optical characteristics of substrates, different phenomena such as heating, roughening, and ablation can be observed according to the energy density

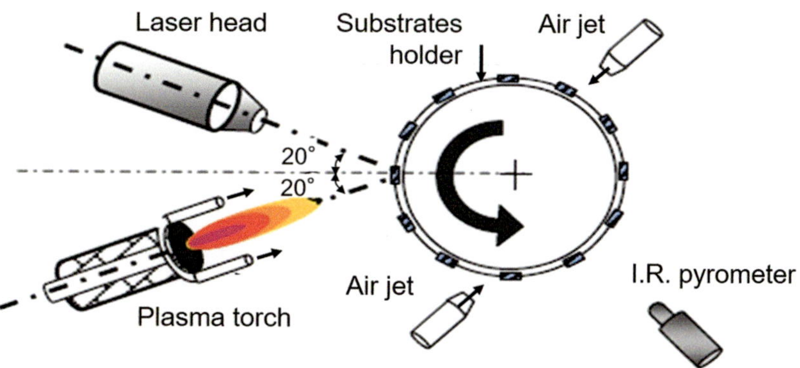

Fig. 14.44 Schematic of the integrated system for the PROTAL process for air plasma spraying. Reprinted with kind permission from Springer Science Business Media [Coddet, G. et al. (1999)], copyright © ASM International

(J/m^2) [Coddet, G. et al. (1999)]. In is essential in this case that the laser and spray gun be rigidly mounted in such a way that the passage of the laser treatment of the surface to be sprayed precedes immediately, or even overlap slightly, the passage of the spray gun for the deposition step. A schematic of a typical integrated system for the PPROTAL process is given in Fig. 14.44. The laser head is maintained near the spray torch in front of the sample holder. Both axes of the laser head and plasma torch make an angle of $20°$ relative to normal axis with respect to the substrate surface.

It is of primary importance that the sprayed spot is in the center of the laser irradiated spot, which must be larger than the spraying spot (see Fig. 14.45). Otherwise, particles would impact on a nontreated surface.

The laser used in the PROTAL process is a Q-switched Nd-YAG laser from Quantel (laser blast 1000), which operates at 1064 μm with an average power output of 40 kW, a maximum pulse frequency of 120 Hz with pulse duration (FWHM) of 10 ns. The laser beam is transferred through SiO_2 optical fibers. Due to present technical limitations, the power delivery is about 10 W per fiber. A specific optical arrangement allows achieving rectangular-shaped beams with a "top-hat" energy distribution that permitted homogeneous irradiation [Coddet, G. et al. (1999)]. The laser head-target distances, defined by the optical system, range from 70 to 120 mm [Landau L. and E. Lifshitz, (1984)]. With this setup, energy densities that can be supplied vary from 5 to 25 kJ/m^2.

14.6.3 Example of Results

First of all, it must be recalled that all laser parameters will have to be changed according to the material of the treated substrates. As per our knowledge, only the following substrate materials and coatings have been studied:

- Aluminum-based alloy 2017 substrates with pure Cu, Ni-20 wt.% Cr, Al_2O_3-13 wt.% TiO_2 coatings [Coddet, G. et al. (1999)], and Ni-5 wt.% Al coatings [Li H. et al. (2006a)].
- Titanium-based alloy Ti-6Al-4V substrate with pure Cu, Ni-20 wt.% Cr, Al_2O_3-13 wt.% TiO_2 [66], Ni-5 wt.% Al coatings [Coddet, G. et al. (1999)], [Li H. et al. (2006a)], [Leung N.P. et al (1992), and Conroy M. et al. (2005)].
- Nickel-based alloy IN718 substrate with Ni-5 wt.% Al coating [Costil S. et al. (2004)].
- Magnesium alloy AZ91 substrate with Ni-20 wt.% Cr coating [Liao H. et al. (2003)].

Moreover the influence of the laser treatment on the surface modification has been studied for the following substrates: pure aluminum (99.99%) [76], Ti-6Al-4V [Bahbou M.F. et al. (2004b) and Li H. et al. (2006b)], aluminum alloy A2017, and iron base alloy A430 [Li H. et al. (2006b)].

In general, it can be stated that the effect of the laser treatment depends on the substrate material, its oxidation (oxide layer thickness, porosity, and crack network), its roughness, and of course on the laser parameters (energy density, number of pulses).

14.6.3.1 Substrate Modifications

For Ti-6Al-4V substrates, the roughness generated by laser treatment varies with the laser energy density and the number of pulses, as indicated in Fig. 14.46. A significant increase of Ra begins at 10 kJ/m^2, corresponding to the emergence of craters. Up to 20 kW/m^2 the craters spread all over the surface and correspond to the maximum roughness. Above that value the surface is smoothened, and the roughness diminishes. Increasing the number of laser pulses smoothens the surface and protects the early-formed craters, thus the R_a decreases [Li H. et al. (2006b)]. Fig. 14.47a presents the initial state of the Ti-6Al-4V surface and Fig. 14.47b shows the craters after exposure to a single laser pulse of 20 kW/m^2.

Craters are systematically created at surface defect locations, defects such as micro-inclusions or small scratches. The effect of the laser energy density and number of pulses is shown in Fig. 14.47. When comparing Fig. 14.47a, b it can be seen that with the increase of the laser energy density, the induced craters become denser with an enlarged size, linked to more ejected material [Li H. et al. (2006b)]. With multiple pulses, the irradiated surface becomes smooth displaying some periodic variation (Fig. 14.47c) that can be attributed

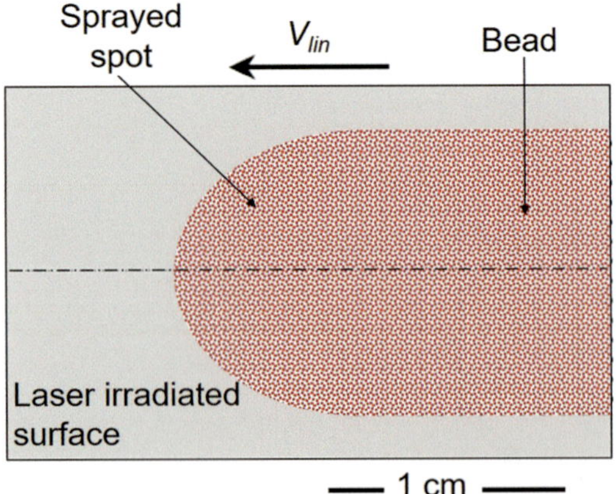

Fig. 14.45 Schematic configuration of the laser-treated zone and of the location of the sprayed particles jet during the PROTAL process. Reprinted with kind permission from Springer Science Business Media [Coddet, G. et al. (1999)], copyright © ASM International

to the laser-induced surface melting at the outermost layer [Li H. et al. (2006b)]. When the number of laser pulses increases to about 20, a "mud-cracked" morphology is formed on the smooth surface and the surface presents a yellowish color (slight oxidation). If more laser pulses are applied, the "mud-cracked" morphology becomes more distinct with a brighter yellow color that is an indication of the laser induced oxides (Fig. 14.47d) [Li H. et al. (2006b)]. Correspondingly, the R_a and the skewness S_K vary, as indicated in Table 14.5 (Fig. 14.48).

Similar results, however with fewer details, have been obtained for other materials [Coddet, G. et al. (1999)], [Costil S. et al. (2004)], [Li H. et al. (2006b)], [Bahbou M.F. et al. (2004b)]: in all cases, craters are formed with laser flux thresholds adapted. Of course, when increasing the number of laser pulses, which can also be done by reducing the relative laser-substrate velocity, the substrate temperature increases [Liao H. et al. (2003)].

14.6.3.2 Splat Formation

For Ni-5 wt.% Al sprayed on Ti-6Al-4V splats tend to be disk shaped when the laser heating of substrate reaches about 165 °C. It is worth noting that with the conventional preheating by the plasma jet, the same result is achieved at 250 °C, while the splat is extensively fingered at 165 °C [Li H. et al. (2006a)].

However, if the substrate has been preheated by the plasma torch at about 200 °C, the difference between splats with and without laser treatment is very small [74]. On non-preheated substrates, splats are close to disk shape when laser treatment is used, and are extensively fingered otherwise [Li H. et al. (2005)]. Of course, the disk-shaped splats are obtained only if the laser energy density is sufficient, that is, roughly 15 kJ/m^2 [Coddet, G. et al. (1999)].

14.6.3.3 Coating Adhesion

Adhesion has been measured either by the interfacial indentation test or the tensile adhesion test. However significant differences were observed between the results (tensile adhesion test) of different treatments according to the method used, the interfacial indentation seeming to be more favorable to the

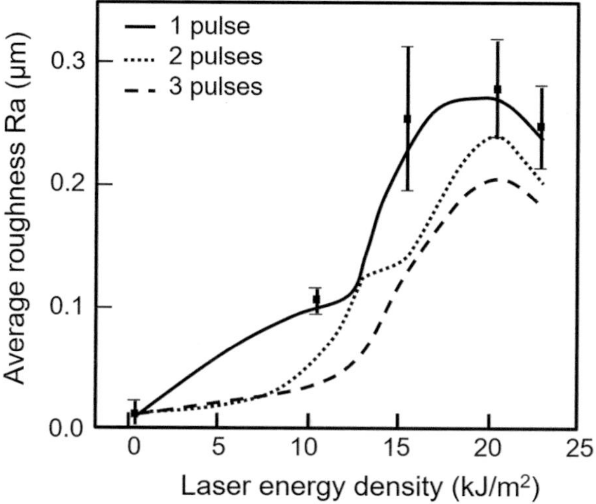

Fig. 14.46 Dependence of a Ti-6Al-4V average surface roughness (R_a) on the laser energy density and number of laser pulses. Reprinted with kind permission from Springer Science Business Media [Li H. et al. (2006b)], copyright © ASM International

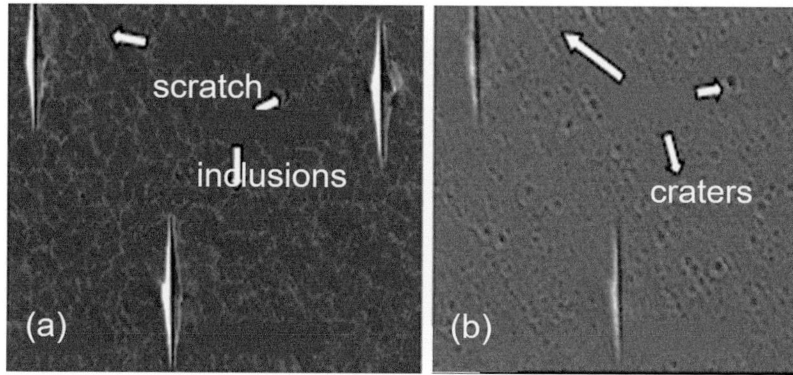

Fig. 14.47 SEM images of the same zone of a Ti-6Al-4V surface: (**a**) initial state, (**b**) surface treated by a single laser pulse at a fluence of 20 kW/m^2. Reprinted with kind permission from Springer Science Business Media [Li H. et al. (2006b)], copyright © ASM International

Table 14.5 Roughness characterization of laser irradiated area on polished Ti-6Al-4V substrate surface (P = pulse). Reprinted with kind permission from ASM International [Li H. et al. (2006b)]

	Reference	10 kJ/m^2, 1 P	20 kJ/m^2, 1 P	20 kJ/m^2, 20 P	20 kJ/m^2, 100 P
R_a (µm)	0.047	0.13	0.35	0.20	0.25
S_K (−)	−0.09	1.24	0.902	0.096	0.445

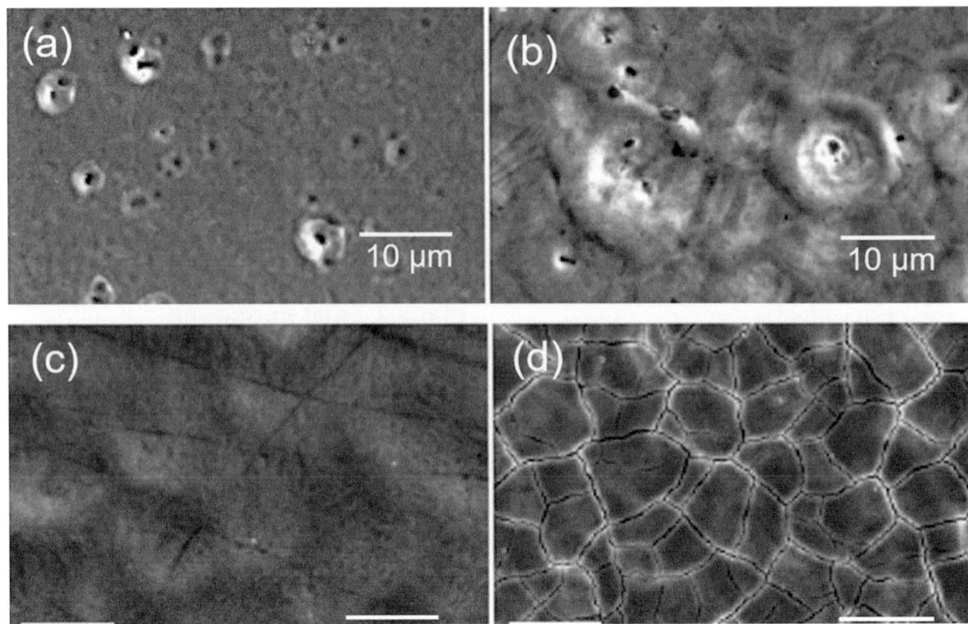

Fig. 14.48 SEM observation of laser irradiated area on Ti-6Al-4V substrate surfaces: (**a**) after a pulse at 10 kJ/m^2, (**b**) 1 pulse of 20 kJ/m^2, (**c**) 10 pulses at 20 kW/m^2, and (**d**) 120 pulses at 20 kJ/m^2. Reprinted with kind permission from ASM International [Li H. et al. (2006b)]

PROTAL process [Coddet C. and T. Marchione (1997a). Compared to the standard preparation (degreasing and grit blasting *but with no preheating*) the PROTAL process applied to a smooth surface is either equivalent (NiCr coating) or worse (pure Cu or Al$_2$O$_3$-TiO$_2$). However, it must be kept in mind that with the PROTAL process and the 7.5 kJ/m^2 flux, the R_a is only about 0.09 µm, and the oxide layer has not been completely removed. Moreover, the size of the laser spot is very close to that of the sprayed spot and thus the adhesion is probably poor in the fringes of the sprayed spot.

However, the process must be carefully optimized for each substrate and sprayed material. For example, with Ni-20 wt.% Cr deposited on magnesium alloy substrates [Li H. et al. (2006b)] the adhesion reaches 50 MPa for an energy density of 10 kJ/m^2, about 20 MPa with 15 kJ/m^2, and close to 30 MPa for 20 kJ/m^2! When preheating the substrate over the transition temperature (250 °C) for Ni-5 wt.% Al sprayed onto Ti-6Al-4V, the adhesion of the coating reaches 30 MPa. On the other hand, when the preheating is followed by one laser pulse at 20 kJ/m^2, it is raised to 40 MPa. To conclude, while this technique seems to be promising because it combines surface preparation with the spray operation, much effort is still necessary to optimize it.

14.7 Case Studies

Bahbou M.F. et al. (2004b) studied the effects of grit-blasting and plasma-spraying angles on the adhesion strength of an alloy (Tribaloy 800) that was plasma sprayed on a titanium-base alloy. Five different spray and grit-blast angles were investigated: 45°, 55°, 65°, 75°, and 90°. The surface texture in different directions was characterized by the classic average roughness and by a fractal analysis number using a two-dimensional fractal analysis method. The grit residue was measured by an X-ray spectrometer. The study showed that the maximum adhesion strength was close to a 90° blasting and spraying angle. However, the grit residue reached its maximum at a 75° blasting angle. From the image analysis of the interface in different directions, it was found that the non-perpendicular grit blasting produces an anisotropic surface. The fractal analysis method showed a rather good correlation with the blasting angle. However, no good correlation between the fractal number and the adhesion strength was found.

According to Zeng Z. et al. (2012), in order to prepare heat-resistant inner layer of hot-forging die, plasma spraying, plasma re-melting, and plasma spray welding were adopted. Cr$_3$C$_2$ coatings of Ni-based were prepared, respectively, with 10%, 20%, and 30% Cr$_3$C$_2$ powder and W$_6$Mo$_5$Cr$_4$V$_2$ substrate. The coating microstructure analysis, the microhardness test, and the measurement of thermal parameters of coating were conducted. The experimental results show that the coating has the better thermophysical property by using plasma spray welding method with the powder ratio of 90% Ni$_{60}$ and 10% Cr$_3$C$_2$, and by this way the microhardness of coating can achieve 1100 HV.

According to Wang J. et al. (2012), in order to produce the hear-resistant inner layer of hot-forging die, the plasma spraying and plasma re-melting and plasma spray welding were adopted. Substrate material was W6Mo5Cr4V2, including 10%, 20%, 30% tungsten carbide (WC) ceramic powder used as coating material to obtain different Nickel- based WC alloys coating. Micro-structure and micro-hardness analysis of the coating layer were conducted, as well as thermophysical properties for the coating layer were measured. The experimental results showed that the coating prepared with $70\%Ni_{60}$, $30\%WC$ powder had the best properties with plasma spray welding, in which the micro-hardness achieved 900 HV, meanwhile it can improve the thermal property of hot-forging die dramatically.

Dong S-J et al. (2013) deposited by atmospheric plasma spraying aluminum coating, as an example of spray coating material with low hardness, while dry ice blasting was applied during the deposition process. The deposited coatings were characterized in terms of microstructure, porosity, phase composition, and the valence states. The results showed that the APS aluminum coatings with dry ice blasting present a porosity of $0.35 \pm 0.02\%$, which is comparable to the bulk material formed by the mechanical compaction. In addition, no evident oxide has been detected, except for the very thin and impervious oxide layer at the outermost layer. Compared to plasma-sprayed Al coatings without dry ice blasting, the adhesion increased by 52% for Al substrate using dry ice blasting, while 25% for steel substrate. Corrosion behavior of coated samples was evaluated in 3.5 wt.% NaCl aqueous using electrochemistry measurements. The electrochemical results indicated that APS Al coating with dry ice blasting was more resistant to pitting corrosion than the conventional plasma-sprayed Al coating.

Tian J.-J.et al. (2018) recall that corrosion of metal plays a detrimental role in service lifetime of parts or systems. Therefore, coating a protective film that is fully dense and defects free on the base metal is an effective approach to protect the base metal from corrosion. In this study, a dense NiCr-20Mo coating with excellent lamellar interface bonding was deposited by plasma spraying of the novel shell–core– structured Mo-clad-NiCr powders, and then post-spray shot peening treatment by cold spraying of steel shots was applied to the plasma-sprayed NiCr-20Mo coating to obtain a fully dense coating through eliminating possibly existed pores and un-bonded interfaces within the NiCr-20Mo coating. Corrosion behaviors of the NiCr-20Mo coatings before and after shot peening were tested to investigate the effect of the post-spray shot peening on the corrosion behavior of the NiCr-20Mo coating. Results showed that a much dense and uniform plasma-sprayed NiCr-20Mo coating with perfect lamellar bonding at most of interfaces was deposited. However, the electrochemical tests revealed the existence of through-thickness pores in the as-plasma-sprayed NiCr-20Mo coating. Through the post-spray shot peening treatment, a completely dense top layer in the coating was formed,

and with the increase in the shot peening intensity from one pass to three passes, the dense top layer became thicker from 100 μm to reach 300 μm of the whole coating thickness. Thus, a fully dense bulk-like coating was obtained. Corrosion test results showed that the dense coating layer resulting from densification of shot peening can act as an effective barrier coating to prevent the penetration of the corrosive medium and consequently protect the substrate from corrosion effectively.

Grishina I. P. et al. (2019) proposed a combined technology for the modification of the surfaces of titanium implants by laser radiation followed by the plasma spraying of biocompatible coatings. They performed comprehensive investigations of the coatings with the use of SEM, optical microscopy, and other methods. The coatings obtained by using the proposed procedure were characterized by the high adhesive strength, structural uniformity, and a sufficiently high degree of hydrophily.

Ding Y. et al. (2019) recall that the adhesive strength to the substrate plays an important role in the performance of thermal-sprayed coatings and could determine the coating's service life. It depends to a large extent on the state of the substrate surface prior to coating deposition. In this work, a novel surface pre-treatment technique was introduced to thermal spraying, namely, bristle blasting. This process is fundamentally a mechanical abrasion process using a rotating brush-like wheel. The adhesive strengths of a plasma-sprayed metallic coating (Ni5Al) and a ceramic coating (Al_2O_3) were examined and compared to that of coatings sprayed with conventional grit blasting surface pre-treatment. A mild steel and an aluminum alloy were selected as substrate materials. The results indicated that bristle blasting could be a practical solution for steel and aluminum alloys when grit blasting is not applicable on site. The adhesion of the sprayed coatings with different pre-treatments increased sequentially from mechanical grinding, bristle blasting to grit blasting. The adhesive strength of the Ni5Al coating deposited on the bristle-blasted substrate reached 60% of the adhesive strength of a coating deposited with the traditional grit-blasting pre-treatment, while for the alumina coating, it was only about 30%. Moreover, the effect of substrate materials should be considered when using bristle blasting as a surface preparation.

14.8 Summary and Conclusions

The proper preparation of the substrate surface prior to spraying is essential for the coating adhesion. It has to be carried out within the shortest time possible prior to spraying. Surface preparation comprises the following four steps:

Cleaning is the first step to removing contaminants such as oil, greases, paint, rust, scale, and moisture. This operation is quite conventional, but it is essential.

Masking is necessary to prevent the deposition of the coating on areas where it is not wanted. It also improves the uniformity of the deposit when coating limited areas.

Roughening is crucial for optimal coating adhesion to the substrate. The degree of roughness has to be compatible with the mean size of splats: the R_t (also called R_z) must be about 2.5 times the mean splat diameter, and the spacing between peaks has to be sufficient for the liquid flattening droplet to penetrate between them. The root mean square value $R_{\Delta q}$ seems to be the roughness parameter that correlates the best with the coating adhesion. Roughening can be achieved by grit blasting or water jet. The former is probably the most commonly used. The degree of roughness, grit residues, and blasting induced stress depends on the blasting system used (suction or pressure-type machines), blasting pressure (important for pressure-type machine), blasting distance that must be optimized, grit material, and size. Since the grit particles can break down during the blasting operation, the grit used must be continuously sieved, and the fines discarded, in order to maintain a close control on the uniformity and degree of roughness generated. The main disadvantage of grit blasting is that grit residues must be removed prior to spraying, which is not necessarily easy. The main advantage of water jet, compared to blasting, is that no residue is formed. The degree of roughening depends strongly on the water jet pressure and mass flux of water per unit surface area (kg/m^2). This parameter is calculated by dividing the water jet mass flow rate (kg/s) at a given pressure by the product of the water jet width (w) multiplied by its transverse velocity (m/s). Because of the rapid wear of the equipment and the liquid jet nozzles, which are rather expensive, this technique is mainly used for sophisticated coatings as in aero-engine industry.

Adsorbates and condensates elimination is performed either by preheating the substrate, generally with the spray torch, to temperatures at which they evaporate, or by blasting the substrate with dry ice just ahead of the spray spot, or by using a laser focused upstream of the spray spot. Substrate preheating is by far the most commonly used on an industrial scale; dry ice blasting has started to be used, while laser pretreatment is a still mainly at the laboratory developed stage.

Nomenclature

Units are indicated in parentheses; when no units are indicated, the parameter is dimensionless.

Latin Alphabet

AA	Arithmetic average (μm), Eq.14.1
c_m	Sound velocity in the substrate (m/s)
c_s	Shock speed in the substrate (m/s)
c_w	Sound velocity in water (m/s)
d	Blasting distance (m)
d_b	Blasting distance (m)
d_o	Sapphire orifice internal diameter for water jet (m)
d_{50}	Particle diameter below which 50 wt.% of the particles are found
Ds	Sample support diameter (m)
E_j	Water jet energy (J)
h_i	Peak height (m)
Ku	Kurtosis, or third moment
L_c	Distance to the beginning of the droplet zone (m)
l_n	Assessment length (m), Fig. 14.5
l_t	Transverse length (m), Fig. 14.5
L_T	Length of the profiled section with the water jet (m)
$\dot{m}_g$	Air mass flow rate (kg/s)
$\dot{m}_w$	Water jet mass flow rate (kg/s)
m_p	Particle mass (kg)
$\dot{m}_p$	Grit particle flow rate (kg/s)
N	Number of measurements (−)
p	Blasting pressure (MPa)
P	Number of passes
r_t	Radius of the target area hit by blasting particles (m)
R_a	Average roughness (μm), Eq. 14.2
R_{sm}	Arithmetic mean of the widths of the profile (μm) Eq. 14.6
R_t	Distance between the highest peak and the deepest undercut (μm)
R_z	Root-mean square roughness (μm) Eq. 14.3
$R_{\Delta q}$	Root mean square value (μm) Eq. 14.4
S_k	Skewness parameter , Eq. 14.5
S_n	Blasting nozzle internal cross section area (m^2)
t_b	Blasting time (s), Eq. 14.7
t_E	Local exposure time ($t_E = L_T/v_t$) to the water jet (s)
t_{EC}	Critical exposure time to a water jet (s)
t_{Es}	Time to saturation point (s) Fig. 14.36
v_g	Air velocity (m/s)
v_j	Water jet velocity (m/s)
v_m	Displacement velocity of the blasting nozzle (m/s)
v_p	Particle velocity (m/s)
v_t	Transverse velocity of the water jet/substrate (m/s)
x	Sampling length (m)
z	Surface height (μm)

Greek Alphabet

δ_{uc}	Characteristic dimension of the undercut (m)
ρ_g	Air specific mass (kg/m^3)
ρ_m	Specific mass of the substrate (kg/m^3)
ρ_p	Specific mass of the particle (kg/m^3)
ρ_w	Specific mass of water (kg/m^3)
σ_p	Surface tension liquid drop
φ	Water jet nozzle efficiency parameter (−)
$\varphi(x)$	Distribution function (−)

References

Abukawa, S., K. Kobayashi, and Y. Kobayashi. 2006. The relationship of work distortion and surface roughness on grit blasting process. In *TS-2006*, ed. B. Marple et al. Materials Park: ASM International.

Al-Kindi, G.A., and B. Shirinzadeh. 2007. An evaluation of surface roughness parameters measurement using vision-based data. *International Journal of Machine Tools and Manufacture* 47: 697–708.

Amada, A., and T. Hirose. 2000. Planar fractal characteristics of blasted surfaces and its relationship with adhesion strength of coatings. *Surf Coat Technol* 130: 158–163.

Amada, S., and A. Satoh. 2000. Fractal analysis of surface roughened by grit blasting. *Journal of Adhesion Science and Technology* 14 (1): 27–41.

Amada, Shigeyasu, and Hiroshi Yamada. 1996. Introduction of fractal dimension to adhesive strength evaluation of plasma-sprayed coatings. *Surface and Coatings Technology* 78: 50–55.

Amada S., T. Hirose, and T. Senda (1999) Quantitative evaluation of residual grits under angled blasting, Surface and Coatings Technology 111 1–9.

ASM Handbook. 1994. Volume 05: Surface Engineering (1056 pages).

ASM Metals Handbook. 1964. ASM International, Materials Park, OH 44073–0002, USA.

Bacova, V., and D. Draganovska. 2004. Analysis of the quality of blasted surfaces. *Materials Science* 40 (1): 125–131.

Bahbou, F., and P. Nylen. 2005. Relationship between surface topography parameters and adhesion strength for plasma spraying. In *ITSC-2005*, ed. E. Lugscheider. Düsseldorf: DVS.

Bahbou, M.F., P. Nylen, and J. Wigren. 2004a. Effect of grit blasting and spraying angle on the adhesion strength of a plasma-sprayed coating. *Journal of Thermal Spray Technology* 13 (4): 508–514.

Bahbou, F., P. Nylen, and G. Barbezat. 2004b. A parameter study of the PROTAL® process to optimize the adhesion of Ni5Al coatings. In *ITSC-2004*. Düsseldorf: DVS.

Bardi, U., L. Carrafiello, R. Groppetti, F. Niccolai, G. Rizzi, A. Scrivani, and F. Tedeschi. 2004. On the surface preparation of nickel superalloys before CoNiCrAlY deposition by thermal spray. *Surface and Coatings Technology* 184: 156–162.

Bellmann, R., and A. Levy. 1981. Erosion mechanism in ductile metals. *Wear* 70: 1–27.

Bradley, C. 2000. Automated surface roughness measurement. *International Journal of Advanced Manufacturing Technology* 16: 668–674.

Cedelle, J., M. Vardelle, and P. Fauchais. 2006. Influence of stainless-steel substrate preheating on surface topography and on millimeter- and micrometer-sized splat formation. *Surface and Coatings Technology* 201 (3–4): 1373–1382.

Celik, E., A.S. Demirkiran, and E. Avei. 1999. Effect of grit blasting of substrate on the corrosion behavior of plasma sprayed Al_2O_3 coatings. *Surface and Coatings Technology* 116–119: 1061–1064.

Coddet, C., and T. Marchione. 1993. *Process for the preparation and coating a surface and apparatus for practicing sand process* (in French), French patent FR-9,209,277, July 21st., (1993).

———. 1997a. *Process for the preparation and coating a surface and apparatus for practicing sand process*, European patent extension EP-0,580,534-A1 April 23rd. (1997).

———. 1997b. *Process for the preparation and coating a surface and apparatus for practicing sand process*, US patent extension, US-5,668,564, Nov 18th. (1997).

———. 1999. *Process for the preparation and coating a surface and apparatus for practicing sand process*, Canadian patent extension, CA-2,101,004 July 13th., (1999).

Coddet, G., S. Montavon, O. Ayrault-Costil, F. Freneaux, G. Rigolet, F. Barbezte, A. Diard Folio, and P. Wazen. 1999. Surface preparation and thermal spray in a single step: The PROTAL process-example of application for an aluminium-base substrate. *Journal of Thermal Spray Technology* 8 (2): 235–242.

Conroy, M., et al. 2005. A comparison of surface metrology techniques. *Journal of Physics Conference Series* 13: 458–465.

Costil, S., H. Li, and C. Coddet. 2004. *New developments in the PROTAL® process, in ITSC-2004*, DVS, Düsseldorf, Germany, proceedings.

Davis, J.R., ed. 2004. *Handbook of thermal spray technology*. Materials Park: ASM International.

Deng, J., and S. Junlong. 2008. Sand erosion performance of B_4C based ceramic nozzles. *International Journal of Refractory Metals and Hard Materials* 26: 128–134.

Ding, Y., H. Li, and Y. Tian. 2019. Bristle blasting surface preparation in thermal spraying. *Journal of Thermal Spray Technology* 28: 378–390.

Dong, S., B. Song, B. Hansz, H. Liao, and C. Coddet. 2011. Improvement in the microstructure and property of plasma sprayed metallic, alloy and ceramic coatings by using dry ice blasting. In *5th. RIPT conference*, CEC Limoges December 2011, e-proceedings.

Dong, S.-J., B. Song, G.-S. Zhou, C.-J. Li, B. Hansz, H.-L. Liao, and C. Coddet. 2013. Preparation of aluminum coatings by atmospheric plasma spraying and dry-ice blasting and their corrosion behaviour. *Journal of Thermal Spray Technology* 22 (7): 1222–1229.

Elbing, F., N. Anagrehb, L. Dorn, and E. Uhlmann. 2003. Dry ice blasting as pretreatment of aluminum surfaces to improve the adhesive strength of aluminium bonding joints. *International Journal of Adhesion and Adhesives* 23: 69–79.

Fukumoto, M., and Y. Huang. 1999. Flattening mechanism in thermal sprayed Ni particles impinging on flat substrate surface. *Journal of Thermal Spray Technology* 8 (3): 427–432.

Fukumoto, M., I. Ohgitani, and T. Yasui. 2004. Effect of substrate surface change on flattening behaviour of thermal sprayed particles. *Materials Transactions* 45 (6): 1869–1873.

Griffiths, B.J., D.T. Gawne, and Guishu Dong. 1996. The erosion of steel surfaces by grit-blasting spraying as a preparation for plasma. *Wear* 194: 95–102.

Grishina, I.P., S.V. Telegin, A.V. Lyasnikova, O.A. Markelova, and O.A. Dudareva. 2019. Development of the combined technology of modification of the surface of titanium implants by laser radiation with subsequent plasma spraying of bio-compatible coatings. *Metallurgist* 63 (1–2): 215–230.

Guessasma, S., G. Montavon, and C. Coddet. 2003. On the implementation of the fractal concept to quantify thermal spray deposit surface characteristics. *Surface and Coatings Technology* 173 (1): 24–38.

He, Jianhong, Bruce Dulin, and Thomas Wolfe. 2008. Peening effect of thermal spray coating process. *Journal of Thermal Spray Technology* 17 (2): 214–220.

Hu, Z., Lei Zhu, Teng Jiaxu, Xuehong Ma, and Shi Xiaojun. 2008. Evaluation of three-dimensional surface roughness parameters based on digital image processing. *International Journal of Advanced Manufacturing Technology* 2008 (January).

Knapp, J.K., and J.A. Taylor. 1996. Waterjet roughened surface analysis and bond strength. *Surface and Coatings Technology* 86–87: 22–27.

Kovacevic, R., M. Hashish, R. Mohan, M. Ramula, T.J. Kim, and E.S. Geskin. 1997. State of the art of research and development in abrasive-water-jetting machining, Transactions of ASME. *Journal of Manufacturing Science and Engineering* 119: 776–785.

Landau, L., and E. Lifshitz. 1984. *Statistical physics*. Moscow: MIR.

Leung, N.P., W. Zupka, and W. Ziemlich. 1992. Cleaning techniques for removal of surface particulates. *Journal of Applied Physics* 73: 513–3523.

Li, H., S. Costil, H.-L. Liao, and C. Coddet. 2005. Role of laser surface preparation on the adhesion of Ni-5 wt%, Al coatings deposited using the PROTAL® process. In *ITSC-2005: Explore its potential!* ed. E. Lugscheider. Materials Park: ASM International.

———. 2006a. Role of the laser surface preparation on the adhesion of Ni-5 wt% Al coatings deposited using the PROTAL® process. *Journal of Thermal Spray Technology* 15 (2): 191–197.

Li, H., S. Costil, S.-H. Deng, C. Coddet, H.-L. Liao, K. Richardt, E. Lugscheider, and K. Bobzin. 2006b. Ni-Cr coatings deposited on a magnesium alloy substrate using the Protal process. In *ITSC-2006*, ed. B. Marple. Materials Park: ASM International.

Liao, H., P. Vaslia, Y. Young, and C. Coddet. 1999. Determination of the residual stress distribution from in situ measurements for thermally sprayed WC-Co coatings. *Journal of Thermal Spray Technology* 6 (2): 235–241.

Liao, H., A. Gammondi, S. Costil, and C. Coddet. 2003. Influence of surface laser cleaning combined with substrate preheating on the splat morphology. In *TS-2003: Advancing the science and applying the technology*, ed. C. Moreau and B. Marple, 883–888. Materials Park: ASM International.

Maruyama, T., K. Akagi, and T. Kobayaski. 2006. Effects of blasting parameters on removability of residual grit. *Journal of Thermal Spray Technology* 15 (4): 817–821.

———. 2007. Effect of the blasting angle on the amount of the residual grit on blasted substrates. In *TS-2007*, ed. B.P. Marple et al. Materials Park: ASM International.

Mellali, M., A. Grimaud, A.C. Léger, P. Fauchais, and J. Lu. 1997. Alumina grit blasting parameters for surface preparation in the plasma spraying operation. *Journal of Thermal Spray Technology* 6 (2): 217–227.

Mohammadi, Z., A.A. Ziaei-Moagyed, and A. Sheik-Mehdi Mesgar. 2007. Grit blasting of Ti-6Al-4 V alloys: Optimization and its effect on adhesion strength of plasma sprayed hydroxyapatite coatings. *International Journal of Materials and Product Technology* 19: 415–423.

Momber, A.W., and R. Kovacevic. 1998. *Principles of abrasives water jet machining*. London: Springer Ltd.

Momber, A.W., Y.C. Wong, R. Ij, and E. Budiharma. 2002. Hydrodynamic profiling and grit blasting of low-carbon steel surfaces. *Tribology International* 35: 271–281.

Montavon G. 2004. *Quality control of thermal sprayed coatings*, Course Thermal Spraying, Aliderte Limoges France.

Ness, E., and R. Zibbell. 1996. Abrasion and erosion of hard materials related to wear in the abrasive water jet. *Wear* 196: 120–125.

Scrivani, A., G. Rizzi, and C. Giolli. 2006. Improved surface preparation of nickel-superalloys for MCrAlY coatings on gas turbine. In *Proceedings of ITSC-2006 conference*, ed. B. Larple et al. Materials Park: ASM International.

Shin, Y.C., S.J. Oh, and S.A. Coker. 1995. Surface roughness measurement by ultrasonic sensing for in-process monitoring. *ASME Journal of Engineering for Industry* 117: 439–447.

Spur, G., E. Uhlmann, and F. Elbing. 1999. Dry-ice blasting for cleaning: Process, optimization and application. *Wear* 233–235: 402–411.

Stratford, S. 2000. Dry ice blasting for paint stripping and surface preparation. *Metal Finishing* 98 (6): 493–499.

Tanaka, Y., and M. Fukumoto. 1999. Investigation of dominating factor on flattening behavior of plasma sprayed ceramic particles. *Surface and Coatings Technology* 12 (121): 124–130.

Taylor, T.A. 1995a. Surface roughening of metallic substrates by high pressure pure waterjet. *Surface and Coatings Technology* 76–77: 95–100.

———. 1995b. Surface roughening of superalloys by high pressure water jets. In *TS-1995: Science and technology*, ed. C.C. Berndt and S. Sampath, 351–358. Materials Park: ASM International.

Thermal Spraying, Practice, Theory and Applications. 1985. American Welding Society, Miami, FL.

Thomas, T.R. 1999. *Rough surfaces*. London: Imperial College press.

Tian, J.-J., Y.-K. Wei, C.-X. Li, G.-J. Yang, and C.-J. Li. 2018. Effect of post-spray shot peening treatment on the corrosion behavior of NiCr-Mo coating by plasma spraying of the shell–core–structured powders. *Journal of Thermal Spray Technology* 27: 232–242.

Tomovich, S.J., and Z. Peng. 2005. Optimized reflection imaging for surface roughness analysis using confocal laser scanning microscopy and height encoded image processing. *Journal of Physics Conference Series* 13: 426–429.

Tonshoff, H.K., H. Janocha, and M. Seidel. 1988. Image processing in a production environment. *Annals CIRP* 37 (2): 579–589.

Varacalle, D.J., Jr., D.P. Guillen, D.M. Deason, W. Rhodaberger, and E. Sampson. 2006. Effect of grit blasting on substrate roughness and coating adhesion. *Journal of Thermal Spray Technology* 15 (3): 348–355.

Vorburger, T.V., and E.C. Teague. 1981. Optical techniques for on-line measurements of surface finish. *Precision Engineering* 3 (2): 61–83.

Wang, L. 2004. Erosion testing and surface preparation using abrasive water-jetting. *Journal of Materials Engineering and Performance* 13 (1): 103–106.

Wang, J., H. Wang, H. Wang, and Z. Zeng. 2012. Preparation of Ni60-WC coating by plasma spraying, plasma re-melting and plasma spray welding on surface of hot forging die. *Journal Wuhan University of Technology, Materials Science Edition* 27 (4): 640–643.

Wigren, J. 1988. Grit blasting as surface preparation before plasma spraying. In *Thermal spray: Advances in coatings technology*, ed. D. Hauck, 99–104. Materials Park: ASM International.

Yang, Y., H.G. Wang, X.P. Cao, and L. Wang. 2006. Influence of substrate surface roughness on adhesive strength. In *Proceedings of the ITSC-2006*, ed. B. Marple et al. Materials Park: ASM International.

Yankee, S.J., R.L. Salsbury, and B.J. Pletka. 1991. Quality control of hydroxylapatite coating: properties, processes and applications. In *TS-1991*, ed. T. Bernecki, 505–513. Materials Park: ASM International.

Zeng, J., and T.J. Kim. 1996. An erosion model for abrasive waterjet milling of polycrystalline ceramics. *Wear* 199: 275–282.

Zeng, Z., H. Wang, Wang Hongfu1, and J. Wang. 2012. *Preparation of Ni60-Cr3C2 coating by plasma spraying, plasma re-melting and plasma spray welding on W6Mo5Cr4V2.*

Abbreviations

CCD	Charge Coupled Device
CCDS	Computer Controlled Detonation Spraying
CS	Cold Spray
CSM	Crystallographic Structure Matching
CS-SEM	Cross-Sectional Scanning Electron Microscopy
CS-TEM	Cross-Sectional Transmission Electron Microscopy
D-Gun	Detonation Gun
DC	Direct Current
DC-APS	DC-Atmospheric Plasma Spraying
DC-VPS	DC-Vacuum Plasma Spraying
EB-PVD	Electron Beam-Physical Vapor Deposition
EL	Elongation Factor
FGM	Functionally Graded Material
FIB	Focused Ion Beam
FS	Flame Spraying
HVAF	High Velocity Air Fuel
HVOF	High Velocity Oxy Fuel
LHS	Left-Hand Side
LPPS	Low-Pressure Plasma Spraying
LV	Laster Velocimetry
OEM	Optical Emission Spectroscopy
OM	Optical Microscopy
PDF	Probability Distribution Function (PDF)
PIC	Particle Impact Condition
PSD	Particle Size Distribution
PS-PVD	Plasma Spraying-Physical Vapor Deposition
PTA	Plasma Transferred Arc
RF	Radio Frequency
RF-ICP	RF-Inductively Coupled Plasma
RF-IPS	RF-Induction Plasma Spraying
RF-VIPS	RF-Vacuum Induction Plasma Spraying
RHS	Right-Hand Side
RT	Room Temperature
SEM	Scanning Electron Microscopy
SF	Shape factor
SOD	Standoff Distances
SOFC	Solid Oxide Fuel Cells
SS	Stainless Steel
SZM	Structure Zone Model
TBC	Thermal Barrier Coating
TEM	Transmission Electron Microscopy
TS	Thermal Spray
VLPPS	Very Low-Pressure Plasma Spray
VOF	Volume of Fluid
WAS	Wire Arc Spraying
XRD	X-ray Diffraction
YPSZ	Yttria Partially Stabilized Zirconia
YZS	Yttria Stabilized Zirconia

15.1 Introduction

As described in the introduction of this book: *"Thermal Spraying comprises a group of coating processes in which finely divided metallic or nonmetallic materials are deposited in a molten or semi-molten condition to form a coating. The coating material may be in the form of powder, ceramic rod, wire, or molten materials"* [Hermanek (2001)].

A block diagram of the principal components of a thermal spray (TS) system is illustrated in Fig. 15.1. The system is centered around the "Spray Torch" in which different sources of energy such as chemical (combustion) or electrical (electric discharge) are used to generate a stream of hot gas in which the material to be sprayed is injected. The material introduced into the torch, whether as powder, wire or cord, is heated, melted/atomized, and entrained by the high-temperature, high-velocity gas stream toward the substrate on which the heated particles or molten droplets impact generating splats that are the building blocks of the coating. The form and properties of these splats depend on the nature of the material sprayed, particle/droplet size distribution, particle/droplet temperature, and velocity prior to their impact on the surface of the substrate as well as on the surface properties of the substrate.

© Springer Nature Switzerland AG 2021
M. I. Boulos et al. (ed.), *Thermal Spray Fundamentals*, https://doi.org/10.1007/978-3-030-70672-2_15

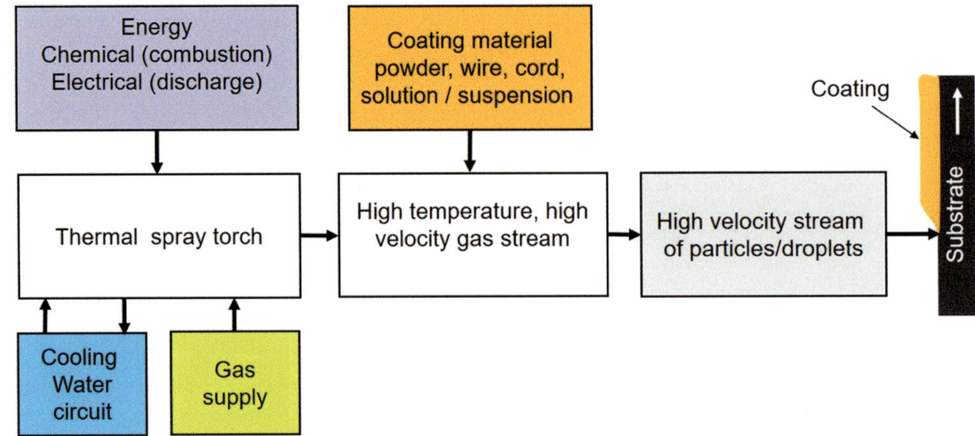

Fig. 15.1 Block diagram of a generalized Thermal Spray (TS) system

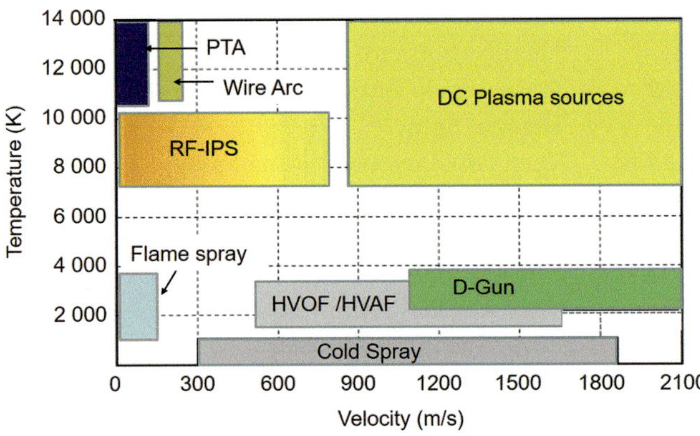

Fig. 15.2 Predominant gas temperatures and velocity mapping associated with different thermal spray processes

A wide range of energy sources have been used in the different thermal spray (TS) technologies. As shown in Fig. 15.2, these varied widely in the predominant gas temperature and velocities associated with each of them. The source design and the main characteristics of each of these technologies have been presented in Part II of this book, Chap. 6 through Chap. 12 dedicated respectively to cold spray (CS), flame spraying (FS) including D- Gun, HVOF, and HVAF, DC atmospheric and vacuum plasma spraying (DC-APS) (DC-VPS), RF induction plasma spraying (RF-IPS), wire arc spraying (WAS), and plasma-transferred arc deposition (PTA).

Cold spray (CS) which is borderline to be a "thermal Spray" technology offers high gas velocities (1900 m/s) with relatively mild to low gas temperatures generally below (1000 K). Flame Spraying, the first thermal spray technology developed in the early twentieth century, is based on the combustion of a gaseous or liquid fuel

generating a stream of hot gases with a temperature generally below 4000 K and gas velocities of 150 m/s. D-Gun, HVOF, and HVAF, which are also combustion-based technologies, achieve considerably higher velocities than standard FS (up to 2000 m/s) with the same temperature limitation of (3000 to 4000 K). DC plasma spraying, on the other hand, makes use of an electric arc to generate a plasma jet with a gas temperature (8000 to 14,000 K) and velocity (900 to 2100 m/s) with the highest velocities being associated with DC-VPS. RF-IPS, WAS, and PTA, which are all plasma spraying technologies, offer gas temperatures comparable with DC-APS or DC-VPS in the range of 8000 to 12000 K, with relatively lower associated gas velocities (100−300 m/s). In the case of RF-ICP, higher gas velocities (up to 1000 m/s or more) can be achieved under vacuum induction plasma spraying (RF-VIPS) conditions. WAS and PTA remain relatively low velocity plasma sources (less than 300 m/s).

With such board variations in the temperature and velocity fields associated with the different energy sources used in thermal spray (TS) technology, it is not surprising that the size and form of the feed material used in each of these technologies vary considerably. The great majority of the spray technologies are capable of receiving the feed material in the form of powder with a PSD in the range of 20–40 µm up to 100–150 µm tha\t are heated and completely or partially melted in-flight prior to their projection on the substrate. Exceptionally, FS can also be operated using wires and cords that are continuously feed into the flame, heating and melting the wire or cord tip and atomizing the molten material formed into a stream of small molten droplets entrained by the gases and projected toward the substrate. WAS, as the name implies, is limited to the use of wires or cords as feed material that are continually feed into the torch with their tip melted by an electric arc struck between the two wires or a wires and a fixed, non-consumable electrode. The molten metal formed at the wire tip is atomized by a high velocity cold gas stream entrain the formed molten metal droplets and projecting them toward the substrate forming the coating. PTA, on the other hand, while being able to use feed material in the form of powder or wires or cords, does not involve the atomization of the feed material with the coating generated through the creation of a molten metal pool on the surface to the substrate, using a transferred arc, in which the feed material is deposited in a molten or partially melted state.

Based on the mechanism of TS coating formation, with the exception of PTA, coatings can be grouped into the following two broad categories:

- *Those sprayed with mainly fully melted particles* (including metals, alloys, ceramics, and cermets introduced within the spray gun as particles, wires, or cords) flattening upon impact on the substrate-forming splats, with mean diameters between 2 and 6 times that of the impacting droplet, which are the building blocks of the coating.
- *Those resulting from the plastic deformation of on-molten or partially melted particles* (metals, alloys, and cermets) upon their impact on the substrate. The formed splat diameter in this case is rather close to that of the initial particle, with a flattening degree generally below 1.5–2.0. The formed coating is further consolidated by peening effect induced by the stream of subsequently impacting particles, obviously with the exception of the last deposited layer.

The microstructure of sprayed coatings is rather complex, exhibiting anisotropic behavior, which is also affected by defects such as pores, cracks, and imbedded nonmelted particles. As discussed in the vast literature published on the subject [Fauchais P, et al. (2001a, b) Fauchais P (2004) Hearley JA, et al. (2000) Sampath S and H Herman (1996) Sampath S, et al. (2003), Davis JR (2004) Azarmi F, et al. (2008) Schmidt T, et al. (2006)] coating microstructure depends strongly on more than 60 independent design and operating parameters that need to be considered for the optimization of the coating quality. Fortunately, the development of sensors able to work in the harsh environment of spray booths [Fauchais P and M Vardelle (2010)] (see Part IV, Chap. 18 Thermal Spray Process Integration has allowed achieving a better understanding of coating generation and the means to control its microstructure. These sensors measure in-flight particle parameters such as temperature, diameter, velocity, and trajectory distributions prior to their impact on the substrate. High-speed images coupled with 3-D transient models of particle impact and splat formation on the surface of the substrate [Mostaghimi J and S Chandra (2018)] have also contributed extensively to the understanding of the basic phenomena involved.

Coating generation starts with particle/molten droplet flattening on impact with the substrate or previously deposited layers, followed by the rapid cooling and solidification of the formed splats which are the building blocks of the coating. In this chapter, we will attempt to address the following general questions:

- What are the physical and chemical characteristics of the substrate surface? In all metals-based substrate surface will be covered with an oxide layer, the thickness and composition of which are dependent on the substrate temperature before spraying.
- Have adsorbates and condensates, present at the substrate surface, been eliminated by preheating just prior to spraying, using dry ice blasting or laser treatment?
- What are the fundamental phenomena linked to the impact of individual particles?
- How do particles adhere first to the substrate and then to the previously deposited layers? (a coating will never protect a substrate if it does not adhere to it!). Of course, melted and unmolten ductile metal, cermet, even ceramic particles, and polymer particles will have to be considered separately.
- How melted or partially melted particles will form splats on smooth substrates or previously deposited layers?
- What happens when melted particles impact on rough substrates or on idealized surfaces with well-defined geometrical patterns?
- How ductile particles impact of on smooth and rough surfaces?
- Coating properties resulting from the impact of melted or ductile particles with:
  - Splat layering with a motionless spray torch?
  - Splat layering with the torch moving in opposite direction to the substrate.
  - Splat layering with beads overlapping to form a pass, the thickness of which is linked to the spray pattern, as well as successive pass layering?
- Formation of polymer coatings?
- Formation of PTA coatings?

Topics covered in this chapter also include;

- How the use of robots moving the spray torch, and a rotating table on which the substrates are fixed allows for a better control of the coating thickness on substrates of complex shapes?
- Temperature control of the substrate and coating during the spraying operation with the impact of hot particles and the jet of hot gases on the substrate/coating.

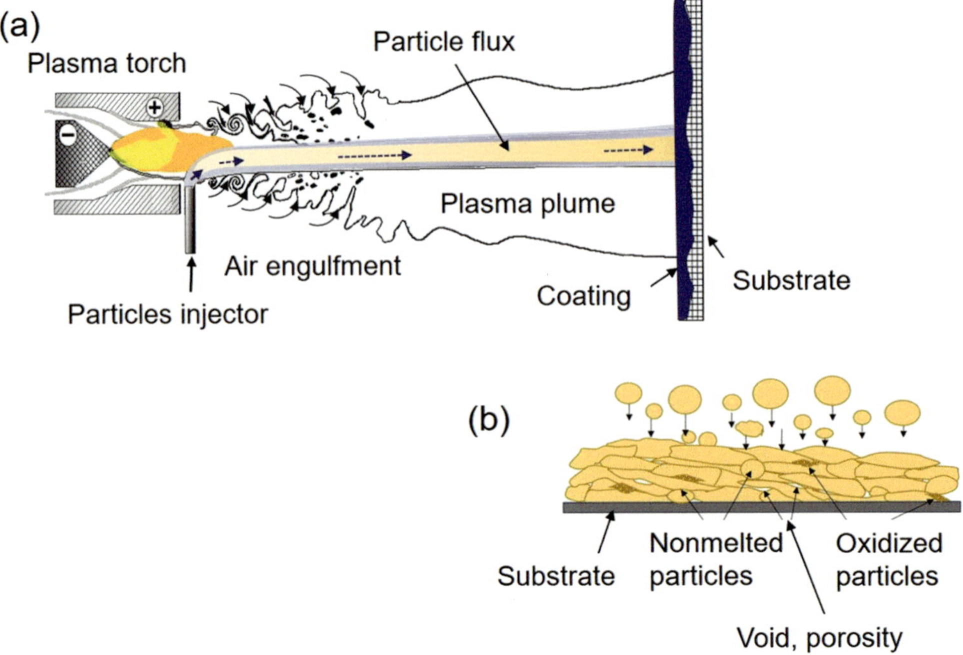

Fig. 15.3 Thermal Spray process design and operating parameters

- Development of residual stresses due to quenching, expansion mismatch, temperature gradient, surface preparation, coating geometry, and phase changes.
- A short review of the influence of the choice of particle morphologies and wires, cored wires, rods, or cords on coating properties is presented.
- Different means to finish coatings and post-treat them is described.

15.2 Basic Concepts

15.2.1 Thermal Spray Parameters

Considering, for example, a typical DC plasma spraying process, as illustrated in Fig. 15.3, the principal design and operation parameters can be summarized as follows:

- *Spray process design parameters*
  - Torch design, nozzle shape and dimensions,
  - Open atmosphere Vs operation in a controlled atmosphere,
  - Nature of the feed material powder, wire, cord, solution/suspension.
  - Axial Vs radial injection, injector dimensions, location and orientation, atomization technique used in the case of solution/suspension spraying.
- *Substrate parameters*
  - Material, shape, size, thickness, and preparation.
  - Substrate handling, Manual Vs robotic
  - Need for masking, pretreatment and cooling during the spraying operation
- *Material parameters*
  - Powder, wire, or cord requirements
  - Material composition, PSD, particle morphology depending on the manufacturing process
  - Wire/cored composition, dimension and ductility
- *Independent operating parameters*
  - Process gas composition and flow rate
  - Ambient operating pressure
  - Combustion/Plasma power
  - Powder/Wire or cord feed rate
  - Powder carrier gas composition and flow rate
  - Spraying distance
  - Spraying patters and relative torch to substrate displacement velocity
  - Substrate temperature control
- *Dependent parameters*
  - Particle velocity and temperature prior to impact on the substrate

- Particle melting
- Impact angle of particle flux on the substrate
- Substrate temperature during preheating, coating and cooling down phases.

In the case of alternate thermal spray technologies, such as cold spaying or combustion spraying, RF induction plasma spraying or wire arc spraying some of these parameters are not applicable while additional parameters can be added with the relative importance of each of these parameters depending strongly on the nature of the thermal spray technology used. The first three groups of parameters (spray process, substrate and material parameters) are essentially design parameters that have to be carefully selected taking into account the functional role expected of the coating and its properties, the nature, material shape, and principal dimensions of the substrate, number of parts to be coated, available coating material in the appropriate form (powder, wire or cord), and, of course, economical constraints. Most of these parameters, however, have a direct or indirect influence on in-flight particle parameters (particle diameter, velocity, temperature, degree of melting and surface chemistry) prior to their impact on the surface of the substrate. These combined with the surface conditions of the substrate (surface finish and chemistry, temperature, and orientation or the substrate surface with respect to the particle trajectories/ angle of attack) constitute the principal parameters

controlling the splat formation on the impact of the particles on the surface of the substrate and consequently the quality of the coating

15.2.2 In-Flight Particle Transformations

15.2.2.1 Particle Trajectory, Velocity, and Temperature History

Once the spray process has been chosen, temperatures, velocities, degree of melting, and surface chemistry of the particles/molten droplets attained in the process prior to their impact on the substrate strongly depend on the spray jet characteristics, particle material, their injection conditions, and trajectories [Vardelle M, et al. (2001), Fauchais P, et al. (2010)]. A detailed discussion presented in Part I, Chap. 4, Plasma-particle momentum and heat transfer, and Chap. 5 Gas and particle dynamics, lays the ground for the computation of the particle trajectories, velocity, and temperature histories for different plasma sources and operating conditions. These are also supported by systematic diagnostic measurements of in-flight particle conditions that are available in literature and served for model validation (see also Chap. 18 Thermal Spray Process Integration.

Typical results reported by [Vardelle et al. (1988)] are given in Figs. 15.4 and 15.5 for atmospheric pressure plasma spraying of fine alumina powder ($d_p = 18 \pm 3$ μm) using DC plasma Jet immerging from 8 mm i.d. anode

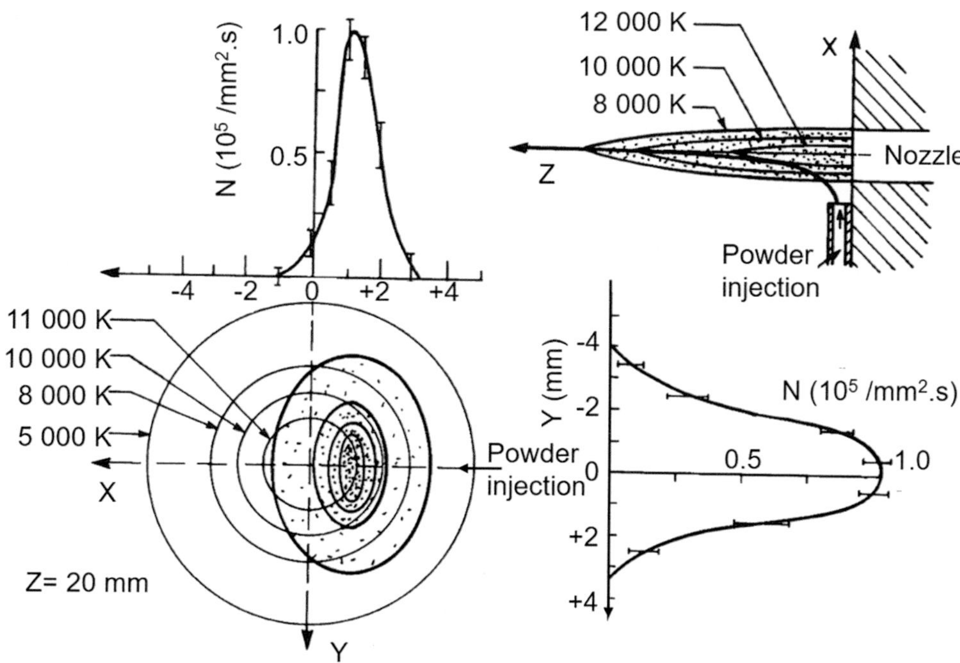

Fig. 15.4 Particle trajectories and number flux density distribution for fine alumina powder, 18 ± 3 μm in 29.2 kW, (Ar/H$_2$) DC plasma jet, 75 slm (Ar) +15 slm (H$_2$), at 75 mm from torch exit with a carrier gas flow rate of 5.5 slm (Ar). [Vardelle et al. (1988)]

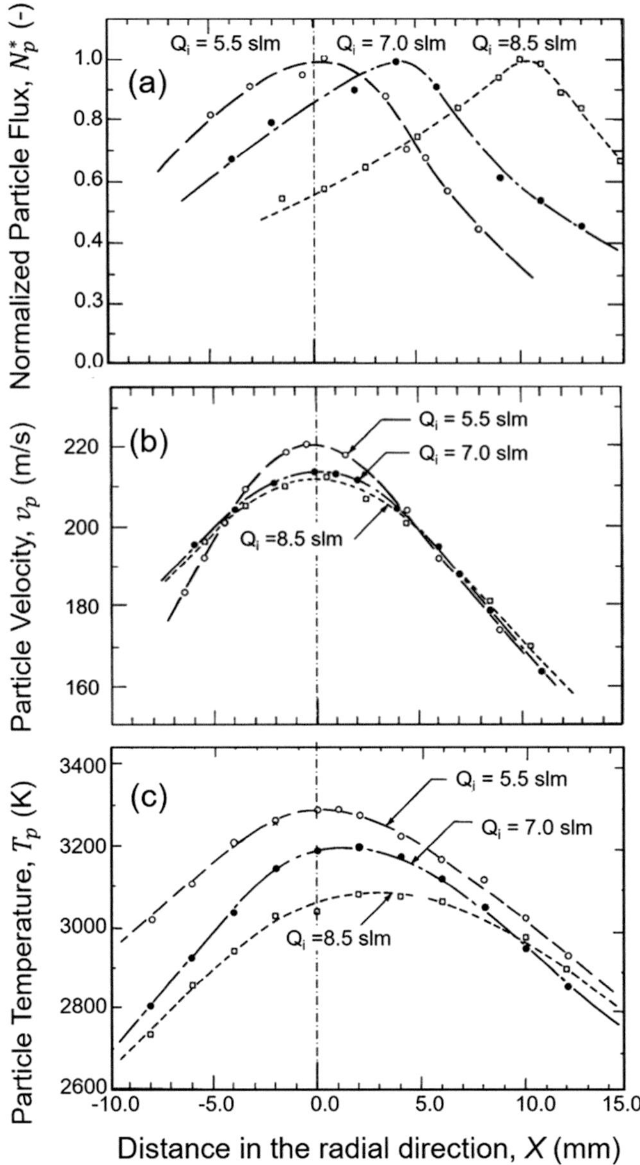

Fig. 15.5 Effect of carrier gas flow rate, Q_i (Ar) on the radial distribution of fine alumina powder $d_p = 18 \pm 3$ μm, in 29.2 kW, (Ar/H$_2$) DC plasma jet, 75 slm (Ar)+15 slm (H$_2$), at 75 mm from torch exit (**a**) Normalized particle number flux distribution, (**b**) particle velocity and (**c**) particle temperature, profiles are in the plane of symmetry with injection from the LHS of the figure. [Vardelle et al. (1988)]

nozzle, operating using Ar/H$_2$ (16.7 vol./H$_2$) as plasma gas at a flow rate of 90 slm, and DC powder of 29.2 kW. The powder was injected at 90° to the axis of the flow using an external injector, 2 mm i.d., and powder carrier gas flow rates varying between 5.5 and 8.5 slm (Ar). The corresponding mean particle injection velocities were $\overline{v_i} = 25$, 30 and 35 m/s, respectively. The 2-D dispersion of the injected powder in the plasma jet at a plane, $z = 20$ mm downstream of the DC torch nozzle exit is given in Fig. 15.4. Superposed on the figure are the 2-D temperature

contours of the plasma. It may be noted that the powder disperses in the plasma stream in both the direction of powder injection (x-axis) and the lateral direction (y- axis) with the number flux density profiles showing a maximum value of $N_p = 1.0 \times 10^5$ particle/mm^2.s in the injection plane (corresponding to 1.2 μg/mm^2.s or a total particle feed rate of about 1.0 g/min, for a mean particle diameter of 18 μm and density of alumina of 3.95 g/cm^3). With the increase in the distance from the plane of symmetry, the particle number flux density drops down to 0.4×10^5 particles/mm^2.s at $y = \pm 2$ mm, and close to zero values at $y > \pm 3$ mm from the central plane of powder injection that essentially defines the cone of powder stream.

Corresponding radial profiles of the normalized particle number flux density, N_p^* particle velocity, v_p, and particle temperature T_p as function of the powder carrier gas flow rates are given in Fig. 15.5.

$$N_p^* = N_p / \left(N_p\right)_{\text{Max}} \qquad (15.1)$$

These were obtained in the plane of symmetry (x-axis) at an axial location $z = 75$ mm from the exit plane of the DC plasma torch, with the powder injection being from the LHS of the figure. It may be noted from Fig. 15.5a that the trajectory of the particle stream is rather sensitive to the powder carrier gas flow rate, with the maximum particle flux density coinciding at this level with the axis of the plasma jet for a carrier gas flow rate of $Q_i = 5.5$ slm, moving further away from the torch axis with the increase in the carrier gas to 7.0 or 8.5 slm. As shown in Fig. 15.5b, c, the corresponding maximum value of the particle velocity and temperature remains centered over the torch axis. It must be emphasized that, as shown in Fig. 15.5a, such values of maximum particle velocity and temperature correspond to a decreasing number of particle flux, down to almost 50% of the maximum particle flux density at $Q_i = 8.5$ slm. Increasing the powder carrier gas flow rate is also responsible for the local cooling of the plasma with a corresponding drop of the maxim particle temperature.

The effect of the particle injection and plasma operating parameters on particle trajectories and their velocity and temperature histories have been studied by [Delluc et al. (2004)] for the case of DC plasma torch with an exit nozzle 7.0 mm i.d. operated using 60 slm Ar/H$_2$ (25 vol./H$_2$), at 43 kW with a thermal efficiency of 60%. Computations were carried out for the injection of 30 μm diameter dense zirconia particle at different velocities between 4.71 and 47.15 m/s. The particles were externally injected 4.5 mm downstream of the nozzle exit, orthogonal to the jet, at a distance of 8.5 mm off the torch axis. The results given in Fig. 15.6a, b show a significant effect of the injection velocity on particle trajectories and velocity with near optimal results achieved with injection velocities between 9 and 20 m/s that gives rise to near axial

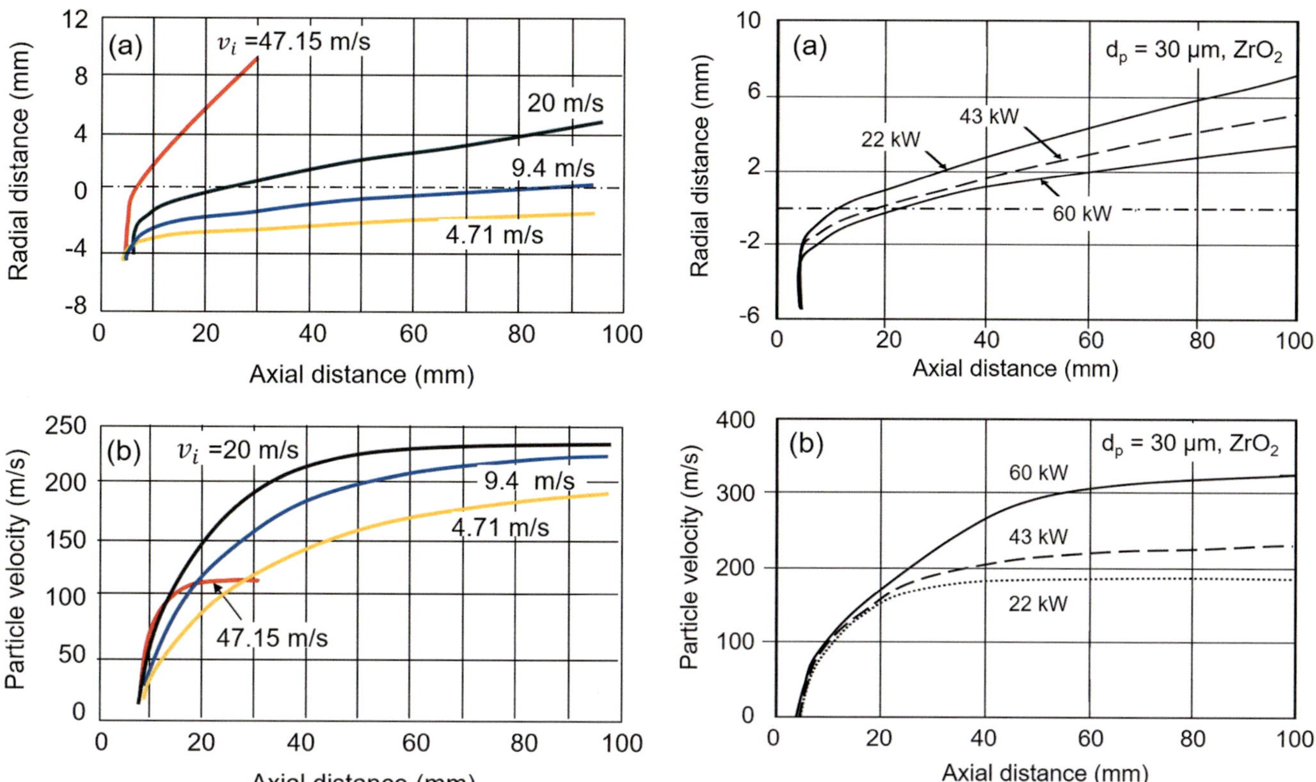

Fig. 15.6 (a) Particle trajectory and (b) particle velocity, for 30 μm Zirconia powder injected externally into DC plasma jet (45 slm Ar+15 slm H$_2$) 43 kW, with different injection velocities. Radial distance = 0 corresponds to torch axis. [Delluc et al. (2004)]

Fig. 15.7 (a) Particle trajectory and (b) particle velocity, for 30 μm zirconia powder injected externally into DC plasma jet (45 slm Ar+15 slm H$_2$) with different DC power ratings 22–60 kW, for a constant powder injection velocity of 20 m/s. Radial distance = 0 corresponds to torch axis. [Delluc et al. (2004)]

particle trajectory crossing the torch axis (radial distance = 0) at an angle of 3.5–4°. Such trajectories provide the particles with a maximum residence time in the core of the plasma jet for maximum heating and acceleration by the plasma. It may be noted that with an injection velocity of 4.71 m/s the particle inertia is too small to penetrate the plasma jet with the particle staying essentially in the fringes of the jet throughout its trajectory. Particle injection velocity of 47.15 m/s seems, on the other hand, as excessive since it gives rise to particle trajectory that crosses the axis of the jet, immerging from its other side in less than 10 mm from its point of injection.

The corresponding effect of plasma torch power on the particle trajectory and velocity is given in Fig. 15.7 [Delluc et al. (2004)]. These were obtained for the same ZrO$_2$ powder and plasma operating conditions as those of Fig. 15.6, with DC plasma power ratings varying between 22 and 60 kW and a constant particle injection velocity of 20 m/s that was optimal for DC plasma power of 43 kW. The results given in Fig. 15.7 show that as the plasma momentum density increases with increasing power level, the particle penetration of the plasma is significantly reduced. As expected, the particle velocity increases with the increase in the power

level providing the injection velocity is adapted to each power level to achieve the optimum trajectory.

As discussed in Chap. 5 5 Gas and Particle Dynamics in Thermal Spray, Sect. 5.5.4.3, Effect of turbulence, for smaller diameter particles of relatively light materials ($d_p < 10$–20 μm, $\rho_p < 2$–3 g/cm^3), turbulent diffusion of particles plays an increasingly important role in the determination of the particle trajectories. Fig. 15.8, after [Pfender E (1989)] shows that for alumina particles injected in a DC argon plasma jet, the ranges of dispersed particle trajectories are obviously larger for the smaller particles (10 μm). This is due to their small mass and consequently their inertia that allows them to follow easier the motion within eddies in the plasma flow. It is also observed that the dispersion becomes larger downstream. This is caused by the accumulated "random walk" influence.

It has also to be noted that particle conditions along their trajectories in any thermal spray source can and will change continuously with distance from the source gas. The effect becomes increasingly more important for refractory metal and ceramic powders, which because of their high melting temperatures, will lose, in-flight, an increasing fraction of their thermal energy by radiation to the surrounding leading to their cooling down and possible solidification while still in-flight entrained by a plasma with a temperature higher than that of the material melting temperature. The effect was demonstrated in Chap. 10, Induction Plasma Spraying, Sect. 10.5.3, dealing with the RF-IPS of tungsten powder. The results given in Fig. 15.9a, b, after [Jiang et al. (1993)], report experimental data for the deposition of W in a 40 kW, Ar/H$_2$ induction plasma on a graphite substrate at different chamber pressures 26.6–53.2 kPa (200–400 Torr). These show a rapid decrease in deposition efficiency and of the relative density of the deposit with the increase in spraying distance that can only be attributed to the cooling and partial solidification of the W-particles in-flight. The interpretation is further supported by the strong dependence of the observed changes on the chamber pressure. The effect being significantly less important at a pressure of 16.6 kPa (200 Torr) that would correspond to the highest particle velocities and the shortest residence time of the particle in-flight prior to its impact on the surface of the substrate.

Finally, attention has to be also given to the fact that most combustion-based and plasma sources used in thermal spray are operating under turbulent flow conditions that also include low-frequency fluctuations and flow oscillations. As demonstrated by the high-speed photographs given in Fig. 15.10 after [Fauchais P (2004)], arc fluctuations can lead to the entrainment of cold surrounding gas into the jet, rapidly reducing its average temperature and velocity and uneven particle heating compared to well-controlled steady-state modeling studies.

15.2.2.2 Inflight Particle Oxidation

As discussed in Chap. 4 Plasma-Particle Momentum and Heat Transfer, Sect. 4.7, Chemical Reactions and Melt Circulation, and later in Chap. 9, DC Plasma Spraying, Sect. 9.3 Controlled Atmosphere Plasma Spraying, the large majority of thermal spray operations, with the exception of "Controlled Atmosphere Plasma Spraying," would involve the exposure of the material to be sprayed to oxygen either as oxidant in a combustion flame or as entrained air in open atmosphere plasma spraying. In either cases, the oxygen can have a detrimental effect of the chemistry of the material being sprayed especially when dealing with reactive metals, alloys or non-oxide ceramics (carbides, nitrides...).

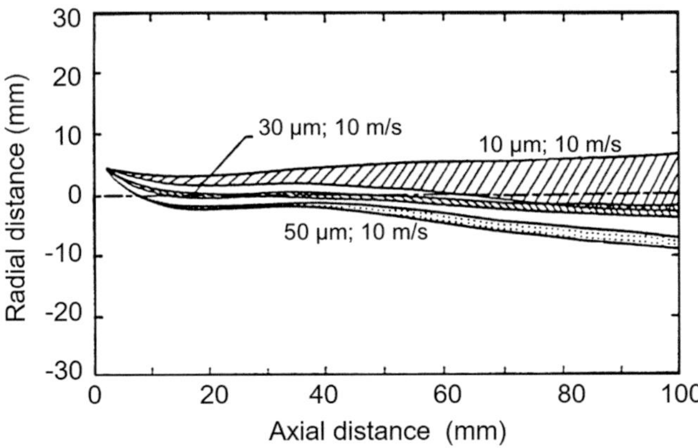

Fig. 15.8 Dispersed 10–50 µm alumina particle trajectories in an argon, DC turbulent plasma jet. (Reprinted with kind permission from Springer Science Business Media Pfender (1989))

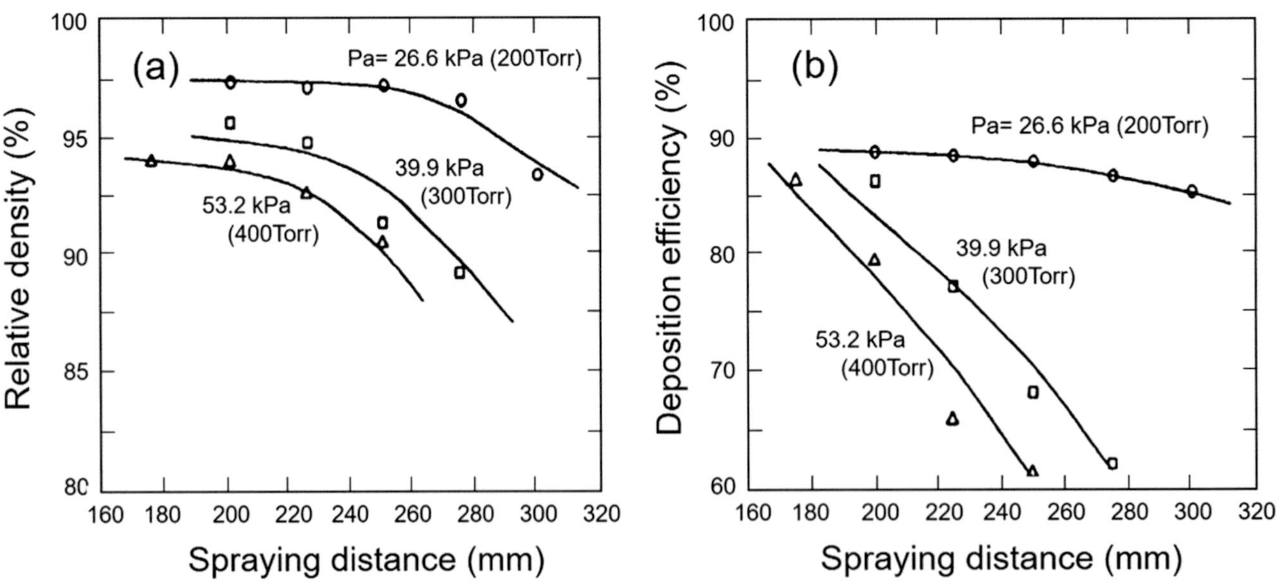

Fig. 15.9 Effect of spraying distance and chamber pressure on (**a**) relative density and (**b**) deposition efficiency for tungsten spraying using an Ar/H$_2$ induction plasma at a plate power 40 kW. ([Jiang et al. (1993)]. Reprinted with permission of ASM International. All rights reserved)

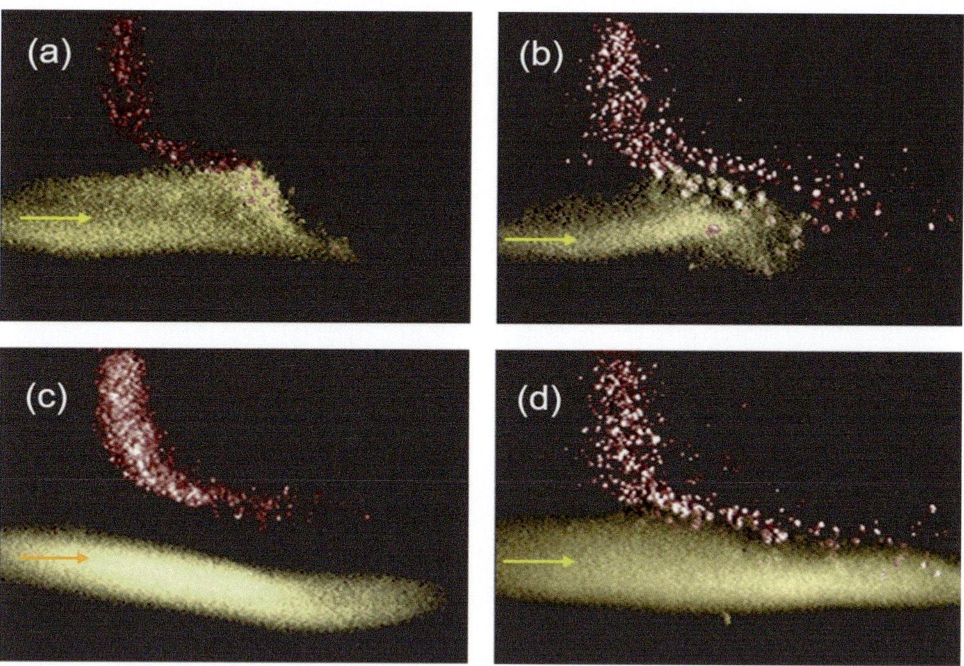

Fig. 15.10 High-speed images showing the fluctuations of the plasma jet and the spray particle fluxes being exposed to varying environments. ([Fauchais (2004)]. Copyright IOP Publishing. Reproduced with permission. All rights reserved)

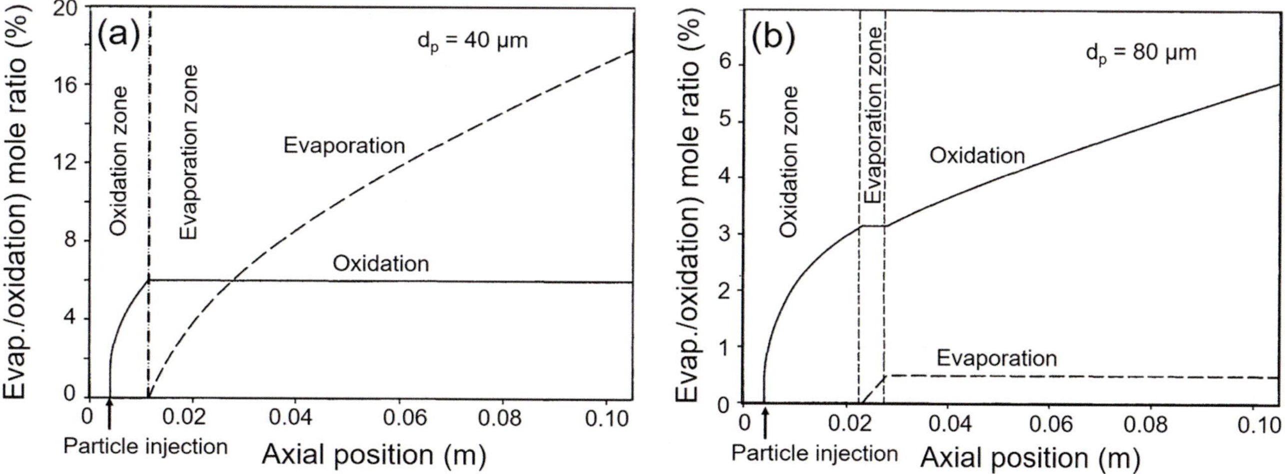

Fig. 15.11 Molar ratio of evaporated and oxidized fractions of an iron particle along its trajectory in an Ar/H$_2$ plasma jet. (**a**) 40 μm, (**b**) 80 μm. (After [Vardelle et al. (1996)])

Under typical plasma spraying conditions, oxidation of the sprayed material can take place either in the vapor phase surrounding the droplet or at the surface of the droplet. In the latter case, the formed oxide layer can result in a considerable reduction of the evaporation rate from the droplet surface. The effect is strongly dependent on the particle diameter as illustrated in Fig. 15.11 [Vardelle et al. (1996)] for iron particles of diameters 40 and 80 μm, injected at 10 m/s into an Ar/H$_2$ DC plasma jet. The results are given in terms of molar ratio of the evaporated to the oxidized fraction of the iron particle as function of its axial position on its trajectory. As expected, the smaller particles, $d_p = 40$ μm, evaporate at a faster rate compared to the larger 80 μm particles, with comparable in-flight oxidation. This implies that if the trajectory of the 40 μm particle is optimum, the 80 μm particle will cross earlier the jet and thus is less heated. This results in a much lower evaporation in the latter case with a strong oxidation, while the opposite is true for the smaller particle.

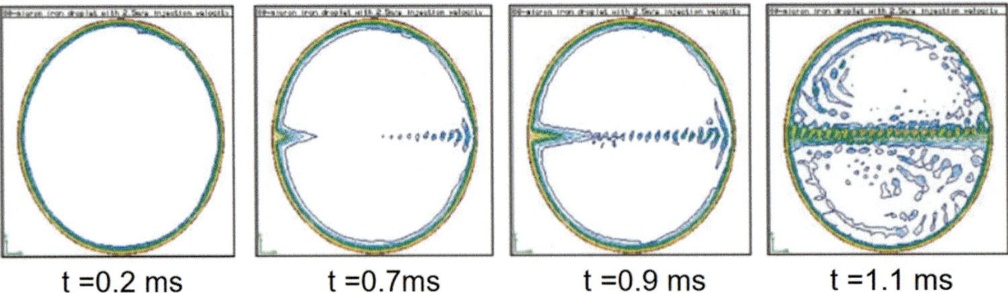

t =0.2 ms   t =0.7ms   t =0.9 ms   t =1.1 ms

Fig. 15.12 Time-resolved modeling of internal convection and oxide penetration by a spherical vortex of Hill in a molten iron particle. [Espié et al. (1999) (2001) (2005)]

The in-flight particle oxidation is controlled by diffusion and convection transfer of oxygen in the particle. The specific reaction rate constant of diffusion is of an Arrhenius type and roughly increases by one order of magnitude when the temperature increases by 100–200 K. It is thus clear that the oxidation of a fully melted particle will be much more important than that of one in a plastic state, which is in turn higher than that of a cold one, as in cold spray.

At the end of the trajectory of the particle in the plasma, the composition of its surface depends on its temperature history, and the reactive species present in the plasma, provided that they can reach the particle surface. Thermodynamic equilibrium calculation can serve for the calculation of the chemical composition of the particles prior to their impact on the substrate. While the plasma conditions often deviate from equilibrium, the knowledge of which phases are thermodynamically stable under given conditions, and how they compare with those present in real systems is very informative. This has been demonstrated, for example, for the oxides in plasma-sprayed chromium steel [Volenik (1997)].

Oxidation can be promoted, under certain conditions encountered in plasma, HVOF and wire arc spraying, by the convective movement induced by the high-energy jet within the liquid drop (fully melted particle). In this case, fresh metal is continuously renewed at the particle surface, while oxide or oxygen is entrained inside the molten particle. Compared to oxidation by diffusion, the oxide content can be multiplied by a factor of 4–6! In the case of DC plasma jets, as with wire-arc spraying the shear stress at the surface of the molten droplet created by the large velocity difference between the molten particle in flight and the plasma flow will induce convective motion within the droplet. Such internal convective currents, as illustrated in Fig. 15.12, will result in the engulfment of some of the surface oxide into the center of the droplet. The phenomena are observed when the ratio of the kinematics viscosities of the plasma and the droplet is greater than 50, and the Reynolds number of the flow relative to droplet is greater than 20 [Neiser et al. (1998)]. The motion inside the droplet can be represented by a Hill vortex for inviscid axisymmetric flow [Espié et al. (1999) (2001)

Fig. 15.13 SEM cross sectional micrograph of a low carbon steel particle collected at z = 100 mm after its flight in DC plasma jet (50 slm (Ar)+10 slm (H$_2$) I=500 A, nozzle i.d. 7 mm. [Espié et al. (1999) (2001) (2005)]

(2005)]. For a 60 μm diameter, low carbon iron droplet, in Ar-H$_2$ DC plasma jet, time dependent computations performed by using commercial code, FIDAP7-62, dedicated to the materials processing field [FIDAP code] indicated the formation of internal convection motion in the molten iron particle with iron oxide engulfment inside the droplet caused by the spherical Hill vortex as shown in Fig. 15.12

The internal circulation in the liquid metal droplet continuously sweeps fresh liquid to its surface and makes it available for oxidation, while in the same time engulfing fragments of the oxide formed at the surface into its core. For a low-carbon iron particle, the oxide formed at the surface is liquid Fe$_x$O. Liquid iron and liquid Fe$_x$O being not miscible due to the significant difference in their surface tension will give rise to the formation of Fe$_x$O nodules within the solidified particle as shown in Fig. 15.13. Such recirculating flow within the droplet gives rise to an increase in the oxygen content in the solidified particle compared to that estimated assuming a pure diffusion model [Espié et al. (1999) (2001) (2005)]. For the iron particles shown in

Fig. 15.13, the captured FeO corresponds to a mass percentage is 15 wt.%, while diffusion calculations estimate the Fe_2O_3 and Fe_3O_4 concentration to be less than 3 wt.%. The oxide shell formed at the surface of particles, with no inside convective movement, is made of Fe_2O_3 and Fe_3O_4. Such oxides are observed in the thin shell surrounding the particle formed at the end of the particle trajectory when the particle velocity is about the same as that of the surrounding plasma plume (no more convective effect). The same effect was observed with stainless steel particles plasma sprayed with Ar-H_2 plasma forming gas [Seyed et al. (2005)]. Similar results have also been observed when spraying Ti-4Al-6V particles in a DC nitrogen plasma under controlled atmosphere or in ambient air. TiN and TiO_2 contaminates are formed observed in this case in the particles with significantly less TiO_2 in controlled atmosphere spraying [Ponticaud et al. (2001)].

In the case of non-oxide ceramic particles such as nitrides and carbides, oxygen present in the spray jet can also react with the sprayed material to modify its composition. This is, for example, the case of tungsten carbides (WC) that can be deeply modified or even destroyed through the transformation of WC to W_2C and then to W, or WO according the following reaction.

$$4WC + O_2 \rightarrow 2W_2C + 2CO \tag{15.2}$$

Such reactions are also controlled by Arrhenius type laws with the reaction rate strongly dependent on the particle temperatures. To limit oxidation problems spraying in a controlled atmosphere or under soft vacuum is an efficient means, but a rather expensive one. Using shielding devices (elongated spray gun nozzle and/or inert gas flow) also reduces oxidation but generally at the expense of high shield inert gas flow rates (up to more than ten times that of the working gas of the gun).

Based on the above it is clear that the choice of the size distribution is a key issue for the coating quality. It must be made taking into account the material to be sprayed, especially its melting difficulty, and the spray process used. The size distribution must also be as narrow as possible to limit the dispersion of the particle trajectories that depend on their injection inertia imparted by the carrier gas flow rate. Once the carrier gas flow rate has been chosen, particle forces are linked to their mass, proportional to the cube of their diameter. Trajectory dispersion within energetic gases will be very important, resulting in broad distributions of heat and momentum transfer to the particles. Correspondingly, the particle hits the substrate or with different velocities and temperatures and of course different melting stages. As pointed out earlier, the dispersion effect will be less important with axial particle injection into the plasma stream compared to radial injection. Particles penetrate within the jet in the case

of radial injection should be adjusted such that the mean particle trajectory forms an angle of 3–4° with the torch axis. Since at the exit of the injector, particle trajectories are not all parallel and the particle flux has a conical shape with a total angle generally below 20° due to interparticle collisions as well as particle-tube wall collisions it is unavoidable that part of the injected particles travel in the jet fringes, where the heat and momentum transfers are lower, and impact the substrate at different angle to the surface of the substrate. In the process of making the selection of the material to be sprayed it is important to keep in mind the dependence of the powder properties on their respective manufacturing process as discussed in great details in (Chap. 13 Powders, wire and cords). For the same material composition, the specific mass can vary by a factor of 5 between atomized and agglomerated particles. The heat transfer is also modified when comparing, for example, dense and agglomerated particles [Ben Ettouil F, et al. (2008)] especially for materials with poor thermal conductivities ($\kappa < 20$ W/m K) where the internal heat propagation phenomenon in the particle is important.

15.2.3 Physical Aspect of Substrate Surfaces

The proper preparation of the substrate surface prior to the thermal spraying operation is a critical step that has a major impact on the adhesion of the coating to the substrate and consequently the functional performance of the coating. A general review on the subject has been reported by [Chandra and Fauchais (2009)]. The topic covered in this section include;

- *Substrate surface topography*
- *Adsorbates and Condensates*
- *Oxide layer management on metallic substrates*

Considering that vast majority of substrates used in thermal spray applications are metallic the question of oxide formation is particularly important since most metals, with the exception of gold and platinum, have an oxidized surface even without preheating them (native oxide). Unless such oxide layer is removed prior to spraying, the bonding of the coating to the surface can be compromised. Moreover, because of the natural reactivity of metallic surfaces, substrate surface preparation has to be carried out as close as possible to the spraying operation to avoid the degradation with time of the substrate surface once exposed to ambient air.

15.2.3.1 Substrate Surface Topography

As discussed in Chap. 14, Surface preparation, proper control of substrate surface topography through roughening is crucial for optimal coating adhesion to the substrate. The degree of roughness has to be compatible with the mean size of splats: the R_t (also called R_z) must be about 2.5 times the mean splat diameter, and the spacing between peaks has to be sufficient for the liquid flattening droplet to penetrate between them. The root mean square value $R_{\Delta q}$ is generally considered as the roughness parameter that correlates best with the coating adhesion. Roughening can be achieved by grit blasting, water jet or the physical machining of roughness pattern on the surface of the substrate. The former is probably the most commonly used. The degree of roughness, grit residues, and blasting induced stress depends on the blasting system used and blasting parameters such as blasting pressure, blasting distance, grit material, and size. The main disadvantage of grit blasting is the grit residues that must be removed prior to spraying, which can be avoided using high pressure water jet blasting.

Several authors have pointed out that mean surface roughness R_a of metals or alloys increases at the nanometer scale with substrate preheating. [Fukumoto et al. (2005)] noted that when a surface was polished to a particular average roughness R_a, coating adhesion was not as high as it was when preheating produced the same roughness. To more accurately characterize the substrate surface, [Fukumoto et al. (2005)] introduced another index of surface topology: skewness (S_K), the definition of which is given in Chap. 14, Surface preparation, Sect. 14.2.3 Measurement of surface roughness, Eq. 14.5. The skewness increases from negative values to positive ones upon preheating a polished surface that has a major impact on splat shape formation on the surface. [Cedelle et al. (2006)] have shown for a stainless steel 304L, changes in the substrate surface topography in the nanometer range can have a large effect on the static wetting behavior of molten metal droplets impacting on its surface. Hence substrate preheating does not modify just the chemical composition and thickness of the oxide layer on the metallic substrate but also its physical aspect, which plays a key role in the splat formation mechanism.

15.2.3.2 Adsorbates and Condensates

Any surface, especially in the presence of an oxide layer, is polluted by adsorbates and condensates. Upon impact of a molten droplet on the substrate, it flattens trapping the adsorbates and any vapors between the flattened droplet surface and the substrate. During a few hundreds of nanoseconds, the local gas pressure can reach a few thousands of MPa. More generally the presence of an evaporable substance on the surface affects significantly the flattening process, through the drop-substrate wettability and the contact resistance created between the flattening particle and substrate [Li C-J et al. (1998), Li C-J et al. (1999), Fukumoto M et al. (2006a, b) and Li H, Danlos Y et al. (2007)].

[Li C-J et al. (2004)] reported a careful study of adsorbates and condensates at the substrate surface. Often, water is the main component at the surface. Its adsorption on an oxide surface is usually the result of one of three possible mechanisms, or combinations of them, depending on the temperature, the intrinsic reactivity of the surface, and the number of defect sites at the surface [Henrich VE and Cox PA (1994)];

- Physisorption of molecular water
- Chemisorption of molecular water
- Chemisorption of molecular water followed by dissociation

Physisorption corresponds to very weak interaction between the substrate and adsorbates, while chemisorption is much stronger and may involve partial charge transfer, which more readily leads to dissociation [Henrich VE and Cox PA (1994)]. Stronger adsorption occurs at steps and defects and new OH_y radicals, resulting from dissociation, bond with a surface metal ion.

When the surface is heated over a critical temperature, adsorbed water, and other adsorbed liquids species, will be desorbed. This desorption process is linked to adsorbed products, surface structural features, and surface materials. Physiosorbed molecular water is generally completely removed by preheating the substrate above 150 °C. However, with chemisorbed water, the thermal desorption temperature depends on adsorbed features and the subsequent desorbed product. For most oxide-covered metal surfaces it occurs at temperatures in the range from 127 to 320 °C as summarized by [Li C-J et al. (2004)] for several oxides and nominal metal surfaces. Further study of the effect of adsorbing and desorbing of organic compounds on splat formation on a stainless-steel surface was reported by [Fukumoto et al. (2005), Fukumoto M et al. (2006a, b Cedelle et al. (2006), McDonald A et al. (2007a, b) and Chandra and Fauchais (2009),].

[Li C-J et al. (2004)] underline that this desorption temperature range is in good agreement with the substrate preheating temperatures reported as transition temperature, as described later in the section on splat formation, above which splats are predominantly disk shaped. [Li C-J et al. (2004)] demonstrated for xylene, ethylene glycol, and glycerol on stainless steel substrates that when the preheating temperature exceeded the boiling point of these liquids by 50 °C, thermal spray molten droplets impacting on the substrate formed regular disk splats. An interesting question addressed was how long it takes after preheating and evaporating all adsorbates and condensates for the surface to be recovered again by the same components as picked up from the surrounding atmosphere? To answer the question [Fukumoto M, et al. (2006a, b)] exposed stainless steel

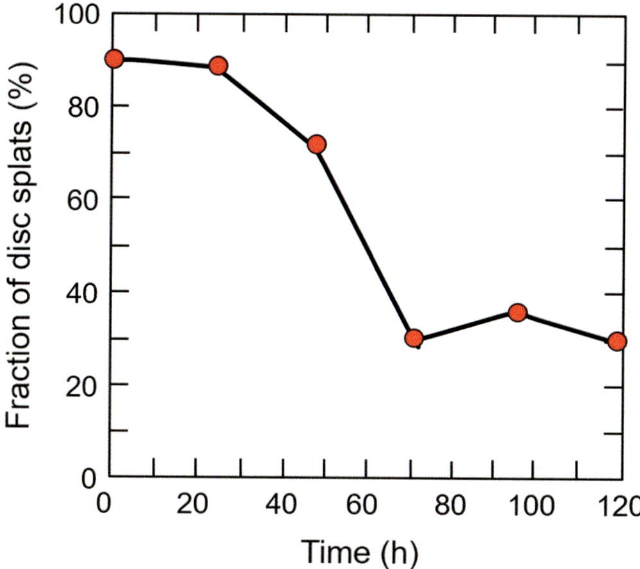

Fig. 15.14 Relationship between elapsed time after heating to 723 K and fraction of disk splats. ([Fukumoto M, et al. (2006a, b)], Reprinted with kind permission from Springer Science Business Media, copyright © ASM International)

304Lsubstrates to ambient air atmosphere at room temperature for different periods of time following their initial heating to a temperature where disk-shaped splats were obtained. The results given in Fig. 15.14, show the fraction of disk-shaped splats remained essentially constant at about 90% for the first 20 h of exposure of the substrate to ambient air. Beyond this initial period, the percentage of disc-shaped splats drops gradually to around 7% at 40 h, and further down to 30% at 70 h beyond which the fraction of disc-shaped splats remains essentially constant. It is clear from the above that substrate surface preparation for thermal spray coating has to be carried out as close as possible to the scheduled spraying time minimizing the exposure of the prepared substrate surfaces to ambient atmosphere to less than 24 h.

[Fukumoto M, et al. (2007a, b)] have also investigated the effect of pressure on desorption of adsorbates and condensates. They studied the particle flattening behavior when reducing the chamber pressure (plasma spraying under soft vacuum) with the substrate kept essentially at room temperature. Indirect measurements (determination of the pressure below which splats became disk shaped on different materials) have shown that desorption probably occurs at pressures between 92 and 26 kPa, depending on substrate material. Alternate approaches used for the elimination of adsorbents and condensates from the substrate surface include blasting the substrate surface with dry ice (CO_2 ice at −79 °C) just before the spray operation [Dong S, et al. (2011)], or scanning the surface by a laser beam provided its energy density is over a certain threshold [Coddet C, et al. (1999) and Costil S, et al. (2005)].

15.2.3.3 Oxide Layer on Substrate and Coating

When preheating the substrate over the transition temperature above which adsorbates and condensates are removed, an oxide layer can be formed on the surface, or if already present, its thickness, composition, and roughness will change depending on the technique used for the substrate preheating. For example, if the substrate is heated by an electrical heating element while being exposed to ambient air, important variations in substrate surface temperature can develop depending on the thermal conductivity of the substrate material. These will result in-turn in variations in the surface oxidation of the substrate and important differences of the thickens of the formed oxide layer. Often, a direct contact between the substrate and the resistance heating element raises excessively its temperature resulting in uneven heating and loss of control on the oxide layer formation.

When a spray torch is used to preheat the surface, even if the mean heating rate and final temperature are well defined, the local variation in the substrate surface temperature will depend on the movement of the torch relative to the substrate. Moreover, with combustion flames (flame spraying, HVOF, HVAF) gas temperatures are below 3000 K and the entrained air contains mainly molecular oxygen. With plasma jets, gas temperatures of up to 5000 K can be achieved at torch-to-substrate distances of 80 mm. Surface heating can vary in these cases between 1 and 5 K/s with the compositions and morphologies of the formed oxide layers different from those obtained by more conventional electrical heating [Haure T (2003), Syed AA et al. (2005), Pech J (1999) and Pech J and Hannoyer B (2000)].

Preheating and/or modifying the substrate surface by a laser may result in the formation of oxide layers with different characteristics depending on the laser used and the substrate treated. [Fukumoto et al. (2006a, b)] reported that for laser power densities below 50 MW/m², the surface roughness does not change, with its skewness remaining negative or close to zero. It is also possible that the laser irradiation results in the removal of part of the oxide layer at the surface if the laser energy density is sufficiently high or if the number of pulses at the same location is increased [Li H et al. (2006a, b)]. It is worth noting that oxide layer elimination requires much higher energy density than that necessary to remove adsorbates and condensates.

Ideally, substrate preheating in an insulated furnace with a controlled atmosphere and well-defined conditions (heating rate, maximum temperature, preheating time) offers the best option, if the size and nature of the substrate allows it. This would provide means for the elimination of adsorbents and condensates and control of the substrate temperature prior to the spray-coating operation, without compromising the surface chemistry of the substrate in terms of oxide layer characteristics such as its composition, thickness and structure.

Table 15.1 Thickness and composition of the oxide layer formed after Ar-H$_2$ (25 vol.%) plasma heating at 500 °C on low carbon steel (C 0.49 wt %) [Pech J (1999)]

Heating rate (K/s)	Oxide layer thickness, δ_{ox} (nm)	wt. fraction of α-Fe$_2$O$_3$		Identified phases
		Grazing XRD	Mössbauer	
1.6	600	0.16	0.29	α-Fe, α-Fe$_2$O$_3$–Fe$_3$O$_4$
2.6	525	0.08	0.27	α-Fe, α-Fe$_2$O$_3$–Fe$_3$O$_4$
4.5	515	0.04	0.14	α-Fe, α-Fe$_2$O$_3$–Fe$_3$O$_4$

Reprinted with kind permission from ASM International

In general, when preheating a metal or an alloy in the presence of either ambient air, oxygen or water vapor, the oxide layer at its surface will change in thickness, roughness, and composition. Thin layers (δ_{ox} < 50 nm) are observed on stainless steel, nickel or titanium and its alloys, eg. for stainless steel 316L [Dong S (2003)], the oxide layer thickness is about 10 nm without any heat treatment and reaches 25 nm when preheated at 250 °C, 40 nm at 400 °C, and 58 nm at 580 °C. At 400 °C a spinel phase can be observed but XPS analysis shows that Cr is present in a very low concentration. At 400 °C and above oxides are hematite and a spinel. The Cr has almost disappeared, and the Ni has increased with Fe/Ni = 3.8 against 6 in the matrix; thus, the spinel should be of the type NiFe$_2$O$_4$. This result is contrary to that obtained using the same preheating conditions with 304L steel, where the spinel detected, mixed with hematite, was Fe$_{2-x}$Cr$_x$O$_3$ [Pech J (1999)].

It is the same for Ti–6Al–4V with a nascent oxide layer of 25 nm before, 32 nm after preheating during 10 min to 250 °C and 34 nm at 400 °C. [Dong S (2003)] also observed that two phases of Ti are present in the sample, α and β, with the latter diminishing when the temperature increases. With heating, Al diffuses toward the surface with a ratio Al/Ti > 1. Oxides formed are TiO, TiO$_2$ and Al$_2$O$_3$ with the thickness of Al$_2$O$_3$ increasing with temperature and, if TiO$_2$ is identified at the surface, TiO is observed at a greater depth [Dong S (2003)].

Thick oxide layers (δ_{ox} > 100 nm) of about 440 nm are observed on low carbon steels (C 0.49-Si 0.19-S 0.02-Mn 0.53-Fe bal. wt.%) when preheated at 450 °C [Dong S (2003) and Syed AA et al. (2005)]. Results presented in Table 15.1 obtained by heating low carbon steel for 10 min at 500 °C with 3 heating rates, [Pech J (1999)], show that when the heating rate is increased while keeping the preheating temperature the same, both the oxide layer thickness and the weight fraction of hematite (α-Fe$_2$O$_3$) decrease. The difference between XRD grazing and Mössbauer results is due to the fact that conversion electron emission comes mainly from the surface and slows down rapidly with the thickness. The oxide layer is heterogeneous, with an upper surface largely constituted of α-Fe$_2$O$_3$ and an inner one of Fe$_3$O$_4$ [Pech J (1999), Valette S (2004) and Pech J et al. (1999)]. It thus becomes possible to monitor the relative thicknesses of Fe$_2$O$_3$ and Fe$_3$O$_4$, the former having poor mechanical resistance. For example, when preheating for 10 min the same substrate at 1.4 K/s up to 520 °C, the hematite fraction measured by Mössbauer is 0.85 against 0.62 when preheating to 440 °C at 4 K/s.

Increased oxide layer thickness has two effects: a larger temperature gradient across the layer and a reduced resistance to thermal shock. For example, on low carbon steel or cast iron, the impact of alumina or zirconia particles results in the fracture of the oxide layer for about 30% of impacts [Pech J (1999)]. Rupture occurs within the thick hematite layer at the substrate surface for both 1040 steel and low carbon steel. The impact of stainless-steel particles on 1040 steel, while does not induce any fracture in the oxide layer, the coating adhesion strength is reduced by a factor of almost 2 compared to that if the substrate was stainless steel preheated under the same conditions.

15.2.4 Splat Adhesion to Substrate

The main adhesion mechanisms for coatings are Diffusion, Crystallographic Structure Matching (CSM) at the interface splat–substrate, Chemical, and Mechanical bonding. The presence of adsorbed and contaminants on the substrate will also affect the strength of adhesion. Preheating the grit-blasted surface before spraying removes adsorbates and condensates and improves coating adhesion by a factor of 2–4 [Fauchais P et al. (1996)]. Thus, the adhesion and cohesion strengths of coatings, as well as other thermomechanical properties, are strongly linked to the quality of contact between splats and the substrate or the previously deposited coating layers. That is why the grit residues must be as low as possible, which depends on the grit-blasting conditions [Fauchais P et al. (1996) and Maruyama T et al. (2006)].

15.2.4.1 Diffusion
Diffusive adhesion of metals and alloys only occurs when the temperature during coating formation is sufficiently high in the absence of any oxide layer on the substrate surface. In a simplified approach the thermal diffusion distance, d_d, is given by;

$$d_d = \sqrt{D_{th}\, t} \qquad (15.3)$$

Where, t is the time during which the contact occurs at temperature T and D_{th} the diffusion coefficient varying as, $e^{-E_A/kT}$ where E_A is an activation energy, T is the absolute temperature, and k is the Boltzmann constant. Good adhesion requires accordingly a high temperature and sufficient contact time. In the case of metals, diffusion bonding can occur only if the solid oxide layer at the substrate and particle surface is removed.

An excellent adhesion between coating and substrate can be achieved, when plasma spraying in soft vacuum. The spraying of a super alloy on a turbine blade made of another super alloy is a typical example in which the spraying is accomplished with the substrate and coating maintained at about 900 °C, but not above to avoid the formation of undesirable phases. In this case, the oxide layer at the substrate surface is removed prior to the spraying operation by establishing a transferred arc between the part to be sprayed, which acts as cathode, and the anode of the plasma torch. Once the plasma jet of the spray torch is established, the medium between the anode nozzle of the torch and the part is electrically conducting. It allows tailoring the arc current of the transferred arc [Itoh A et al. (1990)]. The arc attachment at the cathode (the part to be sprayed) is erratic and cathode spots are continuously moving. The oxide layer is ejected from the cathode spot and the surface is progressively cleaned from its oxides. This is illustrated in Fig. 15.15a showing the cleaning of stainless-steel ASI-304 after 0.15 s with a transferred arc current of 37 A. At crater positions the oxide has disappeared (at the EDS precision). As shown in Fig. 15.15b, following a treatment of a few minutes, and surface polishing to a mirror-polish, the oxygen level at the substrate surfaces is very low and the quantities of Fe and Cr

are the same as those inside a crater. Once the substrate has been cleaned by the reverse transferred arc, it is heated, if necessary, with the help of a direct transferred arc (the part is now the anode), to achieve sufficiently high temperatures for the diffusion to occur during the spraying operation.

Diffusion can also occur at the splat-substrate interface in the few to ten microseconds that are available before splat solidification provided that the initial temperature of the molten splat is sufficiently high. For example [Chraska T and King AH (2002)], at the interface between a YSZ splat and a 304L stainless steel substrate, a 20–30-nm thick oxide layer is found extending uniformly over the length of the splat cross section. The whole splat is in excellent contact with the substrate, except where cracks in the splat intersect with the substrate. The oxide layer interface includes elements from both the ceramic splat (Zr) and the substrate (Cr, Fe), as determined by EDS using a sufficiently small beam spot size. The relatively thick oxide layer grew during substrate preheating to approximately 450 °C carried out in an ambient atmosphere. The heat transferred from the molten ceramic droplet is sufficient to allow zirconia's cations to diffuse into the oxide layer. However, diffusion through a few tens of nanometers is not quite sufficient to achieve a good adhesion.

With a polished Ti-6Al-4V substrate, the presence of alumina diffusing to the substrate surface upon preheating at 400 °C produces excellent adhesion (48 ± 12 MPa) of the sprayed alumina coating. This is probably due to the chemical continuity at the interface substrate coating with the presence of alumina at the surface of the Ti-6Al-4V substrate [Haure T (2003)].

As already emphasized previously [Brossard S et al. (2010)], the substrate heating by the impacting fully melted particle can also promote the diffusion of certain elements of the substrate toward its surface and increase the wettability

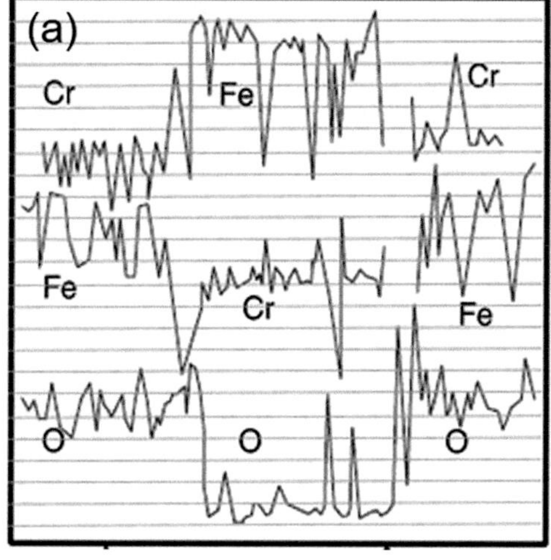

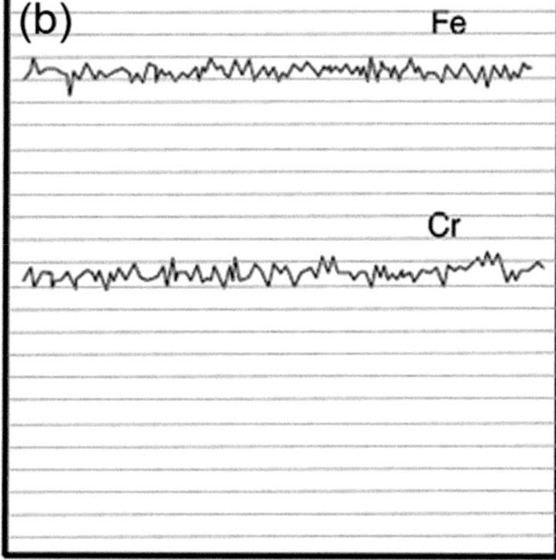

Fig. 15.15 EDS characterization of a US-304 stainless steel. (**a**) Treated by a reverse transferred arc of 37 A. (**b**) Treated by the same arc for a few minutes and mirror polished. ([Itoh A et al. (1990)], Reprinted with kind permission from ASM International)

substrate-flattening particle. With cold-sprayed particles the diffusion is not possible. For example, [Grujicic M (2007)] calculated that with $D_{th} = 10^{-15}$ to 10^{-13} m²/s and the contact time $t_c = 4 \times 10^{-8}$ s, $\sqrt{D_{th}\, t} = 0.004 - 0.1$ nm (fraction of inter-atomic distance), which is by far insufficient for bonding.

15.2.4.2 Crystallographic Structural Matching

[Yang E-J et al. (2012)] have plasma sprayed alumina particles onto single crystalline alumina substrates (α phase) kept at 900 °C. On a cold substrate, due to the fast-cooling rate, alumina crystallizes into the metastable γ phase. When the substrate is preheated at 900 °C, the cooling rate of the splat is much slower and the crystallization occurs in α phase, which is also that of the sintered alumina substrate. Moreover, the growth orientation of alumina splat crystal during rapid solidification was exactly the same as the orientation facets of [001] or [110] of the single crystalline alumina substrate, epitaxial solidification taking place during splat cooling.

Alumina particles were plasma sprayed onto alumina polished ($R_a \approx 0.4$ μm) plasma-sprayed coatings [Denoirjean A et al. (1998)], which were kept at 250 °C before spraying to get rid of adsorbates and condensates. The latter were either as-sprayed (with more than 99 wt% of γ phase) or preheated at 1373 K at a rate of 5 K/min, annealed for 6 h, and cooled at a rate of 5 K/min resulting in a 100% α-columnar structure. Some were also preheated to 1873 K at a temperature ramp of 5 K/min, annealed for 3 h, and cooled at a rate of 5 K/min resulting in a α-granular structure with grains between 3 and 5 μm. As shown in Fig. 15.16a, on the γ-alumina substrate, alumina splats (γ phase) exhibit a columnar and regular structure ~100–150

nm; the adhesion of the alumina coating (300 μm thick) obtained on this smooth substrate is 35 ± 3 MPa! On the columnar α-alumina substrate, splats (γ phase) form a columnar and irregular structure ~150–300 nm (see Fig. 15.16b); the adhesion of the coating is only 3 ± 1 MPa. On the α-alumina substrate with a granular structure, splats (γ phase) achieve a very irregular structure ~100–400 nm (see Fig. 15.16c), and splats have the tendency to peel off and it is impossible to achieve any coating.

It has been shown that the pre-oxidation of smooth ($R_a < 0.05$ μm) low carbon steel substrates in a furnace under a CO_2-rich atmosphere at atmospheric pressure allows the formation of a wüstite (Fe_{1-x} O) layer. The latter improves significantly the adhesion (>55 MPa) of alumina coatings in spite of the rather low roughness (0.10 μm < R_a < 1.00 μm) of the oxidized surface [Maitre A et al. (2002)]. It appears that the plasma preheating of steel samples did not modify the morphology of the pre-oxidized wüstite layer formed ($R_a \approx 0.2$ μm). In spite of maintaining the surface condition, part of the wüstite scale, very probably the most superficial one, is oxidized into magnetite (Fe_3O_4) [Valette S (2004)]. This has already been observed on wüstite and it corresponds to a topotactic phase transformation. This change of composition is liable to occur even inside a grain, so that it may have no harmful consequence on the cohesion of the iron oxide scale. The presence of metastable γ-Al_2O_3 together with α-Al_2O_3 was a little bit surprising because the former phase is generally the only one observed under the spraying conditions used. It is possible that the γ-Al_2O_3 phase was localized in direct contact with magnetite, because they have the same crystallographic structure (spinel) with lattice parameters rather close to 0.793 nm for alumina and of 0.839 nm for magnetite or 0.834 nm for lacunars magnetite.

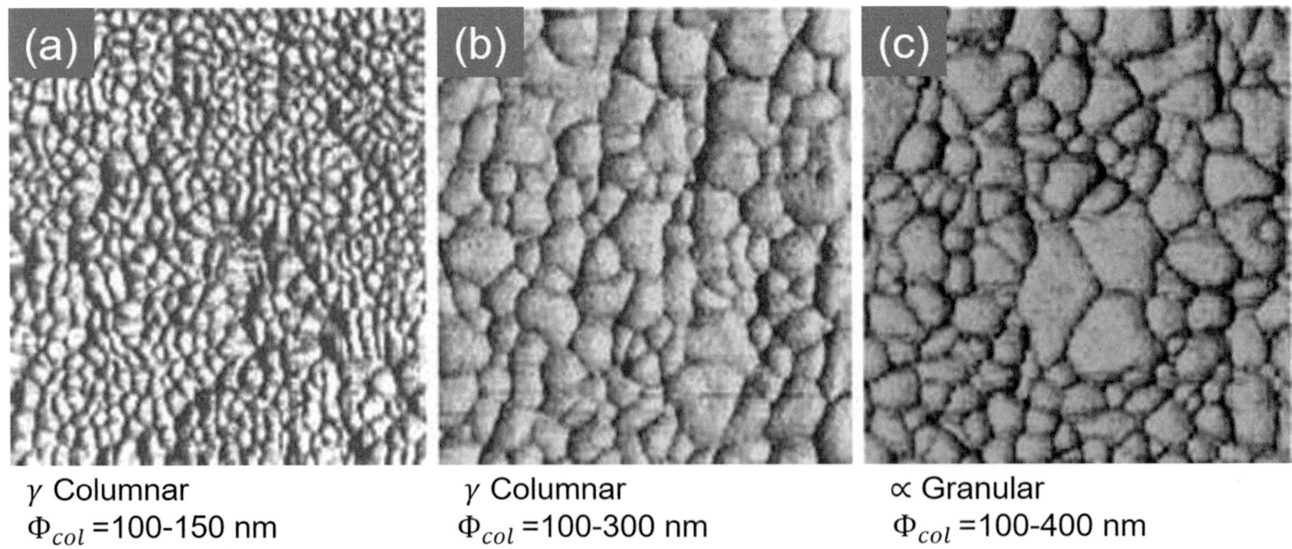

Fig. 15.16 Alumina splats plasma sprayed onto smooth alumina substrates with different microstructures, (**a**) γ columnar, (**b**) α columnar, and (**c**) α granular. ([Denoirjean A et al. (1998)]. Reprinted with kind permission from ASM International)

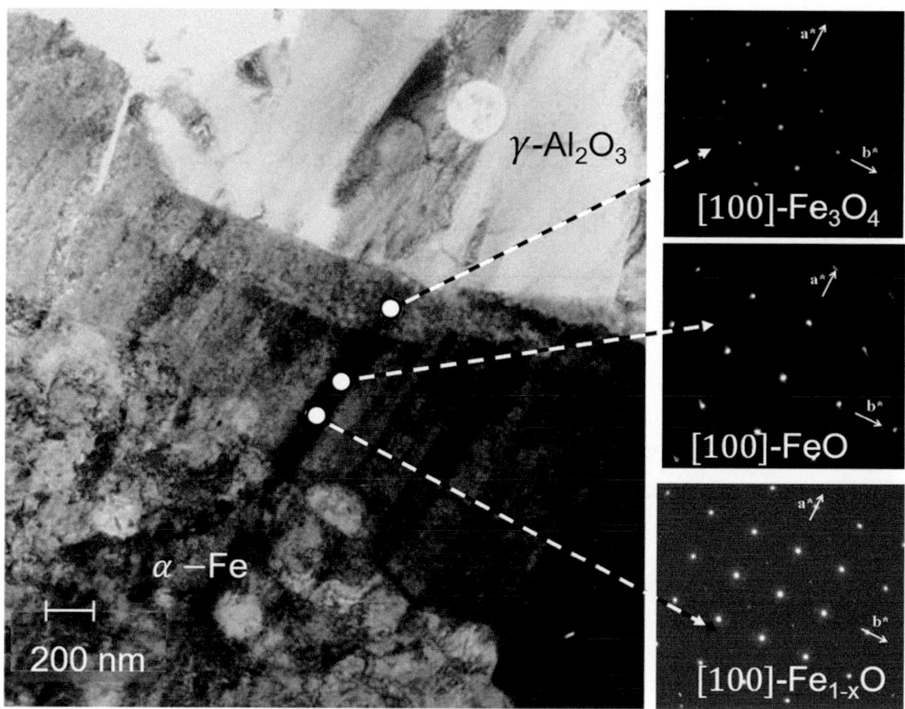

Fig. 15.17 TEM image of the interface between low carbon steel C40E and a plasma-sprayed alumina coating. ([Valette (2004)]. Reprinted with kind permission from Dr. S. Valette)

If such an assumption could be verified, it could explain the good adherence of the coating. Indeed, in these conditions, the coating structure is composed of a steel substrate, the α-Fe phase and the wüstite phase at its surface, with the interface comprising magnetite (topotactic transformation), and finally the γ alumina (epitaxial relationship) coating, as shown in Fig. 15.17. This might justify a physical adhesion, and perhaps also a chemical one if a chemical reaction occurs between both spinel phases, which in principle could form mixed oxides.

15.2.4.3 Chemical Bonding

Chemical adhesion requires that the impacting droplet melts the substrate and a chemical compound of both liquids exists. For example, when Mo particles impact on steel, the melting temperature and effusivity [defined as $(\rho_p.c_p.\kappa_p)^{0.5}$] of the Mo droplet are higher than those of the steel substrate, which melts and reacts to form an intermetallic compound, MoFe$_2$. Mo particle flattening on a stainless-steel substrate results in splats formed of different pieces as shown in Fig. 15.18a, b after [Li et al. (2006a, b, c, d)]. This type of splashing is due to a localized melting of the substrate surface by the impacting droplet. The melted crater on the substrate surface alters the flow direction of the droplet fluid and tends to form a free liquid jet that detaches from the substrate surface Fig. 15.18c, d. Cracking occurs within the Mo splat due to the restraining of the splat contraction during cooling and due

to the low plasticity of the solidified splat material [Li et al. (2006a, b, c, d)]. The liquid film, mainly resulting from the steel substrate material, between the Mo flattening particle and the substrate promotes the floating of Mo pieces initiated from cracks.

A chemical reaction also occurs when a cast iron particle impacts on an Al–Si–Cu aluminum alloy substrate preheated to above 400 K where probably the thin alumina layer is melted or eroded. As it has been observed [Morks MF et al. (2002)], a thin (a few hundreds of nanometers) FeAl$_2$O$_4$ layer is formed between the splat and substrate. Even if the substrate is not melted upon impact, the high temperature induced in the substrate oxide layer can greatly promote adhesion. Deformed substrate ridges, probably due to the slight melting of the substrate surface, are formed immediately adjacent to the splat periphery sprayed at high substrate temperatures. The flattening ratio decreases with preheating, because the ridges act as an obstacle to the splat expansion.

15.2.4.4 Mechanical Bonding

Mechanical adhesion takes place when the substrate is roughened, for example, by grit blasting, but, compared to cold spray, with molten particles the phenomena are quite different. Flattening particles and resulting splats impacting on the rough surface shrink while cooling and adhere to the surface because of the frictional force that develops [Morks MF et al. (2002), Bahbou MF and Nylen P (2005), Bahbou MF and

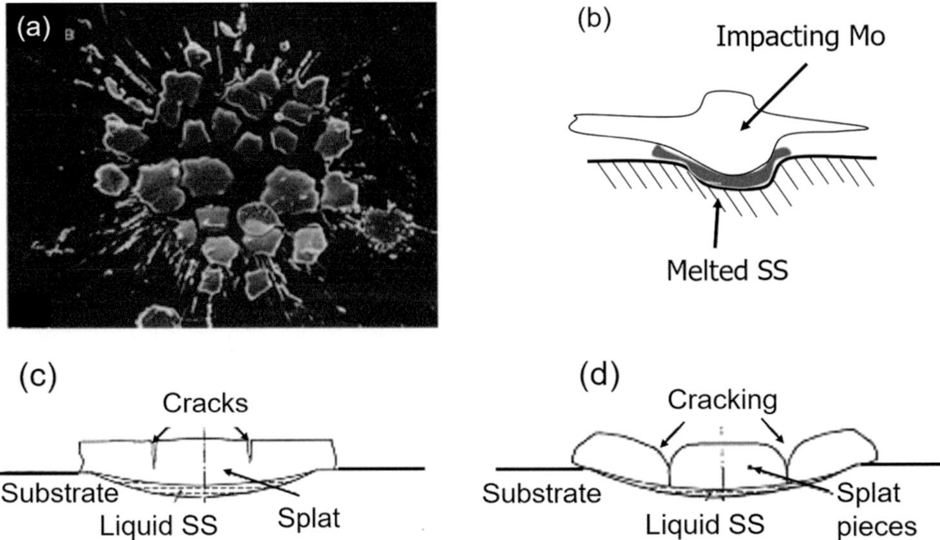

Fig. 15.18 (**a**) Mo splat on stainless steel (SS) substrate. (**b**) Schematic diagram of droplet impact inducing substrate melting. (**c**) Crack initiation in the Mo splat. (**d**) Displacement by floating of the cracked pieces on low melting point liquid. ([Li et al. (2006)]. Reprinted with kind permission from Springer Science Business Media, copyright © ASM International)

Nylén P (2007), and Fauchais P et al. (1996)]. This adhesion depends both on the amplitude of roughness, often characterized by the distance between the highest peak and the deepest undercut (R_t), and on the spacing between peaks and valleys. Bahbou and Nylen [Bahbou MF and Nylén P (2007)] have established that a good correlation exists between the adhesion strength of NiAl (5 wt%) coatings on a Ti-6Al-4V substrate, and the root mean square value $R_{\Delta q}$ of the substrate, which takes into account both the amplitude and spacing of peaks. The correlation is rather poor when variables such as R_a, R_t, or the peak spacing are considered individually. As emphasized previously for the best mechanical adhesion, the peak heights, characterized by R_t, must be adapted to the mean size of the splats, as illustrated in Fig. 14.35.

In the cold spray process, mechanical bonding can occur at the surface of the substrate when no metallic/chemical bonding is feasible. This is the case, for example, of cold spraying aluminum particles on glass or alumina substrates [Zhang D et al. (2005)]. With a polymer substrates, on the other hand, all the damage occurs in the polymer, but the aluminum particles are not deformed at all and no bonding occurs. [Trompetter et al. (2005)] studied the interface between Ni–Cr (20 wt.%) particles HVAF sprayed on an aluminum substrate. The gas temperature was 1300 °C, while the melting temperature of NiCr particles was 1400 °C, resulting in particles being in a plastic state on impact. No evidence of chemical bonding across the interface was observed suggesting that mechanical interlocking was the dominant bonding mechanism.

15.3　Splat Formation

Splat formation is the final and most critical step in the spray coating process on which depends the final properties of the sprayed coating. Splat formation depends, intern, on the particle and substrate surface conditions prior to their impact on the surface of the substrate. Schematic representation of the splat formation step is given in Fig. 15.19 identifying the principal parameters, including the particle impact velocity, v_p, and temperature T_p and the impact angle θ, defined as the angle between the axis parallel to the substrate and the direction of impact. Generally, particle temperature can have a uniform value in the case of metallic particles, or an internal temperature gradient between the surface and core of the particle as with low thermal conductivity ceramics. Depending on the particle heat conduction or more precisely its Biot number [see Chap. 4 Plasma -Particle Momentum and Heat Transfer, and Chap. 5, Gas and Particle Dynamics in Thermal Spray] the particle can be either in a *non-molten* solid state, in a *fully molten* or *partially molten* state, with the surface layers molten while the particle's core still in a solid state at impact, modifying deeply its flattening behavior and consequently splat formation. As discussed in Chap. 14 Surface Preparation, substrate roughness, skewness, oxide layer thickness and composition, and the adsorbates and condensates present on its surface are all elements that can significantly influence the particle flattening and solidification behavior.

Ideally the particle-substrate impact angle, θ should be as close as possible to 90°. Alternately, the effective particle impact velocity, v_{pi} will be the velocity component in the

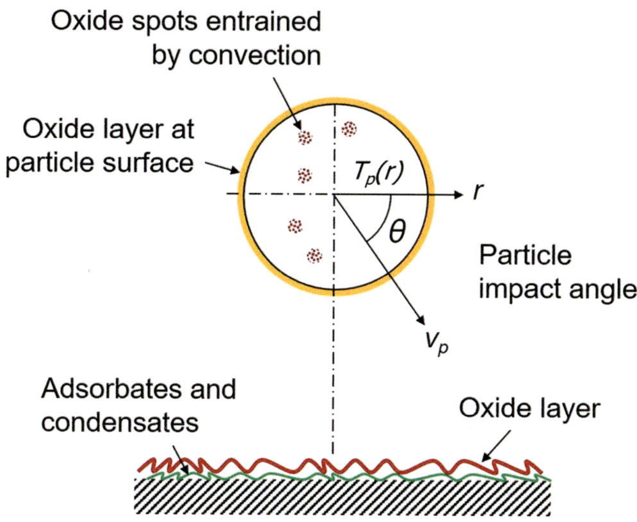

Fig. 15.19 Schematic representation of a single particle impacting on a substrate surface identifying system of coordinates, definition of impact angle, particle parameters and potential presence of oxides, adsorbents and condensates

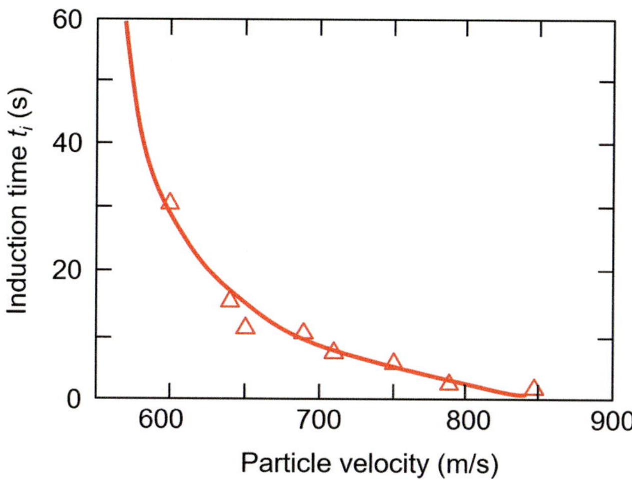

Fig. 15.20 Dependence of induction time, t_i on the impact velocity for aluminum particles impacting on a copper substrate. Powder flux density of 0.06 kg/m^2 s. ([Klinkov and Konsarev (2006)] Reprinted with kind permission from Springer Science Business Media, copyright © ASM International)

perpendicular direction to the surface, that is, $v_p.\sin\theta$. With decreasing angle, v_{pi} decreases with a corresponding drop in deposition efficiency. Generally, the deposition efficiency is low with a rather porous, poor quality coating formation when the spray angle is below 70°. Besides the reduction of the impact velocity, a shear stress is imposed on the flattening of the impacting particle reducing its adhesion.

Depending on the chemical nature of the particle and of the substrate either of them, or both, can be covered by an oxide layer, with the substrate entrapping adsorbents and condensates that can interfere with the splat formation process. On impact with the substrate surface, the particles are flattened forming a "splat" characterized by its "Splat flattening ratio" or "flattening degree," ξ_D defined as the ratio of the splat diameter, D to the diameter of the particle, d_p prior to its impact on the surface;

$$\xi_D = D/d_p \qquad (15.4)$$

The splat flattening ratio will depend strongly on the particle impact velocity, temperature and its state at impact whether solid, molten or only partially molten.

15.3.1 Classification of Splat Formation Mechanisms

15.3.1.1 Splat Formation from Non-molten Solid Particles

As discussed in Chap. 6, Cold Spray, the deposition of coatings through the spraying of solid non-molten particles is possible in the cold spray process through the acceleration of the spray particles to high velocities in the hundreds of m/s, at temperatures essentially 40–50% their melting temperature, T_m [Schmidt T et al. (2006)]. their impact velocity must be higher than the critical velocity, v_{cr}, (depending on particle ductility) but lower than the erosion velocity, v_{er}. Particle deformation in this case is purely plastic with the stress applied to the particle higher than the yield stress. As the yield stress diminishes with the increase in the particle temperature, plastic deformation will be easier with hot particles compared to cold ones. On impact, adiabatic heating occurs in the contact zone with a transition to an adiabatic shear phenomenon only during the dynamic deformation of the solid [Hanson TC et al. (2002)]. While adiabatic heating will result in the local softening of the particle and substrate material in the contact zone, adiabatic melting of the spray particles is not frequently observed in cold spray processes as discussed by [Grujicic M et al. (2003) and Grujicic M (2007)]. The situation is different in HVAF or high power HVOF spraying where particle temperatures prior to impact are much closer to their melting point, a rather low amount of additional energy would be sufficient for the local melting the particles.

Predominant bonding mechanism of solid particles on the substrate have been discussed in Sect. 15.2.4 Splat Adhesion to Substrate. These include Diffusion Bonding, Crystallographic Structural Matching, Chemical and Mechanical Bonding. Most of these are, however, conditional to the availability of a clean, oxide-free contact surfaces between the impacting particle and the substrate that can be insured through proper surface preparation of the substrate prior to the coating operation. Due to the gradual degradation with time of the quality of the substrate surface once exposed to ambient air, surface preparation needs to be carried out as close as possible to the actual coating operation. Even with such a precaution, the Cold Spray process may also require a short surface activation period, generally referred to as "*induction time, t_i*" which is of the order of a few seconds, during which the sprayed particles do not adhere, and bounce off the surface cleaning it further from any remaining, or formed, oxide layers through "*shot pinning.*" Beyond the induction time ($t > t_i$) the particles begin to attach to the surface of the substrate in an avalanche-like manner, rapidly forming a coating. According to [Klinkov and Konsarev (2006)] the induction time t_i decreases rapidly with the increase in the particle impact velocity as shown in Fig. 15.20. for aluminum particles impacting on the copper substrate. It is also function of the mass flux of the powder, with the values given in Fig. 15.20 obtained for a mass flow rate of the powder per unit area of 0.06 kg/m²s.

In order to achieve the necessary plastic deportation of the particle and its proper adhesion to the substrate, the impact velocity of the cold-sprayed particles must be higher than a certain critical velocity, v_{cr}, which depends on the nature and relative hardness of the substrate and of the particle as well as the particle impact temperature as shown in Fig. 15.21 after [Schmidt et al. (2006)]. Too high particle velocity, on the other hand, approaching the velocity range of hydrodynamic penetration, will cause strong erosion of the coating preventing its buildup. Such velocities, called critical erosion velocity, v_{er} that are higher than the critical velocities, v_{cr} define an intermediate particle velocity region for optimal particle adhesion as shown in Fig. 15.21.

As particles are in a solid state their flattening degree, ξ_D is relatively small, at the maximum about 2, depending on particle plasticity and generally below 1.5. The bonding

between particle and substrate requires that the oxide layers are removed, and that adiabatic shear instability and the resultant plastic-flow localization will occur. The average kinetic energy of impacting particles in cold spray, and the relatively short, high pressure, contact time (a few tens of ns) are generally insufficient for the melting of the particle-substrate interface.

According to [Grujicic M (2007)], the adhesion of the impacting particle on the substrate is an atomic length-scale phenomenon. The occurrence of this adhesion is controlled by the quality of the surface (clean with no oxides) and the contact pressures. The strength of the adhesion depends on the attractive or repulsive character of atomic interactions and the crystallographic structure and orientation of the contact surfaces. The adiabatic shear instability and the resultant plastic-flow localization are the key phenomena for the particle-substrate bonding. [Grujicic M (2007)] made a transient non-linear dynamics analysis of the particle-substrate interactions during the cold-spraying of copper particles on aluminum substrate, and aluminum particles on copper substrate. Figure 15.22a represents the particle and substrate morphologies at the end of the impact, that is, 50 ns, of the aluminum particle onto the semi-infinite copper substrate. The particle diameter being 20 μm and its impact velocity 650 m/s, which is slightly over the critical velocity for this combination. [Grujicic M (2007)] reported that on impact, the particle aspect ratio decreases, with the particle creating a crater in the substrate surface with its depth increasing with the increase in the particle inertia. During the impact, a jet of particle and substrate materials are ejected from the crater forming a lip pointing away from the flattened particle as shown in Fig. 15.22a. For the case of a copper particle (same diameter and impact velocity) impacting on an aluminum substrate, shown in Fig. 15.22b, as the particle-substrate contact time increases, the crater depth and width increase while the flattened particle aspect ratio diminishes compared to that of an aluminum particle impacting a copper substrate, Fig. 15.22a. The length of the interfacial jets is also noted to increase.

Similar experimental observations were reported by [Guetta S et al. (2009)] who attributed the difference in the final shapes of impact craters created in each of these two cases (copper particles impacting an aluminum substrate and aluminum particles impacting a copper substrate) to the higher kinetic energy of the copper particles due to their higher specific mass and the strength of Cu compared to that of Al. When the impact velocity is lower than the critical one, the interfacial jet comprises only one material. According to [Grujicic M (2007)] the interfacial jets play a key role in removing oxides and other surface films for both materials. However, it must be emphasized that the "*induction time*" observed at the start of the cold spray process during which sprayed particles do not stick to the surface of

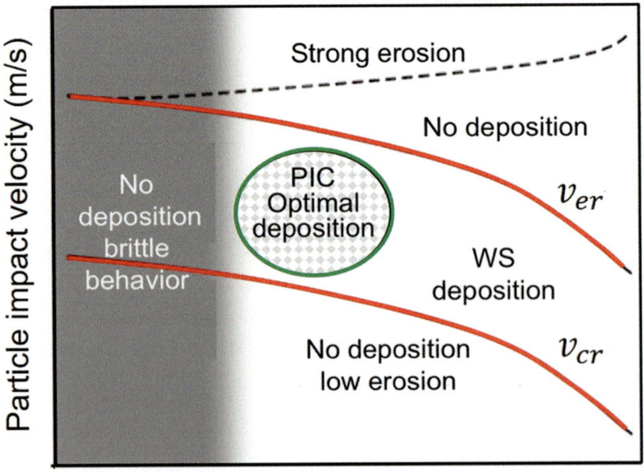

Fig. 15.21 Optimum Particle Impact Conditions (PIC) in the particle velocity Vs particle temperature diagram. ([Schmidt et al. (2006)]. Reprinted with kind permission from Elsevier)

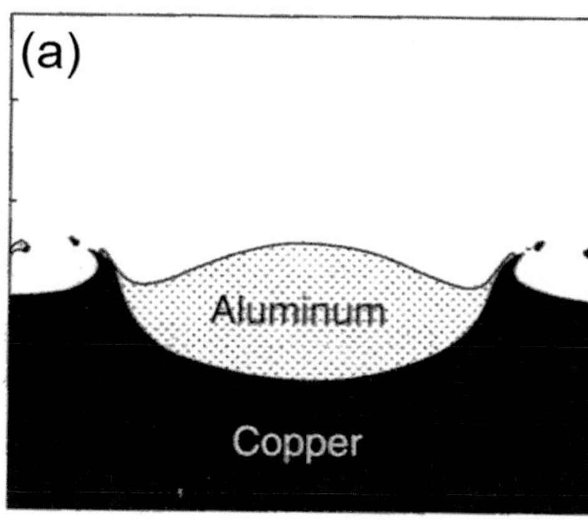

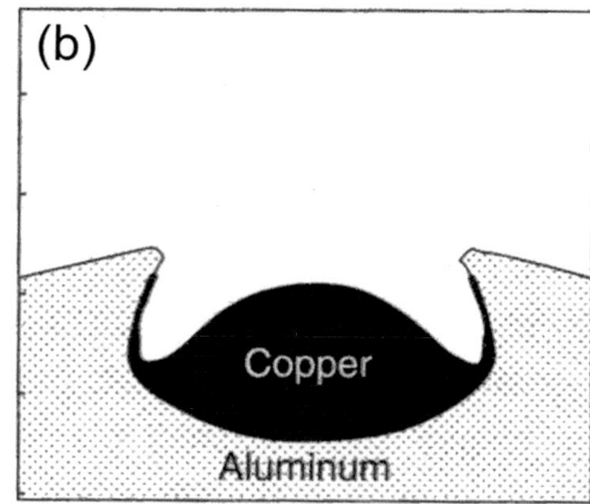

Fig. 15.22 Final calculated shape (impact duration 50 ns) of a 20 µm particle impacting the semi-infinite substrate with impact velocity $v_{pi} = 650$ m/s ($v_{pi} > v_{cr}$) (**a**) Al particle on a Cu substrate (**b**) Cu particle on an Al substrate. ([Grujicic M (2007)]. Reprinted with kind permission from Woodhead publishing)

the substrate, corresponds to the time necessary for the removal of any residual oxide present on the substrate surface. [Grujicic M (2007)] suggested that the formed interfacial jets contribute to particle-substrate bonding through the cleaning of the contact surfaces by some type of nano/micro length-scale mechanical material mixing. They suggested that this phenomenon is due to the instabilities created at the interface, giving rise to interfacial roll-ups and vortices. Of course, the mutual solubility of the two metals plays a key role. Also, the impact velocities during the induction time can begin to form reentrant cavities, creating a rivet-like particle-cavity assembly.

[Fukumoto et al. (2010)] cold-sprayed Cu particles with a mean diameter of 5 µm onto stainless steel SUS-304 substrate with a self-designed cold spray gun working with N$_2$ (1–3 MPa and 673 K) to study the jetting phenomenon. They studied the surface by atomic force microscopy, the cross-section microstructure of an individual particle by focused ion beam and Transmission Electron Microscopy (TEM) for the coating/substrate interface microstructure. The interface microstructure between sprayed particle and substrate revealed an amorphous-like band region at interface during coating formation at high power conditions. For the deposition mechanism of the cold-sprayed particles onto the substrate surface, the deformation of the particles initially induces the destruction of its surface oxide and the appearance of the active fresh surface of the material probably enhances the bonding between particles and substrate. On the other hand, in coating formation at high power conditions, bonding between particle and substrate may be possibly formed via oxygen-rich amorphous like layer at the interface. With high impact velocities, the substrate at the collision center deformed remarkably and metal jetting along the interface between particle and substrate occurred at the periphery of the flattened particle resulting in a good bonding of the particles. Figure 15.23 of a Cu splat on stainless steel

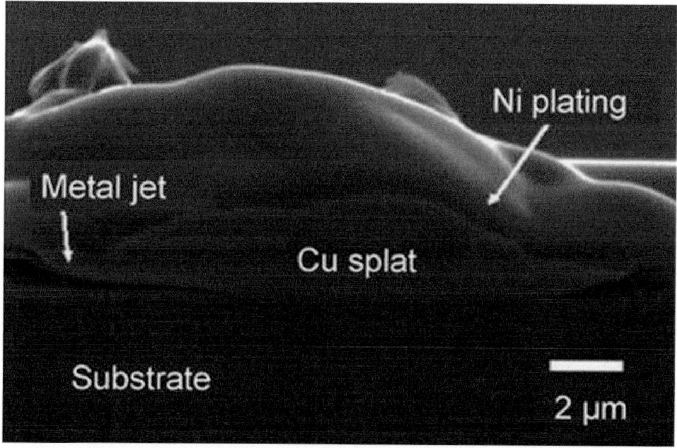

Fig. 15.23 Cross-sectional microstructure of individual deposited Cu particle (mean diameter 5 µm) cold sprayed onto stainless steel SUS304 substrate with N$_2$ (1–3 MPa and 673 K). ([Grujicic M (2007)]. Reprinted with kind permission from Springer Science Business Media copyright © ASM International)

substrate shows the metal jet, similar to that schematically represented in Fig. 15.22a. Increasing the process gas pressure was observed to increase the fraction of metal jet evolution and the deposition efficiency. [Fukumoto M et al. (2010)] conclude that "positive control and high efficiency coating formation are possible by setting the spray conditions."

[Zhang D et al. (2005)] have studied the impact of Al particles on substrates with different strength. They showed that initiation of deposition process onto soft metallic substrates was difficult due to the lack of deformation of the aluminum particles. Deposition onto a tin substrate, for example, resulted in its melting, the strength of the interface was therefore very low, and Al particles rebounded from the surface, only the very small ones being trapped in the tin. The ease of initiation of aluminum deposition increased as the hardness of metallic substrates increased. Metal jetting was observed when

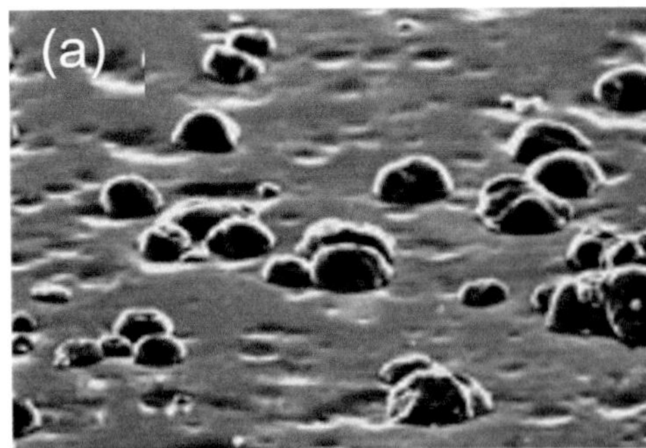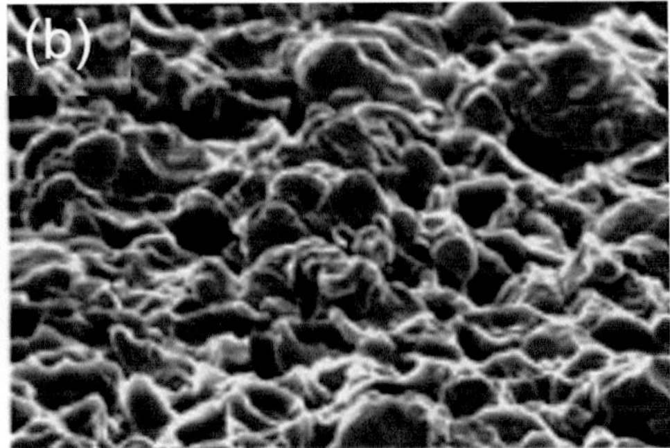

Fig. 15.24 Morphologies of collected copper particles (gas pressure: 0.7 MPa, helium temperature: 723 K, particle size: 5 μm). (**a**) AISI-304 substrate surface. (**b**) A6063 substrate surface. ([Fukumoto et al. (2007a, b)]. Reprinted with kind permission from Springer Science Business Media, copyright © ASM International)

depositing onto steel substrates, where also the shortest *induction time, t*$_i$ was observed. Deposition of aluminum particles on onto aluminum substrate was difficult, due to the tenacious nature of the oxide film on the substrate. Deposition onto nonmetallic substrates (ranging from polymers to ceramics) was difficult because of the lack of metallic bond that can be formed.

[Hussain et al. (2009)] studied the bonding mechanism of copper particles onto aluminum alloy substrates prepared prior to the spraying operation using three different procedures termed as; ground with 58.5 μm SiC-paper, polished, and grit-blasted with ~500 μm grit, respectively. In addition, some substrates were annealed following surface preparation to allow for the separation of the effects of work hardening during preparation from those of changes in surface morphology. They concluded that two types of bonding must be considered: metallurgical bonding between the substrate and the coating, and material interlocking between the material extruded from the substrate during impact of the particles and the coating structure. Of course, as shown by [Huang R and Fukanuma H (2012)] for Cu particles onto Al alloy substrates, the particle velocity (which must be > v_{cr}) plays an important role in improving adhesive strength based on plastic deformation. It is also beneficial to the mechanical interlocking effect resulting in excellent bonding between the coating and substrate.

The situation is somewhat different with HVOF or HVAF when the sprayed particles are at temperatures slightly below their melting temperature. For example, [Trompetter et al. (2006)] HVAF-sprayed NiCr alloy particles onto a variety of substrate materials. Soft substrates predominantly had deeply penetrating solid splats, whereas harder substrates that resisted particle penetration had a higher percentage of molten splats. A possible explanation was that melting occurred in the process of plastic deformation. The percentage of melted particles was greater for harder substrates. The percentage of molten splats increased with the increase in the particle impact velocity. Moreover, as in the cold spray

process, the impact velocity must be sufficiently high ($v_{pi} > v_{cr}$) to break surface oxides and to achieve sufficient plastic deformation for successful bonding.

Of course, the impact velocity must be lower than the erosion velocity (section "(a) Basic Phenomena). The analysis of [Assadi et al. (2003)] suggests that mass density and particle temperature have significant effects on the critical velocity and are thus two of the most influential parameters in cold gas spraying. [Fukumoto et al. (2007a, b)], considering copper particles impacting on stainless steel AISI 304 and aluminum alloy A6063 substrates, have shown that extreme roughening is observed on the A6063 substrate, as illustrated in Fig. 15.24.

[Fukumoto et al. (2007a, b)] have also shown that substrate heating is quite effective for the formation of the first layer resulting in a higher deposition ratio in the cold spray process. This tendency was promoted more effectively using helium instead of air or nitrogen as a working gas. Both higher velocity and temperature of the particles sprayed are the necessary conditions for the higher deposition ratio in the cold spraying. Instead of particle heating, substrate heating may bring about the equivalent effect for particle deposition. The substrate material also plays a key role in the deposit formation. For example, aluminum splat formation has been studied for substrates made of tin, copper, Al alloy, brass, different tool steels, low carbon steel, glass, and alumina (hardness varying from 0.08 to 10.7 GPa) [Li C-J et al. (2006a, b, c, d, e) and, Papyrin AN et al. (2003)]. It has been shown that initiation of deposition onto soft metallic substrates is difficult due to the lack of deformation of the aluminum particles. The initiation of aluminum deposition increases with the substrate hardness. Deposition of aluminum particles onto aluminum is difficult due to the thin aluminum oxide layer on the substrate, which deforms with the underlying aluminum. Of course, deposition onto nonmetallic substrates is difficult due to lack of metallic bond that can be formed [Zhang D et al. (2005)]. To conclude this part more work on the interface between a cold-sprayed particle and the substrate is necessary to improve our understanding of the involved phenomena.

[Klinkov SV and Kosarev VF (2012)] proposed the addition of small fractions of abrasive powder to the metallic cold spray powder with the objective of the continued activation of the substrate surface by the abrasive powder during the deposition process and improve the adhesion of the metal powder to the substrate. A two-probability model was developed to theoretically analyze the influence of substrate/coating surface erosion and activation by the abrasive particles on the total deposition efficiency of a sprayed metal particles. Relations for the metal deposition efficiency derived with regard to the activation effect due to abrasive particles were obtained. The spray gun was working with air (pressure 1.6 MPa, temperature 550–750 K) with the two-component system: copper ($d_{50} = 30$ μm)–silicon carbide with different sizes (8, 14, or 40 μm). The presence of abrasive particles in the mixture enhances the adhesion probability of metal particles to the surface, even if, due to dilution, it decreases the concentration of metal particles in the mixture. The impact of ceramic particles also exerts an erosive action on the surface. The involved process is summarized in Fig. 15.25 involved the following steps;

- Adhesion of metal particle to initially nonactivated substrate surface (transition s_s to s_c)
- Impact of abrasive particle onto nonactivated substrate surface (transition s_s to s_{sa})
- Adhesion of metal particle to activated substrate surface (transition s_{sa} to s_c)
- Impact of abrasive particle onto the formed coating (transition s_c to s_{ca})
- Adhesion of metal particle to activated coating surface (transition s_{ca} to s_c)
- Impact of abrasive particle onto the activated portion of the coating (transition s_{ca} to s_{sa})

where the areas of nonactivated s_c and activated s_{ca} coating surfaces are presented, and also the corresponding areas of nonactivated s_s and activated s_{sa} substrate surfaces. The modeling data were found to be in a good qualitative agreement with experimental data. This suggests that the model developed by [Klinkov SV and Kosarev VF (2012)] provide an adequate description of the kinetics of coating growth from sprayed metal–ceramics mixtures.

[Koivuluoto H. and Vuoristo P (2010)] have shown that Cu + Al$_2$O$_3$ coatings are strongly influenced by the powder characteristics of Cu particles (spherical and dendritic) and volume percentage of Al$_2$O$_3$ (0, 10, 30, and 50 vol.%). Coating denseness and particle deformation level increased with the addition of hard particle. Furthermore, hardness and bond strength increased with increasing Al$_2$O$_3$ fractions. In the comparison between different powder types, spherical Cu particles led to the denser and less oxide-containing coating structure due to the highly deformed particles.

When spraying with high power HVOF guns, where, for example, nitrogen is introduced in the combustion chamber, or HVAF guns, the gas temperature can be lower than the melting temperature of sprayed metals or alloys that can impact in solid phase a few hundred degrees below their melting temperature. For example, [Trompetter WJ et al. (2005)] studied Ni–Cr (Ni80/Cr20, 5–45 μm) alloy particles HVAF sprayed onto Al substrates. For a chamber pressure of 414 kPa, the gas temperature falls to ~1300 °C in the diverging section of the HVAF torch nozzle, which is below the melting point of NiCr powder ($T_m = 1400$ °C). Moreover, the residence time of the spray particles in the gas jet is quite short (~200 μs for a particle velocity of 700 m/s and a spray distance of 150 mm). The particle substrate interface was investigated with focused ion beam microscopy, cross-sectional scanning electron microscopy, and cross-sectional transmission electron microscopy. No evidence of melting or chemical bonding was found in the samples. Instead, evidence of mechanical bonding was found quite similar to that predicted by [Grujicic M et al. (2003) and Grujicic M (2007)] in high-pressure cold spray. Interlocking features were found at certain locations along the particle–substrate interfaces. Vortices of aluminum were found to extend 100–1000 nm from the substrate into thermally sprayed NiCr particles. As illustrated in Fig. 15.26, the size of the turbulent features was proportional to the particle diameter, which is in agreement with the amount of interfacial instability estimated by Reynolds number calculations.

[Trompetter WJ et al. (2010)] have also studied the influence of the substrate surface oxides on NiCr splat–Al substrate bonding sprayed in the previous conditions. No splats were found on the aluminum oxy-hydroxide oxide layer (845 nm), whereas bonding did occur on an alumina layer (650 nm) grown on an aluminum substrate. Cross sections of solid splats showed that the aluminum substrate and surface oxide had been deformed during the particle impact. This has resulted in some locations where the oxide was as thin as 20 nm but was still present between the splat and the substrate even in these very thin local regions. Cross-sectional

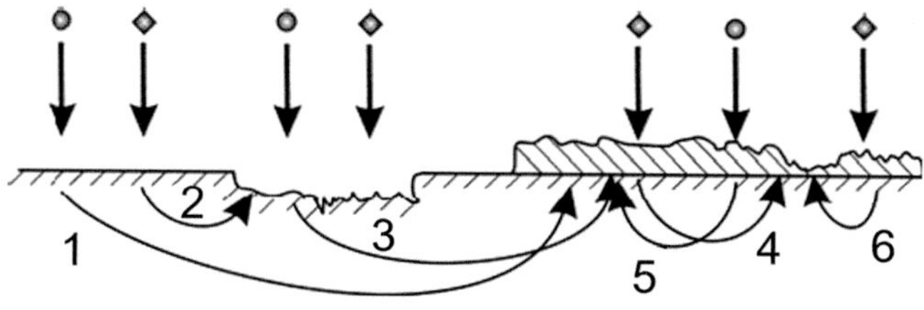

Fig. 15.25 Different phenomena involved in the formation of a coating from a sprayed metal-ceramic mixture. (Involved processes: [Klinkov SV and Kosarev VF (2012)]. Reprinted with kind permission from Springer Science Business Media, copyright © ASM International)

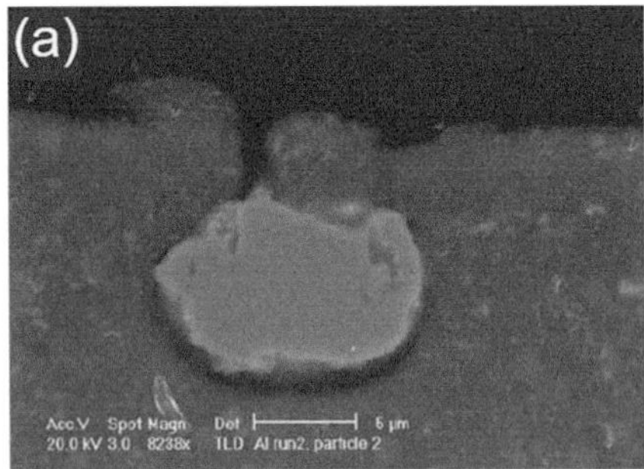  

Fig. 15.26 Cross-sectional SEM images of individual NiCr particles embedded in an Al substrate with varying degrees of interfacial contact and interfacial turbulence: (**a**) low level of interfacial turbulence ($d_p = 14$ µm), (**b**) medium level of interfacial turbulence ($d_p = 21$ µm), and (**c**) high level of interfacial turbulence ($d_p = 36$ µm). (Trompetter WJ et al. (2005). Reprinted with kind permission from Springer Science Business Media, copyright © ASM International)

measurements of molten particles show that the oxide has been penetrated in several locations allowing close proximity of the splat metal to the substrate metal to occur [Trompetter WJ et al. (2010)]. At last, [Trompetter WJ et al. (2006)] have studied the same Ni–Cr (Ni80/Cr20, 5–45 µm) alloy particles HVAF sprayed onto a variety of substrate materials (Al, Cu, Fe, Ti, Ta, glassy C, Si), the hardness's of which varied from 338 to 10,082 MPa. Although the various substrate materials were coated using identical powder material and thermal spray conditions, the type and variation in splat morphologies were strongly dependent on the substrate material. On the Al substrate 52% of solid splats and 45% of semi-molten ones were observed. With the very hard substrates (glassy C and Si) 81–70% of molten splats were obtained, while with Ti, Cu, Fe, Ta substrates semi-molten splat fractions were between 45% and 57% with fully molten ones between 23% and 31%. However, it must be emphasized that percentages do not vary linearly with the hardness. This is illustrated in Fig. 15.27a showing that the solid splat has penetrated deeply into the Aluminum substrate while on the Silicon substrate, Fig. 15.27b, the deformation is negligible. Fig. 15.27c, d illustrate are semi-molten splats obtained on Titanium and glassy C substrates and Fig. 15.27e, f present molten splats on Tantalum and glassy C.

[Trompetter WJ et al. (2006)] found that conversion of particle kinetic energy into plastic deformation and hemain types of splats observed are presented at, dependent on substrate hardness, can make a significant contribution toward explaining the observed behavior. When impacting on soft substrates a low percentage (<10%) of kinetic energy is converted into plastic deformation and heat. Conversely, when particles suffered heavy deformation impacting onto hard substrates most of the particle kinetic energy (e.g., 95%) would have been converted into plastic deformation and heat, as illustrated in Fig. 15.28.

15.3.1.2 Splat Formation from Molten Particles

For *molten particles*, with temperatures either at *their melting point, T_m*, as in flame spraying or HVOF or higher than T_m as

in DC or RF Induction plasma spraying, the mechanism of splat formation is quite different from that of solid particles. Particle splat formation involves in this case a rather complex process involving simultaneously the transient flow of the molten particle material resulting from the high-pressure wave generated by the impact, coupled with the transient cooling and solidification of the material due to its contact with the surface of the substrate. The transformations involved depends on the thermophysical properties of the particle/droplet and the surface properties of the substrate including the liquid-substrate wettability, the effect of any oxide layer present on the surface or adsorbents and condensates as well as the particle impact velocity and angle of impact. The solidification of the molten particle, which often starts before flattening is completed, is closely linked to its thickness and the thermal contact resistance, R_{th}, with the substrate, or the previously deposited layers. These two parameters vary along the radius of the flattening droplet, r and the contact resistance R_{th} is closely correlated to the local pressure diminishing with r.

The observation of the transformation of the shape of the droplet and the monitoring of its transient temperature during the splat formation step has for long been particularly challenging due to the fact that the particle size of most powders used in thermal spray is in the range of 20 and 60 µm, with their impact velocities ranging between 50 and 600 m/s. Typically, a splat formation time can vary depending on the particle velocity, between 50 ns (cold spray) and 10 µs (flame and plasma spraying). Experiments to follow particle impact and splat formation are very complex and require that such investigations are performed on smooth substrates, in order to have the necessary depth of field to observe the particle flattening with a microscopes or special optics. Fast pyrometers (50 ns response time) are used to measure the particle temperature evolution prior to and during splat formation. Impacting droplets can also be photographed with cameras triggered by the pyrometer signal to obtain the corresponding images of the particles at different stages of its flattening and splat formation. Since it is

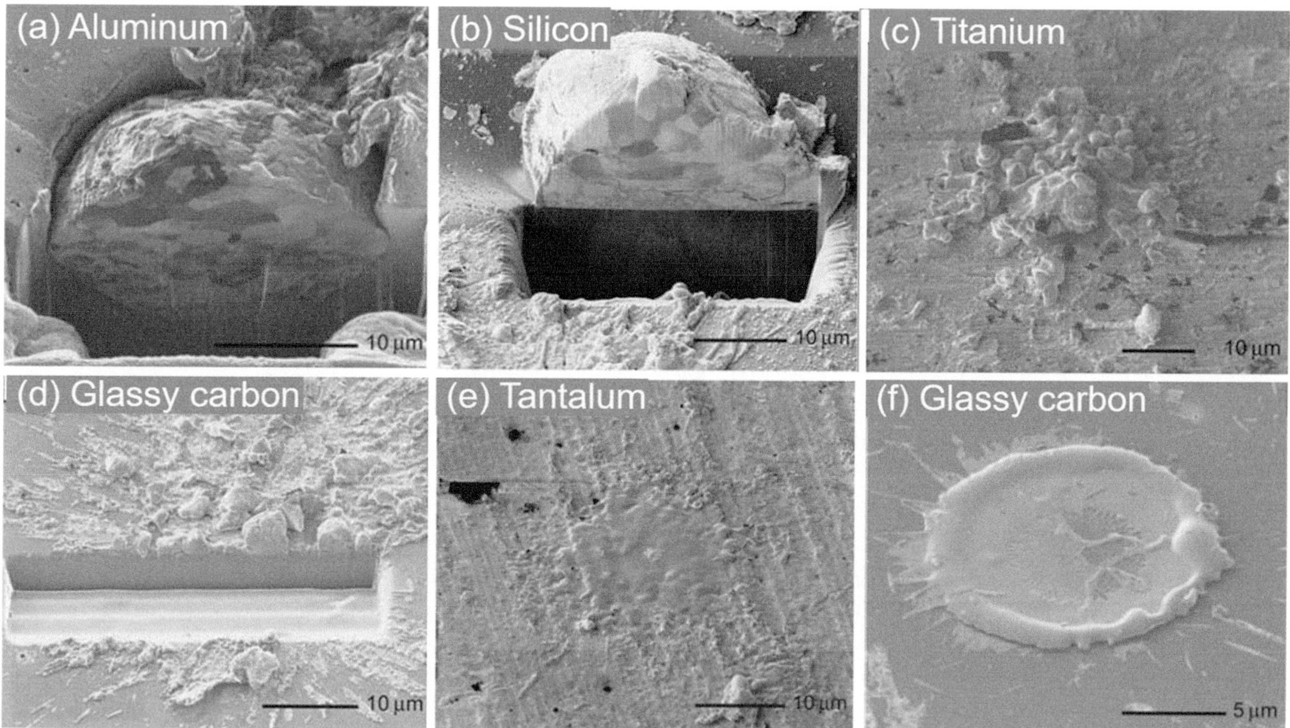

Fig. 15.27 NiCr particles HVAF sprayed and resulting in solid splats on (a) Al and (b) Si (sectioned); semi-molten splats on (c) Ti and (d) glassy carbon (sectioned); (e) molten-splash splat on Ta and molten-disc splat on glassy carbon. ([Trompetter WJ et al. (2006)]. Reprinted with kind permission from Springer Science Business Media, copyright © ASM International)

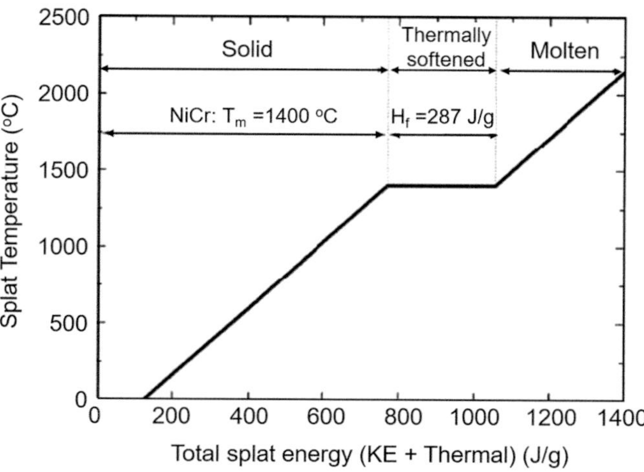

Fig. 15.28 Relationship between splat thermal energy and temperature. ([Trompetter WJ et al. (2006)]. Reprinted with kind permission from Springer Science Business Media, copyright © ASM International)

virtually impossible to follow continuously the flattening of a single molten particles sprayed by plasma or HVOF, the time history of the flattening particle can be determined by measurements made on ensemble of particles, assuming that all impacting particles have the same parameters. Splat characterization can be carried out using Optical Microscopy (OM), Focused Ion Beam (FIB), Cross-Sectional Scanning Electron Microscopy (CS-SEM), and Cross-Sectional Transmission Electron Microscopy (CS-TEM). Moreover, it is also possible to measure particle velocities, temperatures (above about 1400 K), and diameters prior to their impact using Optical Emission Spectroscopy (OEM) and Laster Velocimetry (LV), and for certain experimental conditions, to correlate these measurements to splat shape (see Chap. 18 Process Diagnostics and Online Monitoring).

Alternately, the study of the flattening of millimeter-sized molten particles have been also successfully used to identify some of the basic mechanism involved in the flattening and solidification of droplets [M. Pasandideh-Fard et al. (1996, 1998)]. The main advantage in this case is the relative ease by which such larger particles, combined with their longer flattening time that is of the order of several milliseconds compared to microseconds for micron-sized particles, allows for their observation using high-speed video cameras and infrared pyrometry. Applying these observations, however, to micron-sized thermal-sprayed particles requires matching the values of all the dimensionless numbers (*Reynolds, Weber, Peclet, Stephan, Biot*) that control impact dynamics, which is generally possible when considering one dimensionless number at a time, but rarely possible simultaneously for all of them. Moreover, while, for micrometer-sized particles, the ratio of the thickness of the boundary layer of the flattening particle, to its characteristic dimensions is close to unity,

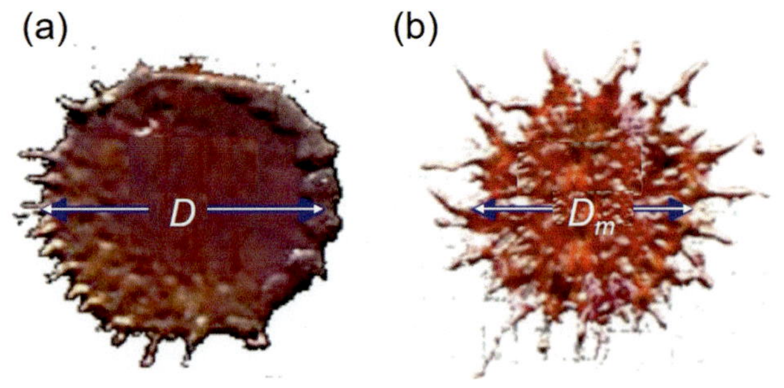

Fig. 15.29 Definition of Yttria Stabilized Zirconia (YSZ) splat with mean diameter: (**a**) Disk-shaped splat and (**b**) Fingered "splashed" splat

it is a few orders of magnitude lower with millimeter-sized particles [Dhiman R et al. (2007)].

On impact of a molten droplet on the substrate, the resultant splat can be broadly classified as either of a "disc shape" with as well-defined diameter, D as shown in Fig. 15.29a, or an "exploded" molten particle, often referred to as "fingered or splashed particle" with a diameter, D_m, rather difficult to determine with precision as shown in Fig. 15.29b.

Two types of impacts are generally considered: normal impact, optimum for coating generation, and off normal one where the impact velocity is reduced. However normal impact cannot always be achieved when the substrate to be sprayed has steep angular areas or convex or concave zones with small radii. Normal impact on the substrate cannot be guaranteed either even if the substrate surface is normal to the spray jet axis, since because of the normal divergence of the spray jet particles entrained in its fringes impact the substrate with an angle generally between 70° and 110°.

[Kudinov et al. (1981)] at the beginning of the 1980s were the first to make broad classification of splat shapes on a smooth surface as shown in Fig. 15.30. These were attributed to the following particle impact conditions;

- (1) and (4) fully melted particle/droplet,
- (6) and (11) fully melted particle with gas inclusions,
- (14) superheating with strong evaporation,
- (17) melted shell surrounding a solid core,
- (20) solid shell with a melted core,
- (27) solidified crown, melted interface and solid core,
- (28) and (29) ductile particles with high velocities

Splat shapes depend on a large number of factors such as the presence of contaminants at the substrate surface, the partial solidification of the particle in-flight resulting in formation of a solid shell around the liquid central part (this occurs especially for low thermal conductivity materials), the degree of in-flight oxidation and the superheating of the particle. According to recent studies where the flattening of

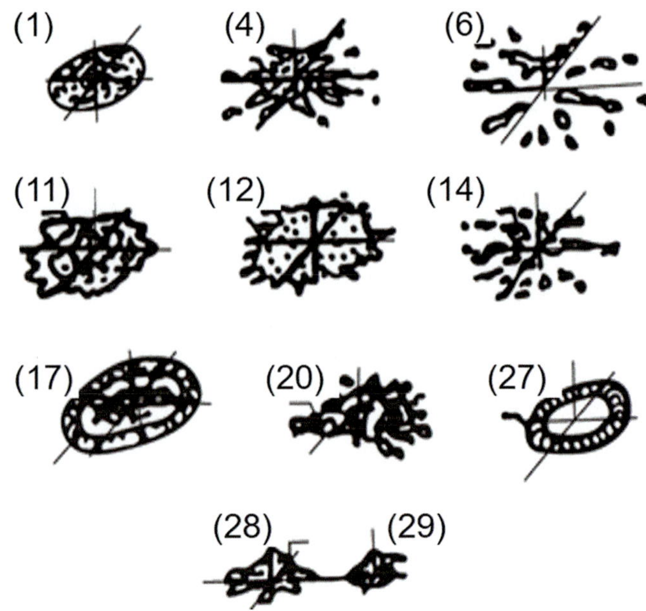

Fig. 15.30 Different morphologies of splats collected on a smooth substrate. (After [Kudinov et al. (1981)])

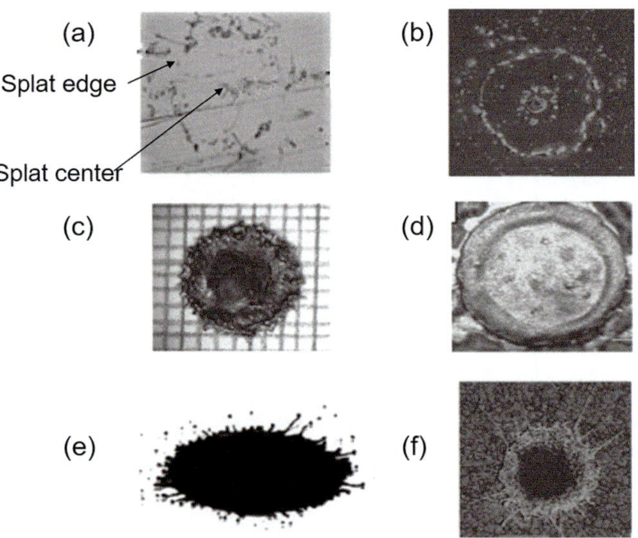

Fig. 15.31 Main types of molten splats: (**a**) alumina on cold 304 stainless steel (SS) [Goutier S et al. (2011) (**b**) Ni on 304 SS pre-oxidized at 150 °C [Dhiman R et al. (2007)] (**c**) Cu on AISI-304 SS [Fukumoto M et al. (2011)] (**d**) alumina on low carbon steel preheated at 200 °C [Sabiruddina K et al. (2011)] (**e**) alumina on SS preheated at 400 °C [Goutier S et al. (2011) and (**f**) Ni on pre-oxidized SS at 650 °C [Dhiman R et al. (2007)]. (Copyright © ASM International, reprinted with kind permission)

droplets has been followed, Examples of the main types of splats observed are presented in Fig. 15.31. [Dhiman R et al. (2007), Goutier S et al. (2011), Fukumoto M et al. (2011) and Sabiruddina K et al. (2011)]

The following discussion will deal exclusively with splat formation from molten or semi-molten particles, with

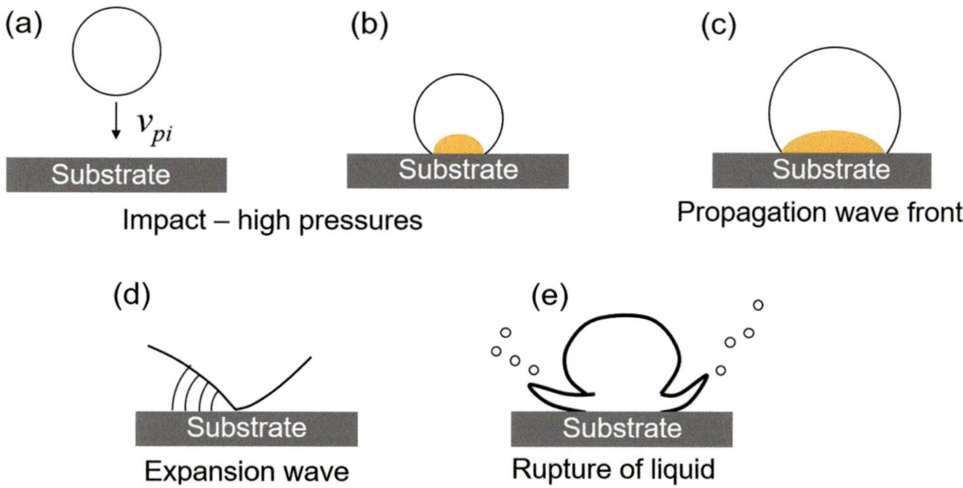

Fig. 15.32 Successive phenomena of impact splashing: (**a**) particle before impact, (**b**) impact generating a very high pressure, (**c**) propagation of a wave front, (**d**) expansion wave at the triple point, and (**e**) rupture of liquid, splashing. ([Armster SQ et al. (2002)]. Reprinted with kind permission from Maney Publishing)

emphasis placed on the type and shapes of splats formed with different materials, spray sources and substrate conditions, followed by an analysis of the basic mechanism involved and a brief review of the numerous modeling studies reported in literature. General reviews on the subject are given by [Fauchais P. (2004), Chandra S and Fauchais P (2009), Fukumoto M et al. (2010), Fauchais P and Vardelle M (2010) and Mostaghimi J and S Chandra (2018)].

15.3.2 The First Instants of Impact

The impact pressure $(\rho_p v_{pi}^2)$ increases rapidly with the particle velocity. If we compare the impact pressure obtained for the same particle at the same temperature, we obtain values for RF plasma spraying with an impact velocity of 50 m/s (Fig. 15.32a) as a base line, p_{ref}. The dynamic pressure obtained with DC plasma spraying with an impact velocity of 200 m/s, $p = 16 \times p_{ref}$, and for spraying with HVOF with an impact velocity of 500 m/s, $p = 100 \times p_{ref}$. The impact pressure, linked to compressibility effects, can be very high (over 100 MPa). The impact generates a pressure pulse distributed along a radius r_{m} smaller than the final splat radius: $r_{m} = d_{p} \cdot v_{pi}/(2\,c_{\ell})$ (Fig. 15.32b), where c_{ℓ} is the sound velocity in the liquid drop [Armster SQ et al. (2002)]. As typical sound velocity in liquid metals is of the order few thousand m/s, r_{m} is about 10% of the impacting particle diameter. Within r_{m} the vaporization of adsorbates and condensates is generally negligible. In general compressibility effects are also neglected. Then the wave front propagates, see (Fig. 15.32c), resulting in an expansion wave (Fig. 15.32d), and induces rupture of the liquid and droplet formation at the triple interface, see (Fig. 15.32e).

Impact splashing occurs immediately after impact (within the first 100 ns) and is driven by fluid instabilities, before any solidification can take place. It depends, for water droplets that have been studied in detail, on the Sommerfeld parameter, Ks, defined as

$$Ks = We^{0.5}\, Re^{0.25} \tag{15.5}$$

where We is the Weber number and Re is the Reynolds number of the particle at impact. Thus

$$Ks = \left(\frac{\rho\, v_{pi}^2\, d_p}{\sigma}\right)^{0.5} \left(\frac{\rho\, v_{pi}\, d_p}{\mu}\right)^{0.25} \tag{15.6}$$

where, d_p and v_{pi} are, respectively, the particle diameter and impact velocity, and ρ, μ, and σ *are* the mass density, the dynamic viscosity, and the surface tension of the impacting droplet material evaluated at the impact temperature of the particle.

A value of $Ks < 3$ corresponds to a rebound; $3 < Ks < 58$ results in deposition, and $Ks > 58$ induces splashing [Escure C et al. (2003)]. High speed imaging of impact splashing of alumina particles (22–44 µm) plasma sprayed with different Sommerfeld parameters, K onto a stainless-steel substrate, is given in Fig. 15.33 after [Escure C et al. (2003)]. The ejected droplets have diameters below one micrometer reaching a distance of few mm above the substrate, outside the dynamic boundary layer. From a practical point of view, these tiny particles reach distances above the substrate that are generally outside the fluid boundary layer. Thus, they are entrained by hot gases far from the substrate and the coating being deposited, creating no defect within coating.

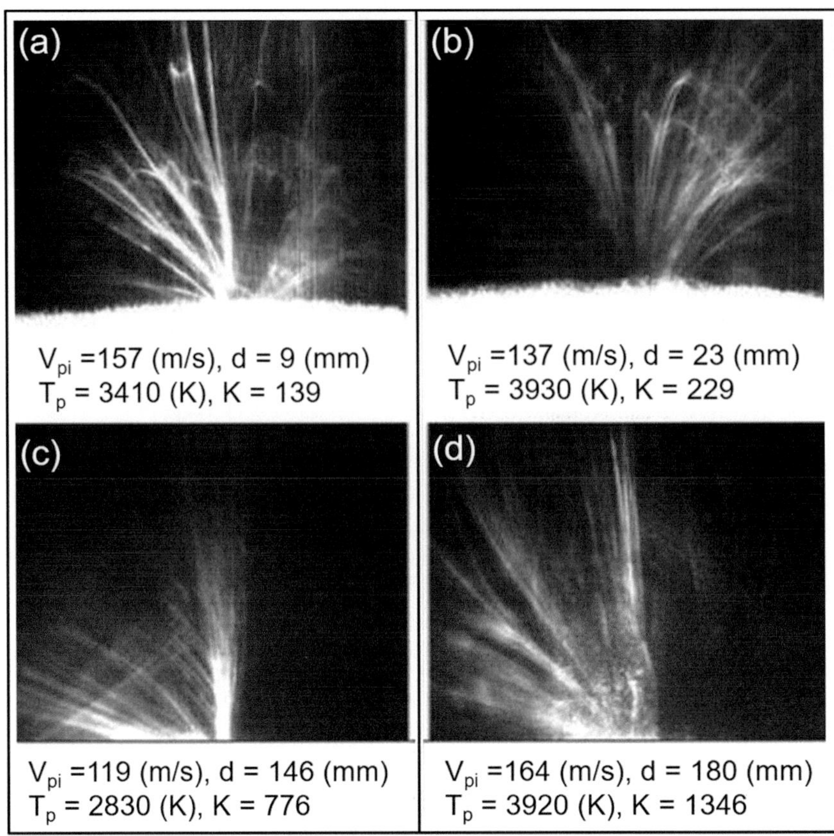

Fig. 15.33 Impact splashing of plasma-sprayed alumina particles with different particle impact velocities, particle temperatures and Sommerfeld parameters, pictures taken with a long exposure time of 1 ms. ([Escure C et al. (2003)]. Reprinted with kind permission from Springer Science Business Media)

15.3.3 Flattening of Molten Particle

15.3.3.1 Transition Temperature and Pressure

On impact of the particle/droplet on the surface of the substrate, a high-pressure wave is generated at the contact point, $r = 0$, as illustrated in Fig. 15.34a, which radiates outwards towards the outer limits of the droplet forcing the melted material to flow laterally flattening the droplet to a disk shape, and possibly deforming the substrate if its material is too ductile. In this process, the kinetic energy of the particle is transformed into work of viscous deformation and surface energy. The contact pressure at $r = 0$ drops rapidly with time as droplet flatten. Figure 15.34b represents the flattening droplet at an intermediate time t_i when the flattening particle reaches the radius r_i, in this figure is also shown the evolution of the pressure with time at the radial position r_i as the radius of the flattening particles continues to grow with time. It is worth noting the appearance of the high-pressure wave at this point in time at $r = r_i$ is at a much lower value compared to its initial level at $r = 0$ and $t = 0$, and for a very short time, corresponds to the further movement of the pressure wave outward with time until it reaches the outer rim of the flattened droplet. Such transient pressure-time variations along the

flattening particle diameter play a key role in its flattening and solidification behavior. As the flattened diameter ratio of the droplet exceeds a value of 1.5–2.0, the pressure in the droplet is not high enough to overcome other forces such as those due to evaporated adsorbates and condensates that could be present on the substrate surface [Chang-Jiu L and Li J-L (2004)]. This results in a lifting of the molten droplet fluid at the outer rim of the splat off the substrate surface, as illustrated schematically in Fig. 15.35. The detachment of the splat rim from the substrate surface results in the decrease in the contact area between the splat and the substrate by a factor of 2–6, which in turn reduces splat cooling rate, and consequently the continuation of the flattening of the splat beyond that could be expected on the same substrate in the absence of adsorbates and condensates. Normally, the quantity of adsorbates and condensates present on the surface depends on the degree of oxidation of the metal substrate and substrate preparation [Li C-J and Li J-L (2004), Withy B et al. (2006), Tran ATT et al. (2008), Birks Net al. (2006), Fukumoto M (2004) and Yang K et al. (2010)]. Heat treatment of the substrate just prior to spraying can reduce the relative amounts of these species and consequent the release of gases from the surface of the substrate during the spray operation.

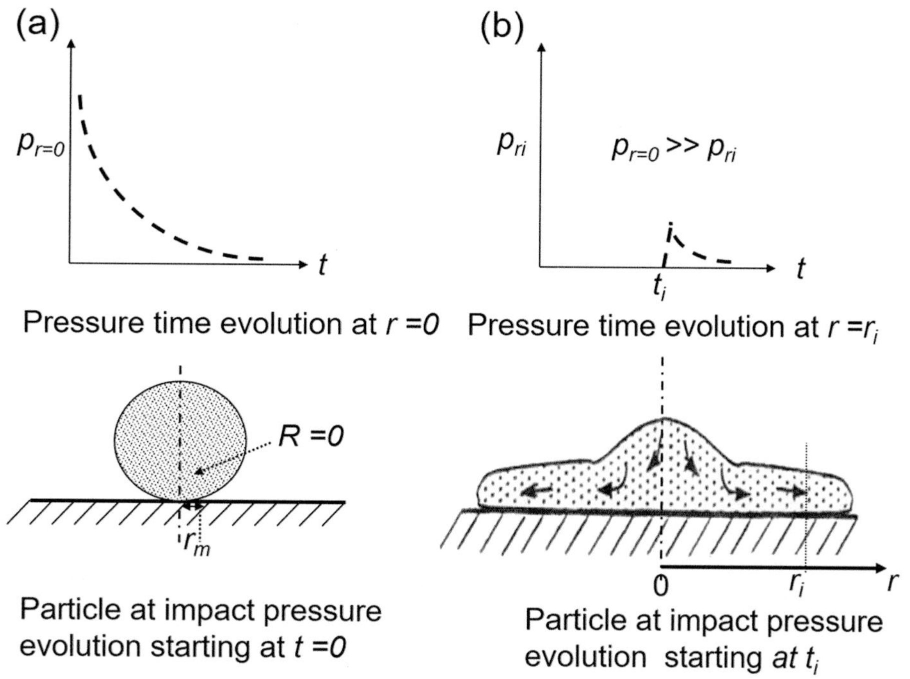

Fig. 15.34 (**a**) Droplet impact at time, $t = 0$ with the impact pressure time evolution at radius 0. (**b**) Droplet at time t_i with the pressure time evolution at radius r_i

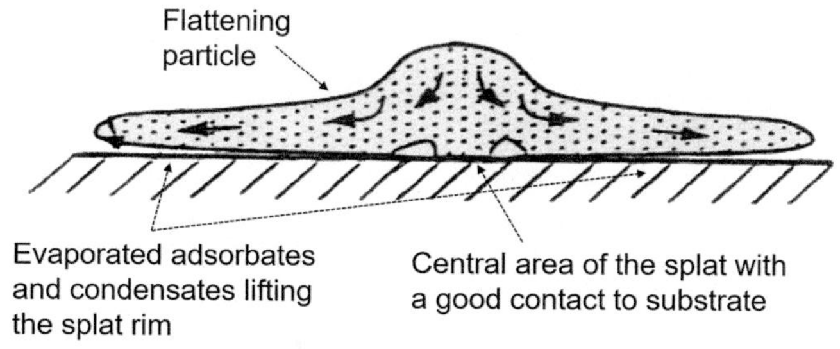

Fig. 15.35 Schematic representation of droplet flattening on a smooth substrate at room temperature where adsorbates and condensates exist

Figures 15.36 and 15.37 present two examples of splats of ceramic and metallic powders collected on smooth ASI-304 stainless steel substrates at Room Temperature (RT = 300 K) and on substrates preheated to (673 K). The Yttria Partially Stabilized Zirconia (YPSZ) splats given on Fig. 15.36 show an extensively fingered "splashed" form of splats with the substrate at RT, Fig. 15.36a, while in contrast, well defined disc shaped splats were obtained on the same substrate when preheated to 673 K Fig. 15.36b [Bianchi L et al. (1977)]. For the YPSZ splats the fraction of the splat surface that was in good contact with the substrate at RT is estimated at about 20%, while on the preheated substrate it is around 60%. Similar observation were reported by [Fukumoto M (2004)] using Nickel particles that showed a strongly fingered "splashed" splats on stainless steel substrate at room temperature, Fig. 15.37a,

and a perfectly circular disc-shaped splat with the substrate at 670 K, Fig. 15.37b. Detailed micrographs giving top and bottom views of the nickel splats obtained by [Fukumoto M (2004)] with RT (300 K) and Preheated (673 K) AISI-304 SS substrates are given in Figs 15.38 and 15.39, respectively. These show that the central part of the splat, which in Fig. 15.38a seems to be in good contact with the substrate, has in fact some sort of lace surface with many holes corresponding to a poor contact area (Fig. 15.38b, c). This is in contrast to the Ni splat on the preheated substrate (Fig. 15.39), where the bottom part shows an excellent contact except in its periphery.

Based on these observations and supported by a wide range of data obtained with a wide range of splat observations of ceramic and metallic materials with different substrate temperatures [Fukumoto M and Y. Huang (1999), Fukumoto (2004) and Fukumoto et al. (2007a) (2007b) (2011)] defined the transition temperature, T_t as the critical substrate temperature over which more than 50% of splats are disk-shaped. Typical results for Ni powder on polished ASI-304 stainless-steel substrate given in Fig. 15.40, [Fukumoto M and Y. Huang (1999)] the fraction of disc-shaped splats and the coating adhesion strength as function of the substrate temperature over the range of 300–800 K. Tabulated values of the transition temperature for different metal and ceramic materials on ASI-304 stainless-steel obtained using APS and HVOF spraying are given in Table 15.2.

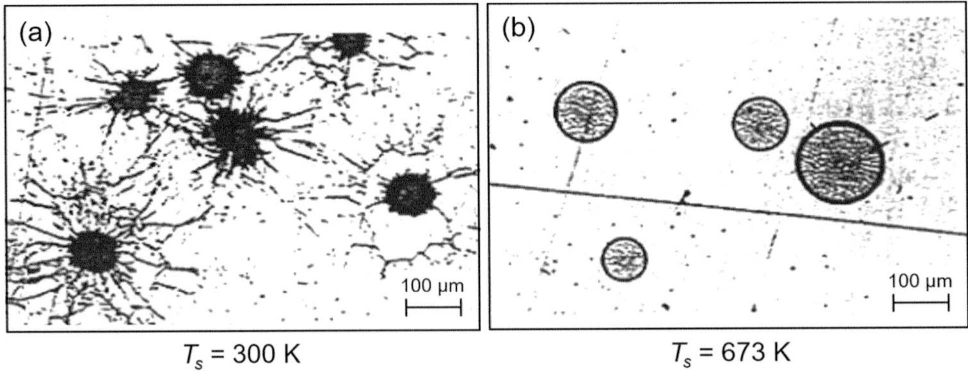

Fig. 15.36 YSZ splats plasma sprayed at atmospheric pressure onto stainless steel AISI-304 substrate. (**a**) At room temperature (300K) and (**b**) preheated to 673 K. ([Bianchi L et al. (1977)]. Reprinted with kind permission from Elsevier)

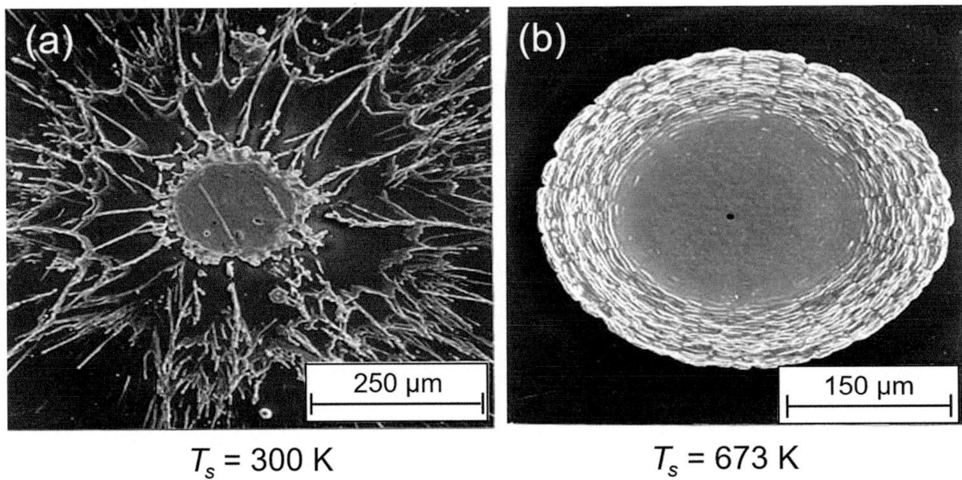

Fig. 15.37 Nickel splats plasma sprayed at atmospheric pressure onto stainless steel AISI-304 substrate. (**a**) At room temperature (300K) and (**b**) preheated to 673 K. ([Fukumoto M and Y. Huang (1999) and Fukumoto M (2004)]. Reprinted with kind permission from ASM International)

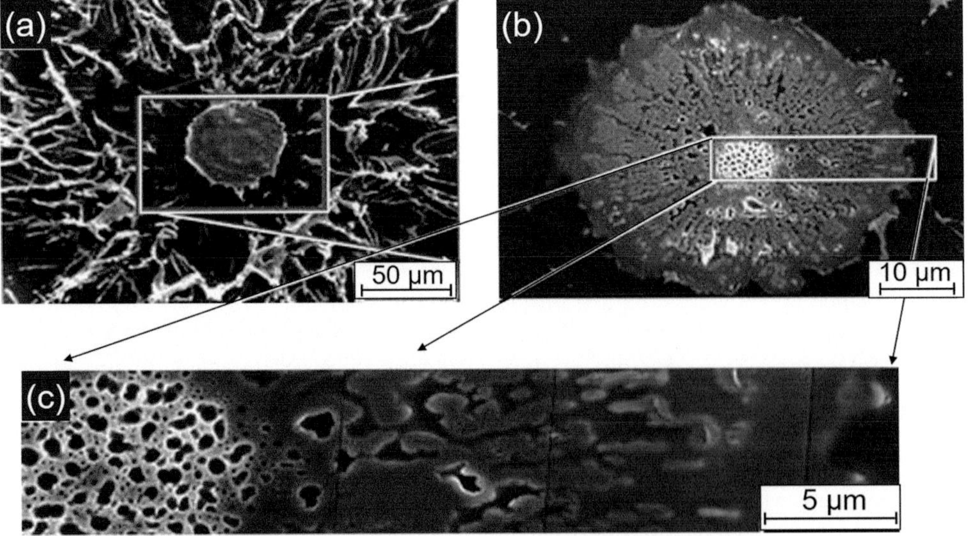

Fig. 15.38 (**a**) Top view and (**b**) bottom view of nickel splats plasma sprayed at atmospheric pressure onto a stainless steel AISI-304 substrate at RT (**c**) enlargement of the rectangular defined area on (**b**). (Reprinted with kind permission from ASM International [Fukumoto M (2004)])

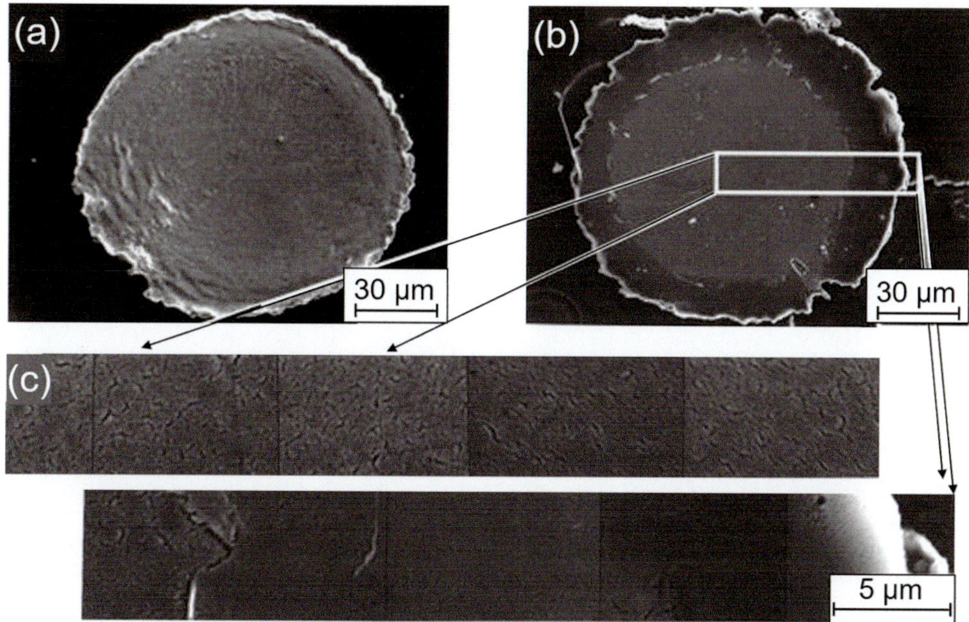

Fig. 15.39 (**a**) Top view (**b**) bottom view of nickel splats plasma sprayed at atmospheric pressure onto a stainless steel AISI-304 substrate at 673 K. (**c**) enlargement of the rectangular defined area on (**b**). ([Fukumoto M (2004)]. Reprinted with kind permission from ASM International)

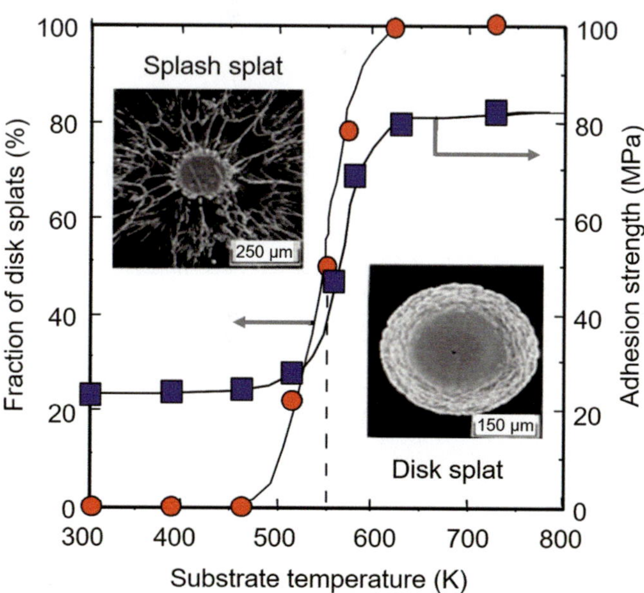

Fig. 15.40 Dependence of the fraction of nickel splats that are in disk form and coating adhesion strength onto AISI-304 stainless steel substrate as function of substrate temperature at atmospheric pressure AISI-304 stainless steel substrate at room temperature (300K) [Fukumoto et al. (2007a)]. (Reprinted with kind permission from ASM International)

According to [Fukumoto M and Y. Huang (1999)], this important change in the splat pattern near the transition temperature is a quite unique phenomenon. Both velocity and temperature of the particle, substrate temperature, and wetting at the particle/substrate interface are dominant factors for the flattening behavior of the particle. In this case, however, as the conditions relating to the particle are constant,

Table 15.2 Transition temperature (K) for splat formation of different metallic and ceramic powders on AISI-304 stainless-Steel substrate sprayed using APS and HVOF [Fukumoto M (2004), Fukumoto et al. (2007a, b)]

Metals				Ceramics	
APS		HVOF		APS	
Ni	610 K	Ni	560 K	Al_2O_3	318 K
Mo	474	Ni-5Al	440	TiO_2	350
Cu	394	Ni-10Cr	400	YSZ	345
		Ni-20Cr	360		
Cu-30Zn	505	Cu-30Zn	455		
Cr	387	Cr	345		

both particle solidification and interface wetting affect the flattening behavior of the particle on changing the substrate temperature. The transition temperature has also been correlated with the vaporization of adsorbates and condensates on the surface of the substrate. Preheating the substrate beyond the transition temperature plays a key role in eliminating contaminants at the substrate surface when spraying in an air atmosphere or in a controlled atmosphere at atmospheric pressure. However, the preheating temperature must not exceed too much the transition temperature since at too high temperatures can modify the oxide layer thickness, its composition and also its Skewness [Cedelle J, Vardelle M, Fauchais P (2006)].

When the pressure is reduced as in soft vacuum spraying, a transition pressure also exists, as shown by [Fukumoto et al. (2007a, b, 2011) and Yang K. et al. (2009)]. Without preheating no chemical modification of the surface occurs when the pressure is lowered, and desorption is the only possible physical change occurring. However, transition

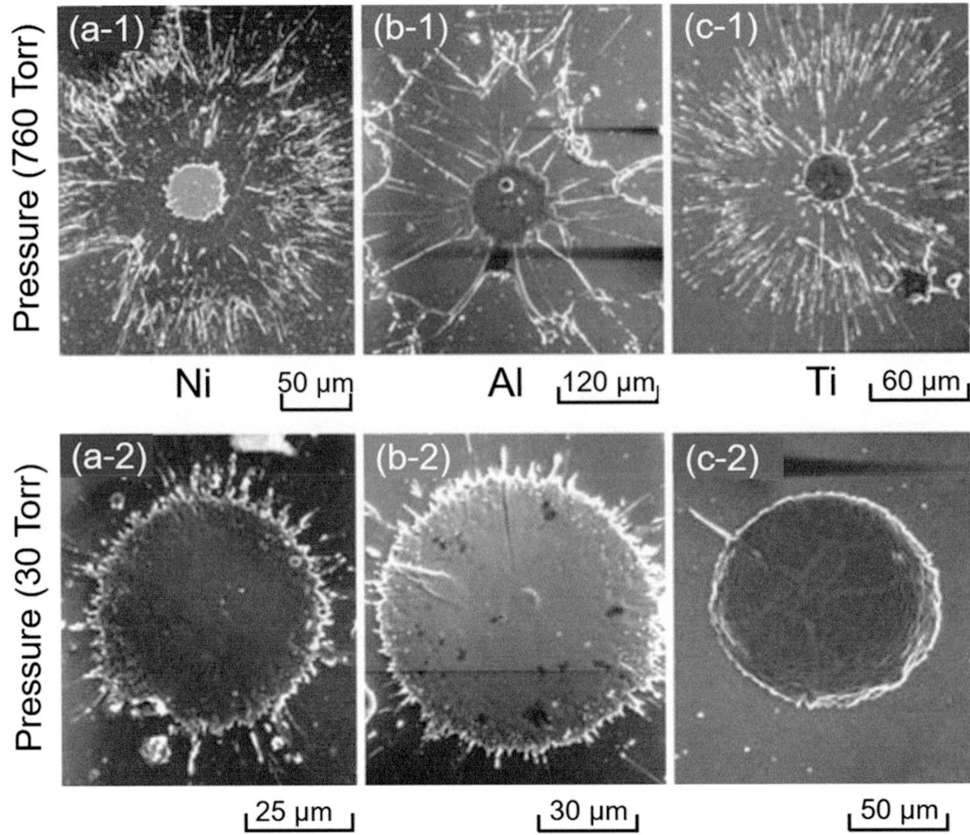

Fig. 15.41 Splat morphologies of metallic particles DC plasma Sprayed on AISI-304 stainless steel substrate at room temperature (300K) and ambient pressures of 760 and 30 Torr (101 and 4 kPa). (**a**) Ni (**b**) Al (**c**) Ti. ([Fukumoto M et al. (2007a)]. Reprinted with kind permission from ASM International)

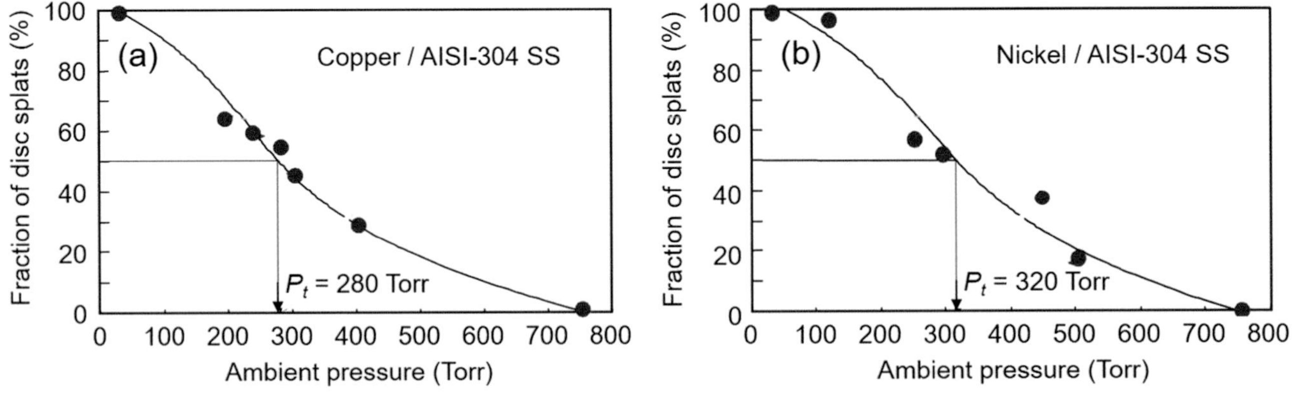

Fig. 15.42 Dependence of fraction of disc splats on ambient pressure for copper and nickel particles DC plasma sprayed on AISI-304 stainless steel substrate at room temperature (300 K). ([Fukumoto M et al. (2007a)]. Reprinted with kind permission from ASM International)

pressures for different materials do not follow the same classification as that obtained for transition temperatures. The differences are probably linked to the native oxide layers of the different substrates, which also depend on surface preparation. Electron micrographs of Ni, Al and Ti splats deposited by [Fukumoto et al. (2007a)] using DC plasma Spraying on polished AISI-304 SS substrate at room temperature at atmospheric, 101 kPa (760 Torr) and low pressure, 4 kPa (30 Torr) conditions are given in Fig. 15.41. These show that the disk splat appears easily by reducing the ambient pressure while the substrate temperature is kept constant at room temperature.

The transition behavior from the splash splat to the disk-shaped splat was recognized in most of the metallic materials by reducing the ambient pressure as shown in *Fig. 15.42* with the pressure at which the fraction of disk splat exceeds 50%

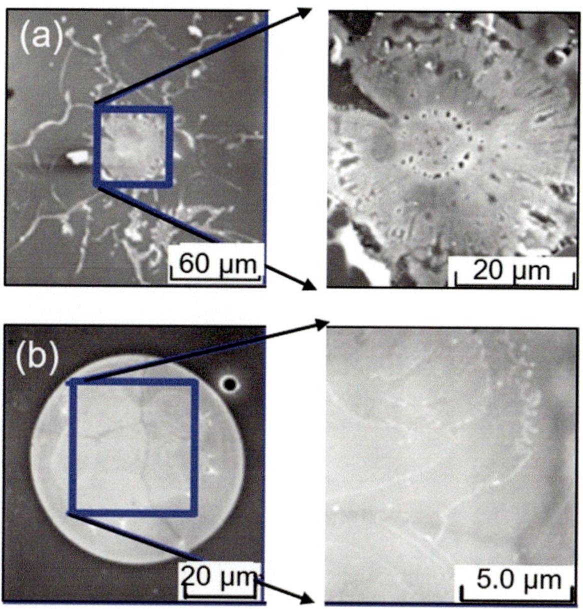

Fig. 15.43 Bottom view of Ni Splat morphologies DC plasma Sprayed on AISI-304 stainless steel substrate at different temperatures and ambient pressures (**a, b**) atmospheric pressure (760 Torr) with substrate temperatures of 300 and 673 K, respectively (**c, d**) room temperature (300 K) with ambient pressures of 760 and 30 Torr, respectively. ([Fukumoto M et al. (2007a)]. Reprinted with kind permission from ASM International)

defined by [Fukumoto et al. (2007a)] as the transition pressure, P_t. While the plasma length is easily changed by the chamber pressure, the effect of spraying distance on the fraction of disk splat appearance under different chamber pressures did not significantly change with the spraying distance in any chamber pressures. Observations of the bottom surface microstructure of Ni splats obtained at different substrate temperatures and ambient pressure are given in Fig. 15.43. These show similar tendency in microstructure changes of the contact surface at the center of the splat, from porous to dense associated with the increase in the surface temperate of the substrate from 300 to 673 K, at atmospheric pressure, Fig. 15.43a, b and the reduction of the ambient pressure from 760 to 30 Torr (101 kPa to 4 kPa) at room temperature, 300 K Fig. 15.43c, d. The effect seems to be related in both cases to the reduction in the vapor pressure of the adsorbates at the surface of the substrate that improves splat-substrate wetting during flattening, resulting in disk-like splats.

The transition mechanism proposed by [Fukumoto M and Y. Huang (1999)] for the modification of the splat patterns around the transition temperature attributes the formation of a splashed splats, when $T_s < T_t$, to the initial rapid solidification of the small central disk that occurs just after the impingement on the substrate as shown in Fig. 15.44a. As the molten liquid film does not spread by flowing on the substrate surface but by jetting away from the initial solidified central disk, the flow speed becomes higher in inverse proportion to the small consumption of the kinetic energy as the viscous one. On the other hand, when $T_s > T_t$,

because of the improved wettability at the splat/substrate interface, the flow tip of the splat spreads steadily and adheres to the substrate, as shown in Fig. 15.44b. In this case, the kinetic energy of the particle is consumed effectively as the viscous energy, and finally the splat flattens to a disk shape.

15.3.3.2 Droplet Flattening Ratio
The first studies on this topic used analytical expressions [Armster SQ et al. (2002)] such as those of [Madejski J (1976), Williams CA and Jones H (1975), Yoshida T et al. (1992) and Fantassi S et al. (1993)]. The analytical expression established by [Madejski J (1976)] assuming that the solidification started only when flattening was completed, related the flattening degree ξ_D defined by Eq. 15.4, to the particle Reynold, Re and Weber, We numbers at impact:

$$\frac{3\xi_D^2}{We} + \frac{1}{Re}\left(\frac{\xi_D}{1.2941}\right)^5 = 1 \tag{15.7}$$

Assuming that $We > 100$, which is the case for most thermal spray processes, at least at the beginning of the flattening process. The splat diameter D_s, assumed to be a disk, is linked to the impacting particle diameter d_p by an expression of the type:

$$D_s/d_p = A.\, Re^{0.2} \tag{15.8}$$

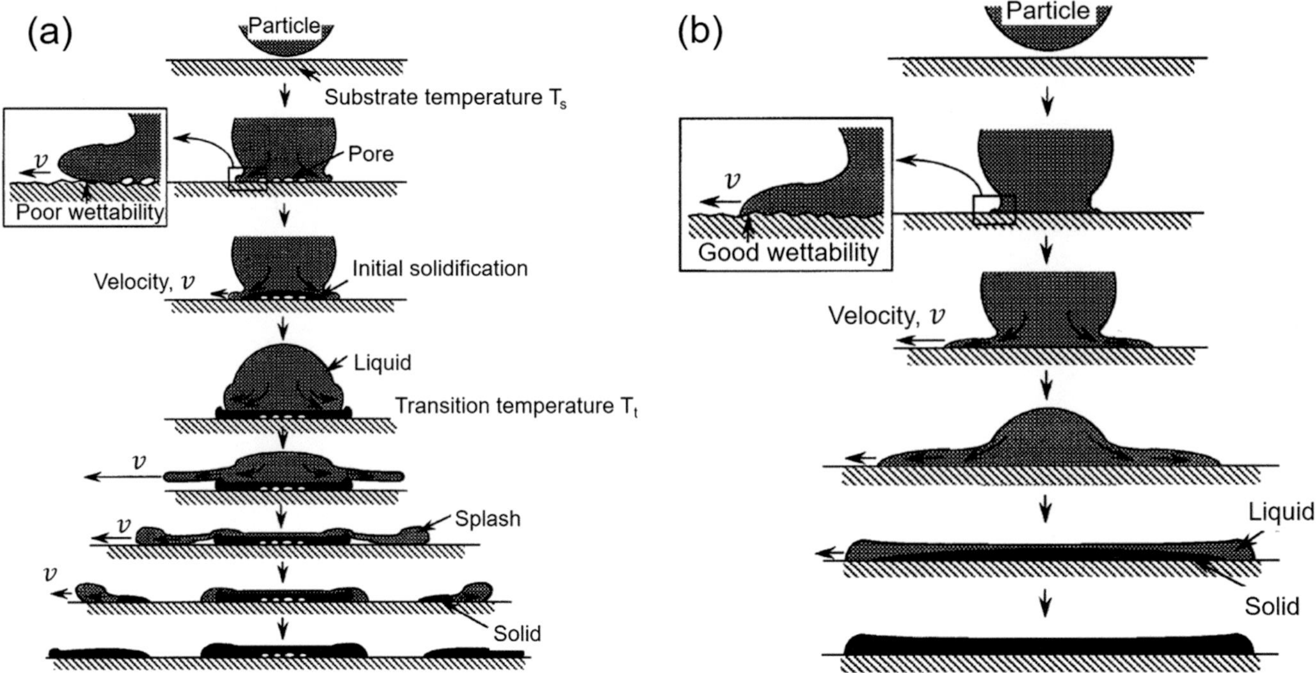

Fig. 15.44 Schematic of proposed splat formation model (**a**) Splash splat ($T_s < T_t$) (**b**) Disc splat ($T_s > T_t$). ([Fukumoto M and Y. Huang (1999)]. Reprinted with kind permission from ASM International)

Table 15.3 Flattening degree and splat thickness of zirconia particles 30 μm in diameter, and temperature of 3200 K, impacting on a smooth substrate with different velocities

Impact velocity, v_{pi} (m/s)	25	135	200	400
Flattening degree, ξ_D	3.6	5.0	5.5	6.3
Splat thickness, δ_s (μm)	2.4	1.2	1.04	0.8

where A is a constant with a value between 1.29 for a perfect disk shape according to [Madejski J (1976)] and 0.83 for extensively fingered splats [Williams CA and Jones H (1975)] where only the central part of the splat is considered being a disk. The flattening degree increases with impact velocity for the same temperature, and the splat thickness, δ_s, diminishes accordingly. The thickness can be calculated by assuming that the liquid spherical particle of diameter d_p is transformed into a perfect disk, with a volume of $\left(\pi \frac{D_s^2 \delta_s}{4} \right)$.

Table 15.3 gives a few values of flattening degree and splat thickness for 30 μm diameter zirconia particles with a temperature of 3200 K impacting with different impact velocities. Impact parameters, however, are not the only parameters that have an influence on particle flattening. Once the droplet has started to flatten, the substrate parameters play a key role. These comprise the substrate material, its mechanical and thermal properties, especially its diffusivity, its oxide layer characteristics (if any), its roughness, and its surface temperature to which reflect the presence of adsorbates and condensates strongly

depends. The particle flattening time varies between a few microseconds to a few tenths of microseconds depending on the particle impact velocity (50 m/s $< v_{pi} <$ 400 m/s).

Assuming that all adsorbates and condensates have been eliminated before the droplet impact, its flattening depends principally on the wettability between droplet and substrate. Generally, wettability of a liquid metals is best on the solid phase of the same material that is the case for a molten droplet impacting on previously deposited layers. Wettability, however, also depends in a complex way on the substrate surface roughness. Generally increasing surface roughness hinders the liquid spreading and splats are smaller than on the same smooth surface [Fauchais P et al. (2004), Fukumoto M et al. (2005), Fukumoto M et al. (2006a, b, 2007a) and Chandra S and Fauchais P (2009)]. The wettability also depends on the substrate thermal conductivity, as shown by [Tanaka and Fukumoto (1999)] for ceramic particles impacting different substrates. The higher the thermal conductivity of the substrate is, the poorer is its wettability.

An oxidized substrate surface offers better wettability to an oxidized particle than a nonoxidized particle. The composition, thickness, and morphology of the oxide layer on the surface of the substrate, which are all dependent on the preheating temperature, preheating time and rate, plays a key role on the wettability of the substrate surface. As shown by [Cedelle et al. (2006)], preheating 304L stainless steel substrate increases its roughness with its skewness becoming positive at the nanometer scale and its wettability is improved. The thickness of the oxide layer can in turn modify its wettability through the increased surface roughness [Chandra S and Fauchais P (2009) and Fukumoto M et al. (2006a, b)].

To account for the substrate-flattening droplet wettability variation [Tanaka Y and Fukumoto M (1999)] proposed the use a new criterion K_f, based on the Sommerfeld parameter, Ks (Eq. 15.6) using the maximum flattening velocity v_f and splat thickness δ_s instead of the impact velocity v_{pi} and droplet initial diameter d_p.

$$K_f = \left(\frac{\rho \cdot v_f^2 \cdot \delta_s}{\sigma} \right)^{0.5} \left(\frac{\rho \cdot v_f \cdot \delta_s}{\mu} \right)^{0.25} \qquad (15.9)$$

The main challenge for the use of this equation is the evaluation of v_f and δ_s for micron-sized particles. [Goutier et al. (2011)] points out that while measurement of the flattening velocity of millimeter-sized drops is relatively accessible, that of micrometer-sized droplets is complex because of the difficulty of following the flattening of such small particles in real time. The proposed approach is to determine the flattening time, through measurement of the time necessary to reach the maximum flattening diameter, assuming negligible cooling of the droplet during its flattening. Knowing the maximum flattening diameter, $D_{s,\,max}$, and flattening time, t_f [Goutier et al. (2011)] it is possible to calculate the mean flattening velocity, $\bar{v}_f$ as;

$$\bar{v}_f = D_{s,\,max} / 2\, t_f \qquad (15.10)$$

Typical results for the flattening of NiAl alloy and Alumina (Al_2O_3) particles are given in Fig. 15.45 for 5 mm drops and 40 μm droplets [Goutier S (2010)]. These are presented in terms of the maximum droplet flattening ratio, $\xi_{D.\,max}$ as function of the Reynolds number and the Weber number, evaluated using the maximum splat radius, $r_{s,\,max} = D_{s,\,max}/2$ and the mean flattening velocity $\bar{v}_f$ where;

$$Re = \left(\rho\, \bar{v}_f\, rs,\, max / \mu \right) \qquad (15.11)$$

$$We = \left(\rho\, \bar{v}_f^2\, rs,\, max / \sigma \right) \qquad (15.12)$$

As noted from Fig. 15.45a, the results obtained with 5 mm drops of metal alloy or ceramic show the maximum splat flattening ratio to have marginal dependence on the Reynold number compared to the corresponding 45 μm droplets. In contrast, the situation is significantly different when the results for the drops and droplets are plotted as function of the Weber number, Fig. 15.45b, where they all follow the same trend. This seems to indicate that the splat flattening is controlled by the competition between the surface tension

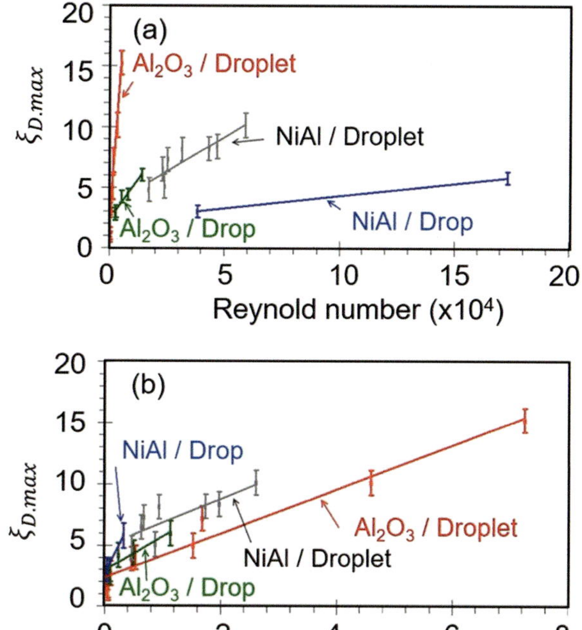

Fig. 15.45 Evolutions of the maximum splat flattening ratio for metal alloy and ceramic, 5 mm drops and 45 μm droplets with (**a**) Reynold number (**b**) Weber number. ([Goutier, S. (2010)]. Reprinted with kind permission)

and inertia forces, and that the effects of the surface tension are preserved during the modification of the particle scale, due to the weak dependence of the surface tension on the liquid temperature during its flattening and cooling stages. This is not the case for the viscous effects, which are strongly dependent on liquid temperature variation in the drop/droplet flattening stag.

Clearly, much effort is still necessary to better understand the flattening phenomena specially to link the previous results to the solidification behavior that was neglected in this study.

15.3.3.3 Droplet Cooling and Solidification

The material solidification depends on splat thickness, thermal diffusivities of both sprayed material and underlying solid layer, and the quality of contact between the latter and the flattening droplet. Unfortunately, the quality of contact at the interface is a function of the particle impact pressure that varies significantly and nonuniformly along the contact surface during impact (see Fig. 15.34). Here again the higher the impact velocity will be, the higher will be the impact pressure and the smaller the thickness of the flattening droplet. Thus, high impact velocities improve the flattening particle solidification. Compared to flattening times, solidification times are

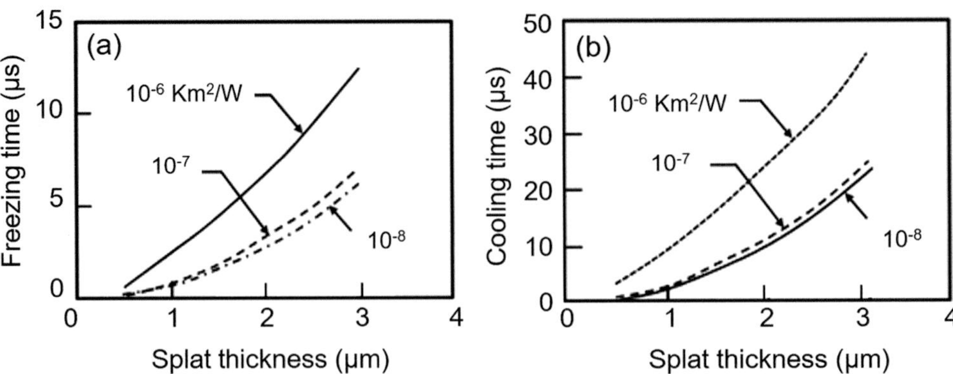

Fig. 15.46 (**a**) Freezing and (**b**) cooling time of disk-shaped zirconia splat deposited on a 304L stainless steel substrate as function of the splat thickness for three different thermal contact resistances. ([Vardelle M et al. (1994)]. Reprinted with kind permission from Springer Science Business Media, copyright © ASM International)

5–10 times longer, but generally solidification starts before flattening is terminated. When solidification starts it can modify the liquid flow at the surface of the flattening droplet. The contact surface of the splat will never represent the whole splat surface, a very good contact corresponds to 60–70%, and a poor one 10–20% [Fauchais P et al. (2004)]. Even when the contact is good, at the splat rim, where the contact pressure is rather low and the surface tension at its maximum, splat curling occurs with the formation of a rounded rim due to the surface tension effects [Xue M et al. (2006)]. The contact of the external rim of the flattening droplet with the substrate is very poor, thereby inducing a slower liquid cooling through the already solidified part of the splat rather than directly toward the substrate or the previously deposited coating layers. Here again, the higher the impact velocity, the smaller the flattening droplet thickness will be due to the impact pressure on the droplet.

To illustrate the influence of two critical parameters, contact resistance between splat and substrate, and splat thickness, results of a simple 1-D transient modeling work [Fauchais P et al. (2004)] are presented. In this study the cooling of a molten liquid splat in direct contact with the substrate is followed through its freezing and subsequent cooling down stages until its temperature reaches that of the substrate. The disk-shaped splat is assumed to have a uniformly contact with the substrate with a thermal contact resistance R_{th} that is either low ($R_{th} = 10^{-7}$ or 10^{-8} m^2 K/W corresponding to a perfect contact, or high, $R_{th} = 10^{-6}$ m^2 K/W corresponding to a relatively poor contact. It must be emphasized that the assumption of a uniform and constant thermal contact resistance to thermally characterize the splat-substrate interface is an over-simplification since, as previously mentioned, R_{th} varies normally with time and the radius of the flattening particle. The evolution of the splat freezing time with the splat thickness is given in Fig. 15.46a as function of the contact resistance (solidification being supposed to occur at melting temperature). The corresponding

cooling time, over which the splat reaches the substrate temperature in given in Fig. 15.46b. As expected, freezing and cooling times of a splat increases rapidly with the splat thickness. The results obtained with $R_{th} = 10^{-7}$ or 10^{-8} m^2 K/W are almost the same, the difference being important for $R_{th} = 10^{-6}$ m^2 K/W. It is important to note that the flattening particle and resulting splat cooling depend essentially on the heat conduction through the substrate or the previously deposited layers. Compared to convective cooling, even when boosted by blowing air jets at the top surface of the splat, conduction accounts for 90% and 95% of the heat loss from the splat.

Because of the relatively low thermal diffusivity of zirconia, important internal temperature gradients can develop between the top surface and the bottom of the splat in direct contact with the substrate. Results obtained in the same study [Vardelle M et al. (1994)] given in Fig. 15.47a, show in the case of a low the thermal contact resistance R_{th} the temperature of the bottom layer of the splat, in close to that of the substrate has essentially the same temperature as that of the substrate that is heated by the flattening particle. On the other hand, in the case of high thermal contact resistance, $R_{th,}$ the substrate is poorly heated and the splat bottom temperature is close to that of its top as illustrated in Fig. 15.47b.

When solidification starts it can, especially for low thermal contact resistances, modify the liquid flow at the flattening droplet surface. It is important to keep in mind that the thermomechanical properties of coatings depend strongly on the surface contact between splats and substrate and between themselves. In areas of good contact, depending on the impacting droplet temperature, substrate diffusivity, oxide layer thickness and composition, and dimension of asperities, some melting can occur at the substrate surface, but the thicknesses involved are relatively small (few tens of nanometers) [Brossard S et al. (2010)]. This phenomenon is generally not sufficient to achieve good adhesion. Heterogeneous nucleation takes place at temperatures lower than the

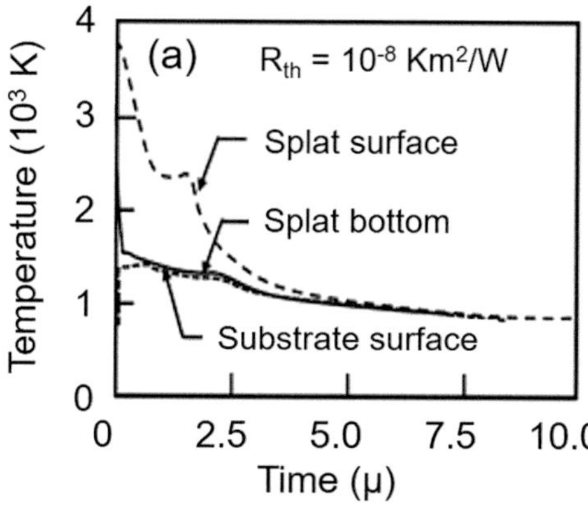

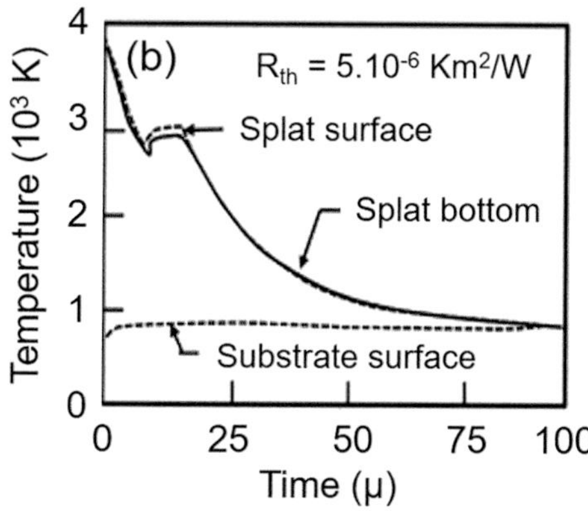

Fig. 15.47 Time dependence of the top and bottom temperatures of disk-shaped zirconia splats plasma sprayed on 304L stainless steel, as well as substrate surface temperature for two thermal contact resistances **(a)** $R_{th} = 10^{-8}$ m² K/W, **(b)** $R_{th} = 5 \times 10^{-6}$ m² K/W. ([Vardelle M et al. (1994)]. Reprinted with the kind permission of Prof. A. Vardelle)

melting one (undercooling) and depends on many parameters, among them the contact angle liquid-substrate (wettability) and the thermal contact resistance. Unfortunately for particle flattening these data are impossible to measure and to the best of our knowledge heterogeneous nucleation has been taken into account only in 1-D models [Lahmar-Mebdoua Y et al. (2009)]. Results show that the cooling curves obtained are not very different from those shown in Figs. 15.46 and 15.47 except for temperatures lower than 100–200 K.

Experimental studies of particles impacting on a surface preheated above T_t have shown that their cooling rates can be as high as two orders of magnitude faster than that on a cold surface. A high cooling rate corresponds to low thermal contact resistance at the splat–substrate interface. For example [McDonald A et al. (2007a, b, c)] plasma sprayed molybdenum and yttria-stabilized zirconia particles onto glass and Inconel 625 substrates held at either room temperature or 400 °C. Samples of Inconel 625 were also preheated for 3 h and then air cooled to room temperature before spraying. The analysis showed that the thermal contact resistance between the heated or preheated surfaces and splats was more than an order of magnitude smaller than that on nonheated surfaces held at room temperature (see Table 15.4). Particles impacting on the heated or preheated surfaces had cooling rates that were significantly faster than those on surfaces held at Room Temperature (RT), which was attributed to smaller thermal contact resistance.

Flattening phenomena are strongly linked to thermal phenomena, including contaminants, controlling the real thermal contact between flattening particle and substrate. Thus [Dhiman et al. (2007)] proposed a single parameter to estimate the importance of freezing during solidification and predict the likelihood of splat break-up. When a molten droplet lands on a solid surface, it spreads into a thin splat of uniform thickness δ_s. If the substrate is at a temperature lower than the melting point of the droplet, a solid layer of

Table 15.4 Thermal contact resistances measured for different particle materials impacting onto different substrates at room temperature or preheated over the transition temperature [McDonald A et al. (2007a, b, c)]

Material/substrate	Substrate temperature (°C)	$R_{th} \times 10^7$ (m² K/W)
Mo/glass	27	490 ± 55
	400	6.5 ± 1.0
Mo/Inconel	27	190 ± 15
	400	12 ± 2.0
	Preheated–cooled to RT	55 ± 10
Zirconia/glass	27	220 ± 30
	400	10 ± 3.0

Reprinted with kind permission from International Journal of Heat and Mass Transfer

thickness δ_{so} grows in the flattening particle from its bottom during the time it takes to reach its maximum spread. The solidification parameter (ϑ_s) is defined as the ratio of the solid layer thickness to splat thickness:

$$\vartheta_s = \delta_{so}/\delta_s \tag{15.13}$$

[Dhiman et al. (2007)] developed an analytical and complex expression to calculate the value of ϑ_s as a function of the dimensionless parameters Re, We, Ste, Pe, and Bi. It is important to note that the splat-substrate thermal contact resistance is included in the Biot number. The magnitude of ϑ_s can be used to predict what the final shape of the splat will be, and what the mechanism of the break-up is, if it occurs. Three outcomes are possible during spreading:

(a) If the solid layer is very thin ($\delta_{so} \ll \delta_s$, or $\vartheta_s \ll 1$), it will have no effect on spreading. The splat will spread into a thin liquid sheet and ruptures internally. This will lead to extensive fragmentation, producing a small central splat surrounded by a ring of debris Fig. 15.48a. It is the case when adsorbates and condensates are present.

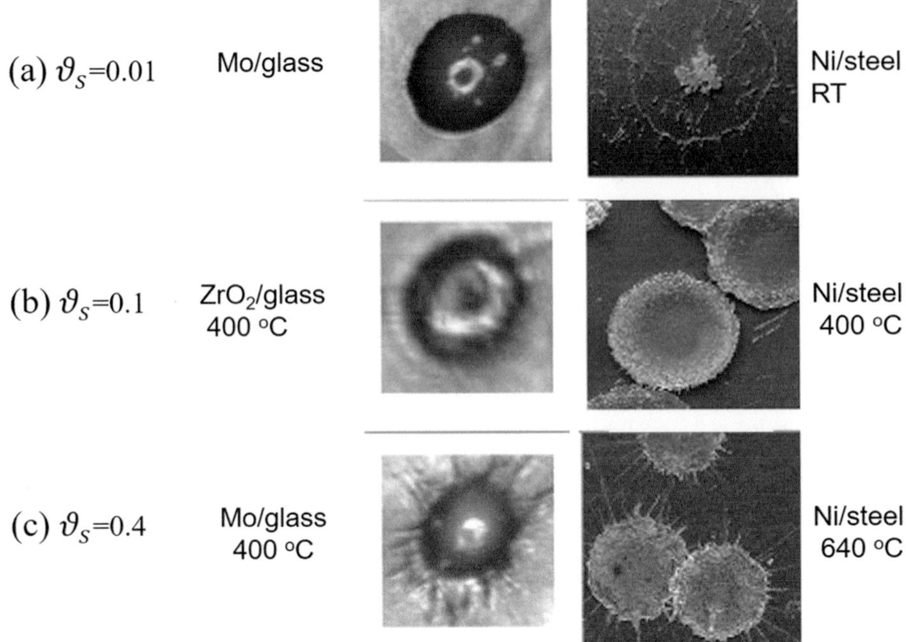

Fig. 15.48 Photographs during and after impact for splats (**a**) Fragmenting during impact, (**b**) forming disk splats, and (**c**) undergoing freezing-induced break-up. ([Dhiman et al. (2007)]. Reprinted with kind permission from Elsevier)

(b) If the solid layer grows by a significant amount during spreading ($\delta_{so} \sim 0.1\,\delta_h$ to $0.3\,\delta_h$ or $\vartheta \sim 0.1$ to 0.3), it will restrain the splat from spreading too far and becoming thin enough to rupture. In this case, a disk-type shape splat will be produced see Fig. 15.48b.

(c) If solidification is very rapid ($\delta_{so} \sim \delta_h$, or $\vartheta \sim 1$), the solid layer will obstruct the outward spreading liquid. A splat with fingers radiating out from its periphery will be produced, see Fig. 15.48c.

Comparison with experimental photographs showed that the value of the solidification parameter gave a reasonably accurate method of predicting the shape of the final splat. Figure 15.48 shows photographs of splats taken both during and after impact, illustrating the three different modes of splat impact. In Fig. 15.48a, for molybdenum and nickel particles landing on substrates at room temperature, thermal contact resistance was high ($\sim 10^{-5}$ m^2 K/W) and $\vartheta_s \sim 0.01$. The splats spread into a thin film that ruptured internally and fragmented. The final splats all showed a central portion at the point of impact that adhered strongly to the substrate, surrounded by a ring.

Increasing the substrate temperature to values above the transition temperature reduces the thermal contact resistance by an order of magnitude, since it evaporated adsorbed contaminants on the surface. Figure 15.48b shows impacts of zirconia and nickel particles on surfaces heated to 400 °C that had $R_{th} \sim 10^{-6}$ m^2 K/W and correspondingly $\vartheta_s \sim 0.1$.

Solidification occurs near the end of droplet flattening, when the spreading liquid has not enough momentum to jet over the solid rim and instead came to rest forming a circular splat with smooth edges.

If ϑ_s is increased further ($\vartheta_s \sim 0.4$, see Fig. 15.48c), the solid layer growth is sufficiently rapid to obstruct the flow of the liquid early during spreading. The liquid has enough momentum that it jetted outward, producing fingers radiating out from the central splat. For nickel particles spreading on a steel substrate oxidized by heating to 640 °C (Fig. 15.48c) the splat is intact and smooth at the center, where solidification is slow. The edges, which solidify very rapidly, have a rough surface since surface tension has not time to level irregularities before solidification occurred. Molybdenum splats on a glass surface heated to 400 °C also have fingers radiating out.

Finally, it must be mentioned that for splats collected on smooth surfaces kept at room temperature, all the splashed material or droplets or fingers collected around their central part (see, for example, Fig. 15.48a) have very poor adhesion and can be removed easily by the tip of the roughness gauge. This means that any new splat deposited on these dispersed materials will have a lower adhesion than it would have had on a clean surface. In contrast, splats obtained on surfaces preheated over the transition temperature have a much better adherence and cannot be separated by the tip of the roughness gauge.

15.3.3.4 Splat Fragmentation

Fragmentation occurs late during impact, when the particle is almost completely flattened, and both flow and solidification

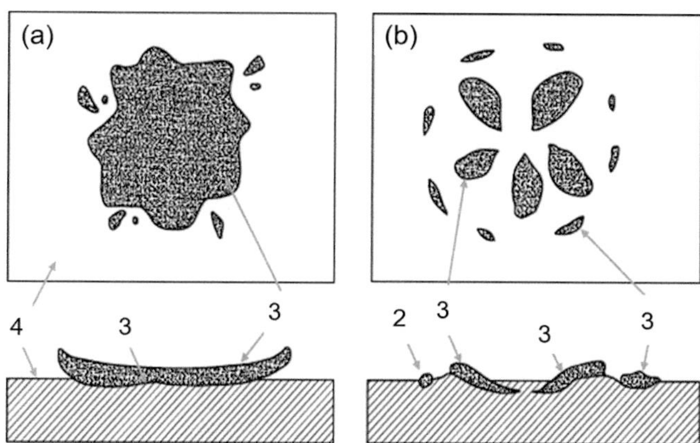

Fig. 15.49 Two morphological forms of lamellae splashed on the substrate: (**a**) pancake, (**b**) flower, (1) crack, (2) deformed substrate, (3) corona, (4) substrate, (5) lamella, (I) top view, (II) X-section view. ([Sidhu TS et al. (2005)]. Reprinted with kind permission from Springer Science Business Media)

phenomena are involved. Generally, particles sprayed onto a substrate held at room temperature form splats that are extensively fragmented, while preheating the surface above the transition temperature, T_t, results in disk-shaped splats. T_t is much lower than the melting temperatures of either the substrate or the impacting particle and seems to correspond to the evaporation of contaminants. The splat fragmentation can be linked to instabilities at the edge of the liquid sheet rapidly decelerating and becoming unstable [Goutier S et al. (2011)].

In the course of the flattening of the droplets on impact with the substrate, cooling and solidification of the liquid phase starts its periphery forcing the outward flowing liquid to jet over the top of the formed solid layer, becoming unstable and breaking up [Dhimam et al. (2007)].

Typically, the flattening of the droplets to their maximum diameter is completed within a few μs after impact, forming a thin sheet with a diameter up to 10x that of their initial droplet, and a thickness of only 0.5 μm [McDonald A et al. (2006, 2007a, b, c)]. The liquid sheet then begins to rupture, initially around the solidified central core and later at other sites. At the end, splats disintegrate almost completely and only a central solidified core remains at the substrate surface.

With HVOF spraying, initially the particle is melted and, during its first contact with the substrate, the impact creates a shock wave inside the lamella and in the substrate. After impact the flattening particle forms either a "pancake-"shaped lamella or splat (associated with moderate particle velocities and moderate heat contents) shown in Fig. 15.49a or a flower-type lamella (connected with an elevated velocity of the particle and an elevated heat content) as shown in Fig. 15.49b [Sidhu TS et al. (2005)]. Although the authors do not discuss at all the substrate preheating effect, Fig. 15.49b represents well the types of splats and the substrate deformation observed with HVOF when particles are

fully melted. [Fukumoto et al. (2004)] have observed the same preheating effect on plasma and HVOF-sprayed splats (see Table 15.4).

Similar results were obtained with D-gun spraying, with which, depending on spray conditions sprayed particles can be fully or partially melted upon impact. [Ulianitsky et al. (2011)] sprayed with Computer Controlled Detonation Spraying (CCDS) device CCD-S2000, working with different gas mixtures, Ni and Cu particles, which velocities and temperatures were measured or calculated in-flight prior to impact. Results are presented in Fig. 15.50. Ni splat in Fig. 15.50b was in a semi-melted state at impact, its temperature being close to the melting one ($T_m = 1726$ K). Figure 15.50a corresponds to an initially solid Ni particle, which becomes melted at the interface upon its impact with the substrate due to kinetic energy release. Figure 15.50c corresponds to splashing for the fully melted Ni splat. With overheated Cu splat ($T_m = 1358$ K, the particle impacting at 2200 K) presented in Fig. 15.50c adsorbates and condensates have resulted in the liquid fast expansion and a result similar to that presented in Fig. 15.31a, b. For YSZ particles, to our best knowledge, no splats were presented, but the fractured YSZ topcoat obtained with D-gun spraying shows that the coating is made of layered splats.

Fully or partially melted cermet particles have a behavior different from that of metal, alloy, or ceramic particles. Upon impact the metal or alloy matrix is melted or in a plastic state, while the ceramic particles imbedded are not melted at all. HVOF or HVAF or D-gun processes are spraying most of the cermet coatings, and the impact of cermet particles will be illustrated with the work of [Watanabe et al. (2006)]. They have sprayed WC-12 wt.% Co on carbon steel (0.45% C) substrates using the HVOF JP-5000 torch from Praxair Technology Inc. (USA). The substrate surface was mirror-finished by mechanical polishing before spraying. Six commercially available WC–Co powders were sprayed with kerosene (0.38 ℓ/min) and oxygen (944 slm). These powders had similar size distributions of 15–45 μm with the same composition but differed in the size distributions of the WC particles contained in one composite WC–12% Co powder particle. Table 15.5 presents their classification.

Splats and coatings of the six types of WC-12 wt.% Co were investigated. The interface fracture toughness was evaluated by a pre-notched four-point bending test, and the microstructure of the splat around the interface was examined by SEM and FIB cross-sectioning. Figure 15.51a, b. shows images of the disc-shaped type A and type E splats, and their respective cross sections along the dotted lines in Fig. 15.51c, d. These images were obtained by sectioning splats with FIB equipment and then observing them with SEM. EDX analysis revealed that splats consisted of the W–Co metal binder phases and the carbides. XRD analysis of the coatings showed the crystalline peaks of WC, W_2C, and W. There is

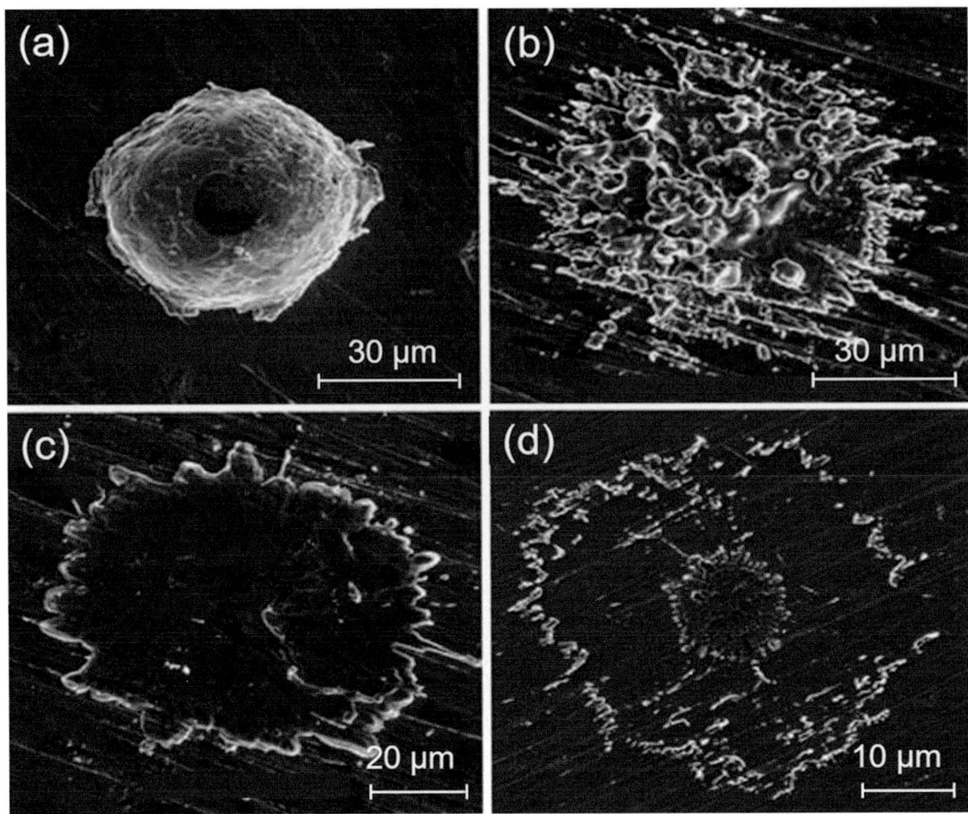

Fig. 15.50 Splat morphology of particles D-gun sprayed at different v_p & T_p collision conditions: Calculated (v_p & T_p) values: (**a**) Ni: $v_p \approx 360$ m/s, $T_p \approx 1450$ K; (**b**) Ni: $v_p \approx 420$ m/s, $T_p \approx 1700$ K; (**c**) Ni: $v_p \approx 460$ m/s, $T_p \approx 2{,}200$ K, (**d**) Cu: $T_p \approx 2200$ K. ([Sidhu TS et al. (2005)]. Reprinted with kind permission from Springer Science Business Media, copyright © ASM International)

Table 15.5 WC particle size distributions in WC-12% Co particles (15–45 μm)

Type	A	B	C	D	E	F
WC size (μm)	0.2	1.0 ~ 1.5	2.0 ~ 2.5	3.0 ~ 4.0	5.0 ~ 7.0	1.0 ~ 5.0

[Watanabe et al. (2006)]. Reprinted with kind permission from Elsevier

a clear tendency of the increase in W_2C peak as the carbide size decreases. Interfaces of both splats display concave peripheries, implying substrate deformation by the high kinetic energy of the molten particles upon impact.

The maximum thickness of the splats was about 4 μm for all of them, but the type A splat had more uniform thickness over the splat. In contrast, a cross section of the type E splat indicates a nonuniform feature due to the existence of large particles. The smaller WC particle at the interface in the type E splats was compressed by the larger WC particle just above it and imbedded into the substrate. In contrast, no such carbides intruded into the substrate in type A and thus the interface is much smoother than that of type E coatings. The WC particle size in the original powders was found to significantly affect the interface fracture toughness, which generally increases as the WC particle size in the original powder increases. The key factors in realizing the excellent adhesion of this coating system is considered to be compression and densification of the surrounding microstructure by the impact

of WC particles, intrusion of the particles into the substrate, and plastic deformation of W–Co alloy phases around a crack tip, which functions as the energy dissipation mechanism for crack propagation in the coating [McDonald A et al. (2007a, b, c)].

15.3.4 Splat Formation of Polymer Particles

When molten polymer droplets hit the surface, as for droplets of conventional materials, they solidify into distinct "frozen" splats. An example of several Nylon-11 ($T_m \approx 190$–260 °C) splats deposited on a glass substrate using the HVOF spray process is shown in Fig. 15.52 [Isocevic et al. (2006a, b)]. The quasi-small spheres observed, probably correspond to small particles traveling in the jet fringes and sufficiently heated to stick on the substrate. The "fried-egg" shaped splats correspond to bigger partially molten particles with a molten shell surrounding its nonmelted core. This feature is a direct

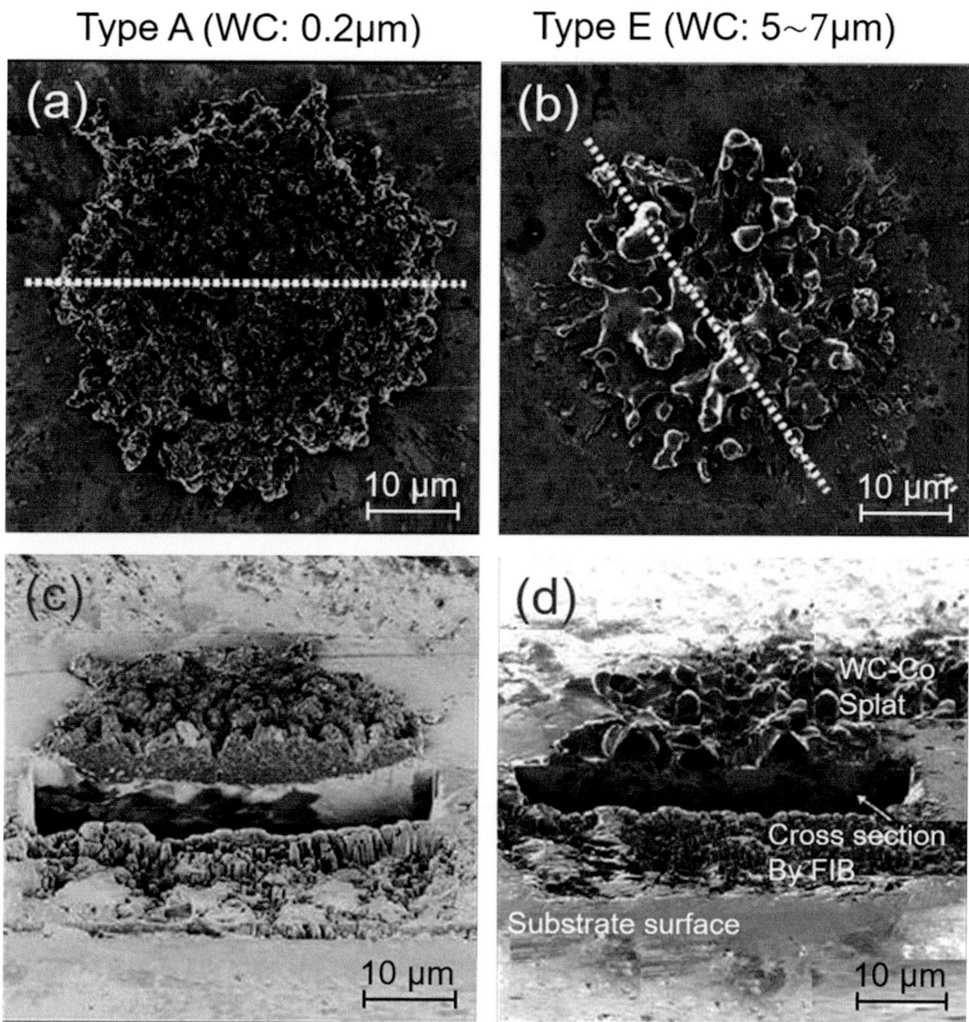

Fig. 15.51 SEM and FIB images of WC–2 wt.% Co HVOF sprayed on carbon steel (0.45% C) substrates: type A and type E splats in top view (**a, b**), and in a cross section of the splats (**c, d**). ([Watanabe et al. (2006)]. Reprinted with kind permission from Elsevier)

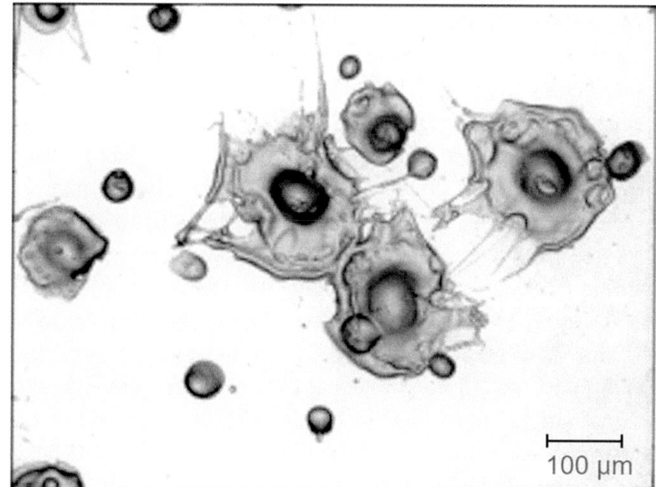

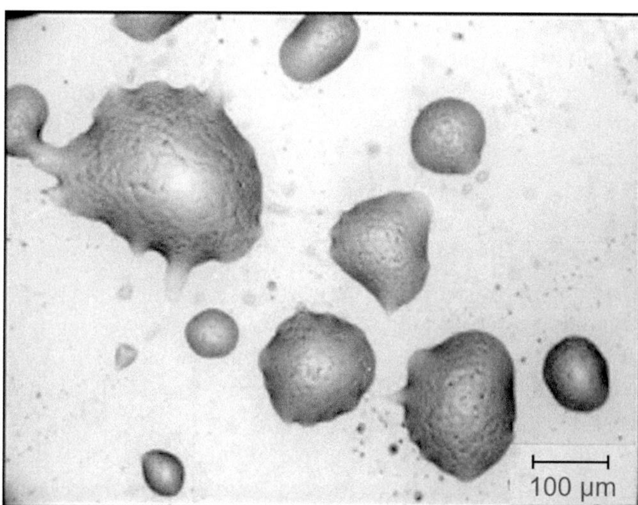

Fig. 15.52 Nylon-11 splats sprayed using HVOF on glass substrate at room temperature. (Reprinted with kind permission from Elsevier [Isocevic et al. (2006a, b)])

Fig. 15.53 Nylon-11 splats sprayed using HVOF on glass substrate preheated to ~190 °C and cooled down to room temperature. (Reprinted with kind permission from Elsevier [Isocevic et al. (2006a, b)])

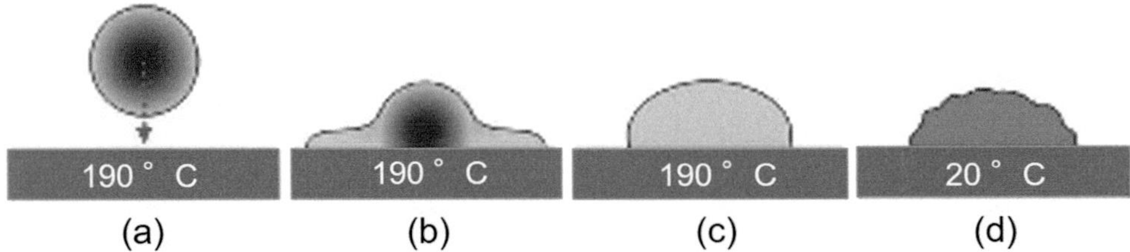

Fig. 15.54 Schematic of partially molten polymer particle transformation following its impact on preheated glass substrate (**a**) particle prior to its impact (**b**) "fried-egg" shaped splat (**c**) postdeposition particle melting and flowing (**d**) droplet shrinkage during its final cooling down to room temperature. (Reprinted with kind permission from Int. J. Heat and Mass Transfer [Isocevic et al. (2006a, b)])

consequence of the poor thermal conductivity of the polymer particles ($\kappa \approx 0.23–0.34$ W/m K) giving rise to important internal temperature gradients in the particle, with a low temperature, high viscosity core, surrounded by the high temperature, low viscosity shell.

When the substrate was preheated to a temperature around the melting temperature of Nylon-11 ($T \sim 190$ °C) splats exhibited a flattened hemispherical shape, as shown in Fig. 15.53 [Isocevic et al. (2006a, b)]. This was attributed to the postdeposition flow activated primarily by surface tension, after the initial "fried-egg" splats were fully melted by the preheated substrate, shown schematically in *Fig. 15.54* [Ivosevic M (2006)].

[Isocevic et al. (2006a, b)] have modeled HVOF spraying of Nylon-11. Their heat and mass transfer equations were identical to those used with conventional particles taking into account the heat propagation phenomenon in the polymer particle. Molten nylon 11 was modeled as a generalized Newtonian fluid with temperature and shear rate dependent viscosity. Predictions, in good agreement with experiments showed that larger polymer particles, 90–120 μm in diameter, develop steep internal temperature gradients between the core and the surface during HVOF spraying. They conclude by that optimal thermal spray process for polymers spraying should utilize a low gas temperature ($\sim 400–500$ °C) to prevent overheating of the polymer particles, a high gas velocity to provide the necessary high kinetic energy for spreading of the high melt viscosity particles, and a sufficiently long dwell time for uniform particle melting.

To the best of our knowledge very few studies have been devoted to polymer splat formation on rough surfaces. [Ivosevic et al. (2006a, b)] studied HVOF-sprayed Nylon-11 particles onto 4140 steel grit blasted with alumina grits of different sizes. As noted in Fig. 15.55 the increase in substrate roughness promoted splat instability, resulting in radial jetting and breakup, and producing more irregularly shaped splat shapes on rougher surfaces.

PolyPropylene (PP) splats were sprayed onto a glass substrate at room temperature using the flame spray process, at various Standoff Distances (SODs) [Alamara K et al. (2011)]. The results of this study indicated that increasing the SOD from 10 cm to 25 cm reduced the splat thickness and

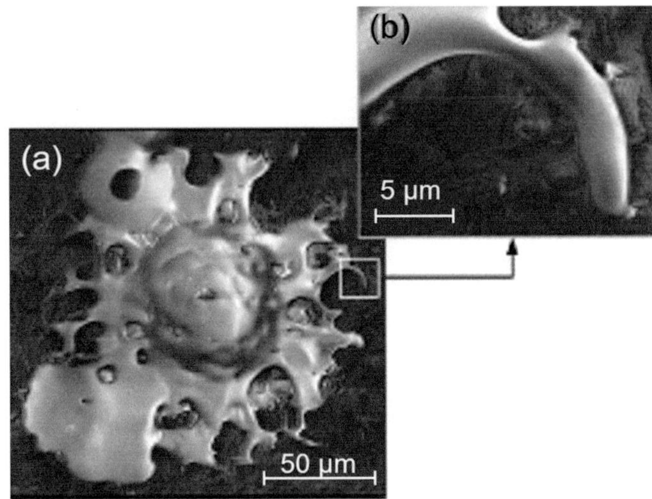

Fig. 15.55 (**a**) Nylon 11 splat on a grit-blasted steel substrate (**b**) Enlarged view of a peripheral splat finger. (Reprinted with kind permission from Springer Science Business Media [Ivosevic et al. (2006a, b)], copyright © ASM International)

deposition efficiency (as in previous cases splats were "fried egg" shaped), though an increase in splash area was demonstrated. Splat circularity was steady around 0.9, indicating that the deposited particles exhibited a circular shape at all SODs.

[Alamara et al. (2010)] flame sprayed PolyEther-EtherKetone (PEEK) ($T_m \approx 341$ °C) with radial injection collecting the molten particles splats on a glass substrate at room temperature. PEEK splats had "fried-egg" shape and, depending on the SOD, presented either disk-shaped or splashed patterns. Decreasing the SOD from 350 to 150 mm produced coherent, integral disk-shaped splats that exhibited minimum splashing behavior. [Withy et al. (2008)] HVAF sprayed PEEK particles on aluminum substrates with six different pretreatments. Surface morphology was the main factor found to influence the splat area density, the area of single splats, and the circularity of single splats on substrates maintained at 323 °C. Substrate surface chemistry was however found to have an influence on the area density of deposited single splats, where surface morphology alone cannot explain the significant differences between the six different pretreated substrates.

The thermal degradation of PEEK particles HVAF sprayed was the subject of study by [Patel et al. (2010a, b)] who observed that flame impingement onto the just deposited coating was partially responsible for the degradation of the molecular structure of deposited PEEK coating. When there is significant flame impingement onto the just deposited coating, as with the coatings produced at a SOD of 200 mm hydrocarbons from the flame react with the coating surface.

15.3.5 Effect of Impact Angle on Splat Formation

With an inclined substrate, the impact velocity, v_{pi} is the velocity component of the particle normal surface of the substrate, $v_{pi} = v_p \cdot \cos\varphi$, while the tangential velocity $v_{pt} = v_p \cdot \sin\varphi$ where φ is the angle between the particle trajectory and the axis normal to the surface of the substrate as shown in Fig. 15.56a. Upon impact, as with substrates normal to the particle velocity vector, impact splashing occurs with the tiny generated droplets centered in the direction symmetrical to the impact velocity relative to the normal, as shown in Fig. 15.55b.

On cold substrates, splats are extensively fingered, especially in the substrate inclination direction as shown in Fig. 15.57a. [Bianchi L et al. (1977), Kang CW and Ng HW (2006), Salimijazi HR et al. (2007) and Kang CW et al. (2006)] have shown that on stainless steel, copper and a few other polished substrates ($R_a < 0.1\,\mu m$) preheated over the transition temperature, when the spray angle, φ, increases from $0°$ to $75°$, splats have an elliptical shape, as shown in Fig. 15.57b, the ratio of the major to the minor axes increasing when the spray angle increases. For different materials (alumina, zirconia, titania, Al, Ni, Astroloy, and Cu), the relationship between the major and minor axes shows a strong linearity over a wide range of splat sizes. This observation implies that the Elongation Factor (EF) defined as the ratio between the major and minor axes of the splat does not depend on particle diameter nor on its impact velocity, but only on spray angle. The splat thickness increases slightly along the inclined surface and it becomes progressively thicker in the direction of the inclined surface when the spray angle increases. This probably explains why, when the spray angle exceeds a critical value, which depends on the sprayed material and the substrate material, splashing occurs in the inclination direction even on substrates preheated above the transition temperature.

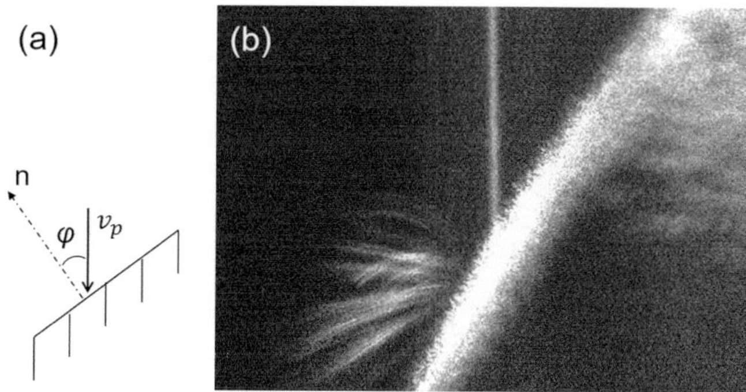

Fig. 15.56 (a) Definition of the impact angle in off-normal impacts and (b) impact splashing on inclined substrates. (Reprinted with kind permission from Springer Science Business Media [Escure C et al. (2003)])

15.3.6 Splat Formation on Rough Surfaces

Experimental studies of the flattening of particles on rough surfaces are very difficult due to the small optical depth of field of optical microscopes in comparison to the level of

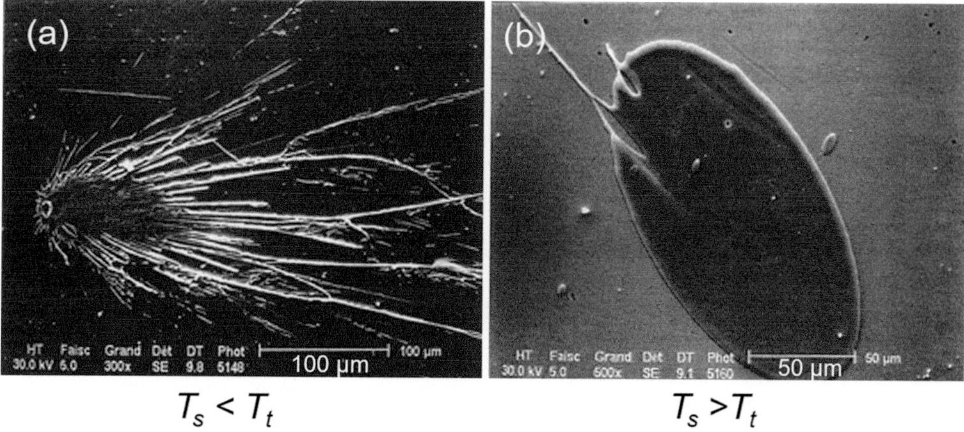

Fig. 15.57 Impact of alumina plasma-sprayed particles ($d_p = 58$ µm, $v_p = 138$ m/s, $T_p = 2400$ K) onto an inclined ASI304 stainless steel (impact angle $30°$). (a) Room temperature substrate. (b) Preheated substrate over transition temperature. (Reprinted with kind permission from Elsevier [Bianchi L et al. (1977)])

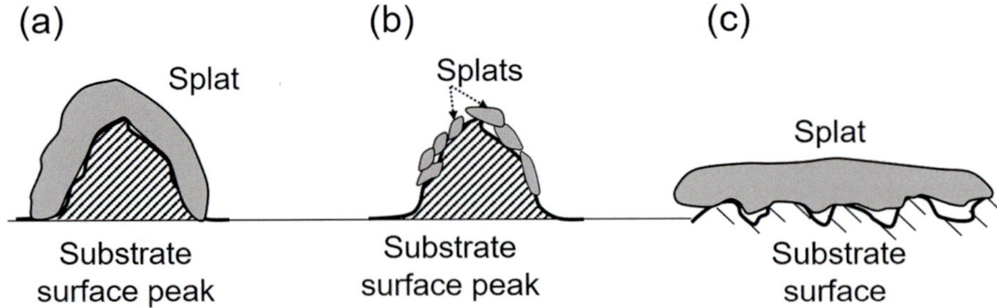

Fig. 15.58 Schematic of splat size relative to grit blasted surface peak sizes. (**a**) Splat size adapted to peak size. (**b**) Too small splat sizes relatively to peak size. (**c**) Too large splat size relatively to peak sizes

surface roughness that makes it impossible to follow the splat formation on the surface of the substrate. This does not impair, however, other diagnostics tools such as on-line measurements of the temperature of the flattening particle and off-line Scanning Electron Micrographs (SEM) of the splat obtained.

It is important, however, to make the distinction between roughened surfaces such as those produced using grit blasting and high pressure waterjet roughening as described in Chap. 14 Surface Preparation, where the roughening is carried out in order to clean the surface and enhance mechanical bonding of the splats to the substrate, and naturally rough surfaces with design patterns related to the functional role of the part. In the former case the level of roughness of the surface that can be characterized by standard parameters such as R_a, R_t, R_{sm}, and $R_{\Delta q}$ (see Chap. 14 Surface Preparation, Sect. 14.2.3 for the definitions). Roughly, the splat dimension must be between 1.5 and $3 \times R_t$ (distance between their highest peak and deepest under-cut) as shown in Fig. 15.58a. If the splat size is too small compared to the roughness level, as shown in Fig. 15.58b, the adhesion will be very poor. Splat adhesion to the substrate will be equally poor if the splat is too big compared to the substrate roughness as illustrated in Fig. 15.58c. However [Bahbou and Nylen (2005)] have studied the influence of R_a, R_t, R_{sm}, and $R_{\Delta q}$ on the adhesion of NiAl (5 wt.% Al) coatings DC plasma sprayed onto Ti-6Al-4V substrates, grit blasted with two brown alumina grit sizes (blasting angle 45°). Results show that the adhesion values measured are poorly correlated with R_a or R_{sm}. However, the coating adhesion seems to be well correlated to $R_{\Delta q}$ (i.e., the average slope: see Chapter 14, Fig 14.7) that is a hybrid parameter containing both amplitude and spacing properties of the surface roughness.

Mehdizadeh et al. (2002), and Raessi M et al. (2006) have shown that for a good adhesion the molten material in the droplet must penetrate into the cavities separating the roughness peaks which require that the impact pressure should be high enough. In the center of the impact region, the pressure is given by $(\rho_p v_{pi}^2)$ where ρ_p and v_{pi} are, respectively, the specific mass and impact velocity of the melted particle, if

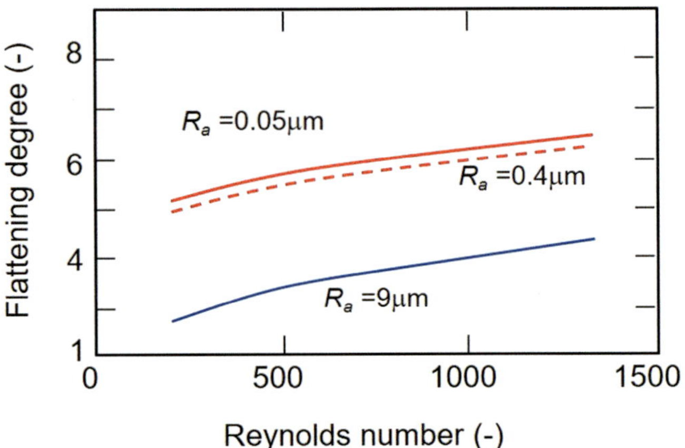

Fig. 15.59 Flattening degree as function of the particle Reynolds number for zirconia particles (22–44 μm) DC plasma sprayed on ASI-304 stainless steel substrates with different roughness levels, $v_{pi} = 200$ m/s, $T_p = 3200$ K. [private communication, University of Limoges, France]

σ_p is the droplet surface tension and δ_{cw} the mean width of the cavity, one must have

$$\delta_{cw} > 4\sigma_p / \rho_p v_{pi}^2 \tag{15.14}$$

If this condition is not fulfilled, the liquid will not penetrate down to the bottom of the cavity. For example, in the splat fringes, where the pressure is low and with the splat curling, the liquid material does not penetrate into the cavities. Here again it must be emphasized that when grit residues block cavities, the adhesion of the splat on it becomes very poor. Moreover, often the grit residues have an expansion coefficient different from that of the splat and the substrate, inducing local stresses upon heating and cooling of the coating/substrate.

The spreading of the droplet and the ability to analyze the oxide layer thickness and composition formed on a rough substrate surface is limited by irregularities in the surface. Roughness in the substrate surface generally results in smaller and thicker splats with a tendency to splash flattening behavior and poorer contact compared to that on smooth substrates. This is illustrated in Fig. 15.59 from University

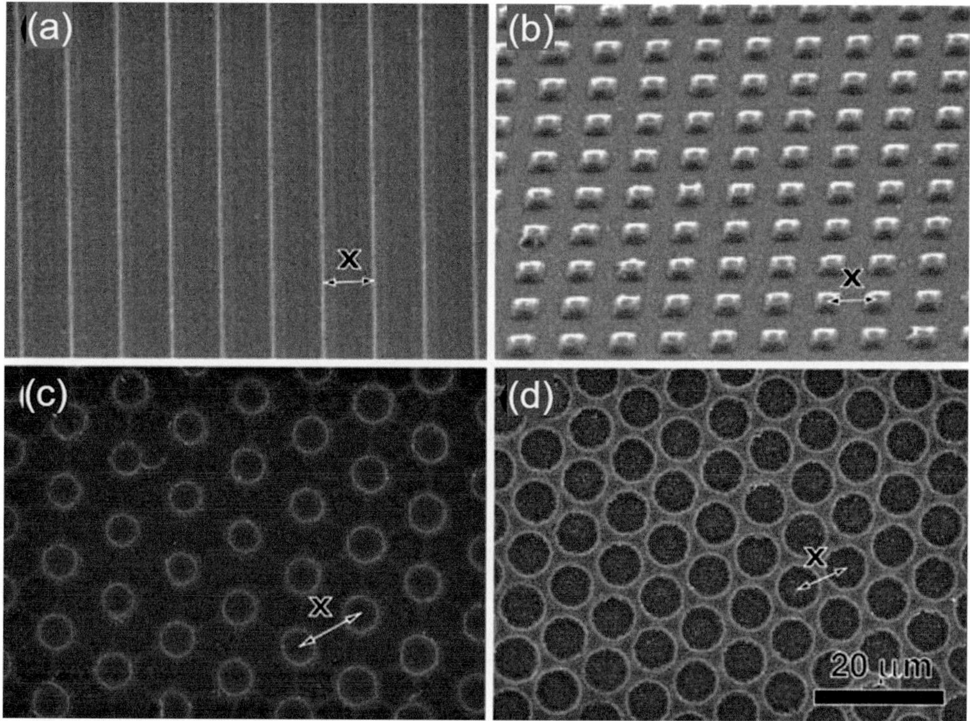

Fig. 15.60 SEM images of four micropatterns: (**a**) line-and-space pattern (**b**) pillar pattern in tetragonal array (**c**) pole pattern in triangular array, (**d**) dimple pattern in triangular array. The x indicates the pattern spacing for each pattern. [Shinoda et al. (2007)]

of Limoges, for zirconia particles $22 < d_p < 44$ μm, impacting at a velocity of $v_{pi} = 200$ m/s, and particle temperature of $T_p = 3200$ K, on smooth ($R_a = 0.05$ μm) or grit-blasted ($R_a = 0.4$ and 9 μm) ASI-304 stainless steel substrates. It can be noted that in both smooth and rough surfaces, the splat flattening ratio, ξ_D increases steadily with the increase in the Reynolds number. Increasing the roughness parameter, R_a, while showing little effect on ξ_D for values of $0.05 < R_a < 0.4$ μm, ξ_D is reduced by a factor of about 2 for $R_a = 9$ μm.

In an attempt to determine the effect of surface roughening pattern on splat formation [Shinoda et al. (2007)] conducted an experimental study in which the surface roughness was artificially micro-patterned and well characterized as shown in Fig. 15.60. These were fabricated on quartz glass substrates by photolithography followed by wet etching. Four micro-patterns were used, their depths/heights were kept at approximately 1 μm (corresponding to $R_a = 0.5$ μm) and their spacing was varied from 4 to 20 μm. The anisotropic deformation of molten alumina droplets (35–65 μm), sprayed with hybrid plasma (DC- RF), with an impact velocity of around 90 m/s was observed on these substrates preheated to 700 K are given in Fig. 15.61. The splats obtained generally showed intense cracking with gas entrapped in each splat off-center, regardless of the type, size, and pattern layout. Droplet spread was affected by a roughness as small as $R_a = 0.5$ μm influencing the fingering

of the splats. The splat given in Fig 15.61a, partially overlapping the linier pattern, show a traditional disk-shaped morphology on the smooth part of the substrate while a deformed "elliptic" splat shape is observed on the pattern part (Fig. 15.61b). Furthermore, the results given in Fig. 15.61c–h for surface convex-concave substrate roughening patterns using poles and dimples with different characteristic length, $x = 5, 8, 10$ and 20 μm, show that in all patterns that droplets deformed fundamentally to disk shapes, and the splat flattening ratio, ξ_D was nearly the same as that of the splats on the smooth substrate. However, the distinct features of splats on these substrates are that they exhibited elongation locally in the form of fingering, rather than the characteristic elongation in one direction observed on the line-and-space pattern.

15.3.7 Modeling Studies of Splat Formation

Many analytical, 2-D and 3-D numerical models have been developed for the prediction of the splat formation on smooth and rough or patterned surfaces have been developed. This section is limited to highlights of some of these modeling studies giving basic concepts typical results. For more information the interested reader can see the review papers by [Amster et al. (2002), Fauchais et al. (2004) and Mostaghimi J and S Chandra (2018)]. These models assume 2-D and 3-D geometry

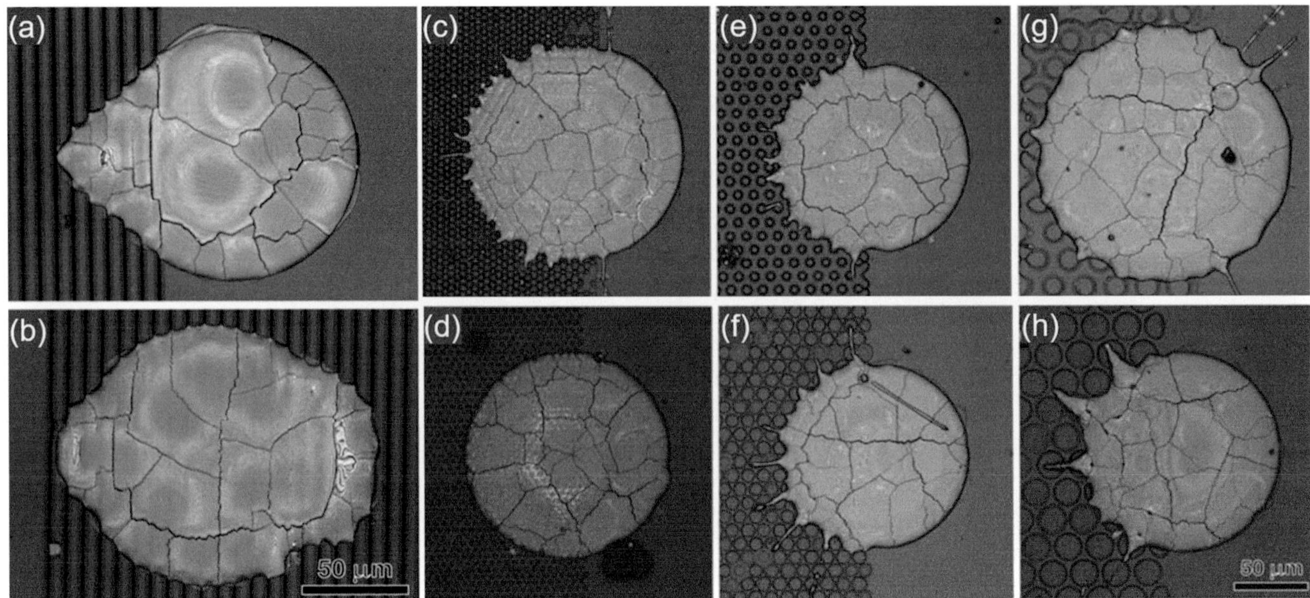

Fig. 15.61 Laser-microscope images of single splats deforming (**a**, **b**) line-and-space pattern with 8 μm trench spacing and 1 μm depth (**c**, **e**, **g**) obstacle-pattern with x = 8,10 and 20 μm, respectively, and (**d**, **f**, **h**) dimple-pattern with x = 5, 10 and 20 μm, respectively. [Shinoda et al. (2007)]

and perfect contact between flattening particle and substrate with a relatively low contact resistance ($R_{th} = 10^{-8}$ m^2 K/W). Splat solidification was often assumed to be delayed until flattening is completed. The particle flattening is characterized by Reynolds, Re, Weber, We, Sommerfeld, Ks, numbers, and its cooling by Stephan, St, Peclet, Pe, and Biot, Bi, numbers. These dimensionless numbers allow predicting the effect of particle parameters on splat formation. However, they cannot predict the breakup or splashing of the flattening particle on the surface. That is why computational fluid mechanics must be used to simulate thermal spray particle impact and gain insight into mechanisms that determine the shape of a splat.

Simulating thermal spray particle impact requires solving simultaneously momentum and energy conservation equations, while accounting for phase change within the solidifying droplet. A transient 3-D model is required to simulate droplet splashing or interactions between multiple droplets, which places significant demands on computing requirement. If gas entrapment under splats is to be modeled a two-fluid model is required, which includes flows in both the droplet and the ambient atmosphere. The model must track the free surface of the particle, which undergoes very large and rapid deformation.

Representing boundary conditions realistically is a challenge. The contact-angle at the liquid–solid contact line must be specified as well as a thermal contact resistance at the interface between the splat and the substrate, if possible varying with time and flattening particle radius. Both are very difficult, if not actually impossible, to determine experimentally. Most models that have been developed assume that thermodynamic equilibrium exists and that particles freeze at

a well-defined solidification temperature. In reality the rate of cooling is very high when a particle impacts a surface and significant undercooling may exist. Results from numerical models are sensitive to fluid properties such as viscosity and surface tension, which vary, sometimes significantly, with droplet temperature and are often not well known for liquid materials such as molten zirconia. Examples of models developed over the past two decades are given in [Pasandideh-Fard M et al. (1996), Bussmann M et al. (2000), Pasandideh-Fard M et al. (2002) and Xue M et al. (2007)].

Numerical codes can simulate fluid flow and heat transfer during particle impacts quite accurately, the main difficulty being quantitatively describing the interface between the splat and the substrate. The interface is defined in models only by specifying values of the thermal contact resistance (R_{th}) and liquid–solid contact angle. During droplet flattening, the flow is mostly driven by fluid inertia, but at the end of its flattening, if it is still liquid, fluid motion becomes sensitive to contact angle balance by viscous and surface tension forces. Thus, the value of thermal contact resistance is very important in the modeling. [Pasandideh-Fard et al. (2002)] modeled the splashing of an impacting nickel particle. Figure 15.62 shows a sequence of computer-generated images showing successive stages of the impact of a 60-μm diameter nickel particle, at a temperature of 1600 °C (150 °C above the melting point of nickel) on a stainless steel substrate initially at 290 °C, R_{th} being assumed constant and equal to 10^{-7} m^2 K/W. Splat breakdown and the formation of splashes was triggered by fluid instability created by the small variations in the liquid velocity around the periphery of the impacting drop, leading to the formation of fingers of liquid around

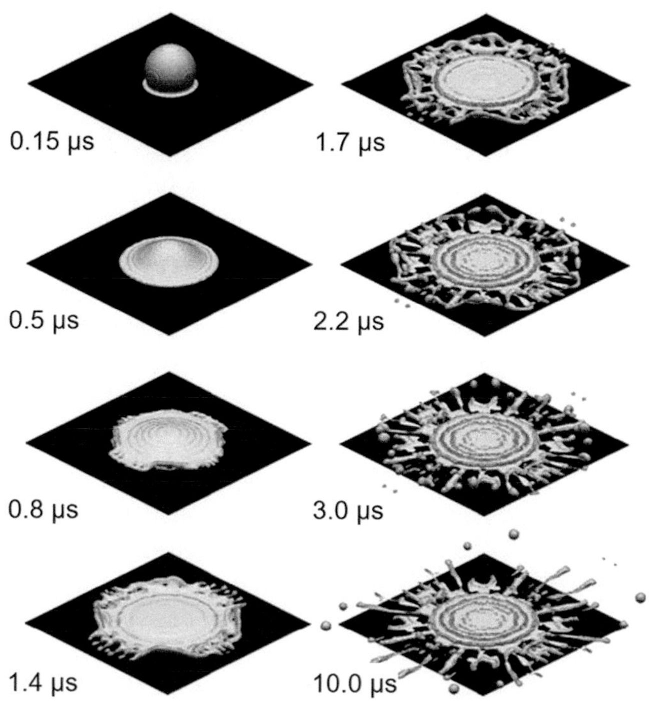

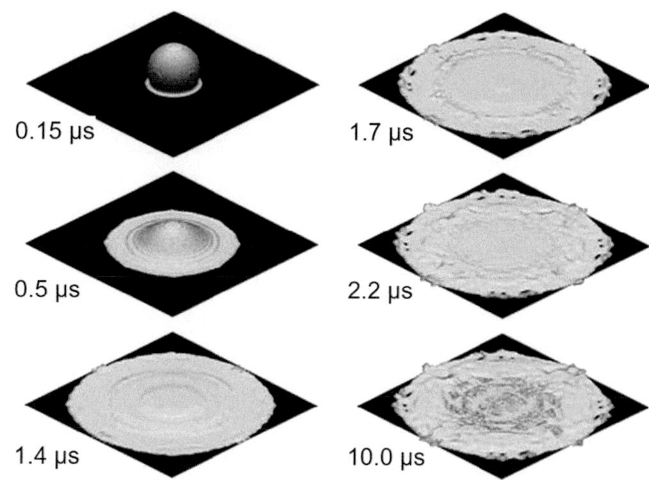

Fig. 15.62 Simulations showing the impact of 60 µm diameter molten nickel particle at 1600 °C with an impact velocity $v_{pi} = 73$ m/s on a smooth stainless-steel plate at a temperature of 290 °C. $R_{th} = 10^{-7}$ m² K/W. (Reprinted with kind permission from Springer Science Business Media [Pasandideh-Fard M et al. (2002)], copyright © ASM International)

Fig. 15.63 Simulations showing the impact of 60 µm diameter molten nickel particle at 1600 °C with an impact velocity $v_{pi} = 73$ m/s on a smooth stainless-steel plate at a temperature of 400 °C. $R_{th} = 10^{-6}$ m² K/W. (Reprinted with kind permission from Springer Science Business Media [Pasandideh-Fard M et al. (2002)], copyright © ASM International)

the drop that grow larger and detached to form small satellite droplets. Simulations in which the substrate temperature was increased from 290 to 400 °C produced little change in the impact process, since the droplet temperature was so high that this relatively small change in substrate temperature did not reduce heat transfer significantly. Increasing R_{th} delayed the onset of solidification until the droplet had finished spreading. Figure 15.63 shows simulation results with all parameters the same as those used in Fig. 15.62, except for the increase in the substrate temperature = 400 °C, and the specific contact resistance $R_{th} = 10^{-6}$ m² K/W. In this case, the droplet spread completely before a significant amount of solidification occurred, and the flow was not destabilized by the solid layer, leaving a disk-shaped splat, looking much like those observed experimentally on surfaces above the transition temperature. [Fauchais et al. (2004)] points out that while these results are consistent with experimental observations regarding the transition of the splat formation from splashing to disc-shaped with the increase in the surface temperature of the substrate, they contradict experimental observation regarding the effect of contact resistance where splat splashing was observed to occur when the contact resistance was high, not when it was low, as in these simulations.

To simulate results obtained with cold surfaces [McDonald et al. (2007a, b, c)] specified a spatially varying thermal contact resistance between a molybdenum splat and glass substrate held at room temperature in a 3-D numerical model. A thermal contact resistance of 10^{-7} m² K/W was specified for the splat central core in contact with the glass. The diameter of the central core was measured from images of the splat after spreading, fragmentation, and solidification. Under the rest of the splat, a thermal contact resistance of 5.0×10^{-5} m² K/W was specified. The results from the numerical simulation show that as the splat spreads, it fragments from the solidified central core. Since the thermal contact resistance in the central core is small compared to the rest of the splat, it solidifies faster and remains attached to the substrate. The rest of the splat, in liquid form, continues to spread and disintegrate. Finally, a central solidified core is left on the substrate, surrounded by a ring of debris (see a similar behavior in Fig. 15.31b). The numerical simulations agree qualitatively with experimental observations. These few results show that models can reproduce the experiments provided the parameters controlling the phenomena at the interface (R_{th} and wet ability) are correctly specified.

Almost at the same time [Parizi et al. (2008)] presented numerical simulation results for molten nickel and zirconia (YZS) droplets impacting on different micro scale-patterned surfaces of silicon. The numerical simulation clearly showed the effect of surface roughness and solidification on the shape of the final splat, as well as the pore creation beneath the sprayed material. They used 3-D transient computational fluid dynamic software, solving the full Navier–Stokes equation, including heat transfer and phase change. A Volume of

Fluid (VOF) tracking algorithm was used to track the droplet-free surface, and the thermal contact resistance at the droplet-substrate interface was also taken into account. Droplet sizes ranged from 15 to 60 μm with impact velocities of 70–250 m/s. The substrate surface pattern was made of regular arrays of cubes 1–3 μm in height, spaced either 1 μm or 5 μm from each other. In the following a few results of this study are presented. Figure 15.64 presents the schematic set up for the patterned surface.

SEM images of top view, and cross section, given, respectively, in Fig. 15.63a, b, are for a nickel splat on a patterned silicon substrate preheated to 350 °C with 1 μm high and 1 μm spacing between the cubes 4 μm in width [Parizi et al. (2008)]. The droplet conditions prior to its impact on the substrate were, $d_p = 50 \pm 3$ μm, $T_p = 2390 \pm 70$ °C, $v_{pi} = 70 \pm 2$ m/s. The results are compared with the predictions of a numerical model under the same operating conditions assuming a thermal contact resistance: $R_{th} = 5 \times 10^{-6}$ m^2 K/W, given in Fig. 15.65c, d. These show a splat diameter of 180 μm predicted by the model that is comparable to the average value of 195 μm that can be measured from the experimental observations (mean value of the splat diameter, excluding all splashes or fingers in splat periphery out of this mean diameter perimeter). When comparing the experimental and simulated splat cross sections (Fig. 15.65b, d), it can be seen that cavities between the textured posts on the substrate are filled with the droplet material in the middle of the splat, whereas close to its boundaries, cavities are partially filled or remain completely open. Close to the impact point of the droplet, where the kinetic energy of the liquid and its normal velocity are the highest, the cavities on the surface are filled up regardless of the size. As the normal velocity of the liquid is reduced and the liquid spreads in radial direction, the liquid energy is not enough anymore to fill cavities.

[Parizi et al. (2008)] have also studied the impact of zirconia particles; Fig. 15.66 shows the calculated top view (Fig. 15.66a) and cross section (Fig. 15.66b) of the splat. The initial velocity of the droplet in this case was more than three times that of the nickel splat (250 m/s compared to 70 m/s). Also, relative to the nickel splat, the thermal contact resistance chosen for the calculation is 200 times less and thus the rate of heat transfer between the substrate and the splat is increased causing a more rapid solidification of the splat on the surface. The shape of the splat in this case is closer to a square than a circle and most of the cavities in between the roughness posts are filled with the material. Such calculations have the potential to be used for characterizing the pore sizes and locations in the coating layers that result from unfilled cavities.

15.4   Coating Formation

As discussed earlier in this chapter, Sects. 15.2.1 Thermal Spray Parameters and 15.2.2 In-flight Particle Transformations, coating formation starts with the injection of the coating material into the Thermal Spray source whether a flame, Wire Arc, DC plasma or RF Induction Plasma source. In most thermal spray devices, the coating material is in the form of powder pneumatically transported and injected into the high temperature flame or plasma stream either internally, radially into the anode nozzle, or axially into some DC plasma torches and inductively coupled plasmas, or externally into the flame or plasma stream at its exit level from the thermal spray torch. In both case, the particle trajectories, their respective temperature history and in-flight transformations depends on the powder characteristics (diameter, shape, porosity, density and chemistry) and their injection conditions (injector diameter, its orientation and carrier gas flow rate) and obviously the flow

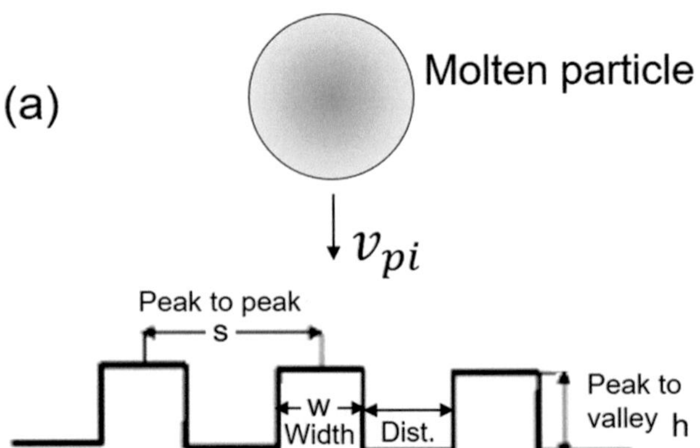

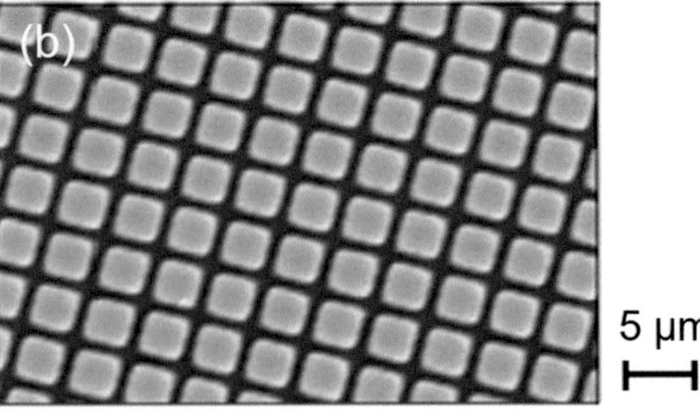

Fig. 15.64 (a) Schematic set up for simulation of the droplet impact on a patterned surface: peak to peak distance (s), width (w), peak to valley distance (h), and distance between two peaks (dist) of the cubes and (**b**) SEM image of a typical textured silicon wafer. (Reprinted with kind permission from Springer Science Business Media [Parizi et al. (2008)], copyright © ASM International)

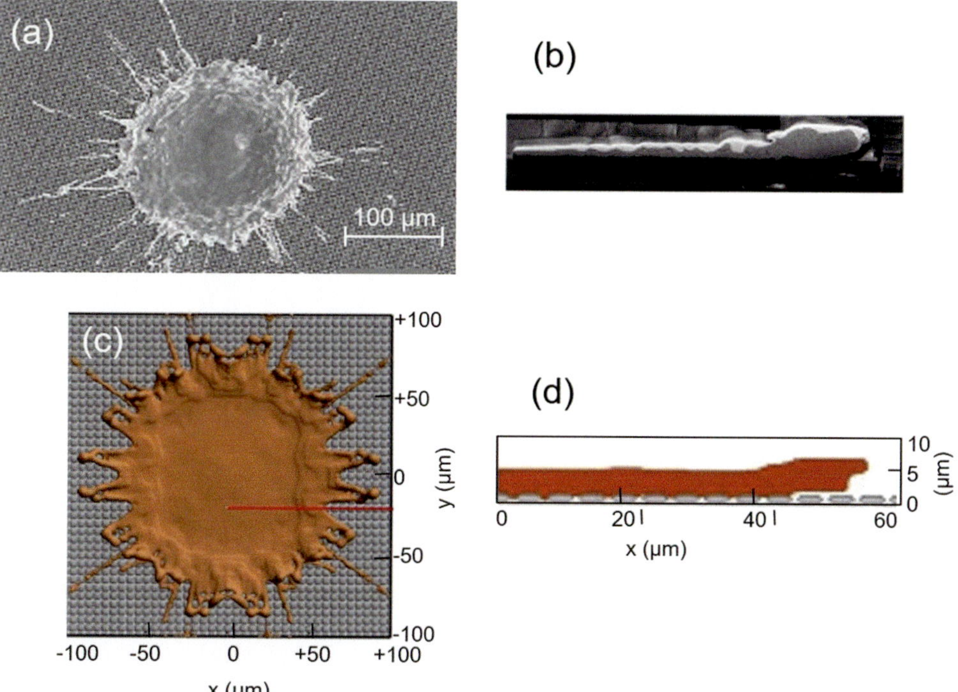

Fig. 15.65 SEM image of the nickel ($d_p = 50 \pm 3$ μm, $T_p = 2390 \pm 70$ °C, $v_p = 70 \pm 2$ m/s) on patterned silicon substrate preheated to 350 °C with 1 μm high and 1 μm spacing between the cubes 4 μm in width, taken from experiments (**a**) top view and (**b**) cross-section and corresponding results of numerical simulation (**c**) top view (**d**) cross section. (Reprinted with kind permission from Springer Science Business Media [Parizi et al. (2008)], copyright © ASM International)

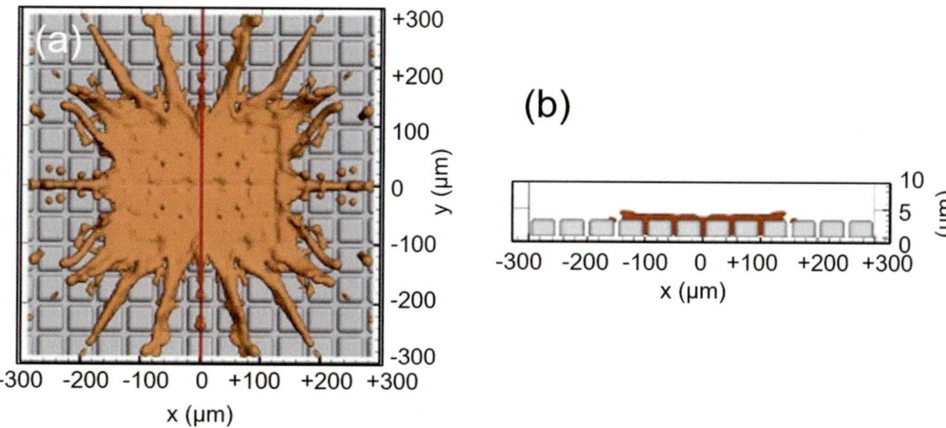

Fig. 15.66 Numerical result for YSZ splat ($d_p = 15$ μm, $T_p = 2900$ °C, $v_{pi} = 250$ m/s) on patterned silicon substrate, preheated to 350 °C with 3.3 μm high and 1 μm spacing between the cubes 4 μm in width, $R_{th} = 10^{-8}$ m^2 K/W. (**a**) Top view and (**b**) Cross section. (Reprinted with kind permission from Springer Science Business Media [Parizi et al. (2008)], copyright © ASM International)

velocity, composition and temperature fields in the high temperature gas stream. A schematic representation of the process in the case of internal powder injection in a DC plasma jet is given in Fig. 15.67a, showing, respectively, the envelope of the plasma gas stream, particle trajectories and the corresponding envelope of the particle flux towards the substrate, and the respective heat flux to the substrate due to the gas and particle streams. It is important to note that,

with the exception of plasma sources with axial powder injection, the maximum of the particle flux is generally off-axis, compared to the maximum of the gas heat flux profile, for all radial injections of powders into the high temperature gas stream.

As the flux of the molten spray droplets impact on the substrate they form splats that pile up on each other as illustrated in Fig. 15.67b forming the coating with potential

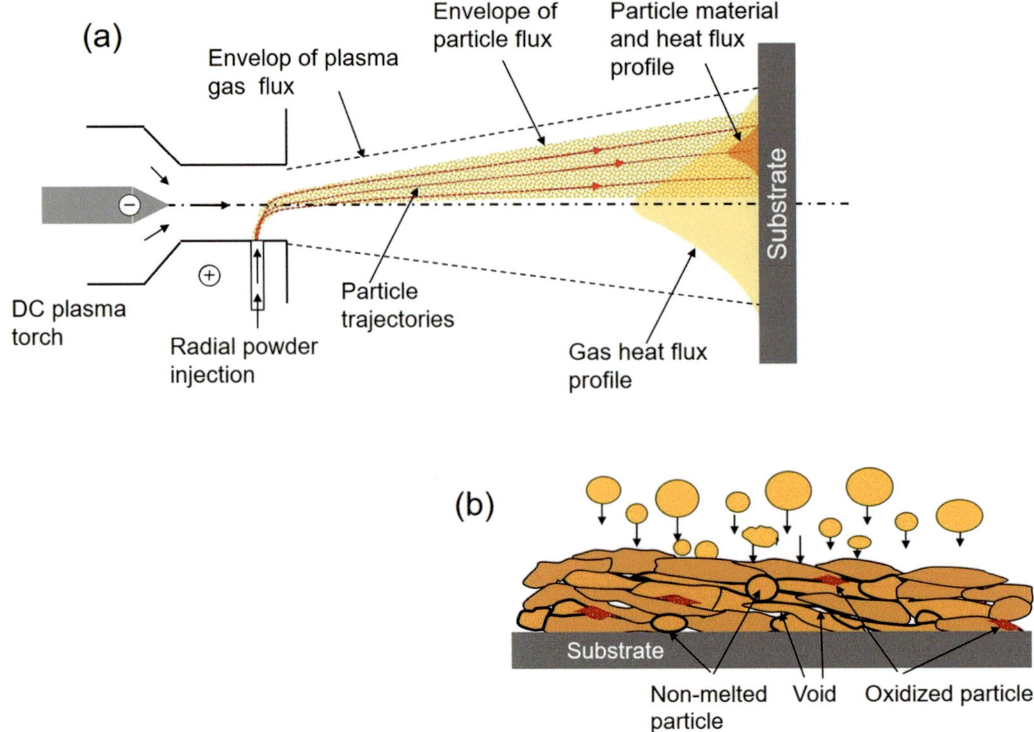

Fig. 15.67 Schematic representation of (**a**) energy and particle flux distributions to the substrate during the thermal spray operation using a DC plasma source (**b**) Coating formation through splat layering on the surface of the substrate

presence of voids/porosity, non-molten particles and oxide particle debris. It is important to point out that, with the exception of the first layer of spats on the surface of the substrate, all subsequent splat layers are essentially formed on earlier splats deposited on the substrate and are not directly affected by the initial substrate preparation. Moreover, the temperature of the earlier deposited coating layers will depend on the equilibrium between the heat flux received by the coating, gas and particle flux, and energy dissipated toward the substrate. The latter depends directly on the microstructure of the formed coating and the contact resistance between different splat layers.

As shown schematically in Fig. 15.68a, the real contact between layered splats determines the coating properties such as thermal conductivity and Young's modulus. Between layered splats, or first splats and the substrate, there exist inter-lamellar pores, the size of which is exaggerated in the figure. The inter-lamellar flat pores are thin voids, orthogonal to the spray direction, filling spaces between splats. These inter-lamellar pores are part of coating porosity and their thickness is a few hundredths to a few tenths of micrometers. The real contact between splats increases with particle impact velocities, provided that they are not either too much super-heated or below their melting temperature. To demonstrate that the splat thickness is not the only parameter controlling coating properties, a 300 μm coating achieved by layering

100 splats, 3 μm thick each, or 300 splats 1 μm thick each, is considered assuming a real contact of 30% between splats whatever maybe their thickness [Renouard-Vallet G et al. (2005)]. The mean real contact, MR, between the 300 layered 1 μm thick splats is $MR_1 = 299 \times 30/299 = 30\%$. When considering 3 μm thick splats that can be considered as three layered 1 μm with perfect contact, one 3 μm thick splat corresponds to two-layered splats with a 100% contact (Fig. 15.68a, b). The same calculation results in: $MR_2 = (99 \times 30 + 200 \times 100)/299 = 76.6\%$. Of course, if it is assumed that the real contact between 1 μm thick splats is 60% and that between 3 μm thick ones is 15%, $MR_1 = 60\%$ while $MR_2 = 71.8\%$. However, the main problem, whatever the mean splat thickness may be, is to know the real surface contact percentage between splats (which is not at all easy to determine).

Splashing of the melted particles during flattening upon impact can drastically affect the coating properties [Gawne DT et al. (1995), Newbery AP and Grant PS (2000) and, Trice RW and Faber KT (2000)]. The lower adhesion of splats deposited on splashed material, as described schematically in Fig. 15.68c, can become more important when spraying metals because splashed material is oxidized rather fast due to the small droplet sizes. Buildup of splashed particles around perturbations on otherwise smooth surfaces was observed under plasma spray conditions [Trice RW and

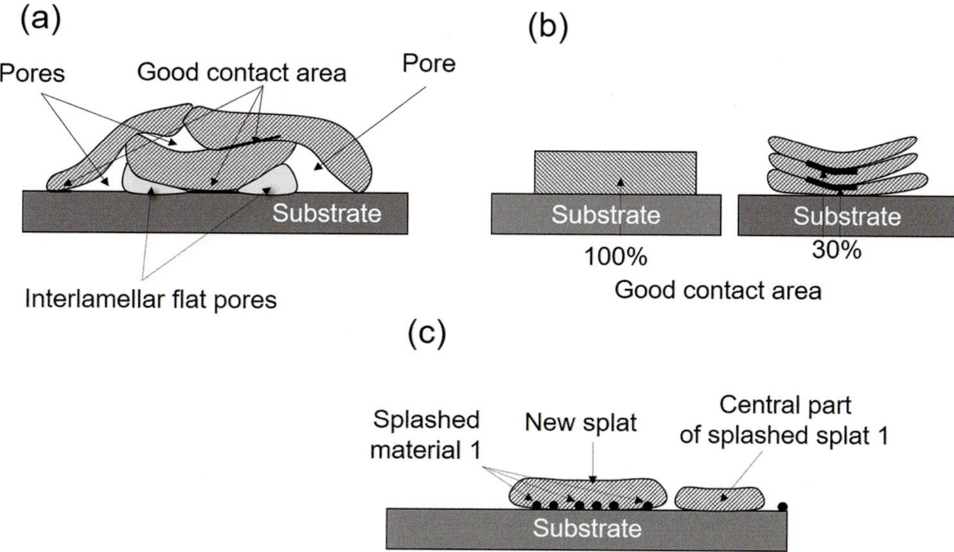

Fig. 15.68 (**a**) Schematically shown contacts between layered splats or first splats and substrate with inter-lamellar contacts, the size of which is drastically exaggerated. (**b**) Same thickness of 3 μm achieved with either a single splat or 3 layered ones. (**c**) New splat deposited in the periphery of a splashed splat with poor contact with substrate due to splashed material

Faber KT (2000)]. It was also shown that the substrate geometry affects the flow of impacting and splashing particles: flat and cylindrical specimens were used by Racek [Racek O (2010)] to investigate the effect of the surface curvature when spraying CrC–NiCr coatings by HVOF. Less splashed particles would be observed on the cylindrical surface, 90 mm in diameter, Fig. 15.69. Due to the surface curvature of the cylindrical surface, the droplets or fragments would travel higher above the surface and fewer protrusions would extend to their path.

Pores, often called globular, which are formed during coating generation, are illustrated in Fig. 15.68. Pores are due to the shadowing effect (Fig. 15.70a), narrow holes in valleys between splats are not completely filled as explained by Eq. 15.13 (Fig. 15.70b), due to unmolten or partially melted particles that create the worst defects within coatings (Fig. 15.70c), or due to exploded particles (Fig. 15.70d). These globular pores are distributed more or less homogeneously and their potential to deteriorate coating properties is proportional to their size [Ctibor P et al. (2006)]. That is why in order to reduce the porosity, the incorporation in the coating of large nonmelted/or partially melted, particles, which are sufficiently heated to stick, must be avoided. This can be done by choosing carefully the particle size distribution and optimizing the particle injection, especially for radial injection. However, sometimes a compromise has to be found between having most particles fully melted but with a rather high oxide content and more nonmelted particles with lower oxide content [Saaedi J et al. (2010)].

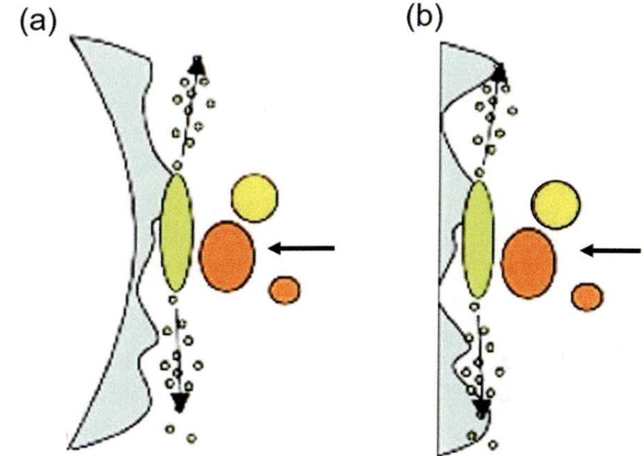

Fig. 15.69 The relation between the substrate curvature and the mechanism of collection of splashed material: (**a**) cylindrical and (**b**) flat surface. ([Racek O (2010)] Reprinted with kind permission from Springer Science Business Media, copyright © ASM International)

Pores can also be generated in the coating as cracks are formed during residual stresses relaxation. They comprise mainly:

- Micro-cracks within splats resulting from quenching stress relaxation, not the worst for coating porosity, observed in ceramic materials, these splats are always parallel to the spray direction.
- Macro-cracks running through layered splats especially at their interfaces; these tend to initiate interconnected porosities. They are often due to the relaxation of an expansion mismatch stress as discussed in Sect. 15.5

Fig. 15.70 Schematic representation of globular pores within coatings, (**a**) shadow effect, (**b**) narrow holes and/or gas in valleys between splats (**c**) nonmelted or partially melted particles, (**d**) exploded particle

Temperature and stress control. They are flat pores that often contribute to the open porosity of coatings. Other stress relaxations can occur, but they can be avoided by optimizing spray or service conditions.

Another important source of defects is linked to the impact angle that reduces the normal impact velocity, resulting in elongated splats. As discussed in the next Sect. 15.4.1 Splat layering, spraying with an angle above a certain value, depending on particle and substrate materials, will promote splashing even on substrates preheated above the transition temperature.

It must be stressed that in most spray conditions, a careful optimization of the spray parameters, including morphology and size distribution of particles, must be made to achieve coatings with properties expected in service conditions. The importance of the choice of sprayed particles to limit their oxidation is, for example, illustrated by the work of [Zeng et al. (2010)] who plasma sprayed (APS) Ni–Cr-based alloy powder with Si (4.3 wt%), B (3.0 wt%), and C (0.8 wt%) additions. They showed that the addition of Si and B to iron effectively reduced the oxygen contents in the coatings,

especially during the in-flight period at higher particle temperature.

15.4.1 Splat Layering

As mentioned earlier, the properties of the coatings obtained depend on the spray pattern generated by the spray device used, its standoff distance, spray angle and the relative movement of the torch with respect to the substrate. For a better understanding of the coating generation, the simple case of flame or plasma spray deposition on a stationary substrate will be discussed first, defining what is commonly referred to as the formation of the "Sweet Spot." Typical plasma spray configurations used are illustrated, respectively, in Fig. 15.71a, b for DC Atmospheric Plasma Spraying (DC-APS) with external radial powder injection, and RF Induction Plasma Spraying (RF-IPS) with internal coaxial powder injection. In both cases, the envelope of the sprayed particles is directed towards the substrate at 90° orientation with respect to the substrate surface, which is optimal in terms of deposition efficiency and density of the coating obtained. This is followed by a discussion of effect of the deposition angle on the quality of the coating obtained. Bead-formation and finally coating formation with overlapping beads are discussed next for different patterns of the relative motion of the plasma torch with respect to the substrate.

The "sweet spot" deposition of refractory metals such as tungsten on a stationary substrate using Radio Frequency Induction Plasma Spraying (RF-IPS) was studied by [Jiang et al. (1993)]. Details of the installation were described earlier in Chap. 10 Induction Plasma Spraying, Fig. 10.75. The spray configuration was similar to that illustrated in Fig. 15.71b with a 100 mm stainless-steel substrate placed coaxially with a Tekna Plasma Systems Inc. PN-50 Induction plasma torch at a spraying distance of 200 mm from the torch nozzle exit. Photographs of the deposit at different stages of the plasma spraying process are given in Fig. 15.72a, b. These were obtained using an Ar/H$_2$ ICP plasma operated at a reduced pressure of 26.6 kPa (200 Torr), power of 40 kW, spraying distance of 200 mm with tungsten powder ($-75+45$ μm) feed rate of 20 g/min. The deposit obtained in each of these two cases has a typical gaussian profile reflecting the particle number flux distribution impacting on its surface. It is represented by standard Gaussian Probability Distribution Function (PDF), $P(x)$ expressed as;

$$P(x) = \frac{1}{\sqrt{\pi}\sigma} \exp\left[-\frac{(x-\mu)^2}{2\sigma^2}\right] \text{for} -\infty < x < \infty \quad (15.15)$$

where x is radial position in the envelop of the particle stream, σ is the standard deviation, and μ is radial position of the maximum particle flux density. Setting $\mu = 0$, the distribution

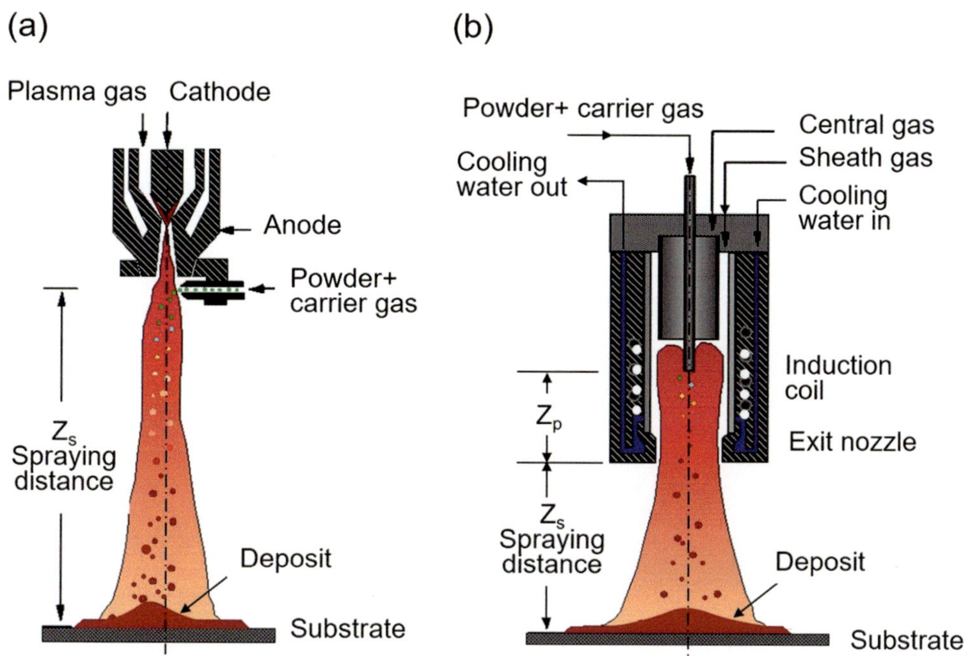

Fig. 15.71 Schematic representation of commonly used powder injecting techniques in (**a**) DC plasma spray torch and (**b**) RF-Induction Plasma Spraying torch

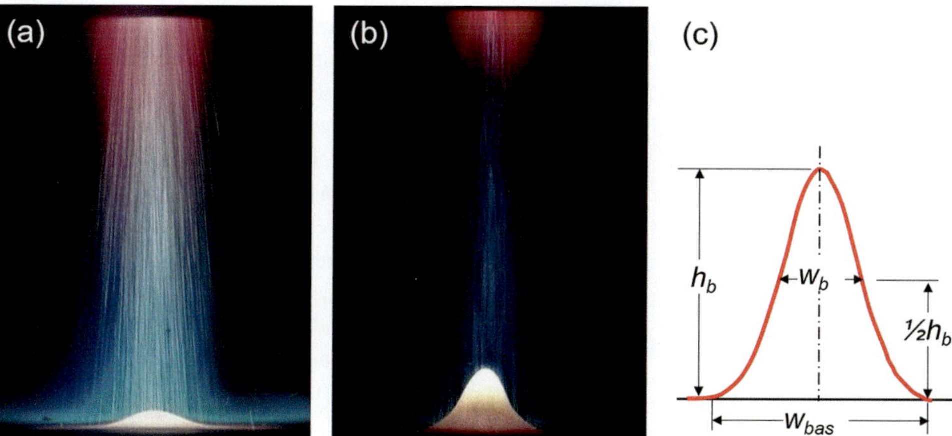

Fig. 15.72 Photographs of the RF-IPS, low pressure (26 kPa), 40 kW spraying of tungsten powder at 20 g/min on a 100 mm diameter stainless-steel substrate at a spraying distance of 200 mm at different instances (and different magnification) during the spraying process. [Jiang et al. (1993)]

function implies that 68.2% of the deposited material will be within the region $\pm\sigma$ and 95.4% with in $\pm 2\sigma$. An estimate of the mean rate, $\bar{n}_p$, of powder particles impinging on the substrate can be obtained simply as;

$$\bar{n}_p = \frac{\dot{m}_p \cdot \eta_d}{m_p} \qquad (15.16)$$

where, $\dot{m}_p$ is the powder feed rate, η_d is particle deposition efficiency, defined as the fraction of the feed material that is deposited on the substrate, and m_p is mass of a single particle. For the case of tungsten powder deposition at a feed rate of 20 g/min, with a deposition efficacy of 90%, assuming a mean particle diameter of 60 μm, and density of tungsten metal of 19,300 kg/m^3, the mean particle number flux rate is estimated at $\bar{n}_p = 1.38 \times 10^5$ particles/s. It is to be pointed out, however, that in the case of powder deposition on a fixed substrate, the deposition efficiency can change in the course of the operation as the deposit builds up. As discussed later in this section, with the increase in the height of the deposited bead h_b, the particles in the fringes of the deposit will be impacting the surface of the deposit at impact angles significantly lower than 90° leading to a decrease in the deposition efficiency. Photograph of a typical "sweet spot" of a tungsten deposit on a fixed substrate is given in Fig. 15.73 with a deposit base diameter of about 60 mm and 30 mm high at its center. The corresponding total weight of the deposit is estimated at 545 g, which would require about 30 min. for its deposition, with a powder feed rate of 20 g/min, and a deposition efficiency of about 90%.

Fig. 15.73 Photograph of a "sweet spot" tungsten deposit on a 100 mm stainless-steel substrate. [Jiang et al. (1993)]

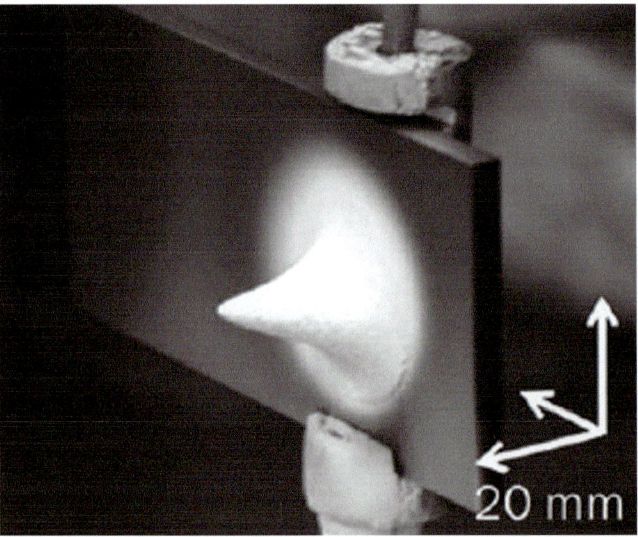

Fig. 15.74 DC plasma-sprayed yttria partially stabilized zirconia on stationary substrate showing 21 mm high "Sweet spot" formation. ([Shinoda et al. (2008)]. Reprinted with kind permission from Springer Science Business Media, copyright © ASM International)

Figure 15.74 from [Shinoda et al. (2008)] shows an example of the deposit obtained by DC atmospheric plasma spraying of Yttria Partially Stabilized Zirconia (YPSZ) on a stationary substrate. The spray configuration used in this case is similar to that schematically illustrated in Fig. 15.71a, with the exception of using an internal radial injection of the powder into the nozzle of the plasma spray torch rather that the illustrated external injection. The formed "sweet spot" has a near Gaussian profile, over a deposition region 36 mm in diameter and height $h_b = 21$ mm. The deposit in this case is, however, not fully axisymmetric, neither it is coaxial with the torch axis due to the radial powder injection configuration giving rise to a 3°–4° deviation of the particle trajectories with respect to the torch axis. Its equivalent base diameter is estimated at $w_{bas} = 20$ mm with diameter at mid-height of $w_b = 8$ mm.

15.4.2 Coating Formation Through Bead Overlapping

A bead, as illustrated in Fig. 15.75a, is obtained when translating the torch relatively to the substrate (1-D movement) at a given relative velocity v_r, for example, in the x direction. The bead parameters, its thickness h_b and its width w_b at mid-height, depend on the spray process, the torch working conditions, the relative torch-substrate velocity v_r, the powder mass feed rate, $\dot{m}_p$ the particle size distribution, the deposition efficiency η_d, the torch nozzle internal diameter d_n, the injection conditions, and the spray distance. The bead height h_b is particularly sensitive to v_r, the deposition efficiency, decreasing with increasing distance above a certain value, and the spray distance. The bead width w_b is linked to the nozzle internal diameter, d_n, the injection conditions, especially the injector i.d., and the spray distance.

To form a uniform coating on a flat surface, the motion of the torch with respect to the substrate should follow a pattern as demonstrated in Fig. 15.76b following a linear motion in one direction along the x-axis, at velocity, v_r shifting at the end of the first travel by a step, Δy for the spraying of the next bead. The value of Δy being dependent on the width of the bead at half height, w_b multiplied by the *overlapping ratio, f,*

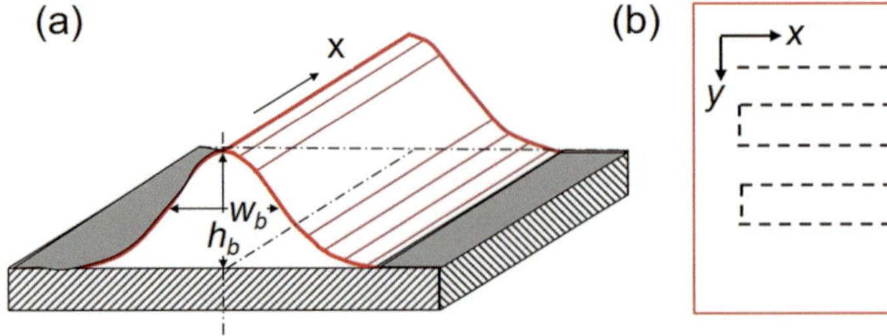

Fig. 15.75 (a) Schematic representation of a bead and (b) Typical spray pattern on a flat substrate

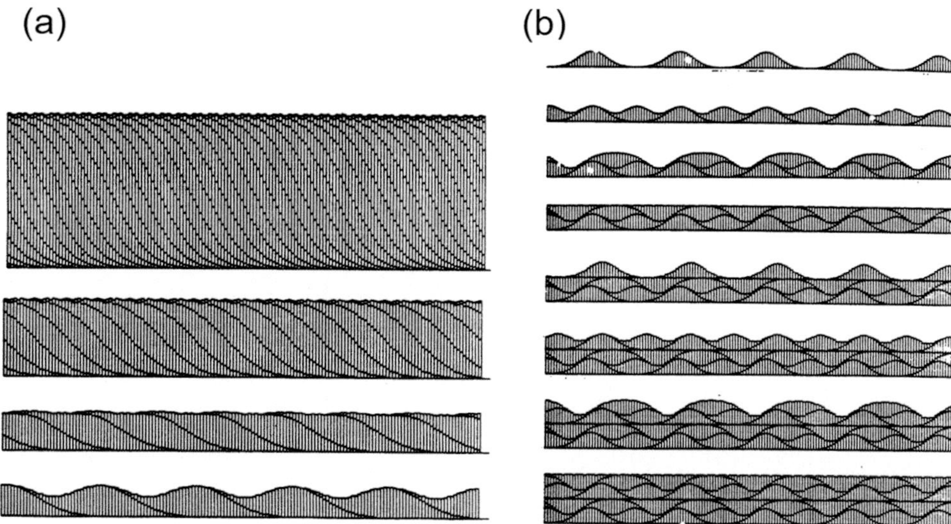

(a)   (b)

Fig. 15.76 Simulations: (**a**) showing the cross section, normal to the spray line, of the first layer of the pass in dependence of the track gap ($\Delta y = f.w_b$) with $f = 0.5, 1.0, 2.0, 3.0$, (**b**) smoothing the surface of a thick coating by starting the next passes with different starting points. ([Fasching MM et al. (1992)]. Reprinted with kind permission from ASM International)

which is typically in the range of 0.33, 0.5, 1.0, 2.0, 3.0. The choice of the beads overlapping ratio depends on the bead thickness and the surface undulation acceptable in the coating. For a medium thick bead of 7–8 μm, an overlapping of 1.0, which implies that $\Delta y = w_b$, results in an undulation of about 3 μm, while with an overlap ratio of 0.33, the undulation is about 1 μm [Montillet D et al. (1999)]. This is also illustrated in Fig. 15.76a after [Fasching MM et al. (1992)] simulating the first layer of the pass in dependence of the track gap ($\Delta y = f.w_b$) with $f = 0.5, 1.0, 2.0, 3.0$. The simulation is based on a symmetric Gaussian distribution and shows the cross section normal to the spray line. As it can be seen the undulation decreases when f increases, but the pass thickness also increases correspondingly, which does not necessarily help to control temperature gradients within passes, especially when spraying low thermal conductivity materials. The undulation increases when passes sprayed with the same procedure are superimposed. To smoothen the surface of thick passes, the starting point of the next pass must be chosen such that the first bead of the next pass is shifted in such a way that it starts in between the beads previously deposited. Figure. 15.76b from [Fasching MM et al. (1992)] illustrates how the evenness diminishes the coating roughness, for an increasing number of layers, by choosing properly the starting point of the next pass.

Another important point that must be emphasized is the spray distance that, especially for plasma spraying, plays a key role, not only in the heat transferred to the substrate or coating under construction but also in the bead thickness and width at mid-height. Finally, the standoff distance results from a compromise between the locations of particle maximum velocities and temperatures, heat transferred to the substrate, which decreases almost exponentially with the spray distance, and particle inertia. For example, in plasma spraying typical stand-off distances are between 100 and 120 mm to achieve heat fluxes below 2 MW/m². The maximum particle velocities and temperatures are obtained between 60 and 80 mm, where the heat fluxes, reach 5–6 MW/m² and are more difficult to control with cooling systems. The final spray distance also depends on particle sizes, that is, particle inertia: a particle 10 μm in diameter is cooling and slowing down much faster than a 50 μm one. It thus becomes important to spray small particles at shorter distances at the expense of a higher heat flux imposed onto the substrate.

15.4.3 Effect of Spray Angle

Special attention has been given to the study of the effect to the spray angle on the coating formation and its characteristics. This is mainly due to the inherent difficulty in insuring that the particle/molten droplet flux envelope is always impinging on the substrate at 90°. This is particularly challenging in the case of coating 3-D parts that would require a continuous manipulation of the torch and/or substrate or both in the course of the spraying operation requiring rather sophisticated robotic tools.

A systematic study of the effect of spray angle on the properties of DC plasma sprayed deposits was reported by [Leigh SH and CC Berndt (1997)] for NiAl ($\overline{d_p} = 60$ μm) and Cr_3C_2-NiCr ($\overline{d_p} = 25$ μm) powders sprayed on a fixed substrate at spray angles between 50° and 90°. A proposed model of off-angle spraying showed that when the spray

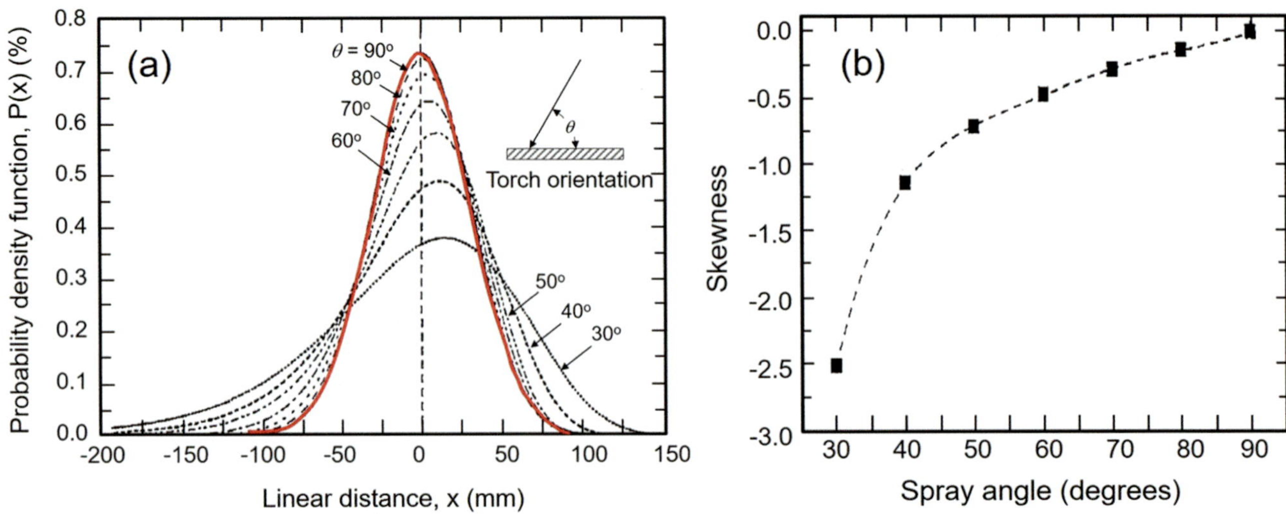

Fig. 15.77 (a) Probability distribution of spray pattern and (b) Skewness factor for different spray angles of molten particle flux towards the substrate. [Leigh SH and CC Berndt (1997)]

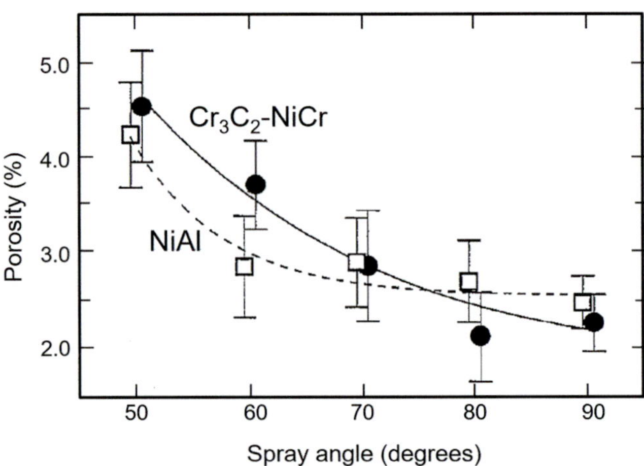

Fig. 15.78 Variation in deposit porosity with spray angle for the DC plasma spraying of NiAl and Cr₃C₂-NCr. [Leigh SH and CC Berndt (1997)]

gun is titled with respect to the substrate, the distribution of the deposited material would deviate from the standard Gaussian distribution as function of the spray angle, θ. Results given in Fig. 15.77 show the probability distribution of spray pattern for angles between 30° and 90° that are increasingly more skewed with the decrease in the spray angle. The deposit profile given in Fig. 15.77a, $\theta = 90°$, represents a standard Gaussian probability distribution function $P(x)$ given by Eq. 15.15

As shown in Fig. 15.78, the decrease in the spray angle also has a significant impact on the porosity of the coating with the lowest porosity reported for a spray angle of 90°, increasing rapidly with the decrease in the spray angle below 60°. The surface roughness of the Cr₃C₂–NiCr deposit, on the other hand, was not sensitive to the spray angle, whereas

NiAl exhibited an increase as the spray angle decreased. Micro-hardness, tensile adhesion strength, and interfacial fracture toughness decreased with spray angle. The change of these properties with spray angle was attributed to the reduction of the normal component of the particle impact velocity on the substrate/or previously deposited coating layers, changes of the splat morphology, and shadowing effects.

Another important point is the risk, when spraying off-normal angles, to have particles intercepted by the bead rebounding and splashing on the other side of it, as shown in Fig. 15.79a. The adhesion of the splashed material being very poor, it will be the same with the next bead deposited on top of it [Kanouff MP et al. (1998)]. Figure 15.79b presents the digitized photograph of the deposit cross section obtained with 15 successive passes along the same bead sprayed with an HVOF gun and a spray angle of 51°. The portion of the deposit near the impact zone of the main spray jet is relatively thick with small roughness elements on it. This portion of the deposit has buckled and separated from the plate due to the thermal stresses created by the transfer of a large amount of heat in the thermal spray material to the plate and deposit. A large amount of splashing occurred when the thermal spray droplets impacted on the target surface (at least 23%). The splashed material, called overspray, redeposit over large areas on the target surface and is composed of isolated columns with large spaces between them.

[Tillmann et al. (2008)] studied the influence of the spray angle (between 90° and 20°) on plasma-sprayed WC–12Co and Cr₃C₂–10(Ni–20Cr) coatings. Each sprayed bead was characterized by a Gaussian fit. Furthermore, metallographic studies of the surface roughness, porosity, hardness, and morphology were carried out and the deposition efficiency

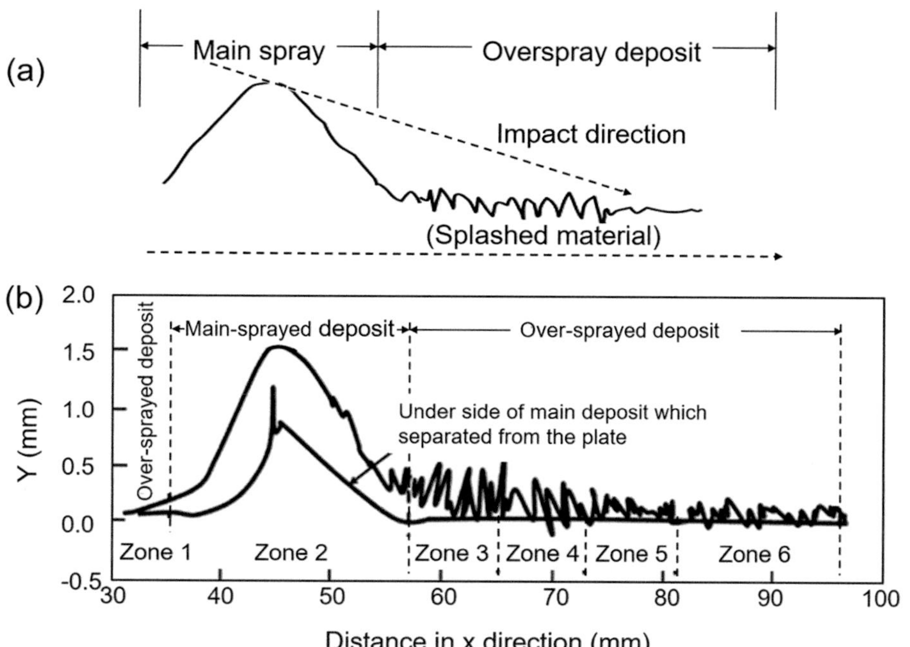

Fig. 15.79 (**a**) Splashed material from inclined spraying where the bead intercepts part of the impacting particles. (**b**) Digitized data for the shape of the deposit cross section obtained from the experiment. ([Kanouff MP et al. (1998)]. Reprinted with kind permission from Springer Science Business Media, copyright © ASM International)

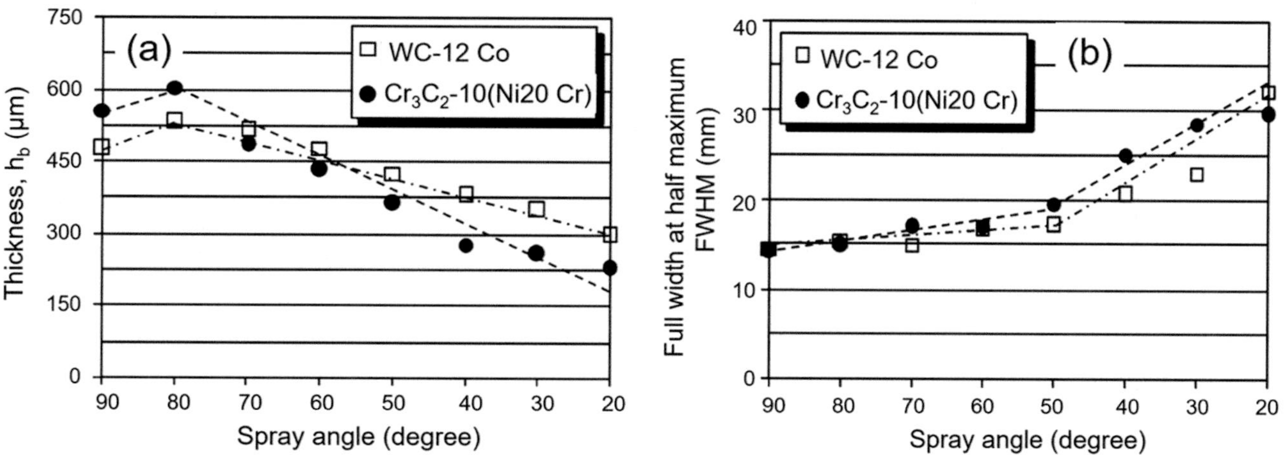

Fig. 15.80 Dependence on the spray angle of plasma-sprayed WC-12Co and Cr_3C_2-10(Ni-20 Cr) beads. (**a**). bead height, h_b. (**b**) Width at mid height w_b. (Reprinted with kind permission from Springer Science Business Media [Tillmann et al. (2008)], copyright © ASM International)

as well as the tensile strength was measured. The thermally sprayed coatings show a clear dependence on the spray angle. A decrease in spray angle changes the thickness, width, and form of the spray beads. Coatings become rougher and their quality decreases. Figure 15.80 represents the evolutions of the thickness h_b of the bead (Fig. 15.80a) and its width w_b at mid-height (Fig. 15.80b). The thickness is maximum for 80° and decreases after that, while the width increases slowly down to 50° and much more afterwards. At smaller spray angles the normal component of the particle velocity regarding the substrate surface diminishes and at the same time the tangential component increases, and, besides the splat elongation, the probability rises to bounce off a sprayed particle from the specimen [Tillmann et al. (2008)]. For the considered materials, the porosity is not really affected up to 50°, but it increases drastically afterwards, being multiplied by a factor of up × 3 at 20°.

[Racek O (2010)] has shown when HVOF spraying Cr_3C_2–NiCr coatings, localized porosity, and high oxide content can be formed due to re-deposition of splashed

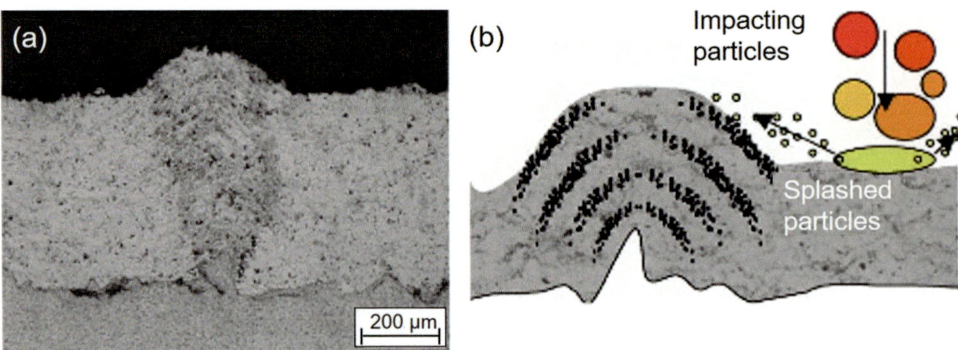

Fig. 15.81 (**a**) Optical micrograph of the cross section of a Cr_3C_2–Ni HVOF coating: (**a**) porosity speckle emanating from a substrate protrusion, (**b**) suggested mechanism of the speckle formation. ([Racek O (2010)]. Reprinted with kind permission from Springer Science Business Media, copyright © ASM International)

metallic particles. Macro-roughness features like speckles were formed when higher particle velocities were used on cold substrates, as illustrated in Fig. 15.81a, showing the coating cross section with a speckle. The speckle grew in size progressively with coating deposition forming an approximately conical volume of high porosity. Its highly porous microstructure consisted of small highly oxidized particles. [Racek O (2010)] has hypothesized that droplets or fragments of splashed or fractured particles are responsible for the effect that leads to a speckle formation. Since the trajectories of the splashed particles are more parallel to the substrate, particles re-deposited primarily on surfaces inclined to the substrate plane such as the sides of a substrate protrusion, Fig. 15.81b. As the speckle or protrusion height over the surrounding coating surface increases, the effect is multiplied due to the increased projected area where the splashed particles deposited, Fig. 15.81b. Fortunately [Racek O (2010)], adjusting the process parameters such as flow of the process gases and the standoff distance can effectively control the splashing phenomenon. Substrate temperature and chemical composition also strongly affects the particle deposition.

It is important to stress, however, spraying at right angle to the surface of the substrate cannot always be guaranteed depending on the geometry of the surface to be sprayed. The study reported by [Henne et al. (1999)] for the near-net shape spraying of metallic bipolar plates for planar Solid Oxide Fuel Cells (SOFC) by Induction Plasma Spraying (RF-IPS) is a typical example where the profile of the substrate shows grooving of the surface with a depth of around 10 mm and width of 20 mm with important variations in the angle of the substrate surface as illustrated in Fig. 15.82c. The objective was to deposit a "near net shaped" plate with different surface profiles through the spraying of the coating material on a sacrificial, graphite substrate with the required surface profile. The spraying system used in this case was Tekna Plasma System RF-IPS torch, with a 70 mm

i.d. plasma confinement tube mounted on a vacuum deposition chamber with central axial powder injection. The powder used was Cr- based alloy (94 wt.% Cr, 5 wt.% Fe and 1 wt.% Y_2O_3) with a particle size distribution ($d_p < 90$ µm). The spraying operation was carried out using Ar/H_2 as plasma gas (5 to 8 vol.% H_2), plate power of 50–75 kW, spraying distance of 150–200 mm and powder feed rate of 40 g/min. The principal challenge is the substrate pattern profile that resulted in the particles impacting the substrate surface at different angles that as shown in Fig. 15.82a, b, resulting in the formation of significant porosity in the coating when the spray angle was below 60°. Micrographs of etched cross sections of samples sprayed under different incidence angles in the range of 90° to 30° are shown in Fig. 15.83a–d. These show a significant increase in porosity and of the pore size with decreasing spray angles reaching more than 20 vol.% for spray angles less than 45° as shown in Fig. 15.84. The phenomena were attributed to droplet shadowing effects as illustrated in Fig. 15.82b.

15.4.4 Polymer Coating Formation

The thermal spraying of polymer coatings is mostly carried out using combustion and flame spray technologies including High Velocity Oxy Fuel (HVOF) and High Velocity Air Fuel (HVAF) technologies. The technique is essentially similar to conventional bead formation with overlapping beads approach used in order to produce a uniform coat on a large substrate surface. One of the main challenges of thermally spraying polymer coatings is the need to exercise a close control on the in-flight particle temperature as well as the substrate/coating temperature during the operation. These need to be kept as close as possible to the polymer melting temperature with limited overheating to avoid polymer thermal degradation. This requires the use of polymer powders with a relatively narrow particle Size Distribution (PSD) in

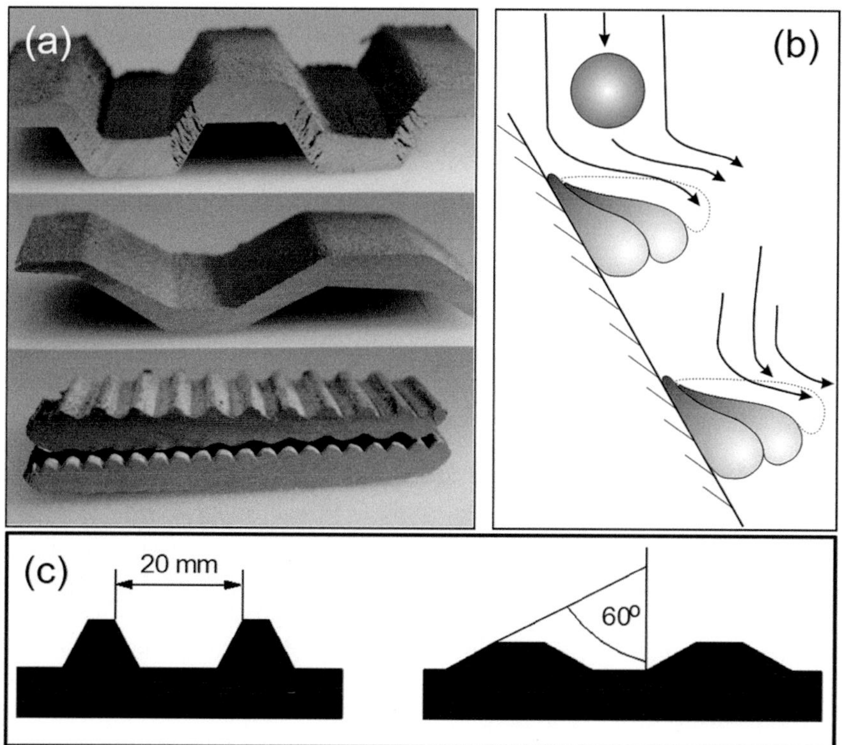

Fig. 15.82 (**a**) RF-IPS of free-standing parts of bipolar plate material made with different substrate profiles (**b**) schematic representation of the showing effect between successive droplet splatting on inclined substrate (**c**) principal dimensions of the graphite substrate used. [Henne et al. (1999)]

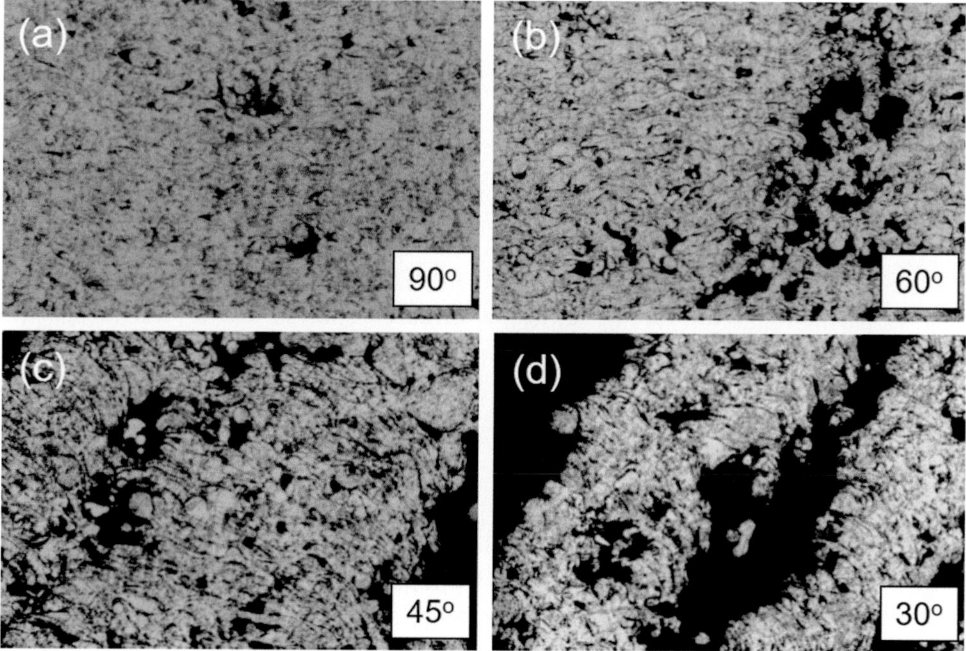

Fig. 15.83 Micrographs of etched cross sections of samples sprayed under different incidence angles. (**a**) 90°. (**b**) 60°. (**c**) 45°. (**d**) 30°. [Henne et al. (1999)]

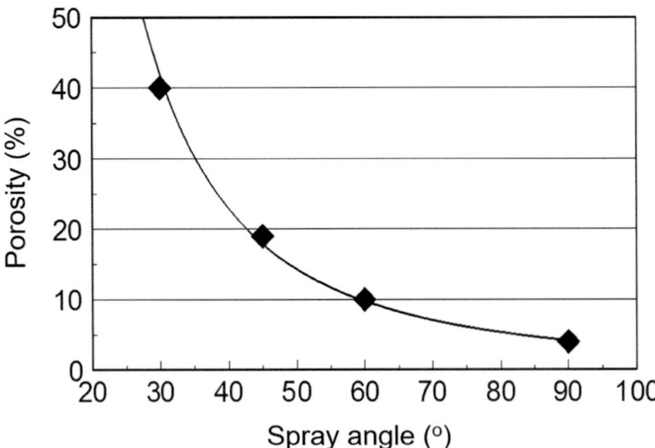

Fig. 15.84 Dependence of the porosity on the spray angle for "enlarged" substrate contours °. [Henne et al. (1999)]

order to avoid overheating the smallest particles while insufficiently melting the larger ones. Substrate preheating to a temperature close to the polymer melting point is also needed to achieve complete melting and flattening of the particles and the uniform flow of the melted polymer. On the other hand, the controlled cooling of the coating following the deposition step, allows for the control of the coating crystallinity. Thermal post-processing of polymer coatings is also often necessary to improve the tribological properties of the coatings as indicated by their lower friction coefficient and lower wear rate.

Numerous examples are available in the open literature for the thermal spraying of a wide range of polymers such as polypropylene (PP) using an oxyacetylene flame process [Alamara et al. (2010)], and Ultra-High Molecular Weight Polyethylene UHMWPE coatings with relative molecular weights up to six million [Bao et al. (2005)]. PEEK coatings using an HVOF process (Stellite Jet Kote II) were sprayed by [Patel et al. (2010a, b)] onto ANSI stainless steel 304 substrates that have been subjected to three different pretreatments to investigate the effect of varying surface roughness and chemistry on the coating's formation. The three surface treatments were as received degreased, chemically etched and steel grit blasted, including substrates with the same R_a but with different skewness (positive, negative, or zero). The formation of surface roughness on these substrates with a positively skewed distribution can be produced using an etching process. A positively skewed roughness with an adapted magnitude promotes mechanical interlocking. PEEK coatings on surfaces with positive skewness (S_k) show better coating adhesion than similar surfaces with negative skewness. [Patel et al. (2010a, b)] also studied the effect of flame impingement on PEEK coating degradation. When there is significant flame impingement onto the just deposited coating, as with coatings produced at 200 mm

spray distance, hydrocarbons from the flame react with the coating surface. The extent of thermal degradation decreases with increasing spray distance. The choice of spray gun nozzle did not have as great an effect as spray distance.

[Normanda et al. (2004)] have flame (acetylene–oxygen)-sprayed PEEK coatings applying two cooling conditions to modify the PEEK crystal structure. A slow cooling, achieved by ambient air, allowed obtaining a partially crystallized structure. The crystallinity was estimated, by DSC measurements, at 36%. A rapid cooling ensured by water quenching permitted to freeze an amorphous state. The same study indicated the presence of $-CH_2-$ groups in the main chain of PEEK. The explanation given by authors relied on the reactions of acetylene with carbon and hydrogen at high temperature. [Li J, Liao H and Coddet C (2002)] and [Zhang et al. (2006)] have also flame-sprayed (acetylene–oxygen) PEEK coating samples. The substrates were preheated to 400 °C (PEEK melting temperature of 343 °C). Different types of coatings were sprayed with the same spray conditions but different cooling rates [Normanda et al. (2004) and Li J, Liao H and Coddet C (2002)]: quenched into iced water immediately after spraying and annealed at different temperatures (180–300 °C) for different holding times (from 1 to 30 min). The annealed samples have a semi-crystalline structure, and the quenched sample crystallinity reached 36%. The analytical results indicated that two crystal entities emerged from the coating during the isothermal treatment following a stepwise crystallization fashion. The semi-crystalline coatings exhibit higher hardness, lower friction coefficients, and wear rates. The increase in the annealing temperature and the holding time in the studied range could lead to a higher coating hardness.

Polymer spraying has a certain number of limitations, particularly on the coating quality such as high porosity and low interfacial adherence. For that reason, a thermal post processing step is often necessary. Poly-ethylene-terephthalates (PETs) obtained from used soda bottles were thermally sprayed by HVOF (kerosene/oxygen/nitrogen) [Branco JRT and Campos SV (1999)]. Heat-treated thermally sprayed coatings delivered a better tribological performance than virgin PET, as indicated by their lower friction coefficient and lower wear rate against a hardened steel pin. [Soveja et al. (2010)] and [Zhang et al. (2009)] have studied the effects produced during a laser beam heat treatment (Nd: YAG, CO_2, or diode lasers) on the morphological structure (compactness) of PEEK and PTFE coatings, which were flame sprayed, and on the mechanical properties (adherence and tribological properties). Whatever laser wavelengths were used, the results showed a good effect of the laser treatment: improvement of both polymer coating compactness and its adherence to the substrate. The coatings' mechanical properties were correlated with their resulting structures.

Using polymer matrix coatings filled with commercial or nanometer-sized ceramic (e.g., silica or alumina) reinforcements or fillers have been reported to improve polymer-coating properties. However, the strong velocity and temperature gradients of thermal spray jets, together with variations in particle size, mass density, and morphology can, result in significant segregation when such dissimilar materials are co-sprayed.

[Ivosevic et al. (2005)] have sprayed functionally graded materials (FGMs) comprising layers of pure thermosetting polyimide, polyimide + WC-Co, and pure WC-Co onto polymer matrix composites (PMCs). For that PMR-II (thermoset polyimide) particles (20–100 µm) were injected axially into the Jet Kote II HVOF gun, working with hydrogen and oxygen, while WC–Co particles (5.6–22.5 µm) were fed externally. The WC–Co topcoat was flame sprayed. The drastic difference of mass densities of polymer (1390 kg/m^3) and WC–CO (12,500 kg/m^3) must be pointed out. Substrates were grit blasted and preheated (230–340 °C). The relative proportions (on a pore-free basis) of polyimide matrix and WC–Co filler were 58 and 42%, respectively. The porosity of the pure polyimide coating was determined to be 26%. The two- and three-layer FGM coatings both failed by the same mechanism: a combination of cohesive failure within the composite layer and a delamination from the substrate.

[Berndt et al. (1998)] have flame sprayed (propane/air mixture) ethylene methacrylic acid copolymer powder (115 µm) mixed either with nickel-chromium alloy particles (mean size 23 µm) or alumina particles (mean size 16 µm). In both cases tensile strength, elongation to break, and energy to break decrease with the filler content.

[Jackson L et al. (2007)] have HVOF (hydrogen/oxygen) sprayed nylon particles (10–180 µm) together with nominal 10 vol.% of nano and multiscale silica reinforcements. Combining Nylon-11 and silica in a 9:1 ratio by volume produced composite feedstock particles. The nanometer-sized particles were made of fumed silica (7 nm mean size), while the multiscale reinforcement comprised equal proportions of each of the five particle sizes ((7, 12, 20, 40 nm, and 10 µm), physically blended prior to ball milling. Each composition was ball-milled for 48 h, using zirconia balls, to ensure sufficient embedding within coating of the softer Nylon-11 with the silica. Nylon-11 coatings reinforced with 7 nm silica exhibited greater sliding wear resistance than pure Nylon-11, but they were better than coatings with the multiscale silica reinforcement. The same study, however, revealed that the crystallinity of the polymer matrix played a significant role in the wear behavior. The advantage of pure Nylon-11 coating is probably due to the occurrence of two additional wear mechanisms: abrasive and fatigue wear.

Another way to spray composite ceramic-polymer coatings is that proposed by [Mateus et al. (2005)]. They have used plasma spraying (Ar-25 vol.% H$_2$ mixture) where

Alumina–Titania (3 wt.%) particles (30 µm in mean diameter) were injected at the anode nozzle exit, while the PTFE (20 µm in mean diameter) and PFA (30 µm in mean diameter) particles were injected 30 mm downstream with the injector oriented toward downstream of the plasma plume flow. The sprayed polymer particles represented only 8 vol.% of the total sprayed materials. A well-melted ceramic matrix in which rounded polymer particles are randomly distributed characterizes the composite coatings. The as-sprayed polymer particles kept at the coatings surface can however be spread, thanks to a thermal treatment at 350 °C for 15 min. Unlike with spraying of PFA, a large material loss was noticed during the spraying of PTFE. Thus, the highest polymer/ceramic ratio is observed in the coatings made with the separate injection of the Al$_2$O$_3$–TiO$_2$ and PFA powders. The friction tests carried out with thermally treated coatings show that the friction coefficient and the wear of the steel 100C6 balls depend on the amount of polymer (PTFE or PFA) at the coatings surface. To keep it high enough, because surface particles reach quickly the decomposition temperature, cooling and spray conditions must be well controlled. The polymer viscosity at the melting state is also a predominant parameter: a low value promotes a good splashing of the melted particles on the surface substrate and also a good particles adhesion. Thus, with PFA a better adhesion is obtained than wit PTFE.

Good results have been obtained with polymer particles doped with nanometer-sized hard particles (silica or diamond). For example, [Schadler et al. (1997)] have HVOF (H$_2$/O$_2$)-sprayed nylon-11 particles ball-milled with either 7 nm or 12 nm silica powders treated to be either hydrophobic or hydrophilic, respectively. Compared to pure nylon coatings, scratch resistance improved by 30% and wear resistance by 50%. The surface chemistry of the silica filler affected the final coating properties. Silica particles with a hydrophobic (methylated) surface resulted in better mechanical properties than those with a hydrophilic (hydroxylated) surface. [Stravato et al. (2008)] have HVOF-sprayed Nylon-11 particles ball-milled with zirconia balls. Qualitative assessment indicated that coating adhesion was improved through the addition of the nanometer-sized diamond to the Nylon-11. Feedstock preparation is an important aspect and further work is clearly needed to investigate and optimize the process for better dispersion and covalent bonding of the nano-diamond particles to the polymer matrix and for characterizing the improvement in properties as a function of nano-diamond content.

15.4.5 Cold Spray Coating Formation

Coating formation by "Cold Spray" technology using ductile particles is quite different from that obtained with conventional thermal spray processes with fully or partially melted

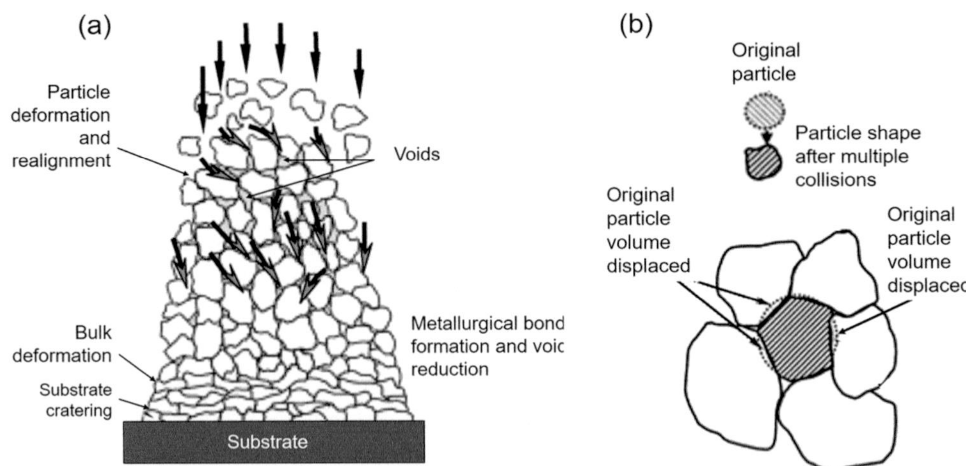

Fig. 15.85 (**a**) Schematic of the progressive plastic deformation of impacting particles when spraying on a stationary substrate. (**b**) Schematic representation of the transformation process. ([Van Steenkiste et al. (2002)]. Reprinted with kind permission from Elsevier)

particles. As discussed earlier in Sect. "15.3.1.1 Splat Formation from Non-molten Solid Particles," the spraying coating using solid non-molten particles is possible through the acceleration of the spray particles to high velocities in the hundreds of m/s, at temperatures essentially in the range of 40–50% their melting temperature, T_m [Schmidt T et al. (2006)]. Their impact velocity must be higher than the critical velocity, v_{cr}, (depending on particle ductility) while lower than the erosion velocity, v_{er} to avoid coating loss by erosion.

As with "sweet spot" deposition of molten particles, cold spray deposition on a stationary substrate, gives rise to the formation of a conical-shaped deposit depending on the torch-nozzle geometry and particle mass flow rate. [Van Steenkiste et al. (2002)] sprayed Al particles (d_p >50 μm) on a stationary brass substrate using Cold Spray (CS) process with pressurized air at 2.0 MPa, heated to temperatures in the range of 208 and 371 °C as process gas. The converging-diverging supersonic gun nozzle used had a rectangular exit aperture of approximately 2.10 mm with an approximate throat diameter of 3 mm and 80 mm length (throat to nozzle exit). The coating cross section (schematically presented in Fig. 15.85a) exhibits distinct areas with very different deposit morphologies can be observed. For a better description of the involved phenomena, [Van Steenkiste et al. (2002)] identified the following four stages in the coating buildup:

- Substrate cratering and buildup of first layer of dense deposit
- Particle deformation and realignment (multilayer coating buildup). As illustrated in Fig. 15.85b, particle rotation and deformation inducing void reduction initiated by the collision of the incident particle with the surface particles, which in turn are forced into their neighbors. The dashed outline in the figure stands for the original contour of an ideal spherical sprayed particle surface, while the solid

line is the contour of the particle's surface after densification and deformation has occurred
- Metallic bond formation between particles (particle-particle bonding) and void reduction.
- Further densification and work hardening of the coating (excess kinetic energy required for this stage)

Coating adhesion is determined by the plastic deformation (adiabatic shear instabilities), which occurs at the particle-particle interfaces when the impact velocity is above the critical velocity [Gärtner F et al. (2003)].

As is the case for fully melted particles, the impact angle also plays an important role. [Li C-J et al. (2005a, b, c, 2006a, b, c, d, e)] studied the critical velocity for particle deposition in cold spraying through numerical simulation and experiments using Cu powders. The critical velocity was linked to particle size distribution parameters and spray angle through the relative deposition efficiency. Results giving the relative spray deposition efficiency, defined as the ratio of the deposition efficiency at any angle to that at 90° incidence on the substrate, as function of particle impact angle, are presented in presented in Fig. 15.86a, b, for copper powder with a mean particle diameter of 56 μm and two nitrogen gas pressures of 1.0 and 2.6 MPa, respectively. The gas temperature was maintained at ≈ 400 °C during the spraying for these two conditions. The average particle velocities for those two conditions were 260 and 325 m/s, respectively, for particles of the same mean diameter. As can be seen, in both cases the relative deposition efficiency is zero for angles less than 50°.

Overlapping beads will make coatings, as with spraying of fully melted particles (thermal spraying). Compared to beads obtained with thermal spraying, the wings of cold-sprayed coatings are more abrupt, particles impacting with an angle

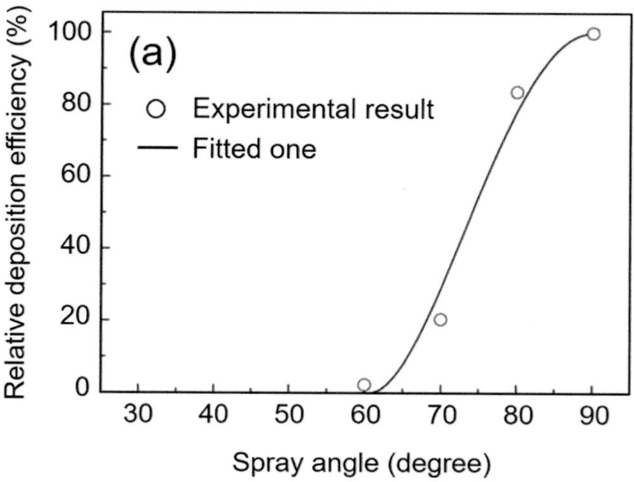

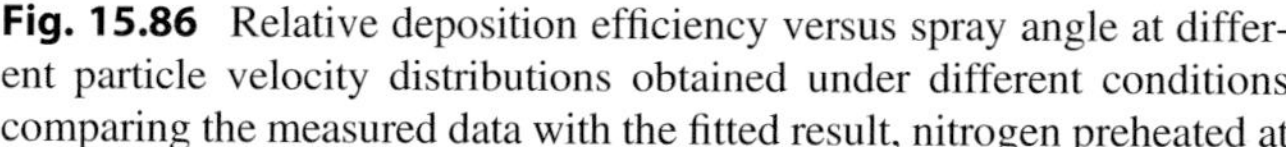

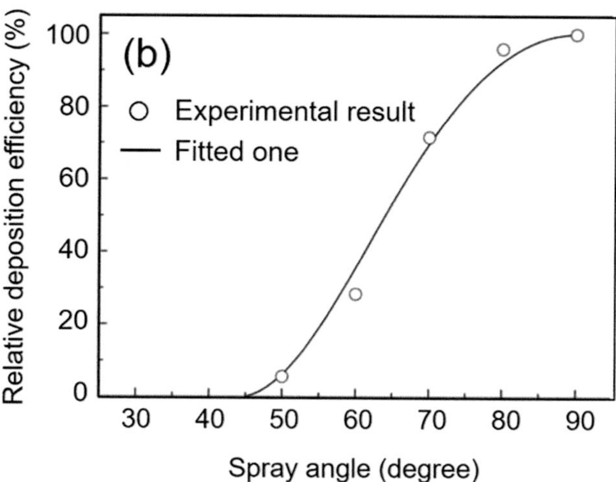

Fig. 15.86 Relative deposition efficiency versus spray angle at different particle velocity distributions obtained under different conditions comparing the measured data with the fitted result, nitrogen preheated at 400 °C pressure, (**a**) 1 MPa; (**b**) 2.6 MPa. ([Li C-J et al. (2006a, b, c, d, e)]. Reprinted with kind permission from Springer Science Business Media, copyright © ASM International)

reaching rapidly a velocity below the critical value. An example of a cold-sprayed copper coating (particle diameter 5–22 μm) is presented in Fig. 15.87 after [Stoltenhoff et al. (2002)]. With a transverse sweep rate speed of the substrate of 125 mm/s and with line spacing of 1.7 mm, a thickness of 200–230 μm per layer was deposited with a deposition efficiency of 72%. Figure. 15.87a shows optical micrographs of the polished coating cross section. In comparison to thermal spray coatings, the coating is very dense and low in oxide content. Microstructure images show no evidence of pores. Differences in the coating density compared with solid Cu can only be determined using leak tests in which a vacuum of 10–12 kPa is produced on one side of the coating and in which the opposite side is exposed to helium. A coating with a thickness of 5.5 mm yielded a leak rate of only 10 kPa L/s, which roughly corresponds to the requirements for high-vacuum components [Stoltenhoff et al. (2002)]. The etched cross section (Fig. 15.87b) shows the deformation of the sprayed spherical particles with a rather good contact between layered particles and few tiny pores. Of course, oxidation (coating oxygen content of less than 0.1 wt%) is that of the original particles [Stoltenhoff et al. (2002)].

[Koivuluoto et al. (2007)] obtained similar results when cold spraying copper, nickel, and nickel–30% copper coatings. SEM analysis of etched coating cross sections showed the particle boundaries, deformation, and microstructure inside the sprayed particles. Coatings seemed to be dense according to the SEM studies. However, corrosion tests showed some through porosity in the cold-sprayed Ni and Ni-30% Cu coatings. The weak points were on the particle boundaries where the bonding is not strong and uniform enough between the particles. Zahiri et al. (2006) have studied the effects of temperature, pressure, and particle size with respect to porosity formation in cold-sprayed copper particles onto aluminum substrates. It is observed that an increase in spray temperature from 300 to 630 °C results in a decrease in the volume fraction of porosity from 44 to 2.5 μm. Similarly, an increase in temperature leads to a decline in average size of pores. These observations are due to the fact that an increase in spray temperature raises the particle velocity and the particle temperature. Furthermore, [Zahiri et al. (2006)] have presented a schematic representation of the particle size distribution effect on porosity formation in Fig. 15.88. Three powder size distributions are considered:

- Powder with a large particle size and a narrow particle size distribution
- Powder with a mixture of small and large particles that represents a broad particle size distribution
- Powder with a small particle size and narrow particle size distribution

In the ideal case, (a) large particles achieve lower velocities than the small ones and, if their velocity is above the critical one, a limited plastic deformation (flattening) occurs. Porosities in the microstructure probably develop due to interparticle gaps and bridging of the particles (Fig. 15.88a). At the opposite case (c) small particles achieve the maximum velocity (in the gas stream) required for particles to significantly deform and eliminate voids between particles. This leads to a considerable reduction in porosity and the average pore size (Fig. 15.88c). For a mixture of large and small particles, case (b), small particles acquire enough velocity (kinetic energy) to deform significantly (Fig. 15.88b). However, the large particles (similar to the schematic in Fig. 15.88a) do not considerably deform. This

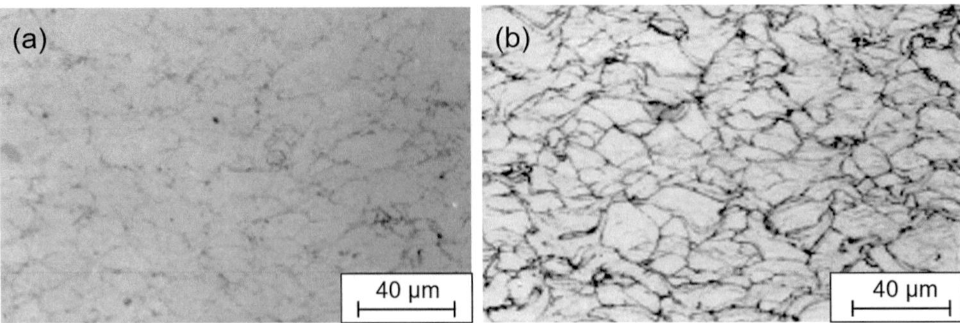

Fig. 15.87 Optical microscopy of the cross section of a cold sprayed Cu coating: (**a**) polished, (**b**) etched. ([Stoltenhoff et al. (2002)]. Reprinted with kind permission from Springer Science Business Media, copyright © ASM International)

results in a heterogeneous porosity formation with small and large cavities [Zahiri et al. (2006)].

15.4.6 PTA Coatings/Deposition

As described in Chap. 12, in Plasma Transferred Arc (PTA) deposition the substrate is part of the electrical circuit that delivers the power for the coating process, and the molten powder of the coating material is mixed with the molten substrate material. Moreover, compared to other coating processes the torch has also an oscillatory movement limiting the local heat flux and producing flatter beads. Figure 15.89 presents a typical cross section of a PTA bead obtained with the part to be coated as anode [Ducos M (1985)]. At can be seen the bead is rather flat, which is due both to the flow of molten-sprayed material and the oscillatory movement. The top of the bead presents irregularities [Wilden J et al. (2006)] linked to the oscillatory movement. According to [Ducos M (1985)]. the deposition distance and the arc current of the pilot arc have no characteristic influence on the bead shape and deposition efficiency. When the transferred arc current increases, the bead becomes wider, less thick, and the deposition efficiency is raised. The increase in the powder flow rate induces thicker beads and lower deposition efficiencies.

However, parameters controlling bead formation also determine the dilution, which increases if the transferred arc current increases, decreases if the powder flow rate is raised or the relative velocity torch–substrate increased. According to [Gatto et al. (2004)] the quality of the deposition is strongly determined by an optimal reciprocal positioning and movement between the torch and the part to be coated, especially for critical geometries. Thus, an accurate design of the process, in particular of the deposit geometry and of the torch tool path is mandatory.

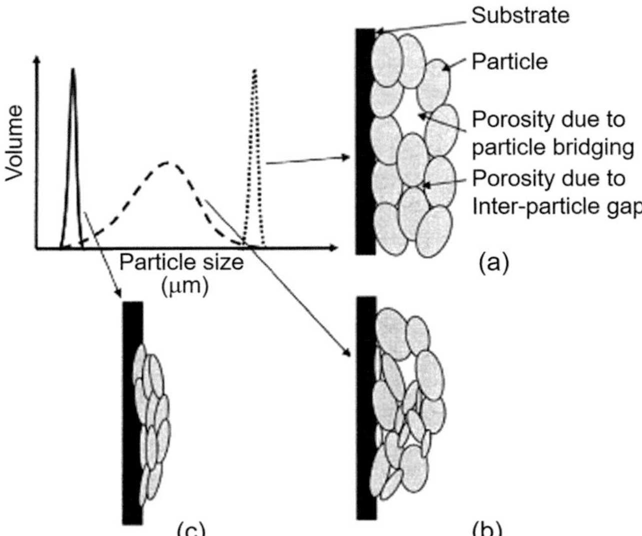

Fig. 15.88 Effect of powder of (**a**) large particle size and a narrow particle size distribution, (**b**) a mixture of small and large particle size with a broad particle size distribution, and (**c**) a small particle size with a narrow particle size distribution on porosity formation in cold spray process. ([Zahiri et al. (2006)]. Reprinted with kind permission from Springer Science Business Media, copyright © ASM International)

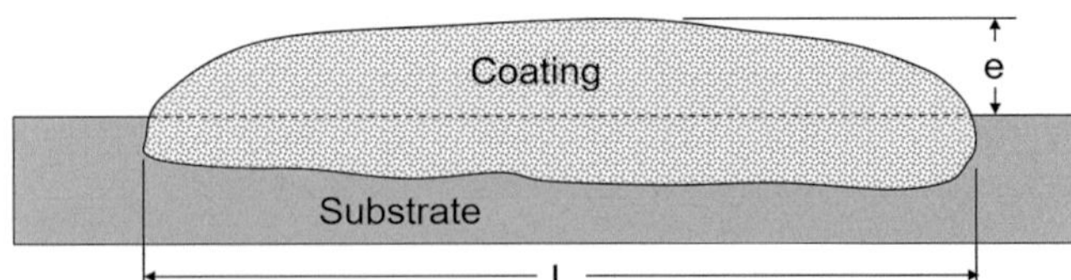

Fig. 15.89 Typical cross section of a PTA bead obtained with the part to be coated as anode. (Reprinted with kind permission from Dr. M. Ducos [Ducos M (1985)])

15.4.7 Coatings Obtained by Very Low-Pressure Plasma Spray

The very Low-Pressure plasma spray (VLPPS) process has been developed with the aim of depositing uniform and thin coatings with large area coverage by plasma spraying

[Mauer G, et al. (2010)]. At typical pressures of 100–200 Pa, the characteristics of the plasma jet change compared to conventional Low-Pressure plasma spraying processes (LPPS or VPS) operating at 5–20 kPa. As described by [Gindrat et al. (2002)] as the ambient pressure is significantly below the exit pressure of the jet (above atmospheric pressure), the plasma jet becomes under-expanded with a hot and dense plume exiting the nozzle. The plasma velocity is almost constant over a large distance from the nozzle exit. Furthermore, at these low pressures the collision frequency is distinctly reduced, and the mean free path is strongly increased, the flow being laminar. Enthalpy probe measurements showed that the axial temperature profile of the plasma jet is at higher levels and more uniform compared to conventional spray processes at higher chamber pressures [Mauer G, et al. (2010)]. Depending on the power level of the plasma torch used (between 40 and 160 kW) the injected powder material is either only molten into liquid droplets or evaporated if the power level is high enough.

15.4.7.1 Dense and Thin Coatings

These coatings are sprayed at pressures in the range of 100–200 Pa with power levels below 60 kW and typical jet dimensions are >1 m in length and >0.2 m in diameter depending on the working parameters. These particular plasma characteristics offer enhanced possibilities to spray thin and dense ceramics. Oxygen ion conducting membrane layers were sprayed using a spray dried $Ba_{0.5}Sr_{0.5}Fe_{0.2}Co_{0.8}O_3$ powder ($d < 45$ μm) with the VLPPS process at 150 Pa working with $Ar/N_2/He$ plasma forming gas mixtures at power levels between 44.7 and 51.2 kW [Mauer G, et al. (2010)]. No hydrogen was used as secondary plasma gas to avoid decomposition due to reduction. The stainless-steel substrate was preheated to 500 °C before spraying. The composition of the plasma gas was very important for coating properties, only a small gap existing between sufficient melting of the perovskite particles and the beginning of their decomposition. Moreover, sufficient gas-tight coatings have not been successfully obtained for gas separating membrane applications. In contrast, VLPPS-sprayed Mg-spinel (−25 + 5 μm) SOFC insulation layers were gas tight and showed adequate electrical resistances at a coating thickness of approximately 40 μm.

[Zotov et al. (2012)] have deposited $La_{0.6}Sr_{0.4}Fe_{0.8}Co_{0.2}O_{3-\delta}$ (LSFC powder spray pyrolyzed) coatings using a pure Ar–He plasma with a power level of 55 kW at a chamber pressure of 200 Pa. Unfortunately, the decomposition of the LSFC powder occurred with loss of oxygen. In addition to the LSFC perovskite, these coatings contained $SrLaFeO_4$, $Fe_{3-x}Co_xO_4$, CoO, and FeCo phases in different amounts. Coating morphologies were strongly dependent on the Ar–He gas mixture [Zotov et al. (2012)]. With Ar-rich plasma without additional oxygen homogeneous coatings with layered microstructures were formed with porosities decreasing with increasing substrate temperature. In contrast, with He-rich plasma more porous columnar microstructures were obtained. When additional oxygen was pumped into the vacuum chamber and the torch power lowered, coatings contained more than 85% perovskite phase and only a few percent $Fe_{3-x}Co_xO_4$ and/or CoO. Morphologies of the LSFC coatings obtained with additional oxygen were linked to the substrate temperature, the deposition rate, and the plasma jet velocities. An increase in the plasma jet velocity leads to the formation of homogeneous coatings [Zotov N et al. (2012)]. These results show that further optimization of the powder injection, the spray parameters, and a better understanding of the growth mechanisms are still necessary to obtain homogeneous gas tight LSFC coatings, suitable for membrane applications [Zotov et al. (2012)].

15.4.7.2 Coatings from Vapor Phase

[Hafiz et al. (2006)] have developed a novel method for generating and depositing nanoparticles to form nanocrystalline films, patterns, and structures. They called it "hypersonic plasma particle deposition." Gas phase products are injected into a reactive thermal plasma that is expanded into a low-pressure chamber via a convergent boron nitride nozzle. The high temperature (∼4000 K) in the injection region of the BN nozzle dissociates the gas-phase reactants (for instance $SiCl_4 + CH_4$ for producing SiC deposits) into their elemental constituents. Then the expansion through the nozzle results in super-saturation of the reactant vapors and consequent nucleation of particles. The substrate is placed 20 mm downstream of the nozzle in the low-pressure chamber (at ∼267 Pa versus ∼60 kPa in the injection region) [Hafiz et al. (2006)]. This distance results in hypersonic velocities at impact due to under-expansion of the flow. Short residence times inhibit particle growth, and most particles remain between 10 and 40 nm in diameter. The particles deposited are directly nucleated from gaseous precursors, eliminating the need for powder feeding and producing dense and hard Si–Ti–N films that are inherently nanostructured [Hafiz et al. (2006)].

The need for better performance from advanced turbine engines requires higher operating efficiencies and longer operating lifetimes. EB-PVD processes operate at low pressures (0.1–5 Pa) [Doering T (2007)] and work well but are very expensive. Because of the lower application cost, and a large installed equipment base of the plasma spray process, Sulzer-Metco Co. has developed thermal plasma torches working at pressures as low as 100 Pa to process the coating material [von Niessen K et al. (2010)]. This involves a technique based on a partial or nearly complete evaporation of the feed material. The powder is injected into the plasma torch (two- or fourfold internal powder injection) to produce a coating either from molten particles or from vapor depending on the powder, torch operating conditions and working pressure. This process is called either plasma spraying-physical vapor deposition (PS-PVD) or very low-pressure plasma spraying (VLPPS). The PS-PVD technology, with a modified plasma gun, works at reduced ambient pressure (down to 100 Pa) and increased power level (total gas flow up to 200 slm and power levels up to 180 kW with 3000 A). Under such a pressure condition, the plasma jet reaches more than 2 m in length and up to 0.4 m in diameter. At this pressure, the plasma jet interaction with the surrounding atmosphere is very weak.

Preheating and cleaning of the substrate can be performed using the plasma or an integrated transferred arc processing to optimize the coating adhesion. Typical substrate temperatures during the spray process can vary between 900 and 1100 °C depending on the spray conditions and substrate material. In general, a higher substrate temperature is advantageous for the formation of a well-defined columnar microstructure. Depending on the working conditions either splat-like or columnar structures are obtained, as shown in Fig. 15.90. The columnar structure is obtained when using a low powder feed rate, a specific mixture of argon and other plasma gases, a high-power level to vaporize the injected particles, and a large spray distance.

[Von Niessen K. and M. Gindrat (2011)], using the very low-pressure plasma spraying process (VLPPS), showed that it was possible to produce from the vapor phase coatings with columnar microstructure. The single columns consist of many fine needles with a high defect density and a high amount of internal porosity, as shown in Fig. 15.90. With

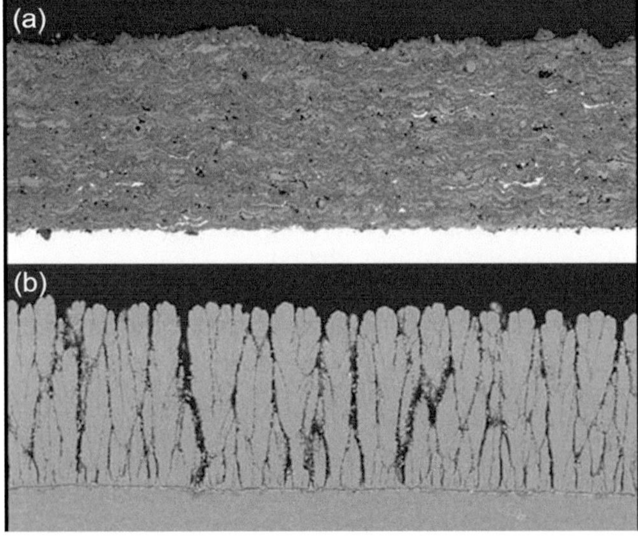

Fig. 15.90 TBC coating of 7 YSZ using different PS-PVD plasma parameters showing (**a**) splat-like structure and (**b**) columnar structure. ([von Niessen K et al. (2010)]. Reprinted with kind permission from Springer Science Business Media, copyright © ASM International)

this morphology, the coatings have proven to be highly strain tolerant during furnace cycling tests and offer a low thermal conductivity. The erosion resistance is lower than that of EB-PVD coatings but at a level of standard APS-TBC coatings with 15% porosity. Moreover, this technology offers a high potential to support the development of new non-line-of-sight coating.

[Hospach et al. (2011)] have obtained columnar-structured YSZ layers between 20 and 750 μm. The geometry and arrangement of the sample and the sample holder showed a big influence on the coating quality. The tested columnar-structured TBCs failed after 1000 cycles, which are fairly close to the already established APS-generated coatings. While this process is promising, a better understanding of phenomena is mandatory to further improve the results. The same authors [Hospach et al. (2012)] have shown that this process offers many opportunities to modify the coating microstructure from dense coatings (droplet deposition) to columnar ones. In between exists a quasi-vapor deposition from clusters resulting in columnar coatings with high porosity and growth rate. With the long and wide plasma plume homogeneous coatings are obtained.

[Mauer et al. (2012)] made a careful study of the way coatings were produced with agglomerated 7YSZ ($d_{50} = 8$ μm) and fused and crushed TiO_2 ($d_{50} = 13$ μm) particles. To ensure an evaporation of the feedstock, high-power density inside the nozzle in proximity to the location of injection together with well-selected feedstock characteristics are required. Even if the jet excitation temperature can reach values in the range of 8000–10,000 K, because of its low density no further particle heating is possible. Besides evaporation, the formation of nanosized clusters is observed. For the coating growth [Mauer et al. (2012)] have used the structure zone model (SZM), initially developed by [Thornton JA (1975)] for magnetron-sputtered coatings. Different growth mechanisms occurred like shadowing and surface and bulk diffusion. However, the vapor deposition of compact columnar structures is only possible if the surface mobility of the "adatoms" is sufficient, an adatom being an atom that lies on a crystal surface and can be thought of as the opposite of a surface vacancy. High substrate temperatures and low deposition rates are the conditions to obtain vapor deposition. These conditions are defined by the ratio $T_s/T_m > 0.6$ and the area specific impingement rates $\dot{n} = \dot{N}/N_A$, where $\dot{N}$ represents the area specific impingement rate (particles/m^2 s) and N_A the Avogadro number (mol^{-1}). Figure 15.91 from [Mauer et al. (2012)] present the SZM in the VLPPS process as far as the zones could be identified. The nomenclature of the zones corresponds to the Thornton SZM. By calculating the homologous temperatures and mole rates, experiments with different materials could be considered in the same diagram. The activation energy for surface diffusion

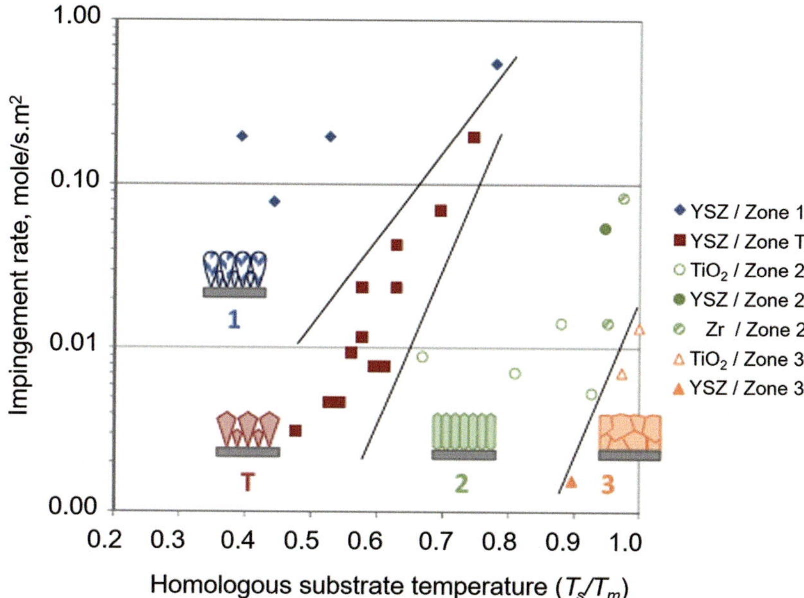

Fig. 15.91 Very low-pressure plasma spraying structure zone model: (1) Porous structure consisting of tapered crystallites separated by voids, (T) Transition structure consisting of densely packed fibrous grains. (2) Columnar grains. (3) Recrystallized grain structure. ([Mauer et al. (2012)]. Reprinted with kind permission from ASM International)

was estimated by [Mauer et al. (2012)] to be 141 kJ/mol for YSZ, which is relatively low.

15.5 Substrate and Coating Thermal Management

Temperature and stress control are intricately interrelated in thermal spray processes with residual stresses in a coated part closely dependent on its mechanical and thermal history during the different stages of the process. With the exception of cold spray, the substrate and coating under formation receive two different heat fluxes during the coating operation: that from the hot gases and that from hot particles. As illustrated earlier in Fig. 15.67, if the particle injection is radial, both heat fluxes have different axes, while they have the same axis with axial powder injection.

Whatever the spray process may be, the torch is moved, more or less rapidly (up to 1.5–2 m/s), relative to the substrate, and thus heat fluxes are transient, or more precisely cyclic with the beads overlapping, forming passes. The heat flux due to hot gases is almost negligible with wire arc spraying and D-gun spraying (substrate heating, without cooling systems, is in the 100 °C range). However, this it is not the case with flame, HVOF, HVAF, and plasma spraying, systems for which external cooling means are generally necessary. In PTA deposition process, on the other hand, heat flux must be sufficient to create a molten pool locally (under the plasma transferred arc) where the particles and the substrate are locally, heated and melted by the arc.

The temperature of the substrate and the coating being deposited are closely linked to the following parameters: the spray process used, the spray distance (due to the mixing with the surrounding atmosphere the hot gas heat flux decreasing almost exponentially with the increase in the spray distance), thermal properties of the substrate and the sprayed material (heat absorption by the substrate, $m.c_p$ kJ/K, is quite different for a 0.5 mm thick substrate compared to that for a 50 mm thick one, with the same other dimensions), relative velocity torch–substrate (the residence time of the torch moved at 1 cm/s is 100 times longer than that of one moved at 1 m/s !), and cooling systems (compressed air blown at the substrate surface is less efficient than a spray of liquid argon, but the cost of the latter is a few orders of magnitude higher).

Temperature monitoring during the different stages of the spraying process, substrate preheating, spray coating, and post deposition cooling, is essential for;

- Coating adhesion: the substrate must be preheated above the transition temperature before performing spraying to get rid of adsorbates and condensates and to improve the coating adhesion and decrease the quantity of splashed material. As mentioned previously, preheating can be replaced by dry ice blasting in front of the sprayed spot or by laser treating in front of the sprayed spot.
- Avoiding temperature gradients within coatings, especially when dealing with low thermal conductivities coating materials, ($\kappa < 30$–40 W/m K), to limit or suppress the resulting stress and eventually undesirable phase changes.

That implies spraying relatively thin passes (<10 μm) for low κ materials,

- The temperature at which the coating is kept during the spraying operation, and post deposition cooling of the coating and substrate is very important to tailor the coating residual stress distribution.

At the splat level for conventional particles, typical times are in the microsecond time range for the particle flattening and a few microseconds for its cooling, the next splat impacting at the same location about 10 μs later. Of course, such measurements, which are necessary for a better understanding of the thermal spray process, are typically performed in laboratories but not in spray booths. In industrial spray processes the important temperature that can be measured rather easily with conventional pyrometers, is the time averaged temperature of the substrate/coating in the tenths of second range, which is responsible for the post deposition residual stress distribution within coating and substrate.

In the following will be presented successively:

- Splat cooling
- Thermal management during the coating operation
- Substrate and coating temperature

This is followed in the next section by an analysis of stresses developed within the substrate/coatings as a result of the different stages of the coating-formation process and during the service of the coating.

15.5.1 Splat Cooling

Besides the heat flux from the spray torch imposed at the splat location, the thermal history during splat layering is closely linked to the number of splats impacting at the same location during the torch movement (a typical splat diameter is in the range 50–150 μm) [Kuroda S et al. (1995), Monnerie-Moulin F et al. (1992), Haddadi A et al. (1997), Fauchais P et al. (2001a, b) and Leger AC et al.(1998)]. This flux depends not only on the powder flow rate, deposition efficiency, and relative torch–substrate velocity but also on the thermal properties of the substrate and the previously deposited layers and the substrate mass and thickness, that is, the ability to conduct away the heat brought by the solidifying splats and the hot gas heat flux. Under conventional spray conditions, a pass thickness is between 3 and 15 μm, corresponding roughly to about 3–15 layered splats in HVOF or DC plasma spraying. [Kuroda et al. (1995)] were the first to propose a method to estimate this time-temperature history. The simplified transient 1-D model of [Leger et al. (1998)] predicts the cooling of layered splats, assuming that;

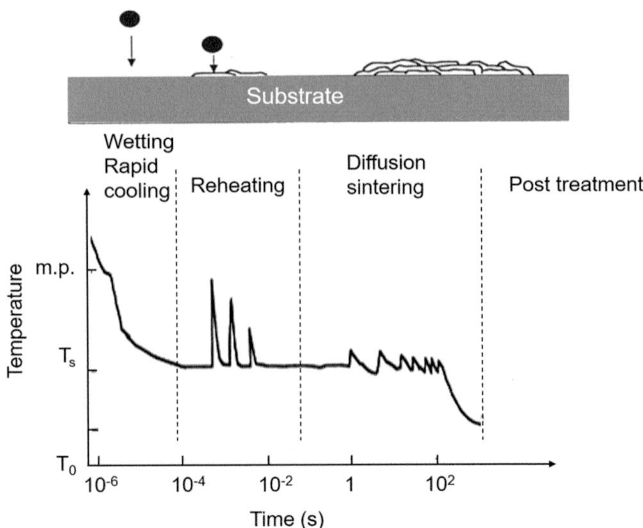

Fig. 15.92 Temperature time history (1-D transient model) of the bottom of the first alumina splat deposited on a low carbon steel XC38 substrate, 5 mm thick. The temperature of the impacting particle is 3000 K, resulting splat thickness 1 μm, time between each splat 15 μs, four splats layered in a sequence and 1.1 ms between each torch passage. ([Leger et al. (1998)]. Reprinted with kind permission from ASM International)

- solidification occurs at the melting temperature,
- the contact splat–substrate and splat-previously deposited layers are characterized by thermal contact resistances,
- time t_i exists between successive impacts, and
- different heat fluxes for the hot gases are imposed.

This model shows that splat cooling is mainly due, about 95%, to heat conduction toward the substrate or previously layered splats. Convective cooling by the cold gases close to the substrate when the torch is moved to another position and radiation from the cooling splats at the coating surface also contributes, by about 5%, to the cooling of the layered splats. When the thermal conductivity of the sprayed coating is higher than 50 W/m K, the gas cooling is more efficient (up to 15%) than for lower thermal conductivity materials. For example, Fig. 15.92 presents the impacts of alumina particles at 3000 K, forming splats 1 μm thick and 120 μm in diameter on low carbon steel XC38. The thermal contact resistance splat–substrate and splat–splat is assumed to be 5×10^{-7} m² K/W, and the time between two successive splats impacting at the same location is 15 μs. During splat layering the heat flux brought by the plasma jet is 2 MW/m². Once 4 splats are layered the torch comes back to the same location 1.1 ms later. The temperature time history presented in Fig. 15.92 is that of the bottom surface of the first splat. When the first splat is formed it cools down very fast, the melting temperature being reached in about 1.1 μs, then it takes around 10 μs to cool down to the substrate preheating temperature of 700 K. When the next splat arrives 15 μs later it reheats the first splat to almost the alumina melting temperature, the solidified first splat insulating partially its bottom surface. The next impacting splats reheat less and less the bottom surface of the first one. Of course, the successively layered four splats (1.1 ms between each passages of the torch) reheat very little the first splat, due to the thicker and thicker layered splats insulating more and more the bottom surface of the first one.

In Fig. 15.93a almost the same result is presented as in the previous figure, except that the initial temperature of the impacting particle is 3200 K, the torch is moved at 1 m/s, and only 3 splats are layered during one passage on the substrate, preheated to 720 K. The time between two successive passages of the torch is identical to the previous one. As can be seen in Fig. 15.93a each splat has plenty of time to cool down between the reheating by the next impacting particle. However, the situation is quite different when the velocity torch–substrate is significantly reduced (by a factor of 60) to layer 180 splats per pass. Once the first pass is deposited, the next one, 2 min later, impacts on a thick insulating material, which has still a surface temperature of 2000 K. Thus, according to the rather low thermal

conductivity of the first alumina layer and its high surface temperature, the successive splats of the second pass are kept in a molten state, as shown in Fig. 15.93b.

Of course, this case is far from those used in most spray protocols, even if it allows achieving ceramic coatings with vertical cracks from bottom to top, thus presenting an excellent compliance. Such calculations also show the necessity to spray thin passes (<10 μm) of materials with low thermal conductivity (<30−40 W/m K). For a very low thermal conductivity material such as zirconia (about 1 W/m K) a 10 μm thick pass creates already a temperature gradient of about 70 K/μm, as shown in Fig. 15.94. To conclude, the lower the thermal conductivity of the sprayed material is, the thinner the pass must be to limit temperature gradients.

15.5.2 Heat Fluxes Contributing to Substrate and Coating

Particle cooling, as described in the previous section, occurs mainly through the substrate and the previously deposited layers. The hot gas heat flux hitting the surface of the substrate during preheating, and then the coating during its formation, contribute to its heating, and cooling gas jets are mainly used to reduce the coating surface temperature. When considering the different spray processes, the following considerations apply to the heat flux imposed onto the substrate and coating:

- The highest heat flux is obtained with suspension and solution plasma spraying where spray distances are in the 30–40 mm for rod type cathode plasma spray torches, and about the same for the tri-cathode Triplex Pro torch (with fluxes slightly higher than for the former), and 60–80 mm for the Axial III torch of Mettech Inc. Heat fluxes under such conditions can be as high as 40–50 MW/m^2.

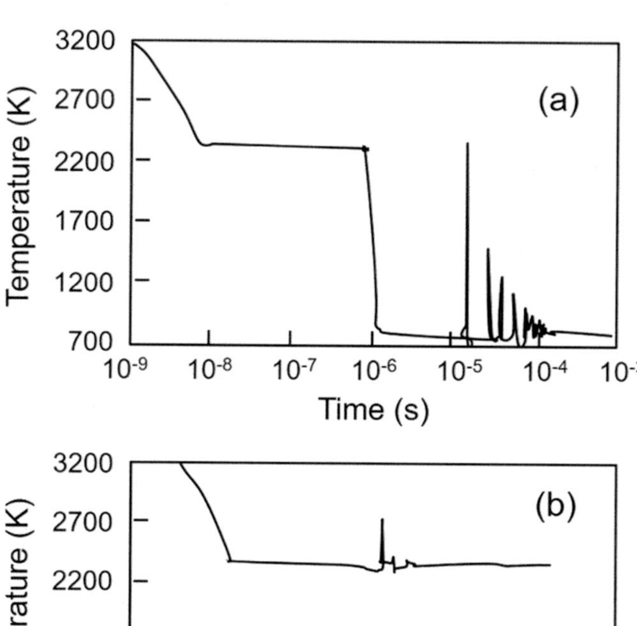

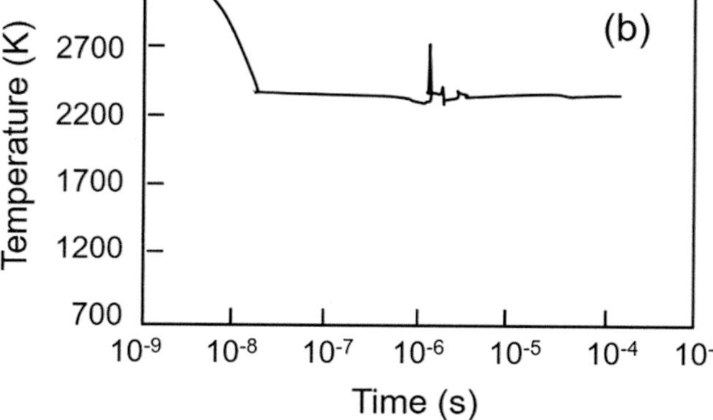

Fig. 15.93 Temperature time history (1-D transient model) of the bottom of the first alumina splat 1 μm thick formed by particles impacting at 3200 K on a low carbon steel XC38 substrate. (**a**) 3 splats layered in one pass. (**b**) 180 splats layered on the previously deposited alumina layer also 180 μm thick. ([Leger et al. (1998)]. Reprinted with kind permission from ASM International)

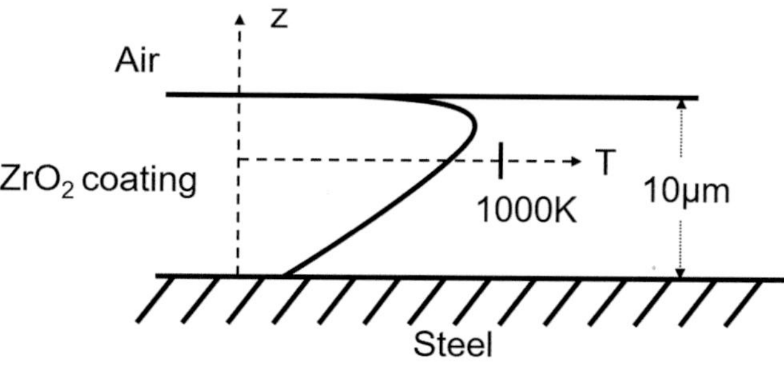

Fig. 15.94 Temperature gradient within a zirconia pass 10 μm thick sprayed on a stainless-steel substrate (1-D transient model). [Leger et al. (1998)]

- When spraying micrometer-sized particles with these torches, spray distances are typically between 80 and 150 mm and heat fluxes are in the 4–2 MW/m² range.
- With HVOF spraying of micrometer-sized particles, heat fluxes at substrate position that have been reported are similar to, or smaller than, those of plasma torches.
- For flame spraying, heat fluxes are below 1 MW/m², lower values being attained when accelerating particles or when atomizing wires with compressed air jets.
- With wire arc spraying the heat flux is below 0.1 MW/m².

Of course, to this gas heat flux must be added that of hot-sprayed particles that depend strongly on the pass thickness and the sprayed material. With DC plasma jets and radial injection both heat flux peaks have different locations, as illustrated in Fig. 15.67. Gas heat fluxes have roughly a Gaussian shape, at least for DC plasma [Monnerie-Moulin F et al. (1992)] or HVOF jets [Honner M et al. (1998)]. This is illustrated in Fig. 15.95 for a JP-5000 HVOF gun working with kerosene and oxygen. The spray distance, as pointed out above, is very important. [Hackett and Settles (1994)] have shown that after a 4 s exposure to the JP-5000 flame at a 270 mm spray distance, the temperature increase in the substrate is already 400 °C, against 100 °C at 350 mm. The coating temperature plays a key role in the coating oxidation (at least for metals, alloys, cermets, and non-oxide ceramics). For example, when spraying aluminum particles, the oxygen

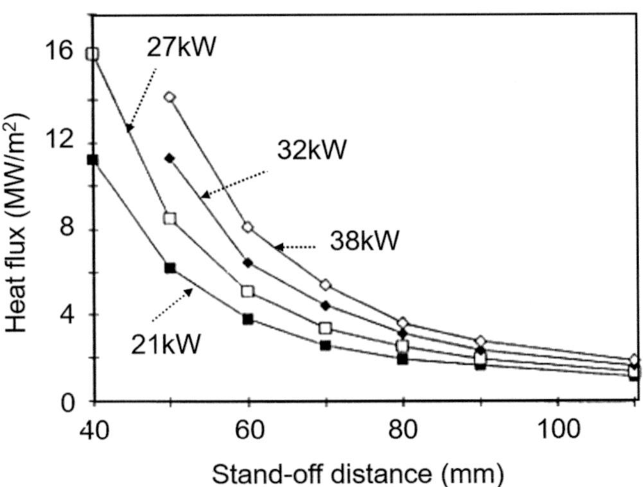

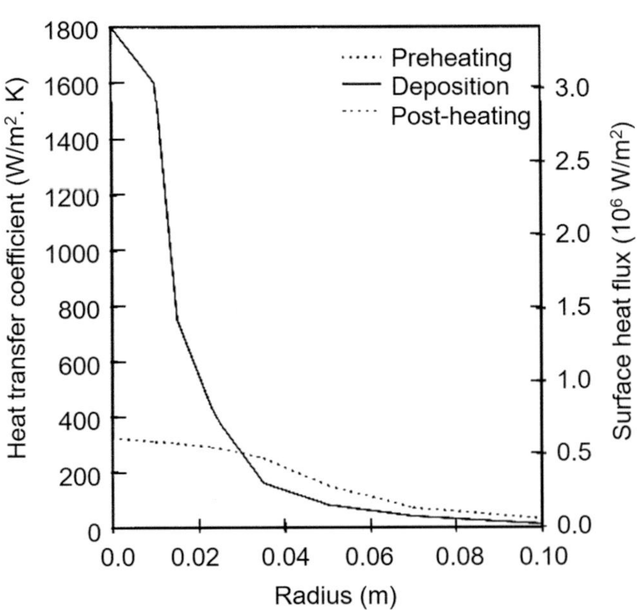

Fig. 15.95 Heat transfer coefficients and heat fluxes over the distance from the axis of the HVOF gun during preheating, deposition and post-heating (JP-5000, TAFA Inc., inside gun diameter 11 mm, kerosene 19.5 l/h, oxygen 893 slm, spray distance 380 mm, sprayed material: WC–Co 8.5 kg/h). ([Honner M et al. (1998)]. Reprinted with kind permission from Elsevier)

Fig. 15.96 Dependence on the spray distance of the maximum heat flux imposed by a DC plasma jet flowing in air at atmospheric pressure (Ar 45 slm+H₂ 15 slm, nozzle internal diameter 7 mm) for four power levels. ([Monnerie-Moulin F et al. (1992)]. Reprinted with the kind permission from Dr. F. Monerie-Moulin)

content of the coating was about 0.23 wt.% if the substrate temperature rise was only 100 °C, against 0.4 wt% for a rise to 400 °C [Hackett and Settles (1994)]. Similar results were obtained by modeling [Li M, Shi D and Christofides PD (2005b)].

The heat flux decreases almost exponentially with the spray distance. With the jet expansion in the surrounding atmosphere, both the jet velocity and temperature decrease very rapidly, as shown, for example, for plasma jets [Coudert JF et al. (1995) and, Fauchais P and Coudert JF (1996)]. An illustration of the maximum heat flux dependence on the spay distance is presented in Fig. 15.96 for an Ar-H₂ plasma jet working with 4 different power levels. If at 80 mm the heat fluxes can be above 4 MW/m², they are around 2 MW/m² or less at 100 mm. This explains why, in spite of the fact that the maximum particle velocities and temperatures are obtained around 60–80 mm, spray distances are often preferred between 100 and 120 mm where the heat flux control is easier. It must be noted that the presence of the substrate intercepting the hot gas jet increases the gas temperature close to the surface by a few hundred degree Celsius [Lapierre D et al. (1994)].

Heat fluxes become higher, as shown in Fig. 15.97 [Brousse E.(2010)], and thus much more difficult to control, with suspension or solution plasma spraying where spray distances are between 30 and 50 mm for a rod type cathode torch. Such heat fluxes also strongly modify the coating microstructures. Of course, these fluxes also depend on the plasma forming gases used and, according to the lower voltage and thermal conductivity of Ar-He mixtures, heat fluxes can be half of those obtained with Ar-H₂ with the same

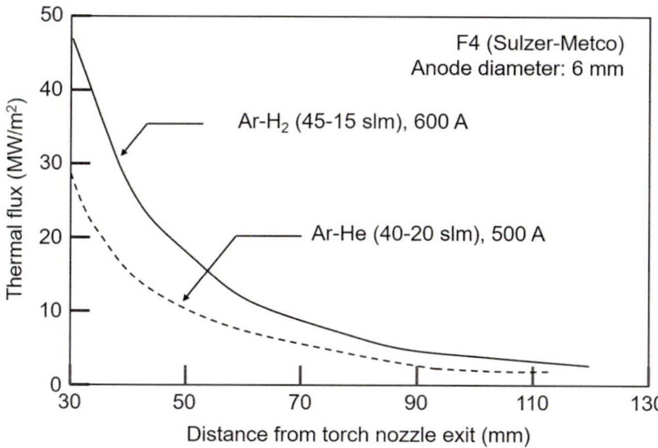

Fig. 15.97 Heat flux evolution along the axis of a rod type cathode torch (PTF-4 torch of Sulzer-Metco) when spraying suspensions or solutions: torch anode nozzle internal diameter 6 mm, (Ar 45 slm+ H_2 15 slm), 600 A, 72 V and (Ar 40 slm+ He20 slm) 600 A. ([Brousse E. (2010)]. Reprinted with kind permission from Dr. E. Brousse)

plasma torch working with the same arc current [Fauchais P et al. (2008)].

The spray pattern and the torch/substrate velocity play also a key role in the hot gas and particle heat transfer. According to [Floristán et al. (2012)], robot trajectory planning should integrate not only an accurate definition and control of process kinematics but also an optimal thermal guidance during deposition based on detailed analysis of heat and mass transfer during deposition process.

15.5.3 Cooling Methods

To monitor and control the surface temperature of the substrate and then that of the coating during deposition, different means can be used: compressed air jets, CO_2 ice/snow, liquid nitrogen, and liquid argon. Whatever the cooling device may be, the most common design of the cooling system is that represented in Fig. 15.98 with two nozzles mounted on the spray torch.

When using air jets for the control of the substrate/coating temperature, the cooling is purely convective and depends strongly on the velocity of the air jet that is a function of the air flow rate, the i.d. of the cooling air nozzle, and the distance between the nozzle and the part to be cooled that must be rather short to avoid air jet expansion. Besides the typical cooling nozzles presented in Fig. 15.98, air-cooling setups as those presented in Fig. 15.99 can be used. They comprise a nozzle, identified as R in Fig. 15.99 fixed or not to the torch with a slot shape (typically 0.8 mm in width ×30 to 40 mm in length), and another slot, called wind jet, identified as B in Fig. 15.99, intercepting the hot spray gas and molten droplet jet about 20 mm before the substrate. Wind jets, to the best of our knowledge, have been used essentially with

Fig. 15.98 Typical cooling nozzles fixed on the spray torch and moving with it, after [LIN-SPRAY Tech Bull]. (Courtesy of Linde Company)

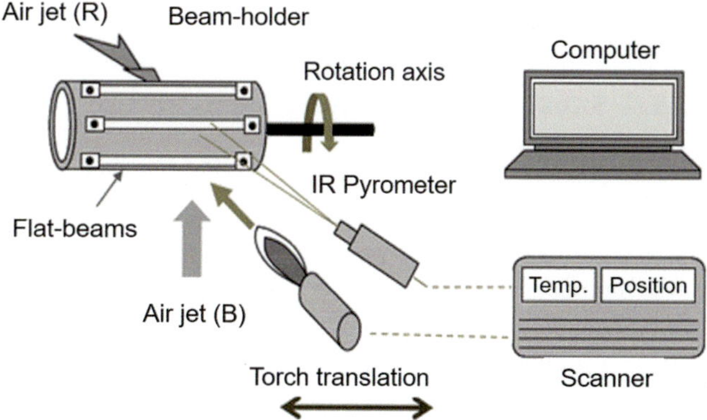

Fig. 15.99 Example of cooling nozzles with a slot shape, (R) nozzle attached to the spray torch and cooling the deposited pass, (B) nozzle, attached to the spray torch, intercepting jets of hot gases and particles in front of the substrate (about 20 mm), called "wind jet"

plasma spraying and, as explained in the next section, their main role is not only cooling but also to eliminate particles poorly heated due to their trajectory in the periphery of the plasma jet. With the ice/snow, liquid nitrogen or liquid argon cooling, the latent heat of vaporization and lower temperatures of the cryogenic fluids contribute much to the cooling. The main thermodynamic properties of these are summarized in Table 15.6.

Cooling systems with liquefied gases consist of a high-pressure tank containing the liquid, withdrawal valves, safety valves, a pressure gauge, a high-pressure hose with a good thermal insulation, valves, filters, atomization nozzle, nozzle holders, and a control panel. Typically, 0.5 to 1.0 Liter of liquid nitrogen is needed per kg of metal to reduce its temperature from +20 °C to −196 °C ($\Delta T = 216$ °C) [Gosh R (2007)]. Liquid argon is used only when spraying in argon-

Table 15.6 Thermal properties of gases used to cool coatings with liquid or frozen gases

Gas	Freezing temperature (°C)	Evaporation/sublimation temperature (°C)	Vaporization enthalpy (kJ/kg)	Specific heat (J/kg K)
Argon	–210	–196	198	1036
Nitrogen	–189	–185	160	520
CO_2	–56	–78	307	

controlled atmosphere to spray refractory materials prone to decompose or react, such as TiB_2.

Carbon dioxide, on the other hand, which does not exist in liquid state at pressures below 0.51 MPa is stored at a pressure either of 1.8 or 5.6 MPa. The gas condenses directly as a solid at temperatures below –78 °C when the pressure is rapidly reduced to 0.1 MPa, and the solid sublimes directly to a gas above –78 °C. In its solid state, carbon dioxide is commonly called known as 'dry ice.' It is generally made of pellets in the few mm size range (typically a diameter of 3 mm for a length of 5 to 15 mm), which sublimes on contact with the part. Depending on the atomization nozzle the quantities of CO_2 vary between 0.165 and 1.2 kg/min. It takes about 0.2 kg of "dry ice" per kg of metal to reduce the temperature by 98 °C [LIN-SPRAY Tech Bull].

15.5.4 Substrate and Coating Temperature Control

The main role of the cooling devices is to keep the coating temperature as constant as possible during spraying, and to avoid temperature gradients, especially when the sprayed material has a low thermal conductivity ($\kappa < 30$–40 W/m K) [Leger AC et al. (1997) and Haddadi A (1998)]. This is illustrated in Fig. 15.100 showing the surface temperature dependence of the substrate during the 150 s of preheating time with the plasma torch, curve (a), followed by a 200 s period of spraying with alumina powder (fused and crushed +22- 44 µm) in the presence of air-cooling nozzles, and finally natural cooling of the coating and substrate for 400 s. The plasma torch used in this case was PTF-4 type with 7 mm nozzle i.d. operated using (45 slm Ar +15 slm H_2), 35 kW and coating thickness 300 µm. The corresponding substrate/coating temperature history is shown in *Fig. 15.100* curve (b), operating under the same conditions with the exception of the absence of the initial substrate preheating period, and of the air cooling during the spraying stage of the operation.

Without preheating and cooling during spraying, the temperature gradient within the alumina coating ($\kappa \approx 18$ W/m K) is about 1600 °C/mm! After preheating the substrate with the plasma torch and the cooling air jets, the substrate temperature is increased by about 100 °C when the coating process starts, but equilibrium is reached in less than 30 s and then the surface temperature of the coating being deposited remains

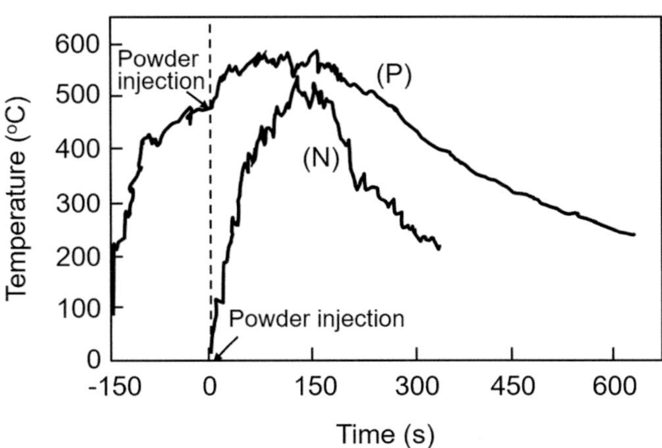

Fig. 15.100 Time evolution of the surface temperature of substrate and alumina coating (during: preheating, spraying and cooling (curve a), and during spraying and cooling cycle in the absence of preheating and air cooling during the spraying operation. (curve b). ([Haddadi A (1998)] Reprinted with the kind permission from Dr. A. Haddadi)

almost constant. Of course, for controlling such temperature dependencies with alumina, the pass thickness must be below 10 µm.

In the presence of wind jet, facing the plasma jet and fixed to the plasma torch, situated 20 mm in front of the substrate, with spray distances of either 80 or 100 mm, the heat flux to the substrate is reduced by a factor of $5-6$ with an airflow rate of 20 m^3/h as shown in Fig. 15.101 [Monerie-Moulin F et al. (1992)].

It is important to note, however, that the main interest in the wind jet approach is not so much for its cooling effect on the substrate and the coating, but rather the possibility to get rid of nonmelted particles traveling in the plasma jet fringes. Unfortunately, also some fully melted small particles are removed since particles removed from the jet as those who have lower momentums than the other ones because of their lower velocity (those traveling in the jet fringes) or their smaller size. Moreover, the wind jet reduces the velocity of the particles crossing the jet by about 20–30 m/s and the temperature by about 200–300 °C [Fantassi S et al. (1993) and Renouard-Vallet G (2004)]. In spite of these drawbacks, the use of wind jets presents some advantages as illustrated below.

[Renouard-Vallet (2004)] pursued spraying of Yttria Stabilized Zirconia (YSZ, 13 wt.% Y_2O_3) particles (fused and crushed with a size distribution $-22+5$ µm) using a vacuum plasma spray process (VPS) to produce about

50 μm thick and dense electrolyte layers for SOFCs. A PTF-4 torch, equipped with a Laval nozzle, was used working with an Ar–H$_2$ mixture (40 slm Ar + 6 slm H$_2$), an arc current of 750 A at a pressure of 8 kPa. The in-flight particle temperature distribution (measured with a DPV-2000) focused at the substrate location on the particle jet trajectory. Coatings were also kept at 550 °C during spraying (with the help of cryogenic cooling), and cross sections are presented in Fig. 15.102a. The porosity was 10.9% (measured by image analysis of the cross sections), and the ionic conductivity was 1.9×10^{-2} S/m (that of sintered YSZ being 1 S/m). Such a low ionic conductivity corresponds to rather poor contacts between layered splats. To get rid of nonmelted particles, a wind jet was used, blowing argon orthogonally to the particle jet 30 mm in front of the substrate. Compared to the temperature distributions without wind jet the inflight particle temperature distribution with wind jets is much narrower, with values between about 2200 and 2950 °C, and the mean value is slightly higher. According to particle velocities, the wind jet has removed both the high velocity small hot particles and the low velocity larger cold particles. The hot particles with an intermediate size were slightly slowed down (less than 20 m/s). Splats collected are rather close to disk shapes with a mean diameter of 30 μm, a mean thickness of 1.5 μm (against 4 μm without wind jet) corresponding to a flattening degree of 4. The coating obtained with the wind jet (also kept at 550 °C during spraying) is presented in Fig. 15.102b. Compared to that without wind jet (Fig. 15.102a), it is denser with 6.1% porosity, and its ionic conductivity was 5×10^{-1} S/m corresponding to a much better contact between layered splats.

At last it must be recalled that preheating the substrate to get rid of adsorbates and condensates can be performed with a laser, attached to the torch and irradiating the substrate surface in advance of the sprayed spot [Li H et al. (2006a, b, c, d, e)], as illustrated in Fig. 15.103.

15.6   Stress Control in Thermal Spraying Operations

Two types of stresses are considered:

Residual Stress commonly defined as the stress that remains in a coated part after its manufacturing and is not being subjected to external forces. Residual stresses can be detrimental or beneficial to the performance of a material. According to [Lyphout C et al. (2008)] residual stresses include;

- *Thermally induced stresses*, resulting from nonuniform heating and cooling during manufacturing process. In a thermally sprayed coating, the final residual stress state through the whole coating/substrate system is determined by the superposition of stresses induced during

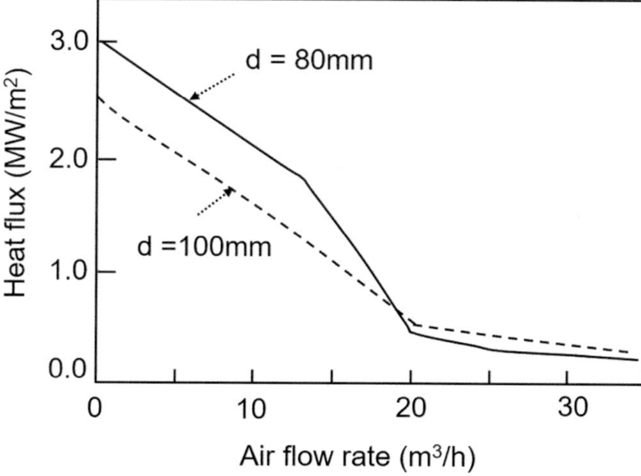

Fig. 101 Dependence of the maximum axial heat fluxes of the DC plasma jet on the cooling air flow rate at, 80 and 100 mm, standoff distances. Wind jet (40 mm in length and 0.8 mm wide) moved with the torch and fixed to be 20 mm in front of the substrate. ([Monerie-Moulin F et al. (1992)]. Reprinted with the kind permission from Dr. F. Monerie-Moulin)

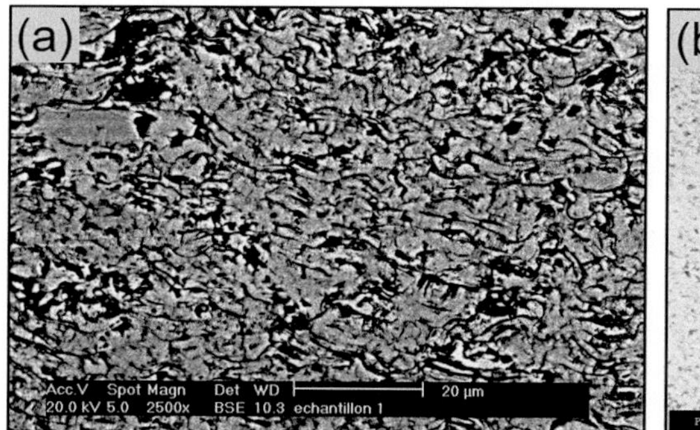

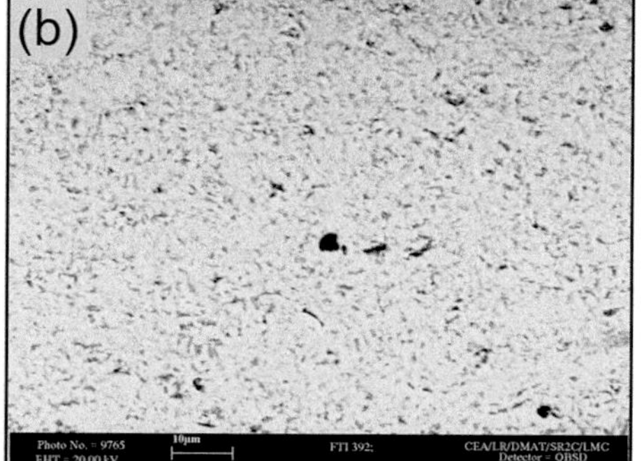

Fig. 15.102 Coatings obtained with (**a**) Spraying performed without wind jet. (**b**) With wind jet. ([Fauchais P and M Vardelle (2010)] Reprinted with kind permission from Springer Science Business Media, copyright © ASM International)

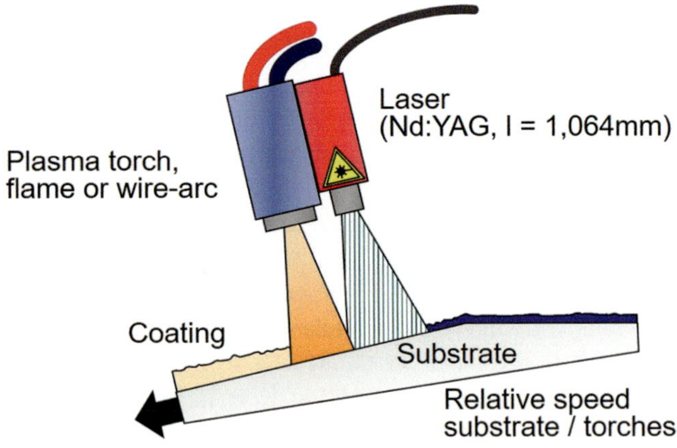

Fig. 15.103 Principle of the substrate preheating with a laser focused just in front of the sprayed spot. ([Li H et al. (2006a, b, c, d, e)], Reprinted with kind permission from Springer Science Business Media, copyright © ASM International)

the spray process: quenching, thermal mismatch, and thermal gradients. The latter stress can be avoided or significantly limited by keeping the time average coating temperature as constant as possible.

- *Mechanically induced stresses* that are generated during the different manufacturing steps (such as grinding, blasting, and machining) and that can produce nonuniform plastic deformation. For spray processes where particles achieve velocities over 500–600 m/s (HVOF, HVAF, D-gun, Cold Spray) their peening effect must be considered because it results in a compressive stress states in the coating, similar to that obtained in the substrate by grit blasting.

Service Stresses

In service conditions, coatings can also be submitted to thermal loads, temperature gradients, and fatigue. Moreover, service conditions can also induce phase or chemical changes resulting in what is called *intrinsic stresses*. Service stresses are mostly due to;

- *Thermal effects*, such as in thermal barrier coatings including cyclic heating and cooling associated with engine operation
- *Constrained oxide growth*, on the bond coat at the ceramic/bond coat interface, or
- *Mechanical effects*, as in thermal spray coatings submitted to fatigue in rolling/sliding contacts.
- *Chemically induced effects*, resulting from volume changes when chemical reactions occur,
- *Precipitations or phase transformations*, occurring, for example, as a result of heating γ-Al_2O_3 coating over 1000 °C inducing its transformation into α-Al_2O_3.

15.6.1 Residual Stresses

15.6.1.1 Thermal Stresses

An excellent overview of the different thermally induced stresses contributing to the residual stress after coating generation is given by [Clyne and Gill (1996)]. As illustrated in Fig. 15.104, a pair of plates bonded together with a misfit strain $\Delta\varepsilon$ in the *x*-direction, assuming a purely elastic behavior, will give rise to stress distribution, $\sigma_x(y)$, and curvature, K', based on simple beam bending theory. The misfit strain is removed by the application of two equal and opposite forces ($-P'$ and P'). When the two plates are joined, an unbalanced moment, M is generated given by;

$$M = P.(H + h')/2 \tag{15.17}$$

where H and h' are the thicknesses of the substrate and deposit, respectively. Balancing this moment generates curvature of the composite plate, K' (equal to the through-thickness gradient of strain) can be expressed as the bending moment divided by the beam stiffness, Γ:

$$K' = M/\Gamma \tag{15.18}$$

Combining Eqs. 15.17 and 15.18, P can be expressed as

$$P = 2\Gamma K'/(H + h') \tag{15.19}$$

Using the expression of the beam stiffness, calculating the misfit strain, $\Delta\varepsilon$ for a change arising from temperature variation according to forces P, it is possible to obtain a complex expression of K' depending upon the Young's moduli of coating and substrate, their respective thicknesses h' and H (for details scc [Clyne and Gill (1996)]. Of course, the curvature depends strongly upon the respective thicknesses of coating and substrate as well as their Young's moduli. For example, it will be negligible when $h'/H \ll 1$. For a given h'/H ratio, the curvature is inversely proportional to the substrate thickness [Clyne and Gill (1996)]. This implies that use of relatively thin substrates is essential to obtain a sufficiently large curvature to achieve a good precision stress measurement through beam curvature [Stokes J and L Looney (2008)].

Quenching Stress During Splat Formation

Quenching stress occurs due to the rapid cooling of flattened fully molten droplets as they solidify on the substrate or the previously deposited layer. They arise from the hindered contraction of individual splats as their temperature drop below their melting temperature T_m where they are no more in a plastic state, until they reach the temperature of the

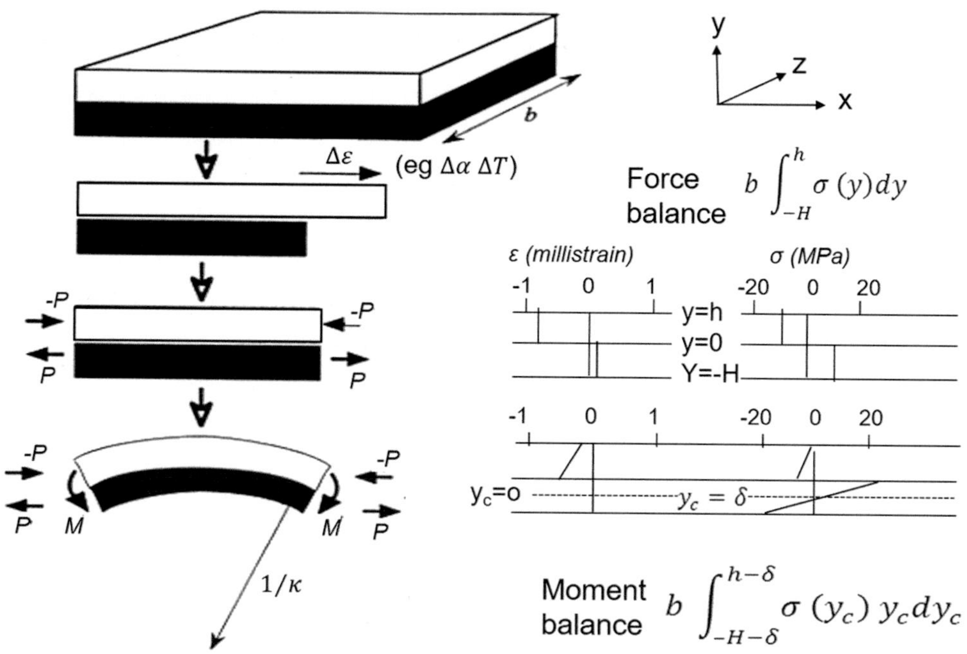

Fig. 15.104 Schematic of the generation of curvature in a flat bimaterial plate as a result of the imposition of a uniform, linear misfit strain, Δε. ([Clyne and Gill (1996)]. Reprinted with kind permission from Springer Science Business Media, copyright © ASM International)

substrate T_s or the previously deposited layers. Quenching stresses are always tensile, with their value depending on the quality of the contact between splat and substrate or the underlying layer on the one hand and the relaxation phenomena taking place on the other hand. Theoretical value of the quenching stress is given by:

$$\sigma_q^o = E_c \times \alpha_c (T_p - T_s) \qquad (15.20)$$

where E_c is Young's modulus (GPa) and α_c (K^{-1}) the expansion coefficient of the splat, σ_q^o being in MPa. Instead of using Young's modulus of the coating, it is recommended to use the effective Young's modulus $E_c' = E_c/(1 - v')$, where v' is Poisson's ratio [Stokes J and L Looney (2008)]. With a perfect contact between splat and the underlying layer, the quenching stress is often much less than σ_q^o because of the relaxation mechanisms that occur, reducing it to a value σ_q, which reflects the degree of relaxation within splats.

Considering the case of alumina, assuming that E_c and α_c values are those of the bulk material, for a 100 °C cooling, Eq. 15.20 gives a value for σ_q^o of 100 MPa, while experimental values for splats cooled down to room temperature are about 10 MPa!. In general quenching stresses of different materials are lower than 100 MPa. As could be expected, the quenching stress is almost independent of the nature of the substrate material [Gill C. (1991), Kuroda S. et al. (1989) and Kuroda S and TW Clyne (1991a, b)]. The relaxation mechanisms involved are summarized in Fig. 15.105 after [Kuroda S and TW Clyne (1991b)]. These show that for

metals or alloys edge relaxation takes place through-thickness yielding, interfacial sliding, and creep, while for ceramics relaxation takes place mainly by the formation of cracks and possible interfacial sliding.

Quenching stress has been studied mainly for Plasma Spraying (APS or VPS) and HVOF-sprayed coatings. [Kuroda and Clyne (1991a, b) and Clyne T.W. and S.C. Gill (1996)]. Typical results are given in Fig. 15.106. giving the quenching stress as function of the Homologous temperature defined as the ratio of the specimen temperature to the melting temperature of the deposit with the data obtained from specimen curvature measurements during spraying. [Clyne T.W. and S.C. Gill (1996)] emphasized that maximum theoretical values for the quenching stress are typically of the order of 1 GPa. This is due to the various stress-relaxation mechanisms, such as interfacial sliding, which can occur during the quenching process. While many ceramic materials undergo extensive micro cracking, metallic deposits are prone to plastic yielding or creep (see *Fig. 15.105*). This explains why, depending on the strength of the metal, reductions in quenching stress occur at relatively high temperatures with the misfit strain tending to zero, as the homologous temperature tends to unity. The results also show that differences between the data for APS and VPS are not significant. For metals, depending on their strengths, reductions in quenching stress occur when moving to relatively high temperatures for substrate preheating with relaxation phenomena (through-thickness yielding, interfacial sliding, and creep). At lower temperatures, all materials

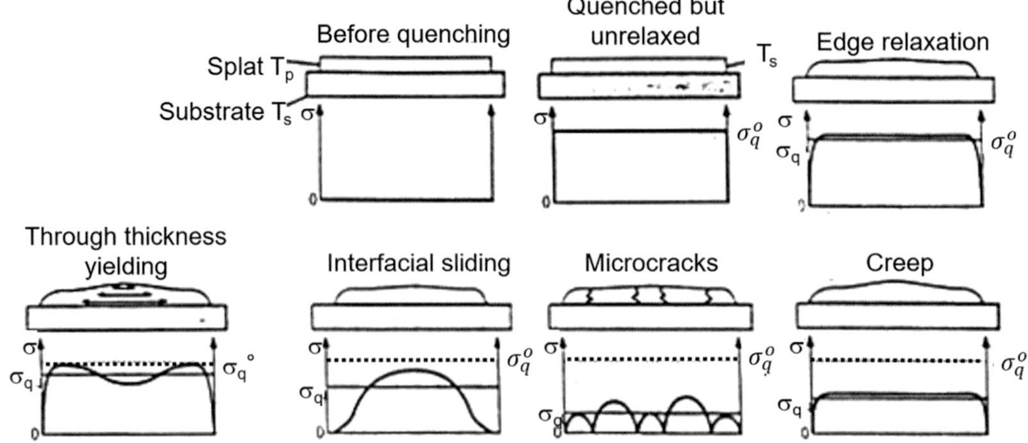

Fig. 15.105 Different mechanisms for quenching stress relaxation. ([Kuroda S and TW Clyne (1991b)]. Reprinted with kind permission from Elsevier)

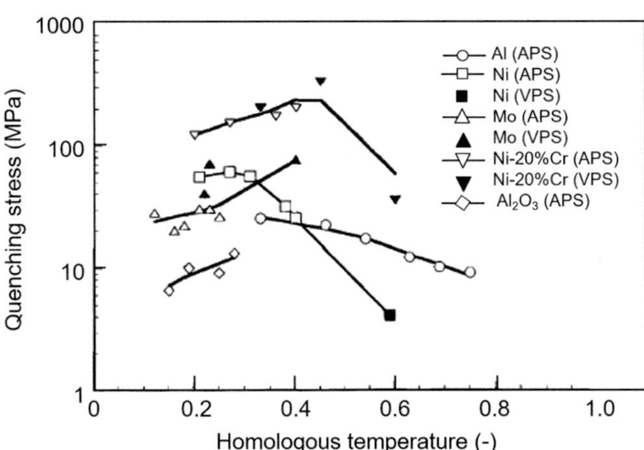

Fig. 15.106 Experimental quenching stresses for plasma spraying in air (APS) or in vacuum (VPS). ([Clyne TW and SC Gill (1996)]. Reprinted with kind permission from Springer Science Business Media, copyright © ASM International)

tend to show gradual increases with temperature, which is attributed to improvements in inter-splat bonding.

Of course, impact conditions and inter-splat bond strength play a key role as well. For example, for HVOF or D-gun spraying, due to the high particle impact velocities, with particle temperatures lower than those obtained with plasma, wire arc, flame spraying, the quenching stress is different, but still tensile. However, a high-temperature shot-peening process also occurs for particle velocities over 500–600 m/s, creating compressive stresses within the coating [Lyphout C et al. (2008), Stokes J and L Looney (2008), Kuroda S et al. (1997), Kuroda S et al. (2001), Sampath S et al. (2004), Bansal P et al. (2006), Bansal P et al. (2007), He J et al. (2008), Bolelli G et al. (2008), Wang T-G et al. (2010), Lima CRC et al. (2006), Jiang Yi et al. (2005) and Clyne TW (2001)].

Expansion Mismatch Stress

Thermal mismatch stress, σ_d, occurs during cooling but after solidification due to difference in thermal expansion coefficients between each deposited layer and the substrate material. Simple elastic analysis of in-plasma stresses of thin coatings attached to a thick substrate results in the simple expression:

$$\sigma_d = \frac{E_c\,(T_s - T_o)\,(\alpha_s - \alpha_c)}{(1 + 2(E_c h'/E_s H))} \tag{15.21}$$

where E_c and E_s are the coating and substrate Young's moduli, α_s and α_c the expansion coefficients of substrate and coating, T_s and T_o are the substrate and room temperature. Of course, if one has a thick substrate and a thin coating ($h'/H \ll 1$) Eq. 15.21 simplifies to;

$$\sigma_d = E_c\,(T_s - T_o)\,(\alpha_s - \alpha_c) \tag{15.22}$$

This equation shows that if $\alpha_c > \alpha_s$ the expansion mismatch is compressive ($\sigma_d < 0$), while if $\alpha_c > \alpha_s$ the expansion mismatch is tensile ($\sigma_d > 0$).

The expansion mismatch thermal stress must be systematically considered for all coating processes except cold spray. However, in most cases the cold spray is performed above room temperature and after deposition the coating and substrate will cool to room temperature generating a thermal misfit [Spencer K et al. (2012)]. Of course, it depends on the cooling range and difference in thermal expansion coefficients between the coating and substrate materials.

Stress Due to Temperature Gradients

The relatively high heat fluxes imposed on coating and substrate during spraying can cause high thermal gradients, especially for sprayed materials with low thermal conductivity ($\kappa_c < 30$ W/m/K), which generate differential thermal expansion at different depths resulting in local misfit strains. Thermal gradients increase also if the coating surface temperature is not kept constant during spraying thanks to adapted cooling devices blowing on the coating surface under construction. Stresses and bending moments will therefore rise. Although they are expected to disappear when the thermal gradient vanishes, they can induce nonelastic effects and thus must be avoided [Clyne TW and SC Gill (1996)]. This can be achieved by adjusting the cooling devices, the spray pattern, and the pass thickness, especially for low thermal conductivity sprayed materials.

15.6.1.2 Mechanically Induced Stresses

Mechanically Induced Stresses can be generated as a result of;

Grit Blasting

During grit blasting [Clyne TW and SC Gill (1996)] compressive residual stresses are generated and distributed within a thin layer beneath the grit-blasted substrate surface. Grit blasting modifies the grain shapes, as in cold working. He et al. (2008) have shown that initial equiaxed grains of 1010 carbon steel in the deformed regions of a substrate are elongated and spirally oriented surrounding the impact spots. There are no visible changes in this microstructure caused by thermal spraying, and the elongated grain regions are completely unchanged in the coated substrate. Figure 14.34 (Chap. 14 Surface Preparation) from [Mellali et al. (1997)] shows the residual stress distribution within a cast iron FT-25 substrate, grit blasted with 0.5 mm white alumina. The area plastically affected in the substrate is about 0.4 mm thick as shown in Fig. 14.34. The maximum stress, corresponding to the depth of the plastically deformed zone, was reached at 0.05 mm from the substrate surface. The magnitude and the depth of the zone affected by the grit blasting are strongly dependent on the blasting conditions (particularly the impact velocity and the size of particles, or more precisely momenta of grit particles) and on Young's modulus of the substrate material. [Lima et al. (2006)] obtained similar results for UNS G41350 steel substrates. When grit blasting thin beams (with thickness below about 4–5 mm), the grit-blasting stress induces a curvature of the beam that depends on blasting conditions and the beam Young's module. A possibility to reduce the beam bending is to grit blast it on both sides, with the same blasting conditions.

The grit blasting stress can be relaxed by heating substrates over the material recovery temperature ($\approx 0.4 T_\mathrm{m}$, T_m being the material melting temperature), in a controlled atmosphere furnace to avoid oxidation.

Peening Effect

For spray processes where particle velocities are above 500–600 m/s a high-temperature shot-peening process occurs resulting in a compressive stress [Kuroda S et al. (1997), Kuroda S et al. (2001), Sampath S et al. (2004), Bansal P et al. (2006), Bansal P et al. (2007), He J et al. (2008), Bolelli G et al. (2008), Wang T-G et al. (2010), Lima CRC et al. (2006), Jiang Yi et al. (2005) and Clyne TW (2001)]. The interaction between impacting particles and substrate depends on the relative hardness value of the impacting particle and the substrate. When a hard particle is sprayed onto a soft substrate, the substrate is deformed (it can even be locally molten), and its surface is penetrated by the impacting particle. In contrast, when a soft particle is sprayed onto a hard substrate, the majority of the impacting particle's kinetic energy is transformed into thermal energy, the particle itself deforms, even melts [He J et al. (2008)]. It is generally assumed that the substrate, which had been already hardened by grit blasting, is not affected and no further mechanical hardening is created. The superposition of quenching and peening stress induces a nonlinear stress that is first tensile, decreasing with the coating thickness and becoming compressive beyond a given thickness [Wang T-G et al. (2010)]. To illustrate the peening effect, [Sampath S et al. (2004)] have compared the stress distribution within Ni–5 wt% Al coatings wire arc sprayed (WAS), air plasma sprayed (APS), HVOF sprayed, and cold sprayed (CS). The steel substrate and coating were air cooled during the spray process. The stress distribution was determined by neutron diffraction. Near-surface effects, when the gauge volume was partially out of the material, were determined on a stress-free reference and subtracted. In the APS deposit, as shown in Fig. 15.107a, the overall stress within the coating is slightly tensile and it increases toward the deposit surface, due to the buildup of deposit layers with tensile quenching stress. To accommodate this coating stress the stress in the substrate (bending effect) will display a gradient from compression, near the interface with the coating, to tension on the back surface. In the case of wire arc spraying, the stress gradient is somewhat higher, while in the substrate it is negligible. [Sampath S et al. (2004)] explain that behavior by the change in quenching stress during the deposit buildup and the low deposit modulus, stemming from higher porosity resulting in insignificant substrate bending. In the HVOF deposit, see Fig. 15.107b from [Sampath S et al. (2004)], the stress is mainly compressive in spite of quenching and thermal stresses, the very high velocity impact of the particles on pre-deposited layers resulting in the peening effect generating compressive residual stress as already emphasized by [Kuroda et al. (2001)]. This results in high tensile stress within substrate close to the interface with coating. Similarly, in the cold-sprayed coating, the stress is compressive [Sampath S et al. (2004)] as well. There is a slight gradient toward compression near the surface. [Sampath S et al. (2004)] explain this effect by the fact that the peening action of impinging particles affects only the immediately adjacent material, while the layers further

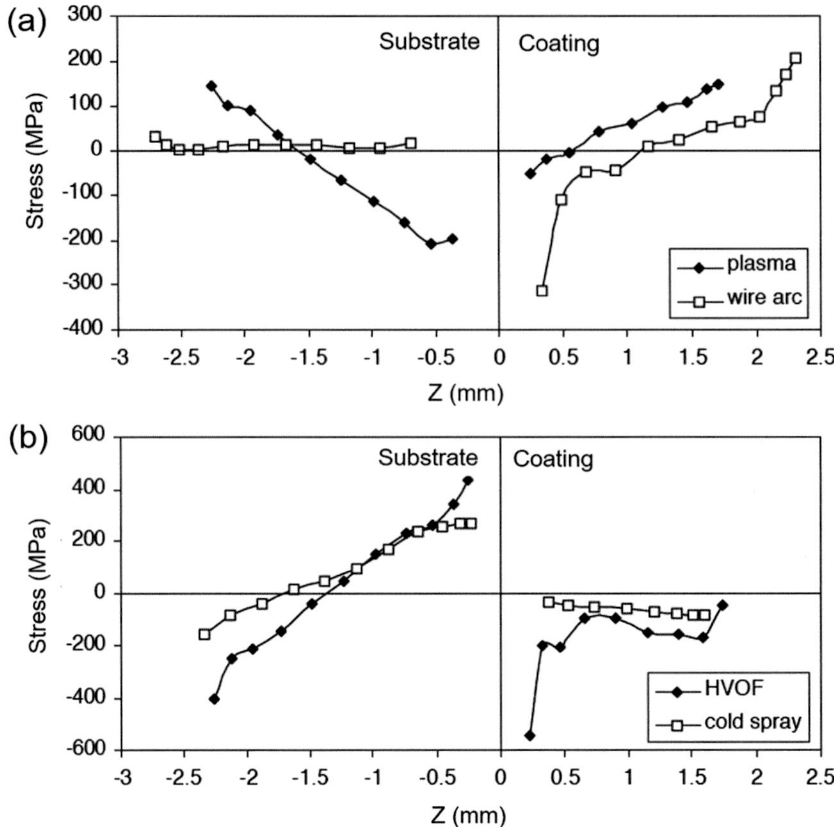

Fig. 15.107 Through-thickness profile of the residual stress in Ni–5 wt% Al deposits on steel substrate sprayed, respectively, by: (a) plasma and wire arc and (b) HVOF and cold spray. ([Sampath S et al. (2004)]. Reprinted with kind permission from Elsevier)

beneath are shifted a little toward tension, to maintain a force balance.

Figure 15.108 after [Wang et al. (2010)] presents the dependence on the coating thickness of the residual stress of D-Gun sprayed WC–Co coatings. In Zone A the residual stress is of tensile nature and decreases dramatically (by $\approx$1.28 MPa/µm) with a coating thickness increase up to 317.4 µm. When the coating thickness is larger than 317.4 µm (Zone B), the residual stress decreases at a lower rate (by $\approx$0.31 MPa/µm) and varies gradually from a tensile nature to a compressive one.

The stress distribution in cold-sprayed coatings is mainly compressive and it seems [Luzin et al. (2011)] that it is determined almost entirely by the plastic deformation process (peening) of the particles of the spray material. [Luzin et al. (2011)] have measured the residual stresses profiles in Cu and Al cold spray coatings by neutron diffraction. The in-plane residual stress profiles are shown in Fig. 15.109a–d. Both Cu coatings on Cu or Al substrates show a significant compressive residual stress at the surface 50–80 MPa, while the surface residual stress of the Al coatings on the same substrates is less than 10 MPa. The overall stress profile of the two Cu coatings is similar, regardless of the substrate material, and the same can be said of the Al coatings. The

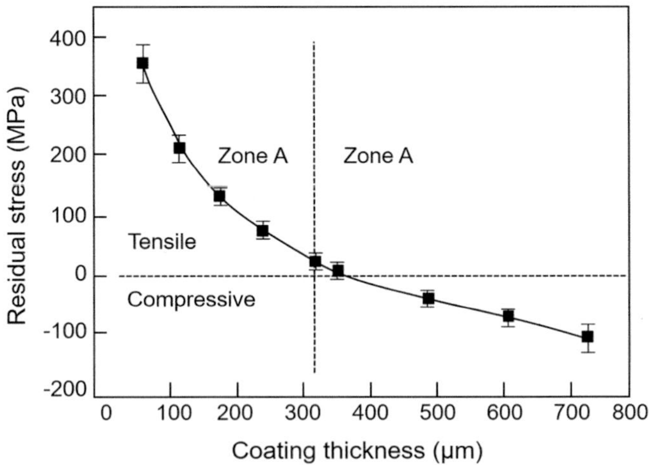

Fig. 15.108 Dependence on the coating thickness of the residual stress of D-Gun sprayed WC–Co coatings. ([Wang et al. (2010)]. Reprinted with kind permission from Elsevier)

experimental residual stress distributions were treated [Luzin et al. (2011)] empirically using Tsui and Clyne's progressive deposition model to demonstrate that the residual stresses accumulate mainly by kinetic and not thermal effects, and these effects were characterized quantitatively. The measured

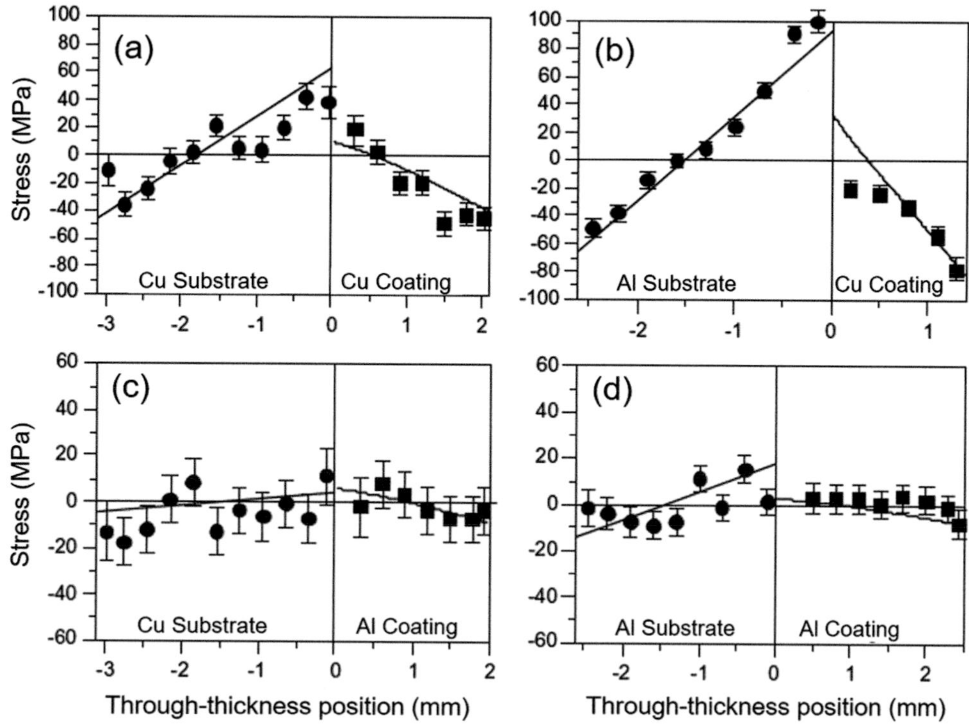

Fig. 15.109 Measurement (symbols) and model fit (solid lines) of the through thickness in-plane stress distributions for (**a**) Cu/Cu sample, (**b**) Cu/Al sample, (**c**) Al/Cu sample, and (**d**) Al/Al sample. ([Luzin et al. (2011)]. Reprinted with kind permission from Elsevier)

compressive residual stresses were determined almost entirely by the plastic deformation process (peening) of the particles of the spray material. An attempt has been made [Luzin et al. (2011)] to explain the higher residual in-plane stress in the Cu coatings by using a physically based model of residual stress accumulation in a shot-peened surface. Calculations within this model in the case of Al and Cu demonstrated that the balance between the elasto-plastic properties of the material and the spraying conditions defines the resultant residual stress.

15.6.1.3 Resulting Residual Stress

At the end of the coating process the residual stress state through the whole coating/substrate system is determined by the superposition of stresses of different nature induced during the spray process: quenching, thermal mismatch, and peening stresses, together with the compressive stress state of the substrate induced during the grit blasting prior to spraying. This is illustrated in Fig. 15.110 from [Lyphout et al. (2008)], it is worth noting that in this figure the grit blasting stress has been assumed negligible. The stress distribution, its intensity and sign in the coating, is strongly dependent on the specific spray process [Sampath S et al. (2004)] and processing conditions [Bansal P et al. (2007), He J et al. (2008) and Bolelli G et al. (2008)]. The residual stress distribution can be measured or calculated. Calculations can be performed either by using analytical expressions such as

those given by [Clyne and Gill (1996), Clyne TW (2001), Stokes and Looney (2008), Wang et al. (2010), and Kroupa F (1997)] who has considered non-homogeneous elastic properties depending on the coordinate z perpendicular to the interface (case of any graded or layered coatings), or by using finite element methods [Lyphout C et al. (2008), Kroupa F (1997), Chang GC et al. (1987), Yokoyama K et al. (2006) and Samadi H and TW Coyle (2009)] with, for example, the ABAQUS code [Bansal P et al. (2006) and Kroupa F (1997)]. [Ghafouri-Azar et al. (2006)] using the experimental spray parameters as inputs to a 3-D stochastic model have simulated coating formation. They used an object-oriented finite element code (OOF) developed at the National Institute of Standards and Technology to calculate residual stresses in the coating. The model, with an adaptive meshing technique, discretized the coating microstructure into a mesh suitable for finite element analysis. The coating geometry used either micrographs of coating cross sections or computer-generated images of coatings.

Whatever the model or calculation method may be, one of the most important difficulties is that the mechanical properties of coatings, depending strongly on spray conditions, can be quite different from those of bulk materials given in tables. In particular, a very difficult problem is the calculation of the quenching stress taking into account the relaxation phenomena.

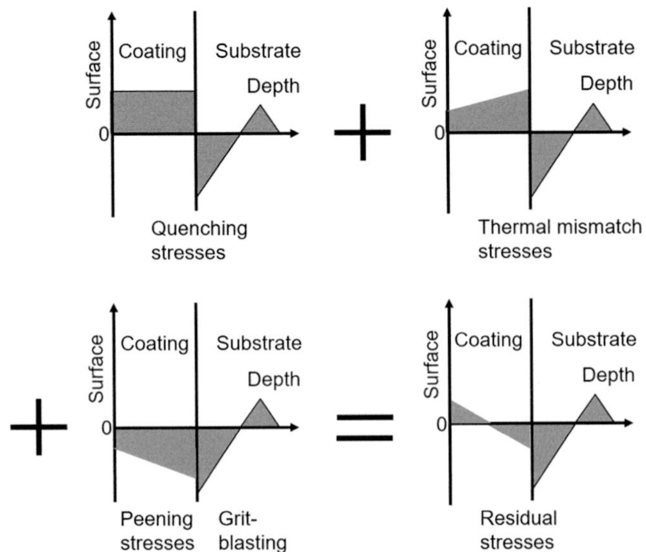

Fig. 15.110 Schematic representation of radial residual stress distribution in HVOF sprayed coating/substrate system. (Reprinted with kind permission from Springer Science Business Media [Lyphout C et al. (2008)], copyright © ASM International)

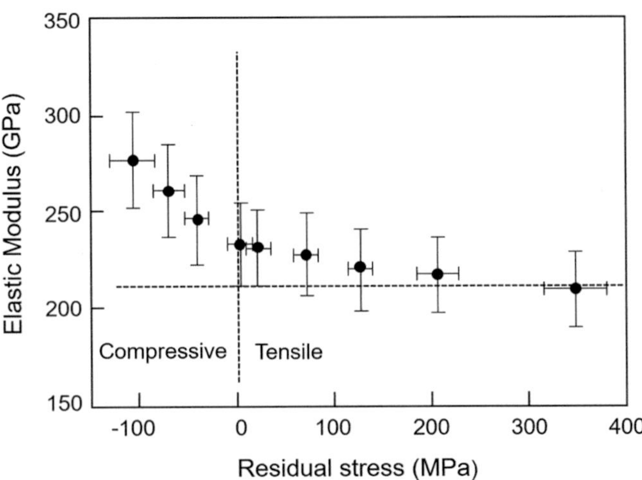

Fig. 15.111 Variations in the elastic modulus with the residual stress evolution through coating thickness see Fig. 15.108 (D-Gun sprayed WC–Co). ([Wang T-G et al. (2010)]. Reprinted with kind permission from Elsevier)

15.6.1.4 Effect of Residual Stress on Coating Adhesion

Coating adhesion is a very important property that is linked to the critical strain energy release rate for the interface, G_{ic}, or the closely related critical interfacial stress-intensity factor, K_{ic} at the interface between coating and substrate [Clyne TW and SC Gill (1996), Clyne W (1996) and Gill SC and TW Clyne (1990)]. In the simplest case of a thin coating with a *uniform* misfit strain, the strain energy release is simply the stored elastic strain energy per unit interfacial area. Assuming that $h' \ll H$ the expression for the strain energy release becomes

$$G = \sigma_c^2 \, h'/2E_c \qquad (15.22)$$

Once the stress is known at the point where debonding occurs, G_{ic} can be taken as equal to the value of G (J/m^2) at this point.

Unfortunately, the analysis is complicated by the fact that the stress field at the crack tip may not represent conditions of a pure crack opening (mode I) [Clyne TW and SC Gill (1996)]. In contrast to a homogeneous medium, the crack is confined to a particular geometric path and, depending on the toughness of the interface and of the two media on either side, crack propagation may continue within the interface also with a shear stress (mixed modes I and II).

Debonding may also occur under combined residual and applied stresses (e.g., generated by the application of an external mechanical load). At last, for an edge crack the driving force near the specimen edge can be reduced during

the initial stages of the propagation. That is why at the edge of the coated substrate proper machining must avoid sharp corners, such as an abrupt stop of the coating at the end of a cylinder as explained in Sect. 12.2. While the problem is rather complex, it is important to estimate stress distribution in the coating through appropriate modeling under different spraying conditions in order to suppress or at least reduce the initiation of spalling which is catastrophic for the coating [Clyne TW and SC Gill (1996)]. Many models have been devoted to the stress distribution calculation [Katanoda H et al. (2007a, b), Stokes J et al. (2008), Bansal P et al. (2007) He J et al. (2008), Clyne TW (2001), Kroupa F (1997), Chang GC et al. (1987), Yokoyama K et al. (2006), Samadi H et al. (2009), Ghafouri-Azar R et al. (2006), Clyne W (1996) and Gill SC and TW Clyne (1990)] all emphasize the importance of having low tensile, or better even compressive stress in the coating near the interface. The residual stress also modifies the coating mechanical properties (elastic modulus, hardness) both increasing when the compressive stress is raised [Lima CRC et al. (2006)], as shown in Fig. 15.111. The increase in these mechanical properties results from the elastic closure of small porosities or micro-cracks. However, it must be kept in mind that coating properties are strongly linked to the real contacts between splats, which do not necessarily depend on the impact energy of particles. This is illustrated in Fig. 15.112 from Sampath et al. [Sampath S et al. (2004)] who have compared the porosity and hardness of Ni–5 wt% Al coatings sprayed by wire arc, APS, HVOF, and cold spray. As shown in Fig. 15.112a, the deposit porosity decreases with increasing particle kinetic energy, but the hardness drops from HVOF deposit to cold-sprayed deposit even though the particle kinetic energy is doubled,

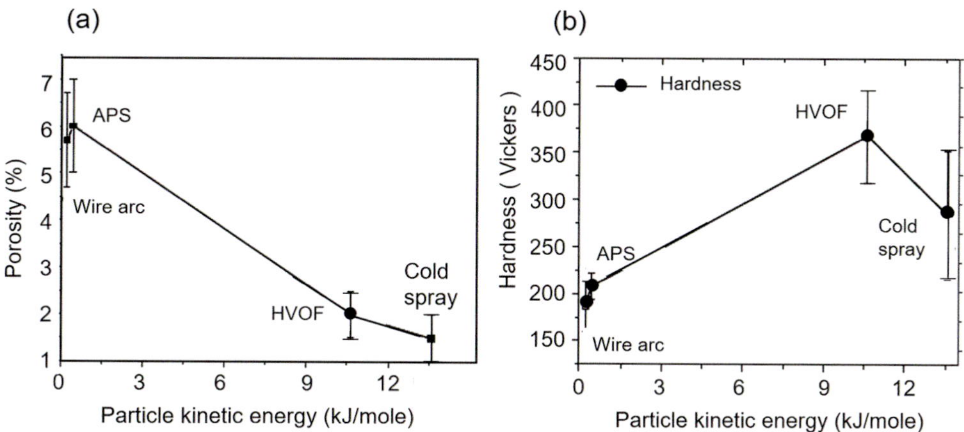

Fig. 15.112 Influence of particle kinetic energy on Ni–5 wt% Al deposits sprayed by wire arc, APS, HVOF, cold spray; (**a**) porosity, (**b**) hardness. ([Sampath S et al. (2004)]. Reprinted with kind permission from Elsevier)

Fig. 15.112b. This implies that deposit properties are strongly linked to the real contacts between splats and not only to the coating porosity or density.

Coating thermal conductivity, closely linked to the real contacts between splats, is the lowest for cold sprayed (11 W/ m K) and APS (10.7 W/m K) coatings, slightly higher for the wire arc deposit (17 W/m K), in spite of the lowest impact velocity, and the highest (22.2 W/m K) for HVOF coatings. By analyzing the microstructure of HVOF, APS, and cold spraying parameters, it is clear that a combination of high kinetic energy and a high thermal energy promotes enhanced splat–splat interaction and a consequent improvement in interlayer/inter-splat adhesion, as reflected by the increase in thermal conductivity. The low porosity associated with dynamic compaction in cold spraying does not automatically provide enhanced inter-splat contact. The wire arc-sprayed material has a coarse porosity and large metallurgical bonded regions with no interfaces. The cold-sprayed material consists of a network of fine interconnected interparticle separations with a lack of metallurgical bonding [Sampath S et al. (2004)].

Tensile stresses within a coating may also induce debonding when the coating thickness increases. For example, [Greving et al. (1994)] have shown that if residual stress measurements indicate that coating thickness for wire arc-sprayed nickel–5 wt% aluminum coatings on AISI 1018 steel minimally influences the residual stress, the degradation of bond strength is raised with increasing coating thickness. They attribute that to the fact that thicker coatings cause higher tensile edge stresses, increase the tendency for debonding, and hence reduce the applied stress required to cause debonding.

Such considerations are especially important for coatings used at high temperatures such as thermal barrier coatings. That is why, especially for ceramic materials, which have lower expansion coefficients compared to those of the metal

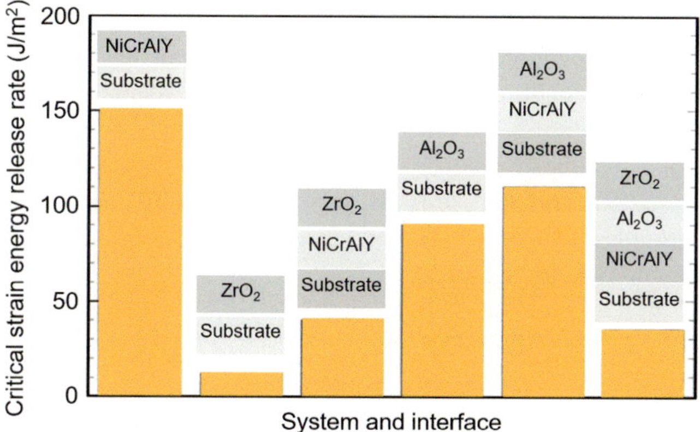

Fig. 15.113 Experimental data for interfacial toughness of various interfaces within multilayer ceramic coating–metal–substrate with or without NiCrAlY bond coat. ([Clyne and Gill (1996)]. Reprinted with kind permission from Springer Science Business Media, copyright © ASM International)

substrate, bond coats (generally about 100 µm thick) are used between the substrate and the ceramic. Their role is to increase the interfacial toughness. For example, [Tsui and Clyne (1996)] studied the evolution of experimentally measured toughness for various interfaces, including some that might form if an alumina oxidation barrier was introduced between the bond coat and the topcoat. The results are presented in Fig. 15.113 from [Clyne and Gill (1996)]. The schematic above each bar in this figure indicates the interface that was measured and the specimen geometry.

Besides superalloys that can work at temperatures up to about 1300 °C, other metals are used as bond coats, especially molybdenum, nickel–aluminum composites or alloys and nickel–chromium alloys. NiAl alloys or composites are rather popular because of their good resistance at high temperatures and their ease of application to various substrates. NiCr and, of course, MCrAlY have the best

Table 15.7 Approximate maximum service temperatures for common bond coatings

Bond coat	°C
Mo	315
80Ni–20Al	620
95Ni–5Al	1010
80Ni–20Cr	1260
94NiCr–6Al	980

Reprinted with kind permission from [American Welding Society [AWS] (1985) Thermal spraying: practice, theory and application. American Welding Society, Miami, FL]

resistance to thermal shocks. Table 15.7 presents the maximum service temperatures of the mostly used bond coats

15.6.2 Service Stresses

In service conditions coatings can be submitted to thermal loads, temperature gradients, fatigue resulting in new stresses that have to be calculated according to the residual stress linked to the spray process. Moreover, service conditions can also induce phase or chemical changes resulting in what is called intrinsic stresses. In the following only two examples will be considered: thermal barrier coatings and fatigue of tribological coatings

15.6.2.1 Intrinsic Stresses

The intrinsic growth stresses are due to chemical reactions, phase transformations, and energetic particle bombardment. For thermally sprayed coatings when chemical reactions or phase changes occur, a change in volume occurs generating stresses that can be compressive when the volume increases, or tensile when it is decreased. Such volume transformations occur generally during service conditions. A typical example is the transformation of plasma-sprayed γ-alumina into α-alumina occurring at about 1000 °C and resulting in an increase in the volume with coating spallation as possible consequence. Thus, coatings must be used at temperatures below those at which such transformations occur.

Thermal Barrier Coatings

Gas turbine engines use nickel- and cobalt-base superalloys in the turbine components, such as airfoils, combustors, transition ducts, and seals [Bose S and J DeMasi-Marcin (1997)]. Increased thermal efficiency demands that the turbine inlet temperature be significantly increased, without structural failure of components from melting, creep, oxidation, thermal fatigue, or other degradation mechanisms. Therefore, the surface temperature of these components must be maintained low enough to retain material properties within acceptable bonds. Due to its low thermal conductivity and good thermocyclic durability when stabilized against

phase transformation, zirconia has become the ceramic of choice that is sprayed on a bond coat.

Thermal stresses induced by high heat fluxes are not only responsible for the delamination of the topcoat but also the oxidation of the bond coat by the hot oxidizing gases diffusing through the porous top coat is also very important [Chang GC et al. (1987), Bose S and J DeMasi-Marcin (1997) and Koolloos MFJ and JM Houben (2000)]. At the rough bond coat, differential thermal contraction induces tensile radial stress in the ceramic above asperities and compressive in-plane stress, thus cracks propagate tangentially from the tip of asperities and stop above valleys (location of state of tri-axial compression). When the bond coat is oxidized, the tensile radial component and the compressive component above valleys decrease resulting in easier crack extension and coating failure. Although many factors control TBC failure, the thermally grown oxide (TGO) layer development and subsequent thickening is generally considered as the most important factor. Calculations by [Koolloos and Houben (2000)] show that pre-oxidation of the bond coat has a beneficial effect on lifetime in the case of a thermal load where oxidation is the main cause of failure. The pre-oxidation time and temperature must be chosen such that a dense alumina layer is formed, and the forming of spinel-type structures is avoided.

Another problem that must be addressed is that the high thermal gradients commonly present in TBCs under service conditions lead to widely varying conditions within the topcoat. Sintering is likely to occur much more rapidly near the free surface, not only because the temperatures are higher there, but because the residual stresses tend to be more compressive [Tsipas SA et al. (2004)]. It is known that tensile stresses retard the progression of sintering.

At the end of the nineties, demands on engine performance have required the implementation of even thicker coatings for increased thermal protection. TBCs with a thickness of more than 1 mm are being developed to fulfill the demand, but these present new problems unseen with traditional thinner TBCs. A major problem concerning the thick coating is a low thermal shock resistance when the coating is subjected to thermal shock loads [Guo HB et al. (2006)]. Segmentation cracks (orthogonal to the substrate and coating) enable TBCs to tolerate greater strains, thereby improving thermal shock behavior of TBCs. By depositing the zirconia (totally or partially stabilized) either by an electron beam evaporation-physical vapor deposition (EB-PVD) process [Strangman T et al. (2007)] or "hot-spraying" particles without cooling during spraying resulting in an increased surface temperature [Guo HB et al. (2006) and Tricoire Au et al. (2005)] it was possible to create segmentation crack network. These techniques allowed achieving compliant microstructures of the top coat accommodating the ceramic–super alloy thermal expansion mismatch strains. However, the drawback of the compliant coating microstructure is an easier access of oxidizing gases to the bond coat surface, the protection of which against oxidation must be improved [Strangman T et al. (2007)].

Unconventional plasma-sprayed thermal barrier coating (TBC) systems were also produced [Smialek JL (2004)]. Approximately 250 µm thick 8YPSZ (8 wt%) coatings were directly sprayed onto grit-blasted surfaces of desulfurized PWA 1484, superalloy with excellent oxidation resistance, without a bond coat. To avoid edge-initiated failure, the surfaces of PWA 1484 were machined with a grid of grooves or ribs (~250 µm wide and high) to break up the shear stresses parallel to the surface. In this case, failure occurred only as independent, single-segment events. For grooved samples, 1100 °C segment life was extended to ~ 1000 h for 5 mm wide segments, with no failure observed out to 2000 h for segments ≤2.5 mm wide. Ribbed samples were even more durable, and segments ≤6 mm remained intact for 2000 h. Larger segments failed by buckling at times inversely related to the segment width.

Functionally Graded Materials

Functionally graded materials (FGMs), whose compositional gradients are designed to have unique properties that are not available from a homogenous material, have great potential for severe thermal loading [Choules BD and Kokini K (1996) and Nemat-Alla M (2009)]. FGMs overcome the drawbacks, such as cracking, of the multilayer composite plates that have been commonly used as thermal barriers [Choules BD and Kokini K (1996)]. For example, [Nemat-Alla M (2009)] has determined, based on the minimum value of the maximum normalized equivalent stresses, the optimum composition of the ZrO_2/6061–T6/Ti–6Al–4V 2D-FGM in x- and y directions, respectively. [Shabana and Noda (2002)] have studied the FGM composite of aluminum and partially stabilized zirconia with volume fractions continuously varying through the thickness. Several functional forms of gradation of the constituents in the FGM layer were examined to reach the optimum profile giving the minimum stress level for the FGM structure under thermo-elasto-plastic behavior.

15.6.2.2 Fatigue of Tribological Coatings

Thermal spray coatings can provide a cost-effective solution for rolling/sliding contacts and other novel tribological applications, if the durability of these coatings is demonstrated. [Ahmed and Hadfield (2002)] have studied various cermet (WC–Co) and ceramic (Al_2O_3) coatings deposited by detonation gun (D-Gun), high-velocity oxy-fuel (HVOF), and high-velocity plasma spraying (HVPS) techniques, in a range of coating thicknesses (20–250 µm) on various steel substrates to determine the different competing failure modes. Four modes of fatigue-wear failure were identified as abrasive, delamination, bulk deformation, and spalling. These failure modes compete

during fatigue failure, and eventual coating failure may be either due to one or a combination of these modes. Coating failures were attributed to micro- and macro-cracking of either the coating material or the coating substrate interface, which also resulted in the attenuation of compressive residual stress.

[Ibrahim A and CC Berndt (2007)] have compared, with a detailed analysis of fatigue and deformation two groups of AISI 4340 steel specimens, one sprayed with HVOF WC–Co coating and the other plated with hard chrome. Rotating beam wear-fatigue tests were performed on the coated and uncoated specimens. The HVOF system produced a near-theoretical dense WC–Co coating that exhibited a high elastic modulus, which imparted high compressive residual stress to the substrate. The hard chrome plating process induced a tensile residual stress in the substrate that allowed the crack initiation process to start at the periphery of the specimen. Thus, the rate of the wear-fatigue process increased, and the wear-fatigue strength decreased [Ibrahim A and CC Berndt (2007)].

15.6.2.3 Conclusions Relative to Residual Stresses

Residual stresses within coatings and substrates, with or without bond coat, play an important role in coating behavior in service conditions. They depend strongly not only on the process and spray conditions used but also on the real contacts between splats, contacts linked to particle velocities and temperatures at impact, and not only to one of these parameters. They can be calculated either by analytical expression or sophisticated Finite Element Methods in order to predict the coating spallation depending on the critical strain energy release rate for the interface. However, stress calculations depend on coating mechanical properties, again linked to real splat contacts, which are a function of the spray process, and these properties must be measured to obtain calculation results close to reality.

15.7 Summary and Conclusions

Coating formation is the final and most critical step in thermal spray coating operation. Its success depends strongly on the thoroughness by which surface preparation was carried out and the use of compatible well calibrated spraying material whether in the form of powder, wire or rod. The spray process is essentially based on the injection of the coating material into the high energy stream whether in the form of high velocity gas as in Cold Spray, high temperature/high velocity combustion products as in flame spraying, HVOF and HVAF spraying, detonation product, as in D-gun spraying, DC arc as in Wire Arc Spraying or Plasma Transferred Arc deposition, DC high temperature and high velocity plasma jet or RF inductively coupled high temperature discharge. With the transfer of the energy to the material to be sprayed it is transformed into a stream of high velocity ductile particles, or molten/partially melted droplets that are projected onto the surface of the substrate. As these impact the substrate, splats are formed depending on the substrate surface properties and its temperature, the particle velocity and temperature and impact angle of the particles on the substrate.

Depending on particle impact conditions that are strongly linked to the spray process, layered splats have different percentages (about 20–60%) of real surface contact between themselves or between them and the substrate. These contacts also strongly depend, at least for those between first splats and substrate, on the substrate preheating temperature and the elimination of adsorbates and condensates, among other parameters. Cold-sprayed particles form coatings by deformation of ductile cold particles when their impact velocity is above the so-called critical velocity. If these coatings are often almost fully dense, the contact between layered deformed particles is generally rather poor (less than that with other thermal spray processes).

Coating properties depend strongly upon;

- The real contact between layered splats or plastically deformed particles
- The different defects generated during spraying: nonmelted particles, pores either globular or elongated (those between splats), micro-cracks resulting from quenching stress in ceramic splats and macro-cracks propagating generally parallel to layered splats
- For cermet coatings the size of ceramic particles (generally WC or Cr_3C_2) imbedded within the sprayed particles
- The substrate roughness and its adaptation to the mean splat size and melting state

Besides depending on the spray process used, coating properties are also strongly dependent on the spray pattern and the coating time averaged (a few tenths of seconds) temperature distribution during the whole spray process. Sophisticated spray robots are required not only to control the spray pattern, especially for parts with complex shapes, but also to control locally the heat flux imposed by hot gases and hot particles. Well-controlled cooling systems must be used during spraying to monitor the coating temperature distribution. The parameters controlling the heat transfer to the substrate and coating under formation are particularly important when spraying polymers. The substrate must be preheated close to the polymer melting temperature, but their decomposition temperature is close to their melting one. As the thermal conductivity of polymers is very low, the precise control of the surface temperature of the coating under formation is mandatory. This is achieved by adapting the spray distance, the cooling system, the thickness of the pass and the duration of the heating by hot gases and particles, both through control of the spray pattern.

Residual stress distribution within the coating, depending on the spray process but also on the coating and substrate temperature monitoring before (preheating), during (coating generation), and after spraying (cooling); for example, a tensile stress within the coating close to the substrate interface is generally detrimental to the coating adhesion under service conditions and must be avoided or limited.

Nomenclature

Units are indicated in parentheses; when no units are indicated, the parameter is dimensionless.

Latin Alphabet

a	Ratio of flattening velocity v_f to droplet impact velocity v_p
a_i	Thermal diffusivity ($i = s$ for solid phase, $i = 1$ for liquid phase) (m^2/s)
A/C	Adhesion/cohesion of a coating (MPa)
b	Splat thickness (m)
Bi	Biot number ($Bi = d_p/R_{th}\,\kappa_d$) ($-$)
c	Sound velocity (m/s)
c_ℓ	Sound velocity within the liquid droplet (m/s)
v_{cr}	Critical particle deposition velocity (m/s)
v_{er}	Erosion velocity (m/s)
c_p	Specific heat at constant pressure (J/kg.K)
d	Distance in the y direction between successive beads sprayed in x direction (m)
d_{50}	Mean powder diameter (m)
d_d	Thermal diffusion distance $d_d = \sqrt{D_{th}\,t}$, Eq. 15.3, (m)
d_p	Particle/droplet diameter (m)
D	Equivalent splat diameter (m)
D_{th}	Diffusion coefficient (m^2/s)
D_s	Splat diameter (m)
e_f	Thermal effusivity $e_f = (\rho_p \cdot c_{pp} \cdot \kappa_p)^{1/2}$ ($J/m^2\,K\,s^{0.5}$)
E_a	Activation energy (J)
E_c	Young's modulus (GPa)
EF	Ratio of the major to the minor axis of an elliptical splat
f	Bead overlapping ratio
G_{ic}	Critical strain energy release rate (J/m^2)
h	Convective heat transfer coefficient ($W/m^2\,K$)
h'	Coating thickness (m)
h_b	Bead height (μm)
h_c	Convective heat transfer coefficient ($\theta = 0$) ($W/m^2\,K$)
H	Substrate thickness (m)
k	Boltzmann Constant ()
Ks	Sommerfeld parameter at impact, $Ks = We^{1/2}Re^{1/4}$, Eq. 15.5 ($-$)
K_f	Flattening splashing parameter, Eq. 15.9
K_{ic}	Critical interfacial stress-intensity factor ($Pa\,m^{1/2}$)
K_{fc}	Critical value of K_f
K'	Beam curvature (m^{-1})
L_p	Latent heat of solidification (J/kg)
m_p	Particle mass (kg)
$\dot{m}_p$	Powder mass flow rate (kg/s)
M	Bending moment of a beam (N.m)
Ma	Impact Mach number; $Ma = v_p/c_l$
MR	Mean real contact between splat layers
N_p	Number flux density (particle /m^2.s)
N_p^*	Normalized number flux density, ($-$), Eq. 15.1
$\bar{n}_p$	Mean rate of particles/droplets impacting the surface (s^{-1})
p	Impact pressure (Pa)
p_h	Water hammer pressure; $p_h = \rho_\ell\,c_\ell\,v_p$ (Pa)

p_t	Transition pressure (Pa)
P	Splat perimeter (m)
P'	Force (N)
Pe	Peclet number ($Pe = v_p\, d_p/a_p$)
Q_i	Powder carrier gas flow rate (slm)
R	Distance in the radial direction (m)
r_s	Splat radius (m)
R_a	Average surface roughness $R_a = \frac{1}{l}\int_0^l z(x).dx$ (µm)
Re	Reynolds number at impact ($Re = v_{pi}\, d_p\, \rho/\mu$)
R_{sm}	Arithmetic mean of the surface roughness profile, $R_{sm} = \frac{1}{m} \times \sum_{i=1}^{m} x_i$ (µm)
R_t	Distance between the highest peak and the deepest undercut (µm)
R_{th}	Thermal contact resistance (K m²/W)
$R_{\Delta q}$	Standard deviation of roughness profile angle, $R_{\Delta q} = \sqrt{\frac{1}{l}\int_0^l \left(\frac{dz(x)}{dx}\right)^2 dx}$
S	Thickness of droplet's solidified layer (m)
$s*$	Dimensionless solid layer thickness ($s* = s/d_p$)
S	Splat surface (m²)
SF	Splat shape factor; $SF = P/4\,\pi\,S$ (m⁻¹)
Ste_l	Stephan number ($Ste = c_p\,(T_p - T_m)/L_p$ (−)
S_k	Skewness factor (−)
t	Time (s)
t_c	Contact time (s)
t_c	Wave propagation time $t_c = d_p\, v_p/4c_\ell^2$ (s)
t_i	Induction time in cold spray process (s)
t_f	Flattening time (s)
t_{ps}	Preheating time (s)
T	Temperature (K)
T_m	Melting temperature (K)
T_{mean}	Coating temperature (K)
T_p	Particle/droplet temperature (K)
T_{ps}	Preheating temperature (K)
T_s	Substrate temperature (K)
T_o	Room Temperature (K)
T_t	Transition temperature (K)
$\overline{v_i}$	Mean particle injection velocity (m/s)
v_f	Maximum flattening velocity (m/s)
v_N	Normal component of the particle impact velocity ($v_N = v_p \cos\varphi$) (m/s)
v_{cr}	Critical deposition velocity (m/s)
v_{er}	Critical erosion velocity (m/s)
v_f	Flattening velocity (m/s)
v_p	Particle /droplet velocity (m/s)
v_{pi}	Particle/droplet impact velocity (m/s)
v_r	The relative velocity spray torch-substrate (m/s)
v_s	Solidification velocity (m/s)
Vm	Heating rate (K/s)
w_b	Bead width at mid-height (µm)
w_{bas}	Bead width at its base (µm)
We	Weber number at impact $We = \rho\, v_{pi}^2\, d_p/\sigma_p$
x	Distance in the plan of powder injection (m))
y	Distance perpendicular to the plan of powder injection (m)
z	Distance in the axial direction (m)
z	Surface roughness (m)

Greek Alphabet

α_c	Thermal expansion coefficient of the coating
α_s	Thermal expansion coefficient of the substrate
Γ	Flexural beam stiffness (Pa m⁴)
δ_{cw}	Mean cavity width (m)
δ_{ox}	Thickness of oxide layer (m)
δ_s	Splat thickness (m)
δ_{so}	Solid layer thickness (m)
$\Delta\varepsilon$	Misfit strain
ε	Strain
φ	Angle between particle trajectory and axis normal to the substrate, Fig. 15.56 (°)
η	Dimensionless radius ($\eta = 2 \cdot r/d_p$)
η_d	Deposition efficiency
θ	Particle/droplet impact angle to substrate (°), Fig. 15.19
ϑ_s	Solidification parameter
κ	Thermal conductivity (W/m K)
κ_p	Particle/droplet thermal conductivity (W/m K)
μ	Viscosity of molten particle (Pa s)
μ_o	Viscosity of molten particle at its melting temperature (Pa.s)
ξ_D	Splat flattening ratio (-)
ν	Kinematic viscosity; $\nu = \mu/\rho$ (m²/s)
ν'	Poisson's ratio
ξ_D	Flattening parameter, $\xi_D = D/d_p$, Eq. 15.4 (−)
ρ_p	Particle/droplet density (kg/m³)
σ	Stress (Pa)
σ_p	Surface tension (J/m²) or (N/m)
σ_p°	Quenching stress (MPa)
σ_q	Quenching stress after relaxation phenomena (MPa)
σ_x	Stress distribution

References

Ahmed, R., and M. Hadfield. 2002. Mechanisms of fatigue failure in thermal spray coatings. *Journal of Thermal Spray Technology* 11 (3): 333–349.

Alamara, K., S. Saber-Samandari, and C.C. Berndt. 2010. Splat formation of polypropylene flame sprayed onto a flat surface. *Surface and Coating Technology* 205: 2518–2524.

———. 2011. Splat taxonomy of polymeric thermal spray coating. *Surface and Coating Technology* 205: 5028–5034.

Armster, S.Q., J.-P. Delplanque, M. Rein, and E.J. Lavernia. 2002. Thermo-fluid mechanisms controlling droplet based materials processes. *International Materials Review* 7 (6): 265–301.

Assadi, H., F. Gartner, T. Stoltenhoff, and H. Kreye. 2003. Bonding mechanism in cold gas spraying. *Acta Materialia* 51: 4379–4394.

AWS. 1985. *Thermal spraying: practice, theory and application.* Miami: American Welding Society.

Azarmi, F., T.W. Coyle, and J. Mostaghimi. 2008. Optimization of atmospheric plasma spray process parameters using a design of experiment for alloy 625 COATINGS. *Journal of Thermal Spray Technology* 17 (1): 144–155.

Bahbou, F., and P. Nylen. 2005. Relationship between surface topography parameters and adhesion strength for plasma spraying. In *ITSC-2005*, ed. E. Lugscheider. Düsseldorf: DVS. e-proceedings.

Bahbou, M.F., and P. Nylén. 2007. On-line measurement of plasma-sprayed Ni-particles during impact on a Ti-surface: influence of surface oxidation. *Journal of Thermal Spray Technology* 16 (4): 506–511.

Bansal, P., P.H. Shipway, and S.B. Leen. 2006. Effect of particle impact on residual stress development in HVOF sprayed coatings. *Journal of Thermal Spray Technology* 15 (4): 570–575.

———. 2007. Residual stresses in high-velocity oxy-fuel thermally sprayed coatings – Modelling the effect of particle velocity and temperature during the spraying process. *Acta Materialia* 55: 5089–5101.

Bao, Y., T. Zhang, and D.T. Gawne. 2005. Influence of composition and process parameters on the thermal spray deposition of UHMWPE coatings. *Journal of Materials Science* 40: 77–85.

Ben Ettouil, F., O. Mazhorova, B. Pateyron, Hélène Ageorges, M. El Ganaoui, and P. Fauchais. 2008. Predicting dynamic and thermal histories of agglomerated particles injected within a d.c. plasma jet. *Surface and Coating Technology* 202 (18): 4491–4495.

Berndt, C.C., J.A. Brogan, G. Montavon, A. Claudon, and C. Coddet. 1998. Mechanical properties of metal- and ceramic-polymer composites formed via thermal spray consolidation. *Journal of Thermal Spray Technology* 7 (3): 337–339.

Bertagnolli, M., M. Marchese, G. Jacucci, I. St. Doltsinis, and S. Noelting. 1997. Thermomechanical Simulation of the Splashing of Ceramic Droplets on a Rigid Substrate. *Journal of Computational Physics* 133: 205.

Bianchi, L., A. Denoirjean, F. Blein, and P. Fauchais. 1977. Microstructural investigation of plasma sprayed ceramic splats. *Thin Solid Films* 299: 125–135.

Birks, N., G.H. Meier, and F.S. Pettit. 2006. *Introduction to the high-temperature oxidations of metals.* Cambridge: Cambridge University Press.

Bolelli, G., L. Lusvarghi, T. Varis, E. Turunen, M. Leoni, P. Scardi, C.L. Azanza-Ricardo, and M. Barletta. 2008. Residual stresses in HVOF-sprayed ceramic coatings. *Surface and Coating Technology* 202: 4810–4819.

Bose, S., and J. DeMasi-Marcin. 1997. Thermal barrier coating experience in gas turbine engines at Pratt & Whitney. *Journal of Thermal Spray Technology* 6 (1): 99–104.

Branco, J.R.T., and S.V. Campos. 1999. Wear behavior of thermally sprayed PET. *Surface and Coating Technology* 120–121: 476–481.

Brossard, S., P.R. Munroe, A.T. Tran, and M.M. Hyland. 2010. Study of the splat microstructure and the effects of substrate heating on the splat formation for Ni-Cr particles plasma sprayed onto stainless steel substrates. *Journal of Thermal Spray Technology* 19 (5): 1100–1114.

Brousse, E. 2010. *Elaboration by thermal spraying of finely structured elements of high temperature electrolyser for hydrogen production: Processes, structures and characteristics.* Ph.D. thesis, University of Limoges, France, Sept 2010 (in French)

Bussmann, M., S. Chandra, and J. Mostaghimi. 2000. Modeling the splashing of a droplet impacting a solid surface. *Physics of Fluids* 12: 3121–3132.

Cedelle, J., M. Vardelle, and P. Fauchais. 2006. Influence of stainless steel substrate preheating on surface topography and on millimeter- and micrometer-sized splat formation. *Surface and Coating Technology* 201: 1373–1382.

Chandra, S., and P. Fauchais. 2009. Formation of solid splats during thermal spray deposition. *Journal of Thermal Spray Technology* 18 (2): 148–180.

Chang, G.C., W. Phucharoen, and R.A. Miller. 1987. Finite element thermal stress solutions for thermal barrier coatings. *Surface and Coating Technology* 32 (1–4): 307–325.

Chang-Jiu, L., and J.-L. Li. 2004. Transient contact pressure during flattening of thermal spray droplet and its effect on splat formation. *Journal of Thermal Spray Technology* 13 (2): 229–238.

Choules, B.D., and K. Kokini. 1996. Architecture of functionally graded ceramic coating against surface thermal fracture. *ASME Journal of Engineering Materials and Technology* 118: 522–528.

Chraska, T., and A.H. King. 2002. Effect of different substrate conditions upon interface with plasma sprayed zirconia—A TEM study. *Surface and Coating Technology* 157: 238–246.

Clyne, W. 1996. Residual stresses in surface coatings and their effects on interfacial debonding. *Key Engineering Materials* 116 (7): 307–330.

Clyne, T.W. 2001. *Residual stresses in thick and thin surface coating, encyclopedia of materials: science and technology,* 8126–8134. Oxford: Elsevier.

Clyne, T.W., and S.C. Gill. 1996. Residual stresses in thermal spray coatings and their effect on interfacial adhesion: a review of recent work. *Journal of Thermal Spray Technology* 5 (4): 401–418.

Coddet, C., G. Montavon, S. Ayrault-Costil, O. Freneaux, F. Rigolet, G. Barbezat, F. Foliot, A. Diard, and P. Wazen. 1999. Surface preparation and thermal spray in a single step: the PROTAL process – Example of application for an aluminium-base substrate. *Journal of Thermal Spray Technology* 8 (2): 235–242.

Costil, S., H. Liao, A. Gammondi, and C. Coddet. 2005. Influence of surface laser cleaning combined with substrate preheating on splat morphology. *Journal of Thermal Spray Technology* 14 (1): 31–38.

Coudert, J.F., M.P. Planche, and P. Fauchais. 1995. Velocity measurements of dc plasma jets based on arc root fluctuations. *Plasma Chemistry and Plasma Processing* 15 (1): 47–69.

Ctibor, P., R. Lechnerová, and V. Beneš. 2006. Quantitative analysis of pores of two types in a plasma-sprayed coating. *Materials Characterization* 56: 297–304.

Ctibor, P., K. Neufuss, F. Zahalka, and B. Kolman. 2007. Plasma sprayed ceramic coatings without and with epoxy resin sealing treatment and their wear resistance. *Wear* 262: 1274–1280.

Davis, J.R. 2004. *Handbook of Thermal Spray Technology.* Materials Park: ASM International.

Delluc, G., L. Perrin, H. Ageorges, P. Fauchais, and B. Pateyron. 2004. Modelling of plasma jet and particle behavior in spraying conditions. In *ITSC-2004, in modeling and simulation V (pub.).* Düsseldorf: DVS, C.D. Rom.

Denoirjean, A., A. Grimaud, P. Fauchais, P. Tristant, C. Tixier, and J. Desmaison. 1998. Splat formation, first step for multitechnique deposition of plasma spraying and microwave plasma enhanced CVD. In *ITSC-1998: meeting in challenges of the 21st century,* ed. C. Coddet, vol. 2, 1369–1374. Materials Park: ASM International.

Dhiman, R., A.G. McDonald, and S. Chandra. 2007. Predicting splat morphology in a thermal spray process. *Surface and Coating Technology* 201: 7789–7801.

Doering, T. 2007. PVD and PACVD (DLC) coating technologies and applications. In *International SAMPE symposium and exhibition, proceedings,* vol 52, 8pp

Dong, S. 2003. *Multifunctional layers obtained by a multi-technique process,* Ph.D. University of Limoges, France, November.

Dong, S., B. Song, B. Hansz, H. Liao, and C. Coddet. 2011. *Improvement in the microstructure and property of plasma sprayed metallic, alloy and ceramic coatings by using dry ice blasting.* In: 5th RIPT conference, CEC, Limoges, Dec 2011, e-proceedings.

Ducos, M. 1985. Industrial applications of arc and RF inductive plasmas of low power. In *Plasmas in industry.* Dopee diffusion, 77210 Avon, France (in French), 251–263

Escure, C., M. Vardelle, and P. Fauchais. 2003. Experimental and theoretical study of the impact of alumina droplet on cold and hot substrates. *Plasma Chemistry and Plasma Processing* 3: 291–309.

Espié, G., P. Fauchais, B. Hannoyer, J.C. Labbe, and A. Vardelle. 1999. Effect of metal particles oxidation during the APS on the wettability, heat and mass transfer under plasma condition. In *Heat and mass transfer under plasma condition*, ed. P. Fauchais, J. Van der Mullen, and J. Heberlein, vol. 891, 143–151. Annals of NY Academy of Sciences.

Espié, G., P. Fauchais, J.C. Labbe, A. Vardelle, and B. Hannoyer. 2001. In *Oxidation of iron particles during APS, ITSC-2001: New surface for a new Millennium*, ed. C.C. Berndt, K.A. Khor, and E. Lugscheider, 821–828. Materials Park: ASM International.

Espié, G., A. Denoirjean, P. Fauchais, J.C. Labbe, J. Dudsky, O. Scheeweiss, and K. Volenik. 2005. In flight oxidation of iron particles sprayed using gas and water stabilized plasma torches. *Surface, and Coatings Technology* 195: 17–28.

Fantassi, S., M. Vardelle, A. Vardelle, and P. Fauchais. 1993. Influence of the velocity of plasma-sprayed particles on splat formation. *Journal of Thermal Spray Technology* 2 (4): 379–384.

Fasching, M.M., L.E. Weiss, and F.B. Prinz. 1992. Optimization of robotic trajectories for thermal spray shape deposition. In *ITSC-1992: advances in coatings technology*, ed. C.C. Berndt, 221–226. Materials Park: ASM International.

Fauchais, P. 2004. Understanding plasma spraying. *Journal of Physics D: Applied Physics* 37: 86–108.

Fauchais, P., and M. Vardelle. 2010. Sensors in spray processes – An invited review. *Journal of Thermal Spray Technology* 19 (4): 668–694.

Fauchais, P., M. Vardelle, A. Vardelle, L. Bianchi, and A.C. Léger. 1996. Parameters controlling the generation and properties of plasma sprayed zirconia coatings. *Plasma Chemistry and Plasma Processing* 16 (1): 99S–125S.

Fauchais, P., A. Vardelle, and B. Dussoubs. 2001a. Quo vadis thermal spraying – invited paper. *Journal of Thermal Spray Technology* 10: 44–66.

Fauchais, P., A. Vardelle, M. Vardelle, A. Denoirjean, B. Pateyron, and M. El Ganaoui. 2001b. Formation and layering of alumina splats: thermal history of coating formation, resulting residual stresses and coating microstructure. In *ITSC-2001: new surfaces for a new millenium*, ed. C.C. Berndt, K.A. Khor, and E. Lugsheider, 865–873. Materials Park: ASM International.

Fauchais, P., M. Fukumoto, A. Vardelle, and M. Vardelle. 2004. Knowledge concerning splat formation: an invited review. *Journal of Thermal Spray Technology* 13 (3): 337–360.

Fauchais, P., R. Etchart-Salas, V. Rat, J.-F. Coudert, N. Caron, and K. Wittmann-Ténèze. 2008. Parameters controlling liquid plasma spraying: solutions, sols or suspensions. *Journal of Thermal Spray Technology* 17 (1): 31–59.

Fauchais, P., G. Montavon, and G. Bertrand. 2010. From powders to thermally sprayed coatings. *Journal of Thermal Spray Technology* 19 (1–2): 56–80.

Floristán, M., J.A. Montesinos, J.A. García-Marín, A. Killinger, and R. Gadow. 2012. Robot trajectory planning for high quality thermal spray coating processes on complex shaped components. In *ITSC-2012*. Materials Park: ASM International. e-proceedings.

Fukumoto, M., and Y. Huang. 1999. Flattening mechanism in thermal sprayed nickel particle impinging on flat substrate surface. *Journal of Thermal Spray Technology* 8 (3): 427–432.

Fukumoto, M. 2004. Particle flattening behavior in thermal spray process. In: *ITSC course*, Osaka, Japan, 7–8 May.

Fukumoto, M., I. Ohgitani, H. Nagai, and T. Yasni. 2005. Effect of substrate surface change by heating on flattening behavior of thermal sprayed particles. In *ITSC-2005*, ed. E. Lugscheider. Düsseldorf: DVS. e-proceedings.

Fukumoto, M., H. Nagui, and T. Yasui. 2006a. Influence of surface character change of substrate due to heating on flattening behavior of thermal sprayed particles. In *ITSC-2006*, ed. B. Marple et al. Materials Park: ASM International.

Fukumoto, M., H. Nagai, and T. Yasui. 2006b. Influence of surface character change of substrate due to heating on flattening behavior of thermal sprayed particles. *Journal of Thermal Spray Technology* 15 (4): 759–764.

Fukumoto, M., T. Yamaguchi, M. Yamada, and T. Yasui. 2007a. Splash splat to disk splat transition behavior in plasma-sprayed metallic materials. *Journal of Thermal Spray Technology* 16 (5–6): 905–912.

Fukumoto, M., H. Wada, K. Tanabe, M. Yamada, E. Yamaguchi, A. Niwa, M. Sugimoto, and M. Izawa. 2007b. Effect of substrate temperature on deposition behavior of copper particles on substrate surfaces in the cold spray process. *Journal of Thermal Spray Technology* 16 (5–6): 643–650.

Fukumoto, M., M. Mashiko, M. Yamada, and E. Yamaguchi. 2010. Deposition behavior of copper fine particles onto flat substrate surface in cold spraying. *Journal of Thermal Spray Technology* 19 (1–2): 89–94.

Fukumoto, M., K. Yang, K. Tanaka, T. Usami, T. Yasui, and M. Yamada. 2011. Effect of substrate temperature and ambient pressure on heat transfer at interface between molten droplet and substrate surface. *Journal of Thermal Spray Technology* 20 (1–2): 48–58.

Gärtner, F., C. Borchers, T. Stoltenhoff, and H. Kreye. 2003. Numerical and microstructural investigations of the bonding mechanisms in cold spraying. In *Thermal spray 2003: advancing the science & applying the technology*, ed. C. Moreau and B. Marple, 1–8. Materials Park: ASM International.

Gatto, A., E. Bassoli, and M. Fornari. 2004. Plasma transferred arc deposition of powdered high performances alloys: process parameters optimization as a function of alloy and geometrical configuration. *Surface and Coating Technology* 187: 265–271.

Gawne, D.T., B.J. Griffiths, and G. Dong. 1995. Splat morphology and adhesion of thermally sprayed coatings. In *ITSC-1995: Current status and future trends, Kobe, Japan*, ed. A. Ohmori, 779–784. Osaka: High Temperature Society of Japan.

Ghafouri-Azar, R., J. Mostaghimi, and S. Chandra. 2006. Modeling development of residual stresses in thermal spray coatings. *Computational Materials Science* 35: 13–26.

Gill, C. 1991. *Residual stresses in plasma sprayed deposits*. Ph.D. thesis, April 1991, Gonville and Caius College, Cambridge.

Gill, S.C., and T.W. Clyne. 1990. Stress distributions and material response in thermal spraying of metallic and ceramic deposits. *Metallurgical Transactions B* 21: 377–385.

Gindrat, M., J.-L. Dorier, C. Hollenstein, M. Loch, A. Refke, A. Salito, and G. Barbezat. 2002. Effect of specific operating conditions on the properties of LPPS plasma jets expanding at low pressure. In *ITSC-2002*, ed. E. Lugscheider and C.C. Berndt, 459–464. Dusseldorf: Verlag für Schweißen und verwandte Verfahren DVS-Verlag.

Goutier, S. 2010. *Etude expérimentale de l'impact et de la solidification de gouttes céramiques et métalliques de tailles micrométrique et millimétrique sur différents types de substrats: compréhension du mécanisme de formation des dépôts par projection thermique*. Ph.D. thesis, University of Limoges, France, 7 Dec.

Goutier, S., M. Vardelle, J.C. Labbe, and P. Fauchais. 2011. Flattening and cooling of millimeter- and micrometer-sized alumina drops. *Journal of Thermal Spray Technology* 20 (1–2): 59–67.

Greving, D.J., J.R. Shadley, and E.F. Rybicki. 1994. Effects of coating thickness and residual stresses on the bond strength of ASTM C633-79 thermal spray coating test specimens. *Journal of Thermal Spray Technology* 3 (4): 371–378.

Grujicic, M. 2007. Particle/substrate interaction in the cold-spray bonding process. In *The cold spray materials deposition process*, ed. V.K. Champagne, 148–177. Cambridge: Woodhead Publishing Ltd.

Grujicic, M., J.R. Saylor, D.E. Beasley, W.S. DeRosset, and D. Helfritch. 2003. Computational analysis of the interfacial bonding between feed-powder particles and the substrate in the cold-gas dynamic-spray process. *Applied Surface Science* 219: 211–227.

Guetta, S., M.H. Berger, F. Borit, V. Guipont, M. Jeandin, M. Boustie, Y. Ichikawa, K. Sakaguchi, and K. Ogawa. 2009. Influence of particle velocity on adhesion of cold-sprayed splats. *Journal of Thermal Spray Technology* 18 (3): 331–342.

Guo, H.B., S. Kuroda, and H. Murakamia. 2006. Segmented thermal barrier coatings produced by atmospheric plasma spraying hollow powders. *Thin Solid Films* 506–507: 136–139.

Hackett, C.M., and G.S. Settles. 1994. Turbulent mixing of the HVOF thermal spray and coating oxidation. In *Proceedings of NTSC-1994*, ed. C.C. Berndt and S. Sampath, 307–312. Materials Park: ASM International.

Haddadi, A. 1998. *Zirconia and alumina coatings plasma sprayed: Columnar growth, residual stress and modeling.* Ph.D. thesis, University of Limoges, France, 22 July (in French).

Haddadi, A., F. Nardou, P. Fauchais, A. Grimaud, and A.C. Leger. 1997. Influence of substrate and coating temperature on columnar growth within plasma sprayed zirconia and alumina coatings. In *UTSC-1997*, ed. C.C. Berndt, 671–680. Materials Park: A united forum for scientific and technological advances. ASM International.

Hafiz, J., R. Mukherjee, X. Wang, P.H. McMurry, J.V.R. Heberlein, and S.L. Girshick. 2006. Hypersonic plasma particle deposition-a hybrid between plasma spraying and vapor deposition. *Journal of Thermal Spray Technology* 15 (4): 822–826.

Hanson, T.C., C.M. Hackett, and G.S. Settles. 2002. Independent control of HVOF particle velocity and temperature. *Journal of Thermal Spray Technology* 11 (1): 75–85.

Haure, T. 2003. *Multifunctional layers obtained by a multi-technique process.* Ph.D. University of Limoges, France, Nov. 2003.

He, J., B. Dulin, and T. Wolfe. 2008. Peening effect of thermal spray coating process. *Journal of Thermal Spray Technology* 17 (2): 214–220.

Hearley, J.A., J.A. Little, and A.J. Sturgeon. 2000. The effect of spray parameters on the properties of high velocity oxy-fuel NiAl intermetallic coatings. *Surface and Coating Technology* 123: 210–218.

Henne, R., M. Müller, E. Pross, G. Schiller, F. Gitzhofer, and M.I. Boulos. 1999. Near-net-shape forming of metallic bipolar plates for planar solid oxide fuel cells by induction plasma spraying. *Journal of Thermal Spray Technology* 8 (1): 110–116.

Henrich, V.E., and P.A. Cox. 1994. *The surface science of metal oxides.* Cambridge: Cambridge University Press.

Hermanek, F.J. 2001. *Thermal spray terminology and company origins.* Materials Park: ASM International.

Honner, M., P. Cerveny, V. Franta, and F. Cejka. 1998. Heat transfer during HVOF deposition. *Surface and Coating Technology* 106: 94–99.

Hospach, A., G. Mauer, R. Vaßen, and D. Stöver. 2011. Columnar-structured thermal barrier coatings (TBCs) by thin film low-pressure plasma spraying (LPPS-TF). *Journal of Thermal Spray Technology* 20 (1–2): 116–120.

———. 2012. Characteristics of ceramic coatings made by thin film low pressure plasma spraying (LPPS-TF). *Journal of Thermal Spray Technology* 21 (3–4): 435–440.

Huang, R., and H. Fukanuma. 2012. Study of the influence of particle velocity on adhesive strength of cold spray deposits. *Journal of Thermal Spray Technology* 21 (3–4): 541–549.

Hussain, T., D.G. McCartney, P.H. Shipway, and D. Zhang. 2009. Bonding mechanisms in cold spraying: the contributions of metallurgical and mechanical components. *Journal of Thermal Spray Technology* 18 (3): 364–379.

Ibrahim, A., and C.C. Berndt. 2007. Fatigue and deformation of HVOF sprayed WC–Co coatings and hard chrome plating. *Materials Science and Engineering A* 456: 114–119.

Itoh, A., K. Takeda, M. Itoh, and M. Koga. 1990. Pretreatements of substrates by using reversed transferred arc in low pressure plasma spray. In *NTSC-1990: research and applications*, ed. F. Bernicki, 245–252. Materials Park: ASM International.

Ivosevic, M., R. Knight, S.R. Kalidindi, G.R. Palmese, and J.K. Sutter. 2005. Adhesive/cohesive properties of thermally sprayed functionally graded coatings for polymer matrix composites. *Journal of Thermal Spray Technology* 14 (1): 45–51.

Ivosevic, M. 2006. *Splat formation during thermal spraying of polymer particles: Mathematical modeling and experimental analysis.* Ph.D. Thesis, Faculty of Drexel University.

Ivosevic, M., R.A. Cairncross, and R. Knight. 2006a. 3D predictions of thermally sprayed polymer splats: modeling particle acceleration, heating and deformation on impact with a flat substrate. *International Journal of Heat and Mass Transfer* 49: 3285–3297.

Ivosevic, M., V. Gupta, J.A. Baldoni, R.A. Cairncross, T.E. Twardowski, and R. Knight. 2006b. Effect of substrate roughness on splatting behavior of HVOF sprayed polymer particles: modeling and experiments. *Journal of Thermal Spray Technology* 15 (4): 725–730.

Jackson, L., M. Ivosevic, R. Knight, and R.A. Cairncross. 2007. Sliding wear properties of HV thermally sprayed nylon-11 and nylon-11/ceramic composites on steel. *Journal of Thermal Spray Technology* 16 (5–6): 927–932.

Jiang, X.L., R. Tiwari, F. Gitzhofer, and M.I. Boulos. 1993. On the induction plasma deposition of tungsten metal. *Journal of Thermal Spray Technology* 2: 265–270.

———. 1995. Reactive deposition of tungsten and titanium carbides by induction plasma. *Journal of Materials Science Letters* 30: 2325–2328.

Kang, C.W., and H.W. Ng. 2006. Splat morphology and spreading behavior due to oblique impact of droplets onto substrates in plasma spray coating process. *Surface and Coating Technology* 200: 5462–5477.

Kang, C.W., H.W. Ng, and S.C.M. Yu. 2006. Imaging diagnostics study on obliquely impacting plasma-sprayed particles near to the substrate. *Journal of Thermal Spray Technology* 15 (1): 118–130.

Kanouff, M.P., R.A. Neiser, and T.J. Roemer. 1998. Surface roughness of thermal spray coatings made with off-normal spray angle. *Journal of Thermal Spray Technology* 7 (2): 219–228.

Katanoda, H., M. Fukuhara, and N. Iino. 2007a. Numerical study of combination parameters for particle impact velocity and temperature in cold spray. *Journal of Thermal Spray Technology* 15 (5–6): 627–633.

———. 2007b. Numerical study of combination parameters for particle impact velocity and temperature in cold spray. In *ITSC-2007: Global coating solutions*, ed. B.R. Marple, M.M. Hyland, Y.-C. Lau, C.-J. Li, R.S. Lima, and G. Montavon, 72–77. Materials Park: ASM International.

Klinkov, S.V., and V.F. Kosarev. 2006. Measurements of cold spray deposition efficiency. *Journal of Thermal Spray Technology* 15 (3): 364–371.

———. 2012. Cold spraying activation using an abrasive admixture. *Journal of Thermal Spray Technology* 21 (5): 1046–1053.

Koivuluoto, H., and P. Vuoristo. 2010. Effect of powder type and composition on structure and mechanical properties of Cu + Al2O3 coatings prepared by using low-pressure cold spray process. *Journal of Thermal Spray Technology* 19 (5): 1081–1092.

Koivuluoto, H., J. Lagerbom, and P. Vuoristo. 2007. Microstructural studies of cold sprayed copper, nickel, and nickel-30% copper coatings. *Journal of Thermal Spray Technology* 16 (4): 488–497.

Koolloos, M.F.J., and J.M. Houben. 2000. Behavior of plasma-sprayed thermal barrier coatings during thermal cycling and the effect of a preoxidized NiCrAlY bond coat. *Journal of Thermal Spray Technology* 9 (1): 49–58.

Kroupa, F. 1997. Residual stresses in thick, non-homogeneous coatings. *Journal of Thermal Spray Technology* 6 (3): 309–319.

Kudinov, V.V., et al. 1981. *High temperature. Dust laden jets*, 381–392. Delft: VSL.

Kuroda, S., and T.W. Clyne. 1991a. The origin and quantification of the quenching stress associated with splat cooling during spray deposition. In *Proceedings of the 2nd. Plasma-Technik Syrup*, ed. H. Eschnauer, P. Hüber, A. Nicoll, and S.B. Sandmeier, vol. 1, 273–284. Wohlen: Plasma-Technik.

———. 1991b. The quenching stresses in thermally sprayed coatings. *Thin Solid Films* 200: 49–66.

Kuroda, S., T. Fukushima, and S. Kitakara. 1989. Generation mechanisms of residual stress in plasma sprayed coatings. In: *Proceedings of the 11th international vacuum congress, 17th conference on solid surfaces*, Sept 1989, Köln, F.R.G

Kuroda, S., T. Dento, and S. Kitahara. 1995. Quenching stress in plasma sprayed coatings and its correlation with the deposit microstructure. J. *Journal of Thermal Spray Technology* 4: 75–87.

Kuroda, S., Y. Tashiro, H. Yumoto, S. Taira, and H. Fukanuma. 1997. Measurement of stress development during HVOF thermal spray. In *UTSC-1997 a united forum for scientific and technological advances*, ed. C.C. Berndt, 805–811. Materials Park: ASM International.

Kuroda, S., Y. Tashiro, H. Yumoto, S. Taira, H. Fukanuma, and S. Tobe. 2001. Peening action and residual stresses in high-velocity oxygen fuel thermal spraying of 316L stainless steel. *Journal of Thermal Spray Technology* 10 (2): 367–374.

Lahmar-Mebdoua, Y., A. Vardelle, P. Fauchais, and D. Gobin. 2009. Splat heat transfer and crystal growth under thermal spray conditions. *High Temperature Materials and Processes* 13 (1): 77–91.

Lapierre, D., R.J. Kearney, M. Vardelle, A. Vardelle, and P. Fauchais. 1994. Effects of a substrate on the temperature distribution in argon-hydrogen thermal plasma jet. *Plasma Chemistry and Plasma Processing* 14 (4): 407–423.

Leger, A.C., A. Grimaud, P. Fauchais, and G. Delluc. 1997. Influence of the torch to substrate velocity and resulting temperature on the residual stresses in alumina plasma sprayed coatings. In *Proceedings UTSC-1997*, ed. C.C. Berndt, 823–829. Materials Park: ASM International.

Leger, A.C., A. Haddadi, B. Pateyron, G. Delluc, A. Grimaud, F. Nardou, and P. Fauchais. 1998. Residual stresses during coating generation: plasma sprayed alumina coating on XC38, measurements and calculations. In *ITSC-1998: meeting the challenge of the 21st century*, ed. C. Coddet, 895–903. Materials Park: ASM International.

Leigh, S.H., and C.C. Berndt. 1997. Evaluation of off-angle thermal spray. *Surface and Coating Technology* 89: 213–224.

Li, C.-J., and J.-L. Li. 2004. Evaporated-gas-induced splashing model for splat formation during plasma spraying. *Surface and Coating Technology* 184: 13–23.

Li, C.-J., J.-L. Li, and W.B. Wang. 1998. The effect of substrate preheating and surface organic covering on splat formation. In *ITSC-1998: meeting the challenges of the 21st century*, ed. C. Coddet, 473–480. Materials Park: ASM International.

Li, C.-J., J.-L. Li, W.B. Wang, A.-J. Fu, and A. Ohmori. 1999. A mechanism of the splashing during droplet splatting. In *UTSC-1999 proceedings, Düsseldorf*, ed. E. Lugscheider and P.A. Kammer, 530–535. Düsseldorf: DVS.

Li, J., H. Liao, and C. Coddet. 2002. Friction and wear behavior of flame-sprayed PEEK coatings. *Wear* 252: 824–831.

Li, C.-J., and J.-L. Li. 2004. Evaporated-gas-induced splashing model for splat formation during plasma spraying. *Surface and Coating Technology* 184: 13–23.

Li, C.-J., W.-Y. Li, Y.-Y. Wang, G.-J. Yang, and H. Fukanuma. 2005a. A theoretical model for prediction of deposition efficiency in cold spraying. *Thin Solid Films* 485: 79–85.

Li, M., D. Shi, and P.D. Christofides. 2005b. Modeling and control of HVOF thermal spray processing of WC–Co coatings. *Powder Technology* 156: 177–194.

Li, C.-J., W.-Y. Li, Y.-Y. Wang, G.-J. Yang, and H. Fukanuma. 2005c. A theoretical model for prediction of deposition efficiency in cold spraying. *Thin Solid Films* 489: 79–85.

Li, C.-J., C.-X. Li, G.-J. Yang, and Y.-Y. Wang. 2006. Examination of substrate surface melting-induced splashing during splat formation in plasma spraying. *Journal of Thermal Spray Technology* 15 (4): 717–724.

Li, H., S. Costil, H.-L. Liao, C.-J. Li, M.P. Planche, and C. Coddet. 2006a. Effect of surface conditions on the flattening behavior of plasma sprayed Cu splats. *Surface and Coating Technology* 200: 5435–5446.

Li, H., S. Costil, S.-H. Deng, H.-L. Liao, C. Coddet, V. Ji, and W.-J. Huang. 2006b. Benefit of surface oxide removal on thermal spray coating adhesion using the PROTAL process. In *ITSC-2006*, ed. B. Marple et al. Materials Park: ASM International. e-proceedings.

Li, C.-J., C.-X. Li, G.-J. Yang, and Y.-Y. Wang. 2006c. Examination of substrate surface melting-induced splashing during splat formation in plasma spraying. *Journal of Thermal Spray Technology* 15 (4): 717–724.

Li, C.-J., W.-Y. Li, and H. Liao. 2006d. Examination of the critical velocity for deposition of particles in cold spraying. *Journal of Thermal Spray Technology* 15 (2): 212–222.

Li, H., S. Costil, H.-L. Liao, and C. Coddet. 2006e. Role of the laser surface preparation on the adhesion of Ni-5wt% Al coatings deposited using the PROTAL® process. *Journal of Thermal Spray Technology* 15 (2): 191–197.

Lima, C.R.C., J. Nin, and J.M. Guilemany. 2006. Evaluation of residual stresses of thermal barrier coatings with HVOF thermally sprayed bond coats using the modified layer removal method (MLRM). *Surface and Coating Technology* 200: 5963–5972.

LINSPRAY CO2 cooling for thermal spraying, pdf document. http://www.linde.com.

Luzin, V., K. Spencer, and M.-X. Zhang. 2011. Residual stress and thermo-mechanical properties of cold spray metal coatings. *Acta Materialia* 59: 1259–1270.

Lyphout, C., P. Nylén, A. Manescu, and T. Pirling. 2008. Residual stresses distribution through thick HVOF sprayed Inconel 718 coatings. *Journal of Thermal Spray Technology* 17 (5–6): 915–923.

Madejski, J. 1976. Solidification of droplets on a cold surface. *International Journal of Heat and Mass Transfer* 19: 1009–1013.

Maitre, A., A. Denoirjean, P. Fauchais, and P. Lefort. 2002. Plasma-jet coating of pre-oxidized XC38 steel: influence of the nature of the oxide layer. *Physical Chemistry Chemical Physics* 4 (15): 3887–3893.

Maruyama, T., K. Akagi, and T. Kobayashi. 2006. Effects of blasting parameters on removability of residual grit. *Journal of Thermal Spray Technology* 15 (4): 817–821.

Mateus, C., S. Costil, R. Bolot, and C. Coddet. 2005. Ceramic/fluoropolymer composite coatings by thermal spraying-a modification of surface properties. *Surface and Coating Technology* 191: 108–118.

Mauer, G., R. Vaßen, and D. Stöver. 2010. Thin and dense ceramic coatings by plasma spraying at very low pressure. *Journal of Thermal Spray Technology* 19 (1–2): 495–501.

Mauer, G., A. Hospach, N. Zotov, and R. Vaßen. 2012. Process conditions and microstructures of ceramic coatings by gas phase deposition based on plasma spraying. In *ITSC-2012 proceedings*, ed. R.S. Lima et al. Materials Park: ASM International. e-proceedings.

McDonald, A., M. Lamontagne, C. Moreau, and S. Chandra. 2006. Impact of plasma-sprayed metal particles on hot and cold glass surfaces. *Thin Solid Films* 514: 212–222.

McDonald, A., C. Moreau, and S. Chandra. 2007a. Effect of substrate oxidation on spreading of plasma-sprayed nickel on stainless steel. *Surface and Coating Technology* 202: 23–33.

———. 2007b. Thermal contact resistance between plasma-sprayed particles and flat surfaces. *International Journal of Heat and Mass Transfer* 50: 1737–1749.

McDonald, A., M. Xue, S. Chandra, J. Mostaghimi, and C. Moreau. 2007c. Modeling fragmentation of plasma-sprayed particles impacting on a solid surface at room temperature. *Comptes Rendus Mecanique* 335: 351–356.

Mehdizadeh, N.Z., S. Chandra, and J. Mostaghimi. 2002. Effect of substrate temperature and roughness on coating formation. In *Proceedings of the ITSC-2002*, ed. E. Lugscheider, 830–837. Düsseldorf: DVS.

Mellali, M., A. Grimaud, A.C. Léger, P. Fauchais, and J. Lu. 1997. Alumina grit blasting parameters for surface preparation in the plasma spraying operation. *Journal of Thermal Spray Technology* 6 (2): 217–227.

Monerie-Moulin, F., F. Gitzhofer, P. Fauchais, M. Boulos, and A. Vardelle. 1992. Heat flux transmitted to a cold substrate by a d.c. Ar-H2 plasma jet. *Journal of High Temperature Materials and Processes* 1: 249–258.

Monnerie-Moulin, F., F. Gitzhofer, P. Fauchais, M. Boulos, and A. Vardelle. 1992. Flux transmitted to a cold substrate by a d.c. Ar-H2 spraying plasma jet. *Journal of High Temperature Materials and Processes* 1 (3): 249–257.

Montillet, D., E. Dombre, F.D. Valentin, and J.M. Goubot. 1999. Modeling, simulating and optimizing the robotized plasma deposition: an expression approach. In *Thermal spray*, ed. E. Lugsheider and P.A. Kammer, 507–512. Düsseldorf: DVS.

Morks, M.F., Y. Tsunekawa, M. Okumiya, and M.A. Shoeib. 2002. Splat morphology and microstructure of plasma sprayed cast iron with different preheat substrate temperatures. *Journal of Thermal Spray Technology* 11 (2): 226–232.

Mostaghimi, J., and S. Chandra. 2018. Droplet Impact and Solidification in Plasma Spraying. In *Handbook of Thermal Science and Engineering*, Chapter 69, ed. F.A. Kulacki, 2967–3008. Cham: Springer.

Neiser, R.A., M.F. Smith, and R.C. Dykhuisen. 1998. Oxidation in wire HVOF-sprayed steel. *Journal of Thermal Spray Technology* 7 (4): 537–545.

Nemat-Alla, M. 2009. Reduction of thermal stresses by composition optimization of two-dimensional functionally graded materials. *Acta Mechanica* 208: 147–161.

Newbery, A.P., and P.S. Grant. 2000. Droplet splashing during arc spraying of steel and the effect on deposit microstructure. *Journal of Thermal Spray Technology* 9 (2): 250–258.

Normanda, B., H. Takenouti, M. Keddama, H. Liao, G. Monteil, and C. Coddet. 2004. Electrochemical impedance spectroscopy and dielectric properties of polymer: application to PEEK thermally sprayed coating. *Electrochimica Acta* 49: 2981–2986.

Papyrin, A.N., S.V. Klinkov, and V.F. Kosarev. 2003. Modeling of particle-substrate adhesive interaction under the cold spray process. In *ITSC-2003: advancing the science & applying the technology*, ed. C. Moreau and B. Marple, 27–35. Materials Park: ASM International.

Parizi, H.B., L. Rosenzweig, J. Mostaghimi, S. Chandra, T. Coyle, H. Salimi, L. Pershin, A. McDonald, and C. Moreau. 2008. Numerical simulation of droplet impact on patterned surfaces. *Journal of Thermal Spray Technology* 16 (5–6): 713–721.

Pasandideh-Fard, M., Y.M. Qiao, S. Chandra, and J. Mostaghimi. 1996. Capillary effects during droplet impact on a solid surface. *Physics of Fluids* 8: 650–659.

Pasandideh-Fard, M., R. Bhola, S. Chandra, and J. Mostaghimi. 1998. Deposition of Tin Droplets on a Steel Plate: Simulations and Experiments. *International Journal of Heat and Mass Transfer* 41: 2929–2945.

Pasandideh-Fard, M., V. Pershin, S. Chandra, and J. Mostaghimi. 2002. Splat shapes in a thermal spray coating process: simulations and experiments. J. *Journal of Thermal Spray Technology* 11: 206–217.

Patel, K., C.S. Doyle, B.J. James, and M.M. Hyland. 2010a. Valence band XPS and FT-IR evaluation of thermal degradation of HVAF thermally sprayed PEEK coatings. *Polymer Degradation and Stability* 95: 792–797.

Patel, K., C.S. Doyle, D. Yonekura, and B.J. James. 2010b. Effect of surface roughness parameters on thermally sprayed PEEK coatings. *Surface and Coating Technology* 204: 3567–3572.

Pech, J. 1999. *Preoxidation generated by blown arc d.c. plasma jets. Relationship between surface, oxidation and adhesion of plasma sprayed coatings.* Ph.D., University of Rouen, France

Pech, J., and B. Hannoyer. 2000. Influence of the oxide layer by d.c. plasma preheating on the adhesion coating and role of the initial surface pretreatment. *Surface and Interface Analysis* 30: 585–588.

Pech, J., B. Hannoyer, O. Lagnoux, A. Denoirjean, and P. Fauchais. 1999. Influence of the preheating parameters on the plasma jet oxidation of a low carbon steel. In *Progress in plasma processing of materials*, ed. P. Fauchais and J. Amouroux, 543–551. New York: Begell House.

Pfender, E. 1989. Particle behavior in thermal plasmas. *Plasma Chemistry and Plasma Processing* 9 (Sup.1): 167S–194S.

Ponticaud, C., A. Grimaud, A. Denoirjean, P. Lefort, and P. Fauchais. 2001. Titanium powder nitridation by reactive plasma spraying. In *Progress in plasma processing of materials*, ed. P. Fauchais, J. Amouroux, and M.F. Elchinger, 527–536. New York: Begell House Inc.

PROTAL process, worldwide patented, original patent: FR9209277.

Racek, O. 2010. The effect of HVOF particle-substrate interactions on local variations in the coating microstructure and the corrosion resistance. *Journal of Thermal Spray Technology* 19 (5): 841–851.

Raessi, M., J. Mostaghimi, and M. Bussmann. 2006. Effect of surface roughness on splat shapes in the plasma spray coating process. *Thin Solid Films* 506–507: 133–135.

Renouard-Vallet, G. 2004. *Elaboration by plasma spraying of dense and thin (a few tens of micro meters) yttria stabilized zirconia electrolytes for SOFCs.* Ph.D. thesis, University of Limoges, France, 8 Feb.

Renouard-Vallet, G., L. Bianchi, P. Fauchais, M. Vardelle, M. Boulos, and F. Gitzhofer. 2005. Influence of spray technology on ionic conductivity of yttria stabilized zirconia. *Journal of High Temperature Material Processes* 9 (2): 195–210.

Saaedi, J., T.W. Coyle, H. Arabi, S. Mirdamadi, and J. Mostaghimi. 2010. Effects of HVOF process parameters on the properties of Ni-Cr coatings. *Journal of Thermal Spray Technology* 19 (3): 521–530.

Sabiruddina, K., P.P. Bandyopadhyay, G. Bolelli, and L. Lusvarghi. 2011. Variation of splat shape with processing conditions in plasma sprayed alumina coatings. *Journal of Materials Processing Technology* 211: 450–462.

Salimijazi, H.R., L. Pershin, T.W. Coyle, J. Mostaghimi, S. Chandra, Y.C. Lau, L. Rosenzweig, and E. Moran. 2007. Effect of droplet characteristics and substrate surface topography on the final morphology of plasma-sprayed zirconia single splats. *Journal of Thermal Spray Technology* 16 (2): 291–299.

Samadi, H., and T.W. Coyle. 2009. Modeling the build-up of internal stresses in multilayer thick thermal barrier coatings. *Journal of Thermal Spray Technology* 18 (5–6): 996–1003.

Sampath, S., and H. Herman. 1996. Rapid solidification and microstructure development during plasma spraying deposition. *Journal of Thermal Spray Technology* 5 (4): 445–456.

Sampath, S., X. Jiang, A. Kulkarni, J. Matejicek, D.L. Gilmore, and R.A. Neiser. 2003. Development of process maps for plasma spray: case study for molybdenum. *Materials Science and Engineering A* 348: 54–66.

Sampath, S., X.Y. Jiang, J. Matejicek, L. Prchlik, A. Kulkarni, and A. Vaidya. 2004. Role of thermal spray processing method on the microstructure, residual stress and properties of coatings: an integrated study for Ni–5 wt.%Al bond coats. *Materials Science and Engineering* A364: 216–231.

Schadler, L.S., K.O. Laul, R.W. Smith, and E. Petrovicova. 1997. Microstructure and mechanical properties of thermally sprayed silica/nylon nanocomposites. *Journal of Thermal Spray Technology* 6 (4): 475–485.

Schmidt, T., F. Gärtner, H. Assadi, and H. Kreye. 2006. Development of a generalized parameter window for cold spray deposition. *Acta Materialia* 54: 729–742.

Seyed, A.A., A. Denoirjean, P. Denoirjean, J.C. Labbe, and P. Fauchais. 2005. In-flight oxidation of stainless steel in plasma spraying. *Journal of Thermal Spray Technology* 14 (1): 117–124.

Shabana, Y.M., and N. Noda. 2002. Geometry effects of substrate and coating layers on the thermal stress response of FGM structure. *Acta Mechanica* 159: 143–156.

Shinoda, K., A. Yamada, M. Kambara, Y. Kojima, and T. Yoshida. 2007. Deformation of alumina droplets on micro-patterned substrates under plasma spraying conditions. *Journal of Thermal Spray Technology* 16 (2): 300–305.

Shinoda, K., H. Murakami, S. Kuroda, S. Takehara, and S. Oki. 2008. In situ visualization of impacting phenomena of plasma-sprayed zirconia: from single splat to coating formation. *Journal of Thermal Spray Technology* 17 (5–6): 623–630.

Sidhu, T.S., S. Prakash, and R.D. Agrawal. 2005. Studies on the properties of high-velocity oxy-fuel thermal spray coatings for higher temperature applications. *Materials Science* 41 (6): 805–823.

Smialek, J.L. 2004. Improved oxidation life of segmented plasma sprayed 8YSZ thermal barrier coatings. *Journal of Thermal Spray Technology* 13 (1): 66–75.

Soveja, A., S. Costil, H. Liao, P. Sallamand, and C. Coddet. 2010. Remelting of flame spraying PEEK coating using lasers. *Journal of Thermal Spray Technology* 19 (1–2): 439–447.

Spencer, K., V. Luzin, N. Matthews, and M.-X. Zhang. 2012. Residual stresses in cold spray Al coatings: the effect of alloying and of process parameters. *Surface and Coating Technology* 206 (19–20): 4249–4255.

Stokes, J., and L. Looney. 2008. Predicting quenching and cooling stresses within HVOF deposits. *Journal of Thermal Spray Technology* 17 (5–6): 908–914.

Stoltenhoff, T., H. Kreye, and H.J. Richter. 2002. An analysis of the cold spray process and its coatings. *Journal of Thermal Spray Technology* 11 (4): 542–550.

Strangman, T., D. Raybould, A. Jameel, and W. Baker. 2007. Damage mechanisms, life prediction, and development of EB-PVD thermal barrier coatings for turbine airfoils. *Surface and Coating Technology* 202: 658–664.

Stravato, A., R. Knight, V. Mochalin, and S.C. Picardi. 2008. HVOF-sprayed nylon-11 + nanodiamond composite coatings: production & characterization. *Journal of Thermal Spray Technology* 17 (5–6): 812–817.

Syed, A.A., A. Denoirjean, B. Hannoyer, P. Fauchais, P. Denoirjean, A.A. Khan, and J.C. Labbe. 2005. Influence of substrate surface conditions on the plasma sprayed ceramic and metallic particles flattening. *Surface and Coating Technology* 200: 2317–2331.

Tanaka, Y., and M. Fukumoto. 1999. Investigation of dominating factors on flattening behavior of plasma sprayed ceramic particles. *Surface and Coating Technology* 120–121: 124–130.

Thornton, J.A. 1975. Influence of substrate temperature and deposition rate on structure of thick sputtered Cu coatings. *Journal of Vacuum Science and Technology* 12 (4): 830–835.

Tillmann, W., E. Vogli, and B. Krebs. 2008. Influence of the spray angle on the characteristics of atmospheric plasma sprayed hard material based coatings. *Journal of Thermal Spray Technology* 17 (5–6): 948–955.

Tran, A.T.T., M.M. Hyland, T. Qiu, B. Withy, and B.J. James. 2008. Effects of surface chemistry on splat formation during plasma spraying. *Journal of Thermal Spray Technology* 17 (5–6): 637–645.

Trice, R.W., and K.T. Faber. 2000. Role of lamellae morphology on the microstructural development and mechanical properties of small-particle plasma-sprayed alumina. *Journal of the American Ceramic Society* 83 (4): 889–896.

Tricoire, Au., P. Fauchais, P. Braillard, A. Malie, and A. Bengtsson. 2005. New concepts for plasma sprayed zirconia TBCs for aeronautical applications. In: *ITSC-2005, Basel, CH*. DVS, Düsseldorf (DVD), e-proceedings.

Trompetter, W.J., M. Hyland, P. Munroe, and A. Markwitz. 2005. Evidence of mechanical interlocking of NiCr particles thermally sprayed onto Al substrates. *Journal of Thermal Spray Technology* 14 (4): 524–529.

Trompetter, W., M. Hyland, D. McGrouther, P. Munroe, and A. Markwitz. 2006. Effect of substrate hardness on splat morphology in high-velocity thermal spray coatings. *Journal of Thermal Spray Technology* 15 (4): 663–669.

———. 2010. The effect of substrate surface oxides on the bonding of NiCr alloy particles HVAF thermally sprayed onto aluminum substrates. *Journal of Thermal Spray Technology* 19 (5): 1024–1031.

Tsipas, S.A., I.O. Golosnoy, R. Damani, and T.W. Clyne. 2004. The effect of a high thermal gradient on sintering and stiffening in the top coat of a thermal barrier coating system. *Journal of Thermal Spray Technology* 13 (3): 370–376.

Tsui, Y.C., and T.W. Clyne. 1996. Adhesion of thermal barrier coating systems and incorporation of an oxidation barrier layer. In *ITSC-1996: practical solutions for engineering problems*, ed. C.C. Berndt, 275–284. Materials Park: ASM International.

Ulianitsky, V., A. Shtertser, S. Zlobin, and I. Smurov. 2011. Computer-controlled detonation spraying: from process fundamentals toward advanced applications. *Journal of Thermal Spray Technology* 20 (4): 791–801.

Valette, S. 2004. *Influence of the preoxidation of a steel substrate on the adhesion of an alumina coating plasma sprayed*. Ph.D. Thesis, University of Limoges, France, 2004, November.

Van Steenkiste, T.H., J.R. Smith, and R.E. Teets. 2002. Aluminum coatings via kinetic spray with relatively large powder particles. *Surface and Coating Technology* 154 (2–3): 237–252.

Vardelle, A., M. Vardelle, P. Fauchais, and M.I. Boulos. 1988. Particle dynamics and heat transfer under plasma conditions. *AICHE Journal* 34: 567–573.

Vardelle, M., A. Vardelle, A.C. Leger, P. Fauchais, and D. Gobin. 1994. Influence of particle parameters at impact on splat formation and solidification in plasma spraying processes. *Journal of Thermal Spray Technology* 4 (1): 50–58.

Vardelle, M., A. Vardelle, K.-I. Li, P. Fauchais, and N.J. Themelis. 1996. Coating generation: Vaporization of particles in plasma spraying and splat formation. *Pure and Applied Chemistry* 68 (5): 1093–1099.

Vardelle, M., A. Vardelle, P. Fauchais, K.-I. Li, B. Dussoubs, and N.J. Themelis. 2001. Controlling particle injection in plasma spraying. *Journal of Thermal Spray Technology* 10: 267–286.

von Niessen, K., and M. Gindrat. 2011. Plasma spray-PVD: a new thermal spray process to deposit out of the vapor phase. *Journal of Thermal Spray Technology* 20 (4): 736–743.

von Niessen, K., M. Gindrat, and A. Refke. 2010. Vapor phase deposition using plasma spray-PVD. *Journal of Thermal Spray Technology* 19 (1–2): 502–509.

Wang, T.-G., S.-S. Zhao, W.-G. Hua, J.-B. Li, J. Gong, and C. Sun. 2010. Estimation of residual stress and its effects on the mechanical properties of detonation gun sprayed WC–Co coatings. *Materials Science and Engineering* A527: 454–461.

Watanabe, M., A. Owada, S. Kuroda, and Y. Gotoh. 2006. Effect of WC size on interface fracture toughness of WC–Co HVOF sprayed coatings. *Surface and Coating Technology* 201: 619–627.

Wilden, J., J.P. Bergmann, and H. Frank. 2006. Plasma transferred arc welding—Modeling and experimental optimization. *Journal of Thermal Spray Technology* 15 (4): 779–784.

Williams, C.A., and H. Jones. 1975. The effect of melt superheat and impact velocity on splat thickness. *Materials Science and Engineering* 19: 293–297.

Withy, B., M. Hyland, and B. James. 2006. Pretreatment effects on the surface chemistry and morphology of aluminium. *International Journal of Modern Physics B* 20 (25–27): 3611–3616.

Withy, B.P., M.M. Hyland, and B.J. James. 2008. The effect of surface chemistry and morphology on the properties of HVAF PEEK single splats. *Journal of Thermal Spray Technology* 17 (5–6): 631–636.

Xue, M., S. Chandra, and J. Mostaghimi. 2006. Investigation of splat curling up in thermal spray coatings. *Journal of Thermal Spray Technology* 15 (4): 531–536.

Xue, M., Y. Heichal, S. Chandra, and J. Mostaghimi. 2007. Modeling the impact of a molten metal droplet on a solid surface using variable interfacial thermal contact resistance. *Journal of Materials Science* 42: 9–18.

Yang, K., K. Tomita, M. Fukumoto, M. Yamada, and T. Yasui. 2009. Effect of ambient pressure on flattening behavior of thermal sprayed particles. *Journal of Thermal Spray Technology* 18 (4): 510–518.

Yang, K., M. Fukumoto, T. Yasui, and M. Yamada. 2010. Study of substrate preheating on flattening behavior of thermal-sprayed copper particles. *Journal of Thermal Spray Technology* 19 (6): 1195–1205.

Yang, E.-J., G.-J. Yang, X.-T. Luo, C.-J. Li, and M. Takahashi. 2012. Epitaxial grain growth during splat cooling of alumina droplets produced by atmospheric plasma spraying. In *ITSC-2012*. Materials Park: ASM International. e-proceedings.

Yi, Jiang, Bin-Shi Xu, and Hai-dou Wang. 2005. Residual stresses within sprayed coatings. *Journal of Central South University of Technology* 12 (Suppl 2): 53–58.

Yokoyama, K., M. Watanabe, S. Kuroda, Y. Gotoh, T. Schmidt, and F. Gärtner. 2006. Simulation of solid particle impact behavior for spray processes. *Materials Transactions* 47 (7): 1697–1702.

Yoshida, T., T. Okada, H. Hamatani, and H. Kumaoka. 1992. Integrated fabrication process for solid oxide fuel cells using novel plasma spraying. *Plasma Sources Science and Technology* 1: 195–201.

Zeng, Z., S. Kuroda, J. Kawakita, M. Komatsu, and H. Era. 2010. Effects of some light alloying elements on the oxidation behavior of Fe and Ni-Cr based alloys during air plasma spraying. *Journal of Thermal Spray Technology* 19 (1–2): 128–136.

Zhang, D., P.H. Shipway, and D.G. McCartney. 2005. Cold gas dynamic spraying of aluminum: the role of substrate characteristics in deposit formation. *Journal of Thermal Spray Technology* 14 (1): 109–116.

Zhang, G., H. Liao, H. Yu, V. Ji, W. Huang, S.G. Mhaisalkar, and C. Coddet. 2006. Correlation of crystallization behavior and mechanical properties of thermal sprayed PEEK coating. *Surface and Coating Technology* 200: 6690–6695.

Zhanga, C., G. Zhang, V. Ji, H. Liao, S. Costil, and C. Coddet. 2009. Microstructure and mechanical properties of flame-sprayed PEEK coating remelted by laser process. *Progress in Organic Coating* 66: 248–253.

Zahiri, S.H., D. Fraser, S. Gulizia, and M. Jahedi. 2006. Effect of processing conditions on porosity formation in cold gas dynamic spraying of copper. *Journal of Thermal Spray Technology* 15 (3): 422–430.

Zahiri, S.H., D. Fraser, and M. Jahedi. 2009. Recrystallization of cold spray-fabricated CP titanium structures. *Journal of Thermal Spray Technology* 18 (1): 16–22.

Zotov, N., A. Hospach, G. Mauer, D. Sebold, and R. Vaßen. 2012. Deposition of La1-xSrxFe1-yCoyO3-δ Coatings with different phase compositions and microstructures by low-pressure plasma spraying-thin film (LPPS-TF) processes. *Journal of Thermal Spray Technology* 21 (3–4): 441–447.

Abbreviations

ALR	Atomizing gas to Liquid mass Ratio
APS	Atmospheric Plasma Spraying
BGCC	Bioactive Glass–Ceramic Coatings
CCD	Charge-Coupled Device
CGDS	Cold Gas Dynamic Spraying
CMAS	Calcium Magnesium Alumino Silicates
CS	Cold Spray
DC	Direct Current
DSD	Droplet Size Distribution
EB-PVD	Electron Beam Physical Vapor Deposition
FCT	Furnace Cycle Test
FE-SEM	Field Emission-SEM
GDC	Gadolinia-Doped Ceria
GFA	Glass Forming Ability
GNP	Glycine-Nitrate Process
GZ	Gadolinium Zirconate
HA	Hydroxyapatite ($Ca_{10}(PO_4)6(OH)$)
HPAL	High-Pressure Acid-Leach
HVAF	High Velocity Air Spraying
HVOF	High Velocity Oxyfuel
HV-SFS	High Velocity-Suspension Flame Spraying
i.d.	Internal diameter
IT-SOFC	Intermediate Temperature SOFC
LSCF	Lanthanum Strontium Cobalt Iron oxide ($La_x Sr_{1-x} Co_y Fe_{1-y} O_{3-\delta}$)
LSM	Lanthanum Strontium Manganese
MIG	Tungsten Inert Gas
MWCNT	Multiwalled Carbon Nanotubes
PAA	Polyacrylic Acid
PEI	Polyethyleneimine
PGDS	Pulsed Gas Dynamic Spraying
PSD	Particle Size Distribution
PTA	Plasma Transferred Arc
R&D	Research and Development
RF-ICP	Radio Frequency Inductively Coupled Plasma
RF-IPS	Radio Frequency Induction Plasma Spraying
RGS	Ratio of Gas to Suspension
RH	Relative Humidity
SBF	Simulated Body Fluid
SD	Spraying Distance
SDC	Samaria-Doped Ceria
SEM	Scanning Electron Microscopy
SHS	Self-propagating High-temperature Synthesis
slm	Standard Liters per Minute
SOFC	Solid Oxide Fuel Cell
SPPS	Solution Precursor Plasma Spraying
SPS	Suspension Plasma Spraying
STS	Solution Thermal Spraying
TBC	Thermal Barrier Coatings
TEM	Transmission Electron Microscopy
TGO	Thermally Grown Oxide
TS	Thermal Spray
TTPR	Triple Torch Plasma Reactor
USNNI	US National NanoTechnology Initiative,
VPS	Vacuum Plasma Spraying
WSP	Water-Stabilized Plasma
XRD	X-Ray Diffraction
YPSZ	Yttria Partially Stabilized Zirconia
YSZ	Yttria Stabilized Zirconia

16.1 Introduction

According to the definition of the US National Nanotechnology Initiative, USNNI, "nanometer-sized" materials, or more commonly referred to as "nanosized" or "nanostructured" materials, are materials with a particle diameter or an internal structure with at least one dimension smaller than 100 nm. From the pioneering works of McPherson in 1973, who was among the first to identify nano-size features in thermally sprayed alumina coatings, the thermal spray community has been actively involved in the development of novel approaches for the manufacturing of nanostructured functional coatings that are at the frontier of materials science

M. I. Boulos et al. (ed.), *Thermal Spray Fundamentals*, https://doi.org/10.1007/978-3-030-70672-2_16

due to their remarkable, and in some cases, novel properties [McPherson R. (1973) and Ashby M.F. et al. (2009)]. This results in particular from the surface-to-bulk ratio, and interface density between features, that is much higher in nanostructured materials compared to those of microsized materials. The physical phenomena involved leading to those remarkable properties lies, at such dimensions, between atom quantum effects (e.g., phonon scattering, surface plasmon) and bulk behavior (e.g., cohesion).

Due to the large volume fraction of the internal interfaces, nanostructured ceramic coatings, for example, exhibit superior dimensional stability, lower thermal diffusivity and hysteresis (due to phonon scattering by boundary defects), higher hardness and toughness (due to small grain sizes), and better wear resistance (due to higher hardness, grain-sliding plasticity) than conventional coatings [Lu Y. and Liaw P.K. (2001) and, Dahotre N.B. and S. Nayak (2005)]. Generally, while nanostructure coatings are not necessarily harder than the conventional ones, they tend to be much tougher. This is explained by the fact that in conventional thermal-sprayed coatings, cracks tends to propagate along splat boundaries due to the weak inter-splat bonding. In contrast, in nanostructured coating because of their ultra-fin, homogeneous structure, with splat sizes in the 100 nm range, and superior splat-to-splat cohesion, crack propagation is significantly limited.

Since the beginning of the mid-1990s, numerous studies have been focused on the use of the thermal spray (TS) technology for the deposition of finely or nanostructured coatings with emphasis on nanostructured ceramics. One of the major challenges for spraying nanosized particles by TS is the problem of injecting them into the core of the high enthalpy high velocity flow, since the particle injection force has to be of the same order of that imparted to them by the flow [Fauchais P. et al. (2006)]. By attempting to use conventional pneumatic powder feeding technique to inject nanometric-sized particles into the flow, the powder carrier gas flow rate has to be significantly increased in order to provide the nanoparticle with the necessary momentum to penetrate and be entrained by the flow. However, this will have a catastrophic impact on the high temperature, high velocity spray flow stream with significant local cooling and disruption of the flow pattern. To circumvent this phenomenon, two approaches have been considered:

- Spraying micrometer-sized particles as in conventional thermal spraying using powders made of agglomerated nano-sized particles
- Spraying sub-micrometer or nano-sized particles using a liquid carrier instead of a gas conventionally used in TS applications

Moreover, when spraying microsized particles made of agglomerated nanopowders, their melting in the thermal spray process will automatically destroy its nanostructure with the exception of cold spray (CS) in which the sprayed particles remain at relatively low temperatures. Cold spraying, on the other hand, requires that the sprayed nanostructured particles be ductile enough to be able to deform and adhere on impact to the surface of the substrate. Alternately, the use of a liquid as a carrier medium for the injection of the nanopowder particles into the spray medium will add a considerable thermal load on the process by locally cooling the spray medium limiting considerably the powder throughput that can be tolerated, and consequently the productivity and economic viability of the approach.

The processing of nanoparticles is indeed not as simple as pointed out in the review of the different processes by [Viswanathan et al. (2006)]. Thermal spray remains, however, a promising approach [Gadow R. et al. (2008a, b)] to achieve nano or finely structured materials with unique properties. Four principal approaches are at different stages of development;

- Formation of *nanocrystalline coatings* through the thermal spraying of complex alloys containing multiple elements that exhibit a glass forming capability when cooled down below their glass transition temperature. Upon the subsequent heating to the 500 to 750 °C range, the deposited metallic glass precursor transforms through a solid-state devitrification into multiple nanocrystalline phases [Branagan D.J. et al. (2005a, b, 2006)].
- Spraying conventional particles (in the size range of 30 to 90 μm) made of *agglomerated nanoparticles*. These can be sprayed using cold spray, or conventional thermal spray technology provided that the operating parameters are set to a narrow window limiting the particle heating to partial melting with the molten part acting as "cement" bonding un-melted nanograins [Lima R.S. and B.R. Marple (2007)]. Coatings exhibiting such a two-scale grain-size architecture of nanograins (unmelted particles) and micrograins (melted material) have found niche applications in thermal barrier coatings and abradable coatings, mostly made of partially stabilized zirconia.
- Thermal spraying submicron-sized or nanosized particles via *suspension plasma spraying* (SPS) in which a liquid carrier is used instead of the conventional carrier gas [Bouyer E. et al. (1997a, b), Gitzhofer F. et al. (1997), Fauchais P. et al. (2005, 2008a, b) and Fazilleau J. et al. (2006)]. Once the suspension has been fragmented and the liquid vaporized by the plasma flow or the High Velocity Oxyfuel (HVOF) flame, particles contained in the droplets are heated, accelerated, and deposited onto the substrate.

They lead to the formation of splats that have equivalent diameters ranging between 0.1 and 2 μm and average thicknesses between 20 and 300 nm. The stacking of splats forms finely structured coatings.

- *Solution precursor plasma spraying* (SPPS) is a natural extension of the suspension plasma spraying approach. In this case, the nanometer-sized particles used for the coating formation are synthesized in-flight through the decomposition of the precursor in the feed solution. This method significantly limits the safety issues associated with the handling of nanometer-sized particles (see, for example, the very complete review of Singh N. et al (2009)) and offers means to avoid many of the drawbacks associated with suspension stabilization, in particular when dissimilar materials (e.g., metallic alloys and oxides) are mixed together.

In this chapter, a review is presented of the basic concepts involved in each of these technological approaches giving highlights of the state of the art, its science bases together with their potential and limitations.

16.2 Spraying of Nanocrystalline Materials

16.2.1 Nanocrystalline Coating Formation

According to [Sergueeva et al. (2008)], crystallization of metallic glasses has been successfully used as one of the methods for the production of nanocrystalline material in various alloy systems, such as iron, nickel, or cobalt-based alloys. This type of transformation involves decomposition of single-phase supersaturated solid solutions into multiphase nanoscale microstructures. To obtain a nanoscale structure, the crystallization process should proceed with the largest nucleation rate possible while suppressing the crystal growth rate. Such conditions can be obtained for compositions of some alloys by applying specific methods of heat treatment. Among metallic glasses those that undergo primary crystallization with a time-dependent, long-range diffusion-controlled growth rate are the most suitable candidates for *nanocrystallization*.

The methodology involves designing alloys that have low-critical cooling rates for metallic glass formation that result in the formation of amorphous coatings when thermally deposited. The amorphous coatings, when heated (after spraying) above their crystallization temperature, are devitrified. Since the diffusion rate in the solid state is very low at the transformation temperature (typically 0.4–0.7 T_m for iron alloys), nanoscale microstructures are formed [Branagan D.J. et al. (2001), Peker A. and W.L. Johnson (1993), Inoue A. (1995) and Sergueeva A.V. et al. (2004)].

In the past two decades, a series of new bulk amorphous alloys with a multicomponent chemistry and high-glass forming ability (GFA) have been developed in zirconium, magnesium, lanthanum, palladium, titanium, and iron-based systems with various rapid solidification techniques [Peker A. and W.L. Johnson (1993) Inoue A. (1995) and Sergueeva A.V. et al. (2004)]. Compared with other amorphous alloy systems, such as zirconium and lead-based metallic glasses, the advantages of iron-based amorphous coatings are their lower materials cost, higher strength, and higher wear and corrosion resistance. However, to be suitable for thermal spray processes, such materials should be produced by gas atomization.

16.2.1.1 Amorphous Alloys Containing Phosphorus

Complex alloys containing phosphorus are known to present amorphous phases and can be produced by gas atomization. Alloy powders of Fe–10% Cr–8% P–2% C(10Cr), Fe–20% Cr–8% P–2% C(20 Cr), and Fe–10% Cr–10% Mo–8% P–2% C(10Mo) compositions (in wt.%) were sprayed by the High Velocity Oxy-Fuel (HVOF) process under different conditions where the oxygen and fuel pressures were changed to vary the flame temperature [Otsubo F. et al. (2000)]. The melting point of particles was under 1273 K, and their particle size distribution (PSD) between 10 and 45 μm. Amorphous coatings with a small percentage of crystalline phases were obtained from the 10Cr and 20Cr alloys and a 100% amorphous coating was formed from the 10Mo alloy. The volume fraction of the crystalline material increased slightly with the rise of the flame temperature. The coatings appeared very dense, although some pores and oxide films were visible in the coatings. The hardness of the 10Cr and 20Cr coatings was 600–700 HV_{5N}. On the other hand, the 10Mo coatings composed of a perfect amorphous phase reveal a constant hardness of 560 HV_{5N} independently of the spray condition. The as-sprayed coatings of the 10Cr and 20Cr alloys exhibited the activation–passivation transition. In contrast, the as-sprayed coating of the 10Mo alloy composed of a 100% amorphous phase structure had excellent corrosion resistance in 1 N H_2SO_4 and 1 N HCl solutions [Otsubo F. et al. (2000)].

16.2.1.2 NiCrB and FeCrB Alloys

Since the mid-1980s, studies were devoted to the production and the characterization of atmospheric and vacuum plasma (APS or VPS) or HVOF-sprayed amorphous/nanometer crystalline NiCrB- and FeCrB-based alloys [Das S.K. et al. (1984) Sampath S. et al. (1993), Kishitake K. et al. (1996a, b) and Dent A.H. et al. (1999)]. For example, NiCrMoB alloys [Dent A.H. et al. (1999)] were sprayed with a HVOF Miller spray system (Praxair Surface Technology, Appleton, WI) using hydrogen as a fuel with a H_2/O_2

volume ratio of 3 to 1. The deposits obtained with two different NiCrMoB alloys, with PSD of 25–63 μm, contained substantial quantities of amorphous/nanocrystalline matrices (as shown by the significant broadening of the fcc nickel peak at $2\theta = 44°$). Hardness values were in excess of 610 ± 10 HV_{3N}. Higher molybdenum and boron levels led to higher hardness values of 810 ± 15 HV_{3N} · The deposited layers were found to contain fine Cr_5B_3 precipitates within the metallic matrix phase. The predominant oxide phase, Cr_2O_3, occurred principally with a columnar-grained morphology at inter-splat boundaries. Anodic polarization curves showed that in 0.5 M H_2SO_4 a passive region occurred between approximately +100 and + 900 mV (SCE). Increased molybdenum and boron levels modified the pre-passivation corrosion current density but had little effect on either passive current density or the onset of trans-passive corrosion. The potentiodynamic corrosion testing also demonstrated that the corrosion behavior of the experimental coatings A and B was similar to that of HVOF-sprayed Inconel 625. However, the presence of oxides at inter-splat boundaries leads to passive current densities higher than those observed in wrought alloys or melt spun ribbons.

In-flight particle oxidation during atmospheric plasma spraying (APS) and (HVOF) spraying is however unavoidable. For example, with NiTiZrSiSn there is a preferential oxidation of Zr and Ti in an in-flight particle [Choi H. et al. (2005)]. As a matter of fact, oxidation triggers the destabilization of the bulk of metallic glass particles because amorphous phase stability and formability are largely affected by the chemical composition depending on the critical cooling rate. With the HVOF process (H_2/O_2 mixture) [Branagan D.J. et al. (2009)] above the 0.20 O_2/H_2 ratio, severe oxidation was observed, inducing destabilization of the amorphous phase.

16.2.1.3 Iron-Based Amorphous Alloys

Nanoscale steels have appeared at the end of the past century and the first advances have boosted the field of hard magnets made of neodymium, iron, and boron. [Branagan et al. (2000)] succeeded to synthesize nanoscale metal matrix composite microstructures in a 9-element modified Nd-Fe-B alloy, optimizing the magnetic behavior of permanent magnets. To obtain amorphous and nanoscale composite thermally deposited steel coatings a specialized iron-based composition ($Fe_{63}Cr_8Mo_2B_{17}C_5SiAl_4$) with a low-critical cooling rate (10^4 K/s) for metallic glass formation was developed. The alloy is processed by inert gas atomization to avoid oxidation and to form micron-sized amorphous spherical particles with a size distribution adapted to thermal spraying. The particle size distribution is particularly important because the particles have to melt to form the amorphous coatings. Collecting and quenching sprayed particles were performed to verify the particle characteristics. The results show that a significant fraction of ($d_p < 50$ μm) particles are amorphous, while particles ($d_p > 75$ μm) are crystalline. Backscattered electron micrographs given in Fig. 16.1 [Branagan D.J. et al. (2001)] indicate that small particles ($d_p < 20$ μm) have an absence of structure, while larger powder particles ($d_p = 50$–75 μm) exhibit a primarily crystalline structure with areas of uncrystallized glass.

A primarily amorphous structure was also formed in the as-sprayed coatings (using DC plasma or HVOF), independent of coating thickness. After heat treatment above the crystallization temperature (568 °C that is less than half the melting temperature), the coatings are devitrified into a multiphase nanocomposite microstructure with 75–125 nm grains containing a distribution of 20 nm second-phase grain-boundary precipitates. A strong effect of annealing conditions on the nature of crystallization products and, as a

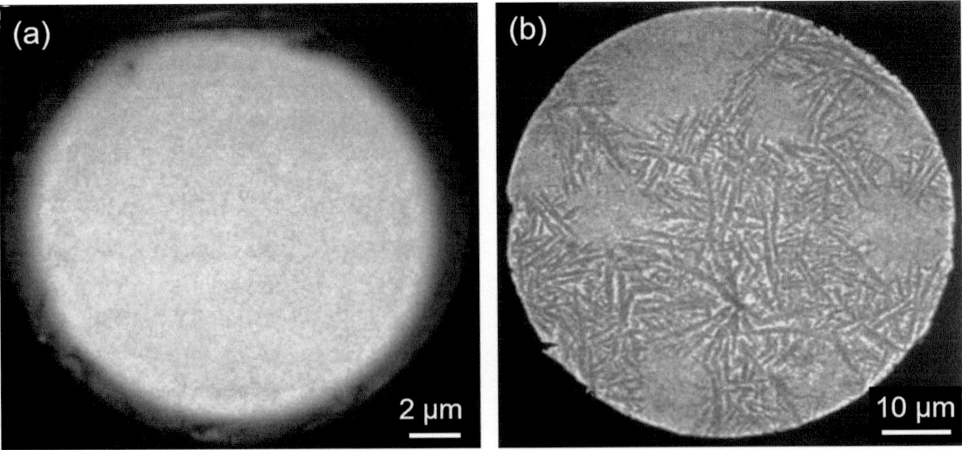

Fig. 16.1 Backscattered electron micrographs of the cross-section of two as-atomized powder particles: (**a**) 11 μm particle with an amorphous structure and (**b**) 59 μm particle with a primarily crystalline structure containing pockets of glass [Branagan D.J. et al. (2001)]. (Reprinted with kind permission from Springer Science Business Media)

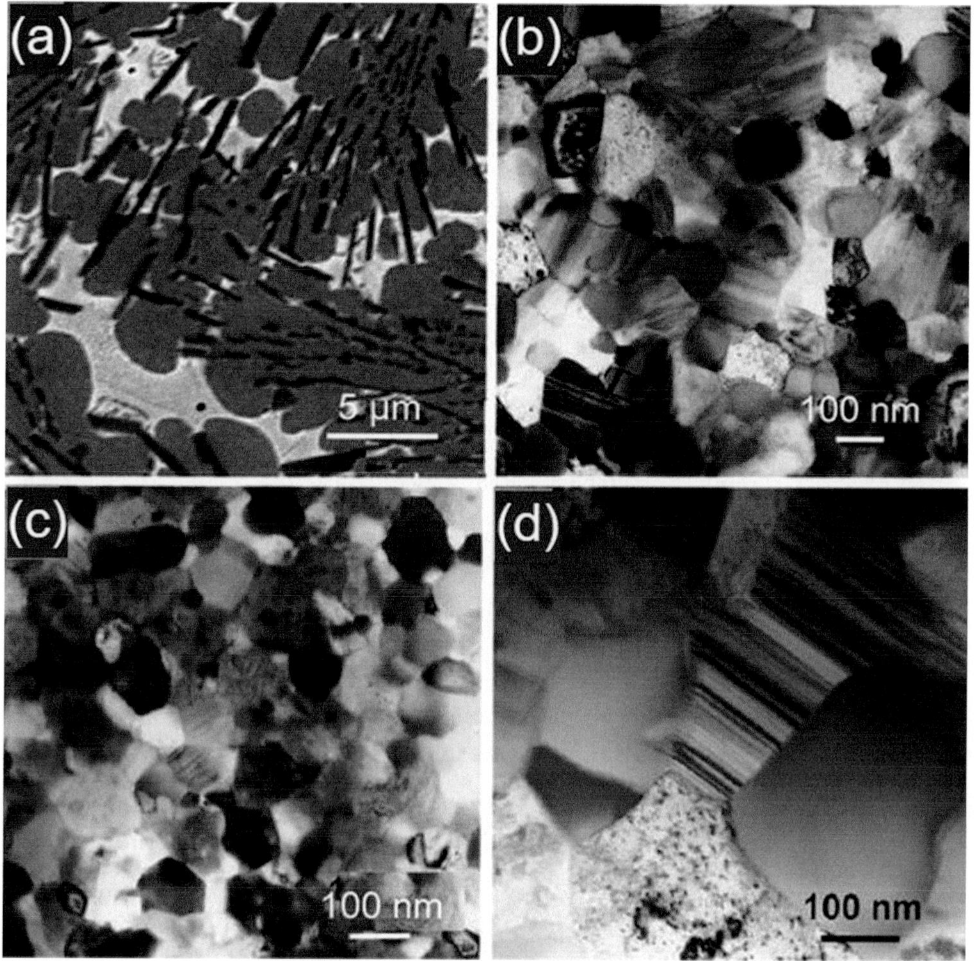

Fig. 16.2 Electron microcopy images of a 5-element steel alloy; (**a**) backscattered electron image of an as-cast ingot, (**b–d**) TEM image of ribbon heat-treated, for 1 h, at (**b**) 700 °C, (**c**) 750 °C, and (**d**) 850 °C [Branagan D.J. (2007)]. (Reprinted with kind permission from Elsevier)

result, on the mechanical behavior of the alloy has been revealed [Branagan D.J. et al. (2001)]. Accordingly, depending on the microstructure produced and the deformation conditions chosen, a wide variety of strength/elongation properties can be generated for various potential applications.

An example of the transformation that takes place in the such materials when reheated after spraying above their glass temperatures are described in the following according to Branagan DJ (2007) for the heat treatment of a five-element glass-forming steel alloy with the stoichiometry composition of $Fe_{63.2}Cr_{15.8}B_{17}W_2C_2$. The as-cast microstructure of the alloy was found to have a dendritic solidification morphology, resulting in a multiphase structure with an average phase scale from 1 to 5 µm as shown in Fig. 16.2a. Phases in the ingot, as determined by X-ray diffraction, included α-Fe, $M_{23}C_6$, and M_3B. The same five-element steel was processed by melt spinning at 15 m/s to form a homogeneous glass (glass-to-crystalline peak at 536 °C). After heat-treating the glass above 536 °C, complete deglazing occurred through the solid–solid-state transformation. TEM micrographs given in Fig. 16.2b, c and d show, respectively, the nanometer-scale

structures obtained after heat treatment for 1 h, at 700 °C, 750 °C, and 850 °C.

The phase size is still considered nanoscale up to at least 750 °C, with only a small change from the average size of 25 nm when heat treated from 700 to 750 °C. At the 850 °C heat treatment, extensive grain growth has clearly occurred with phase sizes approximately double those of the lower temperature heat treated samples [Branagan D.J. (2007)]. When heat-treated at or below 700 °C, the nanometer scale microstructure is found to exhibit a hardness that is approximately 8.83 GPa higher than the micrometer scale microstructure. The nanometer-structured metal hardness exhibits two plateaus with little change in hardness from 600 to 700 °C, followed by a very significant drop in hardness from 700 to 750 °C, and then by a leveling off with finally another significant drop in hardness at 850 °C. Globally, the results indicate that to improve properties in engineering applications, the size of the microstructure alone is not the only important parameter that need to be considered.

Using HVOF deposition, the adhesion strength of the coatings was found to be excellent for a wide variety of metallic substrates. Good wear resistance was obtained in the three-body slurry abrasion tests [Branagan D.J. et al. (2001)]. Due to the high hardness, the amorphous and nanocomposite coatings experienced good wear characteristics. No wear was found in the two-body abrasion tests in the as-sprayed or heat-treated plasma coatings. This was surprising, since the Si_3N_4 pin was much harder than both the as-sprayed and heat-treated coatings. The NanoSteel Corp. was recently created to develop and market a range of patented high-performance Super Hard Steel materials. For example, the new Fe-based commercial alloy ($Fe_{52.3}Mn_2Cr_{19}Mo_{2.5}W_{1.7}B_{16}C_4Si_{2.5}$) demonstrates strength of more than 6 GPa at room temperature [Sergueeva A.V. et al. (2008)], which is much stronger than the commonly used high strength (0.24–0.9 GPa) or even ultrahigh strength steels (0.9–1.5 GPa) of today. Such coatings can be applied with a wide variety of processes: thermal spraying, welding (MIG and PTA), and laser cladding. They are mostly made of Fe–Cr–Mo–W–C–Mn–Si–B or Cr–Mo–B–Si–W–Mn–C–Nb–Fe.

Certain iron-based glass forming materials such as SAM2X5 have an optimized chemistry with a protective adherent oxide layer and in principle a good corrosion resistance. However, [Branagan et al. (2009)] have emphasized that when using optimized HVOF spray parameters to achieve high-density coatings, the development of the glass structure seems to be the dominant factor affecting corrosion performance. All HVOF coatings, regardless of powder lot, were found to be primarily amorphous with only subtle differences in crystallinity. However, because the larger particles in the distribution are not completely re-melted during spraying, the starting crystallinity in the feedstock powder is found to be an important factor. It is believed that the cores of these larger particles cause the coarsening of the microstructure. These pockets or bands of crystallinity can initiate anodic attack and result in reduced corrosion performance. Thus, for a good resistance to corrosion the glass content in the feedstock powder is a paramount factor and must be maximized.

Due to their high abrasion, erosion, and corrosion resistance, hardness and toughness, the main applications of such coatings are boiler tube refurbishment, hard chrome replacement, hard facing, pumps, etc. Generally, HVOF spraying is used because of the high particle velocities resulting in dense coatings. The SAM2XS amorphous steel has been used to replace hard chromium [Branagan D.J. et al. (2006)]. Coatings, sprayed with a JP-5000 HVOF gun, were found to be nanometer-structured, hard (about 1000 HV_{3N} as-sprayed and 1200 HV_{3N} after heat treatment at 700 °C for 10 min, approaching hardness of WC-Co coatings), with an excellent corrosion resistance to sea water and salt fog environment, and a wear resistance far superior to that of stainless steel used for this corrosion resistance. The SHS-7574 powder from NanoSteel was successively sprayed with a D-gun and HVOF [Parco M. et al. (2006)]. In both cases, coatings were dense but those sprayed with a D-gun

showed better adhesion and presented very good wear resistance.

[Branagan et al. (2005a, b)] have introduced a new iron-base cored wire, SHS-7170, which readily forms nanometer composite coatings when sprayed using the wire arc process and exhibits a combination of properties that are superior to the existing available materials for high temperature boiler applications. The oxide content in the coatings is very low and is typically < 1 vol.%. The bond strengths are remarkable for wire arc coatings and represent some of the best-published values, along with high toughness and resiliency. The as-sprayed SHS-7170 wire arc coatings are found to develop an amorphous matrix structure containing starburst-shaped boride and carbide crystallites with sizes ranging from 60 to 140 nm. After heating to temperatures above the peak crystalline temperature (566 °C), a solid/state transformation occurs that results in the formation of an intimate three-phase matrix structure consisting of the same complex boride and carbide phases, along with α-iron inter-dispersed on a structural scale from 60 to 110 nm. The nanometer composite microstructure contains clean grain boundaries, which are found to be extremely stable and resist coarsening throughout the range of temperatures found in boilers. The SHS-7170 coatings exhibit high bond strength along with high toughness and resiliency. The elevated-temperature erosion resistance of the SHS-7170 wire arc coatings was found to be superior based on thickness loss compared with the existing wire arc coatings that have been tested.

Recently a new iron-based alloy SHS-7170 has been developed [Branagan et al. (2005a, b)]. Upon wire arc spraying, it forms an amorphous matrix with very fine dispersed nanometer-sized crystals that were identified as ferrites and $M_2(C,B)_1$ compounds. When exposed to elevated temperatures, the microhardness slightly increases at relatively low temperatures from 316 °C to 482 °C, and it dramatically increases up to ~ 1.77 MPa when the temperature is raised to 649 °C. The wear resistance was improved by a factor of 3.5 with the significant hardening. The SHS-7170 coatings were found to exhibit high bond strength along with high toughness and resiliency. The elevated-temperature erosion resistance of the SHS-7170 wire arc coatings was found to be superior based on thickness loss compared with the existing wire-arc coatings that have been tested. SHS-7170 coatings resisted erosion almost independently of contact angle and at temperatures at least up to 600 °C. [Zhou et al. (2010)] have shown that the SHS-8000 coatings exhibited high bond strength along with high toughness and resiliency. The development of the time–temperature–transformation diagram for the SHS-8000 coating allowed predicting the coating performance as a function of temperature and time during elevated temperature exposure.

Fe-base nanocrystalline SHS-7172 CP1 ($53 < d_p < 150\,\mu m$) wear-resistant materials have been cladded on aluminum alloys using a PTA with a negative work piece polarity process where the heat input into the base material and the

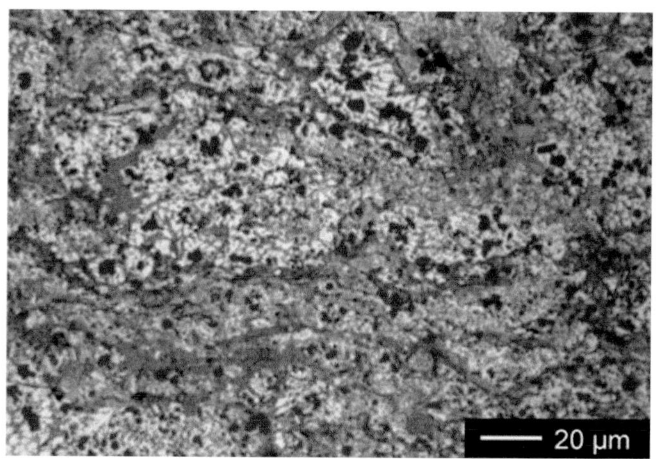

Fig. 16.3 Optical micrograph of the cross-section of HVOF sprayed Al–Si composite structure showing ultrafine primary (2–4 µm) Si and a network of Si in the eutectic Al–Si matrix [Laha T. et al. (2004a, b)]. (Reprinted with kind permission from Springer Science Business Media)

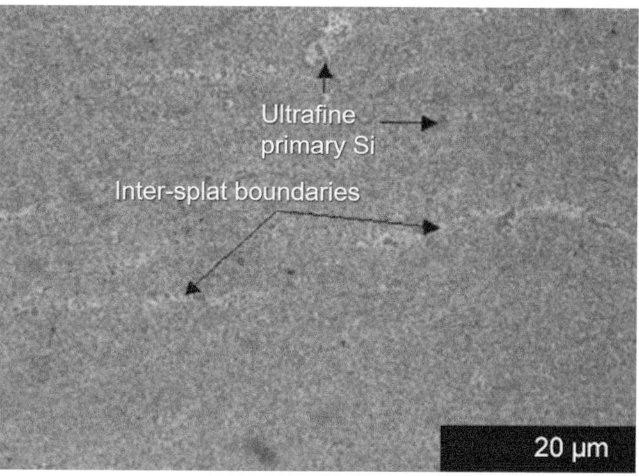

Fig. 16.4 Cross-sectional optical micrograph of the VPS formed Al–Si deposit showing the presence of eutectic Al–Si grains, ultrafine primary silicon particles and a network of silicon particles [Laha T. et al. (2005)]. (Reprinted with kind permission from Elsevier)

formation of spatter can be limited [Wilden J. et al. (2007)]. The dilution is limited to approximately 5%, while the high conductivity of the base material is responsible for a high cooling rate, which induces nanocrystalline solidification.

16.2.2 Spraying Hypereutectic Alloys

When melting alloys that exceed the solubility limit, upon solidification a single solid solution is produced. However, when alloys continue to cool down, a solid-state reaction occurs, permitting a second solid phase to precipitate from the original one and keeping a nanostructure in the coating, if the cooling is fast enough. This effect has been observed both with plasma and HVOF spraying.

Free-standing ring-shaped Al–21Si nanometer composite structures were produced by HVOF spraying of gas atomized hypereutectic Al–21 wt.% Si alloy powders ($15 < d_p < 45$ µm) [Laha T et al. (2004a, b)]. Figure 16.3 presents a cross-sectional micrograph of the coating structure showing a homogenous and highly dense splat structure, consisting of ultrafine primary (2–4 µm) Si grains and a network of Si in the eutectic Al–Si matrix. Formation of a thin oxide layer of uneven thickness can also be observed on the splat surfaces. The module of elasticity (138 ± 17 GPa) shows an intermediate value between those of pure Al (71.9 GPa) and Si (162.9 GPa). The Vickers microhardness value (441 ± 34 N) was also better than that of a conventionally cast eutectic.

A freestanding bulk nanocrystalline structure was vacuum plasma sprayed starting from a hypereutectic Al–21 wt.% Si micron size (15–45 µm) gas-atomized powder [Laha T. et al. (2005)]. The cross-sections of the coating exhibited a dense splat structure containing fine eutectic Al–Si grains with

Fig. 16.5 Low magnification SEM image of fractured surface of a composite cone structure, showing the retention of carbon nanotubes [Laha T. et al. (2004a, b)]. (Reprinted with kind permission from Elsevier)

homogenously distributed ultrafine primary Si (about 1 µm) particles throughout the eutectic Al–Si matrix (Fig. 16.4). The ultrafine primary Si particles precipitated from the Al–Si alloy due to the rapid solidification in VPS process. Further decrease in size of Si particle was due to fragmentation attributed to high velocity impact of the molten powder.

Gas atomized hypereutectic Al–21 wt.% Si alloy powder was blended and mixed with multiwalled carbon nanotubes (MWCNT) (10 wt.%) before being ball milled. The blended powder was plasma sprayed (SG-100 Praxair gun) [Laha T. et al. (2004a, b, 2007]. The multiwalled carbon nanotubes (MWCNT) were successfully retained in the spray formed composite structure (Fig. 16.5). The formation of a β-SiC layer, rather than aluminum carbides at the interface of Al–Si matrix and MWCNT reinforcement, has been confirmed. The β-SiC formation occurred in an ultrathin layer (2–5 nm), attributed to the higher rate of reaction at the triple point of MWCNT/Al–Si alloy/vapor [Laha T. et al. (2007)].

16.3 Spraying of Nanostructured Agglomerated Powders

16.3.1 Powder Characteristics and Coating Formation

In order to spray nanosized particles using regular powder feeders, the nanosized particles need to be agglomerated via spray drying followed by sintering/crushing and/sieving into micro-sized particles. For details, see Chap. 13 Powders, Wires and Cords, Sect. 13.3.2. It is important to point out that for these particles, finding the tradeoff threshold between providing cohesive strength and maintaining the nanostructure character of the feedstock is paramount, that is, too high heat-treatment temperatures and/or long heat-treatment times may cause the partial or total loss of the nanostructural character of the powder due to particle coarsening and sintering effects [Fauchais P. et al. (2011)].

These agglomerated particles can be either dense or rather porous. According to [Fauchais P. et al. (2011)] not all so-called nanostructured agglomerated powders commercially available are formed via the agglomeration of individual nanostructured particles. For example, Fig. 16.6a shows a spray-dried "nanostructured" agglomerated Al_2O_3–13 wt.% TiO_2 (alumina–titania) powder particle (Nanox S2613S, Inframat Corp., Farmington, CT, USA). By viewing this particle at higher magnification (Fig. 16.6b), it is possible to observe that the agglomerate exhibits individual particles varying from about 15 to 300 nm [Killinger A. et al. (2011)]. Therefore, it is suggested that the term "ultrafine" agglomerate is more scientifically rigorous to describe the

morphology of these types of powders, which are formed from a mixture of nanometer and sub-micrometer-sized particles.

[Lima R.S. and B.R. Marple (2007)] have recently published an excellent overview of the spraying of agglomerated nanometer-structured ceramic particles. With the except cold spray, thermal spraying is intrinsically associated with the in-flight melting/partial melting of particles prior to their impact on the substrate. Without some particle melting, it is extremely difficult to produce thermal spray coatings, especially ceramic ones. Thermal spraying nanostructured powders is thus a challenge: if all powder particles are fully molten in the thermal spray jet, all the initial nanostructures will disappear. In order to overcome this challenge, it is necessary to carefully control the temperature and size distribution of the particles in the thermal spray jet to keep part of the particles in a semi-molten state. Based on this fact, the expression "nanostructured Thermal spray coating" is not the most accurate to designate or represent these types of coatings. The expression "bimodal coating," representing the mixture of particles in the coating microstructure that were previously semi-molten and fully molten in the spray jet, is more scientifically rigorous. However, as the term "nanostructured Thermal spray coatings" is the most widely used by the thermal spray community to designate these types of coatings, it will be used in further discussion of the basic phenomena involved.

[Shaw et al. (2000)] were among the firsts to plasma spray Al_2O_3–13 wt.% TiO_2-agglomerated nanosized particles. They showed that the phase content and grain size of the coating exhibit a strong dependency on the spray temperature. Figure 16.7a shows schematically a micro-sized granule containing hundreds of nanosized Al_2O_3 and TiO_2 particles. If, during the short exposure to the plasma jet, the temperature to which particles will be submitted is low enough, the TiO_2 particles (melting temperatures of 1854 °C) will be melted, while the Al_2O_3 nano-grains (melting temperatures 2040 °C) will remain unmeted (see Fig. 16.7b) [Shaw et al.

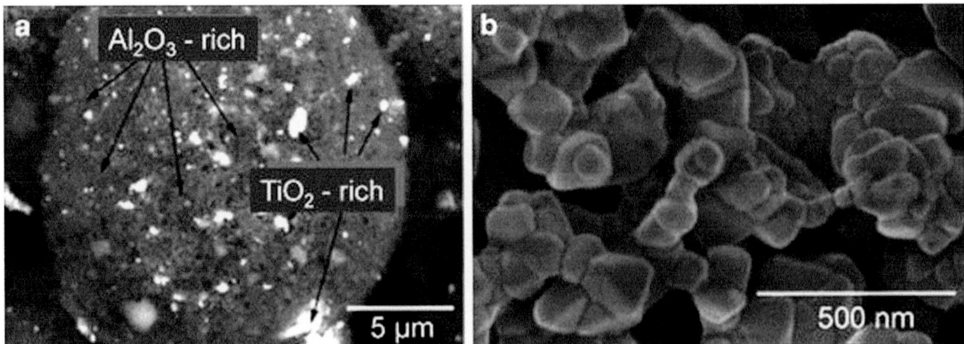

Fig. 16.6 (**a**) Spray-dried agglomerated Al_2O_3–13 wt.% TiO_2 particle. (**b**) Higher magnification showing the "ultrafine" character of the agglomerate [Fauchais P. et al (2011)]. (Reprinted with kind permission from IOP organization)

(2000)]. Besides the temperature and size distribution of the particles, the microstructure of the individual particles that reflects its manufacturing route will have an impact on the outcome of the process [Wang M. and L.L. Shaw (2007)].

With zirconia, the problem is even more complex because particles must be only partially melted forming a molten shell surrounding a nonmelted core nanosized particles as schematically illustrated in Fig. 16.8a. A modeling study of the transient heating and melting of agglomerated nanosized zirconia particles when injected in a DC plasma jet was reported by [Ben-Ettouil F. et al. (2008)]. The evolution of the melting front in agglomerated particles, represented by the ratio (r_m/r_{po}), with r_m the radius of the nonmolten core at any time during the particle trajectory, and r_{po} the initial radius of the agglomerated particle, as a function of the axial position is shown in Fig. 16.8b for different initial particle diameters, 40, 50 and 60 μm, with an initial porosity,

$\alpha_o = 50\%$. The plasma operating conditions were plasma gas flow rate (60 slm, Ar + 25 vol.% H$_2$), nozzle internal diameter 7 mm, and power 18.24 kW. The corresponding evolution of the melting front for a fully dense 50 μm zirconia particle is also given for comparison. For the four cases, the particle trajectory in the plasma jet was assumed the same. It may be noted that the 40 μm diameter particle is completely melted after only a 40 mm trajectory in the plasma, whereas within particles of 50 and 60 μm in diameter, nanostructured solid cores with diameters of 27 and 56 μm, respectively, corresponding to 54 and 71% of their initial diameter are retained at 100 mm from their point of injection. The 50 μm, dense zirconia particle, melts much more rapidly than the porous particles due to their thermal conductively and transient heat propagation in the particle. Computations were also carried out by [Ben-Ettouil F. et al. (2008)] considering particles with different porosities corresponding to different powder manufacturing processes. The results given in Fig. 16.9 were obtained for 60-μm diameter particles with porosities $\alpha = 0.1, 0.3,$ and 0.5 and corresponding effective thermal conductivities, $\kappa = 1.49, 1.16,$ and 0.83 W/mK, respectively. Results are given in terms of the evolution with axial distance in the plasma jet of the non-dimensional radius of the particles, (r_p/r_{po}), and the dimensionless radius of the porous core (r_m/r_{po}), with both referred to the original radius of the particle ($r_{po} = 30$ μm). As noted in Fig. 16.9a, the particle radius shrinks as a result of surface vaporization and densification due to the melting of the outer shells of the particle with the effect increasingly more pronounced for the particles with higher porosity. The same phenomenon is observed in Fig. 16.9b with the radius of the porous core

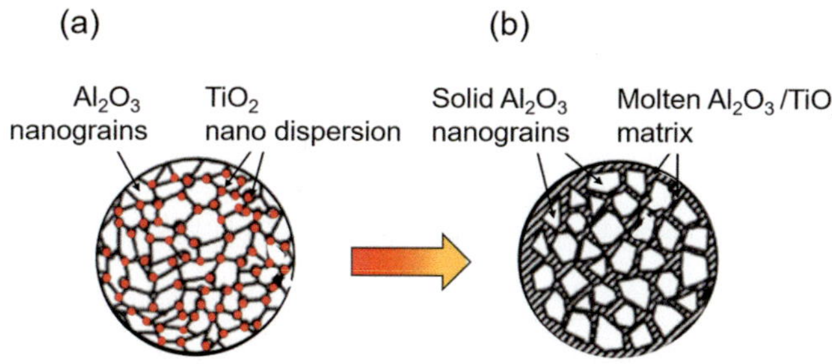

Fig. 16.7 Schematic of (**a**) A reconstituted Al$_2$O$_3$–TiO$_2$ granule and (**b**) the partly melted granule where solid Al$_2$O$_3$ nanometer grains exist within a liquid TiO$_2$. The viscosity of such a "mushy" granule depends on the volume fraction of the liquid TiO$_2$ and the temperature [Shaw et al. (2000)]. (Reprinted with kind permission from Elsevier)

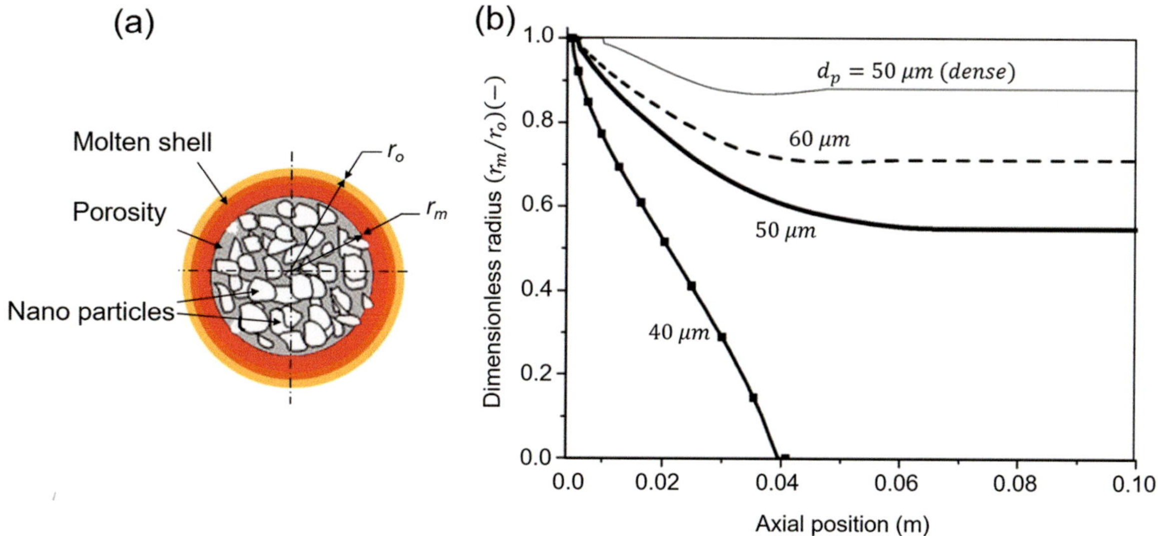

Fig. 16.8 (**a**) Mechanism of nanostructured coating building (**b**) Evolution of the melting front in agglomerated zirconia particles ($d_p = 40, 50,$ and 60 μm) with axial distance from their point of injection in a DC plasma jet [Ben-Ettouil F. et al. (2008)]. (Reprinted with kind permission from Elsevier)

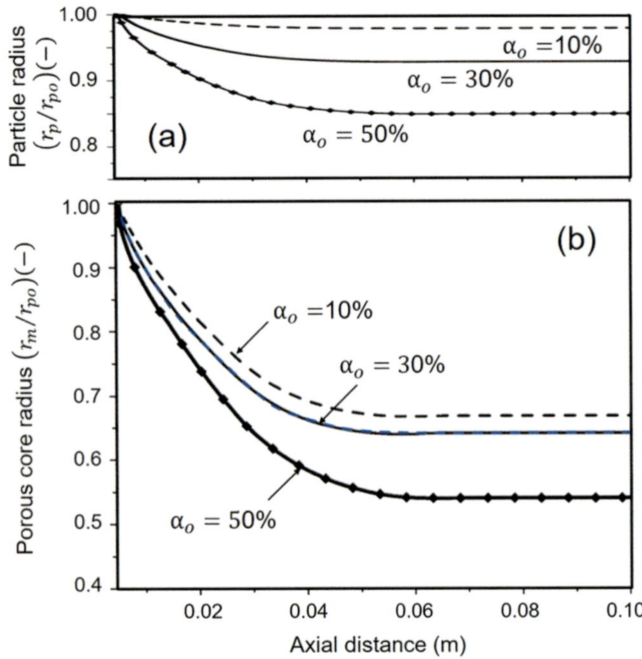

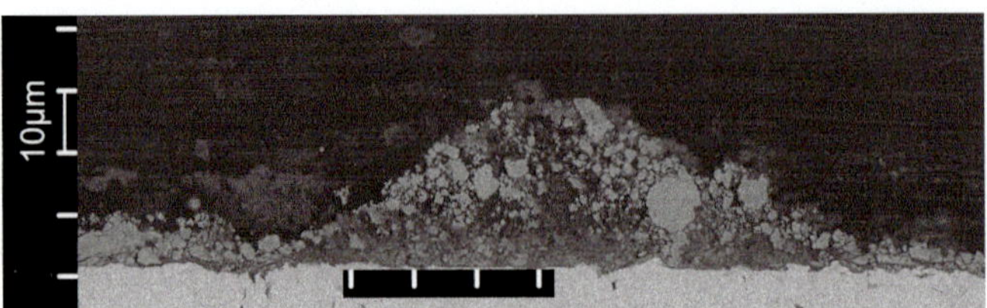

Fig. 16.10 Cross-sectional view of a splat resulting of semi-molten structure of agglomerated nanometer-sized particles of YPSZ (8 wt.% of Y_2O_3) sprayed using triple torch plasma reactor (TTPR) working with Argon as plasma gas [Guru D. and J. Heberlein (2008)]. (Reprinted with kind permission from Prof. J. Heberlein)

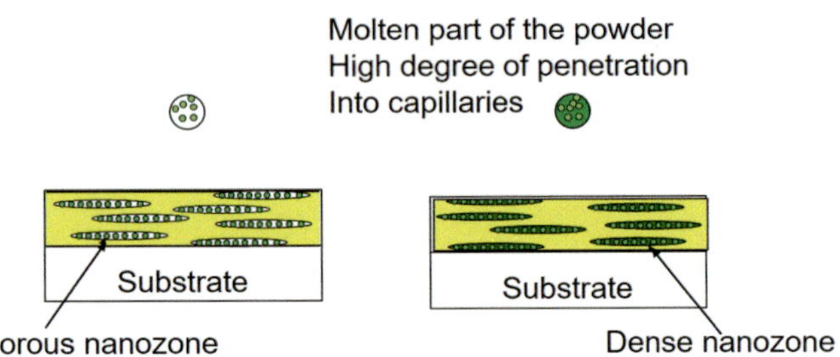

Fig. 16.9 Evolution with axial distance of the ratio of (**a**) particle radius, r_p and (**b**) radius of its porous core, melting front, r_m, to its original radius ($r_{po} = 30$ μm), for 60 μm dimeter, agglomerated zirconia particles of different porosities injected into DC plasma jet [Ben-Ettouil F. et al. (2008)]. (Reprinted with kind permission from Elsevier)

Fig. 16.11 Schematic of the embedding of porous and dense nanozones throughout the coating microstructure during thermal spraying [Fauchais P. et al. (2011)]. (Reprinted with kind permission from IOP organization)

that decreases rapidly as the particle travels along its trajectory and the melting front moves towards the center.

To spray these nanostructured agglomerates, the spray parameters must be optimized to produce conditions (particle temperatures and velocities) that result in only partial melting of agglomerates (to avoid the complete loss of the nanostructure). However, a sufficiently high degree of melting must be obtained to ensure effective deposition on the substrate and the formation of the so-called nanozones [Lima R.S. and B.R. Marple (2007)].

[Guru D. and J. Heberlein (2008)] used the triple torch plasma reactor (TTPR), as described in Chap. 8. DC Plasma Fundamentals, Sect. 8.3.6, Fig. 8.39, along with a custom-designed shutter mechanism to deposit splat samples of agglomerated nanosized particles of YPSZ (8 wt.% of Y_2O_3). The TTPR was working with pure argon as plasma gas and the particle size distribution of the agglomerated powder was in the range of $15 < d_p < 150$ μm. A typical splat, deposited on polished Mo substrate, is given in Fig. 16.10. The photograph shows the features of a semi-molten agglomerate particle having the nanophase structure preserved. A layer of such particles has some plastic deformation properties making it suitable as strain relief intermediate layer between two layers with different thermal expansion characteristics. Many solid particles are present

within the melted matrix. The splat shape shown in Fig. 16.10 is conical because only a shell of the particle was fully melted, and the impact velocity was relatively low.

One of the critical goals of parameter optimization is to control the density of the nanosized zones, that is, the density of the semi-molten nanostructured agglomerates embedded in the coating microstructure [Fauchais P. et al. (2011)]. This is achieved by finding the conditions to adjust the amount of the molten part of each semi-molten particle that penetrates into the capillaries (i.e., the non-molten particle core) of the agglomerates (Fig. 16.8a) during its flight within the thermal spray jet and/or at the impact on the substrate surface and subsequent solidification. It is important to point out that by introducing porous or dense nanosized zones throughout the coating microstructure, it is possible to engineer coatings with very different and even opposite properties for a variety of purposes [Fauchais P. et al. (2011)]. For example, for thermal barrier and abradable seal applications, the presence of a porous nanosized zones is paramount, while for anti-wear applications, dense nanosized zones are absolutely necessary to induce high levels of wear resistance [Lima R.S. and B.R. Marple (2007)]. A schematic of the embedding of porous and dense nanometer-sized zones throughout the coating microstructure is given in Fig. 16.11.

As shown in Fig. 16.12 [Lima R.S. and B.R. Marple (2007)], the coating microstructure is composed of semi-

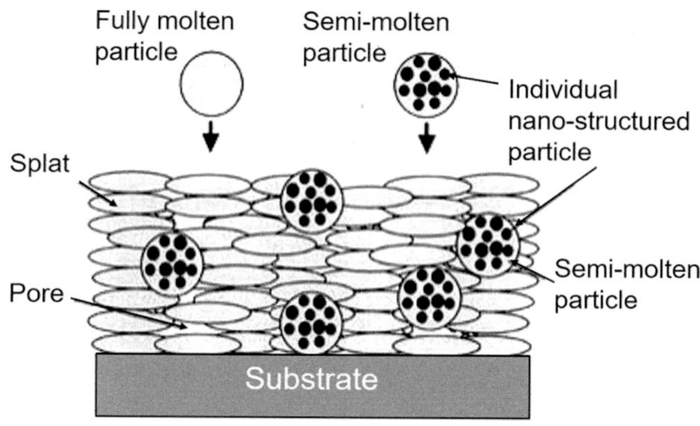

Fig. 16.12 Schematic representation of a typical bimodal microstructure of thermal spray coatings formed by fully molten. and semi-molten nanostructured agglomerated particles [Lima R.S. and B.R. Marple (2007)]. (Reprinted with kind permission from Springer Science Business Media, copyright © ASM International)

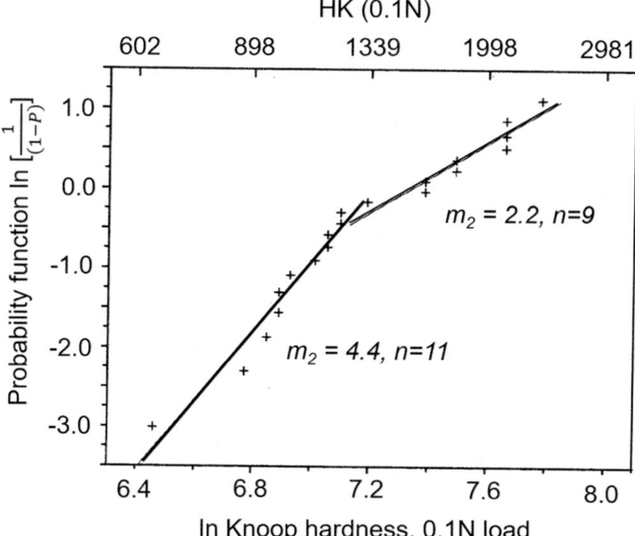

Fig. 16.13 Weibull plot of Knoop microhardness for the nanostructured coating (Nanox S4007) plasma sprayed with the Metco 3MB [Lima R.S. et al. (2002)]. (Reprinted with kind permission from Elsevier)

molten feedstock particles that are spread throughout the coating, surrounded by fully molten particles that act as a cement or binder. Such bimodal nanostructured thermal spray coatings exhibit significantly higher crack propagation resistance and relative toughness when compared to conventional coatings.

[Lima R.S. et al. (2002)] described these coatings as exhibiting "bimodal microstructure" properties as illustrated in Fig. 16.13 for nanostructured YPSZ (ZrO_2–7 wt.% Y_2O_3) experimental feedstock Nanox S4007 (Inframat, North Haven, CT), plasma sprayed in air on low-carbon steel substrates using Metco-3MB torch (GH nozzle, powder port

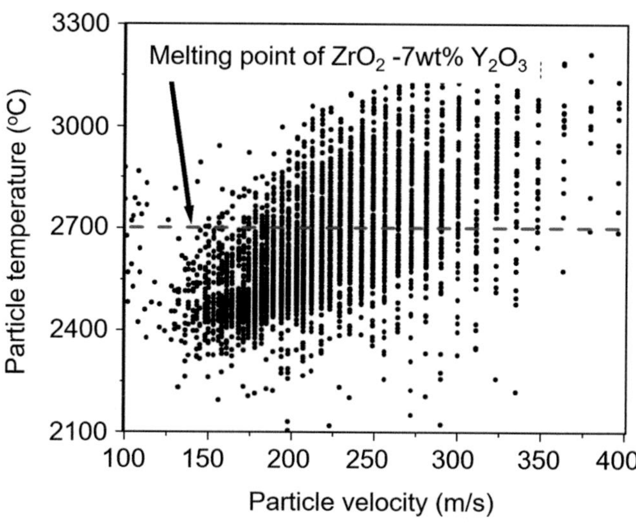

Fig. 16.14 In-flight particle surface temperature and velocity distributions during the DC plasma spraying of nanostructured agglomerated YSZ particles [Fauchais P. et al. (2011)]. (Reprinted with kind permission from IOP organization)

C2) with (40 slm Ar + 12 slm H_2) as plasma gas, 600 A, and 40 kW. The nanostructured coatings show hardness values similar to those of conventional coatings (with the same size distribution) in the low slope regions of the Weibull plots. Since the conventional coating primarily represents the behavior of fully molten particles, this observation indicates that the low slope regions of the Weibull plots reflect microstructural features of the nanostructured feedstock, which are fully molten. On the other hand, the high slope regions, at low hardness values, represent the non-molten feedstock particles in the coating microstructure [Lima R.S. et al. (2002)].

To achieve this bimodal structure, the temperature of the powder particles should be maintained such that it is not significantly higher than the melting point of the material [Fauchais P. et al. (2011)]. In-flight particle temperature and velocity monitoring were used for the optimization of the spray parameters to engineer such coatings produced from nanostructured agglomerated particles. One of the major advantages of this approach is the fact that it can be applied to optimize spray parameters for different thermal spray sources such as plasma spray, flame spray, and high velocity oxy-fuel (HVOF) torches [Fauchais P. et al. (2011)]. An essential element of using in-flight particle monitoring to engineer nanostructured coatings is the knowledge of the melting point of the material to be sprayed [Lima R.S. and B.R. Marple (2007), Lima et al. (2006)] used an atmospheric plasma spray (APS) system, with an Oerlikon-Metco DC plasma torch (F4-MB), for the spraying of the same nanosized feed material (ZrO_2–7 wt.% Y_2O_3 Nanox S4007). Typical results given in Fig. 16.14 show the in-flight particle temperature and velocity distributions of the plasma-sprayed YPSZ particles, monitored using DPV-2000 by Tecnar Automation, Saint-Hubert, QC, Canada, at the same spray distance as that used to deposit the coating.

The melting point of YSZ or YPSZ is around 2700 °C. By controlling and adjusting the main plasma spray parameters (i.e., torch power, gas flow rates, current, and spray distance), it is possible to control the temperature distribution of the particles. If the majority of the particles in Fig. 16.14 are below the melting point of the YSZ, no effective coating deposition will occur. On the other hand, if the majority of the particles are located above the melting point of the material, although a high deposition efficiency could be achieved, most of the nanostructural character of the feedstock would tend to be destroyed during deposition. It can also induce deeper penetration/infiltration of the molten part, formed near the surface of the agglomerates, into their non-molten cores (capillaries) to form dense nanometer-sized zones. At the conditions depicted in Fig. 16.14, the average particle temperature and velocity are 2633 ± 174 °C and 213 ± 52 m/s, respectively. Therefore, distribution of particles within the plasma spray jet is around the melting point of YPSZ, that is, approximately 50% of the particles are above and 50% below 2700 °C. Achieving this type of distribution is an important factor in preserving and embedding part of the original porous nanostructure of the powder throughout the coating microstructure.

However, the particle temperature and velocity distributions are not the only important factors in the processing [Fauchais P. et al. (2011)]. The particle size distribution is also a key factor. For example, the YPSZ agglomerates used are exhibiting a PSD over the range of 40 to 160 μm that is larger than typical PSD values used for thermal spraying, typically in the range of 10–90 μm. The use of large particles is necessary with agglomerated nanosized plasma spraying to avoid a high degree of particle melting in the plasma jet [Fauchais P. et al. (2011)]. As described by [Fauchais P. (1991)], the plasma temperatures a few millimeters downstream of the plasma torch nozzle can be as high as ~15,000 K, which is considerably higher than the melting temperature of any material known in nature. In addition, the particle velocities in the plasma jet, in the order of ~150–300 m/s (Fig. 16.14), are considered to be relatively low compared to those possible for some thermal spray processes. Consequently, the dwell time of these particles in the plasma jet is considered to be "sufficient" to induce a high degree of particle melting. The use of such large agglomerates is essential to engineer architectures that exhibit porous nanometer zones embedded in the coating microstructure. Typical microstructure of the coating produced is given in Fig. 16.15. Lighter and darker zones can be distinguished in the coating microstructure (Fig. 16.15a), that is, the bimodal distribution. In the darker colored zones, at higher SEM magnifications

Fig. 16.15 (a) Microstructure of the nanostructured zirconia–yttria coating made from a nanostructured feedstock. (b) Darker colored region containing the semi-molten feedstock particles. (c) Detail of semi-molten feedstock contained in dark zone of (b) [Lima R.F. et al. (2002)]. (Reprinted with kind permission from IOP organization)

(Fig. 16.15b, c), it is possible to recognize the similarities with the morphology of the nano-YPSZ feedstock. Dark zones shown in Fig. 16.15b, c correspond to semi-molten nanostructured agglomerated YPSZ particles that became embedded in the coating microstructure. This is achieved when the molten part of each semi-molten particle does not exhibit a high level of penetration into the capillaries of the agglomerates during flight and upon impact and solidification. Nanostructured YSZ coatings exhibiting these porous nano-zones may be very good candidates for thermal barrier coatings (TBCs) and pure ceramic abradable coatings for the hot sections of gas turbine engines [Lima R.S. and B.R. Marple (2007, 2008a, b) and Lima R.S. et al. (2006)]. The percentage of semi-molten porous agglomerates embedded in the coating microstructure of Fig. 16.15 was estimated via image analysis to be around 35%.

As pointed out by [Lima R.S. and B.R. Marple (2007)], the key parameter to engineer coatings with very pronounced differences in microstructural characteristics and mechanical performance is the density of the nanosized zones. This density depends on thermal processing, spraying conditions, and feedstock characteristics (e.g., diameter and density). The nanometer-zones that form the coating can be porous, like the original feedstock (see Fig. 16.15b, c), or can be much denser. The porous nanometer-sized zones are expected to appear when the molten part of the agglomerated semi-molten particle does not fully infiltrate its non-molten core during thermal spraying. Using very porous nanostructured agglomerated particles can help obtaining porous nanosized zones. On the other hand, the dense nanosized zones probably occur when the molten part of an agglomerated semi-molten particle fully or almost fully infiltrates into the small capillaries of its non-molten core, either in the spray jet or at impact with the substrate.

Spraying nanostructured powders with less control of temperatures may become possible if compensated by a significant high particle velocity (e.g., HVOF particle velocities) [Lima R.S. and B.R. Marple (2007)]. The control of particles melting is more difficult with plasma spraying [Shaw L.L. et al. (2003)] than with HVOF process, in spite of the fact that spraying ceramic materials by HVOF is a challenge. This is due to the high melting point of ceramic materials and the low flame temperatures of HVOF torches (<3000 °C). HVOF spraying seems to be more adapted to ceramic materials with relatively low melting temperature such as Titania or Hydroxyapatite (HA). Controlling the density of the nanosized zones makes it possible to engineer very distinct properties in thermal spray coatings produced from nanostructured agglomerated powders [Fauchais P. et al. (2011)].

The presence of dense nanosized zones is particularly important to produce high performing anti-wear coatings. These are obtained when the molten part of each semi-molten

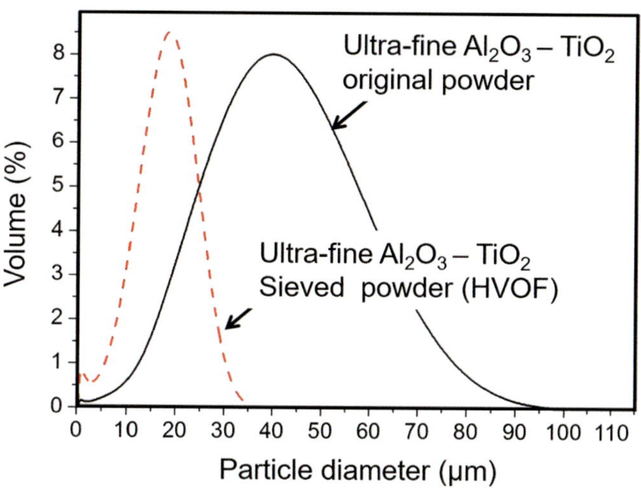

Fig. 16.16 Original and sieved particle size cuts for the ultrafine Alumina–titania powder [Lima R.S. and B.R. Marple (2006)]. (Reprinted with kind permission from IOP organization)

particle exhibits a high level of penetration/infiltration into the capillaries of the agglomerates (non-molten particle core) before and/or at the impact with the substrate surface and subsequent solidification [Lima R.S. and B.R. Marple (2006, 2007) and Fauchais P. et al. (2011)]. To generate such dense nano-sized zones, ultrafine alumina–titania ceramic powder was sprayed using HVOF technology because of its capacity to accelerate the particles to high velocities in the range of 800–1000 m/s, which makes it well suited for the spraying of highly dense coatings with the ultrafine microstructure required for anti-wear applications [Fauchais et al. (2011)]. As previously mentioned, however, HOVF spraying is limited in its ability to heat. As previously mentioned, however, HOVF spraying is limited in its ability to heat the particle to near melting temperatures because of the relatedly low jet temperature and the short dwell time of the particles in the high temperature zone of the spray jet. From a certain point of view, it can be stated that HVOF is a high velocity and low temperature process, in contrast to plasma spray that is a low velocity and high temperature Technology.

To overcome such a limitation, [Lima R.S. and B.R. Marple (2006)] sieved the, as purchased, spray-dried ultrafine alumina–titania agglomerated powder (Al₂O₃–13 wt.% TiO₂) to separate the ~5 to 30 µm fraction from its original PSD of ~5 to 90 as shown in Fig. 16.16. For comparison, a conventional powder (Amperit 744.0, HC. Starck, Goslar, Germany) was also sprayed via HVOF process (DJ-2700-hybrid, Sulzer Metco, Westbury, NY, USA). The powder was constituted from a blend of fused and crushed alumina and titania particles [Lima R.S. and B.R. Marple (2006)]. The particle size distribution for this powder was ~5–25 µm, that is, within the typical particle size range for HVOF torches. The melting point of the Al₂O₃–13 wt.%

TiO_2 composition is ~1900 °C. The in-flight particle characteristics were monitored and adjusted (using DPV-2000, Tecnar Automation, Saint-Hubert, QC, Canada), and the results are shown in Fig. 16.17. It is possible to observe that the overall temperature distributions of particles from both powders are well above the melting point of the alumina–titania composition. As it was the case for the nanostructured YPSZ coatings, the in-flight particle characteristics were being monitored at the same spray distance used to deposit the coatings.

It is important to point out that this type of in-flight particle monitoring is based on pyrometer measurements, that is, the temperature measurements represent the external temperature of the particle. Therefore, even if the temperatures at the outer surface of the particles are above the melting point of the material, it does not necessarily mean that the inner cores of the particles are also fully molten. It has to be stressed that the velocities in the HVOF jet, are in the range of 900–1100 m/s for the ultrafine agglomerates

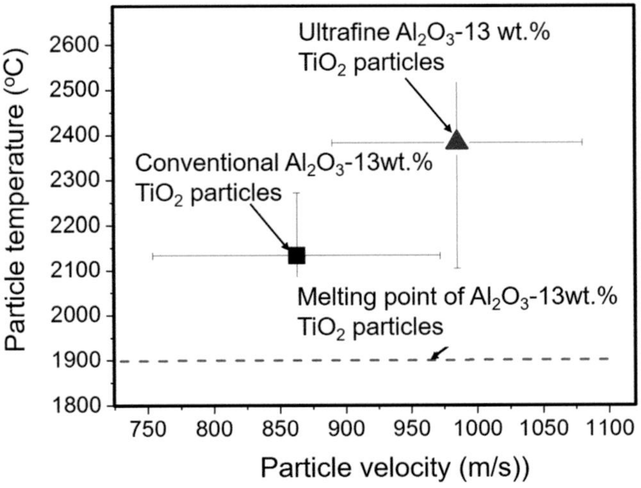

Fig. 16.17 Particle temperature and velocity levels for the ultrafine and conventional Alumina–titania powders [Fauchais P. et al. (2011)]. (Reprinted with kind permission from IOP organization)

(Fig. 16.17), which is considered high for thermal spraying. Consequently, the dwell time of these particles in the HVOF jet is considered to be "short" and higher particle temperatures and smaller particle diameters must compensate for these short-heating times.

[Lima R.S. and B.R. Marple (2005a, b)] have compared coatings sprayed by HVOF (Diamond Jet-2700-hybrid, Sulzer Metco) using nanostructured Titania feedstock (Altair VHP-DCS, 5–20 μm) to plasma-sprayed coatings (APS torch SG-100 Praxair) using conventional Alumina–titania (Al_2O_3–13 wt.% TiO_2) feedstock (Metco 130 nominal particle size range from 15 to 53 μm). The particle temperature and velocity of the feedstock particles in-flight in both spray jets were measured using the DPV-2000 at the spray distance at which the substrates would be positioned when depositing a coating (APS system 64 mm; HVOF system 200 mm). As it could be expected the air plasma-sprayed alumina–titania coating was 33% harder than the HVOF-sprayed nanostructured titania. But the nanostructured titania coating exhibited a superior abrasion wear resistance (27% lower volume loss) when compared with the air plasma-sprayed conventional Alumina–Titania. It also exhibited a crack propagation resistance almost twice that of the air plasma-sprayed conventional alumina–titania coating (i.e., the nanostructured coating is tougher than the conventional one).

To conclude, spraying agglomerated-nanosized particles, to obtain bimodal structures, is not simple and requires staying within a rather narrow spray window. However, as pointed out in the review [Fauchais P. et al. (2011), Lima R.S. and B.R. Marple (2006, 2007)] recommended guidelines based on previous experimental results given in Table 16.1 summarizing the conditions to achieve dense nanosized zones by APS or HVOF or porous nanosized ones by APS. It is important to stress the need to use a sensor measuring the particle temperatures and velocities distributions (particularly the mean temperature $\overline{T}$ and its standard deviation σ_T), and not ensemble measurements. Such measurements should be performed at the same spray distance used to deposit the coating. The particle size

Table 16.1 Spray parameters to achieve dense nanosized zones by APS or HVOF or porous nanosized ones by APS

	Particle size distribution	Particles average surface temperature	Spray distance
APS (porous nanostructured zones)	Large particles with a wide particle size distribution, $(10 < d_p < 150$ μm)	$\overline{T}$ and σ_T must be adjusted (via spray parameters and particle size distribution) to overlap the melting point of the material	Adjusted to be the initial point at which $\overline{T}$ and σ_T overlap the material melting point
APS (dense nanostructured zones)	Typical particle size distribution for APS systems within the range $(5 < d_p < 90$ μm)	$\overline{T}$ and σ_T must be adjusted (via spray parameters and particle size distribution) to overlap the melting point T_M of the material	Compromise between the previous condition and a larger spray distance[a] such as that used with conventional powders
HVOF (dense nanostructured zones)	Small particles exhibiting a narrow particle size distribution $(5 < d_p < 30$ μm)	$\overline{T}$ and σ_T must be adjusted (via spray parameters and particle size distribution) to be over the melting point T_M of the material	Due to the small size of the agglomerates, the spray distance should be around the minimum one required to achieve $\overline{T} > T_M$

[a]Larger spray distances yield longer dwell times and induce a higher degree of particle melting and infiltration into the capillaries of the non-molten cores

distribution is the other key parameter controlling the coating bimodal structure and the nano-zones density.

16.3.2 Applications

Bimodal nanostructured coatings can be engineered to exhibit different properties and microstructures (sometimes with completely opposite performance characteristics such as anti-wear and abradable coatings!) by spraying nanostructured ceramic agglomerated powders via atmospheric plasma spray (APS) or High Velocity Oxy-Fuel (HVOF) [Lima R.S. abd B.R. Marple (2007)].

16.3.2.1 Wear-Resistant Coatings
Numerous studies have been reported on the plasma and HVOF spraying of conventional and nanostructured wear-resistance coatings.

- *Plasma sprayed* [Ahmed I. and T.L. Bergman (1999), Zhu Y. et al. (1999), Gell M. et al. (2001), Ok Chwa S. et al. (2005), Wang Y. et al. (2000), Jiang X.-L. et al. (2003), Song E.P. et al. (2006), Ctibor P. et al. (2006), Ahn Jeehoon et al. (2006), Song Eun Pil et al. (2008), and Lima R.S. and B.R. Marple (2005a)]
- *HVOF sprayed* [Lima R.S. and B.R. Marple (2000, 2003a, b, 2005a, b, 2006), Liu Y. et al. (2003) Turunen E. et al. (2006a, b), Lima R.S. et al. (2007), Varis T. et al. (2007), Ibrahim A. et al. (2007), Lima R.S. et al. (2008) and Lin Xinhua et al. (2005)]

Compared to conventional coatings of titania or alumina–titania, those made with nanostructured agglomerated particles have better abrasion or sliding wear performance. Such coatings exhibit significantly higher crack propagation resistance or relative toughness than conventional ones [Lima R.S. and B.R. Marple (2007)]. The hardness of conventional and nanostructured coatings is generally about the same. The thermal diffusivity of the nanostructured Al_2O_3–3 wt.% TiO_2 coating was higher compared with that of the corresponding

conventional coating at temperatures ranging from 200 to 1000 °C [Lin Xinhua et al. (2005)]. Zirconia nanostructured plasma-sprayed coatings have also been used against wear. The nanostructured coating reduced the wear rate of this coating against stainless steel [Chen Huang et al. (2003), and Shunyan T. et al. (2005)].

[Lima R.S. and B.R. Marple (2007)] compared the abrasion wear behavior of coatings sprayed with conventional Al_2O_3–13 wt.% TiO_2 powder (Amperit 744.0, H. C. Starck, Goslar, Germany) to the nanostructured agglomerated powder depicted in Fig. 16.6 with the size distribution presented in Fig. 16.16. Both were sprayed using HVOF (DJ-2700-hybrid, Sulzer Metco, Westbury, NY, USA). The comparison was achieved through the measurement of volume loss after abrasive wear testing. The ultrafine coating exhibited 4× improvement in wear performance. The superior abrasion wear performance of the ultrafine coating could not be explained based on Vickers microhardness values. Both coatings exhibited average Vickers microhardness number values of ~800$_{3N}$ [Lima R.S. and B.R. Marple (2006)]. Comparable results were reported by [Turunen et al. (2006a, b)] for the abrasion wear resistance of nanostructured and conventional alumina (Al_2O_3) coatings deposited via HVOF in which the nanostructured coating had superior wear resistant properties. According to the study reported by [Lima R.S. and B.R. Marple (2006)], the evolution of the crack propagation resistance for the alumina–titania coatings seemed to be the main reason for the superior wear behavior of the ultrafine coating. The evaluation of the crack propagation resistance, which is a measure of relative toughness, was carried out by indenting the cross-section of the coatings using a Vickers indenter (5 N load) with one of its diagonals parallel to the substrate surface to induce crack propagation between layers. The crack length, from tip-to-tip, of the conventional coating shown in Fig. 16.18a was found to be ~40% longer than that for the ultrafine coating shown in Fig. 16.18b. The observation is consistent with the notion that ultrafine nanostructured coatings have significantly higher crack propagation resistance than conventional coatings.

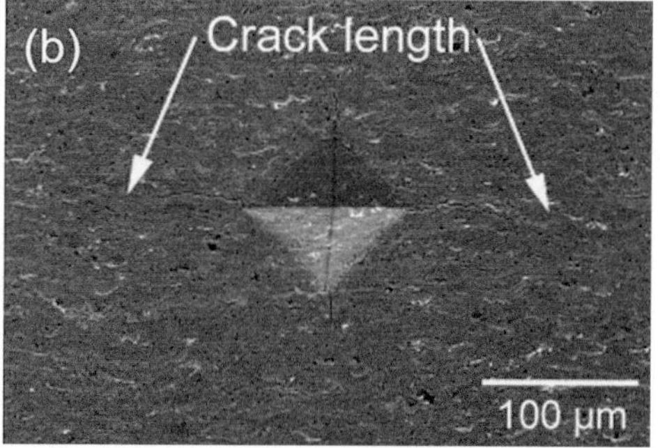

Fig. 16.18 Crack propagation induced by indention on the cross-section of the (**a**) conventional and (**b**) ultrafine alumina–titania coatings [Lima R.S. and B.R. Marple (2006)]. (Reprinted with kind permission from Elsevier)

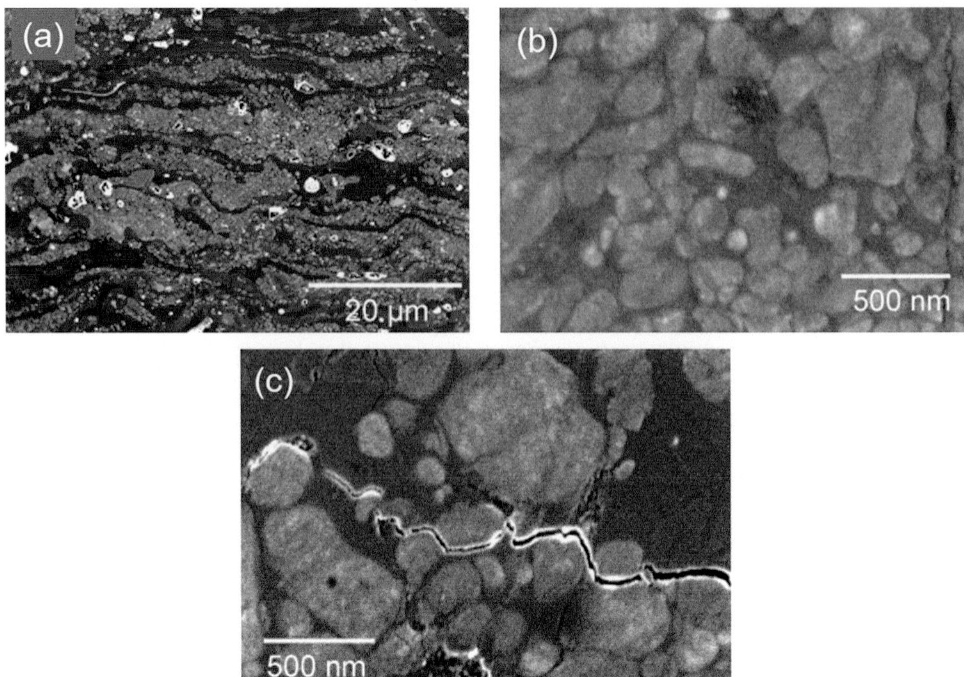

Figs. 16.19 Higher magnification views (cross-section) of the ultrafine coating of Fig. 14.15b. (**a**) Overall view of the dense ultrafine zones. (**b**) Localized view of a dense ultrafine zone. (**c**) Crack tip arresting via path deflection by passing thorough a dense ultrafine zone [Lima R.S. and B.R. Marple (2006)]. (Reprinted with kind permission from Elsevier)

The discussion below follows that is presented in [Fauchais P. et al. (2011)]. The question addressed arises is related to the role of the ultrafine nanostructure of the agglomerated powder in the enhancement of the coating toughness. [Gell M. et al. (2001) and Luo et al. (2003)] observed that the enhanced toughness of these types of coatings was related to crack arresting and deflection when passing through dense nano-zones embedded in the coating microstructure. [McPherson R. (1989)] has shown earlier that for conventional thermal spray ceramic coatings, cracks will tend to propagate between adjacent layers of the coatings (i.e., the splat boundaries), which are the weakest link of the microstructure. When the microstructure of the ultrafine coating given in Fig. 16.18b is analyzed at higher magnifications using SEM, it is possible to observe lamellar zones constituted of a finely dispersed material (Fig. 16.19a) distributed evenly within the coating cross-sectional microstructure [Lima R.S. and B.R. Marple (2006)]. By analyzing one of these finely dispersed regions in more detail (Fig. 16.19b) and comparing it with the microstructure of the ultrafine agglomerated powder (Fig. 16.6), one can realize from the similarities that these finely dispersed zones are semi-molten ultrafine particles embedded in the coating microstructure. However, due to processing conditions (i.e., particle temperature, velocity, diameter, and dwell time), the molten part from the surface of the agglomerates deeply infiltrated into the capillaries of the non-molten cores, thereby creating dense ultrafine microstructure zones after particle impact and solidification, as also reported by [Shaw et al. (2003)]. The percentage of semi-molten dense agglomerates embedded in the coating microstructure of Figs. 16.18b and 16.19 was estimated via image analysis to be ~50% [Lima R.S. and B.R. Marple (2006)]. When the left crack tip of the ultrafine coating of Fig. 16.18b is observed in detail at higher SEM magnifications (Fig. 16.19c), it is noticed that the crack is deflected and arrested when passing thorough a dense ultrafine zone. As in these coatings the microstructure is disrupted periodically by these zones, the crack propagation encounters barriers to propagate through the well-defined layered structure, thereby enhancing coating toughness, as also observed by [Gell et al. (2001) and Luo et al. (2003)].

Improved bond strength values have been reported for these coatings when compared to their conventional counterparts [Lima R.S. and B.R. Marple (2007)]. The explanation for this improved bond strength is similar to that used to explain the enhanced toughness. According to [Bansal et al. (2003)] for coatings produced from nanostructured agglomerated powders, the interfaces between the dense nanometer zones and the substrate surface did not exhibit micrometer cracks or gaps, whereas for those of the conventional coatings, micro-sized cracks or gaps were observed. Therefore, the dense nano-zones would tend to impede crack propagation at the interface, which would enhance interfacial toughness and the bond strength levels of the coating. It has to be pointed out that other authors are attributing these higher toughness levels to an enhanced inter-lamellar strength, and not crack path deflection. [Ahn et al. (2006)] observed that in spray-dried nanostructured agglomerated powders, the individual alumina and titania particles are intimately mixed, as readily observed in Fig. 16.6. The same degree of homogeneous mixture is not found in conventional particles. The melting point of pure alumina is 2050 °C, whereas that of pure titania is 1855 °C. An addition of 13 wt.% of titania into alumina results in a lowering of the melting point of the compound to ~1900 °C. Therefore, Al_2O_3–13 wt.% TiO_2 "well-mixed" agglomerates (Fig. 16.6) would tend to arrive at the substrate exhibiting lower viscosity levels due to the lowering of the melting point, than those of the conventional particles. These lower viscosity levels would translate into an enhanced splat-to-splat cohesion, which would then improve the toughness of the coating. By closely examining Fig. 16.18, it is possible to observe a "lower" degree of homogeneity of the coating produced form the conventional powder (Fig. 16.18a) compared to that produced from the ultrafine powder (Fig. 16.18b). For the conventional coating the alumina-rich (dark) and titania-rich (light) phases are quite well defined, that is, not well mixed. As titania has lower mechanical strength levels than alumina, it probably creates weak-link zones in between layers of the conventional coating. Based on these studies, it is believed that for alumina–titania coatings produced from nanostructured or ultrafine powders, *crack path deflection*, and *enhanced splat-to-splat contact* are acting together to enhance coating toughness.

Such coatings for anti-wear purposes are used in "real world" applications [Fauchais P. et al. (2011)]. Nanostructured alumina–titania coatings deposited by APS are in use to protect the main propulsion shafts of ships of the US Navy. After 4 years of use in naval applications no significant damage was recorded in these types of coatings. This development was based on a research led by the University of Connecticut (USA) [Shaw L.L. et al. (2003), Gell M. et al. (2001), Luo H. et al. (2003) and Bansal P. et al. (2003)]. Nanostructured titania thermal spray coatings have been used with success by Perpetual Technology (Ile des Soeurs, QC, Canada) in ball valves for autoclaves that operate at high temperature (~260 °C), at high pressure (~5.5 MPa), in corrosive sulfuric acid (>95%), at relatively high solids content (>20 wt.%) [Goberman D. et al. (2002)]. The in-field performance of this nanostructured coating was compared to that of a conventional one. After 10 months of service the coatings were inspected. The ball valve coated with the conventional coating exhibited delaminating in several areas and had to be replaced, whereas the ball valve coated with the nanostructured coating was nearly intact and put back into service.

16.3.2.2 Abradable Coatings

By embedding a significant amount of semi-molten porous particles in the coating microstructure, the ceramic coating becomes friable, producing a nanostructured ceramic abradable coating as reported by [Lima et al. (2006a, b, 2007)]. The clearance between the tip of the turbine blades and the turbine case needs to be as tight as possible to improve engine efficiency and performance. It could be stated that abradable coatings are the "opposite" of anti-wear coatings, that is, they must possess a friable structure. It is a challenge to engineer these coatings because they must be at the same time readily abradable and able to withstand the harsh environment of the turbine engine, as described by [Ghasripoor et al. (1997)]. The conventional abradable seal is a composite coating typically made from a metallic-based material, a self-lubricating phase and many pores. Due to environmental and economic issues, there is a significant driving force to increase the combustion temperature of gas turbines. Higher combustion temperatures will increase engine efficiency and reduce pollution. The abradable coatings have also to follow this trend. To achieve this objective two types of high temperature abradables are currently in use. One is based on a powder composite formed from a metallic superalloy (e.g., CoNiCrAlY), a self-lubricating agent (e.g., BN), and a polymer. The second one is based on a powder composite formed from YSZ, BN, and a polymer. For both cases, the polymer is introduced into the coating microstructure during deposition and burned off in a postdeposition heat treatment to create a large porosity network to make both types of coatings friable. This approach has challenges due to the fact it is difficult to engineer homogeneous microstructures when materials of very distinct physical and chemical properties are sprayed together (e.g., metal + polymer and ceramic + polymer). In addition, the postdeposition heat treatment necessary to burn the polymer off the coating microstructure represents an additional issue of cost and time. Therefore, the use of a pure ceramic abradable material could be a major advance in producing the next generation of high temperature abradable coatings. However, the lack of plasticity of ceramic materials is a major barrier impeding progress towards this goal. For a pure ceramic abradable coating design, the concept of the porous nanometer-sized zones, if embedded in the coating microstructure in "sufficient numbers," could be employed to lower the overall stiffness of the coating, allowing the tip of a metallic turbine blade to wear off some of the coating, creating the "seal effect," without damaging the blade [Fauchais P. et al. (2011)]. Figure 16.20a shows the wear scar after rub-rig testing for abradables on the top surface of the nanostructured YSZ coating of Fig. 16.15. The rub-rig is a standard test to evaluate the abradability of a coating. Essentially, a metallic blade is placed at the edge of a spinning wheel, which exhibits rotating speeds and incremental incursion rates similar to those of a real turbine

Fig. 16.20 (a) Wear scar (top surface) of the nanostructured YSZ coating after rub-rig testing. (b) Wear scar (top surface) of a state-of-the-art high temperature metal-based abradable coating after rub-rig testing [Lima R.S. and B.R. Marple (2007)]. (Reprinted with kind permission from Springer Science Business Media, copyright © ASM International)

engine. By rubbing the blade tip against the top surface of the coating and analyzing the blade tip and the coating wear scar after testing, it is possible to infer the degree of abradability of the material.

By examining Fig. 16.20a it is possible to observe the absence of macro-cracks or major damage, delaminating, or debonding in the coating structure after being rubbed off by the tip of a metallic blade. The wear scar is "well-shaped" and smooth, with no wear observed on the blade. By comparison, a state-of-the-art metallic-based high temperature abradable coating (CoNiCrAlY + BN + polyester) was tested under the same conditions (Fig. 16.20b). The volume losses of both coatings were measured, and they exhibited an almost identical value [Lima R.S. and B.R. Marple (2007)]. These results show essentially that the ceramic abradable exhibited a performance level similar to that of the state-of-the-art metallic based abradables and that the use of nanostructured YSZ coatings engineered to contain porous nanometer-sized zones for thermal barriers and abradables may become major applications of this Technology for the next decade.

16.3.2.3 Thermal Barrier Coatings

Thermal barrier coatings (TBCs) are made from low-thermal conductivity (<2 W/m/K) ceramics that also exhibit high mechanical and chemical stability at high temperatures. According to [Padture et al. (2002)], they are employed to protect and insulate the metallic components of the hot sections of gas turbine engines such as, blades, vanes, and combustion chambers, from the hot gas stream provided by the fuel combustion, allowing higher combustion temperatures and improving engine efficiency. These turbine engines are principally employed in aerospace propulsion, power generation, and marine propulsion industries. Considerable attention has been given over the past few decades to the understanding of the basic mechanisms controlling the performance of TBCs and the development of new and improved materials compositions, microstructures and

deposition techniques as reviewed by [Fauchais P. et al. (2011), Lima R.S. and B.R. Marple (2007)] including plasma-sprayed nanostructured TBC coatings [Lima R.S. et al. (2001), Chen H. and C.X. Ding (2002), Chen H. et al. (2003), Chen H. et al. (2004), Liang B. and C. Ding (2005a, b), Liang B. (2005), Liang B. et al. (2006), Gong W.B. et al. (2006), Racek O. et al. (2006), Soltani R. et al. (2006), Racek O. and C.C. Berndt (2007)].

Regarding TBC applications, basically the porous nanometer-sized zones will reduce significantly the increase of the elastic modulus and thermal conductivity values of these coatings when exposed at high temperatures by counteracting densification effects via differential sintering, which are highly desired characteristics for TBC applications. In fact, it has been shown that for specific coatings in some studies the growth rate of the thermal conductivity (Fig. 16.21a) and elastic modulus (Fig. 16.21b) values of conventional YPSZ TBCs at 1400 °C were approximately 5× higher than those of the nanostructured YPSZ TBCs [Lima R.S. and B.R. Marple (2008a, b), Wang W.Q. et al. (2006)]. By examining the micrographs given in Fig. 16.22 it is possible to identify how the differential sintering mechanism works. It may be noted that micrographs of the as-sprayed coating given in Fig. 16.22a exhibits the typical porous nanometer-sized zones of Fig. 16.15, while according to Fig. 16.22b the overall coating porosity in is observed to increase after an exposure at 1400 °C for 20 h, which is a rather counterintuitive effect. Essentially, the nanometer-sized zones sinter at much faster rate than the coating matrix, thereby inducing porosity and reducing the growth rate of thermal conductivity and elastic modulus values, which are governed by sintering. When the conventional YSZ is observed, from the as-sprayed (Fig. 16.23a) to the heat-treated (Fig. 16.23b) coating, the "expected" behavior for a ceramic is noticed, that is, there is a significant reduction of the overall coating porosity due to sintering effects, which yields higher values of thermal conductivity and elastic modulus.

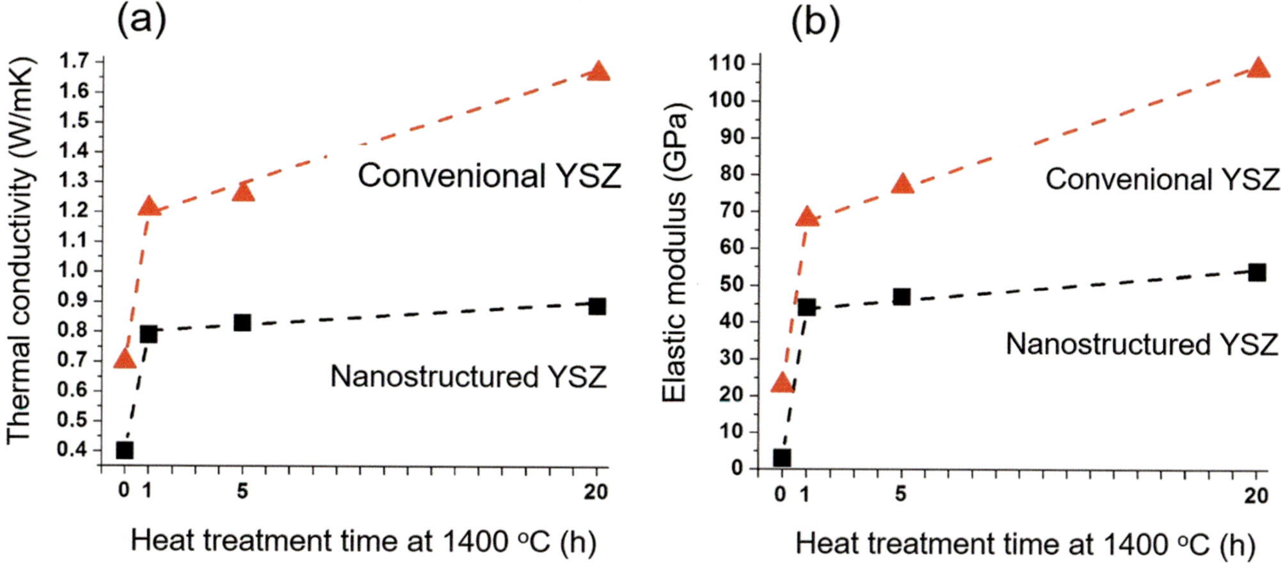

Fig. 16.21 (**a**) Thermal conductivity and (**b**) elastic module values evolution of as-sprayed and heat-treated nanostructured and conventional YSZ coatings at 1400 °C [Lima R.S. and B.R. Marple (2008a, b)]. (Reprinted with kind permission from Springer Science Business Media, copyright © ASM International)

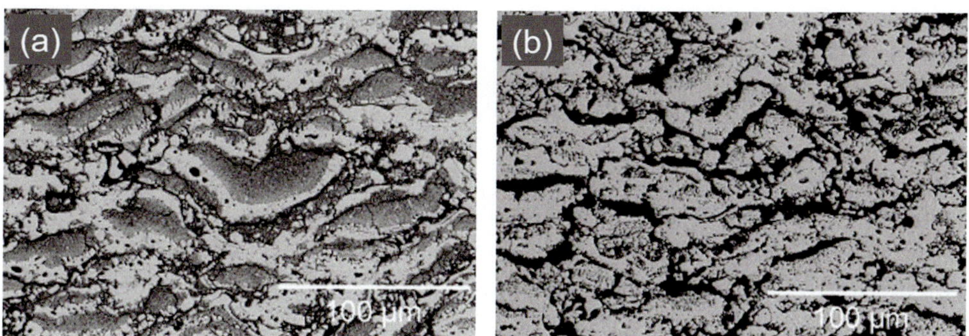

Fig. 16.22 Cross-section of (**a**) as-sprayed and (**b**) heat-treated nanostructured YPSZ coating at 1400 °C for 20 h [Lima R.S. and B.R. Marple (2008a, b)]. (Reprinted with kind permission from Springer Science Business Media, copyright © ASM International)

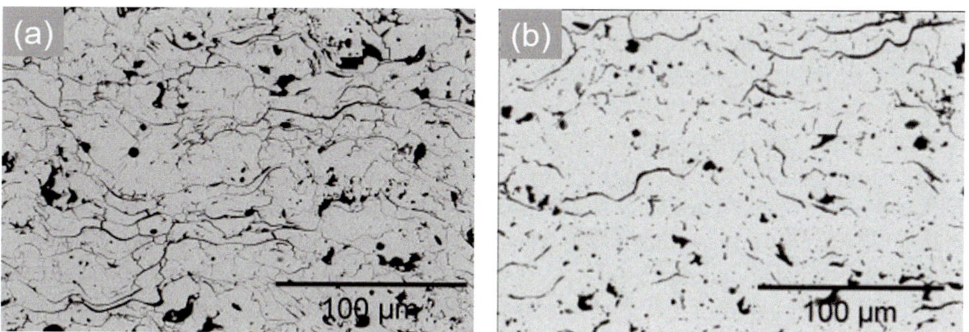

Fig. 16.23 Cross-section of (**a**) as-sprayed and (**b**) heat-treated conventional YPSZ coating at 1400 °C for 20 h [Lima R.S. and B.R. Marple (2008a, b)]. (Reprinted with kind permission from Springer Science Business Media, copyright © ASM International)

More information about the mechanism of differential sintering on TBCs can be found elsewhere [Lima R.S. and B.R. Marple (2008a, b)]. In addition to sintering resistance, [Liang and Ding (2005a), Wang W.Q. et al. (2006), and Liu et al. (2009)] have reported superior thermal shock resistance levels of TBCs produced from nanostructured agglomerated YSZ powders compared to those of TBCs deposited using conventional YSZ powders. [Zhou et al. (2007)] studied the thermal cycling oxidation of nanostructured and conventional YSZ TBCs and reported a superior performance of the nanostructured ones. Therefore, the APS YSZ coatings engineered from nanostructured agglomerated powders have the potential to become the next generation of TBCs. As summarized by [Lima R.S. and B.R. Marple (2007)] nanostructured TBCs have the following advantages: lower thermal diffusivity, higher (2 to 4×) thermal shock resistance, higher creep rate, and healing of the nanoporosity.

As previously discussed, nanostructured YPSZ coatings have demonstrated superior sintering resistance when compared to those of conventional ones [Lima R.S. and B.R. Marple (2008a, b)]. These characteristics are very important because a high increase of elastic modulus and thermal diffusivity/conductivity values caused by sintering hinders the use of these coatings as TBCs. The superior thermal cycle and shock resistances of the nanostructured YSZ coatings [Liang B. and C. Ding (2005a, b), Wang W.Q. et al. (2006), Liu C-B et al. (2009), Zhou C. et al. (2007)] show that they may become an important alternative as TBCs in gas turbines for the next decade.

16.3.2.4 Biomedical Applications

For biomedical applications, mainly two materials are used: hydroxyapatite and titania [Lima R.S. et al. (2005) Li H. et al. (2000) Li H. and K.A. Khor (2006) Yang Y.C. and C. Chang (2003) Li H. et al. (2004) Webster T.J. et al. (1999)]. Hydroxyapatite (HA) ($Ca_5(PO_4)_3(OH)$) is the standard thermal spray ceramic coating applied on implants for load bearing applications, such as acetabular, cups, and hip joints [Sun L. et al. (2001)]. In spite of this successful application, there are concerns regarding the long-term prognostics of these coatings, which may be hindered by coating dissolution and low mechanical performance [Lai K.A. (2002) Yang and Yang (2013)]. Therefore, it is possible to find nanosized zones at the coating/substrate interface, in the internal structure of the coating, as well as on its surface. It has to be pointed out that thermal spray coatings produced from conventional powders do not exhibit a significant presence of nanosized zones on their surfaces. It has been shown, however, that HVOF-sprayed nanosized TiO_2–HA composite coatings exhibit bond strength levels of at least 2.5× that of thermally sprayed HA coatings. Moreover, these coatings exhibited bio-performance levels equivalent or superior to those of HA coatings, which are the current state-of-the-art

material. It was hypothesized that one of the reasons for this enhanced behavior was related to the presence of nanosized zones on the coating surface [Gutwein L.G. and T.J. Webster (2004), Lima R.S. et al. (2006a, b)]. Again, the comments in the review of [Fauchais P. et al. (2011), Lima R.S. and B.R. Marple (2007)] summarize well the results:

- Osteoblast cell culture (in vitro) shows that the HA coatings with a high content of both crystalline HA and nanostructures are preferred for cell proliferation.
- HVOF-sprayed nanometer-sized TiO_2 coatings have bond strength values at least 2.4× higher than those of HA coatings on Ti–6Al–4V substrates. In vivo tests (with rabbits) showed that these coatings (deposited on Ti–6Al–4V substrates) have a higher degree of bone apposition when compared to those of uncoated Ti–6Al–4V substrate.

16.3.2.5 Other Applications

Nanostructured titania coatings provided superior resistance against abrasive and erosive wear for ball valves destined for High-Pressure Acid-Leach (HPAL) service [Kim G.E. and J. Walker (2007)]. The thermal shock behavior of nanostructured Al_2O_3/13 wt.% TiO_2 plasma-sprayed coatings was much higher than that of conventional coatings [Wang Y. et al. (2007a, b)].

16.3.3 Cold Spraying of Nanostructured Powders

Compared to the other spray processes with hot gases, the only way to achieve nanostructured material by cold spraying is to spray nanopowders or micro-size powders with nanosized grains or to spray composite agglomerates made either of micron- and nanoparticles or of only nanoparticles. However, it must be kept in mind that the sprayed particles must be sufficiently ductile.

16.3.3.1 Alloys

Cold spray deposition of conventional and nanocrystalline (atomized and cryomilled) (Al–Cu–Mg–Fe–Ni–Sc) coatings was successfully achieved [Ajdelsztajn L. et al. (2006a, b, c)]. The conventional cold-sprayed coating showed negligible porosity and an excellent interface with the substrate material. This was not the case for the nanocrystalline coating, in which the porosity level was in the range of 5–10%. The microstructure of the feedstock powder was retained after the cold spray process. The difference in porosity between the conventional and nanocrystalline coatings can be explained by the hardness and microstructure of the corresponding feedstock powder, the extent of deformation of which is much less, resulting in a less dense coating. The hardness of

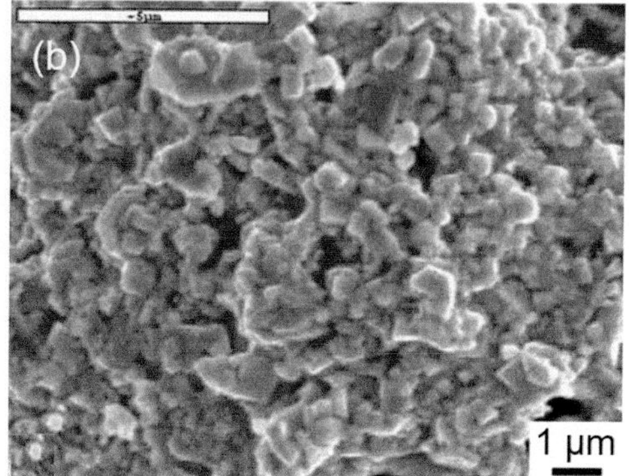

Fig. 16.24 Morphology of nanostructured WC–12Co feedstock (**a**) with detailed surface structure and (**b**) at a high magnification [Li et al. (2007)]. (Reprinted with kind permission from Springer Science Business Media, copyright © ASM International)

the nanocrystalline coating was similar to that of the feedstock powder, suggesting that no work hardening took place [Ajdelsztajn L. et al. (2006a, b, c)].

Nanocrystalline Al-5083 coatings were cold sprayed with helium without preheating of the carrier gas [Ajdelsztajn L. et al. (2005a, b)] onto a grit-blasted aluminum substrate. The critical velocity (over 700 m/s) was successfully achieved. The cold-sprayed coating showed negligible porosity and the interface with the substrate material was excellent. Wang et al. [Wang H-T et al. (2007a, b)] produced by ball milling a metastable Fe (Al) alloy powder with morphology and size distribution suitable for cold spraying. The prepared powder exhibited a lamellar microstructure. The metastable microstructure of the milled Fe(Al) feedstock was completely retained in the coating using cold spraying. The FeAl intermetallic phase was formed during the heat treatment of the as-sprayed coatings at a temperature of 500 °C [Wang H-T et al. (2007a, b)]. Nickel powder was mechanically milled in liquid nitrogen to achieve small non-spherical powder particles with a mean size of 15 μm with an average nanocrystalline grain size in the range of 20–30 nm [Ajdelsztajn L. et al. (2006a, b, c)]. The cryo-milled powder was sprayed with He (also used as carrier gas) onto grit-blasted aluminum substrates using the cold spray process. The Ni coating showed negligible porosity, and the grain structure of the powder was retained. Microhardness values were comparable to the results obtained by electrodeposition. Nanostructured NiCrAlY starting powders were produced using a high-energy milling process and subsequently a coating was deposited on IN738 substrate by cold spraying maintaining the feedstock nanostructure [Zhang Q. et al. (2008a, b)]. The shot-peened coating presented a rather good resistance to oxidation.

16.3.3.2 Composites

The difficulty with cold spraying composites lies in the necessity to spray ductile particles, which is not necessarily

the case for WC–Co particles. The WC–12Co powder sprayed by [Li C.-J. et al. (2007)] exhibited a spherical morphology with nanosized WC particles partially bonded together and voids appearing within particles Fig. 16.24. According to the XRD pattern the powder consisted of two phases of WC and Co. Critical velocities of about 915 m/s were measured. Coating microhardness was about 1800 $HV_{0.3N}$, a value similar to that of sintered bulk material of a similar composition and microstructure. A certain degree of deformation of both impacting particle and deposited layer is required. This is illustrated by [Li et al. (2007)] who cold sprayed with He a powder made from agglomeration of nanosized WC with cobalt followed by partial sintering. The powder particles had a size range from 5 to 44 μm. The nominal WC grain size was 50–500 nm. As shown in Fig. 16.25 small particles (about 10 μm) are almost completely embedded, into the stainless-steel substrate, while large particles penetration is much less. With nanostructured WC–Co, a porous structure, originating from the agglomeration, permits the pseudo-deformation through compaction of particles and the buildup of a thick coating Therefore, it can be considered that a WC–Co powder with WC loosely bonded by the binder is suitable for cold spraying. The annealing treatment at a temperature of 1000 °C had little influence on the microhardness of the coating. However, it seemed that the bonding between deposited WC-Co particles and the toughness of the resulting coatings can be improved by post-annealing treatment [Yandouzi M. et al. (2007)].

The production of cermet coatings with conventional and nanocrystalline feedstock powders by Cold Gas Dynamic Spraying (CGDS) and Pulsed Gas Dynamic Spraying (PGDS) processes was performed by [Yandouzi et al. (2007)]. No degradation of the phase (phase transformation and/or decarburization of WC) was observed. Dense coatings without major defects were difficult to obtain when using the CGDS process. Coatings with low porosity were obtained

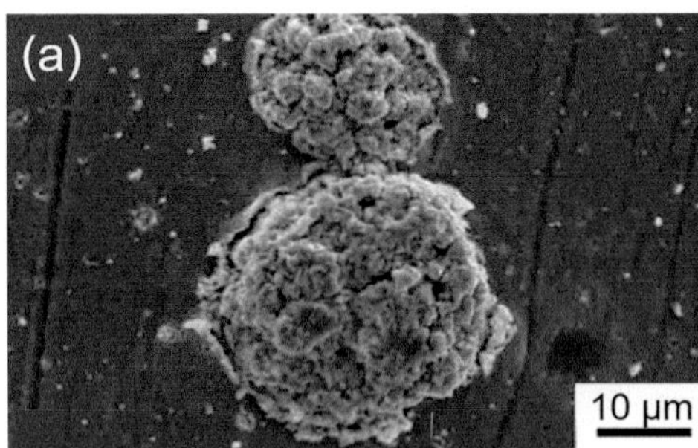

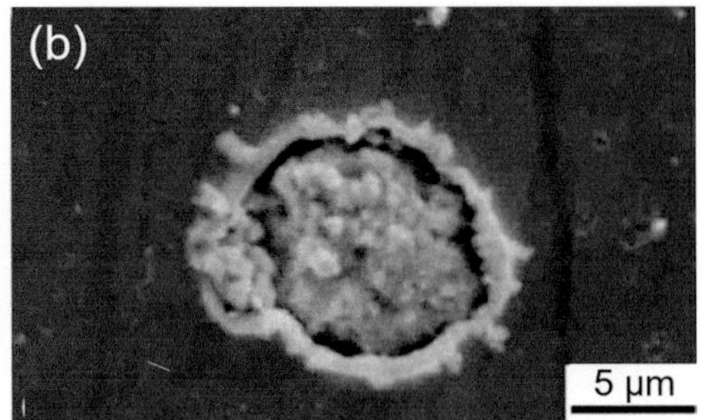

Fig. 16.25 WC–Co particles penetrated into the stainless-steel substrate. (**a**) Two particles of different diameters penetrating to different depths; the large particle penetrated less deeply than the small particle and (**b**) a smaller particle of about 10 µm diameter has almost completely embedded into the substrate [Li et al. (2007)]. (Reprinted with kind permission from Springer Science Business Media, copyright © ASM International)

when the feedstock powder was preheated above 573 K. The nanostructure of the feedstock powder was preserved.

16.3.3.3 Amorphous Alloys

Cold-sprayed Fe-base amorphous alloy coatings were successfully produced. The coatings showed a negligible porosity and excellent interfaces with the substrate material [Ajdelsztajn L. et al. (2006a, b, c)]. The microstructure of the feedstock powder (gas atomized) was retained after the spray process. The use of an amorphous Al–Co–Ce powder to produce nanostructured coatings with high degree of bonding to aluminum substrates alloys was also demonstrated [Sansoucy E. et al. (2007)].

16.3.4 Spraying of Attrition or Ball Milled Powders

Another possibility to achieve coatings with nanosized particles is to spray nanocrystalline powders and melt them only partially.

(a) During thermal spraying of ceramics, even with adapted spray conditions (discussed below), small particles exceed the melting temperature of the alloy or the binder phase for cermets. In contrast larger particles have the melting temperature exceeded only near the surface, whereas particle cores can remain solid. In volumes that remain solid, micrometer-structural changes are negligible (slow kinetics of solid-state diffusion) due to the very short exposure to high temperatures during the spray process [Eigen N. et al. (2005)]. The final crystallite size of the melted alloy or binder phase is determined by the cooling rate during solidification, high cooling rates resulting in much finer crystallites. With cermets, the behavior of the hard phase will strongly depend on the grain size and spatial distributions. In volumes where the binder phase is melted, obvious changes in the microstructure of both the nanostructured and the microstructured cermet material due to the spraying process will be observed. In contrast to the crystallite size of the binder phase, the diameters and distribution of the hard phases depend on their initial size and solubility in the solid and liquid state of the binder phase.

(b) To spray nanostructured alloys or cermets, two techniques (HVOF) or Vacuum Plasma Spraying (VPS) are mainly used to achieve temperatures slightly above the melting temperature of the alloy or cermet binder:

With HVOF, depending on the selection of fuel gas and proportion of the fuel gas to oxygen, the sprayed powder particles experience much lower temperatures (gas temperatures below 3000 °C) compared to plasma spraying. Moreover, the spray velocity in HVOF spraying (700–1400 m/s) is much higher than in plasma spraying, which causes a higher degree of particle impact in the HVOF process. This heavy impact may lead to fragmentation of solidifying particles and thus formation of large number of nucleation sites, which may cause a higher extent of grain refinement in the deposit structure.

With VPS the plasma jet temperature, except in the few first millimeters of the jet, is relatively low (below 5000 K) and the Knudsen effect) reduces the heat transfer. Moreover, particle velocities are higher than those obtained with conventional torches used in atmospheric spraying. Of course, to limit the particle heating, atmospheric argon plasma spraying could be used, but oxidation would be unavoidable. Probably the most important advantage of vacuum plasma spraying is the low oxygen pressure reducing drastically the particle and coating oxidation phenomena.

(c) The last important points are the particle oxidation in flight and the coating oxidation during its deposition. Since the agglomerate size of the nanocrystalline particles is higher than that of the conventional powders, it can be expected that the chemical reactivity of nanocrystalline powders will increase due to the increased surface area. It may result in an increased proportion of oxide phases throughout the grains. For cermets containing materials highly reactive to oxygen, such as WC, some decomposition could occur. It is thus important to control the composition of combustion gases used in HVOF. Vacuum plasma spraying will result in less oxidation than HVOF. This is particularly important for the inter-pass oxidation, which is a consequence of the high temperature that the coating experienced during the spray process. During the time between passes, the coating surface is still at a high temperature and, in the presence of air starts to oxidize (as in HVOF or argon plasma spraying in air).

16.3.4.1 Alloys

Inert gas atomized ASTM F75 Co-based super alloy powders (Cr–28.55 wt.%, Mo–5.958, Si–0.53%, C–0.30%. Fe–0.27% Ni–0.13%, Mn–0.09%, balance Co, Starmet Corp.) have been mechanically milled in methanol environment for 10 h to produce nanocrystalline powders with an average grain size of 12 nm, determined by TEM dark field imaging [Lau M.L. et al. (1998)]. An argon plasma jet (Sulzer Metco 7 M) sprayed these particles onto Ti substrates, and the resultant coating remained nanocrystalline with an average grain size of 21 nm. Moreover, the micro-hardness and porosity of the nanocrystalline Co–Cr coating are found to be higher than those from the conventional Co–Cr coating, indicating that nanocrystalline coatings may be potential coating candidates for implant applications. It was speculated that the marginal increase in micro-hardness of the nanocrystalline Co–Cr coatings might be due to the presence of an increased proportion of oxide phases throughout the grains.

Cryomilled Ni powders were sprayed with Sulzer Metco Diamond Jet DJ-2600 [Ajdelsztajn L. et al. (2002)]. The coating was composed of nanocrystalline grains with an average size of 92.5 ± 41.6 nm (the nanostructure was retained) and with extremely fine NiO particles of 5 nm diameter distributed homogeneously inside the grains. Adjusting the spraying conditions, FeNb microcrystalline powders allowed obtaining amorphous coating, while for the FeSi alloy powder, amorphization was obtained with the addition of boron [Cherigui M. et al. (2005)]. Ball milled Fe-based powder with additions of Si, B, Nb, and Cu elements were HVOF sprayed (Sulzer Metco CDS with a mixture CH_4/O_2 (%) = 0.4) [Cherigui M. et al. (2006)]. The powders sprayed were $Fe_{93.5}$–$Si_{6.5}$, $Fe_{86.5}$–$Si_{13.5}$, Fe_{75}–$Si_{6.5}$–$B_{18.5}$, Fe_{75}–Si_{15}–B_{10}, and $Fe_{73.5}$–$Si1_{3.5}$–B_9–Nb_3–Cu (atomic percentages). Amorphous and nanostructured phases were present in the coatings. Coatings produced had a soft ferromagnetic character, the additions of nonmagnetic elements such as boron, niobium, and copper incorporated in the milled powders having little effect on these properties. However, the nanostructure phase had a significant effect on modifying the magnetic properties of deposits.

Super alloys were also sprayed with an HVOF processes:

- Inconel 718 alloy powder, methanol, or cryo-milled producing nanocrystalline coatings exhibiting thermal stability against grain growth up to 1273 K [Jing H.G. et al. (1998)].
- Oxidation of nano-grain CoNiCrAlY coatings made from cryo-milled powder was studied at 1000 °C [Tang F. et al. (2004a, b)].
- Nanocrystalline NiCrAlY was produced [Ajdelsztajn L. et al. (2005a, b)] by cryo-milling and sprayed with the Diamond Jet DJ-2700 (Sulzer Metco). The final microstructure of the nanostructured coating had a fine grain structure characterized by a "multimodal structure," in which both nanocrystalline regions (approximately 75 nm) and submicron grain regions (from 120 to 550 nm) were observed. The oxidation behavior of nanostructured NiCrAlY coatings was investigated after heat treatments in air, at 1000 °C, for various times. Oxidation led to the formation of a continuous alumina layer, without the presence of other mixed oxides [Ajdelsztajn L. et al. (2005a, b)].

16.3.4.2 Cermets

One of the problems when spraying carbide cermets is the decarburization. Moreover, even under HVOF spraying conditions, the extent of decarburization observed in the nanosized coating material is larger than that in the micro-coating material, because of the high specific surface area of the particles. In particular, heterogeneous melting and localized superheating of the nano-powder, with high surface area is considered to be the controlling factor in decarburization.

WC–Co

[He and Schoenung (2002)] have presented an excellent review of the problems involved with spraying WC–Co. They showed that it was possible to produce nanostructured WC–Co powders without non-WC/Co phases although a high percentage of non-WC/Co phases is frequently reported in conventional WC–Co powders. Nanostructured WC–Co powders experienced higher particle temperatures during spraying than those of the corresponding conventional powder. Therefore, if the same spraying parameters were used, WC particles in nanostructured powders suffered from severe decomposition, degrading the performance of corresponding WC–Co coatings. By controlling agglomerate size of the feedstock powder, fuel chemistry, and fuel–oxygen ratio, near nanostructured WC–Co coatings with a low amount of non-WC/Co phases were successfully deposited. These coatings had increased hardness, toughness, and wear resistance. [Qiao et al. (2000)] showed that when HVOF spraying nanostructured powders having the shape of hollow spheres, particles rapidly reached high temperatures in the various deposition processes and were subject to extensive decarburization. HVOF-sprayed nanostructured WC–12 wt.% Co were compared to Cr_3C_2–NiCr coatings [He J. et al. (2000a, b)]. Results showed that the WC–Co coatings had a nanostructure consisting of nanosized WC carbide particles in an amorphous matrix phase, while in the nanostructured Cr_3C_2–NiCr a few elongated amorphous phases discontinuously distributed in the coating were observed.

Much work has been devoted to nanostructured WC–Co spraying mainly by HVOF [Kear B.H. et al. (1999), He J. et al. (2000a, b), Skandan G. et al. (1999), Ban Z.-G. and L.L. Shaw (2002), Dent A.H. et al. (2001), Qiao Y. et al. (2003), Ban Z.G. and L.L. Shaw (2003), Marple B.R. and R.S. Lima (2005), Bartuli C. et al. (2005), Guilemany J.M. et al. (2006)] and much less by DC plasma spraying [Siegmann S. et al. (2004)]. In nanostructured WC–Co powders manufactured recently; the amount of the non-WC/Co phases was greatly reduced. A grain-growth inhibitor has a strong influence on the microstructure and other properties of the coatings. Preventing WC dissolution in the binder, not only maintains very small grains but also maintains their original shape; it is effective in retarding decarburization, but it decreases the cohesion between WC and the binder [Qiao Y. et al. (2003)]. Superior sliding wear resistance is obtained from coatings deposited with near nano-powders containing an antigrowth additive [Qiao Y. et al. (2003)].

Cr_3C_2–NiCr

Compared to WC–Co system coatings the main shortcomings of Cr_3C_2–NiCr coatings are lower hardness and subsequent low wear resistance, but they are frequently used as protective coatings for application in corrosive environments at elevated temperatures.

Nanostructured Cr_3C_2–25($Ni_{20}Cr$) coatings are synthesized using mechanical milling and sprayed by HVOF or HVAF. Their decomposition seems to be less than that of WC–Co [He J. et al. (2001), Matthews S. et al. (2004), Roy M. et al. (2006)]. Better results are obtained with coatings sprayed by HVAF [Matthews S. et al. (2004)]. After 30 days at 900 °C, the carbide morphology of both HVOF and HVAF coatings was comparable, tending toward an expansive structure of coalesced carbide grains [Matthews S. et al. (2004)]. Nanocrystalline coatings exhibit a 20% increase in hardness, a 40% decrease in surface roughness, and comparable fracture toughness and elastic modulus compared to conventional coatings [Roy M. et al. (2006)].

Other Cermets

A typical example is the Y_2O_3-reinforced milled FeAl sprayed by a Plasma-Technik CDS 100 HVOF torch. The coating displayed the typical dual aspect consisting of fully melted and well-flattened splats together with retained non-melted powder particles, which both contained nanosized grains: columnar nanosized grains formed by rapid solidification in the fully melted splats and equiaxed nanosized grains retained in a deformed matrix at the core of the non-melted powder particles. Oxidation and Al evaporation during thermal spraying led to a modification of the chemistry of the melted zones. The associated Al depletion resulted in the formation of the Fe_3Al phase and more complex structures [Ji G. et al. (2007)].

When cryo-milling Ni powders [He J. and J.M. Schoenung (2003)], the presence of ultrafine AlN particles drastically decreased the dimension of the formed agglomerates and increased their surface roughness. The AlN phase was broken down into ultrafine particles of approximately 30 nm in size. These particles were dispersed in the Ni matrix and enhanced the development of a nanocrystalline structure in the Ni matrix during cryo-milling. A Sulzer Metco DJ-2600 HVOF gun was used to spray these particles. AlN nanometer-sized grains were present in the coatings, and they also led to an increase in the amount of the NiO phase, distributed in the coating in the form of ultrafine, round particles. Indentation fracture indicated that the fine, dispersed AlN particles raised the apparent toughness of the Ni coating. The increase in micro-hardness resulted from both grain refinement and the presence of ultrafine particles.

TiC–Ni-based composites [Eigen N. et al. (2005)] and Al_2O_3 and Al_2O_3–Ni [Turunen E. et al. (2006a, b)] resulted in high quality coatings with optimized HVOF spray parameters. Introduction of nickel alloying decreased hardness and wear resistance of the coatings but increased their toughness.

Using plasma spraying, a coating of Al_2O_3 dispersed in a FeCu or FeCuAl matrix [Basak A.K. et al. (2008)] appeared to offer better wear resistance under sliding and abrasion tests than nanostructured Al_2O_3 coatings, and nanostructured YSZ/NiO anodes provided larger triple phase boundaries for hydrogen oxidation reactions in SOFCs [Hwang C. and Y. Chia-ho (2007)].

16.4 Suspension and Solution Plasma Spraying

Suspension plasma spraying (SPS) and solution precursor plasma spraying (SPPS) are alternat routes for the thermal spraying of protective and functional nanostructured coatings that have emerged over the past two decades as viable coating technology for a wide range of applications. Extensive review papers on the subject have been reported by [Fauchais P. (2004), and Fauchais P. et al. (2008a, b, 2010a, b, 2011, 2015a, b, c, 2016)]. The Technology are based on the use of a liquid as carrier of the coating material in the form of either "nanopowder suspended" or "dissolved solute." On dispersion of the suspension or solution into the high temperature spray stream, whether a combustion flame or a plasma flow, the liquid component is evaporated releasing the suspended solid or solute as fine dispersed grains that on further heating are melted/or partially melted prior to their impact against the surface of the substrate where they form micro splats that are the building blocks of the coating. By replacing the conventional *powder carrier gas* used in thermal spraying, by a *liquid* it is possible to introduce the ultrafine, low inertia, nanopowder into the high temperature spray stream at the expense of significantly increasing the thermal load on the process in the form of energy required for the evaporation of the liquid and further superheating of the generated vapors to the temperature of the spray stream. Whenever applicable, such as in the case of spraying of oxide powders, the use of air/oxygen in the high temperature gas stream, combined with a combustible liquid as solvent or medium for the suspension, such as alcohol, can partially compensate for the increased energy requirement of the process through the released energy of combustion.

To produce finely structured coatings by thermal spray techniques using liquid injection, two routes have been suggested:

Suspension Spraying; Spraying submicron-sized or nanosized ceramic or cermet particles via a suspension [Fauchais P. et al. (2005), Fazilleau J. et al. (2006), Etchart-Salas R. et al. (2007)]. Once drops have been fragmented and vaporized by the plasma flow, in Suspension Plasm Spraying (SPS), or the flame, in High Velocity Suspension Flame Spraying (HVSFS), particles contained in the droplets are heated and sprayed onto the substrate. Splats collected on cold substrates have diameters ranging between 0.1 and 2 μm and average thicknesses between 20 and 300 nm. The stacking of splats or grain shaped particles (resulting from flattened liquid droplets recalescence) form finely structured coatings.

Solution Spraying; Spray solutions of final material precursors. As with suspension, the liquid undergoes rapid fragmentation and evaporation once injected in the plasma jet. This is followed by precipitation or gelation, pyrolysis, and melting to result finally in the impact of molten liquid droplets with average diameters ranging from 0.1 to a few micrometers [Bhatia T. et al. (2001), Gell M. et al. (2004), Jordan E.H. et al. (2004)].

In this section, the basic concepts behind the technology are discussed followed by highlights of recent development and examples typical applications.

16.4.1 Precursor Preparation

Either for SPS or SPPS, the main constituent of the precursor is the solvent, which in most cases is water or a low viscosity organic liquid such as ethanol. As it can be expected, fragmentation of the liquid depends mainly upon its surface tension of the medium, σ_s, and, to a lesser extent, its viscosity, μ_s, while its vaporization is linked to its Thermodynamic properties. For an easier liquid injection, the loading of the solvent with the powder should be relatively low so as to keep the viscosity of the suspension or the solution as close as possible to that of the solvent. Key Thermodynamic and Transport properties for five liquids often used as solvent for SPS or SPPS are summarized in Table 16.2. Water, one of the most widely used, while requiring most energy per unit mass for its vaporization compared to other solvents, has the main advantage of being the only carbon-free solvent avoiding polluting the coating. On the other hand, the combustion of ethanol, pentanol, and triethanolamine, can occur only if oxygen is present and the gas temperature is below the combustion temperature. It means that combustion is possible with HVOF, once the liquids are vaporized, and in the plume of plasma jets run in air but at spray distances higher than those

Table 16.2 Main characteristics of solvents used in suspension or solution spraying

	Molecular weight	Boiling temp.	Viscosity	Surface tension	Specific heat	Energy of vaporization		Energy of combustion
	(g/mole)	(°C)	(mPa. s)	(mN/m)	(kJ/kg K)	(kJ/mole)	(MJ/kg)	(MJ/kg)
Water	18.0	99.98	1.74	71.97	4.18	40.7	2.26	–
Ethanol	46.07	78.0	1.00	22.27	2.44	38.6	0.84	29.0
Triethanolamine	149.19	335.0	–	45.95		67.7	1.85	34.2
Pentanol	88.15	137–139	4.00	25.00		44.6	0.51	35.0
Isopropanol	60.1	82.5	2.37	23.00		44.0	0.73	30.5

commonly used to spray suspensions or solutions, or when the drops are injected with a gas containing oxygen.

16.4.1.1 Solutions

Compared to other spray techniques, solutions offer an excellent chemical homogeneity at the molecular level [Ravi B.G. et al. (2006)]. Several liquid precursors, such as solution/sols and polymeric complexes, have been evaluated for different oxide systems [Gell M. et al. (2008)]. The success in forming the phase required for a given system depends on the decomposition characteristics of the different precursors. These include:

- Mixture of nitrates in water/ethanol solution
- Mixtures of nitrates and organometallic compounds in isopropanol (hybrid sol)
- Mixed citrate/nitrate solution (polymeric complex)
- Co-precipitation followed by peptization (gel dispersion in water/ethanol)

At the molecular level, mixing of constituents allows creating metastable coating phases due to rapid cooling of ultrafine splats during deposition [Karthikeyan J. et al. (1997)]. For example, alumina, YSZ, and zirconia were produced by aluminum isopropoxides, zirconium butoxides, zirconium acetate, and yttrium acetate in isopropanol with n-butanol and distilled water as solvents [Jadhav A.D. et al. (2006), Vasiliev A.L. et al. (2006a, b)]. Other researchers have used aqueous solutions of zirconium, yttrium, and aluminum salts [Vasiliev A.L. et al. (2006a), Chen D. et al. (2008a, b)]. Aqueous solutions allow higher precursor concentrations than organic ones. They are also less expensive to produce and safer to store and handle. According to studies related to spray pyrolysis, volumetric precipitation, when heating droplets within the plasma jet, depends on the solute initial concentration and its value relative to the equilibrium saturation concentration. The latter is determined by concentrating the precursor in an evaporator at room temperature until precipitation occurs [Chen D. et al. (2008a, b)]. This allows for the preparation of several solutions, with different concentrations and different viscosities, surface tensions, and specific masses.

According to [Chen D. et al. (2007)] the precursor concentration in solutions can be varied up to its equilibrium saturation. In their study of several YSZ solutions [Chen D. et al. (2008a, b)] considered two different precursor concentrations: one of high molar concentration (2.4 M) that is 4× higher than that of the low concentration (0.6 M). When the initial YSZ precursor is concentrated four times in water, the solution viscosity increases from 1.4×10^{-3} to 7.0×10^{-3} Pa.s, and the surface tension decreases from 4.82×10^{-2} to 4.663×10^{-2} N/m. Both precursors pyrolyze

below 450 °C and crystallize at ~ 500 °C [Vasiliev A.L. et al. (2006b)]. This indicates that the solution precursor concentration has little effect on the precursor pyrolysis and crystallization temperatures. The concentration of precursors produces almost no variation in the solution specific mass and surface tension, but large variations in the solution viscosity. [Chen D. et al. (2010)] also studied the influence of the properties of the liquid phase. The results are quite similar to those observed with suspensions: droplets with a high surface tension and also high boiling point of the liquid phase experience incomplete liquid phase evaporation in the plasma jet, while droplets created from a low surface tension and low-boiling point liquid phase undergo rapid liquid phase evaporation.

As with suspensions, liquid vaporization and further superheating of the liquid vapors consumes energy that would be necessary to pyrolyze the solutions or melt the particles contained in suspension droplets. Moreover, with solutions often precursors undergo significant endothermic processes in the early stages of heating [Muoto C.K. et al. (2011)]. Endothermic behavior at low temperatures will delay particle heating pushing the location where the melting point is reached further downstream. It appears, from the work of [Muoto C.K. et al. (2011)], that the two characteristics that distinguish the desirable YSZ precursor from others is the ability for the YSZ precursor to form dense oxide particles and the absence of endothermic events during pyrolysis. Dense particles are better at making dense coatings because they will impinge on the substrate at near normal incidence (Stokes number effect), and therefore do not introduce porosity associated with the arriving particle. These authors have plasma sprayed binary mixtures of yttrium nitrate or yttrium acetate, combined with magnesium nitrate or magnesium acetate (the 4 combinations being tested), with water as the solvent to produce Y_2O_3–MgO nanocomposites. Moreover, they have added ammonium acetate (CH_3COONH_4) to their solution of nitrates to increase the exothermic of the decomposition process. Acetates can be oxidized by the entrained air in the jet while the nitrate can act as an oxidizer for the acetate in the acetate–nitrate combinations. Coatings obtained with the ammonium acetate were denser than the four other ones. This relatively high density was attributed to the exothermic decomposition characteristics of the precursor combined with the formation of oxide particles that are not too flakey or fluffy to remain entrained in the core of the plasma jet. However, if the processes are too highly exothermic then the oxide particles may shatter in the plasma, causing them to travel along the colder periphery of the jet; also, the increased heat resulting from excessive exothermic reactions may overheat the substrate [Muoto C.K. et al. (2011)].

16.4.1.2 Suspensions

The easiest way to produce a suspension is to make a simple slurry with particles and a solvent, particle sizes varying from a few tens of nanometers to micrometers. The most frequently used solvents are water or alcohols (ethanol or isopropanol) or a mixture of both [Toma F.L. et al. (2006a, b, c, d), Rampon R. et al. (2006a, b, c)]. After stirring, the suspension stability can be tested by sedimentation. Typical values of slurry stability are a few tens of minutes, the stability increasing with the powder load [Toma F.L. et al. (2006a, b, c, d)]. The pH adjustment is also an important factor to be taken into consideration. However, nanosized particles of oxides have the tendency to agglomerate or aggregate, even when stirring the suspension. The stability problem can be overcome by using a suitable dispersant/ surface acting agent that adsorbs on the particle surface and allows an effective dispersion of particles by electrostatic, steric, or electro-steric repulsions. The concentration of the dispersant must be kept relatively low so as not to affect the viscosity of the suspension or induce a shear-thinning behavior [Fazilleau J. et al. (2006)]. This behavior means that when the shear stress imposed by the plasma flow is low, the suspension viscosity is high, decreasing rapidly when the increase of the shear stress as the particle/drop penetrates more deeply into the plasma flow.

Considering that the increase of the wt.% of powder in the suspension gives rise to the increase of its viscosity different products need be added to the liquid phase to modify its surface tension and/or its viscosity [Rampon R. et al. (2006a, b, c, 2008)]. It is also possible to modify the suspension by adding viscous ethylene glycol (the boiling point of which is 200 °C) at the expense of an additional thermal load on the plasma [Oberste-Berghaus J. et al. (2008)]. The addition of binders also controls the suspension viscosity almost independently of the dispersion. For internal feeding against the pressure in axial injection, especially in combustion chambers (HVSFS), the choice of appropriate rheological properties may be critical in terms of a constant suspension feeding and transport, because any sedimentation of the solid phase will cause clogging in the feeding line [Wang Y. et al. (2012), Arevalo-Quintero et al. (2011)] have investigated the behavior of YSZ, Samaria-Doped Ceria (SDC), and mixtures of YSZ and SDC powders in aqueous suspensions. They have determined the optimum dispersant concentrations of three potential dispersant materials (ammonium Polyacrylic Acid, PAA, polyethyleneimine PEI, and 2-phosphonobutane-1, 2, 4-tricarboxylic acid PBTCA) on the suspension stability.

A key point to achieve good spray conditions is to adapt the size distribution of particles within the suspension to the heat transfer from the hot gases and to limit the width of the particle size distribution [Guignard A. et al. (2012)] (as in conventional spraying). This will allow reducing the trajectory dispersion. At last, powders that have the tendency to agglomerate or aggregate, which is often the case with nanosized particles, especially oxides, when prepared by chemical routes, must be avoided.

Examples of TiO_2 and ZrO_2 slurries have been prepared by [Toma F.L. et al. (2006a, b, c, d), Rampon R. et al. (2006a, b, c), Toma F.L. et al. (2006a, b, c, d), Oberste-Berghaus J. et al. (2005)] as well as Al_2O_3 and ZrO_2–Al_2O_3 mixtures by [Fazilleau J. et al. (2003a, b)], and zirconia using a phosphate ester that provides a combination of electrostatic and steric repulsion [Etchart-Salas R. et al. (2007)]. The problem is more complex with a mixture of WC–Co [Oberste-Berghaus J. et al. (2006a, b)], due to the different acid/base properties of both components: WC or more precisely, WO_3 at its surface is a Lewis acid, while CoO is basic. Thus, a complex equilibrium between the dispersing agent and the suspension pH must be found.

Amorphous powders can also be used in suspensions as demonstrated by [Chen et al. (2009a, b)], who prepared Al_2O_3–ZrO_2 amorphous powders by a chemical solution process. Aluminum nitrate ($Al(NO_3)\cdot 9H_2O)_3$ and zirconium acetate ($ZrO(OOCCH_3)_3$ were dissolved in deionized water based on molar volumes to produce a ceramic composition of Al_2O_3–40 wt.% ZrO_2. The obtained solution was heated at 80 °C and stirred continually to get the sol transformed into a dried gel. The dried gel powders were heated to 750 °C at a heating rate of 10 °C/min and then held for 2 h. The phase composition and microstructure of the as-prepared Al_2O_3–ZrO_2 powders heat treated at 750 °C for 2 h were investigated. In the XRD patterns, no crystalline peaks appeared (amorphous powders). To prepare the suspension, the powder, with an average particle size of ~5 μm, was mixed and ball-milled in ethanol using ZrO_2 balls with a loading rate of 50 wt.% for 24 h [Chen et al. (2009a, b)].

16.4.1.3 Colloidal

A new process called "PROSOL" [French Patent FR0453390] has recently been developed by the CEA Le Ripault to avoid the limitations of suspensions with solid particles. It consists of injecting a stabilized suspension of nanometer-sized particles (1–100 nm) in the form of sol–gel or colloidal prepared by the hydrolysis and condensation of metallic precursors. The inorganic polymerization reaction (nucleation and growth) is controlled by varying the chemical conditions (pH, hydrolysis ratio, etc.) and allows preparing colloidal particles (1–100-nm particles) directly dispersed in the liquid medium [Wittmann-Ténèze K. et al. (2008)]. The method of stabilization permits avoiding the use of additives such as dispersant or adding ultrasound during the deposition phase. Consequently, the main advantages are the purity of the deposited material, a very low level of agglomeration and aggregation, well-structured nanometer coatings featuring sizes below 100 nm, simplification of the process, and the capability of reaching sub-micrometric thickness with high deposition efficiency.

16.4.2 Precursor Injection and Fragmentation

In either of suspension or solution plasma spraying the controlled injection and atomization of the suspension or solution into the high temperature combustion or plasma stream represents the critical step on which the quality of the formed coating strongly depends. Specific reference should be made to a review of the fundamentals governing the basic phenomenon involved given in "Chapter 5, Gas and Particle Dynamics in Thermal Spray, Sect. 5.3 Suspension or solution injection in plasma flows." Basically, the simplest approach is to inject the liquid or suspension directly into the combustion or plasma spray stream where it is atomized in-situ through the shear action of the high velocity, high temperature flow. Alternately, the liquid or suspension is atomized prior to its injection into the spray stream using standard gas or mechanical atomization techniques [Lefebvre A.H. (1989)].

The choice between the two approaches, that is, in-situ *atomization* Vs *pre-atomization*, depends on the characteristics of the solution/suspension and the available mechanical and thermal energy into the spray stream. Generally, direct liquid feeding, also referred to as mechanical feeding, is favored with high velocity, high temperature, spray sources such as HVOF, HVAF or DC plasma spraying. Pre-atomization of the liquid precursor feed is favored in situations where the plasma velocity is relatively low as in some Flame or Combustion Spraying techniques and RF Induction Plasma Spraying (RF-IPS).

It is fairly difficult, however, to extrapolate from the literature the interactions between a liquid and a plasma flow because of the presence of steep temperature gradients and considerably high heat flux conditions which are rarely present in conventional ambient temperature flows. The approach commonly used is to adopted basic dimensionless numbers developed for ambient temperature conditions and combined with empirical correction factors that bring model predictions closer to physical data obtained under plasma conditions. Based on this approach solution or suspension atomization in plasma flows is generally correlated using the Weber number and the Ohnersorge number, defined as follows:

$$\text{Weber number } We = \frac{\rho \Delta v^2 d_s}{\sigma_s} = \left(\frac{\text{Inertia force}}{\text{Surface tension}} \right) \quad (16.1)$$

$$\text{Ohnesorge number } Zo = \frac{\mu_s}{\sqrt{\rho_s d_s \sigma_s}} = \left(\frac{\text{Viscous force}}{\sqrt{\text{inertia} \times \text{surface tension}}} \right) \quad (16.2)$$

Where ρ and ρ_s are, respectively, the specific gravity of the gas and solution/suspension, σ_s surface tension, and μ_s viscosity of the solution/suspension. Physically, the Weber number, We, represents the ratio of [Inertia force/surface tension force], while the Ohnesorge number, Zo, represents the ratio of (Viscous force/(inertia force $\times$ surface tension)$^{1/2}$). Fragmentation of solution or suspension to drops/droplets depends essentially on the Weber number, with the different possible breakup mechanisms classified as:

- *Bag breakup* ($12 < We < 100$), deformation of the drop as a bag-like structure that is stretched and swept off in the direction of the flow.
- *Stripping breakup* ($100 < We < 350$), thin sheets are drawn from the periphery of the deforming droplets.
- *Catastrophic breakup* ($We > 350$) that is a multistage breaking process [Marchand C. et al. (2008)].

Drop/droplet vaporization depends in turn on the thermophysical properties of the liquid phase in the solution/suspension such as, c_{ps} specific heat, T_{vs} vaporization temperature at atmospheric pressure and, L_{vs} Latent heat of vaporization (see Table 16.2).

Whatever the liquid/suspension injection mode used, the main questions that needs to be addressed are; how to control the atomization or fragmentation process? and what happens when the injected liquid droplets interact with the plasma flow?

16.4.2.1 In-Situ Precursor Atomization

In-situ atomization is simply carried out by injecting the liquid solution or suspension precursor directly into the spraying gas stream using a variable speed peristatic pump or simply placed in a pressurized container in which the liquid is stored and forced through a nozzle of specified internal diameter, d_n which typically varies between 50 to 300 μm. At the injector exit, a liquid jet is generated with a diameter of 1.2 to 1.9$\times d_n$ [Delbos C. (2004), Etchart-Salas R. (2007), and Fauchais P. (2008)]. As schematically illustrated in Fig. 16.26a. Rayleigh–Taylor type instabilities develop after a distance of 100 to 150 $\times d_n$, leading to the formation of droplets with a mean drop diameter between 1.5 and 2.0 $\times d_n$ (Fig. 16.26b) generally referred to as "primary atomization." A simple mass balance on the system leads to the following relationships for the mass feed rate of the precursor, $\dot{m}_s$ and the pressure drop across the feed nozzle, Δp.

$$\dot{m}_s = \rho_s v_s \frac{\pi d_n^2}{4} \quad (16.3)$$

$$\Delta p = \frac{1}{2} f \rho_s v_s^2 \quad (16.4)$$

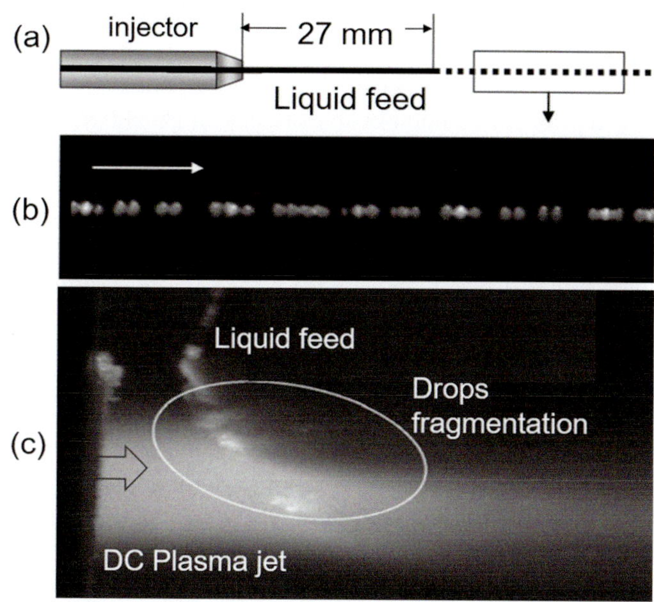

Fig. 16.26 Liquid flow exiting an injector (**a**) general view, (**b**) primary atomization of the liquid stream (**c**) primary and secondary atomization of the liquid in DC plasma jet [Etchart-Salas (2007) and Fauchais P. et al. (2008b)]

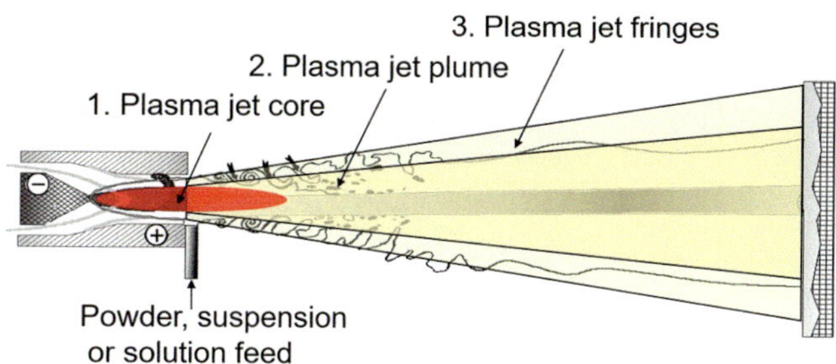

Fig. 16.27 Schematic of the three major zones of DC plasma jet flow

where:

ρ_s specific mass of the solution/suspension
v_s exit velocity of the solution/suspension
d_n injector nozzle i.d.
f friction coefficient, in the range of 0.6 to 0.9 depending on the nozzle shape, length and roughness

For a mass flow rate of 0.47 cm^3/s of water at room temperature using an injection nozzle i.d. of 150 μm, the diameter of the generated drops will be around 300 μm, and a pressure difference of 0.5 MPa, which goes up to 41 MPa when reducing the internal diameter of the injector to 50 μm leading to technological constraints.

Once the formed drops are entrained by the high energy spray stream, they are further fragmented by "secondary atomization" caused by intense shear stresses in the main spray stream. As shown in Fig. 16.26c, for the case of water injection in an Ar/H$_2$, DC plasma jet, this leads to the formation of clouds of materials (liquid and/or solid) composed of a compact "head" of suspension coupled to a "tail" made of small droplets or solid particles. The clouds are equally spaced between themselves, the average distance between them corresponding to the wavelength of the jet instabilities before entering the plasma. This signifies that the initial velocity of the suspension is kept along the injection trajectory after its penetration into the spray stream [Etchart-Salas R. (2007), Etchart-Salas R. et al. (2007) Fauchais P. et al. (2008b)].

Considering DC plasma jet, three major zones can be identified in the flow as shown in Fig. 16.27 [Etchart-Salas R. (2007), Wang Y. et al. (2012)].

- Zone 1, the plasma jet core, where the temperature and velocity of the plasma stream are at their maximum value
- Zone 2 the plasma plume, where the heat and momentum capabilities from the plasma are significantly reduced compared to those in zone 1
- Zone 3 the plasma fringes, where fragmentation can occur though droplet heating will be insufficient

In order to make full use of the high energy level in the core of the plasma jet, it is important to inject the liquid/suspension in zone 1, as close as possible to the torch nozzle exit. When considering liquid injection into a plasma jet, the trajectory and penetration of the injected liquid into the plasma stream is a function of the ratio of the momentum of the injected solution/suspension $\rho_s v_s^2$ to that of the plasma jet ρv^2 [Fazilleau J. et al. (2006), Fauchais P. et al. (2008b), Salas R. (2007), Etchart-Salas R. et al. (2007)]. In general, an "appropriate" liquid penetration is achieved for;

$$\rho_s v_s^2 \gg \rho v^2 \text{ (at least } 10 \times \text{greater)} \qquad (16.5)$$

It is also important to take into account that upon primary fragmentation, formed drops are smaller and thus their velocity has to be sufficiently high to maintain the same level of momentum for further penetration into the plasma flow. Once the liquid/suspension stream or primarily fragmented drops reaches this zone, its further fragmentation into droplets of a few micrometers is almost instantaneous followed by flash vaporization of the liquid component of the drops/droplets [Fazilleau J. et al. (2006), Etchart-Salas R. et al. (2007)].

Fragmentation and vaporization times have been calculated for an Ar/H$_2$ plasma jet (the jet was supposed stationary: that is to say with no arc root fluctuation) versus the drop average diameter [Fazilleau J. et al. (2006)]. Results given in Fig. 16.28 show that fragmentation is about two orders of magnitude faster than vaporization that strongly depends on the mass of each drop/droplet being function of d_p^3. In the

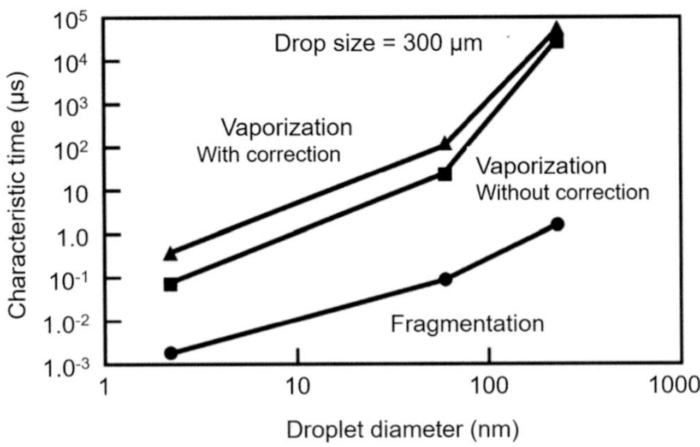

Fig. 16.28 Evolution of the fragmentation and evaporation as function of drop diameter in an Ar/H$_2$ DC plasma jet [Etchart-Salas R. (2007)]

presented case the average size of the drops formed by primary atomization is about 300 µm while the diameter of the fragmented droplets was a few micrometers. The whole transformation (fragmentation & vaporization) is achieved about 10 to 40 mm downstream the injection location of the precursor into the mains spray plasma stream depending on the solvent used, plasma torch power level, and plasma flow velocity [Etchart-Salas R. et al. (2007)]. The in-flight vaporization of the liquid/solvent content of the formed drops or droplets, results in turn in the rapid local cooling of the plasma flow due to thermal loads associated with the liquid vaporization process that is proportional to tis liquid mass and latent heat of vaporization, as well as the energy necessary for the further heating of the evolved vapors up to the main stream gas/plasma temperature. Considering that 20–40% of the energy available in the main spray stream is consumed in the fragmentation and droplet vaporization process, it is not surprising to note a significant local cooling of the spray stream in this process and the need to be shortened the spraying distance, to 30–50 mm compared to 100–120 mm commonly used with conventional DC plasma spraying [Etchart-Salas R. et al. (2007), Vaßen R. et al. (2010)]. Alternately, torch power should be increase by 30 to 40% in solution and suspension plasma spraying in order to compensate for the added thermal load necessary to vaporize the liquid carrier [see Chap. 5, Gas and Particle Dynamics in Thermal Spray, Sect. 5.3.3 Cooling the Plasma Flow by the Liquid]. A description of the sensitivity of liquid injection to arc root fluctuations for plasma jets can also be found in [Chap. 5, Sect. 5.3.2.4 Influence of Arc Root Fluctuations] The effect being far more important than that observed when injecting particles in conventional spraying, which requires use of low fluctuating plasma jets when spraying liquids. Unfortunately, the strong fluctuations with H$_2$ in the plasma forming gas require the use of He as secondary gas, which reduces the voltage and thus also the

power level. Such fluctuations are also observed when using nitrogen as primary forming gas.

For HVOF guns working with C$_2$H$_2$ or liquid fuel, the injection of the solution or the suspension should be made radially, downstream of the nozzle exit level. Using organic solvent does not help, through its combustion, to compensate for the energy lost in its vaporization and transformation into plasma. Indeed, combustion can only occur in areas where temperature is below about 3000 K, and where a sufficient quantity of oxygen has been entrained. Alternately, in high velocity suspension flame spraying (HV-SFS), precursor injection can also be made axially in the combustion chamber which might require the increase of the length of the combustion chamber from its conventional length of 60–90 mm in HVOF spraying to 150–300 mm in order to insure complete evaporation and combustion of the solvent prior to the exiting the combustion chamber [Erne et al. (2012)].

In the following a few examples of in-situ atomization of solution and suspensions precursors are given for radial and axial solution/suspension injection modes.

Radial Injection Mode

To illustrate the droplet formation in the case of radial injection of the liquid stream into a DC plasma flow, examples of shadowgraph images showing the interaction of a water jet with an Ar/H$_2$ DC plasma jet are presented in Fig. 16.29. The image given in Fig. 16.29a is taken with laser illumination and pulse duration of 8 ns, while Fig. 16.29b shows the average of 50 shadowgraph images triggered at the same voltage level at a frequency of 6 Hz, taking into account plasma jet instabilities, allowing the visualization of average trajectories of the drops.

In order to elucidate the dependence of the liquid atomization on the Weber number of the flow, Shadowgraph images of the atomization of water jet with DC plasma jets of pure Ar and of Ar/He/H$_2$, are shown in Fig. 16.30 after [Damiani D. (2012) and Meillot E. et al. (2011)]. The corresponding Weber number for each of these two cases was calculated assuming that all the energy supplied is distributed uniformly in the plasma jet section at the nozzle exit giving rise to a $We = 41$ for the pure Ar plasma jet and $We = 495$ for the Ar/He/H$_2$ flow. It can be readily seen that the primary fragmentation is by far more important with the Ar/He/H$_2$ plasma with a high We (Fig. 16.30b) compared to that with pure Ar, low We (Fig. 16.30a). Moreover, the drops formed by the primary fragmentation are farther fragmented into smaller droplets, 15 mm downstream of the flow, in the Ar/He/H$_2$ plasma case compared to the pure Ar plasma. It can also be noted in Fig. 16.30b that fragmentation increases from the jet fringes to the jet axis.

Detailed diagnostics of the liquid in-situ atomization process with radial injection into a DC plasma jet was reported by [Marchand C. et al. (2008), Fauchais P. and A. Vardelle

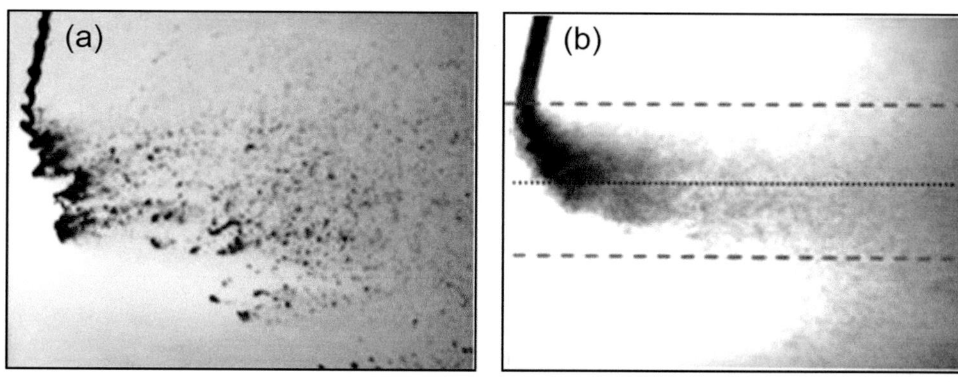

Fig. 16.29 Water jet interaction with (33 slm Ar + 10 slm H$_2$) DC plasma jet (**a**) 8 ns single exposure time (**b**) average of 50-images triggered at the same arc voltage level [Bertolissi G. et al. (2012)]. (Reprinted with kind permission from Springer Science Business Media copyright © ASM International)

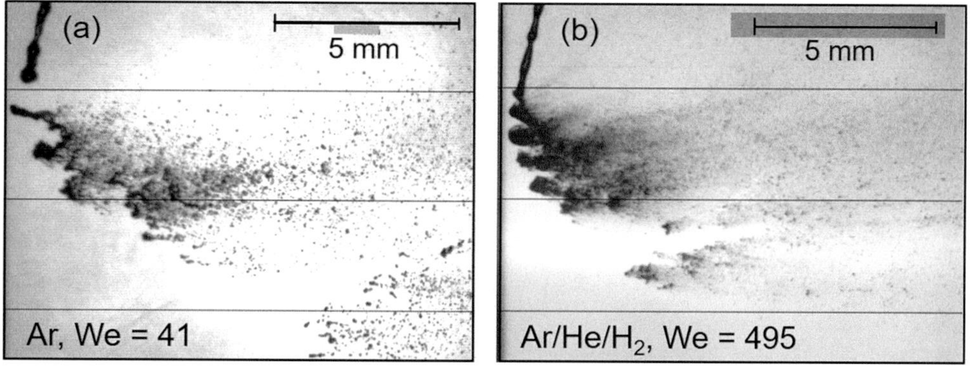

Fig. 16.30 Visualization of a water jet injection within two DC plasma jets with different Weber numbers: (**a**) Ar We = 41, (**b**) Ar/He/H$_2$ We = 495. Horizontal lines correspond to the nozzle axis and limits (torch i.d. 6 mm) [Damiani D. (2012)]. (Reprinted with kind permission from Dr. Damiani)

(2011) and Fauchais P. et al. (2013)] using laser shadow-graph imagery techniques with a double-pulsed Nd:YAG laser (532 nm wavelength with 8 ns pulse duration) with a beam diffuser for the backlight illumination of the liquid material. The detection system consisted of two Charge-Coupled Device (CCD) cameras with 1376 × 1040 pixels and 12 bits resolution and, a programmable hardware timing unit controlled the synchronization of the laser with the cameras [Fauchais P. and A. Vardelle. (2011)]. This system made it possible to discriminate droplets with diameters of about 2–3 μm at the best, and measure their number, velocity and diameter distributions in a relatively small volume (2.25 × 1.8 × 1 mm). Results given in Fig. 16.31 show the penetration, fragmentation and dispersion of a water-based, submicron, YPSZ suspension (10 wt. %, 0.5–1.5 μm) into an Ar/H$_2$ DC plasma jet. Figure 16.31b shows a blow-up of small area indicated by a red square in Fig. 16.31a. It should be noted that suspension droplets observed in these images would not be detectable by this technique once their liquid content was evaporated due to the considerably smaller grain size that would be generated.

Figure 16.32 presents the evolution of the corresponding number droplet size distribution (DSD) along the plasma jet axis, showing their fragmentation. However, as previously underlined, the real understanding of different phenomena controlling suspensions or solutions spraying requires that droplets or particles could be followed at least down to a few tenths of micrometers. It is also possible with this setup to visualize the influence of the plasma enthalpy and that of solvent on the drops and droplets fragmentation [Fauchais P. et al. (2013) and Fauchais P. et al. (2013)]. Figure 16.33 shows the droplet size distribution (DSD) by number, for ethanol-based suspension injected into DC plasma jets, with an anode nozzle diameter of 6 mm, and different gas compositions and specific enthalpies (a) 35 slm Ar, specific enthalpy = 9.1 MJ/kg (b) 35 slm Ar +10 slm H$_2$, specific enthalpy = 21 MJ/kg (c) 45 slm Ar + 3 slm H$_2$ + 45 slm He, specific enthalpy = 19 MJ/kg). It can be noted that with the increase of the specific enthalpy of the plasma the number of droplets with a diameter > 5 μm drops rapidly with the evaporation of the ethanol solvent diminishes in about 15 mm downstream of the point of injection of the

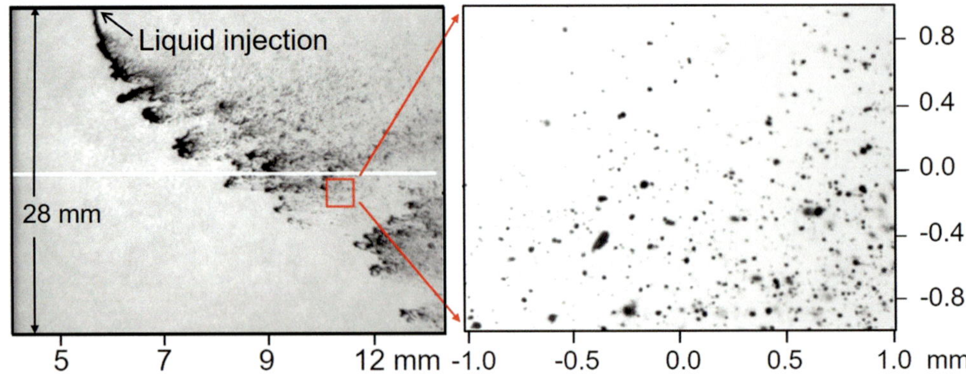

Fig. 16.31 Shadowgraph images of the in-situ fragmentation water-based YPSZ suspension of into an Ar/H_2 DC plasma jet (33 slm Ar + 10 slm H_2, anode–nozzle i.d. 6 mm, arc current 600 A), (**a**) general view, 8 ns exposure (**b**) Zoom on (2.25 × 1.18 mm) area as indicated by red square [Fauchais P. and A. Vardelle (2011)]

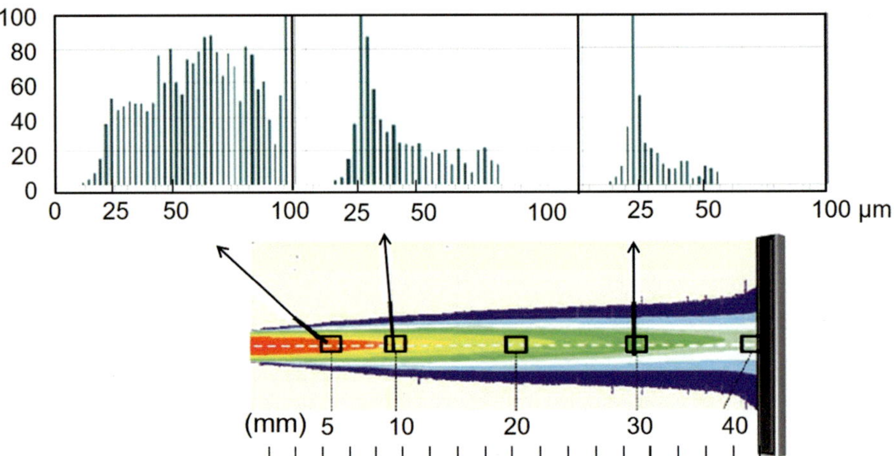

Fig. 16.32 Schematic of the plasma jet with the locations of measurements by shadowgraph imagery and the corresponding Droplet Size Distributions (DSD) obtained (spray conditions as in Fig. 16.31a) [Fauchais P. and A. Vardelle. (2011)]

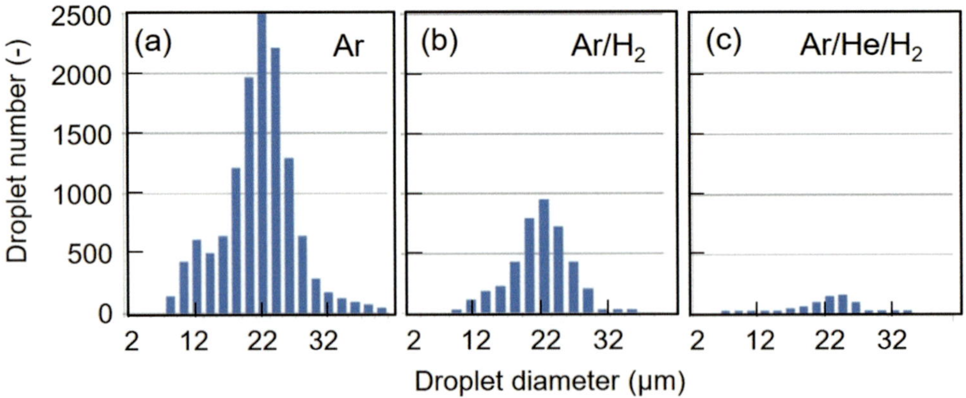

Fig. 16.33 Droplet Size Distribution of ethanol-based droplet diameter (>5 µm) 15 mm downstream of their injection for (**a**) Ar (**b**) Ar/H_2 (**c**) Ar/He/H_2 DC plasma jets [Fauchais P. et al. (2013)]

suspension. When using water instead of ethanol as solvent, droplets formed from water still exist 40 mm downstream of the injection as can be observed in Fig. 16.34 for a (45 slm Ar + 3 slm H$_2$ + 45 slm He) DC plasma with specific enthalpy = 19 MJ/kg. All visible ethanol-based droplets (>5 μm) are observed to vanish at a distance of 20 mm from the point of injection.

Axial Injection Mode

Axial injection of solutions and suspensions in DC plasma torches is limited to small number of multi-electrode torch designs such as the Mettech Axial III plasma torch shown in Fig. 16.35 (see also Chap. 8, Sect. 8.3.6 Multi-electrode DC Plasma Torch, Figs. 8.34 and 8.35). In this case, the liquid or suspension, need to be injected into the plasma stream at a relatively low velocity sufficient for its penetration into the high-energy flow. As the injected stream mixes with the three converging plasma jets of the Axial III torch it is entrained by the high velocity plasma flow, fragmentated and its liquid component partially or completely evaporated [Fauchais P. et al. (2011)]. Considering that the power level in these torches can be up to 2.5 to 3× that of the conventional stick-type cathode DC plasma guns, no significant local cooling of the plasma flow is observed with plasma velocities reaching 500 and 770 m/s [Oberste-Berghaus J. et al. (2008)].

Generated nano particles and molten droplets entrained by the flow are projected towards the substrate on which they are deposited forming the coating.

The injection of the liquid precursor, whether solution or suspension, in HVOF or HVAF spraying torches can be carried out either radially into the high velocity spray stream at the exit level of the combustion chamber, or axially into the combustion chamber, see Chap. 7, Combustion Spraying, Sect. 7.3.4 High Velocity Suspension Flame Spraying (HV-SFS).

In the case of radial injection, as the liquid/solvent component of the solution or suspension, evaporates it significantly cools down the flame resulting in strong disturbance of the free expanding hot gas stream. Torches such as the DJ-2700 (Sulzer Metco, Wohlen, Switzerland) or Top-Gun (manufactured by GTV, Düsseldorf, Germany) with modified combustion chambers have been used with axial injection of the solution or suspension in the HV-SFS process [Gadow R. et al. (2008a, b) and Domongo E. et al. (2009)]. A schematic of the schematic of the system used by [Gadow R. et al. (2008a, b, 2009)] is given in Fig. 16.36. It consists essentially of a modified TopGun-G torch in which the suspension is feed axially and concentrically into the combustion chamber from a pressurized holding tank. Provision is also made for a second container holds pure solvent to flush the suspension line and injection nozzle when shutting down the process. The principal modifications are in the design of the suspension injection nozzle that is capable to evenly inject the suspension into the combustion chamber of the HV-SFS torch.

Propane and ethene were preferentially chosen as combustion fuels as they allow stable operation. Basically, acetylene can also be used mostly when spraying oxide materials, as it delivers higher flame temperatures. However, due to the insertion of liquid fuel, the combustion chamber pressure rises significantly, causing instabilities in the acetylene flow. Therefore, further improvement of the spray equipment is necessary to ensure a stable process with acetylene. Isopropanol was chosen as an alternate organic solvent as it

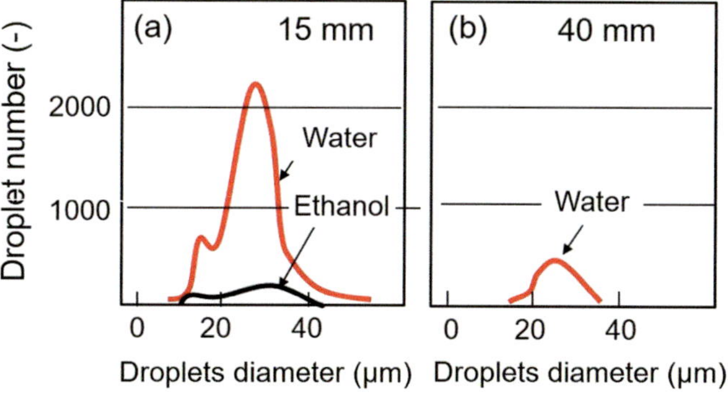

Fig. 16.34 Droplet Size Distribution of ethanol-based droplet diameter (>5 μm) in DC plasma jet (45 slm Ar + 3 slm H$_2$ + 45 slm He), specific enthalpy = 19 MJ/kg at (**a**) 15 mm and (**b**) 40 mm downstream of their point of injection [Fauchais P. et al. (2013)]

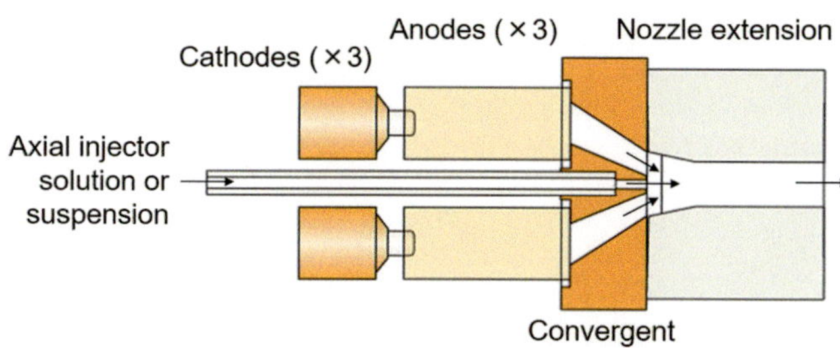

Fig. 16.35 Schematic of the Northwest Mettech Axial III DC plasma torch with central injection. (Courtesy of Northwest Mettech)

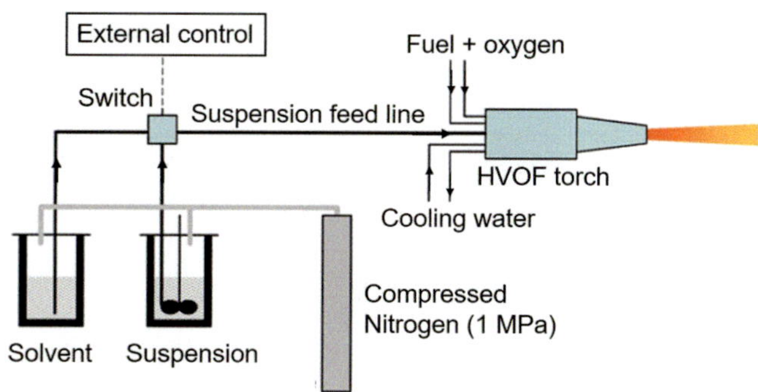

Fig. 16.36 Scheme of the high velocity suspension flame spraying (HVSFS) system used by [Gadow R. et al. (2008a, b)]

delivers fairly high enthalpy values during the combustion process. In most cases some water (10–50 wt.%) was added to facilitate the chemical stabilization of the suspension. However, even small quantities of water decrease the flame temperature significantly and show a strong impact on resulting coating properties. Several suspensions consisting of an organic solvent and a solid phase consisting of a micron or a nanopowder titanium oxide (n-TiO$_2$), chromium (III) oxide (n-Cr$_2$O$_3$), yttrium stabilized zirconia (n-YSZ) and a n-hydroxyapatite (n-HAP) have been prepared and sprayed using HV-SFS Technology. According to [Gadow R. et al. (2008a, b)] direct spraying of suspensions shows much higher flexibility in combining and processing different materials and is far less expensive than alternative approach using agglomerated (nano- and micron-sized) powders.

16.4.2.2 Precursor Pre-Atomization

A wide range of liquid and suspension atomization techniques are available as described in standard textbooks on the subject such as Lefebvre A.H. (1989) "Atomization and Sprays." Essentially two broad approaches are commonly used depending on the rheological properties of the feed material.

- *Spray Atomization, or gas atomization* through the use of an atomizing medium such as a compressed gas that supplies the energy required for the atomization of the liquid or suspension. Also referred to as pneumatic or gas or two-fluid atomization. In specific cases such as DC plasma spraying, the plasma gas stream can be used as the high energy atomizing medium in which the solution or suspension is feed as a liquid jet injected into the high velocity, high shear zone of the plasma stream in which it is atomized by the plasma gas.
- *Mechanical Atomization* where the liquid or suspension is pressurized and pushed through an appropriate nozzle resulting its fragmentation into small droplets. This

approach is also used to provide the liquid/suspension with the necessary momentum energy in order to penetrate the combustion or plasma flow. On reaching to core high velocity region of the flow the liquid is fragmented before vaporization using the kinetic energy of the plasma. Mechanical atomization can also be accomplished using a vibrating device or ultrasound source.

In SPS and SPPS gas atomization is mostly used injecting a low-velocity liquid/or suspension inside a nozzle where it is fragmented high velocity gas stream expanding within the body of the nozzle. For liquids with viscosity between a few tenths to tens of mPa.s, their breakup to drops depends, as discussed earlier, on the Weber number that physically represents the ratio of the forces exerted by the gas flow on the liquid, to the surface tension forces. Atomization also depends, to a lesser degree on the Ohnesorge number, Zo, including the effect of viscosity of the liquid/suspension. Size distribution of the generated droplets depends to a large extent on the Ratio of the Gas to Liquid/Suspension volume feed rats (RGS) that is generally >100, or the Atomizing gas to Liquid mass Ratio (ALR) that is generally <1.0 increasing RGS leads to a decrease of the PSD of the generated droplets at the expense of increased perturbation and local cooling of the thermal spray stream, that is, plasma flow. While increasing the weight percentage of the solid in the suspension broaden the PSD.

In the following examples are given of the adaptation of pre-atomization approach for the with DC and RF-ICP plasma sources.

Gas-Atomized Liquid/Suspension Injection in DC Plasma Jet

An example of the interaction between a gas-atomized, ethanol liquid jet radially injected into Ar/H$_2$ DC plasma jet (45 slm Ar/15 slm H$_2$), 7 mm nozzle i.d. and 600 A arc current is given in Fig. 16.37. This shows the wide spread

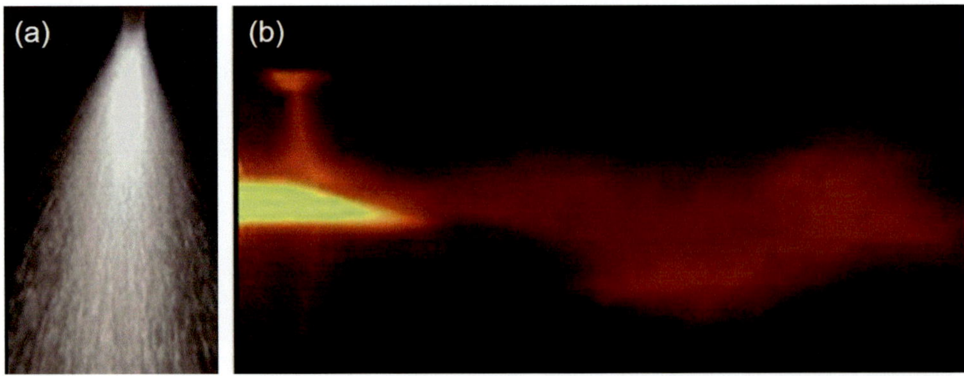

Fig. 16.37 Photograph of (**a**) gas-atomized ethanol jet and (**b**) its interaction with the flow of) DC plasma jet, in which it is radially injected near the nozzle exit [Fauchais P. et al. (2008a, b)]

of the gas atomized stream (Fig. 16.37a) and its significant perturbation of the DC plasma flow with most of the spray droplets entrained in the fringes of the jet as shown in Fig. 16.37b.

Gas-Atomized Liquid/Suspension Injection in RF Induction Plasmas

With RF plasmas, flow velocities in the discharge cavity are generally <100 m/s that is too low for fragmentation to be carried out efficiently by the plasma gases. Solution and suspension spraying require in this case the use of a gas atomization probes for the feeding of the liquid/suspension the discharge stream. Schematic representation of some of the concepts used for gas atomization are given in Fig. 16.38a, b These show, respectively, internal probe designs in which the solution/suspension comes in contact with the high-pressure atomizing gas either internally, or externally to the probe. Such probe designs have been equally used for the feeding and dispersion of dry fine and ultrafine powders using gas atomization concepts. Photographs of the dispersed stream generated by some of these probes are given in Fig. 16.38c–e that show a close control on the "half-angle, α" of the dispersion cone generated at the exit of the probe tip that ensures an adequate spread of the dispersed particles or generated droplets into the plasma with dispersion angles ($\alpha \leq 15°$) to avoid spraying the particles/droplets on the inner walls of the plasma confinement tube in the RF Induction Plasma Torch. As a reference, the spray half-angle shown in Fig. 16.38e, $\alpha = 8.5°$.

In contrast to the case of radial injection of an atomized liquid stream into a DC plasma jet, the injection of an atomized liquid or suspension into an RF-ICP source is much simpler due to the accessibility of axial injection mode and the ability to have the atomization probe tip centrally located into the discharge zone as shown in Fig. 16.39a, b. These two figures represent, respectively, the streamlines and the temperature field in the discharge region of an RF-ICP torch in which a carrier/atomizing gas in axially injected into the center of the discharge. Details of the modeling study are given in Chap. 10 Induction Plasma Spraying, Sect. 10.4.5 Plasma and Particle Dynamics Modeling, Fig. 10.55. These show the carrier/atomizing gas being fulling entrained and heated by the annular plasma stream. A corresponding photograph, though at different operating conditions, pure Ar plasma at low power, 7 kW, given in Fig. 16.39c show similar confinement and entrainment of the probe gas and powder feed into the central region of the discharge.

A good example for the use of the SPS Technology with axial injection into an RF-ICP torch is the production of Hydroxyapatite (HA) (Ca_{10} (PO_4)$_6$ (OH)$_2$) coatings or micrometer-sized powders described by [Bouyer E. et al. (1995, 1997a, b) and Gitzhofer et al. (1997)]. HA material in the form of colloidal water-based suspension was synthesized using "wet chemistry" following the method of [Tagai and Aoki (1980)] involving the reaction;

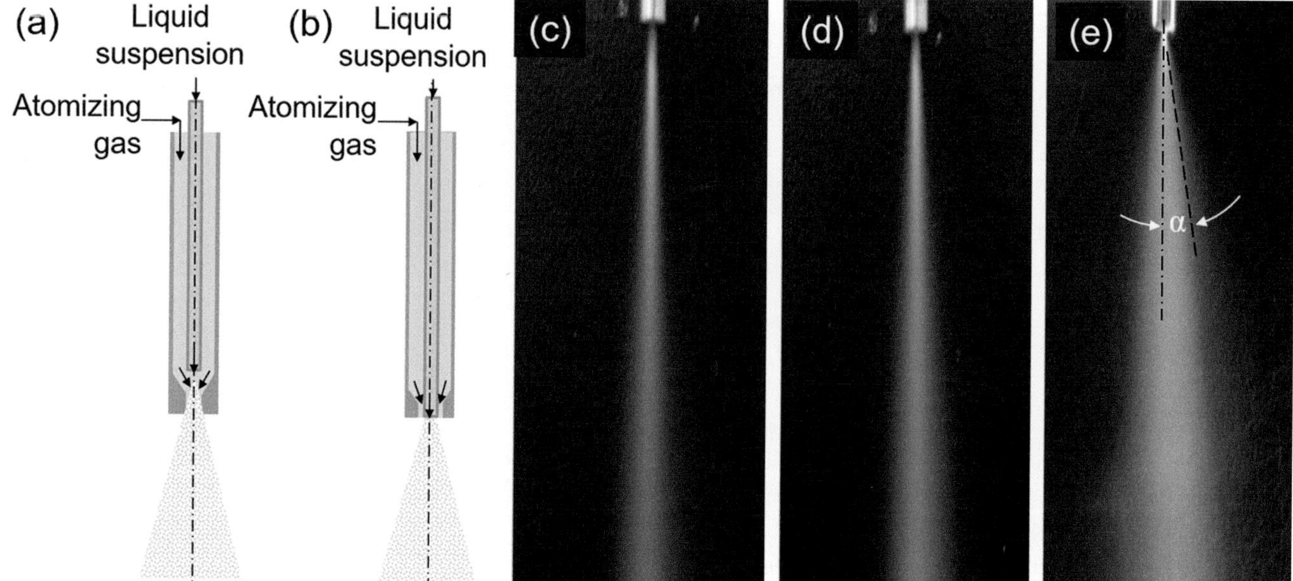

Fig. 16.38 Schematics of gas atomization nozzle designs (**a**) internal mixing (**b**) external mixing (**c–e**) photographs of particle/droplet jet dispersion reflecting different probe design and dispersion gas flow rates, angle $\alpha = 8.5°$. (Results supplied by Tekna Plasma Systems Inc. 2012)

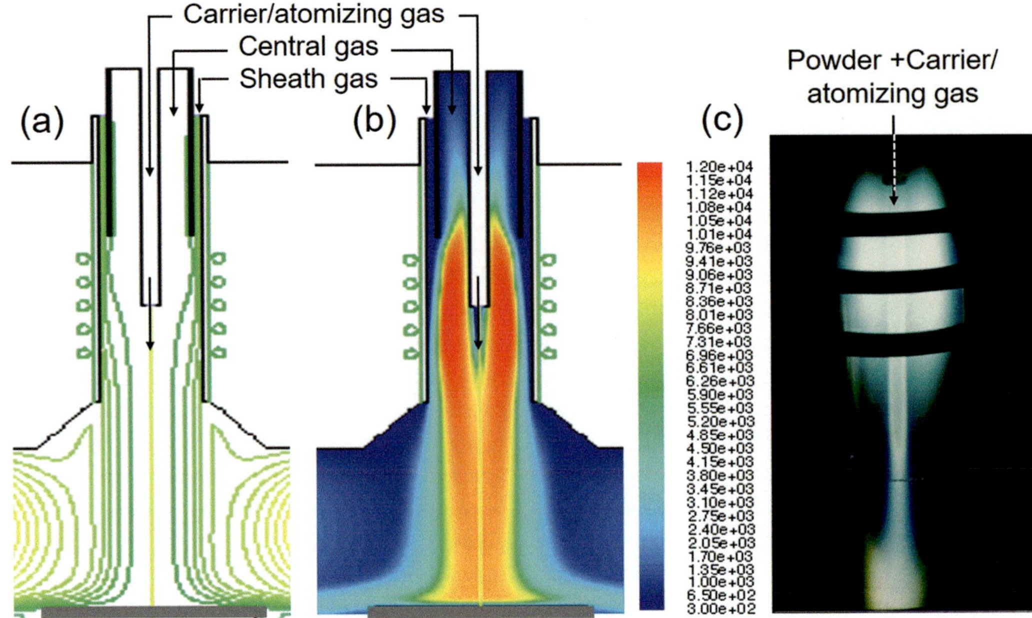

Fig. 16.39 CFD modeling of (**a**) flow (**b**) temperature fields in an RF-IPS torch system with central axial injection of atomizing gas and (**c**) photograph of a similar set up in a quartz tube showing the ICP discharge and he central injection of ultrafine zirconia powder

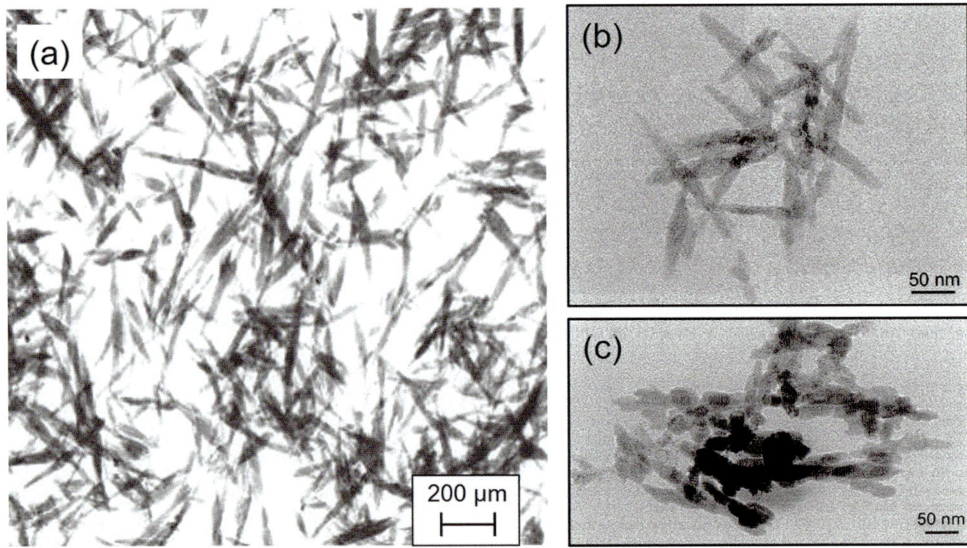

Fig. 16.40 TEM images of HA as synthesized crystals morphology (**a**) as synthesized (**b** and **c**) higher magnification images [Bouyer et al. (1997a, b, 2000)] with kind permission

$$10Ca(OH)_2 + 6H_3PO_4 \rightarrow Ca_{10}(PO_4)_6(OH)_2 + 18H_2O$$

$$(16.6)$$

TEM photograph of the microcrystalline HA given in Fig. 16.40a. show the characteristic "needle-shaped" HA crystals, with high magnification images of the crystal nanostructure given in Fig. 16.40b, c. The corresponding XRD pattern of the "as-synthesized" powder given in Fig. 16.41a showing typical (002) and (211) HA peaks. The

suspension with a solid content of around 40 wt.% had an apparent viscosity that was too high to atomize. The addition of a polymeric surface active agent, Darvan-7 (<1 vol.%) allowed for the stabilization of the suspension and a reduction of its room temperature viscosity to 0.4 Pa.s with an acceptable flowability through the atomizing probe.

As illustrated in Fig. 16.42a, a peristatic pump was used for the controlled feeding of the suspension (5 g/min. HA Dry-basis) to the atomization probe axially located in the RF-ICP torch with its atomization tip in the center of the

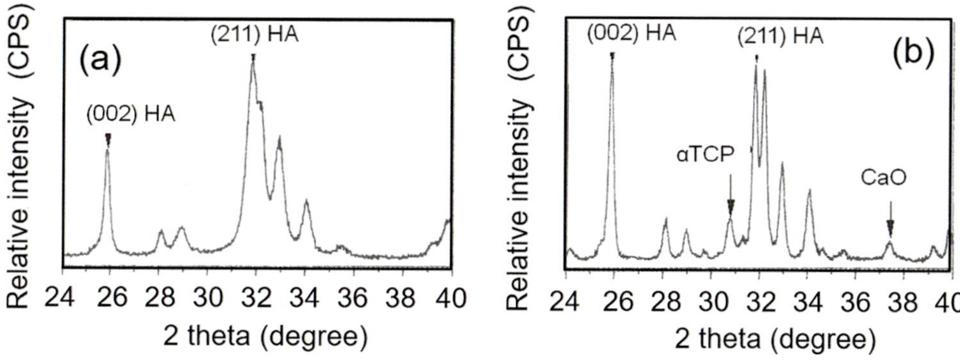

Fig. 16.41 XRD pattern of HA crystals (**a**) as synthesized (**b**) HA coating made by SPS RF-ICP torch. [Bouyer et al. (1997a, b)] with kind permission

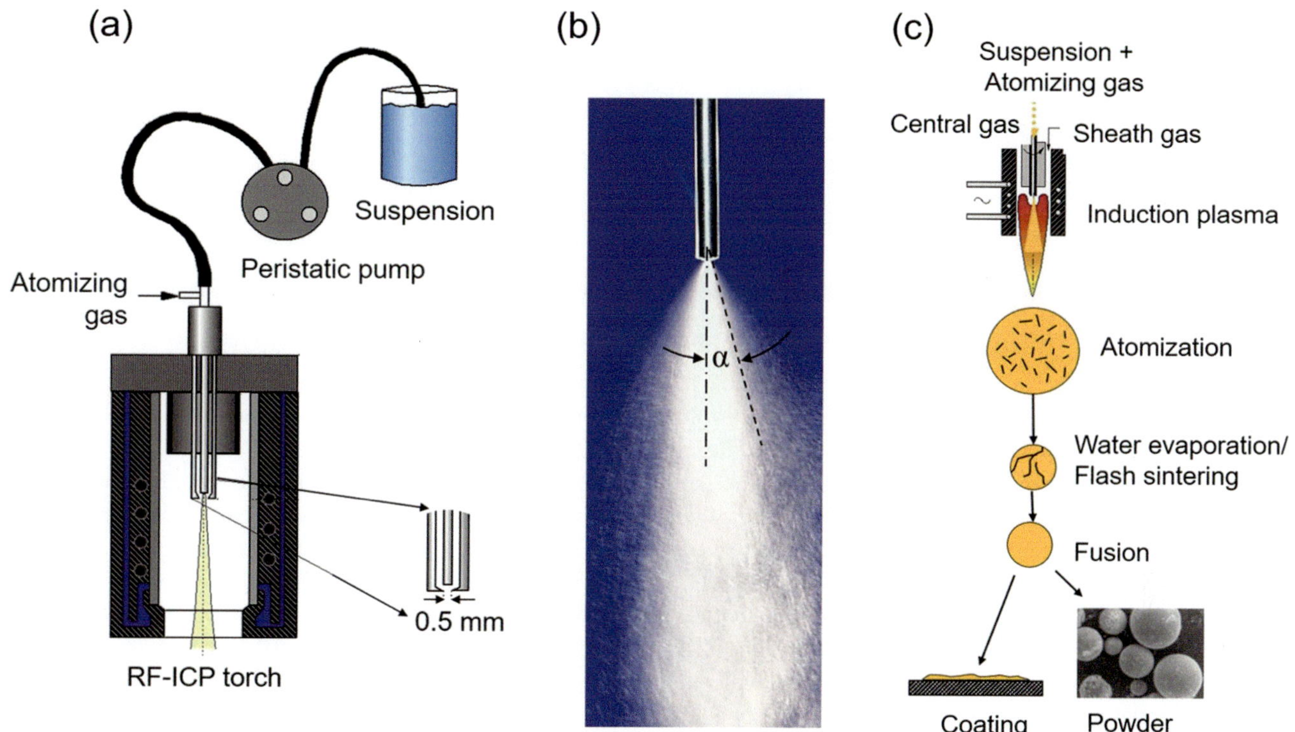

Fig. 16.42 (**a**) Schematic of the setup used for SPS of hydroxyapatite (HA) axially dispersed into the discharge zone of RF-ICP torch. (**b**) Photograph of the gas atomized HA suspension. (**c**) Steps for SPS coating or spherical powder production. [Bouyer et al. (1997a, b)] with kind permission

induction coil region. Pure argon was used as atomizing gas at a flow rate of 5 slm. Figure 16.42b shows a typical image of the atomized droplet flux from the probe tip in a cold simulation in the absence of the plasma discharge with a dispersion half-angle of 15°. The torch central gas flow rate was 45 slm (Ar) and Sheath gas 90 slm (Ar/H_2) or Ar/O_2). Torch power was set at 38.5 kW, and reactor chamber pressure 40 kPa. As illustrated in Fig. 16.42c, once the individual suspension droplets, with a mean droplet size around 80 µm are entrained by the plasma, their water content is evaporated generating a porous gran of the HA of a diameter in the 40 µm

range and a porosity of around 30 vol.%. As indicate in the lower part of Fig. 16.42c, depending on the plasma operating conditions, the outcome of the process can be recovered as porous powder, or if allowed a longer dwell time in the plasma, as a sintered or fused powder with a mean particle diameter in the 20 to 30 µm depending on the on the level of sintering or fusion of the particles.

Electron micrographs of the formed HA powder particles are given in Fig. 16.43a with a wide size range below 40 µm. High magnification photograph given in Fig. 16.43b shows the particles to be spherical, fully melted, with occasional

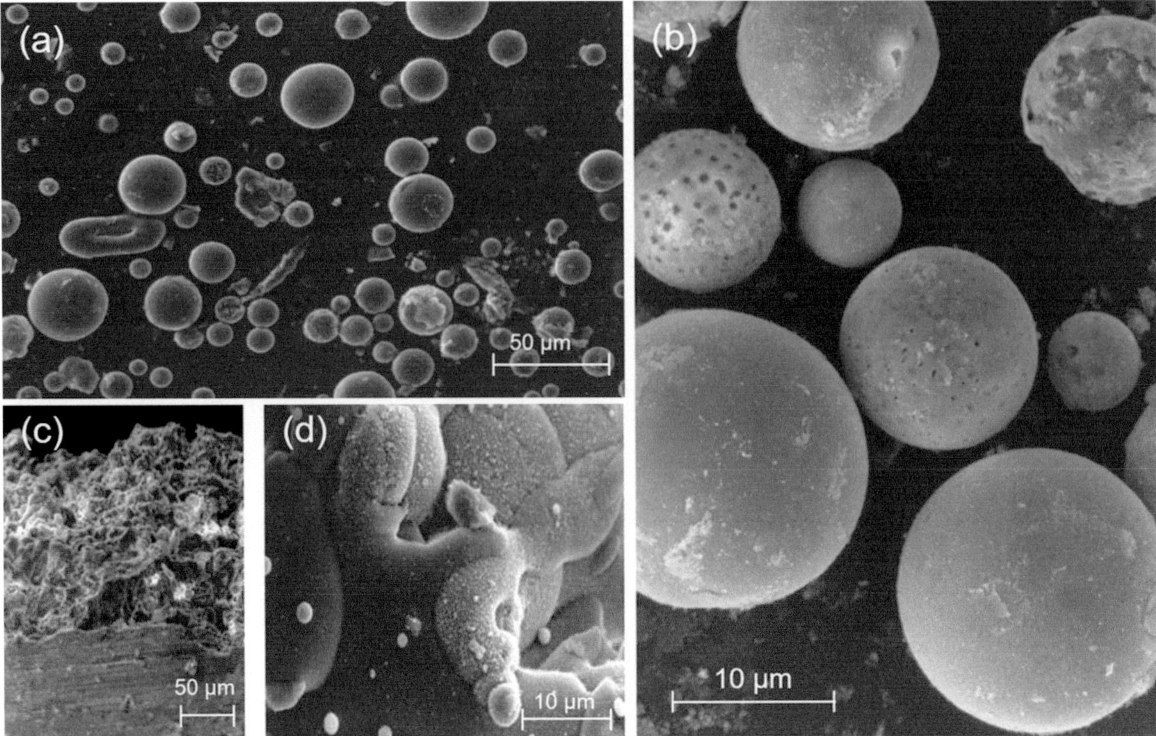

Fig. 16.43 Electron micrographs of hydroxyapatite (HA) powders and coating prepared by SPS using RF-ICP torch. (**a**) HA powder morphology prepared by SPS. (**b**) High magnification HA powder. (**c**) Coating cross-section. (**d**) Coating surface morphology. [Bouyer et al. (1997a, b)] with kind permission

pore structure on its surface that could result from the evolution of residual water vapor escaping the particle. By placing a substrate coaxially with the flux of the sprayed droplets they impact on its surface forming micro splats building a rather porous micro/nanostructured coating. Electron micrographs showing a cross-section of the coating is given in Fig. 16.43c. with a top view of the surface of the coating showing the splat overlapping given in Fig. 16.43d. The corresponding XRD pattern of the coating, given in Fig. 16.41b, reveals the appearance of two new relatively small peaks corresponding to α-Tri-Calcium Phosphate ($Ca_3(PO_4)_2$) and calcium oxide (CaO) indicating a small degree of decomposition of the HA during the plasma treatment which can be described by the following reaction:

$$Ca_{10}(PO_4)_6(OH)_2 \rightarrow 3Ca_3(PO_4)_2 + CaO + H_2O \quad (16.7)$$

Oxygen as sheath gas helps in limiting the decomposition of HA during plasma treatment. As a general rule, this decomposition can be either avoided or at least minimized by using an appropriate choice of plasma gas, namely, a plasma–sheath gas mixture with moderate enthalpy and high oxidizing potential. In addition, the presence of water in the suspension contributes to the resistance of the apatite structure to decomposition during the high-temperature treatment by maintaining a high partial pressure of water vapor in the deposition reactor.

RF-ICP torch systems have been equally used for the spraying of coatings using Suspension Plasma Spraying (SPS) or Solution Precursor Plasma Spraying (SPPS) as described in the example given by [Jia L. and F. Gitzhofer (2010)] for the deposition of a gas-tight ceramic membrane of Gadolinia-Doped Ceria (GDC) for Solid Oxide Fuel Cells (SOFC). The solution precursor of GDC was prepared by dissolving separately cerium nitrate hexahydrate (Alfa Aesar, purity: 99.99%) and gadolinium nitrate hexahydrate (Alfa Aesar, purity: 99.99%) in distilled water according to stoichiometric compositions. The solution concentration was 600 g/L. A magnetic stirrer was used to fully mix the starting precursors.

For SPS deposition of the ceramic membrane, the precursor used was a powder of composition Ce0.8Gd0.2O1.9 (GDC) prepared by the glycine-nitrate process (GNP) method of Xia and Liu (2002). Stoichiometric amounts of Ce(NO3)3 6H2O (Alfa Aesar, 99.99%) and Gd(NO3)3 6H2O (Alfa Aesar, 99.99%) were dissolved in distilled water, to which 0.5 mol glycine per mole nitrate was added. Combustion of the metal nitrate-glycine solution was performed with about 20 mL of the solution (0.1 mol with respect to metal ions) burned at a time. The precursor solution turned to a brown-red gel as the solvent was evaporated and

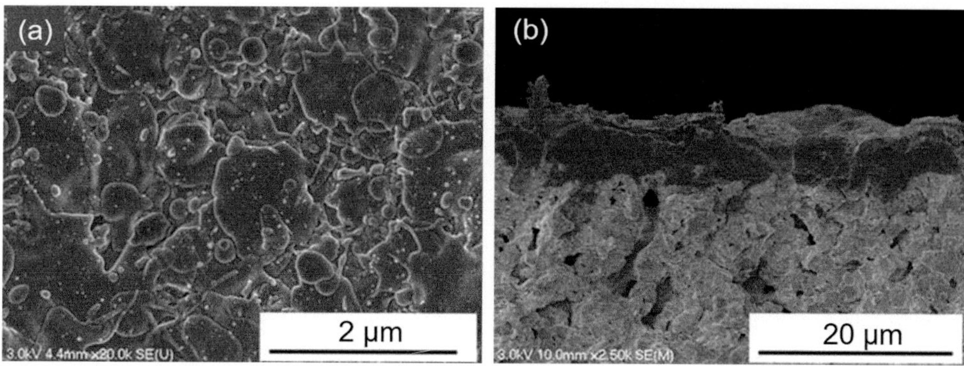

Fig. 16.44 FESEM micrographs of RF Suspension Plasma Sprayed (SPS) GDC coating (**a**) top surface view (**b**) fractured cross-section [Jia L. and F. Gitzhofer (2010)]. (Reprinted with kind permission from Springer Science Business Media, copyright © ASM International)

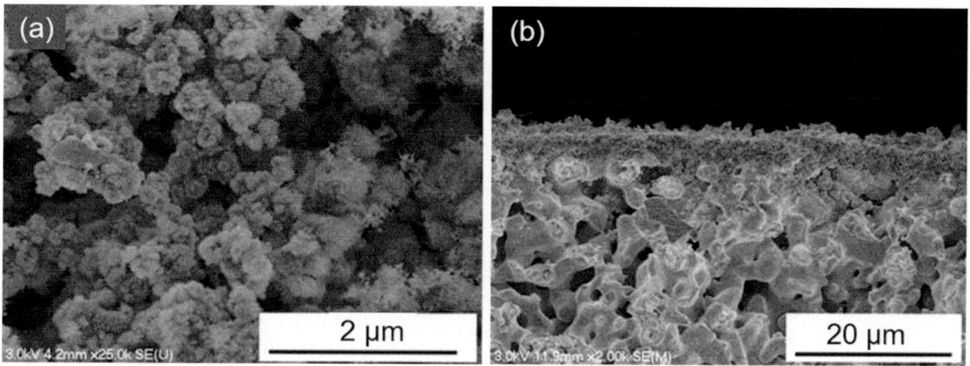

Fig. 16.45 FESEM micrographs of RF Solution Precursor Plasma Sprayed (SPPS) GDC coating (**a**) top view (**b**) fractured cross-section [Jia and Gitzhofer (2010)]. (Reprinted with kind permission from Springer Science Business Media, copyright © ASM International)

then spontaneous combustion occurred, leading to pale-yellow ash. The resultant ash was then collected and calcined in air at 600 °C for 2 h to remove any carbon residues remaining in the oxide powder. The Gadolinia-Doped Ceria (GDC) suspension precursor was prepared according to [Jia L. and F. Gitzhofer (2010)] by dispersing the GNP-GDC powders (with particle mean sizes of 0.6 μm) were directly dispersed into ethanol. The weight ratio of GDC to ethanol was set to 10 wt.%. The suspension viscosity was adjusted by adding Darvan No. 7 (dispersing agent, R.T. Vanderbilt Company, Inc., Norwalk, CT) to the suspension. The mass percentage of the dispersing agent selected for use was 2% of the GDC powder mass to obtain the lowest viscosity for adequate feeding. Before deposition, the suspension mixture was ultrasonically dispersed to break apart the large agglomerates. During injection, a stirrer was used to prevent particles from settling.

Field emission scanning electron microscopy (FE-SEM) photographs of the resulting coating given in Fig. 16.44a, b show, respectively, top view and fractured cross-section of the ceramic membrane formed by SPS with visible layered splats 0.2 to 2 μm diameters, corresponding to a flattening ratio below 2.7 [Jia L. and F. Gitzhofer (2010)]. RF-ICP was

also used for solution precursor plasma spraying (SPPS) as described by [Jia L. and F. Gitzhofer (2010)]. As GDC solution precursor is atomized into the discharge, the liquid component of formed droplets was completely vaporized in the plasma forming nanostructured GDC particles which on impact with the substrate build up a coating with high porosity, as illustrated in the top view and fractured cross-section given in Fig. 16.45. The formation of globular GDC coatings was observed over the entire experimental range of this study.

A similar approach was used by [Shen Y. et al. (2011)] for the RF-ICP spraying of lanthanum strontium cobalt iron oxide ($La_x\,Sr_{1-x}\,Co_y\,Fe_{1-y}\,O_{3-\delta}$) (LSCF) and GDC solutions to produce composite nanopowders, which were used in a next step to produce cathode coatings for intermediate temperature solid oxide fuel cells (IT-SOFCs) by suspension plasma spray (SPS), It has been possible to achieve a homogeneously mixed nanometer-sized composite GDC/LSCF powder without using a prolonged period of mechanical mixing which exhibited perovskite and fluorite structures as well as separated GDC and LSCF phases. By applying a low feeding rate, it was easy to synthesize nanometer-sized powders with different compositions by only adjusting the

Fig. 16.46 SEM micrograph of the surface of SPS cathode coating [Shen Y. et al. (2011)]. (Reprinted with kind permission from Springer Science Business Media, copyright © ASM International)

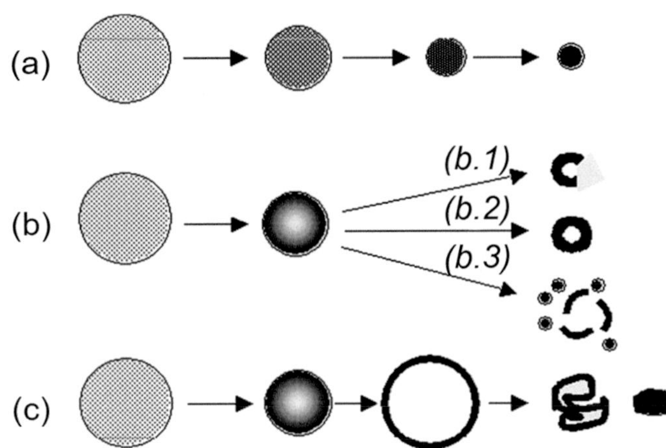

Fig. 16.47 Different routes for droplet vaporization and solid precipitations; (**a**) solid particle resulting from volume precipitation. (**b**) Super-saturation near the surface leading to the formation of a solid dense shell (b.1), highly permeable shell (b.2), or impermeable shell leading to breakup (b.3). (**c**) Elastic shell [Saha et al. (2009)]. (Reprinted with kind permission from Elsevier)

metal nitrate concentrations in the precursor solution. All synthesized powders were spherical with a diameter between 10 and 60 nm regardless of their composition. The deposited coating had a homogeneous nanocauliflower structure as shown in Fig. 16.46, with an average porosity of 51%.

16.4.3 Droplets-Spray Stream Interactions

16.4.3.1 Solutions

As solution drops are formed and further heated inflight the solute starts to precipitate forming a shell around its surface. Modeling studies by [Ozturk A. and B. Cetegen (2005) and Basu et al. (2006)] showed that, compared to the suspension droplets, where the diameter is controlled by the solvent evaporation, the solution droplet size is now fixed by the outer diameter of the precipitate shell formed through the droplet heating. The interior of the droplet is divided into three zones: solid shell, liquid core, and a vapor annulus between them. Depending on the solute and solvent characteristics and thermophysical conditions controlling the precipitation, the shell will be more or less porous. The pressure of the vapor inside it, depending on the rate of vapor leaving it, that its porosity can be calculated as well as the critical pressure over which the shell will fracture (if its failure stress is known). The precipitate formation depends on droplet sizes and solute initial mass fraction. In small droplets (5 μm in diameter), precipitation, whatever may be the initial solute level, encompasses the whole droplet, thus creating solid particles [Oberste Berghaus J. et al. (2007)]. For larger particles (only particles up to 40 μm were modeled) a shell precipitate is formed and subsequently fragments, depending on the vapor pressure within particle.

[Saha et al. (2009)] proposed that, as illustrated in Fig. 16.47, different particle morphologies result from the different conditions to which the drops or droplets, are exposed during their trajectories in hot spray stream. These include solid particles, hollow shells, and fragmented shells. Small droplets with high solute diffusivity exhibit a propensity to precipitate volumetrically to form solid particles as shown in Fig. 16.47a. Rapid vaporization combined with low solute diffusivity and large droplet sizes can lead to significant increase in solute concentration near the droplet surface resulting in surface precipitation to form a crust around the liquid core of the droplet. The crust/shell may have varied levels of porosity. Shells having low porosity usually rupture due to internal pressurization to form shell fragments (Fig. 16.47b, path b.1). For shells with a high level of porosity, the internal pressure rise is counterbalanced by the vapor venting through the pores resulting in hollow shells (Fig. 16.47b, path b.2). Shells that are completely impervious rupture and secondary atomization having a trapped liquid core may be observed (Fig. 16.47b, path b.3). For particular precursors, elastic inflation and subsequent collapse and rupture of the shell can also be observed (Fig. 16.47c).

Essentially, particle morphology resulting from droplet processing is sensitive to the solute chemistry, mass diffusivity, solute solubility, droplet size, thermal history, injection type, and velocity. According to the calculations of [Saha et al. (2009)], the final coating microstructure depends mainly upon the size of the droplet rather than their respective trajectory being on the torch axis or a deviated path after primary precipitation. Globally, droplets in the size range 5 or 10 μm get pyrolyzed completely before reaching the substrate, while those with a diameter larger than 20 μm remain partially pyrolyzed [Ozturk A. and B. Cetegen (2005)].

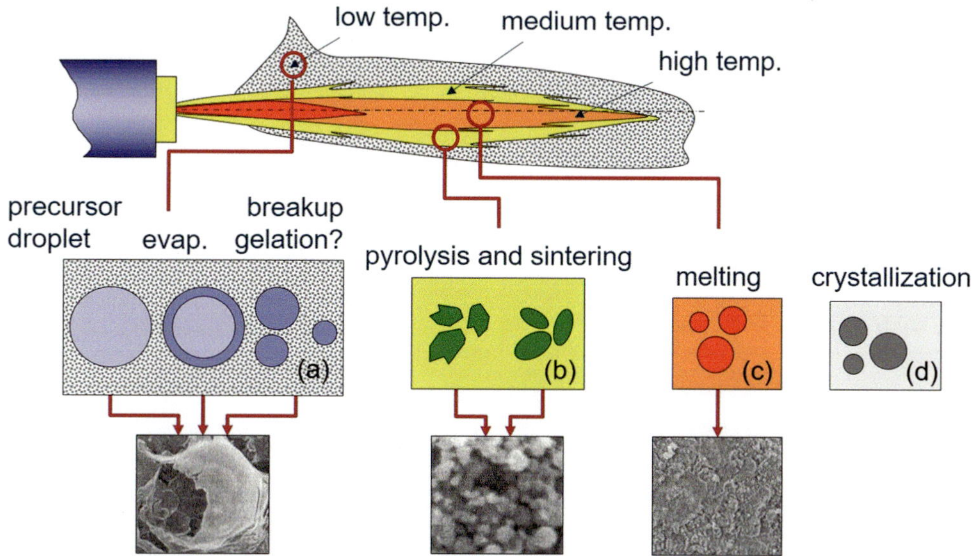

Fig. 16.48 Schematic representation of the different transformations of solution droplets within DC plasma jet [Jordan et al. (2004)]. (Reprinted with kind permission from Springer Science Business Media, copyright © ASM International)

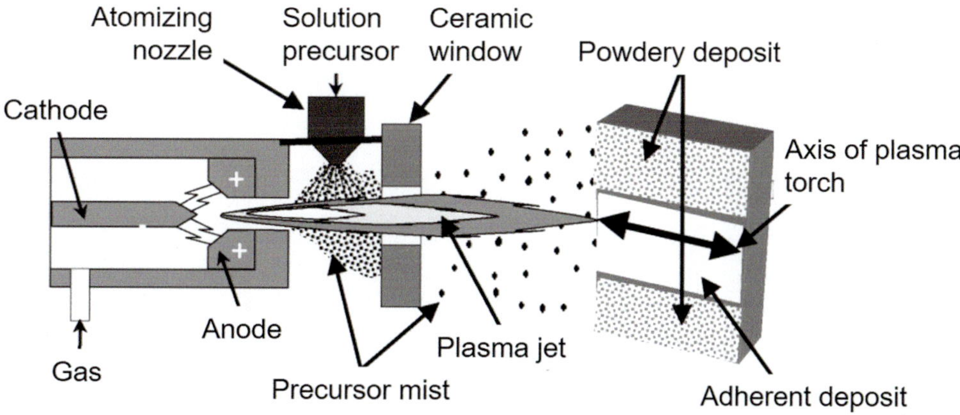

Fig. 16.49 Schematic of the window-shield fixed scan spray setup [Xie et al. (2004a)]. (Reprinted with kind permission from Elsevier)

As pointed out by [Jordan et al. (2004)] the different transformations to which the droplets are exposed during their trajectory in DC plasma jet, illustrated in Fig. 16.48, include precursor solute precipitation, pyrolysis, sintering, melting, and crystallization. Particles traveling in the hot core of the jet exhibit mechanisms (a), (b), and (c) of Fig. 16.48. If the standoff distance increases, molten particles may solidify and crystallize before impact on the substrate (D mechanism). When droplets travel in the jet fringes but still in the hot area ($T < 4000$–5000 K), (a) and (b) (Fig. 16.48) mechanisms are the most probable while for droplets traveling in the low temperature regions of the jet, some precursor solution can reach the substrate in liquid form.

Numerous studies have been reported on the study of the deposition mechanisms and the parameters controlling solution spraying [Jordan E.H. et al. (2004), Chen D. et al. (2008a, b), Saha A. et al. (2009) Xie L. et al. (2004a, b, c)]. The spray patterns produced using solution spraying under various processing conditions were investigated by [Xie et al. (2004a)] using the experiment setup illustrated schematically in Fig. 16.49. In this configuration the movement of the plasma torch was limited to one direction while the substrate held stationary to obtain a bead of a thickness depending on the number of passes of the torch with respect to the substrate. Due to the vast differences in the temperature histories experienced by the precursor droplets traveling in the core of the plasma jet compared to those in its fringes, the microstructure of the central part of the bead was reported to be rather different compared to those its fringes. To study only the central part of the bead, [Xie et al. (2004a)] placed a ceramic mask with a 20-mm diameter hole in its center, between the torch and the substrate, and aligned with its

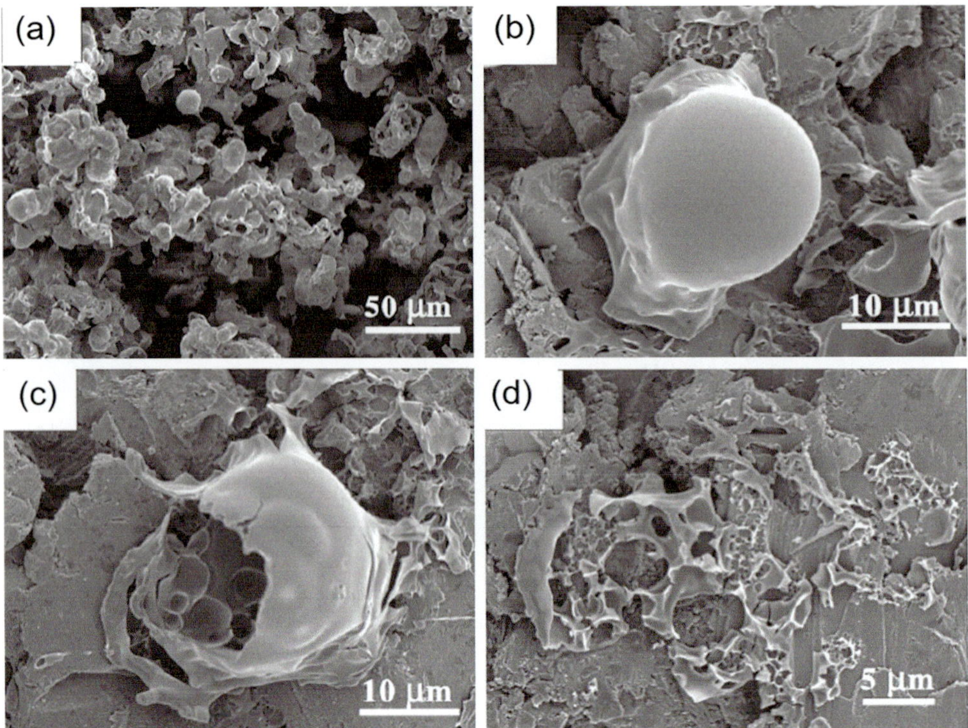

Fig. 16.50 SEM micrographs of deposit from wet-precursor (30 mm offset and 500 °C substrate): (**a**) overview; (**b–d**) detailed views of the individual deposit, [Xie et al. (2004b)]. (Reprinted with kind permission from Elsevier)

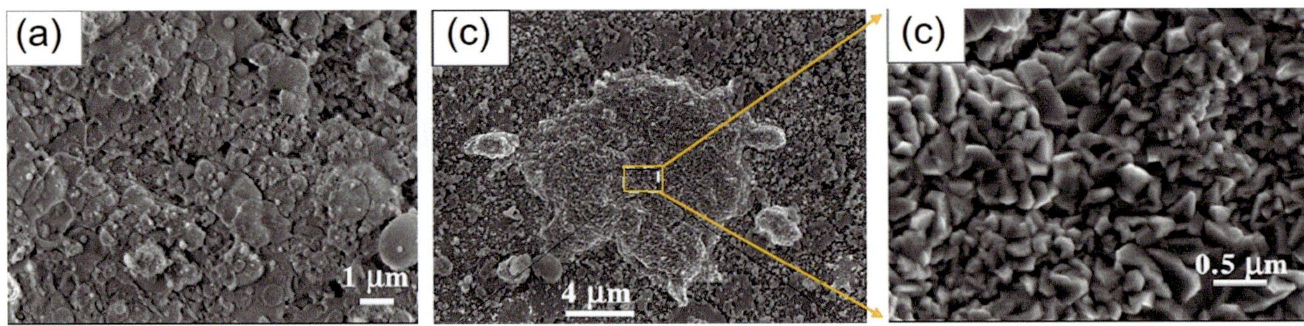

Fig. 16.51 SEM micrograph of the top view of the deposit collected on mask-shielded substrate: (**a**) splats, (**b**) splat-like large deposit, and (**c**) higher magnification SEM image, [Xie et al. (2004b)]. (Reprinted with kind permission from Elsevier)

centerline. The coating material traveling in the outer region of the plasma jet was accordingly intercepted by the mask allowing only those traveling along the central part of the jet to reach the substrate. An aqueous solution of a chemical precursor feedstock was used for the spray deposition of ZrO_2–7 wt.% Y_2O_3 ceramic solid solution coating, using a Metco 9 MB plasma torch (35–45 kW) working with a mixture of Ar and H_2, on a substrate preheated to 500 °C. The results showed that deposits formed by precursor injected into the hot region of the plasma jet were mostly crystalline 7-YSZ, while those from the precursor droplets traveling along the colder fringe region of the plasma jet were amorphous containing significant amount of water

Fig. 16.49a. Higher magnification images showing individual features of the deposit are given in Figure 16.49b–d show the formation of a hollow shell, Fig. 16.49b. Partially broken and fully broken shells were also observes as shown, respectively, in Fig. 16.49c, d. These types of deposits were rarely observed on substrate with no-preheat, reinforcing the notion that evaporation of the wet-precursor occurs after the deposition of the precursor on the hot substrate (Fig. 16.50).

In a subsequent study [Xie et al. (2004b)] reported different splat morphologies in the SPPS coatings obtained. The first comprises splat-like features, as shown in Fig. 16.51a. with a typical splat diameter in the range of 0.5 to 5 μm. The second morphology is the fine spherical particles, which can

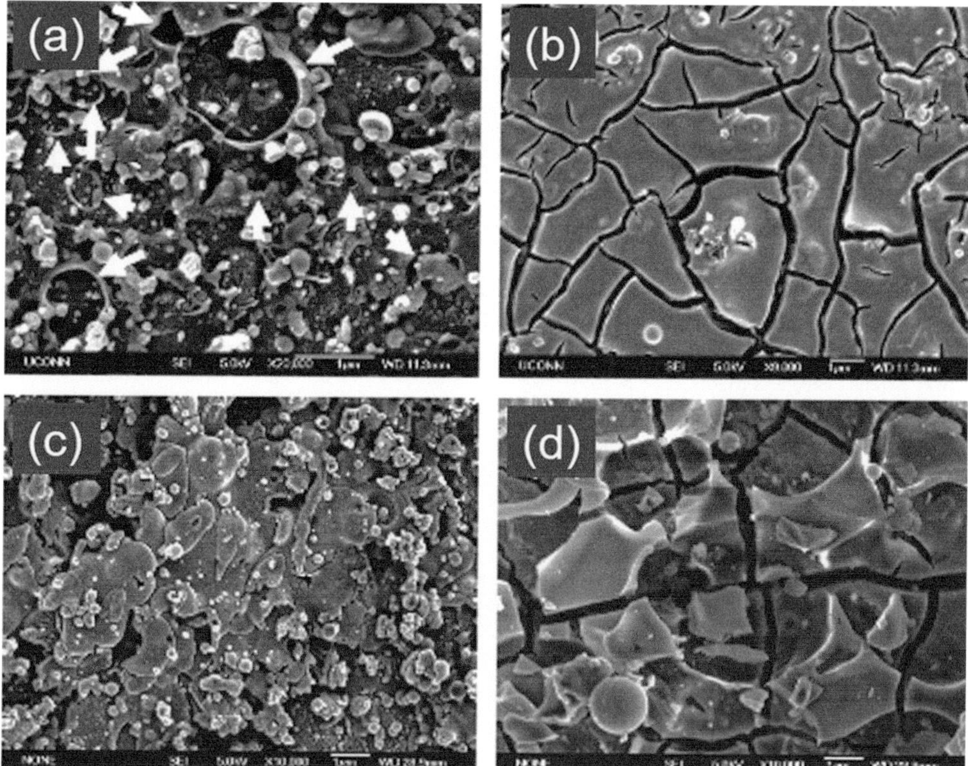

Fig. 16.52 Deposited beads collected on room temperature substrate by single scan experiment. (**a** and **b**) From diluted precursor with (**a**) central zone of deposited bead (**b**) deposited bead fringes. (**c** and **d**) From high concentration precursor with (**c**) central zone of deposited bead (**d**) deposited bead fringes [Chen et al. (2008a, b)]. (Reprinted with kind permission from Elsevier)

also be identified in Fig. 16.51a. The third "splat" like morphology shown in Fig. 16.51b, c, with (Fig. 16.51c) being at higher magnification shows the deposit relatively "rough" consisting of angular grains of sizes ranging from tens to hundreds of nanometers.

The initial solution concentration also plays a key role in the droplet pyrolysis. Once the solvent is selected (water, ethanol, isopropanol, etc.), the solution concentration can be varied up to the equilibrium saturation concentration. A high concentration close to the equilibrium saturation concentration tends to produce volume precipitation. Figure 16.52 after [Chen et al. (2008a, b)] shows typical SEM microstructures of the collected solution coatings of Y-PSZ on substrates at room temperature from a low concentration (Fig. 16.52a, b) and high concentration solutions (Fig. 16.52c, d). In each of these two cases, the micrographs given on (Fig. 16.52a, c) were made for the central zone of the deposit, with low and high concentration feed, while those given in (Fig. 16.52b, d) were taken from the fringes of the deposit at Low and high solution concentrations. As can be noted in (Fig. 16.52a) no splats are observed in the central zone of the deposited bead with the low concentration solution precursor, with the coating mainly composed of ruptured bubbles and a small volume fraction of solid spheres (<0.5 µm). The deposited bead in the central zone made from the high concentration solution

(Fig. 16.52c) is mainly composed of overlapping splats, with an average diameter ranging from 0.5 to 2 µm, and a small amount of nonmelted solid spheres (<0.5 µm). The deposited at the fringe of the bead for both dilute and high concentration solutions shown in (Fig. 16.52b, d) are made of un-pyrolyzed precursor containing significant amounts of water. The mud-like cracks presented at the edges are the result of shrinkage due to liquid phase evaporation.

16.4.3.2 Suspensions

The selection of the particles (morphology and size distributions) and their mass load in the suspension are key parameters. Typical mass loads vary between 5 and 20 wt.%. However, with a plasma spray torch with a rod-type cathode, coatings manufactured with a suspension containing particle mass load over 10% are less cohesive [Fauchais P. et al. (2008a, b)] (due very likely to the loading effect on the plasma flow), unless the enthalpy of the spray stream is increased. The size distribution and particle morphologies are also important parameters to consider. Results are also observed to be strongly dependent on the ability of solid particles contained in the liquid to agglomerate, or worse, to aggregate. Schematic representation given in Fig. 16.53 [Delbos C. et al. (2006)] show the in-flight transformations taking place in suspension droplets formed of Yttria Partially

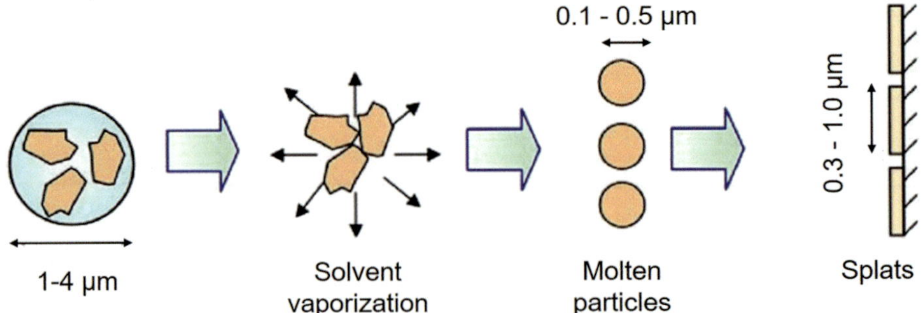

Fig. 16.53 Schematic of suspension droplet treatment when containing attrition-milled particles [Delbos C. et al. (2006)]. (Reprinted with kind permission from Springer Science Business Media)

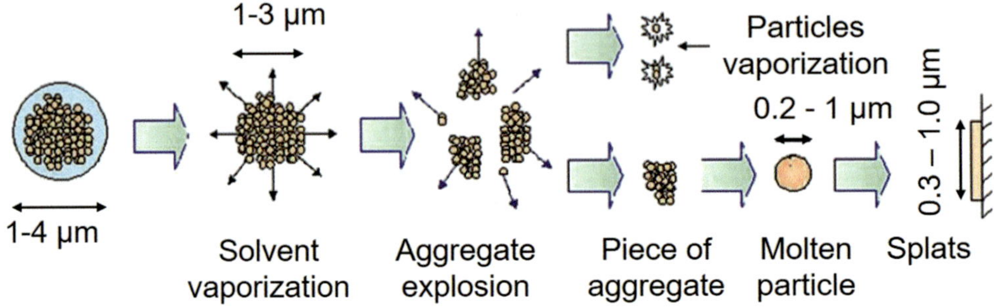

Fig. 16.54 Scheme of the suspension droplet treatment in the plasma jet for the Tosoh nanometric powder containing aggregates and agglomerates resulting in a size distribution between 0.1 and 3 µm [Delbos C. et al. (2006)]. Reprinted with kind permission from Springer Science Business Media

Stabilized Zirconia (YPSZ) particles made by attrition milling, following the solvent evaporation, acceleration (their initial velocities being those gained by droplets before their complete vaporization), heating, and melting by the plasma prior to the impact on the substrate.

When using nanosized particles obtained by a chemical process (commercial YPSZ Tosoh powder), where the size distribution is between 0.1 and 3 µm, due to the agglomeration and aggregation of the initial grains (mean size of 25 nm), the behavior upon solvent vaporization is different. It seems that the bigger agglomerates or aggregates explode upon solvent evaporation resulting in molten particles and smaller particle evaporation as illustrated in Fig. 16.54 [Delbos C. et al. (2006)].

16.4.3.3 Concluding Remarks

Based on the above discussion for solution and suspension spraying, it is to be underlined that small particles contained in suspensions or formed in solutions are poorly accelerated by the plasma flow. That is why finally the solid particle velocity contained in suspensions is that achieved by its carrier droplet just prior its vaporization. For solutions, droplets vaporization is stopped by pyrolysis of the solvent followed by sintering and thus the final size of particles formed is bigger than those contained in suspensions (typically a few µm).

When the liquid jet or drops penetrate radially the plasma jet, they are either progressively fragmented in droplets if the Weber number is below 200–300 or very rapidly if $We > 350$. The value of the Weber numbers along drop and droplet trajectories will condition their fragmentation. To delay fragmentation without modifying the spray and injection conditions, the Weber number can be reduced when shifting from ethanol to water or when reducing the size of initial drops.

With axial injection in HVOF guns, the solvent is mainly evaporated in the combustion chamber, if it is long enough, the solid particles contained in suspensions, being accelerated by the flow downstream of the nozzle throat. Compared to plasma jets, their acceleration is much better than in conventional plasma jets because, trajectories in hot gases are longer, the Knudsen effect is much lower and the gas mass density higher. To the best of our knowledge no solutions have been sprayed that way. For axial injection with the Mettech Axial torch no clear explanation of the phenomena can be given because no measurements can be performed where drops are injected. However, drops are probably fragmented when entrained in the converging section of the three plasma jets and it can be supposed that drops vaporization plays a key role. As the high temperature plasma core is much longer than with conventional plasma jets, formed particles are also accelerated to higher velocities.

16.4.4 Coating Formation

16.4.4.1 Splats

To collect solid sub-micrometric particles, after their plasma treatment, one way is to intercept the particles on a polished stainless steel or glass substrate positioned at different distances from the nozzle exit and traversing the jet at about 1 m/s [Wittmann K. et al. (2002), Delbos C. et al. (2004, 2006), Fazilleau J. et al. (2006)]. As the melted particles flatten and form splats, those re-solidified or not melted rebound or form some sort of gel phase for poorly treated droplets. Similar results are be obtained by moving rapidly (1 m/s) the torch relatively to the substrate positioned at different distances. Of course, to collect such tiny splats the substrate surface has to be polished. Besides, as in conventional plasma spraying [Cedelle J. et al. (2006)]. As discussed in Chap. 15, Conventional Coating Formation preheating the substrate over the transition temperature allows obtaining close to disk-shaped splats while on cold substrates splats with irregular shape are collected, as shown in Fig. 16.55.

Solutions

The surface of spray beads can be divided into adherent deposits (bead central part) and powdery deposits (bead edges) that correspond to the hot and cold regions of the plasma jet, respectively. Four deposition mechanisms, corresponding to splat formation or particle deposition, have been identified [Xie L et al. (2004b), Gell M. et al. (2008)].

- Smaller droplets that undergo further heating to a fully molten state solidify upon impact to form ultrafine splats with an average diameter of 0.5–2 µm, as illustrated in Fig. 16.56a.
- At certain spray distances, droplets undergo re-solidification and crystallization prior to their impact upon the substrate forming fine crystallized spheres, as illustrated in Fig. 16.56b.

- Droplets entrained in the cold regions of the spray jet where they experience sufficient heating to cause solute evaporation leading to the formation of a gel phase, deposited on the substrate. Some droplets also form a pyrolyzed shell containing un-pyrolyzed solution that fracture during deposition, as illustrated in Fig. 16.56c.
- Some precursor solution droplets can reach the substrate in liquid form, having undergone none of the aforementioned processes, as illustrated in Fig. 16.56d.

Suspensions

Splats, obtained when spraying a suspension of attrition-milled YSZ particles with a distribution of 0.2–3 µm, using a DC plasma torch, Ar–H$_2$ plasma (45–15 slm, and mean specific enthalpy $h = 17.9$ MJ/kg) are presented in Fig. 16.57 after [Delbos et al. (2006)]. On the whole the flattening degree is below 2 and the splat distribution follows rather well the particle size distribution. However, some splats are larger than those expected with such flattening degree. These probably correspond to droplets that have contained a few particles that have fused together. This was later confirmed when spraying suspensions of two different powders [Tarasi F. et al. (2011)]. When spraying a solution of alumina and YPSZ particles, splats of alumina and YSZ were obtained and the resulting coating was made of layered alumina and YPSZ splats [Delbos C. et al. (2004)]. Due to the high thermal conductivity of hydrogen, particles below 0.4 µm were either vaporized and afterwards re-condensed or, for those only melted in-flight, re-solidified at the spray distance.

For YPSZ powder made of nanometer-sized particles, which agglomerate easily, even with a dispersant, sprayed with a DC plasma torch Ar–H$_2$–He plasma (40–10–30 slm, mean specific enthalpy $h = 22.3$ MJ/kg) splats collected had a rather wide size distribution, as shown in Fig. 16.58. Most splats collected are below 1 µm, with few with diameters up to 2 µm.

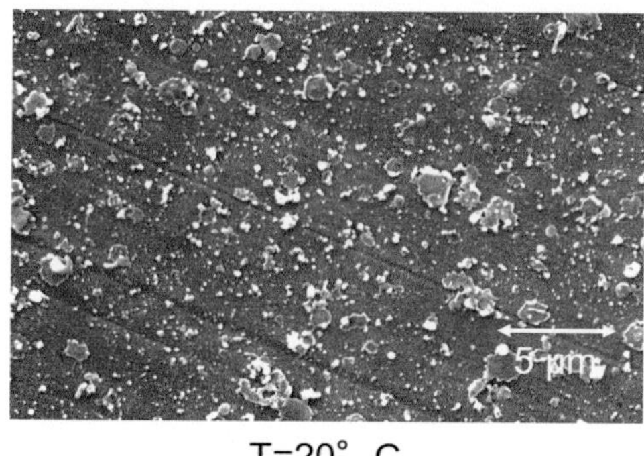
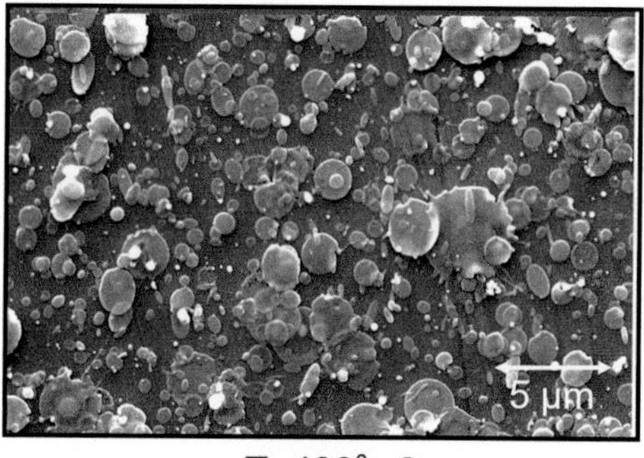

Fig. 16.55 Micrographs of YSZ splats obtained with ethanol-based suspension 10 wt.% Tososh powder, collected on mirror-polished stainless-steel substrates (**a**) cold and (**b**) preheated. [Fauchais P. et al. (2006)]. (Reprinted with kind permission from Springer Science Business Media, copyright © ASM International)

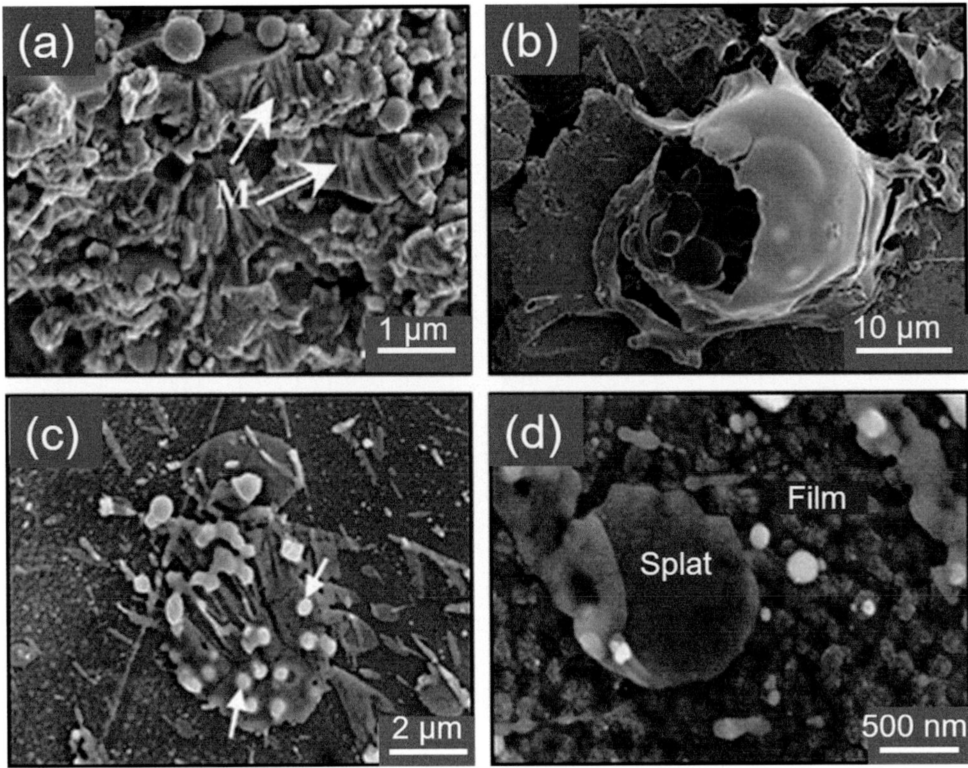

Fig. 16.56 Several deposition modes of SPPS. (a) Fine splats, (b) crystallized spheres, (c) ruptured shell, and (d) vapor deposited film [Gell M. et al. (2008)]. (Reprinted with kind permission from Springer Science Business Media, copyright © ASM International)

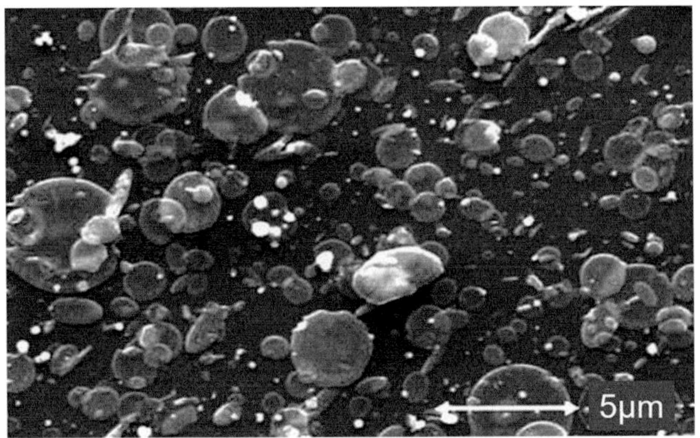

Fig. 16.57 Splats collected on a glass substrate when spraying an ethanol-based suspension of attrition-milled YSZ particles with a distribution of 0.2–3 μm using DC plasma torch [Delbos et al. (2006)]. (Reprinted with kind permission from Springer Science Business Media)

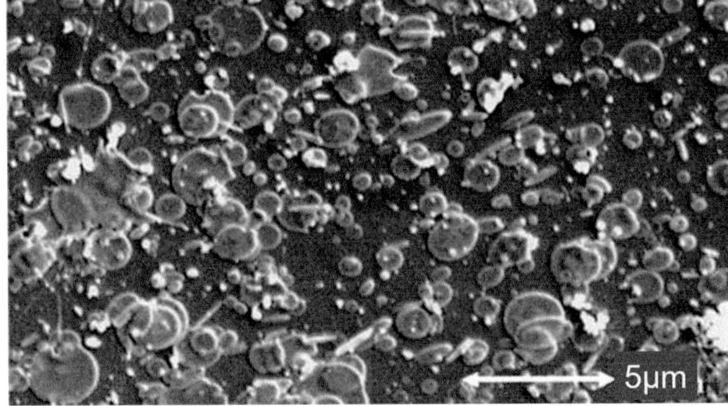

Fig. 16.58 Splats collected on a glass substrate when spraying an ethanol-based suspension of attrition-milled YSZ particles with a distribution of 0.2–3 μm sprayed using a DC plasma torch [Delbos et al. (2006)]. (Reprinted with kind permission from Springer Science Business Media)

According to [Fazilleau J. et al. (2006)] agglomerations of such particles must be avoided or at least limited which can be achieved, for example, by attrition or ball milling the suspension after its preparation. It is also possible to collect splats and particles on a substrate protected by a water-cooled shield, positioned at the end of an actuator and intercepting particles and plasma fluxes. The substrate is exposed to the particle flux only during a few tenths of seconds. To enlarge the collecting zone of splats and particles in one direction, the line-scan test, as developed by [Bianchi et al. (1997)] for conventional spraying, can also be used. [Siegert et al. (2005)] have also used a similar device. Such systems allow collecting splats and almost spherical particles distributed as shown schematically in Fig. 16.59. Particles that have traveled in the hot zones of the jet (zone 1 of Fig. 16.27) are well melted and form splats the layering of which will result in dense coatings, while those that have traveled in the jet fringes (Zone 2 and 3 in Fig. 16.27) form tiny spheres sticking all around the central zone (Fig. 16.59) forming the powdery part of coatings. With solutions similar results are obtained with a mud-like cracked film in the periphery. Under the repeated scan of the high temperature plasma jet, the mud-like film in situ evaporates, pyrolyzes, and crystallizes on the substrate resulting in the formation of a porous coating.

16.4.4.2 Spray Beads

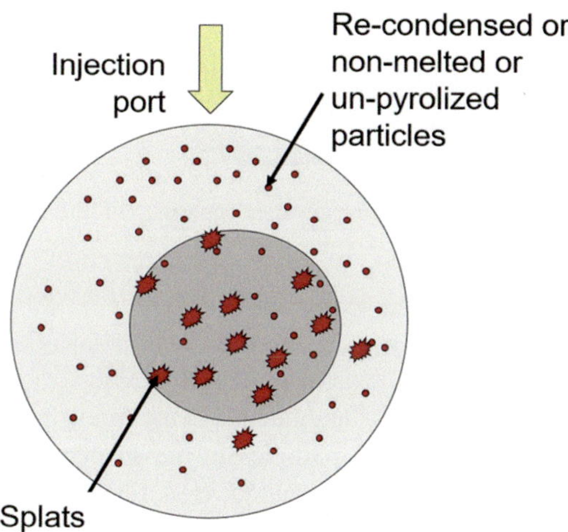

Fig. 16.59 Schematic distribution of splats and particles collected on a disk-shaped substrate: PT-F4 torch nozzle 6 mm i.d., Ar–H$_2$ (45–15 slm), I = 500 A, zirconia suspension 7 wt.%, particle size (0.1–3 μm) Etchart-Salas R. (2007). (Reprinted with kind permission from Dr. Etchart-Salas)

The degree of melting of particles can be evaluated with a line-scan-spray experiment considering either a simple bead in one passage or overlapped beads with successive passes [Ozturkand A. and B. Cetegen (2005), Xie L. et al. (2004b)]. The bead thickness depends upon the relative torch/substrate velocity, the number of passes, the suspension or solution flow rates and the injection parameters, the mass load of powder particles in suspension or the solution concentration and, of course, the torch-operating conditions. A typical bead, manufactured by suspension plasma spraying, is presented in Fig. 16.60 together with the schematic of the torch movement. It can be seen that the bead can be represented rather well with a Gaussian profile. [Xie et al. (2004a)] have reported similar beads with solutions.

Solutions

[Xie et al. (2004a, b) and Chen et al. (2008a, b)] have studied beads obtained with YSZ solutions and conventional plasma spray torch under various operating conditions. The beads can be divided into "*adherent deposit*" and "*powdery deposit*" corresponding, respectively, to particles traveling in warm or cold regions of the plasma jet. Adherent deposits seem to result from precursors traveling in both the warm and cold regions, the poorly treated precursor issued from cold regions being incorporated into the adherent deposit. Powdery deposits in the edges of the spray bead correspond to precursors traveling in the low temperature regions (fringes). Figure 16.61a is a representative macro-photo of samples produced in the fixed scan spray experiments, and Fig. 16.61b–d the top view of the illustrated regions of the sample. As shown in Fig. 16.61b, d, the powdery deposits consist of loosely connected powders, while the structure of the adherent deposits is dense as shown in Fig. 16.61c.

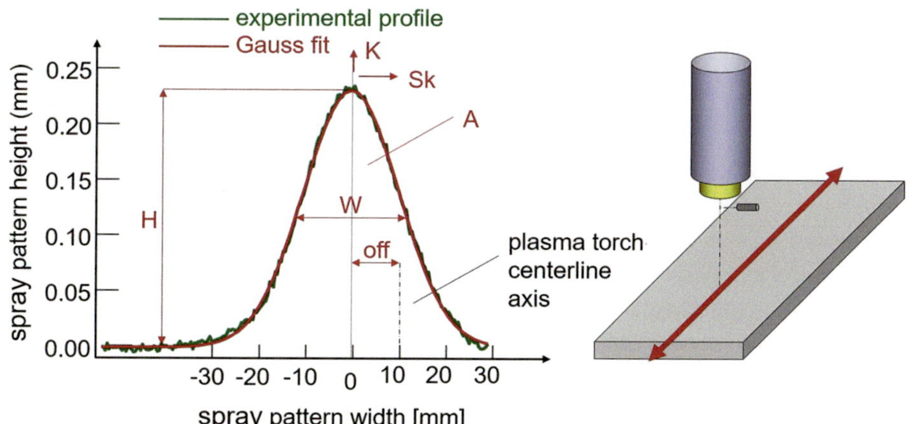

Fig. 16.60 Typical spray bead manufactured by suspension plasma spraying with the schematic of the corresponding torch movement

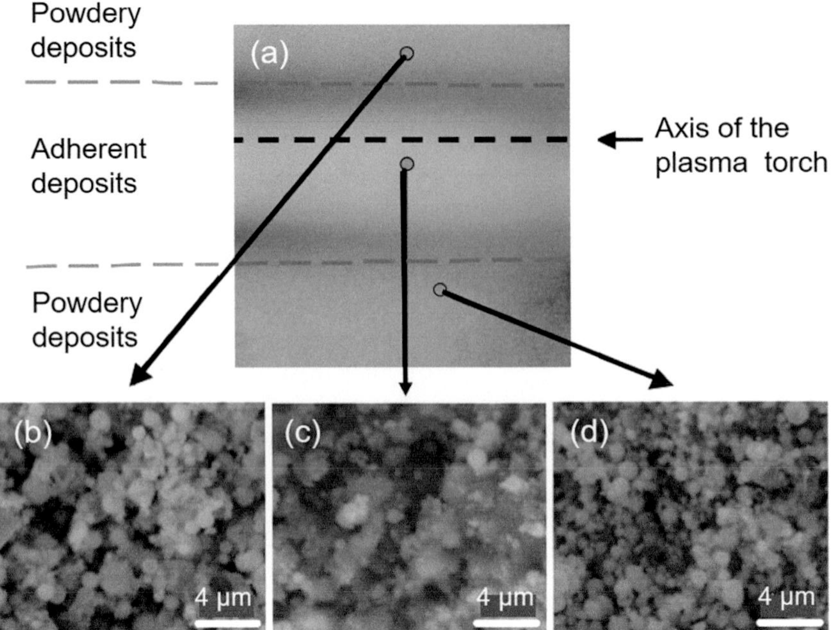

Fig. 16.61 Top view of the samples deposited using SPPS. (**a**) Macrophoto, (**b** and **d**) micrograph of the powdery deposits, and (**c**) micrograph of the adherent deposits [Xie et al. (2004a). (Reprinted with kind permission from Elsevier)

[Chen et al. (2010)] have studied the influence of the precursor concentrations. The central part of the spray bead deposited on a cold substrate consists of semi-pyrolyzed material that will form soft and porous coatings. Upon impacting the substrate at room temperature, the shells on the droplet surfaces fracture and form bubbles. However, bubbles evolve to spongy deposits when the substrate temperature is raised to 450 °C. In contrast, the high concentration precursor experiences volume precipitation leading to a fully melted lamella microstructure and resulting in a dense coating. In both cases, non-pyrolyzed material will in situ pyrolyze and form aggregates when the substrate temperature is above the precursor pyrolysis temperature. This pyrolysis also occurs when reheating the coating by the successive torch passes [Gell M. et al. (2008)].

Suspensions

A similar observation can be made when considering beads obtained with SPS though radial injection of the suspension in a DC plasma jet. As shown in Fig. 16.62 after [Tingaud O. et al. (2008, 2009)] the central part of the bead is relatively dense and mainly consists of a few splats, spherical particles, and nonmelted ones corresponding to particles that have been well treated in the plasma core plus those that have traveled in colder zones. At mid-height of the bead, the coating is less dense with many nonmelted particles while in the edges of the bead, the coating is fully powdery. To our knowledge, no

corresponding studies were reported with axial injection of the suspension in DC plasma or HVOF torches.

Figure 16.63, shows the influence of the spray distance and the solid loading of the particles in the suspension on the coating microstructure [Tingaud O. et al. (2008, 2009)]. The densest coating is obtained at a spray distance of 30 mm (when the plasma heat flux reaches 15 MW/m^2). The density seems to be slightly lower with the 10 wt.% mass load compared to that obtained with 5 wt.% solid loading of the suspension. The coating thickness, when the spray time and the mass load are doubled, is also slightly diminished, which is probably due to the loading effect. When spraying at 40 mm, beads are thicker than at 30 mm but less dense (due to the incorporation of more untreated particles) and it is worst at 50 mm.

As discussed in Chap. 15 Conventional Coating Formation, Sect. 15.4.2 Coating Formation Through Splat Layering, to form a uniform coating on a flat surface, the motion of the torch with respect to the substrate should follow a pattern as demonstrated in Fig. 15.76b following a linear motion in one direction along the x-axis, at velocity, v_r shifting at the end of the first travel by a step, Δy for the spraying of the next bead. The value of Δy being dependent on the width of the bead at half height, w_b multiplied by the *overlapping ratio*, f_b, which is typically in the range of 0.33 to 2.0 or 3.0. The same principle applies in solution or suspension plasma spraying with special attention given to the variation of the coating microstructure across the formed bead.

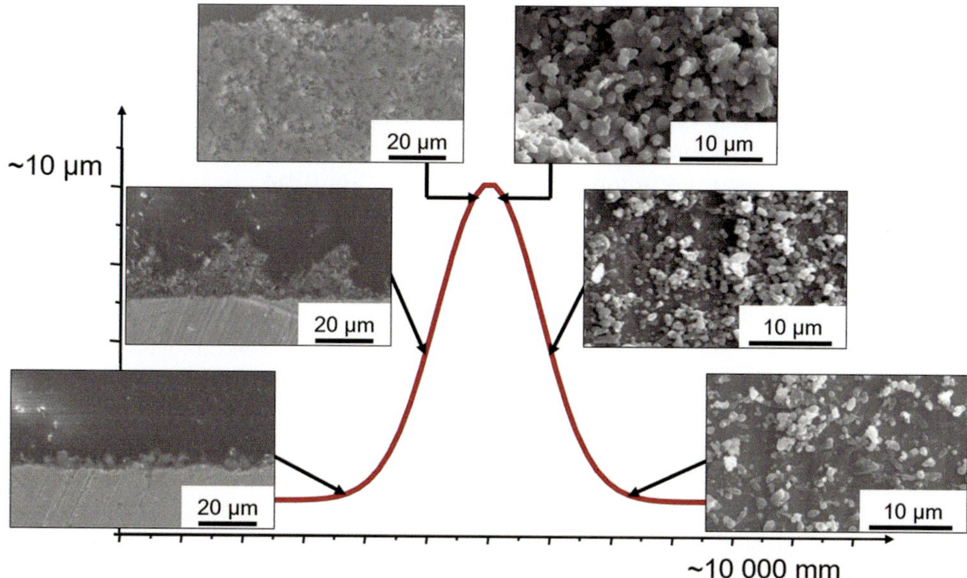

Fig. 16.62 Structure of an alumina suspension sprayed bead (particles size 0.02–1.0 µm, [Tingaud O. et al. (2009)]. (Reprinted with kind permission from Elsevier)

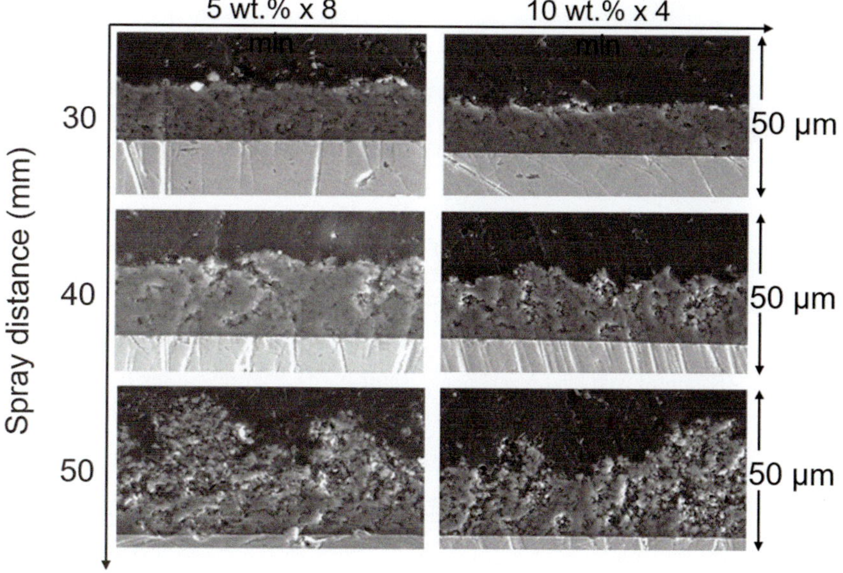

Fig. 16.63 Alumina suspension plasma-sprayed microstructures obtained for three different spray distances (30, 40, 50 mm) and two powder mass loads (5 and 10 wt.%) [Tingaud O. et al. (2008)]. (Reprinted with kind permission from Springer Science Business Media copyright © ASM International)

16.4.5 Spray Parameters and Coating Microstructure

As with "Conventional Thermal Spraying" solution or suspension plasma spraying is no exception in terms of the large number of "Spray parameters" that can have a significant impact on the quality of the coating. These can be grouped in the following categories according to [Fauchais P. et al. (2008a, b)].

- *Thermal source used*, whether combustion (HVOF or HV-SFS) or Plasma (DC plasma torch, Multielectrode axial injection DC plasma, or RF-IPS). As described in earlier chapters (Part II Thermal Spray Technology) corresponding operating parameters for each of these devices controls the flow and temperature fields in the generated high temperature stream. These include total *gas flow rate*, *generated thermal energy* (kW), *specific enthalpy of the thermal spray stream*, and *accessibility for radial or axial injection* of the spray precursor into the spray stream.

- *Precursor stream used*, whether *solution* or *suspension*. These include the choice of the liquid component, *water* or *combustible solvent*, and the rheological properties of the solution or suspension which depends in turn, in the case of *solution* on the solute concentration with respect to saturation level, solution viscosity and surface tension. In the case of *suspensions*, these include particle morphology and the PSD of the solid content, its loading ratio as well as the presence of additives such as surface-active agents often needed to stabilize the suspension.

- *Substrate preparation*, this includes cleanliness and surface roughness as well as substrate preheating to get rid of adsorbents and condensates, and substrate/coating temperature during the spraying operation.

- *Spraying conditions*, solution or suspension *feed rate* and *spraying distance*, also referred to as *stand-off distance* and whether the spraying operation is carried out in the *open atmosphere* or in *controlled environment*.

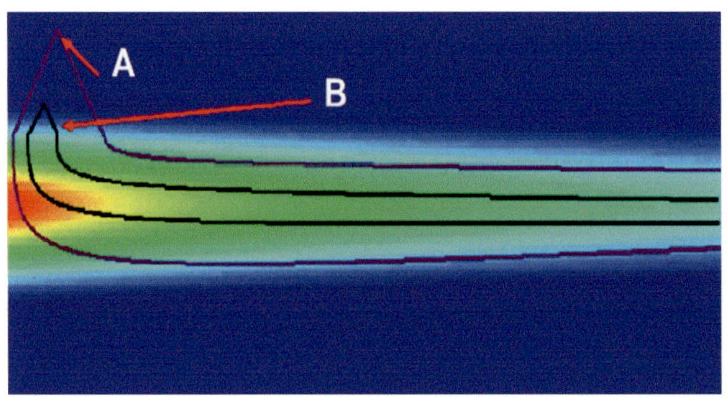

Fig. 16.64 Possible liquid droplet trajectory distributions in a DC plasma jet [Jordan E.H. et al. (2007)]. (Reprinted with kind permission from ASM International)

In the following a few examples are given on the dependence of the microstructure of the deposited coating on some of these parameters for each of solution and suspension spraying approaches.

16.4.5.1 Solution

In *solution spraying* the amount of non-decomposed precursor is of critical importance for the quality of the coating. It can be controlled by the spray parameters, primarily droplet injection momentum density, spray droplet sizes, and trajectories dispersions, as well as precursor concentration [Gell M. et al. (2008)]. Figure 16.63 illustrates the necessity to adjust the liquid droplet trajectories to achieve a distribution as narrow as possible in the hot core of the plasma jet and obtain uniform dynamic and thermal treatments.

When overlapping beads contain un-pyrolyzed particles (mainly those in the jet fringes) embedded within the layers during the spraying procedure, upon further heat exposure either by subsequent plasma torch passes or postdeposition treatment, their pyrolysis generates tensile stress leading to the formation of vertical cracks as illustrated in Fig. 16.64 after [Xie L. et al. (2006), Jordan E.H. et al. (2007)]. Coatings are made of micron-sized lamellae, which have grain sizes smaller than 30 nm, and they exhibit a porous architecture at two scales (nano- and micro-scales) made of voids and through-thickness stress relieving intra-lamellar cracks. Unique intra-lamellar cracks could also result from shrinkage as a result of solvent pyrolysis. Porosities, as those shown in Fig. 16.65b, could also result from the entrapment of poorly treated droplets in the coating. The fraction of un-pyrolyzed material has hence to be carefully controlled through the drop atomization (size and trajectory distributions), the spray pattern, and the solution concentration. To manufacture a dense (and hard) coating under the same operating conditions, the fraction of poorly treated material in the jet fringes needs to

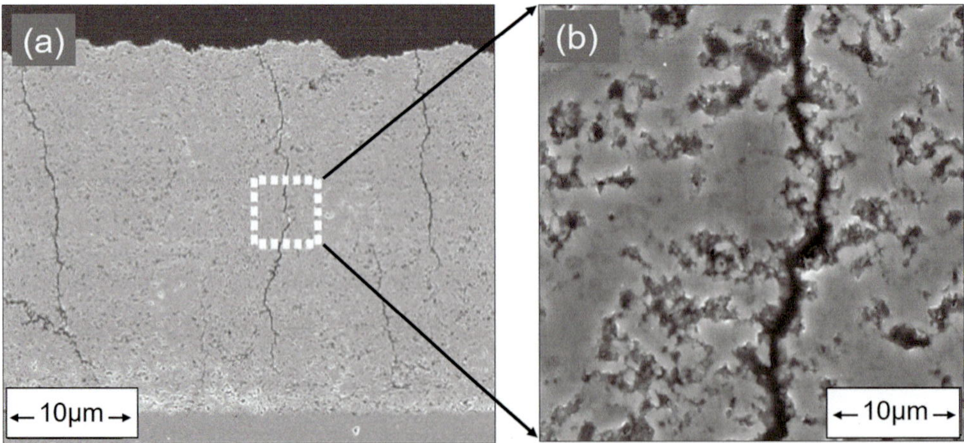

Fig. 16.65 Intra-lamellar cracked zirconia coatings deposited by solution plasma spraying with an Ar–H$_2$ plasma produced by a 9-MB Sulzer Metco DC plasma torch (**a**) general view (**b**) detail [Jordan E.H. et al. (2007)]. (Reprinted with kind permission from ASM International)

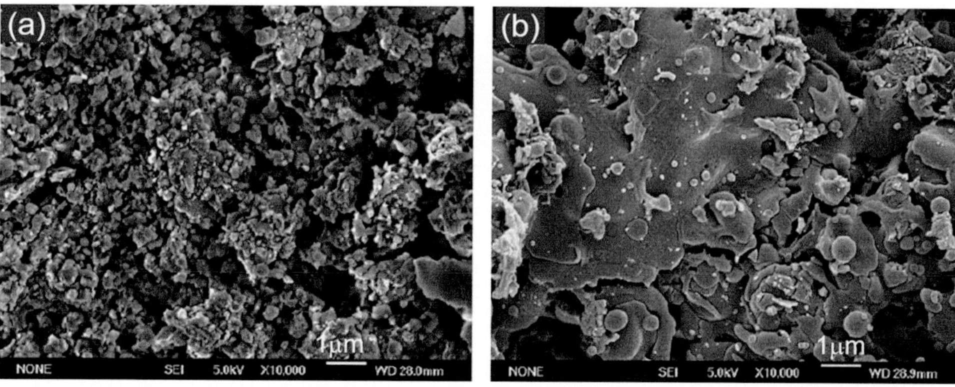

Fig. 16.66 Coating surface morphologies from solution precursors. (**a**) Low concentration. (**b**) High concentration [Chen D. et al. (2008a, b)]. (Reprinted with kind permission from Elsevier)

Fig. 16.67 Dense and hard zirconia coatings deposited by solution plasma spraying with an Ar–H$_2$ plasma using a capillary atomizer [Jordan E.H. et al. (2007)]. (Reprinted with kind permission from ASM International)

be significantly reduced, if not suppressed through the use of appropriate masks. Besides the optimization of the drop injection, dense coatings are linked to solutions with high concentrations [Jordan E.H. et al. (2007)]. Figure 16.66 shows that many more splats are observed at the surface of a coating sprayed with a high solution concentration than with a low concentration. Such layers are reasonably dense (88% total pore level), hard (1023 VHN, average value), and exhibit no apparent intra-lamellar cracking as shown in Fig. 16.67.

16.4.5.2 Suspension

In the following attention is given to the effect of the properties of the powder used in Suspension Plasma Spraying (SPS) and the effect of "powder loading" of the suspension on the quality of the coating obtained. Results are given using different, commercially available, Partially Yttria-Stabilized Zirconia (PYSZ) (8 mol Y$_2$O$_3$%), A summary of the PSD of the powders and the loading ratio of the solid in the suspension used is given in Table 16.3 [Delbos C. (2004), Etchart-Salas R. (2007), Brousse-Pereira E. (2010)]. All suspensions were sprayed using either Ar–H$_2$, Ar–H$_2$-He, or Ar–He, DC plasmas torches of conventional design at the same spray distance of 40 mm.

SEM micrographs of the Tosoh TZ-8Y powder and its PSD showing its nanometer and aggregate character with a broad particle size distribution is given in Fig. 16.68. The specific surface area was 13 m^2/g reflecting its heavy content of loosely agglomerated nanometer sized particles. The powder was SPS onto a 316 L stainless-steel substrate with a Ra of 0.05 μm as ethanol-based suspension injected radially into the plasma jet at the exit level of the torch nozzle (6 mm i.d.) operating using Ar–H$_2$ as plasma gas (45–15 slm), arc current of 500 A, with a mean specific enthalpy of the plasma jet $h_o = 17.9$ MJ/kg. Figure 16.69a shows the very porous nature of the coating with a pronounced columnar structure. It is likely that the columnar structure is due to the high voltage fluctuations (between 40 and 80 V for a mean voltage of 60 V), the impact velocity of particles injected at low voltage being too low to achieve Stokes numbers (> 1), thus promoting the shadow effect. Moreover, the agglomerate fragmentation creates a large dispersion of trajectories and nonuniform treatment of particles resulting in coatings with poor cohesion. Tiny particles were also probably evaporated and then condensed on the coating during its formation creating defects. The agglomerate fragmentation creates a large dispersion of trajectories and nonuniform treatment of particles resulting in coatings with poor cohesion.

Thin coating cross-sections (below a few μm) were analyzed to compare the initial particle distribution to those of the splats and spheres within the coating and results are presented in Fig. 16.69b. All splats have sizes over 0.2 μm, which means that the smallest fully melted particles

Table 16.3 Summary of PSD, grain size and solid loading (wt.%) of the PYSZ powder used in SPS by [Delbos C. (2004) and Etchart-Salas (2007)]

	$d_{0.1}$ (µm)	$d_{0.5}$ (µm)	$d_{0.9}$ (µm)	Mean grain size (nm)	Powder (wt.%)
Tosoh-TZ-8Y	0.10	1.50	3.50	30	7
Medipur attrition milled	0.15	1.00	3.00	>150	7
Marion	0.03	0.06	0.76	20–40	20
UC-001 h attrition milled	0.03	0.05	0.29	30	20
UC-002 h attrition milled	0.26	0.39	0.70	260	20

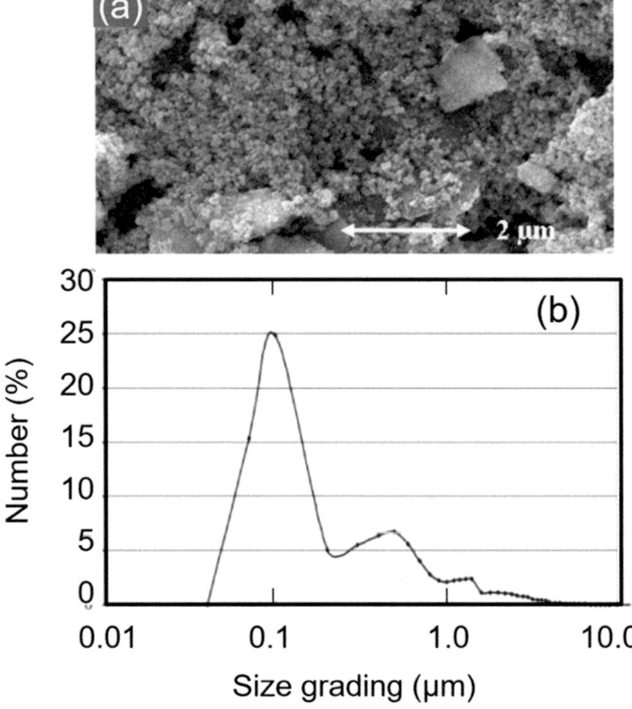

Fig. 16.68 (a) SEM view of Tosoh powder. (b) PSD of the powder by number [Delbos C. (2004)]

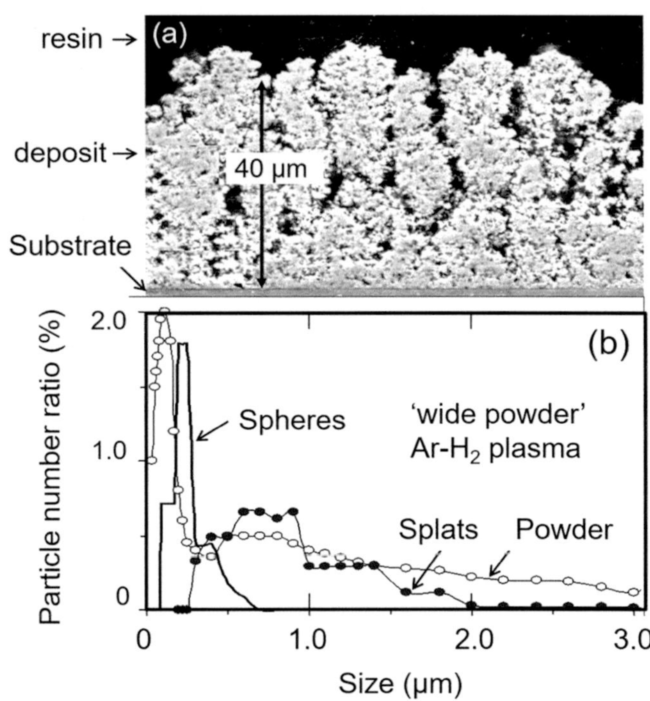

Fig. 16.69 (a) Coating cross-section obtained with SPS of Tosoh powder, ethanol 10 wt.% YSZ, 2 vol.% dispersing agent, sprayed with Ar–H₂ plasma. (b) Size distributions of initial powder, splats, and spheroidized particles [Delbos C. et al. (2006)]. (Reprinted with kind permission from Springer Science Business Media)

impacting the substrate are about 0.1 µm in diameter. The spheres correspond probably to small particles, bigger than 0.1 µm, which were melted rapidly but have started to solidify before reaching the substrate. It means that coatings are mostly formed; thanks to particles over 0.1 µm.

To clarify the role of the different particle sizes, the small particles were separated from the big ones (aggregates and agglomerates in the few micrometer-sized range). It was achieved by forming an ethanol suspension with no dispersant and by its sedimentation during a sufficient time (42 h 25 min) to achieve particles separation. Figure 16.70 represents the PSD of the micron-sized agglomerates and aggregates sprayed with the same spray conditions as those used for (Fig. 16.59) together with resulting splats and spheres. These two distributions are rather similar to those presented in Fig. 16.69 obtained with the initial powder, confirming that coating result mainly of particles with sizes over a few tenths of microns. At last the small particles

(<0.3 µm) have been sprayed in the same conditions, and Fig. 16.71 shows that the splat distributions of both, the sub-micron- and micro-sized particles, are quite similar, coatings formed with splats result mainly of micro-sized particles.

A suspension of Tosoh powder, where most aggregates and agglomerates were taken away, was also sprayed with the same torch using Ar–He (40–80 slm) as plasma gas with a current of 500 A, specific enthalpy of 13 MJ/kg and a mean gas velocity of 1425 m/s, against 17.9 MJ/kg and 1312 m/s for the Ar–H₂ (45–15 slm) plasma. SEM micrographs of the coating cross-section and its fracture are presented in Fig. 16.72a, b. With this more stable plasma ($V_\mathrm{m} = 62$ V and $\Delta V = 15$ V) no more columnar structure, as that shown in Fig. 16.69a, is achieved. Figure 16.72a, b shows that the coating is relatively dense with mostly granular microstructure, $d_p \geq 0.05$ µm as noted in the fraction cross-section given

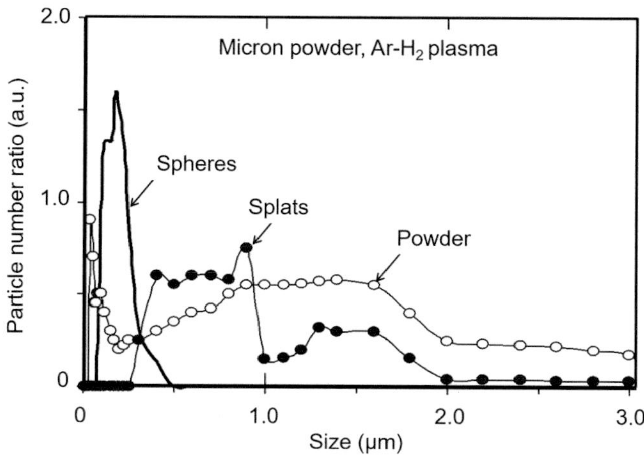

Fig. 16.70 Size distributions of initial powder, splats, and spheroidized particles obtained when spraying suspension of agglomerates and aggregates [Delbos C. et al. (2006)]. (Reprinted with kind permission from Springer Science Business Media)

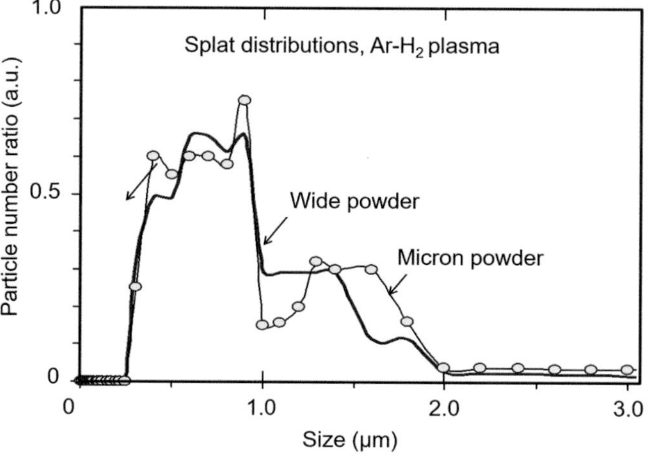

Fig. 16.71 Size distributions of splats obtained when spraying suspensions on the one hand of agglomerates and aggregates (micron particles) and on the other of the Tosoh powder [Delbos C. (2004)]. (Reprinted with kind permission)

in Fig. 16.72b. It must also be noted that the deposition of the same coating thickness is about twice longer than that obtained with initial Tosoh powder sprayed with the Ar–H$_2$ (45–15 slm), 500 A, plasma. It seems that with this faster plasma jet with low voltage fluctuations the small particles reach sufficient velocities and temperatures to be imbedded within the coating under formation.

[Delbos C. (2004)] has sprayed attrition milled (during 13 h), fused, and crushed particles of Medipur (initial size distribution 5–25 µm). SEM micrograph of the resulting particles and their corresponding PSD are given, respectively, in Fig. 16.73a, b.

The suspension of attrition-milled Medipur particles was sprayed first with the same conditions as those used for Tosoh powder suspension (Ar–H$_2$ plasma) and the corresponding micrographs of the coating cross-section is presented Fig. 16.74a. For this coating, many tiny particles, almost spherical, in the range of 0.05–0.3 µm, are contained in the pores. As with Tosoh powder most splats (studied when spraying layers a few µm thick) have sizes over 0.2 µm (see Fig. 16.75). The coating is denser and less porous (6%) than those obtained with Tosoh powder.

In order to improve the heat transfer, this suspension was also sprayed with Ar–H$_2$–He (40-10-50 slm) plasma, which enthalpy was 22.3 MJ/kg against 17.9 for the Ar–H$_2$ one, the corresponding coating cross-section being presented in Fig. 16.72b. In spite of the fact that, with this plasma forming gases containing 10 vol.% of H$_2$ the voltage fluctuations are not negligible ($V_{\mathrm{m}} = 75.1$ V and $\Delta V = 30$ V), the coating presented in Fig. 16.72b is rather dense (about 7% porosity). Compared to the coating presented in Fig. 16.72a, with the same suspension flow rate and number of coating passes, the coating thickness is increased by almost 50% due probably to the better melting of particles.

[Etchart-Salas (2007)] has studied 3 powders, Marion powder made by a "soft chemistry process" and two Unitec Ceramic powders fused and crushed and then milled: UC-001h and UC-002h. The PSD of the powder used is presented in Table 16.3. In all cases the injection with liquid jets of suspensions, all containing 20 wt.% of particles, has been optimized. As shown in Fig. 16.76 the Marion powder

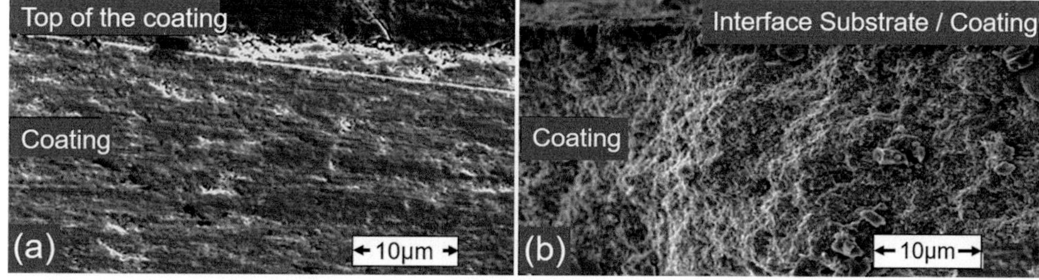

Fig. 16.72 Coating achieved with Ar–He plasma (40–80 slm), anode nozzle i.d. 6 mm, $I = 500$ A, $h_{\mathrm{o}} = 17.9$ MJ/kg: (**a**) Cross-section. (**b**) Fractured cross-section [Delbos C. et al. (2006)]. (Reprinted with kind permission from Springer Science Business Media)

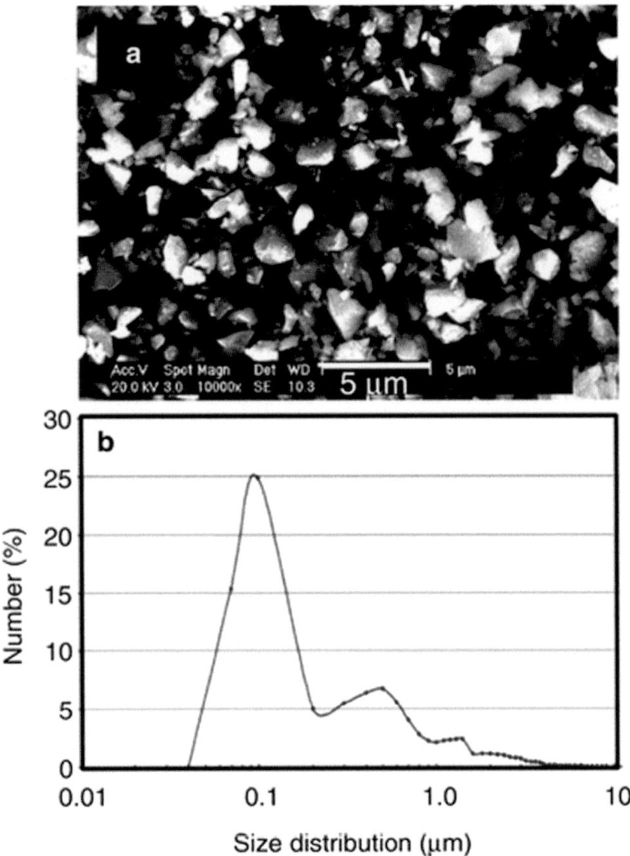

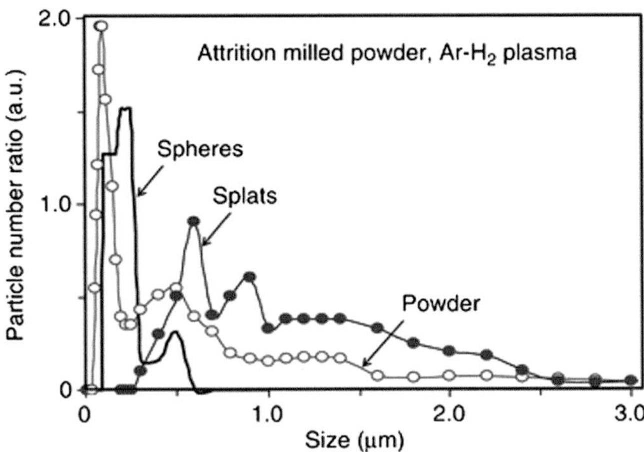

Fig. 16.75 Size distributions of initial powder, splats, and spheroidized particles obtained when spraying suspension of attrition milled Medipur powder [Delbos C. (2004)]. (Reprinted with kind permission)

Fig. 16.73 (a) SEM view of attrition milled Medipur powder. (b) Powder particle size distribution [Delbos C. et al. (2004)]. (Reprinted with kind permission)

Fig. 16.76 SEM micrograph of the Marion powder [Etchart-Salas (2007)]

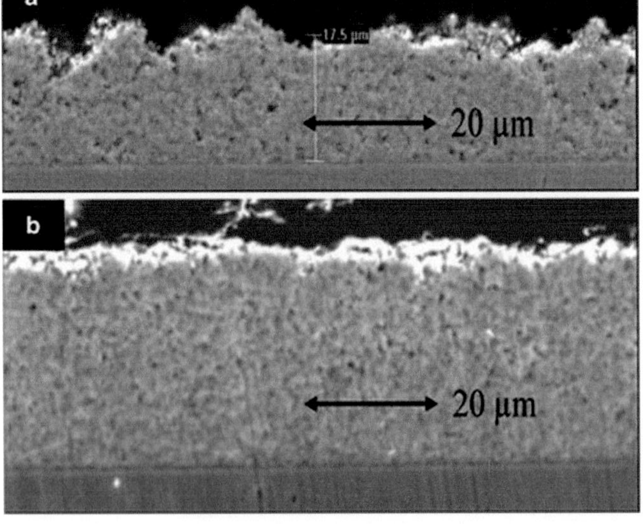

Fig. 16.74 Attrition-milled Medipur powder suspension sprayed: (a) with the Ar–H₂ plasma jet used to spray Tosoh powder suspension (b) with a more powerful jet: Ar–H₂–He (40–10–50 slm), anode nozzle i.d. 6 mm, $h_\mathrm{o} = 22.3$ MJ/kg [Delbos C. et al. (2004)]. (Reprinted with kind permission)

used is made of nanometer-sized grains (20–40 nm) and presents much less agglomerates and especially aggregates than the Tosoh powder, and its size distribution is between 0.03 and 0.9 μm. When sprayed with Ar–He (30–30 slm) with a relatively stable plasma, low arc voltage fluctuations: $V_\mathrm{m} = 50.9$ V and $\Delta V = 18$ V), the coating, as shown in Fig. 16.77, is rather dense (porosity of 4.7%) and its morphology is close to that of the coating obtained with Medipur attrition-milled powder and the Ar–H₂–He plasma (see Fig. 16.72b).

When spraying suspension using attrition-milled powder (UC.001h), shown in Fig. 16.78 with PSD in the range of 0.03 and 0.29 μm, with few agglomerates between 0.3 and 1 μm, slightly denser coatings, shown in Fig. 16.79, are obtained (porosity 4%) compared to those obtained using Marion powder [Etchart-Salas R. (2007)]. Similar results are obtained when larger particles with sizes between 0.26 and 0.7 μm (Unitec UC-002h), also attrition milled, are used.

When suspension plasma spraying of coatings over a large surface requiring the use of a torch movement pattern over

Fig. 16.77 YSZ coating obtained with ethanol-based suspension, 20 wt.% of Marion powder (nanograins) injected in Ar–He (30–30 slm) plasma jet (700 A) specific enthalpy $h_o = 17.9$ MJ/kg [Fauchais P. et al. (2008a, b) (Reprinted with kind permission from Springer Science Business Media, copyright © ASM International)

the substrate surface as shown in Fig. 16.80a, tiny particles traveling in jet fringes have a strong tendency to stick to the hot zirconia sprayed layers ($T > 1000$ °C) and create defects, especially between successive passes. This is particularly the case when injecting the suspension at the end of each pass and starting to spray the next one, particles stick on the previously deposited and still very hot pass. This is illustrated in Fig. 16.80b, c showing the deposition of successive pass with the defects created between them by this deposition of nonmelted tiny particles.

The result is illustrated in Fig. 16.81 with coatings obtained with Unitec UC-002 h powder (attrition milled with a rather narrow size distribution, see Fig. 16.78c). In Fig. 16.81a the fluctuating Ar–H$_2$ (45–8 slm), $I = 500$ A, $h_o = 14.5$ MJ/kg, DC plasma jet was used with the injection conditions optimized. The formation of columnar structure can be observed, but smoothed compared to that obtained when spraying with the same conditions Tosoh powder (obtained by a chemical route and with a broad size distribution, as shown in Fig. 16.69b). When using Ar–He (30/30 slm) DC plasma, I = 700 A, $h_o = 19.6$ MJ/kg, DC plasma with optimized injection much more stable than the Ar–H$_2$ one, the deposition of the nonmolten particles between passes is very clearly seen, as shown in

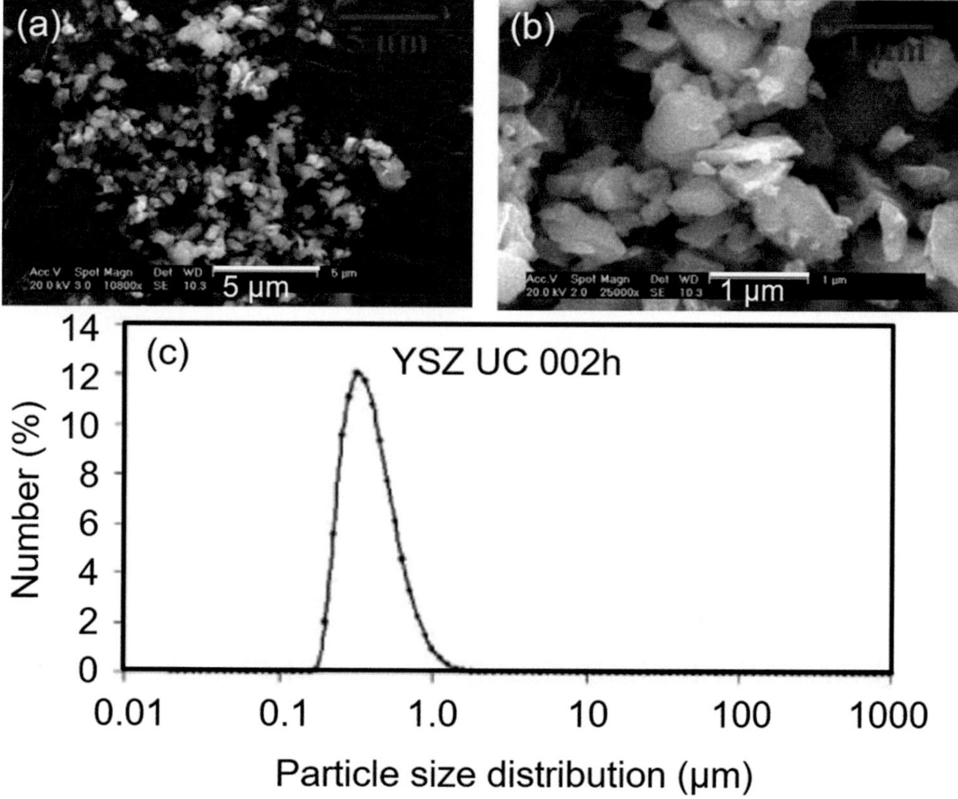

Fig. 16.78   Unitec powder 001 h attrition milled: (a and b) SEM micrographs with different magnification (c) PSD of powder by number [Fazilleau J. (2006)]. (Reprinted with kind permission from Springer Science Business Media)

Fig. 16.81b. When reducing the suspension injection velocity, leading to reduced suspension penetration in the plasma jet with particles traveling more in the jet fringes, where the

Fig. 16.79 YSZ coating obtained with SPS 20 wt.% of UC-001h attrition-milled powder in ethanol injected into Ar–He (30–30 slm) DC plasma, specific enthalpy $h_o = 17.9$ MJ/kg [Fauchais P. et al. (2008)]. (Reprinted with kind permission from Springer Science Business Media, copyright © ASM International)

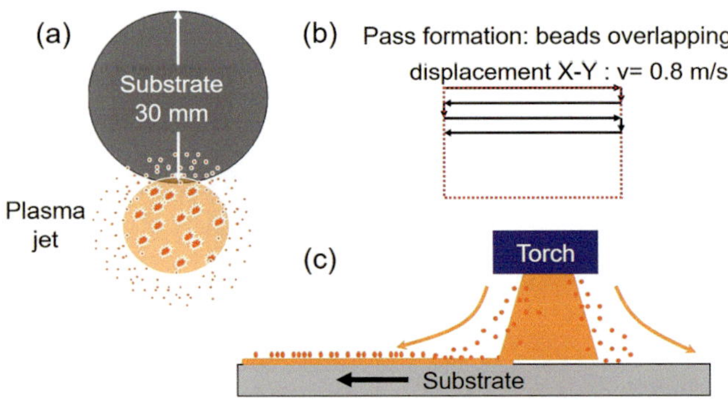

Fig. 16.80 Schematic of successive pass deposition. (**a**) Substrate and particle distribution (well heated in the central part, poorly heated in the periphery). (**b**) Spray pattern. (**c**) Torch movement with the poorly heated particles in the jet fringes [Delbos C. (2004)]. (Reprinted with kind permission)

plasma jet fluctuations have more influence on particles with lower temperatures and velocities, and the columnar structure is emphasized, as shown in Fig. 16.81c. On the cold stainless-steel substrate few non-melted particles have been deposited and the first pass is dense even close to the substrate. It is no more the case for the new successive passes that are more porous close to the previously deposited one. During pass deposition in an x–y pattern with beads overlapping by 1/20, the phenomenon also occurs but the tiny particles are included within beads and less perturb the coating cohesion.

The formation of a powdery layer between successive passes can be avoided by changing the spray pattern and the suspension mass load, as is illustrated in Fig. 16.82 representing the same coating as that of Fig. 16.81b. The spray pattern must be such that beads overlapping limit as much as possible the inclusion of particles traveling in jet fringes and the overheating of successive passes thus also reducing the chances of poorly heated particles to stick to the surface of the coating.

When collecting ceramic sub-micron- or nanosized molten YSZ particles onto smooth alumina, stainless steel or superalloys substrate, preheated above the transition temperature ($\sim$500 K), splats are identical to those obtained with micro-sized sprayed particles. The splat diameters are generally between 3 and 0.2 µm with thicknesses between 300 and 60 nm, corresponding to a flattening degree of about 2 [Fazilleau J. et al. (2006)]. This low flattening degree results from lower impact velocities together with larger surface tension due to the small size of the particles. Moreover, with these splat sizes, for ceramic materials, no cracks during quenching appear.

When using high enthalpy jets (Ar–H$_2$–He) $h_o = 22.3$ MJ/kg, with a heat flux of 18 MW/m^2, the structure close to the stainless-steel substrate is columnar, resulting from splats layering. The columns develop through a thickness of about 4 µm corresponding to the layering of 5 passes, each of them being 0.8 µm thick. It seems that, after a few layers (a few µm thick), the thermal insulation together with the high heat flux from the plasma delays the splat solidification, allowing the

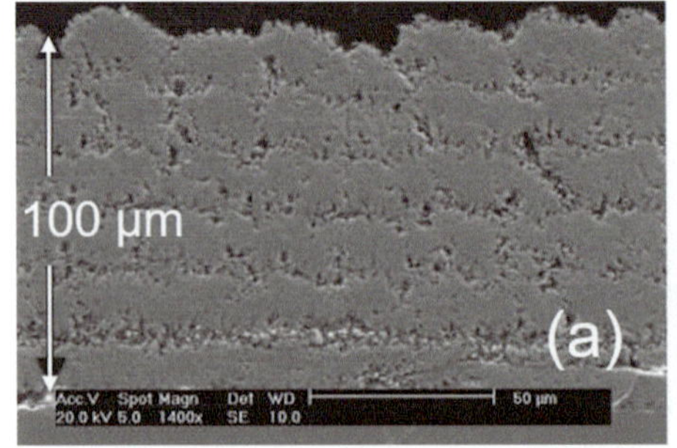

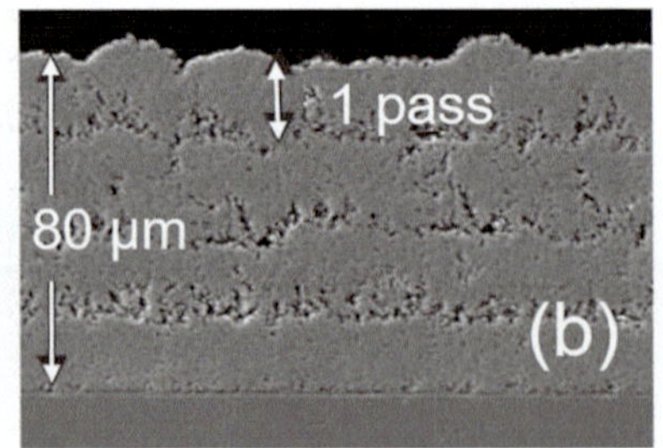

Fig. 16.81 Coatings formed with a PTF-4 DC plasma torch, with 6 mm i.d. anode nozzle, onto a stainless steel smooth substrate (Ra = 0.1 µm) with a SD = 40 mm, using SPS (20 wt.% of powder) of Unitec UC-002h: (**a**) Ar–H$_2$ (**b**) Ar/He (**c**) same spray condition as (**a**), but with a reduced injection velocity [Fauchais P. et al. (2011)]. (Reprinted with kind permission from IOP organization)

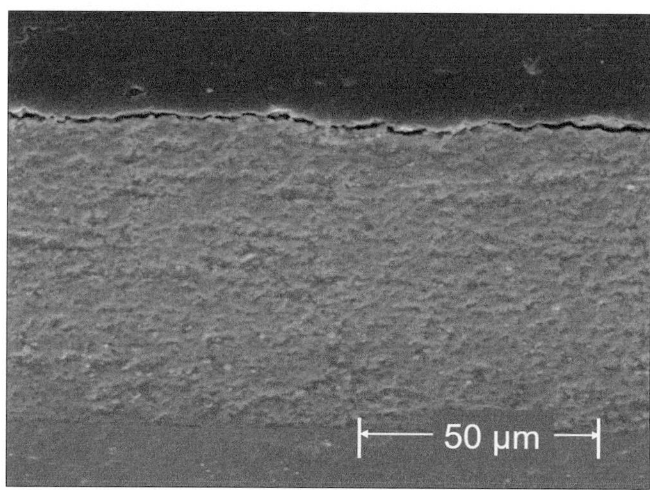

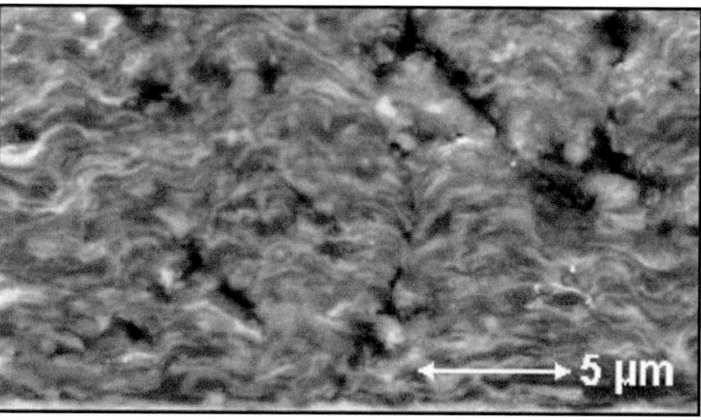

Fig. 16.84 14.76 Cross-section of an alumina–YSZ coating obtained with suspension spraying with the spray conditions of Fig. 14.75 caption, but with a torch nozzle i.d. of 5 mm instead of 6 mm [Delbos C. (2004)]. (Reprinted with kind permission of Dr. Delbos [222])

Fig. 16.82 Cross-section of the same coating as that presented in Fig. 14.73c but with an optimized spray pattern [Fauchais P. et al. (2011)]. (Reprinted with kind permission from IOP organization)

Fig. 16.83 Fractured cross-section of a SPS coating sprayed on a 316 L stainless steel substrate 40 mm downstream of the nozzle exit with a plasma torch 6 mm in i.d., an attrition milled powder and an (Ar–H_2–He) DC plasma [Delbos C. (2006)]. (Reprinted with kind permission from Springer Science Business Media)

surface tension force to take over resulting in quasi-spherical particles in a molten state sticking together before their solidification to achieve the granular structure as observed in Fig. 16.83.

[Delbos et al. (2004)] used YSZ Tosoh powder, with a specific surface area of 13 m^2/g and alumina powder (Baïkowski CR125) with a specific area of 105 m^2/g to prepare a suspension. They were mixed with the same wt.% in ethanol used as the suspension solvent. Other spray parameters were similar to those used for the Tosoh powder, with the exception of the nozzle i.d. which was reduced to 5 mm to increase the plasma jet velocity [Delbos et al. (2004)]. Figure 16.84 shows that the structure of the coating is made up of YSZ and alumina splats alternatively layered (there is a difference of contrast between both oxides: gray for alumina and white for YSZ) with thickness lower than 100 nm, as observed with AFM. Moreover, it could be noticed that the contact between splats seems to be excellent.

[Oberste-Berghaus et al. (2005)] obtained similar results with a Mettech Axial III torch. A follow-up detailed study of alumina–yttria-stabilized zirconia composite SPS using Axial III torch was reported by [Tarasi et al. (2011)]. They used either agglomerates of nanoparticles, or suspensions made using loose nano or micron sized, powder mixtures. Through the study of the in-flight collected particles and coatings produced from the two approaches, it was possible to compare fragmentation, melting and mixing phenomena involved. They showed that both methods can be used in the production of high amorphous coatings provided that the appropriate parameters for each process are utilized. The components in the composite materials sprayed by plasma processes may appear in different forms. They may form crystalline structure of alumina or YSZ with no additional solute atoms. They also can dissolve the solute atoms of the second component and form crystalline solid solutions (even to exceptionally high levels of solubility), and/or form amorphous phase [Tarasi et al. (2011)]. In-flight melting followed by mixing are crucial processes in amorphous coating formation. This melting is strongly controlled by the particle velocity, a lower one resulting in higher amorphous content, in spite of the lower cooling rates.

An example of HV-SFS spraying results is presented with the spraying of TiO_2 nanometer-sized (grain size 12 nm) particles onto aluminum substrates after [Gadow R. et al. (2008a, b)]. In SEM micrographs of titania nanometer-sized powder used appears agglomerated with a primary grain size of approximately 10 nm and agglomerates of about 100 nm. Using XRD analysis both rutile and the anatase phases could be detected in the powder. The suspension used contained 10 wt.% TiO_2 with isopropanol/water mixture as solvent with a volume ratio of 90/10 and DOLAPIX as additive. The

Fig. 16.85 Light microscope image of HVSFS-sprayed TiO$_2$ coating [Gadow R. et al. (2008a, b)]. (Reprinted with kind permission from Elsevier)

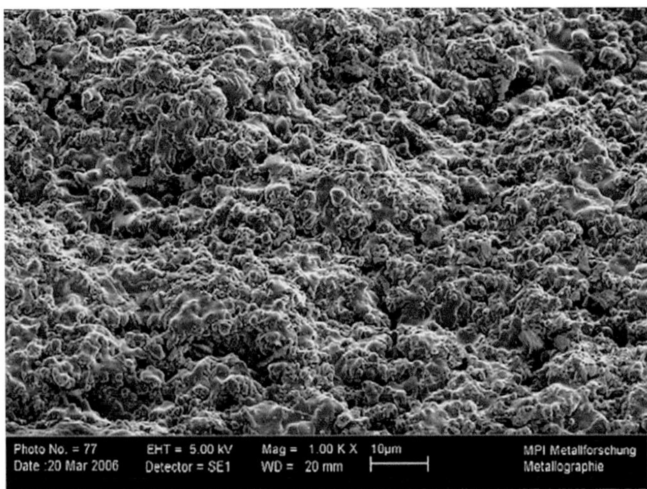

Fig. 16.86 SEM image of HVSFS-sprayed TiO$_2$ coating surface [Gadow R. et al. (2008a, b)]. (Reprinted with kind permission from Elsevier)

addition of water in the solvent facilitated the chemical stabilization of the suspension at the expense of cooling the flame temperature on its evaporation. The HV-SFS torch used was based on a TopGun-G system (manufactured by GTV) modified to meet suspension spray requirements. Propane and ethene were preferentially chosen as combustion fuels as they allow stable operation of the HV-SFS process. The HVSFS torch was mounted on a six-axes robot system using a simple meander kinematics with a 2-mm offset. Sample cooling was performed using two air nozzles attached to the torch body. The suspension was fed at a rate of about 8 g/min using a conically shaped 0.3 mm i.d. nozzle. The thickness of coating, deposited on aluminum substrate, was around 30 µm, the deposition rate being around 4 µm/cycle. As shown in Fig. 16.85, the TiO$_2$ coating appears totally dense, homogeneous, with no visible internal structure, and no detectable non-molten particles. The SEM surface image (Fig. 16.86) reveals fully molten splats ranging from a few tenths to about 10 µm. From the XRD data, the main fraction of the coating consists of anatase (approx. 75%), the rest of the material being rutile. Traces of a sub-stoichiometric species, Ti$_3$O$_5$, could also be identified in the coating.

[Toma et al. (2012)] compared alumina coatings HV-SFS sprayed by SPS with particles in the range of 0.4–3.5 µm to those obtained with HVOF and particles in the size range of 5–25 µm. They used a Top-Gun working with ethylene–oxygen (90–270 slm) with a spray distance of 150 mm for the HVOF spraying. The same gun was also used for the HV-SFS with a modified combustion chamber working with same gases (75–230 slm) and a standoff distance of 80 mm. The results showed that the HV-SFS coating approach presents a specific microstructure: partially melted fine particles with elongated shapes (of about 5 µm in length and 150 nm in thickness), nearly spherical µm-sized particles and clusters of agglomerated tiny sub-µm-sized grains embedded in the matrix of well-melted ones (Fig. 16.87). The main crystalline phase is the α-phase representing 60%. The HVOF coating is more conventional (Fig. 16.88) with layered splats, γ-alumina phase being the main one. The suspension-sprayed Al$_2$O$_3$ coatings present the better electrical resistance stability that can be explained by their specific microstructure and retention of a higher content of α-Al$_2$O$_3$.

[Bolelli et al. (2010)] also HV-SFS-sprayed alumina coatings, using seven different Al$_2$O$_3$-based suspensions prepared by dispersing two nano-sized Al$_2$O$_3$ powders (having analogous size distribution and chemical composition but different surface chemistry), one micron-sized powder and their mixtures in a water + isopropanol solution. HV-SFS coatings were deposited using these suspensions as feedstock and adopting two different sets of spray parameters (propane–oxygen either 55–325 slm or 60–350 slm). As with plasma spraying, the characteristics of the suspension, particularly its agglomeration behavior, have a significant influence on the coating deposition mechanism and, hence, on its properties [Bolelli et al. (2010)]. Dense and very smooth (Ra ~ 1.3 µm) coatings, consisting of well-flattened lamellae having a homogeneous size distribution, were obtained with micro-sized (~1–2 µm) powders. Spray parameters favoring the breakup of the few agglomerates present in the suspension enhance the deposition efficiency (up to 50%), as no particle or agglomerate larger than ~2.5 µm can be fully melted. Nanosized powders, by contrast, generally form stronger agglomerates, which cannot be significantly disrupted by adjusting the spray parameters. When the chosen nanopowder forms small agglomerates

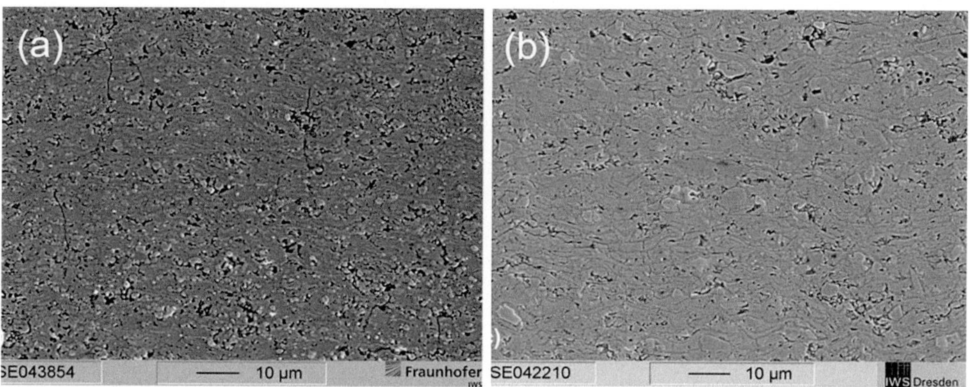

Fig. 16.87 SEM micrographs of cross-sections of thermally sprayed alumina coatings. (**a**) Conventional HVOF-sprayed with particles in the range of 5–25 μm. (**b**) HVSFS sprayed with particles in the range of 0.4–3.5 μm [Toma et al. (2012)]. (Reprinted with kind permission from Springer Science Business Media, copyright © ASM International)

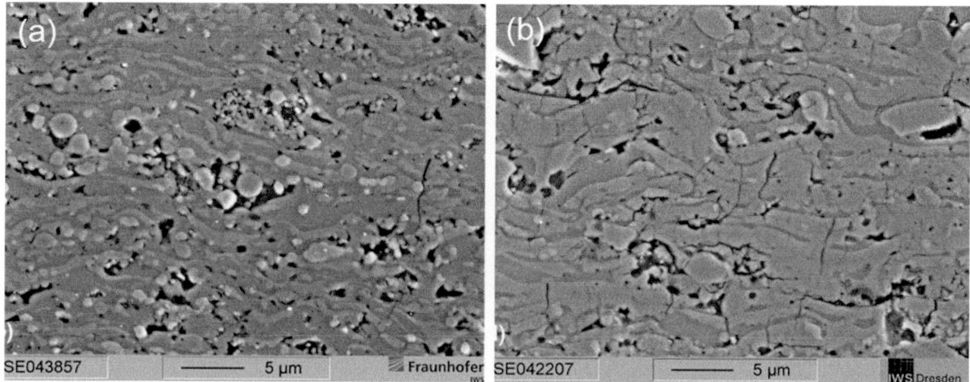

Fig. 16.88 SEM micrographs with higher magnifications of cross-sections of HVOF sprayed coatings depicted in Fig. 16.87. (**a**) Conventional and (**b**) Suspension [Toma et al. (2012)]. (Reprinted with kind permission from Springer Science Business Media, copyright © ASM International)

(up to a few microns), the deposition efficiency is satisfactory, and the coating porosity is limited. If the nanopowder forms large agglomerates (on account of its surface chemistry), poor deposition efficiencies and porous layers are obtained [Bolelli et al. (2010)].

[Müller et al. (2012)] have used two different spray processes, suspension plasma spraying (SPS) and HV-SFS. They used alumina particles: MR52 (0.33–4.07 μm) and APA-0.5 (0.26–1.19 μm) and suspensions were made either with water or isopropanol. For HV-SFS they used a Top-Gun system equipped with a 22-mm long combustion chamber and working with ethene–oxygen (410 slm) with different oxygen to fuel ratios and a spray distance of 120 mm. For SPS they used a F6-Gun of GTV working with Ar–H$_2$ (55–12 slm) and standoff distances of 55 mm with water and 75 mm with isopropanol (to benefit from its combustion).

Figure 16.89 represents particle distributions before spraying and collected in water after HV-SFS spraying. Collected spray particles produced by HV-SFS, exhibited a spherical shape, corresponding to their good melting during their in-flight phase. Those from the course MR52 powder

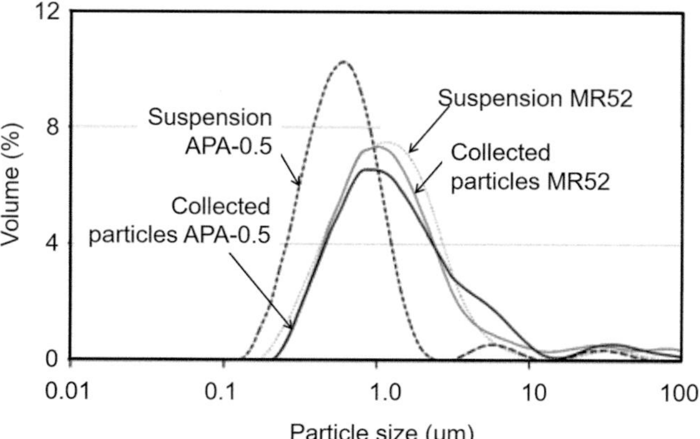

Fig. 16.89 Size distribution of the particles in the used suspensions and of the collected particles from HVSFS spray process for the two used powders [Müller et al. (2012)]. (Reprinted with kind permission from Springer Science Business Media, copyright © ASM International)

exhibit the same size distribution after the spray process as those in the used suspension. The finer APA-0.5 powder has its initial d_{50} value increased from 0.55 μm to a value of

1.23 μm for the collected particles. The size distribution also broadened from a narrow size distribution in the suspension to a broader distribution of the collected particles after spraying. This could be explained by the agglomeration phenomenon. Using water for the suspension results in higher amount of α-alumina phase.

With the SPS process, particle melting was better with isopropanol than with water due to the combustion phenomenon and compared to HV-SFS process splat sizes were smaller. When comparing coatings obtained by HV-SFS and SPS, α-alumina phase is dominant in the HV-SFS coatings, which confirms the better melting of the particles. Using the SPS process, high coating porosities, as high as 40%, were reported, while with HV-SFS low coating porosities (down to 2%) were obtained.

16.4.6 Present and Potential Applications

Compared to conventional thermal sprayed (TS) coatings using micron-sized dry powders those manufactured with Solutions or Suspensions have been recognized to exhibit interesting features the most important of which is their fine nanosized grains, the absence of lamella boundaries and cracks as well as their porous microstructures. While the SPPS and SPS Technology is still in its early stages of development, numerous areas of potential industrial applications have been recognized and are attracting increasing attention. The approach is generally viewed as complimentary rather than competitive to conventional thermal spray Technology and to the best of our knowledge, it is still not being used in large-scale industrial applications.

In the following examples are given of the leading present and potential applications. These are mostly grouped in terms of either applications and/or materials highlighting their relative advantages and limitations compared to conventional TS coating techniques.

16.4.6.1 Biomedical Applications

The use of SPS for the deposition of Hydroxyapatite (HA) using RF-ICP Technology was one of the first attempts in this area reported for biomedical applications by [Bouyer E. et al. (1996a, b, 1997a, b, 2000), Gitzhofer F. et al. (1997)]. As discussed earlier in [Sect. 16.4.2 Precursor Injection and Fragmentation] the approach used involved either SPS Technology to produce micrometer-sized Hydroxyapatite (HA) particles that were easy to spray by conventional means, or direct spraying of the suspension for coating formation.

Subsequent efforts in this area were reported by [Tomaszek R. et al. (2007a, b), Łatka L. et al. (2010), D'Haese R. et al. (2010), Kozerski S. et al. (2010a, b), Podlesak H. et al. (2010), Stiegler N. et al. (2010, 2012)]

including the development of development of multilayer coatings of hydroxyapatite (HA) and TiO_2 onto titanium substrates using SPS [Tomaszek R. et al. (2007a, b)]. [Jaworski et al. (2008)] presented the recent developments in suspension DC plasma-sprayed HA coatings with rather low thickness (10–50 μm). Two types of coatings were tested: duplex and gradient coatings. The density of coatings was improved as well as their cohesion when increasing the torch power level, at the expense of increased HA partially decomposition of HA in the coating.

[Łatka et al. (2010)] plasma sprayed homemade HA powder. The coatings presented two types of zones: (i) dense ones corresponding to splats, as in conventionally sprayed coatings; (ii) sintered ones containing fine hydroxyapatite grains, corresponding to the fine solids from the initial suspension. The latter disappeared after soaking the coating in Simulated Body Fluid (SBF), with the pores filled by the re-precipitated calcium phosphates.

[Gadow R. et al. (2008a, b)] reported that a dense, 50 μm thick, coating structure with a porosity of about 1% was obtained by HV-SFS an ethylene glycol-based suspension with 13 wt.% HA powder. The coatings consisted of about 90% of HA with a small amount of tricalcium phosphate.

[Stiegler et al. (2012)] used the HV-SFS technique to deposit HA coatings onto titanium plates. The use of different dispersion media to prepare the HA suspensions affected the microstructure and mechanical properties of the resulting coatings more than the process parameters. However, coatings containing substantial amounts of amorphous phase tended to dissolve very rapidly in the SBF solution and to be replaced by a precipitation layer consisting of crystalline HA.

Phosphorous containing biocompatible glass coatings prepared using High-Velocity Suspension Flame Spraying (HV-SFS) by [Bolelli G. et al. (2009a, b)] displayed a level of porosity consisting mainly of closed pores with some transverse micro-cracks. The same Technology (HV-SFS) was used by [Altomare et al. (2011)] to deposit 45S5 bioactive glass coatings onto titanium substrates, using a suspension of micron-sized glass powders dispersed in a water–isopropanol mixture as feedstock. Depending on the spray parameters, coatings with different porosities and thicknesses were obtained. The sprayed coatings were entirely glassy but exhibited a through-thickness microstructural gradient and the glass network structure was analogous to that of bulk annealed 45S5 bioglass.

The interaction mechanisms between coatings and the simulated body fluid (SBF) were also analogous to those of the bulk bioglass. They involved controlled dissolution of the glass, polymerization of an amorphous silica layer on the glass surface itself, and growth of a carbonated hydroxyapatite layer on top of the silica layer. Thus it seems that bioglass poses no trouble related to decomposition and phase alteration and the reactivity of the HV-SFS-deposited bioglass coatings seems to be particularly fast, most of the original coating being replaced by the products of the interaction with the simulated body fluid (SBF) in only one week. [Xiao et al. (2011a, b)] plasma sprayed P_2O_5–Na_2O–CaO–SiO_2 Bioactive Glass–Ceramic Coatings (BGCCs), using solutions and suspension spraying. Results indicated that coatings with higher crystallinity were obtained using the solution precursor, while nanostructured coatings predominantly consisting of amorphous phase were synthesized using the suspension precursor. Overall, the liquid precursor plasma spraying process seems to be a promising technique to synthesize nanostructured BGCCs with good in vitro bioactivities. The same authors [Xiao Y. et al. (2011a, b)] plasma sprayed a solution to produce the same BGCC on a Ti substrate. The as-deposited coating consisted of two phases, with a predominant amorphous phase and a crystalline phase of $Na_2Ca_2Si_3O_9$. The in vitro bioactivity of the as-deposited coating was studied by SBF soaking. Results reviled the formation of hydroxide carbonate apatite (HCA) which is likely due to the existence of nanostructure in the as-sprayed coatings.

16.4.6.2 Thermal Barrier Coatings (TBC)

The potential use of solutions (SPPS) or suspension (SPS) for the deposition of thermal barrier coatings (TBC) has been at the center of intense R&D efforts in this area over the last two decades [Padture N.P. et al. (2001), Gell M. et al. (2004, 2008), Jordan E.H. et al. (2004), Xie L. et al. (2004b, c, 2006), Low A.D. et al. (2006), Madhwal M. et al. (2004), Vaßen R. et al. (2009a, b)].

Compared to APS conventional TBCs, coatings produced by *Solution Precursor Plasma Spray (SPPS)* demonstrate interesting properties related to their specific features, including ultrafine splats (which form dense coating regions), through thickness vertical cracks, embedded un-pyrolyzed particles, and voids. The adhesive bond strength, measured according to ASTM C633-79 standard, was 24.2 MPa, while that of conventional APS-deposited coatings was found to be 19.9 MPa [Gell M. et al. (2004)]. The improved bond strength probably results not only from the finer splat size but also may be due to an interfacial thin oxide layer that would form on the substrate and would promote adhesion. The vertical crack density increases the in-plane TBC fracture toughness, improving the coatings cyclic durability in the thermal cycling. [Gell et al. (2008)] have shown that, during thermal cycling, the spallation life was improved by a factor of 2.5 compared to APS coatings on the same bond coat and substrate, and by a factor of 1.5 compared to EB-PVD coatings. The apparent thermal conductivity, as measured by laser-flash technique from 100 to 1000 °C, has been found to be approximately 1.0–1.2 W/m.K, a value lower than that of EB-PVD coatings, but higher than that of

conventional APS coatings. Probably this high conductivity (relative to APS coatings) is due to the increased internal contact area of solution-sprayed coatings. The thermal cyclic stability of these coatings was found to show no significant microstructural or phase changes during 1090 cycles of 1 h each at 1121 °C. The critical features of solution coatings, vertical cracks, and ultrafine splats remained stable throughout the test. Vertical cracks were retained or reformed, even after exposure to 1500 °C. APS and solution sprayed TBCs exhibited similar grain growths, density and hardness increase between 1200 and 1400 °C. Above 1400 °C, APS TBCs exhibited a faster rate of grain growth and transformation into monoclinic phase. The microstructural features of these coatings are retained in very thick coatings (up to 4 mm), the spallation lives being significantly less thickness sensitive and demonstrating superior performance.

Systematic studies of TBCs obtained using SPS and HV-SFS are reported by [Killinger A. et al. (2006, 2011), Kassner H. et al. (2008), Rampon R. et al. (2008)]. Of these, [Kassner et al. (2008)] obtained segmented coatings (7 cracks/mm) for 300-μm thick coatings using plasma spraying. [Ben-Ettouil et al. (2009)] showed, on the other hand, that resistances to Furnace Cycle Test (FCT) of YPSZ-SPS coatings were higher for coatings exhibiting lower crack density. The thermal diffusivity of as-sprayed coatings measured at atmospheric pressure was noted to change from 0.015 to 0.025 mm^2/s for a temperature rise from room temperature to 250 °C. Such low values, which are about 10 times lower than commonly measured values on coatings manufactured with micro-sized agglomerates of nanosized particles, was attributed to the peculiar void network architecture (80% of voids smaller than 30 nm) of such coatings. According to [Vaßen R. et al. (2009a, b)] besides phonon scattering phenomena in the smallest grains of the coating structure, numerous thermal resistances in the structure, thermal resistance is increased by rarefaction effects due to small grain volume. Such nanosized coatings were also observed to exhibit enhanced optical absorption and strong IR scattering compared to conventional micro-sized YPSZ coatings, due to the number and size of the voids, and due to the density of interfaces in between grains and lamellae forming such coatings [Vaßen R. et al. (2009a, b)].

In the recent thermal spray of Suspensions & Solutions Symposium (TS-4) held Sept. 13–14 (2017) in GE Niskayuna, NY, USA, the development of higher temperature TBCs with yttrium aluminum garnet (YAG) using SPPS was reported by [Gell M. (2017)]. These had the potential to be used at temperatures 200 °C higher than conventional YSZ APS-TBCs, due to increase phase stability, sintering and Calcium Magnesium Alumino Silicates (CMAS) infiltration resistance. These coatings also exhibited 50% reduction in high temperature (1000 to 1300 °C) thermal conductivity and increased thermal cyclic durability compared to APS

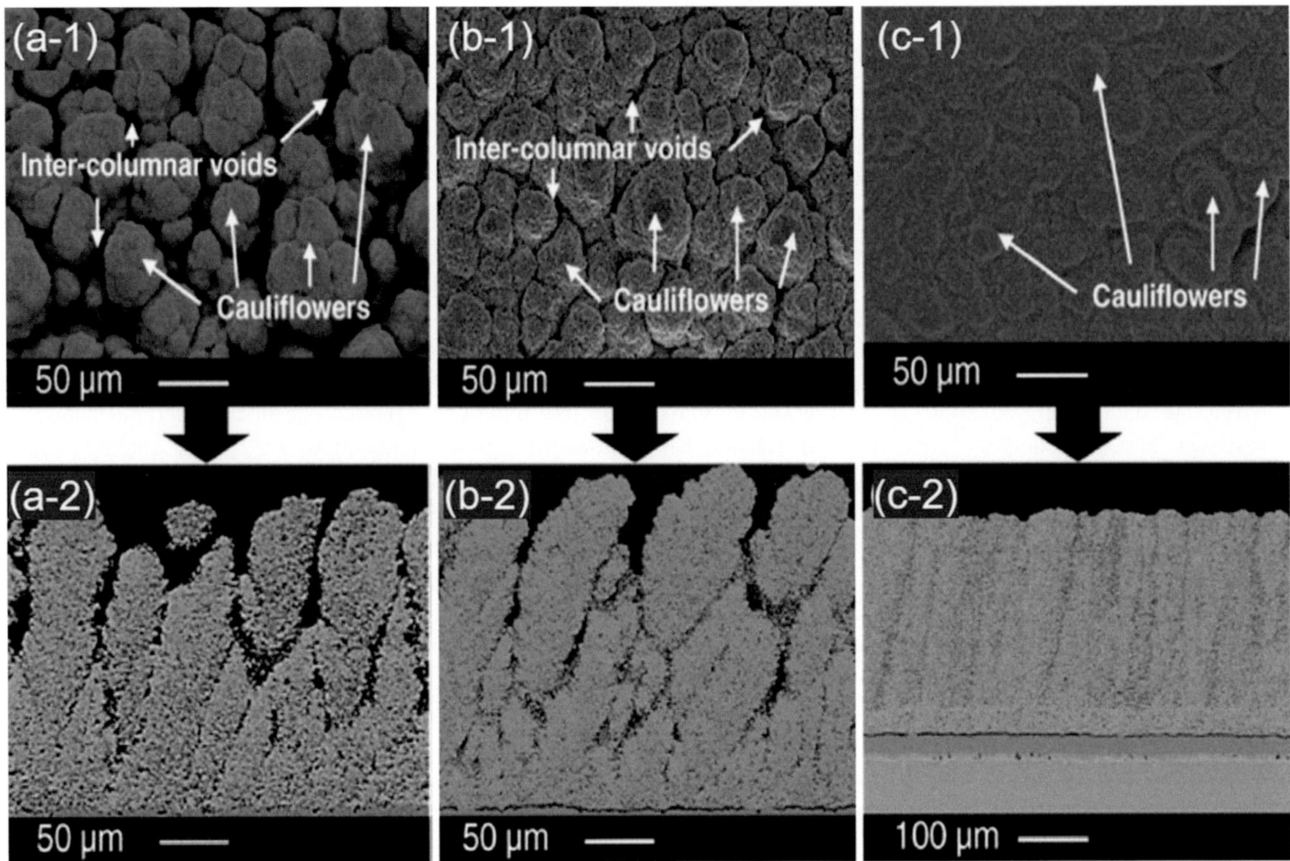

Fig. 16.90 SEM micrographs of SPS-YSZ TBCs, top views (a-1, b-1 and c-1) and cross-sections (a-2, b-2 and c-2) [Bianchi L. et al. (2017)]

YSZ. Subsequent activities in this area was undertaken by Solution Spray Technology Inc. for the further development and commercialize the SPPS Technology such as the deposition of thin (5–25 µm) dense coatings (>95%) for a variety of applications, including environmental barrier coatings, oxygen permeable membranes, fusion reactors and corrosion and contamination protection applications.

Developments of SPS spraying of TBCs by [Kitamura J. et al. (2017), Bianchi L. et al. (2017), Bernard B. et al. (2017)] showing the potential of producing a wide range of coatings with columnar microstructure as shown in Fig. 16.90 after [Bianchi L. et al (2017)]. SEM Micrographs shown in Fig. 16.90a-2 and Fig. 16.90b-2 represents typical SPS columnar structures separated by inter-columnar voids. Figure 16.90a-2 made using a lower transverse speed than that shown in Fig. 16.90b-2 presents an enlarged column size distribution, particularly for the higher coating thickness. The compact structure of coating shown in Fig. 16.90c-2 could be attributed to the use of a lower plasma enthalpy and a higher suspension feed rate.

[Mahade S. et al. (2015)] proposed double layer TBC comprising Solution Spray Technology (GZ)/YSZ and triple layer TBC (GZdense/GZ/YSZ) comprising relatively denser GZ top layer on GZ/YSZ. These materials were deposited by

suspension plasma spray. The multilayered GZ/YSZ TBCs were shown to have lower thermal conductivity than the single layer YSZ. The as-sprayed TBCs were also subjected to thermal cyclic testing at 1300 °C. The double and triple layer TBCs had a longer thermal cyclic life compared to YSZ. In a subsequent study [Mahade S. et al. (2017)] reported two different TBC architectures deposited using the axial suspension plasma spray: double layer TBC comprising gadolinium zirconate (GZ)/YSZ and triple layer TBC (GZdense/GZ/YSZ) comprising relatively denser GZ top layer on GZ/YSZ. The first variation was a triple layered TBC comprising of thin YSZ base layer beneath a relatively porous GZ intermediate layer and a dense GZ top layer. The second variation was a composite TBC architecture of GZ and YSZ comprising of thin YSZ base layer and GZ + YSZ top layer. TBCs after hot corrosion test revealed the infiltration of the molten salts through the columnar gaps. The composite TBC showed a lower hot corrosion induced damage compared to the layered TBC where a considerable spallation was observed. Using pyrochlores due to their excellent properties such as lower thermal conductivity and better CMAS attack resistance compared to YSZ, unfortunately have inferior thermalos-mechanical properties compared to YSZ. [Gupta M. et al. (2017)] studied a bilayer

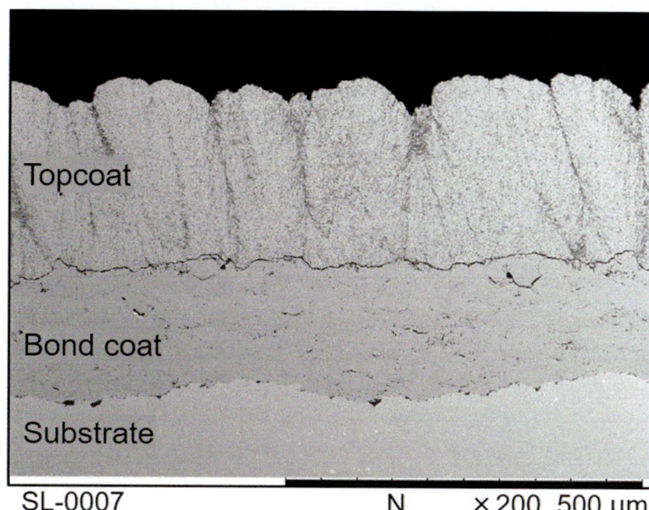

Fig. 16.91 14.21 SEM micrograph of, as-sprayed, single-layer, TBC [Gupta M. et al. (2017)]

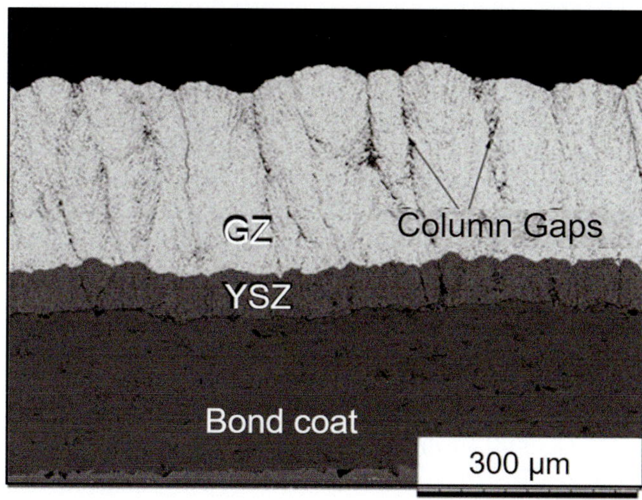

Fig. 16.93 14.23 SEM micrographs of, as-sprayed, cross-section, of double layer GZ/YSZ TBC [Mahade S. et al. (2015)]

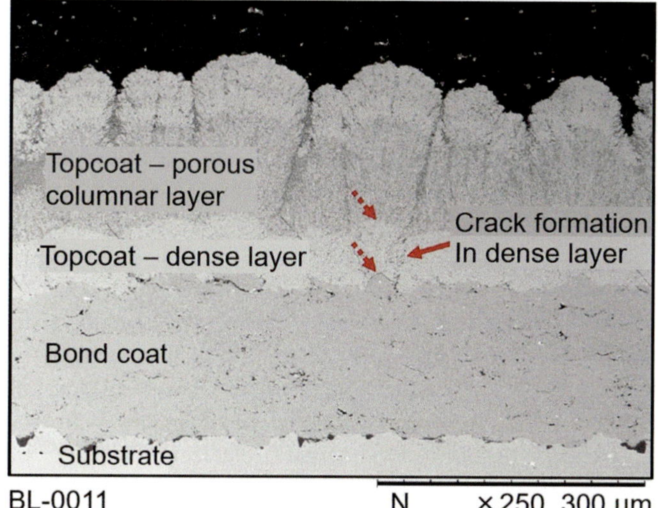

Fig. 16.92 14.22 SEM micrographs of, as-sprayed, bilayer, of TBC [Gupta M. et al. (2017)]

topcoat with a dense layer close to the topcoat-bond coat interface followed by a porous columnar layer fabricated by SPS (see Fig. 16.91) and its lifetime under thermal shock testing was compared to the single layer porous columnar SPS topcoat in TBCs.

They showed that graded coating with denser microstructure near the TC-BC interface and porous microstructure toward the surface with column gaps/vertical throughout the topcoat (Fig. 16.92) could be beneficial to further extend TBC lifetime as it would not contain a clear interface between the porous and dense layer and thus preventing failure in the interface.

[Curry N. et al. (2015)] studied within a TBC system the surface topography of bond coats modified by the combination of polishing and surface grit blasting. YSZ coatings were deposited using an axial feed suspension plasma spray gun. Samples were tested for thermalo-cyclic fatigue lifetime at 1100 °C during 1 h cycles. Thermal shock performance was evaluated using the burner rig test and thermal conductivity evaluated using the laser flash analysis. The results indicate that columnar SPS coating microstructure is strongly influenced by surface topography. Test results suggest that control of surface topography may be an important factor to improve the performance of SPS coatings. [Mahade S. et al. (2015)] plasma sprayed suspension multilayered (double and triple layer) TBCs (Fig. 16.93). Their functional performance (thermal conductivity and thermal cyclic fatigue) were compared with the current standard YSZ single layer TBC. A denser third layer on top of the columnar coatings was created by spray parameter alterations. Lower thermal conductivity and improved thermal cyclic fatigue life were obtained compared to YSZ single layer TBC.

The same authors with R. Vassen [Mahade S. et al. (2017)] compared functional performance (thermal conductivity and thermal cyclic life) of 8YSZ single layer and multilayered Gadolinium Zirconate/YSZ TBCs. multilayered GZ/YSZ TBCs was shown to have a lower thermal conductivity compared to the single layer YSZ coating, despite its higher porosity content. The GZ/YSZ multilayered TBCs also had a lower thermal diffusivity and presented a longer thermal cycling life at 1300 °C compared to the YSZ single layer TBC. To improve both the CMAS and the erosion resistance of the TBC a denser top layer in the case of triple layer TBC could be created using a different suspension of same composition with higher median particle size and higher solid load content. [Zhou D.et al (2017)] used axial injection to spray TBCs by SPS and compared results with those obtained with EB-PVD. In their tests all columnar structure TBCs showed relatively short lifetimes compared to porous APS coatings. They found a correlation between mechanical properties and lifetime for the SPS coatings. [Shen Q. et al. (2017)] presented two models for describing the thermally grown oxide (TGO) layer growth. They investigated the effects of interface roughness on the evolution of TGO thickness and growth stress.

[Chang H. et al. (2017)] investigated CMAS corrosion induced microstructure evolution and composition development in YSZ thermal barrier coatings prepared by electron-beam physical vapor deposition (EB-PVD). [Petorak C. (2017)] examined variations in SPS microstructures and composite SPS coatings in thermal conductivity and CMAS resistance. [Li X.-H. (2017)] compared SPS coatings to APS and EB-PVD ones for possible applications in industrial Siemens gas turbines. [Ganvir A. et al. (2017)] studied, with axial injection, the effect of the feedstock, spray conditions, solid loading, particle size distribution and solvent type to produce different coating microstructures with 8 wt. % YSZ while keeping the same deposition conditions. Based on this work, they determined various factors needed for designing an optimized columnar TBC for gas turbine applications.

[Tang Z. and N. Young (2017)] developed SPS process with high spray rate to achieve comparable deposition rate similar to those obtained with APS and tens of micrometer-sized particles. [Sokolowski P. et al. (2017)] studied the thermal transport properties of suspension-and-solution precursor plasma sprayed YSZ coatings. They used different slurries (water/ethanol) with various solid contents for SPS. The solution was prepared to achieve 8YSZ. They studied the influence of the use of various feedstocks. Burgess [2017] has presented results obtained by Spray-werks Technology Inc. when spraying with Axial III and the new micrometer-sized powder dispenser spraying with a gas solid particle in the nanometer sized. They sprayed 7 wt. % YSZ. An Accuraspray system was used to monitor in-flight particles temperatures and velocities according to spray parameters. The correlation between spray parameters and coating microstructures and thermal conductivities were presented and discussed. [Hazel B. (2017)], from Pratt and Whitney, pointed out that SPS strain tolerant vertically segmented and columnar structures are of particular interest for application of TBC engines. These strain tolerant structures are expected to have the ability to withstand the high temperatures and thermal cycling inherent to gas turbine operation. [Musalek et al. (2017)] compared zirconia-based coatings deposited by two types of high-energy plasma torches: (i) Axial III; and, (ii) hybrid version of Water-Stabilized Plasma (WSP) torch. The suspensions were formulated using solid dispersed phase of: (i) zirconia stabilized with 14 wt.% of Y_2O_3 and (ii) zirconia stabilized with 24 wt.% of CeO_2 + 2.5 wt.% of Y_2O_3 and continuous phase of water with ethanol. The EDS/EDX and XRD studies showed that there was no significant change in chemical/phase composition of zirconia material before and after spraying. Coatings were characterized by sub-micrometric microstructure what corresponded to the size of powder particles used to formulate suspension.

16.4.6.3 Wear-Resistant Coatings

Materials commonly used for the thermal spraying of Wear-resistant Coatings include; Tungsten Carbide-Cobalt, Alumina, Zirconia-alumina, Alumina–Zirconia–Yttria, and Titania-Titanium carbide.

Tungsten Carbide-Cobalt

WC-Co coatings were deposited using an axial injection plasma torch and an ethanol-based suspension of sub-micrometer-sized feedstock powders [Oberste-Berghaus J et al. (2006a, b)]. With a proper selection of spray parameters, porosities below 0.2% were achieved with hardness over 700 HV_{3N}. The best coatings corresponded to the highest particle velocities (~800 m/s) and the lowest particle temperatures (at the minimum, 2200 °C). It seems that the measured carbon loss in the carbide is principally associated to deposition of layers of highly oxidized overspray. The properties of the feedstock powder and feed suspension play a critical role in the resulting coatings quality.

Alumina

[Toma F.-L. et al. (2010a, b)] provided an overview of SPS and HV-SFS of Al_2O_3 coatings using water or ethanol-based, alumina powder suspensions with a mean particle diameter of 0.3 μm. Coatings obtained with ethanol-based suspension axially injected in DC plasma spray gun operating with Ar–H_2 plasma-forming gas mixture had a porosity of 18.9 ± 2.5%, against 11.6% porosity when operating with Ar–He. With HV-SFS (acetylene–oxygen) and aqueous-based suspension, the porosity was 7.6 ± 2.8%. A comparative study of the characteristics (microstructure, phase composition, and mechanical properties) and electrical properties (insulating resistance, electrical resistivity, and dielectric strength) of Al_2O_3 coatings produced by conventional HVOF (fed with micrometer-sized dry powder) and HV-SFS fed with a suspension of particles, with a PSD in the range of 0.4 and 3.5 μm was reported by [Toma F.-L. et al. (2012)]. The results showed that at low levels of Relative Humidity (H < 40%) the electrical resistivity was on the order of 10^{11} Ω.m for both coatings. At a very high humidity (97% RH) the electrical resistivity values for the HV-SFS coatings were in the range of 10^7–10^{11} Ω.m, up to five orders of magnitude higher than those recorded for the HVOF coating (10^6 Ω m). The specific microstructure of the suspension-sprayed coating is probably due to its much lower pore sizes. However, at RH = 97% the electrical properties were found

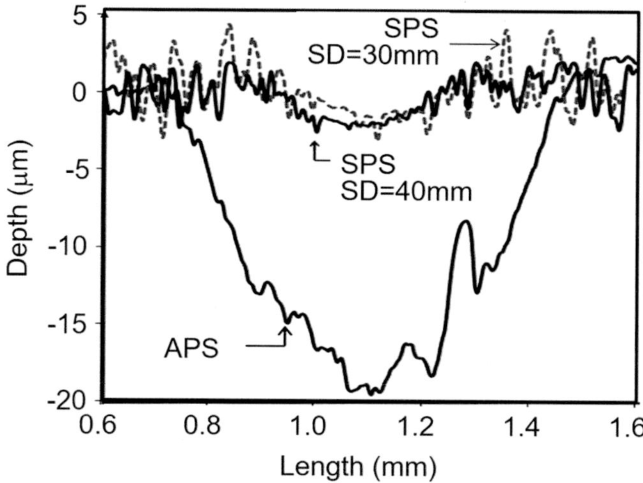

Fig. 16.94 Wear profiles for α-Al_2O_3 micrometer-sized APS coating, and sub-micrometer-sized SPS coatings, spray at distances of 30 and 40 mm [Darut et al. (2008)]. (Reprinted with kind permission from Springer Science Business Media, copyright © ASM International)

to degrade significantly (by up to 4–5 orders of magnitude) in these coatings after long-term exposure. The dielectric strength of suspension-sprayed coatings was found to be 19.5–26.8 kV/mm for coating thicknesses ranging from 60 to 200 µm, which is probably due to its microstructure.

[Darut et al. (2008)] compared the tribological properties of alumina coatings structured at two different scales, a micrometer one (d_{50} of particle size distribution of 36 µm) manufactured by APS, and a sub-micrometer one (d_{50} of particle size distribution of 0.4 µm) manufactured by SPS. Coating architectures were analyzed and their friction coefficient and wear rate in dry sliding mode measured in ball-on-disk friction conditions using α-Al_2O_3 ball with a diameter of 6 mm, a load of 2 N, relative speed of 0.1 m/s and sliding distance of 1500 m. The test was operated in the dry mode and the wear scraps were constantly removed by an air jet located behind the contact point. The friction coefficient of Al_2O_3 coatings dropped by a factor of two (0.4 for SPS layers to be compared to 0.8–0.9 for APS ones) when the characteristic scale was divided by two orders of magnitude. Moreover, wear rate was 30 times lower for SPS layers compared to the one of APS layers, as shown in Fig. 16.94.

[Qiu and Chen (2009)] manufactured by suspension plasma spraying nanostructured fine alumina coating using nanoscale alumina powders (d ~ 20 nm), ethanol being used as liquid phase. The quality of the coatings strongly depended upon the plasma power, spray distance, liquid phase, solid phase concentration, inner diameter of the suspension injector, and suspension flow rate.

Nanostructured Al_2O_3–13 wt.% TiO_2 coatings were manufactured [Zhang J. et al. (2008a, b), Sodeoka S. et al. (2006)] by plasma spraying with nanocrystalline powders.

The TEM analysis revealed that partially melted Al_2O_3 particles, in the size range of 20–70 nm, were embedded in a TiO_2-rich matrix. The mechanism of the substructure formation was explained in terms of melting and flattening behaviors of the powder particles during plasma spray processing.

[Oliker et al. (2009)] found that SPS alumina coatings exhibited better hardness and mainly toughness than conventionally sprayed coatings.

Zircconia-Alumina

Nanometer-sized composites made of *alumina–zirconia* can potentially exhibit higher hardness, fracture toughness, slower crack growth, and lower thermal conductivity than alumina or zirconia alone. Both solution [Vasiliev A.L. et al. (2006a)] and suspension spraying [Gadow R. et al. (2008a, b), Oberste-Berghaus J. et al. (2008), Chen D. et al. (2009a, b), Tingaud O. et al. (2010), Manzat A. et al. (2010)] were used. [Gadow et al. (2008a, b)] have sprayed by HV-SFS a mixture of Al_2O_3 (150 nm) and YPSZ (12 nm) dispersed in 30/70 wt.% isopropanol/water mixtures with a 20 wt.% solid content. The coatings reveal that nanometer-sized particles of zirconia are embedded in an alumina matrix. The spray conditions were tailored to achieve complete alumina melting with the zirconia particles remaining solid. As the time between melting and reconsolidation is extremely short, the two components are almost immiscible. [Oberste-Berghaus J. et al. (2008)] have also deposited Al_2O_3–ZrO_2 coatings by SPS and HV-SFS, both with axial injection. With optimized spray conditions, the smaller nanosized particles create fine laminates of alumina and zirconia layers, while slightly larger particles promote pseudo-alloyed alumina–zirconia amorphous phase components. A large variety of microstructures and properties of suspension-sprayed Al_2O_3–ZrO_2 composites was introduced by the different feedstock, spray systems and operating conditions.

Also an alumina/8 wt.% YPSZ was deposited by axial injection SPS process by [Tarasi et al. (2008)]. The effects of main deposition operating parameters on the microstructural features were evaluated using the Taguchi design of experiment. The results indicate that thermal diffusivity of the coatings, an important property for potential thermal barrier applications, is barely affected by the changes in porosity content during annealing treatments. At last, [Chen et al. (2009b)] have sprayed with a rod-type cathode DC plasma torch an Al_2O_3–ZrO_2 amorphous powder feedstock [Chen et al. (2009a)].

Solution plasma spraying was used to spray 75% dense coatings of metastable ceramics in the ZrO_2–10 mol.% Al_2O_3 binary system [Vasiliev A.L. et al. (2006a)]. The coating microstructure contains a rich variety of features and phases. Coatings were predominantly nanostructured (grain size < 100 nm) and made of tetragonal phase of ZrO_2 with Al^{3+} in solid solution.

[Tingaud et al. (2010)] manufactured by suspension plasma spraying (SPS) several alumina ($d_{50} = 0.6$ μm) and alumina–zirconia (YSZ $d_{50} = 0.4$ μm) composite coatings, varying the operating conditions to achieve dense and cohesive structures. For alumina they measured particle temperatures around 2000 °C and velocities as high as 450 m/s, resulting in coatings with high amount of α-phase. The preliminary results of the SPS wear behavior show rather low friction coefficient and good wear resistance, which is even better with the addition of ZrO_2 in the alumina matrix.

[Darut et al. (2008)] showed that the Al_2O_3-ZrO_2 composite coatings showed higher wear resistance than the Al_2O_3 coatings.

Alumina–Zirconia–Yttria Coatings

Alumina–Zirconia–Yttria coatings were Solution plasma spraying [Vasiliev A.L. et al. (2006a)] was used to process the ternary systems (mol%) $10Al_2O_3$–$86.4ZrO_2$–$3.6Y_2O_3$ and $20Al_2O_3$–$76.4ZrO_2$–$3.6Y_2O_3$. The nanostructures (10–40 nm) in both coatings were made predominantly of tetragonal ZrO_2 phase with some cubic phase. The sub-micrometer-sized regions contain, in addition to t-ZrO_2, small amounts of crystalline Al_2O_3 phases. Thus, solution plasma spraying allows producing metastable ceramics with unusual structures.

[Manzat et al. (2010) and Killinger A. et al. (2010)] have used HV-SFS to spray TiO_2/TiC coatings for automotive engine applications. The suspension was composed of a nanosized titania and a sub-micron-sized TiC dispersed in isopropanol. They succeeded to achieve spraying onto the inner surfaces of light metal and gray cast cylinder liners. Tests performed under dry sliding and lubricated conditions demonstrated an improved wear resistance compared to gray cast iron.

16.4.6.4 Corrosion Resistant Coatings

A CaO–ZrO_2–SiO_2 glass frit was processed by HV-SFS [Bolelli G. et al. (2008)]. Such glass, if dense, should have excellent anticorrosion properties on metal or ceramic substrates. The coatings obtained were much denser than those formerly investigated, and plasma sprayed. Some problems were connected to the formation of molten glass deposits in the combustion chamber and their subsequent embedment in the coating. Work is in progress to overcome this problem and to achieve dense coatings with strong inter-lamellar cohesion.

[Waltz et al. (2012)] have developed a synthesis for the preparation of nanocrystalline magnesium fluoride suspensions, which delivers nearly mono-disperse nanoparticles. These suspensions were deposited with a plasma torch onto TIG welded seams of the magnesium alloy AZ31, thus producing a protective magnesium fluoride layer. The corrosion properties of the magnesium fluoride layer on the welded seams were studied by means of potentiodynamic potential measurements and showed a significant lower corrosion rate.

SnO₂ Layers

An aqueous $SnCl_4$ solution feedstock was used to integrate the spray pyrolysis and sintering of nanometer-sized SnO_2 particles in nanometer-sized porous coating production [Chien and Coyle (2007)]. The coating consisted of interpenetrating pores separating agglomerated particles that themselves exhibited a layered structure with micropores between the layers. Each layer was comprised of SnO_2 nanometer-sized particles. The structure was apparently formed from overlapping viscous hollow spheres of Sn solution, which were pyrolyzed and annealed in situ. Sensors constructed directly from the porous coatings have good ethanol gas sensitivity at 200 °C.

Inconel Coatings

A HV-SFS process has been developed to conduct spraying of small Inconel alloy 625 particles in suspension [Ma X.Q. et al. (2006)]. The coatings exhibited several interesting characteristics for potential applications, including full density, uniform microstructure, and high bond strength. Unfortunately, the preliminary erosion tests, carried out at 90° impact angle, indicated that the conventional HVOF coating was superior to the HV-SFS coatings.

Antierosion Coatings

[Kitamura et al. (2011)] have sprayed yttrium oxide (Y_2O_3) coatings by axial suspension plasma spraying with fine powders ($d_{50} = 1$ μm). Coatings had high hardness (640 $HV_{1.96N}$), low porosity (0.3–0.4%), high erosion resistance (about 1.5 times that of bulk yttria) against CF_4-containing plasma and retention of smooth eroded surfaces. This suggested that the axial suspension plasma spraying of Y_2O_3 is applicable to fabricating equipment for electronic devices, such as dry etching.

Adhesive Layer on Smooth and Thin Substrate

[Vert et al. (2010)] have studied the use of suspension plasma spraying to manufacture 1-mm thick zirconia coatings on smooth and thin (1 mm) substrates. The spraying conditions were optimized for the deposition of coatings on stainless steel AISI 304 L substrates, and the process then adapted for Haynes 230 nickel-based alloy substrates. Special attention was paid to coating adhesion that was investigated by using a Vickers indentation cracking method. The adhesion of various coatings was studied and the effect of the substrate temperature on adhesion confirmed.

16.4.6.5 Titania Photo-Catalytic Coatings

Titania photo-catalytic layers were sprayed by Suspension Plasma Spraying [Tomaszek R. et al. (2006, 2008), Yang G.-J. et al. (2007), Garcia E. et al. (2009), Jaworski R. et al. (2008), Toma F.-L. et al. (2006a, b), Podlesak H. et al. (2008), Znamirowski and Ladaczek (2008), Kozerski S. et al. (2010a, b), Bannier E. et al. (2011), Yuan J. et al. (2012), Toma F.-L. et al. (2009), Vaßen R. et al. (2009a, b), Moroz N.A. et al. (2010)] or HV-SFS [Killinger A. et al. (2006), Bolelli G. et al. (2009), Rauch J. et al. (2009), Gutzmann H. et al. (2010)]. TiO_2 coatings are mostly designed for photo-catalytic applications. As pointed out by [Toma et al. (2006a, b, 2008, 2009)], it is generally assumed that the metastable phase, anatase, presents higher photo-catalytic activity than the stable one, rutile. Coatings sprayed with ethanol base suspensions contained only 23% of anatase ratio (against 82 vol.% in the feedstock powder) and ensured a very low photo-catalytic decomposition of nitrogen oxides. In contrast, with water base suspensions, the anatase phase and crystallites size were preserved and the conversion rate of pollutant reached 40% against 32% for the starting powders [Podlesak H. et al. (2008)].

Besides, [Tomaszek et al. (2007a, b)] measured the field emission characteristics of the obtained TiO_2 deposits. The influence of conditioning on emission properties was observed. The lighting segment excited by electrons from the field titania cathode has also been studied by [Znamirowski and Ladaczek (2008)]. The titania cathode of the segment was manufactured with suspension plasma spraying. Operating tests reveal effective lighting properties of the segment, with very intense light emitted.

[Kozerski et al. (2010a, b)] studied the influence of the suspension solvent and injection on photo catalytic properties of plasma-sprayed titania suspension. The main conclusion of the study is that the coatings containing mainly rutile are also photo catalytically active for the degradation of aqueous solution of dye methylene blue. The photo-catalytic activity was not found to be correlated with the anatase content; thus further studies should clarify the role of suspension plasma spraying processing parameters, substrate material, suspension formulation, and coating microstructure (porosity, content of hydroxyl groups on the coatings surface, gap energy of the coatings, etc.) on the photo-catalytic activity. [Bannier et al. (2011)] prepared TiO_2 coatings by suspension plasma spraying from a TiO_2 nanoparticle suspension using two different substrates (a standard stainless steel and a Pyrex glass). They studied the influence of spray parameters and plasma torch design. A large amount of anatase was obtained, ranging from 32 to 72 wt.%. Coatings displayed a bimodal microstructure with completely fused areas and non-molten ones with agglomerated anatase nanoparticles. The main parameters controlling this bimodal structure were the spray distance and the cooling. [Yuan et al. (2012)] flame sprayed

mixtures of suspensions and powder feedstock. They tailored over a wide range the anatase–rutile ratio and the proportion of nanostructures in the coatings, and up to 30 wt.% of anatase in the coatings was obtained. The hybrid micro−/nanostructured TiO_2 coatings exhibited super-hydrophilic performances ($0°$ contact angle). Results were closely linked to the concentration of nano TiO_2 particles (10–30 nm in size) in the starting suspension.

The HV-SFS process leaves a fairly large degree of freedom to tailor TiO_2 coating characteristics (thickness, porosity, anatase content, hardness, etc.) according to the required functional properties by adjusting operating spray parameters [Bolelli G. et al. (2009a, b)]. For example, coatings with higher anatase content and higher specific surface can be produced to achieve higher photo catalytic efficiency, better than conventional APS and HVOF TiO_2.

Porous titania coatings were plasma sprayed from an aqueous solution containing titanium iso-propoxide [Chen D. et al. (2008a, b)]. It was shown that when increasing the power level, the anatase phase content decreased and the rutile one increased. In fact, the formation of crystalline anatase and rutile in the as-sprayed coatings was the result of plasma flow heat treatment on the gel like deposit (many non-pyrolized droplets).

16.4.6.6 Solid Oxy-Fuel Cell Components

Numerous works are in progress in the world for the SOFC [Brousse E. et al. (2008), Gadow R. et al. (2008a, b), Vaßen R. et al. (2010), Wang Y. et al. (2012), Bacciochini A. et al. (2010), Fauchais P. et al. (2007), Waldbilling D. et al. (2007), Stöver D. et al. (2006), Brinley E. et al. (2007), Oberste Berghaus J. et al. (2008), Monterrubio-Badillo C. et al. (2006), Wang X.-M. et al. (2010), Bouaricha S. et al. (2005), Waldbillig and Kesler (2009a, b), Marchand O. et al. (2010), Michaux P. et al. (2010), Waldbillig and Kesler (2009a, b), McCoppin J. et al. (2011)] For the cathode and electrolyte, [Kassner et al. (2008)] have plasma sprayed, using two injectors and a Triplex torch, LSM (($La_{0.65}Sr_{0.3}$) MnO_3) and YPSZ to produce a functional layer for SOFC cathodes. Works have also been devoted to deposition of $LaMnO_3$ perovskite thin films by suspension plasma spraying [Monterrubio-Badillo C. et al. (2006)].

According to the small particle size, such coatings should provide a higher amount of triple phase boundaries. The cathode manufacturing by thermal spraying is challenging because $LaMnO_3$ doped with SrO, CaO, etc., is prone to decompose easily above the melting temperature. [Monterubio et al. (2006)] have shown that pure $LaMnO_3$ (the easiest to decompose) can be sprayed, with less than 10% of La_2O_3 formation, if the feedstock particles were doped with 10 mol.% of MnO_2. In this condition, particles with sizes between 1 (d_{10}) and 5 (d_{90}) μm (mean size d_{50} of 3 μm) were sprayed with a pure Ar DC plasma jet to limit, as much as possible, the heat transfer, and consequently the particle temperature. [Waldbillig and Kesler (2009a, b)] reported that the properties of $La_{0.8}Sr_{0.2}MnO_3$ (LSM) cathode prepared were strongly influenced by the spray distance used.

Thin (~10–20 μm) electrolyte coatings of Y-PSZ by SPS are supposed to reduce Ohmic losses, but these layers must be impervious to gases. Efforts were mainly devoted to the study of the influence of various solid loadings in suspensions together with their state of dispersion, the stability of the suspension (by adjusting the amount of dispersant), the suspension tailored viscosity, the atomization effect (gas-to-liquid mass ratio (ALR), influence of Weber number), the effect of mechanical injection, the solid particle size distribution, and the manufacturing process [Etchart-Salas R. et al. (2007), Brousse-Pereira E. (2010), Vaßen R. et al. (2010), Kassner H. et al. 2008), Delbos C. et al. (2004), Waldbillig and Kesler (2011), Fauchais P. et al. (2007), Waldbilling D. et al. (2007), Stöver D. et al. (2006), Brinley E. et al. (2007), Oberste Berghaus J. et al. (2008), Monterrubio-Badillo C. et al. (2006), Wang X.-M. et al (2010), Bouaricha S. et al. (2005), Waldbillig and Kesler (2009a, b), Marchand O. et al. (2010), Michaux P. et al. (2010), Waldbillig and Kesler (2009a, b), McCoppin J. et al. (2011), Xiaoming W. et al. (2011)]. By adapting the spray conditions, it became possible to achieve a leakage rate of 0.5 MPa L/s m^2 for a 15 μm thick electrolyte, which is not yet the requirement of 0.08 MPa L/s m^2 [Brousse-Pereira E. (2010)].

With the Prosol process [Wittmann-Ténèze K. et al. (2008)] (sol suspension), the structure and the proportion of the crystalline phases in the coatings are typically the same as those encountered in the injected sol. Coatings are dense when the sol is processed with high enthalpy plasma flows (20 MJ/kg) and consist mainly of very fine particles. [Bouaricha et al. (2005)] sprayed, with the Axial II torch, suspension of nanosized particles of Samarium-Doped Ceria (SDC) in order to produce thin electrolyte layers for SOFC applications with reduced both Ohmic losses and operating temperature. Original phase composition was retained and, independently of thermal spray conditions, the SDC coatings retained the nanostructure of the sprayed feedstock. Depending upon plasma spray parameters and substrate temperature, the process can produce a wide variety of structures ranging from porous to dense coatings, with very low open void content.

A method for manufacturing metal supported SOFCs with atmospheric plasma spraying (APS) is presented in [Waldbillig and Kesler (2009a, b)], making use of aqueous suspension feedstock for the electrolyte layer and dry powder feedstock for the anode and cathode layers. Even with optimized spray parameters, three main defect types were detected in such suspension plasma sprayed electrolyte layers: (1) through-thickness cracks (perpendicular to the substrate surface) caused by thermal stresses developing during coating deposition, (2) voids or defects caused by non-melted particles embedded within the coating, and (3) small inter-splat voids. Three different types of dispersant were also considered for ensuring feedstock dispersion, and the effects of solid particles loading, dispersant type, and dispersant concentration on suspension properties, such as viscosity and feed ability, and layer characteristics such as microstructure and deposited thickness were also examined [Waldbillig and Kesler (2009a, b)].

Ni–YSZ anode and YSZ electrolyte half cells were successfully deposited by SPS with the Axial III torch from Mettech on porous Hastelloy X substrates by [Wang et al. (2012)]. The thickness of the anode and electrolyte layers was ~ 40 μm and 10–20 μm, respectively. The NiO–YSZ coating deposition process was optimized by a design of experiment approach. The YSZ electrolyte spray process was examined by changing one parameter at a time. The maximum power density reached 0.610 W/cm^2 at 750 °C.

16.4.6.7 Other Applications

[Tesar et al. (2019)] investigated deposition of aluminum oxide (Al_2O_3) coatings from liquid feedstocks using a high enthalpy hybrid water-stabilized plasma torch. The most durable coating with the highest density microstructure prepared from ethanol-based suspension showed outstanding mechanical properties, for example, high values of hardness (up to 1211 HV$_{0.3}$), tensile adhesion strength (up to 51 MPa), and wear resistance (material removal rate as low as $1.94 \cdot 10^{-5}$ mm^3/N·m as measured by pin-on-disc test). Moreover, deposition from water-based solution of aluminum nitrate resulted in deposition of highly porous coating composed predominantly of stable α-alumina phase. [Curry et al. (2015)] have also presented new applications in fields like wear, erosion, and corrosion resistance with dense Al_2O_3 and Y_2O_3 ceramics. [Kato N. (2017)] focused their presentation on electric properties of dense Al_2O_3 and Y_2O_3 ceramic coatings. They discussed the influence of spray conditions, plasma power, gas flow rate, and stand-off distance on coating properties, especially mechanical and electric properties.

[Sharifi N. et al. (2007a, b)] employed APS and SPS for developing micro/nano morphologies of superhydrophobic coatings with high water repellence and mobility. Experimental results show that, although coatings developed by the APS process may reach water contact angles as high as 145°, the water mobility of these coatings is low due to relatively large morphological features resulting from the micron-sized feedstock powder. Coatings developed by SPS show superior water repellence (manifested through water contact angles as high as 167°) as well as improved mobility (displayed through water sliding angles as small as 1.3°) due to dual-scale submicron/nano (hierarchical) roughness attributed to the submicron size particles in the feedstock. The same authors [Sharifi N. et al. (2017a)] presented new applications focusing on surface textured coatings including super-hydrophobic coatings. They demonstrated that by carefully designing and controlling the process parameters, one can generate relatively fine and uniform dual-scale (hierarchical) surface textured coatings that after treatment for lowering their surface energy show significantly improved water repellence and water mobility with water contact angles as high as 170° and sliding angles as low as 1.3°. It is also demonstrated that both scale levels of surface textures (i.e., micron-scale and nanoscale) are essential for having simultaneously improved water repellence and mobility. Furthermore, it is established that producing finer, more uniformly distributed and packed surface features lead to more consistent and desirable wetting properties. The results show the significant influence of pre-deposition surface roughness, precursor suspension rheology, and plasma power on the structure and performance of the developed coatings. The same authors [Sharifi N. et al. (2017b)] developed the promising applications of SPS in this field and the necessity to achieve that by a carefully designing and controlling of process parameters.

[Toma F.-L. (2017)] presented tailored compositions of Cr_2O_3-TiO_2-Al_2O_3 in SPS sprayed by plasma and also by HVOF. She discussed the criteria for the selection of raw materials as well as the aspects of the development of binary/ternary suspensions. In most of cases denser microstructure, higher mechanical properties and superior corrosion performances were obtained for these coatings by SPS and suspension with HVOF when compared to coatings produced from feedstock powder in the tens of micrometers range.

[Gadow R. (2017)] presented the last developments in high velocity flame spraying (HV-SFS). His presentation gave an overview about current research activities in hardware component and process development and the new coating product development and industrial applications were discussed

16.5 Summary and Conclusions

For about a decade, the interest in manufacturing on large surfaces "thick" finely structured or nanostructured layers has been growing. Vapor condensation routes such as CVD, PE-CVD, PVD, and EB-PVD allow manufacturing of nanostructured architectures; their thicknesses can hardly be higher than a few micrometers, except for EB-PVD. In contrast, conventional thermal spraying while offering means of depositing coating with thicknesses of 50 µm up to a few millimeters do not allow for the deposition of nanostructed coatings, at least not through the in-flight particle melting approach. The objective of this chapter is to present alternate

thermal spray technological approaches under development over the past two decades for the deposition of nanocrystalline or nanostructured coatings using a wide range of novel precursors. Topics covered included

- Spraying of nanocrystalline materials
- Spraying of nanostructured agglomerated powders
- Suspension and solution thermal/plasma spraying of nanostructured coatings

These techniques seem promising to manufacture dense or porosity-controlled coatings and functionally graded layers.

Some industrial applications are already using coatings made from powders with a size range in the tens of micrometers, especially those from amorphous powders. Nevertheless, coatings made from suspensions or solutions injected into hot gases, are by far more complex because the chemistry (choice of solvent, dispersant, precursors), the liquid fragmentation, and vaporization control the processes. This is probably why publications on first commercial applications of such techniques are still missing.

Moreover, present in-flight particle diagnostic techniques are not easily adapted to measure in-flight droplets below 5 µm, which would permit getting a much better understanding of the phenomena involved. According to papers published during the past decade, the interaction plasma jet and liquid jet or drops is critical for their fragmentation to sizes below around 10 µm and velocities imparted to particles contained in droplet suspensions or formed from solutions. However, to optimize the process first the injection means of the liquid must be improved and new plasma torches developed. Injection setups producing liquid jets or drops with controlled narrow trajectories dispersion, narrow distributions of sizes (if possible, around or below 50 µm), and velocities, both varying independently must be developed. To spray them, plasma torches with higher power levels (more than 100 kW), longer plasma jets, and low voltage fluctuations must be designed. The only one presently available is the Axial III of Mettech plasma torch. At last, the deposition rate, which is about 10–30% that of conventional coatings, should also be improved. For suspensions HVOF spraying guns, called HV-SFS, with longer combustion chambers have been developed to achieve combustion of both combustion gases and organic solvent. Dense and very smooth (Ra ~ 1.3 µm) coatings, consisting of well-flattened lamellae having a homogeneous size distribution, were obtained when micron-sized (~1–2 µm) powders. Both for plasma and HV-SFS suspension spraying, the size distribution of particles used in suspension should be as narrow as possible. Mixtures in suspensions with a wide size distribution because of agglomerates and aggregates result in poor coatings.

Presently SPS and SPPS are in the same state of development as conventional spraying was in the 1980s and the beginning of the 1990s. However, over the past decade, numerous papers in international reviews and conference proceedings have been published on SPS and SPPS, with coatings obtained (mostly ceramics) present very interesting features, especially a much better toughness than conventional coatings. Numerous studies are still necessary to reach a better understanding of the basic phenomena involved. These will require, however, the development of advanced diagnostic techniques to measure the in-flight temperature and velocity of small droplets or particles (d_p < 10 μm), and/or visualize their interactions with the hot gases.

No standards have been yet defined both for the spray process and solution or suspension preparation. It is not surprising that, to our best knowledge, no industrial application of these techniques has been achieved. While many potential applications have been described, it is difficult to predict which one will be the first industrially implemented. However, a spray equipment manufacturer has recently proposed an industrial liquid-based feedstock feeder prototype that might be an indication of the potential interest of industry.

As pointed out by Killinger et al. (2011), "to promote industrial implementation, an urgent concern of future research effort is to qualify liquid feedstock Thermal spray coatings in a more application-oriented way and to communicate the results convincingly." As suspension and solution spray are to be clearly differentiated from classical thermal spray processes, also new industrial standards have to be defined concerning suspension and solution preparation, storage, feeding, as well as qualification and handling of raw materials.

Nomenclature

Units are indicated in parentheses; when no units are indicated, the parameter is dimensionless.

Latin Alphabet

c_{ps}	Specific heat at constant pressure (kJ/kg K)
d_l	Drop or droplet or liquid jet diameter (m)
d_n	Nozzle internal diameter (m)
d_p	Particle diameter (m)
f	Friction coefficient
f_b	Overlapping ratio
h_o	Mean specific enthalpy (J/kg)
l_{BL}	Boumdasry layer thickness (m)
L_{vs}	Latent heat of vaporization (kJ/mole) or (MJ/kg)
$\dot{m}_s$	Mass feed rate (kg/s)
r_m	Radius of the nonmolten core (m)
r_{po}	Initial radius of the agglomerated particle (m)
St	

	Stokes number $St = \rho_p d_p^2 v_p / \mu_g \ell_{BL}$
$\overline{T}$	Mean temperature (°C or K)
T_m	Material melting temperature (°C or K)
T_{vs}	Vaporization/boiling temperature (°C or K)
u_r	Relative velocity gas-drop or droplet (m/s)
v	Gas velocity (m/s)
v_l	Liquid velocity (m/s)
v_p	Particle velocity (m/s)
v_s	Exit velocity of solution/suspension (m/s)
$\overline{V}$	Mean voltage (V)
We	Weber number, $We = \rho \Delta v^2 d_s / \sigma_s$
Zo	Ohnesorge number, $Oh = \mu_s / \sqrt{\rho_s d_s \sigma_s}$

Greek Alphabet

α	Atomized jet half-angle (°)
α_o	Initial particle porosity
ΔV	Amplitude of voltage fluctuations (V)
Δp	Pressure drop across the feed nozzle (Pa)
κ	Thermal conductivity (W/mK)
μ_g	Molecular viscosity (Pa. s)
ρ or ρ_g	Gas specific mass (kg/m^3)
ρ_l	Liquid specific mass (kg/m^3)
ρ_p	Particle specific mass (kg/m^3)
ρ_s	Specific mass of solution/suspension (kg/m^3)
σ_l	Liquid surface tension (N/m)
σ_T	Standard deviation of temperature distribution (°C or K)

References

Ahmed, I., and T.L. Bergman. 1999. Thermal modeling of plasma spray deposition of nanostructured ceramics. *Journal of Thermal Spray Technology* 8 (2): 315–322.

Ahn, J., B. Hwang, E.P. Song, S. Lee, and N.J. Kim. 2006. Correlation of microstructure and wear resistance of Al_2O_3-TiO_2 coatings plasma sprayed with nano-powders. *Metallurgical and Materials Transactions A: Physical Metallurgy and Materials Science* 37A: 1851–1861.

Ajdelsztajn, L., J. Lee, K. Chung, F.L. Bastian, and E.J. Lavernia. 2002. Synthesis and nanoindentation study of high-velocity oxygen fuel thermal-sprayed nanocrystalline and near-nanocrystalline Ni coatings. *Metallurgical and Materials Transactions A: Physical Metallurgy and Materials Science* 33A: 647–655.

Ajdelsztajn, L., B. Jodoin, G.E. Kim, and J.M. Schoenung. 2005a. Cold spray deposition of nanocrystalline aluminum alloys. *Metallurgical and Materials Transactions A: Physical Metallurgy and Materials Science* 36A: 657–666.

Ajdelsztajn, L., F. Tang, G.E. Kim, V. Provenzano, and J.M. Schoenung. 2005b. Synthesis and oxidation behavior of nanocrystalline MCrAlY bond coatings. *Journal of Thermal Spray Technology* 14 (1): 23–30.

Ajdelsztajn, L., A. Zúñiga, B. Jodoin, and E.J. Lavernia. 2006a. Cold-spray processing of a nanocrystalline Al-Cu-Mg-Fe-Ni alloy with Sc. *Journal of Thermal Spray Technology* 15 (2): 184–190.

Ajdelsztajn, L., B. Jodoin, and J.M. Schoenung. 2006b. Synthesis and mechanical properties of nanocrystalline Ni coatings produced by cold gas dynamic spraying. *Surface & Coatings Technology* 201: 1166–1172.

Ajdelsztajn, L., B. Jodoin, P. Richer, E. Sansoucy, and E.J. Lavernia. 2006c. Cold gas dynamic spraying of iron-base amorphous alloy. *Journal of Thermal Spray Technology* 15 (4): 495–500.

Altomare, L., D. Bellucci, G. Bolelli, B. Bonferroni, V. Cannillo, L. De Nardo, R. Gadow, A. Killinger, L. Lusvarghi, A. Sola, and

N. Stiegler. 2011. Microstructure and in vitro behaviour of 45S5 bioglass coatings deposited by high velocity suspension flame spraying (HVSFS). *Journal of Materials Science. Materials in Medicine* 22: 1303–1319.

Arevalo-Quintero, O., D. Waldbillig, and O. Kesler. 2011. An investigation of the dispersion of YSZ, SDC, and mixtures of YSZ/SDC powders in aqueous suspensions for application in suspension plasma spraying. *Surface & Coatings Technology* 205: 5218–5227.

Ashby, M.F., P.J. Ferreira, and D.L. Schodek. 2009. Nanomaterials properties. In *Nanomaterials, nanotechnology and design*, 199–255. Kidlington: Elsevier.

Bacciochini, A., F. Ben-Ettouil, E. Brousse, J. Ilavsky, G. Montavon, A. Denoirjean, S. Valette, and P. Fauchais. 2010. Quantification of void network architectures of as-sprayed and aged nanostructured yttria-stabilized zirconia (YSZ) deposits manufactured by suspension plasma spraying. *Surface & Coatings Technology* 205: 683–689.

Ban, Z.-G., and L.L. Shaw. 2002. Synthesis and processing of nanostructured WC-Co materials. *Journal of Materials Science* 37: 3397–3403.

Ban, Z.G., and L.L. Shaw. 2003. Characterization of thermal sprayed nanostructured WC-Co coatings derived from nanocrystalline WC-18wt.%Co powders. *Journal of Thermal Spray Technology* 12 (1): 112–119.

Bannier, E., G. Darut, E. Sánchez, A. Denoirjean, M.C. Bordes, M.D. Salvador, E. Rayón, and H. Ageorges. 2011. Microstructure and photocatalytic activity of suspension plasma sprayed TiO$_2$ coatings on steel and glass substrates. *Surface & Coatings Technology* 206: 378–386.

Bansal, P., N.P. Padture, and A. Vasiliev. 2003. Improved interfacial mechanical properties of Al$_2$O$_3$-13wt%TiO$_2$ plasma-sprayed coatings derived from nanocrystalline powders. *Acta Materialia* 51: 2959–2970.

Bartuli, C., T. Valente, F. Cipri, E. Bemporad, and M. Tului. 2005. Parametric study of an HVOF process for the deposition of nanostructured WC-Co coatings. *Journal of Thermal Spray Technology* 14 (2): 187–195.

Basak, A.K., S. Achanta, J.-P. Celis, M. Vardavoulias, and P. Matteazzi. 2008. Structure and mechanical properties of plasma sprayed nanostructure ed alumina and FeCuAl–alumina cermet coatings. *Surface & Coatings Technology* 202: 2368–2373.

Basu, S., E.H. Jordan, and B.M. Cetegen. 2006. Fluid mechanics and heat transfer of liquid precursor droplets injected into high temperature plasmas. *Journal of Thermal Spray Technology* 15 (4): 576–581.

Ben-Ettouil, F., O. Mazhorova, B. Pateyron, H. Ageorges, M. El-Ganaoui, and P. Fauchais. 2008. Predicting dynamic and thermal histories of agglomerated particles injected within a D.-C. plasma jet. *Surface & Coatings Technology* 202: 4491–4495.

Ben-Ettouil, F., A. Denoirjean, A. Grimaud, G. Montavon, and P. Fauchais. 2009. Sub-micrometer-sized Y-PSZ thermal barrier coatings manufactured by suspension plasma spraying: Process, structure and some functional properties. In *ITSC-2009*. Materials Park: ASM.

Bernard, B., A. Quet, L. Bianchi, A. Joulia, A. Malié, V. Schick, and B. Rémy. 2017. Thermal insulation properties of YSZ coatings: Suspension plasma spraying (SPS) versus Electron beam physical vapor deposition (EB-PVD) and atmospheric plasma spraying (APS). *Surface & Coatings Technology* 318: 122–128.

Bertolissi, G., C. Chazelas, G. Bolelli, L. Lusvarghi, M. Vardelle, and A. Vardelle. 2012. Engineering the microstructure of solution precursor plasma-sprayed coatings. *Journal of Thermal Spray Technology* 21 (6): 1148–1158.

Bhatia, T., A. Ozturk, L.D. Xie, E.H. Jordan, B.M. Cetegen, M. Gell, X.C. Ma, and N.P. Padture. 2001. Mechanisms of ceramic coating deposition in solution precursor spray. *Journal of Materials Research* 17 (9): 2363–2372.

Bianchi, L., A. Denoirjean, F. Blein, and P. Fauchais. 1997. Microstructural investigation of plasma sprayed ceramic splats. *Thin Solid Films* 299: 125–135.

[Bianchi L. et al (2017)] Bianchi, L., B. Bernard, A. Quet, A. Joulia, and A. Malié 2017. *Effect of coating microstructure and bond coat surface preparation on columnar suspension plasma sprayed columnar YSZ Thermal barrier coatings*. TS4, September, 13–14 2017, in GE Niskayuna, NY, USA.

Bolelli, G., J. Rauch, V. Cannillo, A. Killinger, L. Lusvarghi, and R. Gadow. 2008. Investigation of high-velocity suspension flame sprayed (HVSFS) glass coatings. *Materials Letters* 62: 2772–2775.

Bolelli, G., V. Cannillo, R. Gadow, A. Killinger, L. Lusvarghi, and J. Rauch. 2009a. Properties of high velocity suspension flame sprayed (HVSFS) TiO$_2$ coatings. *Surface & Coatings Technology* 203: 1722–1732.

———. 2009b. Microstructural and in vitro characterization of high-velocity suspension flame sprayed (HVSFS) bioactive glass coatings. *Journal of the European Ceramic Society* 29: 2249–2257.

Bolelli, G., V. Cannillo, R. Gadow, A. Killinger, L. Lusvarghi, J. Rauch, and M. Romagnoli. 2010. Effect of the suspension composition on the microstructural properties of high velocity suspension flame sprayed (HVSFS) Al$_2$O$_3$ coatings. *Surface & Coatings Technology* 204: 1163–1179.

Bouaricha, S., J. Oberste-Berghaus, J.-G. Legoux, C. Moreau, and D. Ghosh. 2005. Production of doped-ceria plasma sprayed nanocoatings using an internal injection of a suspension containing nanoparticles. In *Thermal spray connects, explore its surfacing potential!* ed. E. Lugscheider. Düsseldorf: DVS-Verlag GmbH. e-proceedings.

Bouyer, E., F. Gitzhofer, and M.I. Boulos. 1996a. Parametric study of suspension plasma sprayed hydroxyapatite. In *9th national thermal spray conference – NTSC-96, Cincinnati, OH, USA*, 683–691.

———. 1996b. Experimental study of suspension plasma spraying of hydroxyapatite. In *4th European conference on thermal plasma processes – TPP-4, Athens, Greece*.

———. 1997a. Suspension plasma spraying for for hydroxyapatite powder preparation by RF plasma. *IEEE Transactions on Plasma Science* 25 (5): 1066–1072.

———. 1997b. The suspension plasma spraying of bioceramics by induction plasma. *Journal of Materials Science Letters* 49 (2): 58–62.

———. 2000. Morphological study of hydroxyapatite nanocrystal suspension. *Journal of Materials Science: Materials in Medicine* 11: 523–531.

Branagan, D.J. 2007. Engineering structures to achieve targeted properties in steels on a nanoscale level. *Comput Coupling Phase Dia Thermalochem* 31: 343–350.

Branagan, D.J., M.J. Kramer, Y. Tang, R.W. McCallum, D.C. Crew, and L.H. Lewis. 2000. Engineering magnetic nanocomposite microstructures. *Journal of Materials Science* 35: 3459–3466.

Branagan, D.J., W.D. Swank, D.C. Haggard, and J.R. Fincke. 2001. Wear-resistant amorphous and nanocomposite steel coatings. *Metallurgical and Materials Transactions A: Physical Metallurgy and Materials Science* 32A: 2615–2621.

Branagan, D.J., M. Breitsameter, B.E. Meacham, and V. Belashchenko. 2005a. High-performance nanoscale composite coatings for boiler applications. *Journal of Thermal Spray Technology* 14 (2): 196–204.

Branagan, D.J., M.C. Marshall, B.E. Meacham, J. Branagan, M.C. Marshall, and B.E. Meacham. 2005b. Formation of nanoscale composite coatings via HVOF and wire-arc spraying. In *Thermal spray conference*, ed. E. Lugscheider. Düsseldorf: DVS. e-proceedings.

Branagan, D.J., M.C. Marshall, B.E. Meacham, L.F. Apriliano, R. Bayler, E.J. Lemieux, T. Newbauer, F.J. Martin, J.C. Farmer, J.J. Haslan, and S.D. Day. 2006. Wear and corrosion resistant amorphous/nanostructured steel coatings for replacement of electrolytic hard chromium. In *ITSC-2006*, ed. B.R. Marple et al. Materials Park: ASM International. e-proceedings.

Branagan, D.J., W.D. Swank, and B.E. Meacham. 2009. Maximizing the glass fraction in iron-based high velocity oxy-fuel coatings. *Metallurgical and Materials Transactions A: Physical Metallurgy and Materials Science* 40A: 1306–1313.

Brinley, E., K.S. Babu, and S. Seal. 2007. The solution precursor plasma spray processing of nanomaterials. *JOM July*: 54–59.

Brousse, E., G. Montavon, P. Fauchais, A. Denoirjean, V. Rat, J.-F. Coudert, and H. Ageorges. 2008. Thin and dense yttria-partially stabilized zirconia electrolytes for IT-SOFC manufactured by suspension plasma spraying. In *Thermal spray crossing borders*, ed. E. Lugscheider, 547–552. Düsseldorf: DVS.

Brousse-Pereira, E. 2010, September. *Elaboration by thermal spray of finely structured elements of a high temperature electolyser for hydrogen production: processes, structures, and characteristics.* Ph.D. thesis, University of Limoges, France.

Burgess, A. 2017. *Low κ SPS TBCs using nanostructured YSZ powder.* TS4, September, 13–14, in GE Niskayuna, NY, USA.

Cedelle, J., M. Vardelle, and P. Fauchais. 2006. Influence of stainless steel substrate preheating on surface topography and on millimeter and micrometer-sized splat formation. *Surface & Coatings Technology* 201 (3–4): 1378–1382.

Chang, H., C. Cai, Y. Wang, Y. Zhou, Li Yang, and G. Zhou. 2017. Calcium-rich CMAS corrosion induced microstructure development of thermal barrier coatings. *Surface & Coatings Technology* 324: 577–584.

Chen, H., and C.X. Ding. 2002. Nanostructured zirconia coating prepared by atmospheric plasma spraying. *Surface & Coatings Technology* 150: 31–36.

Chen, X., and E. Pfender. 1983. Effect of the Knudsen number on heat transfer to a particle immersed into a thermal plasma. *Plasma Chemistry Plasma Processing* 3: 97–113.

Chen, H., X. Zhou, and C. Ding. 2003. Investigation of the thermomechanical properties of a plasma-sprayed nanostructured zirconia coating. *Journal of the European Ceramic Society* 23: 1449–1455.

Chen, H., S.W. Lee, C.H. Choi, B.Y. Hur, Y. Zeng, X.B. Zheng, and C.X. Ding. 2004. Plasma sprayed nanostrucutred zirconia coatings deposited from different powders with nano-scale substructure. *Journal of Materials Research* 39: 4701–4703.

Chen, D., E. Jordan, and M. Gell. 2007. Thermal and crystallization behavior of zirconia precursor used in the solution precursor plasma spray process. *Journal of Materials Science* 42 (14): 5576–5580.

Chen, D., E.H. Jordan, and M. Gell. 2008a. Effect of solution concentration on splat formation and coating microstructure using the solution precursor plasma spray process. *Surface & Coatings Technology* 202: 2132–2138.

———. 2008b. Porous TiO₂ coating using the solution precursor plasma spray process. *Surface & Coatings Technology* 202: 6113–6119.

———. 2009a. Microstructure of suspension plasma spray and air plasma spray Al₂O₃-ZrO₂ composite coatings. *Journal of Thermal Spray Technology* 18 (3): 421–426.

———. 2009b. Suspension plasma sprayed composite coating using amorphous powder feedstock. *Applied Surface Science* 255: 5935–5938.

———. 2010. The solution precursor plasma spray coatings: Influence of solvent type. *Plasma Chemistry and Plasma Processing* 30: 111–119.

Cherigui, M., N.E. Fenineche, and C. Coddet. 2005. Structural study of iron-based microstructured and nanostructured powders sprayed by HVOF thermal spraying. *Surface & Coatings Technology* 192: 19–26.

Cherigui, M., N.E. Fenineche, A. Gupta, G. Zhang, and C. Coddet. 2006. Magnetic properties of HVOF thermally sprayed coatings obtained from nanostructured powders. *Surface & Coatings Technology* 201: 1805–1813.

Chien, K., and T.W. Coyle. 2007. Rapid and continuous deposition of porous nanocrystalline SnO₂ coating with interpenetrating pores for gas sensor applications. *Journal of Thermal Spray Technology* 16 (5–6): 886–892.

Choi, H., J. Kim, C. Lee, and K.H. Lee. 2005. Critical factors affecting the amorphous phase formation of Ni-Ti-Zr-Si-Sn bulk amorphous feedstock in vacuum plasma spray. *Journal of Materials Science* 40: 3873–3875.

Ctibor, P., K. Neufuss, and P. Chraska. 2006. Microstructure and abrasion resistance of plasma sprayed titania coatings. *Journal of Thermal Spray Technology* 15 (4): 689–694.

Curry, N., Z. Tang, N. Markocsan, and P. Nylén. 2015. Influence of bond coat surface roughness on the structure of axial suspension plasma spray thermal barrier coatings -thermal and lifetime performance. *Surface & Coatings Technology* 268: 15–23.

D'Haese, R., L. Pawlowski, M. Bigan, R. Jaworski, and M. Martel. 2010. Phase evolution of hydroxyapatite coatings suspension plasma sprayed using variable parameters in simulated body fluid. *Surface & Coatings Technology* 204 (8): 1236–1246.

Dahotre, N.B., and S. Nayak. 2005. Nanocoatings for engine application. *Surface & Coatings Technology* 194 (1): 58–67.

Damiani, D. 2012. *Elaboration, fitting in shape and characterization of materials for energy by the interaction radiation-matter.* HdR University of Limoges, France, 13 December 2012.

Darut, G., H. Ageorges, A. Denoirjean, G. Montavon, and P. Fauchais. 2008. Effect of the structural scale of plasma-sprayed alumina coatings on their friction coefficients. *Journal of Thermal Spray Technology* 17 (5–6): 788–797.

Das, S.K., E.M. Norin, and R.L. Bye. 1984. Ni-Mo-Cr-B alloys: Corrosion resistant amorphous hardfacing coatings. In *Materials research society symposium proceedings*, ed. B.H. Kear and M. Cohen, 233–237. Amsterdam: Elsevier Science.

Delbos, C. 2004. *Contribution to the understanding of liquid injection of ceramics (Y.S.Z., Perovskite…) or metals (Ni,…) in blown arc plasma jets to achieve finely structured coatings for SOFCs.* Ph.D. thesis, University of Limoges, France. (in French).

Delbos, C., J. Fazilleau, J.F. Coudert, P. Fauchais, and L. Bianchi. 2004. Finely structured ceramic coatings elaborated by liquid suspension injection in a DC plasma jet. In *Thermal spray solutions: Advances in technology and application.* Düsseldorf: DVS.

Delbos, C., J. Fazilleau, V. Rat, J.F. Coudert, P. Fauchais, and B. Pateyron. 2006. Phenomena involved in suspension plasma spraying. Part 2: Zirconia particle treatment and coating formation. *Plasma Chemistry and Plasma Processing* 26: 393–414.

Dent, A.H., A.J. Horlock, D.G. McCartney, and S.J. Harris. 1999. The corrosion behavior and microstructure of high-velocity oxy-fuel sprayed nickel-base amorphous/nanocrystalline coatings. *Journal of Thermal Spray Technology* 8 (3): 399–404.

Dent, A.H., S. DePalo, and S. Samph. 2001. Examination of the wear properties of HVOF sprayed nanostructured and conventional WC-Co cermets with different binder phase contents. *Journal of Thermal Spray Technology* 11 (4): 551–558.

Dongmo, E., R. Gadow, A. Killinger, and M. Wenzelburger. 2009. Modeling of combustion as well as heat, mass, and momentum transfer during thermal spraying by HVOF and HVSFS. *Journal of Thermal Spray Technology* 18 (5–6): 896–908.

Eigen, N., F. Gartner, T. Klassen, E. Aust, R. Bormann, and H. Kreye. 2005. Microstructures and properties of nanostructured thermal sprayed coatings using high-energy milled cermet powders. *Surface & Coatings Technology* 195: 344–357.

Erne, M., D. Kolar, C. Hübsch, M. Möhwald, and Fr-W Bach. 2012. Synthesis of tribologically favorable coatings for hot extrusion tools by suspension plasma spraying. *Journal of Thermal Spray Technology* 21 (3–4): 668–675.

Etchart-Salas, R. 2007. D.C. plasma spraying of suspensions of submicronic particles. Experimental and analytic approaches of phenomena implied in the reproducibility and quality of coatings. Ph.D. thesis, University of Limoges, France. (in French).

[Etchart-Salas R.et al. (2007)], Etchart-Salas R., V. Rat, J.-F. Coudert, P. Fauchais (2007) Thermal Spray 2007: Global Coating Solutions. B.R. Marple, M.M. Hyland, Y.-C. Lau, C.-J. Li, R.S. Lima, G. Montavon CD-Rom, ASM International, Materials Park. ISBN:0-87170-809-4

Etchart-Salas, R., V. Rat, J.F. Coudert, P. Fauchais, N. Caron, K. Wittman, and S. Alexandre. 2007. Influence of plasma instabilities in ceramic suspension plasma spraying. *Journal of Thermal Spray Technology* 16 (5–6): 857–865.

Fauchais, P. 1991. Molten particle deposition. In *Engineered materials handbook, vol 4, ceramics and glasses*, ed. S.J. Schneider Jr. Materials Park: ASM International.

Fauchais, P., and A. Vardelle. 2011. Innovative and emerging processes in plasma spraying: From micro- to nanostructured coatings. *Journal of Physics D: Applied Physics* 44: 194011. (14pp).

Fauchais, P., V. Rat, C. Delbos, J.-F. Coudert, T. Chartier, and L. Bianchi. 2005. Understanding of suspension D.C. plasma spraying of finely structured coatings for SOFC. *IEEE Transactions on Plasma Science* 33 (2): 920–930.

Fauchais, P., G. Montavon, M. Vardelle, and J. Cedelle. 2006. Developments in direct current plasma spraying. *Surface & Coatings Technology* 201 (5): 1908–1921.

Fauchais, P., R. Etchart-Salas, C. Delbos, M. Tognovi, V. Rat, J.-F. Coudert, and T. Chartier. 2007. Suspension and solution plasma spraying of finely structured layers for SOFCs. *Journal of Physics D: Applied Physics* 40 (8): 2394–2406.

Fauchais, P., R. Etchart-Salas, V. Rat, J.F. Coudert, N. Caron, and K. Wittmann-Ténèze. 2008. Parameters controlling liquid plasma spraying: solutions, sols, or suspensions. *Journal of Thermal Spray Technology* 17 (1): 31–59.

Fauchais, P., R. Etchart-Salas, V. Rat, J.-F. Coudert, N. Car, and K. Wittmann-Ténèze. 2008a. Parameters controlling liquid plasma spraying: Solutions, sols or suspensions. *Journal of Thermal Spray Technology* 17 (1): 31–59. https://doi.org/10.1007/s11666-007-9152-2.

Fauchais, P., V. Rat, J.-F. Coudert, R. Etchart-Salas, and G. Montavon. 2008b. Operating parameters for suspension and solution plasma-spray coatings. *Surface & Coatings Technology* 202: 4309–4317.

Fauchais, P., G. Montavon, R.S. Lima, and B.R. Marple. 2011. Engineering a new class of thermal spray nano-based microstructures from agglomerated nanostructured particles, suspensions and solutions: An invited review. *Journal of Physics D* 44: 093001.

Fauchais, P., A. Joulia, S. Goutier, C. Chazelas, M. Vardelle, A. Vardelle, and S. Rossignol. 2013. Suspension and solution plasma spraying. *Journal of Physics D: Applied Physics* 46: 224015.

Fauchais, P., M. Vardelle, A. Vardelle, and S. Goutier. 2015a. What do we know, what are the current limitations of suspension plasma spraying? *Journal of Thermal Spray Technology* 24 (7): 1120–1129.

Fauchais, P., M. Vardelle, S. Goutier, and A. Vardelle. 2015b. Specific measurements of in-flight droplet and particle behavior and coating microstructure in suspension and solution plasma spraying. *Journal of Thermal Spray Technology* 24 (8): 1498–1505.

Fazilleau, J., C. Delbos, M. Violier, J.F. Coudert, P. Fauchais, L. Bianchi, and K. Wittmann-Teneze. 2003a. Influence of substrate temperature on formation of micrometric splats obtained by plasma spraying liquid suspension. In *ITSC 2003*, ed. B.R. Marple and C. Moreau, 889–895. Materials Park: ASM International.

Fazilleau, J., C. Delbos, M. Violier, J.F. Coudert, P. Fauchais, L. Bianchi, and K. Wittmann-Ténèze. 2003b. In *ITSC-2003: Advancing the science and applying the technology*, ed. C. Moreau and B. Marple, 889. Materials Park: ASM International.

Fazilleau, J., C. Delbos, V. Rat, J.-F. Coudert, P. Fauchais, and B. Pateyron. 2006. Phenomena involved in suspension plasma spraying. Part 1: Suspension injection and behaviour. *Plasma Chemistry and Plasma Processing* 26 (4): 371–391.

Gadow, R. 2017. *High Velocity Flame Spraying (HVSFS) of nanostructured coatings and related industrial applications.* TS-4, September, 13–14, 2017, in GE Niskayuna, NY, USA.

Gadow, R., F. Kern, and A. Killinger. 2008a. Manufacturing technology for nanocomposite ceramic structural materials and coatings. *Materials Science and Engineering* B148: 58–64.

Gadow, R., A. Killinger, and J. Rauch. 2008b. New results in high velocity suspension flame spraying (HVSFS). *Surface & Coatings Technology* 202: 4329–4336.

Ganvir, A., N. Markosan, M. Gupta, S. Björklund, R. Calinas, and R. Vassen. 2017. *Understanding the functional performance of advanced axial suspension plasma sprayed TBCs.* TS-4, September, 13–14, 2017, in GE Niskayuna, NY, USA.

———. 2017. *The solution precursor plasma spray process-development & applications.* TS-4, September, 13–14, 2017, in GE Niskayuna, NY, USA.

Gell, M., E.H. Jordan, Y.H. Sohn, D. Goberman, L. Shaw, and T.D. Xiao. 2001. Development and implementation of plasma sprayed nanostructured ceramic coatings. *Surface & Coatings Technology* 146–147: 48–54.

Gell, M., L.D. Xie, X.C. Ma, E.H. Jordan, and N.P. Padture. 2004. Highly durable thermal barrier coatings made by the solution precursor plasma spray process. *Surface & Coatings Technology* 13 (1): 97–102.

Gell, M., E.H. Jordan, M. Teicholz, B.M. Cetegen, N. Padture, L. Xie, D. Chen, X. Ma, and J. Roth. 2008. Thermal barrier coatings made by the solution precursor plasma spray process. *Journal of Thermal Spray Technology* 17 (1): 124–135.

Ghasripoor, F., R. Schmid, and M.R. Dorfman. 1997. Abradable coatings increase gas turbine engine efficiency. *Materials World* 5 (6): 328–330.

Gitzhofer, F., E. Bouyer, and M.I. Boulos. 1997. *Suspension plasma spray deposition.* United States Patent, 5,609,921, March 11, 1997.

Goberman, D., Y.H. Sohn, L. Shaw, E.H. Jordan, and M. Gell. 2002. Microstructured development of Al_2O_3-13wt%TiO_2 plasma sprayed coatings derived from nanocrystalline powders. *Acta Materialia* 50: 1141–1152.

Gong, W.B., C.K. Sha, D.Q. Sun, and W.Q. Wang. 2006. Microstructures and thermal insulation capability of plasma-sprayed nanostructured ceria stabilized zirconia coatings. *Surface & Coatings Technology* 201: 3109–3115.

Guignard, A., G. Mauer, R. Vaßen, and D. Stöver. 2012. Deposition and characteristics of submicrometer-structured thermal barrier coatings by suspension plasma spraying. *Journal of Thermal Spray Technology* 21 (3–4): 416–424.

Guilemany, J.M., S. Dosta, and J.R. Miguel. 2006. The enhancement of the properties of WC-Co HVOF coatings through the use of nanostructured and microstructured feedstock powders. *Surface & Coatings Technology* 201: 1180–1190.

Gupta, M., C. Kumara, and P. Nylen. 2017. Bilayer suspension plasma-sprayed thermal barrier coatings with enhanced thermal cyclic lifetime: Experiments and modeling. *Journal of Thermal Spray Technology* 26: 1038–1051.

Guru, D., and J. Heberlein. 2008. Plasma spray processing of nanostructured partially stabilized zirconia for a strain accommodating inter-layer – Splat characteristics. In *ITSC-2008: Thermal spray crossing borders*, ed. E. Lugsheider. Düsseldorf: DVS. e-proceedings.

Gutwein, L.G., and T.J. Webster. 2004. Increased viable osteoblast density in the presence of nanophase compared to conventional alumina and titania particles. *Biomaterials* 25: 4175–4183.

Gutzmann, H., J.-O. Kliemann, R. Albrecht, F. Gärtner, T. Klassen, F.-L. Toma, L.-M. Berger, and B. Leupolt. 2010. Evaluation of the photocatalytic activity of TiO_2-coatings prepared by different thermal spray techniques. In *ITSC-2010: Global solutions for future application*, 182–186. Düsseldorf: DVS-Berichte.

Hazel, B. 2017. *Processing effects on microstructure in spraying suspensions*. TS-4, September, 13–14, 2017, in GE Niskayuna, NY, USA.

He, J., and J.M. Schoenung. 2002. A review on nanostructured WC–Co coatings. *Surface & Coatings Technology* 157: 72–79.

———. 2003. Nanocrystalline Ni coatings strengthened with ultrafine particles Ni-AlN HVOF. *Metallurgical and Materials Transactions A: Physical Metallurgy and Materials Science* 34A: 673–683.

He, J., M. Ice, and E.J. Lavernia. 2000a. Synthesis of nanostructured Cr_3C_2-25(Ni20Cr) coatings. *Metallurgical and Materials Transactions A: Physical Metallurgy and Materials Science* 31A: 555–564.

He, J., M. Ice, S. Dallek, and E.J. Lavernia. 2000b. Synthesis of nanostructured WC-12 wt. % Co coating using mechanical milling and high velocity oxygen fuel thermal spraying. *Metallurgical and Materials Transactions A: Physical Metallurgy and Materials Science* 31A: 541–553.

He, J., M. Ice, J.M. Schoenung, D.H. Shin, and E.J. Lavernia. 2001. Thermal stability of nanostructured Cr_3C_2-NiCr coatings. *Journal of Thermal Spray Technology* 10 (2): 293–300.

Huang, Chen, Chuanxian Ding, Pingyu Zhang, Peiqing La, and Soo Wohn Lee. 2003. Wear of plasma-sprayed nanostructured zirconia coatings against stainless steel under distilled-water conditions. *Surface & Coatings Technology* 173: 144–149.

Hwang, C., and Y. Chia-ho. 2007. Formation of nanostructured YSZ/Ni anode with pore channels by plasma spraying. *Surface & Coatings Technology* 201: 5954–5959.

Ibrahim, A., R.S. Lima, C.C. Berndt, and B.R. Marple. 2007. Fatigue and mechanical properties of nanostructured and conventional titania (TiO_2) thermal spray coatings. *Surface & Coatings Technology* 201: 7589–7596.

Inoue, A. 1995. Thermal properties of Zr-TM-B and Zr-TM-Ga (TM = Co, Ni, Cu) amorphous alloys with wide range of supercooling. *Materials Transactions, JIM* 36: 1411–1419.

Jadhav, A.D., N.P. Padture, E.H. Jordan, M. Gell, P. Miranzo, and E.R. Fullu. 2006. Low thermal conductivity plasma sprayed thermal barrier coatings with engineered microstructure. *Acta Materialia* 54 (12): 3343–3349.

Jaworski, R., L. Pawlowski, F. Roudet, S. Kozerski, and A. Le Maguer. 2008. Recent developments in suspension plasma sprayed titanium oxide and hydroxyapatite coatings influence of suspension plasma spraying process parameters on TiO_2 coatings microstructure. *Journal of Thermal Spray Technology* 17 (1): 240–247.

Jeehoon, Ahn, Byoungchul Hwang, Eun Pil Song, Sunghak Lee, and Nack J. Kim. 2006. Correlation of microstructure and wear resistance of Al_2O_3-TiO_2 coatings plasma sprayed with nanopowders. *Metallurgical and Materials Transactions A: Physical Metallurgy and Materials Science* 37A: 1851–1861.

Ji, G., T. Grosdidier, and J.-P. Morniroli. 2007. Microstructure of a high-velocity oxy-fuel thermal-sprayed nanostructured coating obtained from milled powder. *Metallurgical and Materials Transactions A: Physical Metallurgy and Materials Science* 38A: 2455–2463.

———. 2010. Induction plasma synthesis of nanostructured SOFCs electrolyte using solution and suspension plasma spraying: A comparative study. *Journal of Thermal Spray Technology* 19 (3): 566–574.

Jiang, X.-L., E. Jordan, L. Shaw, and M. Gell. 2003. Mechanism of deformation of nanostructured ceramic coatings. *Materials Science* 39 (2): 305–306.

Jing, H.G., M.L. Lau, and E.J. Lavernia. 1998. Grain growth behavior of nanocrystalline Inonel 718 and Ni powders and coatings. *Nanostructured Mater* 10 (2): 169–178.

Jordan, E.H., L.D. Xie, X.C. Ma, M. Gell, N.P. Padture, B. Cetegem, J. Roth, T.D. Xiao, P.E.L. Bryant, J. Roth, T.D. Xiao, and P.E.C. Bryant. 2004. Superior thermal barrier coatings using solution precursor plasma spray. *Journal of Thermal Spray Technology* 13 (1): 57–65.

Jordan, E.H., M. Gell, P. Benzani, D. Chen, S. Basu, B. Cetegem, F. Wa, and X.C. Ma. 2007. Making dense coatings with the solution precursor plasma spray process. In *ITSC-2007: Global coating solutions*, ed. B.R. Marple et al. Materials Park: ASM International. e-proceedings.

Karthikeyan, J., C.C. Berndt, J. Tikkanen, S. Reddy, and H. Herman. 1997. Plasma spray synthesis of nanomaterial powders and deposits. *Materials Science and Engineering A* 238 (2): 275–286.

Kassner, H., R. Siegert, D. Hathiramani, R. Vassen, and D. Stoever. 2008. Application of suspension plasma spraying (SPS) for manufacture of ceramic coatings. *Journal of Thermal Spray Technology* 17 (1): 115–123.

Kato, N. 2017. *Electric properties of dense ceramic coating prepared by SPS process*. TS4, September, 13–14, in GE Niskayuna, NY, USA.

Kear, B.H., R.K. Sadangi, M. Jain, R. Yao, Z. Kalman, G. Skandan, and W.E. Mayo. 1999. Thermal sprayed nanostructured WC/Co hardcoatings. *Journal of Thermal Spray Technology* 9 (3): 399–406.

Killinger, A., M. Kuhn, and R. Gadow. 2006. High-velocity suspension flame spraying (HVSFS), a new approach for spraying nanoparticles with hypersonic speed. *Surface & Coatings Technology* 201: 1922–1929.

Killinger, A., R. Gadow, A. Rempp, and A. Manzat. 2010. Advanced ceramic tribological layers by thermal spray routes. *Advances in Science and Technology* 66: 106–119.

Killinger, A., R. Gadow, G. Mauer, A. Guignard, R. Vaßen, and D. Stöver. 2011. Review of new developments in suspension and solution precursor thermal spray processes. *Journal of Thermal Spray Technology* 20 (4): 677–695.

Kim, G.E., and J. Walker. 2007. Successful application of nanostructured titanium dioxide coating for high-pressure acid-leach application. *Journal of Thermal Spray Technology* 16 (1): 34–39.

Kishitake, K., H. Era, and F. Otsubo. 1996a. Characterization of plasma sprayed Fe-10Cr-10Mo-(C, B) amorphous coatings. *Journal of Thermal Spray Technology* 5 (2): 145–153.

———. 1996b. Characterization of plasma sprayed Fe-17Cr-38Mo-4C amorphous coatings crystallizing at extremely high temperature. *Journal of Thermal Spray Technology* 5 (3): 283–288.

Kitamura, J., Z. Tang, H. Mizuno, K. Sato, and A. Burgess. 2011. Structural, mechanical and erosion properties of yttrium oxide coatings by axial suspension plasma spraying for electronics applications. *Journal of Thermal Spray Technology* 20 (1–2): 170–185.

Kitamura, J., K. Fujimori, B. Callen, C. Bauer, and R. Rocchio-Heller. 2017. *Suspension spraying-recent activities of material, equipment and coating development*. TS-4, September, 13–14, 2017, in GE Niskayuna, NY, USA.

Kozerski, S., F.-L. Toma, L. Pawlowski, B. Leupolt, L. Latka, and L.-M. Berger. 2010a. Suspension plasma sprayed TiO_2 coatings using different injectors and their photocatalytic properties. *Surface & Coatings Technology* 205: 980–986.

Kozerski, S., L. Pawlowski, R. Jaworski, F. Roudet, and F. Petit. 2010b. Two zones microstructure of suspension plasma sprayed hydroxyapatite coatings. *Surface & Coatings Technology* 204 (9–10): 1380–1387.

Laha, T., A. Agarwal, and T. McKechnie. 2004a. Forming nanostructured hypereutectic aluminum via high velocity oxyfuel spray deposition. *JOM, The Journal of The Minerals, Metals & Materials Society* 56 (1): 54–56.

Laha, T., A. Agarwal, T. McKechnie, and S. Seal. 2004b. Synthesis and characterization of plasma spray formed carbon nanotube reinforced aluminum composite. *Materials Science and Engineering A* 381: 249–258.

Laha, T., A. Agarwal, T. McKechnie, K. Rea, and S. Seal. 2005. Synthesis of bulk nanostructured aluminum alloy component through vacuum plasma spray technique. *Acta Materialia* 53: 5429–5438.

Laha, T., S. Kuchibhatla, S. Seal, W. Li, and A. Agarwal. 2007. Interfacial phenomena in thermally sprayed multiwalled carbon nanotube reinforced aluminum nanocomposite. *Acta Materialia* 55: 1059–1066.

Lai, K.A. 2002. Failure of hydroxyapatite-coated acetabular cups. *The Journal of Bone and Joint Surgery. British Volume* 84-B (5): 64646.

Łatka, L., L. Pawlowski, D. Chicot, C. Pierlot, and F. Petit. 2010. Mechanical properties of suspension plasma sprayed hydroxyapatite coatings submitted to simulated body fluid. *Surface & Coatings Technology* 205: 954–960.

Lau, M.L., E. Strock, A. Fabel, C.J. Lavernia, and E.J. Lavernia. 1998. Synthesis and characterization of nanocrystalline Co43 coatings by plasma spraying. *NanoStructured Materials* 10 (5): 723–730.

Lefebvre, A.H. 1989c. *Atomization and sprays*, 421 pages. New York: Hemisphere Publishing Corporation.

Li, X-H. 2017. *Development of suspension plasma sprayed TBCs for Siemens industrial gas turbines*. TS-4, September, 13–14, 2017, in GE Niskayuna, NY, USA.

Li, H., and K.A. Khor. 2006. Characteristics of the nanostructures in thermal sprayed hydroxyapatite coatings and their influence on coating properties. *Surface & Coatings Technology* 201: 2147–2154.

Li, H., K.A. Khor, and P. Cheang. 2000. Effect of the powders' melting state on the properties of HVOF sprayed hydroxyapatite coatings. *Materials Science and Engineering A* 293: 1–80.

Li, H., K.A. Khor, R. Kumar, and P. Cheang. 2004. Characterization of hydroxyapatite-nano-zirconia composite coatings deposited by high velocity oxy-fuel (HVOF) spray process. *Surface & Coatings Technology* 182: 227–236.

Li, C.-J., G.-J. Yang, P.-H. Gao, J. Ma, Y.-Y. Wang, and C.-X. Li. 2007. Characterization of nanostructured WC-Co, deposited by cold spraying. *Journal of Thermal Spray Technology* 16 (5–6): 1011–1020.

Liang, B., and C. Ding. 2005a. Thermal shock resistances of nanostructured and conventional zirconia coatings deposited by atmospheric plasma spraying. *Surface & Coatings Technology* 197: 185–192.

Liang, B., and Chuanxian Ding. 2005b. Phase composition of nanostructured zirconia coatings deposited by air plasma spraying. *Surface & Coatings Technology* 191: 267–273.

Liang, B., C. Ding, H. Liao, and C. Coddet. 2006. Phase composition and stability of nanostructured 4.7wt.% yttria-stabilized zirconia coatings deposited by atmospheric plasma spraying. *Surface & Coatings Technology* 200: 4549–4556.

Lima, R.S., and B.R. Marple. 2000. From APS to HVOF spraying of conventional and nanostructured titania feedstock powders: A study on the enhancement of the mechanical properties. *Surface & Coatings Technology* 200: 3428–3437.

———. 2003a. High weibull modulus HVOF titania coatings. *Journal of Thermal Spray Technology* 12 (2): 240–249.

———. 2003b. Optimized HVOF titania coatings. *Journal of Thermal Spray Technology* 12 (3): 360–369.

———. 2005a. Superior performance of high-velocity oxyfuel-sprayed nanostructured TiO_2 in comparison to air plasma-sprayed conventional Al_2O_3-13TiO_2. *Journal of Thermal Spray Technology* 14 (3): 397–404.

———. 2005b. Enhanced ductility in thermally sprayed titania coating synthesized using a nanostructured feedstock. *Materials Science and Engineering A* 395: 269–280.

———. 2006. From APS to HVOF spraying of conventional and nanostructured titania feedstock powders: A study on the enhancement of the mechanical properties. *Surface & Coatings Technology* 200: 3428–3437.

———. 2007. Thermal spray coatings engineered from nanostructured ceramic agglomerated powders for structural, thermal barrier and biomedical applications: A review. *Journal of Thermal Spray Technology* 16 (1): 40–63.

———. 2008a. Nanostructured YSZ thermal barrier coatings engineered to counteract sintering effects. *Materials Science and Engineering A* 485: 182–193.

———. 2008b. Toward highly sintering-resistant nanostructured ZrO_2-7wt%Y_2O_3 coatings for TBC applications by employing differential sintering. *Journal of Thermal Spray Technology* 17 (5–6): 846–852.

Lima, R.S., A. Kucuk, and C.C. Berndt. 2001. Evaluation of microhardness and elastic modulus of thermally sprayed nanostructured zirconia coatings. *Surface & Coatings Technology* 135: 166–172.

———. 2002. Bimodal distribution of mechanical properties on plasma sprayed nanostructured partially stabilized zirconia. *Materials Science and Engineering A* A327: 224–232.

Lima, R.S., K.A. Khor, H. Li, P. Cheang, and B.R. Marple. 2005. HVOF spraying of nanostructured hydroxyapatite for biomedical applications. *Materials Science and Engineering A* 396: 181–187.

Lima, R.S., B.R. Marple, A. Dadouche, W. Dmochowski, and B. Liko. 2006a. Nanostructured abradable coatings for high temperature applications. In *Proceedings ITSC-2006*, ed. B.R. Marple, M.M. Hyland, Y.-C. Lau, R.S. Lima, and J. Voyer. Materials Park: ASM International. e-proceedings.

Lima, R.S., H. Li, K.A. Khor, and B.R. Marple. 2006b. Biocompatible nanostructured high velocity oxyfuel sprayed titania coating: Deposition, characterization, and mechanical properties. *Journal of Thermal Spray Technology* 15 (4): 623–627.

Lima, R.S., C. Moreau, and B.R. Marple. 2007. HVOF-sprayed coatings engineered from mixtures of nanostructured and submicron Al_2O_3-TiO_2 powders: An enhanced wear performance. *Journal of Thermal Spray Technology* 16 (5–6): 866–872.

Lima, R.S., S.E. Kruger, and B.R. Marple. 2008. Towards engineering isotropic behaviour of mechanical properties in thermally sprayed ceramic coatings. *Surface & Coatings Technology* 202 (15): 3643–3652.

Lin, Xinhua, Yi Zeng, Xuebin Zheng, and Chuanxian Ding. 2005. Thermal diffusivity of plasma sprayed monolithic coating of alumina–3 wt.% titania produced with nanostructured powder. *Surface & Coatings Technology* 195: 85–90.

Liu, Y., T.E. Fischer, and A. Dent. 2003. Comparison of HVOF and plasma-sprayed alumina/titania coatings – Microstructure, mechanical properties and abrasion behavior. *Surface & Coatings Technology* 167: 68–76.

Liu, C.-B., Z.-M. Zhang, X.-L. Jiang, M. Liu, and Z.-H. Zhu. 2009. Comparison of thermal shock behaviors between plasma-sprayed nanostructured and conventional zirconia thermal barrier coatings. *Transactions of Nonferrous Metals Society of China* 19: 99–107.

Low, A.D., N.P. Jadhav, E.H. Padture, M. Jordan, P. Gell, and E.R. Miranzo. 2006. Fuller Jr., Low-thermal-conductivity plasma-sprayed thermal barrier coatings with engineered microstructures. *Acta Materialia* 54: 3343–3349.

Lu, Y., and P.K. Liaw. 2001. The mechanical properties of nanostructured materials. *Journal of Metals* 53 (3): 31–35.

Luo, H., D. Goberman, L. Shaw, and M. Gell. 2003. Indentation fracture behavior of plasma-sprayed nanostructured Al_2O_3-13wt%TiO_2 coatings. *Materials Science and Engineering A* 346: 237–245.

Ma, X.Q., J. Roth, D.W. Gandy, and G.J. Frederick. 2006. A new high-velocity oxygen fuel process for making finely structured and highly bonded Inconel alloy layers from liquid feedstock. *Journal of Thermal Spray Technology* 15 (4): 670–675.

Madhwal, M., E.H. Jordan, and M. Gell. 2004. Failure mechanisms of dense vertically-cracked thermal barrier coatings. *Materials Science and Engineering A* 384: 151–161.

Mahade, S., N. Curry, S. Björklund, N. Markocsan, and Per Nylén. 2015. Thermal conductivity and thermal cyclic fatigue of multilayered $Gd_2Zr_2O_7$/YSZ thermal barrier coatings processed by suspension plasma spray. *Surface & Coatings Technology* 283: 329–336.

Mahade, S., N. Curry, S. Björklunda, N. Markocsan, P. Nylén, and Robert Vaßen. 2017. Functional performance of $Gd_2Zr_2O_7$/YSZ multi-layered thermal barrier coatings deposited by suspension plasma spray. *Surface & Coatings Technology* 318: 208–216.

Manzat, A., A. Killinger, and R. Gadow. 2010. Application of supersonic flame spraying for next generation cylinder liner coatings. In *Sustainable automotive technology*, ed. J. Wellnitz, 175–181. Berlin: Springer.

Marchand, C., A. Vardelle, G. Mariaux, and P. Lefort. 2008. Modelling of the plasma spray process with liquid feedstock injection. *Surface & Coatings Technology* 202: 4458–4464.

Marchand, O., P. Bertrand, J. Mougin, C. Comminges, M.-P. Planche, and G. Bertrand. 2010. Characterization of suspension plasma-sprayed solid oxide fuel cell electrodes. *Surface & Coatings Technology* 205 (4): 993–998.

Material production process., Handbook of thermal spray technology, ed. J.R. Davis. Materials Park: ASM International.

Matthews, S., M. Hyland, and B. James. 2004. Long-term carbide development in high-velocity oxygen fuel/high-velocity air fuel Cr_3C_2-NiCr coatings heat treated at 900 °C. *Journal of Thermal Spray Technology* 13 (4): 526–536.

McCoppin, J., D. Young, T. Reitz, A. Maleszewski, and S. Mukhopadhyay. 2011. Solid oxide fuel cell with compositionally graded cathode functional layer deposited by pressure assisted dual-suspension spraying. *Journal of Power Sources* 196: 3761–3765.

McPherson, R. 1973. Formation of metastable phases in flame and plasma-prepared alumina. *Journal of Materials Science* 8: 851–858.

———. 1989. A review of microstructure and properties of plasma sprayed ceramic coatings. *Surface & Coatings Technology* 39–40 (1): 173–181.

Meillot, E., R. Vert, C. Caruyer, D. Damiani, and M. Vardelle. 2011. Manufacturing nanostructured YSZ coatings by suspension plasma spraying (SPS): Effect of injection parameters. *Journal of Physics D: Applied Physics* 44: 194008. (8pp).

Michaux, P., G. Montavon, A. Grimaud, A. Denoirjean, and P. Fauchais. 2010. Elaboration of porous NiO/8YSZ layers by several SPS and SPPS routes. *Journal of Thermal Spray Technology* 19 (1–2): 317–327.

Monterrubio-Badillo, C., H. Ageorges, T. Chartier, J.-F. Coudert, and P. Fauchais. 2006. Preparation of $LaMnO_3$ perovskite thin films by suspension plasma spraying for SOFC cathodes. *Surface & Coatings Technology* 200: 3743–3756.

Moroz, N.A., H. Umapathy, and P. Mohanty. 2010. Synthesis and microstructure evolution of nano-titania doped silicon coatings. *Journal of Thermal Spray Technology* 19 (1–2): 294–302.

Müller, P., A. Killinger, and R. Gadow. 2012. Comparison between high-velocity suspension flame spraying and suspension plasma spraying of alumina. *Journal of Thermal Spray Technology* 21 (6): 1120–1128.

Muoto, C.K., E.H. Jordan, M. Gell, and M. Aindow. 2011. Identification of desirable precursor properties for solution precursor plasma spray. *Journal of Thermal Spray Technology* 20 (4): 802–816.

Musalek, R., J. Medricky, T. Tesar, J. Kotlan, Z. Pala, F. Lukac, K. Illkova, M. Hlina, T. Chraska, P. Sokolowski, and N. Curry. 2017. Controlling microstructure of Yttria-stabilized zirconia prepared from suspensions and solutions by plasma spraying with high feed rates. *Journal of Thermal Spray Technology* 26: 1787–1803.

Oberste Berghaus, J., J.-G. Legoux, C. Moreau, F. Tarasi, and T. Chráska. 2007. Mechanical and thermal transport properties of suspension thermal sprayed alumina-zirconia composite coatings. In *ITSC- 2007: Global coating solutions*, ed. B.R. Marple, M.M. Hyland, Y.-C. Lau, C.-J. Li, R.S. Lima, and G. Montavon. Materials Park: ASM International. e-proceedings.

Oberste Berghaus, J., J.-G. Legoux, C. Moreau, R. Hui, C. Deces-Petit, W. Qu, S. Yick, Z. Wang, R. Maric, and D. Ghosh. 2008. Suspension HVOF spraying of reduced temperature solid oxide fuel cell electrolytes. *Journal of Thermal Spray Technology* 17 (5–6): 700–707.

Oberste-Berghaus, J., S. Boccaricha, J.G. Legoux, C. Moreau, and T. Chraska. 2005. Suspension plasma spraying of nanoceramics using an axial injection torch. In *ITSC 2005*. Dusselörf: DVS. e-proceedings.

Oberste-Berghaus, J., B. Marple, and C. Moreau. 2006a. Suspension plasma spraying of nanostructured WC-12Co coatings. *Journal of Thermal Spray Technology* 15 (4): 676–681.

———. 2006b. Suspension plasma spraying of nanostructured WC-12Co coatings. In *ITSC-2006*, ed. B.R. Marple et al. Materials Park: ASM International. e-proceedings.

Oberste-Berghaus, J., J.-G. Legoux, C. Moreau, F. Tarasi, and T. Chraska. 2008. Mechanical and thermal transport properties of suspension thermal-sprayed alumina-zirconia composite coatings. *Journal of Thermal Spray Technology* 17 (1): 91–104.

Ok Chwa, S., D. Klein, F.L. Toma, G. Bertrand, H. Liao, C. Coddet, and A. Ohmori. 2005. Microstructure and mechanical properties of plasma sprayed nanostructured TiO2–Al composite coatings. *Surface & Coatings Technology* 194: 215–224.

Oliker, V.E., A.E. Terentev, L.K. Shvedova, and I.S. Martsenyuk. 2009. Use of aqueous suspensions in plasma spraying of alumina coatings. *Powder Metallurgy and Metal Ceramics* 48 (1–2): 115–120.

Otsubo, F., H. Era, and K. Kishitake. 2000. Formation of amorphous Fe-Cr-Mo-8P-2C coatings by the high velocity oxy-fuel process. *Journal of Thermal Spray Technology* 9 (4): 494–498.

Ozturkand, A., and B. Cetegen. 2005. Modeling of axially and transversely injected precursor droplets into a plasma environment. *International Journal of Heat and Mass Transfer* 48 (21–22): 4367–4383.

Padture, N.P., K.W. Schlichting, T. Bhatia, A. Ozturk, B. Cetegem, E.H. Jordan, M. Gell, S. Jiang, T.D. Xiao, P.R. Strutt, E. Garcia, P. Miranzo, and M.I. Osendi. 2001. Towards durable thermal barrier coatings with novel microstructures deposited by solution precursor plasma spray. *Acta Materialia* 49: 2251–2257.

Padture, N.P., M. Gell, and E.H. Jordan. 2002. Thermal barrier coatings for gas turbine engine applications. *Science* 296: 280–284.

Parco, M., I. Fagoaga, K. Bobzin, E. Lugsheider, J. Zwick, and G. Hildago. 2006. Developement and characterization of nanostructured iron-based coatings by HFPD and HVOF. In *Proceedings ITSC-2006, e-proceedings*, ed. B. Marple. Materials Park: ASM International.

Patent. 2004. Revêtement Nanostructure´ et Procédé de Revêtement, French patent No. 04 52390, 2004.

Patent FR0453390. Revêtement nanostructuré et procédé de revêtement, CEA Le Ripault, Monts, 37, France. (in French).

Peker, A., and W.L. Johnson. 1993. A highly processable metallic-glass $Zr_{41.2}Ti_{13.8}Cu_{12.5}Ni_{10.0}Be_{22.5}$. *Applied Physics Letters* 63: 2342–2344.

Petorak, C. 2017. *Thermal conductivity and CMAS resistance of SPS composite coatings for use in TBC systems*. TS-4, September, 13–14, 2017, in GE Niskayuna, NY, USA.

Pil, Song Eun, Jeehoon Ahn, Sunghak Lee, and Nack J. Kim. 2008. Effects of critical plasma spray parameter and spray distance on wear resistance of Al2O3–8 wt.%TiO2 coatings plasma-sprayed with nanopowders. *Surface & Coatings Technology* 202 (15): 3625–3632.

Podlesak, H., L. Pawlowski, J. Laureyns, R. Jaworski, and T. Lampke. 2008. Advanced microstructural study of suspension plasma sprayed titanium oxide coatings. *Surface & Coatings Technology* 202: 3723–3731.

Podlesak, H., L. Pawlowski, R. dHaese, J. Laureyns, T. Lampke, and S. Bellayer. 2010. Advanced microstructural study of suspension plasma sprayed hydroxyapatite coatings. *Journal of Thermal Spray Technology* 19 (3): 657–664.

Qiao, Y., Y. Liu, and T.E. Fischer. 2000. Sliding and abrasive wear resistance of thermal-sprayed WC-Co coatings. *Journal of Thermal Spray Technology* 10 (1): 118–125.

Qiao, Y., T.E. Fischer, and Andrew Dent. 2003. The effects of fuel chemistry and feedstock powder structure on the mechanical and tribological properties of HVOF thermal-sprayed WC–Co coatings with very fine structures. *Surface & Coatings Technology* 172: 24–41.

Qiu, C., and Y. Chen. 2009. Manufacturing process of nanostructured alumina coatings by suspension plasma spraying. *Journal of Thermal Spray Technology* 18 (2): 272–283.

Racek, O., and C.C. Berndt. 2007. Mechanical property variations within thermal barrier coatings. *Surface & Coatings Technology* 202: 362–369.

Racek, O., C.C. Berndt, D.N. Guru, and J. Heberlein. 2006. Nanostructured and conventional YSZ coatings deposited using APS and TTPR techniques. *Surface & Coatings Technology* 201: 338–346.

Rampon, R., G. Bertrand, F.L. Toma, and C. Coddet. 2006a. Liquid plasma sprayed coatings of yttria stabilized for SOFC electrolyte. In *ITSC 2006*. Materials Park: ASM International. e-proceedings.

Rampon, R., F.-L. Toma, G. Bertrand, and C. Coddet. 2006b. Liquid plasma sprayed coatings of yttria-stabilized zirconia for SOFC electrolytes. *Journal of Thermal Spray Technology* 15 (4): 682–688.

Rampon, R., G. Bertrand, F.-L. Toma, and C. Coddet. 2006c. Building on 100 years of success. In *Proceedings of ITSC-2006, CD-Rom*, ed. B.R. Marple, M.M. Hyland, Y.C. Lau, R.S. Lima, and J. Voyer. Materials Park: ASM International.

Rampon, R., C. Filiatre, and G. Bertrand. 2008. Suspension plasma spraying of YPSZ coatings: Suspension atomization and injection. *Journal of Thermal Spray Technology* 17 (1): 105–114.

Rauch, J., N. Stiegler, A. Killinger, and R. Gadow. 2009. High velocity suspension flame spraying (HVSFS): Process development and industrial applications. In *Thermal spray 2009: Expanding thermal spray performance to new markets and applications*, ed. B.R. Marple et al., 150–155. Materials Park: ASM International.

Ravi, B.G., S. Sampath, R. Gambino, P.S. Devi, and J.B. Parise. 2006. Plasma spray synthesis from precursors: Progress, issues and considerations. *Journal of Thermal Spray Technology* 15 (4): 701–707.

Roy, M., A. Pauschitz, J. Bernardi, T. Koch, and F. Franek. 2006. Microstructure and mechanical properties of HVOF sprayed nanocrystalline Cr_3C_2-25(Ni20Cr) coating. *Journal of Thermal Spray Technology* 15 (3): 372–381.

Saha, A., S. Seal, B. Cetegen, E. Jordan, A. Ozturk, and S. Basu. 2009. Thermalo-physical processes in cerium nitrate precursor droplets injected into high temperature plasma. *Surface & Coatings Technology* 203: 2081–2091.

Sampath, S., R.A. Neiser, H. Herman, J.P. Kirkland, and W.T. Elam. 1993. A structural investigation of a plasma sprayed Ni-Cr based alloy coating. *Journal of Materials Research* 1: 78–86.

Sansoucy, E., G.E. Kim, A.L. Moran, and B. Jodoin. 2007. Mechanical characteristics of Al-Co-Ce coatings produced by the cold spray process. *Journal of Thermal Spray Technology* 16 (5–6): 651–660.

Sergueeva, A.V., N.A. Mara, J.D. Kuntz, D.J. Branagan, and A.K. Mukherjee. 2004. Shear band formation and ductility of metallic glasses. *Materials Science and Engineering A* 383 (2): 219–223.

Sergueeva, A.V., D.J. Branagan, and A.K. Mukherjee. 2008. Microstructure/properties relationship in Fe-based nanomaterials. *Materials Science and Engineering A* 493: 237–240.

Shan, Y., T.W. Coyle, and J. Mostaghimi. 2007a. Numerical simulation of droplet break-up and collision in solution precursor plasma spraying. *Journal of Thermal Spray Technology* 16: 698–704.

———. 2007b. 3D modeling of transport phenomena and the injection of the solution droplets in the solution precursor plasma spraying. *Journal of Thermal Spray Technology* 16 (5–6): 736–743.

Sharifi, N., F. Ben Ettouil, C. Moreau, A. Dolatabadi, and M. Pugh. 2017a. Engineering surface texture and hierarchical morphology of suspension plasma sprayed TiO2 coatings to control wetting behavior and superhydrophobic properties. *Surface & Coatings Technology* 329: 139–148.

Sharifi, N., F. Ben Ettouil, A. Dolatabadi, M. Pugh, and C. Moreau. 2017b. *Engineering surface morphology of suspension plasma sprayed TiO2 coatings to control wetting behavior and superhydrophobic properties.* TS4, September, 13–14, in GE Niskayuna, NY, USA.

Shaw, L.L., D. Goberman, R. Ren, M. Gell, S. Jiang, Y. Wang, T.D. Xiao, and P.R. Strutt. 2000. The dependency of microstructure and properties of nanostructured coatings on plasma spray conditions. *Surface & Coatings Technology* 130: 1–8.

———. 2003. The dependency of microstructure and properties of nanostructured coatings on plasma spray conditions. *Surface & Coatings Technology* 130: 1–8.

Shen, Y., V.A.B. Almeida, and F. Gitzhofer. 2011. Preparation of nano-composite GDC/LSCF cathode material for IT-SOFC by induction plasma spraying. *Journal of Thermal Spray Technology* 20 (1–2): 145–153.

Shen, Q., L. Yang, Y.C. Zhou, Y.G. Wei, and N.G. Wang. 2017. Models for predicting TGO growth to rough interface in TBCs. *Surface & Coatings Technology* 325 (2017): 219–228.

Shunyan, T., B. Liang, C. Ding, H. Liao, and C. Coddet. 2005. Wear characteristics of plasma-sprayed nanostructured yttria partially stabilized zirconia coatings. *Journal of Thermal Spray Technology* 14 (4): 518–523.

Siegert, R., J.E. Doring, J.L. Marque's, R. Vassen, D. Sebold, and D. Stöver. 2005. Influence of injection parameters on the suspension plasma spraying coating properties. In *ITSC-2005*, ed. E. Lugsheider. Düsseldorf: DVS. e-proceedings.

Siegmann, S., O. Brandt, and M. Dvorak. 2004. Thermally sprayed wear resistant coatings with nanostructured hard phases. *Journal of Thermal Spray Technology* 13 (1): 37–43.

Singh, N., B. Manshian, G.J.S. Jenkins, S.M. Griffiths, P.M. Williams, T.G.G. Maffeis, C.J. Wright, and S.H. Doak. 2009. *Biomaterials* 30: 3891–3914.

Skandan, G., R. Yao, R. Sadangi, B.H. Kear, Y. Qiao, L. Liu, and T.E. Fischer. 1999. Multimodal coatings: A new concept in thermal spraying. *Journal of Thermal Spray Technology* 9 (3): 329–331.

Sodeoka, S., M. Suzaki, and T. Inone. 2006. Mechanical properties of plasma sprayed alumina-zirconia nano-composite film. In *ITSC-2006*, ed. B.R. Marple et al. Materials Park: ASM International. e-proceedings.

Sokolowski P., P. Carpio, and L. Pawlowski. 2017. *The thermal transport properties of suspension-and-solution precursor plasma sprayed YSZ coatings*, TS-4, September, 13–14, 2017, in GE Niskayuna, NY, USA

Soltani, R., E. Garcia, T.W. Coyle, J. Mostaghimi, R.S. Lima, B.R. Marple, and C. Moreau. 2006. Thermomechanical behavior of nanostructured plasma sprayed zirconia coatings. *Journal of Thermal Spray Technology* 15 (4): 657–662.

Song, E.P., J. Ahn, S. Lee, and N.J. Kim. 2006. Microstructure and wear resistance of nanostructured Al2O3–8wt.%TiO2 coatings plasma-sprayed with nanopowders. *Surface & Coatings Technology* 201: 1309–1315.

Stiegler, N., A. Killinger, and R. Gadow. 2010. Hydroxyapatite coatings for biomedical applications deposited by different thermal spray techniques. *Surface & Coatings Technology* 205: 1157–1164.

Stiegler, N., D. Bellucci, G. Bolelli, V. Cannillo, R. Gadow, A. Killinger, L. Lusvarghi, and A. Sola. 2012. High-velocity suspension flame sprayed (HVSFS) hydroxyapatite coatings for biomedical applications. *Journal of Thermal Spray Technology* 21 (2): 275–287.

Stöver, D., D. Hathiramani, R. Vaßen, and R.J. Damani. 2006. Plasma sprayed components for SOFC applications. *Surface & Coatings Technology* 201: 2002–2005.

Sun, L., C.C. Berndt, K.A. Gross, and A. Kucuk. 2001. Materials fundamentals and clinical performance of plasma-sprayed hydroxyapatite coatings: A review. *Journal of Biomedical Materials Research* 58: 570–592.

Tagai, H., and H. Aoki. 1980. Chapter 39: Preparation of synthetic hydroxyapatite and sintering of apatite ceramics. In *Mechanical properties of biomaterials*, ed. G.W. Hastings and D.F. Williams. New York: Wiley.

Tang, Z., and N. Young. 2017. *Axial suspension plasma spraying with high spray rate and wide process window*, TS-4, September 13–14, 2017, in GE Niskayuna, NY, USA.

Tang, F., L. Ajdelsztajn, G.E. Kim, V. Provenzano, and J.M. Schoenung. 2004a. Effects of surface oxidation during HVOF processing on the primary stage oxidation of a CoNiCrAlY coating. *Surface & Coatings Technology* 185: 228–233.

Tang, F., L. Ajdelsztajn, and J.M. Schoenung. 2004b. Influence of cryomilling on the morphology and composition of the oxide scales formed on HVOF CoNiCrAlY coatings. *Oxidation Metals* 61 (3/4): 219–238.

Tarasi, F., M. Medraj, A. Dolatabadi, J. Oberste-Berghaus, and C. Moreau. 2008. Effective parameters in axial injection suspension plasma spray process of alumina-zirconia ceramics. *Journal of Thermal Spray Technology* 17 (5–6): 685–691.

———. 2011. Amorphous and crystalline phase formation during suspension plasma spraying of the alumina–zirconia composite. *Journal of the European Ceramic Society* 31: 2903–2913.

Tesar, T., R. Musalek, J. Medricky, and J. Cizec. 2019. On growth of suspension plasma-sprayed coatings deposited by high enthalpy torch. *Surface & Coatings Technology* 371: 333–343.

Tingaud, O., A. Grimaud, A. Denoirjean, G. Montavon, V. Rat, J.F. Coudert, P. Fauchais, and T. Chartier. 2008. Suspension plasma-sprayed alumina coating structures: Operating parameters versus coating architecture. *Journal of Thermal Spray Technology* 17 (5–6): 662–670.

Tingaud, O., A. Bacciochini, G. Montavon, N.A. Denoirjean, and P. Fauchais. 2009. Suspension DC plasma spraying of thick finely-structured ceramic coatings: Process manufacturing mechanisms. *Surface & Coatings Technology* 203: 2157–2161.

Tingaud, O., P. Bertrand, and G. Bertrand. 2010. Microstructure and tribological behavior of suspension plasma sprayed Al_2O_3 and Al_2O_3–YSZ composite coatings. *Surface & Coatings Technology* 205: 1004–1008.

Toma, F.-L. 2017. *New developments on suspension spraying-from hardware to coating solutions*, TS4, September, 13–14, in GE Niskayuna, NY, USA.

Toma, F.L., G. Bertrand, R. Rampon, D. Klein, and C. Coddet. 2006a. Relationship between the suspension properties and liquid plasma sprayed coating characteristics. In *ITSC 2006*. Materials Park: ASM International. e-proceedings.

Toma, F.L., D. Sokolov, G. Bertrand, D. Klein, C. Coddet, and C. Meunier. 2006b. Comparison of the photocatalytic behavior of TiO_2 coatings elaborated by different thermal spray processes. *Journal of Thermal Spray Technology* 15 (4): 576–581.

Toma, F.-L., G. Bertrand, S. Begin, C. Meunier, O. Barres, D. Klein, and C. Coddet. 2006c. Microstructure and environmental functionalities of TiO_2-supported photocatalysts obtained by suspension plasma spraying. *Applied Catalysis B* 68: 74–84.

Toma, F.-L., G. Bertrand, S.O. Chwa, C. Meunier, D. Klein, and C. Coddet. 2006d. Comparative study on the photocatalytic decomposition of nitrogen oxides using TiO_2 coatings prepared by conventional plasma spraying and suspension plasma spraying. *Surface & Coatings Technology* 200: 5855–5862.

Toma, F.-L., L.-M. Berger, T. Naumann, and S. Langner. 2008. Microstructures of nanostructured ceramic coatings obtained by suspension thermal spraying. *Surf Coating Technology* 202: 4343–4348.

Toma, F.-L., L.-M. Berger, D. Jacquet, D. Wicky, I. Villaluenga, Y.R. de Miguel, and J.S. Lindeløv. 2009. Comparative study on the photocatalytic behavior of titanium oxide thermal sprayed coatings from powders and suspensions. *Surface & Coatings Technology* 203 (15): 2150–2156.

Toma, F.-L., L.-M. Berger, C.C. Stahr, T. Naumann, and S. Langner. 2010a. Microstructures and functional properties of suspension-sprayed Al_2O_3 and TiO_2 coatings: An overview. *Journal of Thermal Spray Technology* 19 (1–2): 262–274.

———. 2010b. Microstructures and functional properties of suspension-sprayed Al_2O_3 and TiO_2 coatings: An overview, coatings technology. *Surface & Coatings Technology* 202: 4318–4328.

Toma, F.-L., L.-M. Berger, S. Scheitz, S. Langner, C. Rödel, A. Potthoff, V. Sauchuk, and M. Kusnezoff. 2012. Comparison of the microstructural characteristics and electrical properties of thermally sprayed Al_2O_3 coatings from aqueous suspensions and feedstock powders. *Journal of Thermal Spray Technology* 21 (3–4): 480–488.

Tomaszek, R., Z. Znamirovski, L. Pawlowski, and A. Wojnakowski. 2006. Impedance spectroscopy of suspension plasma sprayed titania coatings. *Surface & Coatings Technology* 201 (5): 2099–2102.

Tomaszek, R., Z. Znamirowski, L. Pawlowski, and J. Zdanowski. 2007a. Effect of conditioning on field electron emission of suspension plasma sprayed TiO_2 coatings. *Vacuum* 81: 1278–1282.

Tomaszek, R., L. Pawlowski, L. Gengembre, J. Laureyns, and A. Le Maguer. 2007b. Microstructure of suspension plasma sprayed multilayer coatings of hydroxyapatite and titanium oxide. *Surface & Coatings Technology* 201: 7432–7440.

Turunen, E., T. Varis, S.-P. Hannula, A. Vaidya, A. Kulkarni, J. Gutleber, S. Sampath, and H. Herman. 2006a. On the role of particle state and deposition procedure on mechanical, tribological and dielectric response of high velocity oxy-fuel sprayed alumina coatings. *Materials Science and Engineering A* 415: 1–11.

Turunen, E., T. Varis, T.E. Gustafsson, J. Keskinen, T. Falt, and S.-P. Hannula. 2006b. Parameter optimization of HVOF sprayed nanostructured alumina and alumina-nickel composite coatings. *Surface & Coatings Technology* 200: 4987–4994.

Varis, T., J. Knuuttila, E. Turunen, J. Leivo, J. Silvonen, and M. Oksa. 2007. Improved protection properties by using nanostructured ceramic powders for HVOF coatings. *Journal of Thermal Spray Technology* 16 (4): 524–532.

Vasiliev, A.L., N.P. Padture, and X. Ma. 2006a. Coatings of metastable ceramics deposited by solution precursor plasma spray: I-binary ZrO_2-Al_2O_3 system. *Acta Materialia* 54 (19): 4913–4920.

Vasiliev, A.L., N.P. Padture, and X.C. Ma. 2006b. Coatings of metastable ceramics deposited by solution-precursor plasma spray: II. Ternary ZrO_2–Y_2O_3–Al_2O_3 system. *Acta Materialia* 54 (19): 4921–4936.

Vaßen, R., A. Stuke, and D. Stöver. 2009a. Recent developments in the field of thermal barrier coatings. *Journal of Thermal Spray Technology* 18 (2): 181–186.

Vaßen, R., Z. Yi, H. Kaßner, and D. Stöver. 2009b. Suspension plasma spraying of TiO_2 for the manufacture of photovoltaic cells. *Surface & Coatings Technology* 203 (15): 2146–2149.

Vaßen, R., H. Kaßner, G. Mauer, and D. Stöver. 2010. Suspension plasma spraying: Process characteristics and applications. *Journal of Thermal Spray Technology* 19 (1–2): 219–225.

Vert, R., D. Chicot, C. Dublanche-Tixier, E. Meillot, A. Vardelle, and G. Mariaux. 2010. Adhesion of YSZ suspension plasma-sprayed coating on smooth and thin substrates. *Surface & Coatings Technology* 205: 999–1003.

Viswanathan, V., T. Laha, K. Balani, A. Agarwal, and S. Seal. 2006. Challenges and advances in nanocomposite processing techniques. *Materials Science and Engineering* R54: 121–285.

Waldbillig, D., and O. Kesler. 2009a. Characterization of metal-supported axial injection plasma sprayed solid oxide fuel cells with aqueous suspension plasma sprayed electrolyte layers. *Journal of Power Sources* 191: 320–329.

———. 2009b. The effect of solids and dispersant loadings on the suspension viscosities and deposition rates of suspension plasma sprayed YSZ coatings. *Surface and Coatings Technology* 203: 2098–2101.

———. 2011. Effect of suspension plasma spraying process parameters on YSZ coating microstructure and permeability. *Surface and Coatings Technology* 205: 5483–5492.

Waldbilling, D., O. Kesler, Z. Tang, and A. Burgess. 2007. Suspension plasma spraying of solid oxide fuel cell electrolytes. In *Thermal spraying 2007, global coating solutions*, ed. B.R. Marple et al. Materials Park, OH: ASM International. e-proceedings.

Waltz, F., M.A. Swider, P. Hoyer, T. Hassel, M. Erne, K. Möhwald, M. Adlung, A. Feldhoff, C. Wickleder, F.-W. Bach, and P. Behrens. 2012. Synthesis of highly stable magnesium fluoride suspensions and their application in the corrosion protection of a magnesium alloy. *Journal of Materials Science* 47: 176–183.

Wang, M., and L.L. Shaw. 2007. Effects of the powder manufacturing method on microstructure and wear performance of plasma sprayed alumina–titania coatings. *Surface & Coatings Technology* 202: 34–44.

Wang, Y., S. Jiang, M. Wang, S.T.D. Wang Xiao, and P.R. Strutt. 2000. Abrasive wear characteristics of plasma sprayed nanostructured alumina/titania coatings. *Wear* 237: 176–185.

Wang, W.Q., C.K. Sha, D.Q. Sun, and X.Y. Gu. 2006. Microstructural feature, thermal shock resistance and iso-thermal oxidation resistance of nanostructured zirconia coating. *Materials Science and Engineering A* 424: 1–5.

Wang, Y., W. Tian, and Y. Yang. 2007a. Thermal shock behavior of nanostructured and conventional $Al_2O_3/13$ wt% TiO_2 coatings fabricated by plasma spraying. *Surface & Coatings Technology* 201: 7746–7754.

Wang, H.-T., C.-J. Li, G.-J. Yang, C.-X. Li, Q. Zhang, and W.-Y. Li. 2007b. Microstructural characterization of cold-sprayed nanostructured FeAl intermetallic compound coating and its ball-milled feedstock powders. *Journal of Thermal Spray Technology* 16 (5–6): 669–676.

Wang, Y., J.-G. Legoux, R. Neagu, S. Hui, and B.R. Marple. 2012. Suspension plasma spray and performance characterization of half cells with NiO/YSZ anode and YSZ electrolyte. *Journal of Thermal Spray Technology* 21 (1): 7–16.

Webster, T.J., R.W. Siegel, and R. Bizios. 1999. Osteoblast adhesion on nanophase ceramics. *Biomaterials* 20: 1221–1227.

Wilden, J., J.P. Bergmann, S. Reich, S. Schlichting, and T. Schnick. 2007. Cladding of aluminum substrates with nano crystalline solidifying wear resistant iron-based materials. In *ITSC-2007: Global coating solutions*, ed. B.R. Marple, M.M. Hyland, Y.-C. Lau, C.-J. Li, R.S. Lima, and G. Montavon. Materials Park: ASM International. e-proceedings.

Wittmann, K., F. Blein, J. Fazilleau, J.F. Coudert, and P. Fauchais. 2002. A new process to deposit thin coatings by injecting nanoparticles suspensions in a DC plasma jet. In *ITSC-2002*, ed. E. Lugsheider. Düsseldorf: DVS. e-proceedings.

Wittmann-Ténèze, K., K. Vallé, L. Bianchi, P. Belleville, and N. Caron. 2008. Nanostructured zirconia coatings processed by PROSOL deposition. *Surface & Coatings Technology* 202: 4349–4354.

Xia, C.R., and M.L. Liu. 2002. Microstructures conductivities, and electrochemical properties of $Ce_{0.9}Gd_{0.1}O_2$ and GDC-Ni anodes for low-temperature SOFCs. *Solid State Ionics* 152–153: 423–430.

Xiao, Y., L. Song, X. Liu, Y. Huang, T. Huang, J. Chen, Y. Wu, and F. Wu. 2011a. Bioactive glass-ceramic coatings synthesized by the liquid precursor plasma spraying process. *Journal of Thermal Spray Technology* 20 (3): 560–568.

Xiao, Y., L. Song, X. Liu, H. Yi, T. Huang, Y. Wu, J. Chen, and F. Wu. 2011b. Nanostructured bioactive glass–ceramic coatings deposited by the liquid precursor plasma spraying process. *Applied Surface Science* 257: 1898–1905.

Xiaoming, W., L. Chengxin, L. Changjiu, T. Lihui, and Y. Guanjun. 2011. Microstructure and electrochemical behavior of $La_{0.8}Sr_{0.2}MnO_3$ deposited by solution precursor plasma spraying. *Rare Metal Materials and Engineering* 40 (11): 1881–1886.

Xie, L., X. Ma, A. Ozturk, E.H. Jordan, N.P. Padture, B.M. Cetegen, D.T. Xiao, and M. Gell. 2004a. Processing parameter effects on solution precursor plasma spray process spray patterns. *Surface & Coatings Technology* 183 (1): 51–61.

Xie, L., X.C. Ma, E.H. Jordan, N.P. Padture, D.T. Xiao, and M. Gell. 2004b. Deposition mechanisms of thermal barrier coatings in the solution precursor plasma spray process. *Surface & Coatings Technology* 177–178: 103–107.

Xie, L., E.H. Jordan, N.P. Padture, and M. Gell. 2004c. Phase and microstructural stability of solution precursor plasma sprayed thermal barrier coatings. *Materials Science and Engineering A* 381: 189–195.

Xie, L., D. Chen, E.H. Jordan, N.P. Padture, A. Ozturk, F. Wu, X.C. Ma, B.M. Cetegem, and M. Gell. 2006. Formation of vertical cracks in solution-precursor plasma-sprayed thermal barrier coatings. *Surface & Coatings Technology* 201: 1058–1064.

Yandouzi, M., E. Sansoucy, L. Ajdelsztajn, and B. Jodoin. 2007. WC-based cermet coatings produced by cold gas dynamic and pulsed gas dynamic spraying processes. *Surface & Coatings Technology* 202: 382–390.

Yang, Y.C., and C. Chang. 2003. The bonding of plasma-sprayed hydroxyapatite coatings to titanium: Effect of processing, porosity and residual stress. *Thin Solid Films* 444: 260–275.

Yang, Y.-C., and C.-Y. Yang. 2013. Mechanical and histological evaluation of a plasma sprayed hydroxyapatite coating on a titanium bond coat. *Ceramics International* 39: 6509–6516.

Yang, G.-J., C.-J. Li, X.-C. Huang, Y.-Y. Wang, and C.-X. Li. 2007. Influence of silver doping on photocatalytic activity of liquid flame sprayed nanostructured TiO2 coating. In *Thermal spray 2007: Global coating solutions*, ed. B.R. Marple et al. Materials Park: ASM International. e-proceedings.

Yuan, J., Q. Zhan, Q. Lei, S. Ding, and H. Li. 2012. Fabrication and characterization of hybrid micro/nanostructured hydrophilic titania coatings deposited by suspension flame spraying. *Applied Surface Science* 258: 6672–6678.

Zhang, Q., C.-J. Li, C.-X. Li, G.-J. Yang, and S.-C. Lui. 2008a. Study of oxidation behavior of nanostructured NiCrAlY bond coatings deposited by cold spraying. *Surface & Coatings Technology* 202 (14): 3378–3384.

Zhang, J., J. He, Y. Dong, X. Li, and D. Yan. 2008b. Microstructure characteristics of Al_2O_3-13wt.%TiO_2 coating plasma spray deposited with nanocrystalline powders. *Journal of Materials Processing Technology* 197 (1–3): 31–35.

Zhou, C., N. Wang, and H. Xu. 2007. Comparison of thermal cycling behavior of plasma-sprayed nanostructured and traditional thermal barrier coatings. *Materials Science and Engineering A* 452–453: 569–574.

Zhou, J., J.K. Walleser, B.E. Meacham, and D.J. Branagan. 2010. Novel in situ transformable coating for elevated-temperature applications. *Journal of Thermal Spray Technology* 19 (5): 950–957.

Zhou, D., O. Guillon, and R. Vassen. 2017. *Development of columnar YSZ Thermal barrier coatings with suspension plasma spraying*, TS-4, September 13–14, 2017, in GE Niskayuna, NY, USA.

Zhu, Y., M. Huang, J. Huang, and C. Ding. 1999. Vacuum-plasma sprayed nanostructured titanium oxide films. *Journal of Thermal Spray Technology* 8 (2): 219–222.

Znamirowski, Z., and M. Ladaczek. 2008. Lighting segment with field electron titania cathode made using suspension plasma spraying. *Surface & Coatings Technology* 202: 4449–4452.

Coating Characterizations

Abbreviations

AAS	Atomic Absorption Spectroscopy
AC	Alternative Current
AE	Acoustic Emission
AES	Atomic Emission Spectroscopy
AFM	Atomic Force Microscopy
AP	Archimedean Porosimetry
BSE	Back Scattered Electrons
CBN	Cubic Boron Nitride
CMAS	Calcium–Magnesium–Aluminosilicate
DB	Double Bar method
DC	Direct Current
DCB	Double Cantilever Beam
D-gun	Detonation Gun
DOF	Depth of Field
DPH	Diamond Pyramid Hardness
DSC	Differential Scanning Calorimetry
DTA	Differential Thermal Analysis
EB-PVD	Electron Beam Physical Vapor Deposition
EDS	Energy-Dispersive X-ray Spectroscopy
EIS	Electrochemical Impedance Spectroscopy
EN	Electrochemical Noise
EPMA	Electron Probe Microanalysis
ESCA	Electron Spectroscopy for Chemical Analysis
EXAFS	Extended X-Ray Absorption Fine Structure
FBR	Fluidized Bed Reactor
FCT	Furnace Cycle Test
FESEM	Field Emission Scanning Electron Microscope
FFT	Fast Fourier Transform
FGM	Functional Gradient Material
FIB	Focused Ion Beam
FTIR	Fourier Transform Infrared Spectroscopy
FWHM	Full-Width at Half-Maximum
GP	Gas Permeation
HA	Hydroxyapatite
HIP	Hot Isostatic Pressing
HK	Knoop Hardness
HR	High resolution
HRTEM	High-Resolution Transmission Electron Microscopy
HU	Universal Hardness
HV	Vickers Hardness
HVOF	High-Velocity Oxy-Fuel
i.d.	Internal Diameter
IA	Image Analysis
ICP	Inductively Coupled Plasma
IRS	Infrared Spectroscopy
JCPDS	Joint Committee Powder Diffraction Standard
JETS	Jet Engine Thermal Shock
LASAT	Laser Adhesion Test
MIP	Mercury Intrusion Porosimetry
MS	Mass Spectrometry
MSANS	Multiple Small Angle Neutron Scattering
ND	Neutron Diffraction
NDT	Nondestructive Technique
NEXAFS	Near Edge X-Ray Absorption Fine Structure
NS	Neutron Scattering
OM	Optical Microscopy
OOF	Object-Oriented Finite Element Analysis
P	Pycnometry
PS	Porod Scattering
PT	Pulsed Thermography
R&D	Research and Development
RC	Resistive/Capacitive Circuit
RCF	Rolling Contact Fatigue
REV	Representative Elementary Volume
RFPPS	RF Precursor Plasma Spray Synthesis
RPM	Rotation Per Minute
RTS	Reactive Thermal Spraying
RVE	Representative Volume Element
SANS	Small-Angle Neutron Scattering
SAW	Surface Acoustic Waves
SAXS	Small-Angle X-ray Scattering
SB	Single Bar Method

© Springer Nature Switzerland AG 2021
M. I. Boulos et al. (ed.), *Thermal Spray Fundamentals*, https://doi.org/10.1007/978-3-030-70672-2_17

SCE	Standard Calomel Electrode
SEM	Scanning Electron Microscopy
SRV	Sliding, Reciprocating, and Vibrating friction
ST	Stereological Protocols
STF	Strain to Fracture
TAT	Tensile Adhesion Test
TBC	Thermal Barrier Coating
TEM	Transmission Electron Microscopy
TG	Thermo Gravimeter
TG-DTA	Thermo Gravimeter-Differential Thermal Analysis
TGO	Thermally Grown Oxide
TSR	Thermal Shock Resistance
TSS	Thermal Spray Society
URCAS	Ultrasonic Reflection Coefficient Amplitude Spectrum
USAXS	Ultrasmall Angle X-Ray Scattering
VH	Vickers Hardness
XANES	X-Ray Absorption Near Edge Structure
XAS	X-Ray Absorption Spectroscopy
XPS	X-Ray Photoelectron Spectroscopy
XRD	X-Ray Diffraction
XRF	X-Ray Fluorescence
YAG	Yttrium Aluminum Garnet
YPSZ	Yttria Partially Stabilized Zirconia
YSZ	Yttria-Stabilized Zirconia

17.1 Introduction

Coatings as most industrial products must be tested.

At the research and development scale, coatings can be characterized using techniques with varying degree of sophistication and complexity, such as nondestructive testing, metallography coupled with image analysis, materials characterization, void content, and network architecture. In a production environment, on the other hand, testing is mostly oriented toward trouble shooting and the quality control, such as adhesion–cohesion, mechanical properties, thermal properties, wear and corrosion resistance, related to coating performance under service conditions. In this chapter, a review is presented of commonly used coating characterization techniques and testing methods for R&D and industrial production scale in the thermal spray coatings industry. These are grouped in terms of techniques used and properties evaluated. Extensive references relevant to ASTM standards are given in Appendix A, where the reader can find detailed description of these techniques.

17.1.1 Differences Between Coatings and Bulk Materials

As described in Chaps. 15 and 16, thermal-sprayed coatings are made of layered splats; the contacts between splats or between splats and substrate surface correspond to small percent of the total surface of the splats, from 15% to 60%. The response of these coatings to any external solicitation will depend strongly on the mean value of these contacts. Moreover, coatings have an anisotropic behavior: for example, hardness measured at coating surface is different from that measured on its cross-section. Within splats, because of the fast quenching of the flattening melted droplets, material grain or column sizes are rather small (below 0.5 μm in diameter). The coating adhesion is either mechanical, as in most cases, or chemical for few specific sprayed materials associated with specific substrates, or metallurgical, that is, controlled by diffusion when metal coatings are plasma sprayed in soft vacuum on hot metal substrates where the oxide layer has been removed. The coating adhesion to substrate is generally below 70 MPa for mechanical adhesion and much higher for diffusion-controlled adhesion. Sprayed coating properties are also lowered by various defects such as pores and the inclusion of unmelted particles.

Wrought, forged, cast, or sintered metal materials have, on the other hand, quite different properties. For materials prepared by metallurgical processes, grain sizes are larger (a few micrometers, that is, 10–100 times that of coatings) due to much lower cooling rates and also heat treatments to achieve specific precipitates. Tensile strengths are generally much higher than that of coatings, from a few hundreds to thousands or more MPa. For sintered engineering ceramic materials, grain sizes (about 0.6–14 μm) are larger than those of thermal-sprayed coatings and if tensile strengths are lower (hundreds MPa) than those of bulk materials, they are still higher than those of coatings.

17.1.2 Characterization and Testing Methods Used for Coatings

The first characteristics that are important and necessary are those related to the coating microstructures at different depths. According to Pawlowski L. (1995), they involve:

- Chemical composition at macro- and microscales
- Grain or column morphology and orientation (texture)
- Defects such as voids, unmelted particles, micro- and macrocracks, and dislocations with their number and distribution

For such characterizations, in most cases, the critical step is the metallographic examination. However, the heterogeneous or composite nature of the thermal spray coating can render the choice of a metallographic preparation protocol to be used rather difficult. According to Riggs W (2004), and Wigren J and K Täng (2007), the observed coating microstructure can vary significantly depending on the protocol used for sample preparation. For example, they cited the evaluation of plasma-sprayed WC/Co coatings sprayed by one supplier and the corresponding metallographic preparation by 27 laboratories that resulted in a wide range of observed properties. Sample mounting technique, whether hot and cold under vacuum, may be the single most important step in the whole metallographic procedure that can have a significant impact on the observed sample microstructure. Improper grinding and polishing can also be the source of many problems as well as the measurement system analysis using image analysis [Wigren J and K Täng (2007)].

Testing methods make it possible to determine properties such as coating adhesion–cohesion, mechanical and thermal properties, or service properties such as wear and corrosion resistance. However, the test results can also depend on the sample preparation techniques. For instance, the adhesion–cohesion test results depend strongly on the penetration of epoxy and pressure used during curing, e.g., the tensile strength of Ni–5Al/Alumina system was found to vary from 60 down to 15 MPa [Wigren J and K Täng (2007)]. However, when the coating was dense enough to prevent the penetration of the glue, this variation in tensile results was not observed. Such tests must be as close as possible to real-world coating service conditions to be used at the production stage. In addition, they should be rapid to operate and easy to be used by less skilled personnel. Generally, test methods consist in observing or measuring the response of coatings to various external solicitations: loads, stress, applied electric signals, or other stimuli [Wigren J and K Täng (2007)]. The coating response depends on its physical, chemical, and structural properties that in turn depend to a great extent on the real contacts between splats. To characterize a specific coating property, the coating response to the applied stimulus must be sufficiently important and clear to be measured. For example, when applying a load, the response of the coating can be purely elastic or plastic or both and it must be known to interpret correctly the response signal.

The result of every test is practically a number: size of a print for hardness, percent of porosity, and thermal signal propagation for thermal diffusivity. The numbers obtained from tests are supposed to be representative of the studied property. However, numbers are useful only if a certain degree of confidence in them is achieved [Riggs W (2004), Wigren J, Täng K (2007) and Wigren J, Johansson J (2011)]. The degree of confidence is usually expressed statistically based on the results of a significant number of tests, being highest if a minimal variance or standard deviation from the average value is obtained. For example, a hardness of 400 HV under a load of 5 N is meaningless: the design engineer will be satisfied if this value is 400 ± 20 HV$_5$ and very disappointed if it is 400 ± 150 HV$_5$.

17.1.3 Statistical Methods

Samuel Clemens wrote: "there are three kinds of lies: lies, damned lies, and statistics." It underlines that statistical methods to characterize sprayed coating properties must be used properly. The complexity of coating structures, their anisotropy implying to consider the measurement directions, their strong dependence on spray conditions and powder characteristics, and the generally wide scatter in different property values make it necessary to perform statistical analysis when determining coating properties. This will be illustrated for microhardness, which is a basic mechanical property as elastic modulus. They are used to characterize the performance of coatings as they affect the erosion wear-resistance performance, stress–strain behavior, contact stress field, coating delaminating, coating fracture, and residual stress state within coatings [Li J, Ding C (2001)]. However, hardness data are often presented in the literature as a single numerical statistic value without any details of how such a value was obtained, which is meaningless.

The measuring equipment, a factor that can affect hardness measurement, must be, as for any type of measuring equipment, tested regularly in order to avoid systematic errors and biases. The coating surface or cross-section preparation is a nonnegligible parameter. The load also is very important because the indenter penetration increases nonlinearly with it (plastic deformation increases with load). To achieve comparable results, the same load must be used with the same time of application. The indentation size effect is of little significance [Factor M, Roman I (2000a, b)] when comparisons are made between samples while keeping the load constant. Indents should only be discounted if there is reason to believe that the measurement is invalid, such as if the testing equipment suffered vibration during the test. Disqualification of outliers to get a "statistically more meaningful result" is not legitimate if, as is usually the case, these represent true variation of the coating structure. Microhardness indentation is also subject to systematic biases between operators that limit the reproducibility of the technique. As it could be expected, these biases increase significantly when the indent size is decreased [Factor M, Roman I (2000a, b)]. Since they may be more important for low loads and high hardness materials, the hardness data reported in the literature for ceramics materials and cemented carbides are not necessarily sound.

Statistically, Normal, Lognormal, and Weibull distributions are widely used to fit experimental data with

"Normal" and "Weibull" distributions being mostly used to characterize thermal-sprayed coatings [Walpole RE, Myers RH (1978), 17.S1].

17.1.3.1 Normal Distribution

In laboratories when considering N events, x_i, such as for example, indents, the largest and the smallest values are discarded, before averaging the N-2 remaining values. The standard deviation, σ, is the square root of the average value of $(x - \eta)^2$, where η is the arithmetic mean of the measured values. In the case where x takes random values from a finite data set x_1, x_2, ..., x_N, with each value having the same probability, the standard deviation σ, is given by;

$$\sigma = \sqrt{\frac{1}{N} \sum_{i=1}^{N} (x_i - \eta)^2} \qquad (17.1)$$

However, the central limit theorem says that the distribution of a sum of many independent, identically distributed random variables tends toward the famous bell-shaped "normal distribution" with a probability density function proportional to $\exp(-(x - \eta)^2/2\sigma^2)$, which is a Gaussian distribution. Therefore, the standard deviation is simply a scaling variable that reflects how broad the curve will be. When the data distribution is approximately normal then about 68% of the data values are within $\pm$ one standard deviation of the mean value (mathematically, $\eta \pm \sigma$, where η is the arithmetic mean), about 95% are within $\pm$ two standard deviations ($\eta \pm 2\sigma$), and about 99.7% lie within ± 3 standard deviations ($\eta \pm 3\sigma$).

However, the assumption of normal distribution is not always fulfilled for nonhomogeneous materials, such as thermal-sprayed coatings, where the dimensions of the microstructural features have a similar order of magnitude than that of the indent depth. In general, flaws such as subsurface porosity, brittleness, and trapped fine particles will result in larger indents, thus producing right- or positively skewed hardness distributions [Walpole RE, Myers RH (1978)]. For an asymmetrical dataset, Weibull distribution is more appropriate than Gaussian distribution. Lin CK and CC Berndt (1993) were among the first to demonstrate that for yttria partially or totally stabilized thermal spray coatings, Weibull statistic is more appropriate.

17.1.3.2 Weibull Statistic

The Weibull distribution function is given by the following equation:

$$F(x.\lambda, m, \delta) = \frac{m}{\lambda} \left(\frac{x - \delta}{\lambda}\right)^{m-1} \exp\left(-\left(\frac{x - \delta}{\lambda}\right)^m\right) \qquad (17.2)$$

where x is the variable, $m > 0$ is the shape parameter, $\lambda > 0$ the scale parameter, and δ the offset.

In materials science, m, the shape parameter is called Weibull modulus. By plotting $[\ln(\ln(1/(1 - F(x - \delta))))]$ versus $[\ln(x - \delta)]$ using a linear regression least-squares fit to calculate the slope and intercept of the best fit straight line, the parameters of the distribution function can be determined [Factor M, Roman I (2000a, b)]. Very often the offset δ is assumed to be zero and the three-factor Weibull distribution is simplified to a two-factor one. "However with modern spreadsheets allowing linear regression and giving the coefficient of determination (r^2 statistic) automatically, finding the offset δ that gives the best fit is trivial, as is the subsequent calculation of the other parameters for this best-fit case" [Factor M, Roman I (2000a, b)]. When measuring Vickers microhardness HV $= x$, and (HV$-$dH) is the characteristic hardness with 63.2% of data points expected to be below this value, and m describes the shape of the distribution. The Weibull distribution is equal to or approximates several other distributions, depending on the value of m. For $m = 1$ exponential distribution, $m = 2$ Rayleigh distribution, $m = 2.5$ lognormal distribution, $m = 3.6$ normal distribution, and $m = 5$ peaked normal distribution. It must be underlined that Factor M, Roman I (2000a, b) are not favorable to the discounting of lowest and highest values since it affects measures of spread such as the range and standard deviation. This also tends to increase the calculated averages, as the data for most coatings are not normally distributed.

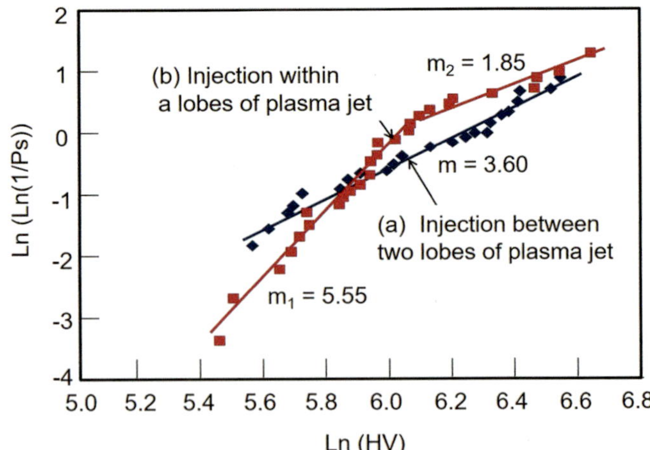

Fig. 17.1 Statistical analysis by microhardness Weibull distribution for lanthanum manganite doped with strontium (La$_{0.8}$Sr$_{0.2}$MnO$_3$: LSM) coatings plasma spraying using Triplex II Pro torch anode nozzle i.d. 6.5 mm, $I = 400$ A, 50% He (**a**) injection between two lobes of plasma jet and (**b**) injected within a lobe of plasma jet [Brousse E (2010)]. (Reprinted with kind permission from Dr. E. Brousse)

An example of Weibull distributions from Brousse E (2010) is presented in Fig. 17.1 for lanthanum manganite doped with strontium, $La_{0.8}Sr_{0.2}MnO_3$, (LSM), coating plasma sprayed using Triplex II Pro torch with an anode nozzle i.d. 6.5 mm, $I = 400$ A, 50% He. The powder used was in the form of micrometer-sized particles made of agglomerated nanometer-sized grains. With this type of particles when the spray conditions are properly adjusted, coatings are made of nanometer-sized areas imbedded within micrometer-sized ones. Thus, the coating hardness depends strongly on locations where measurements are performed: low values in the nanometer-sized zones and high ones in the others. In Fig. 17.1a, the spray conditions were not optimized and only one Weibull modulus was obtained, most sprayed particles being fully melted, while in Fig. 17.1b two Weibull distributions were obtained with the adapted spray conditions resulting in unmelted zones imbedded within fully melted ones. It must be underlined that Weibull distribution is the only way to characterize properly coatings with bimodal distributions.

17.1.3.3 Variance

An important issue when calculating a statistical distribution is the number of data to be considered to have a reliable distribution. It can be determined by calculating the variance according to the number of events, N, taken into consideration. The variance is defined by the mean value of $(x - \eta)^2$, which is very simply calculated, if all events have the same probability, by $(x - \eta)^2/N$.

For the above given example of plasma spraying of lanthanum manganite doped with strontium, $La_{0.8}Sr_{0.2}MnO_3$ (LSM), coating using a Triplex II Pro, the Vickers hardness results given in Fig. 17.2 after Brousse E (2010) were

obtained for different spray conditions [(P1): $d_a = 6.5$ mm, Ar 60 slm, (P2): $d_a = 6.5$ mm, Ar 58 slm + He 19 slm, (P3): $d_a = 6.5$ mm, Ar 54 slm + He 54 slm, and (P4): $d_a = 9.0$ mm, Ar 60 slm]. These show that with the spray conditions P4 (torch with a 9-mm internal diameter anode nozzle) all particles were melted corresponding to the highest hardness 773 ± 55 HV$_3$, and the variance was stabilized with a low number, 5, of measurements. On the contrary with the other spray conditions, the coating hardness was lower (in a ratio of almost twice) with rather large standard deviation and, as shown in Fig. 17.2, the variability was more important (about 0.5 against 0.2 for conditions P4) and a stable value was reached only for 12–14 indents.

17.2 Nondestructive Methods

Simple, fast, reproducible nondestructive coating test methods are one of the dreams of sprayers. If significant progress was achieved at the R&D level, relatively few techniques have been implemented in coating industry. This is probably due to the very complex structure of thermally sprayed coatings that is very difficult to interpret through the signals of nondestructive techniques (NDT). The different devices developed in R&D and industry are summarized in the ASM handbook about "Nondestructive Evaluation and Quality Control of Materials and Manufactured Parts" [ASM International (2007)] and the ASTM reference [17.ND1]. The few devices that can be used for thermal spray coatings are summarily presented below.

17.2.1 Visual Inspection

Visual inspection allows detecting and examining a variety of surface flaws due to corrosion, contamination, delaminating, quality of surface finish, and discontinuities. It comprises image sensors for visual records, magnifying systems, dye and fluorescent penetrants, and magnetic particles for enhancing the observation of cracks or defects. Flexible or rigid boroscopes for illuminating and observing internal, closed, or otherwise inaccessible areas [ASM International (2007)] while indispensable for a variety of applications are not very useful for coatings. A boroscope is an optical device consisting of a rigid or flexible tube with an eyepiece on one end and an objective lens on the other linked together by a relay optical system in between. The optical system is usually surrounded by optical fibers used for illumination of the remote object. An internal image of the illuminated object is formed by the objective lens and magnified by the eyepiece.

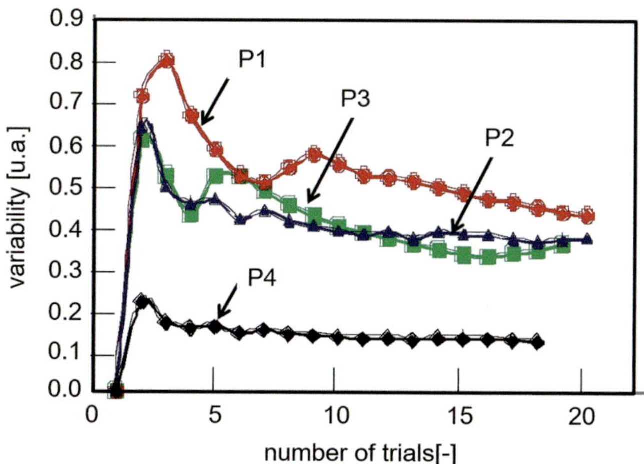

Fig. 17.2 Evolution of the variability of the Vickers hardness for lanthanum manganite doped with strontium, $La_{0.8}Sr_{0.2}MnO_3$ (LSM), coating sprayed using Triplex II Pro, 400A, with different spray conditions [Brousse E (2010)]. (Reprinted with kind permission from Dr. E. Brousse)

17.2.2 Laser Inspection

Laser-based measurement systems are used for dimensional measurements and surface inspection of coating. The following techniques are commonly used for dimensional measurements:

- Scanning laser gage with a precision as high as ± 0.25 μm for parts with dimensions in the range of 10–50 mm. Lower accuracy is to be expected for larger parts.
- Laser triangulation sensors. They measure the standoff distance between a surface and a microprocessor-based sensor providing quick measurements of deviations due to changes in the surface. With two sensors the part thickness or the inside diameter of bores can be measured.
- Surface inspection to detect surface flaws and roughness.

17.2.3 Coordinate Measuring Machines

These provide accurate and flexible 3-D inspection, in-process, before coating, and finished parts, for example, with the coating after machining or rectification. They are normally used in production scale manufacturing facilities, see [ASM International (2007)] for details about the method.

17.2.4 Machine Vision and Robotic Evaluation

A machine vision, also computer vision, includes both visual-sensing and interpretive capabilities allowing identifying shapes, measuring distances and ranges, gaging of sizes and dimensions, determining orientation of parts, and detecting surface shading [ASM International (2007)].

17.2.5 Acoustic Emission

Acoustic emission (AE) is often produced by sudden movement in stressed material due to crack growth and/or plastic deformation. In most cases, the rapid release of energy from localized source within coating or bulk material [Noone MJ, Mehan RL (1974)] produces a stress wave radiated into the structure that can be detected by a sensitive piezoelectric transducer judiciously positioned. It is generally difficult, however, to separate the AE events related to elastic and plastic deformation, and to directly correlate the nature of cracking with the level of AE energy. Multilinear regression analysis of the number of AE events can be helpful to correlate to the spray parameters [Kucuk A et al. (2000b)]. The Weibull modulus of the energy released during four-point bend test can be calculated [Lin CK, Berndt CC (1998)].

AE analysis has been successfully used to monitor cracking in ceramic coatings during mechanical testing such as in four-point bending [Kucuk A, et al. (2000b), Lin CK, Berndt CC (1998) and Lin CK, et al. (1997)] or under thermal cycling conditions [Berndt CC (1985), Voyer J, et al. (1998), Andrews DJ, Taylor JAT (2000) and Robin P, et al. (2010)]. When studying the effect of thermal cycling, the AE behavior monitoring during the heating and cooling periods of a thermal cycle must be separated. These periods show different stress ranges that correspond to different crack lengths and, therefore, correspond to different stress intensity factor ranges that Andrews and Taylor (2000) and Robin P et al. (2010)have used AE as a quality control test for TBCs, as the test is relatively fast and the data can be handled easily by a microprocessor. The test specimens were commercially sprayed straps. The data showed that differences in spraying parameters and microstructure were clearly visible in the emissions during the first thermal cycle.

Recently, Zhao et al. (2010) have extended the ultrasonic coefficient amplitude spectrum (URCAS) to obtain the coating thickness and its longitudinal velocity at the same time on thick substrates. A model was set up first to represent the ultrasonic waves reflected from a coating system at normal incident. Then, an inverse algorithm based on the Gauss–Newton method was introduced to determine the thickness and velocity by comparing the theoretical and measured URCAS. Experimental validation was performed on the inhomogeneous ZrO_2–7 wt% Y_2O_3 (YSZ) coatings on super alloy substrates. The relative errors of the thickness and velocity measurement were in the ranges of 5.33–5.96% and 8.95–9.66% for YSZ coatings. It seems that the URCAS combined with inversion technique can be applied to obtain the thickness and longitudinal velocity of coatings simultaneously.

17.2.6 Laser Ultrasonic Techniques

Nondestructive ultrasonic techniques have been used to determine the elastic constants, such as Young's modulus and Poisson's ratio, of a variety of bulk materials [Boccaccini DN, Boccaccini AR (1997), Asmani M, et al. (2001)]. It is difficult, however, to measure the elastic properties of thermal spray coatings by conventional ultrasonic techniques due to the porosity and extensive micro cracking. The laser ultrasonic technique uses surface acoustic waves (SAWs) or Rayleigh waves, propagating along the surface of a material, to obtain information about it. SAWs are generated by the instantaneous local thermal expansion, laser ablation, or high-intensity pulsed ion beam. As the wave motion is concentrated near the surface, it makes SAWs interesting for testing coatings [Ma XQ, et al. (2001a, b)] or thin films. The surface waves are generated and detected at lengths much greater than the dimension of the coating thickness, and the coating thickness-to-wavelength ratio is the key parameter for the penetration depth of the material [Lima RS, et al. (2003a, b)]. A laser pulse can generate simultaneously many wave modes. The laser and its wave detection are performed on the same surface, but at different places, two types of ultrasonic waves, surface longitudinal and Rayleigh wave velocities, permit measuring Young's modulus, E, and Poisson's ratio, ν. E and ν values measured by Knoop indentation were in good agreement with those obtained with laser ultrasonic technique for titania coatings plasma and high-velocity oxy-fuel (HVOF) sprayed coatings [Lima RS, et al. (2003a, b)]. The titania coatings exhibited significantly different Poisson ratio, ν, values, although they had very similar E values. The E values measured on the cross-section of WC–Co coatings agreed well with those measured via laser ultrasonic.

Laser (nanosecond pulsed) ultrasonic has also been used for the nondestructive characterization of ceramic coatings (alumina) and the estimation of their adhesion strength on metallic substrates [Rosa G, et al. (2001)].

17.2.7 Thermography

Measuring the coating thickness is not necessarily straight-forward technique that is used to detect spatial variations in the measured surface temperature pattern. Thermography reveals flaws by searching anomalous hot spots after thermal excitation. Two types of thermography inspections are used: passive and active. Passive thermography measures the variation of surface radiation to identify anomalous regions. Active thermography uses a controllable thermal source to excite the testing object and reduces the environmental influence such as ambient conditions and emissivity variations [Hung YY, et al. (2009)]. The most efficient thermal tests are the dynamical or active ones, which can detect the presence of subsurface detachments by monitoring the evolution of the coating temperature during a thermal transient pulse. Pulsed thermography (PT) is used today for the inspection of the manufacturing quality of coatings immediately after their deposition. Because of its speed and simplicity, PT allows the design of an automatic inspection procedure for factory quality control [Maldague X, Marinetti S (1996), Shepard SM (2001)].

When performing in-field tests on thermal barrier coatings (TBCs) deposited on gas turbine blades, the main problem encountered is the difficulty to correctly interpret the experimental data. Marinetti S et al. (2007) have defined a procedure to reliably discriminate thickness changes and real defects, and presented and discussed preliminary results. Their approach was based on the analysis of the apparent effusivity profile.

17.2.8 Coating Thickness

Measuring the coating thickness is not necessarily straight-forward and the method for thickness evaluation can also be questioned. Wigren J and K Täng (2007) have considered the thickness evaluation of Ni–5Al plasma-sprayed coating. For that they compared two microscope techniques (average and maximum readings) and two types of micrometer (flat and ball end). Thicknesses were given stroke by stroke (two passes) during the coating buildup. A 50 μm thick metallic coating measured with flat micrometer did not have full metallographic coverage in the above investigation, whereas 50 μm measured with a ball end micrometer did. Important differences in thickness measurements were observed. The measurement tool (flat or ball point micrometer) used for thickness measurements had a significant effect on the reading as well as the microscope techniques [Wigren J, Täng K (2007)].

17.3 Metallography and Image Analysis

Metallography is a critical step to characterize coatings. Sometimes, surfaces are rougher than the original grit-blasted surface, which is especially the case for porous and loose structures such as those observed with certain flame-sprayed coatings. Compared to bulk materials, coating preparation for metallographic examination is rather complex. This is due to coating microstructure that is made of layered splats with porosities and the occasional inclusion of unmelted particles. Ceramic coatings are more brittle than bulk-sintered ceramic materials, mixtures of hard and soft materials in cermets, and mixtures of many phases. All these features offer a challenge for conventional metallographers. In the chapter "Metallography and Image Analysis" by Riggs W (2004), an excellent illustration of the metallographic procedure for a WC–Co coating is given. Pullout of particles during coating preparation for metallographic examination represents one of the main problems that is met especially for cermet coatings. According to Van der Voort GF (1999), George Vander Voort (ed) (2004), Geels K (2007) and 17.Me1–17.Me3 properly prepared metallographic surfaces must meet the following criteria:

- Removal of deformation zone produced during rough polishing
- Be flat and free from scratches, stains, and other imperfections
- Keep all nonmetallic inclusions intact
- Show no chipping or galling of hard and brittle intermetallic compounds
- Be free from all traces of disturbed metal

17.3.1   Coating Preparation

17.3.1.1   Sectioning

The first step of metallographic preparation is sectioning either to reduce the specimen size or to examine its cross-section or both. Of course, the process must not alter the microstructure through generated heat and deformation. Because of the generally good resistance of the coatings to compressive loads and their poor resistance to tensile ones, cutting must be performed from the outermost layer of coating, inward, and then through the coating and not the reverse [Riggs W (2004)]. To limit the flaws created by cutting, the specimen can be first encapsulated in a cold mount-type epoxy before sectioning, which is an excellent technique though rather time consuming.

(a) Abrasive Cutting

This is probably the best solution to eliminate or limit heat generation and deformation during the cutting of the sample. The cutoff machine and abrasive wheels must be matched. Diamond, alumina, and silicon carbide are among the most commonly used abrasives depending on the material to be cut. Independently of bond hardness, the coarser grit size produces the harder action, while finer grits result in a softer action and a smoother surface. The bonding material holding abrasive grains in place is made of either resinoid for dry cutting or rubber for wet cutting. Softer bonds are used to cut hard materials, while harder ones are used to cut soft materials. As already emphasized, the proper cooling of the coating during the cutting process is very important and can be achieved using high volume jets or submerged cutting. The cutting speed is also an important parameter and must be adapted to the wheel dimension used.

For thermally sprayed coatings, Antou G, Montavon G (2007) recommended:

- Cutting speed, that is, the speed of the cutting abrasive surface, should be of the order of a few hundred meters per minute.
- Transverse speed should vary between 0.01 and 0.1 mm/s. It is suggested, when possible, to select the lowest possible transverse speed and to use a device in which the cutting speed can be adjusted rather than the cutting load, since in this latter case the cutting speed varies along the sample.
- Water or oil has to be used as lubricant.

Sauer JP (2005) on behalf of the Thermal Spray Society (TSS) Committee on Accepted Practices has also presented the best practices recommendations for sectioning sprayed coatings.

(b) Precision Sectioning

For small parts with a diameter in the range of 75–125 mm, diamond or CBN (*Cubic Boron Nitride*) rimmed wheels are used. Their speed range extends from a few to 1000 rpm and the load range, if possible controlled electronically, varies from a few tenths to ten Newton. The technique gives excellent results for most coatings, but it is rather slow.

To conclude this part applying the following motto is recommended: "Better spend more time for cutting the sample, this will result in reduced polishing time and in less damage to the structure." For more details about cutting, see [Riggs W (2004), George Vander Voort (ed) (2004)] as well as the recommendations of cutoff machine manufacturers.

17.3.1.2   Mounting

Metallographic samples are mounted to facilitate their manipulation, protect the coating and preserve all of its features during its sectioning and polishing preparation. According to Riggs W (2004), the role of the mount is to:

- Not only ease their gripping by hand or automatic devices but also firmly hold the specimen.
- Avoid damaging the specimen, for example, by a too high temperature or pressure.
- Penetrate and fill nonoccluded (surface-connected pores) without modifying their original size and shape and also minimizing pullout during grinding and polishing.

Mounting involves essentially the encapsulation of the coated sample in polymeric mounts. Two principal techniques are used:

- Hot mounting process: The compression molding technique consumes the minimum amount of time. It uses thermosetting and thermoplastic materials. The first one requires heat and pressure during the molding cycle and can be ejected at maximum molding temperature. Thermoplastic materials remain fluid at maximum molding temperatures and become dense and transparent with a decrease in temperature and an increase in pressure.
- Cold mounting process: Neither heat nor pressure is applied. The sample is placed in a plastic or rubber cup with epoxies, polyesters, and acrylics all comprising a resin and a hardener. An exothermic reaction occurs during polymerization and thus the mixing by volume or weights of both components is critical. Epoxies and polyesters are transparent, whereas acrylics are opaque. Epoxy can fill pores if the gas they contain is evacuated. It can be achieved by, if necessary, heating the samples to get rid of water vapor and then by placing the samples in soft vacuum at a pressure such that the boiling of the epoxy is avoided and during 10 min at least. Vacuum

impregnation with a suitable liquid epoxy produces non-porous samples well consolidated and rigid. With this technology, it becomes possible to distinguish between oxide stringers and delaminations.

According to Wigren J and K Täng (2007), "The choice between hot and cold (vacuum) mounting may be the single most important step in the whole metallographic procedure." True microstructure features (cracks and porosities) that never were seen previously are now revealed with cold mount techniques using low viscosity epoxies. The penetration of epoxy during mounting is also crucial for the following steps in the metallographic procedure, especially the grinding step. Mounting defects such as cracking at the corners, bulging in phenolics, soft mounts, and also porous and friable areas (hot mounts), or bubbles (cold mounts) must be strictly avoided.

For more details, see George Vander Voort (ed) (2004), Geels K (2007), Antou G, Montavon G (2007), Sauer JP (2005), Sauer JP, Blann G (2006), Puerta DG (2005) (2006) as well as the recommendations of cutoff machine manufacturers.

17.3.1.3 Grinding

The prepared metallurgical mount is then subjected to a series of grinding and polishing steps in order to achieve a surface suitable for observation at both low and high magnification, see [Riggs W (2004)]. Of course, the grinding procedure depends on the grinded material properties such as hardness and ductility and it must be adapted to each one. The role of grinding, then followed by polishing, is to produce samples with:

- True and undisturbed microstructure.
- Scratch (polishing artifact) must be limited to dimension below those observable at a given magnification.
- Flat specimen with edge rounding: using a polishing surface with high resilience will result in material removal from both the sample surface and the sides. The effect of this is edge rounding and can be seen with mounted specimens if the resin wears at a higher rate than the sample material.

Of course, the process parameters, including the coolant lubricant, must be such that overheating does not occur; the high temperatures generated in the grinding zone can cause different types of thermal damages. Wigren J and K Täng (2007) have underlined that many problems can arise from improper grinding/polishing. By far, one of the biggest issues in grinding is the smearing caused by the SiC papers. The most critical are the unfilled pores and cracks in a metallic coating. Long polishing, while sometimes necessary because of damage created by the grinding steps, can also bring

forward another problem of edge retention, which causes metallic coatings with oxides and pores to be distorted. The width of pores and oxides can also be enlarged. Ceramic coatings can cause other grinding/polishing issues as discussed by Wigren J and K Täng (2007).

The first grinding is the course, one, the purpose of which is to remove the deformation produced during sectioning and achieves a flat surface. It also removes gross amounts of surface material for the micro-sample preparation. The process is performed with abrasive belts or disc-covered rotating wheels and generally water as coolant, which also flush away the surface removal products. In most cases, the abrasive used is alumina or silicon carbide with grit sizes between about 350 and 80 µm. The abrasive action, depending also on the grinding speed, is very aggressive. For more details, see [Puerta DG (2006)] on behalf of TSS Committee on Accepted Practices.

The second grinding is fine grinding performed, generally in wet conditions, with bonded grains of alumina, silicon carbide, and emery. Prepolishing diamond structured discs with diamonds in the size range 3–15 µm can also be used.

A detailed description of grinding materials and process parameters is given by [W. Riggs (2004)] including grinding rates, grinding deformation, and finally material response to (pressure, abrasive type, abrasion fluid, and grinding time).

17.3.1.4 Polishing

The main objective is to remove the abrasion damage layer (plastically deformed material, scratches, slip/twin/shear damage layer immediately beneath the surface) produced by the grinding process. The material removal is achieved with either rolling abrasive particles or fixed abrasive particle. To minimize the time and labor, interchangeable revolving disks are used, each of which covered with emery paper, cloth, or parchment, according to the stage of polishing for which it is required. The abrasive used should permit accurate sizing, which is the case of diamond abrasives that also have a high hardness, a low coefficient of friction, and excellent inertness. They produce uniform and high rate of material removal with very few induced surface damage. Typically, 6-µm mean size is used to achieve the highest removal rates for most materials (rough polishing), the removal rate decreasing very fast for smaller sizes. The type of cloth used is very important for the end result. The final polishing stage removes any deformation zone resulting from the rough polishing. For this polishing, a wide variety of abrasive materials can be used: alumina, chromium oxide, magnesium oxide, colloidal silica, and diamond, with sizes generally around or below 1 µm. The coating materials that are trickier for grinding and polishing are ceramics and cermets. For details about polishing see the Metallography chapter of W. Riggs (2004) or Van der Voort GF (1999), George Vander Voort (ed) (2004), Geels K (2007), Antou G,

Montavon G (2007), Sauer JP (2005), Sauer JP, Blann G (2006), Puerta DG (2005) (2006) or the recommendation of polishing machine manufacturers.

Antou G, Montavon G (2007) recommended to:

- Use semi- or automatic-polishing machine for a higher reproducibility of the operation.
- Adjust independently the load on each sample to polish a given coating nature at the appropriate pressure regardless of the sample size.
- Select identical directions of rotation for the samples and the abrasive support to avoid damage by spallation.

Wigren J and K Täng (2007) underlined that very often long polishing times are necessary, because of the damage created by the grinding steps, especially for ceramic coatings.

17.3.1.5 Etching

Etching is the operation to reveal the microstructural features (grain boundaries, phases, precipitates, and other microstructure constituents) of the polished specimen. Etching is achieved through selective chemical attack of the surface. To prevent uneven attack and stains, especially after polishing, the specimen must be washed and degreased and then cooled in running water before being etched. Etching is performed by immersing the clean specimen in the etching reagent, for a given time, followed by its rapid rinsing in running water and drying. The etching reagent is chosen according to the material to be etched. For example, Nital reagent (2 vol.% of nitric acid in alcohol) is used for steels and cast iron. Electrolytic etching involves the enhancement of standard acid etching by electric current. Many etchants have been developed to reveal the structure of metals and alloys, ceramics, carbides, and nitrides with some etchants revealing the general microstructure of the coating, while others may be selective to certain phases or constituents. For details, see George Vander Voort (ed) (2004), Geels K (2007).

17.3.1.6 Focused Ion Beam

The use of focused ion beam (FIB) is a relatively "high end" surface preparation technique developed in the semiconductor industry. It is similar to SEM, but instead of using electrons it works with a focused ion beam. Ions, being far heavier than electrons, have a high momentum though are much slower than electrons [Giannuzzi LA, Stevens FA (2004)]. As the ion beam hit the surface of the coating, it will remove atoms from the coating and/or substrate. Depending on the beam position, dwell time, and size material, removal is locally controlled in a highly controlled manner, down to the nanometer scale. As with conventional cutting and polishing techniques, ion beams can also produce surface damage and implantation that can be minimized by FIB milling with lower voltage. For thermally sprayed coatings, FIB milling is essentially the only mean allowing studying the contact between a splat and the substrate since the conventional means described earlier are too aggressive and, in most cases, lead to the separation of the splat from the substrate. FIB can also be used to prepare thin TEM samples in the 100th nm range.

17.3.1.7 Examples of Conventional Coatings Preparation

The Thermal Spray Society (TSS) has carried out round Robin tests for the metallographic preparation of two conventional thermal-sprayed coatings [Puerta DG (2008)] (http://asmcommunity.asminternational.org/portal/site/tss/):

- Molybdenum coating with some oxide content after spraying in air.
- Ni–4Cr–4Al/Bentonite coatings containing metallic and nonmetallic phases with a high porosity level (up to 40%).

Mo coatings are intensively used in aerospace industry and also automotive, marine, and heavy industries. For these coatings, vacuum impregnation with a low-viscosity cold mount epoxy is the recommended mounting method. The amount of material, which must be removed, depends on the sectioning methods employed. All accepted practice preparation procedures utilized combined diamond steps totaling a minimum of 5 min.[1] For abradable coating, the possibility to pullout during metallographic preparation is very high as to the elevated porosity with nonmetallic phases generally loosely bonded. Most laboratories identified vacuum impregnation of a castable epoxy as a critical step to ensure such a coating integrity during preparation.[2]

17.3.2 Microscopy

17.3.2.1 Optical Microscopy

Optical microscopy (OM) examination is the first instrument to be used for metallographic observations because it is faster and much less expensive than scanning electron microscopy (SEM). OM covers a large area; the contrast at magnification below 500 is better than that of SEM and can give the natural color of the specimen. In most cases, magnifications are between 50 and 1000×. However, with some microscopes, it becomes possible to perform examination at higher magnifications, for example, 2000×, and even higher, as

---

[1] ASM Thermal Spray Society (TSS), Accepted Practices Committee on Metallography Accepted Practice–Molybdenum Thermal Spray Coatings, see [Puerta DG (2008)].

[2] ASM Thermal Spray Society (TSS), Accepted Practices Committee on Metallography Accepted Practice–NiCrAl/Bentonite Abradable Coatings, see [Puerta DG (2008)].

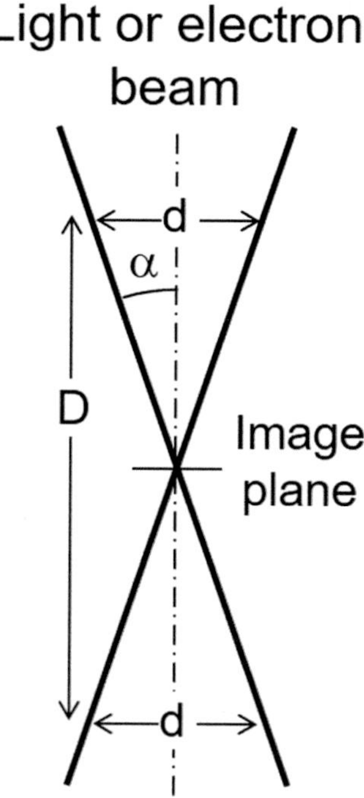

Fig. 17.3 Scheme of the depth of field in optical microscopy

long as diffraction fringes are not present to distort the image. The resolution limit of the OM is at the best about 0.2–0.3 μm. If necessary, special objectives allow using OM at magnifications below 50×. One of the key parameters is image contrast, which depends, on one hand, on the quality of the optics, coatings on lenses, and reduction of flare and glare, and, on other hand, on proper specimen preparation and, if necessary, good etching techniques. The important characteristic of OM is the depth of field, D, area in which the specimen that will be in acceptable focus. Figure 17.3 presents a scheme of the depth of field, D, which can be calculated simply by

$$D = \left(\frac{d}{\tan \alpha}\right) \qquad (17.3)$$

where α is the beam half angle and d its diameter at the depth of field (DOF) position.

Using optical microscopy, the depth of field is the weak point, with D values typically in the micrometers range or below, for example, $D = 5$ μm for a magnification of 20 corresponding to a resolution of 5 μm, and $D = 0.7$ μm for a magnification of 200 corresponding to a resolution of 0.5 μm.

Most observations are performed with bright field illumination where the image of any flat feature perpendicular to the incident light path is bright or appears to be white. Dark field provides higher contrast images and greater resolution than bright field. The light from features perpendicular to the optical axis is blocked and appears dark, while the light from features inclined to the surface appear bright. Polarized light is also used when studying the structure of metals with noncubic crystal structures. Tint-etched surfaces, where a thin film (such as a sulfide, molybdate, chromate, or elemental selenium film) is grown epitaxial on the surface to a depth where interference effects are created, when examined with bright field, produce color images. The process can be improved with polarized light. Another imaging mode is differential interference contrast that converts minor height differences on the plane of polish, invisible in bright field, into visible detail.

In many circumstances, the ability to capture, display, and preserve specimen images is very important. For that purpose, optical microscopes have increasingly built-in digital cameras transmitting high resolution images viewed within the microscope onto a TV or monitor. Adapted software allows performing image analysis. For details about optical microscopes, see George Vander Voort (ed) (2004), Geels K (2007).

17.3.2.2 Scanning Electron Microscopy

In a scanning electron microscope (SEM), an electron beam with an energy ranging from 0.5 to 40 keV is thermionically emitted from an electron gun. The electron beam is focused by one or two condenser lenses to a spot about 0.4–5 nm in diameter. The beam, generally in the final lens, passes through pairs of scanning coils or deflector plates, which deflect it in the x and y axes so that it scans over a rectangular area of the sample surface. The electron beam loses energy by repeated random scattering and absorption within the interaction volume of the specimen, which extends from less than 100 nm to around 5 μm into the surface. The size of the interaction volume depends on the electron's landing energy, the atomic number of the specimen, and the specimen's mass density. The analysis requires high vacuum conditions and specimens must be vacuum compatible (low vapor pressure) and electrically conductive. The charge-up effects on non-conductive samples can be compensated by reduced probe current, by reduced acceleration voltage or, in most cases, by coating them with a thin metal film (Au or Pt) or carbon layer.

Compared to OM, the depth of field of SEM is excellent, for example, at a magnification of 200 the DOF is 100 μm (against 0.7 μm for OM) and at a magnification of 10,000 the DOF is still 2 μm. This allows studying fractured coating surfaces, showing well the splat or grain microstructure. The fractured surface is cut to a suitable size, cleaned of any organic residues, and mounted on a specimen holder for viewing in the SEM. Actually, SEM can reach magnifications up to 500,000. As scanning electron microscopes have evolved, the electron beam cross-section has become smaller and smaller (they reach now a few nanometers in diameter) increasing magnification several fold. The last generation of

SEM, field emission scanning electron microscope (FESEM), has a beam cross-section close to 1 nm in diameter. The field emission tip is made up of a sharply etched piece of monocrystalline tungsten. A field applied to the tip causes electrons to tunnel out of the tip and accelerate down the column.

In SEM and FESEM, the electrons and photons generated in the excitation volume carry different types of information from the analyzed specimen: secondary electrons, backscattered electrons, absorbed specimen current, and cathodoluminescence. Secondary electrons are produced when the incident electrons from the beam interact with the atoms in the surface region of specimen. The impact (collision cascade) causes a path change for the incident electron and ionization of several specimen atoms. The ejected electrons leave the atom with a very small kinetic energy (< 50 eV) and permit a high resolution (about 3.5 nm) of fine surface morphology. As these electrons are sensitive to the orientation of different surface features, they create an image contrast for evaluating the sample's surface topography. SEM is routinely used to observe:

- Single splats
- Surface of as-sprayed coatings
- Metallographic polished surface of coatings or that of their cross-section to see layered splats, structure (columnar or granular), pores, voids from pullouts, and cracks
- Fractured surface of as-sprayed coatings
- Back scattered electrons (BSE) are high-energy electrons produced by the elastic collision of the incident electron beam with the electron cone of the sample atoms. Imaging with them provides elemental composition variation and surface topography (resolution about 5.5 nm). The yield is proportional to the atomic number (Z contrast) and depends on the beam energy and incidence angle. They also allow mapping of individual elements

An example of SEM images from Ingo GM et al. (2005) is presented in Fig. 17.4 for reactive plasma-sprayed composite Ti–TiN–TixNy coating starting from SP700 powder (Ti–4.5Al–3 V–2Mo–2Fe). Figure 17.4a shows the coating surface with high roughness as well as melted and unmolten particles, while Fig. 17.4b presents the coating cross-section by back scattered electrons (BSE) allowing distinguishing nitrided and non-nitrided particles.

The electron beam specimen interaction results in the emission of characteristic X-rays, which can be detected and analyzed (electron probe microanalysis, EPMA). For details about scanning electron microscopes, see [George Vander Voort (ed) (2004), Geels K (2007)].

17.3.2.3 Image Analysis

Image analysis is extensively used to assess the area percentages of some attributes such as porosity; see the chapter of [Riggs W (2004)] and the book of [Russ JC (2011)] or to filter images of spray jets, for example, to follow liquid drops injection in a DC plasma jet [Etchart-Salas R, et al. (2007)]. When a digital image is formed, the digitization process divides it into a horizontal grid of very small regions called "picture elements" or "pixels." In the computer, this digital grid or "bitmap" represents the image. Each pixel is identified by its position in the grid, as referenced by its row (x) and column (y) number. Each pixel has a different color or gray scale value and together they form a representation of the image. Depending on whether pixels are black and white, gray scale, or color, pixels have different bit depths. Bit depth refers to the amount of information allocated to each pixel. For example, in digital color, each color occupies 8 bits (1 byte) while in a gray scale image each picture element has an assigned intensity ranging from 0 to 255. In image analysis, resolution refers to the number of pixels used to represent the image: more pixels correspond to a higher resolving power. With computers, information from image analysis can be easily modified and/or improved:

- Gray scale adjustment to improve the contrast

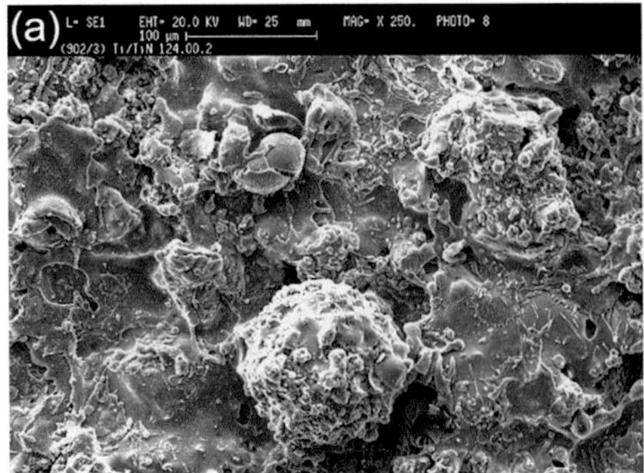
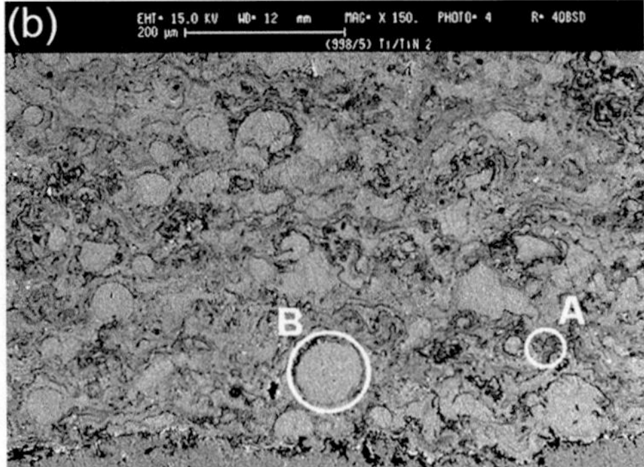

Fig. 17.4 Reactive plasma-sprayed composite Ti–TiN–TixNy: (**a**) SEM image of the surface. (**b**) Cross-section of the coating in BSE micrograph (A) nitrided particles and (B) non-nitrided particles [Ingo GM et al. (2005)]. (Reprinted with kind permission from Elsevier)

- Spatial filtering, which generally results in an image with only edges of high contrast visible, fine detail with low contrast being usually lost to the background
- Noise reduction
- Edge enhancement
- Shape measurement
- Thresholding to select features of interest which is controversial since it is strongly operator dependent

For more details, see the paper of [Antou G and G Montavon (2007)] presenting, from a practical viewpoint, some key points to assess when implementing image analysis coupled to stereological protocols to quantify statistically the architecture of thermal spray coatings and their relevant features (pores, lamellae, and so forth).

Image analysis must be used with care and requires a good knowledge of how information is extracted from the image to achieve reliable data, for example, porosity. At last one must keep in mind that image analysis cannot give more precision than that given by the microscope observation of the prepared sample.

17.3.2.4 Atomic Force Microscopy

Atomic force microscopy (AFM) is based on the specimen surface scan by a cantilever with a sharp tip (probe) at its end. The cantilever is typically made of silicon or silicon nitride with a tip radius of curvature on the order of nanometers. When the tip is brought into proximity of a sample surface, forces between the tip and the sample lead to a deflection of the cantilever according to Hooke's law. Forces implied in AFM include mechanical contact force, van der Waals forces, capillary forces, chemical bonding, electrostatic forces, and magnetic forces. The cantilever deflection is generally measured using a laser spot reflected from its top surface onto an array of photodiodes.

A feedback mechanism makes it possible to adjust the tip-to-sample distance to maintain a constant force between the tip and the sample. For that the sample is moved in the z direction by a piezoelectric tube and in the x and y directions for scanning the sample. See [Eaton P, West P (2010)] for details about the different AFM modes: topographic, nontopographic, and surface modification.

Compared to SEM, AFM has several advantages: it provides a 3-D surface profile against a 2-D one for SEM, it works perfectly well in ambient air or liquid environment, and it can provide higher resolution than SEM. However, it also has some disadvantages: in one pass, SEM can image an area in the order of square millimeters with a depth of field in the order of millimeters, while AFM can only image a maximum height in the order of 10–20 µm and a maximum scanning area of about 150 × 150 µm. The scanning speed of AFM is slower than that of SEM that is capable of scanning at near real time, although at relatively low quality.

In thermal spraying, AFM has been used especially to study splat formation and characterize the variation with oxidation of smooth substrate surface topography at the nanometer scale.

17.3.2.5 Transmission Electron Microscopy

Transmission electron microscopy (TEM) works on the same principle as an old-fashioned slide projector, except that an electron beam replaces light. In the projector, the light passing through the film interacts with it. The light going through the film hits the lenses disposed on the other side and the resulting image is projected onto a screen. In a TEM, a beam of electrons is focused on a single, pinpoint spot or element on the sample being studied [Williams DB, Carter CB (2009)]. The electrons interact with the sample and only those that go through unobstructed hit the phosphors screen on the other side. Electrons that hit the screen are converted to light and so an image is formed. The dark areas of the image correspond to areas on the specimen where fewer electrons were able to pass through (either absorbed or scattered upon impact); the lighter areas are where more electrons pass through, although the varying amounts of electrons in these areas enable the user to see structures and gradients. In a TEM, lenses made of electromagnetic devices focus the electron beam to the desired wavelength or size. The amount of power used to generate electrons controls magnification that can be very high, permitting to see objects a few angstroms in size. Of course, TEM works under high vacuum that can be very low (10^{-7} to 10^{-9} Pa) in certain parts. High-resolution TEM (HR-TEM) is the ultimate tool in imaging defects. In favorable cases, it shows directly a 2-D projection of the crystal with defects and its magnification can reach 10^6.

The drawback of TEM is that specimens studied have to be sliced very thinly (a few micrometers) to ensure that they are electron transparent; they must also be placed in a vacuum. Thus, preparation of specimens is often time consuming. Moreover, it requires expert handling to limit the risk of inadvertent damage during the process, especially for ceramic materials. For example, a wedge-polishing technique was used to prepare samples of YSZ single splats [Chraska T, King AH (2001)]. Wedge-polishing device, supplied by different manufacturers, reduces the need for ion milling by mechanically thinning the area of interest to 0.1 µm or less. These polishing kits are recommended for material applications involving microelectronics, metals, ceramics, minerals, and composites. Electron transparency thickness can be reached for TEM analysis. Focus ion beam (FIB) systems, described earlier in Sect. 17.3.1.6, operate in a fashion similar to a scanning electron microscope (SEM) are also used for TEM sample preparation.

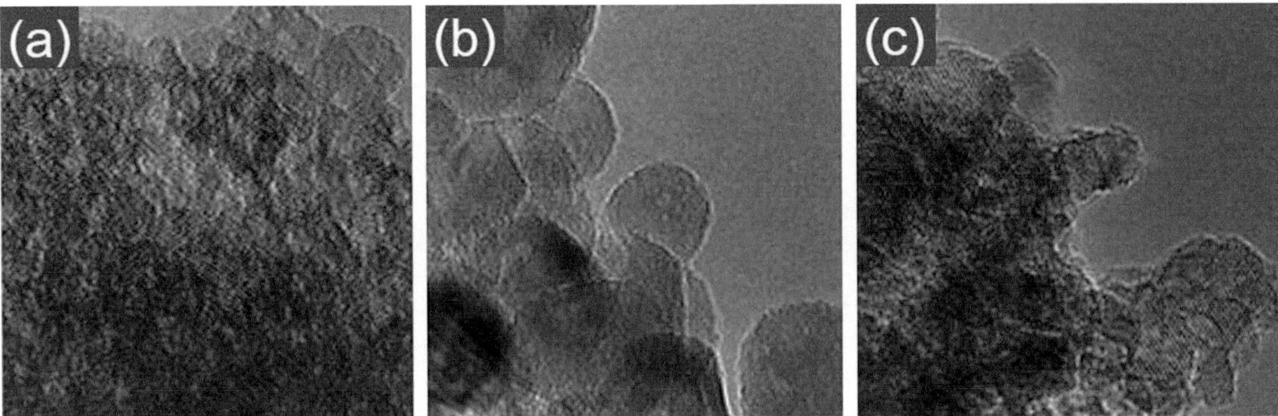

Fig. 17.5 TEM images of (**a**) as-synthesized, (**b**) annealed, and (**c**) hydrothermal treated TiO_2 powders [Tjitra Salim N et al. (2011)]. (Reprinted with kind permission from Elsevier)

TEM allows obtaining information about morphology, crystal size, and chemistry, if equipped as SEM with an EDS (energy-dispersive X-ray spectroscopy) probe. Two observation modes are used: image mode to observe the specimen morphology with a very high magnification and diffraction mode giving information about the crystalline structure. For example, TEM was used by Tjitra Salim N et al. (2011) to characterize TiO_2 powders synthesized by hydrolysis of titanyl sulfate ($TiOSO_4$) in distilled water with a small addition of inorganic salt. At a relatively low hydrolysis temperature, pure anatase TiO_2 could be obtained and post-synthesis treatments (annealing or hydrothermal treatment) did not alter this phase structure. The powder was then agglomerated with fine nano-primary particles, with different post-synthesis treatments leading to different TiO_2 nanostructures. Figure 17.5a represents the powder as synthesized, while Fig. 17.5b shows the annealed powder, and Fig. 17.5c shows the hydrothermal treated powder with its unique oriented agglomeration structures. The hydrothermal treatment consisted in soaking the TiO_2 powder in distilled water and heat treating it for 5 h in an autoclave at 150 °C.

Among the numerous papers about thermal-sprayed coatings where TEM was used to characterize either powders or coatings or both, Chen H et al. (2010) characterized nanostructured zirconia particles, Yan D et al. (2011) studied the structure of coating prepared by reactive plasma spraying Fe_2O_3/Al composite powders, Ma XQ et al. (2008) observed grain structures of suspensions of Ni–14.55Cr–3.22B–4.7Si–4.78Fe sprayed by HVOF, and Chraska T (1999) studied microstructures and interfaces of zirconia coatings produced by plasma spraying. For more details, see [Williams DB, ASM International (2008)].

17.4 Materials Characterization

The techniques described below are limited to nondestructive analytical techniques, which reveal information about the crystallographic structure, chemical composition, and physical properties of coatings. These are based on the observation of the scattered intensity of X-rays, infrared, neutron, and nuclear gamma ray beams hitting a specimen as to incident and scattered angles, polarization, and wavelength or energy.

17.4.1 X-Ray Fluorescence

When a primary X-ray excitation source strikes a sample, the X-rays can be either absorbed by the atom or scattered through the material. The photoelectric effect occurs when the atom absorbs X-ray, the energy of which is all transferred to an innermost electron. If the primary X-rays have sufficient energy, electrons are ejected from the inner shells, creating vacancies. As the atoms return to stable condition, electrons from the outer shells are transferred to the inner shells. This process creates characteristic X-rays whose energy is the difference between the two binding energies of the corresponding shells. Each element having a unique set of energy levels produces X-rays at a unique set of energies. The theoretical relationship between the measured X-ray intensities and concentrations of elements in the sample allows the nondestructive measure of the elemental composition of a sample. This process is called "X-ray Fluorescence" or XRF. A typical X-ray spectrum from an irradiated sample displays multiple peaks of different intensities. The penetration depth of X-ray depends on the coating material and the wavelength of the beam, but it is generally a few micrometers at the maximum 10–15 µm. For details, see ASM International (2008).

17.4.2 Infrared Spectroscopy

In infrared spectroscopy (IRS) molecules absorb specific frequencies that are characteristic of their chemical bonds and organization. The frequency of the absorbed radiation matches the frequency of the bond or group that vibrates. The shape of the molecular potential energy surfaces, masses of atoms, and the associated vibration coupling determine the energies. The frequency of the vibrations can be associated with a particular bond type. For example, the characteristic wave numbers of oxides are between 250 and 1000 cm^{-1}.

Fourier transform infrared spectroscopy (FTIR) refers to a recent development in which infrared light is guided through an interferometer and then through the sample (or vice versa). The data are collected and converted from an interference pattern to a spectrum. The sample's spectrum is then compared to standards as described in ASM International (2008) and 17.Ma1.

In thermal spraying, FTIR has been used to characterize the oxide layer formed at the smooth surface of a metal or alloy substrate subjected to the preheating conditions prior to spraying [Chandra S, Fauchais P (2009)]. In this case, the analysis was performed with a quasi-ranking incidence of the IR light and its depth was between a few nanometers and few micrometers.

17.4.3 Mössbauer Spectroscopy

Mössbauer spectroscopy (MS) is a spectroscopic technique based on the recoil-free, resonant absorption, and emission of gamma rays in solids. Typically, three types of nuclear interaction may be observed: an isomer shift, also known as a chemical shift; quadrupole splitting; and magnetic or hyperfine splitting, also known as the Zeeman effect (splitting of a spectral line into several components in the presence of a static magnetic field) [Volenžk K, et al. (1999)].

Mössbauer spectroscopy gives local information on the nucleus affected, its vibration state, electron density, and magnetic momentum. These data inform about the valence state of the corresponding atoms, bonds with their neighbors, and positions within the crystalline network. The technique is well adapted to the study of iron and its oxides as it allows characterizing phases and their relative percentages. It has been used to characterize the oxidation of steel substrates according to the preheating conditions [Chandra S, Fauchais P (2009)]. It has also been used to characterize the in-flight oxidation of plasma-sprayed alloy steel particles [Volenžk K, et al. (1999)].

17.4.4 X-Ray Diffraction

X-ray diffraction (XRD) is a versatile, nondestructive technique that reveals detailed information about the chemical composition and crystallographic structure of natural and synthetic materials, including coatings [ASM International (2008) and 17.Ma2–17.Ma5]. When a monochromatic X-ray beam is projected onto a crystalline material at an angle, θ, diffraction occurs only when the distance traveled by the rays reflected from successive crystal planes differs by an integer number, n of wavelengths. Bragg developed in 1913 the following relationship between the order of reflection, n, of the crystallographic planes and the angles of incidence (θ) of the X-ray beam.

$$n\lambda = 2d_{hkl} \sin \theta \qquad (17.4)$$

Where, d, is the distance between atomic layers with the same Miller indices (h, k, l), in a crystal, and λ, is the wavelength of the incident X-ray beam. By varying the angle theta, the Bragg's law conditions are satisfied by different d-spacings (characterized by Miller indeces) in polycrystalline materials. Plotting the angular positions and intensities of the resultant diffracted peaks of radiation produces a pattern, which is characteristic of the sample. When the sample is a mixture of different phases, the resultant diffractogram is formed by addition of the individual patterns. The identification of crystalline materials is based on the comparison of experimental values of the reticular distances with reference data banks compiled by the Joint Committee Powder Diffraction Standard (JCPDS). XRD is extensively used in thermal spraying, for example, to characterize structural changes between the sprayed powder and coating. It must be underlined that the penetration distance of the X-ray beam into the sample can vary from few to several micrometers depending on the sample characteristics and wavelength of the beam.

XRD is also used to estimate the average grain size of the feedstock and coating. The applicability of this method was confirmed by comparing results from TEM and XRD techniques [Chraska T (1999), Lima RS et al. (2001a, b)]. The XRD method assumes that the overall broadening of XRD peaks comprises two effects: one arising from the small coherent grain size and one arising from the atomic level micro-strain, that is, ($\Delta d/d$), where d is the atomic spacing. The peak width resulting from a small grain size effect alone can be described by the Scherrer equation [ASM International (2008)]:

$$B(2\theta) = \frac{K\lambda}{D \cos \theta} \qquad (17.5)$$

where $B(2\theta)$ is the true broadening of the diffraction line measured at full width at half-maximum (FWHM), λ is the wavelength of the X-ray radiation, D is the mean dimension of the grains, θ is the Bragg's angle, and K is the Scherrer constant with a commonly used value of 0.9, depending on how the width is determined, the shape of the crystal, and the size distribution. For more information about the K constant, see [Langford JI and AJC Wilson (1978)].

Hugo Rietveld, for the characterization of crystalline materials, showed that the neutron and X-ray diffractions of powder samples result in a pattern characterized by reflections (peaks in intensity) at certain positions. The height, width, and position of these reflections can be used to determine many aspects of the materials structure (see [ASM International (2008)]). X-ray diffraction also makes it possible to measure the angular lattice strain distributions resulting from residual stress. A reflection at high 2-Theta is chosen and the change in the d-spacing with different orientations of the sample is measured. Using Hooke's law, relating stress and strain, the stress can be calculated from the strain distribution.

In amorphous material, the atoms are not arranged in a periodic fashion such that observed with crystals and the scattering intensity is then the summation of each individual atom. Thus, weak X-ray scattering, scattering spreading throughout reciprocal space, and a detailed analysis are required to obtain real-space information characterizing amorphous coatings. Spectra with amorphous elements are characterized by a broad halo as illustrated in Fig. 17.6 after Liu XQ et al. (2009) who sprayed an FeCrMoMnWBCSi amorphous metallic coating with thickness of about 1 mm,

and porosity less than 0.1%, onto mild steel, using high-velocity plasma gun with axial powder injection. Comparing the XRD pattern of the powder used to that of an as-spun ribbon obtained by melt spinning, Fig. 17.6 reveals a broad halo of fully amorphous phase in the ribbon samples. For the powder, it shows a similar broad halo with some crystalline peaks superposed due to Fe_2C, Cr_2B, and $M_{23}C_6$, reflecting that the cooling rate of gas atomization is apparently lower than that of melt spinning. These XRD patterns were also compared to those of coatings obtained with different spray conditions. Similar XRD patterns can be found in many references such as, for example, Kishitake K, Era H, Otsubo F (1996), Zhao XB, Ye ZH (2012).

17.4.5   Small-Angle X-Ray Scattering

Small-angle X-ray scattering (SAXS) is used for the structural characterization of solid and liquid materials in the nanometer (nm) range. It probes inhomogeneities of the electron density on a length scale of typically 1–100 nm, thus yielding complementary structural information to standard XRD data. SAXS is applicable to crystalline and amorphous materials alike. Measurements are commonly performed in transmission geometry, using a narrow, well-collimated, and intense X-ray beam. Scattering angles typically range between 0.1° and 5°. The smallest accessible angle determines the largest resolvable feature size. Some typical applications comprise the determination of nanometer-sized particle and pore size distributions of specific surface areas. For more details, see the book of [Guinier A and G Fournet (1955)].

For coatings, ultra-small-angle X-ray scattering (USAXS) is used as a nondestructive characterization technique recording elastic scattering of X-rays induced by compositional and structural inhomogeneities [Ilavsky J, et al. (2009a, b)]. USAXS has been successfully implemented in quantifying void size distribution in YPSZ-EB-PVD coatings [Flores A, et al. (2007)] and also in YPSZ ($d_{50} = 50$ nm) suspension plasma-sprayed coatings [Bacciochini A, et al. (2010a, b)]. For the EB-PVD coatings, computer modeling was used to obtain values of stereometric (geometrical and spatial) characteristics of the pore populations in terms of their volume fraction, shape (aspect ratio), size, and orientation, each representing statistical average values based on Gaussian distributions. For the nanometer-sized structured coating, about 80% of voids, in number, exhibited characteristic dimensions smaller than 30 nm, the largest voids in the coatings having characteristic dimensions of a few hundreds of nanometers. Such results are difficult, if not impossible, to obtain from other characterization techniques, since it detects with a very high resolution to the whole set of scatters (voids

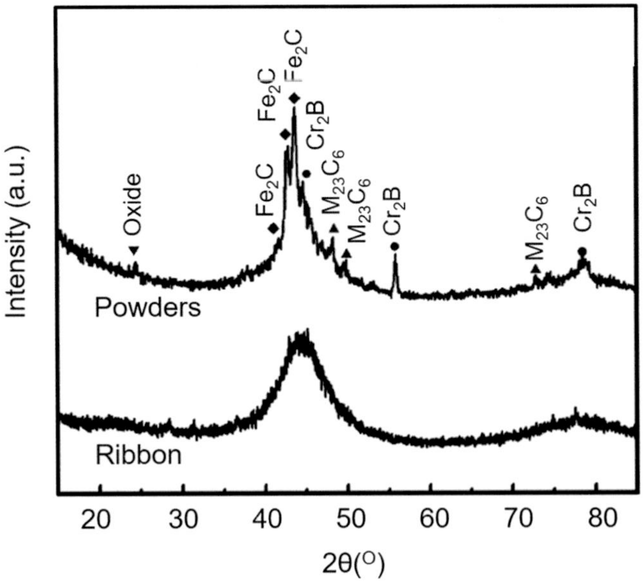

Fig. 17.6 XRD patterns of the as-spun ribbon and the atomized powders of Fe-based alloy [Liu XQ et al. (2009)]. (Reprinted with kind permission from Elsevier)

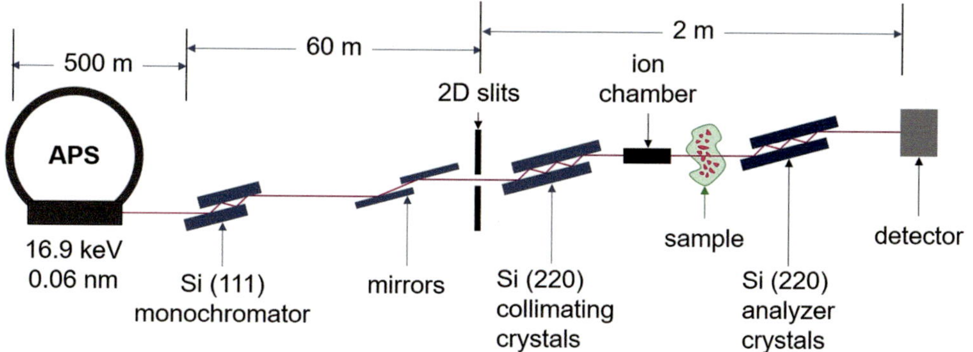

Fig. 17.7 Scheme of the advanced photon source (Argonne National Laboratory, Argonne, IL, USA) (http://usaxs.xor.aps.anl.gov/, see also [Ilavsky J, et al. (2009a, b)])

in this case) regardless of their characteristics (open, connected, or closed).

For both examples cited above, USAXS experiments were conducted on beam line 32-IDat Advanced Photon Source (Argonne National Laboratory, Argonne, IL, USA). Measurements were performed in transmission geometry as shown in Fig. 17.7 that shows the characteristic dimensions of this source. The interested reader can find further details on the Web site of Argonne National Laboratory (http://usaxs.xor.aps.anl.gov/, see also [Ilavsky J, et al. (2009a, b)].

The USAXS instrument uses Bonse–Hart crystal diffraction optics, which allows recording of small-angle scattering curves using a photodiode detector. q is the scattering vector, ranges from 10^{-4} to 1 Å^{-1} with an angular resolution of 10^{-4} Å^{-1}. The scattering vector $|q|$ [Kishitake K, Era H, Otsubo F (1996)] is a typically used quantity in small-angle scattering and relates to the diffraction angle (2θ), as known from X-ray diffraction, via the relationship:

$$|q| = \frac{4\pi \sin\theta}{\lambda} \qquad (17.6)$$

where λ is the X-ray wavelength. Combining this equation with Bragg's law, the length scale, L, probed at a given q range follows the general inverse relationship:

$$L \approx \frac{2\pi}{|q|} \qquad (17.7)$$

This setup delivers approximately 10^{13} photons per second in about 1 mm^2 area at the sample position, for incident photons with energy around 16.9 keV and corresponding to a wavelength of 0.775 Å. The uncertainty estimation for USAXS results is challenging, and absolute intensity of USAXS calibration was estimated, from repeated measurements, to present an uncertainty of about $\pm 5\%$ for the total volume content [Bacciochini A, et al. (2010a, b)].

17.4.6 Small Angle Neutron Scattering

Small angle neutron scattering (SANS) is a laboratory technique, similar to the complementary techniques of small angle X-ray scattering (SAXS) and light scattering. The technique provides valuable information over a wide variety of scientific and technological applications, especially for coatings defects in materials, the data analysis giving information on size and shape. [Allen AJ, et al. (2001a, b), Ilavsky J, et al. (1994) (1999a, b), Kulkarni A, et al. (2003) (2005)].

With SANS experiment a beam of neutrons is directed at a sample, which can be for coatings, a solid or a powder. The neutrons are elastically scattered by changes of refractive index on a nanometer scale inside the sample, which is the interaction with the nuclei of the atoms present in the sample. In this experiment, a monochromatic beam of thermalized neutrons passes through the specimen in transmission geometry and the scattered neutrons are recorded on a 2-D detector. Figure 17.8a shows the principle of the SANS instrument [Kulkarni A, et al. (2003)]. Neutrons are capable of interacting strongly with all atoms, in contrast to X-ray techniques where the X-rays interact weakly with hydrogen, the most abundant element. Small angle neutron scattering is caused by fluctuations of scattering length density $\rho(r)$ on a size scale 1 nm to 5 μm in the studied material. These fluctuations are connected with compositional and/or structural inhomogeneity such as pores. The fluctuations of scattering length density result in the scattering contrast $\Delta\rho(r) = \rho(r) - \overline{\rho}$, where r is the coordinate in the real space and $\overline{\rho}$ the average scattering length density of the sample. The scattering contrast gives rise to the coherent elastic scattering of neutrons to small magnitudes of the scattering vector $|q|$; see Eq. 17.6, ($q = k - k_0$, where k is the wave vectors of the incident and scattered neutron, respectively, $|k| = |k_0| = 2\pi/\lambda$, λ being the incident neutron wavelength).

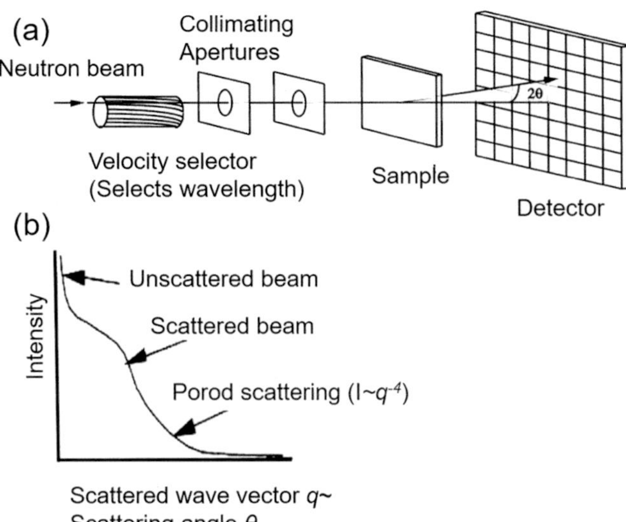

Fig. 17.8 (**a**) Scheme of the SANS instrument. (**b**) Evolution of signal intensity with scattered wave vector q [Kulkarni A, et al. (2003)]. (Reprinted with kind permission from Elsevier)

In this experiment [Kulkarni A, et al. (2003)], a monochromatic beam of thermalized neutrons passes through the specimen in transmission geometry and the scattered neutrons are recorded on a 2-D detector, as shown in Fig. 17.8a. Scattering occurs at the void–grain interface due to differences in scattering length density between the material and the pores. The first measurement is the anisotropic Porod scattering (see Fig. 17.8b), where the orientation averaging of the Porod scattering from the sample gives the total void surface area per unit sample volume, independent of the precise void morphology [Kulkarni A, et al. (2003)]. The second type of measurement is anisotropic multiple SANS (MSANS), involving the measurement of the beam broadening due to anisotropic multiple scattering by long wavelength neutrons. In Porod scattering, the scattering intensity, $I(q)$, is a function of the magnitude and direction of the scattering vector, q, and its orientation average, as shown in Fig. 17.8b. More complete microstructural information is obtained by combining MSANS measurements for different sample orientations, anisotropic Porod surface area distributions, and the total porosity determined from precision density measurements [Kulkarni A, et al. (2003)]. The MSANS beam broadening shows much greater sensitivity to the coarse globular and irregular porosity than does the Porod scattering analysis. For interpretation of the anisotropic MSANS beam-broadening data, the interlamellar pores and intra-splat cracks are considered to comprise two space-filling networks of oblate spheroids, each with a fixed aspect ratio, and the globular pores are considered spheres.

Allen et al. (2001a, b) were among the first to use SANS to determine the porosity of ceramic plasma-sprayed coatings. The voids and the grains within the samples have different scattering length density, ρ, causing some of the neutrons to be scattered at the void/grain interfaces. In the SANS experiment, a beam of "cold" neutrons with wavelength λ, ranging from 5 to 18 Å and controlled by a neutron velocity selector, is passed through parallel-sided sample. Scattered neutrons are recorded in directions orthogonal and parallel to the coating with a 2-D area detector. The intensity of the scattering vector varies with the volume fraction of porosities, the scattering contrast $(\Delta\rho)^2$ (square of the difference in scattering length densities between the grains and the void), and on the distribution of void sizes [Ilavsky J et al. (1999a, b)]. The scattering curve of intensity versus scattering angle can be divided into two regions as shown in Fig. 17.8b. The multiple small angle neutron scattering (MSANS) region is dominated by neutrons scattered many times by large pores inside the sample. The theory of MSANS is useful for pores with sizes in the 0.08–10 μm range. Larger scattering angles are dominated by scattering from the surfaces of pores in a region, which is termed the Porod region [Ilavsky J, et al. (1999a, b)]. The anisotropy in the Porod scattering is strongly amplified by the shape of the scattering material. The anisotropy depends not only on the individual pore shapes and preferred orientation but also on the polydispersity of pore shapes and sizes and to some extent on the surface roughness [Kulkarni A, et al. (2003)].

SANS has the major advantages of not requiring sample preparation and its ability to provide quantitative information concerning the separate crack and pore systems, including their distinctive anisotropies [Ilavsky J, et al. (1999a, b)]. However, the relationship between the SANS results and the underlying structure is more complex and less intuitive than for image analysis. The availability of the SANS technique is limited by the need to have access to a powerful neutron source. Kulkarni et al. (2003) described the quantitative characterization of the microstructure of plasma-sprayed partially stabilized zirconia (PSZ) coatings by means of X-ray and neutron-scattering imaging techniques. Petorak et al. (2010) have used SEM and Porod's specific surface area analysis of SANS results to determine which void systems, either interlamellar pores or intra-lamellar cracks, contributed to the observed relaxation of stress in plasma-sprayed coatings when simulating TBC service conditions. [Allen et al. (2001a, b) explored the relationships between the feedstock or spray process conditions, anisotropic void and crack microstructures of the deposits, and technologically important deposit properties such as thermal conductivity. Kulkarni et al. (2003) described the quantitative characterization of the microstructure of plasma-sprayed partially stabilized zirconia (PSZ) coatings by means of X-ray and neutron-scattering imaging techniques.

17.4.7 X-Ray Absorption Spectroscopy

X-ray absorption spectroscopy (XAS) is widely used for determining the local geometric and/or electronic structure of matter. Usually, it is performed using synchrotron radiation sources providing intense and tunable X-ray beams. The photon energy is tuned using a crystalline monochromator to a range where core atom electrons can be excited (0.1–100 keV photon energy) with the principal quantum numbers $n = 1$, 2, and 3 corresponding to the K-, L-, and M-shells, respectively. For instance, excitation of a 1 s electron occurs at the K-edge, while excitation of a 2p electron occurs at an L-edge. When a monochromatic X-ray beam is directed through a sample and the energy of the X-ray is gradually increased such that the X-ray beam crosses an absorption edge of one of the elements of interest in the sample, the transmitted X-ray light will contain small variations in absorbance, on the high energy side of the absorption edge. These absorbance variations provide information about the structural environment of the atoms surrounding the element whose absorption edge is being examined. An X-ray absorption spectrum is generally divided into four sections:

- Pre-edge ($E < E_0$);
- X-ray absorption near edge structure (XANES), where the energy of the incident X-ray beam is $E = E_0 \pm 10$ eV;
- Near edge X-ray absorption fine Structure (NEXAFS), in the region between 10 and 50 eV above the edge; and
- Extended X-ray absorption fine structure (EXAFS), which starts approximately from 50 eV and continues up to 1000 eV above the edge.

One of the main interests of XAS is its tunability allowing probing the environments (molecular structure) of different elements in the sample by selecting the incident X-ray energy. For details, see the book of Bunker G. (2010).

17.4.8 Electron Probe X-Ray Microanalysis

Electron probe microanalysis (EPMA) associated with SEM is generally considered as microanalytical technique able to image or analyze materials. Elemental analysis is accomplished at a microscale (about a few μm^2). It makes it possible to identify the elements present and quantify them, down to about 100 ppm with a rather good accuracy. Elements lighter than atomic number 8 cannot be measured without reservations.

EPMA implies that the element densities in the probed volume are uniformly distributed. Thus, to obtain information on a large area, the over-scanning must be avoided, and the large area sampled at many discrete points. For details, see [George Vander Voort (ed) (2004) and 17.Ma6].

17.4.9 Auger Electron Spectroscopy

Auger electron spectroscopy (AES) is used specifically in the study of surfaces. The Auger effect is based on the analysis of energetic electrons emitted from an excited atom after a series of internal relaxation events. Auger effect is produced whenever incident radiation—photons, electrons, and ions—interacts with an atom with an energy exceeding that to remove an inner-shell electron (L, M, . . .) from the atom. This interaction leaves the atom in an excited state with a core hole, that is, a missing inner-shell electron. These excited atoms have a limited life and de-excitation occurs very rapidly with the emission of an X-ray or an electron termed Auger electron. SEMs are equipped to detect such electrons and AES is now a practical and straightforward characterization technique for probing chemical and compositional surface environments [George Vander Voort (ed) (2004) and 17.Ma7].

17.4.10 X-Ray Photoelectron Spectroscopy

X-ray photoelectron spectroscopy (XPS), also known as electron spectroscopy for chemical analysis (ESCA), is a quantitative spectroscopic technique that measures the elemental composition, chemical state, and electronic state of the elements that exist within a material. The material is irradiated with a beam of X-rays and the kinetic energy and numbers of electrons that escape from the top (1–10 nm) are measured. XPS works exclusively in ultrahigh vacuum; it allows analyzing the surface chemistry of a material in its "as-received" state or after some treatment. XPS detects all elements with atomic number (Z) of 3 (lithium) and above. Detection limits for most of the elements are in the parts per thousand ranges. For details, see [George Vander Voort (ed) (2004) and 17.Ma8–17.Ma10].

17.4.11 Other Techniques

The characterization means described above are those mostly used for thermal-sprayed coatings. Complementary techniques such as inductively coupled plasma atomic emission spectroscopy (ICP-AES) or inductively coupled plasma mass spectrometry (ICP-MS) can be used for elemental analysis in the, part per million (ppm) or lower concentration range. Atomic absorption spectrometry (AAS) can be used for measuring relatively low concentrations of metallic or semi-metallic elements in a solution of the sample containing the metals, for details, see [ASM International (2008)].

17.5 Void Content and Network Architecture

The thermomechanical behavior of coatings and their thermal insulation performances (e.g., thermal barrier coatings) are mostly related to the void architecture typified usually as void network, see for example Rice RW (1996). Its quantification is hence fundamental and usually includes several characteristics. According to Ilavsky J (2010), one can identify among the principal characteristics:

- **Void global content** (called in some cases as **porosity** or porosity content or apparent density): It is the most commonly used descriptor of the pore microstructure.
- **Void size distribution**: Most manufacturing processes result in a wide range of void sizes rather than a single size. Establishing the extent of sizes and distribution of pores can be of major importance in understanding and controlling material properties as depending on the application and material, large pores or small pores can be wished.
- **Discrimination of voids by shape**: Globular voids corresponding to features exhibiting low elongation ratio, cracks corresponding to features exhibiting high elongation ratio and preferentially oriented perpendicularly to the substrate surface (from 45° to 135° from substrate surface), and delamination's corresponding to features exhibiting high elongation ratio and preferentially oriented, as for them, parallel to the substrate surface (from 0° to 45° from substrate surface).
- **Void network connectivity** to the coating surface on one hand (open void content) and to the substrate and coating surface on other hand (connected void content). Voids being not connected are identified as closed voids. The question arises as to the fraction of pores that can be accessed from the surface.
- **Pore surface area**: Surface area of pores, expressed as specific surface area per weight or volume of sample.
- **Pore anisotropy**: The anisotropy of the void system can be of major importance especially for highly anisotropic microstructures such as the TBCs microstructures obtained by EB-PVD.

Often the pores in the material are modeled as spherical voids, nicely separated from each other. However, this simple

Table 17.1 Conventional techniques implemented to quantify voids in a porous medium such as a thermal-sprayed coating [Fauchais P et al. (2011)]

Technique	Method	Void global content	Void size distribution	Void morphology — Globular voids	Void morphology — Cracks	Void morphology — Delamination	Void network connectivity — Connected void content	Void network connectivity — Open void content	Void network connectivity — Closed void content
Archimedean Porosimetry (AP)	Physical								■
Electrochemical Impedance Spectroscopy (EIS)	Electro-chemical						■		
Gas Permeation (GP)	Physical						■		
Mercury Intrusion Porosimetry (MIP)	Physical		■					■	
Pycnometry (P)	Physical							■	
Small-Angle Neutrons Scattering (SANS)	Physical	■	■	Via void network anisotropy					
Ultrasmall-Angle X-ray Scattering (USAXS)	Physical	■	■						
Stereological protocols (ST) coupled with Image analysis	Stereological	■	■	■	■	■			

model is wrong for the majority of engineered materials. For more complicated pore microstructures, one can imagine pores that are better approximated by other geometrical shapes, such as ellipsoids or platelets [Ilavsky J (2010)].

Several methods are commonly implemented to quantify voids network in a porous medium at the micrometer scale. Reviews on the subject have been presented by Ilavsky J. (2010), Andreola F et al. (2000) and Fauchais P et al. (2011). A listing of conventional techniques used to qualify voids in thermal spray coatings is given in Table 17.1 after Fauchais P et al. (2011) identifying the void characteristics and basic concepts used. In the following, a brief description is presented of some of these techniques highlighting their main characteristics, potential and limitations when applied to nanometer-sized pores in thermal spray coatings.

17.5.1 Archimedean Porosimetry

Archimedean porosimetry (AP) makes it possible to quantify the nonconnected "closed" porosity [Andreola F et al. (2000), Mancini CE, et al. (2001), Matějíček J et al. (2006)]. The substrate is generally removed by chemical etching in acid medium, followed by the weighing of a "freestanding" sample of the coating in ambient air and with the sample immersion into de-mineralized water. Knowing the specific mass of water w_{water} (i.e., depending upon temperature), the volume of the coating sample used is calculated, according to Archimedean, by measuring:

- The dry weight of the sample, w_{dry}.
- The weight of the sample, w_{sat} after complete impregnation of open porosities.
- The weight of water-saturated sample while it is immersed in water, w_{wet}.

The weight of water displaced, w_{wd}, can be calculated as;

$$w_{wd} = w_{sat} - w_{wet} \tag{17.8}$$

Knowing the water temperature and accordingly its specific mass, ρ_w the bulk volume of the sample, V_b is given by;

$$V_b = w_{wd}/\rho_w \tag{17.9}$$

The open pores volume, V_p is obtained by calculating the weight of water they contain, w_{wp} dividing it by the specific mass of water ρ_w

$$w_{wp} = w_{sat} - w_{dry} \tag{17.10}$$

$$V_p = w_{wp}/\rho_w \tag{17.11}$$

the apparent porosity, ϕ_a is given by

$$\phi_a = V_p/V_g \tag{17.12}$$

A value for porosity can alternatively be calculated from the bulk mass density ρ_{bulk} and particle mass density ρ_{part}:

$$\phi_a = 1 - \rho_{bulk}/\rho_{part} \tag{17.13}$$

Closed or sealed pores are not included in the apparent porosity value. The latter is usually close to the total porosity as long as the closed porosity in the sample is low. The first limitation of Eq. 17.13 is that the coating bulk-specific mass must be known. It depends on the coating phase content that can be analyzed by X-ray diffraction analysis. Equation 17.12 on the other hand is limited by the size of the accessible of pores. Percolation test of deionized water drops through the coating can be used to determine the smallest open pore diameter into which the water can penetrate. This is governed by the contact angle, θ, which is for zirconia and deionized water about 59° and the surface energy, γ, of deionized water 72.8 mN/m at room temperature. At atmospheric pressure (about 10^5 Pa), pure water percolates, according to Lucas–Washburn's equation [Washburn EW (1921)], into open voids of equivalent diameter equal or larger than 1.5 μm, while most of open voids in nanometer-sized coatings exhibit characteristic dimensions smaller than micrometer.

17.5.2 Mercury Intrusion Porosimetry

Mercury intrusion porosimetry (MIP) is based on the concept that nonwetting mercury percolates through the open voids according to the applied impregnation pressure. This results in the intrusion versus pressure curve, which can be normalized with respect to sample weight or volume [Ilavsky J, et al. (1997), Van Brakel J, eta al (1981), Siebert B, et al. (1999), Mauer G, et al. (2009)]. The experimental data are analyzed on the basis of the Washburn equation:

$$d = \frac{-4\gamma \cos\theta}{p} \tag{17.14}$$

where d represents the equivalent pore diameter (m), γ is the surface tension of mercury (N/m), θ is the contact angle between the pore walls and mercury, and p is the applied pressure (Pa). The result is converted into equivalent pore sizes on the assumption of tubular pores. Often the mercury porosimetry consists of two parts: the low pressure part (0.1–400 kPa) for the measurement of large pores up to a maximum dimension of 60–100 μm, depending on how much mercury is used, and the high pressure part (0.1 ± 400 MPa) detecting pores and micro-cracks down to the size of 1.8 nm.

However, while MIP techniques cover a large void size range, they are limited to the measurement of accessible open-void networks that combined with the necessity to apply high pressure of mercury can lead to sample deformation and even structural failures [Zhu S, et al. (1995)] and the distortion of the volume distribution versus pore characteristic dimension relationship.

17.5.3 Gas Permeation and Pycnometry

Gas permeation (GP) is determined by measuring the pressure drop across the sample of the coating of known thickness, as function of the gas (air) flow rate. This is usually achieved by applying a gas pressure to one face of the sample while the other face is kept at a reference pressure (usually atmospheric). After checking the linear evolution of the pressure drop versus the gas flow rate, the permeability can be determined from the Darcy's law [Wittmann-Ténèze K et al. (2008)]:

$$Q = k \frac{A.\Delta p}{\mu.e} \qquad (17.15)$$

The leakage rate is then calculated from the linear regression of the curve, $(Q/A)\mu$ versus $(\Delta p/e)$, Q is the imposed gas flow rate (m³/s), A sample measured area (m²), μ gas dynamic viscosity (Pa.s), e sample thickness (m), Δp, pressure drop (Pa), and k permeability of the material.

The ability of measuring the gas permeation of the coating at elevated temperatures (corresponding usually to coating operating temperature) and the possible use of different gases are the two major advantages of this technique [Li C-J, et al.

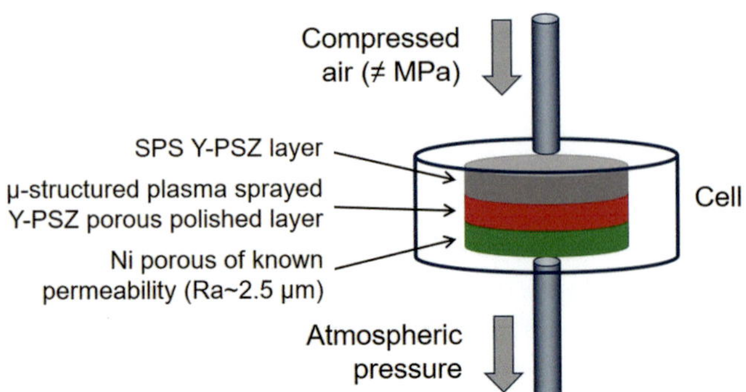

Fig. 17.9 Principle of the permeation measurement of an YSZ coating suspension plasma sprayed [Brousse E (2010)]. (Reprinted with kind permission from Dr. E. Brousse)

(2005), Fox AC, Clyne TW (2004), Golosnoy IO, et al. (2008), Reed JS (1995), Curran JA, Clyne TW (2006), Bacciochini A, et al. (2010a, b), Ilavsky J et al. (1999a, b)]. For thin nanostructured YSZ coatings resulting from thermal-sprayed suspension or liquid precursors, multiple-layered coatings as shown in Fig. 17.9 after [Brousse E (2010)] are used for gas permeation measurement. The under-layer, made of a material of known permeation rate, can influence the coating void architecture, which would have not been the case if the under-layer had been dense as a conventional substrate.

Helium pycnometry is used to determine the skeletal density of (non) porous solid and gives indirect information about pore volume and closed porosity. The simplest type of gas pycnometer (no moving parts) consists of two chambers: one (with a removable gas-tight lid) to hold the sample and a second chamber of fixed, calibrated internal volume, referred to as the reference volume or added volume. A valve allows admitting a gas under pressure to one of the chambers. The pressure in the first chamber is measured with a transducer, the two chambers can be connected, and the second chamber can be vented. The working equation of a gas pycnometer wherein the sample chamber is pressurized first is as follows:

$$V_s = V_c + \frac{V_r}{(1 - p_1/p_2)} \qquad (17.16)$$

where V_S is the sample volume, V_C is the volume of the empty sample chamber (known from a prior calibration step), V_r is the volume of the reference volume (also known from a prior calibration step), p_1 is the first pressure (i.e., in the sample chamber only), and p_2 is the second (lower) pressure after expansion of the gas into the combined volumes of sample chamber and reference chamber.

While pycnometers of any type are recognized as density measuring devices, they are in fact devices for measuring volume only. The volume measured in a gas pycnometer is the amount of 3-D space, which is inaccessible to the gas used, that is, the volume within the sample chamber from which the gas is excluded. Therefore, the volume measured depends on the atomic or molecular size of the gas. Helium is most often the prescribed gas, because of the atom small size and also because it is inert. Indeed, gas atoms form a single layer on the total surface of material and penetrate into open and connected porosity. This technique has already been successfully employed to determine the mass density of plasma-sprayed coatings [Golosnoy IO, et al. (2008)]. By comparing with the theoretical mass density, this measurement gives the percentage of closed porosity of the sample. It has been used for nanometer-sized suspension plasma-sprayed coatings [Reed JS (1995)].

17.5.4 Small Angle Neutrons Scattering

A quantitative characterization of the microstructural features has been demonstrated successfully in the case of plasma-sprayed YPSZ coatings, using a combination of anisotropic Porod scattering (PS) and multiple small angle neutron scattering (MSANS) techniques. The experiment is the same as that described in Sect. 17.4.6. The scattering occurs at the void–grain interface due to differences in scattering length density between the material and pores. Two types of measurements are performed: anisotropic Porod scattering (PS) and anisotropic MSANS [Ilavsky J, et al. (1999b)]. The anisotropic Porod scattering can be divided into two anisotropic contributions: one from interlamellar pores that are predominantly parallel to the substrate and the other from intra-splat cracks that are predominantly perpendicular to the substrate. Orientation averaging of the Porod scattering from the sample makes it possible to obtain the total void surface area per unit sample volume.

Two examples of results are presented in Fig. 17.10 after Ilavsky et al. (1999a, b). Water-stabilized plasma-sprayed alumina SANS results, presented in Fig. 17.10a, show high anisotropy of the apparent Porod surface area distribution dominated by intralamellar cracks. Apparent Porod surface area can be defined as the Porod surface area, which would be viewed in any particular direction by an observer standing in the center of the sample. Knowledge of this apparent surface area in all directions (over 4π) allows calculation of the specific surface area in the sample, and the identification of more surface systems can be distinguished.

Allen AJ et al. (2001a, b), and Kulkarni A. et al. (2003) used this method to study the effect of partially stabilized zirconia feedstock characteristics (particle density, size, and shape) on the anisotropic void structure of plasma-sprayed coatings. They correlated the results with the thermal conductivity and elastic modulus of coatings; properties that are most sensitive to these microstructural features. Ilavsky J. et al. (1999b) used SANS to study the effect of heating on the pores and cracks in YSZ coatings. Strunz P et al. (2004) also studied the effect of heating on plasma-sprayed TBCs using SANS and compared the results with those obtained by mercury porosimetry. Kulkarni A et al. (2006) have also used SANS to study EB-PVD thermal barrier coatings and quantified the voids in terms of component porosities, anisotropy, size, and gradient through the coating thickness.

17.5.5 Ultra Small Angle X-Ray Scattering

Ultra small angle X-ray scattering (USAXS) has been used to characterize EB-PVD thermal barrier coatings [Flores Renteria A, et al. (2007)]. Two data collection methods

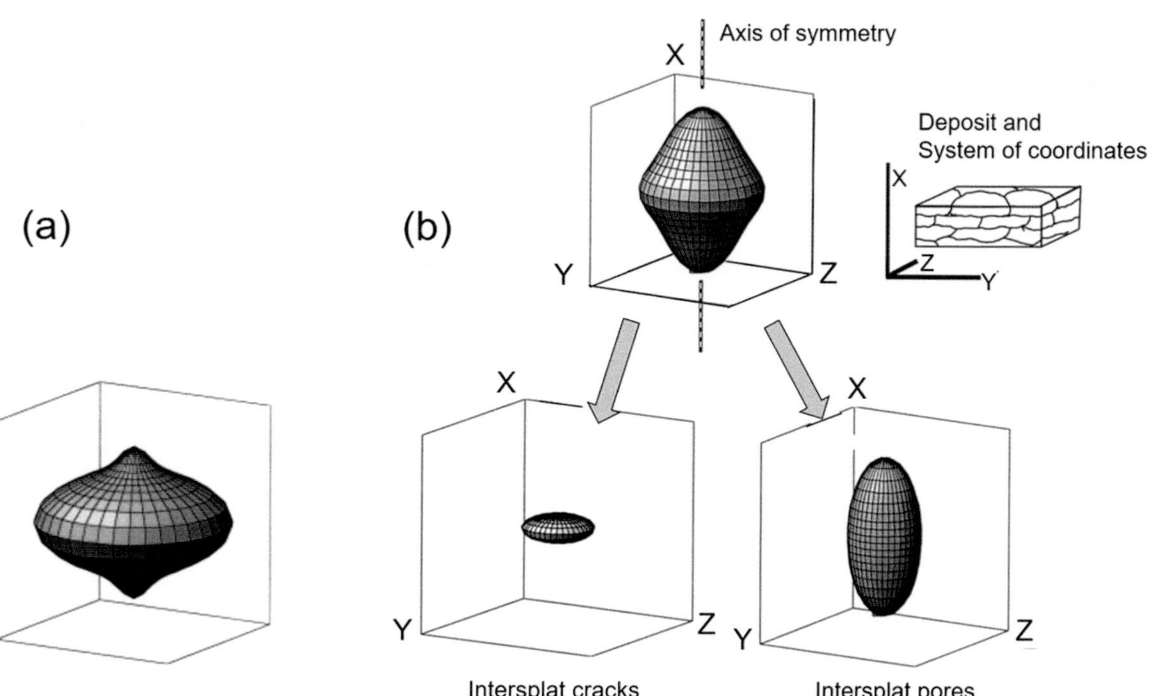

Fig. 17.10 Apparent Porod surface area distribution of (**a**) water-stabilized plasma sprayed alumina and (**b**) gas-stabilized plasma-sprayed alumina separated into two void surface areas: Left (prolate ellipsoid) representing intra-lamellar cracks and right (elongated ellipsoid) representing interlamellar pores [Ilavsky et al. (1999b)]. (Reprinted with kind permission from Elsevier)

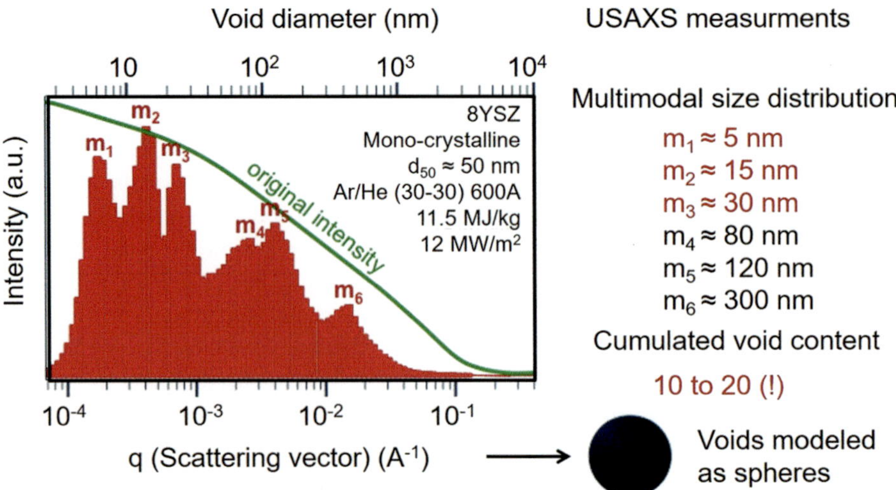

Fig. 17.11 Evolution of the USAXS signal intensity with the scattering vector, q (in *Ängstrom*$^{-1}$) and the calculated corresponding pore size distributions for YSZ suspension plasma-sprayed coating with particle mean size: 50 nm, Ar–He plasma, 30-mm standoff distance [Bacciochini A et al. (2010a, b)]. (Reprinted with kind permission from Elsevier)

were combined. In the first, the scattered intensity was measured as a function of $|q|$ (see Eq. 17.6) for each orientation of the sample azimuthal angle α. In the second, the scattered intensity at a fixed $|q|$ was measured as a function of α by rotating the sample in the beam. Variations in the scattering anisotropies observed at different $|q|$ were associated with the variation in the microstructure anisotropy at different length scales and gave, thus, a quantitative map as a function of the sizes of the scattering populations. The anisotropies in the scattered intensity at different $|q|$ values were presented as a function of azimuthal angle with respect to the substrate-normal direction [Kulkarni A, et al. (2006)]. Flores Renteria A, et al. (2007) have studied, via USAXS-measurement, the stereometric characteristics of the pores within EB-PVD TBCs. A subsequent use of the appropriate statistically representative values made it possible to predict the thermal conductivity by a noninteracting scheme. The projected values agreed well with the measured values.

USAXS measurement was reported by Bacciochini A et al. (2010a, b) to characterize suspensions plasma-sprayed YSZ (50-nm average particle diameter, Ar–He plasma gas mixture, 30-mm spray distance). Figure 17.11 shows the evolution of the signal intensity with the $|q|$ scattering vector (in Angström) and the calculated corresponding pore size distribution, assuming that the pores have spherical shapes. Rotating the analyzer and recording the scattered photons received by the detector enabled the measurement of the X-ray scattering from the sample. USAXS data were fully corrected for instrument effects and analyzed using Igor Pro1software from Wave Metrics Inc. (Oswego, OR, USA) coupled to Irena1 package for analyzing small angle scattering data. The total void content was about 14.3%. The

void distribution was multimodal and characterized by several modes' diameters (i.e., average diameter of an equivalent monomodal distribution) contrary to micrometer sized YSZ coatings, which usually exhibited a bimodal distribution of voids [Bacciochini A et al. (2010a, b)]. In SPS coatings, the typical void mode diameters that were identified are $m_1 < 5$ nm, $m_2 < 15$ nm, $m_3 < 30$ nm, $m_4 < 80$ nm, $m_5 < 120$ nm, and $m_6 < 300$ nm. The most numerous population of voids corresponded to voids exhibiting mode diameter $m_5 < 120$ nm. It is important to underline that commonly applied porosity characterization methods, based on analysis of cross-sectioned coatings or on liquid impregnation, are not suitable to address the void content of SPS nanometer-sized coatings. On the other hand, combination of helium pycnometry and ultra small angle X-ray scattering proved to be suitable for quantifying SPS deposits void contents with sufficient details. Large populations of nanometer and larger voids have been detected. Their characteristic dimensions cover four orders of magnitude (i.e., from 1 to 10,000 nm). About 90% of voids (by number) in the coatings exhibit characteristic dimensions smaller than 50 nm. The photon scattering phenomena and thermal resistance effects are most likely responsible for the significantly lower thermal diffusivity values (about 0.025 mm^2/s) measured in the temperature range of 20–300 °C for these SPS coatings compared to typical values for micrometer-sized conventional APS coatings.

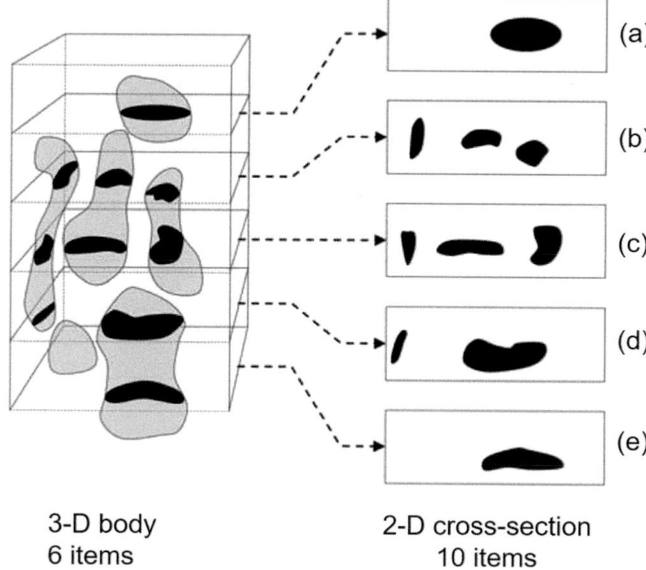

Fig. 17.12 Aanalyses of the Wicksell corpuscular problem [Antou G, et al. (2004a, b) and Montavon G and G Antou (2007)]. (Reprinted with kind permission from Springer Science Business Media, copyright © ASM International)

17.5.6 Stereological Protocols (Coupled to Image Analysis)

Assuming that resolution limits are considered, porosity within a microstructure can be easily detected by image analysis due to the high degree of contrast between the dark pores (voids) and the most highly reflective coating material [Deshpande S, et al. (2004)]. However, image analysis gives only 2-D pore distributions of the coating cross-section (the intersection of the structure with this specific plane). If the cross-section images can show the effect of the coating densification, for example, after laser remelting [Antou G, et al. (2004a, b)], it does not mean that such images are representative of the real 3-D structure. It is illustrated in Fig. 17.12 after [Montavon G and G Antou (2007)] and representing the Wicksell's corpuscle problem demonstrating that "an accurate estimation of the number of objects of interest cannot be obtained from profile counts on individual cross sections." This figure illustrates that the three bodies of the 3-D structure cannot be represented by the ten objects resulting from 2-D cross-sections (a, b, c, d, e).

According to Montavon G and G Antou (2007), the procedure to analyze different 2-D cross-sections of the 3-D coating comprises successively

- sample preparation,
- image acquisition,
- image pretreatment,
- image treatment, and finally,

- stereology implementation to achieve the quantitative interpretation of the coating architecture.

The image acquisition is made either with either, optical microscopy (OM) equipped with cameras for image analysis, but the resolution is low, or by SEM. For SEM, BSE imaging, which is sensitive to the average atomic number of the sample, gives sufficient contrast between dissimilar phases for automatic threshold determination due to the electron–matter interaction. With OM magnification is generally below 500, while with SEMs magnification values can be up to 1000 or more. If high magnification allows seeing many details (pores and cracks), the small surface (between about one-tenth and one square millimeter) analyzed does not necessarily represent the whole coating. Experience indicates that image characteristic dimension should be between 10 and 15 times larger than the objects of interest (voids) to be analyzed to account for the representative elementary volume (REV) of the structure. Thus, statistical measurements (in general 10–15 locations within the coating) are necessary to ensure a representative sample for qualitative analysis. Moreover low-magnification images ($100\times$) can also be taken to reveal larger cross-sections of the coating for porosity analysis as did, for example, Hanson TC, Settlese GS (2003) to analyze 316 L stainless steel HVOF-sprayed coatings.

Several mathematical filters can be implemented to reveal specific structural data from the initial image [Russ JC (1995)]. The objective of this pretreatment is to remove singularities and to reveal feature outlines prior to morphological analysis. It consists of altering the pixel values (i.e., pixel intensities) using mathematical treatments. Four main types of filters are usually implemented at this stage of the treatment [Antou G, Montavon G (2007)]:

- Filters that make it possible to modify the image contrast (linear or nonlinear filters).
- Arithmetic and logic filters.
- Spatial filters (linear or nonlinear filters).
- Frequency filters (fast Fourier transform, FFT, applicable only on square images).

An example of such a processing is presented in Fig. 17.13 after Antou G, Montavon G (2007) showing the different treatments to discriminate globular pores from cracks of the SEM image of Al_2O_3–$13TiO_2$ coating plasma sprayed in air.

As coating architectures are heterogeneous, they must be characterized at several scales and the size of the representative elementary volume (REV) must be defined to describe the overall structure of the material [Antou G, Montavon G (2007)]. The REV must be small enough to resolve the thinner details of the structure: high magnification and image resolution. But the REV must be large enough to

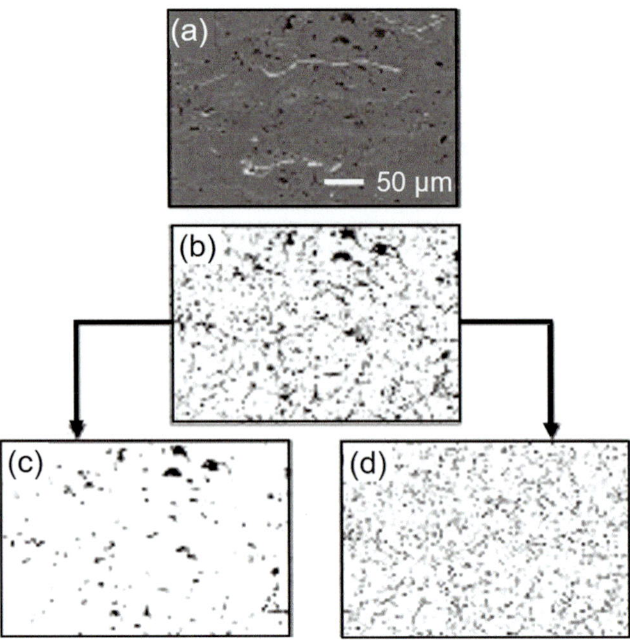

Fig. 17.13 SEM images of Al_2O_3–$13TiO_2$ plasma-sprayed coating in air (**a**) initial SEM image, (**b**) image treated by an image analysis software, (**c**) globular pores, and (**d**) cracks isolated after several filtering and morphological protocols [Antou G, Montavon G (2007)]. (Reprinted with kind permission from Springer Science Business Media, copyright © ASM International)

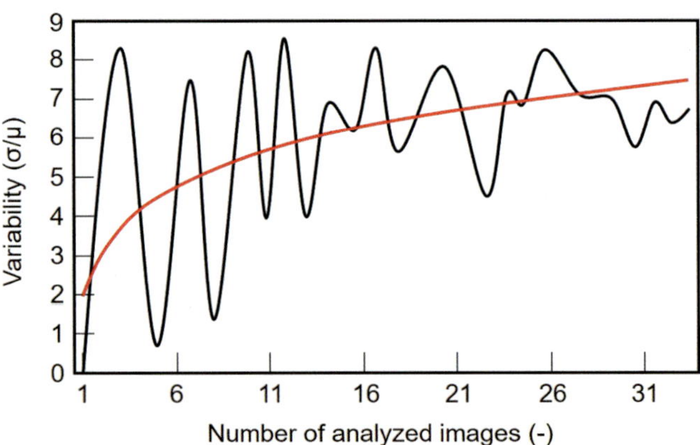

Fig. 17.14 Determination of the number of images to be analyzed [Antou G, Montavon G (2007)]. (Reprinted with kind permission from Springer Science Business Media, copyright © ASM International)

describe the whole material structure: low magnification. To find the compromise, the evolution of the result variability (standard deviation over average value) must be followed as a function of the considered number of images (Fig. 17.14). The appropriate number of images corresponds to the value where this function tends toward an asymptote.

For example, Zhang XC et al. (2009b, c) have proposed the following procedure divided into four steps, that is, gray level transformation, fuzzy enhancement of image, binary

segment of image, and removing the impurities and identification of pores and micro-cracks. Venkataraman R et al. (2007) have processed and analyzed the digitally stored images with image analysis software such as ImageJ and Adobe Photoshop to identify the pores and their distribution. Elsebaei A et al. (2010) have used the Matlab code to identify the pores and their distribution. The measured data of porosity can be highly scattered. The adjustment of the porosity dataset almost does not lead to the variation of the mean value of porosity calculated from the Gaussian distribution, as shown by Zhang XC et al. (2009b). However, the adjustment of the porosity dataset by subtracting the two largest and two smallest data greatly decreases the variance. It was also observed that the measured data of coating porosity might follow the Weibull distribution [Venkataraman R et al. (2007)]. The first limitation of this method is the limited resolution, which makes difficult to take into consideration features smaller than 0.1 μm in average value. So, the technique cannot be used to characterize nanometer-structured coatings. The second limitation is the artifacts (i.e., pullouts, scratches) that result from the cutting and polishing steps [Sauer JP (2005), Puerta DG (2006), Russ JC (1995), Nolan DJ, Samandi M (1997)]. This can be particularly the case for cermet coatings where the hard-ceramic particles imbedded within the soft metal matrix can be pulled out easily if the polishing step is not made carefully. For example, Nolan DJ, Samandi M (1997) have shown that the potential for smearing over (or filling in of) pores is important and insufficient polishing can result in considerable underestimation of porosity, thus providing incorrect representation of coating structure.

This method of image analysis is extensively used to characterize the coatings void structure, for example, according to the spray process [Kang H-K, Bong Kange S (2004), Saravanan P, et al. (2000), Yang G, et al. (2001)], the spray conditions [Scrivani A, et al. (2008)], the spray angle [Tillmann W, et al. (2008a, b)], and the heat treatment [Siebert B, et al. (1999)].

The coating porosity is often measured using the Delesse principle [Underwood EE (1970), Russ JC, DeHoff RT (1999)]: if the porosity is randomly distributed throughout the coating, then the percentage of porous area in a coating cross section is identical to the percentage of porous volume in the entire coating. Fowler DB et al. (1990) have shown that image analysis can reproducibly detect and measure microstructural features (pores, cracks, etc.) within thermal spray coatings. They have statistically tested the reliability of these methods for specific experiments and they assume 95% confidence level. However, coatings are nonuniform in all spatial directions and the 2-D approach is not necessarily representing well the structure. Several quantitative tools allowing the quantification of complex structures exist. They are known as stereological protocols [Montavon G,

Antou G (2007)] and have been used to quantify some specific features of thermal spray coatings [Montavon G, et al. (1998), Leigh S-H, Berndt CC (1999), Antou G, et al. (2006)]. Stereological protocols of the first order aim at statistically quantifying 3-D structures from 2-D randomly oriented plane sections.

NIST was the first to develop, Object-Oriented Finite (OOF) Element Analysis of Microstructures (http://www.ctcms.nist.gov/oof/), a program designed to help materials scientists to calculate macroscopic properties from images of real or simulated microstructures. OOF reads an image, assigns material properties to features in the image, and conducts virtual experiments to determine the macroscopic properties of the microstructure. At the beginning two separate versions existed, one only solving elasticity problems and the other solving coupled elasticity and thermal diffusion problems. Now the current version of OOF is OOF2, which improves on OOF1 in a number of ways. OOF2 can solve a much larger variety of physics problems and can be easily extended to cover even more. This program was, for example, used to characterize and predict properties of plasma-sprayed YSZ coatings [Kulkarni A, et al. (2003), Wang Z, et al. (2003), Michlik P, Berndt CC (2006), Jadhav AD, et al. (2006)], and arc-sprayed WC–FeCSiMn coating [Tillmann W, et al. (2011)].

University of Technology of Belfort-Montbéliar (France) also developed a software package TS2C focused on estimation of the thermal conductivity of thermally sprayed coatings [Bolot R, et al. (2005), Costil S, et al. (2007), Bolot R, et al. (2011)]. Bobzin K et al. (2012) showed that when using 3-D models, the results of the virtual testing and asymptotic homogenization methods match each other more closely compared with the conventional 2-D models. Homogenization methods are based on the assumption that a specific small part of the microstructure can be considered to be representative for the entire material. The dimensions of the representative volume element (RVE) are much smaller than those of the entire microstructure. It requires that the RVE contain a sufficient number of microstructural inhomogeneities, so its morphology is equivalent to that of the entire microstructure [Bobzin K et al. (2012)]. Figure 17.15 presents the way the 3-D representation of the RVE has been derived.

The statistical 3-D reconstruction approach has been used by Bobzin K et al. (2012) to calculate the effective properties of a typical YSZ coating in 3-D domain. The calculated (with this method) effective values of Young's module and the thermal conductivity of the coating agree with the results of micro-indentation and laser flash measurements. The definition of the distance between 2-D images (h), used to reconstruct the 3-D representation, is critical for the correctness of the calculated effective properties [Bobzin K et al. (2012)].

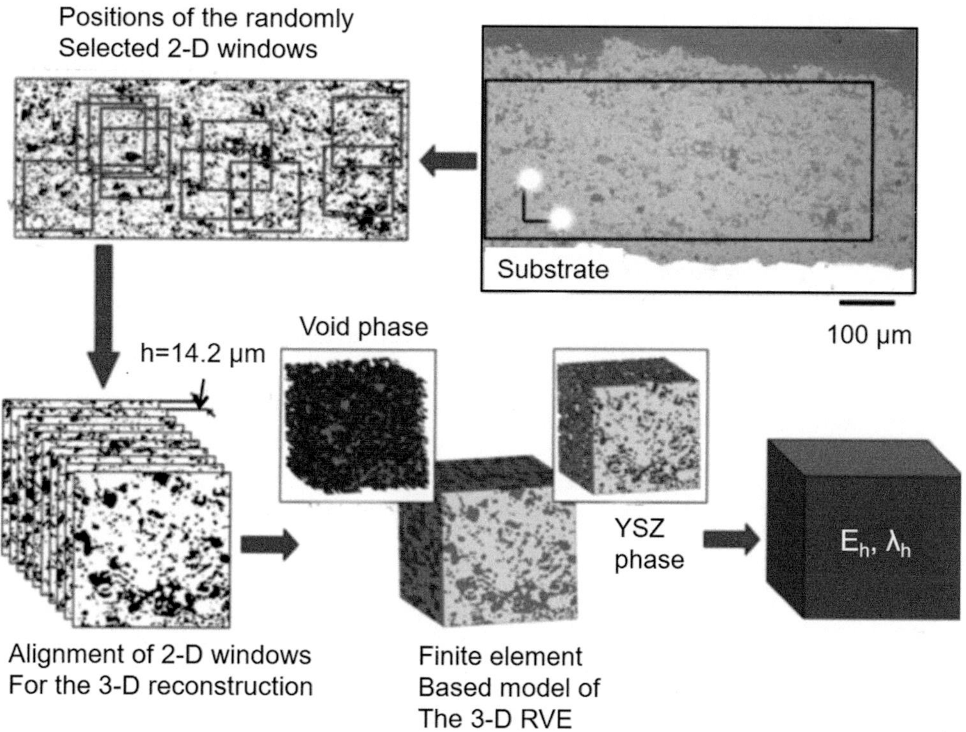

Fig. 17.15 Homogenization of thermally sprayed YSZ coatings obtained by statistical 3-D reconstruction approach [Bobzin K et al. (2012)]. (Reprinted with kind permission from Springer Science Business Media, copyright © ASM International)

17.5.7 Electrochemical Impedance Spectroscopy

The percolation of an electrolyte inside the interconnected void network allows quantifying the void fraction connected to the substrate by analyzing the chemical reaction (passivation most of the time) at the substrate/electrolyte interface. The electrochemical impedance spectroscopy (EIS) technique consists hence in measuring the impedance of the electrochemical cell. In order to access the charge transfer resistance or polarization resistance, EIS results have to be interpreted with the help of a model of the interface. To make an EIS measurement, a small amplitude signal, usually a voltage between 5 and 50 mV, is applied to a specimen over a range of frequencies of 0.001–100,000 Hz. The EIS instrument records the real (resistance) and imaginary (capacitance) components of the impedance response of the system. Depending upon the shape of the EIS spectrum, a circuit model or circuit description code and initial circuit parameters are assumed and input by the operator. The program then fits the best frequency response of the given EIS spectrum to obtain the fitting parameters. The quality of the fitting is judged by how well the fitting curve overlaps the original spectrum. By fitting the EIS data, it is possible to obtain a set of parameters, which can be correlated with the coating condition.

A typical EIS for measuring connected porosity in plasma sprayed TBCs is presented in Fig. 17.16 from Antou G et al. (2006). They have used a flat cell with a volume of 300 mL at room temperature. The gray alumina sample was used as the working electrode (exposed area of 1 cm^2). A platinum mesh was used as a counter electrode, and an Ag/AgCl electrode was used as a reference. A 0.01 mol/L $K_3Fe(CN)_6/K_4Fe$(CN)$_6$ aqueous solution was selected as electrolyte due to its highly reversible electrochemical exchange current density and its minimal interference with the system. The three-electrode system was connected to EG&G Parc Model K0235 EIS measurement system. A sinusoidal voltage perturbation of 10 mV amplitude and a frequency range of 1 mHz–100 kHz was input to the system, and the system response was recorded as a Bode plot.

The immersed coating surface behaves as the working electrode [Zhang J, Desai V (2005), Jayaraj B, et al. (2004)]. The connectivity of a pore network is related to the quantity of voids, which connects the substrate to the surrounding atmosphere.

A similar system was used by Zhang J and V Desai (2005), Jayaraj B et al. (2004), Antou G and G Montavon (2007) and Anderson PS et al. (2004) to study plasma sprayed TBCs. Bolelli G et al. (2009) have also characterized the connected porosity of high-velocity suspension flame sprayed (HVSFS) Al$_2$O$_3$ coatings. Kawakita J et al. (2003) have studied by EIS the through-porosity of Hastelloy-C High-Velocity Oxy-Fuel (HVOF) sprayed coatings by quantitatively analyzing dissolved substances derived from coated steel during immersion in HCl solution. Furthermore, Koivuluoto H et al. (2008) investigated the denseness of cold-sprayed Cu, Ni, and Zn coatings with corrosion tests to get more information about existing through-porosity (open-porosity). The corrosion tests used were open-cell potential measurement and salt spray fog test.

As previously mentioned, the test of deionized water droplet percolation through the coating indicated that the smallest void diameter into which a liquid such as water percolates through the coating void network was in the order of the micrometer. So, the EIS technique is not adapted to most of the nanometer-sized thermal spray coatings.

17.6 Adhesion–Cohesion

17.6.1 Introduction

The quality and performance of thermal spray coatings are strongly dependent on the adhesion between the coating and substrate because the debonding of coating will result in the collapse of the sprayed system. However, the prediction and control of coating adhesion are complex as it depends on:

- Spray process and operating conditions used.
- Feedstock, its particle size distribution and morphology.
- Substrate material: oxidation stage (oxide composition and thickness); roughness (peaks height in comparison with the splat mean size, and peak separation distance, characterized by the root mean square roughness $R_{\Delta q}$, that is, the root mean square average of the roughness profile

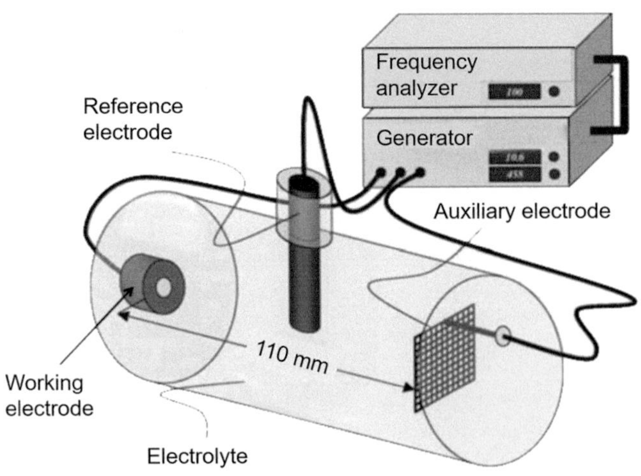

Fig. 17.16 Schematic representation of electrochemical impedance measurement system [Antou G et al. (2006)]. (Reprinted with kind permission from Springer Science Business Media, copyright © ASM International)

ordinates); cleanliness (removal of surface pollution and grit residue); and preheating prior to spraying at a temperature sufficient to get rid of adsorbates and condensates.

- Residual stresses: they depend on the substrate-sprayed material couple, spray conditions, and substrate mean temperature.
- Environmental conditions such as the surrounding atmosphere temperature and humidity, externally generated vibration transmitted into the specimen at the time of testing.

Moreover, the structure of thermal-sprayed coatings is different from that of bulk materials; it generally involves layered splats with a contact surface between successive layers of 15–60% of their surface, unmolten particles, globular pores, cracks. Thus, the basic bonding mechanisms between thermal spray coatings and substrate strongly depend on real contacts between successive splats as well as between individual splats and the substrate. They are classified into three major groups:

- mechanical interlocking or anchoring,
- metal-to-metal bonding (diffusion phenomenon), and,
- chemical bonding (formation of an intermetallic compound with a substrate).

The service failure mode can be described as interfacial (adhesive), cohesive, or mixed interfacial/cohesive [Lin CK, Berndt CC (1994)]. Adhesion can be defined through fracture mechanics [Berndt CC, McPherson R (1981)] that consider the energy required to initiate or propagate cracks and evaluate the adhesion of the coating system in terms of fracture toughness [Heintze GN, McPherson R (1988)]. The experiment must establish the equilibrium condition where the elastic energy provided by an external force (as defined by the geometry of the specimen and applied load) is balanced by the propagation of a stable crack [Berndt CC, McPherson R (1981)]. Over a critical value of the strain energy release rate, G_c (stated in J/m^2) crack propagation occurs and thus failure.

17.6.2 Simple Tensile Adhesion Test

The tensile adhesion test (TAT) is used as a routine quality control tool for thermal spray coatings. The TAT arrangement is illustrated in Fig. 17.17a after [Lin CK, Berndt CC (1994)]. In this test, the coated sample (a cylinder $\varnothing$ 25.4 mm $\times$ 25.4 mm long) is glued to an uncoated similar counterpart that is just grit blasted and then tested in tension in a universal testing machine. The value of the tensile load, at which the separation of the coated–uncoated parts occurs, is registered and transformed in an adhesion value "bond

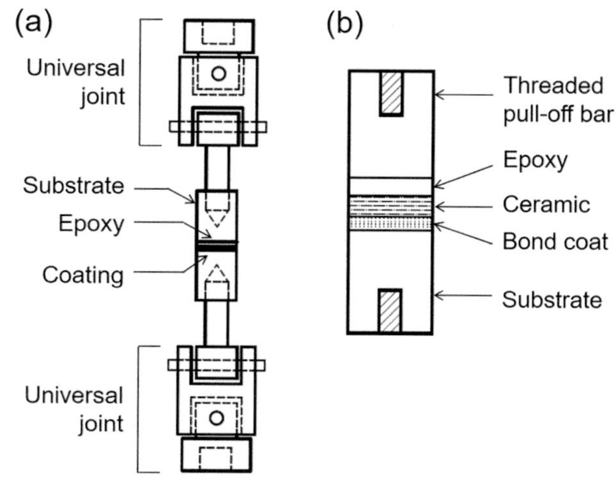

Fig. 17.17 (**a**) Tensile adhesion test configuration specified by ASTM C633–79 and (**b**) tensile adhesion test with cohesive and interfacial ruptures [Lin CK, Berndt CC (1994)]. (Reprinted with kind permission from Springer Science Business Media, copyright © ASM International)

strength, by calculating the load/area relation [Lima CRC, Guilemany JM (2007)]. Four main standards are used in industry or research laboratories [Lin CK, Berndt CC (1994)]: DIN 50160-A (Germany), AFNOR NF A91–202-79 (France), JIS H8666–80 (Japan), and ASTM C633–01 (2008) (USA) [17.A1]. These standards differ in their specimen geometry (e.g., cylinder diameter: 25.2 mm for ASTM against 40 mm for DIN and cylinder length 25.4 mm for ASTM and 50 mm for DIN), test methodology, and failure analysis [Lin CK, Berndt CC (1994)].

The failure region indicates the type and characteristic of the failure, according to Fig. 17.17b, that shows a simple coating system. The failure modes of a coating under TAT conditions are:

- interfacial failure, which occurs along the coating/substrate interface,
- cohesive failure within the coating, and,
- mixed-mode failure, which is a combination of the first two modes.

If the coating is a duplex system, as in a TBC, the coating cohesive failure could be located within the bond coat, within the ceramic top-coating or at the bond coat/ceramic interface. Some failures may occur in a combined way, starting in a region and then expanding to other ones [Lima CRC, Guilemany JM (2007)].

TAT measurement, whatever maybe the standard used, presents some shortcomings the most important of which is glue diffusion within the coating, which can artificially improve the apparent tensile resistance value [Wigren J, Täng K (2007)]. It appears that the tensile strength achieved on a thermal sprayed coating is very dependent on the penetration of the epoxy and the pressure being used during curing. They showed that the tensile strength of a Ni–5Al/ Alumina system can vary between 15 and 60 MPa. Such variations in the tensile resistance of the coating are not observed in cases where the coating is dense enough to prevent epoxy penetration (such as WC/Co). The alignment of test fixtures is also very important because it can create nonuniformly distributed stress or stress singularities within the coated sample, such that the failure is not only controlled by the magnitude of the applied tensile force. That is why, as shown in Fig. 15.17a, universal joints are used to center the TAT arrangement.

Han W, et al. (1993a, b) used finite-element analysis to study the stress distribution along the coating/substrate interface; they found that the stress distribution was nonuniform and the average adhesion strength underestimated. A modified specimen, 50% longer than the standard size, was proposed so that a uniform stress can be obtained along the interface. In their tests, they found that the determined mean tensile strength with the elongated specimen was 21% higher than that determined with the ASTM C633–79 standard specimen. So, they concluded [Han W, et al. (1993a, b)] that elongated specimens provide more accurate estimates of bond strength than ASTM C633–79 standard specimens.

TAT has been used intensively for many applications, for example, to investigate the effect of grit blasting [Białucki P, Kozerski S (2006), Amada S, Hirose T (1998), Staia MH, et al. (2000)], substrate preheating prior to spraying [Pershin V, et al. (2003)], coating thickness [Krishnamurthy N, et al. (2009)], spray angle [Bahbou MF, et al. (2004)], sprayed particle size distribution, and morphologies [Li C-J, Wang Y-Y (2002)].

17.6.3   Other Types of Tensile Tests

Different tests have been defined [Leigh SH, Berndt CC (1994), Rickerby DS (1988), Du H, et al. (2005)] and few examples are given below.

The modified configuration of the TAT, called single bar (SB) method, according to Leigh SH, Berndt CC (1994) has been designed for self-alignment and to accommodate flat and wide thermal sprayed-coated specimens, see Fig. 17.18, with sizes up to 152 × 152 × 152 mm. The specimen is positioned between two plates that are 178 × 178 mm wide. The bottom plate has a 25.4 mm diameter hole at the center, through which the glued pull-off bar passes and provides tensile stress to the specimen [Leigh SH, Berndt CC (1994)]. This bottom plate and the universal joints at the

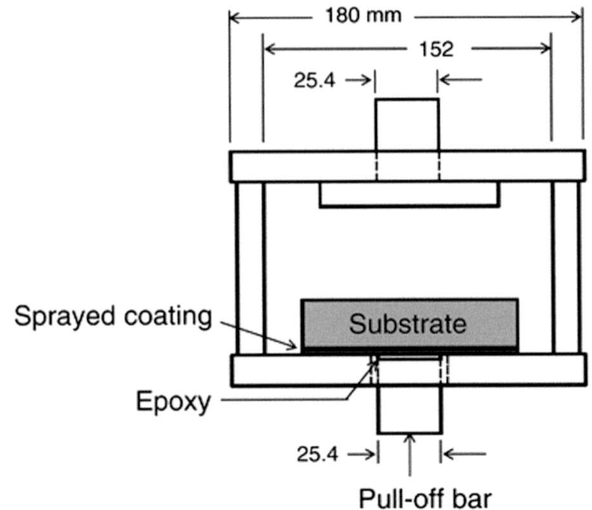

Fig. 17.18 Schematic of TAT assembly for single bar (SB) method [Leigh SH, Berndt CC (1994)]. (Reprinted with kind permission from Springer Science Business Media, copyright © ASM International)

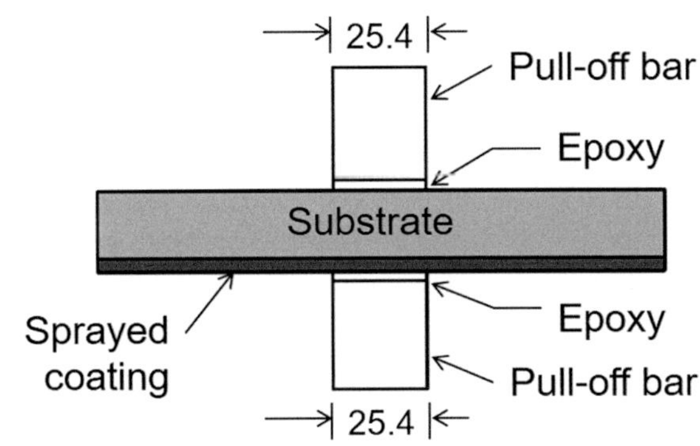

Fig. 17.19 Schematic of TAT assembly for double bar (DB) method [Leigh SH, Berndt CC (1994)]. (Reprinted with kind permission from Springer Science Business, copyright © ASM International)

top and bottom of the fixture result in a self-aligning TAT system. Several tests can be performed on each flat substrate, since the pull-off bar can be repositioned for each TAT.

Another modified TAT, called double bar (DB) method, is shown in Fig. 17.19 from [Leigh SH, Berndt CC (1994)]. This method does not require a special testing fixture except two pull-off bars and two universal joints, which are attached to both sides of the flat specimen. As for SB, several tests can be performed on each flat substrate, since the pull-off bar can be repositioned for each TAT, the problem being to maintain the alignment between the two bars.

For coatings with a strong adhesion (> 60 MPa), such as those obtained with D-gun or high-power HVOF, the adhesions are measured according to the pull-off method [Rickerby DS (1988)] with the application of a force normal to the coating–substrate interface. A schematic diagram of one of these tests is presented in Fig. 17.20 from [Du H, et al. (2005)]. The substrate comprises two parts with conical shapes: one female and one male made of the same material as that of the used substrate. The surface of joined both parts is continuous, and grit blasted to anchor the sprayed coating. Du H, et al. (2005) have sprayed WC–Co powder using a D-gun spray system, onto stainless steel substrate; the tester was also made of stainless steel. They found adhesions values ranging between 120 and 150 MPa, depending on the spray conditions.

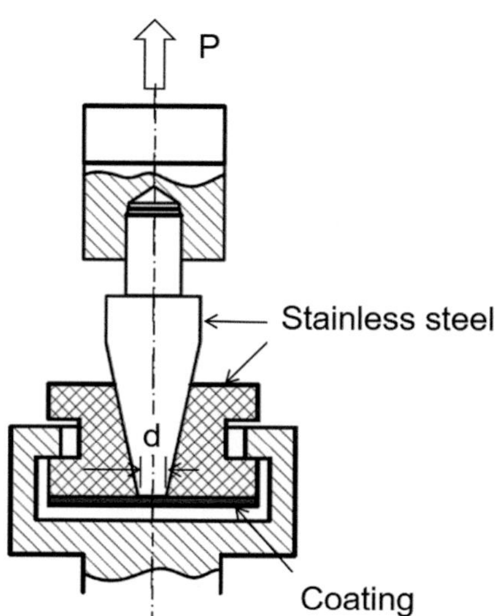

Fig. 17.20 Schematic diagram of the direct pull-off test [Du H, et al. (2005)]. (Reprinted with kind permission from Elsevier)

17.6.4 Shear Stress

In many cases shearing plays a key role in coating peeling off. The principle of a shear stress test is very simple, as shown in Fig. 17.21a. The coating is deposited onto either plane (Fig. 17.21a) or cylindrical (Fig. 17.21b) substrate and a force parallel to the coating is applied at its extremity as shown schematically in Fig. 17.21a, b. While the shear test piece is rather complex to prepare, it needs no adhesive agent. Moreover, the adhesion strength can be measured using specimens, which are cut out from the products [Era H, et al. (1998)].

The main disadvantage of this shear test is a possible stress concentration [Era H, et al. (1998)] at the corner of the protruded step of a test piece, indicated by a dotted arrow in Fig. 17.21a. It results in a lower value than the true shear strength. That is why Era H, et al. (1998) proposed a shear test piece with a semicircular notch at the corner of a protruded step. Calculations and measurements have demonstrated that the reduction of the stress concentration in a notched test piece (radius over 0.3 mm) is effective for a coating that has a Young's modulus equal or larger than that of substrate. The sprayed specimen (Cr_3C_2 cermet coating on mild steel) was cut using a precision machining tool into conventional shear test pieces assuming that no damage occurs to the coating by this machining. Figure 17.21c shows the shear test jig they used and Fig. 17.21d shows the appearance of the shear test. The shear test jig was precisely made of sintered WC-Co cermet. The clearance between the punch and die was measured to be about 20 µm, corresponding to the interface zone between coating

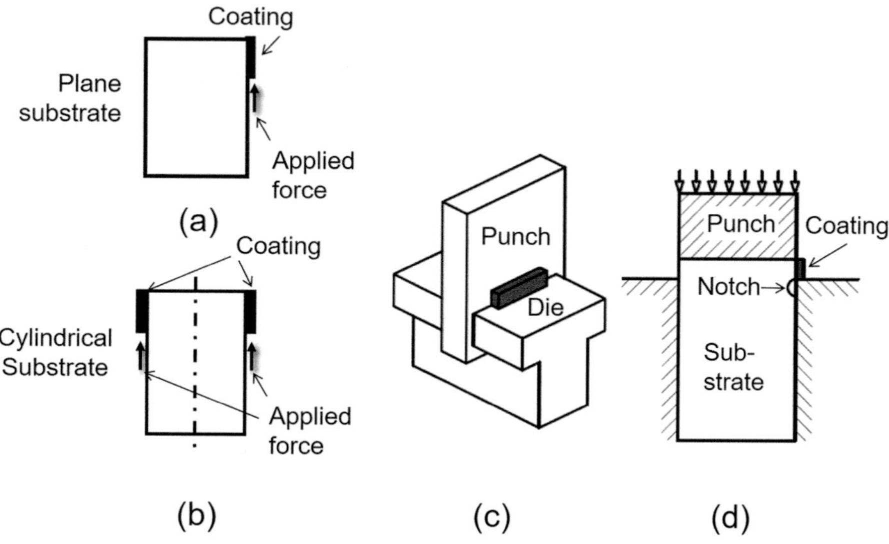

Fig. 17.21 Schematic illustration of shear test: (**a**) Plane substrate, (**b**) cylindrical substrate (**c**) shear test jig, and (**d**) schematic illustration of shear test [Era H, et al. (1998)]. (Reprinted with kind permission from Elsevier)

and roughened substrate. The critical shear stress, τ_c is calculated by

$$\tau_c = \frac{F}{wl} \qquad (17.17)$$

where F is the force applied, l is the length of the coating, and w is the coating thickness − 20 μm.

17.6.5 Fracture Mechanics Approach

The fracture mechanics approach consists in establishing the equilibrium condition where the elastic energy provided by an external force (as defined by the geometry of the specimen and the applied load) is balanced by the propagation of a "stable" crack [Lin CK, Berndt CC (1994)]. One form of this energy-balance criterion derives the strain energy release rate, G (in J/m^2) defined as:

$$G = \frac{\partial(W_e - U)}{\partial A} \qquad (17.18)$$

where W_e is the work done by external forces (J), U is the elastic energy stored in the system (J), and A is the crack surface area (m^2). It is convenient to write G as:

$$G = \frac{P^2}{2w} \frac{dC}{da_x} \qquad (17.19)$$

where P is the force required to extend a crack (N), a_x is the crack length (m), w is the coating thickness (m), and C is the compliance, which is the reverse of the slope of the curve load displacement (m/N). The critical value G_{ic} corresponds to the coating adhesion limit. The strain energy release rate can be related to the fracture toughness, characterizing the interface, K (stated in N/m$^{3/2}$), by

$$K = \frac{\sqrt{E' \times G}}{1 - v^2} \qquad (17.20)$$

where E' is the elastic modulus (MPa), and v is the Poisson's ratio. The toughness represents the ability of a material to deform plastically and absorb energy in the process before fracture occurs.

The TAT can be analyzed according to the fracture mechanics concepts, where the specimen configuration for the TAT geometry is considered as a circumferentially cracked bar (Fig. 17.22). The average fracture strength can be converted into fracture toughness [Lin CK, Berndt CC (1994)] by using Eq. 17.21.

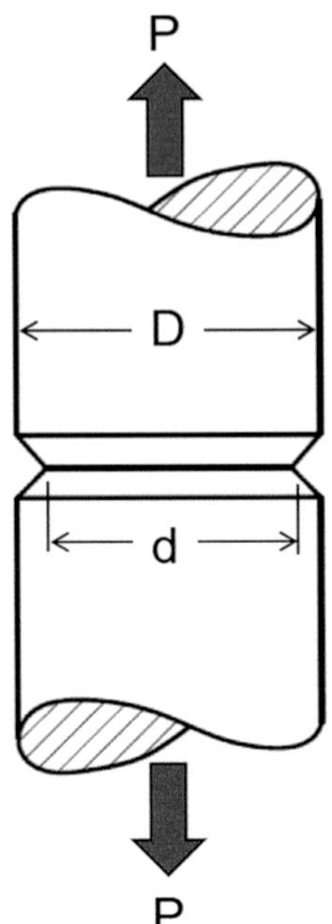

Fig. 17.22 Tensile adhesion test specimen of a circumferentially cracked bar for K_{1c} testing [Lin CK, Berndt CC (1994)]. (Reprinted with kind permission from Springer Science Business Media copyright © ASM International)

$$K_{ic} = P\left[-1.27 + 1.72 \left(\frac{D}{d}\right)\right] D^{-1.5} \qquad \left(0.4 \le \frac{d}{D} \le 0.9\right) \qquad (17.21)$$

where K_{ic} is the critical fracture toughness (N/m$^{1.5}$), P is the fracture force (N), D the outside diameter of the bar (m), and d is the inside diameter in the circumferentially notched bar (m) (see Fig. 17.22). The average reduced-area ratios range between 0.74 and 0.86. This reduction in area corresponds to a reduced failure stress of about 80%, which correlates reasonably well with the underestimation factor of approximately 83% found by Han W, et al. (1993a, b).

For HVOF-sprayed coatings, the tensile adhesion test is widely used. However, the edge of the substrate is heavily deformed and rounded due to the high impact energy of the sprayed particles. This deformation causes a large scatter of the adhesion test results [Watanabe M, et al. (2008)]. To avoid that a pre-crack was formed at the interface of a conventional tensile adhesion test specimen, the coating was sprayed onto the cylindrical substrate used in the ASTM C633-79 test. Before spraying, a carbon layer was placed along the substrate edge with 3 mm width from the edge in order to introduce "a weak bonded region" at the interface as indicated with the black rings in Fig. 17.23. A pencil of 2B grade was used to place the carbon layer. The

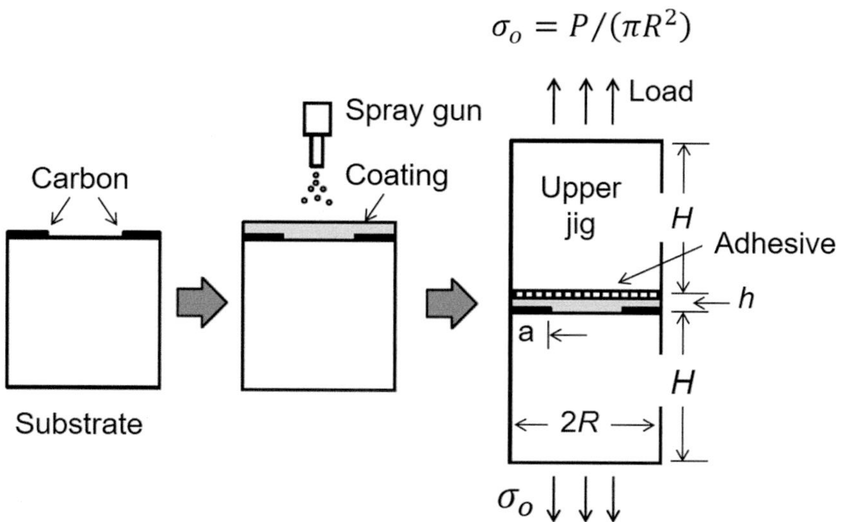

Fig. 17.23 Procedure of sample preparation for the modified adhesion test and schematic of a sample configuration [Watanabe M, et al. (2008)]. (Reprinted with kind permission from Elsevier)

substrate was set in a lathe-turning machine and rotated. The sharpened pencil was touched on the surface by the precise position controller in order to achieve a uniform carbon layer. This system was used to determine the interfacial toughness, according to the fracture mechanics concepts.

According to Eq. (17.19), the strain energy release rate corresponds to crack opening. Clyne TW and SC Gill (1996) have shown, however, that the analysis of interfacial debonding is complicated by the fact that the stress field at the crack tip may not represent conditions of pure crack opening (mode I). "Depending on the toughness of the interface and of the two media on either side," crack propagation may continue within the interface, rather than seeking a mode I path in some other direction, even though the interfacial path is heavily "mixed mode" (i.e., has a substantial shear component at the crack tip) [Clyne TW and SC Gill (1996)]. The so-called phase angle, ϕ, is related to the crack-tip stress-intensity factors, K_i (see Eq. 17.22), for mode I and mode II loading by:

$$\phi = \tan^{-1}\left(\frac{K_{\mathrm{II}}}{K_{\mathrm{I}}}\right) \qquad (17.22)$$

so that $\phi \sim 0$ represents pure opening conditions (mode I) and $\phi \sim 90$ represents pure shear (mode II). The critical value of critical strain energy release rate G_{ic} rises as a shear component is introduced, since this will encourage dissipation of energy at and immediately behind the crack tip as a result of frictional rubbing between asperities. It is expected that roughening of the substrate surface will give rise to a more substantial increase in interfacial toughness if there is a large shear component to the crack-tip driving force. As the dependence of G_{ic} on ϕ is significant, especially over 45°, the interfacial fracture energy values should always be quoted

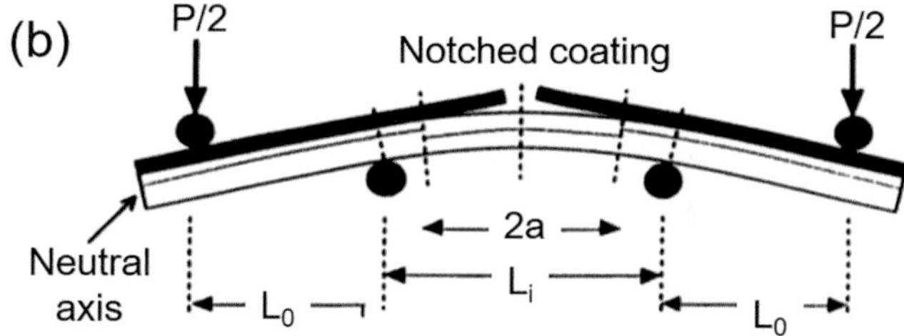

Fig. 17.24 Loading arrangement and specimen geometry used to measure coating adhesion (**a**) three-point adhesion test [Rammo NN et al. (2009)] and (**b**) four-point adhesion test [Lin CK, Berndt CC (1994)]. (Reprinted with kind permission from Springer Science Business Media, copyright © ASM International)

together with ϕ, which values vary widely between different test procedures and are also affected by the presence of residual stresses. Thus, all the adhesion and toughness experiments presented afterwards should be considered with special attention in their interpretation. The same experiment should be used to compare the effect of the different spray parameters on coating adhesion and toughness. Moreover, the results obtained under the same spray conditions with two different measurements devices are not comparable.

17.6.6 Bending Toughness Measurements

"Bending tests" are carried out by either three-point measurements where a rectangular coupon is placed on two rolls and a third roll applies a load at the mid-span location as in Fig. 17.24a from [Rammo NN, et al. (2009)] or four-point measurements where four rolls are used to more homogeneously distribute stresses over the test volume, as in Fig. 17.24b from [Lin CK, Berndt CC (1994)]. During the three-point test, the upper half of the specimen is under compression, while the complementary half is under tension. With the four-point bend test, it is the reverse. This test is more currently used because it is less sensitive to the defects within coatings. However, it is a mixed mode system with ϕ comprised between 35 and 60°. The test must enable a symmetric crack to propagate from a perpendicular notch, see Fig. 17.24b, along the weakest interface of a multilayer specimen.

In principle, stable crack propagation should occur at a constant load between the two inner loading points. Théry P-Y et al. (2007) performed adhesion tests on specimens, either pre-cracked, as shown in Fig. 17.25a or not. Without preset pre-crack, a large load is needed to initiate the crack and the specimen could consequently be overloaded so that catastrophic crack propagation occurs, and the force–displacement diagram exhibited no plateau (see Fig. 17.25b), providing no reliable information on the adhesion energy. A pre-crack has been successfully realized by introducing a 1-mm large band at the center of the specimens that were not grit blasted prior to spraying [Théry P-Y et al. (2007)]. Under these conditions, a stable crack propagation occurs, with the force–displacement diagram exhibiting a plateau after the peak (see Fig. 17.25c) providing reliable adhesion measurements. The length of the pre-crack might be sufficient to make the critical strain energy release rate, G_{ic}, be independent of it [Clyne TW, Gill SC (1996)].

AE is often used to follow crack propagation [Ma XQ, Cho S, Takemoto M (2001b)]. The distribution of AE responses combined with the analysis of crack source parameters is useful for determining the failure modes and the damage progression [Ma XQ, Cho S, Takemoto M (2001b)].

This method can also be used to determine the coating toughness. For that the critical strain energy release rate G_{ic} (see Eq. 17.19) must be calculated either from simple equilibrium equation, assuming an elastic behavior, or using more complex expressions such as those deduced from Euler–Bernoulli beam theory [Bradai MA, et al. (2008), Li H, Khor KA, Cheanga P (2002)]. In order to determine G_{ic}, both the applied load and the displacement of the loading points are continuously monitored and recorded. The specimen is loaded until both cracks have propagated out to the supporting points. The strain is monitored, and any cracking detected by AE so that the strain to fracture (STF) is determined. Unfortunately, the STF parameter is dependent on the residual stress. For example, Howard and Clyne, as cited by

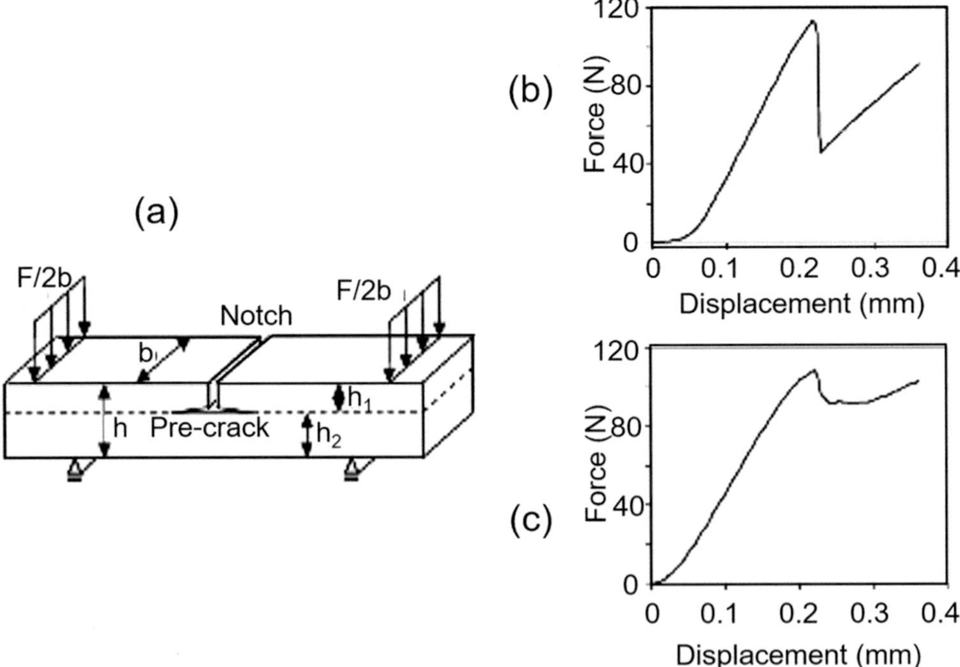

Fig. 17.25 (a) Four-point bending specimen with symmetrical interfacial cracks; (b) force–displacement diagrams recorded during the adhesion test, without any pre-crack; and (c) with an efficient pre-crack [Théry P-Y et al. (2007)]. (Reprinted with kind permission from Elsevier)

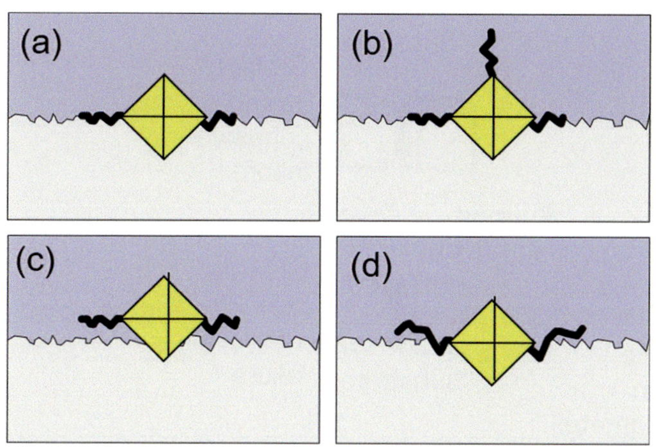

Fig. 17.26 Schematic aspect of the crack generated by interfacial indentation: (a) accepted test, (b) multiple cracks, (c) misalignment, and (d) deviated cracks [Montavon G (2004)]. (Reprinted with kind permission of Prof. G. Montavon)

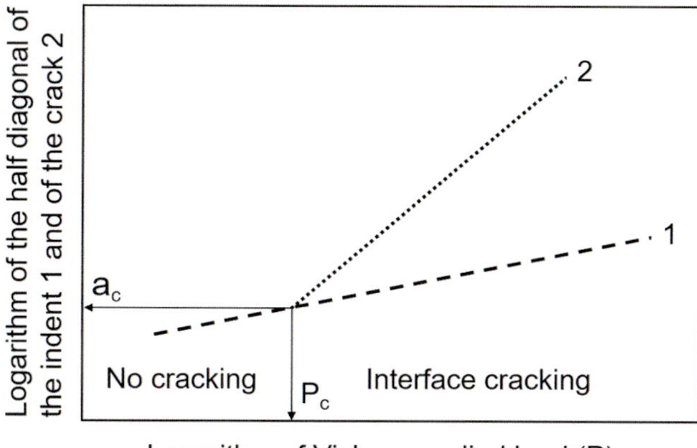

Fig. 17.27 Typical results obtained by performing the interface indentation test: curve 1: logarithm of Vickers hardness indent half diagonal length and curve 2: logarithm of the interface crack length. Below P_c and a_c no interface cracking is observed [Lesage J, et al. (2000)]. (Reprinted with kind permission from Elsevier)

Lin CK and CC Berndt (1994), used the bi-material notched four-point bending specimen to study the interfacial fracture toughness of vacuum plasma-sprayed titanium coatings. G_{ic} was 6.79 ± 0.35 J/m² for coatings exhibiting residual stress and 2.98 ± 0.19 J/m² for coatings without residual stress.

The four-point test has been used to characterize, for example, plasma-sprayed TBCs [Choi SR, et al. (2005)], 100Cr6 steel coatings thermally sprayed on a 35CrMo4 steel substrate [Bradai MA, et al. (2008)], plasma-sprayed ceramics [Westergård R, et al.(2000)], high velocity oxy-fuel-sprayed bioceramic coatings [Li H, Khor KA, Cheanga P (2002)], and detonation-sprayed MoB–CoCr alloy coatings on 2Cr13 stainless steel substrate [Heping L, et al. (2010)].

17.6.7 Indentation Toughness Measurement

Compared to the sophisticated adhesion and/or toughness measurements described above, the indentation test is simple as it consists essentially in obtaining polished cross-sections of coating and substrate and performing Vickers hardness (VH) indentation tests on the interface using different applied loads. Upon indenting the interface, a plastic deformation zone is created by sharing the combined local properties of the coating and the substrate. When the fracture toughness of this composite interface material coating is reached, a local crack occurs [Westergård R, et al. (2000), Heping L, et al. (2010)] (see Fig. 17.26a) [Montavon G (2004)]. The purpose of the interface indentation test is to give a quantitative measure of the apparent fracture toughness of the above mentioned "interface material." Compared to most of the above tests, it does not require the use of a bonder and it may be used for a large range of coating thicknesses, provided they are over 100 µm. However, the test needs to

be carefully performed and avoid multiple cracks (Fig. 17.26b), misalignment (Fig. 17.26c), and cracks deviation (Fig. 17.26d).

Lesage J. et al. (2000) have developed a procedure for coated materials; it consists of the following:

- Applying the Vickers indent with its diagonal coincident with the coating substrate interface.
- For each indentation test, measuring the value of the half diagonal of the indent, a_c, and the length of the crack l, both at the interface.
- Plotting these data as a function of the applied load in bi-logarithmic scale, as represented schematically in Fig. 17.27.
- Determining the coordinates of the critical point indicated in Fig. 17.27 (load P_c, half diagonal of crack length a_c), underneath which no crack is formed.

To achieve a quantitative measure of the apparent fracture toughness, the mechanical properties, strength, and elasticity of the interface materials must be known. Neglecting the effect of residual stresses, according to Lesage J, et al. (2000) the apparent interface fracture toughness K_{ca} may be defined as

$$K_{ca} = 0.015 \, \frac{P_c}{a_c^{3/2}} \left(\frac{E'}{H}\right)_i^{1/2} \tag{17.23}$$

The quantity $(E/H)_i$ may be expressed as follows:

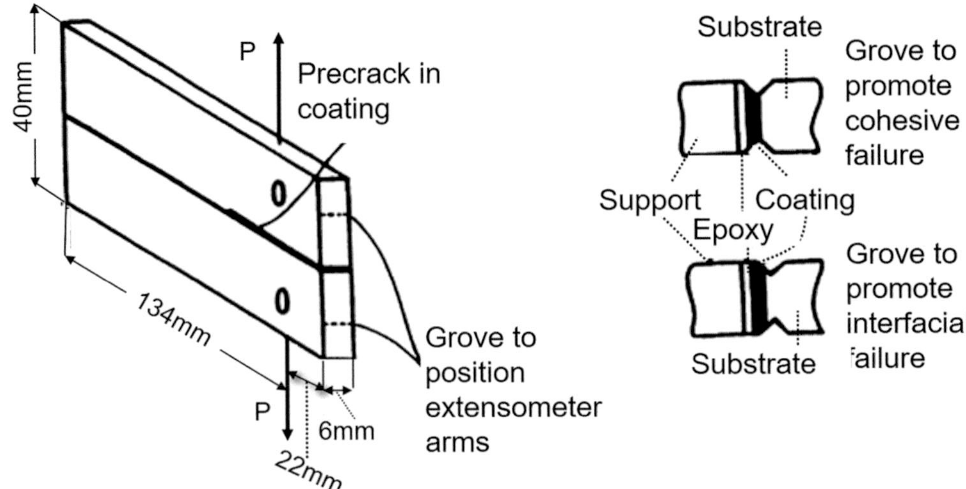

Fig. 17.28 Double cantilever beam specimens used for adhesion measurement. Detail shows the grooving procedure to promote either interfacial or cohesive failure [Lin CK, Berndt CC (1994)]. (Reprinted with kind permission from Springer Science Business Media, copyright © ASM International)

$$\left(\frac{E'}{H}\right)_i^{1/2} = \left[\frac{\left(\frac{E'}{H}\right)_S^{1/2}}{1+\left(\frac{H_s}{H_c}\right)^{1/2}} + \frac{\left(\frac{E'}{H}\right)_C^{1/2}}{1+\left(\frac{H_c}{H_s}\right)^{1/2}}\right] \qquad (17.24)$$

where H is the Vickers hardness and E' the Young's modulus. The subscripts i, S, and C stand for interface, substrate, and coating, respectively.

This method was used, for example, to characterize the effect of thermal treatment on the adhesion of thermal-sprayed NiCr coatings [Lesage J, et al. (2000)], the adhesion of plasma-sprayed cermet Cr_3C_2/Ni–Cr coatings [Hivart P, Crampon J (2007), Marcano Z, et al. (2008)], and that of NiCr–Cr_3C_2 VPS-sprayed coatings [Marot G, et al. (2008)].

17.6.8   Other Methods

17.6.8.1   Double Cantilever Beam Test

In the simple double cantilever beam (DCB) test on coatings (Fig. 17.28), a tension force is applied to the specimen assembly and displacement is measured by an extensometer placed on the arms [Lin CK, Berndt CC (1994)]. When cracking initiates (decreasing load with increasing extension), the DCB is unloaded. Several loading/unloading sequences are performed until the specimen completely fails, at which point the morphology can be examined by optical microscopy and SEM. Of course, the fracture toughness of the adhesive must be greater than that of the coating. As illustrated in Fig. 17.28, it is possible to promote either cohesive or interfacial failure. In this test, both fracture modes occur and the ϕ angle is between 0° and 50°.

Expression of the fracture toughness can be found in the paper of Lin CK, Berndt CC (1994).

17.6.8.2   Double Torsion Test

The double torsion test [Lin CK, Berndt CC (1994)] has been applied to thermally sprayed coatings. The prime advantage of this test is that the crack length need not be measured, because cracking occurs at a constant strain energy release rate or stress-intensity factor.

17.6.8.3   Scratch Test

The scratch test is mostly used to characterize thin, hard coatings (for example, TiN and TiC). In this test, a loaded Rockwell C diamond stylus (see Sect. 17.7.1.1) is drawn across the coating surface under either constant or gradually increasing load. Often the AE is also monitored during the scratching procedure to determine precisely the critical load, P_c, at which failure takes place [Lin CK, Berndt CC (1994)]. Erickson LC et al. (1998) have compared scratch test with other tests and Xie Y and Hawthorne HM (2001) have modeled it. For example, Ma XQ et al. (2008) have used this technique to characterize ultrafine-structured alloy HVOF sprayed using a liquid suspension/slurry containing small alloy powders. In their test, 0.2 mm tip radius Rockwell diamond indenter tip was drawn across the coating with a loading rate of 300 N/min, start load of 0.9 N, and end load of 150 N. Using optical microscopy examinations, the critical loads were measured. Sundararajan G et al. (2010) used this test to get an estimate of the critical load (P_c) at which cracks started appearing at the base of the scratch formed on the cold-sprayed coating surface. Scratch tests were performed on polished and un-etched top surfaces of Cu, Ag, Zn, and SS 316 L coatings with different heat treatments.

17.6.8.4 Laser Shock

(a) Coating Adhesion

Based on the work of Bolis C et al. (2002), laser adhesion test (LASAT), which is a laser shock method, has been used to characterize the interfacial resistance of thermal-sprayed coatings. The pulsed laser was focused on the substrate side on a few mm^2 surfaces, permitting to run different tests on the same sample. Irradiating the surface of a substrate with a high-power laser generated a very intense pressure pulse propagating as a shock wave [Barradas S et al. (2005)]. This shock wave propagated and reflected in the material, creating release waves, the crossing of which led to tensile stresses. Depending on the laser source, the shock pressure can reach few tens GPa during a few ns. To determine the adhesion threshold, shocks were applied to samples with different levels of the incident laser power flux. The debonding limit, that is, when exceeding the interface resistance, was determined from systematic metallographic observations of cross-sections of all the laser-shocked specimens. Below and above the laser shock adhesion threshold, two types of velocity profiles of the coating surface were respectively obtained [Barradas S, et al. (2002)].

- The first type of velocity signal was for a laser power flux of 18 GW/cm^2, that is, below the adhesion threshold. The corresponding curve showed two peaks in the velocity amplitude, which corresponded to the interaction of the shock wave with the coating surface. The time between two peaks was that for the shock wave to propagate through the whole coated substrate material and go back. Therefore, the wave could go through the interface, which was the sign there was no interface damage.
- When increasing the power flux, beyond a certain value the distance between two velocity peaks was shorter and the time between the two peaks corresponded to the time necessary for the wave to cross two times the coating only. This signal was typical of de-bonding at the interface as the shock wave reflects on the void, which was created at the first interaction with the interface.

Barradas S et al. (2005) used an Nd: YAG laser delivering 10-ns Gaussian pulses focused on a 2-mm-diameter spot at the surface of the substrate with a wide range of laser power densities (5–100 GW/cm^2). Laser interferometry allowed measuring the substrate top surface velocity to calibrate LASAT testing numerical simulations in order to calculate the corresponding traction at interface. For the first tests, the tensile stresses in the tested sample were determined using "SHYLAC" finite element code, which allowed 1D simulation only. It corresponded to assuming that the laser beam had a top-hat shape and not a Gaussian one. The first experiments [Barradas S et al. (2005)] on cold-sprayed Cu

onto Al showed that Cu–Cu interlamellar strength was about of the same order as that of Cu–Al coating–substrate adhesion and demonstrated the feasibility of the LASAT testing as a powerful tool for the measuring of adhesion, including in local zones. Rolland G et al. (2011) showed that the laser test was reproducible provided that a black overlay was deposited onto the surface prior to testing to achieve a good and constant absorption of the laser pulse. The optimization of the laser test consisted in determining the maximum laser energy without any surface melting. This requirement came from the fact that the laser absorption was not stable as a function of surface temperature and dramatically increased with melting. Rolland G et al. (2011) carried out finite element calculations using Zebulon® in two steps: that is, thermal calculations then mechanical calculations. The thermal step consisted in applying a thermal flow on the laser-irradiated zone. The laser power profile was modeled with a spherical approximation [Jeandin M et al. (2010)] with which the minimum laser energy needed to melt the surface was calculated. This power profile was used to involve the defocusing effect of the laser, which evolved from a top hat profile to a Gaussian profile.

LASAT has been used to evaluate the feasibility of metallization of thermoplastic polyamide 66 (PA66) using cold spray of aluminum powder [Giraud D, et al. (2012)]. The need of careful interpretation of LASAT when applied to this kind of deposits was highlighted. However, the potential of the test was demonstrated and criteria for future use established.

Guipont has studied the characterization with LASAT of porous ceramic coatings (YSZ and HA) plasma sprayed or deposited by a physical process on metal substrates [Guipont V (2013)]. He showed that, when the laser power corresponding to the adhesion level was reached, a white spot appeared in the ceramic coating at the opposite of the laser impact. This signature of the cracked surface that can be detected with eyes is very convenient.

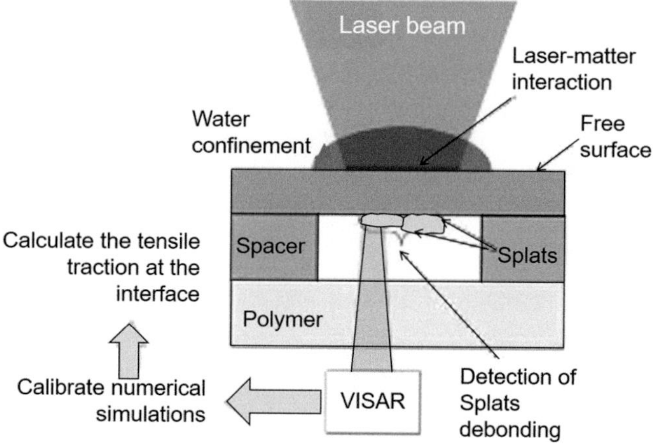

Fig. 17.29 Schematic illustration of LASAT testing setup for splats [Guetta S, et al. (2009)]. (Reprinted with kind permission from Springer Science Business Media, copyright © ASM International)

(b) Splat Adhesion

In cold spray, particle adhesion onto the substrate is influenced by particle impact velocity. Guetta S et al. (2009) developed an original experimental setup (Fig. 17.29) to discriminate the particles as a function of the levels of impact velocity to investigate the influence of this parameter on the resulting splat adhesion. The sample, the top of which surface showed splats, is placed in front of a polymer plane. The upstream laser shock results in the debonding of splats, which are collected at the polymer surface [Guetta S, et al. (2009)]. For splats, laser interferometry (VISAR in Fig. 17.29) resulted in measuring the substrate top surface velocity. This was used to calibrate LASAT testing numerical simulations before the corresponding traction at interface was calculated. Because splat debonding highly depended on traction at the interface, modifying the laser beam intensity decreased the latter. When the traction was low enough, splats no longer debonded. The level of adhesion was therefore reached. Mechanical adhesion resulted from both solid material jetting, which kept the particle in the substrate and molten material jetting, which overlaid the particle and improved adhesion. TEM observation, simulation, and LASAT experiments allowed correlating nanoscale phenomena to macroscopic parameters such as impact velocity [Christoulis DK, et al. (2010)].

17.7 Mechanical Properties

Mechanical properties of materials including hardness, adhesion, Young's modulus, toughness, and residual stress are very important for materials and coatings in service conditions. Some tests of these properties are summarized in [17.M1–17.M4].

17.7.1 Hardness and Indentation Test

17.7.1.1 Hardness

Hardness is generally not considered as an intrinsic property of a material or a coating and its value depends on the testing method used [Chandler H (ed) (2004)] as well as, for coatings, on the direction where the test is performed: at coating surface, orthogonal to splats, or at a cross-section parallel to splats. In the latter case, the coating must be imbedded in resin, which is not necessarily the case for surface measurements. Cross-sections are generally polished, whereas it is not necessarily the case for surfaces. However, on nonpolished rough surfaces, the dispersion of results is more important and thus polishing is recommended. For thin coatings, the substrate may influence the results. It is, however, generally accepted that for coatings thicker than 250 μm the substrate is no more a problem. As previously underlined,

hardness measurements must be statistical and thus can be performed only with multiple readings, at least 10–15.

The principle of the method to measure hardness is the "depth of penetration" (in general for nano-hardness) or the "width of print" (microhardness or hardness) approach left by an indenter applied with a given load, P (between 0.1 mN and tens of N) during a given time (usually a few tens of seconds both for loading and unloading). The load used for nano-hardness measurement, $P < 0.05$ N, microhardness $0.05 < P < 5$–10 N, and hardness >10 N.

Wigren J and K Täng (2007) have pointed out that microhardness is generally performed as per ASTM E384 [17.M7]. A first general misconception is the number of indentations. It is easily shown that 10 indentations are by far too few. A second issue is the metallographic preparation of the specimen before measurement. A preparation routine that tends to smear the coating (i.e., the use of SiC paper) can lead to an underestimation of the microhardness by about 100 Vickers.

When measuring the depth of penetration, a motorized vertical carriage supplies a loading force (from 0.1 mN to kN). The depth of indentation is measured for nano-hardness with a 3-plate capacitance sensor (from 0.1 nm to few μm) and for microhardness with a traditional capacitance sensor (from 0.05 to 250 μm).

For microhardness and hardness measurements, the diameter or diagonals (from about 5 μm to 0.5 mm) of the indents are measured with a digital optical microscope. However, here again to read prints that can be only a few micrometers long, the surface preparation also plays a non-negligible role [Riggs W (2004)]. It is also important to keep a certain distance, depending on the indenter of the indenter used, between the measurement location and the interface coating–substrate or bond coat.

A comparison of hardness testers is given in [17.M5]. For coating nano-, microhardness, and hardness measurements, four types of indenters are used:

- **Berkovitch indenter** is used for nano-hardness measurements. It is a three-sided pyramid, which is geometrically self-similar and has an excellent penetration. It has a very flat profile and a tip radius of about 150 nm, a total included angle of 142.3° and a semi-apex angle of 65.03°. The Berkovich tip has the same projected area to depth ratio as Vickers indenter. For example, Dey A et al. (2009) used this method to characterize hydroxyapatite coatings. It is also possible to use a cone of semi apical angle 30°, which penetration depth is higher than that of the Berkovich indenter.
- **Vickers indenter.** This standard indenter uses a square-based diamond pyramid with an angle between opposite faces of 136° (see Fig. 17.30). Sometimes, this test is called diamond pyramid hardness (DPH). The hardness is calculated by

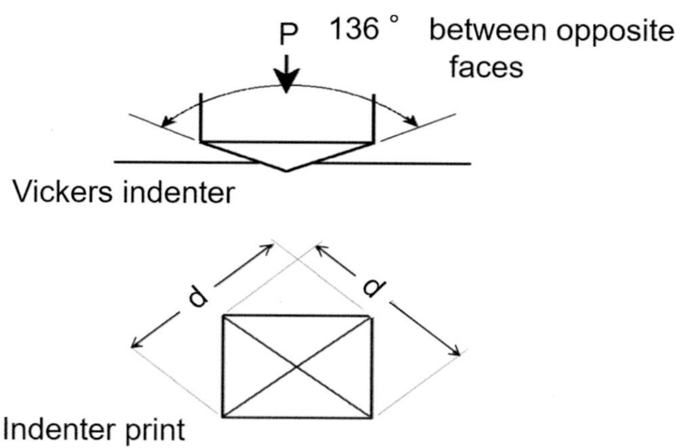

Fig. 17.30 Schematic of a Vickers indenter tip and of its print on the tested coating

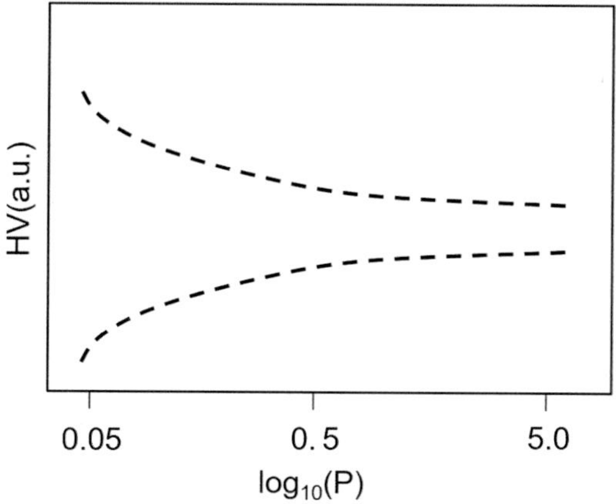

Fig. 17.31 Schematic variation of distributions of maximum and minimum of the Vickers hardness of a thermal sprayed coating for different loads

$$HV = \frac{2P \sin \alpha}{L^2} \, (MPa) \qquad (17.25)$$

where P is the applied load (N), L is the average length of diagonals (m), and α is the angle between opposite faces of diamond (136°). The measurement is acceptable only if the length difference between both diagonals (see Fig. 17.30 representing the indenter print) is less than 5% and if the print is at least half a diagonal length apart from any side of the coating and for measurements on cross-sections if the distance from interface or top is larger than 1.5 times the diagonal length.

In principle, because the impressions made by the pyramid indenter are geometrically similar whatever their size, HV should be independent of the load. However, localized microstructural variations decrease when the applied load increases. This is due to the particular structure of the sprayed coatings that involve splats more or less in good contacts, voids, unmolten particles, the indenter with a higher load being in contact with more splats, voids, and giving a response less and less localized. Thus, when the load increases, the dispersion of results diminishes. If curves representing the minimum and maximum values measured for different loads are plotted, they would look as those schematically presented in Fig. 17.31.

It should be stressed that when hardness measurement is presented it is meaningless unless the load at which measurement has been made is given as well as the measurement dispersion: for example, 600 ± 60 MPa HV_5. This is illustrated in Fig. 17.32 for titania (TiO$_2$) coatings HVOF sprayed on low-carbon steel substrates using a DJ 2700 HVOF torch by [Lima RS, Marple BR (2003)]. Hardness measurements on the surface and cross-section of the titania

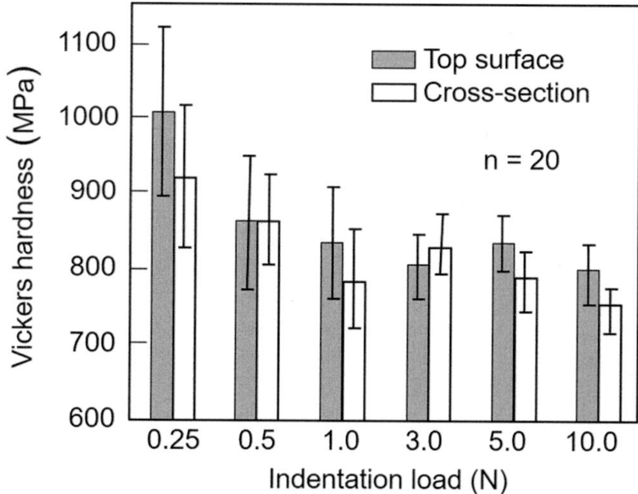

Fig. 17.32 Vickers hardness on the top surface and cross section of HVOF titania coatings for different indentation loads [Lima RS, Kucuk A, Berndt CC (2002)]. (Reprinted with kind permission from Springer Science Business, copyright © ASM International)

coating are noted to decreased from <900 to 1000 Vickers at an indentation load of 0.25 N, to <750–800 Vickers with the increase of the indentation load to 10 N, for both the top surface and cross-section. At lower loads, the indentation diagonal sizes were quite small indicating that only a small region of the coating was being sampled. For example, at an indentation load of 0.25 N, the indentation diagonal length was approximately 7 μm. This minimized the extent of defects, such as coarse pores, splat boundaries, and micro-cracking, enclosed within the indentation zone. As the indentation load increased, so will the volume of coating be being analyzed. At an indentation load of 10 N, the indentation diagonal length was approximately 50 μm. As a consequence, porosity, splat boundaries, and micro-cracking played a significant role on hardness, lowering its value. As

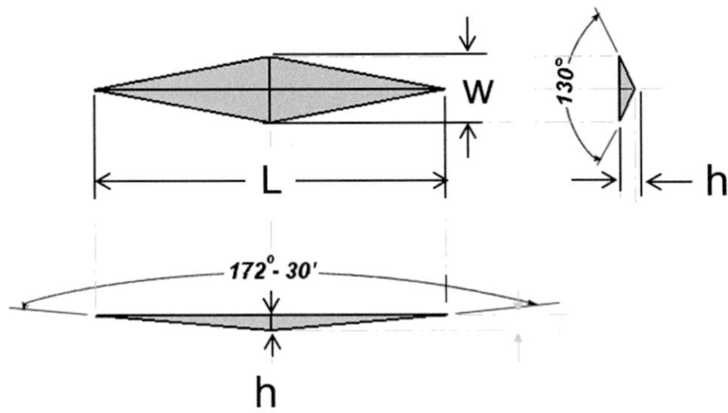

Fig. 17.33 Knoop hardness prints

mentioned earlier, the dispersion of the hardness values decreased with the increase of the indentation load. Hardness measurements made on the cross section of the coating were generally lower than those made on the surface for the coating for the same indentation load [Lima RS, Marple BR (2003)].

For very brittle materials, where only a small indentation may be made for testing purposes, Knoop indenter is often used. The geometry of this indenter is an extended pyramid with a length to width ratio of 7:1 and with face angles of 72°30′ for the long edge and 130° for the short edge, respectively (see Fig. 17.33). The depth of the indentation can be approximated as 1/30 of the long dimension. The Knoop hardness HK or KHN is then given by the formula:

$$HK = \frac{P}{c_f L^2} \, (\text{MPa}) \qquad (17.26)$$

where L is the length of indentation along its long axis and, c_f correction factor related to the shape of the indenter, ideally 0.070279. For measurements close to the interface or the top for coatings cross-sections, the distance must be higher than 0.35 times the length of the longer diagonal. Often Knoop is preferred to measure hardness of ceramic materials because of the longer print of the long diagonal compared to that of the Vickers indenter under the same load. For example, Lima RS, et al. (2002) have measured the Knoop hardness of plasma-sprayed nanostructured partially stabilized zirconia coatings. ASTM tests for Knoop and Vickers harnesses are given in [17.M6–17.M9].

Rockwell C indenter is a diamond cone with an angle of 120°. The test consists in measuring the additional depth to which an indenter is forced by a heavy load beyond the depth of a previously applied light load. The light load eliminates backlash in the load train: the indenter breaks through slight surface roughness and crush particles of foreign matter [Chandler H (ed) (2004)]. Compared to the two previous techniques, Rockwell C technique is faster due to its minimal surface preparation and its ease of use.

17.7.1.2 Indentation

Upon penetration of the indenter into the material, under a load, P, elastic and plastic deformation phenomena occur resulting in a print which is linked to the indenter geometry and the applied load. For example, a spherical indenter with low loads results generally in elastic deformation, while a sharp indenters gives rise to plastic deformation even with very low loads. The penetration depth, h, depends on the geometry of the indenter and the load. The curve $P(h)$ shown in Fig. 17.34a is obtained by first increasing progressively the load (loading step) and then by reducing it to zero (unloading step). When the maximum load, P_{max}, is reached the maximum depth, h_{max}, is obtained. The unloading curve is different from the loading one (see Fig. 17.34a) because of the plastic deformation induced by the load. Figure 17.34b illustrates the different states induced by the loading and unloading steps. The depth h_{max} results from the sum of the contact depth h_p and displacement of the material surface along the contact perimeter as shown in Fig. 17.34c and d for a rounded-cone [Bucaille JL, et al. (2002)] and a sharp-cone indenter. Once the indenter is removed, only the elastic component is recovered, leaving a residual deformation h_r (see Fig. 17.34d). It is worth noting that the length of the Vickers indenter diagonal, observed after unloading, is that corresponding to the residual deformation. This is different for the Knoop print: upon unloading elastic deformation reduces the length of the shorter diagonal as well as its depth, while the elastic recovery of the longer diagonal is much less. It gives the possibility to determine the Young's modulus [Marshall DB, et al. (1982)].

Instrumented indentation method reveals much more information about a thermal spray coating than conventional Vickers or Knoop microhardness methods. It gives information about mechanical properties of the coating such as elastic modulus, elastic/plastic work of indentation, and creep or cyclic behavior [CSM Instruments (2010)]. Thanks to its force–displacement recording capability, the instrumented indentation method is able to sense when an indentation is performed on a void: the depth suddenly increases without the corresponding increase in force (Fig. 17.35). The most significant advantage is probably the automation of measurements. For example, this efficient cyclic indentation method allows automatic increase of maximum load on each subsequent cycle. Dataset of indentation depth and hardness or elastic modulus is obtained within 10 or 20 min depending on the number of cycles [CSM Instruments (2010)]. When the coating hardness increases, the loading and unloading curves are shifted to the left as the indenter penetrates less for the same load.

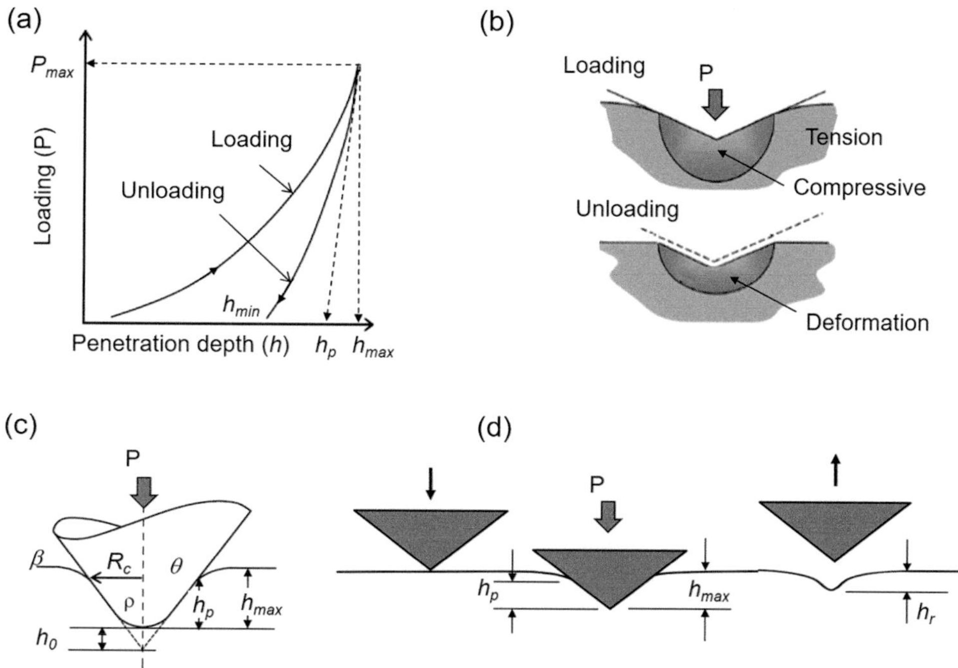

Fig. 17.34 (a) Characteristic curves $P(h)$ during loading and unloading, P being the load, and h the indenter penetration depth, (b) different states induced by loading and unloading steps, and (c) geometrical parameters for a rounded cone indenter under load and (d) definitions of $h_{\max}$, h_{p} and h_{r} [Bucaille JL, et al. (2002)]. (Reprinted with kind permission from Springer Science Business Media)

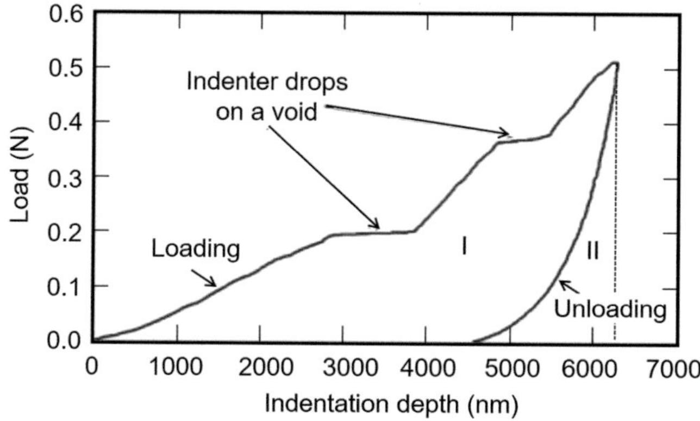

Fig. 17.35 Loading–unloading indentation curve presenting two "indenter drops" due to indentation on voids or other defects, Areas I and II represent respectively plastic deformation and elastic recovery [CSM Instruments (2010)]. (Reprinted with kind permission of CSM Instruments application bulletin)

The area between the loading and unloading curve represents the energy dissipated in the coating due to plastic deformation, and the area under the unloading curve represents the elastic energy for deformation. The normal load, P, and the depth of indentation, h, are measured during the indenter penetration [CSM Instruments (2010)]. The zero-point, defined as $P = 0$ and $h = 0$, is determined by a change in the force signal as the indenter approaches the test

surface. When the indenter is withdrawn the hardness, H can be calculated for the conical indenter by

$$H = \frac{P}{\pi R_{\mathrm{c}}^2} \, (\mathrm{MPa}) \qquad (17.27)$$

The hardness can also be calculated from the depth of the indenter, h_{p} in Fig. 17.36 (plastic component of the coating deformation only) or from $h_{\max}$ (corresponding to both plastic and elastic components) defining the Universal Hardness, HU [Musil J, et al. (2002)], which is lower than H. For example, with a Vickers indenter one has Leigh S-H, Berndt CC (1999):

$$H = \frac{L_{\max}}{\left(26.43 \, h_{\mathrm{p}}\right)^2} \quad \text{and} \quad HU = \frac{L_{\max}}{\left(26.43 \, h_{\max}\right)^2} \qquad (17.28)$$

Musil J et al. (2002) have shown that the area between the loading/unloading curve and the value of, h decrease with increasing hardness H, effective Young's modulus $E^* = E'/(1 - v^2)$ and universal hardness HU, where E' and v are the Young's modulus and the Poisson ratio, respectively. While there is no simple relation between the mechanical response of the coating and H or E alone, this response is strongly dependent on the ratio H/E. More details will be given about indentation in sections related to Young's modulus and toughness measurements.

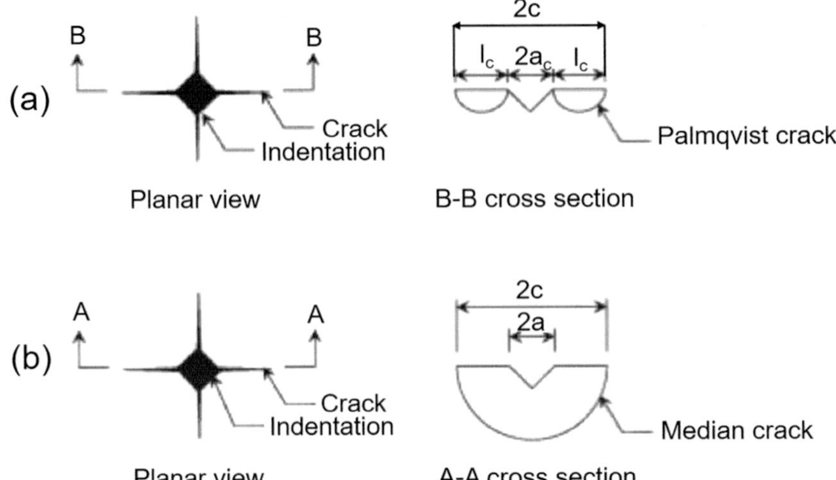

Fig. 17.36 Planar and cross-sectional views of (**a**) the Palmqvist model and (**b**) median crack model [Meacham BE, et al. (2006)]. (Reprinted with kind permission from Elsevier)

17.7.2 Young's Modulus

The elastic constants of thermal spray coatings are strongly dependent on microstructures (depending upon the real contacts between layered splats, cracks, and porosities), which are sensitive to the material processing condition. For example, Leigh S-H, Berndt CC (1999) assumed that the voids within the deposit were ellipsoidal in shape. The structure of the deposit was considered to be transversely isotropic with respect to the spray direction, which required five independent elastic constants of a stiffness tensor. Solid mechanics models containing ellipsoid-shaped voids were applied to obtain the five independent elastic constants of the deposits. The calculated elastic constants were compared to the experimentally determined values. The predicted Young's modulus values from the theoretical calculations were comparable to the experimental values. However, the thin voids and micro-cracks contributed only to a small part of the total porosity; but they had the major influence on the decrease of Young's moduli by small elastic openings and partial closings of their faces. The inter-splat thin voids led to the decrease of Young's module E_3 in the spraying direction, while the intra-splat micro cracks led to the decrease of Young's module E' in the directions orthogonal to the spray direction [Landa M, et al. (2003)]. A few ASTM tests for Young's modulus are given in 17.M10–17.M13. Experimental values of the Young's modulus can be obtained from the following measurements:

17.7.2.1 Indentation

As already pointed out in Sect. 17.7.1.2, the loading/unloading curves of Berkovitch or conical indenters are related to the ratio H/E and different ways are used to determine the Young's modulus. Roy M et al. (2006) working with a Berkovitch indenter have used the following expression

$$\frac{1}{E_r} = \frac{(1 - v^2)}{E'} + \frac{(1 - v_i^2)}{E'_i} \tag{17.29}$$

with

$$E_r = \frac{0.89 \times S}{\sqrt{A}}$$

where S is the slope of the initial part of the unloading curve (see Fig. 17.34a), A is the contact area between the indenter and the substrate, E' and E'_i are the Young's modulus, and v and v_i are the Poisson ratio of the coating and the indenter, respectively.

Ma D et al. (2009) have used the same approach to investigate a set of approximate relationships between H/E' and W_e/W for different types of Berkovich tip geometries with different degree of bluntness. W_e and W are the elastic work and total work corresponding to the areas under the unloading and loading curves recorded in an indentation test.

Feng C, Kang BS (2006) (2008) have used a spherical transparent indenter that can directly measure the indentation-induced out-of-plane deformation as well as the indented surface. Coupling with a multiple partial unloading testing procedure, material's Young's modulus was determined.

Wallace JS, Ilavsky J (1998) have used a spherical indenter to measure the elastic constant: Hertzian indentation. They found that planar defects, which contributed only insignificantly to the porosity of the material, had a very strong effect on the elastic modulus. It has been further shown that there are periodic variations in the elastic modulus throughout the thickness that are correlated with the spray parameters of the initial deposit.

17.7.2.2 Four-Points Bending

The setup used for the four-point bending test is that presented in Fig. 17.24b with the coating placed in tension. All the expressions (see, for example, [Wallace JS, Ilavsky J (1998), Kucuk A, et al. (2000a), Beghini M, et al. (2001a, b) and Podrezov YN, et al. (1999)]) relate the coating Young's modulus, E'_c, to that of the substrate, E'_s, the respective thicknesses of substrate and coating, the relative beam deformation.

Musalek R, et al. (2010) have used different types of loadings: loading up to stepwise increasing loads, with full unloading, loading up to the maximum load, with full unloading, and loading up to stepwise increasing loads, with partial unloading. This allowed observing the coating behavior at different load levels and hysteresis of the stress–strain curves. Coating secant modulus was separated from the total stiffness [Harok V, Neufuss K (2001)]. Bare substrates were tested to assess their elastic limits. The maximum load applied was then selected accordingly.

17.7.2.3 Knoop Hardness

The elastic modulus of coatings can be determined via Knoop microhardness tests. As already emphasized, the elastic modulus is determined by measuring the major and minor diagonals of a Knoop indenter, respectively [Lima RS, et al. (2001a, b)].

17.7.2.4 Ultrasound Propagation

Landa M, et al. (2003) have studied the dependence of the velocity of the longitudinal ultrasound waves on uniaxial pressure for three types of plasma-sprayed ceramics in two directions. The measurement consisted in recording the AE activity (count rate dN/dt) as AE is an important tool for detecting the irreversible changes in brittle materials. The Young's modulus determination requires the knowledge of the coating thickness, including during the pressure application. Acoustic emission was monitored to find the maximum pressure, which did not cause any detectable damage. The instantaneous elastic stiffness c33 and c11, calculated from the measured ultrasound velocities, grew with increasing pressure between 0 and 300 MPa in different materials and in different directions 1.4 up to 4.7 times [Landa M, et al. (2003)]. These results were explained by elastic closing of the inter-splat thin voids and intra-splat micro cracks under pressure. The loading and unloading curves slightly differed, which was explained by relative displacements of the faces of thin voids and micro-cracks oblique to the compression axes, leading to the appearance of friction forces.

17.7.3 Toughness

Among the different mechanical methods that can be used to measure coating toughness, the Palmqvist fracture toughness is popular because no specialized test specimens are necessary and small laboratory samples can be easily tested. Depending on the load applied on a Vickers indenter [Meacham BE, et al. (2006)], either Palmqvist cracks appear for a low load regime, while median cracks show up for high load regime. When increasing loads, Palmqvist cracks form first followed by median cracks. Both types of cracks are represented in Fig. 17.36 from Meacham BE, et al. (2006). Unlike the median crack model, also called halfpenny model, Palmqvist cracks tend to be only as deep as the indenter penetration and are independent of each other.

The critical indentation and crack-related dimensions are the indentation half-diagonal length, a_c, the radial surface crack length, c, and the Palmqvist surface crack length, l_c (Fig. 17.37) [Emiliani ML (1993)]. Equations to determine fracture toughness take the following general form for halfpenny (Eq. 17.30) and Palmqvist (Eq. 17.31) shaped cracks [Emiliani ML (1993)].

$$K_c = k\frac{P}{a_c c^{1/2}} \qquad (17.30)$$

$$K_c = k(E/HV)^{2/5}\frac{P}{a_c l_c^{1/2}} \qquad (17.31)$$

where according to [Emiliani ML (1993)], k is a constant determined typically between 0.001 and 0.5, P is the applied indentor load (N), HV is the Vickers microhardness (GPa), E is the elastic modulus (GPa), and a_c, c, l_c are the indentation and crack-related dimensions (m) respectively that of indentor a and that of Palmqvist, as shown in Fig. 17.36. However, it must be kept in mind that cracks must have a minimum length, depending on the test used. The AE recorded during macro-indentation shows three distinct stages [Emiliani ML (1993)]. The onset of the second stage marked the critical indentation size, for the specific-analyzed coating where radial cracks begin to form. The last stage was associated with the continuation of all mechanisms of accommodation as the indenter descends further into the coating, and therefore changes significantly with load [Faisal NH, et al. (2009)].

Vickers indentation with Palmqvist cracks is extensively used for ceramic coatings, see, for example, [Xie XY, et al. (2010), Richard CS, et al. (1995) and Berka L, Murafa N (2004)]; cermet coatings, see, for example, [Faisal NH, et al. (2009), Lima MM, et al. (2003a, b) (2004), Chivavibul P, et al. (2010) and Prudenziati M, et al. (2010)]. It is also used for certain metal coatings as, for example, those manufactured by PTA with gas-atomized powder and wire arc with cored wire, both with a nominal composition of <25% Cr, <5% Mn, <10% Mo, <10% W, <5% B, <5% C, and < 5% Si, the balance being Fe [Meacham BE, et al. (2006)]. A few ASTM standards for toughness measurements are presented in [17.M14–17.M19].

17.7.4 Residual Stress

The paper of Clyne TW, Gill SC (1996) presents an overview of residual stresses associated with thermally sprayed coatings together with the corresponding measurements.

17.7.4.1 X-Ray Diffraction

A very popular method for direct measurement of residual stresses is the monitoring of the shift of selected XRD peaks [Bunker G (2010), 17.M20]. The method is applicable only to materials with well-defined crystal structures. It becomes difficult, as a result of peak broadening, when the grain size is very fine or if unpredictable variations in composition are likely to occur. The lattice strain is obtained from the shift of *hkl* peak (where *h*, *k*, and *l* stand for Miller indices of the investigated lattice plane), when compared with that for a corresponding unstrained specimen. The drawbacks of the method are the following:

- The limited penetration depth (10–50 μm) of X-rays through most coating materials depending on the source wavelength and coating material.

- The penetration depth is usually about that of the surface roughness of as-sprayed coatings.

To overcome the penetration problem, successive layer-removal can be performed but it changes the stress distribution, if not inducing deformation or damage in the underlying material [Clyne TW, Gill SC (1996)].

Although the lattice strain, and hence stress, in a selected in-plane (*x*) direction can in principle be estimated via the Bragg's equation from a single measurement, it is more accurate to examine the variation in peak shift as a function of φ, the angle between the normal to the coating surface (*y* direction) and the normal to the diffracting planes (which lies in the *x–y* plane) [Clyne TW, Gill SC (1996)]. The stress can then be obtained from

$$\sigma_x = \left(\frac{E}{1+v}\right)_{hkl} \frac{1}{d_0} \left(\frac{\partial d}{\partial \sin^2 \varphi}\right) \qquad (17.32)$$

where *d* is the measured inter-planar spacing and d_0 is the value for an unstrained sample. The Young's modulus, *E*, and Poisson's ratio, v, for the *hkl* planes must be accurately known. In general, they are obtained by experimental measurements, using a sample with a known stress state [Ma XQ, et al. (2001a, b)]. As shown in Fig. 17.37, the lattice strain is obtained from the shift of one *hkl* peak, when compared with that for a corresponding unstrained specimen.

For coatings with different components such as, for example, WC–Co coating, the diffraction peaks of the dominant

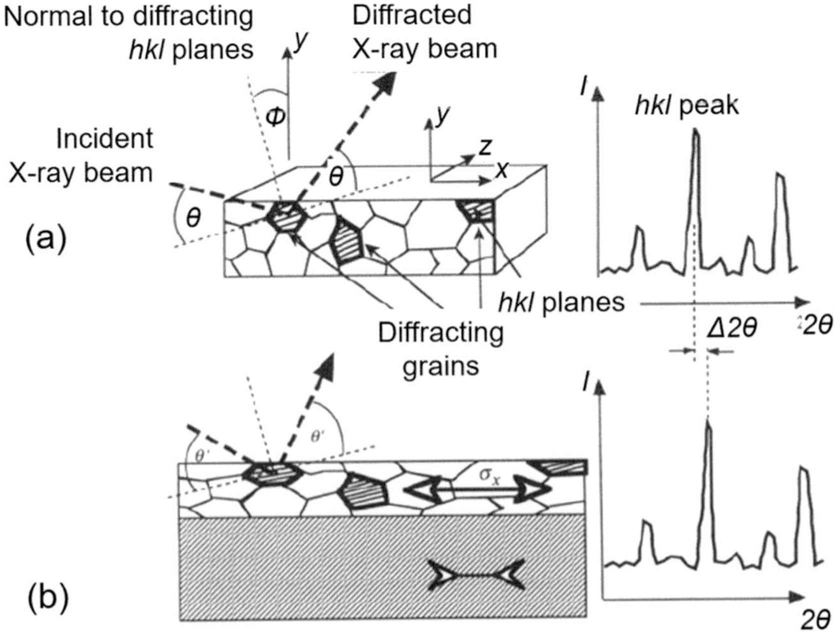

Fig. 17.37 Scheme of the XRD technique for the measurement of residual stresses (**a**) coating detached from its substrate (un-strained specimen) and (**b**) coating with its substrate (strained specimen) [Clyne TW, Gill SC (1996)]. (Reprinted with kind permission from Springer Science Business Media, copyright © ASM International)

phase are used [Santana YY et al. (2008), Taha-al ZY, et al. (2009)].

17.7.4.2 Neutron Diffraction

Compared to low-energy X-rays, the main advantage of working with neutrons is the possibility to analyze greater depths, that is, higher coating thicknesses. Neutrons have the following advantage over X-ray photons: for wavelengths comparable to the atomic spacing, their penetration into engineering materials is in the range of several millimeters, due to their interaction only with nuclei instead of electrons [17.M21].

For example, Lyphout C et al. (2008) used a monochromatic neutron beam of a cross-section of 0.6×0.6 mm^2, in a spatially resolved mode, to determine the residual stress distribution within HVOF-sprayed Inconel 718 Coatings. The Bragg's peaks (220) and (111) for Ni-alloy were measured to $2\theta = 86°$ and $47°$, using a wavelength, λ, of 0.147 and 0.165 nm, respectively. From the obtained diffracted peaks, they used Bragg's law (Eq. 17.4) with the corresponding lattice strain (Eq. 17.30) defined as a function of the "stress-free" lattice parameter d_0.

$$\varepsilon_{hkl} = \frac{(d_{hkl} - d_o)}{d_o} \tag{17.33}$$

where d_{hkl} is the lattice spacing and ε_{hkl} is the lattice strain. Measurements were performed only in radial and axial directions, by scanning the sample from the top to the bottom, using, respectively, the reflection and transmission modes. Sampath S et al. (2004) have also used neutrons to measure the residual stress distribution within Ni–5 wt.% Al bond coats sprayed by different spray techniques.

17.7.4.3 Material Removal

There are two basic methods involving material removal: either a hole is drilled in the specimen or layers are removed by polishing.

(a) Layer Removal Method

This method was adapted to coatings by Greving DJ et al. (1994a, b); it is based on the concept that removing a layer from the surface of a plate or beam with residual stresses releases a force and moment acting on the remaining piece. It is presumed that the remaining piece is large enough and the layer removed small enough so that the change in strain through the thickness of the remaining piece is linear. A strain rosette (gage) on the remaining piece records the change in strain on the surface opposite the face where the layer was removed. The stresses in the layer removed and the change in stresses of the remaining piece can be calculated from force and moment equilibrium, the linear strain change assumption, the strain rosette readings, and the stress–strain

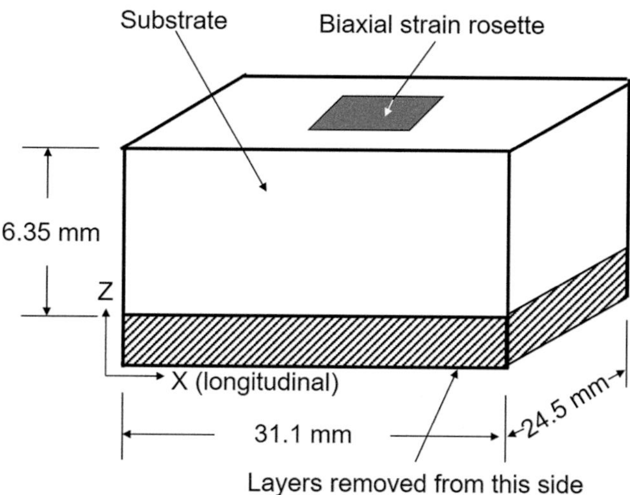

Fig. 17.38 Residual stress specimen ready for layer removal [Greving DJ et al. (1994b)]. (Reprinted with kind permission from Springer Science Business Media, copyright © ASM International)

properties of the material [Greving DJ et al. (1994a, b)]. To calculate the residual stress after removal of one layer, one must know the substrate dimensions, the coating thickness before starting removal, its new thickness after removal, the substrate and coating, Young's modulus, and Poisson ratios. A typical specimen for layer removal is presented in Fig. 17.38. The method was tested for different metal, cermet, or ceramic coatings plasma or HVOF sprayed, with and without bond-coats and with substrate grit blasting [Greving DJ et al. (1994a, b), Pejryd L, et al. (1995), Lima CRC, et al. (2006)].

(b) Hole Drilling Method

The hole drilling strain gage technique [Rendler NJ, Vigness I (1966), Richard CS, et al. (1996) and Schajer GS (1988)] is a semi-destructive method for the determination of residual stresses in the superficial layers of a solid body. The technique simply involves the removal of stressed material by drilling a small hole (about 2–4 mm diameter) on the surface of the body followed by measurement of the relaxation strains occurring around the hole by means of an extensimetric rosette [Valente T, et al. (2005)]. Residual stresses are then calculated from the measured deformations by means of an analytical model based on the solution of the problem of the loaded wide plate with hole [Rendler NJ, Vigness I (1966), Richard CS, et al. (1996) and Schajer GS (1988), Valente T, et al. (2005)].

In the case of nonuniform through thickness stresses, the test must be divided into several small depth increments, and the produced relaxation must be determined for every single incremental step [Valente T, et al. (2005)]. Results may be affected by various errors [Schajer GS, Altus E (1996)] such as alignment [Wang H-P (1979)]. The drilling in brittle ceramics with diamond drills requires an adapted cooling [Balykov AV (2006)].

17.7.4.4 Bending

A rather powerful and nondestructive residual stress measurement technique is the in-situ curvature method, where the curvature of a thin plate sample is continuously monitored during the deposition process [Clyne TW, Gill SC (1996), Kuroda S, et al. (1988) (2001), Gill SC, Clyne TW (1994)]. Kuroda S, et al. (1988) were the first to describe such a continuous measurement procedure during thermal spraying. They used lightly contacting knife-edges and recorded the changing curvature during spraying. Later, Bolelli G et al. (2008) used this method to determine the peening action and residual stresses in HVOF spraying of 316 L. Gill SC, Clyne TW (1994) then further developed this technique and made two advances.

- First, a noncontacting method of curvature measurement was employed with simple video recording, eliminating interference with free bending of the specimen, and allowing the study of a wide range of substrate temperatures.
- Second, a numerical model was developed that allowed the curvature history to be predicted from thermomechanical boundary conditions and material property including the quenching stress [Clyne TW, Gill SC (1996)].

The main advantage of this technique is that various stress contributions can be separately evaluated, especially the quenching stress, which, up to now, cannot be modeled without many assumptions. However, the use of a thin plate with macroscopic curvature generates very different experimental conditions compared to most industrially relevant applications. The latter involves rather thick, medium- or large-sized components like paper or steelmaking rolls, petrochemical industry ball valves, turbine shafts, train axles [Bolelli G et al. (2008)]. The main problems of this technique are the different substrate stiffness that can alter the stress distribution, compared to a real application, and the average deposition temperature and thermal history of the system that can vary significantly with the substrate heat capacity and torch spray pattern. For example, in the experiment of Hugot F et al. (2007), a metallic beam ($110 \times 15 \times 2$ mm) was fixed on a water-cooled rotating cylinder (Fig. 17.39a). The tangential velocity could be varied between 0.2 and 1.8 m/s, the torch being moved up and down at velocities ranging between 2 and 30 cm/s. The metallic plate was supported onto thin edge knifes and held in position by two springs (Fig. 17.39c). The central deflection was measured with a LD-500-5 gauge transducer, the precision of which was 1 μm (Fig. 17.39c). The macroscopic preheating temperature was measured on sample back side by K-type thermocouple

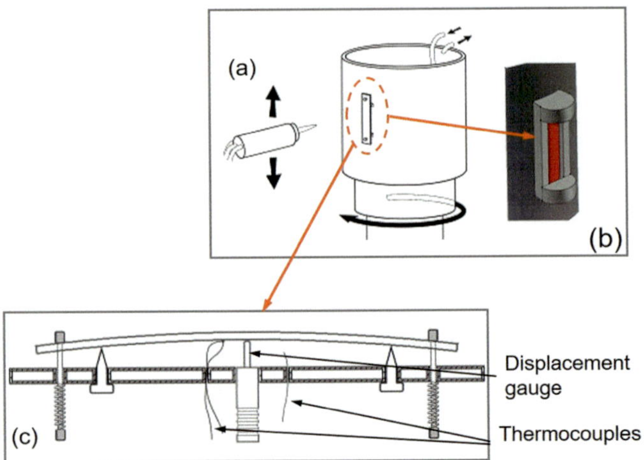

Fig. 17.39 (**a**) Scheme of the experimental stress evaluation device based on beam deflection measurement, (**b**) mask protecting the beam from excessive heating, and (**c**) cross-section of the metallic sample and its measurement device [Hugot F et al. (2007)]. (Reprinted with kind permission from Elsevier)

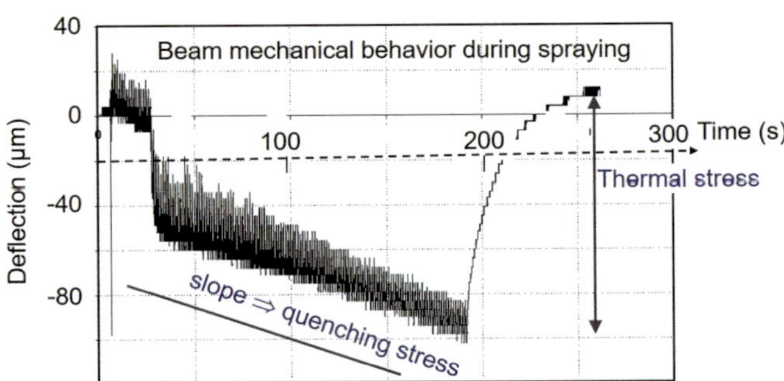

Fig. 17.40 Example of the evolution of the signal measured by the beam displacement gauge (see Fig. 15.40c) during plasma spraying of YSZ on grit-blasted substrate preheated at 200 °C during the first 250 s

(chromel–alumel). This measurement was performed 1 cm ahead of the displacement measurement (Fig. 15.39c). In order to protect the sensors from too high temperatures, a mask was fixed around the substrate (Fig. 17.39b). Two liquid CO_2 atomizing nozzles were positioned around the rotating cylinder, at 90° and 135° from the plasma torch respectively in order to control the substrate mean temperature. A typical signal obtained is presented in Fig. 17.40 for YSZ coating deposited on a grit-blasted stainless-steel substrate. The first 250 s corresponds to the preheating of the substrate up to 200 °C and a beginning of relaxation of the compressive stress induced by grit blasting can be observed. Then, during spraying (about 155 s) the slope of the beam deflection increases linearly (the slope corresponds to the quenching stress). At last, when the plasma jet and particle feeding are stopped, as well as the liquid CO_2 atomization, the coated beam cools down to room temperature in about 700 s. The corresponding gauge signal represents the stress due to the mismatch between substrate and coating expansion coefficients.

Stokes J, Looney L (2008) have listed all the analytical expressions related to quenching and expansion mismatch stress, often used when measuring them, and compared them with the predictions of finite elements model.

17.8 Thermal Properties

The thermal properties of coatings are of primary importance for engineering. The linear expansion coefficient, α_L (K^{-1}) of the coating plays a key role in the residual stress developed in the coating due to differences of expansion coefficients of coating and that of the substrate. In coating service conditions, the service temperature must be such that the stress generated by the expansion mismatch is below that generating the failure of the coating–substrate interface. The removal of the heat flux imparted to coating during service conditions depends strongly upon its thermal conductivity, κ_c (W/m K), which value also controls the temperature gradient generated within coating. Coatings are extensively used as thermal barriers in turbines, automotive engines, and their thermal conductivity, or thermal diffusivity in transient conditions, control their performances. Thermal diffusivity, a (m^2/s), is linked to the thermal conductivity κ_c, specific heat at constant pressure, c_p (J/K kg), and mass density, ρ (kg/m^3), by the expression:

$$a = \kappa_c / \rho \, c_p \qquad (17.34)$$

In the following measurements of the density, thermal expansion coefficient and thermal diffusivity, measurements of the thermal conductivity are also presented. The section ends up with a review of measurements of thermal shock resistance of coatings.

17.8.1 Mass Density

The mass density of the coating, ρ_{300}, at room temperature (300 K) is mostly determined by Archimedean porosimetry [Andreola F, et al. (2000)] (see Sect. 17.5.1. and Eq. 17.8). To obtain its value at temperature T, the expansion coefficient, α_L must be known. If it is linear and isentropic, that is, the same value in all three directions, $\rho(T)$,T being expressed in Kelvin, is given by the following expression:

$$\rho(T) = \frac{\rho_{300}}{1 + 3\alpha_L(T - 300)} \qquad (17.35)$$

In most coatings, however, the linear expansion coefficient of coatings with lamellar structure is not the same in the all three directions since α_L in the spray direction, perpendicular to the surface of the substrate, is different from the linear expansion coefficient in the directions parallel to the

substrate. In most measurement systems of expansion coefficients, the sample length required is a few millimeters. Such a length is easy to cut within the coating, but unfortunately, in most cases the coating thickness is a few hundreds of micrometers and samples in that direction are not suitable for the measurement.

17.8.2 Coefficient of Thermal Expansion

Different techniques include mechanical dilatometers, optical imaging and interference systems, X-ray diffraction methods, and electrical pulse heating techniques are used. For details, see the paper of James JD, et al. (2001); see also the ASTM standards [17.T1, 17.T2]. For thermal-sprayed coatings, the mechanical dilatometry is mostly used. The quantity that must be determined is

$$\alpha_L = \frac{1}{L}\frac{dL}{dT} \text{ or sometimes } \alpha_{L_o} = \frac{1}{L_o}\frac{dL}{dT} \qquad (17.36)$$

In the first expression L is the sample length at temperature T, while in the second the measurement is carried out at room temperature. An alternative means of quantifying the sample expansion is in terms of the fractional increase in volume with temperature:

$$\alpha_V = \frac{1}{V}\left(\frac{dV}{dT}\right)_p \qquad (17.37)$$

If the three linear expansion coefficients α_L are the same: $\alpha_V = 3\alpha_L$.

Thermal expansion measurements can also provide information about deviate from linearity, which happens when phase transformation or oxidation occurs and when stresses relax [Ilavsky J, Berndt CC (1998)]. To avoid the oxidation phenomena the dilatometer furnace must be maintained under controlled atmosphere when studying metals, alloys, or cermets.

17.8.3 Thermal Conductivity and Thermal Diffusivity

Thermal conductivity, κ_c, depends on three mechanisms [Pawlowski L (1995), Wang H, Dinwiddie RB (2004)]:

- Electronic conduction, κ_e, which is the main mechanisms for materials having free electrons (metals and alloys). In general, κ_e varies linearly with temperature at low temperatures and is not temperature dependent at higher temperatures.

- Lattice conduction (phonon conduction), κ_{ph}, which is the main mechanism for materials without free electrons, such as ceramics. At low temperatures, it depends exponentially upon temperature and the scattering of phonons by geometrical defects. At higher temperatures, the "Umklapp process" occurs and κ_{Ph} varies as the reverse of temperature.
- Photon conduction, κ_{ph}, is important in optically absorbing dielectric materials (emission, absorption, and re-emission phenomena). κ_{ph} is proportional to T^3. At high temperatures photons contribute to the conduction through pores, the thermal conductivity of the air or gas contained in the pores being low (~0.026 W/m K).

The thermal conductivity of thermal-sprayed coatings depends strongly on the coating structure, defects, porosity, real contacts between layered splats, and oxide impurity for metals or alloys, which are linked to the spraying process. For example, the low thermal conductivity of YPSZ plasma-sprayed coatings (about 1–2 W/m K) is due to the material but also to the poor contacts between layered splats that reduce phonons propagation. The sintering of coatings at high-temperature service conditions increases the thermal conductivity or diffusivity of the coating by improving the contacts between layered splats [Cernuschi Fet al. (2005), Gitzhofer F, et al. (1986)]. The results of Chi W, et al. (2006) also pointed out the important role of interlamellar porosity in both room temperature and temperature-dependent thermal conductivity. The modeling of thermal conductivity can be achieved using a finite element-based approach coupled with SEM image analysis of the coating microstructure for both through-thickness and in-plane directions for a single image [Kulkarni A, et al. (2003), Tan Y, et al. (2006)].

Most thermal conductivity measurements are performed through thermal diffusivity measurement, which implies measuring the mass density (see Sect. 17.8.1) and the specific heat at constant pressure (see Sect. 17.8.4) to calculate thermal conductivity according to Eq. 17.34. See ASTM references [17.T3, 17.T8]. Baba T, Ono A (2001) have developed an advanced laser flash system for thermal diffusivity measurements. The initial and boundary conditions for measurements are well defined and the transient temperature of the specimen is measured with a calibrated radiation thermometer, the entire curve of the measured temperature history being analyzed. A schematic representation of a Xenon flash diffusivity system used by Chi W, et al. (2006) is shown in Fig. 17.41a. Infrared (IR) detector is used for the monitoring of the temperature of the back surface of the sample (typically a small disk-shaped specimen) when heated by a short Xenon lamp pulses (pulse energy 12 J and duration 6–8 ms, over an 8 mm diameter area of the sample surface). The temperature history on its rear surface being detected by

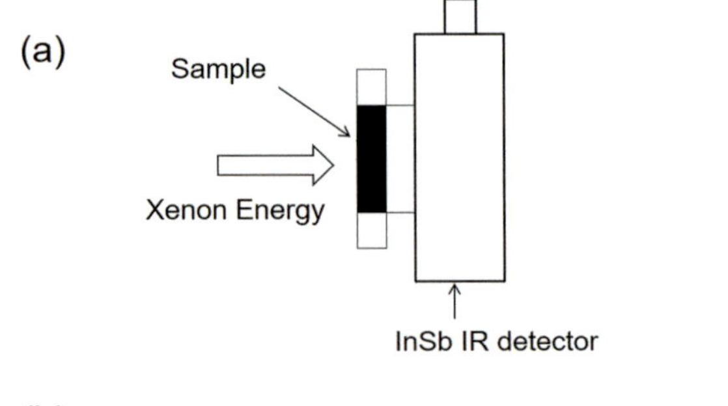

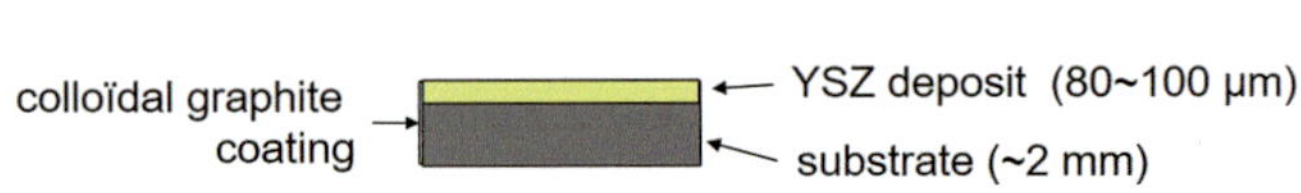

Fig. 17.41 (**a**) Schematic of the laser flash thermal diffusivity measurement system and (**b**) sample preparation using colloidal graphite painted to make surfaces opaque [Chi W, et al. (2006)]. (Reprinted with kind permission from Springer Science Business Media, copyright © ASM International)

the IR detector combined with a TV camera monitoring beam profiles. The infrared detector and the sample holder are completely shielded from the flash source by multiple shielding layers and addition of reflecting surfaces. The detector output signal is monitored from the beginning of the test and in between each shot, which assures that the following flash pulse starts only if the temperature of the sample is steady. When the material studied is translucent in the near infrared, as most oxide ceramics (for example, YPSZ), a thin layer of colloidal graphite paint is used as shown in Fig. 17.41b to make the surfaces opaque prior performing the thermal diffusivity measurement, on both the front and the rear faces of the sample.

For evaluating the thermal diffusivity, α, the simple solution proposed by Parker et al. can be used:

$$\alpha = 0.1388 \, \frac{L^2}{t^{1/2}} \qquad (17.38)$$

where $t^{1/2}$ and L are the time corresponding to the half-maximum increase of the temperature and the sample thickness, respectively. The main advantages of this method are the simplicity and rapidity of measurement and the possibility to measure the thermal diffusivity on a wide range of materials over a wide range of temperature [Cernuschi F, et al. (2005)]. For more details about the method, see the paper of Baba T, Ono A (2001), and, for the reliability of the method, see the paper of Wang H, Dinwiddie RB (2000). The coating diffusivity can be measured with the coating detached from the substrate or with the coating attached to substrate, provided the thermal properties of the substrate are known [He GH, et al. (2000), Rätzer-Scheibe H-J, et al. (2006)]. This technique can also be used for multilayer coatings [Jennifer SY, et al. (2001)] as well as to characterize nanometer-structured, solution plasma-sprayed, coatings. The lower thermal conductivity obtained compared to conventional coatings sprayed with micrometer-sized particles has been explained by the distribution and size of pores [Jadhav AD, et al. (2006)].

Instead of measuring the diffusivity, it is also possible to measure directly the thermal conductivity [Slifka AJ, et al. (1998)] using axial flow with hot plate according to ASTM C 177-85, or comparative cut bar, with guarded or unguarded heat flow meter method.

17.8.4 Specific Heat at Constant Pressure

The heat capacity (c_p) of coatings can be measured with a precision of about 1% [17.T9] using a wide range of calorimeters. These mostly comprise either one cell design, the adiabatic one being the simplest, or two cells where the studied sample is compared to a reference material [17.T10].

17.8.5 Thermal Shock Resistance

Thermal shock is a failure mechanism that occurs in materials, especially ceramics that exhibit a significant temperature gradient when they are subjected to a sudden change in temperature. The thermal shock resistance (TSR) describes the ability of the material (for example, a thermal barrier coating) to resist to rapid temperature variations without failure. Most methods used to characterize TSR involve heating the material to high temperature and then quenching it using water, oil, or air as cooling medium. The result depends both on the heating conditions: steady state flow, transient heating by convection, radiation, and the temperature drop. Quench tests generally must be repeated and in that case the number of cycles necessary to cause a defined damage or weight loss can be used as a measure of the thermal shock resistance [Aksel C, Warren PD (2003)]. According to Hasselman DPH (1970), for steady heat flow and an infinite slab symmetrically heated or cooled, the TSR parameter is defined as

$$\mathrm{TSR} = \frac{\sigma_f\,(1 - \nu)}{\alpha_L\,E'} \tag{17.39}$$

where σ_f is the fracture stress, ν the Poisson ratio, E' the Young's modulus, and α_L the mean expansion coefficient. Four more shock resistance parameters can be defined [Bolcavage A, et al. (2004)]. TSR of coatings is measured with the whole system: coating, bond coat, and substrate. It is thus important to keep in mind that the thermal shock must be adapted to the resistance of all components. For example, if the YPSZ topcoat can sustain temperatures up to 1200 °C and in some cases for relatively short periods 1300–1400 °C (military engines), it is not the case of the bond-coat and substrate superalloys. If possible, characterization means must reproduce service conditions, which is not always the case. Again, for a thermal barrier coating three tests are commonly used: Jet Engine Thermal Shock (JETS), Furnace Cycle Test (FCT), and Fluidized Bed Reactor (FBR) [Bolcavage A, et al. (2004)]:

- With the JETS the flame of a burner rig heats the coating surface, while the opposite side of the substrate is air cooled. It creates a thermal gradient of several hundred degrees Celsius across the TBC. The front surface temperature can reach 1400 °C and can initiate sintering. This test mainly stresses the ceramic layer and bond coat interface thermo-mechanically. Bond coat oxidation is rather low in this case. This test is reasonably quick and typically results are available within 2 days for a 2000-cycle test [Bolcavage A, et al. (2004)]. Figure 17.42a represents schematically the JETS process.

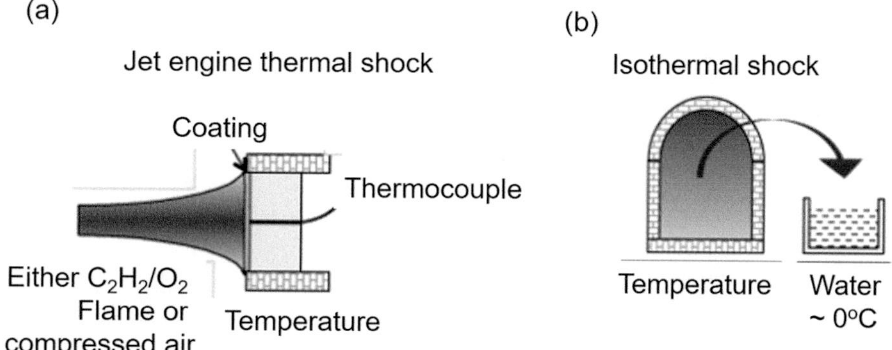

Fig. 17.42 Scheme of two thermal shock setups: (**a**) engine thermal shocks (JETS) heating with a burner rig and cooling with air jet and (**b**) furnace cycle test (FCT) using furnace heating and water soaking cooling [Montavon G (2004)]. (Reprinted with kind permission from Prof. G. Montavon)

- In the FCT test, the TBC system is oxidized and cycled in a furnace between room temperature and temperatures ranging from 1100 to 1200 °C. Thermal gradients across the TBC are moderate during the heat-up phase and significant during the cool-down phase. The sample is exposed for a long time to elevated temperatures, which leads to significant bond coat oxidation. The TBC temperature during soaking is too low to cause pronounced sintering in the ceramic. Figure 17.42b represents schematically the FCT process.
- FBR test consists of a standard fluidized bed furnace using alumina. It is typically operated at 1000 °C. Test samples are transported via a pneumatic system into a cooling zone in which compressed air is directed onto the TBC surface.

According to [Bolcavage A, et al. (2004)]: The JETS test provides performance-ranking data on the ceramic itself thanks to the high thermal gradients within the ceramic layer. The FCT reflects the actual engine condition because it also degrades the bond coat through severe oxidation, though it does not really address TBC degradation in the ceramic layer, which typically initiate above 1300 °C. The FBR used as a quality control tool as the standard 1000-cycle TCF test, does not discriminate differences well enough to provide quantitative data for a performance ranking. ASTM references for thermal shock testing are given in [15.T11–15.T15].

Other means used to measure TSR of thermally sprayed coatings include laser irradiation [Qi ZM, et al. (1988)] and infrared irradiation [Richard CS, et al. (1995)]. To increase the thermal gradient after heating, samples can be quenched in iced water instead of air jets [Irisawa T, Matsumoto H (2006)]. Nusair Khan A, Lu J (2003) have compared TBCs delamination when quenched using water or forced air jets:

- In both cases samples delamination starts from their extreme edges.
- In the case of the water-quenched samples, the thickness of the topcoat was only half of its original value, after 220 cycles, whereas in the case of the forced air-quenched samples, the spallation was only confined to the edges and the thickness of the topcoat remained the same even after 1000 cycles.
- In both cases more thermally grown oxide (TGO) was observed at the edges compared with the central portion of the samples.

To obtain more information than those resulting from visual inspection after a given number of cycles, AE can be used [Robin P, et al. (2010), Bolcavage A, et al. (2004)]. As a crack is initiated and propagates through a material,

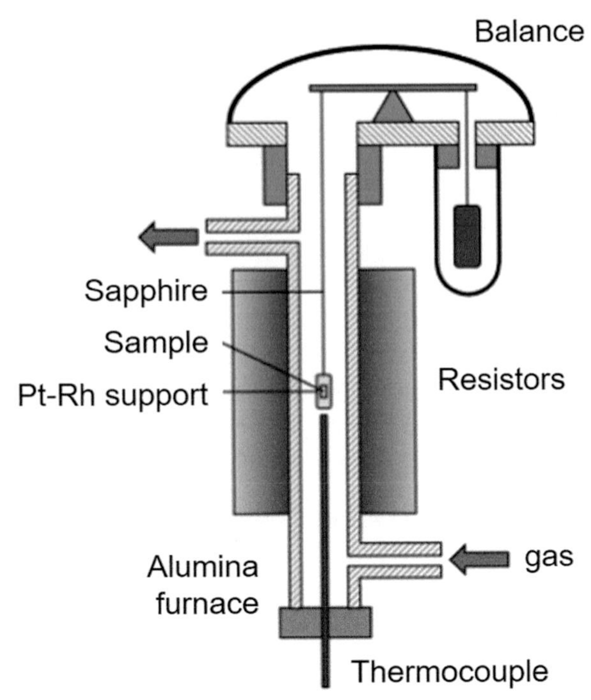

Fig. 17.43 Principle of a thermogravimetric balance [Montavon G (2004)]. (Reprinted with kind permission from Prof. G.Montavon)

measurable characteristics of the AEs from this event reveal details about the microstructure [Andrews DJ, Taylor JAT (2000)]. It is to be noted that a continuous AE signal corresponds to plastic deformation mechanisms, while burst-type AE results from micro-cracking in brittle materials, fracture of hard inclusions in alloys, and phase transformations. Such measurements allow determining when first cracks propagation occurs and the different fracture modes [Ma XQ, Takemoto M (2001)].

17.8.6 Thermal Analysis

Thermal analysis (TA) is a measure of physical property of a substance as a function of temperature [Sorai M, Gakkai NN (2004), Speyer R (1993)]. One of their main advantages of is that they require rather small samples. For more details about these techniques, the reader can see the books of Sorai M, Gakkai NN (2004) or Speyer R (1993). In the following, a few examples are presented of these techniques used in the thermal spraying area to characterize phase changes in powders or coatings, reactions at high temperatures and oxidation. TA techniques mainly used for characterizations of sprayed coatings or sprayed powders include:

Thermogravimetry (TG): In which the mass of a sample is recorded, in a controlled atmosphere or air (see Fig. 17.43), as a function of temperature. On the one hand, mass is lost if the substance contains a volatile fraction and TG

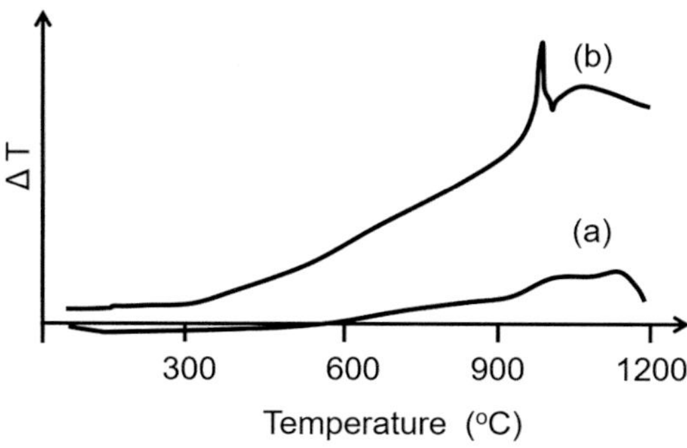

Fig. 17.44 DTA obtained from (a) initial mullite powder and (b) as-sprayed mullite coating at substrate temperature of 300 °C [Salimijazi H, et al. (2012)]. (Reprinted with kind permission from Springer Science Business Media, copyright © ASM International)

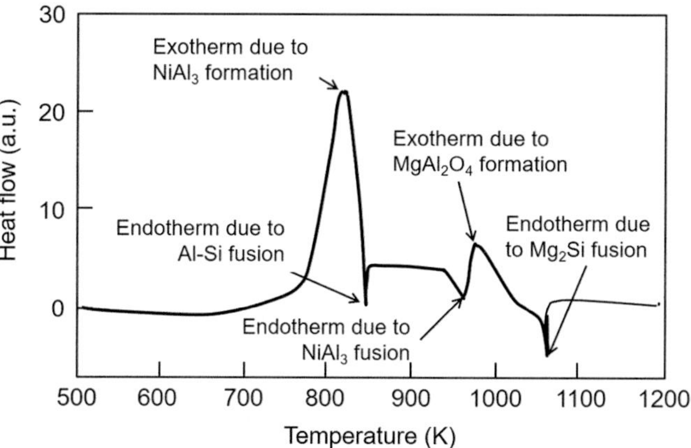

Fig. 17.45 DTA scan of granulated SiO$_2$/Ni/Al–Si–Mg particles heated to 1273 K with heating rate of 0.17 K/s in argon atmosphere [Ozdemir I, et al. (2005)]. (Reprinted with kind permission from Springer Science Business Media, copyright © ASM International)

measurement can be coupled with a device permitting the analysis of volatile products, and on the other, mass is gained with oxidation.

Differential scanning calorimetry (DSC): The difference in the heat flow absorbed by the test sample is compared with the heat flow absorbed by a reference material, as a function of the temperature. Experiments are performed in a controlled atmosphere. DSC provides information about thermal changes that do not involve a change in sample mass such as phase changes or changes in chemical composition. It also allows checking the material purity through changes in heat of fusion and melting point and studying sample oxidation in oxidizing atmospheres.

Differential thermal analysis (DTA): In which the temperature difference between a substance and a reference material is measured as a function of temperature, both materials being subjected to the same programmed temperature profile. Experiments are performed in a controlled atmosphere. The reference material is a known substance, generally inactive thermally within the temperature range of interest.

17.8.6.1 Phase Changes

Salimijazi H, et al. (2012) have compared plasma-sprayed coatings made with commercially available mullite powder particles and a mixture of mechanically alloyed alumina and silica powder particles. DTA was used to study the phase transformations in the coatings. Figure 17.44 represents the DTA curves obtained for mullite initial powder and the as-sprayed mullite coating on stainless steel substrate, at temperature of 300 °C. Neither exothermic nor endothermic peaks were observed for the mullite powder. A peak around 980 °C could be observed in the as-sprayed structure that was attributed to the crystallization of amorphous aluminum silicate.

Garcia E et al. (2011) have air plasma sprayed with an Axial III plasma torch plain mullite, 75 vol.% mullite–

25 vol.% Y-ZrO$_2$, and 50 vol.% mullite–50 vol.% Y-ZrO$_2$ powders. DTA of as-sprayed plain mullite coatings showed no endothermic or exothermic event, supporting the fact that as-sprayed coating was fully crystallized, which was not the case with the two other powders. The authors stated that mullite crystallization was hindered in the composites during the spray process. Liu X, Ding C (2002) studied phase changes of plasma-sprayed wollastonite (CaSiO$_3$) coatings. The DTA curves showed that the wollastonite powder had no chemical reactivity from room temperature to 1000 °C, whereas an obvious exothermic reaction occurred in the wollastonite coatings at about 882 °C. Crystalline wollastonite, glassy phase, and tridymite (SiO$_2$) were observed in the coating. Tridymite (SiO$_2$) likely reacted with other compounds such as CaO and glassy phase to form crystalline wollastonite when the coating was heated at about 882 °C.

17.8.6.2 Reactive Thermal Spraying

Reactive thermal spraying (RTS) in which thermodynamically stable compounds are formed by in-process reactions, has attracted considerable attention leading to the wide availability of in-situ synthesized composite coatings. Ozdemir I et al. (2005) have HVOF sprayed a composite powder of SiO$_2$/Ni/Al–Si–Mg onto an aluminum substrate. To identify the reactions which were taking place in flight, they heated the same materials in a DTA over the temperature range of 298–1273 K under argon atmosphere with a heating rate of 0.17 K/s. Figure 17.45 presents the DTA curve.

Deevi SC, et al. (1997) have studied the reactive spraying of nickel aluminide when nickel and aluminum powders were plasma sprayed onto carbon steel substrates. To illustrate the exothermic reaction between nickel and aluminum, a mixture of nickel and aluminum in a 1:1 atomic ratio was heated in a

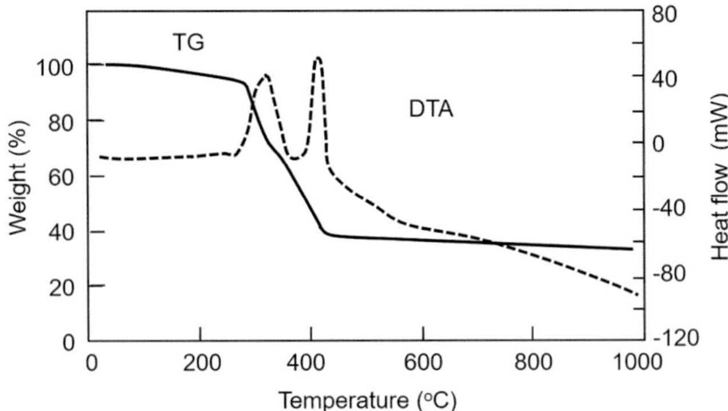

Fig. 17.46 Thermal gravimetric–differential thermal analysis curves of dried MgO–Y$_2$O$_3$ precursor at a heating rate of 10 °C/min [Wang J, et al. (2010)]. (Reprinted with kind permission from Springer Science Business Media, copyright © ASM International)

DTA unit under argon atmosphere. Wang J, et al. (2010) have synthesized by a sol–gel combustion process MgO–Y$_2$O$_3$-nanostructured composite powder (volume ratio of 50:50) with crystal sizes in the 10–20 nm ranges. The powder was used for plasma spraying. The formation process was studied by TG-DTA of the 200 °C dried MgOY$_2$O$_3$ powder precursor as a function of temperature at a heating rate of 10 °C/min. As illustrated in Fig. 17.46, the sample weight decreased continuously with the increasing of temperature up to 1000 °C, with a total weight loss of about 67%. Two exothermic peaks observed at 326 and 419 °C were ascribed to the pyrolysis of the precursor (see the DTA curve). The TG curve showed that after 600 °C, the pyrolysis process was completed.

Ravi BG et al. (2008) studied the RF precursor plasma spray (RF-PPS) synthesis process. It starts with liquid inorganic precursors and directly produces ceramic coatings as yttrium aluminum garnet (YAG). The sols used were either regular or reverse ones [Ravi BG et al. (2008)]. To gain a better insight in the process, thermal analysis (DTA-TG curves) was performed on the coatings. The TG plots showed that the as-sprayed coatings had substantial amount of volatile matter. However, subsonic-sprayed coatings lost less weight than the supersonic-sprayed coatings, probably due to the relatively longer residence times of particles in subsonic spraying allowing for the improved pyrolysis of the sols.

17.8.6.3 Oxidation Resistance

Guilemany JM, et al. (2009) have explored transition metal aluminide in HVOF-sprayed coatings of two different aluminide (FeAl and NbAl$_3$) feedstock powders in terms of resistance to oxidation and mechanical behavior. They have studied the heat flow and mass gain-temperature dependence for the as-sprayed coatings during the oxidation test in DTA–TGA equipment. The oxidation analysis revealed that despite the high melting point of NbAl$_3$, it suffers from intergranular

disintegration at moderate temperatures. The high level of oxidation and coating defects that facilitated oxygen diffusion inwards and Al depletion in the as-sprayed coating itself favored this phenomenon. Fan X, Ishigaki T (2001) used induction plasma spray processing to produce freestanding parts of Mo$_5$Si$_3$-B composite and MoSi$_2$ materials. Oxidation resistance of the specimens was evaluated through the TG tests. The specimen temperature was increased at 20 °C/min to the preset temperature, 1210 °C, and held for 24 h, during which time the mass changes and temperature of the specimens were continuously recorded. The performance of Mo$_5$Si$_3$-B composites at high temperature was found to depend strongly on the composite boron content, for example, at boron contents greater than 1 wt.%; composites demonstrated encouraging oxidation resistance in the high-temperature environment.

17.9 Wear Resistance

Wear is generally classified into six distinguishable types [Pawlowski L (1995)] as described below according to Craig BD (2005) and Chattopadhyay R (2001). Information about standard terminology relating to wear and erosion can be found in 17.W1.

The wear volume, V_w, defined by the following equation, characterizes wear [Chattopadhyay R (2001)]:

$$V_w = K_w \frac{F_N}{H} \times d_w \qquad (17.40)$$

where d_w is the sliding distance, F_N is the applied normal force, H is the hardness or yield stress of the softer surface, and K_w is the wear coefficient. The depth of wear δ_w is expressed as

$$\delta_w = K_w \frac{F_N}{H} \cdot \frac{d_w}{A_w} \qquad (17.41)$$

where A_w is the area of contact, $V = \delta_w A_w$

The wear coefficient K_w depends on the wear mode, materials and lubrication system used, type of motion, stress distribution. Chattopadhyay R (2001)] has summarized typical values of K_w.

For adhesion wear the following equation is used:

$$\frac{V_w}{d_w} = K_{adh} \frac{F_N}{H} \qquad (17.42)$$

where $K_{adh} = K_w/3$.

In erosive wear, the volume loss is given by

$$V_w = \frac{\frac{1}{2}\, mv^2}{H} f(\alpha) \qquad (17.43)$$

where H is the hardness or yield strength of the surface material, $\frac{1}{2}\, mv^2$ is the kinetic energy of the impacting particles, and $f(\alpha)$ is a function of the impingement angle α, which plays a key role in the erosion. For brittle materials, V_w increases with higher angles, while it is the reverse for ductile materials.

It must be underlined that any kind of wear produced with the adapted testing machine is often measured with the use of optical methods giving the wear track profile as well as the debris distribution.

17.9.1 Abrasive Wears

Abrasive wear occurs when a solid surface experiences the displacement or removal of material as a result of a forceful interaction with another hard surface. Both free-flowing particles and abrasive attached to the counter body cause wear. Two types of abrasive wear are defined based on the degree of stress in the component surface or the number of components involved. In the two-body abrasion, under low stress conditions, abrasive particles (asperities or hard particles) sliding against the component surface cause scratches, but the stress induced does not cause fragmentation of abrasives. Three-body or high stress abrasion occurs when the abrasive particles are forced between two mating surfaces leading to loss of materials from both component surfaces [Chattopadhyay R (2001)]. A few abrasive tests proposed by ASTM are given in 17.W2–17.W5.

For the abrasive wear tests, in laboratories two main tests are performed [Kennedy DM, Hashmi MSJ (1998),

Bolelli G, et al. (2006), Cockeram BV, Wilson WL (2001), Wielage B, et al. (1999) Schmidt G, Steinhäuser S (1996)]:

- The pin-shaped material rotates on an abrasive disk, or an abrasive cloth, or paper mounted on a flat disk. The debris can be blown out with a gas jet to determine if their presence improves or not the wear resistance.
- Free or loose abrasive is interposed in between the two surfaces pressed against each other. One is in sliding or rotating motion while the other is stationary.

Many results are described in the literature and a few of them illustrated in the following references [Gawne DT, et al. (2001), Li C-J, et al. (2011), Rainforth WM (2004), Kim HJ, et al. (1994), Tillmann W, et al. (2008a, b), He D-Y, et al. (2008) and Magnani M, et al. (2009)], tests being performed at room or high temperature, depending on the coating application.

17.9.2 Adhesive Wears

Adhesive wear occurs between two surfaces in relative motion because of the inherent roughness of material surfaces. It is also called sliding wear [17.W6]. In adhesive wear, the peaks on the adjacent surfaces that do come into contact plastically deform under pressure and form atomic bonds at the interface and in some cases this is considered solid-phase welding. As the relative motion between the surfaces continues, the shear stress at the bonded contact point increases until the shear strength limit of one of the materials is reached and the contact point is broken, bringing with it a piece of the opposing surface. The broken material can then either be released as debris or remain bonded to the other material's surface [Craig BD (2005) and

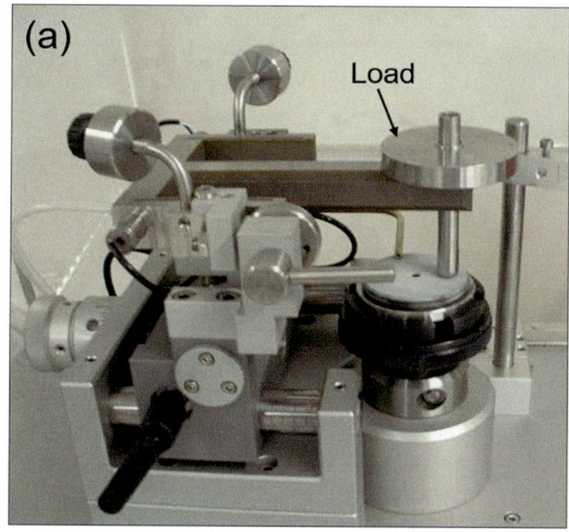

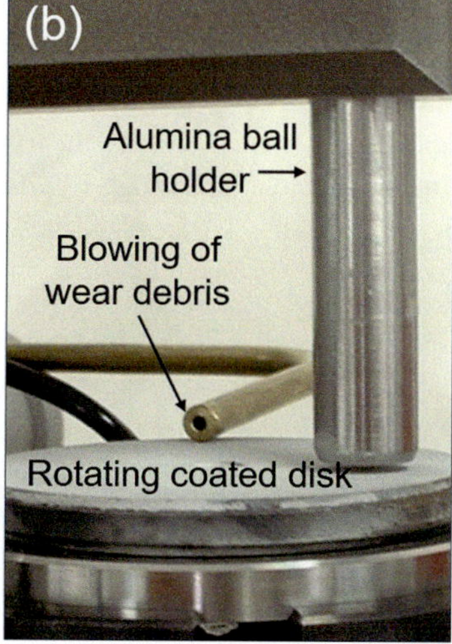

Fig. 17.47 Pin-on disk test with a rotating coated disk, 6 mm diameter ball made of sintered alumina (**a**) general view and (**b**) detailed view

Chattopadhyay R (2001)]. The most used tests are Chattopadhyay R (2001), Kennedy DM, Hashmi MSJ (1998), Bolelli G, et al. (2006), Cockeram BV, Wilson WL (2001), Wielage B, et al. (1999) and Schmidt G, Steinhäuser S (1996):

- The pin-on-disk [17.W7, 17.W8] where a pin specimen, with an extremity either flat or spherical, is tightly pressed against a flat circular-coated disk with a specified load. During the test either the pin or the disk rotates, see Fig. 17.47, and the sliding path forms a circle on the rotating disk surface. The debris can be blown out with a gas jet to determine if their presence improves or not the wear resistance. It is also possible to have the disk vertical and the pin horizontal to entrain the debris by gravity or achieve tests with lubrication.
- The pin-on-plate where the coated plate reciprocates at 50–300 cycles per minute.
- The pin or block-on-ring where the coated sample slides on the edge of a disk or a ring mounted vertically. This test reproduces the type of motion found in engine cam or internal combustion systems.

Few equipment can work at high temperatures to test the evolution of the wear. This is for example the case of the experiment of Yin B, et al. (2010) who studied the sliding wear behavior of HVOF-sprayed Cr_3C_2–NiCr/CeO_2 composite coatings at temperatures up to 800 °C. Friction and wear tests were carried out in a ball-on-disk contact configuration using an SRV (sliding, reciprocating, and vibrating) friction and wear tester.

Many results are described in the literature and a few of them illustrated in the following references [Fernandez JE, et al. (1995), Liu Y, Wang HM (2010), Tiana W, et al. (2009), Lin JF, et al. (1996), Yanga Q, et al. (2006) and Guilemany JM, et al. (2001)].

The extreme form of adhesive wear is galling that involves excessive friction between the two surfaces, resulting in localized solid-phase welding, and subsequent spalling of the mated parts [Chattopadhyay R (2001)]. This process causes significant damage to the surface of one or both

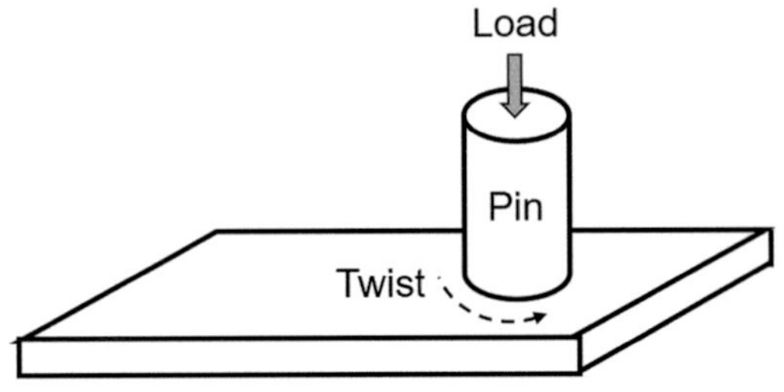

Fig. 17.48 Principle of galling test setup [Chattopadhyay R (2001)]. (Reprinted with kind permission from ASM International)

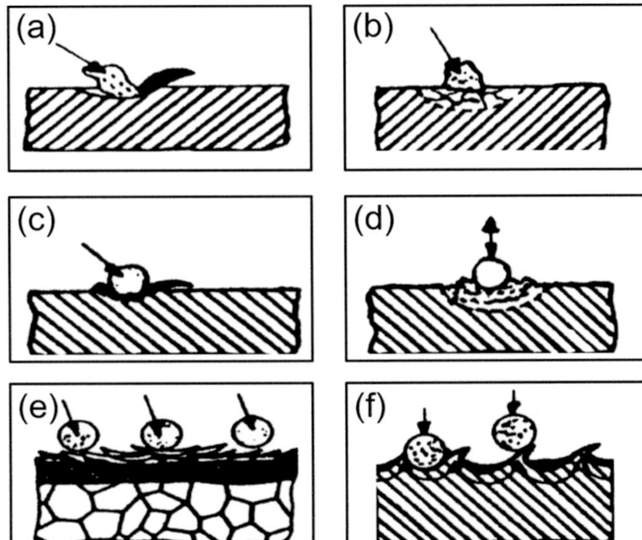

Fig. 17.49 Erosive wear due to particle impact: (**a**) microcutting and microploughing; (**b**) surface cracking; (**c**) extrusion of material at the exit end of the impact craters; (**d**) surface and subsurface fatigue cracks; (**e**) formation of thin platelets due to extrusion and forging; and (**f**) formation of platelets by a backward extrusion process [Stewart S, et al. (2005)]. (Reprinted with kind permission from Elsevier)

materials. The typical test is shown in Fig. 17.48: the pin specimen, gradually load, against a rotating counter plate material. The load at which the pin tends to gall onto the plate is the galling load [Wielage B, et al. (1999)]. The ASTM test is referenced [17.W8].

17.9.3   Erosive Wear

Erosive wear is the continuous deterioration of a material by fluid carrying solid particles [17.W6]. When the fluid is traveling in a direction that is normal to the surface of the material, it can be considered as impact wear. The repeated impact by particles that are very small relative to the size of the material being impacted, results in erosive wear. Whatever may be the size of impacting particles, they can cause deformation to the material being impacted that can result in the ejection of particles from the material's surface or the formation of near-surface cracks that under repeated impact can cause pieces of the surface to fracture. Figure 17.49 from Kennedy DM, Hashmi MSJ (1998) summarizes the different phenomena implied in erosive wear. One conventional test is described in 17.W10.

Cavitation wear, occurring during liquid impingement by the collapse of bubbles on the coating solid surface, is also considered as erosive wear. The deformation of solids by the impact of liquid occurs because of the interaction of stress waves generated by the collapsing bubbles with the surface.

For dry erosion wear, the test mostly used consists of repeated impact erosion by a flow of 50 µm mean diameter

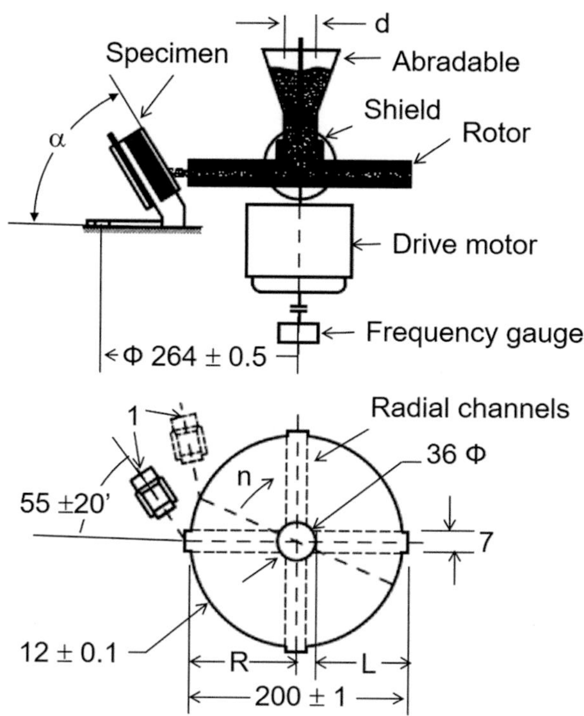

Fig. 17.50 Impact erosive wear tester for materials in abradant particles jet [Kulu P, Pihl T (2002)]. (Reprinted with kind permission from Springer Science Business Media, copyright © ASM International)

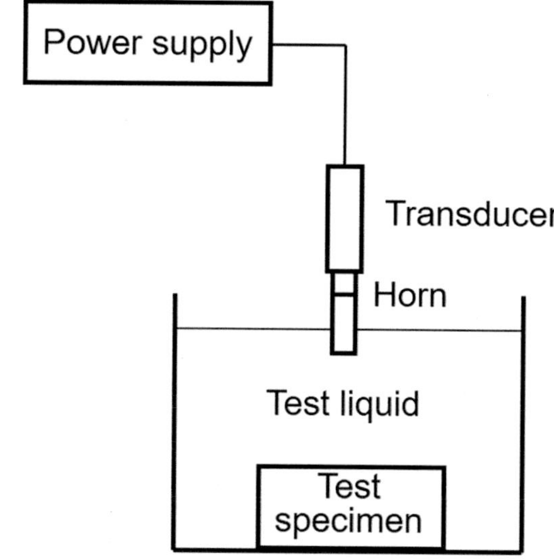

Fig. 17.51 Cavitation erosion test [Chattopadhyay R (2001)]. (Reprinted with kind permission from ASM International)

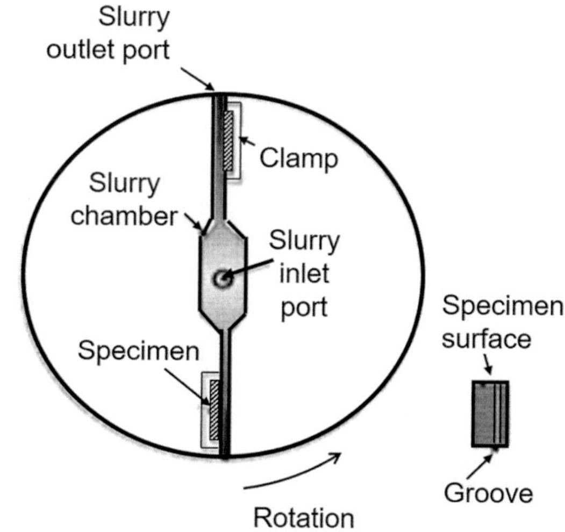

Fig. 17.52 Schematic of the Coriolis erosion tester [Clark HMI, et al. (1999)]. (Reprinted with kind permission from Elsevier)

angular alumina particles exiting a tungsten–carbide nozzle (50 mm in length and 1.5 mm in internal diameter) with a mean velocity of 30 m/s. Crushed silica with grain size below 0.1 mm is also used for erosion tests. As for grit blasting the nozzle wears and must be replaced as soon as its internal diameter has increased by 10%. Blasting time is generally 10 min and the nozzle axis must be kept orthogonally to coating. When dividing erosion rate (mg/min) by the abrasive flow rate (g/min) one obtains the average erosion rate after dividing the result by the specimen mass density. The averaged erosion rate is thus expressed in mm^3/g.

To characterize the impact erosion in severe but cold conditions, Kulu P, Pihl T (2002) have developed an impact erosion wear tester (Fig. 17.50), based on centrifugal force. It produces a stream of abrading-quartz sand with abrasive particle size of 0.1–0.3 mm and abrasive hardness 1100–1200 HV. The velocity of abrasive particles is 80 m/s. It can also use abrasives of different hardness (from 120 to 2000 HV) such as limestone, glass, iron oxide, quartz, and corundum. The measurements of weight loss allow for the calculations for weight and volume losses as a measure for wear rate; that is, the loss of mass or volume per 1 kg of abrading material in mg/kg and mm^3/kg, respectively. The impact angle, α, can be varied easily, see Fig. 17.50.

In liquid cavitation erosion, tensile stresses are generated during liquid impingement by the collapse of bubbles on a solid surface. It results in solid deformation because of the interaction of stress waves generated by the collapsing bubbles. Two ASTM tests are referenced in [17.W11, 17.W12]. According to Chattopadhyay R (2001), a simple testing method of cavitation is that developed by Shal'nev and represented in Fig. 17.51. The material to be tested is exposed to sound waves, produced by a magneto-striction vibrator with a nickel pipe 20 mm in internal diameter, oscillation frequency of 8000 Hz, and amplitude of 0.066 mm.

Liquid erosion with slurry, generally an alumina slurry, combines high abrasive conditions with cavitation. For example, the test rig [Clark HMI, et al. (1999)] showed in Fig. 17.52 uses a steel wheel, which rotates against a flat-coated specimen in slurry containing sharp alumina particles, subjecting the samples to combined impact erosion. The extent of erosion depends on the composition, size, and shape of the eroding particles, their velocity and angle of impact, and the composition and microstructure of the surface being eroded. The method utilizes the combination of centrifugal and Coriolis accelerations in a revolving rotor to pass slurry rapidly across a test surface such that the solid particles are forced against the surface, producing wear during their passage. The device allows using flat plate specimens and rotation speeds up to 7000 rpm, resulting in very short wear test times of 1 or 2 min. For example, a wear scar formed in a ductile material during a Coriolis erosion test

shows a progressive increase in depth with distance from the rotation center of the device.

Different examples of erosive wear for coatings sprayed with various processes are given in the following references [Ji G-C, et al. (2007), Dallaire S (2001), Shivamurthy RC, et al. (2010), Hejwowski T, et al. (2000), Hejwowski T (2009) Mann BS, Arya V (2001) Prasad Sahu S, et al. (2010)].

17.9.4   Surface Fatigue

Surface fatigue wear not only results in material loss from the surface, but it can also reduce the working life of the engineering component. It is the process leading to progressive localized permanent structural change in a material subjected to fluctuating stresses and strains. The cumulative effects of progressive changes in service conditions may result in cracks and/or complete fracture after a certain number of cycles. For coatings, for example, it happens in contact with a rolling motion. Fatigue is discussed according to two categories based on the number of cycles to failure. In high-cycle fatigue situations, materials performance is commonly characterized by an S–N curve, also known as a Wöhler curve. In this test constant cyclic stress amplitude S is applied to a specimen and the number of loading cycles N until the specimen fails is determined. It results in a graph of the magnitude of a cyclic stress (S) against the logarithmic scale of cycles to failure (N), see Fig. 17.53. The coupon with its coating manifests variation in its number of cycles to failure, and the S–N curve should more properly be an S–N–P curve capturing the probability of failure after a given number of cycles of a certain stress. Probability distributions that

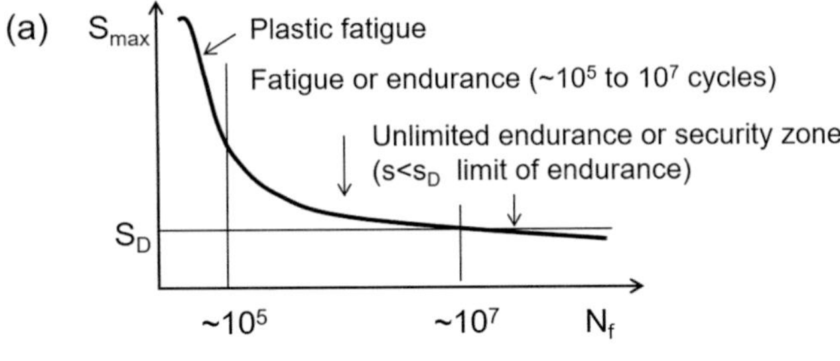
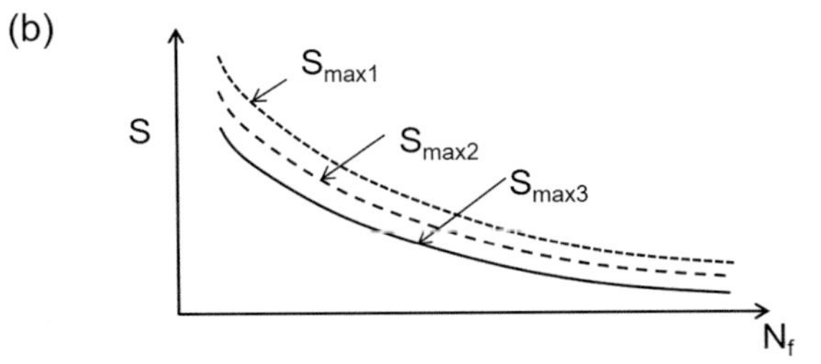

Fig. 17.53 S–N diagram plots: nominal stress amplitude S versus cycles to failure N: (**a**) $S - N$ curve for fatigue and (**b**) evolution of the S–N curve for different $S_{\max}$ [Chattopadhyay R (2001)]. (Reprinted with kind permission from ASM International)

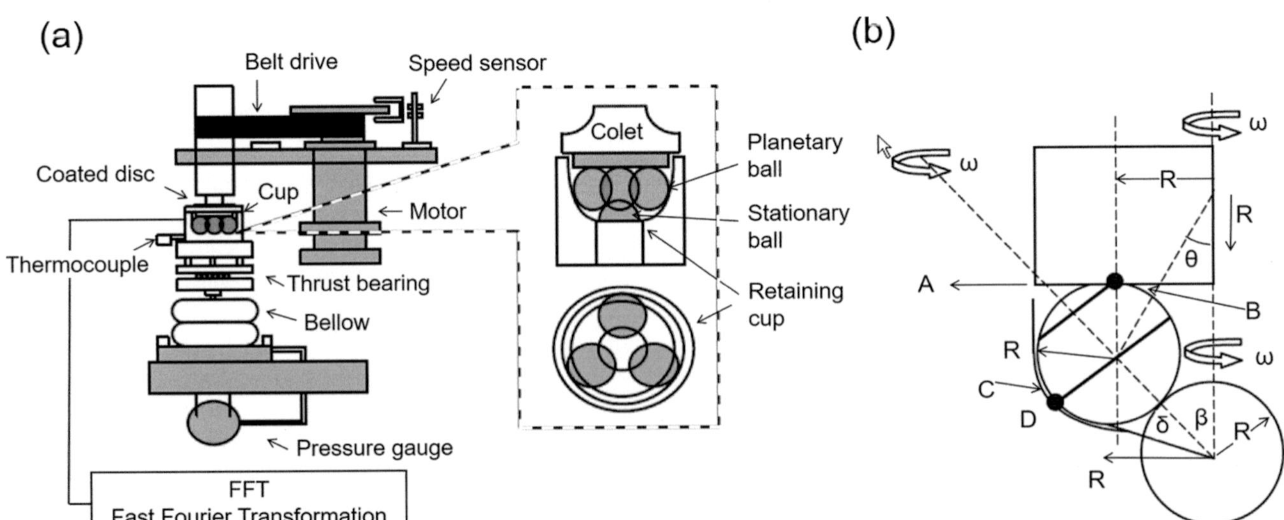

Fig. 17.54 (**a**) Schematic of the modified four-ball machine (**b**) schematics of ball and disk kinematics in cup assembly: $R_i = 7.75$ mm, $R_p = 6.35$ mm, $\theta = 36.83°$, $\beta = 37.61°$, $\delta = 31.16°$, $\omega = 4000 \pm 10$ rpm [Ahmed R, Hadfield M (2002)]. (Reprinted with kind permission from Elsevier)

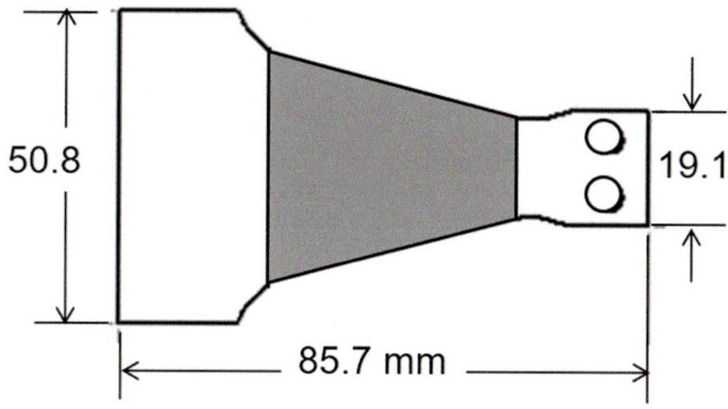

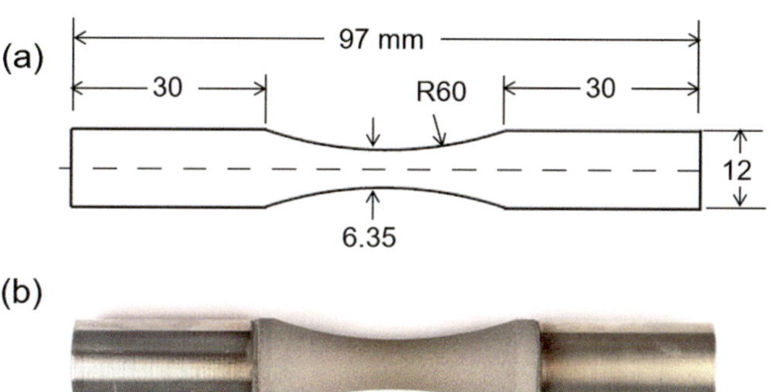

Fig. 17.55 Bending fatigue strength specimen. The gray region corresponds to the coated area. Dimensions are in mm [McGrann RTR, et al. (1998)]. (Reprinted with kind permission from Springer Science Business Media, copyright © ASM International)

Fig. 17.56 Axial fatigue specimens. All the dimensions are in mm. (**a**) Drawing and (**b**) as WC–CoCr sprayed [Agüero A, et al. (2011)]. (Reprinted with kind permission from Springer Science Business Media, copyright © ASM International)

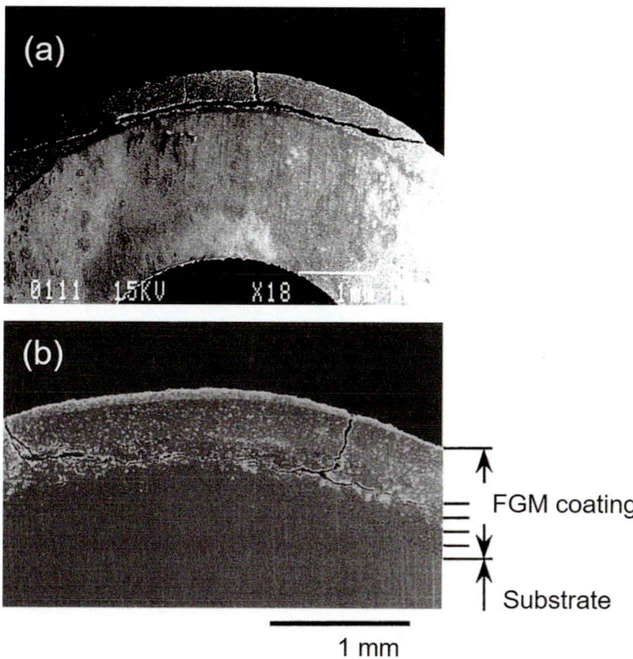

Fig. 17.57 (**a**) SEM micrographs showing interface delamination cracking for non-FGM coating subjected to six thermal fatigue cycles and (**b**) corresponding micrographs showing interface delamination cracking for FGM coating subjected to 10 thermal fatigue cycles, where laser power was 34 W. [Zhou YC, Hashida T (2002)]. (Reprinted with kind permission from Elsevier)

are common in data analysis and in design against fatigue include lognormal distribution, Weibull distribution and others. High-cycle fatigue occurs under low stress S_{max} and fracture occurs at $N = 10{,}000$ cycles or greater ($N \geq 10^4$).

Low-cycle fatigue is represented by the E–N curves ($E = $ strain $= E_p + E_e$, E_p being the plastic strain and E_e the elastic one) and failure occurs below 10,000 cycles ($N < 10^4$). To improve the fatigue resistance, compressive residual stresses can be introduced in the surface by, for example, shot peening to increase fatigue life. On the contrary, tensile stress decreases the fatigue resistance. Of course, fatigue life, as well as the behavior during cyclic loading, varies widely with the material or coating considered. Extreme high or low temperatures can decrease fatigue strength as well as environmental conditions resulting in erosion, corrosion, or gas-phase embrittlement.

Various means are used to characterize coating's fatigue:

Cyclic fatigue and fracture mechanisms [17.W13], thermomechanical fatigue testing [17.W14], cyclic fatigue, and fracture mechanisms [17.W15–17.W17].

For highly loaded machine elements such as roller bearings and gears, subjected to a combined rolling/sliding motion, extensive tests have to be performed [Stewart S, et al. (2005)]. Stewart S, et al. (2005) have proposed a modified version of the standard four-ball machine, where steel discs 31 mm in diameter were fabricated, coated, and then HIPed. A number of modifications to the four-ball machine were thus necessary in order to maintain the correct rolling/sliding kinematics. Figure 17.54a, b presents the details of the setup. Test results revealed that performance of the coating was dependent on the microstructural changes due to post-treatment. Ahmed R, Hadfield M (2002) have used a similar setup to study different coatings, but in this case the substrate material was either 440-C, M-50 bearing steel, or mild steel in the shape of rolling element ball or cone.

Zhang XC, et al. (2009a) have used the rolling contact testing machine developed by Yanshan University in China to the Rolling Contact Fatigue (RCF) resistance and failure mechanisms of plasma-sprayed CrC–NiCr cermet coatings.

Much less sophisticated tests are those based on Young's modulus measurement at increasing deformation of up 0.3% [Kovarik O, et al. (2005) (2008)]. For example, Kovarik O, et al. (2005) have loaded the fatigue samples in reversible bend (as a cantilever beam) at room temperature in a special computer-controlled electromagnetic testing device. An alternating electromagnetic field induced by a pair of coils was used to deflect the free end of the specimen, which was equipped by a yoke. Other authors [McGrann RTR, et al. (1998), Sansoucy E, et al.

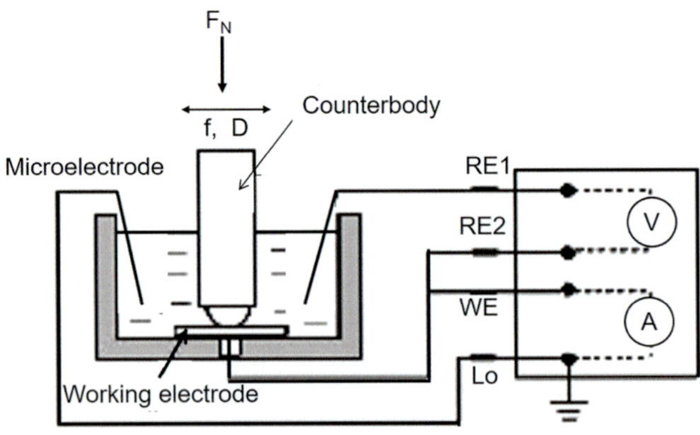

Fig. 17.58 Experimental setup combining a bidirectional sliding test immersed in an electrolyte and electrochemical noise measurements [Basak AK, et al. (2006)]. (Reprinted with kind permission from Elsevier)

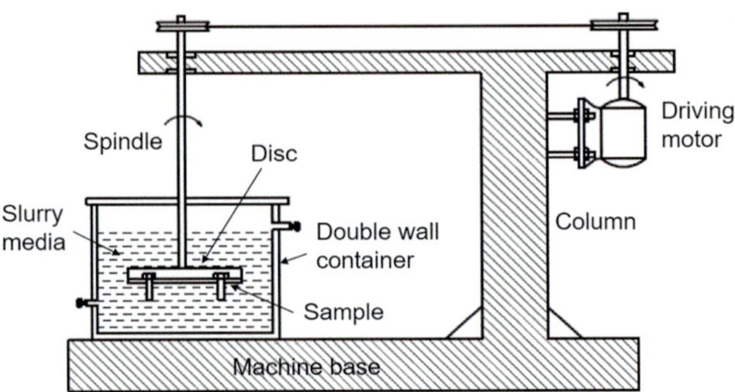

Fig. 17.59 A schematic representation of the slurry wear tester [Prasad BK (2000)]. (Reprinted with kind permission from Elsevier)

(2007)] used specimens supported in the same manner as a cantilevered beam at one end and that were subjected to an alternating force at the other. The fatigue test specimen shown in Fig. 17.55 includes a triangular shape intended to produce a constant stress along the length of the test section.

Other authors [Ibrahim A, et al. (2007), Ibrahim A, Berndt CC (2007)] have used rotating-beam fatigue testing machines, where test specimens are coated bars (12.7 mm in diameter). The fatigue experiments were conducted at room temperature under a rotating beam and stress ratio of $R = -1$ configuration at a load frequency of 50 Hz. The maximum load P_{max} generating a tensile stress, considered positive, is equal to the minimum load P_{min} generating a compressive stress, considered negative. The stress ratio corresponds to the ratio of the minimum stress to the maximum one during one cycle of loading in a fatigue test.

Fatigue resistance has also been tested with a torsional setup [Yan L, et al. (2003)].

For the axial fatigue strength test, a sinusoidal load of 20 Hz with load ratio of $R = -1$, at room temperature ($23 \pm 2\ °C$), and $35 \pm 3\%$ humidity was applied by Agüero A, et al. (2011). Experimental tests considered as fatigue strength the specimen fracture or 5×10^6 load cycles. The test was performed on two MTS computer-controlled servo-hydraulic axial fatigue-testing machines, according to the ASTM E466 standard. Figure 17.56 presents the test sample dimensions and a picture of the coated fatigue specimen.

At last thermal fatigue resistance, generally of TBCs, is tested with furnaces and/or laser [Scrivani A, et al. (2007), Giolli C, et al. (2009), Zhu D, et al. (2004), Zhou YC, Hashida T (2002)]. For example, Scrivani A, et al. (2007) have tested thick (1.8 mm) plasma-sprayed TBCs by introducing them in a furnace, heating them up to 1150 °C in 5 min, letting them stabilize at 1150 °C during 45 min, followed by a 10-min forced

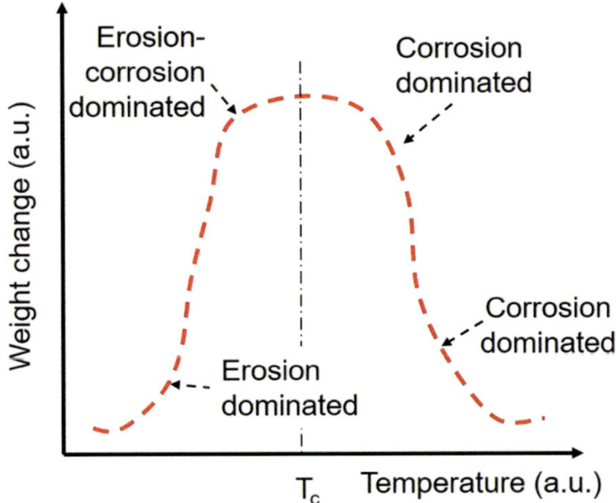

Fig. 17.60 Scheme of corrosion–erosion wear at high temperature [Chattopadhyay R (2001)]. (Reprinted with kind permission from ASM International)

air cooling. Failures occurred after 305–455 cycles, depending on the spray conditions. When looking at the limits of the TBC the thermal fatigue resistance increases with amount of porosity in the topcoat. Following the thermal cycling tests, the compressive in-plane stress increases in the TBC systems and it is less a function of the porosity level of topcoat. [Robin P, et al. (2010)] also observed the modification of the stress distribution. In their setup coatings were heated by radiating lamps and acoustic emission allowed following cracks formation during the heating and cooling stages. The bond coat or the Functional Gradient Material (FGM) play an important role in the coating resistance to thermal fatigue [Zhou YC, Hashida T (2002)]. SEM micrographs given in Fig. 17.57a, show interface delamination along the boundary between PSZ/NiCrAlY (75/25 wt. %) and PSZ/NiCrAlY (50/50 wt.%) for "non-FGM" coating subjected to six thermal fatigue cycles, where the exposed time for every cycle was 70 s and the highest temperature on coating and substrate was 1200 and 600 °C, respectively. The Corresponding micrographs given in Fig. 17.57b obtained

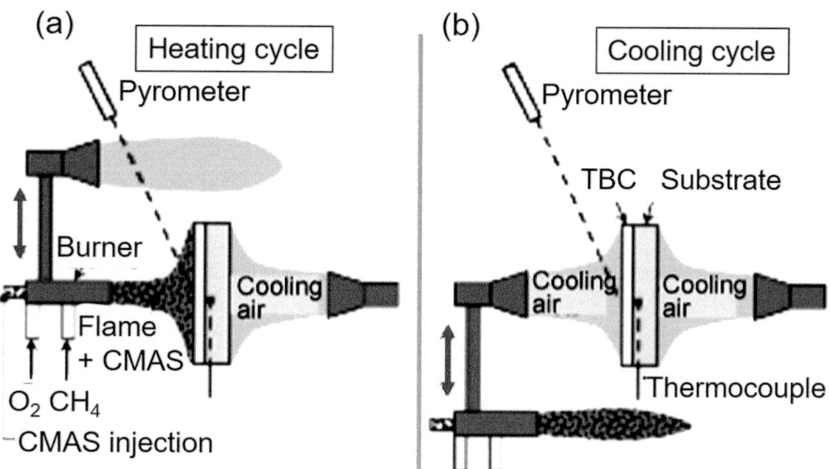

Fig. 17.61 Schematic diagram of thermal-gradient cyclic testing of TBCs with CMAS injection during (**a**) heating and (**b**) cooling cycles [Drexler JM, et al. (2010)]. (Reprinted with kind permission from Elsevier)

with FGM coating subjected to 10 thermal fatigue cycles tests conducted under the simulated advanced gas turbine blade thermal cycling conditions, show higher thermal fatigue resistance compared to non-FGM coating system.

17.9.5 Corrosive Wears

Corrosive wears occur when the effects of corrosion and wear are combined, resulting in a more rapid degradation of the material's surface [17.W18, 17.W19]. A surface that is corroded or oxidized may be mechanically weakened and more likely to wear at an increased rate. Furthermore, corrosion products including oxide particles that are dislodged from the material's surface can subsequently act as abrasive particles. Stress corrosion failure results from the combined effects of stress and corrosion. At high temperatures reactions with oxygen, carbon, nitrogen, sulfur, or flux (chemical cleaning agent) result in the formation of oxidized, carburized, nitride, sulfurized, or slag layer on the surface. Temperature and time are the key factors controlling the rate and severity of high temperature corrosive attack [Chattopadhyay R (2001)].

Three types of tests are mainly used:

Sliding and corrosive test is quite similar to the pin on disk test presented in Fig. 17.47, but a container allows performing the test with lubricant or corrosive solution [Zhang T, Li DY (2001)] or both [Liu R, Li DY (1999)]. For example, 10% H_2SO_4 added to the lubricant with the volume ratio of H_2SO_4 to oil approximately equal to 1:4. It is also possible to perform simultaneously fretting wear test with an electrochemical noise (EN) measurement system, the whole system being in an electrochemical cell, as schematically shown in Fig. 17.58 [Basak AK, et al. (2006)]. Two personal computers control the systems and acquire potential, current, and friction force data. The flat samples are covered with an electrical insulating cover prior

to fretting tests leaving an area of 1 cm^2 exposed to the electrolyte. Corundum balls (Ceratec, The Netherlands) with a diameter of 10 mm, a hardness of 2000 HVN, and a surface roughness $Ra < 0.02$ μm, are used as counter body. Corundum was selected for its high wear resistance, high chemical inertness, and high electrical resistance. Fretting–corrosion experiments are performed at room temperature with a reciprocating ball-on-flat contact configuration. The flat sample is immersed in the Hank's solution for 90 min before starting up experiments in order to reach a stable potential and a low background current.

Pre-weighed samples 15 mm in diameter are fixed on a disk attached with an electric motor. The disk is isolated from the samples by painting appropriately and putting a nonconducting adhesive in between. A schematic representation of the slurry-wear test apparatus is shown in Fig. 17.59 from Prasad BK (2000). Samples are rotated at ambient temperature at a given velocity and the travel distance varies in the range of 15–500 km. Changing the content of the solid sand particles (size 212–300 μm) from 0 to 60 wt.% in the liquid electrolyte modifies the slurry composition. The electrolyte comprises 4 g sodium chloride plus 5 cc concentrated sulphuric acid dissolved in 10 L of water and has a pH of 1.98 highly acidic in nature. Addition of 20, 40, and 60 wt.% sand to the electrolyte changed the pH of the slurry mixture of liquid plus sand to 3.64, 6.85, and 7.12, respectively. Wear rate are computed by weight loss technique. Zhang D-W, Lei TC (2003) conducted a similar experiment.

In high-temperature erosion–corrosion heated fluids containing solid particles lead to material removal from the surface by a specific wear process [Stack MM, et al. (1993)]. The evolution of the weight change of the sample is represented in Fig. 17.60 [Chattopadhyay R (2001)].

Four different behaviors can be observed [Chattopadhyay R (2001), Higuera Hidalgo V, et al. (2001)]:

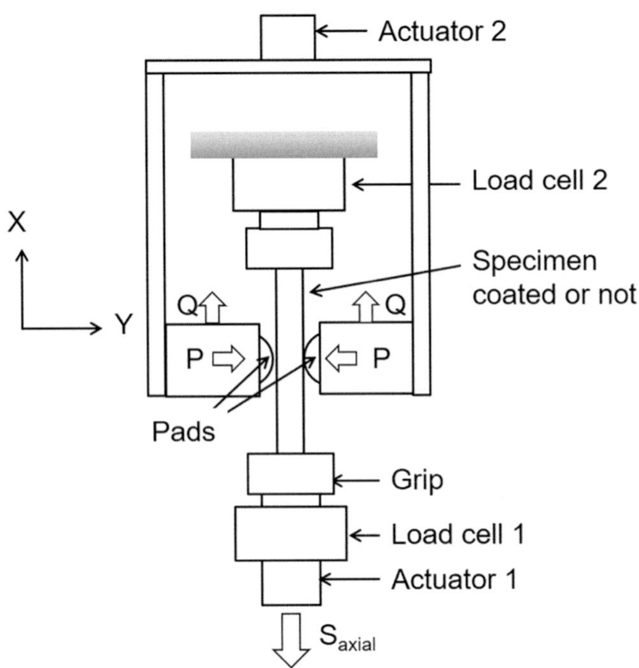

Fig. 17.62 Fretting test apparatus with a servo-hydraulic uniaxial test frame and an additional servo-hydraulic actuator [Lee H, et al. (2005)]. (Reprinted with kind permission from Springer Science Business Media)

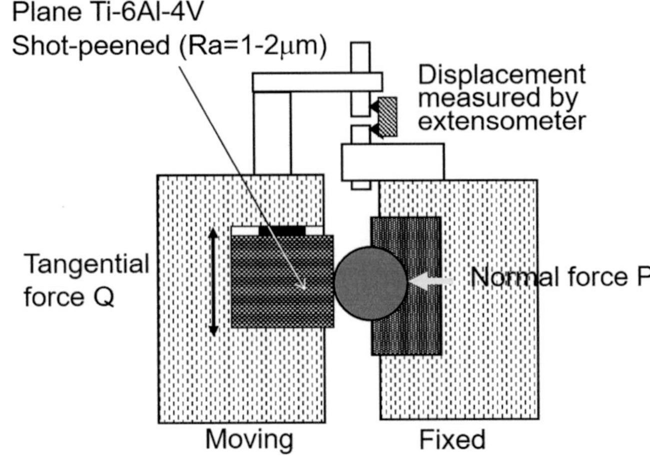

Fig. 17.63 Schematic drawing of fretting-wear rig with a tension-compression hydraulic machine [Fridrici V, et al. (2003)]. (Reprinted with kind permission from Elsevier)

- At low temperature, the material loss is essentially due to erosion.
- Beyond a certain temperature the rate of oxidation increases drastically with temperature. For a temperature in the mid-range, both erosion and oxide scale formation rates remain similar and the wear is erosion–corrosion dominated.
- With the further temperature increase the loss of oxide dominates and over a certain critical temperature, chipping of brittle scale is the main mechanism.
- Above the critical temperature, T_c, the overall weight change from corrosion process tends to zero.

For high temperature tests, for example to reproduce the erosion of fly ash in coal-fired boiler, fly ashes are introduced in a 210 kW laboratory combustion unit using methane as fuel (this type of combustion reproduces well that of coal). Ashes are continuously injected into the combustion chamber by means of a helicoidally feeder [Higuera Hidalgo V, et al. (2001)]. Garcia JR, et al. V (2007) have conducted a similar experiment to study the erosive–corrosive wear of Ni–Cr coatings in boilers atmosphere the high temperatures being provided by a burner using methane as fuel. During erosion wear tests, ashes retrieved from a 900 MW power plant were injected continuously into the combustion chamber by means of a helicoidally feeder, which pumped the ash into a mixing chamber where the ash was fluidized with air before being incorporated into the chamber. This device allows reproducing, at the laboratory scale, the same operating conditions, which exist in the areas of post-combustion gases in industrial coal-fired power plants.

Degradation of thermal barrier coatings (TBCs) in gas-turbine engines by molten calcium–magnesium–aluminosilicate (CMAS) glassy deposits is becoming a pressing issue, as engines are required to operate under increasingly harsh conditions. Drexler JM, et al. (2010) developed a new thermal-cycling test for the evaluation of TBC performance, where a thermal gradient was applied across the TBC, with simultaneous injection of CMAS. The conditions simulated in this new test are closer to actual conditions in engine, as compared to the conventional furnace test without thermal gradient. In a typical test, an equilibrated natural gas/oxygen flame heats the front of the TBC, while compressed air cools the backside (Fig. 17.61a). The front temperature (T_{sur}) is monitored using a long-wavelength (9.6–11.5 μm) pyrometer, and the substrate temperature (TTC) is monitored by a thermocouple (1 mm diameter) placed inside a hole (1.1 mm diameter, 15 mm long) drilled radially into the substrate. The bond coat temperature (T_{sub}) at the ceramic/metal interface is estimated from T_{sur} and TTC. The CMAS suspension is sprayed axially through the gas burner. The heating cycle lasts 5 min. During the cooling cycle (2 min) the cooling air nozzle is brought in front of the TBC (Fig. 17.61b), and the process is repeated.

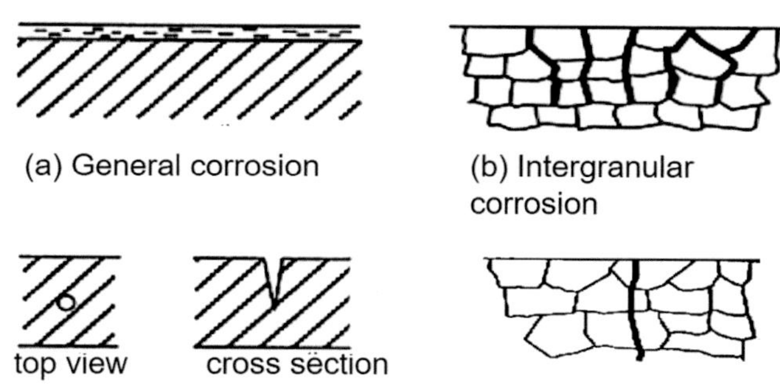

Fig. 17.64 Different types of corrosion: (**a**) general corrosion, (**b**) intergranular corrosion, (**c**) pitting, and (**d**) transgranular corrosion

17.9.6 Fretting

Fretting wear occurs when surfaces, in intimate contact with each other, are subject to a small amplitude relative motion that is cyclic in nature, such as vibration. Fretting wear is normally accompanied by the corrosion or oxidization of the debris and worn surface. If the debris become embedded in the surface of the softer metal, the wear rate may be reduced, but if the debris remain free at the interface between the two materials, the wear rate may be increased. Fatigue cracks also have a tendency to form in the region of wear. ASTM standards are presented in [17.W20–17.W22].

Figure 17.62 shows schematic of a dual actuator fretting test apparatus used by Lee H, et al. (2005) (2007). It allows conducting fretting tests with any prescribed value of relative displacement at an applied cyclic stress to the specimen. The setup comprises a servo-hydraulic uniaxial test frame and an additional servo-hydraulic actuator (actuator 2 in Fig. 17.62), which is directly connected to the fretting fixture so that independent cyclic movement of the fretting fixture is possible under a given contact load. During fretting the specimen is fatigued at a given stress level thanks to actuator 1. The tangential force, Q, depends on the difference between two load cells located at the bottom and top of the specimen. Fretting fixture can be independently controlled in either load or displacement-controlled mode through the actuator 2. The contact load, P, is applied through lateral springs and measured by a pressure gauge.

Another device used by Fridrici V, et al. (2003) is presented in Fig. 17.63. A tension-compression hydraulic machine is used to impose the displacement between the plane and the cylinder. For preselected cycle numbers during a fretting test, the displacement δ, the normal force P, and the tangential force Q are recorded. This enables to plot the fretting loop Q–δ for the preselected cycles.

Other fretting tests have been developed. They involve dovetail geometry [Gean MC, Farris TN (2009)], the temperature of which can be monitored; a tension-compression hydraulic machine [Xu G-Z, et al. (2002)] where the coated flat specimen is stationary and the 52,100 steel ball is vibrated with small reciprocating amplitude; flat-on-flat wear tester designed to simulate the vibrating condition and positioned inside a furnace to obtain an operating condition of 500 °C [Koiprasert H, et al. (2004)]; and a device in which the fluctuating loading is supplied by a variable crank system [Majzoobi GH, et al. (2010)].

17.10 Corrosion Resistance

17.10.1 General Remarks

In corrosion process materials loss occurs through electrochemical or chemical reaction with the surrounding medium. At high temperatures corrosion reactions are oxidation, carburization, nitriding, halogen erosion, sulfidation, and molten-salt corrosion [Chattopadhyay R (2001)], coatings being mainly concerned by the former and the latter.

The dissolution of metallic elements by the formation of ions by electron loss corresponds to anodic reactions such as

$$M(\text{metal}) \rightarrow M^{n+} + ne^{-} \tag{17.44}$$

where M is a metal, M^{n+} a positively charged n times ion, and e^{-} an electron. This electron loss allows the metal ion to bond to other groups of atoms that are negatively charged, for example, the reaction of water or oxygen with electrons from the metal surface, known as cathodic reaction. Let us consider steel rusting where water (H_2O) and oxygen (O_2) are involved:

$$Fe \rightarrow Fe^{2+} + 2e^{-} \tag{17.45}$$

The free electrons produced react with water and oxygen:

$$O_2 + 2H_2O + 4e^{-} \rightarrow 4OH^{-} \tag{17.46}$$

Both reactions (17.44 and 17.45) can then be written as the global reaction:

$$2Fe + O_2 + 2H_2O \rightarrow 2Fe(OH)_2 \tag{17.47}$$

As O_2 dissolves rapidly in water and because there is generally an excess of it, O_2 reacts with the iron hydroxide to form the hydrated iron oxide, $2Fe_2O_3 \cdot H_2O$, often called brown rust:

$$4Fe(OH)_2 + O_2 \rightarrow 2H_2O + 2Fe_2O_3 \cdot H_2O \tag{17.48}$$

It is clear from these equations that the corrosion rate is linked to electrons production, corresponding to a corrosion current flow. However, the presence of an impervious oxide

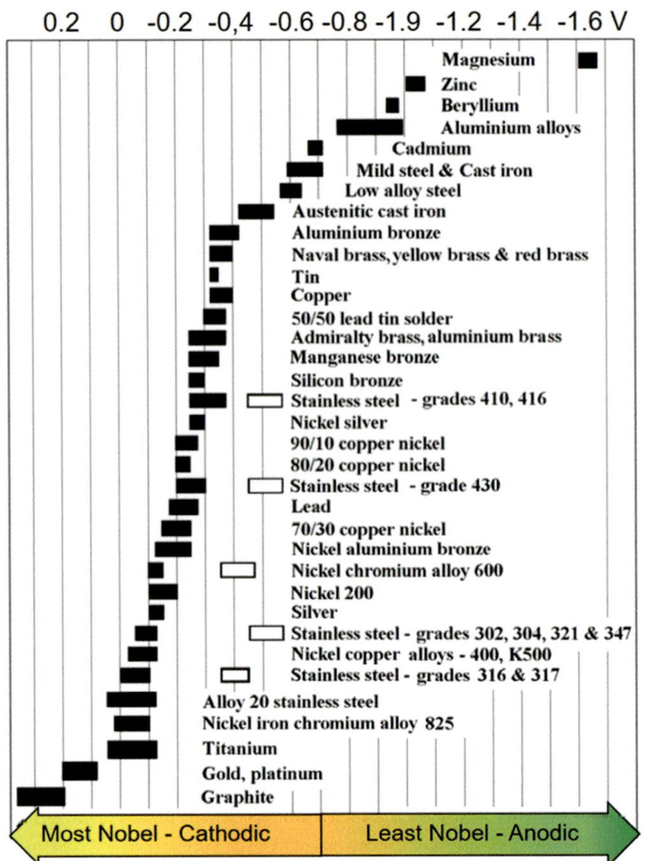

Fig. 17.65 Galvanic corrosion chart: corrosion potentials in flowing sea water at ambient temperature. The more Noble materials at the left side tend to be cathodic and hence protected; those at the right are less noble and tend to be anodic and hence corroded in a galvanic couple [Atlas Steel Technical Note No. 7]. (Reprinted with kind permission from Elsevier)

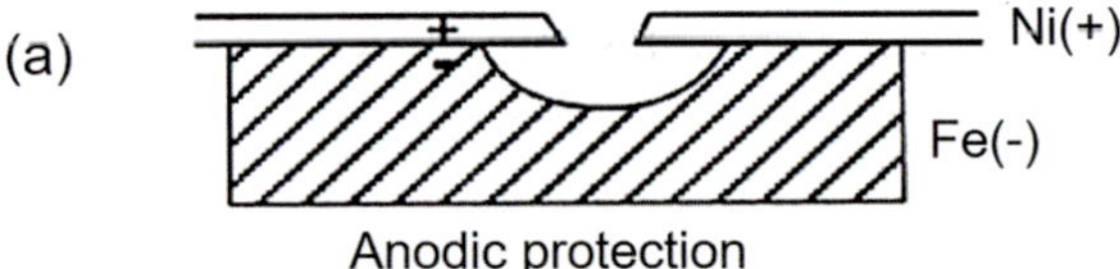

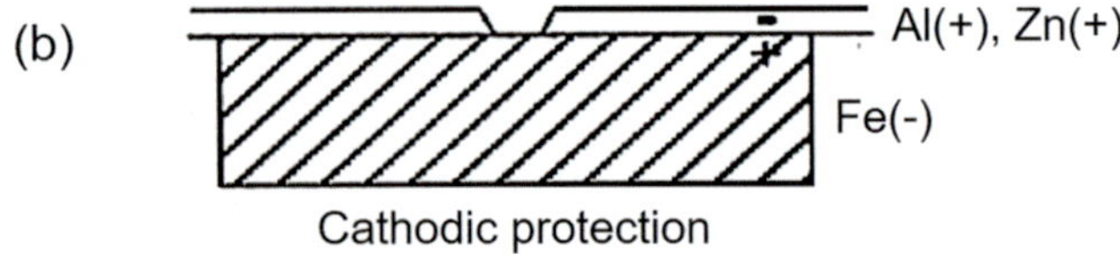

Fig. 17.66 Examples of protective coatings: (**a**) anodic (no discontinuity allowed in the coating) and (**b**) cathodic (discontinuity allowed in the coating, resulting in almost no corrosion of iron)

layer can inhibit the corrosion and the metal is said passivated.

Corrosion rate equation [Chattopadhyay R (2001)] is expressed by

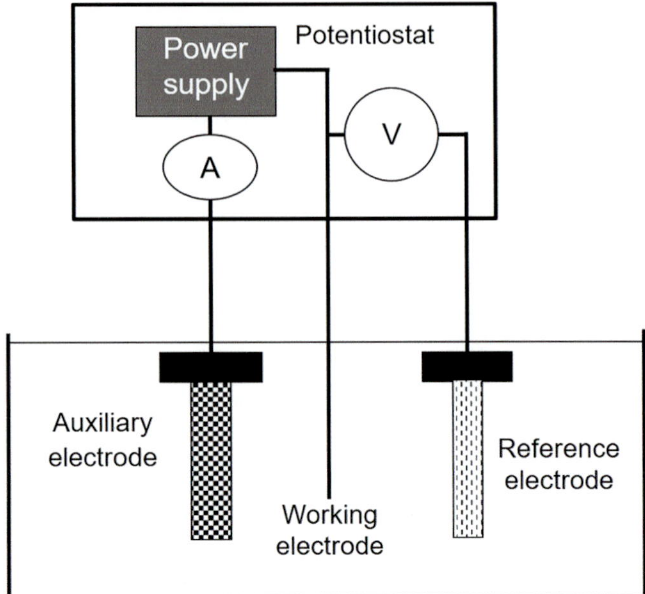

Fig. 17.67 Three-electrode cell used in the electrochemical experiments [Souza VAD, Neville A (2006)]. (Reprinted with kind permission from Springer Science Business Media, copyright © ASM International)

$$C_T = C_0 \left(\Delta G^* / RT \right) \tag{17.49}$$

where C_T is the corrosion rate at temperature $T(\mathrm{K})$, generally expressed in mm/year or mpy, C_0 is the rate at $0\,(\mathrm{K})$, R is the ideal or universal gas constant, and ΔG^* is the activation energy of the corrosion reaction. This equation is transformed in logarithmic form, the energy terms are considered as potentials and rates as currents, resulting in a relationship between corrosion current, I_c, and measured potential, E, of the specimen:

$$E - E_c = \beta \frac{\log I}{\log I_c} \tag{17.50}$$

where E measured potential of the specimen when current flows, E_c corrosion potential (no current flowing), I impressed current, I_c corrosion current (no external current), and β a constant. This equation is used to characterize corrosion by electrochemical measurements, as described later.

The different types of corrosive attack, especially for coatings, are the following:

General corrosion, corresponding to about 30% of failure, where the average rate of corrosion on the surface is uniform as illustrated in Fig. 17.64a.

Localized corrosion, about 70% of failures, comprise of the following:

Galvanic corrosion occurring when two dissimilar metals are in contact with each other in a conductive solution (electrolyte), the more anodic metal being corroded, while the more cathodic one is unaffected. The anodic metal corrodes

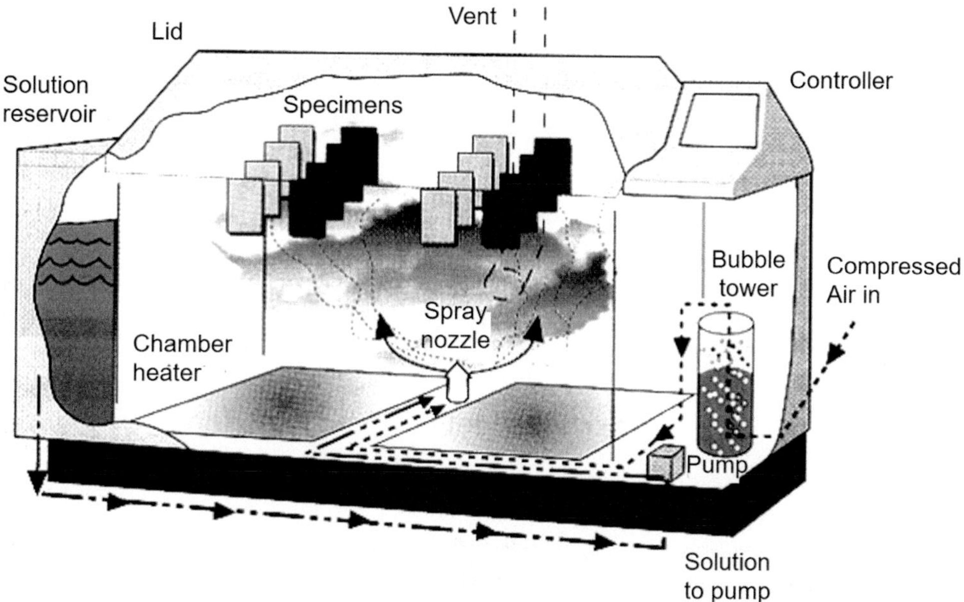

Fig. 17.68 Typical climatic enclosure [Montavon G (2004)]. (Reprinted with kind permission from Prof. G. Montavon)

faster because ions from the metal migrate from anode to cathode. There are three conditions that must exist for galvanic corrosion to occur. First there must be two electrochemically dissimilar metals present. Second, there must be an electrically conductive path between the two metals. And third, there must be a conductive path for the metal ions to move from the more anodic metal to the more cathodic metal. If any one of these three conditions do not exist, galvanic corrosion will not occur. The galvanic or electrochemical series ranks metals according to their potential, generally measured with respect to the Standard Calomel Electrode (SCE). Metals can be arranged in a galvanic series representing the potential they develop in a given electrolyte against a standard reference electrode. The relative position of two metals on such a series gives a good indication of which metal is more likely to corrode more quickly. However, other factors such as water aeration and flow rate can influence the process markedly. The results are often viewed as a galvanic corrosion chart or galvanic corrosion (Fig. 17.65).

This chart says that the "anodic" or "less noble" metals at the negative end of the series, such as magnesium, zinc, and aluminum, are more likely to be attacked than those at the "cathodic" or "noble" end of the series such as gold and graphite. For example, zinc has a negative potential higher than that of iron and it will behave as a cathode relatively to it, while nickel which negative potential is lower than that of iron will behave as an anode relatively to it. The electrolyte, ionic-conducting fluid bridging both metals, plays a key role, as well as the relative surface contact area, the smaller the anodic to cathodic area ratio is the more severe is the anodic metal corrosion. This is very important, for example, to protect low carbon iron from atmospheric corrosion by a coating either anodic (nickel coating) or cathodic (aluminum or zinc). In the first case no discontinuity in the coating can be tolerated, while it has relatively a low importance with cathodic coating, as illustrated in Fig. 17.66. By imposing a voltage between both metals, the current direction can be changed.

Intergranular corrosion, illustrated in Fig. 17.64b, occurring when a chemical element is depleted during the coating or bulk material manufacturing, for example during heat treatment.

Pitting is a localized corrosion characterized by depression or pit formation on the surface as illustrated in Fig. 17.64c. It occurs, for example, when stainless steel is corroded by chloride-containing solutions. The pit cavities act as stress raisers.

Trans-granular corrosion is mainly due to high static tensile stress in the presence of a corrosive environment. It can be intergranular but also trans-granular when cracking occurs, as illustrated in Fig. 17.64d. Of course, the coating material and its microstructure play an important role.

17.10.2 Corrosion Characterization

For terminology and acronyms related to corrosions, see [17. C1].

17.10.2.1 Electrochemical Measurements
ASM 13A book [ASM International (2003)] and the following ASTM references [17.C2–17.C13] give a detailed description of the electrochemical techniques.

In direct current (DC) method potentiodynamic polarization is used. The coating is insulated by epoxy cold resin mounting, except a surface of about 1 cm^2, which is exposed to a liquid electrolyte and the system is allowed to reach a near-electrochemical equilibrium. The equilibrium corroding potential, E_c, is measured relatively to a reference electrode because the absolute potential of the coating sample relatively to the electrolyte is not accessible. The value of either the anodic or cathodic current at E_c, called the corrosion current, I_c, would allow calculating the corrosion rate of the metal, but unfortunately it cannot be measured directly. Thus, additional electrodes are immersed in the solution, and all of them are connected to a device called a potentiostat that allows changing the potential of the metal sample in a controlled manner and measure the current as a function of the potential. Figure 17.67 from [Souza VAD, Neville A (2006)] presents an oversimplified scheme of the three electrodes electrochemical experiment. When the potential of a metal sample in solution is forced away from E_c, it is referred to as polarizing the sample. The response (current) of the sample is measured as it is polarized. The computer controlled potentiostat shift the potential of the working electrode (the specimen) relative to the reference electrode (often saturated calomel electrode) at a predetermined rate (a few mV/min) to potentials more positive than the free corrosion potential (E_c). The current density in the external circuit between the Pt auxiliary electrode and the coating (working electrode) is measured as a function of applied potential until a given current density is reached and then the potential drops at the same rate as on the forward curve. The response is used to develop a model of the sample's corrosion behavior. Various authors achieved similar measurements for cermets or metals either thermal sprayed [Guilemany JM, et al. (2005), Shrestha S, et al. (2001) Ishikawa Y, et al. (2005) Gobinda CS, Khan TI (2010) Paul S, Yadav K (2010) Barletta M, et al. (2010) Luiz de Assis S, et al. (2006)] or cold sprayed [Wang H-R, et al. (2008)].

AC methods are based on the fact that in presence of an electrolyte, the corroding system is equivalent to an electrical circuit comprising a resistive/capacitive (RC) parallel circuit. It is called electrochemical impedance spectroscopy (EIS). The response of the circuit (computer-driven ac impedance device) to frequencies varying from very low frequencies (10^{-2} Hz) to very high ones (10^5 Hz) allows determining the different elements of the coating equivalent electrical circuit [Wang H-R, et al. (2008)].

17.10.2.2 Fog and Salt-Spray Test

The corrosion tester is a box; for details, see [17.C14], where a corrosive solution is atomized by humidified compressed air. Air humidification is achieved by passing it through a bubble tower. The box temperature can be adjusted thanks to heaters. In the box, also called climatic enclosure, temperature, airspeed, and relative humidity are controlled. The linear shrinkage, product mass loss or gain of treated samples, is then measured. The tests are performed either continuously or in a cyclic mode: for example, x hours with the solution followed by x hours without solution. Figure 17.68 represents a typical box.

For example, the corrosion resistance of nickel-base cored wires arc-sprayed coatings was tested by this method and by electrochemical corrosion test [He D, et al. (2007)].

17.10.2.3 Molten Salt

According to Sidhu TS, et al. (2006a, b) (2007), vanadium (V), sulfur (S), and sodium (Na) are common impurities in the low-grade petroleum fuels. Molten sulfate vanadate deposits resulting from the condensation of combustion products of such fuels are extremely corrosive to high-temperature materials in the combustion systems. Further, mixture of Na_2SO_4 and V_2O_5 in the ratio of 40:60 constitutes eutectics with a low melting point of 550 °C and provides a very aggressive environment for hot corrosion to occur. They have applied uniformly a layer of Na_2SO_4–60%V_2O_5 mixture on the warm specimens. Cyclic studies were performed in molten salt for 50 cycles. Each cycle consisted of 1-h heating at 900 °C in a silicon carbide tube furnace in open air atmosphere followed by 20-min cooling at room temperature. Singh H, et al. (2005) have also studied the corrosion of Ni-base superalloy by Na_2SO_4–60%V_2O_5 mixture in a furnace at 900 °C.

Guilemany JM, et al. (2008) have tested the corrosion resistance of Ni-based coatings applied on municipal solid-waste incinerators by using a blowpipe of hot air that expels air at 650 °C, and propels onto the coating drops of aqueous solution 0.05 M of a eutectic mixture KCl:ZnCl (52:48 wt.%).

Li L, et al. (2010) have studied the corrosion of TBCs by CMAS (the acronym of each oxide: CaO, MgO, Al_2O_3, and SiO_2). The CMAS powder was synthesized as tapes of 25.4 mm diameter and glued onto the top of the coating surfaces using cement 520. The infiltration process was accomplished either in a tube furnace or through a flame burner. The coating degradation simulated in this method is similar to reports from field studies of aircraft engines.

17.10.2.4 Oxidation

Matthews S, et al. (2010) have followed the oxidation of Cr_3C_2–NiCr coatings heat treated at 900 °C in still air during different numbers of days. Ye F-X, et al. (2008) performed a similar work on Cr39Ni7C cermet coatings deposited by Diamond jet spray process. For details about oxidation test, see [15.C15].

17.11 Summary and Conclusions

Sprayed coating characterization is very important. However, the needs of research laboratories, often working on relatively small samples, are not at all the same as those of industry working with real part geometry. In laboratories the interest is mostly in the characterization of coating phases, microstructure, and more recently nanostructure, unmolten particles, voids, and cracks distributions, coating–substrate interface or splat interfaces, which are parameters not directly linked to the properties desired by industry in a production scale environment. On the other hand, in the context of a production scale operation, the main interest is measurement of the part geometry after spraying and machining, coating thickness, adhesion/cohesion, hardness, stiffness, Young's modulus, thermal conductivity, and diffusivity.

In both cases, however, the measurement will require using statistical methods that must be carefully achieved to limit the errors to human ones. The dream in production would be that all measurements could be achieved by nondestructive methods. All these measurements impact the reproducibility and reliability of thermal-sprayed coating production. Thus, the measurement uncertainty must be controlled carefully, and statistics used properly while leaving room to detect and accommodate human factor still remains important in coating quality control [Wigren J, Johansson J (2011)].

Nomenclature

Units are indicated in parentheses; when no units are indicated, the parameter is dimensionless.

Latin Alphabet

a	Thermal diffusivity (m^2/s)
a_c	Half-diagonal crack length (µm)
a_k	Crack length (m)
A	Area (m^2)
A_w	Area of contact (m^2)
B	Thickness (m)
c_f	Correction factor related to the shape of the indenter
c_p	Specific heat at constant pressure (J/K kg)
C	Compliance: ratio of the displacement to the load (m/N)
C_T	Corrosion rate at temperature T (mm/year or mpy)
d	Equivalent or inside diameter (m)
d_0	"Stress-free" lattice parameter (nm)
d_{hkl}	Distance between atomic layers with the same Miller indices (h, k, l)
d_w	Sliding distance (m)
D	Outside diameter of the bar (m)
e	Sample thickness (m)
E	Energy (J) or (eV)
E'	Young's modulus (GPa)
E'_c	Young's modulus of the coating (GPa)
E'_s	Young's modulus of the substrate (GPa)
E_c	Corrosion potential (mV)

F	Force applied (N)
$f(\alpha)$	A function of the impingement angle α (−)
F_N	Applied normal force (N)
G_c	Critical value of the strain energy release rate (J/m^2)
h	Penetration depth of an indenter (µm)
H	Hardness or yield stress (MPa)
h_{max}	Penetration depth (µm)
h_p	Plastic component of penetration depth of an indenter (µm)
h_r	Residual depth of the residual deformation once the indenter is removed (µm)
HV_x	Hardness measured with a load of 'x' Newton
I	Impressed current (mA)
I_c	Corrosion current (no external current) (mA)
k	Wave vector of the incident neutron beam
K	Strain energy release rate (related to the fracture toughness) (N/m$^{3/2}$)
k_o	Wave vector of the scattered neutron beam
K_{adh}	Adhesion wear coefficient $K_{adh} = K_w/3$ (−)
K_c	Fracture toughness (N/m$^{3/2}$)
K_{ca}	Apparent interface fracture toughness (N/m$^{3/2}$)
K_w	Wear coefficient (−)
l	Length of the coating (m)
L	Length scale (m)
l_c	Palmqvist surface crack length (µm)
m	Shape parameter in Weibull distribution
p	Pressure (Pa)
P	Force required extending a crack (N)
P_c	Applied load (N) on an indenter
P_{max}	Maximum load on an indenter (N)
q	Scattering vector
q	Modulus of scattering vector
Q	Imposed gas flow (m^3/s)
S	Slope of the initial part of the unloading curve (−)
U	Elastic energy stored in the system (J)
V	Volume (m^3)
V_b	Coating bulk volume (m^3)
V_p	Open pores volume (m^3)
V_w	Wear volume (m^3)
w	Coating thickness (µm)
w_{dry}	The dried coating weight (kg)
w_{sat}	The wet coating weight (kg)
w_{wd}	Weight of water displaced (kg)
W	Total work (J)
W_e	Work done by external forces (J) or elastic work (J)
x_i	Measured value of indent, hardness, Young's module …

Greek Alphabet

α	Impingement angle (°)
φ	Angle with respect to the normal to the substrate surface (°)
ϕ	Phase angle
γ	Surface tension (J/m^2 or N/m)
η	Arithmetic mean of the sample of N events
κ	Thermal conductivity (W/m K)
μ	Viscosity (Pa.s)
λ	Wavelength (nm)
ν	Poisson's ratio
θ	Angle of incidence
ρ	Mass density (kg/m^3)
σ	Standard deviation
$\rho(r)$	Scattering length density
ϕ_a	Apparent porosity (%)
τ_c	Critical shear stress (MPa)
κ_e	Electronic conduction (W/m K)

σ_f	Fracture stress (MPa)
ΔG^*	Activation energy of the corrosion reaction (J)
ε_{hkl}	Lattice strain ($-$)
α_L	Linear expansion coefficient (K^{-1})
κ_c	Thermal conductivity of the coating (W/m.K)
κ_{Ph}	Lattice conduction or phonon conduction (W/m.K)
α_V	Volumetric expansion coefficients
δ_w	Depth of wear (μm)
ρ_X	Specific mass or mass density of X (kg/m^3)
Δp	Pressure drop (Pa)

ASTM Standards

A.1: Adhesion–Cohesion

[17.A1] **ASTM C633** Revision/Edition: 01 Chg: W/REAP Date: 00/00/08 Standard test method for adhesion or cohesion strength of thermal spray coatings

A.2: Corrosion

[17.C1] **ASTM G193** Revision/Edition: 10B Chg: Date: 06/01/10 Standard terminology and acronyms relating to corrosion

[17.C2] **ASTM 03.02** Revision/Edition: 10 Chg: Date: 08/00/10 Corrosion of metals; Wear and corrosion

[17.C3] **ASTM G102** Revision/Edition: 89 Standard practice for calculation of corrosion rates and related information from electrochemical measurements

[17.C4] **ASTM G3** Revision/Edition: 89 Standard practice for conventions applicable to electrochemical measurements in corrosion testing

[17.C5] **BS EN ISO 17475** Revision/Edition: 06 Corrosion of metals and alloys—Electrochemical test methods— Guidelines for conducting potentiostatic and potentiodynamic polarization measurements

[17.C6] **ASTM G5** Revision/Edition: 94 Chg: W/REAP Date: 00/00/04 Standard reference test method for making potentiostatic and potentiodynamic anodic polarization measurements

[15.C7] **ASTM G59** Revision/Edition: 97 Chg: W/REAP Date: 00/00/09 Standard test method for conducting polarization resistance measurements

[17.C8] **ASTM G61** Revision/Edition: 86 Chg: W/REAP Date: 00/00/09 Standard test method for conducting cyclic potentiodynamic polarization measurements for localized corrosion susceptibility of iron-, nickel-, or cobalt-based alloys

[17.C9] **ASTM G102** Revision/Edition: 89 Chg: W/REAP Date: 00/00/10 Standard practice for calculation of corro- sion rates and related information from electrochemical measurements

[17.C10] **ASTM G106** Revision/Edition: 89 Chg: W/REAP Date: 00/00/10 Standard practice for verification of algo- rithm and equipment for electrochemical impedance measurements

[17.C11] **SEMI F77** Revision/Edition: 03 Chg: W/REAP Date: 03/00/10 Test method for electrochemical critical pitting temperature testing of alloy surfaces used in corro- sive gas systems

[17.C12] **ASTM G150** Revision/Edition: 99 Chg: W/REAP Date: 00/00/10 Standard test method for electrochemical critical pitting temperature testing of stainless steels

[17.C13] **ASTM G78** Revision/Edition: 01 Chg: W/REAP Date: 00/00/07 Standard guide for crevice corrosion test- ing of iron-base stainless alloys in seawater and other chloride-containing aqueous environments

[17.C14] **ASTM B117** Revision/Edition: 09 Chg: Date: 07/01/09 Standard practice for operating salt spray (fog) apparatus

[17.C15] **ISO DIS 21608** Revision/Edition: 10 Chg: Date: 10/29/10 Corrosion of metals and alloys—Test method for isothermal-exposure oxidation testing under high- temperature corrosion conditions for metallic materials.

A.3: Mechanical Properties

[17.M1] **ASTM 03.01** Revision/Edition: 10 Chg: Date: 07/00/10 Metals-mechanical testing; elevated and low-temperature tests; Metallography

[17.M2] **ASM METALS HDBK V8** Revision/Edition: 00 Chg: Date: 00/00/00 Mechanical testing and evaluation

[17.M3] **ASTM E6** Revision/Edition: 09B Chg: Date: 05/15/ 09 Standard terminology relating to methods of mechani- cal testing

[17.M4] **BS 7134 P4 S4.2** Revision/Edition: 90 Chg: REAF Date: 00/00/96 Testing of engineering ceramics—thermo- mechanical properties—method for determination of ther- mal diffusivity by the laser flash (or heat pulse) method

[17.M5] **ASTM A833** Revision/Edition: 08A Chg: Date: 11/01/08 Standard practice for indentation hardness of metallic materials by comparison hardness testers

[17.M6] **ASTM C1327** Revision/Edition: 08 Chg: Date: 08/01/08 Standard test method for Vickers indentation hardness of advanced ceramics

[17.M7] **ASTM E384** Revision/Edition: 10 Chg: W/E2 Date: 04/00/10 Standard test method for Knoop and Vickers hardness of materials

[17.M8] **ASTM C1326** Revision/Edition: 08 Chg: W/E1 Date: 09/00/08 Standard test method for Knoop indenta- tion hardness of advanced ceramics

[17.M9] **ASTM C1327-08** Standard test method for Vickers indentation hardness of advanced ceramics

[17.M10] **BS 5600 P4 S4.7** Revision/Edition: 79 Chg: Date: 00/00/79 Determination of the Young's modulus

[17.M11] **JIS Z 2280** Revision/Edition: 93 Chg: W/REAF Date: 10/01/08 Test method for Young's modulus of metallic materials at elevated temperature

[17.M12] **BS DD CEN/TS 1071-7** Revision/Edition: 03 Chg: Date: 10/16/03 Advanced technical ceramics—Methods of test for ceramic coatings—Part 7: Determination of hardness and Young's modulus by instrumented indentation testing

[17.M12] **ASTM C1198** Revision/Edition: 09 Chg: Date: 11/04/09 Standard test method for dynamic Young's modulus, shear modulus and Poisson's ratio for advanced ceramics by sonic resonance

[17.M13] **ASTM C1259** Revision/Edition: 08 Chg: W/E1 Date: 04/00/09 Standard test method for dynamic Young's modulus, shear modulus and Poisson's ratio for advanced ceramics by impulse excitation of vibration

[17.M14] **API TR 6 AM** Revision/Edition: 2 Chg: W/REAF Date: 01/00/03 Material toughness

[17.M15] **ASTM C1421** Revision/Edition: 09 Chg: Date: 05/01/09 Standard test methods for determination of fracture toughness of advanced ceramics at ambient temperature

[17.M16] **BS 7448-1** Revision/Edition: 91 Chg: W/REAF Date: 03/00/07 Fracture mechanics toughness tests—Part 1: Method for determination of K1c, critical CTOD and critical J values of metallic materials

[17.M17] **BS EN ISO 12737** Revision/Edition: 06 Chg: Date: 01/12/06 Metallic materials—Determination of plane-strain fracture toughness

[17.M18] **BS ISO 28079** Revision/Edition: 09 Chg: Date: 08/31/09 Hardmetals—Palmquist toughness test

[17.M19] **ASTM STP381** Revision/Edition: 65 Chg: Date: 00/00/65 Fracture toughness testing and its applications

[17.M20] **BS EN 15305** Revision/Edition: 08 Chg: W/CRGD Date: 06/30/09 Non-destructive testing—Standard test method for determining residual stress analysis by X-ray diffraction

[15.M21] **BS DD CEN ISO/TS 21432** Revision/Edition: 06 Chg: W/REAF Date: 12/01/08 Non-destructive testing—Standard test method for determining residual stresses by neutron diffraction.

A.4: Materials Characterization

[17.Ma1] **ASTM E204** Revision/Edition: 98 Chg: W/REAP Date: 00/00/07 Standard practices for identification of materials by infrared absorption spectroscopy, using the ASTM coded band and chemical classification index

[17.Ma2] **BS EN 13925-1** Revision/Edition: 03 Chg: W/REAF Date: 12/01/08 Non-destructive testing—X-ray diffraction from polycrystalline and amorphous materials—Part 1: General principles

[17.Ma3] **BS EN 13925-2** Revision/Edition: 03 Chg: W/REAF Date: 12/01/08 Non-destructive testing—X-ray diffraction from polycrystalline and amorphous materials—Part 2: Procedures

[17.Ma4] **NBS MONO 25 SEC 16** Revision/Edition: 79 Chg: Date: 10/00/79 Standard X-ray diffraction powder patterns

[17.Ma5] **ASTM F2024** Revision/Edition: 10 Chg: REIN Date: 06/00/10 Standard practice for X-ray diffraction determination of phase content of plasma-sprayed hydroxyapatite coatings

[17.Ma6] **BS ISO 17470** Revision/Edition: 04 Chg: W/REAF Date: 12/01/08 Microbeam analysis—electron probe microanalysis—Guidelines for qualitative point analysis by wavelength dispersive X-ray spectrometry

[17.Ma7] **BS ISO 18516** Revision/Edition: 06 Chg: W/REAF Date: 06/01/10 Surface chemical analysis—Auger electron spectroscopy and X-ray photoelectron spectroscopy—Determination of lateral resolution

[17.Ma8] **AS ISO 15470** Revision/Edition: 06 Chg: Date: 10/20/06 Surface chemical analysis—X-ray photoelectron spectroscopy—description of selected instrumental performance parameters

[17.Ma9] **BS ISO 10810** Revision/Edition: 10 Chg: Date: 12/31/10 Surface chemical analysis—X-ray photoelectron spectroscopy—guidelines for analysis

[17.Ma10] **BS ISO 15470** Revision/Edition: 05 Chg: W/REAF Date: 10/01/10 Surface chemical analysis—X-ray photoelectron spectroscopy—description of selected performance parameters

A.5: Metallography and Image Analysis

[17. Me1] **ASTM E1920** Revision/Edition: 03 Chg: W/REAP Date: 00/00/08 Standard guide for metallographic preparation of thermal sprayed coatings

[17. Me2] **ASTM E3** Revision/Edition: 01 Chg: W/E1 Date: 03/00/09 Guide for metallographic preparation of metallographic specimens

[17. Me3] **ASTM MNL46** Revision/Edition: 07 Chg: Date: 00/00/07 Metallographic and materialographic specimen preparation light microscopy, image analysis and hardness testing

A.6: Non-destructive Methods

[17.ND1] **ESDU 91027** Revision/Edition: 91 Chg: W/AA Date: 11/01/93 Non-destructive examination—Choice of methods

A.7: Statistical Methods

[17.S1] **ASTM C1239-07** Revises ASTM C1239-06a Standard practice for reporting uniaxial strength data and estimating Weibull distribution parameters for advanced ceramics

A.8: Thermal Properties

[17.T1] **ASTM E228** Revision/Edition: 06 Chg: W/REIN Date: 09/01/06 Standard test method for linear thermal expansion of solid materials with a push-rod dilatometer

[17.T2] **ASTM E289** Revision/Edition: 04 Chg: W/REAP Date: 00/00/10 Standard test method for linear thermal expansion of rigid solids with interferometry

[17.T3] **ASTM C177** Revision/Edition: 10 Chg: Date: 06/01/10 Standard test method for steady-state heat flux measurements and thermal transmission properties by means of the guarded-hot-plate apparatus

[17.T4] **ASTM C201-93** (2009) Standard test method for thermal conductivity of refractories

[17.T5] **ASTM E 1461** Revision/Edition: 07 Chg: Date: 11/01/07 Standard test method for thermal diffusivity of solids by the flash method Disk 6–12 mm diameter; 1.5–4 mm thick diffusivity $0.1–1000$ m^2/s

[17.T6] **ASTM C 714** Revision/Edition: 05 Chg: W/REAP Date: 00/00/10 Standard test method for thermal diffusivity of carbon and graphite by a thermal pulse method; 6–12 mm, 2–4 mm thick; $0.04–2.0$ cm^2/s

[17.T7] **ASTM C 1470-06** Standard guide for testing the thermal properties of advanced ceramics

[17.T8] **BS 7134 P4 S4.2** Revision/Edition: 90 Chg: REAF Date: 00/00/96 Testing of engineering ceramics—thermomechanical properties—method for determination of thermal diffusivity by the laser flash (or heat pulse) method

[17.T9] **ASTM D2766-95(2009)** Standard test method for specific heat of liquids and solids

[17.T10] **ASTM E1269-11** Standard test method for determining specific heat capacity by differential scanning calorimetry

[17.T11] **ASTM C1525** Revision/Edition: 04 Chg: W/REAP Date: 00/00/09 Standard test method for determination of thermal shock resistance for advanced ceramics by water quenching

[17.T12] **FORD FLTM BI 107-05** Revision/Edition: 09 Chg: Date: 02/03/09 Thermal shock for coating adhesion

[17.T13] **JDQ149** Revision/Edition: 02 Chg: Date: 06/27/02 Tests for thermal shock resistance of high temperature coatings

[17.T14] **ASTM C 1171-05** Standard test method for quantitatively measuring the effect of thermal shock and thermal cycling on refractories

[17.T15] **ISO DIS 13123** Revision/Edition: 10 Chg: Date: 07/22/10 Metallic and other inorganic coatings—Test method of cyclic heating for thermal barrier coatings under temperature gradient.

A.9: Void Content and Network Architecture

[17.V1] **E2109-01(2007)** Standard test methods for determining area percentage porosity in thermal sprayed coatings

A.10: Wear

[17.W1] **ASTM G40** Revision/Edition: 10A Chg: Date: 07/01/10 Standard terminology relating to wear and erosion

[17.W2] **ASTM C704/C704M** Revision/Edition: 09 Chg: W/E1 Date: 08/00/09 Standard test method for abrasion, resistance of refractory materials at room temperature

[17.W3] **ASTM G105** Revision/Edition: 02 Chg: W/REAP Date: 00/00/07 Standard test method for conducting wet sand/rubber wheel abrasion test

[17.W4] **ASTM G132** Revision/Edition: 96 Chg: W/REAP Date: 00/00/07 Standard test method for pin abrasion testing

[17.W5] **ASTM G65** Revision/Edition: 04 Chg: Date: 11/01/04 10 Standard test method for measuring abrasion using the dry sand/rubber wheel apparatus

[17.W6] **ASTM MNL56** Revision/Edition: 07 Chg: Date: 00/00/07 Guide to friction, wear, and erosion testing

[17.W7] **ASTM G99** Revision/Edition: 05 Chg: W/REAP Date: 00/00/10 Standard test method for wear testing with a pin-on-disk apparatus

[17.W8] **ASTM G133** Revision/Edition: 05 Chg: W/REAP Date: 00/00/10 Standard test method for linearly reciprocating ball-on-disk sliding wear

[17.W9] **ASTM G98** Revision/Edition: 09 Chg: Date: 10/01/09 Standard test method for galling resistance of materials

[17.W10] **ASTM G73** Revision/Edition: 10 Chg: Date: 04/01/10 Standard test method for liquid impingement erosion using rotating apparatus

[17.W11] **ASTM G134** Revision/Edition: 95 Chg: W/REAP Date: 00/00/06 Standard test method for erosion of solid materials by a cavitating liquid jet

[17.W12] **ASTM G32** Revision/Edition: 09 Chg: Date: 05/01/09 Standard test method for cavitation erosion using vibratory apparatus

[17.W13] **ASTM E1942** Revision/Edition: 98 Chg: W/REAP Date: 00/00/04 Standard guide for evaluating data acquisition systems used in cyclic fatigue and fracture mechanisms

[17.W14] **ASTM E2368** Revision/Edition: 10 Chg: Date: 05/01/10 Standard practice for strain controlled thermomechanical fatigue testing

[17.W15] **BS 3518-1** Revision/Edition: 93 Chg: W/REAF Date: 01/01/09 Methods of fatigue testing—Part 1: Guide to general principles

[17.W16] **BS 3518-2** Revision/Edition: 62 Chg: W/REAF Date: 01/01/09 Methods of fatigue testing—Part 2: Rotating bending fatigue tests

[17.W17] **BS 3518-3** Revision/Edition: 63 Chg: W/REAF Date: 01/01/09 Methods of fatigue testing—Part 3: Direct stress fatigue tests

[17.W18] **ASTM G119** Revision/Edition: 09 Chg: Date: 07/15/09 Standard guide for determining synergism between wear and corrosion

[17.W19] **ASTM 03.02** Revision/Edition: 10 Chg: Date: 08/00/10 Corrosion of metals; Wear and corrosion

[17.W20] **ASTM G204** Revision/Edition: 10 Chg: Date: 04/01/10 Standard test method for damage to contacting solid surfaces under fretting conditions

[17.W21] **ASTM STP1159** Revision/Edition: 92 Chg: Date: 00/00/92 Standardization of fretting fatigue test methods and equipment

[17.W22] **ASTM STP1367** Revision/Edition: 00 Chg: Date: 02/00/00 Fretting fatigue: current technology and practices

References

Agüero, A., F. Camón, J. García de Blas, J.C. del Hoyo, R. Muelas, A. Santaballa, S. Ulargui, and P. Vallés. 2011. HVOF-deposited WCCoCr as replacement for hard Cr in landing gear actuators. *Journal of Thermal Spray Technology* 20 (6): 1292–1309.

Ahmed, R., and M. Hadfield. 2002. Mechanisms of fatigue failure in thermal spray coatings. *Journal of Thermal Spray Technology* 11 (3): 334–349.

Aksel, C., and P.D. Warren. 2003. Thermal shock parameters [R, R′′′ and R′′′′] of magnesia–spinel composites. *Journal of the European Ceramic Society* 23: 301–308.

Allen, A.J., G.G. Long, H. Boukari, J. Ilavsky, A. Kulkarni, S. Sampath, H. Herman, and A.N. Goland. 2001a. Microstructural characterization studies to relate the properties of thermal spray coatings to feedstock and spray conditions. *Surface and Coating Technology* 146–147: 544–552.

Allen, A.J., J. Ilavsky, G.G. Long, J.S. Wallace, C.C. Berndt, and H. Herman. 2001b. Micro structural characterization of yttria-stabilized zirconia plasma-sprayed deposits using multiple small-angle neutron scattering. *Acta Materialia* 49: 1661–1675.

Amada, S., and T. Hirose. 1998. Influence of grit blasting pre-treatment on the adhesion strength of plasma sprayed coatings: Fractal analysis of roughness. *Surface and Coating Technology* 102: 132–137.

Anderson, P.S., X. Wang, and P. Xiao. 2004. Impedance spectroscopy study of plasma sprayed and EB-PVD thermal barrier coatings. *Surface and Coating Technology* 185: 106–119.

Andreola, F., C. Leonelli, M. Romagnoli, and P. Miselli. 2000. Techniques used to determine porosity. *American Ceramic Society Bulletin* 79 (7): 49–52.

Andrews, D.J., and J.A.T. Taylor. 2000. Quality control of thermal barrier coatings using acoustic emission. *Journal of Thermal Spray Technology* 9 (2): 181–190.

Antou, G., and G. Montavon. 2007. Quantifying thermal spray coating architecture by stereological protocols: Part II. Key points to be addressed. *Journal of Thermal Spray Technology* 16 (2): 168–176.

Antou, G., G. Montavon, F. Hlawka, A. Cornet, C. Coddet, and F. Machi. 2004a. Evaluation of modifications induced on pore network and structure of partially stabilized zirconia manufactured by hybrid plasma spray process. *Surface and Coating Technology* 180–181: 627–632.

Antou, G., G. Montavon, F. Hlawka, A. Cornet, and C. Coddet. 2004b. Characterizations of the pore-crack network architecture of thermal-sprayed coatings. *Materials Characterization* 53: 361–372.

———. 2006. Exploring thermal spray gray alumina coating pore network architecture by combining stereological protocols and impedance electrochemical spectroscopy. *Journal of Thermal Spray Technology* 15 (4): 765–772.

ASM International. 2003. Corrosion: Fundamentals, testing, and protection. In *ASM handbook*, vol. 13A, 509–513. Materials Park: ASM International.

———. 2007. *ASM handbook: Nondestructive evaluation and quality control*, vol. 17, 9th ed. Materials Park: ASM International.

———. 2008. *ASM handbook*, Materials characterization, vol. 10, 8th ed. Materials Park: ASM International

Asmani, M., C. Kermel, A. Leriche, and M. Ourak. 2001. Influence of porosity on Young's modulus and Poisson's ratio in alumina ceramics. *Journal of the European Ceramic Society* 21: 1081–1086.

Atlas Steel Technical Note No. 7 "Galvanic corrosion," revised August 2010.

Baba, T., and A. Ono. 2001. Improvement of the laser flash method to reduce uncertainty in thermal diffusivity measurements. *Measurement Science and Technology* 12: 2046–2057.

Bacciochini, A., J. Ilavsky, G. Montavon, A. Denoirjean, F. Ben-ettouil, S. Valette, P. Fauchais, and K. Wittmann-Teneze. 2010a. Quantification of void network architectures of suspension plasma-sprayed (SPS) yttria-stabilized zirconia (YSZ) coatings using ultra-small-angle X-ray scattering (USAXS). *Materials Science and Engineering* A528: 91–102.

Bacciochini, A., G. Montavon, J. Ilavsky, A. Denoirjean, and P. Fauchais. 2010b. Porous architecture of SPS thick YSZ coatings structured at the nanometer scale (50 nm). *Journal of Thermal Spray Technology* 19 (1–2): 198–206.

Bahbou, M.F., P. Nylén, and J. Wigren. 2004. Effect of grit blasting and spraying angle on the adhesion strength of a plasma-sprayed coating. *Journal of Thermal Spray Technology* 13 (4): 508–514.

Balykov, A.V. 2006. Main conditions for efficient drilling of holes in brittle nonmetallic materials using diamonds drills. *Material Processing* 63 (5–6): 161–165.

Barletta, M., G. Bolelli, B. Bonferroni, and L. Lusvarghi. 2010. Wear and corrosion behavior of HVOF-sprayed WC-CoCr coatings on Al alloys. *Journal of Thermal Spray Technology* 19 (1–2): 358–367.

Barradas, S., F. Borit, V. Guipont, M. Jeandin, C. Bolis, M. Boustie, and L. Berthe. 2002. Laser shock flier impact simulation of particle-substrate interactions in cold spray. In *ITSC-2002*, ed. E. Lugscheider and C.C. Berndt, 592–597. Düsseldorf: DVS Deutscher Verband für Schweißen.

Barradas, S., R. Molins, M. Jeandin, M. Arrigoni, M. Boustie, C. Bolis, L. Berthe, and M. Ducos. 2005. Application of laser shock adhesion testing to the study of the interlamellar strength and coating-substrate adhesion in cold-sprayed copper coating of aluminum. *Surface and Coating Technology* 197: 18–27.

Basak, A.K., P. Matteazzi, M. Vardavoulias, and J.-P. Celis. 2006. Corrosion-wear behaviour of thermal sprayed nanostructured FeCu/WC–Co coatings. *Wear* 261: 1042–1050.

Beghini, M., L. Bertini, and E. Frendo. 2001a. Measurement of coatings' elastic properties by mechanical methods: Part 1. Consideration on experimental errors. *Experimental Mechanics* 41 (4): 293–304.

Beghini, M., G. Benamati, L. Bertini, and F. Frendo. 2001b. Measurement of coatings' elastic properties by mechanical methods: Part 2. Application to thermal barrier coatings. *Experimental Mechanics* 41 (4): 305–311.

Berka, L., and N. Murafa. 2004. On the estimation of the toughness of surface microcracks. *International Journal of Fracture* 128: 115–120.

Berndt, C.C. 1985. Acoustic emission evaluation of plasma-sprayed thermal barrier coatings. *Journal of Engineering for Gas Turbines and Power* 107: 142–146.

Berndt, C.C., and R. McPherson. 1981. A fracture mechanics approach to the adhesion of flame and plasma sprayed coatings. *Transaction of the Institute of Engineers* 6 (4): 53–58.

Białucki, P., and S. Kozerski. 2006. Study of adhesion of different plasma-sprayed coatings to aluminium. *Surface and Coating Technology* 201: 2061–2064.

Bobzin, K., N. Kopp, T. Warda, and M. Öte. 2012. Determination of the effective properties of thermal spray coatings using 2-D and 3-D models. *Journal of Thermal Spray Technology*. https://doi.org/10.1007/s11666-012-9809-3.

Boccaccini, D.N., and A.R. Boccaccini. 1997. Dependence of ultrasonic velocity on porosity and pore shape in sintered materials. *Journal of Nondestructive Evaluation* 16 (4): 187–192.

Bolcavage, A., A. Feuerstein, J. Foster, and P. Moore. 2004. Thermal shock testing of thermal barrier coating/bondcoat systems. *Journal of Materials Engineering and Performance* 13 (4): 389–397.

Bolelli, G., V. Cannillo, L. Lusvarghi, and T. Manfredini. 2006. Wear behaviour of thermally sprayed ceramic oxide coatings. *Wear* 261: 1298–1315.

Bolelli, G., L. Lusvarghi, T. Varis, E. Turunen, M. Leoni, P. Scardi, C.L. Azanza-Ricardo, and M. Barletta. 2008. Residual stresses in HVOF-sprayed ceramic coatings. *Surface and Coating Technology* 202: 4810–4819.

Bolelli, G., J. Rauch, V. Cannillo, A. Killinger, L. Lusvarghi, and R. Gadow. 2009. Microstructural and tribological investigation of high-velocity suspension flame sprayed (HVSFS) Al_2O_3 coatings. *Journal of Thermal Spray Technology* 18 (1): 35–49.

Bolis, C., et al. 2002. Developments in laser shock adhesion test (LASAT). In *ITSC-2002*, ed. E. Lugscheider et al., 587–593. Materials Park: ASM International.

Bolot, R., G. Antou, G. Montavon, and C. Coddet. 2005. A two-dimensional heat transfer model for thermal barrier coating average thermal conductivity computation. *Numerical Heat Transfer, Part A: Applications* 47 (9): 875–898.

Bolot, R., J.-L. Seichepine, J. Hao Qiao, and C. Coddet. 2011. Predicting the thermal conductivity of AlSi/polyester abradable coatings: Effects of the numerical method. *Journal of Thermal Spray Technology* 20 (1–2): 39–47.

Bradai, M.A., M. Braccini, A. Ati, N. Bounar, and A. Benabbas. 2008. Microstructure and adhesion of 100Cr6 steel coatings thermally sprayed on a 35CrMo4 steel substrate. *Surface and Coating Technology* 202: 4538–4543.

Brousse, E. 2010. Elaboration by thermal spraying of finely structured elements of a high temperature electrolyser for hydrogen production: Processes, structures and characteristics. PhD thesis, University of Limoges, France (in French).

Bucaille, J.L., E. Felder, and G. Hochstetter. 2002. Identification of the viscoplastic behavior of a polycarbonate based on experiments and numerical modeling of the nano-indentation test. *Journal of Materials Science* 37: 3999–4011.

Bunker, G. 2010. *Introduction to XAFS: A practical guide to X-ray absorption fine structure spectroscopy*. Cambridge: Cambridge University Press.

Cernuschi, F., L. Lorenzoni, S. Ahmaniemi, P. Vuoristo, and T. Mäntylä. 2005. Studies of the sintering kinetics of thick thermal barrier coatings by thermal diffusivity measurements. *Journal of the European Ceramic Society* 25: 393–400.

Chandler, H., ed. 2004. *Hardness testing*, 2nd ed. Materials Park: ASM.

Chandra, S., and P. Fauchais. 2009. Formation of solid splats during thermal spray deposition. *Journal of Thermal Spray Technology* 18 (2): 148–180.

Chattopadhyay, R. 2001. *Surface wear: Analysis, treatment, and prevention*, 307 p. Materials Park: ASM International.

Chen, H., Y. Hao, H. Wang, and W. Tang. 2010. Analysis of the microstructure and thermal shock resistance of laser glazed nanostructured zirconia TBCs. *Journal of Thermal Spray Technology* 19 (3): 558–565.

Chi, W., S. Sampath, and H. Wange. 2006. Ambient and high-temperature thermal conductivity of thermal sprayed coatings. *Journal of Thermal Spray Technology* 15 (4): 773–778.

Chivavibul, P., M. Watanabe, S. Kuroda, J. Kawakita, M. Komatsu, K. Sato, and J. Kitamura. 2010. Effect of powder characteristics on properties of warm-sprayed WC-Co coatings. *Journal of Thermal Spray Technology* 19 (1–2): 81–88.

Choi, S.R., D. Zhu, and R.A. Miller. 2005. Fracture behavior under mixed-mode loading of ceramic plasma-sprayed thermal barrier coatings at ambient and elevated temperatures. *Engineering Fracture Mechanics* 72: 2144–2158.

Chraska, T. 1999. TEM studies of microstructures and interfaces of zirconia produced by plasma spraying. Ph.D. thesis, State University of New York at Stony Brook, New York.

Chraska, T., and A.H. King. 2001. Transmission electron microscopy study of rapid solidification of plasma sprayed zirconia—Part I. First splat solidification. *Thin Solid Films* 397: 30–39.

Christoulis, D.K., S. Guetta, E. Irissou, V. Guipont, M.H. Berger, M. Jeandin, J.-G. Legoux, C. Moreau, S. Costil, M. Boustie, Y. Ichikawa, and K. Ogawa. 2010. Cold-spraying coupled to nano-pulsed Nd-YaG laser surface pre-treatment. *Journal of Thermal Spray Technology* 19 (5): 1062–1073.

Clark, H.M.I., H.M. Hawthorne, and Y. Xie. 1999. Wear rates and specific energies of some ceramic, cermet and metallic coatings determined in the Coriolis erosion tester. *Wear* 233–235: 319–327.

Clyne, T.W., and S.C. Gill. 1996. Residual stresses in thermal spray coatings and their effect on interfacial adhesion: A review of recent work. *Journal of Thermal Spray Technology* 5 (4): 401–418.

Cockeram, B.V., and W.L. Wilson. 2001. The hardness, adhesion and wear resistance of coatings developed for cobalt-base alloys. *Surface and Coating Technology* 139: 161–182.

Costil, S., C. Verdy, R. Bolot, and C. Coddet. 2007. On the role of spraying process on microstructural, mechanical, and thermal response of alumina coatings. *Journal of Thermal Spray Technology* 16 (5–6): 839–843.

Craig, B.D. 2005. Material failure modes, Part II: A brief tutorial on impact, spalling, wear, brinelling, thermal shock, and radiation damage. *Journal of Failure Analysis and Prevention* 5 (6): 7–12.

CSM Instruments. 2010. *Advanced mechanical surface testing: Characterization of thermal spray coatings by instrumented indentation and scratch testing. Part II: Indentation of plasma sprayed coatings*, CSM instruments application bulletin, vol. 32. Peseux: CSM Instruments.

Curran, J.A., and T.W. Clyne. 2006. Porosity in plasma electrolytic oxide coatings. *Acta Materialia* 54 (7): 1985–1993.

Dallaire, S. 2001. Hard arc-sprayed coating with enhanced erosion and abrasion wear resistance. *Journal of Thermal Spray Technology* 10 (3): 511–519.

Deevi, S.C., V.K. Sikka, C.J. Swindeman, and R.D. Seals. 1997. Reactive spraying of nickel-aluminide coatings. *Journal of Thermal Spray Technology* 6 (3): 335–344.

Deshpande, S., A. Kulkarni, S. Sampath, and H. Herman. 2004. Application of image analysis for characterization of porosity in thermal spray coatings and correlation with small angle neutron scattering. *Surface and Coating Technology* 187: 6–16.

Dey, A., A.K. Mukhopadhyay, S. Gangadharan, M.K. Sinha, and D. Basu. 2009. Weibull modulus of nano-hardness and elastic modulus of hydroxyapatite coating. *Journal of Materials Science* 44: 4911–4918.

Drexler, J.M., A. Aygun, D. Li, R. Vaßen, T. Steinke, and N.P. Padture. 2010. Thermal-gradient testing of thermal barrier coatings under simultaneous attack by molten glassy deposits and its mitigation. *Surface and Coating Technology* 204: 2683–2688.

Du, H., W. Hua, J. Liu, J. Gong, C. Sun, and L. Wen. 2005. Influence of process variables on the qualities of detonation gun sprayed WC–Co coatings. *Materials Science and Engineering A* 408: 202–210.

Eaton, P., and P. West. 2010. *Atomic force microscopy*. Oxford: Oxford University Press.

Elsebaei, A., J. Heberlein, M. Elshaer, and A. Farouk. 2010. Comparison of in-flight particle properties, splat formation, and coating microstructure for regular and nano-YSZ powders. *Journal of Thermal Spray Technology* 19 (1–2): 2–10.

Emiliani, M.L. 1993. Debond coating requirements for brittle matrix composites. *Journal of Materials Science* 28: 5280–5296.

Era, H., F. Otsubo, T. Uchida, S. Fukuda, and K. Kishitake. 1998. A modified shear test for adhesion evaluation of thermal sprayed coating. *Materials Science and Engineering* A251: 166–172.

Erickson, L.C., R. Westergard, U. Wiklund, N. Axén, H.M. Hawthorne, and L.S. Hogmark. 1998. Cohesion in plasma-sprayed coatings – A comparison between evaluation methods. *Wear* 214: 30–37.

Etchart-Salas, R., V. Rat, J.F. Coudert, P. Fauchais, N. Caron, K. Wittman, and S. Alexandre. 2007. Influence of plasma instabilities in ceramic suspension plasma spraying. *Journal of Thermal Spray Technology* 16 (5–6): 857–865.

Factor, M., and I. Roman. 2000a. Vickers micro-indentation of WC-12%Co thermal spray coating. Part 1: Statistical analysis of microhardness data. *Surface and Coating Technology* 132: 181–193.

———. 2000b. Vickers micro-indentation of WC-12%Co thermal spray coating. Part 2: The between-operator reproducibility limits of microhardness measurement and alternative approaches to quantifying hardness of cemented-carbide thermal spray coatings. *Surface and Coating Technology* 132: 65–75.

Faisal, N.H., J.A. Steel, R. Ahmed, and R.L. Reuben. 2009. The use of acoustic emission to characterize fracture behavior during Vickers indentation of HVOF thermally sprayed WC-Co coatings. *Journal of Thermal Spray Technology* 18 (4): 525–535.

Fan, X., and T. Ishigaki. 2001. Mo_5Si_3-B and $MoSi_2$ deposits fabricated by radio frequency induction plasma spraying. *Journal of Thermal Spray Technology* 10 (4): 611–617.

Fauchais, P., G. Montavon, R.S. Lima, and B.R. Marple. 2011. Engineering a new class of thermal spray nano-based microstructures from agglomerated nanostructured particles. Suspensions and solutions: An invited review. *Journal of Physics D* 44: 093001.

Feng, C., and B.S. Kang. 2006. A transparent indenter measurement method for mechanical property evaluation. *Experimental Mechanics* 46 (1): 91–103.

———. 2008. Young's modulus measurement using a simplified transparent indenter measurement technique. *Experimental Mechanics* 48: 9–15.

Fernandez, J.E., R. Rodriguez, Y. Wang, R. Vijande, and A. Rincon. 1995. Sliding wear of a plasma-sprayed A_2O_3 coating. *Wear* 181–183: 417–425.

Flores Renteria, A., B. Saruhan, J. Ilavsky, and A.J. Allen. 2007. Application of USAXS analysis and non-interacting approximation to determine the influence of process parameters and ageing on the thermal conductivity of electron-beam physical vapor-deposited thermal barrier coatings. *Surface and Coating Technology* 201: 4781–4788.

Flores, A., Saruhan B. Renteria, J. Ilavsky, and A.J. Allenc. 2007. Application of USAXS analysis and non-interacting approximation to determine the influence of process parameters and ageing on the thermal conductivity of electron-beam physical vapor deposited thermal barrier coatings. *Surface and Coating Technology* 201: 4781–4788.

Fowler, D.B., W. Riggs, and J.C. Russ. 1990. Inspecting thermal sprayed coatings. *Advanced Materials and Processes* 11: 41–52.

Fox, A.C., and T.W. Clyne. 2004. Oxygen transport by gas permeation through the zirconia layer in plasma sprayed thermal barrier coatings. *Surface and Coating Technology* 184: 311–321.

Fridrici, V., S. Fouvry, and P. Kapsa. 2003. Fretting wear behavior of a Cu–Ni–In plasma coating. *Surface and Coating Technology* 163–164: 429–434.

Garcia, J.R., J.M. Cuetos, E. Fernandez, and V. Higuera. 2007. Laser surface melting Cr–Ni coatings, in the erosive-corrosive atmosphere of boilers. *Tribology Letters* 28: 99–108.

Garcia, E., J. Mesquita-Guimaraes, P. Miranzo, M.I. Osendi, C.V. Cojocaru, Y. Wang, C. Moreau, and R.S. Lima. 2011. Phase composition and microstructural responses of graded mullite/YSZ coatings under water vapor environments. *Journal of Thermal Spray Technology* 20 (1–2): 83–91.

Gawne, D.T., Z. Qiu, Y. Bao, T. Zhang, and K. Zhang. 2001. Abrasive wear resistance of plasma-sprayed glass-composite coatings. *Journal of Thermal Spray Technology* 10 (4): 599–603.

Gean, M.C., and T.N. Farris. 2009. Elevated temperature fretting fatigue of Ti-17 with surface treatments. *Tribology International* 42: 1340–1345.

Geels, K. 2007. *Metallographic and materialographic specimen preparation, light microscopy, image analysis and hardness testing*, ASTM MNL46. West Conshohocken: ASTM International.

Giannuzzi, L.A., and F.A. Stevens. 2004. *Introduction to focused ion beams: Instrumentation, theory, techniques and practice*. Berlin: Springer.

Gill, S.C., and T.W. Clyne. 1994. Investigation of residual stress generation during thermal spraying by continuous curvature measurement. *Thin Solid Films* 250: 172–180.

Giolli, C., A. Scrivani, G. Rizzi, F. Borgioli, G. Bolelli, and L. Lusvarghi. 2009. Failure mechanism for thermal fatigue of thermal barrier coating systems. *Journal of Thermal Spray Technology* 18 (2): 223–230.

Giraud, D., F. Borit, V. Guipont, and M. Jeandin. 2012. Metallization of a polymer using cold spray: Application to aluminum coating of polyamide 66. In *Thermal spray 2012: Proceedings (ITSC-2012)*, ed. R.S. Lima et al. Materials Park: ASM International. (e-proc).

Gitzhofer, F., D. Lombard, A. Vardelle, M. Vardelle, C. Martin, and P. Fauchais. 1986. Thermophysical properties of zirconia coatings stabilized with calcia or yttria: Influence of spraying parameters and heat treatment. In *Advances in thermal spraying*, 269–275. Oxford: Pergamon.

Gobinda, C.S., and T.I. Khan. 2010. The corrosion and wear performance of microcrystalline WC-10Co-4Cr and near-nanocrystalline WC-17Co high velocity oxy-fuel sprayed coatings on steel substrate. *Metallurgical and Materials Transactions A: Physical Metallurgy and Materials Science* 41A: 3000–3009.

Golosnoy, I.O., S. Paul, and T.W. Clyne. 2008. Modelling of gas permeation through ceramic coatings produced by thermal spraying. *Acta Materialia* 56: 874–883.

Greving, D.J., J.R. Shadley, and E.F. Rybicki. 1994a. Effects of coating thickness and residual stresses on the bond strength of ASTM C633-79 thermal spray coating test specimens. *Journal of Thermal Spray Technology* 3 (4): 371–378.

Greving, D.J., E.F. Rybicki, and J.R. Shadley. 1994b. Through-thickness residual stress evaluations for several industrial thermal spray coatings using a modified layer-removal method. *Journal of Thermal Spray Technology* 3 (4): 379–388.

Guetta, S., M.H. Berger, F. Borit, V. Guipont, M. Jeandin, M. Boustie, Y. Ichikawa, K. Sakaguchi, and K. Ogawa. 2009. Influence of particle velocity on adhesion of cold-sprayed splats. *Journal of Thermal Spray Technology* 18 (3): 331–342.

Guilemany, J.M., J.M. Miguel, S. Armada, S. Vizcaino, and F. Climent. 2001. Use of scanning white light interferometry in the characterization of wear mechanisms in thermal-sprayed coatings. *Materials Characterization* 47: 307–314.

Guilemany, J.M., N. Espallargas, P.H. Suegama, A.V. Benedetti, and J. Fernández. 2005. High-velocity oxyfuel Cr_3C_2-NiCr replacing hard chromium coatings. *Journal of Thermal Spray Technology* 14 (3): 336–341.

Guilemany, J.M., M. Torrell, and J.R. Miguel. 2008. Study of the HVOF Ni-based coatings corrosion resistance applied on municipal solid-waste incinerators. *Journal of Thermal Spray Technology* 17 (2): 254–262.

Guilemany, J.M., N. Cinca, S. Dosta, and I.G. Cano. 2009. FeAl and NbAl3 intermetallic-HVOF coatings: Structure and properties. *Journal of Thermal Spray Technology* 18 (4): 536–545.

Guinier, A., and G. Fournet. 1955. *Small angle scattering of X-rays.* New York: Wiley.

Guipont, V. 2013. Microstructures and interfaces ceramic/metal obtained by physical processes. State thesis, University of Limoges, Limoges.

Han, W., E.E. Rybicki, and J.R. Shadley. 1993a. An improved specimen geometry for ASTM C633-79 to estimate bond strengths of thermal spray coatings. *Journal of Thermal Spray Technology* 2 (2): 145–150.

Han, W., E.F. Rybicki, and J.R. Shadley. 1993b. Application of fracture mechanics to the interpretation of bond strength data from ASTM Standard C633-79. *Journal of Thermal Spray Technology* 2 (3): 235–241.

Hanson, T.C., and G.S. Settlese. 2003. Particle temperature and velocity effects on the porosity and oxidation of an HVOF corrosion-control coating. *Journal of Thermal Spray Technology* 12 (3): 403–415.

Harok, V., and K. Neufuss. 2001. Elastic and inelastic effects in compression in plasma-sprayed ceramic coatings. *Journal of Thermal Spray Technology* 10 (1): 126–132.

Hasselman, D.P.H. 1970. Thermal stress resistance parameters for brittle refractory ceramics: A compendium. *American Ceramic Society Bulletin* 49 (12): 1033–1037.

He, G.H., J.D. Guo, Y.Y. Zhang, B.Q. Wang, and B.L. Zhou. 2000. Measurement of thermal diffusivity of thermal control coatings by the flash method using two-layer composite sample. *International Journal of Thermophysics* 21 (2): 535–542.

He, D., N. Dong, and J. Jiang. 2007. Corrosion behavior of arc sprayed nickel-base coatings. *Journal of Thermal Spray Technology* 16 (5–6): 850–856.

He, D.-Y., B.-Y. Fu, J.-M. Jiang, and X.-Y. Li. 2008. Microstructure and wear performance of arc sprayed Fe-FeB-WC coatings. *Journal of Thermal Spray Technology* 17 (5–6): 757–761.

Heintze, G.N., and R. McPherson. 1988. Fracture toughness of plasma sprayed zirconia coatings. *Surface and Coating Technology* 134: 15–23.

Hejwowski, T. 2009. Erosive and abrasive wear resistance of overlay coatings. *Vacuum* 83: 166–170.

Hejwowski, T., S. Szewczyk, and A. Weronski. 2000. An investigation of the abrasive and erosive wear of flame-sprayed coatings. *Journal of Materials Processing Technology* 106: 54–57.

Heping, L., P. Nie, Y. Yan, J. Wang, and B. Sun. 2010. Characterization and adhesion strength study of detonation-sprayed MoB–CoCr alloy coatings on 2Cr13 stainless steel substrate. *Journal of Coating Technology and Research* 7 (6): 801–807.

Higuera Hidalgo, V., J. Belzunce Varela, A. Carriles Menéndez, and S. Poveda Martınez. 2001. High temperature erosion wear of flame and plasma-sprayed nickel–chromium coatings under simulated coal-fired boiler atmospheres. *Wear* 247: 214–222.

Hivart, P., and J. Crampon. 2007. Interfacial indentation test and adhesive fracture characteristics of plasma sprayed cermet Cr_3C_2/Ni–Cr coatings. *Mechanics of Materials* 39: 998–1005.

Hugot, F., J. Patru, P. Fauchais, and L. Bianchi. 2007. Modelling of a substrate thermomechanical behavior during plasma spraying. *Journal of Materials Processing Technology* 190 (1–3): 317–323.

Hung, Y.Y., Y.S. Chen, S.P. Ng, L. Liu, Y.H. Huang, B.L. Luk, R.W.L. Ip, C.M.L. Wu, and P.S. Chung. 2009. Review and comparison of shearography and active thermography for non-destructive evaluation. *Materials Science and Engineering R* 64: 73–112.

Ibrahim, A., and C.C. Berndt. 2007. Fatigue and deformation of HVOF sprayed WC–Co coatings and hard chrome plating. *Materials Science and Engineering A* 456: 114–119.

Ibrahim, A., R.S. Lima, C.C. Berndt, and B.R. Marple. 2007. Fatigue and mechanical properties of nanostructured and conventional titania (TiO_2) thermal spray coatings. *Surface and Coating Technology* 201: 7589–7596.

Ilavsky, J. 2010. Characterization of complex thermal barrier deposits pore microstructures by a combination of imaging, scattering and intrusion techniques. *Journal of Thermal Spray Technology* 19 (1–2): 178–189.

Ilavsky, J., and C.C. Berndt. 1998. Thermal expansion properties of metallic and cermet coatings. *Surface and Coating Technology* 102: 19–24.

Ilavsky, J., H. Herman, C.C. Berndt, A.N. Goland, G.G. Long, S. Krueger, and A.J. Allen. 1994. In *Thermal spray industrial applications*, ed. C.C. Berndt and S. Sampath, 709–715. Materials Park: ASM International.

Ilavsky, J., C.C. Berndt, and J. Karthikeyan. 1997. Mercury intrusion porosimetry of plasma-sprayed ceramic. *Journal of Materials Science* 32: 3925–3932.

Ilavsky, J., G.G. Long, A.J. Allen, L. Leblanc, M. Prystay, and C. Moreau. 1999a. Anisotropic microstructure of plasma-sprayed deposits. *Journal of Thermal Spray Technology* 8 (3): 414–420.

Ilavsky, J., G.G. Long, A.J. Allen, and C.C. Berndt. 1999b. Evolution of the void structure in plasma-sprayed YSZ deposits during heating. *Materials Science and Engineering A* 272: 215–221.

Ilavsky, J., P.R. Jemian, A.J. Allen, F. Zhang, L.E. Levine, and G.G. Long. 2009a. Ultra-small-angle X-ray scattering at the advanced photon source. *Journal of Applied Crystallography* 42: 1–11.

———. 2009b. Ultra-small-angle X-ray scattering at the advanced photon source. *Journal of Applied Crystallography* 42 (3): 469–479.

Ingo, G.M., S. Kaciulis, A. Mezzi, T. Valente, F. Casadei, and G. Gusmano. 2005. Characterization of composite titanium nitride coatings prepared by reactive plasma spraying. *Electrochimica Acta* 50: 4531–4537.

Irisawa, T., and H. Matsumoto. 2006. Thermal shock resistance and adhesion strength of plasma-sprayed alumina coating on cast iron. *Thin Solid Films* 509: 141–144.

Ishikawa, Y., J. Kawakita, S. Osawa, T. Itsukaichi, Y. Sakamoto, M. Takaya, and S. Kuroda. 2005. Evaluation of corrosion and wear resistance of hard cermet coatings sprayed by using an improved HVOF process. *Journal of Thermal Spray Technology* 14 (3): 384–390.

Jadhav, A.D., N.P. Padture, E.H. Jordan, M. Gell, P. Miranzo, and E.R. Fuller Jr. 2006. Low-thermal-conductivity plasma-sprayed thermal barrier coatings with engineered microstructures. *Acta Materialia* 54: 3343–3349.

James, J.D., J.A. Spittle, S.G.R. Brown, and R.W. Evans. 2001. A review of measurement techniques for the thermal expansion coefficient of metals and alloys at elevated temperatures. *Measurement Science and Technology* 12: R1–R15.

Jayaraj, B., V.H. Desai, C.K. Lee, and Y.H. Sohn. 2004. Electrochemical impedance spectroscopy of porous ZrO_2–8 wt.% Y_2O_3 and thermally grown oxide on nickel aluminide. *Materials Science and Engineering A* 372 (1–2): 278–286.

Jeandin, M., D.K. Christoulis, F. Borit, M.H. Berger, S. Guetta, G. Rolland, V. Guipont, E. Irissou, J.G. Legoux, C. Moreau, M. Nivard, L. Berthe, M. Boustie, K. Sakaguchi, Y. Ichikawa, K. Ogawa, and S. Costil. 2010. Lasers and thermal spray. *Materials Science Forum* 174: 638–642.

Jennifer, S.Y., H. Wang, W.D. Porter, A.R. de Arellano Lopez, and K.T. Faber. 2001. Thermal conductivity and phase evolution of plasma-sprayed multilayer coatings. *Journal of Materials Science* 36: 3511–3518.

Ji, G.-C., C.-J. Li, Y.-Y. Wang, and W.-Y. Li. 2007. Erosion performance of HVOF-sprayed Cr_3C_2-NiCr coatings. *Journal of Thermal Spray Technology* 16 (4): 557–565.

Kang, H.-K., and S. Bong Kange. 2004. Behavior of porosity and copper oxidation in W/Cu composite produced by plasma spray. *Journal of Thermal Spray Technology* 13 (2): 223–228.

Kawakita, J., S. Kuroda, and T. Kodama. 2003. Evaluation of through-porosity of HVOF sprayed coating. *Surface and Coating Technology* 166: 17–23.

Kennedy, D.M., and M.S.J. Hashmi. 1998. Methods of wear testing for advanced surface coatings and bulk materials. *Journal of Materials Processing Technology* 77: 246–253.

Kim, H.J., Y.G. Kweon, and R.W. Chang. 1994. Wear and erosion behavior of plasma-sprayed WC-Co coatings. *Journal of Thermal Spray Technology* 3 (2): 169–178.

Kishitake, K., H. Era, and F. Otsubo. 1996. Characterization of plasma sprayed Fe-17Cr-38Mo-4C amorphous coatings crystallizing at extremely high temperature. *Journal of Thermal Spray Technology* 5 (3): 283–288.

Koiprasert, H., S. Dumrongrattana, and P. Niranatlumpong. 2004. Thermally sprayed coatings for protection of fretting wear in land-based gas-turbine engine. *Wear* 257: 1–7.

Koivuluoto, H., J. Lagerbom, M. Kylmälahti, and P. Vuoristo. 2008. Microstructure and mechanical properties of low-pressure cold-sprayed (LPCS) coatings. *Journal of Thermal Spray Technology* 17 (5–6): 721–727.

Kovarik, O., J. Siegl, J. Nohava, and P. Chraska. 2005. Young's modulus and fatigue behavior of plasma-sprayed alumina coatings. *Journal of Thermal Spray Technology* 14 (2): 231–238.

Kovarik, O., J. Siegl, and Z. Prochazka. 2008. Fatigue behavior of bodies with thermally sprayed metallic and ceramic deposits. *Journal of Thermal Spray Technology* 17 (4): 525–532.

Krishnamurthy, N., S.C. Sharma, M.S. Murali, and P.G. Mukunda. 2009. Adhesion behavior of plasma sprayed thermal barrier coatings on Al-6061 and cast-iron substrates. *Frontiers of Materials Science in China* 3 (3): 333–338.

Kucuk, A., C.C. Berndt, U. Senturk, and R.S. Lima. 2000a. Influence of plasma spray parameters on mechanical properties of yttria stabilized zirconia coatings. I: Four-point bend test. *Materials Science and Engineering A* 284: 29–40.

———. 2000b. Influence of plasma spray parameters on mechanical properties of yttria stabilized zirconia coatings. II: Acoustic emission response. *Materials Science and Engineering A* 284: 41–50.

Kulkarni, A., Z. Wang, T. Nakamura, S. Sampath, A. Goland, H. Herman, J. Allen, J. Ilavsky, G.G. Long, J. Frahm, and R.W. Steinbrech. 2003. Comprehensive microstructural characterization and predictive property modeling of plasma-sprayed zirconia coatings. *Acta Materialia* 51: 2457–2475.

Kulkarni, A.A., A. Goland, H. Herman, A.J. Allen, J. Ilavsky, G.G. Long, and F. De Carlo. 2005. Advanced microstructural characterization of plasma-sprayed zirconia coatings over extended length scales. *Journal of Thermal Spray Technology* 14 (2): 239–250.

Kulkarni, A., A. Goland, H. Herman, A.J. Allen, T. Dobbins, F. DeCarlo, J. Ilavsky, G.G. Long, S. Fang, and P. Lawton. 2006. Advanced neutron and X-ray techniques for insights into the microstructure of EB-PVD thermal barrier coatings. *Materials Science and Engineering A* 426: 43–52.

Kulu, P., and T. Pihl. 2002. Selection criteria for wear resistant powder coatings under extreme erosive wear conditions. *Journal of Thermal Spray Technology* 11 (4): 517–522.

Kuroda, S., T. Kukushima, and S. Kitahara. 1988. Simultaneous measurement of coating thickness and deposition stress during thermal spraying. *Thin Solid Films* 164: 157–163.

Kuroda, S., Y. Tashiro, H. Yumoto, S. Taira, H. Fukanuma, and S. Tobe. 2001. Peening action and residual stresses in high-velocity oxygen fuel thermal spraying of 316L stainless steel. *Journal of Thermal Spray Technology* 10 (2): 367–374.

Landa, M., F. Kroupa, K. Neufuss, and P. Urbanek. 2003. Effect of uniaxial pressure on ultrasound velocities and elastic moduli in plasma-sprayed ceramics. *Journal of Thermal Spray Technology* 12 (2): 226–233.

Langford, J.I., and A.J.C. Wilson. 1978. Scherrer after sixty years: A survey and some new results in the determination of crystallite size. *Journal of Applied Crystallography* 1: 102–113.

Lee, H., S. Mall, J.H. Sanders, and S.K. Sharma. 2005. Wear analysis of Cu–Al coating on Ti-6Al-4V substrate under fretting condition. *Tribology Letters* 19 (3): 239–248.

Lee, H., S. Mall, and K.N. Murray. 2007. Fretting wear behavior of Cu–Al coating on Ti-6Al-4V substrate under dry and wet (lubricated) contact condition. *Tribology Letters* 28: 19–25.

Leigh, S.H., and C.C. Berndt. 1994. A test for coating adhesion on flat substrates – A technical note. *Journal of Thermal Spray Technology* 3 (2): 184–190.

Leigh, S.-H., and C.C. Berndt. 1999. Modelling of elastic constants of plasma spray deposits with ellipsoid-shaped voids. *Acta Materialia* 47 (5): 1575–1586.

Lesage, J., M.H. Staiab, D. Chicot, C. Godoy, and P.E.V. De Mirandad. 2000. Effect of thermal treatments on adhesive properties of a NiCr thermal sprayed coating. *Thin Solid Films* 377–378: 681–686.

Li, J., and C. Ding. 2001. Determining microhardness and elastic modulus of plasma-sprayed Cr_2C_3-NiCr coatings using Knoop indentation testing. *Surface and Coating Technology* 135: 229–237.

Li, C.-J., and Y.-Y. Wang. 2002. Effect of particle state on the adhesive strength of HVOF sprayed metallic coating. *Journal of Thermal Spray Technology* 11 (4): 523–529.

Li, H., K.A. Khor, and P. Cheanga. 2002. Young's modulus and fracture toughness determination of high velocity oxy-fuel-sprayed bioceramic coatings. *Surface and Coating Technology* 155: 21–32.

Li, C.-J., X.-J. Ning, and C.-X. Li. 2005. Effect of densification processes on the properties of plasma-sprayed YSZ electrolyte coatings for solid oxide fuel cells. *Surface and Coating Technology* 190: 60–64.

Li, L., N. Hitchman, and J. Knapp. 2010. Failure of thermal barrier coatings subjected to CMAS attack. *Journal of Thermal Spray Technology* 19 (1–2): 148–155.

Li, C.-J., H.-T. Wang, G.-J. Yang, and C.-G. Bao. 2011. Characterization of high-temperature abrasive wear of cold-sprayed FeAl intermetallic compound coating. *Journal of Thermal Spray Technology* 20: 227–233.

Lima, C.R.C., and J.M. Guilemany. 2007. Adhesion improvements of thermal barrier coatings with HVOF thermally sprayed bond coats. *Surface and Coating Technology* 201: 4694–4701.

Lima, R.S., and B.R. Marple. 2003. High Weibull modulus HVOF titania coatings. *Journal of Thermal Spray Technology* 12 (2): 240–249.

Lima, R.S., A. Kucuk, and C.C. Berndt. 2001a. Integrity of nanostructured partially stabilized zirconia after plasma spray processing. *Materials Science and Engineering* A313: 75–82.

———. 2001b. Evaluation of microhardness and elastic modulus of thermally sprayed nanostructured zirconia coatings. *Surface and Coating Technology* 135: 166–172.

———. 2002. Bimodal distribution of mechanical properties on plasma sprayed nanostructured partially stabilized zirconia. *Materials Science and Engineering* A327: 224–232.

Lima, R.S., S.E. Kruger, G. Lamouche, and B.R. Marple. 2003a. Elastic modulus measurements via laser-ultrasonic and Knoop indentation techniques. In *ITSC-2003: Advancing the science and applying the technique*, ed. C. Moreau and B. Marple, 1369–1378. Materials Park: ASM International.

Lima, M.M., C. Godoy, J.C. Avelar-Batista, and P.J. Modenesi. 2003b. Toughness evaluation of HVOF WC-Co coatings using non-linear regression analysis. *Materials Science and Engineering* A357: 337–345.

Lima, M.M., C. Godoy, P.J. Modenesi, J.C. Avelar-Batista, A. Davison, and A. Matthews. 2004. Coating fracture toughness determined by Vickers indentation: An important parameter in cavitation erosion resistance of WC-Co thermally sprayed coatings. *Surface and Coating Technology* 177–178: 489–496.

Lima, C.R.C., J. Nin, and J.M. Guilemany. 2006. Evaluation of residual stresses of thermal barrier coatings with HVOF thermally sprayed bond coats using the modified layer removal method (MLRM). *Surface and Coating Technology* 200: 5963–5972.

Lin, C.K., and C.C. Berndt. 1993. Microhardness variations in thermally sprayed coatings. In *Proceedings of the national thermal spray conference*, ed. C.C. Berndt and T.F. Bernicki, 561–564. Materials Park: ASM International.

———. 1994. Measurement and analysis of adhesion strength for thermally sprayed coatings. *Journal of Thermal Spray Technology* 3 (1): 75–104.

———. 1998. Acoustic emission studies on thermal spray materials. *Surface and Coating Technology* 102: 1–7.

Lin, J.F., M.H. Liu, and J.D. Wu. 1996. Analysis of the friction and wear mechanisms of structural ceramic coatings. Part 1: The effect of processing conditions. *Wear* 194: 1–11.

Lin, C.K., C.C. Berndt, S.H. Leigh, and K. Murakami. 1997. Acoustic emission studies of alumina-13% titania free-standing forms during four-point bend tests. *Journal of the American Ceramic Society* 80: 2382–2394.

Liu, X., and C. Ding. 2002. Thermal properties and microstructure of a plasma sprayed wollastonite coating. *Journal of Thermal Spray Technology* 11 (3): 375–379.

Liu, R., and D.Y. Li. 1999. Protective effect of yttrium additive in lubricants on corrosive wear. *Wear* 225–229: 968–974.

Liu, Y., and H.M. Wang. 2010. Microstructure and wear property of laser-clad Co_3Mo_2Si/Co_{ss} wear resistant coatings. *Surface and Coating Technology* 205: 377–382.

Liu, X.Q., Y.G. Zheng, X.C. Chang, W.L. Hou, J.Q. Wang, Z. Tang, and A. Burgess. 2009. Microstructure and properties of Fe-based amorphous metallic coating produced by high velocity axial plasma spraying. *Journal of Alloys and Compounds* 484: 300–307.

Luiz de Assis, S., S. Wolynec, and I. Costa. 2006. Corrosion characterization of titanium alloys by electrochemical techniques. *Electrochimica Acta* 51: 1815–1819.

Lyphout, C., P. Nylén, A. Manescu, and T. Pirling. 2008. Residual stresses distribution through thick HVOF sprayed Inconel 718 coatings. *Journal of Thermal Spray Technology* 17 (5–6): 915–923.

Ma, X.Q., and M. Takemoto. 2001. Quantitative acoustic emission analysis of plasma, sprayed thermal barrier coatings subjected to thermal shock tests. *Materials Science and Engineering* A308: 101–110.

Ma, X.Q., Y. Mizutani, and M. Takemoto. 2001a. Laser-induced surface acoustic waves for evaluation of elastic stiffness of plasma sprayed materials. *Journal of Materials Science* 36: 5633–5641.

Ma, X.Q., S. Cho, and M. Takemoto. 2001b. Acoustic emission source analysis of plasma sprayed thermal barrier coatings during four-point bend tests. *Surface and Coating Technology* 139: 55–62.

Ma, X.Q., D.W. Gandy, and G.J. Frederick. 2008. Innovation of ultra-fine structured alloy coatings having superior mechanical properties and high temperature corrosion resistance. *Journal of Thermal Spray Technology* 17 (5–6): 933–941.

Ma, D., C.W. Ong, and T. Zhang. 2009. An instrumented indentation method for Young's modulus measurement with accuracy estimation. *Experimental Mechanics* 49: 719–729.

Magnani, M., P.H. Suegama, N. Espallargas, C.S. Fugivara, S. Dosta, J.M. Guilemany, and A.V. Benedettir. 2009. Corrosion and wear studies of Cr_3C_2-NiCr-HVOF coatings sprayed on AA7050 T7 under cooling. *Journal of Thermal Spray Technology* 18 (3): 354–363.

Majzoobi, G.H., R. Hojjati, M. Nematian, E. Zalnejad, A.R. Ahmadkhani, and E. Hanifepoor. 2010. A new device for fretting fatigue testing. *Transactions of the Indian Institute of Metals* 63 (2–3): 493–497.

Maldague, X., and S. Marinetti. 1996. Pulse phase infrared thermography. *Journal of Applied Physics* 79 (5): 2694–2698.

Mancini, C.E., C.C. Berndt, L. Sun, and A. Kucuk. 2001. Porosity determinations in thermally sprayed hydroxyapatite coatings. *Journal of Materials Science* 36: 3891–3896.

Mann, B.S., and V. Arya. 2001. Abrasive and erosive wear characteristics of plasma nitriding and HVOF coatings: Their application in hydro turbines. *Wear* 249: 354–360.

Marcano, Z., J. Lesage, D. Chicot, G. Mesmacque, E.S. Puchi-Cabrera, and M.H. Staiaa. 2008. Microstructure and adhesion of Cr_3C_2–NiCr vacuum plasma sprayed coatings. *Surface and Coating Technology* 202: 4406–4410.

Marinetti, S., D. Robba, F. Cernuschi, P.G. Bison, and E. Grinzato. 2007. Thermographic inspection of TBC coated gas turbine blades: Discrimination between coating over-thicknesses and adhesion defects. *Infrared Physics & Technology* 49: 281–285.

Marot, G., P. Démarécaux, J. Lesage, M. Hadad, and Staia M.H. St. Siegmann. 2008. The interfacial indentation test to determine adhesion and residual stresses in NiCr VPS coatings. *Surface and Coating Technology* 202: 4411–4416.

Marshall, D.B., T. Noma, and A.G. Evans. 1982. A simple method for determining elastic-modulus-to-hardness ratios using Knoop indentation measurements. *Communications of the American Ceramic Society* 65: C175–C176.

Matějíček, J., B. Kolman, J. Dubský, K. Neufuss, N. Hopkins, and J. Zwick. 2006. Alternative methods for determination of composition and porosity in abradable materials. *Materials Characterization* 57: 17–29.

Matthews, S., B. James, and M. Hyland. 2010. The effect of heat treatment on the oxidation mechanism of blended powder Cr_3C_2-NiCr coatings. *Journal of Thermal Spray Technology* 19 (1–2): 119–127.

Mauer, G., R. Vaßen, and D. Stöver. 2009. Atmospheric plasma spraying of yttria-stabilized zirconia coatings with specific porosity. *Surface and Coating Technology* 204: 172–179.

McGrann, R.T.R., D.J. Greving, J.R. Shadley, E.F. Rybicki, B.E. Bodger, and D.A. Somerville. 1998. The effect of residual stress in HVOF tungsten carbide coatings on the fatigue life in bending of thermal spray coated aluminum. *Journal of Thermal Spray Technology* 7 (4): 546–552.

Meacham, B.E., M.C. Marshall, and D.J. Branagan. 2006. Palmqvist fracture toughness of a new wear-resistant weld alloy. *Metallurgical and Materials Transactions A* 37A: 3617–3627.

Michlik, P., and C.C. Berndt. 2006. Image-based extended finite element modeling of thermal barrier coatings. *Surface and Coating Technology* 201: 2369–2380.

Montavon, G. 2004. *Quality control of thermal sprayed coatings, course thermal spraying*. Limoges: Aliderte.

Montavon, G., and G. Antou. 2007. Quantifying thermal spray coating architecture by stereological protocols: Part I. A historical perspective. *Journal of Thermal Spray Technology* 16 (1): 6–14.

Montavon, G., C. Coddet, C.C. Berndt, and S.-H. Leigh. 1998. Microstructural index to quantify thermal spray deposit microstructures using image analysis. *Journal of Thermal Spray Technology* 7 (2): 229–241.

Musalek, R., J. Matejicek, M. Vilemova, and O. Kovarik. 2010. Non-linear mechanical behavior of plasma sprayed alumina under mechanical and thermal spray loading. *Journal of Thermal Spray Technology* 19 (1–2): 422–428.

Musil, J., F. Kunc, H. Zeman, and H. Polakova. 2002. Relationships between hardness, Young's modulus and elastic recovery in hard nanocomposite coatings. *Surface and Coating Technology* 154: 304–313.

Nolan, D.J., and M. Samandi. 1997. Revealing true porosity in WC-Co thermal spray coatings. *Journal of Thermal Spray Technology* 6 (4): 422–424.

Noone, M.J., and R.L. Mehan. 1974. In *Fracture mechanics of ceramics*, ed. R.C. Bradt, Hasselman DPH, and F.F. Lange, 201–229. Plenum: New York.

Nusair Khan, A., and J. Lu. 2003. Behavior of air plasma sprayed thermal barrier coatings, subject to intense thermal cycling. *Surface and Coating Technology* 166: 37–43.

Ozdemir, I., I. Hamanaka, Y. Tsunekawa, and M. Okumiya. 2005. In-process exothermic reaction in high-velocity oxyfuel and plasma spraying with SiO_2/Ni/Al-Si-Mg composite powder. *Journal of Thermal Spray Technology* 14 (3): 321–329.

Paul, S., and K. Yadav. 2010. Corrosion behavior of surface-treated implant Ti-6Al-4V by electrochemical polarization and impedance studies. *Journal of Materials Engineering and Performance* 20 (3): 422–435.

Pawlowski, L. 1995. *The science and engineering of thermal spray coatings*. New York: Wiley, 2nd (2008), Ed, (656 p).

Pejryd, L., J. Wigren, D.J. Greving, J.R. Shadley, and E.E. Rybicki. 1995. Residual stresses as a factor in the selection of tungsten carbide coatings for a jet engine application. *Journal of Thermal Spray Technology* 4 (3): 268–274.

Pershin, V., M. Lufitha, S. Chandra, and J. Mostaghimi. 2003. Effect of substrate temperature on adhesion strength of plasma-sprayed nickel coatings. *Journal of Thermal Spray Technology* 12 (3): 370–376.

Petorak, C., J. Ilavsky, H. Wang, W. Porter, and R. Trice. 2010. Microstructural evolution of 7 wt.% Y_2O_3–ZrO_2 thermal barrier coatings due to stress relaxation at elevated temperatures and the concomitant changes in thermal conductivity. *Surface and Coating Technology* 205: 57–65.

Podrezov, Y.N., Y.V. Mil'man, D.V. Lotsko, N.L. Lugovoi, D.G. Verbilo, and N.A. Efimov. 1999. A method of determining the mechanical properties of a two-layer composite consisting of a steel matrix and a plasma spray coating based on amorphizing powders. *Powder Metallurgy and Metal Ceramics* 38 (5–6): 224–227.

Prasad, B.K. 2000. Effects of alumina particle dispersion on the erosive–corrosive wear response of a zinc-based alloy under changing slurry conditions and distance. *Wear* 238: 151–159.

Prasad Sahu, S., A. Satapathy, A. Patnaik, K.P. Sreekumar, and P.V. Ananthapadmanabhan. 2010. Development, characterization and erosion wear response of plasma sprayed fly ash–aluminum coatings. *Materials and Design* 31: 1165–1173.

Prudenziati, M., G.C. Gazzadi, M. Medici, G. Dalbagni, and M. Caliari. 2010. Cr_3C_2-NiCr HVOF-sprayed coatings: Microstructure and properties versus powder characteristics and process parameters. *Journal of Thermal Spray Technology* 19 (3): 541–550.

Puerta, D.G. 2005. Metallographic preparation and evaluation of thermal spray coatings: Mounting, TSS committee on accepted practices. *Journal of Thermal Spray Technology* 14 (4): 450–452.

———. 2006. Metallographic preparation and testing of thermal spray coatings: Grinding, TSS committee on accepted practices. *Journal of Thermal Spray Technology* 15 (1): 31–32.

———. 2008. TSS accepted practices committees—Metallography and mechanical testing. *Journal of Thermal Spray Technology* 17 (4): 435–436.

Qi, Z.M., S. Amada, S. Akiyama, and C. Takahashi. 1988. Evaluation of thermal shock strength of thermal-sprayed coatings by the laser irradiation technique. *Surface and Coating Technology* 110: 73–80.

Rainforth, W.M. 2004. The wear behaviour of oxide ceramics—A review. *Journal of Materials Science* 39: 6705–6721.

Rammo, N.N., H.R. Al-Amery, T. Abdul-Jabbar, and H.I. Jaffer. 2009. Adhesion, hardness and structure of thermal sprayed Al/SiC composite, coat on graphite. *Surface and Coating Technology* 203: 2891–2895.

Rätzer-Scheibe, H.-J., U. Schulz, and T. Krell. 2006. The effect of coating thickness on the thermal conductivity of EB-PVD PYSZ thermal barrier coatings. *Surface and Coating Technology* 200: 5636–5644.

Ravi, B.G., A.S. Gandhi, X.Z. Guo, J. Margolies, and S. Sampath. 2008. Liquid precursor plasma spraying of functional materials: A case study for yttrium aluminum garnet (YAG). *Journal of Thermal Spray Technology* 17 (1): 82–90.

Reed, J.S. 1995. *Principles of ceramics processing*. New York: Wiley.

Rendler, N.J., and I. Vigness. 1966. Hole drilling method of measuring residual stresses. *Experimental Mechanics* 6 (12): 577–586.

Rice, R.W. 1996. Porosity dependence of physical properties of materials: A summary review. *Key Engineering Materials* 115: 1–19.

Richard, C.S., J. Lu, G. Béranger, and F. Decomps. 1995. Study of Cr_2O_3 coatings. Part II: Adhesion to a cast-iron substrate. *Journal of Thermal Spray Technology* 4 (4): 347–352.

Richard, C.S., G. Béranger, J. Lu, J.F. Flavenot, and T. Grégoire. 1996. Four-point bending tests of thermally produced WC-Co coatings. *Surface and Coating Technology* 78 (284): 294.

Rickerby, D.S. 1988. A review of the methods for the measurement of coating-substrate adhesion. *Surface and Coating Technology* 36 (1–2): 541–557.

Riggs, W. 2004. Metallography and image analysis. In *Handbook of thermal spray technology*, ed. J.R. Davis, 224–259. Metals Park: ASM International.

Robin, P., F. Gitzhofer, P. Fauchais, and M. Boulos. 2010. Remaining fatigue life assessment of plasma sprayed thermal barrier coatings. *Journal of Thermal Spray Technology* 19 (5): 911–920.

Rolland, G., P. Sallamand, V. Guipont, M. Jeandin, E. Boller, and C. Bourda. 2011. Laser-induced damage in cold-sprayed composite coatings. *Surface and Coating Technology* 205: 4915–4927.

Rosa, G., P. Psyllaki, R. Oltra, S. Costil, and C. Coddet. 2001. Simultaneous laser generation and laser ultrasonic detection of the mechanical breakdown of a coating-substrate interface. *Ultrasonics* 39: 355–365.

Roy, M., A. Pauschitz, J. Bernardi, T. Koch, and F. Franek. 2006. Microstructure and mechanical properties of HVOF sprayed nanocrystalline Cr_3C_2-25(Ni20Cr) coating. *Journal of Thermal Spray Technology* 15 (3): 372–381.

Russ, J.C. 1995. *The image processing handbook*, 2nd ed. Boca Raton: CRC.

———. 2011. *The image processing handbook*, 6th ed. Boca Raton: CRC, Taylor & Francis Group.

Russ, J.C., and R.T. DeHoff. 1999. *Practical stereology*, 2nd ed. New York: Plenum.

Salimijazi, H., M. Hosseini, J. Mostaghimi, L. Pershin, T.W. Coyle, H. Samadi, and A. Shafyei. 2012. Plasma sprayed coating using mullite and mixed alumina/silica powders. *Journal of Thermal Spray Technology* 21 (5): 825–830.

Sampath, S., X.Y. Jiang, J. Matejicek, L. Prchlik, A. Kulkarni, and A. Vaidya. 2004. Role of thermal spray processing method on the microstructure, residual stress and properties of coatings: An integrated study for Ni-5 wt,%Al bond coats. *Materials Science and Engineering* A364: 216–231.

Sansoucy, E., G.E. Kim, A.L. Moran, and B. Jodoin. 2007. Mechanical characteristics of Al-Co-Ce coatings produced by the cold spray process. *Journal of Thermal Spray Technology* 16 (5–6): 651–660.

Santana, Y.Y., P.O. Renault, M. Sebastiani, J.G. La Barbera, J. Lesage, E. Bemporad, E. Le Bourhis, E.S. Puchi-Cabrera, and M.H. Staia. 2008. Characterization and residual stresses of WC–Co thermally sprayed coatings. *Surface and Coating Technology* 202: 4560–4565.

Saravanan, P., V. Selvarajan, M.P. Srivastava, D.S. Rao, S.V. Joshi, and G. Sundararajan. 2000. Study of plasma- and detonation gun-sprayed alumina coatings using Taguchi experimental design. *Journal of Thermal Spray Technology* 9 (4): 505–512.

Sauer, J.P. 2005. Metallographic preparation and testing of thermal spray coatings: Sectioning, TSS committee on accepted practices. *Journal of Thermal Spray Technology* 14 (3): 313–314.

Sauer, J.P., and G. Blann. 2006. Metallographic preparation and testing of thermal spray coatings: Fine grinding and polishing, TSS committee on accepted practices. *Journal of Thermal Spray Technology* 15 (2): 174–176.

Schajer, G.S. 1988. Measurement of non-uniform residual stress using the hole drilling method. Part 1. *Journal of Engineering Materials and Technology* 110 (4): 338–343, Part 2. 110(4):345–349.

Schajer, G.S., and E. Altus. 1996. Stress calculation error analysis for incremental hole drilling residual stress measurement. *Journal of Engineering Materials and Technology* 118 (1): 120–126.

Schmidt, G., and S. Steinhäuser. 1996. Characterization of wear protective coatings. *Tribology International* 29 (3): 207–214.

Scrivani, A., G. Rizzi, U. Bardi, C. Giolli, M. Muniz Miranda, S. Ciattini, A. Fossati, and F. Borgioli. 2007. Thermal fatigue behavior of thick and porous thermal barrier coatings systems. *Journal of Thermal Spray Technology* 16 (5–6): 816–821.

Scrivani, A., G. Rizzi, and C.C. Berndt. 2008. Enhanced thick thermal barrier coatings that exhibit varying porosity. *Materials Science and Engineering A* 476: 1–7.

Shepard, S.M. .2001. Advances in pulsed thermography. In *Proceedings of SPIE, Thermo sense XXIII, Orlando, FL, USA, 16–19 April 2001*, pp. 511–515.

Shivamurthy, R.C., M. Kamaraj, R. Nagarajan, S.M. Shariff, and G. Padmanabham. 2010. Slurry erosion characteristics and erosive wear mechanisms of Co-based and Ni-based coatings formed by laser surface alloying. *Metallurgical and Materials Transactions A: Physical Metallurgy and Materials Science* 41A: 470–486.

Shrestha, S., A. Neville, and T. Hodgkiess. 2001. The effect of post-treatment of a high-velocity oxy-fuel Ni-Cr-Mo-Si-B coating. Part I: Microstructure/corrosion behavior relationships. *Journal of Thermal Spray Technology* 10 (3): 470–479.

Sidhu, T.S., S. Prakash, and R.D. Agrawal. 2006a. Hot corrosion resistance of high-velocity oxyfuel sprayed coatings on a nickel-base superalloy in molten salt environment. *Journal of Thermal Spray Technology* 15 (3): 387–399.

———. 2006b. Characterizations and hot corrosion resistance of Cr_3C_2-NiCr coating on Ni-base superalloys in an aggressive environment. *Journal of Thermal Spray Technology* 15 (4): 811–816.

———. 2007. Study of molten salt corrosion of high velocity oxy-fuel sprayed cermet and nickel-based coatings at 900 °C. *Metallurgical and Materials Transactions A: Physical Metallurgy and Materials Science* 38A: 77–85.

Siebert, B., C. Funke, R. Vaßen, and D. Stöver. 1999. Changes in porosity and Young's modulus due to sintering of plasma sprayed thermal barrier coatings. *Journal of Materials Processing Technology* 92–93: 217–223.

Singh, H., D. Puri, and S. Prakash. 2005. Corrosion behavior of plasma-sprayed coatings on a Ni-base superalloy in Na_2SO_4-60 Pct V_2O_3 environment at 900 °C. *Metallurgical and Materials Transactions A: Physical Metallurgy and Materials Science* 36A: 1008–1015.

Slifka, A.J., B.J. Filla, J.M. Phelps, G. Bancke, and C.C. Berndt. 1998. Thermal conductivity of a zirconia thermal barrier coating. *Journal of Thermal Spray Technology* 7 (1): 43–46.

Sorai, M., and N.N. Gakkai. 2004. *Comprehensive handbook of calorimetry and thermal analysis*, 534 p. New York: Wiley.

Souza, V.A.D., and A. Neville. 2006. Mechanisms and kinetics of WC-Co-Cr high velocity oxy-fuel thermal spray coating degradation in corrosive environments. *Journal of Thermal Spray Technology* 15 (1): 106–117.

Speyer, R. 1993. *Thermal analysis of materials*, Materials engineering. Seattle: Amazon.

Stack, M.M., F.H. Stott, and G.C. Wood. 1993. Review of mechanisms of erosion-corrosion of alloys at elevated temperatures. Part 2. *Wear* 162–164: 706–712.

Staia, M.H., E. Ramos, A. Carrasquero, A. Roman, J. Lesage, D. Chicot, and G. Mesmacque. 2000. Effect of substrate roughness induced by grit blasting upon adhesion of WC-17% Co thermal sprayed coatings. *Thin Solid Films* 377–378: 657–664.

Stewart, S., R. Ahmed, and T. Itsukaichi. 2005. Rolling contact fatigue of post-treated WC – NiCrBSi thermal spray coatings. *Surface and Coating Technology* 190 (1–2): 171–189.

Stokes, J., and L. Looney. 2008. Predicting quenching and cooling stresses within HVOF deposits. *Journal of Thermal Spray Technology* 17 (5–6): 908–914.

Strunz, P., G. Schumacher, R. Vaßen, and A. Wiedenmann. 2004. In situ SANS study of pore microstructure in YSZ thermal barrier coatings. *Acta Materialia* 52: 3305–3312.

Sundararajan, G., N.M. Chavan, G. Sivakumar, and P. Sudharshan Phani. 2010. Evaluation of parameters for assessment of inter-splat bond strength in cold-sprayed coatings. *Journal of Thermal Spray Technology* 19 (6): 1255–1266.

Taha-al, Z.Y., M.S. Hashmi, and B.S. Yilbas. 2009. Effect of WC on the residual stress in the laser treated HVOF coating. *Journal of Materials Processing Technology* 209: 3172–3181.

Tan, Y., J.P. Longtin, and S. Sampath. 2006. Modeling thermal conductivity of thermal spray coatings: Comparing predictions to experiments. *Journal of Thermal Spray Technology* 15 (4): 545–552.

Théry, P.-Y., M. Poulain, M. Dupeux, and M. Braccini. 2007. Adhesion energy of a YPSZ EB-PVD layer in two thermal barrier coating systems. *Surface and Coating Technology* 202: 648–652.

Tiana, W., Y. Wang, T. Zhang, and Y. Yang. 2009. Sliding wear and electrochemical corrosion behavior of plasma sprayed nanocomposite Al_2O_3–13%TiO_2 coatings. *Materials Chemistry and Physics* 118: 37–45.

Tillmann, W., E. Vogli, and B. Krebs. 2008a. Influence of the spray angle on the characteristics of atmospheric plasma sprayed hard material-based coatings. *Journal of Thermal Spray Technology* 17 (5–6): 948–955.

Tillmann, W., E. Vogli, I. Baumann, G. Matthaeus, and T. Ostrowski. 2008b. Influence of the HVOF gas composition on the thermal spraying of WC-Co submicron powders (28 + 1 μm) to produce superfine structured cermet coatings. *Journal of Thermal Spray Technology* 17 (5–6): 924–932.

Tillmann, W., B. Klusemann, J. Nebel, and B. Svendsen. 2011. Analysis of the mechanical properties of an arc-sprayed WC-FeCSiMn coating: Nanoindentation and simulation. *Journal of Thermal Spray Technology* 20 (1–2): 328–335.

Tjitra Salim, N., M. Yamada, H. Nakano, K. Shima, H. Isago, and M. Fukumoto. 2011. The effect of post-treatments on the powder morphology of titanium dioxide (TiO_2) powders synthesized for cold spray. *Surface and Coating Technology* 206: 366–371.

Underwood, E.E. 1970. *Quantitative stereology*. Reading: Addison-Wesley.

Valente, T., C. Bartuli, M. Sebastiani, and A. Loreto. 2005. Implementation and development of the incremental hole drilling method for the measurement of residual stress in thermal spray coatings. *Journal of Thermal Spray Technology* 14 (4): 462–470.

Van Brakel, J., S. Modrý, and M. Svatá. 1981. Mercury porosimetry: State of the art. *Powder Technology* 29 (1): 1–12.

Van der Voort, G.F. 1999. *Metallography principles and practice*, Material science and engineering series. Materials Park: ASM International.

Vander Voort, George, ed. 2004. *ASM handbook*, Metallography and microstructures, vol. 09, 1184 p. Materials Park: ASM International.

Venkataraman, R., G. Das, S.R. Singh, L.C. Pathak, R.N. Ghosh, B. Venkataraman, and R. Krishnamurthy. 2007. Study on influence of porosity, pore size, spatial and topological distribution of pores on microhardness of as plasma sprayed ceramic coatings. *Materials Science and Engineering A* 445–446: 269–274.

Voleník, K., F. Hanousek, P. Chraska, J. Ilavsky, and K. Neufuss. 1999. In-flight oxidation of high-alloy steels during plasma spraying. *Materials Science and Engineering* A272: 199–206.

Voyer, J., F. Gitzhofer, and M.I. Boulos. 1998. Study of the performance of TBC under thermal cycling conditions using an acoustic emission rig. *Journal of Thermal Spray Technology* 7: 181–190.

Wallace, J.S., and J. Ilavsky. 1998. Elastic modulus measurements in plasma sprayed deposits. *Journal of Thermal Spray Technology* 7 (4): 521–526.

Walpole, R.E., and R.H. Myers. 1978. *Probability and statistics for engineers and scientists*, 2nd ed. New York: Macmillan.

Wang, H.-P. 1979. The alignment error of the hole-drilling method. *Experimental Mechanics* 19: 23–27.

Wang, H., and R.B. Dinwiddie. 2000. Reliability of laser flash thermal diffusivity measurements of the thermal barrier coatings. *Journal of Thermal Spray Technology* 9 (2): 210–214.

———. 2004. Development of a lab view based portable thermal diffusivity system. In *Thermal conductivity 27, thermal expansion 15*, ed. H. Wang, 484–492. Lancaster: DESTech.

Wang, Z., A. Kulkarni, S. Deshpande, T. Nakamura, and H. Herman. 2003. Effects of pores and interfaces on effective properties of plasma sprayed zirconia coatings. *Acta Materialia* 51: 5319–5334.

Wang, H.-R., B.-R. Hou, J. Wang, Q. Wang, and W.-Y. Li. 2008. Effect of process conditions on microstructure and corrosion resistance of cold-sprayed Ti coatings. *Journal of Thermal Spray Technology* 17 (5–6): 736–741.

Wang, J., E.H. Jordan, and M. Gell. 2010. Plasma sprayed dense MgO-Y_2O_3 nanocomposite coatings using sol-gel combustion synthesized powder. *Journal of Thermal Spray Technology* 19 (5): 873–878.

Washburn, E.W. 1921. The dynamics of capillary flows. *Physics Review* 17: 273–283.

Watanabe, M., S. Kuroda, K. Yokoyama, T. Inoue, and Y. Gotoh. 2008. Modified tensile adhesion test for evaluation of interfacial toughness of HVOF sprayed coatings. *Surface and Coating Technology* 202: 1746–1752.

Westergård, R., N. Axén, U. Wiklund, and S. Hogmark. 2000. An evaluation of plasma sprayed ceramic coatings by erosion, abrasion and bend testing. *Wear* 246: 12–19.

Wielage, B., S. Steinhäuser, T. Schnick, and D. Nickelmann. 1999. Characterization of the wear behavior of thermal sprayed coatings. *Journal of Thermal Spray Technology* 8 (4): 553–558.

Wigren, J. and J. Johansson. 2011. Improving reliability of thermal spray coatings in production environment. In Presented at 3rd plasma round table, 31 October–4 November 2011, South Africa.

Wigren, J., and K. Täng. 2007. Quality considerations for the evaluation of thermal spray coatings. *Journal of Thermal Spray Technology* 16 (4): 533–540.

Williams, D.B., and C.B. Carter. 2009. *Transmission electron microscopy*, 2nd ed. New York: Springer.

Wittmann-Ténèze, K., N. Caron, and S. Alexandre. 2008. Gas permeability of porous plasma-sprayed coatings. *Journal of Thermal Spray Technology* 17 (5–6): 902–907.

Xie, Y., and H.M. Hawthorne. 2001. A model for compressive coating stresses in the scratch adhesion test. *Surface and Coating Technology* 141: 15–25.

Xie, X.Y., H.B. Guo, and S.K. Gong. 2010. Mechanical properties of $LaTi_2Al_9O_{19}$ and thermal cycling behaviors of plasma sprayed $LaTi_2Al_9O_{19}$/YSZ thermal barrier coatings. *Journal of Thermal Spray Technology* 19 (6): 1179–1185.

Xu, G.-Z., J.-J. Liu, and Z.-R. Zhou. 2002. The effect of the third body on the fretting wear behavior of coatings. *Journal of Materials Engineering and Performance* 11 (3): 288–293.

Yan, L., Y. Leng, and J. Chen. 2003. Torsional fatigue resistance of plasma sprayed HA coating on Ti-6Al-4V. *Journal of Materials Science. Materials in Medicine* 14: 291–295.

Yan, D., Y. Dong, Y. Yang, L. Wang, X. Chen, J. He, and J. Zhang. 2011. Microstructure characterization of the $FeAl_2O_4$-based nanostructured composite coating synthesized by plasma spraying Fe_2O_3/Al powders. *Journal of Thermal Spray Technology* 20 (6): 1269–1277.

Yang, G., H. Zu-kun, X. Xiaolei, and X. Gang. 2001. Formation of molybdenum boride cermet coating by the detonation spray process. *Journal of Thermal Spray Technology* 10 (3): 456–460.

Yanga, Q., T. Senda, and A. Hirose. 2006. Sliding wear behavior of WC–12% Co coatings at elevated temperatures. *Surface and Coating Technology* 200: 4208–4212.

Ye, F.-X., S.-H. Wu, and A. Ohmori. 2008. Microstructure and oxidation behavior of Cr39Ni7C cermet coatings deposited by diamond jet spray process. *Journal of Thermal Spray Technology* 17 (5–6): 942–947.

Yin, B., G. Liu, H. Zhou, J. Chen, and F. Yan. 2010. Sliding wear behavior of HVOF-sprayed Cr_3C_2-NiCr/CeO_2 composite coatings at elevated temperature up to 800 °C. *Tribology Letters* 37: 463–475.

Zhang, J., and V. Desai. 2005. Evaluation of thickness, porosity and pore shape of plasma sprayed TBC by electrochemical impedance spectroscopy. *Surface and Coating Technology* 190: 98–109.

Zhang, D.-W., and T.C. Lei. 2003. The microstructure and erosive–corrosive wear performance of laser-clad Ni–Cr3C2 composite coating. *Wear* 255: 129–133.

Zhang, T., and D.Y. Li. 2001. Improvement in the resistance of aluminum with yttria particles to sliding wear in air and in a corrosive medium. *Wear* 251: 1250–1256.

Zhang, X.C., B.S. Xu, S.T. Tu, F.Z. Xuan, H.D. Wang, and Y.X. Wu. 2009a. Fatigue resistance and failure mechanisms of plasma-sprayed CrC–NiCr cermet coatings in rolling contact. *International Journal of Fatigue* 31: 906–915.

Zhang, X.C., B.S. Xu, F.Z. Xuan, H.D. Wang, Y.X. Wu, and S.T. Tu. 2009b. Statistical analyses of porosity variations in plasma-sprayed Ni-based coatings. *Journal of Alloys and Compounds* 467: 501–508.

Zhang, X.C., B.S. Xu, F.Z. Xuan, H.D. Wang, and Y.X. Wu. 2009c. Microstructural and porosity variations in the plasma-sprayed Ni-alloy coatings prepared at different spraying powers. *Journal of Alloys and Compounds* 473: 145–151.

Zhao, X.B., and Z.H. Ye. 2012. Microstructure and wear resistance of molybdenum based amorphous nanocrystalline alloy coating fabricated by atmospheric plasma spraying. *Surface and Coating Technology* 228: S226–S270.

Zhao, Y., L. Lin, X.M. Li, and M.K. Lei. 2010. Simultaneous determination of the coating thickness and its longitudinal velocity by ultrasonic nondestructive method. *NDT and E International* 43: 579–585.

Zhou, Y.C., and T. Hashida. 2002. Thermal fatigue failure induced by delamination in thermal barrier coating. *International Journal of Fatigue* 24: 407–417.

Zhu, S., R.H. Pelton, and K. Collver. 1995. Mechanistic modelling of fluid permeation through compressible fiber beds. *Chemical Engineering Science* 50 (22): 3557–3572.

Zhu, D., S.R. Choi, and R.A. Miller. 2004. Development and thermal fatigue testing of ceramic thermal barrier coatings. *Surface and Coating Technology* 188–189: 146–152.

Part IV

Process Integration and Industrial Applications

Abbreviations

APS	Atmospheric Plasma Spraying
AWJ	Abrasive Water Jetting
AWS	American Welding Society
BET	Brunauer–Emmett–Teller
CAPS	Controlled Atmosphere Plasma Spraying
CARS	Coherent Anti-Stokes Raman Spectroscopy
CCD	Charge Coupled Device
CS	Cold Spray
CT	Computer Tomography
CW	Continuous Wave
dBA	Decibel
DC	Direct Current
D-Gun	Detonation Gun
DLR	German Aerospace Research Center
EB-PVD	Electron Beam Physical Vapor Deposition
EMI	Electromagnetic Interference
EP	Enthalpy Probe
FS	Flame Spraying
HA	Hydroxyapatite
HD-WAS	High Definition-WAS
HE-APS	High Energy-APS
HE	Heat Exchanger
HEPA	High-Efficiency Particulate Air Class Filters
HIP	Hot Isostatic Pressing
HVAF	High-Velocity Air Fuel
HVOF	High-Velocity Oxy Fuel
ICP	Inductively Coupled Plasma
ICP-AE	ICP-Atomic Emission
ICP-MS	ICP-Mass Spectrometry
IR	Infrared
LC	Laser Cladding
LDA	Laser Doppler Anemometry
LEV	Local Exhaust Ventilation
LHS	Left-Hand Side
LPPS	Low-Pressure Plasma Spraying
LRmC	Laser Remelted Coatings
LTE	Local Thermodynamic Equilibrium
MFC	Mass Flow Controller
MS	Mass Spectrometer
MS-SOFC	Metal-Supported Solid Oxide Fuel Cells
MVE	Molten Volume Ensemble
NIOSH	National Institute for Occupational Safety and Health
NIR	Near Infrared
NRR	Noise Reduction Rating
OCV	Open Circuit Voltage
OEL	Occupational Exposure Limits
OSHA	Occupational Safety and Health Administration
OV	Organic Vapor
PDTS	Pulse Detonation Thermal Spray
PFI	Particle Flux Imaging
PI	Process Integration
PIV	Particle Image Velocimetry
PLC	Programmable Logic Controller
PRPE	Personal Respiratory Protective Equipment
PSD	Particle Size Distribution
PS-PVD	Plasma Spray-Physical Vapor Deposition
PTA	Plasma Transferred Arc
PT-WAS	Plasma Transferred-WAS
PVD	Physical Vapor Deposition
QC	Quality Control
R&D	Research and Development
RCF	Rolling Contact Fatigue
RF	Radio Frequency
RF-IPS	RF Induction Plasma Spraying
RHS	Right-Hand Side
RMS	Root Mean Square
SB	Spray Booth
SDC	Spray and Deposit Control
SOFC	Solid Oxide Fuel Cells
SPS	Spark Plasma Sintering
SWAS	Single-Wire Arc Spraying

© Springer Nature Switzerland AG 2021

M. I. Boulos et al. (ed.), *Thermal Spray Fundamentals*, https://doi.org/10.1007/978-3-030-70672-2_18

TBC	Thermal Barrier Coating
TCP	Tricalcium Phosphate
ToF	Time of Flight
TSC	Thermal Spray Coating
TSS	Thermal Spray Society
TTCP	Tetracalcium Phosphate
ULPPS	Ultralow-Pressure Plasma Spraying
UV	Ultraviolet
VH	Vickers Hardness
VLPPS	Very-Low-Pressure Plasma Spraying
VPS	Vacuum Plasma Spraying
WAS	Wire Arc Spraying
WC	Water Chiller
YPSZ	Yttria Partially Stabilized Zirconia
YSZ	Yttria Stabilized Zirconia

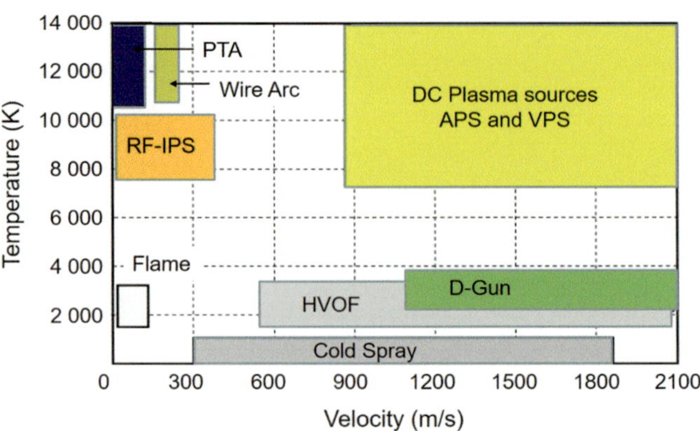

Fig. 18.1 Temperature–velocity mapping showing the range of spray conditions prevailing with different thermal spray coating technologies

18.1 Introduction

Throughout this book, emphasis has been placed on the science based on this wide-reaching, rapidly developing technology of "thermal spray coating" (TSC). The motivation for coating structural parts has been summarized as follows:

- *Improving functional performance* by allowing, for example, tolerance to higher-temperature exposure using thermal barrier coatings
- *Improving component life* by reducing wear due to abrasion, erosion, and corrosion
- *Extending system viability* by rebuilding worn parts to their original dimensions, avoiding the need for replacing the entire component, e.g., a shaft or axle
- *Reducing component cost* by improving functionality of a low-cost material with an appropriate coating

Process integration (PI) is the most important and critical step in setting up an efficient, smooth, flawless, reliable, cost-effective, and safe industrial-scale operation. It has to be conceived with a thorough knowledge of the technology involved, its potential and limitations, required infrastructure, potential hazard, as well as health and safety concerns to the operator. Thermal spray coating (TSC) is no exception in this respect, though some of the process design parameters can be particularly challenging due to the very broad range of conditions met in different thermal spray technologies. As illustrated by the temperature–velocity diagram given in Fig. 18.1, the temperature of the spray medium can vary between 2000 and 14,000 K with spray medium velocities varying between 100 and 2000 m/s or more. Thermal spray coating technologies also vary widely with regard to the energy sources used. For example, cold spray (CS) makes use of high pressure compressed gas as an energy source for the entrainment, acceleration, and projection of the powder to be sprayed on the substrate. Combustion-based technologies,

on the other hand, such as flame spraying (FS), high velocity oxy fuel (HVOF), and detonation-gun (D-Gun), rely on the use of a combustible gas, such as hydrogen, methane, or acetylene, or liquid fuel, such as kerosene, as a source of energy. Plasma-based technologies such as DC atmospheric and vacuum plasma spraying (APS and VPS), RF-induction plasma spraying (RF-IPS), wire arc spraying (WAS), and plasma-transferred arc (PTA) deposition are all electro-technologies using electricity as a sole source of energy for the spraying process.

It is obvious and unrealistic to aim in a single process design to integrate such a broad range of technologies. In a large-scale operation with diverse coating requirements, the integration of different TSC technologies is only possible through the setting up of dedicated parallel production lines designed for the different technologies needed, with common infrastructure and quality control (QC) laboratory facilities. In this chapter, process integration (PI) will be limited to the design of a production line dedicated to a preselected number of key TSC technology. The integration of multiple R&D and industrial production lines in a large-scale production facility is to be considered as a further step of overall plant design as described by numerous textbooks on the subject such as [Peters et al. (2003), Park et al. (1995)].

The basic process flow diagram for a single R&D or production-scale TSC line is schematically represented in Fig. 18.2. The process is split essentially into the following four principal steps:

- Surface preparation of the part to be sprayed involving cleaning, masking, and roughening of the surfaces to be sprayed
- Thermal spraying of the part using appropriate coating technology under optimized operating conditions
- Part post-treatment and finishing whenever necessary
- Quality control and packaging

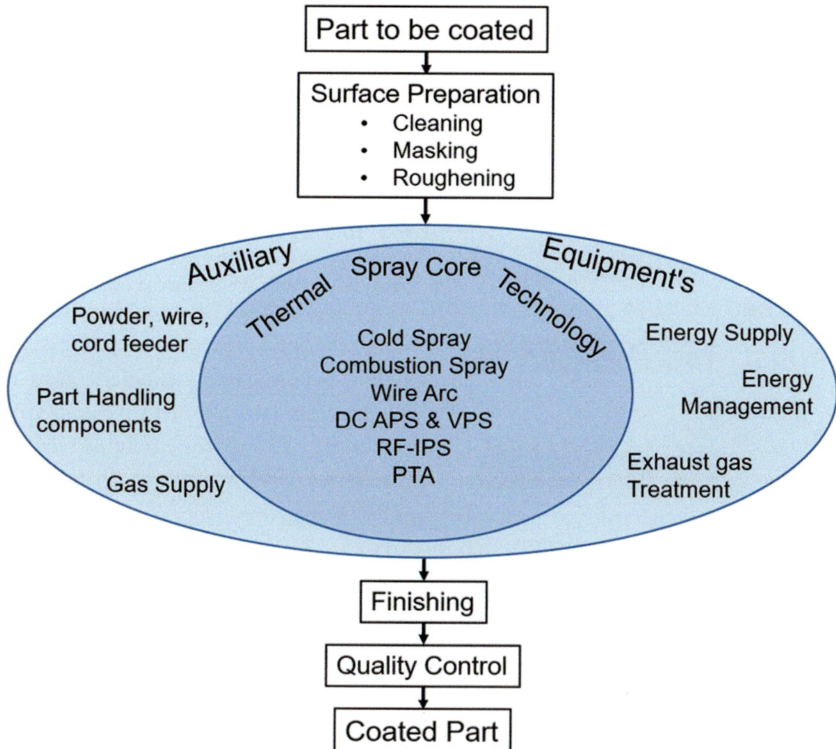

Fig. 18.2 Schematic representation of the principal process steps in a TSC operation

A detailed description of each of these steps is presented in the following sections. It is important to underline however that the successful realization of these steps will depend on the precise control and stability of all processing conditions. For this, a wide range of instrumentation is needed to monitor and control vital surface preparation, thermal spraying, and post-treatment/finishing of the part. A separate discussion is therefore included in this chapter about sensors and instrumentation which need to be integrated in the process design for the monitoring of the operation and control of the different process steps. As can be expected, the extent and level of instrumentation needed can vary significantly depending on the primary use of the installation, whether for R&D development of the spray conditions for a given part or the large-scale spraying of large number of parts.

Health, safety, and environmental impact issues are discussed in the last part of this chapter which need to be kept in mind and fully addressed throughout the design, realization, and operation stages of setting up a thermal spray coating installation independent of its size or scope of the operation.

18.2 Surface Preparation

18.2.1 Substrate Design Considerations

As reviewed in Chap. 14, "Surface Preparation," the preparation of a part for the thermal spray process starts at the design and manufacturing stage where a number of important rules have to be respected in order to avoid creating weakness points in the coating where cracks would initiate once the coated part is exposed to mechanical or thermal stresses. Some of the "Do" and "Not to do" in substrate design recommended by the American Welding Society (1985) [Davis ed. (2004)] are illustrated in Fig. 18.3. Further information on the subject is available in ASM Handbook Vol. 5A Thermal Spray Technology, by Tucker Jr. (ed.) (2013). The principal recommendation for spraying on a flat or cylindrical surface are:

- A coating must never end abruptly at the part extremity. A sharp edge as illustrated on the LHS of Fig. 18.3a may act as a stress raiser where cracks will develop, especially when loads are applied. Coating at the extremity of a part should accordingly present a narrow-feathered band, rather than a sharp edge.

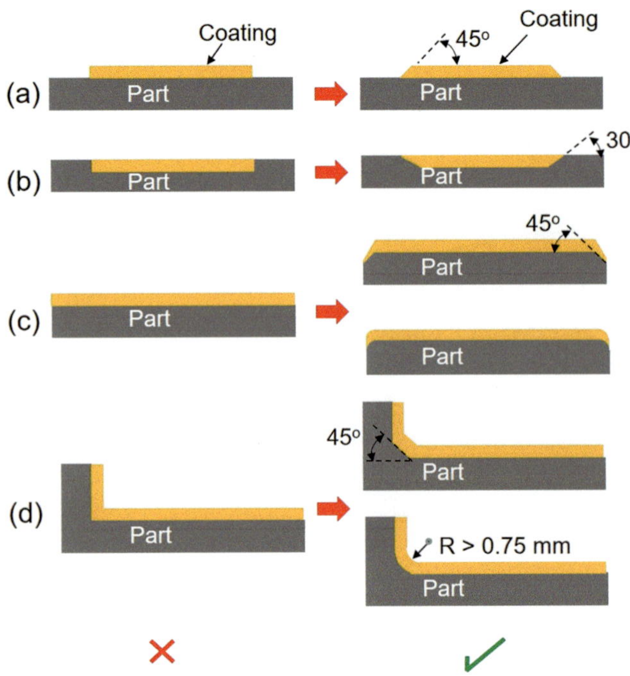

Fig. 18.3 Recommended machining for areas to be sprayed. (Reprinted with kind permission from ASM International [Davis (2004)] and [Thermal Spraying, American Welding Society (1985)]

- When undercutting a substrate area to accept the coating, the coating deposited must never end abruptly. In the undercuts, the corners must be chamfered with an angle of 30° or its cutting edge removed before spraying as illustrated in Fig. 18.3b. Sharp corners capture loose spray particles, dusts, and debris resulting in porous areas.
- For a part with a sharp corner at 90°, the corner must be chamfered at 45° as illustrated in Fig. 18.3c or rounded with radius of >0.75 mm to avoid the development of cracks.
- For a part with a sharp shoulders at 90°, it must be chamfered at 45° as illustrated in Fig. 18.3d or rounded with radius of >0.75 mm to avoid the development of cracks.

For the spraying of thick coatings or coatings on a substrate with rounded profile with a short radius of curvature, it is recommended to machine grooves or threads into the surface to be sprayed in order to enhance the adhesion of the coating to the surface.

18.2.2 Surface Cleaning

This is the first step for the substrate preparation. Before spraying, all contaminants, such as scale, oil, grease, and paint must be removed. After removing all contaminants,

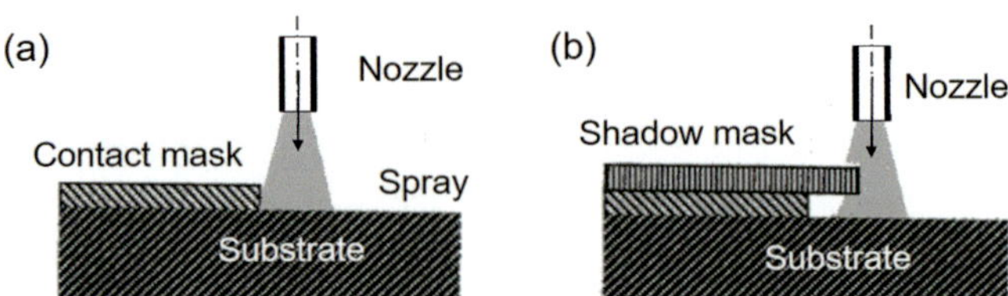

Fig. 18.4 Effect of masking technique on coating profile: (a) Contact masking, (b) shadow masking [Davis (2004)]. Reprinted with kind permission from ASM International

parts should be protected from airborne debris and fingerprints and should be handled with clean fixtures and materials. The main cleaning techniques as described in detail in [ASM Handbook Volume 05: Surface Engineering (1994)] are:

- Solvent degreasing
- Backing
- Ultrasonic cleaning
- Wet or dry blasting
- Acid pickling
- Brushing
- Dray ice blasting

18.2.3 Masking

Masking is necessary to prevent the deposition of the coating on areas where it is not wanted. It also improves the uniformity of the deposit when coating limited areas. A wide range of coating materials are used including self-adhesive tapes or hard masks. In either case, the material to be used should be resistant to the conditions to which it is exposed in the roughening and thermal spray coating stages of the operation. Two type of masks are commonly used as illustrated in Fig. 18.4a and b, contact masks and shadow masks, respectively.

18.2.4 Surface Roughening

After cleaning and masking, several methods are used to produce a surface to which the sprayed coating will adhere. Dry abrasive grit blasting is the most commonly used roughening technique by which dry abrasive particles are propelled toward the substrate at relatively high speeds. On impact, sharp, angular particles act like chisels, cutting small irregularities into the surface creating the so-called roughness depending on the nature and size of the grit used, the substrate material properties, as well as the grit-blasting conditions. Two types of machines are used for the grit-blasting operation:

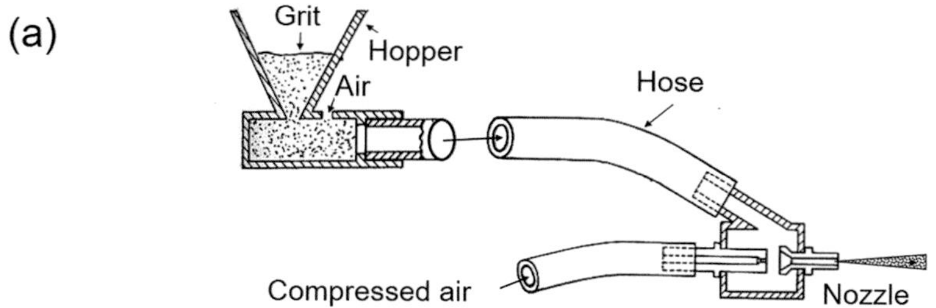

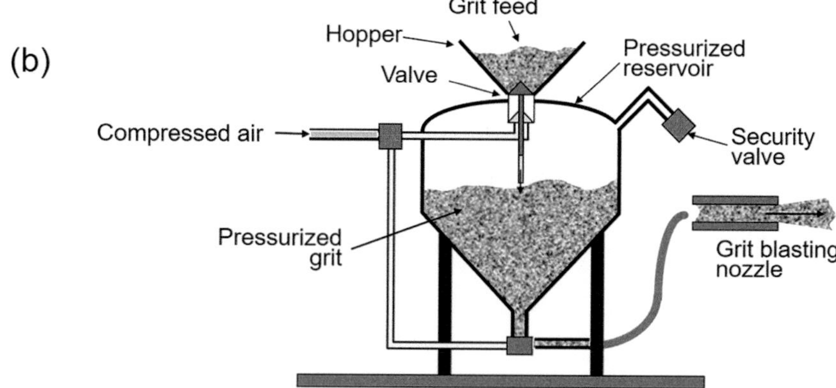

Fig. 18.5 Principle of (**a**) suction-type and (**b**) pressure-type grit-blasting machines

- Blast cabinets with suction-type nozzles
- Pressure machines or blast generators

Schematics of the principle behind the design of such machines are given in Fig. 18.5.

Grit materials commonly used for grit-blasting operations are:

- Aluminum oxide, whether brown alumina containing less than 1.5 wt.% free silica or white alumina, 99.5% pure with a hardness exceeding 9 Mohs
- Silicon carbide
- Silica sand, crushed garnet, crushed slag, or angular chilled iron

The choice of the proper grit for an application depends on the required surface characteristic, "roughness factor," substrate material, and economic and health–safety issues.

Special attention should be given to the problem of the embedment of grit residue in the surface of the substrate which increases with the ductility of the substrate material, becoming a real problem for lightweight alloys. In most cases, surface cleaning procedure after grit blasting, to get rid of as much residue as possible, consists in blowing compressed air at a pressure of 0.4–0.5 MPa with a nozzle i.d. of 4 mm for a few tens of seconds and then immersing the substrate in an acetone bath solution ultrasonically agitated.

High-pressure water jet roughening is an attractive alternate approach to grit blasting that avoids the problem of grit embedment combined with a significantly reduced health hazard and improved surface cleanliness. Abrasive water jetting (AWJ) is another alternative which can improve pertinent surface characteristics with a higher degree of surface roughness due to the abrasion of the water jet and abrasive particles.

It is important to stress that the surface cleaning and roughening steps should be carried out as close as possible to the spraying time in order to avoid the contamination of the surface with environmental dust or debris and/or the reformation of an oxide layer on the surface of the substrate as a result of its exposure to environmental conditions. Alternately, the part to be coated, depending on its size, could be vacuum-sealed rapping.

Laser treatment is another alternate approach for the roughening of the surface of the substrate which consists in coupling one or more Q-switched Nd:YAG lasers to the thermal spray torch. The PROTAL process developed by [Coddet and Marchione (1993, 1997, 1999)], illustrated in Fig. 18.6, combines in a single step the surface preparation with the spraying operation. The purpose of the laser irradiation is to eliminate the contamination films and oxide layers, to generate a surface state enhancing the deposit adhesion,

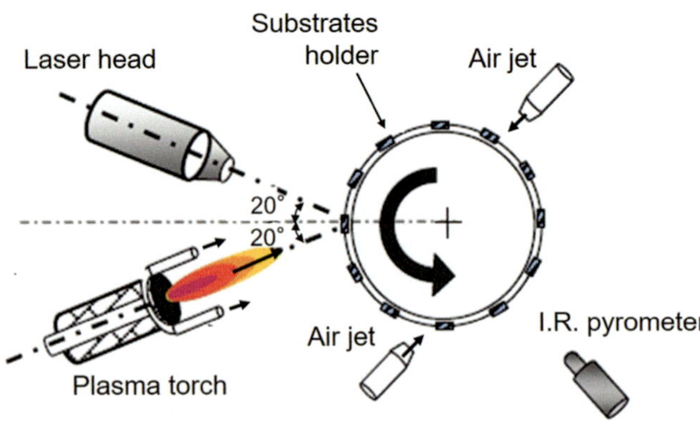

Fig. 18.6 Schematic of the integrated system for the PROTAL® process for air plasma spraying. Reprinted with kind permission from Springer Science Business Media [Coddet et al. (1999)], copyright © ASM International

and to limit the recontamination of the deposited layers by condensed vapors.

18.3 Thermal Spray System Design

The first step in setting up a thermal spray (TS) coating system is to identify the core TS technology to be used in the process which, in turn, closely depends on the objective of the installation whether for R&D or industrial-scale production.

For an R&D installation, diversity of the TS technology is of prime importance combined with the ability to use a wide range of feedstock materials in the form of powder, wire cord, solution, or suspension, the accessibility of reliable adequate instrumentation and control, as well as a well-equipped material characterization laboratory. Operator health and safety issues also have to be of primary concern.

For an industrial-scale production, diversity of the TS technology is less critical provided that the choice of the technology is based on a through comparative analysis of the different options available and their ability to produce a coating with the required functional properties. Productivity, reliability, proper instrumentation, and controls are also key elements to be considered. To this, it is necessary in a production-scale environment to insure operator health and safety, as well as viable economic factors such as capital investment, operating cost, consumables, and quality assurance.

As highlighted earlier in Fig. 18.1, and reviewed in detail in Part II Thermal Spray Technologies of this book, Chaps. 6 through 12, core TS technologies, include:

- Cold spray (CS)
- Flame spray (FS), high-velocity oxygen fuel (HVOF), and detonation gun (D-Gun) spraying

- DC atmospheric plasma spraying (APS), controlled atmosphere plasma spraying (CAPS), vacuum plasma spraying (VPS), and ultralow-pressure plasma spraying (ULPPS)
- RF-inductively coupled plasma spraying (RF-IPS)
- Wire arc spraying (WAS)
- Plasma-transferred arc (PTA) deposition

Of these, CS and PTA deposition are relatively focused technologies that are well adapted to a limited number of applications. CS, for example, is essentially limited to the spraying of ductile metallic powders for electrical and electromagnetic applications. The technology is also used to a lesser extent for repair and restoration of parts especially those made of magnesium alloys and a limited number of corrosion and wear resistance applications.

PTA on the other hand is mostly used for the deposition of thick coatings on metallic substrates mostly in areas of wear and corrosion protection as well as for the rebuilding of worn-out parts to their original dimensions.

A comparative analysis of other TS technologies including combustion and plasma-based processes is presented *in* Table 18.1 *giving* for each technology the:

- Form of the feed material used, powder, nanopowder, solution, suspension, wire, or cord
- The nature of feed materials whether metallic, ceramic, or cermet
- Energy source such as gas and liquid fuels or electricity including DC and radio-frequency (RF) sources

For each, information is given on the most common power range of operation, nature of the spray medium, typical spray gas temperature and velocity at the exit level of the spray gun, range of powder feed rates commonly used, and maximum particle/droplet impact temperature and velocity on the substrate.

While coating properties depend not only on the spray conditions, they also depend on the thermophysical properties of the sprayed material; values are given of typical range of coating density, bulk strength, and, for metallic coating, oxide content either wt.% or parts per million (ppm). Corresponding area of application of each of these technologies has been extensively discussed in Chaps. 6, 7, 8, 9, 10, 11, and 12.

The bottom two lines of the table provide an indicative value of necessary capital investment and operating cost associated with TS technologies. A more detailed discussion of the technoeconomic aspects of the technology and its most widely accepted industrial-scale applications is given in Chap. 19, "Industrial Applications of Thermal Spray Technology."

Table 18.1 Comparative analysis of different thermal spray (TS) technologies

		Combustion spraying			Wire arc spray	DC Plasma spraying			Induction plasma spraying
		FS	HVOF	D-gun	WAS	APS	HE-APS	VPS	RF-IPS
Feed stock	Powder	X	X	X		X	X	X	X
	Nano	X	X	X		X		X	X
	Solution	X				X			X
	Suspension	X				X			X
	Wire	X			X				
	Cord	X			X				
Material	Metal	X	X	X	X	X	X	X	X
	Ceramic	X	X	X	X	X	X	X	X
	Cermet	X	X	X	X	X	X	X	X
Energy	Comp. gas								
	Gas-fuel	X	X	X					
	Liq.-fuel		X						
	Electricity				X	X	X	X	X
Spray medium	Power (kW)	25–75	100–300	100–300	2–6	30–100	100–250	40–120	15–200
	Type of gases	O_2, C_2H_2	CH_4, C_3H_6, H_2, O_2	O_2, C_2H_2	Air, N_2, Ar	Ar, He, H_2, N_2,	Ar, He, H_2, N_2,	Ar, He, H_2, N_2,	Ar, He, H_2
	Gas flow rate (slm)	100–200	400–1100	166.00	500–3000	100–200	280–460	150–250	75–150
	Gas flow rate (m³/h))	6–12	24–66	10	30–180	6–12	17–28	9–15	4.5–9
	Jet temperature (K)	3300	3000	4000	4500	12,000	12,000	12,000	10,000
	Jet Velocity (m/s)	50–100	500–1200	>1000	50–100	300–1000	300–1000	200–600	20–80
Powder	Max spray rate (g/min)	100–150	230	15	250	80	400	150	300
	Max spray rate (kg/h)	7–9	14	1.00	16	5	23	10	20
	Droplet impact Temp. (K)	2500	3300	N/A	>3 800	>3800	>3800	>3800	>3800
	Droplet impact vel. (m/s)	30–180	600–1000	900–1000	250.00	200–300	240–1200	240–610	20–300
Coating	Coating density range (%)	85–90	>95	>95	80–95	90–95	90–95	90–99	95–99
	Bulk strength (MPa)	7–18	68	82	10–40	68	68	>68	>68
	Oxide content (wt.%)	4–6%	0.2	0.1	0.5–3	0.5–1	0.1	ppm	ppm
Cost	Capital investment	$	$$	$$$$	S	$$	$$$	$$$$	$$$$
	Operation cost	$	$	$	$	$	$$	$$	$$

A schematic representation of the different categories of equipment that are necessary for the setting up of thermal spray installation, using, for example, atmospheric plasma spraying (APS), as defined by Oerlikon Metco, one of the leading equipment manufacturers in this field, is given in Fig. 18.7. Comprehensive "Thermal Spray, Equipment Guide" of commercially available equipment was published by Oerlikon Metco, issues 5 and 11 October 2014, and "An introduction to Thermal Spray," issue 6 July 2016.

The basic "system structure" which can be applied to the broad field of thermal spray coating technologies involves the following three groups of *components*:

- *Core components*: including the spray torch/spray gun, power supply, powder feeder, heat exchanger, and materials supply
- *Handling components*: including robotic manipulator for the spray gun, workpiece manipulator, and handling controller

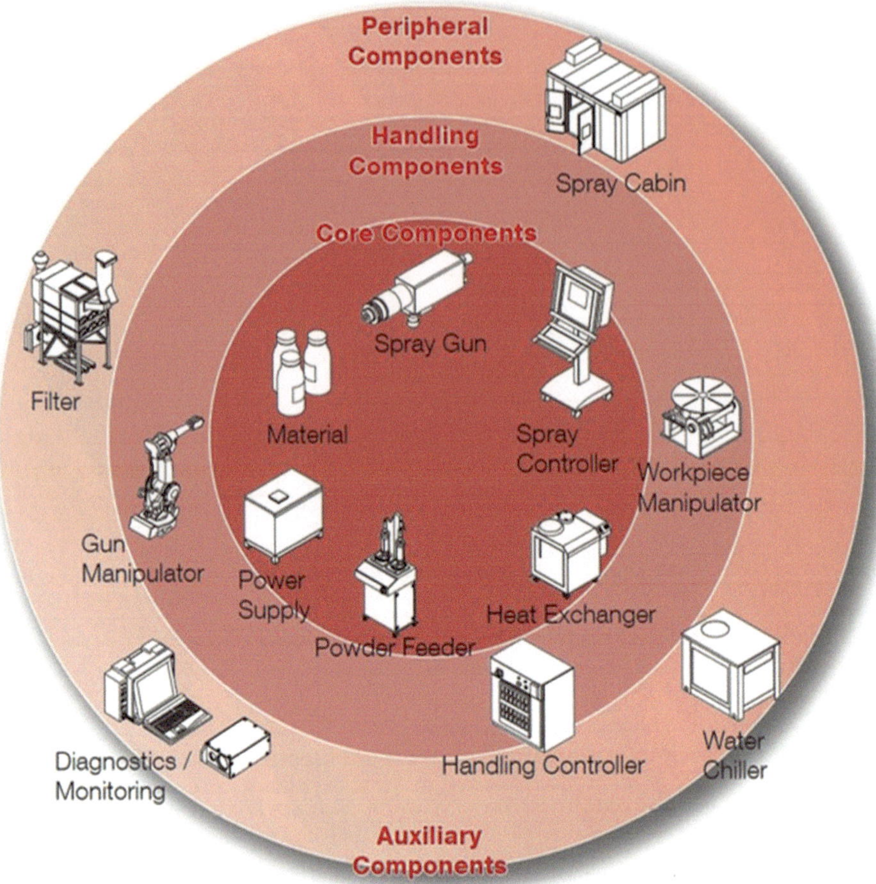

Fig. 18.7 Key elements of an atmospheric plasma spraying (APS) solution. [Oerlikon Metco, Technical Bulletin, #5, October 2014] (Reproduced with kind permission from Oerlikon Metco)

- *Peripheral/Auxiliary Components*: including spray booth, water chiller air filter, or spray chamber, gas-line filters, and vacuum pumping station in the case of vacuum plasma spraying or plasma spraying under controlled atmosphere and process diagnostics/monitoring and control

Obviously, the extent of the equipment needed for any specific application will vary widely depending on whether the technology is "cold-spray based," "combustion-based," or "plasma-based." In this section, a brief general description is given of the main features of each of these categories of equipment. This is followed by a conceptual design of the respective systems for each of the principal thermal spray coating technologies commonly used on an industrial scale.

18.3.1 Spray Torch/Gun

The "spray torch," or commonly referred to as "spray gun," varies widely depending on the technology used and process

requirement. The large majority of spray torches are designed to be "handheld" to be used by the operator to spray coat parts of different sizes and shapes. Their design in this case has to ensure that they are relatively light, with a weight below 5 kg, and integrate essential safety features to avoid operation under unsafe conditions. These will also have to be used in combination with the necessary personal protective equipment to ensure the safety of the operator under all conditions.

With the growing need for higher production capacities and reproducible and precise control of the spraying conditions, torch designs have evolved toward the larger and heavier units operating at power levels of 100 kW or more, which forced the technology to adopt robotic manipulation of the torch and of the workpiece which is becoming increasingly the standard, with exemption made for the TS coating of large infrastructures such as bridges, building, or marine halls, using flame spraying or wire arc spraying, which remain up until now essentially a manually executed operation.

18.3.2 Power Supply

For wire arc spraying and plasma spraying, an adapted DC power supply is needed with current control and an open circuit voltage (OCV) at least double that of the operating voltage of the torch. The "head-room" defined as the difference between the OCV and the operating voltage is needed to avoid arc extinction due to the continuous variation of the arc voltage across the torch associated with the continuous movement of the arc-root position on the surface of the anode. Older power supply units, typically welding-type power supplies, available from tens to hundreds of kW and more, were designed using a saturable core reactor combined with a bulky shock to smoothen out the ripples on the arc current. These are rapidly being replaced by thyristor-controlled units of six or twelve pulse design which are considerably lighter and easier to control at any required current level. While the twelve pulse units generally have lower current ripple than the simpler, six-pulse design, an inductive choke is generally necessary to smoothen the arc current.

It is important to point out that for a WAS, DC-plasma spraying, or PTA deposition, the operating arc voltage will depend on the torch design, as well as the nature of the plasma-forming gas. The lowest voltage requirement is associated with operation with pure argon as plasma gas. In most cases, however, the addition of a molecular gas such as hydrogen, nitrogen, oxygen, or air is necessary to enhance heat transfer between the plasma and the spray material resulting in a significant increase of the arc voltage. The choice of the additive gas nature and concentration, or even the operation with a pure molecular gas such as nitrogen, has to be taken into account during the design phase of the installation, depending on the chemical nature of the material to be sprayed. As a general rule, it is recommended to favor operation in general with a high arc voltage and low arc current since electrode erosion and consequently electrode lifetime are directly related to the arc current rating of the operation.

In a given installation, the electrical power supply for all thermal plasma spraying applications must be placed outside the spray booth (SB). Electrical cables between gun and power source must be as short as possible, and the high-frequency starter for DC arcs must be located as close as possible to the spray gun. This limits the antenna effect of the cables transmitting radio frequency. For the same reason, a certain distance between these cables and the metal conductors such as pipes and structural supports must be maintained. It is also preferable to separate electrical cabinets, consoles, or electrical boxes from an area where combustible gases are present. If a cabinet contains both electrical components and combustible gas-carrying components, it must be designed with strong efficient ventilation.

18.3.3 Gas Supply

Gasses are an essential consumable item of all thermal spray coating operations. For combustion-based technologies such as FS, HVOF, HVAF, or D-gun, gases used provide the energy source used in the process through combustion. Gas fuels used in these case are hydrogen (H_2), acetylene (C_2H_2), methane (CH_4), and propane (C_2H_6). These are combined with oxygen (O_2) and air as oxidants. For plasma-based technologies, most commonly used gases are argon (Ar), nitrogen (N_2), air, hydrogen (H_2), and helium (He).

Combustible gases are either supplied in compressed gas reservoirs or in the case of CH_4 or C_2H_6 could be available on-line. Other gases such as Ar, N_2, and O_2 are available in large cryogenic reservoirs, or smellers bulks, while H_2 and He are generally available in batteries of multiple compressed gas cylinders. Air as process gas is generally generated on-site using an oil-less compressor combined with adequate filtering and air-drying units.

In all cases, gas storage is generally favored outdoors or, if necessary, in an open shed with efficient air circulation. Gas piping should always be metallic, steel, stainless steel, or copper, preferably welded and tested for leaks on a regular basis. Within control cabinets involving gases, piping should preferably be in stainless steel or copper, welded or soldered with reliable fittings for demountable subsystems. More information on the subject can be found in [Dobler and Gifford (2003), Creffield and Cole (1999), Heinrich et al. (1999), Hale et al. (2002)]. In spray booths, one can find sophisticated gas control consoles. The safety precautions are outlined in [Dobler and Gifford (2003)]. The interested reader can also find information on the industrial gases used and quality issues in [Kroemmer and Heinrich (2006)].

18.3.4 Feed Material Supply

Feed materials used for TS application are mostly available in the form of powders, wires, or cords. A detailed review of the standard powder characterization techniques and their respective manufacturing procedures can be found in Chap. 13, "Powders, Wires, and Cords." Special attention has to be given to quality control of the as-received feed materials. Differences can exist between factory-certified particle size distribution (PSD) of powders and that of the as-received powders, especially for agglomerated, friable powders which can deteriorate during the transport process. Particularly critical for micron-sized powders is to avoid the presence of even small percentages of submicron powders in the mix since, once injected into a flame of plasma stream, submicron particles will tend to vaporize, cooling the spray stream and forming soot on condensation which will pollute the coating as well as the environment.

For powders:

- Particle morphology using optical or electron microscopy
- Elemental analysis, using ICP-AE or ICP-MS
- PSD, using screening or light diffusion techniques
- Flowability, using Hall flow
- Apparent and tap density
- Oxygen content for reactive metal powders

For wires or cods:

- Wire diameter and cross section, for cords
- Elemental analysis
- Stiffness and the presence of kinks
- Oxygen content for reactive metal powders

With the increasing use of solution and suspension plasma spraying, feed materials could also be received in the form of bulk chemical or of nanosized particles. In the latter case, their quality control should include the following:

For nanosized particles:

- Particle morphology using electron microscopy
- Elemental analysis, using ICP-AE or ICP-MS
- PSD, using light diffusion techniques
- Specific surface area measurements using BET techniques

Special attention has to be given to the handling and storage of feed materials to avoid the deterioration of their quality and for safety reasons as well. Powders should generally be stored in well-ventilated dry spaces with adequate spacing between different lots and different materials to avoid cross-contamination. Particularly important in the case of metallic powders, especially those of reactive metals, is to take all necessary standard measures to combat fire hazard.

Powder feeding to the spray device is generally carried out using an appropriate, well-calibrated powder feeder of the Archimedes screw or rotating disc type with the powder transferred pneumatically to the spray torch. The powder feed rate must be precisely controllable and stable over long periods of time. Special attention has to be given to feed rate stability with time. Archimedes-type screw powder feeder suffers often from feed rate pulsations depending of the flowability of the powder used. Such pulsations can generally be mitigated by the addition of a feed rate homogenizer unit in the powder transport line. The velocity of the powder carrier gas in the transport line has to be maintained above the "saltation velocity" of the powder, typically above 10 m/s, to avoid the deposition of the powder in the transport line leading to slugging flow conditions. The i.d. of the powder transport line has to be small enough to minimize the total gas flow rate needed for the pneumatic transport of the powder, without chocking it, giving rise to excessive pressure drops between the powder feeder and the injection point of the powder into the spray torch. Three types of powder feeders are commonly used in thermal spray operation [Davis (2004)].

Gravity-feed hoppers (the oldest one): it consists of a simple container, usually with conical-shaped bottoms, with an angle, α, above the "angle of internal friction" of the powder, generally $>60°$, which allows powder to flow freely, assisted by gravity from the bottom of the feeder. They are inexpensive and trouble-free, provided the feed stock powder is free-flowing, but they are not accurate, and it is difficult to control the gas flow independently of the powder rate.

Volumetric powder feeders: packets of powder are delivered by a positive-displacement mechanism into a carburetor where they are picked up by the carrier gas stream. The mechanism involves a screw or a wheel rotating within a bed of powder. Vibrators can be used for powders with poor flow ability. Scrapers can be used to deliver a constant quantity of powder with the wheel system. Using a "cylinder mixer," which consists of a conical chamber with tangential injection and an axial exit, the powder pulsation due to the principle of the system can be damped.

Fluidized beds: the powder is fluidized, and the design is such that a controlled volume of gas powder is sampled with a tube passing through the bed at a uniform rate. Their operation is very much affected by the powder characteristics and particle size distribution.

The most sophisticated powder feeders have a feed rate control, achieved with a weight loss device that uses a load cell to continuously monitor the contents of the hopper. The feed rate is determined by the rate of weight loss averaged over time. The system is capable to adjust for variations, thus keeping the feed rate constant. A limited number of powder feeders are equipped with a hopper heating option that allows to keep the powder at a moderate temperature in order to limit any moisture pickup. The use of preheat powder before their injection was reported by [Czernichowski et al. (2000)] who claimed to notice enhanced contact between grains.

For liquid or suspension injection, attention has to be given to maintaining the feed material at a constant temperature and in constant mild agitation to avoid solid precipitation and insure the uniformity of the solid content of the suspension throughout the process. The addition of a dispersant/surface active agent such as Darvan 7 [Bouyer E. et. (1997)]

is commonly used to stabilize the suspension. Two main injection techniques are used:

Spray atomization where a low-velocity liquid is injected inside a nozzle where it is fragmented by a gas expanding within the body of the nozzle [Filkova and Cedik (1984), Bouyer et al. (1997), Marchand et al. (2011), Cotler et al. (2011)]

Mechanical injection through the use of either a magnetostrictive rod at the back side of the nozzle which superimposes pressure pulses at variable frequencies (up to a few tens of kHz) or a liquid in a pressurized reservoir from where it is forced through an appropriate nozzle [Oberste-Berghaus et al. (2006), Etchart-Salas et al. (2007)]

18.3.5 Spray Gun and Workpiece Manipulators

As mentioned earlier, robotic torch and workpiece manipulation are essential for any large-scale thermal spraying operation. By accurately positioning the spray torch with respect to the surface of the substrate to be sprayed, it is possible to maintain optimal spraying distance throughout the operation and ensure consistent coating quality. Compared to handheld straying operation, the use of robots for torch and substrate manipulation relieved the design engineer, to some degree, from the concern about the weight of the torch and that of the part to be coated. The manipulation equipment is often computer controlled by specific software or hardware, and frequently, an operator in close proximity defines the movement. There is a three-dimensional envelope through which the arm can move and a larger area where a safety hazard may result from unexpected arm movement with the possibility for pinch hazard reaction between the arm and other physical elements of both the interiors. Much work has been devoted to development of robotic trajectories for thermal spray coating operations [Kutay and Weiss (1992), Fasching et al. (1993), Nylen et al. (1996, 1999), Eidelman and Yang (1997), Bonnet et al. (1999), Guessasma et al. (2004), Trifa et al. (2005), Candel and Gadow (2009), Floristán et al. (2012), Deng et al. (2012)].

18.3.6 Control Console

The control console houses the central programmable logic controllers (PLCs) to which the signal from all distributed instrumentation throughout the Thermal Spray Coating systems leads. These include:

- Gas flow rates, whether gaseous fuel in combustion spraying or torch and powder carrier gases in plasma spraying or other auxiliary gas streams. These are usually monitored using mass flow controllers (MFC), rotameters, or other flow monitoring devices.
- Cooling-water flow rates in the different cooling water loops between the spray system and the heat exchanger (HE) as well as between the heat exchanger and the water chiller (WC).
- Cooling-water line and gas pressures though out the systems.
- Cooling-water temperature in and out of the system.
- Power supply vital signs including arc current and arc voltage.
- Robotic system vital signs including the spraying distance and relative velocity between the spray torch and the substrate.
- Auxiliary process instrumentation such as substrate surface temperature and/or in-flight temperature or mean velocity of the particle flux moving toward the substrate.
- Often these are complemented by a recording of the digital imaging of the spray process either for a production record or a closed loop to comeback to clarify any unexpected event.

The first role of the PLC in the control console is to simply archive all of the process information for future examination and analysis and to verify, on-line, that the conditions are within the prescribed safety limits for each of monitored variables. In the event of any deviations from the preset values, the PLC is normally programmed to alert the operator by taking one of the following actions:

- Hi/Lo warning clearly identifying the monitored parameter concerned
- HiHi/LoLo warning indicating a larger deviation from the set value
- Alert combined with a system "EMERGENCY STOP"

PLC programming can also include corrective action to restore the operation to its present value. This role is, however, considerably more demanding since it implies a fundamental understanding of the process dynamics.

Other than direct control of process control, the PLC can be programmed to carry out a number of facilitating tasks for process operation such as:

- Graphical display of the trends for the variation with time of some of the process parameters such as arc current and voltage or some of the in-flight particle parameters such as particle temperature or velocity.
- Energy balance computation based on cooling water flow rates and temperature differences across different system

components. Such analysis can shed light on the energy efficiency of the system and its variation with time.

- By following the dynamics of arc voltage fluctuations including amplitude, and frequency spectrum, the operator can deduce the state of the electrode erosion and take a decision about the timing for electrode replacement in order to restore the system to its original preset condition.

18.3.7 Spray Booth

The use of a well-designed and equipped spray booth is strongly recommended for all thermal spray coating whenever the size of the parts to be coated allows it. Its principal role is to allow carrying out the spraying operation in a confined volume minimizing the exposure of the operator and ambient environment to the intense acoustic noise and ultraviolet (UV) radiation associated running of a combustion flame or plasma torch in an open atmosphere. It also limits the uncontrolled dispersion of overspray dust, which often represents ≈50 wt.% of the powder feed into spray gun, and fumes (mostly oxides with a mean particle diameter below 1 μm) resulting from the condensation of any vaporized metallic or ceramic powder into the atmosphere with severe associated health hazard [Gifford et al. (2003)]. An excellent document on the subject has been prepared by the Thermal Spray Society (TSS), ASN-TSS Safety Committee, "Thermal Spray Booth Design Guidelines, SG003-03," which is available on the TSS website, "Safety Guidelines."

As schematically represented in Fig. 18.8, after [Gifford et al. (2003)], a spray booth is a well-constructed and acoustically insulated chamber with nonflammable insulation (e.g., fiberglass) thick enough (in the range of 15 cm) to damp the sound emitted by the process down to tolerable noise level. Detonation devices which can generate a noise level as high

as 145 dBA will require double-walled booths with a heavier mass of sound absorbent material. The booth should be equipped with wide heavily insulated access doors allowing for the introduction and removal of relatively large pieces of equipment. The doors should be also equipped with relatively large observation windows, preferable with photosensitive coatings that darken when exposed to intense UV radiation, or simply adequate pull-down shades. They should also have an interlock connected to the system's control console to prevent the gun from starting when personnel is inside the booth, and most of all the doors should be airtight once closed.

The booth should be also equipped with an adequate ventilation system allowing the complete renewal of air in the booth at a minimum rate of five (5) exchanges of the entire booth air volume per minute [Gifford et al. (2003)]. The exhaust, dust, and fume-loaded air from the chamber should be channeled to an adequate large-volume air filter for recovery of the dust and fumes before the air is released into the atmosphere. The booth should also have strategically located access ports for the introduction of power and control cable, cooling water in and out as well as process gases and spray feed material.

In the internal booth layout, it is important to streamline the airflow capability to ensure the capture and entrainment of dust particles out to the filters. The respiratory ailments caused by feedstock materials appear as the major risk for operators as described by [Heriaud-Kraemer et al. (2003)]. To get rid of dusts, the use of laminar flows is mandatory; turbulent flows often create areas where the dust is trapped. In most cases, the pressure within the booth should be slightly negative (by about 25–124 Pa) adjusting the dampers on the suction end of the ducting downstream of the filter. This lower pressure results in an inward flow through small leaks, carrying the dust to the filters. Using an entry timer

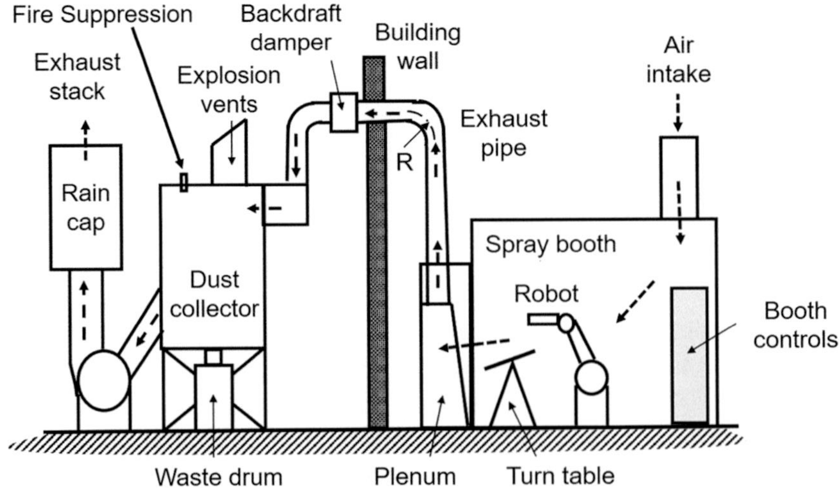

Fig. 18.8 Schematic of the spray booth and the ventilation system. Reprinted with kind permission from ASM International [Gifford et al. (2003)]

to allow the ventilation system to clean the air of hazardous dust and fumes after stopping the spray gun is an extra precaution that would prevent accidental exposure of the operator to polluted air in the spray booth. It is also recommended to use a dual-speed exhaust system maintaining a low-level ventilation mode at all times when the booth door is open and when operators are inside the booth for system maintenance and setup of the spraying operation [Gifford et al. (2003)].

The air inlet can come from the inside of the building or from the outside, but in the latter case, consideration must be given to the heating or cooling of this air. The position of the air intake and exhaust plenums is crucial to achieve laminar flow. It requires using an inlet plenum with many small unevenly distributed inlet air openings and fewer larger openings to create turbulence. As much as possible, obstacles (cooling, parts, fixtures) must be out of the airflow. As shown in Fig. 18.8, the exhaust plenum should be at the opposite end of the spray booth from the air intake to smoothen the flow. The plenum is generally on the wall just beyond the spray area. In this figure, the airflow is from the back of the gun to the front, and the robot is placed near the center of the booth facing the exhaust plenum/spray hood with the part to be sprayed on the turntable between the exhaust plenum and the robot [Gifford et al. (2003)].

18.3.8 Exhaust Gas/Air Evacuation and Filter

The recommended velocity inside the ductwork between the spray booth and the dust collector should be between 17.8 and 20.3 m/s [Gifford et al. (2003)]. The lower-velocity value is chosen to prevent dust deposition on the walls and the higher value to limit the air and dust momentum. The number of bends in the duct work between the booth and the air filter must be minimized to reduce unnecessary pressure drop. The air velocity when entering the dust collector must be reduced to prevent the erosion of the filters and disperse the air across them. Self-cleaning, cartridge, or bag-type filters are used to capture overspray particles and fumes. These must be equipped with a blowback or periodic shaking system to dislodge accumulated dust layers from the surface of the filter medium falling down in the bottom hopper of the filter. It is important also to remember that the position in which a filter cartridge is mounted in the dust collector has a major impact on the effectiveness of the pleated media within the filter. In cases of heavy dust load, cyclone collectors are added upstream of the filter to capture larger size fraction of the dust ($d_p > 5$ μm) before they reach the primary dust collector, while secondary filters (high-efficiency particle air, HEPA)

may be used downstream of the filter for a higher cleaning efficiency depending on local regulation (they capture 99.7% of fine particles down to 0.3 μm). Ductwork must prevent hot particles from reaching the dust collector. A back-draft damper is recommended at the point where the ductwork leaves the building. It prevents the reverse flow of air from outside into the spray booth when the exhaust system is off and also the back flow of smoke in case of a fire in the dust collector [Gifford et al. (2003)].

18.3.9 Cooling Water Chiller and Heat Exchanger

A closed-loop cooling water is an integral part of any plasma spraying, D-gun, and most HVOF spray systems. It has to be adequately designed for the total power of the system, thus providing a safety margin of cooling capacity depending on the thermal efficacy of the system. The cooling capacity being generally backed by an adequate chiller, which cools the hot water against the air, or a cooling tower. Normally two cooling circuits are used. A low-pressure circuit typically has 60–80 psig (400–550 kPa), mostly for cooling auxiliary equipment whenever necessary, while the high-pressure circuit typically has 200–250 psig (1.4–1.7 MPa) for the plasma, D-gun, or HVOF spray torch. Particular attention has to be given to the location of the flow monitoring and controller elements in the circuit which has to be placed downstream of the device being cooled especially for the high-pressure circuit. The justification stems from the need to keep a high backpressure over the critically hot surfaces to avoid local boiling of the water which can result in an almost one or two orders of magnitude drop in the hat transfer coefficient between the surface and the water flow. Such action, however, requires that the spray torches and associated piping and hoses used should be all designed for high-pressure operation. The water quality and temperature also have to be well controlled. Mineralized water is not recommended because of calcia deposits forming in the hottest area, reducing the heat transfer where it should be the highest. With deionized water being somewhat corrosive, the best is to use distilled water, or equivalent, and monitor its electrical conductivity regularly to change it when it becomes too conductive. Cooling water temperature must be above the dew point of the ambient atmosphere to prevent condensation which can be very detrimental in the anode nozzle of a plasma torch or in the chamber of HVOF, while the return water temperature of the cooling water circuit should generally not exceed 40–45 °C.

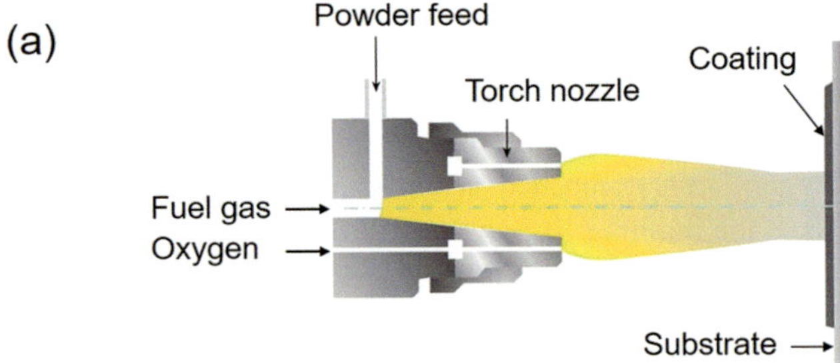

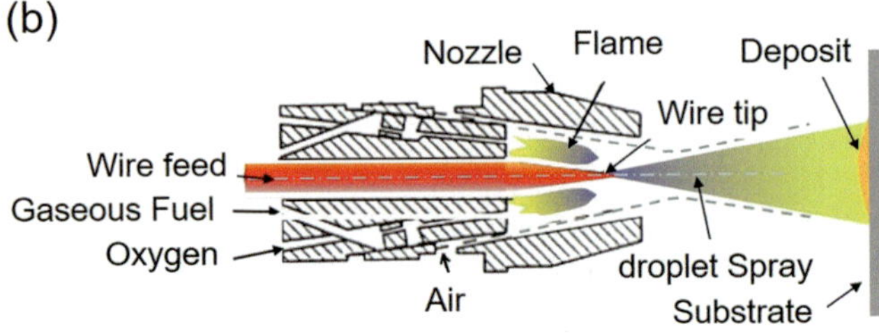

Fig. 18.9 Schematics of typical flame spraying torch used with (**a**) powder and (**b**) wire feed stock. (Reproduced with kind permission from Oerlikon-Metco)

18.4 Examples of Integrated TS Systems

18.4.1 Combustion-Based Spraying

Three subgroups are identified in the area of combustion-based thermal spray technologies.

18.4.1.1 Flame Spraying

Flame spraying (FS) is at the origin of the broad field of thermal spaying which was developed in the early twentieth century and continues to be used on an industrial scale for a wide range of applications mostly for corrosion protection of infrastructure such as bridges and steel structures. The technology as described in Chap. 7, "Combustion Spraying," Sect. 7.2, is based on the feeding of the sprayed material in the form of powder or wire/cord into a gaseous fuel burner as shown in Fig.18.9a, b, respectively. Photographs of such devices are shown in Fig. 18.10a, b with the spraying system shown being for the simplest of such devices with the spray torch, handheld by the operator. In the case of Fig. 18.10a, the powder to be sprayed is placed in a canister above the torch in which it is entrained by the flowing gases, while in Fig. 18.10b, for wire-FS, the wire spool is sparely placed with a wire feeding device. The work peace to be spayed in both case is a small component placed in a handling rig. The full operation can be carried out in this case in a spray booth as illustrated in Fig. 18.11 using a robotic arm for the torch manipulation with respect to the part. The booth is to be also equipped with ambient air evacuation ducts leading to a filter for proper dust collection prior to releasing the air into

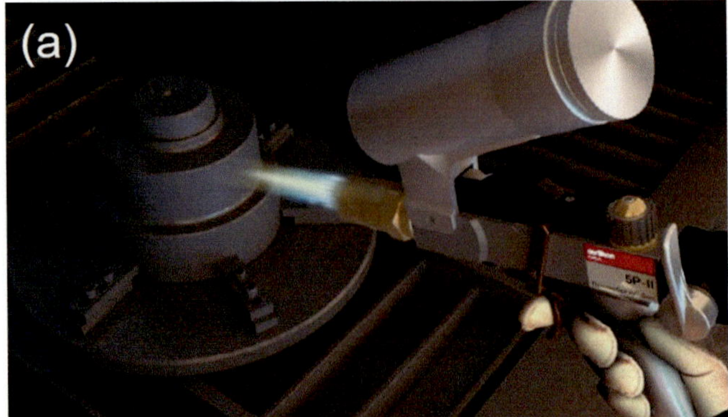

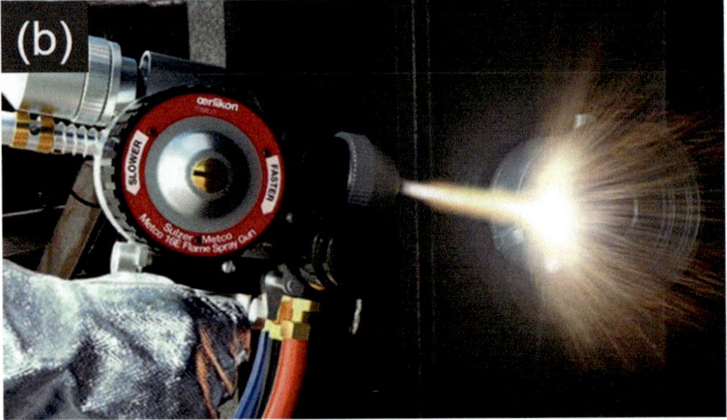

Fig. 18.10 Photographs of typical flame spraying torch used with (**a**) powder and (**b**) wire feed stock. (Reproduced with kind permission from Oerlikon-Metco)

the atmosphere. Robot controller and process control modules and the wire feeder are placed outside the spray booth allowing the operator to run the process with its doors closed protecting him from exposure to the dust-laden environment in the booth. The operator's penetration into the booth will be thus limited to the replacement of the work-piece/pieces and adjustment of the overall setup between spraying runs.

Obviously in the case of using FS technology for the spray coating of large infrastructure parts like buildings or bridges, appropriate canvasing around the region being sprayed has to be used to limit the spread of pollution into the environment in the vicinity of the workplace. Moreover, the operator will have to be equipped with the appropriate personal protection gear including efficient breathing masks, goggles, and face shield.

18.4.1.2 High-Velocity Flame Spraying

High-velocity flame spraying encompasses HVOF which stands for "high-velocity oxygen fuel" in which the spray gun is fed with combustible gas or liquid fuel and oxygen, while HVAF stands for "high-velocity air fuel" where oxygen is replaced by air resulting in lower temperatures and higher gas velocities. Both are very aggressive thermal spray technologies with gas temperature in the 3000–3300 K rang

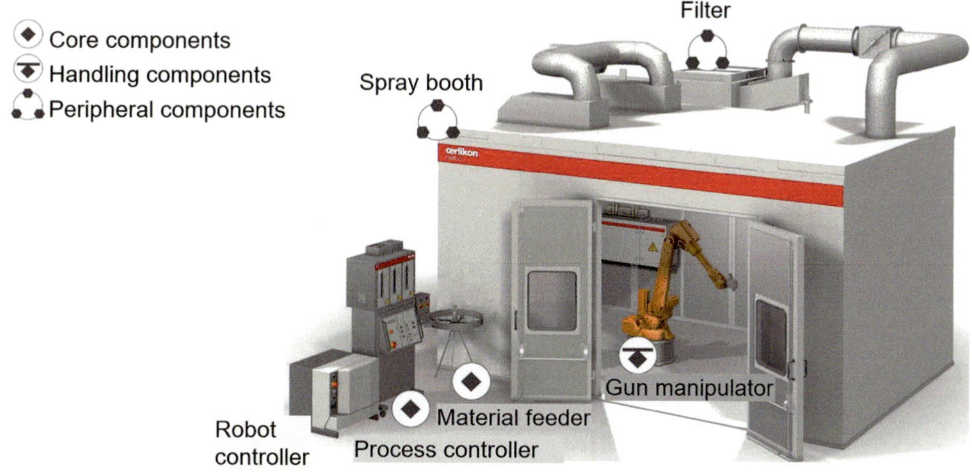

Fig. 18.11 Schematic representation of a typical flame spraying booth showing different components associated with the technology. [Oerlikon Metco, Technical Bulletin, Thermal Spray Equipment Guide, issue 11, p. 32 October 2014] (Reproduced with kind permission from Oerlikon Metco)

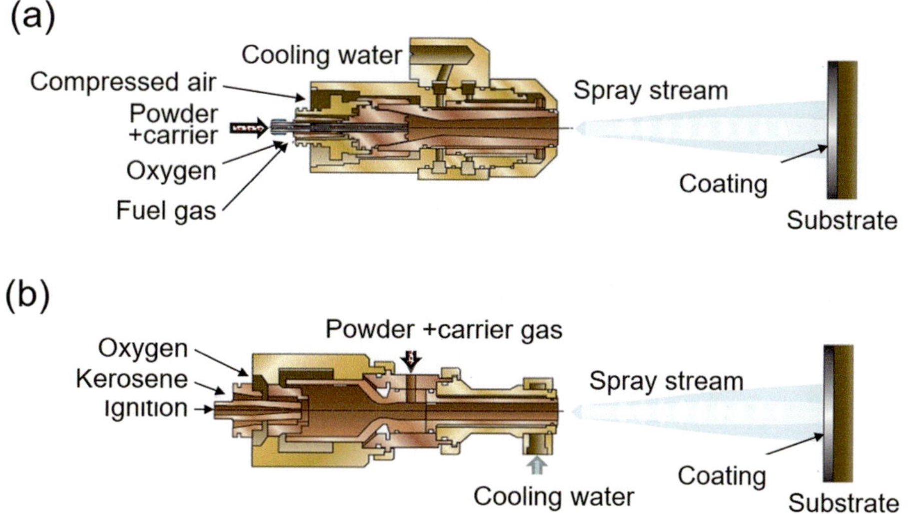

Fig. 18.12 Schematics of typical HVOF, water-cooled, spraying torch for use with (**a**) gaseous fuel and (**b**) liquid fuel. (Reproduced with kind permission from Oerlikon Metco)

and gas velocities varying between 900 and 1600 m/s depending on the fuel/oxygen ratio. As described in Chap. 7, "Combustion Spraying," Sect. 7.3, numerous torch designs have been developed for operation, using gaseous fuel such as propane, propylene, hydrogen, or methane (natural gas) as the fuel source or liquid fuel such as kerosene. Principal design features of the Oerlikon Metco torches are given in Fig. 18.12a, b operating with gaseous and liquid fuels, respectively.

Photographs of the Oelikon Metco WokaStar 610 plasma torch in operation, in the absence of feed powder, are given in Fig. 18.13, while those of the Praxair, JA-8000, are given in Fig. 18.14a, b, in the presence and absence of the feed powder, respectively. Both these torches represent top-of-

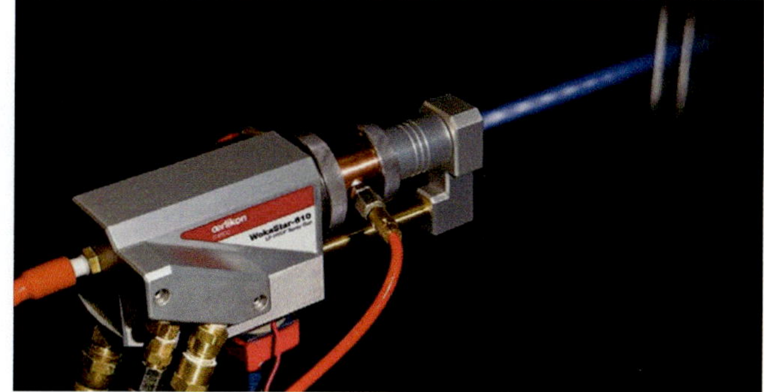

Fig. 18.13 Photograph of the Oerlikon Metco WokaStar 610, liquid fuel HVOF gun, power rating of 293 kW. (Reproduced with kind permission from Oerlikon Metco)

the-line torches operating with kerosene as liquid fuel showing well-defined supersonic jet stream with its characteristic diamond shock wave nodes. A summary of typical operating conditions for some of these torches was reported by [Schwetzke and Kreye (1999)].

With the extremely high noise level associated with the HVOF and HVAF torches in general, typically in the range of 125–135 dBA [Gross (2002)], operation is generally recommended in a well-designed spray booth as illustrated in Fig. 18.15 after Oerlikon Metco. Internal to the booth are the gun and workpiece manipulators, with the process controllers and handling interface placed outside the booth, as well as material feeder. As with other combustion spray

booth, exhaust gases are channeled from the booth through appropriate ductwork to a particulate matter filter before they are exhausted to the atmosphere.

18.4.1.3 Pulse Detonation Thermal Spray

The pulse detonation thermal spray (PDTS) or commonly known as detonation gun (D-gun) was developed in the 1950s in the USA by Union Carbide (Now Praxair) and independently in 1969 at the Institute of Materials Science (Kiev, Ukraine) in the former Soviet Union. As described in Chap. 7, "Combustion Spraying," Sect. 7.4, the technology is closer to cold spray or HVOF/HVAF than conventional thermal spray coating since it relies heavily on the acceleration of particles to be sprayed to high velocities projecting them toward the substrate in essentially a soften solid state where they deform on impact forming the coating. As illustrated in Fig. 18.16, the material to be sprayed is introduced into the gun barrel in the form of small charges, typically of less than 1 g each, of fine powder with a particle size in the range of 10–40 μm. The position of the powder feed in the barrel is a critical parameter in the design and operation of the gun. The powder charge is either axially introduced from the closed end of the barrel using a coaxial probe that can deliver the powder to different locations on the barrel axis, as shown in Fig. 18.16a, or laterally (radially) further downstream at any desired locations as shown in Fig. 18.16b. According to [Kadyrov and Kadyrov (1995)], the process comprises essentially of the following four principal steps schematically represented in Fig. 18.17:

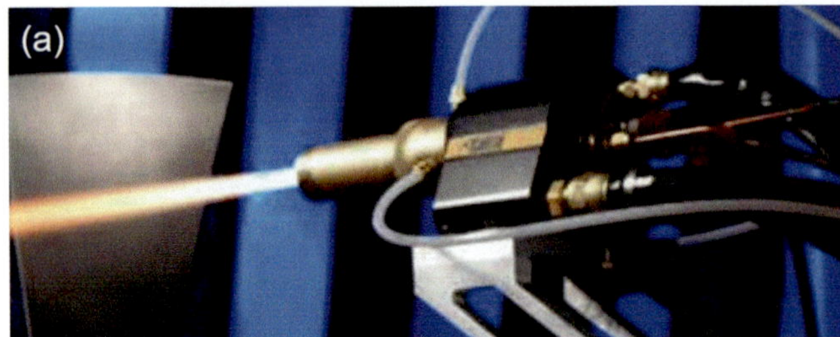

Fig. 18.14 Photograph of the Praxair TAFA JP-8000 liquid fuel HVOF gun showing (**a**) spray jet in operation with powder injection and (**b**) the high-velocity supersonic spray jet in the absence of spray powder showing shock wave diamond nodes

(a) Injection of an oxygen and fuel mixture into the combustion chamber

(b) Injection of the powder "charge" followed by a nitrogen purge between ignition point and the combustion gas

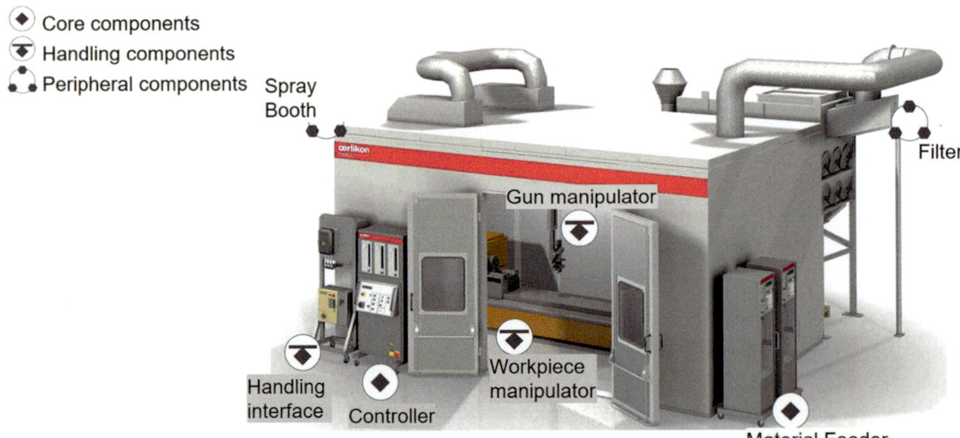

Fig. 18.15 Schematic representation of a typical HVOF spraying booth showing different components associated with the technology. [Oerlikon Metco, Technical Bulletin, Thermal Spray Equipment Guide, issue 11, p. 29 October 2014]. (Reproduced with kind permission from Oerlikon Metco)

mixture to separate the gas supply from the explosive gases and prevent backfiring

(c) Ignition of combustion gas mixture giving rise to detonation and the creation of a detonation wave which entrains, heats, and accelerates the powder toward the open end of the barrel and projecting them on the substrate forming the coating

(d) Nitrogen purge of the barrel at the end of the cycle [Kadyrov and Kadyrov (1995)]

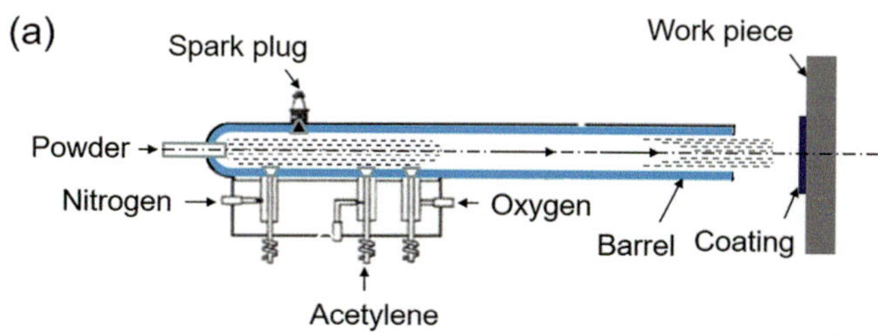

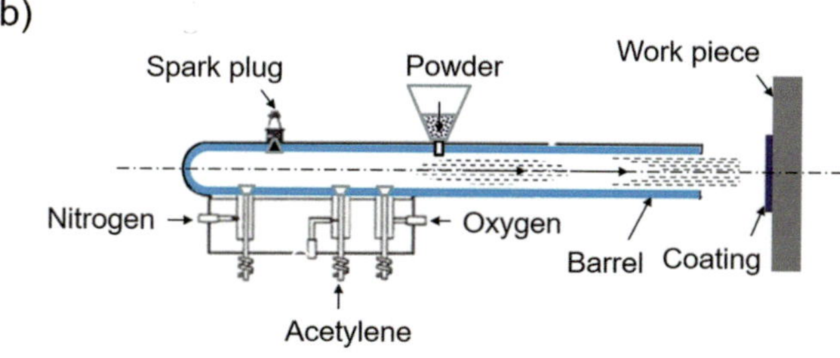

Fig. 18.16 Schematic of detonation gun spray coating process: (**a**) axial powder feed (**b**) lateral (radial) powder feed

The operation of the D-Gun generating one of the highest noise levels among thermal spray operations (145 dBA) has to be carried out in a well acoustic insulated booth comparable with those used with VHOF and VHAF technologies as illustrated earlier in Fig. 18.15.

18.4.2 Wire Arc Spraying

Compared to alternate surface modification technologies, wire arc spraying (WAS) is one of the most competitive technologies because of its high deposition rates up to 25 kg/h, favorable economics, and potential use for a wide range of applications mostly for corrosion protection of infrastructure such as bridges and metallic constructions in general as well as in the marine and automotive industry. The main limitation of the technology is limited, however, to the spraying of metallic ductile wires, such as aluminum, zinc, and steels. The development of the "cord-wire" approach offered an option to circumvent this limitation and expand the use of WAS to nonconductive brittle materials. The feed material in this case is supplied as a metallic tube formed through the folding of a ductile metallic foil as envelop packed with a fine powder of the ceramic or composite cermet-coating material. Cored wires are used on an industrial scale for WAS of carbide-based cermets such as WC/ W_2C + Fe, WC/TiC + Fe, Cr, Ni for wear resistance applications. The technology can be combined with epoxy or silicon polymer coating for sealing open porosity in the

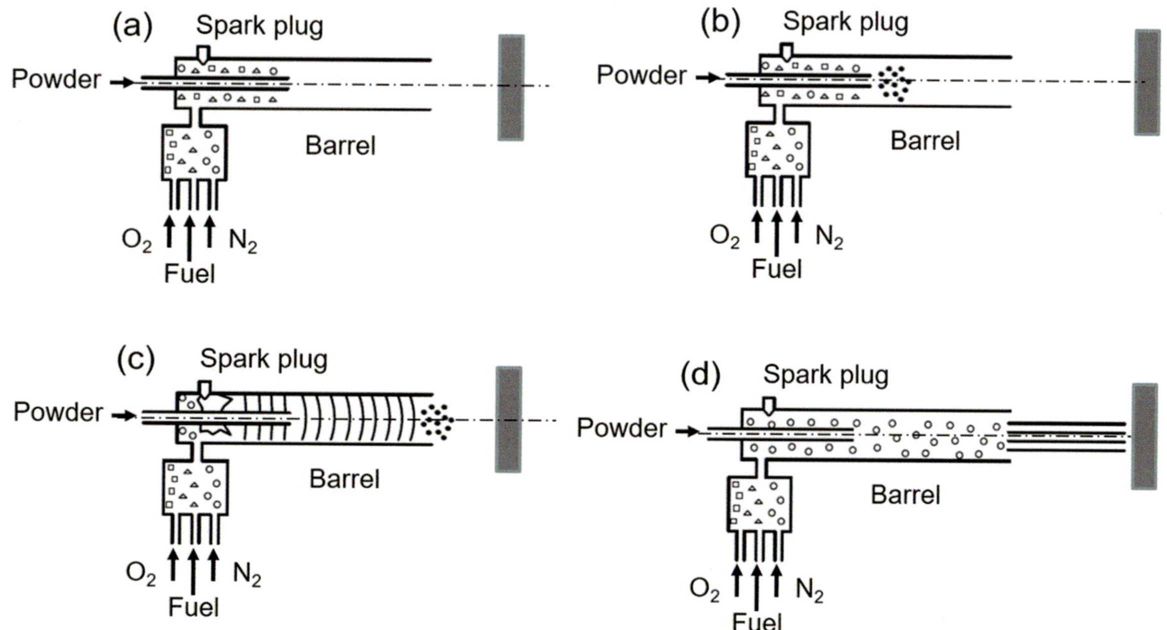

Fig. 18.17 Schematic of the detonation process cycle: (**a**) injection of fuel and oxygen; (**b**) injection of powder charge and nitrogen gas; (**c**) detonation of the combustion mixture, powder acceleration, and projection on the substrate; (**d**) nitrogen purge in preparation for the next cycle [Kadyrov and Kadyrov (1995)]. Reprinted with kind permission from Springer Science Business Media, copyright © ASM International

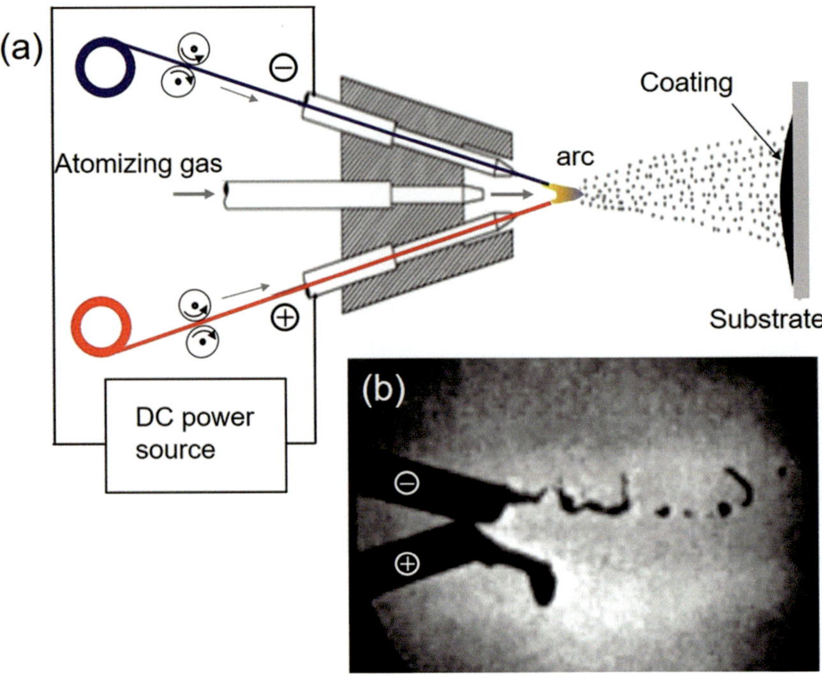

Fig.18.18 (**a**) Principle of wire arc spraying and (**b**) high-speed images of liquid metal droplet atomization recorded at 100 ns exposure time

coating provided that the service temperature is below 200 °C.

As presented in Chap. 11, "Wire Arc Spraying," the basic principle behind the twin wire arc spraying (TWAS) technology, illustrated in Fig. 18.18a, is based on the concept of melting the material to be sprayed in wire form using an electric arc struck between their tips. A high velocity gas stream injected into the gap between the two wires, as shown in Fig. 18.18b, atomizes the molten metal at the tip of the wires, forming small droplets which are entrained by the atomizing gas and projected toward the substrate. As with conventional thermal spraying, as the molten droplets impact on the substrate, they form splats that rapidly solidify building up the coating in successive layers. By electrically connecting the two wires to a DC power supply, and continuously feeding the wires at a closely controlled speed to maintain a constant gap between their tips, the operation can be sustained with an essentially constant arc voltage.

The single wire arc spraying (SWAS) torch illustrated in Fig. 18.18a after [Marantz et al. (1991)], also referred to as plasma-transferred wire arc spraying (PT-WAS) torch, makes use of a nonconsumable thoriated tungsten cathode coaxially placed at the center of an air-cooled copper nozzle, with consumable-wire, anode, placed at 90° to the torch axis downstream of the nozzle exit. The arc is initiated by a high-frequency ignition between the cathode and the nozzle, followed by the transfer of the arc to the consumable-wire anode. The plasma gas is preferably of argon–hydrogen mixture, though other gases have also been used including air and nitrogen, injected in the annular space between the cathode and the copper nozzle. A secondary gas injected as a series of high-velocity micro jets surrounding the central arc nozzle serves to atomize molten metal formed at the tip of the wire [Cook et al. (2003)]. It also serves as shroud gas for narrowing the molten droplet spray pattern [Marantz et al. (1991), Kowalsky et al. (1992)].

An alternate approach schematically illustrated in Fig. 18.19b by Heberlein and his research group at the University of Minnesota, referred to as the high-definition single wire arc spray torch (HDSWAS) [Carlson 918 and Heberlein (2000, 2001, 2002)], reversed the positioning of the wire and the auxiliary nonconsumable cathode. The wire, which still acts as anode, is introduced through the center of the nozzle which was at a floating potential, while an external tungsten rod attached to the downstream face of the nozzle serves as cathode. A constant current welding power supply was used with typical arc currents between 35 and 125 A and voltages between 19 and 22 V. Argon was used as the atomizing gas with backpressures between 100 and 400 kPa.

Most commercially available WAS units, such as the Oerlikon-Metco, Model Flexi 810 Arc™300 shown in Fig. 18.20, are of the twin-wire design. These are of a compact/mobile construction incorporating in a single mobile unit the power supply, wire spools, and wire feeding system as well as the control interface which are used for standard applications with high spraying rates. The Flexi Arc™ 300 features maximum current rating of 300A and use of hand-held spray gun (LD/U2) with a pneumatic push/pull wire feed

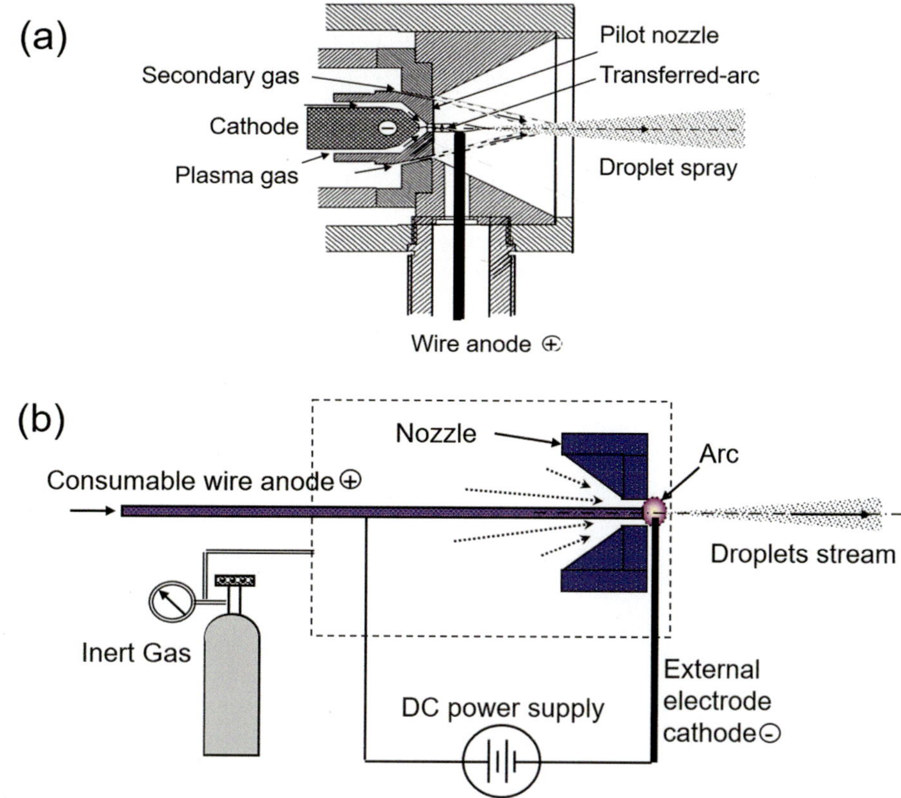

Fig. 18.19 Schematic of (**a**) plasma-transferred wire arc spraying (PTWAS) torch. [Marantz et al. (1991)] and (**b**) high-definition single wire arc spraying torches. [Carlson and Heberlein (2001), Carlson (2005)]

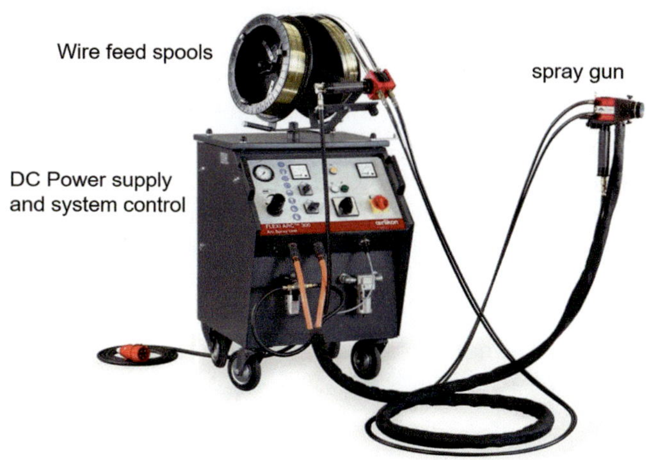

Fig. 18.20 General view of an Oerlikon Metco Flexi Arc™ 300 wire arc spray unit. [Reproduced with kind permission of Oerlikon Metco]

Table 18.2 Typical wire arc operating ranges

Arc voltage	20–40 V
Arc current	100–400 A
Wire feed speed	7–10 m/min
Stand-off distance	100–200 mm
Plasma gases	Air, nitrogen, CO_2
Plasma gas flow rates	800–2400 slm
Gas supply pressure	0.27–0.6 MPa

system. Typical operating parameter ranges of WAS units in general are listed in Table 18.2. It should be noted that high-power twin-wire arc coating installations exist operating at arc currents up to 1500 A. These installations which require water cooling of the contact tips are mainly used for anti-corrosion coatings on large surfaces.

As for the case of flame spraying, while WAS is mostly used for the coating of large infrastructure such as bridges, buildings, or the hulls of large marine structures, it is also widely used for the coating of smaller parts which could be handled in a spray booth such as the one shown in Fig. 18.11 for flame spraying. For larger infrastructure parts such as concrete beams supporting highway overpass, shown in Fig. 18.21, the operator needs to be well equipped with personal protective equipment and whenever possible have the site canvased to limit the spread of the fine dust resulting from the generated fume and overspray.

18.4.3 Atmospheric DC Plasma Spraying

As reviewed in Chaps. 8, "DC Plasma Spraying Fundamentals," and 9, "DC Plasma Spraying Process Technology," a wide range of plasma torch and spray system designs have been developed and used on an industrial scale for a wide range of applications. Plasma torch designs used for atmospheric plasma spraying (APS) have been mostly based on a relatively simple concept using a stick-type thoriated tungsten cathode and an annular coaxial water-cooled anode as schematically illustrated in Fig. 18.22.

Fig. 18.21 Al/Zn/In wire arc spraying of concrete beams and caps of the San Luis Pass Bridge. [Metallisation Ltd. (2006)]

Following arc ignition, using a high-frequency spark generated by a tesla coil, plasma gas is injected into the annular space between these two electrodes, with either an axial or swirling flow pattern, or combination of both, and serves to keep the arc discharge centered in the anode channel (gas stabilization) and the arc root attachment to the anode in continuous movement over its surface, the objective being to distribute the heat load over as larger surface of the anode as possible and consequently reduce anode erosion.

With the continuous variation of the arc length in the torch, the torch voltage will fluctuate reflecting different modes of operation, the restrike, takeover, and steady modes as illustrated in Fig. 18.23a after [Dorier et al. (2001)]. The plasma torch used in this case was an F-4 manufactured by Oerlikon Metco operated with "warn" electrodes and a straight gas injector. The gas flow rate in the restrike mode was 40 slm Ar + 4 slm H_2 with an arc current of 500 A. In the takeover mode, a pure argon plasma was used with a flow rate of 40 slm Ar and arc current of 500 A. The steady mode was achieved at yet a lower gas flow rate of 30 slm Ar and higher current of 600 A. The corresponding energy spectra of the voltage fluctuations are given in Fig. 18.23b, with the same gas flow rate and composition as that used earlier (40 slm Ar + 4 slm H_2) and arc current of 400 A for the restrike mode and pure argon plasma (40 slm Ar), with arc currents of 200 and 600 A for the takeover mode. The effect of the variations of the flow pattern inside the torch was also reported to have a significant impact of the arc voltage fluctuations and the corresponding arc voltage power spectra as shown in Fig. 18.24 for the F-4 torch with an N_2/H_2 plasma gas with vortex and straight gas injectors into the torch [Dorier et al. (2001)].

Some of the leading torches used on a commercial basis for atmospheric plasma spraying (APS) are the F-4 by Oerlikon Metco [Oerlikon Metco, Atmospheric Plasma Spray Solution' Technical bulletin, March 2016)] and

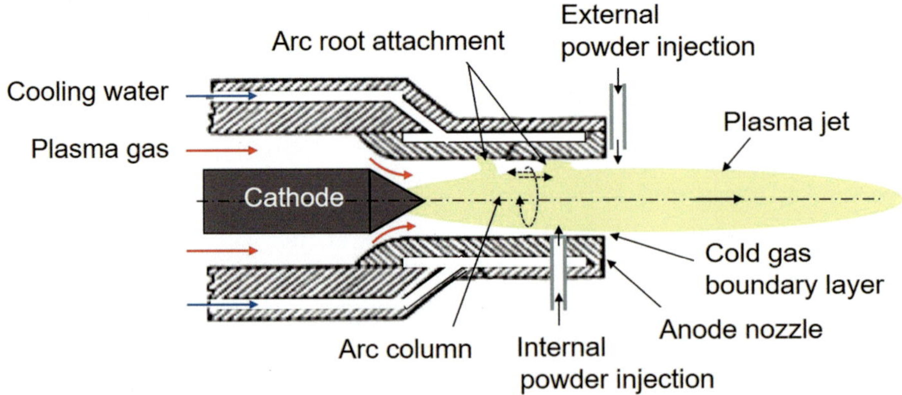

Fig. 18.22 Schematic representation of a DC plasma torch with a stick-type cathode and an annular concentric anode showing internal arc root movement on the anode surface and internal and external powder injection modes

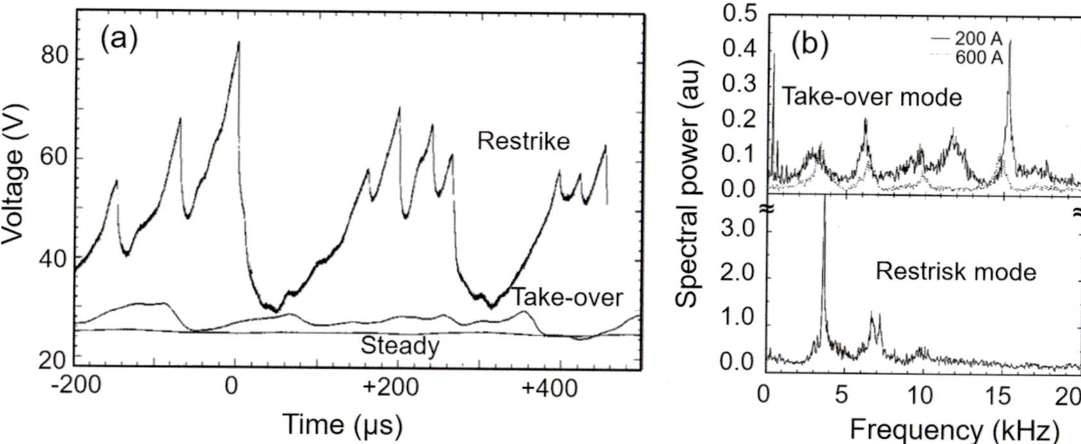

Fig. 18.23 (**a**) Typical voltage traces for an F-4 plasma torch obtained with warn electrodes and a straight gas injector, in the restrike mode (Ar/ H_2), 500 A; in the takeover mode 40 slm Ar, 500 A; and in the steady mode, 600 A, 30 slm Ar (**b**). Corresponding power spectra of the voltage fluctuations in the takeover and restrike modes at at the same gas flow rates and compositions for different currents. [Dorier et al. (2001)]

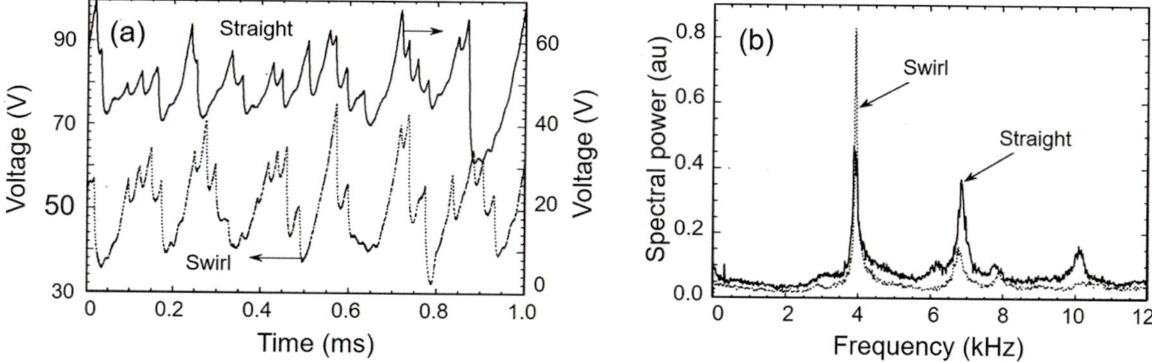

Fig. 18.24 (**a**) Typical voltage traces and (**b**) corresponding power spectra for an F4 torch operating with N_2–H_2 mixture with vortex and straight injection of the plasma-forming gas. [Dorier et al. (2000)]

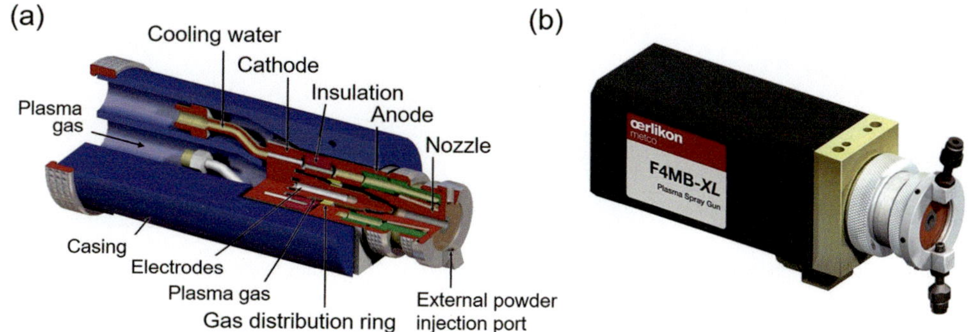

Fig. 18.25 (**a**) Sectional view and (**b**) photograph of the F4-MB-XL DC plasma spray torch. With double-port external powder injection by Oerlikon Metco

SG-100 by Praxair TAFA illustrated in Figs. 18.25 and 18.26, respectively.

In both these cases, the powder to be sprayed is injected radially into the plasma stream either internally in the anode channel with a slight downstream inclination as shown in the SG-100 torch (Fig. 18.26a) or externally at the exit level of the torch nozzle as in the F-4 torch shown on Fig. 18.25b. Few designs were also developed for axial injection of the powder in the plasma stream as is the case of the Triplex torch by Oerlikon Metco (Fig. 18.27) and the Northwest Mettech Axial III torch show in Fig. 18.28.

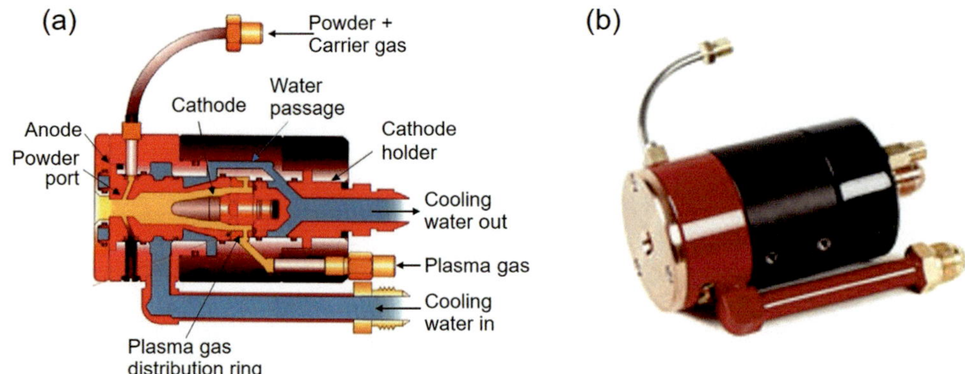

Fig. 18.26 (**a**) Schematic representation and (**b**) photograph of the SG-100, DC plasma spraying torch with internal powder injection by Praxair TAFA

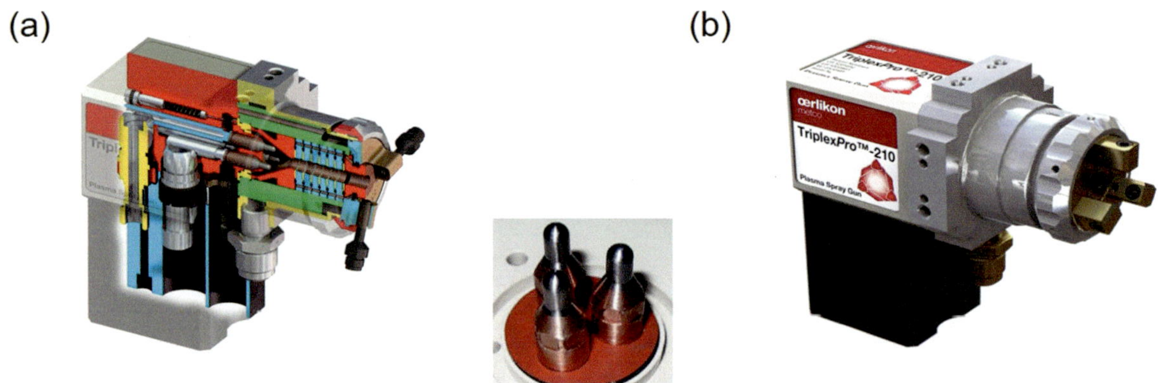

Fig. 18.27 Sectional view of Oerlikon Metco Triplex torch. (Reproduced with kind permission of Oerlikon Metco AG, Switzerland)

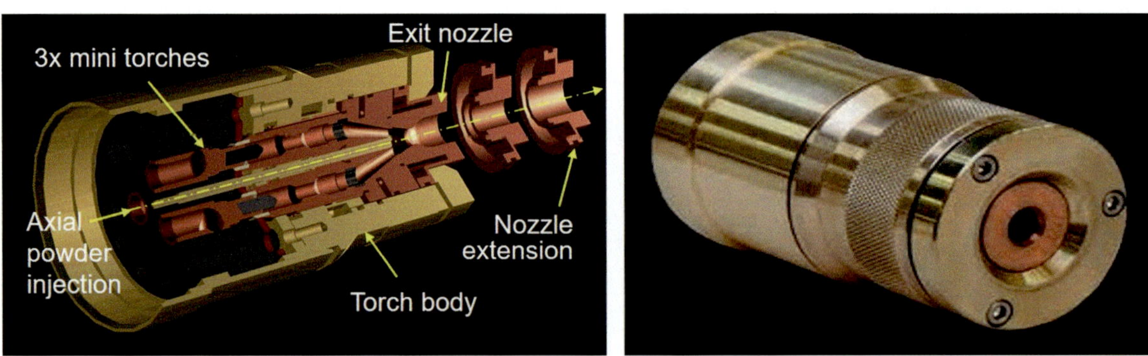

Fig. 18.28 (**a**) Sectional view and (**b**) photograph of the Northwest Mettech Axial III central injection torch. (Courtesy of Northwest Mettech)

A typical integrated DC plasma spraying system, by Oerlikon Metco, is illustrated in Fig. 18.29 showing the principal spray booth equipped with a robotic torch manipulator and a part manipulation lath. Internal to the booth is the "JAMBox" attached to the wall of the booth where the electrical lines and cooling water circuit merge from which the flexible hose connection to the plasma torch supplies the torch with cooling water and electric powder through flexible copper cables immersed into the cooling water hoses. The booth is also equipped with two large sliding doors, shown in wide-open position, with adequate observation windows for process monitoring by the operator. Exhaust gases are conveyed via large ductwork from the booth toward an air filter on the back side of the booth. Outside the booth, the DC power supply and water shield are placed together with the gas management center, powder feeder, robot controller, and process control center. The MultiCoat Operator control panel is placed on a mobile base allowing its displacement by the

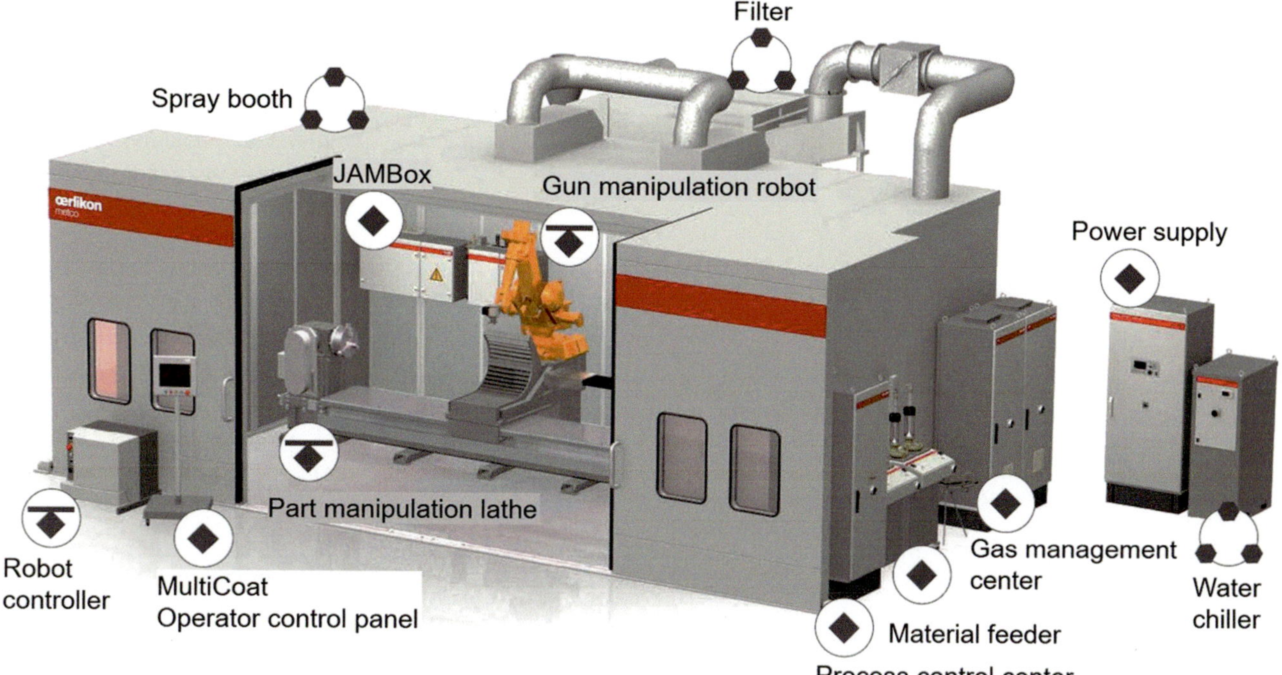

Fig. 18.29 Schematic representation of an integrated DC atmospheric plasma spraying system showing different components associated with the technology. [Oerlikon Metco, Technical Bulletin, Thermal Spray Equipment Guide, issue 11, p. 28 October 2014; Reproduced with kind permission from Oerlikon-Metco]

operator during the running of the system in order to allow visual observation of the process while having complete and unobstructed access to the process control parameters.

18.4.4 Controlled Atmosphere DC Plasma Spraying

As discussed in Chap. 9, "DC Plasma Spray Process technology," Sect. 9.3, DC plasma jets are characterized by the presence of steep temperature and velocity gradients. As the plasma jets immerge into a stagnant ambient atmosphere, a strong shear layer develops in the interface between the jet and the ambient gas. As schematically illustrated in Fig. 18.30 after [Pfender et al. (1991), Russ et al. (1994)], the ring vortex formed around the nozzle exit of the plasma jet due to local intense shear stresses is pulled downstream by the flow, allowing the process to repeat itself again, while adjacently formed vortex rings at the outer edge of the jet have the tendency to coalesce, forming large vortices. The distorted vortex rings start entangling themselves with the adjacent rings, finally resulting in total breakdown of the vortex structure into large-scale eddies and the onset of turbulent flow. This process results in the first large-scale engulfment of the ambient gas into the jet flow. Some entrainment also takes place during the roll-up process of the jet

shear layer. The eddies of cold gas traveling in the axial direction at lower velocities than the flow continuously break down into smaller eddies, while diffusion takes place on the molecular level at all eddy boundaries.

The extent of the impact of ambient gas entrainment on the temperature fields of the plasma jet is demonstrated in Fig. 18.31 after [Roumilhac et al. (1991)], showing the corresponding temperature isocontours for a pure argon DC plasma jet in an open discharge in an ambient atmosphere of different gases. Comparing the temperature fields for the argon plasma jet into an argon ambient atmosphere (Fig. 18.31a) to that in a nitrogen atmosphere (Fig. 18.31b) and in air (Fig. 18.31c) reveals a clear quenching effect resulting from the engulfment of nitrogen or air into the argon plasma flow. The effect is quantitatively noticeable in Fig. 18.32 giving the profile of the plasma temperature and the molar fraction of entrained air in the axial direction along the centerline of a DC pure argon plasma jet in an open discharge into ambient air. The results show that the mole fraction of air on the centerline of the jet reaches 50% only 34 mm downstream of the nozzle exit of the jet.

Other than the above discussed quenching effect of the entrained ambient air entrainment on the temperature field in the plasma jet, the abundance of oxygen mixed with the plasma flow can have a detrimental effect of the chemistry of the material being sprayed especially when dealing with

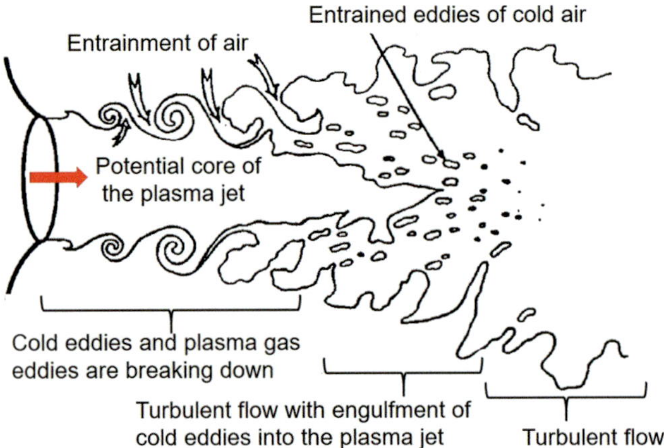

Fig. 18.30 Schematic representation of the main regions of DC plasma jet showing the rolling of the flow around the nozzle exit, ambient cold gas engulfment, and breakdown followed by turbulent flow generation. [Pfender et al. (1991)]

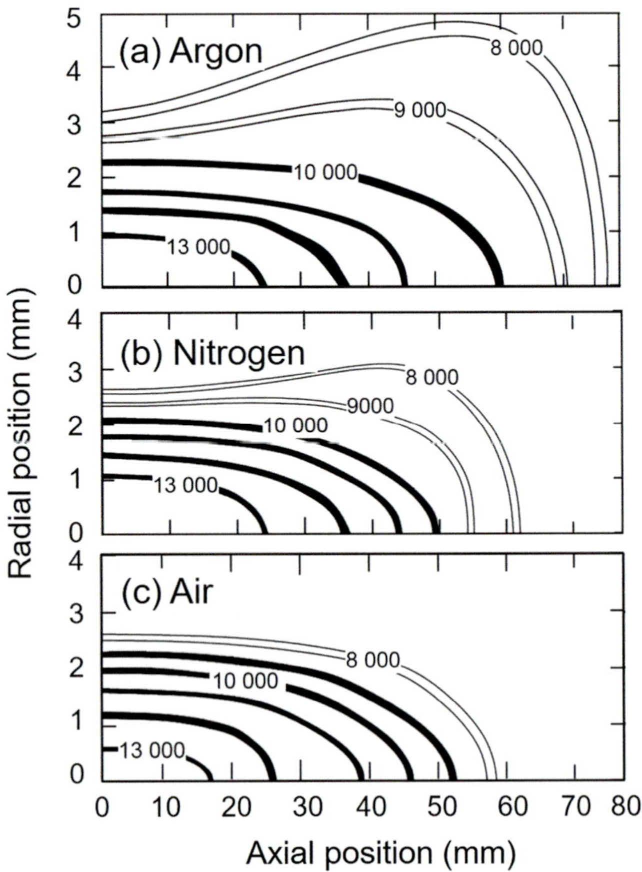

Fig. 18.31 Effect of the ambient gas entrainment on the temperature contours of an argon plasma jet discharged in different ambient gases: (**a**) Argon ambient gas. (**b**) Nitrogen. (c) Air. Derived from data published by [Roumihac et al. (1991)]

pure metals, alloys, or non-oxide ceramics (carbides, nitrides) due to the in-flight oxidation of sprayed material.

Controlled atmosphere plasma spraying (CAPS) was developed in the 1960s to overcome such a problem by enclosing the plasma spray system, including the plasma torch, robotic manipulator, and substrate handling lathe in a gas-sealed large chamber which is evacuated prior to the start of the spray operation to be filled by an inert gas usually argon, to atmospheric pressure. The oxygen content can be lower than 7 ppm if a liquid argon source is used [Freslon (1995)]. Plasma jets in argon atmosphere are broader and longer (about a factor 1.5–2) than in air [Roumilhac et al. (1991)]. Such chambers are used to spray carbides, borides, and refractory metals. Liquid argon sprays allow controlling the coating temperature during the spray process [Freslon (1995)]. If the chamber volume is above 10 m^3, no cooling is necessary for the chamber walls.

A few tests have been performed using chambers at pressures above one atmosphere (up to 300 kPa) [Jäger et al. (1992)]. However, plasma jets are shorter, and electrode erosion increases with the increase of the operation pressure above one atmosphere. This is mainly due to the increased radiation form the arc column and the shrinkage of the cathode attachment with the increase of the operating pressure. The latter gives rise to the increase of the local current density and the thermal load on the electrode and consequently to higher erosion and lower lifetime of the electrodes.

Plasma spraying using CAPS installations was, however, of limited success due to the significant increase of the capital investment necessary and increased operating costs. The use of shrouding gas injected around the plasma torch nozzle exit represents an interesting cost-effective alternative, though it does not ensure the complete elimination of in-flight oxidation of the spray material.

18.4.5 Vacuum DC Plasma Spraying

Vacuum plasma spraying (VPS) is a natural extension of the CAPS approach except for operation under sub-atmospheric pressure "soft vacuum" conditions. The technology was developed on an industrial scale in 1974 by Mulberger (Electro Plasma Inc. acquired later by Sulzer Metco Inc. and presently part of the Oerlikon Metco group), [Muehlberger (1988), and Meyer and Hawley (1991)]. As reviewed in Chap. 9, "DC Plasma Spray Process Technology," Sect. 9.4, one of the key challenges for the development of the technology was the need to redesign the DC torch's nozzle for compressible flow reaching supersonic flow. Under these conditions, speeds inside the nozzle channel reach sonic values, with shock diamonds observed in the plasma jet as schematically represented in Fig. 18.33a. These occur when the pressure inside the jet is different from the surrounding atmosphere leading to sudden velocity drop and the creation of shock waves. The jet is identified as "over-expanded" when the pressure inside the jet is lower than the ambient pressure in the chamber and "under-expanded" with a jet pressure higher than the ambient pressure. As a consequence, through the interaction with the environment, the jet is compressed or expanded, but an over-compensation may occur resulting in another under-expansion/over-expansion of the flow. A complex flow structure evolves with shock waves starting at the nozzle exit being reflected at the discontinuity between the jet and environment, and these reflected shock waves create the shock diamonds. The more shock

diamonds and the clearer they are, the worse the fluid dynamic condition for plasma spray operation.

The use of a properly designed Laval nozzle either as part of the anode or as an attachment to the anode as illustrated in Fig. 18.33b can significantly improve the results by avoiding the shock structures immediately downstream of the anode nozzle exit [Henne and Weber (1982), Henne et al. (1983, 1986, 1988), Meyer and Hawley (1991)]. With a Laval-type nozzle design, the jet will lengthen and enlarges in soft vacuum, as shown in Fig. 18.34. At an absolute pressure of 95 kPa, the jet is only close to 60 mm long, while at 5 kPa, its length reaches 400 mm, extending to more than 1200 mm in length and up to 200 mm in diameter at chamber pressures of 0.1 kPa. The latter case is rather in the "ultralow pressure" region of operation rather than traditional VPS operation. The improved nozzle design and flow dynamics of the plasma jet result in the increase of particle velocities by almost 30–50% which, associated with a particle trajectories parallel to the torch axis, gives rise to higher deposition efficiencies [Meyer and Hawley (1991), Henne et al. (1993), Rahmane et al. (1998)].

An important part of the VPS process is the ability to clean the surface of the substrate and to preheat it to the desired temperature prior to the initiation of the spraying operation. The cleaning and heating processes are carried out with the plasma jet ignited, with no particle injection, and striking a low-current transferred arc, in the range of 20–150 A, between the torch nozzle and substrate, as illustrated in Fig. 18.35. For cleaning step, the substrate is set as cathode (Fig. 18.35a), while for heating, it is connected as anode (Fig. 18.35b). An electrical shield must be used in certain

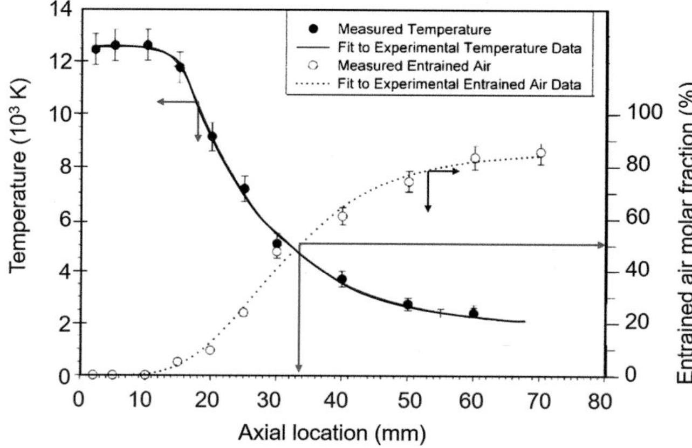

Fig. 18.32 Axial distributions of enthalpy-probe measurements of plasma temperatures, volume fraction of entrained air in an argon DC plasma jet, and CARS temperatures of the entrained oxygen showing the low temperatures of the entrained oxygen. [Finke et al. (2003)]. (Reproduced with kind permission of Elsevier)

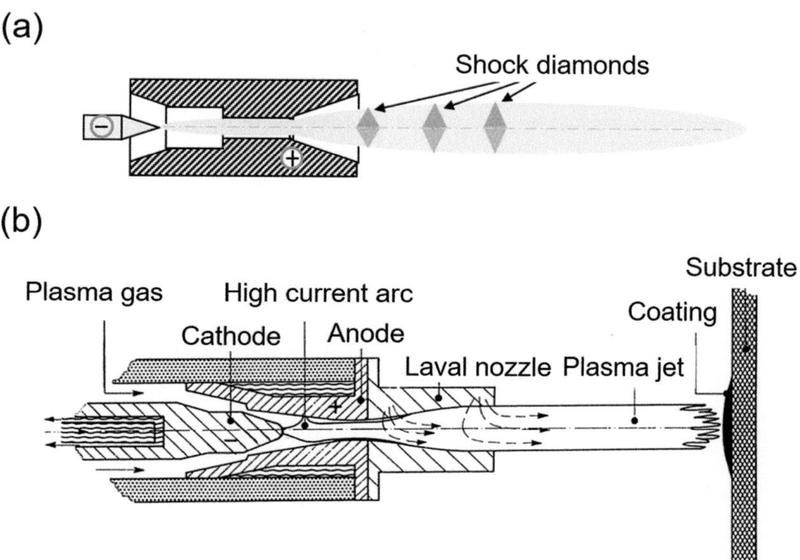

Fig. 18.33 (a) Schematic of shock diamonds and (b) torch nozzle with Laval attachment for improved supersonic flow structure. [Mayr and Henne (1988)]. Reproduced with kind permission of Sulzer Metco AG, Switzerland

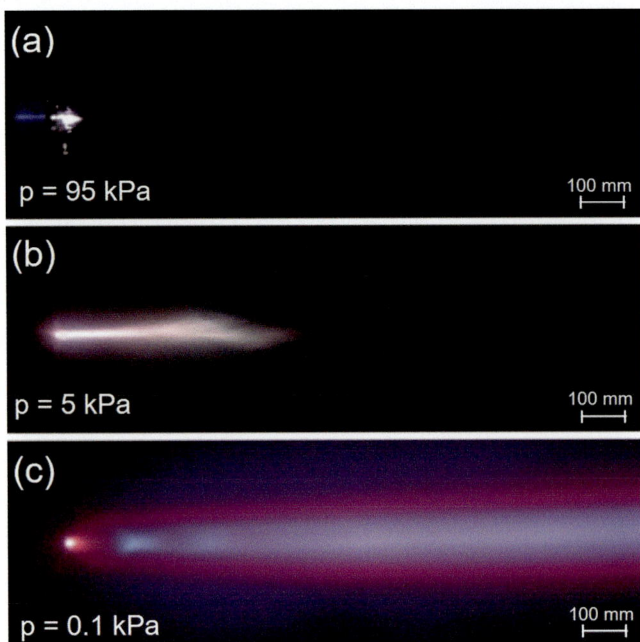

Fig. 18.34 Images of plasma jets in a controlled-atmosphere chamber expanding at different pressures: (**a**) 95 kPa (APS), (**b**) 5 kPa (VPS), (**c**) 0.1 kPa (ULPPS) and deposition process [von Niessen and Gindrat (2010)]

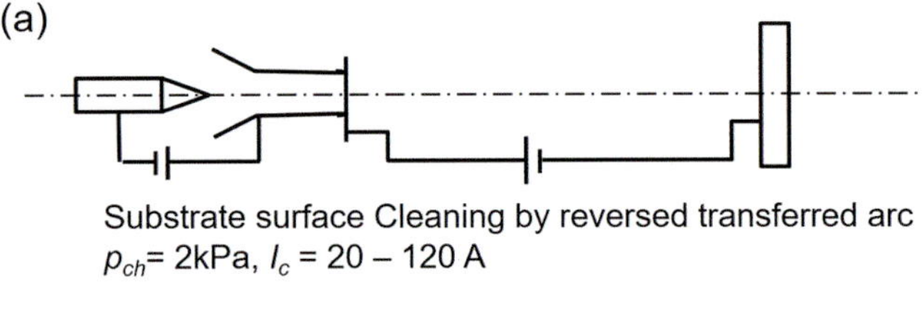

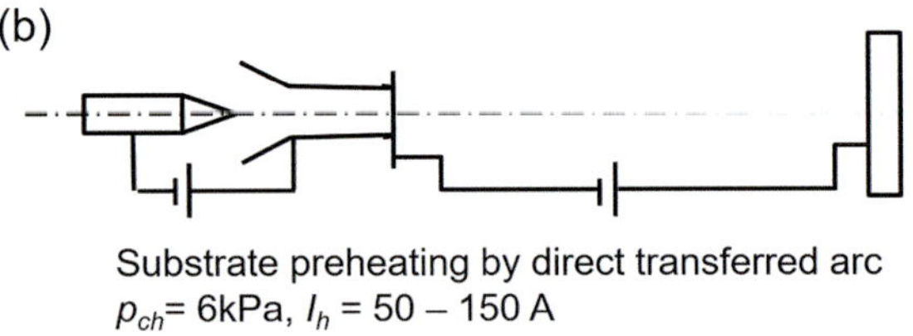

Fig. 18.35 Schematic of the transferred arc cleaning (**a**) and heating (**b**) when vacuum plasma spraying [Itoh et al. (1991)]. Reprinted with kind permission from ASM International

cases to avoid the transferred arc striking on substrate grips, for example. [Itoh et al. (1991)] have cleaned stainless-steel (SUS304) coupons 25 mm in diameter placed 400 mm downstream of the nozzle exit at a chamber of 3.3 kPa. Some stainless-steel coupons were heated in air to 1000 °C and then mirror polished, the oxide layer being 0.45 µm thick. The reversed transferred arc current was 37 A. The arc cathodic spots moved continuously at the substrate surface.

The principle of the industrial installation of Electro Plasma Inc. is shown in Fig. 18.36. The equipment is contained in a water-cooled, vacuum-tight chamber, maintained at an absolute pressure between 10 and 70 kPa. Such systems, known as vacuum plasma spraying (VPS) or low-pressure plasma spraying (LPPS), include vacuum pumping, exhaust cooling, filtration, an inert gas backfill manifold, and robots to move the spray gun and the part, as shown in Fig.18.37. As the temperature inside the chamber can reach values between 150 and 200 °C, and moreover since the interior is rather dusty, the arms of the robots inside the chamber have to be protected against heat and dust. Filters must also be disposed between chamber and pumps to trap the dust.

A schematic representation of a state-of-the-art vacuum spraying system by Oerlikon Metco, model ChamPro™ -VPS System, is given in Fig. 18.38. The unit is designed for operation at absolute pressures of 5 kPa with a modified DC plasma spraying or wire arc spraying (WAS) system components integrating the reverse transferred arc part cleaning. As can be noted in Fig. 18.38, the installation includes vacuum pumping and gas filtration systems as well as specialized handling and system control interface. The ChamPro™-VPS VPS is a batch-processing system in which, once a batch of parts has been loaded, the chamber is pumped down to a low pressure (a few Pa) and backfilled with high-purity dry argon to maintain oxygen levels below 30 ppm. Following the coating step, the parts are allowed to cool down before backfilling the VPS chamber with ambient air, opening the chamber, and unloading the parts. The cycle is repeated with the next batch of parts. Thorough cleaning of the inside of the deposition chamber is necessary on a regular basis for proper operation of the mechanical components. It is also essential to avoid memory effects when switching from one coating material system to another.

A schematic representation of the Oerlikon Metco ChamPro™-HC-LVPS System is shown in Fig. 18.39, equipped with two loading systems and part-preheating antechamber which serves as a buffer space that allows for the continuous introduction, preheating, cooling, and withdrawal of parts from the system without interrupting the spray coating operation in the spray chamber under a constant controlled spray atmosphere.

The main advantages of the VPS process are:

- Coatings are almost oxide-free (a quantity slightly higher than the oxide content of sprayed powder).
- Substrate oxide layer can be eliminated by a reverse transferred arc.
- Substrate can be preheated up to 1000 °C before spraying and maintained at this temperature during the operation.

VPS coatings for metal sprayed on metal substrate show high coating density and excellent coating adhesion by diffusion, once the oxide layer at substrate surface has been

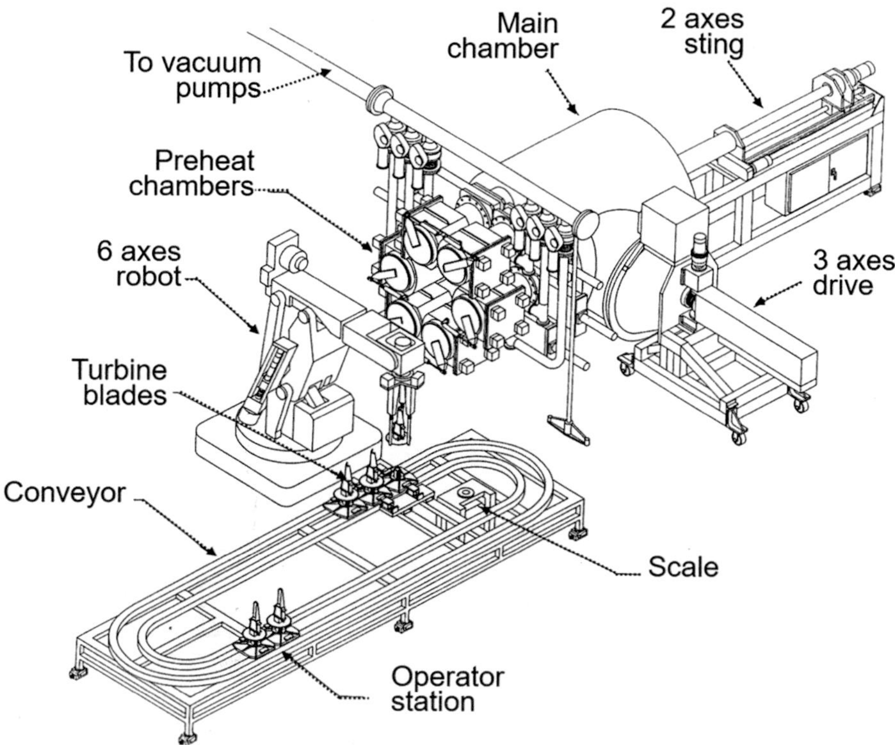

Fig. 18.36 Industrial soft vacuum plasma spray setup to coat turbine blades. Reprinted with kind permission from ASM International. [Meyer and Hawley (1991)]

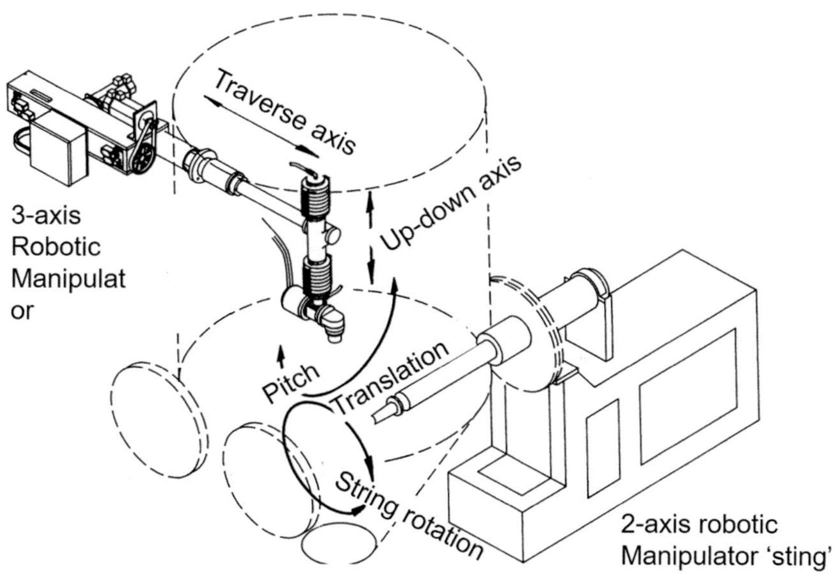

Fig. 18.37 Schematic of the robot manipulating the spray torch and that moving the part to be sprayed. Reprinted with kind permission from ASM International. [Meyer and Hawley (1991)]

removed with the reverse transferred arc. Diffusion requires, however, that during spraying the temperature be kept sufficiently high throughout the full spray cycle.

18.4.6 Ultralow-Pressure Plasma Spraying

As discussed in Chap. 9, "DC Plasma Spray Process technology," Sect. 9.5, the ultralow-pressure plasma spraying (ULPPS) also identified as very-low-pressure plasma

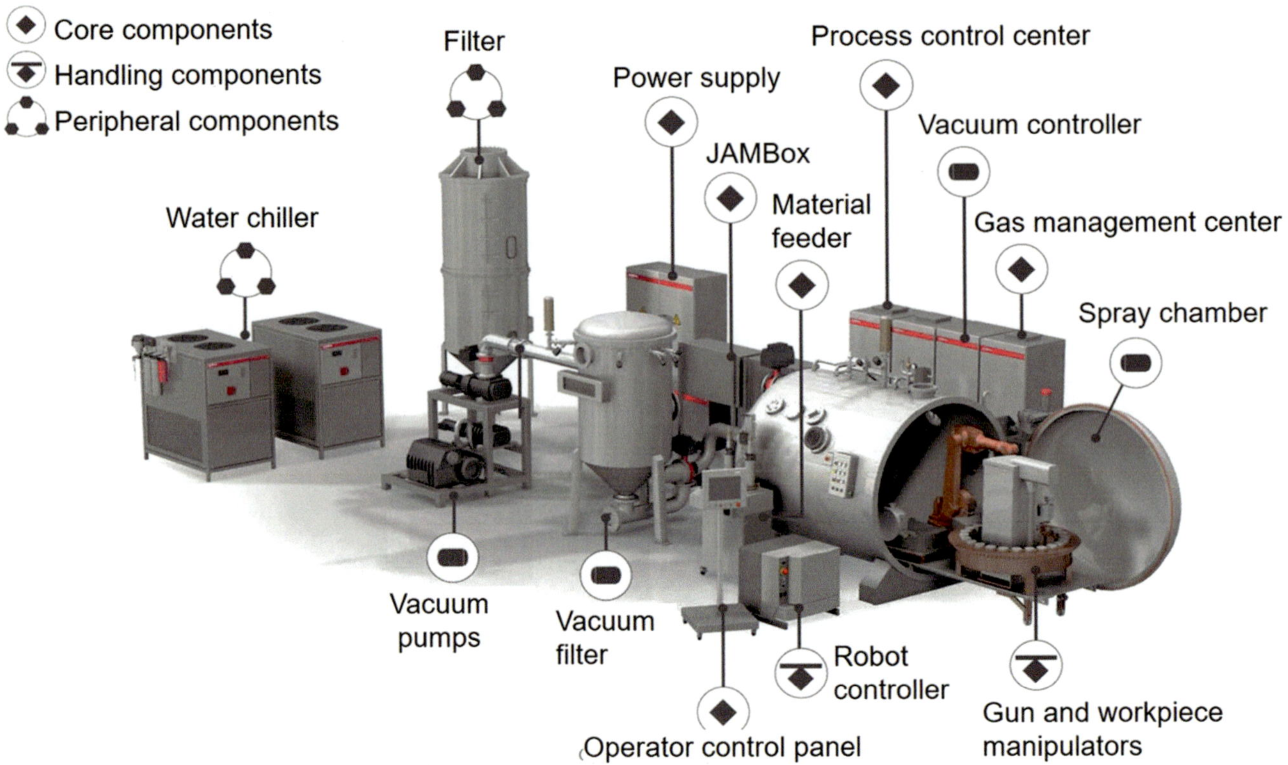

Fig. 18.38 Schematic representation of an Oerlikon Metco ChamPro™-VPS System. [Oerlikon Metco, Technical Bulletin, Thermal Spray Equipment Guide, issue 11, p. 35 October 2014; Reproduced with kind permission from Oerlikon Metco]

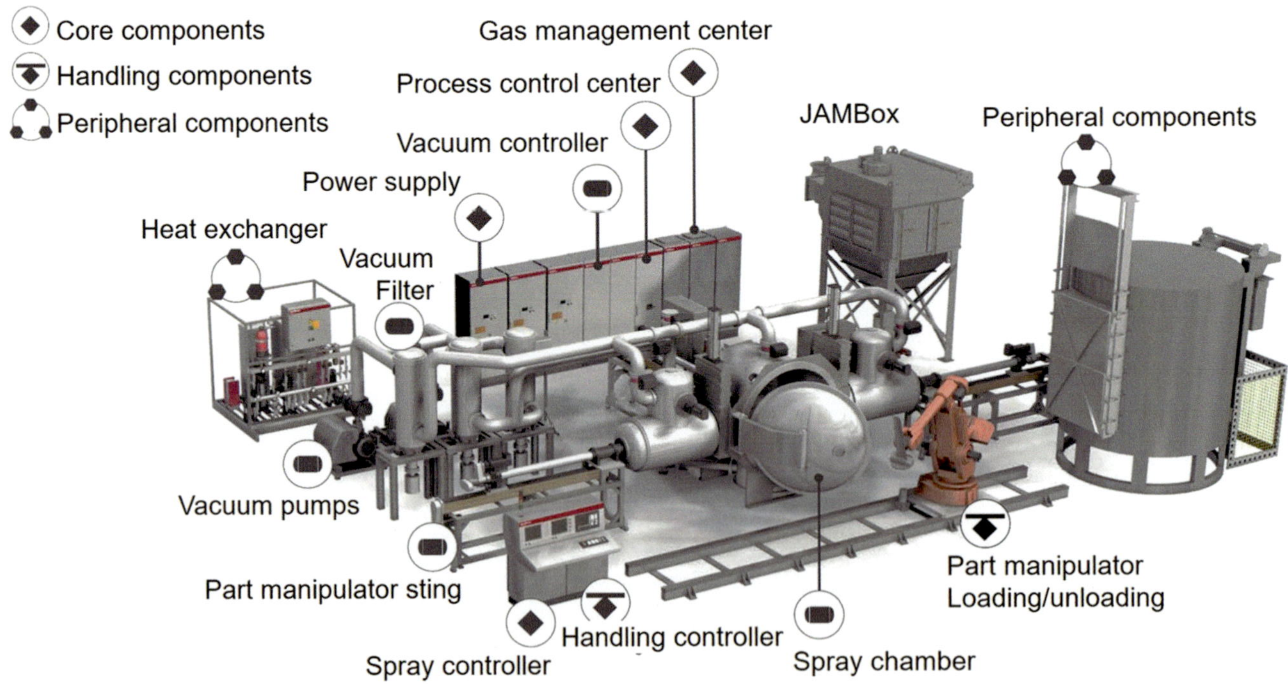

Fig. 18.39 Schematic representation of an Oerlikon Metco ChamPro™-VPS System. [Oerlikon Metco, Technical Bulletin, Thermal Spray Equipment Guide, issue 11, p. 37 October 2014; Reproduced with kind permission from Oerlikon Metco]

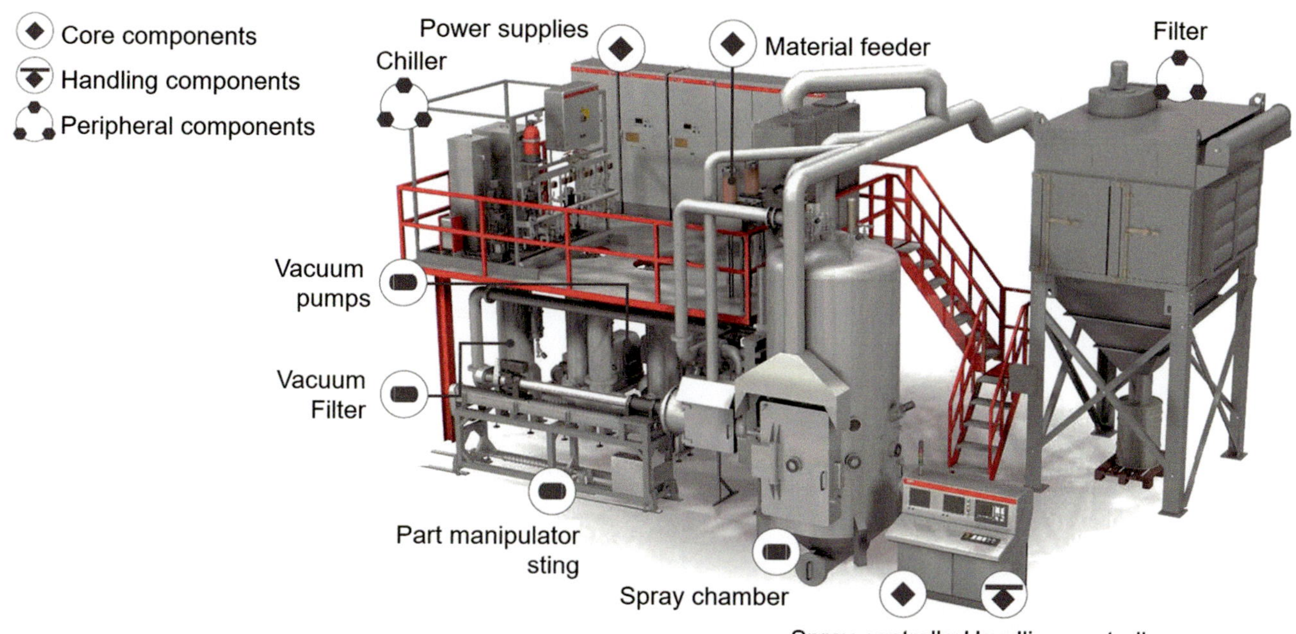

Fig. 18.40 Schematic representation of an Oerlikon Metco ChamPro™-LPPS-Hybrid System, [Oerlikon Metco, Technical Bulletin, Thermal Spray Equipment Guide, issue 11, p. 38 October 2014; Reproduced with kind permission from Oerlikon-Metco]

spraying (VLPPS) and plasma spray-physical vapor deposition (PS-PVD) is a relatively new technology that bridges the gap between conventional VPS spraying processes and physical vapor deposition (PVD) processes. The ULPPS technology is in direct competition to EB-PVD, for the production of ceramic TBC coatings with equivalent quality as EB-PVD technology, at considerably higher deposition rates and lower cost. The two technologies, ULPPS and EB-PVD, differ essentially in the way the precursor is transformed into the vapor phase with ULPPS based on the in-flight heating and evaporation of the precursor in powder form using the plasma jet, while EB-PVD makes use of an electron beam for the precursor evaporation from a solid target or powder. This difference has a significant impact on their respective operating pressure and the deposition rates associated with each of the two technologies, with the ULPPS process operating at pressures < 200 Pa and deposition rates in the range of μm/s, in contrast to the electron beam technology which required considerably lower operating pressures (< 5 Pa), lower deposition rates in the 0.1–100 μm/min range, and significantly higher investment cost.

The process developed originally by Sulzer Metco called plasma spray-physical vapor deposition (PS-PVD) is based on their expertise in low-pressure plasma spraying and involves a high-power plasma spray torch (180 kW–3000 A, gas flow up to 200 slm) working at a pressure as low as 0.1 kPa (1 mbar). Under such a pressure condition, the plasma jet reaches more than 2 m in length and up to 0.4 m in diameter as presented in Fig. 18.34c. The coatings obtained

using a low powder feed rate, adapted spray conditions, and large spray distance exhibited a columnar microstructure similar to that of EB-PVD. The YPSZ columns grow perpendicular to the substrate surface whose roughness (Ra) was lower than 2 μm and temperature in the 1000 °C range [von Niessen and Gindrat (2010)]. Thanks to the versatility of the PS-PVD process, it also makes it possible by adapting the spray parameters, especially when increasing the powder flow rate to limit its evaporation to achieve porous coatings with fine particles and splats imbedded in a matrix resulting from vapor deposition.

A schematic representation of the ChamPro™-LPPS-Hybrid System manufactured by Oerlikon Metco is illustrated in Fig. 18.40, representing a family of processes applied in near vacuum conditions (approx. 0.1 kPa) to produce unique, high-performance functional surfaces. Depending on the feedstock material (liquid or vapor) and the operating parameters, the produced coatings may be thin and dense or thick with unique microstructures.

18.4.7 RF Induction Plasma Spraying

As described in Chap. 10, "Induction Plasma Spraying," Sect. 10.5, RF induction plasma spraying (RF-IPS) is a relatively new technology that evolved rapidly over the past two to three decades. Since its development in the early 1960s, it was mostly used on an industrial scale for the deposition of high-purity materials and crystal growth. It

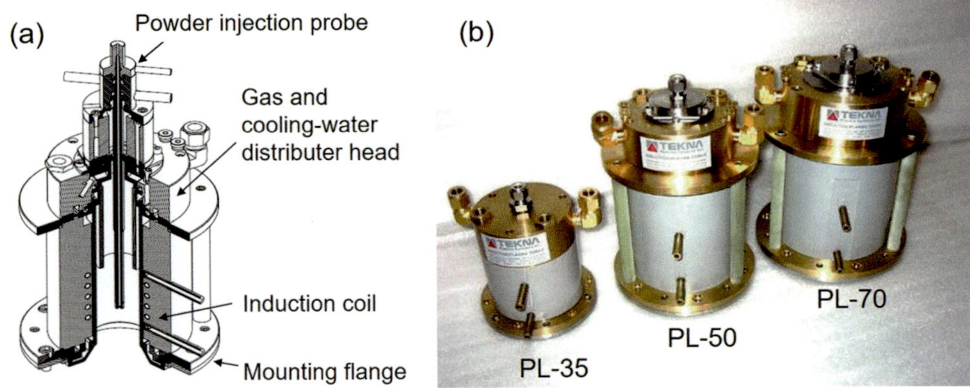

Fig. 18.41 (**a**) Schematic sectional representation of Tekna's ceramic wall induction plasma torch. (**b**) Photograph of the PL-35, PL-50, and PL-70 torches [Boulos and Jurewicz Canadian patent 2,085,133 (1992) and US Patent 5,200,595 (1993)] (Courtesy Tekna Plasma Systems Inc)

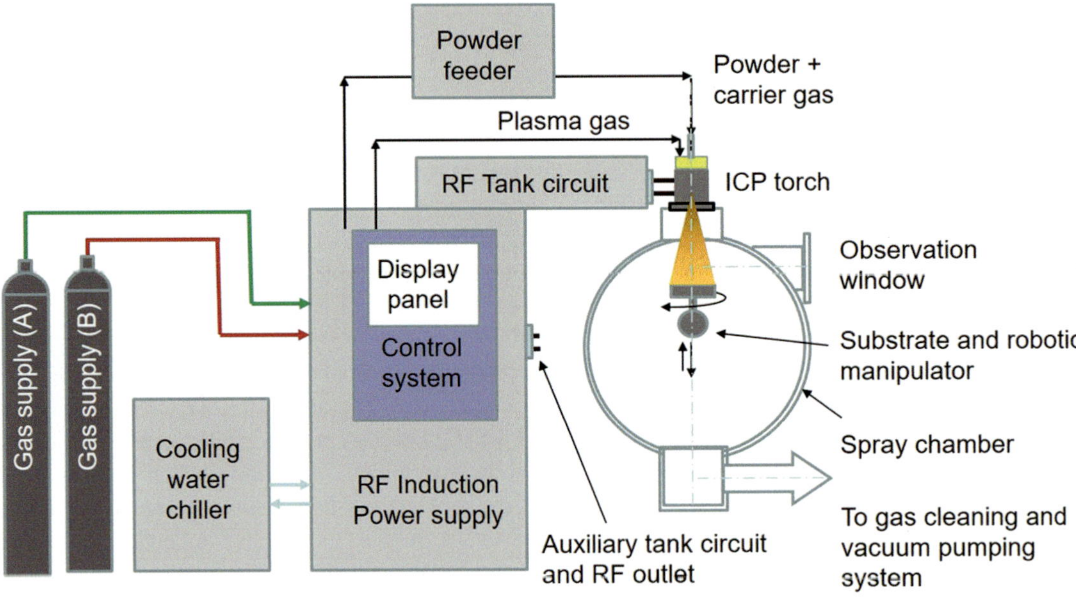

Fig. 18.42 Schematic representations of principal components of an integrated RF-VIPS system

has also been an enabling technology in the fiber-optics industry for the over-cladding of the core of the fiber with high-purity quartz. The subsequent expansion of the technology was carried out by TAFA, in Concord, NH, USA, in the 1970s in connection with NASA's aerospace program and taken over since the early 1990s by Tekna Plasma Systems Inc. in Sherbrooke, Québec, Canada. Tekna's contribution has been mostly in the development of an integrated line of induction plasma torches based on its ceramic plasma confinement tube torch patents [Boulos and Jurewicz (1992, 1993, 1996a, b, 1997, 1999, 2001)] and applications of the technology for in-flight powder melting and spheroidization, synthesis of nanopowder, and induction plasma spraying of metals and ceramics. Cross-section view of the basic torch design is shown in Fig. 18.41a, with photograph of three of these torches, the PL-35, PL-50, and PL-70, given in

Fig. 18.41b. The number associated with each of these torch models reflects the i.d. of the ceramic plasma confinement tube. As discussed in Chap. 10, "Induction Plasma Spraying," Sect. 10.3.4, each of these torches is designed for operation over a given power range with:

- PL-35 rated for operation over at plate power of 10–40 kW
- PL-50 rated for 30–70 kW
- PL-70 rated for 50–120 kW
- PL-100 rated for 80–200 kW

A schematic representation of a typical RF-IPS system given in Fig. 18.42 shows its principal components which center around the RF-induction plasma torch, mounted on top of a water-cooled, vacuum deposition chamber equipped with the centrally located robotic arm with 3 or 4° of freedom for the manipulation of the substrate in the plasma/molten particle stream emerging from the plasma torch. Because of the high ambient temperature in the chamber, the mechanical components of the robotic arm should be water-cooled and adequately protected against radiant heating and dusty environment. Exhaust gases from the chamber loaded with overspray powders and fumes exit the spray chamber from an appropriate outlet at the bottom of the chamber to the gas cleaning and vacuum pumping station. Other components of the system include a cooling water schiller used with a closed loop water-cooling circuit for the cooling of the power supply components including the oscillator vacuum tube, plasma torch, reactor or vacuum deposition chamber, and system accessories. A control system, integrated either in the power supply or in a separate cabinet, serves to monitor vital system parameters including torch ignition, power control, current, voltage, gas and water flow rates, cooling water temperature, and safety interlocks. An appropriate gas and powder feeder supply forms an integral part of the system. The latter is normally placed at a relatively high level above the torch to shorten as much as possible the length of the powder transport line to the entrance of the water-cooled powder injection probe centrally placed in the torch.

A typical medium-sized RF-vacuum induction plasma spraying (RF-VIPS) system is represented schematically in Fig. 18.43. It is composed of a horizontally oriented, water-cooled, stainless-steel cylindrical chamber, 0.5 m in internal diameter and 0.64 m long, on which the ICP torch is mounted on top in a fixed position. A Faraday cage (not shown in the drawing) surrounds the torch to protect the surrounding equipment, instrumentation, and computer-based control modules from electromagnetic interference (EMI) which originated from the induction coil. Coaxial with the torch is the exit port from the chamber through which the plasma gases and powder overspray are evacuated. This port is located directly underneath the torch on the opposite side of the chamber. It is of conical shape ending by a water-cooled cylindrical container "catch pot" in which any over-sprayed powder collects by gravity prior to the exit of the plasma gases from a side port. From this port, the gases are directed toward a filter to recover any entrained fumes and fine powders before being sent to the vacuum pumping station. An access door to the deposition chamber is located on the right-hand side of the figure. It is also water-cooled and equipped with appropriate seal for maintaining the vacuum tightness of the chamber while allowing for quick opening and closing of the chamber for installation of the substrates to be coated and recovery of the coated components. Opposite to the access door is the mechanical module in which the motors and motion control components are located for a three-axis robotic arm, which supports the substrate. The water-cooled arm has to allow for the precise control of the exposure of the substrate to the stream of plasma and molten particles immerging from exit nozzle of the plasma torch. In standard, pilot installations, the robotic arm is designed for the control of the linear motion of the arm in the horizontal direction, the vertical

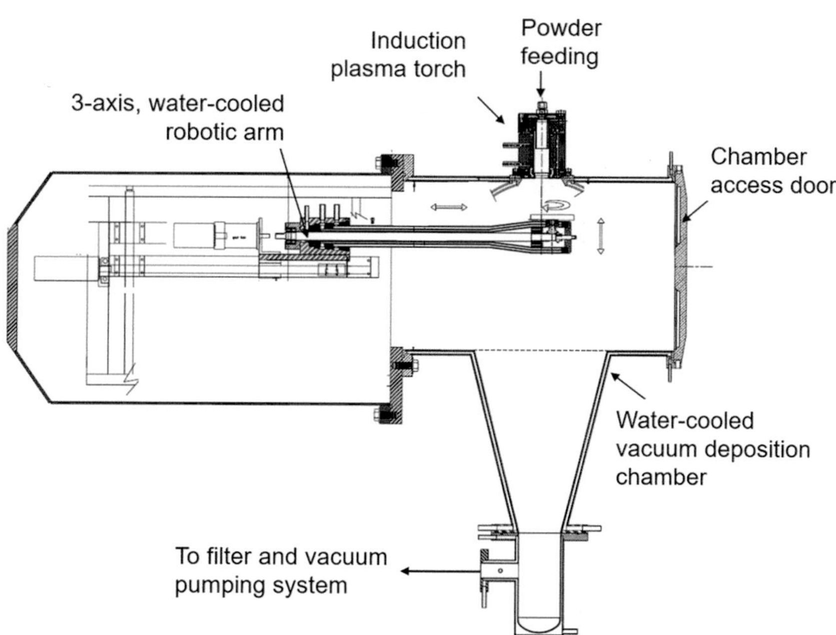

Fig. 18.43   Schematic of a pilot-scale RF-VIPS chamber. [Drawing courtesy of Tekna Plasma Systems Inc. 2012]

Fig. 18.44 Photograph of a 100 kW RF-VIPS system. [Courtesy of Tekna Plasma Systems Inc. 2012]

spraying distance between the substrate and the torch exit, and the speed of rotation of the substrate around a vertical or horizontal axis. In more elaborate and larger units, a fully equipped five-axis robot can be located in the deposition chamber in order to allow for the articulation of more complex parts underneath the plasma torch, at the expense of considerably increased level of complexity and the size of the spraying chamber.

A photograph of an induction plasma vacuum spraying chamber at Tekna Plasma Systems Inc. in Sherbrooke, Québec, Canada, is shown in Fig. 18.44. One notices the powder feeder located on top of the chamber above the induction plasma torch. The latter is not visible since the plasma torch is placed inside the aluminum Faraday cage used to protect system instrumentation and control equipment from electromagnetic interference (EMI) emitted from the induction coil in the torch. On the bottom, right-hand side of the photograph, the RF power supply and control panel for the operation of the system are noticeable.

As with other thermal spray operations, VIPS depends largely on the ability to heat and melt, in-flight, the precursor particles prior to their impact on the surface of the substrate. The splat formation on the substrate is a critical step in building up the coating which required complete melting of the particles prior to their impact on the substrate as well as substrate surface preparation and its temperature. Because of the relatively large volume and low velocity of the RF-ICP discharge, the residence time of the particles in the plasma stream is typically of the order of 10–20 ms, which is 1 order of magnitude larger than the 1–2 ms residence time typical of DC vacuum plasma spraying.

Photographs of VIPS operations for different applications are given in Fig. 18.45. The example given in Fig. 18.45a and b is for the deposition of a refractory metal such as molybdenum or tungsten on a cylindrical graphite mandrel in order to form a near net-shaped part, which is a stand-alone component once the graphite substrate is machined out following the deposition step. A similar setup in operation for the coating of X-ray or a sputtering target using VIPS technology is shown in Fig. 18.45b. The sheath gas used in this case is composed of 120 slm Ar + 20 slm H2. The plasma gas is 30 slm Ar, while the powder gas is 8 slm of He. The chamber pressure was 40 kPa (300 Torr), plate power 80 kW, and spraying distance 100 mm.

An integrated, compact, RF induction plasma spraying (RF-IPS) system was recently developed by Tekna Plasma Systems Inc. (Teck-15) for R&D work in the areas of powder processing and induction plasma spraying. The system illustrated in Fig. 18.46, for 15 kW plate power rating, is housed in a single cabinet with a footprint of 2.0 × 1.0 m, 1.8 m high. The LHS of the cabinet houses the RF power supply, gas control panel, and operator

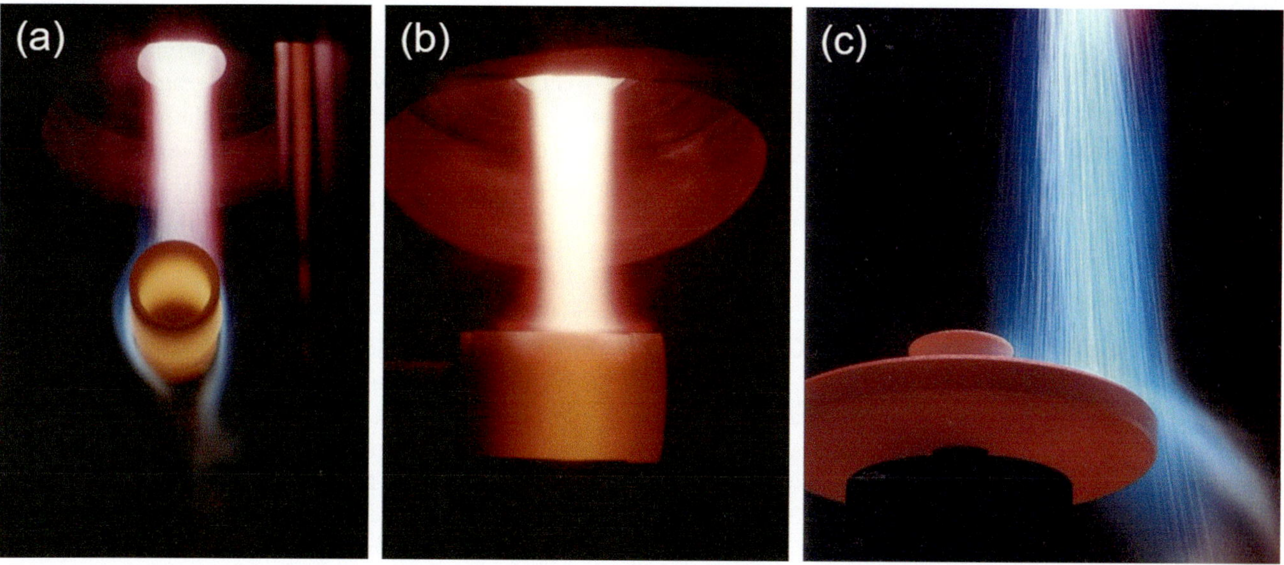

Fig. 18.45 Photograph of VIPS of tungsten on a graphite substrates. (**a**) (**b**) Two views of disposition on a cylindrical substrate. (**c**) W/Re deposition on X-ray graphite test mandrel. [Courtesy of the CRTP, University of Sherbrooke (1995)]

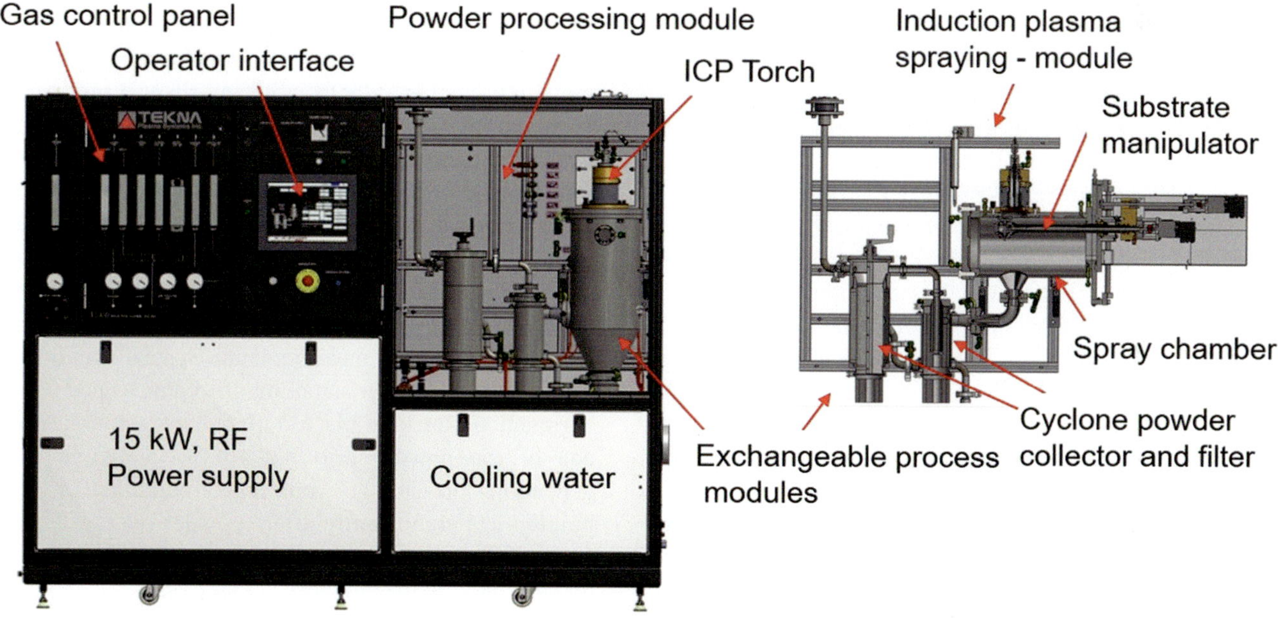

Fig. 18.46 Schematic representation of the Tekna Plasma Systems Inc., Teck-15 integrated induction plasma powder processing and spraying system (15 kW plate power), for R&D and small production scale. [Courtesy of Tekna Plasma Systems Inc.]

interface. The RHS of the cabinet is dedicated to the process part shown in this case for powder processing. It includes the PL-35 induction plasma torch, powder or suspension feeder, injection probe, powder recovery, and exhaust gas cleaning and filtration. The bottom section of the unit is dedicated to the closed-loop cooling water circuit. External to the unit not shown in the figure is the vacuum pumping system which can be a distance from the unit. An important feature of the integrated system design is the ability to interchange the process module. On the RHS of the figure is a water-cooled vacuum plasma spraying module which can be exchanged with the powder processing module. It is composed of a water-cooled vacuum spray chamber 0.245 m in diameter x 0.415 m long, equipped with an access port on its LHS and a top opening for the mounting of the induction plasma torch. The chamber is also equipped with a water-cooled, three-axis robotic arm for the manipulation of the substrate in the spray chamber in a horizontal direction over a span of 125 mm at variable speed 5–50 mm/s. The arm allows also for the rotation of the substrate around a horizontal or vertical axis at variable speed of 10–180 rpm and manual adjustment of the position of the substrate in the vertical direction over a span of 75 mm for the setting of the spraying distance.

18.5 Instrumentation and Process Control

Instrumentation is a key component of any technology that makes all the difference between an "artisanal process" and "scientific-based process." In the former, the outcome is based on the "experienced operator" that can detect based on the color of the flame or the sound of the torch the best operating conditions to obtain the desired product. In the "scientific-based process," on the other hand, the outcome depends on a set of well-defined, measurable, conditions that need to be maintained to achieve the desired result. Over the years, since its development in the early 1960s, thermal spray (TS) has evolved rapidly from its initial empirical-based approach to its present mature science-based technology. This does not to imply either that all the steps in the process are fully understood and controlled, with no room for improvement.

In this section, a review is presented of the role of instrumentation in the advancement of the technology and the level of instrumentation needed in a given TS installation as function of its objectives. However, it is appropriate in the first place to clarify the nuances and interpretations associated with different terms, commonly used in industrial plasma spray operations. The definitions given by [Dwivedi et al. (2010)] related to the variability in industrial TBC production can be used independent of the nature of the coating sprayed.

- *Repeatability* consists in obtaining the same result/coating by repeating a process under the same operating conditions on the same equipment with the same operator.
- *Reproducibility* is defined by the variation in average measurements obtained when two or more people inspect the same parts or items using the same measuring technique. Often this term is used to define variations, for instance, when two or more booths/sites seek to reproduce the same coating through the same or similar processes.
- *Reliability* of a coating is, on the other hand, the probability that the coating will adequately perform its specified purpose for a specified period of time under specified environmental conditions. As such, the term reliability captures all the aspects of repeatability and reproducibility and sets fundamental benchmarks at each step of the process.

Instrumentation and controls do not relieve either the design engineer or operator from their prime responsibility of giving the necessary attention to the hardware elements and environmental conditions in which the coating process is to take place. These include, though not necessarily limited, the following:

- *Hardware design and manufacturing stages*, through double checking of the design computation combined with a through quality check of the critical system components such as spray gun nozzle dimensions and surface finish. Component alignment such as the position of the powder injector with respect to the exit nozzle of the DC plasma torch, or the powder injector probe with the axis of the induction plasma torch. In general, it is strongly recommended at the design stage to have such critical elements self-aligning once placed in its operating position.
- *Substrate preparation*, such as substrate cleaning, masking, and roughening of the surfaces to be coated using standardized procedure to a preset and measurable level. The time laps between the surface roughening stage and the spraying stage should be monitored and kept as short as possible. Alternately, the prepared parts should be vacuum packed in order to avoid any changes with time of the quality of the surfaces to be coated.
- *Sprayed medium quality control* on product lots used in the process. For powders, this can involve microscopic observation of the particle morphology, measurement of particle size distribution (PSD), Hall flow index, apparent and tap density of the powder, and, when dealing with fine powders with a mean diameters below 20–30 μm, BET specific surface area measurements. Special attention has to be given to powder sampling procedures since powder segregation can take place during transportation. Shelf-age of the powder also has to be monitored since, depending on storage conditions, humidity or oxide buildup can significantly affect powder quality. Equivalent measurements and controls are needed for wires and cords used in flame and wire arc spraying.
- *Spray gases quality control*. Special attention is to be given to the purity and oxygen level in the inert spray gases, their storage facility, and stock inventory for the duration of the planned operation. When using compressed air locally generated on site, quality control of the compressed air line for the presence of "oil" is necessary on a regular basis.
- *Booth condition*, such as ambient temperature, humidity, and exhaust gas evacuation, should be monitored on a regular basis. It should be pointed out that for a given setup, the flow rate of ambient air extraction from the spray booth will depend on the cleanliness of the filter elements in the bag house. Periodic calibration of the

powder feeders is also an essential part of booth maintenance to ensure stability of the operating conditions especially when phased with the need to use a new batch of spray powder for the spray coating operation. The proper positioning of the different equipment in the booth is also essential for a reliable and safe operation.

- *Operator training and certification* is also a potential source of process variability for which attention including non-tangible elements such as "day of the week" over which the spray operation is carried out.

18.5.1 Core System Instrumentation

The first level of instrumentation in any spray coating installation should be dedicated to simple reliable instruments monitoring the vital signs of each individual component in the installation as well as the integrated system performance. These are usually accomplished using standard instrumentation involving the measurements of electric current, voltage, gas and water temperature, pressure and flow rates, humidity, weight, and distances. For critical parameters, duplication of instrumentation using two different techniques, for example, using mass flow controllers (MFC) and rotameters, for gas flow measurements, offers an added safety feature in the case of defective performance of any of the measuring instruments.

For example, in a DC plasma spraying system, the following parameters should be continuously monitored and recorded:

- Arc current and voltage. Oscillography observation of the arc voltage and its spectrum analysis are also very useful, though not absolutely necessary, for the following up of the electrode life performance.
- On-line monitoring and recording of cooling water pressure and flow rate to the plasma torch, as well as water temperature in and out of the different equipment components. These are essential to carry out occasional energy balance on the system and evaluate its thermal efficiency under different operating conditions.
- On-line monitoring and recording of line pressure and flow rates of process and auxiliary gases to the plasma torch, powder feeder, as well as any gases used.
- On-line monitoring and recording of ambient temperature and humidity in the spray booth as well as vital signs of the air exhaust system and the filter baghouse.
- Powder feed rate often monitored either directly through continuous weight monitoring of the powder feeder or indirectly through the rotation speed of the disc or screw in the powder feeder. On-line monitoring of the powder feed rate in the powder feed line is also a great asset,

whenever available, especially if it provides dynamic information of the powder flow regime.

By connecting all of the above listed parameters to the system controller (PLC), they can be used in an interlock scheme designed to avoid system operation under unsafe conditions. Equivalent core instrumentation needs to be used to monitor and record for different thermal spray technologies.

18.5.2 Substrate Diagnostics

While it is possible to run the process relying exclusively the system instrumentation, it is a long shot to relate the "coating quality" directly to "'system operating conditions." Process optimization becomes consequently slow and tedious since it would rely on "coating characterization," or even "coating performance" evaluation, which at times can be time-consuming and optimize the operating conditions. Process instrumentation, by providing means for the measurements of intermediate process parameters, such as substrate temperature, spray stream temperature, and velocity, or in-flight particle temperature and velocity, allows to bridge the gap between system operating conditions and coating quality and performance. The effect is particularly significant during R&D development of new coatings, and optimization of coating performance, by providing the necessary information for the understanding of the basic phenomena involved. Process instrumentation is also increasingly being used in a production-scale environment, by allowing for rapid, online detection of the drifting of process parameters, thus avoiding the spraying of multiple "out of spec parts" before realizing the drift in the process conditions, with significant reduction of rejected parts and associated economical savings. In the following examples are given some of the most commonly used process instrumentations in thermal spray operations.

18.5.2.1 Substrate Surface Temperature
As discussed in Chap. 15, "Conventional Coating Formation," Sect. 15.3.3, the shape of splats formed through the spraying of metallic powders on metallic substrates, and their adhesion to the substrate, depends strongly on the substrate temperature. As shown in Fig. 18.47, for nickel powder sprayed on stainless-steel substrate, a critical substrate temperature exists between 500 and 600 K below which formed splats of "splash" shape, while above that temperature they are "disc" shaped. Associated with this transition is a major increase of splat adhesion to the substrate at substrate surface temperatures above the critical value, which is reflected on the adhesion of the coating to the substrate. As indicated in Table 18.3, the transition temperature can vary between

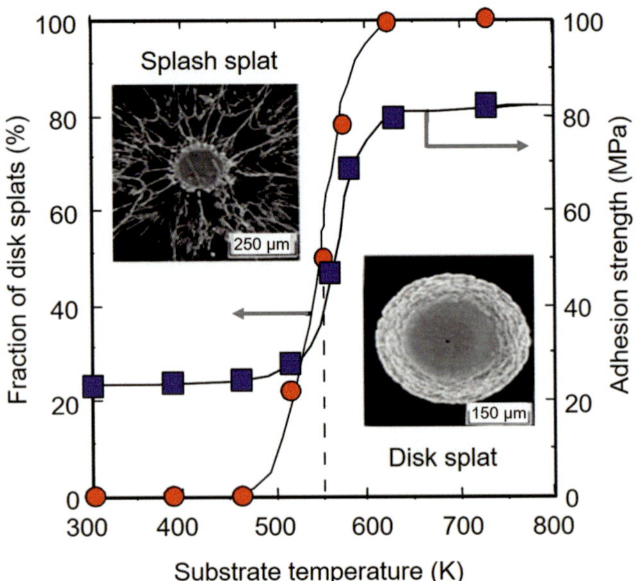

Fig. 18.47 Dependence of the fraction of nickel splats which are in disk form and coating adhesion strength onto AISI-304 stainless-steel substrate as function of substrate temperature at atmospheric pressure AISI-304 stainless-steel substrate at room temperature (300 K) [Fukumoto (2004, 2007)]. Reprinted with kind permission from ASM International

Table 18.3 Transition temperature (K) for splat formation of different metallic and ceramic powders on AISI-304 stainless-steel substrate sprayed using APS and HVOF [Fukumoto (2004, 2007)]

Metals				Ceramics	
APS		HVOF		APS	
Ni	610 K	Ni	560 K	Al$_2$O$_3$	318 K
Mo	474	Ni-5Al	440	TiO$_2$	350
Cu	394	Ni-10Cr	400	YSZ	345
		Ni-20Cr	360		
Cu 30Zn	505	Cu-30Zn	455		
Cr	387	Cr	345		

radiation to the surrounding and conduction to the body of the part being sprayed which depends in turn on the size of the part and its heat capacity.

In a typical thermal spray operation such as flame spraying, DC plasma, or RF induction plasma spraying, the heat flux due to the hot gas jet downstream of the spray torch nozzle exit decreases rather rapidly with the spray distance (almost exponentially). When spraying solutions or suspensions at very short spray distances (30–40 mm with plasma jets and 60–90 mm with HVOF jets), heat fluxes can reach up to a few tens of MW/m^2, especially with plasmas. Special precautions involving the use of mobile thermal protection masks or substrate surface cooling using gas jets during the operation can be effective for the removal of non-molten, low-inertia, fine particles traveling in the fringes of the spray jet and reducing the heat flux during the spray operation. Substrate temperature can also be controlled by variation of the exposure time of the substrate to the spray stream which can be achieved through control of the translation velocity of the spray gun with respect to the substrate surface. The coating temperature can also be controlled through variation of the thickness of the coating deposited by every pass, especially for low thermal conductivity sprayed materials ($\kappa_p < 20$ W/m.K), the spray pattern, the spray distance, and the cooling systems used.

As reported by [Fauchais and Vardelle (2011)], intense research effort has been devoted to the monitoring and control of the substrate/coating temperature during the spray process. The most widely used techniques used are:

- Infrared (IR) pyrometers [Vardelle et al. (2002), Moulla et al. (2005), Doubenskaia et al. (2006)], with wavelengths over 6 μm being less sensitive to radiation of the plasma or hot gases as well as hot particles
- IR thermography [Weisheng et al. (2009), Lugscheider et al. (1998), Friedrich et al. (2001)]
- Embedded and fast response micro-thermocouples in metallic substrate [Verdy et al. (1998), Salimijazi et al. (2007)]

While (IR) pyrometry and thermography are the most favored for such measurements being contactless and offering the ability to monitor the substrate surface temperature at a reasonable distance away from the hot and dust-laden zone of the spraying operation, special care has to be exercised in selecting the type of instrument and wavelength used.

Generally, a "two-wavelength" pyrometer is favored compared to a "single-wavelength, total emission" pyrometer since in the latter case, it is necessary to specify the emissivity of the surface of the substrate and/or coating. In contrast, in "two-wavelength" pyrometry, the surface temperature is deduced based on the ratio of the emission received at the

350 and 600 K depending on the material being sprayed. Moreover, as pointed out by [Fan et al.(1998)], the spraying of ceramic coatings is particularly sensitive to the substrate and coating temperature during the spraying process since too high a temperature either would result in the creation of significant residual stresses during the cooling step of the coating leading to the its severe crackling and spalling.

Substrate temperature is consequently one of the important process parameters that need to be measurement and controlled during the spray process for optimal coating properties. This is not, however, an easy task since the surface temperature of the substrate is dependent on the balance between the heat flux received by substrate/coating during the spraying operation, from impacting hot gases and semi-molten particles [Fauchais and Vardelle (2010)], and heat losses from the surface of the substrate/or coating by

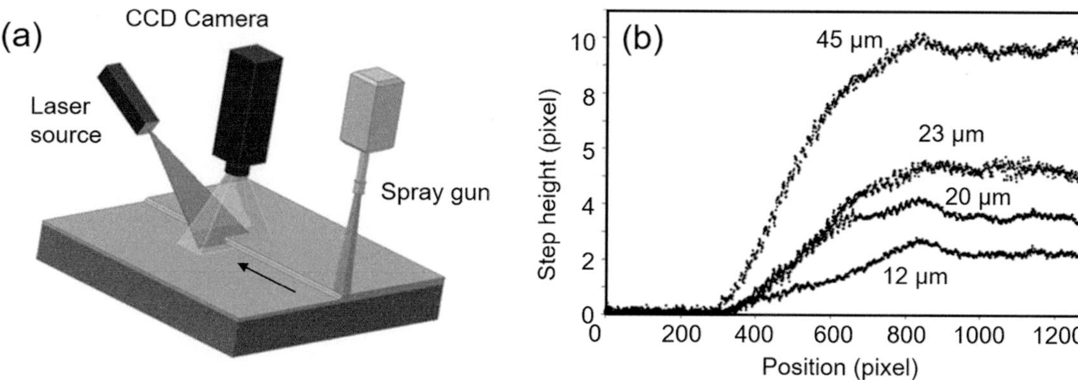

Fig. 18.48 (**a**) Coating thickness measurement setup and (**b**) raw profiles for four-step heights from 12 to 45 µm recorded on a previously coated moving surface. [Nadeau et al. (2006)]

two fixed wavelengths, without the need to specify the surface emissivity of the substate or that of the coating. The measurement implies gray-body emission of the surface of the substrate or that of the coating being measured, i.e., the surface emissivity is independent of the wavelength. Special attention has also to be given to the careful choice of the respective wavelengths to be used to avoid interference by the spectrum of the background emission of the plasma or that of any metallic vapors in the spray stream.

18.5.2.2 Coating Thickness

Coating thickness is an important parameter to monitor and control on-line as much as possible. Unfortunately, most measurements are generally destructive (after spraying) and time-consuming. A novel approach proposed by [Nadeau et al. (2006)], illustrated in Fig. 18.48b, is based on simple optical triangulation to detect the smooth step profile of a pass over the immediately adjacent uncoated (or previously coated) surface. The strategy consists in recording the profile of the coating at the frontier between the new layer and the previous one using a laser line projected across the pass edge and captured with CCD camera as shown in Fig. 18.48b. Measurements obtained using a calibration relating the camera pixels to thickness in (µm) are independent of coating/substrate nature, the surface roughness, or the thermal expansion of the coated part. According to [Nadeau et al. (2006)], the techniques which have a precision of ± 5 µm for coatings on a cylindrical substrate can be used for on-line, real-time individual spray pass thickness during deposition.

18.5.3 Spray Medium Diagnostics

Spray medium diagnostics aim at the identification of the general flow characteristics and determination of temperature, velocity, and composition of the high-energy spray medium carried out principally on plasmas and flames,

including HVOF, HVAF, and D-gun. The following broad range of techniques have been used mostly for R&D and device/process development studies.

- Flow visualization using high-speed photography, shadow graph, and Schlieren techniques
- Emission or laser spectroscopic mostly used for temperature field measurements
- Enthalpy probe techniques, for specific enthalpy, velocity, and gas composition measurements

In the following, a brief review is presented of the approach used and the information that could be obtained. Reference is made to the vast literature on the subject for further information.

18.5.3.1 Flow Visualization

High-speed photography is a powerful tool for the characterization of a thermal spray system operating in the absence or presence of spray particles. It has been effectively used both in an R&D process development stage and in a standard production environment. Different optical setups have been used depending on the objective of the study. These include:

- CCD recording of the spray system
- Shadowgraph
- Schlieren photography

Photography is particularly useful and a relatively simple way to record the effect of operating parameters on the overall characteristics. It is, however, of critical importance to keep in mind the camera setting used due to their impact on the interpretation given to the observed flow structure. The example given in Fig. 18.49 is for an Ar/H$_2$ RF induction plasma jet in an open discharge mode in vacuum spraying chamber in the absence of any powder injection. The plasma torch used is a Tekna PS-70, with a 64 mm i.d. discharge

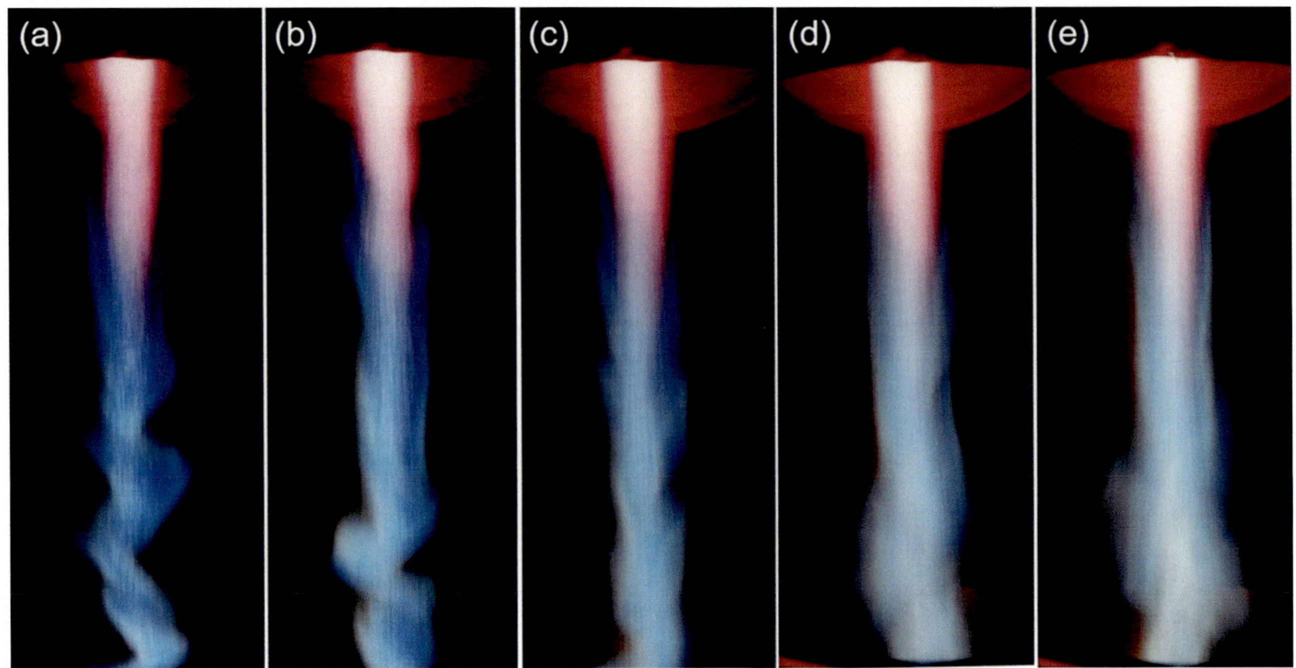

Fig. 18.49 Sequence of photographs taken at different exposure times of low-pressure RF-ICP Ar/H$_2$ plasma jet, immerging from a PS-70 plasma torch, nozzle i.d. = 64 mm, power = 60 kW, chamber pressure = 33 kPa: (**a**) 1/1000 s, (**b**)1/500 s, (**c**) 1/250 s, (**d**) 1/120 s, and (**e**) 1/60 s. [Boulos (2001)]

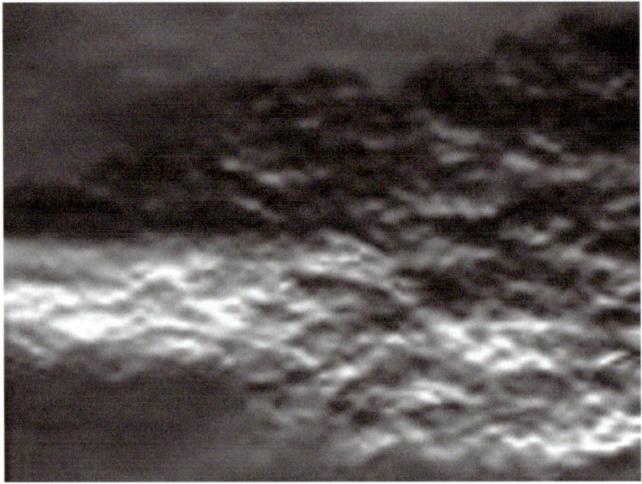

Fig. 18.50 Schlieren photographs of DC plasma jet in ambient air. [Pfender et al. (1991)]

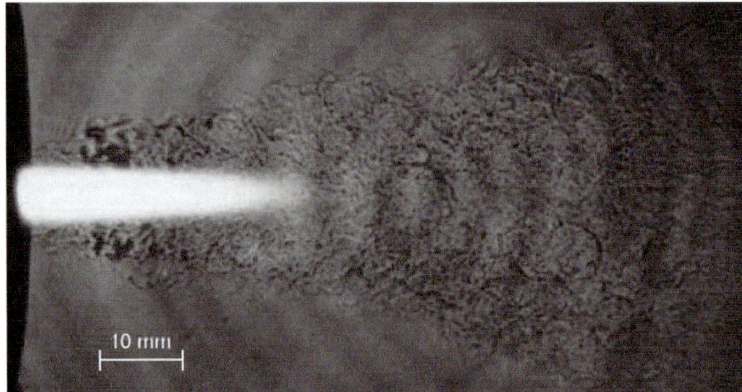

Fig. 18.51 Pulsed laser Schlieren images of turbulent Ar plasma jet, SG-100 torch, nozzle i.d. =8 mm, 35.4 slm (Ar), 900 A. 15.4 V. [Finke et al. (2003)] (Reprinted with permission of ASM International. All rights reserved)

nozzle operating at a plate power of 60 kW and an absolute chamber pressure of 33 kPa. All the photographs given in Fig. 18.49a–e were all for the same operating plasma conditions except for the systematic doubling of the exposure time between every two consecutive photographs with a corresponding reduction of aperture to maintain equivalent total exposure. It may be noted that with the increase of the exposure time, the plasma plume appears to be far more uniform and symmetric compared to that at the shortest

exposure time of 1 ms, in which a 3-D "cork-screw" flow pattern is observed reflecting the swirling velocity component introduced into the flow in this particular case.

Schlieren photography has also been commonly used for the characterization of combustion and plasma flows as shown in Figs.18.50 and 18.51, the latter using pulsed laser Schlieren technique with a Praxair SG-100 DC plasma jet, 8 mm i.d. nozzle, 35.4 slm (Ar), arc current of 900 A, and voltage of 15.4 V [Finke et al. (2003)]. Schlieren photographs of droplet formation in twin-wire arc spraying given in Fig. 18.52 show the diamond shock pattern formed by the atomizing gas flow in a twin-wire arc spraying gun [Hussary

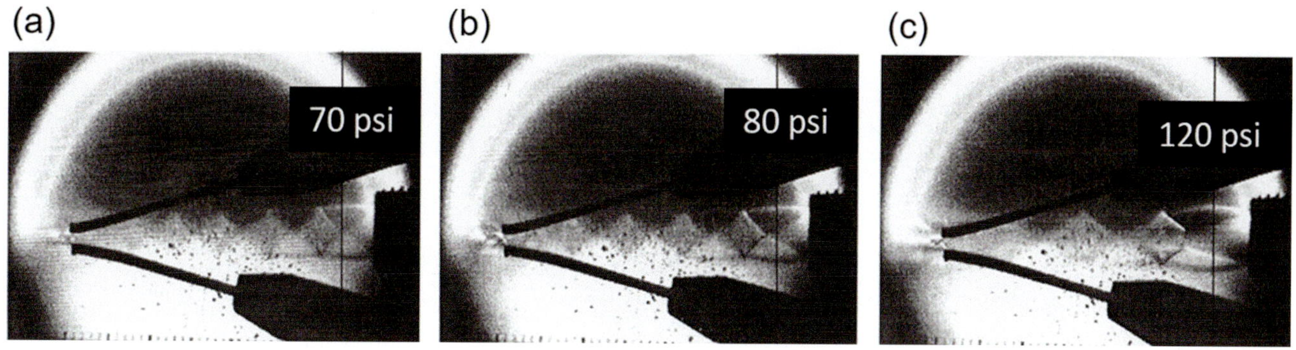

Fig. 18.52 Schlieren photographs of shock pattern outside wire-arc atomizing gas nozzle showing under-expanded flow exiting straight bore nozzle. [Hussary (1999)]

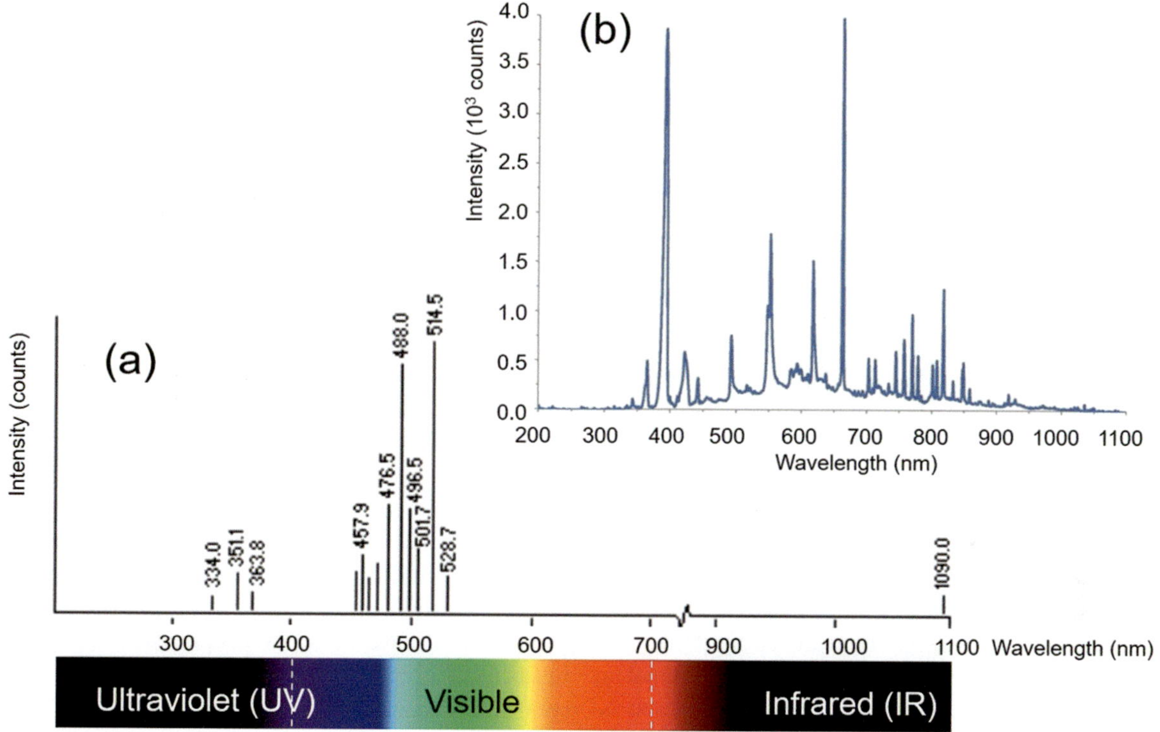

Fig. 18.53 Typical example of emission spectra of (**a**) pure argon plasma and (**b**) Ar/H$_2$ plasma at atmospheric pressure

(1999)]. It should be pointed out that while digital photography has been used for both R&D and industrial production-scale applications, Schlieren has been mostly limited to R&D fundamental studies.

High-speed photography has also been used for the visualization of particle trajectories in a plasma flow. The situation is, however, more complex since in the absence of a powerful flash, the particles, depending on their temperature, might not be visible because of the intense background emission by the plasma. The adaptation of the camera with an appropriate narrow-band filter is an option that has been successfully used provided that the passing band of the filter is chosen to block the strong line emission of the plasma

gas. A typical example of the emission spectra of pure Ar and Ar/H$_2$ plasmas is given in Fig. 18.53. The presence of metal vapor in the plasma medium offers an extra challenge in this case due to the presence of large number of intense metal vapor radiation lines in the spectrum as can be noted in the iron vapor spectrum given as an example in Fig. 18.54.

18.5.3.2 Particle Image Velocimetry

Particle image velocimetry (PIV) is a method for visualizing and analyzing field of flows [Westerweel (1993), Adrian (1988)]. Small particles (in the μm size range) are added to the flow, and measurements are performed downstream of the

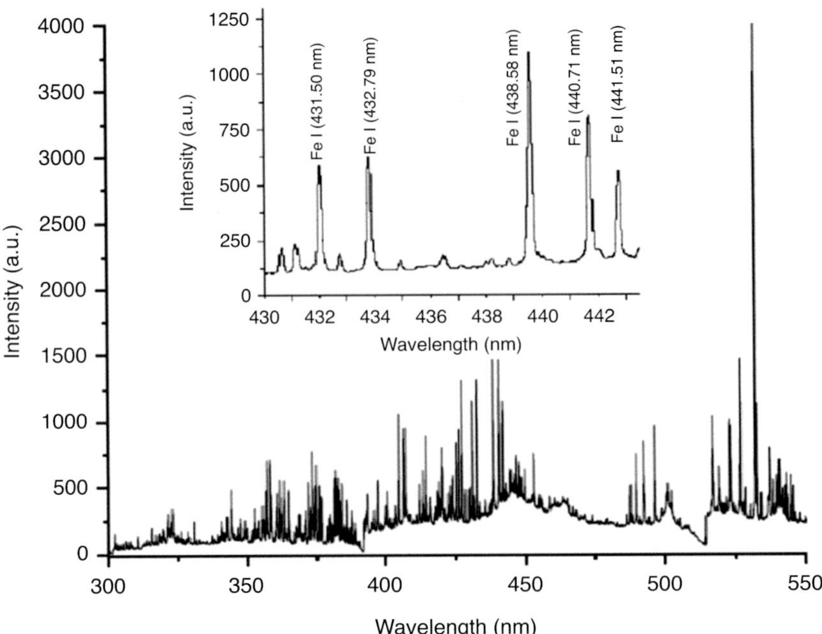

Fig. 18.54 Emission spectrum of neutral iron plasma produced by the second (532 nm) wavelength of the Nd:YAG laser at a distance of 0.05 mm from the target. [Muhammad Salik (2013)]

injection where it is assumed that particles have reached the same velocity as that of the flow (Knudsen effect is generally considered to be negligible below 3000–3500 K). The particle motion is dominated by Stokes drag with Stokes number below 1. Assuming that particles are spherical with low Reynolds number (micrometer-sized particles and low specific mass), their ability to follow the fluid's flow is directly proportional to the difference in mass density between particles and the fluid and directly proportional to the square of the particles' diameters. The smaller they are, the faster they will reach the flow velocity. The scattered light from the particles being dominated by Mie scattering is proportional to the square of the particles' diameters. The particle size needs to be balanced to scatter enough light to accurately visualize all particles within the laser sheet plane but small enough to accurately follow the flow.

In the PIV method, particles in the flow are illuminated by short laser pulses and captured with CCD digital cameras with a few hundred nanoseconds of difference between frames. The frames are split into a large number of interrogation areas or windows. Quantitative information about the velocity of the flow at any one location is obtained by comparing two successive images of the field taken, respectively, at times t and $t + \Delta t$. For this, it is necessary in the first place to identify pairs of small areas (windows), which are the closest to each other in both images. It is then possible to calculate a displacement vector for each window, thanks to signal processing and autocorrelation or cross-correlation techniques. This is converted to a velocity using the time between laser shots and the physical size of each

pixel on the camera. The size of the interrogation window should be chosen to have at least six particles per window on average. For example, this technique has been used by [Alekseenko et al. (2011)] to measure the flow structure and velocities of swirling turbulent propane flames and by [Yilmaz et al. (2009)] to characterize diffusion flame oscillations. PIV has also been by [Zahiri et al. (2009)] to characterize supersonic flow field in cold spraying conditions. Results show that PIV is a practical characterization technique for the optimization and model validation in cold spray applications. [Newbery and Grant (2000)] have used PIV coupled with high-speed video imaging to observe droplet deposition during electric arc spraying, with particular attention given to the behavior of droplets originating from splashing.

18.5.3.3 Emission Spectroscopy

Emission spectroscopic measurements provide information on the population of the excited level of molecules, atoms, and ions [Fauchais et al. (1989, 1992), Coudert et al. (2002)]. Depending on the assumptions made, different temperatures can be deduced. In most cases when measuring plasma jet temperatures in spray conditions, it is generally assumed that the plasma is at local thermodynamic equilibrium (LTE) implying that the temperatures of electrons and heavy elemental particles are equal [Fauchais et al. (1989)].

At equilibrium, the evolution of the volumetric emission coefficient, ε_λ, can be calculated as a function of temperature for a given pressure [Boulos et al. (1994)], and thus temperature can be deduced from ε_λ measurement. However, for atoms, the evolution of ε_λ with temperature varies in ratios of about 10^{16} to 10^{18} when temperature increases from 3000 to 14,000 K! At best, detectors can cover 4–5 orders of magnitude, and thus the temperature range that can be covered is limited and varies roughly from 8000 to 13,000 K for argon or nitrogen atoms which cover well conventional temperatures in the core region of plasma jets. For higher temperatures, such as those obtained in PTA, argon ions, Ar^+, are used and cover values from about 15,000 to 22,000 K, while for lower temperatures (below 8000 K) such as those obtained in the plasma jet plumes or in high-power torches with vortex gas injection and button-type cathodes, molecular spectra must be used. The easiest molecular spectra to be used are those of N_2^+ or OH, though they are by far more complex and time-consuming than atomic spectra [Fauchais et al. (1989), Boulos et al. (1994)].

To record the plasma radiation, an optical setup is used to focus a small measurement volume onto the entrance slit of the monochromator which must cover, with a reasonable accuracy and resolution, the wavelength range from about 3000 to 12,000 Å, depending on the lines or molecular spectra considered. The spectral resolution depends on the width of the entrance slit, the resolution of the detector itself, as well as the accuracy of the diffraction grating, linked to grooves density in the dispersive element. Spectral resolutions better than a few Å are easily achieved. Photodiode arrays are increasingly used to measure the integrated, line-of-sight, volume emission over the measurement volume received by the monochromator. For axisymmetric sources, the Abel's inversion can be used to convert the integrated, line-of-sight, volume emission profile, into emissivity profiles, which in turn can be used to deduce the temperature profile in the measurement volume.

Anodic arc root fluctuations, giving rise to fluctuation of the plasma plum which are characteristic of DC plasma jets, as illustrated by photographs shown in Fig. 18.55a, for an Ar–H_2 plasma jet, offer an important challenge to emission spectroscopic temperature measurements especially because of their relatively high frequencies in the range of 2–10 kHz. While monochromators equipped with CCD detectors cooled with liquid nitrogen and computer controlled allow for a fast acquisition and treatment (a few ms) of the recorded data, their response is still in the milliseconds range. This implies that fluctuations are integrated, and the temperature distribution recorded would "seem" perfectly symmetrical. It is difficult to pretend that the temperature distributions presented in Fig. 18.55b are a good representation of images shown in Fig. 18.55a. Thus, such measurements for highly fluctuating plasma jets obtained with diatomic plasma-forming gases can only give trends about the plasma jet evolution when modifying plasma torches working parameters.

It should be recalled, however, that as discussed in Chap. 8, "DC Plasma Spraying, Fundamentals," Sects. 8.

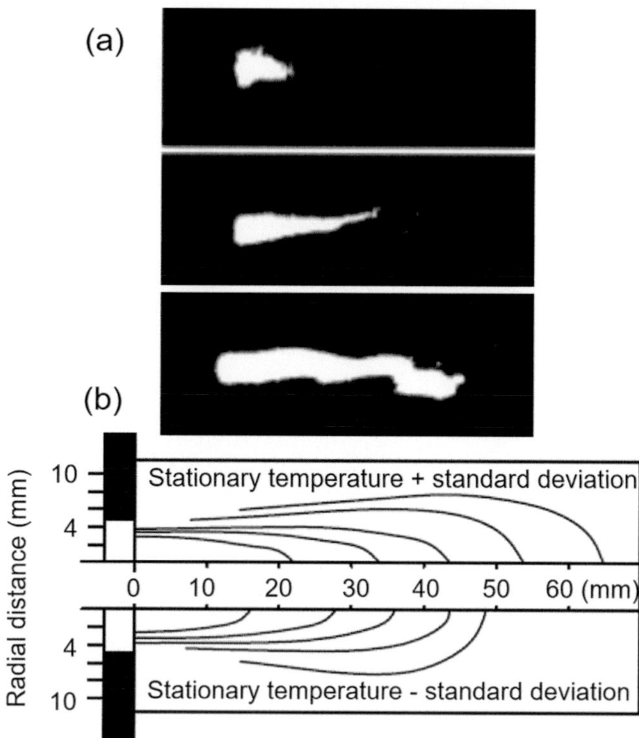

Fig. 18.55 (a) Photographs of DC plasma jet taken with a shutter time of 10^{-4} s, (b) temperature measured with ArI-738.3 nm line ± standard deviation (measuring time 100 ms). Ar–H_2 (45–15 slm) plasma jet; nozzle i.d. 10 mm; I = 632 A; V = 61 V; thermal efficiency of the gun 56%. Reprinted with kind permission from Elsevier for Int. J. Therm. Science. [Fauchais and Vardelle (2000)]

3.3, 8.3.4, and 8.3.6, dedicated efforts have been devoted to the development of laminar flow DC plasma torches, cascade torches, and multi-electrode torches, which have considerably lower voltage fluctuations ($\Delta V/V_m < 20\%$) and consequently a far more stable plasma jet.

For nonsymmetrical spray flow stream, computer tomography (CT) combined with conventional emission spectroscopy offers means of measuring 2-D temperature distributions in a plasma jet through the capture of radial emission distributions of the flow at a given axial location, using a photodiode array, from multiple angles over a sector of 180° in a plane perpendicular to the torch axis. Through the simultaneous deconvolution of the observed integrated line of sight emission profiles, the data can be converted to 2-D local emission intensity distribution, which can be used to compute the corresponding 2-D temperature distribution across the plasma flow. Repeating the procedure at different axial locations allows for the reconstruction of the full 3-D temperature field in the plasma jet. The successful use of the technique is conditional to having a highly stationary emission source (plasma jet) with a well-defined geometrical axis around which the detecting optical system is rotated. It also requires having a high data storage and computation capacity that increases rapidly (to the power 3) with the increase of

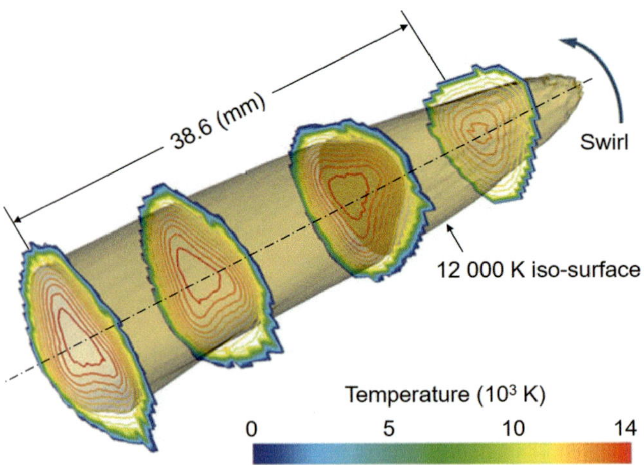

Fig. 18.56 Tomographic reconstruction of gas temperature distribution showing the counterclockwise jet rotation: injector plane cross section, three additional downstream sections, and surface of 12,000 K. Reprinted with kind permission from Springer Science Business Media. [Mauer et al. (2011)], copyright © ASM International

measurement resolution. (For more details on the process, see Landes 2006; Schein et al. 2008; Hlina and Sonsky 2010; Mauer et al. 2011.)

[Mauer et al. (2010, 2011)] reported measurements carried out on Oerlikon Metco Triplex Pro™ torch with a nozzle diameter of 9 mm, plasma gas flow rates 50 slm Ar and 4 slm He, and arc current 540 A, with an effective net power of 30.9 kW. Figure 18.56 shows the tomographic reconstruction of gas temperature distribution with the powder injector plane cross section, three additional downstream sections, and the iso-surface of 12,000 K. The gas distribution ring inside the torch induces the counterclockwise jet rotation. The temperature distribution, over the jet cross section in the plane of particle injection, shows characteristic non-rotationally symmetric profiles with a typical triangular shape.

More sophisticated techniques such as Coherent Anti-Stokes Raman Spectroscopy (CARS) or Rayleigh scattering are also used, essentially for temperatures below 8000 K. The CARS signal is generated by crossing three high-powered, pulsed laser beams, called pump beam at frequency ω_1, pump beam at ω_2, and the Stokes beam at ω_S, in a suitable phase-matched geometry. The beam-crossing region is at the common focus of the three laser beams and defines the CARS probe volume, a 50 μm diameter cylinder that is 1 ± 2 mm in length [Kearney et al. (1999)]. A broadband approach is generally used to generate the entire CARS spectrum on the time scale of the laser pulse. The technique, which works only with molecular spectra, has the advantage of high conversion efficiency, a laser-like coherent signal beam for high collection efficiency, excellent fluorescence and luminosity discrimination, as well as high spatial and temporal resolution. CARS has been used exclusively in an R&D

environment to follow the entrainment of the surrounding air into the fringes of a plasma jet [Pfender et al. (1991)]. Molecular nitrogen was the probe and the temperature determination calibrated against a Pt/Pt-Rd thermocouple [Hall and Eckbreth (1984), Eckbreth and Anderson (1985, 1986), Kearney et al. (1999)].

Integrated Rayleigh scattering, as with CARS, presents an excellent spatial resolution: the measuring volume is the intersection of the laser beam, and the optical path of the detector and the measurement is local. The temporal resolution is obtained by using pulsed lasers. The Rayleigh signal is proportional to the densities of the atomic and molecular species present at the measuring volume and of course to their Rayleigh cross sections. Assuming equilibrium and using Dalton's law, it is possible to determine the temperature [Fauchais and Coudert (1996)]. Such measurements have been used to study the influence of the presence of the substrate on the temperature distribution in the plume of an Ar–H_2 DC plasma jet [Lapierre et al. (1994)].

18.5.3.4 Enthalpy Probe

[Blais et al. (2005)] have presented an excellent historical review of the development of the enthalpy probe (EP) starting from the pioneering work of [Grey et al. (1962, 1963, 1965) and Grey et al. (1964, 1965)] through its first applications to the diagnostics of DC plasma jets by Pfender and his collaborators at the University of Minnesota, [Brossa and Pfender (1988), Capetti and Pfender (1989), and Pfender et al. (1991)]. Subsequent studies by Fink and his collaborators at Idaho National Laboratory EG&G, Idaho Falls, USA [Swank et al. (1993), Fincke et al. (1993, 1994)] and Boulos and his collaborators at the University of Sherbrooke and Tekna Plasma Systems Inc. [Rahman et al. (1995, 1996)] extended their use through their coupling to a mass spectrometer to measure local gas temperature velocity and composition in atmospheric and compressible plasma jets. In 1993, Tekna Plasma Systems Inc., in Sherbrooke, Quebec, Canada, developed a robust enthalpy probe system that is equally suited to R&D development work as well as on-line calibration of industrial DC plasma spraying installations.

As illustrated in Fig. 18.57a, Enthalpy probe systems bridge the gap in plasma temperature measurement between thermocouples for temperatures below 1500 K and emission and laser spectroscopy for temperatures above 10,000 K. A photograph of such a probe placed in a DC plasma jet is shown in Fig. 18.57b. The basic concept used is essentially a calorimetric one including the following steps:

- Measurement of the specific enthalpy of the plasma by an energy balance on a sample of the spray gas extracted from the spray stream.

- Determination of the local equilibrium composition of the gas stream on a sample of the extracted gas using a quadrupole mass spectrometer (MS).
- Computation of the corresponding temperature and local density of the extracted plasma gas using appropriate thermodynamic data.
- Measurement of the dynamic pressure at the tip of the probe; knowing the local plasma temperature and density, it is possible to calculate the local velocity of the plasma stream.

The probe, represented schematically in Fig. 18.58a, consists of double-walled stainless-steel tubing with the outer two annular shells used for its cooling, while the central tube is used for the extraction of the gas sample from the plasma flow. Photographs of the probe tips of two standard probes are given in Fig. 18.58b with the smallest probe 3.17 mm o.d. × o.66 mm i.d. and the largest one, 4.78 mm o.d. × 1.27 mm i.d. Both are coated in this case with an optional thin layer of yttria stabilized zirconia (YSZ) thermal barrier coating in order to reduce the heat flux received from the plasma to the outer shell of the probe. As shown in the circuit diagram given in Fig. 18.59, the enthalpy probe is cooled by a high-pressure (up to 6 MPa) closed loop water circuit which comprises essentially of a variable speed, positive displacement pump with a maximum flow rate of 1.0 ℓ/ min, water-water heat exchanger, and an appropriate turbine water flow meter. Thermocouple probes are used to measure the water temperature at the entrance and exit to the probe. In order to avoid boiling of the water under the intense heat flux to which the probe tip is exposed, a back-pressure booster system is integrated into the circuit in order to raise the overall pressure in the cooling water circuit up to 6 MPa. Gas sampling is carried out through the central tube in the probe connected to a vacuum sampling pump and mass flow controller. From the extracted gas stream, a smaller sample is channeled to a quadrupole mass spectrometer by an appropriate sampling pump for composition analysis.

As described by [Rahman et al. (1995)], the technique is based on a two-step energy balance on the cooling-water circuit of the enthalpy probe. The first step is a "tare" measurement of the heat load to the probe in the absence of gas sampling conditions. The second is a measurement of the heat load under gas sampling conditions. The difference in heat load between the two represents the energy associated with the extracted gas sample. For a known amount of gas extracted from the plasma, the specific enthalpy, associated with the sampled gas, can be calculated.

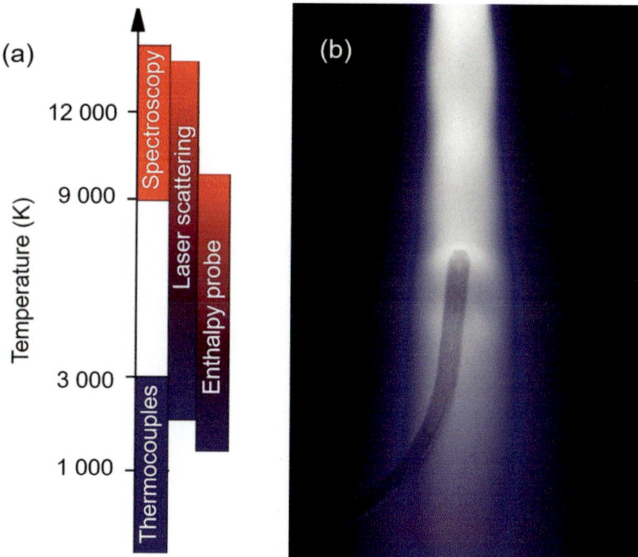

Fig. 18.57 (**a**) Representation of the temperature range covered by enthalpy probe techniques compared to alternate techniques. (**b**) Photograph of an enthalpy probe in a DC plasma jet

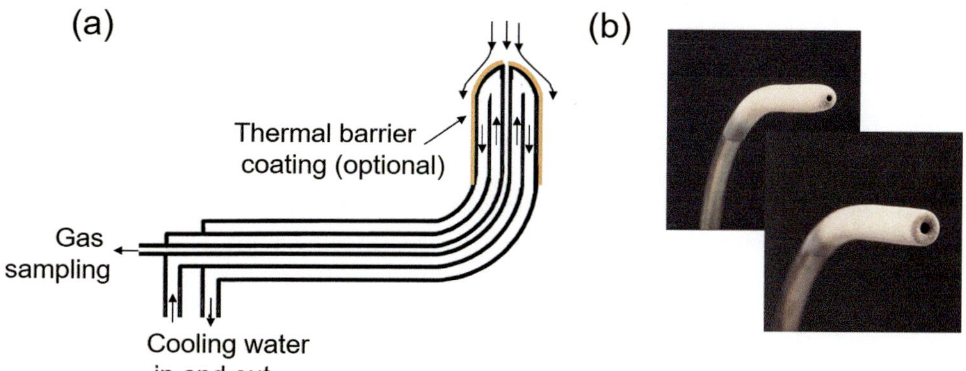

Fig. 18.58 (**a**) Schematic representation of the construction of the enthalpy probe integrating an optional thermal barrier coating (TBC). (**b**) Photographs of two typical enthalpy probe tips plasma spray coated by YSZ-TBC. (Courtesy of Tekna Plasma Systems Inc.) (Reproduced with permission)

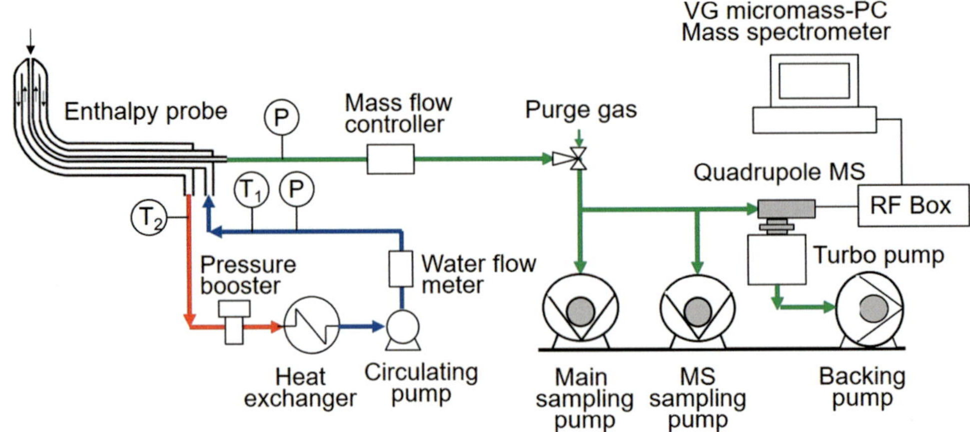

Fig. 18.59 Cooling water and gas sampling circuit diagram for an enthalpy probe system integrating mass spectrometry analysis

$$h_o = h_e + c_p \frac{\dot{m}_w \left(\Delta T_s - \Delta T_t\right)}{\dot{m}_g} \qquad (18.1)$$

where:

$h_o =$ Specific enthalpy of the sampled gas at the location of the probe tip (kJ/kg).

$h_e =$ Specific enthalpy of the gas at reference temperature (25 °C) (kJ/kg).

$\dot{m}_w =$ Cooling water flow rate (kg/s).

$\dot{m}_g =$ Sampled gas mass flow rate (kg/s).

$c_p =$ Specific heat of water (kJ/kg.°C).

$\Delta T_s = (T_2 - T_1)_s$ Temperature difference under gas sampling condition (°C).

$\Delta T_t = (T_2 - T_1)_t$ Temperature difference under "tare" condition (°C).

In systems equipped with a mass spectrometer, as that shown in Fig. 18.59, a small fraction of the sampled gas is used by the mass spectrometer to determine the equilibrium composition of the sampled gas, x_o. Based on the measured specific enthalpy, h_o, and equilibrium composition of the sampled gas, x_o, the local gas temperature, T_o, and the specific density, ρ_o, can be determined using conventional thermodynamic tables.

By measuring the dynamic pressure p_o on the gas sampling line, under "tare" conditions, i.e., in the absence of sampling, the free-stream plasma velocity u_o can be calculated using Bernoulli's equation for subsonic, non-compressible, flow conditions, with Mach number (M < 0.3) as:

$$u_o = \sqrt{\frac{2(p_o - p_a)}{\rho_o}} \qquad (18.2)$$

where:

p_o

Dynamic pressure measured under no sampling condition (Pa)

p_a Ambient pressure surrounding the probe tip (Pa)

ρ_o local gas density at the measurement point

In the Mach number range ($0.3 \leq M \leq 1$), compressibility must be considered, and the velocity expression becomes:

$$u_o = \sqrt{\frac{2\gamma RT}{\gamma - 1}\left[\left(\frac{p_o}{p}\right)^{\frac{\gamma}{\gamma-1}} - 1\right]} \qquad (18.3)$$

where γ is the ratio of the gas specific heat at constant pressure and volume ($\gamma = c_p/c_v$) and R is the ideal gas constant. As for incompressible gas, temperature must be determined from the enthalpy measurement.

In the supersonic regime ($M > 1$), the problem is more complex because of the existence of shock wave and non-equilibrium phenomenon. Calculation procedure and analysis can be found in [Blais et al. (2005)].

Special attention must be given to the precision of the measurement which according to [Grey (1963)] depends on the probe sensitivity σ_s defined as:

$$\sigma_s = \frac{\left(\Delta T_s - \Delta T_t\right)}{\Delta T_s} \times 100 \qquad (18.4)$$

Considering that the probe sensitivity improves with the decrease of ΔT_t, efforts have been made to reduce the "tare" heat load on the probe as much as possible through the coating of the probe tip by an appropriate thermal barrier coating (TBC) as shown in Fig. 18.60b.

Increasing the sampling gas flow rate would also have a favorable effect on the probe sensitivity, as shown in Fig. 18.60a after. The effect is, however, at the expense of

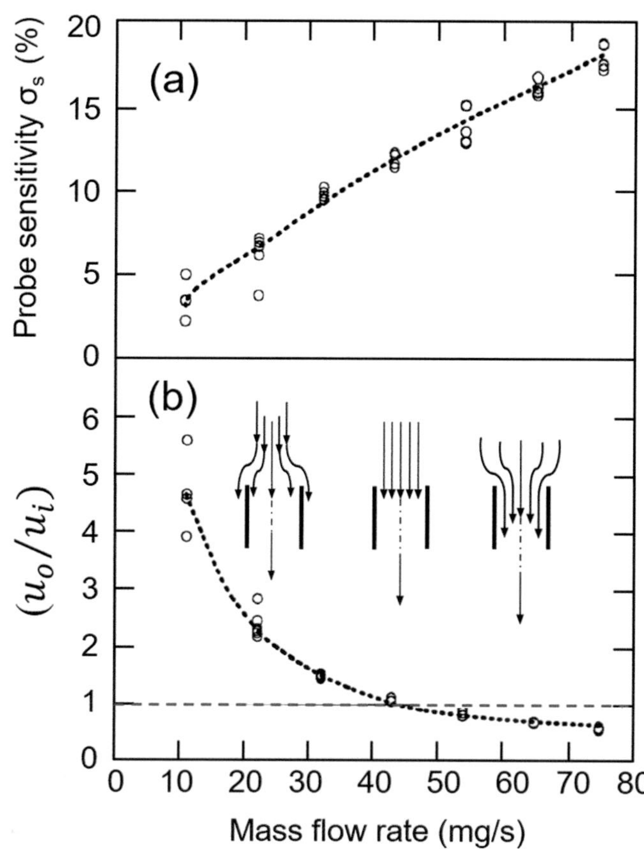

Fig. 18.60   Effect of sampling mass flow rate on (a) probe sensitivity and (b) the ratio of the free stream velocity u_o to the gas velocity at the entrance of the probe. [Rahman et al. (1964)]

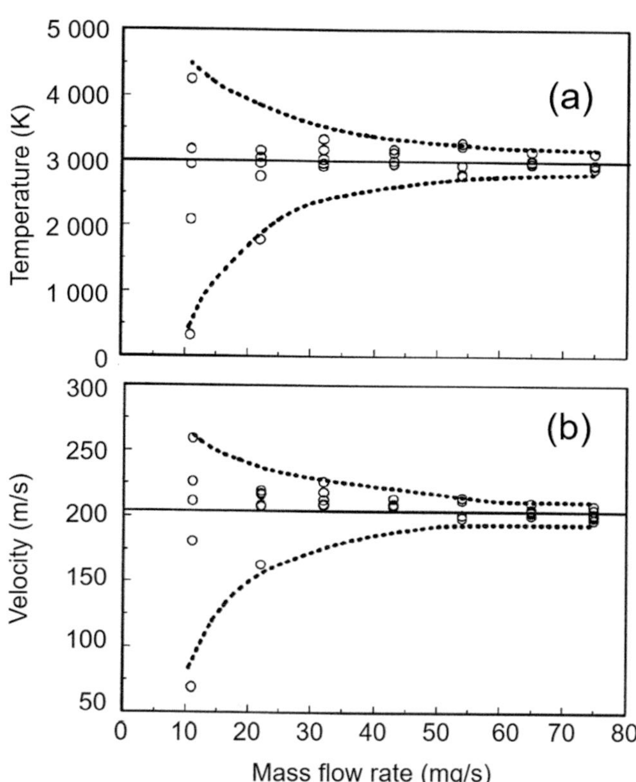

Fig. 18.61   Effect of gas sampling mass flow rate on reproducibility of (a) temperature and (b) velocity measurement in a DC argon plasma jet after [Rahman et al. (1964)]

the deterioration of the spatial resolution of the measurement. As can be noted in Fig. 18.60b, the increase of the sampling gas flow rate $\dot{m}_g$ results in the steady decrease of (u_o/u_i), the ratio of the free stream velocity, u_o, to that of inlet gas velocity in the sampling tube, u_i. As the gas sampling rate is increased, the ratio of the free stream velocity u_o to that in the entrance port of the sampling probe u_i drops below unity $(u_o/u_i < 1.0)$, and the sampled gas will be drawn from adjacent regions to the sampling point, leading to a loss of spatial resolution of the measurement. Ideally, gas sampling should always be carried out near "isokinetic conditions" with the gas velocity at the entrance of the probe equal to that of the free stream velocity $(u_i = u_o)$.

An example of the repressibility of the temperature and velocity measurements for a DC argon plasma stream using an enthalpy probe, with no TBC coating, at different gas sampling flow rates is given in Fig. 18.61, after [Rahman et al. (1964)]. The results show that with the increase of the sampled gas flow rate, the scatter of the data is significantly reduced with the increase of the gas sampling flow rate up to values in the range of 30–40 mg/s. Beyond this point, there is not much improvement in measurement precision, with the

corresponding (u_o/u_i), starting to dip below unity as shown in Fig. 18.61b.

A comparison of enthalpy probe and line-shaped analysis of laser scattering measurements of thermal plasma temperature and velocities was reported by [Fincke et al. (1963)]. The measurements were made on an atmospheric pressure argon DC thermal plasma jet in an open discharge mode in ambient air. Entrained air concentration, temperature, and axial velocity profiles in the radial direction were measured at an axial position of 2 mm downstream of the exit level of the plasma torch nozzle for arc current of 900 A. The measured radial concentration profile of entrained air obtained by the enthalpy probe given in Fig. 18.62 show, at this location, negligible air entrainment by the argon plasma jet avoiding possible complications due to altered composition of sampled gas in subsequent enthalpy and velocity measurements.

A comparison of the radial temperature profiles of the plasma jet as measured using the enthalpy probe and laser scattering presented in Fig. 18.63a show relatively good agreement between both techniques with differences limited to the fringes of the profiles. It is to be noted that the laser scattering data were taken from radial positions of 0 to +4 mm only, while those of the enthalpy probe were taken from −10 to +10 mm. To form the comparisons, the laser scattering data were reflected about the centerline. Corresponding velocity data obtained using laser scattering and enthalpy probe for the same operating conditions, pure argon plasma and 900 A, are given in Fig. 18.63b.

Figure 18.64 shows a comparison of the measured center-line (peak) values of temperature and velocity versus arc current which gives a reasonably realistic comparison of performance of both techniques over a range of operating

conditions. In most cases, the estimated uncertainties (error bars) overlap. The discrepancy in the velocity may be due to the steep gradients present and differences in measurement volume size. Since the disagreement is not large (< 10%), it may also be due to variations of the operating characteristics of the plasma torch.

Further studies of the enthalpy probe diagnostics were reported by [Steffens and Duda (2000)] who studied a 9 MB plasma torch of Sulzer Metco working with an argon/hydrogen mixture and two different anode designs: 5 and 10 mm i.d. anode nozzles. The closest measurements were achieved 40 mm downstream of the nozzle exit to ensure a safe operation. At that distance, as shown in Fig. 18.65a, the maximum temperature was roughly 1500 °C lower with the 10 mm nozzle compared to that obtained with the 5 mm i.d. nozzle, but the temperature distribution was more uniform, thus improving the heating of the particles traveling in the jet periphery. These different jet characteristics were not observed, however, at a distance of 60 mm downstream of the nozzle exit as shown in Fig. 18.65b. The maximum temperature at this measurement location was approximately the same, but the gradient to the fringes of the 5 mm anode was larger compared to that of the 10 mm anode. Excellent reproducibility of enthalpy probe measurements was also reported by [Asmann et al. (2001)] in their study of a plasma jet produced by a triple torch plasma reactor. A more detailed analysis of differences between spectroscopic and enthalpy probe techniques can also be found in [Rajabian et al. (2004)].

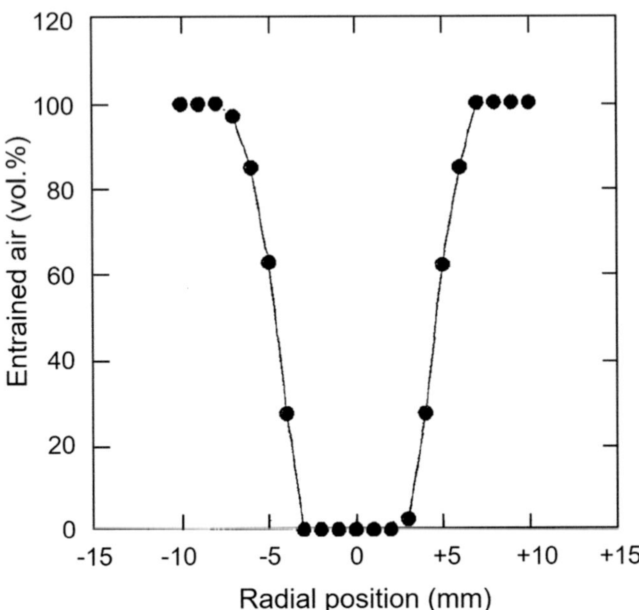

Fig. 18.62 Radial concentration profile of entrained air obtained from enthalpy probe measurements in argon DC plasma jet, argon at 900 A [Fincke et al. (1963)]

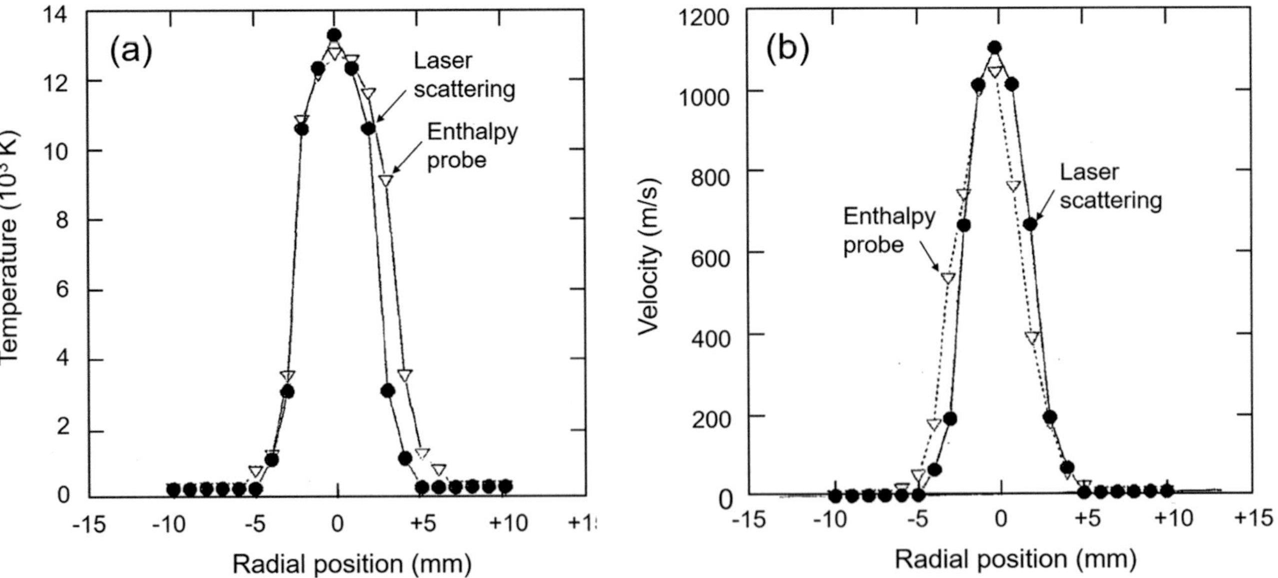

Fig. 18.63 Enthalpy probe and laser scattering measurements of radial profiles of (**a**) temperature and (**b**) velocity, in DC plasma jet, argon at 900 A. [Fincke et al. (1963)]

18.5.4 In-flight Particle Diagnostics

In-flight measurement of individual particle parameters is of primary importance for the understanding and development of thermal spray process. Particularly critical is the follow-up and control of particle trajectories in the spray medium and

measurement of the particle temperature and velocity, prior to their impact on the substrate which have been recognized among the key parameters that have major influence on splat formation and consequently on the microstructure and properties of the coating. Over the past four decades, intensive R&D efforts were devoted to the development of new and sophisticated in-flight particle diagnostic tools and their use for the study of plasma-particle momentum and heat exchange and transient in-flight particle heating and melting. Most of these techniques were, however, too complex and cumbersome for integration in industrial-scale installation used for thermal spray coating. Specific attention was therefore also given to the scaling-down of some of these instruments and packaging them in user-friendly, robust configuration that can be used in a spray booth either as an attachment to the spray torch or as a diagnostic tool in a fixed location monitoring the particle parameters during the spray process or at intermediate time for the calibration of the spray parameters used.

In this section, a review is given of the basic principles used in some of these instruments, in R&D and/or production-scale environment, with typical examples of measurement results. These are grouped into two broad categories dealing with either *individual particle diagnostics* which provide information about the full statistical distribution of the in-flight particle parameters or *ensemble particle diagnostics* which can be observed through mostly the observation of the full in-flight particle cloud through their own emission [Fauchais and Vardelle (2011)].

Measurement volume in this case of *individual particle diagnostics* can be below 1 mm^3, coupled with relatively high-speed detectors and electronics with bandwidths on the order of 0.1–1 MHz or greater allowing for the observation of a single particle [Fauchais et al. (1989, 1992), Coudert et al. (2002), Fincke (2001)]. A sufficient number (several

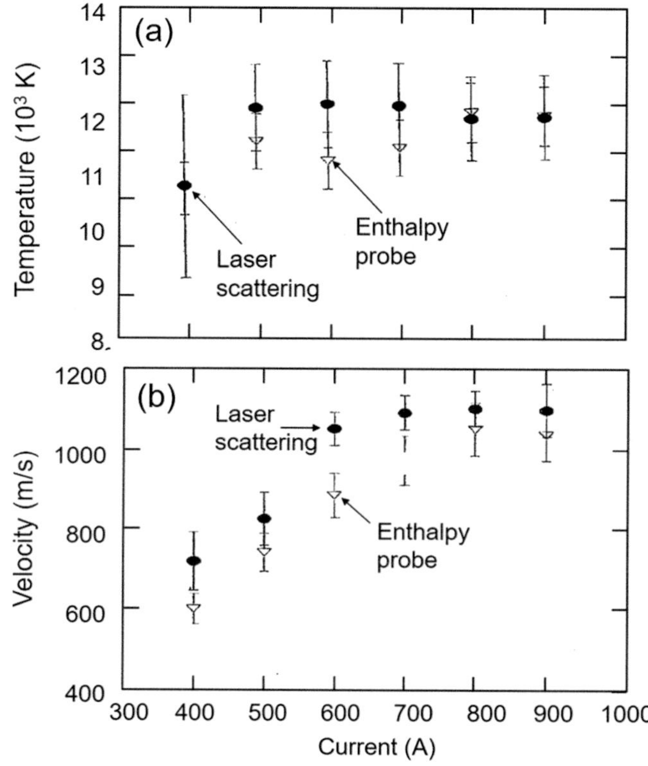

Fig. 18.64 Enthalpy probe and laser scattering measurements of centerline. (**a**) Temperature and (**b**) velocity profiles along the axis of DC plasma jet, argon at 900 A. [Fincke et al. (1963)]

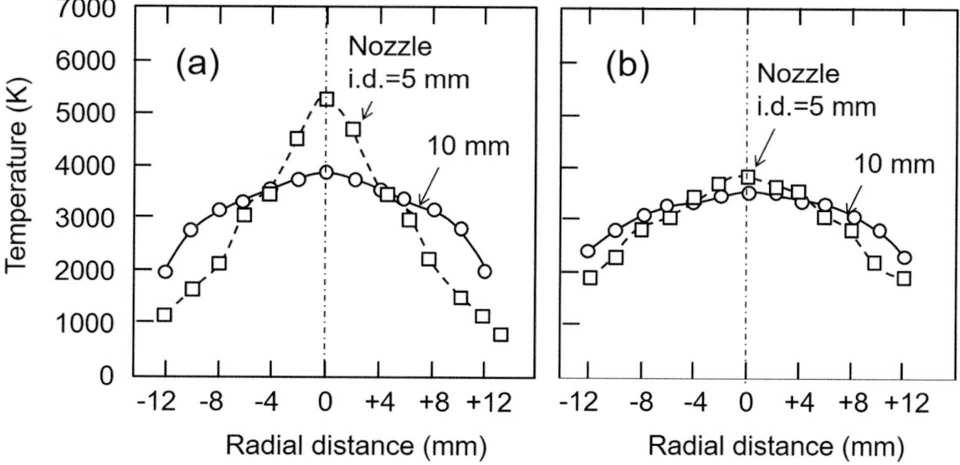

Fig. 18.65 Radial temperature distributions for Ar–H_2 DC plasma jet with different nozzle i.d., measured downstream of the nozzle exit at (**a**) 40 mm and (**b**) 60 mm. [Steffens and Duda (2000)]. Reprinted with kind permission from Springer Science Business Media, copyright © ASM International

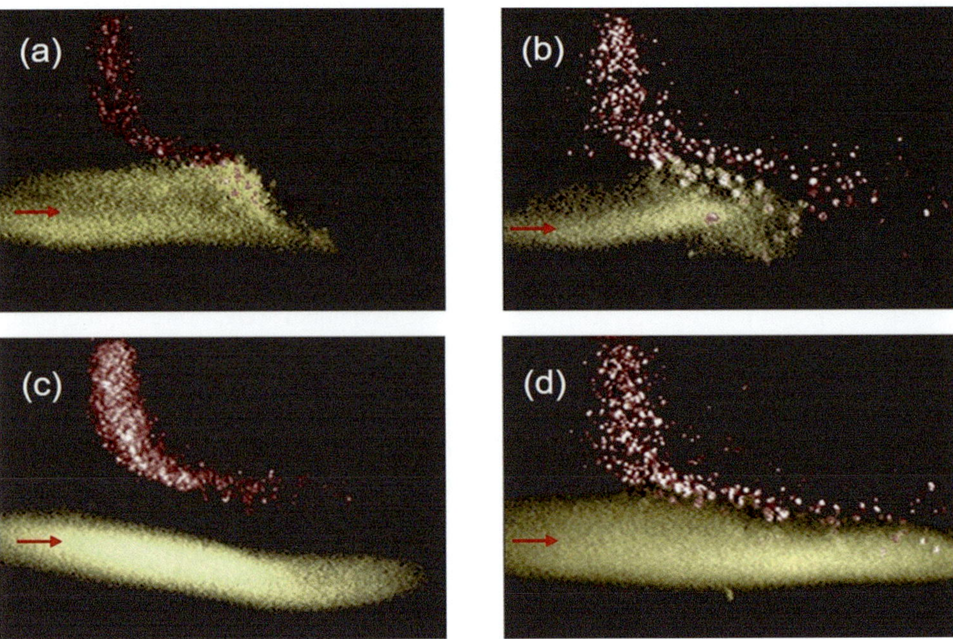

Fig. 18.66 Photographs taken using a Control Vision setup, with a 5 μs nitrogen laser flash, of DC plasma jet in the presence of molybdenum particle injection in the plasma plum. [Fauchais et al. (2004)]

thousands) of individual particles must be observed to achieve an adequate statistical representation of the means and standard deviations of temperatures, velocities, and diameters recorded. Such measurements usually require lightly loaded thermal spray processes and are often performed with particles mass flow rates below 0.5–1 kg/h.

Measurement of *ensemble particle parameters*, on the other hand, applies to a large measurement volume containing an important number of particles at a given time. They do not attempt to distinguish between individual particles. For example, the measurement volume consists of an approximately cylindrical chord, of few tens of mm^3, through the spray pattern. This chord is preferably oriented in the plane of the injector so that the measurement is insensitive to the movement of the spray pattern relative to the measurement volume [Fincke et al. (2001)]. However, such measurements are limited to providing mean values over the particle population in the measurement volume. Ensemble techniques work equally well for heavily loaded processes such as HVOF spraying, and measurement times are a few seconds.

18.5.4.1 Individual Particle Diagnostics

High-end diagnostic tools have been mostly focused on the in-flight measurement of individual particle parameters, coupled with statistical analysis of the data. This includes either monitoring of individual particles illuminated by an intense laser beam/flash, independent of their temperature, or through their own emission which would limit the collected data to particles at sufficiently high temperature to be visible

even in the presence of relatively strong background emission by the spraying medium.

The simplest of such instruments is the *Laser Vision* which is essentially a variable frequency laser strobe, coupled with a fast charge coupled device (CCD) camera which can record the relative position of the particles-in-flight as demonstrated in Fig. 18.66 with 5 μs nitrogen laser flash duration for DC plasma jet in the presence of molybdenum particle injection during a plasma spray operation. Through the rapid acquisition of the photos with a preset frequency, the individual velocity of the particles can be deduced by image analysis, measuring the distance traveled by each particle in the short time between successive laser flashes. Statistical analysis of the data provide means of obtaining full information of the particle velocity distribution, mean value, and standard deviation. The measurement is limited, however, to the velocity component in the plane of observation and by the depth of field of the camera optics.

Corresponding photographs of the entrained droplet flux distribution in a wire arc spraying are shown in Fig. 18.67. These benefit from the fact that the spray atomizing gas medium is essentially at room temperature with no background emission allowing for the clear visibility of the atomized molten metal droplets transported by the gas which shows in this case some periodic variation of droplet flux with time [Hussary and Heberlein (2001)].

Corresponding photograph taken by CCD camera focused to image entrained hot droplet in-flight in the plume of a plasma jet few centimeters to the substrate is given in Fig. 18.68 after Oseir®. The particles in this case are

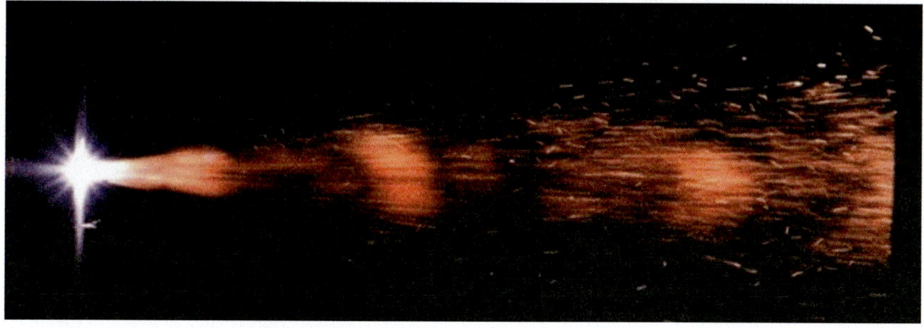

Fig. 18.67   Photograph of a wire arc spray jet showing the periodic variation of droplet flux. [Hussary and Heberlein (2001)]

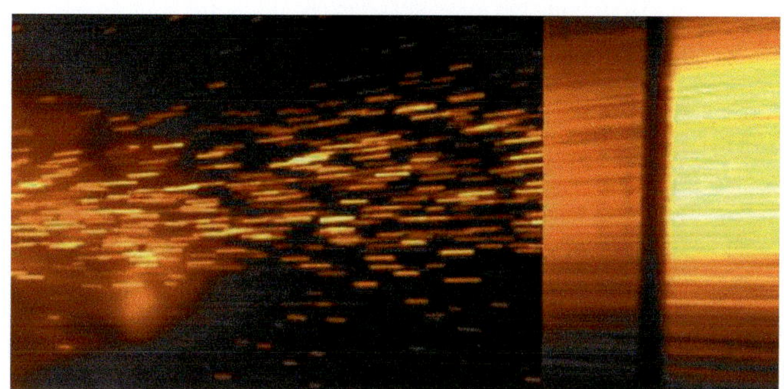

Fig. 18.68   CCD image of particles in-flight in the plume of DC plasma jet. The length of the trace for each particle is function of the particle velocity and exposure time as controlled by the shutter speed. [Courtesy of Oseir®]

observed through their own emission since the radiation of the plum of the plasma jet in this region is low enough not to overcome radiation from hot particles [Fauchais and Vardelle (2011)]. It can be seen that, according to the shutter time, lines represent particles, and their lengths can be linked to their velocities.

Laser Doppler anemometry (LDA) was one of the key diagnostic techniques widely used in combustion studies that was introduced by [Lesinski and Boulos (1988a, b)] for the in-flight measurement of particle velocities in DC and RF induction plasma spraying, respectively. As illustrated in Fig. 18.6a the technique is based on the creation of an interference fringe pattern at the point of intersection of two laser beams of wavelength λ_b and half intersection angle of α_b. The inter-fringe spacing δ_b is given by:

$$\delta_b = \frac{\lambda_b}{2 \sin \alpha_b} \tag{18.5}$$

As a particle of diameter d_p crosses the inter-fringe pattern, it is illuminated by the laser beam emitting a light pulse of a duration depending on its velocity and size, modulated by a frequency $f_o = 1/\tau_f$ function of its velocity, v_p, and inter-fringe spacing, δ_b. The instantaneous particle velocity can be calculated as $v_p = f_o \times \delta_b$ (Fig. 18.69).

[Lesinski and Boulos (1988a)] reported LDA measurements for a DC plasma jet in an open discharge mode in ambient air with a nozzle i.d. =7.1 mm, total gas flow rate of 23.6 slm (Ar/N$_2$, 20 vol. % N$_2$), arc current of 400 A, and DC power of 15.2 kW. Initial measurements were carried out of the plasma flow by seeding the plasma gas using ultrafine zirconia powder, as a tracer, with a particle diameter of 1 μm. Alumina powder with $(\overline{d}_p = 53$ μm, $\sigma_{dp} = 1.13)$ was next injected into the plasma jet using a radial injector with i.d. =2.8 mm and a carrier gas flow rate of 2.7 slm (Ar). The LDA system was set using a He–Ne laser (15 mW), in the forward-scatter mode with an effective measuring volume resolution of 0.3 mm diam × 3.0 mm long. Typical results given in Fig. 18.70a show the radial profiles of the mean plasma velocity and the "intensity of turbulence, T_u %" defined as the ration of the standard deviation of the axial velocity component divided by the mean local velocity, i.e.:

$$T_u = \frac{v'}{\overline{v}} \times 100 \tag{18.8}$$

$$\text{where } v' = \sqrt{\frac{1}{(n-1)} \sum_{i=1}^{i=n} (v_i - \overline{v})^2} \tag{18.9}$$

$$\overline{v} = \frac{1}{n} \sum_{i=1}^{n} v_i \tag{18.10}$$

Corresponding values for the radial velocity profiles of the alumina particles in the plasma jet at different downstream levels from the injection point of the particles are given in Fig.18.70b.

LDA measurements by [Lesinski and Boulos (1988b)] were also carried out for atmospheric pressure RF induction plasma discharge with a plasma confinement tube i.d. of 50 mm operating with a sheath gas flow rate of 63 slm (Ar), central gas flow rate of 11.7 slm (Ar), oscillator frequency of 3 MHz, and plasma power varying between 4.6 and 10.3 kW.

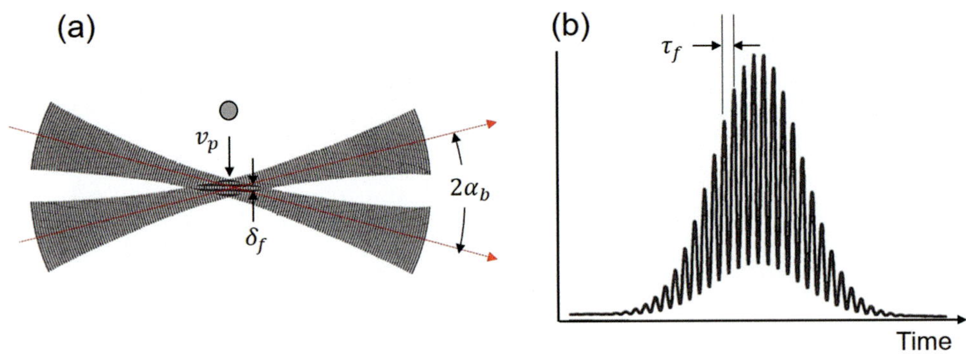

Fig. 18.69 (**a**) Schematic of laser beam intersection and generated interference fringes, (**b**) modulated pulse generated resulting from the passage of s single particle in the measurement volume. [Lesinski and Boulos (1988a, b)]

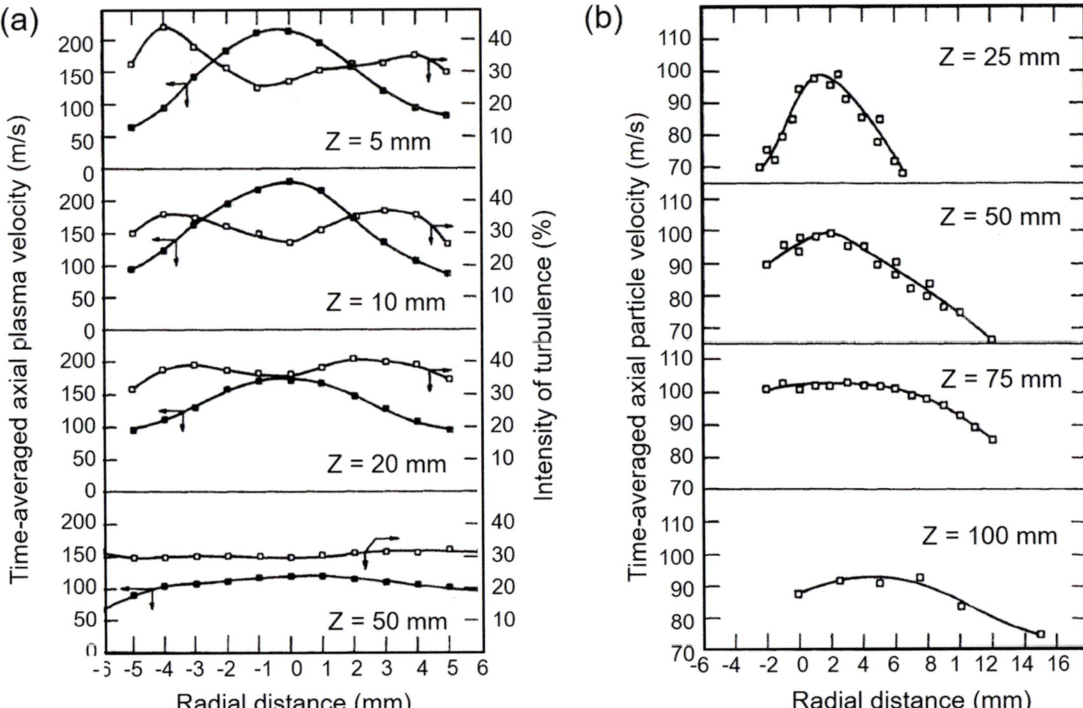

Fig. 18.70 Plasma and particle diagnostic in an open discharge Ar/N$_2$ (20 Vol% N$_2$) DC plasma jet, nozzle i.d. 7.1 mm, current 400 A, power 15.2 kW. (**a**) Radial profiles of the mean plasma velocity and intensity of turbulence (**b**) mean velocity of 53 µm diam alumina particles radially injected into the plasma jets. [Lesinski and Boulos (1988a)]

Alumina powder with a particle diameter of ($\bar{d}_p = 33$ µm, $\sigma_{dp} = 13$ µm) was axially injected into the center of the discharge with carrier gas flow rate of 4.8 and 7.4 slm (Ar). Typical results are given in Fig. 18.71a for the radial profile of the axial plasma velocity over the central region of the discharge as measured using ultrafine zirconia powder seeding (diam. = 1 µm). Corresponding axial velocity of the alumina particles at different axial positions in the discharge tube is given in Fig. 18.71b. In both figures, profiles identified as:

a-1 & b-1 powder gas flow rate = 4.8 slm (Ar) & Power = 4.6 kW,

a-2 & b-2 powder gas flow rate = 7.4 slm (Ar) & Power = 4.6 kW,

a-3 & b-3 powder gas flow rate = 7.4 slm (Ar) & Power = 10.3 kW.

It should be pointed out that using LDA implies measurement of all the particles crossing the measuring volume independent of their temperature.

Simultaneous measurements of particle velocity, temperature, and number flux density in DC plasma jet were reported by [Vardelle et al. (1988)] using the experimental setup shown in Fig. 18.72.

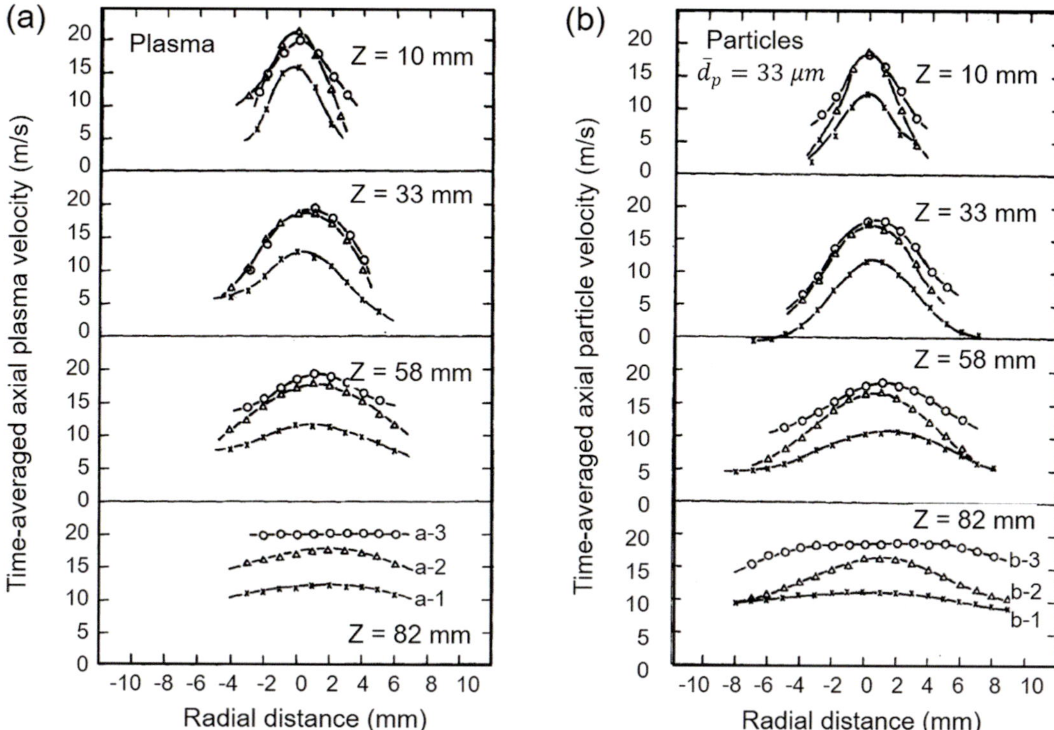

Fig. 18.71 Plasma and particle diagnostic in an atmospheric pressure RF induction plasma discharge in a 50 mm i.d. plasma confinement tube, sheath gas 63 slm (Ar), central gas 11.7 slm (Ar), oscillator frequency 3 MHz, and plasma power varying between 4.6 and 10.3 kW. (**a**) Radial profiles of mean plasma velocity, (**b**) mean velocity of 33 μm diam alumina particles axially injected into the discharge. [Lesinski and Boulos (1988a)]

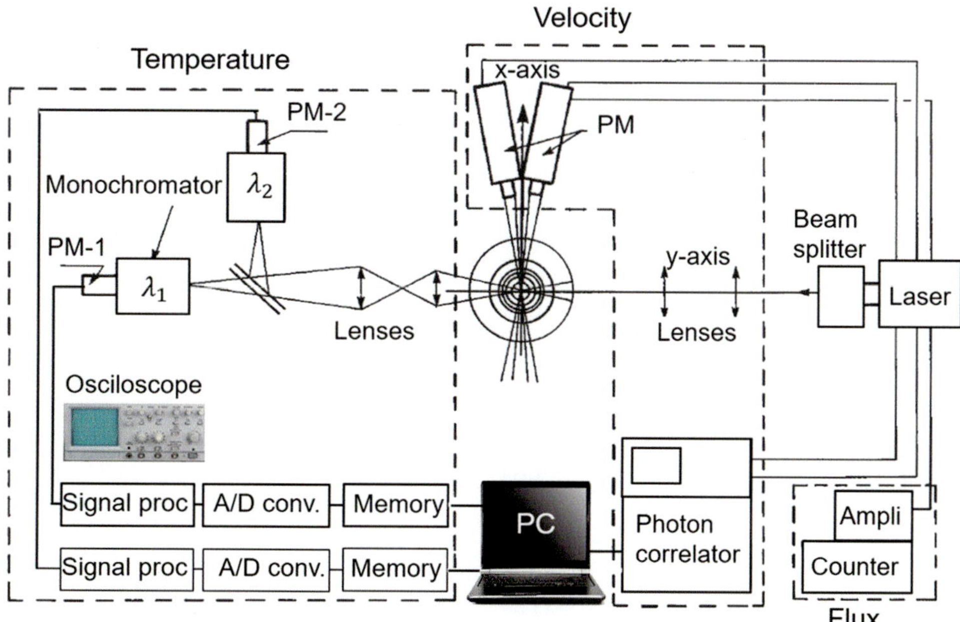

Fig. 18.72 Experimental setup used for measurement of particle number flux, velocity, and surface temperature distributions. [Vardelle et al. (1988)]

Particle velocity measurement was carried out using a 1.2 W, argon ion laser ($\lambda = 514.5$ nm) crossing the plasm jet at the required measurement level. Observation of the scattered light from the particles was carried out off-axis by two photomultipliers directed, respectively, at angles of $81°$ and $99°$ to the optical axis of the laser beam. The response time of each of the photomultipliers was 3 ns. Based on geometrical considerations, the measuring volume was estimated to be in the form of a parallelepiped of dimensions $0.1 \times 0.4 \times 0.5$ mm in the x, y, and z directions, respectively. In order to block the background plasma radiation from reaching the photomultipliers, monochromators with a band-pass width of 2A, centered on the laser wavelength, were placed in front of each of the photomultipliers. These helped to improve considerably the performance of the laser anemometer and allowed the observation of the scattered light from alumina particles as small as 2 μm diameter in the presence of the intense background plasma radiation. The output signal of the two photomultipliers are cross correlated by a photon correlator (Malvern). The uncertainty associated with the reported velocity measurements is estimated to be of the order of 5–7%.

Particle *flux number density* distribution. Measurement was determined by counting the total pulse rate obtained from one of the photomultipliers used in the laser anemometer setup. The measuring volume was therefore identical to that given earlier for the anemometer. The results, however, represent the number of particles per unit time crossing a surface of dimensions 0.1×0.4 mm, in the x–y plane, perpendicular to the axis of the torch.

Particle surface temperature was determined by two-color pyrometry. Details of the experimental technique are given by [Mishin (1985) and Mishin et al. (1987)]. Measurements were made of the particle emission at two wavelengths, 680.0 and 837.0 nm, using a pair of photomultipliers (Hamamatsu R928) of similar transient response characteristics. A monochromator placed in front of each of the photomultipliers was used to define the wavelength of the observed particle emission and to block the background plasma radiation. In order to overcome the problems of coincidence requiring that the two photomultipliers be independently focused on the same particle in space, a beam splitter was used to project simultaneously the image of the particles on the entrance slit of each of the two monochromators. Based on geometrical considerations, the measurement volume was estimated to be in the form of a parallelepiped of dimensions 15×20 mm in the X and Y directions, respectively, and 0.12 mm in the z direction. The particularly long dimension in the y direction is due to the fact that it represented the line of sight of the optical setup used, which had a rather long depth of field. Reduction of the depth of field below this value would have required a larger objective lens that the 20 mm diameter lens used in this study.

The passage of a particle in the measuring volume gives rise to two simultaneous electric pulses generated by each of the two photomultipliers. These are filtered and amplified using an Ortec 113 preamplifier and an Ortec 450 linear amplifier. An Ortec 464 analyzer is used to generate a signal that is proportional to the ratio of the amplitudes of the two pulses. Saturated signals, which generally reflect the presence of more than one particle in the measurement volume, are eliminated using a discriminator prior to statistical analysis of the signal using a multichannel analyzer (Northrop 5300). By proper calibration of the apparatus using a standard tungsten ribbon lamp, the statistical information on the ratio of the particle emission at the two wavelengths can be converted into statistical information about the particle surface temperature distribution. While the measurements were reproducible within 4%, the absolute precision of the temperatures obtained is estimated to be of the order of 300 K at a particle surface temperature of 3000 K. This is mainly due to a number of limiting factors, including uncertainties about the emissivity of the surface of the particles.

The plasma torch used was of standard design with a thoriated tungsten cathode and an 8.0 mm i.d., water-cooled, copper anode. The particles were injected inside or outside the anode nozzle at a distance of 1.6 mm upstream or downstream of the point of discharge of the plasma jet. Measurements were carried out using Ar and Ar/H_2 plasmas. The powder was injected into the flow using a 2 mm i.d. radial powder feed port placed at a radial distance of 8 mm from the torch axis (i.e., 4 mm off the edge of the nozzle). The orientation of the injection was in the positive x-direction. Two types of alumina powders were used. These had a mean particle diameter and the standard deviation, respectively, of 18.0 ± 3 μm and 60 ± 15 μm. The powder feed rate was approximately 4 g/min. Argon was used as a powder carrier gas with a flow rate that was varied between 4.5 and 8.5 slm.

Typical results are given in Fig. 18.73 in terms of the radial distribution of the mean particle axial velocity, mean particle surface temperature, and particle number flux distributions at an axial distance of 75 mm downstream of the torch nozzle exit plane. The plasma gas flow rate was [75 slm (Ar) + 15 slm (H_2)], arc current 400 A, voltage, 73 V, power 29.3 kW, torch efficiency of 60%, and mean specific enthalpy of the plasma at the torch exit of 8.5 MJ/kg. Mean particle size of the powder used was 18.0 ± 3 μm. It may be noted from Fig 18.73a, b that the highest particle velocities and temperatures are obtained at relatively low carrier gas flow rate of 5.5 slm, with the corresponding particle flux density, given in Fig. 18.73c, showing that the particles in this case have an axial trajectory, with the maximum particle flux on the centerline of the jet. It is interesting to note, however, that while increasing the carrier gas flow rate does not seem to affect the maximum velocity and surface

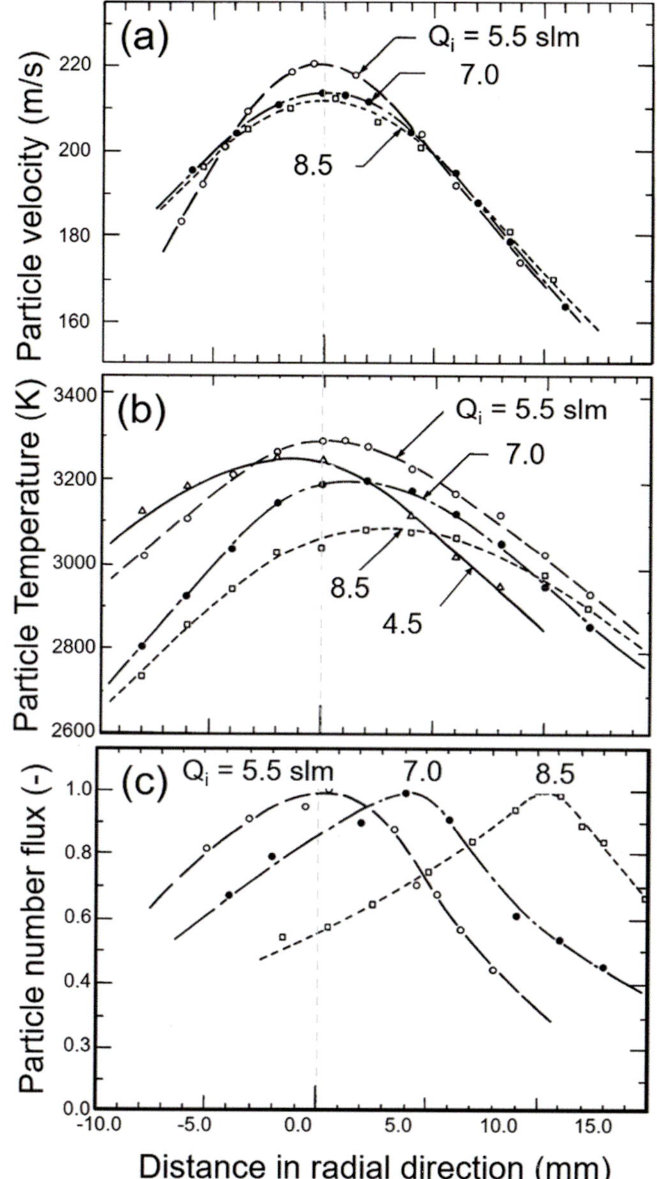

Fig. 18.73 Mean particle parameters at 75 mm downstream the exit plane of the DC torch nozzle operating using 75 slm (Ar) + 15 slm (H$_2$) as plasma gas, arc current 400 A, voltage 73 V, power 29.3 kW, and alumina powder, 18.0 ± 3 μm, injected at 4 g/min using different carrier gas flow rates, Q$_i$. (a) Mean particle velocity, (b) mean particle surface temperature, (c) particle number flux distribution. [Vardelle et al. (1988)]

temperature values attained by the particles along the centerline of the jet, they represent in this case a considerably smaller fraction of the total powder injected into the flow since the peak of the particle flux density distribution, as noted in Fig. 18.73c, has shifted substantially away from the centerline of the jet.

Further development in this area was reported by [Sakuta and Boulos (1988a, b)] who combined the time-of-flight (ToF) approach with two-wavelength pyrometry for the

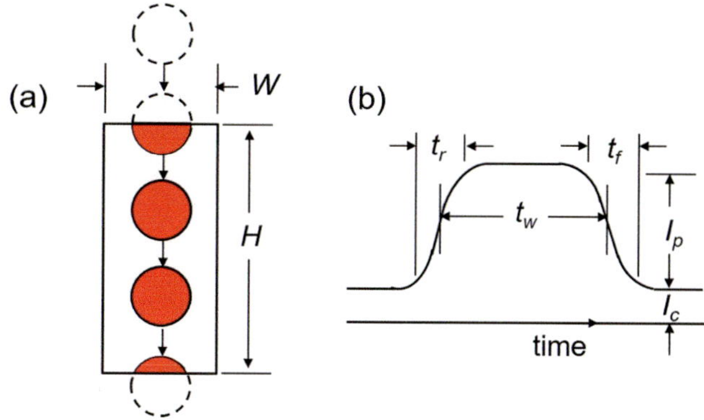

Fig. 18.74 Schematic representation of the basic concept used for ToF measurement of in-flight particle parameters. [Sakuta and Boulos (1988a, b)]

simultaneous measurement of velocity, temperature, and diameter of particles. The approach used, as illustrated in Fig. 18.74a, is based on the observation of an individual particle through its own emission as it crosses a well-defined observation window of high H and width W. Assuming a negligible background emission from the plasma plume, and that the particle number density is sufficiently low not to have more than one particle at a time in the observation window, the total emission from the H × W surface will increase by a value corresponding to the emission from the surface of that particle. Measuring the total emission from the observation window using an appropriate photomultiplier or photodiode array, an electric pulse is generated of trapezoidal form, as shown in Fig. 18.74b, with the base line I_c corresponding to the background emission in the absence of the particle and the top level ($I_c + I_{\lambda i}$) corresponding to the situation where the particle is full in the observation window during the time laps t_w. The rise time of the signal t_r, and the fall time t_f correspond, respectively, to the time taken by the particle to enter and to exit the window.

Based on these data, the in-flight particle parameters can be calculated as follows:

$$\text{Particle velocity}, v_p; v_p = \frac{H}{R \times t_w} \tag{18.11}$$

$$\text{Particle diameter}, d_p; d_p = v_p \times t_r \text{ or } d_p = v_p \times t_f \tag{18.12}$$

Particle temperature, T_p; T_p

$$= \frac{C\,(\lambda_1 - \lambda_2)/\lambda_1 \lambda_2}{[\,ln\,(I_{\lambda 1}/I_{\lambda 2}) + 5 \times ln\,(\lambda_1/\lambda_2)]} \tag{18.13}$$

where:

H Height of observation window
R Image magnification ratio

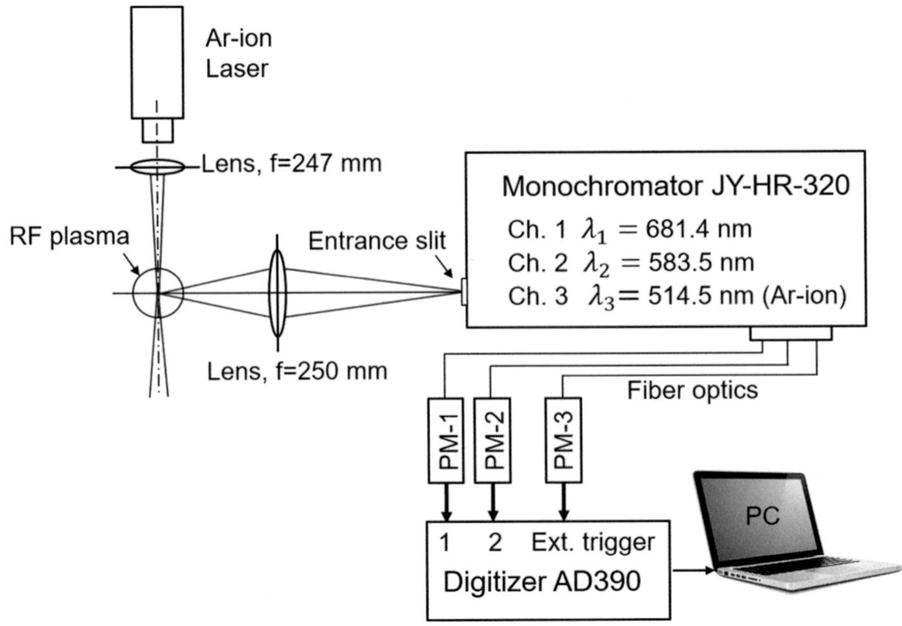

Fig. 18.75 Experimental system setup for ToF particle parameter diagnostics. . [Coulombe et al. (1995)]

t_w Mean time of flight across the observation window, Fig. 18.74

t_r Mean rise time (entry time) of the particle in the observation window, Fig. 18.74

t_f Mean fall time (exit time) of the particle from the observation window, Fig. 18.74

C Calibration constant

λ_i Wavelength of the i observation, with $i = 1$ or 2

$I_{\lambda i}$ Mean pulse height above background level at wavelength λ_i

A schematic diagram of the experimental setup used for the particle diagnostics is presented in Fig. 18.75 after [Coulombe et al. (1995)]. A measurement volume is defined within the plasma stream by the intersection of a focused laser beam with the optical axis of a monochromator. In this way, an image of the particle crossing the measurement volume is focused on the entrance slit of the monochromator. The particle image is enlarged in this configuration by the image magnification ratio, R = 1.93. The entrance slit of the monochromator is set to height H = 2 mm and width W = 1 mm. An argon ion laser (2 W) is used to define the location of the measurement point along the line of observation. At the measurement point, the ($1/e^2$) diameter of the laser beam was 0.14 mm, giving rise to a 0.075 mm^3 measurement volume.

Three photomultipliers (Hamamatsu R1104) focused through optical fibers on the output plane of the monochromator allowed for the simultaneous capture of the light emitted by the measurement volume at three different wavelengths (spectral windows of 2 nm width). The first two wavelengths, λ_1=681.4 nm (PM-1) and λ_2= 583.5 nm (PM-2), are used for the detection of the pulse waveforms associated with the monochromatic emission of the particles. Both wavelengths have been selected by taking into account the presence of atomic lines within the emission spectrum of the plasma, the spectral response of the photomultipliers, and the sensitivity of the surface temperature to the ratio of the detector outputs ($V_{\lambda 1}/V_{\lambda 2}$) [Sakuta and Boulos (1988a, b)]. The third wavelength corresponding to the argon ion laser, λ_2= 514.5 nm (PM-3), is used to detect the scattered laser light and serves as a triggering signal for the pulse acquisition with a 10 bit A/D delay digitizer with 2048 memory addresses. The maximum frequency of the digitizer is 30 MHz for a minimum sampling time of 33 ns per address. The advantage of using the laser wavelength for the triggering of the data acquisition is that only the particles passing through the measurement volume are detected by the optical setup.

Measurements reported by [Coulombe et al. (1995)] were for alumina particles with a mean particle diameter, $\bar{d}_p =$ 49 μm, and standard deviation, $\sigma_{dp} = 4$ μm, injected at a feed rate of 1.6 g/min, in Ar/H$_2$, RF inductively coupled plasma discharge with a powder carrier gas flow rate of 3 slm. Sheath gas was composed of 71.5 slm (Ar) + 5.6 slm (H$_2$), while central gas was 43 slm (Ar). Chamber pressure was 33 kPa (250 Torr), oscillator frequency of 3 MHz, and plate power of 20 kW. The powder injection probe i.d. = 2. 65 mm, and its o.d. = 6.25 mm, with the tip of the injection probe at the center of the coil region. Measurements were carried out in the plasma plum, on the axis of the torch at a distance of 188 mm downstream of the tip of the powder injection probe. Two types of in-flight particle diagnostics were used, time of flight (ToF) for the simultaneous measurement of the velocity and surface temperature of the particles and laser Doppler anemometry (LDA) for the velocity of the particles in the same location. An example of the pulse waveforms captured due to particle emission at λ_1=681.4 nm (PM-l) and λ_2= 583.5 nm (PM-2) is given in Fig. 18.76. These show the typical trapezoidal shape with a rather consistent time of flight in the observation window of 35.4 µs and 35.9 µs at the two wavelengths.

ToF and LDA measurements were each carried out for 1000 particle acquisitions of which in the case of ToF measurements, 821 were accepted, with the balance of 179 rejected either because the emission signal was too weak (too low particle temperature) or significant fluctuations of the intensity of the background radiation from the plasma

stream. Typical results are given in Fig. 18.77. The statistical distribution of the particle surface temperature, Fig. 18.77a, was essentially Gaussian with a mean particle temperature $\overline{T}_p$ = 3017 K and a standard deviation σ_{Tp} = 464 K. Corresponding particle velocity measurement obtained using ToF approach given in Fig. 18.77b (dark bar graph) also shows a Gaussian-type distribution with a mean particle velocity of $\overline{v}_p$ = 23.2 m/s and a standard deviation of σ_{vp} = 5.4 m/s. Superposed on this graph is the corresponding particle velocity measurement obtained using LDA (red bar graph) which shows good agreement with the ToF data, with a mean particle velocity of $\overline{v}_p$ = 22.5 m/s and a standard deviation of 6.5 m/s.

It is important to point out that one of the principal difference between the ToF and LDA approach is that ToF is limited to the measurement of the "hot" particles that are visible through their own emission, typically >1700 K, depending on the operating conditions. [Sakuta and Boulos (1988a, b) and Coulombe et al. (1995)] presented an analysis of particle visibility criteria, identifying critical particle temperatures below which the particle would not be sufficiently visible for a reliable temperature and velocity measurement. LDA on the other hand offers means of measurement of the total number of particles crossing the measurement volume including cold and hot one. The combination of both techniques in single setup, while making the installation rather cumbersome for other than dedicated R&D studies, offers means of obtaining significant fundamental information on the process performance as reported by [Coulombe et al. (1995)]. The experimental setup used in this case is illustrated in Fig. 18.78 comprising both ToF and LDA systems of standard design as described earlier.

Operating conditions were essentially identical to those used in the earlier study [Coulombe et al. (1995)]. Measurements were carried out for alumina powder with a

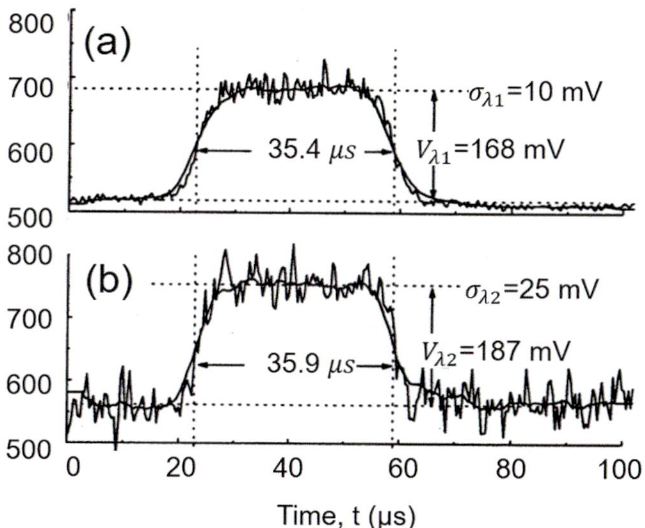

Fig. 18.76 Typical pulse waveforms captured due to particle emission at λ_1=681.4 nm (PM-l) and λ_2= 583.5 nm (PM-2). [Coulombe et al. (1995)]

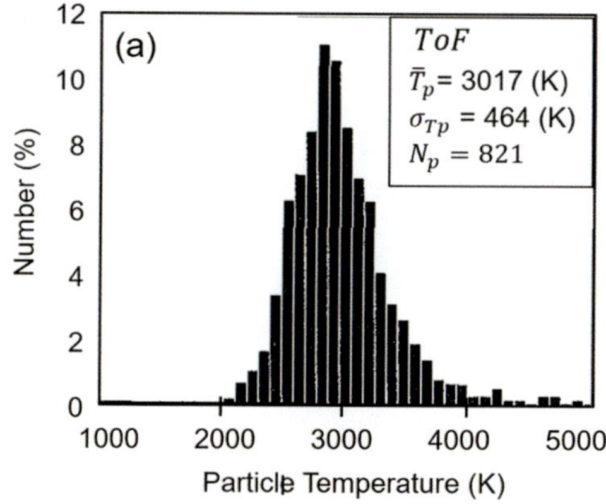

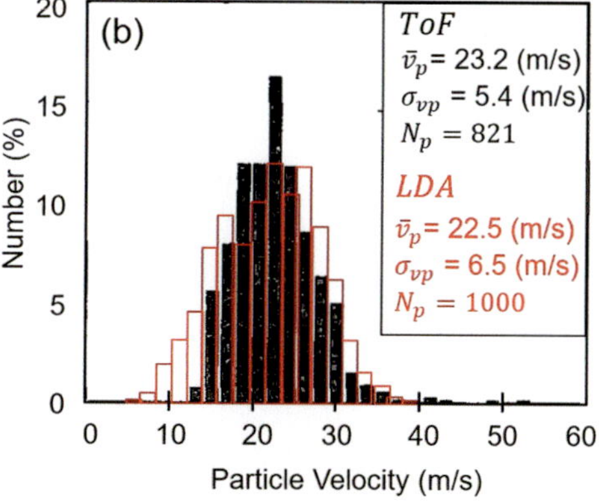

Fig. 18.77 Alumina particle surface temperature and particle velocity distributions in the plum of an Ar/H$_2$ RF induction plasma flow at a chamber pressure of 33 kPa (250 Torr) and a plate power of 20 kW, $\overline{d}_p$=49 µm, σ_{dp}=4 µm. (**a**) ToF particle surface temperature, (**b**) ToF and LDA particle velocity. [Coulombe et al. (1995)]

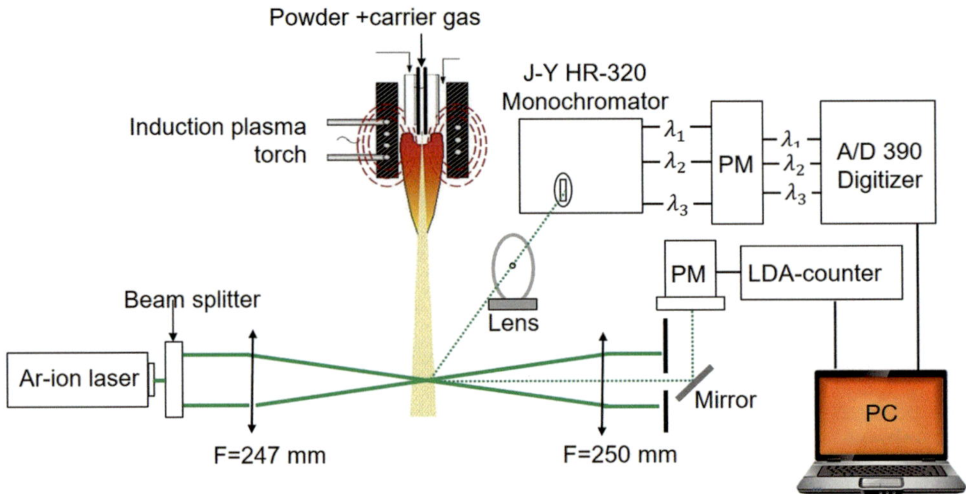

Fig. 18.78 Schematic of the experimental setup comprising ToF and LDA diagnostic techniques used for in-flight alumina particle diagnostics in the plume of an RF induction plasma flow. [Coulombe et al. (1995]

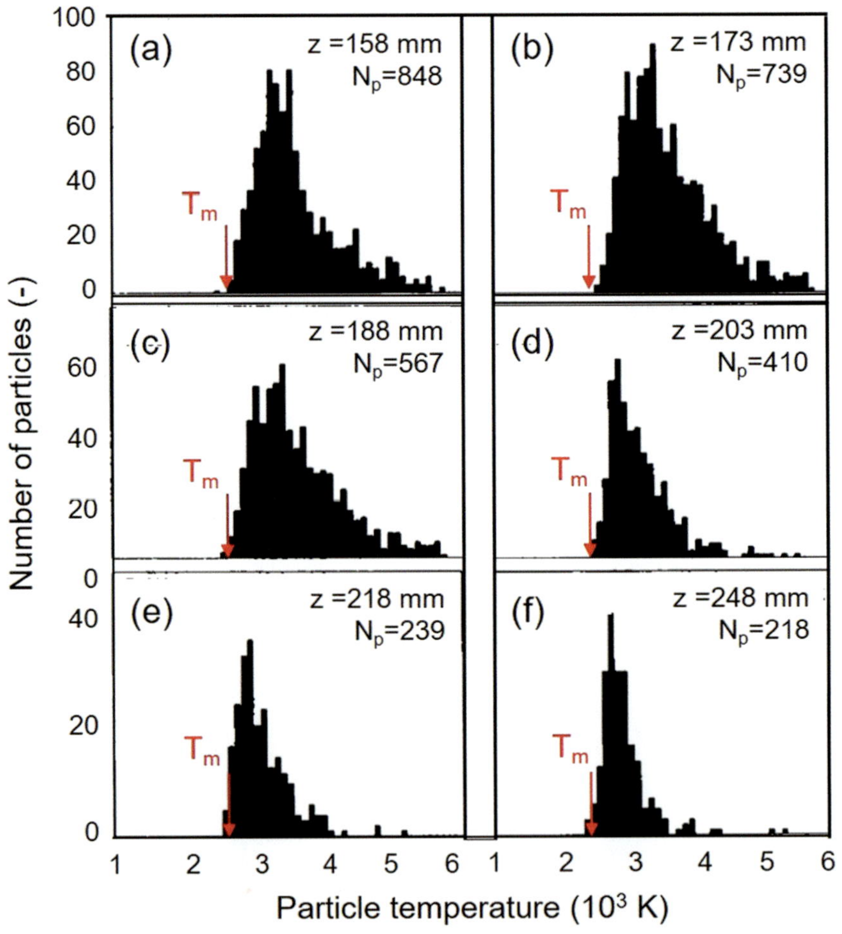

Fig. 18.79 Surface temperature distribution of alumina particles, dp $= 45$–53 µm, axially injected into an Ar/H$_2$ RF induction plasma at a chamber pressure of 33 kPa (250 Torr), measurement made using ToF diagnostics. [Coulombe et al. (1995)]

particle size in the range of 45–53 µm axially injected into the center of the coil region of an RF induction plasma torch at a feed rate of 1.6 g/min. Typical results obtained at a chamber pressure of 33 kPa (250 Torr) are given, respectively, in Figs. 18.79 and 18.80 for the particle surface temperature and particle velocity at different axial positions,

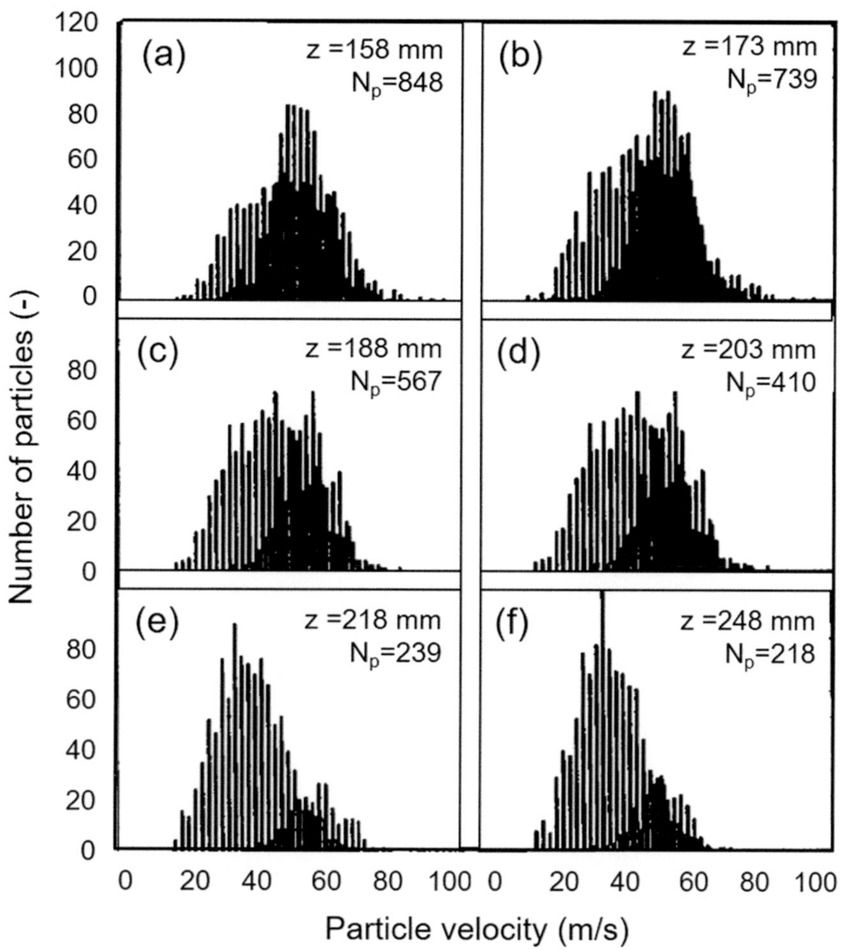

Fig. 18.80 Particle velocity distribution of alumina particles, dp = 45–53 μm, axially injected into an Ar/H$_2$ RF induction plasma at a chamber pressure of 33 kPa (250 Torr), measurement made using ToF and LDA diagnostics. [Coulombe et al. (1995)]

Table 18.4 Compilation of in-flight particle data obtained for 45–53 μm alumina particles axially injected into an Ar/H$_2$ RF induction plasma at a chamber pressure of 33 kPa (250 Torr) and power of 20 kW [Coulombe et al. (1995)]

Ref	z (mm)	TOF					LDA	
		N_p	$\overline{T}_p(K)$	$\sigma_{Tp}(K)$	$\bar{v}_p(m/s)$	$\sigma_{vp}(m/s)$	$\bar{v}_p(m/s)$	$\sigma_{vp}(m/s)$
(a)	158	848	3664	662	53.0	8.7	48.0	13.8
(b)	173	739	3675	675	54.3	8.0	46.7	13.7
(c)	188	567	3534	652	56.0	7.6	42.7	12.0
(d)	203	410	3277	494	57.1	7.8	42.8	12.5
(e)	218	239	3123	417	58.3	8.9	37.2	10.3
(f)	248	218	3024	393	52.9	7.4	36.7	9.6

158–233 mm, downstream of the point of injection. Identified on Fig. 18.79 is the melting temperature of alumina (T_m = 2380 K) which indicates that most of the particles monitored were essentially in a molten state. The observed rapid drop in the number of particles accepted out of the 1000 monitored is an indication that as the particles move further downstream of the plasma source, they freeze and cool down below the lower limit of detectability or that the particle flux is contaminated by cold particles entrained from the ambient atmosphere in the spray chamber. The phenomena are clearly evident from the particle velocity data given in Fig. 18.80, in which measurements are reported by the ToF and LDA techniques, with the ToF technique representing measurement of the high temperature fraction of the particle entrained by the flow and LDA representing the velocity distribution of all of the particles independent of their temperatures. A compilation of the mean temperature and velocity values obtained by these two techniques for this heterogeneous mix of "hot" and "cold" particles is given in Table 18.4 for

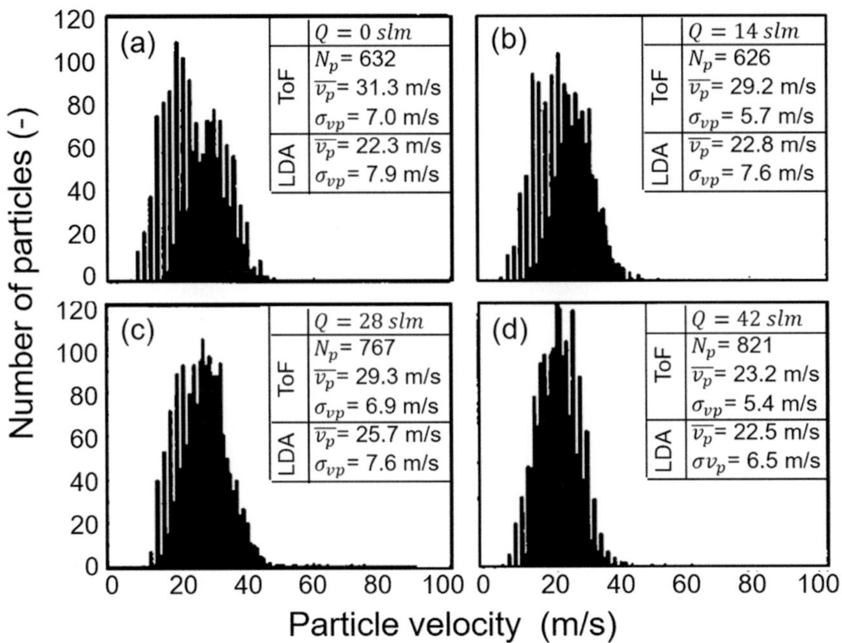

Fig. **18.81** Effect of sheath gas at the exit level of the RF induction plasma torch on "cold" particle entrainment into the main spray jet as determined using ToF and LDA diagnostic techniques, z = 188 mm, p = 33 kPa (250 Torr), and sheath gas flow rate, Q4 = 0–42 slm (Ar). [Coulombe et al. (1995)]

different axial positions downstream of the point of injection of the powder.

In an attempt to sort out the origin of the "cold" particle fraction in the overall particle flux, a dedicated experiment was carried out at a fixed axial distance of 188 mm which is below the exit level of the torch nozzle which was equipped in this case by an annular ring allowing for the injection of a shroud gas around the plasma jet at the torch exit. The results given in Fig. 18.81 represent the alumina particle velocity distribution on the axis of the flow as measured using the ToF and LDA techniques as from the point of injection which corresponds at the same axial location, in the presence of increasing level of shroud gas flow rate, Q. The results clearly showed the progressive overlapping of the two particle velocity distributions signaling the effective shielding of the flow by the shroud gas, avoiding the entrainment of "cold" particles recirculating in the atmosphere of the spray chamber. The effect is also demonstrated by the steady increase of the number of "hot" particles in the plasma jet with the complete overlapping of the particle velocity distributions measured by ToF and LDA techniques.

The principle of the near infrared sensor (NIR Sensor, GTV Verschleißschutz GmbH, Luckenbach, Germany) [Schwenk et al. (2010)] is based on the illumination of single particles in the focus distance by a 10 MHz pulsed infrared laser (1000 nm), the reflected pulsed laser radiation indicating the passage of single particles through the measurement focus spot, which is small enough to detect one particle (local

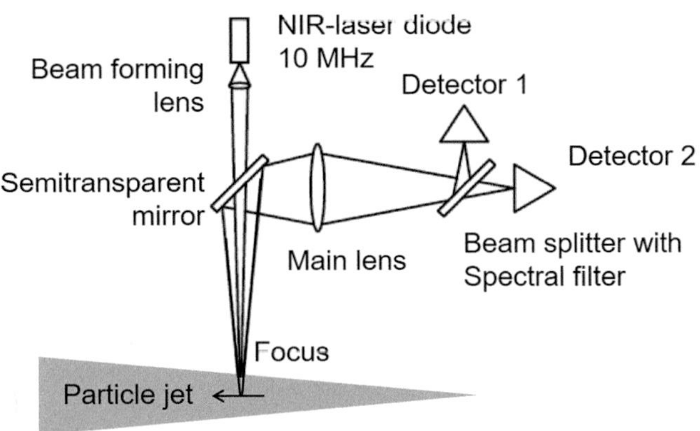

Fig. **18.82** Optical design scheme of the NIR sensor. [Schwenk et al. (2010)] (Courtesy of GTV-MBH D-57629 Luckenbach)

measurement). The in-flight time is inversely proportional to the particle velocity. Thus, the velocity of cold particles can be detected. Furthermore [Mauer et al. (2011)], two fast detectors with different spectral ranges between 900 and 1500 nm as well as between 1500 and 2600 nm, capture, during the time of passage, the emitted thermal radiation of the single particles. The ratio of the two emission signals is used for temperature determination according to the two-color pyrometer principle. Figure 18.82 presents the scheme of the setup. The detection limits are between 700 and 3000 °C for temperature, 30–1200 m/s for velocity, and particle diameters over 5 μm. As up to 300,000 single particles per second are

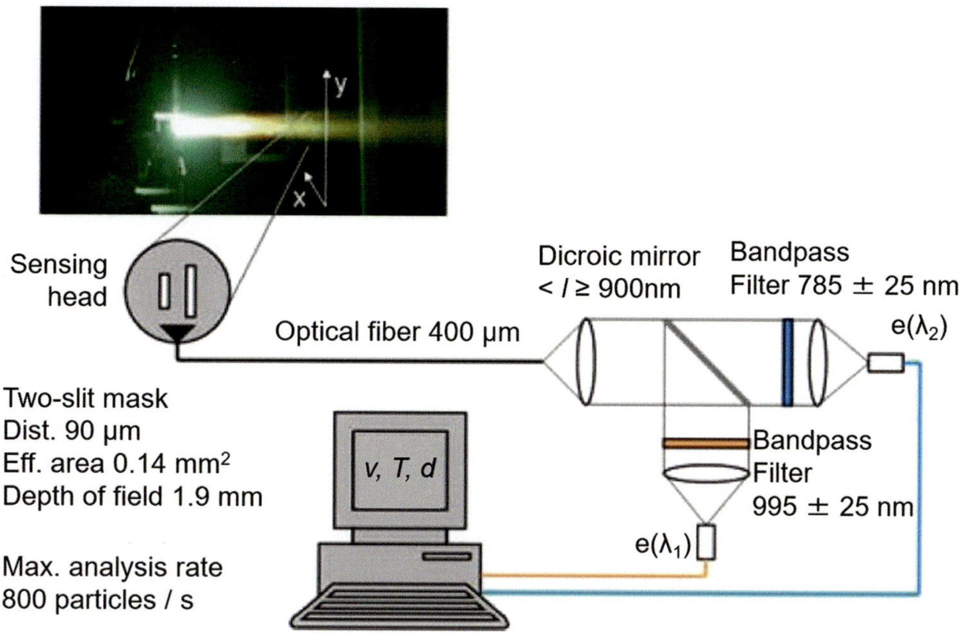

Fig. 18.83 Operation principle of the DPV-2000® diagnostic system. [Mauer et al. (2007)]

detected, statistical distributions and average values can be calculated. Due to operation in the infrared range and the pulsed laser illumination, comparatively cold and slow particles are still detectable.

Most of these techniques were, however, too sophisticated to be used in the harsh environment of a spray booth, but they have permitted a much better understanding of coating generation. Thus, they helped to shift the technology from a trial-and-error approach to a more scientific one progressively transforming the plasma spray process from an art to a science [Moreau et al. (2005)]. They have also been the basis for the development of sensors, less sophisticated but simpler and more robust, able to work in the harsh environment of spray booths [Li et al. (2003), Moreau et al. (1994)].

By the end of the 1990s, a commercially available in-flight particle diagnostic system (temperature, velocity, and diameter), the DPV-2000® (Tecnar Automation, Quebec, Canada), was developed, based on the work of [Moreau et al. (1994)]. As illustrated in Fig. 18.83, after [Mauer et al. (2007)], the DPV-2000® can be used to measure particle velocities, temperatures, and diameters. The velocity is obtained by measuring the time between the two signals which are triggered by a radiating particle passing the two-slit mask of the optoelectronic sensor head. Knowing the distance between the two slits, and the magnification factor of the lens and the time of flight (ToF), the velocity can be calculated. The temperature is acquired using two-color pyrometry, i.e., by calculating the ratio of the energy radiated at two different wavelengths assuming that the surface of the particles act as gray body emitters with the same emissivity at both

wavelengths. The diameter is obtained from the radiation energy emitted at one wavelength assuming that the melted particles are spherical or close to be. Since it is necessary to know the real emissivity of the particle, a prior calibration by means of a powder with known diameter distribution has to be carried out. As the measurement volume is relatively small (<1 mm³), the data is collected for individual particles and can subsequently be analyzed statistically. A certain measurement time is necessary to support the mean and standard deviations by a sufficient number of individual particle data acquisition.

18.5.4.2 Ensemble Particle Diagnostics

Ensemble averages of the particle parameters are equally valuable for production-scale process monitoring with considerable simplified instrumentation. They essentially involve the observation of the particle flux in the spray medium either through their own emission or through their illumination using a laser strobe or laser sheet as discussed earlier with individual particle diagnostic approach. The principal difference is mostly at the level of data acquisition and analysis.

The approach originated with the Spray and Deposit Control (SDC®) approach [Vardelle et al. (2001, 2004)], based on the observation of "hot particles radiation," using either a CCD camera or a photodiode array where the image of a section of the plasma jet plume or the flame is focused. A filter with a 3 nm band pass allows eliminating the most important part of the plasma plume light or that of combustion gases in HVOF. It is possible to record four images, and as the SDC is fixed on the plasma torch or the HVOF gun, the particle trajectories can be continuously monitored. The arrangement is shown in Fig.18.84, where the SDC, fixed orthogonally to the plane defined by the torch and injector axes, measures the flux of hot particles 60 mm downstream of the nozzle exit and, at the same time, the temperature of the coating just behind the sprayed spot.

Figure 18.85 illustrates the basic principle behind the SDC[®] approach in which a schematic of the different mean trajectories of particles as function of the carrier gas flow rate is shown together with the corresponding SDC[®] emission profile for the different carrier gas flow rates, for alumina particle injected into a DC plasma jet operating with a plasma gas flow rate of 45 slm Ar + 15 slm H_2, nozzle diameter of 6 mm i.d., torch power = 20 kW, and injector i.d. = 1.75 mm, placed 3 mm downstream of the nozzle exit with its tip at 8 mm of the torch axis. The maxim of the emission profile is noted to shift in the direction of powder injection with the increase of the carrier gas flow rate. The optimum trajectory observed with a flow rate of 4.5 slm (Ar) corresponds to a mean trajectory of particles, making an angle of 3.5–4° with respect to the torch axis and also induces the best temperature and velocity distributions. When the mean trajectory is not optimal, because the carrier gas flow rate is either too low or too high, particles are less heated, and the SDC[®] signal intensity decreases significantly. The type of measurement achieved by the SDC[®] is an ensemble one: along the lines of sight across the hot gases cross section, defined on the one hand by the camera or photodiode optics and on the other by the pixels, and corresponding to a large measurement volume. All particles contained in this large volume are taken into account during the exposure time.

The Particle Flux Imaging[®] (PFI[®]) from Linspray [Zierhut et al. (1999), Landes (2006)], tested in plasma spraying, records the plasma jet close to the torch and the particle flux (ensemble measurement) in the plasma jet plume and also the sprayed spot in the last version. Using a CCD camera, the very luminous plasma jet close to the torch, the less luminous particle flux in the downstream zone, and the spray spot are imaged simultaneously as shown in Fig. 18.86a.

A PC reduces the information by finding lines of constant radiation intensity in images of the hot plasma jet and the particle flux. These lines can be approximated by ellipses. Their characteristics are typical of the state of the coating

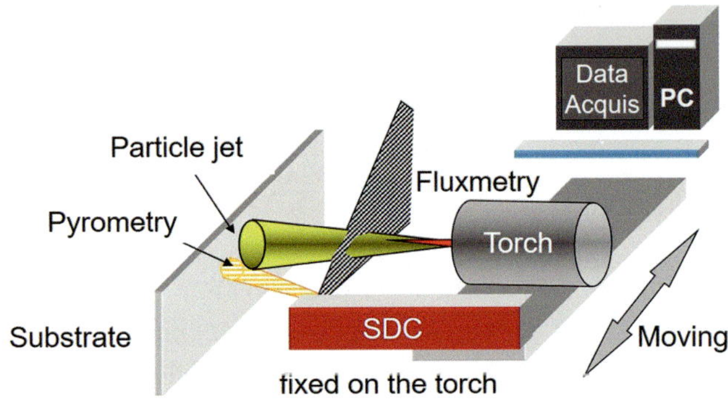

Fig. 18.84 Spray and Deposit Control setup fixed on the spray torch and following the hot particles radiation in the hot gases jet plume (60 mm downstream of the nozzle exit) and the temperature of the coating behind the sprayed spot. [Vardelle et al. (2001)]. Reprinted with kind permission from Springer Science Business Media, copyright © ASM International

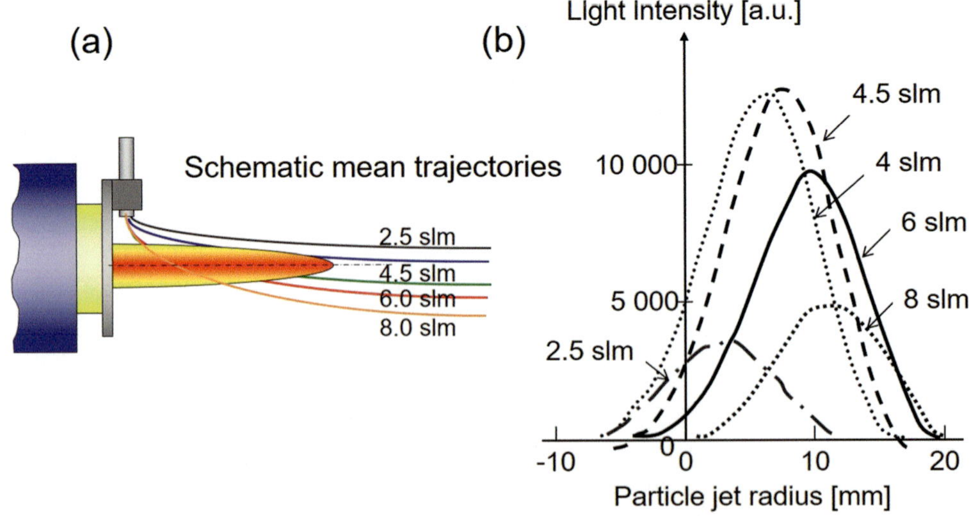

Fig. 18.85 SDC measurement of alumina particles heat flux 60 mm downstream of the nozzle exit of an Ar–H_2 DC plasma jet with different carrier gas flow rates, (**a**) mean trajectories and (**b**) radial distribution of emitted light intensity (Fauchais and Vardelle 2010). Reprinted with kind permission from Springer Science Business Media, copyright © ASM International

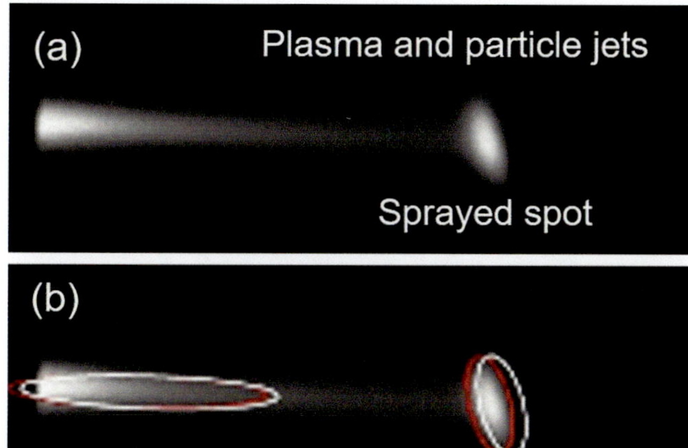

Fig. 18.86 (**a**) Particle Flux Imaging® (PFI-S) image of a running process, (**b**) PFI-S image with two calculated ellipses [Landes (2006)]. Reprinted with kind permission from Elsevier

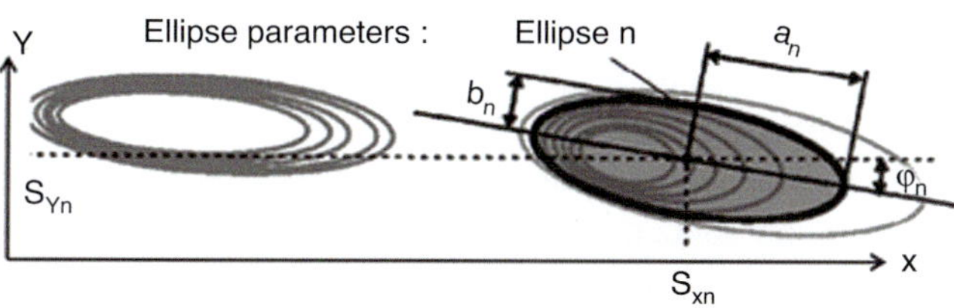

Fig. 18.87 Ellipses characteristics. [Landes (2006)]. Reprinted with kind permission from Elsevier

process, and any variation of macroscopic parameters results in the particles' ellipse displacement. This system also allows detecting the possible perturbation of hot gases jet by the carrier gas when radially injected. In this way variations in the hot plasma jet as well as in the particle flux can be detected without a precise knowledge of exact physical plasma jet or particle parameters. An example of these two ellipses (plasma jet and spray spot) is presented in Fig. 18.86b.

For plasma jets, due to their very high level of radiation, narrow band filters with different spectral transmittance must be used both for plasma and particles. The combined resulting image of the plasma jet and the particle flux represents the actual state of the coating process. The ellipses illustrated in Fig. 18.87 are characterized by the major axis, a_n; the minor axis, b_n; the x-coordinate, S_{xn}, of the center point; the y-coordinate, S_{yn}, of the center point; the angle, φ_n, between the x-axis and major axis; as well as additional information: number of pixels E_N in the contour and the average grayscale value E_{avg} of pixels. The PFI® system, as the SDC®, is moved attached to the gun and images continuously the whole area between the gun and the substrate surface, thus allowing an instant response to any change in

spray conditions. While the PFI® technique does not provide detailed quantitative physical parameters of the process, it is a very useful tool for quality control.

The SprayVIEW® developed by Tecnar (Canada) makes use of CCD camera with adapted filters to observe phenomena such as fine particles bouncing off the plasma and coarse particles flying all the way through the plasma. The good contrast between injected particles and the plasma stream itself permits the visualization of the injection cone and the quantitative characterization of the injection conditions, such as cone width and angle, particle mean velocity, acceleration, and trajectory angle. Being not fixed to the spray gun, the SprayVIEW® can be placed in the spray booth in a fixed location and have the spray gun moved, between spray runs, in front of it on a period basis for validation of consistency of the spray conditions.

Almost in parallel, the SprayWatch® which is an "ensemble" type of diagnostic instrument, developed by Oseir, makes use of an imaging system for in-flight particle temperature, T_p, and velocity, v_p, measurements based on the development work by [Vattulainen et al. (2001)]. As shown in Fig. 18.88, the unit is well suited for the diagnostic of hot particle trajectories in the plume of a plasma jet where the radiation from the hot gases jet is low enough to not overcome radiation from hot particles [Fauchais and Vardelle (2011)].

[Jodoin et al. (2006)] have developed measurements of particles in cold spray process, using the Oseir SprayWatch®, combining a fast shutter CCD camera with a high-power pulsed laser diode (HiWatch) to illuminate particles. The camera exposure time is between 100 ns and 10 ms with a maximum of 7 frames/s. The diode laser produces a laser sheet (at $\lambda = 808$ nm) with a width of 15 mm and a thickness of 1.5 mm, the pulse length being in the 50–2000 ns range. The accuracy in velocity measurement is 1–10 m/s. It is also possible to use the DPV-2000® with particles illuminated by a laser diode (7 W, $\lambda = 830$ nm) [Irissou et al. (2007), Legoux et al. (2007)].

As illustrated in Fig. 18.68, the SprayWatch® system has also been used for the monitoring of liquid droplets or suspensions injected into a DC plasma jet [Etchart-Salas et al. (2007)]. The set used, illustrated in Fig.18.68a, consists of a fast shutter camera (that of the SprayWatch®) coupled with the laser sheet flash at wavelength of 808 nm with photograph of the installation given in Fig. 18.68b. The image is triggered when the voltage reaches a given threshold. Figure 18.89 [Fauchais and Vardelle (2011)] represents the corresponding image obtained with the mechanical injection of an ethanol suspension injected into an Ar/H$_2$ DC plasma jet, image triggered for a 65 V voltage. Plasma gas flow rate was 45 slm Ar + 15 slm H$_2$, nozzle i.d. = 6 mm, and current = 500 A. Liquid was mechanically injected with a nozzle internal diameter of 150 µm. The dashed line at the top

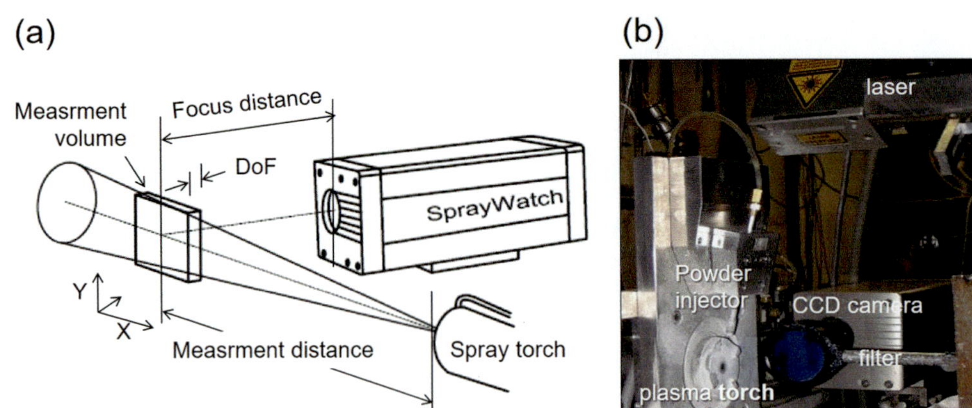

Fig. 18.88 (**a**) Schematic design of the measurement setup where the liquid injection is illuminated by a laser sheet and the corresponding image collected by a CCD camera (that of the SprayWatch® of Oseir (FN)). (**b**) Photograph of the device with the plasma torch, the laser, and the camera. [Etchart-Salas et al. (2007)]

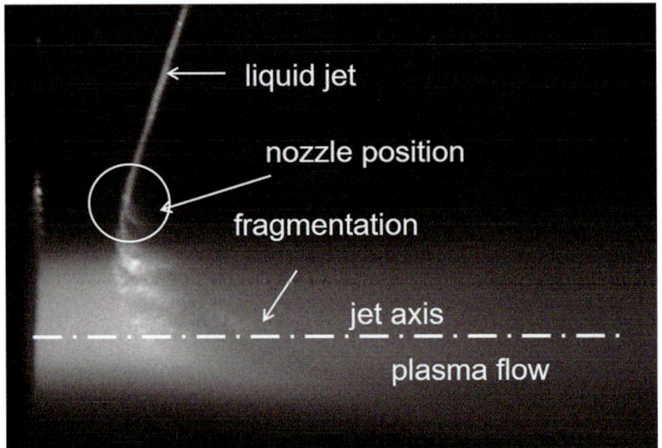

Fig. 18.89 Interaction between ethanol suspension feed stream with a DC plasma jet, image taken when the fluctuating voltage of the Ar–H$_2$ plasma was 65 V. [Fauchais and Vardelle (2011)]. Reprinted with kind permission from Springer Science Business Media, copyright © ASM International

of the figure corresponds to the anode nozzle top position. The distance between the torch axis and the injector is 23 mm, and, just before entering the plasma jet, weak instabilities of the liquid jet can be observed with a wavelength of about 900 nm. Upon its penetration into the jet, and unfortunately also in its fringes, the liquid jet is broken at the neck of its own instabilities by the shear stress produced. Several clouds of material (liquid and/or solid) within the plasma jet are visible in Fig. 18.68. Based on the image size and the number of pixels (600 × 600), one pixel represents about 30 μm^2. It is thus impossible to see fragmented droplets with d$_p$ < 5 μm. Clouds are composed of a compact head of suspension and, behind it, some sort of tail with tiny droplets and/or solid particles resulting from the fragmentation process. The distance between clouds corresponds to 900 μm, implying an initial velocity of the liquid jet in the range of 26.6 ± 2 m/s.

As a complimentary, scaled-down, version of the DPV-2000® Tecnar, Canada, has developed an "ensemble" particle diagnostic device, the AccuraSpray®-G3 presently marketed by Oerlikon Metco for the measurement of ensemble averages of particle velocity and surface temperature in a measurement volume of approximately ϕ 3×25 mm. As illustrated in Fig. 18.90 after [Mauer et al. (2007)], particle velocities are obtained from cross correlation of signals which are recorded at two closely spaced locations. The temperatures are determined by two-color pyrometry. CCD camera integrated into the system enables the analysis of the plume appearance (position, width, distribution, intensity) along a line in spray distance perpendicular to the particle jet. Reported range of measurable parameters are $5 < v_p < 1200$ *m/s* and $1600 < T_p < 4300$ *K*.

18.5.4.3 Individual vs Ensemble Particle Diagnostics

In-flight measurements are not, however, sufficient by themselves to explain coating properties; an example is presented below. [Renouard-Vallet (2004)] sprayed yttria stabilized zirconia, 13 wt.% Y$_2$O$_3$ (YSZ), particles (fused and crushed with a size distribution between 5 and 22 μm) using vacuum plasma spraying (VPS) to produce a 50-μm-thick, dense electrolyte layer for solid oxide fuel cells (SOFC). PTF-4 DC plasma torch, equipped with a Laval nozzle, was used in this study working with an Ar/H$_2$ mixture (40/6 slm) and an arc current of 750 A in a soft vacuum chamber with an absolute pressure of 8 kPa. The in-flight particle temperature distribution (measured with a DPV-2000® focused at the substrate location on the particle jet axis) is presented in Fig. 18.91a. The melting temperature of YSZ (T_m = 3100 K) has been indicated in this figure. As can be seen, the temperature distribution is rather broad: from about 1300 to 4300 K, and, if the absolute calibration of the pyrometer is correct, many particles are unmelted, the mean value, T_{en}, in

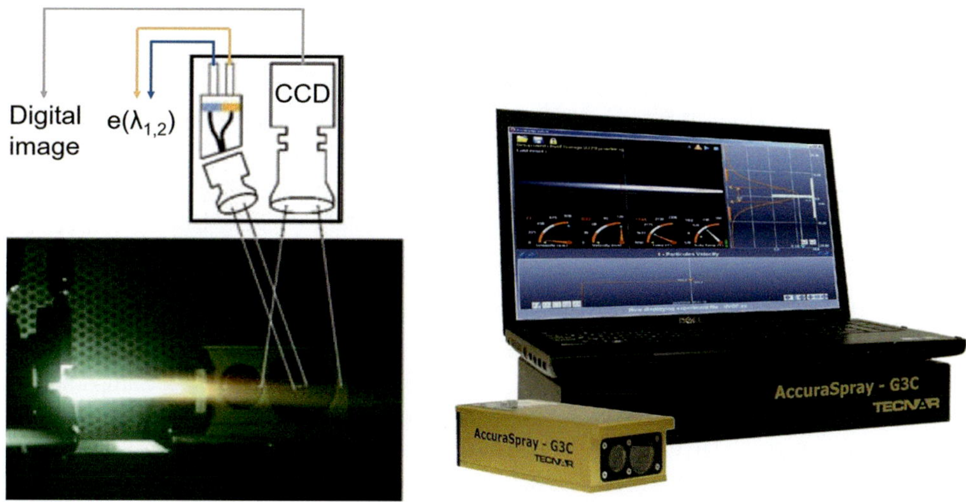

Fig. 18.90 Operation principle and photograph of the AcuuraSpray-G3 diagnostic tool by Tecnar, Canada. After [Mauer et al. (2007)]. Photograph courtesy of Oerlikon Metco with kind permission

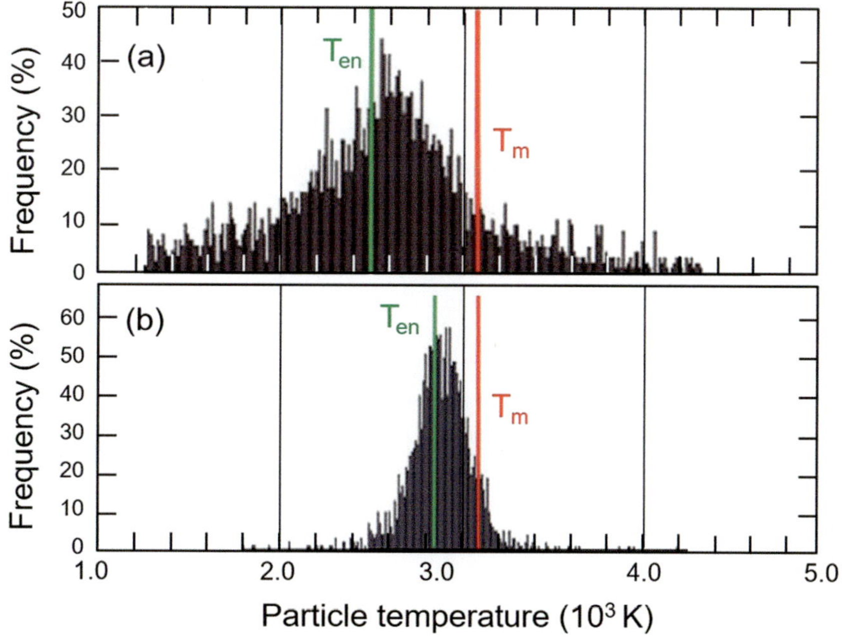

Fig. 18.91 In-flight temperature distribution measured by DPV-2000® and SprayWatch® (ensemble value T_{en}) of YSZ particles sprayed in soft vacuum with a PTF-4 torch, Laval nozzle, Ar–H$_2$ mixture, 750 A. (**a**) Spraying performed without "windjet," (**b**) with "windjet" [Renouard-Vallet (2004)]. Reprinted with kind permission from Springer Science Business Media, copyright © ASM International

Fig. 18.91a, being below T_m. The mean ensemble temperature measurement (T_{en}) with a SprayWatch®, focused at the same location as DPV-2000®, but measuring the whole jet cross section, is in good agreement with the mean temperature resulting from the distribution shown in Fig. 18.91a. However, for more precise comparison, distributions obtained from DPV-2000® all over the measurement volume of the SprayWatch® and weighted by the particle frequency should have been determined [Mauer et al. (2007)]. Particle

velocities (DPV-2000®) were 150 ± 70 m/s [Fauchais et al. (2011)].

To get rid of unmelted particles traveling in the fringes of the jet, a "windjet" has been used, blowing argon orthogonally to the particle jet 30 mm upstream of the substrate. The corresponding particle temperature distribution, presented in Fig. 18.91b, with values between about 2500 and 3200 K, is narrower than that without "windjet." Here again ensemble temperature is in good agreement with the mean value of the

DPV-2000®. The particle velocities (DPV-2000®) are 80 ± 25 m/s: the high-velocity small hot particles and the low-velocity larger cold particles have been ejected by the "windjet."

When looking at the ensemble temperatures, the values obtained with and without the "windjet" are close to the mean values deduced from DPV-2000®. SprayWatch® results correspond to average values representing a wide particle fraction in a comparatively large measurement volume. However, the differences between ensemble results with and without the "windjet" are much less informative than statistical measurements at a given location.

[Mauer et al. (2007)] have compared different measurements performed with DPV-2000® and Accuraspray-G3®. This comparison was made after measuring and then weighting by the local particle flow rates the different local mean values of the particle data (temperatures and velocities) at each DPV-2000® grid point that is contained by the Accuraspray-G3 measurement volume. They found that results obtained with both systems were in good agreement, thus confirming the measurement accuracy of both. They have also identified some application limits for the DPV-2000® and the Accuraspray-G3 diagnostic systems when using a few powder species. It is also reported that measurement of particle temperature close to plasma torch exit is generally more difficult to carry out by the Accuraspray-G3 than by the DPV-2000® system [Bissons et al. (2001)]. That is probably due to the fact that the plasma radiation is dominant relatively to the particle radiation because of the much larger measurement volume of the Accuraspray-G3. The difficulty in identifying the particles through their own emission is more accentuated under conditions where the sprayed particle partially evaporated thus contaminating the spray stream by metal vapor which is known to increase significantly the background radiation of the main spray stream.

Recently, [Colmenares-Angulo et al. (2011)] performed deliberate particle diagnostics using atmospheric plasma spray (APS) and high-velocity oxygen fuel spraying (HVOF) to create a set of first-order process maps. Particle states were measured simultaneously using five in-flight particle sensors: DPV-2000®, AccuraSpray®, SprayWatch®, TDS, and SprayCam. While sensors used similar methods for calculating particle characteristics, absolute values of temperature and velocity were considerably different. While process map trends among sensors are in agreement for the HVOF process, they differ when using plasma spray at high total gas flow conditions. After understanding the stochastic nature of particle detection, they have implemented an open-loop feedback control algorithm to achieve similar particle states with different hydrogen gas flow rates. The resulting particle state window measured by three different sensors under select fixed hydrogen flow rates was significantly narrowed.

18.5.5 Process Control

The DPV-2000®, SprayWatch®, and AccuraSpray-G3® were among the first devices that could work in the harsh environment of spray booths. They allowed monitoring the effect of the conventional spray parameters (gas flow rates, nozzle internal diameter, power level for plasma spray torches, injection conditions, particle size distribution and morphology) on in-flight particle parameters. Extensive research on relationships between in-flight particle parameters and coating properties [Vattulainen et al. (2001), Srinivasan et al. (2006), Planche et al. (2003), Sampath et al. (2009), Prystay et al. (1996), Vardelle et al. (1995), Fauchais et al. (1996), Wang et al. (2008), Legoux et al. (2002)] led to a significant enhancement in the process understanding and to the improvement of coating reproducibility and reliability [Vattulainen et al. (2001), Sampath et al. (2009)]. The linkage to coating performance, however, is still some sort of enigma requiring:

- Real-time process control to improve performance and meet quality requirements
- Close-loop control for direct, real-time monitoring of process performance via real-time sensing of particle spray parameters
- Corrective actions setting process variables and post-process examination

which implies development of controllers that must be able to modify input spray variable in response to:

- The necessity to establish for *each coating* with given service objectives, relationships between macroscopic spray parameters, sensor parameters, coating thermomechanical properties, and service performance
- The development of new robust and easy-to-use sensors that incorporate these relationships

In the following, a brief review is presented on some of the first attempts made to close the control loop in a conventional plasma spray operation.

One of the first results obtained with the DPV-2000® was the demonstration of the particle temperatures and velocity shifts with the plasma torch electrodes wear during a long-term experiment. The investigation by [Leblanc and Moreau (2002)] was carried out using a model F4-MB plasma gun starting with new electrodes and a 1.8 mm i.d. external powder injector, the tip of which was located 10 mm off the axis of the spray gun. The sprayed powder, injected from the top into the plasma plume, was a fused and crushed 7 wt.% yttria partially stabilized zirconia (YPSZ) powder, $-45 + 22.5$ μm. Those operating conditions were used for more than 50 h of spraying, with an average of 2.5 stops/starts per hour. The DPV-2000®

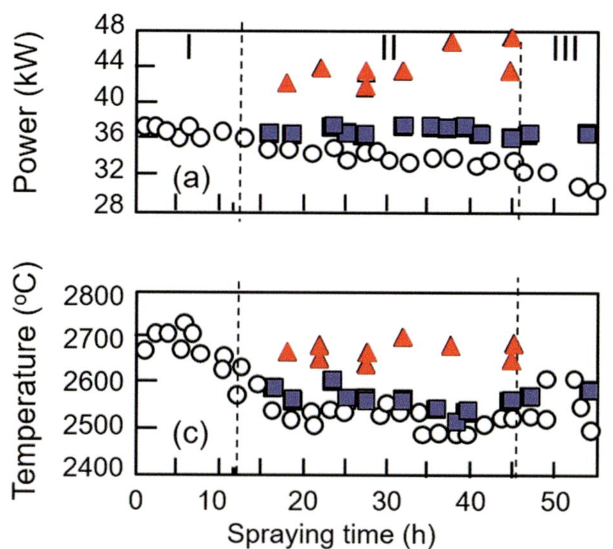

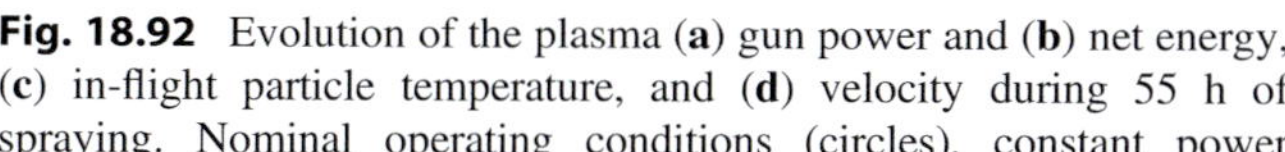

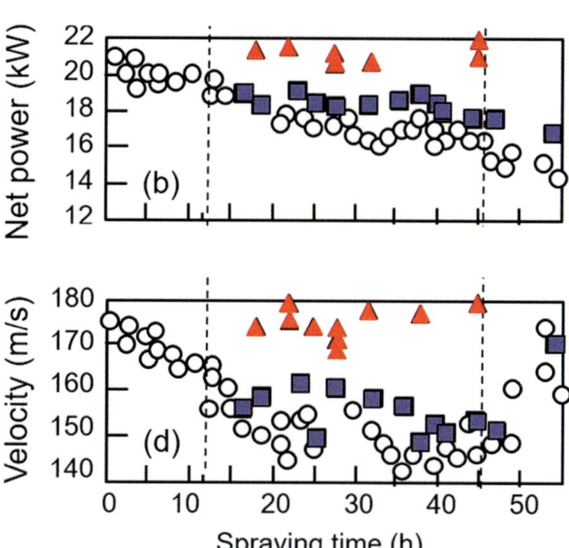

Fig. 18.92 Evolution of the plasma (**a**) gun power and (**b**) net energy, (**c**) in-flight particle temperature, and (**d**) velocity during 55 h of spraying. Nominal operating conditions (circles), constant power (squares), and constant in-flight particle state (triangles) [Leblanc and Moreau (2002)]. Reprinted with kind permission from Springer Science Business Media, copyright © ASM International

was used to monitor the temperature, velocity, size, and trajectory of the sprayed particles at the nominal standoff distance. Results were obtained both under the nominal operating conditions, i.e., at constant arc current (circles) and after adjustment of the arc current (squares and triangles). Figure 18.92 presents the time evolution of the power dissipated, the net energy, the particle temperature, and velocity. They show significant variations in the particle state and gun characteristics with spraying time and demonstrate the necessity to modify the spray parameters to keep the particle temperature and velocity at impact as constant as possible. Deposition efficiencies and coating porosities were compared for different spray gun conditions yielding a similar input power. It has been shown that the same input power obtained by increasing the arc current or by increasing the hydrogen flow rate resulted in different coating properties.

Using AccuraSpray-3G®, [Marple et al. (2007)] compared YSZ coatings sprayed with Ar–H$_2$ and N$_2$/H$_2$ plasmas. With N$_2$–H$_2$ plasma gas mixture, higher in-flight particle temperatures and lower particle velocities were produced as compared with Ar–H$_2$ plasmas. Coatings had similar hardness values; however, Young's modulus and thermal diffusivity following heat treatment were lower. [Planche et al. (2003)] have studied, with a DPV-2000®, alumina particles in-flight and compared measurements with calculations. The calculated particle velocities and temperatures were in good agreement with the experimental results: discrepancies being less than 10%. [Tekmen et al. (2009)], using AccuraSpray-G3®, have optimized the spraying of cast iron to control the

graphite content. A wide range of in-flight particle temperature and velocity values with constant graphite carbon content was determined. [Zhang et al. (2008)] have studied the influence of particle parameters onto the ionic conduction of YSZ. [Teckmen et al. (2008a, b)] have linked the spray conditions to particle parameters and alumina formation when spraying Al–12Si particles. [Yin et al. (2008)] have studied the influence of alumina particle sizes on their in-flight parameters and bonding. [Wang et al. (2004)] have shown the possibility to tailor alumina–zirconia coatings with two powder injection ports. [Fang et al. (2007)] have studied the influence of spray conditions on YSZ particle parameters. [Shinoda et al. (2010)] have studied the powder loading effects for various yttria stabilized zirconia powders under atmospheric DC plasma spraying. Statistical temperature distributions of in-flight particles suggested a rapid increase in the number of semi-molten particles above a certain powder-loading rate. Despite drops in the particle temperature and velocity due to the powder loading effect, the deposition efficiency tends to have increased in some cases.

Such measurements allow drawing what Sampath et al. (2009) call first-order process maps as illustrated in Fig. 18.93. In this figure are identified the state of the particles CoNiCrAlY- (38–57 μm) in response to various torch operating conditions as well as the control vectors identifying the influence of the secondary gas flow and torch current on the particle temperatures and velocities. These quantitative vectors could eventually be used as feedback in a control loop.

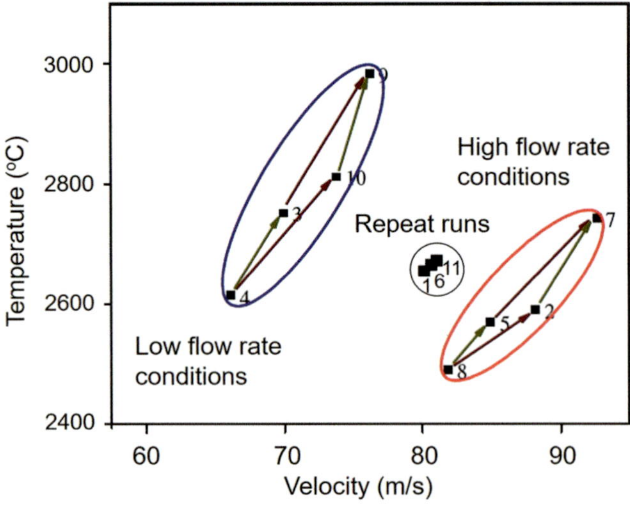

Fig. 18.93 First-order process map for APS of CoNiCrAlY material. Results identify the particle state response for various torch-operating conditions as well as the control vectors identifying the influence of arc current and secondary gas flow rate [Sampath et al. (2009)]. Reprinted with kind permission from Springer Science Business Media, copyright © ASM International

From such measurements, [Srinivasan et al. (2006)] estimate that there exist at least a few control protocols to monitor the particle state (predominantly temperature and velocity) with judicious choice of critical parameters. They have controlled the particle state by varying the critical torch parameters (primary gas flow and arc current) in a narrow range using yttria (8 wt%)–zirconia particles with angular shape. The particle state resulting from averaged individual particle measurements (DPV-2000®) is surprisingly stable with variability in temperature < 1% and variability in velocity of <4%. Ensemble approaches yield a somewhat higher variability (5% in temperature). Despite this, the variability in basic coating attributes, such as thickness and weight, is surprisingly large.

[Zhang and Sampath (2009)] have defined a group of dimensionless parameters, melting index, *MI* and Reynolds number, *Re*, to represent particle in-flight status. The melting index (*MI*) describes the molten state of a given particle by normalizing the measured surface temperature with the dwell time and size. It is assessed as:

$$\mathrm{MI} = \frac{\Delta t_{fl}}{\Delta t_m}$$

$$= \left(\frac{24\kappa_p}{\rho_p h_m}\right)\left(\frac{1}{1 + 4/Bi}\right)\left(\frac{(T_o - T_m)\Delta t_{fl}}{d_p^2}\right) \quad (18.14)$$

where κ_p is the particle thermal conductivity (W/m.K), ρ_p the specific mass of the melted particle (kg/m³), h_m the enthalpy of fusion (J/kg), Bi the Biot number, T_o the gas temperature

near the in-flight particle (K), T_p the measured particle surface temperature (K), d_p the particle size (m), Δt_m the particle melting time (s), and Δt_{fl} the particle in-flight time (s), which has the following expression:

$$\Delta t_{fl} = \frac{2L}{v_p} \quad (18.15)$$

where L is the spray distance (m) and v_p is the particle velocity (m/s).

Similar to MI, they also used particle Reynolds number (Re) as another dimensionless method for describing the kinetic state of a particle.

$$\mathrm{Re} = \frac{\rho_p v_p d_p}{\mu_p} \quad (18.16)$$

where:

ρ_p Mass density of the molten particle (kg/m³)

μ_p Molecular viscosity of the molten particle (Pa/s)

Experimentally and theoretically, Re has a significant effect on particle impact spreading. However, this is true as long as the flattening liquid particle surface tension is not considered, i.e., at the right beginning of the impact. Afterward, flattening is controlled by, on the one hand, the wettability flattening particle–substrate depending on substrate layer surface composition and roughness, as well as adsorbates and condensates existing at its surface, and, on the other hand, the droplet surface tension. Of course, to calculate MI and Re, statistical measurements are mandatory, and DPV-2000® must be used. Figure 18.94 presents a generalized first-order map based on large number of process diagnostic results obtained from multiple materials and processed through a typical air plasma spray torch at one or more process conditions (injection was optimized for each material so as to have maximum heat and momentum transfer for that particular condition). Figure 18.94a is a traditional first-order temperature–velocity representation based on the DPV-generated single particle data. According to their melting temperatures, aluminum is on the bottom of the plot, while molybdenum and tungsten are on top. Moreover, surface temperature is not necessarily linked to particle melting: low-thermal conductivity material can present an important temperature gradient, and metal oxidation also plays a role. The melting percentage of each particle depends on its size, thermal conductivity, latent heat, and residence time in the plasma jet. At last, particle impact velocity is not sufficient to characterize its flattening. According to [Zhang and Sampath (2009)], MI and Re are more appropriate for a more comprehensive description of particle states. The DPV-2000® single-particle data allow calculation of the individual particle MI

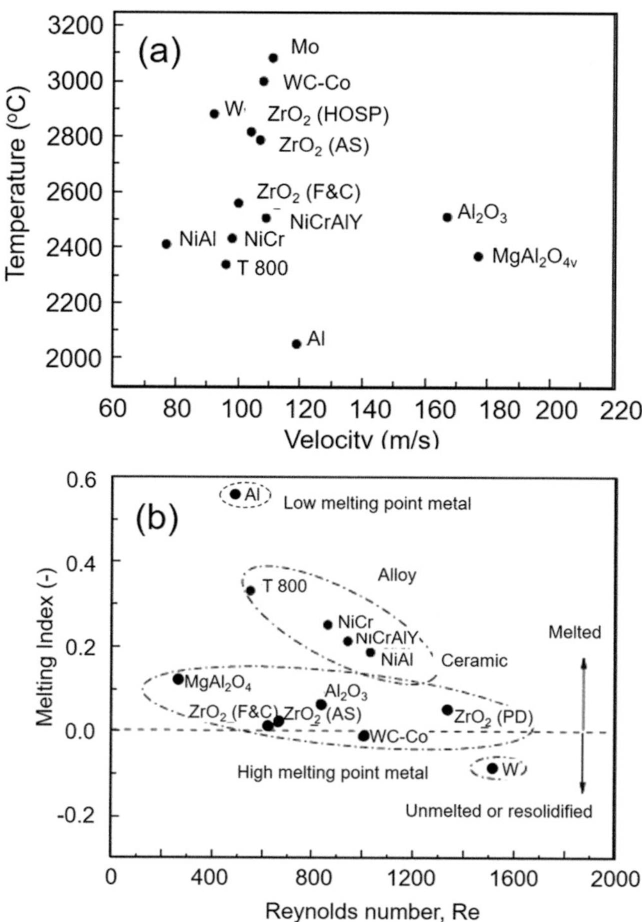

Fig. 18.94 (**a**) T_p–v_p global process map and (**b**) MI–Re global process map. [Zhang and Sampath (2009)]. Reprinted with kind permission from Springer Science Business Media, copyright © ASM International

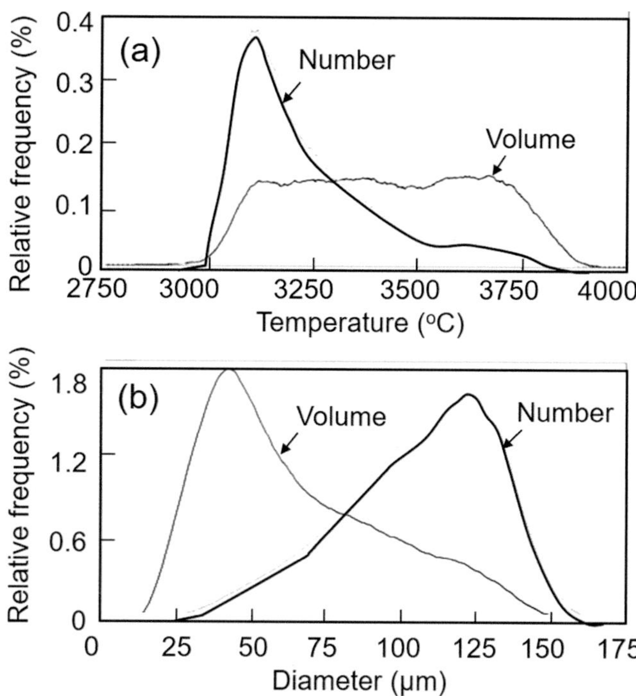

Fig. 18.95 Volume (gray)- and number (black)-weighted distributions of (**a**) temperature and (**b**) diameter of only molten particles (Sulzer Metco 7 MB torch, nitrogen 40 slm, hydrogen 6 slm, I = 600 A, YSZ particles 45–75 µm) [Wroblewski et al. (2010)]. Reprinted with kind permission from Springer Science Business Media], copyright © ASM International

and Re numbers and average values based on these measurements for each of the materials. This is represented in Fig. 18.94b from the T_p, v_p, and diameter data obtained for materials in Fig. 18.94a. In the MI–Re-based first-order process map, it is clearly seen that low-melting point Al shows the highest melting state (value of MI), while W shows the lowest value of MI despite representing a much higher temperature in the T_p–v_p-based first-order process map.

The mass flux of molten particles correlates better with coating thickness than bulk-average temperature or light intensity. However, measurement of molten mass flux for control requires sensors that are capable of sensing particle states at high rates from across a large portion of the full plume. This aim is that of the flux sentinel [Gevelber et al. (2008), Wroblewski et al. (2010)]. Only a subset of particles in the jet plume is incorporated in the coating: primarily those that are molten. The characteristics of this Molten Volume Ensemble, MVE, differ from those of the entire ensemble, with the MVE characterized by smaller, hotter, and faster particles. For [Wroblewski et al. (2010)], the total Molten

Volume Flux, MVF, is a properly weighted metric that correlates well with deposited mass and is defined by:

$$\mathrm{MVF} = \frac{1}{t_M} \sum_{i=1}^{N_{\mathrm{MVE}}} \frac{4}{3} \pi \left(\frac{D_i}{2}\right)^3 \qquad (18.17)$$

where N_{MVE} is the number of particles in the MVF and t_M is the time over which the particles are captured by the sensor. Volume weighting eliminates the bias toward smaller particles inherent with straight number averages, as illustrated in the number and volume-weighted temperature (Fig. 18.95a) and diameter (Fig. 18.95b). Experiments with different torches showed that molten volume flux measured across the entire plume correlated better with deposited mass than other quantities.

Measuring T_p and v_p has also permitted to better understand the working conditions of the Triplex Pro torch. With the three cathodes and the same anode, the plasma jet can be considered as three plasma jets flowing aside. Thus, particles treatment will be different when injecting them in one jet or between two of them. Between two jets, the viscosity is lower and the particle penetration easier, while injection in one jet of higher viscosity requires a higher injection force of particles. To study that, [Mauer et al. (2011)] used a nozzle

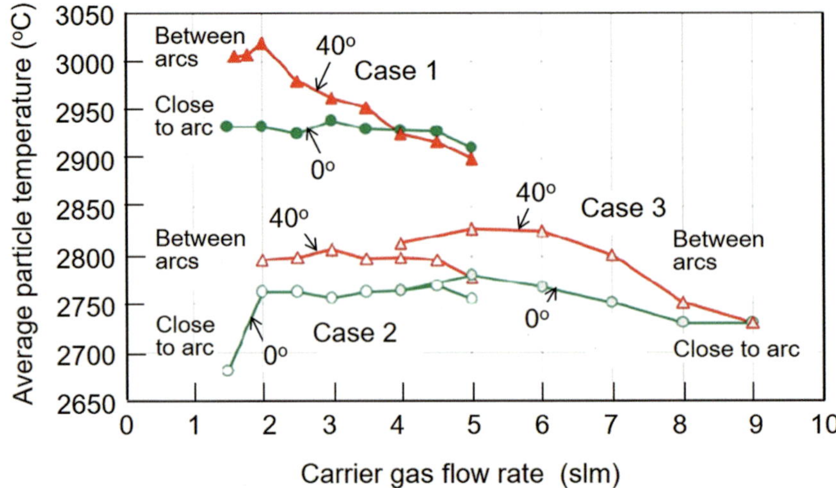

Fig. 18.96 Volume flow-averaged particle temperatures as a function of carrier gas flow and azimuthal injector position (Triplex Pro™ torch, case 1–3, short injector mount) [Mauer et al. (2011)]. Reprinted with kind permission from Springer Science Business Media, copyright © ASM International

internal diameter of 9 mm and an injector diameter of 1.8 mm, with an angle of 90° relative to the torch axis. The azimuthal injector positions were step-adjustable by 20° between 0° and 80°. In Fig. 18.96, injection between two jets is indicated as "between arcs," while that in the jet is called "close to arc." Three spray conditions were used: 40 slm Ar–4 slm He with either $I = 540$ A (case 1) or $I = 470$ A (case 2) and for the third (case 3) 75 slm Ar, 6 slm He, and $I = 540$ A. Results presented in Fig. 18.96 show that the highest temperatures are obtained with case 1 corresponding to a net power of 30.9 kW with gas velocity, v_g, of 1250 m/s, while similar values are obtained for case 2 and 3 corresponding to, respectively, 25.3 kW with $v_g = 1140$ m/s and 37.6 kW with $v_g = 1940$ m/s. In the three cases, the injection between jets results in higher particle temperatures, phenomena called by authors "cage effect."

18.6 Finishing and Post-treatment of Coatings

After spraying, coatings are treated to achieve the final dimension and obtain a smooth surface. However, machining of coatings can be troublesome especially with abrasion-resistant, porous, poorly bonded materials in which particles or splats can easily be pulled out. Coatings with low thermal conductivity (at least lower than that of the bulk material) will also suffer during machining due to reduced heat dissipation from the surface increasing the surface temperature of the coating during grinding. Generally accepted methods, practices, and techniques used to machine wrought material do not apply in this case to the same material when sprayed.

Post-treatments of coatings are also often performed to enhance the bond strength; partially or totally relax residual stress; create compressive residual stress; close or reduce porosity, especially the interconnected one; create a barrier to corrosive or oxidizing products; improve coating homogeneity; improve splat or particle cohesion; obtain hard phase precipitation; or induce chemical modifications. This is achieved by using different techniques, including fusion of self-fluxing alloys, heat treating or annealing, hot isostatic pressing (HIP), austempering, laser glazing, sealing, spark plasma sintering, peening densification, and diffusion. Heat treatment processes have to be moderate and well planned to avoid overheating the surface, especially of ceramic coatings, which could result in the formation of cracks and defects that weaken the coating.

In this section, a brief review is presented of the principal finishing and post-treatment processes commonly used in the industry to achieve final dimension of the sprayed part and a surface finish compatible with its functional role. For more details, the reader is recommended to consult leading references on the subject such as the *Thermal spraying: Practice, Theory, and Application*, American Welding Society, Miami, FL [AWS (1985)], and *Handbook of Thermal Spray Technology*, ASM International, Materials Park, OH [Davis (2004)]. Complimentary references are also integrated in the text whenever necessary.

18.6.1 Machining (Turning, Milling)

For metal coatings, turning and milling can be easily used to establish the final dimension and remove the as-deposited coating roughness. However, cutting lubricants, oils, and coolants will penetrate and remain in coatings with open porosities and remain after machining [AWS (1985)] if the coated part is not treated afterward.

For cermets, especially those with carbide grains, carbide tools are recommended due to the abrasive nature of the coating. For ceramic coatings, grinding is the only option.

18.6.2 Grinding

The required finish and tolerance of ceramic coatings in general require grinding. As with turning and milling, grinding conditions of ceramic-sprayed coatings are distinctly different from those of wrought or cast materials. Individual particles or splats can be pulled out, or even the whole coating can delaminate from the substrate. The ductility of ceramic coatings is very low, and, besides the poor thermal conductivity of coatings, the contact between coating and substrate is never perfect, and heat conduction toward the metal substrate is reduced. Lastly, ceramic coatings are porous, and whatever the quality of the grinding may be, the surface will always have holes after grinding.

Machines commonly used are outside diameter grinders (cylindrical and centerless), surface grinders (horizontal and vertical spindle), and internal grinders [Malkin and Guo (2008)]. However, to work on complex-shaped parts, such as turbine blades, six-axes machines with pressure control are required. To achieve complex machining at the coating surface, such as airflow channels for air-sustained rotating machines, ultrasonic machining can be used. In this machining, a constant stream of abrasive (diamond, boron carbide, silicon carbide, alumina) slurry passes between the tool and the workpiece. The vibrating tool, combined with the abrasive slurry, abrades the material uniformly, leaving a precise reverse image of the tool shape. The tool must have lower hardness and toughness than the ceramic coating to be machined.

Compared to machining of metallic coatings, grinding is rather expensive, and thus the quantity of material to be taken off must be minimized, implying that the coating thickness should be planned as close as possible to the final dimension (± 5 μm at the maximum).

Wet grinding is preferred to dry grinding because of the greater latitude in abrasive and wheel selection; it also allows using harder wheels resulting in better finishes and less dressing [AWS (1985), Davis (2004)]. The selection of the grinding wheel depends on the coating hardness and microstructure. The *Handbook of Thermal Spray Technology* [Davis (2004)] proposes a few rules for this selection.

The grinding variables, wheel speed, work speed, area of contact, grinding fluid filtration, and proper fluid concentration, are also important parameters. For example, for the grinding wheel selection [AWS (1985)], always use the sharpest wheel possible to cut rapidly without overheating and choose wheels with structures and grades to achieve free cutting action. For the wheel structure, the spacing between individual abrasive is one of the key issues as well as the wheel grade, with harder wheels lasting longer but cutting slowly and generating more heat. Moreover, they require frequent dressing, harder grades used on small areas of contact or narrow wheels. Of course, one has to choose the bond type best suited for the operation and equipment: vitrified bonds, which have a high-stock removal and a good precision tolerance, the operating speed being limited to 33 m/s, or resinoid bonds allowing high speeds, rapid stock removal, and fine finishes. Equipment alignment especially wheel mounting (balance) must be systematically checked [AWS (1985), Malkin and Guo (2008)]. The principal grinding parameters are:

- Proper choice of the wheel speed from which depend the performance, finishing, and stock removal
- High work speed corresponding to aggressive wheels
- Low area of contact, a higher pressure corresponding to larger depth of abrasive penetration
- Wet grinding, allowing for the use of harder wheels, giving better finishes, and less dressing

Metal, cermets, and ceramic-sprayed coatings can also be ground and polished using coated abrasive belts which have the advantages of lower part/coating overheating, avoiding the need for tool dressing, and allowing for their rapid replacement when worn. The part is either transverse or rotated under the belt head. Belt speed is generally between 30 and 40 m/s, while the part speed varies between 0.13 and 1 m/s. The grain size of belts can be adapted to rapid stock removal or improve surface finish [AWS (1985)].

Super surface finishing can be achieved using abrasive stones or laps oscillating in front of the moving part. The choice of lubricant is critical with a high viscosity, reducing cutting action, but giving finer finishes. Hand stoning, buffing, and polishing can be accomplished with abrasive stone, stick, wheel, or paper [AWS (1985)]. Hand stoning and buffing give excellent results on soft metals with a bright finish, provided abrasive sizes are progressively decreased from coarse to fine. However, the abrasive device should be rapidly loaded and must be frequently dressed. Carbon and stainless-steel coatings are difficult to hand stone to a good finish. Ceramics or cermets can be hand stoned to a light finish (well suited for parts with unusual contours). However, with these techniques, it is difficult to maintain a uniform thickness.

For large quantities of small parts or when parts have unusual contours, tumbling and burnishing are well adapted. Tumbling is performed in a vibrating or rotating tub with an abrasive medium and water, while burnishing is the same but with hard metal balls instead of the abrasive. These hard balls are used to compress the surface, the final smoothness of which depends on pressure and frequency of ball-to-part contacts (AWS 1985).

18.6.3 Fusion of Self-fluxing Alloys

Coatings are submitted to a post-thermal treatment at a temperature between the solidus and liquidus states of the alloy, where important diffusion processes take place. This temperature is usually in the range 1200–1400 K, meaning that low melting temperature metals, such as aluminum, are excluded of this process. Self-fluxing alloys contain boron and silicon (e.g., CrBFeSiCNi), which are used as deoxidizers. As-deposited self-fluxing alloys are rather porous. Upon coating and reheating, densification occurs (most of the pores are eliminated), and borosilicate ($B_2O_x \cdot SiO_y$) is formed, according to reaction, and diffuses toward the coating surface as slag, which is mechanically removed afterward by turning, milling, etc. The coating shrinkage can reach 20 vol.%, and silicon and boron form complex silicides and borides, dispersed within the coating [AWS (1985), Malkin and Guo (2008)]. Lastly, high fusion temperature allows diffusion between substrate and coating and between the splats, improving the bond strength and the coating cohesion. Of course, the fusing temperature depends on the self-fluxing alloy chosen (between 1020 and 1121 °C), and it is very important to control it. A too high temperature results in excessive fluxing, depleting the alloy of silicon and/or boron and creating more slag. Too low temperatures induce poor bonding and high porosity.

The cheapest fusing method uses a flame torch or torches (depending on the part size) working with oxygen and either acetylene or propane. This process can be well adapted to parts with axial symmetry. A low flame velocity must be used to avoid blowing molten metal. The part and coating are gradually and uniformly heated to 760–870 °C before elevating the coating temperature to its fusing range. The torch movement and the surface temperature must be controlled if possible, with automatic torch fusing equipment [AWS (1985)].

With furnace fusing, the heating, cooling, and working atmosphere can be easily controlled. Vacuum furnaces are also used. Industrially, furnace fusing is used for components with complex shapes or large parts, which are difficult to heat uniformly with flames.

Induction fusing is rather popular due to savings in fusion time and energy cost but limited to parts with cylindrical symmetry. Laser fusing, if the right parameters are selected, can provide good-quality coatings, and the heating can be very well controlled. However, it is generally preferable to preheat the part in a furnace and then let it cool slowly, for example, by placing it in vermiculite [AWS (1985)].

Nevertheless, there are no significant differences in wear behavior of NiCrBSi coatings treated with flame or laser fusing [Gonzalez et al. (2007)].

Crack formation occurs when alloys are brittle and when the expansion mismatch between coating and substrate is high. To limit their formation, temperature gradients must be avoided with a uniform heating and a slow cooling. The latter is achieved with a furnace or by immersing the part in an insulating medium such as vermiculite.

Porosity, as noted previously, occurs with under-fusing (incomplete melting, wetting, and fluxing) or overheating. A problem of non-uniform heating can be observed with large parts. Non-uniform heating can also be observed with parts that are hollow or vary in wall thickness. High sulfur, selenium, or lead in the base metal can also cause porosity problems [AWS (1985)].

18.6.4 Heat Treating or Annealing

By definition, annealing corresponds to a heat treatment that alters the microstructure of a material causing change in properties such as strength, hardness, ductility, etc. For thermally sprayed coatings, annealing or heat treating is performed [AWS (1985)] at high temperatures, but well below the coating melting temperature, T_m. It takes place at atmospheric or reduced pressure, in air or more generally in a controlled atmosphere. It induces changes to its microstructure as well as its thermomechanical and electrical properties enhancing the overall coating toughness. Among the changes that can take place are:

- Formation of amorphous phase
- Relaxation of residual stress (temperature must be over that of recovery $\approx 0.4 \times T_m$)
- Diffusion between substrate and coating improving the bond strength
- Coating densification by increasing inter-splat cohesion
- Recrystallization and grain growth occurring through sintering
- Reduction of coating oxide content when heated in hydrogen atmosphere, or complete oxidation when heated in an oxygen atmosphere
- Carbide precipitation from solid solutions with cobalt

18.6.5 Hot Isostatic Pressing

Hot isostatic pressing (HIP) is the simultaneous application of heat (temperatures below the melting point, about 1100 °C for steel and 975 °C for nickel) and pressure between 50 and 300 MPa in an inert atmosphere to an encapsulated coated substrate. The heat promotes diffusion (enhanced bond strength and splat cohesion), and the pressure collapses and shrinks pores. Of course, at those temperatures, the residual stress is relaxed, and the fracture toughness is enhanced [AWS (1985), Malkin and Guo (2008)]. The main industrial application is for thermally sprayed, high-added value, near-net-shaped components, due to the relatively high cost of the process.

Successful applications of HIP as a post-treatment technology was reported by [Stewart and Ahmed (2003)] who investigated the influence of the HIP on the rolling contact fatigue (RCF) performance of thermal spray (WC–12Co) coatings. By increasing the HIP temperature to 1200 °C and maintaining full film lubrication, it was possible to achieve a fatigue life in excess of 70 million stress cycles without failure in relatively thin (50 μm) cermet coatings. [Stoica et al. (2004)] have HIP treated (850 °C for 1 h) WC–12Co coatings sprayed by a HVOF system onto SUJ-2 bearing steel substrate. The sliding wear performance of test coupons was drastically improved not only in terms of coating wear but also in terms of the total volume loss.

[Khor and Loh (1994)] studied HIP treatment at increasing temperatures and pressures of plasma-sprayed Ni-based alloy coatings. HIP changed the microstructure and increased the microhardness of the coatings. Results showed that, for temperatures below 750 °C, the elastic modulus of HIPed coatings was larger than that of as-sprayed coatings. These changes could be related to plastic flow, interlamellar diffusion, and creep that occur at increased temperatures and pressures. [Abdel-Samad et al. (2006)] have studied the influence of hot isostatic pressing on thermal fatigue resistance of plasma-sprayed coatings onto die-casting molds. HIP treatment was performed at 1100 °C at a pressure of 100 MPa for 6 h using argon gas, heating for 3 h, and slow cooling to room temperature. The hardness did not decrease, residual stresses were attenuated, and the fatigue life of steel die specimens was increased. The HIPed coatings showed low residual tensile stress indicating their high relaxation through the HIPing process.

18.6.6 Austempering Heat Treatment

"Austempering" is a process to promote diffusion at the substrate and coating interface and to collapse the internal pores in a coating. The treatment is performed in molten salt baths as described by [Lenling et al. (1991)] who immersed coated samples in an 870 °C neutral salt for 30 min and then quenched them in a neutral salt bath at 315 °C for 30 min followed by cooling and rinsing the part with room temperature water. Coatings, about 250 μm thick, were composites of WC-Co and Ni base, plasma sprayed onto AISI 5150 steel substrates. The percentage of WC phase was increased by the treatment, and less desirable, carbide phases formed during

plasma spraying were eliminated. The bond to substrate was improved, the coating hardness increased, and residual compressive stress was induced.

18.6.7 Laser Glazing

Laser treatments (auxiliary processes) can be applied before, during, or after thermal spraying (main process), leading to a wide range of coating improvements (microstructure, adhesion, etc.). In this section, the presentation will be limited to laser post-treatments. An excellent review of laser treatments (ablation and texturing) is given by Garcia-Alonso D et al. (2011) covering topics as the use of laser pretreatment to promote coating/substrate adhesion, hybrid laser treatments applied simultaneously during the spraying process and deeply modifying the coatings microstructure as well as laser post-treatments to improve coatings density and adhesion, and induce phase transformations and structure refinement.

The advantage of laser heat treating and glazing is that the thermal energy applied is perfectly controlled with respect to the amount and the location on the surface to be treated. Lasers used are mainly continuous wave (CW), CW-CO_2 lasers at 10.6 μm with power levels up to 5 kW, diode lasers with power levels in the few kW range, CO_2-pulsed lasers with power levels of a few hundred watts, and neodymium-doped/yttrium-aluminum-garnet (Nd/YAG laser) at 1.06 μm. For all treated surfaces, the intensity of the hemispherical reflection coefficient, R, must be considered, the energy or power absorbed by the surface being proportional to $(1-R)$. With a wide range of energy densities possible, accurate temperature profiles can be obtained at precise locations in different depths of the coating. The treatment is limited to heat treating or melting of the surface (laser glazing). In the latter case, the rapid solidification through the coating and substrate allows microstructural refinements and other metallurgical changes modifying coating properties, wear or corrosion resistance, and tribological properties [Crawmer et al. (1987)]. It must be pointed out that for sprayed materials with low thermal conductivities (below 30–40 W/m. K), temperature gradients can induce important stress, which will result in relaxation by macro-cracks orthogonal to the coating.

Plasma-sprayed zircon ($ZrSiO_4$) coatings are generally made of ZrO_2 and SiO_2 due to the fast quenching of dissociated zircon. [Schelz et al. (2008)] have shown that, by using low-energy (8 W) CO_2 laser treatment, the inverse reaction occurred, unfortunately with a low recombination rate. When glassy particles were incorporated in the coating with a large-size distribution from approximately 10 μm to more than 100 μm, the result was the production of a substantial amount of recombined $ZrSiO_4$ when the same laser parameters were used. Thus, the laser treatment with additives may provide technological benefits over other heat treatment techniques where high-temperature annealing (1600–1900 K) must be used for the same result.

The microstructure, electrochemical, and wear performances of HVOF WC 24%, Cr_3C_2 6%, and Ni coatings on Inconel were modified greatly by laser heat treatment [Zhanga et al. (2010)]. The laser beam spot was elliptical 5×4 mm with a power of 400 W. The relative content of Cr_3C_2 and Ni in HVOF-sprayed coatings was increased, as well as the coating hardness, after laser heat treatment, and a dense interface between the coating and substrate was achieved. Laser heat treatment was also helpful to form denser oxide tribo-film and improve the cohesion of WC, Cr_3C_2, and Ni phases, together with better wear performance of the coating.

[Pokhmurska and Ciach (2000)] have remelted (Fe–Cr–B) + Al-based electric arc-sprayed coatings with a CW-CO_2 laser. The surface was covered with a graphite-based antireflection coating. Laser melting markedly improved coating microstructure and density, with much better chemical homogeneity. Results also demonstrated an improvement of mechanical and corrosion properties of coatings.

Al_2O_3–13wt%TiO_2 plasma-sprayed coatings onto a titanium alloy (Ti–6Al–4 V) substrate were remelted with a CW-CO_2 laser [Wang et al. (2009)]. The laser remelted coatings (LRmC) possessed a much denser and more homogeneous structure and excellent metallurgical bonding to the substrate. The average porosity of the LRmC was reduced to 0.9%, compared with 6.2% of the as-sprayed coatings. The metastable γ-Al_2O_3 phase in the as-sprayed coatings transformed to stable α-Al_2O_3 during laser remelting, due to the slow cooling. The microhardness of the coatings was enhanced to 1000–1400 $HV_{0.3}$ after laser remelting, which was much higher than that of the plasma-sprayed coatings.

Aluminum coatings on austenitic stainless-steel substrates were laser remelted (5 kW CW-CO_2 laser) [Song et al. (2000)]. The residual porosity was eliminated, and phases were mainly AlFe Ni, Cr, and CrFeNi. A metallurgical bond between the coatings and the base metal was formed after laser surface remelting.

Continuous wave CO_2 laser and a pulsed laser with cylindrical focal lens have been used to glaze the surface layer of plasma-sprayed ZrO_2–20 wt% Y_2O_3/MCrAlY coatings [Tsai and Tsai (1998)]. Defects, such as bubbles or depressions, occurred usually with a continuous wave laser, but a high-quality glazed layer was produced using a pulsed laser. The surface roughness of the plasma-sprayed ceramic coatings was significantly improved by laser glazing. Surface roughness decreased slightly as the pulse frequency increased for the glazed surface.

Pulsed laser glazing of plasma-sprayed hydroxyapatite (HA) coatings modified the morphology of their surface as well as their phase composition [Dyshlovenko et al. (2006)].

The surface became smoother. The phase content was richer in α-tricalcium phosphate and tetracalcium phosphate (TTCP) than that of as-sprayed deposits; TCP and TTCP were the products of HA decomposition. The fraction of TTCP was larger in all samples treated with a laser.

Ni–20Cr and Stellite-6 powders have been sprayed on boiler tube steels with a shrouded plasma spray process using NiCrAlY as bond coat [Sidhu et al. (2005)]. The coatings showed some porosity (2–4%). Laser remelting of the coating was performed with a pulsed Nd:YAG laser: 300 W with pulse energy of 6 J and pulse width of 20 µs, the repetition rate being 10 Hz. It appeared to be effective in reducing the porosity to less than 0.5%. However, laser remelting further led to a significant dilution of the coating due to surface alloying of the coating with the bond coat and substrate. Laser remelting of the coatings led to a decrease in microhardness.

Laser surface melting was shown to reduce surface roughness of a cordierite glass coating, which was air plasma sprayed [Razavy et al. (2003)]. Subsequent devitrification of the cordierite was found to depend upon the physical and chemical alterations induced by the laser treatment. Crystallization in the as-sprayed coating was controlled by the TiO_2 distribution in the coating microstructure. After laser treatment, the devitrification was dominated by nucleation at micro-cracks, since the inter-splat structure was homogenized both physically by the removal of splat boundaries and chemically by the redistribution of TiO_2 in the coating.

Improvement in copper erosion resistance at temperatures comprised between 25 and 300 °C can be obtained by forming copper coatings containing 45 vol.% boride particles. To resist erosion due to impact of coarse iron ore particles, these coatings must possess microstructural features, which are larger than the size of impacting particles. Laser melting of plasma-sprayed Cu–boride coatings produced large intricate islands of ductile and hard phases [Dallaire et al. (1999)]. The peculiar microstructure architecture developed upon laser melting was found suitable to withstand erosion attacks.

[Ahmed et al. (2010)] have HVOF-sprayed Ni-based superalloy Inconel 625 coatings and laser remelted them using a high-power diode laser. The performance gap between bulk and HVOF-sprayed Inconel 625 was largely due to galvanic corrosion between Cr-depleted re-solidified regions and non-melted material in the coatings. Laser surface melting of HVOF coatings of Inconel 625 largely improved the corrosion resistance due to the elimination of both porosity and localized regions of material depleted in Cr. Several studies have been undertaken recently to adapt yttria partially stabilized zirconia (YPSZ) thermal barrier coating (TBC) characteristics during their manufacturing process. [Antou et al. (2004)] have treated the coating pass by pass, as it was manufactured, with a 3 kW diode laser at 848 nm. The input energy effect was studied by varying the scanning velocity (i.e., between 35 and 60 m/min) and consequently the irradiation time (i.e., between 1.8 and 3.1 ms, respectively). Experiments showed that coating thermal conductivity was lowered by more than 20%, and coating resistance to isothermal shocks was increased significantly.

18.6.8 Sealing

As thermally sprayed coatings are more or less porous with connected porosities, they are not very suitable for corrosion protection. This situation can be drastically improved by curing their interconnected porosities through sealing them.

The advantages of sealing are the following [Davis (2004)]:

- Prevents the corrosive substance (liquids or gases) from penetrating the coating and attacking the coating/substrate interface
- Limits the lodging of wear debris in the coating
- Enhances inter-splat cohesion
- Can provide special surface properties such as nonstick surfaces
- Can extend the life of aluminum or zinc coatings on steel to prevent corrosion

[Knuuttila et al. (1999a)] gave an excellent description of the process presenting various sealant types and impregnation methods, as well as factors influencing impregnation. Impregnation methods can be divided into four categories: atmospheric pressure impregnation, low-pressure impregnation, overpressure impregnation, and a combination of those. The method of choice depends on the size of the coated component, the required penetration depth, and the sealant, the choice of which also depends on the coating material and the application. Generally, low-pressure or overpressure impregnation is suitable for small components. Atmospheric pressure impregnation, on the other hand, is economically viable for larger components, but the result depends strongly upon the sealant-coating wet ability (in fact, it is the dynamic wetting angle that has to be considered) and the sealant viscosity. Sealing requires controlling humidity by keeping the surface temperature above the dew point where water condenses to any surface. The techniques used are brushing, dipping, or spraying.

The advantage of low pressure is the removal of moisture and air from pores and cracks improving the sealant penetration depth. However, the pressure must be adjusted to the sealant vapor pressure to avoid excessive evaporation.

Overpressure impregnation is much less sensitive to the sealant-coating contact angle that can be over 90°.

Sealants for impregnation are either organic or inorganic materials or metals. In sealing, the aim is to seal the porosity to as great a depth as possible. However, there are applications where only a relatively thin sealant layer is considered optimum [Dyshlovenko et al. (2006)]. A low interfacial surface tension between the coating and the sealant is the most important factor dictating the penetration depth. The pore size distribution of the coating is equally important; larger pores might not become filled due to lowered capillary pressure.

18.6.8.1 Organic Sealants

The often-used organic sealants are based on epoxies, phenolics, furans, polymethacrylates, silicones, polyesters, polyurethanes, and polyvinyl esters. Waxes can be used as well. The choice depends upon the service temperature. For example, wax can be used only in cold conditions, phenolic resins up to 150–260 °C, and methacrylate up to 150 °C [Malkin and Guo (2008)]. A good sealing depends on how coatings were treated before impregnation and how long the impregnation time was. These sealants are cured by heat, ultraviolet light, or high-energy radiation [Knuuttila et al. (1999b)]. However, curing the sealants induces some shrinkage, and only low curing shrinkages can prevent the transport of electrolyte through the coating.

[Shrestha et al. (2001)] have studied the erosion–corrosion characteristics of a Ni–Cr–Mo–Si–B coating applied by a high-velocity oxygen fuel (HVOF) process onto an austenitic stainless steel as sprayed, either after vacuum sealing with polymer impregnation or after vacuum furnace fusion. Erosion–corrosion was produced by the impact of a liquid jet of 3.5% NaCl solution at 18 °C at a velocity of 17 m/s, with or without 800 ppm silica sand. The study showed that sealing with polymer impregnation did not significantly modify the erosion–corrosion behavior of the sprayed coating. However, there was a significant improvement afforded by the post-fusion process in coating durability.

[Kim et al. (2001)] have sealed Al_2O_3–13 wt.% TiO_2 using three commercial polymeric sealants. The penetration depth of sealants into the ceramic coating was from 0.2 to 0.5 mm depending on the sealants used. All of the microhardness, two-body abrasive wear resistance, bond strength, and surface roughness of the ceramic coating after the sealing treatment were improved. Three-point bending stress of the ceramic coating after the sealing treatment was decreased. This was attributed to the reduced micro-crack toughening effect since the cracks propagate easily through the splats of the coating without crack deflection and/or branching after the sealing treatment.

[Ctibor et al. (2007)] studied unsealed and epoxy resin-sealed plasma-sprayed specimens prepared from alumina as well as alumina-13 wt. % titania. It was shown that the abrasion wear resistance of the epoxy resin-sealed coatings is significantly better than that of the as-sprayed coatings for both types of applied tests.

18.6.8.2 Inorganic Sealants

They are mostly used for high-temperature applications. Besides aluminum phosphates, sodium, and ethyl silicates, various sol–gel-type solutions and chromic acid have been used for sealing purposes. For example, phosphate acts as refractory glue that forms solid bridges over pores and cracks; it has also a beneficial effect in transforming the residual stress of the coating into compression. Sol–gel processes refer to the formation of a stable sol, the hydrolyzing of the sol to a gel, and the calcinating of the gel at elevated temperature to oxide. The sol includes a variety of metal alkoxides, nitrates, or hydroxides [Knuuttila et al. (1999a)].

One of the most popular sealants, especially for ceramics, is aluminum phosphate. [Leivo et al. (1997)] have studied plasma sprayed Al_2O_3 and Cr_2O_3 coatings sealed by aluminum phosphates. Phosphates were formed throughout the coating, down to the substrate. The sealing increased the hardness of the coatings by 200–300 Vickers hardness (HV) units. Abrasion and erosion wear resistances were increased by the sealing treatment. Sealing also substantially closed the open porosity, as shown in electrochemical corrosion tests.

[Ahmaniemi et al. (2002a, 2004)] have modified thick (1000 µm) thermal barrier coatings (TBC) ($25CeO_2$– $2.5Y_2O_3$–ZrO_2, $22MgO$–ZrO_2, $8Y_2O_3$–ZrO_2) for diesel engines with laser glazing and phosphate-based sealing treatments. In phosphate-sealed coatings, the penetration depth of the sealant was approximately 300–400 µm, and microhardness was at the range of 825–882 $HV_{0.3}$. The open porosity of the phosphate-sealed coatings was reduced by 24–48% depending on the coating material. The optimal thickness of the melted layer in laser-glazed coatings was 50–150 µm. In all cases, the melted zone was significantly densified, and a vertical macro-crack network was introduced.

The same authors, [Ahmaniemi et al. (2005)], have studied several modified thick thermal barrier coatings in three test series, in which the maximum coating temperature was fixed to 1000, 1150, and 1300 °C. The modified coating structures were all segmentation-cracked coatings, and some of these coatings were surface-sealed, using aluminum phosphate or a sol–gel-based sealant. Both sealing treatments reduced the thermal cycling resistance of the segmentation-cracked coatings. The decrease of the thermal cycling resistance of aluminum phosphate sealing was rather high, whereas in sol–gel sealing, it was moderate.

[Ahmaniemi et al. (2002b)] studied the residual stress induced by an aluminum phosphate sealing of plasma-sprayed alumina and chromia coatings. Sealing treatments were performed at 200, 300, and 400 °C with treatment times for these temperatures of 16, 4, and 2 h, respectively. Strengthening of alumina coating is mainly based on chemical reactions between the coating and the sealant, leading to bonding phosphate phases and obviously higher stiffness of the coating. The tensile stress state of the Al_2O_3 coating decreased during the sealing treatment, and the stress became even compressive as measured with the hole-drilling method. The aluminum phosphate sealing treatment temperature seemed to influence strongly the residual stress state of chromia coatings.

[Troczynski et al. (1999)] have sealed zirconia thermal barrier coatings using sol–gel alumina. The overall reduction of porosity was however small (from ~12 to ~11%), preserving the strain and thermal shock tolerance of the coatings. Burner rig tests showed an increase in sealed coating lifetime under thermomechanical fatigue conditions.

Electrodeposition of Ni has been used in an attempt to seal the open porosity of plasma-sprayed alumina coatings [Westergård and Hogmark (2004)]. It is a promising sealing treatment for sprayed alumina coatings intended for tribological applications. The hardness and coating cohesion increased, and the wear resistance was found to increase markedly in all wear tests as a result of the sealing.

Conventional Ni plating was successfully used [Niranatlumpong and Koiprasert (2006)] to seal the surface pores of an arc-sprayed coating. Electroplating of Ni sealed the pores and provided a smooth finish to the sample, enhancing the corrosion resistance of a Hastelloy coating.

The oxidation and hot-corrosion behavior of a Co–Ni–Cr–Al–Y coating produced by HVOF with and without an enamel coating were investigated [Xie et al. (2003)] in air at 900 °C and in molten salt 75 wt.% $NaCl$+25 wt.% Na_2SO_4 at 850 °C. The results showed that the enamel coating possessed excellent hot corrosion resistance in a molten salt, in comparison with the as-sprayed coating. In the hot-corrosion tests, breakaway corrosion did not occur on the samples with the enamel coating, and the composition of the enamel did not significantly change.

For automotive application lightweight so-called metal-supported solid oxide fuel cells (MS-SOFC) have been developed. As usual, these cells are serially stacked to achieve technically suitable voltage and power. Each cell is contained between two metal sheets, where these sheets of ferrite steel have to form a gas-tight cassette. But, to prevent electrical short-circuiting of the electrodes, both sheets have to be insulated from each other at the cassette rim. Gas tightness and electrical insulation at once represent a demanding task to solve, considering the strong operating conditions of about 800 °C and the desired lifetime. DLR-Stuttgart (Germany) has successfully developed a reliable sealing system [Henne and Zerfass (2005), Henne 2007; Arnold et al. 2008, 2011).

The main problem here was to have a ceramic which was at the same time insulating and could be soldered by means of a metal braze, without losing the insulating property. A patent was filed about the procedure consisting in spraying at vacuum conditions to one sheet a powder mixture of $MgAl_2O_4$ (spinel) and fine titanium hydride or another reactive metal hydride. The metal hydride decomposed within the hot plasma jet into the metal and together with the spinel a cermet arised. At high temperature brazing in the atmosphere, the metallic component turned into insulating oxide. Because of the high pyrophoric property of these reactive metals, the required fine fraction cannot be handled, directly, whereas the hydrides were less problematic.

[Caron et al. (2008)] have developed a stable hermetic seal composed of a ceramic matrix charged with glass particles, atmospheric plasma sprayed to obtain efficient air tightness between two SOFC cells. The developed seal was found to be solid, un-distortable, and adhesive to its support at ambient temperature. The sealing properties were acquired when the SOFC was put into service: the glassy phase migrated into the peculiar plasma-sprayed microstructure of the ceramic matrix toward the interface leading to the air tightness of the deposit.

Molten metals, electroplating, and enamel deposition are also occasionally used for sealing and strengthening purposes. The metal-coating wetting is a key issue. For example, zirconia is completely wetted by pure manganese [Knuuttila et al. (1999a)].

18.6.9 Spark Plasma Sintering

Spark plasma sintering (SPS), applied under compressive loading, is a new process for the densification of porous materials. Sparks are generated between the particles and cause the material to be heated rapidly.

[Khor et al. (2003)] have applied SPS for fabricating high-performance yttria stabilized zirconia (YSZ) electrolyte for SOFC starting from detached plasma-sprayed coating. SPS was performed at 1200, 1400, and 1500 °C, each sintering cycle having a holding time of 3 min. A microstructure transition occurred above 1200 °C, where the typical plasma-sprayed lamellar structure transformed to a granular-type structure, but the phase composition remained the same. The thermal conductivity of the as-sprayed YSZ was greatly increased (up to two times) by the SPS post-spray heat treatment, and there was an optimum of multiple SPS cycles for the enhancement of the thermal conductivity. The oxygen ion conduction resistivity was reduced drastically with an increase of the SPS temperature for multiple SPS cycles.

[Prawara et al. (2003)] also used SPS to densify a ZrO_2– 25 wt.% MgO flame-sprayed coating on a steel substrate. In the SPS process, the thermal spray-coated steel was exposed to rapid heating under vacuum at temperatures of 900, 950, and 1000 °C under compressive load for 10 min. If the as-sprayed coating had high porosities of up to 22%, the porosity fell to below 5% after SPS treatment. The bond strength and microhardness of the coating were also increased. These properties could be controlled by adjustment of the SPS temperature and the compressive load.

18.6.10 Peening or Rolling Densification

To close surface-connected pores and improve surface smoothness, shot peening or rolling can mechanically treat metallic-sprayed coatings [60]. This treatment resulted in pore closure and development of compressive residual stress. A few examples are given below.

[Zhou et al. (2007)] have shot peened, with different lengths of operating time, a CoCrAlY bond coat on a disk-shaped Ni-based superalloy. A moderate shot peening processing modified the surface of the bond coat. An EB-PVD (electron beam-physical vapor deposition) YSZ coating was deposited, onto the compacted bond coat, and after 3 cycles of shot peening, with a smooth surface, the coating exhibited a satisfactory thermal cycling lifetime of 1643 cycles at 1293 K.

The shot peening process increased the axial fatigue strength of AISI 4340 steel WC-10Ni HVOF sprayed by [Junior et al. (2010)]. The fatigue limit increased by 13.3% from 750 to 850 MPa.

[Kubiak et al. (2006)] investigated a shot peening treatment and a WC-17% Co coating applied to 30NiCrMo steel against 52,100 steel under plain fretting and complex fretting fatigue loading. Only WC-17% Co decreased the value of the coefficient of friction. Shot peening treatment had very little influence on fretting wear resistance under gross slip conditions and had no influence on the crack nucleation threshold but performed well against crack propagation under plain fretting and the studied fretting fatigue conditions.

[Marin de Camargo et al. (2007)] studied WC-17% Co HVOF-sprayed coatings on an Al 7050-T7451 alloy. The axial fatigue strength of the shot peened coating was improved in comparison to the Al 7050-T7451 substrate HVOF sprayed with WC-17% Co coating. The shot peening process could induce surface modification, such as surface roughening that, considering fatigue damage, accelerated nucleation and fatigue crack propagation.

[Costa et al. (2009)] studied the influence of a WC-10% Co–4% Cr coating deposited by HVOF on the fatigue strength of a Ti–6Al–4 V alloy. Comparison of the fatigue strength of the coated specimens and the base material showed a decrease when the parts were coated. They observed that the effect was more pronounced in high-cycle fatigue tests. The shot peening prior to the thermal spray coating was an efficient surface treatment to improve fatigue resistance of coated Ti-6Al-4 V.

Shot peening of Al wire arc-sprayed coatings was studied by [I'Anson et al. (1991)], the process being optimized, thanks to Taguchi fractional-factorial L4 design parametric study.

Results also depended on the peened layer thickness, which varied from 83 to 207 μm. The porosity of the un-peened Al layer ranged from 4.33 to 13.63%, while that of the peened coating layers varied from 0.16 to 0.83%. The average hardness of the coatings ranged from 61.8 to 74.6 $HV_{0.25N}$ for the peened coatings and 41.6 for the un-peened coating.

18.7 Safety and Environmental Hazards

As underlined by [Montavon and Gross (2004)]: "Safety of both research and production thermal spray facilities is a major interest of managers, workers, and also environmental, health, and safety regulating bodies around the world." Any activity, and of course thermal spraying, presents a certain level of risk that must be carefully assessed. For a safer work environment, a plan of action must be put in place to control unacceptable risks [Russo (2010)]. According to [Davis (2004)], "as of 2004, two safety guidelines have been prepared by the TSS Safety Guidelines Committee and are available as PDF files at the TSS Web Site 'Safety Guidelines for Risk Assessment' describes the necessary steps involved in identifying, documenting and implementing the necessary controls to minimize or eliminate the potential risk in Work Place. 'Safety Guideline of the Handling and Use of Gases in Thermal Spraying' provides practical recommendations for the safe installation, operation, and maintenance of gas equipment used in thermal spraying." An excellent review on "Health and Safety Equipment" is presented by Crawmer DE in the *Handbook of Thermal Spray Technology* by [Davis (2004)]. For more information, visit the TSS website, "Safety Guidelines."

18.7.1 Powders: Respiratory Problems and Explosions

Thermal spray processes involve powders with sizes in the following ranges:

- The *micro-size* (from a few μm to about 200 μm) corresponding to those sprayed
- The *sub-micro-size* resulting mainly from dusts formed by particle evaporation during spraying, thus consisting of oxide particles, or from particles splashing upon flattening

or from particles that after traveling in the jet fringes (poor penetration of the smallest injected ones) rebound on the coating
- *Nano-size* produced when nanometer-sized particles formed from suspensions or solutions travel in the jet fringes and, due to their too low velocity, follow the flow and do not impact on the substrate (Stokes effect)

However, powders used to spray are not the only ones to be considered; other activities linked to thermal spraying can also generate particles and/or fumes such as:

- Grit blasting generates fine particles of the grit medium and of the substrate, the worst case being a cloud of silicon dioxide particles.
- Loading and emptying grit-blast cabinets; loading, unloading, and cleaning hoppers; and cleaning operations within the thermal spray enclosure.
- Vapors from solvents can also be produced when used during the part-cleaning process.

18.7.1.1 Particles and the Pulmonary System

The main risk with particles is their inhalation [Hériaud-Kraemer et al. (2003), Petsas et al. (2007)] from which workers must be protected [Wuest et al.(n.d.)]. Depending on their size, particles penetrate in different regions of the tracheobronchial tree[(Hériaud-Kraemer et al. (2003)]:

- The large particles (average diameters ranging from 5 to 30 μm) are stopped in the pharynx.
- The medium particles (average diameters ranging from 1 to 5 μm) are deposited mostly in the trachea and the bronchi.
- The small particles (average diameters smaller than 1 μm) diffuse into the pulmonary alveoli, from which their harmful effect arises.

Problems could be even worst with nanometer-sized particles even if the properties driving such toxic responses are not yet well understood [Fadeel and Garcia-Bennett (2010), Schulte et al. (2010)].

All particles penetrating at different levels are partially eliminated by phenomena called pulmonary clearance mechanisms, and finally the pulmonary retention level is the deposition level minus the clearance level. Thus, pulmonary retention varies as a function of the particle size and the respiratory frequency. Whatever the mode of clearance, the harmful products operate chemically on the organism, which suffers from clamping; the products also operate physically by destroying specific molecular sequences [Hériaud-Kraemer et al. (2003)].

18.7.1.2 Toxicity of Powders

Powder toxicity is essentially of chemical nature. According to Hériaud-Kraemer et al. (2003), while numerous studies have been made dealing with powder toxicity in a wide range of industries, none has been specifically dedicated to powders commonly used in the thermal spray industry. The large volume of literature on the subject of powder toxicity is, however, most valuable for the setting up on case-by-case bases appropriate guideline for the safe use and handling of powders in thermal spray operations. The powders most susceptible to be dangerous are those containing nickel, cobalt, copper, aluminum, and chromium. Moreover, except aluminum, they are potential occupational carcinogens for humans. The study of [Chadwick et al. (1997)] has shown that levels of exposure to cobalt, chromium, and nickel were highest in plasma sprayers and on occasion exceeded UK Occupational Exposure Limits (OEL). Exposure to metals during detonation gun and electric arc spraying was better controlled, and levels remained below the relevant [Dobler and Gifford (2003)]. This was especially the case when spraying was performed manually or semi-automatically and where control relies on local exhaust ventilation (LEV) and personal respiratory protective equipment (PRPE) [Wuest et al. (n.d.), Chadwick et al. (1997)]. Legislations about the rules related to the exposure of workers to harmful products vary with countries and rely on indexes determining several admitted concentrations of a given product in the atmosphere for a given duration of exposure, permanent or not [Hériaud-Kraemer et al. (2003)].

18.7.1.3 Explosiveness of Powders

Numerous feedstock powders are in an unsteady thermodynamic state [Hériaud-Kraemer et al. (2003)]. An ignition source can induce the release of an enthalpy of reaction. Some powders can react explosively with water vapor or oxygen of the surrounding atmosphere; magnesium and zirconium are reactive with carbon dioxide, while aluminum and titanium are reactive with chlorine. The most critical factors [Hériaud-Kraemer et al. (2003)] are the concentration of particles in the atmosphere (the critical concentration depends on the powder type), their shapes and sizes (small irregular particles are more prone to explosion than large and spherical particles), temperature, pressure, hygrometry, and nature of the ignition source. That is why the local accumulation of powder particles in hazardous areas (spray booth, spray vacuum chamber, etc.) has to be limited by frequent cleanings and appropriate ventilation [Hériaud-Kraemer et al. (2003)].

The spontaneous inflammation of a powder in air at ambient temperature is defined as pyrophoricity, which differs from the autogenous ignition by the absence of temperature increase [Hériaud-Kraemer et al. (2003)]. All metallic powders, whatever their nature, except precious metals, are subjected to pyrophoricity when their sizes are sufficiently small (a few micrometers). Pyrophoricity depends on the size and shape of the particles, their specific surface (below $1 \text{ m}^2/\text{g}$, the potential risk increases drastically), and oxidation enthalpy.

18.7.2 Gases

However, if powders are probably the main risk, spray gases (Ar, N_2, H_2, methane, acetylene) and those resulting from the process (CO_2, CO, NO, etc.) also contribute to hazardous risks.

18.7.2.1 Gases Used for Spray Processes

Problems with gases used for spray processes are discussed below [Dobler and Gifford (2003), Linde catalogue, Creffield and Cole (1999), Heinrich et al. (1999)].

When nonreactive gases at room temperature (nitrogen, argon, and helium), which are inert, colorless, and tasteless, are trapped in a confined area, thus reducing the oxygen normal content (21 vol.%) below 19.5 vol.%, it may result successively in dizziness, unconsciousness, and finally death. When breathing air where oxygen is below 16 vol.%, a person is mentally incapable of diagnosing the situation. The stagnation of argon is more important because it is heavier than air, but that of nitrogen is almost the same, because its weight is close (only slightly lighter) to that of air. Helium, according to its weight, accumulates in the upper parts of the confined area.

Pure oxygen gas, used for combustion, is inert, colorless, and tasteless. It is a local irritant to mucous membranes and can result in destruction of the lung tissue. It lowers drastically the ignition point of flammable materials (an enrichment of a few % considerably increases the risk of fire) and supports and accelerates combustion. Oxygen reacts with most organic materials, especially hydrocarbons. Compounds like grease and oil may ignite spontaneously when subjected to oxygen-enriched atmosphere. Cloths saturated with oxygen are highly flammable and may be very easily ignited.

Combustible gases are flammable and susceptible of explosions. The main hazards with these gases are due to their flammability. Table 18.5 from [Creffield and Cole (1999)] summarizes the properties of fuel gases.

However, the ignition energy is also an important parameter to consider, that of hydrogen and acetylene being very low, as shown in Table 18.6.

Lastly, it must be kept in mind that when combustion occurs in a confined area, it can displace oxygen needed for breathing.

Table 18.5 Properties of fuel gases mostly used in spray processes [Linde catalogue] [Courtesy of Linde]

	Specific gravity (air =1)	Flammability vol. %		Auto-ignition T ($^\circ$C)	
		Air	O_2	Air	O_2
Hydrogen	0.0696	4–75	4–94	572	560
Acetylene	0.908	2–82	2–93	305	296
Methane	0.555	5–15	5–60	580	555
Propane	1.550	2.2–9.5	2.3–45	480	470
Propylene	1.476	2–10.5	2.1–53	460	

Table 18.6 Minimum ignition energies of combustible gases mainly used in spray processes. Reprinted with kind permission of Elsevier [Creffield and Cole (1999)]

Fuel gas	Ignition energy (mJ)
Hydrogen	0.02
Acetylene	0.02
Methane	0.29
Propane	0.25
Propylene	0.27

- *Hydrogen*, which is the lightest gas (twice less weight than helium), is colorless, tasteless, and odorless. It burns with an almost invisible flame, making it difficult to see.
- *Acetylene* (C_2H_2), lighter than air, with an odor close to that of garlic, is a very unstable gas that can easily decompose to the point of a deflagration being started (without the presence of air) that will rapidly progress into an explosive detonation. It is also highly explosive with copper, silver, and mercury as well as their salts and compounds.
- *Natural gas* (CH_4), lighter than air, has a faintly disagreeable odor, and breathing moderate concentrations can cause drowsiness and dizziness. Typically, an identifying odor is added to help in identifying any leakage.
- *Propane* (C_3H_8) heavier than air is a colorless gas, easily liquefied, with a faintly disagreeable odor, but, as for methane, identifying odor is added to help in identifying any leakage. It is mostly transported as liquid.
- *Propylene* (C_3H_6) heavier than air is a colorless gas, easily liquefied, with a faintly sweet odor. At atmospheric pressure and room temperature, propylene is nontoxic. Short-term overexposure to fumes may have anesthetic effects and cause dizziness, nausea, and dryness or irritation of the nose; longer-term exposure may cause liver damage.
- *Kerosene* is a clear, water-white liquid combustible with a mild characteristic odor. A low-energy spark (e.g., static electricity) can ignite very easily its vapor, and thus it must be kept away from sources of ignition.

18.7.2.2 Gases Resulting from the Spray Process

Gases such as nitrogen oxides (NO, NO_2) and phosgene ($COCl_2$) can be produced during the spray process. The nature of the respiratory attack is a function of the spot of action of these harmful substances. Nitrogen oxides and phosgene are insoluble substances penetrating very deeply into the tracheobronchial tree, deep enough to reach the alveoli. The symptoms observed in acute harmful effects range from irritation syndrome to pulmonary edema [Hériaud-Kraemer et al. (2003)].

Carbon oxide gases (CO and CO_2) are combustion products (as water vapor, which is harmless). CO_2, nonreactive gas heavier than air and corresponding to complete combustion of carbon, can cause asphyxiation and death in confined and poorly ventilated local. CO_2 ice is currently used to cool substrate and coating during the spray process, and the resulting CO_2 gas has the same effects as those resulting from combustion products. The story is worst with carbon monoxide CO (resulting from incomplete carbon combustion), which is colorless, odorless, and tasteless, but highly toxic. It combines with hemoglobin to produce carboxylhemoglobin, which is ineffective for delivering oxygen to bodily tissues. Concentrations as low as 667 ppm may cause up to 50% of the body's hemoglobin to convert to carboxylhemoglobin. A level of 50% carboxylhemoglobin may result in seizure, coma, and fatality. For example, in the USA, the OSHA limits long-term workplace exposure levels above 50 ppm.

18.7.2.3 Gas Storage

Five types of spray gas installations are used:

- Cryogenic liquid tanks (mainly argon, nitrogen, oxygen, and hydrogen)
- Pressurized liquid tanks (propane and propylene)
- Cylinders containing gas dissolved in solvent (acetylene)
- High-pressure cylinders (typically argon, nitrogen, oxygen, helium, and hydrogen) that can be interconnected (often 12 bottles)
- Refrigerated liquid tanks for carbon dioxide used to cool down substrate and coating during the spray process

The main risks are discussed next.

Fire and asphyxiation are by far the main ones.

Liquid or cold gas, especially oxygen, can cause severe frostbite to the eyes or skin; of course, it is the same with CO_2 ice. Although some local pain may be experienced initially, frozen tissues are painless, and a casualty might be unaware.

Breathing very cold vapor can cause damage to the lungs. Lastly, unprotected skin coming in contact with un-insulated items of cold equipment (e.g., pipe) may stick fast, and flesh may be torn on removal.

Short-term exposure to propylene fumes may have anesthetic effect.

18.7.3 Prevention and Safety Measures

It must be kept in mind that adequate ventilation can eliminate many health and safety concerns. One of the most important protections for workers is performing the spray process in spray booths, provided that, before entering the spray booth, after coating a part, a sufficient time elapsed until several air exchanges have been completed. In all other cases, either for spraying or cleaning or loading and emptying powders, workers must wear adequate protection [Wuest et al. (n.d.)].

18.7.3.1 Powders

Two cases must be considered: spraying in spray booth and manual spraying.

Automated spray booths are well-ventilated areas where no worker is present during spraying. They are connected to dry filtering systems, and the spray booth as well as the dry filtering system (typically every 3 months) is cleaned regularly. The efficiency of the filtering systems is also annually checked. Depressions are generated in the spray booths to keep possible leaks at the lowest possible level. Non-automated spray booths are connected to a wet filtering system.

In case of manual spraying, whether off-site or in a non-automated spray booth, the operator must use full personal protective equipment including full face masks and possibly dedicated breathing-air line depending on the ambient ail level of dust/fume contamination. When operated in the open air or well-ventilated space, it is possible to limit the personal protective equipment to radiation welding filters and efficient heppa breathing masks. This does not relieve the operator, however, from full compliance to local health and safety regulations.

Air-purifying respirators are typically negative-pressure units in which user's inhalation draws contaminated air in through a filtration medium and exhalation pushes air out through one-way valves back into the atmosphere [Wuest et al. (n.d.)]. Filtering cartridges can protect either from dust or solvent or chemical products. However, the replaceable filter cartridges must be regularly changed. They can be combined for removing suspended particles and organic vapors. Of course, the mask must properly fit and maintained. As a minimum [Howes (2001)], negative-

pressure half-mask respirator equipped with OV/P-100 filters should be used. OV stands for organic vapor, P stands for oil-proof, and the 100 stands for 99.97% efficient against solid or liquid particles, including oil-based particles. These are HEPA-class filters (high-efficiency particulate air) approved for protection against dust, fumes, and mists that have a time-weighted average of less than 0.05 mg per cubic meter or two million particles per cubic foot. These filters generally are color-coded magenta and gray with a black National Institute of Occupational Safety and Health (NIOSH) label. Filters whose elements are encased in a plastic or metal container offer more protection against the open flames and sparks generated by thermal spray processes than low-profile filters with cloth or paper exteriors. If low-profile filters are used, a spark shield should be installed over the element [Howes (2001)]. Lastly, when spraying nanometer-sized particles, filters must be adapted.

If necessary, an air-supplied respirator can be used where the worker wears a half- or full-face mask with breathable air supplied via a pressurized airline. The air must be supplied by a devoted compressor with its own piping system and never, for example, by the air to run the grit blaster [Wuest et al. (n.d.)].

To find related standards and documents, see the guidelines of [Wuest et al. (n.d.)]

18.7.3.2 Gases

For process gas supply, from gas storage to spay torches, through piping, pressure relief valves, vents, valves, regulators, flow meters, connections, and hoses, many rules have been defined for safety and are summarized in [Dobler and Gifford (2003)].

For gases generated during the spray process, again, good ventilation is the key issue.

Gas analyzers and combustible gas detectors are often incorporated in the design of the thermal equipment to provide information to unsafe conditions [Dobler and Gifford (2003)] when preset levels are attained. They can also be installed in gas storage local to detect leakage. In spray booth, they mostly detect un-proper concentration of oxygen and the presence of unacceptable percentages of combustible or other dangerous gas. They can be connected to computerized systems to shutoff valves, ventilation, and/or alarms. They must be disposed according to the weight of the gas relatively to that of air: close to the ground for heavy gases and close to the roof for light ones. Oxygen monitors are usually placed at face level just inside the entrance to spray booths in order to give readings at the operator-breathing zone.

The interested reader will find in the guidelines of Dobler and Gifford (2003) the standards, the gas characteristics and application-specific safety hazards/precautions, and a detailed description and all the recommendations to install safely:

- The bulk gas supplies installation, operation, and maintenance
- The connecting pipes (with especially the specific requirements for oxygen)
- The supply piping to the thermal spray cell
- The bulk supply operation and maintenance considerations
- Regulators, flow meters, and hoses
- Check valves, flashback arrestors, and excess flow valves
- Pressure relief valves and vents

18.7.4 Other Risks

18.7.4.1 Noise

The abrupt expansion of the high-temperature jets exiting the spray gun at atmospheric pressure (e.g., velocities ranging from 400 to 2800 m/s for plasma torches, velocities up to 1800 m/s for HVOF spraying) generates an acoustic level far higher than the maximal permitted level of 85 dBA [Creffield and Cole (1999)] as presented in Table 18.7.

In order to place such values relatively in relation to tolerable noise limit levels, the normal voice level is about 70 dBA, and automobiles on a highway have levels of about 100 dBA. Any exposure over 109 dBA for more than 2 min requires, according to US regulations, hearing protection, and 110 dBA or higher is the case encountered with most spray processes. Over 110 dBA, serious risk of permanent occupational deafness exists if no action is taken to protect the sprayers.

When spraying outside the booth, hearing protection mandates workers to wear either earplugs or earmuffs with the highest Noise Reduction Rating (NRR) available. In areas of very high noise, both earplug and earmuff must be used. For more details, see. Lastly, increasing the distance between the noise source and the workers reduces the sound level: theoretically by 10 dBA at 3 m and 28 at 27 m.

18.7.4.2 Radiation

For plasma spraying in particular, but less so for flame spraying, high-temperature jets generate radiation in a broad range of wavelengths from ultraviolet (UV) to visible and infrared (IR) [Hériaud-Kraemer et al. (2003)]. Plasma systems emitting radiation in the range between 280 and 220 nm are dangerous for the cornea of the eye that absorbs the UV from these regions easily, leading to a condition called flash burn after prolonged exposure. The severity of flash burn depends on the duration of exposure, UV wavelengths, and the energy level at which the luminance and radiance are produced during the process. The eyes can be damaged without discomfort during exposure [Hériaud-Kraemer et al. (2003)]. These UV can affect the exposed skin,

Table 18.7 Sound levels with different spray processes. Reprinted with kind permission from Springer Science Business Media [Hériaud-Kraemer et al. (2003), Gross (2002)], copyright © ASM International; Reprinted with kind permission from ASM [Gifford et al. (2003)]

Thermal spray device	Decibel (dBA)
Detonation gun	145
HVOF (liquid fuel)	133
HVOF (gaseous fuel)	125–135
HVAF	133
Wire flame spray	118–122
Powder flame spray	90–125
Rod flame spray gun	125
Wire arc	105–119
Air plasma spraying (APS)	110–125
Vacuum plasma spraying (VPS)	Ambient[a]
Cold spray	110 Ambient[b]
Water-stabilized plasma	125
RF plasma	Ambient[a]

[a]These processes are carried out in vacuum chambers, and noise from the gun is considerably attenuated. However, the continuous noise from pumps, fans, power supply, etc. is not negligible, and the personnel must be protected

[b]When the cold spray is performed in a closed enclosure in order to recycle gases, especially helium. The same remark can be made for the auxiliary equipment, as for VPS or RF spraying

causing sunburn, sun tanning, and changes in skin cell growth. Repeated exposure to UV may decrease skin elasticity. This can give the appearance of premature aging and can lead to a higher risk of skin cancer. Visible and near-IR radiation can also generate an energy density that reaches the retina. When exposed to a luminous source of excessive energy density without any protection, a subject can suffer from a retinal photo-traumatic lesion known as "arc stroke." It is thus important to install UV dark glass or shades over the windows of spray booths and enclosures. Operator must never view the plume of a spray gun without adequate eye protection.

For eye protection [Wuest et al. (n.d.)], safety glass with safety frames, vinyl-framed goggles of soft pliable body design, and welders' goggles are used, such equipment being available with incremental shades of filtration. Of course, lenses of eye protectors must be kept clean. Face shields should be worn to protect the eyes and face against flying particles, metal sparks, and chemical/biological splash. [Wuest et al. (n.d.)] present a chart for eye and face protection.

18.7.4.3 Thermal Risks

Plasma plumes or flames are directional jets; their thermal effect, caused by extreme energy density and very high temperatures, can be detected as far as 1 m past the exit of the guns [Hériaud-Kraemer et al. (2003)]. The components sprayed reach temperatures from 200 to 500 °C in some cases. Severe burns by contact with warm parts represent major risks encountered by the sprayers. Suitable gloves need be worn when hazards of burns and harmful temperature extremes are present [Wuest et al. (n.d.)]. Gloves must also be used for hazards from chemicals, cuts, lacerations, abrasions, punctures, and biological agents. Glove selection shall be based on performance characteristics of the gloves, conditions, duration of use, and hazards present.

18.7.4.4 Electric Risks

All thermal spray equipment relies to a varying degree on the availability of electric source with proper voltage and current requirements. Electric risks are especially important with DC plasma guns for which a high-intensity current and direct current source (a few hundreds of amperes, up to 1000 A) are used to generate the plasma jet. Grounding of electrical cabinets, booth walls, and other metallic structures must be provided. Electrical cabinets containing electrical equipment for the control of thermal spray installations must follow regulations. The electrical installation must have overload protection and ground fault interrupters, because most spray devices use water-cooling and leaks of coolant are common, especially during maintenance [Hériaud-Kraemer et al. (2003), Petsas et al. (2007)]. In general, the use of electrical gloves is recommended [Wuest et al.(n.d.)].

18.7.4.5 Risks Associated with the Use of Robots

A robot arm can move very fast, up to 2 m/s or even more, and it has no detection capability, resulting in an unavoidable collision with any object or person in its trajectory. That is why numerous national and international standards [Hériaud-Kraemer et al. (2003); Gifford et al. (2003)] apply when a robotic system is used. Numerous emergency stops must be located on the robot control system, on its power delivery system, and near the spray booth. When an anomaly is detected, the system abruptly stops. Anti-intrusion security systems on doors are frequently implemented in thermal spray booths. When the operator is in the booth for maintenance, the robot must be set to manual control (velocity lower than 0.25 m/s) and its control system handled by the operator. Main risks occur during maintenance or when the sprayer intervenes in the booth during production to solve a problem.

18.8   Summary and Conclusions

As presented in this chapter, thermal spray (TS) process integration is critical step that requires bringing together information from a wide range of sources and disciplines in order to achieve the ultimate objective of setting up an efficient, smooth, flawless, reliable, cost-effective, and safe industrial-scale operation. It requires in the first place a clear identification of the following key elements:

- *System objectives*, whether for R&D or industrial-scale production. For an R&D installation, flexibility and the ability to integrate different TS technologies are of prime importance. Emphasis is also to be placed on instrumentation and control for a better understanding of the processing conditions. On the other hand, for a production-scale operation, it is essential to identify in the first place the characteristics of the parts to be coated, in terms of size and material; the nature of the sprayed material whether metallic/alloy, ceramic, or cermet; and any constraints such as sensitivity to contamination/oxidation and the TS technology best adapted for the application.
- *Scale of operation* is particularly important in a production environment in terms of the number of parts to be treated per day or per year and possible and frequency of variations in terms of sprayed parts and materials to be sprayed.
- *Need for surface finishing and post-treatment*, nature of the treatment and necessary equipment.
- *Quality control*, what type of QC is needed and its extent in terms of percentage of sprayed parts to be checked.
- *Safety and environmental hazard*, which in turn depends on the TS technology used and the nature of the spray gases and spray material to be handled.

Each of these elements has an impact on the required:

- *Surface preparation* of the parts to be sprayed
- *Spray system design* involving the proper choice of the TS technology to be used and the sizing and disposition of the different system components in the "spray booth"
- *Auxiliary equipment* needed for supplying the spray material whether in the form of powder, wire, cord, solution, or suspension
- Supply of compressed gases and *electric power*
- *Auxiliary equipment* for the handling and cleaning of exhaust gases
- *Robotics* necessary for the handing of the part to be sprayed
- *System instrumentation and control* necessary for proper on-line diagnostics of key system performance parameters
- *Surface finishing and post-treatment* of the sprayed part
- *Quality control*, system to be adopted and necessary equipment
- *Adequate training* of operators and technical personal responsible for the running of the operation
- *Safety and environmental hazard* analysis

Nomenclature

Units are indicated in parentheses; when no units are indicated, the parameter is dimensionless.

Latin Alphabet

a	Sound velocity (m/s)
Bi	Biot number, $Bi = h\,L_c/\kappa_p$
c_{pw}	The specific heat of water (J/K kg)
C	Calibration constant
d_p	Particle diameter, (m)
f_o	Modulation frequency
h	Heat transfer coefficient (W/m^2 K)
H	Height of observation window (mm)
h_o	Specific enthalpy of the sampled gas (kJ/kg)
h_e	Specific enthalpy of the gas at reference temperature (25°C) (kJ/kg)
h_{fg}	Latent heat of fusion (J/kg)
h_p	Plasma enthalpy (J/kg)
I	Arc current (A)
$I_{\lambda i}$	Mean pulse height above background level at wavelength λ_i
L	Spray distance (m)
L_c	Characteristic length (m)
$\dot{m}_g$	Sampled gas mass flow rate (kg/s)
$\dot{m}_w$	Cooling water flow rate (kg/s)
M	Mach number, $M = v_f/a$
MVF	Molten volume ensemble, $MVF = \dfrac{1}{i_M} \displaystyle\sum_{i=1}^{N_{MVE}} \dfrac{4}{3}\pi\left(\dfrac{D_i}{2}\right)^3$
p_a	Ambient pressure surrounding the probe tip (Pa)
p_0	Stagnation pressure (Pa)
R	Ideal gas constant (8.32 J/K.mole)
Rm	Image magnification ratio (-)
Re	Reynolds number, $Re = \rho v d/\mu$
t_e	Mean exit time of the particle in the observation window (s)
t_r	Mean entry time of the particle in the observation window (s)
t_w	Mean time of flight across the observation window (s)
T_o	Local gas temperature (K)
T_f	Gas temperature near the in-flight particle (K)
T_p	Particle surface temperature (K)
T_m	Melting temperature (K)
T_t	Transition temperature (K)
T_u	Intensity of turbulence $T_u = \dfrac{v'}{\bar{v}} \times 100$ (%)
u_o	Local gas velocity (m/s)
v_f	Flow velocity (m/s)
v_p	Particle velocity (m/s)
$\bar{v}$	Mean velocity $\bar{v} = \dfrac{1}{n}\displaystyle\sum_{i=1}^{n} v_i$ (m/s)
v'	RMS of velocity fluctuation, $v' = \sqrt{\dfrac{1}{(n-1)}\displaystyle\sum_{i=1}^{i=n}(v_i - \bar{v})^2}$ (m/s)
We	Weber number, $We = \rho v^2 d/\sigma$
x_o	Equilibrium composition of sampled gas

Greek Alphabet

α_b	Half intersection angle ($^\circ$)
γ	Specific heat ratio, $\gamma = c_p/c_v$
δ_b	Inter-fringe spacing $\delta_b = \dfrac{\lambda_b}{2\sin\alpha_b}$ (mm)
Δt_m	Particle melting time (s)
Δt_{fl}	Particle in-flight time (s)
ΔT_s	Temperature difference under gas sampling condition ($^\circ$C)
ΔT_t	Temperature difference under "tare" condition ($^\circ$C)
ΔT_f	Cooling water temperature difference during sample flow (K)
ε_λ	Volumetric emission coefficient of the line at the wavelength λ (W/m^3.ster)
κ_p	Particle thermal conductivity (W/m K)
λ_i	Wavelength of the "i" observation, with $i = 1$ or 2
μ_p	Molecular viscosity of the molten particle (Pa/s).
ρ_o	Local gas density at the measurement point (kg/m^3)
ρ_f	Specific mass of the flow (kg/m^3)
ρ_p	Specific mass of the molten particle (kg/m^3)
σ_p	Surface tension (J/m^2)
σ_s	Probe sensitivity $\sigma_s = \dfrac{(\Delta T_s - \Delta T_t)}{\Delta T_s} \times 100$
τ_f	Time between two successive peaks in modulated signal

References

Abdel-Samad, A., E. Lugscheider, K. Bobzin, and M. Maes. 2006. The influence of hot isostatic pressing on plasma sprayed coatings properties. *Surface and Coating Technology* 201: 1224–1227.

Adrian, R.J. 1988. Double exposure, multiple-field particle image velocimetry for turbulent probability density. *Optics and Lasers in Engineering* 9: 211–228.

Ahmaniemi, S., J. Tuominen, P. Vuoristo, and T. Mäntylä. 2002a. Sealing procedures for thick thermal barrier coatings. *Journal of Thermal Spray Technology* 11 (3): 320–332.

Ahmaniemi, S., M. Vippola, P. Vuoristo, T. Mäntylä, M. Buchmann, and R. Gadow. 2002b. Residual stresses in aluminum phosphate sealed plasma sprayed oxide coatings and their effect on abrasive wear. *Wear* 252: 614–623.

Ahmaniemi, S., M. Vippola, P. Vuoristo, T. Mäntylä, F. Cernuschi, and L. Lutterotti. 2004. Modified thick thermal barrier coatings: Microstructural characterization. *Journal of the European Ceramic Society* 24: 2247–2258.

Ahmaniemi, S., P. Vuoristo, T. Mäntylä, C. Gualco, A. Bonadei, and R. Di. 2005. Thermal cycling resistance of modified thick thermal barrier coatings. *Surface and Coating Technology* 190: 378–387.

Ahmed, N., M.S. Bakare, D.G. McCartney, and K.T. Voisey. 2010. The effects of microstructural features on the performance gap in corrosion resistance between bulk and HVOF sprayed Inconel 625. *Surface and Coating Technology* 204: 2294–2301.

Alekseenko, S.V., V.M. Dulin, Y.S. Kozorezov, D.M. Markovich, S.I. Shtork, and M.P. Tokarev. 2011. Flow structure of swirling turbulent propane flames. *Flow, Turbulence and Combustion* 87: 569–595.

Antou, G., F. Hlawka, A. Cornet, G. Montavon, C. Coddet, and F. Machi. 2004. Processing of yttria partially stabilized zirconia thermal barrier coatings implementing a high-power laser diode. *Journal of Thermal Spray Technology* 13 (3): 381–389.

Arnold, J., S.A. Ansar, U. Maier, and R. Henne. 2008. Insulating and sealing of SOFC devices by plasma spraying ceramic layers. In International thermal spray conference & exposition, thermal spray 2008: crossing borders (DVS-ASM), Maastricht, The Netherlands, 2–4 June 2008, pp 83–87.

Arnold, J., S.A. Ansar, R. Ruckdäschel, R. Henne, U. Maier, and H.W. Grünling. 2011. Vacuum plasma sprayed insulating layers suitable for brazing in high temperature fuel cells. In Abstract and manuscript of the ITSC- 2011 in Hamburg, no. 276, 27–29 Sept 2011, pp 143–148.

Asmann, M., A. Wank, H. Kim, J. Heberlein, and E. Pfender. 2001. Characterization of the converging jet region in a triple torch plasma reactor. *Plasma Chemistry and Plasma Processing* 21 (1): 37–63.

AWS. 1985. *Thermal spraying: Practice, theory and application.* Miami: American Welding Society.

Blais, A., B. Jodoin, J.-L. Dorier, C. Gindrat, and C. Hollenstein. 2005. Inclusion of aerodynamic non-equilibrium effects in supersonic plasma jet enthalpy probe measurements. *Journal of Thermal Spray Technology* 14 (3): 342–353.

Bonnet, R., R. Bolot, and C. Coddet. 1999. Simulation of the thermal spray process and off-line programming. In *United thermal spray conference*, ed. E. Lugscheider and P.A. Kammer, 519–526. Düsseldorf: DVS.

Boulos, M.I. 2001. Visualization and diagnostics of thermal plasma flows. *Journal of Visualization* 4 (1): 19–28.

Boulos, M.I., and Jurewicz, J. High Performance Induction Plasma Torch with a Water-Cooled Ceramic Confinement Tube. International Patent Application, PCT/CA92/00156 (April 10, 1992), Canadian Patent, 2,085,133 (April 10, 1992), US Patent, 5,200,595 (April 6, 1993), Chinese Patent, ZL92103380.X (April 11, 1996), European Patent, 533884 (January 22, 1997), Korean Patent, 203,994 (March 25th 1999), Japanese Patent, 3,169,962 (March16th 2001),

Boulos, M.P., P. Fauchais, and E. Pfender. 1994. *Thermal plasmas fundamentals and applications.* Vol. 1. New York: Plenum.

Bouyer, E., F. Gitzhofer, and M.I. Boulos. 1997. The suspension plasma spraying of bioceramics by induction plasma. *JOM* 49: 58–62.

Brossa, M., and E. Pfender. 1988. Probe measurements in thermal plasma jets. *Plasma Chemistry and Plasma Processing* 8 (1): 75–90.

Capetti, A., and E. Pfender. 1989. Probe measurements in argon plasma jets operated in ambient argon. *Plasma Chemistry and Plasma Processing* 9 (2): 329–341.

Candel, A., and R. Gadow. 2009. Trajectory generation and coupled numerical simulation for thermal spraying applications on complex geometries. *Journal of Thermal Spray Technology* 18 (5–6): 981–987.

Carlson, R.R. 2005. *Single wire arc spray.* PhD Thesis, University of Minnesota.

Carlson, R.R., and J.V.R. Heberlein. 2001. Effects of operating parameters on high definition single wire arc spraying. In *Proceedings of the ITSC-2001, Singapore*, ed. C.C. Berndt, K.A. Khor, and E. Lugscheider, 447–453. Materials Park: ASM International.

———, N.A. Hussary, and K. Shi. 2000. High-definition single-wire-arc-spray. In *Proceedings of the 1st ITSC-2000, Montreal, Quebec*, ed. C.C. Berndt, 709–716. Materials Park: ASM International.

Caron, N., L. Bianchi, and S. Méthout. 2008. Development of a functional sealing layer for SOFC applications. *Journal of Thermal Spray Technology* 17 (5–6): 598–602.

Chadwick, J.K., H.K. Wilson, and M.A. White. 1997. An investigation of occupational metal exposure in thermal spraying processes. *Science of the Total Environment* 199: 115–124.

Coddet, C., and T. Marchione. 1993. Procédé de Préparation et de Revêtement de Surface et Dispositif pour la Mise en Oeuvre Dudit Procédé (Process for the Preparation and Coating of a Surface and Apparatus for Practicing Said Process), French Patent FR9209277, 21 July 1993.

Coddet, C., G. Montavon, S. Ayrault-Costil, O. Freneaux, F. Rigolet, G. Barbezat, F. Foliot, A. Diard, and P. Wazen. 1999. Surface preparation and thermal spray in a single step: The PROTAL process – Example of application for an aluminium-base substrate. *Journal of Thermal Spray Technology* 8 (2): 235–242.

Colmenares-Angulo, J., K. Shinoda, T. Wentz, W. Zhang, Y. Tan, and S. Sampath. 2011. On the response of different particle state sensors to deliberate process variations. *Journal of Thermal Spray Technology* 20 (5): 1035–1048.

Costa, M.Y.P., M.L.R. Venditti, H.J.C. Voorwald, M.O.H. Cioffi, and T.G. Cruz. 2009. Effect of WC–10%Co–4%Cr coating on the Ti–6Al–4V alloy fatigue strength. *Materials Science and Engineering A* 507: 29–36.

Cotler, E.M., D. Chen, and R.J. Molz. 2011. Pressure-based liquid feed system for suspension plasma spray coatings. *Journal of Thermal Spray Technology* 20 (4): 967–973.

Coudert, J.F., P. Fauchais, and M. Vardelle. 2002. Diagnostics of plasma spray process and derived on-line control. *High Temperature Material Processes* 6 (2): 247–265.

Coulombe, S., M.I. Boulos, and T. Sakuta. 1995. Simultaneous particle surface temperature and velocity measurements under plasma conditions. *Journal of Measurement Science and Technology* 6: 383–390.

Crawmer, D.E., R.L. Bartoe, and J. Kramer. 1987. Technical note: Improved universal powder mass flow control for thermal spray applications. *Surface and Coating Technology* 33: 353–365.

Creffield, G., and M. Cole. 1999. Safe working practices with thermal spray gases. In *United thermal spray conference*, ed. E. Lugscheider and P.A. Kammer, 397–401. Düsseldorf: DVS.

Ctibor, P., K. Neufuss, F. Zahalka, and B. Kolman. 2007. Plasma sprayed ceramic coatings without and with epoxy resin sealing treatment and their wear resistance. *Wear* 262: 1274–1280.

Czernichowski, A., L. Pawlowski, and B. Nitoumbi. 2000. Injection of hot particles in the plasma flame. *Journal of Thermal Spray Technology* 9 (4): 458–462.

Dallaire, S., D. Dubé, and M. Fiset. 1999. Laser melting of plasma-sprayed copper–ceramic coatings for improved erosion resistance. *Wear* 231: 102–107.

Davis, J.R. 2004. *Handbook of thermal spray technology.* Materials Park: ASM International.

de Camargo, J.A.M., H.J. Cornelis, V.M.O.H. Cioffi, and M.Y.P. Costa. 2007. Coating residual stress effects on fatigue performance of 7050-T7451 aluminum alloy. *Surface and Coating Technology* 201: 9448–9455.

Deng, SiHao, ZhenHua Cai, DanDan Fang, HanLin Liao, and Ghislain Montavon. 2012. Application of robot offline programming in thermal spraying. *Surface and Coating Technology* 206 (19–20): 3875–3882.

Dobler, K., and D. Gifford 2003. Safety guidelines for the handling and use of gases in thermal spraying. Prepared by the ASM-TSS Safety Committee. TSS, ASM International, Materials Park, OH, 27 p.

Dorier, J.L., C. Hollenstein, A. Salito, M. Loch, and G. Barbezat. 2000. Influence of external parameters on arc fluctuations in a F4 DC plasma torch used for thermal spraying. In *Proceedings of the 1st ITSC-2000, Montreal, Quebec, Canada*, ed. C.C. Berndt, 37–43. Materials Park: ASM International.

Dorier, J.L., M. Gindrat, C. Hollenstein, A. Salito, M. Loch, and G. Barbezat. 2001. Time-resolved imaging of anodic arc root behavior during fluctuations of a DC plasma spraying torch. *IEEE Transactions on Plasma Science* 29 (3): 494–501.

Dorier J.L., B. Jodoin, M. Gindrat, A. Blais, C. Hollenstein, and G. Barbezat. 2003. A Novel Approach to Interpret Enthalpy Probe Measurements in Low Pressure Supersonic Plasma Jets, 16th. International Symposium on Plasma Chemistry, Taormina, Italy, 22–27 June 2003.

Doubenskaia, M., P. Bertrand, and I. Smurov. 2006. Pyrometry in laser surface treatment. *Surface and Coating Technology* 201: 1955–1961.

Dwivedi, G., T. Wentz, S. Sampath, and T. Nakamura. 2010. Assessing process and coating reliability through monitoring of process and design relevant coating properties. *Journal of Thermal Spray Technology* 19 (4): 695–712.

Dyshlovenko, S., L. Pawlowski, I. Smurov, and V. Veiko. 2006. Pulsed laser modification of plasma-sprayed coatings: Experimental processing of hydroxyapatite and numerical simulation. *Surface and Coating Technology* 201: 2248–2255.

Eckbreth, A.C., and T.J. Anderson. 1985. Dual broadband CARS for simultaneous multiple species measurements. *Applied Optics* 24: 2731–2736.

————. 1986. Simultaneous rotational coherent anti-Stokes Raman spectroscopy with arbitrary pump-Stokes spectral separation. *Optics Letters* 11: 496–498.

Eidelman, S., and X. Yang. 1997. Three-dimensional simulation of HVOF spray deposition of nanoscale materials. *Nanostructured Materials* 9 (1–8): 79–84.

Etchart-Salas, R., V. Rat, J.F. Coudert, P. Fauchais, N. Caron, K. Wittman, and S. Alexandre. 2007. Influence of plasma instabilities in ceramic suspension plasma spraying. *Journal of Thermal Spray Technology* 16 (5–6): 857–865.

Fadeel, B., and A.E. Garcia-Bennett. 2010. Better safe than sorry: Understanding the toxicological properties of inorganic nanoparticles manufactured for biomedical applications. *Advanced Drug Delivery Reviews* 62: 362–374.

Fan, X., F. Gitzhofer, and M.I. Boulos. 1998. Investigation of alumina splats formed in the induction plasma process. *Journal of Thermal Spray Technology* 7 (2): 197–204.

Fang, J.C., W.J. Xu, Z.Y. Zhao, and H.P. Zeng. 2007. In-flight behaviors of ZrO2 particle in plasma spraying. *Surface and Coating Technology* 201: 5671–5675.

Fasching, M.M., F.B. Prinz, and L.E. Weiss. 1993. Planning robotic trajectories for thermal spray shape deposition. *Journal of Thermal Spray Technology* 2 (1): 45–57.

Fauchais, P., and J.-F. Coudert. 1996. Mesures de température dans les plasmas thermiques. *Revue Générale de Thermique* 35: 324–337.

Fauchais, P., and A. Vardelle. 2000. Heat, mass and momentum transfer in coating formation by plasma spraying. *International Journal of Thermal Sciences* 39: 852–870.

Fauchais, P., and A. Vardelle. 2011. Innovative and emerging processes in plasma spraying: From micro- to nano-structured coatings. *Journal of Physics D: Applied Physics* 44: 194011.

Fauchais, P., J.F. Coudert, and M. Vardelle. 1989. Diagnostics in thermal plasma processing. In *Plasma diagnostics*, ed. O. Auciello and D.L. Flamm, vol. 1, 349–446. New York: Academic.

Fauchais, P., M. Vardelle, A. Vardelle, L. Bianchi, and A.C. Leger. 1996. Parameters controlling the generation and properties of plasma sprayed zirconia coatings. *Plasma Chemistry and Plasma Processing* 16: S99–S126.

Fauchais, P., M. Fukumoto, A. Vardelle, and M. Vardelle. 2004. Knowledge concerning splat formation: An invited review. *Journal of Thermal Spray Technology* 13 (3): 337–360.

Fauchais, P., G. Montavon, R. Lima, and B. Marple. 2011. Engineering a new class of thermal spray nano-based microstructures from agglomerated nanostructured particles, suspensions and solutions: An invited review. *Journal of Physics D: Applied Physics* 44: 093001 (53pp).

Filkova, I., and P. Cedik. 1984. Nozzle atomization in spray drying. In *Advances drying*, ed. A.S. Mujumdar, vol. 3, 181–215. New York: Hemisphere Publishing Corporation.

Fincke, J.R., S.C. Snyder, and W.D. Swank. 1993. Comparison of enthalpy probe and laser light scattering measurement of thermal plasma temperature and velocities. *The Review of Scientific Instruments* 64 (2,3): 711–718.

Fincke, J.R., C.H. Chang, W.D. Swank, and D.C. Haggard. 1994. Entrainment and demixing in subsonic thermal plasma jets: Comparison of measurements and predictions. *International Journal of Heat and Mass Transfer* 37: 1673–1682.

Fincke, J.R., D.C. Haggard, and W.D. Swank. 2001a. Particle temperature measurement in the thermal spray process. *Journal of Thermal Spray Technology* 10 (2): 255–266.

————. 2001b. Particle temperature measurement in the thermal spray process. *Journal of Thermal Spray Technology* 10 (2): 255–266.

Fincke, J.R., D.M. Crawford, S.C. Snyder, W.D. Swank, D.C. Haggard, and R.L. Williamson. 2003. Entrainment in high-velocity, high-temperature plasma jets. Part I: Experimental results. *International Journal of Heat and Mass Transfer* 46: 4201–4213.

Floristán, M., J.A. Montesinos, J.A. García-Marín, A. Killinger, and R. Gadow. 2012. Robot trajectory planning for high quality thermal spray coating processes on complex shaped components. In ITSC 2012. ASM International, Materials Park, OH, e-proceedings.

Freslon, A. 1995. Plasma spraying at controlled temperature and atmosphere. In *Thermal spray: Science and technology*, ed. C.C. Berndt and S. Sampath, 57–63. Materials Park: ASM International.

Friedrich, C.J., R. Gadow, A. Killinger, and C. Li. 2001. IR thermographic imaging-a powerful tool for on-line process control of thermal spraying. In *Thermal spray 2001: New surfaces for a new millennium*, ed. C.C. Berndt, K.A. Khor, and E.F. Lugscheider, 779–786. Materials Park: ASM International.

Fukumoto, M., I. Ohgitani, H. Nagai, and T. Yasni. 2005. Effect of substrate surface change by heating on flattening behavior of thermal sprayed particles. In *ITSC-2005*, ed. E. Lugscheider. Düsseldorf: DVS, e-proceedings.

Fukumoto, M., H. Nagui, and T. Yasui. 2006. Influence of surface character change of substrate due to heating on flattening behavior of thermal sprayed particles. In *ITSC-2006*, ed. B. Marple et al. Materials Park: ASM International.

Fukumoto, M., H. Wada, K. Tanabe, M. Yamada, E. Yamaguchi, A. Niwa, M. Sugimoto, and M. Izawa. 2007a. Effect of substrate temperature on deposition behavior of copper particles on substrate surfaces in the cold spray process. *Journal of Thermal Spray Technology* 16 (5–6): 643–650.

Fukumoto, M., T. Yamaguchi, M. Yamada, and T. Yasui. 2007b. Splash splat to disk splat transition behavior in plasma-sprayed metallic materials. *Journal of Thermal Spray Technology* 16 (5–6): 905–912.

Fukumoto, M., K. Yang, K. Tanaka, T. Usami, T. Yasui, and M. Yamada. 2011. Effect of substrate temperature and ambient pressure on heat transfer at interface between molten droplet and substrate surface. *Journal of Thermal Spray Technology* 20 (1–2): 48–58.

Garcia-Alonso, D., N. Serres, C. Demian, S. Costil, C. Langlade, and C. Coddet. 2011. Pre-/during-/post-laser processes to enhance the adhesion and mechanical properties of thermal-sprayed coatings with a reduced environmental impact. *Journal of Thermal Spray Technology* 20 (4): 719–735.

Gevelber, M., C. Cui, B. Vattiat, Z. Fieldman, D. Wroblewski, and S. Basu. 2005. Real time control for plasma spray: Sensor issues, torch nonlinearites, and control of coating thickness. In *ITSC 2005 proceedings*, ed. E. Lugsheider. Düsseldorf: DVS, e-proceedings.

Gevelber, M., D. Wroblewski, B. Vattiat, O. Ghosh, M. VanHout, and S.N. Basu. 2008. Issues and requirements for developing a plasma spray deposition rate sensor for real-time control. In *Thermal spray 2008: Thermal spray crossing borders*, ed. E. Lugscheider, 912–916. Düsseldorf: DVS.

Gifford, D.J., L. Pollard, G. Wuest, and R.C. Fletcher. 2003. Thermal spray booth design guidelines. Prepared by the ASM-TSS safety committee. ASM International, Materials Park, OH, 47p.

Gonzalez, R., M. Cadenas, R. Fernandez, J.L. Cortizo, and E. Rodriguez. 2007. Wear behaviour of flame sprayed NiCrBSi coating remelted by flame or by laser. *Wear* 262: 301–307.

Grey, J., P.F. Jacobs, and M.P. Sherman. 1962. Calorimetric probe for the measurement of extremely high temperatures. *The Review of Scientific Instruments* 33 (7): 738–741.

Gross, K.A. 2002. Noise emission in thermal spray operations. *Journal of Thermal Spray Technology* 11 (3): 350–358.

Guessasma, S., F.-I. Trifa, G. Montavon, and C. Coddet. 2004. Al2O3–13% weight TiO2 deposit profiles as a function of the atmospheric plasma spraying processing parameters. *Materials and Design* 25 (4): 307–315.

Hale, D.H., K. Dobler, and D. Gifford. 2002. Safety hazards associated with the usage of compressed gases in thermal spraying. In ITSC 2002. DVS, Düsseldorf, Germany, pp 242–246.

Hall, R.J., and A.C. Eckbreth. 1984. CARS: Application to combustion diagnostics. In *Laser application*, ed. J.F. Ready and R.K. Eaf, vol. 5. New York: Academic.

Heinrich, P., C. Penszior, and H. Meinass. 1999. Industrial gases for thermal spraying-from production to application, with the required purity. In *UTSC-1999*, ed. E. Lugscheider and P.A. Kammer, 95–105. Düsseldorf: DVS.

Henne, R. 2007. Solid oxide fuel cells: A challenge for plasma deposition processes. *Journal of Thermal Spray Technology* 16 (3): 381–403.

Henne, R., and W. Weber 1982. Plasmaspritzen im Unterdruck. Ein Verfahren zur Herstellung von dichten Schutzschichten für hochschmelzende Metalle; Proc. 10. Plansee-Seminar, Reutte, A, June 1–5, (1981), which appeared also in High temperature – High pressure, vol 14, pp 237–244.

Henne, R., W. Mayr, and A. Reusch. 1993. Influence of nozzle geometry on particle behavior and coating quality in high velocity VPS. In *Proceedings of the thermal spray conference TS93, Aachen, 7–11*. Düsseldorf: DVS.

Henne, R., W. Schnurnberger, and W. Weber 1983. Low pressure plasma spraying – An interesting method to produce electrodes for water electrolysis. In ITSC 83, 10th international thermal spraying conference, Essen, 2–6 May.

Henne, R., V.M. Bradke, W. Schnurnberger, and W. Weber. 1986. Development and manufacture of electrolyzer components – Applying plasma spraying under reduced pressure. In Proceedings of the 11th international thermal spraying conference, Montreal, Canada, 8–12 Sept 1986. Pergamon Press.

Henne, R, H.R. Zerfass, and J. Arnold. 2005. *Sealing arrangement for a fuel cell stack and process for the production of such a sealing arrangement*. Patent Application DEA 102005045053, 21.09.05

Hériaud-Kraemer, H., G. Montavon, S. Hertert, H. Robin, and C. Coddet. 2003. Harmful risks for workers in thermal spraying: A review completed by a survey in a French Company. *Journal of Thermal Spray Technology* 12 (4): 542–554.

Hlina, J., and J. Sonsky. 2010. Time-resolved tomographic measurements of temperature in thermal plasma jet. *Journal of Physics D Applied Physics* 43: 055202.

Howes, C.P. 2001. Thermal spray safety and OSHA compliance, protecting operators from ultraviolet light, fumes, dust, compressed air, gases, Fabricator Magazine July 12.

Hussary, N.A. 1999. *Fluid dynamic investigations of wire arc spraying process*, MS Thesis, University of Minnesota, Minneapolis.

I'Anson, K., W.L. Riggs, G. Irons, T.J. Steeper, and D.J. Varacalle Jr. 1991. Use of shot peening process for surface modification of thermal spray coatings. In *4th NTSC-1991 proceedings*, ed. T.F. Bernecki, 139–145. Materials Park: ASM International.

Irissou, E., J.-G. Legoux, B. Arsenault, and C. Moreau. 2007. Investigation of Al-Al2O3 cold spray coating formation and properties. *Journal of Thermal Spray Technology* 16 (5–6): 661–668.

Itoh, A., K. Takeda, M. Itoh, and M. Koga. 1991. Pretreatment of substrates by using reversed transferred arc in low pressure plasma spray. In *Thermal spray research and application*, ed. F. Berneki, 245–251. Materials Park: ASM International.

Jäger, D.A., D. Stöver, and W. Schlump. 1992. High pressure plasma spraying in controlled atmosphere up to 2 bars. In *Thermal spray: International advances in coatings technology*, ed. C.C. Berndt, 69–75. Materials Park: ASM International.

Jodoin, B., F. Raletz, and M. Vardelle. 2006. Cold spray modeling and validation using an optical diagnostic method. *Surface and Coating Technology* 200 (14–15): 4424–4432.

Junior, G.S., H.J.C. Voorwald, L.F.S. Vieira, M.O.H. Cioffi, and R.G. Bonora. 2010. Evaluation of WC-10Ni thermal spray coating with shot peening on the fatigue strength of AISI 4340 steel. *Procedia Engineering* 2: 649–656.

Kadyrov, E., and V. Kadyrov. 1995. Gas dynamical parameters of detonation powder spraying. *Journal of Thermal Spray Technology* 4 (3): 280–286.

Kearney, S.P., R.P. Lucht, and A.M. Jacobi. 1999. Temperature measurements in convective heat transfer flows using dual-broadband, pure-rotational coherent anti-Stokes Raman spectroscopy (CARS). *Experimental Thermal and Fluid Science* 19: 13–26.

Khor, K.A., and N.L. Loh. 1994. Hot isostatic pressing of plasma sprayed Ni-base alloys. *Journal of Thermal Spray Technology* 3 (1): 57–62.

Khor, K.A., L.-G. Yu, S.H. Chan, and X.J. Chen. 2003. Densification of plasma sprayed YSZ electrolytes by spark plasma sintering (SPS). *Journal of the European Ceramic Society* 23: 1855–1863.

Kim, H.-J., C.-H. Lee, and Y.-G. Kweon. 2001. The effects of sealing on the mechanical properties of the plasma-sprayed alumina-titania coating. *Surface and Coating Technology* 139: 75–80.

Knuuttila, J., P. Sorsa, and T. Mäntylä. 1999a. Sealing of thermal spray coatings by impregnation. *Journal of Thermal Spray Technology* 8 (2): 251–257.

Knuuttila, J., P. Sorsa, T. Mäntylä, J. Knuuttila, and P. Sorsa. 1999b. Sealing of thermal spray coatings by impregnation. *Journal of Thermal Spray Technology* 8 (2): 249–250.

Kroemmer, W., and P. Heinrich. 2006. Influence of industrial gases on the thermal spray process. In ITSC 2006. e-proceedings, ed. B. Marple et al. Materials Park: ASM International.

Kubiak, K., S. Fouvry, A.M. Marechal, and J.M. Vernet. 2006. Behavior of shot peening combined with WC-Co HVOF coating under complex fretting wear and fretting fatigue loading conditions. *Surface and Coating Technology* 201: 4323–4328.

Kutay, A., and L.E. Weiss. 1992. Economic analysis of robotic operations: A case study of a thermal spraying robot. *Robotics and Computer-Integrated Manufacturing* 9 (3): 279–287.

Lapierre, D., R.J. Kearney, M. Vardelle, A. Vardelle, and P. Fauchais. 1994. Effect of a substrate on the temperature distribution in an argon-hydrogen thermal plasma jet. *Plasma Chemistry and Plasma Processing* 14 (4): 407–423.

Landes, K. 2006. Diagnostics in plasma spraying techniques. *Surface Coating &Technology* 201: 1948–1954.

Leblanc, L., and C. Moreau. 2002. The long-term stability of plasma spraying. *Journal of Thermal Spray Technology* 11 (3): 380–386.

Legoux, J.-G., B. Arsenault, L. Leblanc, V. Bouyer, and C. Moreau. 2002. Evaluation of four high velocity thermal spray guns using WC-10% co-4% Cr cermets. *Journal of Thermal Spray Technology* 11 (1): 86–94.

Legoux, J.G., E. Irissou, and C. Moreau. 2007. Effect of substrate temperature on the formation mechanism of cold-sprayed aluminum, zinc and tin coatings. *Journal of Thermal Spray Technology* 16 (5–6): 619–626.

Leivo, E.M., M.S. Vippola, P.P.A. Sorsa, P.M.J. Vuoristo, and T.A. Mäntylä. 1997. Wear and corrosion properties of plasma sprayed Al_2O_3 and Cr_2O_3 coatings sealed by aluminum phosphates. *Journal of Thermal Spray Technology* 6 (2): 205–210.

Lenling, W.J., M.F. Smith, and J.A. Henfling. 1991. Beneficial effects of austempering post-treatment on tungsten carbide-based wear coatings. In *Thermal spray research and applications*, ed. T.F. Bernecki, 227–232. Materials Park: ASM International.

Lesinki, J., and M.I. Boulos. 1988a. A laser Doppler anemometry under plasma conditions – Part I: Measurements in a DC plasma jet. *Journal of Plasma Chemistry and Plasma Processing* 8: 113–132.

———. 1988b. B laser Doppler anemometry under plasma conditions – Part II: Measurements in an inductively coupled RF plasma. *Journal of Plasma Chemistry and Plasma Processing* 8: 133–144.

Linde catalogue., Acetylene… there is no better fuel gas for Oxy-fuel gas processes. www.linde-gas.com/.

Lugscheider, E., F. Ladru, A. Fischer, and C. Herbst. 1998. Plasma sprayed ceramic coatings for electrical purposes—Necessity of process control. In *Proceedings of the 24th annual conference of the IEEE industrial electronics society, Aachen, Germany*, 2284–2289.

Malkin, S., and C. Guo. 2008. *Grinding technology: Theory and applications of machining with abrasives*. New York: Industrial Press.

Marchand, O., L. Girardot, M.P. Planche, P. Bertrand, Y. Bailly, and G. Bertrand. 2011. An insight into suspension plasma spray: Injection of the suspension and its interaction with the plasma flow. *Journal ofThermal Spray Technology* 20 (6): 1310–1320.

Marantz, D.R., and H. Herman. 1991. *Plasma generating apparatus and method*, US Patent 4982067.

———. 1992. *Plasma spray gun and method of use*, US Patent 5144110.

Marple, B.R., R.S. Lima, C. Moreau, S.E. Kruger, L. Xie, and M.R. Dorfman. 2007. Yttria-stabilized zirconia thermal barriers sprayed using N2-H2 and Ar-H2 plasmas: Influence of processing and heat treatment on coating properties. *Journal of Thermal Spray Technology* 16 (5–6): 791–797.

Mauer, G., R. Vaßen, and D. Stover. 2007. Comparison and applications of DPV-2000 and Accuraspray-g3 diagnostic systems. *Journal of Thermal Spray Technology* 16 (3): 414–424.

Mauer, G., R. Vaßen, D. Stöver, S. Kirner, J.-L. Marqués, S. Zimmermann, G. Forster, and J. Schein. 2010. Improving powder injection in plasma spraying by optical diagnostics of the plasma and particle characterization. In *Thermal spray: global solutions for future applications, 3–5 May 2010, Singapore, DVS-Berichte*, vol. 264, 525–530. Düsseldorf: DVS Media.

Mauer, G., R. Vaßen, and D. Stöver. 2011a. Plasma and particle temperature measurements in thermal spray: Approaches and applications. *Journal of Thermal Spray Technology* 20 (3): 391–406.

Mauer, G., R. Vaßen, D. Stöver, S. Kirner, J.-L. Marqués, S. Zimmermann, G. Forster, and J. Schein. 2011b. Improving powder injection in plasma spraying by optical diagnostics of the plasma and particle characterization. *Journal of Thermal Spray Technology* 20 (1–2): 3–11.

Mayr, W., and R. Henne. 1988. Investigation of a VPS burner with laval nozzle using an automated laser doppler measuring system. In 1st Plasma Technik symposium, Luzern, vol 1, pp 87–97.

Meyer, P.J., and D. Hawley. 1991. Electro plasma Inc., LPPS production systems. In *Thermal spray coatings: Properties, processes and applications*, ed. T.F. Bernecki, 29–38. Materials Park: ASM International.

Metallisation Ltd. 2006. *Technical brochure*. www.mctaallisation.com

Mishin, J. 1985. "Contribution i la Mise au Point d'un Dispositif de Mesure des TempCratures de Surface des Particules en Vol dans un Jet de Plasma d'Arc," Thtse de Doctorat, 3itme cycle, University of Limoges, France (in French).

Mishin, J., M. Vardelle, J. Lesinski, and P. Fauchais. 1987. Two-color pyrometer for the statistical measurement of particulate surface temperature under thermal plasma conditions. *Journal of Physics E: Scientific Instruments* 20: 620.

Montavon, G., and K. Gross. 2004. Safety issue in thermal spraying: The need for a collaborative effort within the community? *Journal of Thermal Spray Technology* 13 (2): 155–157.

Moreau, C., P. Gougeon, M. Lamontagne, V. Lacasse, G. Vaudreuil, and P. Cielo. 1994. On-line control of the plasma spraying process by monitoring the temperature, velocity, and trajectory of in-flight particles. In *Thermal spray industrial applications*, ed. C.C. Berndt and S. Sampath, 431–437. Materials Park: ASM International.

Moreau, C., J.-F. Bisson, R.S. Lima, and B.R. Marple. 2005. Diagnostics for advanced materials processing by plasma spraying. *Pure and Applied Chemistry* 77 (2): 443–462.

Moulla, L., Z. Salhi, M.P. Planche, M. Cherigui, and C. Coddet. 2005. On the measurement of substrate temperature during thermal spraying. In *Thermal spray connects: Explore its surfacing potential*, ed. E. Lugscheider, 679–683. Düsseldorf: DVS.

Muehlberger, E. 1988. Industrial plasma processing technology. In *Proceedings of the 1st plasma Technik symposium*, vol. 3, 105–118. Wohlen: Plasma Technik.

Nadeau, A., L. Pouliot, F. Nadeau, J. Blain, S.A. Berube, C. Moreau, and M. Lamontagne. 2006. A new approach to online thickness measurement of thermal spray coatings. *Journal of Thermal Spray Technology* 15 (4): 744–749.

Newbery, A.P., and P.S. Grant. 2000. Droplet splashing during arc spraying of steel and the effect on deposit microstructure. *Journal of Thermal Spray Technology* 9 (2): 250–258.

Niranatlumpong, P., and H. Koiprasert. 2006. Improved corrosion resistance of thermally sprayed coating via surface grinding and electroplating techniques. *Surface and Coating Technology* 201: 737–743.

Nylen, P., J. Franssen, A. Wretland, and N. Martensson. 1996. Coating thickness prediction and robot trajectory generation of thermal sprayed coatings. In *Thermal spray: Prediction and robot trajectory generation of thermal sprayed coatings. Thermal spray: Practical solutions for engineering problems*, ed. C.C. Berndt, 693–698. Materials Park: ASM International.

Nylen, P., J. Wigren, L. Pejryd, and M.-O. Hansson. 1999. The modelling of coating thickness, heat transfer and fluid flow for a plasma sprayed gas turbine application. *Journal of Thermal Spray Technology* 8 (3): 393–398.

Oberste-Berghaus, J., B. Marple, and C. Moreau. 2006. Suspension plasma spraying of nanostructured WC-12Co coatings. *Journal ofThermal Spray Technology* 15 (4): 676–681.

Oerlikon-Metco. 2014a. Atmospheric plasma spray equipment solutions, issues 5, October.

———. 2014b. Thermal spray equipment guide, issues 11, October.

———. 2016. An introduction to Thermal Spray', issue 6, July.

———. 2017. 'Thermal Spray-Materials Guide', BRO-000. 1.17, April.

Park, C.S., K.C. Porteous, K.F. Sadler, and M.J. Zue. 1995. *Contemporary engineering economics; A Canadian perspective*. Don Mills: Addison-Wesley.

Peters, M.S., K.D. Timmerhaus, and R.E. West. 2003. *Plant design and economics for chemical engineers'*, 988 p. New York: Mc Graw Hill.

Petsas, N., G. Kouzilos, G. Papapanos, M. Vardavoulias, and A. Moutsatsou. 2007. Worker exposure monitoring of suspended particles in a thermal spray industry. *Journal of Thermal Spray Technology* 16 (2): 214–219.

Pfender, E., J.R. Fincke, and R. Spores. 1991. Entrainment of cold gas into thermal plasma jets. *Plasma Chemistry and Plasma Processing* 11 (4): 529–543.

Planche, M.P., R. Bolot, and C. Coddet. 2003. In-flight characteristics of plasma sprayed alumina particles: Measurements, modeling, and comparison. *Journal of Thermal Spray Technology* 12 (1): 101–111.

Pokhmurska, A., and R. Ciach. 2000. Microstructure and properties of laser treated arc sprayed and plasma sprayed coatings. *Surface and Coating Technology* 125: 415–418.

Prawara, B., H. Yara, Y. Miyagi, and T. Fukushima. 2003. Spark plasma sintering as a post-spray treatment for thermally-sprayed coatings. *Surface and Coating Technology* 162: 234–241.

Prystay, M., P. Gougeon, and C. Moreau. 1996. Correlation between particle temperature and velocity and the structure of plasma sprayed zirconia coatings. In *Thermal spray: Practical solutions for engineering problems*, ed. C.C. Berndt, 517–523. Materials Park: ASM International.

Rahman, M., G. Soucy, and M.I. Boulos. 1995. Analysis of the enthalpy probe technique for thermal plasma diagnostics. *The Review of Scientific Instruments* 66 (6): 3424–3431.

———. 1996. Diffusion phenomena of a cold gas in a thermal plasma stream. *Journal of Plasma Chemistry Plasma Processing* 16 (1): 169S–189S.

Rahmane, M., G. Soucy, and M.I. Boulos. 1994. *Mass transfer in induction plasma reactors*. International Journal of Heat and Mass Transfer 37: 2035–2046

Rahmane, M., G. Soucy, M. Boulos, and R. Henne. 1998. Fluid dynamic study of direct current plasma jets for plasma spraying applications. *Journal of Thermal Spray Technolgoy* 7 (3): 349–356.

———. 1995. Analysis of enthalpy probe technique for thermal plasma diagnostics. Journal of Measurement Science and Technology 66 (6): 3424–3431.

Rajabian, M., D.V. Gravelle, and S. Vacquié. 2004. Measurements of temperatures and electron number density in an argon–nitrogen plasma jet generated by a dc torch-operation close to supersonic threshold. *Plasma Chemistry and Plasma Processing* 24 (2): 261–284.

Razavy, F.G., D.C. Van Aken, and J.D. Smith. 2003. Effect of laser surface melting upon the devitrification of plasma sprayed cordierite. *Materials Science and Engineering* A362: 213–222.

Renouard-Vallet, G. 2004. Elaboration by plasma spraying of dense and thin (a few tens of micro meters) yttria stabilized zirconia electrolytes for SOFCs. Ph.D. thesis, University of Limoges France, 8 Feb 2004 (in French).

Roumilhac, P., J.-F. Coudert, and P. Fauchais. 1990. Influence of the arc chamber design and the surrounding atmosphere on the characteristics and temperature distribution of Ar-H$_2$ and Ar-He spraying plasma jets. In *Plasma processing and synthesis of materials*, ed. D. Apelian and J. Szekely, vol. 190, 227–333. Pittsburgh: MRS.

Roumilhac, P.h., J.F. Coudert, and P. Fauchais. 1991. Influence of the arc chamber design and of the surrounding atmosphere on temperature distributions of Ar-H$_2$ and Ar-He spraying plasma jets. In Plasma processing and synthesis of materials III, MRS proceedings, vol 190, pp. 227–238.

Roumilhac, Ph., P. Fauchais, and M. Ducos. 1991. Optical and thermal diagnostics regarding the working conditions of a plasma mini-spray torch. In *1st Plasma Technik symposium*, vol. 1, 121–131. Wohlen: Plasma Technik.

Russ, S., P.J. Strykowski, and E. Pfender. 1994. Mixing in plasma and low-density jets. *Experiments in Fluids* 16: 297–307.

Russo, L. 2010. Safety guidelines for performing risk assessments. ASM Thermal Spray Society, Safety Guidelines, 07/06/2010.

Sakuta, T., and M.I. Boulos. 1988a. A novel approach for particle velocity and size measurement under plasma conditions. *Review of Scientific Instruments* 59: 285–291.

———. 1988b. A novel technique for simultaneous in-flight measurement of particle surface temperature, velocity and size under plasma conditions. *IEEE Transactions on Plasma Science* 108: 389–396.

Salik, Muhammad. 2013. Plasma properties of nano-secod laser iron target in air. *International Journal of Physical Sciences* 8: 1738–1745.

Salimijazi, H.R., L. Pershin, T.W. Coyle, J. Mostaghimi, S. Chandra, Y. C. Lau, L. Rosenzweig, and E. Moran. 2007. Measuring substrate temperature variation during application of plasma-sprayed zirconia coatings. *Journal of Thermal Spray Technology* 16 (4): 580–587.

Sampath, S., V. Srinivasan, A. Valarezo, A. Vaidya, and T. Streibl. 2009. Sensing, control, and in situ measurement of coating properties: An integrated approach toward establishing process-property correlations. *Journal of Thermal Spray Technology* 18 (2): 243–255.

Schein, J., M. Richter, K.D. Landes, G. Forster, J. Zierhut, and M. Dzulkor. 2008. Tomographic investigation of plasma jets produced by multielectrode plasma torches. *Journal of Thermal Spray Technology* 17 (3): 338–343.

Schelz, S., F. Enguehard, N. Caron, D. Plessis, B. Minot, F. Guillet, J.-L. Longuet, N. Teneze, and E. Bruneton. 2008. Recombination of silica and zirconia into zircon by means of laser treatment of plasma-sprayed coatings. *Journal of Materials Science* 43: 1948–1957.

Schulte, P.A., V. Murashov, R. Zumwalde, E.D. Kuempel, and C.L. Geraci. 2010. Occupational exposure limits for nanomaterials: State of the art. *Journal of Nanoparticle Research* 12 (6): 1971–1987.

Schwetzke, R., and H. Kreye. 1999. Microstructure and properties of tungsten carbide coatings sprayed with various high-velocity oxygen fuel spray systems. *Journal of Thermal Spray Technology* 8 (3): 433–439.

Shinoda, K., Y. Tan, and S. Sampath. 2010. Powder loading effects of yttria-stabilized zirconia in atmospheric dc plasma spraying. *Plasma Chemistry and Plasma Processing* 30: 761–778.

Shrestha, S., T. Hodgkiess, and A. Neville. 2001. The effect of post-treatment of a high-velocity oxy-fuel Ni-Cr-Mo-Si-B coating part 2: Erosion-corrosion behavior. *Journal ofThermal Spray Technology* 10 (4): 656–665.

Sidhu, B.S., D. Puri, and S. Prakash. 2005. Mechanical and metallurgical properties of plasma sprayed and laser remelted Ni–20Cr and Stellite-6 coatings. *Journal of Materials Processing Technology* 159: 347–355.

Song, R.G., W.Z. He, and W.D. Huang. 2000. Effects of laser surface remelting on hydrogen permeation resistance of thermally-sprayed pure aluminum coatings. *Surface and Coating Technology* 130: 20–23.

Srinivasan, V., A. Vaidya, T. Streibl, M. Friis, and S. Sampath. 2006. On the reproducibility of air plasma spray process and control of particle state. *Journal of Thermal Spray Technology* 15 (4): 739–743.

Steffens, H.-D., and T. Duda. 2000. Enthalpy measurements of direct current plasma jets used for ZrO2-Y2O3 thermal barrier coatings. *Journal of Thermal Spray Technology* 9 (2): 235–240.

Stewart, S., and R. Ahmed. 2003. Contact fatigue failure modes in hot isostatically pressed WC-12%Co coatings. *Surface and Coating Technology* 172: 204–216.

Stoica, V., R. Ahmed, M. Golshan, and S. Tobe. 2004. Sliding wear evaluation of hot isostatically pressed thermal spray ceramet coatings. *Journal of Thermal Spray Technology* 13 (1): 93–107.

Swank, W.D., J.R. Fincke, and D.C. Haggard. 1993. Modular enthalpy probe and gas analyzer for thermal plasma measurements. *The Review of Scientific Instruments* 64 (1): 56–62.

Tekmen, C., M. Yamazaki, Y. Tsunekawa, and M. Okumiya. 2008a. In-situ plasma spraying: Alumina formation and in-flight particle diagnostic. *Surface and Coating Technology* 202: 4163–4169.

Tekmen, C., Y. Tsunekawa, and M. Okumiya. 2008b. Effect of plasma spray parameters on in-flight particle characteristics and in-situ alumina formation. *Surface and Coating Technology* 203: 223–228.

Tekmen, C., K. Iwata, Y. Tsunekawa, and M. Okumiya. 2009. Controlling graphite content in plasma sprayed cast iron coatings via in-flight particle diagnostic. *Journal of Materials Processing Technology* 209: 5417–5422.

Trifa, F.-I., G. Montavon, C. Coddet, P. Nardin, and M. Abrudeanu. 2005. Geometrical features of plasma-sprayed deposits and their characterization methods. *Materials Characterization* 54 (2): 157–175.

Troczynski, T., Q. Yang, and G. John. 1999. Post-deposition treatment of zirconia thermal barrier coatings using sol-gel alumina. *Journal of Thermal Spray Technology* 8 (2): 229–234.

Tsai, H.L., and P.C. Tsai. 1998. Laser glazing of plasma-sprayed zirconia coatings. *Journal of Materials Engineering and Performance* 7 (2): 258–264.

Tucker, R.C., Jr. 1994. *Thermal spray coating, surface engineering ASM handbook*. Vol. 5. Materials Park: ASM International.

Tucker, R.C., Jr., ed. 2013. *ASM handbook Vol. 5A thermal spray technology*. Materials Park: ASM international.

Vardelle, M., A. Vardelle, P. Fauchais, and M.I. Boulos. 1988. Particle dynamics and heat transfer under plasma conditions. *AICHE Journal* 34 (4): 567–573.

Vardelle, M., A. Vardelle, A.C. Leger, P. Fauchais, and D. Gobin. 1995. Influence of the particle parameters at impact on splat formation and solidification in plasma spraying processes. *Journal of Thermal Spray Technology* 4 (1): 50–58.

Vardelle, M., A. Vardelle, P. Fauchais, K.-I. Li, B. Dussoubs, and N.J. Themlis. 2001. Controlling particle injection in plasma spraying. *Journal of Thermal Spray Technology* 10: 267–286.

Vardelle, M., T. Renault, and P. Fauchais. 2002. Choice of an IR pyrometer to measure the surface temperature of a coating during its formation in air plasma spraying. *High Temperature Material Processes* 6 (4): 469–490.

Vattulainen, J., E. Hämäläinen, R. Hernberg, P. Vuoristo, and T. Mäntylä. 2001. Novel method for in-flight particle temperature and velocity measurements in plasma spraying using a single CCD camera. *Journal of Thermal Spray Technology* 10 (1): 94–104.

Verdy, C., B. Serio, and C. Coddet. 1998. In situ temperature measurement using embedded micro-thermocouples in vacuum plasma sprayed multi-layered structures. In *Thermal spray: Meeting the challenges of the 21st century*, ed. C. Coddet, 821–824. Materials Park: ASM International.

von Niessen, K., and M. Gindrat. 2010. Vapor phase deposition using a plasma spray process. In Thermal spray: global solutions, ITSC 2010. DVS, Düsseldorf, Germany, e-proceedings.

Wang, P., S.C.M. Yu, and H.W. Ng. 2004. Particle velocities, sizes and flux distribution in plasma spray with two powder injection ports. *Materials Science and Engineering A* 383: 122–136.

Wang, H.-T., C.-J. Li, G.-J. Yang, and C.-X. Li. 2008. Cold spraying of Fe/Al powder mixture: Coating characteristics and influence of heat treatment on the phase structure. *Applied Surface Science* 255: 2538–2544.

Wang, Y., C.G. Li, W. Tian, and Y. Yang. 2009. Laser surface remelting of plasma sprayed nanostructured Al_2O_3–13wt%TiO_2 coatings on titanium alloy. *Applied Surface Science* 255: 8603–8610.

Weisheng, X., H. Zhang, G. Wang, and Y. Yang. 2009. A novel integrated temperature investigation approach of sprayed coatings during APS process. *Journal of Materials Processing Technology* 209: 2897–2906.

Westergård, R., and S. Hogmark. 2004. Sealing to improve the wear properties of plasma sprayed alumina by electro-deposited Ni. *Wear* 256: 1153–1162.

Westerweel, J. 1993. *Digital particle image velocimetry – Theory and application*. Delft: Delft University Press.

Wroblewski, D., G. Reimann, M. Tuttle, D. Radgowski, M. Cannamela, S.N. Basu, and M. Gevelber. 2010. Sensor issues and requirements for developing real-time control for plasma spray deposition. *Journal of Thermal Spray Technology* 19 (4): 723–735.

Wuest, G., A. Hall, and D. Crawmer. Guidelines for the use of personal protective equipment (PPE) in thermal spraying. ASM Thermal Spray Society, 27 pages in Safety Guidelines.

Xie, D., Y. Xiong, and F. Wang. 2003. Effect of an enamel coating on the oxidation and hot corrosion behavior of an HVOF-sprayed Co–Ni–Cr–Al–Y coating. *Oxidation Metals* 59 (5/6): 503–516.

Yilmaz, N., R.E. Lucero, A. Burl Donaldson, and W. Gill. 2009. Flow characterization of diffusion flame oscillations using particle image velocimetry. *Experiments in Fluids* 46: 737–746.

Yin, Z., S. Tao, X. Zhoua, and C. Ding. 2008. Particle in-flight behavior and its influence on the microstructure and mechanical properties of plasma-sprayed Al2O3 coatings. *Journal of the European Ceramic Society* 28: 1143–1148.

Zhang, W., and S. Sampath. 2009. A universal method for representation of in-flight particle characteristics in thermal spray processes. *Journal of Thermal Spray Technology* 18 (1): 23–34.

Zahiri, S.H., D. Fraser, and M. Jahedi. 2009. Recrystallization of cold spray-fabricated CP titanium structures. *Journal of Thermal Spray Technology* 18 (1): 16–22.

Zhang, Q., C.-J. Li, X.-R. Wang, Z.-L. Ren, C.-X. Li, and G.-J. Yang. 2008. Formation of NiAl intermetallic compound by cold spraying of ball-milled Ni/Al alloy powder through post-annealing treatment. *Journal of Thermal Spray Technology* 17 (5–6): 715–720.

Zhanga, S.H., J.H. Yoon, M.X. Li, T.Y. Cho, Y.K. Joo, and J.Y. Cho. 2010. Influence of CO_2 laser heat treatment on surface properties, electrochemical and tribological performance of HVOF sprayed WC–24%Cr3C2–6%Ni coating. *Materials Chemistry and Physics* 119: 458–464.

Zhou, Z.-h., S.-k. Gong, H.-f. Li, H.-b. Xu, C.-g. Zhang, and L. Wang. 2007. Effects of shot peening process on thermal cycling lifetime of TBCs prepared by EB-PVD. *Chinese Journal of Aeronautics* 20: 145–147.

Zierhut, J., K. Landes, C. Waas, D. Kutscher, P. Heinrich, and W. Krömmer. 1999. Particle flux imaging. In *ITSC99 proceedings*, ed. E. Lugscheider, 340–344. Düsseldorf: DVS.

Abbreviations

ACP	Amorphous calcium phosphate
ALTM	Air Lock Transition Module
APS	Air plasma spraying
APS	Atmospheric plasma spraying
AS	As-sprayed
BAG	Bioactive glass
BM	Base material
BOF	Basic oxygen furnace
BRT	Burner Rig Test
CAPS	Controlled atmosphere plasma spraying
CFRP	Carbon fiber-reinforced plastic
CMAS	Acronym for CaO, MgO, Al$_2$O$_3$, SiO$_2$
CNT	Carbon nanotubes
CS	Cold Spray
C-SSCS	Composite of Stainless Steel-Carbon Steel
CTE	Coefficient of thermal expansion
CVD	Chemical Vapor Deposition
CW	Corrosion-Wear
dBA	Decibel Authorized
DC	Direct current
D-gun	Detonation-gun
DRC	Diamond-reinforced composite
DWTS	Direct write thermal spray
EAF	Electric arc furnace
EB-PVD	Electron Beam-Physical Vapor Deposition
EC	Erosion-Corrosion
EHC	Electrolytic hard chrome
EIS	Electrochemical impedance spectroscopy
EMI	Electromagnetic interference
EW	Erosive-Wear
FAC	Fe-based alloy coatings
FBC	Fluidized-bed combustor
fcc	Face-centered cubic
FDA	Food and Drug Administration
FG	Functionally graded
FGC	Functionally graded coating
FW	Fatigue wear
GDC	Acronym for, Ce$_{0.8}$ Gd$_{0.2}$ O$_{1.9}$
GS	Gas shroud
HA	Hydroxyapatite Ca$_{10}$ (PO$_4$)$_6$ (OH)$_2$
HAT	HA top coating
HB	Hardness Brinell
HC	Hard chrome
HCC	Hard chromium coating
HEPS	High-energy plasma spray
HIP	Hot isostatically pressed
HPAL	High-pressure acid-leach
HPPS	High-power plasma spray
HTBC	HA/TiO$_2$ (50 vol.% each) Bond Coat
HTH	HA/(HA/TiO$_2$) Bond Coat
HVAF	High-velocity air fuel
HVFS	High-velocity flame spraying
HVLF	High-velocity liquid fuel
HVOF	High-velocity oxy fuel
HVPS	High-velocity plasma spray
HVSFS	High-velocity suspension flame spraying
ICP	Inductively Coupled Plasma
IACS	International Annealed Copper Standard
IPS	Induction plasma spraying
JTST	*Journal of Thermal Spray Technology*
LaMA	Acronym for LaMgAl$_{11}$O$_{19}$
LBT	Land-based turbines
LPPS	Low-pressure plasma spraying
LPPS-TF	Low Pressure Plasma Spray-Tin Film
LSCF	Acronym for, La$_{0.6}$ Sr$_{0.4}$ Co$_{0.2}$ Fe$_{0.8}$ O$_{32-\delta}$
LTA	Acronym for, LaTi$_2$Al$_9$O$_{19}$
LTE	Local thermodynamic equilibrium
MLCC	Multilayer ceramic capacitors
MMC	Metal matrix composite
MSWI	Municipal solid waste incinerator
NTSRS	Net thermal spraying residual stress
ODS	Oxide-dispersion strengthened
OEM	Original equipment manufacturer

© Springer Nature Switzerland AG 2021

M. I. Boulos et al. (ed.), *Thermal Spray Fundamentals*, https://doi.org/10.1007/978-3-030-70672-2_19

PA-12	Polyamide-12
PAH	Progressive abradability hardness
PE-CVD	Plasma Enhanced-CVD
PEEK	Poly-ether-ether-ketone
PEI	Polyetherimide
PGDS	Pulsed gas dynamic spraying
PM	Post-melted
PMC	Polymer matrix composite
PS	Plasma spraying
PSD	Particle size distribution
PS-PVD	Plasma Spray-PVD
PTA	Plasma-transferred arc
PTS	Polymer thermal spray
PVD	Physical vapor deposition
QC	Quality control
RCF	Rolling contact fatigue
RF	Radio frequency
RF-IPS	RF-Induction Plasma Spraying
RH	Relative air humidity
SBF	Simulated body fluid
SER	Specific energy requirement
SFW	Surface fatigue wear
SPS	Spark Plasma Sintering

19.1   Introduction

At its early stages of development, thermal spray (TS) technology was mostly used for surface protection against corrosion and erosion, as well as for the rebuilding, retrofitting, and wear repair [Davis JR (Ed) (2004)]. Up to the early 1980s, the basic phenomena involved in most of thermal spray processes were poorly understood, with the process parameters based more on the operator experience and skills rather than on a solid scientific understanding of the phenomena involved. This often resulted in unsatisfactory process reproducibility and reliability. The wider acceptance of the technology for industrial-scale production started in the late 1980s and early 1990s, with applications limited to high added-value components in the aerospace and nuclear industry. These were mostly driven by the fact that no viable alternate solutions were available, and design engineers and scientists were used to work with rather complex processes.

Since the turn of the century, the range of industrial-scale applications of the thermal spray industry expanded considerably penetrating into new industrial sectors such as biomedical, electric and electronics industry, and high-end automobile industry. These were mostly considered for tribological and wear-resistant applications including lubricity and low-friction surfaces, resistance to corrosion and/or oxidation, thermal protection, electrical and optical components, electromagnetic shielding, electrical insulation, abradable seals, biomedical applications, as well as ornamental applications. The surge was manifested by a major increase in the volume of the scientific literature published in this field as illustrated in Fig. 19.1, which shows the cumulative number of technical manuscripts submitted to the *Journal of Thermal Spray Technology* (JTST), one of the leading scientific journals in this field, during the period 2004–2010. Particularly relevant to note is the very rapid increase of TS

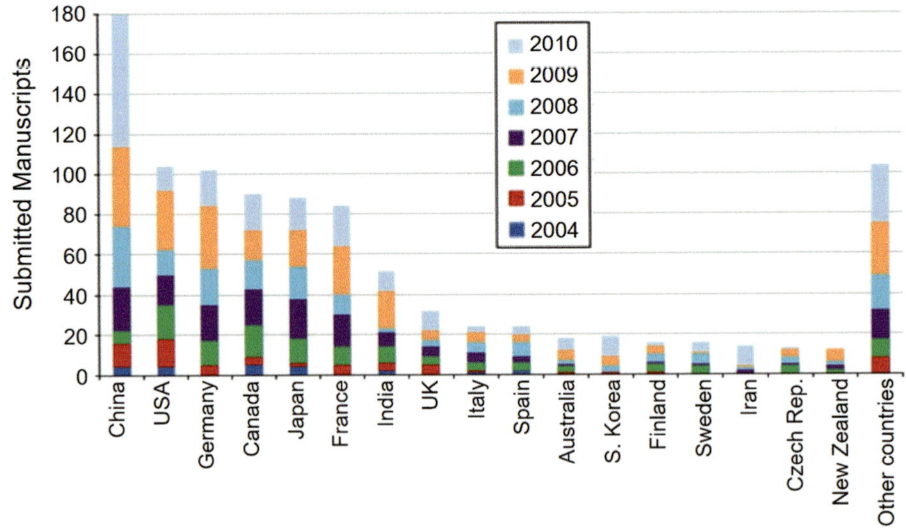

Fig. 19.1 Cumulative number of technical papers submitted to the *Journal of Thermal Spray Technology* over the period 2004–2010. [Dorfman and Sharma (2013a, b)]

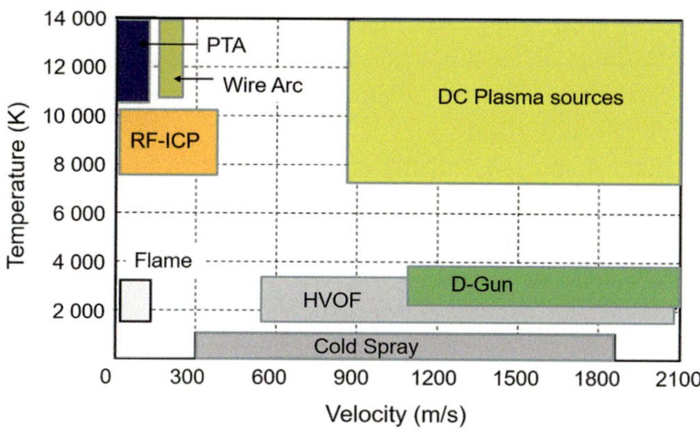

Fig. 19.2 Gas temperatures versus velocity mapping associated with different thermal spray processes

research activities over this period in China and the sustained research and development (R&D) activities in the USA, Germany, Canada, Japan, and France.

The choice of a specific coating and/or thermal spray process technology, for a given service condition, depends, however, on the expectation of the user and the cost that could be tolerated for the application. This affected, in turn, the selection of the material to be applied for the coating and the spray process used, which, as shown in Fig. 19.2, can offer widely different spraying conditions in terms of spray medium velocity and temperature range. The design of functional coatings, however, is often complicated by the fact that components are not always devoted to a single requirement such as wear or corrosion, electrical insulation, or thermal insulation. In most cases, coatings had to satisfy multiple combined needs such as, for example, wear and corrosion resistance.

It is important to point out that, contrary to general perception, because of the specific features of each of these TS technologies, they are mostly complementary, rather than competitive. Any overlap between them in terms of application is not very broad with the majority of TS technologies differing in terms of coating functional properties, and economic factors. To choose between them, it is important to know the advantages and limitation of each of these processes and to make the proper decision of the best process for each application.

In this chapter, a comparative analysis of thermal spray processes is given highlighting their principal characteristics and limitations. This is followed by a classification of the different areas of application of TS technologies for surface modification depending on functional needs and service conditions. Review of the principal TS coating applications by industrial sector and by country is presented next, followed by a simplified economic analysis of the different thermal spray processes. A summary of coating applications by industry and listing of the range of materials that can be used for different applications are provided respectively in Appendices A and B at the end of this chapter.

19.2 Comparative Analysis of Thermal Spray Processes

In this section, a comparative analysis is presented of different thermal spray processes described in detail in Part II, Thermal Spray Technologies, of this book. These are grouped in terms of energy source used in the process, the form of precursor used with each of these processes, and key features of the technology that orients its field of industrial applications.

19.2.1 Cold Spraying

As discussed in Chap. 6, cold spray (CS) is a kinetic process utilizing high-pressure compressed gas (2–4 MPa), such as nitrogen or helium or their mixture, at near room temperature, or preheated up to 700–800 °C, to produce high-velocity supersonic gas jets that act as spray medium to entrain and accelerate the spray material in the form of fin powder to velocities up to 300–1500 m/s. On impact of the particles on the surface of the substrate, they plastically deform creating the coating. The process requires impact particle velocities higher than a so-called *critical velocity*, which depends on the nature of the particle material, its particle size distribution (PSD), and particle morphology.

Cold spray is limited to the spraying of ductile metals, ferrous and nonferrous, metal alloys, composites, and cermets on virtually any substrate. The main advantage of the technology is its ability to avoid the in-flight oxidation of the powder and achieving high coating densities and minimal heating of the substrate during the deposition process, which allows it to be used for the coating of heat-sensitive substrates. With this spray process, the oxide content of the coating reflects essentially that of the sprayed powder (for example, <100 ppm for copper). The mechanical and electrical properties of as-sprayed (AS) coatings are remarkably close to those of bulk materials, especially with very good tensile properties. The ductility of cold-sprayed coatings can, however, be rather low due to the important work hardening inherent to the process.

Other than its principal limitation with respect to the material to be sprayed, the process is rather noisy reaching up to 110 decibels authorized (dBA), thus requiring special protective equipment for the operator. Its need for high gas flow rates can be prohibitively costly especially when using helium, even when mixed with nitrogen. Recycling helium requires that the spray process be carried out in a controlled atmosphere chamber, adapted to the size of the parts to be coated. The chamber should be equipped with an airlock transition module (ALTM) allowing for the introduction and removal of the parts from the chamber without contaminating its gaseous content. Gas recycling also requires important investments in gas recycling pumping system, filters, and high-pressure gas storage tanks. In this case, the noise level associated with the process is reduced to that of the surrounding equipment. Deposition rates are typically in the range of 2–10 kg/h, with deposition efficiencies of 40–90% depending on sprayed materials.

The first and very successful application of CS was for copper metallization. Other industrial-scale applications include its use for metal restoration and sealing, engine blocks, castings, molds and dies, refrigeration equipment, heat exchangers, aluminum piston heads, manifolds, disk brakes, heat sinks for microelectronics (Al and Cu), solid lubricant matrix with base metals, electronically conductive coatings (Cu or Al on ceramic or polymeric components), electromagnetic shielding, as well as localized corrosion protection (Zn or Al coatings).

19.2.2 Combustion-Based Thermal Spraying

As presented in Chap. 7, combustion-based thermal spray technologies are the first to be developed in the early twentieth century and remains one of the most widely used spray-coating processes because of their favorable economics. The principal limitation of the technology is in the nature of the material to be sprayed since the maximum flame temperature is generally below 3500 K. Moreover, because the oxidizing atmosphere predominate in the flame, flame-spray coatings of metals usually contain a few wt.% of oxide (1–4 or 5 wt.% depending on metal reactivity, torch design, and oxygen-to-fuel ratio). Over the years, their wider acceptance of combustion-based thermal spraying driven by scientific and technical improvements as well as growing market demand has led to the development of the following three distinct technologies.

19.2.2.1 Flame Spraying

Flame spraying (FS) is at the origin of all combustion-based thermal spray technologies, making use of mostly oxyacetylene torches achieving premixed combustion temperatures up to about 3000 K. Flame velocities are generally below 100 m/s. Sprayed materials are introduced generally axially into the spray gun either in the form of powder—"powder flame spraying"—or as wire, rod, or cord—"wire flame spraying".

In "powder flame spraying," the sprayed materials are mostly metals or polymers, which are rather easy to melt and spray. Particles are axially injected into the flame. Coatings present a high porosity (>10 vol.%) and a low adhesion (<30 MPa). Powder feed rates are typically between 2 and 10 kg/h, depending on sprayed material and torch used. The deposition efficiency is around 50%. In-flight oxidation of the powder occurs during processing (oxide content up to

6–12 wt.%). They are mostly used for wear-resistance applications under low load. Substrate and coating must be cooled during spraying. Flame spraying is generally rather noisy (90–125 dBA).

Flamed-sprayed, self-fluxing alloys contain Si and B (for example, CrBFeSiCNi) that are used as deoxidizers. When the coating is heat-treated at about 1030 °C, they diffuse toward the coating surface trapping the oxygen. It results in coatings with almost no porosity, an excellent adhesion by diffusion at the substrate–coating interface. It is also very easy to deposit materials of brazing type, and composite coatings are possible. Such coatings are generally limited to substrates that can tolerate the fusing temperature (steel and not aluminum, for example) and possibly any induced distortion. They are mostly used against abrasion (friction, erosion) and corrosion (cold or hot).

Flame spraying can also be used with wires, rods, or cords. In this case, a compressed air jet atomizes the melted tip of the wire (metallic and ductile) or cored wires (containing ceramics or nonductile materials), rod, or cord (for ceramics). Due to the compressed gas atomization, the noise level is not negligible. Compared to powders, the variety of materials that can be sprayed as wires, rods, or cords are larger. Wires of self-fluxing alloys can also be sprayed. Oxides are of course generated during the process, but less than with powders (about 4–8 wt.%), and the deposition efficiency is better (around 70%). Compared to powders the material flow rate varies from 5 to 25 kg/h. With wires the coating adhesion is slightly better than that obtained with powders. The porosity is similar to that obtained with powders except for ceramic materials that are more porous but present an excellent wear resistance. Generally, the metal or alloy coatings are used against wear—especially abrasion or adhesion under low load, for example, in rotating heavy equipment, piston rings, and synchronizing rings—and atmospheric corrosion: bridges, large steel structures, galvanized tubing, or ships. The wire flame-spray noise level is between 118 and 122 dBA, while with rods it is about 125 dBA.

19.2.2.2 High-Velocity Flame Spraying

High-velocity flame spraying (HVFS), or as commonly referred to as "high-velocity oxy fuel" (HVOF) or "high-velocity air fuel" (HVAF), are also combustion-driven processes with internal combustion at pressure below 1 MPa for those with gaseous fuel and slightly higher for those with liquid fuels. They produce very-high-velocity spray medium streams, thanks to a convergent–divergent nozzle design following the combustion chamber. They work mostly with powder spray materials injected axially, or radially, into the flame with some spray gun designs, with a few processes developed to spray wires. Power levels for HVOF guns working with gaseous fuels are in the range of 100–120 kW, with a noise level of 125–135 dBA. For those working with liquid fuels, the power can reach 300 kW with a noise level of 133 dBA or more. The actual trend is to increase particle velocities and reduce their temperatures in order to limit their oxidation, the high kinetic energy of particles compensating for the lower temperature. This was first realized with HVAF guns working with air and where the noise level was 133 dBA or more. In the high-power guns, especially designed for that, the addition of inert gas such as nitrogen in the combustion chamber, at flow rates up to 2000 standard liter per minute (slm), is used to reduce gas temperature and increase its velocity.

Globally this process, working mainly with metals, alloys, and cermets (one of their most successful application), has deposition efficiencies of about 70% at powder feed rates up to 7.2 kg/h for gaseous-fuel guns, and up to 12 kg/h for liquid-fuel guns with deposition efficiencies roughly of 60–80%. Resulting coating porosities are a few vol.%, with a good adhesion (>50–60 MPa) and low oxygen content (between 0.5 and a few wt.%). Due to shot-peening effect, compressive stresses can be achieved in rather thick coatings (up to 6.4 mm). Of course, the process is noisy, dusty, and requires special safety precautions for the handling of large flow rates of explosive gases. HVOF and HVAF spraying are mostly used for wear and corrosion protection. Resistance to wear of the coating, sliding/adhesive wear, fretting, erosion, or cavitation is generally excellent depending on the material and process parameters used. The corrosion resistance is also very good with the high density of coatings and, according to the sprayed material, they are used for hot corrosion, oxidation, and against acidic or alkaline atmospheres and liquids. Substrate and coating must be cooled during spraying.

19.2.2.3 Detonation-Gun Spraying

Detonation-gun (D-gun) spraying, commonly referred to as "D-gun spraying" or "pulsed detonation spraying," was developed in the 1950s. The technology is closer to cold spray or HVOF/HVAF than conventional thermal spray coating since it relies heavily on the acceleration of particles to be sprayed to high velocities, projecting them toward the substrate in essentially a soften solid-state where they deform on impact forming the coating. In contrast to other combustion processes, where the flame is a subsonic wave sustained by a chemical reaction, detonation wave is a shock wave sustained by the energy of chemical reactions in a compressed explosive gas mixture. The ideal detonation wave travels in gases at speeds between 1500 and 3000 m/s, depending on the type and composition of the fuel-oxidizer mixture. The pressure just behind the detonation wave can be as high as 20–30 times the ambient pressure.

This process is pulsed at a frequency between 6 and 100 Hz. Particle velocities are due to the energy of a detonation (explosion). The resulting deposit is extremely hard, dense, and tightly bonded to the substrate. The process is

the noisiest of all the thermal spray processes (more than 145 dBA). Coating porosity is low (below 1 vol.%) and its oxygen content is between 0.1 and 0.5 wt.%. The deposition efficiency is around 90%, but powder flow rates are limited to the 1–2 kg/h range. Sprayed materials in powder form range from metals, alloys, and cermets. Some oxides can also be sprayed by this technique as long as their average particle size is below 20 μm. The main applications are abrasion and adhesion (friction) under low load as well as corrosion. Substrate and coating must be cooled during spraying.

19.2.3 Plasma Spraying

As described in Chaps. 7, 8, and 9, plasma spraying (PS) is based on the concept of generation of a high-temperature, high-velocity spray jet through the striking of an electric arc between two electrodes, or inductive coupling of the electric energy into a discharge that is heated through resistive energy dissipation by the current flowing through the gas at sufficiently high temperatures to have appreciable degrees of ionization and high electrical conductivities. Two distinct features of the technology in general are:

- Electric energy is the sole source of energy in the process.
- Extensive flexibility on the chemistry of the spray medium allowing the process to be carried out in an inert, reducing, or oxidizing atmosphere.

For most plasma-forming gases, the required temperatures at atmospheric pressure are above 8000 K. By definition, in a thermal arc the thermodynamic state of the plasma generally approaches local thermodynamic equilibrium (LTE), which includes kinetic and chemical equilibrium, except for the arc fringes or that close to a cooled wall.

Direct current (DC) plasma spray technology has been the principal driver in this field, leading to the development of a wide range of technologies such as:

- Atmospheric plasma spraying (APS)
- Controlled atmosphere plasma spraying (CAPS)
- Vacuum plasma spraying (VPS)
- Ultra-low-pressure plasma spraying (ULPPS)

Radio frequency induction plasma spraying (RF-IPS), on the other hand, has been developed in parallel mostly for high-end, high-purity applications such as preform cladding with high-purity silica in the fiber optic industry and high-purity crystal growing.

With temperatures over 8000 K, any material, whether metallic or ceramic, can be melted. The technology is essentially based on the in-flight heating and melting of the sprayed material injected into the plasma stream, either radially as in most DC plasma torches or axially as in RF-IPS or a few multi-electrode DC plasma sources. Feed material used is mostly in powder form with a mean particle diameter in the micron size range. Alternately, solution and suspension plasma spraying have been gaining wider acceptance for specific applications such as the spraying of nanostructured coatings.

19.2.3.1 Atmospheric Plasma Spraying

Atmospheric plasma spraying or air plasma spraying (APS) is carried out mostly with DC plasma torches in an open-air atmosphere using Ar, Ar/H_2, Ar/He, or Ar/N_2 as plasma gas operating at power levels ranging from 30 to 90 kW with, in most cases, radial powder injection. For plasma spray torches with one-stick type cathode, and power levels in the range 40–50 kW, the powder flow rate is between 3 and 6 kg/h, the deposition efficiency around 50%, and the torch noise level between 110 and 125 dBA. With high-power torches (150–250 kW), powder flow rates as high as 15–20 kg/h can be reached. With tri-cathode torches such as the Oerlikon-Metco Triplex Torch or the Northwest Mettech Axial III comprising of three plasma jets converging in the torch nozzle, particles are axially injected into the emerging plasma jet.

Typical spraying distance is in the range of 100–120 mm, the heat flux to the substrate is among the highest (≈ 2 MW/m^2), and the substrate as well as the coating must be cooled. To avoid oxidation, spraying must be performed under controlled atmosphere or soft vacuum conditions. APS coating porosities are between 2 and 8 vol.%, the oxygen content of metal or alloy coatings is between 1 and 5 wt. %, and their adhesion is good (>40–50 MPa). They are mainly used to spray oxides. However, they present good resistance to abrasive, adhesive, fretting, or sliding wear. They also produce thermally conductive or resistant surfaces.

19.2.3.2 Controlled Atmosphere Plasma Spraying

Controlled atmosphere plasma spraying (CAPS) is used when oxidation is a problem. Metals with high-melting temperature and non-oxide ceramics can be sprayed only in inert atmosphere (e.g., chamber filled with argon), at atmospheric pressure or slightly over. The technique is mostly used to melt materials with high melting temperature, such as TaC ($T_m \approx 4150$ K). The controlled atmosphere chamber must be water-cooled if its volume is below 10 m^3, and industrial-scale production of such coatings requires using an airlock transition module (ALTM) to introduce and remove parts from the chamber without allowing for air penetrating into the main chamber. The noise level is that of the surrounding pumps, fans, and power supply. The equipment is more complex than APS requiring the use of robots with adequate

thermal protection depending on the ambient temperature in the chamber. With all the equipment required, including that of argon gas recycling, the investment and operating cost is considerably higher than that of APS.

19.2.3.3 Vacuum Plasma Spraying

Vacuum plasma spraying (VPS) is similar to CAPS with the exception that the operation is carried out under sub-atmospheric pressures, in the range of tens of kPa ($5 > p_a > 70$ kPa). VPS is carried out in large vacuum chamber with volumes of a few m^3 up to 10–20 m^3, typically 1.5–2.5 m in diameter by 3–4 m long, which houses the substrate, often on a carousel holding more than one part to be coated, a DC plasma torch, and a robotic torch manipulator adequately protected for operation in a hot dusty atmosphere. The chamber is normally water-cooled with a large access door for ease of servicing, placing the parts to be coated on the carousel substrate holder and retrieving them at the end of the coating cycle. Torch operation is similar to that of APS torches, that is, with similar arc currents, plasma gas selections and flow rates, power levels, and powder flow rates. The torch-to-substrate distance is longer than normally used in APS, typically in the range of 250 and 500 mm. According to the plasma jet expansion, the mean ambient temperature in the chamber is over 200 °C. The noise level is similar to CAPS being essentially that of ambient fans, vacuum pumps, and power supply. The equipment is more complex and consequently is the most expensive of all spray processes, with the exception of ultra-low-pressure plasma spraying (ULPPS). One of the advantages of the process, besides the significant reduction in oxidation, is the possibility to remove the oxide layer from the substrate prior to the spraying operation using the plasma torch in a reversed-polarity-transferred arc mode. The part can also be preheated to a sufficiently high temperature to achieve diffusion bonding between the sprayed metal or alloy and substrate. This step can be carried out using the same transferred arc approach used in surface cleaning from oxides, in straight polarity with the part being then the anode of the transferred arc. VPS is by far one of the most expensive thermal spray installations that is mostly used for coating of high added-value parts, such as turbine blades.

19.2.3.4 Ultra-Low-Pressure Plasma Spraying

Ultra-low-pressure plasma spraying (ULPPS), also identified as very-low-pressure plasma spraying (VLPPS) or plasma spray physical vapor deposition (PS-PVD) and referred to by Oerlikon-Metco as low-pressure plasma spraying–thin film (LPPS®-TF), is a relatively new technology that bridges the gap between conventional VPS and physical vapor deposition (PVD) processes, in which the coating is formed through the deposition of the coating material from the vapor phase at lower temperatures at relatively low deposition rates compared to VPS. ULPPS was developed in direct competition to (EB-PVD), for the production of ceramic TBC coatings with equivalent quality as (EB-PVD) technology, at considerably higher deposition rates and lower cost. The two technologies, ULPPS and (EB-PVD), differ essentially in the way the precursor is transformed into the vapor phase with ULPPS based on the in-flight heating and evaporation of the precursor in powder form using the plasma jet, while (EB-PVD) makes use of an electron beam for the precursor evaporation from a solid target or powder. This difference has a significant impact on their respective operating pressure and the deposition rates associated with each of the two technologies, with the ULPPS process operating at pressures <200 Pa with deposition rates in the range of µm/s, in contrast to the electron beam technology that required considerably lower operating pressures (<5 Pa), has lower deposition rates in the 0.1–100 µm/min range, and significantly higher investment cost.

19.2.3.5 Induction Plasma Spraying

Radio frequency induction plasma spraying (RF-IPS) is usually carried out under controlled atmosphere at atmospheric pressure, or soft vacuum conditions, down to 10 kPa. The process is essentially similar to DC-VPS with the exception that the torch is fixed with respect to the chamber walls, while the substrate has to be translated in a linear or rotating movement across the stream of molten droplets emerging from the torch at the required spraying distance. With torches i.d. in the range of 30–100 mm, against below 10 mm for DC ones, the velocity of the plasma jet is considerably lower (<100 m/s) than that of a corresponding DC plasma torch at a comparable power rating and chamber pressures. It allows spraying bigger particles (up to 200–250 µm) than those sprayed with DC torches. Typical plasma power ratings for radio frequency vacuum induction plasma spraying (RF-VIPS) installations are in the 50–200 kW range, with corresponding powder feed rates of 2–10 kg/h, depending on the sprayed material. Because of the radio frequency (RF) of the supplied power to the induction coil of the torch (mostly in the 1–5 MHz range), special precautions need to be implemented for the electromagnetic interference (EMI) shielding of the plasma torch from the rest of the process and control equipment. RF-IPS is mostly used for the spraying of metals and ceramics on small parts with a high added value and coating densities close to 98% of the theoretical value. Vacuum induction plasma spraying (VIPS) is used on an industrial production scale for the deposition of high-purity silica on fiber optics and refractory metals on X-ray targets. The technology is also extensively used for powder spheroidization and the synthesis of nanostructured or nanosized powders for a wide range of applications beyond the thermal spray industry.

19.2.4 Wire Arc Spraying

Wire arc spraying (WAS) is the oldest of thermal spray processes, which can be traced back to 1915 with the issuing of its first patents in the USA. It was only in the 1960s that the real potential of the technology was recognized, and its applications greatly expanded. WAS is essentially a plasma spray coating process based on the concept of melting the material to be sprayed in wire or cord form using an electric arc struck between the tips of two wires, or a wire and a non-consumable electrode, and atomizing the formed molten metal by a high-velocity gas stream, which projects the droplets toward the substrate. As with conventional thermal spraying, the molten droplets form splats that rapidly solidify on impact with the substrate surface building up the coating in successive layers. The process can be maintained in a continuous mode by electrically connecting the two wires to a DC power supply, and continuously feeding the wires in a closely controlled speed to compensate for the melting of their respective tips such as to maintain a constant gap between them, and consequently a constant arc voltage. A high-velocity gas flow injected between the two wires removes constantly formed molten material from the wire tips, breaks down the larger droplets into smaller ones in a secondary atomization process, and propels them toward the substrate. The maximum arc current can vary from 200 to 1500 A. With this process, the spray rates are between 5 and 30 kg/h, with deposition efficiency of about 80%, higher than most other spray processes. Consequently, it is one of the most economical processes, as long as the materials to be sprayed can be obtained in the form of wires or cored wires. The noise level is similar to that of wire flame process, which is in the range of 118–122 dBA. The oxide content depends strongly on the sprayed materials but is rather high, over 25 wt.% for Al, for example. Using nitrogen instead of air as atomizing gas can reduce the oxidation level in the deposit though, depending on the required gas flow rates (around 1.0 m^3/min or more), the process can become rather expensive. Porosity is usually over 10 vol.% and the coating adhesion is medium, in the 40 MPa range. As the atomizing gas is at room temperature, the process has the added advantage of minimal heating of the substrate during the spraying operation, while the divergence of the spray pattern is a disadvantage. Like other spray processes, it is noisy and dusty. Coatings can be used for abrasion and adhesion (friction) under low load, though their main use is for atmospheric or marine corrosion protection and a broad range of electrical applications.

19.2.5 Plasma-Transferred Arc Deposition

Plasma-transferred arc (PTA) deposition is a combination of welding and thermal spraying process that requires electrically conductive substrates acting mostly as the anode. The feedstock is wires or powder form with particle sizes in 50–150 μm range. Metals, alloys, and cermets are sprayed using PTA techniques. The different guns are characterized by the maximum current varying from 200 to 600 A. The coating is different from those obtained from other thermal spray coating processes because the arc melts to some degree the substrate material, and the molten powder of the coating material is mixed with the molten substrate material. The consequence is that a good metallurgical bond or fusion bond is achieved, and that the porosity is very low; however, the heat penetration region, and in particular the region where the coating material is mixed with the substrate material, can change the properties of both the substrate and the coating. This mixing region is expressed in terms of the "dilution." Torches used are divided into three different categories according to the deposition rate and the power used: micro PTA with deposition rates of 0.1–2 kg/h, used for small components, or components with complex shapes; regular PTA for deposition rates of 2–10 kg/h; and high-power PTA for deposition rates between 10 and 20 kg/h. Coatings are thicker than those of other spray processes; they can be 10 mm thick or more and fused with a metallurgical bond to the substrate. The substrate must be kept as horizontal as possible during spraying. No porosity is observed, and the deposition efficiency is over 90%. The resistance to wear is excellent as well as that to high-temperature corrosion. These coatings are used to achieve high-quality rebuilding of worn surfaces, to coat large and heavy parts without cracks or deformation, and to produce smooth and highly wear-resistant surfaces. PTA coatings are mostly used against wear in different mining, pulp and paper, oil and gas, and power industries.

19.3 Thermally Sprayed Coating Applications

In this section, a review is presented of the principal applications of thermal spray coating technology. The objective is to provide typical examples of present and potential industrial applications of each of these technologies rather than presenting an exhaustive review of each of these topics. The reader must keep in mind, however, that the TS coatings can have mechanical, thermal, and service properties that are distinctly different from those of bulk ones. These depend on the real contacts between layered splats, porosities, crack types and distributions, oxide inclusions, and possible partial modification of the chemistry of the sprayed particles during the deposition process. Coating morphologies and microstructure are also linked to the spray process used, the properties of the sprayed particles, the spray parameters, the substrate used, the coating thickness, and the possible posttreatments of coatings.

19.3.1 Wear-Resistant Coatings

"Wear is the problem" and it occurs in almost all industries where thermally sprayed coatings are used. In all cases, it is a progressive loss of material at the active surface due to the relative movement of another part or particles on this surface. The life time of the coating depends on its resistance to wear and its thickness, which can reach up to 10 mm with PTA coatings. Examples of coatings are presented for the different types of wear described in the following according to American Welding Society (1985), Cartier (2003), Zhum Gahr (1987), Chattopadhyay (2001), Pawlowski (1995), Champagne (2007), and Sobolev et al. (2004).

19.3.1.1 Abrasive Wears

Abrasive wear (AW) corresponds to weight loss with grooves, pits, and scores at the surface due to cutting or deformation, which results from:

- *Two-body abrasion*, where asperities or defects of one of the surface plough or abrade the counter-face.
- *Three-body abrasion*, where hard particles move freely between both surfaces, or are imbedded in one of them.

The phenomena are promoted when temperature, humidity, and aggressiveness of the environment (corrosion) increase. The three-body abrasion depends also on the shape, grain size, and hardness of the abrasive particles and the relative speed of the two bodies.

As a general rule, abrasion increases significantly as soon as the hardness of one metal or alloy becomes equal to that of the abrasive particles. If the abrasive particle belongs to the surrounding medium, the contact must be protected from it and wear debris must be removed or trapped. The roughness of the harder surface must also be reduced to the minimum. It must be emphasized that abrasion wear represents more than 50% of wear. In most cases, wear-resistant coatings are hard with a good resistance to heat and chemical attack. In the following, examples are given of materials commonly used for abrasive wear resistance coatings [American Welding Society (1985)].

Self-Fluxing Alloys

Self-fluxing alloys have generally excellent corrosion resistance. Due to their hardness, not being particularly high, they cannot compete with cermets for sliding wear resistance unless they are either *heat-treated* or sprayed with *hard particles*.

The effect of heat treatment on the wear resistance of self-fluxing alloys was investigated by Bolelli and Lusvarghi (2006) for HVOF-sprayed Co-28%, Mo-17%, and Cr-3% Si coatings. Comparing the as-sprayed coatings with those heat-treated at 200, 400, and 600 °C for 1 h, significant degree of splat boundary oxidation was observed in the as-sprayed coating, because of exothermic oxidative reaction occurring at $T > 810$ °C. The as-sprayed coating had low hardness and toughness, resulting in poor tribological performance. Posttreatment at 600 °C caused the appearance of submicrometric crystalline regions improving hardness and elastic modulus and the sliding wear performance at room temperature was improved. Sakata et al. (2007) flame sprayed Co-based (Co–Cr–W–B–Si) self-fluxing alloy coating on steel substrate followed by a diffusion treatment at 1370–1450 K for 10–100 min under Ar atmosphere. Two types of fine compounds were precipitated in Co-based matrix: a chromium boride dissolving cobalt, and a tungsten boride containing cobalt and chromium. The size of each precipitate became larger with increasing treatment temperature and time. A coating with the proper size borides showed a superior wear resistance with substantially improved abrasion resistance.

The effect of mixing hard metal particles with the self-fluxing alloy during the spraying process on the wear resistance property of the coating was tested by Kulu and Hailing (1998) who flame sprayed NiCrSiB self-fluxing alloy-base composite powders, containing 15–50 wt.% WC–Co hard metal powders. They compared the coating obtained with D-gun-sprayed WC–Co particles. The wear resistance to abrasive particles at *small impact angles* increased with the increase in the hardness of the matrix phase, and the increase in hard phase content in the composite. However, at *straight impact angles* the wear resistance remained relatively low. Further study by Kulu and Pihl (2002) showed for the spraying of self-fluxing alloys, containing tungsten carbide-based hard metal particles (NiCrSiB-[WC-Co]), using the HVOF JP 5000 that a good resistance to oblique and normal impact can be obtained provided the coatings have the following properties: minimum porosity (less than 3 vol.%), hardness higher than that of the abrasive, and a metal matrix structure containing particles of WC–Co granules, or WC–Co particles.

Cermet Coatings

Coating quality is strongly linked to the spray process and parameters, especially with WC particles. If corrosion problems are also to be considered, Cr is important in the metal matrix. For example, Schwetzke and Kreye (1999) have sprayed different WC–Co and WC–Co–Cr powders with various HVOF spray systems such as Jet Kote, Top Gun, Diamond Jet (DJ) Standard, DJ-2600, DJ-2700, JP-5000, and Top Gun-K. Powders exhibited various degrees of phase transformation during the spray process depending on type of powder, spray system, and spray parameters. Phase transformations increased when the injection of the powder occurred in a region where the flame temperature was highest, for example, into the combustion chamber of the

Top Gun system. Decarburization of agglomerated and sintered WC–Co 83–17 powder ranged from 25 to 70% for the various spray systems and fuels. When the carbon loss remained below 60%, the properties of the coatings such as hardness and wear resistance were not significantly influenced. The decarburization seemed to be compensated by the hardening with the formation of a solid solution of cobalt (tungsten, carbon) and hard W_2C and h-phases. While the wear resistance of WC–Co–Cr coatings were comparable to those of WC–Co, the corrosion resistance of WC–Co–Cr was considerably higher.

Kasparova et al. (2011) have evaluated the abrasive wear resistance and adhesive strength of HVOF-sprayed WC–Co and the Cr_3C_2–NiCr coatings. They found that high-stress abrasive conditions change the coating behavior very significantly, particularly that of the Cr_3C_2–NiCr coating. The high plastic deformation and pulling out of entire splats or splat blocks, especially during abrading by the alumina sand, were detected under the high-stress abrasive conditions for both coatings, but much more with Cr_3C_2–NiCr. As pointed out by Houdkova et al. (2010), the spraying angle is one of the deposition parameters that influences the quality of thermally sprayed coatings. They HVOF (TAFA JP-5000) sprayed WC–Co and Cr_3C_2–NiCr coatings with different spray angles. Their results showed that the wear resistance of the coating was not affected up to 30° angle diversion from the normal spray direction for WC–Co, and 15° angle diversion for Cr_3C_2–NiCr coatings. Tillmann et al. (2010) used statistical design of experiments to identify the most relevant factors influencing the HVOF spraying of fine $75Cr_3C_2$–$25(Ni20Cr)$ powders and to find an optimum setting of these factors to produce coatings with improved morphological and mechanical properties. Fine structured coatings obtained showed an extremely dense and finely dispersed structure (porosity < 2 vol.%), a high surface quality (Ra < 2 μm), and a high adhesive strength. These coatings showed a high potential to be used as wear-resistant coatings for large tools without any posttreatment or surface finish.

Considerable emphasis has been placed recently on HVOF thermal spraying of nanostructured WC/Co to achieve high hardness combined with excellent wear resistance. However, it appeared difficult to achieve dense coatings and avoid decarburization. Skandan et al. (2000) have HVOF sprayed homogeneously mixed powders. These consisted of WC/12 Co agglomerates in the range 15–40 μm with a carbide grain size of 2–5 μm (70 vol.% of the mixture) and WC/5 Co, forming particles in the range 0.1–0.5 μm with each particle composed of many WC nanometer-sized crystals (~30 nm in diameter). The coating was dense and had no decarburized phases. The abrasion wear resistance was at least 50% better than that of a pure coarse-grained WC/Co coating. In their review, Lima and Marple (2007) presented superior abrasion and sliding wear performance of nanostructured ceramic coatings (Al_2O_3, Al_2O_3–13 wt.% TiO_2, Al_2O_3–3 wt.% TiO_2, TiO_2, and YSZ) when compared to those of conventional coatings. By employing HVOF and nanosized ceramics, the abrasion wear levels could be reduced by up to 90% in comparison with the wear performance of optimized APS conventional ceramic coatings. It was generally observed that the nanostructured coatings were not harder than the conventional ones; however, they tended to be much tougher. Kim and Walker (2007) have developed nanostructured titania coating specifically for ball valves destined for high-pressure acid-leach (HPAL) service. These coatings were compared to conventional ones. The nanostructured titania coating provided significantly superior resistance against abrasive and erosive wear—the better wear performance attributed to improved toughness.

Gawne et al. (2001) have plasma sprayed ball-milled mixture of glass and alumina powders to produce alumina–glass composite coatings. The alumina raised the mean hardness from 300 Vickers hardness (HV) for pure glass coatings to 900 HV for a 60 wt.% alumina–glass composite coating. The scratch resistance increased by a factor of 3 and the wear resistance by a factor of 5, and maximum value obtained with 40–50 vol.% alumina. This alumina content corresponded to the changeover from a glass matrix to an alumina matrix. Cipri et al. (2007) produced thick aluminosilicate coatings by plasma spray technique. Coatings, efficiently coupling to metallic substrates even after plastic deformation, exhibited very interesting performance in terms of mechanical properties and fracture toughness (elastic modulus of 43 GPa and K_{1c} of about 2 MPa m$^{1/2}$ for 850 μm thick coatings) and compliance. Excellent refractory behavior allows a wide use of such coatings, as wear-resistant thermal barrier coatings (TBCs) in metallurgical and glass plants and in high-temperature heat exchangers. To illustrate the interest of such coatings, Kang et al. (2012) used detonation-gun to spray three different WC–Co–Cr, Cr_3C_2–NiCr, and Stellite-21 coatings on high-tensile steel rotavator blades. The wear rates of Cr_3C_2–NiCr and Stellite-21 coated blades showed significant superiority over the uncoated blade, but not as much as that obtained by WC–Co–Cr coated blades.

19.3.1.2 Erosive Wear

Erosive wear (EW) occurs when hard particles carried by a fluid hit the surface. The impact angle plays an important role on the wear rate depending on the nature of the surface material; ductile, hard metal, ceramic [Zhum Gahr KH (1987), Chattopadhyay R (2001)]. To reduce the wear, one must choose a coating material harder than the abrasive particles, with toughness high enough, especially for impact angles between 30 and 45°, which must be avoided if possible. It is also important to avoid turbulences in service

conditions because they increase erosion and reduce coating roughness to a minimum.

Erosion wear is much more complex than abrasive wear in which the wear resistance is predominantly influenced by the hardness of the coatings. The erosion wear resistance depends on the response of coatings to the impact of erosive particles at high velocity. It is usually considered that erosion wear of cermet coatings is predominately influenced by their microstructures including the splat size, carbide particle size, carbide content, and its distribution within a splat, as well as cohesion between the splats [Kim HJ, et al. (1994), Wang BQ, Luer K (1994)].

Kim et al. (1994) showed that inter-splat cohesive strength of coatings, measured by a simple bonding test, was the most significant factor relating to the wear rate of plasma-sprayed WC-12 wt.% Co coatings. As previously emphasized, the microstructure strongly depended on starting powder, spray system, and spray conditions. Other parameters, such as the influence of time, solid loading, impingement angle, temperature, particle velocity, as well as the erosion–corrosion (E-C) mechanism, must also be considered [Kenichi S, et al. (2005)]. When comparing the influence of dry and slurry erosion on HVOF coatings, it was shown that erosion rates in dry particle impact were about three orders of magnitude higher than those in slurry systems. This difference probably reflects the real erodent target impact velocities, which are mitigated in the slurry test by the water medium [Hawthorne HM, et al. (1999)]. A few examples are presented below.

Ji et al. (2007) HVOF sprayed Cr_3C_2–NiCr coatings and erosion tests were performed at different jet angles of abrasive particles. The erosion occurred dominantly by spalling of splats from the lamellar interfaces, spalling resulting from the propagation of cracks parallel to the interfaces between the lamellae exposed at the surface and underlying coating. The carbide particle size and content in the coating influenced significantly the erosion performance of Cr_3C_2–NiCr coatings. Yang et al. (2008) compared the high-temperature erosion behavior of HVOF-sprayed Cr_3C_2–NiCr coating with that of mild steel for circulating fluidized-bed boiler tubes. The erosion rate of the HVOF-sprayed Cr_3C_2–NiCr coating was not influenced by the temperature in the range of 300–800 °C, while for mild steel at 800 °C the erosion rate was four times that at 300 °C at an erosion angle of 30°.

Osawa et al. (2005) have shown for WC–Co HVOF-sprayed coatings that the substrate also played a role in coating erosion resistance. For the studied variety of steel substrates, the near-surface hardness of the substrate resulting from the work hardening caused by peening during the grit blasting and HVOF spraying played a dominant role. It improved the ability of the substrate to support the coating and thus the integrated coating–substrate performance during impact loading. Kulu et al. (2005) sprayed, by D-gun, HVOF

JP-5000 spraying, and spray fusion processes, tungsten carbide-based hard metal, nickel-based self-fluxing alloy, and composites NiCrSiB. For all of them the erosion was strongly affected by particle impact angle. The erosion rate was five to six times higher at elevated temperature for all materials tested, while in that case, the influence of impact angle had no great effect.

Ramesha et al. (2011) plasma sprayed Inconel-718 (thicknesses of 200 μm and 250 μm) on mild steel. Increased coating thickness decreased the porosity and increased the hardness significantly. Inconel-718 coatings exhibited improved slurry (containing 3.5 vol.% NaCl) erosive wear resistance when compared to uncoated mild steel. The wear resistance also increased with coating thickness. Higuera Hidalgo et al. (2001a, b) compared the behavior of plasma- and flame-sprayed modified nickel–chromium alloy (with small aluminum and titanium additions) subjected to the action of simulated post-combustion gases from a coal-fired boiler combustor. With flame, the adherence was much lower even with a bond coat. High-temperature oxidation rate of nickel–chromium coatings in post-combustion gases from coal-fired boilers (atmospheres with 3–3.5 vol.% of free oxygen) was low. The embedment of fly ash particles on the surface of coatings was especially important at higher temperatures (800 °C). The erosion behavior of nickel–chromium flame and plasma-sprayed coatings at medium temperatures (500 °C) was characterized by low erosion rates and a ductile erosion mechanism. At 800 °C the wear was essentially corrosive.

Krishnamurthy et al. (2012) studied the erosion resistance of plasma-sprayed alumina and calcia-stabilized zirconia coatings on Al-6061 substrate. It was found that erosion of coating systems occurred through spalling of lamella exposed on coating surface resulting from cracking along the lamellar interface. Erosion wear was more at 45° angle of impact. Pores seemed to act as stress concentrators and decreased the load-bearing surface.

The promising behavior of unconventional nanoscale composite coatings must be underlined. Branagan et al. (2005)] have wire arc sprayed (WAS) amorphous SHS-7170. Coatings were found to develop an amorphous matrix structure containing starburst-shaped boride and carbide crystallites with sizes ranging from 60 to 140 nm. The performance of the SHS-7170 coatings in boiler environments was measured via elevated temperature erosion experiments conducted at 300, 450, and 600 °C using bed ash from a circulating fluidized-bed combustor (FBC) boiler, and the results were compared with those for existing boiler coatings. The elevated temperature erosion resistance of the SHS-7170 wire arc coatings was found to be superior, based on thickness loss, compared with the existing wire arc coatings that have been tested. SHS-7170 coatings have resisted to erosion almost independently of contact angle.

19.3.1.3 Friction and Adhesive Wear

Friction and adhesive wear occur when particles are transferred from one interacting surface to the other. When different materials are in contact, particles are mainly transferred from the softer or weaker material onto the harder one. This type of wear is promoted by the increase in the load and/or the temperature, under dry friction or poor lubrication. It depends on the structure, composition, hardness, and melting temperature of the material.

To reduce it, dry friction must be avoided and, if not possible, coatings containing solid lubricants or retaining lubricants must be used. The compatibility of materials is also important since the friction coefficient, f, defined by Eq. 19.1, depends on the couple of materials rubbing against each other.

$$f = \frac{T}{N} \tag{19.1}$$

Where,

T tangential force
N normal force

The friction coefficient is also linked to the load, P (or more precisely the pressure, p), and the relative velocity, v, between both parts (in principle the product, $p \cdot v$, must be below 1 MPa m/s) [Cartier M (2003)]. The roughness of surfaces in contact must also be as low as possible and from a rough surface (Ra of a few μm) to a smooth one (Ra < 0.1 μm), f, can be reduced by 60%.

Dry friction usually results in high local heating and, even with a very low relative velocity, the friction coefficient, f, increases with temperature. Low friction coefficients can be achieved with solid lubricants ($f = 0.001$–0.05), but the relative velocity must be low. Using materials with a high thermal conductivity (both coating and substrate) reduces the heating due to friction. As wear increases with the energy of adhesion, it is also important to avoid micro welding, that is, to use materials with a low solubility and a narrow range of solid solution (e.g., Fe/Al, Fe/Cu). The work of adhesion decreases with the following couples: metal–metal, metal–ceramic, ceramic–ceramic, metal–polymer, ceramic–polymer, and polymer–polymer. A few examples are presented in the following.

High stress, due to too high average pressure or very high local pressure that exceed the elastic limit, might be favorable when it strengthens the material by work hardening, but it might result in surface embrittlement of surface layers reducing their fatigue strength. It is promoted by an increase in the friction coefficient or temperature. Here again the worst occurs when cracks propagate.

Ceramic Materials

Pantelis et al. (2000) plasma sprayed a 450 μm thick Al_2O_3 coating deposited on cast iron substrate. Wear tests were carried out with normal force varying from 50 to 160 N, sliding speed of 1.40 m/s, and average relative humidity of 60%, using as a counter-body a quenched D-2 tool steel (D-type tool steels contain between 10% and 18% chromium; D-2 steel is very wear resistant but not as tough as lower alloyed steels). The coating wear rate presented three stages.

- During the first one, the wear rate was decreasing rapidly, and the wear of the coating proceeded via adhesion mechanism.
- During the second stage, wear rate remained almost constant and the wear of the coating was taking place via a combined "polishing/abrasion/fatigue" wear mechanism.
- During the third and last stage, wear rate increased rapidly due to the "easy" removal of the remaining completely cracked ceramic coating.

Sanchez et al. (2008) plasma sprayed Al_2O_3–13 wt.%TiO_2 coatings on stainless steel substrates from conventional and agglomerated nanostructured powders. The wear resistance of conventional coatings was shown to be lower than that of nanostructured coatings. As already emphasized in Chap. 16, Nanocrystalline and Nanostructured Coatings, Darut et al. (2008) found that the friction coefficient of Al_2O_3 suspension plasma sprayed (SPS) coatings decreased by a factor of 2 and wear rates were 30 times lower for SPS layers compared to coatings obtained with micrometer-sized particles.

Bolelli et al. (2009a, b) HVOF sprayed alumina suspensions and compared them to coatings obtained with conventional powders sprayed with APS and HVOF. With the suspension, porosity is much lower, and pores are smaller than in conventional coatings. Moreover, few inter- or intra-lamellar cracks exist, resulting in reduced pore interconnectivity (evaluated by electrochemical impedance spectroscopy [EIS]). Such strong interlamellar cohesion favors much better dry sliding wear resistance at room temperature and at 400 °C.

Ahn and Kwon (1999) studied the tribological behavior of plasma-sprayed chromium oxide coatings against cast iron both in dry, at 450 °C, and lubricated wear tests at room temperature and 200 °C. Under dry sliding conditions, dispersed smooth surface films were formed by plastic deformation of compacted debris particles that adhered to the surface and these films strongly influenced the friction coefficient and wear rate of the coating. Considerable quantity of CrO_3 was detected at room temperature, whereas CrO_2 (more favorable than CrO_3 in reducing friction) was detected at 450 °C. Under lubricated sliding conditions, tribo-chemical reaction films of different species of carbon–oxygen bond units were formed depending on the test temperature and they

appeared to be effective in reducing friction and preventing wear. Pratap Singh et al. (2011) studied the tribological behaviors of plasma-sprayed conventional and nanostructured (agglomerated nanoparticles) Cr_2O_3–3% TiO_2 ceramic coatings. Samples coated with nanostructured powder exhibited better properties such as higher hardness and less porosity as compared to conventional powder-coated samples. Dry sliding wear resistance of nanostructured powder under 60 N load was better than the wear resistance offered by conventional powder under 50 and 60 N load for similar testing conditions. In general, nanostructured powder exhibits a better wear resistance than conventional powder.

Tao et al. (2010) deposited Al_2O_3 and Cr_2O_3 coatings on stainless steel by atmospheric plasma spraying and tested their dry sliding tribological properties against copper alloy using a block-on-ring configuration at room temperature. The wear resistance of Al_2O_3 coating was superior to that of the Cr_2O_3 coating under these conditions. This was mainly attributed to the better thermal conductivity of Al_2O_3 coating (about 2.8 W/m K for alumina against 2.4 for chromia at 400 °C), which was considered to effectively facilitate the dissipation of tribological heat and alleviate the reduction of hardness due to the accumulated tribological heat. Bolelli et al. (2006a, b, c) plasma sprayed ceramic coatings (Al_2O_3, Al_2O_3–13 wt.% TiO_2, Cr_2O_3) and tested them through pin-on-disk and dry sand–steel wheel tests. The toughest coating (Al_2O_3) displayed the highest wear resistance, which in fact overcame HVOF-sprayed cermets and Cr electroplating, when a low number of wheel revolutions was considered. Against the alumina ball, Al_2O_3 and Al_2O_3–TiO_2 coatings showed high wear rates and friction coefficients (due to chemical affinity), while Cr_2O_3 had better wear resistance, lower friction coefficient, and inflicted less wear on the counterpart. Cr_2O_3 wear scar consisted in plastically deformed splats and debris forming a quite adherent protective tribo-film. In pin-on-disk tests, no coating underwent wear loss against the 100Cr6 ball that possessed lower hardness. Ramachandran et al. (2012) studied the friction and wear behaviors of yttria-stabilized zirconia (YSZ) coatings, lanthanum zirconate (LZ) coatings, and Inconel-738 Base Material (BM) sliding under unlubricated conditions, against a sintered tungsten carbide surface in a pin-on-disk configuration. They found that the wear resistance of the ceramic coatings got deteriorated with the increase in the percentage volume of porosity.

Cermets

Dallaire (2001) studied a cored wire, referred to as Alpha 1800, developed to produce tailored arc-sprayed coatings that were tough enough to resist particle impacts at 90° and sufficiently hard to deflect eroding particles at low-impact angles. Results showed that coatings produced with the new cored wire are at least five times more erosion resistant and ten times more abrasion resistant than coatings produced by arc spraying commercial cored wires.

Yang et al. (2006a, b) sprayed by an HVOF system three-agglomerated WC-12 wt.% Co powders with different carbide grain size distributions. Dry sliding friction and wear behavior of the WC-12 wt.% Co coatings were investigated using sintered alumina (Al_2O_3) as the mating material at 200 °C, 300 °C, and 400 °C. The specific wear rate of the coatings increased when increasing the carbide grain size at a given testing temperature. It decreased when increasing the temperature for a given carbide grain size. The specific wear rate was reduced by more than one order of magnitude when the test temperature was increased from room temperature to 400 °C. The tribo-film formed at higher temperature was denser and more adhesive to the underlying surface, thus providing more surface protection against wear. The results showed that the formation of dense and well-adhered tribo-films played an important role in the low sliding wear rate of the coatings at elevated temperatures.

Yandouzi et al. (2012) sprayed by pulsed gas dynamic spraying (PGDS) process both cryo-milled particles made of an Al matrix reinforced with B_4C powders (Al-5356 + 20% B_4C) and conventional composite. The presence of homogeneously distributed fine B_4C reinforcement particles embedded within the nanostructured Al-5356 matrix significantly improved the dry sliding wear resistance of the coating. Li et al. (2008) cold sprayed dense Al5356/TiN composites with TiN particles uniformly dispersed in the matrix (50 wt.% TiN). The coating porosity was less than 1 vol.%. The deposited composite coating presented an excellent performance compared to that of the composite sprayed without using the ball-milled powder.

Qiao et al. (2001) studied the resistance to abrasive and unlubricated sliding wear of 40 WC/Co coatings applied by HVOF, high-energy plasma spray (HEPS), and high-velocity plasma spray (HVPS), using commercial and nanostructured experimental powders. Phase analysis by X-ray diffraction (XRD) revealed various amounts of decarburization in the coatings, some of which contained WC, W_2C, W, and h-phase. The wear resistance was lowered by decarburization, which produced a hard but brittle phase.

Jacobs et al. (1999) investigated the microstructural properties of WC–Co–Cr and WC–Co coatings deposited by HVOF and HVAF processes. The HVAF-sprayed coatings showed better sliding wear resistance compared to the HVOF coatings. The prime wear mechanism in the WC–Co HVAF coatings was adhesive wear; the lubricious cobalt matrix resulted in very low wear rates and low debris generation. The WC–Co–Cr HVOF coating wear was linked to "pullouts" that became trapped in the contact zone and acted as a third-body abrasive. The HVAF/WC–Co–Cr coatings exhibited better resistance.

Bolelli et al. (2012a, b) studied the tribological performance of two Colferoloy Fe–Cr–Ni–Si–B–C coatings HVOF sprayed and they compared them to other coatings. Under sliding wear conditions, these coatings were a good alternative to Ni-based alloy coatings, but they could not replace Cr_3C_2–NiCr cermets (best sliding wear). These coatings were unsuitable for dry particle abrasion conditions. At 400 °C, all coatings were softened, and their sliding wear behavior was dominated by more severe abrasive grooving and all differences were reduced.

Alam et al. (2001) studied the tribological characteristics of low-pressure plasma spraying (LPPS) aluminum bronze coatings against steel ball. Under optimum operating conditions, the test samples exhibited a dense microstructure with high hardness, low coefficient of friction, and high wear resistance.

Ouyang et al. (2001) studied the tribological properties of VPS ZrO_2–CaF_2–Ag_2O composite coating (ZFA). At 300–700 °C, the ZFA coating exhibited lower friction and wear than at room temperature, 200 °C, or 800 °C. Ag_2O and CaF_2 acted as solid lubricants effectively at 300–400 °C and 600–700 °C, respectively. But with the increase in temperature up to 800 °C, the severe adhesive sliding caused more material transfer and tearing out of coating, and finally led to a high friction and wear.

Metals

Ahn et al. (2005) studied the wear resistance, on a low-carbon steel substrate, of plasma-sprayed molybdenum blend coatings consisting of powders of bronze and aluminum–silicon alloy powders mixed with molybdenum. The wear test results revealed that the wear rate of all coatings increased with rising wear load and that the blended coatings had better wear resistance than the pure molybdenum coatings, although its hardness was lower. The molybdenum coating blended with bronze and aluminum–silicon alloy powders exhibited excellent wear resistance because hard phases such as $CuAl_2$ and Cu_9Al_4 formed inside the coating.

Bolelli et al. (2012a, b) studied the tribological performance of two Fe–Cr–Ni–Si–B–C (Colferoloy) alloy coatings manufactured by HVOF thermal spraying, through rubber-wheel dry particle abrasion test and ball-on-disk sliding wear tests. Colferoloy coatings were validated as alternatives to Ni-based alloys and electroplated chromium under sliding wear conditions but appeared unsuitable for particle abrasion resistance. The different sliding wear behaviors of HVOF-sprayed coatings were explained by coupling micro- and nano-hardness to scratch testing, which reflected cracking resistance and plastic deformability.

Rodriguez et al. (2003) compared NiCrBSi alloys sprayed by either plasma or flame but then fused for the latter. Four sets of specimens were tested varying the composition (with the presence or not of WC in the powders). Coatings were tested in a reciprocating pin on plate wear machine able to select loads ranging from 50 to 200 N and temperatures up to 500 °C. NiCrBSi alloys, deposited by thermal spray techniques, maintained their wear-resistant performances up to bulk temperatures of 500 °C. The alloys deposited by flame spraying with fusion had better wear resistance than those deposited by plasma spraying. The presence of tungsten carbide in the powders was not a beneficial factor on the wear performance of thermally sprayed NiCrBSi coatings. PTA is used to increase wear resistance mainly in steel industry.

Polymers

At low temperatures polymers can be used against friction. Niebuhr and Scholl (2005) sprayed with high-energy plasma polymer–steel coatings for high-contact pressure rolling/sliding systems. Polymers were applied as a thin film (75–100 μm) over a thicker (250–750 μm) steel coating. Twin roller rolling/sliding tests were performed at 5 and 35% creep and contact loads of 1700 N on a 5 mm contact face. A lower coefficient of friction (0.10–0.15) with increased durability compared with that of AISI-1080 steel thermally sprayed coating (coefficient of friction of 0.46) was observed under these rolling–sliding contact conditions.

Li et al. (2002) deposited, using the flame spray process, poly-ether-ether-ketone (PEEK) coatings with three kinds of crystallinities. Investigations were performed under dry sliding conditions against a 100C6 counter-body. The average friction coefficients appeared to decrease while increasing the sliding velocity but were insensitive to the applied load in the range of investigation. The higher was the crystallinity of the coating, the lower was its average friction coefficient. The wear mechanisms of the different coatings were explained in terms of plastic deformation, plough marks, and fatigue tearing.

Li et al. (2007) flame sprayed polyamide1010 (PA-1010) and its composite with nanometer-sized silica (PA-1010/n-SiO_2). The dry friction and wear behaviors of both coatings were investigated under dry sliding conditions. The results showed that the addition of nanometer-sized silica increased the crystallinity of the coatings and reduced friction and wear compared to pure PA-1010. The nano-composite coatings, containing 1.5 wt.% nanosized silica, displayed better properties.

19.3.1.4 Cavitation Wear

Cavitation wear characterized by the formation of pits and pores occurs when the velocity difference between the liquid and solid surfaces is important. They result from sudden variations of pressure in the fluid leading to the creation of shock waves due to implosion combined with fatigue phenomenon. Cavitation wear can be significantly reduced or

eliminated by simply redesigning the solid surface profiles and reducing low-pressure areas. Surface polishing is also very important since a low Ra can reduce the wear by a factor of 5. Alternately, the use of high-toughness alloys with a good resistance to micro-crack formation is essential.

Wire arc spraying (WAS) has been extensively used against cavitation with results depending on surface hardness and internal coherency of the coating. According to Kim and Lee (2010), among ten different coating materials tested, stainless steel with the highest hardness showed the best performance against the cavitation erosion. With Al-based coatings, the addition of silicon to the alloy (12 wt.% Si) improved both the hardness and bond strength, resulting in significant improvement in cavitation erosion resistance by almost 40 times. Unfortunately, the addition of Si to Al-based coatings caused surface pitting similar to those of Zn-based coatings. In both the Cu- and Fe-based coatings, interfacial oxidation occurred in between the coating and steel substrate.

Hahn and Fischer (2010) tested the cavitation resistance of WAS- and HVOF-sprayed $FeC_{0.8}$ coating, and $FeCr_{19}C_{0.1}B_{1.6}$ coating deposited by plasma-transferred arc (PTA) fed with wire. They showed that for all coatings the weight loss immediately starts with the beginning of the test. The major acting wear mechanism within cavitation was surface fatigue with splat delamination and crack propagation along splat interfaces, the worst being when coating had a very low ductility. Kumar et al. (2005) reported that out of 21 different TS coatings, HVOF-sprayed coatings generally exhibited lower cavitation wear rates than plasma-sprayed coatings. The lowest cavitation rate was for TS Stellite-6 coating, which had a cavitation erosion rate of 11.7 mg/h. The results, however, were still considerably higher than that obtained with weld deposited 308 stainless steel, which was 3.2 mg/h. In contrast, slurry erosion wear testing showed that the volume loss for Stellite-6 coatings was 5.33 mm^3/h, which was considerably lower than the volume loss of 11.17 mm^3/h reported for 304 stainless steel.

Factor and Roman (2002) have subjected a selection of WC–Co and Cr_3C_2–25% NiCr coatings produced by PS and HVOF deposition techniques to various wear tests designed to simulate abrasion, cavitation, sliding, and particle erosion type wear mechanisms. Cr_3C_2–25 wt.% NiCr outperforms the WC–Co samples, and it appeared that the WC–12 wt.% Co sample performed better than the WC–17 wt.% Co samples. For all samples the wear loss mechanism was clearly that of the delamination of flakes of material. With WC–Co samples, corrosion mechanisms played an important role, possibly through enhanced corrosion resulting from the cavitation mechanisms causing pit corrosion or similar behavior. Wu et al. (2012a, b) studied a WC–Co–Cr coating deposited by HVOF onto a $CrNi_{18}Ti_9$ stainless steel substrate to increase its cavitation erosion resistance. The microstructural analysis of the coating after the cavitation erosion tests indicated that most of the corruptions took place at the interface between the unmolten or half-melted particles and the Co–Cr matrix, the edge of the pores in the coating, and the boundary of the twin and the grain in the stainless steel $CrNi_{18}Ti_9$.

Santa et al. (2009) studied the slurry and cavitation erosion resistance of six thermal spray coatings and compared them to that of an uncoated martensitic stainless steel. The results showed that the slurry erosion resistance of the steel could be improved by up to 16 times through the application of the thermally sprayed coatings. On the other hand, none of the coated specimens showed better cavitation resistance than the uncoated steel in the experiments.

Lima et al. (2004) deposited the following four different coatings onto an AISI- 1020 steel substrate:

- WC–12%Co
- As-sprayed (AS) 50%WC–12%Co + 50%NiCr
- Post-melted (PM) 50%WC–12%Co + 50%NiCr
- A duplex system comprising WC–12%Co top layer and NiCrAl interlayer

The worst performance in cavitation erosion tests was observed for the WC–12%Co coating, which showed the highest mass loss throughout the test. Conversely, the PM 50%WC–12%Co + 50%NiCr coating exhibited the best cavitation resistance and a correlation between coating toughness and cavitation resistance could be established.

Ding et al. (2011) reported that for conventional, submicron, and multimodal WC–12Co cermet HVOF-deposited coatings, the multimodal WC–12Co coating exhibited the best cavitation erosion resistance among the three coatings tested. The erosion rate was approximately 40% that of the conventional coating, and the cavitation erosion resistance of multimodal WC–12Co coating was enhanced by >150% in comparison with the conventional coating. Dense nanostructure, high microhardness, and strong cohesive strength of WC–12Co coating contributed to the increase in the cavitation erosion performance of multimodal WC–12Co coating.

19.3.1.5 Fatigue Wear

Fatigue wear (FW) is a form of material wear that can result from either surface fatigue, thermal fatigue, or thermal chock fatigue. In the following, highlights of these different fatigue-wear mechanisms are discussed identifying their main features, probable causes, and the role thermal spray coating can play to control them.

Surface Fatigue Wear

Surface fatigue wear (SFW) results in the development of surface defects such as pits and pores that can extend to depths of several tenths of millimeters due to cyclic loading contacts, with stresses induced by rolling, shocks, or sliding

in a lubricated regime. They depend on material properties such as structure, cohesion, elastic limit, toughness, and residual stresses. If tangential stress is predominant, fatigue will start in the "skin" while if shear stress is more important, fatigue will occur essentially in the subsurface. The worst SFW occurs when cracks propagate. The best materials to limit SFW are hard ones with a high toughness with smooth surfaces with no irregularities where cracks are initiated.

Stewart and Ahmed (2002) made, in 2002, an extensive review of the thermally sprayed coatings used in rolling contact fatigue (RCF). The following coatings were applied to steel substrate using different thermal spray technologies:

- WC–12%Co, WC–17%Co, and WC–10%Co–4%Cr were sprayed using APS or HVOF
- NiBSiCrFeC, Al_2O_3–TiO_2, Cr_2O_3–SiO_2–TiO_2, and Mo were sprayed by APS
- WC–Co and WC–Cr–Ni were sprayed by HVOF
- WC–12%Co, WC–15%Co, and Al_2O_3 were sprayed by D-gun

The HVOF-deposited coatings presented superior RCF performance, probably because of their dense microstructure and high cohesive strength combined with a minimum number of detrimental brittle phases. Cermet coatings displayed superior RCF performance followed by ceramic and metallic coating, respectively. Thermal and mechanical mismatch between the coating and the substrate influenced the RCF performance of the coating through the control of the degree of compressive residual stress. Coatings thicker than 200 µm displayed superior RCF performance over thinner ones.

Ahmed and Hadfield (2002) studied cermet (WC–Co) and ceramic (Al_2O_3) coatings deposited by D-gun, HVOF, and high-velocity plasma-spraying techniques, in a range of coating thicknesses (20–250 µm) on various steel substrates to deliver an overview of the various competing failure modes in rolling/sliding contact. Four modes of fatigue failure, competing during fatigue failure and that can be combined, were identified: abrasion, delamination, bulk deformation, and spalling. The authors concluded that *abrasion* can be controlled by appropriate selection of contacting pair and lubrication conditions while *delamination* and *catastrophic* failure mode can be avoided by appropriate selection of coating thickness and fracture toughness. Controlling the hardness of substrate and also increasing the coating thickness can avoid bulk failure. Coating failures were attributed to micro- and macro-cracking of either the coating material or the coating–substrate interface, which also resulted in the attenuation of compressive residual stress.

Savarimuthu et al. (2001) studied the possible replacement of chromium electroplating with WC–Co-sprayed coatings for commercial aircraft applications. Decarburization, volume percent and distribution of hard particles, and porosity might more strongly influence wear behavior than residual stresses. Zhang et al. (2008) addressed the fatigue behavior and failure mechanisms of plasma-sprayed CrC–NiCr cermet coatings in rolling contact. As emphasized previously, failure modes of coatings could be classified into four main categories, that is, surface abrasion, spalling, and delamination within the coating, and at the coating/substrate interface. The interfacial delamination was the main failure mode of the coatings at high contact stress. The failure mechanisms of coatings were associated with the microstructures and the bonding strengths of coatings, the depths of the orthogonal shear stress, and the maximum shear stress.

Berger et al. (2011) studied the dependence on substrate hardness of the rolling contact fatigue of HVOF-sprayed WC–17%Co coatings used for gears and other components with high load-bearing capacity. The durability of HVOF-sprayed WC–17 wt.% Co coatings under rolling contact fatigue loading was significantly improved through the use of substrates of different hardnesses. The highest endurable Hertz pressure for a probability of survival of 50% at 5×10^7 cycles was obtained with quenched and tempered substrates of intermediate hardness of 400 Hardness Brinell (HB). Crack formation and subsequent delamination appeared to be the main causes of failure.

Bolelli et al. (2012a, b) studied the tribological performance of two coatings of Fe–Cr–Ni–Si–B–C (Colferoloy) alloy HVOF sprayed. They compared results to those obtained on Ni–Cr–Fe–Si–B–C and Cr_3C_2–NiCr layers (also HVOF sprayed), hard chromium electroplating, and bulk tool steel. Colferoloy coatings were validated as alternatives to Ni-based alloys and electroplated chromium under sliding wear conditions but appeared unsuitable for particle abrasion resistance. The different sliding wear behaviors of HVOF-sprayed coatings could be explained by coupling micro- and nano-hardness to scratch testing, which reflected cracking resistance and plastic deformability.

Ganesh Sundara Raman et al. (2007) D-gun sprayed Cu–Ni–In coatings on two substrate materials: Ti–6Al–4V and Al-alloy AA-6063. The Cu–Ni–In coating was found to be beneficial on the Ti-6-4 alloy, while deleterious on the Al-alloy substrate under both plain fatigue and fretting fatigue loading. The results were explained in terms of differences in the values of surface hardness, surface roughness, surface residual stress, and friction stress.

Akebono et al. (2008) investigated the factors that most strongly influenced fatigue properties of steel flame sprayed with Ni-based self-fluxing alloy and prepared three types of thermally sprayed specimens, which differed in heating time in the fusing process. The sprayed specimens fused for longer time exhibited lower fatigue strength, because of the segregation of the chromium compound in the coated microstructures. Performing the fusion for a shorter time

was effective for producing sprayed materials of enhanced fatigue-resistant properties.

Thermal Fatigue Wear

Thermal fatigue occurs in components subjected to high temperature gradients resulting from friction or operating environment (for example, external surface subjected to hot gases). It results in high stresses, surface fatigue, and mechanical transfer (transfer of the softer body to the harder one). When thermal fatigue tests are performed in air, very often oxidation problems take an important part in coating performance. For thermal fatigue testing, the heat flux to the sample is applied using a flame, followed by air cooling, with or without convection. As it could be expected, most studies were devoted to thermal barrier coatings (TBCs). The following two examples are not related, however, to TBCs.

Aoh and Chen (2001) investigated high-temperature (700 °C) wear characteristics of Stellite-6 alloy containing Cr_3C_2 after thermal fatigue and oxidation treatment at 700 °C. The coating was deposited by plasma-transferred arc (PTA) process. Thermal fatigue cracks initiated from the surface of Stellite-6 with Cr_3C_2 and propagated into the coating along carbide boundaries. No thermal fatigue crack was observed on sprayed Stellite-6 layer. The Cr_3C_2 content in the alloy enhanced the formation of the oxide layer. Surface oxidation was beneficial to the improvement of wear resistance and the coating exhibited the lowest wear loss if an oxidation treatment at 700 °C/100 h was achieved prior to the wear testing.

According to D'Ans et al. (2011), the major wear mode of steel in the aluminum foundry industry is thermal fatigue, especially affecting die-casting dies. Steel is also intensively corroded by molten aluminum and, sometimes, sliding wear occurs when extracting the molded parts from the die. D'Ans et al. (2011) conducted thermal fatigue tests with samples of hot work tool steel, either simply borided, or protected by a multilayer of YSZ, followed by a nickel superalloy and then a borided layer. The multilayer coating resistance to thermal fatigue (heating between 500 and 800 °C) turned out to be lower than for a single-treatment solution (boriding) and for plain steel.

Eriksson et al. (2011) subjected air plasma-sprayed TBC to isothermal and cyclic heat treatments (thermal cycling fatigue: TCF). Specimen subjected to TCF and isothermal oxidation gave very similar interface between thermally grown oxide (TGO) thickness and TGO composition. The heat treatment influences the adhesion of TBCs: isothermal oxidation up to 290 h has been shown to have a beneficial effect on adhesion compared to the as-sprayed condition, while cyclic heat treatment has been shown to decrease adhesion. Kwon et al. (2006) studied a TBC with topcoat deposited by APS and bond coat by HVOF. Thermal fatigue did not affect the stress–strain curves, except for thermal fatigue for 500 h. However, the thermally grown oxide (TGO) layer thickness was dependent on the exposure time under thermal fatigue, showing a nominal thickness of approximately 4 μm after thermal fatigue for 500 h, independent of the number of thermal fatigue cycles. As the exposure time was increased in the thermal fatigue experiments, the damage to the topcoat was inhibited with less crack coalescence. The higher stiffness in the bond coat induced cracking or delamination at the interface between the bond coat and the substrate, whereas thermal fatigue increased the mechanical properties, especially the elastic modulus, of the bond coat and enhanced the damage tolerance of the TBC system.

Thermal Shock Fatigue Wear

Thermal shock occurs when the thermal gradient causes different parts of an object to expand by different amounts, generating stress or strain. At some point, this stress can exceed the strength of the material, causing a crack to form. If nothing stops this crack from propagating through the material, it will cause the object's structure to fail. The thermal shock resistance is tested by alternate high-temperature exposure and cooling cycles until failure. The heating is achieved by a furnace (moderate thermal gradients), followed by cooling using cold water (high thermal gradients). Alternately, the heating is made with a flame on one side of the sample, while the opposite side of the sample is air-cooled (high thermal gradients). The cooling step can also be achieved through replacing the flame by an air jet (moderate thermal gradients). The heating step can also be achieved with a fluidized bed or a laser.

Luo et al. (2011) deposited by high-velocity wire arc spraying using cored wires FeMnCrAl/Cr_3C_2 and FeMnCrAl/Cr_3C_2–Ni9Al coatings onto low-carbon steel substrates. The specimens were placed inside a furnace at 800 °C, held for 15 min, and then quenched into a water bath (25 °C). The first type of coating consisted of layered splat of mainly Fe solid solution phases mixed with oxide phases, unmelted particles, and pores. The second type of coating exhibited higher bonding strength; Ni and Fe formed the diffusion layer that improved the thermal shock resistance. The cracks in the coatings were mainly initiated and propagated along the oxide phases during the thermal shock test. The uniformly shaped pores in the coatings helped prevent crack initiation and propagation.

Pan et al. (2012) studied the effect of SiC particles on thermal shock resistance of Al_2O_3–20 wt.% 8YSZ coatings air plasma sprayed. The thermal shock resistance of Al_2O_3/8YSZ coatings was increased greatly from 5 to 83 cycles with addition of SiC particles at 1000 °C. Some silicate fibers were formed on the coating surface and through thickness after thermal shock. The $Al_2O_3 \cdot xSiO_2$ silicate fibers increased greatly the thermal shock resistance of ceramic coatings.

Gu et al. (2011) plasma sprayed NiCoCrAlY/8YPSZ coating on aluminum alloy (5A06). The failure after thermal shock tests (furnace heating at 400 °C for 0.5 h, then quenched with cold water, and dried by warm air) was mainly due to the stress caused by thermal expansion mismatch between the bond coat and the substrate as well as the galvanic corrosion of the aluminum alloy. Ni–P, Ni–W–P, and Ni–Cu–P as interlayers were electrolessly deposited on the substrate in order to mitigate the thermal stress. Diffusion layers mainly composed of AlNi, Al_3Ni_2, and Al_3Ni were observed between the interlayers and the substrate after thermal shock test. The oxidation of the substrate was effectively inhibited. Ni–P interlayer was superior to the other two interlayers and enhanced the thermal shock life from 38 to more than 200 cycles.

Kokini et al. (2002) studied plasma-sprayed yttria-stabilized zirconia (YSZ)–NiCoCrAlY bond coat alloy compositionally graded TBC onto Inconel steel subjected to a thermal shock loading. The thermal loading was applied using a continuous CO_2 laser. The results showed that for a given thermal resistance, as the compositional gradation was increased, the maximum surface temperature at which horizontal cracks initiated in the TBC increased.

Liang and Ding (2005) deposited by APS nanostructured and conventional zirconia coatings and studied, by the water quenching method, the thermal shock resistances of as-sprayed coatings. The results showed that the nanostructured as-sprayed coating possessed better thermal shock resistance than the conventional one. The difference in microstructure and microstructural changes occurring during thermal shock cycling explained those results. Gilbert et al. (2008) investigated the reduction in time-dependent deformation through the addition of mullite to YSZ-based TBCs. Fracture mechanics analyses were performed for three coating architectures: monolithic YSZ, monolithic mullite, and a mullite–YSZ composite. The effect of coating architecture and surface crack morphology on interface fracture was investigated. They found that mullite and YSZ combined to influence the thermal shock behavior of the composite coatings. Composite coatings experienced a reduced driving force for interface fracture compared to monolithic YSZ coatings while having a higher thermal resistance compared to monolithic mullite coatings.

19.3.1.6 Other Forms of Wear

Wear by Very High Stresses

Wear by very high stresses is mostly characterized by dry sliding with a pin-on-disk test and depends on the high load applied and sliding distance, the wear response being dominated by subsurface cracking and removal of materials when sliding. Martials used include APS-formed ceramic coating, and HVOF, or D-gun with cermets. Bolleli et al.

(2006a, b, c) have investigated through pin-on-disk and dry sand–steel wheel tests the wear resistance of APS ceramic coatings (Al_2O_3, Al_2O_3–13%TiO_2, Cr_2O_3). These have their best performance in dry particle abrasion conditions, because they never undergo abrasive grooving but only splat detachment. In the pin-on-disk test, the ability to form a smooth and compact surface film by local plastic deformation was the key property determining coating performance. Al_2O_3 and Al_2O_3–13%TiO_2 did not form an adequately compact tribo-film and therefore showed unfavorable properties in terms of sample wear rate, pin wear rate, and friction coefficient. Cr_2O_3, on the other hand, had the ability to form a compact tribo-film by splat plastic deformation giving rise to performances comparable to HVOF-sprayed ceramics. However, its rather low deposition efficiency (below 45%) and its propensity to possibly form Cr_6^+ by in-flight thermal alteration represented serious shortcomings that must be considered. Bolleli et al. (2006a, b, c) also found that Al_2O_3 coatings, manufactured by the high-velocity suspension flame spraying (HVSFS) technique using a nanosized powder suspension, resulted in few interlamellar or interlamellar cracks, inducing reduced pore interconnectivity. Such strong interlamellar cohesion favored much better dry sliding wear resistance at room temperature and at 400 °C.

Ahmaniemi et al. (2002a, b) showed that the aluminum phosphate sealing treatment changed stress states of the alumina and chromia coatings toward compression. Tensile stress state of the Al_2O_3 coating decreased during the sealing treatment and was even compressive when measured with hole-drilling method. Dry abrasion resistance increased correlatively. Stress state of sealed Cr_2O_3 coating was more sensitive to sealing treatment temperature. Higher sealing treatment temperature led to higher compressive stress state and better abrasion wear resistance.

Chen and Hutchings (1998) plasma sprayed coatings of tungsten carbide with 9, 12, and 17 wt.% Co by both air (APS) and low-pressure (VPS) methods. VPS coatings had a lower porosity, better inter-pass bonding, and higher WC contents; these correlated with greater abrasion resistance. All coatings showed poor resistances to high-stress abrasive wear, which was associated with widespread fracture of the carbide constituents of coatings under these conditions. Under low-stress abrasion, however, all coatings showed a substantially greater wear resistance than the steel control. The presence of hard carbides was beneficial, and the VPS coatings all showed a greater wear resistance than the APS ones.

Venkateswarlu et al. (2009) sprayed diamond-reinforced composite (DRC) coatings deposited with both oxyacetylene flame and HVOF techniques. The feedstock material used for the coating consisted of tungsten carbide (WC; average particle size: ~2 μm), bronze (average particle size: ~30 μm), and diamond (average particle size: ~25 μm) powders. In addition to the high hardness of HVOF-sprayed coating, the uniform distribution of diamond particles in the matrix was considered important in view of reduced matrix deformation. HVOF-sprayed specimen, owing to its lower porosity and higher bond strength between reinforced particulates and the matrix, led to substantial reduction in the deformation of matrix, which in turn facilitated higher wear resistance. To improve the resistance of HVOF-sprayed cermets, different means were used. For example, Stoica et al. (2005) have HVOF (JP-5000) sprayed functionally graded (FG) WC–NiCrBSi coatings and heat-treated them at 1200 °C in argon environment. It was possible to achieve higher wear resistance, both in terms of coating wear and the total wear of the test couples. This was attributed to the improvements in the coating microstructure during the heat treatment, which resulted in an improvement in coating's mechanical properties through the formation of hard phases, elimination of brittle W_2C and W, and the establishment of metallurgical bonding within the coating microstructure and at the coating–substrate interface.

Valarezo et al. (2010) tested the effective damage tolerance of a functionally graded coating (FGC) deposited by HVOF. The thick FGC ($\approx$1.2 mm) consisted of six layers with a stepwise change in composition from 100 vol.% ductile AISI-316 stainless steel (bottom layer) to 100 vol.% hard WC–12Co (top layer) deposited onto an AISI-316 stainless steel substrate. The FGC structure showed the ability to reduce cracking with increased compliance in the top layer during static and dynamic normal contact loading, while retaining excellent sliding wear resistance (ball-on-disk tests). One of the main problems seemed to be the poor characteristics of HVOF-sprayed stainless steel. It suffered from extensive oxidation and phase alteration during spraying, which compromised its mechanical strength and probably impaired its adhesion to the substrate.

Bolelli et al. (2012a, b) manufactured HVOF-sprayed functionally graded coating, consisting of two NiAl/WC–Co composite layers with increasing cermet content and a pure WC–Co topmost layer. Results showed that the WC–Co/NiAl FGC could provide good wear and corrosion protection to the substrate under very severe conditions, when very thick layers were demanded. This architecture also solved many of the problems highlighted by the previously considered WC–Co/stainless steel FGC.

Stoica et al. (2004) compared the sliding wear performance of as-sprayed and hot isostatically pressed (HIP) thermal spray cermet (WC–12Co) coatings. Results indicate that HIP technique can be successfully applied to posttreat thermal spray cermet coatings for improved sliding wear performance, not only in terms of coating wear but also in terms of the total volume loss for test couples. The process induced phase transformations: elimination of secondary phase W_2C and metallic tungsten W, and alteration of amorphous binder phase through recrystallization of Co leading to precipitation of the η carbides. It also developed the metallurgical bonding at the interface between the constituent lamellae of the coating, thereby increasing the coating's Young's modulus after posttreatment with HIP technique.

Wear by Fretting and Fretting-Corrosion

Fretting occurs when the two interacting mating surfaces are subjected to an oscillatory motion of small amplitude. It can be more important when corrosion or oxidation occurs, producing for example oxides with ferrous materials (fretting corrosion). In the latter case, the extent and kinetics of phenomena are linked to the balance between the formation of products and their removal through wear. Besides the material properties, the design of the machine must eliminate the micro-movement, trap or remove debris, and limit the contact with the environment. Jin et al. (2006) have underlined that surface coating may protect the contact area from the fretting damage; however, specific contact condition would govern the effectiveness of the coating. They proposed to classify coatings into two major types: hard and soft. Hard coatings are used in the contact conditions where fretting wear is present or dominant, while soft coatings are used to reduce the fretting fatigue-induced failure since they could act as a self-lubricant and reduce the coefficient of friction resulting in improvement in fretting fatigue life. Of course, fretting resistance depends on coating and substrate on the one hand and contact conditions (lubricant, pressure, and amplitude of the movement) on the other. For example, Kim and Korsunsky (2011) studied a pad and a substrate made of Ti-6Al-4V alloy on which was plasma sprayed a metallic interlayer of Cu59–Ni36–In5. A dry film lubricant layer was then applied on the metallic interlayer and on the pad in the form of epoxy matrix loaded with MoS_2 particles. The coated specimen was clamped between two pads that were pressed against the specimen surface with a contact pressure of ~ 125 MPa, conditions simulating the geometric and loading conditions experienced in aerospace components. Each fretting test was performed at a frequency of 2.5 Hz. The durability of the coating was increased when increasing its thickness. Jin et al. (2006) studied Cu–Al coating on Ti-6Al-4V substrate under fretting fatigue condition. At the low applied-stress amplitudes, the coating survived up to one million cycles. At the high applied-stress amplitudes, either the coating was separated from the substrate due to insufficient interfacial strength or the specimen fractured at locations away from the contact area.

Mary et al. (2011) performed fretting wear tests up to 500 °C. A representative punch (Ti17)/plane (CuNiIn plasma coating) interface was investigated under air conditions. Wear was studied by varying pressure, sliding amplitude, and temperature. A threshold pressure $p_{th} \approx 85$ MPa was identified for the transition from a low- to a high-wear regime. Temperatures in the range of 20–450 °C did not modify the respective wear processes and consequently did not affect the related wear rates.

Hager Jr et al. (2008) conducted, for titanium compressor bladed disk assemblies, an in-depth wear analysis on the gross slip fretting wear of Ti-6Al-4V-mated surfaces as well as Ti6Al4V worn against plasma-sprayed CuNiIn, Al–bronze, Mo, and Ni. Tests were performed at room temperature and 450 °C. To eliminate the stress concentrations, the test geometry was an ellipsoid on a flat plate. They determined that all coatings caused significant damage to the mating Ti6Al4V surfaces and that the wear mechanisms were all similar to those of the uncoated baseline case. In gross slip fretting, these soft thermal spray coatings did not effectively protect the mating Ti6Al4V surfaces without solid lubrication.

Koiprasert et al. (2004) HVOF sprayed four types of coating (≈ 300–400 μm), namely, WC–17% Co ($T < 500$ °C), Cr_3C_2–25%NiCr ($T < 800$ °C), AlSi graphite, and CoMoCrSi ($T < 800$ °C) on stainless steel substrates. They used an in-house flat-on-flat fretting wear tester at 500 °C. At that temperature, WC–Co did not perform well, and Cr_3C_2–25%NiCr outperformed WC-Co; but the formation of Cr_2O_3 by oxidation caused rapid wear, and AlSi graphite coating did not provide additional wear resistance to the stainless steel. Only the CoMoCrSi coating, with its low porosity, moderate-to-high hardness, and good high-temperature corrosion resistance, exhibited the best wear property.

Hager Jr et al. (2009) plasma sprayed nickel graphite composite coatings with 5–20% graphite and studied mitigated fretting wear within Ti-6Al-4V contacts at room temperature and 450 °C. The embedded graphite particles reduced the friction of the nickel thermally sprayed coatings during both low- and high-temperature fretting wear experiments. The wear reduction on the mated Ti-6Al-4V surfaces was due to the formation of uniform transfer film graphite based at room temperature and NiO based at 450 °C.

Carrasquero et al. (2008) investigated the fretting wear performance of a Ni–Cr-based alloy, containing B and C, HVOF sprayed onto a SAE-1045 steel substrate. Tests were conducted on both the uncoated and coated substrate, under unlubricated dry conditions, at different applied normal loads, cycles, and amplitudes. They showed that at constant wear amplitude, the wear volume increased with the applied normal load and that at under constant load conditions, the wear volume decreased as the wear test amplitude also decreased, becoming insignificant when it was less than ~ 100 μm.

Tian et al. (2008) studied the fretting wear behavior, against a steel ball, of conventional and nanostructured

Al_2O_3–13 wt.% TiO_2 plasma spray. The improved fretting wear resistance of nanostructured coatings was attributed to the nanometer-sized grains, reduced lamellar structures, and amorphous phases. Fretting wear often occurs in hip joints that must be replaced after about 10 years of use.

Sathish et al. (2011) reported on the sliding wear performance of the nanostructured Al_2O_3–13TiO_2, ZrO_2, and the bilayered (ZrO_2/Al_2O_3–13TiO_2) coated biomedical Ti–13Nb–13Zr alloy in simulated body fluid (SBF) condition. The bilayered coating exhibited 200- and 500-fold increase in the wear resistances when compared with that of the nanostructured Al_2O_3–13TiO_2 and ZrO_2 coatings. This substantial improvement was attributed to the lower porosity and higher adhesion strength of the bilayered coating.

19.3.1.7 Replacement of Hard Chromium

Electrolytic hard chromium plating is extensively used against wear and sometimes corrosion. However, the chromium plating baths contain chromic acid, in which the chromium is in hexavalent state, which is known to be carcinogenic. Considering the harmful effects of the technology on the environment and public health, combined with some intrinsic technical limitations, research efforts were devoted since the beginning of the 1990s to its replacement by sprayed coatings, especially HVOF-sprayed cermets. In the following a few examples are given.

Savarimuthu et al. (2001) studied sliding wear characteristics of HVOF-sprayed WC coatings and electroplated chrome tested against Al–Ni–Bz blocks, Cu–Be blocks, and against themselves. The fatigue resistance of WC coatings increased in the presence of compressive residual stresses.

Sahraoui et al. (2004) investigated and compared microstructural properties, wear resistance, and potentials of HVOF-sprayed Tribaloy©-400 (T-400), Cr_3C_2–25 wt.% NiCr, and WC–12 wt.% Co coatings for a possible replacement of hard chromium plating in gas turbine shaft repair. HVOF carbide-based coatings exhibited higher hardnesses and superior performances in wear resistance than hard chromium coatings. In the case of shaft repair, hard chromium plating process proved to be time consuming and to involve high costs.

Ishikawa et al. (2005) used a commercial HVOF and a gas shroud (GS) attachment HVOF to prepare WC/CrC/Ni cermet coatings under various combustion conditions. The density of coatings improved as the velocity of sprayed particles increased. With the gas-shrouded HVOF, high density was achieved with a lower degree of WC degradation. Such coatings could be alternative candidates of hard chrome (HC) plating.

Guilemany et al. (2005) compared wear and corrosion properties of Cr_3C_2–NiCr (CC-TS) HVOF sprayed and hard chromium coatings (HCCs) obtained on a steel substrate.

Three orders of magnitude lower volume loss was found for CC-TS (HVOF) after friction tests compared with HC. The best corrosion resistance was also obtained for the CC-TS by HVOF.

Picas et al. (2006) compared the mechanical and tribological properties of HVOF CrC75–(NiCr20)25 coatings sprayed from three different agglomerated feedstock powders with various powder size distributions and compared results with conventional hard chromium plating. The CrC–NiCr coating, obtained with the lowest feedstock powder size, presented the best wear resistance under all studied conditions and demonstrated superior performance to hard chrome with regard to mechanical and tribological properties.

Bolelli et al. (2006a, b, c) studied mechanical properties and tribological behavior (abrasion and unlubricated sliding wear resistance) of various kinds of electrolytic hard chrome (EHC) coatings and of metallic and cermet HVOF-sprayed coatings (WC–17Co, WC–10Co–4Cr, Co–28Mo–17Cr–3Si). HVOF-sprayed cermet coatings were harder but less tough than EHC ones. Therefore, they underwent a comparable or even higher mass loss when subjected to three-body abrasion conditions. However, their two-body sliding resistance definitely overcame that of EHC coatings, because they form a tough and uniform surface film protecting them from further damage.

Deng et al. (2007) used high-velocity oxygen/air fuel (HVO/AF) to spray WC-17Co and WC-10Co4Cr on 300M ultrahigh-strength steel. WC-17Co coating exhibited higher fracture toughness than that of WC-10Co4Cr coating. WC-17Co coating demonstrated better impingement resistance than WC-10Co4Cr, and the impingement resistance of chrome electroplating fell between that of WC-17Co and WC-10Co4Cr coatings.

Abdi and Lebaili (2008) flame sprayed NiCrBSiCFe coatings for a possible replacement of hard chromium plating in mechanical parts' repair. Coatings presented a high hardness after cyclic heat treatment, an excellent resistance to wear and friction, a good corrosion resistance, and an excellent adhesion to steel. Tests have been also positive for the replacement of hard chromium in landing gear by HVOF-sprayed WC–CoCr coatings [Krishnan N, et al. (2008), Matthäus G, et al. (2009)].

Heydarzadeh and Ghadami (2010) plasma deposited WC–12 wt.% Co using APS onto steel substrates. Heat treatment of the coatings in an inert atmosphere above 900 °C promoted the formation of Z-carbides from the amorphous phase. Microhardness of coatings for different heat-treating conditions was relatively higher than that of the conventional hard chromium electro deposit.

Lu et al. (2011) compared the microstructure and corrosion behavior of hard chromium coating (HCC) and plasma-sprayed Fe-based alloy coating (FAC) on low-carbon steel, using the salt-spray test and electrochemical corrosion test.

After 60 days of salt-spray test, the weight loss of the FAC was only about 10% of HCC. After sealing treatment, the defects of pores and micro-cracks existing in FAC disappeared. The sealing treatment further decreased the weight loss of FAC to be about 4% of HCC.

Staia et al. (2012) investigated the tribological behavior of a VPS sprayed Cr_2C_3–25 wt.% NiCr chromium carbide coating both in the as-deposited and heat-treated conditions. The samples were subsequently annealed for 2 h at 600 °C, 800 °C, and 900 °C in Ar. The wear values indicated a satisfactory behavior from the tribological point of view. However, the heat treatment had no important consequences on the tribological performance of these coatings. The only problem is that VPS coating is about four times more expensive than HVOF.

19.3.2 Corrosion- and Oxidation-Resistant Coatings

Corrosion and oxidation correspond to a chemical or electrochemical reaction with the surrounding atmosphere. As a general rule, except for sacrificial coatings, those used against corrosion must be as dense as possible and, in most cases, a posttreatment such as fusion of self-fluxing alloys, heat-treatment, sealing, austempering, or laser glazing is necessary for optimal performance.

19.3.2.1 Room Temperature Corrosion

Corrosions at room temperature can roughly be categorized into atmospheric or marine corrosion, and chemical or parachemical corrosion.

Atmospheric and Marine Corrosion

Large steel structures such as bridges, pipelines, oil tanks, towers, radio and television masts, overhead walkways, and large manufacturing facilities in the metallurgical, chemical, energy, and other industries must be protected from corrosion. It is the same, but with more difficulties, with structures exposed to moist atmospheres and seawater such as ships, offshore platforms, and seaports [Evdokimenko Yu I, et al. (2001)]. Sørensen et al. (2009) have presented a review of anticorrosive coatings for containers, offshore constructions, wind turbines, storage tanks, bridges, rail cars, and petrochemical plants and "marine corrosion" corresponding to coatings for ballast tanks, cargo holds and cargo tanks, decks, and engine rooms on ships.

Flame and wire arc spraying meet such requirements. However, coatings obtained by these methods are relatively porous (up to 20%). Sacrificial coatings (for example, Zn or Al on steel) are mainly used. Such coatings must have a cathodic behavior relatively to the ions of the metal to be protected, steels in almost all cases. The cathodic protection can be porous without any corrosion of the underneath metal. Zinc performs better than aluminum in alkaline conditions, while aluminum is better in acidic conditions. Zinc–aluminum (Zn–15 wt.% Al) seems to combine advantages of both materials. Against marine biofouling, undesirable accumulation of marine organisms on artificial surfaces that are immersed in the sea, Murakami and Shimada (2009) have studied flame-sprayed aluminum–copper alloy powders, aluminum–copper blend powders, aluminum–zinc blend powders, and a zinc powder. Coatings were immersed in the sea and their corrosion and marine fouling behaviors were examined. The aluminum–copper coatings had poor anticorrosion and antifouling properties. The aluminum–zinc coating with high zinc content and the zinc coating possessed the anticorrosion and antifouling properties.

Of course, other treatments are possible such as plastic deformation. For example, the initial porosities of 4–14% (depending on spray conditions) of aluminum coatings, wire arc sprayed, were reduced to 0.16–0.83% after being shot-peened with SiC glass beads of 0.21–0.3 mm [Pacheo da Silva C et al (1991)]. Sealers can also present antifouling properties. Chun-long et al. (2009) have arc sprayed aluminum on steel panels and then sealed coatings with nano-composite epoxies especially developed for sealing them. In order to test their performance, some panels were tested in the East China Sea. Test panels were mounted respectively for 3 years, in the marine atmosphere zone, seawater splash zone, tidal zone, and full-immersion zone. Tests included marine atmospheric outdoor exposure test, seawater corrosion test, and coating adhesion test. It was found that the appearance of coating panels was as fine as original but with a little sea species adhering to panels on tidal zone and full-immersion zone. Basically, no change in the morphology and bond strength, and no visible coating crack, blister, rust, and break-off was observed. Schmidt et al. (2006) investigated the corrosion protection performance of 17 different Zn and Al sacrificial coating system configurations during marine atmospheric exposure (20-month exposure time) at Kure Beach, NC, USA. The coating systems incorporated several conversion coating layers, primers, and organic topcoats. The sacrificial Zn coatings provided sacrificial protection at the defects. Of the two thermal spray Zn coatings, flame spray coatings were more protective than arc spray.

Han et al. (2009) wire arc sprayed aluminum wire on special treatment steel (STS) 304 base metal. The electrochemical experiment was performed in natural seawater. The coating with the higher thickness represented good corrosion resistance in seawater. Esfahani et al. (2012) deposited aluminum coating on mild steel by arc spraying. The corrosion behavior was evaluated by electrochemical impedance spectroscopy (EIS) and polarization tests in 3.5 vol.% NaCl solution. The as-coated samples were also subjected to a

1500-h salt-spray assay. EIS measurements showed that the corrosion performance of the coating was improved during a long-time immersion and exposure to saline mist. This could be due to plugging of pores by corrosion products, which hinder further penetration of the electrolyte through the coating. The twin wire arc sprayed aluminum coatings could reliably protect steel structures against corrosion in chloride-containing aqueous solutions.

Gorlach (2009) has developed wire (Zn–Al) spraying by HVAF process. Coatings showed considerably higher bond strength than those obtained by the conventional methods, an advantage in areas where the adhesion strength was critically important. The highly dense coating structure also eliminated the need for a top paint coat, usually applied on metal-sprayed coatings to extend service life. Concrete bridge structural deterioration in coastal marine environments or in areas where deicing salts are used for snow and ice removal is a problem.

To limit atmospheric or marine corrosion, if anodic protection is used (for example, nickel is anodic to steel and the cathodic area of steel exposed controls the rate), the coating should be as dense as possible (if necessary sealed). Cramer et al. (1999), as well as Holcomb et al. (1997), have proposed a cobalt-catalyzed, non-consumable, thermally sprayed titanium anode as an alternative to thermally sprayed zinc anodes for impressed current cathodic protection (ICCP). Ti was wire arc sprayed [Cramer SD, et al. (1999)], the atomization being performed with nitrogen to limit as much as possible the formation of γ-TiO$_2$. The titanium anode had a porous heterogeneous structure composed of α-titanium containing interstitial oxygen and nitrogen, and a face-centered cubic (fcc) phase thought to be Ti(O,N). The final titanium anode thickness was 50–150 μm. Electrochemical aging was studied using catalyzed titanium anodes thermally sprayed on concrete slabs containing a steel mesh cathode. Cobalt catalyst, applied to the anode as an aqueous cobalt nitrate–amine complex, penetrated the anode and accumulated at the anode–concrete interface and in cracks within the concrete adjacent to the interface. Anodic polarization during and after application converted the cobalt to the active catalyst, Co$_3$O$_4$. With aging, cobalt catalyst dissolved in the increasingly acid environment of the interface and dispersed into the Ca-deficient, silica-rich reaction zone to reprecipitate and form a more diffused site for the anode reactions. Stable operation of catalyzed anodes was maintained for a period equivalent to 23 years' service at Oregon Department of Transportation (DOT) bridge ICCP conditions with no evidence that operation would degrade with further aging. Early results from the field experiment at the Depoe Bay Bridge suggested anodes may age more slowly at low current densities with lesser impact from acidification.

To improve the marine corrosion resistance of stainless steel coatings fabricated by high-velocity oxy fuel (HVOF) spraying with a gas shroud attachment, Kawakita et al. (2005) increased the molybdenum (Mo) content of stainless steel with a chemical composition of Fe balance-18 wt.% Cr-22 wt.% Ni-2–8 wt.% Mo. The corrosion rate in sulfuric acid was improved when increasing the molybdenum content. The pitting corrosion resistance in artificial seawater was shifted toward the noble direction with increases in the molybdenum content. The number of rust spots formed by crevice corrosion in artificial seawater decreased after the addition of molybdenum to the coatings.

Chemical or Parachemical Corrosion

Non-sacrificial coatings will never protect the substrate if connected porosities and oxide networks exist, which is the case of most thermally sprayed coatings. The substrate protection requires using a protective bond coat or producing dense coatings or sealing them, which is generally possible if the service temperature is below few hundreds of Celsius degrees. Austenitic stainless steels, aluminum bronze, nickel-base alloys, MCrAlY, cermets with WC, Cr$_2$C$_3$, and matrices containing chromium or nickel, or both, are used against corrosion, often associated with wear. A few examples will be presented below.

Moskowitz (1993) pointed out that the use of vacuum chambers or posttreatments can eliminate most defects, but these methods are costly and impractical on a large scale. Thus, he proposed using modified HVOF process using unique inert gas shrouding to achieve highly dense, low-oxide coatings of metallic alloys. Coatings of corrosion-resistant alloys for severe corrosion applications in petroleum industry, such as type 316L stainless steel and Hastelloy C-276, were shown to act as true corrosion barriers. The oxide content also played a role. For example, 316L stainless steel coatings formed by HVAF and HVOF were applied on carbon steel panels and their resistance to salt spray was tested as sprayed or sealed [Zeng Z, et al. (2008)]. When coatings were sealed, corrosion was less on the HVAF coatings than that on the sealed HVOF coating, and almost no corrosion was observed on the sealed coating sprayed with powder of the largest particles (highest porosity) after even 500 h of salt-spray testing. While the amount of through pores dominated the corrosion resistance of as-sprayed coatings, the degree of oxidation of the coatings (much less with HVAF) determined the corrosion resistance when sealing was applied.

For applications with a severe wear in oil and gas industry, cermets are used, with however some problems for offshore installations. According to Meng (2010), the choice between the wide varieties of tungsten carbides with different alloying binders is not simple. The corrosion resistance must be improved by the proper choice of binder. For example,

Souza and Neville (2007) have shown that WC–CrNi exhibited passive behavior, as stainless steel, and would be compatible for use as a coating/substrate system when exposed to seawater, which was not the case for WC–CrC–CoCr. As previously mentioned, the coating porosity and its oxide content must be as low as possible, which implied using HVOF guns resulting in higher impact velocities (low coating porosity) and lower particle temperatures (less oxidation). This was illustrated by the work of Ishikawa et al. (2005) who used a commercial HVOF with a gas shroud attachment (GS-HVOF) to prepare WC–CrC–Ni coatings. The density of coatings improved as the velocity of sprayed particles increased. Results of corrosion test indicated that through porosity was eliminated at velocities above 770 m/s with a lower degree of WC degradation. Wear resistance and hardness of coatings prepared by GS-HVOF were superior to those prepared by the conventional HVOF. Fedrizzi et al. (2007) showed that Cr_3C_2–NiCr coatings, in sodium chloride solution under sliding wear, presented good barrier properties, and substrate corrosion was never observed. Moreover, when chromium was added to the metal matrix of WC–Co-based systems, tribo-corrosion behavior was enhanced, and lower tribo-corrosion rates were measured. Plasma-sprayed Cr_2O_3–8 wt.% TiO_2 coating was used on hydraulic cylinder piston rods and rolls, but the substrate was rapidly corroded by the diffusion of the corrosive solution through pores: the bond coating was destroyed by the aggressive solution and ceramic coating flakes dropped off. Zhang et al. (2011) have sealed coatings with epoxy and silicone resins. The sealing treatment improved the corrosion resistance of coatings significantly by blocking the open pores and cracks of the coating. Sealed by silicone resin, the coating gained remarkable anticorrosion properties, as after 1200-h salt-spray test, there was still no rust on the silicone resin-sealed coating.

For chemical corrosion, the coating material must be adapted to the corrosive medium. Coatings are often used to reduce costs, especially when solutions have been found in the use of sophisticated materials that are oxidation- and corrosion-resistant at high and low temperatures with high strength such as nickel-base superalloy Inconel 625. However, coatings of these materials on cheaper metals would be cheaper than bulk super alloy. Tuominen et al. (2000) have shown that Inconel 625 HVOF sprayed presented mechanical and corrosion properties typically inferior to wrought materials due to the chemical and structural inhomogeneity of the thermally sprayed coating material. Laser remelting, with high-power continuous wave Nd:YAG laser equipped with large beam optics, resulted in homogenization of the sprayed structure. This strongly improved the performance of the laser-remelted coatings in adhesion, wet corrosion, and high-temperature oxidation testing.

A rudder is one of the essential ship components for navigation and safety and services in the complex conditions of cavitation erosion and marine corrosion [Tuominen J, et al. (2000)]. Kim and Lee (2011) have investigated ten different coatings arc sprayed with Al-, Zn-, Cu-, and Fe-based wire feedstock for a rudder application. In terms of marine corrosion resistance, aluminum coating was the best, while stainless steel coating presented the highest resistance against cavitation erosion. Si addition improved both the hardness and bond strength of Al-based coatings, resulting in drastic increment of cavitation erosion resistance by about 40 times when adding 12 wt.% Si. With Cu- and Fe-based coatings, interfacial oxidation occurred in between the coating and steel substrate, leading to interfacial cracks and coating delamination. The same authors, also for rudders, wire arc sprayed Al–Zn and then sealed them with F–Si sealer, spread with the brush at room temperature for several times. The corrosion rate of the sealed specimen after coating was lower and the hardness higher and they presented good cavitation resistance characteristics.

Dent et al. (1999) studied the corrosion characteristics in 0.5 M H_2SO_4 of two Ni–Cr–Mo–B alloy powders HVOF sprayed. Both coatings exhibited comparable resistance to corrosion. Microstructural examination of samples revealed the prevalence of degradation at splat boundaries, especially where significant oxidation of the deposit occurred. Ishikawa et al. (2001) wire arc sprayed a duplex coating composed of aluminum on 80Ni–20Cr alloy undercoat. The duplex coating presented an excellent performance in a hot, near-neutral aqueous environment. The role of the undercoat layer of the sprayed 80Ni–20Cr alloy was to decrease the surface area of the steel substrate to be protected and reduce the sacrificing load for the aluminum and also to increase the adhesion of the aluminum topcoat by its roughness.

Magnesium alloys present low density and excellent physical and mechanical properties. They are particularly suitable for their use in automotive and aerospace applications, where replacement of steel and aluminum components by magnesium components would contribute to energy savings and reduced environmental impact. However, their drawback is their relatively high corrosion susceptibility and low wear resistance. That is why many studies of thermal spray coatings have been performed to improve their corrosion resistance. Pardo et al. (2009) wire arc sprayed aluminum/silicon carbide composite coatings on AZ31, AZ80, and AZ91D magnesium–aluminum alloys. Presence of SiC particles provides higher hardness and wear resistance but not necessarily a better corrosion resistance. Corrosion was investigated in 3.5 wt.% NaCl solution at 22 °C. Al/SiC composite coatings in the as-sprayed state revealed high level of porosity with poor bonding at the Al/SiC and coating/substrate interfaces, facilitating the degradation of the magnesium substrates by galvanic corrosion. Cold-pressing posttreatment produced more compact coatings with improved corrosion performance. Arrabal et al. (2010) evaluated also the corrosion behavior of wire arc sprayed Al/SiC coatings on AZ31, AZ80, and AZ91D Mg–Al–Zn alloys in neutral salt fog (ASTM B 117) and high relative humidity (98% relative air humidity [RH], 50 °C) environments. Pardo et al. (2009) found that the application of a cold-pressing posttreatment improved the corrosion performance of the coatings. In high-humidity atmosphere, corrosion signs were only visible at the Al/SiC interfaces in the outermost surface of the coatings and in salt fog environment the galvanic corrosion of the substrates was delayed.

Pokhmurska et al. (2008) deposited on magnesium substrates (AM20, AZ31, AZ91) by wire arc spraying Zn, ZnAl4, and ZnAl15 solid wires and cored wires in aluminum core with powder filling containing different hard particles, such as boron, silicon, and tungsten carbide or titanium oxide. Remelting of thermal spray coatings was carried out by means of continuous irradiation of CO_2 laser in nitrogen or argon atmosphere, electron beam in vacuum, and focused tungsten halogen lamp line heater in atmosphere. Electrochemical behavior of modified surface layers was mostly improved due to alloying, homogenization of element distribution, and strong decrease in as-sprayed coating porosity.

19.3.2.2 High-Temperature Corrosion

Hot corrosion degradation of metals and alloys is a serious problem for many high-temperature aggressive environment applications, such as thermal barrier coatings (TBCs), boilers, fluidized-bed combustors, and industrial waste incinerators. At high temperatures, corrosion reactions can take place through oxidation, carburization, nitriding, sulfidation, molten-salt corrosion, and halogen erosion. They are generally addressed using appropriate gas-tight thermally sprayed coatings adapted to the predominant corrosion mechanism as briefly listed below.

- *Against oxidation*, nickel and/or cobalt-based alloyed coatings are used.
- *Against carburization*, nickel and chromium-containing alloys are often used.
- *Against nitriding*, austenitic stainless steel and nickel-base alloys present a good resistance.
- *Against sulfidation*, certain Hastelloy and CoCrAlY are used.
- *Against molten glass*, among the most corrosive media, high-chromium-containing alloys (NiCrMo: Hastelloy or Nistelle, CoCrSiMo: Tribaloy and MCrAlYs) are used.
- *Against molten salt corrosion* resulting from sodium sulfates (such as Na_2SO_4), sodium chlorides (such as NaCl), and vanadium oxides (such as V_2O_5), no protection exists.

In the following, examples are reviewed of corrosion-resistant coating for some of the listed high-temperature corrosion situations and mechanisms commonly met in an industrial environment.

As summarized by Mohan et al. (2010) and Jones (1997) for TBCs, when cost-effective alternative fuels containing appreciable levels of elemental impurities such as V, Na, S, P, and Ca are used, corrosive compounds such as vanadates and sulfates might arise as combustion by-products. In addition, TBCs are also increasingly susceptible to degradation by air-ingested oxide deposits such as CaO, MgO, Al_2O_3, and SiO_2 (CMAS) [Hitchman LN, Knapp J (2010)], especially in aircraft engines that operate in a dust-laden environment.

Mohan et al. (2010) studied the degradation of thermal barrier coatings (TBCs) by fuel impurities and CMAS (CaO, MgO, Al_2O_3, SiO_2) at temperatures up to 1400 °C. The study was carried out on yttria-stabilized zirconia (YSZ) and CoNiCrAlY coatings (300 µm) on free-standing grit-blasted graphite substrates. The degradation by V_2O_5 and a laboratory-synthesized CMAS was due to the destabilizing effect of molten deposits on the YSZ leading to the formation of YVO_4, and reacting with the thermally grown oxide (TGO) forming various reaction product and phase transformations. CMAS melt readily dissolved the YSZ and then reprecipitated ZrO_2 with a composition based on local melt chemistry that deviates in the content of the Y_2O_3 stabilizer. Extensive dissolution of the TGO α-Al_2O_3 from freestanding APS CoNiCrAlY coatings by CMAS melt was observed. Enriching CMAS composition with Al promoted the crystallization of anorthite platelets and $MgAl_2O_4$ spinel and mitigated CMAS ingression [Hitchman LN, Knapp J (2010)]. Li et al. (2010a, b) investigated, in laboratory scale, the effect of CMAS. They found that the porous nature of the thermally sprayed TBCs made them vulnerable to CMAS attack even before discernible chemical reaction started. A denser coating had the potential to resist the CMAS attack for a longer time than a more porous coating.

Habibi et al. (2012) compared the hot corrosion performance of YSZ, $Gd_2Zr_2O_7$, and YSZ + $Gd_2Zr_2O_7$ composite coatings in the presence of molten mixture of Na_2SO_4 + V_2O_5 at 1050 °C. For YPSZ, results were similar to those of Mohan et al. (2010). For the second composite coating, by the formation of $GdVO_4$, the amount of YVO_4 formed was significantly reduced. Molten salt also reacted with $Gd_2Zr_2O_7$ to form $GdVO_4$ but under 1050 °C, these coatings were more stable, both thermally and chemically, than YPSZ and exhibited a better hot corrosion resistance. $LaTi_2Al_9O_{19}$ (LTA) exhibited [Xie X, et al. (2012)] promising potential as a new kind of TBC material, with its excellent high-temperature capability and low thermal conductivity. Xie et al. (2012) demonstrated its good chemical stability in molten salt of Na_2SO_4 and NaCl at 1000 °C. However, the molten salt infiltrated to the bond coat, dissolving the TGO and resulting in hot corrosion of the bond coat.

Mei et al. (2012) studied the corrosion kinetics and mechanisms of NiCoCrAlTaY cold-sprayed coating and uncoated Mar-M247 superalloy in molten Na_2SO_4 salt vapor. Mass gain and microstructures after 200 h indicated that the NiCoCrAlTaY coating provided good oxidation and corrosion resistance in this vapor at 900 °C. Sidhu et al. (2006a. b) HVOF sprayed NiCrBSi, Cr_3C_2–NiCr, Ni–20Cr, and Stellite-6 coatings on a nickel-base superalloy and tested them at 900 °C in the molten salt (Na_2SO_4–60%V_2O_5) environment under cyclic oxidation conditions. Among them, Ni–20Cr-coated superalloy imparted maximum hot corrosion resistance, whereas Stellite-6 coated superalloy indicated minimum resistance. The hot corrosion resistance of all the coatings was attributed to the formation of oxides and spinels of nickel, chromium, or cobalt.

Chatha et al. (2012a, b) deposited by HVOF process 80Ni–20Cr and 75Cr_3C_2–25(Ni–20Cr) coatings on T91 boiler tube steel. Hot corrosion was studied on bare and HVOF-coated steel specimens after exposure to a molten salt (Na_2SO_4–60%V_2O_5) environment at 750 °C under cyclic conditions. The 80Ni–20Cr coating was found to be more protective than the cermet coating. Tani and Harada (2007) examined the corrosion resistance of Ni–50Cr alloy coating, plasma sprayed onto the fireside of steam-generating tubes in a heavy oil-fired boiler. This coating, applied on steam-generating tubes at Tocalo Co. Ltd., had remained for more than 20 years (about 140,000 h) demonstrating the corrosion-resistant effect. Bala et al. (2010) studied the corrosion resistance of Ni–50Cr coatings cold sprayed on two boiler steels SA-213-T22 and SA 516 (Grade 70). The aggressive environment consisted of Na_2SO_4–60%V_2O_5 under cyclic conditions at 900 °C. The Ni–50Cr coated steels showed lesser weight gains than the bare metals and the oxide scales remained intact till the end of the experiment. According to Matsubara et al. (2007), nickel-based self-fluxing alloy coating extended the service life of furnace wall tubes at waste incineration plants due to its excellent corrosion resistance and heat resistance. Fusing of such coatings by induction heating offered improved efficiency and reliability of products. A successful experimental application of 11 units in a waste incinerator revealed virtually no corrosion on the exposed surfaces and showed an improved water heating efficiency over that of the original tubes.

Chatha et al. (2012a, b) investigated the corrosion resistance of HVOF-sprayed Cr_3C_2–NiCr coating on ASME-SA213-T91 boiler steel without and with posttreatment of the coating by sealing and heat treatment. Corrosion tests consisted of exposure to a molten salt (Na_2SO_4–60% V_2O_5) environment at 900 °C under cyclic conditions. Posttreated coating resulted to be more effective than as-deposited coating in enhancing the corrosion resistance of T91 steel. Tao et al. (2009) HVAF sprayed conventional and nanostructured NiCrC coatings. Hot corrosion was conducted in the Na_2SO_4–30%K_2SO_4 environment, in the temperature range of 550–750 °C for periods up to 160 h. Both types of dense coatings possessed high corrosion resistance, especially the nanostructured NiCrC coating. The enhanced grain boundary diffusion in the nanostructured coating not only promoted the formation of a denser Cr_2O_3 scale with a higher rate but also helped to mitigate the Cr depletion at the metal/scale interface. Ceramic coatings are also used against corrosion.

Jansen et al. (2002) compared the performance of dicalcium silicate-based coatings plasma sprayed to that of YPSZ coatings. Coatings were exposed to a V_2O_5–15 wt.% Na_2SO_4 slag at 700 and 900 °C. At 700 °C, gaseous sulfidation was stimulated by an addition of 0.5 vol.% sulfur dioxide in air. The dicalcium silicate withstood combined attack without debonding, while the YPSZ coating deteriorated and spalled. Hirata et al. (2003) investigated the influence of impurities of Al_2O_3 on the hot corrosion resistance against V_2O_5–Na_2SO_4 molten salt. Thickness of damage zone in Al_2O_3 ceramics, whose purity was high, depended linearly on the holding time. On the other hand, thickness of damage zone in Al_2O_3 whose purity was relatively low depended on square root of holding time. In the impurities, SiO_2 especially affected the corrosion rate influencing the diffusion rate of corrosive elements through grain boundaries and thus the corrosion rate.

Two examples of mold protection are presented. For the protection of aluminum for injection mold tools, Gibbons and Hansell (2008) HVOF or plasma sprayed 100 ± 20 μm thick coatings of WC–10Co–4Cr, Cr_2C_3–20(Ni20Cr), 316 Stainless Steel, or Al_2O_3–40TiO_2. Substrates were made of three grades of aluminum: 2014 (90.4–95.0Al 3.9–5.0Cu 0.5–1.2Si 0.4–1.2Mn), 5083 (92.4–95.6Al 4.0–4.9Mg 0.4–1.0Mn), and 6082 (94.7–98.8Al 0.7–1.3Si 0.4–1.0Mn 0.06–1.2Mg). HVOF, except for the alumina–titania coating only plasma sprayed, was the most suitable of the two types of thermal spray systems evaluated, since APS was unable to provide high bond strengths. Furthermore, the processing by HVOF was not found to be detrimental to the fatigue life of the aluminum substrate. Mizuno and Kitamura (2007) HVOF sprayed on AISI-316L substrate MoB/CoCr, a novel cermet material with high durability in molten alloys for aluminum die-casting parts and for hot continuous dipping rolls in Zn and Al–Zn plating lines. This coating presented a much higher durability without dissolution in the molten Al–45 wt.% Zn alloy. Using undercoat was effective to reduce the influence of large difference in thermal expansion between the MoB/CoCr topcoat and stainless-steel substrate.

Combinations of topcoat and undercoat thickness were optimized to obtain intrinsic performance.

19.3.2.3 Oxidation

A variety of oxides can form on the surface and the further oxidation depends on the oxide layer forming a stable and continuous film. Of course, the oxidation depends on the oxygen affinity of the elements forming the alloy. In most cases Cr and Al additions to the coating's matrix are used to form the oxidation barrier. Their quantity must be important enough to provide the renewal of the oxide layer that must be tough, well adhered, and self-healing. Up to 430 °C, chromium–molybdenum steel presents a good resistance to oxidation. Up to 980 °C, the high-chromium-containing (16–28 wt.%) alloys—NiCr, Inconel (Ni–Cr alloys), stellites (Co–Cr alloys), and stainless steel—present a good resistance. Over that temperature superalloys with aluminum (up to 5 wt.%) are used, because aluminum forms an alumina film, which is more adherent than Cr_2O_3, thin enough to be ductile, and nonvolatile at temperatures $\geq 1000\,°C$.

Many studies were devoted to the bond coat and the thermally grown oxide (TGO) layer formation. The problem is that MCrAlY coatings contain only limited aluminum content because high values can lead to brittleness and potential crack. For example, Koolloos and Houben (2000) have shown that pre-oxidation of the bond coat had a beneficial effect on lifetime when the oxidation was the main cause of failure. The pre-oxidation time and temperature were chosen such that a dense alumina layer was formed, and the forming of spinel-type structures avoided. Fossati et al. (2012) sprayed by VPS CoNiCrAlY coatings onto CMSX-4 single-crystal nickel superalloy disk substrates. As-sprayed samples were annealed at high temperatures in low vacuum. Three kinds of finishing processes were carried out producing three types of samples:

- As-sprayed
- Mechanically smoothed by grinding
- PVD coated by using aluminum targets in an oxygen atmosphere

Samples were tested under isothermal conditions, in air, at 1000 °C, and up to 5000 h. Several differences were observed: grinding operations decreased the oxidation resistance, whereas the PVD process can increase the performances over longer time with respect to the as-sprayed samples.

Jiang et al. (2010) designed gradient coating that could provide a balance between high Al content and high stress-bearing ability. The gradient coating showed better performance of re-healing alumina β scale due to its possession of more β phase as Al reservoir. The improved high-temperature performance of the gradient coating was due to the enrichment of Al in the outer layer.

Yuan et al. (2008) sprayed the MCrAlY bond coat by HVOF and D-gun processes. The isothermal oxidation rate at 1100 °C of the TBC system with the HVOF bond coat was two times lower than that of the TBC system with the detonation-sprayed bond coat. The rough surface of the detonation-sprayed bond coat created a large specific surface area and unfavorable oxides on the bond coat.

Pint et al. (2000) investigated by furnace cycling several different oxidation-resistant substrates with dry, flowing oxygen at temperatures from 1000 to 1200 °C. They determined the cyclic lifetimes of YSZ topcoat plasma sprayed or EBPVD deposited. For NiAl + Zr (0.05–0.08 wt.% Zr) substrate, the testing revealed that a more adherent alumina scale may extend the lifetime of commercial EBPVD and PS coatings. Oxide-dispersion strengthened (ODS) FeCrAl substrates with YSZ topcoat seemed promising for the metallic skin of the next-generation space shuttle.

Limarga et al. (2005) produced Al_2O_3/ZrO_2 functionally graded (FG) thermal barrier coating system incorporating an oxygen barrier layer between bond coat and topcoat. Alumina has been highlighted as a potential material as oxygen barrier to improve the oxidation resistance of the bond coat since it has very low oxygen diffusivity. The results showed that FG systems exhibited superior mechanical properties and oxidation resistance at the expense of a slightly lower thermal insulating effect. Thin interlayer was preferred in order to minimize the detrimental effect of phase transformation of γ-Al_2O_3 to α-Al_2O_3 that resulted in tensile residual stresses at the interface.

High-temperature oxidation and corrosion phenomena in boilers are the source of many problems. Both the HVAF-sprayed conventional and nanostructured NiCrC coatings possess good properties against the oxidation and hot corrosion attacks [Chatha SS, et al. (2012a, b)]. Compared with its conventional counterpart, nanostructured NiCrC coating exhibited better performance, which was believed to result from the faster formation of Cr_2O_3 scale with a denser structure. Even after thermal exposure at 650 °C for over 100 h, the nanostructured coating still kept its mean grain size less than 100 nm and seemed to be stable. Consequently, a large number of grain boundaries in the nanostructured coating served as "short circuit" channels for Cr diffusion [Chatha SS, et al. (2012a, b)].

Sidhu et al. (2007) HVOF sprayed WC-NiCrFeSiB coating on Ni-based superalloy (Superni 75) and Fe-based superalloy (Superfer 800H). The coated as well as uncoated specimens were exposed to air and molten salt (Na_2SO_4–25% NaCl) environment at 800 °C under cyclic conditions. The oxides of active elements of the coatings, formed in the surface scale as well as at the boundaries of nickel and tungsten rich splats, contributed to the oxidation and hot corrosion resistance. These oxides acted as barriers for the diffusion/penetration of the corrosive species through the coatings.

Ramesh et al. (2011) HVOF sprayed Ni-based hard facing NiCrFeSiB alloy powder on three kinds of boiler tube steels. Thermo-cyclic oxidation studies were performed in static air at 900 °C. NiCrFeSiB-coated steels showed slow oxidation kinetics and considerably lower weight gains than uncoated steels. The superior performance was attributed to continuous and protective thin oxide scale of amorphous SiO_2 and Cr_2O_3 formed on the surface of the oxidized coatings.

Kaur et al. (2011) deposited with D-gun Cr_3C_2–NiCr coating on T22 boiler steel. High-temperature oxidation in air and oxidation–erosion in actual boiler environment at 700 °C under cyclic conditions were carried out. While the uncoated boiler steel suffered from a catastrophic degradation in the form of intense spalling of the scale, the Cr_3C_2–NiCr coating showed good adherence to the boiler steel during the exposures with no tendency for spallation of its oxide scale.

Singh et al. (2006) deposited by a shrouded plasma spray process Stellite-6 on two Ni-base superalloys, Superni-601 and Superni-718, and one Fe-base superalloy, Superfer 800H. Oxidation studies were conducted on the coated superalloys in air at 900 °C under cyclic conditions for 50 cycles. The rate of oxidation was observed to be high in the early cycles of the study for the coated superalloys, which may be attributed to the interconnected network of pores and splat boundaries present in the coating structure. The Stellite-6 coating after exposure to air oxidation showed the presence of mainly oxides in its surface scale. The phases revealed where oxides of Cr and Co and spinels containing CoCr mixed oxides were present, oxides reported to be protective in nature against the high-temperature oxidation. Bala et al. (2011) deposited by cold spray Ni–20Cr and Ni–50Cr coatings on two boiler steels. The coatings, in general, were found to follow the parabolic rate of oxidation and were successful in maintaining surface contact with their respective substrate steels.

19.3.2.4 Corrosive Wear

Corrosive wear occurs when the effects of corrosion and wear (C-W) are combined, resulting in a more rapid degradation of the material's surface. A surface that is corroded or oxidized may be mechanically weakened and more likely to wear at an increased rate. Furthermore, corrosion products including oxide particles that are dislodged from the material's surface can subsequently act as abrasive particles. Stress corrosion failure results from the combined effects of stress and corrosion. At high temperatures, reactions with oxygen, carbon, nitrogen, sulfur, or flux result in the formation of oxidized, carburized, nitrided, sulfidized, or slag layer on the surface. Temperature and time are the key factors controlling the rate and severity of high-temperature corrosive attack [Chattopadhyay R (2001)].

The effect of fluids containing solid particles is complex and depends on the surface temperature. At low temperature, the main phenomenon is erosion; with increasing temperature both erosion and oxide scale formation are similar. The loss of oxide prevails with further increase in temperature (chipping of the brittle scale). It is thus necessary to use a compatible matrix with hard reinforcing phases such as carbides and oxides. It is also necessary to limit residual stresses within coating for example by using graded layers.

The problem of erosion–corrosion of materials at elevated temperatures is that the extent of wastage in such environments is dependent on a wide variety of parameters, which include properties of the impacting particles (angularity, for example), impact angle, target material, and nature of the corrosive environment, as well as the composition of the oxide scale formed during the erosion–oxidation process.

Low Temperature

Hill et al. (2012) investigated HVOF-sprayed coatings of Fe–Cr–C and Fe–Cr–V–C on high-chromium high-carbon tool steels in the as-processed state and after a quenching and tempering treatment. Coatings were also achieved by super solidus liquid phase sintering (SLPS). In comparison, sintered steels were analyzed in the quenched and tempered conditions only. The performed heat treatment was able to improve the properties of both steels leading to an increase in hardness due to martensite formation (SLPS and HVOF) and also carbide precipitation (HVOF) and an assumed increase in wear. The lower corrosion resistance of the high-vanadium tool steel X420CrVMo18-16 in comparison to X190CrVMo20-4 was due to a lower concentration of dissolved chromium in the metal matrix. The HVOF coatings showed a decrease in corrosion resistance in relation to the sintered material because of smaller precipitates and formation of chromium oxides. Kembaiyan and Keshavan (1995) deposited, on drill bits in mining, and oil and gas drilling, 82W–15Co–3C by Super-D-gun, 80Ni–10Cr–4Fe–4Si–1.8B–0.4C by spray fused and laser cladded, Fe–Cr–Ni by wire feed and fuse welded, WC-based by plasma, and 36Ni–44W–6Cr–6Co–1.8Si–1.9Fe–1.35B by flame and laser cladded, and all were heat-treated afterward. The D-gun coating offered beneficial compressive residual stresses on the inserts and cone shell, thereby being the best coating protecting three cone drill bits from corrosion and fluid erosion. Uozato et al. (2003) plasma sprayed Fe–C powders, with or without nickel added (up to 14 wt.%), on a cast AA383 aluminum alloy plate. Corrosion and wear tests in engine oil with or without sulfuric acid water solutions added were performed as well. The corrosion performance was dependent upon the nickel content.

Souza and Neville (2007) studied the influence of microstructure on the overall material loss in erosion–corrosion environments for WC–Co–Cr coatings applied by HVOF and D-gun processes. The work was devoted to the enhancement of erosion due to corrosion effects. HVOF coatings had a slightly lower corrosion resistance than the D-gun coatings but higher overall erosion–corrosion resistance. The formation of different tungsten species, η phases, and a higher Cr amount reduced the toughness of WC–Co–Cr coatings and the corrosion rates under erosion–corrosion. Stainless steel components coated with Inconel-625 are very common in the oil/gas industry. That is why Al-Fadhli et al. (2006) HVOF sprayed Inconel-625 on three different metallic surfaces:

- Plain stainless steel (SS)
- Spot-welded stainless steel (SW-SS)
- Composite surface of stainless steel and carbon steel welded together (C-SS-CS)

The erosion experiment was carried out using a jet impingement rig through which natural seawater (pH = 8.3) or sand slurry was circulated. The coating exhibited excellent erosion–corrosion resistance, as it was not highly affected by the type of substrate material. The coating was found to be highly sensitive to the presence of sand particles in the impinging fluid. As the period of coating exposure to the flow of slurry fluid increased, weight loss increased significantly, the increment being dependent on the type of substrate material.

Polymer coatings have received increased attention as protection against corrosion and wear for several environmental conditions mainly in the automotive, petrochemical, oil, and gas industries. Lima et al. 2012) flame sprayed (Terodyn-System-2000 gun of Eutectic) polyetherimide (PEI), poly-ether-ether-ketone (PEEK), and polyamide-12 (PA-12) polymers onto carbon steel substrates. The obtained coatings had formed a dense and smooth deposit with good adhesion and no significant defects. The abrasive wear test showed a better performance for PA-12 samples, almost twice better than PEEK and 20% better than PEI. PA-12 and PEEK coatings showed no corrosive attack in a very aggressive H_2SO_4 40 vol.% solution after 2000 h of testing. Under the experimental conditions used in their work [Lima CRC, et al. (2012)], PEEK and PA-12 coatings had a better corrosion performance, but PA-12 had much better abrasive wear performance.

As the presence of sand and mineral particles flowing in a corrosive fluid produced a high degradation rate of the hydro-transport equipment, Flores et al. (2009) studied the erosion–corrosion behavior of two tungsten carbide cermet coatings obtained by PTA process. Two variations of the CNiCrFeBSi matrix, called soft and hard, with different concentration of Cr, C, and B were evaluated with and without reinforcing hard phase. Important microstructural differences were identified between the two metal matrix composites

(MMCs). Elongated precipitates rich in W and Cr were identified near the WC grains and in the matrix phase of the WC-hard overlay. Needle-like precipitates with a lower concentration of W were identified in the WC-soft overlay. For both MMCs the γ-Ni phase and silicides were present in the matrix phase. The MMC with the harder matrix showed the lower mass loss and the erosive degradation was predominant in the overall erosion–corrosion process at the lower temperature. A different behavior was observed when the temperature of the testing solution was increased, and a low sand concentration was used. PTA coatings are excellent, when wear and corrosion are present at low temperature.

High Temperature

Sidhu and Prakash (2006) evaluated the erosion–corrosion (E-C) behavior of as-sprayed plasma and laser remelted Stellite-6 coatings on boiler tube steels (three different ones) in the actual coal-fired boiler environment at 755 °C. Coating has been found to be effective in increasing the E-C resistance of boiler steels in the coal-fired boiler environment. Among the as-coated steels the highest E-C resistance has been observed for T11 steel, probably due to the presence of a continuous chromium-rich band lying at bond coat–substrate interface. A less porous structure obtained after laser remelting was found to be effective for increasing E-C resistance to the given environment for this coated T11 steel.

Wang (1996) HVOF sprayed Cr_3C_2–NiCr coating and tested them in conditions simulating erosive conditions in tubular heat exchangers of fluidized-bed combustor boilers. These coatings showed excellent E-C behavior as compared with 1018 steel and A213 T22 steel. Coating high E-C resistance was attributed to its high compactness, fine grain size structure, and the homogeneous distribution of the skeletal network of hard carbides within a ductile, corrosive-resistant metal binder. Uusitalo et al. (2002) sprayed various coatings by HVOF, wire arc, and plasma and simulated the E-C conditions in the superheater section of a circulating fluidized-bed combustor by a burner-rig type working at elevated temperature. The gas temperature was 850 °C and the specimen temperature was adjusted to 550 °C by cooling the specimen internally with pressurized air. In E-C tests in presence of chlorine, nickel-based HVOF coatings performed the best, whereas carbide containing HVOF coatings and diffusion coatings wore away.

Kaur et al. (2011) D-gun sprayed Cr_3C_2–NiCr coating on T22 boiler steel. Coated and not coated boiler steel were subjected to high-temperature oxidation in air and oxidation-erosion in actual boiler environment at 700 °C. The uncoated boiler steel suffered from a catastrophic degradation in the form of intense spalling of the scale in both environments. The Cr_3C_2–NiCr coating showed good adherence to the boiler steel during the exposures with no tendency for spallation of its oxide scale.

Kim and Walker (2007) developed a nanostructured titania coating for ball valves destined for high-pressure acid-leach (HPAL) service. The nanostructured titania coating provided dramatically superior resistances against abrasive and erosive wear compared to titania conventional coatings. The enhanced wear performance seemed to be due to improved toughness. The nanostructured titania coating has been applied onto ball valve components in ten locations around the world and has performed very well. PTA coatings are also successful against corrosion and wear at high temperatures.

19.3.3 Thermal Barrier Coatings

Thermal barrier coatings (TBCs) help to protect the base metal/alloy from high-temperature exposure, allowing for the running of engine parts at higher temperatures and achieving higher energy efficiencies. The successful use of TBCs in a given application depends, however, on the ability to overcome the often-significant mismatch between the expansion coefficient of the coating and that of the substrate. The challenge lies in the fact that good heat flux protection requires the use of low thermal conductivity materials that are generally either refractory metals (W, Mo, Ta, Nb) or oxides (Al_2O_3, or ZrO_2), which generally has low thermal expansion coefficients compared to those of the metals or alloy substrate. On thermal cycling, such a mismatch of thermal expansion coefficient between the coating and the substrate would result in the development of cracks in the coating followed by its spalling and catastrophic failure. One of the options used to mitigate such a problem is the spraying of an intermediate "bond coat" to bridge the expansion coefficient mismatch. Moreover, because of the inherent porosity of oxide-based TBCs, bond coats contribute to the added protection of the substrate metal or alloy from oxidation by the high-temperature oxidizing environment to which it is exposed. In this section, a review is presented of the range of materials commonly used as TBCs and the influence of their stabilization chemistry and the thermal spray technique used for their deposition on their performance. Examples of their industrial applications is given in the next section: 19.4 Thermal Spray Coating by Industry.

Among oxide-based TBCs, alumina and zirconia are among the most widely used materials. Alumina can generally be used up to 900 °C since at higher temperature the γ phase of plasma-sprayed alumina coatings transform into α phase, with an important volume change resulting in coating cracking and peeling off. Zirconia, on the other hand, while being by far the most accepted TBC material, must be stabilized to avoid phase transformations with temperature during its service life. The problem is then the associated cost of the stabilizer material used. The lowest-cost stabilizers,

being CaO or MgO, mostly used for TBC coatings in diesel engines, are acceptable for exposure to temperatures below 600 °C above which the stabilization becomes inefficient [Billah BM, et al. (2012)]. For TBCs used in higher-temperature environment, yttria is widely used either at a concentration of 7–8 wt.% in yttria partially stabilized zirconia (YPSZ, tetragonal phase) or 12 wt.% in yttria stabilized zirconia (YSZ, cubic phase). Their maximum service temperature in this case is around 1200 °C and the thermal conductivity of plasma-sprayed coatings is around 1 W/m K. When using powder manufactured from chemical route (sol–gel technique with an intimate mixing of zirconia and yttria at the nanometer level) the maximum service temperature reaches 1350 °C, at the expense of higher cost for the powder.

Other stabilizers, such as CeO_2, GeO_2, Nb_2O_5, To_2O_5, and Sc_2O_3, have also been used in spite of their high cost compared to Y_2O_3, and the need for up to 24 wt.% concentration of the stabilizer oxide in the mix for zirconia stabilization. Even if their thermal conductivities are generally slightly lower than that of ZrO_2-7 wt.% Y_2O_3, their maximum service temperature is about the same. Zirconia stabilized with CeO_2 [Hamacha R, et al. (1996), Ma B, et al. (2009)] offers low thermal conductivity, less phase transformation than YSZ, but increased sintering rate compared to YSZ. Zirconia partially stabilized with dysprosia [Markocsan N, et al. (2007), Curry N, Donoghue J (2012)]. On the other hand, while being more resistant to sintering than standard YSZ coatings up to 200 h at 1150 °C, their thermal properties are significantly influenced by the operating conditions with lowest thermal conductivity reached at high temperatures with the 4DyPSZ coatings (0.70 W/m K at 700 °C and 0.76 W/m K at 1000 °C). Zirconia has also been doped by different rare-earth cations with the formation of dopant clusters as in the ZrO_2–Y_2O_3–Nd_2O_3(Gd_2O_3, Sm_2O_3)–Yb_2O_3(Sc_2O_3) [Gupta M, et al. (2012)]. Pyrochlore structure ($A_2B_2O_7$) materials such as La_2ZrO_7, Nd_2ZrO_7, and $Gd_2Zr_2O_7$ that melt at temperatures over 2000 K without phase change offer properties comparable to YSZ with a lower thermal conductivity, especially $La_2Zr_2O_7$, which has excellent thermal stability [Cao XQ, et al. (2004)].

Hexa-aluminates such as (La,Nd)MAl$_{11}$O$_{19}$, where M could be Mg and Mn to Zn, Cr, and Sm, present excellent longtime sintering resistance and structural stability up to 1800 °C. For example, Chen et al. (2011a, b, c) have sprayed YSZ-based composite coatings with the addition of LaMgAl$_{11}$O$_{19}$ (LaMA) as the secondary phase. Unfortunately, composite coatings showed much shorter thermal cycling lifetime than the monolithic YSZ coating due to the presence of as-sprayed LaMA (La MgAl$_{11}$O$_{19}$) coating with a large amount of amorphous phase reducing the thermal insulating efficiency and accelerating the oxidation and degradation of bond coat. Chen et al. (2012) also sprayed five-layer functionally graded TBCs based on LaMA/YSZ system. The amorphous LaMA phase underwent recrystallization processes during high-temperature aging, resulting in the formations of LaMA platelet-like grains and imparting the TBC with improved thermal and mechanical properties and a higher strain tolerance. Such coatings presented a good thermal cycling lifetime at the surface testing temperature above 1350 °C.

Perovskites (ABO_3 crystal structure such as $BaZrO_3$ or $SrZrO_3$ or $CaZrO_3$) have also been extensively studied [Cao XQ, et al. (2004)]. Jarligo et al. (2009) synthesized $Ba(Mg_{1/3}Ta_{2/3})O_3$ and $La(Al_{1/4} Mg_{1/2} Ta_{1/4})O_3$ compositions and plasma sprayed them as ceramic topcoats of TBC either in single layer or in double-layer combination with conventional YSZ. The low value of fracture toughness for the complex perovskites and the thermally grown oxide at the topcoat–bond coat interface of the TBCs were, however, the major factors that led to the coating failure on thermal cycling at about 1250 °C. Similar results were obtained by Ma et al. (2008). Yu et al. (2011) tested $Sm_2Zr_2O_7$ coatings that exhibited low microhardness and elastic modulus, which increased with rise in aging temperature because of the microstructure reconfiguration.

Other ceramics were also tested. $LaTi_2Al_9O_{19}$ (LTA) was synthesized by solid-state reaction at 1773 K by Xie et al. (2010), and the mechanical properties of the LTA bulk were evaluated. A double-ceramic layer LTA/YSZ TBC structure was proposed, and the TBC sprayed by plasma spraying. The LTA/YSZ TBC remained intact even after 3000 cycles, exhibiting a promising potential as new TBC materials. Guo et al. (2009) produced $BaLa_2Ti_3O_{10}$ (BLT) by solid-state reaction and plasma sprayed it. The coating contained segmentation cracks and had a porosity of around 13%. Thermal cycling result showed that the BLT-TBC had a lifetime of more than 1100 cycles of about 200 h at 1100 °C and the failure occurred by cracking at thermally grown oxide (TGO). Ceramic mixtures were also proposed: Liu et al. (2008) showed that the La_2O_3 addition can effectively alleviate the grain growth of YSZ coatings under high temperatures.

Besides the materials choice, attention was also given to the effect of powder properties such as purity and particle morphology on the TBC coatings obtained using different thermal spray technologies. Paul et al. (2007) reported that improvement in the purity of the YSZ powder used in the process, through a reduction in its free alumina and silica content, from 0.1–0.2 wt.% down to 0.01–0.05 wt.%, gave rise to significant reduction in the sintering rates of produced TBC. Tan et al. (2012) showed that with proper process mapping strategies and spraying hollow powders, it was feasible to produce tailored microstructures with substantially lower thermal conductivity than generally reported for YSZ-TBCs. It also allowed enhancing thermo-structural compliance and reducing propensity to sintering during

elevated temperature exposure. Guo et al. (2004) studied the segmentation cracks density of thick TBCs (1.5 mm) for thermal protection of combustor. A segmentation crack density of 3.65/mm exhibited a thermal cycling lifetime of more than 1500 cycles at 1250 °C for the surface/950 °C for the bond coat, indicating an excellent thermal shock resistance.

Zhu et al. (2004) tested conventional TBCs (APS sprayed ZrO_2–8 wt.% Y_2O_3 with a NiCrAlY VPS bond coat) sprayed onto 25.4-mm diameter and 3.2-mm thick nickel base superalloy specimens. A high-power CO_2 laser was used to test the TBC specimens under high-temperature (1287 °C) and high-thermal-gradient cyclic conditions. If the initial average crack propagation rates were in the range of 3–8 μm/cycle, the crack propagation rates increased to 30–40 μm/cycle at later stages of testing and the critical spalling crack size ranged from 3 to 5 mm. According to Shin et al. (2011), the failure of TBC is directly connected to the failure of the blades because the spallation of the ceramic layer accelerates the local corrosion and oxidation. They performed thermal fatigue tests at 1100 °C and 1151 °C for coin-type specimens. Delamination cracks occurred at the edge of a specimen first and then the area of the edge delamination gradually increased toward the center. A thermally grown oxide (TGO) was formed and grew at the interface between the topcoat and the bond coat. The crack propagated with increase in cycle number, growth resulting from the TGO formation, and the high normal stress at the edge. Kim et al. (2010) also investigated the failure mechanisms of coin-shaped plasma-sprayed TBCs for gas turbine blades due to cyclic thermal fatigue. Hernandez et al. (2009) quantified, through finite element analyses, the thermomechanical response of the bond coat (NiCoCrAlY) and the thermally grown oxide (TGO) for a YPSZ. The models they developed include nonlinear and time-dependent behavior such as creep, TGO growth stress, and thermomechanical cyclic loading. Eriksson et al. (2011) for the same type of TBCs studied the effect of three different heat treatments—isothermal oxidation, thermal cycling fatigue (TCF), and Burner Rig Test (BRT)—using a thermal shock rig (TSR). The specimens subjected to BRT gave much thinner interface TGO due to their shorter exposure time to high temperature, and the TGO consisted almost exclusively of Al_2O_3. TCF-subjected specimens had areas of cracked interface TGO, while the thin interface TGO in the BRT-subjected specimens did not contain any cracks. Jang et al. (2006) studied TBC bond coats (0.08, 0.14, and 0.28 mm) HVOF sprayed. The TGO layer with the thickness of 2.8–3 μm was formed after thermal fatigue tests, showing the same effects in both the temperature and the dwell time. The bond coat thickness did not affect the formation of the TGO layer. As the thickness of the bond coat increased, the damage zone in the substrate decreased, the bond coat acting as a buffer layer under applied loads. Re-sintering of the top coating during the thermal fatigue tests was evident from the fatigue damage and the observed increase in the hardness values. Li et al. (2010a, b) examined through thermal cyclic test TBCs' behavior with the NiCoCrAlTaY bond coats deposited by cold spraying and low-pressure plasma spraying. The surface of the cold-sprayed bond coat presented a smoother configuration than that of the LPPS-sprayed bond coat. Ni/Cr oxides easily formed on the LPPS bond coat during oxidation. In contrast, TGO on the cold-sprayed bond coat was uniform in both thickness and composition. The uniform protective TGO enhanced the thermal cyclic behavior of the TBCs. These results suggested that cold spraying is a promising process to deposit MCrAlY bond coat for high performance of TBCs.

The evolution of TBCs for aero engines is well summarized when comparing review papers published at the end of 1990s [Musil J, et al. (1997), Soechting FO (1999), Beele W, et al. (1999)] with more recent ones [Cao XQ, et al. (2004), Vaßen R, et al. (2010), Gupta M, et al. (2012)] and also those related to coating materials [Haynes JA, et al. (2000), Hamacha R, et al. (1996), Ma B, et al. (2009), Markocsan N, et al. (2007), Curry N, Donoghue J (2012), Chen X, et al. (2011a, b, c), Chen X, et al. (2012), Jarligo MO, et al. (2009)]. In 1999, Miller (1997) said that TBCs have evolved from the laboratory to low-risk turbine section applications and then on to an integral part of engine design. Beele et al. (1999) underlined that if partially stabilized zirconia became the standard material very early on, thermal spraying and electron beam physical vapor deposition (EBPVD) in the early 1990s were even considered as competing technologies. In 2008, Feuerstein et al. (2008) emphasized that "the most advanced thermal barrier coating (TBC) systems for aircraft engine and power generation hot section components consisted of electron beam physical vapor deposition (EB-PVD)-applied yttria-stabilized zirconia and platinum-modified diffusion aluminide bond coating. Thermally sprayed ceramic and MCrAlY bond coatings, however, are still used extensively for combustors and power generation blades and vanes. However, as for any industrial production, the coating cost is a key parameter. For thermal spray processes, the labor and material costs are the major ones, while for EBPVD, the equipment cost is substantial (up to 50% of the cost is for depreciation interest) [Feuerstein A, et al. (2008)] and its utilization must be effective. During the last decade many works were devoted to bond coats: see Toscano et al. (2006). Yttria partially stabilized zirconia (YPSZ) was identified as the best ceramic topcoat material and has been established as standard for the last 30 years. However, correlatively, especially the last decade, other ceramic materials have been studied.

Plasma spray physical vapor deposition (PS-PVD), recently developed by Sulzer Metco [von Niessen K, et al. (2010)], is a low-pressure plasma spray technology to deposit TBC coatings out of the vapor phase. PS-PVD is a part of the family of new hybrid processes. It is then possible to deposit a coating by vaporizing the injected material. Coatings show a unique columnar microstructure, so far only known from other vapor phase deposition processes like EB-PVD. But economical evaluations for columnar coatings on turbine components under production conditions show a clear saving potential when using the PS-PVD process. Hospach et al. (2012) showed that quasi-vapor deposition from clusters resulted in columnar coatings with high porosity and growth rate. The low working pressure (about 100 Pa) increased the size of the plasma plume and enabled homogeneous coatings, which made this process interesting, in particular for large-scale applications.

New developments in suspension or solution plasma spraying seem also very promising (see Chap. 16, Nanocrystaline and Nanostructured Coatings). Vaßen et al. (2010) pointed out that suspension spraying offers the manufacture of strain-tolerant, segmented TBCs with low thermal conductivity. In addition, highly reflective coatings, which reduce the thermal load of the parts from radiation, can be produced. Gell et al. (2008) showed that solution-sprayed TBCs have a microstructure characterized by: strain relieving through thickness vertical cracks, and fine, dispersed porosity ultra-fine splats with enhanced splat-to-splat bonding. First tests showed that such coatings have equal or greater durability, lower thermal conductivity, higher in-plane toughness and bond strength, and costs comparable to APS. However, to our best knowledge such coatings have not yet been tested in industry

19.3.4 Electrical and Electronic Coatings

One of the main problems of electrically conductive thermally sprayed materials, except those cold sprayed or plasma sprayed in soft vacuum VPS, is oxidation during the process, which increases the electrical resistance of the coating. Some thermally sprayed coatings are used to produce resistance heating panels. The choice of the resistance material must be such that the variation of its electrical resistance with temperature is as low as possible because otherwise electronic devices must be used to adapt the voltage to the resistance variation.

Alumina is probably the most used material. The α and γ phases present different dielectric constants and loss tangent at high frequency. The γ phase resulting from the spray process can be avoided either by keeping the substrate at 1000 °C, which is not easy to do with most substrates and requires a careful temperature control during cooling to avoid cracks, or by stabilizing it with elements such as MgO or Cr_2O_3. If the γ phase is not the phase suitable for high-frequency applications, as the α phase, the voltage breakdown is close for both phases. The dielectric properties of alumina coatings depend strongly upon their microstructure and spray conditions [Pawlowski L (1995), Beauvais S, et al. (2005)]. One of the drawbacks is that the alumina coating's electrical resistivity is strongly dependent on the air humidity due to the increase in water adsorption in the porous coating, generating ionic conductivity. Toma et al. (2011) made the comparative study of the electrical properties and characteristics of thermally sprayed alumina and magnesium spinel coatings sprayed by HVOF and plasma. At low humidity levels, the electrical resistivity of alumina and spinel coatings was comparable (on order of magnitude of 10^{11} Ω m). At a very high humidity level (95% relative air humidity [RH]), a dramatic decrease in resistivity of about five orders of magnitude for alumina coatings and about four orders of magnitude for spinel coatings was observed, water adsorption generating ionic conductivity. HVOF spinel coatings presented the best dielectric strength values (E_d > 30 kV/mm against 20 kV/mm for sprayed alumina) and were less sensitive to moisture. Thus, these coatings can be considered as potential candidates for use in insulating applications. Toma et al. (2012) studied microstructural characteristics and electrical insulating properties of thermally sprayed alumina coatings produced by suspension HVOF (S-HVOF) and conventional HVOF spray processes. The better electrical resistance stability of the suspension-sprayed Al_2O_3 coatings could be explained by their specific microstructure and retention of a higher content of α-Al_2O_3. Kim et al. (2001) studied, for plasma-sprayed Al_2O_3–13% TiO_2 coatings, the effect of two commercial sealants based on polymers on the electrical insulation properties before and after the impregnation treatment. Cipri et al. (2007) studied the interesting electromagnetic properties (complex permittivity) of Al_2O_3–SiO_2 compounds plasma sprayed. For microwave and nuclear fusion applications, BeO is sprayed by inert plasma-spraying process but its toxicity requires very stringent spray conditions.

Prudenziati et al. (2006, 2008a, b) have studied Ni, Ni20Cr, or Ni5Al coatings plasma sprayed onto alumina films deposited onto steel plates. They compared electrical properties of resistors prepared with Ni and Ni–20Cr powders by thermal spray processes (APS and HVOF). The resistivity of Ni resistors decreased after annealing at temperatures in the range 200–400 °C, which may be due to healing of structural defects. On the contrary, the observed increase in resistivity of Ni–20Cr-based resistors were possibly due to ordering of the atoms. The temperature dependence of resistance for Ni-based resistors was invariably the same as for Ni bulk, regardless of sample origin. On the other hand, a large spread in temperature coefficient of resistance (TCR) values of Ni–20Cr-based resistors was observed. Ni-based resistors had an excellent reproducibility of the resistance versus the temperature curves, making them

excellent candidates for temperature sensors and self-controlled high-temperature heaters.

To limit oxidation in flight, cold spray can be used. In spite of the low flattening degree of cold-sprayed (CS) coatings, the anisotropy does not disappear even after annealing [Champagne VK (2007)]. However, when compared to thermally sprayed coatings, the electrical resistivity of cold-sprayed ones is lower (three to five times) [Gärtner F, et al. (2006), Marx S, et al. (2006), Wu X-K, et al. (2012a, b)].

For electronics, the following applications and development areas of cold spray process can be indicated [Gärtner F, et al. (2006)]:

- Deposition of electric screening coatings on plastics
- Generation of conducting structures on nonmetals
- Deposition of brazing and soldering alloys

For example, Gui et al. (2004) plasma sprayed, onto a graphite substrate, Al matrix composites with high-SiC volume fraction. Al-55SiC and Al-75SiC powders were milled by stainless steel and ZrO_2 balls in a conventional rotating ball mill for 1–10 h. The SiC particles exhibited a reasonably uniform distribution in the composites sprayed from the milled powders. The Fe contamination occurred to the milled powders when stainless steel balls were used. Waveguide devices for microwave applications were achieved by spraying ferrites and dielectrics.

Yamakawa et al. (2009) have studied ceramic trays (a type of kiln furniture) used for firing multilayer ceramic capacitors (MLCC) that are used in a large number of electronic appliances. They are being required in more compact sizes, larger capacities, and with reduced costs year by year. High thermal-shock resistance and reaction resistance are desired properties for these ceramic trays. They showed that the application of plasma-sprayed ZrO_2 topcoat and an Al_2O_3-sintered basecoat made it possible to enhance longevity and reduce cost.

Thermal spray coatings cannot compete with PVD techniques to manufacture electronic devices, but wire arc sprayed coatings are used as shielding material to eliminate electromagnetic and radio frequency interference and dissipate static discharge sparks. Zn and Al are currently used to protect computers, electronic office equipment, medical monitoring devices, housing constructed of temperature-sensitive plastic, and rooms containing military computers. Donner et al. (2011) have cold sprayed copper coatings onto previously thermally sprayed Al_2O_3 coatings. They either applied a bond coat on the ceramic layer or used heated substrates during the cold spray process. By both alternatives, dense copper layers were produced. The electrical conductivity reached 98% of International Annealed Copper Standard (IACS) in the as-sprayed condition on heated substrates, or

90% of the IACS value after spraying on cold substrates with aluminum bond coat and additional heat treatment, thus meeting the requirements for electronic applications. Lin et al. (2011) plasma sprayed surfaces of copper plates with a thick alumina layer to fabricate a composite with a dielectric performance suitable as substrate in electronic devices with high thermal dissipation. They concluded that an alumina layer thickness of 20 μm provided low surface roughness, low thermal resistance, and highly reliable breakdown voltage (38 V/μm). Most telecommunication systems and wireless networks operate at radio and microwave frequencies. However, due to the ever-increasing exploitation of these frequencies, both electromagnetic interference (EMI) pollution and interference occur [Osbond P (1992)]. Various plastic- and foam-layered instruments and computer equipment cases require shielding. That can be achieved by wire arc spraying zinc coating on the case—zinc because of its low melting temperature and wire arc because of the low heat flux of the process. Of course, it is better using nitrogen atomization to keep the electrical conductivity of the coating as low as possible with as less as possible oxide within coating. For military applications, the whole room can be shielded with a zinc or aluminum film wire arc sprayed outside of it.

Broadband electromagnetic wave absorbers usually correspond to spinel ferrites and hexa-ferrites operating at 45–75 GHz. Unfortunately, these materials are usually decomposed when sprayed. Lisjak et al. (2011) found a solution by plasma spraying atomized particles of hexa-ferrite and polymer. The hexa-ferrite crystalline structure was preserved during the spraying, while the polyester partly melted and resolidified during cooling. The coupling of the hexa-ferrite magnetic and dielectric losses with the polyester dielectric losses resulted in the superior properties of the composite coating with respect to the pure single-phase coating of the constituent phases.

Nd–Fe–B permanent magnets have been sprayed by vacuum plasma spraying [Rieger G, et al. (2000)]. The review by Sampath (2010) presents a few ferrites and soft magnetic materials plasma and HVOF sprayed that could be interesting for industrial applications.

According to the review paper by Sampath (2010), interesting prospective seems to be the innovative direct write thermal spray (DWTS), which is a new and exciting manufacturing technology capable of depositing a large number of electronic materials on a wide range of substrates, enabling direct write fabrication of multilayer thick film electronic devices. One of the exciting prospects of this technology is the fabrication of integrated, embedded sensors directly onto thermal spray-coated components. Sampath (2010) also presents the past, present, and future of thermal spray applications in electronics and sensors. The different dielectrics that are used or that have been used are presented: alumina, beryllia, $MgO–3Al_2O_3$ spinel, alumina–titania, cordierite ($2MgO–2Al_2O_3–5SiO_2$), and forsterite ($MgO–SiO_2$);

perovskite-based dielectric systems including lead zirconate titanate (PZT), $BaTiO_3$, and strontium-doped $BaTiO_3$, and the like, deposited by plasma or HVOF.

Cold-sprayed nickel or further ferromagnetic metals and alloys are used as induction heating coats on cooking appliances or cooking pots [Marx S, et al. (2006)]. Superficial metallization on polymer improves the corrosion-resistant and antiaging property but also achieves some special features such as electrical conductivity, thermal conductivity, electromagnetic shielding, and radial protection [Wu X-K, et al. (2012a, b)].

Electrically conductive and flexible aluminum coatings (flame sprayed with powder or wire) were deposited onto diverse polyester textiles for wearable electronics, lighting, or communication in medical techniques [Voyer J, et al. (2008)]. Coatings produced using wire as raw material had much better morphologies than those produced using powder as starting material. Successful preliminary test results for Al–polyester composites used as current collectors showed that application-specific electrically conductive composites with adjustable specific surface conductivity values and microstructures can be produced without inducing any thermal or chemical damages to the fabric material [Voyer J, et al. (2008)].

19.3.5 Medical Applications

Orthopedic and dental market is developing very fast with either bioinert coatings (Ti-6Al-4V, Ti-6Al-7Nb, Ti-13Nb-13Zr) or bioactive ones (hydroxyapatite [HA], tricalcium phosphate) [Yang Y, et al. (2006a, b), Ong JL, et al (2006)]. Two types of coatings are used [Davis JR (ed) (2004)]:

First, bioinert ones, with no activity between them and the bone or soft tissues, the most used materials being Ti and Ti-6Al-4V. These coatings must be porous (at least 30%).
Second, bioactive coatings, interacting with bone to promote bonding interfaces, are ceramic coatings with material compositions similar to that of bone tissue (calcium phosphate). The most used ceramic is hydroxyapatite (Ca_{10} $[PO_4]_6$ $[OH]_2$) exhibiting a strong activity to join the bone. Examination of four retrieved HA-coated orthopedic prostheses by Gross et al. (2004) has indicated bone attachment, along with coating removal, from a range of areas on the prosthesis surface. Coating removal occurred by dissolution provided by the higher bone-remodeling rate accompanied by the release of calcium and phosphate. Coatings dissolved faster on elevated areas or those subjected to a higher level of loading. Coatings located in less loaded areas provided a higher longevity. In fact,

very different methods have been tested to spray HA [Khor KA, et al. (1997a, b)].

RF induction plasma-sprayed coatings [Roy M, et al. (2011)], where the HA coatings, prepared with supersonic plasma nozzle, were highly crystalline in nature with insignificant phase decomposition. Both the crystallinity and purity of HA decreased when sprayed with normal plasma nozzle.

Air plasma-sprayed coatings and HA- and HVOF-sprayed nano-titania coatings on Ti-6Al-4V and fiber-reinforced polymer composite substrates [Legoux J-G, et al. (2006)] were compared. The surface cell coverage after 7 days of incubation was more complete on nano-TiO_2 than HA. Preliminary results indicated that osteoblast activity after 15 days of incubation on nano-TiO_2 was equivalent to or greater than that observed on HA. Chang et al. (1998) studied hydroxyapatite coatings sprayed with a vacuum plasma spray system at different power levels. They showed that the spray power greatly affected the crystallinity, chemical composition, and microstructure of as-sprayed hydroxyapatite coatings, which were linked to the melting state of hydroxyapatite powder. Prevéy (2000)], to overcome the complexities of characterizing plasma-sprayed HA coatings, has developed an external standard method of XRD quantitative analysis that can be applied nondestructively. Khor et al. (1997a, b) have studied the formation of a composite made of HA and Ti-6Al-4V, APS or HVOF sprayed, to enhance the poor mechanical properties of hydroxyapatite (HA). The composite presented excellent bond strength due to the superior interfacial bond between the Ti-6Al-4V rich splats and the substrate. Wang et al. (1998) have studied the preparation of a functionally graded bio-ceramic coating composed of essentially calcium phosphate compounds.

HVOF-sprayed HA compared to plasma-sprayed one: The HVOF system has proven to be a novel method for HA deposition with maximum crystallinity and purity values of 93.81 and 99.84%, respectively. With U.S. Food and Drug Administration (FDA)-approved plasma spray technique, values of only 87.6 and 99.4% crystallinity and purity, respectively, were found [Hasan S, Stokes J (2011)].

HVOF-sprayed nanostructured titania (n-TiO_2) and 10 wt. % hydroxyapatite (n-TiO_2-10 wt.% HA) powders have been engineered by Lima et al. (2010) as possible future alternatives to HA coatings deposited via air plasma spray (APS).

The n-TiO_2-10 wt.% HA coatings exhibited bond strength levels higher than 77 MPa, that is, at least twice those of APS HA coatings deposited on Ti-6Al-4V substrates. In addition, due to the high stability of TiO_2 in the human body, longevity-related concerns of HA coatings, such as

dissolution and osteolysis, were unlikely to occur. HA coatings with a high content of both crystalline HA and nanostructures were preferred for cell proliferation [Lima RS, Marple BR (2007)].

Coatings resulting from agglomerated nanostructure coatings: It was demonstrated by using an osteoblast cell culture (in vitro) that the type of HA coating phase is more important than the nanostructure character of the coating; however, HA coatings with a high content of both crystalline HA and nanostructures were preferred for cell proliferation [Lima RS, Marple BR (2007), Lima RS, et al. (2006, 2010)]. Based on cell culture (in vitro) results observed for bulk nanostructured ceramics, the nanostructure zones being found on the TiO_2 coating surface may have played an important role in producing good biocompatibility results. However, up to this point there is no experimental evidence to prove this [Lima RS, Marple BR (2007)].

ZrO_2/SiO_2 composite coatings have high-abrasive wear resistance compared with pure HA and HA/ZrO_2 coatings because they exhibit high hardness and dense structure. The in vitro test of the composite coatings in simulated body fluid showed a growth of nanometer-sized particles apatite after the immersion of the coatings for 20 days [Morks MF, Kobayashi A (2008)].

Well-adherent, flawless, glass–ceramic (SiO_2-CaO-K_2O) coatings on alumina and Ti6-Al-4V substrates were realized [Vitale-Brovarone C, Verné E (2005), Xie Y, et al. (2009)]. The obtained results were reproducible and the applied techniques were low cost and not complex. When tested in vitro, the coatings showed an extensive precipitation of a thick HA layer, well adherent to the coatings.

SiO_2-CaO-P_2O_5-based bioactive glasses (BAGs) and glass–ceramics are attractive materials for biomedical applications, because of the excellent levels of bioactivity, which can be achieved by formulating and selecting appropriate compositions [Bolelli G, et al. (2009a, b)]. Preliminary results with high-velocity suspension flame spraying bioactive glasses seemed to be promising.

HA coatings deposited by HVOF spraying showed similar advantages as coatings obtained by APS, but with higher crystallinity. Heat treatment of the coating allowed crystallization of the amorphous calcium phosphate (ACP) present in the coatings [Fernandez J, et al. (2007)]. XRD analysis confirmed that the ACP transforms directly to HA and not to other calcium phosphate phases. Similarly, functionally graded calcium phosphate coatings were produced on Ti-6Al-4V substrates by plasma spraying [Fernandez J, et al. (2007)]. The microstructure of the coating was dense with the typical lamellar structure. The top layer of the coating was mainly composed of tricalcium phosphate (TCP), which suggested high bio-resorbability. After post-spray heat treatment, the TCP phase, from either the decomposition of HA or the TCP feedstock, and the tetracalcium phosphate (TTCP) phase were no longer detected by XRD, but the CaO phase remained.

Yu et al. (2003) plasma sprayed HA coatings and post spray treated them by the Spart Plasma Sintering (SPS) technique at 500 °C, 600 °C, and 700 °C for duration of 5 and 30 min. The HA coatings treated in SPS for 5 min revealed rapid surface morphological changes during in vitro incubation (up to 12 days), indicating that the surface activity is enhanced by the SPS treatment.

Bellucci et al. (2012) coated titanium plates by high-velocity suspension flame spraying (HVSFS) technique using a novel bioactive glass composition based on the K_2O–CaO–P_2O_5–SiO_2 composition ("Bio-K"). On half of the samples, an atmospheric plasma-sprayed (APS) TiO_2 bond coat was preliminarily deposited; suspensions of attrition-milled micron-sized glass powders, dispersed in a water and isopropanol mixture, were then sprayed onto both bare and bond-coated plates using five different process parameter sets. All coatings exhibited analogous behavior when soaked in a simulated body fluid (SBF) solution. Their interaction with the SBF involved ion release from the glass, conversion of its surface into an amorphous and hydrated silica layer followed by nucleation, and growth of a carbonated hydroxyapatite film on top of the latter. This interaction seemed particularly fast, as the hydroxyapatite film started to appear after 3 days of soaking.

Other materials were tested for their potential application in biomedicine: Liang et al. (2010) plasma sprayed three kinds of powders composed of Ca_2SiO_4, ZrO_2, and $CaZrO_3$ with different Ca_2SiO_4. Results showed that the chemical stability of the coatings was significantly improved compared with pure calcium silicate coatings and increased with the increase in Zr contents. Results indicated that plasma-sprayed coating with 40 wt.% of Ca_2SiO_4 had medium dissolution rate and good biological properties, suggesting its potential use as bone implants.

Preliminary results showed that thermally sprayed nanostructure TiO_2 coatings exhibited photocatalytic bactericidal activity with *Pseudomonas aeruginosa* [Jeffery B, et al. (2010)].

The antibacterial behavior of HA–Ag (silver-doped hydroxyapatite) nanometer-sized powder and their composite coatings were investigated against *Escherichia coli* (DH5a). HA–Ag nanometer-sized powder and PEEK (poly-ether-ether-ketone)-based HA–Ag composite powders were synthesized using in-house powder processing techniques. These nanometer-sized composite powders were cold sprayed. The results indicated that the antibacterial activity increased with increasing HA–Ag nanometer-sized powder concentration in the composite powder feedstock and cold-sprayed coating [Sanpo N, et al. (2009a)].

The antibacterial behavior of chitosan–copper complex (CS–Cu) powder and their composite coatings were investigated against *E. coli* (DH5a). The cold-sprayed coatings retained the antibacterial properties of the original feedstock powders [Sanpo N, et al. (2009b)].

19.3.6 Clearance Control Coatings

In compressors, gas turbines, and turbochargers, dimensional changes take place between the rotor and stator components because of thermal and mechanical effects during operation. These dimensional changes affect sealing and clearance control systems, consisting of a sacrificial element and a cutting component, that are used. Thermal spray coatings, called abradables, and honeycomb seals form effective sacrificial systems. Abradable coatings are machined in situ and consist of a soft metal with polymer particles in cold sections and Ni–graphite or MCrAlY with polyester or BN particles in hot areas [Davis JR (ed) (2004), Ma X, Matthews A (2009), Johnston RE (2011)]. Additives provide the necessary friability, as well as aid in dry lubrication. Other thermally sprayed coatings can also be used on the "cutting" side of the clearance control system, and when the dynamic member of the system is too soft to cut without a coating.

Abradable materials have a strong impact on the turbine efficiency and fuel consumption [Rajendran R (2012)]. At rotating speeds of the order of 10,000 rpm, rotating blade tip may rub against the stationary casing, due to either thermal expansion or misalignment or rotation, inducing strains [Ma X, Matthews A (2007)]. Abradable seals act as sacrificial layers between the blades and the casing and are soft enough to avoid significant wear to blade tips, thus allowing much smaller clearances. They also offer a wear protection to the shroud material and rubbing blades. High-temperature abradable seals are used in high-pressure gas turbine of jet engines [Rajendran R (2012)].

Abradable materials are rather complex because they must be abradable with a good resistance and strength as well as with a good resistance to oxidation and corrosion. Their porosity also plays a key role and is often achieved by spraying also polymers, whose size and volume percentage are critical for optimizing both the "abradability" and erosion-resistance performance. The polymer is then removed by heating the sprayed coating. At last, the coating composition must be such that particle debris, released from the coating into the engine, are kept to a minimum. Their composition must also be adapted to the blade materials: titanium or nickel or steel. They are classified in low- and mid- (<540 °C) and high-temperature (540–980 °C) abradables. Low-temperature abradables are aluminum based with graphite or polyester or polyimide or boron nitride, or nickel based with graphite or calcinated bentonite clay [Sun F, et al. (2012), Puranen J, et al. (2011)]. In the cold compressor section, aluminum base is preferred due to the high risk of moisture. At high temperatures one uses super alloys (MCrAlY with M=Ni or Co or both) with polyester and/or boron nitride or bentonite or graphite. Recently, ceramic mixed with softer material (BN and polyester) abradable coatings have been developed to reduce slightly the alloy surface temperature. Other seals, which do not interact with the blades, are coatings composed of Babbit (tin–copper–lead), bronze, and AlSi-polyimide. They are used against labyrinth seals along the engine shafts in the compressor and turbine, sealing either gas or oil paths [Davis JR (ed) (2004)].

A few examples are presented below. For high temperatures, Bardi et al. (2008) have tested, for the first stage of industrial gas turbines, the behavior of composite coatings: plasma-sprayed CoNiCrAlY/Al$_2$O$_3$ and laser-cladded CoNiCrAlY/graphite coatings. After 1000 h at 1100 °C both coatings did not show relevant microstructural modifications. Steinke et al. (2010) showed that the hardness could be used as an indication for the abradability. It should be in the range of 500 HV$_{0.5N}$. A hardness of approximately 600 HV$_{0.5N}$ seemed to be the upper end at which a good cut-ins was no longer reliably reproducible. Sporer et al. (2007) performed abradability tests of YSZ abradables with the Sulzer Innotec test rig that can operate up to 1200 °C with blade tip speeds up to 410 m/s.

At intermediate temperatures (<450–480 °C), AlSi-hexagonal BN or NiCrAl–bentonite is popular [Johnston RE (2009, 2011), Stringer J, Marshall MB (2012), Faraoun HI, et al. (2006), Bounazef M, et al. (2004)]. For example, the AlSi–hBN coating was characterized by a proportion of about 40% of hBN "lubricant" particles trapped in the metallic matrix [Faraoun HI, et al. (2006)]. The amount of porosity was about 5%. The NiCrAl–bentonite coating contained about 25% of relatively spherical bentonite particles together with about 30% porosity constituted mainly of large pores [Faraoun HI, et al. (2006)]. Comparatively to the hBN particles, the majority of the large bentonite particles and pores in this coating are distributed without significant preferred orientation [Faraoun HI, et al. (2006)].

Johnston (2011) studied the net thermal spraying residual stress (NTSRS) and showed that it was most sensitive to changes in substrate thickness, abradable deposit thickness, and deposit modulus. Such results could be used for the creation of new abradable materials with optimum stress profiles and greater mechanical integrity. Stringer and Marshall (2012) studied the wear at high speed: impact velocities between 100 and 200 m/s and incursion rates between 3.4 and 2000 μm/s. They suggested that the adhesion of abradable material was due to plucking out of complete phases from the abradable coating, rather than a conventional cutting mechanism. Ma and Matthews (2007) suggested using "progressive abradability hardness" (also

called "specific grooving energy"), abbreviated as "PAH," to measure abradability in the scratch test. At low temperatures, Ma and Matthews (2007) studied, for abradable seals to centrifugal compressors and steam turbines, the development of a new abradable silicon rubber adhesively bonded to a metal substrate.

Lima and Marple (2007) showed that a material that was generally considered as being hard and stiff, as YSZ, could be engineered to produce a nanostructured 100% ceramic coating with a friable structure possessing attributes required for an abradable coating. In order to engineer a friable ceramic coating, the molten part of a semi-molten agglomerate particle must not fully infiltrate into the capillaries of its nonmolten core during thermal spraying. The porous nano-zones spread throughout the coating microstructure acted as weak links, thereby making the coating friable. Abrasive wear-resistant coatings are applied to blade tips and labyrinth seal teeth. They are often made of alumina, alumina–titania, nickel–aluminum cermet, and Ni–Cr–Cr$_3$C$_2$ cermet.

19.3.7 Bond Coatings

Bond coats are very important to:

- Improve the coating adhesion by forming an anchoring sub-coat or adhering to clean and smooth surfaces that cannot be grit blasted properly
- Form a buffer layer with an expansion coefficient between that of substrate and topcoat
- Provide an oxidation- or corrosion-resistant barrier to the substrate

According to the oxidation or corrosion barrier, and the working temperature of the coated part, the bond coat must be properly chosen.

For example, Mo can be used up to 315 °C [American Welding Society (1985)]. NiAl bond coats are oxidation resistant up to 650 °C with Ni–Al (20 wt.%) and 800 °C with Ni–Al (5 wt.%), but they can also be used against corrosion. For example, Lekatou et al. (2008) have investigated the corrosion behavior of an HVOF Ni–5Al/WC–17Co coating on Al-7075 in 0.5 M H$_2$SO$_4$. In the temperature range of 25–45 °C, the coating exhibited pseudo-passivity that effectively protected from localized corrosion. In case of surface film disruption, the bond coat successfully hindered corrosion propagation into the Al alloy. Yılmaz (2009) has shown that the application of NiAl (5 wt.%) bond coat layer in the plasma spraying of Al$_2$O$_3$ and Al$_2$O$_3$–13 wt.% TiO$_2$ on pure titanium substrate has increased the hardness and bonding strength of coatings. Ni–Cr (20 wt.%) bond coats can be used up to 980 °C and are designed to resist oxidation and corrosive gases. Cho et al. (2009) coated micron-sized WC–Co powder onto a 420J2 steel substrate and bond coats of Ni, NiCr, and Ni/NiCr using HVOF spraying. The fracture locations were at interfaces with top coatings of WC–Co and NiCr indicating that adhesion between metal and similar metal is much stronger than between metal and cermet (WC–Co) because of the easy diffusion between similar metal atoms. Bond coats for TBCs have been extensively studied. They are of the MCrAlY type with M = Ni, Co, and the like that can be doped with Pt, Hf, and the like and can support temperatures of 1100–1200 °C. A few examples are presented below. Schulz et al. (2008a, b) have studied NiPtAl bond coats as well as NiCoCrAlY(X) deposited by LPPS and EBPVD underneath conventional EBPVD yttria-stabilized zirconia topcoats on three different substrate alloys. The longest lifetimes were achieved on a novel Hf-doped EBPVD NiCoCrAlY bond coat owing to a differing TGO formation and failure mechanism. Zhao and Xiao (2008) investigated the effect of the Pt content on the durability of thermal barrier coatings (TBCs) with a Pt-enriched $\gamma + \gamma'$ bond coat. During oxidation, impurities such as S, C, and refractory elements segregating to the interface would degrade the TGO adherence to the bond coat. However, a higher content of Pt can inhibit this segregation and thus improve the TBCs life. Tang and Schoenung (2005) observed multilayered accumulation of thermally grown oxide (TGO) locally on the tops of the roughness protrusions of the bond coat in thermal barrier coatings. Chen et al. (2011a, b, c) have compared TGO growth and cracking behaviors in TBC systems with APS-, HVOF-, and CGDS-CoNiCrAlY bond coats during thermal exposure. Results pointed the potential advantage of using CGDS technique to produce TBC bond coats, for its significantly improved durability over the commercial air plasma spray bond coat, as well as its fast deposition rate and low deposition temperature. The TBCs with HVOF bond coats had an extended durability, compared to that obtained by APS but lower than that by CGDS. Other materials than superalloy were tested for TBC bond coat, such as glass–ceramics [Das S, et al. (2009)]. The TBC with YSZ topcoat, glass–ceramic bond coat, and nickel base superalloy substrate was subjected to static oxidation test at 1200 °C for 500 h in air. No TGO layer was found between the bond coat and the topcoat in the case of glass–ceramic bonded TBC system, while the conventional TBC system exhibited a TGO layer of about 16 μm thickness at the bond coat–topcoat interface region.

Bond coats are also used for other applications, discussed below.

Lu et al. (2004) plasma sprayed two-layer hydroxyapatite (HA)/HA + TiO_2 bond coat composite coating (HTH coating) on titanium. The HA + TiO_2 bond coat (HTBC) consisted of 50 vol.% HA and 50 vol.% TiO_2 (HT). The positive effect of the HTBC on the decrease of residual stress in HA top coating (HAT coating) was evidenced through the observation on the surface cracking. The toughening and strengthening of HTBC was thought to be mainly due to TiO_2. Goller (2004) plasma sprayed bioglass, known as 45S5 (45% SiO_2, 6% P_2O_5, 24.5% CaO, and 24.5% Na_2O, all in weight percent), onto a titanium substrate with and without Amdry 6250 (60% Al_2O_3–40% TiO_2) as bond coating layer. The adhesive bonding observed at the bioglass metal interface turned into cohesive bonding by application of the bond coating layer.

Guanhong et al. (2011) have deposited on polymer matrix composites (PMCs; carbon-fiber-reinforced unsaturated polyester) the Al bond coat and Al_2O_3 top coating by APS. The highest shear adhesion strength of the bond coating was 5.21 MPa and the hardness of the surface was much improved by the alumina coating. Liu et al. (2006) arc sprayed Al and Zn and plasma sprayed Al, Zn, Ni_3Al, and Cu on carbon-fiber-reinforced polyimide substrate as bond coats for erosion and thermal-resistant coating. Ni_3Al or Cu damaged the substrate. Arc-sprayed and plasma-sprayed Al and Zn could be used as bond coat materials. For Zn as bond coat material, depositing method had little influence on shear adhesion strength. For Al as bond coat material, plasma spray was superior to arc spray, preheating improving shear adhesion strength.

19.3.8 Freestanding Spray-Formed Parts

The freestanding bodies are produced by plasma spraying for refractory or ceramic materials [Geibel A, et al. (1996), Devasenapathi A, et al. (2002), Chráska P, et al. (1997), Neufuss K, et al. (1997), Shi S, Hwang J-Y (2003), Patel RR, et al. (2010), Brožek V, et al. (2009)], by HVOF spraying for thin alloys or cermets [Helali M, Hashmi MSJ (1996)], and also by cold spray [Pattison J, et al. (2007)]. Of course, rotationally symmetrical shapes are preferred: the core is easy to produce, the stress in the sprayed material is low and evenly distributed, and at last separating the core from the coating is easier. In most cases the core is made of smooth graphite or other material coated by atomization of a product preventing the substrate–coating adhesion. It is also possible to achieve very complex shapes, such as manifolds of racecar engines, by spraying on special materials that can be afterward dissolved. It becomes then possible to coat externally the sprayed ceramic thermal barrier by an alloy to achieve the hot gases tightness. This technique, having a high cost, is hardly suitable for large-scale production, but it offers advantages as far as cost and fast delivery are concerned. For example, sprayed ceramic bodies offer good mechanical strength at high temperatures, and excellent electrical insulation at temperatures where glass insulators give-up chemical resistance against corrosive gases or liquids. For example, in rocketry nozzle inserts are achieved that way. Ceramic tubing is used in furnace construction as well as in chemistry. The production of crucibles for the melting of materials is also an important application. Such crucibles are made of partially (YPSZ) or totally (YSZ) stabilized zirconia, strontium zirconate, alumina, tungsten, and molybdenum. A relatively new application of PTA deposition is the fabrication of freestanding shapes, or solid free form fabrication.

A few examples of freestanding coatings are presented below.

Chen et al. (1993) vacuum plasma sprayed dense, oxide-free, and pore-free freestanding forms of NiAl and NiAl-B. The deposits retained the same phase structure as the starting powders. The as-sprayed freestanding deposits exhibited a large grain size due to self-annealing during vacuum plasma spray processing. The hardness of NiA1-B was about 10% higher than that of NiAI.

Saeidi et al. (2009) produced freestanding VPS- and HVOF-sprayed CoNiCrAlY coatings. The as-sprayed HVOF coating retained the γ/β microstructure of the feedstock powder, and the VPS coating consisted of a single γ phase. A 3 h, 1100 °C heat treatment in vacuum converted the single-phase VPS coating to a two-phase γ/β microstructure and coarsened the γ/β microstructure of the HVOF coating. Oxidation of the freestanding as-sprayed coatings produced a dual-layer oxide consisting of an inner layer of α-Al_2O_3 and an outer spinel layer.

Fan and Ishigaki (2001) used induction plasma spraying (IPS) to produce freestanding parts of Mo_5Si_3-B composite and $MoSi_2$ materials. The oxidation resistance, up to 1210 °C, of the former composite was compared with that of the latter, known to be resistant to high-temperature oxidation. At boron contents greater than 1 wt.%, composites demonstrated encouraging oxidation resistance and with 2 wt.% boron it was comparable to $MoSi_2$.

Waki et al. (2009) sprayed freestanding circular tube specimen of CoNiCrAlY by VPS, HVOF, and APS. They studied the effect of post-spray thermal treatments, in vacuum and in air, on the mechanical properties in the 400–1100 °C temperature range. High-temperature thermal treatment in air was effective in increasing the bending strength and Young's modulus. It was especially effective on the APS coatings, which were produced using powders with average size of 60 µm, and on HVOF coatings, those bending strengths increased by approximately three times.

Plasma-sprayed freestanding zircon pipes (inside diameter 83 mm, wall thickness 1.5–2 mm, lengths up to 2000 mm) are used in furnaces as shields for heating elements [Chráska P, et al. (1997)]. The plasma deposits contained glassy silica and various modifications of zirconia in addition to a small amount of the amorphous phase. This combination of zirconia and silica exhibited properties such as a high thermal-shock resistance, good corrosion resistance, and low wettability.

Chen et al. (2011a, b, c) obtained by plasma spraying freestanding $La_2Zr_2O_7$ coatings, using an amorphous La–O–Zr precursor as the feedstock. During thermal spraying, the amorphous powder crystallized, and fine grains were formed, while with crystallized feedstock powder the average grain size was 750 nm. The thermal conductivity of the as-sprayed coating with the amorphous feedstock powder was approximately 0.42 W/m K and its average coefficient of thermal expansion (CTE) was 11.1×10^{-6}/K, value similar to that of metallic bond coatings.

Henne et al. (1999) processed by induction plasma spraying the material of metallic bipolar plates for solid oxide fuel cells, a chromium alloy with the composition 94% Cr, 5% Fe, and 1% Y_2O_3. Freestanding parts had a density above 95% of the theoretical density of the material. From the deposits obtained, it was positively concluded that, as long as the deposit thickness did not exceed about half the characteristic substrate dimension, the contour of the free side of the deposit sufficiently followed the contour of the substrate.

Agarwal et al. (2003) plasma sprayed a mixture of commercially available aluminum oxide powder (99.8% pure), in the 15–45 µm particle size range, with nanosized aluminum oxide powder (99.9% pure) thoroughly mixed in a rotating ball (alumina ones) mill for 24 h. Coatings were sprayed on conical mandrels. Spray parameters were controlled with an innovative proprietary cooling technique to retain a large fraction of nanosize Al_2O_3 powder particles in the spray deposit. Nanosize particles were partially melted and trapped between the fully melted coarser (micrometer size) Al_2O_3 grains.

Laha et al. (2004) plasma sprayed freestanding structures, hollow conoid (100 m taper-length, 62 mm diameter, 2 mm thickness), of Al-based nanostructured composite with carbon nanotubes (CNT) as second phase. Carbon nanotubes (CNT) were successfully retained in the spray-formed composite structure. The CNTs were observed largely between the consecutive Al–Si splats with dangling structure.

Hussain et al. (2011) cold sprayed titanium freestanding coatings, some of them being vacuum heat-treated to further decrease porosity levels. Cold spraying using N_2 as a process gas heated to 800 °C deposited titanium with less interconnected porosity than N_2 gas at 600 °C due to a higher degree of particle deformation on impact. Pores above 1 µm were reduced to 0.2 vol.% and the total interconnected porosity to 1.8 vol.% after heat treatment of deposits produced at a process gas temperature of 800 °C.

Herman et al. (1994) sprayed by the VPS process dense deposits of unreinforced and composite matrices of Ni_3Al alloy and $MoSi_2$. Geibel et al. (1996) showed that plasma spray deposition was a suitable technique to produce freestanding, near-net-shaped components of the difficult-to-shape NiAl intermetallic compound.

Weiss et al. (1992) described a new spray-forming process based on thermal spray shape deposition. Shape-deposition processes built three-dimensional (3-D) shapes by incremental material build-up of thin, planar, cross-sectional layers. These processes did not require preformed mandrels and can be used to directly build three-dimensional structures of arbitrary geometrical complexity. The basis for the thermal spray approach was to spray each layer using a disposable mask that had the shape of the current cross section. Masks could be produced from paper rolls, for example, with a CO_2 laser.

19.3.9 Emerging Thermal Spray Applications

19.3.9.1 Solid Oxide Fuel Cells

A solid oxide fuel cell (SOFC) is an electrochemical conversion device that produces electricity directly from oxidizing fuel. They are made of four layers comprising three ceramic layers, which are not electrically and ionically active below about 600 °C. Such high operating temperatures result in mechanical and chemical compatibility issues. To achieve an affordable manufacturing cost, the whole SOFCs have been manufactured by plasma spraying: cathode, electrolyte, and anode. One of the disadvantages of the process is the thickness of the electrolyte that must be below a few tens of micrometers and impervious. However, it seems that thin and impervious electrolyte (around 10 µm) can be achieved by suspension plasma spraying. Henne (2007) presented in 2007 an excellent overview of thermally sprayed coatings to make SOFC components and also entire cells. Among the numerous studies, essentially with plasma spraying, published before that year, "only few have already proven their potential for a reliable and efficient production of relevant layers and components and promised the status for a technical breakthrough for mass production where a continuous operation with high throughput and yield producing the desired quality is required" [Henne R (2007)]. Henne recommends for further developments to improve plasma sources, increase deposition efficiency, work on less costly materials, and combine synthesis and deposition of cell materials based on low-cost precursors. He also evokes the possibility of the recycling of "plasma gas" and overspray [Henne R (2007)]. A few examples are given below.

Takenoiri et al. (2000) developed a seal-less planar SOFC consisting of a cell supported with a porous metallic substrate and a metallic separator. The anode and electrolyte were fabricated using flame spraying and APS, respectively, and APS was also used to form a protective coating of the separator. It was shown that the electrical resistance of the separator could be kept lower than 25×10^{-3} Ω cm^2 for at least 3000 h by the application of $(LaSr)MnO_3$ protective coatings.

Barthel et al. (2000) deposited by thermal spray processes (VPS and flame spraying) porous composite cathodes of $(La_{0.8} Sr_{0.2})_{0.98}MnO_3$ (LSM) and ZrO_2–12% Y_2O_3 (YSZ). The electrochemical performance of the cathodes, evaluated by impedance spectroscopy, indicated significant improvements, especially for the bilayer technique, whereas the concentration profiles of the multilayer gradients still must be optimized. Flame spraying seemed promising.

La_{10} $Mg_{0.2}$ $Si_{5.8}$ $O_{26.8}$ was prepared by solid-phase sintering and also plasma sprayed by Sun et al. (2012). The amorphous transformation of lanthanum silicates happened during APS process. The heat treatment effectively ameliorated phase composition of as-sprayed coatings, complete recrystallization being achieved after 1000 °C for 4 h. However, it decreased the permeability of coatings.

Puranen et al. (2011) plasma sprayed Mn–Co–Fe spinel coating on the thin metallic interconnectors in SOFCs. They optimized spray conditions to produce coatings with low thickness and low amount of porosity. The original spinel structure decomposed because of the fast transformation of solid–liquid–solid states but was partially restored by using post-annealing treatment.

Harris et al. (2012) manufactured by axial-injection APS (Mettech Axial torch) single-phase and composite cathodes based on $La_{0.6}$ $Sr_{0.4}$ $Co_{0.2}$ $Fe_{0.8}$ $O_{32-\delta}$ (LSCF) and $Sm_{0.5}$ $Sr_{0.5}$ Co O_3 (SSC). An average surface temperature of approximately 2200 °C was needed for adequate melting of LSCF particles and approximately 2100 °C to melt SSC. But too much particle heating resulted in coarser and denser microstructures.

Suspension plasma spraying [Fauchais P, et al. (2011)] seems to be promising for the SOFC cathode deposition, as coatings with high porosity levels in combination with fine pore size can be produced. Shen et al. (2011), using induction plasma technology and solution plasma spraying, achieved a homogeneously mixed nanosized composite $Ce_{0.8}$ $Gd_{0.2}$ $O_{1.9}$ (GDC)/LSCF powder without using a prolonged period of mechanical mixing. The nano-powders exhibited a perovskite structure and a fluorite structure as well as separated GDC ($Ce_{0.8}$ $Gd_{0.2}$ $O_{1.9}$) and LSCF phases. The coatings were homogeneous and porous (51% porosity) with cauliflower structures. Wang et al. (2012) successfully deposited with Axial III Mettech torch Ni–YSZ anode and YSZ electrolyte half-cells on porous Hastelloy X substrates by SPS. The thickness of the anode and electrolyte layers was ~40 µm and 10–20 µm, respectively. The density of YSZ electrolyte coatings increased as plasma torch input power increased. The anodes of the deposited half-cells were porous and electrolytes dense, with no interconnected pores. Michaux et al. (2010) prepared and characterized porous anode layers with homogeneous nickel distribution and nanometer-sized microstructure for SOFC application. They investigated the effects of some spray parameters on the layer architecture and composition.

19.3.9.2 Sensors

For oxygen permeation membranes, Zotov et al. (2012) prepared $La_{1-x}Sr_xFe_{1-y}Co_yO_{3-\delta}$ (LSFC) coatings by low-pressure plasma spraying–thin film processes using different plasma spray parameters. The microstructures of the investigated LSFC coatings depended sensitively on the oxygen partial pressure, the substrate temperature, the plasma jet velocities, and the deposition rate. Coatings deposited with Ar-rich plasma and higher plasma jet velocities were most promising.

Oxygen sensors based on ionic conduction of plasma-sprayed ZrO_2 (measurement of the small electrical current caused by oxygen ions diffusing through the layer) are now extensively used in catalytic car exhaust pipes [Fedtke P, et al. (2004)].

19.3.9.3 Decorative Coatings

Thermal spraying is not widely adapted to decorative coatings, and the main efforts were achieved on relatively large structures such as bridges, water tower, chimneys, and so on that were sprayed with zinc or aluminum for corrosion protection. They can be left as-sprayed or sealed with plain or colored sealers. The liquid flame spraying process has been developed to uniformly color hot glass objects [Gross KA, et al. (1999)]. Thermal spraying of concrete or bricks have been performed with glazing materials. Arcondéguy et al. (2007, 2008) manufactured glaze layers by flame spraying onto substrates that decomposed when heated and for which the traditional glazing process was not appropriate. Adjusting the chemical composition permitted to adjust the transition temperature of the materials. Adjusting their morphology by posttreatment permitted to increase the deposition efficiency and reduce the pore content of the coatings. Of course, the decorative value of such coatings is important for the aesthetic but it has a cost.

19.3.9.4 Spent Nuclear Fuel

Many works are now devoted to the containment of spent nuclear fuel that poses a challenge from the perspective of materials science. These applications require to safely provide neutron absorption and corrosion resistance at reasonable costs for very long periods. Identifying a single material that can meet all these requirements has been quasi-impossible and the introduction of coatings seems promising. Alternate coating materials such as amorphous metals and ceramics have been developed and studied to better resist corrosion. Amorphous coatings with boron or other neutron-absorbing element have demonstrated promising results against corrosion [Blink J, et al. (2009), Farmer JC, Choi J-S (2007), Blink J, et al. (2007), Farmer J, et al. (2009), Branagan D (2004)]. The absence of

long-range structure (grain boundaries, dislocations, and segregations), in corrosion-resistant materials further enhances the corrosion resistance of amorphous alloys. One of the most promising formulations was found to be $Fe_{49.7}Cr_{17.7}Mn_{1.9}Mo_{7.4}W_{1.6}B_{15.2}C_{3.8}Si_{2.4}$ (SAM2X5), which included chromium (Cr), molybdenum (Mo), and tungsten (W) for enhanced corrosion resistance, and boron (B) to enable glass formation and neutron absorption [Branagan D (2004)]. The parent alloy for this series of amorphous alloys, which is known as SAM40 and represented by the formula $Fe_{52.3}Cr_{19}Mn_2Mo_{2.5}W_{1.7}B_{16}C_4Si_{2.5}$, has less molybdenum than SAM2X5 and was originally developed by Branagan (2004). Coatings, HVOF sprayed, have demonstrated phase stability well above 500–600 °C and at high neutron dose (equivalent to 4000 years inside the container) [Blink J, et al. (2007)].

Alumina plasma-sprayed coatings, manufactured with graded alumina–titania coatings, and phosphate-sealed alumina coatings were investigated to improve the properties of metallic substrates operating in extreme environments of spent nuclear fuel. The effects of particle size distribution, phosphate sealant, and graded titania additions on the dielectric strength of the as-sprayed, thermally cycled, and thermally aged coatings were investigated [Berard G, et al. (2008)]. Spinel ($MgAl_2O_4$) coatings were also investigated [Haslam JJ, et al. (2005)]. All studies demonstrated the necessity to have impervious coatings.

19.3.9.5 Combined Cycle Power Plant Combinations

Hardwicke and Lau (2013) emphasized that recent power generation company announcements centered on integrating renewable resources onto the power grid for cleaner, more efficient energy generation. They think that these technical breakthroughs can certainly utilize the new developments in the thermal spray industry.

19.3.9.6 Future Nuclear Fusion Reactor

While thermonuclear fusion is a potential source of cleaner and safer energy for the future, its technological realization depends on the development of materials able to survive and function under extreme conditions. According to the review by Matejicek et al. (2007), materials to be applied in a fusion reactor will be subjected to extreme and complex loading conditions and have to fulfill very complex and sometimes contradicting requirements. A number of the demands on fusion materials can be fulfilled by the application of coatings. Thermal spraying is just one of the available coating technologies. Thermal spray (especially plasma spray) coatings were developed and tested for a variety of applications in fusion environments, including plasma facing components, tritium permeation barriers, and electrical insulation.

19.4   Thermally Sprayed Coatings by Industry

In his paper [Longo FN (1992)], based on the 1990 multiclient study prepared by Gorham Advanced Materials Institute, entitled "The Expanding Business Opportunities and Challenges in Thermal Sprayed Coatings," Frank Longo, who is a thermal spray industry consultant and is a principal author of this study, identified 34 industrial sectors where thermal spray and other coatings were used. The broad base of end-user activity with thermal spray coatings for 12 industrial sectors is profiled in Tables 19.1, 19.2, and 19.3. These show that the main activities have been in the gas turbine and aircraft industry, turbine and steam, engines and automotive, engines and diesel, transportation, oil and gas exploration, chemical processing, paper and pulp, defense and aerospace, medical and dental, and electric/electronic industry. Longo (1992) estimated that the total market was projected to reach 1.8 to 2.0 billion US dollars by the end of the decade (2000) with strong growth in the powder/consumable business, and standard equipment such as APS and VPS systems and a growing demand for contract coating services which represented close to 50% of the (1990) TS market.

Two decades later, Dorfman and Sharma (2013a, b) estimated that the annual thermal spray industry is worth approximately US$6.5 billion [Hanneforth P (2006)] with the majority of revenue generation in coating services, thus underlying the growth of thermal processes. In terms of market segmentation, approximately 60% of the total TS market at that time belongs to the turbine industry, 15% to automotive, and the remaining 25% is distributed over a large number of other industries.

Dorfman and Sharma (2013a, b) also point out that while future growth of the TS market will continue to depend on the growth of turbine industry, a greater growth potential exists in the "other" industries where currently the penetration of TS is low such as in the area of energy generation using conventional or renewable resources, oil and gas, pulp and paper, metal processing, and biomedical and electronics industries.

The last *ASM Handbook, Volume 5A: Thermal Spray Technology* [Tucker RC (ed) (2013)], which is a replacement for the *Handbook of Thermal Spray Technology* [Davis JR (ed) (2004)], provides a good insight of developments in terms of thermal spray industrial-scale applications including electronics and semiconductors, automotive, energy, and biomedical. Traditional thermal spray market sectors such as aerospace and industrial gas turbines as well as important areas of growth such as advanced thermal barrier materials, wear coatings, clearance control coatings, and oxidation/hot corrosion-resistant alloys are reviewed.

19.4.1 Aerospace

Aero engines are probably those where thermally sprayed coatings, especially APS, VPS, or HVOF sprayed, were extensively developed since the 1970s–1980s. The fan, compressor, combustion chamber, and turbine must be protected with coatings against oxidation, hot gas corrosion, erosion, and wear. As illustrated in Figs. 19.3, and Fig. 19.4, hundreds of components are presently sprayed in aero engines used both for military and traditional industrial gas turbine (IGT) engine units worldwide. Their main role is to reduce friction and wear, corrosion, clearance control, and for high-temperature protection (TBCs) to improve engine efficiency, and extend service life. In aero engines, almost all types of sprayed materials are used (carbides, iron and steel, nickel alloys, cobalt alloys and superalloys, nonferrous metals) except the self-fluxing alloys. HVOF coatings have also been used to replace hard chromium in landing gear. For more details see [Davis JR (ed) (2004), Thintri Inc. (2013)].

TBCs were first successfully tested in the turbine section of a research gas turbine engine in the mid-1970s. By the early 1980s they had entered revenue service on the vane platforms of aircraft gas turbine engines [Billah BM, et al. (2012), Miller RA (1997)], and today they are flying in revenue service on vane and blade surfaces [Miller RA (1997)]. Thermal insulation benefits provided by TBCs and the resulting impact on component creep and thermomechanical fatigue life have made them enablers of high-thrust gas turbine engines. As underlined by Pratt and Whitney, of particular importance is the EB-PVD-based TBC—presenting an excellent compliance upon thermal cycles were then anticipated to improve blade life by a factor of 3 [Schulz U, et al. (2008a, b)].

The aging of TBC's topcoat strongly depends upon the spray conditions and powder morphologies used to spray or deposit it, conditions acting on its sintering properties [Bose S, de Masi-Marcin J (1997), Cipitria A, et al. (2009)]. The second problem is the oxidation of bond coat with the formation of thermally grown oxide (TGO) [Markocsan N, et al. (2009), Golosnoy IO, et al. (2009), Hernandez MT, et al. (2009), Vaßen R, et al. (2009a, b), Toscano J, et al. (2006), Pint BA, et al. (2010)] as well as the bond coat corrosion with oxides such as CMAS (acronym for CaO, MgO, Al_2O_3, and SiO_2) [Mohan P, et al. (2010), Jones RL (1997), Li L, et al. (2010a, b), Hernandez MT, et al. (2009)] or vanadium oxide [Chen Z, et al. (2009)]. Feuerstein et al. (2008) have shown that the most advanced thermal barrier coating (TBC) systems for aircraft engine and power generation hot section components (combustors, blades, and

Table 19.1 Thermal spray processes used by industry [Longo FN (1992)]

	O_2/fuel powder low velocity	Spray and fuse torch	HVOF	Air plasma	Vacuum plasma	Inert gas shroud plasma	Plasma transfered arc	Laser-assisted plasma	D-gun	Gator gard	O_2/fuel wire spray	Electric arc	Electric arc with inert cover or vacuum	Rokide
Gas turbine, aircraft	X	…	X	X	X	X	X	X	X	X	X	X	…	…
Gas turbine, industrial	…	…	…	X	…	X	X	…	…	…	X	…	…	…
Turbine, steam	…	X	…	X	…	X	X	…	…	…	…	…	…	…
Engines, automotive	X	…	…	X	…	…	X	…	…	…	X	X	…	X
Engines, diesel	X	…	…	X	…	…	X	…	…	…	X	X	…	X
Transportation	X	…	…	X	…	…	X	…	X	…	…	X	…	…
Oil and gas exploration	X	X	X	X	…	…	X	…	X	…	X	X	…	X
Chemical processing	X	…	…	X	X	…	…	…	X	…	…	X	X	…
Paper and pulp	X	…	X	X	…	…	X	…	…	…	X	X	…	X
Defense and aerospace	X	…	X	X	X	…	X	…	X	…	X	X	…	…
Medical and dental	…	…	…	X	X	…	…	…	X	…	…	…	…	…
Electric and electronic	…	…	…	X	X	…	…	…	X	…	…	…	…	…

Table 19.2 Thermal spray coating function by industry [Longo FN (1992)]

Industries	Adhesive wear resistance	Adhesive wear resistance tribology	Antifretting	Erosion resistance	Cavitation resistance	Thermal barriers	Clearance control abradability	Restoration of dimension and repair	Corrosion resistance	Corrosion resistance, iron and steel	Near-net shape forming	Electric resistance	Electric conductivity	Impact	Oxidation
Gas turbine, aircraft	X	X	X	X	…	X	X	X	X	X	X	…	…	…	X
Gas turbine, industrial	X	X	X	X	…	X	X	X	X	…	…	…	…	…	X
Turbine, steam	X	X	…	X	X	…	…	X	…	X	…	…	…	…	…
Engines, automotive	X	X	X	X	…	X	X	X	X	X	X	X	…	X	X
Engines, diesel	X	X	…	X	…	X	…	X	…	X	X	…	…	X	X
Transportation	X	X	…	…	…	…	X	X	X	X	…	X	X	…	…
Oil and gas exploration	X	X	…	X	…	…	…	X	X	X	…	…	…	X	…
Chemical processing	X	…	…	X	…	…	…	X	X	X	…	…	…	…	X
Paper and pulp	X	…	…	…	X	…	…	X	X	X	…	…	…	X	…
Defense and aerospace	X	…	…	…	…	X	X	X	…	…	X	…	…	…	…
Medical and dental	…	…	X	…	…	…	…	…	X	…	…	…	…	…	…
Electric and electronic	…	X	X	…	…	…	…	…	…	…	…	X	X	…	…

Table 19.3 Thermal spray powder materials used by industry [Longo FN (1992)]

Industries	Cemented Co-WC	Chromium carbides	Mixed WC, TiC or Cr$_3$C$_2$	Oxide ceramics	Non-oxide ceramics	Fluxed alloys	Iron and steel alloys and composites	Nickel and nickel alloys	Super-alloys, MCRALY's	Cobalt and cobalt alloys	Refractory metals	Plastics	Special composites	Glasses	Cermets	Non-ferrous metals
Gas turbine, aircraft	X	X	…	X	X	…	X	X	X	X	X	X	X	…	X	X
Gas turbine, industrial	X	X	X	X	…	X	X	X	X	X	X	…	X	…	X	…
Turbine, steam	X	X	…	…	…	X	X	X	…	X	…	…	…	…	…	X
Engines, automotive	…	…	…	X	…	…	X	X	…	X	X	…	…	…	…	…
Engines, diesel	X	…	…	X	…	…	X	X	…	X	X	…	…	…	X	…
Transportation	…	…	…	…	…	…	X	X	…	…	X	…	…	…	…	X
Oil and gas exploration	X	…	…	X	…	X	X	X	…	X	…	X	…	…	…	X
Chemical processing	X	…	…	X	…	…	X	X	…	X	X	X	…	X	…	…
Paper and pulp	X	…	…	X	…	X	X	X	…	…	X	…	…	…	…	…
Defense and aerospace	X	X	…	X	…	…	X	…	X	X	X	…	…	X	X	…
Medical and denial	…	…	…	X	…	…	…	…	…	X	…	…	…	…	…	X
Electric and electronic	…	…	…	X	X	…	…	…	…	…	…	…	…	…	…	X

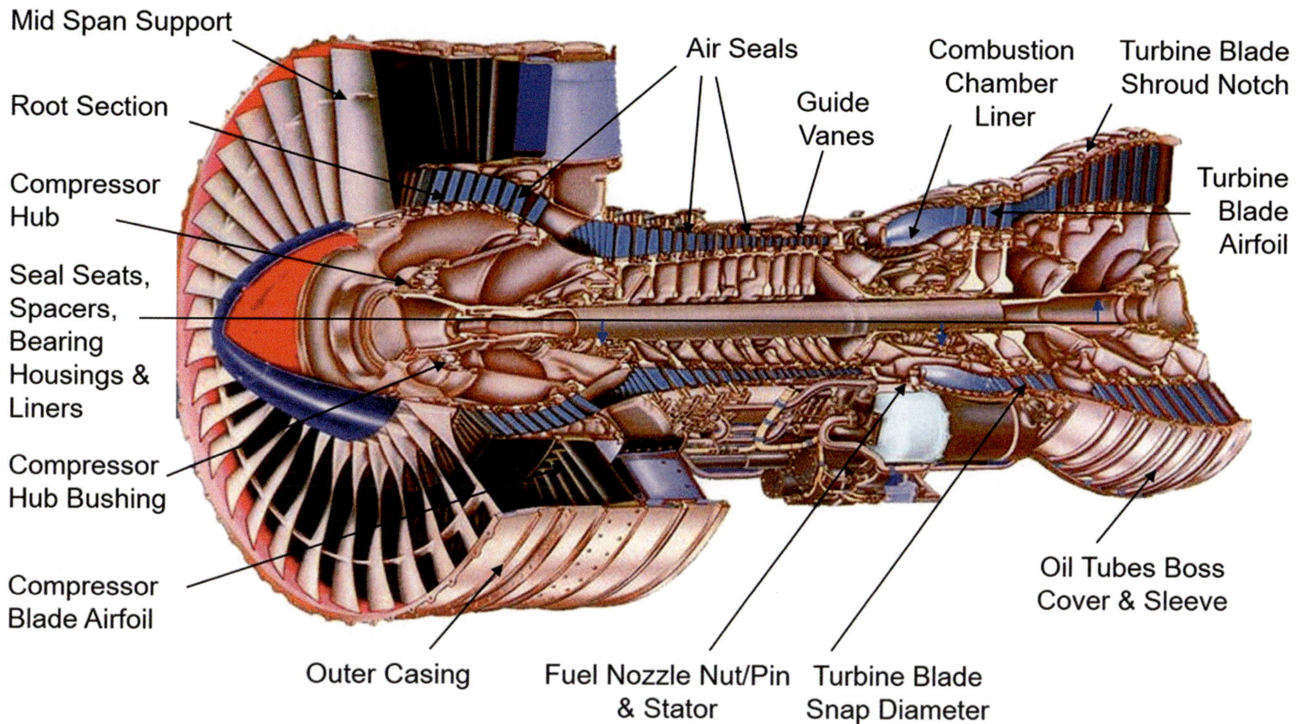

Fig. 19.3 Thermally sprayed coatings on aircraft turbine engine parts. (Reproduced with kind permission from Oerlikon-Metco Corp.)

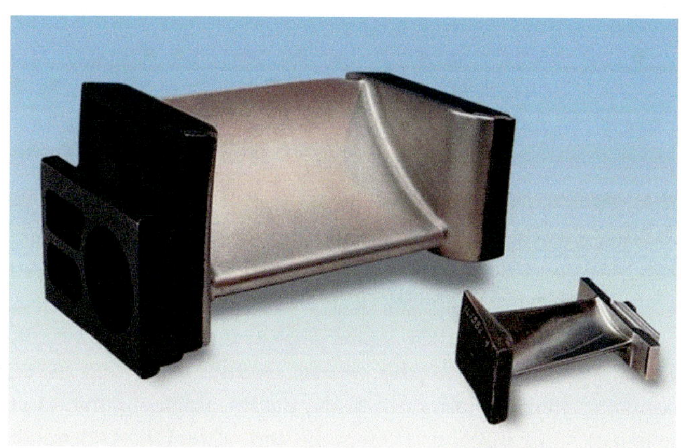

Fig. 19.4 TBC-coated gas turbine vanes. (Courtesy of Sulzer-Metco)

vanes) consist of electron beam physical vapor deposition (EB-PVD)-applied YSZ and platinum-modified diffusion aluminide bond coating. Thermally sprayed ceramic and MCrAlY bond coatings, however, are still used extensively for combustors and power generation blades and vanes. Feuerstein et al. (2008) have reviewed and compared the state-of-the-art processes for the deposition of TBC systems: shrouded plasma and HVOF for MCrAlY bond coat, plasma for low-density YSZ, and dense vertically cracked Zircoat, platinum aluminide diffusion coatings, and EB-PVD-TBC. They outlined and compared key features of coatings and their cost elements actually used in industry. Vaßen et al. (2009a, b, 2010) have presented last developments using advanced processing: methods allowing obtaining highly segmented TBCs [Richer P, et al. (2010)] and the important recent directions of development for TBC systems, including improved processing routes and advanced TBC materials are discussed. Numerous studies were devoted to the resistance of the bond coat [Toscano J, et al. (2006)] and the way to improve it [Billah BM, et al. (2012), Hernandez MT, et al. (2009)] by bond coat treatment, or the choice of the spray process [Richer P, et al. (2010), Karger M, et al. (2011)]. For the TBC topcoat, promising new technologies such as low-pressure plasma spraying–thin film (LPPS-TF) [Hospach A, et al. (2012), Muehlberger E, Meyer P (2009)] or solution or suspension or nanometer-sized agglomerated particles plasma sprayed will probably play an increasing role in future applications [Fauchais P, et al. (2011)].

Against fretting wear, occurring in both lower- and higher-temperature sections of engines, carbides and refractory metals are used, while for clearance control CuNiIn, Nickel–Aluminum-base coatings, where rotating parts need some sealing to reduce bypass of hot combustion or cold compressor gases (through spaces between blade tips and stationary housing), abradable seals are used. For example, in cold sections they are made of polymers and soft metals, while in the hot nickel–graphite, superalloy–polyester or boron carbide, YPSZ with polyester or boron carbide, NiCrAl, and bentonite are used. Seal coatings are also used in the labyrinth seal knife-edges.

For the high-temperature sections, thermal barrier coatings are used for protection against high-temperature metal fatigue, as well as for reduction in the temperature of the metallic substrate to improve component's durability and increase fuel efficiency by higher turbine inlet temperatures. Besides the fuel consumption reduction, emissions of nitrogen oxide, hydrocarbon gases, and smoke are reduced, which is important to have environmental-friendly engines.

Against corrosion, materials of the family MCrAlYs are used with M being nickel, cobalt, molybdenum, iron, or alloys of these elements (NiCo, CoNi, etc.). These alloys are VPS, HVOF, and for some of them cold sprayed. It is important for these materials to reduce or suppress oxidation during spraying and achieve excellent adhesion. The latter is obtained by keeping the substrates clean from any oxide layer, and at high temperature to achieve diffusion bonding between coating and substrate, which can be achieved using VPS.

Thermal barrier coating (TBC) systems are also widely used in modern gas turbine engines to lower the metal surface temperature in combustor and turbine section hardware. The engine components exposed to the most extreme temperatures are the combustor and the initial rotor blades and nozzle guide vanes of the high-pressure turbine. Metal temperature reductions of up to 165 °C are possible when TBCs are used in conjunction with external film cooling and internal component air cooling [Feuerstein A, et al. (2008)]; TBCs are applied on vane bases, burner cans, after burners, and also on other small engine components. An example of coated gas turbine vane is presented in Fig. 19.4 and that of a TBC-sprayed combustion chamber in Fig. 19.5.

Rotating vane assemblies in aircraft engines require limiting the bypass flow of either hot combustion gases or cold compressor gases through the spaces between the blade tips and the stationary housing. These seals can improve significantly the engine efficiency. Two types of seals are used: abradable ones that are machined in situ by the rotating components (blades, for example), and labyrinth ones to match stator and rotor components with an intermeshing saw tooth geometry.

Against wear, chrome plating has been intensively used because of its durability, ease of application, and low cost. It suffers, however, from pitting, spalling, and other failures under stress. Moreover, for commercial airlines the downtime in which the aircraft is out of service, for the replacement of traditional industrial chromium plating, used in landing gear, is simply too long and prohibitively expensive.

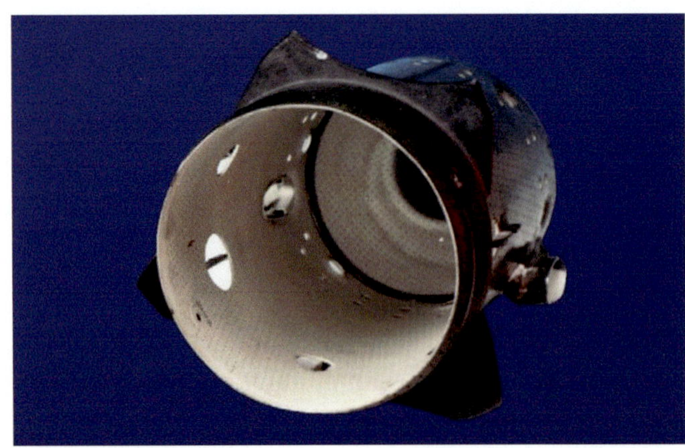

Fig. 19.5 TBC-coated combustion chamber. (Courtesy of Sulzer-Metco)

Alternative coating solutions had also to be developed for the replacement of chrome plating because of serious environmental and health problems associated with hexavalent chromium emissions. Hard trivalent chrome coating, and HVOF- or plasma-sprayed tungsten carbide coatings, present viable alternatives, each with their respective advantages and limitations such as limited coating thickness for trivalent chrome, and only line-of-sight geometries for thermally sprayed coatings [Thintri Inc. (2013)]. Coatings of tungsten carbides especially developed for landing gears and HVOF sprayed have considerably increased the service life of exposed cylinders and seals used in landing gears. Figure 19.6 presents a robot coating a landing gear.

19.4.2 Land-Based Turbines

Compared to aero engines, land-based turbines (LBT) such as the one shown in Fig. 19.7 work in different conditions: the external environment might range from cold (−40 °C) to high temperature (55–60 °C) and corrosive and erosive contaminants due to the environment and fuel are present. Coatings are applied onto bearing journals, bearing seals, sub-shaft journal, labyrinth seals, blades, tip seals, inlet and exhausts, and housing [Davis JR (ed) (2004)]. A detailed description of the different land-based turbines can be found in the paper by Lebedev and Kostennikov (2008). Pomeroy (2005) gives a good description of coatings for gas turbine materials and long-term stability issues. Due to the demand to increase turbine inlet temperatures and thus cycle efficiencies, ceramic insulating coatings can be applied to decrease the temperature of the hottest parts of the turbine components by up to 170 °C. A review of the next generation of TBCs for gas turbines is presented by Curry et al. (2011).

Coatings are applied on bearing journals, bearing seals, stub shaft journals, labyrinth seals, blade nozzles, tip seals, inlets and exhausts, and housings [Davis JR (ed) (2004), Tucker RC (ed) (2013)]. The overview by Rajendran (2012) presents the different coatings used, methods of application and characterization, and degradation mechanisms, and indicates future directions, which are of use to a practicing industrial engineer. Coatings are applied to the compressor blades and vanes for enhanced erosion resistance. Anti-fretting coating protects the contact area of the dovetail part of

Fig. 19.6 Robot coating a landing gear. (Courtesy of Oerlikon-Metco)

the compressor blade root from fretting fatigue failure. Abradable coatings offer close clearance control thereby increasing the engine efficiency. Wear-resistant coatings give extended life to the parts that undergo rotary and reciprocating rubbing motion. Oxidation and corrosion-resistant coatings are applied through either diffusion or overlay process. YPSZ (7 wt.%) thermal barrier coatings offer increased component life with a decrease in operating temperature of the metal. Hardwicke and Lau (2013), in their review paper about advances in thermal spray coatings for gas turbines and energy generation, describe functional coatings widely used in energy generation equipment in industries such as renewables, oil and gas, propulsion engines, and gas turbines. They present the current status of materials, equipment, processing, and properties' aspects for key coatings in the energy industry, especially the developments in large-scale gas turbines. In addition to the most recent industrial advances in thermal spray technologies, future technical needs are also highlighted.

Coating life for land-based turbines (time to refurbish aircraft turbines) should be approximately 24,000 h, with a majority of all service time at maximum conditions, against 8000 h for aircraft turbines with only 5–15% at maximum conditions. Thus, problems of bond coat oxidation, interdiffusion of bond coat and substrate, coating densification, and

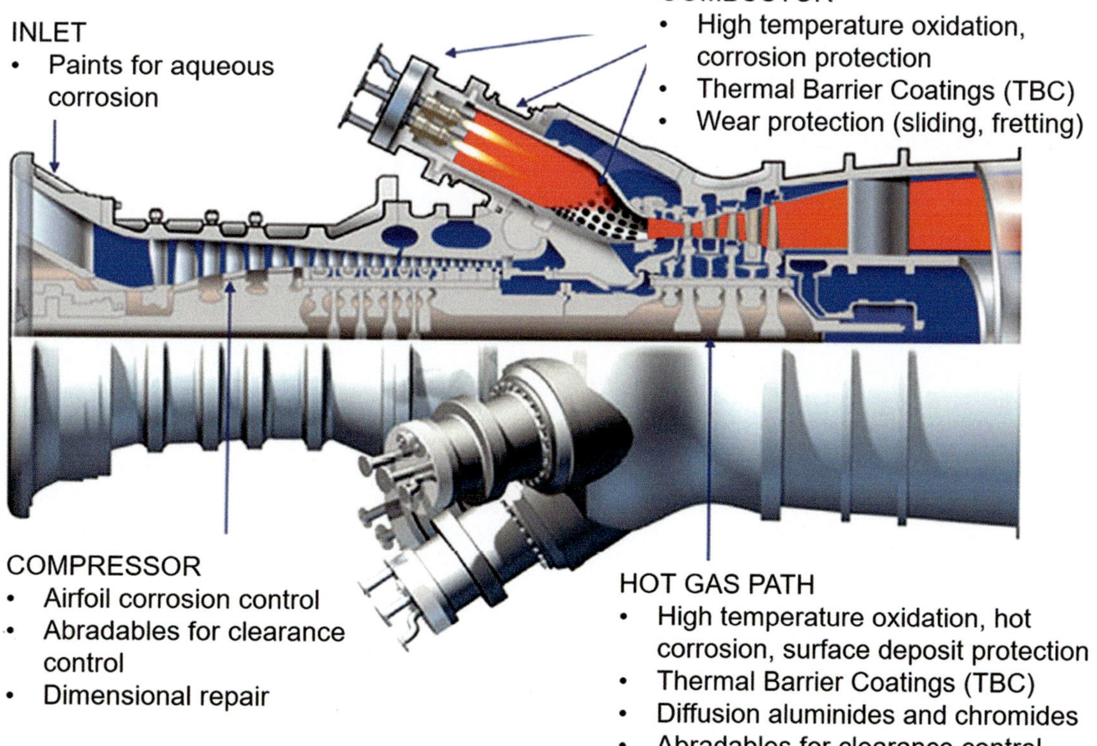

Fig. 19.7 Challenges in traditional industrial gas turbine (IGT) applications for thermal spray coatings with respect to turbine locations. (Schilke 2004; Singh et al. 2019)

changes in thermal or mechanical properties of the coating are more important. Accessibility for inspection, repair, and refurbishment is also more difficult for power generation machines than for aircraft engines. Component size is also more important (a factor of 10 for turbine blades), thus requiring adapted spray processes. At last, the quality of the combustible is far from that used in aircraft engines [Mutasim Z, Brentnall W (1997), Nelson WA, Orenstein RM (1997), Parks WP, et al. (1997), Tamura M, et al. (1999), Wright IG, Gibbons TB (2007)].

Turbines are also exposed to excessive amounts of moisture and chlorides, giving rise to development of two types of hot corrosion and oxidation:

- Type II hot corrosion, occurring at temperatures in the range 500–800 °C and involving the formation of base metal (nickel or cobalt) sulfates that require a certain partial pressure of sulfur trioxide for their stabilization.
- Type I hot corrosion, observed in the range 750–950 °C, involving the transport of sulfur from a sulfate deposit (generally Na_2SO_4) across a preformed oxide into the metallic material with the formation of the most stable sulfides. Once stable sulfide formers (e.g., Cr) are fully reacted with the sulfur moving across the scale, then base metal sulfides can form with catastrophic consequences as they are molten at the temperatures at which Type I hot corrosion occurs.

Sprayed materials with a good corrosion and oxidation resistance are nickel- and/or cobalt-based alloyed coatings [Davis JR (ed) (2004)], such as NiCrMo (Hastelloy or Nistelle), CoCrSiMo (Triballoy), and MCrAlYs plasma or HVOF sprayed. Oxidation- and corrosion-resistant coatings are applied on air inlets, combustor liners, injectors, turbine tip shoes, nozzles, and exhausts [Davis JR (ed) (2004)]. On blades and vanes, MCrAlY coatings are used as bond coats and for corrosion–oxidation protection.

For thermal barrier coatings, mainly ZrO_2–Y_2O_3 (6–8 wt. %), ceria, or dysprosia-stabilized zirconia are used [Wright IG, Gibbons TB (2007)] and also ceria and yttria-stabilized zirconia or for certain applications calcium titanate [Davis JR (ed) (2004)]. It is worth underlying that for abradable and seals in the low-temperature areas, where moisture is important, porous aluminum-base coatings containing polyester, polyimide, or BN, as well as nickel–graphite coatings are used. For higher temperatures (over 450 °C), abradables are made of MCrAlY with BN or polyester [Wright IG, Gibbons TB (2007)]. Ceramic abradables have also been introduced but the expansion mismatch with the metal substrate must be accounted for with the cooling of the superalloy.

19.4.3 Automotive

Thermal spraying has been rather slow in penetrating the automotive industry due to its extreme sensitivity to manufacturing cost and to strict demand on coating reliability. Nevertheless, early signs of acceptability for high-end luxury and racing cars started around the turn of the century. As illustrated in Fig. 19.8, after Oerlikon-Metco Corp., an increasing number of car parts is being treated by different surface modification technologies (from sprayed coatings to PVD or plasma-enhanced chemical vapor deposition [PECVD]). The different coating applications are related to [Davis JR (ed) (2004), Vetter J, et al. (2005), Barbezat G (2003, 2005, 2006)] the power train components, thermal barriers protecting certain components from overheating, brake disks, cylinder bore in diesel engines, aluminum alloy cylinder bore, exhaust valves, crankshaft, transmission and rear end gear clusters, body and chassis, synchronizer rings, shifter forks, valve seats, and connecting rods. Examples of some of these components are given in Fig. 19.9.

Against hot corrosion: Exhaust system components (exhaust headers, exhaust pipes) are protected from hot corrosion by aluminum coatings wire arc sprayed.

Against wear: Piston rings are commonly coated with molybdenum and molybdenum-bearing compounds using the plasma spray process in general. The HVOF spray process is used in the case of high-performance piston rings for diesel engines. Molybdenum coatings are also used on aluminum piston skirts to prevent galling. Some aluminum piston tops are coated with molybdenum to limit erosion due to the impingement of fuel from injectors. Cylinder bore in diesel engines are coated with molybdenum plasma sprayed or WC–Co and WC–Co–Cr HVOF sprayed. Such coatings, slightly porous, retain oil and improve friction. Coated transmission parts such as synchronizing rings and shift forks are widely coated using wire flame spraying with molybdenum to provide a constant coefficient of friction and prevent scuffing.

In order to reduce the weight and size of the engine, iron cylinder blocks are replaced by aluminum ones, usually equipped with cast-iron cylinder linings. Instead of them, Sulzer-Metco has been the first to propose aluminum cylinder bores coated with a special iron alloy, to replace the cast-iron cylinder lining. This plasma coating acts as a direct running surface for the similarly coated piston rings, which reduces diesel or gasoline consumption and the amount of oil needed. The process called "Rota-plasma" developed by Sulzer-Metco was introduced in 2000 for automobile production. It allows coating aluminum alloy (cast aluminum–silicon) cylinder bores to provide a good wear resistance. Coating materials can be carbon steel with its own oxides wustite and magnetite as solid lubricant, composite of carbon tool

Fig. 19.8 Selected parts to be treated by surface technologies in cars. (Reproduced with kind permission of Oerlikon Metco Corp)

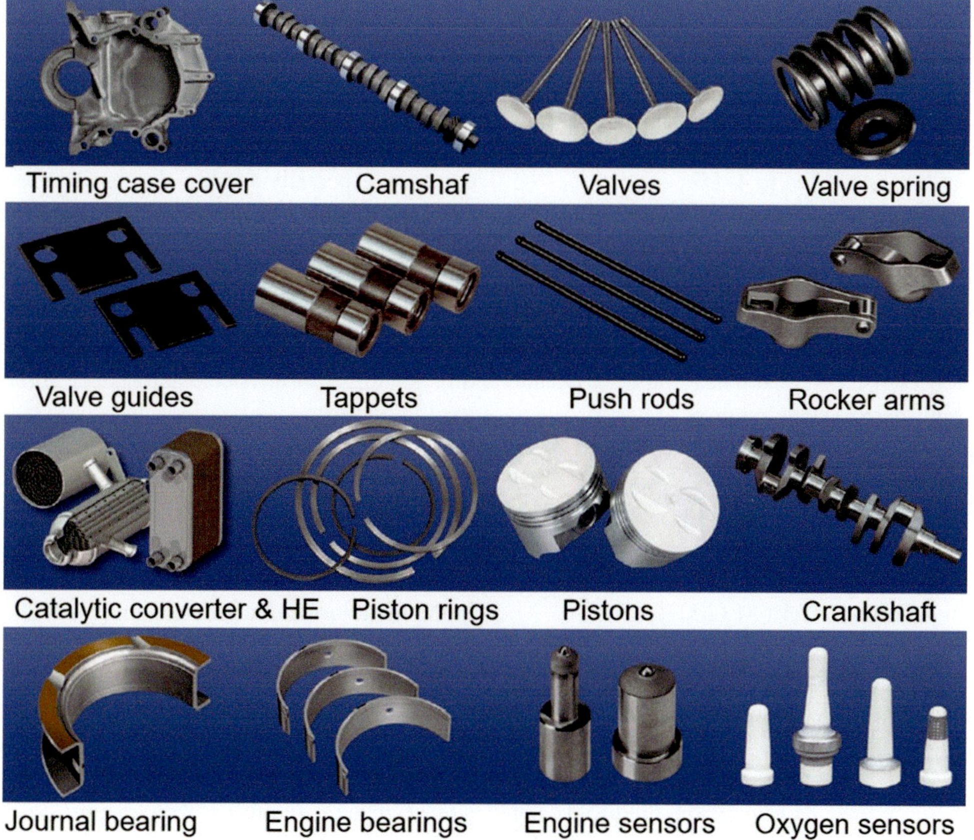

Fig. 19.9 Examples of some of the plasma-coated automobile parts for enhanced longevity and improved performance. ([After Oerlikon-Metco, automotive solutions kit, 2008] Reprinted with kind permission)

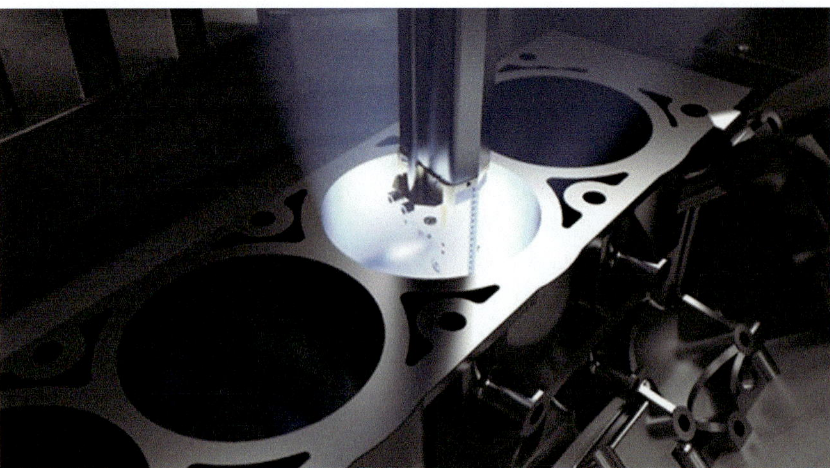

Fig. 19.10 Photo of the Rota plasma process. (Courtesy of Oerlikon-Metco Corp.)

steel and molybdenum, corrosion-resistant steel (Cr and Mo alloyed), and metal matrix composite with ceramic particles. The process is entirely automated and lasts about 1 minute for a bore of conventional car engine (see Fig. 19.10).

The process advantages are the following:

- The pitch distance between the cylinders can be reduced allowing a further reduction in the engine block weight.
- The friction between the piston rings and the liner surface can be significantly reduced (more than 30% in gasoline engines).
- Considering that the majority of the energy losses in the engine is due to thermodynamic cycle efficiency, the potential fuel savings is limited to 4–5%.
- Very low surface wear rate (about 5 nm/h of service).
- Transmission and rear end gear clusters receive often Mo thin coatings plasma sprayed to lessen the friction and produce smoother movement.

TBCs have also been used on piston heads and valves and exhaust ports of high-performance engines. TBC for diesel engines either thick (0.5–1 mm) [Beardsley MB (1997)], or eventually phosphate sealed [Ahmaniemi S, et al. (2002a, b)], or thin [Hejwowski T, Weronski A (2002)] are used to:

- Increase the piston head temperature and improve the fuel combustion [Buyukkaya E, Cerit M (2007)]
- Reduce NO_x emissions [Buyukkaya E, Cerit M (2007)]
- Reduce the heat rejection [Parlak A, et al. (2005)]
- Improve the usability of vegetable oils as a fuel [İşcan B, Aydın H (2012)]

They can also be applied externally on manifolds and other exhaust systems when the air cooling becomes insufficient. They increase the efficiency of turbochargers.

Metal matrix composites are used in brake disks to tailor the mechanical resistance and the thermal conductivity. The wire arc spray provides low-cost filler for weld seam and is also used for dimensional restoration. Ceramic overlays are sprayed for oxygen sensor protection, the sensor itself using zirconia-sprayed coating.

19.4.4 Land-Based and Marine Applications

When metal coatings are needed to enhance the properties of a given surface, wire combustion spray systems or wire arc spray systems are very good solutions to spray protective coatings, especially sacrificial ones. The investment is rather low and coating adherence is generally good (over 20 MPa), with almost no heating of the substrate. However, coatings obtained by these methods are relatively porous (up to 20%), which is not a problem for sacrificial ones. These coatings are extensively used throughout the maritime, paper/pulp/printing, manufacturing, steel, aerospace, automotive, and railroad industries.

19.4.4.1 Sacrificial Coatings

The corrosion protection of large steel structures such as bridges, pipelines, oil tanks, towers, radio and television masts, overhead walkways, and large manufacturing facilities in the metallurgical, chemical, energy, and other industries is a key issue. The protection of structures exposed to moist atmospheres and seawater such as ships, offshore platforms, and seaports is even more difficult [Evdokimenko Yu I, et al. (2001), Sørensen PA, et al. (2009)]. In most cases the surface to be protected is thousands and even tens of thousands of square meters, requiring that coating costs be competitive with those of traditional painting methods. The coating rate must be equal or higher than 10 m^2/h and if possible

Fig. 19.11 Photograph of typical suspension bridge normally WAS with zinc coat followed by painting for environmental protection against corrosion

deposited in one unique pass. The equipment must be mobile and autonomous for operation in field conditions and must work under manual control, automation being generally difficult for large-scale operations, and, at last, the spray gun can be up to 30 m from its power supply and control center [Evdokimenko Yu I, et al. (2001), Sørensen PA, et al. (2009)]. Flame wire arc spraying meets such requirements.

Sacrificial coatings (cathodic behavior relatively to ions, for example, Zn or Al on steel) are mainly used: the thicker they will be the longer will be the protection. Typical thicknesses vary from 50 to 500 μm; the most frequent ones being around 230 μm. Such coatings must have a cathodic behavior relatively to the ions of the metal to be protected, in almost all cases steels. Metals used are then zinc, aluminum, and zinc–aluminum. Zinc performs better than aluminum in alkaline conditions, while aluminum in acidic conditions. Zinc and aluminum replace more and more painting and galvanizing because their lifetime is predictable, they have good erosion resistance, and require one application with no drying problems. Sluice gates and canal lock gates of the St. Denis Canal in France, that have been zinc coated in the early 1930s, have remained in perfect condition with virtually no maintenance for decades. According to literature [Davis JR (ed) (2004), Tucker RC (ed) (2013) *ASM Handbook, Volume 5A*], the lifetime of a 255 μm thick zinc or zinc aluminum coating is about 25 years and it can be extended 15 years by sealing it with vinyl paint. Painting is not the only sealer used; impregnation with special compositions (epoxy resin, silicon resin, etc.) is also being intensively used. As soon as sealing is considered, the porosity of these coatings becomes an advantage for the adhesion of the sealer.

Figure 19.11 represents a typical suspension iron bridge which is normally protected against environmental corrosion by WAS with a zinc coat followed by a paint finish.

Zinc–aluminum (Zn–15 wt.% Al) seems to combine advantages of both materials. If resistance to wear must be improved, aluminum coatings can be sprayed with alumina particles, for example, using cored wires.

In case of the protection of steel reinforcement in concrete, zinc is generally used, but titanium has also been used in spite of the fact it is an anodic protection. In that case the coating is applied directly on the concrete substrate [Davis JR (ed) (2004), Tucker RC (ed) (2013) *ASM Handbook, Volume 5A*]. Aluminum must be avoided where thermite sparkling may occur. That is due to the reaction of rusted steel and aluminum smears when this combustible mix is ignited by an impact [Davis JR (ed) (2004), Tucker RC (ed) (2013) *ASM Handbook, Volume 5A*]. Another interest of such coatings is their antifouling properties. Marine biofouling is the undesirable accumulation of marine organisms on artificial surfaces that are immersed in the sea. For example, when marine biofouling occurs on ship hulls, it leads to an increase in weight of the ship and friction to sail.

In the protection from corrosion and/or erosion of industrial gas turbine compressor airfoils, a sacrificial aluminum-filled metallic coating has been recently developed by Sulzer-Metco. The aluminum-filled metallic coating is made conductive via mechanical abrasive finishing. The thickness of these coatings is typically between 25 and 75 μm and the maximum operating temperature 870 °C. Figure 19.12 presents the coated blades of the compressor airfoil.

Fig. 19.12 Coated blades of the compressor airfoil of an industrial gas turbine compressor. (Courtesy of Oerlikon-Metco Turbo)

19.4.4.2 Non-sacrificial Coatings

Austenitic stainless steels, aluminum bronze, nickel-base alloys, MCrAlY, cermets with WC, Cr_2C_3, and matrices containing chromium or nickel, or both, are used against corrosion, often associated with wear. However, such coatings, presenting no galvanic protection, will never protect the substrate if connected porosities and oxide networks exist, which is the case of most thermally sprayed coatings. The substrate protection requires using a protective bond coat or producing dense coatings or sealing them, which is not always possible if the service temperature is over a few hundreds of Celsius degrees.

Coatings of corrosion-resistant alloys for severe petroleum industry corrosion applications, such as type 316L stainless steel and Hastelloy C-276, were shown to act as true corrosion barriers. The oxide content also plays a role (see Sect. 19.3.2.1.2: Chemical or Parachemical Corrosion).

For applications with a severe wear in oil and gas industry, cermets are used, with however some problems for offshore installations. According to Meng (2010), the corrosion resistance is improved by the proper choice of binder between the wide varieties of tungsten carbides.

19.4.5 Electrical and Electronics Industry

First of all, thermal spray coatings that allow achieving thick coatings cannot compete with the thin film technology (chemical vapor deposition [CVD], PECVD, sputtering, ion plating, etc.) used in electronic devices. However, they can be very useful in capacitors, insulators, resistors, and inductors. Coatings of oxide ceramics (alumina, titanate, beryllia) are used as dielectrics, while metals and alloys are used for current carrying and bonding to ceramic substrates, metals for electromagnetic shielding, and so on [Davis JR (ed) (2004), Tucker RC (ed) (2013) *ASM Handbook, Volume 5A*]. A few examples are presented below.

Alumina plasma-sprayed coatings are currently used in ozonizers. Standard ozonizer tubes consist of a borosilicate glass tube serving as dielectric with a metal coating applied to the inner surface of the tube serving as the high-voltage electrode. In order to increase the ozone production efficiency, a material with a higher permittivity is needed. Alumina coatings with a thickness up to 1000 µm and serving as dielectric are deposited together with the metal electrode on top of the glass tube in ozonizers. Alumina is also used for high-temperature strain gages and insulation of induction heating coils.

Multilayer ceramic chip capacitors (MLCC) are used in a large number of electronic appliances. According to Yamakawa et al. (2009), they are being required in more compact sizes, larger capacities, and reduced costs year by year, and ceramic trays (a type of kiln furniture) are used for firing them. Normally, this ceramic tray consists of an Al_2O_3–SiO_2 body, an Y_2O_3-stabilized ZrO_2 topcoat, and Al_2O_3 basecoat. The application of plasma-sprayed ZrO_2

topcoat and an Al_2O_3-sintered basecoat makes it possible to enhance longevity and reduce cost.

Pure beryllia (BeO) presents excellent thermal and dielectric properties and is used in many high-performance semiconductor parts for applications such as radio equipment. Some power semiconductor devices use beryllia between the silicon chip and the metal mounting base of the package in order to achieve a lower value of thermal resistance than the alumina currently used. It is also used as a structural ceramic for high-performance microwave devices, vacuum tubes, magnetrons, and gas lasers. Vacuum plasma-sprayed beryllia coatings present excellent properties. Unfortunately, their handling, especially that of fine powders, is highly toxic resulting in very stringent safety conditions such as keeping the rooms where spraying is achieved at pressure lower than atmospheric one. These conditions are often economically prohibitive and such coatings are essentially used for army and nuclear applications [Davis JR (ed) (2004)), Tucker RC (ed) (2013) *ASM Handbook, Volume 5A*].

According to Smyth and Anderson (1975), for electrical heaters, arc-sprayed coatings are used, masking permitting to define the geometry. The choice of the material (NiCr, NiAl, etc.) depends on its resistance variation with temperature that must be as low as possible because the electricity source is a constant voltage one. The resistance depends on the coating oxide content that can also modify the alloy composition, chromium being preferentially oxidized compared to nickel, for example. Most sprayed resistors are used at relatively low temperatures (<250 °C). The resistance value can also be increased when adding alumina particles. Coatings can be sprayed on ceramic material, even polished ones, with a very good adhesion, when eutectics are formed or diffusion occurs. It requires heating the substrate to temperatures close to the sprayed alloy or metal temperature.

Cold-sprayed copper coatings are used to improve the heat conductivity between electronic devices, an underlying copper plate, and a soldered copper-coated heat sink, which is fabricated out of aluminum [Gärtner F, et al. (2006)]. Cold spray also permits the generation of solderable surfaces on materials with poor wettability (for example, heat sinks, such as copper on aluminum) [Marx S, et al. (2006)]. Similarly, PTA allows spraying very thick (up to 10 mm or more) copper coatings that are used in powerful computer systems to improve their cooling. These copper layers, Fig. 19.13, achieve a good heat distribution and remove the heat more easily.

Alumina films plasma sprayed are more and more used as insulators in electronic devices. Plasma-sprayed alumina is used for electronic package mount in automotive industry.

Wire arc sprayed coatings are used as shielding material to eliminate electromagnetic and radio frequency interference and dissipate static discharge sparks. Zn and Al are currently

Fig. 19.13 Example of flame wire spraying: copper layer on a computer cooling system. (Ducos 2006a, b)

used to protect computers, electronic office equipment, medical monitoring devices, housing constructed of temperature-sensitive plastic, and rooms containing military computers.

The review paper by Sampath (2010) points out the industrial success of the lambda sensor for exhaust gas oxygen measurement for engine fuel control. This sensor uses a porous plasma-sprayed zirconia membrane coating between electrodes. When subjected to a differential partial pressure of oxygen, a rapid variation in electric resistance related to the oxygen partial pressure can be monitored through an electrical circuit [Sampath S (2010)].

Aluminum coatings, wire arc sprayed, are used in the production of electrolytic condensers. A thin film of copper is sprayed on carbon and ceramic resistors and carbon brushes to provide the electrical connection of high electrical conductivity. Thermally sprayed coatings are also used to produce thick films' electrical circuits that can carry higher current than printed ones. The most used materials for these applications are Cu, Al, Zn, and Ag.

19.4.6 Medical Applications

Orthopedic and dental market is developing very fast with either bioinert coatings (Ti–6Al–4V, Ti–6Al–7Nb, Ti–13Nb–13Zr, etc.) or bioactive ones (hydroxyapatite, tricalcium phosphate). Both orthopedic and dental implants receive coatings designed to aid fixation in bone tissues. For example, according to Yang et al. (2006a, b) and Ong et al. (2006), the report published in *Implant Dentistry* estimated that 910,000 dental implant procedures were performed in 2000, and the numbers grew at an annual rate of 18.6% through 2005. By 2005, annual sales exceeded half a billion dollars. The implant success is dependent on quality bone formation, optimization of the local surface structure, chemistry, morphology, and use of bioactive factors that are equally important [Ong JL, et al. (2006)]. Coated implants

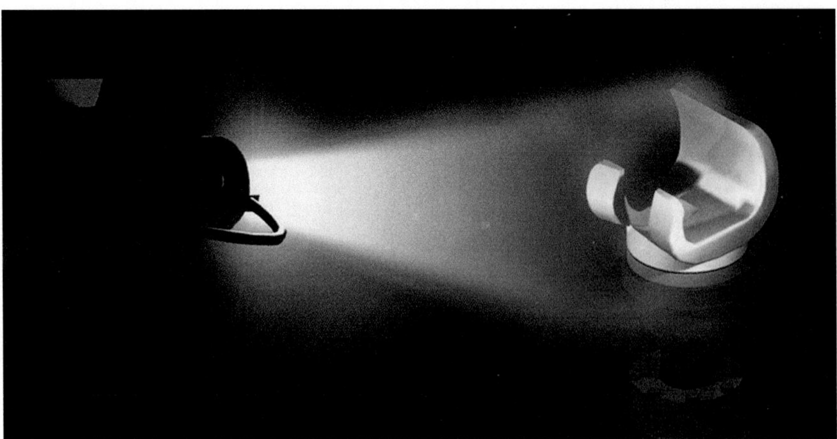

Fig. 19.14 Vacuum plasma spraying of HA coating on a knee prosthesis

are now currently used for tooth, hip, knee, elbow, and shoulder. Two types of coatings are used [Davis JR (ed) (2004), Tucker RC (ed) (2013) *ASM Handbook, Volume 5A*]:

First, bioinert ones, with no activity between them and the bone or soft tissues, the most used materials being Ti and Ti–6Al–4V. These coatings must be porous (at least 30%). Plasma spraying allows preparing porous-graded coatings with dense coating close to substrate to achieve a strong adhesion with substrate and very porous outermost layer (with pore sizes over 100 up to 400 μm). To achieve that, the plasma power was gradually reduced, and the particle sizes increased while regulating the degree of powder melting layers to achieve porous graded coatings [Yang Y, et al. (2006a, b), Ong JL, et al. (2006), Tucker RC (ed) (2013)].

Second, bioactive coatings, interacting with bone to promote bonding interfaces, are ceramic coatings with material compositions similar to that of bone tissue (calcium phosphate). The most used ceramic is hydroxyapatite (HA), $Ca_{10}(PO_4)_6(OH)_2$, exhibiting a strong activity to join the bone. The provision of a high calcium- and phosphorus-rich environment promotes rapid bone formation within the vicinity of the implant. Unfortunately, when HA is plasma sprayed in air, calcium oxide (CaO) is usually formed, thereby affecting the integrity of plasma-sprayed HA coatings. Plasma-sprayed HA coatings also contain other bioresorbable phases, such as tricalcium phosphate (TCP) α or β forms, tetracalcium phosphate (TTCP), and amorphous calcium phosphate (ACP) [Prevéy PS (2000)]. Gross et al. (2004) investigated four HA-coated hip components recovered from patients during revision surgery. Analysis of the coating surface indicated dissolution, osteoclastic resorption, and carbonate apatite precipitation identical to observations from previous in vitro studies. The coating microstructure differed between three coatings that remained on the prosthesis surface, ranging from completely crystalline coatings made by vacuum plasma spraying to less crystalline coatings manufactured by air plasma spraying. Increasing the amorphous and TCP concentration predisposes the coatings to higher dissolution rates [Gross KA, et al. (2004)]. Nelson et al. (2011) have deposited by flame spraying powders of titanium alloy (Ti–6Al–4V) and bioactive glass (45S5) to fabricate composite porous coatings for potential use in bone fixation implants. Bioactive glass and titanium alloy powder were blended and deposited in various weight fractions and with two sets of spray conditions, which produced different levels of porosity. The HA formation on the alloy–bioactive glass composite coating suggested that the addition of bioactive glass to the blend may greatly increase the bioactivity of the coating through enhanced surface mineralization. Drnovšek et al. (2012) studied the ability of nanoparticle bioactive glass (BAG) to infiltrate into the porous titanium (Ti) layer on Ti-based implants to promote osseointegration. Juhasz and Best (2012) reviewed bioactive implants, coatings, and scaffolds made of ceramics, glasses, glass–ceramics, and composites that were able to form a chemical interfacial bond with tissue and can be resorbable or non-resorbable.

The ratio of HA to TCP has been reported to be crucial for bone regeneration. The dissolution rate of HA coating is dependent on the biochemical calcium phosphate phase of the coating as well as coating crystallinity. With adapted spray conditions, it is possible to achieve up to 75% crystallinity for APS, 80% for HVOF spraying, and 95% for VPS [Gross KA, et al. (2004), Khor KA, et al. (1997a, b), Chang C, et al. (1998), Lima RS, et al. (2010), Lima RS, et al. (2006)]. The result also depends on powder preparation, best results being obtained with dense HA particles. In Fig. 19.14 is presented the vacuum plasma spraying (VPS) of HA coating on a knee prosthesis and in Fig. 19.15 biocompatible coating on the lower part of hip joints.

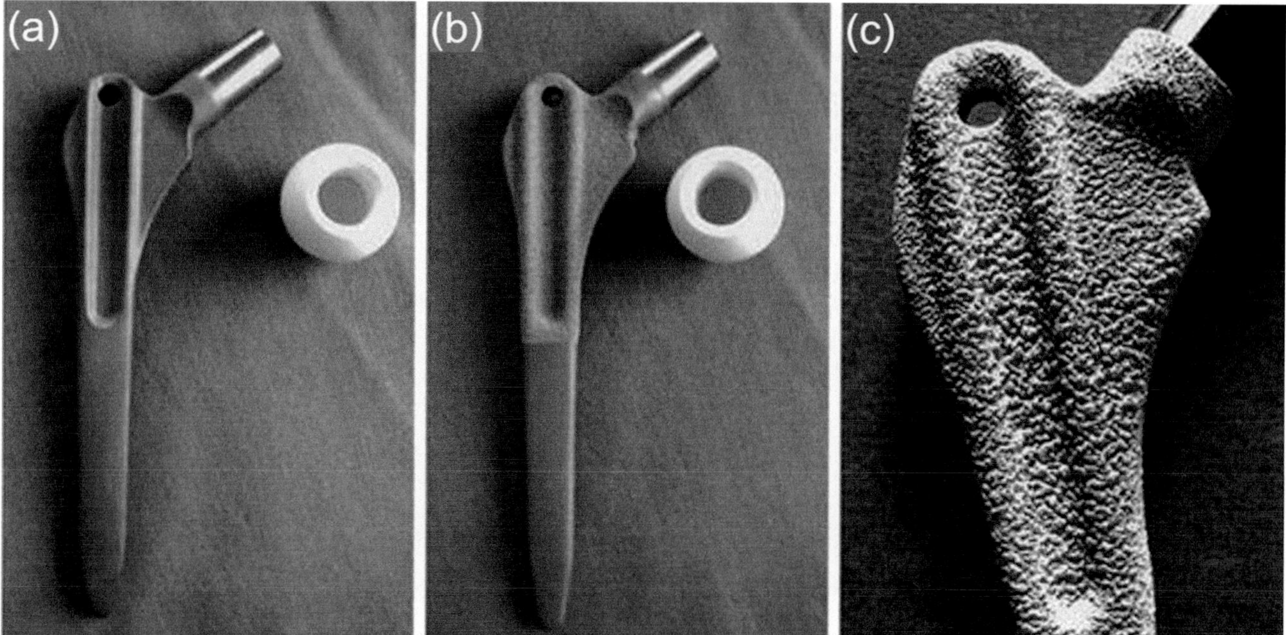

Fig. 19.15 (**a**) Femoral stem Walter, without coating. (**b**) Femoral stem with titanium alloy/hydroxyapatite coating. (**c**) Magnified view of the coating at the proximal end of the porous coated femoral stem. (Landor et al. 2007)

19.4.7 Ceramic and Glass Manufacturing

For the production of a variety of glasses, platinum is currently used thanks to its high melting point, strength, and resistance to corrosion that allows withstanding the abrasive action of molten glass. Rhodium is often alloyed with platinum to increase the strength of the alloy and extend the life of the equipment. According to the prices of these metals, instead of using them as plates or sheets, they are often plasma sprayed in inert atmosphere chambers where over-sprayed powder can be collected. A ceramic coating, that is reapplied regularly according to its wear [Davis JR (ed) (2004), Tucker RC (ed) (2013) *ASM Handbook, Volume 5A*], can protect these precious metal coatings. The electric glass melting for homogenizing, feeding, and shaping was developed with molybdenum electrode materials, since stirrers, mixing paddles, and some mold surfaces are plasma coated with molybdenum and its alloys [Davis JR (ed) (2004), Tucker RC (ed) (2013)].

Mold glass material must have sufficient strength, hardness, and accuracy (no deforming process) at high temperature and pressure. Of course, its oxidation resistance must be good, its thermal expansion low, and its thermal conductivity high. Therefore, the mold material choice depends critically on the transition temperature of the glass material. For low-temperature transition glasses, steel molds with a nickel alloy coating can be used. For example, self-fluxing alloy NiCrBSi flame sprayed and refused. For higher transition temperatures, NiCr–Cr_2C_3 or TiC cermets are used.

Hamashima (2007) has developed a new boride cermet coating consisting of Mo-system ternary boride and iron-group metal alloys. MoCoB–Co/Cr cermets have a very low friction coefficient with glass at high temperature. The cermet coating has been applied to the lower mold on an industrial scale and the results were very satisfactory. In particular, the production efficiency of the forming process was improved substantially.

19.4.8 Printing Industry

According to Döering et al. (2008), all printing presses need an ink transferring unit, a print form, and an impression cylinder. They differ in the process of transferring the print image to the substrate, leading to additional cylinders for generating the inverted image—in case of offset printing the so-called offset cylinder—or even a shortcut from the ink transferring roller to the print form and directly to the substrate if gravure printing is regarded. Additionally, dampening units require other surface energy than ink leading rollers. Moreover, the precision of shape and surface roughness of these core components is rather high. Therefore, not only the requirements for an appropriate coating in terms of the function but also the requirements for the grinding and finishing step after the coating process are rather high. This will be illustrated with few examples.

The ink transporting system in offset printing machines consists of several different rollers, providing a homogenous

Fig. 19.16 SUME™CAL coating after super-finishing on calendar roll. (Courtesy of Oerlikon-Metco)

film of ink for the print form. For example, the duct-roller elevates the ink from the ink box to the ink conditioning system. Its surface roughness R_z must be better than 2 µm and tolerances must be better than 2/1000 mm in concentricity. Such conditions are achieved with chromia-rich coating consisting of a bond coat (Ni, NiCr, NiAl) and the ceramic topcoat [Pawlowski L (1996)].

Rollers in the paper industry are subject to various operating environments resulting in wear, chemical attack from dyes, thermal stress on heated rollers, and mechanical stress from doctor blades. They also must exhibit a high surface finish lasting as long as possible. Figure 19.16 presents a calendar roll coated with the so-called Oerlikon-Metco SUME™CAL developed to meet these requirements.

In contrast to common offset inking units, the anilox unit consists of a laser-engraved roller, taking the ink from a chamber doctor blade system [Döring J-E, et al. (2008)]. The gravure procedure is performed after finishing the roller surface. Producing pure chromia coatings implies using carefully adapted plasma spray conditions and adapted powders. For more details, see the works of Pawlowski [Pawlowski L (1995, 1996)].

Many other rolls are used in printing machines [Pawlowski L (1996)], where different oxides (chromia, alumina–titania [3 or 13 wt.%]) are plasma sprayed. Lima and Marple (2005) have proposed to replace the alumina–titania conventionally plasma sprayed by HVOF-sprayed nanostructured titania coating exhibiting a very dense (nearly pore free) and uniform isotropic microstructure with an excellent wear and corrosion resistance. Plasma-sprayed chrome oxide coatings for inking rollers exhibit very fine microstructures, which can then be laser engraved with a very small and tight pattern (Fig. 19.17).

Other rolls such as blanket cylinders are coated with Hastelloy-C by either APS or HVOF and at last draw rolls are coated with nickel–chromium [Davis JR (ed) (2004), Vetter J, et al. (2005)].

19.4.9 Pulp and Paper

Machines producing paper and cardboard comprise many parts that can be rather important (for example, rolls over one meter in diameter and ten meters in length) with high wear and corrosion problems. Coatings are used in several types of rolls and cylinders, including, for instance, center press rolls, dryer cylinders, calendar rolls, traction rolls, and Yankee cylinders [Vuoristo P, Nylén P (2009)]. Coating materials used are iron- and nickel-base alloys, nickel–chromium self-fluxing alloy, carbides, oxide ceramics, and various multilayers, again depending on the application. For example, rolls are coated by NiCrBSi coatings flame sprayed and others by WC–Co coatings HVOF sprayed. Figure 19.18 presents calendar roll of a paper machine protected by thermal spraying.

Functionally graded (FG) coatings are considered as an option to increase compatibility between ceramic coatings and metallic substrates. In this concept material properties such as coefficient of thermal expansion (CTE) and elastic modulus are designed to gradually change in order to reduce the thermal stresses within the coating. Posttreatments such as coating with fluoropolymers or sealing coatings against corrosive process environments are also used. A few examples are presented below.

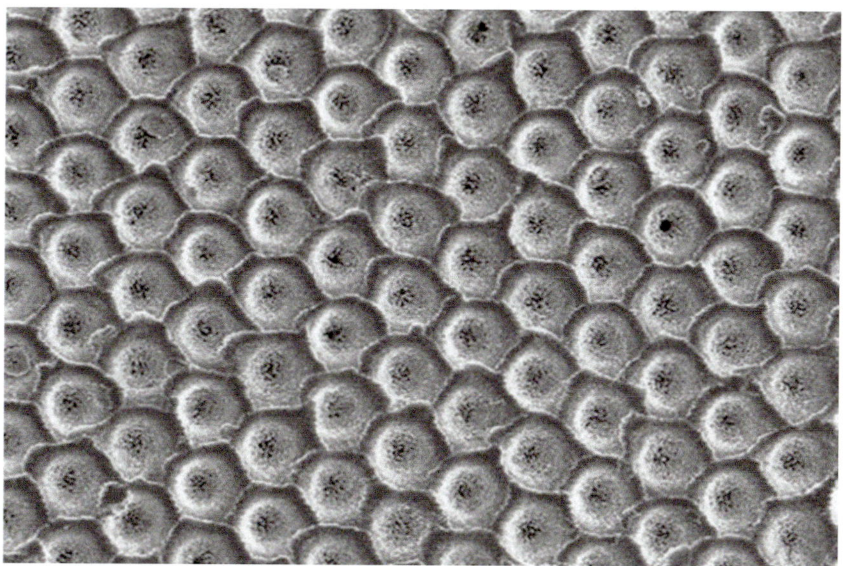

Fig. 19.17 Anilox printing roll with a laser-engraved chromium oxide coating. (Courtesy of Orelikon-Metco)

Fig. 19.18 Plasma spray applications in the paper and printing industry. (Reproduced with kind permission from Oerlikon-Metco Corp.)

Press rolls are used to remove as much water as possible from the sheet using mechanical forces. Initially they were made of granite rock and were very expensive with limited rotation velocity. Now one uses cast iron or steel roll bodies with coatings that are application specific and tailored to perform optimally in various different types of paper production conditions. Factors such as wear and corrosion resistances, and the functionality of the roll surface in the paper manufacturing process, are key properties in these applications [Vuoristo P, Nylén P (2009)]. In most cases alumina–titania and chromia coatings are used with HVOF-deposited bond coat [Davis JR (ed) (2004), Tucker RC (ed) (2013)]. According to the size of the coated cylinders, high-power plasma torches, such as Plazjet, are used for deposition.

Dryer cylinders of paper machines and large Yankee drying cylinders of tissue paper machines are HVOF thermal spray coated with cermets containing various carbides. According to their weight and size, these components are commonly coated on-site (surface preparation, coating, surface finish) in the paper factory. Ceramic-coated center press rolls are, nowadays, used to replace conventional granite rolls [Vuoristo P, Nylén P (2009)]. Ceramic coatings are based on

Fig. 19.19 Paper machine center press roll with a thermally sprayed ceramic coating (Metso Paper Inc.). (Reproduced with kind permission from ASM [Vuoristo P, Nylén P (2009)])

aluminum and chromium oxides with an underlying bond coat and corrosion barrier. They are application specific and tailored to perform optimally in various different types of paper production conditions. According to the size of the components, Metso Paper is the world's leading manufacturer, both in Finland and elsewhere. Figure 19.19 represents a ceramic center press roll of a paper machine after finishing.

Sprayed coatings can also be top-coated with a fluoropolymer layer to improve the release properties of the roll surfaces [Vuoristo P, Nylén P (2009)]. Steels with high molybdenum contents (better corrosion resistance and thermal conductivity) HVOF sprayed are also used [Davis JR (ed) (2004), Tucker RC (ed) (2013) *ASM Handbook, Volume 5A*]. Machine knives (slitters, guillotine, blades) are coated with HVOF-sprayed WC–Co coatings and then sharpened [Davis JR (ed) (2004), Tucker RC (ed) (2013) *ASM Handbook, Volume 5A*].

Carbon-fiber-reinforced plastic (CFRP) rolls, exhibiting excellent characteristics including lightweight, high stiffness, and low flexure, have been increasingly employed in industrial manufacturing fields. Compared to conventional metal rolls, CFRP roll is lighter and stiffer and exhibits lower inertia moment [Yoshiya A, et al. (2009)]. Unfortunately, carbide-cermet coatings failed because of their poor thermal shock resistance on CFPR rolls. Ni-base composite porous coating including ceramic particles developed by Nagai et al. (2009) showed high thermal-shock resistance. Five coated rolls were installed in the actual papermaking line. They achieved successful results with 10% improvement in the line speed, whereby outstanding performance and maintenance-free functioning have been confirmed even after 4 years of use. Yoshiya et al. (2009) have developed a 5.4-m long thermally

sprayed carbon roller, installed in a paper slitter/winder, that is moving stably at ultrahigh speed, 2300 m/min, with no abrasive wear. The roller has a three-layer structure, especially a tungsten carbide cermet layer on grooved metal sleeve, which covers the CFRP substrate roller shell. The thermal spray process is the finishing process because no after-grinding is performed.

Blow tanks, that are storage bins, are also on-site HVOF sprayed with WC–Co coatings to improve their wear resistance.

19.4.10 Metal Processing Industries

Many parts in metal working industries are submitted to severe wear and corrosion; however, its consequences can be reduced by thermally sprayed coatings. The main parts that can be sprayed are [Davis JR (ed) (2004), Vetter J, et al. (2005), Iyengar RK (2009)] components of electric arc furnace (EAF) and basic oxygen furnace (BOF), molds, casting dies, casting salvage, molten metal containment and delivery, and steel mill rolls working in both wet and dry mill environment, for example, entrance and exit rolls of steel processing line, or rolls for galvanized and aluminized steel sheets. PTA coatings are also used in processing industries.

19.4.10.1 Components of Furnaces or Boilers
Boilers are large and expensive installations that suffer enormously from wear caused by corrosion and erosion, aggravated by very high temperatures. The exact type of wear experienced varies from one part of a boiler to another and is influenced by the overall design of the boiler and the

Table 19.4 Different parts of the boiler that can be affected with the corresponding wear phenomena [http://www.castolin.com]

Part of boiler affected	Principal wear phenomena
Combustion chamber waterwalls	Corrosion/abrasion/erosion
Secondary superheater	Corrosion/oxidation/erosion (ash)
Economizer	Erosion (ash)/corrosion
Primary superheater	Erosion (ash)/corrosion
Reheater	Corrosion/oxidation/erosion (ash)
Superheater soot blowers	Erosion (steam)
Combustion chamber soot blowers	Erosion (steam)

Reproduced with kind permission from Castolin, copyright © Castolin Eutectic

type of combustible fuel. For example, Table 19.4 from www.Castolin.com presents the different parts of the boiler that can be affected with the corresponding wear phenomena.

A wide variety of components associated with electric arc furnace (EAF) and basic oxygen furnace (BOF) are under severe attack from heat, particulate, and acidic gases. Water-cooled components, in the off-gas duct systems such as pans, roofs, boxes, and panels, are subjected to high-velocity combustion gases that contain a number of corrosive chemicals that condense and attack the heat transfer surfaces [Kaushal G, et al. (2011)]. Coatings used are those developed for high-temperature wear and corrosion resistance; see for example [Iyengar RK (2009), Kaushal G, et al. (2011), Espallargas N, et al. (2008), Wang B-Q, Verstak A (1999), Higuera HV, et al. (2001a, b), Sidhu TS, et al. (2005), Sidhu HS, et al. (2006a, b), Sanz A (2001), Gross KA, Kovalevskis A (1996)].

19.4.10.2 Molds

In continuous casting the cast shell in the lower half of the mold abrades and wears the bottom of the mold. Diffusion of the copper substrate from the mold into the surface of the cast product leads to a quality defect called "star cracking" [Iyengar RK (2009)]. Chrome- and nickel-based coatings protect copper molds from wear and also enhances caster product quality by greatly reducing cast product contamination and star cracking problems [Iyengar RK (2009)]. Thermal barrier coatings are also used to control the heat flow and retard rapid chilling [Davis JR (ed) (2004), Tucker RC (ed) (2013) *ASM Handbook, Volume 5A*, Sanz A (2001)]. For very corrosive melts, pure yttria is used instead of zirconia partially stabilized with yttria. Of course, bond coats are necessary, and sometimes multilayer coatings, to achieve a good compliance between the expansion coefficients of mold and topcoat. Sanz (2001) has studied different coatings to protect the mold wall. Casting salvage is also achieved by filling the voids, after grinding of porosity or wear zones, with plasma or wire arc coatings that are then re-machined

[Davis JR (ed) (2004), Tucker RC (ed) (2013) *ASM Handbook, Volume 5A*]. Gross and Kovalevskis (1996) have shown that metallic molds from iron, nickel, Ni–Al, and Ni–Cr–B–Si can be produced by air plasma spraying onto steel and chrome-plated steel models. The main processing criterion was the mold temperature that must be heated above 400 °C to avoid coating warpage and fragmentation. Heating to 600–700 °C is required to remove coating porosity and reduce coating pullout. Weiss et al. (1994) have demonstrated the feasibility of making sprayed steel-faced tooling. Kim and Kweon (1996) investigated various flame- and plasma-sprayed coatings to extend the life of these molds. Coating materials studied include plasma-sprayed ceramic coatings with bond coats as well as flame for casting pig iron ingots. Cyclic furnace tests from room temperature to 1100 °C in air, simulating the thermal cycle in casting, indicated that failure occurred along the interface between the bond coat and the gray iron substrate because of iron oxidation, and not at the interface between the ceramic top coating and the bond coating. The field test results indicated that plasma-sprayed alumina coatings with 200 μm top coating thickness are the most promising materials for pig iron casting.

19.4.10.3 Die Casting

Gibbons and Hansell (2006) have shown that two thermal spray materials, one Cr_2C_3–25(Ni–20Cr) and one WC–10Co–4Cr, deposited using JP-5000 HVOF hardware, offered properties that could enable low-cost, low-volume production aluminum injection mold tooling to be upgraded to higher-volume production tooling. Hot dipping rolls—MoB/CoCr, a novel cermet material for thermal spraying, with high durability in molten alloys—have been developed to utilize for aluminum die-casting parts, and for hot continuous dipping rolls in Zn and Al–Zn plating lines [Gibbons GJ, Hansell RG (2008)]. The tests revealed that the MoB/CoCr coating had much higher durability without dissolution in the molten Al–45 wt.% Zn alloy. Using undercoat was effective to reduce the influence of large difference in thermal expansion between the MoB/CoCr topcoat and substrate of stainless steel of AISI-316L, widely used for the hot continuous dipping [Mizuno H, Kitamura J (2007)]. MoB-based cermet feedstock powders (MoB/NiCr and MoB/CoCr) were deposited on SKD61 (AISI H-13) substrates used as a preferred die (mold) material [Khan FF, et al. (2011)]. The durability of these coatings on cylindrical specimens against soldering also has been investigated by immersing in molten aluminum alloy (ADC-12) for 25 h at 670 °C and, subsequently, compared with that of NiCr and CoMoCr coatings. Both types of MoB-based cermet coatings have shown high soldering resistance as negligible intermetallic formation occurred during the immersion test [Khan FF, et al. (2011)]. Weiss et al.

(1994) have used arc-sprayed steel-faced tooling to create matched die sets for injection molding applications.

19.4.10.4 Entrance and Exit Rolls of Steel Processing Line

When coating bridle and accumulator rolls in entrance and exit ends of a steel processing line with tungsten carbide coating surface damage on the roll is eliminated and proper grip provided, and slippage prevented. The surface coating is properly textured to provide the required characteristics or profile on the strip surface [Iyengar RK (2009)]. Multicomponent white cast iron is a new alloy that belongs to system Fe–C–Cr–W–Mo–V, HVOF sprayed, and seems to be promising for rolls [Khan FF, et al. (2011)].

19.4.10.5 Galvanized and Aluminized Steel Sheets

They require very high surface quality, particularly in exposed panels. In continuous galvanizing and aluminizing, the steel strip is dipped in the molten bath through a series of rolls that control the speed and tension of the strip and guide the steel strip through the molten metal bath. The rolls operating in the molten Zn–Al alloy are subjected to severe corrosive environment and require frequent change and repair [Davis JR (ed) (2004), Tucker RC (ed) (2013) *ASM Handbook, Volume 5A*]. Seonga et al. (2001) have shown that WC–Co coatings were not very good with molten Zn–Al. By coating the sink and stabilizer rolls with molybdenum boride, tungsten carbide, and other materials, the rolls remained smoother and produced an improved strip surface.

19.4.11 Petroleum and Chemical Industries

The main problem is to prevent corrosion and wear, all forms of wear, especially those linked to corrosion arising at low, intermediate, and high temperatures. As previously pointed out, against corrosion dense (non-sacrificial) coatings (no pores or cracks and very low oxide level) of the appropriate material are mandatory. For porous coatings, especially ceramic ones, a dense bond coat with a good resistance to the corrosive media is mandatory. Porous coatings can be sealed but the seal must be adapted to the service temperature. Moreover, many components have sizes that require on-site spraying. Usually, dense coatings are obtained with vacuum plasma spraying or D-gun or HVOF, but soft vacuum plasma and D-gun spraying are not adapted to big parts. Thus, HVOF spraying, that can be used on-site, is extensively used against corrosion and abrasive and erosive wear mainly with carbide-based cermets [Davis JR (ed) (2004), Zeng Z, et al. (2008), Souza VAD, Neville A (2007), Ishikawa Y, et al. (2005), Uusitalo MA, et al. (2002), Henne R, et al. (1999), Iyengar RK (2009), Kaushal G, et al. (2011), Espallargas N, et al. (2008), Wang B-Q, Verstak A (1999), Higuera HV, et al.

Fig. 19.20 Typical coating against corrosion sprayed on-site onto refinery pipe. (Courtesy of Oerlikon-Metco)

(2001a, b), Bolelli G, et al. (2008), Godoya C, et al. (2004), Choa JE, et al. (2006)]. Of course, for the protection of external steel structures, pipes, tanks, etc., Al, Zn, or Zn–Al wire arc sprayed or flame sprayed are used [Evdokimenko Yu I, et al. (2001), Sørensen PA, et al. (2009), Murakami K, Shimada M (2009), Pacheo da Silva C et al (1991), Chun-long Y, et al. (2009), Schmidt DP, et al. (2006), Han M-S, et al. (2009)], as well as polymers sprayed by flame, HVOF, or plasma according to the polymer melting temperature [Petrovicova E, Schadler LS (2002)]. For example, Fig. 19.20 presents an aluminum protection flame sprayed on a refinery pipe.

Coatings in chemical industry are used for pressure and storage vessels with Hastelloy B or C, Inconel 600, for heat-affected zones where solutions found in gas turbines are often used. In some chemical reactors, submitted to strong acids in combination with organic solvents, glass lining are used that are repaired by APS-spraying tantalum, bonding well to the glass, and followed by an overlay of chromium oxide [Davis JR (ed) (2004), Tucker RC (ed) (2013)]. For example, Moskowitz (1993) has tested HVOF process using unique inert gas shrouding to produce highly dense, low-oxide coatings of metallic alloys, which were tested in laboratory and plant, with exposures as long as 5 years. Coatings of corrosion-resistant alloys, such as type 316L stainless steel and Hastelloy C-276, were shown to act as true corrosion barriers.

For the oil, gas, and petrochemical industries, the following components are coated using thermal spray technologies: mud drill rotors, pump impellers, plunger, turbine, rotor shaft of centrifugal compressor/pump, pump shafts, boiler tube, thermo-well, mixing screw, mandrels, actuator shafts and housings, housings and valves, valve gates and seats, ball valve with large diameter, progressive cavity mud motor rotors, rock drill bits, riser tensioner rods, impeller/blade drilling and production risers, sub-sea piping, wellhead

Fig. 19.21 General view of practical spraying of ball valve by the MET-JET III system. (Reprinted with kind permission from ASM [Huang XO, et al. (2007)])

Fig. 19.22 Wear band on drilling pipe PTA coated. (Courtesy of Castolin)

connectors, fasteners, compressor rods, mechanical seals, pump impellers, tank linings, external pipe coatings, structural steel coating, etc. [Davis JR (ed) (2004), Tucker RC (ed) (2013) *ASM Handbook, Volume 5A*]. It also seems that wear-coating applications have started to replace hard chrome [Thintri Inc. (2013)].

Two examples of coatings used in oil, gas, and petrochemical industries are presented below. Stainless steel ball valves must be wear resistant, and have a low friction coefficient, a good erosion and heat resistance, good fatigue strength and seal performances, etc.; these are significantly improved by applying HVOF coatings [Huang XO, et al. (2007)] and loping trend of these ball valves for highly harsh working conditions. Figure 19.21 shows the practical spraying of a ball valve having a diameter of over 500 mm. Wear bands on drilling pipes, as shown in Fig. 19.22, are PTA coated.

19.4.12 Electrical Utilities

Coatings against corrosion and wear (C–W) are used in fluidized-bed combustor (FBC) and conventional coal-fired boilers.

19.4.12.1 For Fluidized-Bed Combustor Boilers

The problem is linked to the finely divided mixtures of coal and limestone particles eroding and corroding steam pipes and boiler walls, as well as the high sulfur content of coals or low-grade combustibles resulting in corrosion at high temperatures. Different coatings are used:

- Cr_2O_3 (20 wt.%)–Al_2O_3 on NiCrAlY bond coat (against corrosion through the porous ceramic coating) [Davis JR (ed) (2004), Tucker RC (ed) (2013) *ASM Handbook, Volume 5A*], the addition of approximately 20 wt.% chromia resulting in the formation of one solid solution of $(Al–Cr)_2O_3$ in the α-modification (working temperatures can reach 1000 °C and the transformation of γ phase starts around 900 °C).
- HVOF-sprayed Cr_3C_2–NiCr coatings with high compactness and fine grain size [Wang B (1996)], the wear resistance being due to hard carbide particles homogeneously distributed within coating, the ductile matrix being corrosive resistant.

19.4.12.2 For Coal-Fired Boilers

Plasma-sprayed Stellite-6 coating has been found to be effective in increasing the erosion–corrosion resistance of boiler steels in the coal-fired boiler environment. A less porous structure obtained after laser remelting was found to be effective in increasing erosion–corrosion resistance [Sidhu BS, Prakash S (2006)]. Inconel systems or high-chromium alloy or chromium–nickel alloy coatings, presenting a good resistance to sulfur, have also been used, sprayed with plasma, wire arc, or HVOF [Davis JR (ed) (2004), Tucker RC (ed) (2013)]. Petrovicova and Schadler (2002) have proposed low-cost and high-hardness plasma-sprayed coatings, developed by addition of carbon and hardening elements to high-chromium cast iron (C–Si–Mn–Cr–Mo–V–other–Fe) used for wear-resistant material. Nitrogen gas atomization was applied to manufacture the powder in order to prevent oxidation of particles. These coatings showed the same or more erosion resistance than Cr_3C_2–NiCr cermet coating and had higher reliability for long-period operation and higher practicality.

19.4.13 Textile and Plastic Industries

The abrasive qualities of man-made fibers, particularly nylon and polyester, and the corrosive characteristics of additives, such as fiber finishes and lubricants, combine to deteriorate yarn contact surfaces. High rotation speed of grooved rollers results in considerable abrasion wear. Due to the very high rotational speed, the roller has to be made from aluminum, and therefore, the reduction in wear can only be achieved by the deposition of a coating that also must be as light as possible. For years, thermal spray coatings have been used against wear and corrosion on textile machinery: steel thread guiders, thread breaks, stretch roll, grooved roll, separator roll, oiling roll, draw roll, feed roll, conditioning rollers, snick pins, collets, disk grip, cutter base, feed drum, rotor for open-end, extruder dies, knives, heating bars, heater/hot plate, etc. [Davis JR (ed) (2004), Tucker RC (ed) (2013)]. The high-velocity rotating parts are often coated with alumina or alumina–titania coatings. TiO_2 addition lowers significantly the microhardness of the alumina coating but increases its toughness [Yilmaz R, et al. (2007)]. Parts such as knives and extruder dies are coated by HVOF-sprayed tungsten carbides, while textile rolls are plasma sprayed with alumina–titania (13 wt.%) as illustrated in Fig. 19.23.

Lima and Marple (2005) showed that nanostructured titania feedstock HVOF sprayed exhibited a superior abrasion wear resistance (27% lower volume loss) when compared with an air plasma-sprayed conventional alumina–titania coating, in spite of the fact that the latter is 33% harder than the HVOF-sprayed titania. The higher wear resistance of the HVOF-sprayed nanostructured titania was provided by the nanostructured zones embedded in the dense and uniform coating microstructure acting as crack arrester. They are probably worth to be tested in textile industry.

19.4.14 Polymers

Thermal spraying of the polymers [Davis JR (ed) (2004), Tucker RC (ed) (2013), Petrovicova E, Schadler LS (2002), Lathabai S, et al. (1998), Lins VFC, et al. (2007), Berndt CC, et al. (1998), Leivo E, et al. (2004), Zhang T, et al. (1997), Chen H, et al. (1999), Zhang G, et al. (2007), Zhang G, et al. (2006), Zhang C, et al. (2009), Sweet GK (1993), Brogan JA, et al. (1995), Henne RH, Schitter C (1995), Ivosevic M, et al. (2009)] is one-coat process that acts as both the primer and the sealer, with no additional cure time processes. Polymer thermal spraying is ideally suited for large structures that otherwise could not be dipped in a polymer suspension. Moreover, it seems that functionalized polyethylene polymers such as ethylene methacrylic acid copolymer (EMAA) and ethylene acrylic acid (EAA) can be applied in high humidity. Of course, the use of polymer coatings depends strongly on its service conditions. Especially, it must be kept in mind that melting temperatures vary from 40 to 60 °C for ethylene methacrylic acid copolymer (EMAA) to 300 °C for polyimide. They are deposited onto metals, ceramics, cermets, and composites. As they present a high chemical resistance, a high impact, and abrasion resistance at low temperatures, they are used in many industries, especially in food industry.

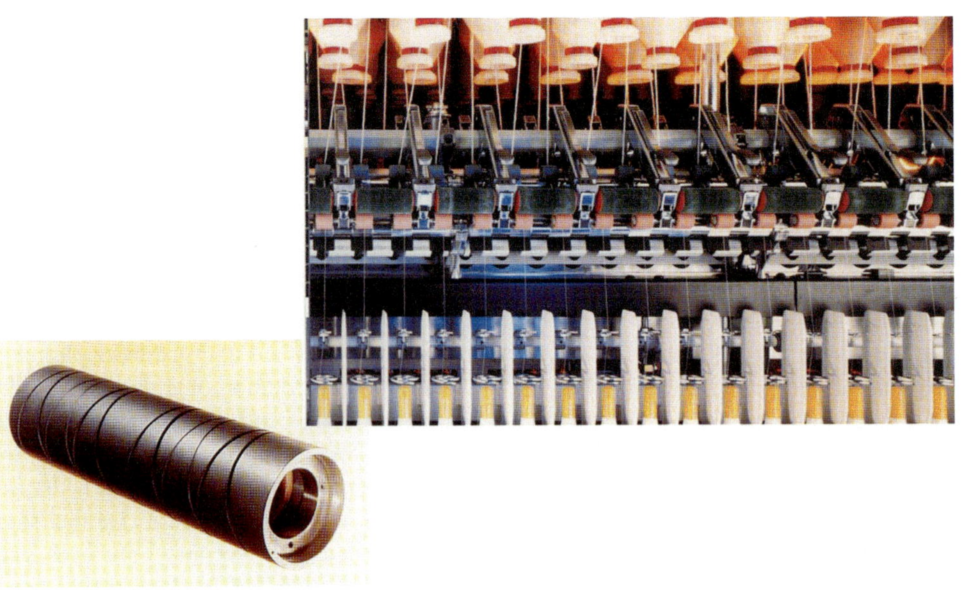

Fig. 19.23 Applications of plasma spray technology for the coating of spools and high-wear parts in the textile industry. (Reproduced with kind permission from Oerlikon-Metco Corp.)

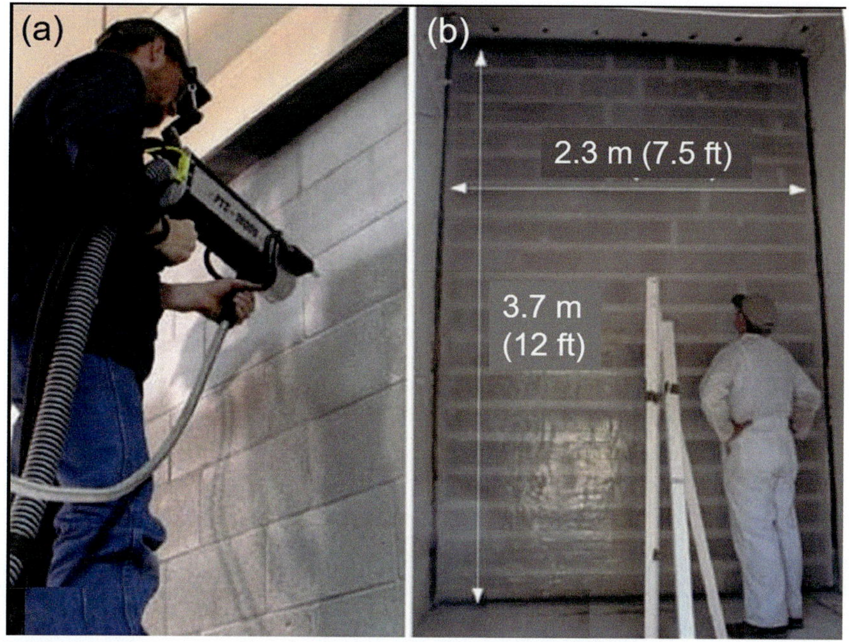

Fig. 19.24 (**a**) Thermal spraying of blast mitigation polymer membrane to the internal side of test walls; (**b**) wall dimensions and appearance of a completed membrane. (Reprinted with kind permission from ASM [Ivosevic M, et al. (2009)])

In food industry, polymer coatings replace paints on the wall because they have a much better resistance to the chemical products used for cleaning (about 1 week for paint against about a quarter for the polymer coating). They are even used on the floors where polymer coatings doped with alumina particles provide an excellent anti-slip lining, the alumina particles rippling out when people walk on it.

A novel coating process (polymer thermal spray, PTS) has been developed by Ivosevic et al. (2009). It utilizes an electro-resistive heating element for rigorous temperature control and for heating the main process gas that could be air, nitrogen, inert, or other gases. They have deposited advanced polymer coatings and structures such as:

- Blast mitigation coatings for protecting civil structures against external explosions.
- Thermal spray forming of syntactic foams based on polyimide micro-balloons.

For blast mitigation, Fig. 19.24 shows thermally sprayed polymer (thermoplastic elastomer) membranes applied to the internal side of the test walls measuring 3.7 × 2.3 m. Figure 19.25 presents the test setup for evaluation of thermally sprayed blast mitigation treatments. The explosion impulse reached 215 kPa and lasted about 50 ms. Both walls successfully passed the test and remained standing. The total wall deflection toward the inside of the test room was 8 in. with no visible damage to the membrane and no flying debris or wall segments penetrating the room.

Spraying syntactic foams based on polyimide microballoons was developed [Ivosevic M, et al. (2009)] in the frame of a research project funded by NASA Langley Research Center. The coating was supposed to be used as high-temperature insulation (up to 325 °C) and also satisfy strict flammability regulations relating to the aerospace industry. As shown in Fig. 19.26a, the polyimide syntactic foam consisted of two NASA Langley-licensed materials, polyimide microspheres (PerFoma-H® from GFT Corporation), and a binder. Polyimide microspheres had a bulk density of 0.043 g/cm^3, a size range of 400–800 μm, and service temperature between −253 °C and 315 °C. The binding powder and microspheres were thermally Co-sprayed using the PTS coating system. The thermally sprayed foam did not support a flame and formed a stable isolative char without smoke at higher temperatures. The foam deposited over various complex shapes is shown in Fig. 19.26 and seemed to be interesting for thermal insulation in airspace applications [Ivosevic M, et al. (2009)].

19.4.15 Reclamation

The effects of wear and corrosion can only be retarded, but not be stopped forever and wear parts must be replaced or resurfaced. It is the same for undersize parts due to manufacturing error. The ability to apply coatings by thermal

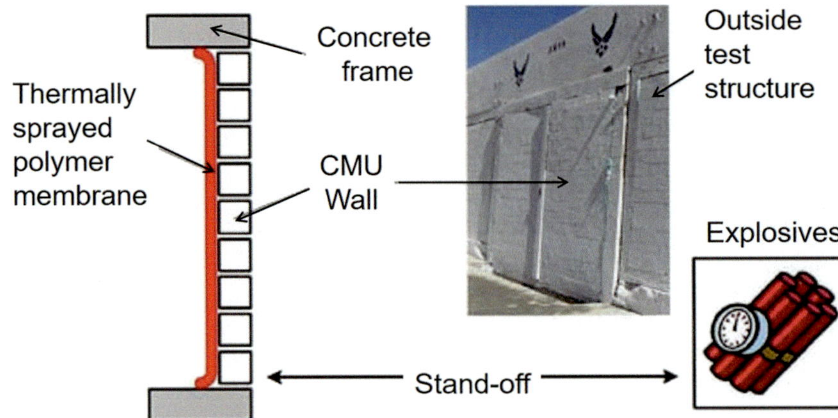

Fig. 19.25 Test setup for evaluation of thermally sprayed blast mitigation treatments. (Reprinted with kind permission from ASM. [Ivosevic M, et al. (2009)])

Fig. 19.26 Thermally sprayed syntactic foam based on polyimide micro-balloons: (**a**) foam microstructure; (**b**) thermally sprayed on pipe T joint; (**c**) on a gas cylinder; and (**d**) on a flat aluminum plate. (Reprinted with kind permission from ASM [Ivosevic M, et al. (2009)])

spraying with a broad spectrum of thickness, finish, and composition requirements makes the different thermal spray processes ideal solutions to resurface components, avoiding their costly replacement. Restoration is performed either with the same material as the base metal or with a more corrosion- and/or wear-resistant material. However, it is better to choose coatings that can be machined. In a wide range of industrial applications, thermal coatings are used to renew components by restoring their specified dimensions and matching and sometimes improving their original performance. Very often it is possible to design a coating that meets the functional requirements of the component, and moreover offers better resistance to corrosion, oxidation, and mechanical wear than the original product. Of course, the cost of the resurfacing must be lower than that of a new part (a fraction of it). For example, this technique is particularly used for roller faces and journals, dryer drums for papermaking, pump seals (shafts and sleeves), pump housings, compressor rods, rotary airlocks and feeders, and conveyer screws [Davis JR (ed) (2004), Tucker RC (ed) (2013)]. PTA also is used for refurbishing components, even with a complex shape as turbine blades.

A few examples are given below to illustrate the use of sprayed coatings in resurfacing components.

The use of aluminum alloy for injection mold tools, especially for the automotive industry, is not new and has been utilized as bridge tooling to support the manufacture of typically up to 10,000 components. Two thermal spray materials, one Cr_2C_3–25(Ni–20Cr) and one WC–10Co–4Cr, deposited using JP-5000 HVOF hardware, have been shown to offer properties that could enable low-cost, low-volume production aluminum injection mold tooling to be upgraded to higher-volume production tooling [Weiss LE, et al. (1994)].

The thermally sprayed Tribaloy T-800 coatings deposited by the HVOF process exhibited lower cavitation wear rates than the stainless-steel bulk material [Hahn M, Fischer A (2010)].

Massive turbines and the valves used to regulate water flow at hydroelectric dams are often exposed to the wearing effects of high-velocity water containing abrasive particulate. As mentioned by Sulzer-Metco, coatings can replace lost material and, at the same time, provide a more wear-resistant surface, increasing useful life by up to 20 times over the original.

Hard chromium plating is usually used to restore to original dimensions the worn surfaces of gas turbine shafts. In the case of shafts repair, hard chromium plating process proved to be time consuming and to involve high costs. High hardness and good wear resistance performances of WC–12%Co HVOF-sprayed coatings support their candidature for the replacement of hard chromium plating in this field [Bolelli G, et al. (2006a, b, c)].

Song et al. (2005) have studied the effect of repair of NiCrAlYSi coating after longtime use on the coherence of coating to substrate and mechanical properties of Ni_3Al-base alloy IC6. For secondary coating repair, the room temperature tensile properties of base alloy had no obvious change, while stress rupture lives decreased, but still were rather long, compared to the alloy with first repair coating. Therefore, they deduced that NiCrAlYSi coating repair is feasible to prolong the service lives of IC6 turbine vanes.

Ducos (1988) has presented a few industries where PTA is used for resurfacing: extruder screw, vanes, turbine blades, engine valves (especially, big diesel engines), and rail train car wheels.

Isakaev et al. (1999) have shown that plasma-sprayed coating of railway frogs, working under contact fatigue and wear-out, provides a good solution for repairing them.

The surfaces of printing press cylinders wear quickly. Coatings of nickel/chrome alloy, Cr_2O_3, and others have

supplanted chrome plating as the preferred method of reclaiming worn cylinders because thermal spray processing is far faster.

It has been shown that cold spray is a promising, cost-effective, and environmentally acceptable technology to impart surface protection and restore dimensional tolerances to aluminum alloy [Tapphorn R, et al. (2009)] or magnesium alloy components on helicopters and fixed-wing aircraft [de Botton O (1988)]. Kashirin et al. (2007) have shown that using a portable cold spray system working with air at pressures below 0.8 MPa made possible to restore on-site aluminum alloy molds for plastics, corrosion defects on car aluminum engines, casting defects for various aluminum alloys, antique art objects such as copper sculptures of Isaac Cathedral in Saint-Petersburg, etc.

19.4.16 Other Applications

As previously described, thermal spray coatings provide superior wear (abrasive, erosive, fretting, etc.) resistance and corrosion protection with coatings having low porosity, high hardness, good toughness for cermets, and great flexibility of composition [Davis JR (ed) (2004), Tucker RC (ed) (2013)]. Thus, they are also used in the following:

Mining industry, where good abrasive and/or adhesive resistance is very important. For example, HVOF-, PTA- (alloys doped with ceramic particles), and wire arc-sprayed cored wire coatings are used on conveyor belt idlers, steel oscillating tub, fan components such as removable leading edges, bolt-covers, blades, deflectors and large inlet entrance cone ducts, scraper or excavator teeth, etc., as illustrated in Fig. 19.27.

Nuclear industry, where the use of cobalt-based alloys (stellite or cermet matrix) is limited, thus coatings against wear and corrosion rely on nickel alloy-based coatings (NiCr–WC or Cr_3C_2). Cermets with hafnium carbide have been developed that present a large neutron cross section. However, sprayed coatings are essentially used in pumps, turbines, heat exchangers, vanes, etc.

In numerous applications developed at CEA-DEN, French Atomic Agency, Atomic Energy Department, particularly those encountered in the processing of nuclear wastes, metallic components are subjected to extreme environments in service, in terms, for example, of aging at moderated temperature (several months at about 300 °C) coupled to thermal shocks (numerous cycles up to 850 °C for a few seconds and a few ones up to 1500 °C) under a reactive environment made of a complex mixture of acid vapors in the presence of an electric field of a few hundred volts and a radioactive activity. Berard et al. (2008) have tested alumina plasma-sprayed coatings manufactured with feedstock of different particle size distributions, graded alumina–titania coatings, and

Fig. 19.27 (**a**) Tooth of excavator PTA-coated with a nickel-based composite (**b**) coating microstructure. (Curtesy of Castolin)

phosphate-sealed alumina coatings to improve the properties of metallic substrates operating in such extreme environments. The effects of particle size distribution, phosphate sealant, and graded titania additions on the dielectric strength of the as-sprayed, thermally cycled, and thermally aged coatings were investigated. Thermal aging test was realized in furnace at 350 °C for 400 h and thermal shocks tests resulted from cycling the coating between 850 and 150 °C using oxyacetylene flame and compressed air cooling. Aluminum phosphate impregnation appeared to be an efficient posttreatment to fill connected porosity of these coatings. Alumina as-sprayed coatings manufactured with +22–45 μm and +5–20 μm particle size distributions exhibited good dielectric strengths after thermal solicitations compared to those manufactured with larger size distributions or to graded titania coatings.

Cement industry: Again, the main role of sprayed coatings is against wear and corrosion in mechanical seal, sleeve, burner tip, boiler tube, thermo-well, kiln support roll, pinion shaft, coating for cement preheat tower, impeller blade, calendar roll, cone crusher, and hydraulic rams with acid-resistant coatings.

Drawing machine: Guide roller and ceramic disk, rod breakdown drawing machine, fine drawing machine, and other accessories.

Waste treatment: Steel tube oxidation causes an important problem in municipal solid waste incinerator (MSWI) plants due to burned wastes containing high concentrations of chemically active compounds of alkali, sulfur, phosphorus, and chlorine. Ni-based HVOF coatings are a promising alternative to the MSWI conventional protection against chlorine environments [Guilemany JM, et al. (2007)].

Sheet metal forming dies: Conventionally, mold and dies are manufactured by machining from bulk metallic materials. Tooling by using arc spray process to spray metal directly onto a 3-D master pattern is an alternative method to manufacture mold and dies for plastic injection molding and other applications [Davis JR (ed) (2004), Tucker RC (ed) (2013)].

19.5 Thermally Sprayed Coatings by Country

Dorfman and Sharma (2013a, b) in their "Commentary; Challenges and Strategies for Growth of Thermal Spray Markets," based on the report by Hanneforth (2006), presented interesting statistics, given in Fig. 19.28, about the regional distribution of the thermal spray (TS) markets around the world in 2006, which they estimated at a total value of US$6.5 billion. According to their report, two-thirds of these markets is split evenly between North America and Europe/Middle East. Of the balance, US$1.0 billion was estimated for the Japanese market, US$0.5 billion for each of China and Asia Pacific, and US$0.3 billion for the rest of the world. The growth rate of the industry has been, however, the highest in Asia and South America [Fukumoto M, (2008a, b), Lee C, (2009a, b), Sundararajan, G, et al. (2009a, b), Nakahira A (2009a, b), Valarezo A (2012)]. In particular, there has been a significant increase in research and development activities in China as indicated earlier in Fig. 19.1, giving the number of technical manuscripts submitted to the *Journal of Thermal Spray Technology* (JTST) during the period (2004–2010).

In the following, a brief description is given on the development of the thermal spray (TS) industry in different regions around the world. These are given based on the scarce information available in the open literature on techno-economic developments in this field. Limited information about thermal spraying in Nordic region of Europe [Vuoristo P, Nylén P (2009)], Korea [Lee C (2009a, b)], India [Sundararajan G, et al. (2009a, b)], Japan [Tani K, Nakahira H (1992), Nakahira A (2009a, b)], and China [Fukumoto M (2008a, b)] was available in connection with the Thermal Spray Conference in 2009 (ITSC-2009). The analysis, and observations made, provides general technological trends, and indications of emerging technologies in this rapidly developing field.

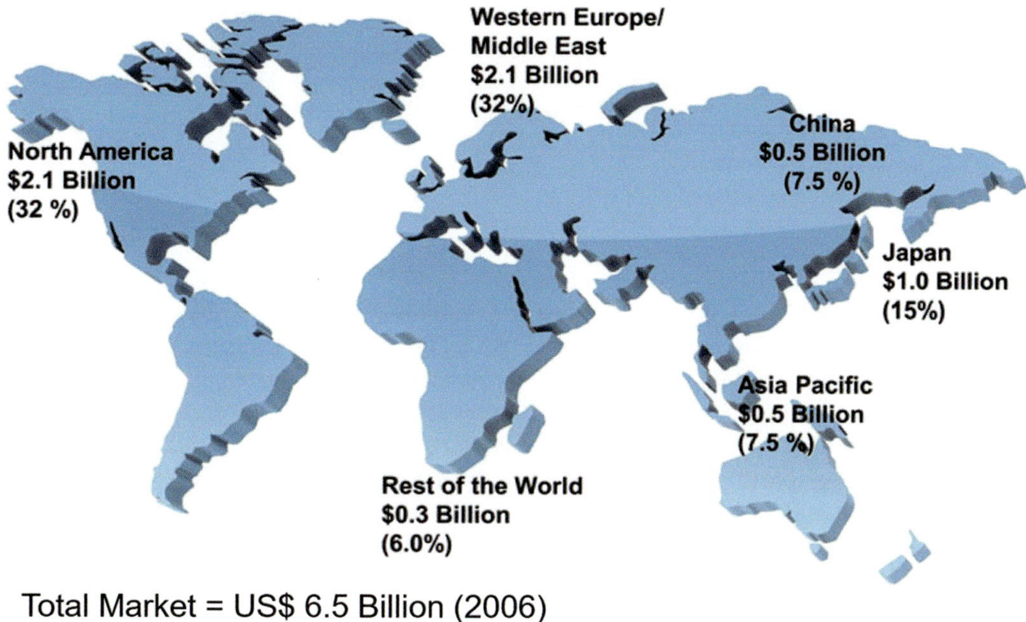

Fig. 19.28 International breakdown of annual thermal spray markets. [Dorfman M, Sharma A (2013a, b)]

19.5.1 North America

Thermal spraying has been in rapid development since the 1970s and 1980s, evolving from an art to a science over the period 1980–2000 [Wigren J, Kristina Täng (2007)]. Extensive research and development efforts were devoted in leading laboratories in this field, which resulted in the emergence of new spray technologies such as HVOF, HVAF, wire arc spraying, and, more recently, solution and suspension plasma spraying. This is illustrated in Fig. 19.29, after TAFA, representing percentages of the different spray techniques in 2000. Flame spraying is now reduced to 50% and the rest of the market comprises all other techniques except of course cold spray that had not yet fully integrated on an industrial scale at that period.

Besides new products and marketing issues, the development and integration of TS technologies at the design stage of new products has been linked to a better reproducibility and reliability of substrate surface preparation, spraying, and posttreatment of coatings. Particularly consistent effort has been made in the following areas:

- Better control of surface preparation with a grit-residue level as low as possible and a roughness adapted to sprayed particles and a desired grit blasting residual stress level.
- Better control of the spray process parameters thanks to computer-control process equipment coupled with adequate online sensors.

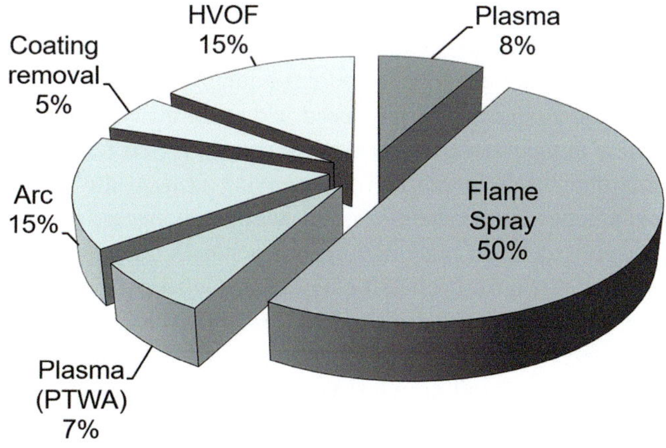

Fig. 19.29 Percentages of the different spray processes in industry in 2000. (After M. L. Thorpe, Oral presentation ITSC)

- Development of robust sensors to follow the key parameters of coating manufacturing.
- Advances in fundamental understanding of correlations between coating properties, particle impact parameters, as well as coating and substrate temperature during spraying.
- Better control of ambient dust fume, especially that resulting from the condensation of vaporized material, in the spray booth and close to the sprayed part.
- Standardization of coating characterization methods.

- Education at the operators and designers' levels coupled with the development of expert systems and proper engineering and material science tools for quality control (QC).

19.5.2 Europe

According to Ducos and Durand (2001), the rapid development of TS technologies in Europe is well illustrated by the breadth of technologies used on an industrial scale with a total annual market value of 615 M€ for 1998 as illustrated in Fig. 19.30 based on the [MAGETEX] study. Particularly significant is the developments of HVOF, D-gun, and wire arc spraying in this period. The spread of the use of TS coating in different industrial sectors in Europe in 1998 is illustrated in Fig. 19.31. As can be seen, these are almost split with one-third mostly devoted to high-end, high-value coating applications, used in aircraft, industrial gas turbines, printing rolls, or medical and dental fields. The balance is equally split between a wide range of applications in the chemical process industries such as primary metals, polymer processing and food, glass, pulp and paper, chemical, oil and gas, textile, and "low end" applications such as mechanical repair and metallic coatings for heavy wear or corrosion protection (including pipes and gas bottles).

Thermal spraying is also used widely in many industrial sectors in the Nordic countries [Vuoristo P, Nylén P (2009)]. Important areas where thermal spraying is used are in the manufacture of products for the petroleum, paper, metals, transport, defense, and high-tech machinery industries. In Finland, thermal spray technology has wide areas of application in the pulp and paper industries. In Sweden, thermal spray technology is of great importance in the manufacture of aero engines and in industrial gas turbine applications. In Norway, thermal spraying is widely used in various offshore applications, including sub-sea oil drilling enterprises.

Unfortunately, if the paper indicates that all of the major spray processes are currently used (from flame and arc spraying, to plasma, HVOF, and HVAF), depending on the application and type of coating applied, no details are given about their distribution between different industrial sectors.

19.5.3 Japan

According to Tani and Nakahira (1992) and Nakahira (2009a, b), the growth rate of the thermal spray (TS) market over the period 2004–2006 is estimated to be more than 5%. Thermal spraying output for fiscal year 2004 for the three sectors of the Japanese market were contract job shop productions just under 40 billion Yen ($\approx$ US\$0.4 billion), in-house production by large enterprises about 45 billion Yen ($\approx$ US\$0.45 billion), and consumables and spray equipment manufacturing 10–15 billion Yen ($\approx$ US\$0.1–0.15 billion). Figure 19.32 represents the repartition of coating services by

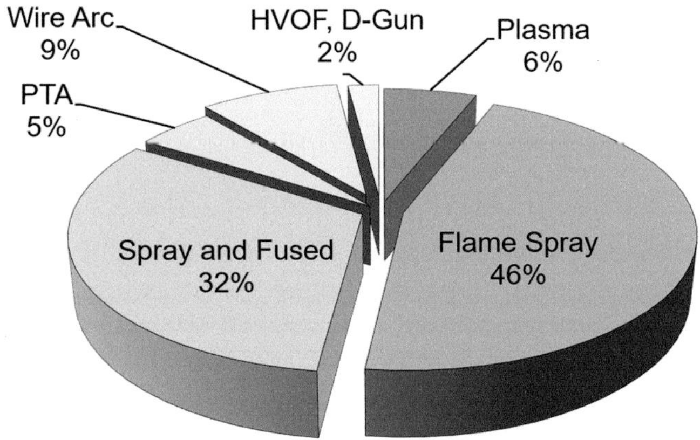

Fig. 19.30 Percentages of the different spray processes in European industry in 1998 (After Ducos and Durand). (Reprinted with kind permission from ASM [Ducos M, Durand JP (2001)])

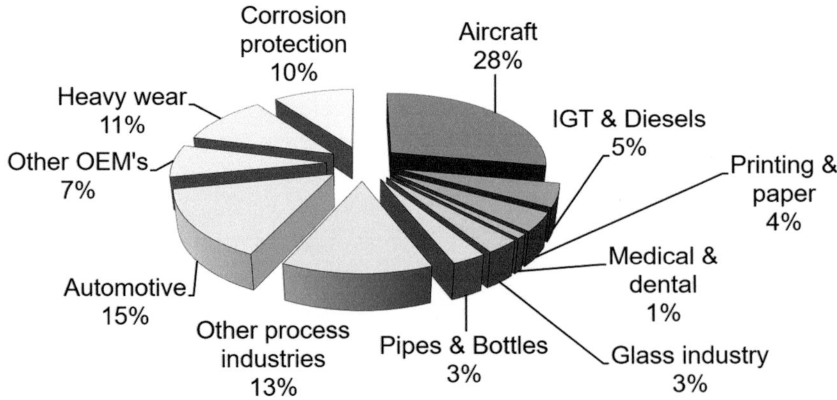

Fig. 19.31 Repartition of the 615 M€ coating activities across end-use sectors in Europe in 1998. (Reprinted with kind permission from ASM [Ducos M, Durand JP (2001)])

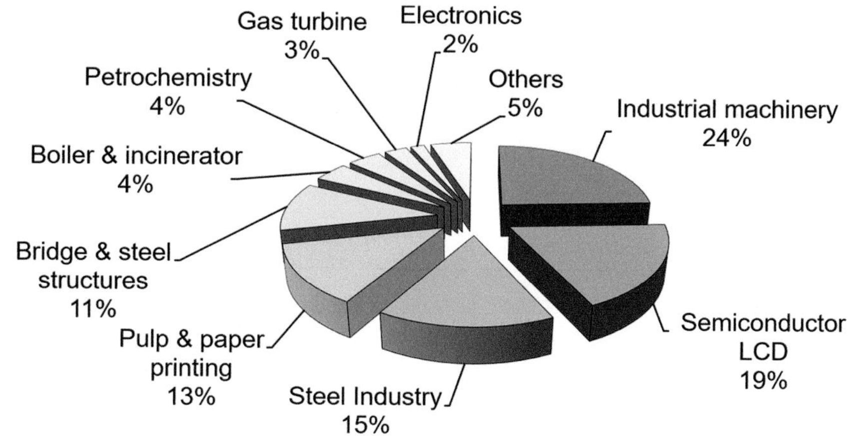

Fig. 19.32 Coating services by job shops in Japan in 2004: total market size, 45 billion Yen ($\approx$ US$0.45 billion). (Reprinted with kind permission from Springer Science Business Media [Nakahira A (2009a, b)], copyright © ASM International)

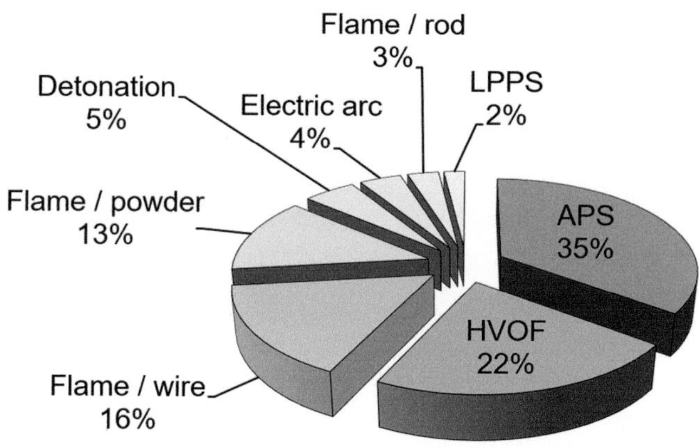

Fig. 19.33 Segmentation of spray process/turnover of contract job shops in fiscal year 2004 in Japan. (Reprinted with kind permission from Springer Science Business Media [Nakahira A (2009a, b)]. copyright © ASM International)

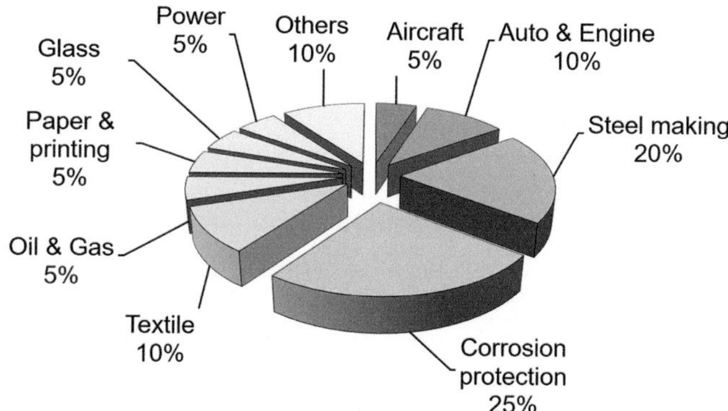

Fig. 19.34 Estimated share of different thermal spray application fields in China. (Reprinted with kind permission from Springer Science Business Media [Fukumoto M (2008a, b)], copyright © ASM International)

job shops in 2004, with total amount of 45 billion Yen ($\approx$ US$0.45 billion) [Nakahira A (2009a, b)]. It is worth noting that the growth rate in the semiconductor/liquid crystal display (LCD) field is very high. Figure 19.33 shows the corresponding distribution among spray processes with the important contribution of APS and HVOF, at the expense of a reduction in flame-sprayed coatings.

19.5.4 China

Statistical data [Fukumoto M (2008a, b)] show that according to Professor Chang-Jiu Li, of the School of Materials Science and Engineering, Xi'an Jiaoton University in China, the total output of the thermal spray industry in China increased from about 1.2 billion RMB (US$0.14 billion) in 2002 to 2.1 billion RMB (US$0.24 billion) in 2005. While this amount corresponds to an annual growth rate of about 20%, which is

significantly higher than the growth rate of 6.1% in North America, the total thermal spray industry output in China in 2005 was only 4% of the North American market, and 20% of the corresponding market in Japan. According to Professor Chang-Jiu Li, the fast and sustained growth of the Chinese economy in the last three decades at an average rate of 10% has provided unique opportunities for China's thermal spray industry to grow the fastest worldwide, a trend that is expected to continue over the next few years.

In terms of thermal spray applications in China, the technology has been widely used in different industrial fields, including steelmaking, textiles, paper and pulp, energy, petroleum and chemicals, aeronautics, and others (Fig. 19.34). The application of thermal spray coatings in the steel industry has been the fastest growing field in the last several years. The energy industry mainly includes coatings for coal-fired power plants and hydroelectric power plants. Applications of thermal spray technology to the aeronautics industry in China are mainly performed in five large job shops that share 5% of the total thermal spray output in

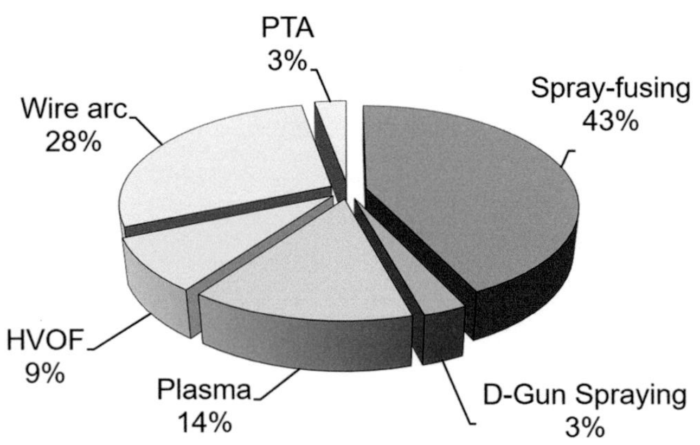

Fig. 19.35 Estimated share of different spray processes in China. (Reprinted with kind permission from ASM International [Lee C (2009a, b)])

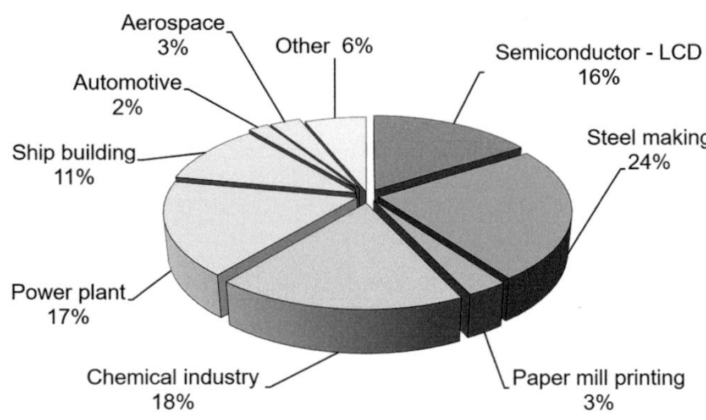

Fig. 19.36 Estimated annual sales of different spray application fields in Korea. (Reprinted with kind permission from ASM International [Lee C (2009a, b)])

China. The national program to develop different types of aircraft is expected to promote further fast expansion of R&D and thermal spraying applications in this field at an annual rate greater than 30%, including domestic and foreign job-shop contracts.

Statistical data also show that Chinese applications in 2003 was mainly driven by flame spraying/spray fusing, which constituted around 43% of the TS market as shown in Fig. 19.35. Wire arc spraying accounted for 28% of the market, plasma 14%, HVOF 9%, and each of PTA and D-gun spraying equally representing 3% of the markets. It is believed that the market share of HVOF and plasma spraying has increased markedly in the last 3 years, surpassing the corresponding figures shown in Fig. 19.35 based on 2003 data.

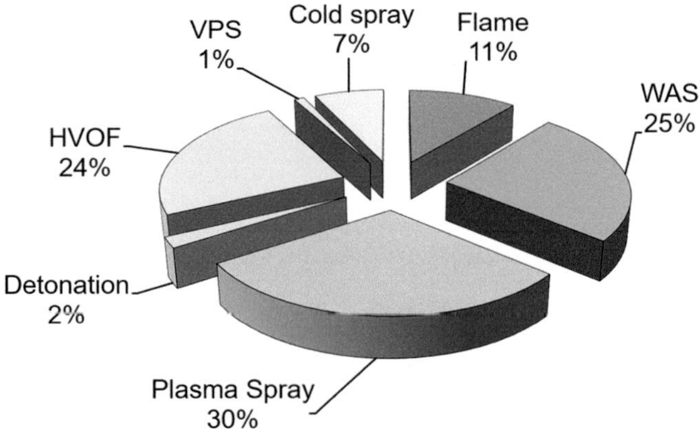

Fig. 19.37 Estimated share of different spray application fields in Korea. (Reprinted with kind permission from ASM International [Lee C (2009a, b)])

19.5.5 South Korea

According to Fukumoto (2008a, b), thermal spraying in Korea began about 40 years ago during an industrialization period in the 1970s. A few small companies started spraying zinc and aluminum during this initial period. Major developments in thermal spraying technology were initiated by companies related to aerospace and steel production in 1980s. Korean Air started flame spraying for commercial airplane repair. Samsung Aerospace Company (now Sermatech Korea) started original equipment manufacturer (OEM) coating for military aircraft-related repair and maintenance. Union Carbide Coating Service (now Praxair Surface Technologies) and Daeshin Metallizing Company initiated industrial applications of thermal spraying such as hearth roll coatings. During a growth period in the 1990s, numerous new thermal spraying companies were established related to steel industries, textile industries, paper industries, and power generation plants. The general thermal spray

market increase has been steady with the exception of aerospace applications because of limited activities in this field.

Figures 19.36 and 19.37, after Lee (2009a, b). show the distribution in 2008 of the thermal spray markets in S. Korea in terms of industrial sectors, for an estimated market of US $0.1 billion, and TS technologies used respectively. As noted, chemical and power plants clearly dominate the market share. Semiconductor and shipbuilding industries also have significant share of the pie. In terms of revenue, semiconductor/LCD, steel making, and power plant industries are the dominant sectors, based on the current fiscal reports.

19.5.6 India

In India the perceptible change in the thermal spray industry occurred in the 1990s and a good indicator of its success is the growth in sale of spray-grade powders [Sundararajan G, et al. (2009a, b)] that has multiplied by a factor of 3 between 2000 and 2008. The number of systems, as shown in

Fig. 19.38, is still low for a country of India's size but it is an important change compared to that prevailing not so long ago. The distribution of the country's thermal spray system installations among the various segments is presented in Fig. 19.39. The data clearly reveal that the highest number of thermal spray units operate as "job-shops." They play a crucial role both in addressing the needs of a vast majority of industrial users and running a crusade to demonstrate "new" applications to the industry and facilitating further growth [Sundararajan G, et al. (2009a, b)].

19.6 Techno-Economic Analysis

19.6.1 Different Cost Contribution Factors

In spite of its fundamental importance toward the acceptance of a new technology, few papers have been devoted to the economic analysis of thermal spray processes [American Welding Society (1985), Feuerstein A, et al. (2008), Ducos M (2006a, b), de Botton O (1988), Ducos M, Durand JP

(2001), MAGETEX, Lee C (2009a, b), Sundararajan G, et al. (2009a, b), Tani K, Nakahira H (1992), Nakahira A (2009a, b), Fukumoto M (2008a, b), de Munter AJ, et al. (2002), Celotto S, et al. (2007), Molz R, Hawley D (2007)]. This is due to the generally confidential nature of the relevant information in an industrially competitive environment. Moreover, in our rapidly evolving global economy, process economics changes rapidly with time and location. In the context of this book, it is more important to focus on the methodology of carrying out a detailed economic analysis, rather than on the absolute values. These can and will change with time. Emphasis is placed on understanding of the relative importance of the different cost factors in the overall process economics and means of controlling them. Typical examples are provided at the end of this section for the illustration of the methodology and a comparative analysis of the different thermal spray technologies. For a more precise economic analysis, it is best to work with your local provider of thermal spray equipment and materials supplier.

The first step in an overall cost analysis is to identify the different steps of the operation, which can involve a single or multiple process technologies. A typical coating process often involves all or most of the following process steps:

(a) Substrate preparation
- Machining
- Masking
- Degreasing
- Grit blasting
- Ultrasonic cleaning
(b) Coating
- Part setting-up
- Preheating
- Coating
- Online posttreatment (whenever necessary)
- Part removal

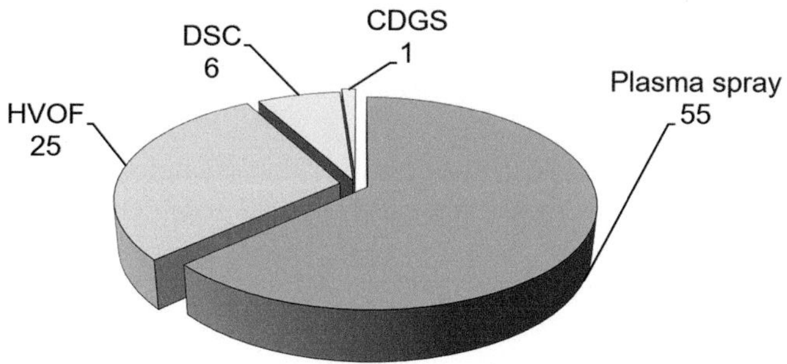

Fig. 19.38 Thermal spray installations in India—distribution in terms of spraying technology used, for a total number of 87 installations (excludes flame, arc, and wire spray systems). ([Sundararajan G, et al. (2009a, b)]. Reprinted with kind permission from ASM International)

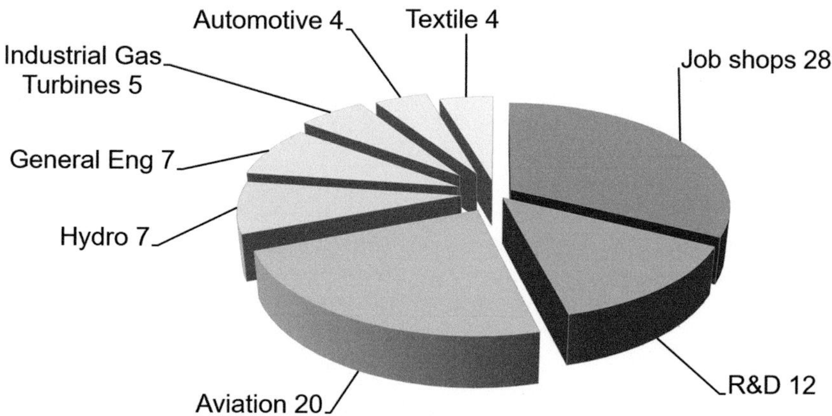

Fig. 19.39 Number of thermal spray installations in India used by different industrial sectors, for a total number of 87 installations (excludes flame, arc, and wire spray systems). ([Sundararajan G, et al. (2009a, b)]. Reprinted with kind permission from ASM International)

(c) Post-coating operations
  - Off-line posttreatment, annealing, austempering, laser glazing, hot isostatic pressing (whenever necessary)
  - Surface machining, grinding, and polishing
  - De-masking
  - Cleaning
  - Quality control
  - Packing and labeling

The overall cost for the coating of a given part constitutes essentially of two broad categories of cost contributing factors:

Direct costs (S_{di}), which includes all cost items directly related to each of the different process steps. These depend directly on the complexity of the technology used, the number, size, shape, and complexity of the parts to be coated, and the nature of the coating material. Direct costs are usually calculated on a per hour basis of the operation. Their conversion to cost per part depends on the number of parts that could be produced in 1 h. The principal factors that contribute to the direct cost of a coating operation are:
  - The cost of material to be sprayed (powder, wires, rods, etc.)
  - The cost of gases, electricity, and consumables supplied for the operation
  - Direct labor cost for each of the different stages of the operation
  - Quality control, packing, and labeling

Indirect costs (S_{in}), often referred to as *fixed cost* or *overhead costs*, include all other expenses, which are related to the investment needed and general running cost of the coating site. The indirect cost factors are usually calculated on a yearly basis of the operation. Their conversion to a per hour basis depends on the number of hours of operation of the coating center per year (N_{hy}). They include, though not limited to, the following:
  - Equipment acquisition and amortization of the capital investment
  - Building, ancillary equipment, and necessary infrastructure
  - Maintenance workshop
  - Quality control laboratory
  - General administration and services
  - Marketing and financial expenses

The total cost of the coating of a single part, S_T, is the sum of all direct, S_{di}, and indirect, S_{in}, cost factors, on a per hour basis, multiplied by the number of hours needed per part (N_{hp}).

$$S_T = \left[S_{di} + \frac{S_{in}}{N_{hy}} \right] \times N_{hp} \qquad (19.2)$$

Where;

S_T Total cost per part (\$/part)
S_{di} Direct cost per hour (\$/h)
S_{in} Indirect cost per year (\$/y)
N_{hy} Operating hours per year (h/y)
N_{hp} Number of hours per part (h/part)

19.6.2 Direct Cost Factors

19.6.2.1 Cost of Materials

This is often the most important single cost item of the coating operation. It varies widely depending on the nature of the coating material, its purity, and its form. As presented in Chap. 13, Powders, Wires, and Cords, coating material are available as powders, wires, cords, or rods depending on the coating technology used. Powders are by far the most common form of materials used in coating operations with particle size ranging from 10 µm up to a few hundred microns in diameter or more. Generally, the cost of a powder increases rapidly with the following parameters:

  - *The purity level of the material*: For example, a 5N Si powder would cost almost one order of magnitude higher than a 3N Si.
  - *Oxygen level*: For example, a "grade 1" titanium powder with an oxygen level lower than 1200 ppm will cost considerably more than a "grade 2" powder of the same particle size but with an oxygen level in the 2500 ppm or "grade 4" with an oxygen level of 4000 ppm.
  - *Particle size distribution*: The narrower the particle size distribution of the powder, the higher will be the cost of the powder on a \$/kg basis.
  - *Powder morphology*: Generally, spherical, free-flowing, dense powders will have a higher cost than a standard spray dried or fused and crushed powder.
  - *Specialty alloys*: When custom made, with tightly controlled specification of its elemental analysis, the powder will cost considerably more than standard materials.

It is important to underline that these parameters are also interactive between them. For example, a titanium powder with a particle size ≤45 µm and an oxygen content of less than 1300 ppm (grade 1) will be more difficult to obtain and would be significantly more expensive than a titanium powder with the same oxygen level (grade 1), but a coarser particle size range (45–106 µm). Delivery time and volume of annual consumption can also have a significant influence on the cost of powders. The same reasoning equally applies for other materials in the form of cords, wires, or rods.

The key parameters that influence the quantity of material needed to spray a part are:

- *The surface to be coated*, A_c, taking into account its shape complexity, its holes distribution that must be protected, and the sprayed material lost at edges.
- *The average thickness of the coating*, δ_c (including the mean thickness that has to be removed from the as-sprayed part to achieve the final specified dimensions, sprayed coatings' dimensional precision being in the range of tenths of millimeters).

These two quantities, together with the feedstock specific mass, ρ_p (kg/m^3), and the deposition efficiency, η_c (%), which depends strongly on the spray process and the geometry of the part to be coated, allow calculating the powder quantity necessary for each part, m_{part} (kg):

$$m_{part} = \frac{\delta_c\, A_c\, \rho_p}{\eta_c} \qquad (19.3)$$

It is important to stress the very important impact of the deposition efficiency, η_c (%), on the overall cost of materials in the coating process. This is because overspray powders, which do not end up in the coating, are collected from the exhaust gases or ambient air strictly for environmental reasons and are never recycled in the coating operation because of the risk of introducing major contaminants or defects in the coating. Moreover, scrap/over-sprayed powders, while they represent a loss of value based on their procurement cost, would also often require additional costs, as a waste material, for their discharge in an environmentally safe way.

Automation is one way of reducing powder waste, and accordingly reduce powder consumption in the spray process, through close control of the spray process. For example, if a multi-axis robot handles the arc spray gun, the operation of powder feeder can be synchronized with the robot, thus stopping the powder feeding during substrate transfer or repositioning, to reduce feedstock losses and coating costs. It is be noted, however, that stopping the powder feeding does not imply stopping the powder carrier gas flow since, in the absence of powder carrier gas, any powder in transient in the powder feeding line would sediment in the transport line and could end blocking it when the powder feeding is resumed.

19.6.2.2 Cost of Gases, Electricity, and Consumables

The cost of gases can represent an important component of the operating cost of the coating operation and often the next cost item in a spray operation after the cost of the spray material. This depends significantly on the nature of the gases used, their purity level, and overall volume flow rates required for the coating operation. The unit cost of gases ($/m^3) also depends on the overall size of the operation and the associated gas consumption. A compilation of typical cost of gases used in thermal spray and cold spray coating operations in North America and Europe is given in Table 19.5. These values are based on 2020 prices and should be adjusted for inflation and price changes in subsequent years.

It is important to note that the decision on which size of gas packaging to use is essentially based on the volume of the operation. For example, the cost for industrial purity argon gas in North America can be more than three times higher when in the form of compressed gas cylinders compared to cryogenic form in industrial-scale Dewar's. The use of cryogenic gases, on the other hand, while generally less expensive per m^3 of gas used, can be completely offset by the constant loss factor that has to be bled to the atmosphere from cryogenic containers for any prolonged period of time in which the gas is not in use. Compressed gas cylinders, or bulk packs, required manual handling and rental fee for the cylinders, which in large-volume operation can be important especially if the operation requires a wide range of different specialty gases. Generally, the cost of the gas varies with its consumption and on-site storage linked to the corresponding delivery mode [Celotto S, et al. (2007)] (from cylinder, vehicles delivering liquid gases with different storage volumes and pressures, trailer for He, bulk vehicle).

The total cost of the gas component in the spray operation depends obviously on the nature of the spray operation, and the operating conditions. For example, with cold spray process, which is one of the highest gas consumer spray operations, the gas consumption rate, as illustrated in Fig. 19.40, depends on the process variables, the type of gas, operating pressure and temperature, and internal diameter of the spray nozzle; see Table 19.6 for system configurations associated with Fig. 19.40 [Celotto S, et al. (2007)]. The use of helium in cold spray requires flow rates that are 4.2 times higher than those for N_2, due to the large

Table 19.5 Range of cost of spray gases and electric energy in North America and Europe

Gas	Cost range
Argon (Ar)	0.8–2.4 US$/m^3
Nitrogen (N$_2$)	0.2–1.5 US$/m^3
Oxygen (O$_2$)	0.2–1.3 US$/m^3
Hydrogen (H$_2$)	6.0–14 US$/m^3
Helium (He)	15–80 US$/m^3
Natural gas (CH$_4$)	0.2–1.5 US$/m^3
Propane (C$_3$H$_6$)	0.2–2.0 US$/m^3
Electric energy	0.09–0.20 US$/kWh

Table 19.6 System configurations associated with Fig. 19.40

System configuration	1	2	3	4	5	7	7	8
Nozzle i.d. (mm)	2.0	2.0	2.0	2.7	2.0	2.7	2.0	2.0
Gas used	N_2	N_2	N_2	N_2	He	He	He	He
Gas pressure (MPa)	2.0	3.0	3.0	3.0	3.0	2.0	2.0	2.0
Gas temperature (°C)	25	25	350	350	25	25	25	350

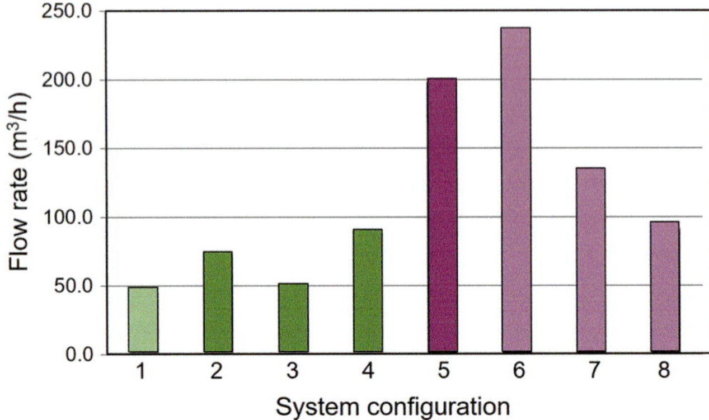

Fig. 19.40 Gas consumption rates for different cold spray system configurations. [Celotto S, et al. (2007)]

difference in sonic velocity. Increasing the gas temperature, on the other hand, allows for the decrease in the gas consumption.

Gas recycling and on-site gas generation are two of the common means of reducing the cost of gases in the coating operation. Gas recycling is generally limited to operation in a closed environment such as in vacuum plasma spraying (VPS), whether DC or RF induction plasma. More recently, gas recycling has been increasingly introduced in cold spray operations, which requires often the use of helium at relatively large volume flow rates with a significant financial burden on the operation. Gas recycling in this case has to be associated with a gas cleaning and recompression operation. On-site gas generation on the other hand is only used in large volume operation, which justifies the capital investment whether for the generation of oxygen or nitrogen on site through cryogenic distillation or hydrogen through water electrolysis.

The cost of energy, whether in the form of fuels in combustion spraying, or electrical power, in plasma spray operations, varies considerably with type of technology used and the scale of operation. A comparison of the power requirement by different spray technologies is given in Fig. 19.41. This shows that HVOF and high-velocity liquid fuel (HVLF) technologies are among the largest energy consumers in which the supplied fuel is in the form of liquid or gas. Comparative data on the energy efficiency and the associated specific energy requirement (SER) of different

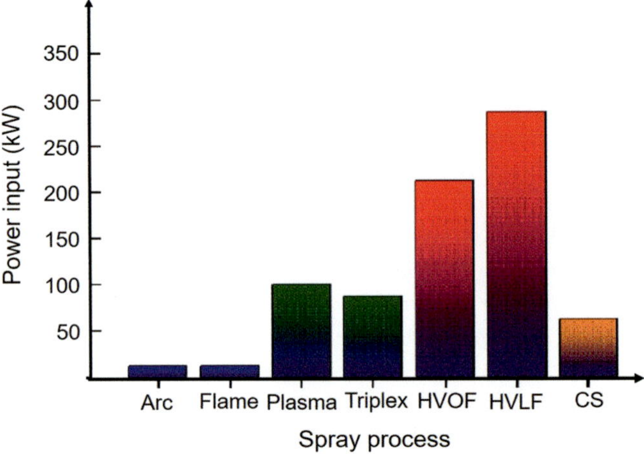

Fig. 19.41 Total power consumption by different spray processes. ([Molz R, Hawley D (2007)]. Reprinted with kind permission from ASM International)

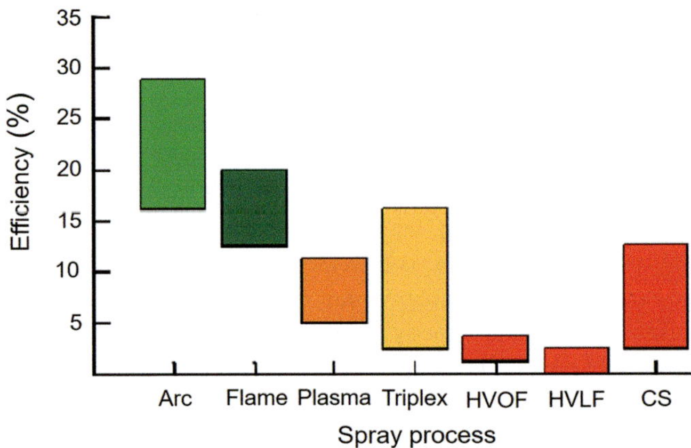

Fig. 19.42 Spray process efficiencies for different spray processes. ([Molz R, Hawley D (2007)]. Reprinted with kind permission from ASM International)

spray processes are provided in Figs. 19.42 and 19.43, respectively.

Molz and Hawley (2007) pointed out that the comparison of the effective performance of alternate thermal spray processes for a given application is not straightforward and often hampered by detailed particularities of each process. They consequently proposed a more generic and global method, based on deriving a unified process efficiency formula that takes into account all energy inputs and energy outputs of a process in the same energy units. They made a direct comparison of process efficiency for each process and specific coating conditions. For example, the energy input to the "system" could either be already included in the supply to the thermal spray gun or created (as in combustion guns) within the gun itself. Two sources exist, one for the gas itself

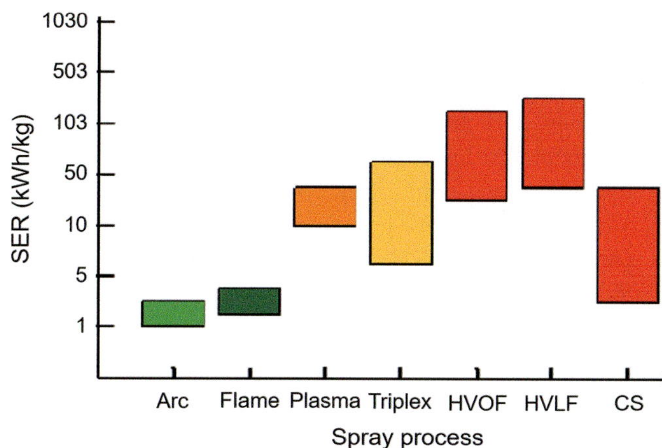

Fig. 19.43 Specific energy requirement (SER) in kWh/kg of coating deposited for different spray processes. ([Molz R, Hawley D (2007)]. Reprinted with kind permission from ASM International)

being used (in raw form prior to any reaction) and one for the material input into the system. This is illustrated in Fig. 19.41, after Molz and Hawley D (2007), representing the total energy required (input) for a few thermal spray processes. As expected, flame and wire arc spraying exhibit the lowest energy consumptions.

Figure 19.42, after Molz and Hawley (2007), represents the process efficiency for each thermal spray process based on the measured power output in terms of particle temperatures and velocities. The drawback of these calculations is that they are based on the particle temperature measured assuming that it is representative of the total thermal energy state of the sprayed powder. As noted in Fig. 19.42, wire arc spraying, while having the highest energy efficiency of all TS processes, barely manages to achieve 30% efficiency. The lowest efficiency ($\approx$2% range) is found to be with the HVOF and HVLF processes.

Figure 19.43, after Molz and Hawley (2007), represents the specific energy requirement (SER) defined as the energy consumed per kg of coating deposited. It is to be noted that the scale in this case is logarithmic in order to fit the entire range of values. The typical high-velocity coatings of carbides, with lower deposit efficiency and higher energy input, require considerably more energy to deposit per unit mass, almost one order of magnitude higher than that to spray pure ceramics or metals.

Electric energy-based technologies, such as plasma spraying and wire arc spraying, use typically industrial electricity rates, which are composed of two components:

- *Power (kW)*, which is set by an annual agreement with the electric energy utilities reflecting the overall size of the operation and its peak energy demand.

- *Energy (kWh)*, which is the electrical energy consumption by the operation for running the plasma system as well as all of the ancillary equipment.

Electricity cost for building illumination, heating, and other operations is not included under direct cost of the operation. These constitute part of the indirect costs. Electricity rates vary widely between locations and period of operation over the range 0.09–0.20 US$/kWh or more.

The cost of consumables and spare parts also varies widely with the nature of the spray process and the complexity of the part. For example, in DC plasma-spraying operation, whether atmospheric or under vacuum, a close monitoring of the state of the electrodes is essential for ensuring a consistent quality of the coating. Generally, the allowed life of the electrodes is based on the recommendations of the equipment manufacturer and/or the operator experience. It is generally accepted practice to replace consumable parts in a setup after a well-defined lifetime, rather than wait for a disruption of the operation and a failure of the quality of the coating. It has to be recognized that the cost of a failed coating is not only limited to the loss of the coating martial and associated labor and expenses, but also to the damage caused to the part being coated, which can be order of magnitude higher than that of the coating operation itself. The cost of consumables is not limited to torch components; it can also include certain important components of the plasma generating equipment. In induction plasma spraying, for example, the RF power supply has a triode tube that needs to be replaced depending on its projected lifetime, which can be in the 10,000–20,000 h. The triode life also depends on the mode of operation and the number of starts and stops per hour of unit. In a 100-kW installation, the cost of such a component can be in the US $10,000–15,000 range.

The cost of consumables also includes the cost of masking tapes used to protect the surface, which should not be exposed to sand blasting or coating, the cost of the grit used in the sand blasting operation prior to coating, and cost of grinding and polishing materials and supplies used for surface finishing of the part.

19.6.2.3 Direct Labor Cost

In thermal spray operations, direct labor cost is inversely proportional to the degree of automation of the operation. The higher is the level of automation and the larger is the scale of the operation, the lower will be the direct cost of labor per part. In a typical coating operation, direct labor is needed for:

- Preparation of the part to be coated, including its masking, sandblasting, and cleaning

- Setting-up of the part in the coating booth: starting of the operation, preheating, and coating, followed by removal of the part from the rig
- Cleaning of the part, surface machining, grinding, and polishing
- De-masking and cleaning
- Quality control
- Packing and labeling

The total cost of labor for the coating operation is also relatively sensitive to the overall management of the facility and the efficient definition of the operation steps and the definition of responsibilities.

19.6.2.4 Direct Cost for Quality Control, Packing, and Labeling

This depends on the maturity of the technology used for the coating operation and on the nature of the parts to be coated. The most stringent quality control rules are generally applied for medical and aerospace applications, where very rigorous quality control (QC) procedures are needed and have to be strictly enforced. In other applications, with generally lower-value items, such as in automobile industry, QC on individual parts would be too expensive. The industry has to rely in this case on well-developed, mature, and reliable technologies.

The overall cost of quality control includes direct and indirect cost factors. Basic investment in QC in instrumentation and laboratory equipment will go under *indirect cost*, while the labor cost associated with the QC procedures will go under *direct cost* and will depend on whether it is applied on all of the coated parts or on randomly selected parts. For obvious reasons, QC procedures tend to use nondestructive testing as much as possible in order to avoid sacrificing the part being tested. In the development stage of a new part, both nondestructive and destructive testing procedures are generally needed.

19.6.3 Indirect or Fixed Cost Factors

19.6.3.1 Capital Investments

This covers the total investment needed for the setting-up of the operation. It includes:

- *Cost of land and building*, including building permits and environmental studies necessary for the setting-up of the operation
- *Ancillary equipment and infrastructure*, including spray booth, dust collection and disposal, air cleaners, sound control, and operator safety equipment

- *Production equipment*, including spray system and associated operations, instrumentation, and data acquisition
- *Equipment for posttreatment and finishing*, including heat treatment furnaces, laser glazing, hot isostatic pressing, machining, and grinding and polishing equipment
- *Maintenance workshop*
- *Quality Control laboratory*
- *Storage space* for materials, supplies, spare parts, and finished part
- *Office space*

Direct investment for the acquisition of the coating production equipment rarely exceeds 15–20% of the total investment needed for the setting-up of a new production facility. The investment structure needed for the expansion of an existing facility can be different with a considerably higher percentage needed for the acquisition of the production equipment.

Capital investment contributes to indirect production cost through the amortization of the investment made. It is important to note that the amortization rules vary from country to country as well as between the different types of investments. For example, investments for the building and infrastructure are generally depreciated over a 25-year period, production equipment, over a 5–7-year period, while computers, automation, and information technology (IT) equipment may need to be amortized over a 3–5-year period due to the rapid evolution of the technology. The simplest method of calculating amortization cost per hour of operation, S_{ih}, is based on a linear amortization over the projected number of years considered as useful life of the equipment, N_y, divided by the number of operating hours per year, N_{hy}, as follows:

$$S_{ih} = \frac{P_e}{N_y \times N_{hy}} \tag{19.4}$$

where P_e is the procurement cost of the equipment. For more information on investment cost calculation, the interested reader can go through the paper by de Munter et al. (2002). Typical investment cost estimate for coating equipment and ancillary equipment are given in Tables 19.7 and 19.8, respectively, based on a 2006 study by MAGETEX and presented by Ducos (2006a, b).

19.6.3.2 Other Indirect or Fixed Costs

These include essentially the cost of general administration and services, insurances, marketing, and financial services. These are strongly dependent on the scale of the operation and its location.

Table 19.7 Investment costs of equipment for spray processes in 2006 [Ducos M (2006a, b)]

	Value in Euro (€)	Value in US$
Manual powder or wire flame	About 5000	About 6500
Automated powder or wire flame	5000–10,000	6500–13,000
HVOF–HVAF	50,000–100,000	65,000–130,000
Wire arc spray	9000–22,500	11,700–29,250
APS	75,000–185,000	97,500–240,500
VPS/CAPS	600,000 to >2 M	780,000 to >2.6 M
PTA	50,000–75,000	65,000–97,500

Reprinted with kind permission from Dr. M. Ducos

APS air plasma spraying, *CAPS* controlled atmosphere plasma spraying, *HVAF* high-velocity air fuel, *HVOF* high-velocity oxy fuel, *PTA* plasma-transferred arc, *VPS* vacuum plasma spraying

Table 19.8 Investment costs of ancillary equipment in 2006 [Ducos M (2006a, b)]

	Value in Euro (€)	Value in US$
Grit blasting	10,000–40,000	13,000–52,000
Linear movements	35,000–75,000	45,500–97,5000
Robot	75,000–150,000	97,500–195,000
Spray booth	12,000–25,000	15,600–32,500
Ventilation and filtering	30,000–45,000	39,000–58,500
Machining and grinding	200,000–1 M	260,000–1.3 M
Quality control	75,000–200,000	97,500–260,000

Reprinted with kind permission from Dr. M. Ducos

19.6.4 Few Examples

19.6.4.1 Cost of DC Atmospheric Plasma Spraying of YPSZ

To illustrate the relative importance of each of these cost factors, de Botton (1988), in his Master of Science Study, calculated the costs for DC atmospheric plasma spraying (APS) of yttria partially stabilized zirconia (YPSZ) coatings on a flat 1 m^2 surface. The coating thickness was 300 μm. The spray rate was 3 kg/h, with a deposition efficiency of 60% and a coating relative density of 0.9. Losses at holes and edges were assumed to be of 10%.

For the cost calculations, he assumed:

- 5 years to recover capital investment (cost of 10% per year)
- Maintenance cost of 4% of capital investment
- 250 production days, 2 shifts per day, for a total of 4000 h/year production time

Results are summarized in Figs. 19.44 and 19.45, which represent the different factors of prices per sprayed part. It is obvious from this figure that the most important one is the powder, followed by labor (automation can reduce it notably) and capital cost. Figure 19.45 reveals that after the coating cost (81.9%), the grinding process is the most important

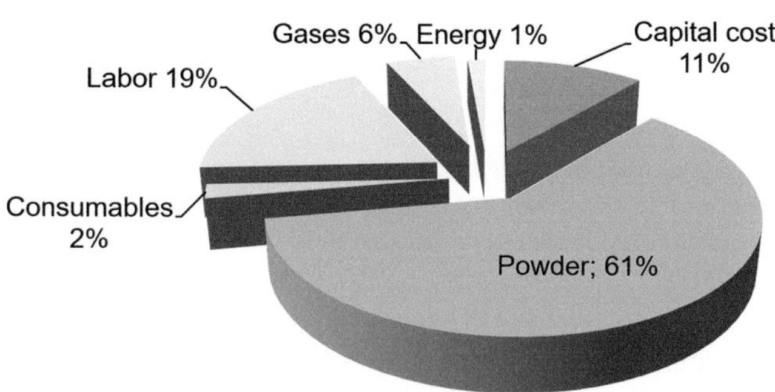

Fig. 19.44 Spray costs distribution of YPSZ coatings. ([de Botton O (1988)]. Reprinted with kind permission from MIT)

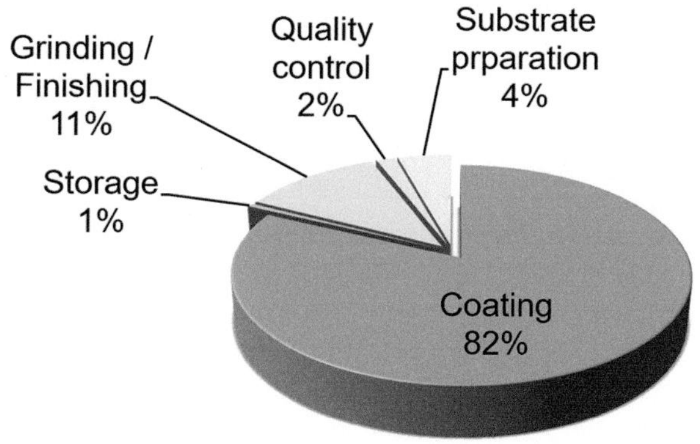

Fig. 19.45 Total costs distribution of YPSZ coatings. ([de Botton O (1988)]. Reprinted with kind permission from MIT)

(11.5%) followed by the surface preparation (4.4%). The quality control and storage represent only 2.1%.

19.6.4.2 MCrAlY Coatings Sprayed Using HPPS and Wire Arc Spray

In a study by Sacriste et al. (2001), the bond coat of TBCs was deposited on a nickel-based super alloy, with a surface area of 100×50 mm^2 and a thickness of 2 mm. As the thickness of the topcoat was limited to 200 μm for this application, a relatively low roughness of the bond coat was

Table 19.9 Cost analysis[a] for the production of NiCrAlY coating using Plazjet-7070 and twin-wire arc spray (Arc Jet-9000) processes operating using nitrogen or air as atomizing gas [Sacriste D, et al. (2001)]

Parameter	Plazjet-7070	Arc Jet-9000	
		Nitrogen	Air
System cost (US$)	198,000	23,800	23,800
Amortization time (years)	5	5	5
Material feed rate (kg/h)	12	12	12
Spraying time per year (h)	278	278	278
Amortization (US$/kg of coating)	19.83	2.38	2.38
NiCrAlY powder and wire (US$/kg)	90	63	63
Powder cost (US$/kg of coating)	1.26	0.11	0.11
Electric power (kW)	115	10	10
Electric energy cost (US$/100 kWh)	7.90	7.90	7.90
Energy cost (US$/kg of coating)	1.26	0.11	0.11
Plasma and atomization gas flow rates (m^3/h)	9.6	105	65
Gas cost (US$/100 m^3)	3.40	3.40	0.50
Gas consumption (m^3/year)	2670	29,200	18,000
Electrodes and tips costs (US$)	635	6	6
Electrodes and tips lifetime (h)	12	6	6
Consumable cost (US$/kg of coating)	7.39	0.63	0.18
Production cost (US$/kg of coating)	162.10	154.80	107.30
Labor cost (US$/kg of coating)	8.45	8.45	8.45
Production cost without amortization (US$/kg)	158.65	105.74	105.29
Total production cost (US$/kg of coating)	178.48	108.12	107.67

Reprinted with kind permission from Springer Science Business Media, copyright © ASM International

[a]The production cost was calculated per kilogram of coating at a deposition rate of 12 kg/h and a deposition efficiency of 60%. The total production was set to two tons per year. Coatings were sprayed without substrate cooling

needed (Ra <10 μm). Two different TS technologies were used for the deposition of the bond coat:

- The high-power plasma spray (HPPS) TAFA, PlazJet-7070 equipped with a 120-mm long anode and working with a mixture of nitrogen and hydrogen, the powder being injected at feed rates up to 12 kg/h.
- The TAFA twin-wire arc spray system used in this study was Arc Jet-9000 model. Wires were fed with the push–pull system capable to work with wires and cored wires 7.50 m long. The NiCrAlY material was available as cored wire with a NiCr envelope containing particles of Al, Cr, and Y. Air and nitrogen were used as atomization gases with different air caps.

Coatings obtained by arc spraying exhibited a higher surface roughness due to a relatively low particle velocity compared to that obtained with the Plazjet-7070 process. The deposition efficiency was the same as that obtained with the Plazjet-7070 gun, for both air and nitrogen atomizing gas. The use of nitrogen atomization resulted in a 10 wt.% reduction in the oxygen content, but no changes were observed in the bond strength (higher than 40 MPa). The composition of

the arc-sprayed NiCrAlY coatings was very heterogeneous, which was attributed to the wire manufacturing method.

A summary of the cost analysis of the optimized results using the three coating conditions (Plazjet-7070, and Arc Jet-9000 operating with N_2 and Air) is presented in Table 19.9, after Sacriste et al. (2001).

Table 19.9 shows that the Arc Jet-9000 process, using air or nitrogen atomization, is about 40% less expensive than the Plazjet-7070 process. As with the majority of spraying systems, the material feedstock is the most expensive parameter for both systems. In this study, it represents about 84% of the total coating price for Plazjet-7070 and 97% for Arc Jet-9000. Because of the high deposition rates, labor cost was relatively low with both systems, representing less than 5% for Plazjet-7070 and of the order of 7% for Arc Jet-9000. One of the major differences between the two systems is in the capital investment, which, for the Plazjet-7070 system, was more than eight times higher than the Arc Jet-9000, representing 11% of total production cost for the Plazjet-7070 compared to about 2% for the Arc Jet-9000 system. The difference in the electric power shows that Arc Jet-9000 is more thermally efficient than plasma spraying, as it requires a power of only 10 kW to spray 12 kg of NiCrAlY

Table 19.10 Different costs for manually wire arc-spraying zinc on 1 m² of steel [Ducos M (2006a, b)]

Parameter	Cost in €/m²
Surface preparation	7.3
Spraying (air atomization)	6.9
Amortizing	1.0
Total	15.2 €/m²

Reprinted with kind permission from Dr. M. Ducos

material per hour, in sharp contrast to 110 kW needed for the Plazjet-7070 system.

19.6.4.3 Manual Wire Flame Zn Coating

According to Ducos (2006a, b) and calculations of MAGETEX, the different costs for manually wire arc-spraying zinc on steel substrate are summarized in Table 19.10. These were based on a coating thickness of 120 µm, labor cost at 40 €/h., and system amortization over 3 years of an operation of 1200 h/y. The cost of zinc wire consumption was estimated at 2 €/h. With this spray process, which is one of the cheapest of surface modification technologies, the surface preparation becomes slightly more important than the spray cost. Because of the low investment cost, amortizing is also rather low as can be noted in Table 19.10.

19.6.4.4 Cost Analysis for NiAl Coatings Using APS Versus WAS

Ducos (2006a, b) has evaluated the cost of a NiAl coating, 1.8 mm thick, for aeronautic application, using atmospheric plasma spraying (APS) in comparison with wire arc spraying (WAS). Coatings are performed with robotized setups according to conditions and costs summarized in Table 19.11.

Figure 19.46 presents the cost distribution of the different items of the spray process for APS. It can be seen that the most important cost factor is the powder (51%), followed by the investment amortization (22%) and labor (18%). Gases, electricity, and consumable parts represent only 9%. The distribution of other costs (surface preparation, grit blasting and cleaning, finishing, and quality control) is presented in Fig. 19.47. It can represent 50–80% of the spray costs. Figure 19.48 represents the cost distribution per sprayed hour for the APS and WA coatings. The WA coating cost per hour is about 60% that of the APS coating, which is mainly due to the investment and labor costs.

19.6.4.5 Cost Comparison for Hard Chromium Replacement

Based on same principles, Ducos and Durand (2001) calculated the costs per m² of Al₂O₃–TiO₂ (13 wt.%) APS sprayed, WC–Co HVOF sprayed, and NiCrBSi HVOF sprayed to

Table 19.11 Spray conditions and costs of APS and WAS deposition of NiAl (5 wt.% Al) 1.8 mm thick coatings for aeronautic application [Ducos M (2006a, b)]

Parameter	APS	WAS
Deposition rate (kg/h)	2.5	11.2
Deposition efficiency (%)	60	70
Material costs (€/kg)	35 (powder)	45 (wire)
Spray time (h)	2.93	0.56
Material used (kg)	7.33	6.29
Consumable parts (€)	13.49	0.45
Gases and electricity (€)	30.26	1.4

Reprinted with kind permission from Dr. M. Ducos

APS air plasma spraying, *WAS* wire arc spraying

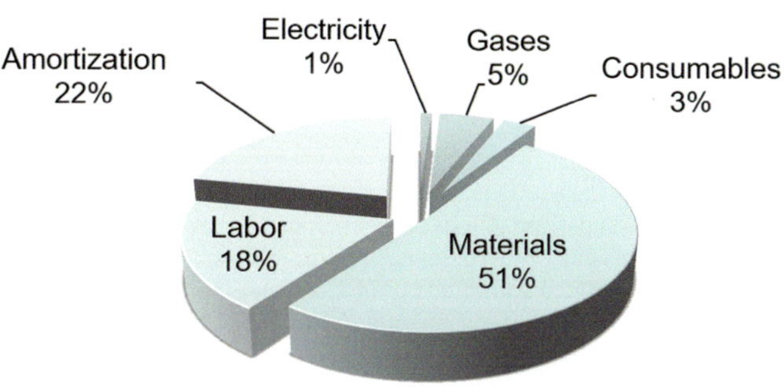

Fig. 19.46 Distribution of direct spray costs for a 1.8 mm thick NiAl (5 wt.% Al) coating plasma sprayed in air. ([Ducos M (2006a, b)]. Reprinted with kind permission from Dr. M. Ducos)

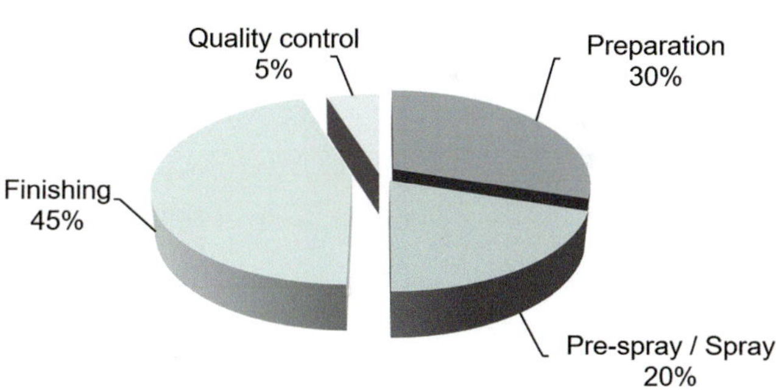

Fig. 19.47 Distribution of pre- and post-spray costs for a 1.8 mm thick NiAl (5 wt.%) coating plasma sprayed in air. ([Ducos M (2006a, b)]. Reprinted with kind permission from Dr. M. Ducos)

replace electrolytic hard chromium. Results are presented in Fig. 19.49, which shows that the cheapest coating is that of NiCrBSi HVOF sprayed at a cost 21.6 € higher than that of hard chromium (18 €). APS ceramic coating comes next to which the cost of the bond coat must be added (46.6 + 20.0 = 66.6 €), followed by the HVOF-sprayed WC–Co coating (122.5 €), which is at a considerably higher cost due to the cost of the powder.

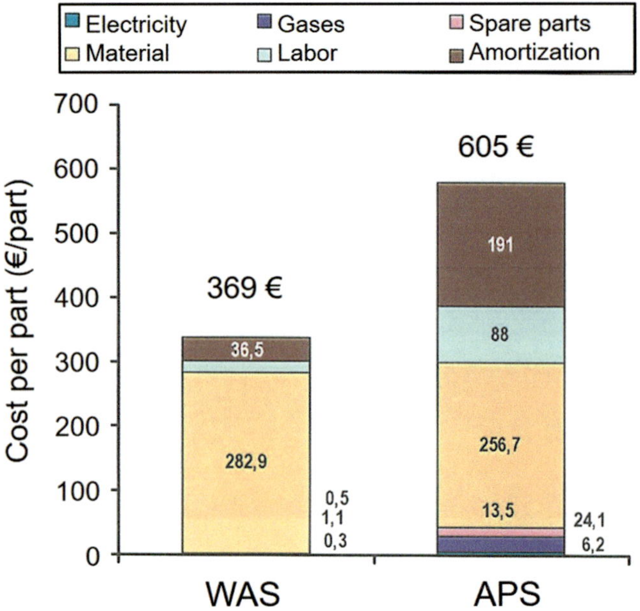

Fig. 19.48 Comparison of the cost per part of 1.8 mm thick NiAl (5 wt. %) coatings plasma sprayed in air or wire arc sprayed. ([Ducos M (2006a, b)]. Reprinted with kind permission from Dr. M. Ducos)

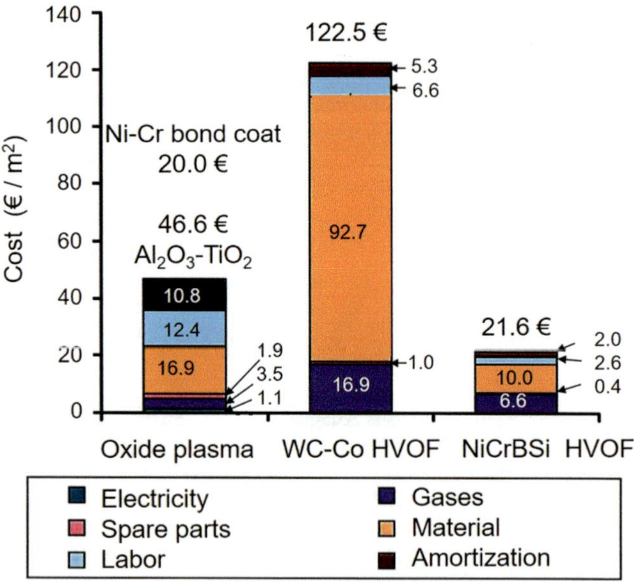

Fig. 19.49 Comparison of the costs per m² of Al₂O₃–TiO₂ (13 wt.%) APS sprayed, WC–Co HVOF sprayed, and NiCrBSi HVOF sprayed to replace hard chromium. ([Ducos M, Durand JP (2001)]. Reprinted with kind permission from ASM International)

19.7 Summary and Conclusions

In this chapter, a brief comparative analysis of different thermal spray processes is presented, grouping them in into five subgroups: cold spray, combustion-based thermal spray, plasma spray, wire arc spray, and plasma-transferred arc deposition. In this section, emphasis is placed on the unique features of each of these technologies, its advantages, and limitation. As demonstrated in the different parts of this book, these technologies are essentially complimentary rather than competing for the same application. The choice of a specific coating and/or thermal spray process, for a given service condition, depends on the service needs, expectation of the end user, and the cost that could be tolerated for the application.

This is followed by a review of different TS coating applications described in terms of surface modification for the enhancement of its resistant to wear, corrosion, and oxidation; providing thermal protection; clearance control; good bonding; electrical and electronic properties; and the deposition of freestanding spray-formed parts. The coating design process is often complicated by the fact that in practice components are not always devoted to a single requirement such as wear or corrosion or electrical insulation or thermal insulation. In most cases, coatings must resist to different combined needs: for example, wear is often linked to corrosion.

The subsequent section provides a review of present and potential industrial applications of the technology in a wide range of industrial sectors including aerospace, land-based turbines and power generation, automotive, land-based and marine applications, application in the electrical and electronic industries, medical applications, ceramics and glass industries, and chemical process industries. This is followed by a brief overview of evolution of this industry on an international scale by country.

In the last section of this chapter, techno-economic analysis is presented discussing different cost contributing factors with typical examples of cost estimation for a few applications. Due to the dependence of the economic parameters on the regional infrastructure and local cost factors, emphasis is placed more on the methodology and the identification of the principal cost factors affecting the process economics rather than their absolute value that can change rapidly with time depending on the local economic context. On this point, it is important to stress that the evaluation of the cost of a coating has to be placed in contrast to cost savings that it could generate either by extending the lifetime of a part, refurbishing a worn part avoiding the extra cost of its replacement, or sometimes simply by reducing the "downtime" of a rather expensive installation simply for repair or the replacement of a worn part.

The following tables prepared by the International Thermal Spray Association [http://www.thermalspray.org] summarize the different TS technologies used in various industrial segments (Table A.1), coating application according to industry served (Table A.2), and materials sprayed in various industrial sectors (Table A.3).

Table A.1 Thermal spray technologies used by various industrial segments

Industry	Oxy fuel	Spray/fuse	HVOF	D-gun	Air plasma	Vacuum plasma	Shroud plasma
Aero gas turbine	✓		✓	✓	✓	✓	✓
Stationary gas turbine	✓		✓	✓	✓	✓	
Hydro-steam turbine	✓		✓	✓	✓	✓	✓
Automotive engines	✓		✓		✓		
Diesel engines	✓		✓		✓		
Transportation non-engine	✓			✓	✓		
Agriculture implementations	✓	✓			✓		
Railroad	✓		✓		✓		
Iron and steel manufacture	✓	✓			✓		
Steel rolling mills	✓	✓	✓		✓		
Iron and steel casting			✓		✓		
Forging	✓		✓				
Copper and brass mills	✓						
Ship and boat manufacture and repair	✓						
Oil and gas exploration	✓	✓	✓	✓	✓		
Mining, construction, and dredging	✓	✓	✓		✓		
Rock products	✓	✓	✓		✓		
Screening							
Cement and structural clay	✓	✓	✓				
Chemical processing	✓	✓	✓	✓	✓		
Rubber and plastic manufacture	✓	✓	✓		✓		
Textile	✓		✓	✓	✓		
Food processing	✓	✓	✓	✓	✓		
Electrical utilities			✓		✓		
Pulp and paper	✓		✓		✓		
Printing equipment			✓	✓	✓		
Defense and aerospace	✓		✓	✓	✓	✓	
Nuclear			✓	✓	✓		
Medical			✓	✓	✓	✓	✓
Business equipment			✓	✓	✓		
Electrical and electronics			✓	✓	✓	✓	
Architectural	✓	✓			✓		
Glass manufacture		✓		✓	✓		

Reprinted with kind permission from *SPRAYTIME*—a publication of the International Thermal Spray Association, a standing committee of the American Welding Society

Table A.2 Thermal spray applications in various industrial segments

Industry	Wear						TBC	Clearance control		Restoration	Corrosion/ Oxidation	Electrical	
	Abrasive	Adhesive	Fretting	Erosion	Cavitation	Impact		Abradable	Abrasive			Resistance	Conductivity
Aero gas turbine	✓	✓	✓	✓			✓	✓	✓	✓	✓		
Stationary gas turbine	✓	✓	✓	✓			✓	✓	✓	✓	✓		
Hydro-steam turbine	✓	✓	✓	✓	✓					✓	✓		
Automotive engines	✓	✓		✓		✓	✓	✓	✓	✓	✓	✓	
Diesel engines	✓	✓		✓		✓	✓			✓	✓		
Transportation non-engine	✓	✓					✓			✓	✓	✓	
Agriculture implementations	✓			✓		✓				✓			
Railroad	✓	✓				✓				✓	✓		✓
Iron and steel manufacture	✓			✓		✓				✓	✓		
Steel rolling mills	✓	✓				✓				✓	✓		
Iron and steel casting	✓			✓		✓				✓	✓		
Forging	✓	✓				✓				✓	✓		
Copper and brass mills	✓									✓	✓		
Ship and boat manufacture/ repair	✓			✓						✓	✓		
Oil and gas exploration	✓	✓		✓		✓				✓	✓		
Mining, construction, and dredging	✓			✓	✓	✓				✓	✓		
Rock products	✓					✓				✓	✓		
Screening	✓					✓				✓	✓		
Cement and structural clay	✓					✓				✓	✓		
Chemical processing	✓			✓						✓	✓		
Rubber and plastic manufacture	✓			✓		✓				✓	✓		
Textile	✓									✓			
Food processing	✓									✓			
Electrical utilities	✓	✓		✓	✓	✓				✓	✓		
Pulp and paper	✓				✓	✓				✓	✓		
Printing equipment	✓	✓								✓			
Defense and aerospace	✓	✓	✓	✓	✓	✓	✓			✓			
Nuclear											✓		
Medical	✓		✓								✓		
Business equipment	✓	✓	✓										
Electrical and electronics										✓	✓		
Architectural	✓					✓							
Glass manufacture	✓	✓								✓	✓		

Reprinted with kind permission from *SPRAYTIME*—a publication of the International Thermal Spray Association, a standing committee of the American Welding Society

Table A.3 Thermal spray-coating materials used in various industrial sectors

Industry	Chrome carbide	Self-fluxing	Iron and steel	Nickel alloys	Superalloys	MCrAIY	Cobalt alloys	Nonferrous
Aero gas turbine	✓		✓	✓	✓	✓	✓	✓
Stationary gas turbine	✓		✓	✓	✓	✓	✓	✓
Hydro-steam turbine	✓	✓	✓	✓	✓		✓	✓
Automotive engines	✓			✓	✓	✓	✓	
Diesel engines	✓		✓	✓	✓	✓	✓	
Transportation non-engine			✓	✓				✓
Agriculture implementations		✓	✓	✓				✓
Railroad		✓	✓	✓				✓
Iron and steel manufacture		✓	✓	✓	✓		✓	✓
Steel rolling mills		✓	✓	✓	✓		✓	✓
Iron and steel casting		✓	✓					✓
Forging		✓	✓	✓	✓		✓	
Copper and brass mills							✓	
Ship and boat manufacture/repair			✓	✓				✓
Oil and gas exploration		✓	✓	✓			✓	✓
Mining, construction, and dredging		✓	✓					✓
Rock products		✓	✓					✓
Screaming			✓					
Cement and structural clay		✓	✓					✓
Chemical processing			✓	✓	✓		✓	
Rubber and plastic manufacture		✓	✓	✓			✓	
Textile			✓					
Food processing		✓	✓					
Electrical utilities		✓	✓	✓			✓	✓
Pulp and paper		✓	✓	✓				✓
Printing equipment								✓
Defense and aerospace	✓	✓	✓	✓	✓	✓	✓	✓
Nuclear								
Medical								✓
Business equipment								✓
Electrical and electronics								✓
Architectural	✓							✓
Glass manufacture	✓	✓	✓					

Appendices

Appendix A: Processes and Coating Applications by Industry

Appendix B: Use of the Different Spray Materials

In this appendix only basic information about the most frequently sprayed materials are presented. For more detailed information, the reader must consult the different powder and wire suppliers. It must also be kept in mind that the coating resulting from the spray process has properties (thermal and electrical conductivities, for example) different from those of powder or wire properties. Such differences result from the increase in the oxide content, the pores generated during spraying, the real contacts between layered splats, etc. Properties also depend on the way the sprayed material has been manufactured. In the following, X-20Y-10Z means 20 wt.% of Y, 10 wt.% of Z, the balance being X.

B.1 Metals

Aluminum: Al (99 wt.%) $T_m = 660\ ^\circ$C:

- Good corrosion resistance in industrial atmospheric conditions
- Good electrical and thermal conductivity
- Soft and ductile $\rightarrow$ repair Al alloys
- Nonmagnetic (electromagnetic shielding)
- Sprayed by flame (powder or wire), wire arc, plasma, and cold spray

Aluminum base: Al–Si (95/5 or 12 wt.%):

- Salvage of parts made of Al, Mg, and their alloys—excellent finish
- Sprayed by flame (powder or wire), wire arc, plasma, and cold spray
- Cobalt-based powders (pure cobalt $T_m = 1495\ ^\circ$C)

Stellite$^\circledR$ (Co–xCr–yW–zC):

- Against galling and cavitation
- For metal friction, high-angle erosion, high-temperature hardness, good resistance to abrasion, and rather good wettability
- Very good oxidation resistance
- Replace WC in high-temperature applications

- Repair of cobalt-based parts
- Sprayed by HVOF, flame, PTA, and plasma

Triballoy$^\circledR$ (Co–xCr–yMo–zSi) or (Co–xCr–yMo–zSi–tNi):

- Against galling for metal-to-metal friction
- High-temperature hardness
- Good resistance to corrosion and oxidation
- Very good wear properties from room temperature to 860 $^\circ$C
- Sprayed by HVOF, flame, PTA, and plasma

Co–25.5Cr–10.5Ni–7.5W–0.5C:

- High abrasive, sliding fretting, and cavitation wear resistance up to 800–850 $^\circ$C
- Good oxidation resistance
- Behave well between 540 and 840 $^\circ$C
- Sprayed by HVOF, flame, PTA, and plasma

Co–28Mo–8Cr–2Si:

- Up to 760 $^\circ$C low coefficient of friction
- Good corrosion resistance
- Sprayed by HVOF, flame, PTA, and plasma

Co–28Mo–17Cr–3Si:

- Excellent sliding wear resistance
- Good hot corrosion resistance and moderate oxidation resistance up to 800 $^\circ$C
- Sprayed by HVOF, flame, PTA, and plasma

Copper Cu (99 wt.%) ($T_m = 1084\ ^\circ$C):

- Very good electrical and thermal conductivities
- Good resistance to inks (paper and printing)
- Used to repair Cu-base alloys
- Nonmagnetic (electromagnetic shielding)
- Sprayed by flame (powder or wire), wire arc, plasma, HVOF, and cold spray

Copper-based powder: Cu–36Ni–5In ($T_m = 1150\ ^\circ$C):

- Very dense coatings with good resistance to galling and fretting
- Sprayed by flame (powder or wire), wire arc, plasma, HVOF, and cold spray

Aluminum bronze: Cu–9.5Al–1Fe:

- For pumps (against cavitation)

- Piston guides (soft-bearing surfaces)
- Shifter forks and compressor air seals (friction)
- Strength and hardness twice that of other bronzes
- Sprayed by flame (powder or wire), wire arc, plasma, HVOF, and cold spray

Supra bronze: Cu–40Zn–0.8Sn–0.75Fe–0.24Mn:

- For metallizing work
- Sprayed by flame (powder or wire), wire arc, plasma, HVOF, and cold spray
- Iron-based powders (pure iron $T_m = 1538$ °C)
- Low-carbon steels (C < 0.25 wt.%)
- An inexpensive low-carbon steel powder
- Corrosion resistant $\rightarrow$ repair + good wear resistance in lubricated service
- May contain martensitic phases
- Produces machinable coatings
- Sprayed by flame, wire arc, plasma, HVOF, and cold spray

High-carbon steels (C > 0.8 wt.%):

- Reclamation
- Wear and erosion resistance
- Sprayed by flame, wire arc, plasma, and HVOF

Stainless steels (SS): Fe-13 or 14Cr–1Ni:

- Good resistance to wear and corrosion
- Best for all purposes (SS)
- Sprayed by flame, wire arc, plasma, and HVOF

Stainless steels (SS): Fe–18Cr–8Mg–5Ni:

- Reclamation
- Corrosion protection
- Low shrinkage and good machinability
- Sprayed by flame, wire arc, plasma, and HVOF

Stainless steels (SS): Fe–17Cr–12NI–2.5Mo–1Si–0.1C (ASI 316):

- Corrosion protection
- Dimensional restoration
- Cavitation and low-temperature erosion resistance
- Sprayed by wire or powder flame, wire arc, and HVOF

Cored wires (wire arc sprayed):

- Some of them (Fe–Cr–P–C–, etc.) form amorphous phases upon spraying
- Rather good resistance to corrosion (for example, with H_2SO_4)
- Good resistance to abrasion

Molybdenum: Mo (pure molybdenum $T_m = 2623$ °C):

- Self-bonding to most metallic surfaces, especially steels
- Natural lubricity and high hardness: good wear properties
- Maximum service temperature 316 °C

Salvage and build-up of Ni-base alloy components:

- High-density coatings
- Fretting resistant
- Used for pump parts, diesel engine fuel, injectors, piston rings, synchronized ring, press fits, valves, gears, cam followers, etc.
- Sprayed by HVOF, flame (powder or wire), wire arc, and plasma

Self-fluxing alloys: Mo+25 (Ni–Cr–B–Si–Fe):

- High wear resistance, low friction coefficient
- Against steel, good scuff resistance
- Can be used for hard-facing, hard-bearing surfaces
- Against abrasion

Nickel: Ni (99.5 wt.%) (pure nickel $T_m = 1455$ °C):

- Good bonding to steel
- Good corrosion and oxidation resistance up to 980 °C
- Resist heat and prevent scaling of carbon and low-alloy steels in hot atmospheres

Salvage and build-up of Ni-base alloys:

- Easily machined
- Sprayed by HVOF, flame (powder or wire), wire arc, and plasma

Ni–20Cr:

- Good surface appearance
- Good machinability
- Protective coatings against oxidizing gases at high temperature (up to 980 °C)
- Electrical conductors
- Surfacing

Other different types of NiCr (Cr 10–17 wt.%) but also addition of Fe, Mo with specific applications:

- See supplier guides for example

Ni–16Cr–8Fe:

- Machinable "stainless" coatings for salvage and build-up applications on corrosion-resistant steels
- Sprayed by HVOF, flame (powder or wire), wire arc, and plasma

Ni–18Cr–6Al (composite or clad):

- Good oxidation resistance
- Good machinability
- Bonding and surfacing layers
- Self-bonds to most metallic surfaces
- Sprayed by HVOF, flame (powder or wire), wire arc, and plasma

Ni–20Cr–10W–9Mo–4Cu–1C–1B–1Fe:

- Wear and corrosion protection
- Coatings contain some amounts of glassy phases (due to addition of refractory metals and metalloids)
- Sprayed by HVOF, flame (powder or wire), wire arc, and plasma

Ni–5Al (clad):

- Good hardness and refractoriness (formation of nickel aluminide)
- Oxidation and abrasion resistant
- Adhere very well to smooth substrates
- Bond coat
- Resizing of over-machined parts or of worn-out parts
- Possible: Al 20 wt.% (clad) → self-bonds to most metal surfaces
- Sprayed by HVOF, flame (powder or wire), wire arc, and plasma

Nickel-based self-fluxing alloys:
Ni–10Cr–2.5B–2.5Fe–2.5Si–0.15C:

- The only one producing machinable-fused coating
- Resistance to abrasive wear, fretting, cavitation, and erosion up to 840 °C

Ni–17Cr–4Fe–4Si–3.5B–1C:

- Dense coating good corrosion resistance
- Ni–17Cr–4Fe–4Si–3.5B–1C

- Dense, hard, and oxide-free coating
- Piston rings, cylinder liners, and utility exhaust fan

Superalloys: M.Cr.Al.Y with M = Ni, Co, Fe, Ni–Co:

- Composition optimized: for substrate compatibility and environmental resistance
- Protection against corrosion and oxidation in high-temperature applications
- Aero gas turbines
- Marine turbines
- Stationary gas turbines
- Design criteria—avoid phase transformation during engine start up and shut down; avoid brittle phases (e.g., μ, σ, αCr)
- Limit brittleness increasing with oxide content
- Form a stable and adherent oxide scale of Al_2O_3 with the addition of active elements (0.5% wt. Y)
- Increase corrosion resistance (by increasing Cr content)
- Control thermal expansion coefficient: increasing with Cr and decreasing with Al
- Keep ductile coatings (ductile to brittle temperature influenced by Cr and Al contents)
- Many compositions exist
- Sprayed by HVOF and plasma (if possible, under soft-vacuum to limit oxidation)

B.2 Ceramics

For most oxides the thermal expansion coefficient is low compared to that of most metals and it has to be taken into account because they are often used at high temperature.

Alumina: Al_2O_3 (T_m = 2050 °C):

- Main problem: Starting from α phase, melting and fast cooling (spraying) result in γ phase. Unfortunately, around 1000 °C γ phase transforms into α phase with a volume increase of about 4% resulting in coating peeling off. Thus, coatings should be used below 900 °C.
- They are not too sensitive to oxygen losses.
- They have a good resistance to abrasive, sliding, and friction wear up to approximately 800 °C.
- Poor resistance to shock or impact loading.
- High dielectric strength and good electrical insulating coatings.
- Reactive with molten salts.
- Porous coatings.
- Sprayed mainly by plasma, sometimes by HVOF (particles below 22 µm in diameter), and also by flame (rods or cords).

Titanium dioxide: TiO_2 (T_m = 1843 °C):

- Very sensitive to oxygen losses resulting in strong modifications of coating properties: color from white to black and especially electrical properties
- Loss and gain of oxygen reversible
- Very good wettability
- Excellent surface finish
- Excellent adhesion
- Rather low porosity
- Sprayed by plasma, HVOF, and also by flame (rods or cords)

Alumina–Titanium dioxide: Al_2O_3–$xTiO_2$:

- TiO_2 in the range 2–50 wt.% (most common: 3–13–40 wt.%): lowers Al_2O_3 coating porosity
- Al_2O_3–$3TiO_2$ can be used with most acids and alkalis
- Good for abrasion, erosion, and sliding wear
- Maximum service temperature, 840 °C
- Less brittle but lower dielectric strength than pure Al_2O_3 coatings sprayed by plasma

Al_2O_3–$13TiO_2$:

- Applications similar to those of Al_2O_3–$3TiO_2$ but with lower hardness and dielectric strength and less resistance to chemical attack
- Maximum service temperature, 540 °C
- Sprayed by plasma

Al_2O_3–$40TiO_2$:

- Formation of Al_2TiO_5
- Softer and less resistant to chemicals
- Excellent finishing properties
- Sprayed by plasma
- Other compositions are used, for example, with SiO_2 below 2 wt.%

Chromium oxide: Cr_2O_3 ($T_m = 2435$ °C):

- Its stoichiometry depends strongly upon spray conditions (high oxygen pressure needed), and sub-stoichiometric coatings have a metallic behavior with a poor corrosion resistance
- Coatings have high hardness (1900–2000 HV_{5N})
- Excellent wear resistance
- Low porosity
- Excellent finish
- Used on sliding surfaces
- Good wear resistance
- Insoluble in acids, alkalis, and alcohol
- Maximum service temperature, 540 °C

- Excellent engraving properties
- Sprayed by plasma, HVOF, and also by flame (rods or cords)

Cr_2O_3–$3TiO_2$–$5SiO_2$:

- TiO_2 limits oxygen losses
- Resists better than Cr_2O_3 to impacts
- High wear and corrosion resistance
- Hydroxyapatite: $Ca_5 (PO_4)_3$ OH
- For coating medical and dental implants
- Biocompatible and bioactive (main constituent of bones)
- Sprayed by plasma

Zirconia: ZrO_2:

- Interesting mechanical properties.
- Good wear resistance.
- Low thermal conductivity (1–5 W/m K).
- Three phases: monoclinic (m), tetragonal (t), and cubic (c). Upon cooling around 1000 °C, cubic or tetragonal phases transform into monoclinic with volume increase of about 10% and coating peeling off. Thus, only totally (c phase) or partially (t' non-transformable t phase) stabilized zirconia can be sprayed.
- Most used stabilizers are CaO, MgO, Y_2O_3, and CeO_2. CaO and MgO are rather cheap but the maximum service temperature is about 500 °C. Very good results are obtained with Y_2O_3; with about 8 wt.% t' phase is obtained, while with 13 wt.% c phase is obtained, as with 24–25 wt.% CeO_2. The maximum service temperature (about 1350 °C at the best) depends not only on the phase (best results with c phase) but also on the way the powder is manufactured (see Sect. 11.1.2.9), best results being obtained when zirconia and stabilizer particles are very small and uniformly distributed.
- Coatings have excellent thermal shock resistance and good oxidation and corrosion resistance.
- They are mainly used as thermal barriers.
- Sprayed by plasma and also by flame (rods or cords).

Zircon: $ZrSiO_4$ (infusible):

- Dissociated during spraying (65% ZrO_2, 35% SiO_2)
- Not wetted by liquid metals: used for casting
- Good resistance to liquid glass
- Good resistance to combustion gases

B.3 Cermets

They are made of a metal matrix, to achieve a good toughness, in which are embedded ceramic particles either of oxides or carbides for the hardness and wear resistance.

Here again the manufacturing process plays a key role, for example, sintered particles behaving very differently than blended ones.

With oxides: In most cases they are blends. For example: Al_2O_3–30(Ni–20Al):

- Denser coatings than pure ceramic, more abrasion, and shock resistant, hard, and smooth
- Addition of alumina particles modifies the electrical resistance of the metal matrix

MCrAlY + Al_2O_3 (<3 μm):

- Hardness increases with Al_2O_3 content
- Electrical resistance decreases with Al_2O_3 %

With carbides (the most used):

- Most used ones are: WC, Cr_3C_2, and also sometimes TiC
- If all of them have melting temperatures over 1900 °C (for example, T_m = 2870 °C for WC), they are relatively sensitive to oxidation.
- WC oxidation starts at 500–600 °C and oxidation produces W_2C decomposing into W over 1300 °C.

Cr_3C_2:

- Is not the only chromium carbide: Cr_7C_3 (T_m = 1782 °C), $Cr_{23}C_6$ (T_m = 1518 °C). Cr_3C_2 is mostly used in spraying and its oxidation starts at 800–900 °C; however, $Cr_{23}C_6$ has an excellent wear resistance.

TiC:

- Has a unique cubic phase (T_m = 3170 °C), in which oxidation starts at 800–900 °C.
- At last, carbides dissolve more or less in the liquid metal or alloy matrix, the dissolution increasing with temperature over the matrix melting temperature. The metal matrix lowers the wear resistance of carbides but increases resistance to mechanical or thermal shock. Phase changes of metal matrix must be avoided during the service (for example, that of Co occurs at about 480 °C). To conclude, chemical changes occur during spraying especially in air atmosphere: oxidation, decomposition, and dilution. Thus, microstructural properties of sprayed cermets depend strongly on spray conditions (VPS, IPS, APS, HVOF, HVAF, etc.), particle morphology and manufacturing process, and ceramic mean grain size. Only a few examples are given below:

WC–8Co:

- Dense, hard, and wear-resistant coating

- Sprayed by plasma, HVOF, and HVAF

WC–12Co:

- Excellent low-temperature wear resistance
- Sprayed by plasma, HVOF, and HVAF

WC–17Co:

- Higher Co level improves toughness and fretting resistance
- Cannot be used in corrosive media

Cr_3C_2–25(Ni–20Cr):

- Oxidation resistant up to 900 °C
- Good corrosion resistance
- Excellent for high-temperature cavitation, abrasion, and sliding wear

Cr_3C_2–7(Ni–20Cr):

- Very good resistance to high-temperature fretting and wear (higher carbide content increases hardness)

B.4 Abradables

They are designed to wear preferentially upon contact with mating part in order to automatically establish clearance. They comprise a metal matrix and nonmetallic filler such as graphite, polyester, polyimide, boron nitride, and friable material, the role of filler being to weaken the matrix integrity. The metal matrix is made of Ni, Al, Cu, Co bases, and superalloys.

The main difference in the base material is related to service temperature:

- Al–Si with C: up to 315–425 °C
- Al–Si with polyester: up to 350 °C

The filler content can be varied:

- Both are used for the compressor section of jet engines

Co–polyester–BN:

- They are used up to 700 °C

Cu: Aluminum bronze alloy–polyester or Cu–14polyester–8Al–1Fe–5Binder

- Maximum service temperature: 650 °C

Ni–graphite:

- They are used up to 480 °C
- Self-lubricating

MCrAlY–polyester and/or BN:

- Temperatures up to 1200–1300 °C

Nomenclature

Units are indicated in parentheses; when no units are indicated, the parameter is dimensionless.

Latin Alphabet

A_c	Surface to be coated (m^2)
f	Friction coefficient
m_{part}	Mass of coating required for each part (kg/part)
N_{hy}	Operating hours per year (h/y)
N_{hp}	Number of hours per part (h/part)
p	Pressure (Pa)
Pc	Cost per kg powder deposited (US\$/kg or €/kg)
Pcc	Cost of spare parts (US\$ or €)
Pcp	Powder cost per hour (US\$/h or €/h)
Pe	Cost of equipment's (US\$ or €)
Pee	Energy cost per 100 kW (US\$ or €)
Pen	Energy cost per deposited kg powder (US\$/kg or €/kg)
Pep	Component cost per deposited kg powder (US\$/kg or €/kg)
Pg	Gas cost per 100 m^3 (US\$ or €)
Pgp	Gas cost per deposited kg powder (US\$/kg or €/kg)
Pt	Plasma torch power (kW)
qp	Powder quantity necessary for each part (kg)
Qp	Powder spraying rate (kg/h)
Ra	Surface roughness (μm)
S_T	Total cost per part (\$/part)
S_{di}	Direct cost per hour (\$/h)
S_{in}	Indirect cost per year (\$/y)
S_{ih}	Amortization rate per hour (US\$/h or €/h)
T	Tangential force (N)
tc	Mean component lifetime (h)
tp	Time necessary to spray the part (h)
v	Velocity (m/s)

Greek Alphabet

η_e	Percentage corresponding to effective spray due to loss at holes and edges (%)
η_c	Powder or wire deposition efficiency (%)
ρ_p	Feedstock specific mass (kg/m^3)
δ_c	Coating thickness (mm)

References

Abdi, S., and S. Lebaili. 2008. Alternative to chromium, a hard alloy powder NiCrBCSi (Fe) coatings thermally sprayed on 60CrMn4 steel. Phase and comportements. *Physics Procedia* 2: 1005–1014.

Agarwal, A., T. McKechnie, and S. Seal. 2003. Net shape nanostructured aluminum oxide structures fabricated by plasma spray forming. *Journal of Thermal Spray Technology* 12 (3): 350–359.

Ahmaniemi, S., M. Vippola, P. Vuoristo, T. Mäntylä, M. Buchmann, and R. Gadow. 2002a. Residual stresses in aluminum phosphate sealed plasma sprayed oxide coatings and their effect on abrasive wear. *Wear* 252: 614–623.

Ahmaniemi, S., J. Tuominen, P. Vuoristo, and T. Mäntylä. 2002b. Sealing procedures for thick thermal barrier coatings. *Journal of Thermal Spray Technology* 11 (3): 320–332.

Ahmed, R., and M. Hadfield. 2002. Mechanisms of fatigue failure in thermal spray coatings. *Journal of Thermal Spray Technology* 11 (3): 333–349.

Ahn, H.-S., and O.-K. Kwon. 1999. Tribological behaviour of plasma-sprayed chromium oxide coating. *Wear* 225–229: 814–824.

Ahn, J., B. Hwang, and S. Lee. 2005. Improvement of wear resistance of plasma-sprayed molybdenum blend coatings. *Journal of Thermal Spray Technology* 14 (2): 251–257.

Akebono, H., J. Komotori, and M. Shimizu. 2008. Effect of coating microstructure on the fatigue properties of steel thermally sprayed with Ni-based self-fluxing alloy. *International Journal of Fatigue* 30: 814–821.

Alam, S., S. Sasaki, and H. Shimura. 2001. Friction and wear characteristics of aluminum bronze coatings on steel substrates sprayed by a low pressure plasma technique. *Wear* 248: 75–81.

Al-Fadhli, H.Y., J. Stokes, M.S.J. Hashmi, and B.S. Yilbas. 2006. The erosion–corrosion behaviour of high velocity oxy-fuel (HVOF) thermally sprayed inconel-625 coatings on different metallic surfaces. *Surface and Coating Technology* 200: 5782–5788.

American Welding Society. 1985. *Thermal spraying, practice, theory and application.* Miami: American Welding Society.

Aoh, J.-N., and J.-C. Chen. 2001. On the wear characteristics of cobalt-based hardfacing layer after thermal fatigue and oxidation. *Wear* 250: 611–620.

Arcondéguy, A., A. Grimaud, A. Denoirjean, G. Gasgnier, C. Huguet, B. Pateyron, and G. Montavon. 2007. Flame-sprayed glaze coatings: effects of operating parameters and feedstock characteristics onto coating structures. *Journal of Thermal Spray Technology* 16 (5–6): 978–990.

Arondéguy, A., G. Gasgnier, G. Montavon, B. Pateyron, A. Denoirjean, A. Grimaud, and C. Huguet. 2008. Effects of spraying parameters onto flame-sprayed glaze coating structures. *Surface and Coating Technology* 202: 4444–4448.

Arrabal, R., A. Pardo, M.C. Merino, M. Mohedano, P. Casajús, and S. Merino. 2010. Al/SiC thermal spray coatings for corrosion protection of Mg–Al alloys in humid and saline environments. *Surface and Coating Technology* 204: 2767–2774.

Bala, N., H. Singh, and S. Prakash. 2010. Accelerated hot corrosion studies of cold spray Ni–50Cr coating on boiler steels. *Materials and Design* 31: 244–253.

———. 2011. Characterization and high-temperature oxidation behavior of cold-sprayed Ni-20Cr and Ni-50Cr coatings on boiler steels. *Metallurgical and Materials Transactions A* 42A: 3399–3416.

Barbezat, G. 2003. Low-cost high-performance coatings produced by internal plasma spraying for the production of high efficiency engines. In *International thermal spray conference 2003*, ed. C. Moreau and B. Marple, 139–142. Materials Park: ASM International.

———. 2005. Advanced thermal spray technology and coating for lightweight engine blocks for the automotive industry. *Surface and Coating Technology* 200: 1990–1993.

———. 2006. Application of thermal spraying in the automobile industry. *Surface and Coating Technology* 201: 2028–2031.

Bardi, U., C. Giolli, A. Scrivani, G. Rizzi, F. Borgioli, A. Fossati, K. Partes, T. Seefeld, D. Sporer, and A. Refke. 2008. Development and investigation on new composite and ceramic coatings as possible abradable seals. *Journal of Thermal Spray Technology* 17 (5–6): 805–811.

Barthel, K., S. Rambert, S. Barthel, S. Rambert, and S. Siegmann. 2000. Microstructure and polarization resistance of thermally sprayed composite cathodes for solid oxide fuel cell use. *Journal of Thermal Spray Technology* 9 (3): 343–347.

Beardsley, M.B. 1997. Thick thermal barrier coatings for diesel engines. *Journal of Thermal Spray Technology* 6 (2): 181–186.

Beauvais, S., V. Guipont, M. Jeandin, D. Juve, D. Treheux, A. Robisson, and R. Saenger. 2005. Influence of defect orientation on electrical insulating properties of plasma-sprayed alumina coatings. *Journal of Electroceramics* 15: 65–74.

Beele, W., G. Marijnissen, and A. van Lieshout. 1999. The evolution of thermal barrier coatings—Status and upcoming solutions for today's key issues. *Surface and Coating Technology* 120–121: 61–67.

Bellucci, D., G. Bolelli, V. Cannillo, R. Gadow, A. Killinger, L. Lusvarghi, A. Sola, and N. Stiegler. 2012. High velocity suspension flame sprayed (HVSFS) potassium-based bioactive glass coatings with and without TiO_2 bond coat. *Surface and Coating Technology* 206: 3857–3868.

Berard, G., P. Brun, J. Lacombe, G. Montavon, A. Denoirjean, and G. Antou. 2008. Influence of a sealing treatment on the behavior of plasma-sprayed alumina coatings operating in extreme environments. *Journal of Thermal Spray Technology* 17 (3): 410–419.

Berger, L.-M., K. Lipp, J. Spatzier, and J. Bretschneider. 2011. Dependence of the rolling contact fatigue of HVOF-sprayed WC–17%Co hard metal coatings on substrate hardness. *Wear* 271: 2080–2088.

Berndt, C.C., J.A. Brogan, G. Montavon, A. Claudon, and C. Coddet. 1998. Mechanical properties of metal- and ceramic-polymer composites formed via thermal spray consolidation. *Journal of Thermal Spray Technology* 7 (3): 337–339.

Billah, B.M., F. Ahmad Khalid, and A. Nusair Khan. 2012. Behavior of calcia-stabilized zirconia coating at high temperature, deposited by air plasma spraying system. *Journal of Thermal Spray Technology* 21 (1): 121–131.

Blink, J., J. Choi, and J. Farmer. 2007. Applications in the nuclear industry for thermal spray amorphous metal and ceramic coatings, UCRL-CONF-232603. *Lawrence Livermore National Laboratory* 9: 1–14.

Blink, J., J. Farmer, J. Choi, and C. Saw. 2009. Applications in the nuclear industry for thermal spray amorphous metal and ceramic coatings. *Metallurgical and Materials Transactions A* 40A: 1344–1354.

Bolelli, G., and L. Lusvarghi. 2006. Heat treatment effects on the tribological performance of HVOF sprayed Co-Mo-Cr-Si coatings. *Journal of Thermal Spray Technology* 15 (4): 802–810.

Bolelli, G., V. Cannillo, L. Lusvarghi, and T. Manfredini. 2006a. Wear behaviour of thermally sprayed ceramic oxide coatings. *Wear* 261: 1298–1315.

Bolelli, G., V. Cannillo, L. Lusvarghi, and S. Ricco. 2006b. Mechanical and tribological properties of electrolytic hard chrome and HVOF-sprayed coatings. *Surface and Coating Technology* 200: 2995–3009.

Bolelli, G., R. Giovnardi, L. Lusvarghi, and T. Manfredini. 2006c. Corrosion resistance of HVOF-sprayed coatings for hard chrome replacement. *Corrosion Science* 48: 3375–3397.

Bolelli, G., L. Lusvarghi, and R. Giovanardi. 2008. A comparison between the corrosion resistances of some HVOF-sprayed metal alloy coatings. *Surface and Coating Technology* 202: 4793–4809.

Bolelli, G., J. Rauch, V. Cannillo, A. Killinger, L. Lusvarghi, and R. Gadow. 2009a. Microstructural and tribological investigation of high-velocity suspension flame sprayed (HVSFS) Al_2O_3 coatings. *Journal of Thermal Spray Technology* 18 (1): 35–48.

Bolelli, G., V. Cannillo, R. Gadow, A. Killinger, L. Lusvarghi, and J. Rauch. 2009b. Microstructural and in vitro characterisation of high-velocity suspension flame sprayed (HVSFS) bioactive glass coatings. *Journal of the European Ceramic Society* 29: 2249–2257.

Bolelli, G., B. Bonferroni, J. Laurila, L. Lusvarghi, A. Milantia, K. Niemi, and P. Vuoristo. 2012a. Micromechanical properties and sliding wear behaviour of HVOF-sprayed Fe-based alloy coatings. *Wear* 276–277: 29–47.

Bolelli, Cannillo V., L. Lusvarghi, R. Rosa, A. Valarezo, W.B. Choi, R. Dey, C. Weyant, and S. Sampath. 2012b. Functionally graded WC–Co/NiAl HVOF coatings for damage tolerance, wear and corrosion protection. *Surface and Coating Technology* 206: 2585–2601.

Bose, S., and J. de Masi-Marcin. 1997. Thermal barrier coating experience in gas turbine engines at Pratt & Whitney. *Journal of Thermal Spray Technology* 6 (1): 99–104.

Bounazef, M., S. Guessasmaa, and B. Ait Saadi. 2004. The wear, deterioration and transformation phenomena of abradable coating BN–SiAl–bounding organic element, caused by the friction between the blades and the turbine casing. *Materials Letters* 58: 3375–3380.

Branagan, D. 2004. *Properties of amorphous/partially crystalline coatings.* US Patent 20040253381.

Branagan, D.J., M. Breitsameter, B.E. Meacham, and V. Belashchenko. 2005. High-performance nanoscale composite coatings for boiler applications. *Journal of Thermal Spray Technology* 14 (2): 196–204.

Brogan, J.A., S. Margolies, H. Sampath, C.C. Herman, and S.D. Berndt. 1995. Adhesion of combustion-sprayed polymer coatings. In *Thermal spray science and technology*, ed. C.C. Berndt and S. Sampath, 521–526. Materials Park: ASM International.

Brožek, V., P. Ctibor, D.I. Cheong, and S.-H. Yang. 2009. Plasma spraying of zirconium carbide-hafnium carbide-tungsten cermets. *Powder Metallurgy Progress* 9: 1–49.

Buyukkaya, E., and M. Cerit. 2007. Thermal analysis of a ceramic coating diesel engine piston using 3-D finite element method. *Surface and Coating Technology* 202: 398–402.

Cao, X.Q., R. Vaßen, and D. Stöver. 2004. Ceramic materials for thermal barrier coatings. *Journal of the European Ceramic Society* 24: 1–10.

Carrasquero, E.J., J. Lesage, E.S. Puchi-Cabrera, and M.H. Staia. 2008. Fretting wear of HVOF Ni–Cr based alloy deposited on SAE 1045 steel. *Surface and Coating Technology* 202: 4544–4551.

Cartier, M. 2003. *Handbook of surface treatments and coatings*, 412 p. New York: ASME Press.

Celotto, S., J. Pattison, J.S. Ho, A.N. Johnson, and W. O'Neill. 2007. The economics of the cold spray process. In *Cold spray materials deposition process - fundamentals and applications*, ed. V. Champagne. Sawston: Woodhead.

Champagne, V.K. 2007. *The cold spray materials deposition process; fundamental and applications*, 362 p. Cambridge: Woodhead.

Chang, C., J. Shi, J. Huang, Z. Hu, and C. Ding. 1998. Effects of power level on characteristics of vacuum plasma sprayed hydroxyapatite coating. *Journal of Thermal Spray Technology* 7 (4): 484–488.

Chatha, S.S., H.S. Sidhu, and B.S. Sidhu. 2012a. High temperature hot corrosion behaviour of NiCr and Cr_3C_2–NiCr coatings on T91 boiler steel in an aggressive environment at 750 °C. *Surface and Coating Technology* 206: 3839–3850.

———. 2012b. The effects of post-treatment on the hot corrosion behavior of the HVOF-sprayed Cr_3C_2–NiCr coating. *Surface and Coating Technology* 206: 4212–4224.

Chattopadhyay, R. 2001. *Surface wear: Analysis, treatment, and prevention*, 307 p. Materials Park: ASM International.

Chen, H., and I.M. Hutchings. 1998. Abrasive wear resistance of plasma-sprayed tungsten carbide–cobalt coatings. *Surface and Coating Technology* 107: 106–114.

Chen, J.Z., H. Herman, and S. Safai. 1993. Evaluation of NiAl and NiAl-B deposited by vacuum plasma spray. *Journal of Thermal Spray Technology* 2 (4): 357–361.

Chen, H., H. Zhao, J. Qu, and H. Shao. 1999. Erosion-corrosion of thermal-sprayed nylon coatings. *Wear* 233–235: 431–435.

Chen, Z., J. Mabon, J.-G. Wen, and R. Trice. 2009. Degradation of plasma-sprayed yttria-stabilized zirconia coatings via ingress of vanadium oxide. *Journal of the European Ceramic Society* 29: 1647–1656.

Chen, X., B. Zou, Y. Wang, H. Ma, and X. Cao. 2011a. Microstructure and thermal cycling behavior of air plasma-sprayed YSZ/LaMgAl11O19 composite coatings. *Journal of Thermal Spray Technology* 20 (6): 1328–1338.

Chen, W.R., E. Irissou, X. Wu, J.-G. Legoux, and B.R. Marple. 2011b. The oxidation behavior of TBC with cold spray CoNiCrAlY bond coat. *Journal of Thermal Spray Technology* 20 (1–2): 132–138.

Chen, H., Y. Gao, H. Luo, and S. Tao. 2011c. Preparation and thermophysical properties of $La_2Zr_2O_7$ coatings by thermal spraying of an amorphous precursor. *Journal of Thermal Spray Technology* 20 (6): 1201–1208.

Chen, X., L. Gu, B. Zou, Y. Wang, and X. Cao. 2012. New functionally graded thermal barrier coating system based on LaMgAl11O19/YSZ prepared by air plasma spraying. *Surface and Coating Technology* 206: 2265–2274.

Cho, T.Y., J. Hong Yoon, J. Young Cho, Y. Kon Joo, J. Ho Kang, S. Zhang, H. Gon Chun, S. Young Hwang, and S. Chol Kwon. 2009. Surface properties and tensile bond strength of HVOF thermal spray coatings of WC-Co powder onto the surface of 420J2 steel and the bond coats of Ni, NiCr, and Ni/NiCr. *Surface and Coating Technology* 203: 3250–3253.

Choa, J.E., S.Y. Hwang, and K.Y. Kim. 2006. Corrosion behavior of thermal sprayed WC cermet coatings having various metallic binders in strong acidic environment. *Surface and Coating Technology* 200: 2653–2662.

Chráska, P., K. Neufuss, and H. Herman. 1997. Plasma spraying of zircon. *Journal of Thermal Spray Technology* 6 (4): 445–448.

Chun-long, Y., A. Yun-qi, and S. Ya-tan. 2009. Three years corrosion tests of nanocomposite epoxy sealer for metalized coatings on the East China Sea. In *Thermal spray 2009: Proceedings of the international thermal spray conference*, ed. B.R. Marple, M.M. Hyland, Y.-C. Lau, C.-J. Li, R.S. Lima, and G. Montavon, 1090–1093. Materials Park: ASM International.

Cipitria, A., I.O. Golosnoy, and T.W. Clyne. 2009. A sintering model for plasma-sprayed zirconia TBCs. Part I: Free-standing coatings. *Acta Materialia* 57: 980–992.

Cipri, F., C. Bartuli, T. Valente, and F. Casadei. 2007. Electromagnetic and mechanical properties of silica-aluminosilicates plasma sprayed composite coatings. *Journal of Thermal Spray Technology* 16 (5–6): 831–838.

Cramer, S.D., B.S. Covino Jr., G.R. Holcomb, S.J. Bullard, W.K. Collins, R.D. Govier, R.D. Wilson, and H.M. Laylor. 1999. Thermal sprayed titanium anode for cathodic protection of reinforced concrete bridges. *Journal of Thermal Spray Technology* 8 (1): 133–145.

Curry, N., and J. Donoghue. 2012. Evolution of thermal conductivity of dysprosia stabilised thermal barrier coating systems during heat treatment. *Surface and Coating Technology* 209: 38–43.

Curry, N., N. Markocsan, X.-H. Li, A. Tricoire, and M. Dorfman. 2011. Next generation thermal barrier coatings for the gas turbine industry. *Journal of Thermal Spray Technology* 20 (1–2): 108–115.

D'Ans, P., J. Dille, and M. Degrez. 2011. Thermal fatigue resistance of plasma sprayed yttria-stabilised zirconia onto borided hot work tool steel, bonded with a NiCrAlY coating: experiments and modeling. *Surface and Coating Technology* 205: 3378–3386.

Dallaire, S. 2001. Hard arc-sprayed coating with enhanced erosion and abrasion wear resistance. *Journal of Thermal Spray Technology* 10 (3): 511–519.

Darut, G., H. Ageorges, A. Denoirjean, G. Montavon, and P. Fauchias. 2008. Effect of the structural scale of plasma-sprayed alumina coatings on their friction coefficients. *Journal of Thermal Spray Technology* 17 (5–6): 788–797.

Das, S., S. Datta, D. Basu, and G.C. Das. 2009. Glass–ceramics as oxidation resistant bond coat in thermal barrier coating system. *Ceramics International* 35: 1403–1406.

Davis, J.R., ed. 2004. *Handbook of thermal spray technology. Sections introduction to applications for thermal spray processing and selected applications.* Materials Park: ASM International.

de Botton, O. 1988. *Master of Science in Technology and Policy.* Cambridge, MA: MIT.

de Munter, A.J., A. Bult, and J.A. de Jong. 2002. On the economical and environmental aspects of TSA coatings. In *International thermal spray conference 2002*, ed. E. Lugscheider. Düsseldorf: DVS. e-Proc.

Deng, C., M. Liu, C. Wu, K. Zhou, and J. Song. 2007. Impingement resistance of HVAF WC-based coatings. *Journal of Thermal Spray Technology* 16 (5–6): 604–609.

Dent, A.H., A.J. Horlock, D.G. McCartney, and S.J. Harris. 1999. The corrosion behavior and microstructure of high-velocity oxy-fuel sprayed nickel-base amorphous/nanocrystalline coatings. *Journal of Thermal Spray Technology* 8 (3): 399–404.

Devasenapathi, A., H.W. Ng, S.C.M. Yu, and A.B. Indra. 2002. Forming near net shape free-standing components by plasma spraying. *Materials Letters* 57: 882–886.

Ding, Z.-X., W. Chen, and Q. Wang. 2011. Resistance of cavitation erosion of multimodal WC-12Co coatings sprayed by HVOF. *Transactions of the Nonferrous Metals Society of China* 21: 2231–2236.

Donner, K.-R., F. Gärtner, and T. Klassen. 2011. Metallization of thin Al_2O_3 layers in power electronics using cold gas spraying. *Journal of Thermal Spray Technology* 20 (1–2): 299–306.

Dorfman, M.R., and A. Sharma. 2013a. Challenges and strategies for growth of thermal spray, keynote lecture presented at ITSC-2012, Houston, TX, USA. *Journal of Thermal Spray Technology* 22 (5): 559–563.

Dorfman, M., and A. Sharma. 2013b. Commentary challenges and strategies for growth of thermal spray markets: the six-pillar plan. *Journal of Thermal Spray Technology Comment* 22 (5): 559–563.

Döring, J.-E., F. Hoebener, and G. Langer. 2008. Review of applications of thermal spraying in the printing industry in respect to OEMs. In *Thermal spray conference: Crossing the border*, ed. E. Lugscheider. Düsseldorf: DVS. e-Proc.

Drnovšek, N., S. Novak, U. Dragin, M. Čeh, M. Gorenšek, and M. Gradišar. 2012. Bioactive glass enhances bone ingrowth into the porous titanium coating on orthopaedic implants. *International Orthopaedics* 36: 1739–1745.

Ducos, M. 1988. Plasma transferred arc reclamation. In *Plasmas in industry. Dopee Diffusion, France*, ed. G. Laroche and M. Orfeuil, 251–262. (in French).

———. 2006a. *Evaluation des coûts de projection thermique* (Costs evaluation in thermal spraying). Cours, ALIDERTE, Limoges.

———. 2006b. *Evaluating the costs of thermal spraying*, ALIDERTE course. ALIDERTE, Limoges (in French)

Ducos, M., and J.P. Durand. 2001. Thermal coatings in Europe, a business perspective. In *Thermal spray 2001*, ed. C.C. Berndt, K.H. Khor, and E. Lugscheider, 1267–1276. Materials Park: ASM International.

Eriksson, R., H. Brodin, S. Johansson, L. Östergren, and X.-H. Li. 2011. Influence of isothermal and cyclic heat treatments on the adhesion of plasma sprayed thermal barrier coatings. *Surface and Coating Technology* 205: 5422–5429.

Esfahani, E.A., H. Salimijazi, M.A. Golozar, J. Mostaghimi, and L. Pershin. 2012. Study of corrosion behavior of arc sprayed aluminum coating on mild steel. *Journal of Thermal Spray Technology*. https://doi.org/10.1007/s11666-012-9810-x.

Espallargas, N., J. Berget, J.M. Guilemany, A.V. Benedetti, and P.H. Suegama. 2008. Cr_3C_2–NiCr and WC–Ni thermal spray coatings as alternatives to hard chromium for erosion–corrosion resistance. *Surface and Coating Technology* 202: 1405–1417.

Evdokimenko Yu, I., V.M. Kisel', V.K. Kadyrov, A.A. Korol', and O.I. Get'man. 2001. High-velocity flame spraying of powder aluminum protective coatings. *Powder Metallurgy and Metal Ceramics* 40 (3–4): 121–126.

Factor, M., and I. Roman. 2002. Use of microhardness as a simple means of estimating relative wear resistance of carbide thermal spray coatings: Part 2. Wear resistance of cemented carbide coatings. *Journal of Thermal Spray Technology* 11 (4): 482–495.

Fan, X., and T. Ishigaki. 2001. Mo_5Si_3-B and $MoSi_2$ deposits fabricated by radio frequency induction plasma spraying. *Journal of Thermal Spray Technology* 10 (4): 611–617.

Faraoun, H.I., T. Grosdidier, J.-L. Seichepine, D. Goran, H. Aourag, C. Coddet, J. Zwick, and N. Hopkins. 2006. Improvement of thermally sprayed abradable coating by microstructure control. *Surface and Coating Technology* 201: 2303–2312.

Farmer, J.C., and J.-S. Choi. 2007. Criticality-control applications in the nuclear industry for thermal spray amorphous metal and ceramic coatings, UCRL-TR-234171. *Lawrence Livermore National Laboratory* 31: 1–7.

Farmer, J., J. Choi, C. Saw, J. Haslam, D. Day, P. Hailey, T. Lian, R. Rebak, J. Perepezko, J. Payer, D. Branagan, B. Beardsley, A. D'Amato, and L. Aprigliano. 2009. Iron-based amorphous metals: high-performance corrosion-resistant material development. *Metallurgical and Materials Transactions A* 40A: 1289–1305.

Fauchais, P., G. Montavon, R.S. Lima, and B.R. Marple. 2011. Engineering a new class of thermal spray nano-based microstructures from agglomerated nanostructured particles, suspensions and solutions: an invited review. *Journal of Physics D: Applied Physics* 44: 093001.

Fedrizzi, L., L. Valentinelli, S. Rossi, and S. Segna. 2007. Tribocorrosion behaviour of HVOF cermet coatings. *Corrosion Science* 49: 2781–2799.

Fedtke, P., M. Wienecke, M.-C. Bunescu, T. Barfels, K. Deistung, and M. Pietrzak. 2004. Yttria-stabilized zirconia films deposited by plasma spraying and sputtering. *Journal of Solid State Electrochemistry* 8: 626–632.

Fernandez, J., M. Gaona, and J.M. Guilemany. 2007. Effect of heat treatments on HVOF hydroxyapatite coatings. *Journal of Thermal Spray Technology* 16 (2): 220–228.

Feuerstein, A., J. Knapp, T. Taylor, A. Ashary, A. Bolcavage, and N. Hitchman. 2008. Technical and economical aspects of current thermal barrier coating systems for gas turbine engines by thermal spray and EBPVD: a review. *Journal of Thermal Spray Technology* 17 (2): 199–213.

Flores, J.F., A. Neville, N. Kapur, and A. Gnanavelu. 2009. An experimental study of the erosion–corrosion behavior of plasma transferred arc MMCs. *Wear* 267: 213–222.

Fossati, A., M. DiFerdinando, U. Bardi, A. Scrivani, and C. Giolli. 2012. Influence of surface finishing on the oxidation behaviour of VPS MCrAlY coatings. *Journal of Thermal Spray Technology* 21 (2): 314–324.

Fukumoto, M. 2008a. The current status of thermal spraying in Asia. *Journal of Thermal Spray Technology* 17 (1): 5–13.

———. 2008b. The current status of thermal spraying in Asia; Hwang SY Status of thermal spraying in Korea; Li C-J The current state of thermal spray activities in China; Tani K, Nakahira A The current status of thermal spray in Japan; Khor MKA Thermal spray activities in Singapore. *Journal of Thermal Spray Technology* 17 (1): 5–13.

Ganesh Sundara Raman, S., B. Rajasekaran, S.V. Joshi, and G. Sundararajan. 2007. Influence of substrate material on plain fatigue and fretting fatigue behavior of detonation gun sprayed Cu-Ni-In coating. *Journal of Thermal Spray Technology* 16 (4): 571–579.

Gärtner, F., T. Stoltenhoff, T. Schmidt, and H. Kreye. 2006. The cold spray process and its potential for industrial applications. *Journal of Thermal Spray Technology* 15 (2): 223–232.

Gawne, D.T., Z. Qiu, Y. Bao, T. Zhang, and K. Zhang. 2001. Abrasive wear resistance of plasma-sprayed glass-composite coatings. *Journal of Thermal Spray Technology* 10 (4): 599–603.

Geibel, A., L. Froyen, L. Delaey, and K.U. Leuven. 1996. Plasma spray forming: an alternate route for manufacturing free-standing components. *Journal of Thermal Spray Technology* 5 (4): 419–430.

Gell, M., E.H. Jordan, M. Teicholz, B.M. Cetegen, N. Padture, L. Xie, D. Chen, X. Ma, and J. Roth. 2008. Thermal barrier coatings made by the solution precursor plasma spray process. *Journal of Thermal Spray Technology* 17 (1): 124–135.

Gibbons, G.J., and R.G. Hansell. 2006. Down-selection and optimization of thermal-sprayed coatings for aluminum mould tool protection and upgrade. *Journal of Thermal Spray Technology* 15 (3): 340–347.

———. 2008. Thermal-sprayed coatings on aluminium for mould tool protection and upgrade. *Journal of Materials Processing Technology* 204: 184–191.

Gilbert, A., K. Kokini, and S. Sankarasubramanian. 2008. Thermal fracture of zirconia–mullite composite thermal barrier coatings under thermal shock: a numerical study. *Surface and Coating Technology* 203: 91–98.

Godoya, C., M.M. Lima, M.M.R. Castro, and J.C. Avelar-Batista. 2004. Structural changes in high-velocity oxy-fuel thermally sprayed WC–Co coatings for improved corrosion resistance. *Surface and Coating Technology* 188–189: 1–6.

Goller, G. 2004. The effect of bond coat on mechanical properties of plasma sprayed bioglass-titanium coatings. *Ceramics International* 30: 351–355.

Golosnoy, I.O., A. Cipitria, and T.W. Clyne. 2009. Heat transfer through plasma-sprayed thermal barrier coatings in gas turbines: a review of recent work. *Journal of Thermal Spray Technology* 18 (5–6): 809–821.

Gorlach, I.A. 2009. A new method for thermal spraying of Zn–Al coatings. *Thin Solid Films* 517: 5270–5273.

Gross, K.A., and A. Kovalevskis. 1996. Mold manufacture with plasma spraying. *Journal of Thermal Spray Technology* 5 (4): 469–475.

Gross, K.A., J. Tikkanen, J. Keskinen, V. Pitkänen, M. Eerola, R. Siikamaki, and M. Rajala. 1999. Liquid flame spraying for glass coloring. *Journal of Thermal Spray Technology* 8 (4): 583–589.

Gross, K.A., W. Walsh, and E. Swarts. 2004. Analysis of retrieved hydroxyapatite-coated hip prostheses. *Journal of Thermal Spray Technology* 13 (2): 190–199.

Gu, L., X. Chen, X. Fan, Y. Liu, B. Zou, Y. Wang, and X. Cao. 2011. Improvement of thermal shock resistance for thermal barrier coating on aluminum alloy with various electroless interlayers. *Surface and Coating Technology* 206: 29–36.

Guanhong, S., H. Xiaodong, J. Jiuxing, and S. Yue. 2011. Parametric study of Al and Al_2O_3 ceramic coatings deposited by air plasma spray onto polymer substrate. *Applied Surface Science* 257: 7864–7870.

Gui, M., S.B. Kang, and K. Euh. 2004. Al-SiC powder preparation for electronic packaging aluminum composites by plasma spray processing. *Journal of Thermal Spray Technology* 13 (2): 214–222.

Guilemany, J.M., N. Espallargas, P.H. Suegama, A.V. Benedetti, and J. Fernández. 2005. High-velocity oxyfuel Cr_3C_2-NiCr replacing hard chromium coatings. *Journal of Thermal Spray Technology* 14 (3): 335–341.

Guilemany, J.M., M. Torrell, and J.R. Miguel. 2007. Properties of HVOF coating of Ni based alloy for MSWI boilers protection. In *Thermal spray 2007: Global coating solutions*, ed. B.R. Marple, M.M. Hyland, Y.-C. Lau, C.-J. Li, R.S. Lima, and G. Montavon, 1115–1119. Materials Park: ASM International. e-Proc.

Guo, H.B., R. Vaßen, and D. Stöver. 2004. Atmospheric plasma sprayed thick thermal barrier coatings with high segmentation crack density. *Surface and Coating Technology* 186: 353–363.

Guo, H., H. Zhang, G. Ma, and S. Gong. 2009. Thermo-physical and thermal cycling properties of plasma-sprayed $BaLa_2Ti_3O_{10}$ coating as potential thermal barrier materials. *Surface and Coating Technology* 204: 691–696.

Gupta, M., N. Curry, P. Nylén, N. Markocsan, and R. Vaßen. 2012. Design of next generation thermal barrier coatings—Experiments and modeling. *Surface and Coating Technology* 220: 20–26.

Habibi, M.H., Li Wang, and S.M. Guo. 2012. Evolution of hot corrosion resistance of YSZ, $Gd_2Zr_2O_7$, and $Gd_2Zr_2O_7$ + YSZ composite thermal barrier coatings in Na_2SO_4 + V_2O_5 at 1050 °C. *Journal of the European Ceramic Society* 32: 1635–1642.

Hager, C.H., Jr., J.H. Sanders, and S. Sharma. 2008. Un-lubricated gross slip fretting wear of metallic plasma-sprayed coatings for Ti6Al4V surfaces. *Wear* 265: 439–451.

Hager, C.H., Jr., J. Sanders, S. Sharma, and A.A. Voevodin. 2009. The use of nickel graphite composite coatings for the mitigation of gross slip fretting wear on Ti6Al4V interfaces. *Wear* 267: 1470–1481.

Hahn, M., and A. Fischer. 2010. Characterization of thermal spray coatings for cylinder running surfaces of diesel engines. *Journal of Thermal Spray Technology* 19 (5): 866–872.

Hamacha, R., P. Fauchais, and E. Nardou. 1996. Influence of dopant on the behaviour under thermal cycling of two plasma sprayed zirconia coatings, Part 1: Relationship between powder characteristics and coating properties. *Journal of Thermal Spray Technology* 5 (4): 431–438.

Hamashima, K. 2007. Application of new boride cermet coating to forming of glass sheets. *Journal of Thermal Spray Technology* 16 (1): 32–33.

Han, M.-S., Y.-B. Woo, S.-C. Ko, Y.-J. Jeong, S.-K. Jang, and S.-J. Kim. 2009. Effects of thickness of Al thermal spray coating for STS 304. *Transactions of the Nonferrous Metals Society of China* 19: 925–929.

Hanneforth, P. 2006. The Global Thermal Spray Industry—100 Years of Success: So What's Next?. iTTSe, Vol 1, No 1, ASM International, Materials Park, May 2006, 14–16

Hardwicke, C.U., and Y.-C. Lau. 2013. Advances in thermal spray coatings for gas turbines and energy generation: a review. *Journal of Thermal Spray Technology* 22 (5): 564–576.

Harris, J., M. Qureshi, and O. Kesler. 2012. Deposition of composite LSCF-SDC and SSC-SDC cathodes by axial-injection plasma spraying. *Journal of Thermal Spray Technology* 21 (3–4): 461–468.

Hasan, S., and J. Stokes. 2011. Design of experiment analysis of the Sulzer Metco DJ high velocity oxy-fuel coating of hydroxyapatite for orthopedic applications. *Journal of Thermal Spray Technology* 20 (1–2): 186–194.

Haslam, J.J., J.C. Farmer, R.W. Hopper, and K.R. Wilfinger. 2005. Ceramic coatings for a corrosion-resistant nuclear waste container evaluated in simulated ground water at 90 °C. *Metallurgical and Materials Transactions A* 36A: 1085–1095.

Hawthorne, H.M., B. Arsenault, J.P. Immarigeon, J.G. Legoux, and V.R. Parameswaran. 1999. Comparison of slurry and dry erosion behavior of some HVOF thermal sprayed coatings. *Wear* 225–229: 825–834.

Haynes, J.A., M.K. Ferber, and W.D. Porter. 2000. Thermal cycling behavior of plasma-sprayed thermal barrier coatings with various MCrAlX bond coats. *Journal of Thermal Spray Technology* 9 (1): 38–48.

Hejwowski, T., and A. Weronski. 2002. The effect of thermal barrier coatings on diesel engine performance. *Vacuum* 65: 427–432.

Helali, M., and M.S.J. Hashmi. 1996. Production of free-standing objects by high velocity oxy-fuel (HVOF) thermal spraying process. *Journal of Materials Processing Technology* 56: 431–438.

Henne, R. 2007. Solid oxide fuel cells: a challenge for plasma deposition processes. *J. Thermal Spray Technology* 16 (3): 381–403.

Henne, R.H., and C. Schitter. 1995. Plasma spraying of high performance thermoplastics. In *Thermal spray science and technology*, ed. C.C. Berndt and S. Sampath, 527–532. Materials Park: ASM International.

Henne, R., M. Müller, E. Proß, G. Schiller, F. Gitzhofer, and M. Boulos. 1999. Near-net-shape forming of metallic bipolar plates for planar solid oxide fuel cells by induction plasma spraying. *Journal of Thermal Spray Technology* 8 (1): 110–116.

Herman, H., S. Sampath, R. Tiwari, and R. Neiser. 1994. Plasma spray forming of intermetallics and their composites. *Journal of Thermal Spray Technology* 3 (3): 295–296.

Hernandez, M.T., A.M. Karlsson, and M. Bartsch. 2009. On TGO creep and the initiation of a class of fatigue cracks in thermal barrier coatings. *Surface and Coating Technology* 203: 3549–3558.

Heydarzadeh, S.M., and F. Ghadami. 2010. Comparative tribological study of air plasma sprayed WC–12% coating versus conventional hard chromium electro deposit. *Tribology International* 43: 882–886.

Higuera, H.V., J. Belzunce Varela, A. Carriles Menéndez, and S. Poveda Martiꞁnez. 2001a. High temperature erosion wear of flame and plasma-sprayed nickel–chromium coatings under simulated coal-fired boiler atmospheres. *Wear* 247: 214–222.

Higuera, H.V., F.J. Belzunce Varela, A. Carriles Menéndez, and S. Poveda Martinez. 2001b. A comparative study of high-temperature erosion wear of plasma-sprayed NiCrBSiFe and WC–NiCrBSiFe coatings under simulated coal-fired boiler conditions. *Tribology International* 34: 161–169.

Hill, H., S. Weber, U. Raab, W. Theisen, and L. Wagner. 2012. Influence of processing and heat treatment on corrosion resistance and properties of high alloyed steel coatings. *Journal of Thermal Spray Technology* 21 (5): 987–994.

Hirata, T., S. Ota, and T. Morimoto. 2003. Influence of impurities in Al_2O_3 ceramics on hot corrosion resistance against molten salt. *Journal of the European Ceramic Society* 23: 91–97.

Hitchman, L.N., and J. Knapp. 2010. Failure of thermal barrier coatings subjected to CMAS attack. *Journal of Thermal Spray Technology* 19 (1–2): 148–155.

Holcomb, G.R., S.D. Cramer, S.J. Bullard, B.S. Covino Jr., W.K. Collins, R.D. Govier, and G.E. McGill. 1997. Characterization of thermal-sprayed titanium anodes for cathodic protection. In *Thermal spray: A united forum for scientific and technological advances*, ed. C.C. Berndt, 141–150. Materials Park: ASM International.

Hospach, A., G. Mauer, R. Vaßen, and D. Stöver. 2012. Characteristics of ceramic coatings made by thin film low pressure plasma spraying (LPPS-TF). *Journal of Thermal Spray Technology* 21 (3–4): 435–440.

Houdkova, S., M. Kasparova, and F. Zahalka. 2010. The influence of spraying angle on properties of HVOF sprayed hardmetal coatings. *Journal of Thermal Spray Technology* 19 (5): 893–901.

Huang, X.O., R.J. Wang, T.J. Zhang, H.J. Luo, and Y.F. Lü. 2007. Several application cases of thermal spraying technology on industrial components and its considerations. In *Thermal spray 2007: Global coating solutions*, ed. B.R. Marple, M.M. Hyland, Y.-C. Lau, C.-J. Li, R.S. Lima, and G. Montavon. Materials Park: ASM International. e-Proc.

Hussain, T., D.G. McCartney, P.H. Shipway, and T. Marrocco. 2011. Corrosion behavior of cold sprayed titanium coatings and free standing deposits. *Journal of Thermal Spray Technology* 20 (1–2): 260–274.

Isakaev, E., A. Yablonsky, A. Kogan, V. Katarzhis, V. Kutnov, and P. Ivanov. 1999. The repair of railway frogs using plasma sprayed coatings, heat and mass transfer under plasma conditions. *Annals of the New York Academy of Sciences* 891: 231–235.

İşcan, B., and H. Aydın. 2012. Improving the usability of vegetable oils as a fuel in a low heat rejection diesel engine. *Fuel Processing Technology* 98: 59–64.

Ishikawa, K., T. Suzuki, S. Tobe, and Y. Kitamura. 2001. Resistance of thermal-sprayed duplex coating composed of aluminum and 80Ni-20Cr alloy against aqueous corrosion. *Journal of Thermal Spray Technology* 10 (3): 520–525.

Ishikawa, Y., J. Kawakita, S. Osawa, T. Itsukaichi, Y. Sakamoto, M. Takaya, and S. Kuroda. 2005. Evaluation of corrosion and wear resistance of hard cermet coatings sprayed by using an improved HVOF process. *Journal of Thermal Spray Technology* 14 (3): 384–390.

Ivosevic, M., S.L. Coguill, and S.L. Galbraith. 2009. Polymer thermal spraying: a novel coating process. In *Thermal spray 2009: Proceedings of the international thermal spray conference*, ed. B.R. Marple, M.M. Hyland, Y.-C. Lau, C.-J. Li, R.S. Lima, and G. Montavon, 1078–1083. Materials Park: ASM International.

Iyengar, R.K. 2009. *Thermal spray coating for steel processing*. Littleton: Technovations International.

Jacobs, L., M.M. Hyland, and M. De Bonte. 1999. Study of the influence of microstructural properties on the sliding-wear behavior of HVOF and HVAF sprayed WC-cermet coatings. *Journal of Thermal Spray Technology* 8 (1): 125–132.

Jang, H.-J., D.-H. Park, Y.-G. Junga, J.-C. Jang, S.-C. Choi, and U. Pai. 2006. Mechanical characterization and thermal behavior of HVOF-sprayed bond coat in thermal barrier coatings (TBCs). *Surface and Coating Technology* 200: 4355–4362.

Jansen, F., W. Xi, M.R. Dorfman, J.A. Peters, and D.R. Nagy. 2002. Performance of di-calcium silicate coatings in hot-corrosive environment. *Surface and Coating Technology* 149: 57–61.

Jarligo, M.O., D.E. Mack, R. Vaßen, and D. Stöver. 2009. Application of plasma-sprayed complex perovskites as thermal barrier coatings. *Journal of Thermal Spray Technology* 18 (2): 187–193.

Jeffery, B., M. Peppler, R.S. Lima, and A. McDonald. 2010. Bactericidal effects of HVOF-sprayed nanostructured TiO_2 on Pseudomonas aeruginosa. *Journal of Thermal Spray Technology* 19 (1–2): 344–349.

Ji, G.-C., C.-J. Li, Y.-Y. Wang, and W.-Y. Li. 2007. Erosion performance of HVOF-sprayed Cr3C2-NiCr coatings. *Journal of Thermal Spray Technology* 16 (4): 557–565.

Jiang, S.M., H.Q. Li, J. Ma, C.Z. Xu, J. Gong, and C. Sun. 2010. High temperature corrosion behavior of a gradient NiCoCrAlYSi coating II: oxidation and hot corrosion. *Corrosion Science* 52: 2316–2322.

Jin, O., S. Mall, J.H. Sanders, and S.K. Sharma. 2006. Durability of Cu–Al coating on Ti–6Al–4V substrate under fretting fatigue. *Surface and Coating Technology* 201: 1704–1710.

Johnston, R.E. 2009. The sensitivity of abradable coating residual stresses to varying material properties. *Journal of Thermal Spray Technology* 101318 (5–6): 1004–1013.

———. 2011. Mechanical characterization of AlSi-hBN, NiCrAl-Bentonite, and NiCrAl-Bentonite-hBN freestanding abradable coatings. *Surface and Coating Technology* 205: 3268–3273.

Jones, R.L. 1997. Some aspects of the hot corrosion of thermal barrier coatings. *Journal of Thermal Spray Technology* 6 (1): 77–84.

Juhasz, J.A., and S.M. Best. 2012. Bioactive ceramics: processing, structures and properties. *Journal of Materials Science* 47: 610–624.

Kang, A.S., J.S. Grewal, D. Jain, and S. Kang. 2012. Wear behavior of thermal spray coatings on rotavator blades. *Journal of Thermal Spray Technology* 21 (2): 355–359.

Karger, M., R. Vaßen, and D. Stöver. 2011. Atmospheric plasma sprayed thermal barrier coatings with high segmentation crack densities: spraying process, microstructure and thermal cycling behaviour. *Surface and Coating Technology* 206: 16–23.

Kashirin, A., O. Klyuev, T. Buzdygar, and A. Shkodkin. 2007. DYMET technology evolution and application. In *Thermal spray 2007: Global coating solutions*, ed. B.R. Marple, M.M. Hyland, Y.-C.

Lau, C.-J. Li, R.S. Lima, and G. Montavon, 141–145. Materials Park: ASM International.

Kasparova, M., F. Zahalka, and S. Houdkova. 2011. WC-Co and Cr_3C_2-NiCr coatings in low- and high-stress abrasive conditions. *Journal of Thermal Spray Technology* 20 (3): 412–424.

Kaur, M., H. Singh, and S. Prakash. 2011. Surface engineering analysis of detonation-gun sprayed Cr3C2–NiCr coating under high-temperature oxidation and oxidation–erosion environments. *Surface and Coating Technology* 206: 530–541.

Kaushal, G., H. Singh, and S. Prakash. 2011. High-temperature erosion-corrosion performance of high-velocity oxy-fuel sprayed Ni-20Cr coating in actual boiler environment. *Metallurgical and Materials Transactions A* 42 (7): 1836–1846.

Kawakita, J., S. Kuroda, T. Fukushima, and T. Kodama. 2005. Improvement of corrosion resistance of high-velocity oxyfuel-sprayed stainless steel coatings by addition of molybdenum. *Journal of Thermal Spray Technology* 14 (2): 225–230.

Kembaiyan, K.T., and K. Keshavan. 1995. Combating severe fluid erosion and corrosion of drill bits using thermal spray coatings. *Wear* 186–187: 487–492.

Kenichi, S., S. Nakahama, S. Hattori, and K. Nakano. 2005. Slurry wear and cavitation erosion of thermal-sprayed cermets. *Wear* 258: 768–775.

Khan, F.F., G. Bae, K. Kang, H. Na, J. Kim, T. Jeong, and C. Lee. 2011. Evaluation of die-soldering and erosion resistance of high velocity oxy-fuel sprayed MoB-based cermet coatings. *Journal of Thermal Spray Technology* 20 (5): 1022–1034.

Khor, K.A., P. Cheang, and Y. Wang. 1997a. The thermal spray processing of HA powders and coatings. *JOM* 49: 51–57.

Khor, K.A., C.S. Yip, and P. Cheang. 1997b. Ti-6Al-4 hydroxyapatite composite coatings prepared by thermal spray techniques. *Journal of Thermal Spray Technology* 6 (1): 109–115.

Kim, K., and A.M. Korsunsky. 2011. Effects of imposed displacement and initial coating thickness on fretting behaviour of a thermally sprayed coating. *Wear* 271 (7–8): 1080–1085.

Kim, H.-J., and Y.-G. Kweon. 1996. The application of thermal sprayed coatings for pig iron ingot molds. *Journal of Thermal Spray Technology* 5 (4): 463–468.

Kim, S.-J., and S.-J. Lee. 2011. Effects of F−Si sealer on electrochemical characteristics of 15%Al−85%Zn alloy thermal spray coating. *Transactions of the Nonferrous Metals Society of China* 21: 2798–2804.

Kim, G.E., and J. Walker. 2007. Successful application of nanostructured titanium dioxide coating for high-pressure acid-leach application. *Journal of Thermal Spray Technology* 16 (1): 34–39.

Kim, H.J., Y.G. Kweon, and R.W. Chang. 1994. Wear and erosion behavior of plasma-sprayed WC-Co coatings. *Journal of Thermal Spray Technology* 3 (2): 169–178.

Kim, H.-J., S. Odoul, C.-H. Lee, and Y.-G. Kweon. 2001. The electrical insulation behavior and sealing effects of plasma-sprayed alumina-titania coatings. *Surface and Coating Technology* 140: 293–301.

Kim, D.-J., I.-H. Shin, J.-M. Koo, C.-S. Seok, and T.-W. Lee. 2010. Failure mechanisms of coin-type plasma-sprayed thermal barrier coatings with thermal fatigue. *Surface and Coating Technology* 205: S451–S458.

Koiprasert, H., S. Dumrongrattana, and P. Niranatlumpong. 2004. Thermally sprayed coatings for protection of fretting wear in land-based gas-turbine engine. *Wear* 257: 1–7.

Kokini, K., J. DeJonge, S. Rangaraj, and B. Beardsley. 2002. Thermal shock of functionally graded thermal barrier coatings with similar thermal resistance. *Surface and Coating Technology* 154: 223–231.

Koolloos, M.F.J., and J.M. Houben. 2000. Behavior of plasma-sprayed thermal barrier coatings during thermal cycling and the effect of a preoxidized NiCrAlY bond coat. *Journal of Thermal Spray Technology* 9 (1): 49–58.

Krishnamurthy, N., M.S. Murali, B. Venkataraman, and P.G. Mukunda. 2012. Characterization and solid particle erosion behavior of plasma sprayed alumina and calcia-stabilized zirconia coatings on Al-6061 substrate. *Wear* 274–275: 15–27.

Krishnan N, Vardelle A, Legoux JG (2008) A life cycle comparison of hard chrome and thermal sprayed coatings: a case example of aircraft landing gears. In: Lugscheider E (ed) Thermal spray conference: Crossing the border. DVS, Dûsseldorf, e-Proc

Kulu, P., and J. Hailing. 1998. Recycled hard metal-base wear-resistant composite coatings. *Journal of Thermal Spray Technology* 7 (2): 173–178.

Kulu, P., and T. Pihl. 2002. Selection criteria for wear resistant powder coatings under extreme erosive wear conditions. *Journal of Thermal Spray Technology* 11 (4): 517–522.

Kulu, P., L. Hussainova, and R. Veinthal. 2005. Solid particle erosion of thermal sprayed coatings. *Wear* 258: 488–496.

Kumar, A., J. Boy, R. Zatorski, and L.D. Stephenson. 2005. Thermal spray and weld repair alloys for the repair of cavitation damage in turbines and pumps: a technical note. *Journal of Thermal Spray Technology* 14 (2): 177–182.

Kwon, J.-Y., J.-H. Lee, H.-C. Kim, Y.-G. Jung, U. Paik, and K.-S. Lee. 2006. Effect of thermal fatigue on mechanical characteristics and contact damage of zirconia-based thermal barrier coatings with HVOF-sprayed bond coat. *Materials Science and Engineering A* 429: 173–180.

Laha, T., A. Agarwal, T. McKechnie, and S. Seal. 2004. Synthesis and characterization of plasma spray formed carbon nanotube reinforced aluminum composite. *Materials Science and Engineering A* 381: 249–258.

Landor, I., P. Vavrik, A. Sosna, D. Jahoda, H. Hahn, and M. Daniel. 2007. Hydroxyapatite porous coating and the osteointegration of the total hip replacement. *Archives of Orthopaedic and Traumatic Surgery* 127 (2): 81–89.

Lathabai, S., M. Ottmuller, and I. Fernandez. 1998. Solid particle erosion behaviour of thermal sprayed ceramic, metallic and polymer coatings. *Wear* 221: 93–108.

Lebedev, A.S., and S.V. Kostennikov. 2008. Trends in increasing gas-turbine units efficiency. *Thermal Engineering* 55 (6): 461–468.

Lee, C. 2009a. Market direction and application opportunities for TS growth in Korea. In *Thermal Spray-2009: Proceedings (ITSC-2009)*, ed. B.R. Marple, M.M. Hyland, Y.-C. Lau, C.-J. Li, R.S. Lima, and G. Montavon, 505–510. Materials Park: ASM International.

———. 2009b. Market direction and application opportunities for T/S growth in Korea. In *Thermal spray 2009: Proceedings of the international thermal spray conference*, ed. B.R. Marple, M.M. Hyland, Y.-C. Lau, C.-J. Li, R.S. Lima, and G. Montavon, 505–510. Materials Park: ASM International. e-Proc.

Legoux, J.-G., F. Chellat, R.S. Lima, B.R. Marple, M.N. Bureau, H. Shen, and G.A. Candeliere. 2006. Development of osteoblast colonies on new bioactive coatings. *Journal of Thermal Spray Technology* 15 (4): 628–633.

Leivo, E., T. Wilenius, T. Kinos, P. Vuoristo, and T. Mäntylä. 2004. Properties of thermally sprayed fluoropolymer PVDF, ECTFE, PFA and FEP coatings. *Progress in Organic Coating* 49: 69–73.

Lekatou, A., E. Regoutas, and A.E. Karantzalis. 2008. Corrosion behaviour of cermet-based coatings with a bond coat in 0.5 M H_2SO_4. *Corrosion Science* 50: 3389–3400.

Li, J., H. Liao, and C. Coddet. 2002. Friction and wear behavior of flame-sprayed PEEK coatings. *Wear* 252: 824–831.

Li, Y., Y. Ma, B. Xie, S. Cao, and Z. Wu. 2007. Dry friction and wear behavior of flame-sprayed polyamide1010/n-SiO_2 composite coatings. *Wear* 262: 1232–1238.

Li, W.-Y., G. Zhang, C. Zhang, O. Elkedim, H. Liao, and C. Coddet. 2008. Effect of ball milling of feedstock powder on microstructure and properties of TiN particle-reinforced Al alloy-based composites fabricated by cold spraying. *Journal of Thermal Spray Technology* 17 (3): 316–322.

Li, L., N. Hitchman, and J. Knapp. 2010a. Failure of thermal barrier coatings subjected to CMAS attack. *Journal of Thermal Spray Technology* 19 (1–2): 148–155.

Li, Y., C.-J. Li, G.-J. Yang, and L.-K. Xing. 2010b. Thermal fatigue behavior of thermal barrier coatings with the MCrAlY bond coats by cold spraying and low-pressure plasma spraying. *Surface and Coating Technology* 205: 2225–2233.

Liang, B., and C. Ding. 2005. Thermal shock resistances of nanostructured and conventional zirconia coatings deposited by atmospheric plasma spraying. *Surface and Coating Technology* 197: 185–192.

Liang, Y., Y. Xie, H. Ji, L. Huang, and X. Zheng. 2010. Chemical stability and biological properties of plasma-sprayed CaO-SiO_2-ZrO_2 coatings. *Journal of Thermal Spray Technology* 19 (6): 1171–1178.

Lima, R.S., and B.R. Marple. 2005. Superior performance of high-velocity oxyfuel-sprayed nanostructured TiO_2 in comparison to air plasma-sprayed conventional Al_2O_3-13TiO_2. *Journal of Thermal Spray Technology* 14 (3): 397–404.

———. 2007. Thermal spray coatings engineered from nanostructured ceramic agglomerated powders for structural, thermal barrier and biomedical applications: a review. *Journal of Thermal Spray Technology* 16 (1): 40–63.

Lima, M.M., C. Godoy, P.J. Modenesi, J.C. Avelar-Batista, A. Davison, and A. Matthews. 2004. Coating fracture toughness determined by Vickers indentation: an important parameter in cavitation erosion resistance of WC–Co thermally sprayed coatings. *Surface and Coating Technology* 177–178: 489–496.

Lima, R.S., H. Li, K.A. Khor, and B.R. Marple. 2006. Biocompatible nanostructured high-velocity oxyfuel sprayed titania coating: deposition, characterization, and mechanical properties. *Journal of Thermal Spray Technology* 15 (4): 623–627.

Lima, R.S., S. Dimitrievska, M.N. Bureau, B.R. Marple, A. Petit, F. Mwale, and J. Antoniou. 2010. HVOF-sprayed Nano TiO_2-HA coatings exhibiting enhanced biocompatibility. *Journal of Thermal Spray Technology* 19 (1–2): 336–343.

Lima, C.R.C., N.F.C. de Souza, and F. Camargo. 2012. Study of wear and corrosion performance of thermal sprayed engineering polymers. *Surface and Coating Technology* 220: 140–143.

Limarga, A.M., S. Widjaja, and T.H. Yip. 2005. Mechanical properties and oxidation resistance of plasma-sprayed multilayered Al_2O_3/ZrO_2 thermal barrier coatings. *Surface and Coating Technology* 197: 93–102.

Lin, K.H., Z.H. Xu, and S.T. Lin. 2011. A study on microstructure and dielectric performances of alumina/copper composites by plasma spray coating. *Journal of Materials Engineering and Performance* 20 (2): 231–237.

Lins, V.F.C., J.R.T. Branco, F.R.C. Diniz, J.C. Brogan, and C.C. Berndt. 2007. Erosion behavior of thermal sprayed, recycled polymer and ethylene–methacrylic acid composite coatings. *Wear* 262: 274–281.

Lisjak, D., M. Bégard, M. Bruehl, K. Bobzin, A. Hujanen, P. Lintunen, G. Bolelli, L. Lusvarghi, S. Ovtar, and M. Drofenik. 2011. Hexaferrite/polyester composite coatings for electromagnetic-wave absorbers. *Journal of Thermal Spray Technology* 20 (3): 638–644.

Liu, A., M. Guo, J. Gao, and M. Zhao. 2006. Influence of bond coat on shear adhesion strength of erosion and thermal resistant coating for carbon fiber reinforced thermosetting polyimide. *Surface and Coating Technology* 201: 2696–2700.

Liu, Y., Y.F. Gao, S.Y. Tao, X.M. Zhou, W.D. Li, H.J. Luo, and C.X. Ding. 2008. Microstructure of plasma sprayed La_2O_3-modified YSZ coatings. *Journal of Thermal Spray Technology* 17 (5–6): 603–607.

Longo, F.N. 1992. Industrial guide–markets, materials, and applications for thermal-sprayed coatings. *Journal of Thermal Spray Technology* 1 (2): 143–145.

Lu, Y.-P., M.-S. Li, S.-T. Li, Z.-G. Wang, and R.-F. Zhu. 2004. Plasma-sprayed hydroxyapatite-titania composite bond coat for hydroxyapatite coating on titanium substrate. *Biomaterials* 25: 4393–4403.

Lu, W., Y. Wu, J. Zhang, S. Hong, J. Zhang, and G. Li. 2011. Microstructure and corrosion resistance of plasma sprayed Fe-based alloy coating as an alternative to hard chromium. *Journal of Thermal Spray Technology* 20 (5): 1063–1070.

Luo, L., S. Liu, J. Li, and Y. Wu. 2011. Thermal shock resistance of FeMnCrAl/Cr3C2–Ni9Al coatings deposited by high velocity arc spraying. *Surface and Coating Technology* 205: 3467–3471.

Ma, X., and A. Matthews. 2007. Investigation of abradable seal coating performance using scratch testing. *Surface and Coating Technology* 202: 1214–1220.

———. 2009. Evaluation of abradable seal coating mechanical properties. *Wear* 267: 1501–1510.

Ma, W., M.O. Jarligo, D.E. Mack, D. Pitzer, J. Malzbender, R. Vaßen, and D. Stöver. 2008. New generation perovskite thermal barrier coating materials. *Journal of Thermal Spray Technology* 17 (5–6): 831–837.

Ma, B., Y. Li, and K. Su. 2009. Characterization of ceria–yttria stabilized zirconia plasma-sprayed coatings. *Applied Surface Science* 255: 7234–7237.

MAGETEX. n.d. Thermal coatings in Europe: a business prospective. MAGETEX, Les bureaux de Sèvres, 2 rue Troyon, 92316 Sèvres.

Markocsan, N., P. Nylén, J. Wigren, and X.-H. Li. 2007. Low thermal conductivity coatings for gas turbine applications. *Journal of Thermal Spray Technology* 16 (4): 498–505.

Markocsan, N., P. Nylén, J. Wigren, X.-H. Li, and A. Tricoire. 2009. Effect of thermal aging on microstructure and functional properties of zirconia-base thermal barrier coatings. *Journal of Thermal Spray Technology* 18 (2): 201–208.

Marx, S., A. Paul, A. Köhler, and G. Hüttl. 2006. Cold spraying: innovative layers for new applications. *Journal of Thermal Spray Technology* 15 (2): 177–183.

Mary, C., S. Fouvry, J.M. Martin, and B. Bonnet. 2011. Pressure and temperature effects on Fretting Wear damage of a Cu–Ni–In plasma coating versus Ti17 titanium alloy contact. *Wear* 272: 18–37.

Matejicek, J., P. Chraska, and J. Linke. 2007. Thermal spray coatings for fusion applications—review. *Journal of Thermal Spray Technology* 16 (1): 64–83.

Matsubara, Y., Y. Sochi, M. Tanabe, and A. Takeya. 2007. Advanced coatings on furnace wall tubes. *Journal of Thermal Spray Technology* 16 (2): 195–201.

Matthäus, G., J. Henry, and D. Ackermann. 2009. Further developments in internal diameter HVOF application of WC-CoCr for hard chrome replacement in critical applications such as landing gear. In *Thermal spray 2009: Proceedings of the international thermal spray conference*, ed. B.R. Marple, M.M. Hyland, Y.-C. Lau, C.-J. Li, R.S. Lima, and G. Montavon, 722–724. Materials Park: ASM International.

Mei, H., Y. Liu, L. Cheng, and L. Zhang. 2012. Corrosion mechanism of a NiCoCrAlTaY coated Mar-M247 superalloy in molten salt vapor. *Corrosion Science* 55: 201–204.

Meng, H. 2010. The performance of different WC-based cermet coatings in oil and gas applications–A comparison. In *ITSC 2010 Thermal spray: Global solutions, future applications*. Düsseldorf: DVS. e-Proc.

Michaux, P., G. Montavon, A. Grimaud, A. Denoirjean, and P. Fauchais. 2010. Elaboration of porous NiO/8YSZ layers by several SPS and SPPS routes. *Journal of Thermal Spray Technology* 19 (1–2): 317–327.

Miller, R.A. 1997. Thermal barrier coatings for aircraft engines: history and directions. *Journal of Thermal Spray Technology* 6 (1): 35–42.

Mizuno, H., and J. Kitamura. 2007. MoB/CoCr cermet coatings by HVOF spraying against erosion by molten Al-Zn alloy. *Journal of Thermal Spray Technology* 16 (3): 404–413.

Mohan, P., T. Patterson, B. Yao, and Y. Sohn. 2010. Degradation of thermal barrier coatings by fuel impurities and CMAS: thermochemical interactions and mitigation approaches. *Journal of Thermal Spray Technology* 19 (1–2): 156–167.

Molz, R., and D. Hawley. 2007. A method of evaluating thermal spray process performance. In *Thermal spray 2007: Global coating solutions*, ed. B.R. Marple, M.M. Hyland, Y.-C. Lau, C.-J. Li, R.S. Lima, and G. Montavon. Materials Park: ASM International. e-Proc.

Morks, M.F., and A. Kobayashi. 2008. Development of ZrO_2/SiO_2 bioinert ceramic coatings for biomedical application. *Journal of the Mechanical Behavior of Biomedical Materials* 1: 165–171.

Moskowitz, L.N. 1993. Application of HVOF thermal spraying to solve corrosion problems in the petroleum industry—an industrial note. *Journal of Thermal Spray Technology* 2 (1): 21–29.

Muehlberger, E., and P. Meyer. 2009. LPPS – thin film processes: overview of origin and future possibilities. In *Thermal spray 2009: Proceedings of the international thermal spray conference*, ed. B.R. Marple, M.M. Hyland, Y.-C. Lau, C.-J. Li, R.S. Lima, and G. Montavon, 737–740. Materials Park: ASM International.

Murakami, K., and M. Shimada. 2009. Development of thermal spray coatings with corrosion protection and antifouling properties. In *Thermal spray 2009: Proceedings of the international thermal spray conference*, ed. B.R. Marple, M.M. Hyland, Y.-C. Lau, C.-J. Li, R.S. Lima, and G. Montavon, 1041–1044. Materials Park: ASM International.

Musil, J., M. Alaya, and R. Oberacker. 1997. Plasma-sprayed duplex and graded partially stabilized zirconia thermal barrier coatings: deposition process and properties. *Journal of Thermal Spray Technology* 6 (4): 449–455.

Mutasim, Z., and W. Brentnall. 1997. Thermal barrier coatings for industrial gas turbine applications: an industrial note. *Journal of Thermal Spray Technology* 6 (1): 105–108.

Nagai, M., S. Shigemura, and A. Yoshiya. 2009. Thermal-sprayed CFRP roll with resistant to thermal shock and wear - for papermaking machine. In *Thermal spray 2009: Proceedings of the international thermal spray conference*, ed. B.R. Marple, M.M. Hyland, Y.-C. Lau, C.-J. Li, R.S. Lima, and G. Montavon, 607–611. Materials Park: ASM International.

Nakahira, A. 2009a. Current status and future prospect of thermal spray coating applications and coating service market of job shops in Japan. In *Proceedings (ITSC-2009) Conference*, ed. B.R. Marple, M.M. Hyland, Y.-C. Lau, C.-J. Li, R.S. Lima, and G. Montavon, 499–504. Materials Park: ASM International.

———. 2009b. Current status and future prospect of thermal spray coating applications and coating service market of job shops in Japan. In *Thermal spray 2009: Proceedings of the international thermal spray conference*, ed. B.R. Marple, M.M. Hyland, Y.-C. Lau, C.-J. Li, R.S. Lima, and G. Montavon, 499–504. Materials Park: ASM International. e-Proc.

Nelson, W.A., and R.M. Orenstein. 1997. Land based gas turbines TBC experience in land-based gas turbines. *Journal of Thermal Spray Technology* 6 (2): 176–180.

Nelson, G.M., J.A. Nychka, and A.G. McDonald. 2011. Flame spray deposition of titanium alloy-bioactive glass composite coatings. *Journal of Thermal Spray Technology* 20 (6): 1339–1351.

Neufuss, K., P. Chráska, B. Kolman, S. Sampath, and Z. Trávnícek. 1997. Properties of plasma-sprayed freestanding ceramic parts. *Journal of Thermal Spray Technology* 6 (4): 434–438.

Niebuhr, D., and M. Scholl. 2005. Synthesis and performance of plasma-sprayed polymer/steel coating system. *Journal of Thermal Spray Technology* 14 (4): 487–494.

Ong, J.L., M. Appleford, S. Oh, Y. Yang, W.-H. Chen, et al. 2006. The characterization and development of bioactive hydroxyapatite coatings. *JOM* 58 (7): 67–69.

Osawa, S., T. Itsukaichi, and R. Ahmed. 2005. Influence of substrate properties on the impact resistance of WC cermet coatings. *Journal of Thermal Spray Technology* 14 (4): 495–501.

Ouyang, J.H., S. Sasaki, and K. Umeda. 2001. Microstructure and tribological properties of low-pressure plasma-sprayed ZrO_2–CaF_2–Ag_2O composite coating at elevated temperature. *Wear* 249: 440–451.

Pacheo da Silva, C., et al. 1991. *2nd Plasma Technik Symposium 1*, 363–373. Wohlen: Plasma Technik.

Pan, Z.Y., Y. Wang, C.H. Wang, X.G. Sun, and L. Wang. 2012. The effect of SiC particles on thermal shock behavior of Al2O3/8YSZ coatings fabricated by atmospheric plasma spraying. *Surface and Coating Technology* 206: 2484–2498.

Pantelis, D.I., P. Psyllaki, and N. Alexopoulos. 2000. Tribological behavior of plasma-sprayed Al_2O_3 coatings under severe wear conditions. *Wear* 237: 197–204.

Pardo, A., M.C. Merino, M. Mohedano, P. Casajús, A.E. Coy, and R. Arrabal. 2009. Corrosion behaviour of Mg/Al alloys with composite coatings. *Surface and Coating Technology* 203: 1252–1263.

Parks, W.P., E.E. Hoffman, W.Y. Lee, and I.G. Wright. 1997. Thermal barrier coatings issues in advanced land-based gas turbines. *Journal of Thermal Spray Technology* 6 (2): 187–192.

Parlak, A., H. Yasar, and O. Eldogan. 2005. The effect of thermal barrier coating on a turbo-charged diesel engine performance and exergy potential of the exhaust gas. *Energy Conversion and Management* 46: 489–499.

Patel, R.R., A.K. Keshri, G.S. Dulikravich, and A. Agarwal. 2010. An experimental and computational methodology for near net shape fabrication of thin walled ceramic structures by plasma spray forming. *Journal of Materials Processing Technology* 210: 1260–1269.

Pattison, J., S. Celotto, R. Morgan, M. Bray, and W. O'Neill. 2007. Cold gas dynamic manufacturing: a non-thermal approach to freeform fabrication. *International Journal of Machine Tools and Manufacture* 47: 627–634.

Paul, S., A. Cipitria, I.O. Golosnoy, L. Xie, M.R. Dorfman, and T.W. Clyne. 2007. Effects of impurity content on the sintering characteristics of plasma-sprayed zirconia. *Journal of Thermal Spray Technology* 16 (5–6): 798–803.

Pawlowski, L. 1995. *The science and engineering of thermal spray coatings*. New York: Wiley.

———. 1996. Technology of thermally sprayed anilox rolls: state of art, problems, and perspectives. *Journal of Thermal Spray Technology* 5 (3): 317–334.

Petrovicova, E., and L.S. Schadler. 2002. Thermal spraying of polymers. *International Materials Review* 47 (4): 169–190.

Picas, J.A., A. Forna, and G. Matthäus. 2006. HVOF coatings as an alternative to hard chrome for pistons and valves. *Wear* 261: 477–484.

Pint, B.A., I.G. Wright, and W.J. Brindley. 2000. Evaluation of thermal barrier coating systems on novel substrates. *Journal of Thermal Spray Technology* 9 (2): 198–203.

Pint, B.A., J.A. Haynes, and Y. Zhang. 2010. Effect of superalloy substrate and bond coating on TBC lifetime. *Surface and Coating Technology* 205: 1236–1240.

Pokhmurska, H., B. Wielage, T. Lampke, T. Grund, M. Student, and N. Chervinska. 2008. Post-treatment of thermal spray coatings on magnesium. *Surface and Coating Technology* 202: 4515–4524.

Pomeroy, M.J. 2005. Coatings for gas turbine materials and long-term stability issues. *Materials and Design* 26: 223–231.

Pratap Singh, V., A. Sil, and R. Jayaganthan. 2011. Tribological behavior of plasma sprayed Cr_2O_3–3%TiO_2 coatings. *Wear* 272: 149–158.

Prevéy, P.S. 2000. X-ray diffraction characterization of crystallinity and phase composition in plasma-sprayed hydroxyapatite coatings. *Journal of Thermal Spray Technology* 9 (3): 369–376.

Process Industries: Power, http://www.castolin.com

Prudenziati, M., G. Cirri, and P. Dal Bo. 2006. Novel high-temperature reliable heaters in plasma spray technology. *Journal of Thermal Spray Technology* 15 (3): 329–331.

Puranen, J., J. Lagerbom, L. Hyvärinen, M. Kylmälahti, O. Himanen, M. Pihlatie, J. Kiviaho, and P. Vuoristo. 2011. The structure and properties of plasma sprayed iron oxide doped manganese cobalt oxide spinel coatings for SOFC metallic interconnectors. *Journal of Thermal Spray Technology* 20 (1–2): 154–159.

Qiao, Y., Y.-R. Liu, and T.E. Fischer. 2001. Sliding and abrasive wear resistance of thermal-sprayed WC-CO coatings. *Journal of Thermal Spray Technology* 10 (1): 118–125.

Rajendran, R. 2012. Gas turbine coatings – an overview. *Engineering Failure Analysis* 26: 355–369.

Ramachandran, C.S., V. Balasubramanian, P.V. Ananthapadmanabhan, and V. Viswabaskaran. 2012. Understanding the dry sliding wear behaviour of atmospheric plasma-sprayed rare earth oxide coatings. *Materials and Design* 39: 234–252.

Ramesh, M.R., S. Prakash, S.K. Nath, Pawan Kumar Sapra, and N. Krishnamurthy. 2011. Evaluation of thermocyclic oxidation behavior of HVOF-sprayed NiCrFeSiB coatings on boiler tube steels. *Journal of Thermal Spray Technology* 20 (5): 992–1000.

Ramesha, C.S., D.S. Devaraja, R. Keshavamurthya, and B.R. Sridharb. 2011. Slurry erosive wear behavior of thermally sprayed Inconel-718 coatings by APS process. *Wear* 271: 1365–1371.

Richer, P., M. Yandouzi, L. Beauvais, and B. Jodoin. 2010. Oxidation behaviour of CoNiCrAlY bond coats produced by plasma, HVOF and cold gas dynamic spraying. *Surface and Coating Technology* 204: 3962–3974.

Rieger, G., J. Wecker, W. Rodewald, W. Sattler, F.W. Bach, T. Duda, and W. Unterberg. 2000. Nd-Fe-B permanent magnets (thick films) produced by a vacuum-plasma-spraying process. *Journal of Applied Physics* 87 (9): 5329–533

Rodriguez, J., A. Martin, R. Fernández, and J.E. Fernández. 2003. An experimental study of the wear performance of NiCrBSi thermal spray coatings. *Wear* 255: 950–955.

Roy, M., A. Bandyopadhyay, and S. Bose. 2011. Induction plasma sprayed nano hydroxyapatite coatings on titanium for orthopaedic and dental implants. *Surface and Coating Technology* 205: 2785–2792.

Sacriste, D., N. Goubot, J. Dhers, M. Ducos, and A. Vardelle. 2001. An evaluation of the electric arc spray and (HPPS) processes for the manufacturing of high power plasma spraying MCrAlY coatings. *Journal of Thermal Spray Technology* 10 (2): 352–358.

Saeidi, S., K.T. Voisey, and D.G. McCartney. 2009. The effect of heat treatment on the oxidation behavior of HVOF and VPS CoNiCrAlY coatings. *Journal of Thermal Spray Technology* 18 (2): 209–216.

Sahraoui, T., N.-E. Fenineche, G. Montavon, and C. Coddet. 2004. Alternative to chromium: characteristics and wear behavior of HVOF coatings for gas turbine shafts repair (heavy-duty). *Journal of Materials Processing Technology* 152: 43–55.

Sakata, K., K. Nakano, H. Miyahara, Y. Matsubara, and K. Ogi. 2007. Microstructure control of thermally sprayed Co-based self-fluxing alloy coatings by diffusion treatment. *Journal of Thermal Spray Technology* 16 (5–6): 991–997.

Sampath, S. 2010. Thermal spray applications in electronics and sensors: past, present, and future. *Journal of Thermal Spray Technology* 19 (5): 921–949.

Sanchez, E., E. Bannier, V. Cantavella, M.D. Salvador, E. Klyatskina, J. Morgiel, J. Grzonka, and A.R. Boccaccini. 2008. Deposition of Al_2O_3-TiO_2 nanostructured powders by atmospheric plasma spraying. *Journal of Thermal Spray Technology* 17 (3): 329–337.

Sanpo, N., M. Lu Tan, P. Cheang, and K.A. Khor. 2009a. Antibacterial property of cold-sprayed HA-Ag/PEEK coating. *Journal of Thermal Spray Technology* 18 (1): 10–15.

Sanpo, N., S. Ming Ang, P. Cheang, and K.A. Khor. 2009b. Antibacterial property of cold sprayed chitosan-Cu/Al coating. *Journal Thermal Spray Technology* 18 (4): 600–608.

Santa, J.F., L.A. Espitia, J.A. Blanco, S.A. Romo, and A. Toro. 2009. Slurry and cavitation erosion resistance of thermal spray coatings. *Wear* 267: 160–167.

Sanz, A. 2001. Tribological behavior of coatings for continuous casting of steel. *Surface and Coating Technology* 146–147: 55–64.

Sathish, S., M. Geetha, S.T. Aruna, N. Balaji, K.S. Rajam, and R. Asokamani. 2011. Sliding wear behavior of plasma sprayed nanoceramic coatings for biomedical applications. *Wear* 271: 934–941.

Savarimuthu, A.C., H.F. Taber, I. Megat, J.R. Shadley, E.F. Rybicki, W.C. Cornell, W.A. Emery, D.A. Somerville, and J.D. Nuse. 2001. Sliding wear behavior of tungsten carbide thermal spray coatings for replacement of chromium electroplate in aircraft applications. *Journal of Thermal Spray Technology* 10 (3): 502–510.

Schilke, P.W. 2004. *Advanced Gas Turbine Materials and Coatings.* GER-3569G, General Electric Company, August. 2004.

Schmidt, D.P., B.A. Shaw, E. Sikora, W.W. Shaw, and L.H. Laliberte. 2006. Corrosion protection assessment of sacrificial coating systems as a function of exposure time in a marine environment. *Progress in Organic Coating* 57: 352–364.

Schulz, U., O. Bernardi, A. Ebach-Stahl, R. Vaßen, and D. Sebold. 2008a. Improvement of EB-PVD thermal barrier coatings by treatments of a vacuum plasma-sprayed bond coat. *Surface and Coating Technology* 203: 160–170.

Schulz, U., K. Fritscher, and A. Ebach-Stahl. 2008b. Cyclic behavior of EB-PVD thermal barrier coating systems with modified bond coats. *Surface and Coating Technology* 203: 449–455.

Schwetzke, R., and H. Kreye. 1999. Microstructure and properties of tungsten carbide coatings sprayed with various high-velocity oxygen fuel spray systems. *Journal of Thermal Spray Technology* 8 (3): 433–439.

Seonga, B.G., S.Y. Hwanga, M.C. Kima, and K.Y. Kimb. 2001. Reaction of WC-Co coating with molten zinc in a zinc pot of a continuous galvanizing line. *Surface and Coating Technology* 138: 101–110.

Shen, Y., V. Alexandra, B. Almeida, and François Gitzhofer. 2011. Preparation of nanocomposite GDC/LSCF cathode material for IT-SOFC by induction plasma spraying. *Journal of Thermal Spray Technology* 20 (1–2): 145–153.

Shi, S., and J.-Y. Hwang. 2003. Plasma spray fabrication of near-net-shape ceramic objects. *Journal of Minerals and Materials Characterization and Engineering* 2 (2): 145–150.

Shin, I.-H., J.-M. Koo, C.-S. Seok, S.-H. Yang, T.-W. Lee, and B.-S. Kim. 2011. Estimation of spallation life of thermal barrier coating of gas turbine blade by thermal fatigue test. *Surface and Coating Technology* 205: S157–S160.

Sidhu, B.S., and S. Prakash. 2006. Erosion-corrosion of plasma as sprayed and laser remelted Stellite-6 coatings in a coal fired boiler. *Wear* 260: 1035–1044.

Sidhu, T.S., S. Prakash, and R.D. Agrawal. 2005. Studies on the properties of high-velocity oxy-fuel thermal spray coatings for higher temperature applications. *Materials Science* 41 (6): 805–823.

———. 2006a. Hot corrosion resistance of high-velocity oxyfuel sprayed coatings on a nickel-base superalloy in molten salt environment. *Journal of Thermal Spray Technology* 15 (3): 387–399.

Sidhu, H.S., B.S. Sidhu, and S. Prakash. 2006b. Comparative characteristic and erosion behavior of NiCr coatings deposited by various high-velocity oxyfuel spray processes. *Journal of Materials Engineering and Performance* 5 (6): 699–704.

Sidhu, T.S., A. Malik, S. Prakash, and R.D. Agrawal. 2007. Oxidation and hot corrosion resistance of HVOF WC-NiCrFeSiB coating on Ni- and Fe-based superalloys at 800 °C. *Journal of Thermal Spray Technology* 16 (5–6): 844–849.

Singh, H., D. Puri, S. Prakash, and V.V. Rama Rao. 2006. On the high-temperature oxidation protection behavior of plasma-sprayed stellite-6 coatings. *Metallurgical and Materials Transactions A* 37A: 3048–3056.

Singh, H., A. Ang, S. Matthews, H. DeVilliers-Lovelock, and B. Singh Sidu. 2019. Thermal spray for extreme environments, editorial. *Journal of Thermal Spray Technology* 28: 1339–1345.

Skandan, G., R. Yao, R. Sadangi, B.H. Kear, Y. Qiao, L. Liu, and T.E. Fischer. 2000. Multimodal coatings: a new concept in thermal spraying. *Journal of Thermal Spray Technology* 9 (3): 329–331.

Smyth, R.T., and J.C. Anderson. 1975. Production of resistors by arc plasma spraying. *Electrocomponent Science and Technology* 2: 135–145.

Sobolev, V.V., J.M. Guilemany, and J. Nutting. 2004. *High velocity oxy-fuel spraying.* London: Maney for the Institute of Materials, Minerals and Mining.

Soechting, F.O. 1999. A design perspective on thermal barrier coatings. *Journal of Thermal Spray Technology* 8 (4): 505–511.

Song, J.X., Y.F. Han, S.S. Li, and C.B. Xiao. 2005. Repair of NiCrAlYSi overlay coating on Ni3Al base alloy IC6. *Intermetallics* 13: 351–355.

Sørensen, P.A., S. Kiil, K. Dam-Johansen, and C.E. Weinell. 2009. Anticorrosive coatings: a review. *Journal of Coating Technology and Research* 6 (2): 135–176.

Souza, V.A.D., and A. Neville. 2007. Aspects of microstructure on the synergy and overall material loss of thermal spray coatings in erosion–corrosion environments. *Wear* 263: 339–346.

Sporer, D., M. Dorfman, L. Xie, A. Refke, I. Giovannetti, and M. Giannozzi. 2007. Processing and properties of advanced ceramic abradable coatings. In *Thermal spray 2007: Global coating solutions*, ed. M.R. Marple, M.M. Hyland, Y.-C. Lau, C.-J. Li, R.S. Lima, and G. Montavon, 495–500. Materials Park: ASM International.

Staia, M.H., M. Suárez, D. Chicot, J. Lesage, A. Iost, and E.S. Puchi-Cabrera. 2012. Cr$_2$C$_3$–NiCr VPS thermal spray coatings as candidate for chromium replacement. *Surface and Coating Technology* 220: 225–231.

Steinke, T., G. Mauer, R. Vaßen, D. Stöver, D. Roth-Fagaraseanu, and M. Hancock. 2010. Process design and monitoring for plasma sprayed abradable coatings. *Journal of Thermal Spray Technology* 19 (4): 756–764.

Stewart, S., and R. Ahmed. 2002. Rolling contact fatigue of surface coatings—A review. *Wear* 253: 1132–1144.

Stoica, V., R. Ahmed, M. Golshan, and S. Tobe. 2004. Sliding wear evaluation of hot isostatically pressed thermal spray cermet coatings. *Journal of Thermal Spray Technology* 13 (1): 93–107.

Stoica, V., R. Ahmed, and T. Itsukaichi. 2005. Influence of heat-treatment on the sliding wear of thermal spray cermet coatings. *Surface and Coating Technology* 199: 7–21.

Stringer, J., and M.B. Marshall. 2012. High speed wear testing of an abradable coating. *Wear* 294–295: 257–263.

Sun, F., N. Zhang, H. Liao, and J. Li. 2012. Effect of heat treatment temperature on performance of plasma-sprayed apatite-lanthanum silicate coatings as electrolytes for IT-SOFC. *Journal of Thermal Spray Technology* 21 (6): 1257–1267.

Sundararajan, G., Y.R. Mahajan, and S.V. Joshi. 2009a. Thermal Spraying in India: Status and Prospects. In *Proceedings (ITSC-2009)*, ed. B.R. Marple, M.M. Hyland, Y.-C. Lau, C.-J. Li, R.S. Lima, and G. Montavon, 511–516. Materials Park: ASM International.

———. 2009b. Thermal spraying in India: status and prospects. In *Thermal spray 2009: Proceedings of the international thermal spray conference*, ed. B.R. Marple, M.M. Hyland, Y.-C. Lau, C.-J.

Li, R.S. Lima, and G. Montavon, 511–516. Materials Park: ASM International.

Sweet, G.K. 1993. Applying thermoplastic/thermoset powder with a modified plasma system. In *Proceedings of the 1993 national thermal spray conference*, ed. C.C. Berndt and F. Bernicki, 381–384. Materials Park: ASM International.

Takenoiri, S., N. Kadokawa, and K. Koseki. 2000. Development of metallic substrate supported planar solid oxide fuel cells fabricated by atmospheric plasma spraying. *Journal of Thermal Spray Technology* 3639 (3): 360–363.

Tamura, M., M. Takahashi, J. Ishii, K. Suzuki, M. Sato, and K. Shimomur. 1999. Multilayered thermal barrier coating for land-based gas turbines. *Journal of Thermal Spray Technology* 8 (1): 68–72.

Tan, Y., V. Srinivasan, T. Nakamura, S. Sampath, P. Bertrand, and G. Bertrand. 2012. TBC optimizing compliance and thermal conductivity of plasma sprayed thermal barrier coatings via controlled powders and processing strategies. *Journal of Thermal Spray Technology* 21 (5): 950–962.

Tang, F., and J.M. Schoenung. 2005. Local accumulation of thermally grown oxide in plasma-sprayed thermal barrier coatings with rough top-coat/bond-coat interfaces. *Scripta Materialia* 52: 905–909.

Tani, K., and Y. Harada. 2007. Enhancement of service life of steam generating tubes in oil-fired boiler for power generation employing plasma spray technology. *Journal of Thermal Spray Technology* 16 (1): 111–117.

Tani, K., and H. Nakahira. 1992. Status of thermal spray technology in Japan. *Journal of Thermal Spray Technology* 1 (4): 333–339.

Tao, K., X.-L. Zhou, H. Cui, and J.-S. Zhang. 2009. Oxidation and hot corrosion behaviors of HVAF-sprayed conventional and nanostructured NiCrC coatings. *Transactions of the Nonferrous Metals Society of China* 19: 1151–1160.

Tao, S., Z. Yin, X. Zhou, and C. Ding. 2010. Sliding wear characteristics of plasma-sprayed Al_2O_3 and Cr_2O_3 coatings against copper alloy under severe conditions. *Tribology International* 43: 69–75.

Tapphorn, R., J. Henness, and H. Gabel. 2009. Kinetic metallization-a repair process for damaged IVD-Al coatings, Mg, and Al alloy components. In *Thermal spray 2009: Proceedings of the international thermal spray conference*, ed. B.R. Marple, M.M. Hyland, Y.-C. Lau, C.-J. Li, R.S. Lima, and G. Montavon, 261–266. Materials Park: ASM International.

Thintri Inc. 2013. Thermal spray wear coatings find growing markets and greater competition. *Spraytime* 20 (1): 1–36.

Tian, W., Y. Wang, and Y. Yang. 2008. Fretting wear behavior of conventional and nanostructured Al2O3–13 wt%TiO2 coatings fabricated by plasma spray. *Wear* 265: 1700–1707.

Tillmann, W., E. Vogli, I. Baumann, G. Kopp, and C. Weihs. 2010. Desirability-based multi-criteria optimization of HVOF spray experiments to manufacture fine structured wear-resistant 75Cr$_3$C$_2$-25(NiCr20) coatings. *Journal of Thermal Spray Technology* 19 (1–2): 393–408.

Toma, F.-L., S. Scheitz, L.-M. Berger, V. Sauchuk, M. Kusnezoff, and S. Thiele. 2011. Comparative study of the electrical properties and characteristics of thermally sprayed alumina and spinel coatings. *Journal of Thermal Spray Technology* 20 (1–2): 195–204.

Toma, F.-L., L.-M. Berger, S. Scheitz, S. Langner, C. Rödel, A. Potthoff, V. Sauchuk, and M. Kusnezoff. 2012. Comparison of the microstructural characteristics and electrical properties of thermally sprayed Al_2O_3 coatings from aqueous suspensions and feedstock powders. *Journal of Thermal Spray Technology* 21 (3–4): 480–488.

Toscano, J., R. Vaßen, A. Gil, M. Subanovic, D. Naumenko, L. Singheiser, and W.J. Quadakkers. 2006. Parameters affecting TGO growth and adherence on MCrAlY-bond coats for TBC's. *Surface and Coating Technology* 201: 3906–3910.

Tucker, R.C., ed. 2013. *ASM handbook, Vol 5A: Thermal spray technology*. Materials Park: ASM International.

Tuominen, J., P. Vuoristo, T. Mäntylä, M. Kylmälahti, J. Vihinen, et al. 2000. Improving corrosion properties of high-velocity oxy-fuel sprayed inconel 625 by using a high-power continuous wave neodymium-doped yttrium aluminum garnet laser. *Journal of Thermal Spray Technology* 9 (4): 513–519.

Uozato, S., K. Nakata, and M. Ushio. 2003. Corrosion and wear behaviors of ferrous powder thermal spray coatings on aluminum alloy. *Surface and Coating Technology* 169–170: 691–694.

Uusitalo, M.A., P.M.J. Vuoristo, and T.A. Mäntylä. 2002. Elevated temperature erosion–corrosion coatings in chlorine containing environments of thermal sprayed. *Wear* 252: 586–594.

Valarezo, A. 2012. Latin America: An Emerging and Growing Market for Thermal Spray. In *Proceedings (ITSC-2012), Conference Houston, TX, 2012*

Valarezo, A., G. Bolelli, W.B. Choi, S. Sampath, V. Cannillo, L. Lusvarghi, and R. Rosa. 2010. Damage tolerant functionally graded WC–Co/stainless steel HVOF coatings. *Surface and Coating Technology* 205: 2197–2208.

Vaßen, R., S. Giesen, and D. Stöver. 2009a. Lifetime of plasma-sprayed thermal barrier coatings: comparison of numerical and experimental results. *Journal of Thermal Spray Technology* 18 (5–6): 835–845.

Vaßen, R., A. Stuke, and D. Stöver. 2009b. Recent developments in the field of thermal barrier coatings. *Journal of Thermal Spray Technology* 18 (2): 181–186.

Vaßen, R., M.O. Jarligo, T. Steinke, D.E. Mack, and D. Stöver. 2010. Overview on advanced thermal barrier coatings. *Surface and Coating Technology* 205: 938–942.

Venkateswarlu, K., V. Rajinikanth, T. Naveen, and Dhiraj Prasad Sinha. 2009. Abrasive wear behavior of thermally sprayed diamond reinforced composite coating deposited with both oxy-acetylene and HVOF techniques. *Wear* 266: 995–1002.

Vetter, J., G. Barbezat, J. Crummenauer, and J. Avissar. 2005. Surface treatment selections for automotive applications. *Surface and Coating Technology* 200: 1962–1968.

Vitale-Brovarone, C., and E. Verné. 2005. SiO$_2$-CaO-K$_2$O coatings on alumina and Ti6Al4V substrates for biomedical applications. *Journal of Materials Science. Materials in Medicine* 16: 863–871.

von Niessen, K., M. Gindrat, and A. Refke. 2010. Vapor phase deposition using plasma spray-PVD™. *Journal of Thermal Spray Technology* 19 (1–2): 502–509.

Voyer, J., P. Schulz, and M. Schreiber. 2008. Electrically conductive flame sprayed aluminum coatings on textile substrates. *Journal of Thermal Spray Technology* 17 (5–6): 818–823.

Vuoristo, P., and P. Nylén. 2009. Industrial and research activities in thermal spray technology in the Nordic region of Europe. In *Thermal spray 2009: Proceedings of the international thermal spray conference*, ed. B.R. Marple, M.M. Hyland, Y.-C. Lau, C.-J. Li, R.S. Lima, and G. Montavon, 517–522. Materials Park: ASM International. e-Proc.

Waki, H., T. Kitamura, and A. Kobayashi. 2009. Effect of thermal treatment on high-temperature mechanical properties enhancement in LPPS, HVOF, and APS CoNiCrAlY coatings. *Journal of Thermal Spray Technology* 18 (4): 500–509.

Wang, B. 1996. Erosion-corrosion of thermal sprayed coatings in FBC boilers. *Wear* 199: 24–32.

Wang, B.Q., and K. Luer. 1994. The erosion-oxidation behavior of HVOF Cr$_3$C$_2$-NiCr cermet coating. *Wear* 174: 177–185.

Wang, B.-Q., and A. Verstak. 1999. Elevated temperature erosion of HVOF Cr$_3$C$_2$/TiC– NiCrMo cermet coating. *Wear* 233–235: 342–351.

Wang, Y., K.A. Khor, and P. Cheang. 1998. Thermal spraying of functionally graded calcium phosphate coatings for biomedical implants. *Journal of Thermal Spray Technology* 7 (1): 50–57.

Wang, Y., J.-G. Legoux, R. Neagu, S. Hui, and B.R. Marple. 2012. Suspension plasma spray and performance characterization of half cells with NiO/YSZ anode and YSZ electrolyte. *Journal of Thermal Spray Technology* 21 (1): 7–15.

Weiss, L.E., F.B. Prinz, D.A. Adams, and D.P. Siewiorek. 1992. Thermal spray shape deposition. *Journal of Thermal Spray Technology* 1 (3): 231–237.

Weiss, L.E., D.G. Thuel, L. Schultz, and F.B. Prinz. 1994. Arc-sprayed steel-faced tooling. *Journal of Thermal Spray Technology* 3 (3): 275–281.

Wigren, J., and Kristina Täng. 2007. Quality considerations for the evaluation of thermal spray coatings. *Journal of Thermal Spray Technology* 16 (4): 533–540.

Wright, I.G., and T.B. Gibbons. 2007. Recent developments in gas turbine materials and technology and their implications for syngas firing. *International Journal of Hydrogen Energy* 32: 3610–3621.

Wu, Y., S. Hong, J. Zhang, Z. He, W. Guo, Q. Wang, and G. Li. 2012a. Microstructure and cavitation erosion behavior of WC–Co–Cr coating on 1Cr18Ni9Ti stainless steel by HVOF thermal spraying. *International Journal of Refractory Metals and Hard Materials* 32: 21–26.

Wu, X.-K., J.-S. Zhang, X.-L. Zhou, H. Cui, and J.-C. Liu. 2012b. Advanced cold spray technology: Deposition characteristics and potential applications. *Science China Technological Sciences* 55 (2): 357–368.

Xie, Y., X. Zheng, C. Ding, W. Zhai, J. Chang, and H. Ji. 2009. Preparation and characterization of CaO-ZrO_2-SiO_2 coating for potential application in biomedicine. *Journal of Thermal Spray Technology* 18 (4): 678–685.

Xie, X.Y., H.B. Guo, and S.K. Gong. 2010. Mechanical properties of $LaTi_2Al_9O_{19}$ and thermal cycling behaviors of plasma-sprayed $LaTi_2Al_9O_{19}$/YSZ thermal barrier coatings. *Journal of Thermal Spray Technology* 19 (6): 1179–1185.

Xie, X., H. Guo, S. Gong, and H. Xu. 2012. Hot corrosion behavior of double-ceramic-layer $LaTi_2Al_9O_{19}$/YSZ thermal barrier coatings. *Chinese Journal of Aeronautics* 25: 137–142.

Yamakawa, O., H. Nihonmatsu, M. Morisasa, and H. Hotta. 2009. Plasma sprayed ceramic tray members for firing ceramic capacitor. In *Thermal spray 2009: Proceedings of the international thermal spray conference*, ed. B.R. Marple, M.M. Hyland, Y.-C. Lau, C.-J. Li, R.S. Lima, and G. Montavon, 624–627. Materials Park: ASM International.

Yandouzi, M., H. Bu, M. Brochu, and B. Jodoin. 2012. Nanostructured Al-based metal matrix composite coating production by pulsed gas dynamic spraying process. *Journal of Thermal Spray Technology* 21 (3–4): 609–619.

Yang, Q., T. Senda, and A. Hirose. 2006a. Sliding wear behavior of WC–12% Co coatings at elevated temperatures. *Surface and Coating Technology* 200: 4208–4212.

Yang, Y., N. Oh, Y. Liu, W. Chen, S. Oh, M. Appleford, S. Kim, K. Kim, S. Park, J. Bumgardner, W. Haggard, and J. Ong. 2006b. Enhancing osseo-integration using surface-modified titanium implants. *JOM* 58: 71–76.

Yang, G.-J., C.-J. Li, S.-J. Zhang, and C.-X. Li. 2008. High-temperature erosion of HVOF sprayed Cr_3C_2-NiCr coating and mild steel for boiler tubes. *Journal of Thermal Spray Technology* 17 (5–6): 782–787.

Yılmaz, S. 2009. An evaluation of plasma-sprayed coatings based on Al_2O_3 and Al_2O_3–13 wt.% TiO_2 with bond coat on pure titanium substrate. *Ceramics International* 35: 2017–2022.

Yilmaz, R., A.O. Kurt, A. Demir, and Z. Tatli. 2007. Effects of TiO_2 on the mechanical properties of the Al_2O_3–TiO_2 plasma sprayed coating. *Journal of the European Ceramic Society* 27: 1319–1323.

Yoshiya, A., S. Shigemura, M. Nagai, and M. Yamanaka. 2009. Advances of thermal sprayed carbon roller in paper industry. In *Thermal spray 2009: Proceedings of the international thermal spray conference*, ed. B.R. Marple, M.M. Hyland, Y.-C. Lau, C.-J. Li, R.S. Lima, and G. Montavon, 601–606. Materials Park: ASM International.

Yu, L.-G., K.A. Khor, H. Li, and P. Cheang. 2003. Effect of spark plasma sintering on the microstructure and in vitro behavior of plasma sprayed HA coatings. *Biomaterials* 24: 2695–2705.

Yu, J., H. Zhao, X. Zhou, S. Tao, and C. Ding. 2011. Effect of thermal aging on microstructure and mechanical properties of plasma-sprayed samarium zirconate coatings. *Journal of Thermal Spray Technology* 20 (5): 1056–1062.

Yuan, F.H., Z.X. Chen, Z.W. Huang, Z.G. Wang, and S.J. Zhu. 2008. Oxidation behavior of thermal barrier coatings with HVOF and detonation-sprayed NiCrAlY bond coats. *Corrosion Science* 50: 1608–1617.

Zeng, Z., N. Sakoda, T. Tajiri, and S. Kuroda. 2008. Structure and corrosion behavior of 316L stainless steel coatings formed by HVAF spraying with and without sealing. *Surface and Coating Technology* 203: 284–290.

Zhang, T., D.T. Gawne, and Y. Bao. 1997. The influence of process parameters on the degradation of thermally sprayed polymer coatings. *Surface and Coating Technology* 96: 337–344.

Zhang, G., H. Liao, H. Yu, V. Ji, W. Huang, S.G. Mhaisalkar, and C. Coddet. 2006. Correlation of crystallization behavior and mechanical properties of thermal sprayed PEEK coating. *Surface and Coating Technology* 200: 6690–6695.

Zhang, G., H. Liao, M. Cherigui, J. Paulo Davim, and C. Coddet. 2007. Effect of crystalline structure on the hardness and interfacial adherence of flame sprayed (poly-ether–ether–ketone) coatings. *European Polymer Journal* 43: 1077–1082.

Zhang, X.C., B.S. Xu, F.Z. Xuan, S.T. Tu, H.D. Wang, and Y.X. Wu. 2008. Rolling contact fatigue behavior of plasma-sprayed CrC–NiCr cermet coatings. *Wear* 265: 1875–1883.

Zhang, C., G. Zhang, V. Ji, H. Liao, S. Costil, and C. Coddet. 2009. Microstructure and mechanical properties of flame-sprayed PEEK coating remelted by laser process. *Progress in Organic Coating* 66: 248–253.

Zhang, J., Z. Wang, P. Lin, W. Lu, Z. Zhou, and S. Jiang. 2011. Effect of sealing treatment on corrosion resistance of plasma-sprayed NiCrAl/Cr_2O_3-8 wt.%TiO_2 coating. *Journal of Thermal Spray Technology* 20 (3): 508–513.

Zhao, X., and P. Xiao. 2008. Effect of platinum on the durability of thermal barrier systems with a $\gamma+\gamma'$ bond coat. *Thin Solid Films* 517: 828–834.

Zhu, D., S.R. Choi, and R.A. Miller. 2004. Development and thermal fatigue testing of ceramic thermal barrier coatings. *Surface and Coating Technology* 188–189: 146–152.

Zhum Gahr, K.H. 1987. *Microstructure and wear of materials.* Amsterdam: Elsevier.

Zotov, N., A. Hospach, G. Mauer, D. Sebold, and R. Vaßen. 2012. Deposition of La12xSrxFe12yCoyO32d coatings with different phase compositions and microstructures by low-pressure plasma spraying-thin film (LPPS-TF) processes. *Journal of Thermal Spray Technology* 21 (3–4): 441–447.

Index

© Springer Nature Switzerland AG 2021
M. I. Boulos et al. (ed.), *Thermal Spray Fundamentals*, https://doi.org/10.1007/978-3-030-70672-2

GPSR Compliance

The European Union's (EU) General Product Safety Regulation (GPSR) is a set of rules that requires consumer products to be safe and our obligations to ensure this.

If you have any concerns about our products, you can contact us on ProductSafety@springernature.com

In case Publisher is established outside the EU, the EU authorized representative is:

Springer Nature Customer Service Center GmbH
Europaplatz 3
69115 Heidelberg, Germany

Batch number: 10167334

Printed by Printforce, the Netherlands